MANUAL OF THE PLANTS OF COLORADO

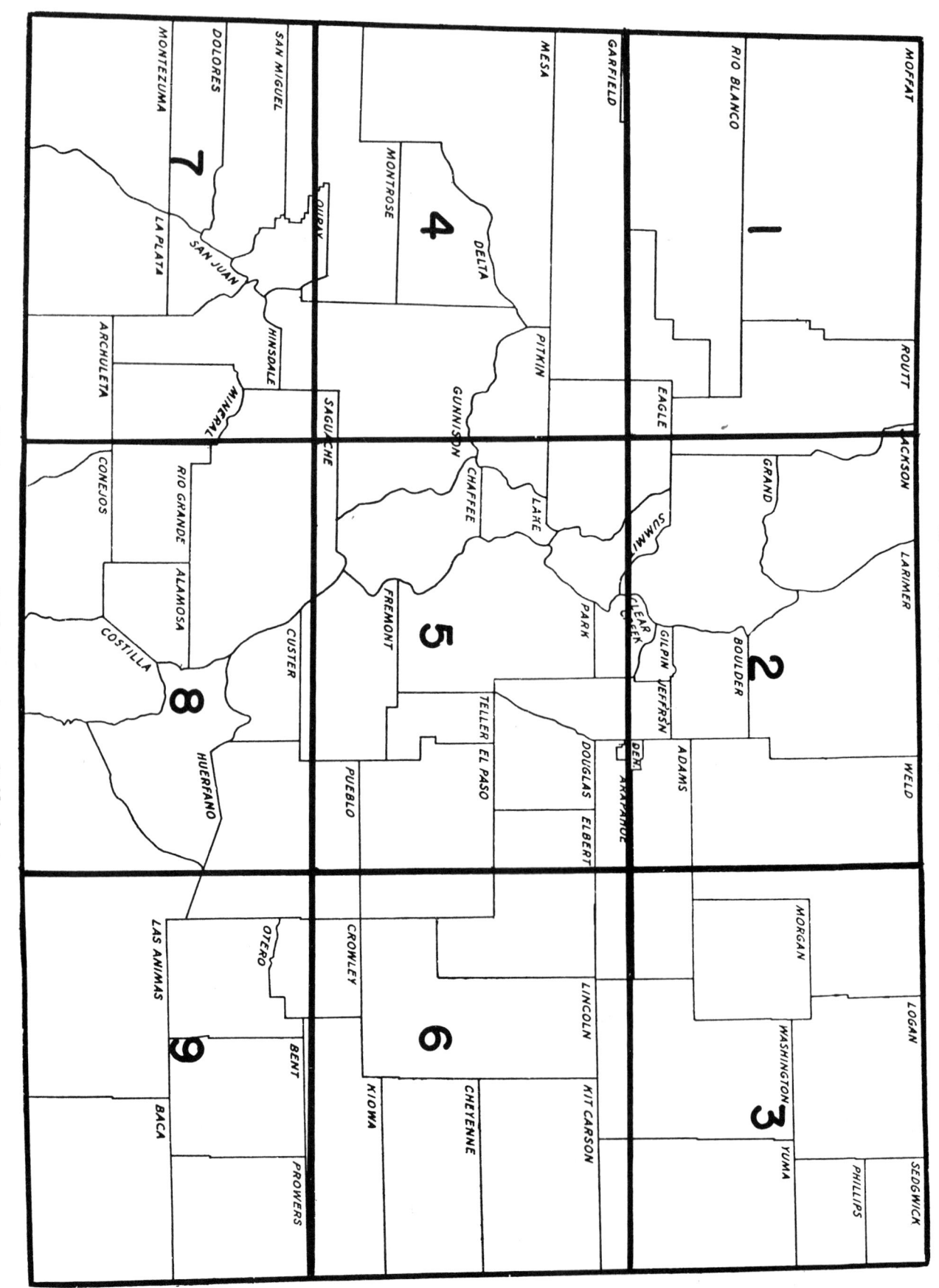

MANUAL OF THE PLANTS OF COLORADO

For the Identification of the Ferns and Flowering Plants of the State

H. D. HARRINGTON

Basalt Regional Library
P. O. Box BB
Basalt, Colo. 81621

*Authorized by the Colorado State Board of Agriculture
and prepared with the cooperation of
Colorado State University*

SAGE BOOKS

Copyright, 1954, by H. D. Harrington

Second Edition, 1964

Sage Books are published by
The Swallow Press Inc., 1139 South Wabash Avenue, Chicago, Illinois 60605

INTRODUCTION

The primary purpose of this book is to provide a means of identifying the plants of Colorado. In order to attain this objective within the limits of reasonable size some important taxonomic information of interest to the professional botanist had to be omitted, such as a more complete list of synonyms and the exact citation of widely distributed specimens. The main foundation for the work was the plant collections in the general Colorado area, but many plants were borrowed from more distant herbaria. Colorado has been reasonably well botanized, at least for a western state, but certain sections have been neglected. Most of the specimens encountered in this study were collected in the vicinity of colleges and universities, or along railroads and highways. In general the plants of the mountainous areas of the state are better represented in herbaria than those from the eastern plains.

About 1 out of every 30 species listed here constitutes a new record for the state, or at least is not credited to Colorado in the monographs and manuals. These may be plants of recent introduction like the weed Chorispora tenella DC., or ones that apparently have always been present but for one reason or another have been missed by plant collectors in the past. Many species not included will undoubtedly be found in the future--in fact, many have come to light since this treatment was originally prepared. In some cases the section concerned was revised to include these additions, but often a memorandum or note was added following a related species or one to which the new entity would naturally be keyed. There had to be a deadline in this respect. The most fertile areas for turning up new additions to the flora in the future probably will be the four corners of the state, and any of the less accessible sections. Our alpine flora needs to be studied more thoroughly. Many circumboreal or northern plants have been located in limited areas at higher altitudes. Examples would be Saussurea alpina DC., Armeria labradorica Wallr. and Cystopteris montana (Lam.) Bernh., each known to the writer in only one general locality in Colorado. Other similar plants will surely be found on intensive search of Colorado's higher altitudes especially the non-glaciated areas.

About 350 plants are included in the keys and lists although actual specimens from this state were not seen. These are plants that have either been actually reported for Colorado or listed from adjacent areas and may be expected here. No specific descriptions are ordinarily given for these, but it is hoped that their presence in the keys will aid in locating and identifying them in the future.

The latest monographs and treatments have usually been followed, and an attempt has been made to bring the nomenclature up to date. This has resulted in unavoidable changes in many familiar names. In some cases these later treatments did not appear to aid in straightening out our local plants of the group. Therefore they were partly or completely ignored; in case of doubt the older or established name has been retained. So-called "common names" are given for the most of the genera. These are the ones in general use in this state or the ones used by the various manuals and floras covering the area. The generic common names from the Standardized Plant Names, second edition 1942, have also been included for the benefit of the many workers in this area who regard that publication as official. These names are copied as listed without any attempt to make the form of the word uniform. When no common name is given it is suggested that the generic name be used. No attempt has been made to collect, or coin, such names for the species.

In general the sequence of the families follows the system of Engler and Prantl. This is the one in widest use both in the manuals and herbaria of this country, and the only system with a world-wide treatment for most of the groups. It is followed here mostly as a matter of convenience and not to indicate that it is considered to be superior in all details to other systems.

Citations to the place of publication of the species are given merely for convenience and not all of them have been actually checked or verified in the original. Diacritical markings have been deliberately omitted. Trinomials have been used when they have appeared to the writer to have particular value. The use of the terms "variety" and "subspecies" follows the particular one used by the author who was followed in that group. No attempt has been made to formally transfer one to another in order to secure uniformity. In most cases the two words apparently refer to about the same type of entity. The abbreviations "var." and "ssp." are listed following the author of the trinomial, where they are available if needed but not unduly conspicuous. In many cases the variety or subspecies is not formally listed but appears as a note following the specific description. In some cases it is ignored altogether. This is rather arbitrary but in general expresses the writer's opinion of the value of the particular entity in this area.

In most cases the synonyms given are limited to those appearing in regional manuals. No new species are proposed, and new combinations have been kept at a minimum. These new combinations are listed together at the back of this book. Extensive revisional work on any group of a local region should involve close comparison with related plants over a wide area and a critical study of all the types concerned, a procedure outside the scope of the present work. However, the special situations and difficulties encountered with local plants are often given in the notes following the particular genus or species, in the hope that such information may be of aid and interest to those who may be revising the group in the future.

The diagnostic keys are often artificial and commonly cut across sectional lines. Supplementary characters have been freely used, making the keys considerably longer but it is hoped more usable. The main purpose of these keys is to help someone identify an unknown plant rather than to present some possible line of evolutionary development. An ideal key would, of course, attain both objectives but this is seldom possible to achieve.

The descriptions to families, genera and species apply to Colorado plants only. The specific descriptions have been made as complete as possible and have been checked as carefully as time permitted against local material, especially in matters of measurement. Information as to habitat and general distribution is usually copied from other sources. The distribution of the species within the state is taken from the specimens studied. Sometimes this is given by listing the counties in which the plant was collected. More commonly local distribution is listed by dividing the state into 9 zones, which are outlined on the map of Colorado accompanying this introduction, and listing the zones from which the specimens were taken. Such data must be used with caution since every new collecting trip adds new information in this respect. That is one of the reasons why the writer's maps for the distribution of each species are not reproduced. Such maps have a visual and therefore a deceptive finality about them.

Altitude is listed in feet instead of meters since it has been found that few if any of the botanists in this area are accustomed to think in terms of the metric system in this respect. The altitudinal range given for each species is incomplete and again is a story that will never be finished. In some cases it is undoubtedly inaccurate due to error or carelessness in recording collection data. Sometimes the base of operations was recorded as the locality, but the plants were apparently collected above or below it without exact information on the matter. Obvious errors in this respect have been corrected, but not all of them could be caught. If used with due caution, however, such data on geographical and altitudinal range should be of definite help in identifying local material.

It was finally decided to omit giving the month of anthesis. In some cases this information has value, but in a state like Colorado with so much variation in altitude the same plant may vary widely in time of blossoming at different levels. For example, Sedum stenopetalum Pursh, flowers in late spring at an altitude of 5000 feet but not until late summer or fall at timberline in the mountains.

In some genera like Lupinus, Rosa, Agoseris, Chrysopsis, and Cirsium, the treatment given here is arbitrary and designed merely to give a temporary convenient system for naming and filing our material. Only comprehensive monographs can give even a reasonably satisfactory understanding of such groups. Again, this is outside the scope of this work.

Many specialists in various groups of plants have aided the project by identifying critical specimens or by giving helpful advice. Their names are too numerous to mention here but their services have been greatly appreciated. Certain genera were contributed in whole or in part by qualified experts. Credit is given in the appropriate place. The writer would like to thank the curators of the various herbaria consulted, who have gone to a great deal of trouble to make their Colorado specimens available at all times. Special mention in this respect should be made to Dr. C. L. Porter, University of Wyoming, Dr. William A. Weber, University of Colorado, and Dr. C. William T. Penland, Colorado College.

Special acknowledgments are due to Dr. L. W. Durrell of Colorado A & M College for initiating this project and securing support for its continuance.

Finally, a great deal of credit must be given to the writer's wife for her constant help throughout the work, especially in checking the manuscript and proof.

Corrections and Additions to the Manual of the Plants of Colorado
(In general limited to those that may affect the identification; only a few of the additions are mentioned.)

Page	
2	First 18 in Key delete "green and"
3	First 5 in Key add (on same or separate plants)
5	First 42 in Key add (at least below)
6	Second 16 in Key change "stamens 4 to 8 in. to stamens 2, 4, or 8
7	First 40 in Key change "Ovules 1 or 4 in. to Ovules 1 to 4
8	Second 58 in Key change "plants not aquatic" to plants seldom aquatic
8	Second 66 in Key change "leaflets toothed" to leaves or leaflets toothed
9	First 74 in Key add often before "borne"
11	Second 28 in Key change "attached to the ground" to attached to the host plant below the ground
11	First 38 in Key add (perhaps not trailing in Ipomoea leptophylla)
27	After species number 5 add The correct name for this plant seems to be P. murrayana)
29	After Pseudotsuga taxifolia description add Recently called P. menziesii (Mirb.) Franco.
34	Species number 17 change "leaves 1.5-5 cm. long" to leaves 1.5-8 cm. long
35	Description of Zannichellia palustris change "flowers of 2-10 carpels" to flowers 2-10
57	Species 27 add P. sanbergii Vasey may be the correct name
66	Species number 2 change "usually 1-3 mm. long" to usually 1-3 cm. long
79	Species number 2 change "ligules 1-3 mm. long" to ligules 1-5 mm. long
123	Second 85 in Key change "Long stolons present" to Long stolons absent
140	First 3 in Key change "Spikes 3-10" to Spikes 3-20
141	In species number 2 change "3 to many spikes" to many spikes
146	Second number 5 in Key change "usually shorter than the bracts" to usually longer than the bracts
151	After description of Liliaceae add The cultivated day lily (Hemerocallis fulva L.) sometimes escapes to ditches but this usually occurs near a dwelling
153	Description of Yucca glauca change "about 3 cm. long" to about 3-6 cm. long
159	After description of genus Iris add Cultivated Iris often occurs along roadsides and persist for a time
166	After species number 4 add Many of the broadleaved cottonwoods now along streams and banks in Colorado are probably escaped P. deltoides Marsh (with glabrous bud scales).
182	In description of family Moraceae add (at least below) after "leaves opposite"
182	After family Ulmaceae add The cultivated Chinese or Siberian elm may occur away from dwellings and appear native.
191	In species number 20 change "stems 10-30 mm. tall" to stems 10-30 cm. tall
196	Second number 32 in Key change "Styles united almost to the middle" to Styles united above the very base
208	Locality for Suckleya suckleana change "northwestern quarter" to northeastern quarter.
218	In description of Nyctaginaceae after "ovary superior" insert but appearing inferior especially in older flowers
223	In description of Portulacaceae after "styles 2-5 in. insert or 2- to 5-branched
236	To common names of Ceratophyllum add COONTAIL
237	In first number 7 in Key after "lower petals" insert short or
242	In species 5 after "12-18 mm. long" insert or even longer
250	In species 22 change "beaks 0.8-0.9 mm. long" to beaks 0.3-0.9 mm. long
252	In Key (and on p. 253 - 2 places) change "Papavera" to Papaver
255	In Key number 1 (both) change "leaflets" to leaves
256	After description to Cruciferae add A new species, Eutrema penlandii Rollins was recently described from Colorado
256	In Key first number 2 change "separated by constrictions" to separated by lateral constrictions
256	In Key second number 2 add (may be notched at apex)
257	In Key first number 24 after "Anthers" insert strongly
257	In Key second number 24 after "Anthers not" insert strongly
257	In Key second number 32 after "sessile stigma;" insert leaves sometimes not at all pinnatifid
269	Both numbers 5 in Key after "Pedicels" insert in fruit
271	Second number 17 of Key change "petals 5-14 cm." to petals 5-14 mm.
273	In species number 14 change "fruit 2.5-4 mm. long" to fruit 2.5-4 cm. long
280	In description of species 2A change "having a shorter stipe to the fruit, this not over 1 mm. long" to having a longer stipe to the fruit, this over 1 mm. long
283	In description to Barbarea orthoceras after "glabrous throughout" insert or a few long hairs present on petiole and edge of leaf lobes
285	In generic description to Ribes add stamens 5
295	To first number 5 of Key after "Branches" insert (at least some)
295	After first number 12 of Key add (however may be deeply toothed)
295	In second number 17 in Key change "follicles" to a capsule
295	In first number 18 in Key change "blade over 3 cm. long" to blade often over 3 cm. long
299	In description to Cercocarpus ledifolius change "up to 18 cm. in diameter" to up to 46 cm. in diameter
308	In first number 2 in Key change "Calyx lobes tomentose" to Calyx lobes hairy at least on the margins
315	In first number 16 in Key change "spines present" to spines usually present
320	To common names of Thermopsis add GOLDEN BANNER
322	In species number 6 change "stems 20-60 cm. tall" to stems 20-100 cm. tall
326	In species number 7 change "stems 10-25 cm. tall" to stems 10-40 cm. tall
328	In species Strophostyles leiosperma change "petals 6-7 mm. long" to petals 6-8 mm. long
341	Change the name "Astragalus plumbeus" to Astragalus molybdenus Barneby in Key and heading
345	In second number 3 of Key insert strigose, after "pods"
349	In second number 6 of Key change "pods 3-4 mm. wide," to pods 3-4 mm. wide, not dorsally flattened
369	To family description of Malvaceae add The cultivated hollyhock (Althaea rosea Cav.) may sometimes be found along ditches and roadsides especially near dwellings.

Corrections and Additions to the Manual of the Plants of Colorado (Continued)

Page	
370	In Key to Malva, second 1 change "Corolla 10-15 mm. long" to Corolla 6-15 mm. long
370	In generic description to Sidalcea change "leaves palmately lobed or cleft" to leaves palmately divided, lobed or cleft
372	In Key to Sida first number 1 change "bractlets present" to bractlets usually present
373	After description Elatinaceae add The genus Bergia has recently been found in Colorado.
376	In species number 5 change "sepals 2.5-3 mm. long" to sepals 2.5-7 mm. long
385	In Key to Echinocactus, second number 2 change "usually over 3 cm. long" to often over 2.5 cm. long
386	In species number 3 change "flowers about 3.5 cm. long" to flowers 2-3.5 cm. long
403	In distribution of Podistera eastwoodae change "southeastern quarter" to southwestern quarter
413	Second number 8 in Key change "flowers purple" to flowers white, purple
419	Second number 2 in Key change "leaves usually 2 cm. long" to leaves usually over 2 cm. long
422	After description to Primula add P. egalikgensis Wormsk. has been reported from Colorado.
425	In family description to Menyanthaceae after 3-foliate insert , sometimes 5-foliate
439	In Key second number 1 add and possibly Leptodactylon
442	In Key first number 10 change "10-30 cm. long" to 10-30 mm. long
443	In locality for species number 11 change "southeastern" to northeastern
458	After description to Mertensia add The corolla tube is below the expanded portion which is called a limb.
461	After second number 2 of Key add ; nutlets rarely over 4 mm. long
463	In first number 17 change word "lower" to upper
463	In second number 17 change word "lower" to upper and "upper" to lower
467	In locality for number 24 change "northwestern" to northeastern
475	In first number 1 of Key add 5 mm. long or more
479	In description to Lamium amplexicaule delete the word "glabrous"
484	In second number 3 of Key after "stellate" insert stems and leaves without spines or prickles
485	In description of family Scrophulariaceae change "central placentae" to central or axile placentae
493	In first 38 of Key add (see also P. hallii)
503	In species number 38 change "calyx glabrous, 3-5 cm. long" to calyx glabrous, 3-5 mm. long
515	After description for Galium add Galium verum L. with flowers yellow seems abundant in Gunnison County.
523	In second number 2 of key change "1/2 to the base" to 3/4 to the base
525	After description to Campanula add Campanula rapunculoides L. is cultivated in Colorado and often escapes into roadsides and lawns.
540	First number 135 should read Receptacle narrowly conical
540	Second number 135 should read Receptacle flat, low-conical or merely convex
547	Species number 2c change "leaves 4-6 mm. long" to leaves 4-6 cm. long
548	Species number 7F change "with subapical spots" to without subapical spots
556	In Key second number 10 change "Plants 5-20 (rarely to 30) mm. tall" to Plants 5-20 (rarely to 30) cm. tall.
585	In species number 20 change "stems 10-15 cm. tall" to stems 10-50 cm. tall
589	In description to Pericome change "pale yellow" to yellow
602	In Key number 9 add ; outer involucre bracts 3-5 mm. long
607	In species number 10 change "stems 30-100 cm. tall" to stems 30-200 cm. tall
639	In species number 4 change "heads 1.5-2.5 mm. high" to heads 1.5-2.5 cm. high
648	In Perigynous change "and pistils on calyx tube" to and petals on the calyx tube
659	Add Linum, 354 to index
663	Add Secale, 69 to index

VEGETATION ZONES IN COLORADO

By David F. Costello*

In Colorado, where the native **vegetation** still remains, one can stand almost anywhere and see from 1 to 50 species of plants. Hardly a square mile exists where the richness of flora will not provide from 150 to 200 species. In the major plant **communities** the number may reach 300 to 500. Many of these species are indigenous to certain elevational zones. Others are widespread, occurring in many sections of the state. A few, such as Taraxacum officinale, Potentilla fruticosa, and Achillea lanulosa occur from the lowest elevations in the eastern plains to the alpine meadows above timberline.

This multiplicity of plant species is related largely to diversity in topography and resulting climate in which many habitats are found. All gradations of environment are found from dry semi-desert and nearly barren rock outcrops to cool mesophytic forests, lakes, and stream sides.

In addition to the major factors of topography and climate, human influences have contributed to the innumerable combinations that intrigue the taxonomist and fascinate the ecologist who is interested in the stages of vegetation development that always follow disturbance by animals and man.

The topographic, geologic, climatic, and watershed diversity of Colorado have induced past and continuing factors that affect the vegetation. Mining has created bare areas, talus slopes, and streamside changes through diversion of channels and water pollution. The demand for timber, beginning in the early 1880's, has changed whole mountain sides and vegetation zones. For example, large areas of the virgin Pinus ponderosa forests in the San Juan Basin have been replaced through the years with extensive stands of Quercus gambellii and an accompanying herbaceous flora atypical of the original conifer stand.

Everywhere grazing has had its effect in changing the flora. Native grasses, such as Festuca thurberi and F. arizonica, have disappeared in many areas. Invaders have appeared or increased in other localities so that today we have communities dominated by Bromus tectorum, Hymenoxys richardsoni, Salsola kali tenuifolia, Cardaria draba, Convolvulus arvensis, and others where they were formerly scarce or absent. Other species which have always been widespread, such as Artemisia tridentata and A. frigida, have increased their ranges over thousands of acres since the coming of white men.

Continued heavy grazing over nearly three-quarters of a century has reduced some types to disclimaxes where subdominants, particularly the short-grasses in the Great Plains, constitute the principal species. Misuse by grazing nearly always depletes the flora, reduces the number and variety of species present, and virtually exterminates the rare ones that are palatable. The class of livestock also influences the flora through selective use. Cattle prefer grasses and consequently favor the development of a flora rich in unpalatable forbs, while sheep prefer certain forbs and favor the **development** of grasses and unpalatable forbs. Game animals, particularly deer, prefer shrubs and on winter concentration areas may destroy many or all of the browse plants in mountain valleys and foothills. Under proper management, livestock and game animals are not detrimental to the flora in general.

Fire, the great enemy of man, is considered to be one of our most destructive elements. Yet, fire contributes much to floristic diversity. It introduces new successions. In Colorado, aspen types, with astounding variety in the shrubby and herbaceous understory, follow burned spruce-fir and **lodgepole pine** forests. Burned sagebrush flats frequently blossom with multitudes of showy forbs and nutritious grasses where competition from shrubs and heavy grazing formerly had left the ground bare. Even on the plains and in meadows, where fire kills the shrubs, the after effects are new and varied floristic combinations.

Many other factors have markedly influenced the vegetation, at least locally, in Colorado. Pocket gophers on western mesas and in the higher mountains have infested more than 4 million acres. Their effect on floristic composition has not been measured adequately, but it is sufficient to be considered an economic factor in the livestock industry. Irrigation farming and extensive plowing of native sod in the Great Plains and in the sagebrush country in the western part of the state have been major factors in the destruction of millions of **acres** of the original vegetation.

---

*Range Conservationist, Rocky Mountain Forest and Range Experiment Station, maintained by U. S. Department of Agriculture, Forest Service, in cooperation with Colorado A & M College, Fort Collins, Colorado (**now located at the Pacific Northwest Forest and Range Experiment Station, Portland, Oregon**).

THE VEGETATION ZONES

The elevational range in Colorado varies from 3386 feet at the lowest point in the southeastern plains to 14,431 feet at the top of Mount Elbert in the Rockies. The difference in mean temperature within the state is equal to the difference between southern Florida and Greenland. Annual precipitation varies from less than 8 inches in the low western valleys to more than 40 inches in the higher mountains.

Topographically, approximately two-fifths of Colorado is level and rolling plains, while the western three-fifths is a mixture of mountains, high mesas, narrow valleys, and large upland parks. The San Luis Valley in the southcentral portion of the state is an extensive level basin, once covered by an inland sea.

Because of the extreme topographic relief in Colorado, vegetation zones are sharply delimited and easily recognized. The eastern plains are characterized in general by short-grasses, but are interrupted by extensive sand hills in which Artemisia filifolia is dominant. The foothills at the western edge of the plains support a shrubby mixture, predominantly Cercocarpus montanus. Coniferous forest occurs generally between 6500 and 11,000 feet. Above the upper timberline, bluegrasses, sedges, and various depauperate willows give aspect to the alpine flora. Between the lowest elevations and the mountain tops, these plant communities can be observed:

    Plains Area
        Mixed prairie (mostly short-grass disclimax)
        Sand sagebrush
    Semi-desert Area (mostly in western Colorado)
        Greasewood
        Saltbush
        Sagebrush
    Mountain and Plateau Area
        Pinon-juniper
        Mountain shrub (Quercus-Cercocarpus-Amelanchier)
        Ponderosa pine-Douglas fir (including the following)
            Mountain parks, aspen, sagebrush
        Spruce-fir (including the following)
            Mountain parks, sagebrush, aspen, lodgepole pine
        Alpine

The Plains Area

The Great Plains area in Colorado, extending eastward from the foothills of the Rockies, varies in elevation from approximately 3400 feet in the southeast to more than 6000 feet on its western border. Originally the area was almost uninterrupted mixed prairie, excepting the sand hills, fringes of broadleaved trees, and fragments of tall-grass prairie along the eastern stream courses. Thousands of acres of the plains are now devoted to irrigation farming along the South Platte and Arkansas Rivers and to dryland farming on the adjoining uplands.

Mixed Prairie--The natural vegetation of the plains is mixed prairie in which Bouteloua gracilis and Buchloe dactyloides persist beneath an overstory of taller grasses such as Agropyron smithii, Andropogon scoparius, Sporobolus cryptandrus, Stipa comata, and Aristida longiseta. South of the Arkansas River, Hilaria jamesii also occurs in the overstory. Most of the potential mixed prairie, however, now is maintained by grazing animals as a disclimax in which short-grasses are dominant.

In Colorado, Bouteloua gracilis either alternates with Buchloe dactyloides or the two are intermingled. The taller grasses are scattered throughout on moderately grazed areas but are more vigorous and abundant in swales, bottomlands, and drainages. Carex filifolia and C. eleocharis are common constituents of the short-grass sod, particularly in northeastern Colorado, while Muhlenbergia torreyi is locally important in the vicinity of Trinidad. In the localities adjacent to the sand hills, Aristida longiseta is a conspicuous component of the vegetation.

The occurrence of forbs is variable from one season to the next, and from one locality to another, but the following are almost constantly present: Sphaeralcea coccinea, Aster tanacetifolius, Bahia oppositifolia, Gaura coccinea, and Salsola kali tenuifolia. Species which fluctuate in abundance from year to year include Lepidium densiflorum, Lappula redowski, Plantago purshii, Cryptantha crassisepala, and Oenothera albicaulis. Among perennials which are locally abundant or occur in colonies are Mirabilis linearis, Dyssodia papposa, Psoralea tenuiflora, P. lanceolata, Sophora sericea, Eriogonum effusum, and Lygodesmia juncea. Yucca glauca is common on sandy soils. Opuntia polyacantha is abundant on the drier uplands.

The most frequent shrubs in the plains are Chrysothamnus nauseosus and Gutierrezia sarothrae. Atriplex canescens, A. nuttallii gardneri, and Eurotia lanata are common on the more alkaline soils in well-drained bottomlands, while Rosa woodsii, R. arkansana, Rhus trilobata, Salix spp., and Populus sargentii are found along stream banks of the major water courses.

Opuntia arborescens lends a distinct aspect in the country between Colorado Springs and Pueblo and southward.

Meadow types occur throughout the mixed prairie association in alkaline swales, on flood plains and around intermittent ponds. In the less alkaline areas Agropyron smithii frequently forms pure stands. On areas where ground water is near the surface Distichlis stricta and Sporobolus airoides are generally dominant. Common associates in these types are Calamovilfa longifolia, Hordeum jubatum, Juncus balticus, Carex spp., and Atriplex canescens.

On thousands of acres in eastern Colorado, abandoned cultivated fields exhibit well-marked stages of natural revegetation. Secondary succession proceeds through an initial stage characterized by annuals such as Salsola kali tenuifolia, Chenopodium album, Lepidium densiflorum, Helianthus petiolaris, Amaranthus retroflexus, A. graecizans (A. blitoides), and Polygonum aviculare. A forb stage consisting of many annual and perennial forbs and a few grasses persists for several years and then is succeeded by a short-lived perennial grass stage in which Hordeum jubatum, Sporobolus cryptandrus, Sitanion hystrix, and Schedonnardus paniculatus are usually abundant. An Aristida stage then occurs and may persist for many years, but ultimately the natural transition is to mixed prairie.

Sand Sagebrush--The sand areas are scattered in many districts, particularly in Weld, Washington, Morgan, Logan, Yuma, Lincoln, Cheyenne, and Kiowa counties. These are characterized by loose sands, without pebbles or rocks, which frequently develop dunes and craters. On loose sands, pioneer species include Muhlenbergia pungens, Redfieldia flexuosa, Oryzopsis hymenoides, Psoralea lanceolata, Heliotropium convolvulaceum, and Polanisia trachysperma. On stabilized sand hills the more frequent species are Aristida longiseta, Agropyron smithii, Bouteloua gracilis, Panicum virgatum, Andropogon hallii, A. scoparius, A. gerardi, Calamovilfa longifolia, Buchloe dactyloides, Yucca glauca, and numerous perennial forbs. Artemisia filifolia is prominent especially on level and gently rolling areas and dominates the vegetation over many square miles.

In the Artemisia filifolia communities the following forbs usually are conspicuous: Tradescantia occidentalis, Helianthus petiolaris, Gilia longiflora, Eriogonum annuum, Mirabilis linearis, Rumex venosus, Chrysopsis villosa, and Opuntia humifusa.

Wet and moist habitats in ponds and small streams commonly support Potamogeton natans, Sagittaria cuneata, Alisma plantago-aquatica, and Scirpus americanus. In swampy areas Typha latifolia, Spartina gracilis, Carex lanuginosa, and Scirpus validus are common. Other species in moist habitats in the sand hill areas include Agrostis alba, Puccinellia nuttalliana, Panicum virgatum, Andropogon gerardi, Distichlis stricta, Bidens cernua, Teucrium canadensis occidentale, Mimulus glabratus, and Sisyrinchium montanum.

Throughout the Artemisia filifolia communities in eastern Colorado, especially on the more compact soils, Bouteloua gracilis now provides the bulk of the herbage cover. Andropogon scoparius, which formerly dominated many square miles, now persists mostly on the borders of borrow pits along highways and in cemeteries where grazing is not allowed. Under moderate grazing use a mixture of short-grasses occurs with the taller species, Andropogon hallii, Calamovilfa longifolia, Stipa comata, Sporobolus cryptandrus, and Aristida longiseta. On many thousands of acres, cultivated crops are now grown where Artemisia filifolia and associated species once covered the landscape.

The Semi-desert Area

The vegetation of this area occurs principally on the lower mountain slopes, plateaus, and dry mountain valleys at elevations below 7000 feet. Precipitation is 15 inches or less and the soils are characteristically alkaline.

Greasewood--Greasewood, characterized by Sarcobatus vermiculatus, is distributed throughout the state on heavy soils with high alkali content. It occurs in localities where the annual precipitation is usually less than 10 inches and the water table is high. Extensive areas of greasewood are found in the San Luis Valley, on the level plains adjoining the foothills north of Pueblo, and in practically all the arid drainage courses in the valley floors of western Colorado.

Grasses and other herbaceous species are typically infrequent in greasewood stands. Where underground seepage occurs, Distichlis stricta and Sporobolus airoides are commonly present. In southern Colorado, Bouteloua gracilis, Agropyron dasystachyum, Hordeum jubatum, and Kochia scoparia are frequent associates of greasewood. Salsola kali tenuifolia, Cardaria draba, and Iva axillaris also are common. On less alkaline soils greasewood is occasionally found in mixture with Artemisia tridentata, A. spinescens, Atriplex canescens, and A. confertifolia. Rabbitbrush (Chrysothamnus spp.) frequently borders dense greasewood stands and may be invaded by the greasewood when the water table becomes elevated by subirrigation as for example in the San Luis Valley.

Saltbush--This type is widespread in the lower valleys of the western three-fifths of the state. It occurs on drier and better drained soils that are slightly less alkaline than those supporting greasewood stands. Extensive areas are found in western Moffat County, on the desert north of Mack, and on the alkaline stretch of country between Hotchkiss and the West Elk Mountains. This type covers a large area between Montrose and Grand Junction from where it continues westward into Utah.

The shrubby component varies from almost pure stands of widely separated plants of Atriplex nuttallii gardneri to uneven mixtures of A. confertifolia, Artemisia spinescens, Eurotia lanata, and Kochia americana.

In southwestern and western Colorado, Hilaria jamesii, Oryzopsis hymenoides, Sphaeralcea coccinea dissecta, Eriogonum cernuum, and Stanleya pinnata bipinnata are components of the vegetation. In northwestern Colorado Agropyron dasystachyum and A. smithii are usually present. Locally, pricklypears (Opuntia spp.) occur in extensive patches.

Saltbush communities are usually bordered on their upper edges by sagebrush stands, or the transition may be to grassland at the edges of foothills as in the area between Colorado Springs and Walsenburg.

Sagebrush--Thousands of acres in western Colorado, particularly in Moffat County, in the valleys of the Gunnison, Colorado, and Rio Grande Rivers, and in the San Juan Basin, support dense stands of Artemisia tridentata with varying mixtures of herbaceous plants in the understory. The sagebrush type occurs on moderately alkaline to neutral soils that are moist to a considerable depth.

In southwestern and middle western Colorado the type is best developed below 7000 feet, although it extends to more than 9500 feet in warm valleys and on level plateaus. In northwestern Colorado the upper limits occur generally between 6000 and 7000 feet, although they reach 10,000 feet in a few localities. These high-elevation areas of sagebrush are extensions into the spruce-fir zone and are not characteristic of semi-desert. Extensive sagebrush stands also are found in North Park and Middle Park.

At its lower border the type merges with the saltbush and greasewood communities. The transition at the upper limits is variable. The boundary between sagebrush and mountain shrub is usually clear-cut. The contact with pinon-juniper is frequently diffusive, with indications in many localities, as on the north slope of the Uncompahgre plateau, that the junipers are invading the sagebrush from above. In southern Colorado the transition is commonly mosaic-insular where ponderosa pine, oak, and sagebrush occupy the same zone. With increase in altitude the sagebrush patches decrease in size and number until the forest associations become completely dominant.

The commonest grasses in the sagebrush type include Agropyron smithii, A. dasystachyum, A. spicatum, Oryzopsis hymenoides, Sitanion hystrix, Poa fendleriana, P. nevadensis, Koeleria cristata, and Hilaria jamesii. Associated with these are many weedy species such as Salsola kali tenuifolia, Arenaria hookeri, Hymenoxys richardsoni, Balsamorrhiza sagittata, Leptodactylon pungens, and Chrysopsis villosa. Many sagebrush areas recently have been invaded by Bromus tectorum.

It is not commonly recognized that the sagebrush association, when subjected to proper grazing management, can support an almost luxuriant stand of herbaceous and woody species. Under continued moderate livestock use, especially at medium altitudes where **rainfall is fairly dependable**, the association is typified by an abundance of such plants as Agropyron spp., Koeleria cristata, Festuca arizonica, F. idahoensis, Carex spp., Purshia tridentata, and Symphoricarpos oreophilus. The denudation in this type, however, is so universal that we are prone to accept a scarcity of plants in the understory as the natural condition.

Mountain and Plateau Area

Pinon-Juniper--Below the ponderosa pine-Douglas fir zone in southern and southwestern Colorado are extensive stands of Pinus edulis intermingled with Juniperus scopulorum, J. mono**sperma**, or J. utahensis. The elevational range is variable, extending from 5500 to 7000 feet **in southcentral** Colorado and up to 8500 feet in the southwestern and middle western parts of the state. The stands vary from open to dense and are modified locally by the intrusion of Artemisia tridentata, Cercocarpus montanus, Symphoricarpos oreophilus, Purshia tridentata, Chrysothamnus spp., and other shrubby communities. In dry western valleys the type, at its lower limits, frequently blends into the desert type.

Pinon-juniper stands are found on the slopes in the lower drainages of nearly all western Colorado rivers--San Miguel, Dolores, San Juan, Animas, Uncompahgre, Gunnison, Colorado, Yampa, White; and Piceance and Yellow Creeks. At higher elevations, 7000 to 8500 feet, the type is typical in Glade Park, Mancos, Bayfield, and much of Mesa Verde National Park. Narrow belts occur on both sides of the Uncompahgre Plateau and the Black Canyon of the Gunnison. The upper edge of the zone frequently contacts the mountain brush community and contains Quercus gambelli, **Amelanchier** alnifolia, and Cercocarpus montanus. At the southeastern edge of the mountains in

El Paso, Pueblo, and Las Animas counties, pinon-juniper merges with the short-grass plains and is characterized by an understory of Bouteloua gracilis and Hilaria jamesii. The herbaceous understory is sparse throughout the type owing to extensive overgrazing in the past. Only a few small relict areas give indications of the nature of the original vegetation. The commonest grasses are Oryzopsis hymenoides, Poa fendleri, Koeleria cristata, and Bouteloua gracilis. The last is mostly limited to the southern portion of the type. Festuca arizonica is abundant in localized areas. Agropyron inerme is frequent in the northern portion and on the Roosevelt National forest; Hesperochloa kingii is common in the type. Stipa comata, Sporobolus cryptandrus, Sitanion hystrix, and Agropyron smithii occur sporadically throughout the type. Oryzopsis micrantha is a constant in the southern part of the pinon-juniper zone on the Eastern Slope, while Poa fendleriana is the outstanding herbaceous subclimax grass on the Western Slope. Stipa scribneri is important locally in the mountains south of Trout Creek Pass.

Forbs are not abundant in the pinon-juniper zone, partly because of the dry site conditions. Species commonly encountered are Chrysopsis villosa, Sphaeralcea coccinea, Salsola kali tenuifolia, Erigeron spp., Grindelia squarrosa, Hymenoxys richardsoni, and Helianthus petiolaris.

Shrubs, other than Artemisia tridentata, are only locally abundant. Winter grazing by deer in many places has reduced the abundance of species such as Cercocarpus ledifolius, Purshia tridentata, Cowania mexicana, and Amelanchier alnifolia. Other shrubs found in the type include Rhus trilobata, Holodiscus dumosus, Artemisia frigida, and Atriplex canescens. **Gutierrezia sarothrae** is common on some areas where most of the palatable plants have long been overused by grazing animals.

Mountain Shrub--Brushlands are dominant in the transitions between grassland and semi-desert at their lower limits and woodland or conifer forest at their upper borders. These thickets form a narrow interrupted belt variously characterized by Quercus gambelli, Prunus virginiana melanocarpa, Cercocarpus montanus, Purshia tridentata, Crataegus spp., Amelanchier spp., and other shrubby forms.

On the eastern edge of the Front Range, Cercocarpus montanus is conspicuous on the hogbacks and rocky slopes north of Denver. Other components of the shrubby layer are Chrysothamnus lanceolatus, Rhus trilobata, Prunus virginiana melanocarpa, Rosa woodsi, and Crataegus **ssp.** The lower contact is with mixed prairie in which Stipa comata, Agropyron smithii, Bouteloua gracilis, B. curtipendula, Poa fendleriana, Sphaeralcea coccinea, Helianthus petiolaris, Chrysopsis villosa, and Artemisia frigida are prominent. South of Denver, Quercus gambelli determines the aspect with an understory composed of many sedges and broadleaved herbs. Festuca arizonica, Muhlenbergia **montana**, and Bouteloua gracilis are abundant on areas which have not been severely grazed.

In southern and southwestern Colorado brushland alternates with pinon-juniper, sagebrush, and ponderosa pine. Quercus gambelli has replaced many ponderosa pine forests which have been removed by fire or logging.

In western Colorado, the oaks are commonly associated with Symphoricarpos oreophilus, Amelanchier alnifolia, Purshia tridentata, and Chrysothamnus viscidiflorus. The oak type does not occur in North, Middle, and South Parks.

On mesophytic slopes the mountain brush understory usually includes Poa pratensis, Stipa comata, S. lettermanii, Orthocarpus luteus, Ceanothus spp., Solidago spp., Viguiera multiflora, Gayophytum ramosissimum, Artemisia carruthii, Erigeron flagellaris, Wyethia amplexicaulis, and Gaura coccinea. In drier situations common components of the ground flora are Stipa scribneri, Oryzopsis hymenoides, **Gutierrezia** sarothrae, Penstemon torreyi, Brickellia grandiflora, Gaillardia aristata, and Phlox multiflora.

In northwestern Colorado, oak and buckbrush thickets occur chiefly on steep slopes at elevations of 6000 to 8000 feet. The lower contact is usually with sagebrush while the upper border frequently touches the aspen and lodgepole pine. In some sections the mountain brush consists almost entirely of Amelanchier alnifolia which grows in extensive patches alternating with **Artemisia** tridentata.

Ponderosa Pine-Douglas Fir--This association occurs on ridges and slopes at elevations varying from 6000 feet in the foothills to 8500 feet in the mountains where the upper contact is with lodgepole pine, Engelmann spruce, and alpine fir. In the southern part of the state, the lower border frequently is in contact with oak or pinon-juniper. In the San Luis Valley the upper limits of the zone extend to 9500 feet.

Pinus ponderosa is distributed principally down the Front Range from the Wyoming line to Trinidad, west to Mesa Verde, and north to the Uncompahgre Plateau. In the remaining portion of the mountainous area Pseudotsuga taxifolia is characteristic, particularly in the northern and western mountains. Douglas fir occurs typically on north slopes and ponderosa pine on south slopes.

Several aspects are common in the ponderosa pine phase of this zone. The open timber stands permit an abundance of light and the development of an extensive herbaceous understory. Grasslands, devoid of trees, are frequent. Meadows and stream banks within the zone support a luxuriant vegetation. In contrast, where closed forest occurs, as under Douglas fir on north slopes, herbaceous plants are scarce and the needle litter constitutes the principal ground cover.

In parks and open timber stands Festuca arizonica and Muhlenbergia montana are the dominant grasses. Festuca idahoensis and Hesperochloa kingii take the place of F. arizonica in the northern part of the state. Other grasses of frequent occurrence are Andropogon scoparius, Blepharoneuron tricholepis, Koeleria cristata, Danthonia intermedia, and D. parryi. Common forbs are Geranium fremontii, Astragalus agrestis, Achillea lanulosa, Oxytropis lambertii, Potentilla fissa, Senecio fendleri, and Campanula rotundifolia. Arctostaphylos uva-ursi is a common shrub throughout, while A. patula is locally important on the Uncompahgre Plateau.

On thousands of acres, heavy grazing has caused the replacement of bunchgrasses by such species as Poa pratensis, Bouteloua gracilis, Stipa robusta, Muhlenbergia filiformis, M. filiculmis, Sitanion hystrix, Hordeum jubatum, Antennaria rosea, Arenaria fendleri, Chrysopsis villosa, Hymenoxys richardsoni, Erigeron flagellaris, and Artemisia frigida. These species are also abundant on abandoned fields. Also locally abundant in treeless parks and areas denuded of their original cover are Gutierrezia sarothrae, Tetradymia spinosa, Solidago petradoria, Salsola kali tenuifolia, Schedonnardus paniculatus, and Bromus tectorum.

Meadow and stream-bank communities in this zone support a rich flora in the absence of overgrazing. Deschampsia caespitosa, Calamagrostis canadensis, Ranunculus spp., Caltha leptosepala, Potentilla fruticosa along with a variety of sedges and rushes are characteristic of the moister soils. Salix spp., **Betula occidentalis,** B. glandulosa, and Cornus stolonifera are common on stream banks. On drier sites and in meadows that have been drained by erosion, Poa pratensis, P. compressa, Iris missouriensis, Trifolium repens, and Taraxacum officinale are common invaders.

Shrub and broadleaved tree communities are common throughout the zone. Some of these represent stages in secondary succession following fire or other disturbances. Aspen stands are one example. Many of the oak communities in southwestern Colorado have developed in the wake of fire and of former logging operations which removed the rich virgin stands of ponderosa pine and Douglas fir. Other shrub communities frequently encountered are local concentrations either of single species or mixtures of Cercocarpus montanus, Ceanothus fendleri, Ribes cereum, Prunus virginiana melanocarpa, Artemisia tridentata, Amelanchier alnifolia, and Holodiscus dumosus.

Spruce-Fir--This is the broad zone of coniferous forest that extends upward from the ponderosa pine-Douglas fir forests to the upper timber line. The climax trees are Engelmann spruce (Picea **engelmanii**) and alpine fir (Abies lasiocarpa). Within this zone extensive stands of aspen (Populus tremuloides) and lodgepole pine (Pinus contorta latifolia) occur. These two species also occur at lower altitudes, being common in the upper reaches of the ponderosa pine zone. Less frequent than any of the above, but common in this zone are white fir (Abies concolor), corkbark fir (A. lasiocarpa arizonica), bristlecone pine (Pinus aristata), and blue spruce (Picea pungens).

The undergrowth in dense tree stands is sparse. Characteristic species are Vaccinium scoparium, Stipa **lettermani**, Galium boreale, Arnica cordifolia, Wyethia amplexicaulis, Aquilegia elegantula, Chimaphila umbellata, Pyrola elliptica, Moneses uniflora, and Carex geyeri. In more open stands the rank herbaceous growth usually includes Trisetum spicatum, Festuca ovina, Bromus anomalus, Phleum alpinum, Carex festivella, C. phaeocephala, Senecio serra, Lathyrus leucanthus, Ligusticum porteri, and Fragaria ovalis. Helenium hoopesii is frequent in spruce-fir forests west of the Continental Divide.

Aspen stands in this zone, when not overgrazed, support a lush understory of dicotyledonous herbs. The following are usually abundant: Thalictrum fendleri, Vicia americana, Thermopsis montana, Achillea lanulosa, Delphinium barbeyi, Pseudocymopterus montanus, Heracleum lanatum, Ligusticum porteri, Pedicularis racemosa, Aquilegia coerulea, Pteridium aquilinum pubescens, and Swertia radiata. Commonly associated with these are such grasses as Elymus glaucus, Festuca thurberi, Bromus carinatus, Agropyron trachycaulum, and Koeleria cristata. Numerous sedges are nearly always present. Usually the most important shrubs are Symphoricarpos oreophilus and Prunus virginiana melanocarpa.

Aspen stands exhibit various subtypes in the understory. On drier sites, Festuca thurberi is frequently dominant or mixed with Bromus spp., Elymus glaucus, Agropyron trachycaulum, and various forbs, including Senecio serra, S. triangularis, Wyethia amplexicaulis, and Hackelia floribunda. On moister sites, where the tree canopy is denser, the herbaceous understory commonly includes the following genera: Heracleum, Ligusticum, Conioselinum, Carum, Osmorrhiza, and Angelica, along with species of Delphinium, Aconitum, and Rudbeckia. Occasionally Pteridium aquilinum pubescens appears in pure stands or in mixture with the genera listed above. In aspen stands on the eastern slope and in southern Colorado, Thermopsis montana is frequently the major herbaceous species. A sheep grazing disclimax commonly found includes Festuca thurberi and Bromus spp. With further deterioration from overgrazing, this degenerates into a mixture of Stipa **lettermani**, Taraxacum officinale, Polygonum douglasii, and various unpalatable **forbs.**

Lodgepole pine communities are widespread in the northern half of the state, especially in the spruce-fir zone. They are found less frequently in the upper limits of the ponderosa pine zone. When the lodgepole pine canopy is dense, a needle layer constitutes virtually the only ground cover. In more open stands the usual understory includes Carex geyeri, Arnica cordifolia, Lathyrus leucanthus, Vaccinium scoparium, Juniperus communis, and Shepherdia canadensis.

Grassland parks of various sizes occur throughout the spruce-fir zone. These derive their aspect from a mixture of sedges and grasses, including Stipa lettermani, Poa reflexa, Agrostis scabra, Phleum alpinum, Trisetum spicatum, and many forbs. The extensive stands of Festuca thurberi which formerly grew on the White River Plateau, Grand Mesa, and in the San Juan Mountains have largely been replaced by inferior forage species, including the above grasses, and many weedy species such as Lupinus wyethii, Wyethia amplexicaulis, Potentilla pulcherrima, Helenium hoopesii, Achillea lanulosa, Polygonum aviculare, Collomia linearis, Eriogonum spp., and Madia glomerata.

Sagebrush extensions break the zonal pattern of both the ponderosa pine-Douglas fir and the spruce-fir belts. Many mountainous areas where sagebrush is now established formerly were grassland parks. Artemisia tridentata is the dominant species. One of the most extensive areas of big sagebrush occurs in North Park where it blends into meadows along moist margins and at its upper limits contacts the coniferous forests. Artemisia cana is locally dominant, as for example, on the west end of Grand Mesa and on the White River Plateau where it grows at elevations of 10,000 feet. It grows on sites more mesophytic than those occupied by Artemisia tridentata. Other shrubby components in sagebrush communities at lower elevations in the spruce-fir zone are Symphoricarpos oreophilus, Purshia tridentata, and Chrysothamnus viscidiflorus lanceolatus. The herbaceous understory contains many of the grasses, sedges, and forbs characteristic of the grassland parks, the borders, and the open spaces of the surrounding spruce-fir forests.

Alpine Zone--The alpine zone is the area above the upper limits of tree growth. It extends to the highest mountain tops where many of the exposed rocks are bare. The lower border is a transition zone merging with the spruce-fir forests below. Frequently this transition or "sub-alpine" belt is characterized by decumbent or scrubby trees extending above the true forests. The lower edge of the true alpine zone varies from 11,500 feet in northern Colorado to 12,000 feet in the southern part of the state, with local differences due to topography.

Floristic composition is remarkably uniform throughout the alpine zone. Essentially the same combinations of species are found, whether one examines the vegetation in the San Juan Mountains or in the Mummy Range. Strong affinities exist between the Colorado **alpine** flora and those of Labrador, Greenland, Norway, and the European Alps. Many of the same species are found in these widely separated localities.

On exposed areas where soils are shallow, one notices especially such pulvinate forms as Arenaria sajanensis, Phlox caespitosa pulvinata, Saxifraga bronchialis, Silene acaulis, Paronychia pulvinata, Trifolium dasyphyllum, and T. nanum. In more favorable sites there may be an abundance of Geum turbinatum, Achillea lanulosa, Stellaria borealis, Gentiana parryi, Podistera eastwoodae, Sibbaldia procumbens, and Castilleja rhexifolia. Over large areas the dominant climax species is Kobresia bellardi. Commonly the Kobresia stands are intermingled with Poa alpina, **Trisetum** spicatum, and Deschampsia caespitosa, resulting in a grassland aspect. Alpine willows are frequent to abundant.

A number of easily distinguished plant communities occur in this zone. The alpine summit community occurs on exposed mountain tops, ridges, and steep slopes. In the more favorable habitats, Kobresia bellardi, Poa alpina, P. lettermani, P. pattersoni, P. rupicola, Trisetum spicatum, Carex drummondiana, C. chimaphila, and Luzula spicata are found in combination with Draba oligosperma, Podistera eastwoodae, Arenaria rossii, Sibbaldia procumbens, Dryas octopetala, Salix petrophila, S. nivalis, and S. nivalis saximontana.

Alpine meadow communities are found on more sheltered benches, slopes, and level areas, where soils are well developed. The grassy aspect results from the presence of such species as Agrostis **humilis**, Avena mortoniana, Deschampsia caespitosa, Phleum alpinum, Trisetum spicatum, Poa alpina, P. arctica, P. lettermani, P. rupicola, Carex atrata, **C. chalciolepis**, C. nigricans, C. nova, C. **phaeocephala**, C. vernacula, Juncus drummondii, and J. parryi. Associated with these are Achillea lanulosa, Hymenoxys grandiflora, Agoseris aurantica, Arnica mollis, Epilobium alpinum, Geum turbinatum, Mertensia alpina, Penstemon hallii, P. harbourii, Polygonum bistortoides, P. viviparum, Senecio taraxacoides, S. soldanella, and Thalictrum alpinum. This community frequently extends downward to the edge of the spruce-fir forest, the transition zone being characterized by a greater abundance of Festuca thurberi, F. ovina, Danthonia intermedia, Ligusticum porteri, Mertensia alpina, Swertia perennis, Caltha leptosepala, and Trollius laxus albiflorus.

Willow communities consisting of the taller shrubby species often make almost impenetrable stands near the lower border of the alpine zone. Salix brachycarpa, S. planifolia nelsoni, S. planifolia monica, and S. pseudolapponum are common. Where former stands of these willows have been killed by grazing the community has been replaced by the characteristic grass-sedge mixture of the alpine meadow communities. The strictly alpine willows, S. cascadensis which is rare in Colorado, S. anglorum antiplasta (S. petrophila), and S. nivalis and its variety saximontana are all prostrate plants of low stature.

Alpine marshes occur where ponds originally existed or where springs and melting snowbanks contribute to a continuously moist habitat. Sedges are the characteristic plants with willows on the drier margins. The most common sedges are *Carex albo-nigra*, *C. arapahoensis*, *C. bella*, *C. illota*, and *C. physocarpa*. Also found with these sedges are *Eleocharis macrostachya*, and *Eriophorum angustifolium*. *Caltha leptosepala* is generally present and *Isoetes bolanderi* may be found as one of the first pioneers in ponds. Distinctive societies of *Pedicularis groenlandica* are frequent in boggy areas.

KEY TO THE FAMILIES OF COLORADO PLANTS

(Carried to unusual genera in some cases)

A. Plants not producing seeds or true flowers, reproducing by spores; fernlike, mosslike, rushlike or floating aquatic plants

Division or Phylum I. PTERIDOPHYTA

1. Plants floating on water, the small rootlets not attached to the soil, body of plant minute, not over 25 mm. wide or long
   -4. Salviniaceae p. 22
1. Plants terrestrial or if aquatic then rooting on the bottom, body of plant not minute, over 25 mm. long or wide
   2. Leaves palmately 4-foliate (like a 4-leaf clover); sporangia borne in short-stalked sporocarps from the very base of the stem
   -3. Marsileaceae p. 21
   2. Leaves various but not palmately 4-foliate; sporangia not enclosed in basal sporocarps (basal only in Isoetaceae)
      3. Stems conspicuously jointed, hollow; leaves minute, united into sheaths at the nodes; sporophylls aggregated into a terminal conelike structure
      -5. Equisetaceae p. 22
      3. Stems not jointed, solid; leaves various but not united into sheaths at the nodes; sporophylls seldom borne in terminal cones
         4. Leaves very numerous, spirally arranged on branching, more or less creeping, rather elongated stems, the leaves not over 1.5 cm. long
            5. Sporangia all bearing minute, uniform spores; leaves over 5 mm. long
            -7. Lycopodiaceae p. 24
            5. Sporangia of 2 kinds, some containing very minute spores (microspores) and others with few large spores (megaspores); leaves less than 5 mm. long
            -8. Selaginellaceae p. 24
         4. Leaves 1 to many (but not very numerous), over 1.5 cm. long, from elongated underground rootstocks or from short rootstocks underground or nearly so (do not mistake the leaf petiole for a stem)
            6. Leaves grasslike or onionlike, undivided; plants partly or completely submerged in water; sporangia borne within the expanded hollow bases of the leaves, both small spores (microspores) and large spores (megaspores) present
            -6. Isoetaceae p. 23
            6. Leaves not grasslike or onionlike, all ours more or less lobed or compound; plants rarely even partially submerged; sporangia not borne in the leaf bases, the spores uniform
               7. Sporangia sessile, very large, in ours borne in a terminal panicle, the sterile leaf blade appearing lateral on a common stalk with it
               -1. Ophioglossaceae p. 12
               7. Sporangia mostly long-stalked (use a strong lens), minute, borne in clusters (sori) on the back or near the margins of ordinary leaves, or on completely separated fertile sporophylls
               -2. Polypodiaceae p. 14

AA. Plants producing seeds and with flowers (these may be inconspicuous); plants of various aspect

Division or Phylum II. SPERMATOPHYTA

B. Ovules and seeds not borne in a closed cavity, borne instead on the face of a scale or bract; leaves needlelike or scalelike, not deciduous in the fall; stigmas none; plants woody

- Class 1. GYMNOSPERMAE

1. Stems jointed, the leaves reduced to scales, these distant, opposite or in whorls; fruit a small thin-scaled cone
   -10. Gnetaceae p. 29
1. Stems not jointed; leaves needlelike and crowded; fruit a woody cone or berrylike
   -9. Pinaceae p. 25

BB. Ovules and seeds borne in a closed ovary; leaves various but seldom needlelike or scalelike, rarely evergreen; stigma or stigmas present; plants herbaceous or woody

- Class 2. ANGIOSPERMAE

C. Embryo with 1 cotyledon; flower parts mostly in 3's (or multiples thereof); leaves mostly parallel-veined, usually entire, rarely toothed or lobed; vascular bundles not in a ring; never woody

- Subclass 1. MONOCOTYLEDONEAE

1. Plants without true stems and leaves, thalluslike, free-floating, aquatic
   -20. Lemnaceae p. 144
1. Plants with stems and leaves, terrestrial to aquatic but not thalluslike and free-floating
   2. Perianth lacking or inconspicuous, of bristles or scales usually arranged in a single whorl, nothing resembling petals present
      3. Aquatic plants, the stems and leaves flaccid, either submerged or floating
         -13. Najadaceae p. 31
      3. Terrestrial plants or lower part in water but the stem or leaves rigid enough to support the part above the water level
         4. Flowers all sessile in the axils of chaffy, imbricated bracts; grasslike plants with leaves sheathing the stem at base; fruit 1-seeded
            5. Flower immediately subtended by only 1 bract (or a second one when present usually entirely enclosing the ovary except at apex); perianth of bristles sometimes present; flowers often unisexual; stems usually solid and triangular; leaves usually 3-ranked with closed sheaths; fruit an achene
               -18. Cyperaceae p. 116
            5. Flowers immediately subtended by 2 bracts (lemma and palea) neither of which completely encloses the ovary; perianth of bristles never present; flowers rarely unisexual; stems usually hollow and round; leaves usually 2-ranked with open sheaths; fruit a caryopsis
               -17. Gramineae p. 38
         4. Flowers not all sessile in the axils of chaffy bracts; sometimes grasslike but often not; fruit 1-seeded or more than 1-seeded
            6. Flowers in globose heads
               -12. Sparganiaceae p. 30
            6. Flowers in elongated spikes or racemes
               7. Flowers monoecious, on a thick spike, the staminate above; numerous bristles present at base of pistillate flower
                  -11. Typhaceae p. 29
               7. Flowers perfect, in a spike or raceme; no bristles present
                  8. Flowers in dense thick spikes these appearing to come from the edge of a leaf (the peduncle leaflike); carpels not separated
                     -19. Araceae p. 144
                  8. Flowers in narrow spikelike racemes on ordinary peduncles (not at all leaflike); carpels almost separated, or at least separating when ripe
                     -14. Juncaginaceae p. 35
   2. Perianth usually of 2 distinct series, at least the inner petaloid (not in Juncaceae, Sparganiaceae and Juncaginaceae) and conspicuous
      9. Plant aquatic with submerged stems and leaves; leaves whorled
         -16. Hydrocharitaceae p. 37
      9. Plant terrestrial, or semiaquatic but then leaves never whorled
         10. Perianth divisions greenish or brownish, not showy; plants rather grasslike in aspect
            11. Carpels 3-6, almost separate, or at least soon separating; flowers in narrow spikelike racemes
               -14. Juncaginaceae p. 35
            11. Carpels 1-3, united, not separating (except as valves in the dehiscent fruit); inflorescence various but never of narrow spikelike racemes
               12. Flowers unisexual, in dense globose heads; staminate flower lacking perianth; pistillate flower with irregular divisions
                  -12. Sparganiaceae p. 30
               12. Flowers perfect, sometimes in globose heads; flowers with regular, rather large perianth divisions
                  -23. Juncaceae p. 146
         10. Perianth divisions (at least the inner) showy and petallike; plants seldom really grasslike
            13. Pistils several to many in a head or ring, each ovary of 1 carpel
               -15. Alismaceae p. 36
            13. Pistils 1 to a flower, more than 1 carpel to each
               14. Stamens 3 or fewer
                  15. Perianth very irregular; ovary inferior; stamens 1 or 2; flowers not subtended by spathes
                     -27. Orchidaceae p. 160
                  15. Perianth regular or nearly so; ovary superior or inferior; stamens 3; flowers subtended by spathelike bracts
                     16. Ovary superior; plants aquatic or semiaquatic
                        -22. Pontederiaceae p. 146
                     16. Ovary inferior; plants terrestrial (often in wet places)
                        -26. Iridaceae p. 159
               14. Stamens more than 3 (usually 6)
                  17. Ovary completely inferior
                     -25. Amaryllidaceae p. 159
                  17. Ovary superior (or only partly inferior in Zygadenus elegans)
                     18. Perianth sharply differentiated into 2 sets, the outer green and sepallike, the inner petallike and different in size or shape from the outer
                        19. Inner perianth segments fugaceous, blue or rose, not bearing glands, less than 2 cm. long; plant not from a bulblike corm
                           -21. Commelinaceae p. 145

    19. Inner perianth segments not fugaceous, white, green, yellow, rose or purple-tinged, often bearing large conspicuous glands, often over **2 cm.** long; plants often from a bulblike corm
                        -24. <u>Liliaceae</u> p. **151**
   18. Perianth parts of the 2 whorls alike or nearly so, not sharply differentiated
                           -24. <u>Liliaceae</u> p. **151**

CC. Embryo with 2 cotyledons; flower parts mostly in 4's or 5's (or multiples thereof); leaves mostly netted veined, sometimes entire but often toothed, lobed or compound; vascular bundles in a ring; plants woody or herbaceous

  - Subclass 2. <u>DICOTYLEDONEAE</u>

D. Perianth none or of a single set, the parts all much alike in color, size and texture (no corolla present)

    (Series) <u>APETALAE</u>

 1. Plants parasitic on the branches of trees, not rooting in the soil, usually lacking chlorophyll
                         -36. <u>Loranthaceae</u> p. **183**
 1. Plants not parasitic on branches of trees, attached to the soil, with chlorophyll present
  2. Plants woody vines, shrubs or trees
   3. Flowers (of one or both sexes) in catkins; plants monoecious or dioecious
    4. Leaves pinnately compound; fruit a large nut enclosed by a husk (walnut), no cup at base
                        -29. <u>Juglandaceae</u> p. **179**
    4. Leaves simple (can be lobed); fruit various but if a nut then without a husk but with a cup at base (acorn)
     5. Both staminate and pistillate flowers in catkins
      6. Plants monoecious; staminate flowers with 4 calyx lobes; fruit indehiscent, not a capsule; seeds not hairy
                     -30. <u>Corylaceae</u> p. **179**
      6. Plants dioecious; perianth absent on all flowers; fruit a dehiscent capsule; seeds hairy
                     -28. <u>Salicaceae</u> p. **165**
     5. Only 1 kind of flowers in catkins
      7. Vines with stems retrorsely prickly; leaves palmately lobed; pistillate flowers only in catkinlike inflorescences   -(Genus)
                     <u>Humulus</u> p. **182**
      7. Trees or shrubs, the stems not retrorsely prickly; leaves entire to pinnately lobed; staminate flowers only in catkins
       8. Fruit a small utricle; leaves entire; shrubs seldom over 2 m. tall
                    -39. <u>Chenopodiaceae</u> p. **203**
       8. Fruit a rather large nut or acorn; leaves toothed or lobed; shrubs or small trees, often over 2m. tall
        9. Staminate catkins present over winter; pistillate flowers remaining in the winter buds at anthesis (only the stigmas protruding); ovary 2-celled; fruit a nut; leaves serrate or somewhat lobulate; plants rarely over 200 cm. tall   -(Genus)
                    <u>Corylus</u> p. **179**
        9. Staminate catkins not present over winter; pistillate flowers out of the buds at anthesis; ovary 1-celled; fruit an acorn; leaves various but often definitely lobed or parted; plants often over 200 cm. tall
                    -31. <u>Fagaceae</u> p.**180**
   3. Flowers not in catkins, perfect or **unisexual**.
    10. Flowers several to many, sessile in a head surrounded by a circle of bracts; ovary inferior
                    -117. <u>Compositae</u> p.**528**
    10. Flowers solitary to variously clustered (rarely in an involucrate head and then ovary superior); ovary superior (in all but first number 11)
     11. Ovary inferior (or appearing so)
      12. Flowers in corymbose, terminal cymes, perfect; ovary usually 2-celled; stamens usually 5
                    -89. <u>Cornaceae</u> p. **414**
      12. Flowers solitary or in axillary clusters, often **unisexual**; ovary 1-celled; stamens 4 or 8
                    -83. <u>Elaeagnaceae</u> p. **386**
     11. Ovary superior
      13. Pistils several to many to each flower; vines with opposite compound leaves
              -(Genus)   <u>Clematis</u> p. **242**
      13. Pistils 1 to a flower (may be lobed); plants various but if vines then the leaves not both opposite and compound
       14. Vines, with climbing or scrambling stems
        15. Leaves alternate; tendrils usually present; stems not retrorsely prickly; fruit a berry
                    -73. <u>Vitaceae</u> p.**368**

15. Leaves opposite; no tendrils present; stems retrorsely prickly; fruit a dry achene
-(Genus) Humulus p. 182
14. Trees or shrubs, stems upright or nearly so
  16. Ovary of 1 carpel (1 cell, 1 placenta, 1 style, 1 stigma)
-56. Rosaceae p. 294
  16. Ovary of 2 or more carpels (more than 1 cell, placenta, style or stigma--more than 1 of any makes more than 1 carpel)
    17. Leaves opposite, lobed or compound; fruit a double samara
-70. Aceraceae p. 365
    17. Leaves opposite or alternate, entire to toothed (rarely compound in Oleaceae); fruit various, if a samara then single
      18. Flowers perfect, subtended by a gamophyllous involucre; stamens 6-9; fruit an achene
-(Genus) Eriogonum p. 185
      18. Flowers perfect or **unisexual** but not subtended by a gamophyllous involucre; stamens 1-5; fruit a samara, utricle or **drupelike**
        19. Leaves toothed; no style present at all (stigmas 2)
-34. Ulmaceae p. 182
        19. Leaves entire; 1-3 styles present
          20. Perianth minute or none; styles and stigmas usually 1; fruit a drupe or samara; leaves always opposite
-93. Oleaceae p. 424
          20. Perianth definite (at least on the pistillate flower); styles and stigmas usually 2 or 3; fruit a utricle; leaves mostly alternate
-39. Chenopodiaceae p. 203
2. Plants herbaceous, obscurely woody (if at all) only at very base of plant
  21. Ovary inferior, adnate to the perianth or hypanthium (or appearing so as in Nyctaginaceae)
    22. Plants aquatic; leaves entire and in whorls or the immersed ones filiform-dissected
-86. Haloragidaceae p. 398
    22. Plants not aquatic; leaves various, rarely as above
      23. Ovary only partly inferior, 1-celled, with many ovules; fruit a dehiscent capsule
-(Genus) Chrysosplenium p. 292
      23. Ovary completely inferior (or appearing so), more than 1-celled or if 1-celled with only 1 or 2 ovules; fruit indehiscent (although lobes may pull apart)
        24. Ovary 2-celled, 1 ovule in each cell; fruit 2-seeded
          25. Perianth parts united; leaves opposite or whorled; flowers in cymes, not umbellate
-109. Rubiaceae p. 515
          25. Perianth parts separated; leaves alternate or basal; flowers in umbels
-88. Umbelliferae p. 399
        24. Ovary 1-celled, 1- to 2-ovuled (or 1- to 3-celled but only 1 cell with ovule); fruit 1-seeded
          26. Perianth really double but appearing single and **corollalike**; calyx limb may be obsolete or reduced to a mere border (these 2 families properly belonging to Gamopetalae series)
            27. Flowers (at least the staminate) sessile in dense heads surrounded by an involucre; stamens 5, the anthers sometimes connate; leaves opposite or alternate
-117. Compositae p. 528
            27. Flowers not sessile in involucrate heads; stamens usually 3, anthers not connate; leaves opposite
-115. Valerianaceae p. 526
          26. Perianth evidently in one series, no evidence of an outer series at all
            28. Leaves alternate; ovules 2 in young ovary (1 developing); inflorescence corymbose
-37. Santalaceae p. 185
            28. Leaves opposite; ovules **only 1 to an** ovary; inflorescence rarely corymbose (often in involucrate heads)
-41. Nyctaginaceae p. 218
  21. Ovary superior
    29. Pistils several to many to a single flower; stamens usually many
-47. Ranunculaceae p. 236
    29. Pistils 1 (may be lobed) to a flower; stamens 1 to many (usually not over 10 in most families)
      30. Submerged aquatic plants with no perianth present at all
        31. Ovary 1-celled with 1 style; fruit 1-seeded; stamens many, leaves whorled, dissected
-46. Ceratophyllaceae p. 236
        31. Ovary 4-celled with 2 styles; fruit 4-seeded; stamens 1; leaves opposite and entire
-67. Callitrichaceae p. 363
      30. Plants usually terrestrial with perianth present (a few terrestrial plants lack perianth and a few semiaquatic plants are present with perianth), none both aquatic and lacking perianth
        32. Ovary 2- or more-celled
          33. Inflorescence of small flowers lacking any perianth, in a dense cylindrical spike, subtended at base by a conspicuous involucre of whitish, petaloid bracts (the whole resembling a single flower)
-32. Saururaceae p. 181
          33. Inflorescence various but not like the above (perianth present or absent)
            34. Flowers **unisexual**; milky juice often present
-66. Euphorbiaceae p. 358
            34. Flowers perfect, milky juice absent

35. Leaves opposite or whorled, entire; stamens 1 to many (rarely 2); flowers axillary, solitary or in small clusters
                         -42. Aizoaceae p. 222
35. Leaves alternate (may be crowded basal) usually toothed; stamens 2; flowers in terminal spikes or racemes
  36. Perennial plants with rootstocks; flowers in spikes; fruit several-seeded
                        -(Genus) Besseya p.489
  36. Annual plants; flowers in racemes; fruit 2-seeded (1 in each cell)
                        -(Genus) Lepidium p.260
32. Ovary 1-celled
 37. Ovary with several to many ovules; fruit a capsule, several to many-seeded
  38. Ovary with free central placentae; leaves opposite
   39. Styles 2-5; calyx of distinct segments
                        -44. Caryophyllaceae p.225
   39. Styles 1; calyx of united segments -(Genus) Glaux p.421
  38. Ovary with 2 or 3 parietal placentae; leaves alternate
   40. Ovary with 2 parietal placentae, half inferior; styles 2; no petals present on any flower; leaf blades reniform, crenate or lobed
                -(Genus) Chrysosplenium p.292
   40. Ovary with 3 parietal placentae, not at all inferior; styles 1 (stigma can be 3-lobed); petals present on upper flowers of plant; leaves oblong-lanceolate, entire
                -(Genus) Helianthemum p. 386
 37. Ovary with only 1 ovule; fruit a 1-seeded achene or utricle
  41. Leaves with stipules present and united around the stem in a sheath just above the nodes (ocreae)
                        -38. Polygonaceae p.185
  41. Leaves without stipules or when present not sheathing the stem
   42. Conspicuous, persistent stipules present; leaves opposite
    43. Leaves palmately divided or lobed
                        -33. Moraceae p.182
    43. Leaves simple, entire or toothed, not at all lobed
     44. Stipules scarious; plants small with spreading or prostrate or densely caespitose stems rarely over 30 cm. tall; stinging hairs not present
              -(Genus) Paronychia p. 228
     44. Stipules not especially scarious; plants with upright stems usually over 30 cm. tall; stinging hairs present on plant
              -(Genus) Urtica p. 183
   42. Stipules lacking; leaves alternate or opposite
    45. Base of perianth closely investing the ovary and fruit, these appearing inferior; perianth large, over 5 mm. long; styles and stigmas 1 (with leaves opposite and entire)
                    -41. Nyctaginaceae p. 218
    45. Perianth base not so closely investing the ovary and fruit as to make them appear inferior; perianth usually less than 5 mm. long; styles or stigmas 2 or more (or if 1 then leaves not both opposite and entire)
     46. Flowers perfect, subtended by a gamophyllous involucre; stamens 6-9; fruit an achene
            -(Genus) Eriogonum p. 185
     46. Flowers perfect or **unisexual** but not subtended by a gamophyllous involucre; stamens 1-5; fruit an achene or utricle
      47. Bracts and perianth more or less scarious or hyaline
                 -40. Amaranthaceae p.216
      47. Bracts and perianth herbaceous to fleshy, not scarious or hyaline
       48. Styles and stigmas 1 (leaves alternate and entire); fruit an achene; plants annual
            -(Genus) Parietaria p. 183
       48. Styles 1 to 3 (but if 1 the leaves toothed); fruit a utricle; plants annual to perennial
                -39. Chenopodiaceae p. 203

DD. Perianth present, evidently double, the outer segments (calyx) and inner segments (corolla) usually conspicuously different in texture, size and color

 E. Petals separate or united only at the very base

   - (Series) POLYPETALAE

 1. Ovary inferior, at least the lower part adnate to the hypanthium or calyx tube
  2. Petals many; stems very thick and succulent, usually spiny; leaves none or greatly reduced and early deciduous
                    -81. Cactaceae p. 382
  2. Petals not over 10; stems not succulent or only moderately so, not spiny (sometimes in Rosaceae); **leaves** present and conspicuous
   3. Plants trees with pinnately compound leaves; fruit a nut (walnut)
                  -29. Juglandaceae p. 179
   3. Plants herbaceous to woody but if trees then the leaves not pinnately compound; fruit various, not a nut
    4. Aquatic plants with whorled, finely dissected leaves
              -(Genus) Myriophyllum p. 398

4. Terrestrial plants or if semiaquatic the leaves not whorled and dissected
  5. Stamens as many as the petals and opposite them
    6. Ovary 2- to 3-celled; plants shrubs or small trees   -72. Rhamnaceae p. 366
    6. Ovary 1-celled; plants herbaceous   -43. Portulacaceae p. 223
  5. Stamens fewer or more than the petals, or if the same number then alternate to them
    7. Stems definitely woody, trees or shrubs
      8. Fruit a drupe, 1- or 2-seeded; calyx limb minute; flowers many in a flat-topped cyme; leaves opposite and entire
          -89. Cornaceae p. 414
      8. Fruit various but not a drupe, 3 or more seeds present; calyx limb well developed; flowers usually in relatively few-flowered racemes or corymbs (if many-flowered the leaves rarely opposite and entire); leaves opposite or alternate, entire, toothed, lobed or compound
        9. Leaves opposite; fruit a dry capsule   -55. Saxifragaceae p. 284
        9. Leaves alternate; fruit a more or less fleshy pome or berry
          10. Leaves palmately veined and lobed; fruit a berry; petals little if any longer than the sepals; stamens not over 5
              -(Genus) Ribes p. 285
          10. Leaves pinnately veined, pinnately lobed or pinnately compound; fruit a pome; petals usually much longer than the sepals; stamens often over 5
              -56. Rosaceae p. 294
    7. Stems herbaceous or slightly woody at the base only
      11. Stamens 11 or more
        12. Plants rough-pubescent from usually barbed hairs; ovary 1, 1-celled
            -80. Loasaceae p. 378
        12. Plants not rough-pubescent; ovaries more than 1, or if appearing as 1 then 2- or more-celled
            -(Genus) Paeonia p. 245
      11. Stamens 10 or less than 10
        13. Plants climbing by means of coiled tendrils; anthers 1-3, more or less united
            -113. Cucurbitaceae p. 523
        13. Plants not climbing, no tendrils present; anthers usually over 3, never united
          14. Ovules and seeds more than 1 in each cell; ovary 1- to 4-celled
            15. Only 1 style present (stigmas may be lobed)
              16. Stamens 10; plants rough-pubescent, usually from barbed hairs
                  -80. Loasaceae p. 378
              16. Stamens 4 or 8; plants not rough-pubescent, at least not from barbed hairs
                  -85. Onagraceae p. 388
            15. Two or more styles present
              -55. Saxifragaceae p. 284
          14. Ovules and seeds only 1 to each cell; ovary 2- to 6-celled
            17. Leaves verticillate and entire; ovary usually 2-celled with 1 style
              -89. Cornaceae p. 414
            17. Leaves alternate, basal or opposite, toothed or entire (never verticillate); ovary 4-celled or if 2-celled with 2 styles present (except in 1 genus in Onagraceae)
              18. Stamens 2, 4 or 8; petals 2 or 4; ovary with 1 style, usually 4-celled (except in Circaea)   **-85. Onagraceae p. 388**
              18. Stamens 5; petals usually 5 (rarely 4); ovary with 2 or more styles, 2- to 6-celled
                19. Ovary with 4-6 cells; fruit a several-seeded berry
                  -87. Araliaceae p. 399
                19. Ovary 2-celled; fruit dry, 2-seeded
                  -88. Umbelliferae p. 399
1. Ovary completely superior
  20. Corolla definitely irregular
    21. Leaves pinnately or palmately compound or completely dissected so as to appear compound
      22. Stamens many, usually over 14; petals much smaller than the sepals; ovaries often 2 or more, or of carpels joined only at base
          -47. Ranunculaceae p. 236
      22. Stamens not over 14; petals as large or larger than the sepals; ovaries 1, if of 2 carpels then these completely joined
        23. Ovary with 1 placenta; petals 5 (or rarely 1), the flowers usually sweetpea-shaped
          -57. Leguminosae p. 314
        23. Ovary with 2 placentae; petals 4, flowers never sweetpea-shaped
          24. Sepals 2; corolla spurred
            -50. Fumariaceae p. 253
          24. Sepals 4 or more; corolla never spurred
            -51. Capparidaceae p. 254
    21. Leaves simple, entire to deeply lobed or pinnatifid, but never truly compound
      25. Petals irregularly but conspicuously cleft; stamens usually many, inserted on 1 side of the flower
          -53. Resedaceae p. 283
      25. Petals entire or only slightly lobed; stamens less than 10, or if many then not 1-sided

26. Stamens many; carpels usually over 1, separate or separating in fruit and becoming follicles; leaves palmately cleft or parted
　　　　　　　　　　　　　　　　　　　　　　　　　　　　　　　　　-47. Ranunculaceae p.236
26. Stamens 10 or less; carpels 1 or if more than 1, united into a compound ovary; fruit a capsule or legume; leaves simple to compound, rarely palmately cleft or parted
　　27. One of sepals spurred, only 3 present; ovary 5-celled, stigmas sessile
　　　　　　　　　　　　　　　　　　　　　　　　　　　　　　　　　-61. Balsaminaceae p.355
　　27. None of sepals spurred (petals may be), 5 present; ovary 1- to 2-celled, style present
　　　　28. Lower petal spurred or saccate at base; stamens 5; ovary 1-celled with 3 placentae
　　　　　　　　　　　　　　　　　　　　　　　　　　　　　　　　　-79. Violaceae p.375
　　　　28. None of petals spurred or saccate at base; stamens 3, 4, 6, 8 or 10, never 5; ovary 2-celled or 1-celled with 1 placenta
　　　　　　29. Ovary 2-celled; stamens 6-8; fruit a capsule
　　　　　　　　　　　　　　　　　　　　　　　　　　　　　　　　　-65. Polygalaceae p.357
　　　　　　29. Ovary 1-celled; stamens 3, 4 or 10; fruit a legume
　　　　　　　　30. Stamens 10; petals longer than the sepals; fruit long and narrow
　　　　　　　　　　　　　　　　　　　　　　　　　　　　　　　　　-57. Leguminosae p.314
　　　　　　　　30. **Stamens 3 or 4;** petals shorter than the sepals; fruit globose
　　　　　　　　　　　　　　　　　　　　　　　　　　　　　　-(Genus) Krameria p.316
20. Corolla regular or nearly so
　　31. Stamens of the same number as the petals and opposite them
　　　　32. Sepals, petals and stamens each 6 in number; either stems or leaf margins spiny
　　　　　　　　　　　　　　　　　　　　　　　　　　　　　　　　　-48. Berberidaceae p.251
　　　　32. Sepals, petals and stamens 2-5 in number (sepals rarely 6); branches and leaves spineless (or spiny in Rhamnaceae)
　　　　　　33. Ovary 2- to 6-celled; woody vines, shrubs or trees
　　　　　　　　34. Leaves palmately veined, palmately lobed or compound
　　　　　　　　　　35. Plants vines, or stems prostrate, usually with tendrils; leaves alternate
　　　　　　　　　　　　　　　　　　　　　　　　　　　　　　　　　-73. Vitaceae p.368
　　　　　　　　　　35. Plants upright, shrubs or trees, no tendrils present; leaves opposite
　　　　　　　　　　　　　　　　　　　　　　　　　　　　　　　　　-70. Aceraceae p.365
　　　　　　　　34. Leaves simple and pinnately veined, not at all lobed
　　　　　　　　　　　　　　　　　　　　　　　　　　　　　　　　　-72. Rhamnaceae p.366
　　　　　　33. Ovary 1-celled; plants herbaceous
　　　　　　　　36. Styles and stigmas 1; sepals usually 5
　　　　　　　　　　　　　　　　　　　　　　　　　　　　　　　　　-92. Primulaceae p.420
　　　　　　　　36. Styles and stigmas 2 or more; sepals usually 2
　　　　　　　　　　　　　　　　　　　　　　　　　　　　　　　　　-43. Portulacaceae p.223
　　31. Stamens fewer or more than the petals, or if the same number then alternate to them
　　　　37. Ovary 1 and 1-celled (may be incompletely 2- or more-celled at base but then 1-celled above)
　　　　　　38. Stamens 13 or more
　　　　　　　　39. Ovary simple (1 placenta, 1 style, 1 stigma)
　　　　　　　　　　40. Ovules 1 or 4; stamens on the calyx tube; plants usually woody
　　　　　　　　　　　　　　　　　　　　　　　　　　　　　　　　　-56. Rosaceae p.294
　　　　　　　　　　40. Ovules many; stamens on the receptacle; plant herbaceous
　　　　　　　　　　　　　　　　　　　　　　　　　　　　-(Genus) Actaea p.244
　　　　　　　　39. Ovary compound (2 or more placentae, styles or stigmas)
　　　　　　　　　　41. Ovary with 2 placentae; plant usually viscid and ill-smelling
　　　　　　　　　　　　　　　　　　　　　　　　　　　　　　　　　-51. Capparidaceae p.254
　　　　　　　　　　41. Ovary with 3 or more placentae; plant not viscid or ill-smelling
　　　　　　　　　　　　42. Petals conspicuously cleft; stamens 1-sided on the flower
　　　　　　　　　　　　　　　　　　　　　　　　　　　　　　　　　-53. Resedaceae p.283
　　　　　　　　　　　　42. Petals entire or obscurely toothed to emarginate; stamens not 1-sided
　　　　　　　　　　　　　　43. Sepals 2-3, caducous; leaves alternate, **not punctate**
　　　　　　　　　　　　　　　　　　　　　　　　　　　　　　　　　-49. Papaveraceae p.252
　　　　　　　　　　　　　　43. Sepals 4 or 5, usually persistent; leaves opposite, punctate with translucent minute dots　　-78. Hypericaceae p.374
　　　　　　38. Stamens 12 or less
　　　　　　　　44. Ovary simple (1 placenta, 1 style, and 1 stigma)
　　　　　　　　　　45. Stamens and petals on the calyx tube; leaves not bipinnate
　　　　　　　　　　　　　　　　　　　　　　　　　　　　　　　　　-56. Rosaceae p.294
　　　　　　　　　　45. Stamens and petals not on the calyx tube; leaves usually bipinnate
　　　　　　　　　　　　　　　　　　　　　　　　　　　　　　　　　-57. Leguminosae p.314
　　　　　　　　44. Ovary compound (more than 1 placenta, style or stigma)
　　　　　　　　　　46. Woody shrubs, trees or vines
　　　　　　　　　　　　47. Leaves scalelike, minute, crowded and imbricate on the twigs; seeds usually hairy at the ends
　　　　　　　　　　　　　　　　　　　　　　　　　　　　　　　　　-76. Tamaricaceae p.373
　　　　　　　　　　　　47. Leaves not scalelike, not minute, not crowded on the twig; seeds not all hairy
　　　　　　　　　　　　　　48. Leaves opposite or whorled, simple and entire; ovary with more than 1 ovule
　　　　　　　　　　　　　　　　　　　　　　　　　　　　　　　　　-77. Frankeniaceae p.374
　　　　　　　　　　　　　　48. Leaves alternate, pinnately or palmately compound; ovary with 1 ovule
　　　　　　　　　　　　　　　　　　　　　　　　　　　　　　　　　-68. Anacardiaceae p.363
　　　　　　　　　　46. Herbaceous plants, woody if at all, only at the very base
　　　　　　　　　　　　49. Petals conspicuously several-cleft; stamens 1-sided on the flower
　　　　　　　　　　　　　　　　　　　　　　　　　　　　　　　　　-53. Resedaceae p.283

49. Petals entire, obscurely toothed or deeply emarginate (this may be deep enough to form a 2-lobed petal); stamens not 1-sided on the flower
    50. Petals inserted on the throat of a bell-shaped or tubular calyx
        -84. <u>Lythraceae</u> p. 387
    50. Petals on the receptacle; calyx of separate or united sepals
        51. Stamens 5, alternating with 5 clusters of gland-tipped staminodia; leaves all basal but one
            -(Genus) <u>Parnassia</u> p. 291
        51. Stamens various but not alternating with 5 gland-tipped staminodia; more than 1 or no cauline leaves present
            52. Ovules 1
                53. Petals 4; stamens 2, 4 or 6; stipules absent or inconspicuous
                    -52. <u>Cruciferae</u> p. 256
                53. Petals 5; stamens 5; stipules usually conspicuous
                    -44. <u>Caryophyllaceae</u> p. 225
            52. Ovules more than 1
                54. Ovules attached to base of ovary or to a free central placenta (not parietal)
                    -44. <u>Caryophyllaceae</u> p. 225
                54. Ovules on 2 or more parietal placentae, rarely axile, never free central
                    55. Ovary with 2 parietal placentae; sepals and petals 4 each
                        -51. <u>Capparidaceae</u> p. 254
                    55. Ovary with 3-5 placentae, these axile or parietal; sepals and petals usually 3 or 5 (rarely 4)
                      56. Styles more than 1; leaves opposite, punctate with minute translucent dots
                        -78. <u>Hypericaceae</u> p. 374
                      56. Styles 1; leaves alternate, not punctate
                        -82. <u>Cistaceae</u> p. 386
37. Ovaries more than 1, or if 1 then 2- or more-celled
    57. Stamens 13 or more
        58. Petals many; plants aquatic with large floating leaves
            -45. <u>Nymphaeaceae</u> p. 235
        58. Petals less than 10; plants not aquatic
            59. Stamens monadelphous
                -74. <u>Malvaceae</u> p. 368
            59. Stamens distinct or in several sets
                60. Leaves opposite, punctate with minute translucent dots
                    -78. <u>Hypericaceae</u> p. 374
                60. Leaves usually alternate or basal, not punctate
                    61. Stamens and petals on the calyx tube (flower perigynous)
                      -56. <u>Rosaceae</u> p. 294
                    61. Stamens and petals on the receptacle (flower hypogynous)
                      -47. <u>Ranunculaceae</u> p. 236
    57. Stamens 12 or less
        62. Plants definitely woody
            63. Leaves simple, not at all lobed
                64. Anthers opening by pores at the 2-horned base (appearing at apex); stamens not on a disk
                    -(Genus) <u>Chimaphila</u> p. 416
                64. Anthers not opening by pores, not 2-horned; stamens on a disk
                    -69. <u>Celastraceae</u> p. 364
            63. Leaves palmately lobed or variously compound
                65. Leaves opposite
                    66. Ovary 5-celled; leaflets entire, usually 2; fruit not a samara
                      -(Genus) <u>Larrea</u> p. 356
                    66. Ovary 2-celled; leaflets toothed or lobed, over 2; fruit a double samara
                      -70. <u>Aceraceae</u> p. 365
                65. Leaves alternate
                    67. Leaflets 3; fruit a samara; shrub
                      -64. <u>Rutaceae</u> p. 357
                    67. Leaflets more than 3; fruit berrylike; shrubs or small trees
                      -71. <u>Sapindaceae</u> p. 366
        62. Plants herbaceous or slightly woody at base only
            68. Ovaries 2, united above by a fleshy common stigma; anthers united with each other and to this stigma; juice milky
                -97. <u>Asclepiadaceae</u> p. 432
            68. Ovaries 1 to many, united if at all only at the very base; anthers not united to each other or to a fleshy stigma; juice rarely milky
                69. Ovaries more than 1, wholly separated or united only at base, each with its own style
                    70. Ovules 1 to an ovary; fruit a 1-seeded achene; ovaries 2 to many (often over 5), not at all united
                        71. Petals and stamens inserted on the calyx tube or on a disk that lines the calyx tube
                          -56. <u>Rosaceae</u> p. 294
                        71. Petals and stamens inserted on the receptacle
                          -47. <u>Ranunculaceae</u> p. 236
                70. Ovules more than 1 to an ovary; fruit a several- to many-seeded follicle; ovaries 2-5, often united at base
                    72. Leaves fleshy and entire; 3-5 ovaries (or carpels) present
                      -54. <u>Crassulaceae</u> p. 283

    72. Leaves not especially fleshy, entire, toothed or lobed; 2 ovaries (carpels) present
    -55. <u>Saxifragaceae</u> p. **284**
69. Ovary only 1, carpels may be lobed but not separating until dehiscence of fruit and with only 1 style (this may branch above) to the whole ovary
    73. Ovules only 1 or 2 to each cell
        74. Flowers **unisexual**, borne in a **calyxlike** involucre, with petaloid glandular appendages often present; plants mostly with milky juice
            -66. <u>Euphorbiaceae</u> p. **358**
        74. Flowers perfect, not borne in a **calyxlike** involucre; plants without milky juice
            75. Sepals and petals 3 each
                -62. <u>Limnanthaceae</u> p. **355**
            75. Sepals and petals 4 or 5 each
                76. Sepals and petals 4 each; stamens 2, 4, or 6
                    -52. <u>Cruciferae</u> p. **256**
                76. Sepals and petals 5 each; stamens 5 or 10
                    77. Leaves alternate, simple and entire; stamens 5
                        -60. <u>Linaceae</u> p. **354**
                    77. Leaves opposite or basal, compound or palmately lobed to divided; stamens usually over 5
                        78. Styles separating at maturity from the central column; leaves palmately **lobed or divided,** or pinnate but the main divisions always toothed or lobed
                            -58. <u>Geraniaceae</u> p. **351**
                        78. Styles not separating from a central column; leaves pinnately (or palmately with 2 leaflets) compound, the leaflets entire
                            -63. <u>Zygophyllaceae</u> p. **356**
    73. Ovules 3 or more to a cell
        79. Leaves palmately compound with 3 entire approximately equal leaflets; ovary 5-celled and 10 stamens present, these stamens united somewhat at base
            -59. <u>Oxalidaceae</u> p. **353**
        79. Leaves simple or if compound then not as above; ovary 2- to 10-celled and stamens 2-10 (but seldom ovary 5-celled and 10 stamens present), the stamens not united below (in all but <u>Hypericaceae</u>)
            80. Stamens 2-3; plants aquatic or creeping on mud, the erect stems not over 10 cm. long
                -75. <u>Elatinaceae</u> p. **373**
            80. Stamens 4-12; plants not aquatic and usually not creeping in the mud, stems usually over 10 cm. long
                81. Plants white (in age blackish), no green chlorophyll present
                    -(Genus) <u>Monotropa</u> p. **416**
                81. Plants with green chlorophyll
                    82. Anthers opening by pores at one end, this end (appearing apical) more or less 2-horned
                        -90. <u>Ericaceae</u> p. **415**
                    82. Anthers opening longitudinally, not by pores at end, not at all 2-horned
                        83. Styles 2-3, distinct
                            84. Ovary 3- to 5-celled; 3 styles present; leaves punctate with minute translucent dots
                                -78. <u>Hypericaceae</u> p. **374**
                            84. Ovary 2-celled; 2 styles present; leaves not punctate
                                -55. <u>Saxifragaceae</u> p. **284**
                        83. Style 1 (at least at base, may be lobed above)
                            85. Stamens inserted on the receptacle, usually 6 (with 2 shorter); sepals and petals 4; leaves various but often toothed or lobed
                                -52. <u>Cruciferae</u> p. **256**
                            85. Stamens inserted on the calyx tube or on a fleshy disk, 4-12 in number but rarely 6 and then all nearly equal in length; sepals and petals 4-6; leaves simple and entire
                                86. Petals and stamens inserted on the throat of a campanulate or tubular calyx tube; flowers purple; leaves opposite (at least at base of plant)
                                    -84. <u>Lythraceae</u> p. **387**
                                86. Petals and stamens inserted on or near a fleshy disk, not on the throat of a campanulate or tubular calyx tube; flowers greenish-yellow or whitish or merely purplish-tinged; leaves alternate
                                    -(Genus) <u>Thamnosma</u> p. **357**

EE. Petals united well above the base

(Series) <u>GAMOPETALAE</u>

1. Ovary partly or completely inferior
    2. Stamens many (over 12); petals over 6; stems fleshy, usually spiny with leaves absent or small and early deciduous
        -81. <u>Cactaceae</u> p. **382**
    2. Stamens not over 12; petals 4-6; stems not conspicuously fleshy, not spiny, leaves definite
        3. Flowers (at least staminate when flowers are **unisexual**) sessile in dense heads surrounded by an involucre; calyx often highly modified

4. Stamens united by their anthers into a ring or tube around the style
-117. Compositae p. 528
4. Stamens with anthers distinct
5. Outer involucre bracts reflexed, united and forming a sheath around the stem below the head; stamens opposite the corolla lobes
-91. Plumbaginaceae p. 420
5. Outer involucre bracts not both reflexed and united; stamens alternate to the corolla lobes
6. Flowers perfect; stems prickly; heads over 3 cm. long
-116. Dipsaceae p. 527
6. Flowers **unisexual**; stems usually not prickly; heads not over 1 cm. long
-117. Compositae p. 528
3. Flowers not in heads (may be in headlike spikes but not involucrate at base), not closely subtended by an involucre; calyx usually normal
7. Plants bearing tendrils; herbaceous vines with scrambling or climbing stems
-113. Cucurbitaceae p. 523
7. Plants without tendrils; not vinelike (stems upright or occasionally scrambling)
8. Stamens the same number as the corolla lobes and opposite to them; ovaries 1-celled
9. Flowers in heads, these involucrate; **seeds 1 to an ovary**
-91. Plumbaginaceae p. 420
9. Flowers in racemes or panicles, not involucrate at base; seeds many to an ovary
-(Genus) Samolus p. 421
8. Stamens fewer or more than the corolla lobes or if the same number then alternate; ovaries usually over 1-celled
10. Stamens borne on the corolla tube; leaves opposite or whorled
11. Stamens twice as many as the corolla lobes; leaves ternately compound
-112. Adoxaceae p. 522
11. Stamens as many as the corolla lobes or less; leaves simple, entire to pinnately divided or pinnately compound, never ternate
12. Stamens 3, always fewer than the corolla lobes; calyx usually reduced to bristles, these elongate and plumose in fruit; ovary with 1 ovule (although 3-celled) and fruit 1-seeded and achenelike
-115. Valerianaceae p. 526
12. Stamens usually 4 or more but in any case just as many as the corolla lobes; calyx definite or reduced but not bristlelike or plumose in fruit; ovary usually with more than 1 ovule; fruit of 2 nutlets, berrylike, capsulelike or drupelike, commonly more than 1-seeded
13. Fruit 2-lobed, dry, separating into 2 one-seeded nutlets, usually bristle-bearing; plants herbaceous or woody only at base (and then flowers not in pairs); leaves usually appearing whorled
-109. Rubiaceae p. 515
13. Fruit not 2-lobed, either fleshy with 1 to several seeds or dry and 1-seeded (never dry and 2-seeded); plants woody shrubs (or woody at base with flowers in pairs on a long peduncle); leaves opposite
-111. Caprifoliaceae p. 518
10. Stamens free from the corolla or very nearly so; leaves alternate
14. Plants shrubby; anthers opening by terminal pores or chinks; stamens 8-10, twice as many as the corolla lobes; fruit berrylike
-(Genus) Vaccinium p. 418
14. Plants herbaceous; anthers splitting longitudinally; stamens 5, just as many as the corolla lobes; fruit a dry capsule
15. Corolla regular; anthers and filaments distinct
-114. Campanulaceae p. 524
15. Corolla irregular; anthers and filaments united (**Lobelioideae**; **Lobeliaceae**)
-114. **Campanulaceae** p. 524
1. Ovary completely superior
16. Stamens more than the lobes of the corolla, 6 to many
17. Stamens many, monadelphous
-74. Malvaceae p. 368
17. Stamens 6 to 10, distinct or diadelphous (rarely monadelphous)
18. Petals united at apex only, free below; ovary 1-celled with 2 parietal placentae
-50. Fumariaceae p. 253
18. Petals united below apex (connate to united at very base); ovary more than 1-celled or 1-celled with 1 parietal placenta
19. Ovary 4- to 10-celled; corolla regular; stamens 10, the anthers commonly opening by terminal pores
-90. Ericaceae p. 415
19. Ovary 1- to 2-celled; corolla usually very irregular; stamens 4-10, anthers opening by longitudinal slits
20. Ovary 1-celled (rarely appearing 2-celled by the intrusion of a suture); petals commonly 5; leaves almost always compound; stamens rarely 6 or 8
-57. Leguminosae p. 314
20. Ovary 2-celled; petals commonly 3; leaves simple; stamens 6 or 8
-65. Polygalaceae p. 357
16. Stamens as many as the corolla lobes or fewer
21. Ovaries 2, completely separate below but with a common stigma at apex; plants with milky sap
22. Stamens with anthers connivent; a crown present between the corolla and the stamens, this usually of hoodlike structures
-97. Asclepiadaceae p. 432

22. Stamens with anthers distinct; no crown present
-96. Apocynaceae p. 430
21. Ovary only 1 (may be lobed); plants seldom with milky sap
  23. Flowers with a spurlike nectary present
    24. Aquatic or semiaquatic plants with leaves dissected into capillary divisions and usually bearing small bladders
-106. Lentibulariaceae p. 513
    24. Terrestrial plants; leaves not dissected and no bladders present
      25. Ovary 5-celled; one of sepals spurred
-61. Balsaminaceae p. 355
      25. Ovary 2-celled; corolla spurred
-105. Scrophulariaceae p. 485
  23. Flowers without a spurlike nectary
    26. Stamens 3-4, more or less united to each other but free from the corolla tube; sepals larger and much more conspicuous than the petals; fruit large, 1-seeded, spiny
-(Genus)   Krameria p. 316
    26. Stamens 2-5, more or less attached to the corolla tube (if 3 or 4 then not at all united to each other); sepals usually smaller and less conspicuous than the petals; fruit various, not as above
      27. Plants lacking green chlorophyll; leaves reduced and scalelike
        28. Stems twining, slender, attached to the stems of some host plant, soon losing contact with the ground; flowers regular; ovary with 2 styles and 2 cells
-(Genus)   Cuscuta p. 435
        28. Stems not twining, short and rather thick, attached to the ground; flowers irregular; ovary with 1 style and 1 cell
-107. Orobanchaceae p. 514
      27. Plants with green chlorophyll; leaves normal (may be rather small)
        29. Stamens of the same number as the corolla lobes and opposite them; ovary 1-celled with basal or free central placentae; flowers regular
          30. Flowers in a headlike inflorescence, this involucrate, the outer bracts reflexed and united into a sheath below the head
-91. Plumbaginaceae p. 420
          30. Flowers in various inflorescences but never headlike and involucrate
-92. Primulaceae p. 420
        29. Stamens of the same number as the corolla lobes and alternate to them or fewer; ovary more than 1-celled or if 1-celled then the placenta rarely basal or free central; flowers regular or irregular
          31. Stamens 5, all anther bearing; corollas regular or nearly so
            32. Ovary developing into 4 (or by abortion less) 1-seeded nutlets, ovary usually 4-lobed
-101. Boraginaceae p. 453
            32. Ovary not developing into nutlets, fruit a capsule or berry usually several- to many-seeded, ovary not 4-lobed
              33. Plants with leaves basal, long petioled, palmately compound with 3 large (over 5 cm. long), entire leaflets
-94. Menyanthaceae p. 425
              33. Leaves simple to compound but not as above (never palmately compound with 3 large, entire leaflets)
                34. Ovary 1-celled
                  35. Leaves opposite or whorled, entire; styles 1 or none; plants mostly glabrous; inflorescence not scorpioid
-95. Gentianaceae p. 425
                  35. Leaves usually alternate (if opposite then not entire); styles 2, or single and 2-cleft above; plants mostly hairy; inflorescence commonly scorpioid
-100. Hydrophyllaceae p. 449
                34. Ovary with 2 or more cells
                  36. Stigmas 3-lobed or style 3-cleft; ovary 3-celled
-99. Polemoniaceae p. 439
                  36. Stigmas 1- or 2-lobed or style 2-cleft; ovary usually 2-celled
                    37. Flowers yellow to whitish on a very dense, thick terminal spike or spikelike raceme over 20 cm. long; flowers slightly irregular
-(Genus)   Verbascum p. 486
                    37. Flowers variously colored, but never whitish or yellow and in spikes or racemes over 20 cm. long; flowers usually strictly regular
                      38. Stems trailing or twining
-98. Convolvulaceae p. 435
                      38. Stems not trailing or twining
                        39. Styles 2, distinct, each one again 2-cleft; ovules 2 in each cell: flowers solitary and axillary
-(Genus)   Evolvulus p. 437
                        39. Styles 1 or if 2 rarely separate to the base, never again 2-cleft; ovules usually over 2 to each cell; inforescence various (sometimes solitary and axillary)

11

40. Style 1, stigma entire or 2-lobed; fruit a capsule or berry
-104. <u>Solanaceae</u> p. **480**
40. Styles 2 or definitely 2-branched before reaching the stigmas; **fruit a capsule**
-100. <u>Hydrophyllaceae</u> p. **449**
31. Stamens, with anthers 2 or 4 (rudimentary stamen may be present in addition); corollas regular or commonly irregular
    41. Ovary with 1 (or possibly 2) ovules to a cell, appearing 4-celled and often 4-lobed; fruit separating at maturity into 4 (or by abortion less) 1-seeded nutlets
        42. Ovary 4-lobed or 4-parted, the style arising in the center between the lobes; corolla usually definitely irregular; inflorescence various (can be heads or spikes)
-103. <u>Labiatae</u> p. **471**
        42. Ovary entire or longitudinally grooved, the style apical; corolla regular to slightly irregular; inflorescence a spike or head
-102. <u>Verbenaceae</u> p. **469**
    41. Ovary with more than 1 ovule to a cell, usually not 4-celled, not 4-lobed (can be 2-lobed); fruit not separating into 1-seeded nutlets
        43. Corolla scarious; flowers small, in dense terminal spikes and with leaves basal; fruit a circumscissile capsule
-110. <u>Plantaginaceae</u> p. **517**
        43. Corolla not scarious; flowers if small seldom in dense terminal spikes combined with leaves all basal; fruit not circumscissile (except in genus <u>Menodora</u>)
            44. Ovary 2-lobed; capsules deeply 2-parted, circumscissile
-(Genus) <u>Menodora</u> p. **424**
            44. Ovary not or only slightly 2-lobed; capsules not 2-parted, not circumscissile
                45. Stigmas 3-lobed; ovary 3-celled
-99. <u>Polemoniaceae</u> p. **439**
                45. Stigmas entire or 2-lobed; ovary 1- or 2-celled
                    46. Ovary 2-celled
-105. <u>Scrophulariaceae</u> p. **485**
                    46. Ovary 1-celled
                        47. Corollas definitely irregular; fruit with a long hooked beak; leaves both suborbicular and over 5 cm. wide
-108. <u>Martyniaceae</u> p. **515**
                        47. Corollas regular or nearly so; fruit without a long hooked beak; leaves not suborbicular and over 5 cm. wide
                            48. Stems creeping and rooting; leaves all basal; corollas small, not over 3 mm. long; ovary 2-celled at base
-(Genus) <u>Limosella</u> p. **511**
                            48. Stems not creeping and rooting; some leaves on the stem, these opposite or whorled; corollas usually large and showy, over 3 mm. long; ovary not at all 2-celled at base
-95. <u>Gentianaceae</u> p. **425**

Division I. <u>Pteridophyta</u> FERNS; FERN ALLIES

Plants with life cycle containing 2 distinct phases, the conspicuous one (sporophyte generation) differentiated **into roots**, stems and leaves and containing vascular tissue. This phase produces spores, these all alike or of 2 kinds, small ones called microspores and large ones called megaspores, the leaf producing them called a sporophyll. These spores on germination give rise to inconspicuous **thalloid plants** (gametophyte generation) called prothallia which develop spermatozoids in antheridia and eggs in archegonia. The eggs after fertilization develop into the asexual, leafy conspicuous plants. No true seeds, fruits or flowers produced.

Family 1. <u>Ophioglossaceae</u> ADDERSTONGUE FAMILY

Plants with short, fleshy erect rootstocks; fronds consisting of a single sterile leaf blade and a compound (in ours) sporophyll both on a common stalk; bud of next year's stalk borne in the hollow base of the old stalk, vernation of the young leaf erect or bent over; sterile blade simple and pinnatifid to variously compound; sporangia large, globose; spores all alike, numerous, yellow in color.

1. <u>Botrychium</u> Sev. GRAPEFERN; MOONWORT

<u>Characters of the Family</u>

1. Buds (in older stalk) hairy; sterile blades usually over 6 cm. long
    2. Sterile blades thin, nearly sessile on the common stalk, usually over 10 cm. long; buds partly exposed
-1. <u>B. virginianum</u>
    2. Sterile blades thick and rather fleshy, with stalks 1-4 cm. long, the blades 6-9 cm. long; buds completely concealed
-2. <u>B. multifidum</u>
1. Buds glabrous; sterile blades not over 6 cm. long
    3. Sterile blades inserted above the middle of the common stalk, commonly near the apex, the blade usually bipinnate or bipinnatifid, the ultimate segments not flabelliform or fan-shaped; sterile blades in the bud usually bent downward near their bases
        4. Sterile blades deltoid in shape; fertile and sterile blades both reflexed in the buds
-3. <u>B. lanceolatum</u>

4. Sterile blades ovate-oblong; fertile blades not reflexed in the buds -4. B. matricariaefolium
 3. Sterile leaf blades inserted at or below the middle of the common stalk, the blade usually pinnate or simple, ultimate segments often flabelliform or fanshaped; sterile blades in the buds erect or bent only near their tips
  5. Sterile blades stalked, simple or of irregularly shaped divisions, sometimes appearing ternate at base, the blade stalks usually inserted near the base of the common stalks -5. B. simplex
  5. Sterile blades sessile, once-pinnate, the segments about the same shape, the blades inserted at or somewhat below the middle of the common stalk
    6. Sterile blades inserted near the middle of the common stalks, broadly oblong, about twice as long as wide -6. B. lunaria
    6. Sterile blades inserted below the middle of the common stalks, over 3 times longer than wide -6A. B. lunaria minganense

1. Botrychium virginianum (L.) Sw., Journ. Bot. Schrad. 1800(2): 111. 1801.
    Plants erect, 5-70 cm. tall, glabrous or sparsely pubescent; buds pilose and partly exposed by the base opening on one side, at least late in the season, the fertile and sterile segments both completely reflexed in vernation; sterile blade sessile, thick-membranous, deltoid, large for the genus, 4-21 cm. long, 5-36 cm. wide, divided, the pinnae once or twice divided but not completely to the rachis, the ultimate segments oblong-lanceolate, variously toothed or incised, often overlapping somewhat; fruiting part pinnately decompound. Our plant is ssp. europaeum (Angstrom) Clausen.---Typically found in forests or moist meadows. The subspecies from Labrador and Newfoundland, west to British Columbia, south to New England, Michigan, Wisconsin, Minnesota, Colorado and Oregon; also in Asia and Europe. Reported from this state from along the eastern slope from Boulder to Custer County. One specimen seen, from El Paso County at about 9500 feet.
2. Botrychium multifidum (S.G.Gmel.) Rupr., Beitr. Pfl. Russ. Reich. 11:40. 1859.
    B. coulteri Underw.---Plants stout and decidedly fleshy, 15-23 cm. tall; buds hairy, the fertile and sterile segments bent back or reflexed; common stalk 2-5.5 cm. long; sterile blade stalk 1-4 cm. long, inserted near the base of the common stalk, the blade 6-9 cm. long, 7-15 cm. wide, ternately decompound, ultimate segments small and crowded; fertile part 6-17 cm. long, in a dense spike; sporangia about 1.2 mm. in diameter. Our plant is ssp. coulteri (Underw.) Clausen.---Grassy meadows. The subspecies from Washington to California, Colorado, Wyoming, and western Montana. Northern Colorado from Larimer to Routt Counties at 6700-9900 feet.
3. Botrychium lanceolatum (S.G.Gmel.) Angstrom, Bot. Not. 1854:68. 1854.
    Plants rather stout and fleshy, 5-25 cm. tall; buds glabrous, the sterile and fertile parts both reflexed in vernation; sterile blade sessile or nearly so, 3-10 cm. long, deltoid, once or twice pinnately or ternately divided, segments lanceolate, the lower pair longest, entire or toothed, the whole blade inserted near the summit of the common stalk; fertile part paniculate, 1-5 cm. long, mostly extending above the sterile blade. Our plant has been called ssp. typicum Clausen.---Dry or damp hillsides in open places or in forests. The subspecies from Newfoundland to Alaska, south to Arizona, Colorado and Maine; also in Greenland, Iceland, northern Europe, and Asia. This plant has been reported from Larimer County south to Chaffee and El Paso Counties at 8000-12,000 feet. One collection seen, from Summit County at 9700 feet.
4. Botrychium matricariaefolium A. Br. ex Koch, Syn. Fil. Germ. ed. 2. 972. 1845.
    Plants stout and rather fleshy, 5-20 cm. tall; buds glabrous the fertile segments erect or slightly inclined, the sterile bent down and clasping the fertile; common stalk 3-11 cm. long; sterile blades with a stalk 1-15 mm. long, inserted well above the middle of the common stalk or near the apex, the blade 1-5 cm. long, 6-40 mm. wide, ovate-oblong, pinnately divided into 6 or more lateral divisions, these entire or coarsely lobed, the basal divisions sometimes again divided; fertile part 3-60 mm. long, paniculate and much divided; sporangia 0.6-1 mm. in diameter. Our plant is ssp. hesperium Max. & Clausen.---Dry slopes, grassy open places or wooded areas. The subspecies from Colorado and Utah. Reported for central and northcentral Colorado at 8000-11,000 feet. From El Paso, Boulder and Larimer Counties at 8500-9100 feet according to specimens seen.
5. Botrychium simplex Hitchc., Am. Journ. Sci. 6:103. 1823.
    Plants rather fleshy, 3-16 cm. tall; buds glabrous with both sterile and fertile segments erect or nearly so; sterile blades with stalks 2-20 mm. long, inserted almost basal to near the middle of the common stalks, the blades 5-40 mm. long, simple, lobed or pinnately divided, the divisions oblong to reniform; fertile spikes simple or compound, 3-50 mm. long. Our plant has been called ssp. typicum Clausen.---Pastures, meadows and gravelly slopes of open places. The subspecies from Newfoundland to British Columbia, south to Pennsylvania, Wisconsin, New Mexico, and California. Has been reported in El Paso, Boulder, and Gilpin Counties. One specimen seen from northcentral Colorado at 9500 feet.
6. Botrychium lunaria (L.) Sw., Journ. Bot. Schrad. 1800(2):110. 1801.
    Plants rather stout, 4-25 cm. tall; buds glabrous, both fertile and sterile segments straight; sterile blade 1-6 cm. long, broadly oblong or deltoid-ovate, once-pinnately divided, the segments flabelliform or fan-shaped to lunate or reniform, often overlapping, entire or sometimes incised, the blade inserted about at the middle of the common stalk; fruiting part racemose or paniculate, 4-80 mm. long. This plant has been called var. typicum Clausen.---Open places such as meadows, slopes, and banks, also in woods. The variety from Newfoundland to Alaska, south to Maine, Michigan, Minnesota, Colorado, Arizona, and California. Reported from the mountainous part of Colorado up to 12,000 feet. Specimens seen from Teller and Clear Creek Counties at 9800-10,800 feet.
6A. Botrychium lunaria minganense (Victorin) Dole, (var.) in Fl. of Vermont ed. 3 p.l, 1937.
    B. lunaria (L.)Sw. in part.---This plant differs from var. typicum in - sterile blade longer, over 3 times as long as wide, often inserted a little below the middle of the common stalk; spores larger, 30-40 microns in diameter (instead of 25-35 microns).---Meadows, eastern

Canada to Alaska, south to Michigan, Wisconsin, Colorado, and California. Reported in scattered areas in the mountainous part of the state up to 12,000 feet. One specimen seen, from north-central Colorado at 9500 feet.

Family 2. Polypodiaceae FERN FAMILY

Plants with underground stems (rootstocks) these horizontal and creeping or shorter, oblique to vertical; sterile blades (fronds) leaflike, fertile leaflike or highly modified, borne separately; sporangia stalked, commonly in clusters called sori, these sori naked or with a special covering called an indusium.

1. Sterile and fertile fronds markedly dissimilar
    2. Ultimate segments of fertile fronds balllike and round in shape; sterile fronds often over 30 cm. long
    -1. Onoclea
    2. Ultimate segments of fertile fronds linear to lanceolate-oblong, not balllike; sterile fronds not over 30 cm. long
    -2. Cryptogramma
1. Sterile and fertile fronds alike or essentially so
    3. Leaves simple (but pinnatifid); indusia absent
    -3. Polypodium
    3. Leaves once or twice pinnate, not simple and pinnatifid; indusia present or absent
        4. Sori each provided with a separate indusium (at least when young) this not connected with the margins of the fronds
            5. Sori oblong or linear (at least twice as long as wide) or indusia horse-shoe or U-shaped (may appear round if middle is packed in with sporangia)
                6. Ferns once-pinnate
                -4. Asplenium
                6. Ferns 2- or 3 times-pinnate
                    7. Fronds less than 15 cm. long; stipes purplish-brown   -4. Asplenium adiantum-nigrum
                    7. Fronds 20 cm. long or more; stipes straw-colored or reddish-brown near base   -5. Athyrium
            5. Sori round or nearly so (may have a narrow sinus or notch at one edge)
                8. Indusia attached by central stalks arising from the center of the sori, indusia usually conspicuous
                    9. Indusia with a sinus or notch at one side; fronds (in ours) with pinnae deeply pinnatifid or fronds thrice-pinnate
                    -6. Dryopteris
                    9. Indusia without a notch or sinus; fronds (in ours) once-pinnate
                    -7. Polystichum
                8. Indusia attached below at one side of the sori or underneath it, often not very conspicuous
                    10. Indusia attached below the sori, the divisions spreading out on all sides   -8. Woodsia
                    10. Indusia attached to one side of the sori, hood-shaped over it when young   -9. Cystopteris
        4. Sori without special indusia except as provided (in some) by the inrolled margins of the segments of the fronds (some genera that lose their indusia readily are repeated here)
            11. Sori covered, at least when young, by the inrolled margins of the fronds which function as indusia
                12. Sporangia borne on the under side of sharply reflexed lobes, these reflexed areas not continuous, ultimate segments of the fronds over 5 mm. wide   -10. Adiantum
                12. Rolled margins of the fronds continuous, or if interrupted then the ultimate segments of the fronds not over 5 mm. wide
                    13. Plants large, usually over 30 cm. tall; sporangia borne on veinlike receptacles connecting the ends of the veins; fronds solitary from long-creeping rootstalks
                    -11. Pteridium
                    13. Plants smaller; sporangia not borne on special transverse receptacles; fronds usually tufted from short or vertical rootstalks
                        14. Under surface of fronds densely and conspicuously white or yellow powdery
                        -12. Notholaena
                        14. Under surface not densely powdery (may be yellowish from hairs)
                            15. Pinnules either tomentose or scaly beneath; ends of veins thickened; inrolled margins of fronds not continuous, the sori not in continuous bands
                            -13. Cheilanthes
                            15. Pinnules glabrous below (may be scaly on rachis and rachilla); ends of veins not thickened; inrolled margins and sori continuous   -14. Pellaea
            11. Sori not covered by the inrolled margins of the fronds
                16. Sori elongate, along the margins of the segments, these latter densely white or yellow powdery below
                -12. Notholaena
                16. Sori round or nearly so, not marginal, the segments not white or yellow powdery below
                    17. Blades triternate, broadly triangular
                        18. Lower pinnules very deeply divided, the segments incised toothed; indusia present when young   -9. Cystopteris montana
                        18. Lower pinnules not very deeply divided or if so then the segments entire or crenately toothed; indusia absent from the first   -15. Phegopteris
                    17. Blades not triternate, more elongated in shape than triangular
                        19. Fronds large, some over 25 cm. long; indusia absent from the first
                        -5. Athyrium alpestre
                        19. Fronds smaller; indusia present in young sori
                            20. Rachises, rachillas or lower surfaces of segments pubescent to glandular-puberulent   -8. Woodsia
                            20. Fronds glabrous   -9. Cystopteris

1. Onoclea L. SENSITIVE FERN

Rootstocks slender and creeping; fronds dimorphous, the sterile foliaceous; fertile fronds becoming brown and indurated, bipinnate but with rigid, contracted, balllike ultimate divisions; sori borne in these balllike structures which are modified inrolled pinnules, sori partly covered by delicate hoodlike indusia which soon wither.

1. Onoclea sensibilis L., Sp. Pl. 1082. 1753.
Sterile fronds 30-90 cm. long, deltoid-ovate, deeply pinnatifid or pinnate near base, the pinnae (or primary segments) lanceolate-oblong, entire, undulate to sinuate-pinnatifid; fertile fronds 30-70 cm. tall.---Damp places, wet meadows and edges of thickets. Newfoundland to Saskatchewan, south to Kansas and Florida. Collected in Douglas County at 7000 feet where it apparently was native.

2. Cryptogramma R. Br. ROCKBRAKE

Rather small ferns of rocky places; fronds dimorphous, the sterile foliaceous, the fertile brownish with fewer, narrower segments; sori marginal or nearly so, continuous or confluent; indusia continuous, formed of the revolute or reflexed margins of the segments.

1. Fronds closely tufted, usually many, the blades herbaceous or thicker, not at all translucent -1. C. acrostichoides
1. Fronds scattered, from creeping rhizomes, blades translucent when fresh -2. C. stelleri

1. Cryptogramma acrostichoides R. Br. in Richards, Bot. App. Frankl. Journ. 754. 1823.
C. crispa var. acrostichoides (R. Br.) C. B. Clark---Rootstocks short, often in massive tufts; fronds tufted, numerous, fertile ones 5-30 cm. long, surpassing the sterile; stipes straw-colored, deciduously scaly below; sterile blades ovate or ovate-lanceolate, 3-12 cm. long, 2-3 times pinnate, pinnae few, rather close, pinnules crenate or incised, segments obtuse, glabrous, leaf tissue herbaceous or chartaceous; fertile blades simpler with fewer segments these linear or linear-oblong, 6-12 mm. long, margins revolute, at first nearly meeting from the side.---Among rocks especially on rock slides. Ontario to Alaska and south to California, New Mexico, Nebraska, and Michigan. Widely distributed in Colorado at altitudes of 7000-12,000 feet.

2. Cryptogramma stelleri (Gmel.) Prantl. in Engler's, Bot. Jahrb. 3:413. 1882.
Rootstocks slender and creeping; fronds scattered, few together, pale or bright green, the blades translucent, 5-25 cm. long, the fertile much longer; stipes brownish; sterile blades ovate to triangular-ovate, tripinnate or sometimes bipinnate, pinnae divided into obovate or cuneate segments, the margins crenate or undulate, often unsymmetrical; fertile blades with linear-oblong or lanceolate-oblong segments with margins strongly revolute, entire, sometimes slightly hyaline. (Descriptions not checked with Colorado material.) Labrador to Alaska, south to Washington, Colorado, Iowa, and New Jersey. Has been reported in Colorado from Summit, Gunnison, Ouray, and La Plata Counties. No specimen seen from the state.

3. Polypodium L. POLYPODY

Ferns rather small; fronds pinnatifid but simple; sori round, large, lacking indusia from the very first.

1. Polypodium hesperium Maxon, Biol. Soc. Wash. Proc. 13:200. 1900.
Rootstocks creeping, densely scaly, rather sweet-tasting; fronds rather close together, variable in size, mostly 8-25 cm. long; stipes rather long sometimes as long as the blade, stramineous, naked; blades linear-oblong to narrowly oblong, sometimes ovate-oblong, deeply pinnatifid or somewhat pinnate below, 5-20 cm. long, 1-4 cm. wide, subcoriaceous, the segments almost always alternate, spreading, oblong to oval, usually rounded and obtuse at ends, margins obscurely to evidently crenate; sori large, about 1/2 way in from margin to midrib of segment, subglobose. Fernald (Rhodora 24:125-142. 1922) maintained P. vulgare for our western plant and listed 2 varieties for Colorado.---Cliffs and rocky slopes. British Columbia to South Dakota, south to California and New Mexico. Scattered over the western two-thirds of Colorado, our specimens from northcentral, central, and southwestern parts at 6300-8700 feet.

4. Asplenium L. SPLEENWORT

Fronds once- to twice- or more-pinnate; sori oblong or linear, borne on the veins; indusia present, attached laterally, oblong to linear, when young arching over the sori, but thin-membranous and often pushed aside and concealed by sporangia at maturity or withering away.

1. Pinnae only 2 to 5, these narrowly linear; sori very elongate -1. A. septentrionale
1. Pinnae many, not narrowly linear; sori not very elongate
  2. Fronds bipinnate to tripinnate -2. A. adiantum-nigrum
  2. Fronds once-pinnate
    3. Stipes brown below, the rachises green -3. A. viride
    3. Stipes and rachises purplish-brown
      4. Pinnae auricled at base on at least the upper side -4. A. platyneuron
      4. Pinnae not at all auricled at base -5. A. trichomanes

1. Asplenium septentrionale (L.) Hoffm., Deutsch. Fl. 2:12. 1795.
Fronds densely tufted, 5-20 cm. long; stipes slender, brown-purple at base, naked for a long ways; blades irregularly forking with 2-5 narrowly linear, rather rigid segments, these tapering both ways, entire or with a few long narrow teeth near end; sori elongate, 2-3 to a segment, usually in pairs near each margin; indusia entire or sparingly short-ciliate.---On rocks. South Dakota and Wyoming south to western Oklahoma, Arizona, and Lower California; also in Europe and Asia. Our specimens from northcentral, central and southcentral Colorado at 5000-8200 feet but reported also from the southwest and southeast corners.

2. Asplenium adiantum-nigrum L., Sp. Pl. 1081. 1753.
A. andrewsii A. Nels.---Rootstocks with matted roots the fronds tufted or a few together; stipes chestnut-brown to blackish below, green above when mature; blades ovate-deltoid to elongate-deltoid, 3-10 cm. long, bipinnate or ternate, pinnae deltoid below to lanceolate above, the segments ovate-lanceolate, incised-serrate; sori short but almost connected in a continuous chain on the pinnae; indusia straight, entire or nearly so.---On rocks and cliffs. Arizona, Utah, and Colorado; also in Eurasia and Africa. Our specimens all from Boulder County at about 5500 feet but the fern should be expected anywhere in the mountainous or rocky areas of the state.

3. **Asplenium viride Huds., Fl. Angl. 385. 1762.**
Rootstocks short-creeping, fronds tufted, 2-20 cm. long; stipes as long or only slightly shorter than the blades, rather weak, reddish-brown near base, straw-colored to green above, the rachises green, sparingly fibrillose; blades linear-lanceolate to oblong-linear, once-pinnate, with 5-25 pairs of pinnae, these round-ovate to rhombic, obtuse or rounded at apex, broadly cuneate at base or obliquely truncate, stalked, crenate to crenately lobed except on the base; sori remote from the margins, 2-4 pairs to a segment, oblong-linear, becoming confluent in age; indusia entire or denticulate.---On rocks or crevices of shaded cliffs. Newfoundland to Alaska and south to Washington and Colorado; also in Europe and Asia. Our few specimens from southwestern Colorado at 10,000-12,000 feet but should be found anywhere in the mountainous portions of the state.

4. Asplenium platyneuron (L.) Oakes in Eaton, Ferns N.A. 1:24. 1878.
Rootstocks short, fronds densely tufted, the fertile **ones 10-50 cm. long;** stipes shining, purplish-brown; blades once-pinnate with 20-40 pair of pinnae, these lanceolate, sessile, crenate, serrate or incised, auricled on upper and sometimes on lower side, lower pinnae gradually smaller, oblong or triangular; sori 5-12 on each side of the midrib of the pinna, oblique and becoming crowded at maturity.---Among rocks and on rocky cliffs. Ontario to Florida and west to New Mexico and Colorado. Our few records from the southeastern corner of the state at about 4500 feet.

5. Asplenium trichomanes L., Sp. Pl. 1080. 1753.
Rootstocks short, the fronds tufted, 5-25 cm. long; stipes and rachises dark brown or purplish-brown, shining; blades simply-pinnate, the rachises very narrowly winged, with 12-35 pairs pinnae these opposite or alternate, the lower ones distant, roundish-oval, roundish-deltoid or oval-oblong the bases cuneate or obtusely truncate, subsessile, margins crenate or entire rarely incised, falling before the rachises; sori 3-5 pairs to a pinna, linear-oblong; indusia subentire.---On rocks and crevices of cliffs. Nova Scotia to Alaska and south to Georgia, Texas and Arizona. Our specimens from central and northcentral Colorado at 5000-8700 feet but should be more widely distributed.

5. Athyrium Roth

Rather large ferns; blades erect-spreading, long-stiped, twice-pinnate or more; sori on the veins, either narrowly oblong, recurved or lunate with indusia laterally attached (at least when young) or else sori round and lacking indusia from the first.

1. Fronds less than 15 cm. long; stipes purple-brown (another genus, keyed out here)   -Asplenium adiantum-nigrum
1. Fronds 20 cm. long or more; stipes straw-colored, reddish or brown near base but not purplish-brown
   2. Pinnules cuneate at base, appearing short-stalked; indusia absent from the first; sori round
        -1. A. alpestre
   2. Pinnules sessile, not cuneate at base; indusia present (at least when young); sori elongated or horseshoe-shaped (may appear round)
      3. Rootstocks horizontal, the young growth at the ends of the old frond bases; indusia often over 1 mm. long   -2. A. angustum
      3. Rootstocks erect or ascending, the young growth surrounding the old frond bases; indusia usually less than 1 mm. long   -3. A. filix-femina

1. Athyrium alpestre (Hoppe) Rylands ex Moore, Ferns Gr. Brit. and Irel. Nat. Print. pl. 7. 1857.
A. americanum (Butters) Maxon ---Rootstocks short, erect or decumbent, branching and forming large tufts; fronds tufted, 20-95 cm. long, blades oblong-lanceolate, 2- to 3-pinnate or pinnatifid on final division; stipes short, stramineous with darker base; pinnae elongate-deltoid, acuminate, their rachises usually very narrowly winged; pinnules oblong-ovate to elongate-deltoid or even narrower, incised or pinnatifid but ultimate margins serrate, base cuneate and appearing stalked; sori round, numerous; indusia lacking. Our plant is var. americanum Butters.---Moist ravines, meadows or thickets. The variety from Alaska, south to California, Nevada, and Colorado. Our few specimens mostly from northcentral Colorado at 10,000-11,000 feet.

2. Athyrium angustum (Willd.) Presl, Rel. Haenk. 1:39. 1825.
Resembles number 3 in general appearance. Rootstocks horizontal with young growth at the end and in advance of the old frond bases, which crowd upon and conceal the rootstock; indusia supposed to be over 1 mm. long (but less than 1 mm. in our one specimen). Our plant is var.

rubellum (Gilbert) Butters.---The variety from Quebec and Ontario to the Black Hills of South Dakota and south to Missouri, Illinois, Ohio, and Pennsylvania. Our one collection from Douglas County at about 7000 feet.
3. Athyrium filix-femina (L.) Roth ex Mertens, Rom. Arch. Bot. 2(1):106. 1799.
Asplenium filix-femina of Manuals; Athyrium cyclosorum Rupr.---Rootstocks short, erect or ascending, the scales light brown, thin (see variety); fronds 20-125 cm. long, erect-arching, the blades oblong-lanceolate, 2- to 3-pinnate; stipes straw-colored to brownish-red; pinnae short-stalked or sessile above, lanceolate, acuminate, many; pinnules lanceolate-oblong to broadly elliptical, serrate or incised; sori short-oblong, lunate or horseshoe-shaped; indusia subentire to ciliate; spores yellowish, less than 40 microns in diameter, smooth or somewhat papillose (see variety).---In woods, fields, and thickets. British Columbia and south to California and Colorado. Our specimens widely distributed over the mountainous two-thirds of the state at 5500-9500 feet. In western Colorado this species shades into a type with the scales of the rootstock dark brown and the spores averaging about 43 microns, blackish in color and reticulated. Further west this form has been held distinct and has been named A. filix-femina var. californicum Butters, Rhodora 19:201. 1917.

6. Dryopteris Adans. WOODFERN

Mostly woodland ferns; fronds large, 1- to 3-pinnate; stipes not jointed to the rootstocks; sori round, on the veins; indusia present (in ours), roundish-reniform, attached at its sinus.

1. Fronds bipinnate or bipinnatifid; sori large   -1. D. filix-mas
1. Fronds tripinnate; sori small, dotlike   -2. D. dilatata

1. Dryopteris filix-mas (L.) Schott, Gen. Fil. Pl. 67, 1834.
Rootstocks stout, erect or decumbent, chaffy; fronds in a close crown, ascending, 25-100 cm. long; stipes stout, brown-scaly, the scales especially large near base; blades oblong-lanceolate, somewhat narrowed below, nearly twice-pinnate; pinnae narrowly deltoid-lanceolate, somewhat stalked, more or less acuminate, tapering from a broad base, the segments somewhat oblique, oblong, obtuse, somewhat decurrent, usually serrate with curved teeth (approaching crenate); sori rather large, round, usually confined near base of segments, nearer the midnerve than the margins of the pinnules; indusia orbicular-reniform, glabrous or glandular.---Rocky damp woods. Newfoundland to British Columbia and south to California, New Mexico, Texas, and Vermont; also in Greenland, Iceland, and north Europe. Reported from the mountainous two-thirds of the state. Our few specimens from 5300-8500 feet.
2. Dryopteris dilatata (Hoffm.) A. Gray, Man. ed. 1. 631. 1848.
Rootstocks stout, creeping or ascending, chaffy; fronds on a crown, 30-100 cm. long or more; stipes stout with brownish often darker centered scales; blades tripinnate or nearly so, triangular, ovate or broadly oblong, acuminate; basal pinnae broadly and unequally ovate or triangular, the upper lanceolate or oblong, segments toothed or pinnatifid, the teeth mucronate; sori mostly subterminal on the veinlets; indusia glabrous or somewhat glandular.---Rocky woods. Newfoundland to Alaska and south to California, Colorado, and North Carolina; also in Greenland and Eurasia. Our one specimen from Larimer County at 10,000 feet. (The eastern records in the U. S. are doubtful.)

7. Polystichum Roth HOLLYFERN

Woody rootstocks present; fronds rather rigid and large, the blades simply pinnate (in ours) with sharply toothed margins; sori round, large; indusia large, round, centrally peltate.

1. Polystichum lonchitis (L.) Roth in Rom., Arch. Bot. 2(1):106. 1799.
Rootstocks erect or decumbent; fronds in a crown, rather rigidly ascending, simply pinnate, 10-40 cm. long, linear to narrowly linear-oblanceolate, tapering at the base, rachises stout, very scaly; pinnae oblong-lanceolate to deltoid, auriculate on upper margin, unevenly serrate-dentate, the teeth conspicuously spreading-spinulose; sori large, round, contiguous, usually in two rows; indusia large, entire or finally toothed, peltate.---Moist cliffs and shaded rock slides. Newfoundland to Alaska south to California, Arizona, New Mexico, and the Great Lakes Region. Scattered over the mountainous area of Colorado at 7000-11,000 feet, no specimens as yet from the southcentral part.

8. Woodsia R. Br. WOODSIA

Fronds tufted, small, usually on rocks, 1- to 2-pinnate, hairy or sub-glabrous; sori round, borne on the veins, separate at least when young; indusia inferior, attached below the sporangia, spreading like a cup on all sides but with narrowly scalelike or filiform divisions, these often partly concealed by the mature sporangia.

1. Lower leaf surface (and often rachilla) bearing flexuous, hyaline, septate hairs   -1. W. scopulina
1. No such hairs present (may be short-glandular--do not mistake the moniliform segments of the indusia for the hairs mentioned above)
    2. Segments of indusia shorter than or not much longer than the sporangia; frond segments without white-crustose, ciliate margins   -2. W. oregana
    2. Segments of indusia long, greatly exceeding the sporangia, segments of fronds with white-crustose margins, the teeth ciliate   -3. W. mexicana

1. **Woodsia scopulina** D. C. Eaton, Canad. Nat. II.2:90. 1865.
    Rootstocks short, forming large tufts; fronds tufted, numerous, 6-30 cm. long; stipes stramineous or brown, the base chaffy; blades oblong-lanceolate to lanceolate, once- to twice-pinnate; pinnae oblong or deltoid-lanceolate or ovate, with 5-9 pairs of segments, these crenate-serrate, the surfaces especially the lower with glandular-puberulence, bearing few to many flat, septate, rather long hairs these often on the rachises; sori round; indusia deeply cleft into narrow segments, the tips hairlike, these may be jointed.---On rocks and in crevices. Quebec to Alaska south to California, Arizona, Oklahoma, Tennessee, and North Carolina. To be expected anywhere in the western two-thirds of the state; our specimens from northcentral, central, and southcentral Colorado at 5000-10,000 feet.
2. **Woodsia oregana** D. C. Eaton, Canad. Nat. II. 2:90. 1865.
    Rootstocks short; fronds tufted, 5-25 cm. long, the fertile longer, lanceolate or oblong-lanceolate in outline; stipes yellowish or straw-colored but darker at base, smooth; pinnae ovate-oblong or triangular, obtuse, glabrous; pinnules crenulate, often lobed or cleft near base, the margins often somewhat recurved; sori round; indusia split into slender hairlike segments, these often enlarged near base, little if any longer than the sporangia. (The form with the rachis and blade glandular-puberulent may be forma glandulosa Taylor.)---On rocks. Quebec to British Columbia and south to California, Arizona, New Mexico, and Oklahoma. Our specimens mostly from along the eastern mountains at 5500-9500 feet but a few in western Colorado. Should be expected anywhere in the western two-thirds of the state.
3. **Woodsia mexicana** Fee, Mem. Fam. Foug. 66. 1854.
    This species is supposed to differ from W. oregana by having longer, narrower segments of the indusia; segments of the pinnae have white-crustose margins, the teeth ciliate, the surface more or less short-glandular. Recently, Taylor (Am. Fern Journ. 37:86. 1947) called the above W. pusilla var. mexicana (Fee) Taylor.---On rocks. Texas, Mexico, and Arizona to Colorado and Utah. This species should be expected in the southern part of the state but many of our northern and central Colorado specimens show all types of intergradation with W. oregana.

9. Cystopteris Bernh. BLADDERFERN

Small and rather delicate ferns the fronds 1- to 4-pinnate; sori round, separate; indusia membranous, attached laterally and arching over the sori in a hoodlike effect when young but finally withering or pushed back and concealed by the sporangia.

1. Frond blades deltoid-ovate, basal pinnae much longer than the others   -1. C. montana
1. Frond blades lanceolate-ovate or oblong-lanceolate, basal pinnae smaller than or no longer than the others
                                                                              -2. C. fragilis

1. **Cystopteris montana** (Lam.) Bernh., Schrad. Nev. Journ. Bot. 1(2):26. 1806.
    Filix montana (Lam.) Underw.--Rootstocks slender and creeping; fronds scattered, 10-40 cm. long; stipes slender, as long or longer than the blades; blades ternate or nearly so, deltoid-ovate, 3-4 times pinnate; pinnae deltoid-lanceolate the basal pair much longer than others; pinnules deeply divided, the segments toothed or incised, glabrous; sori round, small, numerous; indusia hooded, acute, soon withering.---Rocky soil. Circumboreal and south to British Columbia, Montana, Colorado and Labrador. Reported from Central Colorado. Our few specimens from Summit and Ouray Counties at 10,500-11,000 feet.
2. **Cystopteris fragilis** (L.) Bernh., Schrad. Nev. Journ. Bot. 1(2):26. 1806.
    Filix fragilis (L.) Underw. of Manuals ---Rootstocks creeping but rather short; fronds few or several, erect-spreading, glabrous, variable in length, 5-30 cm. long; stipes slender, stramineous or brownish below with very deciduous scales near base; blades variable in shape, mostly lanceolate-oblong to ovate-lanceolate, nearly or quite twice-pinnate; pinnae deltoid-lanceolate or deltoid-ovate; pinnules incised or pinnatifid, decurrent, the segments toothed; sori small, roundish, not marginal; indusia delicate, convex, arching over the young sori, soon deciduous, the apex often toothed or laciniate.---Rocky places, usually in rock crevices. Labrador and Newfoundland to Alaska south to New England, Missouri, Texas, Arizona, and California; also in Greenland, Iceland, and Eurasia. Widely distributed over the western two-thirds of Colorado at 5000-11,500 feet. Rather common.
    A related species, C. bulbifera (L.) Bernh. with leaves 30-100 cm. tall, often bearing bulblets, has been reported for New Mexico, Arizona, and Utah. It is to be expected in southern or southwestern Colorado.

10. Adiantum (Tourn.) L. MAIDENHAIR; LADY FERN

Delicate ferns; fronds compound; stipes slender, dark colored and lustrous; sori appearing marginal on the underside of sharply reflexed indusiform margins of the segments, these margins interrupted, however, by incised lobes.

1. **Adiantum capillus-veneris** L., Sp. Pl. 1096. 1753.
    A. modestum Underw.---Rootstocks creeping, with light brown scales; fronds 8-60 cm. long, laxly ascending or pendulous; stipes slender, purplish-black; blades 6-40 cm. long, 2-3 pinnate at base, more simply pinnate above; pinnae spreading; pinnules stalked, obovate to rhombic or obliquely oblong, 5-30 mm. long, narrowly to broadly cuneate at base, outer edge deeply lobed or incised and denticulate; sori solitary on the lobes, mostly oblong-lunate.---Moist or wet rocky places. British Columbia to South Dakota and south to Florida, Texas, Arizona, and California; also in subtropics of both hemispheres. Our few specimens from the southwestern part of Colorado at 5400-7800 feet.

11. Pteridium Scop. BRACKEN

Rootstocks creeping; fronds large and coarse, borne singly; stipes stout; blades decompound; sori linear, borne on a marginal receptacle connecting the ends of veins; indusia double, the outer one formed of the reflexed margins of the segments, the inner one obscure, delicate, on the inner side of the sporangia.

1. Segments of fronds pubescent beneath all over; sterile and fertile indusia ciliate (often pubescent also); pinnules nearly at right angles to the rachis of the pinnae; blades of fronds ovate-triangular, fairly evenly pinnate -1A. P. aquilinum pubescens
1. Segments of fronds glabrous beneath or pubescent only on the midnerves; sterile and fertile indusia glabrous; pinnules are at an oblique angle to the rachis of the pinnae; blades of fronds broadly triangular -1B. P. aquilinum latiusculum

1A. Pteridium aquilinum pubescens Underw., (var.) Nat. Ferns. ed. 6. 91. 1900.
P. aquilinum (L.) Kuhn. in part; Pteris aquilina L. in part ---Rootstocks creeping, the growing tips usually with a tuft of dark hairs; fronds 30-200 cm. long; stipes usually shorter than blades; blades about 30-100 cm. long, ovate-triangular, not very ternate, usually 3-pinnate; pinnae subacute to obtuse; pinnules subacute to obtuse, usually nearly or quite at right angles to the rachises of the pinnae, pinnatifid or even pinnate, upper surfaces often pubescent, lower surfaces usually densely pubescent; sterile and fertile indusia pubescent or villous, sometimes ciliate on inrolled edge. Fernald (Rhodora 37:247. 1935) maintained that under the Rules the above plant should be P. aquilinum var. lanuginosum (Bong.) Fernald.---Open woods or slopes in damp or dry places. This form of the species ranges from Quebec to Alaska south to the Great Lakes region, South Dakota, Texas, Arizona and California. Our specimens scattered over the western two-thirds of the state at 5500-9100 feet.

1B. Pteridium aquilinum latiusculum (Desv.) Underw. ex Heller, (var.) Cat. N. Am. Pl. ed. 3, 17. 1909.
P. aquilinum (L.) Kuhn. in part; probably Pteris aquilina L. in part.---Differs from var. number 1 as follows: Rootstocks with growing tips usually naked; blades broadly triangular and often ternate; pinnules usually at an oblique angle to the rachises of the pinnae, upper surfaces of segments glabrous or nearly so, lower surfaces pubescent only along the midnerve; fertile and sterile indusia usually glabrous. Found with the var. pubescens in Colorado but apparently less common. Several specimens show intergradations between the varieties. For that reason the writer can see no justification for considering the second variety a separate species, in Colorado.

12. Notholaena R. Br. CLOAKFERN

Small ferns usually on rocks; fronds pinnately compound, farinose below, this white or yellow (in ours); sori marginal or nearly so, round or oblong but more or less confluent; proper indusia wanting but margins of segments revolute and concealing the sporangia especially at first.

1. Blades 4- to 5-pinnate below, the rachises zigzag and widely divaricate; segments white-powdery below -1. N. fendleri
1. Blades 2- to 3-pinnate below, the rachises nearly straight, not widely divaricate; segments yellow-powdery below -2. N. standleyi

1. Notholaena fendleri Kunze, Farnkr. 2:87. 1851.
Rootstocks densely scaly; fronds densely tufted, 5-25 cm. long; stipes dark brown as are the rachises; blades broadly deltoid-ovate, 4- to 5-pinnate below, the rachis branches divaricate and zigzag; ultimate segments oval or elliptical, entire or 3-lobed, white farinose below and less densely so above; sori marginal, elongated, covered somewhat at first and when dried out by the rolled-in margins of the segments; indusia lacking.---In dry crevices of cliffs and rocks. Wyoming to New Mexico and Arizona. Our specimens scattered over the western two-thirds of Colorado at 5500-10,000 feet except the extreme western counties.

2. Notholaena standleyi Maxon, Am. Fern Journ. 5:1. 1915.
N. hookeri Eaton ---Rootstocks short-creeping, with rigid dark brown scales; fronds clustered, 7-25 cm. long; stipes wiry, reddish-brown, shiny; blades pentagonal in outline, bipinnatifid to bipinnate, with 3 main divisions, the lower pairs longer and bearing elongated pinnatifid segments below near bases, middle divisions raised on a short stalk; segments yellow or yellowish-white powdery below, dull green above, curling in dry periods; sori marginal, partly covered by the rolled margins of the segments.---Dry rocky slopes and ledges. Western Oklahoma and Texas to Colorado, Arizona, and Mexico. It has been found in the extreme southeastern corner of the state at 4000-4400 feet.

13. Cheilanthes Sw. LIPFERN

Small rather xerophytic ferns; fronds tomentose or scaly, 1- to 4-pinnate, the ultimate segments small, often inrolled and beadlike; sori distinct or confluent; indusia formed of the revolute or recurved margins of the segments.

1. Fronds densely tomentose at least below, the segments hairy above; rootstocks erect or decumbent but not creeping
   2. Rachises of pinnae with numerous scales; stipes usually scaly   -1. C. eatoni
   2. Rachises of pinnae without scales; stipes usually villous when young   -2. C. feei
1. Fronds glabrous (except for scales), segments glabrous above; rootstocks slender-creeping
   3. Scales of leaf rachises ciliate; rootstock scales persistent   -3. C. wootoni
   3. Scales not ciliate; rootstock scales finally deciduous   -4. C. fendleri

1. **Cheilanthes eatoni** Baker in Hook. & Baker, Syn. Fil. 140. 1867.

Rootstocks thick, short, the scales often with dark brown or blackish bases; fronds tufted, 6-30 cm. long; stipes brown, with long narrow scales and with long villous, septate hairs both the above subappressed (the stipes in ours as long as or longer than the blades); blades 3-pinnate, oblong-lanceolate in shape; pinnae ovate-oblong, rather distant especially below, all rachises with scales and hairs like those on the stipe; pinnules divided into segments 1.5-2 mm. long or terminal one longer, tomentose above with whitish hairs, densely rusty-tomentose below, the hairs entangling the segments and hiding the margins; indusia of inrolled margins of the segments, sometimes membranous at margins.---Rocky places and crevices. Colorado and Utah to Arizona, Mexico, and Texas. Our specimens from the southeastern quarter of Colorado at 3500-6500 feet.

2. **Cheilanthes feei** Moore, Ind. Fil. 38. 1857.

Rootstocks multicipital, divisions short, the scales cinnamon-brown or darker; fronds tufted, erect-spreading, 5-20 cm. long; stipes slender, brown or blackish, shiny; blades ovate-lanceolate to linear-oblong, acuminate, 5-13 cm. long, 3-pinnate or 2-pinnate with pinnatifid pinnules; pinnae ovate to oblong-ovate; pinnules crenately pinnatifid or divided into rounded, crowded segments, lower surfaces densely whitish to brown-tomentose, thinly villous above, no scales; indusia of reflexed or incurved margins.---On or among rocks. Illinois and Minnesota to British Columbia, south to California, Mexico, and Texas. Our specimens scattered over the western two-thirds of the state at 5500-8500 feet but also from Baca County.

3. **Cheilanthes wootoni** Maxon, Biol. Soc. Wash. Proc. 31:146. 1918.

This species resembles **C. fendleri** but the scales of the leaf rachises are ciliate and the scales of the rootstocks persistent.---This fern has been reported from the extreme southeastern corner of Colorado but no specimens of it seen.

4. **Cheilanthes fendleri** Hook., Sp. Fil. 2:103. 1858.

Rootstocks creeping, with tawny scales; fronds mostly scattered or few close together, 6-25 cm. long; stipes brown, chaffy with slender scales, these entire or somewhat denticulate, not ciliate; leaf blades ovate-lanceolate or oblong-lanceolate, 3-pinnate, rachises with scales like the stipes; ultimate segments rounded and entire or 2- or 3-lobed, with 1 or more scales below, glabrous above; indusia of recurved margins.---Shaded rocks and ledges. Western Texas to Colorado and Arizona. Our specimens from Montrose and Las Animas to Larimer County at 5000-7500 feet.

14. Pellaea Link   CLIFFBRAKE

Small ferns mostly on rocks, with short thickened rhizomes; fronds compound; stipes dark; sori confluent, marginal or submarginal; margins of the segments indusiform and rolled or reflexed over the sporangia.

1. Scales of the rootstocks bicolored, the sharply defined midribs blackish, the margins rusty-brown; pinnules long-mucronate   -1. P. longimucronata
1. Scales of the rootstocks uniformly rusty-brown; pinnae or pinnules obtuse or at most acutish
   2. Stipes, rachises and rachillas with long multicellular hairs (especially on the rachillas), also scabrous; stipes dark purplish   -2. P. atropurpurea
   2. Stipes, etc., glabrous or with a few scattered hairs, smooth; stipes usually reddish-brown
      3. Simple pinnae usually linear-oblong; compound pinnae with stalked terminal pinnules; midnerves of pinnae or pinnules rather definite above   -3. P. suksdorfiana
      3. Simple pinnae usually ovate, deltoid or rhomboidal; compound pinnae when present with terminal **pinnules** sessile or nearly so; midnerves usually scarcely visible above
         4. Stipes not corrugated, not especially brittle; some of lower pinnae usually compound; plants apparently limited to the eastern part of Colorado   -4. P. glabella
         4. Stipes marked with transverse corrugations at least near base and breaking very readily at these joints; lower pinnae usually variously lobed to parted but not compound; plants limited to western Colorado   -5. P. breweri

1. **Pellaea longimucronata** Hook., Sp. Fil. 2:143. 1858.

P. wrightiana & P. mucronata of Manuals at least in part---Rootstocks short, the scales linear with blackish midrib and tawny margins; fronds 8-30 cm. long, twice-pinnate; stipes and rachises dark-brown or purplish-brown; blades triangular-ovate or broadly lanceolate; pinnae deltoid-lanceolate with 6-10 pairs of sessile or short-stalked pinnules; pinnules oval (on sterile) to linear-oblong (on fertile), margins revolute, often to the middle on fertile segments, grayish-pruinose otherwise glabrous, truncate at base (on our **specimen**), long-mucronate or very narrowly pointed; sori under the rolled margins which serve as indusia.---Among rocks and on cliffs. Colorado to Nevada, New Mexico, and Arizona. Our one specimen from Fremont County at 5200 feet.

2. **Pellaea atropurpurea** (L.) Link, Fil. Sp. Hort. Berol. 59. 1841.

Rootstocks short-thickened, scales long, rusty-brown, not bicolored; fronds tufted, 4-30 cm. long; stipes dark-purple, wiry, rather shiny but more or less covered with slender flaccid, jointed hairs, also somewhat scabrous; blades rather leathery, lanceolate or ovate-lanceolate, pinnate or usually bipinnate below; sterile segments entire, commonly oval or oblong but variable; fertile segments narrow, mostly linear, entire, sometimes auricled, stalked, obtuse or acute, glabrous above and below except for jointed hairs along midribs (in ours); segment margins rolled over the sporangia, the edges membranous and somewhat erose.---On rocks. Ontario to South Dakota, south to Florida, the Gulf states, New Mexico and Arizona; also in Mexico and Central America. Reported from the mountainous part of southern Colorado, our specimens from El Paso County at about 7200 feet, and Baca County at 4000 feet.

3. Pellaea suksdorfiana Butters, Am. Fern Journ. 11:40. 1921.
P. glabella var. simplex Butters.---Rootstocks tufted, the scales crowded, cinnamon-colored, narrowly linear, 5-8 mm. long, distinctly denticulate; fronds erect, 8-20 cm. long; stipes 2-6 cm. long, castaneous, with a few capillary scales; pinnae 3-8 pairs, simple or compound, linear or linear-oblong, basal commonly 3-divided, rarely 5, the terminal segments stalked; **segments** with thin repand-denticulate margins, the sporangia partially concealed, midnerves rather prominent above; spores averaging about 65 microns long.---Cliffs and crevices of rocks. British Columbia and Washington, south to Utah, Arizona and New Mexico. Our one record from Moffat County at 6000 feet.

4. Pellaea glabella Mett. in Kuhn, Linnaea 36:87. 1869.
P. occidentalis (E. Nels.) Rydb.; P. pumila Rydb.---Rootstocks short and tufted, scales linear, brown-cinnamon colored, 2-5 mm. long, denticulate; fronds 2-10 cm. long; stipes 1-4 cm. long, chestnut-brown, commonly with a few capillary scales; pinnae 2-5 pairs, usually simple, ovate or often narrower, compound pinnae when present with terminal pinnules not distinctly stalked; segments with thin repand-denticulate margins, midnerves scarcely visible above; spores averaging less than 55 microns long. Our plant is var. occidentalis (E. Nels.) Butters.---Crevices in cliffs and ledges. Quebec to British Columbia, south to Virginia and Oklahoma. Our record from northern Larimer County at 6000 feet but also reported from southcentral and west-central Colorado.

5. Pellaea breweri D. C. Eaton, Proc. Am. Acad. 6:555. 1865.
Rootstocks thick and short, scales long, rusty-brown, not bicolored; fronds 8-20 cm. long, tufted; stipes reddish-brown, transversely corrugated at base and brittle, more or less scaly below; blades oblong to **oblong-lanceolate**, pinnate; pinnae deltoid or rhomboidal, becoming lobed to parted especially near base of leaf but not compound, glabrous; margins inrolled except in the sinuses.---Crevices of cliffs and ledges. Wyoming to Washington, south to Utah and California. Our one record from Moffat County at 9300 feet.

15. Phegopteris Fee OAKFERN

Rootstocks slender-creeping; fronds 1- to 3-pinnate or ternate; sori round, small; indusia lacking (in ours). This genus is often merged with Dryopteris.

1. Phegopteris dryopteris (L.) Fee, Gen. Fil. 243. 1850-1852.
Thelypteris dryopteris (L.) Slosson; Dryopteris disjuncta (Ledeb.) C. V. Mort.---Rootstocks slender-creeping; fronds scattered, 6-50 cm. long; stipes slender, elongated, chaffy at least near base; blades broadly triangular, nearly or quite glabrous, 8-20 cm. wide, with 3 stalked primary divisions, these pinnate or twice-pinnate; segments usually pinnatifid into oblong, obtuse, entire or toothed divisions; sori round, small, rather near the margins; indusia lacking.---In moist woods and thickets. Newfoundland to Alaska, south to Oregon, Idaho, Colorado, South Dakota, and Vermont; also in Greenland and Europe. Our specimens from central and north-central Colorado at 7300-9200 feet.

Family 3. Marsileaceae PEPPERWORT FAMILY

Perennial plants, aquatic or semiaquatic with slender creeping rootstocks; leaves long-petioled, 4-foliate (the leaf cloverlike); spores borne in hardened sporocarps, these peduncled, the stalks attached to the lower part of the leaf petioles (stipes) or to the rootstocks near it, the body 2-celled and bearing both megasporangia and microsporangia, with 2 kinds of spores.

1. Marsilea L. PEPPERWORT

Characters of the Family

1. Marsilea vestita Hook. & Grev., Icon. Fil. 2:pl. 159. 1831.
Rootstocks creeping, with dense tufts of reddish-brown hairs at the nodes; leaves few to many, arising from the rootstocks; petioles 5-15 cm. long, slender, with deciduous hairs; leaflets broadly cuneate, spreading or folded, 4-15 mm. long, entire or nearly so, with deciduous hairs; sporocarps solitary, bean-shaped, 4-8 mm. long, on short peduncles, rather pendulous **with** 2 teeth near base (upper side), separated by a sinus, hairy, with 7-10 sori in each of the 2 cells, about 10-20 megasporangia to a sorus. The northern species M. **oligospora Gooding** probably is a synonym of the above.---Edges of ponds, ditches and rivers. Saskatchewan, British Columbia and California to Texas, Arkansas, and South Dakota. Should be expected anywhere in the state at lower altitudes but our specimens scattered over the eastern half at 4000-7500 feet.

Family 4. Salviniaceae SALVINIA FAMILY

Plants small, floating on water, rather mosslike; stems short but branched, bearing rootlets below; leaves borne alternately in 2 rows on the branches, very small (not over 2 mm. long in ours); spores of 2 kinds, large ones (megaspores) and small ones (microspores) in sporocarps borne on the leaves.

1. Azolla Lam. AZOLLA; MOSQUITO FERN

Characters of the Family

1. Azolla mexicana Presl, Abh. Bohm. Ges. Wiss. V.3:150. 1845.
   A. caroliniana of Manuals in part ---In ponds and ditches. Wisconsin to British Columbia, south to Mexico and South America. No records known from Colorado. A related species, A. filiculoides Lam. has been reported from Arizona.

Family 5. Equisetaceae HORSETAIL FAMILY

Rushlike plants from rootstocks; aerial stems unbranched or verticillately branched, mostly erect, cylindrical, hollow, externally fluted, the surface often with small tubercles of silica; leaves reduced, whorled, united to form cylindrical or dilated sheaths, prolonged upward into persistent or deciduous teeth; sporophylls **peltate**, the sporangia below, crowded together in terminal cones; spores all alike, bearing 4 hygroscopic elongated bands.

1. Equisetum L. HORSETAIL; SCOURING RUSHES

Characters of the Family

1. Aerial stems dimorphic, the fertile flesh-colored, succulent; sterile stems green, with numerous, slender side branches — -1. E. arvense
1. Aerial stems uniform, all green, simple or sometimes bearing a few irregular branches
   2. Spikes blunt or nearly so (sometimes somewhat acute); aerial stems annual — -2. E. kansanum
   2. Spikes apiculate; aerial stems evergreen, persisting for 2 or more seasons (sometimes annual in E. variegatum nelsoni)
      3. Stems 5- to 12-angled, central cavity 1/3 to 1/2 the total diameter of stem; leaves and teeth not sharply differentiated, the latter persistent
         4. Ridges of stem each with 2 rows of tubercles — -3. E. variegatum
         4. Ridges of stem each with 1 row of tubercles (or solid cross-bands) — -3A. E. variegatum nelsoni
      3. Stems 16-angled or more, central cavity more than 1/2 the total diameter of the stem; leaves and teeth sharply differentiated, the latter usually deciduous
         5. Sheaths definitely longer than broad, dilated upward, mostly green, each with 1 apical band especially when young; stems smooth, not over 8 mm. thick — -4. E. laevigatum
         5. Sheaths about as wide as long, nearly cylindrical, at maturity ash-colored with 2 dark bands; stems usually rough, some usually over 8 mm. thick
            6. Ridges of stem each with 1 row of tubercles or a cross band of silica; leaves with 3 keels (use strong lens) — -5. E. prealtum
            6. Ridges of stem each with 2 rows of tubercles; leaves with 4 keels — -6. E. hyemale

1. Equisetum arvense L., Sp. Pl. 1061. 1753.
   Rootstocks felted, often bearing tubers; stems dimorphic, sterile later, prostrate to erect, 10-50 cm. tall, 2-3 mm. thick, 10- to 14-furrowed with numerous, verticillate branches, these usually simple, 3- to 4-angled, teeth of sheaths lanceolate, acuminate; fertile stems earlier, 5-25 cm. tall, 3-5 mm. thick, smooth, succulent, pale-brown or tan, sheaths loose with 8-12 teeth; cones lanc-ovoid or oblong-ovoid, 2-4 cm. long, pedunculate. Occasionally the sterile stems will bear small cones during the summer or fall. Such plants have been called E. arvense forma serotinum (C. F. Schutze) Klinge.---Newfoundland to Alaska and south throughout nearly all the United States; Greenland; Eurasia. In Colorado our specimens are from the western borders east to Weld and Las Animas Counties at 4800-10,500 feet. Probably grows in all the eastern counties.
   A related species E. pratense Ehrh. has been consistently reported for Colorado but no specimens have been located authenticating this record.
2. Equisetum kansanum Schaffn., Ohio Nat. 13:21. 1912.
   E. laevigatum of our Manuals at least in part ---Rootstocks black; stems annual, single or clustered, erect, 30-60 cm. tall, 2-8 mm. thick, usually unbranched, light green, 15- to 30-grooved, smooth or nearly so; sheaths green, long, dilated upward, each with a narrow black terminal band and rarely with a fainter one below, leaves keeled on the lower part, teeth partly deciduous, cohering in groups, the persistent triangular bases rigid; cones ovoid or ovoid-oblong, 1-3 cm. long, 7-8 mm. thick, blunt or nearly so at apex, sessile or short-peduncled. This species may not be different from E. laevigatum A. Br.---British Columbia to Ontario, south to southern California, Arizona, New Mexico, Missouri, and Ohio. Widely distributed over Colorado at 3500-9000 feet.
3. Equisetum variegatum Schleich., Cat. Pl. Helvet. 27. 1807.
   Stems perennial, tufted from a branched base, 15-50 cm. tall, 1.5-4 mm. thick, 5- to 12-angled, rough or smooth, the ridges each with 2 rows of tubercles, rarely branched, central cavity 1/3 to 1/2 the total diameter of stem; sheaths enlarged upward, green below, each with a dark ring above, teeth blackish in middle, whitish on connivent margins, prolonged into filiform

deciduous tips, the rest of the teeth persistent; cones 5-10 mm. long, apiculate, sessile.---
Alaska to Labrador and south to California, Colorado, and New Jersey. Apparently rare in this state, having been reported in Gunnison County and near Denver, collected in La Plata County at 10,000 feet.

3A. Equisetum variegatum nelsoni A. A. Eaton, (var.) Fern. Bull. 12:24, 41. 1904.

E. nelsoni (A.A. Eat.) Schaffner ---Stems sometimes annual, ridges are rounded, not biangulate.---Our few specimens from northcentral and southcentral Colorado at 4700-8800 feet.

4. **Equisetum laevigatum** A. Br., Am. Jr. Sc. & Arts 46:87. 1844.

E. intermedium (A.A. Eaton) Rydb.---Stems tufted, erect, usually persisting 2 or more seasons, 20-100 cm. tall, 2-8 mm. thick, simple or sparingly branched, 20- to 30-grooved, smooth or nearly so (but branches somewhat rough); sheaths widening upward, the upper ones green with an apical border, the lower ones especially, with basal fainter bands, leaves keeled below, teeth cohering in groups, partly deciduous; cones 1-2 cm. long, about 5 mm. thick, rather bluntly to sharply apiculate, sessile or nearly so.---British Columbia to New York and south to California, Texas, Missouri, and Illinois. Scattered over the western three-fourths of Colorado according to reports but apparently uncommon. Our collections from 4600-11,000 feet.

5. Equisetum prealtum Raf., Fl. Ludovic. 13. 1817.

E. robustum A. Br.---Stems evergreen, rigidly erect, 50-200 cm. tall, 7-12 mm. thick or even wider at times, 16- to 48-angled, the very rough ridges each with 1 row of tubercles or a cross band not interrupted by a ridge; sheaths cylindrical, about as broad as long, tight, not dilated upward, often split, finally ashy-gray, each with a black basal and an apical band, leaves 3-carinate, teeth cohering in groups; cones oval or ovoid-oval, 1-2.5 cm. long, apiculate, sessile. Similar to the next and may not be distinct.---Nova Scotia to British Columbia, south to southern California and Florida; Mexico. Uncommon in Colorado our specimens from the western two-thirds of the state at 4700-9500 feet.

6. **Equisetum hyemale** L., Sp. Pl. 1062. 1753.

Stems evergreen, erect, rather rigid, 50-100 cm. tall and 5-12 mm. thick, 20- to 40-angled, rough, the ridges usually with 2 rows of tubercles but these sometimes indistinct, central cavity from 1/2 to 2/3 the total diameter of stem, rarely branched; sheaths cylindrical, little if any dilated above, at maturity usually ashy with basal and apical black bands, teeth brown, obscurely 4-carinate, soon deciduous; cone 1-3 cm. long, sharply apiculate, sometimes exserted.---Widely distributed throughout North America but surely uncommon in Colorado. Our few specimens from the western half of the state at 5000-8500 feet. Some believe that the species is Eurasian and represented in our area by the variety listed below. This plant is closely related to E. prealtum in this state. Specimens from southern and western Colorado may have the teeth tardily deciduous and the 2 rows of tubercles on each **ridge** of the stem more distinct. This is E. hyemale var. **californicum** Milde, Monogr. Eguis. 517. 1865.

Family 6.   Isotaceae   QUILLWORT FAMILY

Small plants growing in water or in bogs, submerged or partially emersed; stems short, cormlike, crowned by numerous, crowded onionlike leaves, these not flattened; sporangia in the hollow leaf bases, producing megaspores and microspores. Inconspicuous plants often overlooked by collectors.

1. Isoetes L.   QUILLWORT

Characters of the Family

1. Megaspores spiny; velum covering the sporangia 1/2 or more   -1. I. muricata
1. Megaspores tuberculate or reticulated, not spiny; velum narrower, seldom covering the sporangia over 1/2
    2. Stomata absent on leaves; megaspores with confluent crests becoming somewhat reticulated, over 0.5 mm. in diameter   -2. I. occidentalis
    2. Stomata present; megaspores with separated tubercles, less than 0.5 mm. in diameter   -3. I. bolanderi

1. Isoetes muricata Durieu., Bull. Soc. Bot. Fr. 11:101. 1864.

I. braunii Durieu.---Corm 2-lobed; leaves 10-35, erect or recurved, 7-20 cm. long or rarely longer, tapering to the tip, stomata present but usually few, ligules deltoid, the velum 1/2 to completely covering the sporangia; sporangia oblong, spotted, 4-7 mm. long; megaspores white, about 0.5 mm. in diameter, with numerous spines; microspores smooth to slightly roughened. Sometimes considered to be a variety of I. echinospora Dur.---Throughout most of North America especially the northern part. Our few specimens from El Paso County at around 11,500 feet but should be more widely distributed over Colorado.

2. Iosetes occidentalis Hend., Bull. Torr. Club 27:358. 1900.

I. lacustris paupercula Engelm., I. paupercula (Engelm.) A. A. Eaton ---Corms 2-lobed; leaves 9-30, rarely to 60, 5-20 cm. long, rigid, gradually tapering to apex, ridged, stomata lacking, ligules short-triangular, the narrow velum covering about 1/3 of the sporangia; sporangia orbicular, 5-6 mm. long; megaspores cream-colored, 520-650 microns in diameter, marked with low but conspicuous irregular crests, these chiefly simple on the apical face, branching to form an irregular network on the basal face; microspores spinulose.---Colorado to Wyoming and west to California and Washington. Our 1 record from northcentral Colorado at about 8200 feet.

3. Isoetes bolanderi Engelm., Am. Nat. 8:214. 1874.

Corms lobed; leaves 6-25, 5-10 cm. long, rarely longer, tapering to a fine point, stomata present, but usually few, ligules small, cordate, the velum narrow, covering the sporangia about 1/4 to 1/3; sporangia orbicular or oval, 3-4 mm. long, sometimes to 6 mm.; megaspores white to slate-blue, 300-440 microns in diameter, with low tubercles or short crests; microspores more or less spinulose.---British Columbia to Wyoming, Colorado, Arizona, and California. Our few specimens scattered over the western half of Colorado at 8000-11,500 feet.

Family 7. **Lycopodiaceae** CLUBMOSS FAMILY

Perennial rather mosslike plants; stems mostly trailing; leaves simple, small, 1-nerved, many, imbricate in several ranks; sporangia reniform, 1-celled, solitary in the axils of ordinary or modified leaves, these sometimes in specialized strobili; spores numerous, uniform in size.

1. Lycopodium L. CLUBMOSS; GROUND PINE

Characters of the **Family**

1. Sporophylls in definite spikes (strobili); main stems long (up to 100 cm.); leaves spreading or reflexed, not hollow near base   -1. L. annotinum
1. Sporophylls arranged among the normal leaves; main stem short; leaves usually appressed or ascending, hollow near base   -2. L. selago

1. Lycopodium annotinum L., Sp. Pl. 1103. 1753.
Main stems prostrate, up to 100 cm. long, **sparsely** leafy, giving rise to many erect or ascending branches, these 2-40 cm. tall, simple or sparingly branched; leaves in 5 or more ranks, 5-10 mm. long, linear-lanceolate, serrulate, rigid, spreading or reflexed, acute or tipped with a rigid point; spikes sessile but definite, usually solitary, oblong-cylindric, 1-3 cm. long; sporophylls broadly ovate, erose, abruptly acuminate.---Moist woods or thickets. Alaska to Newfoundland, south to Oregon, Colorado and West Virginia; also in Greenland and Eurasia. Our collections all from the central and northcentral part of Colorado at 8000-11,000 feet.
2. Lycopodium selago L., Sp. Pl. 1102. 1753.
Main stem very short, dichotomously branching into upcurved or erect shoots, these often forking, forming a tuft 2-20 cm. tall; leaves crowded, uniform or nearly so, ascending or appressed, deltoid-lanceolate to linear-acuminate or subulate, acute, usually entire, 5-7 mm. long, hollow near base; sporophylls somewhat shorter than the ordinary leaves but not arranged in spikes; plants often with gemmae.---Moist rocky places. Alaska to Newfoundland, south to Washington, Colorado and North Carolina; also in Greenland and Eurasia. Our few specimens from northcentral Colorado, in Larimer and Clear Creek Counties at 11,000 feet or above.

Family 8. **Selaginellaceae** SELAGINELLA FAMILY

Low terrestrial plants of mosslike habit; leaves numerous, rather scalelike and small, all nearly or quite alike, imbricated in several ranks; sporangia solitary in the axils of ordinary leaves or in specialized sporophylls arranged in spikes at the end of the branches, some sporangia bearing 1-4 large megaspores, others with numerous microspores.

1. **Selaginella** Beauv. SELAGINELLA

Characters of the **Family**

1. Stems rooting only at the base; spikes long, often over 3 cm.   -1. S. selaginoides
1. Stems mostly rooting for their entire length; spikes shorter, not over 3 cm. long
   2. Leaves obtuse or acutish at apex but no apical bristles present, the whole leaf less than 2 mm. long   -2. S. mutica
   2. Leaves with apical bristles, the leaf (counting bristle) usually over 2 mm. long
      3. Stems over 6 cm. long, often over 10 cm.; leaves usually sparsely if at all ciliate   -3. S. underwoodii
      3. Stems usually less than 6 cm. long and never over 10 cm.; leaves definitely ciliate
         4. Spikes 7-11 mm. long; leaves less than 2.5 mm. long, the older usually turning bronze-colored   -4. S. standleyi
         4. Spikes 10-30 mm. long, leaves usually over 2.5 mm. long, the older not at all bronze-colored
            5. Leaves tapering to awns 0.3-0.6 mm. long, the cilia 4-8 on each side; plant usually above 10,000 feet in Colorado   -5. S. scopulorum
            5. Leaves not tapering to the awns, these awns 0.6-1.5 mm. long, cilia 2-12 on each side; plant usually below 10,000 feet in Colorado   -6. S. densa

1. Selaginella selaginoides (L.) Link, Fil. Hort. Berol. 158. 1841.
This is a characteristic species and has consistently been reported for Colorado. However, the writer has never been able to locate a specimen from this state. Wherry (Am. Fern Journ. 28:125-140. 1938) puts it in the "excluded list."---Labrador to British Columbia, south to New Hampshire, Michigan, and Colorado (?); also in Greenland and Eurasia.
2. Selaginella mutica D. C. Eat. in Underw., Bull. Torr. Club 25:128. 1898.
Stems creeping, 7-10 cm. long, sometimes to 15 cm., rooting along the length; leaves 1-2 mm. long, oblong-ovate to narrower, obtuse or very short mucronate, no apical bristles, ciliate with about 8-15 setae on each side, pale green or glaucescent in color; spikes up to 1.5 cm. long; sporophylls ovate-deltoid, almost as long as the leaves.---On rocks and dry cliffs. Western Texas, Colorado, New Mexico, Arizona, and southern Utah. Rather generally distributed in the western 2/3 of Colorado, at 4000-7600 feet according to our specimens but reported from above timberline.
3. Selaginella underwoodii Hieron. in Engler & Prantl., Nat. Pflanzenf. I, 4:714. 1901.
Stems elongate but mostly less than 15 cm. long, creeping and rooting, rather sparsely branched especially in sheltered places; leaves 2-2.5 mm. long, oblong-linear, bases rounded or

cuneate, tips **obtusish** or acute with apical bristles 0.3-1 mm. long, marginal cilia may be present but more commonly reduced or absent, brilliant dark green in color when young; spikes 5-20 mm. long; sporophylls ovate-deltoid, about 2 mm. long, with apical bristles about like leaves, margins from definitely ciliate to sparingly so.----On rocks. Western Texas to Colorado, Wyoming, Arizona, and Utah. Our specimens from central and northcentral Colorado at 5300-9800 feet but to be expected anywhere in our mountainous area. Recently located in Baca County.

A related species has recently been described and recorded from Boulder and El Paso Counties. It is S. weatherbiana Tryon, Am. Fern Journ. 40:69. 1950, and differs from the above in having both subterranean and aerial stems, the latter rooting only at the base. The vegetative leaves in the new species are wider, being 0.4-0.5 mm. wide (instead of 0.3-0.4 mm.)

4. Selaginella standleyi Maxon, Smithsonian Misc. Coll. 72, No. 5, p. 9. 1920.
 Stems prostrate, 6 cm. long or less, densely short-branched; leaves 2-2.5 mm. long, oblong-linear, 10-14 cilia present on each side, the apex obtushish and with abrupt, stout yellowish **bristles**; spike 7-11 mm. long.----Moist rocky slopes or on gravelly soil. Alberta to Montana, but reported from Routt and Gunnison Counties in Colorado by Wherry (Am. Fern Journ. 28:125-140. 1938), at 11,000-14,000 feet. Specimens have been seen from that area at 10,400-12,300 feet.

5. Selaginella scopulorum Maxon, Am. Fern Journ. 11:36. 1921.
 Stems prostrate, short-creeping, 3-6 cm. long, branched, sub-caespitose and often forming large mats; leaves 2.3-3.3 mm. long counting the apical bristles which are 0.3-0.6 mm. long, linear to lanceolate-subulate, appressed, crowded, margins ciliate with 4-8 setae on each side; spikes 1-2.5 cm. long, suberect; sporophylls broader than the leaves.----On rocks. British Columbia and Oregon, to Colorado and Montana. Has been reported as thinly scattered over the western 2/3 of Colorado. Our few specimens from Routt, Larimer, Summit, and La Plata Counties at 8500-12,500 feet.

6. Selaginella densa Rydb., Mem. N. Y. Bot. Gard. 1:7. 1900.
 Stems prostrate, often rooting, densely caespitose and branched, 4-6 cm. long, or rarely to 10 cm.; leaves 2-3.5 mm. long, linear-oblong, abruptly bristled at tips the bristles 0.8-1.5 mm. long, densely crowded, marginal cilia short, 2-12 on each side, pale-glaucous when young; spikes 1-3 cm. long, nearly erect; sporophylls ovate-triangular, rather like the leaves but the apical bristles a little shorter.----On rocky or gravelly slopes. Alberta to Washington and south to Utah, New Mexico and South Dakota. Our specimens scattered over the western two-thirds of Colorado at 5800-10,000 feet, also in Baca County at 4000 feet.

Division II. Spermatophyta SEED PLANTS; FLOWERING PLANTS

Plants of various general aspect. Alternation of generations not readily apparent, the gametophyte plants much reduced and hidden in the flower parts. Plants with flowers, these simple to complex but containing stamens or pistils or both, and producing seeds. Pollen grains (microspores) borne in anther sacs (microsporangia). Ovules (macrosporangia) enclosed in an ovary (except in Gymnospermae), each ovule containing an embryo sac (macrospore). The pollen grain reaches the stigma and grows down to the embryo sac where fertilization occurs. The ovule develops into the seed containing an embryo, this usually consisting of one or more leaves (cotyledons), a radicle and plumule. Food is stored somewhere in the seed for the developing embryo.

Class 1. Gymnospermae NAKED-SEEDED PLANTS

Plants woody. Leaves needlelike or scalelike, not all deciduous at the end of the growing season (evergreen). Each ovule (and later seed) borne on the face of a scale or bract, not enclosed in an ovary. Stigmas none. Fruit a cone or the scales fleshy and the fruit berrylike.

Family 9. Pinaceae PINE FAMILY

Resinous trees or shrubs; leaves evergreen, needle-shaped, narrowly linear, or scalelike and imbricated; flowers unisexual, perianth lacking; pistillate inflorescences with scales spirally arranged or opposite in 4 ranks, the ovules naked; fruits of cones but these fleshy and berrylike in some.

1. Leaves opposite or whorled, less than 15 mm. long, scalelike or narrowly-triangular in juvenile forms
  (J. communis has leaves rather needlelike); fruit berrylike   -1. Juniperus
1. Leaves alternate or fascicled, over 15 mm. long, needlelike or narrowly linear; fruit a dry cone
  2. Leaves in clusters of 2-5, sheathed at the base at least when young; cone scales very thick and woody
                                                                                                -2. Pinus
  2. Leaves single on the twigs, not sheathed at the base; cone scales not very thick or woody
    3. Cones erect, the scales finally deciduous; leaves sessile and flattened   -3. Abies
    3. Cones pendulous, the scales persistent; leaves stalked, flattened or 4-sided
      4. Leaves 4-sided in section (very deciduous in dried specimens); branchlets roughened by the hard peg-like leaf bases; bracts of cones shorter than the scales   -4. Picea
      4. Leaves flattened (rather persistent in drying); branchlets not roughened by the persistent leaf bases; bracts longer than the cone scales   -5. Pseudotsuga

1. Juniperus L. JUNIPER; CEDAR

Trees or shrubs; leaves small, opposite or alternate, subulate and spreading in the juvenile forms, scalelike, appressed and imbricated in the mature forms (except in J. communis), often with darker glands on the back; monoecious or dioecious; pistillate aments with few scales and becoming a fleshy berrylike fruit, the scales not separating at maturity except occasionally near the apex.

1. Leaves all subulate or awl-shaped, not imbricated, 6-15 mm. long, the upper surface whitened; prostrate shrubs -1. J. communis
1. Leaves on mature branches scalelike, closely imbricated, less than 6 mm. long, not whitened above (leaves on juvenile twigs or vigorous shoots may be awl-shaped); prostrate or upright shrubs or trees
   2. Creeping shrubs; berries on recurved peduncles -2. J. horizontalis
   2. Upright trees or shrubs; berries on straight or nearly straight peduncles
      3. Leaves entire (under strong lens), appressed at tips; seeds 1-3 (usually 2); heartwood red or purple-red; branchlets flattened -3. J. scopulorum
      3. Leaves minutely denticulate, appressed or somewhat spreading at tips; seeds 1-2 (usually 1); heartwood brown; branchlets not flattened
         4. Fruit reddish-brown or bluish from a bloom, mealy or fibrous, 7-18 mm. in diameter; limbs usually arising above the ground level; usually monoecious; leaves obtuse to acute, tips closely appressed -4. J. utahensis
         4. Fruit copper-colored or blue, succulent when fresh, 4-7 mm. in diameter; limbs usually arising from below the ground level; usually dioecious; leaves acuminate to acute, tips often spreading -5. J. monosperma

1. Juniperus communis L., Sp. Pl. 1040. 1753.
   J. sibirica Burgsd.---Shrubs rarely over 100 cm. tall, branches depressed and decumbent forming dense usually circular patches; leaves 6-15 mm. long, subulate-triangular, in whorls of 3, prickly-pointed, green below with white central band above, usually bent at base and ascending; berries globose, 7-9 mm. in diameter, dark blue with a glaucous bloom; seeds 1-3. Represented in Colorado by the depressed form which has been variously called var. montana Ait. and var. saxatilis Pallas.---Dry rocky or sandy slopes. Cooler parts of North America and Eurasia. Our records scattered in the western third of Colorado at 5000-7000 feet.
2. Juniperus horizontalis Moench., Meth. 699. 1794.
   J. sabina of Manuals; Sabina horizontalis (Moench.) Rydb.---Banks and hillsides. Nova Scotia to British Columbia, south to New York, Minnesota and Wyoming. Has been reported from Colorado but no specimens seen; the plant may be in the northern mountains.
3. Juniperus scopulorum Sarg., Gard. & Forest 10:420. 1897.
   Sabina scopulorum (Sarg.) Rydb.---Bushy shrub to a tree 12 m. tall; trunk short, often divided near ground, crown usually irregular, bark red-brown to gray-brown, divided by shallow furrows, ridges scaly between; leaves 1-4 mm. long, opposite (very rarely in 3's), scalelike on older twigs, closely appressed, acute or acuminate, usually glandular-dotted on the back, margins smooth, often silver-green or glaucous, juvenile shoots and seedlings with awl-shaped, acerose leaves; flowers dioecious, the staminate with about 6 stamens; fruit blue, glaucous, 4-6 mm. wide; seeds 2 (rarely 1), bony-coated.---Rocky ridges and bluffs. Alberta to British Columbia, south to Texas, Arizona and Nevada. Scattered over the state at 4000-8000 feet.
   Morton (Rhodora 43:344-348. 1941) maintained that our western cedar is a variety of the eastern species and suggested the correct name should be J. virginiana var. scopulorum (Sarg.) Lemmon.
4. Juniperus utahensis (Engelm.) Lemmon, Rept. Calif. State Board For. 3:183. 1890.
   Sabina utahensis (Engelm.) Rydb.---Tree usually not over 6 m. tall; trunk single or with many stems from just above the ground, bushy-shaped, bark thin, ashy-gray to whitish, scaly; leaves 2-3 mm. long, on mature twigs scalelike, opposite (or rarely in 3's), subacute to acute and appressed at apex, usually without dark glandular spot on the back, margins denticulate, juvenile shoots and seedlings with acerose leaves; flowers usually monoecious, staminate with 18-24 stamens; fruit reddish-brown, glaucous, 7-18 mm. wide, the flesh mealy or fibrous; seeds 1 (or rarely 2), bony-coated. The valid name for this species may be J. osteosperma (Torr.) Little.---Dry rocky or sandy slopes. Idaho to California and south to Arizona and New Mexico. Scattered in the western third of Colorado at 5000-9000 feet.
5. Juniperus monosperma (Engelm.) Sarg., Silva N. A. 10:89. 1896.
   Sabina monosperma (Engelm.) Rydb.---Spreading shrub or small tree, the larger branches usually leaving the trunk at the root collar below the ground, bark thin, ashy, ridged and scaly; leaves 1-3 mm. long, rarely longer, opposite (rarely in 3's), scalelike on mature twigs, acute to acuminate and often spreading at the apex, usually with a dark glandular dot on back, margins denticulate, on juvenile shoots and seedlings acerose; flowers dioecious, staminate with 8-10 stamens; fruit copper to blue-colored, glaucous, 4-7 mm. wide, flesh succulent when fresh; seeds 1 (rarely 2). Cory (Rhodora 38:182-187. 1936) maintained this plant is a variety, J. mexicana var. monosperma (Engelm.) Cory.---Dry rocky plains and slopes. Kansas to Nevada and south to Mexico. Scattered over Colorado at 4000-7600 feet but mostly in the southern part.

2. Pinus L. PINE

Trees; primary leaves chafflike, deciduous and inconspicuous, secondary leaves from axils of the primary, conspicuous in fascicles of 2-5 (rarely reduced to 1), needle-shaped or very narrowly linear; monoecious, pistillate cones finally with thick woody scales, these variously thickened or unbonate, cones maturing in 2 or more years, opening to release seeds or remaining closed for years; seeds winged or wingless.

1. Needles in fascicles of 5
    2. Scales of cones with long prickles (especially when young); seeds with definite wings; needles seldom over 4 cm. long -1. P. aristata
    2. Scales of cones lacking such prickles; wings of seeds absent or rudimentary; needles usually over 4 cm. long -2. P. flexilis
1. Needles in fascicles of 2 or 3
    3. Cone scales without prickles; seeds wingless; leaf sheaths deciduous -3. P. edulis
    3. Cone scales with prickles; seeds winged; leaf sheaths persistent
        4. Leaves 8-25 cm. long, often in fascicles of 3 as well as 2; cones 7-15 cm. long -4. P. ponderosa
        4. Leaves less than 8 cm. long, in fascicles of 2; cones 2-5 cm. long -5. P. contorta latifolia

1. Pinus aristata Engelm., Am. Journ. Sc. II. 34:331. 1862.
    Tree 5-15 m. tall, crown dense and bushy or a shrub at high altitudes, bark thin and gray-white at first but red-brown and shallowly furrowed and ridged on mature trunks; leaves 2.5-4 cm. long, in 5's, dark green, stout and curved, long-persistent, usually showing conspicuous resin exudations; staminate flowers orange-red; pistillate cones short-stalked, ovoid or narrowly ovoid, dark purplish-brown, 6-9 cm. long, scales thick with long, fragile, bristlelike, incurved prickles; seeds about 6-8 mm. long, with terminal wings.---Exposed areas on dry rocky soils. Colorado to California, south to Arizona and New Mexico. Northcentral, central and southcentral Colorado at 7000-13,000 feet.
2. Pinus flexilis James, Longs Exped. 2:34. 1823.
    Apinus flexilis (James) Rydb.---Tree 10-15 m. high, with a round-topped head or shrublike at higher elevations, bark of old trunks dark brown to blackish, divided by fissures into ridges, these broken into square plates covered by small closely appressed scales; leaves 3.5-7.5 cm. long, in 5's, dark green, rigid, stout; staminate flowers reddish; pistillate cones short-stalked, ovoid to subcylindrical, horizontal or somewhat declined, 8-25 cm. long, scales thickened at ends, unarmed; seeds 7-12 mm. long, dark brown or blackish, wingless or very narrowly winged (rudimentary).---Various sites but often on summits, ridges or rocky foothills. Alberta to California, south to Texas and Arizona. Northcentral, central and southcentral Colorado at 5000-11,000 feet. One station in eastern Weld County.
3. Pinus edulis Engelm. in Wisliz., Mem. North. Mex. 88. 1848.
    Caryopitys edulis (Engelm.) Small---Tree 5-15 m. tall, trunk rather short, forming a broad compact pyramid and later a dense low round-topped head, bark divided into ridges and small scales; leaves 2-5 cm. long, in clusters of 2 or occasionally 3, sometimes even 1, rigid, incurved, sharp-pointed, margins entire; staminate flowers dark-red when fresh; pistillate cones very short-peduncled, short-ovoid, 2-5 cm. long and almost as wide, tips of scales thickened but without awns or prickles; seeds large, brown, ovate, wingless, edible. Considered by some to be a variety of P. cembroides Zucc.---Rocky soil of foothills, mesas and slopes. Wyoming to Utah, south to western Oklahoma, Texas, Arizona and lower California. Scattered in Colorado, mostly in the southern part but in Larimer and Moffat Counties, at 4000-9000 feet.
4. Pinus ponderosa Dougl. ex P. Lawson, Agri. Man. 354. 1836.
    P. scopulorum (Engelm.) Lemmon; P. brachyptera Engelm.---Tree 15-45 m. tall, sometimes broad and round-topped, bark thick, brown, separating into cinnamon-brown scales or flat plates; leaves 8-25 cm. long, in 3's or 2's, yellowish-green, stout; staminate flowers yellow; pistillate cones subsessile, ovoid or ellipsoid, about 7-15 cm. long sometimes longer, scales thickened at apex with short recurved prickles; seeds ovate, dark purplish, body about 6 mm. long and with a wing about 25 mm. long.---Slopes, valleys and mesas. Widely distributed in the United States from Rocky Mountains to the Pacific coast. Scattered over the **western two-thirds of** Colorado at 5000-9000 feet. The species has been divided into varieties and other species segregated from it but the various entities intergrade and have not been uniformly accepted.
5. Pinus contorta latifolia Engelm., (var.) in S. Wats. in King, Rept. U. S. Geol. Expl. 40th Par. 5:331. 1871.
    P. contorta murrayana (Balf.) Engelm.; P. murrayana Balf.---Tree 20-30 m. tall or more, in dense stands and then slender and branching near top, bark thin, orange-brown to gray, scaly but not ridged or furrowed; leaves 2.5-7 cm. long, in 2's, stout, often twisted; staminate flowers orange-red; pistillate cones subsessile, ovoid to subcylindrical, unsymmetrical, 2.5 cm. long, opening or remaining closed for years on the tree, scales thickened at end with a slender, subpersistent prickle; seeds with a broad oblong wing 10-12 mm. long, the body about 4 mm. long.---Hills and mountains. Saskatchewan to Alaska, south to California and Colorado. Mountainous area of Colorado in the north and central portions, reported in southern part, at 6000-11,000 feet.

3. Abies Link   FIR

Trees, young bark with horizontally elongated resin pockets; leaves flat, narrowly linear, inserted separately and leaving circular scars when fallen, with 2 longitudinal resin ducts (seen in section); monoecious, the pistillate cones erect, with thin, deciduous scales and shorter bracts; seeds winged.

1. Cones yellow to green-purple; scales usually wider than long; bracts with short triangular tips; resin ducts of leaves next to the epidermis
-1. <u>A. concolor</u>
1. Cones dark purple; scales usually longer than wide; bracts with a long subulate tip; resin ducts of leaves in-a-ways from the epidermis
   2. Old bark only moderately thick, light gray to gray-brown; cone scales usually cuneate at base
-2. <u>A. lasiocarpa</u>
   2. Old bark very thick and spongy, yellowish-white; cone scales usually halberd-shaped at base
-2A. <u>A. lasiocarpa arizonica</u>

1. Abies concolor (Gordon) Coopes, Book of Evergreens 220. 1868.
   Tree 20-35 m. tall, with a dense irregular crown, rather rounded in age, bark thin and gray at first, with many resin blisters but deeply furrowed in age; leaves 5-7 cm. long on lower branches, 2-3 cm. on fertile branches, straight or curved especially on fertile branches, acute or acuminate, silver-blue or blue-green, rather obscurely ranked, resin ducts near the epidermis; staminate flowers rose to dark red; pistillate cones oblong to oblong-cylindrical, yellow-green to purple, 7-12 cm. long, scales broader than long and twice as long as the short-tipped bracts; seeds broadly winged, 8-12 mm. long.---Mountain slopes. Wyoming to Oregon, south to lower California and northern Mexico. Scattered over the mountainous part of the state, ours from central and southern part at 7500-10,000 feet.

2. Abies lasiocarpa (Hook.) Nutt., N. A. Sylva 3:138. 1849.
   Tree 20-40 m. tall, with a narrowly pyramidal spirelike crown, or a prostrate shrub at timberline, bark thin, gray, smooth except for resin blisters when young, becoming thickened and shallowly fissured and gray-brown to gray; leaves 2.5-4.5 cm. long on lower branches, often no longer than 1.2 cm. on upper, deep blue-green, blunt or notched at apex, becoming erect by curving at base, resin ducts in-a-ways from the epidermis; staminate flowers dark blue; pistillate cones dark purple, oblong to oblong-cylindrical, 5-10 cm. long, scales a little longer than wide, mostly cuneate at base, bracts 1/3 as long as the scales, laciniate on margins; seeds about 6 mm. long with a broad wing.---In cool, moist mountainous sites. Alberta to Alaska, south to Oregon, Arizona and New Mexico. Scattered over the mountainous part of Colorado at 8500-12,000 feet.

2A. Abies lasiocarpa arizonica (Merr.) Lemm., (var.) Sierra Club Bul. 2:167. 1898.
   <u>A. arizonica</u> Merr.---Rather small tree 10-20 m. tall, bark elastic, corky, irregularly ridged, white to ashy-gray; leaves 2-3 cm. long or shorter on fertile branches; pistillate cones with scales nearly twice as long as the bracts, often halberd-shaped at base.---Colorado, Arizona and New Mexico. In the southern mountains of our state at 8500-11,000 feet.

4. <u>Picea</u> Link   SPRUCE

   Trees; leaves narrow, needle-shaped, 4-sided or nearly terete, blue-green or glaucous, very deciduous in dried specimens, when falling leaving peglike bases on the twigs, many-ranked; monoecious, the staminate aments terminal or from the axils of the older leaves; pistillate cones pendulous, maturing in one season, ovoid or oblong-cylindrical, the scales thin, leathery and persistent; seeds winged.

1. Cones less than 6 cm. long, persistent for one year; twigs pubescent; needles flexible, not very sharp to the touch; bark thin and scaly on mature trees
-1. <u>P. engelmanni</u>
1. Cones over 6 cm. long, persistent for two years; twigs glabrous; needles rigid, sharp to touch; bark of mature trees thick and furrowed
-2. <u>P. pungens</u>

1. Picea engelmanni Parry in Engelm., Trans. St. Louis Acad. Sc. 2:212. 1863.
   Tree 20-40 m. high with a narrowly pyramidal crown, or a depressed shrub at high elevations, bark thin, cinnamon-red or dark brown, scaly; leaves 2-3 cm. long, blue-green or sometimes glaucous, rather flexible the lower ones curving upwards, tip blunt or acute; staminate flowers dark purple; pistillate cones oblong-ellipsoid or oblong-cylindric, sessile or short-stalked, 2.5-6 cm. long, scales flexible, entire or erose at ends; seeds 3-4 mm. long.---Mountains and ravines especially where moist. Alberta to British Columbia, south to California, Arizona and New Mexico. Scattered in the mountains of Colorado at 8500-12,000 feet.

2. Picea pungens Engelm., Gard. Chron. n. ser. 11:334. 1879.
   <u>P. parryana</u> of Manuals---Tree 25-30 m. tall, crown dense and conical at first, ragged-pyramidal in age, bark pale to gray, tinged with cinnamon-red, scaly; leaves 2-3 cm. long, blue-green, frequently glaucous, rigid, acuminate and sharp-pointed at apex; staminate flowers yellow or tinged with red; pistillate cones oblong-cylindrical, sessile or short-stalked, 6-10 cm. long, scales flexuose on margins and apex erose, rather thin; seeds about 2-3 mm. long.---Rich moist soils especially along streams and in valleys. Wyoming, Colorado, New Mexico, Arizona and probably Utah. Scattered in the mountains, our specimens mostly from central, westcentral and southwest parts at 7000-9500 feet. This is our state tree, often called the Colorado Blue Spruce.

5. <u>Pseudotsuga</u> Carr.   DOUGLAS FIR

   Evergreen trees; leaves narrowly linear, single, flattish, appearing 2-ranked by the twisting of the short petioles, not very persistent, leaving a rounded, sessile scar; monoecious; staminate aments from axils of preceding season's leaves; pistillate cones maturing in 1 season, pendulous, scales thin and persistent, shorter than the conspicuous bracts, these with 2 short lobes on sides and caudate-acuminate or aristate between the lobes; seeds winged.

1. **Pseudotsuga taxifolia** (Poir) Britt. ex Sudworth, U. S. D. A. For. Div. Bul. 14:46. 1897.
P. mucronata (Raf.) Sudworth---Tree 20-60 m. tall with a compact pyramidal crown, bark brown, thick, with deep furrows; leaves 2-3 cm. long dark yellow-green, obtuse; staminate flowers orange-red; pistillate flowers green or more commonly reddish-purple; pistillate cones ovoid-oblong, maturing in 1 season, 5-10 cm. long; seeds about 7 mm. long or less.---Hills and mountains, often on deep soils of northern slopes. Alberta to British Columbia, south to California, Arizona and western Texas. Widely distributed in the mountains of Colorado at 6000-11,000 feet.

Family 10. Gnetaceae JOINTFIR FAMILY

Shrubs with opposite or whorled branches, these jointed and striated (resembling those of Equisetum); leaves scalelike, opposite or in whorls of 3, more or less connate; **usually dioecious**, the flowers with persistent bracts; stamens monadelphous, 2-8 stamens subtended by a membranous, 2-lobed, calyxlike perianth; ovulate flowers composed of 1 erect ovule enclosed in an urn-shaped perianth that hardens in fruit, bracts opposite or whorled.

1. Ephedra L. JOINTFIR; BRIGHAM TEA

Characters of the Family

1. Leaves and cone-scales 3 at a node; bracts of ovulate cones clawed, 6-10 mm. wide, scarious -1. E. torreyana
1. Leaves and scales 2 at a node; bracts of ovulate cones not clawed, 3-5 mm. wide, only the margins scarious
    2. Peduncles of ovulate cones wanting or very short; stems never viscid   -2. E. viridis
    2. Peduncles of ovulate cones 5-20 mm. long; stems often viscid (especially when young)
             -2A. E. viridis viscida

1. **Ephedra torreyana** Wats., Proc. Am. Acad. 14:299. 1879.
Shrubs 25-100 cm. tall, branches rigid, terete, up to 3.5 mm. thick, solitary or whorled at nodes; young stems blue-green or glaucous, with many longitudinal furrows; terminal buds less than 4 mm. long, not spinulose; leaves 2-5 mm. long, ternate or whorled, obtusely or sometimes acutely pointed from a brownish-green dorsomedial thickening, connate for 1/3 to 2/3 length but later spreading and recurved; sheath membranous at first, later thickening; staminate spikes solitary or up to 4 in a whorl, ovate, sessile, 6-8 mm. long; bracts ternate in 5-6 whorls, obovate, clawed, hyaline except in center and at base, margins minutely toothed and undulate, 6-9 mm. long; seeds solitary or 2, light brown to yellow-green, scabrous, 7-10 mm. long, equalling or slightly exceeding the bracts.---Dry areas. Colorado to Nevada, south to Arizona and western Texas. Western Colorado, our plants from the westcentral part at 4500-6800 feet.

2. **Ephedra viridis** Coville, Cont. U. S. Nat. Herb. 4:220. 1893.
Erect shrubs 50-100 cm. tall, branches rigid, terete, up to 3 mm. thick, opposite or many at nodes; young stems bright green to yellow-green, later yellowish; terminal buds 1-2 mm. long, conical, obtuse; leaves 1.5-4 mm. long, opposite, dorosomedially thickened, connate for 1/3 to 3/4 their length; sheaths membranous margined, soon falling and leaving a thickened base; staminate spikes paired to numerous, obovate, sessile, 5-7 mm. long; bracts opposite in 4-8 whorls, ovate, 4-7 mm. long; seeds paired, light brown to brown, lightly furrowed longitudinally, 5-8 mm. long, usually about 1/4 longer than the bracts.---Dry areas. Colorado to California, south to Arizona, reported from Wyoming. Western Colorado at 4500-9000 feet.

2A. **Ephedra viridis viscida** (Cutler) Benson, (var.) Am. Journ. Bot. 30:233. 1943.
Differs in having stems thicker and shorter, sometimes viscid especially when young; ovulate cones with peduncles 5-20 mm. long.---With the species in western Colorado at 4500-8500 feet.

Class 2. Angiospermae ENCLOSED-SEEDED PLANTS

Plants woody or herbaceous, mostly the latter. Leaves various but seldom needlelike or scalelike, rarely evergreen. Ovules and seeds borne in a closed ovary. Stigmas present. Fruit various but not commonly conelike.

Subclass 1. Monocotyledoneae

Plants never really woody (with us). Leaves mostly parallel-veined, always simple, usually entire and rarely toothed or lobed. Vascular bundles of the stem irregularly scattered and not arranged in a definite ring. Parts of the flower usually in 3's or multiples of 3. Embryo with 1 cotyledon.

Family 11. Typhaceae CAT-TAIL FAMILY

Perennial marsh or aquatic plants with creeping rhizomes; stems tall and erect; leaves alternate, long, linear, sheathing at base; monoecious, flowers densely crowded in terminal spikes or spikelike racemes, the staminate above; stamens 2-7, the filaments connate; perianth of bristles; ovary superior, stipitate, 1- to 2-celled, styles 1-2; fruit a nutletlike achene.

1. Typha L. CAT-TAIL

Characters of the Family

1. Staminate and pistillate inflorescences usually contiguous, the pistillate flowers commonly without bractlets, pistillate portion more than 2 cm. thick at maturity; leaves commonly over 1 cm. wide   -1. T. latifolia
1. Staminate and pistillate inflorescences usually separated, the pistillate flowers commonly with hairlike bracts dilated at the apex, pistillate portions of inflorescence seldom over 2 cm. thick; leaves usually less than 1 cm. wide   -2. T. angustifolia

1. **Typha latifolia** L., Sp. Pl. 971. 1753.
    Stems stout 1-2.5 m. tall; leaves flat, 6-25 mm. wide; staminate and pistillate spikes usually contiguous, each about 10-30 cm. long, the pistillate portion usually over 2 cm. thick at maturity, the flowers without bractlets; stigmas rhomboid or spatulate; pedicels of mature pistillate flowers 2-3 mm. long; pollen grains often in 4's; fruit reported to burst in water.--- In marshes and shallow lakes. Widely distributed in North America and Eurasia. Scattered over Colorado at 4000-7500 feet.
2. **Typha angustifolia** L., Sp. Pl. 971. 1753.
    Stems rather slender, 1-2.5 m. tall; leaves usually plano-convex, 4-10 mm. wide, sometimes to 15 mm.; staminate and pistillate spikes usually separated, each 10-20 cm. long, the latter usually less than 2 cm. thick; pistillate flowers with bractlets; stigmas linear or linear-oblong; pedicels of mature flowers 1 mm. long or less; fruit reported not to burst in water.---In marshes and shallow lakes. Widely distributed over the northern hemisphere; also in South America. To be expected anywhere in Colorado at lower elevations but our few records from along the eastern foothills at 4700-5500 feet.

Family 12. Sparganiaceae BURREED FAMILY

Perennial marsh or water plants with creeping rhizomes; stems erect or floating; leaves alternate, linear, sheathing at base; monoecious, the flowers in dense, globular heads, the staminate above, usually sessile, the pistillate below, sessile or lower peduncled, often subtended by leafy bracts; perianth reduced to a few chaffy scales; stamens commonly 5, distinct; ovaries superior, 1- to 2-celled, 1- to 2-seeded, 1 style, 1-2 stigmas; fruit a sessile or short-stiped nutlike achene.

1. Sparganium L. BURREED

Characters of the Family

1. Mature achenes with beak less than 1.5 mm. long; mature head of fruit usually about 1 cm. in diameter; anthers about twice as long as wide; usually only 1 staminate head present   -1. S. minimum
1. Mature achenes with beak over 1.5 mm. long; mature head of fruit over 1 cm. in diameter; anthers usually 3 or more times longer than wide; usually over 1 staminate head present
    2. Achenes obpyramidal or obovoid (widest part near apex), mostly truncate at tip, finally 6-8 mm. long; inflorescence often compound; stigmas commonly 2   -2. S. eurycarpum
    2. Achenes fusiform, gradually or abruptly tapering to the beak, about 4 mm. long at maturity; inflorescence not branched (pistillate heads may be peduncled); stigmas usually 1. Note--the next 2 species tend to intergrade.
        3. Leaves 1.5-4 mm. wide, rounded on back, middle and upper with dilated and subinflated sheaths; pistillate heads at maturity 1.2-2 cm. in diameter   -3. S. angustifolium
        3. Leaves 5-12 mm. wide, flat and ribbonlike, slightly if at all dilated or inflated at base; pistillate heads at maturity 2-2.5 cm. in **diameter**   -4. S. multipedunculatum

1. **Sparganium minimum** Fries, Summa Veg. Scand. 2:560. 1849.
    Stems 10-40 cm. long or more, slender, usually floating or in shallow water short and erect; leaves 1-7 mm. wide, flat, narrowly linear, thin; inflorescence simple; staminate heads 1 or sometimes 2; anthers about twice as long as wide; pistillate heads 1-2 or sometimes 3, about 1 cm. in diameter when mature, sessile or lowest peduncled; achenes 2-4 mm. long, short-stipitate, broadly ellipsoid or sometimes obovoid, often constricted below the middle, rather abruptly contracted into a short beak, this seldom over 1 mm. long.---Ponds and streams. Labrador to Alaska and south to Oregon, Colorado and New Jersey; Eurasia. Our few specimens from northcentral Colorado at 8000-8500 feet.
2. **Sparganium eurycarpum** Engelm. in Gray, Man. ed. 2. 430. 1856.
    Stems 50-150 cm. tall, stout, erect; leaves 7-10 mm. wide, linear, flat, slightly keeled below; inflorescence more or less compound; anthers about 1 mm. long, elliptic-clavate; pistillate heads 20-30 mm. in diameter in fruit, 1-4, on the stem or branches, sessile or more commonly peduncled; stigmas usually 2, filiform, about 2 mm. long; achenes 6-8 mm. long at maturity, cuneate-obpyramidal to broadly obovate, 3- to 5-angled, **rounded**, flat, or somewhat depressed at apex, sessile, **beak** about 3 mm. long.---In swamps and along streams. Newfoundland to British Columbia, south to California, Arizona and Virginia. Our few specimens from central, northeastern and southcentral Colorado at 3400-7500 feet.
3. **Sparganium angustifolium** Michx., Fl. Bor. Am. 2:189. 1803.
    Stems 20-50 cm. tall, floating and elongated or erect; leaves 1.5-4 mm. wide, flat but usually rounded on back, not keeled, dilated and scarious near base; inflorescence usually simple; pistillate heads 2-6, the lower 1 or 2 usually peduncled, 10-20 mm. in diameter in fruit; stigmas commonly 1, about 1-1.5 mm. long; achenes about 4 mm. long, gradually or abruptly tapering to a

beak about 2 mm. long, somewhat stipitate, fusiform, brown. Very near S. multipedunculatum and intergrades with it in Colorado.---In water. Newfoundland to British Columbia, south to California and New England. Scattered over the mountainous area of Colorado at 8000-12,000 feet.

4. Sparganium multipedunculatum (Morong) Rydb., Bull. Torr. Club 32:598. 1905.

Stems 30-50 cm. long, usually floating; leaves 5-12 mm. wide, flat, not at all rounded, slightly if at all dilated and inflated at base; inflorescence usually simple; pistillate heads 2-5, the lower 1 or 2 peduncled, 2-2.5 cm. in diameter in fruit; stigmas commonly 1, about 1 mm. long; achenes about 4 mm. long, stipitate, fusiform, tapering to a beak 2.5-3 mm. long.---In shallow water. Canada and south to California and Colorado, and again in northeastern United States. Mostly in northern Colorado at 6000-8000 feet.

Family 13. Najadaceae PONDWEED or NAJAS FAMILY

Immersed aquatic plants; stems leafy; leaves flat or filiform, alternate, opposite or appearing whorled; flowers perfect or unisexual, in axillary spikes or clusters, true perianth lacking (but sepaloid appendages present in Potamogeton); stamens 1-4; pistils 1 or more, ovary superior, 1-celled, ovule usually solitary; fruit mostly druplets or achenes.

1. Flowers perfect; at least the lower leaves alternate; anthers 2 or 4 to a flower
    2. Stamens 4, with sepaloid appendages; flowers sessile in capitate or spikelike inflorescences; fruit sessile; leaves in most species over 0.5 mm. wide   -1. Potamogeton
    2. Stamens 2, no sepaloid appendages; flowers pedicillate, inflorescence corymblike; fruit not sessile; leaves less than 0.5 mm. wide   -2. Ruppia
1. Flowers unisexual; leaves opposite or whorled; anthers solitary
    3. Leaves entire; pistillate flowers 2-5 in a cluster, subtended by scarious cup-shaped involucres; stigmas 1, cup-shaped, sometimes somewhat 2-lobed   -3. Zannichellia
    3. Leaves serrate (though sometimes inconspicuous); pistillate flowers solitary, no involucre present; stigmas 2-4, subulate   -4. Najas

1. Potamogeton L. PONDWEED; FISHWEED

Leaves alternate or upper opposite, often of 2 shapes, floating leaves usually coriaceous and broader, the submerged leaves usually thin and narrower; stipules present, adnate to the base of leaf for part of their length or free, sometimes connate into a tube; inflorescence spicate, axillary; flowers perfect; perianth none; stamens 4, with 4 sepaloid connectives or appendages; ovaries sessile, distinct with sessile or short-styled stigmas; fruit of indehiscent nutlike achenes.

1. Leaves of 2 different shapes, some floating, the other submerged, these usually thinner and narrower, the floating leaves not both clasping the stem and over 1 cm. wide (rarely narrower in P. diversifolius and P. epihydrus). Some species occasionally have only 1 kind of leaves but these are not clasping and are over 1 cm. wide.
    2. Submerged leaves threadlike, or narrowly ribbonlike, the sides parallel, less than 8 mm. wide
        3. Stipules of submerged leaves fused with the bases of the leaves; inflorescences of 2 kinds, in spikes in the axils of the floating leaves and in small heads in the axils of the submerged leaves   -1. P. diversifolius
        3. Stipules not joined with the leaf base; inflorescences uniform
            4. Floating leaves over 2.5 cm. wide, usually cordate or rounded at base; submerged leaves less than 2 mm. wide   -2. P. natans
            4. Floating leaves less than 2.5 cm. wide, usually cuneate at base; submerged leaves over 2 mm. wide
                5. Submerged leaves with a conspicuous, reticulated medial band (hold up to light with lens); fruit 2.5-3.5 mm. long, definitely keeled; floating leaves usually opposite   -3. P. epihydrus
                5. Submerged leaves without such a medial band; fruit 1.7-2.5 mm. long, slightly if at all keeled; floating leaves usually alternate
                    6. Submerged leaves usually less than 5.5 cm. long and less than 6 mm. wide, 5- to 7-nerved   -7A. P. gramineus gramineus
                    6. Submerged leaves usually over 5.5 cm. long and over 6 mm. wide, 7- to 9-nerved   -7B. P. gramineus maximus
    2. Submerged leaves broader, the sides curving in toward the ends, not parallel, over 8 mm. wide (except sometimes in P. gramineus)
        7. Floating leaves with 30-55 nerves; submerged leaves with 30-40 nerves   -4. P. amplifolius
        7. Both floating and submerged leaves with less than 30 nerves
            8. Floating leaves delicate, translucent, the blade tapering into the petiole and not distinct from it, the blade with less than 13 nerves; fruit surface hard and smooth; plant usually reddish
                9. Submerged leaves oblong-linear to linear-lanceolate, 7-25 cm. long, usually more than 8 times as long as wide   -5A. P. alpinus tenuifolius
                9. Submerged leaves oblong to ovate-oblong, 4-10 cm. long, usually less than 8 times as long as wide   -5B. P. alpinus subellipticus
            8. Floating leaves coriaceous, opaque, base rounded or cuneate but distinct from the petiole, the blade with 13 or more nerves; fruit surface soft and porous; plant greenish
                10. Submerged leaves all petioled, the longest petioles over 4 cm. long; fruits 3.5-4 mm. long when mature, usually reddish   -6. P. nodosus
                10. Submerged leaves sessile or some with petioles up to 4 cm. long; fruits 1.7-3.5 mm. long, greenish
                    11. Submerged leaves less than 1.5 cm. wide, sessile, over 5 times longer than wide, nerves usually not over 9; spikes 1-2.5 mm. long

12. Submerged leaves usually less than 5.5 cm. long and less than 6 mm. wide, nerves usually 5-7
-7A. P. gramineus gramineus
12. Submerged leaves usually over 5.5 cm. long and over 6 mm. wide, nerves usually 7-9
-7B. P. gramineus maximus
11. Submerged leaves over 1.5 cm. wide; sessile or on petioles up to 4 cm. long, seldom over 5 times longer than wide, nerves usually over 9; spikes at maturity 3-6 cm. long -8. P. illinoensis
1. Leaves alike and submerged, none over 1 cm. wide unless clasping the stem
13. Leaves threadlike or narrowly ribbonlike, the sides parallel, not over 5 mm. wide (rarely wider in P. robbinsii, reported from Wyoming)
14. Leaves with auricles at base, margins somewhat serrate -9. P. robbinsii
14. Leaves without auricles, margins entire
15. Stipules joined to the base of the leaves making a sheath usually over 1 cm. long; spikes with flowers in whorls
16. Leaves with obtuse or rounded apex; fruit without beak, 2-2.6 mm. long; no style present
17. Leaves to 0.5 mm. wide, retuse to obtuse -10A. P. filiformis borealis
17. Leaves 0.75 mm. or more wide, obtuse to minutely apiculate
-10B. P. filiformis macounii
16. Leaves with a long tapering tip; fruit with a beak, 2.6-4.2 mm. long; style present
-11. P. pectinatus
15. Stipules not joined to the base of the leave; spike continuous or interrupted
18. Leaves 9- to 35-nerved, often over 3 mm. wide -12. P. zosteriformis
18. Leaves 1- to 5-nerved, not over 3 mm. wide
19. Leaves with a pair of small glands where they join the stem; peduncles usually over 1 cm. long; the spike often over 5 mm. long; fruit plump without or with an obscure keel
20. Stipules strongly **fibrous**, becoming whitish and chartaceous; leaves usually revolute; peduncles thickened near apex -13. P. strictifolius
20. Stipules scarious or subherbaceous, not fibrous, usually greenish or brownish; leaves not revolute; peduncles not thickened near apex
21. Stipules connate, forming a tube surrounding the stem just above the node (rupturing in age--best on fresh material); peduncle often over 3 cm. long; spike often over 8 mm. long -14. P. pusillus
21. Stipules not united into a tube (may be rolled but edges not united around the stem); peduncles usually less than 3 cm. long; spike less than 8 mm. long -15. P. berchtoldi
19. Leaves without basal glands; peduncle less than 1 cm. long; spikes 2-5 mm. long; fruit compressed with a thin-toothed keel
22. Leaves over 1.5 mm. broad, 3- to 5-nerved; beak of fruit 0.2-0.4 mm. long
-16A. P. foliosus foliosus
22. Leaves not over 1.5 mm. broad, 1- to 3-nerved; beak of fruit 0.4-0.8 mm. long
-16B. P. foliosus macellus
13. Leaves broader, over 5 mm. wide, the sides often curving in toward the ends (somewhat parallel in P. crispus)
23. Leaves toothed (readily seen without a lens); stem compressed; beak of fruit 2-3 mm. long
-17. P. crispus
23. Leaves entire or very obscurely toothed; stem terete; beak of fruit not over 1.5 mm. long
24. Leaves sessile or only slightly clasping, usually reddish; beak of fruit often over 1 mm. long
25. Submerged leaves oblong-linear to linear-lanceolate, 7-25 cm. long, usually more than 8 times as long as wide -5A. P. alpinus tenuifolius
25. Submerged leaves oblong to ovate-oblong, 4-10 cm. long, usually less than 8 times as long as wide -5B. P. alpinus subellipticus
24. Leaves clasping 1/3 to 3/4 around the stem, rarely if ever reddish; beak of fruit not over 1 mm. long
26. Stipules conspicuous, 3-10 cm. long, persistent; leaves over 10 cm. long, the apex usually boat-shaped; fruit over 4 mm. long and 3 mm. wide at maturity -18. P. praelongus
26. Stipules inconspicuous, not over 2 cm. long, soon reduced to shreds; leaves less than 10 cm. long, the apex not boat-shaped; fruit less than 4 mm. long and 3 mm. wide
-19. P. richardsonii

1. **Potamogeton diversifolius** Raf., Med. Repos. Hex. II. 5:354. 1808.
Stems with many bushy branches or prolonged and remotely branched; submerged leaves 0.5-1.5 mm. wide, linear-setaceous, subobtuse to acute, 3-nerved (but inconspicuously so); floating leaves 0.3-2 cm. wide, 1-4 cm. long, elliptical, oval or narrowly obovate, apex rounded, tapering to a petiole, uppermost 7- to 15-nerved; stipules of submerged leaves mostly adnate to leaf bases for about 1/2 their length; spikes of linear leaves few-flowered, subglobose, subsessile or on short peduncles; upper spikes 5-20 mm. long; fruit 1-1.5 mm. long, suborbicular to semi-reniform, flattened, 3-keeled, beak a minute tooth.---In water. Scattered over the United States and probably in Mexico. Our rather doubtful record from Boulder County at 5400 feet but to be expected almost anywhere in the state.

2. **Potamogeton natans** L., Sp. Pl. 1:126. 1753.
Stems 60-200 cm. long, simple or rarely branched; submerged leaves 0.8-2 mm. wide, 10-20 cm. long, narrowly linear, semiterete, obtuse, with 3-5 obscure nerves; floating leaves 2.5-6 cm. wide, 4-9 cm. long, on long petioles, the blades coriaceous, ovate to oblong-ovate, base cordate, rounded or rarely tapering, apex rounded or obscurely mucronate, with 23-37 nerves; stipules persistent, clasping the stems, whitish, fibrous; peduncles 3-8 cm. long; spikes compact, finally 3.5 cm. long; fruits 3.5-5 mm. long, obovoid, keels none, beak short.---In water. Widely distributed in the Northern Hemisphere. Recorded for northcentral and southcentral Colorado at 5000-9000 feet but to be expected anywhere.

3. **Potamogeton epihydrus** Raf., Med. Repos. II. 5:354. 1808.

Stems 10-200 cm. long, compressed, slender, simple or branched; submerged leaves 2-8 mm. wide, 5- to 7-nerved, linear, flaccid, space between midnerve and inner nerves (or sometimes whole leaf) loosely cellular-reticulated below the middle; floating leaves 4-25 mm. wide, 2-7.5 cm. long, mostly opposite, oblong to oblong-lanceolate, apex rounded or subacute, tapering to a flattened petiole, 11- to 33-nerved; stipules free from leaf, more delicate on submerged leaves; fruiting spikes 8-30 mm. long, from axils of floating leaves; fruits 2.5-3.5 mm. long, flattened, round-obovoid, keeled, beak a mere tooth. Our plants are var. **nuttallii** (C. & S.) Fernald.---In water. Eastern and northeastern United States and scattered in Rocky Mountain area. Our doubtful records from Lake and Boulder Counties at about 9000 feet but to be expected anywhere in the northern part of Colorado.

4. **Potamogeton amplifolius** Tuckerm., Am. Journ. Sci. II, 6:225. 1848.

In water. Eastern United States and west in Arizona, Wyoming, Montana and westward. Sterile material of what appears to be this species has recently been collected in Lake County at 9500 feet.

5A. **Potamogeton alpinus tenuifolius** (Raf.) Ogden, (var.) Rhodora 45:90. 1943.

P. alpinus of our manuals---Stems 1-2 mm. in diameter, terete, simple or rarely branched above; submerged leaves 0.5-2 cm. wide, 4-18 cm. long, oblong-linear to linear-lanceolate, usually with 7 nerves, thin, delicate, translucent, sessile or slightly clasping, obtuse or rarely acutish, entire; floating leaves 1-2 cm. wide, elliptical, oblanceolate to obovate or even oblong-linear, often poorly developed, thin, translucent, tapering with no sharp distinction to a petiole 1-3 cm. long; stipules thin, membranous, free from the leaf; peduncles 3-10 cm. long, as thick as the stem; spikes in fruit cylindrical and crowded, 1.5-3.5 cm. long; fruit 3-4 mm. long (excluding beak), obovate, 1- to 3-nerved, beak curved, 1-1.3 mm. long. Whole plant usually reddish.---Streams and ponds. Nova Scotia to Alaska and south to Colorado, Utah and California; Europe. Our records from the western half of Colorado at 6500-12,000 feet.

5B. **Potamogeton alpinus subellipticus** (Fernald) Ogden, (var.) Rhodora 45:94. 1943.

With broader leaves than the preceding. Recorded from Wyoming and may be in northern Colorado.

6. **Potamogeton nodosus** Poir. in Lam., Enc. Meth. Bot. Suppl. 4:535. 1816.

P. americanus Cham. & Schlecht; P. lonchites Tuckerm. of authors---Stems simple or sparingly branched, terete; submerged leaves 1-3.5 cm. wide, 9-20 cm. long, linear-lanceolate to broadly lance-elliptical, thin, 7- to 15-nerved, acute, tapering to a petiole 2-13 cm. long; floating leaves with blades 2-4 cm. wide, 5-9 cm. long, 13- to 21-nerved, elliptical, apex acute to rounded, cuneate to somewhat rounded at base with long petioles, coriaceous; stipules free, rather delicate; peduncles 3-15 cm. long, usually thicker than stems; spike in fruit 3-6 cm. long; fruit 3.5-4 mm. long, obovate, keels prominent, reddish or brownish-red when mature, beak short.---Ponds and streams. Generally distributed over North America; also in South America, Eurasia and Africa. Not very common in Colorado our records mostly from the northcentral part at 5000-7000 feet but to be expected anywhere in the state. Recently collected in Rio Grande County at 7500 feet.

7A. **Potamogeton gramineus** L. **gramineus**

P. gramineus L.; P. heterophyllus in **part**; P. gramineus var. typicus Ogden---Stems 0.5-1 mm. in diameter, terete, much branched; submerged leaves 0.2-0.6 cm. wide, 1.5-4.5 cm. long, 5- to 7-nerved, apex acute, tapering to a sessile base; floating leaves 1-2 cm. wide, 1.5-5 cm. long, coriaceous, ovate to elliptic, apex obtuse or somewhat mucronate, base cuneate or rounded to a petiole 2-10 cm. long; stipules persistent, free from the leaf bases; peduncles 2-10 cm. long; spikes in fruit 1-2.5 cm. long; fruits 1.7-2.5 mm. long, mostly obovate, nerves usually evident, beak short and recurved.---Lakes and streams. Greenland to Alaska, south to New Jersey, New Mexico and California; Eurasia. Our few records from northcentral and central Colorado at 7500-9000 feet but surely more widely distributed.

7B. **Potamogeton gramineus maximus** Morong ex Bennett, (var.) Journ. Bot. 19:241. 1881.

P. lonchites Tuckerm.---The submerged leaves differ from the preceding in being over 6 mm. wide and over 5.5 cm. long. The nerves run 7 to 9 instead of 5 to 7.---About the same range as the preceding. Appears to be more common in Colorado, the records scattered over the western two-thirds of the state at about 7000-10,000 feet.

8. **Potamogeton illinoensis** Morong, Bot. Gaz. 5:50. 1880.

P. lucens of American authors; P. zizii in part; P. angustifolius in part---Stems simple or branched, terete when fresh; submerged leaves 1.5-4 cm. wide, 5-20 cm. long, blades thin, elliptic or oblong-elliptic to lanceolate, apex acute or mucronate, 9- to 17-nerved, margins with fugaceous 1-celled denticles, sessile or tapering into petioles up to 4 cm. long (sometimes to a midrib in some leaves); floating leaves 2-6.5 cm. wide, 4-13 cm. long, blades elliptic, ovate-elliptic or oblong-elliptic, more or less coriaceous, 13- to 29-nerved, obtuse or bluntly mucronate, base cuneate or rounded to petioles 2-9 cm. long; stipules persistent and conspicuous, free from the leaf bases; peduncles 4-15 cm. long, often thicker than the stems; spikes in fruit 3-6 cm. long; fruits crowded, 2.7-4.2 mm. long, obovate to ovate, nerved and keeled, beak short.---Lakes and streams. Quebec to British Columbia, south to Florida, Texas and California; Mexico and Central America. Our few records from western Colorado at about 8000-9000 feet but to be expected anywhere in the state.

9. **Potamogeton robbinsii** Oakes, Hoveys Magaz. 7:180. 1841.

New Brunswick to British Columbia, south to Delaware, Wyoming and Oregon. To be expected in northern Colorado.

10. **Potamogeton filiformis** Pers., Syn. Pl. 1:152. 1805.

The two varieties intergrade commonly in Colorado.

10A. **Potamogeton filiformis borealis** (Raf.) St. John, (var.) Rhodora 18:134. 1916.

P. interior in part; probably P. filiformis of our manuals---Stems up to 30-40 cm. long in deep water, usually repeatedly branched near base in shallow water; leaves 0.25-0.5 mm. wide,

5-30 cm. long, linear-filiform, retuse or obtuse at apex; stipules united to base of leaf for 1 cm. or more; spikes 0.5-5 cm. long, fruit in verticils, the upper approximate and sometimes the lower; fruit 1-1.5 mm. thick, 2-2.6 mm. long, obovate or obliquely ellipsoid, stigmas sessile, beak lacking.---In water. Newfoundland to Alaska, south to Pennsylvania and Colorado; India, Tibet, and China. Scattered over the western two-thirds of the state at 4800-10,000 feet.

10B. Potamogeton filiformis macounii Morong, (var.) Mem. Torr. Bot. Club 3(2):50. 1893.

Resembles the preceding but leaves are wider, averaging 0.75-2 mm. and their tips are obtuse to apiculate.---About the same range as the preceding. Our few records from the western half of Colorado at 6000-9000 feet.

Recently a similar species, P. vaginatus Turcz., has been located in Park and Larimer Counties. It is a more robust plant with the sheaths of the leaves on the main stem enlarged and conspicuous.

11. Potamogeton pectinatus L., Sp. Pl. 127. 1753.

P. interruptus Kit.---Stems very slender, much branched and very leafy; leaves 0.1-0.3 mm. wide, 2-15 cm. long, very narrowly linear to setaceous, attenuate to almost pungent at tips; stipules adnate to the leaf base for about 1/2 their length, 1-2 cm. long, scarious on margins; spikes 1.5-5 cm. long, interrupted, usually with several whorls; fruit 2.5-4.5 mm. long, with 2 obscure, lateral ridges, but no medial ones, obliquely obovoid, styles about 0.4-0.6 mm. long, stigmas capitate.---In fresh or saline water. New Brunswick to Alaska, south to Florida, Texas and California; Europe. Scattered over the state at 4000-10,000 feet.

12. Potamogeton zosteriformis Fernald, Mem. Am. Acad. Arts & Scs. 17:36. 1932.

Northern United States including Utah and Nebraska. No record or report as yet from Colorado but should be here.

13. Potamogeton strictifolius Ar. Benn., Journ. Bot. 40:148. 1902.

A species of the northern part of the United States. The var. rutiloides Fernald has been reported from western Kansas and Utah and should be in Colorado.

14. Potamogeton pusillus L., Sp. Pl. 127. 1753.

P. panormitanus and varieties of it---Plants of this species resemble P. berchtoldi Fieber but the stipules are connate for at least 2/3 of the way. However, the stipules split on older leaves and the difference is difficult to make out even with fresh material. The peduncles of P. pusillus L. are longer, being 1.5-8 cm. long, while the spikes are 6-12 mm. long, but an overlapping occurs at the lower limits.---Widely distributed throughout the United States. Our records scattered in the state, but few in the extreme eastern part, at 5000-9500 feet.

15. Potamogeton berchtoldi Fieber in Berch. & Fieb., Potamog. 40: 1838.

P. pusillus of Fernald's treatment--Stems very short to 1 m. long, capillary, terete or nearly so, sub-simple to freely branched; leaves 0.3-2.4 mm. wide, 0.8-8.5 cm. long, 3-nerved, linear to linear-setaceous, typically with a pair of small translucent glands at base, midrib bordered on each side by 1 or more rows of lacunae; stipules 3-14 mm. long, faintly nerved, hyaline to subherbaceous, flat or with inrolled but free margins; peduncles 3.5-30 mm. long, filiform and scarcely thickened upwards; spikes in fruit 2-8 mm. long, fruit in 1-3 whorls, these contiguous or but slightly interrupted; fruits 2-2.5 mm. long, little compressed, obliquely obovoid, keel obscure, beak very short.---Water. Widely distributed in North America and Eurasia. Our records from the western half of Colorado at 5000-10,500 feet but to be expected anywhere.

16. Potamogeton foliosus Raf., Med. Rep. Hex. 2. 5:354. 1808.

The following two varieties intergrade commonly in Colorado.

16A. Potamogeton foliosus foliosus

P. foliosus Raf.; P. foliosus var. genuinus Fernald---Stems 6-100 cm. long, compressed-filiform, subsimple to bushy-branched; leaves 1.4-2.7 mm. wide, 4-10 cm. long, 3- to 5-nerved, narrowly linear, acute, subacute or cuspidate, somewhat tapering to base, usually without basal glands; sheaths at first forming delicately fibrous sheaths around the stems above the nodes but soon rupturing and deciduous; peduncles 3-10 mm. long; spikes subcapitate or thick-cylindric; fruits compressed with a thin, acute undulate to dentate keel, beak 0.2-0.4 mm. long.---Ponds and streams. Nearly throughout North America. Our few specimens in northcentral and southcentral Colorado at about 5000 feet.

16B. Potamogeton foliosus macellus Fernald, (var.) Mem. Am. Acad. Arts & Scs. 17:46. 1932.

P. foliosus in part---Differs from the preceding in having somewhat shorter stems; leaves shorter and narrower, not over 1.5 mm. wide; beak of fruit longer, 0.4-0.8 mm. long.---Scattered over the western two-thirds of Colorado at 4500-9500 feet, apparently not very common.

17. Potamogeton crispus L., Sp. Pl. 1:126. 1753.

Stems branched, flattened with the broad sides channelled; leaves 1.5-5 cm. long and 5-15 mm. wide, linear-oblong to oblong, sessile, serrate and crisped at edges, obtuse, midrib prominent and with 2 fine lateral nerves; stipules scarious and fugaceous; spikes about 10-18 mm. long, cylindric, loosely-flowered; fruit about 2.5-6 mm. long, ovoid, long-beaked at apex, sub-sagittate at base. The winter buds are prominent burlike structures with short, indurated leaves.---In water. Nova Scotia to Minnesota, south to Virginia and California; Europe. Our one record from Jefferson County at about 7000 feet.

18. Potamogeton praelongus Wulfen, Arch. Bot. Roem. 3:331. 1805.

Stems simple or branched, whitish or olive-green; leaves 1-3 cm. wide, 10-20 cm. long, ovate-oblong to oblong-lanceolate, with 13-25 nerves, 3-7 being main ones, margins entire, apex cucullate and splitting when pressed, base cordate or rounded, clasping 1/3 to 1/2 way around the stem; stipules 5-10 cm. long, white, persistent and conspicuous; spikes thick in fruit, not very crowded in flower; fruit 4.3-5.5 mm. long, obovate, beak prominent but short and thick.---Deep water. Newfoundland to Alaska, south to New Jersey, Colorado and California. Our few records from northwestern, central and northcentral Colorado at 9000-11,000 feet.

19. **Potamogeton richardsonii** (Bennett) Rydb., Bull. Torr. Club 32:599. 1905.
P. perfoliatus of manuals---Stems 1-2.5 mm. wide, terete, sparingly if at all branched; leaves 1-2 cm. wide, 3-10 cm. long, 15- to 29-nerved, 3-7 more prominent, mostly ovate-lanceolate, basal often ovate and upper lanceolate, margins with fugaceous 1-celled denticles, apex acute or sometimes rounded, cordate and clasping 1/2 to 3/4 around the stems; **stipules** free, whitish, coarsely nerved and soon disintegrating into fibers; peduncles 1.5-3 cm. long, about as thick as the stems; fruit 2.7-3.2 mm. long, obovate, plump, keels rounded or lacking, beak up to 1 mm. long.---Ponds, lakes and rivers. Labrador to Alaska, south to Massachusetts, Colorado, Utah and California. Our few records from the western half of Colorado at 6000-10,000 feet.

2. Ruppia L. WIDGEONWEED

Aquatic, submerged plants; stems slender and branched; leaves alternate, 1-nerved, filiform, sheathing at base; peduncles capillary, elongated finally and coiled after fertilization which occurs at the surface, flowers in a corymbose inflorescence; flowers perfect, of 2 sessile anthers and 4 pistils; perianth absent; stigmas sessile.

1. **Ruppia maritima** L., Sp. Pl. 127. 1753.
R. occidentalis Wats.; R. curvicarpa A. Nels.---Leaves 2-20 cm. long and not over 0.5 mm. wide, basal sheaths 6-40 mm. long; flowers 2-4 together, borne in the sheaths; peduncles elongating finally to 50 cm. long and the individual fruits becoming long-stiped.---North America, South America and Eurasia. Our records from the eastern two-thirds of Colorado at 4000-7500 feet.

3. Zannichellia L. HORNED PONDWEED; POOLMAT

Aquatic plants; leaves opposite or whorled, entire, filiform; stipules membranous; flowers axillary, monoecious; perianth lacking; staminate flower consisting of 1 stamen; pistillate flowers 2-5 in a cluster, subtended by a scarious cup-shaped involucre; stigmas 1, peltate or cup-shaped, sometimes somewhat 2-lobed; fruit an oblong, beaked nutlet.

1. **Zannichellia palustris** L., Sp. Pl. 969. 1753.
Stems slender and capillary, simple or branched; leaves all submerged, 0.5 mm. wide or less, filiform but flat, 1-nerved, acute or pungent at apex; stipules scarious, free from the leaf bases; staminate and pistillate flowers from the same axil, subtended or enclosed by a hyaline bract; pistillate flowers of 2-10 carpels, these sessile at first, smooth or ribbed, styles persistent; fruit 2-4 mm. long.---Ponds, ditches and streams. A cosmopolitan species, generally distributed in the United States. Scattered over the state at 3500-9500 feet.

4. Najas L. (sometimes spelled Naias) NAIAD

Aquatic submerged plants; stems slender and branching; leaves opposite or whorled, narrow, serrulate to spinulose-dentate on margins; stipules sheathing at base; flowers monoecious or dioecious, solitary and axillary; **perianth none; staminate** flowers with a spathelike involucre and 1 stamen; pistillate flower of 1 ovary, stigmas 2-4; fruit a seedlike nutlet.

1. **Najas guadalupensis** (Spreng.) Morong, Mem. Torr. Club 3:60. 1893.
Stems slender, branching; leaves about 0.5-1 mm. wide, 12-25 mm. long, many, narrowly linear, usually tipped with 1 or 2 weak, short spines, margins with 20-40 marginal teeth, these inconspicuous or even wanting; flowers monoecious; anthers 1-celled; pistillate flowers 2-3 mm. long; fruit with 2 or 3 stigmas and usually with 1 or 2 spiny processes, with 16-20 rows of reticulations.---Ponds and lakes. Colorado south to Florida and Texas; Mexico and Central America. Our few records from Larimer County at 5100 feet but should be present in central and southern Colorado also.
N. flexilis (Willd.) R. & S. and N. maritima L. are eastern species but have been reported in Utah. They are to be expected in Colorado.

Family 14. Juncaginaceae ARROWGRASS FAMILY

Herbaceous rushlike plants growing in moist or wet soil; leaves all basal, narrow and grasslike but semi-terete with membranous basal sheaths; flowers perfect in narrow **spikelike** racemes terminating rather long scapes; perianth of 6 segments, the inner 3 inserted higher; stamens 3-6, anthers sessile or nearly so; ovary superior, of 3-6 carpels, these united at first, later separating from base upwards.

1. Triglochin L. ARROWGRASS; PODGRASS

Characters of the Family

1. Carpels 3; fruit linear-clavate, tapering at the base; leaves usually less than 2 mm. wide
-1. **T. palustris**
1. Carpels 6; fruit oblong or ovoid, rounded at base; leaves usually 2 mm. wide or wider
-2. **T. maritima**

1. Triglochin palustris L., Sp. Pl. 338. 1753.
   Rootstocks short, slender fugaceous stolons present; leaves less than 2 mm. wide, narrowly linear, shorter than the scape, sharp-pointed; scapes 20-60 cm. tall; racemes 10-30 cm. long; pedicels in fruit 5-7 mm. long, slender and erect; anthers 6; pistil of 3 carpels, ovary 3-celled and 3-ovuled; fruit 6-7 mm. long, linear or clavate.---Bogs. Greenland to Alaska, south in northeastern United States and scattered in the western states. Our records from the western three-fourths of the state at 4500-10,000 feet.
2. Triglochin maritima L., Sp. Pl. 339. 1753.
   Rootstocks stout, stolons absent; leaves usually 2 mm. wide or wider, linear, shorter than the scapes, rather fleshy; scapes about 10-70 cm. tall; racemes often over 40 cm. long; pedicels in fruit 2-3 mm. long, decurrent; anthers 6; pistil of 6 carpels, ovary 6-celled and 6-ovuled; fruit 5-6 mm. long, oblong or ovoid, rounded below with 6 recurved tips.---Meadows marshes and borders of lakes. Canada south to northeastern United States and widely distributed throughout the western states. Scattered widely over Colorado at 4500-8500 feet.

Howell (Leaflets West. Bot. 5:13-19. 1947) recorded a third species from Colorado which he called T. concinna var. debilis (M. E. Jones) J.T. Howell. It is closely related to the above but the ligules are 2-parted and not entire. No good material has been found in this study.

Family 15. Alismaceae WATERPLANTAIN FAMILY

Herbaceous aquatic or marsh plants; stems scapelike; leaves with usually broad blades and sheathing bases; flowers regular, perfect or unisexual, in racemes or paniculate inflorescences; calyx of 3 sepals; petals 3, separate; stamens 6 or more; pistils several to many in a ring or dense head, ovaries superior, 1-celled, usually 1-ovuled; fruit a compressed achene. (Alismataceae)

1. Leaves sagittate; flowers monoecious, dioecious or sometimes polygamous   -1. Sagittaria
1. Leaves ovate to narrower, never sagittate; flowers perfect   -2. Alisma

1. Sagittaria L. ARROWHEAD; SWAMP POTATO

Perennial plans with rhizomes, these often tuberous; leaves long-petioled, sometimes floating, sagittate (in ours); inflorescence a long verticillate panicle; flowers monoecious, dioecious or polygamous, the staminate above, perianth segments in 2 sets, outer herbaceous, inner white and petaloid; stamens many; pistillate flowers with enlarged receptacles with many ovaries, stigmas persistent; fruit of compressed achenes.

1. Basal lobes of leaves 2 or 3 times longer than the terminal one; beak of achene lateral and minute
    -1. S. longiloba
1. Basal lobes of leaves shorter or little longer than the terminal one; beak of achene various but not both lateral and minute
    2. Lower flowers of inflorescence perfect; leaves usually wider than long; pedicels definitely thickened
        -2. S. calycina
    2. Lower flowers pistillate (only very rarely with a few stamens); leaves usually longer than wide; pedicels not especially thickened
        3. Achene beaks definite (at maturity 1/5 the length of the achene body or more), usually horizontal or strongly oblique; head of fruits at maturity usually over 1.5 cm. wide   -3. S. latifolia
        3. Achene beaks minute, erect, over 1 wing of the body; fruit at maturity seldom over 1.5 cm. wide
            -4. S. cuneata

1. **Sagittaria longiloba** Engelm. in Torr., Bot. U. S. and Mex. Bound. 212. 1859.
   This is a southern species reported from Colorado but no specimens from the state could be located.
2. Sagittaria calycina Engelm. in Torr., Bot. U. S. and Mex. Bound. 212. 1859.
   Lophotocarpus calycinus (Engelm.) J. G. Smith---This plant is in the eastern United States and is reported for New Mexico, Arizona and California. It should be expected in southern Colorado.
3. Sagittaria latifolia Willd., Sp. Pl. 4:409. 1806.
   A variable plant 20-60 cm. tall, rather stout, glabrous; leaves 10-40 cm. long, usually abruptly contracted to an acute apex, basal lobes equalling or shorter than the central one, lobes from broadly ovate to linear-lanceolate, basal often out-curved; bracts 1-3 cm. long, ovate, acute or obtuse; whorls of inflorescence distant; head of fruits 1.5-3 cm. in diameter; achenes about 3 mm. long at maturity, with a broad wing on each side, beak nearly horizontal, definite, about 1/5 to 1/3 the length of the achene body.---Shallow water. Widespread in North America and Central America. Our specimens from northcentral and northeastern Colorado at 3500-6000 feet but probably widespread.
4. Sagittaria cuneata Sheld., Bull. Torr. Club 20:283. 1893.
   S. arifolia Nutt.---Plants about 20-40 cm. tall but varying with water depth, sometimes to 100 cm. tall; leaves 6-18 cm. long, acuminate to acute, basal lobes usually definitely shorter than terminal lobe; bracts ovate-lanceolate to linear-lanceolate; whorls of the inflorescence few to many; head of fruits 1-1.5 cm. in diameter; achenes 2-2.5 mm. long, winged, beak short and erect over the ventral wing. This species is hard to tell from the preceding even when mature achenes are present, and intergradations are to be expected.---Ontario to British Columbia, south to Michigan, Kansas, New Mexico and California. Widely scattered over Colorado at 3500-8000 feet.

2. Alisma L. WATERPLANTAIN

Plants with fibrous roots; leaves basal, erect or floating, not sagittate; flowers in whorled panicles; flowers small, perfect; stamens commonly 6; pistils few to many in a whorl on a flattened receptacle; fruit of achenes.

1. Pedicels slender, ascending; achenes longer than wide, 2-ridged on the back; scapes much longer than the leaves -1. A. plantago-aquatica
1. Pedicels usually stout, recurved in fruit; achenes about as wide as long, 3-ridged on the back; scapes little if any longer than the leaves -2. A. geyeri

1. Alisma plantago-aquatica L., Sp. Pl. 342. 1753.
A. brevipes Greene; A. plantago of manuals---Plants erect, glabrous, with often bulbous or cormlike rootstocks; leaves 3-15 cm. long, oblong or ovate but sometimes narrower, 3- to 9-nerved, usually abruptly acute at apex, cuneate to cordate at base, petioles long, often longer than the blades; scapes 10-100 cm. tall; branches and pedicels of inflorescence divergent in fruit; bracts lanceolate or linear; sepals broadly ovate to suborbicular, obtuse; petals about 3-6 mm. long, white or pinkish; achenes 1.5-2 mm. long, 2-ridged on back, beak near one side of apex.---In water and wet places. Throughout the cooler parts of the Northern Hemisphere, generally distributed throughout the United States. Scattered over the western two-thirds of Colorado at 4500-8000 feet.
Fernald (Rhodora 48:86-88. 1946) claimed our North American plants are different. If so our Colorado plant would probably be A. triviale Pursh.

2. Alisma geyeri Torr. in Nicollet, Rep. Hydrograph. Miss. Riv. 162. 1843.
Plants rather diffuse; leaves 5-9 cm. long, oblong, ovate-lanceolate, elliptic or rarely lanc-linear, acute or acuminate, cuneate at base to a petiole usually longer than the blade; scapes 10-50 cm. long; branches and pedicels of inflorescences divergent in fruit; bracts lanceolate; sepals rounded-ovate; petals 2-4 mm. long, white to pink; achenes about 2 mm. wide, 3-ridged on back, beak erect or nearly so.---Wet places. Throughout the northern part of the United States, south to Nebraska, Wyoming, Utah and California. The new record for Colorado is in the northwest corner at about 6000 feet.

Family 16. Hydrocharitaceae FROGS-BIT FAMILY

Submerged or floating aquatic plants; leaves cauline and opposite or whorled, or basal; flowers dioecious (or apparently polygamous), regular, in 1- to 3-flowered spathes; perianth 3- to 6-parted, the hypanthium tube sometimes elongating; stamens usually 2; ovary inferior, 1-celled with 3 parietal placentae, stigmas 3, nearly sessile

1. Leaves cauline, not over 4 cm. long -1. Elodea
1. Leaves basal, elongated to 100 cm. or more -2. Vallisneria

1. Elodea Michx. WATERWEED

Branching aquatic plants with fibrous roots; leaves whorled or rarely opposite, sessile, entire or finely serrate; flowers 1-3 in axillary, tubular, 2-cleft spathes; sepals 3; petals 3; stamens 3-9; pistillate flower with a hypanthium tube that elongates to 15 cm. long; fruit indehiscent, with 1-5 spindle-shaped seeds. The more familiar name Elodea is retained here.

1. Leaves 1.2-4 mm. wide, averaging 2 mm.; spathes of staminate flower 10-13 mm. or more long, constricted near base to a stalk; staminate flowers about 4-5 mm. long, remaining attached to long threadlike hypanthia; pistillate flowers with sepals about 2.5 mm. long -1. E. canadensis
1. Leaves 0.7-1.8 mm. wide, averaging 1.3 mm.; spathes of staminate flowers globose, about 2-3 mm. long; staminate flowers about 2-2.5 mm. long, sessile, breaking free from spathe and rising to water surface before anthesis; pistillate flowers with sepals 1-1.8 mm. long -2. E. nuttallii

1. Elodea canadensis Michx., Fl. Bor. Am. 1:20. 1803.
Anacharis canadensis (Michx.) Planchon; Philotria canadensis (Michx.) Britt.; P. iowensis Wylie; P. planchonii (Casp.) Rydb.; E. planchonii Casp.; probably Philotria angustifolia (Muhl.) Britt.---Stems slender; leaves 1.2-4 mm. wide, averaging 2 mm., linear; staminate spathes constricted below to form a stalk; the flower 4-5 mm. long, remaining attached to a long threadlike hypanthium; pistillate flower with sepals about 2.5 mm. long, with hypanthium tube 2-15 cm. long.---Lakes, streams and ponds. Present in the northeastern and midwestern states and locally westward. Our few records scattered thinly over Colorado at 5000-8500 feet.
A more robust species, Elodea densa (Planch.) Caspary, with leaves up to 5 mm. wide and flowers more than one in a spathe, is common in cultivation. There is some evidence that it may persist when planted in or escaped to some of our park lakes.

2. Elodea nuttallii (Planchon) St. John, Rhodora 22:29. 1920.
Anacharis occidentalis (Pursh) Marie-Vict.; Philotria minor (Engelm.) Small---Stems slender; leaves 0.7-1.8 mm. wide, averaging 1.3 mm., linear; staminate spathes 2-3 mm. long, globose, apiculate, the flowers 2-2.5 mm. long, sessile, breaking free from the spathe and rising to the surface before anthesis; pistillate flowers with sepals about 1-1.8 mm. long.---Ponds, lakes and streams. Distributed across the northern part of the United States. Our very doubtful records from northcentral Colorado at 5000-5500 feet.

2. Vallisneria L. EEL-GRASS; WILD CELERY

Acaulescent plants rooting in the mud or sand and producing stolons; leaves basal, narrowly linear and very long, entire or nearly so; staminate flowers nearly sessile, breaking away at anthesis and rising to the surface, there opening; stamens usually 2; pistillate flowers on a very long spiral or flexuous scape, the spathe 1-flowered; ovules numerous.

1. Vallisneria spiralis L., Sp. Pl. 1015. 1753.
This is an eastern species that has been reported from this state, probably erroneously. However, the plant is to be expected in eastern Colorado.

Family 17. Gramineae GRASS FAMILY

Herbaceous (in ours) annuals or perennials, with or without rhizomes; stems (culms) usually hollow except at the nodes but sometimes solid; leaves usually narrow, 2-ranked, composed of a sheath enclosing the culm and a free portion called the blade, hairy or membranous appendages (ligules) present on the inside between the blades and the sheaths; inflorescence a spike, raceme or panicle but often modified; flowers usually perfect, arranged in units called spikelets, these consisting of a short axis (rachilla) with 2 or more 2-ranked bracts, the lower 2 empty (glumes) the succeeding ones (lemmas) bearing in their axils a single flower and between the flower and the rachilla another bract (palea); lemma, palea and flower called a floret; ovary superior, 1-celled, 1-ovuled usually with 2 plumose stigmas; fruit a caryopsis. (Poaceae) This treatment includes the cultivated species of grasses.

KEY TO THE TRIBES OF GRAMINEAE

1. Spikelets 3-5, in close groups, these falling entire, sometimes enclosed in involucres
   2. Spikelets in a spiny involucre   -X. (Cenchrus in) Paniceae
   2. Spikelets not in a spiny involucre
      3. Spikelet clusters usually 2, the common peduncle short, the whole partly included in the sheaths, much overtopped by the leaf blades   -VI. (pistillate spikelets of Buchloe in) Chlorideae
      3. Spikelet clusters 3 to many on an elongated rachis overtopping the leaves   -V. (Hilaria in) Zoysieae
1. Spikelets single or if in clusters of 3-5 these not falling entire (a few have 2 spikelets that fall together and Hordeum has 1 perfect and 2 reduced spikelets that fall with the joint of a disarticulating rachis)
   4. Spikelets with 2 or more perfect florets, these essentially similar (or if dioecious with 2 or more similar staminate or pistillate florets in the spikelet); the sterile or reduced florets, if present, above the perfect ones (except in Arrhenatherum)
      5. Spikelets sessile (or very nearly so), on opposite sides or on 1 side of rachis
         6. Spikelets on opposite sides of the rachis, the spike solitary   -II. Hordeae
         6. Spikelets borne on 1 side of the rachis, forming 1-sided spikes, usually over 1 to a culm
                                                                           -VI. Chlorideae
      5. Spikelets in panicles (or rarely in racemes), on distinct and often long pedicels (these panicles may be contracted and spikelike in appearance)
         7. Glumes (at least 1) as long as the lowest floret, usually as long as the whole spikelet; awn from the back of the lemma or appearing so (lacking in a few genera)   -III. Aveneae
         7. Glumes shorter than the first floret; awn when present from the tip of the lemma or from between the teeth of a bifid apex   -I. Festuceae
   4. Spikelets with only 1 perfect floret (except in corn and a few genera where staminate or reduced spikelets are present along with the perfect one), the sterile or reduced florets, if present usually below the perfect one and different from it
      8. Spikelets with 1 or 2 sterile florets below the perfect one, these usually reduced to sterile lemmas; glumes and fertile lemma unlike in texture (1 or both glumes may be absent)
         9. Spikelets all unisexual, the pistillate (in ours) crowded on a thick rachis (cob), and the staminate above (tassel)   -XII. (Corn, Zea in) Tripsaceae
         9. Spikelets bisexual (staminate or reduced spikelets may be present also)
            10. Spikelets laterally compressed, disarticulating above the glumes, leaving the glumes attached to the inflorescence   -VII. Phalarideae
            10. Spikelets round or dorsally compressed, falling entire, the glumes attached to spikelet (however, 1 or both glumes sometimes absent)
                11. Glumes indurated, the sterile and the fertile lemma hyaline or thin-membranous; spikelets in pairs, 1 sessile and perfect, the other pedicellate and usually staminate or reduced (sometimes obsolete or perfect)   -XI. Andropogoneae
                11. Glumes and sterile lemma membranous, the fertile lemma much firmer; spikelets usually not in pairs as above, if so the pedicellate one perfect   -X. Paniceae
      8. Spikelets without any sterile florets below the single perfect one; glumes and fertile lemma usually similar in texture
         12. Spikelets sessile on opposite sides or on 1 side of the rachis
             13. Spikelets on opposite sides of the rachis, the spike solitary; auricles often present on leaves   -II. Hordeae
             13. Spikelets on 1 side of the rachis, forming 1-sided spikes, usually over 1 to a culm; auricles not present   -VI. Chlorideae
         12. Spikelets in panicles, these usually open but often contracted or spikelike but the spikelets definitely pedicellate
             14. Glumes absent; spikelets either round or very much flattened
                 15. Spikelets round, awned (in ours), unisexual   -IX. Zizanieae
                 15. Spikelets flattened laterally, awnless (in ours), bisexual   -VIII. Oryzeae

14. Glumes present (rarely 1 absent); spikelets usually moderately compressed laterally   -IV. Agrostideae

I. Tribe Festuceae

Spikelets with more than 1 perfect floret (or more than 1 similar pistillate or staminate floret in the few dioecious species) the reduced florets, when present, above (in ours); spikelets in open, narrow or spiklike panicles (rarely racemes); glumes seldom as long as the first floret; lemmas awnless or awned from the apex or near the apex; articulation above the glumes, often between the florets.

KEY TO THE GENERA OF FESTUCEAE

(Includes a few related genera in other tribes)

1. Plants dioecious, the sexes dissimilar, the pistillate lemmas with 3 long divergent awns, the staminate lemma awnless or mucronate   -18. Scleropogon
1. Plants with perfect flowers (or if dioecious the sexes similar); lemmas never 3-awned
  2. Rachillas with long silky hairs as long as the lemmas; tall reedlike grasses   -14. Phragmites
  2. Rachillas glabrous or with much shorter hairs (callus may be bearded); grasses rarely reedlike
    3. Inflorescence a few-flowered, capitate head or a short dense raceme, overtopped and partly hidden by several leaves; annual grasses
      4. Leaves sharp-pointed with a definite white marginal band; lemmas 3-nerved, sharp-pointed; ligule made up of hairs; plants often rooting at the nodes   -60. Munroa
      4. Leaves not sharp-pointed, no conspicuous marginal band; lemmas 5- to 7-nerved, obtuse; ligule membranous; plants not rooting at the nodes   -6. Sclerochloa
    3. Inflorescence an exserted open or spikelike panicle not hidden in leaves; annuals or perennials (inflorescence may be partly hidden in the upper sheath especially when immature)
      5. All or most of the spikelets transformed into bulblets; culms bulbous at base   -7. Poa
      5. No bulblets present; base of culms usually not bulbous
        6. Spikelets of 2 kinds, the fertile and staminate mixed on the same inflorescence, the fertile 2- or 3-flowered, the sterile with numerous rigid scales; panicles spikelike   -13. Cynosurus
        6. Spikelets all alike in the same inflorescence; panicles open or contracted but not spikelike
          7. Lemmas 3-nerved, the nerves prominent (except in a few species of Eragrostis); ligules made up of hairs (except in Catabrosa and Leptochloa)
            8. Nerves of lemmas hairy at least near the base, especially the lateral nerves; apex of lemmas usually with a short awn or mucro
              9. Perennial grasses; panicles (in ours) narrow, with short branches   -16. Tridens
              9. Annuals; panicles at maturity with spreading branches
                10. Paleas long-pilose on upper part of keel; ligule a ring of hairs   -17. Triplasis
                10. Paleas not pilose as above (may be pubescent on lower part); ligule membranous   -52. Leptochloa
            8. Nerves of lemmas not pubescent near base, (glabrous or scabrous above and may have hairs on callus); apex of lemmas usually acute
              11. Lemmas 4-6 mm. long; sand plants with large, conspicuous rhizomes; callus villous   -10. Redfieldia
              11. Lemmas less than 4 mm. long; plants not limited to sandy soil, no rhizomes; callus glabrous or short-pubescent
                12. Spikelets with 2 florets; ligules membranous; growing in water or very wet soil   -9. Catabrosa
                12. Spikelets with more than 2 florets; ligules made up entirely or mostly of hairs; seldom growing in water
                  13. Lemmas broadly rounded and emarginate at apex, midnerve excurrent as a short awn; callus short-pubescent; panicle narrow and over 10 cm. long   -16. Tridens
                  13. Lemmas acute or acuminate at apex; callus glabrous; panicle if very narrow then seldom over 10 cm. long   -8. Eragrostis
      7. Lemmas 5- to many-nerved, the nerves usually not conspicuous; ligules membranous (except in Distichlis)
        14. Callus bearded; lemmas awned; sheaths joined in front (unless torn) to very top of ligule   -15A. Schizachne
        14. Callus not bearded (base of lemma may have cobwebby hairs in Poa); lemmas awned or awnless; sheaths split in front at least a short distance
          15. Lemmas awned, the awn 1 mm. or more long
            16. Lateral spikelets nearly sessile on the main rachis, their pedicels less than 1 mm. long (but inflorescence often reduced to a single spikelet)   -2. Brachypodium
            16. Lateral spikelets on longer pedicels; inflorescence rarely reduced to a single spikelet
              17. Spikelets over 12 mm. long; lemmas awned from a bifid apex; sheaths closed (unless torn) for most of their length   -1. Bromus
              17. Spikelets less than 12 mm. long; lemmas with a terminal awn; sheaths open   -3. Festuca
          15. Lemmas awnless or awn-pointed, the point less than 1 mm. long (rarely to 1.1 mm. in Dactylis)

18. Spikelets in dense rather 1-sided clusters toward the end of the panicle branches   -12. Dactylis
18. Spikelets not clustered as above (1-sided in a species of Melica but not clustered)
   19. Lemmas coriaceous (especially on pistillate spikelets); paleas serrate; ligules mostly of hairs; plant dioecious   -11. Distichlis
   19. Lemmas membranous; paleas not serrate; ligules membranous; plant not dioecious (except in Hesperochloa)
      20. Spikelets disarticulating below the glumes, often all turned toward 1 side   -15. Melica
      20. Spikelets disarticulating above the glumes, not 1-sided on the panicle
         21. Culms usually bulbous at base; glumes rather papery; lemmas of 2 or more upper florets often folded together forming 1 structure; sheaths closed   -15. Melica
         21. Culms not bulbous at base; glumes not papery; upper florets not folded closely together; sheaths open or closed (mostly open)
            22. Lemmas 8 mm. long or more, obtuse or mucronate at apex; sheaths closed   -1. Bromus
            22. Lemmas less than 8 mm. long or if approaching that length then acute or awn-pointed; sheaths open or closed
               23. Lemmas with parallel nerves, these not converging toward the tip, hence the apex broadly obtuse or truncate
                  24. Nerves faint; plants usually of low alkaline soil; sheaths open   -4. Puccinellia
                  24. Nerves prominent; plants of fresh water marshes and wet places; sheaths closed for most of the way   -5. Glyceria
               23. Lemmas with nerves converging toward the tip
                  25. Lemmas compressed-keeled   -7. Poa
                  25. Lemmas rounded on the back (glumes can be keeled)
                     26. Lemmas less than 5 mm. long; glumes usually compressed-keeled   -7. Poa
                     26. Lemmas 5 mm. or more long; glumes not compressed-keeled
                        27. Plants dioecious; stigmas hispidulous all around (the species in this genus is also keyed out in Festuca)   -3A. Hesperochloa
                        27. Plants with perfect flowers; stigmas plumose   -3. Festuca

II. Tribe Hordeae

Spikelets with 1 to several florets, sessile on opposite sides of an articulated or continuous rachis, forming a 2-sided spike, this single spike on 1 culm; lemmas awned or awnless; auricles commonly present.

KEY TO THE GENERA OF HORDEAE

1. Spikelets with 1 floret, clustered 3 at a node of the rachis, the lateral 2 pediceled and reduced (except in forms of cultivated barley)   -25. Hordeum
1. Spikelets with 2 or more florets, 1 or more at the node of the rachis but all equal
   2. Spikelets solitary at each node of the rachis
      3. Spikelets with 1 edge turned toward the rachis, the glume on that side wanting (present on the terminal spikelets)   -26. Lolium
      3. Spikelets placed flat-wise to the rachis, **both glumes present**
         4. Spikelets cylindrical, those at the end of the rachis long-awned, those below short-awned or awnless   -21. Aegilops
         4. Spikelets at least somewhat compressed, awnless or about evenly awned all along the spike
            5. Spikelets with a short pedicel (about 0.5 mm. long); inflorescence often reduced to a single spikelet; often over 8 florets to a spikelet   -2. Brachypodium
            5. Spikelets sessile; inflorescence never reduced to a single spikelet, not over 8 florets to a spikelet (except rarely in Agropyron smithii)
               6. Annual plants, cultivated for grain but often growing along fields and roads
                  7. Glumes subulate, 1-nerved   -22. Secale
                  7. Glumes broad, ovate, 3-nerved or more   -20. Triticum
               6. Perennial plants (except 1 introduced species collected only once in Colorado), rarely cultivated and then not for grain   -19. Agropyron
   2. Spikelets 2 or more at all or some nodes of the spike
      8. Rachis readily breaking up at maturity
         9. Awns over 3 cm. long, widely divergent   -24. Sitanion
         9. Awns shorter, not over 2 cm. long, straight   -23. Elymus macounii
      8. Rachis continuous, not breaking up
         10. Rhizomes absent   -23. Elymus
         10. Rhizomes present
            11. Glumes somewhat separated at base, exposing the rachilla below **the lowest floret, the glumes** usually not exactly at the sides of the spikelet   -23. Elymus triticoides
            11. Glumes slightly overlapping at base, hiding the lowest rachilla joint, the glumes at the sides of the spikelet   -19. Agropyron smithii

III. Tribe Aveneae

Spikelets with 2 to several florets, in open or contracted panicles (rarely solitary or in racemes); glumes usually as long or longer than the first floret, commonly longer than all the florets; lemmas usually awned from the back or from between long teeth, the awn usually twisted and bent; articulation usually above the glumes.

KEY TO THE GENERA OF AVENEAE

1. Spikelets with 2 unlike florets, 1 perfect, the other staminate, only 1 of the pair definitely awned (upper sometimes awned in Arrhenatherum)
   2. Lower florets staminate with a twisted, geniculate awn; articulation above the glumes   -32. Arrhenatherum
   2. Upper florets staminate, with a hooked awn; articulation below the glumes   -33. Holcus
1. Spikelets with 2 or more similar florets, all awned or all awnless (except in cultivated oats)
   3. Spikelets large, the glumes over 8 mm. long; awns large and conspicuous (except in cultivated oats)
      4. Lemmas with a flattened awn from between terminal teeth; spikelets usually with 3 or more florets; ligules made up of hairs   -34. Danthonia
      4. Lemmas with the awn not conspicuously flattened, arising from the back (may be absent in cultivated oats); spikelets usually with 2 florets (except in H. hookeri); ligules membranous
         5. Annual grasses; cultivated or as weeds along fields or in waste places; glumes over 15 mm. long; spikelets nodding at maturity   -31. Avena
         5. Perennial grasses; native, not weedy or cultivated; glumes not over 15 mm. long; spikelets erect   -31A. Helictotrichon
   3. Spikelets smaller, glumes less than 8 mm. long; awn, when present, less than 12 mm. long
      6. Lemmas awnless or with a very short, straight awn from just below the apex, the back of the lemmas keeled
         7. Articulation below the glume; glumes very unlike, the first very narrow, the second broadest near tip   -28. Sphenopholis
         7. Articulation above the glumes; glumes nearly alike to somewhat unlike (the first may be somewhat narrower but glumes not as above)
            8. Rachilla at base of lemmas bearded at base; glumes about equal in length, 5-6 mm. long (in ours)   -29. Trisetum
            8. Rachilla at base of lemmas glabrous or short-pubescent, not bearded; glumes unequal in length (usually 1 mm. difference), 2-4 mm. long   -27. Koeleria
      6. Lemmas with a dorsal, often geniculate awn, back of lemmas keeled or rounded
         9. Lemmas rounded on back, awned from the middle or below   -30. Deschampsia
         9. Lemmas keeled, awned from well above the middle (about 1/4 to 1/3 down from apex)   -29. Trisetum

IV. Tribe Agrostideae

Spikelets 1-flowered, usually perfect, no sterile or reduced florets below the perfect one, in open, contracted or spikelike panicles; glumes variable; lemmas awned or awnless; articulation usually above the glumes.

KEY TO THE GENERA OF AGROSTIDEAE

1. Awns of lemmas 3-parted (lateral branches may be short)   -50. Aristida
1. Awns of lemmas simple or absent entirely
   2. Awns of lemma twisted near base and once or twice geniculate
      3. Awns dorsal; lemmas membranous   -35. Calamagrostis
      3. Awns terminal; lemmas indurated
         4. Awns readily deciduous, only slightly twisted or geniculate   -47. Oryzopsis
         4. Awns persistent, definitely twisted and geniculate
            5. Edges of lemmas meeting or overlapping, hiding the paleas completely; callus sharp-pointed; lemmas 4 mm. or more long   -49. Stipa
            5. Edges of lemmas not meeting but exposing part of the paleas; callus short, not sharp-pointed; lemmas 3-4 mm. long   -48. Piptochaetium
   2. Awns of lemma not twisted or geniculate, or absent entirely
      6. Lemmas indurated, terete; callus well developed   -47. Oryzopsis
      6. Lemmas thin or firm but not indurated (rather firm in Calamovilfa and Lycurus but not nearly terete); callus not conspicuous
         7. Florets with a tuft of long hairs at the base on the short callus, these hairs 1 mm. or more long
            8. Lemmas firm and chartaceous, awnless; ligules made up of hairs   -36. Calamovilfa
            8. Lemmas membranous, awned; ligules membranous (may be ciliate)
               9. Lemmas awned from the tip   -44. Muhlenbergia
               9. Lemmas awned from the back (awn often very delicate)   -35. Calamagrostis
         7. Florets without long hairs at base
            10. Lemmas awned, the awn over 1 mm. long (may be delicate)
               11. Glumes awned, the awn 2 mm. long or more
                  12. First glumes 2-nerved, each nerve ending in an awn; spikelets in pairs, 1 perfect, the other staminate or neuter, the pair falling together   -42. Lycurus
                  12. First glumes with 1 awn; spikelets all alike, not as above   -41. Polypogon
               11. Glumes awnless or merely awn-pointed
                  13. Awn of lemmas from the apex or near the apex; paleas over 1/2 the length of the lemmas
                     14. Articulation below the glumes; awns of lemmas from just below the apex and seldom over 1 mm. long   -39. Cinna
                     14. Articulation above the glumes; awns usually terminal and over 1 mm. long   -44. Muhlenbergia
                  13. Awn of lemmas from the middle or below the middle; paleas absent or minute
                     15. Panicle dense and spikelike; glumes united near the base, conspicuously compressed-keeled   -40. Alopecurus

15. Panicle open or contracted but not spikelike; glumes not at all united, not conspicuously compressed-keeled -37. Agrostis
10. Lemmas without awns, if awn-pointed the point less than 1 mm. long
    16. Nerves of lemmas densely silky-hairy -46. Blepharoneuron
    16. Nerves of lemmas not silky-hairy, if lemmas pubescent not especially on the nerves
        17. Glumes awned, the awns 1 mm. or more long
            18. Awn of glumes 4 mm. or more long; spikelets disarticulating below the glumes -41. Polypogon
            18. Awn of glumes not over 3 mm. long; spikelets disarticulating above the glumes (not always plain in Phleum)
                19. Glumes strongly compressed-keeled, hispid-ciliate on the keel; panicle very dense and spikelike -43. Phleum
                19. Glumes not strongly compressed-keeled, not hispid-ciliate on keel; panicle not very dense -44. Muhlenbergia
        17. Glumes awnless, or if awn-pointed the point less than 1 mm. long
            20. Panicles with few spikelets, overtopped by the leaves; glumes minute, less than 1 mm. long, the first often wanting -38. Phippsia
            20. Panicles many-flowered, not overtopped by the leaves; glumes usually over 1 mm. long
                21. Spikelets disarticulating below the glumes -39. Cinna
                21. Spikelets disarticulating above the glumes
                    22. Paleas wanting or if present not over 2/3 of the length of the lemmas; glumes usually as long or longer than the lemmas -37. Agrostis
                    22. Paleas present, about as long as the lemmas; glumes seldom as long as the lemmas
                        23. Lemmas 1-nerved, not apiculate or mucronate (but can be acute); caryopsis at maturity falling free from its lemma and palea; seed loose in the pericarp (when wetted); ligules mostly of hairs -45. Sporobolus
                        23. Lemmas 3-nerved (sometimes inconspicuous), usually apiculate or mucronate; caryopsis at maturity falling with its lemma and palea; seed not loose in the pericarp; ligules membranous (except in M. pungens) -44. Muhlenbergia

V. Tribe Zoysieae

Spikelets 3 to 5, subsessile in short spikes, these falling entire from the axis of a spikelike inflorescence (ours imperfect), the staminate 2-flowered; glumes firmer than the lemmas.
(One genus)-51. Hilaria

VI. Tribe Chlorideae

Spikelets 1- to several-flowered, in 2 rows but these both twisted to 1 side of spikes or spikelike racemes, these spikes solitary, digitate or racemose along an elongated axis; lemmas awnless or awned.

KEY TO THE GENERA OF CHLORIDEAE

1. Inflorescence a few-flowered head hidden among the leaves
    2. Flowers unisexual; spikelets disarticulating below the glumes, several spikelets falling as a unit; leaves not sharp-pointed -61. (pistillate) Buchloe
    2. Flowers bisexual; spikelets disarticulating above the glumes; leaves sharp-pointed -60. Munroa
1. Inflorescence not hidden among the leaves
    3. Spikelets with 1 floret, no rudimentary floret above it (rachilla may be prolonged as a bristle)
        4. Spikes digitate; rachilla prolonged beyond the fertile palea as a bristle -54. Cynodon
        4. Spikes paniculate; rachilla not at all prolonged
            5. Spikelets disarticulating below the glumes; panicle rather narrow, the thick spikes ascending
                6. Glumes unequal, narrow, spikelets very compressed; stout rhizomes present; ligule mostly of hairs -57. Spartina
                6. Glumes equal, broad; spikelets not conspicuously compressed; no rhizomes; ligule membranous -56. Beckmannia
            5. Spikelets disarticulating above the glumes; panicle with long, narrow, spreading spikes (becoming a tumbleweed) -55. Schedonnardus
    3. Spikelets with 2 or more florets (the upper one may be a rudimentary one)
        7. Plants stoloniferous; spikelets with 2 staminate, similar florets -61. (staminate) Buchloe
        7. Plants not stoloniferous; spikelets either with 1 perfect floret and an unlike rudiment above, or with 3 or more perfect, similar florets
            8. Spikelets with 1 perfect floret and 1 or more unlike rudiments above
                9. Spikes aggregated on a very short axis -58. Chloris
                9. Spikes racemose along an elongated axis -59. Bouteloua
            8. Spikelets with 3 or more similar florets
                10. Spikes many, slender, racemose on an elongated rachis; spikelets short-pediceled -52. Leptochloa
                10. Spikes few, rather thick, digitate or nearly so; spikelets sessile -53. Eleusine

VII. Tribe Phalarideae

Spikelets with 1 perfect, terminal floret and a staminate or reduced floret on each side (rarely only 1); glumes membranous (in ours); fertile lemma firm; articulation above the glumes.

KEY TO THE GENERA OF PHALARIDEAE

1. Glumes unequal in length, the first about 1/2 the length of the second; sterile lemmas awned -63. Anthoxanthum
1. Glumes equal; sterile lemma or floret awnless
    2. Lateral florets staminate, as long or longer than the central perfect one; florets brown -62. Hierochloe
    2. Lateral florets (rarely 1) reduced to sterile lemmas not over 2/3 the length of the perfect central one; florets green or pale -64. Phalaris

VIII. Tribe Oryzeae

Spikelets 1-flowered, perfect, strongly compressed laterally, arranged in panicles; glumes wanting (in ours); paleas apparently 1-nerved; stamens 6.   (One genus) -65. Leersia

IX. Tribe Zizanieae

Spikelets unisexual, terete or nearly so, staminate and pistillate borne in the same inflorescence (in ours); glumes absent or very small.   (One genus) -66. Zizania

X. Tribe Paniceae

Spikelets with 1 perfect terminal floret and below this a sterile floret often reduced to a lemma simulating a third glume; inflorescence an open panicle, or of spikelike racemes, these digitate or arranged along an elongated rachis; 2 glumes usually present; fertile lemma and palea firmer than the sterile lemma and glumes; articulation below the glumes.

KEY TO THE GENERA OF PANICEAE

1. Spikelets enclosed in spiny burs, the spines retrorsely barbed -75. Cenchrus
1. Spikelets not in a spiny bur
    2. Spikelets subtended by bristles
        3. Bristles falling with the spikelet at maturity; grasses cultivated for ornament (ours) -74. Pennisetum
        3. Bristles not falling with the spikelet; mostly weeds, or if cultivated, not for ornament -73. Setaria
    2. Spikelets not subtended by bristles
        4. Spikelets awned, awns of at least some sterile lemmas over 1 mm. long; ligules absent entirely -72. Echinochloa
        4. Spikelets awnless, if the sterile lemmas awn-pointed this less than 1 mm. long; ligules present (may be of short hairs or a short membrane)
            5. Spikelets in open panicles, each one on a long pedicel; ligules a ring of hairs -71. Panicum
            5. Spikelets in digitate spikelike racemes or of dense racemes appressed along a main rachis, spikelets sessile or short pediceled; ligules membranous (except in Eriochloa)
                6. First glume and rachilla joint forming a ringlike callus below the spikelet; fertile lemmas with a short deciduous awn; ligules made up of hairs -69. Eriochloa
                6. First glume present or absent but not forming a ringlike callus; fertile lemmas awnless; ligules membranous
                    7. Sterile lemmas conspicuously transversely fluted or wrinkled; cultivated for ornament and rarely persisting -70. Paspalum racemosa
                    7. Sterile lemmas not transversely wrinkled; if present in cultivated ground only as weeds
                        8. Racemes long (usually over 5 cm.) digitate or if on the rachis of a panicle then widely spreading
                            9. Annual weeds; rachis of racemes winged; racemes digitate or nearly so -68. Digitaria
                            9. Perennials, not weedy; rachis of racemes not winged; racemes usually along a main rachis of a panicle -70. Paspalum
                        8. Racemes rather short (rarely over 5 cm.), appressed along the elongated rachis of a panicle
                            10. Spikelets covered with long silky hairs; no long stolons present -67. Trichachne
                            10. Spikelets glabrous; plants developing long stolons -71. Panicum obtusum

XI. Tribe Andropogoneae

Spikelets perfect or imperfect, borne in pairs along a rachis, usually 1 perfect and sessile, the other pedicellate and staminate or reduced; fertile spikelet of 1 terminal floret and below this a staminate or reduced floret; glumes firm; sterile lemma, fertile lemma and palea thin, often hyaline; articulation below the glumes.

KEY TO THE GENERA OF ANDROPOGONEAE

1. Spikelets all perfect, surrounded by a copious tuft of hairs 4 mm. long or more, inserted at base; large grasses cultivated for ornament

2. Panicle rather fan-shaped, less than 20 cm. long, the racemes aggregated at base; rachis of racemes falling entire; both of spikelet pairs pedicellate (but one short)   -76. Miscanthus
2. Panicle oblong, usually over 20 cm. long, the racemes on an elongated rachis; rachis of racemes breaking into sections at maturity; one of the spikelet pairs sessile   -77. Erianthus
1. Spikelets unlike, the sessile one perfect, the **pedicellate** one staminate or reduced (even absent sometimes), not closely surrounded by long hairs (may have hairs on pedicel and rachis but these not tufted around the spikelet and seldom 4 mm. long); native grasses (only Sorghum cultivated and not for ornament)
  3. Racemes solitary or several, these digitate, not in an elongated panicle   -78. Andropogon
  3. Racemes many, disposed along an elongated panicle rachis
    4. Racemes of several to many joints, sessile, the rachis of the raceme thin in the center and thickened at margins; panicle silvery-hairy   -78. Andropogon
    4. Racemes with few joints, these peduncled, rachis of raceme not thicker on margins; panicle, if hairy, not silvery-hairy
      5. Pedicellate spikelets wanting, only the pedicel present; panicle narrow, seldom over 8 cm. wide; plants native; leaves less than 10 mm. wide   -80. Sorghastrum
      5. Pedicellate spikelets present and staminate; panicle usually over 8 cm. wide (narrower in some cultivated sorghums); plants usually cultivated; leaves usually over 10 mm. wide   -79. Sorghum

XII. Tribe Tripsaceae

    Spikelets unisexual, the staminate in pairs or sometimes in 3's, 2-flowered; pistillate spikelets usually single, 2-flowered, the lower floret sterile, embedded in hollows of a thickened axis or enclosed in a thickened involucre or sheath or (as in ours) crowded in rows on a thickened axis (cob); glumes awnless, membranous or rigid; lemmas hyaline, awnless; paleas hyaline; plants **monoecious**.   (One genus) -81. Zea

I. Tribe Festuceae

1. Bromus L. BROMEGRASS

    Annual or perennial plants, rhizomes present or absent; leaf blades flat; sheaths closed for most of the way; ligules membranaceous; inflorescence a panicle, this usually rather open; spikelets large, over 12 mm. long, several- to many-flowered, disarticulating above the glumes and between the florets, the florets perfect or the upper ones reduced and imperfect; glumes unequal, acute; lemmas round or keeled on the back, 5- to 9-nerved, awned from between apical teeth (may appear dorsal) or awnless; paleas shorter or as long as the lemmas, the grain adherent to it; ovary with a hairy cushionlike appendage at the apex with the styles inserted laterally below it, the stigmas plumose.

1. Lemmas compressed-keeled, the spikelets strongly flattened; first glume more than 1-nerved
  2. Awn absent or short, less than 3 mm. long   -1. B. catharticus
  2. Some awns over 3 mm. (B. carinatus complex, with numerous intergradations)
    2A. Lemma glabrous or at most scabrous; sheaths usually glabrous; panicle branches erect or ascending   -2. **B. polyanthus**
    2A. Lemmas pubescent to puberulent; sheaths usually pilose-pubescent; panicle branches spreading or reflexed
      2B. Glumes pubescent; awn less than 7 mm. long; panicle branches spreading but not deflexed   -3. B. marginatus
      2B. Glumes glabrous or scabrous; some awns 7 mm. long or more; panicle branches often deflexed   -4. B. carinatus
1. Lemmas rounded on the back, the spikelet little if any flattened; first glume with 1 or more nerves
  3. Rhizomes present; awn of lemma not over 3 mm. long
    4. Lemmas glabrous to somewhat scabrous (rarely ciliate on the lower margin); escaped from cultivation and most common around fields and along roads   -5. B. inermis
    4. Lemmas pubescent; native in the mountains
      5. Lemmas pubescent along the margins and across the back at the base   -6. B. pumpellianus
      5. Lemmas densely pubescent over the back   -6A. B. pumpellianus tweedyi
  3. Rhizomes not present; awns often over 3 mm. long
    6. Weedy annuals
      7. Awns geniculate, twisted; teeth of lemmas aristate; rare in Colorado   -7. B. trinii
      7. Awns straight to divergent, not twisted (except slightly in No. 12); teeth of lemmas short to long but not aristate
        8. Lemmas narrow with sharp callus and acuminate at apex, the teeth over 2 mm. long; first glume 1-nerved
          9. Awns 2-3 cm. long; second glume over 1 cm. long; rare in Colorado   -8. B. sterilis
          9. Awns less than 2 cm. long; second glume shorter; locally common in Colorado
            10. Spikelets pubescent   -9. B. tectorum
            10. Spikelets glabrous or at most scabrous   -9A. B. tectorum glabratus
        8. Lemmas broad, callus not especially sharp, not acuminate at apex, the teeth rarely over 1 mm. long; first glume 3- to 5-nerved
          11. Awns wanting or very short; lemmas very broad and conspicuously inflated   -10. B. brizaeformis
          11. Awns definite (sometimes short on lower florets of No. 11); lemmas not very broad or conspicuously inflated
            12. Sheaths glabrous or at most puberulent; awn of lemmas mostly less than 5 mm. long   -11. B. secalinus

12. Sheaths (at least the lower ones) pilose-pubescent; awns mostly over 5 mm. long
    13. Awns at maturity becoming somewhat twisted and divergent; panicle open, the branches lax or flexuous -12. B. japonicus
    13. Awns remaining straight and not at all twisted; panicle branches stiffly erect or spreading
        14. Panicle contracted, rather dense, rarely over 10 cm. long, the branches erect or ascending -13. B. racemosus
        14. Panicle open, usually over 10 cm. long, the branches spreading -14. B. commutatus
6. Perennials, native and hardly weedy
    15. Lemmas partially or completely glabrous, the hairs when present confined to the base and the margins
        16. Lemmas pubescent on the margins and at the base -15. B. ciliatus
        16. Lemmas glabrous -15A. B. ciliatus laeviglumis
    15. Lemmas pubescent over the entire back (but often more densely so on the lower part)
        17. Panicle small, usually less than 10 cm. long
            18. First glume 3-nerved (at least on some spikelets); sheaths pilose to glabrous -16. B. anomalus
            18. First glume 1-nerved; sheaths densely pilose to lanate -16A. B. anomalus lanatipes
        17. Panicle larger, over 10 cm. long
            19. Sheaths glabrous or nearly so; leaf blades mostly less than 5 mm. wide; first glume more than 1-nerved -17. B. frondosus
            19. Sheaths pilose, at least the lower; leaf blades mostly over 5 mm. wide; first glume 1-nerved
                20. Sheaths shorter than the internodes, only 4-6 nodes present, sheaths without a ring of pilose hairs, no flanges or auricles at the base of the blade -18. B. purgans
                20. Sheaths longer than the internodes, the nodes 8-20 (sheaths much overlapping), sheaths with a ring of pilose hairs at the base of the blade, flanges or auricles present -19. B. latiglumis

1. Bromus catharticus Vahl., Symb. Bot. 2:22. 1791.
    B. unioloides (Willd.) H. B. K.---Annual or biennial plants; culms 40-100 cm. tall, stout, pale gray-green, glabrous; leaf blades 2-5 mm. wide, glabrous or sparingly pilose; sheaths pubescent or glabrous; ligules 2-4 mm. long; panicle 10-20 cm. long, lower branches often drooping; spikelets 2-3.5 cm. long, about 6- to 12-flowered; first glume 7-10 mm. long, 3- to 5-nerved, glabrous or slightly scabrous; lemmas 12-16 mm. long, 7-nerved, glabrous to scabrous or scabro-puberulent, awnless or with straight, stout awns rarely over 2.5 mm. long.---Cultivated or escaped. Mostly in the southern part of the United States. Apparently rare in Colorado, at about 5000 feet altitude.

2. Bromus polyanthus Scribn., Shear in U. S. D. A. Div. Agrost. Bull. 23:56. 1900.
    B. paniculatus (Shear) Rydb.---Perennial plants without rhizomes; culms 40-100 cm. tall, rather stout, erect, smooth; leaf blades 4-8 mm. wide, scabrous especially above; sheaths glabrous or at most scabrous; ligules about 2 mm. long; panicles elongated, erect; spikelets 2-3.5 cm. long, 7- to 11-flowered; first glume 6-8 mm. long, 3-nerved, glabrous or scabrous; second glume 9-11 mm. long, 3- to 7-nerved, glabrous or scabrous; lemmas 12-14 mm. long, 7-nerved, the awns 4-6 mm. long, straight.---Intergrades with No. 3 and 4 in this area. In meadows, open woods, etc. Montana to Washington and south to Arizona and California. In the mountainous part of Colorado (western two-thirds) from 6000 to 11,000 feet altitude.

3. Bromus marginatus Nees. in Steud., Syn. Pl. Glum. 1:322. 1854.
    B. latior (Shear) Rydb.---Perennial plants (mostly), without rhizomes; culms 30-120 cm. tall, stout, erect, pubescent below the panicles; leaf blades 4-12 mm. wide, pilose-pubescent throughout, sometimes sparsely so; sheaths pilose-pubescent; ligules 2-3.5 mm. long; panicle 10-40 cm. long, erect, branches erect or spreading; spikelets 2-4 cm. long (mostly 2.5-3), 5- to 10-flowered; first glume 7-11 mm. long, 3- to 7-nerved; second glume 9-12 mm. long, 5- to 9-nerved; lemmas 11-14 mm. long, 7- to 9-nerved, pubescent with rather coarse hairs, awns rather stout, straight, 4-6 mm. long.---Intergrades with No. 2 and 4. Open woods, meadows and waste places. In the western part of the United States. Scattered in the western two-thirds of Colorado from 5000 to 10,000 feet altitude.

4. Bromus carinatus Hook. and Arn., Bot. Beech. Voy. Suppl. 403. 1840.
    B. hookerianus Thurb.---Biennial or annual plants (sometimes look like perennial) without rhizomes; culms 50-100 cm. tall, erect; leaf blades 3-8 mm. wide, or rarely more; sheaths retrorsely pilose to nearly glabrous; ligules about 2 mm. long; panicle 18-40 cm. long with spreading often deflexed branches or reduced to a raceme; spikelets 2-3 cm. long, 5- to 10-flowered, the florets not closely overlapping in anthesis, exposing the rachilla; first glume 9-12 mm. long, 5-nerved; second glume 9-12 mm. long, 5-nerved; lemmas 10-20 mm. long, with 5 rather faint nerves, more or less appressed pubescent, rarely glabrous, the awns 7-15 mm. long.---Open ground, open woods or waste places. Western half of the United States to western Nebraska, Kansas and Texas. Apparently rare in Colorado, our one specimen collected at 7500 feet in Routt County.

5. Bromus inermis Leyss., Fl. Hal.16. 1761.
    Perennial plants with creeping rhizomes; culms 40-100 cm. tall, erect; leaf blades 5-14 mm. wide (averaging about 8), glabrous to scabrous; sheaths glabrous; ligules 1.5-2 mm. long; panicle 16-20 cm. long, rather dense, erect but somewhat open, often reddish-tinged; spikelets 2-2.5 cm. long, 5- to 10-flowered; first glume 4-6 mm. long, 1-nerved; second glume 6-9 mm. long, 3-nerved; lemmas 9-12 mm. long, 5- to 7-nerved, mucronate or with awns 1 to 2 mm. long, rarely to 3.---Cultivated and escaped along roads and fields. Especially in the western two-thirds of the United States. Rather generally distributed over Colorado, mostly from the western two-thirds of the state at 4500 to 10,000 feet. This species tends to intergrade with the following one especially at higher altitudes.

6. Bromus pumpellianus Scribn., Bull. Torr. Club 15:9. 1888.
    Bromus pumpellianus melicoides Shear---Perennial plants with creeping rhizomes; culms 30-100 cm. tall, erect; leaf blades 3-10 mm. wide, glabrous or somewhat pubescent especially above; sheaths glabrous or sparsely pilose-pubescent; ligules about 1 mm. long or less; panicle 5-20 cm.

long, erect, rather narrow, green to purplish; spikelets about 2 cm. long, sometimes longer, 7- to 11-flowered; first glume 6-8 mm. long, 1-nerved; second glume 7-9 mm. long, 3-nerved; lemmas 10-12 mm. long, 5- to 7-nerved, the awns about 2-3 mm. long.---In meadows and on hillsides. Colorado to Idaho and western South Dakota. In Colorado, in the mountains in a strip from Larimer to Costillo County and in Montezuma County at 8000 to 10,500 feet.

6A. Bromus pumpellianus tweedyi Scribn. in Beal, (var.) Grass. N. A. 2:622. 1896.

Differs from the species in having the leaves and sheaths usually pilose, the lemmas with densely pubescent or villous margins with some hairs on the back.---Dry rocky slopes from Alberta to Colorado. In our state in the northcentral part from Larimer to Gilpin County at 9000 to 9500 feet.

7. **Bromus trinii** Desv. in Gay, Fl. Chile 6:441. 1853.

Annual plants; culms 30-60 cm. tall, often pubescent at the nodes, erect or spreading below; leaf blades 1-2 mm. wide, sometimes to 5 mm., pilose-pubescent to smooth; sheaths usually pilose but sometimes glabrous; ligules 0.5-1 mm. long; panicle 8-20 cm. long, rather narrow, the branches erect or somewhat spreading; spikelets 1.5-2 cm. long, 3- to 7-flowered, narrow; first glume 6-10 mm. long, mostly 1-nerved, glabrous, acuminate or subulate at apex; second glume similar but 10-16 mm. long and mostly 3-nerved; lemmas 8-14 mm. long, 5- to 7-nerved, pubescent but sometimes sparsely so, awns 1-2 cm. long, bent below the middle.---Plains and slopes. Western United States. Collected once in Colorado on La Veta Pass at 9100 feet, to be expected in the western part of the state.

8. **Bromus sterilis** L., Sp. Pl. 1:77. 1753.

Annual plants; culms 50-100 cm. tall; leaf blades 2-4 mm. wide, scabrous to pilose; sheaths usually pubescent; ligules about 1.5 mm. long; panicle 10-20 cm. long, the branches drooping; spikelets 2.5-3.5 cm. long, 6- to 10-flowered, oblong-lanceolate; first glume 7-9 mm. long, 1-nerved; second glume 11-13 mm. long, 3-nerved; lemmas 17-20 mm. long, 5- to 7-nerved, scabrous to puberulent with apical teeth about 2 mm. long, awns rough.---Along fields, roadsides and in waste places. From British Columbia to California and New Mexico and the eastern states to Illinois and Alabama. From Boulder and Larimer Counties at about 5000 feet but to be expected anywhere near cultivated areas.

9. **Bromus tectorum** L., Sp. Pl. 1:77. 1753.

Annual plants; culms 15-60 cm. tall, erect or spreading; leaf blades 3-7 mm. wide, usually softly pubescent; sheaths pubescent, at least the lower ones; ligules about 2-3 mm. long; panicle about 5-15 cm. long, broad, drooping, rather secund, often purplish; spikelets 12-20 mm. long, 5- to 8-flowered; first glume 4-6 mm. long, 1-nerved; second glume 8-10 mm. long, 3-nerved; lemmas 8-13 mm. long, 7-nerved, villous or pilose above (scant on some), awns 12-16 mm. long.--- Along roadsides, on slopes and in waste land. Throughout the United States except in a few of the extreme southern states. Scattered over the state especially near fields and roads, from 4000-9500 feet.

9A. Bromus tectorum glabratus Spenner, (var.) Fl. Friburg. 1:152. 1825.

Like the species but the spikelets glabrous. All intergradations occur in Colorado.--- About the same range as the species in the United States and Colorado but less common.

10. **Bromus brizaeformis** Fisch. & Mey., Ind. Sem. Hort. Petrop. 3:30. 1837.

Annual plants; culms 15-60 cm. tall; leaf blades 2-5 mm. wide, pilose below less dense above, flat or slightly keeled; sheaths pilose; ligules about 1 mm. long; panicle 3-15 cm. long, drooping, rather secund; spikelets 10-25 mm. long, 5-15 florets, broad, flattened, oblong-ovate; first glume 5 mm. long, 3- to 5-nerved; second glume 6-7 mm. long, 5- to 9-nerved, both glumes broad especially the second; lemmas about 8 mm. long, broad and inflated, obtuse, glabrous or slightly scabrous, awns none or a very short one rarely present on the upper ones.---Sandy fields and waste places. Scattered throughout the United States except the southern states. Has been collected in Colorado in northcentral area from 5000 to 8500 feet but to be expected any place near roads and cultivated fields.

11. **Bromus secalinus** L., Sp. Pl. 1:76. 1753.

Annual plants; culms 30-60 cm. tall, erect; leaf blades 3-7 mm. wide, rather sparsely pilose especially above, often glabrous below; sheaths glabrous or somewhat puberulent; ligules about 1-1.5 mm. long; panicle 7-12 cm. long, nodding, lower branches short and erect; spikelets 10-24 mm. long, turgid at maturity; first glume 4-6 mm. long, 3- to 5-nerved; second glume 6-7 mm. long, 7-nerved, broader than the first; lemmas 6-8 mm. long, 7-nerved, smooth or scaberulent on the back, the margins involute at maturity, broad, awns undulate, from obsolete to 5 mm. long.--- Intergrades with species 12, 13, and 14. In waste places or a weed in fields. Throughout the United States. Apparently rare in Colorado, from the northcentral and central areas from 5000 to 6000 feet.

12. **Bromus japonicus** Thunb., Fl. Japon. 52. 1784.

Annual plants; culms 20-60 cm. tall, erect, geniculate or sometimes decumbent at base; leaf blades 3-7 mm. wide, or sometimes narrower, pilose or pubescent especially above; sheaths pilose; ligules about 1 mm. long or sometimes to 2 mm.; panicle 10-18 cm. long, diffuse, somewhat drooping; spikelets 1.5-2.5 cm. long, 5-9 florets; first glume 4-6 mm. long, 3- to 5-nerved; second glume 6-8 mm. long, 5- to 7-nerved, both glumes rather broad; lemmas 7-9 mm. long, 7- to 9-nerved, glabrous or scaberulent on the back, awns 8-10 mm. long, somewhat twisted and flexuous at maturity.---Intergrades with species 11, 13, and 14. A weed in fields, or in waste places. Throughout the United States except in the southeastern states. In the eastern half of the state at 4000 to 7500 feet.

13. **Bromus racemosus** L., Sp. Pl. ed. 2 1:114. 1762.

Annual plants (has been reported as a biennial); culms 20-60 cm. tall, usually scabrous below the panicle, erect or ascending; leaf blades 1-4 mm. wide, pilose on both sides, but especially below; sheaths pilose especially the lower ones; ligules 1-1.5 mm. long; panicle 5-10 cm. long, erect, dense; spikelets 1-2 cm. long, 5- to 9-flowered; first glume 3-5 mm. long, 3- to 5-nerved; second glume 6-8 mm. long, 5- to 7-nerved; lemmas 5-8 mm. long, about 7-nerved, glabrous to scabrous, broad, awns 5-8 mm. long.---Intergrades with species 11, 12, and 14. Weed in fields and in waste places. In western United States and scattered in the eastern part. Scattered in Colorado mostly in the southern part, apparently not common, the collections from 5000 to 9200 feet.

14. Bromus commutatus Schrad., Fl. Germ. 353. 1806.
Annual plants; culms 30-100 cm. tall, erect or decumbent at base; leaf blades 2-7 mm. wide, sparsely to densely pubescent; sheaths sparsely to densely pubescent or short-pilose, the hairs usually retrorse; ligules about 1-2 mm. long; panicle 10-12 cm. long, open but the branches rather stiffly ascending or drooping; spikelets 1-2.5 cm. long, 6- to 10-flowered, somewhat flattened with rather closely imbricated florets; first glume 4.5-6 mm. long, 3- to 5-nerved; second glume 6-8 mm. long, 7- to 9-nerved, both glumes glabrous to scaberulous; lemmas 7-10 mm. long, 7- to 9-nerved, minutely scabrous, thin and acute, awns usually 7-9 mm. long. Intergrades with species 11, 12, and 13.---A weed in fields and waste places. Throughout the United States but less common in the south. Apparently uncommon in Colorado, from Larimer, Logan and Pueblo Counties at 3900 to 6500 feet.

15. Bromus ciliatus L., Sp. Pl. 1:76. 1753.
B. richardsoni Link---Perennial plants, no rhizomes present; culms 60-120 cm. tall, erect; leaf blades 5-10 mm. wide, glabrous to scabrous; sheaths glabrous to short-pilose; ligules 0.5-1 mm. long; panicle 15-25 cm. long, sometimes about 10 in ours, broad with spreading and drooping branches; spikelets 1.5-2 cm. long, 5- to 11-flowered, first glume 6-8 mm. long, 1-nerved; second glume 8-11 mm. long, 3-nerved; lemmas 10-12 mm. long, 3- to 5-nerved, oblong-lanceolate, pubescent from near the margins in to the outer nerve for 1/2 to 3/4 the length, glabrous or nearly so on the back, awns slender, straight, 3-5 mm. long.---Moist woods, rocky slopes, gulches and stream sides. Northern part of United States but extending south to Western Texas and Arizona in the western half of the country. Common in Colorado in the mountainous part from 6000 to 11,000 feet but occasionally above and below that altitude.

15A. Bromus ciliatus laeviglumis Scribn.; Shear, (var.) U. S. D. A. Div. Agrost. Bul. 23:32.1900.
Differs from the species in the glabrous lemmas. Swallen (Proc. Biol. Soc. Wash. 54:43-46, 1941) transferred this variety to B. purgans.---Has been collected twice in Colorado in Boulder and Gilpin counties at 5300 to 9000 feet.

16. Bromus anomalus Rupr. ex. Fourn., Bull. Acad. Sc. Brux 9(2):236. 1840.
B. porteri (Coult.) Nash---Perennial plants, no rhizomes present; culms 30-60 cm. tall, pubescent at the nodes (some of ours very slightly); leaf blades 2-5 mm. wide, flat, scabrous; sheaths sparsely pilose to glabrous, shorter than the internodes; ligules about 1 mm. long, sometimes to 2; panicle about 10 cm. or less long (some of ours up to 15); spikelets 16-32 mm. long, 5- to 10-flowered; first glume 6-7 mm. long, 3-nerved (not on every glume on some specimens); second glume 8-9 mm. long, 3-nerved; lemmas 11-12 mm. long, lanceolate, evenly and densely pilose on the back, awns 2-4 mm. long, straight.---Open woods, hillsides and meadows. Western United States especially in the Rocky Mountain area. In the mountainous part of Colorado from 5300 to 11,000 feet.

16A. Bromus anomalus lanatipes (Shear) Hitchc., (var.) Journ. Wash. Acad. Sc. 23:449. 1933.
B. porteri lanatipes Shear; B. lanatipes (Shear) Rydb.---More robust than the species; leaf blades 3-6 mm. wide; sheaths densely woolly; first glume 1-nerved; second glume 5- to 7-nerved.--- Open woods and hillsides. Colorado to western Texas and Arizona. Scattered over southern and northcentral Colorado from 5000 to 9500 feet.

17. Bromus frondosus (Shear) Wooton & Stand., N. Mex. Col. Agri. Bul. 81:144. 1912.
Perennial plants, no rhizomes present; culms 60-90 cm. tall, slender, weak, often reclining; leaf blades 3-5 mm. wide, rarely to 10 mm., glabrous; sheaths glabrous to slightly pilose; ligules about 1 mm. long; panicle over 20 cm. long, open, erect but with drooping branches; spikelets 2-2.5 cm. long, 7- to 11-flowered; first glume 5-6 mm. long, 3-nerved (sometimes faint); second glume 6-8 mm. long, 5-nerved; lemmas 8-12 mm. long, 5- to 7-nerved, pubescent or partly glabrous, awns 2-4 mm. long, straight.---Open woods and rocky slopes Colorado, Utah, New Mexico and Arizona. Mountainous part of Colorado from central north and south belt at 5000 to 10,000 feet.

18. Bromus purgans L., Sp. Pl. 1:76. 1753.
Perennial plants, no rhizomes present; culms 50-120 cm. tall, erect; leaf blades 5-10 mm. wide or less, only 4 or 5 present; sheaths pubescent to glabrous, mostly shorter than the internodes; ligules 1-2 mm. long; panicles 15-25 cm. long, large, nodding with slender drooping branches; spikelets 2-3 cm. long, 5- to 10-flowered; first glume 5-7 mm. long, 1-nerved (rarely some 3-nerved); second glume about 7-9 mm. long, 3-nerved; lemmas 10-13 mm. long, 5- to 7-nerved, pubescent but often more densely so near the margins, the awns 4-6 mm. long, slender and straight. ---Moist woods, canyons and rocky slopes. Western two-thirds of the United States. In the mountainous area of Colorado, especially the northcentral part from 6000 to 10,000 feet.

19. Bromus latiglumis Hitchc., Rhodora 8:211. 1906.
Resembling No. 18 but with 8-20 nodes, the sheaths overlapping; sheaths pilose in a ring near the collar; flanges or auricles present.---Banks of streams, meadows and plains. Eastern part of the United States except the southeastern states, west to Montana, Colorado and New Mexico. Our one Colorado record from Huerfano County at 8500 feet.

2. Brachypodium Beauv. FALSEBROME

One introduced species; has been collected once in Colorado.

1. Brachypodium distachyon (L.) Beauv., Ess. Agrost. 101, 155. 1812.
Annual plants; culms 10-35 cm. tall, glabrous, decumbent or sprawling at base; leaf blades 1-3 mm. wide, 2-8 cm. long, flat; sheaths glabrous, somewhat keeled; ligules membranous, 1-2 mm. long; panicles (?) 2-5 cm. long, narrow, often reduced to 1 spikelet, but the 1-4 laterals if

present with very short pedicels or actually sessile; spikelets 1.5-3 cm. long, 5- to 15-flowered, the florets rather remote, somewhat laterally compressed; first glume 4-6 mm. long, 3- to 7-nerved; second glume 5-7 mm. long, 7- to 9-nerved, both glumes stiff and pointed; lemmas 6-8 mm. long, 5- to 7-nerved, rounded on back, hispidulous at least near margins with a straight, stiff, terminal awn 5-12 mm. long; paleas about as long as lemmas.---On ballast. New Jersey, California, Colorado, and Oregon according to Hitchcock. Our record from near Denver at 5200 feet probably in or near a garden, the seed perhaps coming in as an adulterant.

3. Festuca L. FESCUE

Annual or perennial bunch grasses, rarely with rhizomes; leaf blades flat to filiform; ligules membranous, variable in length; inflorescence an open or a narrow panicle; spikelets up to 12 mm. long; few- to several-flowered, rachilla disarticulating between the florets and above the glumes, florets perfect; glumes rather narrow, acute, unequal; lemmas rounded on the back, membranous to somewhat indurated, 5-nerved, these sometimes obscure, usually terminally awned but sometimes awnless, rarely from a minutely bifid apex; **paleas** as long or nearly as long as the lemmas; stigmas plumose. In addition to the species in the Key, F. pacifica Piper may be in western Colorado. Swallen (Flowering Plants and Ferns of Arizona) lists it from British Columbia and Montana, south to New Mexico and Arizona and Baja, California.

1. Leaf blades flat, averaging over 3 mm. wide
  2. Lemmas acuminate or tapering into an awn up to 2 mm. long; lower panicle branches spreading or reflexed
                              -1. F. sororia
  2. Lemmas merely acute, awnless; lower panicle branches ascending
    3. Panicle spikelike, the branches short and spikelet-bearing nearly to the base; plants dioecious; stigmas bearing branches on all sides; no auricles on the leaves; native in the mountains
                    (Genus No. 3A.) -Hesperochloa kingii
    3. Panicle contracted after flowering but not spikelike; plants not dioecious; stigmas with bilateral branches; some of leaves with auricles present; introduced grasses usually found in waste places and along roadsides
      4. Panicle contracted after flowering; spikelets 6- to 12-flowered   -2. F. elatior
      4. Panicle open; spikelets 4- to 5-flowered   -2A. F. elatior arundinaceae
1. Leaf blades involute or if flat, less than 3 mm. wide
  5. Annual; spikelets with 7 or more florets; leaves short, some less than 5 cm. long -3. F. octoflora
  5. Perennial; spikelets with less than 7 florets; leaves averaging well over 5 cm. long
    6. Ligules over 3 mm. long, acute; lower panicle branches long and usually spreading or drooping; lemma awnless or cuspidate   -4. F. thurberi
    6. Ligules shorter, not acute; lower panicle usually contracted; lemmas awned (except in numbers 5 and 8)
      7. Lemmas over 7 mm. long, acute, rarely short-awned   -5. F. scabrella
      7. Lemmas shorter, awned (some florets awnless in No. 8)
        8. Basal sheaths rusty brown, their nerves separating as fibers; culms decumbent at base
          9. Lemmas smooth or scabro-ciliate near the margins   -6. F. rubra
          9. Lemmas pubescent on the back   -6A. F. rubra lanuginosa
        8. Basal sheaths remaining firm and entire; culms erect
          10. Panicle contracted, almost spikelike, the branches bearing spikelets nearly to the base, the panicle mostly less than 10 cm. long; culms not over 30 cm. tall; leaf blades short, generally less than half the length of the culm; lemmas 3-5 mm. long
            11. Culms to 30 cm. tall; panicle over 5 cm. long; usually below timberline -7. F. ovina
            11. Culms to 15 cm. tall; panicle to 5 cm. long; mostly found above timberline
                    -7A. F. ovina brachyphylla
          10. Panicle more open, the lower branches somewhat naked below, the whole usually over 10 cm. long; culms usually over 30 cm. tall; leaf blades usually more than half the length of the culm; lemmas 5-7 mm. long
            12. Awn short, less than 2 mm. long or obsolete; ligule long, often over 1 mm. long; limited to the southern half of Colorado   -8. F. arizonica
            12. Awn over 2 mm. long; ligule short, rarely over 1 mm. long; in both southern and northern Colorado   -9. F. idahoensis

1. Festuca sororia Piper, Cont. U. S. Nat. Herb. 16:197. 1913.
Perennial plants, no rhizomes; culms 60-90 cm. tall or sometimes shorter, glabrous; leaf blades 3-6 mm. wide, 10-25 cm. long, flat, glabrous both sides but scabrous on the margins; sheaths glabrous; ligules short, 1 mm. long or less; panicle 10-15 mm. long, open, drooping, branches mostly solitary, the longest up to 10 cm. long, naked below; spikelets 7-12 mm. long, 3- to 5-flowered; first glume **3-4 mm.** long, 1-nerved, acute; second glume 5-6 mm. long, 3-nerved, scabrous on the keel; lemmas 6-9 mm. long, rather indistinctly 3- to 5-nerved, somewhat keeled, scabrous especially on the nerves, narrow and tapering to a pointed apex or to awns up to 2 mm. long.---Open woods, hillsides, canyons and meadows. Southern Colorado, Utah, Arizona and New Mexico. Our record from Mineral County at 9500 feet but listed from 6000 to 10,000 feet by Hitchcock.

2. Festuca elatior L., Sp. Pl. 75. 1753.
Perennial plants, tufted, rarely with short rhizomes; culms 25-100 cm. tall, glabrous; leaf blades 4-8 mm. wide, 10-60 cm. long, flat, glabrous below, scabrous above (slightly on ours), some auriculate; sheaths glabrous; ligules less than 1 mm. long; panicle 10-20 cm. long, erect or nodding at the tip, contracted after anthesis; spikelets 6-12 mm. long, 6- to 8-flowered; first glume 3 mm. long, 1-nerved or faintly 3-nerved; second glume 4 mm. long, 3- to 5-nerved; lemmas 5-7 mm. long, smooth, coriaceous, apex acute, awnless (rarely short-awned).---Meadows,

roadsides and waste places. Introduced in the northern parts of the United States. Scattered over Colorado from 4000 to 8000 feet, apparently not very common.

2A. **Festuca elatior arundinaceae** (Screb.) Celak., (var.) Prod. Fl. Bohm. 51, 1869.
This variety has more open panicles than the species with 4-5 flowers to the spikelet. Considered a distinct species by some.---May be present in Colorado and has been so reported. Recently collected in the western part of the state.

3. **Festuca octoflora** Walt., Fl. Car. 81. 1788.
F. tenella Willd.---Annual plants; culms 8-30 cm. tall, glabrous, erect and filiform; leaf blades 1-2 mm. wide, 2-10 cm. long, involute; sheaths glabrous to puberulent; ligule less than 1 mm. long; panicle 3-8 cm. long, narrow, branches appressed and rather 1-sided; spikelets 5-9 mm. long, 7- to 13-flowered (often 8); first glume about 2-3 mm. long, 1-nerved; second glume 3- to 4-nerved, 3-4 mm. long; lemmas 3.5-5 mm. long, obscurely 5-nerved, rather firm, involute, from glabrous to scabrous, awns 3-5 mm. long, straight and scabrous. This is a variable species. Fernald (Rhodora 47:107, 1942) segregated several varieties, the above would be F. octoflora var. tenella (Willd.) Fernald. The form with lemmas hirtellous to pubescent would be F. octoflora var. hirtella Piper. The awn of the lemma is wanting or not over 2 mm. long in F. octoflora var. glauca Fernald. Fernald (Rhodora 47:107, 1945) maintained that this species should be segregated in a different genus, Vulpia octoflora (Walt.) Rydb.---Open sterile ground. Throughout the United States. Scattered over Colorado at altitudes of 3500 to 8500 feet.

4. **Festuca thurberi** Vasey in Rothr., Cat. Pl. Survey W. 100th Merid. 56. 1874.
Perennial plants, no rhizomes; culms in dense tufts, 50-90 cm. tall, erect, scabrous to smooth; leaf blades about 1 mm. wide, 6-20 cm. long, rather glaucous, involute; sheaths usually scabrous, usually only slightly so; ligules acute, 3-6 mm. long; panicle 10-15 cm. long, branches rather long, spreading or ascending, spikelet-bearing from above the middle; spikelets 8-10 mm. long, 3- to 6-flowered, pedicels pubescent; first glume about 3 mm. long, 1-nerved; second glume 4-5 mm. long, 3-nerved; lemmas 6-8 mm. long, 5-nerved, glabrous or somewhat scabrous near the margins, acute to somewhat cuspidate, purple to green.---Dry slopes and rocky hills. Wyoming to New Mexico and Utah. In mountainous area of Colorado from 6500 to 12,000 feet.

5. **Festuca scabrella** Torr. in Hook., Fl. Bor. Amer. 2:252. 1840.
F. campestris Rydb.---Perennial plants, no rhizomes; culms densely tufted, 30-90 cm. tall, smooth or scabrous; leaf blades involute and narrow, variable in length, the lower 10-50 cm. long, sharp-pointed, scabrous; sheaths glabrous or scabrous; ligules short, less than 1 mm. long; panicle 5-15 cm. long, narrow, branches in 1's or 2's or the lower in 3's, naked below, appressed or ascending; spikelets 8-12 mm. long, 3- to 6-flowered; first glume 7-8 mm. long, 1-nerved, acute or obtuse, glabrous or slightly scabrous; second glume 8-12 mm. long, 3-nerved, acute, glabrous or slightly scabrous, lanceolate or oblong-lanceolate; lemmas 7-10 mm. long, fairly definitely 5-nerved, scabrous, acute, mucronate or very short-awned.---Prairies, hillsides and open woods. Newfoundland to British Columbia, south to Oregon, North Dakota and Colorado. Our two records from Colorado are from Huerfano County at 11,250 feet and Custer County at 8500 feet.

6. **Festuca rubra** L., Sp. Pl. 1:74. 1753.
F. vallicola Rydb.; F. earlei Rydb.---Perennial plants, rhizomes present or absent; culms 30-100 cm. tall, rarely less, loosely to closely tufted, bent or decumbent at the reddish-rusty brown base, occasionally erect; leaf blades usually folded or involute, 1.5-2.5 mm. wide, 5-15 cm. long, smooth; sheaths glabrous the lower brown and fibrillose; ligules about 0.5 mm. long, longer at the sides than at the middle; panicle 3-20 cm. long, rather contracted and narrow, the branches usually erect and ascending; spikelets 7-10 mm. long, 4- to 7-flowered; first glume 2-3 mm. long, 1-nerved, narrow; second glume 3-6 mm. long, 3-nerved; lemmas 4-7 mm. long, 5-nerved, green or purplish, smooth or scabrous near apex, awns about half as long as the body.---Meadows, hills, bogs and marshes. California, Washington to Arizona, Colorado and Montana. Only locally common in scattered areas of the mountainous area in Colorado from 7000 to 10,000 feet.

6A. **Festuca rubra lanuginosa** Mert. & Koch., (var.) Deut. Fl. ed. 3, 1:654. 1823.
This variety has pubescent lemmas.---Oregon and Wyoming and northward; Michigan, Vermont to Connecticut; Europe. Our only Colorado record from Boulder County at about 8500 feet.

7. **Festuca ovina** L., Sp. Pl. 1:73. 1753.
F. saximontana Rydb.; F. calligera (Piper) Rydb.; F. minutiflora Rydb.---Perennial plants, no rhizomes; culms 20-40 cm. tall (rarely over 30), densely tufted, scabrous near panicle; leaf blades about 1 mm. wide, 5-10 cm. long, involute, scabrous to glabrous; sheaths scabrous to glabrous; ligules about 0.5 mm. long, shorter in the center; panicle 5-10 cm. long, narrow, somewhat spikelike, sometimes 1-sided; spikelets 7 mm. long, 3- to 5-flowered; first glume 2-3 mm. long, 1-nerved; second glume 3-4 mm. long, 3-nerved; lemmas 3-5 mm. long, 5-nerved but these indistinct, firm, puberulent on upper half, the awns 1.5-2.5 mm. long.---Open woods and stony slopes. Rather widely distributed over the United States except in the southeastern part. In the mountainous area of Colorado, rather common at 6500 to 12,000 feet. The species often intergrades with the variety.

7A. **Festuca ovina brachyphylla** (Schult.) Piper, (var.) Cont. U. S. Nat. Herb. 10:27. 1906.
F. brachyphylla Schult.---Culms shorter than in the species, 4-20 cm. tall (seldom over 15), densely tufted; panicle shorter, 1-5 cm. long; spikelet and its parts may be much shorter than in the species.---Rocky slopes mostly above timberline. California, Rocky Mountains to New Mexico and the higher mountains of Vermont, New Hampshire and New York. In higher altitudes of Colorado mountains rarely lower than 10,000 feet.

8. **Festuca arizonica** Vasey, Cont. U. S. Nat. Herb. 1:277. 1893.
Perennial plants without rhizomes; culms 25-90 cm. tall, tufted, pale blue-green, scabrous; leaf blades filiform, involute, 15-25 cm. long, scabrous; sheaths definitely scabrous, somewhat flattened in age; ligules up to 2 mm. long, rather conspicuous; panicle 7-20 cm. long, narrowly contracted, somewhat secund; spikelets 8-12 mm. long, 4- to 6-flowered; first glume 3-5 mm. long, 1-nerved or somewhat 3-nerved; second glume 5-7 mm. long, 3-nerved; lemmas 6-7 mm. long, 5-nerved, scabrous to smooth, the awns lacking or up to 3 mm. long (rarely over 2).---Open pine woods or

rocky slopes, Arizona, Nevada, New Mexico and Colorado. Mostly in the southern part of Colorado in the mountains from 6500 to 10,500 feet.
9. Festuca idahoensis Elmer, Bot. Gaz. 36:53. 1903.
F. ovina var. ingrata Hackel; F. ingrata (Hack.) Rydb.---Perennial plants without rhizomes; culms 30-80 cm. tall, densely tufted, blue-green; leaf blades filiform, involute, usually elongate, over 1/2 the length of the culm, numerous, blue-green, scabrous or rarely smooth; sheaths scabrous to glabrous; ligules very short; panicle 10-15 cm. long, narrow, dense; spikelets 7 mm. long, 4- to 7-flowered; first glume about 3-5 mm. long, 1-nerved; second glume 5-6 mm. long, faintly 3-nerved; lemmas about 5-7 mm. long, 5-nerved, scabrous to glabrous, the awns about 2-4 mm. long.---Open woods, rocky slopes and hillsides. British Columbia to Alberta, south to California and Colorado. In the mountains of Colorado from 6000 to 11,500 feet.

3A. Hesperochloa SPIKE FESCUE

Sheaths open; ligules membranous; spikelets 3- to 5-flowered, disarticulating above the glumes, apparently dioecious; stigmas hispidulous all around.

1. Hesperochloa kingii (S. Wats.) Rydb., Bull. Torr. Club 39:106. 1912.
Festuca kingii (S. Wats.) Cassidy; F. confinus Vasey---Perennial plants, rhizomes absent or short ones present; culms 40-80 cm. tall, erect, tufted, glabrous; leaf blades 3-6 mm. wide, 10-30 cm. long, flat, firm, striated, blue-green somewhat scabrous; sheaths glabrous or somewhat scabrous; ligules 1-2 mm. long, occasionally longer; panicle 7-20 cm. long, narrow, erect, branches short and appressed; spikelets 7-12 mm. long, 3- to 5-flowered; first glume 3-4 mm. long, 1-nerved; second glume 4-5 mm. long, 1-nerved but faintly 3-nerved on some, both subscarious; lemmas 5-8 mm. long, obscurely 5-nerved, rounded on back, awnless, acute, scabrous on back; paleas somewhat shorter than the lemmas.---Dry hills, rocky slopes and open ground. Oregon and California, east to Montana and Colorado. Most of our specimens from northcentral and northwestern Colorado (one in Costilla County) at 5500-10,000 feet.

4. Puccinellia Parl. ALKALI-GRASS

Perennial plants (ours); leaf blades involute or flat; sheaths split; ligules membranous; inflorescence an open panicle; spikelets several-flowered, the rachilla disarticulating above the glumes and between the florets; flowers perfect; glumes unequal, obtuse to acute; lemmas rounded on the back, acute to obtuse, with 5 indistinct parallel nerves; paleas as long or nearly as long as the lemmas; stigma sessile, plumose.

1. Lower panicle branches becoming reflexed at maturity; lemmas holding their width well out toward the end (about as broad 3/4 from base as at middle), 1.5-2 mm. long; ligules seldom over 1.5 mm. long; first glume usually shorter than 1.5 mm. -1. P. distans
1. Lower panicle branches spreading but rarely reflexed at maturity; lemmas narrowed to a truncate or obtuse tip (narrower 3/4 out from base than at middle), 2-3 mm. long; ligules usually over 1.5 mm. long (at least on upper leaves); first glume about 1.5 mm. long -2. P. airoides

1. Puccinellia distans (L.) Parl., Fl. Ital. 367. 1848.
Perennial plants, seldom if ever with rhizomes; culms 20-60 cm. tall, erect or decumbent at base, glabrous; leaf blades 1-3 mm. wide, 2-10 cm. long, involute or flat, glabrous or somewhat scabrous; sheaths glabrous, definitely nerved; ligules to 1.5 mm. long; panicle 5-20 cm. long, loosely pyramidal, the branches fascicled, rather distant, the lower widely spreading or reflexed at maturity, naked below; spikelets 4-5 mm. long, 4- to 6-flowered, little flattened; first glume about 1 mm. long, 1-nerved; second glume 1-1.5 mm. long, 3-nerved but the 2 lateral ones obscure; lemmas 1.5-2 mm. long, obscurely 5-nerved, somewhat pubescent at the base, rather broad even 3/4 out from the base, the apex broadly obtuse or truncate. This grass frequently intergrades with number 2.---Moist more or less alkaline soil, sometimes in waste places. Quebec to Alaska, south to Maryland, Michigan, Wisconsin and North Dakota; Washington, south to New Mexico and California; Eurasia. Has been collected in central, northcentral and southwestern Colorado at 4500 to 8500 feet rarely at higher altitudes, but probably more widely distributed.
2. Puccinellia airoides (Nutt.) Wats. and Coult. in Gray, Man. ed. 6. 668. 1890.
P. nuttalliana (Schult.) Hithc.---Perennial plants, seldom if ever with rhizomes; culms 30-60 cm. tall, rarely taller, usually erect; leaf blades 1-3 mm. wide, (rarely over 2), 3-10 cm. long, usually involute; sheaths glabrous, rather heavily nerved; ligules 1-3 mm. long; panicle 10-20 cm. long (rarely to 30), open-pyramidal, branches fascicled, distant, appressed at first and partly included, then spreading, naked below; spikelets 4-7 mm. long, 3- to 6-flowered, only slightly compressed; first glume about 1.5 mm. long, 1-nerved; second glume about 1.8-2.0 mm. long, 3-nerved the laterals usually indistinct; lemmas 2-3 mm. long, with 5 indistinct nerves, narrowed to obtuse tips. Intergrades with number 1 and the panicle difference given varies with its maturity.---Moist usually alkaline soils. Rather generally distributed throughout the western and the extreme northeastern part of the United States. Scattered throughout the western two-thirds of the state at 4500-9500 feet.

5. Glyceria R. Br. MANNA GRASS

Perennial plants, usually with rhizomes, growing in wet soil; leaf blades flat; sheaths usually closed for most of the way; ligules membranous; inflorescence an open or contracted panicle; spikelets few- to many-flowered, subterete or somewhat compressed, rachilla disarticulating above the glumes and between the florets; glumes unequal, small, obtuse or acute; lemmas broad, rounded on the back, obtuse or truncate and scarious at apex, with 5-9 parallel and prominent nerves; paleas somewhat shorter to slightly longer than the lemmas.

1. Spikelets linear, over 7 mm. long; panicles narrow, with appressed branches   -1. G. borealis
1. Spikelets ovate to oblong, less than 7 mm. long; panicle (at maturity) open with numerous spreading branches
  2. Lemmas with 5 prominent nerves (2 fainter, marginal ones may show); ligules often over 5 mm. long; apex of lemmas with a broad scarious band (may be 1/5 or 1/6 the length), usually with a purple band below this   -2. G. pauciflora
  2. Lemmas with 7 prominent nerves; ligules rarely over 5 mm. long; apex of lemmas with a moderate to narrow scarious band, not as wide as above
    3. Leaf blades narrow, 2-6 mm. wide; first glume 0.5-0.9 mm. long; culms usually less than 100 cm. tall   -3. G. striata
    3. Leaf blades wider; first glume 1 mm. long or more; culms usually over 100 cm. tall
      4. First glume 1.5 mm. long or more; spikelets 5-7 mm. long; lemmas 2.5 mm. long or more; panicle very compound   -4. G. grandis
      4. First glume 1-1.5 mm. long; spikelets 3-5 mm. long; lemmas 2-2.5 mm. long; panicle only moderately compound   -5. G. elata

1. Glyceria borealis (Nash) Batchelder, Manchester Inst. Proc. 1:74. 1900.
Panicularia borealis Nash---Perennial plants; culms 50-100 cm. tall, erect or somewhat decumbent at base, slender, glabrous; leaf blades 2-8 mm. wide (seldom over 4), 8-20 cm. long, flat or folded, somewhat scabrous both sides; sheaths glabrous to scabrous, somewhat keeled; ligules 5-7 mm. long on upper leaves, acute; panicle 10-30 cm. long, narrow, the branches and pedicels appressed; spikelets 6-15 mm. long (ours about 10), 6- to 12-flowered; first glume 2-2.5 mm. long, 1-nerved; second glume 3-4 mm. long, 1-nerved (has been reported as 2-nerved); lemmas 3.5-4.5 mm. long, 7-nerved, rather thin, scaberulent on the nerves, apex obtuse, scarious.---In shallow water, marshy areas or wet meadows. Newfoundland to Alaska, south to Connecticut, Indiana, South Dakota and in the mountains to New Mexico and California. In Colorado in the mountains of northcentral and central parts at 6500-10,000 feet.
2. Glyceria pauciflora Presl., Rel. Haenk. 1:257. 1830.
Panicularia pauciflora (Presl.) Kuntze---Perennial plants with creeping rhizomes; culms 40-100 cm. tall, erect or from a decumbent rooting base, glabrous; leaf blades 4-12 mm. wide, 10-25 cm. long, thin, flat, scaberulous; sheaths split, smooth or scaberulous on the nerves; ligules 5-6 mm. long, bluntly acute; panicle 8-20 cm. long, open or rather dense, nodding, the branches ascending or spreading, rather flexuous, the spikelets on the upper half; spikelets 4-5 mm. long, 5- to 6-flowered; first glume 1-2 mm. long, 1-nerved; second glume similar but 3-nerved, 1.5-2.5 mm. long; lemmas 2-3 mm. long, 5-nerved (with usually 2 faint marginal ones too), scaberulent on the nerves, usually truncate and dentate at tip with a broad scarious band 1/5-1/6 the length of the lemma, usually with a purple band below it. Recently this species has been placed in the genus Puccinellia (Rhodora 54:42-45. 1952).---Shallow water, edge of marshes or wet meadows. Alaska to South Dakota, south to California and New Mexico. In the western mountainous half of Colorado at 7500 to 10,500 feet.
3. Glyceria striata (Lam.) Hitchc., Proc. Biol. Soc. Wash. 41:157. 1928.
Panicularia nervata (Willd.) Kuntze; G. nervata (Willd.) Trin.---Perennial plants with short rhizomes usually present; culms 20-100 cm. tall, in small or large clumps, erect and slender; leaf blades 2-6 mm. wide, 5-30 cm. long, flat or folded, glabrous or somewhat scabrous; sheaths glabrous or somewhat scabrous; ligules 2-3 mm. long; panicle 5-20 cm. long, ovoid, open, erect or finally drooping, branches in 1's to 3's, naked below; spikelets 2.5-4 mm. long, 3- to 7-flowered; first glume 0.5-0.9 mm. long, 1-nerved; second glume 1-1.5 mm. long, 1-nerved; lemmas 1.5-2.5 mm. long, conspicuously 7-nerved, obtuse or rounded at apex and with a narrow scarious band at apex. Fernald (Rhodora 21:47-49, 1929) maintained that all our Colorado plants are variety stricta (Scribn.) Fernald.---Moist meadows, swamps and wet places in general. Generally distributed over the United States. Widely spread in Colorado at 3500-10,500 feet.
4. Glyceria grandis S. Wats. in A. Gray, Man. Ed. 6, 667. 1890.
Panicularia grandis (S. Wats.) Nash; P. americana (Torr.) MacM.---Perennial plants with thick rhizomes; culms 100-150 cm. tall, erect or sometimes decumbent at base, thick (often 1 cm. in diameter), leaf blades 6-12 mm. wide, 15-40 cm. long, flat or rarely folded, glabrous or scaberulous above; sheaths loose, rather flattened and keeled, glabrous or lower scabrous; ligules about 3 mm. long, sometimes longer; panicle 20-40 cm. long, very compound, open at maturity, nodding at summit; spikelets 5-7 mm. long, 4- to 7-flowered; first glume about 1.5 mm. long, 1-nerved; second glume 2 mm. long, 1-nerved; lemmas about 2.5 mm. long (sometimes to 2.8), 7-nerved, purplish, obtuse or truncate and with a moderately wide scarious band at apex.---Banks of streams, marshes and lakes in very wet places. Generally distributed over the northern part of the United States and reaching New Mexico in the Rocky Mountains. Widely distributed over Colorado but has been collected most often in the mountainous part. Elevation 4000-9000 feet, but reported up to 10,000 feet.
5. Glyceria elata (Nash) Hitchc. in Jepson, Flora Calif. 1:162. 1912.
Panicularia elata Nash---Perennial plants, with rhizomes usually present; culms 60-150 cm. tall, glabrous; leaf blades 6-12 mm. wide (ours about 6-7), 20-40 cm. long, flat, thin and lax, scabrous; sheaths smooth and scaberulous; ligules 3-4 mm. long on upper leaves; panicle 15-30 cm. long, oblong, branches naked at base, spreading or reflexed at maturity; spikelets 3-5 mm. long, 5- to 8-flowered; first glume 1-1.5 mm. long, 1-nerved; second glume 1.5-2 mm. long, 1-nerved; lemmas 2-2.5 mm. long, 7-nerved, with a medium to narrow hyaline border at apex. This species is rather hard to tell from immature or depauperate specimens of number 4.---Wet meadows, edges of marshes, lakes and moist woods. Montana to British Columbia, south in the mountains to New Mexico and southern California. Our Colorado records scattered, mostly from the northcentral part of the state at 6000-10,500 feet.

6. Sclerochloa Beauv. HARDGRASS

Spikelets 3- to 4-flowered, the upper sterile, disarticulating below the glumes.

1. Sclerochloa dura (L.) Beauv., Ess. Agrost. 98, 174, 177. 1812.
Annual plants; culms 2-8 cm. tall, erect or spreading; leaf blades 1-3 mm. wide, 7-40 mm. long, folded, glabrous but scaberulous on keel and margin, the upper exceeding the inflorescence; sheaths glabrous the upper inflated; ligules membranous obtuse or truncate, varies from very short to 2 mm. long; racemes 1-2 cm. long, dense and spikelike, often partly included; spikelets 6-8 mm. long, 3- to 4-flowered, the upper florets somewhat separated; first glume about 2-3 mm. long, 3-nerved the laterals sometimes **obscure**, obtuse; second glume 3-4 mm. long, strongly 5- to 7-nerved, obtuse; lemmas (of lower florets) 5-6 mm. long, 5- to 7-nerved, obtuse, narrowly oblong, margins scarious, glabrous or scabrous on nerves, not particularly keeled; paleas as long or slightly shorter than the lemmas, margins hyaline.---Dry sandy or **gravelly** soil. A European species introduced in a few western states--Washington, Idaho, Colorado and Utah. Our specimens from Garfield County in western Colorado at 5800 feet.

7. Poa L. BLUEGRASS

Annual or perennial plants with rhizomes present or absent; leaf blades flat, folded or involute, usually ending in boat-shaped tips; sheaths usually open; ligules membranous; inflorescence an open or narrow panicle; spikelets 2- to several-flowered, the rachilla disarticulating above the glumes and between the florets; glumes acute, keeled, somewhat unequal; lemmas 5-nerved, the intermediate pair sometimes obscured, definitely keeled to rounded on the back, awnless, somewhat scarious near the apex, glabrous to pubescent, sometimes webbed at base; paleas somewhat shorter to as long as the lemmas; style short, stigmas plumose.

1. Florets usually partly or completely converted into bulblets with dark purple bases; culms rather bulblike at base -1. P. bulbosa
1. Florets normal; culms not bulblike at base
  2. Creeping rhizomes present (Pratenses group - see note at end of second No. 2)
    3. Culms conspicuously flattened -2. P. compressa
    3. Culms terete or slightly flattened (sheaths may be flattened and keeled)
      4. Lemmas glabrous or scabrous, no crisp-pubescent or villous hairs
        5. Lemmas rounded on the back, the keel obscure; rhizomes, when present short; panicle narrow -19. P. ampla
        5. Lemmas keeled and compressed; rhizomes long-creeping; panicle open -3. P. nervosa
      4. Lemmas with crisp-pubescent or villous hairs at least on the nerves below
        6. Base of lemmas webbed (a bunch of long, crinkly hairs attached at the very base)
          7. Lemmas less than 4 mm. long, not pubescent on the internerves near base; the common bluegrass of lawns, fields and roadsides below 10,000 feet -4. P. pratensis
          7. Lemmas over 4 mm. long, pubescent on the internerves below (as well as the nerves); usually growing above 10,000 feet -5. P. arctica
        6. Base of lemmas not webbed
          8. Panicle narrow, the branches ascending; leaves folded or involute, rather stiff
            9. Rhizomes when present, short; spikelets 7-8 mm. long; ligules less than 2 mm. long -30. P. fendleriana
            9. Rhizomes long-creeping; spikelets 5-7 mm. long; ligules over 2 mm. long -6. P. arida
          8. Panicle open, the branches spreading at maturity (rarely rather narrow in some forms); leaves flat or sometimes folded but not stiff
            10. Internerves of lemmas glabrous, the keel and marginal nerves with or without crisp-pubescence; lower sheaths usually retrorsely pubescent -3. P. nervosa
            10. Internerves of lemmas pubescent near base (as well as keel and marginal nerves); lower sheaths glabrous or scabrous (or somewhat pubescent on the innovations)
              11. Panicle less than 10 cm. long, the lower branches mostly in 2's; usually above 10,000 feet; culms not commonly over 30 cm. tall -5. P. arctica
              11. Panicle over 10 cm. long, the lower branches mostly in 3's; below 10,000 feet; culms over 30 cm. tall -7. P. glaucifolia
  2. Creeping rhizomes not present (these long, often scaly structures may be left in the ground if the culms are pulled out; short offset stems where the base of the decumbent culm may take root at the lower nodes do not constitute rhizomes; tillering culms forcing outward and upward through the soil must not be confused with true rhizomes; unless material is available showing clearly the presence or absence of rhizomes the beginner is advised not to attempt the specific identification of a Poa)
    12. Annual grasses (Annuae group, see note at second number 12)
      13. Panicle narrow and contracted; culms usually over 20 cm. tall; sheaths scabrous -8. P. bigelovii
      13. Panicle open, the branches spreading; culms usually less than 20 cm. tall; sheaths glabrous -9. P. annua
    12. Perennial grasses (the root and culm of an annual grass constitutes one unit; a perennial would have an enlarged root or culm base from which the new growth develops in the spring; last year's dried culm bases may still be present on a perennial)
      14. Lemmas webbed at the bases (a cluster of long crinkly hairs, often as long as the lemmas--Palustres group)
        15. Lemmas glabrous except on the keel below, nerves prominent -10. P. trivialis
        15. Lemmas pubescent on the lateral nerves or on the intermediates, the nerves themselves not conspicuous
          16. Lemmas less than 3 mm. long; spikelets less than 4.5 mm. long

17. Lower panicle branches reflexed at maturity; leaves often over 2 mm. wide -11. P. reflexa
17. Lower panicle branches usually ascending, occasionally spreading; leaves seldom over 2 mm. wide
    18. Panicle usually more than 10 cm. long; leaves usually more than 8 cm. long; culms somewhat decumbent at the usually purplish base -12. P. palustris
    18. Panicle usually less than 10 cm. long; leaves usually less than 8 cm. long, at least on the main culms; culms erect at the base, usually not purplish below -13. P. interior
16. Lemmas more than 3 mm. long; spikelets over 4.5 mm. long
  19. Panicle narrow and condensed; culms seldom over 25 cm. tall -31. P. pattersoni
  19. Panicle open, the branches spreading; culms usually over 25 cm. tall
    20. Sheaths somewhat keeled, usually distinctly **retrorsely** scabrous; panicle generally over 18 cm. long, its branches usually in 3's to 5's; culms usually over 50 cm. tall -14. P. occidentalis
    20. Sheaths not keeled (except in pressing) smooth or at most faintly scaberulous; panicle rarely over 18 cm. long, its branches usually in 2's or 3's; culms usually less than 50 cm. tall
      21. Lower panicle branches reflexed at maturity; lemmas rather broad near apex (as wide 2/3 out from the base as at the middle), seldom over 4 mm. long -11. P. reflexa
      21. Lower panicle branches ascending or spreading at right angles to the rachis, not reflexed; lemmas long-acute (wider at middle than 2/3 out from base), lemma often 4 mm. long
        22. Panicle over 11 cm. long, the branches often more than 2 together; culms typically over 50 cm. tall; lemmas with intermediate nerves rather obscure, web scant when present at all -15. P. macroclada
        22. Panicle less than 11 cm. long, the branches rarely over 2 together; culms less than 50 cm. tall; lemmas with intermediate nerves rather definite, web usually definite -16. P. leptocoma
14. Lemmas not webbed at base (basal hairs, if present, not much longer than other hairs)
  23. Lemmas glabrous or at most scabrous, no crisp-pubescent or villous hairs present
    24. Spikelets little compressed, the lemmas rounded on the back, the keel obscure (glumes may be keeled); panicle over 8 cm. long; plants seldom above 10,000 feet (Nevadenses group)
      25. Sheaths scabrous or scaberulous; ligules over 3 mm. long, acute -17. P. nevadensis
      25. Sheaths smooth (rarely scaberulous in No. 19); ligules less than 3 mm. long, truncate or rounded at apex
        26. Leaf blades involute, rather stiff; lemmas 3.5-4 mm. long; panicle narrow, less than 1.5 cm. wide -18. P. juncifolia
        26. Leaf blades flat or sometimes folded; lemmas 4-6 mm. long; panicle wider -19. E. ampla
    24. Spikelets laterally compressed, the lemmas definitely keeled and V-shaped in section; panicle less than 8 cm. long; plants usually growing above 10,000 feet (except No. 20; Epiles group)
      27. Leaf blades all filiform, scabrous; spikelets over 6 mm. long; plants not growing over 10,000 feet -20. P. cusickii
      27. Leaf blades flat (at least on culms), not scabrous; spikelets less than 6 mm. long; plants usually above 10,000 feet
        28. Culms less than 15 cm. tall, in dense tufts; lemmas less than 3.5 mm. long; ligule about 2 mm. long; leaf blades of culms and innovations alike, flat or folded -21. P. lettermani
        28. Culms taller, solitary or few in a tuft; lemmas over 3.5 mm. long; ligules 2.5-3.5 mm. long on culm leaves; leaf blades of culms flat, those of innovations filiform -22. P. epilis
  23. Lemmas with crisp-pubescent, silky or villous hairs, at least near the base
    29. Panicle open, the lower branches spreading and naked below
      30. Panicle about as broad as long; spikelets broad, the bases cordate or subcordate; leaves short, seldom over 5 cm. long; plant usually above 10,000 feet -23. P. alpina
      30. Panicle longer than broad; spikelets not especially broad, the bases not at all cordate; some of leaves over 5 cm. long; plants usually below 10,000 feet
        31. Spikelets little compressed, the lemmas rounded on the back, not keeled (glume may be); some of crisp basal hairs on internerves -25. P. gracillima
        31. Spikelets compressed, the lemmas V-shaped in cross-section, keeled; crisp pubescent hairs, when present, confined to the keel and marginal nerves (internerves glabrous or with short, straight hairs)
          32. Lemmas 4.5-5 mm. long, not long-acute (width 2/3 out from base about the same as at the middle); spikelets 3- to 5-flowered; no web present -24. P. stenantha
          32. Lemmas 4-4.5 mm. long, long-acute (width 2/3 out from base less than at middle); spikelets 2- to 3-flowered; scant web often present -15. P. macroclada
    29. Panicle narrow, contracted, the branches appressed or ascending
      33. Spikelets little compressed, the lemmas rounded on the back, not keeled (glume can be - Scabrellae group)
        34. Sheaths scabrous -26. P. scabrella
        34. Sheaths glabrous or rarely slightly scaberulous
          35. Panicles less than 10 cm. long; culms seldom over 30 cm. tall; leaves less than 10 cm. long; lemmas 3.5-4 mm. long -27. P. secunda
          35. Panicles longer; culms over 30 cm. tall; leaves over 10 cm. long; lemmas 4-5 mm. long -28. P. canbyi
      33. Spikelets compressed, the lemmas V-shaped and keeled in cross-section (Alpinae group)
        36. Panicle about as wide as long; spikelets broad, cordate to subcordate at base -23. P. alpina
        36. Panicle definitely longer than wide; spikelets not particularly broad, not at all cordate at base
          37. Leaves folded or involute, firm and stiff; panicle often more than 5 cm. long; culms usually over 25 cm. tall; spikelets 3- to 6-flowered; plants commonly below 10,000 feet
            38. Ligules over 3 mm. long, noticeable when viewed from the side of sheath -29. P. longiligula

38. Ligules short, seldom over 1.5 mm. long, not noticeable from side -30. P. fendleriana
37. Leaves flat or if involute then soft and lax; panicle less than 5 cm. long; culms usually less than 25 cm. tall; spikelets 2- to 3-flowered (rarely more); plants often above 10,000 feet
    39. Leaf blades 5-10 cm. long, the culms not much longer than the basal leaves; lemmas 3.5-4.5 mm. long
-31. P. pattersoni
    39. Leaf blades 1-5 cm. long, the culms much longer than the basal leaves; lemmas 2.5-3.5 mm. long
-32. P. rupicola

1. Poa bulbosa L., Sp. Pl. 70. 1753.

Perennial plants, no rhizomes; culms 30-60 cm. tall, densely tufted with bulbous bases; leaf blades 1-2 mm. wide, 5-15 cm. long, flat or loosely involute, glabrous; sheaths glabrous; ligules 2-5 mm. long; panicle 5-10 cm. long, ovoid, somewhat contracted, the branches ascending or appressed, some spikelet-bearing to the base; spikelets mostly proliferous, the florets converted into bulblets with dark purple bases, the bracts extending into slender green tips 5-15 mm. long, unaltered spikelets about 5-flowered; first glume 2-3 mm. long, faintly 3-nerved; second glume 2-3 mm. long, 3-nerved; lemmas 2-3 mm. long, webbed at base, silky-hairy on keel and marginal nerves, the intermediate nerves faint, compressed-keeled.---Fields, meadows, roadsides and waste places. Introduced from Europe and scattered over the United States. Irregularly distributed over Colorado, our few collections from the northern half of the state at 5000-8000 feet.

2. Poa compressa L., Sp. Pl. 69. 1753.

Perennial plants with creeping rhizomes; culms 15-50 cm. tall, solitary or few together, often decumbent at base, wiry, glabrous, strongly flattened; leaf blades 1-4 mm. wide, 2-10 cm. long, flat or V-shaped, bluish-green, somewhat scabrous above and on margins; sheaths compressed and keeled, glabrous; ligules 0.8 to 1.5 mm. long, longest in middle but collar-shaped; panicle 3-7 cm. long, narrow or branches sometimes somewhat spreading, branches usually in 2's, spikelet-bearing to base; spikelets 4-6 mm. long, 3- to 8-flowered; first glume 2-3 mm. long, acute; second glume slightly longer than first and broader; lemmas 2-3 mm. long, web scant or lacking, keel and marginal nerves pubescent, glabrous between, compressed-keeled.---Open ground, thin woods, waste places and roadsides. Introduced from Europe into most of the United States. Widely distributed in Colorado but has not been collected in the northwestern and southwestern parts. Altitude 3500-9500 feet.

3. Poa nervosa (Hook.) Vasey, U. S. D. A. Div. Agrost. Bull. 13(2) pl. 81. 1893.

P. subreflexa Rydb.; P. vaseyana Scribn.; P. wheeleri Vasey; P. olneyae Piper---Perennial plants, creeping rhizomes present; culms 30-60 cm. tall, erect; leaf blades 2-4 mm. wide, 2-7 cm. long (on innovations up to 20) flat or folded, mostly basal, somewhat scabrous to smooth; sheaths glabrous or (in ours) mostly retrorsely pubescent and purplish on lower ones, closed for most of the length on most; ligules 1-2 mm. long, collar-shaped but acute in middle; panicle 5-12 cm. long, apex nodding, open, branches mostly in 2's and 3's, naked below, the lower often reflexed; spikelets 4-6 mm. long, 3- to 8-flowered; first glume 3-4 mm. long, 1-nerved; second glume 3.5-4.5 mm. long, somewhat 3-nerved; lemmas 4-5 mm. long (has been reported as 3), rather strongly nerved, glabrous or pubescent on the lower part of the nerves (ours sometimes scabrous), not webbed, compressed-keeled.---Open woods, canyons and near streams or lakes. In the Rocky Mountain states from Montana to New Mexico and westward to the Pacific. Scattered over the mountainous part of Colorado, especially the northcentral area, at 6000-11,500 feet.

4. Poa pratensis L., Sp. Pl. 67. 1753.

Perennial plants with creeping rhizomes; culms 20-60 cm. tall, rather tufted, erect, glabrous, somewhat compressed; leaf blades 2-4 mm. wide, sometimes to 6, 5-25 cm. long, flat or V-shaped in section, glabrous or slightly pubescent; sheaths compressed but little keeled, glabrous or very finely puberulent; ligules 0.2-1.0 mm. long, collar-shaped; panicle 5-15 cm. long, pyramidal, branches in whorls of 3 to 5, open; spikelets 3-6 mm. long, 3- to 5-flowered; first glume about 2-3 mm. long, acute; second glume about 3-3.5 mm. long; lemmas about 3-3.5 mm. long, webbed at base, midnerve and marginal nerves silky-hairy, intermediate nerves glabrous, compressed-keeled. A widely distributed and rather variable species especially in general appearance.---Open woods, meadows and fields. Introduced from Europe into most of the United States but not common in arid portions of the gulf states. Widely distributed in Colorado, but has been little collected in the eastern plains and northwestern corner. Altitude 4000-12,000 feet.

5. Poa arctica R. Br., Sup. App. Parrys Voyage 288. 1823.

P. grayana Vasey; P. longipila Nash; P. alpicola Nash; P. callichroa Rydb.; P. tricholepis Rydb.; P. phoenicea Rydb.; also has been called P. cenisia All. & P. laxa Haenke---Perennial plants with creeping rhizomes; culms 10-30 cm. tall, loosely tufted, erect from a decumbent base; leaf blades 1-4 mm. wide, 2-10 cm. long, flat or folded, keeled, mostly basal but 1-2 cauline, weakly scabrous to smooth; sheaths glabrous to short-pubescent on the innovations; ligules 1.5-3.5 mm. long; panicle 3-10 cm. long, pyramidal, lower branches usually in 2's, sometimes reflexed, spikelet-bearing near tip; spikelets 5-7 mm. long, 3- to 5-flowered; first glume 3-3.5 mm. long, 3-nerved, acute; second glume about 4 mm. long, 3-nerved, wider; lemmas about 4.5 mm. long (one 3.8), purplish near apex, villous on keel and marginal nerves and pubescent on the internerves at base, often but not always webbed, compressed-keeled.---Meadows, mostly above timberline in the Rocky Mountain states from Montana to New Mexico and in Washington and Oregon. Also in the arctic regions. Has been collected in the western half of Colorado at 9000-14,000 feet.

6. Poa arida Vasey, Contrib. U. S. Herb. 1:270. 1893.

P. sheldoni Vasey; P. pseudopratensis Scribn. & Rydb.; P. pratericola Rydb. & Nash; P. pratensiformis Rydb.---Perennial plants with creeping rhizomes; culms 20-50 cm. tall, erect, glabrous or sometimes scabrous below panicle; leaf blades 2-3 mm. wide, 3-15 cm. long, folded or involute, mostly basal, about 1 culm leaf, glabrous or weakly scabrous above and on margins; sheaths glabrous or slightly scabrous above; ligules 2-4 mm. long, acute in middle; panicle 2-10 cm. long, narrow and contracted, branches ascending (although somewhat open in a few specimens); spikelets 5-8 mm. long, 4- to 8-flowered; first glume about 3 mm. long, acute, rather broad;

second glume slightly longer; lemmas 3-4 mm. long, densely villous on keel and on marginal nerves and more or less villous below on internerves, no web present, apex rather obtuse, compressed-keeled.---Prairies, sandy soil, plains and alkaline meadows. In the Rocky Mountain states from Montana and North Dakota to Texas and Arizona. Widely distributed over Colorado but has not been found in the northwest corner. Altitude 3500-9500 feet perhaps higher.

7. Poa glaucifolia Scribn. & Willms., U.S.D.A. Div. Agrost. Cir. 10:6. 1899.
Perennial plants with creeping rhizomes (some authorities state these are sometimes obscure); culms 60-100 cm. tall, loosely tufted; leaf blades 2-3 mm. wide, 3-10 cm. long; flat or nearly so, glaucous, scabrous (in ours); sheaths slightly scabrous (in ours); ligules to 2 mm. long; panicle mostly 10-20 cm. long, narrow but rather open, the branches usually in somewhat distant whorls, mostly in 3's, ascending, very scabrous and naked below; spikelets 6-7 mm. long, 2- to 4-flowered (often to 6 in ours); first glume 3.5-4 mm. long, 1-nerved; second glume 4-4.5 mm. long, 3-nerved, both glumes rather narrow; lemmas 3.5-5 mm. long, villous on lower half of keel and marginal nerves and more or less on the intermediate nerves, not webbed, compressed-keeled.---Moist places, ditches and open woods in the Rocky Mountain and intermountain states. Our records from northcentral and central Colorado at 5000-9000 feet.

8. Poa bigelovii Vasey & Scribn., Vasey Descrip. Cat. Grasses U. S. 81. 1885.
Annual plants; culms 15-40 cm. tall, more or less tufted, erect; leaf blades 1-3 mm. wide (rarely wider) flat and flaccid; sheaths flattened and scabrous; ligules about 2 mm. long; panicle 5-15 cm. long, narrow and interrupted; spikelets 4-6 mm. long, 3- to 6-flowered; first glume 2-3 mm. long, faintly 3-nerved; second glume 2.5-3.5 mm. long, fairly strongly 3-nerved, both glumes rather broad; lemmas 3-4 mm. long, villous on midrib and marginal nerves, sometimes pubescent on the internerves below, webbed but often scantily so, compressed-keeled.---Open ground, often in arid regions. In the southwestern part of the United States from Oklahoma to California. Our record from southcentral Colorado at about 6000 feet.

9. Poa annua L., Sp. Pl. 68. 1753.
Annual plants; culms 5-20 cm. tall, tufted, spreading, sometimes rooting at lower nodes and forming mats, flattened; leaf blades 1-4 mm. wide, 2-10 cm. long, soft, flat but somewhat keeled, somewhat scabrous on the margins; sheaths flattened and somewhat keeled, glabrous; ligules 1.5-2.5 mm. long, rounded in the middle; panicle 1-8 cm. long, open-pyramidal; spikelets 3-5 mm. long, 3- to 6-flowered; first glume 1.5-2 mm. long, 1-nerved; second glume 2-2.5 mm. long, somewhat 3-nerved; lemmas 2.5-3 mm. long, more or less pubescent on lower half of nerves, long basal hairs sometimes weblike, compressed-keeled.---Open ground, lawns, pastures, waste places and moist areas in general. Introduced from Europe into Canada, Alaska and the United States to California and Florida. Our records from northcentral, central, southcentral and southwestern Colorado at 5000-9600 feet but probably more widely distributed.

10. Poa trivialis L., Sp. Pl. 67. 1753.
P. callida Rydb.---Perennial plants, rhizomes lacking; culms 30-100 cm. tall, erect from decumbent, often rooting bases, scabrous below the panicle; leaf blades 2-4 mm. wide, 5-15 cm. long, flat, scabrous; sheaths somewhat keeled, retrorsely scabrous or scaberulous; ligules 4-7 mm. long; panicle 6-15 cm. long, oblong, the branches spreading or ascending, the lower usually in 5's; spikelets 3-4 mm. long, usually 2- to 3-flowered; first glume about 2 mm. long, 1-nerved, narrow, almost subulate; second glume about 2.5 mm. long, with 3 conspicuous nerves, both glumes with scabrous keels; lemmas 2.5-3.0 mm. long, nerves conspicuous, glabrous except the keel silky-pubescent below, web conspicuous (may be sparse but long), compressed-keeled.---**Moist places in general.** Introduced from Europe mostly in the northeastern and northwestern part of the United States. Our Colorado collection from Boulder County at about 5350 feet.

11. Poa reflexa Vasey & Scribn., Contrib. U. S. Nat. Herb. 1:276. 1893.
P. pudica Rydb.---Perennial plants, no rhizomes; culms 20-40 cm. (some of ours to 50) tall, solitary or in small tufts, erect, glabrous; leaf blades 1-4 mm. wide, 4-7 cm. long, flat, glabrous; sheaths glabrous; ligules 1-2 mm. long (a few to 3); panicle 5-15 cm. long, nodding, branches naked below, in 1's, 2's, or 3's, the lower usually reflexed at maturity, spikelet-bearing near the ends; spikelets 3-4.5 mm. long, 2- to 4-flowered; first glume 2.5-3 mm. long, 1-nerved; second glume 3-3.5 mm. long, 3-nerved; lemmas about 3 mm. long (but a few of ours to 4.5), rather broad out near apex, villous on keel and marginal nerves and sometimes between, webbed at base, compressed-keeled.---Open slopes, moist meadows, grassy summits. Montana to British Columbia, south to New Mexico and Arizona. In the higher mountainous area of Colorado at 7000-13,000 feet, occasionally down to 6000 feet.

12. Poa palustris L., Syst. Nat. Ed. 10, 2:874. 1759.
P. triflora Gilib.; P. serotina Ehr.; and perhaps P. crocata Michx.---Perennial plants, no rhizomes; culms 30-80 cm. tall, sometimes more, loosely tufted, glabrous, decumbent at the usually flattened purplish base; leaf blades 1-2 mm. wide (up to 4 on some doubtful specimens), usually some over 8 cm. long, flat to involute; sheaths keeled, glabrous to **scaberulent**; ligules variable but usually 3-5 mm. long on cauline leaves; panicle 10-30 cm. long, pyramidal or oblong, nodding, branches in rather distant fascicles, naked below; spikelets about 4 mm. long, 2- to 4-flowered; first glume 2-3 mm. long, 1- or somewhat 3-nerved; second glume about 3 mm. long, 3-nerved; lemmas 2.5-3 mm. long, villous on the keel and marginal nerves, internerves rather faint, base webbed (rather scant on some), compressed-keeled.---Moist or rather dry meadows and open ground, aspen groves, swamps and hills. Newfoundland and Quebec, south to Virginia, Missouri, New Mexico and California; Eurasia. The Colorado records are from the northcentral, central and southcentral parts of the state at 4500-10,000 feet but probably more widely distributed.

13. Poa interior Rydb., Bull. Torr. Club 32:604. 1905.
P. subtrivialis Rydb.; close to P. crocata Michx.---Perennial plants, no rhizomes; culms 20-50 cm. tall, usually tufted, erect, often scabrous below the panicle; leaf blades 1-2 mm. wide, 2.5-10 cm. long, flat or somewhat involute; sheaths sometimes keeled, often scaberulent; ligules usually less than 1 mm. long (rarely to 1.5); panicle 5-10 cm. long, narrowly pyramidal, branches erect or ascending (occasionally spreading); spikelets 3-4.5 mm. long, 2- to 5-flowered; first

glume about 2.2 mm. long, 3-nerved; second glume about 2.5 mm. long, 3-nerved; lemmas 2.2-2.6 mm. long (in ours), silky-villous below on keel and marginal nerves, webbed at base (sometimes scant), compressed-keeled.---Quebec to British Columbia and Washington, south to Vermont, Michigan, Minnesota, Nebraska, Texas, New Mexico and Arizona. Our records from the western two-thirds of Colorado, generally distributed at 5000-12,000 feet.

14. Poa occidentalis Vasey, Contrib. U. S. Nat. Herb. 1:274. 1893.
P. platyphylla Nash & Rydb.---Perennial plants, no rhizomes; culms 40-100 cm. tall, or taller, erect, stout, few in a tuft, scabrous; leaf blades 3-6 mm. wide, 10-20 cm. long, flat, scabrous; sheaths somewhat keeled, retrorsely scabrous (very little on some); ligules 2-8 mm. long; panicle 15-30 cm. long, open, branches in distant whorls of 3's or 5's, spreading or reflexed and naked below; spikelets 4.5-7 mm. long, 3- to 6-flowered; first glume about 3 mm. long, 1-nerved; second glume about 3.2-3.5 mm. long, 3-nerved; lemmas 4.5-5 mm. long, villous on lower part of keels and marginal nerves, and often sparingly **pubescent** below, intermediate nerves definite, webbed at base (in ours scanty), compressed-keeled.---Open woods, **moist** banks, rich hillsides and along streams and lakes. Colorado and New Mexico. Scattered over the mountainous area of Colorado but has not been collected in the northwest corner of the state. Altitude 6000-11,500 feet.

15. Poa macroclada Rydb., Bull. Torr. Club **32:604.** 1905.
Perennial plants, no rhizomes; culms 50-80 cm. tall, glabrous; leaf blades 1-3 mm. wide, 7-12 cm. long, flat, glabrous; sheaths glabrous or slightly scaberulous, rather strongly striate; ligules 2-3 mm. long, acute; panicle 10-30 cm. long, open, branches spreading or finally reflexed, distant, in 3's and 5's, rarely 2's, naked below; spikelets 5-6 mm. long, 2- to 3-flowered; first glume 3-3.5 mm. long, 1-nerved; second glume 3.5-4.5 mm. long, 3-nerved but the laterals rather **faint**; lemmas 4-4.5 mm. long, pubescent on the keel and marginal nerves, long-acute from about the middle, web scant or wanting, compressed-keeled.---Moist places in the mountains. Colorado, Montana and Idaho. In Colorado known from **Gunnison** County and probably in Mesa County. Altitude 4500-10,500 feet.

16. Poa leptocoma Trin., Mem. Acad. St. Peters.6. VI. Math. Phys. Nat. 1:374. 1830.
Perennial plants, no rhizomes; culms 20-50 cm. tall, solitary or few in a tuft, erect or decumbent at base, glabrous; leaf blades 2-4 mm. wide, 4-10 cm. long, flat and lax; sheaths glabrous to scabrous; ligules 3-4 mm. long on upper leaves; panicle 5-12 cm. long, nodding, branches capillary, ascending or spreading, subflexuous the lower mostly in 2's, with few spikelets; spikelets about 6 mm. long, 2- to 4-flowered, narrow; first glume about 3 mm. long; second glume about 3.5-4.5 mm. long, both glumes narrow and acuminate; lemmas 3.5-4.5 mm. long, narrowed to the acuminate apex, pubescent on the keel and marginal nerves or sometimes nearly glabrous, webbed at base, compressed-keeled.---Bogs, wet meadows, and streamsides. Alaska south to California, Nevada, Utah and New Mexico. In the mountainous area of Colorado except in the northwestern part. Altitude 8500-13,000 feet.

17. Poa nevadensis Vasey in Scribn., Bull. Torr. Club 10:66. 1883.
Perennial plants, no rhizomes; culms 30-80 cm. tall, densely tufted, erect, scabrous below the panicle; leaf blades less than 1 mm. wide (when folded), 3-25 cm. long, shortest on the culm, narrow, involute, scabrous; sheaths scabrous (sometimes slightly); ligules **3.5 mm.** long, acute; panicle 10-15 cm. long, narrow, pale in color, branches appressed; spikelets 6-8 mm. long, 3- to 5-flowered; first glume 3.5-4.5 mm. long, 1-nerved or somewhat 3-nerved; second glume a little longer, fairly definitely 3-nerved, both glumes narrow and scabrous especially on the keel; lemmas 4-5 mm. long, width carried out well beyond the middle, glabrous or scabrous below, not webbed, rounded on the back, not keeled.---Low meadows, hillsides and dry or wet places. Montana to Washington south to Colorado and California; also in Maine. Our collections from the mountainous area of Colorado, rather scattered, from 7000-9000 feet.

18. Poa juncifolia Scribn., U.S.D.A. Div. Agrost. Bull. 11:52. 1898.
P. brachyglossa Piper---Perennial plants, no rhizomes; culms 50-100 cm. tall, tufted; leaf blades less than 1 mm. wide (when folded), to 10 cm. long, involute, rather stiff, smooth; sheaths glabrous and smooth (in ours); ligules about 1 mm. long; panicle 10-20 cm. long, narrow, about 1 cm. wide, branches appressed; spikelets 6-10 mm. long, 3- to 6-flowered; first glume 3- to 3.5 mm. long, 1- or obscurely 3-nerved; second glume about 0.5 mm. longer, 3-nerved; lemmas **3.5-4 mm.** long, carrying width well out near apex, glabrous or scabrous near base, not webbed, rounded on the back, not keeled.---Dry hills, plains and alkaline meadows. Montana to British Columbia, south to California and Colorado. Our specimens from northwest and northcentral Colorado but probably more widely distributed. Altitude 7500-9000 feet, perhaps higher.

19. Poa ampla Merr., Rhodora 4:145. 1902.
P. confusa Rydb.; P. truncata Rydb.---Perennial plants usually lacking rhizomes; culms 60-100 cm. tall, tufted, erect, smooth or rarely scaberulous, sometimes glaucous; leaf blades 1-3 mm. wide, long (the basal often over 20 cm.), flat or sometimes folded, smooth or scabrous; sheaths smooth, rarely scaberulous; ligules up to 2 mm. long, rounded or truncate; panicle 10-15 cm. long, narrow, usually dense and often 2 cm. wide; spikelets 6-10 mm. long, 3- to 8-flowered; first glume 3-4 mm. long, 3-nerved; second glume 4-5 mm. long, 3-nerved; lemmas 4-6 mm. long, glabrous or minutely scabrous, not webbed, carrying width well out near apex, rounded on the back, not keeled.---Meadows, hills, valleys, **moist** or dry ground. Montana to the Yukon Territory, south to New Mexico, Arizona and California. Our records all from the northcentral and northwestern parts of Colorado but probably has a wider distribution. Altitude 5000-9500 feet.

20. Poa cusickii Vasey, Contrib. U. S. Nat. Herb. 1:271. 1893.
P. scaberrima Rydb.; P. nematophylla Rydb.; P. idahoensis Beal; P. subaristata Scribn.---Perennial plants, no rhizomes; culms 20-60 cm. tall, densely tufted, erect, glabrous, often straw-colored; leaf blades 1 mm. wide or less, 5-15 cm. long (on basal), culm leaves 1-5 cm. long, usually only 1 noticeable one present, filiform, erect, scabrous; sheaths smooth; ligules 0.5-2 mm. long; panicle 2.5-8 cm. long, narrow-oblong contracted or somewhat open at anthesis, usually pale- or tawny-colored; spikelets 6-9 mm. long, 3- to **4-flowered**; first glume 3.5-4 mm. long,

1-nerved or rarely 3-nerved; second glume 4-5 mm. long, 3-nerved, both glumes broad, acute or obtuse at apex; lemmas 4.5-6 mm. long, smooth or scabrous especially on the keel, not webbed, compressed-keeled.---Dry hills, rocky slopes and canyons. Alberta to British Columbia, south to Colorado, Nevada and California. Our only specimen from the northwest part of Colorado at 6200 feet.

21. Poa lettermani Vasey, Contrib. U. S. Nat. Herb. 1:273. 1893.

Perennial plants, no rhizomes; culms 3-10 cm. tall, lax, tufted, as long or only slightly longer than the leaves; leaf blades about 1 mm. wide, 3-5 cm. long, flattish and lax, sometimes folded; sheaths glabrous, strongly nerved; ligules about 2 mm. long; panicle 1-3 cm. long, narrow and contracted; spikelets 3-5 mm. long, 3- to 4-flowered; first glume about 3 mm. long; second glume a little longer than the first; lemmas 2.5-3 mm. long, shorter somewhat than the glumes, broadly obtuse and erose at apex, glabrous, no web, compressed-keeled.---Rocky peaks. British Columbia, Washington, Wyoming, Utah and Colorado. Our Colorado collections from Larimer, Clear Creek and El Paso Counties but to be expected on any of the higher mountain summits. Altitude 11,500-14,000 feet.

22. Poa epilis Scribn., U. S. D. A. Div. Agrost. Cir. 9:5. 1899.

P. subpurpurea Rydb.; P. paddensis Williams---Perennial plants, no rhizomes; culms 20-40 cm. tall, solitary or few in a tuft, erect, glabrous; leaf blades 1-3 mm. wide, 3-10 cm. long, about 3 blades to a culm, culm leaves flat, those of innovations usually folded or involute; sheaths smooth; ligules 2.5-3.5 mm. long on culm leaves; panicle 2-6 cm. long, usually condensed and spikelike, usually purplish; spikelets about 5-6 mm. long, 3-4, rarely to 6-flowered; first glume 3-5 mm. long, 1-nerved; second glume somewhat longer, 3-nerved at base; lemmas 4-6 mm. long, rather narrowed to apex, glabrous or minutely scabrous, compressed-keeled.---Mountain meadows and forest openings. Alberta to British Columbia, south to Colorado, Utah, Nevada and California. Scattered over the mountainous area of Colorado, not collected in the northwestern part. Altitude 8000-12,000 feet, mostly above 10,000 feet.

23. Poa alpina L., Sp. Pl. 67. 1753.

Perennial plants, no rhizomes; culms 10-30 cm. tall, tufted from a thick crown, erect, glabrous; leaf blades 2-6 mm. wide, 2-6 cm. long, flat, holding width well out to the keeled apex, glabrous to somewhat scabrous on midnerve and margins; sheaths glabrous; ligules about 2-3 mm. long; panicle 1-8 cm. long, ovoid to short-pyramidal, about as wide as long; spikelets 5-7 mm. long, 3- to 5-flowered, broad, somewhat cordate at base especially when bracts are open; first glume 3 mm. long; second glume somewhat longer, both glumes broad and abruptly acute; lemmas 3-4 mm. long, very broad, strongly villous on the keels and marginal nerves, pubescent on the internerves below, not webbed, compressed-keeled.---Mountain meadows, banks and along streams and lakes. Arctic regions extending south to Michigan, Colorado and Utah, and Oregon. Also in Mexico. Widely distributed in the high mountains of Colorado from 8500 to 13,500 feet but mostly from above 10,000 feet.

24. Poa stenantha Trin., Mem. Acad. St. Petersb. VI. Math., Phys. Nat. 1:376. 1830.

Perennial plants, no rhizomes (short offsets resembling rhizomes on some of ours); culms 30-50 cm. tall, tufted; leaf blades 1-2.5 mm. wide, to 10 cm. long, flat or loosely involute, mostly basal, the uppermost culm leaf below the middle; sheaths glabrous; ligules 2-5 mm. long; panicle 5-15 cm. long, nodding, branches in 2's or 3's, naked below, arcuate-drooping; spikelets 6-8 mm. long, 3- to 5-flowered; first glume about 3 mm. long, 1-nerved; second glume 3.5-4 mm. long, 3-nerved; lemmas 4.5-5 mm. long, pubescent on lower part of keel and marginal nerves, sparsely pubescent on the internerves below, not webbed, compressed-keeled.---Moist open ground or rocky hillsides. Alaska, Alberta and British Columbia, south to Colorado, Idaho and Oregon. Our Colorado specimens from northcentral and westcentral Colorado at 8700-11,000 feet.

25. Poa gracillima Vasey, Contrib. U. S. Nat. Herb. 1:272. 1893.

Perennial plants, lacking rhizomes; culms 20-60 cm. tall, loosely tufted, usually decumbent at base, smooth; leaf blades of culm flat or folded, basal leaves mostly filiform, to 2 mm. wide and 2-5 cm. long, smooth or nearly so; sheaths smooth; ligules 2-5 mm. long on culm leaves; panicle 4-10 cm. long, pyramidal, branches loose, in whorls, the lower in 2's to 6's, spreading or even reflexed, naked below; spikelets 4-6 mm. long, 3- to 5-flowered; first glume 3-3.5 mm. long, 3-nerved; second glume 3.5-4.5 mm. long, 3-nerved; lemmas 3.5-4.5 mm. long, minutely scabrous and crisp-pubescent at base especially on the nerves, not webbed, rounded on back, not keeled.---Cliffs and rocky slopes. Alaska to Alberta, south to Wyoming, Nevada and California. Our records from mountain peaks of central and southcentral Colorado at 11,000-13,000 feet.

26. Poa scabrella (Thurb.) Benth., Vasey Grasses U. S. 42, 1883.

P. buckleyana Nash ---Perennial plants, no rhizomes; culms 20-80 cm. tall, tufted, erect, usually scabrous below the panicle; leaf blades 1-2 mm. wide, 2-10 cm. long, flat or somewhat involute on drying, lax, mostly basal; sheaths scabrous; ligules 3-5 mm. long; panicle 5-12 cm. long, narrow and usually contracted but sometimes rather open near base; spikelets 5-10 mm. long, 3- to 7-flowered; first glume about 3 mm. long, 3-nerved; second glume similar but about 0.5 mm. longer; lemmas 3.5-5 mm. long, crisp-pubescent on the back toward the base, especially on the nerves, puberulent or scabrous above at least on the midnerve, not webbed, rounded on back, not keeled.---Meadows, open woods, rocks and dry hills. Recorded for western Montana to Washington and California. Our Colorado specimens are from the northcentral part of the state at 5300-7800 feet.

27. Poa secunda Presl., Rel. Haenk. 1:271. 1830.

P. sandbergii Vasey---Perennial plants, no rhizomes; culms commonly not over 30 cm. tall, but sometimes to 60 cm., erect from a dense often extensive tuft of short basal foliage; leaf blades 1-3 mm. wide, 3-10 cm. long, soft, flat, folded or involute, glabrous; sheaths glabrous; ligules 2-4 mm. long, long-acute; panicle 2-10 cm. long, narrow, the branches short and appressed (except in anthesis); spikelets 4-6 mm. long, 2- to 4-flowered; first glume about 3-5 mm. long, 1-nerved or somewhat 3-nerved; second glume about 4 mm. long, 3-nerved; lemmas 3.5-4 mm. long, minutely scaberulous, crisp-pubescent near the base especially on the keel, not webbed, rounded on the back.---Plains, dry woods and rocky hillsides. North Dakota to the Yukon Territory, south to Nebraska, New Mexico, Utah, Nevada and California; also in Chile. Generally distributed throughout the mountainous area of Colorado at 4500-12,500 feet.

28. Poa canbyi (Scribn.) Piper, Contrib. U. S. Nat.Herb. 11:132. 1906.
P. lucida Vasey; P. laevigata Scribn.; P. helleri Rydb.---Perennial plants, no rhizomes; culms 50-120 cm. tall, tufted, erect, smooth; leaf blades 1-2 mm. wide (some to 3), 10-15 cm. on longest, mostly flat but some folded, scabrous above; sheaths smooth or scabrulous; ligules 2.5 mm. long, acute; panicle 10-20 cm. long, usually over 1 cm. wide, narrow, compact to rather loose, often pale; spikelets 6-10 mm. long, 3- to 5-flowered; first glume about 3.5-4 mm. long, 1-nerved or somewhat 3-nerved; second glume slightly longer, 3-nerved, both glumes narrow; lemmas 4-5 mm. long, somewhat crisp-pubescent on the back, not webbed, rounded on back, not keeled.---Sandy, dry meadows and hills, sometimes in forests. Michigan to Yukon Territory, south to Nebraska, Colorado, Arizona and Oregon. Generally distributed over the mountainous two-thirds of Colorado except in the extreme southcentral part. Altitude 4500-9800 feet.

29. Poa longiligula Scribn. & Will., U. S. D. A. Div. Agrost. Circ. 9:3. 1899.
Perennial plants, no rhizomes; culms 25-60 cm. tall, erect, tufted, smooth or scabrous below panicle; leaf blades mostly basal, 1-2 mm. wide, 8-12 cm. long, folded or involute, firm and stiff, rather light blue; sheaths scabrous or smooth; ligule 3-7 mm. long, acute, noticeable from the side of the leaf; panicle 4-12 cm. long, contracted; spikelets 4-7 mm. long, 3- to 6-flowered; first glume 3.5-4 mm. long, 1-nerved; second glume about 4-5 mm. long, 1-nerved or sometimes 3-nerved at base; lemmas 4.5-5 mm. long, only 3 nerves showing definite, strongly pubescent on keel and marginal nerves, sometimes slightly pubescent on internerves below, not webbed, compressed-keeled.---Hillsides, plains and canyons. North Dakota to Oregon south to New Mexico and California. Scattered over western three-fourths of Colorado except in southern part at 4600-12,000 feet. This species is closely related to and intergrades with number 30.

30. Poa fendleriana (Steud.) Vasey, Bull. U. S. D. A. Div. Bot. 13(2), pl. 74. 1893.
P. eatoni S. Wats.; P. longipedunculata Scribn.; P. brevipaniculata Scribn. & Williams---Perennial plants, no rhizomes (rarely present on anamolous specimens); culms 30-50 cm. tall, or occasionally not much over 15, tufted, erect, usually scabrous below the panicle; leaf blades 1-4 mm. wide, 3-12 cm. long (except culm leaf which may be less than 1 cm.), folded or involute, stiff and firm, scabrous, mostly basal; sheaths somewhat scabrous; ligules less than 1 mm. long (but on some specimens it intergrades with No. 29), not conspicuous from the side; panicle 2-10 cm. long, oblong-contracted; spikelets 7-8 mm. long, 5-flowered; first glume 3-3.5 mm. long, 1-nerved; second glume about 4 mm. long, 1- or indistinctly 3-nerved; lemmas about 4 mm. long, villous on the keel and marginal nerves below, not webbed, compressed-keeled.---Mesas, open dry woods and rocky hills. Manitoba to British Columbia, south through South Dakota to Texas and California; also in Mexico. Widely distributed in the mountainous two-thirds at 5000-11,500 feet, also in southeastern Colorado at 4000 feet.

31. Poa pattersoni Vasey, Contrib. U. S. Nat. Herb. 1:275. 1893.
Perennial plants, no rhizomes; culms 10-20 cm. tall, (sometimes to 30 according to Rydberg), in dense tufts with numerous basal leaves about as long; leaf blades about 1 mm. wide, 5-10 cm. long, usually folded; sheaths smooth; ligules about 2 mm. long; panicle 1-4 cm. long, narrow and condensed, purplish; spikelets 5-6 mm. long, 2- to 4-flowered; first glume 2.8-3.3 mm. long, somewhat 3-nerved; second glume 3.2-3.5 mm. long, 3-nerved; lemmas about 4 mm. long (3.4 on some), pubescent on the keel and marginal nerves, short-pubescent on the internerves, not webbed or sometimes sparsely so, compressed-keeled.---Mountain summits and slopes. Montana, Wyoming, Utah and Colorado; also Oregon. Scattered in the high mountains of Colorado, apparently not very common. Altitude 11,000-12,500 feet sometimes lower and probably higher.

32. Poa rupicola Nash , Mem. N. Y. Bot. Gard. 1:49. 1900.
Perennial plants, no rhizomes; culms 10-30 cm. tall, densely tufted, erect, rather stiff, longer than the basal leaves, scaberulous below the panicle; leaf blades 0.5-1.5 mm. wide, 1-5 cm. long, flat or involute but lax; sheaths smooth to scaberulous; ligules 1-1.5 mm. long; panicle 2-5 cm. long, narrow, the branches ascending or appressed, purplish; spikelets 3.5-5 mm. long, usually 3-flowered; first glume 2.5-3 mm. long, 1- or sometimes 3-nerved; second glume only slightly longer, 3-nerved, both glumes usually broad; lemmas 2.5-3.5 mm. long, villous on keel and marginal nerves below and sometimes pubescent on the internerves below, not webbed, compressed-keeled.---Rocky slopes, sometimes in dry forests. British Columbia and south in the mountains to New Mexico, Arizona and Oregon. Generally distributed throughout the mountainous part of Colorado at 7500-13,500 feet but rarely below 9000 feet.

8. Eragrostis Host. LOVEGRASS

Annual or perennial plants, no rhizomes; leaf blades flat or sometimes involute; sheaths open; ligules a short ring of hairs; inflorescence an open or contracted panicle; spikelets few- to many-flowered, florets closely crowded, the rachilla disarticulating above the glumes and between the florets, or continuous with the lemma and fruit deciduous, palea persistent; glumes somewhat unequal, usually keeled, acute or acuminate; lemmas 3-nerved, these usually distinct, usually keeled, acute or acuminate; paleas from 1/2 to nearly as long as the lemmas, 2-keeled.

1. Plants with creeping stems, rooting at the nodes, forming mats   -1. E. hypnoides
1. Plants not creeping and forming mats (base of culm usually decumbent, however)
  2. Plants perennial; not generally weedy grasses
    3. Spikelets nearly sessile, the lateral pedicels not more than 1 mm. long
      4. Panicles condensed, rarely over 15 cm. long and 5 cm. wide; lemmas about 3 mm. long -2. E. oxylepis
      4. Panicles open (becoming a "tumble weed") over 15 cm. long and much over 5 cm. wide; lemmas about 1.5-2 mm. long   -3. E. curtipedicellata
    3. Spikelets on pedicels over 1 mm. long

5. Panicles at maturity about as broad as long, lower branches spreading or finally reflexed; ligules 2-4 mm. long; lemmas 1.5-2 mm. long -4. E. spectabilis
5. Panicles longer than broad, lower branches ascending; ligules about 0.5 mm. long (not hairs of collar); lemmas 2.5-3 mm. long -5. E. trichodes
2. Plants annual; weedy grasses, usually around disturbed areas, lawns gardens, fields or roadsides
6. Lemmas with raised "glands" along the keel (these appearing as small bumps, depressed at the ends); spikelets wide, some over 2.5 mm. -6. E. cilianensis
6. Lemmas without such glands (may be scabrous); spikelets seldom over 2 mm. wide
7. Spikelets about 1 mm. wide, not over 5 mm. long (except in number 8); lemmas with lateral nerves often obscure
8. Plants without glandular depressions on the panicle branches
9. Spikelets 3-5 mm. long; lemmas 1-1.5 mm. long; culms not over 50 cm. tall -7. E. pilosa
9. Spikelets 5-7 mm. long; lemmas about 2 mm. long; culms over 50 cm. tall -8. E. orcuttiana
8. Plants with glandular depressions on the panicle branches -9. E. perplexa
7. Spikelets over 1 mm. wide, usually about 1.5 mm., some over 5 mm. long; lemmas with lateral nerves prominent
10. Panicle narrow with branches ascending at maturity, the lower less than 5 cm. long, spikelet-bearing almost to the base; lemmas often over 2 mm. long -10. E. barrelieri
10. Panicle open and **diffuse** at maturity, lower branches often over 5 cm. long (except in depauperate specimens) usually naked below; lemmas rarely over 2 mm. long
11. Primary branches of panicle simple or lower with branchlets bearing 2 or 3 spikelets; spikelets loosely imbricated on the branches, often not overlapping; plant generally less than 30 cm. tall; rather uncommon in Colorado -11. E. pectinacea
11. Primary branches of panicle usually bearing several appressed branchlets these with 2 or more spikelets; spikelets appearing imbricate and overlapping on the primary branches; plants over 30 cm. tall in favorable situations (often less here); common in Colorado -12. E. diffusa

1. Eragrostis hypnoides (Lam.) B. S. P., Prel. Cat. N. Y. 69. 1888.
Annual plants; culms branching, creeping and rooting at the nodes, forming mats; leaf blades 1-2 mm. wide, 3-5 cm. long, flat or involute, scabrous or pubescent above; sheaths glabrous or pubescent; ligules about 0.5 mm. long; panicle 1-5 cm. long, elliptic, loosely-flowered when mature; spikelets mostly 5-10 mm. long when mature, several- to many-flowered, linear; first glume about 1 mm. long, 1-nerved; second glume 1.5 mm. long, 3-nerved; lemmas 1.4-2 mm. long, glabrous, acute, strongly nerved.---Sandy river banks, sandy or gravelly shores. Quebec to Washington, south through Mexico to Argentina. Our collection from Weld County in the bed of an irrigation reservoir at 4600 feet.
2. Eragrostis oxylepis (Torr.) Torr., Rpt. U. S. Expl. Miss. Pacif. 4:156. 1857.
E. secundiflora of Manuals---Perennial plants; culms 15-50 cm. tall, tufted, suberect, simple or branching; leaf blades 2-5 mm. wide (ours run less), 5-15 cm. long, flat or involute near tip, tapering to a fine point; sheaths glabrous but pilose on the collar when young; ligules very short; panicle 4-15, rarely to 40 cm. long, condensed, more or less interrupted, with stiffly ascending branches; spikelets 6-20 mm. long, 3-5 mm. wide, mostly 8- to 40-flowered, strongly compressed, usually reddish-brown; first glume 2-3 mm. long, 1-nerved; second glume somewhat longer, both scabrous on the keel; lemmas about 3 mm. long, acute, scabrous on keel, strongly nerved.---Sandy, usually dry soil. Listed for Florida to Kansas and New Mexico, Mexico and California. Our records from eastern Colorado at 3500-4500 feet.
3. Eragrostis curtipedicellata Buckl., Proc. Acad. Nat. Sci. Phila. 1862:97. 1863.
Perennial plants; culms 20-55 cm. tall, erect, tufted; leaf blades 1-3 mm. wide, about 10-15 cm. long, flat or loosely involute, glabrous or somewhat scabrous; sheaths more or less pilose at the throat; ligules about 0.5 mm. long; panicles 15-20 cm. long at anthesis, open, spreading, pilose in axils of branches, rachis and branches more or less viscid, branches stiffly ascending or spreading, the whole finally elongating and breaking away as a "tumble weed"; spikelets 3-6 mm. long, 6- to 12-flowered, appressed on the branches with pedicels less than 1 mm. long; glumes about 1.5-1.8 mm. long; lemmas 1.5-2 mm. long, strongly nerved, oblong.---Plains, open woods, dry slopes and draws. Kansas, Oklahoma and Texas, west to Colorado and New Mexico. Our records from Baca County at 4000-4500 feet.
4. Eragrostis spectabilis (Pursh) Steud., Nom. Bot. ed 2 1:564. 1840.
Called E. **pectinacea** in Manuals---Perennial plants; culms 20-60 cm. tall, in dense tufts, erect or spreading; leaf blades 3-8 mm. wide, 10-30 cm. long, flat or folded, glabrous or pilose especially above; sheaths glabrous or sparsely pilose, collar long-pilose; ligule 2-4 mm. **long**; panicle 15-40 cm. long, about 2/3 the entire height, diffuse, included at base at first, bright reddish-purple, branches stiffly spreading or lower reflexed, the whole panicle finally breaking off as a "tumbling weed"; spikelets 4-8 mm. long, 1.5-2 mm. wide, 4- to 12-flowered; first glume 1.5-2 mm. long, 1-nerved or slightly 3-nerved; second glume a little longer, 3-nerved, both scabrous on keel; lemmas 1.5-2 mm. long, scabrous on keel near tip, prominently nerved.---Sandy, dry soil. Maine to Minnesota, south to Florida, Kansas, Colorado and Arizona; also Mexico. Our records from Boulder and Baca Counties at 4400-6500 feet but should be found in eastern part of Colorado.
5. Eragrostis trichodes (Nutt.) Wood, Class Book 796. 1861.
Perennial plants without rhizomes; culms 40-120 cm. tall, tufted, erect; leaf blades 2-6 mm. wide, 12-90 cm. long, flat or subinvolute, elongated to a slender point, scabrous above near apex; sheaths smooth but pilose at throat and sometimes on the upper half; ligules about 0.5 mm. long or less; panicle about 1/2 the length of the culm, oblong, diffuse, branches capillary, loosely ascending, not crowded, sparsely pilose in the axils, naked below; spikelets 4-8 mm. long, 4- to 6-flowered, pale or purplish, lanceolate to ovate-lanceolate, on pedicels well over 1 mm. long; first glume 2-3 mm. long, 1-nerved; second glume somewhat longer, 1-nerved or somewhat 3-nerved, both scabrous on keel.---Sandy plains and open woods. Illinois to Colorado and south

to Texas and Arkansas. Our one record from Logan County at 3900 feet.

A closely related species considered by some botanists to be only a form of the above, appears occasionally in plantings of Sand Love Grass. It has been identified as E. pilifera Scheele and has spikelets 8- to 15-flowered, stramineous to bronze in color.

6. Eragrostis cilianensis (All.) Link., Vign. Lut. Malpighia 18:386. 1904.

E. major Host.; E. megastachya Link.---Annual plants; culms 10-45 cm. tall, ascending or spreading, a ring of "glands" below the nodes; leaf blades 2-7 mm. wide, 6-25 cm. long, flat, margin near collar and midrib somewhat papillose, with a disagreeable odor when fresh; sheaths glabrous but collar long-pilose in front; ligules about 0.5 mm. long; panicle 5-20 cm. long, erect, condensed or somewhat open, the branches ascending; spikelets 5-15 mm. long, 2.5-3 mm. wide, 10- to 35-flowered, **compressed**; first glume about 2 mm. long, 1- or 3-nerved, often deciduous, glandular on the keel; second glume 3-nerved, a little longer otherwise the same; lemmas 2.5-3 mm. long, keel glandular and scabrous near apex, strongly nerved.---Cultivated ground, waste places, gardens, fields and roadsides as a weed. Introduced into suitable areas all over the United States. Our records are from the eastern half of Colorado at 4000-9500 feet, but to be expected around cultivated areas any place in the state.

7. Eragrostis pilosa (L.) Beauv., Ess. Agrost. 71, 162, 175. 1812.

Annual plants; culms 10-50 cm. tall, tufted, erect or ascending from a decumbent base; leaf blades 1-3 mm. wide, 3-15 cm. long, flat; sheaths usually pilose on the collar; ligules less than 1 mm. long; panicle 5-20 cm. long, open and becoming somewhat diffuse, often included at first, branches flexuous, ascending or spreading; spikelets 3-5 mm. long, about 1 mm. wide, 3- to 9-flowered, gray to blackish, little compressed; first glume about 0.8 mm. long, 1-nerved, acute; second glume about 1.2 mm. long, 1- or slightly 3-nerved; lemmas 1.2-1.5 mm. long, lateral nerves **obscure**, faintly scaberulous on keel near apex, rather rounded on the back at least near base, acute, rather loose on the rachilla.---Cultivated ground and waste places. Massachusetts to Colorado, south to Florida and Texas and through Mexico to Argentina; also in California. Apparently uncommon in Colorado, our 3 collections from the south and southcentral part of the state at 5200-7500 feet.

8. Eragrostis orcuttiana Vasey, Contrib. U. S. Nat. Herb. 1:269. 1893.

Plants resembling E. pilosa but with coarser culms 60-100 cm. tall; spikelets 5-7 mm. long; lemmas over 1.5 mm. long (ours about 1.8 mm.).---Fields, waste places and sandy river banks. Colorado to California, south to Arizona; also in Oregon. Our one record from Teller County, Colorado at about 9500 feet.

9. Eragrostis perplexa L. H. Harvey, Univ. Microfilms, Publ. 967:194. 1948.

Plants resembling E. pilosa but panicle branches and keels of the sheaths bearing minute glands depressed in the center.---Low alkaline areas and buffalo wallows. North Dakota to Colorado, south to Texas. Our record from southcentral Colorado at 7500 feet.

A related species, E. lutescens Scribn., with panicles narrow and dense instead of open has been reported from Colorado but the specimen on which the record is based is not at all typical.

10. Eragrostis barrelieri Daveau. in Morot., Journ. Bot. 8:289. 1894.

Annual plants; culms 20-50 cm. tall, rarely taller, erect or decumbent at base (may root at lower nodes), tufted, commonly a glandular ring below each node; leaf blades 2-5 mm. wide, 5-15 cm. long or shorter on the upper, flat, glabrous, scabrous or sparsely pilose above near base; sheaths glabrous or pilose on the edges, the collar long-pilose; ligules short; panicle 8-15 **cm.** long, erect, open but narrow, often with 1-3 axillary panicles included at base; spikelets 5-5 mm. long, 1.1-1.5 mm. wide, 10- to 15-flowered, linear; first glume 1-1.3 mm. long, 1-nerved; second glume 1.5-2 mm. long, **1-nerved**, both glumes acute and scabrous on keel; lemmas about 2 mm. long, scabrous on upper part of keel, definitely nerved.---Fields, roadsides and waste places. Introduced from Europe into Kansas, Texas, Arizona and California according to Hitchcock. Our few specimens are from Alamosa, El Paso and Prowers Counties at 3400-7500 feet, but to be expected along the river valleys anywhere in southeastern Colorado.

11. Eragrostis pectinacea (Michx.) Nees., Fl. Afri. Aust. 406. 1841.

E. purshii Schrad.---Annual plants; culms 10-40 cm. tall, erect, ascending or spreading, usually from a tufted, decumbent base; leaf blades 1-2 mm. wide (has been reported to 6 mm. in Texas), 4-15 cm. long, flat or soon involute, scabrous above; sheaths with pilose collar; ligules about 1 mm. long; panicle 8-20 cm. long, open, ovate, often partly included at base, primary branches mostly simple, the spikelets not overlapping; spikelets 3-8 mm. long, about 1.5 mm. wide, 5- to 15-flowered, dark lead-green; first glume about **1.2 mm. long**; second glume about 1.8 mm. long, both glumes acute and scabrous on keel; lemmas 1.5-2 mm. long, nerves rather prominent, acute, scabrous on keel above. This species seems to intergrade with small specimens of number 12.---Fields, roadsides, waste places in open, often sandy ground. Widely distributed over the eastern part of the United States to the Rocky Mountains but rare in the western states. Our specimens scattered over the eastern half of Colorado at 4500-5300 feet, apparently uncommon.

12. Eragrostis diffusa Buckl., Proc. Acad. Nat. Sc. Phil. 1862:97. 1863.

Annual plants; culms usually 30-50 cm. tall (but may be not over 15 in poor soil), tufted, decumbent or geniculate at base, sometimes erect; leaf blades 2-5 mm. wide, 5-15 cm. long, flat, scabrous above; sheaths glabrous, collars of lower leaves long-pilose; ligules short, less than 1 mm. long; panicle 5-20 cm. long, often included at base with branches opening and spreading, freely branched especially the lower, with spikelets imbricated along the primary branches; spikelets 4-9 mm. long, to about 2 mm. wide, 6- to 15-flowered, dark lead-colored; first glume 1.2-1.4 mm. long, 1-nerved; second glume 1.4-1.7 mm. long, both glumes scabrous on keel; lemmas about 2 mm. long, scabrous near the keel, the nerves prominent. Apparently more common than number 11 with which it seems to intergrade.---Weed in fields, gardens, roadsides and waste land especially in sandy soil. Oklahoma and Texas to Nevada and California and introduced in the southern states; also in Mexico. Common in Colorado in the eastern half of the state at 3500-7500 feet but to be expected around cultivated areas anywhere.

9. Catabrosa Beauv. BROOKGRASS

Sheaths closed; ligules membranous; spikelets 2-flowered, the florets rather distant, the rachilla readily disarticulating above the glumes and between the florets; lemmas 3-nerved.

1. **Catabrosa aquatica** (L.) Beauv., Ess. Agrost. 97, 149, 157. 1812.
Perennial plants; culms 10-45 cm. long, slender, weak, creeping and rooting at the nodes, glabrous; leaf blades 2-8 mm. wide, 3-12 cm. long, flat and abruptly acute, glabrous or scaberulous on nerves and margins; sheaths glabrous or scaberulous, somewhat flattened; ligules 5-6 mm. rounded or acute; panicle 10-20 cm. long, erect, oblong or pyramidal, the branches in whorls and spreading; spikelets about 2.5-3.5 mm. long; first glume 1-1.5 mm. long; second glume 1.5-2 mm. long, both truncate, scarious and often toothed at apex; lemmas 2-3 mm. long, nerves parallel and prominent (in fact each forming a keel), brownish, glabrous, with broad scarious often erose apex; paleas about as long as the lemmas.---Mountain meadows, around springs and along streams, usually growing in water. Newfoundland and Labrador to Alberta, south through North Dakota and Oregon to Arizona; also in Eurasia. Scattered over the mountainous two-thirds of Colorado at 5000-10,000 feet.

10. Redfieldia Vasey BLOWOUT GRASS

Sheaths open; ligules consisting mostly of hairs; spikelets mostly 3- to 4-flowered, the rachilla disarticulating above the glumes and between the florets; lemmas 3-nerved; styles long, the stigmas short and plumose.

1. **Redfieldia flexuosa** (Thurb.) Vasey, Bull. Torr. Club 14:133. 1887.
Perennial plants with long-creeping, straw-colored or tawny rhizomes; culms 60-100 cm. tall, erect or sometimes decumbent at base and rooting at the lower nodes; leaf blades 2-5 mm. wide, 15-30 cm. long, involute or folded especially near the long-tapering point; ligule 1-3 mm. long, collar-shaped; panicle 10-50 cm. long, 1/3 to 1/2 the length of the culm, open and diffuse at maturity, the branches slender, naked below; spikelets 5-7 mm. long, (sometimes plump from fungus-infected grain); first glume 3-3.5 mm. long, 1-nerved; second glume a little longer, 1- or faintly 3-nerved; lemmas 4-6 mm. long, somewhat keeled, light brown, scabrous (almost puberulent) on the back and villous at base; paleas somewhat shorter to longer than the lemmas.--- Sandhills and blowouts. South Dakota to Oklahoma, west to Colorado and Arizona. On the eastern plains area of Colorado and also in the San Luis Valley. Altitude 3500-8000 feet. A good sand-binder.

11. Distichlis Raf. SALTGRASS

Sheaths split; ligules mostly of hairs; plants dioecious, spikelets 5- to 18-flowered or the staminate with more florets, disarticulating above the glumes and between the florets in the pistillate ones; styles thickened at base, rather long, the stigmas plumose.

1. **Distichlis stricta** (Torr.) Rydb., Bull. Torr. Club 32:602. 1905.
D. dentata Rydb.---Perennial plants with short or long scaly rhizomes; culms 10-40 cm. long, the staminate longer, erect or slightly decumbent at base, glabrous; leaf blades 1-4 mm. wide, 5-15 cm. long, flat to involute, glabrous to pubescent, numerous above and reduced below to the sheaths only; sheaths overlapping strongly in 2 ranks, glabrous or with some long hairs near the throat; ligules about 0.5 mm. long; panicle 3-6 cm. long, contracted, pale green or straw-colored, the pistillate usually more condensed and with a shorter culm and exceeded by the leaves, the staminate on a longer culm and usually exceeding the leaves; staminate spikelets 1-2.5 cm. long, flat; pistillate spikelets 8-15 mm. long, not so flat; first glume 3-5 mm. long, 1-nerved or the 2 laterals faintly showing; second glume a little longer, with up to 7 faint nerves showing; lemmas 5-6 mm. long, closely imbricated, with several, often very obscure nerves, the staminate flatter, the pistillate firmer; paleas somewhat shorter than the lemmas (especially in pistillate) to somewhat longer (especially in staminate), keels broadly winged and serrulate in the pistillate only.---Alkaline soil of salt marshes, river bottoms and edges of lakes. Saskatchewan to Washington, south to Texas and California; also in Mexico. Widely distributed all over Colorado except in the high mountains at 3500-9000 feet.

12. Dactylis L. ORCHARD GRASS

Sheaths closed for most of the length; ligules membranous; spikelets few-flowered, finally disarticulating above the glumes and between the florets; lemmas 5-nerved; florets perfect or uppermost staminate; styles distinct, stigmas plumose.

1. **Dactylis glomerata** L., Sp. Pl. 71. 1753.
Perennial plants, in large tussocks, no rhizomes (but has been credited as having creeping rhizomes); culms 50-100 cm. tall, erect, glabrous; leaf blades 2-12 mm. wide, 10-40 cm. long, flat, scabrous; sheaths flattened and keeled, glabrous to somewhat scabrous; ligules 2-6 mm. long; panicle 5-20 cm. long, contracted, branches ascending (except in anthesis), lower branches naked below, the spikelets nearly sessile and crowded in 1-sided clusters near the ends; spikelets 5-9 mm. long, to 5 mm. wide, 2- to 5-flowered, flat; first glume 4-6 mm. long, 1- or faintly 2-nerved, not symmetrical; second glume 5-6.5 mm. long, 1- to 3-nerved, both glumes with midnerve keeled and hispid-ciliate at least near apex, margins scarious; lemmas 4-8 mm. long, compressed-keeled, hispiduous on keel and ciliate on margins, mucronate or short-awned; paleas about as long as lemmas.---Fields, gardens, groves, roadsides and waste places. Widely distributed over the

United States and commonly cultivated as a meadow or pasture grass. Widely scattered over Colorado especially around cultivated areas, from 4500-10,000 feet.

13. Cynosurus L. DOGTAIL

Spikelets of 2 kinds, sterile and fertile together, the fertile sessile, the sterile short-pedicelled and partly covering the fertile; rachilla of fertile spikelets disarticulating above the glumes; style short, stigma loosely plumose.

1. Cynosurus cristatus L., Sp. Pl. 72. 1753.
Perennial plants; culms 30-60 cm. tall, tufted, erect or geniculate at base, glabrous; leaf blades 1-4 mm. wide, 4-15 cm. long, glabrous or scabrous especially on margins; sheaths glabrous; ligules membranous, about 1 mm. long, truncate; panicle 3-8 cm. long, erect, dense and spikelike; spikelets of 2 kinds paired, the pair about 5 mm. long; sterile spikelets with glumes 2-3 mm. long, narrow and several narrow somewhat remote lemmas; fertile spikelets 2- to 3-flowered, with subequal glumes, the first 2 mm. long, the second 3 mm. long, 1-nerved, lemmas 3-4 mm. long, rounded on back, scabrous near apex, awn-tipped, rather firm; paleas like lemmas in length.--- Fields and waste places. Introduced from Europe especially in the northeastern and northwestern parts of the United States. Collected twice in Colorado, both in El Paso County at 6000-8000 feet. These specimens may have come from parks or plantings.

14. Phragmites Trin. REED

Sheaths split; ligules about 1/2 of its length consisting of hairs; spikelets several-flowered, the rachilla disarticulating above the glumes and at the base of each joint among the florets; lemmas 3-nerved; style short, stigmas plumose.

1. Phragmites communis Trin., Fund. Agrost. 134. 1820.
P. phragmites (L.) Karst.---Perennial plants with stout, creeping rhizomes; culms 200-400 cm. tall, erect, glabrous; leaf blades 10-40 mm. wide, 15-40 cm. long, flat, narrowed to a fine tip, glabrous or scabrous; sheaths glabrous; ligules about 1 mm. long, collar-shaped; panicle 15-40 cm. long, tawny, branches ascending to nodding; spikelets 12-15 mm. long, 3- to 7-flowered, numerous, rachilla with long silky hairs exceeding the florets and at maturity spreading the parts; first glume 3-5 mm. long, 3-nerved; second glume 6-8 mm. long, 3- to 5-nerved; lemmas 8-12 mm. long, 3-nerved, the laterals indistinct, glabrous, narrow, acuminate; paleas less than half the length of the lemmas.---Marshes, banks of lakes, streams and wet land in general. Widely distributed over the United States and in South America, Eurasia, Africa and Australia. Scattered over the entire state of Colorado except the high mountains at 3500-6500 feet. Often failing to produce inflorescences.

15. Melica L. MELIC GRASS; MELIC; ONIONGRASS

Perennial plants, with rhizomes; culms often bulbous at base; leaf blades flat to subinvolute; sheaths closed for most of the length; ligules membranous; inflorescence a narrow panicle; spikelets rather few- to several-flowered, falling entire or the rachilla disarticulating above the glumes and between the florets, the terminal florets usually reduced and several folded together; glumes somewhat unequal, thin, often hyaline especially on the margins; lemmas membranous, often rather firm, 5- to 11-nerved, scarious-margined, rounded on back, awnless; paleas somewhat shorter than the lemmas; styles distinct, stigmas plumose.

1. Culms not bulbous at base; spikelets falling entire, spreading or reflexed on 1 side of the panicle
  -1. M. porteri
1. Culms usually bulbous at base; spikelets disarticulating above the glumes, mostly ascending not particularly on 1 side of the panicle
   2. Bulb connected to the rhizome direct; first glume longer than 1/2 the length of the spikelet; ligule often over 3 mm. long
  -2. M. bulbosa
   2. Bulb connected to the rhizome by a slender stem; first glume usually less than 1/2 the length of the spikelets; ligules usually less than 3 mm. long
  -3. M. spectabilis

1. Melica porteri Scribn., Proc. Acad. Nat. Sc. Phil. 1885:44. 1885.
M. parviflora Scribn.---Perennial plants with slender rhizomes; culms 50-100 cm. tall, tufted, erect or spreading, glabrous; leaf blades 2-5 mm. wide, 10-25 cm. long, flat, scabrous especially near apex; sheaths retrorsely scabrous or smooth; ligules 2-4 mm. long, wider than blade; panicle 15-20 cm. long, narrow, 1-sided, branches short and appressed with few spikelets, green or tawny; spikelets 8-15 mm. long, 3- to 5-flowered, rachilla prolonged and bearing several smaller, overlapping, sterile lemmas; first glume 5-6 mm. long, 3-nerved; second glume 6-8 mm. long, 5-nerved, both glumes mostly scarious; lemmas 6-8 mm. long, 5 strong nerves and several fainter ones, scaberulent, subacute, scarious-margined near apex, upper lemmas rolled together.---Canyons, hillsides, streambanks and moist places. Missouri, and Nebraska to Texas and Arizona. Our collections from the mountainous area of Colorado scattered in all but the northern one-third of the state at 4500-10,000 feet.

2. Melica bulbosa Geyer, Porter & Coult., Syn. Fl. Colo. 149. 1874.
M. bella Porter---Perennial plants, with rhizomes; culms 30-60 cm. tall, bulbous at base, the bulbs connected directly to the rhizome, tufted, erect; leaf blades 2-4 mm. wide, 10-35 cm. long (or even more), flat or somewhat involute, smooth or scabrous; sheaths smooth or scabrous; ligules 4-6 mm. long; panicle 6-20 cm. long, narrow, branches short, rather stout, appressed; spikelets 7-15 mm. long, 3- to 9-flowered, rather papery in age; first glume 7-8 mm. long,

moderately broad, 3- to 5-nerved; second glume 9-10 mm. long, 5- to 7-nerved, both glumes scabrous, especially on the keel and mostly scarious; lemmas 10-11 mm. long (recorded as 7-8 in Texas) with 7-11 fairly definite or obscure nerves, scarious-margined near apex, puberulent on the back. This species intergrades with number 3 in general appearance, the presence or absence of a slender stem between the rhizome and the bulb apparently the best distinction.---Rocky woods, meadows and open slopes. Montana to British Columbia south to Colorado and California; also in Texas. Our specimens from the northwest quarter of Colorado at 4500-9000 feet.

3. Melica spectabilis Scribn., Proc. Acad. Nat. Sc. Phil. 1885:45. 1885.

Perennial plants with rhizomes; culms 30-90 cm. tall, bulbous near base, this connected to the rhizome proper by a slender stem; leaf blades 2-4 mm. wide, 5-18 cm. long, flat to subinvolute, scaberulent; sheaths glabrous or the lower somewhat scaberulent; ligules about 2 mm. long; panicle 8-15 cm. long, narrow, the branches appressed or flexuous and somewhat spreading, sometimes recurved, spikelet-bearing near ends; spikelets 9-12 mm. long, 4- to 8-flowered, terminal floret often spherical, purple-tinged; first glume 4-5 mm. long, 3- to 5-nerved; second glume 5-6 mm. long, 5-nerved, both with margins tawny-scabrous; lemmas 6-7 mm. long with 7-9 rather distinct nerves, scaberulent on the nerves, margin scarious near apex, then a purple band below, then green. See note at end of number 2.---Rocky open woods, meadows, and thickets. Montana to British Columbia south to Colorado and California. All our specimens are from the northwest quarter of Colorado at 6500-11,000 feet.

15A. Schizachne Hack. FALSE MELIC

Sheaths closed right to the top of the ligule (unless torn); ligules membranous, joined in front; spikelets several-flowered, disarticulating above the glumes and between the florets; lemmas rounded on the back, pilose on the callus.

1. Schizachne purpurascens (Torr.) Swallen, Journ. Wash. Acad. Sc. 18:204. 1928.

Perennial plants, no rhizomes; culms 30-90 cm. tall, erect from a loosely tufted, decumbent base, glabrous; leaf blades 2-5 mm. wide, 5-15 cm. or more long, flat, glabrous or sometimes scabrous; sheaths glabrous, the nerves definite; auricles absent; ligules 0.5-1.5 mm. long; panicle 6-12 cm. long, branches more or less drooping, each with 1 or 2 spikelets; spikelets 1-2 cm. long, excluding awns, 3- to 6-flowered, purplish at base; first glume 5-7 mm. long, 1-nerved or with 2 other short basal ones; second glume 7-8 mm. long, 5-nerved, the laterals short, both purplish and scarious near apex; lemmas 8-10 mm. long, with about 13 ribbed nerves, bidentate, the teeth 1-2 mm. long, scabrous on nerves especially near apex, awned from just below the teeth, the awn 1-1.5 cm. long, as long or longer than the lemma, scabrous, divergent but little twisted; paleas about 2/3 as long as the lemmas.---In woods, Newfoundland to Alaska, south to Pennsylvania, Kentucky, South Dakota and Montana and in the mountains from British Columbia to New Mexico; also in Siberia and Japan. Our few specimens from northcentral, central and southwestern Colorado but probably present in most of the mountainous part of the state. Altitude 6500-11,000 feet.

16. Tridens Roem. & Schult.

Perennial plants, sometimes with stolons or rhizomes present; leaf blades mostly flat but sometimes folded or involute; sheaths open; ligules made up completely or mostly of hairs; inflorescence a dense cluster of spikelets or a narrow almost spikelike panicle; spikelets several-flowered, disarticulating above the glumes and between the florets; glumes subequal; lemmas broad, rounded on back, rounded at apex and emarginate, mucronate or 2-toothed, 3-nerved, these usually pubescent below; paleas broad, 1/2 to nearly as long as the lemmas, the nerves near the margins.

1. Panicle exceeded by a fascicle of leaves; plants stoloniferous; culms low, usually less than 10 cm. tall
 -1. T. pulchellus
1. Panicle exserted, without a fascicle of leaves; plants lacking stolons; culms usually over 10 cm. tall
  2. Lemmas glabrous except for callus hairs; paleas nearly as long as the lemmas   -2. T. albescens
  2. Lemmas long-pubescent at base, at least on the nerves; paleas seldom over 3/4 as long as the lemma
   3. Panicle dense, subcapitate, less than 4 cm. long; leaf blades with a thick white margin; plants seldom over 25 cm. tall   -3. T. pilosus
   3. Panicle narrow, spikelike, over 4 cm. long; leaf margins not thickened; plants often over 25 cm. tall
    4. Second glume with 3, rarely 5 nerves; panicle to 25 cm. long, the branches each with 5-10 spikelets
     -4. T. elongatus
    4. Second glume with 1 nerve; panicle to 12 cm. long, each branch with 2-4 spikelets   -5. T. muticus

1. Tridens pulchellus (H. B. K.) Hitchc. in Jepson, Fl. Calif. 1:141. 1912.

Triodia pulchella H. B. K.---Perennial plants; culms 3-15 cm. tall, tufted, scabrous to puberulent, consisting of a long internode bearing at apex a fascicle of leaves and often stoloniferous; leaf blades about 0.5 mm. thick, 1-3 cm. long, involute, scabrous, often recurved, at base of plant or at apex of culms; sheaths short, scarious on margins and somewhat villous at throat; ligules about 0.5 mm. long; panicle capitate, usually not exceeding the leaf blades of the fascicle, spikelets sessile or nearly so; spikelets about 4-10 mm. long, mostly 5- to 10-flowered, white-woolly; glumes 4-7 mm. long, 1-nerved, keeled, first somewhat shorter; lemmas 4-5 mm. long, pilose below, cleft about half way to base, making 2 narrow lobes with an awn between scarcely exceeding the lobes, awn straight or somewhat divergent.---Mesas, hills and plains especially in dry areas. Texas to Nevada and southern California, south to Mexico. Apparently rare in Colorado, collected in Mesa and Montezuma Counties at about 4500-5000 feet.

2. **Tridens albescens** (Vasey) Woot. & Standl., N. Mex. Col. Agri. Bul. 81:129. 1912.

Triodia albescens Vasey---Perennial plants; culms 30-80 cm. tall, erect, loosely tufted; leaf blades 2-6 mm. wide, 4-20 cm. long, glabrous or somewhat scabrous above; sheaths smooth; ligules about 1 mm. long; panicle 10-21 cm. long, narrow, dense, greenish, white or rarely purplish, branches appressed, the spikelets short-pediceled; spikelets 5-7 mm. long, 6- to 21-flowered, flattened, florets closely imbricated; first glume 3.5-5 mm. long; second glume 4-6 mm. long, both keeled and hyaline; lemmas 3-4 mm. long, broad, rounded at hyaline apex, apex emarginate and the midnerve somewhat excurrent as a short awn, pubescent on callus, otherwise glabrous.---Plains, and open woods. Kansas and Colorado to Texas and New Mexico; also Mexico. Our one specimen labeled merely "Colorado" but to be expected in the extreme southern or southeastern part.

3. **Tridens pilosus** (Buckl.) Hitchc., Contrib. U. S. Nat. Herb. 17:357. 1913.

Triodia pilosa (Buckl.) Merr.; Erioneuron pilosum (Buckl.) Nash; Tricuspis acuminata Munro---Perennial plants; culms 8-25 cm. tall, densely tufted, these easily pulled up, erect or spreading; leaf blades 1-2 mm. wide, 1-5 cm. long, flat or folded, with a thickened whitish margin and midnerve, more or less pilose, margins papillose-ciliate near base; sheaths sometimes with papillose hairs, keeled, collar long-pilose; ligules about 0.5 mm. long; panicle 1-2 cm. long, contracted, dense, almost capitate, of 3-10 short-pediceled, large, pale or purplish spikelets; spikelets 8-15 mm. long, 4-8 mm. wide, 6- to 12-flowered, compressed; first glume 4-5 mm. long; second glume 5-6 mm. long, both glumes 1-nerved, glabrous and scarious; lemmas 5-7 mm. long, margins and rarely the lateral nerves pilose near apex and long pilose at base on the nerves and also between, minutely 2-toothed at apex, the midnerve arising between as a short awn 0.5-1.5 mm. long.---Plains, rocky hills or dry gravelly soil. Kansas to Nevada, south to Texas, Arizona and Mexico. Our specimens from the southern half of the state extending into Mesa County on the west. Altitude 4000-6000 feet.

4. **Tridens elongatus** (Buckl.) Nash in Small, Fl. Southeast. U. S. 143. 1903.

Triodia elongata (Buckl.) Scribn.---Perennial plants; culms 30-80 cm. tall, tufted, erect, scabrous below panicles; leaf blades 2-4 mm. wide, 5-25 cm. long or shorter on culm leaves, flat or soon involute, mostly basal, scabrous sometimes sparsely pilose; sheaths more or less scabrous; ligules about 1 mm. long, mostly of hairs but a triangle of solid tissue extending up in front; panicle 8-20 cm. long, erect, narrow, almost spikelike, branches appressed; spikelets 6-12 mm. long, 6- to 9-flowered, pale or purplish; first glume 4.5-5 mm. long, 1- or sometimes 3-nerved; second glume 6-7 mm. long, 3- to 5-nerved, both acute or somewhat obtuse, glabrous; lemmas 4-6 mm. long, pilose below on the nerves, 2-toothed or mucronate at apex.---Plains, prairies and sandy or rocky slopes. Missouri and Kansas to Texas and Arizona. Most of the collections in southeastern Colorado but found in Larimer County. Altitudes 4200-6800 feet.

5. **Tridens muticus** (Torr.) Nash in Small, Fl. Southeast. U. S. 143. 1903.

Triodia mutica (Torr.) Scribn.---Reported from Texas and Arkansas to California, north to Nevada, Utah and Colorado. The Colorado record was apparently based on a misidentification and no specimens from this state could be located.

17. Triplasis Beauv. SANDGRASS

Ligules a ring of hairs; spikelets few-flowered, rather V-shaped from the side, the florets remote, perfect or the upper staminate, readily disarticulating above the glumes and between the florets, cleistogamous flowers prevalent in sheaths; lemmas 3-nerved, these silky-villous; styles short, stigmas plumose.

1. **Triplasis purpurea** (Walt.) Chapm., Fl. South. U. S. 560. 1860.

Annual plants; culms 25-70 cm. tall, erect to widely spreading, pubescent at nodes, often purplish; leaf blades 1-3 mm. wide, 1-8 cm. long, flat to involute, scabrous, often papillose-ciliate near base; sheaths glabrous or pilose near base and often villous near throat; ligules 1 mm. long or less; panicle 3-7 cm. long, with few spreading branches bearing few spikelets, often with small axillary inflorescences hidden in the sheaths (in lower sheaths, late in season an inflorescence reduced to 1 spikelet may be present); spikelets 5-8 mm. long, 2- to 5-flowered, short-pediceled, usually purplish; first glume about 2 mm. long, 1-nerved; second glume from slightly to 1 mm. longer than first, 1 or somewhat 2-nerved, both acute or subacute and erose (on ours) at apex; lemmas 3-4 mm. long, 2-lobed at apex these often erose, midnerve excurrent as a short awn and callus short-villous; paleas silky-villous on upper part of the keels.---Dry sandy places. New Hampshire to Minnesota and Nebraska, south to Florida and Texas. Our four specimens are from northeastern Colorado as far west as central Weld County. Altitude 4200-4600 feet.

18. Scleropogon Phil. BURROGRASS

Plants monoecious or dioecious; ligules made up of hairs; staminate spikelets with rachilla not disarticulating; pistillate spikelets disarticulating above the glumes but not between the florets, falling together as a cylindrical, many-awned fruit, the lower floret forming a sharp callus.

1. **Scleropogon brevifolius** Phil., An. Univ. Chile. 36:206. 1870.

Perennial plants; culms 10-20 cm. tall, erect or spreading, tufted, producing wiry stolons with internodes 5-15 cm. long; leaf blades 1-2 mm. wide, 1-4 cm. long, flat or sometimes folded, sharp-pointed, somewhat scabrous (ours papillose-ciliate near the base); sheaths often pilose near or on collar; ligules 0.5-1 mm. long.(Pistillate parts) panicle 1-5 cm. long excluding awns, narrow, with few spikelets; spikelets 2.5-3 cm. long, 3- to 7-flowered; first glume 7-12 mm. long, 3- to 5-nerved; second glume 12-17 mm. long, 3- to 5-nerved; lemmas 6-11 mm. long, the bearded callus about 1.5 mm. long, body long and narrow, margins rolled, the 3 nerves produced into 3 twisted divergent but rather straight subequal awns 2-10 cm. long, a membranous lobe present on each side; paleas about as long as lemmas. (Staminate parts) panicle narrow; spikelets 1-4 cm. long, 8-18 mm. long; first glume 7-9 mm. long, 3-nerved; second glume 7-9 mm. long, 3- to 5-nerved; with a mucro at apex; paleas about as long as the lemmas. The mature pistillate spikelets break away and form "tumbleweeds," the pointed callus penetrating clothing or wool.--- Semi-arid plains, rocky ridges and open valleys. Texas to Colorado and Arizona; also in Mexico and Argentina. Our specimens from the southern part of the state from Pueblo and Las Animas Counties at 4700-5500 feet.

II. Tribe Hordeae

19. Agropyron Gaertn. WHEATGRASS

Perennial plants, (1 rare species annual), rhizomes present or absent; leaf blades flat to involute; sheaths open, with auricles present (small or rudimentary on some); ligules membranous; inflorescence a terminal spike, the spikelets solitary (rarely in pairs), sessile, placed flatwise on 2 sides of a continuous (or disarticulating in 2 species) rachis; spikelets several-flowered, mostly disarticulating above the glumes and between the florets; glumes usually rather firm, equal or subequal, usually acute or awned; lemmas rounded on the back, rather firm, 5- to 7-nerved, these often indistinct, acute or awned from apex; paleas nearly as long or slightly longer than the lemmas; style very short, stigmas plumose.

Recently 3 species of this genus not included here have been introduced in this area. When sown on rangeland they may appear to be a part of the native flora. They are A. trichophorum (Link.) Richt. (Hairy Wheatgrass), A. intermedium (Host.) Beauv. (Intermediate Wheatgrass) and A. elongatum (Host.) Beauv. (Tall Wheatgrass).

1. Awns long, some over 1 cm. and longer than their lemmas
  2. Awns straight or somewhat curved but not strongly divergent
    3. Rachis of spike disarticulating after maturity; awn of glumes 1-4 cm. long; spikelets often in pairs near middle of spike   -1. A. saundersii
    3. Rachis continuous, not disarticulating; awn of glumes (not lemmas) less than 1 cm. long; spikelets rarely in pairs
      4. Culms usually over 50 cm. tall; spike over 7 cm. long; awn of lemmas straight or very nearly so; glumes broadest below the middle   -2. A. subsecundum
      4. Culms usually less than 50 cm. tall; spike usually less than 7 cm. long; awn often somewhat divergent; glumes broadest above the middle   -2A. A. subsecundum andinum
  2. Awns definitely divergent (at least when dry)
    5. Culms decumbent or prostrate; spike, counting awns, almost as wide as long, the rachis usually finally disarticulating (must be old)   -3. A. scribneri
    5. Culms erect (usually geniculate at base in number 2A); spike, counting awns, much longer than wide, the rachis not disarticulating
      6. Lemmas definitely pubescent; rhizomes present   -4. A. albicans
      6. Lemmas glabrous or scabrous; rhizomes present or absent
        7. Glumes broadest above the middle; culms seldom over 50 cm. tall; spike usually less than 7 cm. long; no rhizomes present   -2A. A. subsecundum andinum
        7. Glumes broadest at or below the middle; culms usually over 50 cm. tall; spike usually over 7 cm. long; rhizomes present or absent
          8. Spikelets distant on the rachis, each overlapping the one above by less than 1/3 its length, often not at all; awns very divergent, some at right angles to rachis   -5. A. spicatum
          8. Spikelets overlapping by 1/3 of their length or more; awns less divergent, never at right angles
            9. Long creeping rhizomes present; plants below 8000 feet in Colorado   -6. A. griffithsi
            9. No rhizomes present, or rarely very short ones; plants above 8500 feet in Colorado   7. A. bakeri
1. Awns absent or less than 1 cm. long, shorter than their lemmas
  10. **Rhizomes present**
    11. Lemmas pubescent over most or all of back
      12. Sheaths definitely pilose, at least on lower   -9A. A. smithii palmeri
      12. Sheaths glabrous, scabrous or sometimes puberulent but not pilose
        13. Glumes rigid, gradually tapering from near the base into a short awn; spikelets compressed; leaf blades 2-6 mm. wide (when unrolled)   -9B. A. smithii molle
        13. Glumes not rigid, long-acute but broad at the middle, not tapering from the base; spikelets terete or somewhat compressed; leaf blades 1-3 mm. wide (when unrolled)   -8. A. dasystachyum
    11. Lemmas glabrous, scabrous or pubescent only at base
      14. Glumes rigid, tapering from near base into a short awn, the first usually 1- to 3-nerved; plants glaucous; spikelets 6- to 10-flowered   -9. A. smithii
      14. Glumes not particularly rigid, long-acute but rather broad at middle, not tapering up from near base, the first often 5-nerved; plants green or occasionally glaucous; spikelets 3- to 7-flowered
        15. First glume less than 7 mm. long; leaf blades involute, less than 3 mm. wide; spikes less than 10 cm. long   -10. A. riparium
        15. First glume 7 mm. long or more; leaf blades flat (or sometimes involute in number 12), some usually over 3 mm. wide; spikes often over 10 cm. long

16. Under surface of leaf smooth, upper surface often pilose, flat; glumes not much over 1/2 the length of spikelet; rachilla glabrous or scabrous; an introduced weed   -11. *A. repens*
16. Under surface of leaf scabrous or scaberulous; upper surface not pilose (may be pubescent), flat or involute; glumes nearly as long as the spikelet; rachilla villous or sometimes appressed-pubescent; native grass, not a weed
    17. Culms rarely over 100 cm. tall; spikes not 1-sided, moderate in thickness; leaves less than 20 cm. long, rarely over 6 mm. wide   -12. *A. pseudorepens*
    17. Culms stout, usually over 100 cm. tall; spikes somewhat 1-sided, stout; leaves usually over 20 cm. long, some over 6 mm. wide   -12A. *A. pseudorepens magnum*
10. Rhizomes not present
    18. Spikelets very compressed, closely imbricate and divergent
        19. Spike 4-7 cm. long; perennial grass not uncommon in cultivation for hay or pasture
                -13. *A. desertorum*
        19. Spike 1-1.5 cm. long; annual grass, not cultivated and very rare   -14. *A. triticeum*
    18. Spikelets terete or somewhat compressed, distant or if imbricated not closely so, not divergent
        20. First glume 4-7 mm. long, about half as long as the spikelet; rachilla internodes glabrous or scabrous; leaves involute, 1-2 mm. wide; spikelets distant, little if any overlapping on the rachis
                -15. *A. inerme*
        20. First glume 8-15 mm. long, almost as long as the spikelet; rachilla internodes villous or short-pubescent; leaves flat or involute, 2-8 mm. wide; spikelets usually somewhat overlapping on the rachis
            21. Lemmas glabrous (except sometimes at base); leaves glabrous or scabrous; spikes 7 cm. long or more; glumes not conspicuously scarious-margined; plants not usually alpine
                -16. *A. trachycaulum*
            21. Lemmas pubescent all over; leaves pubescent both sides; spikes 3-6 cm. long; glumes conspicuously scarious-margined; plants alpine   -17. *A. latiglume*

1. **Agropyron saundersii** (Vasey) Hitchc., Biol. Soc. Wash. Proc. 41:159. 1928.

Elymus saundersii Vasey---Perennial plants, no rhizomes; culms 40-100 cm. tall, erect, glabrous; leaf blades 1.5-4 mm. wide, 5-20 cm. long, flat or becoming involute, scabrous both sides; sheaths glabrous, auricles moderate in size; ligules 1 mm. long or less; spike 8-15 cm. long, erect or somewhat nodding, tardily disarticulating at the rachis joints, the spikelets sometimes in pairs near the middle, somewhat imbricate; spikelets 1-1.5 cm. long, 2- to 5-flowered; first glume about 6 mm. long; second glume about 7 mm. long, both glumes variable, narrow with 2 nerves or wider with 3-5 nerves, tapering into an awn 1-4 cm. long (in some a short lateral awn present near base of main one); lemmas 8-10 mm. long, nerves indistinct rather scabrous, with a straight or nearly straight awn 1.2-5 cm. long; rachilla glabrous or scabrous.---Dry slopes and mountains. Colorado and Utah. Our collections from Jackson, Denver and probably Gunnison Counties of Colorado. Recorded by Hitchcock from "Veta Pass." Altitude 5200-9100 feet.

2. **Agropyron subsecundum** (Link) Hitchc., Amer. Journ. Bot. 21:131. 1934.

A. richardsoni (Trin.) Schrad.; *A. caninoides* (Ramaley) Beal and as *A. caninum* (L.) Beauv. by Rydberg.---Perennial plants, no rhizomes; culms 30-100 cm. tall, erect, tufted, glabrous, green or glaucous; leaf blades 3-8 mm. wide, 12-25 cm. long, flat or involute, scabrous (in our specimens); sheaths glabrous or rarely pubescent, auricles short; ligules about 0.5 mm. long; spike 6-15 cm. long, erect or somewhat nodding, sometimes 1-sided, rachis sometimes disarticulating, spikelets rather closely imbricated; spikelets 1-2 cm. long, few-flowered (ours 2-4); first glume almost as long as the spikelet; second glume as long or a little longer, both glumes 4- to 7-nerved, tapering to an awn, the 2 glumes often both in front of the spikelet; lemmas 7-12 mm. long, 5-nerved, scabrous, the awn straight or nearly so, usually 1-3 mm. long, callus very short-pilose; rachilla villous. Tends to intergrade some with number 16.---Moist meadows, open woods, among bushes in foothills and mountains. Newfoundland to Alaska, and south to Maryland, Indiana, Nebraska, New Mexico, Arizona and California. Scattered in the foothills and mountainous western two-thirds of Colorado but has not been collected in the northwestern part. Altitude 5000-10,500 feet.

2A. **Agropyron subsecundum andinum** (Scribn. & Smith) Hitchc., (var.) Amer. Journ. Bot. 21:132.1934.

A. andinum (Scribn. & Smith) Rydb.---Culms less than 50 cm. tall, usually geniculate at base; sheaths usually papery on lower; spike 3-8 cm. long; glumes broadest above the middle; lemmas with awns often less than 1 cm. long, often somewhat divergent.----Alpine, in mountain meadows and high mountain tops. Montana to Washington, south to Colorado and Nevada. Our specimens from the northwest quarter of the state at 7200-13,500 feet.

3. **Agropyron scribneri** Vasey, Bull. Torr. Club 10:128. 1893.

Perennial plants, no rhizomes; culms 20-50 cm. long, tufted, decumbent-spreading or geniculate, even prostrate at base; leaf blades 2-5 mm. wide, 3-8 cm. long, flat or involute on the innovations, mostly basal, rather rigid, puberulent on both sides; sheaths glabrous or short-pubescent, auricles rather small; ligules usually less than 1 mm. long, short-ciliate; spike 3-7 cm. long, often nodding or flexuous, dense, the rachis usually disarticulating at maturity; spikelets 8-15 mm. long, 2- to 5-flowered; first glume 4-7 mm. long, 1-nerved or obscurely more, ending in a divergent awn; second glume like the first but a little longer and nerves more definite; lemmas 8-10 mm. long, nerved toward tip, more or less scabrous, tapering to a divergent awn 1.5-3 cm. long; rachilla minutely scabrous on internodes.---Alpine slopes and mountain tops. Montana to New Mexico, Nevada and Arizona. Scattered through the high mountains of Colorado at 10,700-13,000 feet.

4. **Agropyron albicans** Scribn. & Smith, U. S. D. A. Div. Agrost. Bull. 4:32. 1897.

Perennial plants with rhizomes; culms 40-80 cm. tall, erect; leaf blades 1-2 mm. wide, 10-20 cm. long, mostly involute, smooth or scaberulous; sheaths scaberulous, auricles present; ligules 1 mm. long or less; spike 7-10 cm. long, spikelets distant or loosely imbricated; spikelets 10-15 mm. long, 4- to 8-flowered; glumes 5-8 mm. long, 1- to 3-nerved, glabrous or pubescent, awn-pointed; lemmas about 1 cm. long, pubescent, with a divergent awn 10-15 mm. long.---Plains and dry hills. South Dakota to Alberta and Colorado. Our two specimens from southern Colorado but to be expected anywhere. Not common. Altitude 6000-7000 feet.

5. Agropyron spicatum (Pursh) Scribn. & Smith, U. S. D. A. Div. Agrost. Bul. 4:33. 1897.
A. vaseyi Scribn. & Smith---Perennial plants, no rhizomes (see note below); culms 40-70 cm. tall, tufted, often in large bunches, erect, glabrous, green or glaucous; leaf blades 1-3 mm. wide, 5-20 cm. long, flat or loosely involute, glabrous or pubescent above; sheaths glabrous, auricles small; ligules about 0.5 mm. long; spike 6-15 cm. long, rather narrow and lax, spikelets distant, little if any imbricated; spikelets 10-20 mm. long, 4- to 8-flowered; first glume 5-8 mm. long; second glume 6-10 mm. long, both glumes 3- to 4-nerved, obtuse, acute or short-awned; lemmas 8-10 mm. long, nerves rather indistinct, glabrous or scabrous, awn divergent (usually at right angles) and 1-2 cm. long; rachilla scaberulous. Some typical looking specimens do have rhizomes and would intergrade with number 6. Daubenmire (Bull. Torr. Club 66:327-329, 1939) stated that rhizomes occur under favorable conditions. He also pointed out the intergradations that exist between A. spicatum and A. inerme.---Plains, dry slopes, canyons, rocky hills and open woods. Michigan to Alaska, south to South Dakota, New Mexico and California. Our specimens from central, northcentral and northwestern Colorado at 5000-8000 feet.

6. Agropyron griffithsi Scribn. & Smith; Piper, Biol. Soc. Wash. Proc. 18:148. 1905.
Perennial plants, with creeping rhizomes; culms 50-80 cm. tall, glabrous, erect; leaf blades to 3.5 mm. wide, 8-12 cm., sometimes to 20 cm. long, flat to involute, smooth or scabrous; sheaths glabrous, auricles small, often rudimentary; ligules 0.6-1 mm. long; spike 7-15 cm. long, spikelets loosely to closely imbricate; spikelets 12-20 mm. long, 3- to 7-flowered; first glume 6-8 mm. long; second glume 7-10 mm. long, both glumes 3- to 5-nerved, tapering into an awn 2-3 mm. long; lemmas 8-12 mm. long, nerves stronger near apex, glabrous or scabrous above, tapering to an awn 1-2 cm. long, this divergent usually about 45 degrees or less but sometimes at right angles; rachilla long- to short-pubescent, even scabrous.---Open dry, sandy or alkaline soil and rocky slopes. North Dakota, South Dakota, Wyoming and Colorado. Our specimens mostly from northcentral and central Colorado, one in Crowley County. Altitudes 4200-8000 feet.

7. Agropyron bakeri E. Nels., Bot. Gaz. 38:378. 1904.
Perennial plants, no rhizomes, or rarely short ones present; culms 40-70 cm. tall, erect, loosely tufted, glabrous; leaf blades 2-5 mm. wide, 10-30 cm. long, flat, scabrous both sides; sheaths glabrous, auricles small to rudimentary; ligules 1 mm. long or less; spike 5-12 cm. long, rather slender, spikelets loosely to densely imbricated; spikelets 10-15 mm. long, 3- to 6-flowered; first glume about 9 mm. long, 3- to 5-nerved; second glume about 10 mm. long, 5- to 7-nerved, both awn-pointed; lemmas 9-12 mm. long, nerves indistinct in center, scabrous to glabrous, awn 6-35 mm. long, divergent when dry; rachilla short-pubescent to scabrous.---Open slopes and woods. Michigan and Alberta to Washington and New Mexico. Scattered over the high mountainous area of Colorado, no specimens from southcentral and northwestern parts. Altitudes 8500-11,500 feet.

8. Agropyron dasystachyum (Hook.) Scribn., Bull. Torr. Club 10:78. 1883.
A. lanceolatum Scribn. & Smith; A. subvillosum (Hook.) E. Nels.---Perennial plants with creeping rhizomes; culms 40-90 cm. tall, erect or somewhat decumbent at base, glabrous, often glaucous; leaf blades 1-3 mm. wide, 5-20 cm. long, flat to involute, smooth or scabrous above; sheaths glabrous to puberulent, auricles rather small; ligules less than 1 mm. long (ours about 0.5); spike 6-15 cm. long, spikelets loosely to closely imbricated; spikelets 1-2 cm. long, 3- to 8-flowered; first glume 5-8 mm. long; second glume 6-10 mm. long, both 3- to 5-nerved, scarious-margined, scabrous or pubescent on most, long-acute but broad in middle; lemmas 8-12 mm. long, obscurely nerved, densely to sparsely pubescent, acute or rarely mucronate or short-awned; rachilla from short-puberulent to pubescent.---Plains, often in sandy or gravelly soil. Michigan to British Columbia, south to Illinois, Nebraska, Colorado, Arizona, Nevada and Oregon. Our specimens mostly from the northcentral, central and southcentral parts of Colorado at 5000-10,000 feet.

9. Agropyron smithii Rydb., Mem. N. Y. Bot. Gard. 1:64. 1900.
A. occidentale Scribn.---Perennial plants, with creeping rhizomes; culms 30-80 cm. tall, erect, glaucous; leaf blades 2-6 mm. wide, 10-25 cm. long, flat but somewhat involute in drying, scabrous to short-villous above, glaucous; sheaths glabrous to scaberulous, glaucous; ligules 0.5-1 mm. long; spike 7-15 cm. long, spikelets imbricate and usually divergent, occasionally 2 at a node; spikelets 1-2 cm. long, 6- to 10-flowered; first glume 8-10 mm. long; second glume 10-12 mm. long, both glumes rigid, gradually tapering from near base into a short awn, 1 strong nerve and 4 others often showing, asymmetric; lemmas 8-12 mm. long, glabrous or pubescent near base, obscurely-nerved, acute, acuminate, mucronate or sometimes short-awned. The 2 varieties listed intergrade with each other and with the species.---Moist, usually alkaline soil on hills and plains. Colorado to Utah, south to New Mexico and Arizona. Widely distributed in Colorado at altitudes of 3500-10,000 feet.

9A. Agropyron smithii palmeri (Scribn. & Smith) Heller, (var.) Cat. N. Amer. Pl. ed. 2:3. 1900.
A. palmeri (Scribn. & Smith) Rydb.---Differs from the species in the lower sheaths being pubescent; lemmas scabrous or pubescent (all ours pubescent).---Sandy soil of dry grasslands, hillsides and mountains. Colorado to Utah, south to New Mexico and Arizona. Our specimens from central and southern Colorado but to be expected anywhere in the state. Altitude 4000-10,500 feet.

9B. Agropyron smithii molle (Scribn. & Smith) Jones, (var.) Contrib. West. Bot. 14:18. 1912.
A. molle (Scribn. & Smith) Rydb.---Differing from the species in having lemmas and sometimes the glumes pubescent.---Dry valleys, plains, hillsides and meadows. About the same range as the species in the United States. In Colorado widely distributed over the entire state at 4000-8500 feet.

10. Agropyron riparium Scribn. & Smith, U. S. D. A. Div. Agrost. Bul. 4:35. 1897.
Perennial plants, rhizomes present; culms 40-80 cm. tall, tufted, not a real sod-former, glabrous; leaf blades 1-3 mm. wide, 5-20 cm. long, involute, scabrous to puberulent above; sheaths

glabrous, auricles small and rudimentary; ligules less than 1 mm. long (ours not over 0.5); spike 5-10 cm. long, the spikelets imbricated but not strongly so; spikelets 8-15 mm. long, 5- to 7-flowered; first glume 4-6 mm. long; second glume 5-8 mm. long, both glumes rather broad but long-acute, green with scarious margins, 3- to 5-nerved; lemmas 7-9 mm. long, glabrous, scabrous or somewhat pubescent on the lower edges, nerves faint, acute or very short-awned; rachilla glabrous or scabrous.---Dry or moist meadows, hills and stream banks. North Dakota to Alberta and Washington, south to Oregon, Arizona and Colorado. Not very common in Colorado, our 7 specimens scattered over the western two-thirds of the state except in the southwestern corner. Altitude 4500-9000 feet.

11. Agropyron repens (L.) Beauv., Ess. Agrost. 102, 146, 180. 1812.

Perennial plants, creeping rhizomes present; culms 30-100 cm. tall, erect or somewhat decumbent at base, sometimes glaucous; leaf blades 4-8 mm. wide; 10-30 cm. long, flat, thin, dark green especially early in the season, smooth below, scabrous above and on margins, often pilose above, especially near base; sheaths glabrous or lower pilose, auricles moderately large, often over 1 mm. long; ligules less than 1 mm. long; spike 5-15 cm. long; spikelets 8-15 mm. long, 3- to 7-flowered; glumes subequal, 8-10 mm. long, not much over 1/2 the length of spikelet, 3- to 7-nerved, awn-pointed; lemmas 8-10 mm. long, glabrous to scabrous, merely pointed or with an awn up to 1 mm. long; rachilla glabrous to scabrous, lower internodes with side widening upward. This species is very variable and our specimens often appear distinct from those of the eastern states.---Waste places, fields, roadsides, meadows and pastures. Introduced from Eurasia and now widely distributed in the United States. Our specimens from the northcentral, central and southcentral parts of Colorado but to be expected anywhere around cultivated areas. Altitude 5000-9500 feet.

12. Agropyron pseudorepens Scribn. & Smith, U. S. D. A. Div. Agrost. Bull. 4:34. 1897.

Perennial plants, with creeping rhizomes; culms 30-100 cm. tall, glabrous; leaf blades 2-6 mm. wide, 12-20 cm. long, flat or involute when dry, scabrous (weakly so below on some), sometimes short-pilose above; sheaths glabrous, or somewhat scaberulent, auricles short, often reduced; ligules less than 1 mm. long (rarely to 1.5); spikes 7-25 cm. long, spikelets not strongly imbricated; spikelets 10-16 mm. long, 3- to 7-flowered; first glume 7-10 mm. long, 5-nerved; second glume 8-11 mm. long, 5- to 7-nerved, both glumes awn-pointed, rather broad and nearly as long as the spikelet; lemmas 9-11 mm. long, acuminate or with a short awn up to 2 mm. long; rachilla usually villous, sometimes pubescent, the hairs somewhat appressed.---Meadows, thickets, canyons, slopes and open woods. South Dakota to Washington, south to New Mexico and Arizona; also Michigan. Scattered over the mountainous areas of Colorado, none of our specimens from the western or northwestern parts. Altitude 5000-11,500 feet.

12A. Agropyron pseudorepens magnum Scribn. & Smith,(var.) U. S. D. A. Div. Agrost. Bull. 4:34. 1897.

Culms 100-150 cm. tall, larger and more robust than in the species; leaf blades usually over 6 mm. wide and 20 cm. long; spike stout, somewhat 1-sided.---A large robust form of the species. Our only specimen from Gilpin County at about 9000 feet.

13. Agropyron desertorum (Fisch.) Schult., Mantissa 2:412. 1824.

A. cristatum of Manuals---Perennial plants, no rhizomes (a rhizomatous strain in cultivation); culms 35-100 cm. tall, densely tufted, erect, glabrous; leaf blades 2-7 mm. wide, 5-20 cm. long, flat, short-pilose above (in ours), glabrous below; sheaths glabrous or sometimes pubescent on lower, auricles rather small; ligules about 0.5 mm. long; spikes 4-7 cm. long, dense, the spikelets closely imbricate and spreading; spikelets 5-15 mm. long, about 8-flowered; first glume 4-5 mm. long, 1- or faintly 3-nerved; second glume 5-6 mm. long, 1- to 3-nerved, both glumes strongly keeled, tapering to a short awn; lemmas 4-6 mm. long, nerves indistinct, narrowed to an awn 2-4 mm. long. True A. cristatum (L.) Gaertn. has longer awns on the glumes.---Cultivated fields, sometimes persisting. Cultivated especially in North Dakota, South Dakota, Wyoming and Colorado; native to Eurasia. Our specimens scattered over Colorado at 5000-8500 feet.

14. Agropyron triticeum Gaertn., Nov. Comm. Petrop. 14:540. 1770.

Annual plants; culms mostly 10-30 cm. tall, erect, geniculate or decumbent at base, pubescent below the inflorescence; leaf blades 1-3 mm. wide, 4-8 cm. long, flat, scabrous to pubescent above; sheaths glabrous, rather inflated, auricles small; ligules less than 1 mm. long; spike 1-1.5 cm. long, to 1 cm. wide, oval or ovate, spikelets crowded and divergent; spikelets 5-8 mm. long; glumes subscarious-margined, keel scabrous, wide but tapering to an awn-pointed tip; lemmas 5-6 mm. long, indistinctly nerved, scabrous, firm, awn-pointed; rachilla scabrous.--- Introduced from Russia into Montana and Idaho according to Hitchcock. Our one specimen probably an accidental introduction into Colorado, in Arapahoe County at 5200 feet.

15. Agropyron inerme (Scribn. & Smith) Rydb., Bull. Torr. Club 36:539. 1909.

Perennial plants, no rhizomes; culms 30-70 cm. tall, tufted, smooth; leaf blades 1-2 mm. wide, 10-20 cm. long, involute, scabrous and pubescent above; sheaths glabrous, auricles small; ligules 0.5-1.5 mm. long; spike 7-15 cm. long, narrow, the spikelets little if any imbricated; spikelets 1-2 cm. long, 4- to 8-flowered; first glume 4-7 mm. long; second glume 6-9 mm. long, both glumes awnless, 3- to 5-nerved, rather wide and about 1/2 the length of the spiklet; lemmas about 1 cm. long, faintly nerved, acute, awn-tipped or sometimes obtuse to truncate; rachilla glabrous or merely scabrous. Daubenmire (Bull. Torr. Club 66:327-329, 1939) found all intergradations between this and A. spicatum and concluded A. inerme would be, at most, a variety of it. However, in Colorado the two seem about as distinct as many other pairs of species in this puzzling genus.---Dry plains, hills, meadows and slopes. Montana to British Columbia, south to Utah, Wyoming, Nebraska and Oregon. Our specimens scattered throughout the mountainous west two-thirds of Colorado but none from the southcentral part. Altitude 4500-8500 feet.

16. Agropyron trachycaulum (Link) Malte, Nat. Mus. Canada Ann. Rept. 1930:42. 1932.

A. pauciflorum (Schwein) Hitchc.; A. tenerum Vasey and as A. violaceum (Hornem.) Lange and A. biflorum (Brign.) Roem. & Schult. in Rydberg---Perennial plants, no rhizomes; culms 40-100 cm. tall, tufted, erect; leaf blades 2-8 mm. wide, 7-20 cm. long, flat or involute on drying, glabrous to scabrous; sheaths glabrous or the lower somewhat pubescent, auricles small, often rudimentary; ligules 0.5-1 mm. long; spike 7-25 cm. long, usually slender, spikelets distant or imbricated (usually little); spikelets 10-20 mm. long, 3- to 7-flowered; glumes subequal, 3- to 7-nerved, often asymmetrical, rather broad, acute or awn-pointed, almost as long as the spikelet; lemmas 8-13 mm. long, obscurely nerved, glabrous except sometimes at base, acuminate to awn-pointed, the awn to 4 mm. long; rachilla short, pubescent to villous.---Hillsides, meadows and banks usually in dry soil. Labrador to Alaska and over the United States except in the southeastern states. Our specimens widely distributed over the western three-fourths of Colorado but should be in the eastern one-fourth too. Altitude 4500-12,000 feet.

17. Agropyron latiglume (Scribn. & Smith) Rydb., Bull. Torr. Club 36:539. 1909.
Perennial plants, no rhizomes; culms 20-50 cm. tall, erect or more commonly curved or geniculate at base; glabrous or short-pubescent below the nodes; leaf blades 2-5 mm. wide, 3-10 cm. long, the culm leaves shortest, flat or somewhat involute when dry, pubescent both sides; sheaths glabrous or pubescent on lower ones; ligules about 0.5 mm. long; spikes 3-6 cm. long, dense, the spikelets closely imbricated; spikelets 10-15 mm. long, 3- to 5-flowered; glumes subequal, about 1 cm. long, broad, 3- to 5-nerved, short-awned (ours to 1.5 mm. long), scarious-margined; lemmas about 1 cm. long, pubescent all over, margins somewhat scarious, acute or short-awned; rachilla appressed-pubescent.---High mountains. Listed by Hitchcock from Montana to Labrador and Alaska. Our one specimen is from Lake County, Colorado, at 12,000 feet.

20. Triticum L. WHEAT

Ligules membranous; spikelets several-flowered, solitary and placed flatwise to the continuous or articulated rachis, the rachilla disarticulating above the glumes and between the florets; grain pubescent at apex.

1. Triticum aestivum L., Sp. Pl. 85. 1753.
Annual or winter annual plants; culms 60-100 cm. tall, tufted, erect; leaf blades 8-18 mm. wide, 10-25 cm. long, flat, glabrous or somewhat scaberulous above; sheaths glabrous or occasionally scabrous or pubescent, auricles small to moderate, scattered long hairs often present on them; ligules about 1 mm. long; spike 4-12 cm. long, dense; spikelets 9-12 mm. long, 2- to 5-flowered; glumes subequal, 6-8 mm. long, usually keeled on 1 side near apex, this keel extending into a mucro, outer side broader, hence glume broad and asymmetrical; lemmas broad, keeled, asymmetrical, more or less 3-toothed at apex, the middle extending into an antrorsely scabrous awn (or awnless in beardless types); paleas about as long as the lemmas.---Fields and waste places but hardly established. Found in cultivation all over Colorado at lower elevations.

21. Aegilops L. GOATGRASS

Ligules membranous; spikelets 2- to 5-flowered, solitary and flatwise on 2 sides of the rachis, sunken partly in it, finally disarticulating at the base of the internode of the rachis but spike apt to fall entire first.

1. Aegilops cylindrica Host., Icon. Gram. Austr. 2:6. 1802.
Annual or winter annual plants; culms 30-60 cm. tall, erect, glabrous or pubescent; leaf blades 2-3 mm. wide, 5-10 cm. long, flat, scabrous or sparsely pilose; sheaths glabrous, auricles usually present, small to medium sized; spike 5-9 cm. long, spikelets few; spikelets 8-10 mm. long, 2- to 4-flowered, glabrous or hispid; glumes equal or subequal, several-nerved, keeled near side nearest rachis, keel extending into an awn from 1-2 mm. long on lower spikelets to 3 cm. on the terminal ones; lemmas 2-lobed at apex, these lobes unequal, awned between the lobes, awn of terminal spikelet up to 5 cm. long, then 3-4 mm. on next and shorter to basal spikelets; paleas about as long as lemmas.---Weed in fields and waste places. Missouri, Kansas, Oklahoma, Colorado and New Mexico, introduced from Europe. Our specimens scattered in cultivated areas of Colorado at 3500-5500 feet.

22. Secale L. RYE

Ligules membranous; spikelets solitary and placed flatwise on the rachis, usually 2-flowered, rachilla disarticulating above the glumes and between the florets; grain pubescent at apex.

1. Secale cereale L., Sp. Pl. 84. 1753.
Annual or winter annual plants; culms 60-150 cm. tall, tufted, erect; leaf blades 6-13 mm. wide, 10-25 cm. long, scabrous or sometimes smooth; sheath glabrous or occasionally pubescent, somewhat glaucous, auricles rather small; ligules about 1 mm. long; spike 8-15 cm. long, dense; spikelets 1-1.6 cm. long, usually 2-flowered or a third rudimentary one above; glumes subequal, 7-9 mm. long, narrow, nearly subulate, 1-nerved, acuminate or awn-pointed; lemmas 5-nerved, keeled, the keel terminating into a long awn up to 25 mm. long, this antrorsely scabrous, lanceolate, unsymmetrical; paleas about as long as the lemmas.---Cultivated and escaped into old fields, roadsides and waste places. To be expected in Colorado anywhere around cultivated areas, our specimens up to 8500 feet.

23. Elymus L. WILDRYE

Perennial plants, usually without rhizomes; leaf blades flat to sometimes involute; sheaths open, small or large auricles present; ligules membranous; inflorescence a spike, the spikelets in pairs (sometimes 3 or 4 or even solitary at some nodes); spikelets 2- to 6-flowered; glumes equal, rather rigid, sometimes indurated, twisted in front of the spikelet, usually narrow or subulate, acute to aristate, somewhat asymmetrical; lemmas obscurely 5-nerved, rounded on back, acute or awned from the tip; paleas somewhat shorter than the lemmas; styles very short, stigmas plumose.

1. Some awns over 5 mm. long
   2. Awns straight or nearly so
      3. Glumes terete, bowed out at base   -1. E. virginicus
      3. Glumes flat at base, not bowed out
         4. Rachis breaking up at maturity; leaf blades involute or subinvolute; glumes linear; lemmas 7-10 mm. long (excluding awns); awn of glumes over 5 mm. long   -2. E. macounii
         4. Rachis not disarticulating; leaf blades flat; glumes lanceolate; lemmas 9-12 mm. long; awn of glumes less than 5 mm. long   -3. E. glaucus
   2. Awns divergent on drying
      5. Awns very divergent, making the spike bristly; glumes indurated at base (E. canadensis and intergrading varieties)
         6. Lemmas villous or hirsute   -4. E. canadensis
         6. Lemmas glabrous to hispidulous - scabrous
            7. Spike robust; leaf blades often over 10 mm. wide   -4A. E. canadensis robustus
            7. Spike not robust; leaf blades less than 10 mm. wide   -4B. E. canadensis brachystachys
      5. Awns only slightly spreading, the spike not bristly; base of glumes slightly if at all indurated   -3. E. glaucus
1. Awnless or with awns less than 5 mm. long
   8. Spikes thick, the spikelets 2-4 at a node; some leaf blades over 8 mm. wide; plants often over 100 cm. tall   -5. E. cinereus
   8. Spikes slender, the spikelets paired near the middle of the rachis, solitary above and below; leaf blades less than 8 mm. wide; plants less than 100 cm. tall
      9. Long creeping rhizomes present; glumes subulate to lanceolate (E. triticoides and varieties)
         10. Spikelets paired near the middle; spike mostly over 10 cm. long; plant usually over 60 cm. tall
            11. Sheaths glabrous or scabrous   -6. E. triticoides
            11. Sheaths pubescent   -6A. E. triticoides pubescens
         10. Spikelets mostly solitary; spike mostly less than 10 cm. long; plant usually less than 60 cm. tall   -6B. E. triticoides simplex
      9. Rhizomes lacking; glumes subulate
         12. Awn of lemmas 2-5 mm. long; leaves scattered along the culm, some often flat   -7. E. ambiguus
         12. Lemmas awnless or awn-tipped, these less than 2 mm. long; leaves mostly basal, almost always involute   -8. E. salinus

1. Elymus virginicus L., Sp. Pl. 84. 1753.
E. striatus Willd.---Perennial plants, no rhizomes; culms 50-100 cm. tall, tufted, erect, sometimes glaucous; leaf blades 4-15 mm. wide, 10-30 cm. long, flat scabrous especially toward the tip; sheaths glabrous or lower pubescent, auricles present but may be small; ligules 1 mm. long or less; spike 5-15 cm. long, usually erect, often partly included, spikelets crowded, usually 2 at a node; spikelets about 10-16 mm. long, 3- to 5-flowered; glumes about 12 mm. long, strongly nerved, linear-lanceolate, firm, rounded, yellow-indurated at base, broadened above, tapering into an awn 4-18 mm. long. E. virginicus var. australis (Scribn. & Ball) Hitchc., was reported by Meserve from Boulder County, but the writer cannot locate a specimen collected in Colorado.---Moist ground, along streams, ponds and in low woods. Widely distributed throughout the United States except California, Washington, Nevada and Utah. Rare in Colorado, our three specimens from Larimer and Pueblo Counties at 5000-6000 feet.

2. Elymus macounii Vasey, Bull. Torr. Club 13:119. 1886.
Perennial plants, lacking rhizomes (very rarely present); culms 40-80 cm. tall, densely tufted, erect, slender, glabrous; leaf blades 2-5 mm. wide at base, 10-20 cm. long, firm and subinvolute, scabrous; sheaths glabrous or rarely pubescent, auricles small; ligules 0.5-1 mm. long; spike 4-12 cm. long, slender, erect or somewhat nodding, the rachis finally disarticulating, spikelets often 1 at a node for most of the spike, 2 at a node below, imbricate and appressed; spikelets mostly 2-flowered; glumes subequal, 5-7 mm. long, 3-nerved, scabrous, narrowly linear, tapering to an awn 3-5 mm. long; lemmas 7-10 mm. long, glabrous or scabrous near apex, the awn straight, 7-15 mm. long.---Meadows and open ground. Minnesota to Alaska and Washington, south to Iowa, Nebraska, New Mexico and California. Our specimens mostly from northcentral, central and southcentral Colorado, largely in the mountains at 4500-9500 feet.

3. Elymus glaucus Buckl., Acad. Nat. Sc. Phila. Proc. 1862:99. 1863.
E. nitidus Vasey; E. marginalis Rydb.---Perennial plants, no rhizomes; culms 40-100 cm. tall, loosely or densely tufted, erect or somewhat decumbent at base; leaf blades 8-15 mm. wide (one form in our area 3-6), 10-25 cm. long, flat, usually scabrous both sides; sheaths scabrous to smooth, auricles small, sometimes absent; ligules 1 mm. long or less; spike 5-20 cm. long, erect to nodding somewhat, often purplish, usually dense, spikelets usually 2 at a node; spikelets to 15 mm. long, several-flowered; glumes 8-15 mm. long, with 2 to 5 scabrous nerves, lanceolate, the bases flat and somewhat indurated, acuminate or awn-pointed, this to 4 mm.; lemmas 9-12 mm. long, glabrous or scabrous with an erect or somewhat spreading awn 10-20 mm. long.---Open woods, copses, dry hills and meadows. Ontario and Michigan to southern Alaska, south through South Dakota and Colorado to New Mexico and California; also in Missouri and Arkansas. Our specimens from the mountainous, western two-thirds of Colorado at 6300-11,000 feet.

4. Elymus canadensis L., Sp. Pl. 83. 1753.
Perennial plants, lacking rhizomes; culms 60-150 cm. tall, tufted, erect or decumbent at base, green or glaucous; slender to rather stout; leaf blades 6-20 mm. wide, 10-30 cm. long, flat, somewhat scabrous above or sparsely hispid; sheaths glabrous, rarely pubescent, auricles rather large; ligules 0.5-1 mm. long; spike 10-25 cm. long, thick and bristly, drooping, often interrupted below, spikelets imbricated, commonly in 3's or 4's at a node; spikelets 12-18 mm. long, 3- to 5-flowered; glumes 14-30 mm. long, including the awns which are about 1/2 as long as the body,

mostly 2- to 4-nerved, narrow, rigid, linear or lanc-linear, bases thickened and indurated but not bowed out; lemmas 7-15 mm. long, scabrous-hirsute to hirsute-pubescent, with a divergently curved awn 10-30 mm. long. The species intergrades with the two varieties.---River banks, fence rows and open forests. Quebec to Alaska and south to North Carolina, Missouri, Texas, Arizona and California. Widely distributed in Colorado, probably in every county. Our specimens at 3500-9000 feet.

4A. Elymus canadensis robustus (Scribn. & Smith) Mackenz. & Bush, (var.) Man. Fl. Jackson Co. 38. 1902.

E. robustus Scribn. & Smith---Differs from the species in the stouter, denser spikes only slightly nodding; lemma may be more scabrous than hispid. Intergrades with the species and var. 4B.---Prairies and river banks. Massachusetts to Montana, south to Kentucky, Missouri, Texas and Arizona. Our specimens from scattered areas in Colorado at 4500-7500 feet.

4B. Elymus canadensis brachystachys (Scribn. & Ball) Farwell, (var.) Mich. Acad. Sc. Rpt. 21:327. 1920.

E. brachystachys Scribn. & Ball---Leaf blades narrower than in species; lemmas glabrous or nearly so (merely scabrous not hirsute). Intergrades with species and variety 4A.---Moist open or partly shaded ground. Arkansas, Oklahoma, Texas and New Mexico. Scattered in Colorado mostly in the eastern half of the state at 3500-8000 feet.

5. Elymus cinereus Scribn. & Merr., Bull. Torr. Club 29:467. 1902.

E. condensatus of Manuals---Perennial plants, rhizomes when present short and thick; culms 75-200 cm. tall, tufted, glabrous or puberulent below the nodes; leaf blades 8-20 mm. wide, 15-40 cm. long, flat, firm, scabrous above; sheaths glabrous or scabrous, auricles moderate to large; spike 12-30 cm. long, erect, usually dense, the spikelets 3-5 at each node of the rachis; spikelets 12-17 mm. long, 3- to 4-flowered; glumes to 12 mm. long, subulate, awn-pointed, 1-nerved or nerveless; lemmas 8-10 mm. long, glabrous to strigose, awnless, mucronate or with a short awn rarely to 2 mm. long.---Dry plains, slopes, sand hills, along gullies, ditches and valleys. Minnesota to British Columbia, south to Colorado, New Mexico and California. Our records from the western two-thirds of Colorado but none from the extreme southern part, at 5300-10,000 feet.

6. Elymus triticoides Buckl., Acad. Nat. Sc. Phila. Proc. 1862:99. 1862.

Perennial plants with creeping rhizomes; culms 50-100 cm. tall, slender, erect, usually glaucous; leaf blades 2-5 mm. wide (unrolled), 15-25 cm. long, flat or soon involute especially toward tip, glabrous or scabrous, sometimes pubescent; sheaths glabrous or scabrous, auricles rather small; ligules 1 mm. long or less; spike 8-20 cm. long, erect (most of ours slender and spikelets little imbricated), spikelets 1-2 at a node; spikelets 12-20 mm. long, 3- to 6-flowered, the florets rather distant on most; glumes subequal, 6-12 mm. long, nerveless or 1- to 3-nerved, varying from subulate to lanceolate, tapering to an awn point; lemmas 6-10 mm. long, about 7-nerved, these more distinct near apex, glabrous or scabrous above, long-acute or awn-pointed, the awn up to 3 mm. long.---Meadows and hillsides in moist or alkaline soils. Montana to Washington, south to Texas and California. Apparently not very common in Colorado, our specimens from the northcentral, central, southcentral and westcentral portion of the state at 4500-8500 feet.

6A. Elymus triticoides pubescens Hitch. in Jepson, (var.) Fl. Calif. 1:186. 1912.

Leaf blades pubescent; sheaths pubescent.---Reported by Meserve from along the Gunnison River in western Colorado but no specimen from this state could be located by the writer.

6B. Elymus triticoides simplex (Scribn. & Will.) Hitchc., (var.) Am. Jr. Bot. 21:132. 1934.

Culms usually less than 60 cm. tall; leaf blades short, involute; spikes usually less than 10 cm. long; spikelets mostly solitary at the rachis nodes. Not a separate species in our material.---Wyoming and Colorado to California and Oregon. Our four specimens from southcentral Colorado at 7500-9500 feet.

7. Elymus ambiguus Vasey & Scribn., Contrib. U. S. Nat. Herb. 1:280. 1893.

Perennial plants, no rhizomes; culms 30-70 cm. tall, loosely tufted, erect, glabrous; leaf blades 2-6 mm. wide, 10-30 cm. long, flat to somewhat involute, usually scabrous, especially above; sheaths glabrous, auricles moderate in size; ligules about 1 mm. long; spike 5-15 cm. long, erect, rather dense, spikelets solitary at the base and apex of the rachis; spikelets 12-15 mm. long, 2- to 4-flowered; glumes 8-12 mm. long, subulate, scabrous near awned apex; lemmas about 1 cm. long, glabrous or scabrous on back, short-awned, these 2-5 mm. long (some lemmas may be awnless on same spike). The variety E. ambiguus strigosus (Rydb.) Hitchc., with strigose or pubescent lemmas is listed for Colorado.---Open slopes, canyons and rocky hillsides. Colorado, Montana and Utah. Probably scattered over the mountainous part of Colorado but no specimens available from the southwestern and westcentral parts. Ours from the southcentral, central, northcentral and northwestern portions of the state at 5200-8500 feet.

8. Elymus salinus Jones, Calif. Acad. Sc. Proc. II 5:725. 1895.

Perennial plants, no rhizomes (one atypical specimen has short ones); culms 30-80 cm. tall, in large tufts, erect, smooth or scabrous below the inflorescence; leaf blades 2-3 mm. wide (unrolled), 10-15 cm. long, involute, rather basal, the culm leaves shorter, firm, scabrous or puberulent near base; sheaths somewhat scabrous, often fibrous at base, auricles present or absent; ligules 0.5-1 mm. long; spike 5-12 cm. long, erect, rather slender, spikelets mostly solitary sometimes rather distant on the rachis; spikelets 10-15 mm. long, 5- to 9-flowered; glumes 4-10 mm. long, subulate, often falcate, sometimes reduced; lemmas 8-10 mm. long, glabrous, scabrous, rarely sparsely strigose, nerves obscure, awnless or awn-tipped, this not over 2 mm. long.---Rocky slopes, sagebrush hills and saline soils. Wyoming, Idaho, Utah and Arizona. Our two specimens from western Colorado, Garfield and Montezuma Counties. The former one, however, is not typical of the species.

24. Sitanion Raf. SQUIRREL TAIL

Sheaths split; ligules membranous; spikelets 2- to few-flowered, 2 at a node of the rachis, this disarticulating at the base of each joint which remains as an awn-pointed stipe to the spikelets above it; style short, stigmas plumose.

1. Sitanion hystrix (Nutt.) J. G. Smith, U. S. D. A. Div. Agrost. Bull. 18:15. 1899.
S. longifolium J. G. S.; S. brevifolium J. G. S.; S. montanum J. G. S.; S. molle J. G. S.; S. pubiflorum J. G. S.; S. cinereum J. G. S.; S. insularis J. G. S.; S. strigosum J. G. S.; S. rigidum J. G. S.; S. glabrum J. G. S.; S. marginatum Scribn. & Merr.; S. elymoides Raf.; S. ciliatum Elmer; S. basalticola Piper---Perennial plants, no rhizomes; culms 10-50 cm. tall, tufted, erect or spreading, glabrous or glaucous; leaf blades 1-5 mm. wide, 5-20 cm. long, flat or involute, glabrous or pubescent; sheaths glabrous, scabrous or pubescent, the upper often inflated, auricle long and narrow, rudimentary or absent; ligules less than 1 mm. long; spike 2-10 cm. long, including divergent awns, about as broad, erect, often partly included, green or purplish, later pale; spikelets 10-15 mm. long, 2- to 6-flowered; glumes very narrow, 1- to 2-nerved, each nerve extending into a scabrous, divergent awn 5-9 cm. long, sometimes bifid to the middle or bearing a bristle along 1 margin; lemmas 7-10 mm. long, obscurely nerved, scabrous, strigose or glabrous, with a terminal antrorsely scabrous widely divergent awn 4-9 cm. long; paleas about as long as the lemma, keels strigose-ciliate.---Dry hills, plains, open woods, meadows, grasslands, and rocky slopes. South Dakota to British Columbia, south to Missouri, Texas, and California; also in Mexico. Our specimens from all over Colorado at 4000-10,500 feet.

25. Hordeum L. BARLEY

Annual or perennial plants, no rhizomes; leaf blades usually flat; sheaths open, auricles absent to very large; ligules membranous; inflorescence a spike, 3 spikelets to a node, the middle one sessile, the laterals sessile or pedicelled and often sterile and reduced, rachis usually disarticulating; spikelets 1-flowered (rarely 2), rachilla disarticulating above the glumes and between the florets, prolonged beyond the last palea as a bristle or a minute floret; glumes rigid, narrow, often subulate and awned, standing in front of the spikelet; lemmas 5-nerved, these usually obscure, rounded on back, terminally awned at least in the central spikelet (except in a variety of cultivated barley); paleas about as long as the lemmas; styles short, stigmas plumose.

1. Lateral spikelets sessile, usually not reduced or imperfect; cultivated annuals, sometimes escaping along roads and fields; awnless or with awns often over 10 cm. long
   2. Awns present on lemmas   -1. H. vulgare
   2. Awns absent, replaced by irregular lobes or teeth   -1A. H. vulgare trifurcatum
1. Lateral spikelets stalked, usually reduced and sterile; annuals or perennials, not cultivated but acting as weeds; awns present but rarely over 6 cm. long
   3. Glumes of the middle spikelet bowed out at base, dilated upward into a lanceolate shape; annuals   -2. H. pusillum
   3. Glumes of middle spikelet not bowed out, narrow and awnlike from the base; perennials
      4. Awn of lemma of middle spikelets 15-60 mm. long, spike at maturity bushy with very divergent awns, at maturity spreading over 45 degrees from the rachis
         5. Awn of lemma of middle spikelets 3-6 cm. long; spike usually over 5 cm. long (unless partly broken away)   -3. H. jubatum
         5. Corresponding awn 1.5-3 cm. long; spike often less than 5 cm. long   -3A. H. jubatum caespitosum
      4. Awn of lemma of middle spikelets 6-12 mm. long, the awns at maturity seldom diverging over 45 degrees from the rachis   -4. H. brachyantherum

1. Hordeum vulgare L., Sp. Pl. 84. 1753.
Annual or winter annual plants; culms 60-120 cm. tall, coarse, erect; leaf blades 5-18 mm. wide, 10-30 cm. long, flat, somewhat scabrous; sheaths glabrous, auricles large, whitish; ligules to 3 mm. long; spike 2-10 cm. long, dense, erect or nearly so, not disarticulating; spikelets 3 at a node, all perfect (except in 2-row barley); glumes equal, about 8 mm. long, narrow, divergent at base with an awn 7-10 mm. long; lemmas 10-12 mm. long, fusiform or lanceolate, narrowed into a scabrous awn 10-15 cm. long; paleas about as long as the lemmas.---Cultivated for grain and along road shoulders, sometimes spontaneous in waste places but not persistent. To be expected anywhere in cultivated areas of Colorado, our specimens from 4000-9000 feet.
1A. Hordeum vulgare trifurcatum (Schlecht.) Alefeld, (var.) Landw. Fl. 341. 1866.
Lemmas with irregular short lobes or teeth instead of awns.---Sometimes cultivated in Colorado.
2. Hordeum pusillum Nutt., Gen. Pl. 1:87. 1818.
Annual plants; culms 10-35 cm. tall, tufted, erect or decumbent at base; leaf blades 1-4 mm. wide, 1-7 cm. long, flat or involute when dry, lower pubescent; sheaths from glabrous on upper to pubescent on lower, auricles absent or very small; ligules less than 1 mm. long; spike 2-7 cm. long, erect, the rachis disarticulating; spikelets with central sessile and perfect, laterals reduced and pedicelled; first glume of lateral spikelets and both glumes of central with indurated, bowed out bases, dilated above and rather lanceolate; second glume of lateral spikelets bristlelike, attenuate into an awn 5-15 mm. long; lemmas of middle spikelet 6-8 mm. long with an awn about 3-5 mm. long, rarely longer, lemmas of lateral spikelets smaller and only awn-pointed.---Plains, valleys, mesas, hillsides especially in dry or alkaline soil. Widely distributed in the United States except the northeastern part. Also widely distributed in Colorado, no specimens as yet from the northwestern part but surely there. Altitude 3400-6500 feet.

3. Hordeum jubatum L., Sp. Pl. 85. 1753.
Perennial plants, no rhizomes; culms 20-70 cm. tall, tufted, erect or decumbent at base; leaf blades 2-5 mm. wide, 3-15 cm. long, usually flat, more or less scabrous; sheaths glabrous or the lower pubescent, auricles absent; ligules less than 1 mm. long; spike 5-10 cm. long, nodding when mature, often as wide as long (counting spreading awns), green or purple, but pale in age, rachis disarticulating; spikelets with middle sessile and perfect, laterals pedicelled, sterile and reduced to 1-3 spreading awns; glumes of fertile spikelets awnlike, 2.5-6 cm. long, spreading; lemmas of fertile spikelet 5-8 mm. long with an awn 2.5-6 cm. long, obscurely 5-nerved, scabrous at apex. Intergrades with its variety in Colorado.---Open ground, meadows, prairies, along streams, ditches and in waste places as a weed. Now widespread over the United States except in the southeastern states. Widely distributed over Colorado, no specimens from the northeast corner but surely there. Altitude 3500-9500 feet, probably to 10,000.

3A. Hordeum jubatum caespitosum (Scribn.) Hitchc., (var.) Biol. Soc. Wash. Proc. 41:160. 1928.
H. caespitosum Scribn.---Spike usually 3-5 cm. long; lemmas of middle spikelet 1.5-3 cm. long. This variety intergrades with the species.---North Dakota to Alaska, south to Kansas and Arizona. Scattered with the species over Colorado, no specimens yet from the extreme eastern part. Altitude 5000-9000 feet.

4. Hordeum brachyantherum Nevski, Acta Inst. Bot. Acad. Sci. U.R.S.S. I.2:61. 1936.
H. nodosum of Manuals---Perennial plants, no rhizomes; culms 10-60 cm. tall, tufted, erect or spreading, sometimes geniculate and decumbent at base; leaf blades 2-5 mm. wide, 4-12 cm. long, flat, scabrous; sheaths glabrous to puberulent, auricles none; ligules 1 mm. long or less; spike 2-8 cm. long, slender, rachis disarticulating; spikelets with central one sessile and perfect, laterals pedicelled and sterile; glumes all setaceous or awnlike, about 1 cm. long; lemmas of middle spikelet 6-8 mm. long, scabrous, the awn 6-12 mm. long, this somewhat divergent. Sometimes hard to tell from forms of the preceding.---Meadows, grasslands, streamsides in moist or dry ground. Montana to Alaska and south to New Mexico and California; introduced at various places in the eastern states. Our specimens widely distributed in the western mountainous two-thirds of Colorado at 5000-10,500 feet.

26. Lolium L. RYEGRASS

Short-lived perennial plants; leaf blades flat; sheaths open, auricles usually present; ligules membranous; inflorescence a spike, the rachis continuous; spikelets several-flowered, solitary and placed edgewise to the rachis; rachilla disarticulating above the glumes and between the florets; first glume wanting (except on terminal spikelet); second glume on the outside; lemmas 5- to 7-nerved, rounded on the back, obtuse to awned; paleas about as long as the lemmas; styles short, stigmas plumose.

1. Awns present on some lemmas, 1-8 mm. long; spikelets often over 10-flowered; lemmas sometimes over 7 mm. long
-1. L. multiflorum
1. Awns absent or lemmas merely awn-pointed; spikelets rarely over 10-flowered; lemmas typically less than 7 mm. long
-2. L. perenne

1. Lolium multiflorum Lam., Fl. Franc. 3:621. 1778.
Perennial plants but short-lived, no rhizomes; culms 50-100 cm. tall, tufted, erect or often geniculate at base; leaf blades 3-8 mm. wide, 10-30 cm. long, flat, smooth or scaberulous especially above; sheaths smooth or scabrous, auricles usually present; ligules about 1 mm. long; spike 15-25 cm. long (or even longer), spikelets distant or somewhat imbricated above; spikelets 12-15 mm. long, 8- to 20-flowered; glumes about 6 mm. long, usually about as long as lowest floret, 5-nerved, obtuse; lemmas 6-8 mm. long, awn 1-8 mm. long (but some lemmas may be awnless).---Fields, meadows, lawns and waste places. Newfoundland to Alaska and south to Virginia and California, occasionally further south. Our specimens from Otero, Boulder and Larimer Counties but to be expected anywhere around cultivated or inhabited areas. Altitude 4000-7000 feet.

2. Lolium perenne L., Sp. Pl. 83. 1753.
Short-lived perennial plants, no rhizomes; culms 30-60 cm. tall, tufted, smooth, erect or decumbent at base; leaf blades 2-6 mm. wide, 5-20 cm. long, flat, smooth or rough above; sheaths glabrous or sometimes scaberulous, auricles present on most of the leaves; ligules about 1 mm. long, sometimes to 2; spikes 10-25 cm. long, spikelets distant to somewhat imbricated; spikelets 10-15 mm. long, 6- to 10-flowered; glume 6-12 mm. long, usually about as long as the first floret, 3- to 5-nerved, obtuse; lemmas 4-7 mm. long, awnless or nearly so. This species intergrades freely with number 1.---Meadows, pastures, along ditches, in lawns and waste places. Newfoundland to Alaska and south to Virginia and California, occasionally further south. Our specimens from El Paso, Boulder and Larimer Counties in Colorado but to be expected anywhere in inhabited areas. Altitude 5000-6500 feet.

III. Tribe Aveneae

27. Koeleria Pers. JUNEGRASS

Sheaths split; ligules membranous; spikelets 2- to 4-flowered, compressed, the rachilla disarticulating above the glumes and between the florets, prolonged behind the last palea as a bristle or much reduced floret; lemmas 5-nerved, keeled.

1. Koeleria cristata (L.) Pers., Syn. Pl. 1:97. 1805.
K. gracilis Pers.---Perennial plants, no rhizomes (has been recorded as having short, inconspicuous ones); culms 15-60 cm. tall, tufted, erect, puberulent below the inflorescence; leaf

blades 1-3 mm. wide, 5-25 cm. long, flat or involute, mostly basal, more or less rough or puberulent especially above; sheaths retrorsely pubescent or scabrous especially on lower ones; ligules 0.5-1 mm. long; panicle 4-5 cm. long, dense and spikelike (opening some in anthesis and closing again), often lobed or interrupted below, pale green or purplish; spikelets 4-6 mm. long; first glume 2-3 mm. long, 1-nerved, rather narrow, scarious except for nerves; second glume 3-4 mm. long, 3- to 5-nerved, wider than first and about as long as spikelet, scabrous; lemmas 3-4 mm. long, middle nerve plain, other 4 faint, somewhat scarious at margins, acute or short-awned, this borne just below the apex; paleas nearly as long as the lemmas.---Prairies, open woods, rocky hillsides, often in sandy soil. Widely distributed except in southeastern United States. Widely distributed all over Colorado at 3500-11,500 feet.

28. Sphenopholis Scribn. WEDGEGRASS; WEDGESCALE

Perennial plants without rhizomes, except very rarely; leaf blades flat; sheaths split; ligules membranous; inflorescence a rather narrow to spikelike panicle; spikelets 2- to 3-flowered, the pedicel disarticulating below the glumes; glumes very unlike, the first very narrow, the second broadly obovate; lemmas firm, keeled, obscurely nerved, obtuse to mucronate; paleas somewhat shorter than the lemmas, hyaline, exposed; style short, stigma plumose.

1. Panicle dense, usually spikelike, erect or nearly so; second glume much wider than the lemma, the keel rounding into an obtuse apex (subcucullate) -1. S. obtusata
1. Panicle lax and nodding, not spikelike; second glume not much wider than lemma, keel nearly straight out to an acute or obtuse apex -2. S. intermedia

1. Sphenopholis obtusata (Michx.) Scribn., Rhodora 8:144. 1906.
   Perennial plants, no rhizomes (but reported as sometimes present); culms 25-80 cm. tall, tufted but also sometimes solitary, sometimes decumbent at base, often scabrous below panicle; leaf blades 2-6 mm. wide, sometimes up to 9, 3-20 cm. long, flat, glabrous to pubescent, often scabrous; sheaths glabrous or pubescent; ligules 1.5-4 mm. long; panicle 5-20 cm. long, erect, dense and spikelike, sometimes interrupted and lobed, light green to purplish; spikelets 2.5-3 mm. long, 2 florets close together, sometimes a third; first glume 1.5-2 mm. long, 1-nerved, about 0.2-0.3 mm. wide as seen from side; second glume 2-2.5 mm. long, very broad, 0.6-1 mm. as seen from side, 3- to 5-nerved, subcucullate, both scabrous on keel; lemmas 2-3 mm. long, narrower than second glume, scabrous on back, minutely papillose, obscurely nerved, obtuse or rarely mucronate or with a short awn from just below the apex.---Open woods, old fields, moist meadows along streams and ditches often in sandy soil. Maine to British Columbia and south to Florida, Arizona and California; also in Mexico and Dominican Republic. Our specimens are scattered over the eastern two-thirds of Colorado but reported from the western part of the state. Altitude 3500-8500 feet.

2. Sphenopholis intermedia (Rydb.) Rydb., Bull. Torr. Club 36:533. 1909.
   Probably S. pallens (Spreng.) Scribn. and Eatonia pennsylvanica (D.C.) A. Gray in Rydberg--- Perennial plants, no rhizomes; culms 30-100 cm. tall, in small tufts, erect; leaf blades 2-6 mm. wide, 8-15 cm. long, flat, lax, scaberulous to sparsely pilose; sheaths glabrous to pubescent; ligules 1-3 mm. long; panicle 8-20 cm. long, from rather dense to open, nodding, branches spikelet-bearing to the base; spikelets 2.5-4 mm. long, 2- to 3-flowered; first glume 1.5-2 mm. long or sometimes longer, narrow; second glume 2.3-3.3 mm. long, 4 to 5 times broader than first (to 0.7 mm. wide as seen from the side), keel almost straight to apex, not cucullate, both glumes scabrous on keel; lemmas 2.3-3 mm. long, obscurely nerved, somewhat narrower than the second glume, subacute or rarely mucronate, scaberulous on the middle nerve.---In meadows or damp rocky woods. Widely distributed over the United States. Our specimens from northcentral, central and southwestern Colorado, but probably scattered over the state. Altitude 6500-8000 feet.

29. Trisetum Pers. TRISETUM

Annual or perennial plants, usually without rhizomes; leaf blades flat or somewhat involute; sheaths split; ligules membranous; inflorescence a spikelike or somewhat open panicle; spikelets usually 2-flowered, rachilla villous, articulated above the glumes and between the florets (except in 1 species); glumes somewhat unequal, awnless, acute, second longer than the first floret; lemmas indistinctly 3- to 5-nerved, keeled, 2-toothed at apex, usually awned from the back just below this cleft apex, awn usually geniculate; paleas from shorter to about the length of the lemmas; styles distinct, stigmas plumose.

1. Awn less than 2 mm. long, included within the glumes or else wanting -1. T. wolfii
1. Awns over 2 mm. long, divergent, not hidden by glumes
   2. Articulation below the glumes (a small jointlike swelling on the pedicel marks the spot); plants annual; first glume 3-nerved -2. T. interruptum
   2. Articulation above the glumes; plants perennial; first glume 1-nerved
      3. Panicle yellowish from golden spikelets; spikelets mostly 3- to 4-flowered; introduced grass, rare in Colorado and then in waste places or along fields -3. T. flavescens
      3. Panicle pale green or purplish, not yellowish; spikelets 2-flowered (a reduced third floret sometimes present); native grasses
         4. Panicle dense, more or less spikelike (although sometimes interrupted below); plants densely tufted; culms seldom over 50 cm. tall; often found above 10,000 feet in Colorado -4. T. spicatum
         4. Panicle narrow, loose to rather dense but not spikelike; plants loosely tufted; culms often over 50 cm. tall; seldom found over 10,000 feet in Colorado -5. T. montanum

1. **Trisetum wolfii** Vasey, Monthly Rep. U. S. D. A. Feb. Mar. 156. 1874.
Graphephorum wolfii Vasey; G. brandegii (Scribn.) Rydb.; G. muticum in Rydberg.---Perennial plants, no rhizomes (short ones present on a few of ours); culms 40-80 cm. tall; leaf blades 2-6 mm. wide, 5-15 cm. long, flat, scabrous, rarely pilose above; sheaths scabrous or rarely pilose on lower; ligules 1-2 mm. long; panicle 7-12 cm. long, rather dense but not spikelike, erect, green, pale or sometimes purplish; spikelets 5-7 mm. long, 2- or sometimes 3-flowered, rachilla villous, disarticulating above the glumes; first glume 5-6 mm. long, 1-nerved; second glume 5-6 mm. long, 3-nerved; lemmas 4-5 mm. long, often a little shorter than the glumes, obtuse, scaberulous, awnless or with a short awn 0.5-2 mm. long.---Meadows and moist ground. Montana to Washington, south to New Mexico and California. Our specimens from the western half of the state, but none in the northwest corner as yet. Altitude 7200-11,000 feet.

2. **Trisetum interruptum** Buckl., Acad. Nat. Sc. Phila. Proc. 1862:100. 1862.
Annual plants; culms 10-40 cm. tall, erect or spreading, tufted; leaf blades 1-4 mm. wide, 3-12 cm. long, flat or involute near tip, often scabrous, sometimes pubescent; sheaths usually scaberulous or the lower pubescent; ligules 1.5-2.5 mm. long; panicle 5-15 cm. long, narrow but not spikelike, interrupted, pedicels with a swelling or joint below the glumes, later disarticulating there; spikelets 4-6 mm. long, 2-flowered or a third rudimentary one present; first glume 4-5 mm. long, 3-nerved, rather narrow; second glume 4-5 mm. long, 5-nerved, about twice as wide as the first, both glumes almost or as long as the lemmas; lemmas glabrous, awned from 1/3 to 1/4 down from apex, the awn twisted and flexuous, 2-7 mm. long, but the awn of the first and third floret sometimes straight and short.---Open dry or moist ground, prairies and hills. Texas to Colorado and Arizona. Our one specimen from western Colorado but probably in the southern part of the state also.

3. **Trisetum flavescens** (L.) Beauv., Ess. Agrost. 88, 153. 1812.
Perennial plants, no rhizomes; culms 30-80 cm. tall, solitary or few in a tuft, erect or decumbent at base; leaf blades 2-6 mm. wide, 4-12 cm. long, flat, scabrous at least on margins; sheaths glabrous or lower sparsely pilose; ligules 1-2 mm. long; panicle 5-15 cm. long, narrow but rather loose, yellowish; spikelets 5-8 mm. long, 3- to 4-flowered, usually golden-yellow, disarticulating above the glumes; first glume 2.5-4 mm. long, 0.2-0.3 mm. wide, as seen from side, 1-nerved; second glume 5-6 mm. long, 1 mm. wide, as seen from side, 3-nerved, both glumes scabrous on keel; lemmas 4-6 mm. long, glabrous or scaberulous, awn inserted 1/4 to 1/3 below apex, twisted and flexuous on lower part, then geniculate, usually 5-10 mm. long.---Waste places. Vermont, New York, Missouri, Colorado, Washington, California; introduced from Europe and to be expected anywhere. Our one specimen was probably cultivated in a grass plot.

4. **Trisetum spicatum** (L.) Richt., Pl. Eur. 1:59. 1890.
T. majus (Vasey) Rydb.; T. subspicatum (L.) Beauv.---Perennial plants, no rhizomes; culms 10-50 cm. tall, densely tufted, glabrous or puberulent; leaf blades mostly 1-3 mm. wide sometimes to 4.5, 3-15 cm. long or longer on basal, flat or loosely involute, glabrous or puberulent; sheaths puberulent or sometimes glabrous; ligules 1-2 mm. long; panicle 2-15 cm. long, variable but dense and usually spikelike, often interrupted at base, pale or purplish; spikelets 4-6 cm. long, 2- or sometimes 3-flowered, disarticulating above the glumes; first glume 5-7 mm. long, 1-nerved, rather narrow; second glume 6-8 mm. long, somewhat wider than first, 3- or rarely 5-nerved, both glumes glabrous or scabrous on keel, sometimes equal in length; lemmas 5 mm. long, glabrous or scaberulous, the first lemma slightly longer than the glumes, awn 4-6 mm. long, geniculate, flexuous at base, attached about 1/3 below apex of lemma.---Meadows, ridges, slopes and grasslands. Arctic America, south to Connecticut, Pennsylvania, Michigan and Minnesota and in the mountains to New Mexico and California; also in Mexico, South America and the Old World. In the high mountains of Colorado, rather widely distributed at 8000-12,500 feet.

5. **Trisetum montanum** Vasey, Bull. Torr. Club 13:118. 1886.
Graphephorum shearii (Scribn.) Rydb.---a short-awned form.---Perennial plants, no rhizomes; culms 30-80 cm. tall, loosely tufted, erect or decumbent at base; leaf blades 2-4 mm. wide, 10-20 cm. long, flat, rather scabrous; sheaths glabrous, scabrous or retrorsely pubescent on the lower; ligules 2-4 mm. long; panicle 6-15 cm. long, narrow, fairly dense but not spikelike, pale or purplish; spikelets 5-6 mm. long, 2-flowered or sometimes a reduced third present, disarticulating above the glumes; first glume about 4 mm. long, about 0.5 mm. wide from side, 1-nerved, hyaline except near nerve; second glume about 5 mm. long, about 1 mm. wide from side, 3-nerved but laterals short, hyaline near apex and margin, both glumes scabrous on keel; lemmas 4-5 mm. long, margins hyaline, awn 5-6 mm. long, spreading from base, attached just below the base of the terminal teeth.---Meadows, gulches, among bushes and moist places. Colorado, Utah, Arizona, and New Mexico. Our specimens scattered in the mountainous three-fifths of the state but none in the extreme west or northwestern part. Altitude 6000-10,500 occasionally to 11,500 feet.

30. Deschampsia Beauv. HAIRGRASS

Perennial plants, no rhizomes; leaf blades flat or involute; sheaths split; ligules membranous; inflorescence an open panicle; spikelets 2-flowered, disarticulating above the glumes and between the florets, the hairy rachilla prolonged beyond the uppermost paleas as a bristle; glumes about equal, acute; lemmas rounded on back, thin, truncate and 2- to 4-toothed at apex, bearded at base and awned from about the middle to near the base, the awn straight or geniculate; paleas somewhat shorter than lemmas; styles distinct, stigmas plumose.

1. Glumes longer than the terminal floret, the first about 1 mm. wide (as seen from side); callus hairs over 1 mm. long; lemmas 2-3 mm. long, the awn attached near the middle; leaves usually flat and thin -1. **D. atropurpurea**
1. Glumes equaling or shorter than the terminal floret, the first about 0.5 mm. wide (as seen from side); callus hairs less than 1 mm. long; lemmas 3-4 mm. long, the awn attached near the base; leaves usually folded
-2. **D. caespitosa**

1. Deschampsia atropurpurea (Wahl.) Scheele, Flora 27:56. 1844.
   Culms 15-80 cm. tall, loosely tufted, erect, glabrous, purplish at base; leaf blades 3-5 mm. wide, 5-10 cm. long, flat (rarely folded or involute according to report), glabrous or scaberulous; sheaths glabrous; ligules 3-4 mm. long; panicle 5-10 cm. long, open, loose, branches capillary, naked below, often drooping; spikelets length of glumes; first glume 5-6 mm. long, about 1 mm. wide (as seen from side), 1-nerved; second glume about as long, over 1 mm. wide, 3-nerved, both scabrous on keel; lemmas 2-3 mm. long, obscurely nerved, scabrous or glabrous, truncate-erose at apex, callus hairs about 1/2 the length of lemma, over 1 mm. long, awned from about middle, the awn 2.5-3.5 mm. long, twisted somewhat near base, geniculate or straight especially on lower florets.---Woods, wet meadows and around lakes. Newfoundland and Labrador to Alaska, south to New Hampshire, Colorado and Oregon; also in Eurasia. Our specimens all from northcentral Colorado at 9000-10,500 feet.
2. Deschampsia caespitosa (L.) Beauv., Ess. Agrost. 91, 149, 160. 1812.
   D. curtifolia Scribn.; D. alpicola Rydb.; D. confinis (Vasey) Rydb.; D. pungens Rydb.---Culms 25-100 cm. tall, in dense tufts leafy at base, erect; leaf blades 1.5-4 mm. wide, 5-20 cm. long, flat, folded or more or less involute, lower elongate especially at lower altitudes, scabrous especially above; sheaths glabrous, somewhat keeled; ligules 4-11 mm. long; panicle 10-30 cm. long, open, loose and nodding or erect and contracted in some high altitude forms, branches naked and capillary below; spikelets 3-5 mm. long, 2-flowered, the florets distant in anthesis, pale, tawny or purple; first glume 3.5-4 mm. long, 0.4 mm. wide (as seen from side), 1-nerved; second glume 4-5 mm. long, 0.7 mm. wide (as seen from side), 3-nerved the laterals rather indistinct, both glumes with margins broadly hyaline; lemmas 3-4 mm. long, smooth, callus hairs short, apex truncate and 4-toothed, hyaline margined, awned from near base, this awn from straight and included to weakly geniculate and twice the length of the spikelet. Our plant has been called D. caespitosa spp. genuina (Reichenb.) Lawrence (Am. Journ. Bot. 32:302. 1945).---Bogs, wet places, wet meadows and swamps. Greenland to Alaska, south to New Jersey, West Virginia, Illinois, North Dakota, New Mexico and California; also in Old World. Our specimens widely distributed over the western mountainous half of Colorado, none from the northwest corner. Altitude usually 7000-13,000 feet but occasionally down to 5000 and probably upward to 14,000 feet.

31. Avena L. OAT

Annual plants; leaf blades flat; sheaths open; ligules membranous; inflorescence an open panicle; spikelets over 2 cm. long on most, 2-flowered or a third reduced floret present, the rachilla bearded (except sometimes in cultivated oats), disarticulating above the glumes and usually between the florets; glumes with 7 or more nerves, 15 mm. long or more, usually exceeding the lowest floret at least, membranous or papery; lemmas 5- to 9-nerved, these not very conspicuous, rather indurate, rounded on back, bidentate at apex, with a dorsal twisted and geniculate awn (cultivated oats may be awnless or with a short straight awn); paleas somewhat shorter than the lemmas; styles distinct.

1. Awn twisted, geniculate; lemmas white- or tawny-hairy at least near the base; florets rather easily separated, often with 3 to a spikelet; a weed -1. A. fatua
1. Awn absent or if present not twisted or geniculate (may be somewhat curved); lemmas glabrous; florets not easily separated, the second usually enclosed in the lemma of the first and falling with it, usually only 2 florets present; cultivated -2. A. sativa

1. Avena fatua L., Sp. Pl. 80. 1753.
   A. fatua var. glabrata Peterm.---Annual plants; culms 30-75 cm. tall, erect, glabrous, rather stout; leaf blades 5-10 mm. wide, 10-30 cm. long, flat, scabrous, often pilose on margins near base; sheaths glabrous or pubescent on lower ones; ligules 3-4 mm. long; panicle 15-30 cm. long, loose and open, often partly included, spikelets drooping; spikelets 2-2.5 cm. long, mostly 3-flowered but some with 2, the florets easily separating; glumes subequal, the length of the spikelet, 9-, 10- or 11-nerved (in ours), glabrous; lemmas about 15 mm. long, rather indistinctly nerved, long-pilose on the back with white or tawny hairs, these may be almost absent except near base, awned from near middle, this twisted at base, then geniculate, 25-40 mm. long.---Cultivated ground, waste places, fence rows and roadsides. Introduced from Europe especially in the northern and western parts of United States. Our specimens from along the eastern foothills in Colorado but to be expected anywhere around cultivated areas. Altitude 5000-6500 feet.
2. Avena sativa L., Sp. Pl. 79. 1753.
   Annual plants; culms 60-100 cm. tall, tufted, erect, stout; leaf blades 5-20 mm. wide, 10-40 cm. long, flat, glabrous to scabrous; sheaths somewhat keeled above, glabrous or sparsely pubescent; ligules 3-4 mm. long; panicle 15-40 cm. long, open, spikelets nodding; spikelets 2-2.5 cm. long, mostly 2 florets present, these tending to fall together; glumes subequal, the length of the spikelet, several- to many-nerved; lemmas 1.5-2 cm. long, rather indurated below, glabrous, awnless or with a straight or curved awn this usually short (but up to 2 cm. long on some).---Commonly cultivated and occasionally escaped. To be expected anywhere in Colorado where cultivated grains are grown. We have one specimen from Lake County at 10,000 feet.

31A. Helictotrichon Besser OAT

Perennial plants; leaf blades flat, folded or involute; sheaths open; ligules membranous; inflorescence a rather contracted panicle; spikelets less than 1.5 cm. long, 2- to 6-flowered, the rachilla bearded, disarticulating above the glumes and between the florets; glumes 1- to 5-nerved, not over 15 mm. long usually exceeding the lowest floret, membranous or papery; lemmas about 5- to 9-nerved but these inconspicuous, somewhat indurated, rounded on back, bidentate at apex with a dorsal twisted and geniculate awn; paleas somewhat shorter than the lemmas; styles distinct. Formerly a part of the genus Avena.

1. Glumes longer than the florets; spikelets mostly 2-flowered; lemmas not brownish; culms usually less than 20 cm. tall; panicle usually less than 5 cm. long -1. H. mortoniana
1. Glumes shorter than the florets (unless the terminal ones are shattered); spikelets 3- to 6-flowered; lemmas typically brown; culms usually over 20 cm. tall; panicle seldom less than 5 cm. long (except in high alpine plants) -2. H. hookeri

1. **Helictotrichon mortoniana** (Scribn.) Henr., Blumea 3:429. 1940.
 Avena mortoniana Scribn.---Perennial plants; culms 5-20 cm. tall, densely tufted, glabrous; leaf blades 1-2 mm. wide (unrolled), 1-12 cm. long, involute, scaberulous, mostly basal or on innovations; sheaths glabrous or puberulent, thin; ligules about 1 mm. long or less; panicle 2-5 cm. long, narrow, the short branches erect and bearing a solitary, erect spikelet, pale or purplish; spikelets 8-12 mm. long, mostly 2-flowered; glumes subequal, the length of the spikelet, margins hyaline, scabrous on keel near tip, acuminate, first 1-nerved, the second 3-nerved; lemmas 7-9 mm. long, firm, rather indistinctly nerved, usually with 4 long bristlelike teeth at apex, callus bearded with hairs 1-2 mm. long; awn attached near middle of lemma, twisted and flexuous, geniculate, 1-1.5 cm. long. The second floret is reduced and sterile on a few of our plants.---Meadows and mountain tops. Colorado, Utah and New Mexico. Our specimens all from northcentral Colorado at 11,000-13,500 feet but should be in the high mountains in the southern part.

2. **Helictotrichon hookeri** (Scribn.) Henr., Blumea 3:429. 1940.
 Avena hookeri Scribn.; A. americana Scribn.---Perennial plants; culms 15-45 cm. tall, densely tufted, leaves mostly basal, erect, glabrous; leaf blades 2-4 mm. wide, 4-20 cm. long, flat, or scaberulous below and twisted in age, with a thick white nerve on margins; sheaths keeled, older ones papery; ligules 2-3 mm. long; panicle 5-10 cm. long (a few high altitude ones less than 5), narrow, branches erect, mostly with a single spikelet; spikelets about 1.5 cm. long, with 3-6 florets, these extending beyond the glumes; glumes subequal, 8-14 mm. long, thin and mostly scarious, 3-nerved, the laterals not definite; lemmas 10-12 mm. long, firm, brown, scaberulous, callus short-bearded, the hairs less than 1 mm. long, apex with narrow teeth, awned from a little above the middle; awn twisted below, geniculate, sometimes bent twice, 10-18 mm. long.---Dry slopes, prairies, ridges, hillsides and mountain tops. Manitoba to Alberta, Montana and New Mexico. Our specimens in northcentral, central and southcentral Colorado at 9000-12,000 feet.

32. Arrhenatherum Beauv. OATGRASS

Sheaths split; ligules membranous; spikelets 2-flowered, disarticulating above the glumes, the rachilla produced or prolonged beyond the upper floret; styles short, distinct, stigmas plumose.

1. **Arrhenatherum elatius** (L.) Presl., Fl. Cech. 17. 1819.
 Perennial plants, often with short rhizomes; culms 70-150 cm. tall, loosely tufted, erect, glabrous; leaf blades 3-8 mm. wide, 5-25 cm. long, flat, glabrous or scabrous; sheaths glabrous, rather loose; ligules 1-2 mm. long; panicle 15-25 cm. long, narrow (but spreading in anthesis), the branches short and usually spikelet-bearing from near base, pale or purplish; spikelets 7-9 mm. long, 2-flowered, lower floret staminate, awned, the upper perfect, awned or awnless, rachilla hairy; first glume about 6 mm. long, 1-nerved; second glume 8-9 mm. long, about the length of the spikelet, 3-nerved, both glumes hyaline, glabrous or scabrous on the keel; lemmas about 8 mm. long, scabrous or pilose, that of first floret bearing near base a twisted, geniculate awn about 13-17 mm. long, lemma of second floret awnless or with a straight awn from near apex up to 4 mm. long. A variety, the culms bulbous at base, may be present in Colorado but no specimens are available. It is Arrhenatherum elatius bulbosum (Willd.) Spenner.---Meadows, fields, open ground and waste places. Introduced from Europe and escaped from cultivation especially in the northern and eastern United States. Our specimens all from along the eastern foothills of Colorado at 5000-6700 feet.

33. Holcus L. VELVETGRASS

Sheaths open; ligules membranous; spikelets 2-flowered, the pedicel disarticulating below the glumes; styles distinct, stigmas plumose.

1. **Holcus lanatus** L., Sp. Pl. 1048. 1753.
 Perennial plants, rhizomes probably either present or absent; culms 30-60 cm. tall, tufted, erect from a decumbent or curved base, pubescent; leaf blades 4-10 mm. wide, 2-20 cm. long, flat, velvety gray-green; sheaths velvety gray-green; ligules 1-2 mm. long; panicle 8-15 cm. long, contracted except in anthesis, 2-4 cm. wide, densely flowered, often interrupted below, pale or purplish; spikelets about 4-5 mm. long, upper floret staminate, lower perfect; glumes the length of the spikelet, subequal; first glume 1-nerved, less than 1 mm. wide (as seen from side), acute or obtuse, hirsute on nerves; second glume 3-nerved, over 1 mm. wide (from side) acute or short-awned, hirsute on nerves; lemma of lowest floret 2-2.5 mm. long, smooth, nerves not noticeable, awnless, lemma of upper floret somewhat smaller, with a short, hooked awn inserted below apex; paleas nearly as long as the lemmas.---Open ground, meadows and moist places occasionally cultivated. Maine to Iowa, south to Georgia and Louisiana; Pacific Coast; British Columbia and Idaho to Arizona and California; introduced from Europe. Not common in Colorado, our only specimen from El Paso County at 6000 feet.

34. Danthonia Lam. & DC. OATGRASS

Perennial plants, lacking rhizomes; leaf blades flat or involute; sheaths split; ligules a short ring of hairs; inflorescence a contracted or open panicle or raceme (or of 1 spikelet); spikelets several-flowered, rachilla disarticulating above the glumes and between the florets, some cleistogamous spikelets often present hidden in the sheaths; glumes subequal, papery, usually longer than the uppermost floret; lemmas rounded on back, obscurely nerved, hairy at least on the margins, 2-toothed at apex, the teeth usually awned, lemmas awned from between the teeth with a stout, flat, twisted, geniculate awn which appears dorsal because the teeth are so long; paleas somewhat shorter than lemmas, papery; styles distinct, stigmas plumose.

1. Panicle usually reduced to a single spikelet (when 2 are present, lower usually reduced in size); sheaths long-pilose -1. D. unispicata
1. Panicle of 2 to several spikelets; sheaths glabrous or sometimes pilose near base (may be pilose near collar or on margins)
    2. Panicle open, the slender branches spreading or reflexed; lemmas pilose near margins, glabrous on back -2. D. californica
    2. Panicle narrow, the branches erect or ascending; lemmas variously pilose (see below)
        3. Lemmas glabrous on back, pilose only near margins, 7-9 mm. long; glumes 12-16 mm. long -3. D. intermedia
        3. Lemmas sparsely to densely pilose on back; lemma and glumes typically shorter or longer than above (see below)
            4. Glumes over 15 mm. long (16-23); lemmas over 9 mm. long (11-15), awn over 9 mm. long -4. D. parryi
            4. Glumes less than 15 mm. long (9-12); lemmas less than 9 mm. long (4-5), awn less than 9 mm. long -5. D. spicata

1. **Danthonia unispicata** (Thurb.) Munro in Macoun, Cat. Can. Pl. 2(4):215. 1888.
   Culms 15-25 cm. tall, tufted, often in mats or clumps to 30 cm. in diameter; leaf blades 2-3 mm. wide, 5-15 cm. long, flat, somewhat pilose especially on margins, mostly in radical tufts; sheaths long-pilose, the hairs spreading or reflexed and somewhat papillose at base; ligules about 1 mm. long; panicle usually reduced to a single spikelet, if 2, the lower usually reduced in size, very rarely 3 present; spikelets 10-20 mm. long, 3- to 7-flowered; first glume length of spikelet, glabrous, with 1 strong nerve and 4 indistinct laterals, somewhat hyaline and purplish on margins; second glume similar but a little shorter; lemmas about 10 mm. long, pilose below on margins, otherwise glabrous, callus pilose, the 2-awned teeth at apex 2-3 mm. long, awn somewhat flattened and twisted, 6-8 mm. long.---Open or rocky ground, meadows and hills. Montana to British Columbia, south to Wyoming and California. Our two Colorado collections from the northwest corner of the state at 7800 and 8600 feet.

2. **Danthonia californica** Bolander, Proc. Calif. Acad. 2:182. 1863.
   Culms 30-80 cm. tall, erect or somewhat geniculate at base, often disarticulating at the nodes, rather slender, glabrous; leaf blades 1-4 mm. wide, 5-30 cm. long, flat or involute especially on innovations, scabrous; sheaths glabrous, somewhat pilose near inside of collar; ligules 0.5-1 mm. long; panicle mostly with 2-5 spikelets, pedicels slender, spreading or reflexed, often flexuous, about 1-4 cm. long; spikelets 15-20 mm. long, several-flowered; glumes equal or subequal, length of spikelet, 5-nerved, the 2 laterals indistinct, margins hyaline; lemmas 10-15 mm. long (including the 2-awned teeth at apex--these teeth about 4 mm. long), pilose near margin below, glabrous on back, callus pilose, awn with flat twisted base, 5-10 mm. long.---Wet meadows, along ditches and in open woods. Montana to British Columbia, south to Colorado and California. Our Colorado specimens are from Routt, Jackson and Teller Counties at 6500-8500 feet. To be looked for in the mountains especially in the northern half of the state.

3. **Danthonia intermedia** Vasey, Bull. Torr. Club 10:52. 1883.
   Culms 10-50 cm. tall, densely tufted; leaf blades 2-4 mm. wide, 5-25 cm. long, flat or involute, glabrous or sparsely pilose; sheaths glabrous or lower rarely pilose, collar pilose; ligules 0.5-1 mm. long; panicle 2-6 cm. long, narrow, few-flowered, branches appressed, bearing a single spikelet, purplish (in ours); spikelets 12-16 mm. long, 3- to 8-flowered; glumes subequal, the length of spikelet, 5- or obscurely 7-nerved, papery; lemmas 7-9 mm. long, pilose on or near margins below and on callus, glabrous on back, terminal teeth awned, main awn flattened, twisted and divergent, 7-11 mm. long.---Meadows, bogs, mountain slopes and dry forests. Newfoundland and Quebec to Alaska, south to Michigan, New Mexico and California. Our Colorado collections are from the mountainous western two-fifths of the state except in the northwestern corner. Altitude 7500-12,500 feet.

4. **Danthonia parryi** Scribn., Bot. Gaz. 21:133. 1896.
   Culms 20-60 cm. tall, in tough clumps, erect, glabrous; leaf blades 1-4 mm. wide, 10-25 cm. long, involute or flat, rather filiform; sheaths somewhat keeled, glabrous or ciliate on margins near throat and pilose on collar; ligules 0.5-1 mm. long; panicle 3-7 cm. long, contracted with ascending branches, usually with 3-8 spikelets; spikelets 16-23 mm. long, 5- to 7-flowered; glumes subequal, length of spikelet, 5-nerved but only the middle distinct, papery; lemmas 11-15 mm. long (including the aristate teeth 4-5 mm. long), sparsely to densely pilose on back, margins and callus, awn flattened, twisted and divergent, 11-15 mm. long.---Open grasslands or woods, rocky hillsides and valleys. Alberta to New Mexico. Our Colorado specimens from the mountainous western two-thirds of the state but none from the northwest and westcentral parts. Altitude 6000-10,000 feet.

5. **Danthonia spicata** (L.) Beauv.; Roem.& Schult., Syst. Veg. 2:690. 1817.
   D. thermale Scribn.---Culms 20-70 cm. tall, tufted, slender, rather weak; leaf blades to 2 mm. wide, 5-15 cm. long, flat to involute, glabrous or sparsely pilose, curled or flexuous, mostly basal; sheaths glabrous or sparsely pilose, a tuft of hairs at throat at least on lower; ligules about 0.5 mm. long; panicle 2-4 cm. long, contracted, the short branches erect or ascending with

1 or 2 spikelets; spikelets 10-12 mm. long, several-flowered; glumes subequal, the length of the spikelet, acuminate, papery especially on margins, 1 plain midnerve and about 4 indistinct laterals; lemmas 4-5 mm. long, including the apical teeth, pubescent or villous on the back, awn flat, twisted and divergent, 5-8 mm. long. Fernald (Rhodora 45:242, 1943) concluded our plant would be D. spicata var. pinetorum Piper.---Dry or sterile rocky soil of open slopes or mesas. Newfoundland to British Columbia, south to Florida, Texas, Kansas and in the mountains to New Mexico and Oregon. Our specimens from the northcentral and central parts of Colorado at 5300-8100 feet.

IV.   Tribe Agrostideae

35.   Calamagrostis Adans.   REEDGRASS

Perennial plants with rhizomes; leaf blades flat or involute; sheaths open; ligules membranous; inflorescence an open or contracted panicle; spikelets small, numerous, 1-flowered, perfect, rachilla disarticulating above the glumes and prolonged beyond the palea as a minute bristle; glumes about equal, acute or acuminate, longer than the lemma; lemmas 5-nerved, rounded on back, rather thin, dorsally awned, the awn straight or geniculate, callus hairy or pilose; paleas as long or somewhat shorter than lemmas, hyaline; styles short, stigmas plumose.

1. Awns over 6 mm. long, longer than the glumes, geniculate   -1. C. purpurascens
1. Awns less than 6 mm. long, shorter than or rarely as long as the glumes (may or may not be geniculate)
    2. Awns geniculate and usually twisted below, some protruding sidewise from glumes; callus hairs seldom over 1/2 as long as the lemma
        3. Collars of sheaths glabrous; plants less than 40 cm. tall; leaves involute, not over 2 mm. wide; callus hairs about 1/2 as long as the lemma   -2. C. montanensis
        3. Collars pubescent (sparse on some); plants over 40 cm. tall; leaves flat or involute near tip, some usually over 2 mm. wide; callus hairs about 1/3 as long as lemma   -3. C. rubescens
    2. Awns straight or nearly so, little if any twisted; callus hairs 1/2 as long as the lemma or longer
        4. Panicle rather loose and open; callus or rachilla hairs about as long as lemma
            5. Collars pubescent; callus hairs about 2/3 as long as the lemma (rachilla hairs may be as long)
                                                                                -4. C. scribneri
            5. Collars glabrous; callus hairs as long as the lemma   -5. C. canadensis
        4. Panicle contracted; callus hairs 1/2 to 3/4 as long as the lemma
            6. Collars pubescent   -4. C. scribneri
            6. Collars glabrous (next 3 intergrade some)
                7. Ligules over 3 mm. long (check on culm leaves and be sure ligule is not partly broken); leaves stiff and scabrous
                    8. Leaf blades remaining flat; plants pale green or glaucous   -6. C. scopulorum
                    8. Leaf blades involute on drying; plants green (sometimes pale green) but not glaucous
                                                                                -7. C. inexpansa
                7. Ligules 1-3 mm. long; leaves lax and smooth   -8. C. neglecta

1. Calamagrostis purpurascens R. Br. in Richards, Bot. App. Franklin Journ. 731. 1823.
    C. vaseyi Beal---Rhizomes, when present, short; culms 20-60 cm. tall, tufted, erect, glabrous, puberulent or scabrous below the panicle; leaf blades 2-4 mm. wide, 4-10 cm. long, mostly involute but sometimes flat, scabrous; sheaths usually scabrous, the old ones persistent; ligules 3-5 mm. long or more; panicle 5-12 cm. long, dense and more or less spikelike, often interrupted below, pale green or purplish; spikelets 5-7 mm. long; glumes subequal, length of spikelet, scabrous on keel, acuminate, 1-nerved on first, somewhat 3-nerved on second; lemmas 4-5 mm. long, the apex with 4 setaceous teeth, callus hairs about 1-1.5 mm. long, awn from near base, twisted and straight until near end of lemma, then geniculate, 7-8 mm. long; rachilla hairs 1-1.5 mm. long, sometimes shorter.---Rocky hills, cliffs and open land. Greenland to Alaska, south to Quebec, South Dakota, Colorado and California. Our Colorado collections from the mountainous western two-thirds of the state except from the northwest and extreme southwest corner. Altitude 8500-13,000 feet.
2. Calamagrostis montanensis (Scribn.) Scribn.; Vasey, Contrib. U. S. Nat. Herb. 3:82. 1892.
    Rhizomes creeping; culms 15-40 cm. tall, stiffly erect, scabrous below panicle; leaf blades 0.5-2 mm. wide, 5-15 cm. long, more or less involute, tapering to a fine point; sheaths glabrous or upper scaberulous, rather papery on lower; ligules 1-3 mm. long; panicle 4-10 cm. long, dense, erect, more or less interrupted, branches short and crowded, pale or sometimes purplish; spikelets 3.5-5 mm. long; glumes equal or nearly so, length of spikelet, first 1-nerved, second 3-nerved but laterals faint; lemmas from 1 mm. shorter than to as long as the glumes, 4-toothed at apex, the teeth short, scaberulous, callus hairs about 1/2 as long as the lemma, awned at base, awn about length of lemma but twisted at base on some (for about 1.5 mm.) and geniculate, the terminal segment exserted sidewise from glume (about 1.5 mm. long); rachilla hairs from 2/3 to as long as lemma.---Plains and dry open ground. Manitoba to Alberta, south to South Dakota, Wyoming and Idaho. Our two Colorado collections are from Larimer County at 8000-8500 feet.
3. Calamagrostis rubescens Buckl., Acad. Nat. Sc. Phila. Proc. 1862:92. 1862.
    C. suksdorfii Scribn.; probably C. cusickii Vasey in Rydb.---Rhizomes slender and creeping; culms 45-100 cm. tall, tufted, glabrous; leaf blades 2-5 mm. wide, 12-30 cm. long, flat or somewhat involute especially near tip, scabrous; sheaths glabrous but pubescent on collar, sometimes obscurely so; ligules 1-4 mm. long; panicle 7-15 cm. long, narrow and spikelike or sometimes looser and interrupted, pale or purple; spikelets 3.6-5 mm. long; glumes equal or subequal, the length of the spikelet, long-acute, scabrous on midnerve, first 1-nerved, second 3-nerved; lemmas about as long as the glumes to 1 mm. shorter, thin, pale, callus hairs about 1/3 as long as lemma, awn near base, twisted for about 2 mm., geniculate, the last segment 1-2 mm. long.---Open

woods, prairies and banks. Manitoba to British Columbia, south to Colorado and California. Our one specimen from Routt County, Colorado, at 9100 feet.

4. Calamagrostis scribneri Beal, Grasses N. A. 2:343. 1896.

Rhizomes creeping; culms 60-100 cm. tall, tufted, erect, glabrous or scaberulous below nodes; leaf blades 3-5 mm. wide, 10-25 cm. long, flat, scabrous, long-pointed; sheaths scabrous, pubescent on collar, lower loose and papery-brown; ligules 5-7 mm. long; panicle 10-15 cm. long or more, rather narrow but lax and not dense or spikelike, pale or purple; spikelets 3-4 mm. long; glumes subequal, length of spikelet, scabrous on back with 1 or 2 green nerves, acute; lemmas somewhat shorter than glumes, thin and hyaline, with sharp teeth at apex, glabrous except callus hairs, these 1/2 to 2/3 as long as lemma, awn from about middle of lemma, delicate, slightly bent, somewhat longer than lemma; rachilla hairs about as long as lemma. Seems to intergrade some with number 5.---Moist meadows and open marshes. Montana and Washington to Colorado and Oregon. Our collections rather thinly scattered over the mountainous western two-thirds of Colorado at 7500-11,500 feet.

5. Calamagrostis canadensis (Michx.) Beauv., Ess. Agrost. 15, 152, 157. 1812.

C. blanda Beal; C. canadensis var. acuminata Vasey---Rhizomes numerous and creeping; culms 50-100 cm. tall, tufted, erect or suberect; leaf blades 2-8 mm. wide, 10-30 cm. long, flat, lax, often glaucous, scabrous both sides; sheaths glabrous to scabrous; ligules 3-5, sometimes to 7 mm. long; panicle 10-25 cm. long, from open to rather dense, nodding, pale or purple; spikelets 3-4 mm. long; glumes about equal, length of spikelet, scabrous on keel, acute or acuminate; lemmas nearly as long as glumes, thin, scabrous, apex with 2-4 teeth, callus hairs about as long as lemma, awn at or below middle, delicate, straight extending to about the tip of lemma; rachilla sparsely pilose, hard to make out in callus hairs.---Marshes, wet places, open woods, meadows, wet thickets and streamsides. Greenland to Alaska, south to Maryland, North Carolina, Missouri, Kansas, Colorado, Arizona and California. Rather widely distributed in the western two-thirds of Colorado but no collections yet from the extreme corners. Altitude 5000-11,000 feet.

6. Calamagrostis scopulorum M. E. Jones, Proc. Calif. Acad. II. 5:722. 1895.

C. scopulorum var. lucidula Kearney---Rhizomes short; culms 40-80 cm. tall, glabrous, pale or glaucous, erect; leaf blades 3-7 mm. wide, 10-30 cm. long, flat, scabrous; sheaths glabrous; ligules 3-5 mm. long; panicle 7-15 cm. long, contracted, sometimes spikelike, pale or purplish; spikelets 4-6 mm. long; glumes equal, as long as spikelets, acuminate to long-acute, scabrous on keel, first 1-nerved, second 3-nerved; lemmas as long or a little shorter than glumes, with 4 narrow teeth at apex, callus hairs about 2/3 as long as lemma, awn from middle or just above middle, straight and delicate, about as long as lemma; rachilla pilose.---Moist soil in gulches and among rocks. Wyoming, Colorado and Utah. Our three specimens all from western Colorado at 9000-10,500 feet.

7. Calamagrostis inexpansa A. Gray, N. Amer. Gram. and Cyp. 1:20. 1834.

C. americana Scribn.; C. elongata (Kearney) Rydb.; C. hyperborea Lange; C. h. var. stenodes Kearney; C. h. var. americana (Vasey) Kearney; C. h. var. elongata Kearney---Rhizomes slender; culms 30-100 cm. tall, tufted, erect, scabrous below panicle; leaf blades 1.5-4.5 mm. wide, 10-30 cm. long, flat or involute on drying, rigid, scabrous; sheaths glabrous or somewhat scabrous, the basal ones numerous, withering but persistent; ligules 4-7 mm. long; panicle 5-15 cm. long, narrow, dense, branches mostly erect, more or less lobed, spikelet-bearing to base, pale green or purplish; spikelets 3.5-5 mm. long; glumes subequal, length of spikelet, scabrous especially on midnerve, the first glume 1- or slightly 3-nerved, second 3- or slightly 5-nerved; lemmas as long or nearly as long as glumes, scabrous, acutish, callus hairs 1/2 to 4/5 as long as lemma, awn attached about at middle, delicate, straight, about as long as lemma; rachilla hairs about length of lemma.---Meadows, marshes and wet places, sometimes among rocks. Greenland to Alaska, south to Maine, New York, Illinois, Missouri, Nebraska, New Mexico and California. Our specimens in the mountainous western two-thirds of Colorado except in the extreme west and the northwest part. Altitude 6000-10,000 feet.

8. Calamagrostis neglecta (Ehrh.) Gartn.; Mey and Scherb., Fl. Wett. 1:94. 1799.

C. micrantha Kearney---Rhizomes slender; culms 30-80 cm. tall, erect; leaf blades filiform to 2 mm. wide on culms, 10-20 cm. long, involute especially on basal, to flat, lax, smooth; sheaths glabrous; ligules 1-3 mm. long; panicle 5-13 cm. long, narrow, more or less lobed; spikelets about 4 mm. long; glumes equal or nearly so, length of spikelet, acute, scabrous especially on nerves; lemmas about 3 mm. long, 1/4 to 1/5 shorter than glumes, acute, callus hairs 1/2 to 2/3 as long as lemma, awn from near middle, straight and shorter than lemma.---Marshes, open meadows and wet places. Greenland to Alaska, south to Maine, Vermont, New York, Michigan, Wisconsin, Colorado, Utah, and Oregon; also in Eurasia. Our two specimens from Colorado in Clear Creek and Saguache Counties at about 8500 feet.

36. Calamovilfa Hack. SANDREED

Perennial plants with long scaly rhizomes; leaf blades long, involute and tapering to tip; sheaths split; ligules made up of hairs; inflorescence an open or contracted panicle; spikelets rather flattened, 1-flowered, numerous, rachilla disarticulating above the glumes, not prolonged beyond the palea; glumes unequal, chartaceous, 1-nerved, acute or acuminate, the second almost as long as the lemma; lemmas chartaceous, 1-nerved, awnless, pilose or glabrous, callus bearded; paleas similar to lemmas in length and texture; styles distinct, stigmas plumose.

1. Lemmas glabrous on back (disregard callus hairs); spikelets 5-7 mm. long; panicle usually contracted
-1. C. longifolia
1. Lemmas pubescent on back near base (as well as on callus); spikelets 7-9 mm. long; panicle open at maturity
-2. C. gigantea

1. **Calamovilfa longifolia** (Hook.) Scribn. in Hack., True Grasses. 113. 1890.
    Culms 50-150 cm. tall, mostly solitary, glabrous; leaf blades 4-10 mm. wide, 15-30 cm. long or even longer, stiff, glabrous or scabrous especially near tip and margins; sheaths glabrous or sometimes pubescent, long-pilose at collar, crowded; ligules 0.5-2 mm. long; panicle 15-35 cm. long, usually contracted, the slender, smooth branches erect or ascending, whitish or yellowish; spikelets 5-7 mm. long, crowded; first glume 4-6 mm. long, shorter than lemma; second glume 5-7 mm. long, as long or longer than the lemma, both rigid, glabrous; lemmas 4-7 mm. long, glabrous, callus hairs about 1/2 as long as the lemma.---Sand hills, open woods and sandy prairies, often on sand dunes. Michigan to Alberta, south to Indiana, Colorado and Idaho. From the foothills to the eastern border of Colorado at 3500-7000 feet.
2. **Calamovilfa gigantea** (Nutt.) Scribn. & Merrill, U. S. D. A. Div. Agrost. Cir. 35:2. 1901.
    Culms 100-200 cm. tall, mostly solitary, glabrous, erect; leaf blades 5-12 mm. wide at flat base, tapering to a long involute tip, up to 80 cm. long, glabrous; sheaths glabrous but more or less villous at throat especially on innovations; ligules 0.5-2 mm. long; panicle 30-60 cm. long, open at maturity, branches ascending or spreading, naked at base, whitish to purplish; spikelet 7-9 mm. long; first glume 4.5-6.5 mm. long, shorter than lemma; second glume 7-9 mm. long, as long or a little longer than lemma; lemmas about 6.5-9 mm. long, pubescent on lower half especially near middle, callus hairs about 1/2 the length of lemma.---Sand dunes. North Dakota to Texas and west to Arizona. Our three Colorado specimens from Cheyenne and Baca Counties at 4400 feet.

37. Agrostis L. BENTGRASS

Perennial plants, rhizomes present or absent (stolons in 1 species); leaf blades flat to involute; sheaths split; ligules membranous; inflorescence an open or contracted panicle; spikelets small, numerous, 1-flowered, rachilla disarticulating above the glumes, usually not prolonged beyond the palea; glumes equal or subequal, membranous, keeled, usually longer than lemma; lemma obtuse, thin-membranous, awnless or dorsally awned; paleas to about 2/3 as long as the lemma or absent altogether; styles short, stigmas plumose.

1. Paleas evident, 1/2 or more as long as the lemmas, 2-nerved
    2. Glumes scabrous all over; paleas over 3/4 as long as the lemmas   -1. A. semiverticillata
    2. Glumes scabrous on the keel only, if at all; paleas not over 3/4 as long as the lemmas
        3. Rhizomes or stolons absent; plant above 9000 feet in Colorado; culms seldom over 30 cm. tall
            4. Rachilla prolonged behind the palea; panicle from somewhat narrow to lax   -2. A. thurberiana
            4. Rachilla not prolonged; panicle very narrow   -3. A. humilis
        3. Rhizomes or stolons present; plants usually below 9000 feet in Colorado; culms over 30 cm. tall
            5. Panicle contracted, narrow; long stolons present on isolated plants; leaves 1-3 mm. wide (sometimes to 4)   -4. A. palustris
            5. Panicle open at maturity; no stolons present; leaves 3-8 mm. wide   -5. A. alba
1. Paleas lacking, or very small and nerveless, not over 1/3 as long as lemma
    6. Panicle narrow, some branches bearing spikelets to near bases
        7. Culms over 20 cm. tall; leaf blades over 1.5 mm. wide; ligules 2-6 mm. long   -6. A. exarata
        7. Culms 10-20 cm. tall; leaf blades less than 1.5 mm. wide; ligules 1-2 mm. long   -7. A. variabilis
    6. Panicle open, the branches spreading or reflexed at maturity, naked below
        8. Lemmas awned (only rarely some awnless); spikelets 2.5-3 mm. long   -8. A. borealis
        8. Lemmas awnless (occasionally some awned); spikelets 1.5-2.5 mm. long (rarely to 3)
            9. Panicle branches branching below the middle; basal leaves flat or loosely involute, not filiform; culms seldom over 25 cm. tall; spikelets 1.2-1.8 mm. long   -9. A. idahoensis
            9. Panicle branches branching at or above the middle; base with a tuft of filiform, capillary leaves; culms often over 25 cm. tall; spikelets 2 mm. long or more
                10. Panicle branches long, usually over 4 cm. long, ascending or spreading   -10. A. scabra
                10. Panicle branches shorter, often divaricate   -10A. A. scabra geminata

1. **Agrostis semiverticillata** (Forsk.) C. Chr., Dansk. Bot. Arkiv. 4(3):12. 1922.
    A. verticillata Vill. of Manuals---Perennial bunchgrass plants but culms usually decumbent at base, sometimes with long creeping and rooting stolons; culms 15-60 cm. tall; leaf blades 3-10 mm. wide, 4-14 cm. long, flat, firm, scabrous at least above and on margins; sheaths shorter than internodes, smooth or scabrous; ligules 1.5 mm. long (reported to 4 mm.), scabrous outside; panicles contracted, 3-6 cm. long, oblong to somewhat pyramidal, lobed or verticillate especially at base; spikelets 1.5-2 mm. long, rachilla not prolonged; glumes equal, length of spikelet, 1-nerved, acute, scabrous on back; lemmas about 1 mm. long, awnless, truncate and toothed at apex, hyaline; palea hyaline, nearly as long as the lemmas.---Moist or wet ground along streams and ditches. Washington south to California and Texas; Old World. Southeast and southwest corners of Colorado at 4000-4600 feet.
2. **Agrostis thurberiana** Hitchc., Bull. U. S. D. A. Pl. Ind. 68:23. 1905.
    A. atrata Rydb.---Rhizomes absent; culms 10-30 cm. tall, slender, in small tufts, erect or somewhat decumbent at base, glabrous; leaf blades 0.5-2 mm. wide, 2-10 cm. long, flat, lax, smooth or scaberulous, mostly basal; sheaths glabrous; ligules 0.5-2 mm. long; panicle 3-7 cm. long, rather narrow but lax and somewhat drooping, branches 2-4 to a whorl, ascending or somewhat spreading; spikelets about 2 mm. long, pale or purple, the rachilla prolonged beyond the palea for about 0.3 mm. (sometimes bearing a reduced floret); glumes about equal, the length of the spikelet, somewhat scabrous above on the keel; lemmas nearly as long as the glumes, hyaline, awnless; paleas about 2/3 the length of the lemmas.---Bogs, meadows and moist places. Colorado to British Columbia and south to California. Our six specimens from Mineral, Gilpin, Boulder, Larimer and Jackson Counties, so to be expected anywhere in the higher mountains of Colorado. Altitude 9500-12,000 feet.

3. **Agrostis humilis** Vasey, Bull. Torr. Club 10:21. 1883.

Rhizomes absent; culms up to 20 cm. tall, tufted; leaf blades 1 mm. wide or less, 2-10 cm. long, flat or folded, glabrous below, minutely pubescent above, mostly basal, usually 1 to a culm; sheaths glabrous; ligules 1-2 mm. long; panicle 1-4 cm. long, narrow, purple; spikelets 1.5-2.5 mm. long, rachilla not prolonged; glumes about equal, length of spikelet, 1-nerved, acuminate, purple, somewhat scabrous on keel; lemmas 1.5-2 mm. long, shorter than glumes, purplish (in ours), awnless; paleas about 2/3 to 3/4 as long as lemmas.---Bogs and alpine meadows. Wyoming and Colorado to Washington and Oregon. Our few specimens from northcentral Colorado at 9000-12,000 feet.

4. **Agrostis palustris** Huds., Fl. Angl. 27. 1762.

A. depressa Vasey; A. reptans Rydb.---Culms 15-50 cm. tall, usually decumbent and rooting at base, developing long stolons in isolated plants; leaf blades 1-4 mm. wide, 3-10 cm. long, flat, scabrous especially above; sheaths glabrous; ligules 1-4 mm. long; panicle 5-12 cm. long, condensed, rather narrow, green or purplish; spikelets 2-2.5 mm. long, rachilla not prolonged; glumes nearly equal, length of spikelet, 1-nerved, glabrous except scabrous keel, acute; lemmas about 1.5 mm. long, somewhat shorter than the glumes, often toothed at apex, rarely awned; paleas 1/2 to 3/4 as long as the lemmas.---Cultivated for lawns and golf courses and now found in meadows, along streams and in marshes. Newfoundland to Maryland; British Columbia to California; introduced at various points in the United States including Utah, Colorado, New Mexico and Arizona. Our few specimens from northcentral and southeastern Colorado but to be expected anywhere in the state. Altitude 4500-8500 feet.

5. **Agrostis alba** L., Sp. Pl. 63. 1753.

Rhizomes creeping; culms 30-100 cm. tall, tufted, erect or somewhat decumbent at base, glabrous; leaf blades 3-8 mm. wide, 4-20 cm. long, flat, usually scabrous; sheaths glabrous to scaberulous; ligules 3-6 mm. long; panicle to 20 cm. long, open especially in anthesis, the lower branches usually verticillate and contracted, green to reddish; spikelets 2-2.5 mm. long, rachilla not prolonged; glumes nearly equal, length of spikelet, acute, scabrous on keel; lemmas 1.5-2 mm. long, somewhat shorter than glumes, obtuse, rarely awned; paleas 1/2 to 2/3 as long as the lemmas.---Cultivated for meadows, pastures and lawns and escaping in wet meadows, swamps and streamsides. Now present in all the cooler parts of the United States; also in Eurasia. Widely distributed in Colorado but no specimens yet from the southeast and northwest corners. Altitude 3500-9000 feet.

6. **Agrostis exarata** Trin., Gram. Unifl. 207. 1824.

A. grandis Trin.; A. asperifolia Trin.---Short-lived perennial plants, no rhizomes; culms 20-100 cm. tall, tufted, somewhat geniculate at base, glabrous; leaf blades 2-8 mm. wide, 5-15 cm. long, flat, scabrous; sheaths smooth or scabrous; ligules 2-6 mm. long; panicle 5-20 cm. long, contracted, sometimes spikelike, rarely somewhat open, often interrupted or lobed below, erect or slightly nodding, green or purplish, branches spikelet-bearing to base; spikelets 2.2-3.5 mm. long, the rachilla not prolonged; glumes about equal, length of spikelet, acuminate or awn-pointed, scabrous on keel, often on back; lemmas 1.5-2.2 mm. long, about 2/3 as long as the glumes, acute or obtuse, callus glabrous or sparingly-pilose, awnless or sometimes with a straight or bent awn; paleas less than 0.5 mm. long, about 1/4 or 1/3 as long as the lemmas.---Moist ground in meadows and streamsides or rather dry open ground. Nebraska to Alberta and Alaska, south to New Mexico and California; also in Mexico. Our specimens scattered in the mountainous two-thirds of the state except in the extreme western and northwestern parts. Altitude 5000-11,000 feet.

7. **Agrostis variabilis** Rydb., Mem. N. Y. Bot. Gard. 1:32. 1900.

A. rossae Vasey of Manuals---Rhizomes absent; culms 10-25 cm. tall, densely tufted, erect, glabrous; leaf blades about 1 mm. wide, 2-7 cm. long, flat or somewhat involute, glabrous or nearly so; sheaths glabrous or somewhat scabrous; ligules 1-2 mm. long; panicle 2-6 cm. long, contracted, at least some of lower branches spikelet-bearing at base, usually purple; spikelets 2-2.5 mm. long, rachilla not prolonged; glumes equal or nearly so, as long as spikelet, acuminate, glabrous except scabrous on keel; lemmas 1.5-2 mm. long, shorter than glumes, awned or awnless; paleas small, about 1/4 as long as the lemmas.---Rocky creeks and mountain slopes. Colorado and Utah to Alberta, Washington and California. Our few specimens from northcentral Colorado at 9000-12,000 feet.

A. rossae Vasey is an annual grass restricted to Yellowstone Park area (Lfts. West. Bot. 5:123-124. 1948).

8. **Agrostis borealis** Hartm., Handb. Skand. Fl. ed. 3, 17. 1838.

A. bakeri Rydb.---Rhizomes absent; culms 20-35 cm. tall, tufted, erect or slightly geniculate below, glabrous; leaf blades 1-2 mm. wide, 5-10 cm. long, flat, smooth or nearly so; sheaths smooth or scaberulous; ligules 1.5-2.5 mm. long; panicle 6-10 cm. long, open, branches ascending, branched above the middle, lower 3-5 cm. long; spikelets 2.5-3 mm. long, the pedicels thickened just below, rachilla not prolonged; glumes about equal, length of spikelet, acuminate, purplish; lemmas about 3/4 as long as the glumes, obtuse, bearing above the middle a straight awn (in ours) not much longer than the glumes; paleas none or very minute.---Rocky slopes and moist meadows. Newfoundland and Greenland to Alaska, south to New York, West Virginia and North Carolina, the western form (A. bakeri) in Colorado, Wyoming, Alberta and Washington. Our only specimen from the type locality in Mineral County, Colorado, at 10,500 feet.

9. **Agrostis idahoensis** Nash, Bull. Torr. Club 24:42. 1897.

Rhizomes absent; culms 10-30 cm. tall, tufted, erect, glabrous; leaf blades mostly less than 1 mm. wide (but sometimes to 2), 4-10 cm. long, flat or somewhat involute, glabrous below, scaberulous above, mostly basal; sheaths glabrous; ligules 1-2 mm. long; panicle 5-10 cm. long, with capillary, scabrous, loosely spreading branches, naked at base; spikelets 1.5-2 mm. long (but some of ours to 3), rachilla not prolonged; glumes equal or nearly so, length of spikelet, acuminate, purplish and scabrous on the 1 nerve; lemmas 1.2-1.6 mm. long (a few to 2), shorter than the glumes, glabrous, awnless; paleas minute.---Grasslands and meadows. Montana to Washington, south to New Mexico and California; also in Alaska. Our few specimens from northcentral, northwestern and southcentral Colorado at 9500-11,500 feet.

10. Agrostis scabra Willd., Sp. Pl. 1:370. 1797.
A. hiemalis (Walt.) B. S. P. in Manuals---Rhizomes absent; culms 20-60 cm. tall, tufted, erect, glabrous, slender; leaf blades 0.7-2 mm. wide, 2-12 cm. long, usually flat but basal ones usually involute-filiform, usually scabrous, mostly basal; sheaths glabrous or scaberulous; ligules 1-4 mm. long; panicle up to 30 cm. long, large, diffuse, purplish, scabrous branches whorled, bearing spikelets near ends, the basal ones not branching until 2/3 from base; spikelets 1.5-2.5 mm. long (rarely to 3), rachilla not prolonged; glumes nearly equal, length of spikelet, long-acuminate, scabrous on the 1 nerve; lemmas 2/3 to 3/4 the length of the glumes, awnless or very rarely awned; paleas wanting.---Meadows, hillsides, streamsides in moist and dry ground. Newfoundland to Alaska, south to Maryland, Illinois, Nebraska, New Mexico, Arizona and California, rarely in the southeastern states. The true A. hiemalis is confined to the southeastern part of the United States. Widely distributed over the western two-thirds of the state at 4500-11,000 feet.

10A. Agrostis scabra geminata (Trin.) Swallen, (var.) Proc. Biol. Soc. Wash. 54:45. 1941.
A. hiemalis var. geminata (Trin.) Hitchc.---Differs from the species in having panicle branches short, seldom over 4 cm. long, divaricate, branching sometimes at the middle or farther out; lemmas awnless or perhaps awned.---Our Colorado specimens from the northcentral and central parts of the state at 9500-12,000 feet.

38. Phippsia (Trin.) R. Br. ICEGRASS

Spikelets 1-flowered, the rachilla disarticulating above the glumes; lemmas thin, somewhat keeled.

1. Phippsia algida (Phipps)R. Br., Chlor. Melv. 27. 1823.
Perennial plants, no rhizomes; culms 2-10 cm. tall, densely tufted, glabrous; leaf blades 0.5-2 mm. wide, 5-25 mm. long, flat with boat-shaped tips, soft; sheaths thin, loose, glabrous; ligules membranous, 0.5-2 mm. long; panicle 5-35 mm. long, narrow, the short appressed branches few-flowered; spikelets 1-2 mm. long; first glume sometimes wanting, if present very short; second glume less than 1/2 the length of lemma, rather truncate and erose at apex, glumes unequal; lemmas length of spikelet, 3-nerved the 2 laterals short and not distinct, abruptly acute, awnless; paleas a little shorter than lemmas, rather dentate.---Mountain summits. Arctic regions; also in Colorado. Our few Colorado specimens from Clear Creek County at about 11,000 to 14,000 feet.

39. Cinna L. WOODREED

Sheaths split; ligules membranous; spikelets 1-flowered, the pedicel disarticulating below the glumes, rachilla forming a stipe below the floret and prolonged beyond the palea as a minute bristle; styles short, stigmas plumose.

1. Cinna latifolia (Trev.) Griseb. in Ledeb., Fl. Ross. 4:435. 1853.
Perennial plants, no rhizomes; culms 50-150 cm. tall, erect or decumbent at base, scaberulous below the nodes; leaf blades 4-15 mm. wide, 10-20 cm. long, flat, scabrous; panicle 10-30 cm. long, oblong, branches capillary, spreading or drooping, usually green; spikelets 3-4 mm. long; glumes subequal, length of spikelet, first 1-nerved, second faintly 3-nerved; lemmas a little shorter than glumes, 3-nerved, scabrous on keel, usually with a short straight awn just below the apex, this awn as much as 1 mm. long; paleas about 3/4 as long as the lemmas, the two keels very close together, appearing as one.---Moist woods and streamsides. Newfoundland and Labrador to Alaska, south to Connecticut, North Carolina, New Mexico, Utah and California; also in Eurasia. All our specimens from northcentral Colorado but may well be farther south in the state. Altitude 7600-10,300 feet.

40. Alopecurus L. FOXTAIL

Mostly perennial plants; leaf blades flat or folded; sheaths split; ligules membranous; inflorescence a spikelike panicle; spikelets 1-flowered, laterally compressed, disarticulating below the glumes; glumes equal, usually united below, about as long as lemmas; lemmas 5-nerved, thin, with a slender awn from just below the middle, the margins of lemma somewhat united at base; paleas wanting; style short, stigmas long and sub-plumose.

1. Panicle oblong or ovoid, about 1 cm. wide; glumes densely woolly or villous all over -1. A. alpinus
1. Panicle cylindrical, less than 1 cm. wide; glumes villous only on keel and nerves
  2. Awns included or exserted less than 1.5 mm. beyond the glumes -2. A. aequalis
  2. Awns exserted 2 mm. or more beyond the glumes
    3. Perennial plants; anthers usually over 0.75 mm. long; often growing in water -3. A. geniculatus
    3. Annual plants; anthers less than 0.75 mm. long; growing in moist, open ground (usually not in water)
        -4. A. carolinianus

1. Alopecurus alpinus Smith, Engl. Bot. Pl. 1126. 1803.
Perennial plants with slender, creeping rhizomes; culms 10-80 cm. tall, erect or often decumbent at base, glabrous; leaf blades 2-5 mm. wide, 5-15 cm. long, flat but often folded at tip, usually scabrous, leaves few; sheaths glabrous or scaberulous; ligules 1-3 mm. long; panicle 1-4 cm. long, woolly, ovoid or oblong, about 1 cm. wide or more; spikelets 2.8-4 mm. long, woolly; glumes 3-nerved, woolly, length of spikelet; lemmas about as long as glumes, glabrous below,

pubescent above especially on the veins, awn attached near base, twisted below, then geniculate, from slightly exserted beyond the glumes to exserted 5 mm.---Mountain meadows, along brooks and lakes, Greenland to Alaska, south to Colorado; Arctic regions and Eurasia. Our few specimens from the northcentral and central parts of Colorado at 8500-11,500 feet.

2. Alopecurus aequalis Sobol.,Fl. Petrop. 16. 1799.

A. aristulatus Michx.; A. caespitosus Trin.---Perennial plants, no rhizomes; culms 15-50 cm. tall, erect or spreading, sometimes rooting at nodes; leaf blades 1-4 mm. wide, 3-12 cm. long, flat, scabrous; sheaths glabrous; ligules 2-5 mm. long; panicle 3-7 cm. long, cylindrical, about 4-5 mm. thick; spikelets about 2 mm. long; glumes length of spikelet, ciliate, pilose on nerves, pubescent or scabrous on back, 3-nerved; lemmas about as long as the glumes, glabrous, awn from below middle, this straight or slightly geniculate, included or exserted to 1 mm. beyond glumes.--- In water, wet places, wet meadows, streams and lakes. Greenland to Alaska, south to Pennsylvania, Illinois, Kansas, New Mexico and California; also in Eurasia. Widely distributed over the western two-thirds of Colorado at 4500-11,500 feet.

3. Alopecurus geniculatus L., Sp. Pl. 60. 1753.

Perennial plants, no rhizomes (in ours); culms 15-45 cm. tall, decumbent, often geniculate and rooting at nodes; leaf blades 1-4 mm. wide, 3-15 cm. long, flat, scaberulous; sheaths glabrous; ligules 1-4 mm. long; panicle about 2-7 cm. long, 4-5 mm. wide, cylindrical; spikelets 2-3 mm. long; glumes length of spikelet, 3-nerved, pilose on nerves and sometimes pubescent on the back, often purplish near tip; lemmas about as long as glumes, glabrous or pubescent above on keel, awn from below middle, this usually twisted at base and geniculate, exserted for 2-3 mm. giving the panicle a bristly look.---In water, marshes and wet places. Newfoundland to Saskatchewan and British Columbia, south through New England to New York, New Jersey, and Virginia and through Washington to California; Montana, Colorado and Arizona; also in Eurasia. Our one specimen from Jefferson County, Colorado, at about 5500 feet.

4. Alopecurus carolinianus Walt., Fl. Carol. 74. 1788.

Widely distributed over the United States and reported from Colorado but not seen within the state by the writer.

41. Polypogon Desf. POLYPOGON

Sheaths split; ligules membranous; spikelets 1-flowered, numerous, the pedicels disarticulating a short distance below the glumes leaving a sharp-pointed callus; styles short, stigmas plumose.

1. Polypogon monspeliensis (L.) Desf., Fl. Atlant. 1:67. 1798.

Annual plants; culms 15-50 cm. tall, tufted, erect or decumbent at base, striate, glabrous or scaberulous; leaf blades 3-6 mm. wide, 4-15 cm. long, flat, scabrous; sheaths striate, scabrous; ligules 3-6 mm. long; panicle 2-15 cm. long, 1-2 cm. wide, spikelike, dense, bristly, pale green at first, maturing to a tawny-yellow, sometimes interrupted below; spikelets about 2 mm. long; glumes equal or nearly so, length of spikelet, hispidulous, bearing an awn from below each apex, this 4-8 mm. long, straight or nearly so; lemmas about 1/2 as long as the glumes, smooth and shining, usually toothed at apex, awnless or with a delicate awn up to 1.7 mm. long just below apex; paleas about 2/3 as long to almost as long as lemmas.---Waste places, around the margins of lakes and swamps, and along streams (ours). New Brunswick to Georgia, west to Alaska and California, more common in the western states; also in South America and Europe. Scattered over all of Colorado, no specimens from the northwest or southwest corner as yet. Altitude 4000-7000 feet.

42. Lycurus H. B. K. WOLFTAIL

Sheaths split; ligules membranous; spikelets 1-flowered, borne in pairs, the lower sterile, the pair falling together (articulating below the glumes); styles distinct, short, stigmas plumose.

1. Lycurus phleoides H. B. K., Nov. Gen. & Sp. 1:142. 1815.

Perennial plants, no rhizomes; culms 20-60 cm. tall, tufted, erect or decumbent at base, flattened, scaberulous; leaf blades 1-3 mm. wide, 4-10 cm. long, flat or folded, scabrous above and on midrib below; sheaths flat, keeled, glabrous or scaberulous, blue-green; ligules 2-5 mm. long; panicle 3-6 cm. long, slender, about 5 mm. wide, spikelike, dense, terminal and from upper axils; spikelets 3-5 mm. long, exclusive of awns; first glume 1-1.5 mm. long, 2-nerved, the nerves each ending in an awn 2-5 mm. long; second glume as long or slightly longer, 1-nerved and 1-awned, this awn to 3-6 mm. long; lemmas 3-5 mm. long, pubescent especially on nerves and margins, firmer than glumes, 3-nerved, acuminate and the apex with an awn 2-4 mm. long; paleas about as long as the lemmas, similar in texture.---Plains, rocky hillsides and gulches. Colorado to Texas and Arizona; also in Mexico. Our rather few specimens along the eastern foothills from Larimer to Pueblo County and in southeastern Colorado. Altitude 4400-8000 feet.

43. Phleum L. TIMOTHY

Perennial plants, lacking rhizomes or if present then inconspicuous; leaf blades flat; sheaths split; ligules membranous; inflorescence a spikelike panicle; spikelets 1-flowered, laterally compressed, reported as disarticulating above the glumes but often appearing below; glumes equal, membranous, keeled, short-awned; lemmas shorter than glumes, hyaline, 3- to 5-nerved, truncate; paleas hyaline, narrow, about as long as lemmas; styles long, stigmas sub-plumose.

1. Panicle cylindrical, over 5 times longer than wide; sheaths not inflated; base of culms swollen or bulbous, the culm erect; awns of glumes rarely over 2 mm. long -1. P. pratense
1. Panicle oblong or ovoid, less than 5 times longer than wide; middle and upper sheaths inflated; base of culms not swollen or bulbous, the culms decumbent at base; awns of glumes often over 2 mm. long -2. P. alpinum

1. **Phleum pratense L.**, Sp. Pl. 59. 1753.
Culms 40-90 cm. tall, tufted, erect from a swollen or bulblike base, glabrous or slightly scaberulous; leaf blades 4-8 mm. wide, 5-20 cm. long, scabrous especially above; sheaths glabrous, little inflated except sometimes the lower ones; ligules 2-4 mm. long; panicle 3-12 cm. long, dense, cylindrical, 5-9 mm. wide; spikelets 2.5-3.5 mm. long exclusive of awns; glumes length of spikelet, 3-nerved, hyaline except keel, awn 1-2 mm. long; lemmas about 1/2 to 2/3 as long as the glumes, truncate to acute, scabrous on keel above.---Cultivated and commonly escaped along roadsides, fields, meadows, aspen groves and waste places. Throughout the United States, also in Eurasia. Our specimens scattered over the western two-thirds of the state but probably in the eastern third too. This grass is sometimes found at high altitudes especially along pack-horse trails. Altitude 4500-11,500 feet, mostly below 9000 feet.

2. **Phleum alpinum L.**, Sp. Pl. 59. 1753.
Culms 20-60 cm. tall, densely tufted, erect from a decumbent, somewhat creeping base; leaf blades 3-8 mm. wide, 2-10 cm. long, glabrous to scabrous; sheaths glabrous, mostly inflated; ligules 1.5-3 mm. long; panicle 1-4 cm. long, sometimes to 5 cm., 8-13 mm. wide, ovoid or oblong, dense; spikelets 3-4 mm. long, exclusive of awns; glumes length of spikelet, 1- to 3-nerved, ciliate on keel, hyaline except keel, the awn 1.5-3 mm. long; lemmas 2-2.5 mm. long, scaberulent on keel and often on nerves.---Meadows, bogs, forest borders and along streams. Greenland to Alaska, south to Maine and New Hampshire; Michigan; South Dakota and Washington south to California and New Mexico; also in Eurasia and Arctic regions of the Southern Hemisphere. Our specimens widely distributed over the mountainous two-thirds of the state at 8000-13,000 feet.

44. Muhlenbergia Schreb. MUHLY

Annual or perennial plants, rhizomes present or absent; leaf blades flat to involute; sheaths split; ligules variable, usually membranous but often ciliate at apex (1 species mostly of hairs except near sides); inflorescence an open or contracted panicle; spikelets 1-flowered (occasionally spikelets have 2 flowers), perfect, rachilla disarticulating above the glumes; glumes membranous, convex or keeled, obtuse to awned, usually shorter than lemma; lemmas firm-membranous, 3-nerved (these often obscure), apex acute, awned from the apex or from between 2 short teeth; paleas usually about as long as lemmas; styles distinct, stigmas plumose.

1. Panicle open at maturity, the long branches naked at base, the whole inflorescence usually over 4 cm. wide, often as wide as long
    2. Lemmas awnless or mucronate, not tapering to an awn
        3. Plants annual, no rhizomes present
            4. Pedicels filiform or capillary, elongated, much longer than the spikelets; glumes sparingly pubescent; panicles over 5 cm. long -1. M. minutissima
            4. Pedicels relatively stout and short, at least the lateral ones shorter than the spikelets; glumes glabrous; panicles less than 5 cm. long -2. M. wolfii
        3. Plants perennial, rhizomes present
            5. Leaf margins and midnerve white-cartilaginous (use lens); ligules usually splitting at sides forming 2 lateral auriclelike structures; plants rarely over 20 cm. tall; leaf blades rarely over 3 cm. long -3. M. arenacea
            5. Leaf margins and midnerve not white-cartilaginous; no auriclelike structures on ligules; plants often over 20 cm. tall; leaf blades often over 3 cm. long -4. M. asperifolia
    2. Lemmas tapering into an awn 1 mm. long or longer
        6. Leaf blades stiff, sharp-pointed (really needlelike); long-creeping rhizomes present; awn of lemmas about 1 mm. long; ligules made up mostly of hairs (except right near sides) -5. M. pungens
        6. Leaf blades not stiff nor very sharp-pointed; long-creeping rhizomes absent; awn of lemmas 2 mm. long or more; ligules membranous or short-ciliate
            7. Plants diffusely branched, prostrate or spreading; leaves not in a close basal cluster, the blades usually flat and early deciduous from the sheaths; awn of lemmas 5-10 mm. long -6. M. porteri
            7. Plants branching only at very base, erect (may be decumbent at base); leaves mostly in a close basal cluster, involute, blades not readily deciduous; awn of lemmas less than 4 mm. long
                8. Leaf blades 1-4 cm. long, recurved or flexuous, leafy portion of culm 1/8 to 1/6 the entire length; panicle seldom over 20 cm. long; common in eastern Colorado -7. M. torreyi
                8. Leaf blades 5-13 cm. long, erect or somewhat spreading, leafy portion of culm 1/3 to 1/2 the entire length; panicle often over 20 cm. long; collected twice in southeastern Colorado -8. M. arenicola
1. Panicle narrow and contracted at maturity, the short branches rarely spreading, the whole inflorescence less than 4 cm. wide, much longer than wide
    9. Lemmas tapering into a definite awn usually 2 mm. long or more
        10. Hairs near base of lemma long, as long or nearly as long as the lemma; strong scaly rhizomes present -9. M. andina
        10. Hairs near base of lemma much shorter or absent; rhizomes lacking (culms may root at nodes in number 12)
            11. Second glume 3-nerved, each nerve usually ending in a tooth; rather common in Colorado
                12. Awn of lemmas 6-20 mm. long; ligules 7 mm. long or more; culms usually over 30 cm. tall; panicle 5-15 cm. long; leaf blades 5-15 cm. long -10. M. montana
                12. Awn of lemmas 2-5 mm. long; ligules 3-5 mm. long; culms usually less than 30 cm. tall; panicle 1-5 cm. long; leaf blades seldom over 5 cm. long -11. M. filiculmis

11. Second glume 1-nerved, not 3-toothed at apex; rare in Colorado
   13. Glumes minute, the first obsolete or seldom over 0.5 mm. long, both rounded or truncate at apex; culms spreading and usually rooting at the nodes; awn of lemmas less than 5 mm. long
   -12. M. schreberi
   13. Glumes larger, the first usually over 1 mm. long, acute to awn-tipped; culms mostly erect, not rooting at nodes; awn of lemmas 5 mm. long or more
      14. Annual plants; culms less than 20 cm. tall; leaf blades less than 5 cm. long, with white-thickened margins
      -13. M. brevis
      14. Perennial plants; culms over 20 cm. tall; leaf blades over 5 cm. long, margins not white-thickened
      -14. M. pauciflora
9. Lemmas acute, mucronate or merely awn-pointed, this point not over 1 mm. long (glumes may be awned)
   15. Leaf blades flat, 2-6 mm. wide; plants often branching throughout; culms often over 50 cm. long; glumes tapering into an awn or awn point; strong creeping rhizomes present
      16. Awn of glumes about 2 mm. long, this extending beyond the lemmas giving the panicle a bristly appearance; common in Colorado
      -15. M. racemosa
      16. Awn of glumes less than 2 mm. long and about as long as the lemma, the panicle not bristly; apparently rare in Colorado
      -16. M. mexicana
   15. Leaf blades flat to involute but less than 2 mm. wide; plants branching only at base; culms less than 50 cm. long; glumes acute or mucronate (except number 19); rhizomes absent (except number 17)
      17. Numerous creeping rhizomes present, the culms not densely tufted; culms usually nodulose-roughened
      -17. M. richardsonis
      17. Rhizomes lacking, culms in dense tufts; culms not nodulose-roughened
         18. Glumes short, less than 1 mm. long and less than 1/2 as long as the lemma; annual plants
         -18. M. filiformis
         18. Glumes over 1 mm. long, over 1/2 the length of the lemma; perennial plants
            19. Second glume 3-nerved, each nerve usually ending in a tooth; ligules 3-5 mm. long
            -11. M. filiculmis
            19. Second glume 1-nerved, the apex never 3-toothed; ligules less than 3 mm. long
               20. Panicle dense, usually over 5 mm. wide; glumes awn-pointed, the awn often to 1 mm. long; ligules 1-3 mm. long
               -19. M. wrightii
               20. Panicle rather loosely flowered, usually less than 5 mm. wide; glumes not awn-pointed (may be cuspidate); ligules 0.5 mm. long
               -20. M. cuspidata

1. Muhlenbergia minutissima (Steud.) Swallen, Contrib. U. S. Nat. Herb. 29:207. 1947.
   Sporobolus microspermus and S. confusus of Manuals for Colorado plants.---Annual plants; culms 7-30 cm. tall, tufted often in large patches, erect or spreading, branched below, smooth or nearly so; leaf blades 0.5-2 mm. wide, 1-10 cm. long, flat, lax, scaberulous; sheaths glabrous or scaberulous; ligules membranous, 1-2 mm. long; panicle 3-20 cm. long, open, oblong, 1/2 to 2/3 the entire length, purplish, branches and branchlets spreading; spikelets 1-1.8 mm. long, the pedicels enlarged just below; glumes subequal, 1/2 to 4/5 as long as the lemmas, obtuse, sparsely pilose; lemmas length of spikelet, pubescent especially on margins and nerve, obtuse.---Sandy or rocky open ground. Montana to Washington, south to Nebraska, Texas and New Mexico, west to California and south to Costa Rica. Widely distributed over the western two-thirds of Colorado except in the northwest and westcentral parts. Altitude 5000-9000 feet.
2. Muhlenbergia wolfii (Vasey) Rydb., Bull. Torr. Club 32:600. 1905.
   Sporobolus ramulosus of Manuals for Colorado plants---Annual plants; culms 10-20 cm. tall, occasionally even less, spreading, branching below, glabrous, angled; leaf blades 1.2 mm. wide or less, 1-3 cm. long, flat, lax, glabrous below, scabro-pubescent above; sheaths glabrous, striate; ligules membranous, truncate or erose, 0.5 mm. long or less; panicle 2-5 cm. long, oblong, open, branches ascending; spikelets 1-1.5 mm. long, on short, rather appressed pedicels; glumes subequal about 1/2 as long as the lemma, broad, glabrous, obtuse; lemmas length of spikelet, glabrous or scaberulous, acutish, rounded on back.---Open dry ground. Colorado, New Mexico, Arizona and south to Guatemala. Our one specimen from Teller County at 9400 feet but should be any place in southern Colorado.
3. Muhlenbergia arenacea (Buckl.) Hitchc., Biol. Soc. Wash. Proc. 41:161. 1928.
   Perennial plants with creeping or caespitose and vertical rhizomes; culms 10-20 cm. tall, tufted, erect, ascending or somewhat decumbent at base, often scabrous; leaf blades about 1 mm. wide, 1-3 cm. long, flat at base with an involute rather sharp point, usually glabrous below, scaberulous above, midnerve and margins white-thickened; sheaths glabrous or scaberulous; ligules membranous, 0.5-1 mm. long in center but margins usually splitting and forming 2 erect auriclelike structures up to 2 mm. long; panicle 5-12 cm. long, open and diffuse, sometimes included at base, branches capillary; spikelets 1.8-2.4 mm. long, rarely to 3, usually purplish, rarely 2-flowered, on rather long pedicels; glumes 1-1.3 mm. long, about 1/2 as long as the lemma, the second a little longer than the first, 1-nerved, glabrous or scaberulous, subacute to apiculate; lemmas length of spikelet, glabrous or with a few hairs near base, abruptly mucronate or rarely very short-awned.---Mesas and plains often in low places. Texas to Arizona. Present in southeastern Colorado but very local, our two specimens from Pueblo and Otero Counties at about 4500 feet.
4. Muhlenbergia asperifolia (Nees. & Mey.) Parodi, Revista Fac. Agron. Buenos Aires 6:117. 1928.
   Sporobolus asperifolius (Nees. & Mey.) Nees.---Perennial plants with slender scaly rhizomes; culms 10-40 cm. tall, branching at base, spreading, slender, compressed, glabrous and pale or glaucous; leaf blades 1-3 mm. wide, 2-5 cm. long, flat or involute near tip, scabrous especially on margins; sheaths glabrous, somewhat compressed-keeled especially the lower; ligules membranous but margins ciliate, 0.5-1 mm. long; panicle 5-15 cm. long and about as wide, open and diffuse, branches scaberulous, the inflorescence breaking away at maturity; spikelets 1.5-2 mm. long, on long pedicels, sometimes 2-flowered, often swollen from the grain infected with a fungus; glumes subequal, from 1/2 to as long as the lemmas, 1-nerved, scabrous on keel, acuminate or awn-pointed; lemmas length of spikelet, thin, rather broad, some usually mucronate from an obtuse

apex, nerves often very obscure.---Damp, marshy often alkaline soil in moist prairies, around lakes or along streams. Illinois and Alberta to British Columbia, south to Texas and California; also in Mexico and South America. Widely scattered over all of Colorado at 3500-8000 feet.

5. **Muhlenbergia pungens** Thurb. in A. Gray, Acad. Nat. Sci. Phila. Proc. 1863:78. 1863.

Perennial plants with strong creeping rhizomes, often forming patches or rings of growth; culms 20-40 cm. tall (rarely taller), more or less tufted, erect from a decumbent base, pubescent (on ours); leaf blades 1-3 mm. wide, 3-5 cm. long (or longer sometimes), flat at base but soon involute, spreading, often curved, sharp-pointed, puberulent to scabrous; sheaths glabrous to scabrous and woolly-hairy on lower ones, hyaline margins wide and extending up into ligules as a triangular flap; ligules nearly all of hairs, 0.5-1 mm. long; panicle 5-15 cm. long (occasionally longer), open at maturity, oblong, branches in fascicles; spikelets 2.5-4 mm. long, purple or brownish; glumes subequal, about 1/2 the length of the lemmas, scabrous, short-awned from midrib; lemmas length of spikelet, terete, nerves indistinct, tapering to an awn about 1 mm. long.---Dry hills, sandy plains and sand dunes. South Dakota, Colorado and Utah to New Mexico and Arizona. Our specimens from the sand hills of eastern Colorado and the San Luis Valley but one specimen from Moffat County. Altitude 3500-8000 feet.

6. **Muhlenbergia porteri** Scribn.; Beal, Grasses N. Amer. 2:259. 1896.

Perennial plants, no rhizomes; culms 30-100 cm. tall, loosely tufted from a knotty, woody base, slender, wiry, widely spreading and branching, scaberulous; leaf blades 1-1.5 mm. wide, 2-5 cm. long, usually flat, early deciduous from the sheath; sheaths smooth, soon spreading away from culm; ligules membranous, 1-2 mm. long; panicle 5-10 cm. long, about as wide, open, branches long and brittle; spikelets 3-4 mm. long, on very long pedicels; glumes subequal, about 1/2 to 3/4 the length of lemmas, rarely almost as long, 1-nerved, scabrous on keels, acute or almost awn-tipped; lemmas length of spikelet, sparsely pubescent, purplish, tapering to an awn 5-10 mm. long.---Dry hills, mesas, and plains. Texas to Colorado, Nevada and California, south to Mexico. Our one specimen from Fremont County but to be expected in southern Colorado. Altitude 6000 feet.

7. **Muhlenbergia torreyi** (Kunth) Hitchc.; Bush, Am. Midl. Nat. 6:84. 1919.

M. gracillima Torr.---Perennial plants; culms 10-25 cm. tall, in loose tufts, usually gregarious and forming large patches often bare in the center, bases decumbent, slender but rather stiff; leaf blades 1-3 mm. wide, 1-4 cm. long, involute-setaceous, crowded at base, recurved or flexuous forming a curly cushionlike structure, somewhat scabrous; ligules membranous, 2-5 mm. long; panicle usually about 1/2 the length of the culm, open, branches spreading, usually purplish; spikelets about 3 mm. long, on pedicels as long or longer; glumes subequal, 1/2 to 2/3 as long as the lemmas, 1-nerved, acuminate and awn-pointed, scabrous on nerve; lemmas length of spikelet (2.5-3.2 in ours), glabrous or nearly so, tapering to a delicate awn about 2-3 mm. long.---Plains, mesas and dry hills often in rocky or sandy soil. Kansas and Colorado to Texas and Arizona. Scattered over the eastern half of Colorado but specimens from Chaffee, Saguache and probably Rio Grande Counties. Altitude 4000-8000 feet.

8. **Muhlenbergia arenicola** Buckl., Acad. Nat. Sci. Phila. Proc. 1862:91. 1862.

Perennial plants, no rhizomes (possibly short ones present); culms 30-50 cm. tall, tufted, erect or ascending from a decumbent base, slender, glabrous or puberulent near nodes; leaf blades about 1 mm. wide, 5-13 cm. long, involute, erect, scabrous; sheaths glabrous to scabrous; ligules membranous, 4-8 mm. long; panicle 15-40 cm. long, open, branches spreading or ascending, branchlets appressed, pale green or purplish, sometimes included at base; spikelets 2.5-3.5 mm. long, on pedicels as long or longer; glumes subequal, 1/2 to 2/3 as long as the lemma, scabrous on the 1 nerve, acuminate or awn-pointed, rather scarious; lemmas length of spikelet, nerves rather distinct (in ours), nearly glabrous, tapering to a delicate awn about 3 mm. long.---Sandy plains, mesas and hills. Kansas and Colorado to Arizona and Mexico. Our two specimens from southeastern Colorado at 4000-4500 feet.

9. **Muhlenbergia andina** (Nutt.) Hitchc., U. S. D. A. Bull. 772:145. 1920.

M. comata (Thurb.) Benth.---Perennial plants, with numerous scaly rhizomes; culms 30-80 cm. long, erect or sometimes spreading, puberulent or scaberulent below nodes; leaf blades 1-5 mm. wide, 3-15 cm. long, flat or involute at the tip, scabrous; sheaths keeled, smooth or scaberulous; ligules membranous but short-ciliate, about 1 mm. long; panicle 6-12 cm. long, narrow, almost spikelike but usually interrupted or lobed below, grayish-silky, often purple-tinged; spikelets 3-4 mm. long; glumes subequal, 3-4 mm. long, 1-nerved, long-acute or acuminate, scabrous on keel; lemmas 2-3 mm. long, with copious hairs near base, nearly as long as the lemma, tapering into a delicate awn 3-7 mm. long.---Meadows, moist thickets and river beds, especially in sandy soil. Montana to Washington south to New Mexico and California. Apparently uncommon, our specimens from central and southcentral Colorado at 6000-10,800 feet.

10. **Muhlenbergia montana** (Nutt.) Hitchc., U. S. D. A. Bull. 772:145, 147. 1920.

M. subalpina Vasey; *M. gracilis* var. *breviaristata* Vasey; *M. gracilis* of Manuals---Perennial plants, no rhizomes; culms 30-60 cm. tall, densely tufted, erect, glabrous or scabrous below panicle; leaf blades 1-2.5 mm. wide, 5-15 cm. long, flat to involute, firm, scabrous; sheaths glabrous or scaberulous, lower flat, pale and loose; ligules membranous, 7-10 mm. long, sometimes even longer; panicle 5-15 cm. long, narrow, erect or nodding, dense or loose; spikelets 3-4 mm. long, pedicels scabrous, the lateral ones very short; first glume about 2-2.5 mm. long, 1-nerved, scabrous, mucronate, apex often erose; second glume somewhat longer, 3-nerved, each nerve usually ending in a tooth or short awn; lemmas length of spikelet, pilose below, scabrous above, blotched with purple (in ours), tapering to an awn 6-20 mm. long.---Canyons, mesas, rocky hillsides often in dry ground. Montana to Utah and California, south to Texas and Mexico. Our specimens from the western, mountainous two-thirds of Colorado except in the northwestern part. Altitude 5500-10,000 feet.

11. **Muhlenbergia filiculmis** Vasey, Contrib. U. S. Nat. Herb. 1:267. 1893.

Perennial plants, no rhizomes; culms 10-30 cm. tall, densely tufted, erect, filiform; leaf blades 2-5 cm. long, rarely to 10, involute-filiform, often curved, scabrous, mostly basal; sheaths glabrous, the lower flattened, pale and loose; ligules membranous, 3-5 mm. long; panicle

1-5 cm. long, narrow with appressed, scabrous branches, erect; spikelets 2.5-3.5 mm. long; first glume 1.5-2 mm. long, 1-nerved, mucronate or erose at apex; second glume somewhat longer than first, 3-nerved, each nerve usually ending in a tooth or short awn; lemmas length of spikelets, pubescent below, scabrous above, usually tapering to an awn 2-5 mm. long, a few with shorter awns.---Open sandy or rocky soil, dry glasslands or hillsides, Wyoming, Colorado and New Mexico. Our Colorado specimens from the northcentral, central and southcentral parts at 8500-10,000 feet.

12. **Muhlenbergia schreberi** Gmel., Syst. Nat. 2:171. 1791.

Perennial plants, no rhizomes; culms spreading, decumbent at base and rooting at nodes, flowering branches ascending, 10-30 cm. long or more, lax, glabrous; leaf blades 2-4 mm. wide, 5-10 cm. long, flat, scabrous and pilose on margins near base; sheaths keeled, glabrous or slightly scabrous; ligules membranous, short-ciliate, 1 mm. long or less; panicles 3-15 cm. long, slender, loosely flowered, lax, nodding, the branches appressed or ascending, terminal and axillary present; spikelets about 2 mm. long; glumes minute, first often absent, less than 0.5 mm. long, not over 1/4 as long as the lemma, rounded or truncate at apex; lemmas length of spikelet, narrow, sparsely hairy below, scabrous especially on the nerves, tapering to a scabrous, delicate awn 2-5 mm. long.---Damp shady places, hills and woods. New Hampshire to Wisconsin and Nebraska, south to Florida and Texas; also in Mexico. Our one specimen from a yard near Denver, Colorado, and probably was a casual introduction, possibly resulting from adulterated lawn seed.

13. **Muhlenbergia brevis** C.O. Goodding, Journ. Wash. Acad. Sc. 31:505. 1941.

M. depauperata Scribn. of Manuals, at least in part---Annual plants; culms 3-20 cm. tall, tufted, slender, erect, branched below, scabrous to hispidulose below the nodes; leaf blades 1-2 mm. wide, 0.5-4 cm. long, flat to involute, scabrous or puberulent above with white-thickened margins and midrib; sheaths compressed-keeled, glabrous; ligules membranous, 1-3 mm. long; panicle 1-12 cm. long, often included at base, narrow, rather densely flowered, branches erect; spikelets 4-5 mm. long; first glume 1-3 mm. long, 2-nerved, minutely to deeply bifid; second glume 1.5-4 mm. long, 1-nerved, acuminate to setaceous, both glumes scabrous, shorter than lemmas; lemmas length of spikelet, nerves prominent, scabrous especially on nerves, and appressed-pubescent between nerves at base, awn 10-20 mm. long, scabrous and somewhat flexuous.---Open ground. Colorado and Texas to Arizona; also in Mexico. Our few specimens from Conejos County in southern Colorado at 9000 feet.

14. **Muhlenbergia pauciflora** Buckl., Acad. Nat. Sc. Phila. Proc. 1862:91. 1862.

Perennial plants, no rhizomes; culms 30-50 cm. tall, loosely tufted, wiry, erect, branching at lower nodes, glabrous; leaf blades 1 mm. wide or less, 5-15 cm. long, flat, or involute, scaberulous; sheaths glabrous or scaberulous; ligules membranous, 1-3 mm. long; panicle 5-15 cm. long, narrow, contracted but interrupted, branches erect or ascending, scaberulous, spikelet-bearing from base; spikelets 3.5-4.5 mm. long; glumes subequal, 1/2 to 3/4 as long as lemmas (counting awn tips), narrow, 1-nerved, acuminate to awn-tipped; lemmas length of spikelet, nerves rather plain, scaberulous on keel and sparsely pilose near base, purplish (in ours), tapering to an awn 5-12 mm. long (has been reported to 20 mm.).---Rocky hills and canyons. Texas to Colorado and Arizona, south to Mexico. Our one specimen from La Plata County at 7300 feet but to be expected in extreme southern Colorado.

15. **Muhlenbergia racemosa** (Michx.) B. S. P., Prel. Cat. N. Y. 67. 1888.

Perennial plants, with stout, creeping rhizomes; culms 40-80 cm. tall, erect or reclining, often branched with erect branches, somewhat roughened below nodes; leaf blades 2-6 mm. wide, 5-15 cm. long, flat, scabrous; sheaths keeled, glabrous; ligules membranous, 0.5-2 mm. long; panicle 3-12 cm. long, narrow, compact or lobed, erect or nearly so, bristly from awned glumes; spikelets 4-6 mm. long, subsessile; glumes subequal, 4-6 mm. long, including awn of about 2 mm., 1-nerved, scabrous especially on keel; lemmas 3-4 mm. long, acuminate or very short awn-tipped, pilose below, scabrous above.---Moist meadows and low ground along creeks and ponds. Newfoundland to British Columbia, south to Maryland, Kentucky, Oklahoma and Arizona; also in Virginia. Fernald (Rhodora 45:231-232, 1943) limits this plant to the interior of the United States. Our specimens widely distributed over Colorado, none as yet from the northwestern part, at 3500-9000 feet.

16. **Muhlenbergia mexicana** (L.) Trin., Gram. Unifl. 189. 1824.

M. foliosa of Manuals---Perennial plants with creeping scaly rhizomes; culms 30-100 cm. tall, erect or ascending, freely branching, leafy, scaberulous and strigose below the nodes; leaf blades 2-6 mm. wide, 5-15 cm. long, flat, scabrous; sheaths keeled, glabrous or scaberulous; ligules membranous, 0.5-1 mm. long; panicle 9-15 cm. long, narrow with numerous short appressed branches, dense to rather loose; spikelets 2-3.5 mm. long; glumes subequal, about as long as lemmas, narrow, scabrous especially on keel; lemmas length of spikelet, scabrous above, pilose below, acuminate or awn-tipped.---Moist thickets, low woods and low open ground. Quebec and Maine to Montana, south to North Carolina, Indiana, Kansas, New Mexico and Arizona. Our specimens from northcentral and eastcentral Colorado but it should be more widely distributed. Altitude 4000-6500 feet.

17. **Muhlenbergia richardsonis** (Trin.) Rydb., Bull. Torr. Club 32:600. 1905.

M. squarrosa Rydb.---Perennial plants, with numerous creeping rhizomes; culms 5-50 cm. long, slender, wiry, nodulose-roughened especially in age; leaf blades 0.5-1 mm. wide, 1-5 cm. long, flat or involute, glabrous or scaberulent below, scabrous above; sheaths glabrous; ligules membranous, 1-3 mm. long; panicle 2-10 cm. long, narrow, interrupted to almost spikelike, branches short and appressed; spikelets 2.5-3 mm. long, on short glabrous pedicels; glumes subequal, about 1/2 as long as the lemma, rather broad, acute; lemmas length of spikelet, nearly terete, scaberulous near apex, rather purplish or lead-colored, acute or mucronate.---Dry or moist, open ground of prairies and meadows often in sandy or alkaline soil. New Brunswick and Maine to Alberta, south to South Dakota, and in the mountains to New Mexico, to California and Washington. Our specimens widely distributed over the western two-thirds of Colorado at 6500-9500 feet.

18. **Muhlenbergia filiformis** (Thurb.) Rydb., Bull. Torr. Club 32:600. 1905.
    M. simplex (Scribn.) Rydb.; **M. aristulata** Rydb.---Annual plants; culms 5-30 cm. tall; tufted, erect or somewhat spreading, rather lax, glabrous, filiform; leaf blades 0.5-2 mm. wide, 1-4 cm. long, flat, folded or involute near tip, scabro-pubescent especially above; sheaths glabrous; ligules membranous, 1-2.5 mm. long; spikelets 2-3 mm. long; glumes about equal, less than 1 mm. long, about 1/3 as long as lemma, obtuse or acute, rather scabrous on the 1 nerve; lemmas length of spikelet, minutely puberulent below, somewhat scabrous above, acute and mucronate.--- Open woods, meadows and along streams. South Dakota and Kansas to British Columbia, south to New Mexico and California. Our specimens from central and northcentral Colorado but should be more widely distributed. Altitude 6000-10,000 feet.
    A variety **M. filiformis fortis** E. H. Kelso, Rhodora 38:298. 1936, has been described differing by having longer culms, ligules and lemmas. Some specimens from central and northcentral Colorado agree fairly well but too many intergradations occur. The specific description includes that of the variety.

19. **Muhlenbergia wrightii** Vasey in Coulter, Man. 409. 1885.
    Perennial plants, no rhizomes; culms 25-50 cm. tall, densely tufted from a rather hard crown, erect, slender, puberulent or scabrous; leaf blades 1-2 mm. wide, 4-15 cm. long, flat, folded or involute, scabrous especially below; sheaths glabrous or scaberulous between nerves; ligules membranous, 0.5-1 mm. long; panicle 4-15 cm. long, spikelike, dense but interrupted below; spikelets 2.5-3.5 mm. long; glumes about equal, rather variable but usually about 2 mm. long, scabrous (in ours) acuminate to awn-pointed; lemmas length of spikelet, scabrous or minutely pubescent below, blotched with purple (in ours), acuminate or mucronate, almost awn-pointed.---Plains, foothills and open slopes, sometimes in wet places. Colorado, Utah, New Mexico and Arizona; also in Mexico. Scattered over the western two-thirds of Colorado except in the extreme western part at 5000-8000 feet.

20. **Muhlenbergia cuspidata** (Torr.) Rydb., Bull. Torr. Club 32:599. 1905.
    Perennial plants, no rhizomes; culms 20-40 cm. tall, densely tufted, erect from a hard bulb-like scaly base, wiry, glabrous or scabrous; leaf blades 0.5-2 mm. wide, 3-10 cm. long, flat or involute-setaceous when dry; sheaths glabrous or puberulent especially at base; ligules membranous, about 0.5 mm. long; panicle 5-10 cm. long, narrow, contracted and somewhat spikelike but loosely flowered; spikelets 2.5-3 mm. long; glumes subequal, about 2/3 to 3/4 as long as lemmas, scabrous on keel, acuminate-cuspidate but not definitely awned; lemmas length of spikelet, scaberulent or minutely pubescent on back, apex acuminate-cuspidate.---Prairies, plains and gravelly or stony slopes. Michigan and Wisconsin to Alberta, south to Ohio and New Mexico. Our specimens from the northcentral part of Colorado at 5000-6800 feet. Should be found throughout the eastern plains area.

45. Sporobolus R. Br.   DROPSEED

Annual or perennial plants; leaf blades flat to involute; sheaths split; ligules usually made up of hairs but sometimes membranous at base only; inflorescence an open or contracted panicle; spikelets 1-flowered, the rachilla disarticulating above the glumes; glumes usually unequal and shorter than the lemmas, 1-nerved; lemmas 1-nerved, awnless, membranous; paleas similar to lemmas in texture and length, often splitting in 2 parts at maturity; grain free from the lemma and palea, the pericarp free from seed and slipping off it when moistened; styles short, stigmas plumose. In addition to the species in the key, S. nealleyi Vasey has been reported from Colorado probably on the basis of a misidentified specimen.

1. Spikelets over 3.5 mm. long (4-6)
  2. Second glume less than 5/6 of the length of the lemmas; panicle contracted, rather spikelike -1. S. asper
  2. Second glume as long or longer than the lemmas; panicle open at maturity, not at all spikelike
                                                                                      -2. S. heterolepis
1. Spikelets less than 3.5 mm. long (1-3)
  3. Plants annual; glumes subequal in length   -3. S. neglectus
  3. Plants perennial; glumes definitely unequal in length
    4. Lower panicle branches in distinct whorls; spikelets less than 1.6 mm. long   -4. S. pyramidatus
    4. Lower panicle branches more scattered, not in whorls; spikelets over 1.6 mm. long
      5. Pedicels capillary, widely spreading, mostly over 5 mm. long   -5. S. texanus
      5. Pedicels shorter, usually somewhat appressed along the branchlets
        6. Panicle narrow and spikelike, the branches closely appressed and spikelet-bearing well to the base
          7. Spikelets 2.5-3 mm. long; culms robust, 1-2 m. tall   -6. S. giganteus
          7. Spikelets 2-2.5 mm. long; culms not especially robust, less than 1 m. tall
                                                                                      -7. S. contractus
        6. Panicle open, its branches spreading at maturity, naked near base (**in number 9 the panicle may be partly or entirely included in the sheath and appear spikelike. But the branches are naked at base and any exserted portion will soon begin to spread**)
          8. Second glume and lemmas rounded and glabrous on back (use strong lens); panicle branches often over 8 cm. long, the branchlets at maturity not appressed to the branches; collar of leaf pilose on margin or on sides near margins, the hairs usually ascending
                                                                                      -8. S. airoides
          8. Second glume and lemmas keeled, scabrous on the keel; panicle branches seldom over 8 cm. long, the branchlets appressed along the branches (compare fig. 610 and 616 in Hitchcock's Manual, 1951); collar usually pilose in a line from margin well toward the back, the hairs usually widely divergent   -9. S. cryptandrus

1. **Sporobolus asper** (Michx.) Kunth., Rev. Gram. 1:68. 1829.
   Perennial plants, no rhizomes, or rarely short ones present; culms 50-100 cm. tall, solitary or in small tufts, erect, rather stout, glabrous; leaf blades 1-4 mm. wide, 10-50 cm. long, flat or becoming involute especially toward the long tip, lower elongate, usually sparingly pilose above near base and often below; sheaths glabrous, more or less pilose at margins of collars; ligules of hairs 1/2 to 1/3 its length, seldom over 0.5 mm. long; panicle 5-25 cm. long, contracted and rather spikelike, pale or purplish, terminal and axillary, usually included at base, often entirely so; spikelets 4-6 mm. long, on short pedicels; first glume 1/2 to 2/3 the length of spikelet, keeled; second glume 2/3 to 3/4 the length of spikelet, keeled, both glumes scabrous on keel; lemmas length of spikelet, about 1-2 mm. longer than second glume, scabrous on keel.---Prairies, sandy meadows and fields. Vermont to Michigan, North Dakota and Utah, south to Louisiana and New Mexico. Our specimens from northcentral and southeastern Colorado at 4000-5000 feet. To be expected anywhere on the eastern plains.

2. **Sporobolus heterolepis** (A. Gray) A. Gray, Man. 576. 1848.
   Perennial plants, no rhizomes; culms 30-70 cm. tall, densely tufted, erect, slender, glabrous; leaf blades 2-4 mm. wide, 10-25 cm. long, flat, becoming involute near tip, glabrous but scabrous on margins; sheaths glabrous, or lower somewhat pilose on back, sparsely pilose on edge of collar; ligules a ring of hairs less than 1 mm. long; panicle 5-25 cm. long, open, the branches ascending or spreading, naked below, few-flowered above; spikelets 4-6 mm. long, grayish; first glume 2-3 mm. long, rather subulate; second glume as long or longer than lemma, acuminate or awn-pointed, both scabrous on the 1 nerve; lemmas 4-6 mm. long, obtuse or acute.---Prairies, dry hillsides and plains. Quebec to Saskatchewan, south to Wyoming, Texas, Illinois, Arkansas and Connecticut. Our six specimens from along the eastern foothills of Colorado from Boulder to Pueblo County at 5300-7200 feet. Probably further east on the plains.

3. **Sporobolus neglectus** Nash, Bull. Torr. Club 22:464. 1895.
   Annual plants; culms 5-35 cm. tall, tufted, freely branching, erect or decumbent at base, flattened, even channeled on 1 side below nodes; leaf blades 0.5-2 mm. wide, 3-9 cm. long or sometimes longer, involute toward the tip, scabrous especially on margins, often pilose above near base; sheaths glabrous or sparsely pilose near collar, upper often inflated; ligules mostly of hairs, less than 0.5 mm. long; panicle 2-6 cm. long, spikelike, usually more or less included; spikelets 1.8-3 mm. long, pale, plump; glumes subequal, nearly as long as lemmas, acute, glabrous, rather broad, white-scarious with 1 greenish nerve; lemmas length of spikelet, glabrous, whitish, acute.---Dry open ground and sandy fields. Quebec and Maine to North Dakota, south to Maryland, Tennessee and Texas; also Washington and Arizona. Our two specimens from eastern Colorado, from Prowers and Morgan Counties at 3600-4200 feet. To be looked for in all of the eastern part of the state.

4. **Sporobolus pyramidatus** (Lam.) Hitchc., U. S. D. A. Misc. Pub. 243:84. 1936.
   S. argutus (Nees.) Kunth of Manuals---Perennial plants; culms 10-40 cm. long, in spreading tufts, glabrous; leaf blades 1-4 mm. wide, less than 10 cm. long, flat, sparsely ciliate near base; sheaths glabrous, keeled; ligules of hairs about 1 mm. long; panicle 3-7 cm. long, pyramidal, branches naked below, lower in distinct whorls, pale; spikelets 1-1.5 mm. long, glabrous; first glume small, about 0.5 mm. or less, thin; second glume length of lemmas, thin; lemmas length of spikelet, thin.---Sandy or gravelly soil on plains, often in alkaline soil. Kansas and Colorado to Louisiana and Texas; Arizona and Florida; also in tropical America. The Colorado record is from a specimen in the New York Botanical Garden collected by George Vasey in 1886 from "Colorado." If present in the state then most likely in the eastern or southeastern part.

5. **Sporobolus texanus** Vasey, Contrib. U. S. Nat. Herb. 1:57. 1890.
   Perennial plants, no rhizomes (in ours, but reported present in Kansas and Texas); culms 30-50 cm. tall, in close tufts, erect or spreading, slender, glabrous; leaf blades 1-4 mm. wide, 4-15 cm. long, flat but involute on drying, smooth below, scabrous above; sheaths glabrous or lower pilose, sometimes papillose-pilose, pilose on edge of collar; ligules of hairs about 0.5 mm. long; panicle 12-30 cm. long, almost as wide, open, diffuse, branches and branchlets stiffly spreading, the whole breaking away at maturity; spikelets 2-2.5 mm. long on long capillary pedicels; first glume 1/3 to 1/2 as long as spikelet, narrow, acute, glabrous; second glume about as long as spikelet, scabrous on the 1 nerve, acute to acuminate; lemmas as long as spikelets, scabrous on the nerves.---Mesas, valleys, plains and salt marshes. Kansas to Colorado, Texas and New Mexico. Our few specimens from southern Colorado from Mineral to Prowers County at 3500-10,800 feet, mostly below 5000 feet.

6. **Sporobolus giganteus** Nash, Bull. Torr. Club 25:88. 1898.
   This grass resembles the next species in having a spikelike panicle. The culms are more robust, 1-2 m. tall and the spikelets are larger, about 2.5-3 mm. long.---Mesas and sandy ground. Oklahoma to Colorado, south to Texas and Arizona. Our one record is from the extreme southeastern corner of the state at 3700 feet.

7. **Sporobolus contractus** Hitchc., Am. Journ. Bot. 2:303. 1915.
   S. strictus (Scribn.) Merr.---Perennial plants, no rhizomes; culms 30-100 cm. tall, tufted, erect, glabrous; leaf blades 1.5-5 mm. wide, 3-25 cm. long, mostly flat, smooth below, somewhat scabrous above; sheaths glabrous but pilose on collar from side to back, strongly nerved; ligules of hairs less than 1 mm. long (ours 0.4); panicle up to 50 cm. long (ours to 15), spikelike, usually erect, often partly or rarely entirely included, green or pale; spikelets 2-2.5 mm. long; first glume about 1/2 the length of lemma, narrow; second glume somewhat shorter than lemmas, scabrous on keel; lemmas length of spikelet, scabrous on keel, flattened, hyaline.---Mesas, dry bluffs and sandy fields. Colorado to Nevada, south to Texas, California and Sonora; also in Maine. Our three specimens from southcentral and southwestern Colorado at 6500-7500 feet.

8. **Sporobolus airoides** (Torr.) Torr., Pacific R. R. Rep. 7(3):21. 1856.
   Perennial plants, no rhizomes; culms 50-100 cm. tall, in large, tough bunches, robust, erect or spreading, glabrous; leaf blades 2-6 mm. wide, 5-30 cm. long, flat at first, soon becoming

involute, scabrous above and on margins; sheaths glabrous but ciliate near collar, collar pilose, the hairs on margin or near margin and ascending; ligules mostly of hairs, about 0.5 mm. long; panicle 10-40 cm. long, often 1/2 the length of the culm, about 1/2 to 2/3 as wide as long at maturity, open, the branches spreading, naked at base, branchlets spreading; spikelets 1.8-2.2 mm. long; first glume about 1 mm. long, rather deciduous; second glume about length of spikelet, both 1-nerved, rounded and glabrous on back; lemmas length of spikelet, glabrous, rounded on back, abruptly acute.---Meadows, valleys and prairies, especially in dry alkaline soil. South Dakota to Washington, south to Texas and California. Widely scattered over Colorado although no specimens as yet from the northeastern or northwestern corners. Altitude 4000-8000 feet.

9. Sporobolus cryptandrus (Torr.) A. Gray, Man. 576. 1848.
Perennial plants; rhizomes lacking or short ones rarely present; culms 20-80 cm. tall, in tufts, erect, spreading or sometimes prostrate, glabrous; leaf blades 2-5 mm. wide, 5-15 cm. long, flat but more or less involute on drying especially near tip, scabrous or glabrous above, scabrous on margins; sheaths ciliate-margined, collar pilose from margin well around to back, hairs usually divergent and conspicuous; ligules mostly of hairs, 0.5 mm. long or less; panicle as much as 25 cm. long, usually included at base, sometimes entirely, branches finally spreading or reflexed, naked at base, branchlets and pedicels appressed along the branches; spikelets 1.6-2.5 mm. long; first glume about 1 mm. long, 1-nerved, glabrous, readily deciduous; second glume about as long as lemmas, rather definitely keeled, scabrous especially on keel (with strong lens); lemmas length of spikelet, keeled and finely scabrous especially on keel.---Sandy open ground of prairies, plains and hillsides. Maine and Ontario to Alberta and Washington, south to North Carolina, Indiana, Louisiana, Arizona and Mexico. Very widely distributed over all of Colorado at 3500-8500 feet.

46. Blepharoneuron Nash

Sheaths split; ligules membranous; spikelets 1-flowered, the rachilla disarticulating above the glumes; styles distinct and stigmas plumose.

1. Blepharoneuron tricholepis (Torr.) Nash, Bull. Torr. Club 25:88. 1898.
Perennial plants, no rhizomes; culms 20-60 cm. tall, densely tufted, erect, slender, smooth; leaf blades 1-2 mm. wide, 5-15 cm. long, soon involute, crowded near base of culm, scabrous on margins and upper nerves; sheaths inflated somewhat on basal; ligules erose, 1 mm. long or less; panicle 5-15 cm. long, ovate, elliptical or narrower, but rather open, grayish; spikelets 2.2-3 mm. long, on capillary longer pedicels; first glume about 0.5 mm. shorter than lemma, 1- or faintly 3-nerved, acute or obtuse, thin, lead-colored; second glume similar but a little longer, sometimes as long as lemmas and definitely 3-nerved; lemmas length of spikelet, 3-nerved, these densely white-pubescent except toward the apex, obtuse or acute or the midnerve excurrent as a mucro; paleas about as long as lemmas, silky-hairy between the nerves.---Rocky slopes, valleys and dry woods. Colorado to Utah, south to Texas, Arizona and Mexico. Scattered over the western two-thirds of Colorado but no specimens from the northwest corner. Altitude 6500-12,000 feet, most common at 7500-9500 feet.

47. Oryzopsis Michx. RICEGRASS

Perennial plants, no rhizomes; leaf blades flat to involute; sheaths split; ligules membranous but sometimes ciliate; inflorescence an open or contracted panicle; spikelets 1-flowered, rather large in most species, disarticulating above the glumes; glumes equal or nearly so, about as long as spikelet; lemmas indurated, broad, nearly terete, usually pubescent, callus short and blunt, a deciduous awn, variable in length, from apex; paleas enclosed by edges of lemmas; styles distinct, stigmas plumose.

1. Pubescence of lemmas long and silky, the hairs 1-4 mm. long
  2. Panicle at maturity diffuse, the branchlets dichotomous and widely spreading (see fig. 636 in Hitchcock's Manual, 1951); lemmas seldom over 3.5 mm. long; very common -1. O. hymenoides
  2. Panicle narrow, the branches appressed or ascending; lemmas usually over 3.5 mm. long; very rare in Colorado
    3. Awns 7-12 mm. long; panicle somewhat open, many-flowered, over 6 cm. long; plants often over 30 cm. tall -2. O. bloomeri
    3. Awns about 6 mm. long; panicle very narrow, few-flowered, less than 6 cm. long; plants 15-30 cm. tall -3. O. webberi
1. Lemmas glabrous or with short, appressed hairs
  4. Spikelets, excluding awns, 6-8 mm. long; leaf blades mostly flat -4. O. asperifolia
  4. Spikelets 3-5 mm. long; leaf blades usually involute
    5. Lemmas 2-2.5 mm. long, usually glabrous; panicle usually over 6 cm. long; awns of lemmas 5-10 mm. long, not geniculate -5. O. micrantha
    5. Lemmas 3-5 mm. long, always somewhat pubescent; panicle less than 6 cm. long; awn of lemmas less than 5 mm. long or if longer then at least somewhat geniculate
      6. Awn 1-2 mm. long, or wanting; panicle branches closely appressed -6. O. pungens
      6. Awn 4-6 mm. long (but readily deciduous); panicle branches loosely ascending or spreading -7. O. exigua

1. Oryzopsis hymenoides (R. & S.) Ricker; Piper, Contrib. U. S. Nat. Herb. 11:109. 1906.
Eriocoma hymenoides (R. & S.) Rydb.; E. cuspidata Nutt.---Culms 20-60 cm. tall, rather densely tufted, erect, glabrous; leaf blades 1-2 mm. wide, 10-30 cm. long, involute or sometimes flat, stiff, mostly glabrous; sheaths glabrous or scaberulous; ligules 3-7 mm. long, acuminate; panicle 10-30 cm. long, diffuse, the many slender branches in pairs, branchlets dichotomous, spreading and flexuous at maturity; spikelets 6-8 mm. long, on long capillary pedicels; glumes

length of spikelet, 3- to 5-nerved, glabrous, puberulent or scabrous especially on margins, papery on margins, ovate, long-acuminate the tips spreading; lemmas about 3 mm. long, densely pilose, the hairs about 3 mm. long, brown to blackish when mature, awn nearly straight to somewhat twisted, 3-6 mm. long.---Deserts, plains, canyons, hillsides and sand dunes, Manitoba to British Columbia, south to Texas, California and Mexico. Very widely distributed and common in Colorado at 4000-9500 feet.

2. Oryzopsis bloomeri (Boland) Ricker; Piper, Contrib. U. S. Nat. Herb. 11:109. 1906.

Culms 30-60 cm. tall, tufted, glabrous; leaf blades slender-involute, 10-20 cm. long, tapering to a fine point, glabrous or scaberulous, mostly crowded toward base; sheaths glabrous; ligules 1 mm. long (but varies from 1-6 mm. on same plant); panicle 7-15 cm. long, somewhat open, the branches slender and rather stiffly ascending, somewhat scabrous; spikelets 8-10 mm. long (ours 6-7), rather long-pedicelled; glumes length of spikelet, broad, indistinctly 3- to 5-nerved, pale or purplish, usually puberulent near base especially on nerves; lemmas about 5 mm. long (ours 3-4), villous with hairs over 1 mm. long, pale brown beneath the hairs, awn weakly geniculate or nearly straight, twisted below, about 12 mm. long (ours 7-9). This is a puzzling species and as can be noted from the comments in parenthesis, our specimens vary from the published descriptions. It has been suggested that the plant is a hybrid or a variety of number 1 and is easily confused with immature specimens of that species.---Dry, open ground. Montana to Washington south to New Mexico and California. Our two specimens from Larimer County in northcentral Colorado at 6500-8000 feet.

3. Oryzopsis webberi (Thurb.) Benth.; Vasey, Grasses U. S. 23. 1883.

This species is closely related to the preceding, differing from it in the characters given in the key. Our one record is from westcentral Colorado at about 5500 feet.

4. Oryzopsis asperifolia Michx., Fl. Bor. Am. 1:51. 1803.

Culms 20-70 cm. tall, tufted; innovations erect, fertile culms spreading or prostrate, glabrous above, scaberulous below; leaf blades 3-8 mm. wide, upper not over 1 cm. long, basal 10-30 cm. long, flat or somewhat rolled, margins and upper surface scabrous, scabrous and rather glaucous below; sheaths glabrous, lower purplish and short; ligules short-ciliate, about 0.5 mm. long; panicles 5-8 cm. long, narrow, nearly simple, few-flowered; spikelets 6-8 mm. long, excluding awns; glumes subequal, length of spikelet, broad, green but whitish and apiculate at apex, 7- to 9-nerved; lemmas length of spikelet, sparsely pubescent with appressed hairs, a dense tuft of hairs at base, pale or yellowish at maturity, oblong but tapering to an acute apex, the awn green, scaberulous, angled and flexuous, 5-12 mm. long.---Dry banks, wooded slopes and aspen groves. Newfoundland to British Columbia and Montana, south to Connecticut, Indiana, South Dakota and New Mexico. Our eight specimens mostly along the eastern mountain slopes from Larimer to Huerfano County but one from La Plata County. Apparently uncommon in Colorado. Altitude 7000-9200 feet.

5. Oryzopsis micrantha (Trin. & Rupr.) Thurb., Acad. Nat. Sci. Phila. 1863:78. 1863.

Culms 15-70 cm. tall, densely tufted with numerous innovations, erect, slender, scaberulous; leaf blades 0.5-2 mm. wide, 5-15 cm. long, flat or involute especially on innovations, scabrous; sheaths glabrous; ligules about 0.5-1 mm. long; panicle 5-15 cm. long, open at maturity, branches distant, single or in pairs, finally spreading or reflexed, the spikelets appressed near ends; spikelets 3-4 mm. long, excluding awns on short appressed pedicels; glumes subequal, length of spikelet, thin and hyaline, 5- to 6-nerved, acute or acuminate, the first broader; lemmas 2-2.5 mm. long, glabrous or appressed pubescent, shining, yellow-brown at maturity, awn rather flexuous and scabrous, 5-10 mm. long.---Open dry woods, rocky slopes, gulches and hillsides often under trees or bushes. Saskatchewan to Montana, south to New Mexico and Arizona. Our specimens scattered over the western two-thirds of Colorado but none from the westcentral part as yet. Altitude 5000-9500 feet.

6. Oryzopsis pungens (Torr.) Hitchc., Contrib. U. S. Nat. Herb. 12:151. 1908.

This species is related to the next and differs from it in the characters given in the key. Our one record is from southcentral Colorado at 10,000 feet.

7. Oryzopsis exigua Thurb.; Wilkes, U. S. Expl. Exp. Bot. 17:481. 1874.

Culms 15-25 cm. tall, densely caespitose with numerous innovations, stiffly erect, scabrous; leaf blades involute-filiform, 5-10 cm. long the culm blades shorter, stiffly erect, scabrous, mainly basal; sheaths smooth to scaberulous, strongly nerved; ligules acute, 2-3 mm. long; panicle 3-6 cm. long, narrow, the branches appressed, the lower to 2 cm. long; spikelets about 4 mm. long, short-pedicelled, pale in color; glumes length of spikelet, acute, thin, faintly nerved, glabrous; lemmas about length of glumes, appressed-pilose, awn geniculate, somewhat twisted below, about 5 mm. long.---Dry open ground or open woods. Montana to Washington, south to Oregon, Nevada and Colorado. Our few specimens from Larimer and Routt Counties at about 9500 feet.

48. Piptochaetium Presl.

Sheaths split; ligules membranous; spikelets 1-flowered, articulating above the glumes; lemmas indurated, awned from apex. This genus is closely related to Oryzopsis and might well be united with it.

1. Piptochaetium fimbriatum (H. B. K.) Hitchc., Journ. Wash. Acad. Sc. 23:453. 1933.

Perennial plants, no rhizomes; culms 30-80 cm. tall, densely caespitose, erect, slender, glabrous; leaf blades less than 1 mm. wide, involute-filiform or rarely flat, culm leaves usually less than 6 cm. long, basal from 6 cm. to as long as the culm; sheaths glabrous; ligules 1-3 mm. long, acute; panicle 5-15 cm. long, open at maturity, the branches slender and spreading, a few spikelets near end; spikelets 3-6 mm. long (usually 5); glumes subequal, length of spikelet, glabrous, pale or purplish, obtuse or acuminate, margins hyaline, first 3-nerved, second 5-nerved but nerves not distinct; lemmas 3-4 mm. long, oblong but asymmetric near apex, edges not meeting (as in Stipa), dark brown at maturity, appressed-pubescent especially on callus and a short-ciliate ridge around awn base, awn weakly twice-geniculate, twisted below, scaberulous, green especially near base, 1-2 cm. long, paleas narrow, indurated in center, central keel of 2 nerves and sulcus between, exposed between edges of lemmas, the point projecting above the lemmas.---Open rocky woods, Colorado to Texas, Arizona and Mexico. The Colorado record is based on a specimen in the U. S. National Museum collected by George Vasey from "Colorado." Mr. Paul Ginter, formerly of the U. S. Forest Service, told the writer that he saw this plant growing in Colorado west of Raton Pass in Las Animas County. To be expected anywhere in the southern part of the state.

49. Stipa L. NEEDLEGRASS

Perennial plants without rhizomes; leaf blades usually involute; sheaths split; ligules membranous but usually short-ciliate; inflorescence an open or contracted panicle; spikelets 1-flowered, disarticulating above the glumes, perfect; glumes membranous or papery, pointed, usually long and narrow; lemmas narrow, terete, indurated, callus sharp, terminating in a persistent, geniculate awn twisted at base; paleas enclosed in the rolled lemma; styles short, stigmas plumose. The hard, sharp-pointed fruits of some species are injurious, especially to sheep, working through the wool and into the skin.

1. Awns partly or completely plumose
  2. Awns 12-18 cm. long, plumose only on the terminal segment; lemmas over 12 mm. long at maturity
                                                                                     -1. S. neomexicana
  2. Awns less than 4 cm. long, plumose below (plumose or not plumose above); lemmas less than 12 mm. long
    3. Plumose hairs of awn 5-8 mm. long; lemmas over 7 mm. long   -2. S. speciosa
    3. Plumose hairs about 2 mm. long or less; lemmas less than 7 mm. long   -3. S. porteri
1. Awns glabrous, scabrous or appressed scabro-pubescent, not plumose
  4. Lemmas 8-25 mm. long, glabrous or sparsely pubescent; awns usually over 7 cm. long; glumes over 15 mm. long
    5. Last segment of awn curved and flexuous and over 7 cm. long; lemmas 8-12 mm. long; glumes usually less than 23 mm. long; panicle usually included at base   -4. S. comata
    5. Last segment of awn almost or quite straight and less than 7 cm. long; lemmas 12-25 mm. long; glumes usually over 23 mm. long; panicle usually not included at all
      6. Lemmas 12-15 mm. long; leaves 1-3 mm. wide; glumes seldom over 3 cm. long
                                                                            -4A. S. comata intermedia
      6. Lemmas 16 mm. long or more; leaves 3-5 mm. wide; glumes over 3 cm. long   -5. S. spartea
  4. Lemmas less than 7 mm. long, or if 7-9 mm. long, then distinctly pubescent at apex; awns less than 7 cm. long; glumes less than 15 mm. long
    7. Panicle open, branches distant, spreading or drooping; lemmas about 4 mm. long   -6. S. richardsoni
    7. Panicle narrow, the branches appressed; lemmas 4-9 mm. long
      8. Sheaths, at least the lower, velvety-pubescent   -7. S. williamsii
      8. Sheaths glabrous, scabrous or rarely sparsely pubescent (may be ciliate on 1 margin or pilose near collar)
        9. Awns over 4 cm. long, the terminal segment flexuous; apex of lemma glabrate   -8. S. arida
        9. Awns less than 4 cm. long, the terminal segment straight or nearly so; apex of lemma pubescent or villous
          10. Sheaths villous at the throat and 1 margin usually ciliate; lemmas over 6 mm. long (5-6 in number 11)
              11. Hairs at apex of lemma over 2 mm. long; glumes 10-15 mm. long; awns less than 2 cm. long   -9. S. scribneri
              11. Hairs at apex of lemma less than 2 mm. long; glumes 7-11 mm. long; awns over 2 cm. long
                  12. Lemmas 6-8 mm. long; ligules over 1.5 mm. long on culm leaves; leaves often over 3.5 mm. wide; panicle large and compact; glumes rather firm, the nerves not especially conspicuous   -10. S. robusta
                  12. Lemmas 5-6 mm. long; ligules less than 1.5 mm. long; leaves seldom over 3.5 mm. wide; panicle narrow but looser; glumes with conspicuous green nerves (when young), translucent between, almost to base   -11. S. viridula
          10. Sheaths not villous at the throat or only very slightly so, usually not ciliate; lemmas less than 6 mm. long (except 5-7 mm. in number 12 and 12A.)
              13. Lemmas 5.5-7 mm. long; culm leaves often flat; awns 20-40 mm. long
                  14. Awns less than 2.5 cm. long; panicle usually less than 15 cm. long; culms seldom over 80 cm. tall   -12. S. columbiana
                  14. Awns over 2.5 cm. long; panicle usually over 15 cm. long; culms often over 80 cm. tall   -12A. S. columbiana nelsoni
              13. Lemmas 4-5.5 mm. long; culm leaves involute-filiform; awns usually not over 20 mm. long
                  15. Hairs near apex of lemma 2 mm. long or more; glumes about 8.5 mm. long or longer   -13. S. pinetorum
                  15. Hairs near apex of lemma shorter; glumes 6-8 mm. long   -14. S. lettermani

1. Stipa neomexicana (Thurb.) Scribn., U. S. D. A. Div. Agrost. Bull. 17:132. 1899.
Culms 20-60 cm. tall, densely tufted, erect; leaf blades 1-4 mm. wide when unrolled, 10-30 cm. long, folded or involute, glabrous below, scabrous above; sheaths glabrous or pubescent; ligules short-ciliate, 0.5-1 mm. long; panicle 3-8 cm. long, narrow, bearing a few erect branches with 1 or 2 spikelets; spikelets 3-5 cm. long (counting tips of glumes), pale, on more or less pubescent pedicels; glumes subequal, length of spikelet, mostly 5-nerved or sometimes 7, narrowed into an awn point almost as long as body; lemmas 13-18 mm. long, the callus 4-5 mm. long,

5-nerved, sparsely villous, rather tuberculate above, brown at maturity, awn 12-18 cm. long, the lower segment 1/4 to 1/3 the entire length, rather straight and twisted, appressed-villous, middle segment 1-2 cm. long, flexuous or geniculate and somewhat twisted, the terminal segment rather flexuous, not twisted, plumose with pale or tawny hairs 2-3 mm. long.---Mesas, canyons, rocky slopes, and dry hills. Texas and Colorado to Utah and Arizona. Apparently not very widespread in Colorado, our specimens from Las Animas to La Plata County extending north to Mesa and Larimer County at 5000-6500 feet. Apparently this grass is spreading northward as it has become rather common around Fort Collins in the last 10 years.

2. **Stipa speciosa** Trin. & Rupr., Mem. Acad. St. Petersb. VI 5(1):45. 1842.

Culms 30-60 cm. tall, densely tufted, erect, glabrous; leaf blades 0.5-1 mm. thick (when rolled up), 10-30 cm. long, closely involute, glabrous or scaberulous below, densely scabrous above, rather stiff; sheaths glabrous or lower felty-pubescent, pilose at collar; ligules variable (in ours) lower ones 1-2 mm. long, long-ciliate, upper completely membranous, to 10 mm. long; panicle 10-15 cm. long, narrow, compact, pale or tawny, branches short and appressed; spikelets 14-20 mm. long (counting awn-tips of glumes); glumes subequal, length of spikelet, narrowing to a soft point 5-6 mm. long, first 3-nerved, second 5-nerved; lemmas 7-10 mm. long, callus about 1.5-2 mm. long, body soft-pubescent or glabrate near apex, awn once-geniculate, first segment 1-2 cm. long twisted, strongly plumose on lower part, the hairs 5-8 mm. long, upper segment 15-20 mm. long, straight, scabrous, whole awn about 15-30 mm. long.---Deserts, canyons and rocky hills. Colorado to California; South America. Our two specimens from Montezuma County in southwest Colorado at 5200-8500 feet.

3. **Stipa porteri** Rydb., Bull. Torr. Club 32:599. 1905.

S. mongolica Turcz. of Manuals---Culms 20-40 cm. tall, tufted, glabrous; leaf blades less than 1 mm. wide when unrolled, 2-12 cm. long, involute, sulcate, scaberulous; sheaths glabrous; ligules somewhat 2-lobed, 1.5-3 mm. long; panicle 5-10 cm. long, open branches slender, flexuous, few-flowered, distant; spikelets about 5 mm. long; glumes equal, length of spikelet, rather broad, obtuse or notched at apex, glabrous below, pubescent near apex, faintly 5-nerved; lemmas about length of spikelet, callus short, appressed-pilose below, scaberulous above, awn once-geniculate, twisted at base and entirely plumose, the hairs 1-2 mm. long, whole awn 12-15 mm. long.---Mountains of Colorado. Our few specimens collected in or near Lake County, Colorado, at 9500 feet or above.

4. **Stipa comata** Trin. & Rupr., Mem. Acad. St. Petersb. VI. Sc. Nat. 5(1):75. 1842.

Culms 30-75 cm. tall, tufted, erect, glabrous or pubescent at nodes; leaf blades 1-2 rarely to 3 mm. wide, 5-30 cm. long, basal longest, mostly involute-filiform, occasionally flat on culm leaves, scabrous or puberulent above, usually smooth below; sheaths glabrous or scaberulous, the upper usually inflated, no pilose hairs on collar; ligules 2-4 mm. long, sometimes shorter on innovations or longer on culms; panicle 10-20 cm. long, rather narrow, the branches erect, naked below, usually included at base; spikelets 15-20 mm. long on long pedicels; glumes subequal, length of spikelet with soft awnlike tips, 5-nerved; lemmas 8-12 mm. long, slender, pale or light brown, callus about 3 mm. long, villous with short white hairs or glabrate near apex, awn scabrous, more or less strongly twice-geniculate, first segment twisted, 1-4 cm. long, second segment loosely twisted, flexuous, usually 1-2 cm. long, terminal segment curved or sinuous, 7-12 cm. long, whole awn 12-18 cm. long.---Prairies, plains, and dry hills often in sandy soil. Indiana to Yukon Territory, south to Texas and California. Our specimens widely distributed over all of Colorado at 3500-8500 feet.

4A. **Stipa comata intermedia** Scribn. & Tweedy, (var.) Bot. Gaz. 11:171. 1886.

S. tweedyi Scribn.----Differs from the species in having panicle usually exserted; glumes over 23 mm. long; lemmas 12-15 mm. long (instead of 8-12 mm.), last segment of awn nearly or quite straight and less than 7 cm. long.---Plains and valleys. Montana to Washington, south to New Mexico and California. Our specimens thinly scattered over the western half of the state, none as yet from the westcentral part but undoubtedly present there. Altitude 5500-9000 feet.

5. **Stipa spartea** Trin., Mem. Acad. St. Petersb. VI. 1:82. 1830.

Culms 50-120 cm. tall, tufted, erect, glabrous or scaberulous beneath panicle; leaf blades 3-5 mm. wide, 20-30 cm. long, flat or somewhat involute on drying, scabrous above; sheaths glabrous, not villous at throat; ligules 3-5 mm. long, sometimes longer; panicle 10-20 cm. long, narrow, branches few, distant, ascending, with 1 or 2 spikelets near ends, the whole finally exserted; spikelets 3-4 cm. long (counting awn-tips of glumes); glumes subequal, 5-nerved (with 2 intermediate short ones near base), with soft awnlike tips, pale, glabrous; lemmas 16-25 mm. long, callus 7 mm. long, pubescent on callus and decreasingly so upward on body, brown, awn stout, twice-geniculate, first segment scabrous, twisted, 4-8 cm. long, second segment loosely twisted, flexuous, 1.5-3 cm. long, third segment straight, 6-9 cm. long, whole awn 12-20 cm. long.---Prairies. Ontario to British Columbia south to Pennsylvania, Indiana, Kansas and New Mexico. Our few specimens from along the eastern foothills of Colorado from Larimer to El Paso County but should be expected anywhere in the eastern part of the state. Altitude 5300-7200 feet.

6. **Stipa richardsoni** Link, Hort. Berol. 2:245. 1833.

Culms 50-100 cm. tall, tufted, erect, glabrous; leaf blades less than 2 mm. wide when unrolled, 5-15 cm. long, involute, scabrous or glabrous; sheaths glabrous or lower sometimes puberulent, collar glabrous or occasionally with a few pilose hairs; ligules 1 mm. long or less; panicle 10-20 cm. long, open, the branches spreading or drooping, distant, naked below, mostly in pairs, 7-15 cm. long; spikelets 6-10 mm. long; glumes subequal or the first longer, length of spikelet, acute, rather obscurely 3-nerved, glabrous or scaberulous near apex; lemmas about 5 mm. long, fusiform, dark brown, villous or glabrate near apex, awn twice-geniculate, first bend often obscure, twisted and appressed-pubescent to the second bend, first and second segments totaling 7-9 mm. long, third segment straight, scabrous, whole awn 20-30 mm. long.---Bottom lands, rocky or wooded slopes and hillsides. Saskatchewan to Colorado and British Columbia. Our few specimens from Larimer and Jackson County in northcentral Colorado at 8000-8600 feet.

7. **Stipa williamsii** Scribn., U. S. D. A. Div. Agrost. Bull. 11:45. 1898.

Culms 60-100 cm. tall, tufted, erect, glabrate; leaf blades 1-4 mm. wide, 10-30 cm. long, flat or somewhat involute on innovations, more or less pubescent; sheaths pubescent especially on lower, some of upper glabrate; ligules about 0.5 mm. long; panicle 10-25 cm. long, narrow, rather compact; spikelets 6-10 mm. long; glumes equal, length of spikelet, narrow, acuminate or short-awned, 3-nerved, scabrous or glabrous; lemmas about 6-7 mm. long, callus sharp, body narrowly fusiform, appressed villous, the apical hairs scarcely longer, awn twice-geniculate, 2.5-5 cm. long, twisted and scabro-pubescent to second bend, third segment straight or nearly so. Not a very well-marked species.---Dry hills and plains. Montana to Washington, south to Colorado and California. Our one specimen from the U. S. National Herbarium collected in Routt County, Colorado, at 8000 feet.

8. **Stipa arida** M. E. Jones, Proc. Calif. Acad. II. 5:725. 1895.

Culms 30-80 cm. tall, densely tufted, erect, smooth or scaberulous below nodes; leaf blades 1-2 mm. wide or less, 5-20 cm. long, involute or sometimes flat, scabrous, often glaucous or pale; sheaths glabrous or scaberulous, no hairs on collar; ligules 0.5-1 mm. long, short-ciliate; panicle 8-15 cm. long, narrow, compact, the branches short and appressed, somewhat nodding, pale or silvery; spikelets 7-12 mm. long; glumes subequal, length of spikelet, scaberulous, acuminate, first 3-nerved, second 3- to 5-nerved; lemmas 4-5 mm. long, pale or brown, callus sharp, about 0.5 mm. long, white-pilose, body narrow, appressed-villous except near apex with white hairs, awn 4-6 cm. long, twisted at base for 1-2 cm., not twisted but flexuous to end, often once or twice geniculate.---Rocky slopes and dry hillsides. Rare. Colorado, Utah and Arizona. Our one Colorado specimen, in the U. S. National Herbarium, from La Plata County at 6500 feet.

9. **Stipa scribneri** Vasey, Bull. Torr. Club 11:125. 1884.

Culms 30-70 cm. tall, tufted, erect, glabrous or pubescent below the nodes; leaf blades 1-2 mm. wide, 10-30 cm. long, flat or involute especially toward the tip and base, glabrous or scabrous especially above; sheaths glabrous but villous near collar, and 1 margin somewhat ciliate; ligules short-ciliate, 1 mm. long or less; panicle 10-25 cm. long, narrow, rather compactly flowered, branches appressed; spikelets 10-15 cm. long; glumes subequal to unequal, length of spikelet, long-acuminate, pale, 3-nerved, scaberulous; lemmas about 7-8 mm. long, pale, callus about 1 mm. long, body very narrowly fusiform, short appressed-villous, hairs at apex longer, 2-3 mm. long, awn 14-20 mm. long, twice-geniculate, scabrous, twisted to second bend, third segment straight.---Mesas, rocky slopes of mountains and foothills. Colorado, Utah, New Mexico and Arizona. Our specimens from along the eastern range of foothills in Colorado from Larimer to Fremont County. Probably more widely distributed. Altitude 5000-9500 feet.

10. **Stipa robusta** (Vasey) Scribn., U. S. D. A. Div. Agrost. Bull. 5:23. 1897.

S. vaseyi Scribn.---Culms 60-160 cm. tall, tufted, erect, robust, glabrous; leaf blades 4-8 mm. wide on culm or narrower on innovations, 20-50 cm. long, flat or involute especially on innovations, mostly glabrous but scaberulous on margins; sheaths glabrous or scattered-pubescent, ciliate on 1 margin (at least on younger leaves), pilose on lower margins and sides of collar; ligules 2-4 mm. long on culm leaves, may be shorter on innovations; panicle 15-30 cm. long, often to 2 cm. or more wide, narrow, compact, pale or green, often interrupted below, branches appressed; spikelets 8-11 mm. long; glumes equal or nearly so, length of spikelet, acuminate or tapering to a fine soft point (hardly awned), 3-nerved on first, 5-nerved on second, these not especially conspicuous, rather firm between nerves, not translucent; lemmas 6-8 mm. long, dark brown at maturity, callus about 0.5-1 mm. long, body narrow-fusiform, villous with appressed hairs at apex, these 1-2 mm. long, awn 2-3 cm. long, finally somewhat twice-geniculate, twisted to second bend, scabrous.---Dry plains, hills and open woods. Colorado, Arizona and New Mexico; also in Mexico. Our specimens from northcentral, central and southcentral Colorado at 5000-9000 feet.

11. **Stipa viridula** Trin., Mem. Acad. St. Petersb. VI.2(1):39. 1836.

Culms 50-100 cm. tall, tufted, erect, smooth or scabrous above; leaf blades 1-3 or sometimes to 5 mm. wide, 6-30 cm. long, involute or sometimes flat, scabrous on margins, often on upper and lower surfaces; sheaths ciliate on 1 margin (at least when young), long-pubescent or pilose on collar; ligules 0.5-1 mm. long; panicle 10-20 cm. long, erect, narrow with appressed branches, green or pale at maturity; spikelets 7-11 mm. long; glumes subequal, length of spikelet, with a hyaline, narrow tip about 2-3 mm. long, with 3 prominent nerves, these greenish in young spikelets and translucent between, scaberulous on nerves; lemmas 5-6 mm. long, callus short 0.5-1 mm. long, body appressed-pubescent, the hairs at apex a little longer, fusiform, rather plump and brownish at maturity, awn 2-3 cm. long, finally twice bent, scabrous and twisted to the second bend. This species is hard to tell from depauperate forms of **number 10**. The glumes of young spikelets here with **conspicuous** nerves and translucent between, is the most characteristic difference.---Plains, dry slopes, dry prairies and hills. New York, Wisconsin to Alberta, south to Kansas and New Mexico. Our specimens scattered over most of Colorado often in waste places as a weed. Altitude 3500-10,000 feet.

12. **Stipa columbiana** Macoun, Cat. Can. Pl. 2(4):191. 1888.

S. minor (Vasey) Scribn.---Culms 30-80 cm. tall, tufted, erect, glabrous; leaf blades 0.5-1 mm. wide (sometimes to 2 mm.), 10-20 cm. long, mostly involute especially on innovations, glabrous or scaberulous; sheaths naked at throat, not ciliate (in ours); ligules 1-2 mm. long or rarely more or less; panicle 5-15 cm. long, narrow, the branches short and appressed, from compact to rather loose, often purplish; spikelets 8-10 mm. long; glumes subequal, length of spikelet, glabrous or somewhat scaberulous, 3-nerved, these not very conspicuous; lemmas 6-7 mm. long (reported as 5 mm.), rarely shorter, callus rather short, body appressed-villous, the hairs not longer near apex, awn 20-25 mm. long (sometimes to 35 mm.), more or less twice-geniculate, scabrous. Related to and intergrades some with number 14.---Dry plains, meadows and open woods. Wyoming to Yukon Territory, south to Texas and California. Our specimens from the western two-thirds of Colorado, none however from the extreme western part. Altitude 7000-10,000 feet.

12A. Stipa columbiana nelsoni (Scribn.) Hitchc., (var.) Contrib. U. S. National Herb. 24:254. 1925.
S. nelsoni Scribn.---Differs from species--culms often over 80 cm. tall; leaf blades often wider, over 2 mm. wide; panicle larger and denser, often over 15 cm. long; lemmas with awns usually over 25 mm. long, often to 35 or even 40 mm.---Plains, hills and canyons. Alberta to Washington, south to Colorado and Baja California. Our specimens from the western half of Colorado at 5500-9500 feet.

13. Stipa pinetorum M. E. Jones, Proc. Calif. Acad. Sc. II.5:724. 1895.
Culms 30-60 cm. tall, tufted, sometimes dying out in center of tuft, erect, puberulent below nodes, otherwise glabrous, slender; leaf blades hardly 0.5 mm. thick (when rolled), 5-12 cm. long, involute-capillary, more or less flexuous and scabrous, mostly basal; sheaths glabrous, throat not villous; ligules about 0.5 mm. long; panicle 8-15 cm. long, narrow, branches short, appressed or ascending, few-flowered; spikelets 8.5-13 mm. long; glumes equal, length of spikelet, acuminate long-pointed, faintly nerved, glabrous or slightly scabrous; lemmas 4.5-5.5 mm. long, narrowly fusiform, pale or brown, villous especially on upper half, these hairs forming a tuft at apex 2 mm. long or more, apex of lemmas with 2 slender teeth, awn twice-geniculate, first segment twisted, about 6 mm. long, second segment somewhat twisted, 4 mm. long, third segment 8-10 mm. long, the whole awn 6-21 mm. long.---Open pine woods. Rare. Colorado, Utah, Nevada and California. Our two specimens from near Denver and in Lake County, Colorado, but should be more widely distributed. Altitude 5200-9200 feet.

14. Stipa lettermani Vasey, Bull. Torr. Club 13:53. 1886.
Culms 30-60 cm. tall, tufted, erect, glabrous or minutely scaberulous; leaf blades about 0.5 mm. wide (when rolled), 5-20 cm. long, involute-filiform, glabrous or scaberulous; sheaths glabrous, not ciliate or villous at collar; ligules 0.5-1.5 mm. long, sometimes longer; panicle 10-18 cm. long, narrow, rather loosely flowered, pale or greenish; spikelets 6-8 mm. long; glumes equal or subequal, length of spikelet, rather firm, 3-nerved, these rather inconspicuous glabrous or minutely scaberulous; lemmas 4-5 mm. long, rather narrow, appressed-villous, the hairs at apex longer, 1-1.5 mm. long, awn slender, scabrous, obscurely twice-geniculate, 15-20 mm. long.---Open ground, hills, plains, meadows or open woods. Wyoming to Montana and Oregon, south to New Mexico and California. Our specimens distributed over the western half of Colorado at 6500-11,500 feet.

50. Aristida L.   THREE AWN

Annual or perennial plants without rhizomes; leaf blades flat or more commonly involute; sheaths split; ligules with 1/2 or more of length consisting of hairs; inflorescence an open or contracted panicle; spikelets 1-flowered, rachilla disarticulating above the glumes, usually leaving a sharp callus on the lemmas; glumes membranous, equal or unequal; lemmas narrow, terete, indurate, terminating in a trifid awn (the lateral branches short on some); paleas covered entirely by the edges of the lemmas; styles distinct, stigmas plumose.

1. Central awn spirally coiled at base at maturity; plants annual
  2. Lateral awns over 5 mm. long, 1/2 to 2/3 as long as the central, somewhat spreading   -1. A. basiramea
  2. Lateral awns 1-3 mm. long, much less than 1/2 as long as the central, straight and erect -2. A. curtissii
1. Central awn not spirally coiled; plants perennial (in all but number 3)
  3. Plants annual, much branched just above the base; first glume about 2/3 to 3/4 the length of the second; awns 1-1.5 cm. long; callus obtuse
                                                                                -3. A. adscensionis
  3. Plants perennial, little if at all branched above the base; first glume either about equal to second or about 1/2 as long; awns 1-8 cm. long; callus sharp-pointed
    4. Glumes equal or subequal in length; lemmas narrowed into a twisted neck or "beak" 2-5 mm. long
      5. Panicle open, the long branches spreading or deflexed -4. A. divaricata
      5. Panicle narrow, the branches short and appressed -5. A. arizonica
    4. Glumes unequal, the first about 1/2 as long as the second; lemmas with a short, little-twisted beak if any present at all
      6. Awns less than 3 cm. long
        7. Panicle over 10 cm. long; outside of collar usually villous; rare in Colorado -6. A. wrightii
        7. Panicle less than 10 cm. long; outside of collar not villous (may be pilose inside near margins); common -8. A. fendleriana
      6. Awns over 3 cm. long
        8. Branches of the nodding panicle flexuous and slender (the panicle over 8 cm. long); rare in southern Colorado only -7. A. purpurea
        8. Branches of the erect panicle stiff and appressed (or the lower flexuous in plants with panicle less than 8 cm. long) panicle of others 2-20 cm. long; common in all of Colorado
          9. Leaf blades short, mostly in a curled, basal cluster; awns seldom over 4 cm. long; second glume less than 18 mm. long -8. A. fendleriana
          9. Leaf blades about as long as culm, not in conspicuous basal clusters; awns over 4 cm. long; second glume usually over 18 mm. long
            10. Panicle branches (especially the lowest) rather flexuous, panicle small, about 3-7 spikelets present -9A. A. longiseta rariflora
            10. Panicle branches not flexuous (lower sometimes curved), panicle larger, over 7 spikelets present
              11. Culms 30-50 cm. tall; awns 4-7 cm. long; panicle often over 15 cm. long -9B. A. longiseta robusta
              11. Culms 15-30 cm. tall; awns 6-8 cm. long; panicle seldom over 15 cm. long -9. A. longiseta

1. Aristida basiramea Engelm.; Vasey, Bot. Gaz. 9:76. 1884.
   Annual plants; culms 30-50 cm. tall, branching at base, erect or ascending, slender, glabrous or scaberulous; leaf blades 0.5-1.5 mm. wide, 3-12 cm. long, involute or sometimes flat on basal, scabrous above, usually glabrous below; sheaths glabrous or with a few pilose hairs; ligules less than 1 mm. long; panicles terminal and axillary, the latter mostly enclosed by the sheaths, terminal 5-10 cm. long, narrow with erect branches; spikelets 12-14 mm. long (second glume); first glume 8-10 mm. long, acuminate; second glume 12-14 mm. long, keeled, mucronate, acuminate or short-awned; lemmas 8-12 mm. long, callus about 1 mm. long and pubescent, body firm, narrow, scabrous on keel or near keel, central awn coiled at base 10-15 mm. long, laterals not coiled or very slightly, about 1/2 to 2/3 the length of the central, over 5 mm. long.---Open barren, rocky or sandy soil. New Hampshire, New York, Michigan and North Dakota to Illinois and Kansas; also in Maine. Our one specimen from Boulder County, Colorado, at 5200 feet but might be found anywhere in eastern Colorado.
2. Aristida curtissii (A. Gray) Nash, in Britt. Man. 94. 1901.
   Annual plants; culms 15-50 cm. tall, tufted, sparsely if at all branching; leaf blades 1-2 mm. wide, 4-16 cm. long, mostly involute, at least on drying, sparingly long-villous above near base (in ours); sheaths glabrous, usually loose, with a few long-villous hairs on collar; ligules short, not over 0.5 mm. long; panicle 5-10 cm. long, narrow and spikelike, branches short and erect or else spikelets sessile and partially enclosed in sheaths (ours with spikelets at very base of plant); spikelets 10-12 mm. long (second glume), few; first glume 7-8 mm. long; second glume 10-12 mm. long, both 1-nerved, awn-pointed or acute, scarious; lemmas 6-11 mm. long, glabrous except on callus, purple-blotched (in ours), lateral awns 2-4 mm. long, straight, central awn spirally coiled at base, geniculate above coil, 10-15 mm. long.---Open, dry ground, often on rocky slopes. Maryland to West Virginia, Florida; Illinois to Wyoming and Oklahoma. Our one specimen on a rocky slope just west of Fort Collins, Larimer County, at 5200 feet.
3. Aristida adscensionis L., Sp. Pl. 82. 1753.
   A. fasciculata Torr.; A. bromoides H. B. K.---Annual plants; culms 10-80 cm. tall, tufted and much branched at base, erect or prostrate-spreading, glabrous; leaf blades 2 mm. wide or less, 5-15 cm. long, varying from flat to involute, scabrous above, glabrous below; sheaths glabrous; ligules about 1 mm. long or less, 1/2 or more of hairs; panicle 5-17 cm. long, narrow and rather compact, branches appressed at first, later somewhat spreading; spikelets 8-10 mm. long; first glume 5-7 mm. long; second glume 8-10 mm. long, narrow, 1-nerved, scaberulous on keel, blunt or acute at apex; lemmas 6-10 mm. long, about as long as second glume, callus short and obtuse, scabrous on keel and rather compressed above, awns about equal, divergent, about 10-15 cm. long.---Dry open ground. Mississippi, Kansas to Texas, west to Nevada and California; also in the Old World. Our three specimens from the southeast corner of Colorado at 4200-4800 feet.
4. Aristida divaricata Humb. & Bonpl.; Willd., Enum. Pl. 1:99. 1809.
   A. humboldtiana Trin. & Rupr.---Perennial plants; culms 25-60 cm. tall, tufted, erect or prostrate-spreading, glabrous or retrorsely scaberulous; leaf blades less than 3 mm. wide often about 1 mm., 5-20 cm. long, flat or involute especially on basal leaves, scaberulous above and on margins, often pilose near base and glabrous or nearly so below; sheaths glabrous or scaberulous on upper, usually pilose near collar and ciliate; ligules 0.5-1 mm. long; panicle large, usually about 1/2 the length of the plant or more, diffuse, the branches spreading or deflexed, naked below, scabrous, as much as 10-15 cm. long, the panicle sometimes included at base; spikelets about 12 mm. long, on somewhat appressed pedicels; glumes nearly equal, 10-12 mm. long, narrow, 1- or obscurely 3-nerved, scabrous on keel, acuminate or very short-awned; lemmas about 10 mm. long, pubescent callus about 1 mm. long, body slender, narrowed to a twisted beak 2-5 mm. long, awns somewhat divergent, scabrous, 10-15 mm. long, the laterals slightly shorter.---Dry hills, plains and hillsides. Kansas to California south to Texas and Central America. Our few specimens from southeastern Colorado as far northwest as El Paso County at 4200-6500 feet.
5. Aristida arizonica Vasey, Bull. Torr. Club 13:27. 1886.
   Perennial plants; culms 30-120 cm. tall, tufted, erect, glabrous; leaf blades 1-4 mm. wide, 6-30 cm. long, flat or involute near tip or sometimes throughout, scaberulous above, glabrous below, older ones curled back or flexuous; sheaths glabrous, sometimes villous in throat; ligules about 0.5 mm. long; panicle 8-25 cm. long, narrow with appressed branches, exserted; spikelets 10-15 mm. long, appressed; glumes equal or nearly so, 1-nerved, often scabrous on keel, awn-pointed or with an awn to 2 mm. long; lemmas 10-15 mm. long including the apical twisted beak of 3-5 mm., callus pubescent, about 1 mm. long, awns about equal, ascending or spreading, 1-2 cm. long.---Dry plains, stony hillsides and open forests. Colorado, Texas, New Mexico and Arizona; also in Mexico. To be expected anywhere in southern Colorado, our one specimen from Rio Grande County at 9000 feet.
6. Aristida wrightii Nash in Small, Fl. S. E. U. S. 116. 1903.
   Perennial plants; culms 30-60 cm. tall, caespitose, erect; leaf blades about 1 mm. wide, 8-20 cm. long, involute, curved and flexuous, scabrous above and often below; sheaths glabrous, usually villous on a line outside collar; ligules about 0.5 mm. long; panicle 15-20 cm. long, narrow, branches rather distant but appressed and ascending; spikelets 12-16 mm. long; first glume 6-8 mm. long; second glume 12-16 mm. long, both 1-nerved and pointed; lemmas 10-13 mm. long, callus sharp, body glabrous below, scabrous above, awns nearly equal, divergent, about 2 cm. long (ours to 3.3 cm.).---Dry plains and hills. Texas, Colorado and Utah to California and Mexico. Our one specimen, from the U. S. National Herbarium, from Fremont County, Colorado, at 5300 feet but to be expected from the southern part of the state.
7. Aristida purpurea Nutt., Trans. Am. Phil. Soc. II.5:145. 1837.
   Perennial plants; culms 30-50 cm. tall, tufted, spreading especially on outer culms, branched at base and often above, sometimes scabrous below panicle; leaf blades 1-1.5 mm. wide, 3-10 cm. long, involute, scabrous especially above; sheaths scaberulous on lower, villous at throat; ligules less than 1 mm. long, 1/2 or more of hairs; panicle 10-20 cm. long, nodding, narrow but

loose, the branches capillary, flexuous or curved; spikelets 15-20 mm. long; first glume 6-10 mm. long, scabrous on keel; second glume about twice as long as first, not scabrous, both acuminate or with a short awn from between two teeth; lemmas 9-10 mm. long, the pubescent callus less than 1 mm. long, body scabrous near apex in lines, awns nearly equal, 3-5 or sometimes to 7 cm. long, finally spreading. Our plants apparently are not typical.---Dry hills and plains. Arkansas and Kansas to Utah and California, south to Mexico. Our three Colorado specimens from Lincoln County at 5400 feet, Baca County at 4500 feet, and Archuleta County at 6800 feet.

8. Aristida fendleriana Steud., Syn. Pl. Glum. 1:420. 1855.

Perennial plants; culms 15-30 cm. tall, tufted, often forming large bunches, erect, glabrous or glabrate; leaf blades involute-filiform, mostly less than 10 cm. long, often less than 5 cm., mostly basal and curled back, glabrous to scabrous; sheaths glabrous or scaberulous, pilose near collar with hairs 2 mm. long; ligules less than 1 mm. long; panicle 2-6 cm. long, narrow, strict, almost a raceme; spikelets 13-16 mm. long; first glume 7-8 mm. long or sometimes to 10 mm.; second glume about twice as long as first, both glumes narrow, 1-nerved, long-acute but not awned, often scabrous on keel; lemmas 9-12 mm. long, rigid, gradually narrowed to apex, body glabrous except minutely scabrous in lines on upper half, callus pubescent, 1-1.5 mm. long, awns about equal or the laterals a little shorter, 2-5 cm. long, divergent.---Dry plains and hills often in sandy soil. South Dakota to Montana south to Texas, Utah and California. Scattered over all of Colorado but apparently not as common as number 9. Altitude 3500-9500 feet.

9. Aristida longiseta Steud., Syn. Pl. Glum. 1:420. 1855.

Perennial plants; culms 15-30 cm. tall, tufted often in large bunches, erect, glabrous; leaf blades less than 1.5 mm. wide (unrolled), 2-12 cm. long, involute, scabrous above, scaberulous below, often curved and flexuous; sheaths glabrous to scaberulous, pilose near collar especially on innovations; ligules about 0.5 mm. long; panicle 5-15 cm. long, erect but not stiff, narrow, branches ascending or appressed or the lower somewhat curved, bearing 1-3 spikelets near ends; spikelets about 2 cm. or more long, pale or purplish; first glume 8-10 mm. long; second glume about twice as long as first, usually over 2 cm., both glumes 1-nerved, long-acute or mucronate but not awned at apex, somewhat scabrous on keel; lemmas 11-14 mm. long, pubescent callus about 1 mm. long, body glabrous or upper part scaberulous but not in lines, awns about equal, 6-10 cm. long, divergent, sometimes contorted near base.---Plains, foothills, mesas and rocky slopes. North Dakota to Montana, south to Texas, Arizona and Mexico. Our specimens widely distributed over Colorado but no specimens as yet from the northwest corner. Altitude 3500-8000 feet.

9A. Aristida longiseta rariflora Hitchc., (var.) Contrib. U. S. National Herb. 22:565. 1924.

Differs from species--panicle branches more flexuous, few-flowered, about 3-7 spikelets on the entire panicle; plant usually small.---Dry plains, canyons and hillsides. Texas to Colorado to Arizona. Our specimens scattered over Colorado with the species but not as abundant. Altitude 4200-6000 feet.

9B. Aristida longiseta robusta Merr., (var.) U. S. D. A. Div. Agrost. Circ. 34:5. 1901.

Differs from species--culms 30-50 cm. tall, taller and more robust; leaf blades longer; panicle longer, the branches more stiffly ascending; lemmas with shorter awns, mostly 4-7 cm. long.---Plains, mesas and hills. Minnesota, Washington to Texas and Arizona. Our specimens are from the eastern half of Colorado, rather widely distributed and fairly abundant. Altitude 3500-6800 feet.

V. Tribe Zoysieae

51. Hilaria H. B. K.   HILARIA

Sheaths split; ligules membranous; spikelets sessile in groups of 3, the groups falling from the axis entire; central spikelet fertile, 1-flowered; lateral spikelets staminate, 2-flowered (occasionally 3-flowered); glumes coriaceous, those of 3 spikelets forming a false involucre; paleas about equal or slightly shorter than lemmas.

1. Hilaria jamesii (Torr.) Benth., Journ. Linn. Soc. 19:62. 1881.

Pleuraphis jamesii Torr.---Perennial plants, scaly creeping rhizomes present; culms 15-60 cm. tall, in small tufts, erect from a decumbent rhizomelike base, branching below, glabrous or somewhat hairy at nodes; leaf blades 2-4 mm. wide, 1-15 cm. long (mostly less than 8 cm.), mostly involute, almost pungent at tip, scabrous especially below and on raised nerves, blue-green, mostly on lower part of stem; sheaths strongly nerved, scabrous on sides of nerves, margins rather thick, collar usually pilose; ligules 2-3.5 mm. long, toothed and truncate at top, nerved from sides of sheath; spike 3-10 cm. long, usually exserted, purplish or pale; spikelets about 7-9 mm. long, with long pilose hairs at base; glumes of lateral spikelets 4-5 mm. long, about 3-nerved, linear-oblong, awns from about middle on 1, above on 1, these as long or somewhat longer than glumes, hispidulous, keeled, very unequally 2-lobed at apex; glumes of central spikelets ciliate, divided at apex into 4-8 awned lobes, midnerve excurrent below apex as an awn; lemmas of central spikelet 3- to 5-nerved, awned from just below lobed apex with a short hispid awn.---Deserts, canyons, dry plains and hillsides. Wyoming and Utah to Texas and California. Widely distributed in Colorado except in central and northcentral parts (one specimen from Eagle County), as far north as Moffat County on the west, to Yuma County on the east. Altitude 3600-7500 feet (one specimen recorded from 10,805 feet).

VI. Tribe Chlorideae

52. Leptochloa Beauv. SPRANGLETOP

Annual plants; leaf blades flat or involute; sheaths split; ligules membranous; inflorescence a panicle, somewhat open; spikelets 2- to several-flowered, short-pedicelled on 1 side of rachis (but this obscure), rachilla disarticulating above the glumes and between the florets; glumes unequal, 1-nerved, keeled; lemmas awned or awnless, 3-nerved, the nerves pubescent at least below; paleas about as long as lemmas, pubescent below on keels; styles distinct, stigmas plumose.

1. Lemmas acute at tip, midnerve projecting between 2 narrow teeth as an awn, this up to 3 mm. long on most, lemmas 3-4.5 mm. long; second glume 3-4 mm. long — 1. L. fascicularis
1. Lemmas truncate or broadly obtuse at tip, midnerve projecting as a mucro, not as an awn, lemmas 2-3 mm. long; second glume 2-3 mm. long — 2. L. uninervia

1. **Leptochloa fascicularis** (Lam.) A. Gray, Man. 588. 1848.
Culms 20-100 cm. tall, tufted, erect or decumbent-spreading at base, simple or freely branched, somewhat succulent, glabrous, somewhat flattened; leaf blades 1-5 mm. wide, 10-50 cm. long (sometimes shorter), soon involute, rough on nerves and margins; sheaths glabrous or scabrous especially near summit, rather loose, often purplish; ligules 3-5 mm. long; panicle about 15-30 cm. long when mature, of many racemelike branches bearing appressed overlapping short-pedicelled spikelets (so twisted it hardly looks 1-sided); spikelets 7-10 mm. long, 6- to 12-flowered, on pedicels about 1 mm. long; first glume 2-2.5 mm. long; second glume 3-4 mm. long, both glumes 1-nerved, rather acute, scabrous on keel, somewhat scarious and often purplish near margins; lemmas 3-4.5 mm. long, acute, narrow, midnerve projecting as an awn between the bifid apex, the awn from very short on some to up to length of lemmas.---Wet ground, waste places, marshy areas and sandy river bottoms. New Hampshire to Florida and Texas; Illinois and South Dakota through Colorado and New Mexico to California and Washington; south to Argentina. Our Colorado specimens from the eastern half of Colorado (one in Delta County, one in Alamosa County) at 3500-7500 feet.

2. **Leptochloa uninervia** (Presl.) Hitchc. & Chase, Contrib. U. S. National Herb. 18:383. 1917.
Culms 20-90 cm. tall, densely tufted, erect or geniculate-decumbent at base, somewhat branched, glabrous, little compressed; leaf blades 1-4 mm. wide, 10-40 cm. long, flat or loosely involute, firm; sheaths glabrous or scabrous, little compressed; ligules 2-4 mm. long; panicle 10-30 cm. long, branches solitary or fascicled, ascending or appressed; spikelets 3-8 mm. long, 3- to 10-flowered, lead-colored, appressed and usually overlapping, on pedicels about 1 mm. long; first glume about 1.5-2 mm. long, acute; second glume 2-3 mm. long, obtuse or mucronate, both glumes scabrous on the 1 nerve; lemmas 2-3 mm. long, scarcely narrowed to the truncate apex, oblong, midnerve ending in a short mucro between 2 short teeth.---Swamps, wet ground and wet roadside ditches. Mississippi to Colorado and California; south to Argentina. Our one Colorado specimen from El Paso County at 6300 feet so to be expected in the southern part of the state.

53. Eleusine Gaertn. GOOSEGRASS

Sheaths split; ligules membranous; spikelets few- to several-flowered, compressed, the rachilla disarticulating above the glumes and between the florets.

1. **Eleusine indica** (L.) Gaertn., Fruct. et Sem. 1:8. 1788.
Annual plants; culms 15-60 cm. tall, tufted, erect or decumbent at base, flattened, glabrous; leaf blades 3-8 mm. wide, 5-25 cm. long, flat or folded, glabrous or scabrous especially above and on midnerve below; sheaths flattened and keeled, glabrous or somewhat scabrous, margin near collar may be pilose; ligules short-ciliate, 1 mm. long or less; spikes 2-8 cm. long, 2 to 10, crowded and whorled or 1 inserted lower, the spikelets sessile on 1 side of a rachis, this not prolonged beyond; spikelets 3-5 mm. long, 3-6 florets, crowded; first glume 2-2.5 mm. long, 1-nerved; second glume about 3 mm. long, 3- to 9-nerved, both glumes acute, scabrous on keel, scarious; lemmas 3-4 mm. long, 3 strong nerves close together forming a keel and another pair near margins, obtuse or nearly so, rarely mucronate, glabrous except scabrous on keel; paleas shorter than lemmas.---Waste places, fields and open ground. Massachusetts to South Dakota and Kansas; south to Florida and Texas, also in California and Oregon; a common weed in both hemispheres. Our two Colorado specimens from Garfield and Pueblo Counties at 4700-5800 feet. Also seen by the writer near Rocky Ford. To be expected anywhere near cultivated areas especially in the southeastern part of the state.

A plant resembling the above but with the rachis of the spike extending beyond the spikelets has been reported from this state. It is Dactyloctenium aegyptium (L.) Richt., but the only specimen seen was collected in 1873, and the locality data are incomplete.

54. Cynodon Rich. DOGTOOTHGRASS

Sheaths split; ligules of hairs; inflorescence of several slender spikes digitate at apex of culm; spikelets sessile in 2 rows on 1 side of the flattened rachis, numerous, imbricated, compressed, with 1 perfect floret, the second reduced to a bristle or sometimes with a small rudiment at apex, articulation above the glumes; lemmas 3-nerved, the 2 laterals near margins; paleas about as long as lemmas.

1. **Cynodon dactylon** (L.) Pers., Syn. Pl. 1:85. 1805.
Perennial plants, with creeping scaly rhizomes and long flat stolons; culms 10-30 cm. tall, flattened, glabrous, erect or ascending; leaf blades 1-4 mm. wide, 2-5 cm. long, flat, smooth but scabrous on margins and often below, villous near base, often strictly 2-ranked, especially on stolons; sheaths crowded, glabrous or villous near collar; ligules to 2 mm. long; spikes 3-8; spikelets 1.8-2.4 mm. long, purplish; glumes subequal, 1.5-2 mm. long, about 2/3-3/4 length of lemmas, narrow, acuminate, 1-nerved, scabrous on keel; lemmas length of spikelet, boat-shaped, acute, silky-hairy on keel and near margins below, otherwise glabrous.---Open ground, grassland, waste places, lawns and golf courses. Maryland to Oklahoma, south to Florida and Texas, west to California; occasionally northward. This grass has been planted several places in Colorado. Our records from Mesa, Denver, Cheyenne and Bent Counties, where it seems to persist year after year.

55. Schedonnardus Steud. TUMBLEGRASS

Sheaths split; ligules membranous; spikelets 1-flowered, the rachilla disarticulating above the glumes; styles distinct, stigmas plumose.

1. **Schedonnardus paniculatus** (Nutt.) Trel. in Bran. & Cov., Rpt. Geol. Surv. Ark. 1888:236.1891.
Annual plants; culms 20-40 cm. tall, erect or decumbent at base, scabrous or glabrous, at maturity the axis elongates and the culm turns over toward the ground, the panicle breaking away as a tumbleweed; leaf blades 1-3 mm. wide, 2-6 cm. long, flat or v-shaped, mostly basal, glabrous or scabrous below, wavy or twisted at maturity; sheaths somewhat scabrous especially near keel, flat or keeled near base; ligules 1.5-3 mm. long; panicle 1/2 or more the length of plant, axis long and usually falcate, with 3-13 slender spikes, solitary and remote, spikelets sessile and appressed on 2 sides of a 3-angled rachis--hence rather 1-sided; spikelets 3-6 mm. long; first glume 2-4 mm. long, including awn-point; second glume 3-6 mm. long, both glumes awn-pointed, narrow, rigid, hispidulous or scabrous on the 1 nerve; lemmas 3-5 mm. long, 3-nerved, narrow, rigid, acuminate, rounded on back, pubescent below, rather scabrous above; paleas narrow, about as long as lemmas.---Often sandy soil, prairies and plains. Illinois to Saskatchewan and Montana, south to Texas and Arizona; also in Argentina. Our specimens from the eastern half of Colorado (one specimen from Mesa County) at 3500-7500 feet.

56. Beckmannia Host. SLOUGHGRASS

Sheaths split; ligules membranous; spikelets 1-flowered, disarticulating below the glumes; styles distinct, stigmas plumose.

1. **Beckmannia syzigachne** (Steud.)Fernald, Rhodora 30:27. 1928.
B. erucaeformis of Manuals---Annual plants but sometimes almost appearing perennial; culms 20-100 cm. tall, erect, rather coarse, glabrous or scaberulous; leaf blades 3-10 mm. wide, 8-10 cm. long, flat, firm, scabrous; sheaths rounded but somewhat keeled and scabrous; ligules 5-8 mm. long, sometimes smaller, long-acute to acuminate; inflorescence of numerous, short, appressed or ascending spikes in a narrow more or less interrupted panicle, branches 1-5 cm. long, naked at base, the panicle 6-20 cm. long, spikelets nearly sessile in 2 rows on rachis but to 1 side; spikelets 2.8-3.5 mm. long, orbicular, flat; glumes subequal, about length of spikelet, obscurely 3-nerved, wrinkled, definitely keeled, margin scarious, apex rounded but apiculate; lemmas about length of glumes, acute, mucronate or awn-pointed, the point usually protruding beyond glumes, 3- to 5-nerved, glabrous; paleas nearly as long as lemmas.---Marshes, ditches, wet meadows and seepage areas. Manitoba to Alaska, south to Illinois, Kansas, New Mexico and California; New York, Ohio; also in Asia (the European B. erucaeformis (L.) Host. has 2-flowered spikelets). Scattered over the western two-thirds of Colorado but no specimens yet from the extreme western edge. Altitude 4500-10,500 feet apparently rare above 9500 feet.

57. Spartina Schreb. CORDGRASS

Perennial plants with stout creeping rhizomes; leaf blades flat or involute on drying; sheaths split; ligules mostly of hairs; inflorescence a panicle of 1-sided spikes; spikelets 1-flowered, much flattened laterally, sessile and closely imbricated in 2 rows but 1-sided on the rachis, disarticulating below the glumes, rachilla not prolonged beyond the palea; glumes keeled, 1-nerved, acute or short-awned; lemmas keeled with obscure lateral nerves, firm, awnless; paleas similar to lemmas in texture and length, sometimes longer; styles elongate, stigmas threadlike, papillose or short-plumose.

1. Second glume awnless; leaves 2-5 mm. wide, usually scabrous above; spikes 1-5 cm. long; culms less than 100 cm. tall
-1. S. gracilis
1. Second glume awned on most spikelets, this awn 1-7 mm. long; leaves 5-12 mm. wide, seldom scabrous above; spikes 4-10 cm. long; culms usually over 100 cm. tall
-2. S. pectinata

1. **Spartina gracilis** Trin., Mem. Acad. St. Petersb. VI. Sci. Nat. 4(1):110. 1840.
Culms 30-100 cm. tall, usually solitary, erect, glabrous; leaf blades 2-5 mm. wide, 15-20 cm. long, flat or becoming involute, glabrous below, very scabrous above; sheaths glabrous; ligules about 1 mm. long; spikes 4-8, 1-5 cm. long, closely appressed, usually naked at base, the panicle 6-20 cm. long, spikelets on a hispidulous-ciliate rachis; spikelets 6-8 mm. long; first glume about 1/2 as long as lemmas, narrow, acute or obtuse; second glume a little longer than lemmas, acute or somewhat aristate but not awned, both glumes rather firm, ciliate on keel; lemmas 5-7.5 mm. long, flat, apex rather truncate, hispid-ciliate near apex on keel.---Alkali

meadows and plains. Saskatchewan to British Columbia, south to Colorado, Arizona and to Washington. Our specimens from the western three-fourths of Colorado except in the northwestern and southwestern corners. However, to be expected anywhere in the state. Altitude 4500-9000 feet.
2. **Spartina pectinata** Link, Jahr. Gewachsk. 1(3):92. 1820.
S. cynosuroides of Manuals---Culms 100-200 cm. tall, stout, erect, glabrous; leaf blades 6-12 mm. wide, 20-60 cm. long, flat but involute on drying, margins very scabrous, glabrous or somewhat scabrous above; sheaths glabrous; ligules 1-3 mm. long; spikes 5-30, appressed or ascending, naked below, the lower 4-10 cm. long, whole panicle 15-30 cm. long; spikelets 7-10 mm. long; first glume as long as lemmas or sometimes to half as long on lower spikelets, acuminate or short-awned; second glume length of spikelet, 12-15 mm. long including a scabrous awn 1-7 mm. long, keel may be of several nerves crowded, both glumes hispid-ciliate on keel; lemmas 7-9 mm. long, glabrous except the hispidulous keel, apex rather 2-lobed.---Marshes, sloughs and wet meadows. Newfoundland and Quebec to Washington, south to North Carolina, Kentucky, Illinois, Arkansas, Texas and New Mexico; also along the eastern coast. Our Colorado specimens from the eastern half of the state at 3500-7000 feet.

58. Chloris Sw. CHLORIS; WINDMILL GRASS; FINGERGRASS

Annual or perennial plants, no rhizomes; leaf blades flat or folded; sheaths split, flattened and keeled; ligules membranous, short-ciliate; inflorescence of 7-16 spikes, these somewhat whorled but tending to cluster in digitate fashion at end of culm; spikelets in 2 one-sided rows, 2 florets (or rarely 3) present, the lower fertile and the upper staminate or rudimentary, rachilla disarticulating above the glumes; glumes unequal; lemmas 3- to 5-nerved, these pubescent, apex bifid, awned from just below apex, sterile lemmas truncate at apex; paleas about as long as lemmas; styles distinct, stigmas plumose.

1. Lemmas merely short-pubescent on the nerves, the spike not at all feathery; perennial plants
-1. C. verticillata
1. Lower lemmas conspicuously long-ciliate near the lateral nerves above, the hairs usually over 2 mm. long, the spikes rather feathery; annual plants
-2. C. virgata

1. **Chloris verticillata** Nutt., Trans. Am. Phil. Soc. II.5:150. 1837.
Perennial plants, no rhizomes; culms 10-40 cm. tall, tufted, erect or sometimes decumbent at base and sometimes rooting at nodes; leaf blades 1-4 mm. wide, 3-7 cm. long, folded or flat, margins scabrous, surfaces scabrous or glabrous, light green; sheaths glabrous, margins hyaline, often pilose near collar: ligules less than 1 mm. long; spikes 8-13 in number, 5-11 cm. long, in several whorls when mature and spreading widely, the axis and rachis pubescent, finally breaking away forming a "tumbleweed"; spikelets 2.5-3 mm. long, flattened, 1 or more rudimentary truncate floret present; first glume about 2 mm. long; second glume about 3 mm. long, both glumes 1-nerved, narrow, scabrous on keel, acuminate or awn-pointed; lower lemmas 2-3 mm. long, 3-nerved, these pubescent, obtuse, awn 2-8 mm. long.---Plains, roadsides and waste ground. Missouri to Colorado, south to Louisiana and New Mexico; introduced in Maryland, Illinois, Indiana and California. Our specimens all from southeastern and eastern Colorado at 3500-6000 feet.
2. **Chloris virgata** Sw., Fl. Ind. Occ. 203. 1797.
Annual plants; culms 25-60 cm. tall, erect or decumbent at base, sometimes branching and rooting at nodes; leaf blades 2-7 mm. wide, 4-25 cm. long, flat, margins scabrous, scaberulous both surfaces to nearly smooth or often pilose above near base; sheaths glabrous or with a few hairs near throat; ligules 1 mm. long or less; spikes 7-16 in number, 2-8 cm. long, aggregated, sometimes included, sessile, erect or somewhat spreading, pale green or tawny, silky; spikelets 2.5-4 mm. long, flattened, 2-flowered the upper staminate or sterile (rarely with 2 perfect and 1 rudimentary present); first glume 1.5-2.5 mm. long; second glume 3-3.5 mm. long, awn-pointed, both glumes 1-nerved, narrow, scabrous on keel; lower lemmas 3-3.5 mm. long, 3- to 5-nerved, long-ciliate on upper part near lateral nerves, the hairs 2-4 mm. long, short-ciliate below, callus bearded, awn 5-10 mm. long.---Open ground, roadsides, fields and waste places. Nebraska to Texas and California; Maine and Massachusetts and a few other eastern states; also in tropical America. Our one Colorado specimen from Prowers County in southeastern Colorado at 3400 feet.

59. Bouteloua Lag. GRAMA

Annual or perennial plants, seldom with rhizomes; leaf blades flat to involute; sheaths split; ligules usually composed of hairs for 1/2 the length or more; inflorescence of 1-sided spikes, these 1 to many; spikelets 1-flowered but 1 or more rudimentary florets above, sessile in 2 rows on 1 side of a rachis, disarticulating at the base of the spike (in 1 species) and above the glumes in the rest; glumes at least somewhat unequal, 1-nerved; lemmas as long or longer than glumes, 3-nerved, the nerves extending into short awns or mucros; paleas about as long as lemmas; styles distinct, stigmas plumose.

1. Over 10 spikes to a panicle; spikelets less than 12 to 1 spike; articulation below the spike
-1. B. curtipendula
1. Less than 10 spikes present to a panicle; spikelets 12 or more on 1 spike; articulation above the glumes
    2. Rachis of spike prolonged beyond spikelets, usually for 5 mm. or more; leaf margins papillose-hairy near base; rachilla below rudiment glabrous or short-pubescent; second glume very conspicuously dark-tuberculate
-2. B. hirsuta
    2. Rachis of spike little if any prolonged beyond the spikelets; leaf margins not papillose (rarely sparsely so in number 4); rachilla pilose-tufted below rudiment; second glume not tuberculate (or sparsely so in number 4)

3. Culms felty-pubescent; spikes 3-8 to a culm — -3. B. eriopoda
3. Culms glabrous or nearly so; spikes usually 1 or 2 (except in number 6)
    4. Perennials, often with rhizomes; spikes usually over 25 mm. long, often 2 to a culm (may be 1 - 3); very common in Colorado — -4. B. gracilis
    4. Annuals; spikes usually less than 25 mm. long, either 1 or over 3 to a culm; only locally common, usually in southern Colorado
        5. Spikes single; spikelets 4-6 mm. long (including awns) — -5. B. simplex
        5. Spikes usually 4 or more to a culm; spikelets 3-4 mm. long (including awns) — -6. B. barbata

1. Bouteloua curtipendula (Michx.) Torr. in Emory, Notes Mil. Rec. 154. 1848.
    Antheropogon curtipendula (Michx.) Fourn.---Perennial plants with scaly rhizomes, these rather short; culms 25-80 cm. tall, tufted, erect, glabrous; leaf blades 2-4 mm. wide, 5-20 cm. long, flat or involute especially near ends and widest at middle, scabrous on margins and upper surface, often papillose-pilose near base; sheaths glabrous or papillose-pilose, collars pilose near margins; ligules membranous with short-ciliate margins, about 0.5 mm. long; panicle 5-30 cm. long, of many (usually over 20) spikes 1-sided, on short peduncles, spreading or pendulous, often twisted and falling entire; spikelets 5-10 mm., divergent, 2-12 on a spike; first glume 3-5 mm. long; second glume 5-8 mm. long, both glumes narrow, scabrous on the 1 nerve, acute or awn-pointed; lower lemmas 5-8 mm. long, narrow, smooth or somewhat scabrous on the nerves, awns short; rudiment reduced to a small scale with 1 awn or almost a full-formed lemma with 2 awns, sometimes enclosing a second rudiment.---Plains, hillsides, prairies, mesas and valleys. Maine and Ontario to Montana, south to Maryland, Alabama, Texas, Arizona and California; introduced in South Carolina. Widely distributed over the eastern half of Colorado and in the extreme south extending west to La Plata County. Altitude 3500-7500 feet.

2. Bouteloua hirsuta Lag., Var. Cienc. 2(4):141. 1805.
    Perennial plants, rhizomes absent or rarely present, often sod-forming; culms 10-50 cm. tall, tufted, erect or geniculate at base, glabrous; leaf blades 1-2 mm. wide, 2-12 cm. long, flat or involute, margins papillose-hairy near base, rather flexuous; sheaths glabrous but collars papillose-hairy at or near margins; ligules almost completely of hairs about 0.5 mm. long or less; panicle with 1-4 spikes, these 1-5 cm. long, the rachis prolonged for 5-8 mm. beyond the spikelets; spikelets 5-7 mm. long, many to a spike; first glume 2-3 mm. long, narrow, acuminate, hyaline, 1-nerved; second glume 3-5 mm. long, broader, acuminate or short-awned, rather conspicuously tuberculate-hispid on back near 1 nerve, these tubercles blackish; lower lemmas 5-6 mm. long, 3-cleft to near middle and awned from lobes, pilose on margins and midnerve below, callus sparingly pubescent; rudiment with broad lobes and 3 awns, rachilla glabrous or nearly so (in ours).---Plains, rocky hills, mesas and open ground. Wisconsin and South Dakota to Texas, Colorado, Arizona and California; also in Mexico and Florida. Our specimens from the eastern half of Colorado at 3500-6500 feet. Not as widespread or as common as B. gracilis.

3. Bouteloua eriopoda (Torr.) Torr., Pacif. R.R. Rep. 4:155. 1857.
    Perennial plants, no rhizomes but plants sparsely stoloniferous; culms 20-60 cm. tall, tufted, geniculate, branching and straggling, felty-pubescent, wiry from a rather hard base; leaf blades 1-2 mm. wide, 3-15 cm. long, flat or involute, glabrous; sheaths glabrous or the lower felty-pubescent; ligules about 1/2 or more of hairs, less than 1 mm. long; panicle 5-15 cm. long, with 3-8 spikes 1.5-3.5 cm. long, these erect, spreading, somewhat curved, on short hairy peduncles; spikelets about 6-10 mm. long, 12-20 to a rachis; first glume 2.5-3.5 mm. long; second glume 5-7 mm. long, both glumes keeled, narrow, acuminate; lower lemmas 5-8 mm. long, sometimes puberulent on nerves, narrow, mid-awn 1-2 mm. long, laterals shorter; rudiment of 3 awns 4-5 mm. long, united at base on a stipe which is pubescent in tufts.---Mesas, hills and dry open ground. Texas to Utah and Mexico. Our specimens from the southeastern corner of Colorado at 4400-4800 feet but might be expected in the southwestern corner also.

4. Bouteloua gracilis (H. B. K.) Lag.; Steud., Nom. Bot. ed. 2. 1:219. 1840.
    B. oligostachya (Nutt.) Torr.---Perennial plants, often with rhizomes and forming a sod; culms 15-60 cm. tall, densely tufted, erect but sometimes geniculate at nodes, glabrous; leaf blades 1-2 mm. wide, 3-15 cm. long, flat or involute, usually glabrous but sometimes sparsely pilose or papillose-pilose above, often scabrous; sheaths glabrous but collar pilose on margins, the hairs usually papillose; ligules mostly of hairs less than 1 mm. long; spikes 1-3, usually 2, each 2-5 cm. long, straight or arcuate, the rachis not prolonged beyond spikelets; spikelets about 6 mm. long including awn; first glume 2.5-3.5 mm. long, scabrous on keel; second glume 5-6 mm. long, often papillose-hispid on keel, sometimes strongly so, both glumes 1-nerved, awn-pointed or acuminate, persistent, pale or purplish; lower lemmas about 6 mm. long, awns about 1 mm. long or less, nerves appressed-pilose especially the middle one, pilose at base; rudiment of 3 awns and lobes, the awns about 3 mm. long, sometimes additional rudiments above, the rachilla tufted-pilose.---Plains, mesas, prairies and sand hills. Wisconsin to Manitoba and Alberta, south to Missouri, Texas and California; also in Mexico. Widely distributed over Colorado but no specimens yet from the northwestern corner. Altitude 3500-10,500 feet.

5. Bouteloua simplex Lag., Var. Cienc. 2(4):141. 1805.
    B. procumbens (Durand) Griffiths; B. prostrata Lag.---Annual plants; culms 5-20 cm. tall, tufted often densely so, branching, decumbent-spreading, glabrous; leaf blades 0.5-1.5 mm. wide, 1-5 cm. long, flat or involute, slender-pointed, glabrous, scaberulent or somewhat pubescent above; sheaths glabrous or lower pubescent, often pilose near collar; ligules 1/2 or more of hairs, about 0.5 mm. long; spike 1, 1-3 cm. long, recurved to end and not prolonged; spikelets 4-6 mm. long, including awns, many to a spike; first glume 2-3 mm. long, narrow, glabrous; second glume 3.5-5 mm. long, broader, scabrous on nerve, both glumes 1-nerved, scarious, acute; lower lemmas 3-5 mm. long, broad, central awn about 1 mm. long, appressed-pilose on each side of keel, and at base especially on nerves; rudiment of 3 awns, these scabrous and thick, the rachilla tufted-pilose.---Open ground of plains and hills. Texas to Colorado, Utah, Arizona and Mexico; also in South America. Our specimens thinly scattered over the eastern half of the state and west in the extreme south to La Plata County. Altitude 4000-8500 feet.

6. <u>Bouteloua barbata</u> Lag., Var. Cienc. 2(4):141. 1805.
   <u>Annual plants</u>; culms 5-30 cm. tall, tufted, erect from a geniculate base or decumbent-spreading, slender, branching; leaf blades 1-3 mm. wide, 1-6 cm. long, flat, scabrous above, glabrous below; sheaths glabrous, short; ligules 1/2 or more of hairs, less than 1 mm. long; spikes 4-7 (sometimes 3) each 1-3 cm. long, distant, short-peduncled, ascending or spreading, rather curved; spikelets 3-4 mm. long, including awns, 25-40 to a spike; first glume 1.5-2 mm. long; second glume 2.5-3 mm. long, both glumes with a short awn between 2 teeth or acuminate, hyaline, scabrous on 1 nerve, often purplish; lower lemmas 2-4 mm. long, densely villous on margin and often on back, divaricate awns about as long as lemmas; rudiment with 3 divergent awns, lobed between, rachilla tufted-villous.---Open ground, mesas, rocky hills and waste places. Texas, New Mexico, Utah, Arizona and Mexico. Our two specimens from southern Colorado in Otero County at 4250 feet and Montezuma County at 6500 feet.

60. <u>Munroa</u> Torr. FALSE BUFFALOGRASS

Sheaths split; ligules of hairs; spikelets in 2's or 3's, the lower 1 or 2 larger, 2- to 4-flowered, the upper 2- to 3-flowered, the group enclosed in broad sheaths forming a cluster or head, overtopped by the leaf blades, rachilla disarticulating above the glumes and between the florets; styles elongate, stigmas barbellate or short-plumose.

1. <u>Munroa squarrosa</u> (Nutt.) Torr., U. S. Rept. Expl. Miss. Pacif. 4(5):158. 1857.
   <u>Annual plants</u>; culms up to 10 cm. long, prostrate or spreading, branching freely, often rooting at the nodes, forming mats, the leaves fascicled and the internodes scabrous; leaf blades 1-3 mm. wide, 1-3 cm. long, stiff and pungent, glabrous or somewhat scabrous below, margin weakly scabrous with a band of white; sheaths crowded, somewhat keeled, margins ciliate, collar pilose on margins; ligules 0.5-1 mm. long; spikes short, usually about 3 together; spikelets about 8-10 mm. on longest; glumes subequal, 3-5 mm. long, somewhat shorter than lemmas, 1-nerved, acute or awn-pointed, scabrous on keel, rest scarious; lemmas about 3-5 mm. long, 3-nerved, coriaceous on lower spikelets, a tuft of hair near middle on each side, acuminate or awn-pointed (to 1 mm. long), the end spreading, lemmas of upper spikelets membranous. Some plants are woolly-hairy from the old egg cases of aphids. The variety floccuosa Vasey was described from such a specimen.---Open ground, plains, hillsides and sandy flats, often in ground recently disturbed. Alberta and North Dakota to Montana, south to Texas and Arizona. Our specimens widely scattered over all of Colorado except the extreme western part. Altitude 3500-9000 feet rarely to 10,000 feet.

61. <u>Buchloe</u> Engelm. BUFFALOGRASS

Sheaths split; ligules of hairs; spikelets dioecious or rarely monoecious, the staminate 2-flowered, pistillate 1-flowered, disarticulating below a cluster of 4 or 5 spikelets.

1. <u>Buchloe dactyloides</u> (Nutt.) Engelm., Trans. Acad. St. Louis 1:432. 1859.
   <u>Bulbilis dactyloides</u> (Nutt.) Raf.---Perennial plants, rhizomes lacking but stolons present forming a dense sod, staminate and pistillate types dissimilar; culms commonly 10-20 cm. long, pistillate shorter, in small tufts or large patches, erect or decumbent at base, stolons 5-60 cm. long, often rooting at nodes with the leaves in clusters; leaf blades 1-3 mm. wide, 4-10 cm. long, flat, usually curled, sparsely pilose on margins, the hairs usually papillose; sheaths loose, glabrous, collar pilose on margin; ligules less than 1 mm. long; (staminate plant)-spikes 1 or 2 or rarely 3-4, each 5-15 mm. long, spreading, nearly or quite sessile, with about 10 spikelets, the spikes on a slender culm 5-20 cm. tall; spikelets about 4 mm. long, 2-flowered in two 1-sided rows on the rachis; first glume about 2 mm. long; second glume about 3 mm. long, both glumes rather broad, 1-nerved, obtuse, whitish; lemmas 4-5 mm. long, 3-nerved, awnless; paleas like lemmas in length and texture; (pistillate plant)-spikes 1 or 2 in a head, enclosed in the sheaths and overtopped by the leaf blades, each spike with 4-5 spikelets; spikelets in a cluster, this falling entire, 3-4 mm. thick and about 7 mm. long, the indurated rachis and second glumes forming a sort of involucre; first glume inside, narrow, mucronate, minute or well-developed; second glume rigid, rounded on back, expanded above, apex with 3 rigid, narrow lobes; lemmas indurate, obscurely nerved, 3-toothed, central tooth longest, 3-nerved; paleas about as long as lemmas.---Dry plains, prairies and hills. Minnesota to Montana south to Iowa, Texas, Louisiana; Arizona and Mexico. Our specimens widely distributed over the eastern half of Colorado at 3500-6000 feet.

VII. Tribe <u>Phalarideae</u>

62. <u>Hierochloe</u> R. Br. SWEETGRASS

Sheaths split; ligules membranous; spikelets 3-flowered, the 2 laterals staminate, the center perfect, disarticulating above the glumes, the 3 florets falling together; styles distinct, stigmas plumose.

1. <u>Hierochloe odorata</u> (L.) Beauv., Ess. Agrost. 62, 164. 1812.
   <u>Savastana odorata</u> (L.) Scribn.; Torresia odorata (L.) Hitchc.---Perennial plants with slender creeping rhizomes; culms 20-60 cm. tall, erect, glabrous; leaf blades 2-5 mm. wide, 1-3 cm. long on few culm leaves, 10-20 cm. long on numerous basal leaves, flat, scabrous on margins; sheaths glabrous; ligules about 2 mm. long; panicle 4-12 cm. long, pyramidal, from compact to loose; spikelets 4-6 mm. long, golden-brown, mostly on short pedicels; first glume 3-5 mm. long, 1-nerved

or faintly 3-nerved; second glume 4-6 mm. long, 3-nerved the laterals short and faint, both glumes subequal, hyaline, glabrous, acute or erose at apex; lemmas of lateral florets 3-5 mm. long, broad, scaberulous on back, pubescent at margins and near apex, somewhat mucronate, faintly 5-nerved, rather indurated; lemmas of central florets 3-5 mm. long, pubescent only near apex, nerves very faint but probably 5, indurate; paleas about as long as lemmas. This grass is more or less fragrant on drying.---Meadows, bogs and moist places. Labrador to Alaska, south to New Jersey, Indiana, Iowa, Oregon and in New Mexico and Arizona; also in Eurasia. Our Colorado specimens from the northcentral, central and southcentral parts of the state in the mountainous areas. Altitude 7500-11,500 feet.

63. Anthoxanthum L. VERNAL GRASS

Sheaths split, short auriclelike structures sometimes present; ligules membranous; spikelets with 3 florets, middle perfect, 2 laterals reduced to sterile lemmas, rachilla disarticulating above the glumes, the 3 florets falling together; styles distinct, stigmas elongated, plumose.

1. Anthoxanthum odoratum L., Sp. Pl. 28. 1753.
Perennial plants, no rhizomes; culms 30-60 cm. tall, tufted, erect, glabrous; leaf blades 2-6 mm. wide, 2-15 cm. long, flat, short-pilose to glabrate; sheaths glabrous; ligules 1-5 mm. long; panicle 3-8 cm. long, spikelike, loosely cylindrical, acute at apex, soon brownish-yellow; spikelets 7-10 mm. long, somewhat compressed laterally; first glume about 1/2 as long as spikelet, with 1 green nerve, the rest mostly scarious, long-acute to acuminate, long-pubescent or sparingly pilose especially on keel; second glume length of spikelet, 3-nerved, margin widely scarious, sparingly short-pilose, acuminate; first sterile lemma 2-3 mm. long, 2-lobed at apex, appressed-pilose with golden hair, awned about 1/3 down from apex with a straight, untwisted awn 2-4 mm. long; second sterile lemma like first but awned from near base, the awn geniculate and twisted near base, somewhat longer than spikelet; lemma of fertile floret about 2 mm. long, shorter than sterile lemmas, smooth and shining, brown, coriaceous; paleas like lemmas in length and texture and enclosed by edges of lemmas. This grass is more or less fragrant on drying.---Meadows, pastures and waste places, rarely cultivated in gardens. Greenland and Newfoundland to Louisiana and Michigan, and on the Pacific Coast from British Columbia to California; introduced from Eurasia. Our two Colorado collections were from El Paso County at 6000-8000 feet and may have been from plants growing in a garden.

64. Phalaris L. CANARYGRASS

Annual or perennial plants with rhizomes; leaf blades flat; sheaths split; ligules membranous; inflorescence a narrow or a spikelike panicle; spikelets laterally compressed with 1 perfect terminal (central) floret and 1 or 2 sterile lemmas on sides, these not conspicuous, rachilla disarticulating above the glumes, all the florets falling attached; glumes equal or nearly so, keeled and often winged on the keel, longer than lemmas; lemmas coriaceous; paleas similar to lemmas in length and texture and enclosed by its sides.

1. Rhizomes present; culms over 60 cm. tall; panicle narrow but not spikelike
  2. Leaves normal   -1. P. arundinacea
  2. Leaves striped with white   -1A. P. arundinacea picta
1. Annual plants, no rhizomes; culms less than 60 cm. tall; panicle dense and spikelike
  3. Glumes 7-9 mm. long, broadly winged above; fertile lemmas 4-6 mm. long; sterile lemmas lanceolate
    -2. P. canariensis
  3. Glumes 4-6 mm. long, narrowly winged if at all; fertile lemmas less than 4 mm. long; sterile lemmas linear
    4. Only 1 sterile lemma present on a spikelet; panicle rounded both ends; glumes moderately winged
     -3. P. minor
    4. Two sterile lemmas present; panicle tapering somewhat both ends; glumes wingless or very nearly so
     -4. P. caroliniana

1. Phalaris arundinacea L., Sp. Pl. 55. 1753.
Perennial plants with stout, creeping rhizomes; culms 60-150 cm. tall, erect, stout, glabrous; leaf blades 6-16 mm. wide, 10-30 cm. long, flat, glabrous or weakly scabrous on margins especially near long-elongated apex; sheaths glabrous or scaberulous; ligules 2-5 mm. long; panicle 7-16 cm. long, narrow and dense but not spikelike (spreads some in anthesis); spikelets 4-6 mm. long, flattened, crowded and pale, sterile lemmas 2; glumes subequal, 4-6 mm. long, rather narrow, 3-nerved, acute, flattened, the keel scabrous and very narrowly winged; fertile lemmas 3-4 mm. long, coriaceous and shining, pale or gray-brown, with a few appressed hairs, indistinctly nerved; sterile lemmas about 1 mm. long or less, very narrow and inconspicuous, hairy.---Marshes, river banks, wet meadows and along ditches. New Brunswick to Alaska, south to North Carolina, Kentucky, Oklahoma, New Mexico, Arizona and California; also in Eurasia. Our specimens scattered over the western two-thirds of Colorado at 4500-9000 feet.
1A. Phalaris arundinacea picta L., (var.) Sp. Pl. 55. 1753.
Leaf blades striped with white.---Grown in gardens and lawns for ornament. Commonly planted in Colorado.
2. Phalaris canariensis L., Sp. Pl. 54. 1753.
Annual plants; culms 30-60 cm. tall, erect or sometimes spreading from base, glabrous or scaberulous; leaf blades 4-10 mm. wide, 4-20 cm. long, flat, scabrous; sheaths scaberulous or scabrous, the upper inflated; ligules 3-6 mm. long; panicle 1-4 cm. long, dense and spikelike, ovate to oblong-ovate; spikelets 7-9 mm. long, flattened, with 2 sterile lemmas; glumes subequal, length of spikelet, 2-3 mm. wide, with 3 green nerves and whitened between, abruptly

pointed, glabrous or sparingly pubescent, scabrous keel prominently winged upward; fertile lemmas 4-6 mm. long, obscurely nerved, coriaceous, acute, appressed pubescent; sterile lemmas about 1/2 as long as fertile, lanceolate, glabrous or sparingly hairy especially near apex. This species produces the commercial "canary seed."---Waste places and gardens. Nova Scotia to Alaska, south to Virginia, Kansas, Wyoming, California and occasionally southward. Our specimens scattered over Colorado mostly in the thicker populated areas at 4000-7200 feet.

3. Phalaris minor Retz., Obs. 3:8. 1783.

Annual plants; culms 30-60 cm. tall, tufted, erect or sometimes spreading, glabrous; leaf blades 3-6 mm. wide, 5-30 cm. long, flat, scabrous; sheaths smooth or scabrous, upper often inflated; ligules 2-5 mm. long; panicle 2-5 cm. long, dense and spikelike, ovate to cylindrical, spikelets crowded, imbricated; spikelets 4-6 mm. long, flattened, pale with green stripes, 1 sterile floret present; glumes subequal, length of spikelet, first may be narrower, 3-nerved, rather scarious between nerves, rather abruptly acute, scabrous on the moderately winged keel; fertile lemmas about 3 mm. long, obscurely 3- to 5-nerved, coriaceous, appressed-pubescent (sparse on ours) acute; sterile lemma 1/3 to 1/2 as long as the fertile, narrow, pubescent at apex.---Fields and waste places. New Brunswick to New Jersey, Louisiana, Texas, Colorado; Oregon and California; introduced from the Mediterranean region. Our one Colorado specimen from Larimer County at 5100 feet but to be expected anywhere around cultivated areas.

4. Phalaris caroliniana Walt., Fl. Car. 74. 1788.

Annual plants (reported as perennial by Silveus); culms 25-60 cm. tall, tufted, erect or occasionally decumbent at base, glabrous or scabrous; leaf blades 2-10 mm. wide, 5-20 cm. long, flat, smooth or slightly scabrous, sometimes glaucous, cauline leaves reduced upward; sheaths glabrous, the upper somewhat inflated; ligules 2-6 mm. long; panicle 2-6 cm. long, dense and spikelike, oblong, usually tapering at both ends; spikelets 4-6 mm. long, flattened, with 2 sterile lemmas; glumes subequal, length of spikelet, with 3 green prominent nerves, rather abruptly narrowed to an acute tip, very narrowly winged on keel; fertile lemmas 3-4 mm. long, nerves obscure, coriaceous and shining, lanceolate-ovate, acuminate, pubescent with long, appressed hairs; sterile lemmas 1.2-1.7 mm. long, narrow, hairy, closely appressed to fertile floret.--- Old fields, sandy soil and moist or wet ground. Virginia, Colorado, south to Florida and Texas, west to Arizona, California and Oregon. Our few specimens from southeastern Colorado at 5000 feet.

VIII. Tribe Oryzeae

65. Leersia Sw. CUTGRASS

Sheaths split; ligules membranous; spikelets 1-flowered, strongly compressed laterally, disarticulating from the pedicels, the glumes absent; styles distinct, stigmas plumose with branched hairs.

1. Leersia oryzoides (L.) Swartz., Prodr. Veg. Ind. Occ. 21. 1788.

Homolocenchrus oryzoides (L.) Poll.---Perennial plants with slender creeping rhizomes; culms 50-150 cm. tall, erect or decumbent at base, often rooting at nodes, branching, glabrous except pilose-hispid at nodes; leaf blades 4-12 mm. wide, 6-25 cm. long, flat, very scabrous with small prickles on margins, sometimes pubescent above near base; sheaths retrorsely-scabrous; ligules about 1 mm. long; panicle 10-20 cm. wide, branches flexuous and spreading at maturity, bearing short racemes of imbricated spikelets, lateral included panicles often present; spikelets 4-5 mm. long, 1.5-2 mm. wide, elliptic or oblong-elliptic, very flat on very short pedicels; glumes wanting; lemmas length of spikelet, 3-nerved, very flat, abruptly tipped, short-hispidulous especially on keel; paleas about as long as lemmas and similar in pubescence, the keel especially hispid.---Marshes, riverbanks, ponds, lakes and wet places in general. Quebec and Maine to Washington, south to Florida, Texas, New Mexico, Arizona and California; also in Europe. Our rather few specimens scattered over Colorado, mostly in the northern two-thirds at 4500-5500 feet.

IX. Tribe Zizanieae

66. Zizania L. WILDRICE

Sheaths split; ligules membranous; spikelets unisexual, both on the same inflorescence, 1-flowered, disarticulating from the pedicels, the glumes wanting.

1. Zizania aquatica L., Sp. Pl. 991. 1753.

Annual plants; culms less than 150 cm. tall, thick and robust, striate, glabrous except on nodes; leaf blades 6-12 mm. wide, 20-60 cm. long, flat, scabrous especially on margins; sheaths glabrous but the collar puberulent; ligules 8-15 mm. long, acute; panicle 30-60 cm. long, lower branches staminate and ascending or spreading, upper branches pistillate and ascending at maturity; pistillate spikelets 10-20 mm. long, terete, linear; staminate spikelet 5-10 mm. long, lanceolate; glumes obsolete, represented by a collarlike ridge below the lemmas; pistillate lemmas length of spikelet, 3-nerved, chartaceous, somewhat scabrous, tapering to a scabrous awn 3-7 cm. long; staminate lemmas 5-nerved, membranous, pubescent near apex, acuminate or awn-pointed or with an awn up to 6.5 mm. long (in ours); pistillate paleas 2-nerved, clasped by edges of lemma; staminate paleas 3-nerved, about as long as lemmas. Our plant is var. angustifolia Hitchc.---Marshes and borders of streams or ponds, usually in shallow water. Quebec and New Brunswick to North Dakota, south to New York and Nebraska. Our one specimen from El Paso County, Colorado. The grass was sown some years ago in a pond to provide food for wild fowls and the collection was made the first year. The wildrice has not persisted in the area.

X. Tribe Paniceae

67. Trichachne Nees. COTTONTOP

Sheaths split; ligules membranous; spikelets in pairs, short-pedicelled, in 2 rows along 1 side of a **slender** rachis, copiously silky, 1-flowered with a sterile lemma, articulation below the glumes.

1. Trichachne californica (Benth.) Chase, Journ. Wash. Acad. Sc. 23:455. 1933.
   T. saccharata (Buckl.) Nash; Valota saccharata (Buckl.) Chase---Perennial plants, no creeping rhizomes (in ours); culms 30-80 cm. tall, glabrous, erect from a knotty, felty-pubescent base, freely branching; leaf blades 2-5 mm. wide, 5-25 cm. long, flat or soon involute, glabrous or densely puberulent; sheaths keeled, the upper glabrous to sparsely hairy especially near collar, lower densely villous; ligules 2-4 mm. long, erose at apex; panicle 8-20 cm. long, narrow, erect or finally nodding, branches appressed, silvery-white or purplish from hairs; spikelets 3-4 mm. long, 1 of the pair very short-pedicelled, the hairs 2-3 mm. long; first glume minute, less than 0.7 mm. long, sometimes obsolete, glabrous; second glume almost as long as spikelet, copiously hairy especially on the margins; sterile lemmas slightly longer than second glume, 5-nerved, almost glabrous on back but long-villous near margins; fertile lemmas slightly striate, lanceolate to ovate, abruptly narrowed to a short membranous tip, margins hyaline in a narrow band, rest of lemma cartilaginous and brown at maturity; paleas similar to lemmas in color and texture.---Plains, hillsides and dry open ground. Texas to Colorado, Arizona and Mexico. Our few records from southeastern Colorado at 5500-5800 feet. Should be looked for in the southern part of the state.

68. Digitaria Heist. CRABGRASS; FINGERGRASS

Annual plants; culms decumbent at base often rooting at nodes; leaf blades flat; sheaths open; ligules membranous; inflorescence of several (2-10) spikelike racemes in whorls or approximate at the ends of culms; **spikelets** in 2 rows, 1-sided on a winged rachis, in pairs, 1 sessile or nearly so, 1 short-pedicelled, with 1 perfect floret but with a sterile lemma present, disarticulating below the glumes; first glume minute or wanting; second glume from 1/2 to as long as spikelet; fertile lemmas cartilaginous with pale-hyaline margins; paleas similar to lemmas in length and texture, enfolded by its edges; styles distinct, stigmas plumose.

1. Second glume about 1/2 as long as spikelet, the first small but present; spikelets 2.5 mm. long or more; sheaths papillose-hairy at least on lower; pedicels sharply 3-angled or winged   -1. D. sanguinalis
1. Second glume as long as spikelet, the first usually absent entirely; spikelets about 2 mm. long; sheaths glabrous or rarely a few hairs on lower; pedicels terete or somewhat 3-angled   -2. D. ischaemum

1. Digitaria sanguinalis (L.) Scop., Fl. Carn. ed. 2, 1:52. 1772.
   Syntherisma sanguinalis (L.) Dulac.---Culms 20-100 cm. tall, erect or ascending from a decumbent base, rooting at nodes, freely branching; leaf blades 4-10 mm. wide, 4-15 cm. long, occasionally purplish, pilose both sides especially near base, the hairs often papillose; sheaths somewhat keeled above, pilose, usually papillose-pilose especially on lower ones and on collars; ligules 1-3 mm. long; inflorescence of 2-10 racemes each 5-18 cm. long, the rachis winged with wings wider than midrib; spikelets about 2.5-3.5 mm. long, elliptic-lanceolate, somewhat planoconvex, on 3-angled pedicels, the angles somewhat winged; first glume minute, triangular, glabrous; second glume about 1/2 as long as spikelet, 3- to 5-nerved, ovate-lanceolate, appressed-ciliate; sterile lemmas length of spikelet, 5- to 7-nerved, lateral nerves appressed-pubescent, purplish; fertile lemmas length of spikelet, pale gray or purplish-tinged, somewhat longitudinally striate, acutely apiculate.---Fields, gardens, lawns and waste places, a troublesome weed. Throughout the United States, more common in the east and south; temperate and tropical areas of the world. Our specimens from the eastern half of Colorado at 4000-5500 feet but to be expected anywhere in cultivated areas at least at lower altitudes.
2. Digitaria ischaemum (Schreb.) Muhl., Descr. Gram. 131. 1817.
   Syntherisma humifusa (Pers.) Rydb.; S. ischaemum (Schreb.) Nash---Culms 15-50 cm. long, erect or soon prostrate and rooting at the nodes; leaf blades 2-6 mm. wide, 2-10 cm. long, glabrous or with scattered hairs above near base; sheaths somewhat keeled, glabrous or with a few scattered hairs on basal ones; ligules 1-2.5 mm. long; inflorescence of 2-6 racemes each 3-10 cm. long, widely spreading at maturity, rachis with wings wider than midrib; spikelets about 2 mm. long, lanceolate-elliptic, on pedicels round or somewhat triangular in sections; first glume wanting or very minute; second glume length of spikelet, 3- to 5-nerved, short-villous, often with capitellate hairs; sterile lemmas length of spikelet, 5- to 7-nerved, like second glume but less hairy; fertile lemmas length of spikelet, dark brown when mature, finely longitudinally striate.---Waste places, lawns and gardens, often a troublesome weed. Quebec to North Dakota, south to South Carolina, Tennessee and Arkansas, occasionally westward; introduced from Eurasia. Our specimens from northcentral and central Colorado at 4500-6500 feet but to be expected anywhere in cultivated areas, at least at lower altitudes.

69. Eriochloa H. B. K. CUPGRASS

Sheaths split; ligules mostly of hairs; spikelets 1-flowered with a sterile floret below, disarticulating below the glumes, a ringlike callus just below the spikelet, back of fertile lemma turned away from rachis.

1. **Eriochloa contracta** Hitchc., Biol. Soc. Wash. Proc. 41:163. 1928.
Annual plants; culms 30-70 cm. tall, freely branching, erect or decumbent at base, often rooting at lower nodes, pubescent at least near nodes; leaf blades 2-5 mm. wide, 8-20 cm. long, flat or folded in drying, pubescent or sometimes nearly glabrous; sheaths glabrous or short-pubescent; ligules about 1 mm. long; panicle 8-15 cm. long, contracted, composed of spikelike, overlapping racemes 1-2 cm. long, the axis pubescent; spikelets 3.5-4.5 mm. long, awn-tipped, in 2 rows but 1-sided on rachis, the pedicels short; first glume reduced to a minute sheath adnate to the callus; second glume length of spikelet, 5- to 7-nerved, appressed-villous, awn-tipped, somewhat indurated at base; sterile lemmas somewhat shorter than second glume, acuminate almost awn-tipped, appressed-villous, about 5-nerved, often with a hyaline palea; fertile lemmas indurated, minutely papillose, margins somewhat inrolled, mucronate or awn-tipped, this awn antrorsely scabrous and readily deciduous; paleas similar to lemmas in length and texture, enclosed by its edges.---Open ground, ditches, low fields and moist ground. Kansas to Louisiana and New Mexico; introduced in Missouri and Virginia. Our 3 specimens from southeastern Colorado in Otero, Bent and Prowers Counties at 3600-4200 feet.

70. Paspalum L. PASPALUM

Perennial plants with rhizomes present or absent, or annual; leaf blades flat; sheaths split; ligules membranous but with hairs back of it at base of blade making it look hairy; inflorescence a panicle of spikelike racemes; spikelets 1-flowered, a sterile lemma below, plano-convex, subsessile in pairs or solitary on 1 side of a rachis, the back of the fertile lemma turned toward it; first glume usually wanting or at least minute; second glume and sterile lemmas about equal; fertile lemma chartaceous-indurated with inrolled margins; palea similar to lemma in texture and length and enclosed by its edges; styles long, stigmas plumose.

1. Rachis of raceme flat and winged, about 1 mm. wide; annual plants; sterile lemmas transversely wrinkled
 -1. P. racemosum
1. Rachis of raceme not winged or very narrowly winged in number 2; perennial plants; sterile lemmas not wrinkled
 2. Spikelets with a fringe of long hairs near margins, the spikelet 3-4 mm. long; axillary racemes lacking
 -2. P. dilatatum
 2. Spikelets not ciliate at margins, about 2-2.5 mm. long; axillary racemes usually present (at least in sheaths)
 -3. P. stramineum

1. **Paspalum racemosum** Lam., Tabl. Encycl. 1:176. 1791.
Annual plants; culms up to 100 cm. long, widely spreading or sprawling, often rooting at the nodes, branching, nodes brown; leaf blades 6-20 mm. wide, 4-12 cm. long, tapering to a rounded or subcordate base, glabrous or scabrous on margins, sometimes glaucous below; sheaths glabrous, rather loose; ligules 1-3 mm. long; panicle 10-20 cm. long, pale or purple, of numerous (often over 20) racemes, these 1-2 cm. long with a flat rachis about 1 mm. wide; spikelets 2.5-3 mm. long, 1-1.3 mm. wide, elliptic, abruptly pointed, solitary on short, pubescent pedicels; first glume wanting; second glume length of spikelet, thin, loose, rugulose on back below, fluted or wrinkled near margins; sterile lemmas length of spikelet, loose, thin, transversely fluted or wrinkled on either side of midnerve; fertile lemmas 1.5-1.8 mm. long, narrowly obovate, smooth and shining.---Sometimes cultivated for ornament. Native of Peru. Our one Colorado specimen from a garden or lawn near Denver perhaps the seed coming in as an adulterant.
2. **Paspalum dilatatum** Poir, in Lam., Encycl. 5:35. 1804.
Perennial plants with a knotted base of short rhizomes; culms 30-125 cm. tall, few to several, ascending to erect from a curved or decumbent base, nodes often pubescent; leaf blades 3-12 mm. wide, about 10-25 cm. long, margins scabrous, sparsely ciliate at base near or on collar, long-hairy near ligules; sheaths flattened, usually glabrous on upper and pubescent on lower ones, ciliate on collar, flattened; ligules 2-4 mm. long; panicle of 3-8 ascending or drooping, rather broad, spikelike racemes commonly 4-8 cm. long, rachis narrowly winged, about 1.2 mm. wide with numerous long hairs at base of rachis; spikelets 3-4 mm. long, about 2 mm. wide, ovate, pointed, in pairs, short-pedicelled and overlapping; first glume wanting; second glume length of spikelet, 5- to 7-nerved, sparsely pubescent on back but bearing near margins a fringe of long hairs, abruptly pointed; sterile lemmas 3- to 5-nerved, abruptly pointed, somewhat pubescent, sometimes ciliate; fertile lemmas broadly elliptic, pale, minutely papillose-striate.---Low ground, dry prairies and marshy meadows. New Jersey to Tennessee and Florida, west to Arkansas and Texas; also in Oregon, Colorado, Arizona and California; native to South America. Our one Colorado specimen from Delta County. Should be looked for in the southern part of the state.
3. **Paspalum stramineum** Nash in Britt., Man. 74. 1901.
Perennial plants, no rhizomes; culms 20-100 cm. tall, tufted, erect or rarely spreading, glabrous or appressed-pubescent at nodes, more or less purplish; leaf blades 5-15 mm. wide, 5-25 cm. long, yellowish-green, basal short and brownish, margins crinkly and papillose-ciliate, somewhat puberulent above and below with scattered long hairs especially along the midrib; sheaths flattened, pilose on margins and throats and lower ones densely pubescent, often brownish; ligules about 1 mm. long, hairs back of it about 2 mm. long; racemes 2 or 3 in number, 6-14 cm. long, arching, rachis flattened but little if at all winged; spikelets 2-2.3 mm. long, about 2 mm. wide, pubescent to glabrous, suborbicular and plano-convex, in pairs, short-pedicelled; first glume lacking or very minute; second glume length of spikelet, 3-nerved, glabrous to pubescent; sterile lemmas 2-nerved, the midnerve usually lacking, glabrous to sparsely hairy; fertile lemmas obtuse.---Open ground or open woods especially in sandy soil. Indiana to Minnesota, Texas, Arizona and Mexico. Our specimens from northeastern and southeastern Colorado at 3500-4500 feet. Should be expected in sandy soil anywhere in the eastern part of the state.

71. Panicum L. PANICUM; WITCHGRASS

Annual or perennial plants, rhizomes and stolons sometimes present; leaf blades flat but sometimes involute toward tip; sheaths split; ligules made up of hairs (except 1 species); inflorescence an open panicle or a panicle of narrow densely-flowered racemes; spikelets 1-flowered with a sterile lemma (sometimes staminate) below, disarticulating below the glumes, back of fertile lemma turned toward rachis (this sometimes obscure); first glume shorter than spikelet, often minute; second glume about the length of spikelet; sterile lemma resembling the second glume in length and texture; fertile lemmas chartaceous-indurated, nerves obsolete; paleas similar to lemmas in texture and enclosed by their inrolled edges.

1. Plants with long stolons (herbarium specimens may not show these); ligules membranous; first glume obtuse or broadly acute, at least 3/4 as long as spikelet   -1. P. obtusum
1. Plants lacking stolons; ligules made up entirely or mostly of hairs; first glume less than 1/2 the length of spikelet (or if 2/3 as long then due to an acuminate point)
  2. Spikelets less than 2 mm. long; plants forming a winter rosette of wider basal leaves
    3. Vernal leaves glabrous or nearly so above; autumnal phase spreading or decumbent, often forming matlike structures   -2. P. tennesseense
    3. Vernal leaves pubescent above sometimes pilose near base and margins; autumnal phase stiffly erect or ascending   -3. P. huachucae
  2. Spikelets 2 mm. long or more; plants not forming a winter rosette (except in numbers 4 and 5)
    4. Plants forming basal winter rosettes of wider leaves; leaves seldom over 8 cm. long, less than 15 times longer than wide; spikelets about 3 mm. long
      5. Leaves glabrous above (may be ciliate on margins), 6-12 mm. wide; panicle 2/3 to 3/4 as wide as long   -4. P. scribnerianum
      5. Leaves pubescent above, 3-6 mm. wide; panicle 1/2 to 2/3 as wide as long   -5. P. wilcoxianum
    4. Plants not forming winter rosettes; some of leaves over 8 cm. long, over 15 times longer than wide; spikelets various lengths (about 3 mm. long only in numbers 6, 7, 9 and 11A)
      6. Plants perennial and without rhizomes; leaves not over 5 mm. wide
        7. Panicles 3-6 cm. long; spikelets sparingly pilose, never over 3.2 mm. long; first glume not over 1/3 length of spikelet   -6. P. perlongum
        7. Panicles 6-20 cm. long; spikelets glabrous, often over 3.2 mm. long; first glume over 1/3 length of spikelet   -7. P. hallii
      6. Plants either annual or perennial with conspicuous rhizomes; leaves usually over 5 mm. wide
        8. Perennial plants with rhizomes; not a weed nor cultivated as a crop; sheaths glabrous (except on margins); sterile paleas almost as long as sterile lemmas   -8. P. virgatum
        8. Annual plants; usually weeds in waste places or cultivated as a crop plant; sheaths glabrous or pilose to papillose-hirsuite; sterile paleas much shorter than sterile lemmas or lacking
          9. First glume not over 1/4 length of spikelet, truncate or triangular-tipped   -9. P. dichotomiflorum
          9. First glume 1/2 as long as the spikelet or more, acute or acuminate
            10. Spikelets over 4 mm. long; cultivated as a crop plant, rarely escaping   -10. P. miliaceum
            10. Spikelets less than 4 mm. long; not cultivated but usually a weed
              11. Spikelets 2-2.5 mm. long, ovate or elliptic; leaf blades not crowded toward base of plant   -11. P. capillare
              11. Spikelets 3-3.5 mm. long, lanceolate; leaf blades usually crowded near base (this variety intergrades with species)   -11A. P. capillare occidentale

1. Panicum obtusum H. B. K., Nov. Gen. & Sp. 1:98. 1815.
Perennial plants, rhizomes if present short and clustered, forming a knotty crown, but with stolons often over 100 cm. long, these with long internodes and swollen, geniculate, villous nodes; culms 20-80 cm. long, tufted, erect or decumbent at base, compressed, glabrous, strongly nerved; leaf blades 2-7 mm. wide, 3-30 cm. long, flat but involute near tip, those of stolons shorter and narrower, glabrous, somewhat scabrous or pubescent above near base; sheaths glabrous or the lower pubescent; ligules membranous, 0.8-1.5 mm. long; panicle 3-12 cm. long, short-exserted or included at base, narrow with a few appressed racemelike branches, these densely flowered; spikelets 3-4 mm. long, obovoid, green at first but turning brown at maturity, usually in pairs, 1 pedicelled, 1 nearly sessile; first glume about 3/4 to 4/5 as long as spikelet, 5-nerved or sometimes 3-nerved; second glume about length of spikelet, 7- to 9-nerved; sterile lemmas about length of spikelet, 5- to 7-nerved, staminate; sterile paleas as long as sterile lemmas, hyaline; fertile lemmas smooth, shining, pale, subacute, minutely pubescent near apex.--- Sandy or gravelly soil mostly along rivers, roadsides and ditches. Missouri to Colorado south to Texas and Arizona; also in Mexico. Our specimens from southeastern Colorado as far north as El Paso County at 3500-6000 feet.

2. Panicum tennesseense Ashe, Journ. Elisha Mitchell Sci. Soc. 15:52. 1898.
Perennial plants, no rhizomes; (Vernal phase) - culms 25-60 cm. tall, suberect or spreading, slender, papillose-pilose or glabrous above, nodes pubescent or pilose especially the lower; leaf blades 2-8 mm. wide, 4-10 cm. long, ascending or spreading, with a thin white marginal band, glabrous above or rarely a few scattered hairs near base, lower surface appressed-pubescent to nearly glabrous, sometimes ciliate near base; sheaths papillose-pilose, papillose or nearly glabrous; ligules of hairs 2-5 mm. long; panicle 3-7 cm. long, nearly as wide, lower branches widely ascending; spikelets 1.5-1.7 mm. long, obovate, obtuse, pubescent, often purplish; first glume about 1/4 as long as spikelet, pubescent or rarely glabrous, acute, faintly 1-nerved; second glume somewhat shorter than spikelet, 5- to 9-nerved, these rather faint, pubescent; sterile lemmas as long or sometimes slightly shorter than spikelet, 5- to 9-nerved, these faint, pubescent; sterile paleas less than 1/2 as long as the sterile lemmas, hyaline; fertile lemmas

about length of spikelet, exposed at apex, elliptic, obtuse or slightly apiculate, smooth and shining; (Autumnal phase)--culms widely spreading or prostrate, with numerous flabellate branches, often forming mats; leaves reduced, usually ciliate at base, exceeding the reduced panicles.---Open rather moist ground and borders of woods. Quebec to Minnesota, south to Georgia and Texas and a few places west to Utah and Arizona. Our few records from Yuma and Boulder Counties at 3500-5500 feet.

3. Panicum huachucae Ashe, Journ. Elisha Mitchell Sci. Soc. 15:51. 1898.
Perennial plants, no rhizomes; culms tufted, erect or ascending at first then becoming geniculate at lower nodes and much branched; (Vernal phase)--culms 20-60 cm. tall, densely tufted, usually stiffly upright, papillose-pubescent, rather harsh, nodes bearded; leaf blades 5-8 mm. wide, 4-10 cm. long, erect, firm, upper surface copiously short-pilose especially toward base, lower surface densely pubescent; sheaths usually papillose-pilose at least on upper ones, the hairs often to 2 mm. long but sometimes short and sparse; ligules of hairs 3-4 mm. long; panicle 4-6 cm. long, nearly as wide, exserted, usually purplish, branches ascending or spreading, in fascicles; spikelets 1.5-1.8 mm. long, 1 mm. wide, elliptic, turgid, short-pubescent often papillose-pubescent; first glume about 1/2 length of spikelet; second glume almost as long as spikelet, pilose, 7-nerved; sterile lemmas about length of second glume, 7-nerved, pilose; sterile paleas small or absent; fertile lemmas length of spikelet, elliptic, somewhat apiculate, smooth and shining; (Autumnal phase)--culms stiffly erect or ascending, often papillose only, branches fascicled; leaves reduced, crowded, ascending, the blades 20-30 mm. long but much exceeding the reduced panicles.---Prairies, meadows and open ground. Nova Scotia to Montana westward here and there to California. Our one Colorado specimen from Larimer County at 5100 feet.

4. Panicum scribnerianum Nash, Bull. Torr. Club 22:421. 1895.
Perennial, no rhizomes; (Vernal phase)--culms 15-60 cm. tall, tufted few to many, erect or spreading, often geniculate at base, glabrous, harshly pubescent or ascending pilose; leaf blades 5-12 mm. wide, 5-10 cm. long, firm, rounded and ciliate at base, glabrous above short-pubescent to glabrous below; sheaths ciliate-pilose, some of hairs usually papillose, rarely glabrous, pilose at throat; ligules of hairs about 1 mm. long; panicle 4-6 cm. long, 2/3 to 3/4 as wide, often included at base, branches widely spreading or ascending; spikelets 3.2-3.3 mm. long (but most of ours just make 3 mm.), obovate, turgid, sparsely pubescent to nearly glabrous; first glume about 1/3 as long as spikelets, acute, 1-nerved; second glume a little shorter than spikelet, 7- to 9-nerved, glabrous or more commonly pubescent; sterile lemmas about as long as spikelet, 7- to 9-nerved, glabrous or more commonly pubescent; sterile paleas about 1/3 as long as lemmas, hyaline; fertile lemmas about as long as spikelet, broadly elliptic, somewhat apiculate, pale; (Autumnal phase)--culms branching from middle or upper nodes, late in season developing crowded branchlets with ascending leaf blades and small partly included panicles.---Meadows, dry prairies, rocky slopes or sandy soil. Maine to British Columbia south to Maryland, Tennessee, Texas and Arizona. Our specimens from along the eastern foothills of Colorado from Larimer to Las Animas County at 5000-7200 feet.

5. Panicum wilcoxianum Vasey, U. S. D. A. Div. Bot. Bull. 8:32. 1889.
Perennial plants, no rhizomes; (Vernal phase)--culms 10-25 cm. tall, tufted, erect, copiously papillose-hirsute; leaf blades 3-6 mm. wide, 5-8 cm. long, flat but involute-acuminate, firm, erect or ascending, hirsute above and below; sheaths copiously papillose-hirsute; ligules of hairs about 1 mm. long; panicle 2-5 cm. long, about 1/2 to 2/3 as wide, finally exserted; spikelets 2.7-3 mm. long, obovate-elliptic, pubescent, on long pedicels; first glume about 1/3 as long as spikelet, acute, 1-nerved, pubescent; second glume slightly shorter than spikelet, 7- to 9-nerved, pubescent; sterile lemmas length of spikelet, 7- to 9-nerved, pubescent; sterile paleas about 1/3 as long as lemmas, hyaline; fertile lemmas about as long as spikelet, sometimes a little shorter, elliptic, glabrous, shiny; (Autumnal phase)--culms branching from all the nodes, forming bushy tufts with rigid, erect leaf blades overtopping the reduced panicles.---Prairies. Manitoba and North Dakota to Illinois, Kansas and New Mexico. Our one Colorado specimen from Pueblo County at 6900 feet, but probably another from Rio Grande County at 7800 feet belongs to this species.

6. Panicum perlongum Nash, Bull. Torr. Club 26:575. 1899.
Perennial plants, rhizomes lacking (vernal and autumnal phase similar); culms 15-40 cm. tall, tufted with few in a tuft, unbranched; leaf blades 2-5 mm. wide, 8-15 cm. long on upper ones, lower shorter, glabrous or puberulent above, short-pilose below, pilose on margins near base some of hairs usually papillose; sheaths pilose, some of hairs papillose; ligules of hairs about 1 mm. long; panicle 3-6 cm. long, exserted, branches ascending or nearly erect; spikelets 2.7-3.2 mm. long, oval, turgid, blunt, sparingly pilose; first glume 1/4 to 1/3 as long as spikelet, 1-nerved or nerveless, acute, ovate or triangular; second glume as long or somewhat shorter than spikelet, 7- to 9-nerved, sparsely pilose; sterile lemmas about length of spikelet, 7- to 9-nerved, sparsely pilose; sterile paleas 1/3 to 1/2 as long as lemmas, hyaline; fertile lemmas about length of spikelet, oval-obovate, pale, minutely striated longitudinally, rounded and minutely umbonate at apex.---Prairies, rocky slopes, usually in dry soil. Indiana to Manitoba and North Dakota, south to Colorado and Texas. Our three Colorado specimens from along the eastern foothills from Larimer to Pueblo County at 5300-5700 feet.

7. Panicum halli Vasey, Bull. Torr. Club 11:61. 1884.
Perennial plants without rhizomes (vernal and autumnal phases alike); culms 15-60 cm. tall, erect, tufted; leaves 2-5 mm. wide, 4-15 cm. long, glabrous to sparsely ciliate near base, the hairs often papillose, usually crowded near base of plant and often curled in age like shavings; sheaths glabrous to sparsely papillose-hispid; ligules about 1-1.5 mm. long, mostly of hairs; panicles 6-20 cm. long, exceeding the leaves, branches stiffly ascending and naked at base, bearing short-appressed branchlets; spikelets 3-3.7 mm. long, green or purplish; first glume about 1/2 to 2/3 as long as the spikelet, 3- to 7-nerved, obtuse or acutish at apex; second glume length of spikelet, 5- to 9-nerved, acutish; sterile lemmas like second glume, its sterile palea about 1/2 as long; fertile lemmas about 3/4 as long as the spikelet, smooth and shining,

becoming brown at maturity.---Dry prairies and fields, rocky hills and slopes. Texas to Arizona and south to Mexico. Comes into Colorado in the southeast corner at 4000 feet.

8. Panicum virgatum L., Sp. Pl. 59. 1753.

Perennial plants with numerous creeping rhizomes (vernal and autumnal phases alike); culms 40-150 cm. tall, single or somewhat tufted, robust and tough, green or glaucous, usually glabrous; leaf blades 3-12 mm. wide, 10-50 cm. long, flat, margins somewhat scabrous, often scattered-hairy above near base; sheaths glabrous but margins ciliate especially near collar; ligules mostly of hairs 1.5-3.5 mm. long; panicle 15-40 cm. long, about 1/3 to 1/2 as wide as long, open, ovate, branches ascending or spreading, bearing spikelets mostly on outer half; spikelets 3-6 mm. long (usually 4-5 mm.), elliptic-ovate, turgid, glabrous, acuminate, on short pedicels; first glume about 2/3 as long as spikelet, 5-nerved, acuminate or cuspidate; second glume as long as spikelet, 5- to 7-nerved, acuminate; sterile lemmas somewhat shorter than second glume, 5-nerved, acuminate or pointed, empty or staminate; sterile paleas somewhat shorter than lemmas, hyaline; fertile lemmas about 3/4 as long as spikelet, smooth and shining, narrowly ovate, abruptly acute. Specimens with shrunken, sterile spikelets are occasionally found.---Prairie, open ground, open woods, meadows, marshes and riverbanks, often in sandy soil. Quebec and Maine to Montana, south to Florida, Nevada and Arizona; also in Mexico and Central America. Our Colorado specimens distributed over the eastern half of the state at 3500-7000 feet. However, it has been reported in La Plata County.

9. Panicum dichotomiflorum Michx., Fl. Bor. Am. 1:48. 1803.

Annual plants; culms 50-100 cm. tall, ascending or spreading from a geniculate base; leaves 3-20 mm. wide, 10-50 cm. long, flat or sometimes folded, the white midrib prominent, sometimes sparsely pilose above; sheaths somewhat flattened, ciliate on margins near summit otherwise glabrous; ligules 1-2 mm. long, membranous for about 1/2 up, then of hairs; panicles 10-40 cm. long, terminal and axillary, often included at base, main branches ascending (rarely reflexed), naked at base, short branches bearing short-pedicelled rather crowded spikelets; spikelets 2-3.2 mm. long, glabrous; first glume not over 1/4 the spikelet length, 1- to 3-nerved, truncate or broadly triangular; second glume length of spikelet, somewhat pointed, rather faintly 7-nerved; sterile lemmas like second glume, its palea wanting or as long as the fruit; fertile lemmas about 2-2.3 mm. long, elliptic, smooth and shining.---Moist ground and as a weed in waste places or fields. Maine to Nebraska, south to Florida and Texas, occasionally westward. Found in Larimer County, Colorado, at 5100 feet.

10. Panicum miliaceum L., Sp. Pl. 58. 1753.

Annual plants; culms 20-100 cm. tall, stout, erect or decumbent at base, glabrous or hispid below the pubescent nodes; leaf blades 8-20 mm. wide, 7-30 cm. long, flat, margins scabrous, pubescent both sides or sometimes glabrous, hairs often papillose; sheaths pilose or papillose-pilose especially near throat; ligules mostly of hairs 0.7-2 mm. long; panicle 10-30 cm. long, rather dense, nodding at maturity, usually included somewhat at base, scabrous branches ascending, spikelet-bearing near ends; spikelets 4.5-5 mm. long, ovoid, acuminate, glabrous; first glume 1/2 to 2/3 as long as spikelet, acuminate, strongly 5-nerved; second glume as long as spikelet, strongly 11- to 13-nerved, acuminate; sterile lemmas slightly shorter than second glume, strongly 11- to 13-nerved, acuminate; sterile paleas small and hyaline; fertile lemmas about 3 mm. long, elliptic, smooth and shining, straw-colored or reddish-brown.---Waste places, escaped from cultivation. Cultivated in the cooler parts of the United States for forage or sometimes as a grain for hogs. Not common in Colorado.

11. Panicum capillare L., Sp. Pl. 58. 1753.

Annual plants; culms 15-60 cm. tall, tufted, erect or ascending from a decumbent base, simple or sparingly branched, papillose-hispid to nearly glabrous, pubescent at nodes; leaf blades 8-20 mm. wide, 6-25 cm. long, flat, midnerve prominent, margins papillose-pilose, surfaces pilose to nearly glabrous; sheaths papillose-pilose to **hirsute; ligules** almost entirely of hairs 0.5-1.5 mm. long; panicle 7-35 cm. long, often half the length of the plant, nearly as wide as long at maturity, very diffuse, branches capillary, long and scabrous, panicle breaking off and becoming a "tumbleweed"; spikelets 2-2.5 mm. long, elliptic, somewhat acuminate, glabrous, green or purplish (sometimes with staminate flowers only and the lemmas more membranous); first glume about 1/2 as long as spikelet, with 1 strong nerve, and 2-4 fainter ones sometimes present, acute; second glume length of spikelet, 7-nerved, the nerves distinct, somewhat acuminate at apex; sterile lemmas length of spikelet, 7-nerved, these not as distinct as on second glume; fertile lemmas smooth and shining, pale or brownish.---Open ground, fields, gardens and waste places. Maine to Montana, south to Florida and Texas, occasionally westward. Our specimens widely distributed over Colorado at 3500-9500 feet.

11A. Panicum capillare occidentale Rydb., (var.) Contrib. U. S. Nat. Herb. 3:186. 1895.

P. barbipulvinatum Nash---Differs from species--culms shorter; leaf blades shorter and tending to crowd near base of plant, less pubescent; panicles more exserted, the branches early divaricate; spikelets about 3 mm. long (2.5-3.3 mm.).---Open ground, moist areas and waste places. Prince Edward Island and Quebec to British Columbia, south to New Jersey, Missouri, Texas and California. Our few specimens are from northern Colorado at 4200-6700 feet. The variety seems to intergrade with the species in this state and most of our specimens are intermediates.

72. Echinochloa Beauv. COCKSPUR

Annual plants, often weedy; leaf blades flat; sheaths split; ligules absent entirely; inflorescence a panicle of several 1-sided rather narrow racemes; spikelets 1-flowered, a sterile floret below, plano-convex, hispid, subsessile, solitary or in irregular clusters, disarticulating below the glumes; first glume about 1/3 to 1/2 as long as spikelet, pointed; second glume length of spikelet, sometimes short-awned; sterile lemmas often short-awned; fertile lemmas smooth, shining, chartaceous; fertile paleas similar in length and texture to lemmas, enclosed by its edges except at apex; styles distinct, stigmas plumose. The species and the following 3 varieties intergrade in Colorado and can be **distinguished** only arbitrarily.

1. Awns 5-10 mm. long on some of the spikelets of the panicle; racemes usually ascending or spreading
 -1. **E. crusgalli**
1. Awns absent or less than 3 mm. long; racemes appressed (except 1A)
   2. Panicle open, the racemes ascending or spreading -1A. **E. crusgalli mitis**
   2. Panicle closed, the racemes appressed
      3. Racemes thick, incurved; spikelets (and whole plant) usually purplish -1B. **E. crusgalli frumentacea**
      3. Racemes not especially thick, not incurved; spikelets (and plant) green or sometimes slightly purplish
 -1C. **E. crusgalli zelayensis**

1. Echinochloa crusgalli (L.) Beauv., Ess. Agrost. 53, 161. 1812.
 Culms 20-100 cm. tall, erect or somewhat decumbent at base, often branching from base, rather succulent, glabrous; leaf blades 5-15 mm. wide, 10-30 cm. long, sometimes keeled, glabrous or somewhat hairy near base, margins weakly scabrous, midnerve rather conspicuous above; sheaths glabrous to sparingly pubescent; panicle 10-20 cm. long, erect or nodding, deep purple to pale green, racemes erect or spreading, the lower more distant; spikelets 2.5-3.5 mm. long, excluding awns, in 2 rows on 1 side of rachis; first glume 3-nerved, hispidulous; second glume 5-nerved, hispid on nerves, pubescent between, mucronate or short-awned; sterile lemmas length of spikelet, 5- to 7-nerved, hispid, often tuberculate-hispid, awn length variable, 5-10 mm. long; sterile paleas thin and papery; fertile lemmas finely striated, pale or brownish.---Moist open places, old fields, gardens, yards, ditches and waste ground. New Brunswick to Washington, south to Florida and California. Our specimens from northcentral, central and southcentral Colorado at 4500-7500 feet, but to be expected anywhere around cultivated areas at lower altitudes.
1A. Echinochloa crusgalli mitis (Pursh) Peterm., (var.) Fl. Lips. 82. 1838.
 Differs from species--panicle branches usually ascending or spreading; spikelets all awnless or with awns less than 3 mm. long.---Scattered over the state in cultivated areas, the specimens more abundant than those of the species. Altitude 4000-7500 feet.
1B. Echinochloa crusgalli frumentacea (Roxb.) Wight, (var.) Cent. Dict. Sup. 810. 1909.
 Differs from species--panicle branches thick, appressed, incurved; spikelets awnless, glumes and sterile lemmas mostly purple; fertile lemmas pale, usually somewhat exposed, making a strong color contrast to the outer bracts; whole plant usually reddish-purple in color.---Occasionally cultivated as a forage grass and escaping here and there throughout the United States. Our 3 specimens from Otero and El Paso Counties and near Denver at 4200-5200 feet.
1C. Echinochloa crusgalli zelayensis (H. B. K.) Hitchc., (var.) U. S. D. A. Bull. 772:238. 1920.
 Culms less succulent than any of the others; panicle branches more or less appressed; spikelets awnless or with awns less than 3 mm. long.---Moist often alkaline soil. Oklahoma to Oregon, south to Texas and California; also in Mexico to Argentina. Apparently not very common in Colorado, our four specimens from El Paso, Boulder, Weld and Washington Counties, at 4000-6000 feet.

73. Setaria Beauv. BRISTLE GRASS; MILLET

Annual plants (or 1 species perennial); leaf blades flat; sheaths split; ligules 1/2 or more of hairs; inflorescence a spikelike panicle; spikelets 1-flowered but a sterile floret below, subtended by 1 to several bristles, usually disarticulating below the glumes but above the bristles, awnless; glumes unequal; sterile lemmas as long as spikelets; fertile lemmas coriaceous-indurated, smooth or transversely rugose; fertile paleas similar to lemmas in texture; styles elongated, stigmas plumose.

1. Bristles downwardly barbed, the panicle sticky and burlike -1. **S. verticillata**
1. Bristles upwardly barbed, the panicle not burlike
   2. Perennial plants -2. **S. macrostachya**
   2. Annual plants
      3. Second glume about 2/3 as long as spikelet; fertile lemmas strongly transversely wrinkled; bristles 5-16 below each spikelet, these bristles yellowish at maturity -3. **S. lutescens**
      3. Second glume 3/4 to as long as the spikelet; fertile lemmas very finely wrinkled; bristles 1-3 directly below each spikelet (may branch and be more at sides), bristles usually green or purple (yellow only in a cultivated variety of number 5)
         4. Panicle cylindric, tapering somewhat to apex, less than 10 cm. long and 12 mm. wide; spikelets about 2 mm. long, falling entire; bristles green or pale; leaves seldom over 1 cm. wide; a common weed -4. **S. viridis**
         4. Panicle often lobed or interrupted, large and heavy, over 10 cm. long and over 12 mm. wide; spikelets about 3 mm. long, fertile lemmas falling from outer bracts; bristles purple or occasionally yellow; leaves often over 1 cm. wide; not common except as cultivated -5. **S. italica**

1. Setaria verticillata (L.) Beauv., Ess. Agrost. 51, 178. 1812.
 Annual plants; culms 30-100 cm. tall, tufted, often much branched at base and geniculate-spreading, smooth but scabrous below the panicle; leaf blades 3-10 mm. wide, 8-20 cm. long, scabrous, rarely glabrous, often somewhat hairy; sheaths flattened and keeled, margins ciliate, glabrous or lower pubescent; ligules about 1/2 hairs, 0.7-1.5 mm. long; panicle 5-15 cm. long, spikelike, dense, cylindrical, more or less interrupted at base, burlike; spikelets about 2 mm. long, oblong-elliptic, usually with 1 retrorsely barbed bristle below each, this 1-3 times as long as spikelet; first glume about 1/3 as long as spikelet, 3-nerved, triangular; second glume length of spikelet, 5- to 7-nerved, glabrous; sterile lemmas length of spikelet, 5- to 7-nerved but 2 short, glabrous; sterile paleas small, hyaline; fertile lemmas weakly transversely rugose.---Cultivated ground, gardens and waste places. Massachusetts to North Dakota, south to Alabama and Missouri, occasional west to California; also in tropical America; introduced from

Europe. Only four specimens from Colorado, from Larimer to El Paso County at 5000-6300 feet. Apparently uncommon but to be expected anywhere around cultivated areas.

2. Setaria macrostachya H. B. K., Nov. Gen. & Sp. 1:110. 1815.

Chaetochloa composita of Manuals---Perennial plants, no rhizomes; culms 40-120 cm. tall, tufted, erect or geniculate at base, usually pale or glaucous, branching at base, flattened, nodes pubescent, internodes scabrous; leaf blades 2-4 mm. wide, 10-30 cm. long, sometimes folded, scabrous above, smooth or scabrous below, rarely pubescent both sides; sheaths flattened, ciliate, pubescent at collar; ligules 1/2 or more of hairs, 1-3 mm. long; panicle 10-25 cm. long, mostly spikelike (ours are) cylindrical but tapering to tip, often lobed or interrupted below; spikelets 2-2.5 mm. long, pale green, with 1 bristle below each, these 10-15 mm. long, upwardly barbed; first glume about 1/2 as long as spikelet, 1- to 3-nerved, glabrous, broadly acute; second glume about 3/4 as long as spikelet, 5- to 7-nerved, glabrous, obtuse or apiculate; sterile lemmas length of spikelet, 5-nerved, acute; sterile paleas short; fertile lemmas finely transversely wrinkled.---Open dry ground or dry woods. Texas to Colorado and Arizona; also in Mexico. Our three Colorado specimens from Fremont and Baca Counties at 4000-5300 feet.

3. Setaria lutescens (Weigel) F. T. Hubb., Rhodora 18:232. 1916.

Chaetochloa glauca (L.) Scrib.---Annual plants, culms 25-100 cm. tall, erect, ascending or even sometimes prostrate, much branched at base, flattened, glabrous but scaberulent below panicle; leaf blades 4-10 mm. wide, 5-20 cm. long, scabrous above, often pilose near base, glabrous below, sometimes twisted in a loose spiral; sheaths glabrous, compressed-keeled; ligules mostly of hairs 0.5-1 mm. long; panicle mostly 3-10 cm. long, about 1 cm. wide (counting bristles), spikelike, dense, cylindrical, yellow at maturity; spikelets about 3 mm. long, broadly ovate, with 5-20 yellowish, antrorsely-scabrous bristles just below each, the longest 2-3 times longer than spikelet; first glume about 1/2 as long as spikelet, 3-nerved, glabrous, broadly acute; second glume about 2/3 as long as spikelets, 5-nerved, acute or mucronate, glabrous; sterile lemmas about as long as spikelet, 5-nerved; sterile paleas small; fertile lemmas with rugose, transverse ridges, these rather conspicuous, greenish or **tawny**.---Cultivated soil of gardens or fields and waste places in general. New Brunswick to South Dakota, south to Florida and Texas; occasional from British Columbia to California and New Mexico. Our few Colorado specimens from the northcentral and central parts of the state at 4700-6000 feet. Not as common as number 4.

4. Setaria viridis (L.) Beauv., Ess. Agrost. 51, 178. 1812.

Chaetochloa viridis (L.) Scribn.---Annual plants; culms 20-50 cm. tall, sometimes taller, tufted, erect or geniculate-spreading, usually branched at base, usually glabrous but scabrous below panicle; leaf blades 4-10 mm. wide, 5-25 cm. long, glabrous or scabrous; sheaths slightly if at all keeled, margins ciliate, glabrous, scabrous or lower ones pubescent and **long-hairy** near collar; ligules about 1/2 of hairs, 1-1.5 mm. long; panicle 3-8 cm. long, spikelike, dense, cylindrical but tapering somewhat upward; spikelets 2-2.5 mm. long, with 1-3 bristles directly below each (but may appear to be more because of side branches) these antrorsely barbed, 3-4 times as long as spikelet; first glume about 1/3 as long as spikelet, 3-nerved, **triangular**; second glume as long or nearly as long as spikelet, 5-nerved, glabrous; sterile lemmas as long as spikelet, 5-nerved, glabrous; sterile paleas absent or very short; fertile lemmas pale, faintly transversely-rugose.---Cultivated ground of fields and gardens and waste places in general. Newfoundland to British Columbia, south to Florida and California. A very common weed, widely scattered over Colorado, our specimens at 3500-8500 feet.

5. Setaria italica (L.) Beauv., Ess. Agrost. 51, 170, 178. 1812.

Chaetochloa italica (L.) Scribn.---Annual plants; culms 50-150 cm. tall, simple or branching at base, stout, erect, scabrous below panicle; leaf blades 5-25 mm. wide, sometimes wider, 10-40 cm. long, scabrous on margins and often above and below, often pubescent near collar; sheaths glabrous, scabrous or pubescent, margins ciliate especially near collar, this pilose; ligules almost all hairs, 0.5-2 mm. long; panicle 8-20 cm. long, to 3 cm. thick (smaller on escaped plants) dense and spikelike, cylindrical but often interrupted, yellowish, greenish or purplish; spikelets about 3 mm. long, with 1-3 bristles below each (may appear more because of branching), these upwardly barbed, 1-3 times as long as spikelet or longer; first glume about 1/3 as long as spikelet, 3-nerved, glabrous, triangular-ovate; second glume about 3/4 as long as spikelet, 5- to 7-nerved, glabrous; sterile lemmas length of spikelet, 5- to 7-nerved, glabrous; sterile paleas small or wanting; fertile lemmas finely wrinkled transversely, deciduous from the 3 outer bracts, tawny, **reddish**, brown, pale or black.---Cultivated in the warmer parts of the United States and escaped into waste places. Our specimens from northeastern Colorado at 3500-5100 feet, but might be expected anywhere in cultivated areas at rather low altitudes.

74. Pennisetum L. Rich. PENNISETUM

Annual or perennial plants; leaf blades flat or folded; sheaths split; ligules mostly or all of hairs; inflorescence a dense or spikelike panicle; spikelets 1-flowered, a sterile floret below, solitary or in **groups** of 2 or 3, surrounded by an involucre of bristles, these not united except at base, often plumose, falling with the spikelet; glumes unequal, the first sometimes wanting, second shorter or length of spikelets; sterile lemmas as long as spikelets; fertile lemmas chartaceous, smooth, the margins thin; paleas similar to lemmas in length and texture. In addition to the 2 species listed below, Pearl Millet (P. glaucum (L.) R. Br.) may be grown in Colorado. Dr. Robertson of the Colorado A. and M. College stated to the writer that Pearl Millet probably will not mature in this state.

1. Panicles 10 cm. long or less, oval, tawny or pale from the bristles; first glume usually over 1 mm. long
 -1. **P. villosum**
1. Panicles over 10 cm. long, cylindrical, bristles red or purplish; first glume minute or absent
 -2. **P. setaceum**

1. **Pennisetum villosum** R. Br. (Salt. Voy. Abyss. 62. 1814 - nom. nud.), Fresen. Mus. Senckenb. Abh. 2:134. 1837.
Perennial plants from a base of branching, knotty rhizomes; culms 30-60 cm. tall, usually in tufts, rather flattened, pubescent below the panicle; leaf blades 3-6 mm. wide, 10-60 cm. long, scabrous on margins and near long apex; sheaths ciliate especially toward the top and villous near collar; ligules 1-2 mm. long; panicle 3-15 cm. long, 1-5 cm. wide, dense, ovoid or sometimes oblong, feathery, pale or white-tawny; spikelets 8-10 mm. long, bristles numerous, plumose below, about 2-5 cm. long; first glume 1-2 mm. long; triangular or truncate, 1-nerved or nerveless, glabrous or nearly so; second glume 3-4 mm. long, 1-nerved this scabrous, lanceolate, long-acute; sterile lemmas slightly shorter than spikelet, long-acute to acuminate, 7- to 9-nerved, scabrous especially near tip; sterile paleas well developed often with a staminate flower; fertile lemmas as long as spikelet, 5- to 9-nerved, these not strong, scabrous toward apex, long-acute to acuminate.---Cultivated for ornament, sparingly escaping in dry ground. Introduced from Africa. Occasionally cultivated in Colorado.

2. **Pennisetum setaceum** (Forsk.) Chiov., Bull. Soc. Bot. Ital. **1923**:113. 1923.
P. ruppelii Steud.---Perennial plants; culms about 100 cm. tall, tufted, erect, pubescent below panicle; leaf blades 2-4 mm. wide, up to 60 cm. long, long-pointed mostly folded, scabrous especially on margins, often with scattered long hairs; sheaths ciliate, especially toward the villous collar; ligules **0.5-1 mm.** long; panicle 15-35 cm. long, cylindrical, spikelike, erect or nodding, pink or purplish, bristly; spikelets 6-7 mm. long, purplish from numerous bristles, these plumose below, variable but often to 4 cm. long; first glume minute or wanting; second glume 2-3 mm. long, 1-nerved, acuminate, thin, glabrous; sterile lemmas almost as long as spikelet, 3- to 5-nerved, acuminate or awn-pointed; sterile paleas absent (in ours); fertile lemmas as long as spikelet, 5-nerved, somewhat scabrous near apex, acuminate or short awn-pointed.--- Cultivated for ornament especially as a border plant or around fountains. Occasionally found in Colorado especially in parks or in large lawns.

75. Cenchrus L. SANDBUR

Sheaths split; ligules made up of hairs; spikelets surrounded by spines composed of numerous coalescing bristles forming a bur, disarticulating below this, 1-flowered but with a sterile floret below; fertile lemmas indurated; fertile paleas similar to lemmas in length and texture; styles often connate, stigmas plumose.

1. **Cenchrus pauciflorus** Benth., Bot. Voy. Sulph. 56. 1840.
C. tribuloides and C. carolinianus of Manuals---Annual plants; culms 20-90 cm. long, tufted, often forming mats, erect or decumbent, branching, flattened, scabrous or pubescent below the inflorescence; leaf blades 2-7 mm. wide, 6-15 cm. long, flat or sometimes folded, sometimes scabrous above and on margins, widest at base and tapering to apex; sheaths flattened and keeled, sometimes ciliate especially near collar, rather loose; ligules 0.5-1 mm. long; racemes 3-8 cm. long, of 6-20 burs, often partly included, usually 2 spikelets to a bur, burs 3-6 mm. wide (excluding spines), pubescent, spines 3-4 mm. long, spreading or reflexed, flat at base; spikelets 5-7 mm. long, about 2 mm. wide; first glume not over 1/3 as long as spikelet, 1-nerved; second glume about 3/4 as long as spikelet to almost as long, 3- to 5-nerved; sterile lemmas about as long as second glume, 3- to 5-nerved; sterile paleas thin; fertile lemmas length of spikelet, falcate-acuminate, visible at apex, 5-nerved.---Sandy open ground, often a weed. Maine to Oregon, south to Florida, Texas and California; also to South America. Our specimens mostly scattered over the eastern half of Colorado in sandy or gravelly soil (one in Mesa County) at 3500-6400 feet.

XI. Tribe Andropogoneae

76. Miscanthus Anderss. SILVERGRASS

Sheaths split; ligules membranous but ciliate; spikelets alike but in pairs, 1-flowered with a sterile floret below, surrounded by a tuft of hairs; fertile lemmas hyaline, awned.

1. **Miscanthus sinensis** Anderss., Oefv. Sv. Vet-Akad. Forh. 12:166. 1856.
Perennial plants; culms 100-300 cm. tall in large dense clumps, erect, robust; leaf blades 6-12 mm. wide, to 100 cm. long, flat, margins scabrous, midrib conspicuous above; sheaths somewhat flattened, glabrous but ciliate and often pilose near collar; ligules 1-3 mm. long; panicle 10-20 cm. long, fanlike of many continuous, silky, aggregated, spreading racemes; spikelets in pairs, 1 short-pedicelled, 1 longer, 4-5 mm. long, tuft of hairs at base 4-7 mm. long; glumes subequal, about length of spikelet, somewhat coriaceous, indistinctly 3-nerved, long-acute, sparingly pilose, the second ciliate; sterile lemmas hyaline, somewhat shorter than glumes, 2-toothed; fertile paleas absent (in ours); fertile lemmas as long as sterile, ciliate, deeply bifid with a twisted awn 5-7 mm. long from between the lobes; fertile paleas hyaline, shorter than lemmas.---Cultivated for ornament and growing wild in some eastern states; native of Asia. Occasionally cultivated in Colorado.

77. Erianthus Michx.

Sheaths split; ligules membranous at base; spikelets 1-flowered, a sterile floret below, in pairs, densely villous at base and on pedicels; fertile lemmas hyaline, awned.

1. **Erianthus ravennae** (L.) Beauv., Ess. Agrost. 14, 162, 177. 1812.
Perennial plants; culms up to 400 cm. tall and 1.5 cm. thick at base, tufted, erect, glabrous or hairy at nodes; leaf blades 5-12 mm. wide, 30-80 cm. long or basal ones longer, flat above middle, long-pointed, usually scabrous, pilose above and below near base; sheaths striate, scabrous with long hairs at throat, often pubescent all over; ligules about 1 mm. long, backed by long hairs; panicle 20-60 cm. long, oblong, silky, silvery-brown, branches rather appressed, finally bearing numerous short racemes with villous rachis joints; spikelets 4.5-5.5 mm. long, in pairs, 1 sessile, 1 pedicelled, densely villous at base and on pedicel, the hairs about as long as spikelet, rachis joint falling with spikelet; glumes equal, as long as spikelet, 3-nerved, coriaceous, awn-pointed, usually pilose especially near base with spreading hairs; sterile lemmas about 3/4 as long as spikelet, hyaline, long-pointed; fertile lemmas 3/4 as long as spikelet, 3-nerved, hyaline, awnless or with a slender awn at tip 3-5 mm. long; fertile paleas shorter than lemmas, hyaline.---Cultivated for ornament, established in Arizona; native to Europe. Occasionally cultivated in Colorado.

78. **Androgopon** L. BLUESTEM

Perennial plants; culms solid; leaf blades flat; sheaths split; ligules membranous; inflorescence of spikelike racemes, these solitary, or 2-6 in approximate digitatelike clusters or many on a main rachis; spikelets in pairs, 1 sessile and perfect, the other pedicellate and either staminate, neuter or reduced; glumes of fertile spikelet coriaceous, awnless; sterile lemmas shorter than glumes, hyaline; fertile lemmas narrow, hyaline, awnless or bearing a bent or twisted awn from apex or from between apical lobes; fertile paleas hyaline, smaller than lemmas; styles distinct, stigmas plumose.

1. Racemes solitary on long peduncles or if several on a culm these widely separated　　-1. **A. scoparius**
1. Racemes sessile or on very short peduncles, 2 or more clustered together
   2. Racemes more than 6, on an elongated axis; rachis of racemes and pedicels flattened with center thin and margins thicker; pedicellate spikelets rudimentary and much smaller than sessile ones　-2. **A. saccharoides**
   2. Racemes 2-6, digitate or closely clustered on the short axis; rachis of racemes and pedicels not thinner in center; pedicellate spikelets about as large as sessile ones
      3. Rhizomes short or wanting; hairs of rachis joints and pedicels less than 3 mm. long; awns of sessile spikelets 7-20 mm. long; plants not limited to sandy soil　　-3. **A. gerardi**
      3. Rhizomes long and creeping; hairs of rachis joints and pedicels over 3 mm. long; awns of sessile spikelets rarely over 5 mm. long, often lacking; plants limited to sandy soil in Colorado
      　　　　-4. **A. hallii**

1. Andropogon scoparius Michx., Fl. Bor. Am. 1:57. 1803.
Schizachyrium scoparium (Michx.) Nash---Rhizomes absent or sometimes short ones present; culms 40-120 cm. tall, tufted, erect, compressed, glaucous to purple, freely branching above; leaf blades 3-7 mm. wide, 5-25 cm. long, flat or sometimes folded, margins scaberulous, surfaces glabrous or pubescent; sheaths flattened, glabrous, scabrous or pubescent; ligules 1-2 mm. long; racemes 2-6 cm. long, each on a long peduncle, usually of 4-10 joints, apex of joints cup-shaped, rachis pilose; spikelets with pedicellate one reduced or rarely staminate, sessile one 6-8 mm. long, the pedicel pilose with spreading hairs 1-3 mm. long; first glume flattened; nerved near margins, scabrous, firm; second glume keeled, scabrous on keel, thinner than first, ciliate on hyaline margins, both glumes length of spikelet; sterile lemmas almost as long as spikelet, hyaline, ciliate, acuminate; fertile lemmas almost as long as spikelet, hyaline, 2-toothed at apex and awned between, the awn twisted and bent, 8-18 mm. long.---Prairies, dry hills, fields and open woods, often in sandy soil. Quebec and Maine to Alberta and Idaho, south to Florida and Arizona. Our specimens from the eastern half of Colorado as far west as Larimer, Chaffee and Archuleta Counties. Altitude 3500-8000 feet.

2. Andropogon saccharoides Swartz., Prod. Veg. Ind. Occ. 26. 1788.
Amphilophus saccharoides (Sw.) Nash; *Amphilophus torreyanus* (Steud.) Nash---Rhizomes lacking; culms 50-100 cm. tall, tufted, erect or ascending, often branched below, green or glaucous, nodes glabrous to appressed-hispid; leaf blades 3-6 mm. wide, 5-20 cm. long, upper shorter, commonly glaucous, glabrous or somewhat scabrous; sheaths somewhat keeled above, glabrous or sometimes pubescent on lower; ligules 1-2 mm. long; panicle 5-15 cm. long, silvery-white, silky, dense, oblong, of numerous racemes 2-4 cm. long, rachis and pedicels flat with thin, scarious center and thicker margins, the latter villous with hairs 5-6 mm. long; spikelets with pedicellate one rudimentary, reduced to a scale 2-3 mm. long, sessile one 3.5-4.5 mm. long, callus bearded; first glume length of spikelet, flat in section, villous on lower half, firm, nerves not prominent, scabrous toward apex; second glume about length of spikelet, keeled, rather U-shaped in section, scabrous near apex, firm; sterile lemmas shorter than spikelet, hyaline; fertile lemmas shorter than spikelet, narrow, the awn 10-20 mm. long, twisted below.---Prairies, plains and rocky slopes. Missouri to Colorado and Alabama to Arizona and California; also in Mexico to Brazil. Our specimens all from southeastern Colorado as far west as Fremont County at 3500-5300 feet.

3. Andropogon gerardi Vitman, Summa Pl. 6:16. 1792.
A. furcatus Muhl.; A. provincialis Lam.---Rhizomes absent or sometimes short thick ones present; culms 80-200 cm. tall, in large tufts, rather stout, often glaucous, purplish; leaf blades 4-10 mm. wide, 10-45 cm. long, flat or somewhat folded, margins scarious, glabrous or papillose-pilose on upper surface near base, sometimes on margins; sheaths somewhat flattened, glabrous or pubescent; ligules 1-3 mm. long, margin short-toothed, may be backed by hairs 6-7 mm. long; racemes 4-10 cm. long, 2-6 digitate, spikelike, usually purplish, rachis flat but not particularly thin in center, ciliate and fringed at top of internodes, the hairs usually 1-2 mm. long; spikelets 7-10 mm. long, pedicellate spikelet similar to sessile one but not awned, pedicel

flat and ciliate; glumes subequal, about length of spikelet, first one flattened somewhat grooved in middle, usually scabrous, lateral **nerves** more prominent, second glume rounded-keeled, scabrous on keel; sterile lemmas slightly shorter than spikelet, hyaline, flat; fertile lemmas hyaline, with a bent twisted awn 7-18 mm. long.---Prairies, meadows and sandy hills. Quebec and Maine to Saskatchewan and Montana, south to Florida, Wyoming, Utah and Arizona; also in Mexico. Our specimens from the eastern half of Colorado mostly from along the foothills. Might be expected anywhere in the state. Altitude 3500-9500 feet but usually below 6500 feet.

4. Andropogon hallii Hack., Sitzungsb. Akad. Wiss. 89:127. 1884.

Rhizomes long, robust and yellow; culms 60-200 cm. tall, more or less glaucous, glabrous; leaf blades 3-10 mm. wide, 20-30 cm. long, margins often scabrous, sometimes pubescent above near base; sheaths somewhat keeled near top, glabrous; ligules 2-5 mm. long, villous just behind; racemes 2-5, approximate or digitate, spikelike, 5-10 cm. long, lateral ones often partly included at base, rachis copiously villous-ciliate, the hairs 3-5 mm. long, flattened and widest in center; spikelets with pedicellate one staminate, the pedicel long-ciliate with hairs 3-5 mm. long, sessile spikelet 8-11 mm. long, often awned, hairs all light gray to pale yellowish; first glume flat and somewhat grooved in center; second glume round-keeled, both glumes length of spikelet, subequal, glabrous below, pubescent near apex; sterile lemmas shorter than glumes, hyaline, awnless; fertile lemmas shorter than glumes, membranous or hyaline, awnless or with an awn less than 7 mm. long.---Sand hills and sand dunes. North Dakota and Montana to Texas, Wyoming, Utah and Arizona; also in Iowa. Our specimens scattered in the sandy areas of the eastern half of Colorado at 3500-5200 feet.

79. Sorghum Moench. SORGHUM; JOHNSONGRASS

Annual or perennial plants with rhizomes; culms stout; leaf blades flat; sheaths split; ligules membranous but ciliate, the hairs less than 1/2 the total length; inflorescence an open panicle; spikelets in pairs, 1 sessile and perfect, the other pedicellate, well developed and usually staminate, end of rachis with 3 spikelets instead of 2, disarticulating below the glumes; glumes about length of spikelet, chartaceous; sterile lemmas shorter than glumes, hyaline; fertile lemmas shorter than glumes, hyaline, with a deciduous, flat, twisted and bent awn; paleas small and narrow.

1. Stout creeping rhizomes present; plants seldom if ever cultivated, often a weed; sessile spikelet breaking away from the rachis evenly, leaving a small cuplike structure -1. S. halepense
1. Annual plants, no rhizomes; plants cultivated as a field crop; sessile spikelet breaking away unevenly, carrying a small portion of the rachis with it -2. S. vulgare

1. Sorghum halepense (L.) Pers., Syn. Pl. 1:101. 1805.

Holcus halepensis L.---Perennial plants with long, stout, scaly rhizomes; culms 50-150 cm. tall, erect; leaf blades 5-30 mm. wide, 20-50 cm. long, rather narrowed to base, margins sometimes scaberulous, glabrous rarely somewhat hairy, midnerve conspicuous below; sheaths glabrous; ligules 2-4 mm. long, ciliate, the hairs 1/4-1/3 the total length; panicle 15-50 cm. long, finally open, oblong to oval, branches whorled, branchlets with racemes; spikelets with pedicellate narrower, usually staminate, sessile one 4.5-5.5 mm. long, pale to purplish; (sessile spikelet)--first glume about as long as spikelet, 5- to 7-nerved these not conspicuous, sparsely to densely pubescent, rather dorsally flattened, 3-toothed at the obtuse apex; second glume about same as first but narrower, keeled toward the pointed apex; sterile lemmas somewhat shorter than glumes, ciliate; fertile lemmas oval, ciliate, 2-lobed, the awn 7-15 mm. long.---Open ground, fields, roadsides and waste places as a weed. Massachusetts to Iowa and Kansas, south to Florida and Texas, west to California; native to the Mediterranean region. Our specimens from southeastern Colorado (one doubtful one from Yuma County) at 3500-4200 feet.

2. Sorghum vulgare Pers., Syn. Pl. 1:101. 1805.

Differs from number 1 in being an annual plant, with spikelets averaging a little longer, the sessile spikelet breaks away with part of the rachis and leaves an uneven break.---There are many varieties of this cultivated species, a key to which is given in "U. S. D. A., Bur. Pl. Ind. Bul. 175, April 8, 1910." Some of the common names of these varieties are Sorghum, Sudangrass, Broomcorn, Kafir Corn and Milo.

80. Sorghastrum Nash INDIANGRASS

Sheaths split; ligules membranous; spikelets borne in pairs (theoretically), 1 sessile and perfect, 1 pedicellate and reduced or absent entirely, rachis disarticulating just below the pairs of spikelets.

1. Sorghastrum nutans (L.) Nash in Small, Fl. S. E. United States 66. 1903.

Perennial plants with short, scaly rhizomes; culms 40-100 cm. tall, erect, rather robust, nodes villous; leaf blades 3-10 mm. wide, 10-30 cm. long, flat, scabrous below and often above; sheaths glabrous or pilose, hairs especially long near collar; ligules 2-4 mm. long, veins of sheaths extending up into it in front; panicle 12-30 cm. long, narrowly oblong, rather dense, apex usually nodding, bronze-yellow, rachis joints and pedicels gray-hirsute; spikelets 6-8 mm. long, nearly terete, yellowish or golden-brown, the pedicel pilose and without a spikelet; first glume length of spikelet, 5- to 9-nerved these indistinct, narrowed to obtuse apex, leathery-cartilaginous, pilose, rather flat on back; second glume similar but 5- to 7-nerved, and short-hairy only near margins above; sterile lemmas somewhat shorter than glumes, hyaline, ciliate; fertile lemmas shorter than glumes, hyaline, ciliate, 2-toothed at apex and between them an awn 10-16 mm. long, flat, bent, tightly twisted to bend, loosely twisted to end; paleas very short or obsolete.---Prairies, open woods, dry slopes, meadows and bottom lands. Quebec and Maine to Manitoba and North Dakota south to Florida and Arizona; also in Mexico. Our specimens scattered over the eastern half of Colorado except in the extreme south at 3500-6800 feet.

XII. Tribe Tripsaceae

81. Zea L. MAIZE; INDIAN CORN

Sheaths split; ligules membranous; spikelets imperfect, the staminate at top of culm (tassel), the pistillate crowded on a thickened axis (cob), the whole included in many bracts or spathes (husks), the stigmas protruding at the end (silks), the whole forming a structure called an "ear."

1. Zea mays L., Sp. Pl. 971. 1753.
Annual plants; culms 100-500 cm. tall; leaf blades broad, channelled, crinkly-edged; sheaths overlapping; ligules hyaline, short; plant monoecious; (staminate)--terminal of several spike-like racemes mostly 15-30 cm. long; spikelets 2-flowered, in pairs, 1 sessile, the other pedicillate; glumes membranous, acute; lemmas and paleas hyaline; (pistillate)--inflorescence in axils of leaves, 8-16 or more rows on the axis; spikelets sessile, in pairs, of 1 fertile floret and 1 sterile floret, this latter sometimes developed; glumes broad, rounded or emarginate at apex; sterile lemmas hyaline; fertile lemmas and paleas hyaline; grains at maturity much exceeding the bracts; style long and slender.---In cultivation and sometimes escaping but not persisting. Several races, including sweet corn and popcorn are grown in Colorado.

Family 18. Cyperaceae SEDGE FAMILY

Grasslike or rushlike herbaceous plants; stems (culms) usually solid and triangular; leaves when present 3-ranked, mostly narrowly linear with closed sheaths; flowers perfect or imperfect each subtended by 1 or sometimes 2 (when a perigynium is present) scales, these 2-ranked or spirally imbricated and forming 1 to many spikes (often called spikelets); perianth when present of hypogynous bristles; stamens 1-3; ovary superior, 1-celled and 1-ovuled, stigmas 2-3; fruit an achene.

1. Scales 2-ranked, the spikes often strongly flattened; perianth bristles none; flowers perfect   -1. Cyperus
1. Scales spirally imbricated, the spikes terete or sometimes slightly flattened; perianth bristles present or absent; flowers perfect or imperfect
  2. Achene enclosed by an inner scale (called a perigynium--this is a saclike, often beaked envelope with an apical opening or split down 1 side) in addition to the outer scale; flowers imperfect; perianth bristles wanting
    3. Perigynium split to the base, enclosing both an achene and a staminate flower or flowers; inflorescence a single spike
          -2. Kobresia
    3. Perigynium closed except for the apical orifice (may be partly split), enclosing only the achene; inflorescence of 1 or more than 1 spike
          -3. Carex
  2. Achenes subtended by 1 outer flat scale only and not enclosed by it; flowers perfect (at least some in each spikelet); perianth bristles present or absent
    4. Perianth bristles many, long-silky, appearing at maturity like conspicuous tufts of cotton many times longer than the scales
          -4. Eriophorum
    4. Perianth bristles none or few and short, not at all cottonlike
      5. Perianth of 1 very thin hyaline scale (less than 1 mm. long); stamens 1; achenes 0.8 mm. long or less
          -5. Hemicarpha
      5. Perianth none or bristlelike (these often over 1 mm. long); stamens usually over 1; achenes over 0.8 mm. long at maturity
        6. Spikes 1 to a culm
          7. Outer scales (lower) bracteate (this bract 2-3 mm. long); bases of styles not tuberculate or different in texture from the tips of the achenes   -6. Scirpus
          7. Outer scales not bracteate (lower scales may be empty but not enlarged); bases of styles tuberculed and different in texture from the tips of the achenes (this may be obscure in a few)
            -7. Eleocharis
        6. Spikes 2 or more to a culm
          8. Style-bases swollen, these styles and their 2 branches rather flattened; perianth bristles none
            -8. Fimbristylis
          8. Style-bases not swollen or very little so; styles and their 2 or 3 branches terete; perianth bristles usually present
            -6. Scirpus

1. Cyperus L. FLATSEDGE

Annual or perennial plants; culms simple, triangular, leafy near base; leaf blades narrow; bracts 1 or more; inflorescence of more than 1 spike, these sessile or on long rays (peduncles), the central one commonly sessile; scales 2-ranked, sometimes decurrent; perianth none; flowers perfect; stamens 1-3; styles deciduous, stigmas 2-3; achenes lenticular or trigonous, not tuberculate.

1. Plants annual; culms not tuberous-thickened at base nor bearing tubers at end of rootstocks
   2. Styles 2-cleft; achenes lenticular
      3. Scales 2.5-3 mm. long; style branches long-exserted, divided nearly to the achenes -1. C. diandrus
      3. Scales 2-2.3 mm. long; style branches short-exserted, divided less than 1/3 from the base
         -2. C. rivularis
   2. Styles 3-cleft; achenes 3-angled
      4. Awns of scales over 0.5 mm. long -3. C. aristatus
      4. Awns of scales absent or apex cuspidate or mucronate, the point never over 0.5 mm. long
         5. Spikes in globose heads; rachillas wingless or nearly so -4. C. acuminatus
         5. Spikes in ovate to cylindric clusters; rachillas bearing freely-deciduous, scalelike wings
            -5. C. erythrorhizos
1. Plants perennial; culms from tuberous-thickened bases or with narrow rootstocks bearing tubers at ends
   6. Spikes in globose or subglobose heads -6. C. filiculmis
   6. Spikes in oblong or narrower heads
      7. Rachillas wingless or obscurely winged; scales with mucros over 0.3 mm. long
         8. Rays well-developed, some usually over 3 cm. long; mucros of scales straight or incurved
            -7. C. schweinitzii
         8. Rays wanting or rarely very short; mucros of scales excurved -8. C. fendlerianus
      7. Rachillas winged, this wing over 0.4 mm. wide; scales acute or mucronulate, the point less than 0.3 mm. long
         9. Plants with slender rootstocks ending in tubers; rachillas persistent, the scales deciduous from it; leaves over 6 to a culm -9. C. esculentus
         9. Plants without tuber-bearing rootstocks (base of culms tuberlike); rachillas finally deciduous; leaves seldom over 6 to a culm -10. C. strigosus

1. Cyperus diandrus Torr., Cat. Pl. N. Y. 90. 1819.
This eastern species has been recorded from as far west as Colorado but no specimens could be located from the state.

2. Cyperus rivularis Kunth, Enum. 2:6. 1837.
Plants annual; culms 4-40 cm. tall, caespitose; leaves 0.5-3 mm. wide, mostly flat, 1-3 to a culm; sheaths light brown; rays none or to 5, with 2-11 spikes; bracts 2-4, unequal, leaflike and longer than the inflorescence; spikes 3-20 mm. long, 2-3.5 mm. wide, 6- to 32-flowered; scales 2-2.3 mm. long, 1-3 mm. wide, about equal, obtuse and little if any mucronate, keels green, sides chestnut-brown to dark brown, not hyaline; stamens 2; style branches 2, divided 1/3 up from base or less; rachillas more or less persistent, about 0.5 mm. wide.---Along streams and borders of ponds. Maine to Kansas and south to North Carolina. The few Colorado records from the eastern plains of Colorado at 3500-5500 feet.

3. Cyperus aristatus Rottb., Descr. et Icon. 4:23. 1773.
C. inflexus Muhl.---Plants annual; culms 1-20 cm. tall, caespitose, somewhat odoriferous on drying; leaves 2-10 cm. long, 0.5-3 mm. wide, 2-3 to a culm; sheaths purplish-brown especially near base; inflorescence of 1-3 congested heads, often with 1 or more rays, the heads with 2-5 spikes; bracts 2-4, leaflike and longer than the inflorescence; spikes 4-14 mm. long, 0.8-2 mm. wide, 5- to 20-flowered; scales 1-1.8 mm. long with an awn 0.5-1 mm. long, this somewhat recurved, oblong-lanceolate with several prominent nerves; stamens 1; rachillas wingless, commonly deciduous after the scales; 3 style branches present.---In moist often sandy ground. Throughout most of North America; in South America and the Eastern Hemisphere. Scattered in the eastern half of Colorado at 3500-9500 feet.

4. Cyperus acuminatus Torr. & Hook., Ann. Lyc. N. Y. 3:435. 1836.
Annual plants; culms 5-35 cm. tall, tufted, very slender; leaves usually less than 2 mm. wide, light green; bracts much elongated; umbels 1- to 4-rayed, the rays short, spikes in capitate clusters; spikes 4-8 mm. long, over 2 mm. wide, many-flowered, flat and ovate-oblong; scales oblong, pale-green or yellowish, 3-nerved, cellular with a strong lens, with short, sharp, more or less recurved tips less than 0.5 mm. long; stamens 1; styles 3-cleft; achenes 3-angled; oblong, about 1/2 as long as the scales.---Moist or wet ground. Illinois to California, south to Florida and Arizona. Our one record from Boulder County at 5500 feet.

5. Cyperus erythrorhizos Muhl., Gram. 20. 1817.
Massachusetts to California south to Florida and Arizona. Should be in Colorado, but no records known to the writer.

6. Cyperus filiculmis Vahl., Enum. Pl. 2:328. 1806.
C. bushii Britt.---Perennial plants with short thick rootstocks; culms 10-45 cm. tall, usually glabrous, tuberous-thickened at base; leaves 5-25 cm. long, 2-4 mm. wide, flat, 2-4 on a culm; sheaths becoming tan or yellowish; inflorescence a solitary sub-globose spike or with 2-4 rays up to 7 cm. long; bracts 3-4, leaflike, unequal, longer or shorter than the inflorescence; spikes 5-12 mm. long, 1-2 mm. wide, 5- to 11-flowered, oblong-lanceolate to nearly linear; scales 1.8-3.5 mm. long, 7- to 11-nerved, readily deciduous, oblong, mucronulate or with a mucro up to 0.3 mm. long; stamens 3; style branches 3; rachilla wingless, more or less deciduous. This species tends to intergrade with C. schweinitzii in Colorado.---Dry often sandy ground. Minnesota to Washington south to Colorado, Texas and Missouri. Our records from the eastern half of the state at 4000-6000 feet.

7. Cyperus schweinitzii Torr., Ann. Lyc. N. Y. 3:276. 1836.
Perennial plants with short, rather thick rootstocks; culms 6-70 cm. tall, trigonous, usually scabrous above, from tuberous-thickened base; leaves 2-6 mm. wide, 7-35 cm. long, flat, about 3-5 to a culm; sheaths brownish or yellowish below; inflorescence with 1-6 unequal rays, longest up to 10 cm. long, ascending, the spikes at the ends; bracts 3-6, leaflike, unequal, shorter or longer than the inflorescence; spikes 5-20 mm. long, 2-3 mm. wide, ovate or linear-lanceolate; scales 2-3.5 mm. long, ovate-elliptic, equal, deciduous, 9- to 15-nerved, keel about straight, ending in a mucro less than 1 mm. long this somewhat incurved; stamens 3; style branches 3;

rachilla wingless, more or less readily deciduous.---Rather dry often sandy soil. Ontario to Saskatchewan, south to New Mexico and Missouri. Apparently rather common in the eastern half of Colorado at 3500-5500 feet.

8. Cyperus fendlerianus Boeckel., Linnaea 35:520. 1868.

Plants perennial from short, rather thick rootstocks; culms 20-50 cm. tall, usually scabrous above, from tuberous-thickened bases; leaves 1-2.5 mm. wide, 5-30 cm. long, flat, 2-3 to a culm; sheaths yellow or yellowish-brown; bracts leaflike, shorter or longer than the inflorescence; inflorescence congested and headlike; spikes 4-10 mm. long, 3- to 8-flowered, oblong or narrower; scales 2-2.5 mm. long, 9- to 13-nerved, membranous or somewhat thicker, somewhat greenish at least on midrib, with an excurved mucro up to 0.5 mm. long; stamens 3; style branches 3; rachilla wingless, more or less deciduous but usually not before the scales.---Moist places. Texas to Colorado and south to Arizona and New Mexico. Our few records from central Colorado at 6000-7500 feet but to be expected anywhere in the southern part.

9. Cyperus esculentus L., Sp. Pl. 45. 1753.

Plants perennial with scaly rootstocks bearing terminal tubers; culms 25-80 cm. tall, rather stout; leaves 4-8 mm. wide, often over 6 to a culm; bracts 3-6, the longer much exceeding the inflorescence; umbel 4- to 10-rayed, often compound; spikes 8-25 mm. long, about 3 mm. wide, many-flowered, numerous, straw-colored or yellowish; scales ovate-oblong, subacute, 3- to 5-nerved; achenes 3-angled, styles 3-cleft; rachilla winged and persistent.---Moist ground. Ontario to Alaska and south to tropical America; eastern hemisphere. Our one record from the northeastern corner of Colorado at 3500 feet.

10. Cyperus strigosus L., Sp. Pl. 47. 1753.

This species is widespread in the United States and is to be expected in Colorado. However, no records could be found by the writer.

2. Kobresia Willd.

Perennial herbaceous plants; culms slender, leafy below; inflorescence terminal and spikelike; bracts absent or very small; flowers monoecious, usually 2 together, the inner pistillate with a perigynium but the margins of this free to base or united only below, outer staminate with 3 stamens also enclosed by the perigynium; perianth none; styles slender and continuous to achene, stigmas 3.

1. Inflorescence 2-4 mm. wide; scales 2-4 mm. long; perigynia 3-3.5 mm. long; culms usually exceeding the leaves
-1. K. bellardi
1. Inflorescence 4-5 mm. wide; scales 4-5 mm. long; perigynia about 5.5 mm. long; culms exceeded by the leaves
-1A. K. bellardi macrocarpa

1. Kobresia bellardi (All.) Degland in Loisel., Fl. Gall. 626. 1807.

Elyna bellardii (All.) Koch---Culms 3-45 cm. tall, slender, densely tufted; leaves narrow-filiform, revolute; old leaf sheaths brown and fibrillose; inflorescence densely-flowered or interrupted below, 15-30 mm. long, 3-4 mm. in diameter; scales 2-4 mm. long, obtuse; perigynium 3-3.5 mm. long; achenes shorter than the perigynia.---Moist open places. Greenland to Alaska south to Colorado and Oregon; Europe. Scattered on the high peaks in Colorado at 9000-12,500 feet.

1A. Kobresia bellardi macrocarpa (Clokey) comb. nov. (var.)

K. macrocarpa Clokey---Culms 3-10 cm. tall, densely tufted, stiff and smooth; leaves 0.25-0.5 mm. wide; leaf sheaths brown and conspicuous; inflorescence 10-20 mm. long, 4-5 mm. wide; scales 4-5 mm. long, obtuse; perigynium 5.5 mm. long; achenes shorter than the perigynia.---Rather moist open ground. Known only from central Colorado at about 13,000 feet.

3. Carex L. SEDGE

Perennial grasslike herbaceous plants with long-creeping or short rootstocks (called stolons in some manuals), sod-forming or bunch type; culms mostly triangular, phyllopodic or aphyllopodic; leaves 3-ranked; sheaths closed; bracts present or absent; flowers monoecious or sometimes dioecious, in 1 or more spikes (when small often called spikelets), the spikes pistillate, staminate, androgynous or gynaecandrous; flowers solitary in the axils of scales; staminate flowers of 3 stamens; pistillate flower an ovary with 2 or 3 stigmas, completely surrounded except at apex by a perigynium, styles jointed to achene or continous; perianth lacking; achenes trigonous, lenticular or plano-convex. This is a difficult genus, partly because of the large number of species involved. Mackenzie's treatment (N. A. Flora 18:1-478. 1931-1935.) is followed in general although many "species" probably should be considered as varieties.

1. Spikes 1 to a culm
  2. Stigmas 2; achenes lenticular
    3. Heads orbicular to ovoid-orbicular, dense (really made up of several crowded spikes), over 7 mm. wide
      4. Leaf blades not over 1.5 mm. wide; perigynia ovate-elliptical, less than twice as long as wide, inflated
-43. C. perglobosa
      4. Leaf blades 2-4 mm. wide; perigynia ovate-lanceolate, over twice as long as wide, not inflated
-44. C. vernacula
    3. Heads more elongated, less than 7 mm. wide (only 1 true spike present)
      5. Spikes entirely staminate or pistillate (or nearly so); plants not caespitose but with long rootstocks; perigynia oblong-ovate
-1. C. gynocrates
      5. Spikes androgynous; plants loosely to closely caespitose; perigynia ovate, elliptical-lanceolate to lanceolate
        6. Perigynia ovate, 2-3 mm. long, ascending or spreading
-2. C. capitata

             6. Perigynia lanceolate or elliptical-lanceolate, 3-4.5 mm. long, erect-appressed   -3. C. hepburnii
    2. Stigmas 3 (rarely 4); achenes triangular
       7. Perigynia pubescent or puberulent (may be very sparse and only near base of beak)
          8. Spikes entirely staminate or pistillate or nearly so
             9. Fertile culms with 2-4 well developed leaves, not clothed at base with the dried leaves of the
                previous year; scales somewhat shorter and narrower than the perigynia; beaks of perigynia
                about 0.25 mm. long; plants aphyllopodic   -4. C. scirpoidea
             9. Fertile culms with 5-10 well developed leaves, clothed at the base with last year's dried
                leaves; scales longer and wider than the perigynia; beaks of perigynia about 0.5 mm. long;
                plants phyllopodic   -5. C. pseudoscirpoidea
          8. Spikes androgynous
             10. Leaf blades flattened or somewhat folded at base, 1.5-2 mm. wide; culms stoutish; perigynia
                 usually over 3.7 mm. long, the beaks about 1 mm. long   -6. C. oreocharis
             10. Leaf blades involute at base (resembling culms), 0.25-0.5 mm. wide; culms filiform; perigynia
                 less than 3.7 mm. long, the beaks about 0.5 mm. or less long
                 11. Beak of perigynia 0.2-0.4 mm. long, margins of body rounded on the angles, the whole
                     perigynium often over 3 mm. long; pistillate scales with bright white hyaline margins;
                     basal sheaths usually strongly filamentose; below 8000 feet in Colorado -7. C. filifolia
                 11. Beak of perigynia about 0.5 mm. long, margins of body rather sharply angled, the whole
                     perigynium seldom over 3 mm. long; pistillate scales with dingy white hyaline margins;
                     basal sheaths little if at all filamentose; above 8000 feet in Colorado -8. C. elynoides
       7. Perigynia glabrous
          12. Spikes entirely staminate or pistillate, some of heads usually with 2 or more spikes
                                                                                                 -78. C. hallii
          12. Spikes androgynous, only 1 to a culm
             13. Perigynia subulate to linear-lanceolate; styles continuous with the achenes; lower perigynia
                 strongly reflexed   -9. C. microglochin
             13. Perigynia lanceolate or broader; styles jointed to the achenes; perigynia not strongly
                 reflexed (except in the next 2 species)
                 14. Pistillate scales soon deciduous; perigynia stipitate (the stipe sometimes to 0.5 mm.
                     long), the lower usually reflexed at maturity, lanceolate or ovate-lanceolate
                     15. Leaf blades 0.25-1.5 mm. wide; staminate flowers few, occupying less than 1/3 of
                         the spike, this 3-5 mm. wide   -10. C. pyrenaica
                     15. Leaf blades 1.5 mm. wide or more; staminate flowers many, usually occupying about
                         1/2 the length of the spike, this 6-9 mm. wide   -11. C. nigricans
                 14. Pistillate scales not deciduous; perigynia sessile or substipitate, the lower not re-
                     flexed, variously shaped but not usually lanceolate or ovate-lanceolate
                     16. Perigynia definitely inflated, often 15 or more to a spike, ovoid in shape
                                                                                                 -12. C. engelmannii
                     16. Perigynia not inflated, less than 15 to a spike, variously shaped but rarely ovoid
                         17. Beaks of perigynia definite, about 0.5 mm. long or more
                             18. Leaves flat or channeled, 1 mm. wide or more; perigynia coriaceous,
                                 oblong-ovoid, their beaks with hyaline tips, 1-6 to a spike; plants with
                                 long-creeping rootstocks   -13. C. obtusata
                             18. Leaves setaceous, less than 0.5 mm. wide; perigynia membranous, lanceolate,
                                 elliptical-lanceolate or obpyramidal, the beaks without hyaline tips,
                                 5-15 to a spike; plants caespitose
                                 19. Perigynia 3.5-4.5 mm. long, lanceolate or elliptical-lanceolate;
                                     staminate flowers occupying less than 1/2 the length of the spike
                                                                                                 -3. C. hepburnii
                                 19. Perigynia 2.5-3 mm. long, obpyramidal; staminate flowers occupying
                                     1/2 or more of the spike   -8. C. elynoides
                         17. Beaks of perigynia absent or minute, not over 0.2 mm. long
                             20. Perigynia large, over 5 mm. long and 2 mm. wide at maturity, 1-3 to a
                                 spike   -14. C. geyeri
                             20. Perigynia shorter and narrower than above, 1-15 to a spike (some usually
                                 over 3)
                                 21. Culms 15-60 cm. tall, usually definitely overtopping the leaves;
                                     leaves less than 1.5 mm. wide, about 2 to a fertile culm
                                                                                                 -15. C. leptalea
                                 21. Culms 6-12 cm. tall, overtopped by the leaves; leaves 2-3 mm. wide,
                                     8-12 to a fertile culm   -16. C. drummondiana
 1. Spikes 2 or more to a culm
    22. Stigmas 2; achenes lenticular
       23. Lateral spikes sessile, short, terminal spikes usually not entirely staminate
          24. Perigynia narrowly wing-margined (at least on upper part) to broadly so; spikes (at least some)
              gynaecandrous
             25. Culms arising singly or few together from long-creeping rootstocks; lateral spikes usually
                 pistillate or staminate   -17. C. siccata
             25. Culms caespitose, rootstock if present very short; lateral spikes usually bisexual
                 26. One or more bracts long, conspicuously exceeding the heads   -18. C. athrostachya
                 26. Bracts shorter, not exceeding the heads
                     27. Scales about as long and as wide above as the perigynia, concealing the latter
                         above or nearly so
                         28. Perigynia oblong-lanceolate, more than 3 times longer than wide (counting
                             beak), 5.5-8 mm. long and ventrally nerved   -19. C. petasata

119

28. Perigynia usually ovate or obovate, less than 3 times longer than wide, if over 5.5 mm. long then nervelessly ventrally
  29. Beaks of perigynia short, less than 1 mm. long, terete, not at all winged or margined; achenes rarely over 1.5 mm. long   -20. C. phaeocephala
  29. Beaks of perigynia 1 mm. long or more, at least somewhat flattened, winged or margined especially toward the base; achenes usually over 1.5 mm. long
    30. Perigynia finely nerved ventrally, broadest above the middle; culms less than 40 cm. tall; spikes closely approximate; scales obtuse   -21. C. arapahoensis
    30. Perigynia nerveless ventrally, usually broadest below the middle; culms often over 40 cm. tall; spikes from somewhat approximate to moniliform; scales acute (or sometimes obtuse in C. praticola)
      31. Beaks of perigynia flattened, little raised in center, somewhat bidentate at ends, 1-1.3 mm. long; perigynia ovate or elliptical-ovate, 4-5.5 mm. long; achenes 2.5 mm. long   -22. C. xerantica
      31. Beaks of perigynia raised and rounded in the center (although somewhat winged) little bidentate at end, 1.25-2 mm. long; perigynia ovate-lanceolate to narrowly ovate, 4.5-6.5 mm. long; achenes 1.5-2 mm. long   -23. C. praticola
27. Scales shorter than the perigynia and narrower above, the perigynia exposed near the apex
  32. Perigynia 3.5 mm. long or less, plano-convex, the margins very narrowly winged (hardly more than sharp-edged in C. illota)
    33. Perigynia merely sharp-edged, not really winged; scales obtuse (sometimes almost acute), blackish-brown, wider than perigynia; culms 10-35 cm. tall   -24. C. illota
    33. Perigynia narrowly winged; scales brown, acute or acuminate, narrower than perigynia; culms 20-80 cm. tall   -25. C. bebbii
  32. Perigynia over 3.5 mm. long, definitely wing-margined, often flat and scalelike
    34. Perigynia narrowly lanceolate to narrowly ovate-lanceolate, usually over 3 times longer than wide (counting beak)
      35. Scales brownish-black or blackish; perigynia brownish-black or blackish-tinged
        36. Perigynia narrowly lanceolate, 6-7 mm. long, the beaks appressed; culms erect   -26. C. ebenea
        36. Perigynia usually broader than lanceolate, 4.5-6 mm. long, the beaks conspicuous in the spike, not appressed; culms ascending or decumbent   -27. C. haydeniana
      35. Scales brown; perigynia greenish-white or straw-colored
        37. Scales light brown, center green; perigynia often greenish-white; lower bracts often setaceous; lower spikes often somewhat remote   -28. C. scoparia
        37. Scales chestnut-brown with light brown center; perigynia never greenish-white; lower bracts scalelike (may be short-awned from a broad tip); spikes all crowded   -29. C. stenoptila
    34. Perigynia ovate or broader, less than 3 times longer than wide
      38. Perigynia broadly ovate to almost orbicular; scales yellowish-brown   -30. C. brevior
      38. Perigynia ovate (rarely broadly ovate in C. haydeniana with blackish-brown or black scales); scales chestnut-brown to black
        39. Perigynia 6-7 mm. long, the beak flattened and little raised in center; scales chestnut-brown; spikes 10-14 mm. long   -31. C. egglestoni
        39. Perigynia 3.75-6 mm. long, the beak (although winged) somewhat raised in center; scales chestnut-brown to blackish; spikes 5-12 mm. long
          40. Perigynia 3.75-5 mm. long, ovate; culms 30-100 cm. tall, rather stiffly erect or ascending; scales chestnut-brown to brownish-black
            41. Perigynia plano-convex, copper-colored at maturity, narrowly winged   -32. C. pachystachya
            41. Perigynia flat, green or straw-colored (may be brown at edges) at maturity, definitely winged   -33. C. festivella
          40. Perigynia 4.5-6 mm. long, lanceolate to broadly ovate (can be ovate); culms 10-40 cm. tall, often curving or decumbent at base; scales brownish-black or blackish   -27. C. haydeniana
24. Perigynia not wing-margined, at most sharp-edged; spikes gynaecandrous, androgynous or unisexual
  42. At least some of the spikes gynaecandrous (the others may be staminate or pistillate) or heads dioecious
    43. Culms arising singly or a few together from long creeping rootstocks; perigynia over 3.5 mm. long
      44. Heads strictly dioecious; perigynia not over 4 mm. long, merely sharp-edged; culms 6-30 cm. tall   -34. C. douglasii
      44. Heads various but not dioecious; perigynia usually over 4 mm. long, very narrowly winged; culms 20-90 cm. tall   -17. C. siccata
    43. Culms caespitose, the rootstocks short if present; perigynia not over 3.5 mm. long (except in C. deweyana)
      45. Perigynia over 4 mm. long (4.5-5.5); scales often short-awned   -35. C. deweyana
      45. Perigynia less than 4 mm. long (1.8-3.5); scales never short-awned
        46. Spikes closely crowded in a dense suborbicular head; scales blackish-brown   -24. C. illota
        46. Spikes not closely crowded, the head somewhat elongated; scales brown or lighter in color
          47. Perigynia lanceolate to ovate-lanceolate, widest near base and tapering upward, beaks 1 mm. long or more, over 1/3 the length of the whole   -36. C. angustior
          47. Perigynia ovate or broader, widest at the middle or somewhat below the middle, beaks less than 1 mm. long, less than 1/3 the length of the perigynia
            48. Perigynia white-punctate, ascending to loosely spreading, beaks very short, not over 0.5 mm. long; spikes usually 5 or more to a head; bracts of lower spikes often prolonged to as long or longer than the spikes

49. Leaves glaucous, the blades 2-4 mm. wide; spikes with 10-30 perygynia on each, these appressed-ascending at maturity   -37. C. canescens
49. Leaves not glaucous, the blades 1-2.5 mm. wide; spikes with 5-10 perygynia, these loosely spreading at maturity   -38. C. brunnescens
48. Perygynia not white-punctate, squarrose at maturity, beaks 0.5-0.8 mm. long; spikes usually less than 5; bracts of lower spikes usually short   -39. C. interior
42. At least some of the spikes androgynous (the others usually staminate)
  50. Perygynia widest at very base, tapering up to end of beak in a nearly straight line on each side, strongly nerved above and below
    51. Perygynia 3-4 mm. long, the beak less than 1.5 mm. long, definitely shorter than the body; scales often dark-tinged; inflorescence rather capitate
      52. Leaves all clustered near base of culm; sheaths not green-mottled dorsally, not cross-rugulose; leaf blades less than 3 mm. wide   -40. C. jonesii
      52. Leaves scattered on the culm; sheaths green-mottled dorsally and cross-rugulose; leaf blades usually over 3 mm. wide   -41. C. neurophora
    51. Perygynia 4-6 mm. long, the beak over 1.5 mm. long, about as long as the body; scales not dark-tinged; inflorescence elongate   -42. C. stipata
  50. Perygynia widest at the middle or below the middle (but not very near base), the sides usually abruptly contracted into the beak (hence not straight on each side from widest part to end of beak), usually not strongly nerved especially ventrally
    53. Spikes so closely crowded into a round or ovoid-orbicular head that the individual ones cannot be discerned (appears like 1 thick spike)
      54. Leaf blades not over 1.5 mm. wide; perygynia ovate-elliptical, less than twice as long as wide, inflated at maturity; scales brown   -43. C. perglobosa
      54. Leaf blades 2-4 mm. wide; perygynia ovate-lanceolate, over twice longer than wide; scales dark brown to blackish   -44. C. vernacula
    53. Spikes not so densely crowded, usually somewhat separated, in any case the lower ones readily discernable
      55. Sheaths cross-rugulose ventrally; pistillate scales awned, the awn about 1/3 the entire length; leaves longer than the culm; most of the spikes with long bracts (nearly as long or longer than their spikes)   -45. C. vulpinoidea
      55. Sheaths not cross-rugulose ventrally (usually thin-hyaline); pistillate scales from obtuse to short-awned; leaves usually shorter than the culms; all or most of the spikes with short bracts
        56. Culms loosely to densely caespitose, the rootstocks short and stout
          57. Sheaths brown-spotted ventrally especially near mouth; spikes often over 10 to a head; perygynia ovate (widest part definitely nearer the base), 2-2.75 mm. long   -46. C. diandra
          57. Sheaths whitish, not brown-spotted ventrally (may be green and white mottled); spikes 4-10 to a head; perygynia ovate-elliptical, oblong-elliptical to oval (widest part at or near the middle), 2.5-5 mm. long
            58. Perygynia 3.5-5 mm. long, not really stipitate, serrate from the middle to end of beak; head of spikes orbicular to ovoid-oblong
              59. Sheaths dorsally septate-nodulose; scales tawny; beak of perygynium about 1/4 as long as the body   -47. C. gravida
              59. Sheaths not at all septate-nodulose; scales chestnut-brown; beak of perygynium about 1/3 as long as the body   -48. C. hoodii
            58. Perygynia 2.5-3.5 mm. long, or if longer then definitely stipitate, serrate at apex only; heads oblong to linear-oblong
              60. Beaks obliquely cleft dorsally, the perygynia not definitely stipitate; inflorescence usually less than 6 mm. wide; scales shorter than their perygynia   -49. C. vallicola
              60. Beaks nearly equally bidentate, the perygynia stipitate; inflorescence usually over 6 mm. wide; scales about as long as their perygynia   -50. C. occidentalis
        56. Culms arising singly or a few together from long-creeping, slender (long and stout in C. praegracilis) rootstocks
          61. Perygynia usually over 4 mm. long, usually over 2½ times longer than wide, narrowly wing-margined, the beaks over 1.5 mm. long; spikes various but seldom androgynous   -17. C. siccata
          61. Perygynia 1.75-4 mm. long, less than 2½ times longer than wide, not at all wing-margined, the beaks less than 1.5 mm. long; spikes usually androgynous
            62. Upper sheaths green-striated ventrally (see note in second 62); perygynia 10-20 to a spike; spikes often over 10 to a head   -51. C. sartwellii
            62. None of sheaths green-striated (do not mistake nerves of inner sheaths showing through hyaline outer sheaths for ventral striations); perygynia less than 15 to a spike; spikes usually less than 10 to a head
              63. Heads slender, 3-5 mm. wide; perygynia white-punctate; scales shorter than their perygynia   -52. C. disperma
              63. Heads 5-10 mm. wide; perygynia not white-punctate; scales as long or longer than their perygynia
                64. Perygynia 1.75-2.25 mm. long (rarely to 2.5), the beaks less than 0.5 mm. long, the whole chestnut-brown at maturity   -53. C. simulata
                64. Perygynia 2.5-4 mm. long, the beaks longer, the whole becoming blackish or blackish-brown at maturity

65. Culms sharply-angled above and normally rough on the angles; rootstocks long and stout; perigynia 3-4 mm. long, the beaks 1 mm. long or more   -54. C. praegracilis
65. Culms obtusely angled and smooth above; rootstocks long and slender (seldom over 2 mm. thick); perigynia 2.5-3 mm. long, the beaks seldom over 1 mm. long   -55. C. eleocharis
23. Lateral spikes peduncled or if sessile then elongated, terminal spike usually staminate (or if not then lateral spikes peduncled)
  66. Perigynia 4-5 mm. long; styles continuous with the achenes and of the same bony texture, not withering (bent back in ours)   -56. C. physocarpa
  66. Perigynia 1.5-3.5 mm. long; styles jointed to the achenes, not of the same structure, at length withering
    67. Perigynia beakless or nearly so; bracts long-sheathing at base (at least 3 mm.); scales reddish-brown or light-brown (except margins and nerves)
      68. Mature perigynia whitish-puberulent, not fleshy or translucent, obscurely ribbed; scales usually appressed   -57. C. hassei
      68. Mature perigynia golden-yellow or brown, fleshy and translucent (at least when fresh), coarsely ribbed; scales usually widely spreading at maturity   -58. C. aurea
    67. Perigynia with at least a short beak present; bracts sheathless or very short-sheathing; scales purplish-brown to black (occasionally light brown in C. emoryi)
      69. Perigynia strongly many-ribbed, beaks 0.5 mm. long or more, the apex bidentate   -59. C. nebraskensis
      69. Perigynia usually nerveless (except for the 2 side ribs) but sometimes lightly few-nerved, beaks less than 0.5 mm. long, the apex entire or nearly so
        70. Lowest bracts short, usually less than 5 cm. long, seldom much longer than their spikes and definitely shorter than the inflorescence; spikes usually clustered somewhat above; scales black
          71. Old leaves at base of culm short and inconspicuous; lower leaves reduced to bladeless sheaths on fertile culms, only about 3-6 normal leaves present; old sheaths strongly purple colored; scales ovate to oblong-ovate   -60. C. gymnoclada
          71. Old leaves at base long and conspicuous; over 6 normal leaves present, none reduced on fertile culms; old sheaths brown or purplish-brown, seldom strongly purplish; scales obovate, lanceolate or oblong-lanceolate
            72. Perigynia finally squarrose-spreading, beaks over 0.2 mm. long; scales obovate, shorter than the perigynia   -61. C. scopulorum
            72. Perigynia spreading-ascending, beaks not over 0.2 mm. long; scales lanceolate or oblong-lanceolate, longer than the perigynia   -62. C. chimaphila
        70. Lowest bracts long (over 5 cm.) about equal to or exceeding the inflorescence; spikes not at all clustered; scales light brown to purplish-brown (blackish in C. aquatilis)
          73. Fertile culms with lower leaves (of season's growth) reduced to sheaths, only 3-4 normal leaves present, the culms arising laterally from rootstocks, not enveloped at base with the old leaves of the previous year; scales light brown to purplish-brown   -63. C. emoryi
          73. Fertile culms with all leaves blade-bearing, 5-12 to a culm, the culms arising from the center of the previous year's tufts of leaves; scales purplish-brown to black
            74. Perigynia ovate, widest part below the middle, lightly nerved above and below; leaves 1.5-2.5 mm. wide   -64. C. kelloggii
            74. Perigynia oval to obovate, widest part at or above the middle, nerveless above and below; leaves 2.5-5 mm. wide   -65. C. aquatilis
22. Stigmas 3; achenes triangular in section
  75. Perigynia pubescent to long-hairy
    76. Spikes with 25-75 perigynia, the lower spikes 15-50 mm. long; lowest bract over 8 cm. long; plants often over 40 cm. tall   -66. C. lanuginosa
    76. Spikes with 1-25 perigynia, the lower spikes 3-12 mm. long; lower bract less than 8 cm. long; plants less than 40 cm. tall
      77. Perigynia punctate, not truly pubescent, 2.5-3 mm. long, beak truncate, not tapering into the body (base of beak little if any wider)   -67. C. torreyi
      77. Perigynia definitely puberulent to long-hairy at least above, 3-4.5 mm. long, beak at least weakly bidentate, tapering into the body
        78. Lower spikes very widely separated from others (usually near base of culm) and very long-peduncled
          79. Upper pistillate spikes with 3-15 perigynia; scales usually shorter than their perigynia; beaks of perigynia often serrated on margins   -68. C. rossii
          79. Upper pistillate spikes with 1-3 perigynia; scales usually as long as their perigynia; beaks of perigynia slightly if at all serrated on their margins   -69. C. pityophila
        78. Lower spikes contiguous to or but somewhat separated from the upper ones (not anywhere near base of culm) and sessile or nearly so
          80. Beaks of perigynia very short, less than 0.5 mm. long, body hirsute; scales about 1/2 as long as the perigynia; lower bracts sheathing (to as much as 7 mm.)   -70. C. concinna
          80. Beaks of perigynia about 0.75 mm. long, the body short-pubescent; scales somewhat shorter to longer than perigynia; lower bracts not sheathing   -71. C. heliophila

75. Perigynia glabrous
- 81. Styles continuous with the achenes, of the same bony texture, not withering; leaves septate-nodulose; perigynia definitely inflated at maturity, often over 4.5 mm. long
  - 82. Sheaths long-hairy; perigynia subcoriaceous, beak long-toothed, teeth 1.2-3 mm. long -72. C. atherodes
  - 82. Sheaths glabrous or nearly so; perigynia membranous, teeth of beak not over 1 mm. long
    - 83. Perigynia many-nerved (usually over 14); each scale with a long rough awn much longer than the short body of the scale, this body less than 1/2 as long as its perigynium -73. C. hystricina
    - 83. Perigynia fewer nerved (seldom over 12); scales awnless or short-awned, the awns seldom as long as the body which is over 1/2 as long as the perigynia
      - 84. Bracts very long, several times longer than the inflorescence, the lower usually sheathing; perigynia squarrose-spreading, 7-10 mm. long with beaks 2-3.5 mm. long -74. C. retrorsa
      - 84. Bracts not exceeding the inflorescence over 2 or 3 times at most, lower not sheathing; perigynia ascending-spreading to squarrose-spreading, 3.5-8 mm. long, beaks 1-2 mm. long
        - 85. Long stolons (rhizomelike) present; perigynia squarrose-spreading; basal sheaths not filamentose; culms usually bluntly triangular below the spikes, all of the season's leaves bearing blades -75. C. rostrata
        - 85. Long stolons present; perigynia ascending-spreading; basal sheaths becoming filamentose; culms sharply triangular, some of lower leaves often reduced to sheaths -76. C. vesicaria
- 81. Styles jointed to the achenes, dissimilar in texture, at length withering (hard to make out in C. paupercula); leaves not septate-nodulose; perigynia not inflated or very slightly so, not over 4.5 mm. long (except in C. sprengelii)
  - 86. Pistillate scales purplish-brown to black
    - 87. Perigynia narrowly lanceolate, the beaks over 1 mm. long; at least lower bracts long-sheathing -77. C. misandra
    - 87. Perigynia ovate, obovate or wider, the beaks less than 1 mm. long; lower bracts sheathing only slightly if at all
      - 88. Terminal spikes staminate
        - 89. Perigynia 2-3 mm. long; leaves 2-4 mm. wide; (terminal spikes sometimes pistillate too) -78. C. hallii
        - 89. Perigynia 3.5-4.5 mm. long; leaves 3-8 mm. wide; (terminal spike always staminate) -79. C. raynoldsii
      - 88. Terminal spikes gynaecandrous (or pistillate sometimes in C. hallii)
        - 90. Terminal spikes pistillate or sometimes staminate at the tip; rootstocks long -78. C. hallii
        - 90. Terminal spikes gynaecandrous; rootstocks short (except in C. buxbaumii)
          - 91. Fertile culms lateral from outside the tufts of leaves of the previous year; many of the lower leaves of the season reduced to sheaths; rootstocks long; scales often short-awned -80. C. buxbaumii
          - 91. Fertile culms central from the middle of the previous season's old leaves; all of leaves with blades; rootstocks short; scales seldom if ever short-awned
            - 92. Pistillate scales 1.5-2.5 mm. long; leaves not over 3 mm. wide; perigynia obtusely-triangular in cross-section, seldom over 3 mm. long -81. C. media
            - 92. Pistillate scales over 2.5 mm. long; leaves usually over 3 mm. wide; perigynia flattened (subinflated-triangular in C. nelsonii), usually over 3 mm. long
              - 93. All the spikes sessile or very nearly so, aggregated into a dense head; beaks of perigynia 0.5-1 mm. long
                - 94. Perigynia subinflated-triangular, appressed on the spikes, oblong-obovoid, over 2½ times as long as wide -82. C. nelsonii
                - 94. Perigynia flattened (except where distended by achene), spreading-ascending or becoming squarrose, ovate-suborbicular to obovoid, about twice as long as wide -83. C. nova
              - 93. At least the lower spikes on short or long peduncles and somewhat remote from the others; beaks of perigynia seldom ever 0.5 mm. long
                - 95. Lower spikes on short, erect peduncles (shorter than the spikes)
                  - 96. Scales purplish-black with rather conspicuous white margins and tips, ovate, shorter than the perigynia; lateral spikes about 4 mm. wide -84. C. albo-nigra
                  - 96. Scales copper or brownish-black, the margins narrow, lanceolate to ovate-lanceolate, longer than the perigynia; lateral spikes 6-7 mm. wide -85. C. chalciolepis
                - 95. Lower spikes on long, somewhat drooping peduncles (some as long or longer than their spikes)
                  - 97. Scales shorter than their perigynia, their margins conspicuously white-hyaline; perigynia not glandular roughened, the beaks 0.3 mm. long; lateral spikes linear or oblong-linear (less than 5 mm. wide) -86. C. bella
                  - 97. Scales (at least near base of spike) longer than their perigynia, their margins very narrow; perigynia glandular roughened (under lens), the beaks about 0.5 mm. long; lateral spikes ovate to narrowly-oblong (over 5 mm. wide) -87. C. atrata

86. Pistillate scales not so dark, varying from greenish-white, castaneous to chestnut-brown
    98. Perigynia 5-6 mm. long, the beak 2 mm. long or more; rootstocks and bases of culms densely fibrillose
        -88. C. sprengelii
    98. Perigynia 2-4.5 mm. long, the beak seldom over 1 mm. long; rootstocks and culm bases not densely fibrillose
        99. Perigynia 2-5 to a spike; spikes all androgynous; perigynia about 4 mm. long; lower 2 or 3 scales leaflike and exceeding the inflorescence
            -89. C. saximontana
        99. Perigynia more than 5 to a spike; terminal spikes staminate or gynaecandrous, the laterals pistillate; perigynia seldom over 3 mm. long (except in C. chalciolepis and C. limosa); none of scales leaflike (bracts may be)
            100. Pistillate scales copper-colored or dark purple; perigynia dark purple (at least tinged with)
                -85. C. chalciolepis
            100. Pistillate scales reddish-brown, castaneous to chestnut; perigynia green, yellowish-green or greenish-brown, never dark purple
                101. Lower spikes on slender drooping peduncles longer than the spikes
                    102. Perigynia broadly ovoid or elliptical, less than twice as long as wide, nearly beakless; lower bracts not long-sheathing (seldom over 2 mm.)
                      103. Pistillate scales lanceolate to ovate-lanceolate, over twice as long as wide, usually deciduous before the perigynia; culms loosely caespitose; leaves deep green
                          -90. C. paupercula
                      103. Pistillate scales ovate to suborbicular, rarely over twice as long as wide, persistent; culms not at all caespitose; leaves more or less glaucous
                          -91. C. limosa
                    102. Perigynia ovoid-lanceolate, over twice as long as wide, with a beak up to 1 mm. long; at least the lower bracts long-sheathing (for over 2 mm.)
                      -92. C. capillaris
                101. Lower spikes erect on short pedicels rarely as long as the spikes
                    104. Beaks of perigynia short, less than 0.4 mm. long and less than 1/5 the length of the whole, beak cylindrical, not enlarging at base to enter the body; bracts not long-sheathing; sheaths of leaves pubescent
                      -67. C. torreyi
                    104. Beaks of perigynia over 0.4 mm. long, more than 1/5 the total length, beak enlarging at base to enter the body; bracts long-sheathing; sheaths not pubescent
                      -93. C. viridula

1. Carex gynocrates Wormsk; Drejer, Nat. Tidssk. 3:434. 1841.
C. redowskyana C. A. Mey.---Culms 3-30 cm. tall, single or a few from slender rootstocks, smooth; leaves 0.5 mm. wide, involute or folded; sheaths hyaline ventrally; inflorescence of 1 spike, this 5-15 mm. long, 2-4 mm. wide, (or to 6 mm. when perigynia are spreading), pistillate, staminate or androgynous, with 4-10 perigynia these erect-ascending or spreading (even somewhat reflexed at maturity); bracts none; scales somewhat wider but shorter than the perigynia, oblong-ovate to ovate, short-cuspidate to acute or acuminate, light reddish-brown to brown with narrow margins and nerves poorly defined; perigynia 2.5-3.5 mm. long, 1.5-2 mm. wide, biconvex, oblong-ovate or ovate, substipitate, abruptly contracted to a beak 0.5 mm. long, yellowish or brownish-black at maturity, finely-nerved but this obscure ventrally; styles jointed; stigmas 3.---Moist ground. Greenland to Alaska, south to New York and Colorado. Our few records from central and northcentral Colorado at 11,500 feet.

2. Carex capitata L., Syst. Nat. ed. 10. 1261. 1759.
Greenland to Alaska, south to New Hampshire, Mexico and California. This plant should be in Colorado but no records are known to the writer.

3. Carex hepburnii Boott. in Hook., Fl. Bor. Am. 2:209. 1839.
C. nardina Fries in part---Culms 2-15 cm. tall, very densely caespitose, very slender and wiry, smooth; leaves 0.25 mm. wide, 1-10 cm. long, setaceous, 1-2 to a culm; inflorescence of 1 spike, this 5-15 mm. long, 2-4 mm. wide, oblong, androgynous, with 5-15 erect-appressed perigynia; bracts none; scales wider but about the length of the perigynia, ovate or obovate, obtusish or acute, reddish-brown or brownish with narrow whitish-hyaline margins and straw-colored centers; perigynia 3-4.5 mm. long, 1.25-1.75 mm. wide, biconvex or plano-convex, lanceolate, substipitate, round-tapering to a beak about 0.5 mm. long, glabrous or upper part weakly serrate, obscurely striated both sides, whitish except near apex; styles jointed; stigmas 2 or 3.---Rather dry ground. Alberta to Alaska, south to Colorado and Washington. Our records from central, northcentral and western Colorado at 10,500-12,500 feet.

4. Carex scirpoidea Michx., Fl. Bor. Am. 2:171. 1803.
Culms 10-40 cm. tall, 1 to few, from creeping scaly rootstocks, somewhat roughened above; leaves 1-3 mm. wide, flat or somewhat canaliculate; sheaths puberulent ventrally; inflorescence of 1 spike, this staminate or pistillate, 1.5-3 cm. long, 2.5-5 mm. wide, linear, with many erect-appressed perigynia; bracts present or absent; scales somewhat narrower and shorter than the perigynia, obovate-oblong, obtusish or acutish, ciliate, puberulent, chocolate-brown to blackish with lighter midribs and narrow margins; perigynia 2-3 mm. long, about 1 mm. wide, ovoid or **oblong-ovoid**, compressed-triangular in cross-section, 2-ribbed, obscurely few-nerved near base, short-pubescent, abruptly short-beaked, this about 0.25 mm. long; styles obscurely jointed; stigmas 3.---Open places. Greenland to Alaska, south to Michigan and Colorado; Eurasia. Our one good record from Park County, probably at high altitudes.

5. Carex **pseudoscirpoidea** Rydb., Mem. N. Y. Bot. Gard. 1:78. 1900.
Culms 15-35 cm. tall, loosely caespitose from stout elongated rootstocks, 1-several together, smooth or somewhat roughened above; leaves 2-3 mm. wide, flat; sheaths short-pubescent ventrally; inflorescence of 1 spike, this 12-36 mm. long and 2-5 mm. wide, linear, staminate or pistillate, the latter with many appressed perigynia; bracts absent or present; scales wider and longer than the perigynia, broadly ovate, acute or obtuse but erose-ciliate at tips, margins white-hyaline,

brownish-black, more or less pubescent; perigynia 2.5 mm. long, about 1.25 mm. wide, obovoid, somewhat triangular in cross-section, tapering to base, abruptly beaked, this about 0.5 mm. long, reddish-brown above, obscurely nerved, strongly pubescent; styles jointed; stigmas 3 or 4.---Dry mountain slopes. Montana to Washington, south to Colorado and Nevada. Our one record from central Colorado at about 11,000 feet.

6. Carex oreocharis Holm, Am. Journ. Sc. IV. 9:358, 357. 1900.

Culms 10-35 cm. tall, densely caespitose, stiff, smooth to roughened; leaves 1.5-2 mm. wide, flattened-canaliculate near base and channeled above, 2-3 to a culm; sheaths thin ventrally; inflorescence of 1 spike, this 2-4 cm. long, androgynous, staminate flowers conspicuous, pistillate with 3-10 erect-appressed perigynia; bracts short; scales wider but shorter than the perigynia, broadly ovate or orbicular, obtuse or cuspidate, light yellowish with conspicuous white-hyaline margins and several-nerved green centers; perigynia 4-4.5 mm. long, 2-2.2 mm. wide, oblong-obovoid, obtusely triangular in cross-section, membranous, substipitate, beak 1 mm. long, cylindrical, hyaline-tipped and obliquely cut, yellow-green, obscurely 2-ribbed, otherwise nerveless, short-pubescent above; styles short, jointed to achenes; stigmas 3.---Rather dry slopes. Central and northcentral Colorado at 7500-10,600 feet.

7. Carex filifolia Nutt., Gen. 2:204. 1818.

Culms 8-30 cm. tall, very densely caespitose, filiform, smooth or nearly so; leaf blades about 0.25 mm. wide, involute and wiry; sheaths short; inflorescence of 1 spike, this 1-3 cm. long, 3-5 mm. wide, linear, erect, androgynous, with the upper half staminate and with 5-15 erect-ascending perigynia present; bracts none; scales longer and wider than the perigynia, very broad, obtuse to mucronate, light reddish-brown with conspicuous white-hyaline margins; perigynia 3-3.5 mm. long, about 2 mm. wide, obovoid or obovoid-orbicular, obtusely triangular in section, abruptly narrowed to a beak 0.2-0.4 mm. long, membranous, obscurely 2-ribbed, puberulent above, dull-whitish to greenish straw-colored tinged with yellowish-brown; styles obscurely jointed to achenes and finally deciduous; stigmas 3.---Plains and foothills. Western Canada south to Texas and New Mexico. Our records from central and northern Colorado at 5000-7500 feet.

8. Carex elynoides Holm, Am. Journ. Sc. IV. 9:356, 357. 1900.

Culms 5-12 cm. tall, rarely taller, densely caespitose, smooth, filiform, wiry; leaves about 0.25 mm. wide, wiry and involute, 2-3 to a culm; sheaths thin ventrally; inflorescence of 1 spike, this 8-15 mm. long, 2-4 mm. wide, linear, erect, upper half staminate, lower half with 4-8 erect-ascending perigynia; bracts none; scales wider and longer than the perigynia, oblong-obovate, obtuse to acute, reddish-brown with dingy-white margins; perigynia 2.5-3 mm. long, about 1.5 mm. wide, obpyramidal, triangular, abruptly narrowed to a beak about 0.3-0.5 mm. long, obscurely 2-ribbed, lower part whitish to green, upper yellowish-brown, slightly puberulent at base of beak; styles obscurely jointed but deciduous; stigmas 3. Doubtfully separated from C. hepburnii Boott. in Colorado.---Mountain tops and slopes. Montana to Utah and south to Colorado. Our records from the western half of Colorado at 9000-13,500 feet, being more common at the higher altitudes.

9. Carex microglochin Wahl., Sv. Vet.-Acad. Nya Handl. 24:140. 1803.

Culms 2-25 cm. tall, few, from very slender long rootstocks, smooth; leaves about 0.5 mm. wide, involute; sheaths hyaline ventrally; inflorescence of 1 spike, this 7-14 mm. long, 6-9 mm. wide, androgynous with 3-10 perigynia these soon reflexed; bracts none; scales wider but about 1/2 as long as the perigynia, oblong-obovate, obtuse, early deciduous, light chestnut-brown, often with hyaline margins and lighter midribs; perigynia 4-5 mm. long and about 1 mm. wide, subulate to linear-lanceolate, suborbicular in cross-section, substipitate, gradually long-tapering into the beak, the orifice oblique, light green to straw-colored, obscurely many-striated; styles continuous; stigmas 3.---Open places, often in moist soil. Greenland to Alaska, south to Quebec and Colorado; Eurasia. Our two records from central and southcentral Colorado at 10,500-12,000 feet.

10. Carex pyrenaica Wahl., Sv. Vet.-Acad. Nya Handl. 24:139. 1803.

Culms 3-20 cm. tall, densely caespitose, about the length of or exceeding the leaves, smooth and wiry; leaves 0.25-1.5 mm. wide, channeled, 2-4 to a culm but bladeless sheaths may be present at base; sheaths tight; inflorescence of 1 spike this 5-20 mm. long and 3-5 mm. wide, linear-oblong, erect, androgynous, the perigynia 10 to many; bracts none; scales shorter but somewhat wider than perigynia, ovate, obtuse, brownish or blackish-chestnut, with narrow, hyaline margins and lighter centers, soon deciduous; perigynia 3-4 mm. long, 0.8-1.5 mm. wide, lanceolate or ovate-lanceolate, compressed-triangular in cross-section, erect at first but spreading or reflexed at full maturity, glabrous, brownish to straw-colored, nerveless, membranous, stipitate at base and tapering to a beak about 0.5 mm. long; styles jointed; stigmas 3 or rarely 2.---Open slopes in the mountains. Northwest Territories to Alaska, south to Colorado and Oregon; also in Eurasia. Scattered in the high mountains of Colorado at 10,500-12,500 feet.

11. Carex nigricans C. A. Meyers, Mem. Acad. St. Petersb. Sav. Etr. 1:211. 1831.

Culms 5-30 cm. tall, loosely caespitose with short rootstocks, smooth; leaves 1.5-2 mm. wide, flat or canaliculate, usually shorter than the culms; sheaths yellowish- or brownish-tinged; inflorescence of 1 spike, this 8-15 mm. long and 6-9 mm. wide, oblong or oblong-ovoid, erect, androgynous, staminate part conspicuous and over 1/2 the total length, perigynia 10-25, possibly more; bracts none; scales much shorter than the perigynia, broadly ovate, dark brown, deciduous; perigynia 3.5-4.5 mm. long and 1-1.5 mm. wide, lanceolate or ovate-lanceolate, compressed-triangular in cross-section, not inflated, glabrous, membranous, yellowish-brown or brownish, early deciduous and reflexed, stipe about 0.5 mm. long, tapering to a beak about 0.5 mm. long; styles jointed; stigmas 3 or rarely 2.---Alaska south to Colorado and California; Asia. Scattered in the high mountains of Colorado at 9500-13,500 feet.

12. Carex engelmannii L. H. Bailey, Proc. Am. Acad. 22:132. 1886.

Culms 5-20 cm. tall, from slender rootstocks, erect, smooth or nearly so; leaves 0.3-0.5 mm. wide, involute-filiform, on lower part of culm; sheaths hyaline ventrally; inflorescence of 1 spike, this 1-1.5 cm. long and 6-10 mm. wide, ovoid, androgynous the upper 1/4 staminate, with 15-40 ascending perigynia; bracts none; scales mostly slightly shorter than the perigynia but wider near base, ovate, acute to cuspidate, reddish-brown to **straw-colored**, margins hyaline, centers lighter; perigynia 4.5-5 mm. long, possibly shorter, and 1.8-2.25 mm. wide, ovoid, at least somewhat inflated, nerveless, straw-colored or brown, somewhat stipitate, beak 0.5 mm. long or less, slightly if at all bidentate; styles jointed; stigmas 3.---Open slopes in the higher mountains. Wyoming to Washington, south to Colorado and California. Our records from central and westcentral Colorado at 9000-12,000 feet.

13. Carex obtusata Lilj., Sv. Vet.-Acad. Nya Handl. 14:69. 1793.

Culms 6-20 cm. tall, 1-3 together, from purplish-black, long-creeping rootstocks, roughened above, exceeding the leaves; leaves 1-1.5 mm. wide, channeled and stiff, from near base of culm; sheaths tight; inflorescence of 1 spike this 5-12 mm. long and 3-6 mm. wide, androgynous the upper 1/2 to 2/3 staminate, with 1-6 perigynia, these ascending or spreading-ascending at maturity; bracts none; scales shorter and narrower than the perigynia, ovate to ovate-**lanceolate**, acuminate or cuspidate, light brownish with wide, whitish, hyaline margins and lighter midnerves; perigynia 3-3.5 mm. long and 1.75-2 mm. wide, oblong-obovoid, suborbicular-triangular in cross-section but not inflated, glabrous, dark chestnut or blackish-brown, shining and coriaceous, broadly stipitate, beaks conspicuously flaring at tips which are white-hyaline and obliquely bidentate; styles jointed; stigmas 3.---Dry hills and slopes. Yukon and British Columbia, south to New Mexico; Eurasia. Our rather few records from northcentral, central and southwestern Colorado at 7500-10,500 feet but probably more widely distributed.

14. Carex geyeri Boott., Trans. Linn. Soc. 20:118. 1846.

Culms 10-40 cm. tall, loosely caespitose with elongated woody rootstocks, erect, rough above; leaves 2-3.5 mm. wide, flat or canaliculate; sheaths tight, hyaline ventrally; inflorescence of 1 spike, this 5-25 mm. long and 1.5-3 mm. wide, linear-oblong, androgynous, with 1-3 perigynia, these erect-ascending, overlapping to rather separate; bracts none; scales wider and usually longer than the perigynia, ovate or broader, lower short-awned, upper acutish to obtusish, fulvous with broad hyaline margins; perigynia about 6 mm. long and 2.5 mm. wide, oblong-obovoid, triangular in cross-section, glabrous, 2-ribbed, otherwise nerveless, greenish-straw or brownish tinged, membranous, short-stipitate, rounded to a very short beak; styles jointed and deciduous; stigmas 3.---Open slopes and woods in the mountains. Alberta to British Columbia, south to Colorado and California. Our records scattered over the western two-thirds of Colorado, except the southcentral part, at 6000-11,000 feet.

15. Carex leptalea Wahl., Sv. Vet.-Akad. Nya Handl. 24:139. 1803.

Culms 15-60 cm. long, caespitose with short rootstocks, smooth or roughened below the spike; leaves 0.5-1.25 mm. wide, flat or canaliculate, much shorter than the culms; sheaths tight and short; inflorescence of 1 spike, this 4-16 mm. long and 2-3 mm. wide, androgynous, with 1-10 erect-ascending perigynia; bracts none; staminate scales **connate** at bases; pistillate scales from very short to 5 cm. long and exceeding the spike, ovate-orbicular and obtuse to lanceolate and cuspidate, yellowish-green, often red-dotted or red-tinged; perigynia 2.5-5 mm. long, oval-elliptic to oblong-elliptic, orbicular or flattened-orbicular in cross-section, finely striated, yellowish-green to light green, substipitate with spongy bases, beakless, the tips entire or emarginate; styles jointed; stigmas 3.---Wet meadows and bogs. Labrador to Alaska, south to Florida, Texas and California. Our one record from Lake County at 9000 feet.

16. Carex drummondiana Dewey, Am. Journ. Sc. 29:251. 1836.

C. rupestris of Manuals---Culms 4-12 cm. long, loosely caespitose with scaly rootstocks, rather stout, rough above, usually exceeded by the leaves; leaves 1.5-3 mm. wide, flat and canaliculate, 8-12 to a fertile culm; inflorescence of 1 spike, this 1-2 cm. long, 3-4.5 mm. wide, linear or oblong-linear, androgynous, the upper 1/2 or less staminate, with 6-15 erect-ascending perigynia; bracts none; scales wider and usually longer than the perigynia, orbicular-ovate, obtushish to acute, deep chestnut-brown with broad white hyaline margins and lighter centers; perigynia about 4 mm. long and 1.75 mm. wide, oblong-obovoid, triangular-flattened in cross-section, membranous, shining, glabrous, greenish-white below to brown above, 2-keeled and obscurely many-nerved, substipitate, tapering to a beak 0.2 mm. long, this sparingly serrate-ciliate on edges; styles jointed; stigmas 3. O'Neill and Duman (Rhodora 43:417-418. 1941) concluded the correct name for this plant is C. rupestris Bell.---Higher mountains. Canadian mountains to Colorado. Our few records from westcentral, central and northcentral Colorado at 11,500-13,000 feet.

17. Carex siccata Dewey, Am. Journ. Sci. 10:278. 1826.

Culms 20-90 cm. tall, 1-3 together from long-creeping rootstocks, roughened on the angles above; leaves 2-3 mm. wide, flat or somewhat channeled; sheaths hyaline ventrally; inflorescence of 6-12 spikes in a linear-oblong head, upper aggregated, lower somewhat separated; lower spikes usually pistillate with few flowers, middle staminate and upper gynaecandrous, perigynia appressed-ascending; bracts scalelike, the lower short-cuspidate; scales as wide as but shorter than the perigynia, ovate-lanceolate, obtushish or acute, light reddish-brown or yellow-brown with often conspicuous silvery-white hyaline margins and green 3-nerved centers; perigynia 3-6 mm. long and 1.75-2 mm. wide, oblong-lanceolate, plano-convex and narrowly wing-margined, pale green or ferruginous-tinged, many-nerved both faces, substipitate, somewhat contracted to a beak 2-3 mm. long, more or less bidentate; styles jointed; stigmas 2. Considered by some not to be specifically distinct from C. foenea Willd.---Dry open ground. Maine to Alberta, south to New Jersey, Arizona and Washington. Our records from northcentral, central and southcentral Colorado at 5000-13,000 feet.

18. Carex athrostachya Olney, Proc. Am. Acad. 7:393. 1868.

C. tenuirostris Olney---Culms 5-60 cm. tall, caespitose, rough beneath the head; leaves 1.5-4 mm. wide, flat; sheaths thin ventrally; inflorescence of aggregated spikes; spikes 4-20, 4-10 mm. long, closely aggregated or head somewhat lobed below, gynaecandrous with 15-40 appressed-ascending perigynia; bracts with 1-3 strongly developed, 1.5-6 cm. long, longer than the inflorescence; scales somewhat narrower and shorter than the perigynia, oblong-ovate, acute or short-cuspidate, brownish or reddish-brown with hyaline margins and green 3-nerved centers; perigynia 3-4.5 mm. long and 1.25-1.5 mm. wide, lanceolate-ovate, flat and narrowly wing-margined, light green or straw-colored, lightly nerved dorsally, little if at all nerved ventrally, substipitate, ciliate-serrulate from the middle to the tapering beak which is about 1 mm. long; styles jointed; stigmas 2.---Moist meadows. Saskatchewan to Alaska, south to Colorado and California. Our records from central and northcentral Colorado at 5500-11,500 feet.

19. **Carex petasata** Dewey, Am. Journ. Sci. 29:246. 1836.
Culms 30-80 cm. tall, caespitose, smooth; leaves 2-3 mm. wide, flat; sheaths white-hyaline ventrally; inflorescence of 3-6 spikes aggregated into an erect head; spikes 8-16 mm. long and 6-9 mm. wide, oblong-ovoid or terminal often clavate, gynaecandrous with many appressed perigynia; bracts scalelike or occasionally short-prolonged; scales about as long and as wide as the perigynia, ovate, light reddish-brown with broad hyaline margins and lighter 1- to 3-nerved centers; perigynia 5.5-8 mm. long, oblong-lanceolate, plano-convex and narrowly wing-margined to base, brownish-green to brown, many-striated on both faces, substipitate, tapering to a beak about 2 mm. long, this rather terete; styles jointed; stigmas 2. Some of our plants intergrade somewhat with a recently described western species, C. eastwoodiana Stacey.---Meadows and open woods. Saskatchewan to British Columbia, south to Colorado and Oregon. Our few records from northcentral Colorado at 8000-10,000 feet.

20. **Carex phaeocephala** Piper, Contrib. U. S. Nat. Herb. 11:172. 1906.
Culms 9-30 cm. tall, very densely caespitose, stiff and more or less rough; leaves 1.5-2 mm. wide, canaliculate to somewhat involute; sheaths white-hyaline ventrally; inflorescence of 2-5 (rarely 7) spikes aggregated in a head but the lower somewhat separated; spikes 6-12 mm. long and 5-8 mm. wide, ovoid or oblong-obovoid, gynaecandrous with 10-20 appressed-ascending perigynia; bracts shorter than the head; scales as long and as wide as the perigynia, ovate, acute, brownish-black to reddish-brown with broad white hyaline margins and lighter 1- to 3-nerved centers; perigynia 4-6 mm. long and 1.75-2.5 mm. wide, oblong-ovate to obovate, concavo-convex with rather conspicuous winged margins, straw-colored to brownish-black, substipitate, rather abruptly contracted into an almost terete beak 1 mm. long; styles jointed; stigmas 2.---Mountains. Alberta to British Columbia, south to Colorado and California. Our records from central and northcentral Colorado at 8500-12,500 feet.

21. **Carex arapahoensis** Clokey, Rhodora 21:83. 1919.
Culms 15-40 cm. tall, densely caespitose, roughened below the inflorescence; leaves 1.5-3 mm. wide, flat or somewhat channeled above; inflorescence of 3-6 spikes aggregated into a more or less globose head; spikes 7-11 mm. long and 4-8 mm. wide, obovoid, gynaecandrous with numerous appressed-ascending perigynia; bracts short, rarely the lower short-prolonged; scales about as long and as wide as the perigynia, ovate, obtuse, dark chestnut-brown, conspicuously white-hyaline at apex and upper margins, midveins inconspicuous or obsolete; perigynia 4.5-5.25 mm. long, obovate, flat and wing-margined to base, dark brown at maturity, finely nerved on both faces, sessile, abruptly beaked, this flat and 1 mm. long; styles jointed; stigmas 2.---Dry mountainsides. Apparently limited to Colorado, our records from northcentral, central and south-central parts of the state at 11,500-12,500 feet.

22. **Carex xerantica** L. H. Bailey, Bot. Gaz. 17:151. 1892.
Culms 30-60 cm. long, caespitose, roughened below the inflorescence; leaves 2-3 mm. wide, flat; sheaths white and hyaline ventrally; inflorescence of 3-6 spikes, approximate but distinct; spikes 8-14 mm. long or terminal even longer, 5-8 mm. wide, ovoid-elliptic, gynaecandrous with numerous closely appressed perigynia; bracts short-prolonged below to scalelike above; scales about as wide and long as the perigynia, ovate, acute, light reddish-brown with wide hyaline margins and 3-nerved green centers; perigynia 4-5.5 mm. long and 2-2.5 mm. wide, ovate, strongly winged to base, greenish to yellowish at base, lightly nerved dorsally and nerveless or nearly so ventrally, tapering to a flat beak this 1-1.3 mm. long; styles jointed; stigmas 2.---Open ground. Manitoba to Alberta, south to New Mexico. Our few records from northcentral Colorado at 6000-9500 feet but the plant should be in central and southern parts of the state.

23. **Carex praticola** Rydb., Mem. N. Y. Bot. Gard. 1:84. 1900.
Culms 20-70 cm. tall, caespitose, roughened below head; leaves 1-3.5 mm. wide, flat; sheaths white and hyaline ventrally; inflorescence of 2-7 spikes somewhat approximate but each well defined, rather moniliform; spikes 6-16 mm. long and 4-6 mm. wide, ovoid-oblong to oblong or the upper almost clavate, gynaecandrous, staminate flowers few, perigynia 6-20, loosely to closely appressed; lower bracts sometimes prolonged but seldom the length of the inflorescence; scales as wide and as long as the perigynia, ovate, obtuse or acute, reddish-brown with broad silvery-hyaline margins and green or tan centers; perigynia 4.5-6.5 mm. long and 1.5-2 mm. wide, ovate-lanceolate to narrowly ovate, flattened plano-convex, wing-margined to base, light green to whitish, many-nerved dorsally, nerveless or nearly so ventrally, short-stipitate, tapering to a beak 1.25-2 mm. long, this terete and obliquely cut at apex, slightly if at all bidentate; styles jointed; stigmas 2.---Meadows and open woods. Greenland to Alaska, south to Quebec, Colorado and California. Our records from central and northcentral Colorado at 7000-9500 feet.

24. **Carex illota** L. H. Bailey, Mem. Torr. Club 1:15. 1889.
C. bonplandii minor of Manuals---Culms 10-35 cm. tall, caespitose, roughened on the angles; leaves 1.5-3 mm. wide, flat or somewhat canaliculate; sheaths hyaline ventrally; inflorescence capitate, of 3-6 closely aggregated spikes; spikes 4-6 mm. long and 3-5 mm. wide, suborbicular, gynaecandrous with 5-15 ascending perigynia; lowest bract dilated at base and usually somewhat setaceous-prolonged; scales wider than but about the length of the bodies of the perigynia, broadly ovate, obtuse, blackish with very narrow hyaline margins and yellow-brown centers; perigynia about 3 mm. long and 1.25 mm. wide, ovate-lanceolate, plano-convex and sharp-edged to base, blackish-brown above and straw-colored below, obscurely nerved both faces, substipitate, tapering to a beak 0.75 mm. long, this bidentulate; styles jointed; stigmas 2.---Higher altitudes in the mountains. Wyoming to British Columbia, south to Colorado and California. Our

records thinly scattered in northcentral, central, southcentral and possibly southwestern Colorado at 8500-12,000 feet.

25. Carex bebbii Olney (name only); Fernald, Proc. Am. Acad. 37:478. 1902.

Culms 20-80 cm. tall, densely caespitose, roughened above; leaves 2-4.5 mm. wide, flat; sheaths white-hyaline ventrally; inflorescence of 3-12 spikes in an oblong or linear-oblong head, ours not at all crowded; spikes 4-9 mm. long and 3-6 mm. wide, subglobose to ovoid, gynaecandrous with many ascending perigynia; bracts of lower spikes setaceous but shorter than the inflorescence; scales narrower (at least above) and shorter than the perigynia, oblong-lanceolate, acute or short-acuminate, brown, sometimes with narrow hyaline margins, with lighter centers; perigynia 3-3.5 mm. long and 1-2 mm. wide, ovate, plano-convex and narrowly winged to base, finely nerved dorsally, nerveless or nearly so dorsally, substipitate, tapering to a beak 0.75-1 mm. long, this flat and winged below; styles jointed; stigmas 2.---Moist ground. Newfoundland to British Columbia, south to New Jersey, Colorado and Washington. Our one record from Routt County at 6500 feet.

26. Carex ebenea Rydb., Bull. Torr. Club 28:266. 1901.

Culms 25-50 cm. tall, caespitose, roughened above; leaves 2-3 mm. wide, flat; sheaths white-hyaline ventrally; inflorescence of 5-10 spikes, these very closely aggregated; spikes 7-12 mm. long and 6 mm. wide, gynaecandrous with many perignia their beaks appressed; lower bracts 1 cm. long or less, shorter than the heads; scales about as wide below as the perignia but exceeded in length by their beaks, ovate-lanceolate, acute or obtuse, brownish-black or black with narrow hyaline margins and midribs nearly obsolete; perigynia 6-7 mm. long and about 1.75 mm. wide, narrowly lanceolate to ovate-lanceolate, flat and narrowly winged to base, brownish-black or blackish, finely many-nerved both faces, substipitate, narrowed into a beak 2 mm. long; styles jointed; stigmas 2.---Mountains. Wyoming to Utah, south to New Mexico and Arizona. Scattered in the mountainous two-thirds of Colorado, except the extreme western part, at 9000-12,000 feet.

27. Carex haydeniana Olney in S. Wats., Bot. Kings Exp. 366. 1871.

C. nubicola Mack.---Culms 10-40 cm. tall, densely caespitose, smooth, ascending or decumbent; leaves 1.5-4 mm. wide, flat; sheaths white-hyaline ventrally; inflorescence ovoid or globose, of 4-7 closely aggregated spikes; spikes 5-9 mm. long and 4.5-8 mm. wide, ovoid or subglobose, gynaecandrous with many perigynia whose beaks are not appressed but still conspicuous; bracts scalelike or lower short-prolonged; scales shorter and narrower than the perigynia, ovate, acute, brownish-black with the hyaline margins almost obsolete and the centers somewhat lighter in color; perigynia 4.5-6.5 mm. long and 1.75-2.75 mm. wide, broadly to narrowly ovate, sometimes lanceolate, flat and winged to base, greenish to straw-colored and blackish or brownish-tinged, faintly nerved dorsally, nerveless or nearly so ventrally, somewhat substipitate, with a rather terete beak 1.5-2 mm. long; styles jointed; stigmas 2.---Mountains. Alberta to Oregon, south to Colorado and California. Our few records thinly scattered in the mountains of the state at 6500-11,000 feet.

28. Carex scoparia Schkuhr.; Willd., Sp. Pl. 4:230. 1805.

Culms 15-100 cm. tall, densely caespitose, roughened below the inflorescence; leaves 1.5-3 mm. wide, flat or canaliculate; sheaths white-hyaline ventrally; inflorescence of 3-12 spikes aggregated into an oblong or linear-oblong, even globose head, the spikes well defined, sometimes the head moniliform; spikes 6-16 mm. long and 3-9 mm. wide, short-oblong to subglobose, gynaecandrous with many appressed or erect-ascending perigynia; lower bracts usually setaceous and prolonged but shorter than the heads; scales nearly as wide as perigynia but shorter than their beaks, ovate or oblong-ovate, acute, light brownish with narrow white hyaline margins and a 3-nerved green center; perigynia 4-6.5 mm. long and 1.2-2 mm. wide, lanceolate to narrowly ovate-lanceolate, flat and wing-margined to base, greenish-white to straw-colored at maturity, somewhat nerved on both faces, the beak flat and 1.2-2 mm. long; styles jointed; stigmas 2.---moist or wet ground. Newfoundland to British Columbia, south to South Carolina, New Mexico and Oregon. Our few records from central and northcentral Colorado at 7000-9500 feet, but to be expected anywhere in the state.

29. Carex stenoptila Hermann, Leaf. West. Bot. 4:194. 1945.

Culms 17-75 cm. tall, densely caespitose with stout rootstocks, stiff, triangular and scaberulous above; leaves 1-3.5 mm. wide, 2-4 to a fertile culm; sheaths loose, thin and white-hyaline ventrally; inflorescence of 7-10 spikes in an oblong or suborbicular head 12-19 mm. long; spikes about 5-8 mm. long and 3-5 mm. wide, oblong-elliptic, gynaecandrous with 10-25 ascending perigynia; bracts scalelike; scales not concealing the perigynia, cuneate-oblong or oblong-lanceolate, obtuse to acute, pale to dark chestnut-brown with paler conspicuous midribs; perigynia 4.5-5 mm. long and about 1.3 mm. wide, oblong-lanceolate to narrowly ovate-lanceolate, plano-convex and very narrowly wing-margined, membranous, straw-colored, 5- to 7-nerved on each face, base sessile, margins serrulate above tapering to a beak about 1.5 mm. long, this shallowly bidentate; styles jointed and deciduous; stigmas 2.---Dry or moist ground. Known only from the type collection in Montrose County, Colorado, at 9500 feet.

30. Carex brevior (Dewey) Mack.; Lunell, Am. Midl. Nat. 4:235. 1915.

C. straminea Willd.; C. festucacea of Manuals---Culms 30-100 cm. tall, caespitose, smooth or rough below the inflorescence; leaves 1.5-4 mm. wide, flat; sheaths white-hyaline ventrally; inflorescence of 3-10 spikes aggregated into a short or moniliform head; spikes 7-15 mm. long and 5-9 mm. wide, subglobose, ovoid or oblong, gynaecandrous with 8-20 ascending-spreading perigynia; bracts scalelike or the lower short-prolonged; scales shorter and narrower than the perigynia, ovate, obtuse or acutish, yellow-brown with margins narrowly **hyaline** and with 3-nerved green centers; perigynia 4-5.5 mm. long and 2.5-3.5 mm. wide, very broadly ovate to suborbicular, plano-convex to concavo-convex, strongly winged to base, strongly nerved dorsally, nerveless or nearly so ventrally, abruptly contracted into a flattened beak 1 mm. long; styles jointed; stigmas 2.---Open moist ground. Quebec to British Columbia, south to Tennessee, New Mexico and Oregon. Our records from the eastern half of Colorado at 3500-7000 feet.

31. **Carex egglestoni** Mack., Bull. Torr. Club 42:614. 1915.
    C. straminiformis L. H. Bailey---Culms 40-80 cm. tall, in dense clumps, roughened below head; leaves 2-5 mm. wide, flat; sheaths white-hyaline ventrally; inflorescence of 3-6 spikes, these closely crowded but distinguishable; spikes 10-14 mm. long and 6-10 mm. wide, ovoid or ovoid-oblong, gynaecandrous with many appressed-ascending perigynia; lower bracts short-prolonged or scalelike; scales narrower and shorter than the perigynia, ovate or ovate-lanceolate, acute or short-acuminate, chestnut-brown with narrow but conspicuous white, hyaline margins and lighter midribs; perigynia 6-7.5 mm. long and about 2 mm. wide, ovate, concavo-convex, strongly winged to base the edges somewhat wavy, green or brownish in age, striate both faces at least when young, abruptly contracted to a somewhat bidentate beak 1.5 mm. long; styles jointed; stigmas 2.---Dry open ground. Wyoming, Utah and Colorado. Our records from northcentral, central and southwestern Colorado at 9000-12,000 feet.

32. **Carex pachystachya** Cham.; Steud., Syn. Cyp. 197. 1855.
    C. festiva var. pachystachya Bailey---Culms 30-100 cm. tall, densely caespitose, smooth or slightly rough above; leaves 2-4 mm. wide, flat; sheaths white-hyaline ventrally; inflorescence of 4-12 spikes closely aggregated into a suborbicular or oblong head; spikes 5-8 mm. long and 4-6 mm. wide, suborbicular or oblong, gynaecandrous, staminate flowers inconspicuous, with 10-30 ascending or ascending-spreading perigynia; bracts shorter than the head, usually very short; scales about as wide as but shorter than the perigynia, ovate, acute, chestnut-brown or blackish, margins scarcely hyaline, with lighter midribs; perigynia 3.5-5 mm. long and 1.25-2.25 mm. wide, ovate, plano-convex and narrowly wing-margined to base, copper-colored at maturity, several-nerved dorsally, rather abruptly tapering to a beak 1-1.5 mm. long; styles deciduous and jointed; stigmas 2.---Mountains. Alberta to Alaska, south to Colorado and California. Our rather few records scattered over the mountainous two-thirds of Colorado at 6000-10,000 feet.

33. **Carex festivella** Mack., Bull. Torr. Club 42:609. 1915.
    C. festiva of Manuals at least in part---Culms 30-100 cm. tall, caespitose, somewhat roughened below the head; leaves 2-6 mm. wide, flat; sheaths hyaline ventrally; inflorescence of 5-20 densely aggregated but distinguishable spikes in an ovoid or oblong-ovoid head; spikes 5-12 mm. long and 4-8 mm. wide, oblong-ovoid, gynaecandrous with 15-30 perigynia, these appressed but with erect tips; lower bracts sometimes prolonged but shorter than the head; scales narrower and shorter than the perigynia, ovate, obtuse or acute, dark chestnut to brownish-black at least at maturity, with narrow, hyaline margins and with ill-defined lighter midnerves; perigynia 3.75-5 mm. long and 1.5-2 mm. wide, ovate, thin and wing-margined to base, light green or straw-colored on edges and beak dark brown, lightly nerved dorsally, somewhat nerved near base ventrally, beak 1.25-1.7 mm. long, this rather terete; styles jointed; stigmas 2.---Meadows and slopes. Manitoba to British Columbia, south to Mexico and California. Scattered in the mountainous part of Colorado at 5500-12,000 feet.

34. **Carex douglasii** Boott. in Hook., Fl. Bor. Am. 2:213. 1839.
    Culms 6-30 cm. tall, 1-few from long-creeping slender rootstocks, smooth, obtusely triangular; leaves 1-2.5 mm. wide, involute, flat or canaliculate below; sheaths thickened at the mouth; dioecious; staminate inflorescence of many spikes aggregated into an oblong-orbicular to linear oblong head; pistillate inflorescence of many aggregated but distinguishable spikes; pistillate spikes 5-15 mm. long and 4-8 mm. wide, oblong-elliptical; bracts inconspicuous; pistillate scales wider and longer than the perigynia, lanceolate or ovate-lanceolate, acuminate to cuspidate, straw-colored with margins hyaline and centers green and 3-nerved; perigynia 3.5-4 mm. long and about 1.75 mm. wide, ovate-lanceolate, plano-convex and only sharp-edged above, straw-colored or finally light brown, somewhat striate both faces, stipitate, tapering into a serrated, bidentulate beak about 1.75 mm. long; styles long and jointed; stigmas 2.---Open rather dry ground. Manitoba to British Columbia, south to New Mexico and California. Our records scattered in the western two-thirds of Colorado at 4500-10,500 feet.

35. **Carex deweyana** Schw., Ann. Lyc. N. Y. 1:65. 1824.
    Culms 20-120 cm. tall, caespitose usually no elongated rootstocks present, slender, roughened below head; leaves 2-5 mm. wide, flat, thin, light green or yellowish-green; sheaths hyaline ventrally; inflorescence of 3-4 spikes, lower strongly separated; spikes 5-12 mm. long and 3.5 mm. wide, the terminal larger and usually gynaecandrous the perigynia appressed-ascending, lateral spikes usually staminate; lower bracts prolonged to 1-4 cm.; scales about as wide as but shorter than the perigynia, ovate or oblong-ovate, obtuse to awned, thin and hyaline, whitish or sometimes brownish-tinged, with 3-nerved green centers; perigynia 4.5-5.5 mm. long and 1.5-2 mm. wide, oblong-lanceolate, plano-convex and sharp-edged above, light green, nerveless or obscurely nerved dorsally, sessile or nearly so but the stalk often attached above the very base, tapering to a shallowly bidentate beak 2 mm. long; styles jointed; stigmas 2.---Dry often shaded ground. Labrador to British Columbia, south to Pennsylvania, Iowa and Colorado. Apparently rare in Colorado, known from western Las Animas County at 8000 feet and reported from Larimer County.

36. **Carex angustior** Mack.in Rydb., Fl. Rocky Mountains 124. 1917.
    Culms 10-60 cm. tall, caespitose, somewhat roughened above; leaves 0.75-2 mm. wide, flat or canaliculate; sheaths hyaline and sometimes red-dotted ventrally; inflorescence of 3-5 spikes, these approximate or nearly so but not capitate; terminal spikes gynaecandrous, lateral ones pistillate, 4-6 mm. long, suborbicular or short-oblong with widely spreading perigynia; bracts scalelike or lowest short-prolonged; scales narrower and shorter than the perigynia, ovate, short-cuspidate, yellowish with inconspicuous margins and definite green midribs; perigynia 2.5-3.5 mm. long and about 1.25 mm. wide, lanceolate to ovate-lanceolate, plano-**convex** and sharp-edged, yellowish-brown, obscurely if at all nerved, tapering or slightly contracted to a bidentate beak 1-1.75 mm. long; styles jointed; stigmas 2.---Moist meadows or swamps. Newfoundland to Washington, south to North Carolina and California. Our few records thinly scattered in the mountainous part of Colorado at 8000-11000 feet.

37. <u>Carex</u> <u>canescens</u> L., Sp. Pl. 974. 1753.
Culms 10-80 cm. tall, densely caespitose, erect, smooth except just below the head; leaves 2-4 mm. wide, flat, glaucous-green; sheaths white-hyaline ventrally; inflorescence of 4-8 erect or flexuous spikes, the upper rather approximate, the lower separated; spikes 3-12 mm. long and 3-5 mm. wide, suborbicular to oblong, gynaecandrous with 10-30 appressed-ascending perigynia; lower bracts cuspidate-prolonged; scales about the width of but shorter than the perigynia, broadly ovate, obtuse or acute, hyaline but usually brownish-tinged at maturity with green 3-nerved centers; perigynia 1.8-3 mm. long and 1.25-1.75 mm. wide, oval-ovate, plano-convex and somewhat sharp-edged, gray-green or yellowish-brown, nerved on both sides although these sometimes obscure, somewhat stipitate, contracted into a very short beak; styles jointed; stigmas 2.---Swamps. Greenland to Alaska, south to Virginia, Arizona and California; Eurasia; South America; Australia. Our records from the mountainous two-thirds of Colorado at 8500-11,500 feet.
Some of our plants have the spikes all approximate in the inflorescence and perigynia less than 1.8 mm. long. They may be C. praeceptorium Mack., a species of the northwestern United States.

38. Carex brunnescens (Pers.) Poir. in Lam., Encyc. Suppl. 3:286. 1813.
Culms **7-70 cm.** tall, caespitose, smooth or somewhat roughened above; leaves 1-2.5 mm. wide, flat, not especially glaucous; sheaths thin ventrally; inflorescence of 5-10 spikes the upper closely approximate, lower more or less separated; lateral spikes 3-7 mm. long and 3-4 mm. wide, suborbicular to short-oblong, usually gynaecandrous, with 5-10 ascending or loosely spreading perigynia, terminal spikes longer, gynaecandrous; lower bracts setaceous-prolonged; scales about the width of but shorter than the perigynia, ovate, obtusish or acute, white-hyaline with 3-nerved centers, often brownish-tinged; perigynia 2-2.5 mm. long and 1-1.5 mm. wide, oval-ovate, plano-convex and somewhat sharp-edged, greenish or brownish, lightly several-nerved both faces, substipitate, tapering into a somewhat bidentate beak 0.5 mm. long; styles jointed; stigmas 2.---Wet ground. Greenland to Alaska, south to North Caroline, Colorado and Washington; Eurasia. Our few records from central and northcentral Colorado at about 8000-11,000 feet.

39. Carex interior L. H. Bailey, Bull. Torr. Club 20:426. 1893.
Culms 15-50 cm. tall, very densely caespitose, somewhat roughened below head; leaves 1-2 mm. wide, flat or canaliculate; sheaths hyaline ventrally; inflorescence of 2-4 (rarely to 6) spikes somewhat but not closely crowded into an oblong head; lower spikes about 4 mm. long and wide, suborbicular, usually pistillate with 1-10 perigynia, these radiating at maturity; terminal spike gynaecandrous to staminate; bracts scalelike or lower **short-prolonged**; scales as wide and about half as long as the perigynia, broadly ovate, very obtuse, yellowish-brown with broad white-hyaline margins and broad 3-nerved green centers; perigynia 2.25-3.25 mm. long and 1.5-2 mm. wide, oblong-ovoid or ovoid, plano-convex and narrowly sharp-edged, green, straw-colored or light brown, several-nerved at least dorsally, substipitate, abruptly beaked, this 0.5-0.8 mm. long and shallowly bidentulate; styles jointed; stigmas 2.---Swampy meadows and bogs. Labrador to British Columbia, south to Pennsylvania, Mexico and California. Our few records from central and northcentral Colorado at 7000-11,000 feet.

40. Carex jonesii Bailey, Mem. Torr. Club 1:16. 1889.
Culms 20-60 cm. tall, caespitose but with slender rootstocks, roughened below the head; leaves 1.5-2.5 mm. wide, flat, from near base of plant; inflorescence of 4-8 spikes closely aggregated into an orbicular or short-oblong head 8-18 mm. long and 8-10 mm. wide, the spikes poorly defined; spikes androgynous with 5-10 ascending or finally spreading perigynia; bracts none; scales about as wide but shorter than the perigynia, ovate-triangular, acute, brownish or blackish with hyaline margins and inconspicuous lighter midribs; perigynia 3-4 mm. long and 1.25-1.5 mm. wide, ovate-lanceolate, plano-convex, straw-colored or brown-tinged, definitely nerved both faces, rather spongy below and substipitate, widest near base and tapering to a bidentate beak **about 0.75-1 mm.** long; styles jointed; stigmas 2.---Mountains. Montana to Washington, south to Wyoming and California. Comes into northern Colorado, our record at 9500 feet.

41. Carex neurophora Mack.in Abrams, Ill. Flora Pacif. St. 1:298. f. 706. 1923.
C. nervina of Manuals---Culms 30-70 cm. tall, caespitose with short rootstocks, smooth or roughened above; leaves 2.5-3.5 mm. wide, flat, crowded on lower part of culm; sheaths more or less green- or white-mottled dorsally, cross-rugulose ventrally at least on lower ones; inflorescence of 5-10 spikes closely crowded and not readily distinguishable, in an ovoid or oblong head 1.5-2.5 cm. long and 8-12 mm. wide; spikes androgynous, with several to many perigynia, these finally widely spreading; bracts none; scales about as wide as but about 1/2 as long as the perigynia, ovate, obtuse or acute, brown, margins and nerves not conspicuous; perigynia 3.25-4 mm. long and about 1.5 mm. wide, lanceolate to lanceolate-ovate, plano-convex, **light** brown to deep chestnut-brown, nerved on both faces, short-stipitate, somewhat spongy and widest at base, tapering into a shallowly bidentate beak about 1-1.75 mm. long; styles jointed; stigmas 2.---Mountains. Montana to Washington, south to Wyoming and Oregon. Comes into northern Colorado, our record from 9500 feet.

42. **Carex stipata Muhl.;** Willd., Sp. Pl. 4:233. 1805.
Culms 30-120 cm. tall, densely caespitose, serrulate on angles above; leaves 4-8 mm. wide, flat; sheaths septate-nodulose dorsally and cross-rugulose ventrally; inflorescence of many spikes, these hard to distinguish, crowded into a compound oblong-linear or ovoid head; spikes androgynous the staminate flowers inconspicuous, with 4-10 ascending or spreading perigynia; lowest bract often bristleform and prolonged; scales narrower and only as long as the body of the perigynia, ovate-triangular, acuminate or cuspidate, brownish-hyaline with 3-nerved green centers; perigynia 4-5 mm. long and about 1.5 mm. wide, lanceolate, plano-convex and sharp-edged to base, yellow at maturity, strongly nerved both faces, substipitate, widest at very base and tapering to a serrated, bidentate beak 2-2.5 mm. long; styles jointed; stigmas 2.---Swamps and wet places. Newfoundland to Alaska, south to North Carolina, New Mexico and California; Japan. Our records from northcentral, central and southcentral Colorado at 5000-8000 feet.

43. **Carex perglobosa** Mack., Bull. Torr. Club 34:606. 1908.
Culms 6-15 cm. tall, loosely caespitose with slender rootstocks, erect, smooth; leaves 0.75-1.5 mm. wide, flattened at least on lower; sheaths thin ventrally; inflorescence of 6-15 entirely indistinguishable spikes crowded into a globose head about 1 cm. in diameter; spikes androgynous with ascending or spreading perigynia; bracts absent; scales somewhat wider than and about as long as the perigynia, ovate-orbicular, obtusish or acute, brownish with silvery-hyaline tips and margins and lighter midribs; perigynia about 4 mm. long and 2.25 mm. wide, ovate-elliptic to ovate-oval, plano-convex and somewhat sharp-edged but rather inflated, straw-colored or yellowish-brown at maturity, finely and often obscurely nerved both faces, substipitate, tapering to a bidentulate beak 1-1.25 mm. long; styles jointed; stigmas 2.---Open slopes. Colorado and Utah. Our records from central and northcentral Colorado at 12,000-13,500 feet.

44. **Carex vernacula** L. H. Bailey, Bull. Torr. Club 20:417. 1893.
Culms 5-20 cm. tall, loosely caespitose with long-creeping slender rootstocks, smooth; leaves 2-4 mm. wide on fertile culms, flat or somewhat channeled; sheaths hyaline ventrally; inflorescence of many entirely indistinguishable spikes crowded into an ovoid-orbicular head about 1 cm. in diameter; spikes androgynous, staminate flowers not conspicuous, with several appressed perigynia; bracts absent or lower somewhat prolonged; scales about as long and as wide as the perigynia, ovate-lanceolate, acute to acuminate-aristate, dark brown or blackish with narrow hyaline margins and green midnerves; perigynia 3.5-4.5 mm. long and about 1.5 mm. wide, ovate-lanceolate, plano-convex and sharp-edged, not inflated, yellowish straw-colored to brownish, nerved but obscurely so ventrally, stipitate, contracted to a bidentulate beak 1-1.5 mm. long; styles jointed; stigmas 2.---Open slopes. Wyoming to Washington, south to Colorado and California. Our records from the mountains of Colorado at 10,000-13,500 feet.

45. **Carex vulpinoidea** Michx., Fl. Bor. Am. 2:169. 1803.
Culms 20-90 cm. tall, caespitose, usually shorter than leaves, roughened above; leaves 2-4 mm. wide, flat or channeled; sheaths cross-rugulose especially on the lower; inflorescence of many spikes aggregated into a greenish-yellow, compound head 3-10 cm. long, crowded above, somewhat separated below; spikes hard to distinguish, androgynous with several to many spreading or ascending-spreading perigynia; bracts setaceous-prolonged, conspicuous throughout the head; scales narrower than the perigynia and with a rough awn as long or longer, ovate, yellow-brown with hyaline margins and green centers; perigynia 2-3 mm. long and 1-1.5 mm. wide, body ovate-orbicular to orbicular, plano-convex, yellowish-green or straw-colored, several-nerved on dorsal face, contracted into a bidentate beak about 1 mm. long; styles jointed; stigmas 2.---Swampy ground. Newfoundland to British Columbia, south to Florida, Arizona and Oregon. Apparently rare in Colorado, our two records from Las Animas and Boulder Counties at 5000-5600 feet.

46. **Carex diandra** Schrank, Cent. Bot. Anmerk. 57. 1781.
C. teretiuscula Gooden.---Culms 30-70 cm. tall, caespitose, slender, erect, strongly roughened above; leaves 1-2.5 mm. wide, flat or canaliculate; sheaths tight and thin ventrally, strongly dotted with reddish-brown, prolonged beyond base of blade; inflorescence of many spikes in a head 2.5-5 cm. long and up to 1 cm. thick, upper spikes hardly distinguishable; spikes androgynous, staminate flowers inconspicuous, with several ascending or spreading perigynia; bracts awl-shaped and long-cuspidate, sometimes 1 cm. long but often absent; scales as wide at base as but usually narrower above than the perigynia, from shorter to longer, acute to somewhat cuspidate, brownish with hyaline margins and lighter midribs; perigynia 2-2.75 mm. long and 1-1.25 mm. wide, ovate, unequally biconvex, brown and shining, strongly few-nerved dorsally, little-nerved ventrally, stipitate, tapering or abruptly contracted into a green or whitish, serrulated beak; styles jointed; stigmas 2.---Wet meadows Newfoundland to Alaska, south to New Jersey and Colorado; Eurasia. Our few records from southwestern and northcentral Colorado at 7400-9000 feet.

47. **Carex gravida** L. H. Bailey, Mem. Torrey Club 1:5. 1889.
C. lunelliana Mack.---Culms 30-60 cm. tall, caespitose from stout, woody, fibrillose rootstocks, rather stout; leaves 4-8 mm. wide, flat; sheaths loose, green and white-mottled, septate-nodulose dorsally; inflorescence of 6-10 spikes rather crowded into an oblong or oblong-ovoid head 1-3 cm. long; spikes androgynous with 5-15 ascending, or at maturity spreading perigynia; at least lower bracts setaceous but shorter than the head; scales narrower and usually shorter than the perigynia, ovate, acuminate, cuspidate or short-awned, rather tawny-colored with 3-nerved green centers; perigynia 4-4.25 mm. long and 2.5-3.5 mm. wide, suborbicular to oval, plano-convex, serrulate almost to middle, greenish to straw-colored, several-nerved dorsally, beak about 1-1.5 mm. long strongly bidentate with teeth about 0.5 mm. long; styles jointed; stigmas 2. Our plant is var. **lunelliana** (Mack.) Hermann.---Dry prairies. Illinois to Colorado, south to Texas and New Mexico. Our few collections from extreme eastern Colorado at 3700-4400 feet.

48. **Carex hoodii** Boott. in Hook., Fl. Bor. Am. 2:211. 1839.
Culms 25-80 cm. tall, densely caespitose, roughened above; leaves 1.5-3.5 mm. wide, flat or channeled at base; sheaths thin and hyaline ventrally; inflorescence of 4-8 spikes aggregated into an orbicular or ovoid, dense head 1-2 cm. long, the lower spikes rarely somewhat separated; spikes androgynous, staminate flowers not conspicuous, with 5-10 ascending or finally spreading perigynia; bracts absent or occasionally a narrow one present at base of head; scales about as long and as wide as the perigynia, triangular-ovate, acute, acuminate or cuspidate, chestnut-brown with conspicuous hyaline margins and green midribs; perigynia 3.5-5 mm. long and 1.75-2 mm. wide, ovate-elliptic, flat or somewhat concave, green or soon brownish in center, nerveless or nearly so on both sides, tapering or somewhat contracted into a bidentate beak 0.9-1.25 mm. long; styles jointed; stigmas 2.---Meadows and slopes. Alberta to British Columbia, south to Colorado and California. Our records from northcentral Colorado at 8000-11,000 feet.

49. **Carex vallicola** Dewey, Am. Journ. Sci. II. 32:40. 1861.
Culms 25-60 cm. tall, caespitose, roughened below the heads; leaves 1-2 mm. wide, flat, light green; sheaths thin ventrally; inflorescence of several spikes closely aggregated into an

oblong to oblong-linear head about 1.5-3 cm. long, the individual spikes poorly defined; spikes androgynous with 2-10 ascending or somewhat spreading perigynia; bracts absent or lower present; scales shorter than the perigynia, broadly triangular, acute to short-cuspidate, hyaline with centers brownish or straw-colored; perigynia about 3.5 mm. long and 2 mm. wide, oblong-elliptic, plano-convex, greenish, obscurely nerved dorsally and nerveless ventrally, beak about 1 mm. long, obliquely cut dorsally and sometimes slightly bidentulate; styles jointed; stigmas 2.---Dry slopes. South Dakota to Oregon, south to Wyoming and Nevada. Our few records from north-central and northwestern Colorado at 7500-8000 feet.

50. Carex occidentalis L. H. Bailey, Mem. Torr. Club 1:14. 1889.

Culms 25-70 cm. tall, caespitose, roughened above; leaves 1.5-2.5 mm. wide, flat; sheaths thin ventrally; inflorescence of 5-10 spikes aggregated into an oblong or linear-oblong head 1.5-3 cm. long and 6-8 mm. thick, upper spikes almost entirely indistinguishable, lower a little separated; spikes androgynous the staminate flowers inconspicuous, with several loosely-ascending to somewhat spreading perigynia; bracts absent or prolonged in some; scales somewhat wider than and about the length of the perigynia, ovate-triangular, acuminate to short-awned, brownish, margins not hyaline, with 3-nerved green centers; perigynia 2.5-4 mm. long and about 1.75 mm. wide, oblong-elliptic, plano-convex and sharp-edged, light green to brownish, obscurely nerved both sides, stipitate, abruptly narrowed to a bidentate beak 0.5-0.9 mm. long; styles jointed; stigmas 2. Some of our Colorado plants intergrade with C. hookerana Dewey, a species not credited to the state.---Dry ground. Wyoming to Utah, south to New Mexico and Arizona. Our records well scattered over the western two-thirds of Colorado at 5500-10,500 feet.

51. Carex sartwellii Dewey, Am. Journ. Sci. 43:90. 1842.

Culms 40-80 cm. tall, single or in small clumps from long slender rootstocks, roughened above; leaves 2.5-4 mm. wide, flat, septate-nodulose; sheaths green-striate ventrally, septate-nodulose both ventrally and dorsally; inflorescence of many spikes densely aggregated into an elongated head; spikes 6-9 mm. long and 4-6 mm. wide, ovoid, androgynous or the middle and upper staminate, with 15-20 appressed perigynia; bracts short or lower sometimes setaceous-prolonged; scales slightly shorter and narrower than the perigynia, ovate-triangular, obtuse to short-cuspidate, dull reddish-brown with hyaline margins and green midribs; perigynia 2.5-3 mm. long and 1.5-1.75 mm. wide, ovate to broadly ovate, plano-convex, straw-colored to light brownish, somewhat nerved both faces, substipitate, abruptly beaked this 0.5 mm. long; styles jointed; stigmas 2.---Marshes and bogs. Ontario to British Columbia, south to New York and Colorado. Our three Colorado records from Larimer, Douglas and Ouray Counties at 7000-8500 feet.

52. Carex disperma Dewey, Am. Journ. Sc. 8:266. 1824.

C. tenella Schkuhr.---Culms 15-60 cm. tall, loosely caespitose with long slender rootstocks, roughened below the heads; leaves 0.75-2 mm. wide, thin and flat; sheaths thin ventrally; inflorescence of 2-4 spikes in a head not over 3-5 mm. wide, upper closely aggregated and the lower separated; spikes androgynous, the 1-2 staminate flowers very inconspicuous, with 1-6 ascending perigynia; bracts bristleform and somewhat enlarged at base or rudimentary; scales narrower and usually shorter than the perigynia, ovate-triangular, acuminate to short-mucronate, white-hyaline with green centers; perigynia 2.25-3 mm. long and about 1.5 mm. wide, oblong-ovoid, unequally biconvex, light green to yellowish-green, finely many-nerved both faces, white-punctate, short-stipitate, abruptly contracted to a short beak 0.4 mm. long or less; styles jointed; stigmas 2.---Bogs and swamps. Newfoundland to Alaska, south to New Jersey, New Mexico and California; Eurasia. Widely scattered in the mountainous part of Colorado at 6000-11,000 feet.

53. Carex simulata Mack., Bull. Torr. Club 34:604. 1908.

C. marcida Boott.---Culms 15-50 cm. tall, 1 to a few from thick long-creeping rootstocks, roughened above; leaves 2-4 mm. wide, flat or canaliculate; sheaths thin ventrally; inflorescence of 5-15 spikes densely aggregated into an oblong to linear-oblong head, the spikes not readily distinguishable; spikes mostly about 7 mm. long and 3.5 mm. wide, androgynous, the staminate flowers inconspicuous, with about 10 perigynia present but some spikes often almost completely staminate; bracts absent or lower short-prolonged; scales wider and longer than the perigynia, ovate-triangular, cuspidate or short-awned, brown with narrow hyaline margins and prominent lighter midveins; perigynia 1.75-2.25 mm. long and about 1.5 mm. wide, broadly ovate, unequally biconvex, not winged, yellowish-brown to chestnut, coriaceous, nerveless or nearly so, short-stipitate, abruptly narrowed to a short beak 0.25-0.5 mm. long; styles jointed; stigmas 2.---Wet meadows and swamps. Montana to Washington, south to New Mexico and California. Reported from all the mountainous parts of Colorado but our rather few records from the northern and central parts at 6200-10,000 feet.

54. Carex praegracilis W. Boott., Bot. Gaz. 9:87. 1884.

C. marcida Boott.---Culms 20-75 cm. tall, single or in small clumps from long-creeping stout rootstocks, usually roughened above; leaves 1.5-3 mm. wide, flat or somewhat channeled; sheaths hyaline ventrally; inflorescence of 5-15 spikes aggregated into a linear-oblong or oblong-ovoid head, the lower sometimes separated; spikes 4-8 mm. long and 4-6 mm. wide, androgynous, the staminate flowers inconspicuous, with about 10 erect-ascending perigynia; bracts absent or lower somewhat enlarged at base and upper scalelike; scales wider and longer than the perigynia, ovate, lower acuminate or upper cuspidate, light chestnut-brown with conspicuous hyaline margins and lighter midribs; perigynia 3-4 mm. long and about 1.5 mm. wide, ovate to ovate-lanceolate, plano-convex and sharp-margined, straw-colored to brownish-black at maturity, lightly nerved dorsally only, short-stipitate, tapering to a beak 1 mm. long or more; styles jointed; stigmas 2.---Moist open ground. Manitoba to Yukon, south to Mexico and California; South America. Our records scattered over Colorado, although none from the northwest and southeast corners, at 3500-9500 feet.

55. Carex eleocharis L. H. Bailey, Mem. Torr. Club 1:6. 1889.

C. stenophylla of Manuals---Culms 2.5-20 cm. tall, 1 to several from slender long-creeping rootstocks; leaves 1-1.5 mm. wide, canaliculate-flattened at base, involute above; sheaths thin ventrally; inflorescence of 3-7 spikes densely aggregated into an ovoid or linear head 5-20 mm. long and 5-10 mm. wide, the spikes not very distinguishable; spikes orbicular to ovoid, androgynous, the staminate flowers rather conspicuous, with 1-8 appressed-ascending perigynia, the upper spikes mostly staminate; bracts rudimentary or sometimes as long as the lower spikes; scales somewhat wider and about as long as the perigynia, ovate-orbicular, obtuse to cuspidate, chestnut- to reddish-brown with white-hyaline margins; perigynia 2.5-3 mm. long and 1.5-1.75 mm. wide, ovate-orbicular, plano-convex with ventral margins somewhat elevated, straw-colored to blackish at maturity, nerveless or nearly so ventrally, many-striated dorsally, substipitate, contracted to a beak 0.5-0.75 mm. long; styles jointed; stigmas 2.---Dry open ground. Manitoba to Yukon, south to Iowa, New Mexico and Oregon. Our records scattered in northern, central and southcentral Colorado at 4000-10,000 feet.

56. Carex physocarpa Presl., Rel. Haenk. 1:205. 1828.

C. pulla and C. saxatilis of Manuals---Culms 20-80 cm. tall, 1 to several from long-creeping slender rootstocks; leaves 1.5-5 mm. wide, flat with the margins often revolute; sheaths hyaline ventrally; staminate spike 2-4 cm. long and about 2.5 mm. wide, linear, usually solitary, terminal and peduncled; pistillate spikes 1.5-3.5 cm. long and 6-12 mm. wide, oblong-cylindrical, 1 to 3, separate, slender-peduncled, spreading or drooping, 20- to 75-flowered; lowest bract leaflike, from shorter than to exceeding the inflorescence, little if any sheathing; scales somewhat narrower and usually shorter than the perigynia, ovate or ovate-lanceolate, acute to acuminate, purplish-black with hyaline margins and lighter midribs; perigynia 4-5 mm. long and 1.75-2.25 mm. wide, ovoid-lanceolate, suborbicular in cross-section but not inflated, shiny, dull greenish-yellow to blackish, obscurely few-nerved dorsally, nerveless or nearly so ventrally, substipitate, abruptly very short-beaked, this about 0.5 mm. long; styles persistent and continuous, usually bent down against the achenes at maturity; stigmas 2.---Mountains. Western Canada south to Colorado and Utah. Our rather few records scattered in the mountains of Colorado at 10,000-12,000 feet.

57. Carex hassei L. H. Bailey, Bot. Gaz. 21:5. 1896.

C. garberi var. bifaria Fernald---Culms 5-70 cm. tall, loosely caespitose with long rootstocks, slender, somewhat exceeding the leaves; leaves 2-4 mm. wide, flat; sheaths concave at mouth; terminal spike staminate or gynaecandrous, 6-20 mm. long; pistillate spikes 7-25 mm. long, linear-oblong, 3-5, approximate or lower much separated often nearly basal to culms, with 7-20 appressed-ascending perigynia; lower bract leaflike, sheathing for about 3 mm. and the blade longer than the spikes; scales usually a little shorter and narrower than the perigynia, ovate-suborbicular, obtuse to acuminate, reddish-brown tinged with hyaline margins and lighter broad midribs; perigynia 2-2.5 mm. long and about 1.5 mm. wide, elliptic-obovoid, suborbicular in cross-section, grandular or whitish-puberulent, not fleshy or translucent, rather obscurely ribbed, short-stipitate at base, rounded and beakless or nearly so at apex; styles jointed; stigmas 2.---Moist ground. New Brunswick to Yukon, south to Pennsylvania, Utah and California. Our few records from central and probably northwestern Colorado at about 8000 feet.

58. Carex aurea Nutt., Gen. 2:205. 1818.

Culms 5-55 cm. tall, loosely caespitose with long rootstocks, more or less roughened above; leaves 2-4 mm. wide, flat or channeled at base; sheaths thin ventrally; staminate spike 3-10 mm. long and 1.5-3 mm. wide, linear, terminal, erect; pistillate spikes 4-20 mm. long and 3-5 mm. wide, oblong or linear-oblong, 2-5, erect, upper short-peduncled or sessile, lower widely separated or even basal to culms on peduncles up to 4 cm. long, with 4-20 ascending or finally spreading perigynia; bracts leaflike, conspicuously sheathing at base, usually much exceeding the inflorescence; scales shorter and narrower than the perigynia, ovate or ovate-orbicular, short-cuspidate to obtusish, widely spreading at maturity, reddish-brown with white-hyaline margins and green or yellowish-green 3-nerved center; perigynia 2-3 mm. long and about 1.5 mm. wide, orbicular-obovoid, flattened-oval in cross-section but not inflated, golden-yellow or yellow-brown and rather fleshy when fresh, coarsely ribbed, minutely if at all beaked; styles jointed; stigmas 2.---Moist ground. Newfoundland to British Columbia, south to Connecticut, New Mexico and California. Our many records scattered in the mountainous parts of Colorado at 6500-10,500 feet.

59. Carex nebraskensis Dewey, Am. Journ. Sci. II. 18:102. 1854.

Culms 25-120 cm. tall, caespitose with long stout rootstocks; leaves 3-12 mm. wide, flat or channeled, more or less septate-nodulose; sheaths smooth; staminate spike usually solitary or with smaller ones at base, terminal; pistillate spikes 1.5-6 cm. long and 5-9 mm. wide, oblong to cylindrical, erect, upper nearly sessile the lower short- or even long-peduncled, rather contiguous, 2-5, with many ascending perigynia; lowest bract leaflike, sheathless, usually longer than the inflorescence; scales narrower and from longer to shorter than the perigynia, lanceolate, obtusish to acuminate, purplish, brown or brownish-black, often with narrow hyaline margins, lighter centers 1- to 3-nerved; perigynia 3-3.5 mm. long and about 2 mm. wide, oblong-ovate, plano-convex or unequally biconvex, flattened, straw-colored, coriaceous, strongly many-ribbed, abruptly contracted to a bidentate beak 0.4-1 mm. long; styles jointed; stigmas 2.---Wet meadows and swamps. South Dakota to British Columbia, south to Kansas, New Mexico and California. Widely scattered over Colorado at 3500-9000 feet, apparently more abundant in the northern part.

60. Carex gymnoclada Holm, Am. Journ. Sci. IV. 14:424. 1902.

C. rigida of Manuals---Culms 20-60 cm. tall, single or in small clumps from stout but rather long rootstocks, rough or smooth, with bladeless sheaths at base; leaves 2.5-5 mm. wide, septate-nodulose; sheaths thin and hyaline ventrally, the older sheaths strongly purplish; staminate spike solitary; pistillate spikes 5-25 mm. long and 4-6 mm. wide, oblong to cylindric, 1-3, contiguous or somewhat separated, erect, sessile or short-peduncled, with many ascending or somewhat spreading perigynia; lower bract sheathless, shorter than the inflorescence; scales narrower and shorter to longer than the perigynia, ovate or oblong-ovate, obtuse to acute, black with somewhat hyaline margins and midrib often whitish; perigynia 2.25-3.5 mm. long and 1.5-2 mm.

wide, broadly obovate to suborbicular, plano-convex or biconvex, grandular, pale below and purplish-black above, 2-ribbed and nerveless otherwise, slightly stipitate, abruptly contracted to a beak 0.1-0.25 mm. long; styles jointed; stigmas 2.---Wet ground. Colorado to Washington, south to California. Our rather few records from central and northcentral Colorado at 10,500-12,000 feet.

61. Carex scopulorum Holm, Am. Journ. Sci. IV. 14:422. 421. 1902.

Culms 10-40 cm. tall, 1 to few from stout horizontal rootstocks, stout; leaves 3-7 mm. wide, flat with revolute margins; sheaths usually tinged with yellowish-brown ventrally; staminate spike solitary and terminal; pistillate spikes 1-2.5 cm. long and 6-7 mm. wide, 2-6, aggregated or lower separate and short-peduncled, the upper sessile, erect, with many squarrose-spreading perigynia; lower bracts sheathless, shorter than the inflorescence; scales narrower than but shorter to as long as the perigynia, obovate, usually obtuse, black, sometimes with narrow-hyaline margins and lighter midribs; perigynia 2.5-3.5 mm. long and 1.5-2 mm. wide, orbicular to obovoid, biconvex and turgid, splotched above with purplish-black, with 2 marginal ribs, abruptly contracted to a short beak 0.2-0.5 mm. long; styles jointed; stigmas 2.---Mountains. Wyoming to California, south to Colorado. Our records from central and northcentral Colorado at 8500-13,000 feet.

62. Carex chimaphila Holm, Am. Journ. Sci. IV. 16:33, 32. 1903.

Culms 20-50 cm. tall in small clumps from long stout rootstocks, stout; leaves 3-6 mm. wide, flat with slightly revolute margins; sheaths tinged with yellowish-brown ventrally; staminate spike 1, sometimes partly pistillate; pistillate spikes 1-3 cm. long and 7-10 mm. wide, oblong, 2-4, contiguous or the lower remote, erect, lower on peduncles shorter than the spikes, with 20-40 spreading-ascending perigynia; bracts sheathless, none as long as the inflorescence; scales narrower and usually longer than the perigynia, lanceolate or oblong-lanceolate, acute or acuminate, black but sometimes with lighter margins and midribs; perigynia 2.5-3.5 mm. long and 1.5-2 mm. wide, broadly obovoid to suborbicular, somewhat flattened but biconvex and rather turgid, purplish-blotched above, with 2 marginal ribs, abruptly contracted to a minute beak 0.1-0.2 mm. long; styles jointed; stigmas 2. This species is closely related to the preceding.---Mountains. Wyoming and Colorado. Our records from the high mountainous area of Colorado at 11,000-12,000 feet.

63. Carex emoryi Dewey in Torr., Bot. Mex. Bound. Surv. 230. 1859.

Culms 40-100 cm. tall, loosely caespitose from long stout rootstocks, arising laterally and not enveloped at base by the leaves of the previous year; leaves 3-5 mm. wide, flat with slightly revolute margins; sheaths whitish-hyaline and thin ventrally, the lower septate-nodulose; staminate spikes terminal and 1 or 2 lateral; pistillate spikes 2-10 cm. long and 4-6 mm. wide, linear, 2-4, sessile or lower short-peduncled, erect, with many appressed-ascending perigynia; lower bract leaflike and somewhat longer or shorter than the inflorescence, sheathless; scales narrower and from shorter to longer than the perigynia, lanceolate, obtusish to acuminate, light brown or purple-brown with broad lighter midribs; perigynia 2.25-2.75 mm. long and 1.5-1.75 mm. wide, broadly ovate to obovate, unequally biconvex, light green to straw-colored, with 2 marginal ribs and few-nerved dorsally, abruptly contracted to a beak 0.2-0.3 mm. long; styles rather indistinctly jointed; stigmas 2.---Wet meadows and swamps. New Jersey to Manitoba, south to Virginia and New Mexico. Our few records from central and northcentral Colorado at 5000-6000 feet.

64. Carex kelloggii W. Boott. in S. Wats., Bot. Calif. 2:240. 1880.

Culms 10-60 cm. tall, caespitose with short to somewhat elongated rootstocks; leaves 1.5-2.5 mm. wide, flat or channeled toward the base; sheaths thin and dotted with reddish-brown ventrally and obscurely if at all septate-nodulose; staminate spikes 1; pistillate spikes 1.5-3.5 cm. long and about 4.5 mm. wide, linear-cylindric, 3-5, approximate or but little separated, erect, upper sessile and lower short-peduncled, with many appressed-ascending perigynia; lower bract leaflike, longer than the inflorescence, sheathless or nearly so; scales narrower than and from much shorter to as long as the perigynia, oblong-ovate, obtuse to acute, dark purplish-brown with narrow-hyaline margins and broad lighter midribs; perigynia 1.5-3 mm. long and about 1.25 mm. wide, ovate, flattened biconvex and 2-edged, light green, lightly nerved both faces, stipitate, abruptly contracted to a beak 0.1-0.25 mm. long; styles jointed; stigmas 2. This species is very difficult to distinguish from C. aquatilis Wahl. in this state.---Wet meadows and swamps. Alberta to Alaska, south to Colorado and California. Our few records from central and northcentral Colorado at 5000-11,500 feet but some of these are doubtful.

65. Carex aquatilis Wahl., Sv. Vet.-Akad. Nya Handl. 24:165. 1803.

C. variabilis L. H. Bailey; C. substricta (Kukenth.) Mack.---Culms 20-80 cm. tall, caespitose in small or large clumps from long slender rootstocks, erect, smooth or roughened above; leaves 2.5-5 mm. wide, flat or channeled at base; sheaths thin ventrally, often purplish-dotted, more or less septate-nodulose; staminate spikes 1 or 2; pistillate spikes 1-4 cm. long and 2.5-4 mm. wide, linear to oblong, 2-4, upper approximate and short-peduncled the lower strongly separated and on peduncles shorter than the spikes, with many erect-appressed perigynia; lower bracts leaflike, usually longer than the inflorescence, sheathless; scales narrower and from shorter to longer than the perigynia, ovate to oblong-ovate, short-acuminate to obtuse, blackish with obscure margins and lighter midribs; perigynia 2.5-3 mm. long and 1.25-1.75 mm. wide, oval to obovate, flattened and unequally biconvex, green, 2-ribbed otherwise nerveless or nearly so, substipitate, abruptly beaked, this 0.1-0.3 mm. long, usually more or less dark colored; styles jointed; stigmas 2. The var. altior (Rydb.) Fernald intergrades too much with the species to attempt to keep separate in this state.---Wet meadows and swamps. Greenland to Alaska, south to Quebec, New Mexico and California. Our records scattered in the mountains of Colorado but few from the extreme west, at 5000-11,500 feet.

66. Carex lanuginosa Michx., Fl. Bor. Am. 2:175. 1803.

Culms 30-100 cm. tall, caespitose with long rootstocks, roughened above; leaves 1.5-5 mm. wide, flat with revolute margins, septate-nodulose; sheaths thin and often purplish-tinged ventrally; staminate spikes usually 2; pistillate spikes 1.5-5 cm. long and 5-8 mm. wide, oblong-cylindrical, 2-3, widely separated, erect, sessile or short-peduncled, with many ascending perigynia; lower bracts leaflike usually longer than the inflorescence, sheathless or nearly so; scales narrower and from shorter to longer than the inflorescence, lanceolate, long-acuminate, mucronate or awned, reddish-brown with hyaline margins and broad green centers; perigynia 2.5-3.5 mm. long and 1.75-2 mm. wide, broadly obovoid to ovoid, suborbicular in cross-section and somewhat inflated, dull brownish-green, densely hairy, with numerous ribs, abruptly contracted to a deeply bidentate beak 1 mm. long; styles jointed; stigmas 3.---Wet meadows and swamps. New Brunswick to British Columbia, south to Tennessee, New Mexico and California. Our records scattered over the state, but few from the extreme western and eastern parts, at 3500-10,000 feet.

67. Carex torreyi Tuck., Enum.Caric. 21. 1843.

C. abbreviata of Manuals---Culms 25-40 cm. tall, caespitose, rough and short-pubescent above; leaves 1.5-3 mm. wide, flat; sheaths soft-pubescent, light brown ventrally; staminate spikes usually 1; pistillate spikes 6-12 mm. long and 4-7 mm. wide, short-oblong, 1-3, erect, approximate or the lower somewhat separated, sessile or short-peduncled; bracts sheathless, from shorter to somewhat longer than the inflorescence; scales about as wide as but only about 1/2 the length of the perigynia, ovate-orbicular, lower acuminate and upper acute, reddish-brown or brownish-yellow with broad white-hyaline margins and green centers; perigynia 2.5-3 mm. long and about 2 mm. wide, broadly ovoid or broadly obovoid, obscurely triangular in cross-section, punctate and not truly hairy, yellowish-green, finely ribbed, very short-stipitate, abruptly contracted to a cylindrical beak 0.3-0.4 mm. long; styles jointed to an apiculate achene; stigmas 3.---Rather dry ground. Manitoba to Alberta, south to South Dakota and Colorado. Our one record from northcentral Colorado at 6800 feet.

68. Carex rossii Boott. in Hook., Fl. Bor. Am. 2:222. 1839.

C. deflexa var. farwellii Britt.; C. deflexa var. rossii (Boott.) Bailey---Culms 5-30 cm. tall, caespitose, slightly roughened above; leaves 1-2.5 mm. wide, slightly channeled above and margins slightly revolute; sheaths thin and finely striate ventrally, often minutely hairy dorsally; staminate spike erect; pistillate spikes 3-5 mm. long and 3-4 mm. wide, suborbicular to short-oblong, upper 1-2 erect, sessile or short-peduncled, somewhat approximate, but the lower widely separate, nearly basal on the culms, erect on slender often very long peduncles, with 3-15 ascending perigynia; lower bract leaflike, usually longer than the inflorescence, sheathless or nearly so; scales wider and shorter than the perigynia, ovate, acute, acuminate or even awned, greenish, reddish or purplish-brown with hyaline margins and green centers; perigynia 3-4.5 mm. long, about 1-1.5 mm. wide, elliptic or ovate, not flattened, pale green, 2-keeled, short-pubescent, stipitate, abruptly contracted to a bidentate beak 0.75-1.5 mm. long; styles jointed; stigmas 3.---Dry soil. Michigan to Yukon, south to Colorado and California. Our records scattered over the western two-thirds of Colorado, except from the extreme western border, at 5500-12,000 feet.

Recently a related species, Carex geophila Mack., known before only from New Mexico and Arizona, has been collected in Boulder County. The bracts of the lower non-basal pistillate spikes are not leaflike but are shorter than the inflorescences.

69. Carex pityophila Mack., Bull. Torr. Club 40:545. 1913.

Culms 5-20 cm. tall, in dense clumps, very rough; leaves 0.75-1.5 mm. wide, flat with slightly revolute margins; sheaths thin ventrally; staminate spike 1; pistillate spikes 2-5, upper approximate or little separated, 1- to 3-flowered, lower basal or nearly so, long-peduncled, erect, 2- to 5-flowered; lower bract leaflike, usually longer than the inflorescence, sheathless or nearly so; scales nearly as long and as wide as the perigynia, ovate, acute to short-cuspidate, more or less purplish-tinged with hyaline margins and green centers; perigynia 3.5-4.5 mm. long and about 1.5-1.75 mm. wide, oval to elliptic, not flattened, 2-keeled, short-pubescent especially above, stipitate, abruptly beaked, this 0.75-1 mm. long; styles jointed; stigmas 3. Our plants are closely similar to C. rossii and would best be considered a variety of that species.---Dry ground. Colorado and Utah, south to New Mexico. Our few records from central and southern Colorado (with a doubtful one from northcentral) at 6000-11,500 feet.

70. Carex concinna R. Br., Richards in Frankl. Journ. 751. 1823.

Culms 5-20 cm. tall, loosely caespitose from slender elongated rootstocks; leaves 2-2.5 mm. wide, flat or somewhat involute at base; sheaths hyaline ventrally; staminate spike 3-6 mm. long, very slender; pistillate spikes 4-8 mm. long and 3-4 mm. wide, short-oblong to suborbicular, 2-3, erect, sessile to peduncled, aggregated or lowest somewhat remote, rarely 1 long-peduncled spike from about the middle of the culm, with 5-12 ascending perigynia; bracts with blades reduced to absent, sheathing to 7 mm.; scales narrower and about 1/2 the length of the perigynia, orbicular-ovate, obtuse or slightly cuspidate, short-ciliate, dark reddish-brown with hyaline margins and straw-colored or obsolete midribs; perigynia 3-3.5 mm. long and 1-1.25 mm. wide, oblong-obovoid, obtusely triangular in cross-section, whitish, greenish or yellowish-white, 2-ribbed, hirsute, stipitate, short-tapering to a short beak not over 0.4 mm. long, styles jointed; stigmas 3, short.---Dry soil. Newfoundland to Yukon, south to Michigan and Colorado. Our one record from central Colorado at 10,300 feet.

71. Carex heliophila Mack., Torreya 13:15. 1913.

C. pennsylvanica of Manuals; C. pennsylvanica var. vespertina Bailey---Culms 5-35 cm. tall, in rather small clumps from long slender rootstocks, rough above; leaves 1-2.5 mm. wide, flat and channeled near base with margins slightly revolute; lower sheaths often filamentose; staminate spike usually 1; pistillate spikes 4-6 mm. long, suborbicular, 1 to 2 rarely 3, contiguous to somewhat separated, sessile or nearly so, with 5-15 ascending perigynia; bracts shorter than the inflorescence, sheathless; scales slightly narrower and from somewhat shorter to longer than the perigynia, ovate, obtuse or cuspidate, reddish-brown to tawny with hyaline margins and lighter centers; perigynia about 3.5 mm. long and 2-2.25 mm. wide, oval, suborbicular in cross-section, dull green, 2-keeled, short-pubescent, stipitate, abruptly contracted to a beak 0.75

mm. long; styles jointed; stigmas 3. Possibly not specifically distinct from the eastern C. pennsylvanica Lam.---Plains and hills. Manitoba to Alberta, south to Missouri and New Mexico. Our records from northcentral, central, southcentral and southwestern Colorado at 5000-9000 feet.

72. Carex atherodes Spreng., Syst. 3:827. 1826.

C. aristata R. Br.---Culms 3-150 cm. tall, with long slender rootstocks, smooth; leaves 3-12 mm. wide, flat, septate-nodulose; sheaths densely to rather sparsely soft-hairy; staminate spikes 2-6, erect, upper contiguous and the lower separated; pistillate spikes 5-12 cm. long and 8-15 mm. wide, narrowly cylindrical, 2-4, erect, sessile to short-peduncled, separated widely, with 30-100 ascending-spreading perigynia; bracts leaflike, the lowest strongly sheathing; scales narrower and from longer to shorter than perigynia, ovate, abruptly aristate, dull reddish-brown with hyaline margins and green centers; perigynia 7-10 mm. long and about 2 mm. wide, lanceolate to ovoid-lanceolate, somewhat inflated and suborbicular in cross-section, yellowish-green or light brownish, strongly ribbed, short-stipitate, tapering to a bidentate beak, the teeth 1.2-3 mm. long; styles continuous; stigmas 3.---Swamps and wet ground. Ontario to Yukon, south to New York, Colorado and Oregon; Eurasia. Scattered in the mountainous part of Colorado, most of our records from the northcentral part, at 5000-8500 feet.

73. Carex hystricina Muhl.; Willd., Sp. Pl. 4:282. 1805.

Culms 15-100 cm. tall, caespitose but with long very slender rootstocks, more or less roughened above; leaves 2-10 mm. wide, thin, flat with somewhat nodulose margins, septate-nodulose; sheaths hyaline ventrally; staminate spike terminal; pistillate spikes 1-6 cm. long and 10-15 mm. wide, oblong or oblong-cylindric, 1-4, approximate to separated, upper erect on short peduncles, lower nodding on long peduncles, with many spreading perigynia; bracts leaflike, the lower sometimes sheathing; scales with bodies very narrow and shorter than the perigynia, ovate or oblanceolate, with long rough awns longer than the bodies, light reddish-brown but awns and centers green; perigynia 5-7 mm. long and 1.5-2 mm. wide, narrowly ovoid to elliptic, inflated and round in cross-section, light green or straw-colored, closely many-ribbed, stipitate, tapering to a beak 2 mm. long, this bidentate with teeth 0.5 mm. long; styles continuous; stigmas 3.--- Wet meadows and swamps. New Brunswick to Washington, south to Virginia, Arizona and California. Our records in the northeast, northcentral and western parts of Colorado at 3500-6000 feet.

74. Carex retrorsa Schw., Ann. Lyc. N. Y. 1:71. 1824.

Culms 20-100 cm. tall, caespitose, stout, smooth or slightly rough above; leaves 3-10 mm. wide, flat, septate-nodulose; sheaths tinged with yellow-brown ventrally; staminate spikes usually 1 or 2; pistillate spikes 1.5-8 cm. long and 14-20 mm. wide, oblong-cylindrical to short-oblong, 3-8, erect, aggregated and short-peduncled or lower separate and long-peduncled, with many perigynia, these wide-spreading or reflexed especially below; bracts very long, several times longer than the inflorescence, lower more or less long-sheathing; scales narrow and about as long as the perigynia bodies, lanceolate, acute to cuspidate, brown or purplish-brown with green centers; perigynia 7-10 mm. long and about 3 mm. wide, ovoid, inflated and suborbicular in cross-section, yellowish-green or brownish-tinged, coarsely nerved, very short-stipitate, contracted to a beak 2-3.5 mm. long, this bidentate with teeth 0.5 mm. long; styles continuous and twisted or bent; stigmas 3.---Swamps and wet places. Quebec to British Columbia, south to New Jersey, Colorado and Oregon. Our two records from northwestern and southwestern Colorado at 6500 feet.

75. Carex rostrata Stokes, With. Brit. Pl. ed. 2, 2:1059. 1787.

C. utriculata Boott.---Culms 30-120 cm. tall, caespitose with stout whitish rootstocks, stout, smooth or somewhat roughened above; leaves 2-12 mm. wide, flat or channeled at base, the margins often revolute, septate-nodulose; sheaths thin ventrally; staminate spikes 2-4, the lower sessile; pistillate spikes 1-1.5 cm. long and 6-20 mm. wide, cylindric to oblong, 2-5, erect, more or less separated, upper sessile or short-peduncled, lower short-peduncled, with many ascending or finally squarrose perigynia; bracts leaflike, slightly if at all sheathing; scales narrower and from shorter to longer than the perigynia, oblong, ovate to linear-lanceolate, blunt to acuminate or even short-awned, light brown or purplish-brown with hyaline margins and green centers; perigynia 3.5-8 mm. long and 2.5-3.5 mm. wide, oval-ovoid to ovoid, inflated and suborbicular in cross-section, yellow-green to light brownish, strongly several-nerved, substipitate, the beak 1-2 mm. long and bidentate with teeth 0.5-0.75 mm. long; styles continuous; stigmas 3. The variety utriculata (Boott.) Bailey cannot be separated in Colorado plants.---Swamps and wet ground. Greenland to Alaska, south to West Virginia, New Mexico and California. Our records scattered over the western mountainous part of Colorado at 6500-11,000 feet.

76. Carex vesicaria L., Sp. Pl. 979. 1753.

C. monile Tuck.---Culms 30-100 cm. tall, caespitose, no long horizontal rootstocks, rough or smooth above; leaves 2-7 mm. wide, flat with margins more or less revolute, at least somewhat septate-nodulose; sheaths thinner ventrally, often yellowish-brown, the basal becoming filamentose; staminate spikes 2-4, approximate or separated; pistillate spikes 2.5-7.5 cm. long and 5-15 mm. wide, oblong-cylindric to narrowly cylindric, 1-3, more or less strongly separated, erect, sessile or short-peduncled, with many ascending perigynia; bracts leaflike, lowest longer than the inflorescence, not sheathing; scales narrower than and from 1/2 as long as the perigynia, ovate to lanceolate, acute to awned, purplish-brown to yellowish-brown with hyaline margins and lighter centers; perigynia 4-8 mm. long and about 3 mm. wide, ovoid to globose-ovoid, inflated and suborbicular in cross-section, yellowish-green to brownish-tinged, several- to many-nerved, substipitate, beak 2 mm. long, bidentate with erect teeth 0.5-1 mm. long; styles continuous and often bent or flexuous; stigmas 3.---Swamps or wet ground. Newfoundland to British Columbia, south to Delaware, New Mexico and California; Eurasia; Africa. Our few records from northcentral and southcentral Colorado at 7500-9500 feet.

77. Carex misandra R. Br., Chlor. Melv. 25. 1823.

Culms 10-30 cm. tall, densely caespitose, smooth or slightly rough above, leaves clustered near base; leaves 1.5-3 mm. wide, flat or canaliculate; sheaths thin ventrally, usually brownish-tinged; terminal spike gynaecandrous or rarely staminate, drooping; pistillate spikes 7-20 mm. long and 4-6 mm. wide, oblong or linear-oblong, 2- or 3, approximate or separate, lower nodding on long peduncles, with many appressed-ascending perigynia; at least lower bract long-sheathing, blade short or wanting; scales wider but definitely shorter than the perigynia, ovate, obtuse to acuminate, white-hyaline at apex, with slender midrib; perigynia 3.5-5 mm. long and about 1 mm. wide, narrowly lanceolate, rather flattened, purplish-black above, greenish to straw-colored below, 2-edged but obscurely if at all nerved, short-stipitate, long-tapering to a beak 1.7-2.5 mm. long, this ciliate-serrate on the margins; styles jointed; stigmas 3.---Mountains. Greenland to Alaska, south to Colorado; Eurasia. Our few records from northcentral Colorado at 11,000-12,500 feet.

78. Carex hallii Olney; Porter, Rep. U. S. Geo. Surv. Terr. 5:496. 1872.

Culms 10-60 cm. tall, loosely caespitose with long slender rootstocks, smooth or nearly so; leaves 2-4 mm. wide, flat with revolute margins; sheaths white-hyaline ventrally; spikes 1-5, often all pistillate or terminal one staminate; pistillate spikes (often reduced to 1) 6-30 mm. long and 3-5 mm. wide, contiguous or lower separated, erect, sessile or short-peduncled, with many appressed-ascending perigynia; bracts all shorter than the inflorescence, upper reduced, slightly if at all sheathing; scales wider and slightly longer than the perigynia, orbicular to broadly ovate, obtuse to mucronate, brownish-purple with conspicuous white hyaline margins and lighter centers; perigynia 2-3 mm. long and about 2 mm. wide, broadly obovoid or suborbicular, plano-convex, light green and often dark tinged toward apex, 2-ribbed, nerveless or slightly nerved dorsally, short-stipitate, beak short, 0.1-0.2 mm. long, faintly bidentulate; styles jointed; stigmas 3.---Mountain meadows. Hudson Bay to Alberta, south to North Dakota and Colorado. Our few records from northcentral, central and possibly southcentral Colorado at 7000-10,500 feet.

79. Carex raynoldsii Dewey, Am. Journ. Sc. II. 32:39. 1861.

Culms 20-75 cm. tall, loosely caespitose with stout creeping rootstocks, stiff and stout, smooth or nearly so; leaves 3-8 mm. wide, flat with revolute margins; sheaths thin and yellow-tinged ventrally; staminate spike terminal, nearly or quite sessile; pistillate spikes 1-2 cm. long and 6-8 mm. wide, oblong, 2-4, approximate or lowest slightly separated, erect, upper sessile or nearly so, lower more or less strongly peduncled, with many ascending-spreading perigynia; bracts leaflike, sheathless, about the length of the inflorescence; scales about as wide as but shorter than the perigynia, broadly ovate, short-acute to cuspidate, dark purplish-black, with very narrow hyaline margins and often with lighter midribs; perigynia 3.5-4.5 mm. long and 1.75-2 mm. wide, oblong-oval, oblong-obovoid to elliptic, suborbicular in cross-section, yellowish-green to yellowish-brown, several-nerved, substipitate, abruptly contracted to a short beak 0.3-0.5 mm. long; styles jointed (but achenes apiculate); stigmas 3.---Mountain meadows. Alberta to British Columbia, south to Colorado and California. Our records from the northern half of the mountainous part of Colorado at 7500-12,500 feet.

80. Carex buxbaumii Wahl., Sv. Vet.-Akad. Nya Handl. 24:163. 1803.

Culms 25-100 cm. tall, loosely caespitose with slender scaly rootstocks, rough above; leaves 1.5-4 mm. wide, flat with revolute margins; sheaths yellowish-brown and purple-dotted ventrally; inflorescence of 2-5 spikes, the terminal gynaecandrous, lateral pistillate, approximate or lower separated; pistillate spikes 5-20 mm. long and 6-10 mm. wide, ovoid or oblong-ovoid, sessile, with 10-40 appressed-ascending perygnia; bracts sheathless, the lower shorter than or equalling the inflorescence; scales narrower and usually longer than the perigynia, lanceolate, long-acuminate, aristate to obtusish, purplish-black or purplish-brown with lighter midribs; perigynia 2.5-4 mm. long and 1.5-2 mm. wide, elliptic or obovoid, triangular-biconvex, glaucous-green, minutely papillose, 2-ribbed and finely many-nerved, short-stipitate at base, abruptly contracted to a short, bidentulate beak about 0.2 mm. long; styles jointed but rather obscurely so; stigmas 3.---Swamps and wet ground. Newfoundland to Alaska, south to Georgia and California. Our one record from northcentral Colorado at about 7200 feet.

81. Carex media R. Br.; Richards. in Frankl. Journey 763. 1823

C. vahlii, C. halleri, C. alpina of Manuals---Culms 20-80 cm. tall, loosely caespitose with short slender rootstocks, smooth or slightly roughened above; leaves 2-3 mm. wide, flat; sheaths hyaline ventrally; inflorescence usually of 3 spikes, the terminal gynaecandrous, the lateral pistillate, closely aggregated or approximate; lateral spikes 4-8 mm. long and 3-4.5 mm. wide, short-oblong or suborbicular, erect, sessile or lower short-peduncled, with 8-25 ascending perigynia; bracts slightly if at all sheathing, usually shorter than the inflorescence; scales nearly as wide as but much shorter than the perigynia (1.5-2.5 mm. long), broadly ovate, acute or obtuse, purple-black with white-hyaline margins; perigynia 2-3.5 mm. long and about 1.25 mm. wide, obovoid or oblong-obovoid, rather triangular in cross-section, yellowish-green and becoming yellowish-brown or purplish-tinged above, 2-ribbed, abruptly contracted to a beak 0.3-0.5 mm. long; styles jointed; stigmas 3. Our plants is var. stevenii (Holm) Fernald. Fernald (Rhodora 44:304. 1942) states that C. vahlii Schk., C. halleri Gunn., and C. alpina Lilj., seem to be merely synonyms of C. norvegica Retz., an arctic plant not in Colorado.---Open often rocky ground. The variety from Colorado and Utah. Our records from the mountainous two-thirds of Colorado, none from the northwest corner as yet, at 6000-13,000 feet.

82. Carex nelsonii Mack. in Rydb., Fl. Rocky Mts. 137. 1917.

Culms 10-30 cm. tall, loosely caespitose, smooth; leaves 3-4 mm. wide, flat or channeled at base; sheaths whitish ventrally; inflorescence of 2-3 closely aggregated sessile spikes, the terminal gynaecandrous and the lateral pistillate; pistillate spikes 10-12 mm. long and 5-8 mm. wide, oblong or obovoid, with many appressed perigynia; bracts scalelike or the lowest short-prolonged; scales narrower and shorter than the perigynia, ovate to ovate-lanceolate, obtuse or acutish, black, margins and midribs slightly if at all noticeable; perigynia about 4 mm. long and 1.5 mm. wide, oblong-obovoid, subinflated-triangular in cross-section, yellowish-green but blotched with purple-black, with 2 lateral ribs, grandular-roughened on edges above, substipitate, contracted to a beak 0.6-1 mm. long; styles jointed; stigmas 3. This species seems hardly distinct from C. nova Bailey in Colorado.---Mountain meadows. Wyoming and Colorado. Our rather few records from the mountains of Colorado, but none from the southcentral or northwestern parts, at 8500-13,000 feet.

83. Carex nova Bailey, Journ. Bot. 26:322. 1888.

C. melanocephala Turcz.---Culms 15-60 cm. tall, caespitose, smooth or somewhat roughened above; leaves 2.5-5 mm. wide, flat; sheaths hyaline ventrally; inflorescence of 3-4 spikes, the terminal gynaecandrous, lateral pistillate, sessile and closely aggregated; pistillate spikes 7-12 mm. long and 6-10 mm. wide, suborbicular with many spreading-ascending or squarrose perigynia; bract somewhat below the inflorescence, slightly if at all sheathing, longer to shorter than the inflorescence; scales shorter and narrower than the perigynia, lanceolate or oblanceolate to obovate, obtuse, acute to short-cuspidate, purplish-black with narrow hyaline margins and inconspicuous midribs; perigynia 3-4 mm. long and 2-3.5 mm. wide, ovate-orbicular to obovoid, much flattened, purplish-black with green margins, 2-ribbed, substipitate, abruptly contracted to a beak 0.5-1 mm. long; styles jointed; stigmas 3.---Meadows and along streams. Montana to Idaho, south to New Mexico and Utah. Our records from the mountains of Colorado, none as yet from the northwestern part, at 9000-12,500 feet.

84. Carex albo-nigra Mack. in Rydb., Fl. Rocky Mts. 137. 1917.

Culms 10-30 cm. tall, caespitose, roughened toward the top; leaves 2.5-5 mm. wide, flat; sheaths whitish ventrally; inflorescence usually of 3 spikes, terminal gynaecandrous and the lateral pistillate; pistillate spikes 8-10 mm. long and 4 mm. wide, narrowly oblong, approximate or lower slightly separated and erect on a peduncle shorter than the spike, with 8-20 appressed perigynia; lowest bract leaflike, about as long or somewhat shorter than the inflorescence, subsheathing; scales mostly wider than and nearly the length of the perigynia, ovate, acute or obtuse, purple-black, mostly conspicuously white-hyaline on upper margins; perigynia 3-3.5 mm. long and about 2 mm. wide, broadly ovate or obovate, flattened, purplish-black, 2-ribbed, substipitate, abruptly contracted to a short beak scarcely 0.5 mm. long; styles jointed; stigmas 3.--- Mountains. Alberta to Washington, south to Arizona and California. Our rather few records from northcentral, central and southcentral Colorado at 10,500-13,000 feet.

85. Carex chalciolepis Holm, Am. Journ. Sci. IV. 16:28, 29. 1903.

Culms 20-75 cm. tall, densely caespitose, smooth or slightly roughened above; leaves 2.5-6 mm. wide, flat; sheaths white or yellowish ventrally; inflorescence of 2-4 closely aggregated spikes or the lowest slightly separated, terminal gynaecandrous the lateral pistillate; pistillate spikes 1-2.5 cm. long and 6-10 mm. wide, oblong to broadly ovoid, varying from sessile to short-peduncled these shorter than the spikes, with many appressed-ascending perigynia; lower bract leaflike, from shorter to longer than the inflorescence, slightly if at all sheathing; scales mostly narrower but longer than the perigynia, lanceolate or ovate-lanceolate, acute to short-cuspidate, dark copper-colored to brown-purple or even blackish, margins and midribs obsolete; perigynia 3-5 mm. long and 2.5-5 mm. wide, broadly ovate-suborbicular to obovate, flattened, dark purple often greenish especially at margins, 2-ribbed, abruptly contracted to a short beak about 0.5 mm. long; styles jointed; stigmas 3.---Mountain slopes and meadows. Wyoming to Utah, south to Colorado and Arizona. Scattered in the mountains of Colorado at 8000-13,000 feet.

86. Carex bella L. H. Bailey, Bot. Gaz. 17:152. 1892.

Culms 50-90 cm. tall, caespitose; leaf blades 3-6 mm. wide, flat; sheaths thin ventrally, red-spotted near mouth; inflorescence of 3-4 spikes, the terminal gynaecandrous, lateral pistillate; pistillate spikes 12-25 mm. long and 4-5 mm. wide, oblong-linear, upper erect and short-peduncled, lower more or less drooping with peduncles longer than the spikes, with 15-30 appressed perigynia and sometimes with a few staminate flowers near bases; lowest bract leaflike, seldom exceeding the inflorescence, somewhat sheathing; scales about as wide as but shorter than the perigynia, ovate, obtuse or acute, dark purplish-brown with white-hyaline margins and lighter centers; perigynia 3-4 mm. long and 1.75-2 mm. wide, broadly oval to oblong-oval, flattened, whitish-green, rarely purplish-tinged, 2-ribbed, substipitate, abruptly contracted to a somewhat bidentate beak about 0.3 mm. long; styles jointed; stigmas 3.---Mountains. Colorado to Utah, south to New Mexico and Arizona. Our records from the mountains of Colorado, none as yet from the northwestern part, at 8000-11,500 feet.

87. Carex atrata L., Sp. Pl. 976. 1753.

Culms 15-50 cm. tall, caespitose with short and slender rootstocks; leaves 2-8 mm. wide, flat or channeled above; sheaths hyaline and usually brown-tinged ventrally; inflorescence of 3-7 spikes, the terminal gynaecandrous and lateral pistillate; pistillate spikes approximate or lower separated, upper short-peduncled and weakly erect, lower on peduncles usually as long or longer than the spike and weakly erect or nodding, with 15-50 closely appressed perigynia; lowest bract leaflike, shorter to longer than the inflorescence, sheathless; scales about as wide as and somewhat longer than the perigynia, ovate to oblong-ovate, acute to obtusish, black or brownish-black with narrow margins; perigynia 3-4 mm. long and 1.5-3 mm. wide, broadly oval to obovate, more or less flattened, yellowish-brown and more or less purple-blotched, 2-ribbed, substipitate, abruptly contracted to a beak o.5 mm. long; styles jointed; stigmas 3.---Mountain meadows. Greenland to Alberta, south to Colorado and Nevada; Eurasia. Our few specimens scattered in the mountains of central, northwestern and southwestern Colorado at 9000-12,500 feet.

88. Carex sprengelii Dewey; Spreng., Syst. 3:827. 1826.

C. longirostris Torr.; C. longirostris var. minor Boott.---Culms 30-90 cm. tall, caespitose from fibrillose, elongated rootstocks; leaves 2.5-4 mm. wide, flat; sheaths white-hyaline ventrally, older ones fibrillose; staminate spikes 2-3, approximate and terminal; pistillate spikes 1-3.5 cm. long, and 8-10 mm. wide, oblong-cylindric, 2-4, wide separated, pendulous or upper erect, on peduncles 1/2 to 3 times as long as the spikes, with many spreading-ascending perigynia; bracts leaflike, from shorter to longer than the inflorescence, short-sheathing; scales narrower than the perigynia and shorter than their beaks, lanceolate to ovate-lanceolate, acute to awned, greenish-white to brown-tinged with green centers; perigynia 5-6 mm. long and about 2 mm. wide, body suborbicular to oblong-orbicular, orbicular in cross-section, greenish to straw-colored, 2-ribbed and often nerved near bases, short-stipitate, abruptly contracted to a long, deeply bidentate beak 2.5-3 mm. long; styles jointed to the bent-apiculate apex of the achenes; stigmas 3.---Thickets, banks and rocky places. New Brunswick to Alberta, south to Delaware and Colorado. Our few records from northcentral and central Colorado at 5500-7700 feet.

89. *Carex saximontana* Mack., Bull. Torr. Club 33:439. 1906.

C. backii and *C. durifolia* of Manuals---Culms 5-30 cm. tall, caespitose, narrowly winged and serrate on the angles, enlarged above; leaves 3-5 mm. wide, flat; sheaths thin and hyaline ventrally; spikes androgynous, 1-3, the lower nearly basal on the culm, on long capillary peduncles, staminate flowers few with scales connate to the middle, pistillate part with 2-5 erect perigynia; proper bracts absent but lower scales leaflike, 7-35 mm. long, exceeding the inflorescence; upper scales shorter than the perigynia, usually ovate-lanceolate, acute to acuminate, white-hyaline; perigynia about 4 mm. long and 2 mm. wide, obovoid, 2-keeled, light green, faintly many-nerved, substipitate, abruptly contracted to a somewhat serrulate beak about 1 mm. long; styles jointed; stigmas 3.---Woods and open slopes. Manitoba to British Columbia, south to Nebraska and Utah. Our few records from northcentral Colorado at 5500-7500 feet.

90. *Carex paupercula* Michx., Fl. Bor. Am. 2:172. 1803.

C. magellanica Lam.---Culms 10-80 cm. tall, loosely caespitose, from slender short or long rootstocks; leaves 2-4 mm. wide, flat; sheaths hyaline and often red-dotted ventrally; terminal spike staminate or gynaecandrous, lateral spikes 1-4, pistillate, gynaecandrous or sometimes staminate, approximate, drooping or nearly erect, on peduncles 1 to 4 times as long as the spikes; spikes 4-22 mm. long and 4-8 mm. wide, suborbicular to oblong, with 5-20 ascending perigynia; lowest bract leaflike, as long as or longer than the inflorescence, slightly sheathing; scales narrower but usually longer than the perigynia, lanceolate or ovate-lanceolate, cuspidate or acuminate, castaneous to brown, midribs green or lighter brown near apex; perigynia 2.5-3 mm. long and 1.75-2.25 mm. wide, broadly ovoid to broadly elliptic, compressed-triangular in cross-section, pale green or sometimes purple-blotched near tips, few-nerved on each face, somewhat stipitate, very short-beaked or nearly beakless; styles faintly jointed to the apiculate achenes; stigmas 3.---Bogs and wet places. Newfoundland to Alaska, south to Pennsylvania and Utah; Eurasia. Our rather few specimens from northcentral, central and southcentral Colorado at 7500-11,500 feet.

91. *Carex limosa* L., Sp. Pl. 977. 1753.

Newfoundland to Yukon, south to Delaware, Montana and California. This plant has been reported from Colorado but no specimens could be located by the writer.

92. *Carex capillaris* L., Sp. Pl. 977. 1753.

Culms 3-60 cm. tall, caespitose in small clumps, erect or decumbent, smooth; leaves 0.75-2.5 mm. wide, flat or channeled; sheaths short and tight; the loose terminal spikes usually staminate, slender; lateral spikes pistillate, 5-15 mm. long and 3-4 mm. wide, linear-oblong, 2-3, approximate or widely separated on slender elongated drooping pedicels, with 3-20 ascending perigynia; bracts shorter than the inflorescence, long-sheathing; scales somewhat wider but shorter than the perigynia, orbicular-ovate, obtuse or acute, light chestnut with apex and margins white-hyaline, midribs lighter; perigynia 2.5-3 mm. long and about 0.75 mm. wide, ovoid-lanceolate, obtusely triangular in cross-section, greenish-brown, 2-ribbed but otherwise nerveless, stipitate, beak about 1 mm. long the orifice oblique; styles jointed, stigmas 3.--- Open slopes and meadows. Greenland to Alaska, south to Maine, New Mexico and Nevada; Eurasia. Our few records scattered in the mountains of Colorado, mostly in the central part, at 6000-12,000 feet.

93. *Carex viridula* Michx., Fl. Bor. Am. 2:170. 1803.

Culms 6-30 cm. tall, caespitose, smooth; leaves 1-3 mm. wide, canaliculate; sheaths thin ventrally; terminal spike normally staminate, linear; lateral spikes pistillate, 5-10 mm. long and 4-7 mm. wide, oblong or globose-oblong, 2-6, erect, upper sessile and aggregated, lower little to widely spreading with very short peduncles, with many spreading perigynia; lower bracts longer than the inflorescence, conspicuously sheathing; scales narrower than and about 1/2 as long as the perigynia, obovate, usually short-cuspidate, reddish with narrow margins and green centers; perigynia 2-3 mm. long and about 1.25 mm. wide, obovoid, obtusely triangular in cross-section, yellowish-green, several- to many-ribbed, abruptly contracted to a minutely bidentulate beak 0.5-0.75 mm. long; styles jointed; stigmas 3.---Moist ground. Greenland to Alaska, south to New Jersey, New Mexico and California. Our one record from southwestern Colorado at 6500 feet.

4. *Eriophorum* L. COTTONSEDGE; COTTONGRASS

Bog plants with rootstocks; culms triangular or terete; leaves grasslike or some reduced to sheaths; spikes single or few in a head or umbel subtended by 1 or more involucre bracts; scales spirally imbricated; flowers perfect; perianth of 6 to many filiform smooth long bristles, these much exserted from the scales at maturity forming a cottony mass; stamens 1-3; styles rather long, deciduous; stigmas 3; achenes 3-angled.

1. Spikes several, some distinctly pedicelled; involucre of 1-4 leaflike bracts; bristles white
  2. Leaves less than 2 mm. wide, triangular-channeled throughout; usually only 1 bract present; bristles seldom over 2.5 cm. long -1. *E. gracile*
  2. Leaves over 2 mm. wide, flat below the middle; 2 or more bracts present; bristles at maturity often over 2.5 cm. long -2. *E. angustifolium*
1. Spikelets solitary; involucre wanting (lower scale of spike may be somewhat enlarged); bristles white to reddish-brown -3. *E. chamissonis*

1. Eriophorum gracile Koch. in Roth, Cat. 2:259. 1800.

Culms 20-60 cm. tall, slender, terete or somewhat triangular, rather weak and often reclining; leaves 1-2 mm. wide, triangular-channeled, 2-3 to a culm; upper sheaths long; involucre bract 1-2 cm. long, usually 1; spikes 2-5, the central one often sessile, others on peduncles mostly less than 2 cm. long; scales 2.5-4 mm. long, oblong, lead-colored or black, midveins prominent; achenes linear-oblanceolate to linear-oblong; bristles 1-2 cm. long, white.---Swamps and bogs. Quebec to British Columbia, south to Pennsylvania, Colorado and California; Eurasia. Our few records from northcentral, southcentral and southwestern Colorado at 8000-12,000 feet.

2. Eriophorum angustifolium Roth, Tent. 1, 24. 1788.

E. ocreatum A. Nels.; E. polystachon of Manuals---Culms 30-60 cm. tall, slender, obtusely angled; leaves 2-6 mm. wide, flat toward the base, conduplicate toward the tip; inflorescence of 2-12 spikes on peduncles 0.5-7 cm. long or central sessile; spikes 1-2 cm. long in anthesis, longer in fruit, ovoid or lanceolate; bracts 2-3, dark at base, longer or shorter than the inflorescence; scales 4-10 mm. long, ovate or lanceolate, purple-green, brown or lead-colored, midrib obsolete near the membranous tips; achenes oblong-ovoid; bristles at maturity 2.5 cm. long or longer, white.---Bogs and swamps. Newfoundland to Alaska, south to Maine, New Mexico and Oregon; Eurasia. Scattered in the mountains of Colorado our records from the northcentral, central, southcentral and southwestern parts at 7500-11,500 feet.

3. Eriophorum chamissonis C. A. Mey.; Ledeb., Fl. Alt. 1:70. 1829.

E. russeolum Fries.---Culms 10-80 cm. tall, somewhat triangular; leaves filiform, triangular-channeled in cross-section, usually borne below the middle of the culm, upper leaves inflated; spikes solitary, erect; involucres absent; scales ovate-lanceolate, obtuse to mucronate, thin, purplish-brown with rather broad whitish margins; bristles numerous, white to reddish-brown, 3-5 times longer than the scales.---Bogs and swamps. Labrador to Alaska, south to Wyoming and Idaho; Eurasia. Our two specimens from central and southcentral Colorado at 10,500-12,500 feet.

5. Hemicarpha Nees. & Arn.

Annual plants; culms tufted, glabrous, filiform, terete or compressed; leaves 1-2 to a culm, the lower sheath bladeless; inflorescence of 1 to several spikes in a terminal cluster with 2 or 3 unequal bracts; spikes sessile, many-flowered; glumes spirally imbricated, deciduous; perianth of a scale next to the rachilla; stamens 1; stigmas 2; achenes elliptic to trigonous in cross-section.

1. Hemicarpha micrantha Pax; Engler and Prantl, Nat. Pflzf. 2(2):105. 1887.

H. aristulatus Coville---Culms 5-20 cm. long, longer than the leaves; leaves 0.5-10 cm. long, capillary; spikes 3-8 mm. long, narrowly ovoid; bracts 5-20 mm. long, the longest appearing as a continuation of the culms; scales 0.8-1.3 mm. long, rhombic or rhombic-ovate, each with a spreading mucro nearly as long as the body, brownish; perianth scale about as long as the achene, nerveless, hyaline, obtuse truncate or toothed at apex; achenes 0.5-0.8 mm. long, obovoid, dark brown or black. Our plant has been called var. aristulata Coville.---Sandy banks. The variety from Minnesota to Washington, south to Texas and California. Our two records from northcentral Colorado at about 5200 feet but the plant is easily overlooked and may be widely distributed in the state.

6. Scirpus L. BULRUSH

Annual or perennial herbaceous plants; leaves grasslike or reduced to sheaths; spikes terete or somewhat flattened, solitary or in capitate or umbellate clusters subtended by 1 to several bracts; scales spirally arranged; flowers perfect; perianth of 1-6 bristles, these very small or even lacking in some, smooth, barbed or pubescent; stamens 2-3; styles 2-3 cleft, not much swollen at base, not tubercled, usually wholly deciduous from achenes; achenes triangular or lenticular. Dr. Alan Beetle gave valuable help in outlining the local distribution of this genus.

1. Involucre bracts none at all; spikes single; leaves reduced to sheaths or the blades represented by mucronate
tips   -(species of) Eleocharis
1. Involucre bracts present (the outer scale bracteate in number 4); spikes usually more than 1 (except in number 4); leaves with definite or reduced blades
    2. Involucre bracts 2 or more, leaflike; culms triangular; plants perennial
        3. Spikes 3-10, rather crowded, each 1 cm. long or more; achenes 3-4 mm. long   -1. S. paludosus
        3. Spikes many in compound umbels, each spike less than 8 mm. long; achenes about 1 mm. long
            4. Style branches 2; bristles 4; spikes 3-4 mm. long   -2. S. rubrotinctus
            4. Style branches 3; bristles 6; spikes 3-8 mm. long   -3. S. pallidus
    2. Involucre bracts 1, hardly leaflike (or 2 in number 5, the only annual species); culms terete or triangular
        5. Spikes solitary; involucre bract short, to 3 mm. long   -4. S. pumilus
        5. Spikes 2 or more (rarely 1 on depauperate plants); involucre bract longer
            6. Annual plants; achenes horizontally ribbed; culms less than 30 cm. tall   -5. S. supinus
            6. Perennial plants; achenes not horizontally ribbed; culms usually over 30 cm. tall
                7. Spikes 2-10; leaves not reduced to bladeless sheaths; culms rarely over 100 cm. tall
                    8. Culms definitely triangular; leaves flat or keeled, lax; achenes apiculate, their coats
                       not reticulate   -6. S. americanus
                    8. Culms terete or nearly so; leaves convolute and stiffly erect; achenes not apiculate,
                       somewhat reticulate   -7. S. nevadensis

7. Spikes many; leaves mostly reduced to bladeless sheaths; culms often well over 100 cm. tall
   9. Spikes ovoid; roots fibrous; scales about 2.75 mm. long, little longer than the mature achenes
                                                                                                                        -8. S. validus
   9. Spikes subcylindric; roots thick; scales about 4 mm. long, definitely longer than the mature achenes
                                                                                                                        -9. S. acutus

1. **Scirpus paludosus** A. Nels., Bull. Torr. Club 26:5. 1899.
    S. campestris Britt.---Plants perennial; culms 5-150 cm. tall, from horizontal rootstocks, these thickened and often forming tubers at the nodes, trigonous; leaves about 8 mm. wide; inflorescence of 3 to many spikes, sessile or subumbellate, the primary stalks up to 5 cm. long, spikes 1-2.5 cm. long and about 5 mm. wide, ovate to cylindric, acute; bracts 2-5, leaflike, the outer up to 20 cm. long; scales about 6 mm. long, bifurcate and short-awned this to 2 mm. long, light brown, somewhat pubescent; achene 3-4 mm. long; ovate to obovate, lenticular, apiculate, light to dark brown; bristles 2-6, about 1/2 as long as the achene, minutely barbed; styles 2-branched.---Moist or wet ground. New Brunswick to British Columbia, south to New Jersey, Arizona and California; South America; Australia; Asia. Scattered over Colorado our records from 4000-9500 feet.
2. **Scirpus rubrotinctus** Fernald, Rhodora 2:20. 1900.
    S. microcarpus of Manuals at least in part---Plants perennial, culms 70-150 cm. tall, triangular, often stout; leaves to 10 mm. wide, usually longer than the inflorescence; lower sheaths usually reddish-tinged; inflorescence of 3 to many spikes in glomerules of 3 to many, these on raylets or rays up to 15 cm. long; spikes 3-4 mm. long, ovoid to cylindric; involucre bracts usually 3, leaflike, the longest usually exceeding the inflorescence; scales ovate, obtuse or mucronate, brown with green midribs or suffused with black; achene 0.75-1 mm. long, oblong-obovate, plano-convex, nearly white; bristles 4, as long or longer than the achene, downwardly barbed; style branches 2.---Moist or wet ground. Newfoundland to British Columbia, south to New Jersey, Colorado and Idaho. Our records scattered in the western two-thirds of Colorado at 4500-8000 feet.
3. **Scirpus pallidus** (Britt.) Fernald, Rhodora 8:163. 1906.
    S. atrovirens var. pallidus Britt.---Plants perennial; culms 70-110 cm. tall from short rootstocks, triangular, rather stout; leaves 6-18 mm. wide, pale, somewhat nodulose; inflorescence of many dense heads of spikes, these in compound open umbels, the primary rays up to 10 cm. long; spikes about 3 mm. long or more, ovoid or cylindrical; involucre bracts 2-4, leaflike, often longer than the inflorescence; scales 2-3 mm. long, ovate, greenish to blackish-tinged, the green midrib prolonged as an awn; achene about 1 mm. long, oblong-obovoid, obtusely trigonous, whitish; bristles 6, somewhat shorter than the achene, retrorsely barbed; styles branches 3.---Moist or wet ground. Manitoba, south to Texas and Arizona. Our records from northcentral, central, southcentral and southeastern Colorado at 3500-6500 feet.
4. **Scirpus pumilus** Vahl., Enum. Pl. 2:243. 1805.
    North America and Eurasia. This species has been reported from central Colorado but the record is doubtful.
5. **Scirpus supinus** L., Sp. Pl. 49. 1753.
    S. saximontanus Fernald---Plants annual; culms 10-35 cm. tall, terete or obscurely angled; leaves reduced on lower sheaths at least; inflorescence of 2 or more spikes, both sessile and rayed; bracts 2, the first 3-12 cm. long; scales about 3-3.5 mm. long, hyaline-margined, the green midrib excurrent; achenes 1.5-2 mm. long, trigonous, dark gray-brown, horizontally ridged, apiculate; bristles usually absent; style branches 3. Our plants are var. hallii A. Gray.---Wet or moist ground. The variety from Massachusetts to South Dakota, south to Florida and Mexico. Our one record from Weld County at about 4800 feet.
6. **Scirpus americanus** Pers., Syn. 1:68. 1805.
    Plants perennial; culms 10-100 cm. tall, from horizontal rootstocks, often decumbent or lax, trigonous; leaves mostly 3 or more, narrow, keeled; inflorescence of 2-7 spikes (rarely 1 and sometimes more than 7) sessile in capitate clusters; spikes 5-20 mm. long and about 4 mm. wide; bracts 1, 1.5-12 cm. long, trigonous and appearing to be a continuation of the culm; scales 4 mm. long, mucronate, brown; achene 3 mm. long, obovate; bristles 4, from 1/2 to as long as the achene; style branches mostly 3. Our plant is var. polyphyllus (Bock.) Beetle.---Wet or moist ground. The variety from Ontario to British Columbia, south to the Gulf states, Texas and Mexico; West Indies; South America; New Zealand. Scattered over all of Colorado at 3500-8500 feet.
7. **Scirpus nevadensis** S. Wats., Bot. Kings Expl. 360. 1871.
    Plants perennial; culms 10-50 cm. tall, single or somewhat clustered from long horizontal rootstocks, stout, subterete or very obscurely trigonous; leaves stiff, convolute and appearing terete; inflorescence of 2-8 capitate sessile spikes, these 6-18 mm. long, ovoid or ovoid-oblong; bract solitary, 1-7 cm. long, appearing as a continuation of the culms; scales ovate, obtuse or acute, walnut-brown, strongly nerved; achene about 2 mm. long, broadly ovate to obovate, minutely reticulate; bristles 1-3, shorter than the achene; style branches 2.---Moist or wet ground especially alkaline soil. Saskatchewan to British Columbia, south to Nebraska and California; South America. Our few records from southcentral Colorado at 7600-7800 feet.
8. **Scirpus validus** Vahl., Enum. 2:268. 1806.
    S. lacustris of Manuals in part---Plants perennial; culms about 200-400 cm. tall, from slender rootstocks bearing fibrous roots, thick, terete; leaves reduced to sheaths or the upper with a stiff blade up to 10 cm. long; inflorescence of many spikes in a compound umbel, the primary rays to 4 cm. long, the secondary rays to 2 cm. long; spikes 5-10 mm. long and 4-5 mm. wide, ovate; 1 main bract of inflorescence present but also with smaller secondary ones; scales about 2.75 mm. long, mucronate, pale straw-brown or shaded with red; achene 2.5 mm. long, apiculate, lenticular or plano-convex, deep gray; bristles typically 6, about equalling the achene, retrorsely scabrous; style branches 2.---Wet ground or shallow water. Throughout most of Canada and United States. Our records from all over Colorado at 3500-8000 feet.

9. Scirpus acutus Muhl.; Bigel., Fl. Bost. 15. 1814.
S. occidentalis (S. Wats.) A. Chase; S. lacustres of Manuals in part---Plants perennial; culms about 200-400 cm. tall from thick rootstocks bearing thick spongy roots, thick, terete; leaves reduced to sheaths or upper sometimes with a narrow blade to 8 cm. long; inflorescence of many spikes in a compound umbel, the primary rays to 8 cm. long, the secondary rays to 3 cm. long; spikes 7-20 mm. long and 3-5 mm. wide, ovate-acute to cylindric; 1 main bract present but with shorter secondary ones; scales 4 mm. long, notched and mucronate, brownish; achenes about 2 mm. long, broadly ovate, apiculate, somewhat plano-convex, pale brown, black or dark gray; bristles typically 6, variable in length but never much longer than the achene, minutely scabrous near apex; style branches 2. This species is difficult to distinguish from number 8 with immature material.---Wet soil and shallow water. Widespread in Canada and the United States. Our records scattered over Colorado at 3500-8500 feet.

7. Eleocharis R. Br. SPIKESEDGE; SPIKE-RUSH

Annual or perennial grasslike plants; culms simple, usually terete; leaves reduced to sheaths or the lower with rudimentary blades; spikes solitary, terminal, erect; bracts absent; scales spirally imbricated; perianth of 1-12 bristles, these sometimes short or wanting, often retrorsely barbed; stamens 2 or 3; achene lenticular or triangular, the style base persistent at the apex and differing from it in texture, usually forming a tubercle; style branches 2 or 3.

1. Involucre bracts present, awned or strongly mucronate; leaf blades 0.5-1 mm. long - Scirpus pumilus
1. Involucre bracts absent, outer scales not awned or mucronate; leaf blades absent (sheaths may be mucronate)
  2. Style branches 3; achenes triangular in section; plants perennial
    3. Achenes with conspicuous longitudinal ribs
      4. Culms coarse and flattened, 1 mm. wide, usually over 20 cm. tall -1. E. wolfii
      4. Culms capillary, not flattened, not over 0.5 mm. wide, seldom over 20 cm. tall -2. E. acicularis
    3. Achenes smooth or reticulated, no longitudinal ribs
      5. Style bases confluent with, and not definitely set off as a tubercle plainly distinguishable from the achenes
        6. Culms dwarf, less than 7 cm. tall (rarely over 4); plants often appearing annual-3. E. parvula
        6. Culms over 7 cm. tall; plants perennial
          7. Culms compressed, 1-2 mm. wide; spikes 7-20 mm. long, 12- to 20 flowered; scales more than 2-ranked -4. E. rostellata
          7. Culms 3-angled, less than 1 mm. wide; spikes 4-7 mm. long, 2- to 10-flowered; scales 2-ranked -5. E. pauciflora
      5. Style bases tuberclelike, readily distinguished from the bases of the achenes
        8. Tubercles conical or broadly deltoid, not flattened at tips
          9. Plants perennial with creeping rootstocks; bristles equalling or shorter than achenes -6. E. montevidensis
          9. Plants annual; bristles exceeding the achenes -7. E. obtusa
        8. Tubercles flattened on top (may be apiculate in very center) -8. E. bolanderi
  2. Style branches 2; achenes lenticular, biconvex or plano-convex in cross-section; annual or perennial plants
    10. Perennial from stolons or rhizomes, not caespitose; tubercles spongy, usually higher than wide -9. E. macrostachya
    10. Annual caespitose plants; tubercles not spongy, wider than high
      11. Tubercles small, not over 1/4 the width of the achenes; achenes black or green at maturity -10. E. atropurpurea
      11. Tubercles larger, over 1/2 the width of the achenes; achenes light to dark brown at maturity
        12. Tubercles flattened, very short, less than 1/4 as high as the length of the achene -11. E. engelmanni
        12. Tubercles deltoid, not flattened, over 1/4 as long as the achene body -7. E. obtusa

1. Eleocharis wolfii A. Gray in Paterson, Cat. Pl. Ill.46. 1876.
New York to Canada, south to Tennessee and Kansas. This species has been reported from eastern Colorado but the writer has not seen any material from the state.
2. Eleocharis acicularis (L.) R. & S., Syst. Veg. 2:154. 1817.
Plants perennial; culms 2-20 cm. tall, tufted from slender rootstocks, filiform, somewhat angular and sulcate; sheaths loose, somewhat reddish-striate at base, scarious at apex; spike 2-7 mm. long, ovate to linear, acute, somewhat flattened, 3- to 15-flowered; scales ovate-lanceolate, acute, green with reddish-brown or purplish sides but the margins scarious; achene 0.7-1 mm. long, obovate-oblong, yellow, white to brown, with many longitudinal ribs; style base pyramidal and acute; bristles 3-4 and equalling the achene or lacking; style branches 3.--- Wet or moist ground. North America and Eurasia. Our records scattered in central and eastern Colorado at 3500-10,000 feet.
3. Eleocharis parvula (R. & S.) Link ex Bluff, Comp. Fl. Germ. ed. 2.1(1):93. 1836.
Scirpus coloradensis Britt.---Plants perennial; culms 3-4 cm. tall, forming mats, capillary, erect or arching; spike about 2-3.5 mm. long, ovate, acute, 3- to 9-flowered; scales ovate to ovate-lanceolate, brown to purple, with scarious margins and greenish keels; achenes 1.2-1.5 mm. long, obovate, greenish to light brown; style base small, not readily distinguishable from the achene; bristles present or sometimes absent; style branches 3. Our plants are var. anachaeta (Torr.)Svenson.---Moist soil. The variety from South Dakota to Idaho, south to South America. Our few records from northcentral and northwestern Colorado but the plant is very inconspicuous and is probably generally distributed.
4. Eleocharis rostellata Torr., Fl. N. Y. 2:347. 1843.
Plants perennial; culms 15-100 cm. tall, 1-2 mm. wide, flattened, erect or arching from short vertical rootstocks, the roots thick and whitish; sterile culms often reclining and rooting at the tips; sheaths truncate or oblique; spike 7-20 mm. long, oblong, narrowed above and

below, 12- to 20-flowered; scales ovate, obtuse or the upper acute, rigid, light brown, midrib lighter; achene 2-3 mm. long, obovoid, narrowed above, finely reticulated; style base little distinguishable from the achene, beaklike, about 1/4 to 1/3 the length of the achene; bristles about equalling the achene, toothed; style branches 3.---Moist and wet, often alkaline ground. North America especially in the west. Our one record from Archuleta County at 7000 feet.

5. Eleocharis pauciflora (Lighf.) Link , Hort. Berol. 1:284. 1827.

Scirpus pauciflorus Lighf.---Perennial plants; culms 8-20 cm. tall from slender rootstocks, striate, 3-angled, slender; sheaths truncate, straw-colored or brown; spike 4-7 mm. long, ovate, 2- to 10-flowered; scales dark brown at least along the middle, with pale scarious margins and apices, the 2 lower fertile but enlarged to 1/2 or more the length of the spike; achene 2-3 mm. long obovoid to fusiform, triangular or plano-convex in cross-section, reticulate with rectangular cells; style base small, hardly confluent with the achene beak; bristles variable in length, 2-6, slender, irregularly barbed; style branches 3.---Wet ground. North America; South America; Eurasia. Scattered in the mountains of Colorado at 6500-11,500 feet.

6. Eleocharis montevidensis Kunth., Enum. 2:144. 1837.

E. arenicola Torr.---South Carolina to California and south to South America. This species has been reported from Colorado but no specimens have been located by the writer.

7. Eleocharis obtusa (Willd.) Schultes, Mant. 2:89. 1824.

Plants annual; culms 3-50 cm. tall and to 1.5 mm. in diameter, usually erect, yellowish-green; sheaths purplish at base, apex somewhat oblique; spike 3-13 mm. long, globose-ovoid to ovoid-cylindric, many-flowered; scales ovate-oblong to suborbicular, brown with narrow scarious margins and lighter midribs; achene 1-1.5 mm. long, obovoid, pale to deep brown; style base flattened, deltoid, nearly as wide as the achene but readily distinguishable from it; bristles 6-7, somewhat exceeding the achene, coarse and somewhat retrorsely barbed; style branches 2 or 3.---Mud and wet ground. Rather generally distributed over the United States. Our one record from near Denver at 5300 feet.

8. Eleocharis bolanderi A. Gray, Proc. Am. Acad. 7:392. 1868.

Plants perennial; culms 10-30 cm. tall, caespitose from short, indurated, rather vertical rootstocks, wiry, angular and striate; sheaths straw-colored or purplish, somewhat enlarged near summit; spike 3-8 mm. long, elliptic or ovate, many-flowered; scales 2 mm. long, ovate or lowest orbicular, acute, dark brown or black; achenes 1-1.5 mm. long, trigonous, obovoid to oval, golden-yellow to black; style base tuberclelike, easily distinguishable from the apex of the achene, flattish and narrower than the achene apex; bristles 3-4, 1/2 to 3/4 the length of the achene, retrorsely toothed; style branches 3.---Wet or moist ground. Colorado to Oregon, south to Utah and California. Our few records from northcentral and southwestern Colorado at about 5500 feet.

9. Eleocharis macrostachya Britt. in Small, Fl. S. E. U. S. 184. 1903.

E. calva, E. glaucescens, E. xyridiformis, E. palustris, E. mamillata, E. palustris var. major of authors and Manuals---Plants perennial; culms 10-100 cm. tall, tufted from rootstocks; sheaths reddish-purple below, oblique at apex; spike 6-20 mm. long, lanceolate or cylindrical, acute, many-flowered; scales oblong-ovate to oblong-lanceolate, acute to obtuse, light green, straw-colored to dark brown, 1 or 2 of the lower sterile; achene 1.5-2 mm. long, obovate, lenticular, yellowish or light brown, smooth; style base tuberclelike and readily distinguishable from the achene, rather narrow with constricted apex; bristles as long as the achene or reduced, slender, retrorsely barbed; style branches 2.---Moist or wet ground. West of the Mississippi River from British Columbia to Mexico. Apparently common in all parts of Colorado at 3500-9000 feet.

10. Eleocharis atropurpurea (Retz.) Kunth., Enum. 2:151. 1837.

Widely distributed throughout various parts of the world. This species has been reported for Colorado but no specimens were located.

11. Eleocharis engelmanni Steud., Syn. Pl. Cyp. 79. 1855.

E. monticola Fernald---Plants annual; culms 10-40 cm. tall, caespitose, terete; sheaths pale brown; spike 5-9 mm. long, ovate-lanceolate to oblong-lanceolate, acute; scales obtuse to acute, brown with margins scarious and midnerves green to tan; achene 1-3 mm. long, broadly obovate, biconvex, brown; style branches tuberclelike and readily distinguishable from the achene, almost as broad as long, short and depressed, less than 1/4 as long as the achene; bristles 5-7, from longer than the achene and retrorsely toothed to smooth and rudimentary; style branches 2. Our plant is var. monticola (Fernald) Svenson.---Moist and wet ground. The variety from Colorado to Washington, south to Arizona. Our few records from central Colorado at 5500-7500 feet.

8. Fimbristylis Vahl.

Annual or perennial grasslike herbaceous plants; culms leafy below; spikes in umbellate heads; scales spirally imbricated; perianth none; flowers perfect; stamens 1-3; styles 2-3 cleft, completely deciduous at maturity, swollen at base; achene lenticular to trigonous; bristles none.

1. Fimbristylis puberula (Michx.) Vahl., Enum. 2:289. 1806.

F. interior Britt.; F. castanea and F. thermalis of Manuals---Plants perennial; culms 20-60 cm. tall, loose-tufted, smooth, striate; leaves less than 2 mm. wide (when unrolled), involute; sheaths green to brown; spikes in a compound umbel or sometimes in a simple umbel, the central one sessile; spike 6-10 mm. long, ovoid or ovoid-oblong, obtuse to subacute; scales ovate to oblong, obtuse or mucronate especially on lower ones; achene about 1 mm. long, broadly obovate, biconvex, pale brown or lighter, longitudinally striate and reticulated; bristles lacking; style branches 2.---Meadows. Atlantic Coast states to Colorado. Our few records from northeastern Colorado at 3500-4000 feet.

Family 19. Araceae ARUM FAMILY

Perennial plants of moist habitats, with stout, creeping, aromatic rootstocks; leaves long, 2-ranked; scapes 3-angled, bearing a lateral spadix and foliaceous spathe continuous with the scape; spadix cylindric, yellow-green; flowers perfect, crowded; perianth of 1 whorl with 6 parts; 6 stamens present; ovary solitary, superior, 2- to 4-celled with several ovules; stigma sessile; fruit berrylike, dry but gelatinous inside, 1- to few-seeded.

1. Acorus L. SWEETFLAG

Characters of the Family

1. Acorus calamus L., Sp. Pl. 324. 1753.
Leaves 50-200 cm. long and 1-2 cm. wide, linear, erect, long-attenuate, 2-ranked, ridged down the middle; spathe a leaflike extension of the scape about 20-75 cm. long above the spadix; spadix 5-7 cm. long, about 1 cm. in diameter, lanceolate-cylindric and spikelike, seldom fruiting or even flowering except in shallow water or spring-fed marshes.---Swamps and stream edges. Nova Scotia to Montana, south to Florida, Texas and Colorado; Eurasia. Our one specimen from northcentral Colorado at about 5100 feet.

Family 20. Lemnaceae DUCKWEED FAMILY

Plants small, free-floating or submerged, with an undifferentiated single or aggregated, flat or globular body (frond), no definite stem or leaf; fronds with or without roots, mostly propagating by buds, rarely flowering; flowers monoecious, on upper surface or margins, reduced with perianth lacking, borne in pouches; fruit a utricle.

1. Roots 2 or more to each frond; fronds purple below -1. Spirodela
1. Roots 1 to each frond; fronds may be purplish but usually green or pale below -2. Lemna

1. Spirodela Schleid. DUCKWEED; DUCKSMEAT

Fronds flattened, 5- to 15-nerved, with several roots from 1 frond; flowers from a saclike spathe, 2 or 3 staminate and 1 pistillate; staminate flowers of 1 stamen; pistillate flowers of 1 pistil, this 1-celled and 2-ovuled.

1. Spirodela polyrhiza (L.) Schleid., Linnaea 13:392. 1839.
Fronds 3-10 mm. long, solitary or in colonies of 2-5, round-obovate, flat, dark green above and somewhat convex, generally purplish below, with 5-11 lateral veins and 4-11 rootlets.--- In still water. Almost cosmopolitan. Our few records from northern Colorado at 5200-6500 feet.

2. Lemna L. DUCKWEED

Plants small, fronds free-floating or submerged, flattened, 1- to 3-nerved, usually bearing a single root; flowers 3, borne in a pouch, 2 staminate, composed of a single stamen; 1 pistillate of a single, naked pistil, ovary 1-celled, 1 to several ovules present.

1. Fronds narrowing to stalked bases, 6-12 mm. long; often forming submerged colonies -1. L. trisulca
1. Fronds not stalked, 2-5 mm. long; floating until cold weather
  2. Fronds pale below and mottled or brown-streaked above, usually gibbous below -2. L. gibba
  2. Fronds green or purplish below, uniformly green above, not gibbous
    3. Fronds unsymmetrical at base and apex -3. L. valdiviana
    3. Fronds symmetrical or nearly so
      4. Fronds oblong-obovate, usually indistinctly 3-nerved -4. L. minor
      4. Fronds oblong-elliptical, usually indistinctly 1-nerved or nerveless -5. L. minima

1. Lemna trisulca L., Sp. Pl. 970. 1753.
Fronds 6-12 mm. long, 2-3 mm. wide, usually submerged, oblong with tapering stalked bases, remaining connected and often forming extensive chainlike colonies, denticulate near apex, indistinctly 3-nerved.---In water. Nova Scotia to British Columbia, south to New Jersey, Texas and California, apparently in all continents but South America. Our rather few records from the northern mountains of Colorado as far south as Saguache County at 6500-10,000 feet.
2. Lemna gibba L., Sp. Pl. 970. 1753.
Fronds 2-5 mm. long, solitary or 2-4 in a group, oblong-obovate, sometimes orbicular, outline symmetrical or nearly so, floating, pale below and usually gibbous at base, obscurely 3- to 5-nerved, upper surface mottled or brown-streaked.---In water. Nebraska to California south to Mexico; almost cosmopolitan. Our few records from northcentral Colorado at about 5500 feet.
3. Lemna valdiviana Phillipi, Linnaea 33:239. 1864.
Lemna cyclostasa (Ell.) Chev.---Fronds 2-4.5 mm. long, elliptic-oblong to obovate-oblong, subfalcate and very shortly stalked at base, hence unsymmetrical at both base and apex, indistinctly 1-nerved.---In water. Widely distributed in North America and South America. Our few records from the northern part of Colorado at 4800-8000 feet.
4. Lemna minor L., Sp. Pl. 970. 1753.
Fronds 2-4 mm. long, solitary or a few together, each oblong-obovate, symmetrical or nearly so, thickish, convex, obscurely 3-nerved, upper surface sometimes keeled and with a row of papillae along midrib, green or purplish both sides.---In water. Throughout North America; almost cosmopolitan. Our records scattered over the state at 3500-8500 feet.

5. **Lemna minima** Phillipi, Linnaea 33:239. 1864.
Southern and southwestern United States, Mexico and South America. It is to be expected in Colorado but no specimens have been seen.

Family 21. Commelinaceae SPIDERWORT FAMILY

Annual or perennial, herbaceous, somewhat succulent plants; leaves alternate with sheathing bases; flowers perfect, regular or irregular, subtended by spathes or leaflike bracts; outer perianth segments 3, sepallike; inner segments 3, petallike, showy, fugaceous; stamens 5 or 6, sometimes partly sterile; ovary superior, 2- or 3-celled, styles 1, entire or 2- or 3-lobed at apex; fruit a capsule.

1. Flowers irregular; stamens of 2 types, the filaments naked; inflorescence subtended by 1 spathelike bract -1. Commelina
1. Flowers regular; stamens all alike, the filaments bearded below; inflorescence subtended by several leaflike bracts -2. Tradescantia

1. Commelina (Plum.) L. DAYFLOWER

Stems branching, erect or procumbent; bracts spathelike, compressed and folded; flowers irregular, 2 of the sepals partly united; sterile stamens 2 or 3, smaller than fertile.

1. Spathe 3-6 cm. long, not connate at base, caudate (the tip usually as long or longer than the body), glabrous or pubescent with short hairs only -1. C. dianthifolia
1. Spathe less than 3 cm. long, connate at base, acute or short acuminate at apex, pubescent especially near base with short hairs and with long flaccid hairs -2. C. crispa

1. **Commelina dianthifolia** Delile in Redout., Liliac. 7: pl. 390. 1801.
Perennial plants with tuberous-thickened roots; stems simple or much branched, usually erect, sometimes decumbent; leaves linear-lanceolate; spathe 3-6 cm. long, not connate at base, very long-acuminate or caudate (the tip usually as long or longer than the body), glabrous or pubescent with short hairs; all 3 petals blue, the posterior short-stalked, 10-12 mm. long.---Shaded or open ground. New Mexico and Arizona, south to Mexico. Our few records from southcentral Colorado at 7800-8500 feet.

2. **Commelina crispa** Wooton, Bull. Torr. Club 25:451. 1898.
Perennial plants with thickened but scarcely tuberous roots; stems branched, slender, erect or reclined, finely villous-pubescent; leaves 3-7 cm. long, 4-6 mm. wide; sheaths auricled, these round and ciliate; spathe less than 3 cm. long, with margins connate at base, acute or short-acuminate, pubescent toward base with short, subappressed hairs and long flaccid hairs, orbicular-cordate (if unfolded); sepals unequal, the upper narrower and about 1/2 as long as the 2 united lower; upper petals 1-1.5 mm. long, blue, lower somewhat longer and white. Possibly not specifically distinct from C. erecta L.---Sandy and rocky ground. Indiana to Colorado, south to Texas and Arizona. Our few records scattered in the eastern half of Colorado at 3500-5700 feet.

2. Tradescantia L. SPIDERWORT

Caulescent perennial plants with thickened roots; leaves narrow; inflorescence an umbellate cyme, subtended by 2 or sometimes 3 leaflike bracts; petals showy, ephemeral; stamens 6 with reniform anthers and pilose filaments; style filiform, stigma capitate.

1. Pedicels and sepals glabrous or somewhat barbate-pubescent near tip, not glandular -1. T. canaliculata
1. Pedicels and sepals pubescent (rarely approaching glabrate), at least some of the hairs glandular
    2. Plant bright green or yellowish-green, never glaucous; sepals with long lax glandular hairs, always interspersed with some non-glandular ones -2. T. bracteata
    2. Plants glaucous; sepals with glandular-puberulent, only rarely with non-glandular, hairs present -3. T. occidentalis

1. **Tradescantia canaliculata** Raf., Atl. Journ. 1:150. 1832.
This eastern species has been recorded for western Kansas and Nebraska and may someday be found in eastern Colorado.
2. **Tradescantia bracteata** Small in Britt. & Brown, Ill. Fl. ed. 1, 3:510. 1898.
This is a plant of the Mississippi river area and extends west to western Kansas. It is to be expected in eastern Colorado.
3. **Tradescantia occidentalis** (Britton) Smyth, Trans. Kan. Acad. Sc. 16:163. 1899.
T. laramiensis Goodd.; T. universitatis Cockerell---Stems 15-60 cm. tall, erect or ascending, more or less branching, glabrous; leaf blades 2-20 mm. wide, linear-lanceolate, glabrous; bracts 1.5-2 cm. long; pedicels more or less densely glandular-pubescent, rarely approaching glabrate; sepals more or less densely glandular-pubescent, rarely approaching glabrate; petals 12-16 mm. long, blue to rose. Our material seems to be the var. typica And. & Wood., but var. scopulorum (Rose) And. & Wood. has been reported from Colorado.---Prairies and plains especially in sandy soil. Wisconsin to Montana, south to Texas and Arizona. Our records well scattered over the eastern half of Colorado at 3500-8000 feet.

Family 22. <u>Pontederiaceae</u> PICKEREL-WEED FAMILY

Perennial, herbaceous submerged plants, floating or rooting on muddy shores; stems slender, branched, leafy; leaves alternate, petioled, entire, variously shaped; flowers solitary or a few in the spathe, perfect; perianth tubular of 6 petallike lobes; stamens 3, unequal or equal, on the throat of the perianth; ovary 1- to 3-celled, superior; fruit a several- to many-seeded capsule.

1. <u>Heteranthera</u> Ruiz. & Pav. MUDPLANTAIN

Characters of the Family

1. Leaves linear, grasslike; stamens all alike -1. <u>H. dubia</u>
1. Leaves lanceolate to broader; stamens with unlike anthers -2. <u>H. limosa</u>

1. Heteranthera dubia (Jacq.) MacM., Met. Minn. 138. 1892.
   <u>Zosterella dubia</u> (Jacq.) Small---Quebec to Washington, south to Arkansas, Arizona and California. This plant should be in Colorado but no specimens have been located.
2. Heteranthera limosa (Sw.) Willd., Neue. Schrift. Ges. Nat. Fr. Berlin 3:439. 1801.
   Stems 10-50 cm. long, floating, branched; leaf petioles 3-20 cm. long, blades 1-3 cm. long, ovate, oval, oblong-ovate to lanceolate, rounded or subcordate at base; stamens unequal; perianth white or blue, the tube 1.5-2 cm. long.---In shallow water. Virginia to South Dakota and Colorado, south into South America. Our few records from northcentral Colorado at around 5000 feet.

Family 23. <u>Juncaceae</u> RUSH FAMILY

Annual or perennial grasslike herbaceous plants, commonly in moist places; leaves alternate, sheathing; flowers small, perfect, regular, perianth of 6 distinct, similar scalelike segments; stamens 3 to 6; ovary superior, 1- or 3-celled, stigmas 3; fruit a capsule; seeds few to many.

1. Capsule with numerous seeds, 1- to 3-celled; leaf sheaths open; plants glabrous -1. <u>Juncus</u>
1. Capsule with 3 seeds, 1-celled; leaf sheaths closed; plants glabrous or hairy -2. <u>Luzula</u>

1. <u>Juncus</u> L. RUSH

Annual or perennial glabrous grasslike herbaceous plants of wet habitats; stems usually simple; sheaths open, the blades terete or flattened either way, often septate and often reduced to the sheaths; inflorescence paniculate or corymbose, cymose or capitate; flowers subtended by a bract or sometimes by 2 bractlets; stamens 3-6; ovary 1-celled or by intrusion of the placenta 3-celled; stigmas 3; seeds sometimes caudate. Juncus brachycephalus (Engelm.) Buch. has been collected in El Paso County. It may be a casual since Colorado is so far from its recorded range, and the species is not included in this treatment of the genus.
   Dr. F. J. Hermann gave valuable help in the treatment of this genus, especially in identifying critical material.

1. Lower bracts of the inflorescence terete and not channeled, erect, (exactly simulating a continuation of the stem and the inflorescence appearing lateral); leaves not septate
    2. Seeds caudate at each end; flowers 1-5; plants densely caespitose
        3. Uppermost leaf-sheaths merely bristle-tipped, the blade rudimentary -1. <u>J. drummondii</u>
        3. At least most of upper leaf sheaths bearing well-developed blades
            4. Capsules acute; perianth segments 5-7 mm. long -2. <u>J. parryi</u>
            4. Capsules retuse at apex; perianth segments 4-5 mm. long -3. <u>J. hallii</u>
    2. Seeds merely apiculate but not caudate; flowers usually over 5; plants not densely caespitose
        5. Perianth segments green or straw-colored in age, 2.5-3.5 mm. long; stems striated, usually shorter than the bracts -4. <u>J. filiformis</u>
        5. Perianth segments dark brown or purplish-brown, 4-5 mm. long; stems not striated, usually shorter than the bracts -5. <u>J. balticus</u>
1. Involucre bracts not simulating a continuation of the culms or if so then the bract conspicuously channeled on the upper side (inflorescence usually appearing terminal); leaves septate or not septate
    6. Leaves transversely flattened, the flat surface facing toward and away from the stems (in <u>J</u>. <u>vaseyi</u> leaves are terete but with a shallow groove above), not at all septate
        7. Low annual plants, usually less than 20 cm. tall; inflorescence 1/2 as long as the plant or longer
            8. Perianth 4-6 mm. long, the segments rather appressed to the oblong capsules which are 3-4.5 mm. long -6. <u>J. bufonius</u>
            8. Perianth 3-4 mm. long, the segments usually spreading away from the subglobose capsules which are 2-3 mm. long -7. <u>J. sphaerocarpus</u>
        7. Perennial plants, typically over 20 cm. tall; inflorescence occupying less than 1/3 the height of the plant
            9. Flowers borne in heads, not bracteolate (bracts only at bases of pedicels)
                10. Stamens 3; perianth 2-3.5 mm. long -8. <u>J. marginatus</u>
                10. Stamens 6; perianth 5-6 mm. long -9. <u>J. longistylis</u>
            9. Flowers sometimes congested but not in heads, bracteolate (with bracts inserted directly at base of perianths)
                11. Seeds long-caudate; leaves terete with a shallow groove above -10. <u>J. vaseyi</u>

11. Seeds not caudate; leaves flat or involute
    12. Auricles cartilaginous, opaque and yellowish when dry; perianth segments spreading from mature capsules -11. **J. dudleyi**
    12. Auricles membranous or scarious, whitish (or sometimes finally brown); perianth segments not spreading (except in J. tenuis)
        13. Panicles compact, short, not over 2 cm. long; capsules completely 3-celled -12. **J. confusus**
        13. Panicles looser, over 2 cm. long (except in depauperate plants); capsule 1-celled or incompletely 3-celled
            14. Auricles scarious, extending well upward from point of insertion, 1-3.5 mm. long; bracteoles blunt at apex -13. **J. tenuis**
            14. Auricles membranous, slightly if at all extending upward from point of insertion, hence very short; bracteoles acuminate to aristate -14. **J. interior**
6. Leaves terete or ensiform (sharp edges turned toward and away from stem), always more or less septate
    15. Leaves ensiform, strongly flattened, septa of leaves not conspicuous -15. **J. saximontanus**
    15. Leaves terete or but slightly flattened, septa of leaves conspicuous or inconspicuous
        16. Inflorescence of 1-3 heads or clusters of flowers; septa of leaves usually not conspicuous; seeds often caudate; plants usually above 9000 feet in Colorado
            17. Heads (usually 1) becoming spheroid, over 12 flowers present; leaves somewhat compressed but not channeled above -16. **J. mertensianus**
            17. Heads or clusters not spheroid, 1-12 flowered; leaves somewhat channeled above
                18. Capsules 1½ to 2 times longer than the perianth which is 4-7 mm. long -17. **J. castaneus**
                18. Capsules about as long or slightly exceeding the perianth, which is 3-4 mm. long
                    19. Perianth segments obtuse; plants 5-15 cm. tall; flowers in a capitate cluster -18. **J. albescens**
                    19. Perianth segments acute; plants 20-50 cm. tall; flowers in a cluster but hardly capitate -10. **J. vaseyi**
        16. Inflorescence of 4 or more heads (except in depauperate plants); septa of leaves usually conspicuous; seeds never caudate; plants usually found below 9000 feet in Colorado
            20. Capsules ovoid, ovoid-oblong or obovoid; perianth 2-3 mm. long
                21. Capsules obovoid, truncate or very broadly rounded at apex, shorter than the perianth; inflorescence seldom over 4 cm. long; anthers usually longer than their filaments -19. **J. badius**
                21. Capsules ovoid-oblong, not truncate or very broadly rounded at apex, longer than the perianth; inflorescence often over 4 cm. long; anthers usually shorter than their filaments
                    22. Perianth segments obtuse, mucronate or acute; capsules broadly acute or obtuse below the short tip -20. **J. alpinus**
                    22. Perianth segments acuminate; capsules tapering to the short tip -21. **J. articulatus**
            20. Capsules subulate-pointed; perianth 3-5 mm. long
                23. Leaf blades erect or ascending; perianth 3-4 mm. long, the inner segments equalling or exceeding the outer; heads 6-12 mm. in diameter -22. **J. nodosus**
                23. Leaf blades divaricate or divergent; perianth 4-5 mm. long, the inner segments shorter than the outer; heads 10-15 mm. in diameter -23. **J. torreyi**

1. Juncus drummondii E. Meyer in Ledeb., Fl. Ross. 4:235. 1853.
J. subtriflorus (Mey.) Coville---Perennial plants; stems 10-45 cm. tall, densely tufted from matted rootstocks; basal leaf sheaths bladeless or the upper reduced to a short bristlelike rudiment; inflorescence 1- to 5-flowered; the bract about 2 or 3 cm. long, appearing like a continuation of the stem; perianth 5-7 mm. long with 2 bractlets at base, segments lanceolate, acute or acuminate, about equal, margins dark brown but scarious, lighter near midrib; stamens 6; capsule oblong and retuse before opening, about as long as the perianth, 3-celled; seeds ovate, caudate with rather long white tails at each end.---Mountain slopes especially in damp soil. Alberta to Alaska, south to New Mexico and California; Eurasia. Our records scattered in the mountains of the western half of Colorado at 9000-12,500 feet.

2. Juncus parryi Engelm., Trans. St. Louis Acad. 2:446. 1866.
Perennial plants; stems 10-30 cm. tall, tufted from matted rootstocks, slender, terete, only the uppermost sheath bearing a leaf blade, the others bristlelike or absent; upper leaf 3-8 cm. long, sulcate below, terete above; inflorescence appearing lateral; bract 1.5-6 cm. long, longer than the inflorescence, terete and appearing to be a continuation of the stem; inflorescence 1- to 3-flowered; flowers with a pair of bractlets at base; perianth 5-7 mm. long, inner somewhat shorter and more acute, outer acuminate, margins scarious, center brown; stamens 6; capsule oblong, triangular, acute, 3-celled, somewhat longer than the perianth; seeds ovate or oblong, caudate.---Slopes and meadows. Montana to British Columbia, south to Colorado and California. Our few records scattered in the mountains of Colorado at 9500-13,000 feet.

3. Juncus hallii Engelm., Trans. Acad. St. Louis 2:446. 1866.
Perennial plants; stems 10-30 cm. tall, caespitose, filiform; leaves shorter than the stems, terete and filiform but grooved above at least near the sheath, not septate, sheaths brown; inflorescence a few-flowered (2-5) contracted panicle; bract scarcely exceeding the inflorescence; perianth 4-5 mm. long, sepals slightly longer, brown or with greenish midrib, scarious on margins, acute; stamens 6; capsule oblong, triangular, 3-celled, retuse, about as long as perianth, stigmas subsessile; seeds oblong-linear, caudate.---Mountains. Montana to Colorado. Our few records from central and northcentral Colorado at 8800-10,000 feet.

4. Juncus filiformis L., Sp. Pl. 326. 1753.
Perennial plants; stems 7-50 cm. long, arising from a matted rootstock, erect, slender, finely striate; sheaths purplish tinged, obtuse, with a short bristlelike remnant of a blade often present or this absent entirely; panicle 5- to 10-flowered, 1-3 cm. high; bract terete, appearing like a continuation of the stem, usually longer than the stem proper; perianth 2.5-3.5 mm. long; bractlets obtuse; segments of perianth lanceolate, greenish or stramineous in age, margins hyaline, equal or outer somewhat longer, acute to acuminate, the inner usually

less pointed, sometimes almost obtuse; stamens 6; capsule obovoid, green to stramineous in age, somewhat pointed, 3/4 to nearly as long as the petals.---Moist or wet places. Greenland to Alaska, south to Pennsylvania, Utah and Washington; Eurasia. Our few records from northcentral Colorado at 8500-9000 feet.

5. Juncus balticus Willd., Mag. Gesell. Naturf. Freund Berlin 3:298. 1809.

J. ater Rydb.---Perennial plants with stout horizontal rhizomes; stems 20-80 cm. tall, slender, usually 2-3 mm. in diameter at base; leaves basal and reduced to loose brownish sheaths; bract of inflorescence 3-20 cm. long, terete, little channeled, erect, exactly simulating a continuation of the stem; inflorescence appearing lateral, 2-6 cm. long, of 6 or more flowers on healthy plants, usually rather congested, the branches only about 1-4 cm. long; flowers subtended by bractlets; perianth 4-5 mm. long, purplish-brown, segments lanceolate, acute or acuminate, nearly equal; stamens 6; capsule at maturity as long or longer than the perianth, narrowly obovoid, mucronate; seeds oblong to narrowly obovoid, not caudate. The above description applies to var. montanus Engelm., our most common form. It intergrades in Colorado with var. vallicola Rydb., a form with large inflorescences and with sepals longer than the petals.---Moist places. The variety from Montana to Alaska, south to Kansas, New Mexico and California. Apparently our commonest species, the records scattered all over Colorado at 3500-11,000 feet.

6. Juncus bufonius L., Sp. Pl. 328. 1753.

Annual plants; stems 5-20 cm. tall, branching from base; leaf blades 1-3 on the stem, flat, 1 mm. wide or less; inflorescence about 1/2 the length of the stem with single flowers and leaves at the lower nodes, each flower subtended by 2 bractlets; perianth 4-6 mm. long, segments lanceolate, scarious-margined, acuminate, nearly equal; stamens usually 6; capsule 3-4.5 mm. long, oblong, obtuse, mucronate, 3-celled (perianth not spreading away from capsule at maturity); seeds broadly oblong, minutely reticulate.---Moist places. Nearly throughout North America; almost cosmopolitan. Our records scattered over Colorado, but few from the eastern 1/4 of the state, at 4500-9500 feet.

7. Juncus sphaerocarpus Nees. in Funk., Flora 1:521. 1818.

Annual plants; stems 5-20 cm. tall, branching from base and in inflorescence; leaves less than 1 mm. wide, flat or involute, about 1-3 below the inflorescence; inflorescence over 1/2 the height of plant; flowers single on branches, each subtended by 2 bractlets; perianth 3-4 mm. long, segments with scarious margins, lanceolate, acuminate, spreading in fruit; stamens 6; capsule subglobose to broadly ovoid, obtuse, 3-celled; seeds oblong, minutely reticulate. Very similar to J. bufonis L. in this state.---Moist or wet places. Idaho to Oregon, south to Arizona and California. To be expected in western Colorado. Our one record is from El Paso County at 7500 feet but the specimen intergrades somewhat with number 6.

8. Juncus marginatus Rostk., Monog. Junc. 38. 1801.

J. marginatus var. setosus Coville; J. setosus (Coville)Small---Plants perennial; stems 15-75 cm. tall, tufted from branching rootstocks, somewhat bulbous at base; leaves 1-3 mm. wide (rarely to 5), transversely flattened; sheaths with rounded auricles; bract of inflorescence leaflike (about as long as the inflorescence); inflorescence of 2-20 subspherical heads, these each 4 to 10-flowered; perianth 2-3.5 mm. long, inner segments somewhat longer, margins very hyaline, outer acute to long-acuminate, inner broadly obtuse or aristate; stamens 3; capsule about as long as the perianth, obovate, truncate or retuse, almost completely 3-celled; seeds pointed at each end.---Moist or wet places. Maine to Ontario, south to Florida and Arizona. Our one record from northcentral Colorado at about 5000 feet.

9. Juncus longistylis Torr., Bot. Mex. Bound. 223. 1859.

Perennial plants from short rootstocks; stems 20-50 cm. tall, loosely caespitose; basal leaves flat, with well-developed auricles; stem leaves 1-4 mm. wide, flat or somewhat involute, the flat edges inserted next to the sheaths; sheaths with scarious margins; inflorescence of 2-8 heads, each 3- to 8- (rarely less) flowered; perianth 5-6 mm. long, segments greenish or light brown in the center, brown on the sides and with broad hyaline margins, lanceolate, acuminate; flowers not bracteolate; stamens 6; capsule 3-celled, oblong, from somewhat shorter to somewhat longer than the perianth, mucronate; seeds oblong, apiculate.---Moist ground. Alberta to British Columbia, south to New Mexico and California. Scattered over the state, except in the extreme eastern part, at 4500-10,000 feet.

10. Juncus vaseyi Engelm., Trans. St. Louis Acad. 2:448. 1866.

Perennial plants; stems 20-50 cm. tall, sometimes taller, erect, stiff, more or less tufted; leaves long, rigid, erect, terete, shallowly channeled above; sheaths with short semi-membranous auricles, lower sheaths bladeless, brown or purplish; inflorescence 1-3 cm. long, rather short and crowded, bracts as long or longer than the inflorescence; bracteoles present; perianth 3-4 mm. long, green or straw-colored, sepals lanceolate, acute, scarious-margined, petals as long or slightly shorter; stamens 6; style short; capsule equalling or somewhat exceeding the perianth, oblong-cylindrical, obtuse or retuse, straw-colored to brown, 3-celled; seeds caudate.---Moist ground. Maine to Saskatchewan, south to Colorado and Idaho. Our few records from the mountains of the state, especially in the north, at 8000-10,000 feet.

11. Juncus dudleyi Wiegand, Bull. Torr. Club 27:524. 1900.

J. tenuis var. dudleyi (Wieg.) Hermann---Perennial plants; stems 25-100 cm. tall, tufted, pale green, erect; leaves short, seldom over 1/2 length of stem, basal, narrow, flat or involute, sheaths all blade-bearing, margins not scarious; auricles rounded, thick and cartilaginous, yellowish or yellowish-brown; inflorescence usually less than 5 cm. long, rather congested, relatively few-flowered; perianth 4-5 mm. long, green, pale stramineous or yellow, the segments lanceolate, finally strongly spreading, acute to somewhat acuminate, margins scarious; bracteoles ovate; stamens 6; capsule somewhat shorter than the perianth, ovoid, somewhat apiculate, trigonous, not completely 3-celled; seeds oblong. This species seems to be closely related in this state to the next 3.---Moist ground. Maine to Alberta, south to New York, New Mexico and California. Our records scattered over the state at 3500-8000 feet.

12. Juncus confusus Coville, Proc. Bio. Soc. Wash. 10:127. 1896.
Perennial plants, sparingly tufted; stems 35-50 cm. tall, slender, erect; leaves narrow, almost filiform, flat or involute; auricles produced beyond insertion, scarious; inflorescence 0.5-2 cm. long, short and compact, pale; bract of inflorescence 2-7 cm. long, exceeding the inflorescence; bracteoles present, large, ovate, scarious, obtuse or acutish; perianth 3.5-4 mm. long, parts nearly equal, appressed, segments stramineous with dark stripes on each side, lanceolate, acutish, scarious at margins; stamens 6; capsule oblong, a little shorter than perianth, triangular, retuse at apex, completely 3-celled; seeds oblong, apiculate.---Moist meadows. Saskatchewan to British Columbia, south to New Mexico and California. Our records from the mountainous part of Colorado, none as yet from the southcentral part, at 5000-11,000 feet.

13. Juncus tenuis Willd., Sp. Pl. 2:214. 1799.
J. macer S. F. Gray---Perennial plants; stems 15-60 cm. tall, densely tufted, slender; leaves about 1-2.5 mm. wide, 1/2 to 2/3 as long as the stems, transversely flat or involute, not stiff, all or almost all basal; auricles scarious, extending upwards for 1-3.5 mm. from point of insertion; inflorescence a loose terminal panicle, usually over 2 cm. long, flowers singly inserted near ends of branches; pedicels with a pair of bractlets at base and flowers each with a pair of blunt bracteoles just below; bract of inflorescence exceeding it; perianth 3-5 mm. long, segments lanceolate, green with scarious margins, rather spreading; stamens 6; capsule ovate or oval, shorter than the perianth, rounded at apex, 1-celled but the septa extending about 1/2 way into the center; seeds not caudate. Our Colorado plants seem to be var. multicornis E. Mey.---Meadows, dry or moist ground. Newfoundland to Washington, south to Florida, Arizona and Oregon. (The species almost cosmopolitan.) Our one record from northcentral Colorado at about 6000 feet.

14. Juncus interior Wiegand, Bull. Torr. Club 27:516. 1900.
J. arizonicus Wiegand---Perennial plants; stems 50-100 cm. tall, rather stout, nearly terete; leaves about 1/3 as long as the stem, blades 1-1.25 mm. wide, flat or involute; sheaths loose, auricles short, little projecting, membranous; inflorescence large and open, 3-10 cm. long, many-flowered; bracts usually 2, often exceeding the inflorescence; bracteoles present, acuminate; perianth 3-4 mm. long, segments about equal, lanceolate, very acute, appressed or erect; stamens 6; capsule oblong or acute-oblong, obtuse or apiculate, about as long as the perianth, incompletely 3-celled in center; seeds apiculate.---Dry or moist places. Illinois to Washington, south to Arkansas and Arizona. Our records from the eastern two-thirds of Colorado at 3500-8000 feet.

15. Juncus saximontanus A. Nels., Bull. Torr. Club 29:401. 1902.
J. parous Rydb.; J. brunnescens Rydb.---Perennial plants; stems 30-60 cm. high, from stout creeping rootstocks, compressed and 2-edged; leaves 2-5 mm. wide, flattened laterally (1 edge next the stem)ensiform, incompletely septate, usually auricled; sheaths flattened, edges hyaline; inflorescence a panicle of 2-12 close heads, these usually many-flowered; bracts of inflorescence usually shorter than it and rather leaflike; flowers with bractlets; perianth 2.5-3 mm. long, light to darker brown or greenish especially when young, outer segments lanceolate, acuminate, the inner acute and shorter than outer; stamens 6; capsule oblong, obtuse, acute or mucronate, shorter or as long as perianth; seeds reticulate, not caudate.---Wet ground. Alberta to British Columbia, south to New Mexico and Arizona. Our records scattered over the western two-thirds of Colorado at 4500-10,500 feet.

16. Juncus mertensianus Bong., Mem. Acad. St. Petersburg VI. Math. Phys. Nat. 2:167. 1832.
Perennial plants; stems 10-40 cm. tall, slender, caespitose from slender matted rootstocks; leaves 1-3 mm. wide, terete, 2-3 to a stem, somewhat compressed, septate but often obscurely so; sheaths with scarious margins and bearing auricles; inflorescence usually a solitary head this becoming spherical, about 10-12 mm. in diameter, many-flowered; perianth 3.5-4 mm. long, dark brown to brownish-black, segments about equal, lanceolate, acute to acuminate especially on the outer; stamens 6; capsule trigonous, oval, obtuse to mucronate, equalling or slightly shorter than the perianth, reportedly 1-celled; seeds caudate or not caudate.---Moist or wet places. Alberta to Alaska, south to New Mexico and California. Our records scattered, in the mountains of Colorado at 8500-11,500 feet.

17. Juncus castaneus Smith, Fl. Brit. 1:383. 1800.
Perennial plants; stems 10-40 cm. tall, terete, leaves mostly basal; leaves tapering from an involute tubular base to a slender channeled apex, the general effect being terete, the upper epidermis being membranous, 1-2 mm. thick; sheaths not auriculate; lowest bract usually exceeding the inflorescence; inflorescence of 1-3 heads, few-flowered, no bractlets present; segments of perianth 4-7 mm. long, lanceolate, acute, chestnut-brown, petals somewhat shorter and often almost obtuse; stamens 6; capsule 1½ to 2 times as long as perianth, brown, tapering to an acute apex, narrowly oblong; seeds long-caudate.---Mountains. Greenland to Alaska, south to New Mexico; Eurasia. Our few records from central and northcentral Colorado at 7000-12,000 feet.

18. Juncus albescens (Lange) Fernald, Rhodora 26:202. 1924.
J. triglumis L. of Am. authors---Plants perennial; stems 5-15 cm. tall, loosely tufted from branching rootstocks, erect, terete; leaves 1-7 cm. long, 1-5 in number, terete; sheaths auriculate; inflorescence a capitate cluster of 1-5 (mostly 3) flowers; bracts almost as long as the flowers, the lower acuminate; perianth 3-4 mm. long, brown, segments ovate-lanceolate or oblong-lanceolate, nearly or quite obtuse, about equal; stamens 6; capsule equalling or slightly exceeding the perianth, obtuse or mucronate, 3-angled, imperfectly 3-celled; seeds about 2 mm. long, caudate.---Mountains. Greenland to Alaska, south to New York and New Mexico; Eurasia. Our few records from central and northcentral Colorado at 11,000-12,000 feet.

19. Juncus badius Suksdorf, Deutche Bot. Monatschr. 19:92. 1901.
J. truncatus Rydb.---Perennial plants with slender rhizomes; stems 20-50 cm. tall, nearly or quite terete; leaves 1-2 mm. in diameter, terete or somewhat flattened laterally, septate, 1 short leaf present over 1/2 way up stem; sheaths with scarious margins and rounded rather

scarious auricles; inflorescence open, 2-5 cm. long with 4-12 heads, these 6-9 mm. across and 5- to 10-flowered; bracts 5-30 mm. long; perianth 2-3 mm. long, dark brown, the segments lanceolate, acuminate, about equal; bractlets ovate; capsule slightly shorter than the perianth, obovoid, truncate or broadly rounded at apex; seeds not caudate.---Moist or wet ground. Wyoming to Washington, south to New Mexico and Arizona. Our few records from northcentral Colorado at 8000-9500 feet.

20. Juncus alpinus Vill., Hist. Pl. Dauph. 2:233. 1787.

Perennial plants with creeping rhizomes; stems 15-40 cm. tall, in loose tufts; leaves 1-2 mm. in diameter, terete but often slightly grooved above, definitely septate; auricles rounded and rather scarious; inflorescence rather open, 2-8 cm. long, with 5-25 heads, these 4-7 mm. broad, 3- to 12-flowered; bracts 1-6 cm. long; perianth 2-2.5 mm. long, inner segments oblong and acute, outer segments longer, obtuse, mucronate or acute, brown with scarious margins; bractlets ovate-lanceolate; capsules slightly longer than the perianth, ovoid-oblong, broadly acute or obtuse below a short tip; seeds not caudate.---Borders of lakes and ponds. Greenland to Alaska, south to Pennsylvania, Nebraska, Idaho and Washington; Europe. Our one record from southcentral Colorado at 8000 feet.

21. Juncus articulatus L., Sp. Pl. 327. 1753.

Perennial plants with creeping rhizomes; stems 20-60 cm. tall, in tufts; leaves 1-2 mm. in diameter, terete, definitely septate; auricles rounded and rather scarious; inflorescence open with spreading branches, 5-10 cm. long, with many heads, these 4-7 mm. broad, 6- to 12-flowered; bracts 1-6 cm. long; perianth 2-2.5 mm. long, segments nearly equal, lanceolate and acuminate, brown with scarious margins or greenish on midribs; bractlets ovate-lanceolate; capsules longer than the perianth, ovoid-oblong, acute and tapering to the tip; seeds not caudate.---Moist ground. Circumboreal, south in North America to New York, Indiana and California. Our one record from westcentral Colorado at 6500 feet.

22. Juncus nodosus L., Sp. Pl. ed. 2. 466. 1762.

Perennial plants; stems 10-40 cm. tall, single from slender rootstocks and tuberlike thickenings; stem leaves 2-4, the blades erect; inflorescence a panicle of 1-30 heads, not usually over 6 cm. long; heads spherical, 7-12 mm. in diameter; lowest bract longer than the inflorescence; perianth segments 3-4 mm. long, lanceolate-subulate, the outer usually shorter; stamens 6; capsule lanceolate-subulate, trigonous, 1-celled, exceeding the perianth; seeds oblong, apiculate, reticulate.---Moist or wet ground. Nova Scotia to British Columbia, south to Virginia, New Mexico and Nevada. Our records well scattered over Colorado at 3500-7500 feet.

Recently a similar species, Juncus acuminatus Michx., has been collected from Boulder County. It has 3 stamens instead of 6, and the capsule is wider than lanc-subulate, being described as lanc-ovoid.

23. Juncus torreyi Coville, Bull. Torr. Club 22:303. 1895.

Perennial plants; stems 40-100 cm. tall (rarely as short as 20 cm.), 2-4 mm. thick, from tuberlike thickenings on the slender rootstocks; stem leaves 1-4, the blades usually divergent from the sheath, rather stout, terete, septate; inflorescence a panicle of 1-20 spherical heads, these 10-15 mm. in diameter, rather congested, the panicle usually exceeded by its lowest bract; perianth 4-5 mm. long, segments narrow, lanceolate-subulate, the outer longer than the inner; stamens 6; capsules subulate, 3-sided, 1-celled, exceeding the perianth; seeds oblong, acute at both ends, reticulate. Depauperate specimens tend to intergrade with J. nodosus L.---Moist or wet ground. Massachusetts to Washington, south to Alabama, Texas and California. Our records scattered over the state but mostly from the eastern half, at 3500-8000 feet.

2. Luzula DC. WOODRUSH

Perennial grasslike herbaceous plants; stems leafy; leaves flat to involute; leaf sheaths closed; inflorescence umbellike, capitate or spikelike; flowers subtended by lacerate or dentate bractlets; stamens 6; ovary 1-celled; ovules and seeds 3, not caudate.

1. Flowers 1-3 together on slender pedicels in an open panicle   -1. L. parviflora
1. Flowers many together, congested into globose or oblong spikes (flowers sessile or nearly so)
   2. Inflorescence erect, the spikes borne on peduncles; seeds with a white strophiolelike base
      -2. L. multiflora
   2. Inflorescence nodding, the spikes sessile or nearly so; seeds without such a strophiolelike base
      3. Plants over 40 cm. tall; inflorescence subcapitate; bractlets not ciliate   -3. L. subcapitata
      3. Plants less than 40 cm. tall; inflorescence usually elongate; bractlets ciliate   -4. L. spicata

1. Luzula parviflora (Ehrh.) Desv., Journ. de Bot. 1:144. 1808.

Juncoides parviflorum (Ehrh.) Coville---Stems 30-60 cm. tall, single or few in a tuft from rootstocks, terete, erect; leaves 5-12 mm. wide, glabrous, flat; sheaths closed, sometimes hairy near blade; inflorescence a nodding, decompound panicle often over 7 cm. long; 2 or 3 bracts present, the lowest leaflike, about 1/4 to 1/2 the length of the inflorescence; flowers single (or rarely 2 or 3 together); bractlets ovate, entire or lacerate; perianth 1.75-2.5 mm. long, segments green or dark brown, lanceolate, acute; capsule ovoid, somewhat longer than the perianth, green to reddish-brown or brown; seeds ellipsoid to almost oblong, not narrowed at both ends (may be slightly at one end). As here treated the species is a complex badly in need of revision. The writer cannot separate L. wahlenbergii Rupr., from the above.---Meadows and slopes. Labrador to Alaska, south to New York, New Mexico and California; Eurasia. Our records scattered in the mountainous part of Colorado at 8000-11,500 feet.

2. Luzula multiflora (Retz.) Lejeune, Fl. Envir. Spa 169. 1811.

L. intermedia (Thuill) A. Nels.; Juncoides intermedium (Thuill)Rydb.; prob. called L. campestris in some manuals---Stems 10-50 cm. tall, tufted, slender; leaves 1-6 mm. wide, flat, long-hairy on edges and near base (this sometimes sparse); sheaths rather long and tight, long-hairy at mouth; inflorescence erect, of 2-12 usually oblong (or globose) spikes on erect unequal peduncles up to 5 cm. long; lower bracts leaflike, often exceeding the inflorescence; bractlets often lacerate and hyaline; perianth 2.5-3.3 mm. long, segments pale or tinged with brown, lanceolate, narrowly acuminate, margins scarious; capsule obovate, almost as long as the perianth; seeds dark brown, whitish at one end.---Mountains. Greenland to British Columbia, south to New York, New Mexico and California; Eurasia. Our records from central and northcentral Colorado at 8500-10,000 feet.

Some of our plants have seeds with smaller caruncles (less than 1/6 the length of the entire seed) and a more compact inflorescence. These plants have recently been identified as the northern species, L. sudetica (Willd.) DC.

3. Luzula subcapitata (Rydb.) comb. nov.
Juncoides subcapitatum Rydb.---Stems 30-40 cm. tall, glabrous; leaf blades 5-10 mm. wide, glabrous; inflorescence compact, of 6-10 heads forming a capitate mass, this irregular; perianth parts about equal, 1.5-2 mm. long, ovate, somewhat acuminate, brown but lighter on midrib; capsules broadly obovoid, somewhat shorter than the perianth.---Mountains. Colorado. Our two records from northcentral Colorado at 10,000 and 13,000 feet.

4. Luzula spicata (L.) DC., Fl. Fr. 3:161. 1805.
Juncoides spicatum (L.) Kuntze---Stems 10-40 cm. tall, closely tufted, erect; leaves 1-3 mm. wide, 1-3 on a stem, rather distant, flat, channeled or involute, sparingly webbed near collar; sheaths webbed on margins near collar; inflorescence of nodding glomerules these sessile or nearly so, forming an interrupted spike or compact panicle of spikes, usually 12-25 mm. long; lowest bract rather leaflike, involute, often longer than the inflorescence; bractlets ovate-lanceolate, acuminate, as long as perianth, sparingly lacerate; perianth 2-3 mm. long, brown, segments lanceolate, aristate-acuminate; capsules ovoid, about 2/3 to 3/4 as long as the perianth; seeds brown.---Mountains. Greenland to Alaska, south to New Hampshire, New Mexico and California; Eurasia. Our records scattered in the mountains of Colorado at 9000-13,000 feet.

Family 24. Liliaceae LILY FAMILY

Perennial, usually herbaceous plants from bulbs, corms, rootstocks or woody caudices; flowers solitary or clustered, regular or nearly so, usually perfect, of 6 separate or united segments; stamens commonly 6; ovary superior (partly inferior in Zygadenus elegans), 3-celled, styles 3 and separate or united into one with a capitate, usually 3-lobed stigma; fruit a capsule or a berry.

1. Plants with large woody caudices, these mainly subterranean or largely above ground; leaves numerous in large rosettes at the apex of the caudex or its branches, narrow, elongate, rigid and often spine-tipped
  2. Flowers over 2 cm. long, perfect; seeds numerous in each cell of the capsule; leaves filiferous along the margins -1. Yucca
  2. Flowers less than 1 cm. long, polygamo-dioecious; seeds solitary in each cell; leaves not filiferous along the margins -2. Nolina
1. Plants herbaceous without woody caudices; leaves not in large rosettes, never rigid or spine-tipped
  3. Perianth segments very unlike, the 3 outer much narrower and sepaloid, the 3 inner ones broader and petaloid
    4. Leaves and bracts alternate, linear or linear-lanceolate, grasslike; plants from coated corms; fruit a capsule -3. Calochortus
    4. Leaves or leaflike bracts whorled, broad, not at all grasslike; plants from short rootstocks; fruit a berry -4. Trillium
  3. Perianth segments all alike or nearly so (may be in 2 whorls)
    5. Stems climbing (in ours) by tendrils from the base of the leaf petioles; plants dioecious -5. Smilax
    5. Stems not at all climbing, no tendrils present; flowers perfect
      6. Stamens with their filaments united into a tube, this with toothlike lobes between the anthers -6. Androstephium
      6. Stamens not united by their filaments (may sometimes cohere somewhat by their anthers)
        7. Perianth segments united into a tube 3-8 cm. long, the base of each flower borne below the ground level -7. Leucocrinum
        7. Perianth segments separate or if united, then the tube much shorter, flowers borne above the ground level
          8. Perianth segments united into a tube for at least 1/2 their length; fruit a berry -8. Polygonatum
          8. Perianth segments distinct or united only slightly by their bases; fruit various
            9. Flowers in umbels, borne on scapose peduncles; odor of onions when fresh -9. Allium
            9. Flowers not in umbels on scapose peduncles; no onion odor present
              10. Leaves scalelike, with filiform leaflike branchlets clustered in their axils; stems tall, much-branched (often over 1 m. tall) -10. Asparagus
              10. Leaves not scalelike, no filiform clustered branchlets in their axils; stems simple or sparingly branched (seldom over 1 m. tall except in Veratrum where the stem is unbranched except in the inflorescence)
                11. Plants 1-2 m. tall; leaves 20-30 cm. long; flowers in large panicles (usually over 25 cm. long) -11. Veratrum
                11. Plants less than 1 m. tall; leaves rarely over 15 cm. long; flowers various, if in panicles then in smaller ones
                  12. Underground parts bulbs or corms (both bulb and rootstock in Lloydia); fruit a capsule; leaves linear-lanceolate or narrower (except in Erythronium)

13. Perianth segments over 4 cm. long, orange-red or brownish-red
-12. <u>Lilium</u>
13. Perianth segments less than 4 cm. long, not orange-red or brownish-red (at least when young)
  14. Styles 3, distinct; inflorescence a panicle or raceme, this usually many-flowered; flowers small, seldom over 1 cm. long
-13. <u>Zygadenus</u>
  14. Style 1, entire to divided 3/4 way in to base; flowers solitary or in few-flowered clusters; flowers usually over 1 cm. long
    15. Leaves 2, the widest one over 10 mm. wide attached at about ground level; flowers yellow, over 22 mm. long
-14. <u>Erythronium</u>
    15. Leaves 3 or more, not over 10 mm. wide, some attached well up from ground level; flowers various but if yellow then not over 22 mm. long
      16. Flowers white with purple veins, erect; plants alpine; stems from bulbs attached to rootstocks
-15. <u>Lloydia</u>
      16. Flowers purple, brown, yellow or orange, nodding; plants not alpine; stems from bulbs not attached to rootstocks
-16. <u>Fritillaria</u>
12. Underground parts rootstocks, these usually horizontal; fruit a berry; leaves lanceolate or broader
  17. Flowers axillary or extra-axillary (not terminal), the perianth segments spreading or recurved near their apex
-17. <u>Streptopus</u>
  17. Flowers terminal, the perianth segments not recurved
    18. Flowers in racemes or panicles, not nodding, the perianth not over 7 mm. long; stems simple
-18. <u>Smilacina</u>
    18. Flowers solitary or few in an umbellike cluster, nodding, the perianth 8 mm. long or more; stems branching
-19. <u>Disporum</u>

1. <u>Yucca</u> L. YUCCA; SPANISH BAYONET; SOAPWEED

Large plants with thick, branching, mainly subterranean caudices; leaves firm, narrow, rigid, pointed, commonly with threadlike fibers along the edges; flowers in terminal panicles or racemes, perfect, large; perianth segments nearly or quite distinct, all nearly alike; style rather short and thick; fruit dry or fleshy.

1. Fruit fleshy and berrylike, pendulous; leaves usually over 3 cm. wide; perianth segments usually over 6 cm. long
-1. <u>Y. baccata</u>
1. Fruit dry, dehiscent, upright; leaves seldom over 3 cm. wide; perianth segments seldom over 5 cm. long
  2. Styles (in fresh flowers) white or very pale green (paler than the ovary at anthesis), slender ovoid to oblong-cylindric, stigmas somewhat spreading at anthesis
    3. Capsules 3.5-4.8 cm. long, with thin wall, fragile with smooth epidermis; leaves not rigid, often very flexible; plants limited to southwestern Colorado
-2. <u>Y. angustissima</u>
    3. Capsules about 5 cm. long, wall thick and roughened especially along sutures; leaves rigid and stiff; plants not limited (though may be present) to southwestern Colorado
-3. <u>Y. standleyi</u>
  2. Styles (in fresh flowers) apple green, (darker than ovary at anthesis), slender to thick-ovoid, stigmas erect or spreading at anthesis
    4. Styles oblong-cylindric, about 3 mm. wide near base; leaves often over 1 cm. wide, usually flexible and not very stiff; plants seldom east of the continental divide
-4. <u>Y. harrimaniae</u>
    4. Styles ovoid, over 3 mm. wide near base; leaves rarely over 1 cm. wide, usually stiff; plants east of the continental divide
-5. <u>Y. glauca</u>

1. Yucca baccata Torr., Bot. Mex. Bound. 221. 1859.
Caudices very short, the plants appearing acaulescent or nearly so; stems bearing a tuft of leaves at or near the ground; leaves 40-75 cm. long, about 5 cm. wide at middle (when flattened) pale bluish, concave, rigid, spreading, commonly twisted, with a few coarse flattened, recurved fibers on the margins; panicle 50-90 cm. long; perianth variable in length, 5-15 cm. long (usually about 7 or 8), segments narrowly to broadly lanceolate, the flower oblong-campanulate in shape; styles 5-7 mm. long, not swollen; fruit conical, large, often 15-20 cm. long, tapering from base to apex, sometimes constricted at upper third, fleshy and indehiscent, pendant.---Dry plains and slopes. Colorado to Nevada, south to Texas, New Mexico and California. Our records from southwestern Colorado at 5500-7500 feet, but reported in Las Animas County.

2. Yucca angustissima Engelm. ex Trel., Mo. Bot. Gard. Ann. Rpt. 13:58. 1902.
Caudices short, plants appearing acaulescent or nearly so; stems with leaves clustered near the ground; leaves 20-40 cm. long, to 8 mm. wide, often pungent but ours rather limber, curly-filamentous on the margins; inflorescence 10-150 cm. long, racemose or short-branched below; sepals and petals about 3-5.7 cm. long, greenish-white, ovate-lanceolate, sometimes wider or narrower; style not swollen, whitish or pale green in color; capsule dry, dehiscent, about 3.5-5 cm. long, commonly constricted in middle.---Mesas and slopes. Utah to Nevada, south to New Mexico and Arizona. Our few records from westcentral and southwestern Colorado at 4500-9000 feet.

3. Yucca standleyi McKelvey, Yuccas of S. W. U. S. Part 2. p. 108. 1947.
Plants acaulescent or subacaulescent; leaves rigid, narrowly linear, yellow-green, filamentous on whitish margins; inflorescence 4-60 cm. or more long, a simple raceme; flowers greenish-white, segments 5-6.5 cm. long, elliptic, ovate or obovate; style white, little swollen; capsule dry and dehiscent, not usually constricted, about 5 cm. long, oblong-cylindric or somewhat ovoid.---Plains and hillsides. Colorado, Utah, New Mexico and Arizona. Our few records from central and southern Colorado at 5000-7500 feet, but the plant has been reported for the northcentral, central, southwestern and westcentral parts of the state.

4. Yucca harrimaniae Trel., Rep. Mo. Bot. Gard. 13:59. 1902.
Plants acaulescent, often caespitose; leaves 6-15 (or even to 40) mm. wide, linear to spatulate-lanceolate, thin but rigidly spreading, glaucous or finally green, concave, pungent, bordered with coarse, recurved whitish fibers; inflorescence 25-50 cm. long, flowering from near the base of the scapes; flowers greenish-white with large, broad often obtuse segments about

5-5.5 cm. long; style narrow, green; capsule about 4-5.7 cm. long sometimes constricted, upright, finally dry.---Dry plains and slopes. Colorado, Utah, New Mexico and Arizona. Our records scattered in the southwestern quarter of Colorado at 6000-7500 feet. However, this plant has been reported east of the continental divide in Huerfano and Costilla Counties.

5. Yucca glauca Nutt., Fraser Cat. No. 89. 1813.
Plants subacaulescent or with branching decumbent stems; leaves about 6-12 mm. wide, rigid, spreading, narrowly linear, finely but sparingly filamentous on the white margins, pale in color; inflorescence 30-200 cm. long, simple or occasionally short branches present, nearly sessile; flowers 3-5 cm. long, greenish-white, globose, oblong or campanulate, segments ovate to lanceolate, acute; style swollen, apple-green; capsule dry, dehiscent, upright, oblong, about 3 cm. long, usually not constricted.---Dry plains and slopes. Kansas to Colorado, south to Texas and New Mexico. Our records scattered over the eastern half of Colorado, also in the San Luis Valley, at 4000-8500 feet.

2. Nolina Michx. BEARGRASS

Plants with large woody caudices, these subterranean in our species; leaves numerous, clustered at base, rather rigid and pungently pointed, entire, not filiferous, long and narrow; flowers in open panicles, polygamo-dioecious, numerous; perianth segments in 2 whorls but similar, white, distinct, 1-nerved; stamens 6, reduced to staminodia in fertile flowers, filaments short and thick; ovary with obsolete style; capsule dry, 3-lobed or 3-winged, opening rather irregularly; seeds turgid, 1 in each cell.

1. Nolina microcarpa Wats., Proc. Amer. Acad. 14:247. 1879.
N. greenei Wats.---New Mexico to Arizona, south to Mexico. This plant has been reported for Colorado and has been found in Oklahoma just a few miles south of the state line, but no material actually collected in the state could be located.

3. Calochortus Pursh. MARIPOSA LILY; SEGO LILY

Herbaceous plants from coated corms, subscapose, frequently bulbiferous or swollen about at ground level or from lower leaf axils; leaves few, narrow, basal, alternate; flowers solitary or few in a subumbel, perfect, conspicuous; outer perianth segments sepaloid (called sepals here), greenish, separate, non-glandular; inner petallike (called petals here), separate, with a large gland near base of each; fruit a septicidal, 3-angled or 3-winged capsule.

1. Sepals longer than the petals, long-attenuate at the apex; anthers linear -1. C. macrocarpus
1. Sepals shorter than the petals, obtuse to abruptly acuminate; anthers oblong to lanceolate
   2. Anthers acute; glands on petals oblong; hairs on faces of petals usually branched -2. C. gunnisonii
   2. Anthers obtuse; glands on petals circular to lunate; hairs on faces of petals seldom branched
      3. Stems flexuous, not bulbous near base (in addition to the corm); glands on the petals not depressed, never surrounded by membranes -3. C. flexuosus
      3. Stem straight or nearly so, usually bulbous near base; glands on petals more or less depressed, surrounded by membranes
         4. Petals white to purplish -4. C. nuttallii
         4. Petals yellow -4A. C. nuttallii aureus

1. Calochortus macrocarpus Dougl., Tran. Hort. Soc. Lond. 7:276. 1830.
C. acuminatus Rydb.---Montana to British Columbia, south to Colorado and Nevada. Reported for western Colorado but no specimens could be found and the record is doubtful.

2. Calochortus gunnisonii Watson, Bot. U. S. Geol. Expl. 40th Par. p. 348. 1871.
Stems erect, unbranched; leaves linear; inflorescence 1- to 3-flowered, subumbellate; flowers white to purple, erect, campanulate, a purplish band on petals above gland and a purple dot on claw, sepals with one large spot or similarly marked; sepals shorter than petals, lanceolate, glabrous; petals obovate, obtuse, densely bearded above gland with branching gland-tipped hairs; anthers lanceolate, acute to apiculate, longer than the filaments; ovary linear, tapering to a persistent trifid stigma.---Meadows and open slopes. South Dakota to Montana, south to New Mexico and Arizona. Our records scattered over the western two-thirds of Colorado at 5500-11,000 feet.

3. Calochortus flexuosus Watson, Am. Nat. 7:303. 1873.
Stem erect or often more or less decumbent or climbing, usually branched at least near top, rarely bulbiferous; leaves linear, attenuated; inflorescence 1- to 4-flowered; flowers white to pink, erect, campanulate, each petal with yellow gland at base and usually a purple spot on claw; sepals shorter than petals, similarly marked, glabrous; petals obovate, rounded and obtuse above, sparsely invested with short hairs near the gland, this not depressed, lunate to circular, covered with short processes; anthers oblong, obtuse to umbonate; ovary linear; fruit acute, 3-angled; seeds strongly flattened.---Dry slopes and plains. Colorado to California, south to Arizona. Our few records from southwestern Colorado at 5400-6200 feet.

4. Calochortus nuttallii Torrey in Stansbury, Exp. Utah p. 397. 1852.
Stem erect, usually unbranched, often bulbiferous near base; leaves linear becoming involute; inflorescence 1- to 4-flowered, subumbellate; flowers white or lilac to magenta-tinged, erect, campanulate, petals yellow at base, and with a reddish-brown spot above the gland, sepals marked the same; sepals usually shorter than the petals, lanceolate, short-acuminate; petals obovate, may be short-acuminate, invested near gland with sparse hairs, these unbranched, glands circular, depressed, surrounded by conspicuous fringed membranes; anthers obtuse, equalling filaments; ovary linear, tapering to a persistent trifid stigma.---Dry plains and hillsides. Nebraska to California, south to New Mexico and Arizona. Our records scattered in the western half of Colorado, especially in the extreme west, at 4500-7500 feet.

4A. **Calochortus nuttallii aureus** (Wat.) Ownbey, (var.) Ann. Mo. Bot. Gard. 27:493. 1940.
C. aureus Wat.---Lower than the species, 10-20 cm. tall, our specimen bulbous near base: petals lemon-yellow, each with a maroon blotch over the circular gland.---New Mexico, Utah and Arizona. Our one record from southwestern Colorado at about 5800 feet.

4. Trillium L. WAKEROBBIN

Plants from stout, short rootstocks; leaves in a whorl of 3 near the end of the scape, each one broad and net-veined; flowers perfect, solitary and terminal or rarely in terminal umbels; perianth in 2 unlike whorls of 3 each, outer sepaloid, green, distinct; inner whorl petaloid, pink, white or purplish, distinct; stamens usually 6, with rather short filaments; stigmas sessile; fruit a lobed berry.

1. **Trillium ovatum** Pursh, Fl. Am. Sept. 245. 1814.
Rootstocks rather cormlike; stems 20-40 cm. tall; leaves sessile or nearly so, 6-12 cm. long, oval to broadly elliptic, short-acuminate; flowers on peduncles 3-5 cm. long; sepals linear-lanceolate, 1.5-4 cm. long; petals about as long as the sepals, pink, white or turning purplish.---Montana to British Columbia, south to Colorado and California. Our few records from northeastern Routt County at 9500-11,000 feet.

5. Smilax L. CARRION FLOWER; GREENBRIER

Stems usually climbing by a pair of tendrils at the base of the broad, entire margined leaves, our plants herbaceous; flowers dioecious, in axillary peduncled umbels; perianth of 6 similar, distinct segments; ovary with 3 spreading sessile stigmas; fruit a small berry.

1. **Smilax herbacea** L., Sp. Pl. 1030. 1753.
Nemexia lasioneuron (Hook.) Rydb.; N. herbacea melica A. Nels.---Stem herbaceous, climbing, usually less than 2 meters long, glabrous; leaves 4-10 cm. long, numerous, ovate to cordate, 5- to 9-nerved, entire, short-acuminate, cordate to truncate or widely cuneate at base, glabrous and scabrous-hirsutulose, on petioles 1-5 cm. long, these bearing at base a pair of coiled tendrils; peduncles 3-8 cm. long, many-flowered; flowers carrion- or sweet-scented; perianth greenish, 3.5-4.5 mm. long; berries 6-10 mm. thick, blue-black. Our plant is var. lasioneuron (Hook.) A. DC.---Woods and edges of clearings. The variety from Saskatchewan south to Kansas and Colorado and eastward. Our records from central and northcentral Colorado at 5200-6800 feet.

6. Androstephium Torr. FUNNELLILY

Scapose herbaceous plants with membranous or fibrous-coated corms; leaves basal, narrow; flowers in a bracted umbel on rather short inarticulate pedicels; perianth funnelform, 6-cleft to the middle or below; flowers perfect; stamens 6, adnate to the perianth tube, the filaments united at least to the middle into a tube with erect segments between the anthers; capsule 3-angled.

1. **Androstephium breviflorum** S. Wats., Am. Nat. 7:303. 1873.
Brodiaea paysonii Nels.---Corm ovoid, fibrous-coated; scapes 10-30 cm. tall, often scabrous especially below; leaves several, equalling or exceeding the scapes; bracts lanceolate, scarious; umbel 3- to 12-flowered, pedicels 10-20 mm. long; perianth 15-20 mm. long, light violet-purple, whitish or tinged with blue-rose, segments longer than the tube; lobes of the crown shorter than the anthers.---Dry mesas and slopes. Colorado to California, south to Arizona. Our records from western Colorado at 4500-7000 feet.

7. Leucocrinum Nutt. STARLILY; SANDLILY; MOUNTAINLILY

Low, acaulescent herbaceous plants with deep seated short, vertical rootstocks and fleshy roots; leaves basal, surrounded by scarious sheaths; flowers perfect in umbellike, sessile clusters, the ovaries and pedicels below ground; perianth segments all similar and united into a long tube, rather salverform; ovary sessile; style much elongated, filiform; stigma 3-lobed; fruit a capsule.

1. **Leucocrinum montanum** Nutt.; A. Gray, Ann. Lyc. N. Y. 4:110. 1848.
Leaves 6-15 or more, 2-6 mm. wide, linear, surrounded at base by scarious bracts; flowers 4-8, white, tube 3-8 cm. long, segments 20-25 mm. long, linear-lanceolate; style long, with dilated stigma; capsule 6-8 mm. long, truncate.---Plains and hills. South Dakota to Montana, south to New Mexico and California. Our records scattered over Colorado, none as yet from the extreme western part, at 3500-8000 feet.

8. Polygonatum Adans. SOLOMAN'S SEAL

Caulescent herbaceous plants from long rootstocks; leaves broad, many-nerved, sessile, alternate; flowers perfect, in axillary 1- to few-flowered umbels or racemes, the pedicel jointed; flowers of 6 like segments, these united in a tube for over 1/2 their length; anthers sagittate; style slender, stigma capitate or obscurely 3-lobed; fruit a globular purple or black berry.

1. **Polygonatum commutatum** (Schutesf.) Dietr.; Otto & Dietr., Allg. Gartenz 3:223. 1835.
Ontario to Manitoba, south to Georgia and New Mexico; possibly in Utah. There seems no reason why the plant should not be in Colorado but no specimens from the state could be located.

9. Allium L. ONION

Onion-scented bulbous plants, the bulb sometimes borne from a short rootstock; leaves sheathing, terete to flat, linear; flowers in a scapose, terminal umbel; bracts free or partly connate; perianth segments rather persistent, white to rose-purple, separated or nearly so; stamens adnate to the perianth base; ovary more or less completely 3-celled, often crested at apex; fruit a capsule. This genus is sometimes placed in the family Amaryllidaceae.

The complete bulb is necessary for accurate identification. Partially or completely developed capsules are desirable in order to prove the presence or absence of crests. Some of the species intergrade considerably in Colorado.

1. Umbel at least partially nodding, the scape more or less recurved at apex; perianth segments 4-6 mm. long, often obtuse; stamens exserted beyond the perianth -1. A. cernuum
1. Umbel not nodding, the scape straight or nearly so at apex; perianth segments often over 6 mm. long, obtuse only in A. rubrum; stamens included
   2. Flowers mostly transformed into bulblets -2. A. rubrum
   2. Flowers not transformed into bulblets (or only rarely)
      3. Leaves hollow, terete (may be pressed flat in specimens); pedicels usually shorter than the flowers -3. A. schoenoprasum
      3. Leaves not hollow, flat or channeled (rarely terete); pedicels longer than the mature flowers
         4. Bulbs borne on short, stout rootstocks; leaves over 4 mm. wide; ovary not crested at apex -4. A. brevistylum
         4. Bulbs without rootstocks; leaves not over 4 mm. wide; ovary crested or not crested (crested in most species)
            5. Outer bulb coat transversed by coarse often interwoven fibers; if reticulated then from these loose fibers (the inner coats often have coarse diamond-shaped attached reticulations)
               6. Perianth parts 10 mm. long or more, long-attenuate at the apex; peduncles often 2 or 3 from the same set of loose sheaths -5. A. macropetalum
               6. Perianth parts 4-10 mm. long, acute to acuminate but not long-attenuate; peduncles solitary from the close sheaths
                  7. Ovary without crests at all; involucre bracts ovate and usually 3; perianth usually not over 7 mm. long -6. A. drummondi
                  7. Ovary with crests (may be rather small); involucre bracts usually 2 (or if 3 then lanceolate); perianth often over 7 mm. long
                     8. Leaves usually 2 per stem; tips of inner perianth segments spreading; flowers usually white -7. A. textile
                     8. Leaves usually 3 or more per stem; tips of inner perianth segments erect, flowers usually pink -8. A. geyeri
            5. Outer bulb coats thin and membranous, marked on the surface with fine quadrangular or hexangular reticulations but no coarse loose fibers present
               9. Perianth segments 8-12 mm. long, the inner serrulate near apex; scapes usually over 10 cm. long, usually surpassing the leaves -9. A. acuminatum
               9. Perianth segments 6-8 mm. long, all entire; scape short, not over 10 cm. long, usually surpassed by the leaves
                  10. Ovaries and capsules distinctly 6-crested -10. A. nevadense
                  10. Ovaries and capsules not crested or indistinctly 3-crested -11. A. brandegei

1. Allium cernuum Roth in Roem., Arch. Bot. 1(3):40. 1798.
   A. recurvatum Rydb.; A. neomexicanum Rydb.---Bulbs narrowly ovoid, from elongated horizontal to vertical rootstocks, the bulb on the lower end, bulb coat slightly if at all fibrous; scapes 20-60 cm. tall; leaves variable, averaging 2-4 mm. wide, flat or almost terete, shorter than the scapes; umbel nodding, the scape more or less recurved at the apex; involucre bracts short, usually 2; perianth 4-6 mm. long, the segments obtuse or acutish, white or rose; stamens and style exserted; capsule strongly crested at maturity. According to Ownbey (Research Studies, State College Washington 15:225-228. 1947.) our plant would be var. obtusum Cockerell.---Dry hills, plains and meadows. Widespread throughout the United States. Our records from the western three-fourths of Colorado at 5000-11,000 feet.
2. Allium rubrum Osterh., Bull. Torr. Club 27:506. 1900.
   A. fibrosum Rydb.; A. arenicola Osterh.; A. sabulicola Osterh.---Bulbs single, rarely few, ovoid to more elongated, the outer coats fibrous-reticulated; scapes 20-50 cm. tall; leaves about 3-4 to a scape, up to 7 mm. wide, flat or concave; spathe 2- or 3-bracted; most of the flowers of the umbel replaced by small spherical red bulblets; perianth segments 6-8 mm. long, ovate to ovate-lanceolate, usually obtuse, usually pink, the outer often with a dark midnerve; stamens and style equalling the perianth; ovary not crested.---Meadows and banks. Alberta to British Columbia, south to New Mexico, Arizona and Oregon. Scattered in the mountainous half of Colorado at 6800-11,500 feet.
3. Allium schoenoprasum L., Sp. Pl. 301. 1753.
   A. sibericum of Manuals---Bulbs rather small, often oblique, oblong-ovoid, with membranous usually white coats lacking reticulations; scapes 30-60 cm. long, rather stout; leaf blades shorter than scape, linear, terete, hollow (but may be pressed flat), sheaths long; bracts ovate, usually 2; umbels capitate the pedicels 2-6 mm. long, shorter than the flowers; perianth segments 8-12 mm. long, rose-colored with dark midribs, lanceolate, acuminate; stamens included; capsule not crested. The North American plants have been considered by some to belong to 2 entities, var. laurentianum Fern. and var. sibericum (L.) Hartm.---Moist meadows or plains. Widespread in the northern parts of North America, Europe and Asia. Our few records from north-central Colorado at 7500-8500 feet.

4. **Allium brevistylum** S. Wats. in King, Geol. Expl. 40th. Par. 5:350. 1871.

Bulbs obliquely elongate from stout thick rootstocks, the outer coat only slightly fibrous; scapes 30-60 cm. high, stout; leaves 4-10 mm. wide, flat, shorter to nearly as long as the scapes; bract usually solitary, ovate; flowers few, erect; perianth deep rose colored, the segments 8-12 mm. long, lanceolate, long-acuminate; stamens included; capsule not crested.---Meadows and open woods. Montana to Utah, south to Colorado. Our records from northcentral Colorado at 6500-9000 feet.

5. **Allium macropetalum** Rydb., Bull. Torr. Club 31:401. 1904.

Bulbs ovoid, solitary, coated with loose fibers; scapes 5-20 cm. or more tall, stout, often 2 or 3 from same set of sheaths; leaves almost equalling the scapes, usually 2, 1-4 mm. wide; sheaths very loose and scarious; umbel many-flowered; involucre bracts usually 3, ovate-lanceolate, acuminate; pedicels stout, the outer reflexed in fruit; perianth parts 8-12 mm. long, pink with prominent purplish midvein, lanceolate, long-attenuate, scarious in fruit; capsule with 6 conspicuous crests.---Dry plains and hills. Colorado to Utah, south to Texas and Arizona. Our one record from westcentral Colorado at about 4700 feet.

6. **Allium drummondi** Regel, Acta Horti Petropolitani 3(2):112. 1875.

A. nuttallii Wats.---Plains, hills and prairies. Nebraska south to Texas and New Mexico. Has been reported from Colorado and is to be expected in the eastern or southeastern part.

7. **Allium textile** Nels. & Macbr., Bot. Gaz. 56:470. 1913.

A. reticulatum Fraser---Bulbs usually solitary, ovoid, prominently fibrous-reticulated on outer coats; scapes 10-30 cm. tall; leaves narrow, 1-4 mm. wide, channeled; bracts usually 2, ovate; perianth 4-7 mm. long, segments white to light-rose, often with rose midvein, ovate-lanceolate, acute or acuminate; stamens included; capsule with 6 rounded crests just below the summit.---Plains and dry hills. Saskatchewan to Alberta, south to New Mexico and Arizona. Widely scattered in all parts of Colorado at 4000-10,500 feet, commonest at lower altitudes.

8. **Allium geyeri** Wats., Proc. Am. Acad. 14:227. 1879.

A. funiculosum A. Nels.; *A. dictyotum* Greene; *A. pikeanum* Rydb.---Bulbs usually solitary, ovoid to oblong-ovoid, outer coats fibrous-reticulated; scapes 10-60 cm. tall; leaves usually 3 to a scape, blades shorter than the scape, about 2 mm. wide, (sometimes to 4 mm.); bracts 2-3, ovate to lanceolate; perianth segments 4-10 mm. long, white or rose-colored, ovate or ovate-lanceolate, acute or acuminate; stamens included; capsule with rather inconspicuous crests. Intergrades in Colorado with the preceding species.---Plains, meadows and slopes. Alberta, south to Texas, Arizona and Washington. Our records scattered in the western two-thirds of Colorado at 5000-12,000 feet.

9. **Allium acuminatum** Hook., Fl. Bor.Amer. 2:184. 1840.

Bulbs ovoid or nearly spherical, outer coats gray, with 4-sided or hexagonal-pitted meshes or reticulations, inner coats white; scapes 10-45 cm. long; leaves 2-3 mm. wide, shorter than the scapes; bracts 2; flowers many; perianth dark rose or reddish-purple, the segments 8-15 mm. long, ovate-lanceolate or lanceolate with long recurved tips, somewhat keeled on the back and gibbous below, the inner slightly shorter, serrulate near tip; stamens included; capsule with 3 small, nearly obsolete crests.---Dry plains and hillsides, often in rocky places. Montana to British Columbia, south to Arizona and California. Our records mostly from the western third of Colorado at 4500-9500 feet.

10. **Allium nevadense** S. Wats. in King, Geol. Expl. 40th Par. 5:351. 1871.

Bulbs ovoid to nearly globose, inner coats white to pink, outer gray-brown, no loose fibrous material but outer coats reticulate; scapes 3-10 cm. tall, slender; leaves 1 or 2 to a scape; umbels 7- to many-flowered; involucre bracts 2-3, ovate, about 1-1.5 cm. long; pedicels 10-15 mm. long, slender; perianth segments 6-12 mm. long, white to rose, oblong-ovate to lanceolate, acuminate; capsule prominently 6-crested.---Dry slopes and plains. Idaho to Oregon, south to Arizona and Nevada. Our one record from westcentral Colorado at 6500 feet.

11. **Allium brandegei** Wats., Proc. Amer. Acad. 17:380. 1882.

Bulbs globose-ovoid, rather small, coats all membranous, the outer coats reticulate-veined but not fibrous; scapes up to 10 cm. high, exceeded by the narrowly linear (usually 2) leaves; bracts 2, ovate; flowers many; perianth 6-8 mm. long, segments rose or whitish, broadly lanceolate, acuminate or acute, entire; ovary not crested.---Mountains. Idaho to Oregon, south to Colorado and Utah. Our few records from northwestern Colorado at about 6500 feet.

10. *Asparagus* L. ASPARAGUS

Stems at first simple, scaly in the juvenile state, later becoming tall and much-branched; leaves minute, scalelike, with filiform branchlets in their axils; flowers commonly solitary, perfect, small, nodding; perianth greenish-white, the segments equal, separate or somewhat united at bases; style slender and short; stigma lobes short; fruit a globose berry.

1. **Asparagus officinalis** L., Sp. Pl. 313. 1753.

Rootstocks much branched; young stems succulent, stout, edible, later branching and becoming up to 200 cm. tall, with filiform branchlets 1-2 cm. long, these mostly clustered; flowers mostly solitary at the nodes, greenish and drooping; perianth about 6 mm. long; campanulate; stamens included; berry red.---In cultivation and frequently escaped especially in cultivated areas. Introduced from the Old World.

11. *Veratrum* L. FALSE HELLEBORE

Tall herbaceous plants with leafy stems from thick rootstocks; leaves broad, clasping, strongly veined and plaited; flowers very numerous in large ample panicles, the lower often staminate and upper perfect; perianth whitish to greenish, without glands; styles 3, separate; ovary 3-celled, several-seeded, becoming a capsule.

1. Veratrum californicum Durand, Journ. Acad. Nat. Sc. Phila. II. 3:103. 1855.
 V. speciosum Rydb.; V. tenuipetalum Heller---Stems 50-200 cm. tall, stout; leaves 20-30 cm. long, 8-20 cm. wide, broadly oval to ovate, narrower above, sessile and sheathing at base, acute or obtuse, softly pubescent below; inflorescence often compound, often very large (to 50 cm. long); petals and sepals 6-15 mm. long, greenish-white, ovate to oblanceolate, obtuse or acute; bractlets membranous, not as long as the flowers.---Moist meadows and slopes. Montana to Washington, south to New Mexico and California. Scattered in the mountains of Colorado at 8500-11,000 feet.

12. Lilium L. LILY

Stems tall, leafy, from thick-scaled bulbs; leaves alternate or in whorls, narrow; flowers large, perfect, the segments separate, petals and sepals with a nectar-bearing groove at bases; stamens slightly attached to perianth, anthers versatile; style 1; stigma capitate but 3-lobed; fruit a capsule.

1. Lilium umbellatum Pursh, Fl. Amer. Sept. 229. 1814.
 L. montanum A. Nels.---Bulbs depressed-globose, of thick fleshy scales; stems 30-60 cm. tall; leaves alternate, except uppermost in a whorl or two, linear to lanceolate, glabrous; flowers 1-3, umbellate; petals and sepals 5-6 cm. long, red to orange-red, spotted at base of inner side with purplish-black spots, the blades oval to elliptic-oblong, acute to short-acuminate, long-clawed; capsules cylindric to cylindric-ovoid. Possibly not distinct from L. philadelphicum L.---Slopes and hills, often in open woods. Ohio to Alberta, south to Arkansas and New Mexico. Our records from northcentral, central, southcentral and southwestern Colorado at 6000-10,000 feet.

13. Zygadenus Michx. DEATH CAMAS

Plants glabrous and herbaceous with membranous-coated bulbs; leaves narrow, grasslike, crowded near base; flowers perfect, in racemes or narrow panicles; perianth segments separate or united below, each with a gland near the base, withering-persistent; ovary superior or partly inferior with 3 styles; fruit a 3-lobed capsule. This is a puzzling genus and the treatment of the last 2 species is rather arbitrary (Zigadenus).

1. Ovary partly inferior; glands of perianth obcordate   -1. Z. elegans
1. Ovary superior; glands of perianth obovate or semicircular
    2. Plant stout, over 30 cm. tall (to 70); flowers usually in panicles; claws of petals short, usually less than 1 mm. long; leaves 8-12 mm. wide; perianth segments acute or acuminate at apex   -2. Z. paniculatus
    2. Plant slender, 20-35 cm. tall; flowers usually in racemes; claws of petals long, usually about 1 mm. long; leaves 2-8 mm. wide; perianth segments obtuse or rounded at apex   -3. Z. gramineus

1. Zygadenus elegans Pursh, Fl. Amer. Sept. 1:241. 1814.
 Z. coloradensis Rydb.; Anteclea elegans (Pursh) Rydb.; A. coloradensis Rydb.---Plants 20-60 cm. tall, rather glaucous, glabrous; bulbs ovoid; leaves 3-15 mm. wide, keeled; inflorescence a raceme or sparingly branched below with ascending branches; bracts linear-lanceolate to ovate; flowers greenish-white or yellowish-white, sometimes tinged with purple, the segments 5-10 mm. long (usually 7-10), narrowly obovate or oblong, with an obcordate gland just above the short claw; capsule ovoid.---Moist plains and meadows. Saskatchewan to Alaska, south to New Mexico and Arizona. Our records scattered in the mountains at 5500-12,000 feet.
2. Zygadenus paniculatus (Nutt.) S. Wats. in King, Geol. Expl. 40th Par. 5:343. 1871.
 Plants 30-70 cm. tall, rather stout; bulbs ovoid, about 4 cm. long; basal leaves 8-12 mm. wide, cauline sheathing at base; inflorescence a rather dense panicle, at least at base; bracts lax and scarious; perianth segments 4 mm. long, acute or acuminate, sepals sessile, the petals subcordate to cuneate at base and short-clawed (probably claw definitely less than 1 mm. long).---Hills and plains. Montana to Washington, south to New Mexico and California. Scattered in the western two-thirds of Colorado at 5000-8000 feet. The most of the records are northwest of a line drawn from Montezuma to Larimer County.
3. Zygadenus gramineus Rydb., Bull. Torr. Club 27:535. 1900.
 Z. intermedius Rydb.; Toxicoscordium gramineum Rydb.---Plants 20-35 cm. tall, rather slender; bulb 2-3 cm. long, elongated-ovoid; leaves 3-8 mm. wide, linear, all with sheaths surrounding the stems; flowers in racemes; bracts scarious; perianth parts obtuse, petals long-clawed, the claw about 1 mm. long, subcordate at base. This species is only arbitrarily separated from the preceding.---Hills and plains. Saskatchewan to Idaho, south to New Mexico and Utah. In the western two-thirds of Colorado at 4500-8000 feet, all our records southeast of a line drawn from Montezuma to Larimer County.

14. Erythronium L. DOGTOOTH VIOLET; FAWNLILY

Herbaceous plants from membranous deep-seated corms; stems rather scapiform above the ground; bearing 2 leaves; flowers perfect, solitary or few, nodding; perianth of 6 distinct, similar, recurved segments with a grooved nectary at base of each; style entire, often 3-cleft above; fruit a 3-angled capsule.

1. Erythronium grandiflorum Pursh, Fl. Am. Sept. 1:231. 1814.
 Corms rather elongated; leaves 10-25 cm. long, usually oblong-elliptical, acute or sometimes oblanceolate and obtuse; scapes 15-32 cm. high; flowers 1 to several, rarely to 16, segments 20-40 mm. long, 6-17 mm. wide, golden-yellow but streaked with green near base outside, lanceolate to ovate-lanceolate, acuminate to acute or sometimes obtuse; filaments only slightly wider

below, anthers golden-yellow; style long and slender; stigma with short lobes that tend to recurve. Our plants are ssp. chrysandrum Applegate.---Moist slopes and shaded areas. The subspecies from Wyoming, Colorado, Idaho and Utah. Our records from the mountains of the western half of Colorado, except in the southcentral part, at 6500-11,000 feet.

15. Lloydia Salisb. ALPLILY

Small caulescent herbaceous plants with tunicated bulbs arising from creeping rootstocks; leaves narrowly linear; flowers perfect, solitary or few in a raceme; perianth segments with a transverse foldlike gland near base, nearly alike, distinct; style 1; fruit a capsule.

1. Lloydia serotina (L.) Sweet, Hort. Brit. ed. 2. 527. 1830.
Bulbs oblong, fibrous-coated, ending in creeping rootstocks; stems 5-15 cm. tall; leaves 1-2 mm. wide, several, about as long as the stems; flowers usually 1 but sometimes to 4, erect, about 1 cm. long, broadly turbinate; perianth segments yellowish-white, tinged with rose on the back and purple-veined, oblanceolate, obtuse; capsules obovoid, 8 mm. long.---Mountains at high altitudes. Alberta to Alaska, south to New Mexico and Nevada; Eurasia. Our records from the higher mountains of Colorado at 10,000-13,000 feet.

16. Fritillaria L. FRITILLARY

Caulescent herbaceous plants from bulbs, these made up of 1 or more thick fleshy scales; stems simple, erect; leaves whorled or alternate, narrow; flowers perfect, in racemes or solitary and nodding, perianth campanulate or tubular, the segments in 2 whorls with a gland or nectary near base, this sometimes obscure; stamens included, inserted on base of perianth segments; ovary sessile, style 1, entire or trifid; fruit a 6-angled or winged capsule.

1. Flowers brown or purplish, mottled; style divided over 1/2 way to base into lobes   -1. F. atropurpurea
1. Flowers yellow to orange (when young), not mottled; styles shortly 3-lobed, much less than 1/2 way down
   -2. F. pudica

1. Fritillaria atropurpurea Nutt., Journ. Acad. Phil. 7:54. 1834.
Bulbs of a few to several scales; stems 10-50 cm. tall, naked below; leaves 6-14, 1.5-4 mm. wide, alternate or whorled, scattered at middle of stem and above, narrowly linear; flowers 1-4, brown or purple, more or less spotted with yellow and white, open campanulate, nodding; perianth segments 8-20 mm. long, oblong-rhombic to linear, gland obscure; styles distinct almost to base (2/3 to 3/4).---Plains and slopes, often among bushes. North Dakota to Oregon, south to New Mexico and California. Our rather few records from the western fourth of Colorado at 6500-10,000 feet.

2. Fritillaria pudica (Pursh) Spreng., Syst. Veg. 2:64. 1825.
Ochrocodon pudicus (Pursh) Rydb.---Bulbs of small thick scales; stems 8-30 cm. tall, leafy; leaves 3-10 mm. wide, several, alternate, scattered, linear to lanceolate; flowers usually solitary, nodding, yellow to orange, sometimes brown-veined outside, turning brick-red in age; perianth segments 15-22 mm. long, oblong and obtuse, the basal gland small and obscure; style undivided, the stigmas shortly 3-lobed.---Plains and hillsides. Alberta to British Columbia, south to Wyoming and California. Our one record from northwestern Colorado at 7500 feet.

17. Streptopus Michx. TWISTED-STALK

Caulescent branching, herbaceous plants from horizontal rootstocks; leaves many-nerved, clasping or sessile, broad; flowers 1-2 in lateral, extra-axillary inflorescences; flowers perfect, nodding, perianth greenish-white, segments alike, distinct; style 1; fruit a berry.

1. Streptopus amplexifolius (L.) DC. & Lam., Fl. Franc. 3:174. 1805.
Rootstocks short and stout; stems 30-100 cm. tall, branched, flexuous, glabrous; leaves entire or nearly so, ovate to ovate-lanceolate, glabrous, glaucous below, acuminate at apex, cordate-clasping; flowers 1 to 2, on slender peduncles which are decurrent (though obscurely so) for an internode, the pedicel jointed to it, the pedicel short, rarely over 1 cm. long; flowers 8-12 mm. long, greenish-white, campanulate, the segments widely spreading above; style filiform, stigmas entire or obscurely 3-lobed; berries globose to elliptical, 8-15 mm. long. The above has been divided into 2 varieties, both to be expected here. The var. americanus Roem. et Schult. has leaves lacking papillate projections while the var. chalazatus Fasset is minutely but copiously papillate beneath. Our Colorado plants intergrade between the two.---Moist woods. Greenland to Alaska, south to North Carolina, New Mexico and California. Our records scattered in the western two-thirds of Colorado at 6500-11,000 feet.

18. Smilacina Desf. FALSE SOLOMON'S SEAL; SOLOMON'S PLUME

Plants with spreading rootstocks; stems unbranched, leafy; leaves sessile, several-nerved, alternate; flowers perfect in panicles or racemes; perianth small, white with segments alike, distinct or nearly so, spreading; style 1; fruit a berry.

1. Flowers in panicles; segments of perianth about 2 mm. long; leaves ovate-elliptic or oblong   -1. S. racemosa
1. Flowers in racemes; perianth segments 5-7 mm. long; leaves oblong-lanceolate or narrower   -2. S. stellata

1. **Smilacina racemosa** (L.) Desf., Ann. Paris Mus. d'Hist. Nat. 9:51. 1807.
   Vagnera brachypetala Rydb.; V. racemosa (L.) Morong---Rootstocks rather thick; stems slender or stout, erect or ascending, glabrous to finely pubescent above; leaves ovate, elliptic to oblong, sessile and clasping to short-petioled, acute to acuminate, finely pubescent below; flowers in ovoid to cylindrical panicles; perianth about 2 mm. long; berries red, usually speckled with purple, 4-6 mm. in diameter. Several varieties have been proposed (Am. Midl. Nat. 33:644-666. 1945, and Rhodora 40:405-407. 1938.) for our western plants but too many intergradations occur. S. racemosa var. amplexicaulis (Nutt.) Wats. is the most distinct.---Woods and thickets. Throughout most of temperate North America. Our records scattered in the western two-thirds of the state at 6500-10,000 feet.

2. **Smilacina stellata** (L.) Desf., Ann. Paris Mus. d'Hist. Nat. 9:52. 1807.
   Vagnera leptopetala Rydb.; V. liliacea (Greene) Rydb.; V. stellata (L.) Morong---Rootstocks stout; stems 20-60 cm. tall or sometimes more, glabrous, erect or nearly so; leaves narrowly lanceolate to oblong-lanceolate, sessile, acute to acuminate, minutely pubescent below; flowers in sessile or short-peduncled racemes; perianth 5-7 mm. long, rarely shorter, segments linear to oblong-lanceolate, obtuse to acute; berries purplish-black or at least dark-striped.---Moist shaded or open ground. Throughout most of temperate North America; Europe. Our many records scattered in the western three-fourths of Colorado at 4500-10,000 feet.

19. Disporum Salisb. FAIRYBELLS

Caulescent, branching herbaceous plants from long rootstocks; leaves broad, sessile or clasping; flowers perfect, terminal or few in an umbellike cluster, drooping; perianth of 6 narrow, distinct, equal segments; style 1, slender; fruit a bright red or orange, rarely yellowish berry.

1. **Disporum trachycarpum** (S. Wats.) Benth. & Hook., Gen. Pl. 3:832. 1883.
   Stems 30-60 cm. tall, rather flexuous, more or less pubescent; leaves 3-10 cm. long, ovate to oblong-lanceolate, acute or acuminate, rounded or subcordate at base, sessile; flowers solitary or few together; perianth 8-15 mm. long, narrowly campanulate, segments white or ochroleucous, acute, little spreading; style long, stigma usually with linear lobes; fruit depressed-globose, 8-10 mm. in diameter.---Canyons and hillsides, often in shade. Manitoba to British Columbia, south to New Mexico and Arizona. Scattered in the mountainous part of Colorado at 5500-9500 feet.

Family 25. Amaryllidaceae AMARYLLIS FAMILY

Perennial pubescent, herbaceous plants with corms or short rootstocks; leaves narrow, grasslike, basal; flowers few, solitary or subumbellate, perfect, regular; perianth of 6 segments, these similar, in 2 whorls, distinct above the ovary; stamens 6, with sagittate anthers (in ours); ovary inferior, 3-celled; style 1; stigmas 3 (in ours decurrent on style); fruit a capsule.

1. Hypoxis L. STARGRASS; GOLDSTARGRASS

Characters of the Family

1. **Hypoxis hirsuta** (L.) Coville, Mem. Torr. Club 5:118. 1894.
   Corms subglobose or ellipsoid; leaves linear, 1-5 mm. wide, mostly longer than the scapes, sometimes villous; scapes 5-20 cm. tall, 1- to 7-flowered, villous above, umbellate or subumbellate with short subulate bracts; perianth 6-10 mm. long, segments yellow within when fresh, green and villous without, lanceolate to elliptic; anthers sagittate; stigmas 3, decurrent on the angles of the style.---Meadows and open woods. Maine to Saskatchewan, south to Florida and Texas. Apparently rare on the plains and eastern foothills of Colorado, our two records from 5000-7800 feet.

Family 26. Iridaceae IRIS FAMILY

Perennial herbs from bulbs or rootstocks; leaves narrow, equitant; flowers perfect, regular or nearly so, subtended by spathelike bracts; perianth of 6 segments, in 2 series of 3 each; stamens 3, on the perianth; ovary inferior, 3-celled; style 3-cleft, distinct; fruit a many-seeded capsule.

1. Style branches large and petaloid; flowers large, over 5 cm. wide; sepals and petals unlike -1. Iris
1. Style branches small, not petaloid; flowers smaller; sepals and petals alike -2. Sisyrinchium

1. Iris L. BLUE FLAG; IRIS

Herbaceous plants with creeping, thick horizontal rootstocks; leaves long, narrow, linear; flowers solitary or in terminal racemes or panicles, large and showy, in ours blue or violet; perianth segments united below into a tube, the outer spreading and recurved, the inner narrower and erect; styles petaloid, arching over the stamens, the stigma under and below the tips.

1. **Iris missouriensis** Nutt., Journ. Acad. Nat. Sci. Phila. 7:58. 1834.
   I. pelogonus Goodding---Rootstocks thick; leaves light green, glaucous, linear, acute, usually shorter than the stem and up to 1 cm. wide; stems branched or sometimes simple; spathes usually

opposite, 3.5-7 cm. long, scarious except near base and keel; perianth tube to 1 cm. long; sepals about 6 cm. long and 2 cm. wide, the claw yellowish-white, blade lilac or purple-veined on a paler background; petals somewhat shorter, not veined; capsule oblong, 3-5 cm. long.---Moist meadows and marshes or along streams. North Dakota to British Columbia, south to New Mexico and California. Our records scattered over the western two-thirds of Colorado at 5000-10,000 feet.

2. Sisyrinchium L. BLUE-EYED GRASS

Plants from short rootstocks; stems more or less compressed and winged; leaves linear and grasslike; flowers in terminal clusters from a 2-bracted spathe (also basal in one species); perianth blue or light blue, with segments alike, distinct or nearly so; filaments united at least near base.

1. Outer bract of spathe much longer than the inner one (sometimes 2-3 times longer)
  2. Plants with flowers and fruit in part on basal dwarf stems   -1. S. heterocarpum
  2. All the flowers borne terminally on normal stems   -2. S. montanum
1. Outer bract of spathe equal or subequal to inner, rarely over 1/3 longer
  3. Flowers over 12 mm. long; spathes over 2 cm. long   -3. S. occidentale
  3. Flowers not over 10 mm. long; spathes usually less than 2 cm. long
    4. Capsules glabrous, 4-5 mm. long; stems often flexuous   -4. S. demissum
    4. Capsules scabro-puberulent, 2-3.5 mm. long; stems stiffly erect   -5. S. halophilum

1. Sisyrinchium heterocarpum Bickn., Bull. Torr. Club 26:348. 1899.
Stems 15-30 cm. tall, 1-1.5 mm. wide, erect, narrowly winged; leaves 1-2 mm. wide, half the length of stem or more; spathes of unequal hyaline-margined bracts, inner 1.5-2 cm. long, outer 2.5-4.5 cm. long, sometimes purplish-tinged below; perianth to 10 mm. long, violet-purple, often with a large yellow eye; fruit of terminal spathes 5-7 mm. long, subglobose to somewhat obovoid; many of capsules basal from lower leaf axils on short peduncles, capsules 7-10 mm. long, obovoid-pyriform with a narrowed cuneate base.---Moist meadows. Wyoming and Colorado. Our few records from northcentral Colorado at 5500-8500 feet.
2. Sisyrinchium montanum Greene, Pitt. 4:33. 1899.
S. angustifolium of Manuals---Stems 10-55 cm. tall, 1-3 mm. wide, stiff, glaucous, simple or rarely branched, winged; leaves 1-3.5 mm. wide; spathe usually 1 and terminal, bracts of spathe very unequal, (the outer commonly twice as long as the inner) outer 20-65 mm. long, often purplish-tinged especially near margins; flowers 10-12 mm. long, deep violet; capsule elliptical to globose, 4-6 mm. long.---Moist meadows and banks. Newfoundland to British Columbia, south to Virginia, New Mexico and California. Rather common, the records scattered over the state at 4000-10,000 feet.
3. Sisyrinchium occidentale Bickn., Bull. Torr. Club 26:447. 1899.
Stems 15-35 cm. long, 1.2 mm. wide, erect, winged; leaves 1-2.5 mm. wide; bracts of spathes 20-30 mm. long, green or purplish, slightly scarious-margined, the two sometimes nearly equal but usually subequal, the outer sometimes 15 mm. longer; flowers 3-6, 10-14 mm. long, deep violet-blue; capsule glabrate or nearly so at maturity.---Moist or wet meadows and parks. Montana to Idaho, south to New Mexico and Nevada. Scattered over Colorado but apparently uncommon, at 3500-8000 feet.
4. Sisyrinchium demissum Greene, Pitt. 2:69. 1890.
Stems 15-30 cm. tall, glaucous or glaucescent, slender, often twisted and flexuously curved, narrowly margined; leaves about 1/2 the height of stem, 1-2.5 mm. wide; bracts of spathe 12-20 mm. long, equal or nearly so, scarious-margined and sometimes ciliate above; flowers pale violet-blue, segments 7-10 mm. long, tapering-aristulate or retuse; ovary slightly pubescent but capsule glabrate, 4-5 mm. long.---Moist meadows. Kansas to Arizona. Our two records from southcentral Colorado at 7600-7700 feet.
5. Sisyrinchium halophilum Greene, Pitt. 4:34. 1899.
Stems 10-30 cm. tall, to 2 mm. wide, erect and stiff or nearly so, margined to slenderly winged; leaves 1/2 length of stem or more, 1-3 mm. wide; spathe with the 2 bracts equal or nearly so, outer 12-20 mm. long; flowers 4-8; perianth to 10 mm. long, violet-blue, segments abruptly acuminate; capsule scabrous-puberulent, 2-3.5 mm. long, subglobose.---Moist meadows. Wyoming to Idaho, south to Colorado and Nevada. Two collections, from Moffat and Gunnison Counties at 6500-9000 feet seem to be closest to this species.

Family 27. Orchidaceae ORCHID FAMILY

Perennial herbaceous plants, some without chlorophyll; stems from bulbs, corms or thickened roots; flowers perfect, irregular, solitary, racemose or spiked; perianth of 6 segments, the 3 outer sepallike and similar or nearly so (2 may be united), the three inner with 2 lateral ones similar but lower modified to form a lip, this sometimes saccate or spurred; stamens apparently 1 or 2, united with the style to form a column, pollen grains granular or united into masses (pollinia); ovary inferior, 1-celled, 3 parietal placentae; fruit a 3-valved capsule; seeds minute, very numerous.

1. Plants without green leaves   -1. Coralorhiza
1. Plants with green leaves (these sometimes withered before anthesis)
  2. Lip a large inflated sac (over 1 cm. long - usually over 2 cm.)
    3. Leaves 2 or more; roots coarsely fibrous; fertile anthers 2   -2. Cypripedium
    3. Leaf 1; stems from a corm; fertile anther 1   -3. Calypso

2. Lip not saclike or if so then small (not over 1 cm. long)
    4. Flowers with a distinct spur projecting downward from the base of the lip   -4. Habenaria
    4. Flowers without a spur (the lip may be somewhat saccate at base)
        5. Leaves 1 to a stem   -5. Malaxis
        5. Leaves 2 or more to a stem
            6. Leaves only 2, near middle of stem and apparently opposite   -6. Listera
            6. Leaves over 2, alternate on the stem or crowded near base
                7. Leaves basal (or sometimes appearing to be borne on the stem because of the elongated bases of the petioles), sometimes with conspicuous whitish veins   -7. Goodyera
                7. Leaves cauline, without whitish veins
                    8. Flowers 1-2 cm. long, green or purplish; inflorescence lax, not densely spiraled; sepals and petals separate, spreading   -8. Epipactis
                    8. Flowers less than 1 cm. long, white or ochroleucous; inflorescence a dense spiral; sepals and petals more or less united or connivent   -9. Spiranthes

        1. Corallorhiza (Haller) Chat. CORAL ROOT

Plants lacking chlorophyll, with coralloid roots; leaves reduced to scales; inflorescence a spikelike raceme with minute bracts subtending the flowers; sepals equal or nearly so, the lateral ones often united at base to form a short spur which is adnate to the ovary; petals similar to sepals but usually smaller; lip entire, toothed or lobed; anthers solitary. Species of this genus can be identified best in fresh condition. The flower parts are rather fleshy and tend to shrink in size on drying. The color may also be changed considerably. Albino individuals are not uncommon in some species.

1. Lip with 2 small teeth near base, lip about 5 mm. long; sepals and petals 1-nerved; stems slender usually less than 3 mm. wide   -1. C. trifida
1. Lip with 2 large teeth near base, (the whole 3-lobed) or with entire denticulate or undulate margins, the lip over 5 mm. long; sepals and petals 3-nerved; stems rather stout (often over 3 mm. wide)
    2. Lip 3-lobed because of 2 large teeth near base, usually spotted; spur adnate but conspicuous   -2. C. maculata
    2. Lip not at all lobed, the margin entire or rather evenly denticulate, crenulate or undulate, spotted or not spotted; spur absent or inconspicuous
        3. Lip purple, not spotted, 6-15 mm. long, striate-veined
            4. Sepals and petals 10-15 mm. long   -3. C. striata
            4. Sepals and petals 6-8 mm. long   -3A. C. striata vreelandii
        3. Lip whitish with purplish spots, 4-6 mm. long, not striate-nerved   -4. C. wisteriana

1. Corallorhiza trifida (L.) Chat., Spec. Inaug. Corallorrhiza 8. 1760.
    C. corallorrhiza (L.) Karst---Plants 6-25 cm. tall, slender; stems yellowish; raceme 2-7 cm. long, 4- to 15-flowered; sepals and petals about 6 mm. long, 1 mm. wide, linear, dull purple, 1-nerved; lip about 5 mm. long, whitish, not spotted, oblong or elliptic, 2-lobed or 2-toothed near the base (in ours minutely puberulent); spur very small, adnate to the ovary, often not noticeable at all.---Woods. Nova Scotia to Alaska, south to Georgia and Colorado; Europe. In the mountains of Colorado at 9000-10,500 feet.
2. Corallorhiza maculata Raf., Am. Mo. Mag. 2:119. 1817.
    C. multiflora Nutt.---Plants 20-50 cm. tall, stout; stem purplish (or yellowish in albino forms), with 3 or more reduced scalelike leaves; raceme 5-20 cm. long, 10- to 30-flowered; flowers mainly brownish-purple; sepals and petals 5-9 mm. long and 1.5-3 mm. wide, linear or linear-lanceolate, 3-nerved; lip 6-8 mm. long, 3-lobed, crenulate, 3-nerved, usually spotted with purple or magenta on a whitish background; spur manifest, adnate to the ovary.---Woods. Nova Scotia to Alaska, south to Florida, New Mexico and California. In mountains of Colorado at 6000-9500 feet.
3. Corallorhiza striata Lindl., Gen. & Sp. Orch. 534. 1840.
    C. ochroleuca Rydb.---Scapes 15-50 cm. tall, rather stout; stems purple (or yellowish in albino forms); raceme 5-15 cm. long, 10- to 30-flowered; flowers purplish with dark veins; sepals and petals 2-3 mm. wide, 10-15 mm. long, 3-nerved, narrowly elliptic or linear-lanceolate; lip ovate, elliptic or almost oblong, entire or slightly undulate, somewhat concave and fleshy, about equalling the petals, fleshy short lamellae present near base on upper surface; spur none but petals and sepals somewhat gibbous at base.---Woods. Quebec to British Columbia south to Michigan, Arizona and California. Our few records from the mountains of Colorado at 6500-10,000 feet.
3A. Corallorhiza striata vreelandii (Rydb.) L. Wm., (var.) Ann. Mo. Bot. Gard. 21:343. 1934.
    Doubtfully distinct from C. striata supposed to differ from it as follows: Petals and sepals 10-15 mm. long - C. striata; Petals and sepals 6-8 mm. long - C. striata vreelandii.
4. Corallorhiza wisteriana Conrad, Journ. Acad. Nat. Sci. Phila. 6:145. 1829.
    Plants 10-20 cm. tall, sometimes taller; stems yellowish to purple; raceme 2.5-10 cm. long, rather few-flowered; flowers greenish-yellow to purplish-brown; sepals and petals about 5-9 mm. long, 1-2.5 mm. wide, 1- to 3-nerved; lip 5.5-7 mm. long, margins erose or undulate but not toothed, white but spotted with purple or magenta; spur small. Our plants are difficult to tell from C. maculata Raf. since the shape of the lip is variable and often somewhat lobed. However, the plants are shorter (usually less than 20 cm. tall) and come into blossom earlier (before June).---Woods and meadows. Pennsylvania to Idaho, south to Florida and Arizona. Reported from southwestern Colorado but our record from the northcentral part at 7000 feet.

2. Cypripedium L. LADY'S SLIPPER

Stems from a cluster of thickened, coarse roots, leafy with flat several-nerved large leaves sheathing somewhat at base; flowers solitary or few in a terminal raceme; one of petals an inflated lip; sepals spreading, 2 lateral ones often united below the lip; stamens 3, the upper sterile and somewhat petaloid, covering the summit of the style; stigma broad, obscurely 3-lobed.

1. Leaves 2, opposite or nearly so; flowers dark purple -1. C. fasciculatum
1. Leaves 3 or more, alternate; flowers not purple -2. C. calceolus

1. Cypripedium fasciculatum Kellogg ex S. Wats., in Proc. Am. Acad. 17:380. 1882.
C. knightae A. Nels.---Plants 6-18 cm. tall; stems more or less villous, bearing a single pair of opposite leaves; leaves 4-9 cm. long, broadly ovate to elliptical-lanceolate; flowers 2-3 in a cluster, dark purple; lower sepals united almost to the tips, ovate-lanceolate; petals similar but shorter and a little broader; lip 10-15 mm. long.---Open woods and hillsides. Wyoming, Colorado and Utah, probably to California. Our records from northcentral and northwestern Colorado at 8500-10,000 feet.
2. Cypripedium calceolus L., Sp. Pl. 951. 1753.
C. parviflorum Salisb.; C. pubescens Willd.; C. veganum Ckll.---Plants 20-40 cm. tall, pubescent; leaves 8-15 cm. long, elliptic-lanceolate, acuminate, glandular-pubescent; flowers solitary; sepals about 3-4 cm. long, usually shorter than the lip; petals 4-5.5 cm. long, the lip 3-4 cm. long, yellow, as deep or deeper than broad. Our plants belong to var. pubescens (Willd.) Correll.---Bogs, meadows and deciduous woods. A Eurasian species, the variety from Newfoundland to British Columbia, south to Georgia and Arizona. Our records from northcentral, central and southcentral Colorado at 7000-9000 feet.

3. Calypso Salisb. CALYPSO

Low herbaceous plants with a solid corm; scapes 1-flowered and sheathed by several loose scales, bearing 1 basal leaf; sepals and petals similar; lip large, saccate, 2-parted below near apex; column petaloid, bearing the liplike anther just below the apex; pollinia 2, the lower pair smaller.

1. Calypso bulbosa (L.) Oakes in Thompson, Cat. Vermont Pl. 28. 1842.
Cytherea bulbosa (L.) House---Stems 5-20 cm. tall, smooth, sheathing bracts membranous; leaf usually appearing later than flower, broadly ovate or oval and sometimes cordate, petioled; sepals and petals 10-15 mm. long, lip 2 cm. long, flower rose-colored, variegated with purple and yellow.---Woods. Labrador to Alaska, south to New England, Michigan, Arizona and California. Scattered in the mountains of Colorado at 7000-10,000 feet.

4. Habenaria Willd. BOG ORCHID

Plants with 1 or more leaves from tuberous-thickened roots; flowers small, in elongated, bracted, spikelike racemes; sepals spreading; upper petals erect, the lower with a spreading or drooping lip and with a tubular spur at the base; glands (to which pollen masses are attached) exposed and separate. The species of this genus, perhaps due to extensive hybridization, are difficult to identify. Many specimens will fall midway between the categories of the species key.

1. Stems with but 1 leaf, this basal, stems without bracts; spikes 3-7 cm. long -1. H. obtusata
1. Stems with 3 or more leaves, some cauline (may not be green at flowering time or reduced to bracts); spikes usually over 7 cm. long
    2. Leaves clustered at or near base of stem, usually withering at or before flowering; stems with scalelike bracts; sepals 1-nerved -2. H. unalascensis
    2. Leaves scattered on stem (sometimes mostly near base) usually green at flowering time; stems with leaves or foliaceous bracts; sepals 3-nerved
        3. Bracts leaflike, the lower twice or more than twice the length of the flowers; lip 2- or 3-lobed at the apex -3. H. bracteata
        3. Bracts much less than twice as long as the flowers; lip entire
            4. Lip rhombic-lanceolate, conspicuously and abruptly dilated near the base; flowers white, rarely greenish -4. H. dilatata
            4. Lip linear to lanceolate, not dilated at base; flowers greenish or purplish
                5. Lip lanceolate to elliptic; raceme usually densely flowered and short (seldom over 10 cm. long); spur about equalling the lip -5. H. hyperborea
                5. Lip linear or oblong-linear to lanceolate; raceme usually laxly flowered and elongated (usually over 10 cm. long); spur usually shorter or longer than the lip
                    6. Spur saccate at base, clavate, shorter than the lip -6. H. saccata
                    6. Spur slender, not clavate, usually exceeding the lip -7. H. sparsiflora

1. Habenaria obtusata (Banks ex Pursh) Richards, Frankl. Narr. Journ. App. 750. 1823.
Lysiella obtusata (Pursh) Rydb.---Stems 10-25 cm. tall, slender, scapiform with a single leaf at the base, this 4-12 cm. long, obovate or oblanceolate; inflorescence loosely flowered, 3-7 cm. long; flowers greenish-yellow, upper sepals broad and rounded, lateral sepals and petals lanceolate-oblong, the lip about 6 mm. long, entire, linear-lanceolate, deflexed, about equal to somewhat shorter than the spur.---Damp or wet shaded places. Newfoundland to Alaska, south to

New York and Colorado. Our plants scattered in the mountains, (except southcentral and northwestern parts) at 8500-10,500 feet.
2. Habenaria unalaschensis (Spreng.) S. Wats., Proc. Am. Acad. 12:277. 1877.
    Piperia unalaschensis (Spreng.) Rydb.---Stems 30-50 cm. tall, slender, erect; leaves at or near the base only, these usually oblanceolate, obtuse or acute, usually withering before anthesis; stem leaves bractlike, small, alternate; inflorescence variable, from few to many flowers in a long inflorescence (ours long and lax); bracts shorter than the flowers; flowers greenish or purplish-green; sepals and petals about 2-4 mm. long, lanceolate, 1-nerved; lip oblong, obtuse, sub-hastate near base; spur filiform or slightly clavate, about equalling lip.---Damp woods. Montana to Alaska, south to California and Colorado. Our few records from western and northcentral Colorado at about 8000 feet.
3. Habenaria bracteata (Willd.) R. Br. in Ait., Hort. Kew ed. 2, 5:192. 1813.
    H. viridis var. bracteata (Muhl.) Gray; Coeloglossum bracteatum (Willd.) Parl.---Stems 15-60 cm. tall, leafy even up through inflorescence; leaves 4-12 cm. long, up to 5 or 6 to a stem, the lower oblanceolate, obovate, upper oblong-lanceolate to lanceolate, acute but lower often obtuse; bracts linear-lanceolate, longer than the flowers (except above), the lower commonly 2 or more times as long; flowers greenish, the spur rather whitish; sepals spreading; petals narrow, with an oblong or somewhat spatulate lip about 6-8 mm. long, 2- or 3-toothed at the apex; spur about 1/2 as long as the lip or less.---Moist or wet, usually shaded ground. New Brunswick to British Columbia, south to North Carolina and Arizona; apparently also in Europe. Scattered in the mountains of Colorado at 6500-9500 feet.
4. Habenaria dilatata (Pursh) Hook., Ex. Fl. 2. pl. 95. 1825.
    Limnorchis dilatata (Pursh) Rydb.; L. borealis (Cham.) Rydb.---Stems 10-60 cm. tall, slender, leafy; leaves 4-10 cm. long, several, oblong-lanceolate, obtuse or acute on upper; spike dense or lax; bracts as long or somewhat longer than the flowers; flowers white or occasionally yellowish or greenish; upper sepals ovate, lateral ones narrower; petals lanceolate erect, lip about 7 mm. long, entire, rhombic-lanceolate, conspicuously dilated at base and tapering to apex; spur not clavate, longer or shorter than lip. H. dilatata var. albiflora (Cham.) Correll and H. dilatata var. leucostachys (Lindl.) Ames cannot be maintained for Colorado plants. One collection from North Park has species and both varieties on the same sheet. They were all growing close together and looked alike.---Moist or wet usually shaded ground. Labrador to Alaska, south to New York, Colorado and California. In the mountains of Colorado at 8500-12,000 feet.
5. Habenaria hyperborea (L.) R. Br. in Aiton, Hort. Kew ed. 2, 5:193. 1813.
    Limnorchis viridiflora (Cham.) Rydb.; H. septentrionalis Tidestrom---Stems 10-50 cm. tall, rather stout; leaves 5-15 cm. long, several, lower oblanceolate and obtuse, upper leaves lanceolate and acute; spikes about 5-10 cm. long, rather short and dense; bracts about as long as the flowers; flowers green; upper sepals ovate, shorter than the lanceolate lateral ones; petals erect, lanceolate; lip about 4-5 mm. long, lanceolate, entire not conspicuously dilated at the base; spur about equalling the lip, little if any dilated at end.---Moist or wet, usually shaded soil. Greenland to Alaska, south to New Jersey, Colorado and California. Rather widespread and apparently rather common in the mountains of Colorado at 6000-10,500 feet.
6. Habenaria saccata Greene, Erythea 3:49. 1895.
    Limnorchis stricta (Lindl.) Rydb.; Habenaria neomexicana Tidestrom; L. purpurascens Rydb.; H. purpurascens (Rydb.) Tidestrom---Stems 20-60 cm. tall, leafy; leaves 4-10 cm. long, lanceolate, the lower obtuse; spike usually over 10 cm. long, typically elongated, remotely-flowered; bracts linear lanceolate, the lower longer than the flowers; upper sepals ovate-oblong, lateral oblong-lanceolate, greenish; petals purplish, narrow; lip entire, linear or lanceolate, scarcely if at all dilated at base; spur shorter than the lip (about 1/3 to 2/3 as long), clavate.---Moist or wet, usually shaded ground. Alberta to Alaska, south to New Mexico and Oregon. Scattered in the mountains of Colorado (except the northwest) at 9000-10,500 feet.
7. Habenaria sparsiflora S. Wats., Proc. Am. Acad. 12:276. 1877.
    Limnorchis ensifolia Rydb.; L. sparsiflora (Wats.) Rydb.; L. laxiflora Rydb.---This western and southwestern species has been consistently reported from western and southern Colorado but no specimens could be located by the writer.

5. Malaxis Soland.

Plants from a swollen base; leaves 1; flowers small, in a raceme; sepals spreading; petals linear, lower with a spreading or drooping lip that lacks a spur; pollinia 4.

1. Malaxis monophyllos (L.) Sw., Kongl. Svens. Vetens. Acad. Nya Handl. 21:234. 1800.
    Stems 5-30 cm. tall; leaves ovate, suborbicular-oval to narrower; sepals 2-3 mm. long, 1-nerved; petals 2-2.5 mm. long, the lip enlarged at base, triangular-ovate in outline. Our plant is the var. brachypoda (A. Gray) Morris and Eames.---Moist ground. Newfoundland to Alaska, south to Pennsylvania, Minnesota and California. Our one record from Boulder County.

6. Listera R. Br. TWAYBLADE; LISTERA

Plants with rootstocks and fibrous-fleshy roots; leaves 2, opposite, near the middle of the stem; inflorescence a terminal loose raceme; flowers small, greenish or purplish; sepals and petals similar in color; lip notched or cleft at the apex, often with a tooth near base or auriculate, longer than the rest of the perianth; spur none; anther one, pollinia 2.

1. Lip cleft about 1/2 way to the base, the whole linear -1. L. cordata
1. Lip rarely notched at apex, cuneate or oblong in shape
   2. Lip auriculate at base, 5-6 mm. long; leaves 1.5-3.5 mm. long -2. L. borealis
   2. Lip not auriculate (may have 2 teeth near base), usually over 6 mm. long; leaves 3-7 mm. long
-3. L. convallarioides

1. Listera cordata (L.) R. Br. in Ait., Hort. Kew. ed 2, 5:201. 1813.
    L. nephrophylla Rydb.; Ophrys nephrophylla Rydb.---Stems 10-20 cm. long, slender, glabrous or slightly pubescent above the leaves; leaves about 15-30 mm. long, cordate to rounded-reniform, mucronate; lip 4-5 mm. long, narrow, cleft at apex about 1/2 way to base or more, the segments linear, a short tooth present on margin of lip near base.---Moist woods. Montana to Alaska, south to New Mexico and Oregon. Probably scattered in the mountains, our records from northcentral and southwestern Colorado at 9000-11,500 feet.
2. Listera borealis Morong, Bull. Torr. Club 20:31. 1893.
    Ophrys borealis (Morong) Rydb.---This northern species has been reported for Colorado by several workers but no such material could be located by the writer.
3. Listera convallarioides (Sw.) Nutt., Gen. N. Am. Pl. 2:191. 1818.
    Ophrys convallarioides (Sw.) Wight---Stems 10-25 cm. long, slender, glandular-pubescent above the leaves; leaves 3-6 cm. long, broadly oval or ovate to orbicular, obtuse or abruptly acute; lip about 9 mm. long, narrowly cuneate, 2-lobed at the apex, often with a blunt spreading tooth on each side near the base, but not auriculate.---Moist woods. Newfoundland to Alaska, south to New England, Arizona and California. Our few records from central and northcentral Colorado at 8000-10,500 feet but probably anywhere in the mountains.

7. Goodyera R. Br. RATTLESNAKE PLANTAIN

Stems from creeping rootstocks and thick roots; leaves alternate, basal or nearly so, often white-reticulated along the veins; inflorescence a bracteate, spicate raceme or spike; flowers small; dorsal sepal and petal connate into a hood over the lip; lip saccate or somewhat so, entire; no spur present; anther one; pollinia 2, 1 in each cell.

1. Flowers in 1-sided raceme; lip deeply saccate, the margins recurved or flaring; leaves 1-2 cm. long
-1. G. repens
1. Flowers in a loose spiral; lip not really saccate, the margin not recurved or flaring; leaves 3.5 cm. long
-2. G. oblongifolia

1. Goodyera repens (L.) R. Br. in Ait., Hort. Kew ed. 2. 5:198. 1813.
    Goodyera repens var. ophioides Fernald; Peramium ophioides (Fernald) Rydb.---Scapes 10-20 cm. tall, slender, glandular-puberulent; leaves 1-2 cm. long, ovate, about 5-nerved, dark green with white markings especially along the cross-veins, contracted to a short, winged petiole; flowers about 4-5 mm. long, in a 1-sided raceme, greenish-white; lip saccate at base with an elongated, recurved tip and recurved margins.---Nova Scotia to Alaska, south to South Dakota and New Mexico. Our few records from southwestern and central Colorado at 8500-9500 feet.
2. Goodyera oblongifolia Raf.; Herb. Raf. 76. 1833.
    G. decipiens (Hook.) Hubbard; Peramium menziesii (Lindl.) Morong; P. decipiens (Hook.) Piper ---Scapes 10-40 cm. tall. pubescent; leaves ovate-lanceolate to oblong, often reticulated with white, acute at apex, cuneate at base to a narrowly winged, short petiole; spikes many-flowered, not or only slightly secund, bracts small; flowers 6-9 mm. long, the whitish lip swollen at base with a long narrow, recurved or spreading apex, but not saccate with recurved margins.---Woods. Nova Scotia to Alaska, south to New Hampshire, Arizona and California. Our few records scattered over the mountainous part of Colorado at 6000-9500 feet.

8. Epipactis L. C. Rich HELLEBORINE

Simple-stemmed caulescent herbaceous plants with creeping rootstocks; leaves alternate, sheathing; inflorescence a few-flowered, leafy-bracted raceme; sepals and petals distinct, rather similar; lip constricted near the middle, the upper portion dilated; spur none; anther 1, behind the truncate stigma.

1. Epipactis gigantea Dougl. ex Hook., Fl. Bor. Am. 2:202. 1839.
    Serapias gigantea (Dougl.) A. A. Eat.---Plants 20-100 cm. tall, sparsely pubescent; leaves 5-20 cm. long, from ovate to narrowly lanceolate, scabrous on the veins below; flowers greenish or purplish from the veins; lip with saccatelike base this with short, erect lobes, strongly-nerved, these with callous tubercles near base.---Woods and moist situations. Montana to British Columbia, south to Texas, Arizona and California. Our few records from central and western Colorado at 5500-8000 feet.

9. Spiranthes Rich. LADIES TRESSES

Erect, caulescent plants from a cluster or tuberous roots; leaves alternate, mostly basal but some cauline; inflorescence of terminal spikes, these in 1-3 rows, flowers spiral around the rachis; sepals and petals narrow, more or less connivent; anthers 1; lip oblong or pandurate, often with protuberances or callosities near base on upper surface; spur none.

1. Lip narrowed above the middle, dilated at apex, panduriform; protuberances at base minute or obsolete
 -1. S. romanzoffiana
1. Lip not contracted at middle nor dilated near apex; protuberances at base prominent, nipplelike
 -2. S. porrifolia

1. Spiranthes romanzoffiana Cham. & Schl., Linnaea 3:32. 1828.
   S. stricta (Rydb.) A. Nels.; Ibidium strictum (Rydb.) House---Stems 15-45 cm. tall, glabrous or sparingly pubescent; leaves from oblong-lanceolate to linear-lanceolate; spike dense, flowers about 3-ranked, conspicuously bracteate; lip recurved, contracted below the wavy-crenulate, rounded apex and somewhat pandurate, callosities near base of lip obsolete or obscure.---Moist or wet ground. Newfoundland to Alaska south to New York, New Mexico and California. Apparently rather common, our records from the mountains of Colorado (except northwestern part) at 5500-10,500 feet.
2. Spiranthes porrifolia Lindl., Gen. & Sp. Orchid. 467. 1840.
   S. romanzoffiana var. porrifolia (Lindl.) Ames & Correll; Ibidium porrifolium (Lindl.) Rydb.---This eastern species has been reported for Colorado but no specimens were located by the writer.
   Two additional species have been recorded from adjacent states and may someday be located in Colorado. They are S. cernua (L.) L. C. Rich. and S. vernalis Engelm.

Subclass 2. Dicotyledoneae

Plants woody or herbaceous. Leaves mostly netted-veined with 1 or more main veins and smaller lateral ones but this sometimes obscure, entire, toothed, lobed or compound. Vascular bundles arranged in a ring. Parts of the flower usually in 4's or 5's or multiples thereof. Embryo with 2 cotyledons.

Family 28. Salicaceae WILLOW FAMILY

Trees and shrubs; leaves simple, deciduous, alternate, stipulate; dioecious; flowers in aments, solitary in the axils of scalelike bractlets; perianth absent; stamens 1 to many; ovary superior, solitary, 1-celled with 2-4 parietal placentae; stigmas mostly 2 but each sometimes bifid; fruit a 2-valved capsule; seeds numerous, bearing apical tufts of hair.

1. Winter buds covered by several scales, these buds are at least partly developed most of the summer; bractlets of the aments laciniate or fimbriate; flowers borne on broad or cup-shaped disks; catkins pendulous; trees
 -1. Populus
1. Winter buds with only 1 scale; bractlets of the aments entire or merely dentate; flowers without disks; catkins upright; mostly shrubs (only a few species becoming trees)
 -2. Salix

1. Populus L. COTTONWOOD; POPLAR; ASPEN

Trees with furrowed bark and more or less resinous buds with more than 1 scale; leaves varying in shape, petioled; catkins long and drooping, appearing before the leaves; bractlets lobed or fimbriate; flowers with a broad or cup-shaped disk; stamens usually numerous. In addition to the species listed, P. alba L., a cultivated tree, may persist or escape. The writer has no record of this happening in Colorado, however.

1. Petioles definitely flattened laterally
    2. Leaves suborbicular; bark remaining smooth, whitish or cream-colored; stigmas 2, filiform
 -1. P. tremuloides
    2. Leaves not suborbicular; bark not whitish, furrowed on older trunks; stigmas usually 3 or 4 and dilated
        3. Leaves oblong-ovate, margins finely crenate-serrate and bases rounded or cuneate; petioles only slightly flattened laterally
 -2. P. andrewsii
        3. Leaves deltoid or deltoid-ovate, margins coarsely toothed, bases truncate or cordate; petioles definitely flattened laterally
            4. Pedicels as long or longer than the capsules; leaves with not more than 10 teeth on each side; no glands present at top of petioles
 -3. P. wislizeni
            4. Pedicels shorter than the capsules; leaves with more than 10 teeth on each side; glands usually present at top of petioles
 -4. P. sargentii
1. Petioles terete or slightly flattened on upper surface
    5. Leaves whitish below, the veins conspicuously reticulated
        6. Leaves broadly ovate, at least 2/3 as broad as long; young twigs, petioles and veins of leaf below pubescent; base of leaf cordate or truncate
 -5. P. candicans
        6. Leaves ovate to narrower, rarely over 2/3 as wide as long; twigs, petioles and veins of leaf glabrous or sparingly puberulent; base of leaf rounded or cuneate, not usually cordate or truncate
 -6. P. balsamifera
    5. Leaves green below (may be somewhat lighter green than above)
        7. Leaves lanceolate to ovate-lanceolate; petioles short, not over 1/3 the length of the blade
 -7. P. angustifolia
        7. Leaves rhombic-lanceolate, rhombic-ovate to oblong-ovate; petioles more than 1/3 as long as the blades (usually over 1/2 as long)
            8. Leaves rhombic-lanceolate or rhombic-ovate; petioles not at all flattened laterally
 -8. P. acuminata
            8. Leaves oblong-ovate; petiole often somewhat flattened latterally
 -2. P. andrewsii

1. Populus tremuloides Michx., Fl. Bor. Am. 2:243. 1803.

Trees small to medium sized, up to 30 m. tall; bark smooth, green-white or cream-colored, often marked by dark wartlike protuberances, becoming furrowed at base of old trunks; twigs slender, red-brown to gray; buds sharp-pointed, slightly if at all resinous; leaves semi-orbicular or broadly ovate, apex abruptly acute or acuminate, base rounded or subcordate, crenate-serrate, rather irregularly so, green and glabrous above, dull green or pale below, glabrous, often turning golden-yellow or reddish in autumn; petioles 3-7 cm. long, flattened laterally, sometimes with glands at base of blade and top of petiole; stigmas 2, filiform. Our Colorado tree has been called P. tremuloides var. aurea (Tide.) Daniels, supposed to have more intense autumnal coloration than the species proper. But separate individuals vary from green, yellow or reddish in the autumn and the variety is of doubtful value.---Mountain slopes and valleys. Labrador to Alaska, south to New Jersey, Mexico and Arizona. Our records scattered over the mountainous part of Colorado at 6000-10,000 feet.

2. X. Populus andrewsii Sarg., Trees & Shrubs. 2:212. 1913.

P. sargentii Dode X P. acuminata Rydb.---Trees with light orange-brown or yellow-brown twigs; leaves 9-10 cm. long, usually yellow-green, not very reticulated below, glabrous, oblong-ovate, finely crenate, rounded or cuneate at base; petioles terete or somewhat flattened laterally.---This tree, supposed to be of hybrid origin has been reported to grow native near Boulder and Walsenburg and sometimes appears in cultivation.

3. Populus wislizeni (S. Wats.) Sarg., Silva N. Amer. 14:71. 1902.

Trees to 30 m. tall; bark thick, pale gray-brown, becoming furrowed on old trunks; twigs stout, yellowish or orange-brown, glabrous; buds puberulent, resinous, not aromatic; leaves broadly deltoid-ovate, apex abruptly short-acuminate or sometimes long-acuminate, base truncate or cordate, coarsely and irregularly crenate-serrate with usually less than 10 teeth on each side, yellow-green, lustrous and glabrous; petioles 3-5 cm. long, flattened laterally, glabrous; pedicels recorded as longer than the capsules (as long or even somewhat shorter in most of our Colorado specimens).---Valleys and river banks. Texas to Colorado, south to Mexico. Our records from southern and western Colorado (where it extends north to Moffat County) at 4000-7000 feet.

4. Populus sargentii Dode, Ex. Mon. Ined. du Gen. Populus 40,fig. 46. 1905.

P. occidentalis (Rydb.) Britton of Manuals---Trees to 30 m. tall; bark gray and smooth at first, becoming thick and deeply furrowed; twigs rather stout, glabrous and yellowish; buds resinous, puberulent, not aromatic; leaves broadly deltoid, usually slightly longer than broad, usually abruptly contracted into long slender acuminate tips, base truncate or slightly cordate, coarsely crenate-serrate, glabrous, light green or yellow-green; petioles almost as long as the blade, laterally flattened, 2 small glands usually present at junction of blade and petiole; pedicels shorter than the capsules.---River bottoms and plains. Saskatchewan and Alberta, south to Texas and New Mexico. Our records from the eastern half of Colorado at 3500-6500 feet.

5. Populus candicans Ait., Hort. Kew. 3:406. 1789.

Trees to 30 m. tall; young branches pubescent; buds large, viscid; leaves broadly ovate, at least 2/3 as broad as long, acute or acuminate, cordate or truncate at base, coarsely crenate-serrate, dark green and pubescent above, whitish and sparingly pubescent below, more densely so on the veins; petioles 3-6 cm. long, terete, pubescent or ciliate.---River banks and valleys. Newfoundland to Alaska, south to New Jersey and Alberta. It has been reported from this state but possibly on the basis of a tree escaped from cultivation. We have several records from Colorado but probably all these are from cultivated plants.

6. Populus balsamifera L., Sp. Pl. 1034. 1753.

P. tacamahacca Mill.---Trees to 30 m. tall; young bark smooth, green-brown to red-brown, becoming deeply furrowed; twigs glabrous or pubescent; buds to 2.5 cm. long, long-pointed, resinous and fragrant; leaves ovate to ovate-lanceolate, rather thick and firm, acute or short-acuminate, rounded or broad cuneate, rarely cordate at base, finely crenate-serrate, lustrous dark green and glabrous above, much paler, whitish, with conspicuous reticulated veins below; petioles 2-7 cm. long, terete, glabrous or rarely pubescent; capsules ovoid. The correct name for this plant has caused a great deal of controversy and the issue has not been settled. The writer prefers to use the older and established name.---Along streams and in canyons. Newfoundland to Alaska, south to New York and Nevada. Our records from central and northcentral Colorado at 6000-12,000 feet but some of them are probably from cultivated trees. However, the plants from northcentral part of the state appear native.

7. Populus angustifolia James in Longs, Exp. 1:497. 1823.

Pyramidal trees, medium sized, to 20 m. tall with ascending branches; bark smooth except near base where it is in furrows and ridges; twigs glabrous, yellow-green or orange when young; buds 1-2 cm. long, slender, long-pointed, very resinous and somewhat aromatic; leaves rather variable, mostly narrowly lanceolate to ovate-lanceolate, rarely elliptic or obovate, acute, acuminate, or even obtuse, cuneate or rounded at base; glandular-serrate or finely crenate with margins somewhat revolute, green above, paler green below, glabrous; petioles short, less than 1/3 the length of the blades, terete or somewhat flattened on upper side; aments 2-6 cm. long, densely flowered, glabrous; capsules broadly ovoid.---Along streams. Saskatchewan and Alberta, south to Nebraska, Mexico and Arizona. Our records from lower elevations of the mountainous part of Colorado at 5000-8000 feet.

8. Populus acuminata Rydb., Bull. Torr. Club 20:50. 1893.

Trees to 20 m. high, with round top and spreading or ascending branches; bark smooth and whitish when young, later ridged; twigs slender, often 4-angled, yellow-brown; buds 1-2 cm. long, glabrous, somewhat resinous but not aromatic; leaves rhombic-lanceolate to rhombic-ovate, acuminate, usually cuneate at base, crenate-serrate, rather thick and leathery, green and glabrous both sides; petioles 2-4 cm. long, about 1/2 to 2/3 length of blade, terete, slender; aments mostly 3-5 cm. long, slender, pistillate elongating to 15 cm.; ovaries broad-ovoid or oblong-ovoid on short pedicels.---River banks and valleys. Saskatchewan, south to Texas and -Arizona. Scattered at lower elevations over the mountainous two-thirds of Colorado at 4500-8500 feet.

2. Salix L. WILLOW

(Contributed by E. C. Smith)

Trees or shrubs (in Colorado mostly shrubs); buds with only 1 scale; aments mostly erect, in bloom before, with, or after the appearance of the leaves; bracts of the aments (scales) entire or merely denticulate; stamens 2-10, mostly 2 or 5, anthers yellow or red; ovaries and fruits glabrous or hairy; styles usually entire, rarely notched or cleft; stigmas entire, notched or cleft; glands at base of each flower 1 or 2, at base of pedicel sometimes in pistillate aments. Three specific keys are given here, staminate, pistillate and vegetative.

The European basket willow, Salix purpurea L., not included in the keys, is well established along a creek near Colorado Springs. It is easily recognized by its small oblanceolate or strap-shaped, nearly opposite leaves and its slender cylindrical aments with purple-tipped scales.

I. Key to Plants with Staminate Aments

(Where leaves are used as key characters they are those which accompany the aments and so, often immature; where anther color is used it refers to the color before dehiscence unless otherwise specified.)

1. Scales of aments pale yellow, ochroleucous, rarely pale green; stamens 2 or 5
  2. Stamens 5; trees or shrubs
    3. Trees, 10-15 m. tall; aments slender, 4-6 (8) cm. long, the scales well separated   -1. S. amygdaloides
    3. Erect shrubs 2-5(7) m. tall, individuals of certain species attaining tree size under special conditions; aments 1-4 cm. long, dense
      4. Aments 2-4 cm. long
        5. Leaves glaucous beneath; small trees   -2. S. lasiandra
        5. Leaves not glaucous, only slightly paler beneath; shrubs
          6. Leaves lanceolate, glabrous from first appearance, glands not very prominent; plants common in Colorado   -3. S. caudata and var.
          6. Leaves often ovate-lanceolate, pubescent while young, glands very prominent, especially on the rather persistent stipules, plants rare in Colorado   -4. S. lucida
      4. Aments 1-2 cm. long   -5. S. serissima
  2. Stamens 2; shrubs 2-4 (7) m. tall
    7. Aments 2-4 cm. long, terminating lateral leafy branchlets 1-3 cm. long in early season, up to 10-15 cm. in later season; leaves very narrow, 6-10 times longer than broad
      8. Leaves (with fully developed aments) pubescent above and beneath   -7. S. exigua and vars.
      8. At least some of the leaves glabrous or glabrescent at the above mentioned stage on the upper surface
        9. Only a few leaves glabrous or nearly so   -8. S. interior and var.
        9. All leaves glabrous on the upper surface   -9. S. melanopsis tenerrima
    7. Aments 1-2 cm. long, on very short branchlets; leaves definitely less than 6 times longer than wide   -27. S. bebbiana and var.
1. Scales of aments darker, at least as to the tips; stamens 2
  10. Erect shrubs, 0.5-5 (7) m. tall; plants rarely alpine
    11. Plants 2.5-5 (7) m. tall
      12. Twigs usually pruinose
        13. Aments rarely over 1 cm. long, subsessile, small leaves at base; scales with light-colored base and darker tip, red or brown   -29. S. geyeriana and var.
        13. Aments 1-3 cm. long, sessile, naked at base; scales brown to black
          14. Aments 1-2 cm. long; anthers yellow, but often with red tips before dehiscence; plants mostly below 7500 feet altitude   -10. S. irrorata
          14. Aments 1-3 cm. long; anthers red before dehiscence; plants mostly at or above 7500 feet altitude   -24. S. subcoerulea
      12. Twigs not pruinose
        15. Aments sessile, usually without leaves or bracts at base
          16. Aments rarely more than twice longer than broad (sometimes leaves must be waited for, but often immature leaves are present at tips of sterile branches)
            17. Anthers nearly or quite 1 mm. long, yellow; leaves rarely present   -26. S. scouleriana
            17. Anthers 0.5 mm. long, red to violaceous, drying dark; leaves present or absent
              18. Aments ovoid, 1-2 cm. long, 1-1.5 cm. wide; leaves usually not present   -25A. S. planifolia monica
              18. Aments elliptic-cylindrical, 1.5-2.5 (3) cm. long, 0.8-1 cm. wide; leaves present at tips of twigs   -25B. S. planifolia nelsonii
          16. Aments at least twice longer than broad, mostly three times
            19. Aments ascending, straight; anthers yellow or only lightly margined with red   -13. S. monticola
            19. Aments wide-spreading, often curved; anthers red or purple, drying violet-gray   -14. S. pseudomonticola
        15. Aments subsessile to short-peduncled, with bracts or leaves at base
          20. Leaves at base of aments small and bractlike
            21. Aments rarely more than 1 cm. long   -28. S. petiolaris
            21. Aments 2-3 cm. long

                22. Twigs yellow, later reddish on 1 side; scales light brown, darker in drying   -11. S. lutea
                22. Twigs dark brown; scales bicolored   -11C. S. lutea ligulifolia
            20. Leaves at base of aments like other leaves only smaller, up to 1 cm. long on peduncles, to 2 cm. on
                sterile twigs
                23. Leaves green beneath, not glaucous   -12. S. pseudocordata
                23. Leaves glaucous beneath
                    24. Aments 2-3 (4) cm. long, subsessile or with peduncles 0.5 cm. long; leaves glabrous or
                        nearly so   -14A. S. pseudomonticola padophylla
                    24. Aments 2-4 (6) cm. long, subsessile to "peduncles 1-1.5 cm."; leaves on peduncles and
                        youngest on sterile twigs clothed with long hairs, vestiges on midrib remaining for some
                        time   -15. S. barclayi
    11. Plants 0.3-1 (2) m. tall
        25. Twigs glabrous
            26. Twigs at anthesis yellow or orange; young leaves silky-villous; anthers yellow   -16. S. wolfii
            26. Twigs at anthesis chestnut or dark-brown; leaves (if present) glabrous, glaucous beneath;
                anthers red   -25A. and 25B. S. planifolia vars.
        25. Twigs and leaves of current season pubescent, villous or tomentose at anthesis
            27. Leaves 2.5-6 (8) times longer than broad
                28. Leaves linear-oblong, revolute, floccose-tomentose, sparingly above, dense beneath, veins
                    impressed in upper surface   -21. S. candida
                28. Leaves not as above, elliptical, lanceolate or oblanceolate, wider on shoots and sprouts
                    29. Anthers not over 0.5 mm. long, subglobose; pubescence of leaves short, subtomentose,
                        the hairs twisted; leaves not glaucous beneath   -17. S. brachycarpa
                    29. Anthers 0.5-0.8 mm. long, oblong-ellipsoid or subglobose; pubescence longer, looser,
                        subpilose, the hairs straight; leaves glaucous beneath
                        30. Scales of aments usually broad, obtuse, dark brown to black, mostly black in
                            drying; pubescence on upper side of leaves tardily vanishing; anthers oblong-
                            ellipsoid   -18. S. pseudolapponum
                        30. Scales of aments usually narrow and acute, light brown or with tips darker than
                            base; leaves quickly glabrate above; anthers subglobose
                                                                                                  -19. S. glauca glabrescens
            27. Leaves less than 2.5 times longer than wide   -20. S. cordifolia
10. Prostrate shrubs with stems on or below the surface of the ground; alpine or subalpine
    31. Scales of aments pilose, brown to black
        32. Aments 1-2 cm. long, many-flowered; leaves 1.5-3 cm. long   -22. S. anglorum antiplasta
        32. Aments 5-10 mm. long, few-flowered; leaves 8-15 mm. long   -23. S. cascadensis
    31. Scales of aments glabrous or nearly so, green
        33. Aments 3-5 mm. long, mostly 3-flowered   -30. S. nivalis
        33. Aments 1-2 cm. long, several to many-flowered   -30A. S. nivalis saximontana

II. Key to Plants with Pistillate Aments

1. Capsules glabrous
    2. Scales of aments pale yellow, ochroleucous (or green in S. fragilis)
        3. Trees 10-15 (20) m. tall; aments 4-8 cm. long
            4. Capsules narrowly ovoid; pedicels 2 mm. long; petioles slender, 5-15 (20) mm. long; plants
               native   -1. S. amygdaloides
            4. Capsules long-conic; pedicels less than 1 mm. long; petioles 5-8 mm. long; introduced street trees
               but well established in several places near cities   -6. S. fragilis
        3. Shrubs or small trees 2-4 (7) m. tall; aments usually less than 6 cm. long
            5. Styles of capsules 0.5 mm. long; leaves lanceolate or broader, serrulate, the teeth gland-tipped
                6. Aments 3-6 (10) cm. long; capsules 5-7 mm. long, straw-colored
                    7. Leaves glaucous beneath   -2. S. lasiandra
                    7. Leaves not glaucous, slightly paler beneath
                        8. Twigs reddish-brown; leaves lanceolate (except on vigorous shoots), glabrous; plants
                           rather common in Colorado   -3. S. caudata and var.
                        8. Twigs usually yellowish; leaves lanceolate to ovate-lanceolate, pubescent when young;
                           plants rare in Colorado   -4. S. lucida
                6. Aments 1.5-3.5 cm. long; capsules 6-8 (9) mm. long, at maturity olive-brown or deep reddish-
                   brown   -5. S. serissima
            5. Styles wanting; leaves narrower, linear to linear-elliptic or linear-oblong, remotely denticulate
               or sometimes entire. (In this section accurate determination requires mature leaves and fruits)
                9. Leaves permanently pubescent on both faces   -7. S. exigua and vars.
                9. Mature leaves glabrous or nearly so, at least on upper surface
                    10. Young leaves pubescent, sometimes silky, mature leaves glabrous or nearly so on both
                        faces, veins on under side prominent   -8. S. interior and var.
                    10. Even young leaves glabrous above, veins lightly impressed   -9. S. melanopsis tenerrima
    2. Scales darker, light to dark brown, at least at tips, or with tips some shade of red
        11. Styles 0.3-0.7 mm.; shrubs 2-4 (6) m. tall
            12. Twigs pruinose; aments sessile   -10. S. irrorata
            12. Twigs not pruinose; aments on short leafy branches
                13. Capsules ovoid-conic; twigs yellow or red on one side; leaves lanceolate to ovate-lanceolate
                                                                                                  -11. S. lutea and vars.
                13. Capsules lanceolate; twigs dark brown; leaves ligulate   -11C. S. lutea ligulifolia
        11. Styles 1-1.5 mm. long; shrubs 0.5-4 (6) m. tall
            14. Shrubs 2-4 (6) m. tall
                15. Aments 2-3 cm. long; styles 1 mm. long; leaves green above and below   -12. S. pseudocordata

15. Aments 3-7 cm. long; styles 1-1.5 mm.; leaves glaucous or subglaucous beneath
  16. Aments sessile or sometimes in fruit short-peduncled
    17. Aments 3-5 cm. long; capsules 5-6 (7) mm. long; pedicels 1.25 mm. long, pilose; leaves on fruiting twigs elliptical, or oblong-elliptical, base mostly rounded   -13. S. monticola
    17. Aments 3-7 (8) cm. long; capsules 6-8 mm. long; pedicels 1.5-2.5 mm.; leaves mostly ovate, base subcordate to cordate or rounded   -14. S. pseudomonticola
  16. Peduncles distinct, varying in length with the season and position on twig, 0.3-2 (3) cm. long
    18. Aments 3-4 cm. long; capsules to 6 mm. long; leaves on fruiting twigs mostly oval, but sometimes broadest above the middle, base mostly rounded; margin crenulate
      -14A. S. pseudomonticola padophylla
    18. Aments 4-6 cm. long; capsules 6-8 mm. long; leaves mostly oval, base cuneate or acute, margin crenulate-serrulate with inflexed teeth, often minutely glandular, sometimes entire
      -15. S. barclayi
14. Shrubs 0.5-1 (2) m. tall   -16. S. wolfii
1. Capsules pubescent
  19. Erect shrubs
    20. Shrubs 0.3-1 (2) m. tall, some primarily alpine and subalpine, but growing also at lower altitudes, others less common above timberline and taller at middle altitudes
      21. Twigs glabrous; aments sessile
        22. Twigs usually strongly pruinose; capsules 3-5 mm. long, silvery-silky -24. S. subcoerulea
        22. Twigs rarely faintly pruinose; capsules gray-pubescent, 4-6 mm. long
          -25. S. planifolia vars.
      21. Twigs of current season pubescent, tomentose or pilose; aments subsessile or on peduncles or leafy twigs
        23. Leaves (in Colorado specimens) oblong-linear to oblong-elliptical, 4-7 times longer than broad, permanently densely floccose-tomentose beneath, glabrescent and veins impressed above, margin subentire, revolute   -21. S. candida
        23. Leaves not as above
          24. Aments 1-2 cm. long; capsules rarely over 5 mm. long, scales light yellowish-brown; apex of leaves obtuse to short-acute, base mostly rounded when mature, rarely broadly cuneate   -17. S. brachycarpa
          24. Aments and capsules longer, scales various; leaves not as above
            25. Aments 1.5-2.5 (3) cm. long; capsules 5-7 mm. long; styles entire, 1 mm. long; scales dark reddish-brown to black, at early stage with lighter colored base
              -18. S. pseudolapponum
            25. Aments 2-4 cm. long; capsules 5-8 (9) mm. long; styles entire to cleft; scales mostly light colored, but sometimes drying black
              26. Styles slender, notched, some cleft to middle or deeper; all stigmas bifid; leaves 2.5-3.5 times longer than broad, elliptical to oval
                -19. S. glauca glabrescens
              26. Styles thicker and shorter, entire; stigmas entire, notched or bifid; leaves 1.5-2 (2.25) times longer than broad, broadly oval or obovate to suborbicular   -20. S. cordifolia
    20. Shrubs 2-4 m. tall, of montane zone
      27. Styles 1-1.5 mm. long; aments sessile to subsessile; capsules not rostrate
        28. Capsules 3-5 mm. long, silvery-pubescent; twigs usually pruinose; leaves appressed-pubescent beneath   -24. S. subcoerulea
        28. Capsules 5-6 mm. long, gray-pubescent; twigs rarely faintly pruinose; leaves glabrous, glaucous beneath   -25. S. planifolia vars.
      27. Styles less than 0.5 mm. or wanting; aments sessile or on peduncles or leafy twigs; capsules rostrate
        29. Aments 2-6 (8) cm. long; capsules 6-8 (9) mm. long
          30. Aments dense; styles distinct, 0.3-0.5 mm. long   -26. S. scouleriana
          30. Aments loose; styles wanting   -27. S. bebbiana and var.
        29. Aments 1-2 cm. long; capsules 3-6 mm. long
          31. Twigs not pruinose; mature leaves serrulate   -28. S. petiolaris
          31. Twigs usually pruinose; leaves entire   -29. S. geyeriana
  19. Prostrate shrubs, stems on or below the surface of the ground, branches rising only 2-6 cm. above the ground
    32. Styles 0.5-1 mm. long; scales of aments brown, thinly pilose; leaves subglaucous to green beneath
      33. Aments 1.5-4 cm. long, many-flowered; capsules 4-6 mm. long; leaves subglaucous beneath, mostly elliptical   -22. S. anglorum antiplasta
      33. Aments 1 cm. long, several-flowered; capsules to 4 mm.; leaves green beneath, ligulate to elliptical   -23. S. cascadensis
    32. Styles wanting; scales of aments green, glabrous or nearly so; leaves intensely glaucous and reticulate beneath
      34. Aments 3- to 6-flowered, 1 cm. or less long; leaves 0.7-1.2 cm. long   -30. S. nivalis
      34. Aments many-flowered, 1-2 cm. long; leaves 1.5-2.5 cm. long   -30A. S. nivalis saximontana

III. Key Based on Vegetative Characters

1. All mature leaves with serrate, dentate or crenate margins
  2. Trees 10-15 (20) m. tall
    3. Leaves finely serrulate, glaucous beneath; twigs slender, yellow when young, grayish in age; plants native   -1. S. amygdaloides
    3. Leaves coarsely serrate, not glaucous beneath; twigs green, stouter, brittle at base; introduced street tree established in several places   -6. S. fragilis

2. Shrubs or small trees, 2.5-4 (6) m. tall
    4. Leaves with glandular-serrate margins and sometimes glands on petioles where these join the blades
       5. Leaves glaucous or subglaucous beneath
          6. Leaves 6-10 cm. long, glaucous beneath; plants of southwestern Colorado   -2. S. lasiandra
          6. Leaves 4-6.5 cm. long, subglaucous beneath; plants of northcentral Colorado   -5. S. serissima
       5. Leaves not glaucous beneath, merely paler below
          7. Twigs reddish-brown; leaves lanceolate (except on vigorous sterile twigs), glabrous, glands moderately prominent; plants rather common in Colorado   -3. S. caudata
          7. Twigs yellowish; leaves lanceolate to ovate-lanceolate, young leaves pubescent beneath, glabrous when mature, glands more prominent, especially on the stipules; plants rare in Colorado   -4. S. lucida
    4. Leaves not gland-margined
       8. Leaves linear, linear-lanceolate or linear-oblong, remotely denticulate (except sometimes in S. exigua)
          9. Leaves permanently pubescent on both faces   -7. S. exigua and vars.
          9. Leaves glabrous at maturity
             10. Young leaves usually pubescent or pilose, the mature glabrous on both faces, veins raised on under side   -8. S. interior and var.
             10. Young leaves glabrous above, faintly pubescent beneath, the mature glabrous on both faces, veins lightly impressed beneath   -9. S. melanopsis tenerrima
       8. Leaves broader, glaucous beneath, not remotely denticulate
          11. Leaves narrowly elliptical, less often narrowly lanceolate or oblanceolate, 4-8 cm. by 1-1.5 cm.   -28. S. petiolaris
          11. Leaves elliptical or lanceolate to ovate, oval or obovate, 2-4 cm. wide
             12. Leaves elliptical-oblong, lanceolate or oblanceolate, base mostly rounded, less often cuneate   -13. S. monticola
             12. Leaves ovate, oval or obovate, base cuneate to cordate
                13. Leaves mostly ovate, varying to ovate-elliptical or ovate-oblong, base subcordate to cordate or less often rounded   -14. S. pseudomonticola
                13. Leaves mostly oval to obovate, base cuneate to subcordate
                   14. Bases of leaves mostly rounded, subcordate on large leaves of shoots or sprouts, crenate to crenulate   -14A. S. pseudomonticola padophylla
                 14. Bases of leaves mostly broadly cuneate, on vigorous shoots often rounded or even subcordate, crenate-serrulate, the glandular teeth inflexed, rarely entire   -15. S. barclayi

1. Some or all leaves with entire margins
   15. Some, but not all leaves with entire margins
      16. Twigs usually pruinose
         17. Leaves glabrous, glaucous beneath, most leaves entire, some in late season undulate or sharply serrate   -10. S. irrorata
         17. Leaves sparsely pubescent above, densely appressed silvery pubescent beneath, mostly entire, those of late season (rarely) remotely crenulate   -24. S. subcoerulea
      16. Twigs not pruinose
         18. Leaves glaucous beneath
            19. Leaf bases rounded to subcordate
               20. Leaves lanceolate, oblanceolate or ovate-lanceolate; twigs yellow, red on one side   -11. S. lutea
               20. Leaves oblong-lanceolate or ligulate; twigs dark brown or black   -11C. S. lutea ligulifolia
            19. Leaf bases cuneate or acute
               21. Leaves obovate to oblanceolate, often tomentose beneath   -26. S. scouleriana
               21. Leaves elliptic, oblong-lanceolate to oval or obovate, sparingly hairy beneath on midrib and veins, which are often raised   -27. S. bebbiana
         18. Leaves not glaucous beneath
            22. Leaves green on both faces, subentire to shallowly glandular-serrulate; plants 2-4 m. tall   -12. S. pseudocordata
            22. Leaves floccose-tomentose, loose on upper faces, dense on the lower, margin revolute, entire or rarely glandular-serrulate; plants 0.5-2 m. tall   -21. S. candida
   15. All leaves with entire margins
      23. Erect shrubs over 30 cm. tall
         24. Shrubs 2-4 (6) m. tall
            25. Twigs usually pruinose; leaves subglaucous to hairy beneath
               26. Leaves 2-4 cm. long, subglaucous beneath   -29. S. geyeriana
               26. Leaves 4-6 (8) cm. long, silvery appressed pubescent beneath   -24. S. subcoerulea
            25. Twigs glabrous, rarely faintly pruinose; leaves glaucous beneath
               27. Leaves broadly elliptical or oval, acute apex   -25A. S. planifolia monica
               27. Leaves narrowly elliptical, oblong-elliptical or oblanceolate, apex acute or acuminate   -25B. S. planifolia nelsonii
         24. Shrubs 0.3-1 (2) m. tall
            28. Twigs glabrous; leaves glabrous to hairy
               29. Twigs yellow when young, chestnut color at maturity, slender; leaves silky-villous on both faces when young, dull green, at maturity with vestiges of pubescence, puberulent or glabrate   -16. S. wolfii
               29. Twigs darker, thicker; leaves glabrous at all stages   -25. S. planifolia vars.
            28. Twigs of current season pubescent, tomentose or pilose; leaves pubescent, at least in early stages, but tending in vars. to become glabrous or glabrate
               30. Leaves 2.5-8 (10) times longer than broad, usually linear-oblong to elliptical

31. Leaves linear-oblong or oblanceolate, 4-8 (10) times longer than wide, soon glabrous above with impressed veins, permanently densely white-tomentose beneath -21. S. candida
31. Leaves elliptical, lanceolate or oblanceolate (but wider on vigorous shoots), 2.5-3.5 times longer than broad, tomentose beneath only in S. brachycarpa
    32. Petioles rarely over 3 mm. long; leaves with apex obtusish or abruptly short-acute, pubescence of twisted hairs -17. S. brachycarpa
    32. Petioles longer; leaves with definitely acute apices, pubescence of longer, straight, more pilose hairs
        33. Average mature leaves 2.5-3.5 cm. long, 0.6-1.2 cm. wide, tardily glabrate; stipules wanting -18. S. pseudolapponum
        33. Average mature leaves slightly larger, early glabrous; stipules present -19. S. glauca glabrescens
30. Leaves less than 2.5 times longer than broad, oval, orbicular or oblanceolate, base usually rounded, varying to cordate -20. S. cordifolia
23. Trailing shrubs, the stems on or beneath the ground, the branches rising 3-6 cm.
    34. Leaves in early stage pubescent, soon glabrous, except on margins, paler to subglaucous beneath or same color as above, not strongly reticulate
        35. Leaves paler to subglaucous beneath, elliptic, rarely oval or obovate, 1.5-3 cm. long -22. S. anglorum antiplasta
        35. Leaves same green color above and beneath, ligulate to lanceolate 1-1.8 cm. long -23. S. cascadensis
    34. Leaves glabrous, intensely glaucous beneath, strongly reticulate
        36. Plants scarcely 2 cm. above the ground; leaves orbicular, oblong-obovate or elliptic, 7-12 mm. long -30. S. nivalis
        36. Plants 3-6 cm. above the ground; leaves elliptic-oblong to suborbicular, 1.5-2.5 cm. long -30A. S. nivalis saximontana

1. Salix amygdaloides Anderss., Oef. Svensk. Vet. Akad. Foerh. 15:114. 1858.
    Trees 10-15 m. tall; branchlets slender, yellowish or ashy-gray; leaves lanceolate to ovate-lanceolate, apex acuminate, base acute to rounded (the youngest often oblanceolate and obtuse), margin finely glandular-serrulate, glabrous and green above, paler and somewhat glaucous beneath, mature leaves 5-10 cm. long and 2.5-3 cm. wide, on vigorous shoots ovate, up to 16 cm. long and 6 cm. wide; petioles slender, 1-2 cm. long; aments terminal on lateral leafy twigs; staminate aments 3-5 cm. long and 1 cm. thick; stamens usually 5, filaments villous at base, anthers yellow; pistillate aments lax in fruit, 4-8 cm. long and 1 cm. thick; capsules 4-5 mm. long, glabrous, lanceolate to ovoid, styles 0.3 mm. long; stigmas short, bifid; pedicels glabrous, 1.5-2 mm. long.---Along streams, lake shores and wet places generally. Quebec to Manitoba; Vermont to Oregon; Missouri to Arizona. In Colorado, the common tree willow along streams of the eastern plains, the canyons of the front range, and in Routt, Montrose and Montezuma Counties at 3500-7500 feet.

1A. Salix amygdaloides pilosciuscula Schneider., (forma) Journ. Arnold Arb. 1:11. 1919.
    This form differs from the species by having the twigs and young leaves pilose (rarely the petioles and midribs of the leaves tomentose), branches of the second year yellowish and the aments denser, the fruits close together.---Colorado to Oregon. In Colorado, scattered and not common, but with same range as the species.

1B. Salix amygdaloides angustissima E. C. Smith, (forma) Am. Midl. Nat. 27, No. 1:231. 1942.
    This form differs from the species by its linear-lanceolate leaves, at maturity 1 cm. wide or less. This form was first mentioned by Dr. C. R. Ball in 1909 as found on "high plains east of the Rocky Mountains." A recent collection was made 10 miles south of Cheyenne Wells.

2. Salix lasiandra Benth., Pl. Hartweg. 335. 1857.
    Small trees within our limits; branches rather stout, often shining; leaves lanceolate to broadly-lanceolate or sometimes oblanceolate, apex acuminate, base acute to rounded, margin closely glandular-serrulate, dark green and shining above, glaucous beneath, 6-10 cm. long, 1.5-3.5 cm. wide, on sprouts to 20 cm. long, and 5 cm. wide; petioles stout, glandular near the top; stipules small, acute, glandular-dentate or -serrate; staminate aments 2-6 cm. long and 1.3 cm. thick, terminal on leafy branchlets 1-2 cm. long; filaments glabrous, anthers yellow; pistillate aments 3-7 (10) cm. long, on leafy branchlets 2-4 cm. long; capsules lanceolate, 5-7 mm. long, glabrous, pale straw color or light brown; styles short (0.5 mm.); stigmas short, thick; pedicels 1.5-2 mm. long; scales lanceolate to ovate, usually dentate, sometimes glandular, sparsely pilose.---Stream banks and wet meadows. British Columbia to southern California; Alberta to New Mexico. In Colorado, Archuleta, San Miguel and Montrose Counties, the exact altitude not certain.

3. Salix caudata (Nutt.) Heller, Muhlenbergia 2:186. 1906.
    S. fendleriana Anderss.; S. lasiandra var. caudata (Nutt.) Sudw.---Shrubs 2-6 (10) m. tall; twigs long, rather thick, glabrous, shining, reddish or reddish-brown; leaves lanceolate or sometimes oblanceolate, on vigorous sterile shoots ovate-lanceolate or even ovate, apex acuminate to caudate, base acute to rounded, margin glandular-serrulate, glabrous on both faces, paler, but not glaucous beneath, on flowering branches 2-3.5 cm. long and often merely acute, on vigorous sterile branches 12 cm. long and 3-4 cm. wide, long-caudate; petioles 2-3 mm. long; stipules when present small, lanceolate to reniform, glandular-serrate; staminate aments 3-4.5 cm. long and 1-1.5 cm. thick, on leafy pubescent branchlets; filaments sparsely hairy at extreme base, anthers yellow; pistillate aments 3-6 cm. long, 1-1.5 cm. thick; capsules straw-colored, lanceolate, 5-7 mm. long; styles 0.5 mm. long; stigmas notched; scales lanceolate, oblanceolate or obovate, at first green, then pale yellow, thinly pilose at base, deciduous; pedicels 1-1.5 mm. long.---Beside streams and lakes, sometimes in wet meadows. British Columbia to Oregon, Alberta to New Mexico. In Colorado well distributed at elevations 5500-8500 feet with extremes of 4800-9625 feet.

3A. Salix caudata bryantiana Ball and Bracelin, (var.) Journ. Wash. Acad. Sci. 28:445. 1938.
Differs from the species by branchlets and peduncles glabrous, either from the beginning or very soon after the leaves appear, and by the presence with mature leaves of stipules which are markedly glandular-margined. This variety is more common than the type in Colorado.

4. Salix lucida Muhl., Neue Schrift. Gas. Nat. Fr. Berlin 4:239. 1803.
Shrubs 3-5 (7) m. tall; branchlets glabrous, lustrous, yellowish-brown; leaves lanceolate, ovate-lanceolate or ovate on vigorous vegetative shoots, apex caudate-acuminate, base rounded or narrowed, glandular-serrulate, the youngest pubescent, the mature glabrous and green on both faces, 7 cm. long and 2.5 cm. wide to 15 cm. long and 4.5 cm. wide; petioles 0.5-1.3 cm. long; stipules small, semi-cordate or oblong, very glandular, commonly persistent; staminate aments 2-5 cm. long and 1 cm. thick; stamens 3-5; filaments pilose at base, anthers yellow; pistillate aments 3-5 cm. long and 1-1.5 cm. thick, terminal on leafy twigs; capsules conic-ovoid, glabrous, greenish to pale brown, 4.5-6.5 mm. long; styles 0.5-1 mm. long; stigmas short, bifid, green; scales green at first, then pale yellow, more or less pubescent, deciduous; pedicels 1-1.5 mm. long.---Swamps, beside lakes and streams. Newfoundland to New Jersey, Manitoba to Colorado. In Colorado rare, near the eastern boundary at 4200-5000 feet.

5. Salix serissima (Bailey) Fernald, Rhodora 6:6. 1903.
Shrubs 2-3 m. tall; stems light gray, branches straw-color, ultimate twigs lustrous reddish-brown, in sharp contrast to the branches; leaves elliptic, elliptic-lanceolate or oblanceolate, apex acute or short-acuminate, base cuneiform, margin finely glandular-serrulate, in maturity dark shining green above, pale or subglaucous beneath, on flowering branches 3-4 (5) cm. long and 0.6-1.5 cm. wide, on vigorous sterile shoots up to 9 (10) cm. long and 3 cm. wide; staminate aments widely ellipsoid to almost globular, 1-1.5 (2) cm. long and 1-1.2 cm. thick, on short pilose peduncles terminating leafy, glabrous lateral twigs; filaments pilose on lower half, anthers yellow; scales short, obovate, entire, pale straw color, white-pilose; pistillate aments 1.5-2.5 (3) cm. long and 1.5-1.75 cm. thick; ovaries dark, drying almost black; capsules thick-walled, lustrous, olive or brownish, 7-10 (12) cm. long; styles 0.3 mm. long; stigmas notched; pedicels 1-1.5 mm. long, glabrous, shining; the scales narrow, oblong or oblanceolate, white-pilose.---Eastern Canada to Alberta; Massachusetts, New Jersey to Montana. In Colorado known from only one station, a boggy meadow near Long's Peak Inn, Boulder County, at 9000 feet.

6. Salix fragilis L., Sp. Pl. 1017. 1753.
Introduced street trees with rough gray bark, reaching a maximum height of 25 m. and a diameter of 2.5 m.; twigs slender, green or reddish-green, brittle at base; leaves glabrous, dark green above, somewhat lighter and sometimes subglaucous beneath, at maturity 10-15 cm. long and 2.5-4 cm. wide, lanceolate or elliptical, apex acuminate, base acute in young, more or less rounded in mature condition, margin glandular undulate-serrate; petioles 5-10 mm. long; stipules small or wanting, deciduous; aments slender, at ends of lateral leafy twigs; staminate aments 3-5 cm. long; filaments 2, pilose at base, anthers yellow; pistillate aments 5-7 cm. long; capsules glabrous, subulate-conical, 5 mm. long, subsessile to pedicels 1 mm. long; styles 0.4-0.7 mm. long; stigmas short, bifid; scales straw-colored, mostly lanceolate or oblong, fugacious.---Established along streams outside cities where originally planted. Descended from nursery stock, they sometimes show evidence of characters not belonging to the species. Pistillate forms only. At 4000-6000 feet in Colorado.

7. Salix exigua Nutt., N. Am. Sylva 1:75. 1843.
Shrubs or small trees 2-4 (8) m. tall; stems ash-gray, branches more or less reddish, glabrous, longitudinally wrinkled, ultimate twigs greenish and canescent; leaves linear to narrowly linear-lanceolate, acute at apex and base, entire or remotely denticulate, 2-8 (15) cm. long, 3-10 mm. wide, at first densely silvery-pubescent, later appressed pubescent with very short hairs the color of the leaf, giving a dull grayish-green effect, the pubescence noticeable only with a lens; petioles very short; stipules wanting; aments at ends of leafy branchlets 1-4 (7) cm. long; staminate aments 2-4 cm. long, slender, often 2-3 close together; filaments crisped pilose at base, anthers yellow; pistillate aments 3-6 cm. long and 1-1.3 cm. wide; ovaries more or less hairy; capsules glabrous, 5 mm. or less long, sessile or with pedicels shorter than the gland; stigmas sessile, short, bifid; scales whitish-green at first, later pale yellow or straw color, lanceolate, acute, pilose at base and on inner side.---On bars and shallows of streams which partly submerge the shrubs at certain seasons and on the borders of streams and ditches. Washington and Oregon to Wyoming and Colorado. In Colorado widely distributed at elevations from 5000-7500 (9500) feet.

The following two varieties, first described by Rydberg as new species and later discarded by him, were tentatively restored and given a new status by Schneider. Dr. Ball now redefines and retains var. stenophylla, but discards var. luteosericea. Whatever the final status adopted, both forms as defined by Schneider are found in Colorado.

7A. Salix exigua stenophylla (Rydb.) Schn., (var.) Bot. Gaz. 65:25. 1918.
This variety differs from the species by the glabrous ovaries, pedicels longer than the gland, the less opaque pubescence of the leaves and the longer capsules (6 mm.).---Colorado, Wyoming, New Mexico and Arizona. In Colorado with the species, but more common.

7B. Salix exigua luteosericea (Rydb.) Schn., (var.) Bot. Gaz. 67:334. 1919.
This variety differs from both the species and var. stenophylla by the longer, looser and more permanently silvery pubescence of the leaves, especially on the under side.---Nebraska and Colorado. In Colorado mostly limited to plains and east of the mountains.

8. Salix interior Rowlee, Bull. Torr. Bot. Club 27:253. 1900.
Salix rubra Rich., not Hudson; S. longifolia Muhl., not Lam.; S. fluviatilis Sarg. and other recent authors, in part---Low shrubs to small trees, 1-5 (8) m. high; twigs slender, ascending, smooth, brown with reddish tinge or gray and densely tomentose; leaves linear-elliptical or linear-oblong, apex very acute, often cuspidate, base cuneate, margin usually remotely, but distinctly denticulate, the teeth often spinulose, the young leaves silky-pubescent, the mature leaves 6-10 cm. long, 5-10 mm. wide, often wider on young vigorous shoots,

pure green and veiny on both sides, glabrous; petioles wanting or very short; stipules in Colorado specimens usually absent; aments of early season on short lateral branches which bear 4 to 6 leaves, those of later season on ascending branches sometimes as long as 15 cm.; staminate aments 2-4 cm. long, slender; filaments crisp-hairy at base, anthers yellow; scales usually narrow, but occasionally quite broad, yellowish, hairy to almost glabrous, early deciduous; pistillate aments 2-4 (5) cm. long and 1 cm. wide; ovaries sessile, clothed with appressed silky hairs; capsules nearly smooth at maturity; styles wanting; stigmas cleft; pedicels 0.75-1.5 mm. long, usually fascicled in clusters of 2 or 3 on the axis.---Banks of streams and irrigation ditches. Maine to Maryland, North Dakota to Colorado. In Colorado from eastern boundary to foothills of the Front Range and a few scattered stations near western boundary at 4500-6700 feet.

8A. Salix interior pedicellata (Anderss.) Ball, (var.) Canadian Field Nat. 40:175. 1926.
S. longifolia pedicellata Anderss.---Differs from species in the narrower leaves (3-5 mm. wide), and in the longer pedicels.---Western Canada and U. S. near northern boundary as far east as Minnesota. In Colorado east of mountains and near western boundary.

9. Salix melanopsis tenerrima (Henderson) Ball, (var.) Proc. Nat. Acad. Sci. 21, No. 4:182-183. 1935.
S. linearifolia Rydb.---Low shrubs; stems ashy-gray, twigs dark red, except the youngest, glabrous and shining; leaves on fruiting branchlets linear, 4-7 cm. long and 2-4 (5) mm. wide on seasonal shoots to 10 cm. long and 7 mm. wide, on sprouts to 12-14 cm. long and 10-13 mm. wide, denticulate or spinulose-denticulate, glabrous and shining on both faces except in the youngest stage; petioles wanting or very short; stipules wanting except on vigorous sterile shoots where they are 1-4 mm. long; staminate aments slender, 2-4 cm. long, subsessile, leafy at base; filaments long-pilose at base; anthers yellow; pistillate aments 2-3 (4) cm. long, on leafy twigs 0.5-2.5 (3) cm. long; capsules ovate-lanceolate, 4-5 mm. long, glabrous; stigmas sessile, short, thick, bifid; pedicels 0.5-1 mm. long; scales oblong to obovate, thinly pilose.---Swampy ground and banks of streams. British Columbia to California, Alberta to Utah and Colorado. One collection from Colorado, in North Park, Jackson County at about 8500 feet.

10. Salix irrorata Anderss., Oefv. Vet. Akad. Handl. 15:117. 1858.
Shrubs 2.5-3 m. high; twigs dark reddish-brown or sometimes lighter, glabrous or youngest puberulent, usually pruinose; buds stout, broadly ovoid, light brown or almost black by action of external factors; leaves narrowly oblong-elliptic to oblanceolate, apex acute to short-acuminate, base long-acute or cuneate, margin entire or on late season shoots undulate-serrate, the youngest more or less pubescent, the mature leaves dark green and glabrous above, veiny and green beneath, sometimes glaucous or subglaucous, average mature leaves 5-6 (7) cm. long and 0.8-1.2 cm. wide, on vigorous late season shoots up to 10 cm. long and 1.5-2.5 cm. wide; mature staminate aments sessile, 1.5-2 cm. long and 1 cm. thick; filaments glabrous, united at base, anthers yellow, red-tipped; pistillate aments subsessile, with or without leaflike bracts at base, 2.5-4 cm. long and about 1 cm. thick; capsules ovoid-conic, glabrous, 3-4 mm. long; styles 0.5-0.7 mm. long; stigmas short, entire or emarginate; scales black when dried, ovate to obovate or slight variations from these forms, clothed with long white pilose hairs, the same in staminate and pistillate aments.---Along streams in canyons and a short distance on the plains. Arizona, New Mexico, and Colorado. In Colorado known from the front range, altitudes 5500-7500 (9500) feet, mostly on eastern side, probably more widely distributed.

11. Salix lutea Nutt., N. Am. Sylva 1:63. 1842.
Shrubs or small trees 3-5 (7) m. high; twigs yellow, turning reddish on one side as season advances; leaves lanceolate to ovate-lanceolate, sometimes widest at the middle, apex acute to short-acuminate, base rounded to cordate, margin serrulate to subentire, yellowish-green, glabrous, glaucous beneath, 5-6 (8) cm. long and 2-5 cm. wide, or on vigorous shoots up to 10 cm. long and 4-4.5 cm. wide; petioles 0.5 cm. on flowering branches, up to 1 or 1.5 cm. on vigorous sterile shoots; stipules small except on vigorous shoots where they reach a length of 1-1.5 cm., mostly reniform and glandular-dentate to glandular-serrate; staminate aments sessile to sub-sessile, 2.5 to 4.5 cm. long and 0.8-1 cm. wide; filaments naked at base, anthers red or red-tipped; pistillate aments 2-4 cm. long, less than 1 cm. thick, terminal on short leafy twigs; capsules ovate-conic, 4 mm. long; styles 0.5 mm. long; stigmas short, thick, notched or entire, red; scales bicolored at first with very light base and light reddish-brown tip, both darkening as the season advances.---Oregon to California; Alberta to Arizona and New Mexico; Manitoba to South Dakota. In Colorado, sparingly in the front range between 7000 and 8000 feet, and in several western counties.

11A. Salix lutea platyphylla Ball, (var.) Bot. Gaz. 71:430-431. 1921.
Distinguished from the species by the broader, ovate-lanceolate or elliptic-obovate leaves, 4-6 cm. long and 2-25 mm. wide, on sprouts to 9-11 cm. long and 3.5-4.5 cm. wide, and by the longer pedicels, 2-4 mm. long. Almost impossible to separate at the blossoming stage.---Oregon to Wyoming; south to Nevada and Colorado. In Colorado known only from just south of the Wyoming state boundary.

11B. Salix lutea famelica Ball, (var.) Bot. Gaz. 71:426-427. 1921.
Distinguished from the species by the smaller, narrower and more strongly nerved leaves and the usually shorter aments and pedicels; leaves on a sterile shoot in Colorado A. & M. College herbarium are from 4 cm. long, 6 mm. wide to 7 cm. long, 1 cm. wide, long-acuminate apex, acute or rarely obtuse base, petioles to 1 cm. long, reddish-brown, the color extending into the midrib, growing fainter towards the apex.---Distribution uncertain, not in recent manuals; type from Montana. In Colorado, one station in Yuma County at 3400 feet.

11C. Salix lutea ligulifolia Ball, (var.) Bot. Gaz. 71:428. 1921.
Shrubs with clustered stems 2-3 m. high; bark dark gray; twigs dark brown, glabrous or the youngest pubescent; leaves at blossoming time narrowly elliptic, acute at both ends, 3-5 cm. long and 1-1.3 cm. wide, average mature leaves 6-7 cm. long and 1.2-1.5 cm. wide, on vigorous late season shoots up to 11 cm. long and 3 cm. wide, ligulate or oblong-lanceolate, apex acute to short-acuminate, base rounded to subcordate, margin entire or undulate-serrate, reticulate on

both faces, glabrous, dark green above, pale and at maturity often glaucous beneath; petioles 1-1.5 cm. long; stipules semi-cordate to reniform, dentate; staminate aments 2-4 cm. long and 1 cm. wide, sessile, leafy-bracted; filaments glabrous, united at base; anthers red or purple, drying dark purple; pistillate aments 2-4 cm. long and 1 cm. thick, on leafy peduncles 0.5-1 cm. long; capsules glabrous, lanceolate to lance-ovoid, on glabrous pedicels 1-1.5 (2) mm. long; styles 0.2-0.6 mm. long; stigmas short, red, mostly bifid; scales black or black-tipped, small, ovate to obovate or even orbicular in the staminate aments, lanceolate to oblanceolate in the pistillate, in both clothed with dense cover of long, crinkly white hairs.---Streambanks in mountain canyons or plains near the foothills. California to New Mexico. In Colorado, mountainous counties and adjacent plains at 5000-7500 (9500) feet.

12. Salix pseudocordata (Anderss.) Rydb., Fl. Col. 94. 1906.

S. pseudomyrsinites Anderss. of several American authors---Shrub 2-4 m. high; branchlets slender, mostly short, divaricate, bright chestnut to dark brown, lustrous, glabrous except ultimate twigs of the season, which are pubescent; leaves oval-oblong, sometimes widest slightly above the middle, apex acute to short-acuminate, base somewhat rounded, occasionally acute, 3 cm. long and 1 cm. wide to 6 cm. long and 2 cm. wide, margin minutely serrulate to subentire, glabrous on both faces (except the youngest, which are pubescent), pure green, slightly paler beneath, reticulate-veined; petioles 3-8 mm. long, glabrous; stipules lanceolate to ovate, glandular-serrate; aments on leafy twigs about 0.5 cm. long; staminate aments 2-3 cm. long and 1-1.3 cm. thick; filaments glabrous, anthers yellow; pistillate aments 2-3 cm. long and 1 cm. thick; capsules greenish, 4-5 mm. long, glabrous; styles 0.5-0.7 mm. long; stigmas entire; pedicels 1-1.5 mm. long; scales in early stage light brown, later black, oblong or oblanceolate, acute to truncate, thinly clothed with long white woolly hairs.---Seepage areas and along streams. British Columbia south to Oregon; Alberta to Colorado. In Colorado the northern counties, Routt to Larimer and Grand; south to Lake, Gunnison and El Paso at 5300-9500 feet.

13. Salix monticola Bebb. in Coulter's, Man. Rocky Mt. Bot. 336. 1885.

Shrubs 2.5-5 m. high; twigs at flowering time usually yellow or yellowish-green, rarely dark reddish-brown, drying black; leaves on fertile branches 3-5 cm. long and 1-1.5 cm. wide, oblong-lanceolate, elliptical-lanceolate or oblanceolate, the widest point varying from the middle to below or above that point, apex acute to short-acuminate, base rounded or in early stages narrowed and cuneate, margin crenate-serrulate, upper surface dark green, glabrous, lower surface pale and glaucous, on vigorous sterile shoots reaching 8-9 (10) cm. long and 3 cm. wide, with subcordate base; petioles 3-5 mm. long or on sterile shoots 5-10 mm. long; stipules small or wanting on fertile branches, large (up to 1.3 cm.), semi-ovate, acute, serrate on vigorous sterile branches; staminate aments 2-3 (4) cm. long and 1-1.5 cm. thick, sessile, without basal bracts; filaments not pilose at base, anthers usually yellow, but sometimes red-tipped; pistillate aments 3-4 (5) cm. long and 1-1.5 cm. thick, subsessile or rarely in age with leafy stalks 5 mm. long, leaflike bracts often present; capsules ovoid-rostrate, glabrous, 5-6 mm. long; styles 1-1.5 mm. long; stigmas short, bifid or entire; scales lanceolate to oblanceolate, ovate to obovate, densely villous, somewhat narrower in the staminate aments; pedicels 1 mm. or less, pilose at base.---Streambanks, lake borders and boggy meadows. Rocky Mountains within United States. In Colorado common at elevations between 7000 and 9000 (10,500) feet.

14. Salix pseudomonticola Ball, Contrib. U. S. Nat. Herb. 22:321. 1921.

Shrubs 1-4 m. high; branchlets divaricate, grayish-brown to brown, the youngest pilose-pubescent or glabrate and lighter colored; leaves ovate, oblong-ovate or elliptical-ovate, 4-7 cm. long and 1.8-4 cm. wide, on vigorous sterile branches up to 9 cm. long and 4-4.5 cm. wide, apex acute to abruptly short-acuminate, base mostly subcordate to cordate, but occasionally rounded, dark green to yellowish-green above, glabrous, lighter green or subglaucous beneath, margin crenate-serrate; petioles 5 mm. long (10 mm. on vigorous sterile shoots), glabrous; stipules inconspicuous on branches bearing aments, up to 1 cm. or more on vigorous sterile shoots; staminate aments 2-3 cm. long and 1.5 cm. thick; filaments glabrous, distinct, anthers red or dark purple, at least as to the tips; pistillate aments 3-5 (7) cm. long (or according to Dr. C. R. Ball, in northern localities up to 10 cm. long) and 1.5 cm. thick; capsules glabrous, 6-8 mm. long; styles 0.8-1.3 mm. long; stigmas entire or notched; scales brown, drying black, mostly wide and rounded at apex, pilose with long white hairs, the upper portion on the outside mostly free of the hairs which are massed at the base and margin.---Riverbanks and swampy plains. Alberta to Saskatchewan, south to Montana and South Dakota. In Colorado sparsely scattered over the state 6400-8500 feet.

14A. Salix pseudomonticola padophylla (Rydb.) Ball, (var.) Journ. Wash. Acad. Sci. 28:450. 1938.

Salix padophylla Rydb.---Shrub 3-4 (5) m. high; stem ash-gray, twigs yellowish-green or sometimes brown; leaves commonly oval, apex abruptly acute, base rounded, margin crenate-serrulate, 3-6 cm. long and 1.5-3 cm. wide, on vigorous vegetative shoots up to 9 cm. long and 3.5 cm. wide, with cordate base (in a form provisionally included in this variety leaves are obovate, apex rounded, base cuneate, 2.5 cm. by 1.5 cm. to 5 cm. by 3 cm.), when young sparingly hairy, when mature glabrous above, glabrous and glaucous beneath; petioles 5-8 mm. long; stipules glandular-dentate; staminate aments subsessile, 2-3 cm. long and 1-1.2 cm. wide, with small leaves at base; filaments glabrous, anthers yellow, but often with red tips; pistillate aments 3-4 (5) cm. long and 1 cm. thick, peduncles 0.2-0.5 cm. long at early stage, sometimes 1 cm. when fruit is mature, bearing true leaves, usually with glandular-dentate stipules as in S. barclayi; capsule ovoid-conic, glabrous, 6 mm. long; styles 1-1.5 mm. long; stigmas deeply cleft or entire, red; pedicels 0.5-1 mm.; scales dark brown, usually drying black, pilose with long white hairs twice the length of the scale.---Streambanks, lake borders and bogs. Wyoming, Montana, Colorado and Arizona. In Colorado known from the following counties: Larimer, Grand, Summit, Lake, Park, Chaffee, Gunnison, Teller, El Paso, Hinsdale, San Juan, La Plata, and Conejos, at 7000-10,000 feet.

15. **Salix barclayi** Anderss., Oef. Svensk. Vetensk. Akad. Foerh. 15:125. 1858.

Shrubs 2-3 (4) m. high; branchlets reddish-brown, drying almost black, the youngest often pubescent or tomentose; leaves oval or widest slightly above the middle, apex acute to short-acuminate, base acute to cuneate, margin crenate-serrulate to subentire, sometimes glandular, sparsely to densely pilose on first appearance, soon glabrate or with vestigial hairs on midrib and primary veins, glaucous beneath, strongly reticulate, 3-7 cm. long and 2.5-2.7 cm. wide, on vigorous sterile shoots up to 10 cm. long and 5 cm. wide, with tendency to rounded base; stipules vary from ovate to reniform, often glandular-serrulate to dentate, largest on sterile shoots; staminate aments 2-3 (4) cm. long and 1.5 cm. thick, at ends of leafy twigs 0.3-1 cm. long; filaments naked at base, anthers often red or orange in early stage, usually turning yellow before dehiscence; pistillate aments 2 cm. long at blossoming stage, at maturity 4-5 (7) cm. long and 1.5 (2) cm. wide, the leafy twigs 0.5-1.5(3) cm. long; ovaries slender, tapering from base to tip, 3-4 mm. long; capsules ovoid at base, rostrate, 6-8 mm. long; styles 1.25-1.5(2) mm. long; stigmas cleft, notched or entire, red; scales almost black, lanceolate to ovate or obovate, mostly acute, long-villous; pedicels 0.5-1 mm. long.---Borders of streams and lakes. Alaska and south in Cascades and Sierras to southern California; Yukon to southern Colorado in the Rocky Mountains. In Colorado, known only in the front range from Wyoming boundary to Colorado Springs at altitudes from 7000-10,000 feet.

Salix barclayi var. **hebecarpa** Anderss. differs from type in "ovaries and capsules distinctly hairy on the upper half" and long white hairs of the scales straight, instead of crinkly. One specimen from Larimer County at Joe Wright Creek where it enters the Poudre River, at an elevation of 9500 feet.

16. **Salix wolfii** Bebb, Wheeler Exp. 241. 1878.

Shrubs rarely over 1 m. high; twigs yellow to orange when young, chestnut-brown when older, glabrous and shining except the youngest; leaves oblanceolate, elliptical or rarely obovate, 2-4 cm. long and 0.6-1.3 cm. wide, acute at apex, acute or narrowed and slightly rounded at base, entire, dull green and silvery-pubescent when young, glabrate, rigid and green on both sides when mature; petioles short, pubescent when young, glabrous at maturity; stipules small or wanting; aments on short leafy twigs, the leaves often exceeding the aments; staminate aments 0.5-1.75 cm. long and 0.5-0.7 cm. thick; filaments smooth and free, anthers yellow; pistillate aments 1-2 cm. long, nearly as thick, subglobose to long-oval; capsules conical from ovoid bases, glabrous, 3-4.5 mm. long, subsessile; styles about 1 mm. long; stigmas bifid or entire, red; scales lanceolate or oblanceolate to ovate, obtuse, sparingly villous.---Boggy meadows and slopes where snow remains late in season at altitudes chiefly 7500-10,000 feet, extremes 5900-11,000 feet. Idaho to Wyoming, south to Utah and Colorado. In Colorado, Larimer, Jackson, Boulder, Grand, Lake, Gunnison and Routt Counties.

16A. **Salix wolfii idahoensis** Ball, (var.) Bot. Gaz. 40:378. 1905.

Leaves larger, 3-6 cm. long, 1-1.8 cm. wide, oblanceolate; peduncles in proportion; capsules about 5 mm. long, greenish, thinly tomentose.---With the species, rare in Colorado, common in northern part of the species range. In Colorado, Routt, Grand, Garfield and Boulder Counties.

17. **Salix brachycarpa** Nutt., N. Am. Sylva 1:69. 1842.

Erect shrubs, 0.5-1 m. tall in alpine zone, to 2 m. at middle altitudes; branches typically short, stoutish, sometimes tortuous, with short internodes and prominent nodes, those of the season tomentose or pilose, the older glabrous, often with bark splitting; leaves elliptic-oblong to oblanceolate, apex obtusish or abruptly acute, base rounded, rarely cuneate or subcordate, entire, 2.5 cm. long and 0.6 cm. wide to 3.5 cm. long and 1.3 cm. wide, on sterile twigs or sprouts sometimes oval or ovate, to 4.5 cm. long and 2.5 cm. wide, tomentose, more densely so beneath, nerves on at least some leaves distinctly raised; petioles about 3 mm. long in typical plants; stipules wanting; staminate aments on very short leafy twigs, 10-12 mm. long; filaments glabrous, anthers subglobose, less than 0.5 mm. diameter, often red or reddish at emergence, yellow at maturity; pistillate aments more ellipsoid or even cylindrical, 1-2 (3) cm. long; ovaries 2-3 mm. long, obtuse, densely tomentose or pilose; capsules 4.5-5 mm. long, subsessile, pubescence shorter and less dense; styles about 0.5 mm. long, entire, notched or cleft; stigmas bifid, red; scales at early stage whitish-green, then yellowish-brown, narrow to broad, sometimes black in drying.---Moist slopes above timberline in dense thickets which are heavily browsed; at lower elevations in open meadows near streams. From the Atlantic to the Pacific; from Northwest Territories to Colorado where it is widely distributed from 7500-12,000 feet.

17A. **Salix brachycarpa sansoni** Ball, (var.) Univ. Cal. Pub. Bot. 17:414. 1934.

Differs from the species in the smaller size in all its parts and nearly globular aments.

17B. **Salix brachycarpa antimima** Raup, (var.) Rhodora 33:241. 1931.

Differs from the type by having leaves glabrescent when mature, even beneath; twigs of the season at maturity glabrescent. Both varieties with the species; sansoni rare, antimima common.

18. **Salix pseudolapponum** von Seem., Bot. Jahrb. 29 Beiblatt 65:28-29. 1901.

In Coulter and Nelson's Manual as synonym for S. glaucops Anderss.---Shrub, 0.3-1 (1.5) m. tall; twigs of current season pubescent, the year-old glabrous, brown, the older branchlets yellowish-gray, cracking and forming thin flakes or strips; leaves entire, mostly elliptical, tapering evenly to apex and base, apex long-acute, base acute or less often narrowed and rounded, at anthesis to 2.5 cm. long by 0.8 cm. wide, later in the season to 3.5 cm. long and 1 cm. wide or rarely on vigorous sterile shoots to 4 cm. long and 1 cm. wide, the young leaves loosely pubescent, the older glabrous or nearly so, glaucous beneath; petioles at anthesis 3 mm. long, at maturity 5 mm. long, at all stages longer than the buds; stipules wanting or rare and very small; staminate aments 0.8-1.5 cm. long; filaments pilose at base; anthers oblong-ellipsoid, 0.5-0.8 mm. long, usually red or red-tipped, drying almost black; scales ovate, obovate or suborbicular, at first bicolored, then wholly black, long-pilose, chiefly on margin and tip, clearly visible at all stages of development; pistillate aments 1-2.5 cm. long at anthesis, to 3 cm. long in fruit, on twigs 0.5-1.5 cm. long, these with a few densely silky-villous leaves at anthesis; mature fruits 5-8 mm. long, grayish-villous; styles and stigmas red, sometimes drying yellowish-brown; styles entire or merely notched, 1 mm. long; stigmas bifid; scales as in staminate aments, but narrower.---Moist slopes above timberline and borders of lakes and streams. Rocky Mountains, Alberta to New Mexico; in Colorado widely scattered above timberline to 12,000 feet.

19. Salix glauca glabrescens (Anderss.) Schneider, (var.) Bot. Gaz. 66:329-330. 1918.

S. glaucops Anderss.; S. austinae Rydb. in part---Shrubs, erect or ascending, 0.5-1 (1.5) m. tall; twigs of the season often 5-10 cm. long, straight, with long internodes, pilose at first, the pile growing less as the season advances, the 1- and 2-year-old branchlets glabrous, lustrous, purplish-brown; leaves of fertile twigs at anthesis narrowly to broadly elliptical or a few spatulate, 2 cm. long and 0.3 cm. wide to 4 cm. long and 1.5 cm. wide, the mature, especially on sterile twigs, 5-7 cm. long and 2-2.5 cm. wide, apex acute, sometimes apiculate, base acute, cuneate or rarely obtuse, entire, above glabrous from the first or very quickly as the leaves develop, beneath densely pilose, becoming glabrous or glabrate more slowly than above; petioles 5-10 mm. long, reddish or yellowish; stipules wanting or inconspicuous; staminate aments 2-3 cm. long, on leafy pubescent twigs rarely more than 5 mm. long; filaments crisped-hairy at base, anthers ellipsoid or subglobose, 0.6-0.8 mm. long, often tipped or streaked with red; pistillate aments 2-4 (5) cm. long, on leafy twigs 1-2 (3) cm. long; ovaries conical, subsessile, white-pilose, 3-5 mm. long; capsules short-pedicelled, 6-8 (9) mm. long, covered sparingly with very short erect hairs or subglabrous in part; styles entire, notched, or cleft less than half their length; scales usually oblong or oblanceolate with obtuse apex, but rarely wider and acute, light brown or with tips red or darker brown, at first pilose, at maturity subglabrous, at least in part.---Wet slopes above timberline or boggy places at lower elevations. Alberta to Colorado. In Colorado so far as collected, Trail Ridge in Rocky Mountain National Park at 10,000-11,700 feet.

20. Salix cordifolia Pursh, Fl. Am. Sept. 2:611. 1814.

This species has been reported from Colorado but apparently is represented here only by the following variety.

20A. Salix cordifolia callicarpaea (Trautv.) Fernald, (var.) Rhodora 28:184. 1926.

Salix callicarpaea Trautv.---Erect shrub less than 1 m. high; twigs silky-villous; young leaves mostly elliptic to oblong, varying to oblanceolate or broader, the mature leaves frequently broader and oval, apex and base acute to obtuse, margins entire, silky-villous when young, tending to be glabrate when mature with slight silkiness on the midrib and veins beneath; staminate aments cylindrical, 1.5-2 cm. long and 0.5-0.8 cm. thick, on short leafy peduncles; filaments glabrous; anthers ellipsoid, 0.7-1 mm. long and 0.1-0.2 (0.3) mm. thick, red; pistillate aments 2-4 cm. long and 0.6 cm. thick at blossoming stage, up to 1.5 cm. thick at maturity, the leafy peduncles 2-3 cm. long; capsules sessile, pilose, 5 mm. long; styles 0.5-1 mm. long, red; stigmas deeply lobed, red; scales at early stage light brown, later darker, even black, mostly wide and rounded at tip, long-pilose especially on margin and tip.---Subarctic and alpine. Labrador, Newfoundland and Quebec. Specimens from Rocky Mountain National Park at about 11,300 feet are provisionally placed here.

21. Salix candida Fluegge in Willd., Sp. Pl. 4:708. 1806.

Shrubs 0.5-1 (1.5) m. tall; branches red or reddish, glabrous, those of the current season densely white-tomentose; leaves linear-oblong, often ten times longer than broad, 3-5 cm. long and 3-5 mm. wide, exceptional leaves oblanceolate or broadly elliptical, 4-8 cm. long and 1-1.5 cm. wide or on sprouts to 10 cm. long and 1.5-2 cm. wide, apex and base usually acute, margin entire or rarely crenulate, revolute, above sparsely tomentose, dark green, veins impressed, beneath densely white-tomentose; petioles 2-5 mm. long, tomentose; stipules narrow, acute, tomentose, sometimes infrequent or wanting; staminate aments subsessile, 1.5-2.5 (3) cm. long; filaments glabrous; anthers red or purple; pistillate aments 1-3 (4) cm. long, subsessile with leaves at base; capsules lanceolate, densely tomentose, 5-7 mm. long; styles 1 mm. long; stigmas red, notched or entire; scales long-pilose on apex and margin, sparsely on center and base.---Cold bogs and marshy areas. Labrador to British Columbia; New Jersey to Colorado and Montana. In Colorado, one station in western Park County at 8600 feet.

22. Salix anglorum antiplasta Schn., (var.) Bot. Gaz. 66:134-135. 1918.

Salix arctica var. petraea Anderss.; S. petrophila Rydb.---Creeping alpine plants, the very woody horizontal stems under or on the ground sometimes 60 cm. long and 6-8 mm. thick, the branches ascending, glabrous, brown or yellowish, 3-6 cm. long; leaves elliptical to obovate, apex rounded to acute, often apiculate, base acute, margin entire, deep green and glabrous above, paler and sometimes subglaucous beneath, more or less pilose when young, especially on midrib and veins below, strongly veined, moderately reticulate, 1-4 cm. long and 1 cm. wide; petioles slender, yellowish; stipules small or wanting; staminate aments 1-2 cm. long and 1 cm. thick; filaments not pilose at base, anthers royal purple to dark red; pistillate aments 2-3 cm. long and 1 cm. thick, on pilose leafy twigs 1-1.5 cm. long; capsules sessile, lanceolate, gray-villous to gray-tomentose, 4-5 (6) mm. long; styles about 1 mm. long, red; stigmas entire or deeply cleft; scales oval to obovate, the apex usually rounded, long-villous, black or in early stages brown.---Alpine rocky slopes with thick cover of gravelly soil kept moist by seepage from melting snow above. British Columbia and Alberta to Utah and Colorado. In Colorado widespread at altitudes between 11,000 and 13,000 feet.

23. Salix cascadensis Cockerell, Muhlenbergia 3:9. 1907.

S. tenera Anderss.---Creeping alpine undershrubs, the yellowish or reddish-brown branchlets rising not more than 2 cm. above the ground; leaves elliptical or sometimes ligulate, from 7 mm. long and 2 mm. wide to 20 mm. long and 5 mm. wide, apex and base acute, margin entire, green and shining on both faces, slightly lighter beneath, nerves raised, not reticulate; petioles glabrous, 1-3 mm. long; staminate aments 6-12 mm. long, on leafy twigs 5 mm. or less long; filaments 2, often pink when freshly gathered, entire or rarely cleft, anthers red or purple, drying almost black; pistillate aments 5-12 (15) mm. long, terminating leafy twigs 0.5-1 (1.5) cm. long, the upper portion bare of leaves; capsules sessile, grayish-villous, 3-4 mm. long; styles 0.5-1 mm. long, entire or rarely cleft; stigmas entire or cleft; scales oblanceolate, obovate or broadly oval, black, pilose on inner side and margin, much of outer surface glabrous.---Alpine summits. Washington to California, Montana and Wyoming to Colorado. Rare in Colorado, reported only from Trail Ridge, Rocky Mountain National Park at 11,700 feet.

24. Salix subcoerulea Piper, Bull. Torr. Club 27:400. 1900.

Shrubs 1-3 m. high; twigs glabrous, purplish-brown, pruinose; leaves oblong-lanceolate, elliptical or oblanceolate, apex acute or acuminate, base acute, margin entire or rarely obscurely crenulate, green and minutely pubescent above (sometimes glabrate in aged leaves), densely silvery appressed-pubescent beneath, 2-5 cm. long and 0.5-1.2 cm. wide, but on vigorous sterile shoots up to 10 cm. long and 3 cm. wide; petioles 2-5 mm. long (on the vigorous shoots 1 cm. long); stipules wanting or rarely present on late season shoots; staminate aments precocious, sessile, 1-3 cm. long, when mature 1 cm. thick; filaments united at base, without basal hairs; anthers red or red-tipped; pistillate aments subsessile, often with small immature leaves at base, 2-3 cm. long (up to 4 cm. in fruit) and 0.6-1 cm. thick; capsules ovoid or lance-ovoid, 3.3-5 mm. long, silvery-pubescent with short appressed hairs; styles about 1 mm. long; stigmas thick, two-lobed or entire, red; pedicels 1 mm. or less in length; scales black, ovate to obovate, sparsely pilose, hairs as long as capsule. Some unusual specimens from Joe Wright Creek in Larimer County have more widely spreading, slender branches and oval or obovate leaves shorter and relatively wider than those usually found. Further collections may determine that this should rank as a variety.---Beside streams, pools and in wet mountain meadows. Wyoming to Washington, south to New Mexico and California. In Colorado widely distributed at 7500-11,000 feet.

25A. Salix planifolia Pursh monica (Bebb) Schn., (var.) Journ. Arnold Arbor. 1:78. 1905.

Salix chlorophylla Anderss.---Shrubs less than 1 m. high above timberline, but at lower altitudes 2-3 m. high; branches glabrous, bright chestnut to darker; leaves elliptical to oval, varying to obovate, 2-3 cm. long and 0.8-1.5 cm. wide, apex abruptly acute, sometimes apiculate, base acute to narrowed and slightly rounded, margin usually entire, glabrous above, glaucous beneath (the youngest short silky-pilose beneath), on vigorous sterile shoots relatively broader and up to 4 cm. by 2 cm., or 5 cm. by 3 cm.; petioles 2-5 mm. long, glabrous; stipules none or very small and fugitive; staminate aments 1-2 cm. long and 1-1.3 cm. thick, sessile, not bracted, prococious usually at high altitudes, sometimes coetaneous at lower elevations; filaments glabrous, but sparsely pilose at base, anthers usually red or purple, but sometimes yellow; pistillate aments 2-3 cm. long and 1-1.3 cm. thick, sessile, not bracted; capsules ovate-conic, sessile, gray-pubescent, 4-6 mm. long; styles 1-1.5 mm. long; stigmas entire or cleft, red; scales lanceolate, ovate, oval or obovate, usually obtuse, long-pilose.---In thickets above timberline on slopes kept moist by melting snow, margins of high lakes and streambanks at lower altitudes. Northwest Territories to Manitoba, south to New Mexico and Arizona. In Colorado in the mountainous counties at 8000-13,000 feet.

25B. Salix planifolia nelsonii (Ball) Ball, (var.) Journ. Arnold Arboretum 2:188. 1921.

S. nelsonii Ball---Shrubs, at high altitudes less than 1 m., at middle altitudes up to 3 m. high; twigs shining, bright chestnut or reddish-brown, occasionally faintly pruinose; leaves elliptical, lanceolate or oblanceolate, apex acute, base cuneate or narrowed and rounded, 3-3.5 cm. long and 1-1.5 cm. wide, on vigorous sterile shoots up to 7 cm. long and 2 cm. wide, margin entire, glabrous above, glaucous below (the youngest thinly silky-villous), rather strongly veined on both surfaces, reticulate below; petioles 3-5 mm. long on average leaves, up to 10 mm. on vigorous shoots; stipules wanting; staminate aments 2-3 cm. long and 1-1.5 cm. thick, sessile, usually not bracted; filaments not pilose at base; anthers red, purplish or merely tinged with these colors and yellow at base; pistillate aments 2-4 (5) cm. long and 1 cm. thick, dense, sessile, often with a few bracts at base; capsules sessile, silky-pubescent, 5-6 (7) mm. long; styles 0.5-1 mm. long; stigmas entire, about length of style, red; scales black, mostly wide and rounded at tip, long-pilose.---Wet meadows and along streams. Front Range Wyoming-Colorado. In Colorado, Larimer to Routt Counties and south to El Paso-Lake Counties, and Conejos and Ouray Counties, at elevations from 7000-10,000 rarely to 11,500 feet; probably in mountain counties not visited.

26. Salix scouleriana Barratt in Hook., Fl. Bor. Am. 2:145. 1839.

Shrubs commonly 3-4 m. high, occasionally a tree 10 m. high with trunk diameter of 15 cm.; branches reddish-brown to almost black, twigs glabrous to densely pubescent; leaves obovate, oblanceolate or rarely oval, apex obtuse or abruptly acute, base cuneate or acute, 3-5 cm. long and 1.5-2 cm. wide, on vigorous vegetative shoots up to 10 cm. long and 4.5 cm. wide, margin entire or shallowly crenate-serrulate, dark green and glabrous above, glaucous, strongly reticulate and sometimes densely pubescent beneath; petioles 1-2 cm. long, glabrous or pubescent, matching the leaves; stipules small except on vigorous shoots where they may be 1 cm. long; staminate aments sessile, usually naked at base, precocious, 2-3 cm. long and 1.5 cm. thick; filaments pilose at base; anthers usually yellow, but sometimes tipped with red; pistillate aments sessile or subsessile, 3-6 (8) cm. long and 1.5-2 cm. thick, usually naked at base; capsules subconic, rostrate, tomentose, 7-9 mm. long; styles less than 0.5 mm. long; stigmas 1 mm. long, entire or divided, red; pedicels 1-2 mm. long, tomentose scales at early stage brown with lighter base, later completely black, lanceolate or oblanceolate, ovate to obovate or oval, long-pilose, the hairs twice the length of the scale.---In the usual moist situations and also in drier places, sometimes on well-drained slopes where melting snow from above furnishes adequate moisture for a short growing season. Saskatchewan to British Columbia, south to New Mexico and California. In Colorado widely scattered in the mountains at 8000-10,000 feet, with extremes of 6500-11,000 feet.

27. **Salix bebbiana** Sarg., Gard. and For. 8:463. 1895.
   Salix rostrata Richardson---Shrubs or small trees, usually with a single stem and bushy top even when low (2-3 m. tall), under optimum conditions (when growing with other species of trees) up to 8 m. high; twigs pubescent, puberulent, or glabrous; leaves vary from elliptic or oblanceolate to ovate-oblong or obovate, 3-10 cm. long and 1-4 cm. wide, apex acute (rarely obtuse), base rounded (rarely acute), margin serrate, crenate or subentire, pubescent above, tomentose beneath, reticulate, glabrate and strongly rugose in age; staminate aments sessile, bracted, 1-1.5 (2) cm. long and 0.5 cm. wide; filaments with a few hairs on the bases, anthers yellow; pistillate aments 2-4 (6) cm. long and 1.5-2 cm. wide, on short leafy twigs; capsules 6-8 mm. long, rostrate, pubescent; styles obsolete or very short; stigmas deeply cleft or entire, red; scales of aments lanceolate or oblanceolate to ovate or oval, much wider than in eastern forms, when young very pale, with rose-colored tips, darker in mature specimens, when pressed, pale brown in lower half, darker brown at tip; pedicels 1-2 (3) mm. long, slender, pubescent.---Riverbanks and open hillsides. Widely distributed from Atlantic to Pacific coast and from Alaska to California, New Mexico, New Jersey, etc. In Colorado common in the Front Range at 5000-9000 (9800) feet. Unusual in its single stem, range in size and adaptation to dry as well as moist situations.

27A. **Salix bebbiana perrostrata** (Rydb.) Schn., (var.) Journ. Arnold Arbor. 2:71. 1920.
   Differs from the species in its more fugitive pubescence, looser aments, usually obovate entire leaves and the absence of the rugose lower surface of the old leaves.---The common form in Colorado, occurring in the mountains and foothills of both eastern and western sections of the state at altitudes of 5000-9500 feet.

28. **Salix petiolaris** J. E. Smith, Trans. Linn. Soc. London 6:122. 1802.
   Shrubs 2-4 m. high; branches slender, yellowish or darker, drying black; leaves lanceolate, elliptical or oblanceolate, apex and base acuminate, 4-6 cm. long and 1-1.5 cm. wide on fruiting branches, up to 8 cm. long and 1.5 cm. wide on vigorous sterile shoots, margin finely serrulate, the young leaves silvery-pubescent with entire margin, the mature glabrous and shining above, glaucous beneath, midrib and veins prominent, reticulate; petioles 4-6 mm. long, glabrous, yellow; stipules small, deciduous; staminate aments 1-1.5 (2) cm. long and less than 1 cm. thick, sessile or subsessile, bracted; filaments pilose at base, anthers yellow; pistillate aments 1.5-2 cm. long and 1 cm. or less thick, on short leafy twigs; capsules 3-4 mm. long, pubescent, rostrate, stigmas deeply cleft, sessile, red; pedicels 1-2 mm. long; scales bicolored, in early stage with base pale yellow and apex reddish-brown, the colors gradually deepening until base is light brown and the apex black, the darker portion increasing in relative size.---Swamps and stream banks. New Brunswick to Saskatchewan, south to New Jersey and Colorado. In Colorado along the Front Range at 7000-8000 feet. Only recently recognized, probably more widely distributed.

29. **Salix geyeriana** Anderss. Oef., Vet. Akad. Foerh. 15:122. 1858.
   Shrubs 2.5-4 m. high; twigs slender, glabrous, dark reddish-brown, pruinose; leaves elliptical or widest slightly above the middle, 2-4 (6) cm. long and 0.5-1.2 cm. wide, apex acute to short-acuminate, base acute, silky-pilose on both faces when young, usually glabrous or subglabrous in age, entire; petioles slender, 3-5 mm. long; stipules wanting; staminate aments 1-1.3 cm. long and 0.5-0.8 cm. thick, subsessile, leafy-bracted; filaments pilose at base; anthers yellow or sometimes red-tipped; pistillate aments 1-1.5 cm. long and nearly as thick, on leafy twigs up to 5 mm. long; capsules pubescent, rostrate, 5-6 mm. long; styles short or absent; stigmas short, bifid, red; scales narrow, bicolored, apex red or brown, base light color, the hairs mostly on margin and tip, sometimes almost disappearing in mature aments, somewhat broader on staminate aments.---Middle altitudes beside streams, in wet meadows and bogs. British Columbia to Nebraska. In Colorado, specimens seen from the Front Range; probably grows in most mountain areas from 6500 feet up.

29A. **Salix geyeriana argentea** (Bebb) Schneider, (var.) Journ. Arnold Arbor. 2:74. 1920.
   This differs from the species in the more silvery pilose pubescence which is more permanent and the shorter, relatively broader leaves, 1.5 cm. long, 0.7 cm. wide. Associated with the species.

30. **Salix nivalis** Hook., Fl. Bor. Am. 2:152. 1839.
   Caespitose creeping undershrubs, the slender branches from the buried stem rarely more than 2-3 cm. long, lying flat upon the ground; leaves clustered at the end of these branches, elliptic, oval, obovate or orbicular, apex mostly rounded, but sometimes abruptly short-acute, base rounded to acute, 3-7 (10) mm. long and 3-5 (7) mm. wide, green and glabrous above, glaucous and strongly reticulate beneath, margin slightly revolute; petioles 2-5 mm. long; stipules wanting; staminate aments 2-4 (5) mm. long, ovoid or nearly globular, terminating glabrous leafy twigs 3-5 mm. long; anthers and sometimes the filaments dark red; scales green, sometimes with red tips, glabrous or pilose on margin, spatulate or obovate, the tip rounded or truncate, the base abruptly narrowed into a stalk or gradually into an acute point; pistillate aments 3- to 6-flowered, 3 (4) mm. long, at the ends of leafy twigs; capsules grayish-pubescent, ovoid, 2-3 mm. long; styles very short or wanting; stigmas entire or bifid, dark red; scales as in staminate aments or somewhat narrower.---Alpine areas where the snow lingers late in the season. Alberta to British Columbia, south to New Mexico and Oregon. In Colorado known only from Trail Ridge in the Rocky Mountain National Park at 11,700 feet, probably in many alpine areas.

30A. **Salix nivalis saximontana** (Rydb.) Schn., (var.) Bot. Gaz. 67:47. 1919.
   Salix saximontana Rydb.---Differs from the species by its larger size in all its parts (leaves 1.5-2.5 cm. long and 1-2 cm. wide; aments with 10 to 20 flowers, 1-2 cm. long) at the tips of coarsely pubescent twigs.---Distributed the same as the species and also Nevada and New Mexico. In Colorado widely distributed above timberline at 10,500-12,500 feet.

Family 29. Juglandaceae WALNUT FAMILY

Trees or occasionally shrubs; leaves alternate, pinnately compound, strong-scented; flowers monoecious, the staminate in long drooping catkins, the pistillate solitary or a few in a cluster; staminate perianth 3- to 5-lobed, stamens 8-40; pistillate flower bracted and usually 2-bracteolate, the calyx 4-lobed and with 4 small petals somewhat adnate to the ovary; ovary inferior, 1-celled or incompletely 2- to 4-celled, styles 2; fruit drupelike with a fibrous husk enclosing a nut.

1. Juglans L. WALNUT

Characters of the family in our Area

1. Juglans rupestris Engelm., in Sitgreaves Zuni & Colo. Rept. 171. 1853.
J. major (Torr.) Heller---The var. major Torr. is from Mexico, New Mexico and Arizona and reported for Colorado but the writer has seen no specimens from the state and doubts its occurrence here.

Family 30. Corylaceae BIRCH FAMILY

Trees or shrubs; leaves alternate, simple, deciduous, stipulate; plants monoecious, the flowers in aments (at least the staminate), these appearing with or before the leaves; staminate catkins elongated and pendulous, bracts subtending 2-3 flowers each with a minute 4-parted calyx, or perianth lacking; pistillate catkins conelike, seldom drooping, without a perianth, bracts with 2 or 3 flowers at base, or else the pistillate flowers remaining enclosed in buds at anthesis and only the stigmas protruding; ovary superior or inferior, 1- to 2-celled, style 2-cleft or divided; fruit a 1-seeded nut or samara (Betulaceae).

1. Pistillate flowers few, in a scaly bud at anthesis (only the stigma protruding); fruit a large nut over 1 cm. long enclosed in an involucre -1. Corylus
1. Pistillate flowers many in a catkin; fruit a small nutlet, narrowly or widely winged
   2. Scales of fruiting catkin woody and persistent; stamens 4; some of buds usually stalked with 2 or 3 valvate scales the length of the buds; pith of twigs triangular in section -2. Alnus
   2. Scales of fruiting catkin thin and deciduous; stamens 2, each bifid at apex; none of buds stalked but with several scales, the outer shorter than the buds; pith round in section -3. Betula

1. Corylus L. HAZEL-NUT; FILBERT

Shrubs with smooth bark; buds with several imbricate scales; staminate catkins pendulus, elongated, present over winter, bracts with 4-8 stamens, no perianth; pistillate inflorescence headlike, each enclosed in a scaly bud, only the red stigmas protruding; ovary usually with 1, rarely 2 ovules in each of the 2 cells; stigmas 2: fruit a nut, included or surrounded by a large involucre.

1. Corylus cornuta Marsh., Arbust. 37. 1785.
C. rostrata Ait.---Shrubs 1-2 m. tall; twigs brownish, somewhat pubescent to glabrous; leaves 4-12 cm. long, ovate, obovate to ovate-oblong, subcordate at base, abruptly acuminate, densely serrate and sometimes lobulate, glabrescent above, downy-pubescent, especially on the veins below, petioles 6-15 mm. long; involucre tubular, constricted above the nut into a tubular beak 2.5-4 cm. long, this densely bristly and lacerate at summit; nut ovoid, 1-1.5 cm. long.--- Thickets. Nova Scotia to North Dakota, south to Georgia, Iowa and Colorado. Our records from central and northcentral Colorado at 5400-8000 feet, mostly along the eastern foothills.

2. Alnus Mill. ALDER

Shrubs or trees; buds stalked, with 2 or 3 scales; staminate catkins elongated, each bract with 3 flowers; pistillate catkins short, each bract with 2 flowers, becoming woody, bracts persistent; fruit a winged or wingless nutlet.

1. Alnus tenuifolia Nutt., N. A. Sylva 1:32. 1842.
Large shrubs or small trees up to 10 m. tall, with thin red bark; twigs pubescent especially at first; buds reddish, puberulent; leaves ovate or oblong-ovate, obtuse, acute or short-acuminate at apex, rounded or broadly cuneate to subcordate at base, more or less lobulate and doubly serrate, yellow-green, glabrous or pubescent below on veins and in their axils; cones when ripe 1-2 cm. long, ellipsoid.---Along streams and in canyon floors. Alaska and Yukon, south to New Mexico and California. Scattered over the mountainous two-thirds of Colorado at 5000-10,000 feet.

3. Betula L. BIRCH

Trees and shrubs, outer bark smooth, often separable in sheets; twigs and buds often aromatic; winter buds with several scales; staminate bracts with 3 flowers; pistillate catkins oblong or cylindrical, bracts conspicuous, 3-lobed at apex (in ours), deciduous, 3-flowered; fruit a small samaralike nutlet.

1. Bark white or gray, finally separating into sheets, the outer peeling in shreds; leaves usually over 4 cm. long -1. B. papyrifera
1. Bark dark brown or gray-brown, not peeling or separating; leaves usually less than 4 cm. long
  2. Low shrubs less than 2 m. tall; leaves small, seldom over 2 cm. long, rounded at apex; cones 1-2 cm. long; bracts with lateral lobes equalling the middle; wing of samara narrow -2. B. glandulosa
  2. Treelike shrubs or trees over 2 m. tall; leaves 2-4 cm. long, acute or short-acuminate at apex; cones 2-3 cm. long; bracts with lateral lobes shorter than the middle; wing of samara wide (wider than nutlet body) -3. B. occidentalis

1. Betula papyrifera Marsh., Arb. Am. 19. 1785.
B. andrewsii A. Nels.---Trees up to 25 m. tall but in our range reduced and seeming to grow in stools or clumps of few-several stems up to 10-15 cm. in diameter; bark on older trunks chalky-white to silvery-gray, peeling naturally into thin layers; leaves ovate, abruptly acuminate or long-acute, rounded or cuneate at base, irregularly or doubly serrate with short teeth; staminate aments 4-10 cm. long, 2 or 3 at the ends of the twigs; pistillate aments 2-4 cm. long, peduncled, solitary; fruiting bract 3-lobed, the middle one longer and narrower than the others.---Woods. Labrador to Alaska, south to New Jersey and Colorado. Our only record comes from a small stand in Boulder County.

2. Betula glandulosa Michx., Fl. Bor. Am. 2:180. 1803.
Low shrubs 1-2 m. tall, or procumbent at high elevations; twigs and branches glabrous but densely resinous-glandular; leaves small, 1-2.5 cm. long, suborbicular to obovate, apex rounded, base rounded or broadly cuneate, margin crenate-serrate except near base, glabrous but gland-dotted below with 2-4 pairs of veins, petioles 2-10 mm. long; catkins in fruit 1-2 cm. long, 4-5 mm. thick; bractlets puberulent, with 3 equal lobes, the lateral lobes ascending; samara with a narrow wing.---Moist or wet places. Greenland to Alaska, south to Maine, Colorado and Oregon. Scattered in the mountainous part of Colorado at 7500-11,000 feet but no records as yet from the southcentral part.

3. Betula occidentalis Hook., Fl. Bor. Amer. 2:155. 1839.
B. fontinalis Sarg.---Treelike shrubs to small trees to 12 m. tall but usually forming clumps; bark thin, smooth, dark brown, marked by horizontal lenticels; twigs densely resinous-glandular; leaves 2-4 cm. long, broadly ovate to ovate, acute or short-acuminate, sharply lobed, entire near base, rounded, truncate or broadly cuneate at base, thin, bright green, resinous, glabrous or puberulent, with 3-5 pairs of veins, petiole 5-15 mm. long; catkins in fruit 2-3 cm. long, 5-10 mm. thick; bractlets pubescent, lateral lobes ascending, shorter than the middle lobe; wing of samara much wider than the nutlet.---River banks and plains. Saskatchewan to British Columbia, south to Nebraska, New Mexico and California. Scattered over the mountainous two-thirds of Colorado at 5000-9000 feet.

Family 31. Fagaceae BEECH FAMILY

Shrubs or small trees; leaves alternate, simple, petioled and (in ours) toothed or lobed, deciduous or persistent over winter, usually 2 or 3 crowded out near end of twig; flowers monoecious; staminate flowers in pendulous catkins with a perianth of 4-7 lobes, and 5-10 stamens; pistillate flowers solitary or clustered, in an involucre of flat scales, perianth 6-lobed; ovary 1-celled with 3 styles, superior; fruit a nut (acorn), 1-seeded, partly enveloped by an involucre of scales (cup).

1. Quercus L. OAK

Characters of the Family

Our species are very variable and have been hopelessly split by Rydberg and others. Often two species can be found on the same bush. The mature acorn is always desirable and often necessary for identification. Our species mature their acorns in the fall but during the summer the old acorns may remain on some of the twigs or be found on the ground below the plant.

1. Leaves undulate, dentate or very shallowly lobed (the sinuses less than 1/3 the distance in the midrib), persistent until spring
  2. Acorn 15-20 mm. long, the cup usually somewhat turbinate at base; leaves 1-3 cm. long with reddish tomentum below when young -1. Q. turbinella
  2. Acorn 10-15 mm. long, the cup hemispherical; leaves 3-5 cm. long, without reddish tomentum (may be light brown)
    3. Leaves sinuately lobed, the lobes spinulose-tipped, the leaf strongly crisped and the lobes often appearing twisted -2. Q. pungens
    3. Leaves sinuately dentate, the teeth mucronate, leaf only slightly if at all crisped -3. Q. undulata
1. Leaves lobed (the sinuses extending 1/3 or more the distance to the midrib) usually deciduous in the autumn (except Q. pungens)
  4. Leaves with midvein and lateral veins (hence end of lobes) ending in a mucro or a weak spine
    5. Lobes of leaves mucronate, leaves flat, deciduous before spring -4. Q. fendleri
    5. Lobes of leaves spinulose-tipped, leaves crisped, the lobes often twisted, persistent until spring -2. Q. pungens
  4. Leaves with midrib and lateral veins not ending in a mucro or spine (apex of leaf and end of lobes rounded or sometimes acute)
    6. Leaves oblong to oblanceolate, over $2\frac{1}{2}$ times longer than wide, 3-6 cm. long, usually blue-green above; mature acorns usually less than 10 mm. long -5. Q. venustula

6. Leaves obovate to rather broadly oblong, not over 2½ times longer than wide, 5-10 cm. long, usually bright green above; mature acorns 12-15 mm. long
   7. Acorn cup hemispherical, covering 1/3 or more of the acorn; plants usually over 2 m. tall -6. Q. gambellii
   7. Acorn cup saucer-shaped, covering less than 1/4 the acorn; plants seldom over 2 m. tall -7. Q. vreelandii

1. Quercus turbinella Greene, Illus. W. Amer. Oaks 1:37. 1889.
   Q. subturbinella Trel.---Shrubs 1-3 m. tall; bark dark brown or gray; twigs tomentose; leaf blades 1-3 cm. long, thick, persistent until spring, oblong, elliptical-oblong, oval or rarely ovate, cuneate to cordate at base, acute at apex, sinuate-dentate with spinulose-tipped teeth, upper surface light blue-green and somewhat shiny, lower surface stellate-pubescent to glabrate, often fulvous especially when young, strongly reticulated; cup hemispheric, often somewhat turbinate at base, 8-12 mm. in diameter; acorn 15-20 mm. long.---Dry hills and slopes. Colorado to California, south to Mexico. Our one record from western Montrose County at 5500 feet.
2. Quercus pungens Liebm., Overs. Danske Vidensk. Selsk. Forh. 1854:171. 1854.
   Texas to Arizona. Has been recorded for southern Colorado, but no specimens could be located by the writer.
3. Quercus undulata Torr., Ann. Lyc. N. Y. 2:248. 1828.
   Q. pauciloba Rydb.---Shrubs 1-3 m. tall; bark gray and rough; young twigs sparingly stellate-pubescent; leaf blades firm and persistent until spring, rather blue-green or light-green, glabrate or stellate-pubescent below in age, acute at apex, cuneate at base, oblong or ovate-oblong, margins sinuate-dentate, teeth mucronate (sometimes obscurely so); cup hemispheric, 7-10 mm. in diameter; acorn 10-15 mm. long.---Dry hills and slopes. Colorado to Utah, south to New Mexico and Arizona. Our records mostly from southern Colorado (one from Douglas County) at 4000-6500 feet. It seems to be most common in the southeastern part of the state.
4. Quercus fendleri Lieb., Overs. Dansk. Vid. Selsk. Forh. 170. 1854.
   Shrubs 1-3 m. tall; bark of older branches gray; twigs and buds brownish, puberulent; leaf blades deciduous before spring, oval or elliptic, lobed about halfway to the middle, the lobes somewhat triangular, acute and mucronate, upper surface pale green and glabrate or sparingly stellate, lower stellate-pubescent; cup hemispheric, 10-12 mm. in diameter; acorn 12-18 mm. long.---Dry hills and slopes. Colorado to Texas and New Mexico. Our few records from the southern part of Colorado as far north as El Paso County at 5500-6500 feet.
5. Quercus venustula Greene, N. A. Oaks 2:69. 1890.
   This species is listed for Colorado, and New Mexico but the writer could not locate any good material from this state. The species may have to be merged with Q. gambellii.
6. Quercus gambellii Nutt., Journ. Acad. Phila. ser. 2, 1:179. 1848.
   Q. gunnisonii Rydb., Q. novomexicana (A. DC.) Rydb.; Q. novomexicana var. andrewsii Trel.; Q. utahensis (A. DC.) Rydb.; Q. leptophylla Rydb.---Shrubs or trees, usually 3-5 m. tall but sometimes taller or shorter; leaves 6-10 cm. long, rarely less, deciduous before spring, very variable, usually narrowly or broadly obovate in outline, lobed usually over 1/2 way to the midrib, sometimes almost to it, the lobes acute to obtuse, not mucronate, varying from velvety-pubescent below to glabrate, usually bright green above; cup 10-15 mm. wide covering about 1/3 to 1/2 the acorn; acorn usually 12-20 mm. long. This is a complex and variable species. However, the various segregates that have been proposed are of no value for Colorado plants, often 2 or more being present on the same individual.---Dry hills, slopes and along streams. Colorado to Nevada, south to New Mexico and Arizona, perhaps of wider distribution. Widely scattered over Colorado at 4000-8500 feet except the eastcentral, northeast and northcentral parts.
7. Quercus vreelandii Rydb., Bul. N. Y. Bot. Gard. 2:204. 1901.
   This differs from the preceding only in the short acorn cup which is less than 1/4 as long as the acorn.---Reported from Colorado and New Mexico. A few specimens from southern Colorado seem distinct.

Family 32. Saururaceae LIZARDTAIL FAMILY

Plant perennial, scapelike from creeping-rootstocks; leaves mostly basal, simple, petioled; flowers in a compact spike surrounded at base with a persistent colored involucre of 5-8 bracts; flowers perfect, bracteolate; perianth none; stamens 6-8; ovary superior, of 3-4 carpels, these distinct or united, each 1- or 2-ovuled, the placenta parietal, the ovary sunken in the rachis of the spike, styles 3-4, distinct; fruit a capsule, dehiscent at the apex.

1. Anemopsis Hook. et Arn. YERBA MANSA

Characters of the Family

1. Anemopsis californica Hook., Ann. Nat. Hist. 1:136. 1838.
   Stems 15-50 cm. long, ours glabrous or nearly so, with a broadly ovate clasping leaf above the middle and a fascicle of 1-3 small petioled leaves in its axil; basal leaves petioled, this 10-30 mm. long, blades 5-15 cm. long, elliptic-oblong, rounded above, narrowed somewhat to a cordate base; spikes 1.5-4 cm. long; involucre bracts 12-30 mm. long, white or reddish below; floral bracts whitish, 4-6 mm. long, clawed. Our plant has been called var. subglabrata Kelso.---Saline or moist ground. Colorado to California, south to Texas and Mexico. Apparently rather recently introduced in the eastern half of Colorado, our few records from 3500-5000 feet.

Family 33. Moraceae MULBERRY FAMILY

Trees, shrubs, herbs or vines; leaves opposite, palmately-lobed or divided; flowers small, regular, dioecious; staminate flowers in panicles, perianth parts 5, stamens 5; pistillate flowers in bracted spikes, petals none; ovary 1-celled, 1-ovuled, superior, stigmas 2; fruit an achene. Our genera are often placed in a separate family the Cannabinaceae.

1. Erect annual herbs; stems not prickly; leaves divided to very near base -1. Cannabis
1. Twining vines; stems retrosely (although weakly) prickly; leaves not divided to base (deeply lobed, however) -2. Humulus

1. Cannabis L. HEMP; MARIJUANA

Erect annual herbaceous plants; leaves opposite, petioled, palmately 5- to 11-divided; staminate flowers in long panicled racemes; pistillate in leaf-bracted spikes or glomerules.

1. Cannabis sativa L., Sp. Pl. 1027. 1753.
   Stems 1-4 m. tall, branched or rather simple; leaf blades divided to the bases but cordate or ovate in general outline, the segments 5-15 cm. long, linear, serrate, acuminate at apex, long-cuneate at base, usually pubescent both sides, darker green and often glabrate above.---Waste places, escaped from cultivation. New Brunswick to Minnesota, south to Georgia and Colorado. Our records from northcentral Colorado at 5000-7000 feet., some from cultivated plants. It is illegal to grow the species in this state.

2. Humulus L. HOP

Perennial twining vines; stems and petioles retrorsely prickly (though weak); leaves opposite, palmately-lobed; staminate flowers in loose panicles; pistillate flowers 2 together under a large persistent bract which with the others at maturity forms a large conelike "hop."

1. Humulus americanus Nutt., Journ. Acad. Nat. Sc. Phil. ser. 2. 1:181. 1847.
   H. lupulus var. neomexicanus Nels. & Ckll.; H. neomexicanus (A. Nels. & Ckll.) Rydb.---Vines 5-10 m. long; leaf blades 3- to 7-divided or parted usually over 1/2 way to base, with 3 main lobes, the divisions lanceolate or narrow, terminal one narrower at base than at middle, over twice as long as wide, acuminate or narrowly acute, somewhat scabrous above, lower surface usually copiously glandular, lobes of leaves serrate, petiole about as long or shorter than the blades, rough; bracts of pistillate flowers ovate or lanceolate, acute or acuminate. Our plants have rather variable leaves, some with few obtuse shallow lobes resembling the cultivated species, varying to the typical native form described above. It has been suggested that hybridization may be responsible for this variation.---Usually among bushes. Wyoming to Utah, south to New Mexico and Arizona. Our records scattered over the mountainous part of Colorado at 4000-9000 feet.

Family 34. Ulmaceae ELM FAMILY

Trees or shrubs, bark often becoming corky in age, with pith finely chambered at the nodes; leaves alternate, simple, unequal at base, deciduous, somewhat palmately-veined; flowers perfect or unisexual, axillary, solitary or in small clusters; perianth 4- to 6-parted; stamens 4-5; ovary superior, 1-celled, style none, stigmas 2, spreading or recurved; fruit a drupe with thin flesh and hard-shelled seed.

Various species of elm are in cultivation in Colorado and may possibly escape. Also Ulmus americana L. is native in western Kansas and Nebraska and might be found in Colorado. However, only the genus Celtis is included here.

1. Celtis L. HACKBERRY

Characters of the Family

1. Celtis occidentalis L., Sp. Pl. 1044. 1753.
   Shrubs to trees up to 40 m. tall, the older bark becoming corky and wartlike but the larger branches often remaining smooth; leaves 2-10 cm. long, broadly ovate to ovate-lanceolate, rounded to cordate at base, short-acuminate to acute at apex, serrate to almost entire, smooth to scabrous above, glabrous to hairy below. As here considered a variable complex including what has been called C. crassifolia Lam., C. reticulata Torr., C. douglasii Planch. and C. rugulosa Rydb. The writer cannot separate Colorado material into any of the various species or varieties that have been proposed, in view of the many intergradations encountered. Fernald and Schubert (Rhodora 50:155-162. 1948) limited the use of the species term to what has been passing as C. occidentalis var. crassifolia (Lam.) Gray and took up C. tenuifolia Nutt. for what has been called C. occidentalis var. pumila (Pursh) Gray.---Widely distributed in the United States and Mexico. Our records scattered over Colorado at 4000-7200 feet. The plant with leaves scabrous above seems to be limited to the eastern half of the state.

Family 35. Urticaceae NETTLE FAMILY

Herbaceous plants; leaves alternate or opposite; flowers perfect or unisexual, greenish, axillary; perianth of 2-5 small calyxlike lobes; no petals present; stamens as many as perianth segments and opposite them; ovary superior, 1-celled; style simple; fruit an achene.

1. Perianth parts distinct or nearly so; flower clusters not involucrate; leaves opposite, toothed; plants with stinging hairs -1. Urtica
1. Perianth parts partly united in pistillate flowers; flower clusters involucrate with foliose bracts; leaves alternate and entire; plants without stinging hairs -2. Parietaria

1. Urtica L. NETTLE

Annual or perennial herbaceous plants, the herbage usually hispid with stinging hairs; leaves opposite, toothed, with stipules; perianth parts distinct or nearly so, 2 become enlarged in the pistillate flower; flower cluster not involucrate by leafy bracts.

1. Teeth of leaves broadly triangular, only slightly directed forward; petioles of lower leaves more than 1/2 as long as the blades -1. U. gracilenta
1. Teeth of leaves ovate, strongly directed forward; petioles seldom over 1/2 as long as the blades
  2. Leaves broadly ovate, cordate at base -2. U. dioica
  2. Leaves lanceolate to ovate-lanceolate, rounded to subcordate at base -2A. U. dioica procera

1. Urtica gracilenta Greene, Bull. Torr. Club 8:122. 1881.
   Stems 80-200 cm. tall, slender, simple or sparingly branched from the base, nearly glabrous to somewhat strigose and bristly; leaves thin, petioles long especially on lower leaves, blades 5-15 cm. long, ovate-lanceolate but sometimes wider or narrower, long-acuminate, rounded or broadly-cuneate at base, teeth broadly triangular, not definitely directed forward (more dentate than serrate); flower clusters slender, nearly equalling to slightly exceeding the petioles.---Along streams and in canyons or ditches. New Mexico and Arizona, extending north to Colorado at least. Our one record from southwestern Colorado at 11,400 feet.
2. Urtica dioica L., Sp. Pl. 984. 1753.
   This eastern species has been reported from Colorado and some specimens do tend to key out to it on the basis of lower leaves being somewhat cordate. It is represented in Colorado by the following variety.
2A. Urtica dioica procera (Muhl.) Wedd., (var.) Mon. Fam. Urt. 78. 1856.
   U. gracilis Ait.; U. viridis Rydb.---Stems 50-200 cm. tall, slender, sparsely bristly but glabrous above, or sparingly pilose; leaves lanceolate to ovate-lanceolate, acuminate or rounded to cordate at base, stipulate, glabrous or sparingly pilose beneath, petiole length variable, 1-5 cm. long; flower clusters shorter than the leaves.---Waste places and alluvial ground. Widely distributed in temperate North America. Our records from the western two-thirds of Colorado at 4500-9700 feet, but should be expected anywhere in the state.

2. Parietaria L. PELLITORY

Annual herbaceous plants, without stinging hairs; leaves alternate, entire, 3-veined, without stipules; perianth parts united in fertile flowers; flower-cluster involucrate by leafy bracts.

1. Involucre 2-3 times as long as the flower; leaf blades 2-7 cm. long, commonly lanceolate, narrowly cuneate below to a petiole less than 1/2 the length of the blade; stem simple or sparingly branched from below -1. P. pensylvanica
1. Involucre not over 2 times longer than the flower; leaf blades 0.5-2 cm. long, commonly ovate, rounded or widely cuneate to a petiole usually 1/2 or more the length of the blade; stem diffusely branched from the base -2. P. floridana

1. Parietaria pensylvanica Muhl. ex. Willd., Sp. Pl. 4:955. 1806.
   P. occidentalis Rydb.---Stems 10-40 cm. tall, slender, erect or ascending, simple or branched below, puberulent to villous; leaf blades 2-7 cm. long, lanceolate or ovate-lanceolate, narrowly cuneate at base, obtuse to acuminate at apex, petioles less than 1/2 as long as the blades; bracts of involucre linear, 2-3 times as long as the flowers.---Rocks or shaded places. Widely distributed in temperate North America. Our records from central and northcentral Colorado at 5000-7500 feet but the plant should be expected anywhere in the state.
2. Parietaria floridana Nutt., Gen. Pl. 2:208. 1818.
   P. obtusa Rydb.; P. debilis of Manuals---Stems 5-25 cm. long, very slender, usually diffusely branched from base, finely villous; leaves 5-20 mm. long, oblong to ovate-oblong, rarely lanceolate, base rounded or rather widely cuneate, obtuse to acute, petioles often 1/2 or more longer than the blades; bracts of involucre oblong, not much longer than the flowers.---Shaded ground. Southeastern United States to California. Our records from northcentral Colorado at 4900-6700 feet but the plant is to be expected anywhere in the state.

Family 36. Loranthaceae MISTLETOE FAMILY

Plants parasitic on trees and shrubs, absorbing food through specialized roots (haustoria) and with or without chlorophyll; stems swollen-jointed, brittle when dry; leaves opposite, mostly reduced to scales; flowers dioecious; perianth calyxlike with 2-5 lobes or teeth these coherent at base; stamens as many as perianth lobes and inserted on the tube; ovary inferior, 1-celled, style and stigma 1; fruit a berry.

1. Anthers 1-celled; perianth of pistillate flower 2-lobed; fruit compressed, green, blue or purplish; parasitic on evergreen conifers other than the genus Juniperus   -1. **Arceuthobium**
1. Anthers 2-celled; perianth of pistillate flower normally 3-lobed; fruit globose or subglobose, whitish or reddish; parasitic on the genus Juniperus or on deciduous trees   -2. **Phoradendron**

1. Arceuthobium M. Bieb.   SMALL or DWARF MISTLETOE

Parasitic on conifers; plants without chlorophyll, the leaves reduced to connate scales; flowers axillary but often crowded into apparent spikes or panicles; perianth of staminate flowers 3-lobed, the pistillate 2-lobed; anthers 1-celled; berry compressed, at maturity ejecting the seeds forcibly several yards.

1. Accessory branches arising from collateral buds forming a whorled arrangement; staminate flowers at the end of the branches on distinct pedicels in a paniculate cluster; host almost exclusively Pinus contorta   -1. **A. americanum**
1. Accessory branches arising from superimposed buds in fanlike arrangement (flat, 2-ranked sprays); staminate flowers nearly all axillary in the scales of a simple or compound spike; host rarely if ever P. contorta
   2. Stems slender, about 1 mm. or less in diameter at base, seldom over 2 cm. long; plants scattered along the stem of the host plant; host genus Pseudotsuga   -2. **A. douglasii**
   2. Stem relatively stout, the largest over 2 mm. in diameter at base, seldom less than 2 cm. long; plants commonly clustered on the host plant; host not Pseudotsuga
      3. Stems seldom less than 3 mm. in diameter at base; plants yellowish, normally flowering in the early summer; host Pinus ponderosa   -3. **A. vaginatum**
      3. Stems 2-3 mm. in diameter at base; plants usually olive-green or brown; flowering in late atumn; hosts various, not Pinus ponderosa   -4. **A. camylopodum**

1. Arceuthobium americanum Nutt. ex Engelm. in Gray, Bost. Journ. Nat. Hist. 6:214. 1850. Razoumofskya americana Kuntze---Stems usually olive-green; shoots 30-60 mm. long, much branched, the branches from collateral buds to form a whorl; staminate flowers usually yellowish-green, always borne singly on the end of pedicellike segments which arise in whorls or pairs at stem nodes; pistillate flowers in pairs or whorls at the stem nodes; flowers in the spring, April to June.---Parasitic almost exclusively on Pinus contorta. Saskatchewan to British Columbia, south to Colorado and California. Our records scattered in the range of the host in the mountains of Colorado at 8000-10,000 feet.
2. Arceuthobium douglasii Engelm., U. S. Geol. Sur. W. of 100th Meridian 6:253. 1878. Razoumofskya douglasii (Engel.) Kuntze---Stems short, about 8-20 mm. long, rarely to 35 mm., less than 1 mm. in diameter, usually olive-greenish in color, branched, diffuse (rarely in tufts) on younger branches of host; staminate flowers usually green or yellowish-green, mostly axillary in pairs in spikelike inflorescences, but may be single on ends of pedicellike segments; pistillate flowers axillary, mostly in pairs; flowers in the spring, April to June.---Parasitic on the genus Pseudotsuga. Montana to British Columbia, south to Mexico and California. Scattered with the host plant in the mountains of Colorado at 7000-8000 feet.
3. Arceuthobium vaginatum (Willd.) Presl. in Berch., O. Priroz. rostlin aneb. Rost. 2:28. 1825. Stems variable in size, 2-15 cm. long, more than 2 mm. in diameter at base, usually over 3, brownish-yellow or less commonly olive-green, robust, the pistillate longer lived and becoming larger, much branched, the branches arising from superimposed buds in a fanshaped arrangement, usually in tufts on the host but sometimes scattered on the younger twigs; staminate flowers light green to yellow, normal ones axillary, never subtended by pedicellike segments; pistillate flowers axillary, the lateral in pairs; flowering in April to June.---Parasitic on the genus Pinus. Colorado to Utah, south to Texas and Mexico. Our records from plants on Pinus scopulorum, in the mountains of Colorado at 6500-8000 feet.
4. Arceuthobium campylopodum Engelm. in Gray, Bost. Journ. Nat. Hist. 6:214. 1850. Razoumofskya campylopoda (Engelm.) Kuntze---Stems 2-15 cm. long or more, the largest 2-4 mm. in diameter, often much branched, accessory branches arising from superimposed buds in fanlike arrangement, mostly in tufts on various sized branches of the host or sometimes scattered on the twigs; staminate flowers mostly light green to yellow, mostly in pairs at the nodes, never subtended in pedicellike joints; pistillate flowers mostly axillary, the lateral in pairs; blooming in late summer or early autumn, August to September.---This species has been divided into several forma depending on the host. Of these, five may be in Colorado on the following host plants:

1. Pinus flexilis and P. aristata = forma cyanocarpum (A. Nels.) Gill.
2. Pinus edulis and P. monophylla = forma divaricatum (Engelm.) Gill.
3. Abies lasiocarpa and A. concolor = forma abietinum (Engelm.) Gill.
4. Picea engelmanni and P. pungens = forma microcarpum (Engelm.) Gill.
5. Pinus contorta = forma typicum Gill.

The writer has seen Colorado specimens of only the first two of the above list. The forma cyanocarpum is fairly common in the mountains of Colorado, our records from 8000-10,000 feet.

2. Phoradendron Nutt. AMERICAN MISTLETOE

Plants with or without chlorophyll; leaves foliaceous or reduced to connate scales; inflorescence axillary, of several jointed spikes; perianth of staminate flower 3-lobed, the pistillate 3-lobed also; anthers 2-celled; berry globose, not ejecting the seed at maturity.

1. Leaves reduced to scales about 2 mm. long; host plants species of Juniperus -1. P. juniperinum
1. Leaves with well-developed blades, usually over 1 cm. long; host plants various deciduous trees -2. P. cockerellii

1. **Phoradendron juniperinum** Engelm., Mem. Amer. Acad. n. s. 4:58. 1849.
Stems about 10-30 cm. long, stout, ultimate branches somewhat quadrangular, glabrous; leaf scales 1-2 mm. long, spreading, deltoid; spikes solitary, 3 mm. long with either 2 pistillate or 6 to 8 staminate flowers; fruit subglobose, straw- or wine-colored.---Parasitic on species of the genus Juniperus. Colorado to Utah, south to Texas and Mexico. Our records from southwest and westcentral Colorado at 5300-7500 feet.

2. **Phoradendron cockerellii** Trelease, Monog. Phoradend. 38. 1916.
This southern species has been listed for our state and a specimen with the label "Leadville" has been seen by the writer. However, the locality data is open to question. The plant may be expected in the southern part of the state.

Family 37. Santalaceae SANDALWOOD FAMILY

Herbaceous glabrous perennial plants with horizontal rootstocks, these developing narrow processes with holdfast organs at the end and fastened to the underground parts of other perennial plants in the vicinity; leaves alternate, simple, sessile, entire; flowers in small terminal or in crowded axillary clusters appearing terminal, the whole corymbose; flowers regular, perfect, small, rose-tinged, pink, white or green, no petals present; perianth 4- or 5-cleft, somewhat united above ovary, the tube lined with a fleshy disk; stamens 5, anthers with a tuft of hairs at base; ovary inferior, 1-celled, styles 1 if present, stigmas capitate; fruit a 1-seeded drupe or nut.

1. Comandra Nutt. BASTARD TOADFLAX

Characters of the Family

1. **Comandra umbellata** (L.) Nutt., Gen. 1:157. 1818.
C. pallida A. DC.---Stems 10-40 cm. tall; leaves 15-35 mm. long, linear, linear-lanceolate to lanceolate or oblong, acute or obtuse, pale and glaucous; flowers 3-4 mm. long, perianth 5-parted; fruit globose or ovoid, about 4-5 mm. in diameter.---Plains and foothills. Widely distributed over the United States and western Canada. Our western plant has been segregated as C. pallida and may possibly be worthy of varietal status. Scattered widely over Colorado at 3500-9000 feet.

Family 38. Polygonaceae BUCKWHEAT FAMILY

Herbaceous plants or sometimes shrubs or trees; leaves alternate or sometimes opposite or whorled, simple, often all basal, generally entire; stipules when present joined into a tube sheathing the stem; flowers mostly perfect, regular; perianth 2- to 6-parted or cleft, these lobes often partly or completely petaloid, petals as such wanting; stamens 2-9; ovary superior, 1-celled with 1 ovule, styles or stigmas 2 or 3; fruit an achene.

1. Flowers subtended by a campanulate, turbinate or cylindrical, gamophyllous involucre; stipules lacking -1. Eriogonum
1. Flowers not subtended by an involucre; stipules present as sheaths around the stems above the nodes
  2. Sepals 4; styles 2; leaf blades reniform and basal -2. Oxyria
  2. Sepals 5 or 6; styles 2 or 3; leaf blades various but not reniform and basal
    3. Sepals 5 (rarely 4), all similar and erect in fruit
      4. Achenes little if any exceeding the perianth, if exserted the leaf blades narrow and cuneate at base; not cultivated crop plants -3. Polygonum
      4. Achenes long exserted; leaf blades deltoid, hastate or cordate; cultivated crop plants, sometimes escaping -4. Fagopyrum
    3. Sepals 6, in 2 sets, the outer reflexed, the inner erect and usually enlarging in fruit
      5. Sheaths (ocreae) large and prominent; stamens mostly 8-10; garden plants rarely persisting or even escaping -5. Rheum
      5. Sheaths not very prominent; stamens 6; weedy plants, not cultivated -6. Rumex

1. Eriogonum Michx. UMBRELLA PLANT

Annual or perennial herbaceous plants or sometimes suffrutescent or shrubby; leaves basal or cauline, alternate, opposite or whorled, entire and lacking stipules; flowers in involucrate clusters variously disposed; involucres companulate, cylindric or turbinate, 4- to 8-toothed or lobed; flowers perfect; perianth more or less petaloid, parted or deeply cleft, with 6 segments; stamens 6-9; styles 3-parted; achene 3-angled or 3-winged. This is a large and difficult genus in this state and it is badly in need of a thorough revision at this writing. The names used here have been selected very arbitrarily.

1. Plants annual or sometimes biennial but without woody caudices
    2. Plants scapose, the stem leaves all or nearly all reduced to scalelike bracts, large leaves confined to the bases of the stems
        3. Involucres sessile (except sometimes those in the forks of the inflorescence) or on very short peduncles not over 2 mm. long
            4. Involucres sessile; perianth rose-colored; leaves ovate or oblong, not over 1 cm. wide
                -1. E. densum
            4. Involucres usually very short-peduncled; perianth yellowish; leaves suborbicular, 2-6 cm. wide
                -2. E. hookeri
        3. Involucres on peduncles over 4 mm. long
            5. Perianth hairy (usually whitish-hispidulose) on the outside, usually yellow or sometimes fading to pink
                6. Scapes and often the peduncles glandular-pubescent; involucres often cleft to below the middle
                    -3. E. glandulosum
                6. Scapes and peduncles not glandular-pubescent; involucres not cleft to below the middle
                    7. Scapes definitely inflated on most plants (inflated part usually over 1 cm. wide); leaves glabrous to sparsely pilose or hirsute below
                        8. Accessory branches at the lower forks of the inflorescence many and nearly as long as the primary ones, all divaricate; plants annual
                            -4. E. fusiforme
                        8. Accessory branches of the lower forks of the inflorescence few and small or none, branches if present ascending; plants perennial
                            -5. E. inflatum
                    7. Scapes not inflated or only slightly so; leaves sometimes tomentose to lanate below
                        9. Perianth yellow; leaves villous to hirsute below (often sparsely so) -4. E. fusiforme
                        9. Perianth white to rose (may fade to yellowish); leaves tomentose to lanate below
                            -6. E. subreniforme
            5. Perianth glabrous, usually white to rose (sometimes fading to yellowish)
                10. Leaves glabrous or sparingly pilose below
                    -7. E. gordoni
                10. Leaves densely tomentose or lanate below
                    11. Involucres not over 1 mm. long; perianth 1-1.5 mm. long; peduncles filiform
                        12. Scapes more or less villous or tomentose at the nodes, usually over 25 cm. tall; perianth usually white or rose
                            -6. E. subreniforme
                        12. Scapes glabrous, seldom over 20 cm. tall; perianth yellow with reddish veins
                            -8. E. wetherillii
                    11. Involucres 1.5-2 mm. long; perianth usually over 1.5 mm. long; peduncles slender but hardly filiform
                        13. Peduncles more or less reflexed or decurved; involucres often over 10-flowered; outer lobes of perianth not broadly flabellate, usually more or less panduriform
                            -9. E. cernuum
                        13. Peduncles spreading, not reflexed; involucres usually about 5-flowered; outer lobes of perianth broadly flabellate
                            -10. E. rotundifolium
    2. Plants not scapose, the stems leafy, with the lower stem leaves like the basal except smaller
        14. Leaves glabrous or glabrate, not at all hairy below; perianth hispidulous, yellowish to greenish
            -11. E. salsuginosum
        14. Leaves tomentose to short-pilose below; perianth glabrous, minutely glandular to hispidulous, white, rose or yellow
            15. Involucre bracts scalelike; plants over 30 cm. tall
                -12. E. annuum
            15. Involucre bracts appearing leaflike; plants rarely over 30 cm. tall
                16. Involucres all sessile; leaves oblong-elliptical to orbicular; perianth less than 2.5 mm. long
                    -13. E. divaricatum
                16. Involucre peduncled; leaves linear to narrowly oblanceolate; perianth 2.5-3 mm. long
                    -14. E. pharnaceoides
1. Plants perennial, woody below or at least the caudices thickened and woody
    17. Achenes conspicuously 3-winged; perianth not accrescent
        18. Involucres, stems and leaves manifestly hairy; perianth yellow to greenish -15. E. alatum
        18. Involucres, stems and leaves glabrous (leaves may be ciliate); perianth purplish to yellowish
            -15A. E. alatum triste
    17. Achenes not conspicuously winged; perianth accrescent
        19. Perianth attenuated and stipelike at base (this stipe short in a few species); bracts if present at all, leaflike, rather indefinite in number
            20. Perianth externally hairy
                21. Involucre either solitary at the end of the scapes, in umbels, or in compact or subcapitate clusters
                    22. Leaves not over 12 mm. long; flowering stems rarely over 10 cm. high; plants densely matted; inflorescence of 1 involucre
                        23. Involucre with long lobes, these longer than the tube and reflexed; perianth definitely short-stipitate at base
                            -16. E. caespitosum
                        23. Involucre with lobes much shorter than the tube, not reflexed; perianth very inconspicuously stipitate at base
                            -17. E. acaule
                    22. Leaves usually over 20 mm. long; flowering stems often over 10 cm. long; plants may be caespitose but hardly densely matted; inflorescence usually of more than 1 involucre
                        -18. E. flavum
                21. Involucres in branching open cymes
                    24. Perianth white, cream-colored or pink
                        -19. E. jamesii
                    24. Perianth yellow
                        -20. E. arcuatum

20. Perianth externally glabrous
    25. Stems with a whorl of leaves near the middle (in addition to the bracts of the inflorescence); leaves oblanceolate to linear   -21. E. heracleoides
    25. Stems without a medial whorl of leaves; leaves oblanceolate to orbicular
        26. Perianth pale to ochroleucous or in age often purplish-rose   -22. E. subalpinum
        26. Perianth bright yellow (sometimes tinged with red in age)   -23. E. umbellatum
19. Perianth not attenuated or stipelike at base; bracts usually scalelike and in 3's
  27. Involucres in capitate or umbellate clusters (sometimes reduced to 1)
    28. Perianth externally hairy
      29. Scapes wanting, the heads sessile or on very short peduncles less than 2 cm. long   -17. E. acaule
      29. Scapes present, 2 cm. long or longer
        30. Ovaries and fruits pubescent; perianth usually yellowish; flowering stems usually over 10 cm. long; leaves silky-strigose above, tomentose below
          31. Involucres in umbellate or paniculate clusters; perianth about 2.5-4 mm. long   -24. E. lachnogynum
          31. Involucres in capitate clusters; perianth about 6 mm. long   -25. E. tetraneuris
        30. Ovaries and fruits glabrous; perianth white or rose; flowering stems often less than 10 cm. long; leaves equally hairy above and below
          32. Plants acaulescent; leaves 6-10 mm. long, silky-villous both sides   -26. E. villiflorum
          32. Plants leafy-stemmed and suffruticose; leaves 10-40 mm. long, tomentose both sides   -27. E. multiceps
    28. Perianth glabrous on the outside
      33. Perianth segments unequal, the outer 3 much broader than the inner
        34. Leaves linear, revolute; flowering stems seldom over 6 cm. long   -28. E. bicolor
        34. Leaves spatulate to orbicular, not revolute; flowering stems usually over 10 cm. long
          35. Leaf blades broadly oval to suborbicular, about as broad as long; perianth white, yellow to purplish, rarely ochroleucous
            36. Perianth yellow   -29. E. ovalifolium
            36. Perianth white, rose or purple   -29A. E. ovalifolium purpureum
          35. Leaf blades elliptical, oblong or spatulate, definitely longer than wide; perianth ochroleucous when fresh   -30. E. ochroleucum
      33. Perianth segments equal or nearly so
        37. Perianth yellow, 2-3 mm. long   -31. E. chrysocephalum
        37. Perianth white or rose, 3-5 mm. long
          38. Involucres with lobes about as long as the tube; leaves strongly revolute; peduncles usually longer than the leaves   -32. E. pauciflorum
          38. Involucres shallowly toothed, these much shorter than the tube; leaves slightly if at all revolute; peduncles usually shorter than the leaves   -33. E. coloradense
  27. Involucres racemose or in dicotomous or trichotomous cymes (may be corymbose) but not at all in capitate or umbellate clusters
    39. Perianth pubescent on the outside
      40. Most of the scapes conspicuously inflated; leaves hirsute to glabrate both sides; achenes glabrous   -5. E. inflatum
      40. Scapes not inflated or only slightly so; leaves tomentose below; achenes glabrous to pubescent
        41. Ovaries and achenes hairy; plants not very woody at base; involucres 3-4 mm. long; perianth 3-4 mm. long   -24. E. lachnogynum
        41. Ovaries and achenes glabrous; plants definitely woody at base; involucres 2-3 mm. long; perianth about 2.5 mm. long   -34. E. nebraskense
    39. Perianth glabrous outside
      42. Involucres in rather narrow racemose inflorescences (at least racemose near top)
        43. Leaves crowded on the short caudex, flowering stems not leafy; leaves ovate to elliptical   -35. E. racemosum
        43. Leaves scattered on the suffruticose branches, flowering stems leafy (at least near base); leaves various--see below
          44. Leaves obovate to oblanceolate, 5-20 mm. long; plants more or less caespitose   -36. E. wrightii
          44. Leaves linear or linear-oblong, 20-40 mm. long; plants not caespitose   -37. E. leptocladon
      42. Involucres in repeatedly forked cymes, not at all racemose
        45. Leaves ovate to orbicular, not over 1 cm. long; some of involucres on peduncles up to 2 cm. long   -38. E. tenellum
        45. Leaves oval to linear, some over 1 cm. long (except sometimes in E. contortum with narrow leaves); involucres usually sessile or nearly so
          46. Leaves ovate, oval to oblong, obtuse at apex
            47. Involucres 4-5 mm. long, glabrous or sparingly floccose; plants of southeastern Colorado   -39. E. fendlerianum
            47. Involucres 2-4 mm. long, tomentose; plants of western Colorado
              48. Cymes divaricately branched; leaves usually less than 2.5 cm. long   -40. E. divergens
              48. Cymes with ascending branches; leaves usually over 2.5 cm. long   -41. E. corymbosum
          46. Leaves linear to narrowly oblong, often acute at apex
            49. Leaves not over 2 cm. long, at least somewhat revolute
              50. Leaves only moderately to slightly revolute; perianth 2.5-3 mm. long   -42. E. microthecum

50. Leaves strongly revolute (most of lower surface hidden); perianth 2-2.5 mm. long
    51. Stems depressed, not over 10 cm. long; leaves less than 12 mm. long; perianth yellowish   -43. E. contortum
    51. Stems over 10 cm. long; leaves over 12 mm. long; perianth white or rose   -44. E. simpsoni
49. Leaves averaging well over 2 cm. long, revolute or flat
    52. Perianth yellow
        53. Shrubby plants with long leafy stems, usually longer than the inflorescence   -45. E. nudicaule
        53. Undershrubs, leafy at the base or on short stems, these shorter than the inflorescences   -46. E. campanulatum
    52. Perianth white or rose
        54. Leaves averaging 5 cm. long or more, margins not at all revolute
            55. Involucres tomentose; leaves spatulate   -47. E. spathulatum
            55. Involucres glabrous or nearly so; leaves usually narrower than spatulate
                56. Involucres all sessile; leaves linear to lanceolate to narrowly oblanceolate   -48. E. lonchophyllum
                56. Involucres in the forks of the lower branches of the inflorescence definitely peduncled; leaves various but often wider than lanceolate or narrowly oblanceolate -49. E. salicinum
        54. Leaves shorter, averaging less than 5 cm. long, or if longer then margins definitely revolute
            57. Leaf margins revolute; involucres glabrous externally (or slightly floccose in E. simpsoni)
                58. Inflorescence dense and broomlike, with numerous branches, these erect or nearly so; leaves linear   -50. E. leptophyllum
                58. Inflorescence more open; leaves may be linear but usually wider
                    59. Stems and inflorescences tomentose; leaves seldom over 3 cm. long   -44. E. simpsoni
                    59. Stems and inflorescences glabrous or nearly so; leaves often over 3 cm. long
                        60. Involucres all sessile, campanulate   -45. E. nudicaule
                        60. Involucres in the forks of the lower branches of the inflorescence definitely peduncled, campanulate to turbinate
                            61. Involucres broadly campanulate, about as wide as long, 2.5-3 mm. long; leaves 2-5 cm. long   -51. E. scoparium
                            61. Involucres turbinate, definitely longer than wide, 3-4 mm. long; leaves 2-7 cm. long   -52. E. tristichum
            57. Leaf margins plane, not revolute; involucres usually tomentose on the outside
                62. Leaves elliptic to linear-oblong   -53. E. ainsliei
                62. Leaves linear, oblanceolate or spatulate
                    63. Involucres 3-4 mm. long, all sessile; plants woody only at very base; leaves spatulate; perianth over 2.5 mm. long   -47. E. spathulatum
                    63. Involucres 1.5-3 mm. long, some usually pedunculate; plants with woody stems; leaves usually narrower than spatulate; perianth seldom over 2.5 mm. long
                        64. Inflorescence a rather close corymbose cyme seldom over 12 cm. long; branches of the stem few and short; plants seldom over 25 cm. tall
                            65. Leaves narrowly lanceolate to spatulate, glabrate or loosely floccose above; perianth glabrous   -54. E. salinum
                            65. Leaves oblanceolate, densely tomentose on both sides; perianth usually somewhat pubescent   -34. E. nebraskense
                      64. Inflorescence an open cyme repeatedly branched, often over 12 cm. long; branches of stem many; plants 20-50 cm. tall
                        66. Involucres tomentulose, 1.5-2.5 mm. long   -55. E. effusum
                        66. Involucres glabrous, over 2.5 mm. long   -49. E. salicinum

1. Eriogonum densum Greene, Pitt. 3:17. 1896.
    E. vimineum var. densum (Greene) Stokes---Utah and Nevada, south to New Mexico and Arizona. This plant should be expected in western or southwestern Colorado.
2. Eriogonum hookeri S. Wats., Proc. Amer. Acad. 14:295. 1879.
    E. deflexum Torr. subsp. hookeri (Wats.) Stokes---Annual plants; stems 20-60 cm. tall, scapose, glabrous and somewhat glaucous; leaves 2-6 cm. wide, basal, orbicular, white-lanate especially below, on long petioles; inflorescence an open much branched cyme, the branches somewhat flexuous and zigzag; bracts scalelike; involucres 1.5-2 mm. long, hemispheric to very broadly campanulate, teeth broad and about 1/3 the length of the involucre or more, glabrous externally, mostly sessile or on short peduncles less than 2 mm. long, these spreading or reflexed; perianth 1.5-2 mm. long, yellow or pinkish-yellow, glabrous, the outer segments nearly orbicular, cordate at base, inner narrower and shorter; achenes not winged, glabrous.---Canyons and slopes, often in shade. Utah to Nevada, south to Arizona. Our few records from western Colorado at about 4500-5500 feet.
3. Eriogonum glandulosum Nutt.; Benth. in DC., Prodr. 14:21. 1856.
    E. trichopes ssp. glandulosum (Nutt.) Stokes---Annual acaulescent plants; scapes about 10-20 cm. tall, glandular-pubescent on the scapes and often on the branches of the inflorescence although hairs sometimes sparse; leaves 1-3 cm. long, basal, ovate to round or reniform, usually cordate to truncate at base, broadly rounded at apex, hirsute below and on petiole, glabrous or nearly so above, petiole usually longer than the blade; inflorescence a widely branched cyme over 1/2 the length of the scape, the peduncles often abruptly bent above the middle; bracts short, seldom over 1 cm. long, narrowly linear, not at all leaflike; involucre 1.5-2 mm. long, broadly campanulate or broadly turbinate, glabrous, teeth or lobes often 1/2 or more the length of the involucre, on long capillary peduncles; perianth 1.5-2.5 mm. long, externally yellow or fading to white or pink, short-hispidulous, segments oblong or ovate; achenes glabrous, not winged.---Slopes and hills. Colorado, Utah and Arizona. Our few records from western Colorado at 4500-5000 feet.

4. **Eriogonum fusiforme** Small, Bull. Torr. Club 33:56. 1906.

E. trichopes ssp. minus (Benth.) Stokes---Annual scapose plants; scapes 20-60 cm. tall, branching into a longer inflorescence, glabrous and glaucous, lower internodes almost always inflated above the middle; leaves 1-3 cm. long, crowded at base, ovate, orbicular to reniform, truncate to subcordate at base, rounded or subacute at the tip, with a variable petiole, this about as long, longer or shorter than blades, villous, hairs scattered on both sides; inflorescence a wide diffusely branched cyme, the lower branches often inflated; bracts scalelike, in 3's; involucres about 1 mm. long, turbinate, lobed almost to middle, glabrous, on capillary peduncles about 1 cm. long; perianth 2-2.5 mm. long, yellowish or somewhat paler, very hairy outside; achenes glabrous.---Slopes and bluffs often in sandy soil. Colorado, Utah and Arizona. Our records from western Colorado at 4500-6000 feet.

5. **Eriogonum inflatum** Torr. & Frem. in Frem., Exped. Rocky Mt. Rept. 317. 1845.

Perennial plants from stout rather woody roots, scapose; scapes 20-80 cm. tall, glabrous and glaucous, lower internodes usually more or less inflated above the middle; leaves 10-25 mm. long, rosulate and basal, rounded or reniform to oblong-elliptic, truncate to cordate, rounded at apex, on petioles as long or longer than the blades, hirsute especially below to glabrate; inflorescence a much-branched open cyme forming a round head; bracts scalelike, in 3's; involucre 1-1.5 mm. long, broadly turbinate, 5-lobed to about the middle, glabrous outside, peduncled on capillary but straight and rigid peduncles 5-20 mm. long; perianth 1.5-2.5 mm. long, yellowish, conspicuously pubescent, outer segments somewhat wider than the inner; achenes glabrous. This species may not be distinct from E. fusiforme Small.---Dry plains and hills. Colorado to California, south to New Mexico and Arizona. Our few records from western Colorado at 4000-4500 feet.

6. **Eriogonum subreniforme** Wats., Proc. Am. Acad. 12:260. 1877.

Annual scapose plants; scapes 10-30 cm. tall, more or less villous or tomentose at the nodes; leaves 1-2 cm. wide, crowded, round or round-reniform, rounded at apex, truncate or cordate at base, tomentose or short-lanate below, green and sparingly hairy above; inflorescence an open, rather spreading cyme, much branched and occupying most of the plant; bracts scalelike, more or less hairy; peduncles 5-20 mm. long; involucres 0.6-0.8 mm. long, turbinate-campanulate, glabrous, lobes shorter than tube; perianth 1-1.3 mm. long, whitish or light rose in color, outer lobes oblong to elliptic, glabrous, inner subequal; ovary glabrous, achene about 1 mm. long.---Dry ground. Utah, New Mexico and Arizona. Our few records from western Colorado at 4500-5000 feet.

7. **Eriogonum gordoni** Benth. in DC., Prod. 14:20. 1856.

E. trinervata Sm.---Annual plants; scapes 20-40 cm. tall, often branching right above leaves, glabrous; leaves 1-3 cm. long, basal, ovate to orbicular, apex rounded, truncate to broadly cuneate, petiole about as long as the blade, glabrate to sparingly hirsute; inflorescence an open much branched cyme often starting at base of plant; bracts scalelike, in 3's; involucre about 1 mm. long, campanulate, shallowly toothed, glabrous outside, peduncled; perianth 2-2.5 mm. long, white or light rose, glabrous, outer segments somewhat wider than inner; achenes glabrous.---Dry plains and hills. Wyoming, Colorado and Utah. Our records scattered in the western two-thirds of Colorado, mostly in the southern half of the state, at 4500-7000 feet.

8. **Eriogonum wetherillii** Eastw., Proc. Calif. Acad. Sc. II. 6:319. 1896.

Utah, New Mexico and Arizona. This species should be looked for in western and southwestern Colorado.

9. **Eriogonum cernuum** Nutt., Journ. Acad. Nat. Sc. Phila. ser. 2, 1:162. 1848.

Annual plants; scapes 10-40 cm. tall, diffusely di- or trichotomously-branched, often from near the base, glabrous; leaves 1-2 cm. long, all basal or near the base, broadly ovate, oval or orbicular, broadly cuneate, obtuse or acute, on petioles usually longer than the blades, tomentose below, less so and often glabrate above; inflorescence an open branched cyme, often arising from near base of plant, branches usually spreading, forming a rounded head; bracts not leaflike except leaves may be near first branches; involucres turbinate-campanulate, on more or less strongly reflexed peduncles often less than 1 cm. long, glabrous, rather stout, flowers often over 10 to an involucre; perianth 1-2 mm. long, white to rose, glabrous, outer lobes somewhat panduriform; achenes glabrous, not winged.---Dry hills and plains. Saskatchewan to Alberta, south to Nebraska, Arizona and California. Our records widely scattered over the western two-thirds of Colorado at 4500-9500 feet.

10. **Eriogonum rotundifolium** Benth. in DC., Prodr. 14:21. 1856.

E. cernuum ssp. rotundifolium (Benth.) Stokes---Annual acaulescent plants; scapes 10-30 cm. tall, glabrous; leaves 1-2 cm. long, basal, orbicular, tomentose below, less so and often glabrate above, broadly cuneate or truncate at base, obtuse or mucronate at apex, petioles usually longer than the blades; inflorescence an open cyme often arising from near base of plants, branches spreading, forming a roundish head; bracts reduced; involucres turbinate-campanulate, glabrous, the peduncles usually over 1 cm. long, slender, spreading or ascending, seldom reflexed; flowers about 5 to an involucre rarely up to 10; perianth 1-2 mm. long, white to rose, glabrous, outer lobes flabellate, not panduriform; achenes glabrous, not winged.---Dry ground. Colorado south to Texas, Mexico and Arizona. Our few records from western and southern Colorado at 4500-7500 feet.

11. **Eriogonum salsuginosum** (Nutt.) Hook., Journ. Bot. Kew 5:264. 1853.

Annual plants, stems 8-20 cm. tall, branched from near base, glabrous, caulescent; leaves of 2 general kinds, basal leaves 2-4 cm. long, thick, spatulate, glabrous; stem leaves oblanceolate to linear merging with the bracts, glabrous; inflorescence an open branched cyme occupying most of the plant's length; bracts linear, foliose; involucre of nearly distinct lanceolate bracts, 2-3 mm. long, glabrous or nearly so outside, some sessile in the forks but others long-peduncled; perianth about 2 mm. long, yellowish, pubescent, segments oblong-lanceolate, closely appressed to the achene; achene 3-angled, glabrous.---Hills and plains. Wyoming to Utah, south to New Mexico and Arizona. Our few records from western Colorado at about 5000 feet.

12. **Eriogonum annuum** Nutt., Trans. Amer. Phil. Soc. n. s. 5:164. 1837.

Annual plants; stems 30-100 cm. tall, white-tomentose, leafy, simple or branched; leaves 3-5 cm. long, alternate, oblong, oblong-lanceolate or oblanceolate, acute or obtuse, cuneate, entire or with somewhat revolute or crisped margins, tomentose below, tomentose or floccose above, petioled; inflorescence an open and much branched cyme or this more commonly compact and corymbose; final bracts scalelike but main branches often subtended by leaves; involucres 2-3 mm. long, turbinate-campanulate, white-tomentose, mostly peduncled and secund, teeth short; perianth 1-2 mm. long, white to rose sometimes dark reddish, glabrous, outer segments definitely wider than inner; achene glabrous.---Sandy plains and hills. South Dakota to Montana, south to Texas and Mexico. Our records from the eastern half of Colorado at 3500-7000 feet.

13. **Eriogonum divaricatum** Hook., Journ. Bot. & Kew Gard. Misc. 5:265. 1853.

Annual caulescent plants; stems 10-20 cm. long, low and prostrate or nearly so, puberulent to short-appressed pilose; basal leaves orbicular to elliptic-oblong, thick, puberulent to appressed pilose, on petioles longer than the blades, primary cauline leaves reduced to small triangular bracts, usually bearing 2 secondary leaves in their axils, these similar to basal leaves but smaller; involucres sessile in the axils along the stems, 1-2 mm. long, turbinate-campanulate to hemispheric, deeply 5-cleft, appressed pilose; perianth about 1-1.5 mm. long, yellow or possibly rose or white, glandular to hispidulous outside, segments oblong-ovate to oblong-lanceolate.---Dry hills and plains. Colorado, Utah and Arizona. Our one record from the southwestern corner of Colorado at 5000 feet.

14. **Eriogonum pharnaceoides** Torr. in Sitgreaves, Zuni & Colo. Rept. 167. 1854.

Utah, New Mexico and Arizona. This plant is to be expected in western or southwestern Colorado.

15. **Eriogonum alatum** Torr. in Sitgreaves, Zuni & Colo. Rept. 168. 1854.

Plants perennial by long thick taproots; stems 30-100 cm. tall (or less in unfavorable situations), not woody above ground, erect, often 1 to a plant, paniculately branched above in the inflorescence, pubescent to glabrate; leaves 3-10 cm. long, alternate but mostly basal, spatulate to narrowly oblanceolate, margins entire or somewhat undulate, obtuse or short-acute, long-cuneate, pubescent to scattered hairy; inflorescence an elongated, open, long-branched paniculate cyme, branches usually pubescent; bracts leaflike; involucres 2.5-3.5 mm. long, pedunculate, sericeous, usually sparsely so, campanulate or campanulate-turbinate, rather shallowly toothed; perianth about 2 mm. long, yellowish, glabrous; achenes usually nearly orbicular, 6-7 mm. long, over half as long as wide, conspicuously 3-winged, finally exserted.--- Plains and hills. Nebraska to Colorado, south to Texas and Arizona. Our records well scattered over Colorado at 3500-10,000 feet.

15A. **Eriogonum alatum triste** (Wats.) St., (ssp.) Gen. Eriog. 20. 1936.

E. triste S. Wats.---Doubtfully distinct from the species. Supposed to have a shorter involucre, 2-2.5 mm. long, and be almost glabrous on involucre and leaves (may be ciliate on leaves); achenes longer, 7-8 mm. long, elliptic and about 1/2 as long as wide; perianth purplish or yellowish.---About the range of the species. The writer has seen no Colorado material clearly belonging to this subspecies.

16. **Eriogonum caespitosum** Nutt., Journ. Acad. Nat. Sc. Phila. 7:50. 1834.

Perennial plants with base intricately branched forming dense woody mats; scapes 2-10 cm. tall, tomentose; leaves 5-10 mm. long, crowded at base of plant, oval or elliptical, cuneate at base to a short petiole, obtuse or subacute, more or less revolute, white-tomentose often more thinly so above; inflorescence of 1 involucre without bracts; involucre tube about 3 mm. long with longer reflexed lobes, turbinate, tomentose, peduncled; perianth 3-5 mm. long, yellow or purplish-brown in age, short-stipitate at base, pubescent, segments elliptical to oblong; achenes pubescent.---Hills and plains. Montana to Idaho, south to Colorado and Nevada. This plant should be in northern Colorado but our specimens are doubtful.

17. **Eriogonum acaule** Nutt., Journ. Acad. Nat. Sc. Phila. II. 1:160. 1847.

E. caespitosum ssp. acaule (Nutt.) Stokes---Wyoming to Idaho, south to Colorado and Nevada. Should be expected in northern Colorado. Recently a related species keying out here has been found in the extreme southwestern corner of the state. It has been identified as E. pulvinatum Small. The leaves are wider, being narrowly obovate, and are more densely white-tomentose. Also the perianth is white or yellowish-white instead of yellow.

18. **Eriogonum flavum** Nutt. in Fras., Cat. 1813.

E. crassifolium Benth.; E. chloranthum Greene; E. xanthum Small---Perennial plants with branching caudices, these branches short and stout; scapes 10-20 cm. tall sometimes even less, tomentose or floccose; leaves 2-5 cm. long, clustered on the caudex branches, oblanceolate, spatulate to oblong, cuneate at base to a petiole that is sometimes as long as the blade, obtuse or subacute at apex, whitish-tomentose both sides or somewhat glabrate above, the hairs often rusty in age; inflorescence an umbel, with 1 to several involucres; bracts several, leaflike; involucres 4-6 mm. long, turbinate, the teeth very shallow, tomentose, peduncles short; perianth 4-6 mm. long, yellow, pubescent externally, somewhat stipitate below but this often obscure, segments elliptical to obovate; achenes long-villous above.---Dry hills, plains and canyons. Manitoba to Alberta, south to Colorado and Arizona. Our records scattered in Colorado, except the southeastern and northwestern parts, at 5000-12,500 feet.

19. **Eriogonum jamesii** Benth. in DC., Prodr. 14:7. 1856.

Perennial plants, the caudices at least somewhat branched, these suberect or matted; flowering stems 10-30 cm. tall, rather scapose, tomentose; leaves 3-8 cm. long, mostly basal but sometimes scattered, spatulate-obovate to oblong, long-cuneate at base, acute or obtuse at apex, tapering to a petiole about as long as blade, this expanded at base, tomentose below, glabrate above; inflorescence an open cyme, not capitate, 1-many times di- or trichotomous; bracts several and leaflike, all branches bracted; involucres 4-7 mm. long, campanulate or turbinate-campanulate, teeth not very deep, tomentose, central involucre sessile, lateral sessile or short-stalked; perianth 4-6 mm. long, white to rose, pubescent externally, segments broadly oblong or obovate, attenuate and stipelike at base.---Dry plains and hills. Our records scattered, mostly in the southern half of Colorado, at 4000-10,000 feet.

20. Eriogonum arcuatum Greene, Pitt. 4:319. 1901.

E. jamesii var. arcuatum (Gr.) St.; E. jamesii flavescens S. Wats. in Manuals; E. bakeri Greene; E. jamesii ssp. bakeri (Greene) Stokes; E. vegetius A. Nels. is close---Perennial plants with branching, stout, woody caudices; flowering stems 10-30 mm. tall, scapose, tomentose or becoming glabrate; leaves 1.5-5 cm. long, on the caudex, oblong, elliptical-lanceolate to oval or suborbicular, long-cuneate or rounded to slender petioles usually as long or longer than the blades, tip obtuse or subacute, white-tomentose below, thinner above or glabrate; inflorescence a cyme, di- or trichotomous; bracts leaflike; involucres 5-10 mm. long or sometimes slightly less, turbinate or turbinate-campanulate, the teeth short and wide, tomentose, central one sessile, lateral ones sessile or short-pedunculate; perianth 3-6 mm. long, yellow, externally pubescent, segments broadly oblong or obovate, the whole perianth stipelike at base, though sometimes rather shortly so.---Hills and mountains. Wyoming to Utah, south to New Mexico and Arizona. Our records scattered in the western half of Colorado, except the northwestern part, at 5000-11,500 feet.

21. Eriogonum heracleoides Nutt., Journ. Acad. Nat. Sc. Phila. 7:49. 1834.

Perennial plants from loosely spreading caudices, these rather woody and leafy at ends; scapes 20-40 cm. high, tomentose or floccose with a whorl of leaves near middle; leaves 2-4 cm. long, on a petiole of about equal length, oblanceolate or sometimes narrower, long-cuneate at base, obtuse or acute at apex, white-tomentose below, green and glabrate or somewhat floccose above; inflorescence of involucres in a simple or compound umbel; bracts several, foliacious; involucres turbinate, tomentose, tube 3-4 mm. long, lobes about 2 mm. long and reflexed, peduncled; perianth 4-6 mm. long, white or rose, ochroleucous, glabrous, stiped at base.---Dry often rocky plains and slopes. Wyoming and Utah. Our few records from northwestern Colorado at 5500-8000 feet.

22. Eriogonum subalpinum Greene, Pitt. 3:18. 1896.

E. umbellatum ssp. subalpinum (Greene) Stokes; E. rydbergii Greene---Perennial plants, caudices branching, open and depressed, somewhat woody; flowering stems 10-30 cm. tall, scapose, tomentose to floccose; leaves 1-5 cm. long, the petiole as long or longer, on the branches of the caudex, usually clustered at the ends, spatulate-oblong, elliptical to orbicular, cuneate at base, obtuse or acute at apex, tomentose below, glabrous or glabrate above; inflorescence of umbels, the involucres usually in a simple umbel; bracts several, leaflike, sometimes one leaflike one part way down scape; involucres turbinate, tomentose, the tubes about 3 mm. long and with lobes about as long, usually reflexed; involucres peduncled; perianth usually around 6 mm. long but sometimes to 9, pale to ochroleucous or somewhat rose-colored in age, glabrous, stiped at base, lobes obovate, the inner longer.---Dry ground. Alberta to British Columbia, south to Colorado and Nevada. Our records from the northern three-fourths of the mountainous part of Colorado at 5500-11,000 feet.

23. Eriogonum umbellatum Torr., Ann. Lyc. N. Y. 2:241. 1828.

E. biumbellatum Rydb.; E. neglectum Greene; E. umbelliferum Small---Perennial plants with rather short depressed rather woody caudices; scapes 10-30 cm. tall, tomentose to floccose; leaves 1-4 cm. long, the petioles about as long, crowded on the ends of the caudex, obovate, spatulate or oval, cuneate at base, acute to obtuse at tips, usually tomentose below and floccose or glabrate above but sometimes glabrate on both sides; inflorescence of involucres in umbels these usually simple; bracts several, leaflike; involucres turbinate, tomentose, the tube 3-4 mm. long, lobes about as long and reflexed, peduncled; perianth 5-7 mm. long, bright or sulphur yellow, glabrous, stipitate at base, segments spatulate, obtuse.---Dry plains, slopes and hills. Wyoming to Washington, south to Arizona and California. Our records from the western three-fourths of Colorado at 5000-10,500 feet.

24. Eriogonum lachnogynum Torr. ex Benth. in DC., Prodr. 14:8. 1856.

Perennial plants, branches of the caudices short, caespitose and not very woody, forming tufted mats; flowering stems 10-30 cm. tall, scapose, short-tomentose, pubescent or sometimes glabrate; leaves on caudex, in tufts, leaves 1-3 cm. long, lanceolate, oblanceolate or narrowly-elliptical, cuneate at base to a petiole about as long as blade, acute at apex, often revolute, tomentose below, silky above; inflorescence a branching, rather open cyme, sometimes somewhat umbellate; bracts variable, subulate to leaflike; involucres 3-4 mm. long, turbinate-campanulate, teeth wide and shallow, silky-tomentose, peduncled except sometimes the central ones; perianth 2.5-4 mm. long, gray, pink or yellow, pubescent, not stipitate at base, segments unequal, outer orbicular or oblong, inner spatulate; ovary and achenes densely pubescent.---Dry ground. Kansas to Colorado, south to Texas and Arizona. Our records from southeastern Colorado at 3500-5500 feet.

25. Eriogonum tetraneuris Small, Bull. Torr. Club 33:52. 1906.

E. lachnogynum ssp. tetraneuris (Small) Stokes---This species has been recorded for Colorado. However it seems doubtfully distinct from E. lachnogynum Torr.

26. Eriogonum villiflorum Gray, Proc. Am. Acad. 8:630. 1873.

Utah to New Mexico. This species should be in southwestern Colorado but no specimens from the state have been seen by the writer.

27. Eriogonum multiceps Nees. in Pr., Neiwied. Reise Nord Amer. 2:446. 1841.

Perennial plants, caudices depressed, woody, much branched and forming tufts or mats, leaves crowded on the tomentose stems; scapes erect, 3-15 cm. long; leaves 2-4 cm. long, sometimes longer, linear-oblanceolate to spatulate, cuneate to a long slender petiole about as long as the blade, acute or obtuse at apex, densely white-tomentose both sides; inflorescence capitate; bracts reduced or sometimes foliose; involucres 2.5-3.5 mm. long, turbinate to tubular-campanulate, lobes moderately long, tomentose, mostly sessile; perianth 2-5 mm. long, white to rose, pubescent, segments cuneate-oblong; achenes glabrous, not winged.--Dry ground. Saskatchewan to Idaho, south to Nebraska and Colorado. Our few records from northeastern Colorado at 4500-5000 feet.

28. **Eriogonum bicolor** Jones, Zoe 4:281. 1893.

E. microthecum ssp. bicolor (Jones) Stokes---Perennial plants; flowering stems 2-6 cm. tall from a matted, branched, woody base, tomentose; leafy portion of stem usually longer than the inflorescence; leaves 10-15 cm. long, about 2 mm. wide, linear, slightly cuneate at base, acute or obtuse at apex, margin revolute, both sides strongly tomentose; inflorescence umbellate or subcapitate, 1-3 cm. long; involucre bracts few, usually scalelike; involucres 2-4 mm. long, broadly turbinate, with lobes rather short, tomentose, short-pedunculate; perianth 3-4 mm. long, white to reddish, often reddish near base and white above, not stipelike but abruptly enlarged near base, glabrous, outer segments obovate or orbicular when mature, in ours carinate near base, much longer and wider than the elliptic inner segments; achenes glabrous, not winged.---Dry ground. A Utah species coming into extreme westcentral Colorado. Our record from Mesa County at 4500 feet.

29. **Eriogonum ovalifolium** Nutt., Journ. Acad. Nat. Sc. Phila. 7:50. 1834.

Perennial plants, caudices low, caespitose, somewhat pulvinate; flowering stems 10-30 cm. tall, sometimes shorter, scapose; leaves 5-20 mm. long, closely crowded on the caudex, broadly oval to broadly obovate or suborbicular, nearly as wide as long, abruptly cuneate to a short petiole, obtuse or subacute at apex, both sides densely tomentose; inflorescences of 1 capitate head; bracts small, in 3's; involucres 3-5 mm. long or 6-7 in some forms, campanulate with short broad teeth, tomentose, sessile or nearly so; perianth 3-5 mm. long, yellow, glabrous, not stipitate, outer lobes broad, nearly orbicular and often cordate, inner narrower, spatulate.---Dry ground. Alberta to Washington, south to New Mexico, Arizona and California. Our records from northwestern, westcentral and southwestern Colorado at 4500-8500 feet.

29A. **Eriogonum ovalifolium purpureum** (Nutt.) Nels., (var.) Bot. Gaz. 34:23. 1902.

E. orthocaulon Sm.---Differs from the species in having the perianth white, rose-veined or purplish. Often growing in the same patch as yellow-colored type and with the same general distribution in Colorado.

30. **Eriogonum ochroleucum** Small, Mem. N. Y. Bot. Gard. 1:123. 1900.

E. ovalifolium ssp. ochroleucum (Small) Stokes---Perennial plants, caudices short-branched and caespitose, only slightly woody; scapes 10-30 cm. tall, tomentose to loosely floccose; leaves 1-2 cm. long, densely crowded on the caudex, elliptic, oblong or spatulate, cuneate to a long slender petiole longer than the blade, acute to obtuse at apex, both sides tomentose; inflorescence of involucres capitate in a single cluster; bracts small; involucres about 4-5 mm. long, campanulate with short teeth, tomentose, sessile or nearly so; perianth 4-5 mm. long, ochroleucous, or somewhat rose-colored in age, glabrous, outer segments elliptic or oblong, inner spatulate.---Dry places. Montana to Idaho, south to Colorado and Nevada. Our few records from westcentral Colorado at 4500-5000 feet.

31. **Eriogonum chrysocephalum** Gray, Proc. Am. Acad. Sc. 11:101. 1876.

Wyoming to Idaho, south to Utah. This plant has been reported from Colorado but no specimens have been seen by the writer.

32. **Eriogonum pauciflorum** Pursh, Fl. Amer. Sept. 2:735. 1814.

Perennial densely caespitose plants, the branches somewhat woody; scapes 6-10 cm. tall, tomentose or glabrate especially above; leaves 2-6 cm. long, basal on the branches of the caudex, linear to linear-oblanceolate, long-cuneate at base, acute or obtuse at apex, tomentose below, tomentose to glabrate above, margins revolute, petioles short and not set off; inflorescence capitate; involucres turbinate, deeply lobed usually about to the middle, the lobes broad and scarious, glabrous or tomentose especially on midrib of lobes, the tube 2-3 mm. long, sessile; perianth 3-5 mm. long, white or rose, glabrous, segments oval or elliptic; achenes not winged.---Dry especially sandy soil. South Dakota to Wyoming, south to Colorado. Our rather few records mostly from northcentral and central Colorado at 5500-10,000 feet.

33. **Eriogonum coloradense** Small, Bull. Torr. Club 33:53. 1906.

E. multiceps ssp. coloradense (Small) Stokes---Perennial densely caespitose plants, the short branches woody; scapes about 6-10 cm. long, glabrate at least above; leaves 2-5 cm. long, crowded and basal on the branches of the caudex, linear to linear-spatulate, almost as long as the scapes, tomentose below, tomentose to glabrate above, margins weakly if at all revolute, tapering to a short petiole; inflorescence capitate; involucres about 4-5 mm. long, turbinate-campanulate, shallowly toothed, sessile; perianth 3-5 mm. long, white or rose, glabrous, segments oblong to ovate; achenes not winged.---Mountains. Apparently limited to Colorado, our few records from Gunnison and Park Counties at 9500-12,000 feet.

34. **Eriogonum nebraskense** Rydb., Fl. Rocky Mts. 224. 1061. 1917.

E. multiceps ssp. nebraskense (Rydb.) Stokes---Low shrubs 15-30 cm. tall, with rather short woody branched base and tomentose stems; leaves 2-3 cm. long, scattered on lower part of flowering stems, oblanceolate, cuneate to a short petiole, obtuse or acute at apex, margins not revolute, both sides densely tomentose; inflorescence a short corymbose cyme, branches often whorled, the whole inflorescence less than 10 cm. high; bracts reduced; involucres 2-3 mm. long with short teeth, turbinate, tomentose, sessile or short-pedunculate in the forks of the inflorescence; perianth about 2.5 mm. long, rose-colored, pubescent to nearly glabrous, segments oblong or obovate; achenes not winged, glabrous.---Dry plains. Wyoming to Nebraska. Our few records from southeastern Colorado at about 5000 feet.

35. **Eriogonum racemosum** Nutt., Journ. Acad. Nat. Sc. Phila. Ser. 2, 1:161. 1848.

Perennial scapose plants with unbranched or sparingly branched, short woody caudices; scapes 30-80 cm. tall, solitary or few, tomentose especially below; leaves 2-7 cm. long, petioles as long or longer, appearing at the surface of the ground, elliptical to ovate or nearly round, cuneate to subcordate, acute, entire or slightly wavy, tomentose below, floccose above; inflorescence spikelike or branched, the involucres racemose; bracts scalelike, in 3's; involucres 3-5 mm. long, tubular-campanulate, teeth very short, tomentose, sessile or nearly so; perianth about 3 mm. long at first, finally 4-5, white or pink, glabrous; ovary and achenes glabrous.---Dry plains and slopes. Colorado to Utah, south to Texas and Arizona. Our records from the southwestern quarter of the state at 6500-8500 feet.

36. **Eriogonum wrightii** Torr. ex. Benth. in DC., Prodr. 14:15. 1856.
E. trachygonum ssp. wrightii (Torr.) Stokes; E. wrightii ssp. typicum Stokes---Colorado to California, south to Texas and Mexico. The writer has not been able to locate any material from Colorado.
37. **Eriogonum leptocladon** T.&G., U. S. Rept. Expl. Miss. Pacif. 2:129. 1855.
E. ramosissimum Eastw.; E. effusum ssp. leptocladon (T. & G.) Stokes---Utah, Arizona and New Mexico. This species should be looked for in southwestern Colorado.
38. **Eriogonum tenellum** Torr., Ann. Lyc. N. Y. 2:241. 1828.
Perennial plants, caudices woody, depressed, caespitosely branched and spreading, sometimes the very short branches erect; scapes 20-40 cm. tall, glabrous; leaves 0.5-1 cm. long, sometimes to 1.5 cm., petiole about as long or longer, crowded on the caudex, ovate or orbicular, cuneate, obtuse, both sides tomentose; inflorescence an open cyme rather sparingly branched, usually longer than the scape; bracts scalelike, usually in 3's; involucres on peduncles 1-2 cm. long or more or sometimes sessile in the axils of the branches only, about 3 mm. long, turbinate with ovate-triangular lobes, glabrous; perianth 2.5-3 mm. long, white or pink, glabrous, outer segments broader than inner; ovary and achenes glabrous.---Dry, often rocky ground. Colorado to Utah, south to Texas and Mexico. Our records from the southwestern part of Colorado at 4000-5000 feet.
39. **Eriogonum fendlerianum** (Benth.) Small, Bull. Torr. Club 33:55. 1906.
E. effusum ssp. fendlerianum (Benth.) Stokes---Perennial plants with woody branching bases, these leafy for a distance longer than the length of the inflorescence; flowering stems 10-50 cm. tall, tomentose in the leaves, glabrous or glaucous above and in the inflorescence; leaves 3-6 cm. long, scattered, oval or oblong, broadly cuneate to a short petiole, apex obtuse or mucronate, margins undulate or somewhat revolute, tomentose below, glabrous or floccose above; inflorescence corymbose, branching 2 or more times, branches rather stout; bracts scalelike, ternate; involucres 4-5 mm. long, turbinate or turbinate-campanulate, lobes moderate, much shorter than the tube, glabrous or nearly so, sessile or peduncled in axils of inflorescence branches; perianth 3-4 mm. long, white to pink, glabrous, segments elliptic or oval, outer somewhat longer.---Dry plains and slopes. Colorado and New Mexico. Our records from central and southcentral Colorado at 5000-6500 feet.
40. **Eriogonum divergens** Small, Bull. Torr. Club 33:55. 1906.
E. effusum ssp. divaricatum (T. & G.) Stokes---Colorado to Nevada, south to New Mexico and Arizona. The plant should be expected in western Colorado.
41. **Eriogonum corymbosum** Benth. in DC., Prodr. 14:17. 1856.
E. effusum ssp. corymbosum (Benth.) Stokes---Perennial plants, bases definitely woody with many leafy branches, these longer than the inflorescence; flowering stems 30-100 cm. tall, not scapose, tomentose or floccose; leaves 2.5-5 cm. long, scattered, elliptic to ovate, broadly cuneate to a shorter petiole, obtuse or mucronate, margins undulate or partly revolute, tomentose below, thinly tomentose to floccose above; inflorescence corymbose, branched several times but the branches ascending; bracts mostly foliose; involucres 2-4 mm. long, campanulate, teeth moderate but not as long as tube, sessile or short-pedunculate in the axils of the inflorescence; perianth 3-4 mm. long, white or rose, glabrous, segments elliptic to obovate, inner somewhat shorter.---Dry plains and slopes. Colorado to Nevada, south to New Mexico and Arizona. Our records from westcentral and southwestern Colorado at 5000-7500 feet.
Recently E. batemani Jones has been collected in northwestern Colorado. It keys out to the above except that the involucres are glabrous instead of tomentose. Also the plant is acaulescent with the leaves clustered on the ends of the caudex branches.
42. **Eriogonum microthecum** Nutt., Journ. Acad. Nat. Sc. Phila. ser. 2, 1:162. 1848.
Perennial plants with woody branches and leafy tomentose twigs longer than the inflorescence; plants 10-60 cm. tall; stem above leaves tomentose or floccose to glabrate; leaves 1-2 cm. long, scattered on the stem, oblanceolate or oblong-lanceolate, long-cuneate to a short petiole, acute or obtuse at apex, margins often revolute, tomentose below, floccose to glabrate above; inflorescence a corymbose cyme, di- to trichotomus, short with branches becoming divaricate; bracts small and triangular (sometimes almost leaflike); involucres about 2.5-3 mm. long, turbinate to almost tubular, the teeth rather short, tomentose but sometimes sparsely so, sessile or peduncled in the axils of the inflorescence branches; perianth 2-3 mm. long, yellow or white, glabrous, segments round to obovate, not attenuate at base; achenes pointed.---Dry plains and slopes. Montana to Washington, south to Nebraska and California. Our records scattered in the western half of Colorado at 4500-8000 feet.
43. **Eriogonum contortum** Small in Rydb., Fl. Colo. 107. 1906.
E. effusum ssp. contortum (Sm.) Stokes---Dwarf perennial plants, the bases gnarled, woody and short-branched; flowering stems less than 10 cm. long, numerous, slender, erect or nearly so, tomentose; leaves about 1 cm. long, scattered on lower portion of plant, clavate or narrowly linear, thick, acute or obtuse at apex, very strongly revolute, tomentose below and lighter tomentose to glabrate above, petioles short; inflorescence a short cyme sometimes reduced to 1 involucre; bracts short and scalelike; involucres 2-4 mm. long, long-campanulate to turbinate with short teeth, glabrous outside, sessile or short pedunculate; perianth 2-2.5 mm. long, yellow, often rose at base, sometimes pale yellow, glabrous, segments obovate to oblong; achenes glabrous, not winged.---Dry ground. Colorado. Our few records from westcentral Colorado at about 4500 feet.
44. **Eriogonum simpsoni** Benth. in DC., Prodr. 14:18. 1856.
E. effusum ssp. simpsoni (Benth.) Stokes---This species resembles E. microthecum Nutt. and may be a variety of it. It differs in the more strongly revolute leaves, the lower surface almost hidden, and the shorter perianth which is only 2-2.5 mm. long.---Dry ground. Colorado to Nevada, south to New Mexico and Arizona. Our few records from southern Colorado at 4000-7500 feet.

45. Eriogonum nudicaule (Torr.) Small, Bull. Torr. Club 33:54. 1906.

Perennial plants; stems 10-30 cm. tall sometimes more, scapiform, depressed, or suberect, with short woody bases, glabrous or nearly so; leaves closely set at base of scapes, the leafy portion shorter than the inflorescence; leaves 3-5 cm. long, nearly erect, narrowly oblanceolate or linear, long-cuneate to a short petioled base, acute or obtuse at apex, mostly revolute, tomentose below, floccose or glabrate above; inflorescence cymosely-corymbose, branches rather long; bracts scalelike; involucres 3-4 mm. long, campanulate, lobes about 1/4 as long as the tube, the whole glabrous, sessile; perianth 2.5-3.5 mm. long, white or veined with purple, glabrous, segments obovate or oblong; achenes glabrous, not winged.---Dry plains and hills. Kansas to Utah, south to Texas. Our one record from southwestern Colorado at 6500 feet.

46. Eriogonum campanulatum Nutt., Journ. Acad. Nat. Sc. Phila. ser. 2, 1:163. 1848.

E. brevicaule ssp. campanulatum (Nutt.) Stokes---Perennial plants, branches woody below, not imbedded in the ground but shrubby, leafy portion tomentose, rather stout, usually shorter than the inflorescence; flowering stems 10-30 cm. tall, scapose, glabrous; leaves 3-7 cm. long, linear, narrowly oblanceolate or spatulate, cuneate to a petiole shorter than the blade, obtuse or subacute at the apex, both sides tomentose but lighter above, somewhat floccose, margins more or less revolute; inflorescence a branching cyme, this corymbose; bracts mostly triangular, usually small but sometimes longer and foliose; involucres 2-2.5 mm. long, broadly turbinate or oblong-turbinate (hardly campanulate), the lobes moderate to short, glabrous externally, peduncled in the axils of the inflorescence; perianth about 2-2.5 mm. long, yellow, glabrous, segments oblong or somewhat fiddle-shaped.---Dry plains and hills. Montana to Idaho, south to Nebraska, Colorado and Utah. Our records from northcentral and northwestern Colorado at 5000-8500 feet.

A form with base depressed forming mats, often somewhat imbedded in the ground and the leaves usually not revolute, with a petiole as long or longer than the blade, has been called E. brevicaule Nutt. and may be present in this state but is doubtfully distinct from the above.

47. Eriogonum spathulatum Gray, Proc. Am. Acad. Sc. 10:76. 1874.

E. spathulatum var. brandegei (Rydb.) Stokes; E. brandegei Rydb.---Utah and Colorado. Reported for southern Colorado but no material from the state seen by the writer.

48. Eriogonum lonchophyllum T. & G., Proc. Am. Acad. 8:173. 1870.

Perennial plants, caudices with leafy tomentose branches about 10 cm. long or more, the woody basal branches often long; scapes glabrous or nearly so; leaves 5-10 cm. long, linear to lanceolate, cuneate to a short petiole, tip acute to nearly obtuse, margins little if any revolute, densely white-tomentose below, loosely floccose and glabrate above; inflorescence a branching cyme 15 cm. or more long; bracts narrow and reduced; involucre 3-4 mm. long, narrowly campanulate to narrowly turbinate, teeth short, glabrous outside, all sessile; perianth about 3 mm. long, white or rose, glabrous, segments obovate to oval; achenes glabrous, not winged.---Dry ground. Colorado, Utah and New Mexico. Our few records from westcentral and southwestern Colorado at 5500-6500 feet.

49. Eriogonum salicinum Greene, Pl. Baker. 3:16. 1901.

E. effusum ssp. salicinum (Greene) Stokes---Perennial plants, bases woody-branched, the leafy portion longer or shorter than the inflorescence and tomentose; scapes tall, usually well over 30 cm. (counting leafy stems), robust, glabrous or glaucous; leaves often over 5 cm. long, rather basal on woody branches, lanceolate or oblong-lanceolate, cuneate to a petiole shorter than the blade, acutish at apex, tomentose below, floccose or glabrate above, margins not revolute; inflorescence a corymbose, diffusely branched cyme; bracts scalelike and ternate; involucres about 2.5-3, rarely to 4 mm. long, turbinate, with teeth much shorter than the tube, glabrous, mostly sessile but peduncled in the axils of the branches of the inflorescence; perianth 2-3 mm. long, white or rose, segments oblong.---Canyons and slopes. Colorado. Our records from the westcentral part of the state at 5500-7000 feet.

50. Eriogonum leptophyllum (Torr.) Woot. and Standl., Contrib. U.S. Nat. Herb. 16:118. 1913.

Perennial plants definitely woody at base, about 20-60 cm. tall; stems leafy, glabrous or obscurely puberulent; leaves 2-4 cm. long, linear, margins entire and revolute, the glabrous upper surface often completely hiding the white-lanate lower surface; inflorescence a dense broomlike cyme with numerous close branches, the whole usually less than 6 cm. long; bracts small and triangular; involucres 2-3 mm. long, cylindric-turbinate, glabrous or very obscurely ciliolate, sessile, lobes much shorter than the tube; perianth 2.5-3.5 mm. long, white or rose-colored especially on the midveins, glabrous, segments obovate to oblong.---Dry ground. New Mexico and Arizona. Our one record from southwestern Colorado at about 5000 feet.

51. Eriogonum scoparium Small, Bull. Torr. Club 33:54. 1906.

E. nudicaule ssp. scoparium (Sm.) Stokes---Perennial plants; stems from woody bases, leafy stems woody and shorter than the inflorescence, tomentose, the scapes glabrous or glabrate; leaves 2-5 cm. long, linear to linear-spatulate, cuneate at base to a short petiole, apex acute or obtuse, margins revolute, densely tomentose below, floccose above; inflorescence a branched open cyme; bracts scalelike; involucre 2.5-3 mm. long, broadly-campanulate, about as wide as long, teeth wide and short, glabrous outside, mostly peduncled, those in the forks of the inflorescence definitely so; perianth 2.5-4 mm. long, white or rose, glabrous, the lobes oblong or obovate; achenes glabrous, not winged.---Dry plains, valleys and slopes. Colorado and New Mexico. Our records from the western half of Colorado at 6000-7500 feet.

52. Eriogonum tristichum Small, Bull. Torr. Club 33:55. 1906.

E. nudicaule ssp. tristichum (Small) Stokes---Perennial plants with branching woody bases, the woody, leafy portion rather short, usually shorter than the inflorescence; flowering stems 10-30 cm. tall, sometimes more, longer than the leafy portion, rather scapose-looking, glabrous above the leaves; leaves 2-7 cm. long, on woody stems, rather close and erect, linear to linear-oblong, long-cuneate to a short petiole, acute at apex, margins revolute, tomentose below, glabrous to floccose above; inflorescence of cymosely-corymbose involucres 2- to 3-branched; bracts scalelike, rarely leaflike, ternate; involucres 3-4 mm. long, turbinate, lobes

shorter than the tubes, glabrous, mostly sessile but pedicelled in forks of inflorescence; perianth 2.5-4 mm. long, white or rose, glabrous, segments oblong to obovate-oblong.---Dry plains, valleys and hills. Colorado and New Mexico. Our records from western Colorado at 6000-9000 feet.
53. Eriogonum ainsliei Standl., Contrib. U. S. Nat. Herb. 16:117. 1913.
   E. effusum ssp. ainsliei (Woot. & Standl.) Stokes---Colorado and New Mexico. Reported from southeastern Colorado but the writer has seen only one very doubtful specimen from Pueblo County.
54. Eriogonum salinum (A. Nels.), Bull. Torr. Club 31:240. 1904.
   E. effusum ssp. salinum (A. Nels.) Stokes---This appears to be a dwarf form of E. effusum Nutt. growing on exposed ground. It is reported for Wyoming and Colorado but no material from this state has been located.
55. Eriogonum effusum Nutt., Journ. Acad. Nat. Sc. Phila. ser. 2, 1:164. 1848.
   E. effusum ssp.typicum Stokes---Perennial plants, shrubby at base and much branched, the branches many, leafy stems tomentose; flowering stems 20-50 cm. tall, rather scapose, tomentose or floccose; leaves 2-4 cm. long, linear to oblanceolate or narrowly oblong, cuneate below to a short petiole, obtuse or subacute at apex, plane or margins undulate; inflorescence a diffuse, open cyme, usually over 8 cm. long, corymbose and branched 2-3 or more times; bracts small and triangular but occasionally almost foliose; involucres 1.5-2.5 mm. long, narrowly campanulate, lobes moderate, not as long as the tube, usually tomentulose, sessile or short-pedunculate especially in the axils of the inflorescence; perianth 2-2.5 mm. long, white or rose-colored, glabrous, segments obovate to elliptic.---Dry plains and hills. Nebraska, Montana to Utah, south to New Mexico. Our records scattered over Colorado, none as yet from the corners of the state, at 4000-9500 feet.

2. Oxyria Hill MOUNTAIN SORREL

Low perennial herbaceous plants with thick fleshy taproots; stems erect, with acrid juice; leaves alternate but mostly basal, long-petioled, round or reniform; stipules united into a cylinder around the stem; inflorescence a narrow-panicled raceme, the flowers verticillate; flowers greenish, perfect, small; perianth segments 4, the 2 outer larger than the inner; stamens 6; style short, 2-parted; achenes lenticular, flattened, broadly winged.

1. Oxyria digyna (L.) Hill, Hort. Kew. 158. 1768.
   Stems 5-30 cm. tall, simple or few-branched; leaves 1-3.5 cm. in diameter, margin rather undulate; perianth 1.5-2 mm. long, red or greenish, inner segments erect, 4-6 mm. long in fruit, otherwise not modified, the outer becoming reflexed; achenes in fruit broadly winged, these membranous.---In cold often wet places among rocks at high altitudes. Greenland to Alaska, south to New Hampshire and Arizona; Eurasia. Our records scattered in the high mountains of Colorado at 9500-12,500 feet.

3. Polygonum L. KNOTWEED; SMARTWEED

Annual or perennial, terrestrial or aquatic herbaceous plants or rarely shrubby; stems usually with swollen joints, simple or branched; stipules united into a sheath around the stem; leaves alternate, entire, usually narrow; flowers axillary or in spikelike racemes terminating the stem or branches; flowers usually perfect, sometimes heterostyled; perianth 5- or 6-parted, often petaloid, erect in fruit; stamens 3-9; styles and stigmas 2 or 3; achene lenticular or 3-angled.

1. Stems twining; outer perianth segments keeled or winged at maturity
   2. Outer perianth segments conspicuously winged in fruit; achenes smooth and shining -1. P. scandens
   2. Outer perianth segments merely keeled at maturity; achenes grandular and dull -2. P. convolvulus
1. Stems not twining; outer perianth segments wingless and rarely even obscurely keeled
   3. Stems with sharp recurved prickles; leaves sagittate -3. P. sagittatum
   3. Stems glabrous to hairy but no sharp prickles present; leaves not sagittate
      4. Leaves with a hingelike joint at the attachment of the blade and the sheath; flowers in axillary clusters or if in rather spikelike terminal clusters then in the axils of bracts; ocreae 2-lobed or deeply lacerate
         5. Flowers crowded toward the ends of the branches, appearing to be in terminal spikes, bracts leaflike (usually broader than the leaves)
            6. Inflorescence very short and congested, mostly continuous; perianth about 2 mm. long; achenes about 1.5 mm. long
               7. Bracts oblong, very obtuse, with broad white petaloid margins -4. P. polygaloides
               7. Bracts linear or linear-lanceolate, acute, no white margins present
                  8. Leaves 4-10 mm. long; bracts not much different from the leaves; styles wanting, stigmas sessile or subsessile -5. P. kelloggii
                  8. Leaves 10-40 mm. long; bracts definitely shorter and wider than the leaves; styles short but evident -6. P. watsonii
            6. Inflorescence rather long, not congested and usually interrupted; perianth 2.5-4 mm. long; achenes 3-4 mm. long
               9. Upper bracts foliacious, not reduced; flowers deflexed in fruit -7. P. montanum
               9. Upper bracts reduced or subulate; flowers usually erect or spreading in fruit, seldom deflexed -8. P. spergulariaeforme
         5. Flowers scattered along the stems in small axillary clusters
            10. Fruit reflexed
                11. Perianth and achenes 1.5-2.6 mm. long; leaves linear to linear-lanceolate -9. P. engelmannii
                11. Perianth and achenes over 2.6 mm. long; leaves linear-lanceolate to wider

12. Upper bracts leaflike; flowers tending to crowd out near the ends of the branches, often appearing as a terminal spike -7. P. montanum
12. Upper bracts reduced and subulate; flowers not tending to crowd out near ends of branches -10. P. douglasii
  10. Fruit erect or nearly so, not reflexed
    13. Achenes 5-8 mm. long, conspicuously exserted (about twice as long as the perianth) -11. P. exsertum
    13. Achenes not over 3.5 mm. long, included or seldom over slightly exserted
      14. Stems decumbent or prostrate; achenes dark brown or reddish-brown at maturity
        15. Leaves lanceolate or broader -12. P. aviculare
        15. Leaves narrowly linear-lanceolate -12A. P. aviculare angustissimum
      14. Stems erect or ascending from a branching base; achenes black at maturity (except in P. erectum)
        16. Leaves lanceolate, oblanceolate or narrower, the upper reduced in length and width, rather scattered
          17. Upper bracts subulate; achenes smooth and shining; stems branching at base, seldom throughout -13. P. sawatchense
          17. Upper bracts linear-oblong, not subulate; achenes minutely roughened, not shining; stems usually branching throughout -14. P. ramosissimum
        16. Leaves elliptic-lanceolate, ovate-lanceolate or broader, the upper leaves not reduced, rather crowded
          18. Perianth about 3 mm. long; leaves 1.5-6 cm. long; stems 20-90 cm. long, usually branching throughout -15. P. erectum
          18. Perianth about 2 mm. long; leaves 0.5-1.5 cm. long; stems 5-15 cm. long, usually branching only at base -16. P. minimum
4. Leaves without such a joint; flowers in terminal (sometimes also axillary) spikelike clusters, the bracts of the inflorescence scarious and reduced to sheaths; ocreae not commonly deeply 2-lobed or lacerate (may be ciliate)
  19. Stems simple from fleshy rootstocks; basal leaves long-petioled, cauline sessile or short-petioled; inflorescence solitary; ocreae oblique, lobed or lacerate, more or less open on one side; plants of high altitudes
    20. Racemes bearing bulblets below, linear, 5-8 mm. thick; basal leaves usually subcordate or cordate at bases -17. P. viviparum
    20. Racemes lacking bulblets, oblong and over 1 cm. thick; basal leaves cuneate
      21. Basal leaves lanceolate or broader; stems often over 30 cm. tall; perianth over 4 mm. long -18. P. bistortoides
      21. Basal leaves linear; stems not over 30 cm. tall; perianth seldom over 4 mm. long -18A. P. bistortoides linearifolium
  19. Stems branching at base or throughout, the rootstocks when present seldom fleshy; leaves all cauline and similar; inflorescences terminating the branches, seldom solitary; ocreae truncate, not lobed or lacerate (may be ciliate); plants of low and middle elevations
    22. Inflorescences all terminal, usually solitary; plants perennial, usually aquatic or semiaquatic; flowers bright pink or rose; stamens 5
      23. Inflorescence 1-3 cm. long, over 1 cm. wide; peduncles usually glabrous -19. P. amphibium
      23. Inflorescence 3-10 cm. long, less than 1 cm. wide; peduncles usually hairy -20. P. coccineum
    22. Inflorescences several to many, axillary as well as terminal; plants mostly annuals, not really aquatic; flowers bright rose to white or green; stamens usually over 5
      24. Sheaths with marginal bristles
        25. Perianth glandular-punctate, pale green or whitish; inflorescence lax and interrupted
          26. Racemes erect; achenes smooth and shining; leaves lanceolate to linear-lanceolate -21. P. punctatum
          26. Racemes nodding (at least in fruit); achenes grandular to dull; leaves lanceolate to ovate-lanceolate -22. P. hydropiper
        25. Perianth not glandular-punctate or very obscurely so, rose, pink, or purple; inflorescence rather densely flowered, not interrupted (except sometimes at base)
          27. Ocreae conspicuously fringed; plants annual -23. P. persicaria
          27. Ocreae short-fringed, not conspicuously so; plants perennial -24. P. persicarioides
      24. Sheaths without marginal bristles (sometimes short-ciliate when young)
        28. Leaves floccose-tomentose beneath -25. P. incanum
        28. Leaves glabrous or nearly so beneath
          29. Inflorescences erect or nearly so, seldom over 5 cm. long; peduncles bearing stalked glands; perianth 3-4 mm. long
            30. Flowers with only 1 style length, or if 2 these are not segregated on separate plants -26. P. pennsylvanicum
            30. Flowers with 2 style lengths, these segregated on different plants -27. P. longistylum
          29. Inflorescences drooping or erect, sometimes over 5 cm. long; peduncle glands if present sessile; perianth not over 3 mm. long
            31. Inflorescence erect, not over 5 cm. long; some of sheaths usually ciliate; plants perennial -24. P. persicarioides
            31. Inflorescences drooping, often over 5 cm. long; none of sheaths ciliate; plants annual
              32. Styles united only at the base -28. P. incarnatum
              32. Styles united almost to the middle -29. P. lapathifolium

1. Polygonum scandens L., Sp. Pl. 364. 1753.
    Bilderdykia scandens (L.) Greene; Tiniaria scandens (L.) Small---Perennial plants, glabrous or at least somewhat scurfy; stems 50-300 cm. long, branched, twining; leaf blades 1-12 cm. long, ovate to oblong-ovate, cordate to somewhat sagittate at base, short-acuminate, larger

long-petioled; stipule sheaths with entire margins; flowers in axillary or terminal racemes; flowers 10 mm. long in fruit, yellowish-green, outer segments strongly winged (but sometimes 1 or all 3 wanting); achenes 3.5-4.5 mm. long, black, smooth and shining.---Thickets. Nova Scotia to Montana, south to Florida, Louisiana and Colorado. Our few records from northcentral and central Colorado at 5000-6000 feet.

2. Polygonum convolvulus L., Sp. Pl. 364. 1753.
Bilderdykia convolvulus (L.) Dum.; Tiniaria convolvulus (L.) Webb. & Moq.---Annual, glabrous but scurfy plants; stems 20-100 cm. long, branched and twining or trailing, the internodes long; leaves 2-6 cm. long, long-petioled, ovate-sagittate or the upper lanceolate-sagittate, acuminate; stipule sheaths with entire margins; flowers in axillary clusters or racemes; perianth 3.5-4 mm. long, greenish, outer lobes keeled in fruit; achenes 3-angled, black, dull and granular.--- Waste places, along fences and in thickets. Widely distributed in North America; Eurasia. Our records scattered throughout the state, mostly from the eastern half, at 4000-9000 feet.

3. Polygonum sagittatum L., Sp. Pl. 363. 1753.
Tracaulon sagittatum (L.) Small--Annual plants; stems 30-150 cm. long, slender, decumbent or reclining, branched, armed on the 4 angles by sharp recurved prickles; leaves 1-12 cm. long, lanceolate or oblong, sagittate at the base, obtuse or acute, lower petioled, upper sessile; stipule sheaths not ciliate; flowers in rather dense, terminal heads or racemes; perianth segments about 4 mm. long, white, green or rose; stamens usually 8; style 3-parted to below the middle; achenes about 3 mm. long, 3-angled, black or brownish-red, smooth.---Meadows. Newfoundland to Saskatchewan, south to Florida and Texas. Our one record from El Paso County at about 6000 feet.

4. Polygonum polygaloides Meisn. in DC., Prodr. 14:101. 1856.
Annual plants glabrous throughout; stems 5-20 cm. tall, erect, simple or more commonly corymbosely-branched; leaves 5-30 mm. long, linear or narrowly linear, acute; ocreae 2-parted or lacerate; flowers crowded at the ends of the branches in a short raceme; bracts ascending, oblong, with conspicuous white-petaloid margins, apex acute to obtuse; perianth 2 mm. long, white or pinkish or the midrib darker; stamens 8; style evident, 3-parted below the middle; achenes blackish or dark brown, somewhat striated.---Moist or wet ground. Montana to Washington, south to Wyoming and Oregon. Our one record from northwestern Colorado at 6300 feet.

5. Polygonum kelloggii Greene, Fl. Franc. 134. 1891.
Annual glabrous plants; stems 3-8 cm. high, erect and simple, or branched from the base; leaves 5-10 mm. long, linear to linear-lanceolate; flowers in a leafy-bracted spikelike inflorescence, this terminal; bracts similar to leaves but usually somewhat shorter and slightly wider; perianth 1.5-2 mm. long, green with white or cream-colored margins; stamens about 3; style wanting or very short; achenes 1.5 mm. long, 3-angled, ovoid, light brown, dull and granular especially above.---Moist or wet soil. Montana to Washington, south to Colorado and California. Our records from northcentral and southwestern Colorado at 6500-9500 feet.

6. Polygonum watsoni Small, Columbia Univ. Dept. Bot. Mem. 1:138. 1895.
P. confertiflorum Nutt.---Annual glabrous plants; stems 3-15 cm. high, erect, simple or branched; leaves 1-4 cm. long, linear or narrowly linear-lanceolate; flowers in short terminal racemes; bracts similar to leaves but reduced in length and usually wider; perianth 2 mm. long, green, segments with pinkish margins; stamens 3-5; styles 3-cleft to below middle, the stigmas not at all sessile; achenes 3-angled, ovoid, striate-granular, dull black or dark brown.--- Moist or wet places. Saskatchewan to British Columbia, south to Colorado and California. Our few records from northcentral Colorado at 6500-9500 feet.
P. unifolium Small has been reported from Colorado but is probably only a dwarf reduced form of the above.

7. Polygonum montanum (Small) Greene, Pl. Baker. 3:13. 1901.
P. douglasii var. montanum Small; P. douglasii var. latifolium (Engelm.) Greene; P. commixtum Greene---Annual plants; stems 10-30 cm. tall, branched from near base; leaves 10-40 mm. long, elliptic or oblong-lanceolate, acute, cuneate, midnerve prominent; branches floriferous from near base but few flowers below, mostly forming a bracted spikelike inflorescence at end of branches; bracts foliaceous; perianth dark green or purplish, with lighter margins; achenes reflexed in fruit, black, or somewhat striate, smooth and shining.---Mountains. Alberta to California, south to New Mexico. Our records scattered in the western half of Colorado at 7500-12,000 feet.

8. Polygonum spergulariaeforme Meisn. ex Small, Bull. Torr. Club 19:366. 1892.
Annual plants, scurfy throughout but glabrous; stems 10-40 cm. tall, mostly erect, simple or corymbosely branched from near base; leaves 1-3 cm. long, linear-oblong or linear-lanceolate, acute, cuneate to a sessile base; inflorescence axillary, of clusters of several flowers, these confined to ends of branches and appearing as interrupted racemes; upper bracts reduced and subulate; perianth segments 3.5-4 mm. long, whitish or pink, green-nerved; stamens 8; styles 3-parted to middle or near base; achenes about 3.4-4 mm. long, 3-angled, black, smooth and shining except on the angles and near the apex.---Plains and slopes. British Columbia, south to Wyoming and California. Our one rather doubtful specimen from Chaffee County at 9000 feet.

9. Polygonum engelmannii Greene, Bull. Calif. Acad. 1:126. 1885.
Annual plants; stems 5-30 cm. tall, slender, rather scurfy, usually diffusely-branched at base, the branches erect or spreading; leaves 2-15 mm. long, linear to linear-lanceolate or linear-oblanceolate, acute, cuneate but sessile; stipule sheaths at length somewhat lacerate; inflorescence axillary, of clusters of 2-4 flowers extending from base of plant to end of branches; perianth 2-2.5 mm. long, green but the segments with whitish margins; stamens 5-8; achenes ovoid, 3-angled, black, smooth and shining, 2-2.5 mm. long, reflexed in fruit.---Hills and mountains. Montana to British Columbia, south to Colorado and Utah. Our records from central and northcentral Colorado at 7000-9600 feet.

10. Polygonum douglasii Greene, Calif. Acad. Sci. Bul. 1:125. 1885.
Annual plants; stems 10-40 cm. long, simple or sparingly branched, slender, erect or ascending, glabrous except at nodes which may be scabrous, often glaucescent; leaves 1-5 cm. long,

narrowly lanceolate to oblong or oblanceolate, obtuse or acute, cuneate at base, sessile or nearly so; stipules finally lacerate; inflorescence axillary, 1-3 flowers in a cluster, borne in all but the lowest leaf axils; upper bracts reduced and subulate; perianth segments 3-4 mm. long, green with white or rose on the margins; stamens 8; achenes 3-4 mm. long, 3-angled, black, smooth and shining, reflexed at maturity.---Plains and slopes, often in sandy ground. Saskatchewan to British Columbia, south to New Mexico and Arizona. Our records scattered over the western two-thirds of Colorado at 5000-11,000 feet.

11. Polygonum exsertum Small, Bull. Torr. Club 21:172. 1894.

This eastern species has been reported from this area and may be in Colorado. However, Brenckle (Phytologia 2:169-171. 1946.) suggested that the exserted achenes usually used to distinguish this plant may be produced by other related species as a response to seasonal and weather conditions. In any case no specimens from Colorado have been seen by the writer.

12. Polygonum aviculare L., Sp. Pl. 362. 1753.

P. buxiforme Small---Annual or rarely perennial plants with a blue-green aspect; stems 10-100 cm. long, mostly prostrate or ascending, simple or much branched, slender; leaves 6-25 mm. long, lanceolate to oblong-lanceolate, usually acute but sometimes obtuse, cuneate at base; stipule sheaths silvery, becoming lacerate; flowers in axillary clusters of 1-5; perianth 2-3.5 mm. long, sometimes less, green or with pinkish margins; stamens usually 8; achenes dull and minutely granular, dark brown.---Waste places. Widely distributed in North America; Eurasia. Our records scattered in the state, mostly from cultivated areas, at 5000-9500 feet.

An introduced related species P. monspeliensis Thieb. with upright stems and broader leaves may be in Colorado.

12A. Polygonum aviculare angustissimum Meisn., (var.) in DC., Prod. 14:98. 1857.

P. neglectum Besser---This variety has leaves linear to narrowly linear-lanceolate; achenes usually acuminate (instead of acute as in the species).---Most of our records are from northcentral Colorado at 4500-8500 feet.

13. Polygonum sawatchense Small, Bull. Torr. Club 20:213. 1893.

Annual plants; stems 5-30 cm. tall, slender, erect, simple or more commonly branched from near the base, striate; leaves oblanceolate rarely obovate but narrower above, acute, cuneate, often rather revolute, midrib prominent; stipule sheaths 2-parted or finally lacerate; flowers in clusters of 2 or 3 in axils of nearly all the leaves; upper bracts subulate; perianth 2-3 mm. long, green or slightly lighter on the margins; stamens 6-8; achenes 3-angled, erect in fruit, smooth, black and shining.---Plains and slopes. South Dakota to Washington, south to New Mexico and California. Our records scattered over Colorado, except the extreme eastern part, at 4500-8000 feet.

14. Polygonum ramosissimum Michx., Fl. Bor. Amer. 1:237. 1803.

P. rubescens Small; P. prolificum (Small) Robins.---Annual plants; stems 10-80 cm. tall, erect or ascending, nearly simple to much branched throughout, slender but rather rigid, yellowish-green, glabrous; leaves 1-4 cm. long, lanceolate, linear-oblong, sometimes wider, acute, cuneate; stipule sheaths early becoming lacerate; inflorescence of axillary clusters of several flowers; upper bracts somewhat reduced but not subulate; perianth segments about 2-3.5 mm. long, greenish or yellowish; stamens 3-6; achenes about 3 mm. long, erect in fruit, 3-angled, ovoid, black, usually not shining but rather granular.---Roadsides, plains, and slopes, often in sandy ground. Widely distributed in North America. Our records scattered in the eastern half of Colorado at 3500-5500 feet.

15. Polygonum erectum L., Sp. Pl. 363. 1753.

Annual glabrous plants; stems 20-60 cm. tall, rather stout, erect or ascending, simple or much branched, yellowish-green in age; leaves 1.5-6 cm. long, oval, oblong-elliptic or obovate, obtuse or acutish, cuneate to a sessile or subsessile base, little reduced in the inflorescence; flowers in small axillary clusters of several flowers, scattered along the stem; perianth segments 3 mm. long, greenish, paler or whitish on margins; stamens usually 6; achenes about 3 mm. long, 3-angled, dark brown, dull and granular. ---Waste places. Maine to Alberta, south to Georgia and New Mexico. Our one record from Weld County at 4600 feet.

16. Polygonum minimum S. Wats., Bot. King Expl. 315. 1871.

Annual plants; stems 5-15 cm. long, slender, simple or branched from base, leafy throughout; leaves 5-15 mm. long, ovate to ovate-lanceolate, or obovate, sessile or subsessile, cuneate at base, acute or apiculate at apex, sometimes rather crowded at ends of branches and little if any reduced; flowers mostly 2 or 3 in the axils of most of the leaves, the pedicels 2-3 mm. long; perianth 1.5-2 mm. long, greenish, the margins of the segments often rose-colored; achenes 2-2.5 mm. long, not reflexed in fruit, 3-angled, blackish, smooth and shining.---Dry often sandy soil. British Columbia and Alaska, south to Colorado and California. Our few records from northcentral and southcentral Colorado at 7500-11,000 feet.

17. Polygonum viviparum L., Sp. Pl. 360. 1753.

Bistorta vivipara (L.) Gray---Perennial plants from short cormlike scaly rootstocks; stems 10-30 cm. long, simple, erect; basal leaves 2-10 cm. long, sometimes longer, long-petioled, oblong or lanceolate, acute usually cordate or subcordate at base; stem leaves linear to narrowly lanceolate, subsessile or clasping, often rather revolute; stipule sheaths open above; inflorescence a terminal, linear spikelike raceme, 5-8 mm. thick and 2-8 cm. long, at least the lower flowers replaced by sessile bulblets; perianth pale rose or white; stamens exserted; achenes 3-angled, granular and dull.---Moist ground. Greenland to Alaska, south to New Mexico; Eurasia. Our records from the higher mountains of the state at 8500-13,000 feet.

18. Polygonum bistortoides Pursh, Fl. Amer. Sept. 271. 1814.

Bistorta bistortoides (Pursh) Small---Perennial plants with large, fleshy, horizontal chaffy rootstocks; stems 25-70 cm. tall, erect, simple; basal leaves 10-25 cm. long, oblong to oblong-lanceolate or oblanceolate, acute or obtuse at apex, cuneate at base, glabrous or scabrous-pubescent below, on long petioles; cauline leaves sessile, mostly lanceolate, reduced; stipule sheaths oblique at apex; inflorescence a spikelike solitary, terminal, densely-flowered raceme 1-6 cm. long and 1-1.5 cm. thick; perianth about 4 mm. long, light rose to white; stamens exserted; achenes 3-angled, smooth and shining.---Moist ground. Montana to British Columbia, south to New Mexico and California. Our records scattered in the mountainous parts of Colorado at 7500-12,500 feet.

18A. **Polygonum bistortoides linearifolium** (Wats.) Small, (var.) Bull. Torr. Club 19:252. 1892.
Bistorta linearifolia (Wats.) Greene---This variety differs in having the basal leaves linear, and the stems less than 30 cm. tall; perianth is 4 mm. long or less. Intergrades with the species in Colorado.---Montana to Nevada, south to Colorado and Utah. Our records scattered in the mountains of the state at 10,500-13,000 feet.

19. **Polygonum amphibium** L., Sp. Pl. 1:361. 1753.
P. natans A. Eaton; P. hartwrightii Gray; Persicaria hartwrightii (Gray) Greene---Perennial, often aquatic plants; stems emersed or terrestrial, floating to erect, from rootstocks; leaves 7-15 cm. long, elliptic to lanceolate, rather coriaceous, glabrous to somewhat pubescent, base rounded to subcordate to cuneate, apex obtuse to attenuate; stipule sheaths with margins scarious, or herbaceous and reflexed or spreading; inflorescence erect, usually a single spikelike raceme, 1-3 cm. long and over 1 cm. in diameter, densely-flowered; peduncles glabrous; flowers pink to rose, heterostyled; achenes lenticular, nearly orbicular to obovoid, black, shining or granular and dull. Our plant has been called var. stipulaceum (Coleman) Fernald and is found in two forms, an aquatic and a terrestrial. The extremes of these 2 forms are unlike in general appearance but intergradations are very common.---In water or wet ground. Labrador to Alaska, south to New Jersey, New Mexico and California; Europe.

20. **Polygonum coccineum** Muhl. ex Willd., Enum. Pl. 428. 1809.
P. muhlenbergia (Meisn.) Watson; P. emersum (Michx.) Britt.; Persicaria coccinea (Muhl.) Greene---Perennial plants; stems 30-100 cm., from elongated rootstocks, emersed and floating or terrestrial and erect, mostly glabrous below, pubescent above with simple or glandular hairs, stems rooting at base in the submerged form; leaves lanceolate to ovate-lanceolate, subcoriaceous, acute to acuminate, sometimes obtuse, glabrous to pubescent; stipule sheaths entire or short-ciliate; inflorescence usually terminal of spikelike racemes on stout, pubescent peduncles, the inflorescence 3-10 cm. long and seldom over 1 cm. wide; flowers scarlet or pink or deep rose; stamens and styles exserted but heterostyled; achenes lenticular, orbicular or broadly obovate, black, shining or minutely roughened. Comes in 2 habitat forms. The forma natans (Wiegand) Stanford is the aquatic form and the forma terrestre (Willd.) Stanford grows in moist soil. The latter intergrades with the aquatic form in this state.---In water or wet ground. Maine to Alaska, south to Virginia, Mexico and California. Our records scattered over Colorado, few from the extreme western or eastern parts, at 4500-7500 feet.

21. **Polygonum punctatum** Ell., Bot. S. C. & Ga. 1:455. 1817.
Persicaria punctata (Ell.) Small---Annual or perennial plants mostly glabrous throughout; stems 30-100 cm. tall, erect or ascending or rarely prostrate especially at base, simple or branched; leaf blades 2-16 cm. long, lanceolate to narrowly lanceolate, sometimes oblong-lanceolate, acuminate at apex, cuneate at base, short-petioled, usually glabrous, punctate; ocreae with bristly margins, these rather long; racemes 1-6 cm. long, terminal, usually more than 1, each linear, loosely flowered and interrupted at least below; perianth about 2 mm. long, greenish, 5-parted to below the middle, conspicuously glandular-punctate; stamens 8; style 2- or 3-parted to base; achenes lenticular or 3-angled, smooth and shining. Fassett (Brittonia 6:369-393. 1949.) listed the varieties and forms of the above. He gave var. confertiflorum (Meisn.) Fassett and var. parviflorum Fassett for Colorado.---Moist or wet places. Throughout the most of the United States and south to Central America. Our few records from northcentral Colorado at 4500-5500 feet.

22. **Polygonum hydropiper** L., Sp. Pl. 361. 1753.
Persicaria hydropiper (L.) Opiz.---Annual glabrous plants; stems 20-60 cm. tall, erect or assurgent; leaves 1-9 cm. long, lanceolate to oblanceolate, acute or acuminate, cuneate, short-petioled, more or less papillose and punctate; stipule sheaths bristly-ciliate; inflorescence 2-6 cm. long, linear, drooping, interrupted; perianth 2.5-3 mm. long, greenish, definitely glandular-dotted; stamens 6; styles 2- or 3-parted; achenes lenticular or 3-angled, dark brown, granular and dull, not shining.---Wet or moist ground ground. Newfoundland to British Columbia, south to Georgia, Mexico and Central America; Europe. Our few records from northcentral Colorado at about 5000-5200 feet.

23. **Polygonum persicaria** L., Sp. Pl. 361. 1753.
Persicaria persicaria (L.) Small---Annual glabrous or puberulent plants; stems 20-80 cm. tall, erect, usually branched from the base; leaves 2-18 cm. long, lanceolate to linear-lanceolate, acuminate, cuneate, subsessile, punctate, usually with a dark spot near the middle; stipule sheaths ciliate; inflorescence 1-3 cm. long, not over 1 cm. thick, densely-flowered, mostly erect; perianth segments pink or dark rose; stamens mostly 6; styles 2- to 3-parted; achenes 2-2.5 mm. long, lenticular or 3-angled, black, smooth and shining.---Waste places, often in moist ground. Widespread in North America; Europe. Our records scattered in Colorado, mostly in cultivated areas, at 3500-7500 feet.

24. **Polygonum persicariodes** H. B. K., Nov. Gen. 2:197. 1817.
This species is closely related to P. persicaria L. but is a perennial plant.---Nebraska south to Texas and Mexico. Our one somewhat doubtful record is from eastern Colorado at 3700 feet.

25. **Polygonum incanum** F. W. Schmidt., Fl. Boem. 4:90. 1795.
P. lepathifolium incanum (Schmidt) Koch; P. tomentosa Schrank; Persicaria incana (Schmidt) Gray; Persicaria tomentosa (Schrank) Bickn.---Annual plants; stems 10-50 cm. tall, simple or moderately branched, erect, somewhat scurfy; leaves lanceolate or lanceolate-oblong, acute, acuminate or obtuse, cuneate, subsessile, glabrous above, lighter below and the lower ones at least retaining more or less flocculose-tomentum; stipule sheaths not ciliate; inflorescence 1-3 cm. long, erect or only slightly nodding, dense; peduncles with sessile glands; perianth

2-2.5 mm. long, green to pinkish-white; achenes lenticular, 2.5 mm. long, dark-brown, smooth and shining.---Swamps and wet ground. Newfoundland to British Columbia, south to New York and Colorado. Our few records scattered in the eastern half of Colorado at 4000-5000 feet.
26. Polygonum pennsylvanicum L., Sp. Pl. 362. 1753.
P. omissum Greene; Persicaria pennsylvanica (L.) Small---Annual plants; stems 30-100 cm. tall, rather stout, erect, branching, glabrous below, more or less glandular-hairy above, these glands usually red; leaves 5-20 cm. long, lanceolate, blades cuneate-rounded at base, glabrous or glabrescent, petioled; stipule sheaths usually not ciliate; inflorescence 2-5 cm. long, peduncles copiously covered with glandular hairs, the ends usually red; flowers with same length styles or if 2 style lengths present then these on same plant; perianth segments 3-4 mm. long, pink, deep rose or light purple; stamens about 8; achenes 3-3.5 mm. wide at maturity, lenticular, rather shining. Our plants have been called var. laevigatum Fernald.---Waste ground, especially in moist rich ground. Nova Scotia to Minnesota, south to Florida and Mexico. Our records scattered in the eastern half of Colorado at 3500-5500 feet.
27. Polygonum longistylum Small, Bull. Torr. Club 21:169. 1894.
This plant is recorded for Colorado by Stanford (Rhodora 27:183. 1925.) as var. omissum (Greene) Stanford. It is closely related to P. pennsylvanicum L. but differs in having flowers with 2 style lengths, these on separate plants, not a very practical difference when dealing with individual specimens.---This plant has been recorded as coming into eastern Colorado. We have a doubtful specimen from Weld County.
28. Polygonum incarnatum Ell., Bot. S. C. & Ga. 1:456. 1817.
Persicaria incarnata (Ell.) Small---Annual plants, glabrous or nearly so; stems 60-100 cm. tall, erect; leaves 5-20 cm. long, lanceolate or linear-lanceolate, acute to acuminate, cuneate, short-petioled, sparingly punctate and ciliate; stipule sheaths not ciliate; inflorescence 3-8 cm. long, drooping, rather densely flowered, of spikelike racemes; perianth rose, white or greenish, 5-parted; stamens 6; style 2-parted almost to base; achenes about 2 mm. long, lenticular, rarely 3-angled, brown or black, shining.---Moist or wet ground. Vermont to Idaho, south to Florida and California. Our few records from northcentral and central Colorado at 4500-6000 feet.
29. Polygonum lapathifolium L., Sp. Pl. 360. 1753.
Persicaria lapathifolia (L.) Gray---Annual plants, glabrous or nearly so; stems 30-60 cm. tall, erect, simple or branched; leaf blades 5-20 cm. long, lanceolate or oblong-lanceolate, acuminate, cuneate, short-petioled, punctate and ciliate; inflorescence 2-8 cm. long, drooping or erect, densely-flowered, of spikelike racemes, peduncles with sessile glands often present; stipule sheaths not ciliate; perianth 2-2.5 mm. long, pink, white or greenish; stamens 6; style 2-parted to below middle but not to base; achenes about 2 mm. long, lenticular or rarely 3-angled, brownish to black, shining. A form occurs in eastern Colorado with leaves conspicuously punctate below and racemes dense and erect. It has been called P. lepathifolium var. nodosum (Pers.) Small.---Moist or wet ground. Throughout most of North America; Eurasia. Our records scattered in the eastern half of Colorado at 3500-7500 feet.

4. Fagopyrum (Tourn.) Mill. BUCKWHEAT

Caulescent, annual, glabrous, herbaceous plants; leaves alternate, petioled, with hastate or cordate blades; stipules united as a sheath around the stems; flowers in terminal or axillary cymes or panicles; perianth segments 5, greenish-white and petaloid; stamens 8; styles 3-parted, stigmas capitate; achenes 3-angled.

1. Fagopyrum esculentum(L.) Moench., Meth. Pl. 290. 1794.
Stems 10-90 cm. tall, becoming strongly grooved; leaves 2.5-8 cm. long; stipule sheaths fugaceous; achenes 5 mm. long, with pinnately striate face when mature, acute-angled.---Escaped from cultivation in many parts of the United States but scarcely naturalized. Our few records from eastern Colorado at 4500-7000 feet. The correct name for this plant may be F. sagittatum Gilib.

5. Rheum L. RHUBARB

Stout herbaceous plants; leaves broad and radical; stipule sheaths large and prominent; flowers bisexual, greenish or whitish, small but numerous, in panicled racemes; perianth 6-parted, not much enlarged; stamens mostly 9; fruit a winged achene.

1. Rheum rhaponticum L., Sp. Pl. 371. 1753.
Perennial plants, roots large and fleshy; leaves mostly radical, cordate-ovate, entire, the blades 30-45 cm. or more long; petiole long and thick, edible; flowering stems 1-2 m. long, strict but more or less branched above, hollow.---This is a cultivated species sometimes tending to escape or persist in vacant lots and fields. It has been found growing in the Colorado mountains away from any dwelling but the plants may have been originally set out by someone.

6. Rumex L. DOCK

Herbaceous, mostly perennial plants; stems leafy; leaves simple, alternate, entire or wavy-margined; stipules united as a cylinder around the stems; flowers small, perfect or unisexual, in simple or compound racemes and panicles; perianth 6-parted, the 3 inner segments becoming enlarged and winged in fruit (the valves), often bearing grainlike tubercles on the back; stamens 6; ovary 3-angled.

1. Flowers dioecious; inflorescence slender and leafless; perianth 1.5 mm. long in flower; foliage with a sour acid taste; leaves often hastate
  2. Leaves cuneate at base, never hastate; valves in fruit enlarging to 2.5 mm. long, longer than the achene -1. R. paucifolius
  2. Leaves, at least some, hastate at base; valves in fruit about 1.5 mm. long, shorter than the achene -2. R. acetosella
1. Flowers perfect or polygamous; inflorescence usually with moderate to stout leafy branches; perianth usually over 1.5 mm. long in flower; foliage usually not acid; none of leaves hastate
  3. Valves with 1-3 definite, sharp-pointed teeth on each margin, grains present on 1 or more valves
    4. Basal leaves broadly-ovate (at most 2.5 times longer than broad), cordate at base; valves in fruit 5-6 mm. long, usually only 1 grain present -3. R. obtusifolius
    4. Basal leaves lanceolate or narrower, not cordate; valves in fruit about 2 mm. long, usually 3 grains present -4. R. fueginus
  3. Valves entire or merely crenulate or denticulate, grains present or absent
    5. Valves in fruit over 10 mm. long (14-18), no grains present
      6. Valves in fruit over 20 mm. wide; perianth in flower about 5 mm. long; plants with deep rootstocks -5. R. venosus
      6. Valves in fruit less than 20 mm. wide; perianth in flower less than 5 mm. long; plants with clusters of tuberous roots -6. R. hymenosepalus
    5. Valves in fruit less than 10 mm. long (2-7), grains present or absent
      7. Valves in fruit, at least one of them, bearing a swollen tubercle organ on the back
        8. Leaves crisped on the margins; plants without axillary shoots
          9. Leaves oblong-lanceolate or narrower, broadest at the middle, usually cuneate at base, petioles somewhat canaliculate above; valves 5-6 mm. long in fruit -7. R. crispus
          9. Leaves ovate-lanceolate, often broadest below the middle, somewhat cordate, truncate or broadly cuneate, petioles flat on upper side; valves 6-7 mm. in fruit -8. R. patientia
        8. Leaves not crisped on the margins; plants with axillary shoots
          10. Valves in fruit about 3 mm. long; leaves linear-lanceolate -9. R. triangulivalvis
          10. Valves in fruit 4-6 mm. long; leaves lanceolate or wider -10. R. altissimus
      7. Valves in fruit all lacking tubercles or grains
        11. Stems with axillary shoots; valves in fruit not over 3.5 mm. long; leaves lanceolate or narrower -11. R. utahensis
        11. Stems without axillary shoots; valves in fruit over 3.5 mm. long; leaves oblong-lanceolate or wider
          12. Plants low, less than 30 cm. tall; all leaves rounded at apex, not more than $2\frac{1}{2}$ times as long as broad -12. R. praecox
          12. Plants over 30 cm. tall; some of leaves (the later ones) acute or narrowed at apex, about 3 times longer than broad
            13. Plants with vertical tap roots; valves not abruptly acute at apex -13. R. occidentalis
            13. Plants with creeping horizontal rootstocks; valves usually abruptly acute
              14. Valves over $1\frac{1}{2}$ times as long as broad, base truncate, margin denticulate near base -14. R. pycnanthus
              14. Valves about as long as wide, base subcordate, margin entire or indistinctly-crenulate-denticulate -15. R. densiflorus

1. Rumex paucifolius Nutt. in Wats., Bot. King's Rep. 314. 1871.
Perennial plants with tap roots and short rootstocks, somewhat tufted; stems 20-50 cm. tall, erect; lower leaf blades 5-12 cm. long, lanceolate or oblong-lanceolate sometimes ovate-lanceolate, cuneate to a rather long petiole, acute to obtuse; inflorescence with suberect branches; perianth about 1.5 mm. long, reddish or yellowish-green; valves 2.9-4 mm. long in fruit, margin entire, no grains present; achenes 1.2-1.8 mm. long.---Meadows and parks. Alberta to British Columbia, south to Colorado and California. Our few records from northern Colorado at 6500-10,000 feet.

2. Rumex acetosella L., Sp. Pl. 338. 1753.
Perennial plants with creeping rootstocks; stems 10-60 cm. tall, slender, erect or nearly so, simple or branched, glabrous; leaves 2.5-15 cm. long, hastate (at least lower), the upper linear and not hastate, obtuse or acute, on rather long petioles; panicles narrow, leafless, turning reddish in age; perianth about 1 mm. long, green or purplish; valves about 1-1.5 mm. long in fruit, margins entire, no grain present; achenes somewhat exceeding the valves in length.--Waste places. Almost throughout temperate North America; Europe. Scattered in Colorado but most of our records from the northcentral, central and southcentral parts at 4500-10,000 feet.

3. Rumex obtusifolius L., Sp. Pl. 335. 1753.
Perennial plants from short taproots; stems 50-100 cm. tall, no axillary shoots; basal leaves large, 10-30 cm. long, broadly ovate, cordate, long-petioled, acute or obtuse; upper leaves narrower, margins somewhat undulate; panicle rather open, leafy below; perianth 3 mm. long, greenish or purplish; valves 5-6 mm. long, margins with 2 or 3 definite spreading sharp teeth on each side, usually only 1 valve with a grain; achenes 2 mm. long.---Waste places. Nova Scotia to Alaska, south to Georgia, New Mexico and California. Our records from north-central and northeastern Colorado at 3500-5100 feet.

A related species, Rumex pulcher L., has recently been collected in Boulder County. The inflorescence is one-half to three-fourths the height of the plant (instead of less than one-half) and the leaves are smaller, not over 15 cm. long and 6 cm. wide.

4. Rumex fueginus Philippi, Anal. Univ. Chile 91:493. 1895.
R. maritimus of Manuals; R. persicarioides of Manuals at least in part---Annual (perhaps biennial) plants; stems 15-60 cm. tall, erect to ascending, may be branched but no axillary shoots present, glabrous to papillose-scabrous; leaves 3-20 cm. long, linear-lanceolate to

lanceolate, acute, cuneate to somewhat cordate, often rather sagittate at base; panicle large but flowers in dense whorls along the branches; perianth 1-1.5 mm. long, greenish; valves 1.7-2.2 mm. long, margins with 1-3 bristlelike teeth on each side, all valves with grains; achenes 1.3-1.4 mm. long.---Open plains and slopes often in sandy soil. Throughout most of North America; South America. Our records scattered in Colorado, mostly in the northcentral part, at 4000-8500 feet.

5. Rumex venosus Pursh, Fl. Am. Sept. 2:733. 1814.

Perennial plants from deep seated woody rootstocks; stems 15-40 cm. tall, ascending or erect, stout often flexuous, glabrous and rather pale colored, axillary shoots present; leaves 3-10 cm. long, ovate to oblong-ovate to ovate-lanceolate, cuneate, acute or somewhat acuminate, rather firm and coriaceous; racemes 1 to several, the panicle dense but often interrupted; perianth about 5 mm. long, reddish; valves in fruit 14-18 mm. long or more and 20-30 mm. wide, margins entire, grains none, usually becoming bright red to orange-red at maturity; achenes 5-7 mm. long.---Open banks, ravines and roadsides, often in sandy ground. Saskatchewan and Alberta south to Missouri, Nevada and Washington. Our records scattered in the eastern half of Colorado, also in the northwestern corner, at 3500-8500 feet.

6. Rumex hymenosepalus Torr., Bot. Mex. Bound. 177. 1859.

Perennial plants from clusters of tuberous roots; stems 30-100 cm. tall, erect to ascending, stout, glabrous, no axillary shoots present; leaves oblong to obovate-lanceolate, somewhat fleshy, apex acute, obtuse or acuminate, base cuneate to a thick decurrent petiole, margins somewhat crisped; inflorescence large, branches suberect; perianth 2-4 mm. long, yellowish-green; valves in fruit large, 10-17 mm. long, including margins, their margins entire or nearly so, no grains present; achenes 4-9 mm. long. The above includes 2 varieties based on fruit shape and size of seed. They can be separated according to the following key:

1. Achene becoming 7 mm. or more long; valves in fruit often broader than long
 -R. hymenosepalus var. salinus (A. Nels.) Rech. f.
1. Achene 4-5 mm. long; valves not broader than long
  2. Valves elliptic, longer than wide
 -R. hymenosepalus var. eu-hymenosepalus Rech. f.
  2. Valves round, about as wide as long
 -R. hymenosepalus Torr.

Dry often sandy ground. Wyoming to California, south to Texas and Mexico. Our records from westcentral and southwestern Colorado at 4500-6500 feet.

7. Rumex crispus L., Sp. Pl. 335. 1753.

Perennial plants from taproots; stems 30-100 cm. tall, erect, straight, without axillary branches; leaves 10-30 cm. long, oblong to linear-lanceolate, the basal obtuse or acute at apex, cuneate or rounded to subcordate at base, margins distinctly crisped and wavy; panicle open or strict, elongated, the whorls usually dense; flowers perfect, perianth about 1.5 mm. long, greenish; valves variable in length, usually 2-5 mm., margins entire or minutely erose, usually with 3 grains, rarely one; achenes 2 mm. long.---Waste ground and cultivated fields. Throughout most of temperate North America; Europe. Our records scattered over Colorado at 3500-8500 feet.

8. Rumex patientia L., Sp. Pl. 333. 1753.

Kansas to Utah. This plant can be expected in Colorado.

9. Rumex triangulivalvis (Danser) Rech. f., Repert. Sp. Nov. 40:297. 1936.

R. mexicanus Meisn. of Manuals, at least in part---Perennial plants apparently from taproots; stems 30-80 cm. tall, 1 or more, erect, axillary shoots present; leaves 6-15 cm. long, linear-lanceolate or narrowly lanceolate, pale green, acute and cuneate to short petioles; inflorescence moderate, fairly dense; perianth 1.5-3 mm. long, greenish to purplish; valves in fruit about 3 mm. long, margin entire or minutely crenulate especially near base, usually all grain-bearing; achenes about 2 mm. long.---Valleys and along streams. Quebec to British Columbia, south to Maine, Missouri, New Mexico and California. Our records scattered in the western two-thirds of Colorado at 4500-10,000 feet.

10. Rumex altissimus Wood, Class Bk. ed. 2. 477. 1847.

R. brittanica of Manuals at least in part---Perennial plants from taproots; stems 50-100 cm. tall, erect, with axillary shoots; leaves 5-20 cm. long, lanceolate or oblong-lanceolate sometimes ovate-lanceolate below, acute, cuneate; panicle rather open; flowers perfect, densely whorled, perianth about 2 mm. long, light green; valves in fruit 4-6 mm. long, margin entire or nearly so, with 1 or more bearing grains or tubercles; achene about 3 mm. long.---Open ground often along streams. Eastern and central United States to Colorado and Arizona. Our records from the northeastern quarter of Colorado at 3500-11,500 feet, mostly below 6000 feet.

11. Rumex utahensis Rech. f., Repert Sp. Nov. 40:298. 1936.

R. mexicanus of Manuals at least in part---Perennial plants; stems 15-60 cm. tall, erect or rarely ascending, glabrous, with axillary shoots developing in some of the leaf axils; leaves 5-15 cm. long, lanceolate or narrower, entire or nearly so, glabrous; flowers perfect in terminal panicles of racemes; perianth greenish, the inner lobes 2.5-3.5 mm. long in fruit; fruit deltoid in shape with an acute or obtuse tip and a truncate base, no grain present, valves coarsely reticulated.---Open ground, often along streams and valleys. Alberta south to Colorado and Nevada. Our few records from northcentral and central Colorado at 7000-10,000 feet.

12. Rumex praecox Rydb., Bull. Torr. Club 33:137. 1906.

Perennial plants from short tuberlike but creeping rootstocks; stems not over 30 cm. tall, thick, erect, no axillary shoots present; basal leaves oval, ovate or elliptic, usually about 2 times as long as broad, sometimes to 2½, rather thick, broadly rounded at apex, rounded or broadly cuneate at base; panicle compact; perianth about 2-3 mm. long; valves in fruit to 5 mm. long, margins entire, grains none; achenes to 3 mm. long.---Open ground, often along streams and valleys. Colorado and possibly Wyoming. Reported from northcentral and southcentral Colorado, our one positive record from the former at 10,000 feet.

13. Rumex occidentalis Wats., Proc. Am. Acad. 12:253. 1877.
R. polyrrhizus Greene---Perennial plants from taproots; stems 50-150 cm. tall, stout, erect, glabrous, axillary shoots absent; leaves 10-40 cm. long (on lower), oblong-ovate to lanceolate, obtuse or acute, cordate or subcordate, rarely truncate at base; racemes erect, forming a rather narrow panicle; perianth 3-4 mm. long sometimes shorter, greenish; valves in fruit 5-8 mm. long, margin subentire to somewhat denticulate, no grains present; achenes 3-4 mm. long.---Usually in moist places. Labrador to British Columbia, south to New Mexico and California. Our records scattered over Colorado, few in the extreme east or western parts, at 4000-10,500 feet.

14. Rumex pycnanthus Rech. f., Repert. Sp. Nov. 38:372. 1935.
R. subalpinus Jones---Perennial plants from taproots (but also reported as having creeping rootstocks); stems 1-2 m. tall, erect; basal leaves large, to 40 cm. long, oblong-ovate to oblong-lanceolate, base obliquely truncate or cuneate; cauline leaves not numerous, oblong-ovate; panicle large and dense; perianth 2.5-3 mm. long, greenish; valves in fruit 5-6 mm. long, denticulate near base, longer than wide, abruptly acute, no grains present; achenes 3-3.5 mm. long.---Moist ground. Colorado and Utah. Reported for the western half of Colorado but our one doubtful record from the northcentral part of the state at about 9000 feet.

15. Rumex densiflorus Osterh., Erythea 6:13. 1898.
Perennial plants with horizontal creeping rootstocks; stems 50-100 cm. tall, erect, glabrous, lacking axillary shoots; leaves, especially the lower, to 40 cm. long, elliptical or oblong-lanceolate or ovate, cuneate to obliquely truncate, rounded or cordate, obtuse to acute at apex; upper leaves gradually reduced; flowers polygamous, in dense clusters; perianth 2-3 mm. long, usually reddish; valves in fruit 5-6 mm. long, margins entire or indistinctly crenate-dentate, no grains present; achenes about 3 mm. long.---Often along streams and valleys. Wyoming, Colorado and possibly Idaho. Our records from northcentral, central and southwestern Colorado at 5000-11,500 feet, mostly above 9000 feet.

Family 39. Chenopodiaceae GOOSEFOOT FAMILY

Plants shrubby, or herbaceous annuals or perennials; leaves alternate or opposite, lacking stipules; flowers perfect or unisexual, small, green or greenish, regular or nearly so, usually in cymose glomerules, these arranged in various inflorescences; perianth when present with 1-5 segments, persistent and sometimes variously modified but no petals present; stamens as many as and opposite the perianth segments or fewer; ovary superior, rarely adnate below to the perianth, 1-celled with 1 to 3 stigmas; fruit a 1-seeded utricle.

1. Leaves scalelike; stems and branches fleshy, jointed with short, thick internodes   -1. Salicornia
1. Leaves not scalelike; stems and branches not both fleshy and jointed
    2. Spiny shrub with fleshy, narrowly linear leaves; flowers imperfect, the staminate in catkinlike spikes without a perianth but subtended by peltate bracts; pistillate flowers with a perianth that develops horizontal wings in fruit   -2. Sarcobatus
    2. Herbs, non-spiny shrubs or (as in Atriplex confertifolia) a spiny shrub with leaves oblanceolate,ovate or wider and not fleshy; flowers perfect or if imperfect the staminate flowers with a perianth and the pistillate flowers lacking a definite perianth but enclosed by 2 accrescent bracts
        3. Flowers imperfect, the pistillate lacking a definite perianth and enclosed by 2 accrescent bracts; plants often shrubby
            4. Bracts dorsally compressed;pubescence if any of simple inflated hairs that collapse and become scurfy when dry   -3. Atriplex
            4. Bracts laterally compressed; pubescence if present of stellate hairs at least in part
                5. Fruiting bracts densely long-villous; leaves narrow with revolute margins   -4. Eurotia
                5. Fruiting bracts glabrous or merely scurfy-pubescent; leaves not revolute
                    6. Annual herbs with toothed leaves; fruiting bracts subhastate at base, ovate-rhombic in shape   -5. Suckleya
                    6. Perennial shrubs or undershrubs with entire leaves; fruiting bracts not subhastate, obovate or orbicular   -6. Grayia
        3. Flowers perfect or polygamous, perianth present on all (sometimes small), no flowers subtended by 2 accrescent bracts; plants herbs or at most undershrubs
            7. Fruit included and largely concealed in the perianth lobes; stamens usually over 3
                8. Perianth in fruit developing conspicuous, horizontal scarious wings
                    9. Leaves toothed, lanceolate or broader; flowers in panicled spikes, not in the axils of leaves or bracts   -7. Cycloloma
                    9. Leaves entire,lanceolate to narrower; flowers 1 to several, clustered in the axils of leaves or bracts
                        10. Embryo coiled; leaves and bracts spine- or bristle-tipped; plants often globose and becoming tumbleweeds
                            11. Leaves abruptly narrowed to a weak bristle; seeds vertical; plants present in western Colorado   -8. Halogeton
                            11. Leaves gradually narrowed to a usually spinulose tip; seeds horizontal; plants widely distributed in Colorado   -9. Salsola
                        10. Embryo annular; leaves and bracts never spinulose-tipped; plants seldom globose in outline, never true tumbleweeds   -10. Kochia
                8. Perianth not developing horizontal, scarious wings (may bear a hook, horn or rarely a narrow thick wing)
                    12. Perianth lobes each armed with a slender hooked spine; plants pilose   -11. Bassia
                    12. Perianth lobes not armed with slender hooked spines (may be horned or narrowly winged); plants not pilose (may be glabrous, glandular-pubescent, scurfy or puberulent)

13. Embryo a flattened spiral; leaves fleshy, narrowly linear, often subterete, glabrous or nearly so, not scurfy, always entire -12. Suaeda
13. Embryo annular; leaves not fleshy, sometimes linear but usually wider, flat, usually scurfy or pubescent, often toothed or lobed -13. Chenopodium
7. Fruit largely exposed, perianth lobe 1 or rarely to 3 and then minute; stamens 1-3
   14. Flowers in the axils of bracts wider and different in shape from the foliage leaves; fruit winged or acute; leaves entire; perianth lobes 1-3; stamens 1-3 -14. Corispermum
   14. Flowers in the axils of leaves or reduced leaves or lacking any bracts; fruit not winged or acute; leaves (in our common species) hastate-lobed at base; perianth lobe 1; stamen 1 -15. Monolepis

1. Salicornia L. GLASSWORT; SAMPHIRE

Annual, fleshy, glabrous herbaceous plants; stems jointed with opposite branches; leaves scalelike; flowers perfect or polygamous, in cylindrical spikes sunk in cavities of the internodes, 3-7 together; perianth fleshy, saclike with a truncate or 3- to 4-toothed border; stamens 1 or 2; styles 2 or lacerate above; fruit included in the perianth, pericarp adherent; seed vertical, radicle inferior.

1. Salicornia rubra A. Nels., Bull. Torr. Club 26:122. 1899.
S. herbacea of Manuals---Annual plants with strong taproots; stems 10-25 cm. tall, erect, divaricately-branched throughout; leaves scalelike, short, broadly-triangular, wider than long, subacute; fruiting spikes 2-4 cm. long, at maturity ruby-red, fruit covered with short curved hairs. The above may not be distinct from S. europea L.---Borders of alkaline lakes and ponds. Saskatchewan to British Columbia, south to Kansas and Nevada. Our few records from northcentral, southcentral and northwestern Colorado at 6000-8500 feet.

2. Sarcobatus Nees. GREASEWOOD

Shrubs with spiny, much branched, rigid stems; leaves alternate or opposite, entire, narrow and fleshy; flowers monoecious or dioecious; staminate flowers in terminal, catkinlike spikes, perianth lacking but each flower subtended by a peltate, stipitate bract, 3 or less than 3 stamens to a flower; pistillate flowers solitary or 2 together in the axils of the leaves, with a perianth, this margined by narrow borders which develop into broad membranous, horizontal wings; stigmas 2, recurved; fruit coriaceous, winged, the lower part turbinate, the upper conical; seed erect, orbicular, embryo spirally coiled.

1. Sarcobatus vermiculatus (Hook.) Torr. in Emory's, Notes Mil. Rec. 149. 1848.
S. baileyi Cov.---Shrubs 30-300 cm. tall, erect; leaves 1-4 cm. long, linear or linear-filiform, glabrous or sparsely stellate-pubescent, acute or obtuse at apex, narrowed at base; staminate spikes 6-25 mm. long; wings of fruiting perianth 6-12 mm. wide, sometimes tinged with red.---Usually flat ground. Saskatchewan to Washington, south to Texas and California. Our records well scattered over the western part of Colorado east to Jackson, Pueblo and Las Animas Counties at 4500-8500 feet.

3. Atriplex L. SALTBRUSH; ORACHE

Annual or perennial herbaceous plants or shrubs, more or less scaly, scurfy or mealy; leaves mostly alternate, often opposite below, sessile or petiolate; flowers monoecious or dioecious, in axillary clusters or in terminal panicles or spikes; staminate flowers without bracts, perianth 3- to 5-parted, stamens 3-5, the filaments free or united; pistillate flowers subtended by 2 more or less united compressed bracts, usually without a definite perianth; ovary 1-celled, stigmas 2; fruit a utricle, the pericarp usually free; seed erect or rarely horizontal. Well-matured fruiting bracts are necessary for accurate determination of the species.
Since this treatment was prepared a new and peculiar species of Atriplex has been found in the extreme southwestern corner of the state. It is A. pleiantha W. A. Weber and has the following characteristics: plants annual; perianth present on both staminate and pistillate flowers; pistillate bracts enclosing 2-6 flowers; radicle inferior.

1. Plants perennial shrubs or at least definitely woody at base; dioecious, rarely monoecious; leaves entire or slightly undulated-margined, never hastate at base
  2. Bracts of pistillate flowers with 4 conspicuous longitudinal wings; leaves sessile or nearly so and plant woody throughout
    3. Leaf blades widest at middle, oval or broadly elliptical in shape; shrubs under 40 cm. tall
-1A. A. canescens garrettii
    3. Leaf blades widest above middle, linear-spatulate to oblong-spatulate; shrubs 20-150 cm. tall
      4. Shrubs 20-40 cm. tall; leaves oblong to oblong-spatulate; wings of bracts usually thick
-1B. A. canescens aptera
      4. Shrubs usually over 40 cm. tall; leaves linear-spatulate or narrowly oblong; wings of bracts usually thin
-1C. A. canescens canescens
  2. Bracts of pistillate flowers never with 4 conspicuous wings (irregular flattened appendages may be present); leaves petioled or if sessile the plant woody only at base
    5. Plants spiny, the spines of sharp-pointed twigs, woody throughout; pistillate bracts with entire or nearly entire margins and smooth faces, 6-12 mm. long -2. A. confertifolia
    5. Plants not spiny, woody usually at base only; pistillate bracts with toothed margins or appendaged faces, sometimes both, 4-7 mm. long
      6. Bracts of fruit broader than long, with a truncate, dentate apex, the sides usually smooth or nearly so; leaves obovate or broadly elliptic, short-petioled -3. A. obovata

6. Bracts of fruit longer than broad, the apex often entire, the sides usually appendaged; leaves usually narrower than obovate or broadly elliptic, sessile or petioled
   7. Plants low, seldom over 25 cm. tall; leaves sessile, not over 2 cm. long   -4. A. corrugata
   7. Plants normally over 25 cm. tall, leaves usually short-petioled, usually over 2 cm. long
      8. Leaves broadly elliptic or ovate, usually broadest below the middle   -5A. A. nuttallii cuneata
      8. Leaves oblong-linear, spatulate or sometimes nearly obovate, usually broadest above the middle
         9. Leaves oblong, spatulate or obovate; bracts usually appendaged, widest at or below the middle   -5B. A. nuttallii nuttallii
         9. Leaves narrowly oblong to oblanceolate; bracts usually not appendaged, widest at, above or below the middle
           10. Fruiting bracts widest above the middle   -5C. A. nuttallii gardneri
           10. Fruiting bracts widest at or below the middle   -5D. A. nuttallii tridentata
1. Plants annual; monoecious (or imperfectly dioecious in A. powellii); leaves entire, toothed or hastate
  11. Foliage green or greenish, sparsely mealy or scurfy if at all only when young
    12. Fruiting bracts united only near the base, over 3 mm. long; radicle inferior
      13. Fruiting bracts 8 mm. or more broad; leaves thin, often over 7 cm. long   -6. A. hortensis
      13. Fruiting bracts not over 5 mm. broad; leaves rather thick, often fleshy, not over 7 cm. long
        14. Some of leaves triangular-hastate or rhombic with basal angles or lobes; bracts usually truncate or broadly rounded at base   -7A. A. patula hastata
        14. Leaves lanceolate or oblong not hastate; bracts usually narrowly rounded or broadly cuneate at base   -7B. A. patula patula
    12. Fruiting bracts united to above the middle, about 2 mm. long; radicle superior   -8. A. dioeca
  11. Foliage gray or whitish with a permanent scurf, at least on the lower surface
    15. Fruiting bracts ovate, widest below the middle
      16. Leaves definitely toothed, 2-6 cm. long; bracts 4-6 mm. long; radicle lateral   -9. A. rosea
      16. Leaves entire, 1-2 cm. long; bracts 2-3 mm. long; radicle superior   -10. A. tenuissima
    15. Fruiting bracts widest at or above the middle
      17. Leaves linear or oblong-linear, sessile, 0.5-1.5 cm. long, widest part about at the middle; fruiting bracts 1.5-2 mm. long   -11. A. wolfii
      17. Leaves lanceolate to wider, often petioled, 1-6 cm. long, widest part below the middle; fruiting bracts over 2 mm. long
        18. Leaves cordate at base; fruiting bracts of 2 definite kinds on each plant (one to 3 mm. long with smooth faces, the other to 6 mm. long, the faces heavily appendaged)   -12. A. saccaria
        18. Leaves cuneate, rounded or sometimes truncate at base but not cordate; fruiting bracts all alike on the same plant (may be smaller younger ones and larger older ones present)
          19. Staminate and pistillate flowers in separate clusters, often on separate plants; leaves conspicuously 3-nerved, entire; bracts often panduriform, faces appendaged and crowned at summit with horizontal winglike lobes   -13. A. powellii
          19. Staminate and pistillate flowers intermixed in some clusters especially near the middle of the plant; leaves 1-nerved or very obscurely 3-nerved, entire or toothed; bracts not panduriform, faces smooth or appendaged, no horizontal winglike lobe near apex
            20. Fruiting bracts 2-3.5 mm. long, cuneate at base to the truncate, toothed apex, the sides entire and the face smooth; leaves usually sessile   -14. A. truncata
            20. Fruiting bracts 4-8 mm. long, not cuneate to a truncate apex, the sides usually somewhat toothed and the faces usually sharp-tubercled or appendaged; leaves, at least the lower, usually petioled
              21. Radicle lateral; fruiting bracts becoming hard and bonelike at maturity, mostly widest below the base; leaves remotely sinuate-dentate, never hastate; introduced weeds; branches round or somewhat angled at base   -9. A. rosea
              21. Radicle superior; fruiting bracts not becoming especially hard, widest at or above the middle; leaves repand-dentate to entire, often somewhat hastate at base; native plants; branches definitely angled
                22. Upper leaves short-petioled   -15A. A. argentea argentea
                22. Upper leaves strictly sessile   -15B. A. argentea expansa

1. **Atriplex canescens** (Pursh) Nutt., Gen. Pl. 1:197. 1818.
A. garrettii Rydb.; A. aptera A. Nels.; A. odonoptera Rydb.; A. occidentalis (Torr.) Dietr.---Perennial shrubs, 20-250 cm. tall, woody throughout, loosely to densely branched; stems rather stout, gray-scurfy; leaves 1-5 cm. long, sessile or very nearly so, oval, elliptic, spatulate to oblanceolate, long-cuneate at base, usually obtuse at apex, entire, rather thick, with a gray scurf; plants dioecious or rarely monoecious; staminate flowers in glomerules, in dense spikes of long terminal panicles, these leafy below; body of fruiting bracts little compressed, united to summit, conspicuously 4-winged from sides and back of bracts, faces smooth or with small appendages, the whole bract 4-15 mm. long, rarely longer. The following 3 subspecies are in Colorado.

1A. **Atriplex canescens garrettii** (Rydb.) H. & C., (ssp.) Phylog. Meth. Tax. 344. 1923.
A. garrettii Rydb.---Low shrubs to 40 cm. tall; leaves broadly elliptic or oval, the widest part about at the middle.---Colorado, Utah and Arizona. Our few records from western Colorado at about 8000 feet.

1B. **Atriplex canescens aptera** (Nelson) H. & C., (ssp.) Phylog. Meth. Tax. 243. 1923.
A. aptera A. Nels.---Low shrubs to 40 cm. tall; leaves oblong to oblong-spatulate.---Alberta to Colorado. Should be present in northern Colorado but no specimens seen by the writer.

1C. **Atriplex canescens canescens**
A. canescens ssp. typica H. & C.---Shrubs usually over 40 cm. tall; leaves linear-spatulate to narrowly-oblong.---South Dakota to Washington, south to Texas, Mexico and California. Our records scattered over the entire state, except the northeastern part, at 4000-8000 feet.

2. **Atriplex confertifolia** (Torr. & Frem.) Wats., Proc. Am. Acad. 9:119. 1874.

A. collina Woot. & Standl.---Perennial shrubs, 20-100 cm. tall sometimes slightly more, woody throughout, usually of rounded outline; stems stout, sparsely scurfy at first only, usually forming spines these appearing as the ends of short branches; leaves 1-2 cm. long, crowded, short-petiolate, orbicular-ovate, orbicular-obovate to elliptic, rounded or cuneate at base, obtuse at apex, entire, rather firm but not especially thick, 1- to 3-nerved, permanently scurfy; plants dioecious; staminate flowers in glomerules, sessile in the axils of upper leaves; fruiting bracts 6-12 mm. long, united at base over seed, flat above, orbicular to broadly elliptic, usually entire, the faces not appendaged.---Dry plains and slopes. Montana to California, south to Mexico. Our records from the western two-thirds of Colorado, so far from the extreme western part and in a group in Fremont, El Paso and Pueblo Counties, at 4500-7500 feet.

3. **Atriplex obovata** Moq., Chenop. Enum. 61. 1840.

Perennial plants, woody at least at the base; stems 20-50 cm. tall, rigidly erect from the much branched woody base, gray-furfuraceous; leaves 1-3 cm. long, short-petioled, obovate or broadly elliptic, cuneate at base, obtuse or retuse at apex, entire or undulate, thick, whitish-scurfy; plants dioecious; staminate flowers in glomerules, these sessile on branches of a panicle; fruiting bracts 4-5 mm. long and 5-7 mm. broad, compressed, united at least to the middle, obovate, margins toothed, the sides usually smooth, but sometimes with small or elongated tubercles or crests.---Dry ground. Colorado south to Texas, Mexico and Arizona. Our few records from westcentral and southwestern Colorado at about 4500-5500 feet.

4. **Atriplex corrugata** Wats., Bot. Gaz. 16:345. 1891.

A. nuttallii corrugata (Wats.) A. Nels.---Perennial plants, woody at bases and forming dense leafy mats; stems 10-20 cm. high from the decumbent woody bases, erect, densely furfuraceous; leaves 0.5-2 cm. long, crowded, sessile, broadly linear to linear-spatulate, cuneate below, obtuse at apex, entire, densely white-furfuraceous; plants dioecious or sometimes monoecious; staminate flowers in long glomerules, sessile on terminal spikes that overtop the stems; fruiting bracts 4-6 mm. long, united nearly to the summit, narrowly fan-shaped, widest above the middle, sides with thick wartlike or flattened appendages.---Dry valleys and plains. Colorado, Utah and New Mexico. Our records from western Colorado at 4500-6500 feet.

5. **Atriplex nuttallii** S. Wats., Proc. Am. Acad. 9:116. 1874.

A. cuneata A. Nels.; A. pabularis A. Nels.; A. pabularis eremicola (Osterh.) A. Nels.; A. falcata (Jones) Standl.; A. gardneri (Moq.) Standl.; A. oblanceolata Rydb.; A. tridentata Kuntze; A. neomexicana Standl.---Perennial plants, woody at the bases and much branched; stems 20-50 cm. tall, erect or ascending, often herbaceous, white or gray-scurfy; leaves 1.5-5 cm. long, short-petioled, oblong-linear to ovate or obovate, usually cuneate at base, obtuse at apex, margins entire, rather thick, densely scurfy; plants dioecious or rarely monoecious; staminate flowers in glomerules of terminal spikes or narrow panicles a few centimeters long, leafy in lower part; fruiting bracts 4-7 mm. long, united nearly to the summit, lanceolate, elliptic to cuneate-oblong, more or less dentate especially at summit, the faces smooth, tuberculate or with various appendages. Represented in Colorado by the following 4 subspecies which often intergrade.

5A. **Atriplex nuttallii cuneata** (Nelson) H. & C., (ssp.) Phylog. Meth. Tax. 324. 1923.

A. cuneata A. Nels.---Leaves broadly elliptic or ovate, usually widest below the middle.---Colorado, Utah, New Mexico and Arizona. Our few records from western Colorado at 4500-7500 feet.

5B. **Atriplex nuttallii nuttallii**

A. oblanceolata Rydb.; A. nuttallii ssp. typica H. & C.---Leaves oblong, spatulate or obovate; bracts appendaged, widest below the middle.---Saskatchewan to Idaho, south to Nebraska and Utah. Our records from westcentral and northern Colorado at 4500-6000 feet.

5C. **Atriplex nuttallii gardneri** (Moq.) H. & C., (ssp.) Phylog. Meth. Tax. 324. 1923.

A. erimicola Osterh.---Leaves narrowly oblong or oblanceolate; bracts smooth, widest above the middle.---Wyoming to Idaho, south to Colorado. Our one record from northcentral Colorado at about 8000 feet.

5D. **Atriplex nuttallii tridentata** (Kuntze) H. & C., (ssp.) Phylog. Meth. Tax. 324. 1923.

A. tridentata Kuntze; A. pabularis A. Nels.---Leaves narrowly oblong to oblanceolate; bracts not appendaged, widest at or below the middle.---Wyoming to Utah, south to Colorado. Our few records from northern and western Colorado at 5500-8500 feet.

6. **Atriplex hortensis** L., Sp. Pl. 1053. 1753.

Annual plants; stems 50-200 cm. tall, mostly erect, widely branched from the base, soon glabrous; leaves 4-12 cm. long, sometimes longer, petioled, ovate, triangular to lanceolate-oblong, truncate, cordate or subhastate at base, obtuse or acute at apex, margins entire or sinuate-dentate, thin, bright green and glabrous in age; flowers monoecious, staminate and pistillate usually mixed in the inflorescence; fruiting bracts 8-18 mm. long, nearly orbicular or round-ovate, margins entire or obscurely denticulate, faces smooth but reticulate-veiny; radicle inferior.---A European plant now introduced in various parts of the United States. Our records scattered in the western two-thirds of Colorado at 4500-8000 feet.

7. **Atriplex patula** L., Sp. Pl. 1053. 1753.

A. carnosa A. Nels.; A. hastata L.; A. lepathifolia Rydb.; A. subspicata Rydb.---Annual plants; stems 20-100 cm. tall, erect or decumbent, simple or widely branched; leaves variable as to size, around 2-7 cm. long, petioled or sometimes sessile, typically subdeltoid-lanceolate to ovate but sometimes linear, cuneate to cordate or hastate at base, acute or obtuse at apex, margins entire or coarsely dentate, rather thick, glabrate and green at maturity; plants monoecious, the staminate and pistillate flowers usually together; fruiting bracts 3-12 mm. long, united only near bases, linear to broadly-deltoid, margin entire or sparsely denticulate, face smooth or with small clustered tubercles; radicle inferior.---Alkaline or saline meadows. Represented in Colorado by the following 2 subsepcies.

7A. **Atriplex patula hastata** (L.) H. & C., (ssp.) Phylog. Meth. Tax. 249. 1923.

A. hastata L.---Some of leaves at least triangular-hastate with basal angles or lobes; bracts

usually truncate or broadly rounded at base.---Throughout the United States except the southeastern part. Our few records from northcentral and southwestern Colorado at 5000-7500 feet.
7B. Atriplex patula patula
    A. patula ssp. typica H. & C.---Leaves lanceolate or oblong, not hastate at base; bracts usually narrowly rounded or broadly cuneate at base.---Newfoundland to British Columbia, south to Gulf States and California. Our few records from northcentral and southcentral Colorado at 4500-7500 feet.
8. Atriplex dioeca (Nutt.) Macbride, Contr. Gray Herb. n. s. 53:11. 1918.
    A. suckleyana Rydb.; Endolepis suckleyi Torr.---Saskatchewan to Alberta, south to Nebraska and Wyoming. May be found in northern Colorado.
9. Atriplex rosea L., Sp. Pl. ed 2. 1493. 1753.
    A. spatiosa A. Nels.---Annual plants; stems 10-200 cm. tall, erect, simple or branched from the base; branches terete or little angled, nearly glabrous; leaves 2-6 cm. long, sessile or petioled, ovate or rhombic-ovate to lanceolate, cuneate or rounded at base, acute or obtuse at apex, remotely sinuate-dentate, thinly or densely furfuraceous, rather thick; plants monoecious, the staminate and pistillate flowers usually together near middle; fruiting bracts 4-6 mm. long, united usually to the middle, rhombic or ovate from a broad base, becoming strongly indurated in age, margins dentate, usually with sharp tubercles on the face; radicle lateral.---An introduced weed. Widespread throughout the United States. Our few records from the western two-thirds of Colorado at 4500-7000 feet.
10. Atriplex tunuissima A. Nels., Bot. Gaz. 34:359. 1902.
    A. greenei A. Nels.---Annual plants; stems 15-30 cm. tall, erect, branched throughout, branches slender, terete, moderately white-scurfy; leaves 1-2 cm. long, sessile, lanceolate-ovate to linear, rounded to base, acute at apex, densely gray-scurfy, 1-nerved; flowers monoecious, staminate and pistillate flowers together at least toward middle of plant; fruiting bracts 2-3 mm. long, united nearly to apex, ovate, widest below middle, acute, sparsely-toothed along the edges, faces with a few acute tubercles; radicle superior.---Dry ground. Wyoming and Utah. Our few records from northwestern and westcentral Colorado at 5000-6000 feet.
11. Atriplex wolfii Wats., Proc. Am. Acad. 9:112. 1874.
    Annual plants; stems 10-30 cm. tall, erect, branched from base, branches scurfy, glabrate; leaves 0.5-1.5 cm. long, numerous, sessile, linear or the lowest oblong-linear, obtuse or acutish at apex, entire, rather thin, pale scurfy, 1-nerved, often conduplicate; flowers monoecious, staminate and pistillate mixed at least near middle of plant; fruiting bracts 1.5-2 mm. long, united to apex, cuneate-oblong, truncate at apex and with 3 small teeth on apex, face smooth or with a few small tubercles; radicle superior.---Dry flats. Wyoming, Utah and Colorado. Our rather few records scattered in the western half of Colorado at 4500-8500 feet.
12. Atriplex saccaria Wats., Proc. Am. Acad. 9:112. 1874.
    A. cornuta M. E. Jones---Annual plants; stems copiously branched throughout to form a globose plant 10-50 cm. tall, stout, angled, roughly furfuraceous; leaves 1-3 cm. long, all petiolate or upper ones sessile, broadly cordate-ovate or subreniform, cordate or some broadly-truncate at base, acute at apex, entire, thick (but drying thin), gray or whitish-scurfy; plants monoecious; fruiting bracts of 2 kinds, united to summit, the smaller ones 3 mm. long, cuneate, truncate at apex, toothed only at summit, faces smooth; larger bracts up to 6 mm. long, globoid through development of numerous conspicuous appendages; radicle superior.---Dry alkaline ground. Wyoming to Nevada, south to Texas and Arizona. Our few records from western Colorado at about 5500 feet.
13. Atriplex powellii S. Wats., Proc. Am. Acad. 9:114. 1874.
    A. philonitra A. Nels.---Annual plants; stems 10-100 cm. tall, erect, sparingly branched from the base, obtusely angled, whitish-furfuraceous on younger branches; leaves 1-3.5 cm. long, lower leaves on petioles as long as blade, upper sessile, broadly-ovate or rhombic-ovate, rounded or abruptly cuneate at base, acute at apex, entire, firm, gray especially below with a fine scurf, conspicuously 3-nerved from base; plants imperfectly dioecious, when together on same plant, then seldom on the same cluster; fruiting bracts 3-4 mm. long, united to apex, broadly spatulate or broadly oblong, ending above in a broad flattened terminal lobe, faces usually appendaged; radicle superior.---Alkaline plains. Alberta, south to New Mexico and Arizona. Our records from western Colorado at 4500-7500 feet.
14. Atriplex truncata (Torrey) Gray, Proc. Am. Acad. 8:398. 1872.
    A. subdecumbens M. E. Jones---Annual plants; stems 20-100 cm. tall, usually erect, branched or simple, branches angled, lightly furfuraceous, soon glabrate; leaves 1-4 cm. long, mostly sessile but lower often short-petioled, deltoid, triangular-ovate or rounded-ovate, truncate, subhastate or rarely rounded at base, acute or obtuse at apex, entire or only undulate, rather thin, gray-furfuraceous especially below; flowers monoecious, staminate and pistillate usually together near middle of plant; fruiting bracts 2-3.5 mm. long, united to the apex, broadly-cuneate to the truncate apex, margined only at apex and here 2- to 4-toothed, faces smooth or with 1 or 2 small tubercles; radicle superior.---Alkaline or dry flat ground. Montana to British Columbia, south to New Mexico and California. Our records from western and northcentral Colorado at 4500-8500 feet.
15. Atriplex argentea Nutt., Genera 1:198. 1818.
    A. expansa S. Wats.; A. caput-medusae Eastw.---Annual plants; stems 15-80 cm. tall, erect, branched from base, the plants usually globoid in outline, branches rather stout, angled, furfuraceous when young; leaves 2-5 cm. long, sessile, subsessile or petioled, lanceolate, ovate or deltoid, cuneate or subhastate at base, obtuse or subacute at apex, entire or repand-dentate, moderate in thickness, gray-furfuraceous, sometimes glabrate; flowers monoecious, staminate and pistillate mixed in clusters at least at middle of plant; fruiting bracts 4-8 mm. long and as wide, united to middle or above, obovate to cuneate-orbicular, margins green, subentire to laciniate, faces smooth or appendaged; radicle superior---Alkaline ground. Represented in Colorado by the 2 following subspecies.

15A. Atriplex argentea argentea
   A. argentea ssp. typica H. & C.---The upper leaves are petioled or subsessile, leaf blades triangular-ovate to rounded ovate.---Saskatchewan to Oregon, south to New Mexico and California; introduced in eastern United States. Our records scattered over the state at 4000-8000 feet.
15B. Atriplex argentea expansa (S. Wats.) H. & C., (ssp.) Phylog. Meth. Tax. 284. 1923.
   The upper leaves are strictly sessile, the leaf blades ovate to lanceolate-ovate.---Texas to California, south to Mexico. Our few records from northcentral and western Colorado at 5000-5500 feet but the plant should be expected anywhere in the state especially in the southern part.

4. Eurotia Adans. WINTERFAT; WHITE SAGE

Plants shrubby or suffrutescent, densely stellate-tomentose; leaves alternate, entire; flowers dioecious or polygamous, in axillary clusters or terminal spikelike inflorescences; staminate flowers with a 4-parted perianth, stamens 4; pistillate flower lacking a perianth but with 2 bracts, these united to the apex, somewhat obcompressed, pilose, 2-beaked; stigmas 2; pericarp free; seed vertical with radicle inferior.

1. Eurotia lanata (Pursh) Moq., Chenop. Monog. 81. 1840.
   Plants 30-100 cm. tall, woody only at base, the branches stout, erect, densely and coarsely stellate-tomentose with longer straight hairs intermixed, these white or rufous especially in age; leaf blades 1-4 cm. long, linear to narrowly-lanceolate, short-petiolate or the upper sessile, obtuse at apex, cuneate below, margins revolute; fruiting bracts 4-8 mm. long, lanceolate, densely villous with long white or rufous hairs.---Plains and hills. Saskatchewan to Washington, south to Texas and California. Our records well scattered over the state at 4000-8000 feet.

5. Suckleya A. Gray POISON SUCKLEYA

Annual, sparsely furfuraceous, succulent plants; leaves alternate, petiolate; flowers monoecious, in axillary glomerules; staminate with 3- to 4-parted perianth, 2 of the segments larger, stamens 3 or 4; pistillate flowers without perianth, in two bracts, these obcompressed, carinate, united except at the top, with crenulate, narrow dorsal wings in fruit; fruit enclosed by the bracts; pericarp free or adherent; seed compressed, radicle superior.

1. Suckleya suckleyana (Torr.) Rydb., Mem. N. Y. Bot. Gard. 1:133. 1900.
   Stems 10-30 cm. long, stout, terete, much branched, mostly prostrate, soon glabrate, often reddish and striate, rather brittle; leaves with petioles usually longer than the blades, the latter 1-3 cm. long, suborbicular to rhombic-ovate, apex rounded, base short-cuneate, margin repand-dentate, glabrate; fruiting bracts 4-6 mm. long, ovate-rhombic, subhastate at base, bidentate at apex, glabrous; fruit ovate, about 3 mm. long; embryo subannular. A late fall seedling occurs. The plant may not be over 1 cm. tall, consisting of 2 linear cotyledons and from 1 to several small leaves with flowers and fruit in their axils.---In moist sink holes, around the borders of lakes, ponds and irrigation reservoirs. Montana to Colorado. Our records from the northwestern quarter of Colorado at 4000-5500 feet.

6. Grayia H. & A. HOP-SAGE

Much branched, usually spiny shrubs with stiff, divergent branches, more or less stellate or scurfy; leaves alternate, sessile, entire, rather fleshy; flowers dioecious or rarely monoecious, small, in axillary clusters these forming terminal or axillary spikes or panicles; staminate flowers ebracteate, perianth 4- to 5-parted, stamens 4-5; pistillate flowers with 2 bracts, these obcompressed, in fruit winged, perianth none; stigmas 2; fruit orbicular and compressed, included in the bracts, pericarp free; seed vertical, radicle inferior.

1. Fruiting bracts glabrous at maturity, not carinate, over 6 mm. wide; plants often spiny   -1. G. spinosa
1. Fruiting bracts scurfy-pubescent at maturity, carinate, not over 6 mm. wide; plants not spiny -2. G. brandegei

1. Grayia spinosa (Hook.) Moq. in DC., Prodr. 13(2):119. 1849.
   Erect shrubs, 30-100 cm. tall, much branched, older ones with dark gray bark easily shredding, branchlets usually spinose, sparingly or densely scurfy pubescent at first; leaves 1-3 cm. long, oblanceolate, spatulate or obovate, obtuse or subacute at apex, scurfy-pubescent when young, in age glabrate; fruiting bracts obovate-orbicular to orbicular, 5-12 mm. long, usually over 6 mm. wide, glabrous, not carinate, white or reddish-tinged.---Dry plains or slopes. Wyoming to Washington, south to Arizona and California. Our records from western Colorado at 4500-7000 feet.
2. Grayia brandegei A. Gray, Proc. Am. Acad. 11:101. 1876.
   Shrubs 20-70 cm. tall, with erect branches, these branched below often from gnarled woody bases, not at all spiny, densely scurfy-pubescent; leaf blades 1.5-4.5 cm. long, linear-oblanceolate to obovate, apex obtuse and rounded to acute, scurfy-pubescent especially below; fruiting bracts 5-6 mm. long and not over 6 mm. wide, obovate-orbicular or orbicular, scurfy-pubescent, carinate.---Dry hills, often among rocks. Colorado, Utah and Arizona. Our few records from western Colorado at 5000-6500 feet.

7. Cycloloma Moq. WINGED PIGWEED; RINGWING

Annual herbaceous plants with diffusely branched stems; leaves alternate, petiolate; flowers polygamous in paniculate spikes, without bracts; perianth 5-lobed, these inflexed, carinate, in fruit developing a thin, horizontal irregularly denticulate wing; stamens 5; styles 3, short; fruit depressed-globose, pericarp free; seed horizontal, embryo annular.

1. **Cycloloma atriplicifolium** (Spreng.) Coult., Mem. Torr. Club 5:143. 1894.
Stems 10-60 cm. tall, obtusely angled, somewhat striate, tomentose when young but becoming glabrate; leaves 2-8 cm. long, deciduous in age, petioled, oblong, narrowly oblong to lanceolate, acute at apex, cuneate at base, coarsely and irregularly sinuate-dentate, the teeth mucronate, tomentose when young, glabrate in age; inflorescence broadly paniculate; perianth 2.5-4 mm. wide in fruit, villous or glabrate, covering the fruit.---Sandy ground. Indiana to Manitoba, south to Texas and Arizona. Our records from the eastern half of Colorado at 3500-8000 feet.

8. **Halogeton** C. A. Mey.   HALOGETON

Annual plants, stems branched from the base; leaves alternate, cylindrical and succulent, abruptly bristle-tipped; flowers bracted or ebracteate, polygamous, some perfect others pistillate only, in axillary glomerules or clusters; perianth segments 5, in fruit developing conspicuous scarious horizontal wings; stamens 5 or less, commonly 3; stigmas 2, filiform; seeds vertical with coiled embryo, the radicle (in our material) extending upward and pushing up the pericarp.

1. **Halogeton glomeratus** (Bieb.) Mey. in Ledeb., Fl. Alt. 1:378. 1829.
Stems glabrous, 8-30 cm. tall; leaves 5-10 mm. long; perianth in flower about 1.5 mm. long; perianth wings in fruit 3-4 mm. wide; fruit 1.5 mm. long.---Roadsides, ditches and denuded soil, often on alkaline ground. Introduced from Asia and is spreading out from a focal point in eastern Nevada. The plant is in eastern Utah and has been found in westcentral Colorado in Mesa County. This is the plant that contains an oxalate that has poisoned sheep. It can be expected to spread in Colorado especially in the western part.

9. **Salsola** L.   RUSSIAN THISTLE

Annual plants, much branched, becoming tumbleweeds; leaves narrow, entire, often rigid and spine-tipped; flowers perfect, small, solitary or several in the axils of the leaves, each subtended by 2 bracts; perianth 5- or rarely 4-parted, in fruit with horizontal, scarious dorsal wings; stamens 5, rarely less; styles 2; ovary sometimes rather flattened; pericarp free; endosperm wanting, the embryo coiled in a flattened or conical spiral.

1. **Salsola kali** L., Sp. Pl. 222. 1753.
S. pestifer A. Nels.---Plants about 30-80 cm. high, intricately branched, branches ascending or spreading, forming clumps often 100 cm. across; leaves 3-6 cm. long, filiform to linear-filiform, pungent-pointed; fruiting calyx 3-6 mm. broad, segments with thin transverse wings (or some wingless on lower part of plant). Our plants are var. tenuiflora Tausch. A form with more virgate habit, leaves and bracts less spiny and calyx segments wingless, has been noted in Cheyenne Wells (Leaft. West. Bot. 5:104. 1948) and identified as an Eurasian species S. collina Pall. The state seed laboratory has noted 2 kinds of seeds. In one kind the spiralled embryo is cochleate and symmetrical. In the other the spiral is flat or pushed over to one side. Differences in seed character do not seem to be correlated with the characters given above for S. collina and the typical form.---Dry plains, fields and roadsides. The variety introduced from Eurasia, common throughout western United States and occasional eastward. Our records scattered over Colorado at 3500-8500 feet.

10. **Kochia** Roth.   SUMMER-CYPRESS

Perennial or annual herbaceous plants or undershrubs; leaves alternate or opposite, entire, narrow; flowers mostly perfect or some pistillate, solitary or few in the axils; perianth 5-lobed, persistent, finally developing horizontal wedge-shaped, scarious wings; stamens 3-5; stigmas 2 or 3; fruit depressed-globose; pericarp free from the seed; seed horizontal, embryo annular.

1. Annual plants; leaves petiolate, the blades linear-lanceolate, thin; calyx wings small, less than 1 mm. long
    -1. **K. scoparia**
1. Perennial plants; from woody bases; leaves sessile, the blades narrowly-linear, subterete; calyx wings larger, over 1 mm. long in fruit
    -2. **K. americana**

1. **Kochia scoparia** (L.) Schrad., Neues Journ. Bot. Schrad 3(3):85. 1809.
Annual plants; stems 30-100 cm. tall, erect, branched, leafy, glabrous or sparingly pubescent, often tinged with red; leaves 2-7 cm. long, lanceolate to linear, acuminate, cuneate at base to a short petiole, thin, usually ciliate, often pilose-sericeous; flowers clustered in the leaf axils forming dense leaf spikes, these pilose or glabrate in age; wings of perianth short, 0.6 mm. long or less.---Waste places and fields. A Eurasian plant escaped in many parts of the United States. Our records from the eastern half of Colorado at 4000-9700 feet but probably more widely distributed.

2. **Kochia americana** S. Wats., Proc. Am. Acad. 9:93. 1874.
K. vestita (S. Wats.) Rydb.---Perennial plants; stems erect from stout woody roots and branched woody bases, the season's branches 10-40 cm. tall, mostly simple, glabrous to pubescent; leaves 5-30 mm. long, narrowly linear, erect or ascending, acute, fleshy, sessile, glabrous to sericeous; flowers solitary or in 2's or 3's, hairy; perianth in fruit 2 mm. across, the wings 1.5-2 mm. long, conspicuous.---Plains and foothills. Wyoming to California, south to New Mexico and Arizona. Our rather few records from western Colorado at 4500-5500 feet.

11. Bassia All.

Annual plants; stems much branched; leaves alternate, sessile, narrow, entire; flowers minute, perfect, solitary or glomerate in the axils of the leaves or leaflike bracts, in panicles of short spikes; perianth villous or tomentose, with 5 lobes each armed on the back with a usually hooked spine; ovary with 2 styles; fruit enclosed by the perianth, pericarp free; seed horizontal, embryo annular.

1. Bassia hyssopifolia (Pall.) Kuntze, Rev. Gen. Pl. 1:547. 1891.
   Echinopsilon hyssopifolius (Pall.) Moq.---Pilose throughout; stems 20-50 cm. long, branching from the base; leaves 1-4 cm. long, oblong-lanceolate to linear-lanceolate; perianth lobes each armed with a stout spreading hooked spine.---Waste ground and fields. An Asiatic plant introduced in western United States. Our rather few records well scattered over Colorado at 4000-7500 feet.

12. Suaeda Forsk. SEEPWEED; SEABLITE

Annual or perennial plants, herbaceous or shrubby; leaves alternate, fleshy, terete or subterete, entire; flowers perfect or polygamous, solitary or clustered in the upper axils, bracteate; perianth 5-lobed or 5-parted, keeled, sometimes appendaged or narrowly winged on the back in age; stamens 5; styles usually 2; fruit included, pericarp usually free; seed horizontal or vertical, embryo spiral. A very difficult genus to determine the species, especially with dried material.

1. Perianth lobes (at least some of them) corniculate-appendaged or winged; annual plants
   2. Leaves broadest at the base, the upper ovate or lanceolate; perianth lobes corniculate-appendaged
      3. Stem branched from base, the branches decumbent   -1. S. depressa
      3. Stems simple or branched only above; erect   -1A. S. depressa erecta
   2. Leaves narrowed at base (except sometimes the upper), the upper ones linear or linear-lanceolate; perianth lobes winged   -2. S. occidentalis
1. Perianth lobes not appendaged or winged; annual or perennial plants
   4. Annual or rarely perennial plants but not suffruticose at base; stems erect and slender   -3. S. nigra
   4. Perennial plants, suffrutescent at base; stems erect or ascending
      5. Leaves strongly flattened, those of the inflorescence much reduced; plants never glaucous
                                                                                      -4. S. torreyana
      5. Leaves subterete, those of the inflorescence usually not much reduced; plants often glaucous
                                                                                      -5. S. fruticosa

1. Suaeda depressa (Pursh) S. Wats., King's Geol. Expl. 40th Par. 5:294. 1871.
   Dondia depressa (Pursh) Britt.---Annual or rarely perennial plants, glabrous, green or glaucous; stems 20-50 cm. tall, branched from base, these branches stout, more or less decumbent; leaves 0.7-3 cm. long, narrowly-linear, subterete, usually broadest at base, those of inflorescence shorter and broader, sommonly lanceolate or ovate; calyx lobes acute or obtuse, one or more corniculate-appendaged, carinate or keeled; seed vertical or horizontal, 1 mm. broad.---Saline or alkaline ground. Minnesota to Saskatchewan, south to Texas and California. Our records from northcentral, central and southcentral Colorado at 4500-9000 feet.
1A. Suaeda depressa erecta S. Wats., (var.) Proc. Am. Acad. 9:90. 1874.
    S. erecta (Wats.) A. Nels.; Dondia erecta (S. Wats.) A. Nels.---In this variety the stem is strictly erect, simple or branched above.---Alkaline or saline ground. Saskatchewan south to Nevada and Colorado. Our records from the western two-thirds of Colorado at 4500-8500 feet.
2. Suaeda occidentalis S. Wats., Proc. Am. Acad. 9:90. 1870.
   Dondia occidentalis (S. Wats.) Heller---Annual glabrous and green or somewhat glaucous plants; stems 6-25 cm. tall, erect or spreading, simple or branched; leaves 1-2.5 cm. long, narrowly linear, not widened at base, acute or acuminate, those of inflorescence little reduced and little if any wider than the lower ones; perianth lobes obtuse, transversely-winged in fruit, the wings irregularly lobed; seed horizontal, 1 mm. broad.---Alkaline or saline ground. Wyoming to Washington, south to Colorado and Nevada. Our few records from western and northcentral Colorado at 4900-5100 feet.
3. Suaeda nigra (Raf.) J. F. Macbride, Contr. Gray Herb. II. no. 56:50. 1918.
   S. diffusa S. Wats.; Dondia nigra (Raf.) Standl.; D. diffusa (S. Wats.) Heller---Annual or rarely perennial plants, green, nearly or quite glabrous; stems 20-80 cm. tall, erect, simple or branched, branches slender and usually flexuous; leaves 6-25 mm. long, narrowly linear, usually distinctly narrowed at base, acute, somewhat flattened, those of inflorescence much reduced; perianth lobes obtuse or acute, rounded on the back; seed more often vertical, 1 mm. broad.---Alkaline and saline ground, sometimes in sagebrush plains. Wyoming to Oregon, south to Mexico. Our records thinly scattered over the western two-thirds of Colorado at 4500-7000 feet.
4. Suaeda torreyana S. Wats., Proc. Amer. Acad. 9:88. 1874.
   Dondia torreyana (S. Wats.) Standl.---Perennial plants somewhat woody at base; stems 30-80 cm. tall, erect, branched, glabrous, sparsely leafy; leaves 1.5-3 cm. long, green, linear, strongly flattened, acute or acuminate, those of the inflorescence much reduced; branches of the inflorescence slender and lax; perianth lobes obtuse, rounded on back; seed vertical or horizontal, 1-1.5 mm. broad. Doubtfully distinct from S. fruticosa (L.) Forsk.---Alkaline and saline soil. Wyoming to Oregon, south to New Mexico and California. Our few specimens from southern and western Colorado at 4500-7000 feet.
5. Suaeda fruticosa (L.) Forsk., Fl. Aegypt. 70. 1775.
   S. moquini (Torr.) A. Nels.; Dondia moquini (Torr.) A. Nels. of Manuals---Perennial plants, somewhat woody at base, glaucous, nearly or quite glabrous; stems 30-80 cm. tall, much branched,

rather densely leafy; leaves 1-3 cm. long, narrowly linear, subterete, acute or obtuse, those of inflorescence little reduced; branches of the inflorescence stout; perianth lobes obtuse or acute, rounded on back; seeds mostly horizontal, about 0.8 mm. wide. May not be distinct from S. torreyana S. Wats.---Alkaline and saline soil. Alberta to Mexico; Europe; Asia; Africa. Our rather few records in the western half of Colorado at 4500-5500 feet.

13. Chenopodium L. GOOSEFOOT; PIGWEED

Annual or perennial herbaceous plants, often with mealy-coated or glandular stems and foliage; leaves alternate, usually petiolate, with entire to pinnatifid blades; flowers small, green, perfect, ebracteate, in glomerules, these axillary or forming spikes or panicles; perianth herbaceous or fleshy, 2- to 5-lobed; stamens 1-5; styles usually wanting, stigmas 2-5; utricle erect or depressed.

1. Foliage and inflorescence glandular-pubescent or resinous-glandular with yellow dots, not farinose
   2. Flowers in irregular glomerules, the inflorescences spicate-paniculate and axillary; odor fetid; pericarps with glandular-yellow dots   -1. C. ambrosioides
   2. Flowers solitary or loosely clustered, in dichotomous cymes; odor pronounced but not fetid; pericarps not glandular
      3. Plants densely glandular-pubescent; keel of perianth lobes smooth, not tuberculate   -2. C. botrys
      3. Plants sparsely puberulent, lower leaves and perianth with yellow resinous dots; keel of perianth lobes tuberculate   -3. C. graveolens
1. Foliage and inflorescence glabrous, sometimes puberulent but more commonly farinose, no glandular hairs or resinous dots present
   4. Perianth at maturity conspicuously bright red and very fleshy; flowers in dense glomerules, these in elongated spikes
      5. Leaves hastate and entire; glomerules of flowers about 3-6 mm. in diameter   -4. C. overi
      5. Leaves hastate and dentate; glomerules of flowers about 5-10 mm. in diameter   -5. C. capitatum
   4. Perianth not bright red or very fleshy at maturity; flowers usually in glomerules in paniculate spikes
      6. Leaves all narrowly linear to linear, 1-nerved and entire
         7. Plants densely farinose; seeds usually less than 1 mm. in diameter   -6. C. leptophyllum
         7. Plants glabrous, leaves sparingly farinose below; seeds usually more than 1 mm. in diameter
            8. Pericarp red; perianth lobes not at all carinate; leaves usually somewhat farinose below, 1-1.5 cm. long   -7. C. cycloides
            8. Pericarp white or green; perianth lobes more or less carinate; leaves not at all farinose below, over 1.5 cm. long
               9. Pericarp adherent to the seed; seed surface marked by minute alveolar cells   -8. C. pallescens
               9. Pericarp free from the seed; seed surface smooth   -9. C. subglabrum
      6. Leaves broader, often toothed or lobed, if narrowly lanceolate or oblong then 3-nerved on lower ones
         10. Seeds vertical in the pericarp, only an occasional seed horizontal
             11. Lower leaf surface densely farinose   -10. C. glaucum
             11. Lower leaf surface glabrous or nearly so
                 12. Perianth of flowers with vertical seeds 3-tipped, the lobes almost completely fused   -11. C. chenopodioides
                 12. Perianth of flowers with vertical seeds 3- to 5-lobed, these cleft to the middle or below   -12. C. rubrum
         10. Seeds all or nearly all horizontal in the pericarps
             13. Leaves narrowly lanceolate or narrowly oblong, entire or somewhat lobed at base
                 14. Pericarp adherent to the seed; plants with a rank odor; leaves entire; calyx open in fruit   -13. C. hians
                 14. Pericarp free from the seed; plants without a rank odor; leaves entire or lobed at base; calyx enclosing the fruit   -14. C. pratericola
             13. Leaves from broadly-lanceolate to as wide as long, margins on most of the species toothed or lobed
                 15. Leaf blades broadly ovate, 5-15 cm. long, with 1-3 (rarely 4) large triangular, acute, remote teeth on each side, glabrous, often subcordate at base   -15. C. gigantospermum
                 15. Leaf blades variously shaped, if broadly ovate then less than 5 cm. long, margin entire or variously toothed or lobed, glabrous or farinose, seldom if ever subcordate at base
                     16. Leaves very wide, as broad or nearly as broad as long
                         17. Pericarp adherent to the seed; seed-surface roughened by alveolar cells or ridges
                             18. Plant densely farinose, ill-scented; calyx lobes carinate; leaves 1-4 cm. long   -16. C. watsoni
                             18. Plant glabrous or leaves very sparingly farinose below, not ill-scented; calyx lobes little if any carinate; leaves 3-8 cm. long   -17. C. murale
                         17. Pericarp free from the seed; seed surface smooth
                             19. Plants glabrous or sparingly farinose; leaves thin, often over 2.5 cm. long; plants slender, often over 25 cm. tall   -18. C. fremontii
                             19. Plants densely farinose; leaves thick, seldom over 2.5 cm. long; plants diffusely branched from base, seldom over 25 cm. tall   -19. C. incanum
                     16. Leaves (most of them) 2 to 4 times longer than wide, not as broad as long
                         20. Pericarp free from seed, not adherent
                             21. Plants definitely farinose, on lower leaf surface at least
                                 22. Stems and perianth lobes definitely farinose   -20. C. albescens
                                 22. Stems and perianth lobes not farinose   -10. C. glaucum
                             21. Plants glabrous or very sparingly farinose on the lower leaf surface

211

23. Leaves thin, 3.5-6 cm. long, the upper acute, the lower sinuate-dentate; stems slender
-21. C. standleyanum
23. Leaves thick, 1.5-3 cm. long, the upper usually obtuse, the lower not sinuate-dentate; stems moderate to stout
-22. C. atrovirens
20. Pericarp adherent to the seed
24. Inflorescence mostly of axillary panicles seldom as long as the subtending leaf; perianth not carinate; leaves almost as wide as long; seed surface with minute ridges and acute margins -17. C. murale
24. Inflorescence often axillary but also of elongated terminal panicles much longer than the leaves; perianth lobes definitely carinate; leaves definitely longer than wide; seed surface smooth or with alveolar cells, margin obtuse or rounded
25. Plants bright green, not at all farinose -23. C. paganum
25. Plants blue-green, the leaves, at least on lower surface, sparingly to densely farinose
26. Calyx lobes widely spreading, in fruit, obtusely carinate -13. C. hians
26. Calyx lobes enclosing the fruit, sharply and acutely carinate
27. Seeds smooth and shiny; pericarp smooth or somewhat ridged -24. C. album
27. Seeds minutely alveolate, not very shiny; pericarp roughened by round shallow pits
-25. C. berlandieri

1. Chenopodium ambrosioides L., Sp. Pl. 219. 1793.
Annual plants with fetid odor; stems 30-100 cm. tall, glabrous or puberulent below, usually glandular-villous or tomentulose above and about the inflorescence, simple or branched; leaves 2-12 cm. long, reduced above, the lower usually oblong to oblong-lanceolate in outline but varying from lanceolate to broadly ovate, acute or obtuse at apex, cuneate at base to a short petiole, margins irregularly and coarsely sinuate-dentate or sinuate-pinnatifid, puberulent to glabrous, usually dotted with yellow resinous glands; inflorescence of glomerules, these in dense or interrupted spikes; perianth short, deeply cleft, the lobes glabrous to short-villous, usually gland-dotted; pericarp gland-dotted, free; seed horizontal or vertical, usually the former, 0.6-0.8 mm. wide, smooth and shining, the margin obtuse.---Waste ground. Throughout most of North America; introduced from tropical America. Our one record from northcentral Colorado at 4500 feet.

2. Chenopodium botrys L., Sp. Pl. 219. 1753.
Annual sweet-scented plants; stems 10-60 cm. tall, densely glandular-viscid throughout, much branched; leaf blades 1-5 cm. long, oblong or oval in outline, acute or obtuse at apex, truncate to cuneate at base, the petiole definitely shorter than the blades, margins irregularly sinuate-pinnatifid often 1/2 way to midrib, with rounded lobes, both sides glandular-pubescent; inflorescence of densely many-flowered cymes forming a narrow elongated panicle; perianth cleft nearly to the base, the lobes densely glandular-pubescent, not carinate; pericarp adherent; seed vertical or horizontal, usually the latter, 0.6-0.8 mm. in diameter, dull or shining.--- Waste places. Throughout the United States and Canada; Europe. Our few records scattered over the state at 4500-8000 feet.

3. Chenopodium graveolens Willd., Enum. Pl. Hort. Berol. 1:290. 1809.
C. incisum Poir.; C. cornutum (Torr.) Benth. & Hook.---Annual strong-smelling plants; stems 20-60 cm. tall, glabrate or sparsely puberulent, simple or more commonly branched from the base; leaf blades 2-6 cm. long, deltoid-ovate to narrowly oblong in outline, obtuse to acuminate at apex, truncate to cuneate at base to petioles shorter than the blades, margins sinuate-pinnatifid or laciniate-pinnatifid, the lobes usually entire, glabrous or minutely villous above, with yellow glands below; inflorescence of numerous, loosely flowered cymes these forming a panicle, the flowers sessile in the forks of the branches and solitary at the ends of slender lateral branches, but these latter flowers often abortive and their pedicels spinulose; perianth deeply cleft, the lobes covered with yellow glands and corniculate-appendaged on the back; pericarp adherent; seed horizontal, 0.5-0.8 mm. wide, smooth or nearly so. The correct name for this plant may be C. incisum Poir.---Dry places. Colorado, south to Texas and Central America. Our records from the southern half of Colorado at 7500-8500 feet.

4. Chenopodium overi Aellen, Fedde Report. 26:159. 1929.
Blitum hastatum Rydb.---Annual glabrous plants; stems 10-40 cm. tall, erect or ascending, usually branched from base; leaf blades 2.5-7 cm. long, ovate to lanceolate-ovate, acute, hastate at base or cuneate to a petiole often as long as the blade especially on lower leaves, glabrous; inflorescence in globose glomerules sessile in the axils, and in slender terminal spikes; perianth deeply lobed becoming conspicuously red and fleshy or berrylike at maturity; pericarp free from the seed; seed vertical, 0.8-1 mm. long, slightly compressed.---Dry stony ground. Wyoming to Oregon, south to New Mexico and Nevada. Our few records from central and northcentral Colorado at about 8000-9500 feet.

5. Chenopodium capitatum (L.) Asch., Fl. Brand. 1:572. 1864.
Blitum capitatum L.---Annual glabrous plants; stems 10-60 cm. tall, branched from the base or simple; leaves triangular on lower to narrow above, acute, lower leaves usually hastate-lobed at base, the petiole about as long as the blade, coarsely sinuate-dentate, glabrous; inflorescence of globose glomerules these sessile or short-pedunculate in the upper axils and forming a short spikelike terminal inflorescence; perianth 3- to 5-lobed or deeply cleft, becoming conspicuously red and fleshy in fruit like a berry; pericarp free from the seed; seed vertical about 0.8 mm. long, with acute margins.---In moist or rocky soil, often in shade. Quebec to Alaska, south to New Jersey, New Mexico and Arizona; Europe. Our records well scattered in the mountains of Colorado at 6000-11,000 feet.

6. Chenopodium leptophyllum Nutt. apud. Moq. DC., Prodr. 13:71. 1879.
C. inamoenum Standl.---Annual plants; stems 10-80 cm. tall, farinose to glabrate, simple or more commonly branched above; leaves 1-3 cm. long, linear, lanceolate-linear, obtuse or acute at apex, narrowly cuneate at base to short petioles usually not over 1/2 the length of the blades, margin entire, 1-nerved, both sides farinose or glabrate above; inflorescence of glomerules,

these in paniculate spikes; perianth cleft about to the middle, the lobes carinate and farinose; pericarp free or often somewhat adherent; seed horizontal, 0.8-1.1 mm. wide, smooth and shining, margins obtuse.---Dry ground. Manitoba to Alberta, south to Mexico and California. Scattered in Colorado, our records mostly from the northern and western parts, at 4500-8500 feet.

7. Chenopodium cycloides A. Nels., Bot. Gaz. 34:363. 1902.
Annual plants; stems 25-40 cm. tall, simple to branched, glabrous to sparsely farinose, usually reddish; leaves 1-1.5 cm. long, linear, glabrous above, sparingly farinose below, entire; flowers in dense or interrupted panicles of spikes; perianth farinose, lobes rounded on back, widely spreading in fruit; pericarp adherent, red, minutely tuberculate; seed horizontal, 1-1.5 mm. wide.---Dry ground. Kansas and New Mexico. Our one record from southern Colorado at 5000 feet.

8. Chenopodium pallescens Standley, N. Am. Fl. 21:15. 1916.
Annual plants; stems 25-60 cm. tall, erect, branching, angled and striate, glabrate or sparsely farinose; leaf blades 1.5-3.5 cm. long, linear, 1-nerved, entire, glabrous to sparsely farinose, on petioles 2-5 mm. long; inflorescence of glomerules in interrupted paniculate spikes; perianth slightly farinose, carinate, deeply cleft; pericarp adherent, finely tuberculate; seed horizontal about 1.5 mm. wide, nearly smooth with rounded margins.---Dry ground. Oklahoma and New Mexico. Our one somewhat doubtful record from southeastern Colorado at 6000 feet.

9. Chenopodium subglabrum (S. Wats.) A. Nels., Bot. Gaz. 34:362. 1902.
Montana to Washington, south to Nebraska and Utah. Should be found in the state but no records known to the writer.

10. Chenopodium glaucum L., Sp. Pl. 220. 1753.
Annual plants; stems 10-45 cm. tall, erect or prostrate, branching throughout, glabrous; leaf blades 1-5 cm. long, oblong-obovate to ovate-oblong, obtuse at apex, narrowly to broadly cuneate to a petiole from short to as long as the blade, margin sinuate-dentate or upper entire, glabrous above, farinose below; flowers in dense glomerules, these in short axillary and terminal, often branched spikes; perianth deeply cleft, the lobes not farinose or carinate; pericarp free; seeds of lateral flowers usually vertical but sometimes more horizontal than vertical ones present, 0.6-1 mm. broad, smooth or finely tuberculate, margins acute or obtuse. Represented in Colorado by ssp. eu-glaucum Aellen with the petiolelike leaf bases long, often as long as the blade, and ssp. salinum (Standley) Aellen with the petiolelike leaf bases very short.---Waste places and saline areas. A European species widely distributed in Canada and the United States. Our records widely scattered over Colorado at 4000-7500 feet.

11. Chenopodium chenopodioides (L.) Aellen, Ostenia 98. 1933.
C. humile of Am. Authors, not Hook.; C. humile Am. Auths., not Wats.---Plains and meadows. Saskatchewan to British Columbia, south to Nebraska and California. Reported from Colorado but no plants seen by the writer.

12. Chenopodium rubrum L., Sp. Pl. 218. 1753.
Annual plants, glabrous or glabrate throughout; stems 20-80 cm. tall, simple or branched at base or above; leaves 3-10 cm. long, ovate-rhombic, triangular-hastate to lanceolate, acute or obtuse at apex, broadly or narrowly cuneate at base, often somewhat hastate, petiole about as long or somewhat shorter than the blades, sparingly sinuate-dentate on margins or the upper entire; inflorescence of sessile glomerules, these in short or elongate spikes; perianth lobes cleft to middle or below, slightly fleshy and somewhat red in age, not carinate; pericarp free; seed vertical, 0.8-1 mm. long, smooth with obtuse or somewhat acute margins. The above description applies to the typical and apparently more common plant. A form occurs in western Colorado which has a prostrate or ascending stem less than 20 cm. tall and leaf blades entire except for the hastate lobes at base of some. The glomerules of flowers are solitary or in subspicate inflorescences. It has been called C. rubrum f. humile (Hook.) Asch. et Graeb.---Alkaline or saline soil. Newfoundland to British Columbia, south to New Jersey, New Mexico and Arizona. Our records from northern and western Colorado at 4000-9500 feet.

13. Chenopodium hians Standley, N. Am. Fl. 21:16. 1916.
Ill-scented plants; stems 40-80 cm. tall, erect, usually simple below and sparsely branched above, branches farinose; petioles about 1/2 as long as the blades these 1-3 cm. long, narrowly lanceolate-oblong, oblong or elliptic-oblong, cuneate at base, rounded or obtuse at apex, somewhat hastate at base, margin entire, rarely sinuate-dentate, green and glabrate above, farinose below, rather thick; flowers in glomerules, these rather large, in axillary or paniculate spikes; perianth lobes farinose, obtusely carinate, erect or spreading later and exposing the fruit; pericarp adherent; seed horizontal about 1 mm. broad, nearly smooth, margin obtuse.---Dry ground. Wyoming to Nevada, south to New Mexico and Arizona. Our few records from northern Colorado at 5000-10,000 feet.

14. Chenopodium pratericola Rydb., Bull. Torr. Club 39:310. 1912.
C. desiccatum A. Nels.; C. oblongifolium (S. Wats.) Rydb.; C. leptophyllum var. oblongifolium Wats.---Annual plants; stems 10-80 cm. tall, glabrate rarely farinose, usually branching throughout; leaf blades 1-6 cm. long, narrowly lanceolate to narrowly oblong, sometimes wider, obtuse or acute at apex, cuneate at base to a petiole half as long as the blade or shorter, margins entire but often hastately toothed near base, rather thick and mostly 3-nerved, usually glabrate or sparingly farinose above and often bright green, densely farinose below; inflorescence of glomerules arranged in dense paniculate spikes; perianth rather deeply cleft, the lobes farinose, carinate; pericarp free; seed horizontal, 1 mm. broad, smooth and shining, with obtuse margins. Several subspecific entities have been recorded for Colorado.---Waste places and fields, often in sandy soil. North Dakota to Washington, south to Missouri, Arizona and California. Our records scattered over Colorado, mostly in the northern and eastern parts at 3500-8500 feet.

15. Chenopodium gigantospermum Aellen; Fedde, Rep. Spec. Nov. 26:147. 1929.

C. hybridum of Amer. authors, not of L.; C. hybridum var. gigantospermum (Aellen) Rouleau---Annual plants, glabrous except the inflorescence; stems 30-100 cm. tall, erect, usually much branched; leaves thin, bright green, broadly ovate, rhombic-ovate or triangular-ovate, acuminate at apex, truncate, rounded or often subcordate at base, the petioles about 1/2 as long as the blades, glabrous; inflorescence of small glomerules these in paniculate spikes or loose cymes, the branches more or less farinose; perianth lobes often somewhat farinose, deeply cleft, obscurely if at all carinate; pericarp adherent or free; seed horizontal, 1.6-2 mm. wide, sharp-edged to obtuse, puncticulate or with minute grooves.---Waste places and in shade. Quebec to British Columbia, south to Virginia, New Mexico and California; Europe. Our records scattered in the western two-thirds of Colorado at 5000-8000 feet.

16. Chenopodium watsoni A. Nels., Bot. Gaz. 34:362. 1902.

Ill-scented plants, 10-80 cm. tall, much branched from the base and throughout; stems farinose usually densely so; petioles as long or shorter than the blades, leaf blades 1-4 cm. long and about as wide, broadly rounded-deltoid, rounded-rhombic or ovate, rounded, truncate or broadly-cuneate at base, rounded or obtuse at apex and shortly apiculate, entire or lower with 1 or 2 rounded teeth on each side, hence somewhat hastate, farinose above and below; flowers in rather large glomerules, in axillary or paniculate spikes, the inflorescence leafy; perianth densely farinose, carinate on the lobes; pericarp adherent; seed horizontal, about 1 mm. wide, surface of seed and pericarp with alveolar cells making pits.---Dry places. Montana south to New Mexico and Arizona. Our few records from northeastern Colorado at 4500-5500 feet.

17. Chenopodium murale L., Sp. Pl. 219. 1753.

Annual plants; stems 20-60 cm. tall, glabrous or sparingly farinose above, simple or branched from the base, erect or somewhat decumbent; leaves 3-8 cm. long, ovate or ovate-rhombic, acute or obtuse at apex, cuneate or subtruncate at base to a petiole as long or shorter than the blade, margins sinuate-dentate to laciniate-serrate, glabrate or sometimes farinose below; inflorescence of small glomerules arranged mostly in axillary but also in terminal cymes or panicles, those in the axils shorter than the subtending leaves; perianth deeply cleft, the lobes obscurely if at all carinate, usually sparingly farinose; pericarp adherent, puncticulate; seed horizontal, 1.2-1.5 mm. in diameter, dull, puncticulate or finely ridged, margins acute.---Waste places. A European species established in the most of North America. The species is surely in Colorado but no certain records could be found.

18. Chenopodium fremontii S. Wats., Bot. King's Expl. 287. 1871.

Annual plants; stems 20-80 cm. tall, glabrous or very sparingly farinose, erect or ascending, usually branched throughout; leaf blades thin, triangular-rhombic or broadly triangular, about as wide as long, rounded at apex, rarely acute, truncate or broadly cuneate at base, the petiole 1/2 as long as the blade or longer, hastately lobed at base, these lobes usually broadly rounded, otherwise margin nearly or quite entire, glabrous or nearly so above, sparingly farinose below; inflorescence of small glomerules in slender or stout usually flexuous paniculate spikes; perianth deeply cleft, the lobes usually sparsely farinose; pericarp free; seed horizontal, about 1 mm. wide, smooth or slightly rugulose, shining, margin obtuse.---Canyons and slopes. North Dakota to British Columbia, south to Texas, Mexico and Arizona. Our records scattered in the western two-thirds of Colorado at 4500-8500 feet.

19. Chenopodium incanum (S. Wats.) A. Heller, Pl. World 1:23. 1897.

Annual plants; stems 10-30 cm. tall, usually less than 20, densely branched throughout, farinose; leaf blades 7-20 mm. long, thick, triangular-rhombic to rhombic-orbicular, about as wide as long, obtuse at apex, rounded to broadly cuneate at base, the petiole about 1/2 as long as the blade, the blade hastately lobed, farinose both sides, but more densely so below; inflorescence of large glomerules arranged in crowded paniculate spikes; perianth lobes farinose, carinate; pericarp free; seed horizontal, 0.8-1.2 mm. broad, smooth, shining, margins obtuse or very slightly acute.---Dry plains and slopes. Nebraska to Utah, south to Texas, Mexico and California. Our records from the eastern half of the state at 4500-6000 feet.

20. Chenopodium albescens Small, Fl. S. E. U. S. 385. 1903.

Texas to Arizona. This species has been reported from southern Colorado but the record is doubtful.

21. Chenopodium standleyanum Aellen, Fedde Rep. Spec. Nov. 26:153. 1929.

C. boscianum Moq.---Pennsylvania to Nebraska, south to Florida and New Mexico. This plant should be looked for in eastern, especially southeastern Colorado.

22. Chenopodium atrovirens Rydb., Mem. N. Y. Bot. Gard. 1:131. 1900.

C. wolfii Rydb.; C. aridum A. Nels.---Annual plants; stems 10-50 cm. tall, sometimes less in unfavorable habitats, erect, usually much branched, glabrate; leaves 1.5-3 cm. long, ovate-rhombic to broadly oblong, obtuse to acute at apex, rounded or cuneate at base to a petiole seldom over 2/3 the length of the blade, margins entire or more rarely with 2 rounded lobes near the base, glabrate above, thinly farinose below when young, soon glabrate; inflorescence of glomerules arranged in interrupted paniculate spikes, the branches sparingly farinose; perianth deeply lobed, the lobes carinate, more or less farinose; pericarp free; seed horizontal, about 1 mm. wide, smooth or nearly so, shining, margins obtuse.---Arid ground. North Dakota to California, south to New Mexico. Reported for Colorado and certainly here but no typical specimens could be located by the writer.

23. Chenopodium paganum Riechenb., Fl. Germ. Exc. 579. 1832.

This eastern species should be looked for in eastern Colorado.

24. Chenopodium album L., Sp. Pl. 219. 1753.

Includes C. lanceolatum Muhl.---Annual, pale, more or less farinose plants; stems 30-200 cm. tall, erect usually branched, especially above, usually glabrous; leaf blades 2-8 cm. long, ovate-rhombic, rarely lanceolate, obtuse or acute, broadly cuneate to a petiole about 1/2 as long as the blades, irregularly sinuate-dentate, often 3-lobed, rather thick, pale and glabrate above, more or less farinose below; inflorescence of large glomerules these in rather dense paniculate spikes; perianth lobes farinose, definitely carinate; pericarp adherent, smooth or irregularly furrowed; seed horizontal, 1.3-1.5 mm. wide, smooth and shining, with obtuse margins.---Waste places, fields and thickets. A European species now widespread in North America. Our records from northcentral Colorado at 4500-5000 feet but the plant must be widespread in the state.

25. Chenopodium berlandieri Moq., Chenop. Enum. 23. 1840.
Annual, pale, more or less farinose plants; stems 30-150 cm. tall, erect, usually branched, usually glabrous; leaves 1.5-4 cm. long, ovate, elliptic or rhombic, more rarely lanceolate, acute or obtuse, rounded but more often broadly cuneate to a petiole about 1/2 as long as the blade, margins irregularly sinuate-dentate or sometimes entire, seldom 3-lobed, rather thick, pale and glabrate above, more or less farinose below; inflorescence of glomerules, these in slender or stout paniculate spikes; perianth lobes farinose, definitely carinate; pericarp adherent, with round or shallow pits; seed horizontal about 1.5 mm. wide, minutely roughened by alveolar cells, margins obtuse or rounded. This species is separated with difficulty from C. album L. especially when the seeds and fruit are immature.---Waste ground and fields. Kansas to California, south to South America. Apparently common, our records scattered over the state at 4000-8500 feet.

14. Corispermum L. BUGSEED; TICKSEED

Annual, herbaceous, caulescent plants; stems branching; leaves alternate, sessile, entire, 1-nerved; flowers perfect, solitary or glomerate, arranged in a narrow or loose terminal spike, each flower in the axil of a leaflike, scarious-margined bract, the upper ones wider than the leaves; perianth usually of 1 segment, but sometimes 2 or 3 and then unequal; stamens 1-3, rarely 5, if more than 1 then unequal; ovary exserted, styles 2, persistent, utricle winged or margins acute; pericarp adherent, seed vertical; radicles inferior.

1. Fruit with a distinct wing, this about 0.5 mm. wide
   2. Spikes laxly-flowered, slender; lower bracts definitely narrower than the fruit which is about 3 mm. long
      -1. C. nitidum
   2. Spikes dense, stout; lower bracts usually euqlling or wider than the fruit which is about 4 mm. long
      -2. C. hyssopifolium

1. Fruit acute-margined or very narrowly winged
   3. Plants glabrous   -3. C. emarginatum
   3. Plants villous   -4. C. villosum

1. Corispermum nitidum Kit. ex Schultes, Oestr. Fl. ed. 2. 1:7. 1814.
Stems 20-60 cm. tall, much branched, glabrous or sparsely stellate-villous; leaf blades 1.5-5 cm. long, about 1 mm. wide, narrowly linear to linear-filiform, cuspidate, glabrous or sparsely stellate-pubescent; spikes slender, usually laxly flowered, bracts usually not imbricated; bracts ovate or broadly ovate to linear-lanceolate especially on lower, these narrower than the fruit; fruit 2-3 mm. long, sometimes rather narrowly but still conspicuously winged.---Sandy valleys and slopes. Our few records from eastern Colorado, mostly near the foothills, at 4500-8000 feet.

2. Corispermum hyssopifolium L., Sp. Pl. 4. 1753.
C. marginale Rydb.; C. imbricatum A. Nels.---Stems 15-60 cm. tall, much branched, glabrous or somewhat stellate-villous; leaves 1-7 cm. long and 1-2.5 mm. wide, linear, cuspidate, glabrous or somewhat stellate-pubescent; spikes usually densely-flowered, the bracts often much imbricated; bracts ovate or lanceolate, the upper wider, the lowest longer and narrower, all except the lowest as wide or wider than the fruit; fruit 3.5-4.5 mm. long, conspicuously winged.---Sandy ground. Ontario to Washington, south to New York, Missouri and Mexico; Eurasia. Our few records mostly from the eastern half of Colorado at 4000-8500 feet.

3. Corispermum emarginatum Rydb., Bull. Torr. Club 31:404. 1904.
Stems 30-50 cm. tall, branched from base, glabrous; leaves 1-4 cm. long, 1-2 mm. wide, narrowly linear, cuspidate; spikes rather slender to stout; bracts ovate to broadly ovate or lower longer and narrower, broader than the fruit; fruit 2.5-3 mm. long and about 2 mm. wide, margins acute or narrowly winged.---Sandy ground. Alberta to Nevada, south to Colorado. Our few specimens from central and southcentral Colorado at 6000-7500 feet.

4. Corispermum villosum Rydb., Bull. Torr. Club 24:191. 1897.
Stems 10-40 cm. tall, rather stout and much branched from near the base, villous-stellate when young, glabrate in age; leaves 2-4 cm. long and 1-3 mm. wide, linear, thick, villous-stellate especially when young, acuminate or cuspidate at apex; spikes densely flowered; bracts imbricated, upper ovate-oblong to broadly ovate, lower oblong to linear-lanceolate, all broader than the fruit and scarious-margined, especially on upper ones, stellate-pubescent especially when young; fruit 2-3 mm. long and 2 mm. wide, margin acute or narrowly winged.---Sandy ground. Saskatchewan to Washington, south to New Mexico. Our records from northcentral, central and southcentral Colorado at 4500-8000 feet.

15. Monolepis Schrad. POVERTY WEED

Annual herbaceous plants; stems branching; leaves alternate; inflorescence a panicle of spikes, these axillary or terminal; flowers polygamous, ebracteate, with 1 perianth segment, this persistent; stamen 1; stigmas 2, subulate; pericarp at least somewhat adherent to the erect, laterally compressed seed.

1. Leaves hastately lobed at base, over 12 mm. long; flowers in dense axillary clusters   -1. M. nuttalliana
1. Leaves entire, not over 12 mm. long; flowers paniculate on slender pedicels   -2. M. pusilla

1. **Monolepis nuttalliana** (Schult.) Greene, Fl. Franc. 168. 1891.

Stems 10-30 cm. tall, decumbent or ascending with stout, succulent branches, these somewhat farinose when young becoming glabrate and green; leaves 1-7 cm. long, petioled or the upper sessile, leaf blades triangular to lanceolate, hastately-lobed at base, the lobes divergent, terminal lobe usually entire, obtuse or acutish at apex, blade cuneate to a decurrent base, sparingly farinose when young; flowers in dense axillary clusters, apical portion forming an interrupted spike; perianth acute; pericarp pitted and gray when dry; seed 1-1.5 mm. wide.---Waste places and dry plains. Manitoba to Alberta, south to Texas, Mexico and California. Our many records scattered over the state at 4000-10,000 feet.

2. **Monolepis pusilla** Torr. in S. Wats., Bot. King's Exp. 289. 1871.

Wyoming to Washington, south to Colorado and California. Reported for western Colorado but no specimens could be located by the writer.

Family 40. Amaranthaceae AMARANTH FAMILY

Herbaceous plants; leaves opposite or alternate, simple, usually entire, lacking stipules; flowers perfect, monoecious, dioecious or polygamous, solitary, racemose, spicate or capitate, each flower, or rarely each cluster, subtended by a bract and 2 bractlets, these usually hyaline and never foliaceous; perianth of 2-5 scarious or chartaceous, rarely herbaceous segments; corolla wanting; stamens usually opposite the perianth segments and the same number; ovary superior, 1-celled, styles 1 or 2 or none; fruit a membranous utricle, circumscissle, irregularly dehiscent or indehiscent.

1. Leaves alternate; plants glabrous, puberulent or villous; anthers 4-celled, appearing 2-celled after dehiscence
   2. Pistillate flowers with no perianth; plants dioecious   -1. Acnida
   2. Pistillate flowers with a perianth; plants dioecious or monoecious (our common species monoecious)
             -2. Amaranthus
1. Leaves opposite; plants stellate-pubescent or tomentose; anthers 2-celled but often appearing 1-celled after dehiscence
   3. Plant stellate-pubescent; filaments united only at base; perianth segments separate or united only near base, not tubular and enclosing the utricle, not crested, winged or spiny   -3. Tidestromia
   3. Plant sericeous-tomentose; filaments united into a long tube; perianth segments united into a long tube enclosing the utricle, and bearing wings, crests or spines at maturity   -4. Froelichia

1. Acnida L. WATERHEMP

Annual, glabrous herbaceous plants; stems branched; leaves alternate, petiolate, entire; flowers dioecious, glomerate these axillary or in terminal spikes or panicles; sepals 5 in the staminate flowers, these scarious or membranous; 5 stamens present; perianth lacking entirely in pistillate flowers; styles short or none; fruit indehiscent, circumscissle or irregularly dehiscent; seed erect and compressed.

1. Fruit circumscissle; plants mostly of sandy, damp or dry habitats   -1. A. tamariscina
1. Fruit indehiscent or irregularly dehiscent; plants mostly of muddy habitats   -2. A. altissima

1. **Acnida tamariscina** (Nutt.) Wood, Bot. & Fl. 289. 1873.

This is closely related to A. altissima differing in having circumscissle fruit and perhaps in habitat, recorded for damp sandy or dry ground. The staminate flowers are recorded as being over 3 mm. long.---South Dakota, south to Missouri, Texas and New Mexico. A specimen from Fremont County at 5300 feet may be this species.

2. **Acnida altissima** Riddell; Moq. in DC., Prodr. 13(2):278. 1849.

Stems 40-120 cm. tall, rather stout, erect, usually much branched, the branches ascending, glabrous; petioles slender, as long or shorter than the blades, the latter 2-12 cm. long, lanceolate to rhombic-ovate, cuneate below and narrowed to an obtuse but narrow apex, glabrous; bracts as long or shorter than the sepals, shorter than the utricle; staminate calyx 2-2.5 mm. long; fruit irregularly dehiscent or indehiscent, smooth, or verrucose mostly below the middle.---Swamps and low ground, often in mud. Ontario south to Ohio and Colorado. A specimen from El Paso County seems to be this species.

2. Amaranthus L. AMARANTHUS; PIGWEED

Annual weedy plants; leaves alternate, petiolate, entire or rarely sinuate-dentate or undulate; flowers monoecious, dioecious or polygamous, glomerate, these axillary, spicate or paniculate; bracts and bractlets usually conspicuous; perianth segments 2-5, distinct; utricle dehiscent or indehiscent; seeds erect, compressed, smooth, embryo annular with inferior radicle.

1. Perianth segments of pistillate flowers broadly spatulate, the blade considerably wider than the claw; plants dioecious
   2. Bracts equalling or shorter than the pistillate perianth   -1. A. torreyi
   2. Bracts longer than the pistillate perianth, usually 2 or 3 times longer   -2. A. palmeri
1. Perianth segments of pistillate flowers linear, lanceolate, oblong or sometimes narrowly spatulate but the blade little wider than the claw; plants monoecious, the staminate flowers sometimes few
   3. Flowers all in glomerate axillary clusters; plants low, much branched from the base, these branches prostrate or ascending; stamens 3
      4. Perianth parts 4 or 5; bracts as long or slightly longer than the perianth; seed about 1.5 mm. wide; stems prostrate   -3. A. graecizans

4. Perianth parts 3; bracts 2-4 times longer than the perianth; seed 0.6-0.8 mm. wide; stems erect or ascending
    5. Plant densely viscid-pubescent; leaf blades crispate   -4. A. pubescens
    5. Plant glabrous or sparingly pubescent; leaf blades flat or nearly so   -5. A. albus
3. Flowers, at least the upper in simple or paniculate spikes, some of these terminal; plants usually not branching at very base; stamens 3-5
    6. Plants glabrous or nearly so; stems seldom over 70 cm. tall   -6. A. wrightii
    6. Plants pubescent or villous, at least above and in the inflorescence; stems often over 70 cm. tall
        7. Spikes stout, 8-20 mm. thick; sepals of pistillate flowers over 2 mm. long, usually equalling or longer than the fruit; main bracts usually 4-6 mm. long; plants common in Colorado   -7. A. retroflexus
        7. Spikes slender 6-12 mm. thick; sepals of pistillate flowers seldom over 2 mm. long, usually shorter than the fruit; main bracts usually 2-3.5 mm. long; plants rare in Colorado   -8. A. hybridus

1. **Amaranthus torreyi** (A. Gray) Benth. in S. Wats., Bot. Calif. 2:42. 1880.
   Stems 30-100 cm. tall, stout, erect, glabrous or somewhat pubescent especially above, simple or branched at base; leaves 1-8 cm. long, with slender petioles, blades variable in shape, from oblong-linear or lanceolate to oval-oblong, round to acutish at apex, rounded or cuneate at base, glabrous or nearly so; flowers dioecious, in slender terminal spikes, these in a panicle; bracts lanceolate, spine-tipped, usually shorter than the perianth; perianth parts of staminate flowers oblong, 2-3 mm. long, scarious, obtuse, acute or midrib excurrent; perianth parts of pistillate flower about 2 mm. long, broadly spatulate to spatulate, narrowed to a claw, obtuse and rounded or truncate at apex, midrib usually excurrent; stamens 5; fruit subglobose, circumscissle; seed about 1 mm. wide.---Dry often sandy ground. Iowa to Nevada, south to Texas and New Mexico. Scattered in Colorado, mostly in the eastern half of the state, at 3500-5500 feet.

2. **Amaranthus palmeri** S. Wats., Proc. Amer. Acad. 12:274. 1877.
   Stems 60-100 cm. tall, erect, stout, much branched, glabrous or more or less villous-pubescent above; leaves 1-5 cm. long, slender and long-petioled, blades rhombic-ovate to rhombic-lanceolate, acute, abruptly acuminate or attenuate at apex, cuneate or rounded at base, usually glabrous; flowers dioecious, in slender, terminal, erect or drooping, dense spikes, also in axillary clusters; bracts usually over twice as long as the perianth, tapering to a spine tip; sepals of staminate flowers 2-3 mm. long, oblong, acute or acutish, the nerve usually excurrent; sepals of pistillate flowers narrowly to broadly spatulate, 2-3.5 mm. long, obtuse or truncate, often emarginate, the nerve rarely excurrent; stamens 5; fruit subglobose, circumscissle; seed round, about 1 mm. wide.---River banks, valleys and cultivated areas. Kansas to California, south to Texas and Mexico. Our records for eastern Colorado are uncertain but the plant is surely present in the state.

3. **Amaranthus graecizans** L., Sp. Pl. 990. 1753.
   A. blitoides S. Wats.---Stems 30-70 cm. long, stout, prostrate, much branched, glabrous or sparsely pubescent; leaves 1-3 cm. long, often crowded and enlarged toward the ends of the branches, petioled, oval, spatulate or obovate, cuneate at base, rounded and cuspidate at apex, glabrous, often white-margined; flowers monoecious, in small axillary clusters; bracts equalling or slightly exceeding the perianth, short-pungent at tips; perianth parts 4-5, membranous or scarious; stamens 3; fruit subglobose, circumscissle, smooth or nearly so, equalling or slightly longer than the perianth; seed round, 1.3-1.5 mm. wide.---Roadsides, fields and waste places. Rather generally distributed in North America and Europe as a weed. Our records scattered over Colorado at 3500-8500 feet.

4. **Amaranthus pubescens** (Uline & Bray) Rydb., Bull. Torr. Club 39:313. 1912.
   Colorado to Nevada, south to New Mexico and Arizona. Reported from southern Colorado but no specimens seen by the writer.

5. **Amaranthus albus** L., Syst. ed. 10:1268. 1759.
   A. graecizans of manuals---Stems 20-60 cm. tall, stout and erect but bushy-branched, the branches divaricate or ascending, whitish, glabrous or sparingly puberulent or villous; leaves 1-4 cm. long, slender petioled, oblong, spatulate or obovate, cuneate at base, rounded or mucronate-cuspidate at apex, glabrous and papillose or sparingly puberulent; flowers monoecious or polygamous, in dense or loose axillary clusters; bracts 2-4 times longer than the perianth, pungent-pointed and spreading; sepals 3, membranous; stamens 3; fruit subglobose, circumscissle, rugose, longer than the perianth; seed round, about 0.8 mm. in diameter. Intergrades with A. pubescens in Colorado.---Waste places and cultivated areas. Widely distributed in North America, Europe, Asia, Africa and South America. Our records scattered over Colorado, mostly from the northern half of the state, at 4500-6000 feet.

6. **Amaranthus wrightii** S. Wats., Proc. Am. Acad. 12:275. 1877.
   Colorado, south to New Mexico and Arizona. Reported from southern Colorado but no specimens seen by the writer.

7. **Amaranthus retroflexus** L., Sp. Pl. 991. 1753.
   Stems 30-150 cm. tall, erect, stout, sometimes taller in favorable situations, usually much branched, abundantly villous especially above, often densely so in the inflorescence; leaves with slender petioles, ovate, rhombic-ovate to lanceolate, acute, obtuse or emarginate at apex, cuneate or rounded at base, glabrate above, more or less villous beneath; flowers monoecious, in densely crowded spikes 8-20 mm. wide, these in panicles, or some axillary; bracts about twice as long as the sepals, tapering to stout spinose tips; sepals of staminate flowers ovate-oblong to lanceolate, acute or acutish, nerve excurrent; sepals of pistillate flowers 3 mm. long, linear-oblong, rounded to truncate at apex, usually emarginate, the midrib excurrent; stamens 5; fruit subglobose, circumscissle; seed round, 1 mm. wide. A. powellii S. Wats. appears to be merely a more glabrous form of the above.---Waste places and fields. Widely distributed from southern Canada to Mexico; Eurasia and Africa. Our records scattered over Colorado at 4500-8500 feet.

8. Amaranthus hybridus L., Sp. Pl. 990. 1753.
Stems 30-150 cm. tall, stout erect or ascending, usually branching, glabrous to rough-pubescent and usually somewhat villous above; leaves long-petioled, blades rhombic-ovate, ovate to lanceolate, acute or rounded at apex, cuneate or rounded at base, pubescent or glabrous; flowers monoecious, in spikes, these rather slender and 6-12 mm. wide, in terminal panicles and also axillary; bracts about twice as long as the sepals or less, spinulose-tipped; sepals of staminate flowers narrowly oblong to ovate, acute, the midnerve excurrent; sepals of pistillate flower 1.5-2 mm. long, oblong or linear-oblong, acute the nerve usually excurrent; fruit subglobose, circumscissle; seed round, about 1 mm. wide.---Waste places and fields. A weed now widely distributed in North America and throughout the world. Our one certain record from northcentral Colorado at about 4700 feet.

3. Tidestromia Standl.

Annual or perennial herbaceous plants; stems branched, stellate; leaves opposite, petiolate, entire; flowers perfect, glomerate in the axils of the leaves; bracts and bractlets hyaline and pubescent; perianth 5-parted; stamens 5, filaments connate only at base; fruiting calyx not winged, crested or spiny.

1. Tidestromia lanuginosa (Nutt.) Stand., Journ. Wash. Acad. Sc. 6:70. 1916.
Cladothrix lanuginosa Nutt.---Annual plants; stems diffusely branched, the branches 10-60 cm. long, prostrate or ascending; petioles equalling or shorter than the blades; leaf blades 5-30 mm. long, oval to orbicular or rhombic-ovate, round or obtuse at apex, cuneate to rounded at base, both sides densely stellate or sometimes glabrate; perianth 1-3 mm. long, longer than the bracts.---Dry ground. Kansas to Utah, south to Texas, Mexico and Arizona. Our records from southern Colorado at 4000-5500 feet.

4. Froelichia Moench. SNAKECOTTON

Annual or perhaps biennial herbaceous plants; stems woolly or silky; leaves opposite, and entire; flowers perfect, sessile, in spikes; perianth 5-lobed, the tube lanate, the lobes glabrate, longitudinally crested or tubercled in fruit; stamens 5, filaments united into a tube; fruit indehiscent, included in the tube of the filaments.

1. Stem branched near base, the erect portion little longer than the lower ascending or prostrate branches; wings of the calyx tube at maturity divided to the base, forming distinct spines -1. F. gracilis
1. Stem erect, usually simple at base and sparsely branched above; wings of the calyx tube with a continuous, dentate margin -2. F. floridana

1. Froelichia gracilis (Hook.) Moq. in DC., Prodr. 13(2):420. 1849.
Stems slender, much branched at the base, these branches 20-50 cm. long, ascending or procumbent, densely or sparsely villous-tomentose; leaves more numerous at base of plant, short-petiolate, blades 1-5 cm. long, linear-oblanceolate to lanceolate-elliptic, acute to acuminate at apex, cuneate at base, sericeous or tomentose, the hairs sparse above; spikes 1-3 cm. long; bracts stramineous, fuscous or blackish; calyx tube at maturity with 2 rows of distinct or almost distinct spines.---Sandy valleys and plains. Iowa to Colorado, south to Arkansas, Arizona and Mexico. Our records scattered in the eastern half of Colorado at 3500-6000 feet.
2. Froelichia floridana (Nutt.) Moq. in DC., Prodr. 13(2):420. 1849.
F. campestris Small---Stems 40-100 cm. tall, rather stout, erect, usually simple at base and sparsely branched above, sericeous-tomentose with white or brownish hairs; leaves short-petiolate, blades 3-10 cm. long, oblanceolate, spatulate to oblong, obtuse to acute at apex, cuneate at base, canescent to subscabrous above, sericeous-tomentose below; spikes 1-10 cm. long; bracts fulvous or blackish; calyx tube narrowly winged at maturity, the wings dentate, one or both sides, with 1 or more basal spines. Our plant is var. campestris (Small) Fernald.---Dry often sandy soil. The variety from Illinois to Colorado, south to Missouri and Oklahoma. Our rather few records scattered in the eastern half of Colorado at 4000-6000 feet.

Family 41. Nyctaginaceae FOUR-O'CLOCK FAMILY

Annual or perennial herbaceous plants; stems usually swollen at the joints; leaves simple, usually opposite, entire, lacking stipules; flowers in terminal or axillary clusters, perfect, regular, subtended by bracts, these often united into a calyxlike involucre, perianth 3- to 5-lobed, corollalike, tube short or long, the lower part persistent and closely investing the fruit; petals wanting; stamens 1 to many; pistil 1, ovary superior, 1-celled, 1-ovuled; style short or long; fruit indehiscent, usually angled, ribbed or winged, perianth tube a part of it.

1. Flowers many in a head surrounded by 4-6 separate bracts; stigmas linear or fusiform; anthers included in the perianth tube
    2. Wings of fruit thin, nearly transparent, continuous around the body of the fruit; perianth limb 4- (rarely 5-) parted; annual plants -1. Tripterocalyx
    2. Wings of the fruit thick, opaque, interrupted above and below (may even be absent); perianth limb 5-parted; perennial plants -2. Abronia
1. Flowers in clusters of 1 to 10, surrounded by united bracts or subtended by 3 bracts united at base; stigmas capitate or hemispheric; anthers not included in the perianth tube

3. Fruit strongly compressed, the winglike and usually toothed margins inflexed over the dorsal (outer) face and with 2 rows of tubercles or glands on the dorsal face; flowers subtended by a 3-parted involucre -3. **Allionia**
3. Fruit terete or angled, not strongly compressed, the margins not winglike, not dentate and inflexed over one face, no tubercles or glands present; flowers surrounded by a 4- to 5-lobed involucre   -4. **Mirabilis**

1. Tripterocalyx (Torr.) Hook.   SANDPUFFS

Annual succulent herbaceous plants; stems branched; leaves opposite, those of the pair unequal, long-petioled; flowers capitate, the heads many-flowered, surrounded by an involucre of 4-6 distinct bracts; perianth tubular-funnelform or salverform, the tube elongated and the limb 4- rarely 5-parted; stamens 4 or 5, adnate to the perianth tube; fruit fusiform with 2-4 broad, thin, nearly transparent, conspicuously reticulated wings completely surrounding the body, the whole nearly orbicular. This genus is considered a section of Abronia by many botanists but in our area they seem distinct as considered by Standley (N. A. Flora 21:240-242. 1918).

1. Perianth about 1.5 cm. long or less, the limb greenish, white or pink   -1. **T. micranthus**
1. Perianth 2 cm. long or more, the limb pink outside, white inside   -2. **T. wootonii**

1. **Tripterocalyx micranthus** (Torr.) Hook., Journ. Bot. & Kew. Misc. 5:261. 1853.
   Abronia micrantha of Manuals---Stems 10-40 cm. long, branched at base, ascending or procumbent, viscid-pubescent to glabrate; leaves 2-5 cm. long, elliptic, lanceolate-oblong to ovate, broadly cuneate or truncate at base, rounded or obtuse at apex, often glaucous below, scabrous or glabrate; bracts 4-8 mm. long, lanceolate or ovate, acuminate; perianth about 1.5 cm. long, viscid-puberulent outside to glabrate, the limb greenish-white or pinkish; fruit 1-3 cm. long, usually 3-winged.---Dry often sandy ground. North Dakota to Montana and Nevada, south to Kansas and Arizona. Our records scattered over Colorado at 4000-8000 feet.
2. **Tripterocalyx wootonii** Standl., Contrib. U. S. Nat. Herb. 12:329. 1909.
   Stems 18-50 cm. tall, branched at base, ascending or decumbent, puberulent; leaves 2-5 cm. long, lanceolate-elliptic to oval, cuneate below, rounded or obtusish at apex, more or less pubescent; bracts 8-10 mm. long, ovate to ovate-lanceolate, acuminate; perianth 2.5-3 cm. long, limb about 1 cm. wide, puberulent outside, reddish especially on tube and throat, limb whitish except on broad midrib outside which is greenish-red; fruit 1.5-2 cm. long, usually 3-winged. This species is doubtfully separate from T. cyclopterus (A. Gray) Standley, Contrib. U. S. Nat. Herb. 12:329. 1909.---Dry often sandy ground. New Mexico and Arizona. Our record from the southwestern corner of Colorado at 5500 feet.

2. Abronia Juss.   SANDVERBENA

Perennial herbaceous plants; stems branched, leafy or scapose; leaves opposite, petiolate, the pair usually unequal; flowers capitate, few to many in a head, the heads long-peduncled, surrounded by 5 or more distinct scarious bracts; perianth funnelform to salverform, the tube slender and elongate and the limb 5-lobed; stamens usually 5; fruit leathery to indurate, deeply lobed or winged, these 2-5 in number, but not completely surrounding the body.

1. Perianth 13-15 mm. long; bracts 5-7 mm. long, acute or acuminate at apex; fruit 4-5 mm. long   -1. **A. carletoni**
1. Perianth 15-30 mm. long; bracts 8-20 mm. long, acute, acuminate or obtuse (if shorter than 8 mm. then rounded and sometimes apiculate); fruit 5-10 mm. long
   2. Stems viscid-villous above
      3. Fruit turbinate, truncate and wide at apex, about as wide as long; whitish-straw-colored; plants of western Colorado   -2. **A. salsa**
      3. Fruit biturbinate, narrowing to apex and base, longer than broad, dark olive or brown; plants mostly in eastern part of Colorado   -3. **A. fragrans**
   2. Stems glabrous or viscid-puberulent above, not at all villous
      4. Bracts rounded and very obtuse at apex (may be apiculate also); fruit turbinate, truncate and widest at apex; perianth 15-20 mm. long   -3A. **A. fragrans elliptica**
      4. Bracts acute or acuminate at apex; fruit biturbinate, narrowed to base and apex; perianth 20-30 mm. long   -3B. **A. fragrans glaucescens**

1. **Abronia carletoni** Coult. & Fisher, Bot. Gaz. 17:349. 1892.
   Stems 20-50 cm. tall, few to many, ascending or procumbent, viscid-puberulent; petioles shorter than the blade; leaf blades 1-4 cm. long, linear-lanceolate to oblong-ovate, obtuse to acuminate at apex, rounded to cuneate at base, thick, puberulent or glabrous; bracts 5-8 mm. long, ovate or oval-ovate, acute or attenuated at apex, whitish or rose-colored, viscid-puberulent; perianth 13-15 mm. long, rose-colored to white; fruit biturbinate, 4-5 mm. long, short-villous or puberulent, 5-lobed and reticulate-veined, narrowed but sometimes subtruncate at apex.---Open dry ground. Colorado to Texas. Our few records from eastern Colorado at about 5000-5500 feet.
2. **Abronia salsa** Rydb., Bull. Torr. Club 29:684. 1902.
   Stems 20-50 cm. tall, few to many, erect or decumbent, whitish or straw-colored, short-pubescent or short-villous above; petioles 1-4 cm. long; leaf blades 2-5 cm. long, oblong, oval or oblong-ovate, but sometimes lanceolate or ovate, densely viscid-puberulent below, often above; flowers many in a head; bracts 10-16 mm. long, broadly oval or ovate; perianth 18-25 mm. long, tube greenish, limb white; fruit turbinate, 5-10 mm. long, whitish-stramineous, lobed, the lobes compressed and winglike, truncate and flattened at apex. This species is closely related to the following one.---Open usually sandy ground. Colorado and Utah. Our few records from southwestern and westcentral Colorado at 4500-5500 feet.

3. **Abronia fragrans** Nutt. ex Hook., Journ. Bot. & Kew. Gard. Misc. 5:261. 1853.

Stems 20-100 cm. long, erect, or procumbent, branched, often whitish in color, viscid-puberulent or villous below, or rarely glabrate, densely viscid-villous above; leaves 2-9 cm. long, variable in shape, mostly ovate-oblong or narrowly deltoid-ovate, truncate to rounded or rarely subcordate at base, rounded or acute at apex, viscid-puberulent to glabrous; bracts 1-2 cm. long, oval-ovate, oval, ovate or obovate, acute or acuminate; flowers numerous; perianth 2-3 cm. long, white or sometimes greenish-white; fruit 5-10 mm. long, longer than wide, usually biturbinate, narrowing both ways from middle, lobes narrowly winged or crested. Intergrades in Colorado with its 2 varieties especially in western Colorado.---Dry ground. South Dakota to Idaho, south to Texas and Mexico. Our records scattered over Colorado at 3500-8000 feet.

3A. **Abronia fragrans elliptica** Heimerl., (var.) in Rydb., Bull. Torr. Club 29:684. 1902.

A. elliptica A. Nels.; *A. glabra* Rydb.---Stems usually 10-50 cm. long, sometimes longer, erect, or decumbent, usually branched, whitish or tinged with red, glabrous or viscid-puberulent especially above; leaves 1-5 cm. long, variable mostly oval, oblong-oval or elliptic, rounded or truncate at base, sometimes subcordate, rounded or obtuse at apex, glabrous or minutely puberulent; bracts 8-15 mm. long, broadly oval or obovate-oval, rounded at apex, sometimes apiculate also; flowers numerous; perianth 15-20 mm. long, with greenish-white or pink tube and whitish limb; fruit turbinate, wide and truncate near apex, or biturbinate on outer ones, 5-8 mm. long, with winglike lobes.---Dry ground. Wyoming, south to New Mexico and Arizona. Our records scattered over Colorado at 3500-8000 feet.

3B. **Abronia fragrans glaucescens** A. Nels., (var.) Bot. Gaz. 34:364. 1902.

A. glaucescens (A. Nels.) Standl.---This variety differs from the species in lacking the viscid-villous hairs on the upper part of the stem, which may be glabrous or sparsely puberulent. ---Wyoming and Colorado. Our few records from northcentral and southcentral Colorado at 4500-7000 feet.

3. Allionia L. UMBRELLA-WORT

Annual or perennial plants; stems prostrate, dichotomously branched, pubescent; leaves opposite, the pair unequal in size, petiolate, entire or sinuate; flowers perfect, in axillary, pedunculate clusters of 3, each cluster subtended by a bract which encloses the fruit; perianth short-funnelform, 4- or 5-lobed with an oblique tube; stamens 4-7; stigma capitate, style capillary; fruit coriaceous, compressed, the margin entire or toothed, the dorsal (seemingly the inner) face bearing 2 rows of stipulate glands.

1. Plants perennial; outer margin of fruits with about 3 broadly triangular non-glandular teeth on each side, usually incurved and covering nearly the entire surface -1. *A. incarnata*
1. Plants annual; outer margin of fruit with several relatively slender gland-tipped teeth on each side, these spreading or moderately incurved -2. *A. choisyi*

1. **Allionia incarnata** L., Syst. Nat. ed. 10, 2:890. 1759.

Wedeliella incarnata (L.) Cockerell; *Wedelia incarnata* (L.) Kuntze---Perennial plants; stems 10-50 cm. long, numerous, densely villous-viscid or glandular-puberulent; leaves 1-4 cm. long, petioled, oval, ovate, to oblong, subcordate or rounded at the oblique base, apex rounded to acute, entire or sinuate, paler below, villous-viscid or glandular-puberulent at least when young; involucres on long or short peduncles, the lobes 4-6 mm. long or sometimes to 8, obovate-orbicular; perianth 5 mm. long or more, purplish-red or rarely white; fruit 3-4.5 mm. long, inner side 3-nerved, margins with 3 or rarely to 5 broad teeth on each side, rarely entire, margin incurved, glandular within.---Dry hills and valleys. Colorado to California, south to Texas and South America. Our one record from westcentral Colorado at 4300 feet.

2. **Allionia choisyi** Standl., Field Mus. Nat. Hist. Bot. Ser. 8:310. 1931.

A. glabra (Choisyi) Standl.---Annual plants; stems 20-80 cm. long, slender, puberulent or villous above, little viscid, often glabrate below; petiole 3-18 mm. long; leaf blades 1-4 cm. long, ovate-oval, oval or oblong, subcordate to broadly cuneate at base, rounded or obtuse to nearly acute at apex, whiter below, puberulent or glandular-puberulent at first, usually soon glabrate; involucres 5-7 mm. long, puberulent and usually short-villous, often viscid; fruit 3-4 mm. long, pale brown or olive, 3-costate on the inner surface, shallowly rugose, margin spreading or somewhat incurved each with 5-8 long slender teeth, these gland-tipped.---Dry sandy ground. Texas to Arizona and southward. Our one record from southeastern Colorado at 4200 feet.

4. Mirabilis L. FOUR-O'CLOCK

Plants perennial; leaves opposite, sessile or petiolate; involucres axillary or terminal, 1- to 10-flowered, calyxlike, of united bracts, sometimes becoming enlarged and papery in fruit; perianth campanulate, salverform or funnelform; stamens 3-5; fruit smooth or conspicuously 5-angled or 5-ribbed.

1. Fruits smooth or slightly 5-ribbed, not 5-angled, usually not constricted at base, glabrous or nearly so; involucre in fruit little enlarged, remaining leaflike in texture; leaves usually broadly ovate
    2. Perianth 3-6 cm. long; involucre campanulate, 3- to 10-flowered; stamens 5, the filaments connate at base -1. *M. multiflora*
    2. Perianth less than 1 cm. long; involucre subrotate, 3-flowered; stamens 3, the filaments free -2. *M. oxybaphoides*
1. Fruits strongly 5-angled longitudinally, constricted at base, pubescent in most species; involucres enlarging and becoming membranous in fruit; leaves various but often narrower than ovate
    3. Leaves linear or linear-lanceolate, less than 1 cm. wide, usually sessile or short-petioled

4. Flowers solitary, sometimes 2 in an involucre; perianth about 7 mm. long, white or pale pink; stems and fruits glabrous -3. M. glabra
4. Flowers 3 to an involucre; perianth 8-10 mm. long, pale pink to purplish-red; stems usually puberulent or short-villous at least near the nodes; fruit pubescent or sparsely strigose
   5. Stems glaucous; leaves sessile or gradually narrowed to short, stout petioles, the blades glaucous or glaucescent below -4. M. linearis
   5. Stems green; leaves, at least the lower ones rather abruptly contracted to slender petioles, bright green on both sides -5. M. decipiens
3. Leaves lanceolate or wider, the widest over 1 cm. broad, often long-petioled
  6. Leaves sessile or on short petioles not over 5 mm. long, the blades ovate-oblong to lanceolate (may be wider in M. carletoni)
    7. Stems glabrous or merely puberulent below (may be pilose above); leaves lanceolate to ovate-lanceolate -6. M. lanceolata
    7. Stems densely short-pilose or hirsute below, at least near the nodes and usually throughout; leaves often wider than ovate-lanceolate
      8. Fruit glabrous; stems viscid-pilose below and near the nodes -7. M. carletoni
      8. Fruit pubescent (at least sparsely so); stems hirsute at least below -8. M. hirsuta
  6. Leaves on slender petioles, those of the lower leaves over 8 mm. long, the blades usually deltoid-ovate or broader
    9. Stems densely hirsute below -9. M. rotundifolia
    9. Stems glabrous or puberulent below (may be pilose in the inflorescence)
      10. Involucres glabrous, sparsely strigose or pubescent only at base; stems often numerous from the root -10. M. nyctaginea
      10. Involucres densely viscid-pilose all over; stems solitary or few from the root -11. M. comata

1. **Mirabilis multiflora** (Torr.) Gray in Torr., Bot. Mex. Bound. Surv. 173. 1859.
Quamoclidion multiflorum Torr.---Plants 30-100 cm. high, erect or spreading, forming clumps, usually stout, densely leafy, glaucous or glaucescent, pubescent, often viscid to glabrate; petioles about 1/2 the length of the blade or shorter; blades 3-7 cm. long, broadly ovate or ovate-oblong, cordate or rounded at base, acute to rounded and apiculate at apex, thick, glabrous to pubescent, often glandular; peduncles solitary in axils and cymose at the ends of the branches; involucre 1.5-3.5 cm. long, campanulate, usually 6- to 8-flowered, 5-lobed; perianth 3-5 cm. long, rose-colored to purplish-red, shallowly 5-lobed; stamens 5; fruit 6-8 mm. long, smooth or slightly furrowed at base. The viscid-glandular plant mostly of the southwestern corner of Colorado has been given subspecific rank.---Dry slopes and plains. Colorado to Utah, south to Mexico. Our records mostly from the southern half of Colorado, except in the extreme eastern part, at 4500-8000 feet.

2. **Mirabilis oxybaphoides** Gray in Torr., Bot. Mex. Bound. Surv. 173. 1859.
Allionella oxybaphoides (A. Gray) Rydb.---Stems usually 30-60 cm. tall, ascending or decumbent, much branched, usually forming clumps 40-120 cm. in diameter, viscid-pubescent to glabrate; leaves 1-5 cm. long, deltoid or ovate, often broadly so, usually cordate at base but sometimes truncate, acute to acuminate at apex, entire or undulate, viscid-pubescent to glabrate; inflorescence cymose or axillary; peduncles of involucres solitary, slender, usually longer than the involucres; involucre 5-6 mm. long in anthesis, subrotate, deeply 5-cleft; perianth 7-9 mm. long, sparsely hairy or glabrate; fruit 2.5-3 mm. long, smooth or obscurely ridged transversely or black-spotted.---Dry hillsides and valleys. Colorado to Utah, south to Texas and Arizona. Our records from the southern half of Colorado, except in the extreme eastern part, at 6000-8000 feet.

3. **Mirabilis glabra** (Wats.) Stand., Field Mus. Nat. Hist. Bot. Ser. 8:304. 1931.
Allionia glabra (S. Wats.) Kuntze---Kansas to Utah, south to Mexico. This plant should be in Colorado but no specimens could be located by the writer.

4. **Mirabilis linearis** (Pursh) Heimerl., Ann. Cons. Jard. Geneve 5:186. 1901.
Allionia linearis Pursh; A. linearis bodinii (Holz.) A. Nels.; A. divaricata Rydb.; A. glandulifera A. Nels.; A. decumbens Nutt. of Manuals; A. pilosa decumbens (Nutt.) A. Nels. in part; A. diffusa Heller---Stems 20-100 cm. tall, erect or decumbent, simple or branched below, usually glaucous, often very whitish, glabrous or puberulent below, viscid-puberulent or short-pilose above; leaves 3-10 cm. long, linear or linear-lanceolate, attenuate and sessile or long-cuneate at base to a short petiole, narrowed to an obtuse or acute apex, thick, usually glaucous at least below, glabrous or viscid-puberulent; inflorescence axillary or cymose-paniculate; involucres about 4 mm. long at anthesis, densely viscid-villous, usually with 3 flowers; perianth about 10 mm. long, purple-red to pink; fruit 4-5 mm. long, 5-angled, densely pubescent to sparingly strigose, the sides transversely rugose.---Dry ground. South Dakota to Montana, south to Missouri, Texas and Mexico; sometimes adventive eastward. Our records well scattered over Colorado at 3500-10,000 feet.

5. **Mirabilis decipiens** (Standl.) Standl., Field Mus. Nat. Hist. Bot. Ser. 8:305. 1931.
Colorado to New Mexico and Arizona. This species has been reported for southern Colorado but no specimens could be located by the writer.

6. **Mirabilis lanceolata** (Rydb.) Standl., Field Mus. Nat. Hist. Bot. Ser. 8:305. 1931.
Allionia lanceolata Rydb.; A. sessilifolia Osterh.---Stems 40-100 cm. high, few, erect or ascending, simple or sparsely branched above, glabrate or sparsely puberulent below, short-pilose above; leaves sessile or short-petioled, this not over 4 mm. long; blades 3-10 cm. long, lanceolate or ovate-lanceolate, rarely almost ovate, rounded to cuneate at base, obtuse or acute at apex, glabrous or sparsely short-pilose or strigose; inflorescence paniculate, much branched; involucres about 4 mm. long in anthesis, usually 3-flowered, densely viscid-pilose; perianth about 8 mm. long, pink to purple, sparsely pilose; stamens 5; fruit 4-5 mm. long, 5-angled, pubescent, the side rugose or short-tuberculate.---Dry ground. Wyoming to Colorado, possibly further south. Our records from northcentral Colorado at 4500-8500 feet, but a doubtful specimen from the southeastern part of the state.

7. **Mirabilis carletoni** (Standl.) Standl., Field Mus. Nat. Hist. Bot. Ser. 8:305. 1931.
   Allionia carletoni Standl.---Stems 60-120 cm. tall, few or solitary, simple or sparsely branched above, densely viscid-pubescent throughout; leaves sessile or subsessile, the petioles if present not over 4 mm. long; leaf blades 4-8 cm. long, deltoid-ovate, to broadly ovate or ovate-oblong, sometimes lanceolate, subcordate to rounded at base, obtuse or acute at apex, thick, short-pilose or sometimes glabrate, margins entire or sometimes wavy; inflorescence paniculate, with opposite branches and small bracts; involucres 5-6 mm. long at anthesis, enlarging and reticulate in fruit, viscid-pilose; flowers usually 3 in each involucre; perianth pink and pubescent; stamens usually 3; fruit obovoid, 5 mm. long, glabrous, brown, smooth or rugose or tuberculate on the sides, 5-angled.---Dry plains. Kansas, Colorado and Oklahoma. Our one record from southeastern Colorado at 4200 feet.

8. **Mirabilis hirsuta** (Pursh) Macm., Metasp. Minn. Vall. 217. 1892.
   Allionia hirsuta Pursh; *A. pilosa* (Nutt.) Rydb.---Stems 20-100 cm. tall, solitary or few, erect or decumbent, simple or sparsely branched below the inflorescence, densely hirsute or long-pilose below and usually to inflorescence, sometimes merely puberulent but always hirsute near the nodes; petioles usually wanting, sometimes to 5 mm. long; leaf blades 2-10 cm. long, variable in shape, mostly ovate-oblong, but may be lanceolate to broadly ovate, cuneate to subcordate at base, narrowed to an obtuse or acute apex, rarely rounded, densely hirsute to viscid-puberulent or glabrate; inflorescence both axillary and cymose-paniculate; involucres 4-5 mm. long in anthesis, densely to sparsely viscid-pilose, usually 3-flowered; perianth 8-10 mm. long, pink to purplish-red, sparsely pilose; stamens 3-5; fruit 5-angled, 4-5 mm. long, densely pubescent, the sides rather rugose.---Dry often sandy ground. Manitoba, south to Missouri, Texas and New Mexico. Our records scattered in the eastern half of the state, except in the extreme eastern part, at 5000-8000 feet.

9. **Mirabilis rotundifolia** (Greene) Standl., Field Mus. Nat. Hist. Bot. Ser. 8:305. 1931.
   Allionia rotundifolia Greene; *A. polyatricha* Standl.---This species is endemic to south-central Colorado but no specimens have been seen by the writer.

10. **Mirabilis nyctaginea** (Michx.) MacM., Metasp. Minn. Vall. 217. 1892.
    Allionia nyctaginea Michx.; *A. floribunda* (Choisy) Kuntze---Stems 30-100 cm. tall, ascending, often numerous from a somewhat fleshy root, glabrous or sometimes puberulent, simple or sparsely branched; leaves petiolate, these 1-3 cm. long on lower leaves; leaf blades 2-10 cm. long, mostly deltoid to ovate-deltoid, sometimes ovate-oblong, cordate to rounded at base, acute to acuminate or rarely rounded at apex, green and glabrous on both sides or slightly puberulent; inflorescence axillary or in cymes, these small and dense; involucres 3-flowered, 5-6 mm. long in anthesis, glabrous, sparsely strigose or pubescent near the base; perianth about 10 mm. long, glabrous or sparsely villous, red or pink, rarely white; stamens 3-5; fruit 5 mm. long, densely short-pilose, 5-angled, more or less rugose.---Open ground. Wisconsin to Montana, south to Texas and Mexico; adventive eastward. Our records from the eastern half of Colorado at 4000-6000 feet.

11. **Mirabilis comata** (Small) Standl., Field Mus. Nat. Hist. Bot. Ser. 8:306. 1931.
    Allionia comata Small---Stems 30-100 cm. tall, few or solitary from a woody root, usually erect, usually not branched below the inflorescence, glabrous below or puberulent usually in lines, pilose or short-pilose in the inflorescence; petioles 1-5 cm. long; leaf blades 3-10 cm. long, elongate-deltoid, deltoid-ovate or sometimes ovate-lanceolate, truncate or subcordate at base, obtuse, acute or attenuate at apex, glabrous to sparsely viscid-puberulent; inflorescence cymose-paniculate; involucres 3-5 mm. long at anthesis, densely viscid-pilose, the hairs often blackish and jointed; perianth about 10 mm. long, purplish-red, sparingly pilose; stamens 3-5; fruit 3-5 mm. long, minutely pilose, 5-angled and usually tuberculate.---Meadows and thickets. Texas to Arizona and south into Mexico. Our few records from southern Colorado at 8500-9000 feet.

Family 42. Aizoaceae   CARPET-WEED FAMILY

Annual herbaceous plants; stems rather succulent, prostrate; leaves opposite or verticillate, simple, entire; flowers axillary, solitary or in small clusters, small, perfect and regular; sepals 4-5; corolla none; stamens 1 to many; ovary superior, 3- to 5-celled, styles as many as cells of the ovary; fruit a capsule.

1. Calyx of distinct sepals; leaves whorled; capsule longitudinally dehiscent; stamens 3-5   -1. Mollugo
1. Sepals united at base; leaves opposite; capsule circumscissle; stamens many   -2. Sesuvium

1. Mollugo L.   CARPET-WEED

Plants not very succulent; stems branched; leaves apparently verticillate, mostly in 3's and 6's; sepals 5, persistent, with hyaline margins; stems 3-5; ovary 3- to 5-celled; fruit a longitudinally dehiscent capsule, thin-walled; seeds with curved embryos.

1. **Mollugo verticillata** L., Sp. Pl. 89. 1753.
   Plants glabrous throughout; stems 5-30 cm. long, branched at base and dichotomously so above, prostrate; leaves in whorls of 3-8, 1-2.5 cm. long, spatulate to linear-oblanceolate, obtuse to acute at apex, cuneate below to a short petiole; flowers 2-5 from each node, on slender pedicels 3-14 mm. long; sepals 1.5-2.5 mm. long, oblong; capsule slightly exceeding the sepals; seeds reniform, ridged on the back and sides.---Waste places and cultivated ground. Widely distributed in North America; Eastern Hemisphere. Apparently not common in Colorado, our few records scattered over the state at about 4500-9000 feet.

2. Sesuvium L. SEA PURSLANE

Plants with fleshy stems; leaves opposite, fleshy; flowers solitary, sessile or on short pedicels; calyx tube turbinate, lobes 5, usually horned on the back near the apex; stamens 1 to many; ovary 3- to 5-celled, half-inferior; capsule 3- to 5-celled, circumscissle; seeds with annular embryo.

1. Sesuvium verrucosum Raf., New Fl. 4:16. 1838.
   S. sessile Pers.---Glabrous plants; stems 10-70 cm. long, much branched, prostrate or ascending; leaves 1-3 cm. long, broadly spatulate to linear, rounded or acutish at apex, base narrow but clasping; calyx lobes 5-7 mm. long, ovate-lanceolate, scarious-margined, short-horned near the apex; stamens many.---River banks and valleys. Arkansas to California, south to tropical America. Our few records from southcentral Colorado at 6000-7500 feet.

Family 43. Portulacaceae PURSLANE FAMILY

More or less succulent, annual or perennial herbaceous plants; leaves alternate, opposite or basal, simple, entire; flowers perfect, regular or nearly so; sepals commonly 2 (in Lewisia 6-8); petals distinct, mostly 4 or 5, imbricate; stamens usually of the same number as the petals and opposite them; ovary superior or sometimes partly inferior, 1-celled, styles 2-5 or sometimes more; fruit a loculicidal or circumscissle capsule; seeds 1 to many.

1. Ovary partly inferior; capsule circumscissle, the calyx lobes coming away with the top of the ovary; annual plants -1. Portulaca
1. Ovary superior; capsule valvate or circumscissle but the calyx lobes remaining on the plant; mostly perennial plants
    2. Capsule circumscissle near the base; plants with fleshy roots seldom over 1 cm. in diameter, or with a globose corm and sepals to 4 mm. long and petals to 6 mm. long -2. Lewisia
    2. Capsule splitting longitudinally by 3 valves; underground parts various, if a fleshy root then this often over 1 cm. in diameter, or if a globose corm with sepals and petals longer than the above
        3. Sepals deciduous; seeds many; leaves alternate, crowded at base of stem, terete or semiterete -3. Talinum
        3. Sepals persistent; seeds 1-6; stem leaves opposite, leaves fleshy but flat
            4. One pair of stem leaves present, these not connate; plants perennial by thick taproots or globose corms -4. Claytonia
            4. Over 1 pair of stem leaves present or if 1 pair these partially or completely connate at base; plants annual or perennial by runners ending in bulblets -5. Montia

1. Portulaca L. PURSLANE

Annual more or less succulent plants; stems diffuse or ascending; leaves alternate or opposite; flowers solitary or crowded at the top of the stem and branches; sepals 2, united below and partly adnate to the ovary; petals 4-6, mostly 5, inserted on the calyx; stamens 8 to many, inserted on the calyx; styles 3-8, slender; ovary partly inferior; capsule circumscissle.

1. Petals pink or red-purple; plants with tufts of long-crinkly hairs in the axils of the leaves; leaves linear, subterete -1. P. parvula
1. Petals yellow; plants glabrous or with a few inconspicuous hairs in the axils of the leaves; leaves spatulate or broader, fleshy but flat
    2. Seeds (with a hand lens) conspicuously and sharply tuberculate, almost echinate, 0.9-1 mm. wide; leaves often retuse or emarginate at apex -2. P. retusa
    2. Seeds minutely and not sharply granulate, seldom over 0.8 mm. wide; leaves rounded or truncate at apex -3. P. oleracea

1. Portulaca parvula A. Gray, Proc. Amer. Acad. 22:274. 1887.
   P. pilosa of Manuals---Stems usually less than 10 cm. long, usually many and corymbosely branched, hairy in the axils of the leaves; upper leaves rather persistent on drying, lower usually deciduous; leaf blades 5-14 mm. long, nearly terete, linear, obtuse at apex, petioles very short; inflorescence terminal and axillary, flowers solitary or in clusters, surrounded by long hairs and an involucre of leaves; sepals 2.5-3 mm. long, triangular or triangular-orbicular; petals 3-4 mm. long, red; capsule 1.8-2.5 mm. wide, circumscissle a little below the middle, the lid straw-colored, whitish or rose-tinged; seeds 0.4-0.5 mm. wide, black, often tuberculate.---Sandy soil. Missouri to Colorado, south to Mexico and California. Our one record from southeastern Colorado at 4200 feet.
2. Portulaca retusa Engelm., Bost. Journ. Nat. Hist. 6:154. 1850.
   Arkansas to Utah, south to Texas and Arizona. This plant should be in Colorado but no specimens seen.
3. Portulaca oleracea L., Sp. Pl. 445. 1753.
   Stems prostrate or ascending, fleshy, axillary hairs few and inconspicuous, branches 6-30 cm. long or more; leaves 0.6-3 cm. long, fleshy but flat, obovate-cuneate or spatulate, rounded or truncate at apex; flowers solitary or clustered; sepals 3-4.5 mm. long, ovate to orbicular, keeled; petals 3-4.5 mm. long, yellow, opening in sunshine, soon closing; stamens 6-10; style lobes 4-6; capsule 4-9 mm. long, circumscissle about at the middle; seeds 0.7-0.8 mm. wide (rarely to 1 mm.), granular.---Cultivated ground and waste places. Maine to Washington, south to Florida and Mexico. Our records scattered over Colorado at 4500-8500 feet.

2. Lewisia Pursh   BITTERROOT

Small, perennial, somewhat fleshy, scapose plants with thick fleshy roots and a short caudex or a small corm; leaves mostly basal, narrow; inflorescence a 1-flowered scape or a panicle; flowers usually showy, soon withering; sepals 2-8, persistent; petals 4-18; stamens 5 to many; styles 3-8; ovary superior; capsule circumscissle at base and then splitting upward; seeds usually many.

1. Sepals 6-8, 1.5 cm. long or more; pedicels jointed to the peduncle
1. Sepals 2, 1 cm. long or less; pedicles not jointed to the peduncle   -1. L. rediviva
  2. Basal leaves absent or solitary, stem leaves 2 or 3, opposite or whorled; plants with globular corms
                                                                                          -2. L. triphylla
  2. Basal leaves numerous, stem leaves all crowded at base; plants with thick roots, not corms
    3. Sepals glandular-toothed, suborbicular, 4-5 mm.long; petals 8-10 mm.long; scapes 2-5 cm. long
                                                                                                     -3. L. pygmaea
    3. Sepals entire or nearly so, ovate, 7-10 mm. long; petals 10-15 mm. long; scapes 5-10 cm. high
                                                                                                      -3A. L. pygmaea nevadensis

1. **Lewisia rediviva** Pursh, Fl. Am. Sept. 368. 1814.
Plants with thick fleshy roots and short caudices; leaves 1-5 cm. long, densely clustered at base, linear to clavate, very fleshy and subterete, obtuse at apex; scapes 1-3 cm. long, numerous, 1-flowered with 5-8 whorled subulate scarious bracts at the base of the pedicel, this distinctly jointed to the peduncle; sepals 4-8, 1.5-2.5 cm. long, rose-colored or white; petals 12-18, 2-2.5 cm. long, rose-colored or sometimes white; stamens 35-50; styles 5-8; capsule 5-6 mm. long; seeds 2-2.5 mm. long.---Ridges and slopes often in rocky ground. Montana and British Columbia, south to Colorado and California. Our records from northern Colorado as far south as Grand County, at 7000-9000 feet.

2. **Lewisia triphylla** (S. Wats.) Robinson in Gray, Syn. Fl. N. Amer. 1(1):269. 1897.
Erocallis triphylla (S. Wats.) Rydb.---Plants with deep-seated globose corms 5-8 mm. thick; stems 1-5, slender, 3-10 cm. long; basal leaves 1, but not present at flowering time, cauline leaves 2-4, 1-5 cm. long, verticillate, linear; inflorescence umbellate or corymbosely-paniculate, 2- to 15-flowered; sepals 2, 2.5-4 mm. long, oval, entire; petals 5-8, 4-6 mm. long, white or pink; stamens 4-5; styles 3-5; capsule 3-4 mm. long; seeds about 1 mm. long.---Mountains. Montana to Washington, south to Colorado and California. Our few records from northcentral Colorado at about 10,000 feet.

3. **Lewisia pygmaea** (A. Gray) Robinson in Gray, Syn. Fl. N. Amer. 1(1):268. 1897.
Oreobroma pygmaea (A. Gray) Howell---Subacaulescent plants with fusiform, sometimes branched roots; basal leaves 2-7 mm. long, 1-3 mm. wide, numerous, linear or linear-oblanceolate; scapes 1.5-5 cm. long, bracteate at or below the middle; sepals 2, 4-5 mm. or sometimes to 6 mm. long, suborbicular, rounded or truncate and glandular-toothed at apex; petals 6-8, 8-10 mm. long, pink or white; stamens 5-8; stigmas 3-5; capsule 4-6 mm. long; seeds about 1 mm. long. Intergrades in Colorado with its variety.---Hills and slopes, often in rocky soil. Montana to Washington, south to Arizona and California. Our records scattered over the western half of Colorado at 8500-14,000 feet.

3A. **Lewisia pygmaea nevadensis** (Gray) Fosberg, (var.) Am. Midl. Nat. 27:256. 1942.
L. nevadensis (A. Gray) Robinson; Oreobroma nevadense (A. Gray) Howell---Plants with fleshy taproots about 1 cm. thick; leaves 3-10 cm. long and 2-4 mm. wide, basal, 5-15, linear to linear-spatulate; scapes 5-10 cm. tall, several, usually 1-flowered, bracted usually below the middle, these lanceolate and scarious; sepals 7-10 mm. long, ovate or broadly ovate, acute, entire or minutely denticulate; petals 5-8, 10-15 mm. long, but rather variable in length; stamens 6-12; capsule about 5-10 mm. long, ovoid; seeds many, black and shining.---Colorado to Washington, south to California. Our few records in western Colorado at 7000-12,000 feet.

3. Talinum Adans.   FAME-FLOWER

Glabrous perennial herbaceous plants with cormlike rootstocks; leaves alternate, without stipules; flowers in cymes or solitary and axillary; sepals 2, usually deciduous; petals 5, soon withering; stamens as many or more than the petals; style 3-lobed or cleft; ovary superior; capsule 3-valved; seeds many.

1. Flowers in terminal cymes; leaves 1.5 cm. long or more
  2. Flowers small, petals not over 7 mm. long; stamens 4-8   -1. T. parviflorum
  2. Flowers larger, petals 10-15 mm. long; stamens over 12   -2. T. calycinum
1. Flowers solitary in the axils of the upper leaves; leaves 10 mm. long or less   -3. T. brevifolium

1. **Talinum parviflorum** Nutt. ex T. & G., Fl. N. Amer. 1:197. 1838.
Roots fleshy; stems short, the plants subacaulescent, the scapes 5-20 cm. tall; leaves 1.5-5 cm. long, terete or nearly so, linear and somewhat broadened at base; inflorescence a loose cyme bracted at forks, pedicels slender; sepals 2-3 mm. long, ovate, deciduous; petals 5-7 mm. long or less; stamens 4-8; capsule ellipsoid, 3-5 mm. long.---Open, often rocky ground. Minnesota to North Dakota, south to Arkansas, Texas and Arizona. Our records scattered over Colorado, mostly in the eastern half, at 3500-8500 feet.

2. **Talinum calycinum** Engelm. in Wisliz., Tour. Northern Mex. 88. 1848.
Rootstocks rather thick; stems 3-10 cm. long, erect; scapes 6-20 cm. long; leaves 1-2 mm. wide and 2-7 cm. long, crowded at base of plant, subterete, narrowly linear, acute; inflorescence a terminal cyme, bracted at the forks; sepals 4-6 mm. long, ovate to orbicular, rather persistent; petals 10-15 mm. long, pink, broadly obovate; stamens 30 or more; capsule subglobose,

6-7 mm. long; seeds black and smooth.---Open often sandy ground. Missouri to Nebraska, south to Arkansas, Texas and Mexico. Our few records from northeastern Colorado at about 3500-4000 feet.
3. Talinum brevifolium Torr. in Sitgreaves, Rep. Exp. 156. 1853.
Utah, New Mexico and Arizona. This plant should be in southwestern Colorado.

4. Claytonia L. SPRING BEAUTY

Perennial, glabrous, rather fleshy herbaceous plants with deep-seated corms or fleshy taproots; basal leaves 1 or more, stem leaves 1 pair, usually opposite; inflorescence of terminal racemes, 1-2 bracts present at the base of the racemes; sepals 2, persistent; petals 5; stamens 5, opposite and adnate to the base of the petals; styles 3; ovary superior; capsule 3-valved; seeds 1-6.

1. Plants with fusiform, fleshy taproots; basal leaves numerous   -1. C. megarrhiza
1. Plants with globose corms; basal leaves few or none   -2. C. lanceolata

1. Claytonia megarrhiza (Gray) Parry in Wats., Bibl. Index 118. 1878.
Plants with large thick fleshy, purplish-red taproots, caudex thick and short; stems not over 10 cm. high, several, fleshy; basal leaves 1-5 cm. long on winged petioles, these from short to 15 cm. long, numerous, spatulate to nearly orbicular, sometimes oblanceolate; stem leaves 1-3 cm. long, opposite or alternate, linear to spatulate; inflorescence not exceeding the leaves; sepals 4-7 mm. long, ovate, acute; petals 5-10 mm. long, white to pink, clawed; capsule 4-6 mm. long; seeds 2-2.5 mm. long.---In rock crevices and rock slides in the higher mountains. Montana to Washington, south to Colorado and Utah. Our records scattered in the higher mountains of Colorado at 10,000-13,500 feet.
2. Claytonia lanceolata Pursh, Fl. Am. Sept. 175. 1814.
C. multicaulis A. Nels.; C. multiscapa Rydb.; C. rosea Rydb.---Stems 5-20 cm. tall, from globose corms 1-2 cm. broad, 1 to several stems from each corm; basal leaves present or absent, 5-10 cm. long, petioled, oblanceolate to oblong or ovate; stem leaves 2-6 cm. long, sometimes longer, linear to oblong-lanceolate, sessile, 1- to 5-ribbed; sepals 4-7 mm. long, round-ovate to oval; petals 7-12 mm. long, white or pink, rounded to emarginate at the apex; capsule as long or shorter than the sepals; seeds 2 mm. long.---Open slopes, hills and valleys especially in moist ground. Alberta to British Columbia, south to New Mexico and California. Our records well scattered in the western two-thirds of Colorado at 5000-10,000 feet.

5. Montia L. INDIAN LETTUCE

Annual or perennial, glabrous, rather succulent plants; leaves more or less fleshy, stem leaves opposite, 1 or more pairs present; inflorescence of loose axillary or terminal simple or compound racemes; sepals 2, persistent; petals 5, more or less united at base or free, often unequal in size; stamens opposite the petals; styles 3; ovary superior; capsule 3-valved from the apex; seeds 1-3. Perhaps better united with Claytonia.

1. Stems with 2 or more pairs of leaves, these not at all connate at base   -1. M. chamissoi
1. Stems with 1 pair of leaves, these more or less connate at base   -2. M. perfoliata

1. Montia chamissoi (Ledeb.) Durand & Jackson, Index Kew. Suppl. 1:282. 1903.
M. chamissonis (Esch.) Greene; Crunocallis chamissonis (Ledeb.) Rydb.---Perennial plants with floating or creeping leafy stems 10-30 cm. long, rooting at the nodes and producing runners these bearing globose cormlets at the apex; leaves 1-5 cm. long, of several pairs, oblanceolate or spatulate, sessile or short-petiolate; flowers in axillary or terminal racemes, 3- to 8-flowered, pedicels slender and recurved in fruit; sepals 1.5-3 mm. long, orbicular or obovate, unequal or nearly equal; petals 5-9 mm. long; white or pink; stamens 3-5; capsule obovoid; seeds 1.5 mm. long, muricate.---Moist or wet ground. Minnesota to Alaska, south to Iowa, New Mexico and California. Our many records from northcentral, central and southcentral Colorado at 6000-10,500 feet.
2. Montia perfoliata Donn, Hort. Cantab. 25. 1796.
Limnia depressa (A. Gray) Rydb.; L. humifusa (Howell) Rydb.---South Dakota to British Columbia, south to Colorado and California. No specimens from this state were located by the writer, although the var. depressa (Gray) Jepson is to be expected.

Family 44. Caryophyllaceae PINK FAMILY

Annual or perennial herbaceous plants; stems usually swollen at the nodes; leaves opposite, simple and entire, with or without stipules; flowers usually perfect, regular, usually cymose; sepals 4 or 5, distinct or united; petals distinct, 4 or 5, or rarely fewer or none; stamens usually as many or twice as many as the petals; ovary superior, 1-celled or incompletely 3- to 5-celled at base, ovules basal or on a free central placenta; styles 2-5, rarely 1; fruit a capsule opening by valves at least at the apex, or a 1-seeded achene or utricle. In addition to the genera treated in the key, escaped plants of Gypsophila are to be expected. G. paniculata L. or Baby's breath has been so reported in the state.

1. Sepals united, usually for most of their length forming a definite tube; petals clawed, usually with appendages at the apex of the claw
   2. Calyx lobes foliaceous, over 1 cm. long, longer than the tube; styles 5, alternate to the sepals
                                                                                                        -1. Agrostemma
   2. Calyx lobes not foliaceous, less than 1 cm. long, usually much shorter than the tube; styles 2-5, when 5 opposite to the sepals
      3. Calyx terete or 5-angled, the veins not conspicuous, not over 5; styles 2   -2. Saponaria
      3. Calyx tube with 10 or more conspicuous longitudinal nerves running both to the teeth and to the sinuses; styles 3-5
         4. Styles normally 3; capsule opening by 6, rarely 3 teeth   -3. Silene
         4. Styles normally 5 or 4; capsule opening by 10, rarely 5 teeth   -4. Lychnis
1. Sepals distinct or united only at the very base, petals when present, not definitely clawed or appendaged
   5. Stipules present
      6. Fruit a 1-seeded achene or utricle; petals wanting or very minute   -5. Paronychia
      6. Fruit a several- to many-seeded capsule; petals present and definite
         7. Styles and capsule valves 5; leaves much fascicled in the axils   -6. Spergula
         7. Styles and capsule valves 3; leaves slightly if at all fascicled   -7. Spergularia
   5. Stipules not present
      8. Petals deeply cleft or very deeply emarginate at apex (rarely absent); capsule with twice as many valves as styles
         9. Capsule elongate, cylindric, often curved, opening at the apex with 10 (rarely 8) valves; styles usually 5   -8. Cerastium
         9. Capsule ovoid or oblong, short and straight, splitting into 6 valves; styles usually 3   -9. Stellaria
      8. Petals entire or slightly emarginate at apex, not cleft; capsule with as many valves as styles (but each valve split at apex in some species of Arenaria and hence appearing to be twice as many)
         10. Styles 5, rarely 4; valves of fruit 5 rarely 4   -10. Sagina
         10. Styles 3; valves of fruit 3 (or each split and fruit appearing 6-valved)   -11. Arenaria

1. Agrostemma L. CORNCOCKLE

Annual plants; stems hairy; leaves linear to linear-lanceolate, sessile; flowers solitary on the ends of the branches or on long axillary peduncles; calyx oblong, not inflated, 5-lobed, the lobes linear, elongated and foliaceous; petals 5, shorter than the sepal lobes, clawed but not appendaged; stamens 10; styles 5, alternate to the sepals and opposite to the petals; capsules 1-celled; seeds many.

1. Agrostemma githago L., Sp. Pl. 435. 1753.
Weed in grain fields and waste places. Widely distributed in Canada and the United States. This plant should be in Colorado and has been so reported, but no specimens have been found by the writer.

2. Saponaria L. SOAPWORT; WHEAT COCKLE

Annual or perennial plants; leaves sessile or nearly so and somewhat connate at base; stipules absent; flowers in terminal corymbose or paniculate cymes, showy; calyx of united sepals, 5-toothed, narrowly ovoid or tubular, terete or angled; petals 5, showy, with or without appendages; stamens 10, sometimes becoming petaloid; styles 2; capsule 1-celled or incompletely 2- to 4-celled at base, dehiscent at apex with 4 teeth.

1. Annual plants; calyx strongly 5-angled; petals rose, not appendaged; inflorescence a broad, open, flat-topped corymbose cyme   -1. S. vaccaria
1. Perennial plants; calyx not angled; petals white to pink, appendaged inside with long teeth at the junction of the blade and the claw; inflorescence a rather compact cyme   -2. S. officinalis

1. Saponaria vaccaria L., Sp. Pl. 409. 1753.
Vaccaria vaccaria (L.) Britt.; V. segetalis (Neck.) Garcke---Annual plants; stems 20-100 cm. tall, glabrous, branching above; leaves 3-7 cm. long, lanceolate to ovate, connate at base; inflorescence an open corymbose cyme; calyx tube 10-15 mm. long, becoming ovoid, sharply 5-angled and pale between the angles, much inflated in fruit; petals rose or pale red, not appendaged inside, exserted 5-10 mm. beyond the calyx, blades crenulate.---A weed in grain fields and waste places, often very abundant. Ontario to Alaska, south to Florida and California. Widely distributed in Colorado, our records from 4500-7500 feet.

2. Saponaria officinalis L., Sp. Pl. 408. 1753.
Perennial glabrous plants; stems 30-80 cm. tall, stout, erect, simple or sparingly branched; leaves ovate or oval, sometimes lanceolate, narrowed at base to a short broad petiole, 3- to 5-nerved; inflorescence a dense terminal cyme; calyx tube 15-20 mm. long, cylindrical or nearly so, faintly nerved but not angled; petals pale pink or white with an appendage of filiform teeth at the top of the claw, the blade obcordate; stamens often petaloid.---Roadsides and waste places often escaped from cultivation. Nova Scotia to Colorado, south to Florida and New Mexico. Our records from northern Colorado at 5000-7500 feet.

3. Silene L. CATCHFLY; CAMPION

Annual or perennial plants; no stipules present; flowers usually cymose; calyx of united sepals, more or less inflated, 10- to many-nerved, 5-toothed; petals 5, usually an appendage present at base of blade which is usually cleft or divided; stamens 10; styles 3, rarely 4 or 5, capsule longitudinally dehiscent at apex with 6 or rarely 3 valves.

1. Plants pulvinate-caespitose in a cushionlike mat, stems not over 10 cm. tall, bearing 1 flower each
                                                                                   -1. S. acaulis
1. Plants with elongated stems over 10 cm. long, not forming cushionlike mats; stems over 1-flowered
    2. Annual plants
        3. Plants glabrous or nearly so except for a glutinous band near middle of upper internodes; calyx not
            over 1 cm. long, the nerves not conspicuous   -2. S. antirrhina
        3. Plants viscid-pubescent or hirsute throughout, no glutinous bands present; calyx 12 mm. long or more,
            conspicuously nerved   -3. S. noctiflora
    2. Perennial plants
        4. Flowers solitary in the axils of leaves or terminal, hence inflorescence leafy; flowers not over 10 mm.
            long, calyx 5-8 mm. long   -4. S. menziesii
        4. Flowers in bracted terminal inflorescences; flowers 10 mm. long or more, calyx over 8 mm. long
            5. Petals purple with ciliate claws, and appendages 0.8-3 mm. long; calyx mostly densely glandular-
                pubescent, little inflated in fruit, nerves little if any anastomosing; plants glandular-hairy
                above   -5. S. scouleri
            5. Petals white, claws not ciliate, appendages absent or minute; calyx usually glabrous, definitely
                inflated in fruit, nerves reticulated; plants usually glabrous throughout -6. S. cucubalus

1. Silene acaulis L., Sp. Pl. ed 2:603. 1762.
    Perennial plants; stems 3-12 cm. high, caespitose-pulvinate and matted; glabrous or puberulent; leaves 5-20 mm. long (rarely to 40), linear, sessile, crowded, ciliate-serrulate; flowers solitary at the end of the branches, sessile or on slender peduncles; calyx 5-9 mm. long, oblong-campanulate, glabrous, rather lightly nerved; petals pink or purple, exceeding the calyx, entire or 2-lobed; capsule usually cylindric and equalling or exceeding the calyx. Hitchcock and Maguire (Univ. of Wash. Pub. in Biol. 13:22. 1947) said that our plants are ssp. subacaulescens (F. N. Williams) Hitch. and Maguire.---Alpine or arctic areas. Circumpolar and extending south in the Rocky Mountains to New Mexico. Our records from the high mountains of the state at 10,000-12,500 feet.
2. Silene antirrhina L., Sp. Pl. 419. 1753.
    S. antirrhina depauperata Rydb.; S. antirrhina vaccariifolia Rydb.---Annual plants; stems 15-80 cm. tall, erect, simple or sparingly branched, glabrous or finely puberulent, usually with a viscid belt near middle of upper internodes; leaves 2-6 cm. long, linear to lanceolate or oblanceolate, usually acute, the lower narrowed into a petiole, upper gradually reduced to subulate bracts; inflorescence a cymose panicle; calyx 4-10 mm. long, of united sepals, green or purple-tipped, ovoid in fruit; petals white to rose-tipped, exceeding the sepals and 2-cleft at apex, or shorter than calyx with a truncate apex or even wanting; capsule on ripening expanding the calyx. Several varieties of above have been proposed most of which intergrade in this area.---Fields and waste places. Throughout temperate North America. Our records thinly scattered over Colorado at 4000-7500 feet.
3. Silene noctiflora L., Sp. Pl. 419. 1753.
    Annual plants; stems 30-100 cm. tall, erect, stout, simple or branched, viscid-pubescent at least above; leaves 4-10 cm. long, the lower obovate to oblanceolate, obtuse, viscid-hirsute, narrowed to a broad petiole, upper ovate-lanceolate, acute or acuminate, sessile; inflorescence a few-flowered loose panicle; calyx 12-30 mm. long, cylindrical at first, conspicuously 10-nerved, these anastomosing, expanded in fruit; petals somewhat exceeding the calyx, white or pink, 2-cleft, the appendages denticulate; capsule sessile, ellipsoid.---Cultivated and waste ground. Nova Scotia to Washington, south to Florida and Utah. Our few records from central and northcentral Colorado at 5000-9500 feet.
4. Silene menziesii Hook., Fl. Bor. Amer. 1:90. 1830.
    S. stellaroides Nutt.---Perennial plants with several rootstocks arising from taproots, terminating in erect or ascending stems 10-30 cm. tall, these glandular-pubescent and dichotomously branched above; leaves 2-5 cm. long, many, oblanceolate to obovate to linear-lanceolate, acute or short-acuminate at apex, cuneate at base; flowers in the axils of the upper leaves or terminating the leafy stems; calyx 5-8 mm. long, tubular-campanulate to oblong, inconspicuously nerved, glandular-pubescent; petals 6-10 mm. long, white, 2-cleft at apex and sometimes with small lateral teeth; capsule sessile, ellipsoid, equalling or somewhat exceeding the calyx.---Often among bushes. Saskatchewan to California, south to Missouri and Arizona. Our records scattered over the western half of Colorado at 5500-10,000 feet.
5. Silene scouleri Hook., Fl. Bor. Amer. 1:88. 1830.
    S. hallii S. Wats.---Perennial plants; stems 20-50 cm. tall, pubescent and more or less glandular, several from a stout vertical root; lower leaves 5-15 cm. long, narrowly oblanceolate, long-cuneate to base, pubescent, becoming linear-oblanceolate or linear above on the stem; inflorescence subracemose or narrowly thrysoid, more or less nodding; lower bracts longer than the flowers; calyx 7-16 mm. long, oblong or campanulate, strongly 10-nerved, these green or purplish, with triangular teeth; petals purple, somewhat longer than the calyx, with broad ciliate claws and bifid blades; capsule ovate, short-stipitate. Our plant is ssp. hallii (Wats.) Hitchc. & Maguire. Typical S. scouleri Hook. has white or purple-tinged petals with narrower claws than the subspecies. Some of our plants approach the typical form.---In the mountains. The subspecies from Colorado to Utah, south to New Mexico and Arizona. Our records scattered in the mountains of Colorado at 7000-10,000 feet.
6. Silene cucubalus Wibel, Prim. Fl. Werth. 241. 1799.
    S. vulgaris (Moench) Garcke; S. latifolia (Mill) Britten & Rendle---Perennial plants; stems 20-80 cm. long, spreading or ascending, glabrous and more or less glaucous; leaves 3-5 cm. long, upper reduced to bracts, ovate or lanceolate, acute, sessile, glabrous; inflorescence a loose cymose panicle; calyx campanulate to subglobose, finally 10-15 mm. long, with 15-20 inconspicuous nerves connected by lateral veinlets, inflated in fruit; petals white, no auricles present

but blades deeply bifid, the two lobes entire.---Fields, roadsides and waste places. New Brunswick to Washington, south to New Jersey and California. Our few records from central and northcentral Colorado at 8000-9000 feet.

4. Lychnis L. CAMPION

Perennial or biennial plants; no stipules present; flowers perfect or unisexual, in terminal cymes; sepals united, the tube ovoid, more or less inflated, 5-toothed, 10-nerved; petals 5, narrowly clawed and with rather narrow, entire or 2-cleft blades, usually with a crown at base of blade, the whole often included in the calyx; stamens 10; styles 5, rarely 4; capsule longitudinally dehiscing with 10 or sometimes 5 valves.

1. Tall plants over 25 cm. tall; stems with more than 1 flower, usually over 3; plants seldom growing above 10,000 feet in Colorado
   2. Calyx 12-18 mm. long; plants dioecious or polygamous; leaves ovate-lanceolate to broader   -1. L. alba
   2. Calyx 10-12 mm. long; flowers perfect; leaves linear to oblanceolate   -2. L. drummondii
1. Dwarf plants seldom over 20 cm. tall; stems with a single flower (rarely 2 or 3); plants usually well above 10,000 feet in Colorado
   3. Flowers in anthesis erect; calyx little inflated in fruit   -3. L. kingii
   3. Flowers in anthesis nodding; calyx much inflated in fruit   -4. L. apetala

1. Lychnis alba Mill., Gard. Dict. Ed. 8 No. 4. 1768.
Melandrium album (Mill.) Garcke---Perennial or possibly biennial plants; stems 30-60 cm. high, branching, more or less glandular; leaves 2-8 cm. long, ovate-oblong or ovate-lanceolate, upper sessile, lower short-petioled; flowers dioecious or polygamous; calyx 12-18 mm. long, tubular or oblong-cylindrical; petals 2-3 cm. long, white or pinkish, blade 2-cleft.---Waste places. In the eastern part of Canada and the United States. Our 3 records scattered widely in the western half of Colorado at 7500-9500 feet.
2. Lychnis drummondii (Hook.) S. Wats. in King, Geol. Expl. 40th Par. 5:37. 1871.
Wahlbergella drummondii (Hook.) Rydb.; L. striata Rydb.---Stems 20-50 cm. tall, 1 or more from stout vertical roots, finely glandular-pubescent or viscid above; leaves oblanceolate (basal) to linear (on stem), cuneate below, the basal to a margined petiole, acute or obtuse at apex; flowers few, slender to stout-pedicelled; calyx 10-12 mm. long, tubular or oblong-cylindric, viscid-pubescent, strongly 10-nerved; petals varying from included to exserted 4 or 5 mm., white or purplish, the blade narrow, entire or 2- to 4-lobed; young ovary swollen near tip; seeds tuberculate.---Dry slopes and plains. Manitoba to British Columbia, south to New Mexico and Arizona. Our many records from the mountainous half of Colorado at 5500-12,500 feet.
3. Lychnis kingii S. Wats., Proc. Am. Acad. 12:247. 1877.
Wahlenbergella kingii(S.Wats.)Rydb.---Closely related to the next, differing in the characters given in the key.---Alpine areas. Wyoming, Colorado and Utah. Our few records scattered in the higher mountains at 12,000-14,000 feet.
4. Lychnis apetala L., Sp. Pl. 437. 1753.
Wahlbergella apetala (L.) Fries.---Dwarf plants; stems 5-10 cm. tall, sometimes nearly glabrous below, glandular-pubescent or with jointed reddish villous hairs above; leaves linear-oblanceolate, basal cuneate to a margined ciliate petiole, acute; flowers 1-5 cm. long, nodding, solitary or sometimes to 3; calyx 10-15 mm. long, ovate-campanulate, inflated in fruit, with short, acute teeth, strongly 10-nerved; petals white or purplish, usually included, blade narrow and bifid. Our Colorado plants appear to be ssp. montana (S. Wats.) Maguire which we have been confusing with L. kingii (Rhodora 52:233-243. 1950).---Alpine and arctic regions. Greenland to Alaska, south to Colorado and Utah; Eurasia. Our one rather doubtful record from central Colorado at about 11,000 feet.

5. Paronychia Adans. WHITLOW-WORT; NAIL WORT

Perennial tufted herbaceous plants with woody or nearly woody caudices; leaves crowded, with scarious stipules; flowers clustered, small, essentially apetalous, with scarious bracts; sepals 5, concave or hooded at the awn-tipped apex; petals wanting or very minute; stamens 5, sometimes alternating with as many staminodes; styles partly united; fruit an ovoid or globose utricle included in the calyx.

1. Leaves elliptic or oblong; plants growing above 11,000 feet in Colorado   -1. P. pulvinata
1. Leaves linear or linear-subulate; plants growing below 11,000 feet in Colorado
   2. Flowers solitary (or in pairs); leaves little if any exceeding the bracts, leaves 4-6 mm. long, the stipules almost as long   -2. P. sessiliflora
   2. Flowers more or less clustered; leaves much longer than the bracts, leaves 6-20 mm. long, much longer than the stipules
      3. Stems prostrate and forming mats, seldom over 6 cm. high from the ground, internodes short, seldom over 5 mm. long, the leaves closely crowded   -3. P. depressa
      3. Stems erect or ascending, seldom less than 10 cm. tall, internodes usually well over 5 mm. long especially above, the leaves not closely crowded   -4. P. jamesii

1. Paronychia pulvinata Gray, Proc. Acad. Sci. Phila. 1863. 58. 1864.
Plants arising from thick woody caudices and forming dense cushionlike mats, the stems about 3-5 cm. tall; leaves 3-6 mm. long, oblong or elliptic, obtuse, bright green, flat but thick, nerveless; stipules 2-4 mm. long, silvery, broadly ovate, entire, mostly obtuse, equalling or nearly equalling the leaves and with them densely covering the short stems; flowers mostly solitary and immersed among the leaves; bracts exceeding the flowers and as long or nearly as long

as the leaves; sepals 2-2.5 mm. long, broadly oval, cuspidate from the back, the processes not over 0.5 mm. long; style short.---Mountain tops and upper slopes. Wyoming, Colorado and Utah. Our records from central and northcentral Colorado at 11,500-13,000 feet.

2. Paronychia sessiliflora Nutt., Gen. Am. 1:160. 1818.
P. sessiliflora brevicuspis A. Nels.; P. brevicuspis (A. Nels.) Rydb.; P. brevispina (A. Nels.) Rydb.---Stems 4-10 cm. tall, very densely caespitose from woody caudices, much branched and crowded; leaves 4-6 mm. long, linear-subulate, spinulose-tipped, acute or mucronate, upper recurved-spreading; stipules subulate or oblong-linear, slightly shorter than the leaves; flowers terminal, solitary or sometimes paired; bracts equalling or nearly equalling the leaves; sepals 2-3 mm. long, oblong-linear, pubescent at base, 3-nerved, margins scarious, arched above, with awns often over 1 mm. long; style as long as the sepals.---Dry hills and ridges. Saskatchewan to Alberta, south to Texas and New Mexico. Our records scattered over Colorado, except in the extreme western and eastern parts at 4000-9000 feet.

3. Paronychia depressa Nutt. ex T. & G., Fl. N. Amer. 1:171. 1838.
P. diffusa A. Nels.---Stems 7-16 cm. long, but seldom over 5 or 6 cm. above the ground, from rather woody taproots, prostrate-spreading and forming close mats, the internodes short, rarely over 5 mm. long; leaves 5-10 mm. long or sometimes to 15, crowded on the short branches to the very summit, cuspidate or bristly-pointed; stipules silvery, ovate-lanceolate, shorter than the leaves but conspicuous; inflorescence contracted, flowers crowded and immersed in the leaves; bracts longer than or as long as the flowers, rarely shorter, but definitely shorter than the leaves; sepals 2-3 mm. long, pubescent below, awned from apex, this awn about 0.5 mm. or more in length.---Dry hills and plains. South Dakota to Wyoming, south to Nebraska and Colorado. Our records from central, northcentral and northeastern Colorado at 3500-10,000 feet.

4. Paronychia jamesii T. & G., Fl. N. Amer. 1:170. 1838.
P. wardii Rydb.---Stems 10-30 cm. tall rarely as little as 7 cm., caespitose and freely branched from woody caudices and also above, erect or nearly so; leaves 1-2 cm. long, linear-lanceolate, obtuse or mucronate, sessile, puberulent or minutely scabrous; stipules silvery, linear-lanceolate, much shorter than the leaves, especially the upper; cyme few-flowered with a central subsessile flower in each group; bracts shorter than the leaves; sepals about 2-2.5 mm. long, pubescent near base, linear-oblong, 3-ribbed, the lateral ones obscure, with a short cusp near apex.---Dry plains and hills. Nebraska and Colorado, south to Texas, Mexico and Arizona. Scattered over Colorado, our records mostly from the eastern half, at 3500-7500 feet.

6. Spergula L. SPURRY

Annual plants; leaves subulate, much-fascicled in the axils, rather fleshy; stipules present; flowers in terminal cymes; sepals 5, persistent; petals 5, white, entire; stamens 10 or rarely 5; styles 5; capsule 5-valved, longitudinally dehiscent; seeds acute-margined or narrowly winged.

1. Spergula arvensis L., Sp. Pl. 440. 1753.
Stems 15-60 cm. tall, slender, erect or ascending, branching near the base, glabrous or sparingly pubescent; leaves 2-5 cm. long, linear-filiform, somewhat fleshy, appearing verticillate; stipules small, connate at first; cyme loose; flowers on slender pedicels often reflexed in fruit; sepals 3-4 mm. long, ovate; petals equalling or somewhat exceeding the sepals; capsule ovoid, slightly exceeding the sepals; seeds roughened with minute whitish papillae.---Cultivated ground and waste places. Widely distributed in the United States and eastern Canada. Apparently uncommon here, our 2 records from northcentral Colorado at 5000-5200 feet.

7. Spergularia J. & C. Presl SANDSPURRY

Annual or perennial herbaceous plants; leaves linear, fleshy, with scarious stipules; inflorescence a terminal cyme; sepals 5, persistent; petals 5, rarely fewer or wanting, pink or whitish, entire; stamens 2-10; styles 3; ovary 1-celled; capsule 3-valved to the base; seeds smooth or tubercled.

1. Stamens 9-10; leaves mostly fascicled in the axils
  2. Seeds minutely but definitely papillose, not winged; capsules 3.5-5 mm. long; stipules ovate-lanceolate to lanceolate, definitely longer than wide -1. S. rubra
  2. Seeds smooth or very nearly so, usually winged; capsules 5-8 mm. long; stipules deltoid, about as long as wide -2. S. media
1. Stamens 2-5; leaves not fascicled or occasionally some nodes with 1 or 2 axillary leaves -3. S. marina

1. Spergularia rubra (L.) Presl, Fl. Cech. 94. 1819.
Tissa rubra (L.) Britt.---Annual or short-lived perennial plants; stems 5-25 cm. long, prostrate, branched from base, glabrous or glandular; leaves 4-20 mm. long, fascicled, linear-filiform and more or less flattened; stipules 2.5-5 mm. long, lanceolate-triangular, acuminate; at least lower bracts foliaceous; sepals 3.5-5 mm. long, lanceolate, usually glandular; petals shorter than the sepals, pink; stamens 6-10; capsule 3.5-5 mm. long; seeds dark brown, minutely papillose.---Widely distributed in the United States and Canada; also in South America. Our one record from northwestern Colorado at 8000 feet.

2. Spergularia media (L.) Presl, Fl. Sic. 1:161. 1826.
Annual or short-lived perennial plants; stems 5-40 cm. long, erect or prostrate, glabrous to glandular-pubescent; leaves 10-50 mm. long, often fascicled in the axils, linear-filiform, sometimes short-mucronate; stipules 2-6 mm. long, deltoid, often short-acuminate; bracts foliaceous; sepals 3-6 mm. long, ovate to narrowly ovate, glabrous to glandular-pubescent;

petals about as long as or somewhat shorter than the sepals, white to light rose; stamens 9-10 rarely fewer; capsules 5-8 mm. long; seeds smooth or roughened, usually winged.---Rather widely distributed in Europe, North America and South America. Our few records from northcentral Colorado at about 5000 feet but locally common there.
3. Spergularia marina (L.) Griseb., Spicil. Fl. Rumel et Bith. 1:213. 1843.
S. sparsiflora (Greene) A. Nels.; Tissa salina (Presl) Greene; T. sparsiflora Greene---Annual plants; stems 10-30 cm. tall, usually diffusely branched, glabrous or glandular-pubescent; leaves 5-40 mm. long, not fascicled or with 1 or 2 in the axils, terete, bluntly mucronate, glabrous or glandular-pubescent; stipules 2-3.5 mm. long, deltoid, apex acute or very slightly acuminate; bracts foliaceous; sepals 2.4-5 mm. long, ovate, glabrous or glandular-pubescent; petals white, pink or rose, shorter than the sepals, ovate; stamens 2-5; capsule 3.5-6.5 mm. long; seeds smooth or roughened, sometimes narrowly winged.---Alkaline plains. Widely distributed in the United States and Canada. Our few records from northcentral Colorado at 4500-5000 feet.

8. Cerastium L. MOUSE-EAR CHICKWEED

Annual or perennial, pubescent or hirsute plants; flowers in terminal cymes; sepals 5 or rarely 4, separate or nearly so; petals of the same number as the sepals, rarely lacking, 2-cleft at the apex; stamens 10 or rarely fewer; styles as many as the sepals and opposite them, rarely fewer; capsule cylindrical, often curved, opening at the apex only, by twice as many valves as the style number.

1. Annual plants; mature capsule 2-3 times as long as the calyx; petals shorter or slightly longer than the sepals (sometimes nearly twice as long in C. nutans)
  2. Pedicels in fruit mostly 3 or more times as long as the calyx, usually strongly curved near the apex
                                                                                -1. C. nutans
  2. Pedicels in fruit rarely over twice as long as the calyx, straight or slightly curved
    3. Leaves elliptic, oval or obovate, broadly rounded at apex; petals shorter than the calyx or rarely equalling it   -2. C. viscosum
    3. Leaves oblanceolate or oblong-lanceolate, sometimes oblong, acute or somewhat narrowed to an obtuse apex; petals usually slightly longer than the calyx   -3. C. bractypodum
1. Biennial or perennial plants; mature capsule 1-2 times as long as the calyx; petals about twice as long as the calyx (except in C. vulgatum and sometimes C. beeringianum)
  4. Petals about equalling the sepals in length (if rarely longer than only very slightly so) -4. C. vulgatum
  4. Petals about twice as long as the sepals (often less in C. beeringianum)
    5. Leaves narrow, linear, linear-lanceolate or rarely linear-oblong (bracts may be wider), usually acute at apex   -5. C. arvense
    5. Leaves wider, oblong, ovate or oval, sometimes oblong-lanceolate or oblanceolate, usually obtuse at apex   -6. C. beeringianum

1. Cerastium nutans Raf., Prec. Somiolog. 36. 1814.
C. longipedunculatum Muhl.---Annual plants; stems 20-50 cm. long, weak, ascending or reclining, glabrate to glandular-pubescent; leaves 1.5-5 cm. long, oblong, oblong-lanceolate to spatulate, acute or obtuse, lower petioled, upper sessile; inflorescence loosely cymose, pedicels slender, over 3 times and usually over 5 times as long as the calyx, rather erect but usually strongly curved and bent at the apex; bracts herbaceous; sepals 4-5 mm. long, lanceolate, acute or obtuse; petals longer, sometimes to twice as long as the sepals; capsule twice as long as the calyx or more.---Moist or wet places. Nova Scotia to British Columbia, south to North Carolina and Arizona. Our records from southern Colorado at 6500-9000 feet.
2. Cerastium viscosum L., Sp. Pl. 437. 1753.
Throughout the United States and Canada. No specimens from Colorado seen by the writer.
3. Cerastium brachypodum (Engelm.) Robinson in Britt., Mem. Torr. Club 5:150. 1894.
Annual plants; stems 5-25 cm. tall, tufted, erect, simple or sparingly branched, viscid-pubescent or puberulent; leaves 5-25 mm. long, lower oblanceolate, spatulate or oblong, the upper narrower, obtuse or acute, petioled on lower; cymes terminal; fruiting pedicels short, seldom over twice as long as the calyx, sometimes deflexed but usually straight or gradually curved; bracts herbaceous; sepals about 3-4 mm. long; petals 5-6 mm. long; capsule 2 to 3 times as long as the calyx.---Dry, often sandy ground. South Dakota to Alberta, south to Virginia and Mexico. Our records from northcentral, central, and southcentral Colorado at 5000-8000 feet.
4. Cerastium vulgatum L., Sp. Pl. ed. 2. 627. 1762.
Biennial or perennial plants; stems 10-40 cm. long, viscid-pubescent, usually tufted; leaves 8-25 mm. long, oblong or oblong-ovate to oblong-oblanceolate, acute or obtuse, villous; bracts small but foliaceous though margins may be scarious; flowers in loose cymes, pedicels becoming longer than fruit; sepals 4-6 mm. long, scarious-margined; petals about as long as sepals, 2-cleft; capsule about twice as long as sepals, straight or somewhat curved. Our plant has been called var. hirsutum Fries.---Roadsides, fields and streambanks. Widely scattered in temperate North America. Our records from northcentral, central and southcentral Colorado at 4500-8500 feet, with one specimen marked from 13,000 feet.
5. Cerastium arvense L., Sp. Pl. 438. 1753.
C. scopulorum Greene; C. strictum L.; C. occidentale Greene---Perennial plants; stems 10-30 cm. high, tufted, erect or ascending, simple or branched, glabrous to densely villous, glandless to glandular; leaves 1-4 cm. long, linear-subulate to linear-oblong, rarely to narrowly ovate, acute or obtuse; flowers cymose, rather few, pedicels slender, erect; bracts variable, from linear to broadly ovate, usually definitely scarious-margined; sepals 4-7 mm. long, lanceolate and acute; petals about twice as long as the sepals or longer; capsule equalling or twice as long as the calyx. This is a variable species and the forms have been segregated as species. The one with villous stems has been called C. campestre Greene. The form with bracts broadly

ovate has been called C. oreophilum Greene. However, these 2 forms intergrade with each other and with the typical form in this area.---Dry, usually rocky or sandy ground. Widely distributed in Canada and the United States; Eurasia. Our records well scattered in the western two-thirds of Colorado at 5000-12,000 feet.

6. Cerastium beeringianum Cham. & Schlecht., Linnaea 1:62. 1826.
C. buffumae A. Nels.; C. variabile Goodding; C. pilosum Greene---Perennial plants; stems 5-25 cm. high, densely or loosely matted, spreading or ascending, more or less viscid-pubescent or glandular-pilose; leaves 5-15 mm. long, spatulate to oblong-lanceolate to oblong, rarely linear, mostly obtuse; inflorescence 1- to 15-flowered, pedicels usually slender, mostly ascending; bracts ovate to oblong-lanceolate, acutish, not scarious; sepals 3.5-7 mm. long in anthesis, broadly lanceolate to oblong-ovate, the inner scarious-margined and often tinged with purple; petals longer than the sepals but seldom over 1.5 times as long; capsule 2 to 3 times as long as the calyx. The forms with long showy petals have been called C. earlei Rydb. and C. pulchellum Rydb.---In the mountains, often among rocks. Quebec to Alaska, south to New Mexico; Asia. Our records scattered in the mountains of Colorado at 10,000-12,000 feet.

9. Stellaria L. CHICKWEED; STARWORT

Annual or perennial plants; stems low and spreading; inflorescence usually an open cyme; sepals 5, rarely 4, separate or nearly so; petals white, of the same number as the sepals, rarely wanting, deeply 2-cleft at apex; stamens 10 or less; ovary 1-celled; styles 3 or rarely 4 or 5; capsule globose or oblong, longitudinally dehiscent nearly or quite to the base by twice as many valves as there are styles (usually 6). A rather difficult genus and intergradations between the species are occasionally found. Dissimilar looking plants sometimes key down to the same species.

1. Annual plants; leaves ovate, petioled on lower ones -1. S. media
1. Perennial plants; leaves all sessile, ovate to linear
  2. Plants glandular-pubescent at least above; petals 6-8 mm. long; leaves 5-10 cm. long -2. S. jamesiana
  2. Plants not glandular; petals lacking or less than 6 mm. long; leaves rarely if ever over 5 cm. long
    3. Petals equalling or surpassing the sepals in length
      4. Cyme diffuse, many-flowered, the pedicels finally spreading or deflexed; leaves spreading or ascending
        5. Leaves linear or lanceolate-linear, 2-6 cm. long, widest at or above the middle; seeds smooth -3. S. longifolia
        5. Leaves lanceolate or oblong-lanceolate, sometimes narrowly lanceolate, 2-3 cm. long, widest near the base or at least below the middle; seeds minutely roughened -4. S. graminea
      4. Cyme usually few-flowered, the pedicels erect or ascending; leaves ascending
        6. Bracts of the cyme small and scarious especially the upper, not leaflike
          7. Sepals ovate, obtuse or mucronate; cyme not reduced to a single flower; leaves 1-3 cm. long -5. S. longipes
          7. Sepals lanceolate, acute; cyme often reduced to a single flower; leaves 1-2 cm. long -5A. S. longipes laeta
        6. Bracts of the cyme foliaceous, resembling the upper leaves
          8. Stems less than 15 cm. long; leaves linear to narrowly lanceolate; flowers 1-3 -5A. S. longipes laeta
          8. Stems over 15 cm. long; leaves lanceolate to oblong-lanceolate; flowers usually over 3 -6. S. crassifolia
    3. Petals lacking or minute, not over 1/2 as long as the sepals
      9. Bracts of the inflorescence small and scarious; branches of the inflorescence divaricate or reflexed at maturity; leaves oblong-lanceolate -7. S. umbellata
      9. Bracts foliose, resembling the upper leaves, not scarious; branches of the inflorescence erect or ascending; leaves linear to ovate (may be oblong-lanceolate)
        10. Leaves linear to ovate-lanceolate, some over 12 mm. long; plants with running rootstocks; stems erect -8. S. borealis
        10. Leaves ovate, not over 12 mm. long; plants caespitose, no running rootstocks; stems diffuse -9. S. obtusa

1. Stellaria media (L.) Cyrill., Pl. Char. Comm. 36. 1784.
Alsine media L.---Annual plants; stems 10-40 cm. long, diffusely branched, prostrate or ascending, glabrous except for pubescent lines; leaves 0.5-3.5 cm. long, ovate or oval, upper sessile, the lower petioled; flowers in terminal or axillary leafy cymes, pedicels slender; sepals 4-6 mm. long, oblong, acute; petals 2-parted, shorter than the calyx.---Waste places and cultivated ground especially in shaded portions of lawns. Widely distributed in Canada and the United States; Europe. Our specimens from central and northcentral Colorado at 5000-10,000 feet.

2. Stellaria jamesiana Torr., Ann. Lyc. N. Y. 2:169. 1828.
Alsine jamesiana (Torr.) Heller; A. curtisii Rydb.---Perennial plants with rootstocks; stems 20-50 cm. long, erect or ascending, more or less diffusely branching, glandular-pubescent throughout or glabrate below, sharply angled; leaves 5-10 cm. long rarely shorter, lanceolate to linear-lanceolate, sessile, broadest near the base; flowers in loose, leafy-bracted terminal and axillary cymes; sepals 4-6 mm. long, oblong-lanceolate, acute, margins scarious; petals 6-8 mm. long, deeply lobed but not deeply cleft or parted, often twice as long as the sepals; capsule shorter than the calyx.---Mountains, often in moist shaded areas. Wyoming to Idaho, south to Texas and California. Widely scattered in the western half of Colorado at 5500-9500 feet.

3. **Stellaria longifolia** Muhl. ex Willd., Enum. Hort. Ber. 479. 1809.
  Alsine longifolia (Muhl.) Britton---Plants perennial by running rootstocks; stems 10-50 cm. high, erect or ascending but slender and weak, sharply 4-angled, glabrous or scabro-puberulent on the angles; leaves 2-6 cm. long, rarely less, linear or linear-lanceolate, sessile, acute at apex, cuneate at base, often ciliate below; cyme usually many-flowered, open, the long filiform pedicels spreading or at length reflexed; bracts small and scarious; sepals 2-3.5 mm. long, lanceolate, acute; petals longer than the sepals; capsule longer than the calyx; seeds smooth.---Low meadows. Newfoundland to Alaska, south to Maryland and Arizona. Our records scattered in the mountains of Colorado at 5500-10,000 feet.

4. **Stellaria graminea** L., Sp. Pl. 422. 1753.
  Alsine graminea (L.) Britt.---Perennial plants from running rootstocks; stems 20-50 cm. long, weak, ascending or reclining, sharply 4-angled, glabrous or very nearly so; leaves 20-30 mm. long, narrowly lanceolate to lanceolate or oblong-lanceolate, usually widest below the middle, often ciliate below; cymes terminal, diffuse, the long-filiform pedicels spreading or at length reflexed, especially the lower; bracts small and scarious; sepals about 3-5 mm. long, lanceolate, acute; petals longer than the sepals, sometimes only slightly so; seeds minutely roughened.---Fields, roadsides and low meadows. Newfoundland to Ontario, south to Maryland and Colorado. Our one record from northcentral Colorado at 5300 feet.

5. **Stellaria longipes** Goldie., Edinb. Phil. Journ. 6:327. 1822.
  Alsine longipes (Goldie.) Coville; A. longipes var. stricta (Richardson) Rydb.; A. strictiflora Rydb.---Plants perennial by rootstocks; stems 5-30 cm. tall, erect or ascending, simple or somewhat branched, 4-angled, glabrous or nearly so; leaves 1-3 cm. long, lanceolate or linear-lanceolate, sessile, acute or acuminate at apex, usually broadest below the middle, erect or ascending, rather firm and shining; inflorescence of 1-3 flowers in the axils of leaves, or in a cyme with scarious bracts, the pedicels erect or ascending; sepals 3-5 mm. long, lanceolate or ovate-lanceolate, obtuse, mucronate or acute, scarious-margined; petals somewhat longer than the sepals, 2-cleft; capsule longer than the sepals, at maturity often dark colored.---Usually moist places. Greenland to Alaska, south to New Mexico and California; Asia. Our records scattered in the mountainous half of Colorado, except the extreme western part, at 7000-12,500 feet.

5A. **Stellaria longipes laeta** (Richards.) Wats., Bibl. Index. 112. 1878.
  S. laeta Richards.; Alsine laeta (Richards.) Rydb.---Differs from species in being reduced in size, 3-15 cm. tall (instead of 15-30), and with 1 or very few flowers present, these in the axils of leaves; sepals are lanceolate and acute (not ovate or ovate-lanceolate and obtuse or mucronate). But all the above intergrade in our material and the variety is of rather doubtful value in Colorado.

6. **Stellaria crassifolia** Ehrh., Hannov. Mag. 8:116. 1784.
  Alsine crassifolia (Ehrh.) Britton---Perennial plants with rootstocks; stems 15-40 cm. long, weak, ascending or diffuse, simple or branched, glabrous; leaves 6-8 mm. long, lanceolate to oblong-lanceolate, sessile and narrowed at base, acute or obtuse; cymes terminal, few-flowered or some flowers axillary and solitary, peduncles ascending; bracts foliaceous, none scarious but the upper small; sepals ovate-lanceolate to lanceolate-oblong, acute or acuminate; petals as long or more often slightly longer than the sepals; capsule finally longer than the calyx.---Moist or wet ground. In arctic America extending south to Colorado. Our rather few records from southcentral, central and northcentral Colorado at 7500-10,000 feet.

7. **Stellaria umbellata** Turcz., Bull. Soc. Nat. Mosc. 15:173. 1842.
  Alsine baicalensis Coville---Perennial plants with rootstocks; stems 10-40 cm. tall, slender, usually branched, ascending or decumbent at base, glabrous; leaves oblong-lanceolate, sessile, cuneate below, acute to acuminate at apex, glabrous or ciliate at base; cyme open, many-flowered, the slender pedicels spreading or at length reflexed; bracts scarious; sepals 2-3 mm. long, rarely to 5, scarious-margined, acute; petals wanting or minute; capsule about twice as long as the calyx.---In moist ground. Montana to Oregon, south to Arizona and California; Asia. Our records scattered in the mountainous half of Colorado, none as yet from the northwestern corner, at 8000-14,000 feet.

8. **Stellaria borealis** Bigelow, Fl. Bost. ed 2. 182. 1824.
  Alsine borealis (Bigel.) Britton; A. calycantha (Bong.) Rydb.; A. alpestris (Fries) Rydb.---Perennial plants with running rootstocks; stems 10-40 cm. long, erect or ascending, simple or branching, glabrous or nearly so; leaves 1-4 cm. long, sessile, variable in shape, linear to ovate-lanceolate, rarely ovate, thin, ciliate at base; flowers in leafy-bracted cymes, rarely solitary, on slender pedicels; sepals 3-4 mm. long, lanceolate or ovate-lanceolate, acute, scarious-margined; petals minute or absent; capsule as long or longer than the sepals.---Moist ground. Labrador to Alaska, south to New Jersey and California. Our records from central and northcentral Colorado at 7500-9500 feet.

9. **Stellaria obtusa** Engelm., Bot. Gaz. 7:5. 1882.
  Alsine obtusa (Engelm.) Rose; A. polygonoides Greene---Caespitose perennial plants; stems 4-15 cm. long, decumbent or prostrate, glabrous; leaves 6-12 mm. long, ovate, sessile, acute, glabrous or ciliate at base; flowers solitary on axillary peduncles, no true bracts present; sepals 2-3 mm. long or possibly somewhat longer; petals absent or minute and shorter than the sepals; capsule somewhat longer than the sepals.---Moist or wet places. Alberta to British Columbia, south to Colorado and Washington. Our record from northwestern Colorado, altitude not certain.

10. Sagina L. PEARLWORT

Low matted perennial plants; leaves filiform or subulate; flowers axillary, on slender pedicels; sepals 4 or 5; petals of the same number or wanting, entire or emarginate; stamens usually 4 or 5; styles 4 or 5, alternate with the sepals; capsule finally longitudinally dehiscent to the base, 4- or 5-valved, these opposite the sepals; seeds many, not strophiolate.

1. **Sagina saginoides** (L.) Britton, Torrey Bot. Club Mem. 5:151. 1894.
   S. saginoides var. hesperia Fernald---Stems 2-10 cm. tall, decumbent and often rooting at the nodes forming mats; leaves 4-15 mm. long, linear-filiform or linear-subulate; pedicels 5-20 mm. long, often curved at apex; sepals 5, 1.3-3 mm. long, oval or ovate, obtuse; petals usually little shorter than the sepals; capsule becoming longer than the sepals. S. nivalis Fries may be a long-petaled, high altitude form of the above.---Moist places. Greenland to Alaska, south to New Mexico and California; Eurasia. Our records scattered in the western two-thirds of Colorado at 6000-12,000 feet.

11. Arenaria L. SANDWORT

Annual or perennial plants; leaves sessile, without stipules; inflorescence a terminal cyme or capitate cluster, or sometimes axillary or solitary; sepals 4 or 5, distinct or nearly so; petals 4 or 5, white or rarely yellowish, entire or nearly so; stamens 10, rarely 8; styles usually 3, rarely 2-5; ovary 1-celled or sometimes 3-celled at first, longitudinally dehiscent, opening by 3 entire or 2-cleft valves; seeds sometimes with a broad strophiole.

1. Annual plants; leaves ovate   -1. A. serpyllifolia
1. Perennial plants; leaves rarely ovate, usually much narrower
   2. Leaves lanceolate, oblong-lanceolate to wider, not pungent-tipped; stems finely puberulent; capsule valves 2-cleft at apex
      3. Leaves short, about 5 mm. long, seldom over 10 mm.; stems less than 15 cm. long, rarely over 10 cm.
                                                                                                           -2. A. saxosa
      3. Leaves 10 mm. long or more; stems usually well over 15 cm. long
         4. Seeds with a basal appendage (strophiole) at the hilum (may be deciduous from ripe seeds); plants seldom over 25 cm. tall; petals often longer than the sepals
            5. Leaves oblong or oval; sepals obtuse or very obscurely acute; petals about twice as long as the sepals   -3. A. lateriflora
            5. Leaves lanceolate or oblanceolate; sepals very acute to acuminate; petals variable but often shorter than the sepals   -4. A. macrophylla
         4. Seeds lacking a strophiole; stems usually over 25 cm. long; petals shorter than the sepals
                                                                                                           -5. A. confusa
   2. Leaves narrow, usually subulate, filiform or narrowly linear, rarely to linear-lanceolate, usually pungent-tipped; stems glabrous or glandular-pubescent, rarely pubescent only; valves cleft or entire
      6. Valves of the capsule entire or slightly emarginate; dwarf, caespitose plants with stems less than 10 cm. long; usually in alpine or subalpine zones
         7. Sepals obtuse
            8. Petals 0.5-1 mm. wide, shorter than to barely exceeding the glabrous or puberulent sepals; capsule 4-6 mm. long; leaves of basal shoots obscurely keeled   -6. A. sajanensis
            8. Petals 1.5-2.5 mm. wide, conspicuously exceeding the pilose or hirsute sepals; capsule 6-10 mm. long; leaves of basal shoots keeled   -7. A. obtusiloba
         7. Sepals acute, acuminate, or subulate-tipped
            9. Petals about 2 mm. longer than the sepals, which are 4-5 mm. long; plants glabrous; leaves 1-nerved   -8. A. macrantha
            9. Petals wanting or shorter than the sepals, or not over 1 mm. longer; sepals 2.5-4 mm. long; plants glabrous or glandular-pubescent; leaves 1- or 3-nerved
               10. Leaves linear-subulate, 3-ribbed; sepals with 3 uniform ribs; stems glabrous to glandular-pubescent   -9. A. rubella
               10. Leaves triquetrous (3 sharp angles), 1 angle (rib) much stronger than the others; sepals with 1 rib stronger than the others; stems glabrous   -10. A. rossii
      6. Valves of the capsule 2-cleft at the apex; stems usually over 10 cm. long; plants usually below the subalpine zone in Colorado
         11. Inflorescence 1 or more dense, congested, capitate or subcapitate cymes, the flowers sessile or subsessile
            12. Sepals lanceolate or linear-lanceolate, acuminate, over 5 mm. long; low plants seldom over 10 cm. tall, seldom over 7500 feet in Colorado   -11. A. hookeri
            12. Sepals ovate, acute, less than 5 mm. long; plants usually over 10 cm. tall, often growing above 7500 feet in Colorado   -12. A. congesta
         11. Inflorescence an open cyme, the flowers definitely pedicellate
            13. Petals shorter, equal to or rarely barely exceeding the sepals; sepals ovate-lanceolate to linear-lanceolate
               14. Plant stems glandular above, usually 20 cm. tall or more; leaves erect or nearly so, some of them over 20 mm. long   -13. A. fendleri
               14. Plant stems glabrous, less than 20 cm. tall; leaves more or less spreading, seldom over 20 mm. long   -13A. A. fendleri eastwoodiae
            13. Petals definitely longer than the sepals (over 1 mm.); sepals ovate-lanceolate to ovate
               15. Plants densely glandular-pubescent at least in the inflorescence; leaves not pungent; stems rarely over 10 cm. tall   -13B. A. fendleri tweedyi
               15. Plants glabrous or sparingly glandular-pubescent; leaves distinctly pungent; stems over 10 cm. tall   -14. A. uintahensis

1. **Arenaria serpyllifolia** L., Sp. Pl. 423. 1753.
   Annual plants; stems 5-30 cm. long, much branched from the base, decumbent or ascending, finely pubescent or puberulent; leaves 4-8 mm. long, ovate, acute or short-acuminate, sessile or the lowest short-petiolate; flowers in open leafy-bracted cymes; sepals about 3 mm. long, ovate, acute or acuminate; petals smaller than the sepals; capsule valves 2-cleft at apex; seeds not strophiolate.---Dry often sandy or rocky ground. Quebec to British Columbia, south to Florida, Colorado and Oregon. Our one record from Clear Creek County at 7500 feet.

2. **Arenaria saxosa** A. Gray, Pl. Wright. 2:18. 1853.

A. polycaulos Rydb.---Perennial plants; stems 5-15 cm. tall, usually about 10, many, spreading from stout roots and decumbent at base, puberulent; leaves about 5-15 mm. long, ovate or ovate-lanceolate, rarely narrower, acute or mucronate; inflorescence terminal, solitary or few-flowered; sepals ovate-lanceolate, acuminate; petals as long or somewhat longer than the sepals; capsule valves 2-cleft at apex; seeds not strophiolate.---Dry ground. Colorado, Arizona, New Mexico and Mexico. Our records from central, southcentral and southwestern Colorado at 9000-11,500 feet.

3. **Arenaria lateriflora** L., Sp. Pl. 423. 1753.

Moehringia lateriflora (L.) Fenzl.---Perennial plants; stems 10-25 cm. long, ascending or erect but decumbent at least at the base, simple or sparingly branched, finely pubescent; leaves 1-3 cm. long, oval-oblong or elliptic-oblong, thin, obtuse; inflorescence a terminal or axillary cyme, few-flowered or even solitary; sepals 2-3 mm. long, oblong to ovate, obtuse or obscurely acute; petals longer than the sepals, usually about twice as long; ovary 3-celled below; capsule valves cleft at apex; seeds strophiolate.---Usually moist shaded places. Labrador to Alaska, south to New Jersey and New Mexico; Eurasia. Our records scattered in the western mountainous half of Colorado at 5000-9000 feet.

4. **Arenaria macrophylla** Hook., Fl. Bor. Am. 1:102. 1830.

Moehringia macrophylla (Hook.) Torr.---Perennial plants; stems 10-30 cm. long, decumbent, usually branched, finely puberulent, often angled; leaves 2-8 cm. long, lanceolate or oblanceolate, acute, cuneate; cymes terminal or in upper axils, 1- to 5-flowered; sepals 3-4 mm. long, lanceolate or ovate-lanceolate, very acute to acuminate; petals variable (probably dimorphic), from somewhat longer to shorter than sepals; capsule 3-celled at first, valves cleft at apex; seeds strophiolate. Our plants intergrade somewhat in leaf shape with A. lateriflora L.---Moist usually shaded places. Labrador to British Columbia, south to Vermont, New Mexico and California. Our few records from southwestern Colorado at 10,000-11,000 feet.

5. **Arenaria confusa** Rydb., Bull. Torr. Club 28:275. 1901.

Perennial plants; stems 20-50 cm. tall, diffuse, many from the root, prostrate or decumbent at base; leaves 10-20 mm. long, rarely less, lanceolate, oblong-lanceolate or rarely narrower, acute or mucronate; inflorescence a few- to many-flowered cyme, pedicels slender, somewhat divergent; lower bracts foliose; sepals about 3 mm. long, ovate-lanceolate or narrower, acuminate; petals shorter than the calyx; capsule valves 2-cleft at apex; seeds not strophiolate.---Usually sandy soil. Colorado to California, south to Texas and Arizona. Our records from southcentral and southwestern Colorado at 7000-8500 feet.

6. **Arenaria sajanensis** Willd. in Schlecht, Berl. Gesell. Nat. Fr. Mag. 7:200. 1813.

Perennial, densely caespitose plants with woody much branched caudices; stems 2-6 cm. tall, decumbent, glandular-pubescent; leaves 4-10 mm. long, crowded on the branches, linear, obtusish, obscurely keeled only; flowers solitary or in 2's or 3's; sepals usually 2-3 mm. long, rarely to 4, obtuse, oblong or oval, sometimes narrower, 3-nerved, glabrous or puberulent; petals short and narrow, shorter than or barely exceeding the sepals and about 0.5-1 mm. wide; anthers 0.2-0.3 mm. long; capsule 3-6 mm. long, the valves membranous to firm, entire or merely emarginate at apex, not cleft; seeds not strophiolate.---Arctic and alpine areas. Arctic Region, south to Labrador, Arizona and Oregon; Asia. Our records scattered in the mountains of Colorado at 9500-12,500 feet.

7. **Arenaria obtusiloba** (Rydb.) Fernald, Rhodora 21:14. 1919.

Alsinopsis obtusiloba Rydb.---Perennial, densely caespitose plants, caudices branched and woody; stems 1-6 cm. tall, decumbent, glandular-puberulent; leaves 4-8 mm. long, densely crowded on the branches, these highly marcescent, subulate or narrowly linear, obtuse at apex, thick-ribbed, almost keeled from prominent midvein; flowers solitary or 2 or 3 together; sepals 4-5 mm. long, obtuse, oblong-lanceolate, elliptic or oblong, 3-nerved, long-pubescent, pilose or hirsute, often rather cucullate at apex; petals definitely surpassing the sepals in length, 1.5-2.5 mm. wide; anthers 0.5-1 mm. long; capsule 6-10 mm. long with valves entire or merely emarginate at apex, not cleft, rather firm-walled; seeds not strophiolate. This species may not be distinct from A. sajanensis Willd.---In the mountains. Labrador to Alaska, south to New Mexico and California. Our records scattered in the high mountains of Colorado at 10,000-13,000 feet.

8. **Arenaria macrantha** (Rydb.) A. Nels. ex Coult. & Nels., Man. 186. 1909.

Alsinopsis macrantha Rydb.---Perennial plants; stems 4-10 cm. long, caespitose, spreading, glabrous; leaves 5-10 mm. long, subulate-filiform, 1-nerved, obtuse; inflorescence few-flowered or flowers solitary; sepals 4-5 mm. long, lanceolate, acute or acuminate, 3-nerved; petals definitely exceeding the sepals, usually 2 mm. or more longer; capsule valves entire; seeds not strophiolate. May not be distinct from the next.---In the mountains, often in sandy soil. Colorado. Our records from the southwestern part of the state at 10,000-11,500 feet.

9. **Arenaria rubella** (Wahlenb.) Smith, Eng. Bot. 4:276. 1828.

A. aequicaulis A. Nels.; Alsinopsis propinqua (Richards.) Rydb.; A. quadrivalvis (R. Br.) Rydb.; A. verna of Manuals---Perennial plants; stems 3-10 cm. high, from taproots, branched from the base and tufted, glabrous to glandular-pubescent; leaves 3-10 mm. long, linear-subulate, rather pungent, ascending, the basal ones 3-nerved; flowers in open cymes; sepals 2.5-3.5 mm. long, short-acuminate or strongly acute; petals shorter than the sepals, rarely slightly longer; capsule valves entire; seeds not strophiolate. The glabrous plants with leaves 1-nerved have been called A. filiorum Maguire.---In the mountains especially in sandy or rocky places. Hudson Bay to Mackenzie, south to New Mexico and California; Eurasia. Our records from the higher mountains of Colorado at 9000-13,000 feet.

A caespitose or mat-forming plant, A. nuttallii Pax, similar to the above species, has been reported from Colorado.

10. Arenaria rossii Richardson, Frank. Journ. 738. reprint 10. 1823.
Alsinopsis rossii (Richard.) Rydb.---Perennial densely caespitose, tufted plants; stems 1-8 cm. tall, glabrous, leafy; leaves 5-8 mm. long, linear-subulate to linear, glabrous, rather fleshy, 3-angled in section, 1 angle definitely stronger than the other 2, obtuse; flowers solitary or rarely 2-flowered; sepals 3-4 mm. long, glabrous, the tips spreading; petals usually shorter than the sepals, sometimes wanting; capsule shorter than the sepals, the valves entire; seeds reddish-brown, not strophiolate.---High mountains. Arctic regions, south to Colorado and Oregon. Our one rather doubtful record from southwestern Colorado at 12,000 feet.

11. Arenaria hookeri Nutt. ex T. & G., Fl. N. Amer. 1:178. 1838.
A. pinetorum A. Nels.---Perennial plants; stems densely tufted and branched from a deep woody root, the flowering stems 4-15 cm. long, more or less pubescent; leaves 5-50 mm. long, linear-subulate, rigid, pungent; inflorescence a dense capitate cyme, the flowers sessile or subsessile; bracts lanceolate-subulate, scarious-margined; sepals 5-9 mm. long, lanceolate or narrowly-lanceolate, resembling the bracts, acuminate; petals longer than the sepals, often twice as long; capsule valves 2-cleft at the apex; seeds not strophiolate.---Dry hills and plains. Montana, south to Nebraska and Colorado. Our records from the eastern half of Colorado, and also in the northwestern corner, at 3500-7500 feet.

12. Arenaria congesta Nutt. ex T. & G., Fl. N. Amer. 1:178. 1838.
Perennial plants with short, branched, somewhat woody caudices; stems 10-30 cm. tall, simple, with 2-4 pairs of leaves, glabrous; leaves 1-5 cm. long, filiform-subulate, glabrous, pungent, ascending or erect; inflorescence of 1 or more dense capitate cymes; bracts scarious, ovate; sepals about 4 mm. long, ovate, midnerve evident, the rest scarious; petals longer than the calyx, usually about twice as long; capsule valves 2-cleft at apex; seeds not strophiolate.---Dry plains and hills. Montana to Washington, south to Colorado and California. Our records from the western half of Colorado, mostly in the northwestern quarter of the state, at 6000-12,000 feet.

13. Arenaria fendleri A. Gray, Mem. Am. Acad. ser. 2. 4:13. 1849.
A. fendleri var. diffusa Porter; A. fendleri var. porteri Rydb.---Perennial plants from woody roots; stems variable as to length, 10-30 cm. tall, or even shorter at high altitudes, tufted, erect, usually bluish-green, slender, glandular-pubescent at least above; leaves 2-8 cm. long, filiform or subulate, pungent but not strongly so; inflorescence an open cyme; sepals 4-5 mm. long, linear-lanceolate, acuminate, scarious-margined; petals shorter to slightly longer than the sepals; capsule valves 2-cleft at apex; seeds not strophiolate. This is a puzzling complex in this state but attempts to recognize subspecific entities have not been very successful.---Mountains and hills. Wyoming to New Mexico and Arizona. Our records well scattered in the western two-thirds of Colorado at 6000-13,000 feet.

13A. Arenaria fendleri eastwoodiae (Rydb.) comb. nov. (var.)
A. eastwoodiae Rydb.---Perennial plants with caespitose, rather woody bases; stems 10-20 cm. long, usually not over 15, glabrous; leaves 1-2 cm. long, filiform, sharply pungent; inflorescence an open cyme; bracts subulate to lanceolate, margins scarious; sepals 4-6 mm. long, linear-lanceolate to ovate-lanceolate, acuminate; petals shorter than or barely equalling the sepals or rarely slightly longer, white or yellowish-white; capsule valves 2-cleft at apex; seeds not strophiolate. The variety intergrades with the species to some degree in Colorado.---Colorado to Utah, south to New Mexico and Arizona. Our records from northwestern Colorado at 5000-9000 feet.

13B. Arenaria fendleri tweedyi (Rydb.) Maguire, (var.) Bull. Torr. Club 74:50. 1947.
A. tweedyi Rydb.---This plant intergrades with the species in our area. A few plants of southwestern Colorado may possibly be distinct.

14. Arenaria uintahensis A. Nels., Bull. Torr. Club 26:7. 1899.
Perennial plants; stems 10-20 cm. tall, decumbent at base, nearly glabrous to sparingly glandular-pubescent; leaves 1-3 cm. long, narrowly linear or filiform, somewhat acerose; inflorescence an open cyme; bracts reduced and scarious; sepals 4-5 mm. long, ovate or ovate-lanceolate, acute; petals somewhat longer than the sepals; capsule valves 2-cleft at apex; seeds not strophiolate. Intergrades somewhat with the A. fendleri A. Gray complex in Colorado and related to the western A. kingii (S. Wats.) M. E. Jones.---Dry hills. Wyoming to Oregon, south to Colorado and California. Our records from the western row of counties at 4500-9000 feet.

Family 45. Nymphaeaceae WATER LILY FAMILY

Perennial, acaulescent, aquatic herbaceous plants with horizontal or tuberous rootstocks; leaves entire with elongated petioles and broad floating blades; flowers solitary, long-peduncled, regular, perfect; sepals distinct, 3-12, green to petaloid; petals separate, numerous, often intergrading into staminodia and stamens; stamens many; pistil superior, of many carpels, these united into a compound ovary, stigmas disklike, of 8-24 radiating rays or parts, ovules 2 - many, parietal; fruit a leathery berry.

1. Nuphar J. E. Smith POND LILY; COWLILY

Characters of the Family

1. Nuphar polysepalum Engelm., Trans. St. Louis Acad. 2:282. 1865.
Nymphaea polysepala (Engelm.) Greene---Leaves with long petioles, floating or sometimes raised above the surface, the blades 20-40 cm. long, oval, the basal sinuses narrow or closed; sepals 2-3 cm. long, yellow often tinged with red; petals smaller, mostly concealed by the numerous stamens; fruit ovoid, about 3.5 cm. in diameter, with a constricted neck.---Lakes and slow streams. Montana to Alaska, south to Colorado and California. Our records scattered in the mountains of Colorado at 9000-11,000 feet.

Family 46. Ceratophyllaceae HORNWORT FAMILY

Submerged aquatic herbaceous plants; leaves verticillate and finely dissected; flowers monoecious, regular, inconspicuous, solitary and sessile in the axils, subtended by an 8- to 12-cleft perianthlike involucre, the perianth wanting; stamens many, with short filaments or anthers sessile; ovary superior, of a single carpel, 1-celled and with 1 ovule; fruit a nutlet, the style persistent.

1. Ceratophyllum L. HORNWORT

Characters of the Family

1. Ceratophyllum demersum L., Sp. Pl. 992. 1753.
Stems 20-100 cm. long but varying in length with the depth of the water; leaves 5-12 in each whorl, dichotomously forked into linear or filiform segments, these often toothed; achenes oblong, smooth or tuberculate, the smooth with a basal spur often present on each side, and the tuberculate achenes narrowly to broadly winged, the body about 5 mm. long.---Lakes and ponds. Newfoundland to Washington, south to Florida, Mexico and California; Eurasia. Our records scattered over Colorado at 5000-6500 feet.

Family 47. Ranunculaceae CROWFOOT FAMILY

Annual or perennial plants, herbaceous, or climbing and shrubby; leaves usually alternate, simple or compound, without stipules but the petiole base often dilated; flowers perfect or unisexual, hypogynous, regular or irregular; sepals 3-15, usually green and caducous, usually imbricate, distinct; petals absent, or present and usually the same number as the sepals, distinct; stamens numerous, rarely few; carpels superior (or appearing inferior in Paeonia), few to many, rarely solitary, distinct, 1-celled, with 1 - many ovules; fruit of achenes, follicles or berries. The generic key is frankly artificial since the type of fruit is often difficult to ascertain especially under field conditions. The genus with a berry has only 1 pistil to a flower; those with follicles usually less than 10 and those with achenes usually more than 10.
Since this treatment was prepared a plant with petaloid sepals, small fleshy petals hollow at the apex, 3-9 pistils becoming follicles and with basal trifoliate leaves may have been collected in a bog in Gunnison County. It appears to be Coptis groenlandica (Oeder) Fern.

1. Flowers irregular, commonly blue or white; fruit of follicles
  2. Upper sepal spurred at base   -1. Delphinium
  2. Upper sepal hooded at apex   -2. Aconitum
1. Flowers regular, blue, white, red, yellow or green; fruit of follicles, achenes or a berry
  3. Annual plants with sepals spurred at base; fruit of achenes   -3. Myosurus
  3. Perennial plants, the sepals not spurred (petals are spurred in Aquilegia); fruit of achenes, follicles or a berry
    4. Petals spurred (rarely saclike) at base; fruit of follicles   -4. Aquilegia
    4. Petals not at all spurred at base (may be absent); fruit of achenes, follicles or a berry
      5. Petals lacking, the sepals often petaloid
        6. Fruit of follicles; leaves simple, not at all lobed   -5. Caltha
        6. Fruit of achenes; leaves compound or deeply lobed
          7. Stem leaves whorled or opposite (may be of large opposite leaflike bracts); sepals petaloid and almost always showy, somewhat persistent
            8. Basal leaves long-petioled, the stem naked up to the 1-3 sets of involucrelike leaves near the flowers or inflorescence, these leaves sessile or short-petioled; sepals imbricate; plants never climbing
              9. Styles becoming elongated (to 3 cm.) and plumose; flowers large, the sepals over 2 cm. long   -6. Pulsatilla
              9. Style short, not elongating in fruit, glabrous to pubescent, not at all plumose; flowers smaller, the sepals never over 2 cm. long   -7. Anemone
            8. Several to many pairs of stem leaves present, these essentially alike from base to apex of the stem, no modified involucrelike leaves present; sepals valvate; often climbing plants   -8. Clematis
          7. Stem leaves alternate (most of the leaves often basal); sepals not petaloid, not showy, caducous
            10. Leaves simple, palmately lobed or parted; flowers perfect; anthers oval or ovate, about 1 mm. long   -9. Trautvetteria
            10. Leaves decompound; flowers often imperfect; anthers linear, usually well over 1 mm. long   -10. Thalictrum
      5. Petals present (these are small and strap-shaped in Trollius); sepals are readily deciduous in some
        11. Fruit a fleshy berry; flowers numerous, in narrow terminal racemes, only 1 pistil to a flower   -11. Actaea
        11. Fruit of dry follicles or achenes; flowers various but not numerous in narrow racemes; over 1 pistil present to a flower
          12. Fruit of follicles; flowers large, the sepals over 1 cm. long; petals brownish-red to whitish
            13. Sepals persistent; petals about as long as the sepals; petals and stamens on a fleshy disk   -12. Paeonia

13. Sepals readily deciduous; petals reduced to small linear structures much shorter than the sepals; petals and stamens on the receptacle, no fleshy disk present -13. Trollius
12. Fruit of achenes; flowers smaller, the sepals rarely over 1 cm. long; petals yellow or white -14. Ranunculus

1. Delphinium L. LARKSPUR

Perennial herbaceous plants with erect, simple or branching stems; leaves variously palmately divided or lobed, alternate; flowers in terminal racemes, perfect, irregular; sepals 5, petaloid, the uppermost produced into a spur; corolla of 2 sets of 2 petals each, the 2 lower with a claw, the 2 upper prolonged into the spur; stamens many; pistils 3 or fused into 1; fruit of follicles.

1. Sepals and petals cream-white or merely tinged with blue or with a blue spot; seeds somewhat echinate
   2. Leaves equally distributed on the stems; stems usually over 50 cm. tall -1A. D. virescens penardi
   2. Leaves chiefly basal clusters; stems usually under 50 cm. tall -1B. D. virescens wootoni
1. Sepals or petals (often both) blue, purple or dark colored; seeds not echinate
   3. Leaves all strictly basal; sinuses of lower petals 3-4 mm. deep -2. D. scaposum
   3. Leaves not all in a basal tuft; sinuses of lower petals not over 3 mm. deep
     4. Stems not over 15 cm. tall; alpine plants; racemes little exceeding the leaves; flowers drab and not at all showy -3. D. alpestre
     4. Stems over 15 cm. tall (if sometimes less then plants not found above 10,000 feet); racemes usually well exceeding the leaves; flowers usually bright and showy
        5. Stems low, seldom over 50 cm. tall, easily detached from the tuberlike root or root cluster; lower pedicels usually elongate; typically spring-flowering plants -4. D. nelsoni
        5. Stems over 50 cm. tall, arising from a woody-fibrous root not tuberlike and not easily detachable; lower pedicels usually not elongate; typically flowering in summer
           6. Sepals bicolored, the sepal itself whitish and contrasted with the blue spur and dark violet-blue petals -6B. D. occidentale cucullatum
           6. Sepals not bicolored, the sepals and spurs nearly uniform in color
               7. Stems ashy-puberulent (especially below), not glandular; flowers a lively rich blue; sinuses of lower petals obsolete; plants of northern Colorado seldom growing over 6500 feet altitude -5. D. geyeri
               7. Stems glabrous to densely hirsute, often glandular, not ashy-puberulent; flowers dull blue, dark purple or sometimes bright blue; sinuses of lower petals 1 mm. deep or more; plants not confined to northern Colorado and usually found above 6500 feet
                  8. Flowers narrow, the upper sepal and its spur forming nearly a straight line (even in old flowers); sepals often widest above the middle; stems usually glandular-hairy and the sepals not over 12 mm. long
                     9. Sepals dull gray on back with a close puberulence; racemes 15 cm. long or more -6. D. occidentale
                     9. Sepals dark blue-purple throughout; racemes 8-12 cm. long -6A. D. occidentale quercicola
                  8. Flowers not especially narrow, the upper sepal flaring (at least in older flowers); sepals widest at or below the middle; stems either without glandular hairs or if glandular-hairy then the sepals 14 mm. long or more
                     10. Rachis of raceme and pedicels lustrous, glandular-hirsute; sepals dark purple, often acuminate; stems usually 5-20 from a single root -7. D. barbeyi
                     10. Rachis and pedicels non-glandular; sepals dull or bright blue, not acuminate; stems single or few from a root
                         11. Sepals 8-10 mm. long; lower petals somewhat exserted; seeds strongly wing-angled; stems seldom over 100 cm. tall -8. D. ramosum
                         11. Sepals 11-15 cm. long; lower petals included; seeds narrowly wing-angled; stems usually over 100 cm. tall -9. D. robustum

1. **Delphinium virescens** Nutt., Genera N. Am. Plants 2:14. 1818.
Represented in Colorado by the following 2 subspecies.
1A. **Delphinium virescens penardi** (Huth.) Ewan, (ssp.) U. Colo. Studies Ser. D. 2:167. 1945.
D. penardi Huth.; D. carolinianum var. penardi (Huth.) A. Nels.; D. camporum Greene---Plants tall and stout from a cluster of several woody-fibrous, elongated, deep-seated roots; stems 30-80 cm. tall, sometimes even taller, simple, closely canescent-pubescent or pannose, the hairs short and curled; leaves subbasal or on lower part of stem, palmately parted and parted again into linear segments; racemes moderately compact or spikelike, flowers many; sepals 10-12 mm. long, more or less whitish, the spur 11-13 mm. long and straight or curved at tip; lower petals moderately comose-bearded, the sinuses 3-4 mm. deep; follicles 16-20 mm. long, puberulent when young, nearly glabrous in age; seeds 2 mm. long, black and somewhat echinate.---Dry plains and slopes. Colorado, Kansas and Oklahoma. Our records scattered over the eastern half of Colorado, with one from La Plata County, at 3500-8000 feet.
1B. **Delphinium virescens wootoni** (Rydb.) Ewan, (ssp.) U. of Colo. Studies Ser. D. 2:169. 1945.
D. wootoni Rydb.---This is much like ssp. penardi but the stems are shorter, mostly less than 30 cm. tall; leaves are in a close basal tuft, rarely also cauline but mostly basal; racemes are densely spicate. This subspecies intergrades very commonly with ssp. penardi in Colorado.---Colorado, south to Texas and Arizona. Our records from the southeastern quarter of Colorado at 4500-6000 feet.
2. **Delphinium scaposum** Greene, Bot. Gaz. 6:156. 1881.
Scapose plants from slender woody fibrous roots; scapes 20-50 cm. tall, simple, erect, glabrous; leaves wholly basal or very nearly so (a few reduced leaves rarely on the stem), 3- to 5-parted or divided, the divisions round lobed or toothed; raceme open, 5- to 10-flowered; sepals

10-15 mm. long, usually royal-blue, spurs often bronze-colored near tips, straight, much longer than the sepals; upper petals white or blue-tinged; lower petals dark blue, cleft to a sinus 3-4 mm. deep; follicles 10-20 mm. long, ovate or oblong, glabrous; seeds dark brown with a whitish papery pellicle.---Dry ground. Colorado to Nevada, south to New Mexico and Arizona. Our records from southwestern and westcentral Colorado at 4500-7500 feet.

3. Delphinium alpestre Rydb., Bull. Torr. Club 29:146. 1902.
 Low caespitose plants from slender vertical rootstocks; stems about 10-15 cm. tall, several together, rather weak and slender, puberulent or glandular-hairy above; leaves crowded, palmately divided into 5 divisions, these again lobed, blades small, 2-3 cm. wide; racemes short, little exserted above the leaves, congested at first; flowers dark and drab, not showy, brownish on back; sepals 11-15 mm. long, spur 7-9 mm. long, recurved at apex; lower petals with short sinuses; ovary hirsute (follicles not known).---Among rocks at high altitudes. Apparently limited to Colorado, our 2 records from the central and southcentral parts at 11,500 feet or higher.

4. Delphinium nelsoni Greene, Pitt. 3:92. 1896.
 D. dumetorum Greene; D. pinetorum Tide.---Plants from short fasciculate rootstocks or a cluster of a few tuberlike roots; stems 10-50 cm. tall, easily separating from the roots, simple, rather stout, puberulent to subglabrous; leaves few, blades 3-5 cm. wide, orbicular, palmately deeply parted and these divisions again lobed into narrowly oblong or linear divisions; racemes rather short, 6- to 10-flowered, the flowers rather showy, purple; sepals 11-17 mm. long, about equal, rich blue-purple to pale blue, spur usually 10-15 mm. long, slender and nearly straight; upper petals nearly white, entire or cleft at apex; follicles 13-19 mm. long, densely short-hairy or glabrate; seeds 2 mm. long, dark brown, narrowly wing-margined. A form with whitish flowers is not uncommon in Larimer County.---Plains and slopes often among bushes or trees. South Dakota to Idaho, south to Colorado and Arizona. Our records scattered over the state, none as yet from the extreme eastern part, at 5000-10,500 feet.

5. Delphinium geyeri Greene, Erythea 2:189. 1894.
 Plants with tough woody fibrous, usually vertical rootstocks; stems 20-70 cm. tall, usually several, strict, ashy-puberulent especially below; leaves several, mostly toward the base, divided, the divisions repeatedly dissected, the ultimate divisions straight and linear to linear-filiform; racemes narrow, strict and rather dense, flowers showy; sepals 10-15 mm. long, rarely more, oblong-ovate, acute to obtuse, spur up to $1\frac{1}{2}$ times as long as its sepal, straight or nearly so; lower petals hairy near middle, lower sinuses obsolete or short, closed; upper petals whitish; follicles 12-15 mm. long, finely hirsutulose; seeds 2-3.5 mm. long, flat, body roughened.---Open plains and slopes often among bushes. Wyoming south to Nebraska and Utah. Our records from northern Colorado at 4500-7500 feet.

6. Delphinium occidentale (Wats.) Wats., Bot. Calif. 2:428. 1880.
 D. reticulatum (A. Nels.) Rydb.; D. multiflorum Rydb.---Plants from deep vertical woody roots; stems 60-200 cm. tall, somewhat straw-colored especially at base, somewhat glaucous; leaves divided into 3-7 divisions, these cleft below middle; racemes usually over 15 cm. long, dense, spicate or loosely paniculate in age, rachis thinly to densely glandular-hairy, bracts small; sepals 6-12 mm. long, narrowly ovate-oblong, rounded or acute at apex, usually paler or gray canescent on the back, otherwise blue-purple, spur 9-12 mm. long, horizontal, straight or curved somewhat near tip; lower petals with sinuses 1-2 mm. long; upper petals small and included; follicles 9-12 mm. long, short-oblong, glabrous to glandular-pubescent; seed wing-angled. Intergradations between the species and the following 2 subspecies are very common in Colorado.---Meadows, thickets and open woods. Montana to Idaho, south to Colorado and Nevada. Our records scattered in the western half of Colorado, mostly in the north, at 6500-9500 feet.

6A. Delphinium occidentale quercicola Ewan, (ssp.) U. of Colo. Stud. Ser. D. 2:139. 1945.
 Differs from the species in having the sepals dark blue-purple throughout.---With the species in Colorado but uncommon.

6B. Delphinium occidentale cucullatum (A. Nels.) Ewan, (ssp.) U. of Colo. Stud. Ser. D. 2:138. 1945.
 D. cucullatum A. Nels.---Stems, rachis and pedicels of inflorescence not densely hirsute (as is common in the species); flowers distinctly bicolored, sepals white at least distally, contrasting with light blue spurs and violet-blue petals.---Our few good records from northcentral Colorado.

7. Delphinium barbeyi (Huth.) Huth., Bull. Boiss. Herb. 1:335. 1893.
 D. subalpinum (Gray) A. Nels.; D. cockerelli A. Nels.---Plants from stout woody rootstocks; stems 50-200 cm. tall, hollow, more or less glandular-hirsute, densely so in the inflorescence, often subglabrous near base; leaves ample, little reduced above, broader than long, divided or cleft into 3 primary segments, these entire below and cleft or coarsely toothed in upper half; raceme rather compact, short-oblong, flowers scented and showy with conspicuous bracts below; sepals 14-16 mm. long, narrowly ovate, attenuate or acute or even caudate, rich dark purple, spur about 10 mm. long, curved near tip; lower petals colored like sepals, sinuses about 3 mm. deep; upper petals little exserted, edged with white; follicles 14-17 mm. long, oblong-ovate with purplish veins, subglabrous; seeds smoky-brown, wing-angled.---Meadows and open woods. Wyoming, Colorado and Utah. Our records well scattered in the western half of Colorado at 8000-13,000 feet.

8. Delphinium ramosum Rydb., Bull. Torr. Club 28:276. 1901.
 D. elongatum Rydb.---Plants from stout but short woody roots; stems 50-100 cm. tall or sometimes to 200 cm., hollow, usually glabrous at middle, thinly hairy below and often pubescent in the racemes; leaves divided to near the base into 5-7 segments, these cleft, ultimate divisions oblong or lanceolate; racemes loosely spicate, simple, 16- to 25-flowered, these dark but still a bright blue; sepals 8-10 mm. long, ovate, acute, puberulent on back, spurs 9-10 mm. long, slender, nearly straight; lower petals with sinuses 1.5-2 mm. deep; follicles 10-13 mm. long, short-ovate, puberulent; seeds wing-angled, dark-brown.---Meadows, among bushes

or in open woods. Colorado and possibly northward according to report. Our records scattered in the mountains of Colorado, none as yet from the extreme west, at 6500-10,000 feet.
9. Delphinium robustum Rydb., Bull. Torr. Club 28:276. 1901.
Plants from heavy woody roots; stem 100-200 cm. tall or more, simple, hollow, stout, finely puberulent throughout; leaves divided to the base into 5-7 segments, these again cleft; racemes bracteate, elongate, loosely paniculate on the lower half, with dark azure-blue flowers; sepals 11-15 cm. long, broadly ovate, acute, puberulent to glabrous, spur 9-15 mm. long, straight; lower petals included, the shallow sinuses 1-2 mm. deep; upper petals entire or crisped; follicles 11-13 mm. long, short-oblong, dark veined; seed glabrous, tawny, narrowly wing-margined.---Edges of meadows or in open woods. Colorado and possibly New Mexico. Our few records from southcentral and southwestern Colorado at 7000-8500 feet.

2. Aconitum L. MONKSHOOD

Perennial herbaceous plants; stems usually tall, leafy; leaves alternate, palmately lobed or parted; flowers usually blue-purple, in simple racemes or panicles, perfect, irregular; sepals 5, petaloid, the upper one larger and helmet-shaped or hooded at apex; petals 2-5, small, concealed in the hooded sepal, the upper 2 hooded; stamens numerous; pistils 3-5, many-ovuled, forming follicles in fruit.

1. Front line of hood nearly straight, the beak directed downward -1. A. columbianum
1. Front line of hood curved and concave, the beak porrect and abruptly horizontal -1A. A. columbianum bakeri

1. Aconitum columbianum Nutt. ex T. & G., Fl. N. Amer. 1:34. 1838.
A. insigne Greene---Stems 60-150 cm. tall, stout, glabrous below but pubescent or tomentose above, often viscid; leaves palmately parted into 3-5 divisions, these cleft and toothed, the whole blade 5-15 cm. wide; flowers in a rather loose terminal simple or few-branched raceme, usually blue but not uncommonly pale to white; hood 10-20 mm. long, the front line commonly nearly straight and the beak directed downward but sometimes the beak porrect to some degree; follicles 10-20 mm. long usually more or less pubescent. A. lutescens A. Nels. appears to be only an albino form of the above.---Moist meadows and open woods. Montana to British Columbia, south to New Mexico and California. Our records well scattered over the western two-thirds of Colorado at 7500-11,500 feet.
1A. Aconitum columbianum bakeri (Greene) comb. nov. (var.)
A. bakeri Greene; A. porrectum Rydb.---Seems to differ from A. columbianum Nutt. only in the front line of hood being concave hence the beak porrect and horizontal. Many intergradations occur in our plants but the extremes are different looking.---Our records scattered over the western half of Colorado at 7500-11,500 feet.

3. Myosurus L. MOUSETAIL

Annual, acaulescent plants with fibrous roots; leaves basal, tufted, narrow, entire; flowers small, solitary, regular, perfect; sepals usually 5, spurred at the base, deciduous; petals rudimentary or none; stamens 5 to numerous; pistils numerous on a cylindrical axis which becomes greatly elongated in fruit; fruit of achenes.

1. Sepals 1-nerved; spikes less than 1 cm. long -1. M. aristatus
1. Sepals 3- to 5-nerved (may be very faint on some); longest spikes over 1 cm. long
   2. Back of achenes (not counting beaks) orbicular or suborbicular in outline; small plants with leaves not over 15 mm. long and scapes not over 2 cm. tall; plants limited to southwestern Colorado -2. M. nitidus
   2. Back of achenes not orbicular or suborbicular but definitely longer than wide; plants with leaves and scapes usually longer than above; plants not limited to southwestern Colorado -3. M. minimus

1. Myosurus aristatus Benth. in Hook., Lond. Journ. Bot. 6:459. 1847.
Reported from British Columbia, south to Wyoming, Utah and California. Apparently what we have been calling by this name is a subspecies of M. minimus L. (El Aliso 2(4):389-403. 1952) but the species is to be expected in this state.
2. Myosurus nitidus Eastw., Bull. Torr. Club 32:194. 1905.
M. egglestonii Wooton & Standley---This species is reported from northern Arizona, northern New Mexico and southwestern Colorado.
3. Myosurus minimus L., Sp. Pl. 284. 1753.
M. apetalus of Manuals---Scapes 2-18 cm. tall; leaves 2-8 cm. long, filiform to narrowly linear; sepals 2-3.5 mm. long, 3- to 5-nerved but often faintly so; petals linear, seemingly absent sometimes; fruiting spikes 1-6 cm. long; achenes 1-3 mm. long, beak variable in length, up to 1 mm. long. The plants with beaks less than 0.5 mm. above the achene body have been called ssp. minimus and the ones with longer beaks have been named ssp. montanus Campbell. The latter type is more common in this state.---Moist or wet places. The 2 subspecies from British Columbia, south to Texas and California. Our records scattered over Colorado at 4000-9500 feet.

4. Aquilegia L. COLUMBINE

Plants perennial and herbaceous; leaves 2 or 3 times ternate, rarely once ternate; flowers perfect, regular and showy; sepals 5, petaloid, deciduous; petals 5, the blade (laminae) rather small but produced back between the sepals into a hollow spur, rarely saccate; stamens numerous, the inner reduced to staminodia, these petaloid or scalelike; ovaries commonly 5 but often variable (3-10 on A. caerulea), sessile with many ovules, becoming follicles in fruit.

1. Spurs lacking entirely or reduced to short saclike structures
  2. Sepals 3-4 cm. long, blue   -8B. A. caerulea daileyae
  2. Sepals 1-1.5 cm. long, white to pinkish   -3A. A. micrantha mancosana
1. Spurs present as slender structures
  3. Spur hooked at the tip, rarely over 15 mm. long, shorter or but little longer than the petals
    4. Spurs blue, 3-5 mm. long; sepals blue, 9-12 mm. long; stamens about as long as the petals; plants 5-20 cm. tall   -1. A. saximontana
    4. Spurs yellow. 6-18 mm. long; sepals yellow, 12-22 mm. long; stamens 5-10 mm. longer than the petals; plants 20-70 cm. tall   -2. A. flavescens
  3. Spurs not hooked at the tip, 15-70 mm. long, usually twice or more than twice the length of the petal
    5. Flowers nodding; spurs 15-22 mm. long; petals 6-8 mm. long (A. micrantha is keyed both places)
      6. Leaves viscid-pubescent beneath; spurs whitish rarely reddish; petals whitish   -3. A. micrantha
      6. Leaves glabrous to pilose below but not viscid; spurs red or light red; petals yellowish
        7. Sepals 10-20 mm. long, nearly or quite twice as long as the petals; basal leaves usually triternate; stamens to 10 mm. longer than the petals   -4. A. triternata
        7. Sepals 7-11 mm. long, not much longer than the petals; basal leaves biternate; stamens 4-6 mm. longer than the petals   -5. A. elegantula
    5. Flowers erect; spurs 22-70 mm. long; petals 8-25 mm. long
      8. Leaflets thick, closely clustered, 5-15 mm. wide; plants 5-20 cm. tall   -6. A. scopulorum
      8. Leaflets normal, not crowded, often wider than 15 mm.; plants 20-120 cm. tall
        9. Spurs 15-28 mm. long, mostly whitish; sepals 8-20 mm. long; stamens 3-8 mm. longer than the petals; leaves viscid-pubescent below   -3. A. micrantha
        9. Spurs 30-70 mm. long, mostly blue or yellow; sepals 20-40 mm. long; stamens either about as long as the petals or 8-10 mm. longer; leaves glabrous to pubescent below but hardly viscid
          10. Flowers golden-yellow; stamens 8-10 mm. longer than the petals; petals 8-16 mm. long   -7. A. chrysantha
          10. Flowers blue or white; stamens little if any longer than the petals; petals 15-25 mm. long
            11. Sepals blue   -8. A. caerulea
            11. Sepals whitish or pale   -8A. A. caerulea ochroleuca

1. **Aquilegia saximontana** Rydb. in Robins., Syn. Fl. 1:43. 1895.
Stems 8-15 cm. tall rarely to 25 cm., usually densely tufted, glabrous, hardly longer than the leaves; basal leaves biternate, glabrous, cauline few; leaflets small, 12-16 mm. long; flowers small for the genus, 1.5-2 cm. long and about as wide, nodding; sepals about 8-15 mm. long; petals about 7-8 mm. long, white or whitish; spurs 3-7 mm. long, blue, incurved or hooked; follicles 5-6, about 1 cm. long. Our Colorado plants have incorrectly been called A. brevistyla.---Among rocks at high altitudes. Limited to central and northcentral Colorado. Our records at 10,000-12,000 feet.

2. **Aquilegia flavescens** S. Wats., Bot. King. Expl. 10. 1871.
Stems 20-60 cm. tall, subglabrous to sparsely pilose below, glandular-pubescent to pilose above; basal leaves biternate to almost triternate, rather thin, glabrous to pubescent beneath, somewhat glaucous below; leaflets 1-4 cm. long, broadly obovate, cuneate, mostly cleft; cauline leaves usually present; sepals 12-22 mm. long, spreading to reflexed, usually yellow, acute; petals 6-10 mm. long, cream-colored, broadly rounded; spurs 6-18 mm. long, yellowish, incurved or hooked at apex; stamens 5-10 mm. longer than petals; follicles 15-22 mm. long, glandular-pubescent.---Often in open woods. Alberta to British Columbia, south to Colorado and Utah. Our one record from "Colorado."

3. **Aquilegia micrantha** Eastw., Proc. Calif. Acad. Sc. Ser. 2. 4:559. 1895.
A. pallens Payson; A. lithophila Payson---Stems 30-60 cm. tall, mostly glandular-pubescent throughout or sometimes glabrous below; leaves biternate to triternate, glabrous or pubescent above, viscid-pubescent beneath (though this sometimes sparse); flowers 3-4.5 cm. long, erect or somewhat nodding; sepals 8-20 mm. long, white, yellowish, pale blue or reddish; lamina of petals 5-10 mm. long, yellowish or whitish; spurs 1.5-3 cm. long, whitish or yellowish, rarely reddish, slender; follicles glandular-pubescent.---Usually in canyons and on rocky cliffs. Colorado and Utah. Our records from western Colorado at 5000-8000 feet.

3A. **Aquilegia micrantha mancosana** Eastw., (var.) Proc. Calif. Acad. Sc. Bot. Ser. 3. 1:77. 1897.
A. eastwoodiae Rydb.; A. ecalarata Eastw.---Spurs are lacking or merely saccate.---Apparently limited to southwestern Colorado.

4. **Aquilegia triternata** Payson, Contrib. U. S. Nat. Herb. 20(4):147. 1918.
Stems 20-60 cm. tall, several to many from thick caudices, slender, sparsely pubescent; basal leaves long-petioled, mostly triternate; leaflets rather small, pubescent or glabrous; flowers about 3-4 cm. long and 2-3 cm. wide, nodding; sepals 10-20 mm. long, light red; petals about 7-8 mm. long, yellow or pale red; spurs 18-24 mm. long, light red, stout; follicles 15-20 mm. long.---Moist places often on rocky cliffs. Colorado, New Mexico and Arizona. Our one record from western Colorado at 6400 feet.
Recently a plant from Ria Blanca County has been named A. barnebyi Munz, Leafl. West. Bot. 5:177. 1949. It will key out to the above. Our western Colorado species need to be clarified.

5. **Aquilegia elegantula** Greene, Pitt. 4:14. 1899.
A. elongatula Greene appears to be a misprint in Rydberg's Manual---Stems 10-40 cm. tall, erect, slender, usually glabrous; basal leaves biternate, stem leaves few; leaflets small, glabrous and glaucous beneath; flowers 3-3.5 cm. long and 1-2 cm. wide, nodding; sepals 7-11 mm. long, red or with green or yellow tips; petals 6-8 mm. long, usually shorter than the sepals, yellow; spurs 16-20 mm. long, scarlet, straight; follicles 5, 1.5-1.8 cm. long.---Wooded slopes, often in rocky ground. Colorado, Utah and New Mexico. Our records scattered over the western half of Colorado, except in the extreme north, at 7000-10,000 feet.

6. Aquilegia scopulorum Tidestrom, Amer. Midl. Nat. 1:167. 1910.
Reported from southwestern Wyoming to Utah and Nevada. The species should be looked for in northwestern Colorado.

7. Aquilegia chrysantha A. Gray, Proc. Amer. Acad. 8:621. 1873.
A. thalictrifolia Rydb.---Stems 30-100 cm. tall, glabrous or more or less pubescent above, often viscid; basal leaves mostly triternate; leaflets often pubescent below; flowers 5-9 cm. long and 4-7 cm. wide, clear golden-yellow throughout, erect; sepals 2-3.5 cm. long; petals 8-20 mm. long; spurs 4-7 cm. long, very slender; follicles about 2.5 cm. long.---In the mountains especially along streams or in rocky ravines. Colorado south to Texas, Mexico and Arizona. Our few records from central and southern Colorado at 5500-7000 feet.

8. Aquilegia caerulea James in Long's Exped. 2:15. 1823.
Stems 20-80 cm. tall, usually glabrous below to more or less glandular-pubescent above; leaves mostly basal, long-petioled, biternate; leaflets glabrous or sparsely pubescent especially beneath, glaucous below, often deeply cleft or parted; flowers large, usually 6-8 cm. long and 6-10 cm. wide, smaller at high altitudes, usually erect; sepals 20-40 mm. long, typically a deep blue; petals 15-25 mm. long, white; spurs 30-50 mm. long, mostly blue, slender, straight or spreading; follicles 2-3 cm. long, variable in number but usually 5, 6 or 7. This is the state flower of Colorado.---Woods and slopes, often in aspen groves. Montana to Idaho, south to New Mexico and Arizona. Our records well distributed over the western two-thirds of Colorado at 7000-12,500 feet.

8A. Aquilegia caerulea ochroleuca Hook., (var.) Bot. Mag. 90:pl. 5477. 1864.
A. caerulea var. albiflora Gray; A. caerulea var. leptocera A. Nels.---Like the species but sepals whitish or pale colored. Our few records from northwestern, northcentral and south-central Colorado at 7500-10,500 feet.

8B. Aquilegia caerulea daileyae Eastw., (var.) Proc. Calif. Acad. Sc. Ser. 3. 1:76. 1897.
Spurs lacking. Intergrades with the species to some degree. Our few records from north-central Colorado at about 7000-7500 feet.

5. Caltha L. MARSH MARIGOLD

Fleshy perennial herbaceous glabrous plants; leaves mostly basal, few, simple, with broad, entire or crenate blades; flowers regular, showy; sepals petaloid, early deciduous; petals none; stamens numerous, with short filaments; pistils several to many, sessile, ovules numerous; fruit a head of follicles, each several-seeded.

1. Caltha leptosepala DC., Syst. 1:310. 1818.
C. rotundifolia (Huth) Greene---Stems scapose, 1-flowered or 2-flowered; leaves radical, broadly ovate, oblong, elliptical or orbicular, the blades becoming large, often to 8 cm. or more long, a sinus present at base, crenate to subentire; sepals 5-15, 10-20 mm. long, oblong-obovate to linear-oblong, whitish or somewhat blue especially outside; follicles 12-20 mm. long when mature.---Moist or wet ground. Alberta to British Columbia, south to New Mexico and Utah. Our records scattered in the higher mountains of Colorado at 8500-13,000 feet.

6. Pulsatilla Mill. PASQUE FLOWER

Perennial scapose herbaceous plants with short caudices and thick taproots; leaves basal, palmately divided; stem leaves (or bracts) 3, forming a whorled involucre, sessile and often connate, resembling the basal leaf blades; sepals petaloid, petals wanting; stamens numerous, the outer often sterile; achenes numerous, with persistent, long plumose styles. Closely related to Anemone and our species often placed in that genus.

1. Pulsatilla ludoviciana (Nutt.) Heller, Cat. N. Amer. Pl. ed. 2. 4. 1900.
P. hirsutissima (Pursh.) Britton in the Manuals---Plants densely villous; scapes 10-40 cm. tall; leaves divided and lobed several times into linear segments, appearing later than the flowers; flowers mostly solitary, long-peduncled in fruit; sepals 2-4 cm. long, ovate-oblong, whitish to purple; achenes silky, with styles about 3 cm. long.---Prairies and slopes. Illinois to Alaska, south to Texas and Washington. Our records scattered over the western two-thirds of Colorado at 5000-11,500 feet, mostly below 8500 feet.

7. Anemone L. ANEMONE; WINDFLOWER

Perennial herbaceous plants with erect, scapelike stems; leaves basal, long-petioled, palmately parted or divided; 2 or 3 stem leaves (or bracts) forming an involucre this subtending or rather remote from the flower; flowers terminal, solitary or cymose, often appearing umbellate; sepals 4-20, petaloid; petals none; stamens numerous; fruit a head of many compressed achenes with short glabrous or pubescent but not plumose styles.

1. Stems 10-20 cm. tall, always 1-flowered; basal leaves not over 5 cm. wide; achenes villous -1. A. parviflora
1. Stems usually over 20 cm. tall, some usually over 1-flowered; basal leaves over 5 cm. wide; achenes glabrous, villous or woolly
    2. Achenes villous or woolly; involucre leaves definitely short to moderately long-petioled
        3. Head of achenes cylindrical, 2-4 cm. long; achenes with styles (at least the bases) persistent; lower involucre leaves with a petiole over 2 cm. long; sepals greenish white -2. A. cylindrica
        3. Head of achenes globose or ovoid, not over 15 mm. long; achenes with styles usually deciduous; lower involucre leaves with a petiole variable in length but often less than 2 cm. long; sepals variable in color, greenish-yellow to red -3. A. globosa
    2. Achenes glabrous to pubescent; involucre leaves sessile

4. Achenes glabrous, stipitate, the styles short and curved; flowers umbellate; plants of mountains usually above 9000 feet   -4. A. zephyra
4. Achenes pubescent, at least when young, not stipitate, the styles straight; flowers cymose; plants of plains and foothills below 9000 feet   -5. A. canadensis

1. Anemone parviflora Michx., Fl. Bor. Am. 1:319. 1803.
   Plants with slender rootstocks; stems 10-20 cm. tall, simple, 1-flowered, sparingly villous especially above; root-leaves 3-parted, the divisions cuneate and obtusely lobed or crenate; involucre leaves sessile, with similar lobes; sepals 5-6 present, 8-15 mm. long, white or purple-tinged outside, oval; head of fruit spherical or short-ovoid; achenes densely villous all over or sometimes glabrate on the back.---Rather moist ground. Labrador to Alaska, south to Wisconsin, Colorado and Oregon. Our few records from central and westcentral Colorado at 9500-13,000 feet.
2. Anemone cylindrica A. Gray, Ann. Lyc. N. Y. 3:221. 1836.
   Plants with rather thick rootstocks; stems 30-60 cm. tall, silky-hairy; basal leaves 3- to 5-divided, the divisions cuneate and again cleft, the lobes cut or toothed, strigose-pubescent; involucre leaves similar but shorter petioled, the petiole over 2 cm. long on primary involucres, several pairs of involucre leaves usually present; flowers 1-10; sepals 8-10 mm. long, greenish-white, oblong; head of achenes 2-4 cm. long, cylindrical; achenes woolly, style-tipped.---Meadows and slopes. New Brunswick to British Columbia, south to New Jersey and Arizona. Our records scattered over the western two-thirds of Colorado at 5000-8000 feet.
3. Anemone globosa Nutt. ex Pritz., Linnaea 15:673. 1841.
   Plants with stout rootstocks; stems 10-50 cm. tall, silky-villous, with long somewhat appressed hairs; basal leaf blades deeply parted or divided, the divisions twice cleft or parted, the ultimate segments linear to narrowly lanceolate, more or less villous; involucre leaves similar but short-petioled; stems with 1-3 peduncles; sepals 4-9, 6-12 mm. long, varying from greenish-yellow to red or purplish, oval; head of achenes globose or ovoid, not over 15 mm. long; achenes densely villous.---Open woods or grassy slopes. Saskatchewan to British Columbia, south to New Mexico and California. Our records scattered over the western half of Colorado, except in the extreme western part, at 6000-12,500 feet.
4. Anemone zephyra A. Nels., Bot. Gaz. 42:51. 1906.
   Plants with thick erect caudices; stems 10-40 cm. long, long-pilose, often sparsely so, rather stout; basal leaves 5-8 cm. wide, deeply parted or divided into cuneate, often petiolate divisions, these cleft and again incised into linear-oblong lobes; involucre leaves sessile, somewhat like the lower leaf blades but much smaller; flowers solitary or several and umbellate; sepals 10-15 mm. long, broadly obovate, whitish to lemon-yellow; head of achenes globose; achenes glabrous, stipitate, style short and curved. Critical study may indicate that our plant should be A. narcissiflora L.---Higher mountains. Colorado and Wyoming, possibly north into Alaska. Our records from central, westcentral and northcentral Colorado at 7500-12,000 feet, mostly above 10,000 feet.
5. Anemone canadensis L., Syst. Ed. 12. 3:App. 231. 1768.
   Plants with spreading rootstocks; stems 20-60 cm. tall, appressed-pubescent to strigose, branching above into a cymose inflorescence, the flowers long-peduncled; basal leaves 10-12 cm. wide, 3- to 5-parted, the divisions 3-cleft and toothed, strigose or appressed pubescent beneath; involucre leaves similar but sessile, several sets often present; sepals commonly 4-6, 12-18 mm. long, white (cream-colored when dry), oval, oblong, or obovate; head of achenes globose; achenes pubescent.---Low moist ground, common along streams and ditches. Labrador to Alberta, south to Maryland and New Mexico. Our records from eastern Colorado, mostly east of the Continental Divide, at 4500-9000 feet.

8. Clematis L. VIRGIN'S BOWER

Perennial plants; stems erect or climbing, often woody; leaves opposite, compound, the petioles often twisted and aiding in climbing; flowers solitary and large or clustered and smaller, perfect, polygamo-dioecious or dioecious; sepals 4-5, petaloid, erect or spreading, valvate; petals none; stamens numerous, the outer often with dilated, petaloid filaments; pistils many; fruit a head of achenes, these with elongated plumose styles.

1. Sepals and stamens erect (except at apex), the sepals thick and connivent, at least near the base; plants not climbing, the stems erect or nearly so
   2. Leaves appearing bipinnate, the leaflets bimultifid, ultimate divisions lanceolate to narrowly linear   -1. C. hirsutissima
   2. Leaves pinnate, the leaflets simple or merely lobed, lanceolate to ovate in shape   -1A. C. hirsutissima scottii
1. Sepals and stamens spreading, the sepals not connivent, usually thin; plants half-woody climbers (rarely trailing)
   3. Flowers in corymbose cymes, dioecious or polygamo-dioecious, numerous; sepals white   -2. C. ligusticifolia
   3. Flowers solitary or few, perfect; sepals only rarely white
      4. Flowers yellow; staminodia absent   -3. C. orientalis
      4. Flowers purple, lavender or blue (rarely white); some of outer filaments often petaloid and sterile
         5. Leaves ternate; leaflets ovate, cordate at base (may be toothed or cleft)   -4. C. columbiana
         5. Leaves biternate; leaflets ovate-lanceolate to lanceolate (may be cleft or parted)   -5. C. pseudoalpina

1. Clematis hirsutissima Pursh, Fl. Am. Sept. 2:385. 1814.
   C. douglasii Hook.; C. eriophora Rydb.; Viorna hirsutissima (Pursh) Heller; V. bakeri (Greene) Rydb; V. jonesii (Kuntze) Rydb.; V. eriophora Rydb.---Stems erect, clustered, 20-70 cm. tall (rarely to 150 cm.), unbranched or with sterile branches, white-villous to sparsely villous;

leaves pinnately compound with 7-13 bimultifid leaflets, the ultimate divisions variable but lanceolate to narrowly linear, sometimes the lower leaves simple or nearly so, densely pubescent to nearly glabrous when mature; flowers 2-4.5 cm. long, solitary and terminal, rather campanulate in shape; sepals somewhat spreading at tip, thick, glabrous and deep purple inside, densely hirsute outside; achenes broadly fusiform, hirsute, densely canescent or silky, the styles 4-6 cm. long.---Hillsides, valleys and canyons. Montana to Washington, south to New Mexico and Nevada. Our records scattered over the western two-thirds of Colorado at 5000-9000 feet.

1A. Clematis hirsutissima scottii (Porter) Erickson, (var.) Ann. Mo. Bot. Gard. 30:47. 1943.
C. scottii Porter; Viorna scottii (Porter) Rydb.---Like the species but the leaves pinnate with 5-11 leaflets, these simple or merely lobed, lanceolate to ovate in shape. The variety intergrades somewhat with the species in Colorado.---Apparently with about the same distribution as the species. However, our records from mountainous part of southern Colorado as far north as El Paso County, at 6500-8500 feet.

2. Clematis ligusticifolia Nutt. ex T. & G., Fl. N. Amer. 1:9. 1838.
Woody climbing plants; stems often 4-6 m. tall; leaves pinnately 5- to 7-foliate, the leaflets 3-8 cm. long, petiolate, lanceolate to ovate in the typical form, rounded to cuneate at base, sometimes acute but usually more or less acuminate, sparingly strigose, incisely toothed or cleft; flowers essentially dioecious in corymbose cymes; sepals about 1 cm. long, white; pistillate flowers with sterile stamens, the anthers lacking and the filaments dilated; achenes pubescent with long, straight hairs, styles 4-5 cm. long.---Roadsides and thickets. North Dakota to British Columbia, south to New Mexico and Arizona. Our records well scattered over the western two-thirds of Colorado at 5000-8500 feet.

3. Clematis orientalis L., Sp. Pl. 543. 1753.
Climbing, somewhat woody vines; stems 3-6 m. tall; leaves ternate, biternate or triternate, often appearing pinnate; leaflets thin, coarsely toothed or lobed; flowers solitary or few together, perfect; sepals 1.5-2.5 cm. long, yellow, tinged with green, spreading, usually pubescent both sides; achenes pubescent, styles 2-5 cm. long. The correct name for our plant may be C. aurea Nels. & Macbr.---Among bushes. Ontario to New Mexico. Our few records scattered in the western half of Colorado, probably escaped from cultivation, at 6000-8000 feet.

4. Clematis columbiana (Nutt.) T. & G., Fl. N. Amer. 1:11. 1838.
C. occidentalis Hornem.; Atragene columbiana Nutt.; A. occidentalis Hornem. of Rydb.; A. grosseserrata Rydb.; A. diversiloba Rydb.---Tall half-woody vines; leaves 3-foliate, leaflets 3-6 cm. long, thin, ovate, obliquely cordate at base, acute or short-acuminate, margins entire, sparingly toothed or cleft, glabrous or sparingly hirsute; flowers solitary or elongated peduncles; sepals 2.5-5 cm. long, thin, spreading, purple or blue, rarely whitish; outer filaments often dilated and somewhat petaloid; achene densely pubescent, the styles 3-5 cm. long.--- In woods and thickets. Alberta to British Columbia, south to Colorado and Washington. Our records from northcentral Colorado at 6000-10,000 feet.

5. Clematis pseudoalpina (Kuntze) A. Nels. in Coult. & Nels., New Man. Rocky Mt. 198. 1909.
C. pseudoalpina tenuiloba (Gray) A. Nels; Atragene pseudoalpina (Kuntze) Rydb.; A. tenuiloba (A. Gray) Britt.---Half-woody vines, or stems trailing on the ground; leaves biternate, the leaflets lanceolate or ovate-lanceolate, deeply toothed, cleft or deeply parted; flowers solitary; sepals 3-5 cm. long, purple, blue or rarely white, spreading, thin; outer stamens often sterile, the filaments somewhat petaloid; achenes glabrous to densely pubescent, styles 3-5 cm. long.---Usually in woods or thickets. South Dakota to Montana, south to New Mexico and Arizona. Our records from the western half of Colorado, with few from the extreme northern part, at 6000-10,000 feet.

9. Trautvetteria Fisch. & Mey. FALSE BUGBANE

Perennial herbaceous plants; stems branching above to form the corymbose cymes or panicles; leaves palmately cleft or parted, mainly basal and long-petioled, cauline short-petioled or sessile; flowers perfect, usually numerous; sepals 4 or 5, petaloid, caducous; petals none; stamens many; pistils many, forming a head of achenes, each tipped by a minute, short recurved style.

1. Trautvetteria carolinensis (Walt.) Vail, Mem. Torrey Club 2:42. 1890.
T. grandis Nutt.; T. media Greene---Stems 30-100 cm. tall, stout, glabrous or nearly so; basal leaves 5- to 11-parted, the lobes coarsely and irregularly toothed; filaments clavate, about as wide as the anthers; achenes 3-4 mm. long.---In the mountains often in partial shade. British Columbia, south to New Mexico and California. Our few records from southcentral Colorado at 11,000-12,000 feet.

10. Thalictrum L. MEADOW RUE

Perennial herbaceous plants; stems erect; leaves alternate, ternately decompound; flowers small, perfect, polygamous or dioecious, in panicles or racemes; sepals 4 or 5, green or petaloid; petals none; stamens many, filaments filiform or dilated above; achenes usually few, sessile or stipitate, often ridged or nerved.

1. Flowers perfect; stigmas in anthesis usually shorter than the sepals, these usually 5 in number
  2. Flowers in leafless racemes; stems scapose, seldom over 20 cm. tall; filaments filiform   -1. T. alpinum
  2. Flowers in leafy panicles; stems leafy, over 20 cm. tall; filaments clavate   -2. T. sparsiflorum
1. Flowers dioecious or polygamo-dioecious, only rarely perfect; stigmas in anthesis usually longer than the sepals, these usually 4 in number
  3. Leaflets entire or 3-lobed, the lobes with entire margins, the leaflets longer than wide; filaments white (when fresh), more or less clavate; flowers polygamo-dioecious   -3. T. dasycarpum

3. Leaflets 3-lobed at least some of the lobes crenate-toothed, the leaflets usually wider than long; filaments colored, not at all clavate; flowers usually dioecious (mature fruit is necessary to differentiate the following 2 species)
    4. Achenes distinctly compressed and 2-edged, thin-walled, oblique at base, 1 edge nearly straight -4. **T. fendleri**
    4. Achenes turgid, not compressed, the wall thick, only slightly if at all oblique at base   -5. **T. venulosum**

1. **Thalictrum alpinum** L., Sp. Pl. 1:545. 1753.
Stems 5-30 cm. tall, scapiform, erect, leafless or with a small leaf near base, glabrous; leaves mostly basal, ternate with the primary divisions pinnate; leaflets not over 1 cm. long, orbicular to cuneate, obovate, firm, 3- to 7-lobed, margins more or less revolute; flowers racemose, not leafy, perfect; sepals 1.5-2.3 mm. long, usually 5 in number, ovate or elliptic; stamens 8-15, filaments filiform, about as long as the sepals; ovaries 3-6, the stigmas during anthesis almost always shorter than the sepals; achenes about 3 mm. long, subglobose to oblanceolate, sessile or short-stiped, with thick ribs. Our plants have been listed as var. hebetum Boivin.---Bogs and moist ground. Greenland to Alaska, south to New Mexico and California; Eurasia. Our records scattered in the western half of Colorado, except in the extreme western part at 9000-12,500 feet.

2. **Thalictrum sparsiflorum** Turcz. ex Fisch., Mey. & Lall., Index Sem. Hort. Petrop. 1:40. 1835.
Stems 25-80 cm. tall, leafy, branching above, glabrous or slightly pubescent; leaves 2- to 4-ternate, the lower petioled, the upper subsessile; leaflets variable in shape, cordate or round at base, usually 3-cleft and the divisions 3-lobed or crenate-toothed, pulverulent-glandular beneath; inflorescence paniculate, leafy, flowers perfect; sepals 3-4 mm. long, usually 5 in number, elliptic; stamens with the filaments clavate; stigmas during anthesis shorter than the sepals; achenes 5-8 mm. long, oblique with a straight back, half-obovate, stipitate, glandular-puberulent, with 3-4 faint ribs on each side. Our plants appear to be all var. saximontanum Boivin.---Moist ground. Hudson Bay to Alaska, south to Colorado and California; Asia. Our records scattered in the western half of Colorado, except the extreme western part, at 7000-9000 feet.

3. **Thalictrum dasycarpum** Fisch. & Lall. ex Fisch. May. & Lall., Ind. Sem. Hort. Petrop. 8:72. 1842.
T. purpurascens of Manuals---Stem 50-150 cm. tall, erect, stout, glabrous or more often viscid-pubescent, branching above, leafy; leaves 3- to 5-ternate; leaflets thick, oblong or obovate, margin entire or 3-lobed at apex, the lobes entire; inflorescence paniculate, flowers polygamo-dioecious; sepals about 3-5 mm. long, usually 4 in number; filaments usually white when fresh, more or less clavate; stigmas in anthesis longer than the sepals; achenes 3-5 mm. long, ovoid or oblanceolate with 6-8 thick, almost corky ribs, glabrous or pubescent, short-stipitate.---Meadows and thickets. Ontario to Alberta, south to Ohio, Missouri, New Mexico and Arizona. Our records from northcentral, central, southcentral and southwestern Colorado at 5000-8000 feet.

4. **Thalictrum fendleri** Engelm. ex Gray, Pl. Fendl. 5. 1849.
T. stipitatum Rydb.---Stems 30-60 cm. tall, pubescent or glabrous; leaves 3- to 4-ternate, the leaflets obliquely orbicular to subcordate about as wide as long, 3-lobed, the lobes mostly with 1 or more crenate teeth, glabrous to glandular or puberulent below; inflorescence a leafy panicle, flowers dioecious, only rarely polygamous; sepals of staminate flower 3-5 mm.long, of pistillate flower about 1.5 mm,; filaments filamentose, colored, usually yellow; achenes 4-6 mm. long, compressed, ovate, obovate to lanceolate, short-stipitate, oblique at base, one edge nearly straight, usually more or less pubescent, nerves rather conspicuous, about 3 on each side.---In the mountains often in woods or thickets. Wyoming to Oregon, south to Texas, New Mexico and California. Our records widely scattered in the western two-thirds of Colorado at 6000-10,000 feet.

5. **Thalictrum venulosum** Trelease, Proc. Bost. Soc. Nat. Hist. 23:302. 1886.
T. confine var. greeneanum Boivin---Stems 20-70 cm. tall; leaves 2- to 4-ternate, glabrous; leaflets rounded in outline or cuneate, about as wide as long, thick and veiny below, 3-lobed and the lobes of some at least crenately toothed; sepals of staminate flowers 2-5 mm. long, of pistillate flowers about 1.5-2.5 mm.; filaments filamentous, colored, not white; achenes 4-6 mm. long, narrowly ovoid, not compressed, short-stipitate, slightly if at all oblique at base, nerves fairly conspicuous.---In the mountains, often in open woods. South Dakota to British Columbia, south to Colorado and Oregon. Our records scattered in the western two-thirds of Colorado, except the extreme western part, at 4500-8000 feet. The plant has been reported up to 12,000 feet in Colorado.

11. *Actaea* L.   BANEBERRY

Erect perennial herbaceous plants, with thick rootstocks; leaves large, ternately decompound, basal and cauline; flowers small, perfect, numerous, in terminal racemes; sepals 3-5, fugaceous, petaloid; petals 4-10, small and narrow, clawed; stamens numerous, the filaments flattened and conspicuous; pistil 1, sessile, stigma sessile and broad; fruit a several-seeded berry.

1. **Actaea arguta** Nutt. ex T. & G., Fl. N. Amer. 1:35. 1838.
A. eburnea Rydb.; A. viridiflora Greene---Stems 4-80 cm. tall, rather stout, glabrous or minutely pubescent; leaves few, the lower petioled and the upper nearly sessile, ternate, the divisions long-petioled and pinnate, the leaflets ovate or ovate-lanceolate, serrate or 3- to 5-lobed; raceme elongating in fruit, often to 6-10 cm.; petals shorter than the stamens; filaments white or green; fruiting pedicels 7-20 mm. long; berries 6-8 mm. long, red or white. Our plants may not be distinct from the eastern A. rubra (Ait.) Willd.---Woods, especially in moist ground. South Dakota to Alaska, south to New Mexico and California. Our records scattered in the western two-thirds of Colorado at 7000-11,500 feet.

12. Paeonia L. PEONY

Perennial herbaceous plants with fascicled thickened roots; leaves ternately compound or dissected; flowers large, solitary, perfect, regular; sepals mostly 5 or 6, persistent; petals 5 to 6, borne on a fleshy disc adnate to the base of the sepals; stamens many, on the disc; pistils few, appearing inferior, styles short or none; fruit of follicles.

1. Paeonia brownii Dougl. ex Hook., Fl. Bor. Amer. 1:27. 1829.
 British Columbia, south to Wyoming, Utah and California. This plant should be looked for in northern or northwestern Colorado.

13. Trollius L. GLOBEFLOWER

Perennial herbaceous plants; leaves palmately divided or parted, flowers usually solitary, terminal, regular, perfect; sepals 5-15, petaloid, deciduous; petals 5-8, small, with a nectariferous gland at the base of each; stamens numerous; pistils several to many; fruit of follicles, each many-seeded.

1. Trollius laxus Salisb., Trans. Linn. Soc. 8:303. 1803.
 T. albiflorus (Gray) Rydb.---Stems 20-60 cm. tall, erect but rather weak, with 2-4 leaves; leaves long-petioled below, short-petioled toward the apex, blades palmately parted or divided, nearly orbicular in outline, the lobes cleft and incised; sepals 1.5-2 cm. long, white or cream-colored, ovate, oval or obovate; petals 3-5 mm. long, linear, mostly shorter than the stamens; follicles 8-12 mm. long, usually many. Our plant is var. albiflorus A. Gray, possibly not distinct.---Moist or wet ground. The variety from Alberta to British Columbia, south to Colorado and Washington. Our records scattered in the high mountains of Colorado at 9000-12,000 feet.

14. Ranunculus L. BUTTERCUP; CROWFOOT

Annual or perennial herbaceous plants with fibrous, fascicled roots; stems erect or procumbent; basal leaves entire, lobed or pinnately compound, the petioles dilated at the base; cauline leaves usually alternate; flowers solitary or few; sepals 5, usually highly deciduous; petals yellow or sometimes white with a nectariferous pit usually covered by a scale, the blade rarely reduced; stamens 10 or more, rarely fewer; pistils 5-many; achenes usually beaked. The mature heads of achenes as well as the flowers are necessary for accurate determination of the species. Fortunately these can usually both be found on the same plant when collected at the right time.

1. Petals white or the claw sometimes yellowish, not glossy; achenes transversely ridged; aquatic plants with leaves divided into filiform divisions
    2. Beak of achene less than 0.5 mm. long; leaves usually petioled
        3. Stems 1-2.5 mm. in diameter; stamens usually 10 or more -1A. R. aquatilis capillaceus
        3. Stems less than 1 mm. in diameter; stamens 5-8 -1B. R. aquatilis eradicatus
    2. Beak of achene 0.7-1.1 mm. long; leaves sessile or nearly so -2. R. circinatus
1. Petals yellow (may whiten on drying), usually glossy; achenes not transversely ridged (except sometimes in R. sceleratus), either smooth or longitudinally striate; plants mostly not aquatic, leaf divisions seldom filiform
    4. Sepals persistent in fruit; fruit 3-chambered (the lateral 2 empty) with a lanceolate-shaped beak about 3-4 mm. long -3. R. testiculatus
    4. Sepals deciduous during or soon after anthesis; fruit 1-chambered, the beak not both lanceolate and 3-4 mm. long (rarely over 2 mm. long)
        5. Pericarp of fruit longitudinally striate with 3 or more nerves on each face, the outer layer thin and usually fragile; leaves either bipinnately compound or cordate to reniform with crenate teeth
            6. Leaves simple, all basal; stolons present; fruiting receptacle much enlarged and elongated after anthesis
                7. Scapes 2-5 cm., rarely to 11 cm. long; sepals and petals 3-5 mm. long; stamens usually 15-25; achenes 40-50 in a head 3-8 mm. long, rarely longer -4. R. cymbalaria
                7. Scapes 5-30 cm. tall; sepals and petals 3-8 mm. long; stamens 20-35; achenes 100-300 in a head 5-15 mm. long -4A. R. cymbalaria saximontanus
            6. Leaves compound, some cauline; stolons never present; fruiting receptacle but slightly enlarged after anthesis -5. R. ranunculinus
        5. Pericarp of fruit not longitudinally striated, thick and firm; leaves various but rarely like either of the above
            8. Leaves entire or merely denticulate or wavy, none cleft, divided or deeply toothed
                9. Stems rooting at the nodes; leaves not over 1 cm. wide; petals 2-5 mm. long (rarely to 7)
                    10. Radical leaves with blades markedly broader than the petioles -6A. R. flammula ovalis
                    10. Radical leaves with blades little if any wider than the petioles -6B. R. flammula filiformis
                9. Stems not rooting at the nodes; leaves over 1 cm. wide; petals 7-12 mm. long (rarely 5-15) -7. R. alismaefolius
            8. At least some of the leaves coarsely toothed, lobed, parted or divided (sometimes only at the apex)
                11. Achenes with spines on the margins -8. R. arvensis
                11. Achenes without spines, smooth or hairy
                    12. Styles and achene beaks lacking or very minute, the beak not over 0.2 mm. long
                        13. Submerged aquatic plant, rooting at the lower nodes; head of achenes globose -9. R. natans

13. Terrestrial or growing in mud, not really submerged, not rooting at the lower nodes; head of achenes ovoid to cylindrical
    14. Herbage villous; beak of achene 0.2-0.3 mm. long   -10. R. micranthus
    14. Herbage glabrous or nearly so; beak of achene lacking or to 0.2 mm. long
        15. Some of basal leaves merely crenate; achene smooth   -11. R. abortivus
        15. All of leaves cleft or parted; achenes with minute depressions like pin pricks on each face   -12. R. sceleratus
12. Styles and achene beak longer, well over 0.2 mm. long, usually over 0.5 mm.
  16. Achenes with conspicuous corky thickenings on or along the keels; plants submerged or creeping in mud and rooting at the lower nodes; the submerged leaves dissected into filiform divisions
    17. Achenes corky-thickened on the conspicuous keel; petals 7-15 mm. long   -13. R. flabellaris
    17. Achenes corky-thickened on each side of the inconspicuous keel; petals 4-7 mm. long   -14. R. gmelinii
  16. Achenes without corky thickenings; terrestrial plants, only 2 species (R. repens, R. macounii) ever rooting at the nodes; leaves never dissected into filiform divisions (may be parted into linear divisions)
    18. Achenes turgid, the distance from edge to edge 1-2.5 times the distance from face to face; receptacle in fruit usually ovoid to cylindrical (globose only in R. glaberrimus), usually 3-15 times as long as in anthesis; sepals always tinged with purple, black or lavender
      19. Roots conspicuously tuberous, the tubers truncate, 3-5 mm. in diameter   -15. R. jovis
      19. Roots not distinctly tuberous, if slightly fusiform-tuberous the thickened portion not truncate and seldom over 2 mm. in diameter
        20. Sepals and usually the pedicels, densely covered with blackish to reddish-brown hairs
          21. Radical leaves orbicular, reniform or cordate, palmately lobed or parted   -16. R. nivalis
          21. Radical leaves elliptic to spatulate, shallowly toothed at apex   -17. R. macauleyi
        20. Sepals and pedicels glabrous or if hairy the hairs not blackish or reddish-brown
          22. Fruiting receptacle and head of achenes globose; achenes each with a broad thin stipe or wing at base; some of lower leaves usually entire
            23. Basal leaves round or ovate, 3- to 5-lobed at apex   -18. R. glaberrimus
            23. Basal leaves elliptic or oblanceolate, entire   -18A. R. glaberrimus ellipticus
          22. Fruiting receptacle and head of achenes ovoid to cylindrical (may be almost globose in R. pygmaeus); achenes not stipitate or winged at base; lower leaves either lobed or toothed, not entire
            24. Stems dwarf, not over 5 cm. tall (to base of highest pedicel) at fruiting time
              25. Petals 5 mm. long and 4 mm. wide, twice as long as sepals   -19. R. gelidus
              25. Petals to 3.5 mm. long and 1-2.8 mm. wide, same length as sepals   -20. R. pygmaeus
            24. Stems 5 cm. long or more (to base of the highest pedicel)
              26. Fruiting receptacle 2-4 mm. long, 1-3 mm. in diameter; petals shorter than the sepals   -10. R. micranthus
              26. Fruiting receptacle 4-16 mm. long, usually 3-9 mm. in diameter; petals as long or longer than the sepals
                27. Some of the basal leaf blades merely crenate and not lobed
                  28. Nectary scales ciliate with hairs nearly 1 mm. long, the adjacent petal surface often hairy; petals 5-12 mm. long; achenes usually canescent
                    29. Radical leaf blades cordate at base; petals 8-15 mm. long   -21. R. cardiophyllus
                    29. Radical leaf blades cuneate or rounded at base; petals 5-10 mm. long   -21A. R. cardiophyllus coloradensis
                28. Nectary scales and petals glabrous; petals 3-8 mm. long; achenes usually pubescent   -22. R. inamoenus
                27. All the basal leaf blades lobed, parted or divided
                  30. Sepals and usually the pedicels densely tomentose; achenes pubescent   -23. R. pedatifidus
                  30. Sepals and pedicels glabrous or thinly pilose; achenes usually glabrous
                    31. Basal leaf blades with the 3 primary divisions these again deeply twice divided into linear divisions; head of achenes ovoid   -24. R. adoneus
                    31. Basal leaf blades with the 3 primary divisions only once-lobed or entire, no linear divisions; head of achenes ovoid-cylindrical or cylindrical
                      32. Achene body oblong, the beaks about 1 mm. long, straight; petals 6 mm. long or more, over $1\frac{1}{2}$ times as long as the sepals; leaves cleft more than 1/2 way to the base
                        33. Ultimate basal leaf lobes and sinuses rounded or obtuse   -25. R. eschscholtzii
                        33. Ultimate basal leaf lobes and sinuses usually sharply acute   -25A. R. eschscholtzii eximus
                      32. Achene body obovate, the beaks 0.3-0.5 mm. long, curved or recurved; petals 3-6 mm. long not over $1\frac{1}{2}$ times as long as the sepals; leaves not cleft more than 1/2 way to the base   -22A. R. inamoenus alpeophilus
    18. Achenes compressed, the distance from edge to edge 3-15 times the distance from face to face; receptacle in fruit globose or ovoid (or cylindrical in R. pennsylvanicus), usually 1-3 times its height in anthesis; sepals usually not tinged with lavender or purple

34. Achene beaks 2-4 mm. long, straight or somewhat hooked at the very tip; roots often tuberous
-26. R. fascicularis
34. Achene beaks not over 2 mm. long, usually regularly curved or hooked; roots not tuberous
    35. Stems rooting at the nodes
        36. Petals 8-13 mm. long; stems prostrate; fruiting receptacle globose or hemispherical -27. R. repens
        36. Petals 5-7 mm. long; stems erect or ascending; fruiting receptacle ovoid or oblong -28. R. macounii
    35. Stems not rooting at any of the nodes
        37. Fruiting receptacle at least 3 times its length in anthesis; head of achenes ovoid or cylindrical; petals not over 7 mm. long; terminal segment of basal leaves petioled
            38. Sepals about twice as long as the petals, the latter 2-3 mm. long; head of achenes 11-17 mm. long
-29. R. pennsylvanicus
            38. Sepals as long or a little shorter than the petals, the latter 3-7 mm. long; head of achenes 7-9 mm. long
-28. R. macounii
        37. Fruiting receptacle not over 2.5 times its length in anthesis; head of achenes globose or hemispherical; petals over 7 mm. long (in all but R. unicinatus and its variety); leaves merely divided, the terminal segment not petioled (although appearing so in R. acriformis)
            39. Petals small, not over 6 mm. long; achenes often hispid
                40. Petals minute, not over 3 mm. long; beak of achene about 2 mm. long, body of achene appressed hispidulous
-30A. R. uncinatus parviflorus
                40. Petals 3.5-6 mm. long; beak of achene about 1 mm. long, the body of achene glabrous
-30B. R. uncinatus earlei
            39. Petals 6-18 mm. long; achenes glabrous
                41. Beak of achene short, not over 0.6 mm. long; pubescence of stem spreading -31. R. acris
                41. Beak of achene 1 mm. long or more; pubescence of stem appressed or spreading
                    42. Pubescence of stem appressed; sepals spreading; stem unbranched in first 20 cm.; petals clawless
-32. R. acriformis
                    42. Pubescence of stem spreading; sepals reflexed; stems usually branched in the first 20 cm.; petals with claws
-32A. R. acriformis montanensis

1. Ranunculus aquatilis L., Sp. Pl. 556. 1753.
    Represented in Colorado by the following 2 intergrading varieties.
1A. Ranunculus aquatilis capillaceus (Thuill.) DC., (var.) Prodr. 1:26. 1824.
    Batrachium trichophyllum (Chaix.) Bosch.; B. flaccidum (Pers.) Rupr.; B. drouetii (F. Schutz) Nym. of our Manuals---Aquatic perennial plants; stems 1-2.5 mm. in diameter, branching, submerged, glabrous or hispidulous; leaves 2-4 cm. long and 3-5 cm. broad, all submerged, finely dissected into filiform divisions, petioles 2-30 mm. long; pedicels glabrous; sepals 2-3 mm. long, light green, deciduous; petals 4-8 mm. long, white or yellow at base; stamens usually 10 or more; achenes in a globose cluster, transversely ridged, glabrous or hispid at first, finally glabrous or hispidulous on edge, beaks 0.3-0.5 mm. long.---Ponds and slow streams. Labrador to Alaska, south to Kentucky, New Mexico and California; Eurasia. Our records scattered over Colorado, none as yet from the extreme eastern part, at 4500-10,000 feet.
1B. Ranunculus aquatilis eradicatus Laestad., (var.) N. Act. Reg. Soc. Scient. Ups. 11:242.1839.
    Batrachium confervoides Fries.---Resembles var. capillaceus but the stems are 0.4-1 mm. in diameter; petals 4-6 mm. long; stamens 5-8; achenes glabrous from the beginning.---Pools and lakes. Greenland to Alaska, south to Wyoming; Europe. A few plants from northern and western Colorado are probably this variety.
2. Ranunculus circinatus Sibth., Fl. Oxon. 175. 1794.
    Batrachium longirostre (Godr.) F. Schultz.---Aquatic perennial plants; stems leafy, branched, glabrous, rooting at lower nodes; leaves all submerged, sessile or very short-petioled, finely dissected into filiform divisions; stipular sheaths 12-40 mm. wide, hairy; sepals shorter than the petals, deciduous; petals 5-7 mm. long, white; stamens 10-18; achenes 8-30 in a globose head; achenes about 1.5 mm. long, the beaks about 1 mm. long. Benson (American Midl. Nat. 40:240-245. 1948) made the above var. subrigidus (W. Drew) L. Benson, and separated R. longirostris Godr. as a valid species.---Ponds and slow streams. Quebec to Montana, south to Tennessee, New Mexico and Nevada. Some records from near Denver at about 5200 feet may be this species.
3. Ranunculus testiculatus Crantz, Stirp. Austr. Ed. 1 fasc. 2. 97. 1763.
    Annual terrestrial plants; stems 3-10 cm. tall, scapose, not branched above base, densely pilose to somewhat glabrate; leaves basal, simple, 3-parted, the lateral segments cleft into linear divisions, the petiole somewhat winged; sepals 5, 3-5 mm. long, green, narrowly elliptic, acute, hairy, persistent in fruit but not marcescent; petals 5, 4-6 mm. long, yellow; achenes 30-50 in a cylindrical head; achene beaks 3-4 mm. long, lanceolate, body tomentose, 3-chambered, the lateral 2 empty.---Along roadsides. An introduction from the Old World now present from Washington to Utah. Our few records are from extreme western Colorado at 4500-5100 feet.
4. Ranunculus cymbalaria Pursh, Fl. Am. Sept. 2:392. 1814.
    Represented in Colorado by the following variety. However, intergrades occur especially in the northeastern part of the state.
4A. Ranunculus cymbalaria saximontanus Fernald, (var.) Rhodora 16:162. 1914.
    Halerpestes cymbalaria (Pursh) Greene--as to species---Glabrous, scapose, perennial plants; scapes 5-30 cm. long; stolons becoming long and slender; leaves mostly basal, 15-25 mm. long, cordate, ovate or reniform, shallowly crenately lobed or merely crenate; sepals 3-8 mm. long, glabrous, greenish-yellow, deciduous; petals 5, 3-8 mm. long, yellow; stamens about 20; achenes 100-300 in a cylindrical head 5-15 mm. long; achenes 1.5 mm. long, thin-walled, compressed, each face with about 4 longitudinal nerves or striations, beaks short.---Moist or wet, often sandy soil. Alberta to British Columbia, south to Mexico and California. Our many records well scattered over Colorado at 3500-10,000 feet.

5. Ranunculus ranunculinus (Nutt.) Rydb., Bot. Surv. Nebr. 3:23. 1894.

Cyrtorhyncha ranunculina Nutt.; C. neglecta Greene; C. rupestris Greene ---Glabrous perennial plants; stems 7-35 cm. tall, not stoloniferous; basal leaves on long petioles, biternate, the divisions deeply parted into oblong or linear lobes; stem leaves few, ternately or pinnately parted, short-petioled; flowers few; sepals 3-5 mm. long, yellow and petaloid, spreading and early deciduous; petals about 1/3 longer than the sepals but narrower, pale yellow, 5-9 present; achenes 5-15 in a globose head not conspicuously enlarging in fruit; achenes 2-3 mm. long, 3- to 6-striated on each face, styles recurved.---Hillsides, ravines and cliffs. Wyoming to New Mexico. Our records mostly from northcentral and central Colorado at 5000-7200 feet.

6. Ranunculus flammula L., Sp. Pl. 548. 1753.

Represented in Colorado by the following 2 varieties.

6A. Ranunculus flammula ovalis (Bigel.) L. Benson, (var.) Bull. Torr. Club 69:305. 1942.

Glabrous or slightly appressed-pubescent perennial plants; stems 10-45 cm. long, creeping or reclining, almost always rooting at the nodes; radical leaves 1-5 cm. long and 1.5-7 mm. wide, rarely wider, with the blades wider than the petioles, oblanceolate, linear-spatulate to lanceolate, entire, not lobed; cauline leaves the same or narrower; sepals 2-5 mm. long, greenish-yellow, spreading or reflexed, deciduous, glabrous or pubescent; petals 5 or sometimes 10, 3-6 mm. long; stamens 20-30; achenes 10-25 to a head, this globose; achenes smooth and glabrous, beak short and thick, usually recurved.---Moist or wet ground. Newfoundland to Alaska, south to Colorado and California. Our records from northcentral and central Colorado at 7500-9500 feet.

6B. Ranunculus flammula filiformis (Michx.) Hook., (var.) Fl. Bor. Amer. 1:11. 1829.

R. reptans L.---This variety has leaf blades little if any wider than the petiole, hence very narrowly linear.---Moist or wet ground. Greenland to Alaska, south to Wyoming; Eurasia. Our one record from northcentral Colorado at about 5500 feet.

7. Ranunculus alismaefolius Geyer ex Benth., Pl. Hartw. 295. 1848.

R. calthaeflorus Greene; R. unguiculatus Greene---Glabrous or slightly hairy perennial plants; stems 10-30 cm. long, erect or reclining at base, usually branching above; leaves 2-6 cm. long, some over 1 cm. wide, narrowly elliptic or oval to lanceolate, entire or slightly denticulate; inflorescence rather corymbose; sepals shorter than the petals, yellowish-green, sometimes pubescent on the back, deciduous; petals 10, rarely 7-12 mm. long (rarely 5-12 mm.), usually oblanceolate; achenes 25-60 in a hemispheric or subglobose head; achenes about 2 mm. long, beaks 0.5-1 mm. long, straight or somewhat curved near tip. Our plants have been called var. montanus S. Wats.---Mountain meadows. The variety from Wyoming to Nevada, south to Colorado. Our records scattered over the western half of Colorado at 8500-11,500 feet.

A related species with achenes 3-3.5 mm. long has been considered to be distinct. It is R. oreogenes Greene and has been reported to come into western Colorado from the Southwest.

8. Ranunculus arvensis L., Sp. Pl. 555. 1753.

Waste ground. New York to Oregon, south to Utah and California. This species may be found in Colorado.

9. Ranunculus natans C. A. Meyer in Ledeb., Ic. 2:114. 1830.

R. intertextus Greene---Perennial plants of water or mud; stems creeping, rooting at the nodes, slender; leaves 1-3 cm. wide, petioled, reniform, truncate to cordate, glabrous, palmately 3- to 7-lobed or cleft with rounded, entire lobes; sepals slightly shorter than the petals; petals 3-4 mm. long, yellow, obovate; achene head globose but definitely enlarging after anthesis, 20 or more achenes to a head; achenes turgid, about 1 mm. long, the beaks minute, practically lacking. Our plant has been called variety intertextus (Greene) Benson.---In ponds and lakes. Alberta, south to Colorado and Idaho (for variety). Our records from the western half of Colorado, none as yet from the southwest or westcentral parts, at 5500-10,000 feet.

10. Ranunculus micranthus Nutt. ex T. & G., Fl. N. Amer. 1:18. 1838.

Massachusetts to South Dakota, south to District of Columbia and Arkansas. This plant has been reported for Colorado but no specimens could be located.

11. Ranunculus abortivus L., Sp. Pl. 551. 1753.

Perennial plants; stems 10-60 cm. tall, branched, glabrous or nearly so; radical leaves 1-6 cm. broad, simple, crenate or some 3-lobed, parted or divided or even 3-foliate; cauline leaves alternate and bractlike, deeply once- or twice-parted, short-petioled or the upper sessile; flowers usually many; sepals 2-3 mm. long, longer than the petals; petals yellow; achenes many in an ovoid, fusiform cylindrical (rarely subglobose) short head not over 4 mm. long; achenes glabrous, with minute curved beaks 0.1-0.2 mm. long. Our plant would be var. typicus Fernald.---Moist ground, often in open woods. Labrador to British Columbia, south to Florida, Texas and Colorado. Our records from northcentral, central and southcentral Colorado at 5500-9000 feet.

12. Ranunculus sceleratus L., Sp. Pl. 551. 1753.

R. eremogenes Greene---Perennial plants; stems 15-70 cm. tall, erect, stout and rather fleshy, not rooting at the nodes, glabrous or nearly so; basal leaves 3-10 cm. wide, reniform, long-petioled, 3-parted, the primary divisions again cleft or parted, the lobes obtuse and the sinuses rounded; upper cauline leaves sessile or short-petioled with narrow lobes; flowers several; sepals 3-4 mm. long, shorter than to somewhat longer than the petals; petals 2-4 mm. long, or sometimes to 5, yellow; head of achenes oblong or oblong-ovoid; achenes obovate, 0.8-1 mm. long, glabrous but somewhat thickened on the margin and with very minute pin-prick depressions on the face near this thickening, the beak minute, not over 0.1 mm. long. Our plants seem to be var. multifidus Nutt.---Borders of lakes, ponds and streams. The variety from Alaska, south to Iowa, New Mexico and California. Our records scattered over the state, except in the extreme eastern part, at 4500-10,000 feet.

13. Ranunculus flabellaris Raf. apud Bigel., Amer. Mo. Mag. 3:344. 1818.
    R. delphinifolius Torr.---This species resembles the next but has longer petals, 7-15 mm. long. The achenes are corky on the conspicuous keel, not on each side of it as in the next. Our specimens have 3-lobed petals and intergrade to R. gmelinii DC.---Maine to British Columbia, south to Louisiana, Nevada and California. Our few records from Routt, Lake, and Montrose Counties at 6500-10,000 feet.
14. Ranunculus gmelinii DC., Syst. 1:303. 1818.
    R. purshii Rich.; R. limosus Nutt.---Aquatic or palustrine glabrous perennial plants; stems 10-60 cm. long or more, rooting at the lower nodes; leaves 1.5-2.5 cm. broad, often all cauline, cordate, palmately divided into 3-7 more or less lobed or cleft divisions, petioled, the immersed leaves dissected into filiform divisions; sepals a little shorter than the petals; petals 4-7 mm. long, yellow, circular or obovate; head of achenes globose or ovoid; achenes 1-1.2 mm. long, turgid, with thickened corky keels especially at bases, beaks about 0.5-0.6 mm. long, flattened, straight. Our plant is var. hookeri (D. Don) L. Benson.---Wet ground or in shallow water. The variety from Nova Scotia to Alaska, south to Minnesota, New Mexico and Oregon. Our records scattered over the western half of Colorado, few from the extreme western part, at 5000-10,000 feet.
15. Ranunculus jovis A. Nels., Bull. Torr. Club 27:261. 1900.
    Moist ground of mountain meadows. Wyoming to Idaho, south to Wyoming and Nevada. Reported for Colorado but doubtfully present.
16. Ranunculus nivalis L., Sp. Pl. 553. 1753.
    Labrador to Alaska; Greenland; Iceland. This species was reported from Colorado on the basis of a doubtful specimen.
17. Ranunculus macauleyi A. Gray, Proc. Am. Acad. 15:45. 1879.
    Perennial plants with fleshy-fibrous fascicles of roots; stems 10-15 cm. tall, glabrous or hairy above; basal leaves 5-20 mm. wide, petioled, glabrous or somewhat pilose when young, blades oblong, elliptic or spatulate, 3- to 10-toothed at the truncate apex; stem leaves oblong-cuneate, sessile, also toothed at apex; sepals covered with villous black to reddish-brown hairs; petals about 1 cm. long, much longer than the sepals, yellow, obovate; stamens many; achenes in an ovoid head; bodies of achenes turgid, glabrous, the beaks about 0.5-0.8 mm. long, nearly straight. Benson (Amer. Midl. Nat. 40:123-124. 1948) divided this species into 2 varieties, var. typicus and var. brandegeei but many of our plants seem to intergrade between the two.---Alpine meadows, often among rocks. Colorado and New Mexico. Our records from southcentral and southwestern Colorado at 12,000-12,500 feet.
18. Ranunculus glaberrimus Hook., Fl. Bor. Am. 1:12. 1829.
    Montana to British Columbia, south to Colorado and California. All material seen by the writer has been the variety.
18A. Ranunculus glaberrimus ellipticus Greene, (var.) Fl. Fran. 1:298. 1892.
    R. ellipticus Greene---Perennial glabrous plants; stems 4-10 cm. long, rarely longer, usually reclining; basal leaves elliptic or oblanceolate to round, entire and fleshy, cuneate at base to a long petiole; lower cauline leaves often entire but some of cauline leaves lobed, often into 3 obtuse lobes, the middle larger, but sometimes only into 2 lobes; flowers solitary or few; sepals about 1/2 as long as the petals, lavender-tinged; petals 6-12 mm. long, yellow, broadly obovate; stamens many; achenes many in a globose head; achenes about 1.5 mm. long, turgid, glabrous or finely pubescent, beaks 1/3 to 1/2 as long as the body, slender and often curved, base of achenes stiped, this often winged.---Meadows and hillsides. South Dakota to British Columbia, south to New Mexico and California. Our records from the western half of Colorado, none as yet from the southcentral part, at 5000-10,000 feet.
19. Ranunculus gelidus Kar. & Kar., Bull. Soc. Nat. Mosc. Bot. 15:133. 1842.
    R. grayi Britt.; R. drummondii Greene---Nearly glabrous perennial plants; stems up to 5 cm. long (to base of highest pedicel); basal leaves 8-30 mm. wide, broadly cordate to reniform in outline, 3-parted or 3-divided, the divisions again cleft or lobed, petioles with stipularlike bases; cauline leaves alternate; petals longer than the sepals, often twice as long, about 5 mm. long and 4 mm. wide, yellow; achenes 50-80 in a cylindrical or ovoid head 5-9 mm. long; achenes about 2.5 mm. long, glabrous, beak about 0.6-0.7 mm. long, recurved.---High mountain areas. Arctic regions and Alaska, south to Colorado. Our record from Clear Creek County at 13,500 feet.
20. Ranunculus pygmaeus Wahl., Fl. Lapp. 157. 1812.
    Plants perennial; stems not over 5 cm. long at fruiting time (measured to base of highest pedicel), glabrous or nearly so; radical blades 5-11 mm. wide, simple, semicircular 3-parted or divided or cleft, middle lobe entire, laterals 2- to 3-lobed, truncate or nearly cordate, petioles with stipularlike bases; cauline leaves often appearing opposite; petals about as long as sepals, 1.5-3.5 mm. long and 1-2.8 mm. wide, yellow; achenes 30-50 in an ovoid or rarely subglobose head 2-4 mm. long; achenes about 1 mm. long, glabrous, beak about 0.2-0.7 mm. long, somewhat recurved at tip.---Mountains. Greenland to Alaska, south to Colorado; Eurasia. Our one record from northcentral Colorado at about 11,000-12,000 feet.
21. Ranunculus cardiophyllus Hook., Fl. Bor. Am. 1:14. 1829.
    Perennial plants; stems 15-40 cm. long, erect, not rooting at nodes, more or less pilose; radical leaf blades 1-5 cm. wide, simple, cordate, crenate or sometimes the apex lobed or even parted, on long petioles, pilose; cauline leaves divided into linear lobes especially the upper and the bracts; flowers few to several; sepals shorter than the petals, pilose; petals 8-15 mm. long, yellow, the nectary scale ciliate with hairs about 1 mm. long and the adjacent petal surface often hairy too; achenes many in an oblong or cylindrical head; achenes about 2 mm. long, turgid, oblong or obovate, finely canescent, beaks 0.6-1 mm. long, slender and recurved.---Moist meadows and valleys. Alberta, south to New Mexico. Our records from northcentral, central, southcentral and southwestern Colorado at 5000-11,000 feet.
21A. Ranunculus cardiophyllus coloradensis L. Benson, (var.) Am. Journ. Bot. 27:804. 1940.
    Basal leaf bases are cuneate or rounded; petals are 5-10 mm. long. Possibly a good species.
    Apparently limited to southwestern Colorado, our one record from San Juan County at 9500 feet.

22. Ranunculus inamoenus Greene, Pitt. 3:91. 1896.
R. micropetalus (Greene) Rydb.---Perennial plants; stems 10-30 cm. tall, not rooting at the nodes, glabrate or hirsute; basal leaves 1-3.5 cm. wide, simple and ovate to orbicular, at least some merely crenate, often some 3-lobed or divided, long-petioled; stem leaves sessile, once or twice ternately divided into oblanceolate or linear segments; sepals usually somewhat shorter than the petals, glabrous to pilose; petals 3-8 mm. long, yellow, narrowly elliptical or obovate; achenes many in a cylindrical head; achenes about 1.5 mm. long, turgid, obovate, densely pubescent to rarely glabrous, beaks 0.8-0.9 mm. long, slender, recurved. Our plant has been called var. typicus L. Benson.---Meadows and valleys. The variety from Alberta, south to New Mexico. Our records scattered over the western two-thirds of Colorado, except in the northwestern corner, at 7000-11,000 feet.

22A. Ranunculus inamoenus alpeophilus (A. Nels.) L. Benson, (var.) Bull. Torr. Club 68:651. 1941.
R. alpeophilus A. Nels.---Glabrous or nearly so; radical leaves 3-parted or 3-lobed; achenes glabrous; otherwise like the species.---British Columbia, south to Colorado and Nevada. Our records from the western half of Colorado, none as yet from the extreme western part, at 9500-11,500 feet.

23. Ranunculus pedatifidus J. E. Smith in Rees., Cyclop. 29:No.72. 1819.
R. affinus R. Br.---Perennial plants; stems 12-30 cm. tall, slender, glabrous or sparingly silky; basal leaves 1.5-4 cm. wide, more or less digitately lobed or divided into oblong or linear lobes; stem leaves sessile with linear divisions; flowers usually few; sepals 3-5 mm. long, densely tomentose and usually the pedicels too; petals about twice as long as the sepals, yellow; head of achenes ovate or oblong; achenes 2 mm. long, turgid, pubescent, with rather short recurved beaks.---Wet or moist ground, usually in meadows. Greenland to Alaska, south to Colorado; Asia. Our records from northcentral and central Colorado at 7500-12,000 feet.

24. Ranunculus adoneus A. Gray, Proc. Acad. Phila. 15:56. 1864.
Perennial plants; stems 10-20 cm. tall, often clustered, glabrous; radical leaves 2-3 cm. wide, reniform to semicircular in outline, deeply parted, the primary divisions again twice-lobed into linear segments, long-petioled; cauline leaves alternate, about 2; bracts dissected; flowers 1-2; sepals about 1/2 as long as the petals, villous; petals 8-18 mm. long, yellow, broadly obovate; head of achenes ovoid or somewhat oblong; achenes turgid, oblong, glabrous, the beaks 1.2-1.5 mm. long, curved or hooked. Divided by Benson (Am. Midl. Nat. 40:130-131. 1948) into var. typicus L. Benson and var. alpinus (S. Wats.) L. Benson, and both recorded for Colorado.---Moist ground in the high mountains. Wyoming, Utah and Colorado. Our records mostly from central and northcentral Colorado at 11,000-13,000 feet.

25. Ranunculus eschscholtzii Schlect., Animad. Ranunc. 2:16. 1820.
R. ocreatus Greene---Perennial plants; stems 10-30 cm. tall, little if any branched, glabrous; basal leaves 1-4 cm. wide, simple, semicircular to reniform, deeply 3-parted (more than 1/2 way to the base), the middle lobe again 3-lobed or entire, the lateral lobes 3- to 7-parted, the ultimate lobes and sinuses rounded; petioles rather long; lower stem leaves similar but short-petioled; sepals shorter than the petals, pilose outside; petals 7-10 mm. long, yellow, obovate to oblong; achenes many, in a cylindrical or ovoid-cylindrical head; achenes turgid, somewhat oblong in shape, glabrous, beaks of achenes 1 mm. long or less, straight or nearly so. Our plants have been called var. typicus L. Benson.---Moist meadows or slopes. The variety from Alberta to Alaska, south to New Mexico and California. Our records from the western half of Colorado, except the southcentral and northwestern parts, at 10,000-12,000 feet.

25A. Ranunculus eschscholtzii eximius (Greene) L. Benson, (var.) Bull. Torr. Club 68:654. 1941.
R. eximius Greene---Montana to Idaho, south to Wyoming and Arizona. Reported from Colorado and 2 specimens from northcentral Colorado may be this variety.

26. Ranunculus fascicularis Muhl. ex Bigel., Fl. Bost. ed. 1. 137. 1814.
The eastern var. apricus (Greene)Fernald has been reported from Colorado but the record is open to question.

27. Ranunculus repens L., Sp. Pl. 554. 1753.
Perennial plants; stems 10-50 cm. long, soon prostrate and rooting at the nodes, usually appressed hairy; leaves 2-8 cm. wide, compound, pinnate with 3 sessile or petiolulate leaflets, the middle petiolule sometimes much longer (up to 4 cm.), leaflets toothed or cleft, cuneate at base or sometimes subtruncate; sepals about 1/2 as long as petals; petals 7-13 mm. long; achenes 20-25, in a subglobose head; achenes about 2.5-3.5 mm. long, compressed, beak about 1 mm. long, stout and hooked. Our plants have been called var. typicus G. Beck.---Meadows and roadsides. Newfoundland to Washington, south to Virginia and Utah; Europe. Our few records from northcentral Colorado at 5200-5300 feet.

28. Ranunculus macounii Britt., Trans. N. Y. Acad. Sci. 12:3. 1892.
R. rivularis Rydb.---Perennial plants; stem 20-60 cm. long, finally decumbent and often rooting at the nodes, glabrous to densely hirsute with long hairs; basal leaves 3-divided or pinnate, the segments petiolate, especially the terminal, and 3-parted, the parts again lobed; sepals somewhat shorter than the petals, spreading, pubescent or glabrous; petals 5-7 mm. long, yellow, obovate; stamens many; achenes many in an ovoid to oblong sometimes nearly globose head; achenes about 3 mm. long, compressed, glabrous, smooth or minutely pitted, the beaks short, 1/4 to 1/3 as long as the body of the achene, short and almost straight.---Moist or wet ground. Newfoundland to Alaska, south to Michigan, New Mexico and California. Our records well scattered over the western two-thirds of Colorado at 5000-9500 feet.

29. Ranunculus pennsylvanicus L. f., Suppl. 272. 1781.
Perennial plants; stems 30-90 cm. tall, simple below, branched above, stout, with hirsute or hispid spreading hairs; basal leaves early deciduous; leaves pinnate, the leaflets petiolate especially the terminal, and 3-lobed once or twice again, appressed-hispidulose; flowers several; sepals longer than the petals, reflexed, somewhat hairy; petals 2-4 mm. long, yellow; stamens many; achenes about 2.5 mm. long, compressed, glabrous, beaks about 1 mm. long, broad and flat,

straight.---Moist or wet ground. Newfoundland to Alaska, south to New Jersey, New Mexico and Washington. Our few records from southcentral Colorado at 6500-7500 feet.
30. Ranunculus uncinatus D. Don. in G. Don., Gen. Syst. Gard. 1:35. 1831.
The typical plant, called var. typicus Benson, has been reported for western Colorado but the writer has seen no specimens. The two varieties listed below are more common in the state.
30A. Ranunculus uncinatus parviflorus (Torr.) L. Benson, (var.) Am. Midl. Nat. 39:761. 1948.
R. bongardi Greene---Perennial plants; stems 35-80 cm. tall, moderately stout, hispid, the hairs commonly reddish-brown; basal leaves 3-10 cm. wide, 3-parted, the parts lobed and acutely toothed, appressed-hispidulous, petioles hispid; cauline leaves similar but smaller; sepals about equal to petals or slightly shorter; petals minute, not over 3 mm. long; achenes 8-20 in a globose to hemispherical cluster about 4 mm. in diameter; achenes 1.5-2 mm. wide and 0.6-0.8 mm. thick, compressed, obovate, more or less appressed-hispidulose, smooth or nearly so, beaks 1.7-2 mm. long, slender but broader at base, recurved and hooked at tips.---Shaded, rather moist places. Montana to Alaska, south to Colorado and California. Our records from Routt and Jackson Counties at 6500-9500 feet.
30B. Ranunculus uncinatus earlei (Greene) L. Benson, (var.) Amer. Midl. Nat. 39:761. 1948.
R. earlei Greene---Perennial plants; stems 20-50 cm. tall, sparingly hairy or glabrous; basal leaves larger than the cauline, 3-parted, the parts lobed and toothed, petioles long, pilose; cauline leaves similar but smaller or only bracts present; sepals shorter than the petals, pilose; petals 3.5-7 mm. long, elliptic-obovate; achenes many in a subglobose or ovoid head; achenes about 2-3 mm. long, compressed, glabrous somewhat puncticulate, beaks 0.7-1.2 mm. long, hooked, wide at base.---Along streams. Apparently limited to Colorado, our records scattered in the western third of the state at 5500-11,500 feet.
31. Ranunculus acris L., Sp. Pl. 554. 1753.
This European species has become naturalized in various parts of Canada and the northern United States. It should be looked for in Colorado.
32. Ranunculus acriformis A. Gray, Proc. Am. Acad. 21:374. 1886.
Perennial plants; stems 20-50 cm. tall, not rooting at the nodes, strict and slender, not branching below, appressed-pubescent or appressed-hirsute; leaves 2-6 cm. wide, bi- or ternately-parted into narrow segments, these again 2- to 3-cleft into lanceolate-linear or linear lobes; sepals definitely shorter than the petals, strigose; petals 6-10 mm. long, sometimes longer, yellow, orbicular-obovate, often retuse; achenes many in a globose or hemispherical head; achenes compressed, glabrous, beaks 0.9-1.7 mm. long, about 1/2 as long as the achene body and curved or hooked. Our plants have been called var. typicus Benson.---Moist or wet meadows. Alberta to Colorado. Our records from northcentral to northwestern Colorado at 6500-8000 feet.
32A. Ranunculus acriformis montanensis (Rydb.) L. Benson, (var.) Am. Midl. Nat. 40:43. 1948.
R. montanensis Rydb.---Montana to Wyoming and Idaho. This variety should be looked for especially in northern Colorado.

Family 48. Berberidaceae BARBERRY FAMILY

Shrubs or undershrubs; wood and inner bark yellow; leaves alternate, simple or compound, exstipulate; flowers perfect, hypogynous, regular, yellow, in racemose or subracemose clusters; sepals 6, distinct; petals 6, distinct; stamens 6 and opposite the petals, each anther opening by 2 apical valves; ovary superior, 1-celled, simple, stigma peltate on the style; fruit a few-seeded berry, this sometimes rather dry.

1. Berberis L. BARBERRY

Characters of the Family

1. Leaves unifoliate, apparently simple, deciduous; stems spiny
   2. Second-year twigs gray in color; racemes usually over 10-flowered; fruit 8-12 mm. long; an escape in Colorado -1. B. vulgaris
   2. Second-year twigs reddish-brown; racemes 6- to 10-flowered; fruit 4-6 mm. long, plants native to southwestern Colorado -2. B. fendleri
1. Leaves pinnately compound, evergreen; stems not spiny (edges of leaf are)
   3. Leaflets not over 3 cm. long, with less than 5 teeth on each side (usually 2 or 3), the petiole of the leaf short (1 pair of leaflets crowded close to the base); upright shrubs -3. B. fremontii
   3. Leaflets over 3 cm. long, over 5 teeth on each side (usually over 8), the petiole of the leaf over 2 cm. long; trailing shrubs -4. B. repens

1. Berberis vulgaris L., Sp. Pl. 330. 1753.
Shrubs to 2.5 meters tall or more, branches becoming arched and drooping at ends, gray the second year; spines present at most of the nodes especially on the shoots, 1-2 cm. long, usually 3-parted; leaves 2-4 cm. long, alternate or fascicled, unifoliate and apparently simple, elliptic-obovate to obovate-oblong, obtuse or rarely acutish, cuneate into a petiole, margin bristly-serrate; racemes terminating lateral branchlets, pedicels 5-12 mm. long; berry oblong to elliptic, 8-12 mm. long, red or purple.---Thickets, along roadsides especially in cultivated areas. A European plant now naturalized especially in the eastern and midwestern parts of the United States. Our few specimens from northcentral Colorado at 4500-5600 feet. However, according to the U. S. D. A. records this plant once was scattered over all of Colorado especially near cultivated fields.
2. Berberis fendleri A. Gray, Mem. Amer. Acad. n. ser. 4:5. 1849.
Shrubs 50-200 cm. tall; twigs purplish-brown the second year, somewhat shining, older twigs often turning gray; spines usually 3-parted, 8-15 mm. long, present at most of the nodes;

leaves 2-6 cm. long, alternate or fascicled, unifoliate, apparently simple, blades spatulate, oblanceolate or elliptic-obovate, entire to spinulose-serrate or dentate; racemes terminating lateral branches, about 6- to 10-flowered, the pedicels 4-6 mm. long; berry oval or short-ellipsoid, 5-6 mm. long, red.---Mountains and valleys. Colorado and New Mexico. Our records from southwestern Colorado to Montrose County and the San Luis Valley at 5400-8500 feet.
3. Berberis fremontii Torr., Bot. U. S. & Mex. Bound. 30. 1859.

Odostemon fremontii (Torr.) Rydb.---Erect shrubs 1-3 m. tall, with rigid branches; leaves pinnately compound of usually 3-7 leaflets, a pair usually close to the base of the petiole, leaflets 15-25 mm. long, coriaceous and evergreen, ovate or oblong, sinuately large-toothed, the teeth spiny, usually about 3 on each side; racemes terminating the short stubby branchlets, 3- to 8-flowered; berries blue to blue-black, ovate-globose, 6-14 mm. in diameter, becoming inflated and dry at maturity.---Canyons and dry slopes. Colorado to Utah, south to New Mexico and Arizona. Our records from southwestern and westcentral Colorado at 4900-6500 feet.
4. Berberis repens Lindl., Bot. Reg. 14:pl. 1176. 1828.

B. aquifolium and Odostemon aquifolium of Manuals---Stems rarely more than 25 cm. high, creeping and stoloniferous; leaves pinnately compound of 3-7 leaflets, these 3-9 cm. long, coriaceous and evergreen, oval to ovate or oblong, acute to rounded at apex, margin sinuately bristle-toothed, usually 8 or more on each side, dull green above, paler below; racemes dense, many-flowered; berry ellipsoid-globose, 7-8 mm. long, black or blue with a bloom.---Hills and slopes often in partial shade. Montana to British Columbia, south to New Mexico and California. Our records well scattered over the mountainous part of Colorado at 5500-10,000 feet.

Family 49. Papaveraceae POPPY FAMILY

Annual or perennial herbaceous plants with white, yellow, or colorless sap; leaves alternate, usually lobed, dissected, or compound, exstipulate; flowers perfect, regular, showy, solitary or in clusters; sepals 2-3, free, caducous; petals 4-6, separate, deciduous; stamens numerous; ovary superior, 1-celled; style short or none; stigma simple or divided; fruit a capsule, dehiscent by pores or by 4-6 valves near the apex. The family as treated above is closely related to the Fumariaceae and the 2 families are often united. However, they separate very nicely in our area on the basis of stamen number and regular or irregular corollas.

1. Herbage, sepals and capsules prickly; leaves sinuate-dentate or sinuate-pinnatifid (like a thistle); sepals with hornlike appendages; flowers usually white -1. Argemone
1. Herbage, sepals and capsules not prickly; leaves compound or dissected, not at all thistlelike; sepals without hornlike appendages; flowers yellow
    2. Perianth and stamens borne on the rim of a tubelike expansion of the receptacle; sepals not separating but pushed off by the expanding petals; capsule linear-elongate; leaves dissected into narrowly oblong or linear divisions -2. Eschscholzia
    2. Perianth and stamens strictly hypogynous; sepals not remaining united; capsule obovoid; leaves pinnately compound -3. Papaver

1. Argemone L. PRICKLY POPPY

Annual, biennial or perennial glaucous herbaceous plants with white or yellow sap; leaves pinnatifid or lobed, the divisions spinose-tipped (thistlelike); flowers large and showy; sepals 3, each with a horned apex; petals 4-6, white or yellow; stigma with radiating lobes, ovary with 4-6 nerviform placentae; capsule oblong, opening at the top by 4-6 valves.

1. Petals pale yellow or orange
1. Petals white -1. A. mexicana
    2. Stems and leaves with few or no hairs between the prickles
        3. Horns of the sepals long and slender, dilated only at the base and without lateral spines, or only a few near the base -2. A. intermedia
        3. Horns of the sepals stout, dilated well above the base and usually with several lateral spines, these often extending to the apex of the horn -3. A. platyceras
    2. Stems and leaves hispidulous-pubescent between the prickles -3A. A. platyceras hispida

1. Argemone mexicana L., Sp. Pl. 508. 1753.
Florida to Arizona and southward. This species has been reported from Colorado but no specimens were found by the writer.
2. Argemone intermedia Sweet, Hort. Brit. ed. 2. 585. 1830.
Stems stout, 30-100 cm. tall, more or less prickly, not pubescent between prickles; leaves deeply pinnately cleft, oblong or oblong-obovate, the lobes spinulose and with prickles on the surface especially on the veins, no pubescence between the prickles; sepals more or less spinulose, tipped by a horn 5-8 mm. long, this wide at base only and not spinulose except sometimes at very base; petals white, 2-4 cm. long; capsule about 4 cm. long.---Plains and open slopes. South Dakota to Wyoming, south to Texas, New Mexico and Arizona. Our records scattered over Colorado at 3500-7000 feet, mostly in the northeastern quarter of the state.
3. Argemone platyceras Link & Otto, Icon. Pl. Rar. 1:85. 1828.
Stems 30-80 cm. tall, more or less prickly; leaves deeply pinnately cleft, oblong to oblong-obovate, prickly especially on the veins below; sepals more or less spinose with ascending spines, tipped by a horn about 8-10 mm. long, this somewhat dilated well above the base with lateral prickles present at least near the apex of horn; petals 2-5 cm. long, white; capsule 2-5 cm. long. The latex of the above plant has been reported both as white and orange in color by different observers.---Plains and slopes. Nebraska to Wyoming, south to Mexico. Our records from northern Colorado at 4500-6500 feet.

3A. Argemone platyceras hispida (Gray) Prain, (var.) Journ. Bot. 33:367. 1895.
  A. hispida Gray; Enomegra hispida A. Nels.; probably A. squarrosa Greene---Differs from the species in having the stems and leaves hispidulose-pubescent between the spines or prickles. The variety is reasonably distinctive in Colorado. Our records scattered over the eastern half of Colorado, mostly along the foothills, at 4000-8000 feet.

2. Eschscholzia Cham. CALIFORNIA POPPY

Annual or perennial plants with colorless sap; leaves ternately dissected, smooth and glaucous; flowers solitary on long peduncles; receptacle dilated and with 1 or 2 tubelike expansions above; sepals 2, coherent and pushed off by the expanding petals; petals orange or yellow; capsule elongated, dehiscent from the base.

1. Eschscholzia californica Cham. in Nees, Hor. Phys. Ber. 73. pl. 15. 1820.
  This species is commonly cultivated in this state and has been reported to escape freely in various localities.

3. Papavera L. POPPY

Perennial subscapose herbaceous plants with milky sap; leaves alternate or basal, lobed or dissected; flowers showy, solitary; sepals 2 or rarely 3; petals 4, rarely 6, yellow; stamens numerous; ovary with 4-20 placentae, stigma disclike; capsule dehiscent by pores just under the stigma.

1. Papavera nudicaule L., Sp. Pl. 507. 1753.
  P. alpinum of Manuals---Plants caespitose; scapes 5-15 cm. long, blackish-hirsute; leaves basal, 2-10 cm. long, ovate in outline, deeply lobed or parted, the divisions usually incised-toothed to cleft, rarely entire; calyx densely black-hirsute; petals 1-3 cm. long; capsule about 1 cm. long. Our plants are var. radicatum (Rottb.) DC. ---Exposed summits of hills and mountains. The variety from Greenland to Alaska, south to New Mexico. Our records from the higher mountains of Colorado at 10,000-14,000 feet.

Family 50. Fumariaceae FUMITORY FAMILY

Herbaceous plants with watery juice; leaves alternate or basal, usually finely dissected, exstipulate; flowers perfect, irregular, racemose or paniculate; sepals 2, small and bractlike; petals 4, the outer 2 spreading above, 1 spurred at base, inner 2 smaller and narrower and united at apex over the stigmas; stamens 6, diadelphous in 2 sets of 3 each; ovary 2-carpellate, superior, 1-celled, with 2 parietal placentae or 1-seeded; stigma 2-lobed; fruit a 2-valved capsule.
  In addition to the species of the genus Corydalis, another plant Fumaria officinalis L. may possibly be present as an escape in Colorado. It has a 1-seeded, indehiscent fruit and small purplish flowers not over 1 cm. long.

1. Corydalis Vent. CORYDALIS

Characters of the Family

1. Corolla white, rose or purplish; stems stout, usually over 50 cm. tall — -1. C. caseana
1. Corolla yellow; stems slender, not over 50 cm. tall
   2. Spur 4-5 mm. long, about 1/3 as long as the body of the flower; capsule pendulous and torulose — -2. C. aurea
   2. Spur 5-9 mm. long, usually nearly as long as the body of the flower; capsule erect or ascending and slightly if at all torulose — -2A. C. aurea occidentalis

1. Corydalis caseana A. Gray, Proc. Amer. Acad. 10:69. 1874.
  C. brandegei Wats.; Capnoides brandegei (Wats.) Heller---Perennial, glabrous and usually somewhat glaucous plants; stems 30-150 cm. tall, stout, mostly erect; leaves large, 20-30 cm. long, 2- or 3-pinnately divided, leaflets 2-3 cm. long, oval or obovate to lanceolate, acute or mucronate; corolla about 2 cm. long, the spur longer than the body, white to purplish or rose; capsule reflexed, oblong, 1-1.5 cm. long. Our plant is ssp. brandegei (Wats.) Ownbey.---Moist ground. The variety from Colorado, Utah and New Mexico. Our records from the southwestern quarter of Colorado, none in the extreme corner, at 7500-11,000 feet.
2. Corydalis aurea Willd., Enum. Pl. 740. 1809.
  Capnoides aureum (Willd.) Kuntze---Annual or biennial, glabrous and more or less glaucous plants; stems 10-40 cm. high, diffusely branched and leafy from base; leaves bipinnate, the leaflets pinnatifid; corolla 12-15 mm. long, the spur 4-5 mm. long, about 1/2 as long as the body of the flower (sometimes to 2/3), usually curved downward somewhat; capsules usually pendant, linear, 2-3 cm. long, torulose.---Hillsides, banks and along streams mostly in open ground. Nova Scotia to Alaska, south to Pennsylvania, New Mexico and California. Scattered over Colorado except in the extreme eastern and northwestern parts at 5000-10,500 feet.
2A. Corydalis aurea occidentalis Engelm., (var.) in Gray's Man. 5th Ed. 62. 1875.
  C. montana Engelm.; Capnoides montanum (Engelm.) Britton---Differs from the species in relatively longer spur (5-9 mm. long, usually about as long as the body of the flower); the capsule ascending and only slightly if at all torulose. The variety is reasonably distinct from the species in Colorado but a few intergradations occur.---South Dakota to Wyoming, south to Texas and New Mexico. Our records scattered over the state with the species at 4500-8500 feet.

Family 51. Capparidaceae CAPPER FAMILY

Annual herbaceous plants, often ill-scented; leaves alternate, palmately compound or sometimes simple above, the leaflets nearly or quite entire; flowers usually in racemes, usually regular, perfect; sepals 4, rarely more, distinct or united below; petals distinct, 4 or rarely more, often clawed; stamens 6 to many, on the receptacle; ovary superior, sessile or stipitate, 2-carpellate, 1-celled with 2 parietal placentae, style short, stigma usually single; fruit a 2-valved capsule, the valves separating from the framelike placentae.

1. Capsule short, not over 6 mm. long, usually wider than long; receptacle without appendages or glands
 -1. Cleomella
1. Capsule 6 mm. long or longer, definitely longer than wide; receptacle with tubular appendages or inconspicuous glands
  2. Receptacle with petaloid, tubular appendages; petals unequal, laciniate or fimbriate at apex
 -2. Cristatella
  2. Receptacle with solid glands, these not tubular or petaloid; petals nearly equal, entire, emarginate or 3-toothed at apices, not laciniate
    3. Capsule sessile or very short-stipitate; stamens usually over 10; plant viscid-pubescent
 -3. Polanisia
    3. Capsule stipitate, this stipe 6 mm. long or longer; stamens 6; plant glabrous or nearly so, not viscid-pubescent
 -4. Cleome

1. Cleomella DC.

Glabrous plants; leaves palmately trifoliate, petioled and leaflets petiolate; flowers small, yellow, in terminal bracted racemes and axillary; sepals 4, distinct, deciduous; petals sessile or nearly so, entire; stamens 6, filaments equal, distinct; capsule short, usually wider than long, inflated, stipitate; seeds usually 2 on each placenta.

1. Stipe about 4-7 mm. long, longer than its capsule
  2. Seeds rugose; style about 0.5 mm. long -1. C. angustifolia
  2. Seeds smooth; stipe about 1.5 mm. long -2. C. nana
1. Stipe about 2 mm. long, shorter than its capsule
  3. Capsule much wider than long, widest near apex -3. C. palmerana
  3. Capsule very little if at all wider than long, widest at middle or just below -4. C. montrosae

1. Cleomella angustifolia Torr. in Hook., Kew Journ. 2:225. 1850.
Stems 30-60 cm. tall, diffusely branched; leaflets 1-3 cm. long, linear-lanceolate or oblong-linear, acute; sepals 1-1.5 mm. long; petals 3-5 mm. long; capsule about 5 mm. long and about 6 mm. wide, broadly rhombic with rounded angles, style not over 0.5 mm. long; stipe 4-7 mm. long, longer than capsules, somewhat shorter than pedicels; seeds 2-3 mm. long, transversely rugose.---Plains and valleys. Nebraska to Utah, south to Texas and Colorado. Our one authentic record from northeastern Colorado at 3500 feet.

2. Cleomella nana Eastw., Bull. Torr. Club 30:490. 1903.
Plants 3-15 cm. tall, simple or branching below; leaflets narrowly oblong-lanceolate, 6-15 mm. long; flowers in sessile, rather short racemes the primary ones overtopped by the laterals, pedicels 5-10 mm. long; sepals about 1.5 mm. long, subulate-cristate; petals about 4-5 mm. long, yellow; styles about 1.5 mm. long, stigma more or less 2-lobed; capsule rhomboid at least when young, 4-5 mm. long on a stipe 4-11 mm. long; seeds smooth. Our specimens have longer leaflets and stipes than recorded for the species but otherwise check to the above.---Dry areas. Utah and Colorado. Our one record from southwestern Colorado at 5000 feet.

3. Cleomella palmerana M. E. Jones, Zoe 2:236. 1891.
A species of Utah, reported also from Colorado but no specimens have been seen by the writer.

4. Cleomella montrosae Payson, Bot. Gaz. 60:375. 1915.
Stems 10-30 cm. tall, diffusely branched; leaflets 1-2.5 cm. long, linear-lanceolate to oblong-lanceolate, acute; sepals about 1 mm. long; petals 2-3 mm. long, yellow or orange; capsule 3-4 mm. in diameter, globose or irregularly ovoid, style about 1 mm. long; stipe about 2 mm. long, shorter than capsule and pedicels; seeds 2-3 mm. long, usually 2 mm.---Dry ground. Known only from Montrose County, Colorado, our specimens from 5800 feet.

2. Cristatella Nutt.

Viscid, glandular-pubescent plants; leaves palmately trifoliate, the leaflets entire; flowers small, in terminal bracted racemes; sepals 4, slightly united at base, spreading; petals 4, fan-shaped, laciniate or fimbriate at apex, clawed, the 2 lower smaller than the upper; receptacle short, produced upward into a petaloid, tubular appendage as long as the shorter petals; stamens 6-14; capsule elongate, somewhat flattened, stipitate; seeds many.

1. Cristatella jamesii T. & G., Fl. N. Amer. 1:124. 1838.
Stems 10-40 cm. tall, erect, branching above, glandular-puberulent; leaflets 8-25 mm. long, linear or linear-oblong, nearly sessile; petals (the longest) 3-4 mm. long, whitish to cream-colored; capsule on an ascending pedicel but itself erect, body about 20-25 mm. long and 3-4 mm. wide, glandular; stipe 3-5 mm. long.---Dry often sandy soil. Iowa to South Dakota, south to Arkansas and Colorado. Our records from northeastern Colorado at 3500-5000 feet.

3. Polanisia Raf. CLAMMY WEED

Branching, viscid-pubescent plants with a disagreeable odor; leaves palmately trifoliate or rarely simple; flowers in bracteate terminal racemes, whitish to pale rose; sepals 4, deciduous; petals 4, somewhat irregular, slender or clawed, entire or notched at apex; stamens usually over 10, with conspicuous, long, purple filaments; capsule elongate, somewhat flattened, sessile or nearly so; seeds many.

1. Stamens about 12-16, usually about twice as long as the petals which are 8-12 mm. long   -1. P. trachysperma
1. Stamens about 11, barely exceeding the petals which are 4-5 mm. long   -2. P. graveolens

1. Polanisia trachysperma T. & G., Fl. N. Amer. 1:669. 1840.
Stems 20-80 cm. tall, usually branched; leaflets 2-5 cm. long, oblanceolate to oval; sepals tinged with purple; petals 8-12 mm. long (a few shorter), long-clawed, notched at apex, yellowish-white to white; stamens about 12-16, exerted usually to twice as long as the petals; capsule 3-5.5 cm. long, 5-7 mm. wide, not stipitate, or if so the stipe very short. Seems to intergrade with the next especially in southeastern Colorado.---Canyons and stream beds especially in sandy ground. Saskatchewan to British Columbia, south to Texas and Arizona. Our records scattered over the state at 3500-6500 feet.
2. Polanisia graveolens Raf., Am. Journ. Sci. 1:378. 1819.
This eastern and northern species has been reported for this state and seems to be represented only by a few intergradations with the preceding one.

4. Cleome L. SPIDERFLOWER; BEE-PLANT

Erect branching plants, not glandular-pubescent; leaves palmately 3- to 5-foliate, leaflets entire or serrulate; flowers in terminal racemes, pinkish, white or yellow; sepals 4, distinct or united at base; petals 4, nearly equal, more or less clawed; stamens 6; capsule elongate, long-stipitate, many-seeded.

1. Petals yellow; leaflets (at least lower) 5- to 7-foliate   -1. C. lutea
1. Petals white to rose-purple; leaflets 3-foliate
  2. Leaflets narrowly linear (about 1 mm. wide); leaves short-petioled, this not over 5 mm. long; petals 4-5 mm. long; body of capsule 6-15 mm. long   -2. C. sonorae
  2. Leaflets wider; leaves long-petioled, this over 1 cm. long; petals 8-12 mm. long; body of capsule 2.5-8 cm. long
    3. Body of capsule 2.5-5 cm. long; calyx lobes lanceolate or ovate, acuminate   -3. C. serrulata
    3. Body of capsule 6-8 cm. long; calyx lobes broadly triangular, abruptly acuminate   -3A. C. serrulata angusta

1. Cleome lutea Hook., Fl. Bor. Amer. 1:70. 1830.
Peritoma luteum (Hook.) Raf.---Stems 30-60 cm. tall, glabrous, erect, branching above; leaflets glabrous, 5- to 7-foliate or upper 3-foliate, lower long-petioled, upper short-petioled or sessile; leaflets 2-5 cm. long, oblong to linear-oblong to linear-oblanceolate, entire; racemes elongating in fruit; sepals somewhat united below, rather persistent; petals 5-8 mm. long, yellow, oblanceolate to obovate; pedicels about 10-12 mm. long; filaments around 10-15 mm. long; capsule linear, stipe 1-2 cm. long, longer than pedicel, body of fruit 1-4 cm. long.---Dry open ground. Nebraska to Washington, south to New Mexico and California. All our records from western Colorado at 4500-7000 feet.
2. Cleome sonorae A. Gray, Pl. Wright. 2:16. 1853.
Peritoma sonorae (A. Gray) Rydb.---Stems glabrous, 20-60 cm. tall, erect, simple or branched; leaves short-petioled or the upper sessile; leaflets 3, 1-2 cm. long, narrowly linear, about 1 mm. wide, glabrous; racemes loose, pedicel spreading and filiform; sepals separate or very nearly so, deciduous; petals 4-5 mm. long, white to pink or rose; stamens about as long as petals; stipe of capsule 6-8 mm. long, somewhat reflexed in most, shorter than pedicels, body of capsule 6-15 mm. long, about 2-2.5 mm. thick, acute both ends.---Dry often saline ground. Colorado, New Mexico and Arizona. Our few records from southcentral Colorado at about 7500 feet.
3. Cleome serrulata Pursh, Fl. Amer. Sept. 441. 1814.
Peritome serrulatum (Pursh) DC.; P. inornatum Greene---Stems 30-100 cm. tall, erect and freely branching above, glabrous and somewhat glaucous; leaves 3-foliate, the lower on long petioles, upper nearly or quite sessile, nearly or quite glabrous; leaflets 2.5-8 cm. long, oblanceolate, lanceolate to oblong-obovate, entire or minutely serrulate; racemes elongating, with numerous flowers, the pedicels spreading or recurved; sepals united near base, rather persistent; stamens long-exserted (a few exceptions - perhaps immature); petals 8-12 mm. long, pink, rose, or whitish; stipe of capsule 1-2 cm. long, about as long as the pedicel, body of capsule 2.5-5 cm. long.---Plains, draws, roadsides and waste places in general, especially in sandy soil. Saskatchewan to Oregon, south to Kansas and Arizona. Our records well scattered over the entire state at 3500-8500 feet.
3A. Cleome serrulata angusta (Jones) Tide., (var.) Contrib. U. S. Nat. Herb. 25:249. 1925.
Differs from species in having the sepals broadly triangular and abruptly acuminate, and in the longer pods (6-8 cm. long). Intergrades some with the species in this state.---Utah and Nevada. Our few records from western Colorado, as far east as Jackson and Gunnison Counties, at 6000-8000 feet.

Family 52. Cruciferae MUSTARD FAMILY

Annual, biennial or perennial herbaceous or rarely suffrutescent plants; leaves alternate; flowers perfect, regular or nearly so, in racemes, spikes or corymbs; sepals 4; petals 4 (rarely less or wanting), usually clawed; stamens usually 6 with 2 shorter than the others; ovary superior, of 2 carpels, 2-celled by a septum that stretches across from the placentae, rarely 1-celled; style often wanting, stigma discoid or somewhat 2-lobed; fruit a silicle or silique, rarely indehiscent, the two valves usually falling away from the septum which is often called the replum. The fruit, especially when immature, may be flattened in pressing and the true compression may be obscured. Both flowers and well-matured fruit are desirable and usually necessary in identifying the mustards.

In addition to the following genera there is some evidence that Hesperis matrionalis L., commonly cultivated and known as Dame's Rocket, may sometimes escape and persist. Also a species of Braya, near B. humilis Robinson but possibly new, has been collected recently at Hoosier Pass in Park County.

1. Fruit on a stipe over 1 cm. long at maturity -1. Stanleya
1. Fruit without a stipe or with a short one seldom over 3 mm. long
  2. Fruit didymous, (double) the 2 cells partly separated by constrictions
    3. Fruit strongly flattened; seeds 1 in each cell -2. Dithyrea
    3. Fruit with cells inflated; seeds 2 or more in each cell -3. Physaria
  2. Fruit not didymous
    4. Fruit strongly compressed at right angles to the septum, not over twice as long as broad; petals white, purplish or greenish, never yellow
      5. Seeds 1 in each cell; hairs when present simple
        6. Plants spreading by rootstocks (growing in patches); fruit pubescent or leaves clasping -4. Cardaria
        6. Plants lacking rootstocks (except in L. latifolium with fruit glabrous and leaves not clasping) -5. Lepidium
      5. Seeds 2 to many in each cell; pubescence when present of stellate hairs
        7. Stem leaves auriculate-clasping; fruit 4-15 mm. long
          8. Plants hairy, at least below; fruit triangular in shape, the sides straight -6. Capsella
          8. Plants glabrous; fruit obovate, orbicular or oblong-cuneate but not strictly triangular -7. Thlaspi
        7. Stem leaves not at all auriculate-clasping; fruit 3-4 mm. long -8. Hutchinsia
    4. Fruit (in cross section) round, quadrangular, or compressed parallel to the septum, in most genera over twice as long as broad; petals various in color but often yellow
      9. Fruit not more than twice as long as wide
        10. Fruit flat, definitely compressed parallel to the broad partition
          11. Fruit oval to orbicular; seeds 1-8 (usually 2) in each cell
            12. Pubescence of 2-forked hairs; fruit acute at apex; petals white -9. Koniga
            12. Pubescence stellate, more than 2-branched; fruit notched at apex; petals yellow to whitish -10. Alyssum
          11. Fruit elliptic to longer; seeds many in each cell -11. Draba
        10. Fruit terete or quadrangular, only slightly if at all flattened
          13. Fruit obovoid (widest above the middle), valves with a distinct central nerve extending from base to apex; upper leaves sagittate-clasping -12. Camelina
          13. Fruit broadest at or below the middle, valves nerveless or the nerve not extending to the apex; upper leaves not sagittate-clasping
            14. Plants glabrous or with simple pubescence; seeds plump, not flattened; stigma lobes extending over the placentae -13. Rorippa
            14. Plants silvery-stellate; seeds flattened; stigma lobes extending over the valves -14. Lesquerella
      9. Fruit over twice as long as wide
        15. Plants at least somewhat hairy with some or all of the hairs stellate or branched
          16. Fruit at maturity strongly compressed parallel to the partition (if subterete as in Arabis glabra the pods erect and crowded)
            17. Fruit elliptic to oblong-linear, not over 2 cm. long (rarely over 1.5 cm.), often twisted; petals yellow to white -11. Draba
            17. Fruit elongated-linear, over 2 cm. long, never twisted; petals white, ochroleucous or purplish-pink -15. Arabis
          16. Fruit terete or quadrangular, not strongly compressed (except in pressing); pods seldom strictly erect
            18. Leaves of stem pinnate, bipinnate or deeply pinnatifid (basal leaves may be entire)
              19. Caespitose perennial plants, not over 20 cm. tall; plants growing in the alpine zone of Colorado; petals white to rose -16. Smelowskia
              19. Biennial or annual plants, not caespitose, stems usually over 20 cm. tall; weedy plants of lower altitudes; petals yellow or yellowish -17. Descurainia
            18. Leaves entire to shallowly pinnatifid (rarely more than sinuate-dentate)
              20. Hairs closely appressed; stigma deeply 2-lobed; petals yellow, orange or maroon, often over 8 mm. long -18. Erysimum
              20. Hairs not closely appressed; stigma entire or obscurely lobed; petals white to purplish-pink, not over 8 mm. long
                21. Petals purple-pink, over 5 mm. long; fruit over 4 cm. long -19. Malcolmia
                21. Petals white or whitish, not over 5 mm. long; fruit not over 4 cm. long
                  22. Cauline leaves auriculate-clasping; fruit 2-3 cm. long and 1.5-2 mm. wide; anthers not sagittate or twisted -20. Halimolobos

22. Cauline leaves not auriculate-clasping; fruit 3-4 cm. long, not over 1 mm. wide; anthers sagittate at base, becoming spirally curved   -21. Pennellia
15. Plants glabrous or with simple hairs, no branched or stellate ones present
   23. Sepals in anthesis divaricate or reflexed; anthers sagittate at base and stigma entire -22. Stanleyella
   23. Sepals in anthesis erect or ascending; anthers sometimes sagittate at base but if so then stigma not entire (except in Hesperidanthus)
      24. Anthers sagittate at base; stigma usually with 2 lobes these extending over the valves (except in Sisymbrium, keyed both ways)
         25. Fruit strongly flattened parallel to the septum
            26. Fruit erect or ascending, 4-5 mm. wide; cauline leaves auriculate-clasping   -23. Streptanthus
            26. Fruit pendant, 1-2 mm. wide; cauline leaves not auriculate-clasping   -24. Streptanthella
         25. Fruit terete or only slightly flattened
            27. Calyx urn-shaped, nearly closed at the top in anthesis; petals linear, and channeled, the blades little wider than the claws; stems more or less inflated   -25. Caulanthus
            27. Calyx campanulate in anthesis, open; petals with blades distinctly wider than the claws; stems not inflated
               28. Stigma conic, often pointed, entire or nearly so; outer sepals strongly gibbous at base; petals purple-pink, 12-15 mm. long and stem leaves not clasping   -26. Hesperidanthus
               28. Stigma not conic or pointed, often 2-lobed; outer sepals little if any gibbous at base; petals white, purple or yellow, less than 12 mm. long or if longer then the stem leaves clasping
                  29. Fruit torulose, the septum with a central strip of elongated cells, appearing like a broad midrib under a hand lens; stigma entire or slightly 2-lobed, these extending over the valves   -27. Thelypodium
                  29. Fruit not torulose, the septum not differentiated; stigma usually definitely 2-lobed these extending over the edges of the septum   -28. Sisymbrium
      24. Anthers not sagittate at base; stigma entire or 2-lobed, these extending over the edges of the septum
         30. Fruit at maturity strongly compressed parallel to the septum (if subterete as in Arabis glabra the pods are erect and crowded)
            31. Fruit elliptic to oblong-linear, not over 2 cm. long (rarely over 1.5 cm.), often twisted; petals yellow to white   -11. Draba
            31. Fruit elongate-linear, over 2 cm. long, never twisted; petals white, ochroleucous or purplish-pink   -15. Arabis
         30. Fruit terete, quadrangular or only slightly compressed (except in pressing); pods seldom crowded-erect
            32. Fruit with a stout, indehiscent beak; leaves (at least some) pinnatifid; fruit sometimes indehiscent
               33. Fruit indehiscent, the seeds separated by transverse, spongy partitions
                  34. Stems glandular-viscid; petals deep rose, 7-12 mm. long; fruit breaking at maturity into joints   -29. Chorispora
                  34. Stems not glandular-viscid; petals white, yellow or purplish, 15-20 mm. long; fruit not breaking into joints   -30. Raphanus
               33. Fruit dehiscent, without transverse partitions (the septum longitudinal)
                  35. Beak of fruit flat and often 2-edged, containing 1 seed
                     36. Valves of fruit with 1 strong nerve; seeds in 2 rows in each cell; stems glabrous   -31. Eruca
                     36. Valves of fruit with 3 strong nerves; seeds in 1 row in each cell; stems usually hairy   -32. Brassica
                  35. Beak of fruit elongated-conical, not flattened or 2-edged, seedless   -32. Brassica
            32. Fruit scarcely beaked, merely tipped by a slender style or sessile stigma; fruit dehiscent
               37. Cauline leaves (at least the upper ones) auriculate-clasping
                  38. Fruit 4-angled, the valves keeled, sessile; petals pale yellow   -33. Conringia
                  38. Fruit terete, sessile or stiped; petals usually white (if yellow the fruit definitely stiped)   -28. Sisymbrium
               37. Cauline leaves not auriculate-clasping
                  39. Petals white or purple, not yellow
                     40. Plants aquatic or semiaquatic, rooting at the nodes; fruit terete, with the seeds usually in 2 rows in each cell   -13. Rorippa
                     40. Plants terrestrial (although often in wet soil), not rooting at the nodes; fruit somewhat compressed, the seeds in 1 row in each cell   -34. Cardamine
                  39. Petals yellow
                     41. Fruit 4-angled, the valves keeled; seeds flattened   -35. Barbarea
                     41. Fruit terete; seeds plump and turgid
                        42. Seeds in 2 rows in each cell, fruit not over 14 mm. long   -13. Rorippa
                        42. Seeds in 1 row in each cell, the fruit usually over 14 mm. long (2.5-10 cm. in our common ones)   -28. Sisymbrium

1. **Stanleya** Nutt.   PRINCE'S PLUME; DESERT PLUME

Plants herbaceous, biennial or perennial, rarely subshrubby; stems tall and stout; leaves entire to pinnately compound; flowers large, in elongated terminal racemes; petals yellow to cream-colored, with long claws; stamens about equal, the anthers long and narrow; fruit slender, nearly terete, long-stipitate, this stipe 1-3 cm. long; seeds not at all winged.

1. Middle and upper cauline leaves sessile and sagittate-clasping; inner surface of petal claw glabrous
    -1. S. viridiflora
1. Middle and upper cauline leaves petiolate, not sagittate at base; inner surface of petal claw densely villous
    2. Petal blade pale yellow to white, 4-10 mm. wide; fruit ascending; no woody caudex present
        -2. S. albescens
    2. Petal blade bright yellow, 1.5-3 mm. wide; fruit ascending or widely spreading; woody caudex present
        3. Fruit not definitely torulose and tortuous; leaves rarely bipinnate; plants usually glabrous
            4. Upper cauline leaves oblanceolate or narrower, lower cauline leaves pinnatifid (rarely bipinnatifid)
                -3. S. pinnata
            4. Upper cauline leaves ovate-lanceolate or wider, entire, lower cauline leaves entire or somewhat divided
                -3A. S. pinnata integrifolia
        3. Fruit definitely torulose and tortuous; leaves mostly bipinnate; plants pubescent
            -3B. S. pinnata bipinnata

1. **Stanleya viridiflora** Nutt. in T. & G., Fl. N. Amer. 98. 1838.
    Perennial plants with a simple woody caudex; stems 30-100 cm. tall, glabrous, simple at least at base, angulate or winged; basal and lower leaves petiolate, entire, dentate or divided, obovate to oblanceolate; cauline leaves similar but middle and upper sessile, reduced in size and sagittate at base; inflorescence elongated; sepals 12-16 mm. long, glabrous, greenish; petals lemon-yellow to whitish, glabrous; fruit 4-7 mm. long, glabrous, nearly terete, stipe 15-25 mm. long.---Hills and dry roadsides. Montana to Oregon, south to Wyoming, Utah, and Nevada. Our one record from northwestern Colorado at 6500 feet.
2. **Stanleya albescens** M. E. Jones, Zoe 2:17. 1891.
    Biennial plants branching at base but caudex not woody; stems 1 to several, simple or branched, glaucous; leaves petioled, lyrate-pinnatifid to rarely entire, lower even bipinnatifid, glabrous and glaucous; sepals 10-15 mm. long, glabrous, white, green-tipped; petals white, 10-15 mm. long, blade 4-10 mm. broad, claw villous within; fruit 3-6 cm. long, nearly terete, erect, nearly straight or somewhat curved inward, glabrous, stipe 10-15 mm. long.---Dry hills, plains and banks. Colorado, Utah, New Mexico, and Arizona. Our records from westcentral Colorado at 4500-6000 feet.
3. **Stanleya pinnata** (Pursh) Britton, Trans. N. Y. Acad. 8:62. 1889.
    S. arcuata Rydb.; S. canescens Rydb.; S. glauca Rydb.---Plants subshrubby or sometimes definitely woody at base, with a branching woody caudex; stems several to many, simple or usually branched above, glaucous, glabrous, or sparsely pubescent; leaves petiolate, the lower pinnatifid or rarely bipinnate, upper leaves entire or divided, oblanceolate or narrower; sepals 10-15 mm. long, glabrous; petals 10-16 mm. long, blade glabrous 1.5-3 mm. wide, claw densely villous; fruit nearly terete, 3-8 cm. long, stipe 10-25 mm. long. This plant has been called variety typica Rollins.---Dry hills, valleys and banks. North Dakota to Idaho, south to Texas, New Mexico and California. Our records scattered over Colorado, except in the northeastern part, at 4000-8000 feet.
3A. **Stanleya pinnata integrifolia** (James) Rollins, (var.) Lloydia 2:118. 1939.
    S. integrifolia James---Upper leaves broadly ovate to ovate-lanceolate; lower cauline leaves entire or somewhat divided; fruit not especially torulose or tortulose. This variety intergrades rather commonly with the typical form of the species.---Our records from the western one-fourth of Colorado at 4500-7500 feet.
3B. **Stanleya pinnata bipinnata** (Greene) Rollins, (var.) Lloydia 2:119. 1939.
    S. bipinnata Greene---Plants pubescent; leaves mostly bipinnate; fruit torulose and tortulose.---Our few records from northcentral Colorado.

2. Dithyrea Harv. SPECTACLE POD

Annual or perennial, stellate-pubescent plants; stems branching; leaves entire to sinuate-dentate; flowers racemose; sepals erect or spreading, rather broad; petals clawed, white or purplish; anthers linear and sagittate; fruit didymous, strongly compressed at right angles to the septum, cells suborbicular, stigma sessile; seeds 1 in each cell.

1. **Dithyrea wislizeni** Engelm. in Wisliz., Mem. North. Mexico 96. 1848.
    Annual or possibly biennial plants with taproots; stems erect, 20-60 cm. tall; leaves numerous, the cauline 2-6 cm. long, linear-lanceolate to ovate-lanceolate, sinuate-dentate to nearly entire; petals 5-8 mm. long, white to white-yellow; pedicels 7-10 mm. long, spreading in fruit; fruit cells about 5-6 mm. wide, truncate or shallowly notched above.---Dry places. Colorado and Utah, south to Mexico. This plant has been so definitely reported from southwestern Colorado that a description is included here, checked with a plant from New Mexico.

3. Physaria (Nutt.) Gray DOUBLE BLADDER POD; TWINPOD

Perennial caespitose, stellate plants with taproots; stems simple above the base; basal leaves usually numerous, often in a rosette; cauline leaves usually few; inflorescence a congested or somewhat elongated raceme which elongates in fruit; sepals pubescent; petals usually spatulate, yellow, glabrous; fruit didymous, more or less inflated, pubescent, ovules 2-6 in each cell; seeds wingless.

1. Sinuses of fruit equal above and below; valves nearly orbicular -1. P. australis
1. Sinuses unequal, upper deep, lower shallow or absent; valves variously shaped but not orbicular
    2. Basal leaves rounded at apex; apical sinus of fruit both broad and deep (equalling the replum length in width and depth)
        -2. P. vitulifera

2. Basal leaves acute at apex; apical sinus of fruit either wide and shallow or narrow and deep
   3. Apical sinus of fruit narrow (less than 1 mm. across) and deep (equalling the replum length)
                                                                                                        -3. P. osterhoutii
   3. Apical sinus of fruit broad (at least 2 mm.) and more shallow
      4. Basal leaves entire (or with a single tooth on each side), less than 4 cm. long; plants seldom over 12 cm. tall   -4. P. acutifolia
      4. Basal leaves lobed, more than 4 cm. long; plants often over 12 cm. tall   -5. P. floribunda

1. **Physaria australis** (Payson) Rollins, Rhodora 41:408. 1939.
    P. didymocarpa australis Payson; P. didymocarpa Gray in part---Perennial, caespitose plants, silvery-stellate throughout; stems 5-15 cm. long, numerous, usually somewhat decumbent, simple; basal leaves 2-8 cm. long, numerous, entire or rarely with scattered teeth, blade obovate to orbicular, obtuse; cauline leaves 1-3 cm. long, entire, spatulate to oblanceolate; inflorescence racemose, elongating in fruit; sepals linear-oblong, pubescent; petals about 1 cm. long, spatulate, yellow; pedicels 6-12 mm. long, divaricate; fruit erect, inflated, pubescent, apical sinus deep, narrow or even closed, basal sinus similar; valves 6-10 mm. long and 3-6 mm. wide, suborbicular; replum oblong, constricted, obtuse at top, ovules 2 to a locule; style 4-6 mm. long; seeds 2-3 mm. wide, brown, wingless.---Banks and slopes. Wyoming to Idaho, south to New Mexico and Utah. Our records from the western one-fourth of Colorado, and also in the northcentral part, at 4500-7500 feet.

2. **Physaria vitulifera** Rydb., Bull. Torr. Club 28:278. 1901.
    Perennial, caespitose plants, silvery-stellate throughout; stems 10-20 cm. long, numerous, usually decumbent, lateral, simple; basal leaves 3-6 cm. long and 1-2 cm. broad, numerous, panduriform or merely obovate, obtuse, margins deeply and broadly incised, rarely entire; cauline leaves entire, oblanceolate to spatulate, often somewhat acute; sepals 6-8 mm. long, oblong, pubescent; petals about 1 cm. long, yellow, spatulate; style 5-7 mm. long; fruit often rigid, inflated, pubescent, obtuse or truncate, apical sinus broad, open and deep; valves 5-6 mm. high and 3-4 mm. broad; replum oblong, often constricted, 2-3 mm. long, less than 1 mm. wide, 2 ovules in each cell; 1-2 seeds in each locule, about 2.5 mm. broad, brown, wingless.---Dry valleys and canyons. Apparently limited to Colorado, our records mostly from the northcentral and southwestern parts at 4500-9000 feet.

3. **Physaria osterhoutii** Payson, Ann. Mo. Bot. Gard. 5:146. 1918.
    Perennial, caespitose plants, silvery-stellate throughout, the rays of stellae usually forked; caudex simple or branched; stems 8-15 cm. long, slender, numerous, erect or somewhat decumbent; basal leaves 2-5 cm. long, oblanceolate, often hastate, incised or with broad teeth, rarely entire; cauline leaves 1-2 cm. long, linear-oblanceolate, acute, entire or rarely with a few teeth; inflorescence congested, flowers numerous; sepals 5-7 mm. long, linear-oblong, yellowish, pubescent; petals 8-10 mm. long, yellow, spatulate; replum 2-3 mm. long, oblong, obtuse, ovules 2 to each locule; style 4-5 mm. long; pedicels recurved in fruit; fruit 5-7 mm. long and 4-5 mm. wide, pendant, base truncate or obtuse, apex deeply emarginate; valves somewhat inflated, rather loosely stellate-pubescent; seeds about 2 mm. wide, orbicular, 1-2 in each locule.---Banks and slopes. Apparently limited to Colorado, our records from the northcentral and northwestern parts at 5500-7000 feet.

4. **Physaria acutifolia** Rydb., Bull. Torr. Club 18:279. 1901.
    P. floribunda Rydb. of Manuals in part---Perennial, caespitose plants, silvery-stellate throughout; stems 5-10 cm. tall, several to numerous, decumbent, simple; basal leaves 2-3.5 cm. long, oblanceolate or broader, acute, entire or with 1 or 2 broad teeth; cauline leaves 1-1.5 cm. long, few, oblanceolate, entire, acute; sepals 5-7 mm. long, linear; petals 8-10 mm. long, yellow, spatulate; replum about 3 mm. long, obovate to somewhat longer, obtuse at apex; style 5-7 mm. long; ovules 2; fruit inflated, erect, slightly cordate at base or nearly obtuse, apical sinus broad and deep; valves 4-5 mm. wide, 6-8 mm. high, suborbicular; seeds about 2 mm. broad, orbicular, brown, only slightly flattened, 1-2 to a locule.---Banks and slopes. Apparently limited to Colorado. Our records scattered over the western half of the state, mostly in the central and northcentral parts at 5000-10,000 feet, possibly to 12,500 feet.

5. **Physaria floribunda** Rydb., Bull. Torr. Club 18:279. 1901.
    Perennial, caespitose plants, silvery-stellate throughout; stems 10-20 cm. tall, numerous, simple, decumbent or erect; radical leaves 4-8 cm. long, broadly oblanceolate, pinnatifid or merely dentate, rarely entire, terminal lobes acute or obtuse, but not rounded; cauline leaves 1-3 cm. long, spatulate to linear-oblanceolate, acute, entire; inflorescence loosely racemose; sepals 5-7 mm. long, linear-oblong; petals 9-11 mm. long, yellow, spatulate; pedicels 6-12 mm. long, spreading or somewhat recurved; replum 2.5-4 mm. long, linear-oblong, constricted, obtuse at top, ovules 2 to a locule; style 5-8 mm. long; fruit erect to pendant, somewhat inflated, obtuse or somewhat cordate at base, deeply or broadly notched above; valves 4-6 mm. high, 3-5 mm. wide; seeds about 2 mm. wide, brown, marginless.---Dry banks and slopes. Colorado and New Mexico. Our records scattered over the western two-thirds of Colorado, except in the extreme northern part, at 4500-9000 feet.

4. Cardaria Desr. WHITEWEED; WHITETOP

    Perennial, herbaceous plants, spreading by horizontal roots; leaves not lobed, sessile and auriculate-clasping at base; petals white; fruit a silicle, not winged, more or less inflated, indehiscent; seeds 1 in each cell, but each cell of ovary 2-ovuled. Sometimes united at least in part with Lepidium.

1. Ovary and fruit pubescent   -1. C. pubescens
1. Ovary and fruit glabrous   -2. C. draba

1. **Cardaria pubescens** (Meyer) Rollins, Rhodora 42:305. 1940.
   <u>Hymenophysa pubescens</u> C. A. Meyer---Perennial plants spreading by horizontal roots; stems 20-40 cm. tall, branching, at least above, short-pubescent, leafy; leaves 10-35 mm. long, oblong, acute or obtuse, auriculate-clasping at base, finely serrate-dentate, short-pubescent or puberulent; flowers in corymbose racemes; flowers 2 mm. long; fruits 3-4 mm. in diameter, subglobose, puberulent. Our plants have been called variety <u>elongata</u> Rollins.---As a weed in cultivated fields, roadsides and waste land. Occasionally present in the area from Michigan to Washington, south to Colorado and California. Our rather few records from northcentral, central and southcentral Colorado at 5000-7500 feet.

2. **Cardaria draba** (L.) Desv., Journ. de Bot. Desv. 3:163. 1813.
   <u>Lepidium</u> <u>draba</u> L.; <u>L. repens</u> (Scrank.) Boiss.---Perennial plants from spreading root systems; stems 25-50 cm. tall, erect, sometimes repent however, sparsely pubescent to cinereous-hoary; lower leaves oblanceolate to obovate, denticulate, petioled; upper leaves sessile with clasping bases, usually dentate or denticulate; pedicels about 1 cm. long, slender, terete; sepals 1.5-2 mm. long; petals 3-5 mm. long, clawed; style about 1 mm. long; fruit 3-5 mm. long, variable, obovate to triangular-ovate to ovate, often acute at apex and cordate at base, much inflated, glabrous.---Fields, roadsides and waste ground as one of our most serious weeds. Now rather widespread in the United States. Our records scattered over Colorado, especially in cultivated areas, at 3500-8500 feet.

5. Lepidium L. PEPPERGRASS; PEPPERWEED

Plants annual or perennial, herbaceous or woody at the very base; leaves entire to tripinnate; flowers small, mostly in small dense racemes these elongating in fruit; petals white to greenish, sometimes very small or wanting; stamens 2, 4 or 6; ovary sessile; fruit a silicle, flattened contrary to the septum, more or less winged or emarginate on the margins, little if any inflated, seeds 1 in each cell, style lacking or present; ovules only 1 in each cell.

1. Upper cauline leaves perfoliate or sagittate-clasping at base
   2. Upper cauline leaves perfoliate; basal leaves bipinnatifid into linear divisions -1. <u>L. perfoliatum</u>
   2. Upper cauline leaves merely clasping; basal leaves not lobed or merely pinnatifid -2. <u>L. campestre</u>
1. Upper cauline leaves neither perfoliate nor clasping at base
   3. Style longer than the sinus of the fruit or sinus absent; plants perennial or rarely biennial
      4. Fruit without a sinus at apex; style almost obsolete; plants with spreading roots, growing in patches -3. <u>L. latifolium</u>
      4. Fruit with a definite sinus; style 0.3 mm. long or more; plants not in patches -4. <u>L. montanum</u>
   3. Style shorter than the sinus of the fruit which is always present; plants annual or biennial (in <u>L. ramosissimum</u>)
      5. Pedicels strongly flattened, at least twice as wide as thick
         6. Fruit pubescent -5. <u>L. lasiocarpum</u>
         6. Fruit glabrous -7B. <u>L. densiflorum ramosum</u>
      5. Pedicels terete or if flattened not as much as above
         7. Flowers and fruit congested in numerous, short, axillary racemes as well as in terminal longer racemes; fruit narrowing toward the apex, elliptic in shape; plants biennial -6. <u>L. ramosissimum</u>
         7. Racemes terminal on branches, no short, numerous axillary ones present; fruit not narrowing toward the apex, rotund to oblong or obovate in shape; plants annual
            8. Upper half of fruit averaging greater in width than the lower; petals absent or if present rarely as long as the sepals
               9. Fruit averaging about 2.5 mm. long; pedicels only slightly flattened -7. <u>L. densiflorum</u>
               9. Fruit averaging 3 mm. long or more; pedicels distinctly flattened -7A. <u>L. densiflorum bourgeauanum</u>
            8. Lower half of fruit averaging greater in width than the upper; petals as long or longer than the sepals
               10. Upper part of stems and pedicels glabrous -8. <u>L. virginicum</u>
               10. Entire stems as well as pedicels puberulent to hirsute -8A. <u>L. virginicum pubescens</u>

1. **Lepidium perfoliatum** L., Sp. Pl. 643. 1753.
   Annual plants; stems about 15-40 cm. tall; basal leaves multifid or multipinnate, puberulent to glabrous; cauline leaves variable, lowest dissected, middle auriculate and the upper perfoliate, glabrous; pedicels terete, longer than the fruit; sepals 1 mm. long, ovate, pilose on back; petals slightly longer than sepals, narrow; stamens usually 6; fruit 4 mm. long, rhombic-ovate to rhombic-oval, glabrous or sparsely pubescent, slightly winged at apex, sinus 0.2 mm. deep, style about as long as sinus.---Roadsides and waste places. Now distributed in the United States especially the western and central parts. Our records scattered over the western half of Colorado, none as yet from the southcentral part, at 4500-8500 feet.

2. **Lepidium campestre** (L.) R. Br. in Ait., Hort. Kew. ed. 2, 4:88. 1812.
   Annual, densely short-villous or pubescent plants, simple to profusely branched; leaves oblanceolate to narrowly oblong; basal leaves 4-12 cm. long, pinnatifid, lobed to entire, petioled; cauline leaves denticulate, sessile and sagittate-clasping; pedicels slender, slightly flattened, about equalling the fruit or a little longer; sepals 1.5 mm. long, villous to glabrate; petals about 2 mm. long; stamens 6; fruit 5-6 mm. long, oblong-ovate, pustulose and somewhat hairy to glabrous, margins and apex fairly broadly winged, upper surface concave, apex somewhat emarginate; style about equalling the sinus.---Fields and waste places. Widespread. Our few records from northcentral, central, northwestern and southwestern Colorado at 5500-10,000 feet.

3. **Lepidium latifolium** L., Sp. Pl. 644. 1753.
Perennial plants from widely spreading root systems; stems about 40-60 cm. tall, glabrous or nearly so; leaves all entire to dentate, basal as much as 30 cm. long and 6-8 cm. broad, petioled; cauline leaves 1-4 cm. broad, reduced in size but many, upper nearly sessile; racemes numerous, many-flowered, compounded; pedicels terete, longer than the fruit; sepals less than 1 mm. long, oval, somewhat pilose on back; petals spatulate, white; stamens 6; fruit 2 mm. long, ovate-rotund, sparsely pilose, not emarginate, tipped by stigma and almost obsolete style.---Fields and waste places. Now becoming established at various parts of the United States and Mexico. Our few records from northcentral to southcentral Colorado at 5500-8000 feet.

4. **Lepidium montanum** Nutt.; T. & G., Fl. N. Amer. 1:116, 669. 1838.
L. alyssoides Gray; L. eastwoodiae Wooton; L. jonesii Rydb.; L. crandallii Rydb.; L. spathulatum Vasey; L. vaseyanum Thell.; L. crenatum (Greene) Rydb.; L. brachybotryum Rydb.---Plants biennial or perennial often somewhat frutescent; stems 1 to several, freely branched or simple, glabrous to densely pubescent; basal leaves 3-15 cm. long, parted to variously pinnatifid or sometimes unlobed; cauline leaves reduced, entire to pinnatifid, glabrous or somewhat puberulent; sepals 1-2 mm. long, glabrous to pilose on the back; petals usually twice as long as the sepals, white; stamens 6 or rarely 2; style 0.3-1 mm. long; fruit oval to ovate, glabrous, emarginate. This species is a complex showing great variations in habit, leaf form and pubescence. Hitchcock (Madrona 3:265-320. 1936) gave a key to 6 varieties in or near Colorado. However, only about 2 out of 3 of our plants key down satisfactorily, the others showing all kinds of intergradations especially in relative pubescence. The extremes, however, are certainly different in general appearance.---Plains, canyons, valleys and hills, often in dry ground. Wyoming to Nevada, south to Texas and Arizona. Our records scattered in the western three-fourths of Colorado, except in the northcentral part, at 4500-12,000 feet.

5. **Lepidium lasiocarpum** Nutt.; T. & G., Fl. N. Amer. 1:115. 1838.
Annual plants; stems 5-30 cm. tall, prostrate to erect, rather hispid-hirsute to pubescent; leaves 1-6 cm. long, variable, usually linear to oblanceolate; basal often lobed to pinnatifid; pedicels 1.5-4 mm. long, flattened, pubescent both sides; sepals 1 mm. long; petals no longer than sepals, spatulate to linear or lacking; stamens 2 or 4; style lacking or nearly so; fruit 3-4.5 mm. long, oval to elliptical or oblong-obovate, hirsute-hispid on both surfaces or even glabrous, finely reticulate, apex winged, the sinus 1/10 - 1/5 the length of the fruit. Our plant has been called variety typicum C. Hitchc.---Dry ground. Colorado to Nevada, south to Mexico and California. Our few records from westcentral and southwestern Colorado at 4500-5500 feet.

6. **Lepidium ramosissimum** Nelson, Bull. Torr. Club 26:124. 1899.
L. divergens Osterh.; L. fletcheri Rydb.---Profusely branched biennial or possibly annual plants; stems 15-50 cm. tall, densely but finely puberulent; basal leaves usually pinnatifid, lobes often again toothed; upper cauline leaves entire and linear, lower oblanceolate or oblong and few-toothed; inflorescence of many few-flowered, short racemes, in the axils of upper leaves, terminated by a longer raceme, the whole stem floriferous; pedicels about equalling the fruit, slightly flattened to somewhat wing-margined; sepals 1 mm. long; petals not as long as sepals, linear; stamens 2; style lacking or very short; fruit 2.5-3.5 mm. long, often ciliate or pubescent, elliptic, shallowly notched and winged at apex, sinus 1/6 to 1/8 the length of the fruit.---Dry plains, slopes and mesas. In the Rocky Mountains from Manitoba to New Mexico. Our records mostly from northcentral and central Colorado at 6000-10,500 feet.

7. **Lepidium densiflorum** Schrad., Ind. Sem. n. Gotting. 4. 1832.
L. apetalum in whole or in part---Annual plants; stems about 30-50 cm. tall, sometimes shorter, diffusely branched, puberulent to pubescent; leaves mostly oblanceolate; basal leaves 4-6 cm. long, coarsely toothed to pinnatifid at middle and base; racemes 6-15 cm. long, numerous, many-flowered; pedicels but slightly flattened, scarcely equalling to somewhat longer than the fruit; calyx about 1 mm. long, usually somewhat pilose on back; petals usually lacking or rudimentary, rarely as long as sepals; stamens usually 2; style lacking; fruit about 2.5 mm. long, elliptical-ovate to obovate-rotund (wider nearer tip on the average), glabrous to pubescent. This plant has been called variety typicum Thell.---Plains, roadsides, fields and waste places. Canada and the United States especially in the eastern parts. Our records scattered over Colorado, but few from the extreme western part, at 3500-9500 feet.

7A. **Lepidium densiflorum bourgeauanum** (Thell.)C. Hitch., (var.) Madrono 3:279. 1936.
L. bourgeauanum Thell.---Like the species but the pedicels flattened, chiefly on lower side, but not twice as broad as thick; fruits averaging 3 mm. long or more, inclined to be more oblong-obovate in shape.---Our records scattered over Colorado mostly in the northern half of the state, at 4000-8500 feet.

7B. **Lepidium densiflorum ramosum** (Nelson) Thell., (var.) Bull. Herb. Bois. ser 2. 4:706. 1904.
L. ramosum A. Nels.---Like the species but pedicels definitely flattened on both sides, about twice as broad as thick; fruit about 3.5 mm. long, more oblong, glabrous.---Our records mostly from the westcentral and central parts of the state at 4500-9500 feet.

8. **Lepidium virginicum** L., Sp. Pl. 645. 1753.
L. idahoense Heller; L. medium Greene---Annual plants; stems about 15-60 cm. tall, branched, glabrous above, puberulent below; leaves irregularly toothed to pinnatifid but upper entire and linear; lower leaves spatulate to oblanceolate; racemes numerous and many-flowered; pedicels slender, terete or nearly so, usually longer than the fruit; sepals about 1 mm. long, usually glabrous; petals equal to 3 times longer than sepals, spatulate to obovate, clawed; stamens usually 2; style practically lacking; fruit 2.5-4 mm. long, glabrous, elliptic-rotund to orbicular (widest below the middle). This description is of variety medium (Greene) Hitch.---Banks valleys, plains and hills. Wyoming to Washington, south to Texas and California. Our records mostly from the northcentral, central and southcentral parts of the state at 5000-7500 feet.

8A. **Lepidium virginium pubescens** (Greene) C. Hitch., (var.) Madrono 3:283. 1936.
   L. hirsutum Rydb.---Stems and pedicels puberulent, often almost hispid.---Our records scattered over the western half of Colorado, except for the extreme western part, at 5500-10,000 feet.

6. Capsella Medic. SHEPHERDS PURSE

Annual caulescent plants, with branched hairs; basal leaves in a rosette, usually lyrate-pinnatifid; stem leaves dentate or entire, auricled at base; flowers racemose, perfect, white; fruit triangular, cuneate at base, strongly compressed at right angles to the septum, dehiscent, with short styles; seeds numerous in each cell, not winged.

1. **Capsella bursa-pastoris** (L.) Medic., Pflanzeng. 1:85. 1792.
   Bursa bursa-pastoris (L.) Britt.---Stems 10-40 cm. tall, erect, branching, pubescent below, mostly glabrous above; lower leaves dentate to pinnatifid, usually the latter; pedicels slender, spreading or ascending; petals about 1.5-2 mm. long; fruit 5-8 mm. long, triangular, more or less emarginate at apex.---A common weed in gardens, lawns, fields and waste places. Widespread in North America; naturalized from Europe. Our records scattered over Colorado especially near cultivated areas at 4000-9000 feet.

7. Thlaspi L. PENNY CRESS; WILD CANDY TUFT

Annual or perennial glabrous, herbaceous plants; basal leaves often forming a rosette; stem leaves auriculate-clasping; inflorescence a raceme; petals white or tinged with purple; fruit 2-celled, dehiscent, flattened at right angles to the partition, obovate, orbicular or oblong-cuneate; seeds 2 or more in each cell.

1. Plants annual; stems usually 30 cm. tall or more, usually branched above; fruit orbicular or nearly so, at least 8 mm. wide at maturity; style minute -1. **T. arvense**
1. Plants perennial; stems less than 30 cm. tall, not branched above; fruit obovate to elliptical, not over 6 mm. wide; style 1-3 mm. long -2. **T. alpestre**

1. **Thlaspi arvense** L., Sp. Pl. 646. 1753.
   Annual plants; stems 15-50 cm. tall, erect, branching above; basal leaves oblanceolate, petioled, early deciduous; stem leaves oblong to lanceolate, the upper clasping; pedicels spreading; inflorescence in fruit rather elongated; petals 2-4 mm. long, white; style nearly or quite obsolete; fruit 8-15 mm. wide, suborbicular, broadly winged.---A weed of fields, roadsides and waste places. Introduced from Europe into various parts of North America. Our records scattered over Colorado at 4000-9500 feet.
2. **Thlaspi alpestre** L., Sp. Pl. ed. 2. 903. 1763.
   T. fendleri A. Gray; T. coloradense Rydb.; T. purpurascens Rydb.; T. glaucum A. Nels.; T. nuttallii Rydb.---Perennial plants from short, simple or branched caudices; stems 2-30 cm. tall, 1 to several; basal leaves oval, obovate or spatulate, entire or somewhat toothed, petioled; cauline leaves oblong or ovate, leaves sometimes purplish-tinged; inflorescence crowded to elongated; petals white to purplish; pedicels usually spreading; style 1-3 mm. long; fruit not over 6 mm. wide, obcordate or broadly elliptical, truncate or retuse and narrowly winged. This is a variable complex and the extremes certainly look different. However, all these named forms listed above as synonyms are connected in Colorado by a puzzling series of intergrades and the writer has never been able to separate them in a reasonable fashion. This complex needs critical study.---Hills, valleys and mountains. Montana to Washington, south to New Mexico and California. Our records well scattered over the western two-thirds of Colorado at 5000-13,000 feet.

8. Hutchinsia R. Br.

Annual plants; stems low and spreading; glabrous or stellate-puberulent; leaves entire to pinnately lobed; flowers in racemes; petals minute, white; fruit oval or elliptic, compressed at right angles to the septum, valves strongly 1-nerved; style none or very short; seeds numerous in each cell.

1. **Hutchinsia procumbens** (L.) Desv., 3:168. 1814.
   Stems 5-20 cm. long, branching from the base, slender; lower leaves 10-25 cm. long, short-petioled, entire to pinnatifid; upper leaves sessile or nearly so; pedicels ascending or spreading; sepals and petals about 1 mm. long; fruit about 3-4 mm. long, elliptical or oval, obtuse, usually not emarginate.---Rather moist ground. Labrador to British Columbia, south to Colorado, Arizona and California. Our few records from Grand and Gunnison Counties at 7000-8000 feet.

9. Koniga Adans. SWEET ALYSSUM

Perennial or apparently sometimes annual, herbaceous plants, pubescent or canescent with 2-forked hairs; leaves entire; inflorescence of terminal racemes; petals obovate, entire; filaments slender with 2 glands at base; fruit compressed parallel to partition, dehiscent; seeds 1 in each cell.

1. **Koniga maritima** (L.) R. Br. in Denh. and Clapp., Narr.Exp. Afric. 214. 1826.
   Stems 10-30 cm. tall, ascending or procumbent, freely branching, minutely pubescent with appressed forked hairs; basal leaves oblanceolate, narrowed into a petiole; cauline leaves

lanceolate or linear, nearly sessile; petals 2.5-3 mm. long, white, clawed, fragrant; fruiting pedicels ascending; fruit 3-3.5 mm. long, including the style which is about 0.5 mm. long, oval to orbicular, pointed at the ends.---In waste places, roadsides and along streets. Introduced from Europe. Our one record from northcentral Colorado at about 4600 feet.

10. Alyssum L. ALYSSUM

Annual plants with stellate hairs; stems low, usually branching; leaves entire; flowers racemose; sepals more or less spreading; petals whitish to yellowish; filaments commonly dilated at the base; fruit orbicular in outline, dehiscent, compressed parallel to the septum; seeds 1-8 in each cell, wingless.

1. Alyssum alyssoides L., Syst. Veg. ed. 10. 1130. 1759.
Stems 5-30 cm. tall, erect but usually branching at the base, stellate; leaves 1-3 cm. long, linear-oblong or spatulate, stellate; petals about 2-3 mm. long, white to pale yellow; pedicels spreading; fruit 3 mm. wide, notched at apex.---Fields and waste places. Naturalized from Europe in various parts of North America. Our records from northwestern, northcentral, central and southcentral Colorado at 4500-7000 feet.

11. Draba (Dill.) L. WHITLOW-GRASS

Annual, biennial, or perennial herbaceous plants; stems leafy or scapose, usually pubescent with simple, forked or stellate hairs; racemes corymbose or elongated; petals white or yellow; fruit dehiscent, 2-celled, strongly compressed parallel to the partition, sometimes twisted; seeds numerous, in 2 rows in each cell.

1. Plants annual; style not over 0.2 mm. long or absent entirely
  2. No cauline leaves present, all the leaves in a basal rosette; petals lemon-yellow when fresh
                                                                                                                        -1. D. crassifolia
  2. Flowering stems with 1 or more leaves; petals lemon-yellow to white when fresh
    3. Petals lemon-yellow in anthesis (often fading to whitish); plants usually found above 6000 feet in Colorado; inflorescence and pedicels usually glabrous
      4. Most of the pedicels 1.5 times as long as the fruit or more   -2. D. nemorosa
      4. Pedicels only slightly if at all longer than the fruit
        5. Stems sparsely pubescent below, glabrous above; fruit usually glabrous, 8-22 mm. long, 16-40 seeded   -3. D. stenoloba
        5. Stems definitely pubescent throughout; fruit hispidulous, 6-10 mm. long, 40-80 seeded
                                                                                         -4. D. rectifructa
    3. Petals white in anthesis; plants rarely found above 6000 feet in Colorado; inflorescence and pedicels usually stellate (except in D. reptans)
      6. Leaves and stems with branched but sessile hairs; fruit glabrous or finely stellate, 2-6 mm. long   -5. D. brachycarpa
      6. Leaves and stems with hairs simple or branched and stalked, at least below; fruit glabrous or hispidulous with simple hairs, 4-20 mm. long
        7. Inflorescence, pedicels and upper part of stems glabrous; fruit rarely over 2 mm. wide; leaves usually entire
          8. Fruit hispidulous   -6. D. reptans
          8. Fruit glabrous   -6A. D. reptans reptans
        7. Inflorescences, pedicels and upper part of stems pubescent; fruit usually 2 mm. wide or more; leaves usually dentate   -7. D. cuneifolia
1. Plants biennial or perennial; style over 0.2 mm. long (except in D. fladnizensis)
  9. Flowering stems leafless and scapose
    10. Lower surface of leaves with sessile, doubly-pectinate branched trichomes, the long axis of many parallel to the axis of the leaf (the branched trichomes definitely longer than wide)
                                                                                        -8. D. oligosperma
    10. Lower surface of leaves glabrous or if hairy the trichomes not of the above type
      11. Petals white in anthesis; fruit 1.5-2 mm. wide; style not over 0.5 mm. long
        12. Leaves ciliate with unbranched hairs, branched hairs if present only once or twice forked and sparse, the leaf therefore greenish; style of fruit 0-0.2 mm. long   -9. D. fladnizensis
        12. Leaves with all the hairs branched and densely matted, the surface usually gray; style of fruit 0.2-0.5 mm. long   -10. D. nivalis
      11. Petals lemon-yellow or yellow at anthesis; fruit 2-5.5 mm. wide; style over 0.5 mm. long (except D. crassifolia)
        13. Style 0-0.15 mm. long; petals 2-3 mm. long   -1. D. crassifolia
        13. Style 0.5-1.2 mm. long; petals usually over 3 mm. long
          14. Leaves linear to linear-oblanceolate, ciliate with unbranched hairs, usually glabrous or sparsely stellate below; fruit 2-3.5 mm. wide   -11. D. densifolia
          14. Leaves oblanceolate to ovate, densely stellate, few if any unbranched hairs present; fruit 3.5-5.5 mm. wide   -12. D. ventosa
  9. Flowering stems with 1 or more leaves
    15. Style in fruit averaging less than 1 mm. long
      16. Petals white when fresh (may turn to tan or brownish on drying); plants not over 10 cm. tall
        17. Style lacking or nearly so; leaves long-ciliate with unbranched hairs (surface with branched hairs); fruit not over 6 mm. long   -9. D. fladnizensis
        17. Style over 0.2 mm. long; leaves only rarely with a few long, marginal, unbranched hairs; fruit usually over 6 mm. long
          18. Fruit glabrous or obscurely and sparsely stellate, 4-7 mm. long and 1.5-2 mm. wide; cauline leaves often absent entirely   -10. D. nivalis

                18. Fruit densely pubescent, some of hairs stellate, 4-12 mm. long and 1.5-3 mm. wide; cauline leaves
                    always present   -13. D. lanceolata
            16. Petals lemon-yellow to yellow; plants often over 10 cm. tall
                19. Leaves oblanceolate to broader, over 4 mm. wide
                    20. Fruit about 1/3 as wide as long; leaves glabrous except for a few cilia   -14. D. crassa
                    20. Fruit about 1/4 as wide as long; leaves pubescent on the surfaces
                        21. Petals 4.5-6 mm. long, yellow; sepals 2-3.5 mm. long; style in fruit over 0.2 mm. long
                            22. Fruit pubescent   -15. D. aurea
                            22. Fruit glabrous   -15A. D. aurea leiocarpa
                        21. Petals 2-4.5 mm. long, lemon-yellow; sepals about 1.5 mm. long; style not over 0.1 mm.
                            long   -3. D. stenoloba
                19. Leaves linear to linear-oblanceolate, not over 4 mm. wide
                    23. Style of fruit absent or nearly so, never over 0.2 mm. long; stems usually leafless but some-
                        times with 1 leaf; petals 2-3 mm. long, lemon-yellow; sepals about 1 mm. long
                                                                                                    -1. D. crassifolia
                    23. Style 0.2 mm. long or more; stems with 1-6 leaves; petals 3-5 mm. long, yellow; sepals 1.5 mm.
                        long or more
                        24. Stems and pedicels with appressed, stellate hairs only   -20C. D. spectabilis dasycarpa
                        24. Stems and pedicels glabrous or with slender simple or branched hairs, never appressed
                            stellate
                            25. Leaves ciliate on lower half, otherwise glabrous; flowers bracteate -16. D. graminea
                            25. Leaves ciliate and also more or less pubescent on the surfaces; flowers not bracteate
                                26. Stems glabrous or very sparsely pubescent; basal leaves 10-25 mm. long; fruit
                                    5-14 mm. long   -17. D. exunguiculata
                                26. Stems densely and conspicuously pilose; basal leaves 5-15 mm. long; fruit 4-8
                                    mm. long   -18. D. grayana
    15. Style in fruit 1-3.5 mm. long
        27. Petals white; leaves cinereous with very fine closely interwoven hairs, these all stellate with at least
            5 rays; fruit pubescent   -19. D. smithii
        27. Petals yellow; leaves often with simple hairs but if stellate and interwoven then the rays 4 or less;
            fruit glabrous or pubescent
            28. Leaves linear-oblanceolate, not over 3 mm. wide; sepals and petals about equal; stems glabrous or
                very sparsely pubescent, 2-7 cm. tall   -17. D. exunguiculata
            28. Leaves oblanceolate, lanceolate or broader, over 3 mm. wide; sepals at least 1 mm. shorter than
                petals; stems usually definitely hairy, 2-50 cm. tall
                29. Stellate hairs present (at least on lower leaf surface) these sessile or very short-stalked,
                    unequally 4-rayed, the long rays paralleling the axis of the leaf; fruit plane or nearly so
                    30. Cauline leaves usually over 5, 3-12 mm. wide; fruit narrowly oblong or narrowly lanceolate
                        31. Stems pilose with simple or branched hairs, at least below
                                                                                        -20A. D. spectabilis spectabilis
                        31. Stems with sessile, stellate hairs   -20B. D. spectabilis oxyloba
                    30. Cauline leaves about 3, 1-4 mm. wide; fruit ovate to elliptic-lanceolate
                                                                                        -20C. D. spectabilis dasycarpa
                29. Stellate hairs, when present, stalked and equally rayed; fruit often twisted
                    32. Style of fruit less than 1.5 mm. long
                        33. Fruit pubescent   -15. D. aurea
                        33. Fruit glabrous   -15A. D. aurea leiocarpa
                    32. Style of fruit (at least some) over 1.5 mm. long
                        34. Stems either glabrous or with mostly simple but sometimes branched hairs (not
                            stellate), some over 1 mm. long
                            35. Stems and leaves uniformly hairy   -21A. D. streptocarpa streptocarpa
                            35. Stems glabrous; leaves sparingly hairy   -21B. D. streptocarpa tonsa
                        34. Stems with hairs, these mostly branched but in any case less than 1 mm. long
                                                                                                   -22. D. helleriana

1. Draba crassifolia R. Grah., Edinb. N. Phil. Journ. 182. 1829.
    D. parryi Rydb.---Biennial or perennial, perhaps annual plants; stems 2-20 cm. tall, leaf-
less or rarely with 1 leaf, 1 to several stems but each simple, glabrous above and sparsely
hairy near the base; leaves 10-25 mm. long, basal, many, linear-spatulate to narrowly oblanceolate, ciliate, upper surface usually with few appressed unbranched or forked hairs, entire;
pedicels 2-10 mm. long, usually shorter than the fruit, glabrous; sepals about 1 mm. long, glabrous or pilose; petals 2-3 mm. long, yellow, fading to white; style lacking or not over 0.15
mm. long; fruit 5-12 mm. long and 2-3 mm. wide, glabrous. Our plant has been called variety
typica Hitchc.---Near timberline or above, mostly on rocks and talus. Greenland to Alaska,
south to Colorado, Arizona and Washington. Well scattered in the high mountains of Colorado at
10,000-12,500 feet.
2. Draba nemorosa L., Sp. Pl. 643. 1753.
    Annual plants; stems 5-25 cm. tall, glabrous above, hispidulous below with branched hairs;
leaves 1-3 cm. long, ovate-lanceolate to obovate-spatulate, dentate to denticulate especially
the cauline, hairs simple and branched, rather long; pedicels longer than the fruit, glabrous;
sepals about 1.5 mm. long; petals about 4 mm. long, pale yellow, rarely white; style absent
or very nearly so; fruit 4-11 mm. long and 2-3 mm. broad, hispidulous with fine, short, simple
hairs or glabrous; seeds 25-50.---Hills and plains, mostly in open ground. British Columbia,
south to Colorado and California. Our records from northcentral, central and southcentral
Colorado at 5000-10,000 feet.
3. Draba stenoloba Ledeb., Fl. Ross. 1:154. 1842.
    D. nitida Greene---Plants annual, biennials or short-lived perennials; stems 5-30 cm. tall,
simple or branched, with 1-8 leaves, glabrous or hirsute at base, glabrous above; leaves 10-40

mm. long, mostly basal, obovate to oblanceolate, usually denticulate with simple or forked hispidulous hairs; pedicels usually as long or longer than the fruit, glabrous; sepals about 1.5-2 mm. long, pilose; petals about 3-3.5 mm. long, yellow to cream-colored; style lacking or not over 0.1 mm. long; fruit 8-22 mm. long and 1.5-2.5 mm. wide, usually glabrous. Our plant has been called variety nana (Schulz.)Hitchc.---In the mountains often in partial shade. The variety from Alberta to British Columbia, south to Colorado and California. Our records from the western half of Colorado, except in the extreme southern part, at 8000-11,500 feet.

4. Draba rectifructa C. L. Hitchc., U. of Wash. Publ. in Bot. 11:110. 1941.

D. montana Wats.---Annual plants; stems 10-40 cm. tall, simple or branched, caulescent, grayish from branched and unbranched hairs; leaves 1-3 cm. long, ovate-lanceolate or obovate, denticulate or entire, grayish with branched and unbranched hairs intermixed; inflorescence elongated, longer than rest of stem; pedicels about as long as fruit, spreading or ascending; sepals about 2 mm. long; petals about 3-4 mm. long, yellow (but fading to white or pinkish); style lacking; fruit 6-10 mm. long and 2-2.5 mm. wide, narrowly oblong, hispidulous the hairs unbranched; seeds 40-80, about 0.8 mm. long.---In the mountains, often on gravelly soil. Colorado to Utah, south to New Mexico and Arizona. Our records scattered over the western half of Colorado, except in the extreme northern part, at 7500-10,000 feet.

5. Draba brachycarpa Nutt. ex T. & G., Fl. N. Amer. 1:108. 1838.

Montana to Oregon, south to Arizona. This plant should be looked for especially in western Colorado.

6. Draba reptans (Lam.) Fernald, Rhodora 36:368. 1934.

D. micrantha Nutt.; D. coloradensis Rydb.---Annual plants; stems 3-15 cm. tall, branched at base, leafy below, with a few stalked stellate hairs below, glabrous above; leaves 5-40 mm. long, mostly basal but some definitely cauline, usually entire, ovate to obovate or rarely oblanceolate with unbranched hairs above and stellate below; pedicels 1-6 mm. long, glabrous; sepals about 2 mm. long; petals 1-5 mm. long, white; style absent or less than 0.1 mm. long; fruit 5-20 mm. long, nearly erect, hispidulous, plane, seldom as much as 2 mm. broad. This plant has been called forma micrantha (Nutt.) Hitchc., with forma stellifera (Schulz.) Hitchc. in northwestern Colorado.---Dry plains, hills and sandy flats. In the southwestern part of the United States to Montana and Oregon. Our records in northcentral and western Colorado at 4500-7500 feet.

6A. Draba reptans reptans

D. reptans var. typica Hitchc.---Like forma micrantha but the fruit glabrous. Less common in Colorado. Our few records from the northern part of the state at 5000-6500 feet and possibly up to 10,000 feet.

7. Draba cuneifolia Nutt. ex T. & G., Fl. N. Amer. 1:108. 1838.

Annual plants; stems 5-15 cm. long, branched from the base, usually hirsute on the lower portions, the hairs usually not branched; leaves 1-5 cm. long and 5-20 mm. wide, on lower part of stem, ovate, obovate to oblanceolate, mostly somewhat dentate on upper half at least, hirsute with simple or branched hairs; inflorescence less than 1/2 of the length of the stem, pedicels 2-5 mm. long; sepals 1.5-2.5 mm. long, stellate; petals varying from small or even lacking to 5 mm. long, white; styles lacking; fruit 5-15 mm. long and seldom over 2.25 mm. wide, linear to oblong, hispidulous, densely terminally clustered; seeds 40-200, about 0.7-0.8 mm. long. Our plant has been called var. typica Hitchc.---Plains and dry mesas. The variety from Colorado to California, south to Texas and Mexico. Our few records from western Colorado at 4500-7000 feet.

8. Draba oligosperma Hook., Fl. Bor. Am. 1:51. 1833.

D. saximontana A. Nels.; D. andina (Nutt.) A. Nels.---Low caespitose perennial plants; stems 1-10 cm. tall, scapose, glabrous or with pectinate hairs near base; leaves 3-11 mm. long, crowded-basal, linear-lanceolate to linear-oblanceolate, at least lower surface with pectinate hairs, the long axis of many hairs parallel to the midrib; pedicels 3-10 mm. long; sepals 2-2.5 mm. long; petals 3-4.5 mm. long, yellow (may dry to whitish); styles 0.1-1 mm. long; fruit 2.5-7 mm. long and 2-4 mm. wide, usually with short stiff hairs.---Open ridges and knolls. Alberta to British Columbia, south to Colorado and California. Our few records from central and westcentral Colorado at about 11,000-11,500 feet.

9. Draba fladnizensis Wulfen in Jacq., Miscell. Austra. 1:147. 1778.

Perennial plants from simple or branched caudices; stems 2-6 cm. tall, leafless or rarely with 1 or 2 small leaves present, glabrous or pubescent below with simple or forked hairs; leaves 5-10 mm. long, oblanceolate, ciliate with long, simple hairs, surface glabrous or with forked hairs; pedicels 2-5 mm. long, glabrous; sepals 1-2 mm. long, usually glabrous; petals 2-3 mm. long, white; styles lacking or less than 0.2 mm. long; fruit 3-6 mm. long and 1.5-2 mm. wide, glabrous or rarely pubescent.---Arctic-alpine situations. British Columbia, south to Colorado and Utah; Eurasian. Our few records scattered in the high mountains of Colorado at 11,000-13,000 feet.

10. Draba nivalis Liljebl., Vet. Acad. Handl. Stockl. 208. 1793.

Caespitose perennial plants; stems 1-5 cm. tall, rarely taller, acaulescent, glabrous to short-stellate; leaves linear to obovate, stellate; pedicels usually shorter than the fruits; sepals about 2 mm. long, glabrous or stellate; petals 2.5-5 mm. long, white; style 0.2-0.5 mm. long, rarely to 7 mm.; fruit 4-7 mm. long and 1.5-2 mm. wide or sometimes to 3, often contorted, glabrous or pubescent. Our plant has been called variety exigua (Schulz.) Hitchc.---Mountain peaks. The variety limited to Colorado. Our records from northcentral and central Colorado at 10,000-13,000 feet.

11. Draba densifolia Nutt. ex T. & G., Fl. N. Amer. 1:104. 1838.

This northern species has been reported as far south as western Wyoming and Utah. No records as yet from Colorado.

12. Draba ventosa Gray, Am. Nat. 8:212. 1874.

Reported from western Wyoming and northeastern Utah. May someday be found in northwestern Colorado.

13. Draba lanceolata Royle, Illustr. Bot. Himal. Mts. 1:72. 1839.

D. cana Rydb.---Perennial plants; stems 5-25 cm. tall, with 1-10 leaves, several together, simple to branched, with soft, branched hairs and often with soft, simple hairs nearly 1 mm. long especially near the base; basal leaves 10-30 mm. long, oblanceolate, entire or denticulate; cauline leaves 5-25 mm. long, lanceolate to ovate, usually denticulate, all leaves grayish with soft, stellate, many-branched hairs, or basal leaves sometimes with simple cilia; pedicels ascending or appressed, mostly somewhat shorter than the fruit, soft-pubescent with branched hairs; sepals about 2 mm. long; petals about 4 mm. long, white; style 0.2-0.5 mm. long; fruit 4-12 mm. long and 1.5-3 mm. wide, soft-pubescent, the hairs simple or stellate.---Slopes and meadows often in gravelly or rocky ground. Canada and Alaska, south to Colorado and Nevada; Eurasia. Our few records from northcentral, westcentral and central Colorado at 11,000-12,000 feet.

14. Draba crassa Rydb., Mem. N. Y. Bot. Gard. 1:182. 1900.

D. chrysantha Wats. in part---Perennial plants with fleshy roots; stems 5-15 cm. tall, sometimes decumbent, moderately pubescent with short, simple or branched hairs, 2-6 leaves present; leaves thick, the basal long-petioled, 2-8 cm. long and 0.5-1 cm. wide, bases persistent at the crown, entire or nearly so, glabrous except for a few cilia; cauline leaves 1-2 cm. long, ovate or obovate, entire or nearly so; pedicels 5-10 mm. long, soft-pubescent; sepals about 2.5-3 mm. long, pilose; petals 4-6 mm. long, yellow; style about 0.75 mm. long; fruit 10-16 mm. long and 3-5 mm. wide, undulate or twisted, glabrous.---On talus and rocky slopes. Greenland to Alaska, south to Colorado, Arizona and Washington. Our rather few records scattered in the high mountains of Colorado, mostly from the central and southwestern parts, at 10,000-14,000 feet.

15. Draba aurea M. Vahl. in Hornem., Fors. Dansk. Oecon, Plantel. ed. 2. 599. 1806.

D. luteola Greene; D. aureiformis Rydb.; D. decumbens Rydb.---Perennial or perhaps biennial caulescent plants, the caudex simple or branched; stems 10-50 cm. tall, erect or decumbent, 1 to several, usually simple, pilose-hirsute with hairs to 1.5 mm. long and also with branched hairs; basal leaves 1-5 cm. long, oblanceolate to spatulate, petioled, mostly entire; cauline leaves usually 4 mm. wide or more, 3 or more, ovate to oblanceolate, all leaves with branched or simple hairs; racemes often many-flowered, pedicels 3-20 mm. long; sepals 2-3.5 mm. long, pilose or stellate; petals 4-6 mm. long, pale to bright yellow; styles 0.3 to 1.5 mm. long (about 1 mm. on most); fruit 8-16 mm. long and 2-4 mm. wide, flat or contorted, pubescent.---In the mountains, sometimes in shade. Alaska, south in the Rocky Mountains to New Mexico and Arizona. Our records from the western half of Colorado, except in the northwest, at 9000-12,500 feet.

15A. Draba aurea leiocarpa (Payson and St. John) Hitch., (var.) U. Wash. Pub. in Biol. 11:29. 1941.

British Columbia, south to Colorado and Arizona. Reported from Northcentral Colorado but no specimens from the state were seen by the writer.

16. Draba graminea Greene, Pl. Baker. 3:5. 1901.

Perennial plants; stems 1-5 cm. long, caespitose, caulescent, with slender unbranched hairs; leaves 2-4 cm. long and 0.5-2 mm. wide, linear or linear-oblanceolate, glabrous except long-ciliate on lower half, the hairs simple or sometimes forked; basal leaves many; cauline leaves 1-6; raceme of 3-15 bracteate flowers, pedicels shorter than fruit; sepals about 2 mm. long, glabrous; petals 3-5 mm. long, yellow; style 0.5-0.75 mm. long; fruit 5-10 mm. long and 2.5-4 mm. wide, elliptic-ovate, thick, glabrous; seeds 8-16, about 1.5 mm. long.---Rocky alpine areas. Apparently limited to Colorado. Our few records from the southwestern part of the state at 12,000-13,500 feet.

17. Draba exunguiculata (Schulz) Hitchc., U. Wash. Pub. in Biol. 11:46. 1941.

D. chrysantha Wats. in part---Perennial plants; stems 2-7 cm. tall, caespitose, several, rather decumbent at base, glabrous or rarely sparsely pilose; leaves rather persistent and forming thick tufts; basal leaves 10-25 mm. long and 1-3 mm. wide, linear-oblanceolate, ciliate and sparsely pubescent with few long, simple or sometimes forked hairs; 1-4 cauline leaves present; racemes 5- to 20-flowered, crowded, not bracted; sepals about 3 mm. long, sparsely pilose; petals scarcely exceeding the sepals, yellow; style 0.25-1.25 mm. long; fruit 5-14 mm. long and 1.5-3 mm. broad, plane, glabrous; seeds 20-30.---Rocky alpine areas. Colorado. Our few records from central and northcentral Colorado at about 11,000-13,000 feet.

18. Draba grayana (Rydb.) Hitchc., U. Wash. Pub. in Biol. 11:29. 1941.

D. streptocarpa grayana Rydb.---Perennial plants; stems 2-5 cm. tall, several, pilose with curled simple or branched hairs, from a branched caudex, forming compact tufts, with 2-5 cauline leaves present; leaves 5-15 mm. long, linear-oblanceolate, conspicuously ciliated with simple hairs, surface with long, simple or rarely forked hairs; pedicels 2-5 mm. long, villous; sepals about 2 mm. long, pilose; petals 3-4.5 mm. long, yellow; style 0.5-1 mm. long; fruit 4-8 mm. long and 2-3 mm. wide, plane, glabrous.---Gravelly alpine areas. Colorado. Our few records from northcentral and central Colorado at 11,500-13,000 feet.

19. Draba smithii Gilg ex Schulz, Pflanz. IV. 105:177. 1927.

Perennial plants from branched caudices; stems 10-20 cm. tall, caulescent, cinereous with dense, short, stellate hairs these with 5 or more branches; basal leaves about 10 mm. long, obovate to linear-oblanceolate, entire or nearly so; cauline leaves 3-8, leaves pubescent like the stems; pedicels mostly longer than the fruit; sepals about 1.5-2 mm. long, stellate; petals about 3.5-4 mm. long, white; style 1-2 mm. long; fruit 5-9 mm. long and 2-2.5 mm. broad, usually contorted, finely stellate.-- Mostly in the crevices of rocks. Apparently limited to southern Colorado at 8000-10,000 feet, our one record from Las Animas County at 8000 feet.

20. Draba spectabilis Greene, Pitt. 4:19. 1899.

Perennial plants; stems 10-40 cm. tall, with 3-15 leaves, mostly simple, pilose to stellate with sessile hairs; leaves rather sparsely pubescent on lower surface with sessile or short-stalked cruciform hairs the longer rays usually parallel to the ribs, upper surface often with the same hairs; basal leaves 15-40 mm. long, petiolate, spatulate-oblanceolate or spatulate-obovate; cauline leaves 5-20 mm. long, sessile, ovate to lanceolate, denticulate to dentate; pedicels 10-20 mm. long; sepals about 2-3 mm. long, glabrous or hairy; petals 5-8 mm. long, yellow; styles 1-3.5 mm. long; fruit 5-15 mm. long and 2-3 mm. broad, glabrous to somewhat pubescent, plane or slightly contorted. Represented in Colorado by 3 varieties.

20A. Draba spectabilis spectabilis
D. spectabilis var. typica Hitchc.---Stems pilose below with simple or branched hairs; cauline leaves usually over 5, 3-12 mm. wide; fruit elliptic-oblong to narrowly lanceolate.---In the mountains, often in moist meadows or forests. Colorado and Utah. Our records from southwestern Colorado at 8000-11,500 feet.

20B. Draba spectabilis oxyloba (Greene) Gilg. & Schulz, (var.) Pflanz IV. 105:184. 1927.
D. oxyloba Greene---Stems pilose with simple or branched hairs; cauline leaves usually over 5, 3-12 mm. long; fruit narrowly oblong or narrowly lanceolate.---Mountains, often in meadows or forests. Apparently limited to Colorado. Our records from the southwestern and westcentral parts of the state at 8000-11,000 feet.

20C. Draba spectabilis dasycarpa (Schulz) Hitchc., (var.) U. Wash. Pub. in Bot. 11:44. 1941.
Cauline leaves about 3, 1-4 mm. wide; fruit ovate to elliptic-ovate.---Alpine ridges and slopes. Limited to Colorado. Our records from the central and southwestern parts of the state at 12,000 feet. and above.

21. Draba streptocarpa Gray, Am. Journ. Sc. II. 33:242. 1862.
Perennial plants; stems 2-30 cm. tall, tufted, with the old leaves persistent, glabrous or hirsute, the hairs simple or forked some as much as 1-2 mm. long; basal leaves 10-35 mm. long, many, elliptic-lanceolate; cauline leaves 1-2 cm. long and averaging 3 to 4 mm. wide, 4-20, oblanceolate or elliptic to lanceolate or rarely wider, entire or rarely denticulate; racemes 10- to 60- flowered, pedicels usually somewhat shorter than the fruits; sepals about 3 mm. long, sparsely pilose; petals 6-7 mm. long, yellow; styles 1-2.5 mm. long; fruit about 10 mm. long or somewhat longer, contorted, glabrous, papillose or with marginal hairs. The following 2 varieties are present in this state.

21A. Draba streptocarpa streptocarpa
D. streptocarpa Gray; D. streptocarpa var. typica Hitchc.---Stems and leaves uniformly hairy.---Woods, meadows and slopes. Colorado and New Mexico. Our records from northcentral to southcentral Colorado at 7500-13,000 feet.

21B. Draba streptocarpa tonsa (Woot. & Standl.) Schulz., (var.) Pflanz. IV. 105:193. 1927.
Stems glabrous; leaves sparsely hairy.---Probably limited to Colorado and New Mexico. Our few records from the southcentral part of the state at 9000-10,000 feet.

22. Draba helleriana Greene, Pitt. 4:17. 1899.
Perennial or perhaps biennial plants; stems 5-40 cm. tall, 1 to several, often branched above, with 2-12 leaves, with branched hairs mostly or with simple hairs not over 1 mm. long; basal leaves few; cauline leaves 10-60 mm. long, oval, lanceolate to obovate, entire or dentate, upper surface with cruciform trichomes only; pedicels from subequal to longer than the fruit, usually pubescent; sepals 2.5-3 mm. long, hairy; petals 4.5-8 mm. long, yellow; style 1.5-3.5 mm. long; fruit 4-15 mm. long and 2-3 mm. wide, plane or contorted, glabrous or pubescent.---Open or shaded ground. Colorado and New Mexico. Our records from the southern part of the state at 7000-11,500 feet.

12. Camelina Crantz FALSE FLAX

Annual herbaceous plants; stems erect; leaves entire to denticulate, the cauline auriculate; flowers perfect, small, in elongated racemes; petals yellow or yellowish; fruit short obovoid or pear-shaped, only slightly flattened parallel to the septum, valves 1-nerved; seeds several to many in each cell, wingless.

1. Camelina microcarpa Andrz. in DC., Regni Veg. Syst. 2:517. 1821.
Stems 30-80 cm. tall, hirsute below, simple or branched above; lower leaves lanceolate, entire or nearly so, more or less hirsute; upper leaves sagittate-clasping; fruiting raceme often over 20 cm. long; petals about 3-4 mm. long, yellowish; fruit 4-7 mm. long, strongly margined.---Waste places, fields and roadsides. Naturalized from Europe and now found in various parts of North America especially from British Columbia to Arizona. Our records widely scattered over Colorado at 3500-7000 feet.

A very closely related species, C. sativa (L.) Crantz, with glabrous stems and fruit reported to be 6-8 mm. long, is to be expected in this area. Our plants show some intergradation.

13. Rorippa Scop. CRESS

Annual or perennial herbaceous plants; stems usually branched; leaves simple to pinnate; flowers in terminal or axillary racemes; sepals spreading in anthesis; petals white or yellow; fruit terete or nearly so, globose to narrowly cylindric, sessile; seeds commonly in 2 rows, small.

1. Leaves crenate or sinuate, the basal 15 cm. or more long   -1. R. armoracia
1. Leaves pinnate to pinnatifid, rarely over 15 cm. long
  2. Flowers white and leaves pinnately compound; plants semi-aquatic   -2. R. nasturtium-aquaticum
  2. Flowers yellow (or if whitish then leaves pinnatifid); plants terrestrial though in moist soil
    3. Plants perennial with horizontal rootstocks; petals much exceeding the sepals, 4 mm. long or more   -3. R. sinuata
    3. Plants annual or biennial, no horizontal rootstocks present; petals little if any longer than the sepals, seldom over 2.5 mm. long

4. Stems erect below, branched above, usually over 30 cm. tall; petals about 2 mm. long; plants glabrous or hairy
   5. Stems villous or hispidulous; fruit globose or ovoid, not over 2 times longer than broad   -4. R. hispida
   5. Stems glabrous or nearly so; fruit linear to linear-oblong, 2-6 times longer than broad   -5. R. islandica
4. Stems diffusely branched from the base, rarely over 30 cm. tall; petals 1-1.5 mm. long (2 mm. in R. alpina); plants glabrous
   6. Fruit globose or nearly so, 2-3 mm. long   -6. R. sphaerocarpa
   6. Fruit elliptic to linear, 3-5 mm. long or more
     7. Fruit linear, 7-15 mm. long, style not over 0.5 mm. long   -7. R. lyrata
     7. Fruit elliptic to oblong, 4-8 mm. long, the style about 1 mm. long   -8. R. obtusa

1. Rorippa armoracia (L.) A. S. Hitchc., Spring Fl. Manhattan 18. 1894.
Perennial or biennial plants from deep, thick roots; basal leaves 15-30 cm. long, oblong, crenate to sinuate, on long petioles; cauline leaves sessile at least above, crenate to pinnatifid; pedicels slender and ascending; flowers white and showy, the petals 3-8 mm. long; fruit oblong or nearly globose.---This is "horse-radish," often cultivated and sometimes escaping into moist ground. Apparently this happens rarely in Colorado, our one record from the northcentral part of the state at about 5000 feet.

2. Rorippa nasturtium-aquaticum (L.) Schinz & Thell., Fl. Schweiz, ed. 3. 240. 1909.
Sisymbrium nasturtium-aquaticum L.; Rorippa nasturtium (L.) Rusby---Aquatic or semiaquatic perennial plants; stems floating, creeping or ascending, rooting at the nodes, glabrous; leaves pinnately divided into ovate or oval segments, the terminal larger; sepals about 1/2 as long as the petals; petals 3-4 mm. long, white; fruiting pedicels divaricate; fruit spreading or curved upward, 1-3 cm. long and about 2-3 mm. thick, the beak about 1 mm. long.---In water or very wet ground. A European plant now naturalized throughout temperate North America. Our records widely scattered over Colorado, but few from the extreme eastern part, at 4000-8000 feet.

3. Rorippa sinuata (Nutt.) A. S. Hitchc., Spring Fls. Manhattan 18. 1894.
Radicula sinuata (Nutt.) Greene---Perennial plants with creeping rootstocks; stems 10-40 cm. tall, glabrous, diffuse, numerous; leaves pinnatifid, oblong or lanceolate, the segments entire or toothed and obtuse; pedicels slender, about 6 mm. long, often curved in age; petals about 4 mm. long, yellow; style about 1-3 mm. long; fruit 8-14 mm. long and about 2 mm. wide, curved upward. The writer cannot separate R. columbiae Suksd., R. calycina (Engelm.) Rydb. & R. trachycarpa (A. Gray) Greene from the above in this state.---Valleys and roadsides, especially in somewhat moist ground. Ontario to Saskatchewan, south to Illinois, Texas, Arizona and Oregon. Our records well scattered over the state at 3500-7500 feet.

4. Rorippa hispida (Desv.) Britton, Mem. Torr. Club 5:169. 1894.
Radicula hispida (DC.) Heller in Manuals---Annual or biennial plants; stems 30-100 cm. tall, stout, erect, branched above, main stem and branches more or less hirsute; leaves lyrate-pinnatifid, lower petioled, more or less hirsute on petioles and veins; pedicels slender, spreading or ascending; petals about 2 mm. long, yellow; style about 1 mm. long; fruit globose or ovoid, usually not over twice as long as wide, 4-6 mm. long.---In water or wet ground. New Brunswick to Alaska, south to New Mexico and Arizona. Our records scattered in the western two-thirds of the state, few in the extreme western part, at 4500-9000 feet.

5. Rorippa islandica (Oeder ex Murray) Borbas, Balaton Tavanak es Partmellekenek 392. 1900.
R. palustris (L.) Bess.; Radicula terrestris (R. Br.) Woot.& Standl.---Annual or biennial plants, glabrous or nearly so throughout; stems 30-80 cm. tall, simple below, branched above; lower leaves lyrate-pinnatifid, petiolate; upper leaves dentate to somewhat lobed, nearly sessile; pedicels usually spreading, slender; petals about 2 mm. long, yellow; style about 1 mm. long; fruit 5-7 mm. long, linear or linear-oblong, 2-6 times longer than wide, spreading or curved.---Water or wet ground. Widely distributed in North America; Europe. Our records from the general northcentral area of Colorado at 4500-9500 feet.

6. Rorippa sphaerocarpa (A. Gray) Britton, Mem. Torr. Club. 5:170. 1894.
Radicula sphaerocarpa (A. Gray) Greene---Annual or biennial plants; stems 10-30 cm. tall, erect or decumbent, diffusely branched from the base, glabrous; leaves oblong, lower lyrate-pinnatifid or sinuately lobed, upper sometimes nearly entire; petals yellow, about 1.5 mm. long; pedicels 2-3 mm. long, rarely to 5; style about 0.5 mm. long; stigma not much enlarged; fruit 2-3 mm. in diameter, globose or nearly so.---Often along streams. Illinois to Wyoming and California, south to Texas and Arizona. Our 2 records from central and southcentral Colorado at 5000-9000 feet.

7. Rorippa lyrata (Nutt.) Rydb., Mem. Torr. Club 1:176. 1900.
Radicula lyrata (Nutt.) Greene---Annual plants; stems 10-30 cm. tall, diffusely branched at base, glabrous; leaves deeply pinnatifid; petals about 1 mm. or less long; pedicels 1-3 mm. long, ascending; style about 0.5 mm. long; fruit 7-15 mm. long, linear. This plant is closely related to the next.---Moist or wet places. Montana to Washington, south to Colorado and California. Our few records from northcentral Colorado at 4500-9000 feet.

8. Rorippa obtusa (Nutt.) Britt., Mem. Torr. Club 5:169. 1894.
Radicula obtusa (Nutt.) Greene---Annual or biennial plants; stems 10-30 cm. tall, diffusely branched from the base, glabrous; leaves pinnatifid or sinuate, the lobes obtuse or sometimes acutish; pedicels ascending or spreading, 2-4 mm. long, sometimes longer; petals about 1 mm. long, sometimes to 2 mm., yellow; style about 1 mm. long; fruit 4-8 mm. long and 1-2 mm. thick, oblong or linear, straight or somewhat curved.---Moist or wet places. Michigan to Washington, south to Texas, Arizona and California. Our records scattered over the western two-thirds of Colorado at 4500-11,000 feet.

Two related species have been recorded for this state. R. alpina (S. Wats.) Rydb. has petals 2 mm. long. R. curvipes Greene has the flowers unilateral and the leaves mostly sinuate or if pinnatifid then with acute divisions. All these characters intergrade commonly in Colorado plants.

14. Lesquerella Watson BLADDER POD

Annual, biennial or perennial herbaceous plants with stellate pubescence; leaves simple, entire to pinnatifid, mostly in a basal rosette; flowers in racemes, usually yellow or white, tinged with purple, pedicels straight or recurved; sepals shorter than petals; petals entire; fruit subglobose to oblong, usually inflated, valves nerveless, dehiscent, style persistent, usually long and slender; seeds several to many, flattened to winged.

1. Fruit glabrous
  2. Plants annual; fruit globose, about 4 mm. in diameter   -1. L. gordonii
  2. Plants perennial; fruit globose to elongated, usually over 4 mm. in diameter
    3. Basal leaves linear to linear-oblanceolate, the petiole usually shorter than the blade -2. L. fendleri
    3. Basal leaves broadly oblanceolate, oval or broadly orbicular to rhombic, the petiole about as long as the blade
      4. Fruiting inflorescence contracted and subumbellate, the pedicels erect or ascending   -3. L. ovalifolia
      4. Fruiting inflorescence elongated, the pedicels sigmoid (somewhat S-shaped)   -4. L. pruinosa
1. Fruit stellate-pubescent
  5. Pedicels uniformly recurved, not straight or sigmoid (S-shaped); fruit globose or nearly so
    6. Basal leaves oval or suborbicular; fruit 6-7 mm. in diameter at maturity   -5. L. macrocarpa
    6. Basal leaves linear to oblanceolate, fruit 3-4 mm. in diameter   -6. L. ludoviciana
  5. Pedicels sigmoid or straight; fruit ovate, obovate to oblong (except in L. rectipes)
    7. Pedicels straight, not at all sigmoid
      8. Stems 5-18 cm. tall; fruit 4-6 mm. long, not compressed at apex; radical leaves narrowly linear; plants of southeastern Colorado   -7. L. intermedia
      8. Stems 1-5 cm. tall; fruit 2.5-4 mm. long, compressed especially near apex; radical leaves oblanceolate to wider; plants of northwestern Colorado   -8. L. subumbellata
    7. Pedicels sigmoid
      9. Radical leaves linear to narrowly oblanceolate
        10. Fruit 4-5 mm. long, the styles 2-4 mm. long; pedicels 5-10 mm. long; radical leaves 1-4 cm. long; plants limited to northern Colorado   -9. L. alpina
        10. Fruit 5-9 mm. long, the styles 4-5 mm. long; pedicels 10-15 mm. long; radical leaves 4-10 cm. long; plants of southern Colorado   -10. L. calcicola
      9. Radical leaves oblanceolate to obovate
        11. Fruit globose or nearly so, 3-6 mm. long, with 2-6 ovules in each cell; plants of the western half of Colorado   -11. L. rectipes
        11. Fruit oblong, 6-8 mm. long, with 6-10 ovules in each cell; plants of the eastern part of Colorado   -12. L. montana

1. **Lesquerella gordonii** (Gray) Wats., Proc. Am. Acad. 23:253. 1888.
Oklahoma, Texas to Arizona and Mexico. Should be looked for in southeastern Colorado.
2. **Lesquerella fendleri** (Gray) Wats., Proc. Am. Acad. 23:254. 1888.
L. stenophylla (A. Gray) Wats.---Perennial plants, the caudex usually branched, silvery-stellate throughout; stems 5-30 cm. tall, tufted, erect or spreading, leafy; radical and cauline leaves similar, linear to linear-oblanceolate, entire or toothed; flowers yellow; inflorescence in fruit short or elongated; pedicels 5-12 mm. long, erect or ascending; styles 2-6 mm. long; pods glabrous, globose or elongated, 3-7 mm. in diameter, ovules 8-16 in each cell.---Dry hills and plains. Kansas to Utah, south to Texas and Mexico. Our records from southeastern Colorado, and extending to El Paso and Fremont Counties, at 4500-6000 feet.
3. **Lesquerella ovalifolia** Rydb. in Britt. & Brown, Ill. Fl. 2:137. 1897.
L. engelmanni (A. Gray) S. Wats. of Manuals in part---Perennial plants, densely stellate, caudex often branched; stems 5-20 cm. long, erect or decumbent; radical leaves broadly oblanceolate to orbicular, entire, abruptly narrowed to the petioles; cauline leaves linear-oblanceolate; inflorescence in fruit typically contracted and subcorymbose; flowers yellow; pedicels erect or ascending, 5-15 mm. long; style equalling or exceeding the fruit; pods glabrous, globose or somewhat elongated, 4-5 mm. in diameter, ovules 5-8 in each cell.---Dry plains and hills. Nebraska to Colorado, south to Texas and New Mexico. Our records in southeastern Colorado to Pueblo County, at 3600-6000 feet.
4. **Lesquerella pruinosa** Greene, Pitt. 4:307. 1901.
Perennial plants, stellate throughout, the caudex woody and sometimes branched; stems 10-25 cm. tall, erect or decumbent, simple; radical leaves 3-8 cm. long, blades broadly rhombic to oval, entire or repand, abruptly narrowed to the long petiole; cauline leaves obovate; flowers yellow; fruiting inflorescence elongated but rather crowded; pedicels 5-10 mm. long, sigmoid; styles 4-6 mm. long; fruit 6-9 mm. long, glabrous, ellipsoid, ovules 3-4 in each cell.---Dry hills. Limited to Colorado. All our records from Archuleta County at 7000 feet.
5.. **Lesquerella macrocarpa** A. Nels., Bot. Gaz. 34:366. 1902.
Recorded from southern Wyoming and may be found in northern Colorado someday.
6. **Lesquerella ludoviciana** (Nutt.) S. Wats., Proc. Am. Acad. 23:252. 1888.
L. argentea (Pursh) MacM.---Perennial plants, stellate throughout; stems 15-40 cm. tall, erect or decumbent, usually unbranched; radical leaves 3-10 cm. long, broadly linear to linear-oblanceolate, entire or slightly toothed; cauline leaves similar but smaller; flowers yellow; inflorescence elongated in fruit; pedicels 1-1.5 cm. long, recurved; styles equalling or exceeding the fruit; fruit globose or slightly elongated, 3-4 mm. in diameter, stellate, ovules 4-6 in each cell.---Hills and plains on rocky slopes. North Dakota to Montana, south to Illinois and Arizona. Our records scattered over Colorado, except in the southeastern and southcentral parts, at 3500-7500 feet.

7. **Lesquerella intermedia** (Wats.) Heller, Plant World 1:22. 1897.
   Perennial caespitose plants, stellate throughout, caudex much branched; stems 2-18 cm. tall, erect or decumbent; radical leaves 1-4 cm. long, narrowly linear, involute, cauline similar; flowers yellow; inflorescence crowded and subcorymbose in fruit; pedicels 10-15 mm. long, straight or slightly curved, ascending to nearly horizontal; styles 2-6 mm. long; fruit oval or ovate, 4-6 mm. long, acute but not compressed at the apex, stellate, ovules 2-8 in each cell.---Dry hills and plains. Colorado, Utah, New Mexico and Arizona. Our few records from El Paso, Fremont and Pueblo Counties at 5000-6500 feet.

8. **Lesquerella subumbellata** Rollins, Am. Journ. Bot. 26:420. 1939.
   Perennial plants, stellate throughout; stems 1-5 cm. tall, erect or ascending, unbranched; radical leaves 6-20 mm. long, ovate to oblanceolate, often angular, the blades rather abruptly narrowed to long petioles; cauline leaves oblanceolate or linear; inflorescence corymbose, appearing umbellate in fruit; pedicels 3-6 mm. long, ascending, straight or nearly so; petals 4-7 mm. long, yellow; styles 1.5-2.5 mm. long; fruit 2.5-4 mm. long, stellate, ovoid or obovoid, somewhat compressed parallel to the septum especially at apex, ovules about 2 in each cell.---Gravelly slopes and rocky knolls. Reported only from the type locality in northeastern Utah, but we have several records from the northwestern corner of Colorado at 9000 feet.

9. **Lesquerella alpina** (Nutt.) Wats., Proc. Am. Acad. 23:251. 1888.
   L. parvula Greene---Caespitose perennial plants, stellate throughout, caudex much branched; stems 2-14 cm. long, erect; radical leaves 1-4 cm. long, linear to linear-oblanceolate, entire, cauline leaves similar; sepals about 4-5 mm. long, stellate; petals 5-7 mm. long, yellow; pedicels 5-10 mm. long, sigmoid; style 2-4 mm. long; fruit 4-5 mm. long, erect or ascending, ovate, compressed at apex, stellate, ovules 2-4 in each cell.---Hills and plains, often in the mountains. Montana, Wyoming and Colorado. Our records from northcentral Colorado, south to Lake County, at 7500-10,000 feet.

10. **Lesquerella calcicola** Rollins, Am. Journ. Bot. 26:419. 1939.
    Perennial plants, silvery-stellate throughout, caudex simple or closely branched; stems 10-30 cm. long, simple, erect; radical (or basal) leaves 4-10 cm. long, linear-oblanceolate, numerous, flat, cauline leaves similar; flowers yellow; inflorescence congested; pedicels 10-15 mm. long, sigmoid; styles 3-5 mm. long; fruit 5-9 mm. long, stellate, oblong or oval, ovules 2-4 in each cell.---Dry ground. Colorado and New Mexico. Our records from Pueblo, Huerfano and Las Animas Counties at 5000-6500 feet.

11. **Lesquerella rectipes** Woot. & Standl., Contrib. U. S. Nat. Herb. 16:127. **1913.**
    L. montana of Manuals in part---Perennial plants, stellate throughout; stems 10-40 cm. long, ascending or decumbent, usually simple; radical leaves 2-5 cm. long, oblanceolate to ovate, entire or repand-dentate; cauline leaves linear or narrowly oblanceolate, usually entire; flowers yellow; fruiting inflorescence elongated or somewhat crowded; pedicels 8-10 mm. long, usually sigmoid at maturity; styles 3-5 mm. long; fruit globose or nearly so, 3-5 mm. in diameter, stellate, ovules 3-6 in each cell. Forms resembling L. montana occur in this state.---Dry ground. Colorado to Utah, south to New Mexico and Arizona. Our records from southwestern Colorado (to Montrose and Conejos Counties), at 5000-7500 feet.

12. **Lesquerella montana** (Gray) Wats., Proc. Am. Acad. 23:251. 1888.
    L. curvipes of Manuals in part; L. montana var. suffruticosa Payson---Perennial plants, stellate throughout, caudex sometimes branched and enlarged; stems 10-20 cm. long, decumbent, usually unbranched; radical leaves 1-4 cm. long, rather variable, entire or toothed, oblanceolate to obovate, abruptly narrowed to the petiole; cauline leaves variable in like manner; flowers yellow; fruiting inflorescence elongated, pedicels 8-12 mm. long, definitely sigmoid; styles 3-6 mm. long; fruit ellipsoid to oblong, 6-8 mm. long, rarely longer, stellate-puberulent, ovules 6-10 in each cell.---Dry slopes, plains and mountains. South Dakota to Wyoming, south to New Mexico. Our records from northcentral, central and southcentral Colorado at 4500-9500 feet.

15. Arabis L. ROCKCRESS

Biennial or perennial herbaceous plants (but often appearing annual), glabrous or pubescent with branching hairs; leaves entire, toothed or pinnatifid; inflorescence a raceme, elongating in fruit; petals white to purple; silique sessile, elongated, flattened parallel to the septum (rarely semi-terete); styles present or absent; seeds winged or wingless, uniseriate to biseriate. Mature fruit is necessary in identifying the species and some intergradations are to be expected in Colorado plants.

1. Mature fruit strictly erect, mostly appressed to the rachis; styles sometimes present
   2. Stems hirsute at base; styles present and definite
      3. Fruit semi-terete; seeds biseriate, wingless or narrowly winged; sepals not at all saccate at base; cauline leaves usually glaucous -1. A. glabra
      3. Fruit flattened; seeds uniseriate, rather prominently winged on 1 side; sepals more or less saccate at base; cauline leaves not glaucous
         3A. Petals 3-5 mm. long, white to yellowish-white -2A. A. hirsuta pycnocarpa
         3A. Petals 5-9 mm. long, white to pink -2B. A. hirsuta glabrata
   2. Stems glabrous to pubescent at base, not hirsute; styles absent or very short
      4. Seeds biseriate, prominently winged on 1 side; petals 7-10 mm. long; stems 1 to several, glabrous or sparsely pubescent at base, often over 40 cm. tall -3. A. drummondi
      4. Seeds uniseriate, wingless or nearly so; petals 5-7 mm. long; stems numerous, densely pubescent at base, rarely over 40 cm. tall -4. A. crandallii
1. Mature fruit widely spreading to reflexed (or loosely ascending in A. divaricarpa, A. selbyi and A. fernaldiana), not strictly erect or appressed; styles absent or very short

5. Mature fruiting pedicels ascending, never diverging at right angles to the rachis or fruit ascending in like manner
   6. Basal leaves sparsely and loosely pubescent or glabrous, not grayish; stems 1 or few, 30 to 90 cm. tall, glabrous or sparsely pubescent below   -5. A. divaricarpa
   6. Basal leaves densely gray-canescent; stems usually numerous, 15-50 cm. tall, densely pubescent below
      7. Petals 9-14 mm. long; styles about 1 mm. long   -6. A. fernaldiana
      7. Petals 5-8 mm. long; styles absent or very short
         8. Basal leaves 1.5-3 cm. long and 2-4 mm. wide; fruit 1 mm. wide, straight   -4. A. crandallii
         8. Basal leaves 3-7 cm. long, 6-20 mm. wide; fruit 1.5-2 mm. wide, often curved   -7. A. selbyi
5. Mature fruiting pedicels diverging at nearly right angles to the rachis, or reflexed; fruit diverging or reflexed in like manner
   9. Basal leaves glabrous (petiole may be ciliate); stems entirely glabrous, not over 15 cm. long; fruit 1.5-2.5 cm. long   -8. A. oxylobula
   9. Basal leaves (at least some) hirsute or pubescent; stems hirsute or pubescent at least below, often over 15 cm. tall; fruit usually over 2.5 cm. long
      10. Basal leaves hirsute with the marginal hairs longer, or glabrous; stems hirsute below
         11. Seeds uniseriate; fruiting pedicels 3-7 mm. long; stems caespitose, several to many, 10-30 cm. tall
            12. Basal leaves linear to narrowly lanceolate; cauline leaves lacking auricles
               13. Trichomes on leaves and stem forked; valves of fruit constricted between the seeds   -9A. A. demissa demissa
               13. Trichomes on leaves and stem simple; valves of the fruit not constricted between the seeds   -9B. A. demissa russeola
            12. Outer basal leaves oblanceolate; cauline leaves auriculate   -9C. A. demissa languida
         11. Seeds biseriate; fruiting pedicels 10-20 mm. long; stems not caespitose, 1 to several, 25-60 cm. tall
            14. Basal leaves dentate, oblanceolate, obtuse; petals pink   -10. A. fendleri
            14. Basal leaves entire; linear-oblanceolate, acute; petals white   -10A. A. fendleri spatifolia
      10. Basal leaves pubescent, never hirsute; stems hirsute or not hirsute below
         15. Petals 12-20 mm. long, the limb spreading at right angles; seeds biseriate; fruit densely pubescent; cauline leaves linear   -11. A. pulchra
         15. Petals 4-12 mm. long (rarely to 14 mm.), the limb not spreading; seeds uniseriate or imperfectly biseriate; fruit usually glabrous; cauline leaves lanceolate to wider (in all but A. cobrensis)
            16. Basal and cauline leaves linear or narrower; petals white, about 4 mm. long; seeds widely-winged (the wings over 0.5 mm. wide)   -12. A. cobrensis
            16. Basal and cauline leaves linear-oblanceolate, linear-lanceolate or wider; petals 4-10 mm. long, pink to purple (rarely white and then over 6 mm. long); seeds narrowly-winged (the wing not over 0.3 mm. wide)
               17. Basal leaves not over 2 cm. long, cauline leaves not over 1 cm. long; petals 4-6 mm. long; stems 6-20 cm. tall
                  18. Fruiting pedicel 2-5 mm. long; fruit 2-3.5 mm. wide; basal leaves spatulate, cauline usually ovate   -13. A. lemmonii
                  18. Fruiting pedicel 5-8 mm. long; fruit 1-1.5 mm. wide; basal leaves linear-oblanceolate, cauline oblong   -14. A. gunnisoniana
               17. Basal leaves 2-10 cm. long (rarely less in A. holboellii), cauline leaves 1-8 cm. long; petals 5-14 cm. long; stems usually over 20 cm. tall
                  19. Fruiting pedicels strictly reflexed and appressed to rachis, the fruit appressed in like manner   -15A. A. holboellii retrofracta
                  19. Fruiting pedicels spreading at right angles to the rachis or strongly descending but not appressed to the rachis, hence the fruit not closely appressed
                     20. Stems finely appressed pubescent below; basal leaves linear-oblanceolate, always entire, finely pubescent, the trichomes minute   -16. A. lignifera
                     20. Stems hirsute below with spreading hairs; basal leaves linear-oblanceolate to broader, often toothed, the hairs usually coarse and long
                        21. Fruiting pedicels hirsute; fruit strongly arcuate, widely spreading, 6-12 cm. long; petals 8-14 mm. long   -17. A. sparsiflora
                        21. Fruiting pedicels glabrous to pubescent, not hirsute; fruit straight to curved inward, not strongly arcuate, spreading to pendulous, 3-7 cm. long; petals 6-10 mm. long
                           22. Stems several to many; fruiting pedicels 10-20 mm. long, glabrous; fruit pendulous to widely spreading   -18. A. perennans
                         22. Stems 1 to several; fruiting pedicels 5-15 mm. long, glabrous or pubescent; fruit pendulous, never widely spreading
                             23. Cauline leaves with auricles; plants usually over 20 cm. tall; basal leaves mostly over 3 mm. wide   -15B. A. holboellii pinetorum
                             23. Cauline leaves lacking auricles; plants usually less than 20 cm. tall; basal leaves less than 3 mm. wide   -15C. A. holboellii pendulocarpa

1. **Arabis glabra** (L.) Bernhardi, Syst. Verz. Erf. 195. 1800.
   Turritis glabra L.---Biennial or rarely perennial plants; stems 40-150 cm. long, 1 or few, pubescent, usually hirsute below, glabrous or glaucous above; basal leaves 6-15 cm. long, broadly oblanceolate, spatulate or oblong, repand, dentate or divided, rarely entire, usually coarsely pubescent, petioles usually hirsute; cauline leaves lanceolate to ovate, sessile, auriculate and sagittate, entire or toothed below, glabrous or sparsely pubescent; pedicels erect; petals 5-7 mm. long, yellowish-white to purplish; style short and stout, stigma

enlarging; fruit 4-10 cm. long and 1-1.5 mm. wide, erect, semi-terete, glabrous; seeds wingless or narrowly winged, biseriate or nearly so. Our plant has been called var. typica Hopkins.---Plains and mountains. The variety from Quebec to British Columbia, south to North Carolina and California. Our records scattered over the western half of Colorado at 5500-9000 feet.

2. Arabis hirsuta (L.) Scop., Fl. Carn. Ed. 2., 2:30. 1772.

Biennial or perennial plants; stems 20-70 cm. tall, 1 to several, hirsute or glabrous above; basal leaves 2-8 cm. long, oblong, oblanceolate or broadly spatulate, entire, dentate or repand, petioled, usually hirsute both surfaces; cauline leaves 1-5 cm. long, lanceolate, oblong or spatulate, sessile and auriculate, entire to dentate, hirsute to glabrous; pedicels erect or ascending; petals 3-9 mm. long, white or pink; style 0.5-1.0 mm. long; fruit 3-6 cm. long and 1-2 mm. wide, erect to ascending, glabrous; seeds prominently winged on 1 side, uniseriate. The following 2 varieties are in or near Colorado.

2A. Arabis hirsuta pycnocarpa (Hopkins) Rollins, (var.) Rhodora 43:318. 1941.

A. hirsuta and A. ovata of Manuals at least in part---This variety has small flowers, the petals 3-5 mm. long, white to yellow-white, and the pubescence of stem is spreading.---Mountains and plains. Quebec to British Columbia, south to Missouri and Arizona. Our records widely scattered in the western half of Colorado at 5000-9000 feet.

2B. Arabis hirsuta glabrata T. & G., (var.) Fl. N. Amer. 1:80. 1838.

A. rupestris Nutt. in part---This variety is reported from western Wyoming and Utah and may be found in this state someday.

3. Arabis drummondi Gray, Proc. Amer. Acad. 6:187. 1866.

A. philonipha A. Nels.; A. connexa Greene; A. oxyphylla Greene---Biennial or perennial plants; stems 30-90 cm. tall, 1 to several, glabrous or sparingly appressed-pubescent below; basal leaves 2-8 cm. long, narrowly oblanceolate to broader, entire to dentate, petiolate, glabrous or pubescent; cauline leaves oblong to oblong-lanceolate, sessile and auriculate, glabrous, entire to somewhat dentate; petals 7-10 mm. long, usually white but sometimes pink; pedicels 10-20 mm. long, erect, glabrous; style very short or absent; fruit 4-10 cm. long and 1.5-3 mm. wide, erect and straight, glabrous; seeds prominently winged on 1 side, biseriate.---Mountains. Labrador to British Columbia, south to Delaware and California. Our records widely scattered in the western half of the state at 6000-11,500 feet.

4. Arabis crandallii Robinson, Bot. Gaz. 28:135. 1899.

Perennial plants; stems 15-40 cm. tall, numerous from a branching caudex, densely pubescent to sparsely so above; basal leaves 1.5-4 cm. long, oblanceolate to spatulate, entire or obscurely dentate, densely pubescent and canescent; cauline leaves 8-15 mm. long, entire, sessile, usually auriculate, densely pubescent and canescent; pedicels 5-10 mm. long, erect or slightly spreading, pubescent; petals 5-7 mm. long, white or pink; style very short or absent; fruit 3-6 cm. long and about 1 mm. wide, erect or nearly so, glabrous, somewhat constricted between the seeds; seeds wingless or very nearly so, uniseriate.---Mountains. Limited to Colorado. Our records mostly in or near Gunnison County (one from Grand) at 7000-8000 feet.

5. Arabis divaricarpa A. Nels., Bot. Gaz. 30:193. 1900.

A. oblanceolata Rydb.---Biennial or perennial plants; stems 30-90 cm. tall, 1 or few, glabrous or sparsely pubescent below; basal leaves 2-6 cm. long, broadly oblanceolate to narrowly spatulate, dentate to subentire, loosely pubescent; cauline leaves narrowly oblong to lanceolate, entire to dentate, sessile and sagittate, glabrous or the lower sparsely pubescent; pedicels divaricate to loosely ascending, seldom as much as at right angles to the rachis, glabrous; petals 6-10 mm. long, pink or purple; style very short or absent; fruit 2.5-9 cm. long and 1.25-2.5 mm. wide, loosely ascending to sometimes somewhat pendulous; usually straight, glabrous; seeds narrowly winged, uniseriate or imperfectly biseriate. Our plant has been called variety typica Hopkins.---Hills and mountains. Quebec and Alaska, south to New York, Colorado and California. Our records scattered in the western half of Colorado, few from the extreme western part, at 6500-10,000 feet.

6. Arabis fernaldiana Rollins, Rhodora 43:430. 1941.

Perennial plants; stems 15-40 cm. tall, several to numerous from a matted branching caudex, erect to rather decumbent at base, densely pubescent below, glabrous above; basal leaves 1-4 cm. long and 2-5 mm. wide, entire, spatulate to oblanceolate, narrowly petiolate, densely pubescent with short trichomes and canescent, petioles usually ciliate; cauline leaves sessile and auriculate; petals 9-14 mm. long, white to pink; pedicels ascending, glabrous; style about 1 mm. long; fruit 4-6 cm. long and 1.5-2 mm. wide, erect, glabrous, straight or nearly so; seeds winged or wingless. Our plant would be variety typica Rollins.---Among rocks and on partly shaded slopes. Reported for Nevada but we have two records from northwestern Colorado at 5000-6000 feet.

7. Arabis selbyi Rydb., Bull. Torr. Club 31:557. 1904.

Perennial plants; stems 25-50 cm. tall, several to many from usually branched caudex, pubescent below, usually glabrous above; basal leaves 3-7 cm. long, oblanceolate to broadly spatulate, entire to dentate, densely to sparsely pubescent, grayish; cauline leaves few, linear-oblong to lanceolate, sessile and auriculate, entire, lower pubescent, the upper glabrous; fruiting pedicels ascending or spreading at right angles to the rachis, glabrous to sparsely pubescent, straight or nearly so; petals 6-8 mm. long, pink; style none; fruit 3-6 cm. long and 1.5-2 mm. wide, ascending or spreading at right angles to the rachis, straight or nearly so, glabrous; seeds narrowly winged, uniseriate.---Canyons and slopes. Colorado, Utah, and New Mexico. Our records in western Colorado as far east as Lake County, mostly in the westcentral part, at 5500-9500 feet.

8. Arabis oxylobula Greene, Pitt. 4:195. 1900.

Perennial plants with simple or branching caudices; stems 8-15 cm. tall, caespitose, simple, glabrous; basal leaves 2-3 cm. long, linear to narrowly oblanceolate, entire, glabrous or petioles sparsely hirsute with simple hairs; cauline leaves 8-15 mm. long, few, oblong, remote, entire or minutely denticulate, sessile but not auriculate; inflorescence loose, few-flowered; sepals 2-3 mm. long, glabrous; petals about 5 mm. long, pink; pedicels filiform, 3-6 mm. long,

glabrous, arched or horizontal; fruit 1.5-2.5 cm. long and 1.5-2 mm. wide, glabrous, widely pendulous or spreading, uniseriate; seeds winged all around.---Mountains. Known only from western Colorado. Our one record from Garfield County at 6300 feet.
9. Arabis demissa Greene, Pl. Baker. 3:8. 1901.
Perennial plants; stems 10-30 cm. tall, caespitose, several to many, hirsute below or glabrous throughout; basal leaves 1.5-3.5 cm. long, linear to oblanceolate, entire, hirsute with margins ciliate, or rarely nearly glabrous; cauline leaves 5-10 mm. long, sessile, usually not auriculate, sparsely hirsute to glabrous; pedicels 3-7 mm. long, arching downward, glabrous; petals 4.5-6.5 mm. long, white or pink; style absent; fruit pendulous, 2-4 cm. long and 1.5-2 mm. wide, glabrous, often constricted between the seeds; seeds wingless or very narrowly winged, uniseriate. Represented in Colorado by at least 2 of the following varieties.
9A. Arabis demissa demissa
A. rugocarpa Osterh.; A. aprica Osterh.; A. demissa var. typica Rollins---This variety has the basal leaves linear to narrowly oblanceolate; cauline leaves without auricles; forked trichomes present on stems and leaves; fruit constricted between the seeds.---Apparently limited to Colorado. Our records from a band between Grand and Ouray Counties at 7000-10,500 feet.
9B. Arabis demissa russeola Rollins, (var.) Rhodora 43:387. 1941.
Basal leaves are linear to narrowly lanceolate; cauline leaves lack auricles; all hairs are simple; valves of fruit not constricted between the seeds.---Reported for Wyoming and Utah. Our one record from northwestern Colorado at 7500 feet.
9C. Arabis demissa languida Rollins, (var.) Rhodora 43:388. 1941.
Wyoming and Utah. The plant should be looked for in northwestern and northern Colorado.
10. Arabis fendleri (Wats.) Greene, Pitt. 3:156. 1897.
Plants perennial or appearing biennial; stems 25-60 cm. tall, 1 to several, hirsute below, glabrous above; basal leaves 2-6 cm. long, oblanceolate, dentate, margins ciliate and surface usually coarsely pubescent; cauline leaves oblong to lanceolate, sessile and auriculate, lower pubescent, upper glabrous, entire or rarely dentate; pedicels slender, glabrous, arched downward in fruit; petals 5-8 mm. long, pink; style absent; fruit 3-6 cm. long and 1.5-2.5 mm. wide, pendulous, glabrous, obtuse; seeds wingless or narrowly winged, biseriate. The above would be variety typica Rollins.---Mountains. The variety from Colorado to Nevada, south to Texas and New Mexico. Scattered in the western half of Colorado, except the extreme western part, at 5000-8000 feet.
10A. Arabis fendleri spatifolia (Rydb.) Rollins, (var.) Rhodora 43:394. 1941.
A. spatifolia Rydb.---This variety has basal leaves entire, linear-oblanceolate and acute; petals white. It intergrades somewhat with the typical form of the species in this area.---Wyoming, Colorado, Utah and New Mexico. Our records scattered in the western two-thirds of Colorado, except the extreme western part, at 5000-10,000 feet.
11. Arabis pulchra M. E. Jones ex Wats., Proc. Am. Acad. 22:468. 1887.
A. formosa Greene---Perennial plants with subshrubby bases; stems 20-60 cm. tall, 1 or several, densely pubescent or glabrous above; basal leaves 3-8 cm. long, entire or rarely dentate, linear, petiolate, densely pubescent; cauline leaves linear, sessile but not auriculate, densely pubescent; pedicels pendulous or recurved in fruit, pubescent; petals 12-20 mm. long, white or rarely purple; style very short or absent; fruit 4-7 cm. long and 2.5-3.5 mm. wide, pendulous to somewhat reflexed, densely pubescent, straight; seeds rather prominently winged, biseriate. Our plant is var. pallens M. E. Jones.---Valleys, slopes and flat ground. The variety from Colorado, Utah, New Mexico and Arizona. Our records from western Colorado, mostly from the westcentral part, at 4500-6500 feet.
12. Arabis cobrensis M. E. Jones, Contrib. West. Bot. 12:1. 1908.
A. canescens Nutt.---Wyoming to Oregon and Nevada. It has been recorded in southern Wyoming near to northwestern Colorado.
13. Arabis lemmonii Watson, Proc. Amer. Acad. 22:467. 1887.
A. egglestonii Rydb.---Deep-rooted perennial; stems 6-20 cm. tall, several to many from a branching caudex, slender, simple, glabrous or rather sparsely stellate; basal leaves 1-2 cm. long, spatulate-oblanceolate, entire or rarely with 1-2 teeth; cauline leaves sessile, oblong-lanceolate to ovate, glabrous or lower pubescent; petals 4-6 mm. long, pink or purple; fruiting pedicels 2-6 mm. long, glabrous; style very short or absent; fruit 2-4 cm. long and 2-3.5 mm. wide, usually horizontal or somewhat descending, glabrous; seeds narrowly winged, uniseriate. Our plant has been called variety typica Rollins.---High mountains. The variety from Montana to British Columbia, south to Colorado and California. Our few records from Park and Gunnison Counties at 12,700-13,000 feet.
14. Arabis gunnisoniana Rollins, Rhodora 43:434. 1941.
Perennial plants, the caudex simple or closely branched; stems 10-20 cm. tall, caespitose, simple, densely pubescent below, glabrate above; basal leaves 1-2 cm. long, numerous, linear-oblanceolate, entire or rarely toothed, densely pubescent; cauline leaves 5-8 mm. long, few, remote, oblong, pubescent or upper glabrous, auricles small; sepals 2-3 mm. long; petals 4-6 mm. long, pink or purplish; pedicels widely curved at about right angles to the rachis, pubescent or rarely glabrous; fruit 2.5-4 mm. long and 1-1.5 mm. wide, glabrous, at right angles to stem or somewhat descending, uniseriate; seeds narrowly winged.---Apparently limited to Gunnison County, Colorado. Our few specimens from about 8000 feet.
15. Arabis holboellii Hornem., Fl. Dan. 11, t. 1879. 1828.
Biennial or perennial plants; stems varying from 10-90 cm. tall, 1 to several, pubescent throughout to glabrous above; basal leaves 1-5 cm. long, linear-oblanceolate to spatulate, entire or dentate, densely pubescent, often pannose; cauline leaves oblong to lanceolate, auricles present or absent, entire, lower densely pubescent, the upper glabrous or pubescent; fruiting pedicels 6-16 mm. long, straight to somewhat curved, pubescent or glabrous, loosely descending to strictly reflexed; petals 6-10 mm. long, white to purplish-pink; fruit 3-7 cm. long and 1-2.5 mm. wide, loosely pendulous to strictly reflexed, usually glabrous, straight or nearly so; seeds narrowly winged, uniseriate or imperfectly biseriate. Three varieties are present in Colorado.

15A. *Arabis holboellii retrofracta* (Graham) Rydb., (var.) Contrib. U. S. Herb. 3:484. 1896.
A. retrofracta Graham in part; *A. rhodantha* Greene; *A. exilis* A. Nels.; *A. consanguinea* Greene; *A. caduca* A. Nels.---Plants over 20 cm. tall; basal leaves over 3 mm. wide; cauline leaves with auricles, usually revolute; petals 7-10 mm. long; pedicels pubescent.---Plains, hills and mountains. Quebec to British Columbia, south to Michigan, Colorado and California. Our records scattered over the western half of Colorado at 5500-11,000 feet.
15B. *Arabis holboellii pinetorum* (Tidestrom) Rollins, (var.) Rhodora 43:447. 1941.
Plants are 20 cm. tall; basal leaves over 3 mm. wide; cauline leaves with auricles; fruit pendulous.---Saskatchewan to British Columbia, south to Nebraska and California. Our records scattered in the mountainous western half of Colorado at 5500-10,000 feet.
15C. *Arabis holboellii pendulocarpa* (A. Nels.) Rollins, (var.) Rhodora 43:446. 1941.
A. pendulocarpa A. Nels.---Plants usually less than 20 cm. tall; basal leaves less than 3 mm. wide; cauline leaves lacking auricles.---Cliffs and ridges, often in rocky ground. Montana to British Columbia, south to Colorado and California. Our few records from north-central and northwestern Colorado at 7500-9500 feet.
16. *Arabis lignifera* A. Nels., Bull. Torr. Club 24:123. 1899.
Plants perennial; stems 20-50 cm. tall, 1 or usually few from a simple or branched caudex, erect, densely pubescent below with minute, appressed, stellate hairs, glabrous above; basal leaves 2-5 cm. long, linear-oblanceolate, entire, densely pubescent with minute hairs; cauline leaves 1-3 cm. long, entire, auriculate; sepals 3-4.5 mm. long, pubescent, purplish; petals 5-8 mm. long, pink to purplish; pedicels 5-12 mm. long, glabrous to sparsely pubescent, arching downward; style none or very short; fruit 3-6 cm. long and 1.5-2 mm. wide, laxly pendulous, commonly curved inward, glabrous, valves 1-nerved to middle or above; seeds narrowly winged, uniseriate or somewhat biseriate.---Dry slopes and flats. Wyoming to Idaho, south to Arizona and Nevada. Our records from western Colorado at 6000-8500 feet.
17. *Arabis sparsiflora* Nutt. in T. & G., Fl. N. Amer. 1:81. 1838.
A. perelegans A. Nels.---Montana to Washington, south to southern Wyoming and California. A speciman from northern Colorado is close to this species.
18. *Arabis perennans* Watson, Proc. Am. Acad. 22:467. 1887.
A. eremophila Greene in part; *A. gracilenta* Greene in part; *A. recondita* Greene---Perennial plants; stems 15-60 cm. tall, several to many from a simple or branched caudex; simple or branched above, pubescent below, glabrous above; basal leaves 2-6 cm. long, oblanceolate to broader, petioled, dentate to rarely entire, densely pubescent; cauline leaves lanceolate, auricled and somewhat sagittate at base, entire or sparingly dentate; sepals 3.5-4.5 mm. long, not saccate, pubescent; petals 6-9 mm. long, purple or pink; pedicels 1-2 cm. long, slender, spreading, arched downward, glabrous; style obsolete; fruit 4-6 cm. long and 1.2-2 mm. wide, pendulous to widely spreading, glabrous, curved inward; seeds winged all around, 1-1.5 mm. broad, uniseriate.---Plains and mountains. Colorado to California, south to New Mexico. Our records from western, mostly westcentral Colorado, at 5500-8000 feet.

16. *Smelowskia* C. A. Mey.

Low caespitose perennial plants with suffruticose caudices; leaves pinnatifid, canescent with stellate hairs; flowers racemose, perfect; sepals subequal, somewhat spreading; petals exserted, obovate; stigma sessile or nearly so; fruit lanceolate or ovate, 4-angled, or more or less compressed usually at right angles to the septum, valves keeled.

1. *Smelowskia calycina* C. A. Meyer in Ledeb., Fl. Alt. 3:170. 1831.
S. americana Rydb.; *S. lineariloba* Rydb.---Densely caespitose plants, the caudices with old leaf bases present; stems 5-20 cm. tall, several to many, simple, pubescent with short-branched or long, simple hairs (may be both); basal leaves petioled, pubescent and ciliate near base of petiole, from almost entire to pinnately divided; cauline leaves 1-3 cm. long, pinnatifid, nearly sessile, segments linear to oblong, pubescent; inflorescence congested-corymbose; pedicels 5-10 mm. long, ascending, pubescent with long, simple hairs; sepals 2.5-3.5 mm. long, pubescent; petals 5-7 mm. long, white to rose; style 1.5 mm. long or less; fruit 5-12 mm. long, tapering both ends, usually glabrous. Our plant has been called variety *typica* Rollins and var. *americana* Drury and Rollins.---Mountain peaks, often in rocky ground. The variety from Alberta to British Columbia, south to Colorado, Utah and Washington; Asia. Our records scattered in the high mountains of Colorado at 11,000-14,000 feet.

17. *Descurainia* Webb & Berthel TANSY MUSTARD

Annual or biennial plants with stellate hairs and sometimes glandular and simple hairs; stems leafy, simple or branched; leaves deeply pinnatifid to tripinnate; racemes terminal, elongating on maturity; flowers small, the petals yellow or yellowish, clawed, obtuse, obovate; stigma entire; fruit narrow, usually 5-20 times as long as wide, dehiscent, terete or nearly so; seeds in 1 or 2 series in each cell. Intergradations particularly between the subspecies are not uncommon in our plants.

1. Upper as well as lower leaves bipinnate into linear lobes; septum of fruit with 2-3 longitudinal nerves (these may be faint); fruit 10-30 mm. long
-1. *D. sophia*
1. Upper leaves merely pinnate, the divisions often incised; septum of fruit with 1 longitudinal nerve or none; fruit 3-20 mm. long
   2. Fruit clavate or subclavate (widest nearest the apex and tapering to base); lower leaves usually bipinnate; seeds in 2 rows (or crowded sometimes into 1)
      3. Fruiting pedicel 18-23 mm. long; fruit 14-18 mm. long; leaves canescent; plants limited to extreme western Colorado
-2A. *D. pinnata paysonii*

3. Fruiting pedicel 4-15 mm. long; fruit 4-14 mm. long; leaves not canescent (except in D. pinnata ssp. halictorum); plants not limited (but may be present) to western Colorado
   4. Fruiting pedicels diverging about 75 degrees; leaves canescent   -2B. D. pinnata halictorum
   4. Fruiting pedicels diverging about 45 degrees; leaves not canescent (can be pubescent)
      5. Petals 1.5 mm. long or less; fruiting pedicels 4-6 mm. long; fruit 4-8 mm. long
                                                                                                 -2C. D. pinnata nelsonii
      5. Petals 2-3.5 mm. long; fruiting pedicels 6-15 mm. long; fruit 5-15 mm. long (less than 8 mm. only in ssp. brachycarpa)
         6. Stems glandular and moderately pubescent; fruit 5-10 mm. long   -2D. D. pinnata brachycarpa
         6. Stems rarely glandular, if so then otherwise glabrous or nearly so; fruit 8-15 mm. long
            7. Pedicels 10-15 mm. long, usually longer than the fruit; terminal leaflet usually greatly elongated   -2E. D. pinnata filipes
            7. Pedicels 6-12 mm. long, usually shorter than the fruit; terminal leaflet not greatly elongated   -2F. D. pinnata intermedia
2. Fruit linear or short-fusiform, not clavate; lower leaves simply pinnate (the pinnae may be again pinnatifid); seeds in 1 row (2 rows in D. obtusa doubtful in Colorado)
   8. Seeds in 2 rows in each cell, at least in some of the fruits   -3. D. obtusa
   8. Seeds always in 1 row in each cell
      9. Fruit attenuate at the apex, fusiform, with a prominent style, the whole 3-6 mm. long, 2- to 6-seeded   -4. D. californica
      9. Fruit not attenuate at apex, linear, with a short or obsolete style, the whole 6-15 mm. long, 8- to 28-seeded
         10. Pedicels and fruits closely appressed, the pedicels spreading only about 5-20 degrees from the rachis
            11. Plants canescent   -5A. D. richardsonii richardsonii
            11. Plants nearly glabrous to moderately pubescent   -5B. D. richardsonii procera
         10. Pedicels and fruit ascending to spreading, the pedicels spreading 40-80 degrees
            12. Plants glandular hairy   -5C. D. richardsonii viscosa
            12. Plants not glandular hairy   -5D. D. richardsonii incisa

1. Descurainia sophia (L.) Webb ex Prantl; Engl. & Prantl, Nat. Pflanzenfam. III.2:192. 1892.
    Sophia sophia (L.) Britt.---Annual or biennial plants; stems 25-75 cm. tall, short-branched above, sparsely to densely stellate-pubescent; leaves 2-3 pinnate, the ultimate divisions usually linear; petals 2-2.5 mm. long, yellow, about as long as the sepals; pedicels 8-15 mm. long, spreading about 45°; fruit 10-30 mm. long and about 1 mm. wide, linear, somewhat torulose, often somewhat curved; style short; seeds in 1 row, septum 2- to 3-nerved, these sometimes faint.---Waste places, fields and roadsides. Naturalized from Europe and now widespread throughout most of the United States and Canada. Our records scattered over the state, especially in cultivated areas, at 4000-8000 feet.
2. Descurainia pinnata (Walt.) Britt., Mem. Torrey Club 5:173. 1894.
    Annual plants; stems 10-70 cm. tall, commonly branching, sparsely pubescent to densely canescent; leaves pinnate, the leaflets often pinnatifid especially below, the upper mostly pinnate, sparsely pubescent to densely canescent; petals whitish to yellow; pedicels 3-20 mm. long, spreading 30°-90°; style short; fruit 4-20 mm. long and 1-2 mm. wide, clavate or sub-clavate; seeds usually in 2 rows but sometimes crowded into 1 row, septum 1-nerved if at all. Intergradations commonly occur among the subspecies listed below.
2A. Descurainia pinnata paysonii Detling, (ssp.) Am. Midl. Nat. 22:515. 1939.
    Leaves canescent; fruiting pedicels 18-23 mm. long, diverging 45°-80°; petals 2-3 mm. long, yellow; fruit 14-18 mm. long.---Wyoming to Idaho, south to Arizona. This subspecies should be in western Colorado but no specimens have been seen by the writer.
2B. Descurainia pinnata halictorum (Cockerell) Detling, (ssp.) Am. Midl. Nat. 22:505. 1939.
    Leaves canescent; fruiting pedicel 8-12 mm. long, spreading 65°-90°; petals 1.5-2.5 mm. long, whitish to yellow; fruit 5-10 mm. long.---Arkansas to Oregon, south to Texas and Mexico. Our many records scattered over the state at 4000-8000 feet.
2C. Descurainia pinnata nelsonii (Rydb.) Detling, (ssp.) Am. Midl. Nat. 22:512. 1939.
    Sophia nelsonii Rydb.---Lower leaves sparsely to moderately pubescent, upper leaves even glabrous; fruiting pedicels 4-6 mm. long, diverging about 45°; petals 1.5 mm. long, yellow; fruit 4-8 mm. long.---Montana to Washington, south to Colorado and Utah. Our few records scattered in the western two-thirds of Colorado at 4500-8000 feet.
2D. Descurainia pinnata brachycarpa (Richard.) Detling, (ssp.) Am. Midl. Nat. 22:509. 1939.
    Sophia magna Rydb.---Leaves moderately to densely pubescent, not canescent, glandular-hairy; fruiting pedicels 8-16 mm. long, diverging about 45°; petals 2-3 mm. long, yellow; fruit 5-10 mm. long.---Canada south to Texas and east to New England. Our records scattered in Colorado at 4500-8000 feet.
2E. Descurainia pinnata filipes (Gray) Detling, (ssp.) Am. Midl. Nat. 22:513. 1939.
    Sophia filipes (Gray) Heller---Leaves glabrous to puberulent; fruiting pedicels 10-15 mm. long, diverging 45°-70°; petals 2-3.5 mm. long, light yellow; fruit 10-15 mm. long.---Wyoming to Washington, south to Colorado and Oregon. Our records from central and southcentral Colorado at 7500-9500 feet.
2F. Descurainia pinnata intermedia (Rydb.) Detling, (ssp.) Am. Midl. Nat. 22:511. 1939.
    Sophia intermedia Rydb.---Leaves glabrous to sparsely pubescent; fruiting pedicels 6-12 mm. long, diverging 45°-70°; petals 2-3 mm. long, yellow; fruit 8-12 mm. long.---Alberta to British Columbia, south to Colorado and California. Our records scattered in the western two-thirds of Colorado at 5000-7000 feet.
3. Descurainia obtusa (Greene) O. E. Schulz, Pflanzenreich IV.105:321. 1924.
    New Mexico and Arizona. Reported from areas very close to southwestern Colorado.

4. Descurainia californica (Gray) O. E. Schulz, Pflanzenreich IV.105:330. 1924.
Annual or biennial plants; stems 30-80 cm. tall, branched above, pubescent especially below, to glabrous; leaves simply pinnate with 2-4 pairs of pinnae, these lanceolate, entire, serrate or incised, pubescent; petals 1.5-2 mm. long, yellow, barely exceeding the sepals; pedicels 3-7 mm. long, spreading 15^0-45^0; style prominent, 0.5 mm. or more long; fruit 3-6 mm. long and 1-1.25 mm. wide, tapering to base and acute at apex hence rather fusiform; seeds in 1 row, 2-6 to a fruit, septum 1-nerved if at all.---In the mountains. Wyoming to Oregon, south to New Mexico and California. Our records scattered in the mountainous part of Colorado, except the extreme northern part, at 7000-9500 feet.

5. Descurainia richardsonii (Sweet) O. E. Schulz, Pflanzenreich IV.105:318. 1924.
Biennial or possibly perennial plants; stems 30-100 cm. tall, usually branched above, glabrous to canescent, sometimes glandular; leaves simply pinnate or the lower with pinnae pinnatifid, the segments mostly broad; flowers pale to bright yellow; style short; fruit 5-15 mm. long and 0.5-1.5 mm. wide, narrowly linear; seeds in 1 row, 8-28 to a fruit. Four subspecies are present in Colorado.

5A. Descurainia richardsonii richardsonii
D. richardsonii ssp. typica Detling---Plant canescent, no glands present; fruiting pedicels 3-6 mm. long, diverging 5^0-20^0; fruit 5-10 mm. long.---Yukon, south to Great Lakes, North Dakota and Colorado. Our few records from northcentral Colorado at 7500-9000 feet.

5B. Descurainia richardsonii procera (Greene) Detling, (ssp.) Am. Midl. Nat. 22:491. 1939.
Sophia hartwegiana of Manuals; S. procera Greene; S. brevipes (Nutt.) Rydb.---Plants nearly glabrous to pubescent, not canescent, rarely glandular; fruiting pedicels 2-6 mm. long, diverging 5^0-20^0; fruit 6-12 mm. long.---Montana south to New Mexico. Our records scattered in the western half of the state at 5000-11,500 feet.

5C. Descurainia richardsonii viscosa (Rydb.) Detling, (ssp.) Am. Midl. Nat. 22:492. 1939.
Plants moderately to densely pubescent not canescent, glandular hairs present on the stem and rachis; fruiting pedicels 6-10 mm. long, spreading 40^0-80^0; fruit 9-15 mm. long.---Alberta to Washington, south to New Mexico and California. Our records scattered in the western half of Colorado at 4500-9000 feet.

5D. Descurainia richardsonii incisa (Engelm.) Detling, (ssp.) Am. Midl. Nat. 22:494. 1939.
Sophia incisa (Engelm.) Greene; S. purpurascens Rydb.; S. leptophylla Rydb.---Plants subglabrous to moderately pubescent, not canescent, no glandular hairs present; fruiting pedicels 4-10 mm. long, diverging 45^0-80^0; fruit 8-15 mm. long.---Montana to California, south to Mexico. Our records scattered in the western half of Colorado at 5500-9500 feet.

18. Erysimum L. WALLFLOWER

Annual, biennial or perennial herbaceous plants with short, harsh, appressed 2-branched hairs; leaves entire, toothed or lobed; flowers in racemes; sepals erect, 1 pair gibbous at base; petals large (over 6 mm. long), yellow to orange or maroon, clawed; style short, the stigma 2-lobed, the lobes over the septum; fruit linear, elongated, more or less 4-angled; seeds many, wingless, 1 row in each cell. This is a puzzling genus and the treatment here is very arbitrary.

1. Petals 4-8 mm. long, yellow
  2. Fruit 2-3 cm. long, on slender pedicels over 7 mm. long   -1. E. cheiranthoides
  2. Fruit 2-8 cm. long, pedicels stout, 2-5 mm. long
    3. Fruit divaricately spreading, 5-8 cm. long; plants annual   -2. E. repandum
    3. Fruit erect, 2-5 cm. long; plants perennial or biennial   -3. E. inconspicuum
1. Petals 10-16 mm. long, yellow to orange or maroon
  4. Plants more or less caespitose; stems less than 20 cm. tall; plants growing over 10,000 feet in Colorado; petals yellow   -4. E. nivale
  4. Plants little of any caespitose; stems over 20 cm. tall and typically growing below 10,000 feet in Colorado (if shorter and over 10,000 feet then the petals orange to maroon)
    5. Petals orange or maroon in color   -5. E. wheeleri
    5. Petals yellow   -6. E. asperum

1. Erysimum cheiranthoides L., Sp. Pl. 661. 1753.
Cheirinia cheiranthoides (L.) Link ---Annual plants; stems 20-60 cm. tall, erect, simple or branching, sparingly puberulent, the hairs with 2 or sometimes 3 branches; leaves 2.5-3 cm. long, lanceolate, linear-lanceolate to oblanceolate, entire or obscurely dentate; petals 4-5 mm. long, yellow; style about 1 mm. long; fruit 2-3 cm. long, glabrous or nearly so, nearly erect, on slender pedicels over 7 mm. long.---Newfoundland to Alaska, south to North Carolina, Utah and California; Europe. Our records scattered in the western half of Colorado, except the extreme western part, at 4500-9000 feet.

2. Erysimum repandum L., Amoen. Acad. 3:415. 1756.
Cheirinia repanda (L.) Link ---Annual plants; stems 20-40 cm. tall, simple or much branched; leaves 3-6 cm. long, lanceolate to linear-lanceolate, mostly entire or repand-denticulate; petals 6-8 mm. long, light yellow; fruiting pedicels stout, 2-5 mm. long; fruit 5-8 cm. long and 1.5-2 mm. thick, spreading, beak short and stout.---Waste places and roadsides. Ohio to Washington, south to Arizona. Our few records from northern Colorado at 5000-6500 feet.

3. Erysimum inconspicuum (Wats.) MacM., Met. Minn. 268. 1892.---
E. parviflorum Nutt.; Cheirinia inconspicua (Wats.) Rydb.---Perennial or biennial plants; stems 30-60 cm. tall, erect, simple or branched; leaves 2.5-7.5 cm. long, linear to oblanceolate, entire or sparingly dentate; petals 6-8 mm. long, yellow; fruiting pedicels 4-6 mm. long, stout; fruit 2-5 cm. long and about 2 mm. wide, erect.---Dry ground. Ontario to Alaska, south to Kansas and Nevada. Our records scattered over the western half of Colorado at 6500-7500 feet.

4. **Erysimum nivale** (Greene) Rydb., Bull. Torr. Club 31:558. 1904.

E. radicatum Rydb.; Cheirinia radicata Rydb.---Plants perennial, more or less caespitose from 2 or more caudex branches; stems 5-20 cm. tall; leaves 3-7 cm. long, entire, denticulate or sinuate-dentate, linear to narrowly oblanceolate, somewhat pubescent; petals about 13-20 mm. long; pedicels 4-7 mm. long, ascending; fruit 3-8 cm. long, erect or nearly so.---Mountains. Colorado and Utah. Our records scattered in the higher mountains of Colorado at 10,000-13,000 feet.

5. **Erysimum wheeleri** Rothr. in Wheeler, Rep. U. S. Surv. 100th. Merid. 6:64. 1879.

E. amoenum (Greene) Rydb.; Cheirinia wheeleri (S. Wats.) Rydb.; C. amoena (Greene) Rydb.---This belongs to the E. asperum complex. Perennial or biennial plants; stems 10-60 cm. tall (according to altitude and local conditions); leaves entire to sinuate-dentate; petals orange to maroon but intergrades with yellow; fruit 3-8 cm. long, erect or nearly so.---Mountains. Colorado, New Mexico and Arizona. Scattered in the western two-thirds of the state. No records yet from northwestern Colorado, at 5500-11,000 feet.

6. **Erysimum asperum** (Nutt.) DC., Syst. 2:505. 1821.

Very closely related to, if not synonymous with the following: E. elatum Nutt.; E. aridum A. Nels.; E. asperrimum (Greene) Rydb.; E. oblanceolatum Rydb.; E. bakeri (Greene) Rydb.; E. argillosum (Greene) Rydb.; (and all the above in the genus Cheirinia in Rydbs. Manual)---The following description covers all the above: Biennial or short-lived perennial plants; stems 20-70 cm. tall; leaves entire, denticulate or sinuately-dentate, narrowly linear to oblanceolate; petals yellow; fruit 4-8 cm. long, erect, ascending or rarely divaricate.---Widespread throughout the western half of the United States. Our records well scattered over all of Colorado at 3500-9500 feet.

19. Malcolmia R. Br.

Annual plants; rough pubescent with mostly branched hairs; stems several; leaves petioled, coarsely dentate; flowers in racemes; sepals erect, the lateral ones often somewhat gibbous at base; petals purplish-pink, the blades shorter than the claws; fruit elongated, slender, with a short conical beak.

1. **Malcolmia africana** (L.) R. Br. in Ait. f., Hort. Kew. ed. 2, 4:121. 1812.

Hesperis africana L.---Stems 10-40 cm. tall, rather rigid, branched above; leaves oblong or lanceolate; sepals about 3-5 mm. long, hairy; petals about 6-8 mm. long; fruit about 4-6 cm. long and 1-1.5 mm. thick, hairy, ascending-spreading, rather 4-sided.---Dry slopes and draws. Introduced from the Mediterranean region into Utah, Nevada and Arizona according to report. Our records from westcentral Colorado at 4500-5000 feet.

20. Halimolobos Tausch

Biennial or perennial herbaceous plants; stems erect, ours stellate or branched-pubescent; basal leaves sinuate-dentate, petioled; cauline leaves sessile and auriculate, all leaves pubescent with branched hairs; sepals erect, pubescent, not saccate; petals white with narrow claw; style about 0.5 mm. long; stigma not evidently lobed; fruit elongated, terete or nearly so, erect, glabrous; seeds wingless.

1. **Halimolobos virgata** (Nutt.) O. E. Schulz, Pflanzenreich IV. 105:290. 1924.

Arabidopsis virgata (Nutt.) Rydb.; A. stenocarpa Rydb.; Stenophragma virgata (Nutt.) Greene---Stems 10-40 cm. tall, usually single from a simple caudex, branched above or below, hirsute below, pubescent above; basal leaves 2-6 cm. long, narrowly oblanceolate; cauline leaves lanceolate; sepals 2-3 mm. long; petals 3-4 mm. long; pedicels 5-12 mm. long, divaricately ascending, pubescent; anthers broadly oblong, short (about 0.5 mm. long); fruit 2-4 cm. long and about 1 mm. wide.---Hills and plains often in rather moist ground. Saskatchewan to Alberta, south to Colorado and Utah. Our records from northcentral and central Colorado at 7000-9000 feet.

21. Pennellia Nwd.

Biennial plants, pubescent at least below with stellate or branched hairs; stems slender, erect, simple or sparingly branched with erect branches; basal leaves oblanceolate, more or less toothed; stem leaves linear or lanceolate-linear, entire; calyx with 2 of the sepals somewhat longer, all ascending, greenish; petals broadly clawed, whitish, small; anthers at length spirally curved; stigma entire or slightly 2-lobed; fruit slender, erect, terete, sessile. This genus probably better merged with Thelypodium.

1. **Pennellia micrantha** (A. Gray) Nieuwland, Am. Midl. Nat. 5:224. 1918.

Heterothrix micrantha (A. Gray) Rydb.; Thelypodium micranthum (A. Gray) S. Wats.---Stems 30-70 cm. tall; basal leaves 2-5 cm. long, stellate; cauline leaves often glabrate; raceme elongated, narrow; petals 3-4 mm. long; fruit 2.5-4 cm. long, slender, seldom over 1 mm. thick.---Mountains. Colorado to Utah, south to Mexico. Our one somewhat doubtful record from southwestern Colorado at about 7000 feet.

22. Stanleyella Rydb.

Biennial glabrous plants; stems tall, branched; leaves thin, the lower coarsely toothed or pinnatifid, the upper entire; sepals spreading or reflexed in anthesis, not at all saccate; petals white or pale purplish, short-clawed; style short; filaments linear or filiform, anthers more or less coiled at maturity, sagittate at base; fruit terete, subsessile, very slender; seeds wingless. May not be a distinct genus from Thelypodium.

1. Stanleyella wrightii (A. Gray) Rydb., Bull. Torr. Club 34:435. 1907.
Thelypodium wrightii A. Gray ---Stems 40-150 cm. tall, much branched, glabrous or sparingly pilose at base; basal leaves 10-15 cm. long, lyrate-pinnatifid; stem leaves reduced upward, becoming entire, linear-lanceolate; sepals about 5 mm. long; petals longer than the sepals; inflorescence in flower corymbose, elongating in fruit; pedicels widely spreading or descending; fruit about 4-7 cm. long, not over 1 mm. thick, spreading, torulose.---Mountains and plains, Colorado and Utah, south to Texas and Arizona. Our records mostly from southern Colorado, one from Garfield County, at 5000-7500 feet.

23. Streptanthus Nutt. TWISTFLOWER

Glabrous, perennial, herbaceous plants; cauline leaves cordate-clasping; calyx urn-shaped, lateral sepals saccate at base, all erect, constricted at the throat and with spreading tips; petals narrow, crisped on margin, ochroleucous, yellow or brown-purple; stigma 2-lobed or entire, the lobes developing over the valves; fruit linear, flattened parallel to the septum.

1. Streptanthus cordatus Nutt. ex T. & G., Fl. N. Amer. 1:77. 1838.
Euklisia cordata (Nutt.) Rydb.; Cartiera cordata (Nutt.) Greene; S. coloradensis A. Nels.---Short-lived perennial plants; stems 30-90 cm. tall, mostly unbranched above the base; basal leaves spatulate, variously dentate especially toward the apex, sometimes deeply so; stem leaves entire or dentate toward the apex, pale green, sagittate at base; sepals 7-10 mm. long, greenish or purple; petals 10-15 mm. long, recurved at apex; fruit 5-10 cm. long and about 4-5 mm. wide, strongly flattened.---Dry places. Wyoming to Oregon, south to New Mexico and California. Our rather numerous records well scattered over the western third of Colorado at 4500-8000 feet.

24. Streptanthella Rydb.

Glabrous annual or perhaps biennial plants; stems slender, usually branched; leaves entire to dentate; flowers rather small; sepals saccate at base, especially the outer, erect; petals narrow; anthers subsagittate at base, short; fruit flat, narrowed to a conspicuous beak, dehiscent at base but valves remaining connected at apex, sessile, pendant on recurved pedicels; seeds flattened and narrowly winged.

1. Streptanthella longirostris (S. Wats.) Rydb., Fl. Rocky Mt. 364. 1917.
Streptanthus longirostris S. Wats.; Euklisia longirostris (S. Wats.) Rydb.---Stems 20-60 cm. long; lower leaves narrowly oblanceolate to ovate-spatulate, usually repand or sinuate-dentate, readily deciduous; stem leaves linear to linear-lanceolate, mostly entire; sepals 3-5 mm. long; petals somewhat longer than the sepals, white to yellowish; anthers about 1 mm. long, apiculate; fruit 3-6 cm. long and 1-2 mm. wide, strongly compressed, style short but beak simulates a long style.---Dry often sandy soil. Wyoming to Washington, south to Mexico. Our records from western Colorado at 4500-6000 feet.

25. Caulanthus Wats. WILD CABBAGE

Annual or short-lived perennial herbaceous plants; stems branched or unbranched, sometimes inflated; cauline leaves sessile or petioled; calyx closed at anthesis or nearly so; petals white or purplish, claws not at all conspicuous; style usually short; fruit elongated, terete or slightly flattened; seeds wingless or narrowly winged.

1. Caulanthus crassicaulis (Torr.) S. Wats., Bot. King's Expl. 27. 1871.
Biennial or short-lived perennial plants; stems 30-100 cm. tall, glabrous and glaucous, erect, unbranched, stout, more or less inflated and hollow; radical leaves 5-15 cm. long, rosulate, glabrous or glaucous, varying from entire and oblanceolate to irregularly lyrate or runcinate; upper stem leaves entire, all leaves petioled; sepals 10-15 mm. long, purplish, hirsute, not saccate at base; petals 15-20 mm. long, purplish to brown, linear and channeled, curved outward; pedicels 3-5 mm. long; stout style nearly obsolete and stigma broadly 2-lobed; fruit 10-13 cm. long, erect or ascending, subsessile.---Dry slopes often in rocky or gravelly ground. Wyoming to Idaho, south to Arizona and California. Our records from northwestern Colorado at 5500-8000 feet.

26. Hesperidanthus (B. L. Robinson) Rydb.

Glabrous perennial plants; stems erect, branched above; stem leaves sessile or nearly so; sepals rather firm; erect, the outer ones strongly gibbous at base; petals purple to purple-pink; pedicels in fruit 1-2 cm. long; style conical, little if any lobed; fruit terete, linear, short-stipitate.

1. Hesperidanthus linearifolius (A. Gray) Rydb., Bull. Torr. Club 34:434. 1907.
Thelypodium linearifolium (A. Gray) S. Wats.; Sisymbrium linearifolium (Gray) Payson---Perennial herbaceous plants; stem 40-150 cm. tall; basal leaves 5-10 cm. long, oblanceolate to spatulate, toothed; stem leaves linear and entire; sepals 5-6 mm. long; petals 12-15 mm. long; style conical, slightly if at all lobed; fruit 4-10 cm. long and about 1.5 mm. wide, suberect,

short-stipitate.---Dry places. Colorado to Mexico. Our records from central and southern Colorado at 5000-8000 feet.

27. Thelypodium Endl.

Biennial or perennial herbaceous plants; stems mostly erect, simple or branched; flowers purple, lilac, rose or white; sepals little saccate if at all, erect; petals rather narrow; anthers sagittate at base; style short; stigma entire or slightly 2-lobed, the septum often with a mid-vein; fruit elongated, terete or slightly flattened parallel to the partition, sessile or stipitate, horizontal or erect; seeds not winged.

1. Cauline leaves clasping and auriculate at base -1. T. sagittatum
1. Cauline leaves not clasping, little if at all auriculate at base
  2. Pedicels 3-5 mm. long, conspicuously flattened at base; stipe of fruit 1-3 mm. long; plants west of the Continental Divide in Colorado -2. T. rhomboideum
    3. Stipe 1-2 mm. long
    3. Stipe 2-3 mm. long -2A. T. rhomboideum gracilipes
  2. Pedicels 5-10 mm. long, seldom much flattened at base; stipe of fruit not over 1 mm. long; plants both east and west of the Continental Divide -3. T. lilacinum

1. Thelypodium sagittatum (Nutt.) Endl. in Walp., Rep. 1:172. 1842.
T. torulosum Heller; T. paniculatum A. Nels.---Biennial or short-lived perennial plants, glaucous, glabrous or sparsely hirsute near base; stems 30-60 cm. tall, branched above and often from the base; basal leaves entire, oblanceolate, rather deciduous; stem leaves ovate-lanceolate or narrower, clasping and auriculate at base; sepals 5-7 mm. long; petals 2-3 times as long as the sepals, white to purple, blades oblanceolate, gradually narrowed to slender claws which about equal the blades in length; inflorescence divergent-ascending; style 1-1.5 mm. long; fruit 2-6 cm. long, erect or nearly so, somewhat torulose, subsessile.---Dry ground. Wyoming to Idaho, south to Colorado and Nevada. Our records from northcentral and northwestern Colorado at 7500-9500 feet.

2. Thelypodium rhomboideum Greene, Pitt. 4:314. 1901.
Pleurophragma platypodum Rydb.---Biennial glabrous plants with taproots; stems 40-200 cm. tall, erect, usually simple at base and branched above; radical leaves 4-12 cm. long, oblanceolate, entire or sinuate; cauline leaves linear to linear-lanceolate, sessile or nearly so, not clasping; sepals white or pale purple; petals 6-8 mm. long, white, narrow; inflorescence very dense in fruit but not corymbose, pedicels 3-5 mm. long, horizontal or somewhat reflexed, conspicuously laterally flattened at base; style about 1 mm. long; fruit 1.5-3 cm. long, incurved, irregularly torulose; stipe 1-2 mm. long.---Dry places. Colorado to Nevada. Our records from western Colorado (but reported from Grand County) at 5500-8500 feet.

2A. Thelypodium rhomboideum gracilipes (Robinson) Payson, (var.) Ann. Mo. Bot. Gard. 9:277.1922.
T. gracilipes (Robinson) Rydb.; Pleurophragma gracilipes (Robinson) Rydb.; T. integrifolium var. gracilipes Robinson---Differs from the species in longer stipe to fruit, 2-3 mm. long (instead of 1-2 mm.).---Colorado, Utah and New Mexico. Our few records from southwestern Colorado at 5000-6000 feet.

3. Thelypodium lilacinum Greene, Pl. Baker. 3:9. 1901.
Pleurophragma integrifolium and T. integrifolium of Manuals as to Colorado plants---Biennial, glabrous plants; stems 50-200 cm. tall, erect, simple or branched at base, branched above; radical leaves 4-12 cm. long, oblanceolate, entire, early deciduous; cauline leaves narrowly lanceolate to linear, sessile or very slightly auriculate; sepals white to purple; petals 6-9 mm. long, purple, lilac or nearly white; inflorescence corymbose at first, somewhat elongating in fruit, pedicels 5-10 mm. long, horizontal or ascending, not usually conspicuously flattened at base; style about 1 mm. long; fruit 1.5-3.5 cm. long, irregularly torulose, arcuate-ascending; stipe usually less than 1 mm. long. May not be distinct from the preceding species.---Dry ground. Nebraska to Washington, south to New Mexico and California. Our records scattered over Colorado, except the extreme western part, at 4000-8500 feet.

28. Sisymbrium L.

Annual, biennial or perennial herbaceous plants; stems simple or branched, the hairs when present simple; leaves entire to pinnatifid, the cauline often clasping; sepals slightly if at all saccate at base, erect or spreading; petals yellow, white or purple; style short, stigma nearly entire or 2-lobed with the lobes extending over the partition edges; fruit terete, often torulose, slender, sessile or stipitate; seeds wingless, in 1 or 2 rows in each cell.

1. Cauline leaves entire and auriculate-clasping (sometimes only slightly so in S. salsugineum)
  2. Petals yellow; fruit stipitate -1. S. aureum
  2. Petals white to purple; fruit sessile or stipitate
    3. Fruit with a stipe 2-8 mm. long -2A. S. elegans juniperorum
    3. Fruit sessile or nearly so
      4. Fruit 4 cm. long or more; stigma 2-lobed; petals 8-16 mm. long -2. S. elegans
      4. Fruit not over 3 cm. long, stigma subentire; petals about 2 mm. long -3. S. salsugineum
1. Cauline leaves pinnatifid or if entire then not at all auriculate-clasping
  5. Middle and upper cauline leaves entire (lower may be pinnatifid); plants perennial with long-creeping rootstocks; fruit 2.5-6 cm. long -4. S. linifolium
  5. Middle and upper cauline leaves hastate or pinnatifid; plants annual (or biennial ?) without rootstocks; fruit shorter or longer than 2.5-6 cm. long
    6. Fruit 1-1.5 cm. long, strictly erect; upper leaves hastate -5. S. officinalis
    6. Fruit 7-10 cm. long, spreading; upper leaves deeply pinnatifid -6. S. altissimum

1. **Sisymbrium aureum** (Eastw.) Payson, U. Wyo. Publ. Sc. 1:13. 1922.
   Thelypodiopsis aurea (Eastw.) Rydb.---Biennial plants; stems 30-60 cm. tall, glabrous or sparsely hairy near the base, with long flat hairs; radical leaves lanceolate, blade 2-5 cm. long, dentate with margined petioles; cauline leaves 1-5 cm. long, ovate, entire, glaucous, acute, auriculate-clasping; sepals about 5 mm. long, not saccate, yellow; petals about 7-8 mm. long, bright yellow; pedicels ascending-divergent; style 2-3 mm. long, stigma somewhat 2-lobed; fruit 4-7 cm. long, nearly erect, somewhat torulose, terete, on a slender stipe 2-5 mm. long.---Dry areas. Colorado. Our specimens from the southwestern part at 7000-8000 feet.

2. **Sisymbrium elegans** (Jones) Payson, Univ. Wyo. Publ. Sc. 1:13. 1922.
   Thelypodiopsis elegans (M. E. Jones) Rydb.; *T. bakeri* (Greene) Rydb.; *T. wyomingensis* (A. Nels.) Rydb.; *Streptanthus wyomingensis* A. Nels.---Stems 20-100 cm. tall, glabrous or sparingly hairy at base; basal leaves 2-5 cm. long; cauline leaves ovate, oblong or lanceolate, usually glaucous, auriculate-clasping; petals 6-12 mm. long, white or pinkish-rose; style from very short to about 2 mm. long; fruit 4-9 cm. long, sessile or the stipe short, less than 1 mm. long.---Dry often sandy rocky ground. Wyoming, Colorado and Utah. Our records from the western third of Colorado, except the southwestern corner, at 4500-7000 feet.

2A. **Sisymbrium elegans juniperorum** (Payson), (var.) comb. nov.
   Sisymbrium juniperorum Payson---Differs from the species in having a shorter stipe to the fruit, this not over 1 mm. long.---Dry ground. Colorado. Our specimens from westcentral Colorado at 6000-7000 feet.

3. **Sisymbrium salsugineum** Pallas, Reise 2 app. 114. 1773.
   Arabidopsis glauca (Nutt.) Rydb.---Reported from Montana and central Colorado. No specimens from the state were found by the writer.

4. **Sisymbrium linifolium** Nutt. in T. & G., Fl. N. Amer. 1:91. 1838.
   Schoenocrambe linifolia (Nutt.) Greene; *S. decumbens* Rydb.; *S. pinnata* Greene; *S. linifolia pinnata* (Greene) A. Nels.---Perennial herbaceous plants with long-creeping rootstocks; stems 20-50 cm. tall, simple at first, branching in age, glaucous and glabrous; leaves linear or linear-lanceolate, cuneate at base, entire or lower pinnatifid; racemes elongated in fruit; sepals about 5 mm. long, erect, slightly saccate at base; petals 6-10 mm. long, yellow; pedicels 3-10 mm. long; stigma 2-lobed; fruit 2.5-6 cm. long and about 1 mm. wide, ascending, sessile or subsessile.---Hills, plains often in gravelly soil. Montana to Washington, south to New Mexico and Utah. Our records scattered in the western two-fifths of Colorado, mostly west of the Continental Divide, at 4500-8500 feet.

5. **Sisymbrium officinalis** (L.) Scop., Fl. Carn. ed. 2. 2:26. 1772.
   Erysimum officinale L.---Annual (or biennial) plants; stems 30-60 cm. tall, erect, more or less pubescent with simple hairs or glabrous; basal leaves 7-15 cm. long, petioled, pinnatifid, the upper nearly sessile, subentire or hastate; pedicels 2 mm. long, erect in fruit; sepals short, not saccate; petals about 3 mm. long, yellow; stigma somewhat 2-lobed; fruit 10-15 mm. long, linear, acuminate at apex, strictly appressed.---A weed in fields and waste places. Nova Scotia to British Columbia, south to Florida and California; Europe. Apparently not common in Colorado, our 2 records from northcentral Colorado at 4500-5500 feet.

6. **Sisymbrium altissimum** L., Sp. Pl. 659. 1753.
   Norta altissima (L.) Britt.---Annual (or biennial) plants; stems 60-100 cm. tall, erect, widely branching, glabrous or pubescent especially below; lower leaves petioled, pinnatifid, upper leaves smaller, deeply pinnatifid with narrow divisions; pedicels thickened in fruit; sepals about 5 mm. long, ascending, little if any saccate; petals 6-8 mm. long, yellow to deep cream-colored; stigma 2-lobed; fruit 7-10 cm. long, spreading, very narrow, about 1 mm. thick.---A common weed in fields and waste places. Naturalized from Europe and now widely distributed in the United States. Our records well scattered over Colorado, especially in cultivated areas, at 4500-8500 feet.

29. Chorispora R. Br. ex DC.

Annual plants; stems branched, leafy; leaves entire to pinnatifid; flowers racemose, bractless; sepals erect, the lateral ones gibbous at base; petals clawed, rose or purplish; fruit terete, elongated, breaking up into indehiscent joints, beak long; seeds many, alternately 2-seriate.

1. **Chorispora tenella** DC., Syst. II.435. 1821.
   Stems 20-50 cm. tall, glandular-viscid, branched; lower leaves runcinately-lobed to dentate; middle and upper leaves 2-10 cm. long, dentate, oblong-lanceolate; sepals 4-6 mm. long; petals 7-12 mm. long, purple to deep rose; fruit 2.5-4 cm. long, linear, the beak 8-15 mm. long at maturity.---A weed in fields and waste places. Naturalized from Asia and established at various places in the United States. Apparently spreading rapidly in Colorado. Our records scattered over the state (none as yet from the southeastern part) at 4000-9500 feet.

30. Raphanus L. RADISH

Annual or biennial plants; stems erect, branching; leaves lyrate-pinnatifid; flowers in racemes; sepals erect, the lateral ones somewhat saccate at base; petals showy, large, pale yellow, white or purple; fruit terete, elongated, fleshy or corky, constricted and continuous and spongy between the seeds, indehiscent, tapering above into a long slender style.

1. Fruit with several, longitudinal raised nerves, constricted between the 6-10 seeds; petals not purplish-veined when fresh, usually yellow -1. **R. raphanistrum**
1. Fruit not strongly nerved, only slightly constricted between the 1-4 seeds; petals usually pink or white, when fresh with purplish veins -2. **R. sativus**

1. **Raphanus raphanistrum** L., Sp. Pl. 669. 1753.
   Annual or biennial plants from slender roots; stems 30-70 cm. tall, branching, sparsely hirsute at least below; leaves irregularly dentate or the lower deeply lyrate-pinnatifid; sepals about 6-7 mm. long; petals 13-20 mm. long, usually yellow (rarely purplish) but fading to white; fruit fleshy, about 6-10 seeded, constricted between the seeds when dry, longitudinally ridged, these raised, body about 20-25 mm. long, beak 10-20 mm. long.---Fields and waste ground. Naturalized from Europe especially in the eastern part of the United States. Our two records from Weld and Montezuma Counties at 4500-7500 feet.
2. **Raphanus sativus** L., Sp. Pl. 669. 1753.
   Stems 30-50 cm. tall, from a more or less fleshy root, glabrous or sparsely pubescent with stiff hairs; lower leaves deeply pinnatifid the lobes crenate or dentate, more or less hairy; petals 15-20 mm. long, variable but usually white with purple veins, sometimes yellowish or purplish; pods fleshy, 2- to several-seeded, not longitudinally grooved, only slightly constricted between the seeds, often equalled or exceeded by the long conical beak.---Escaped from cultivation in various parts of North America. Our one record from northcentral Colorado at 4500 feet but it is possible that the plant was cultivated.

31. Eruca (Tourn.) Mill. GARDEN ROCKET; ROCKET SALAD

Annual or biennial plants; stems erect, branching; leaves pinnately lobed or toothed; flowers in racemes; sepals erect; petals large; fruit linear to linear-oblong, turgid, with a long flattened beak; seeds in 2 rows in each cell.

1. **Eruca sativa** Mill., Gard. Dict. ed. 8. no. 1. 1768.
   Waste ground and fields. Native to Europe but introduced in various parts of the United States. It has been reported from Utah and New Mexico but the writer has seen no Colorado material.

32. Brassica L. MUSTARD

Annual or biennial plants; stems erect, branching, glabrous, glaucous or hairy; leaves sessile or petioled, at least the lower lyrate-pinnatifid; sepals slightly if at all saccate at base, ascending or spreading; petals yellow; fruit elongated, commonly torulose, beaked, the beak indehiscent; seeds in 1 row, wingless.

1. Upper cauline leaves with a clasping auriculate base -1. B. campestris
1. Upper cauline leaves petioled or sessile but not auriculate-clasping
   2. Beak half or more as long as the body of the fruit, strongly flattened and usually containing 1 seed in an indehiscent cell
      3. Fruiting pedicels spreading, 8-12 mm. long; beak over 1/2 the length of the fruit, the body hairy -2. B. hirta
      3. Fruiting pedicels ascending, 4-6 mm. long; beak of fruit shorter than the body which is glabrous -3. B. kaber
   2. Beak usually less than half as long as the body of the fruit, conical, seedless
      4. Pedicels in fruit less than 5 mm. long, erect; fruit appressed to the rachis, 1-2 cm. long, not over 2 mm. wide, somewhat quadrangular -4. B. nigra
      4. Pedicels in fruit over 5 mm. long, spreading; fruit not appressed to the rachis, 3-5.5 cm. long, 2-3.5 mm. wide, nearly terete (but 4-nerved) -5. B. juncea

1. **Brassica campestris** L., Sp. Pl. 666. 1753.
   Annual or biennial plants; stems 30-100 cm. tall, glabrous and glaucous or slightly pubescent below; basal leaves lyrate-pinnatifid; upper leaves oblong to lanceolate, entire or dentate, auriculate-clasping at base; petals 6-7 mm. long, yellow; pedicels 15-25 cm. long, spreading; fruit 4-7 cm. long, about 3 mm. wide, beak about 1-2 cm. long.---Fields and waste places. Native of Eurasia and distributed at various places in the United States. Our two records from central Colorado at 7500-8000 feet.
2. **Brassica hirta** Moench., Meth. Pl. Sup. 84. 1802.
   B. alba of Manuals; Sinapis alba L.---Annual plants; stems 30-70 cm. tall, more or less hirsute; lower leaves 10-20 cm. long, lyrate-pinnatifid; upper leaves often entire or merely dentate; petals about 1 cm. long, yellow; fruiting pedicels 8-12 mm. long, spreading; fruit about 3 cm. long, the dehiscent part hairy, as long or shorter than the flattened, sword-shaped beak.---Waste places and fields. Naturalized from Europe into various parts of North America. Reported from northern Colorado and surely is present but no specimens seen to date by the writer.
3. **Brassica kaber** (DC.) L. C. Wheeler, Rhodora 40:306. 1938.
   B. arvensis of Manuals; Sinapis arvensis L.---Annual plants; stems 30-60 cm. tall, branching above, more or less hirsute especially below; lower leaves usually lyrate-pinnatifid with a rounded toothed terminal lobe, usually hispid on the veins below; upper leaves oblong or lanceolate, merely toothed; petals 8 mm. long, yellow; pedicels about 5 mm. long, ascending; fruit 3-4 cm. long, ascending, glabrous, the beak flattened and sword-shaped, about 5-10 mm. long.---Fields and waste places. Naturalized from Europe and now widely distributed in the United States. Our records from the northern and western part of Colorado at 4500-8500 feet.
4. **Brassica nigra** (L.) Koch in Roelh., Deut. Fl. ed. 3. 4:713. 1833.
   Annual plants; stems 50-200 cm. tall, erect, branching, sparsely pubescent or glabrous; lower leaves pinnatifid with a large terminal lobe and a few small lateral ones; upper leaves lanceolate to oblong, entire; petals 7-8 mm. long, bright yellow; pedicels in fruit less than 5 mm. long, appressed to the rachis; fruit appressed, 15-20 mm. long and about 1-1.5 mm. thick,

4-sided, beak slender, 2-4 mm. long.---Fields and waste places. Naturalized from Europe and widely distributed in the United States. Our rather few records from eastern half of Colorado at 4500-8000 feet.

5. Brassica juncea (L.) Cosson, Soc. Bot. France Bul. 6:609. 1859.

Annual or perhaps biennial plants; stems 30-120 cm. tall, erect, glabrous or nearly so, branched above; lower leaves lyrate-pinnatifid, long-petioled, with a terminal lobe, the upper lanceolate or linear, sessile or nearly so, mostly entire; petals 8-10 mm. long, yellow; pedicels more than 5 mm. long, in fruit spreading or ascending; fruit 3-6 cm. long and 2.3-5 mm. wide, erect or ascending but not appressed, nearly terete but the valves with a strong midrib, beak 4-8 mm. long.---Fields and waste places. Naturalized from Europe and widely distributed in the United States. Our records from northcentral Colorado at 4500-7500 feet.

33. Conringia (Heist.) Adans. HARE'S EAR MUSTARD

Plants annual, glabrous; stems leafy; leaves sessile with clasping base at least above; sepals narrow; petals pale yellow, narrow; style rather long, stigma entire or nearly so; fruit elongate and narrow, more or less 4-angled, dehiscent; seeds in 1 row in each cell, not winged.

1. Conringia orientalis (L.) Dumort., Fl. Belg. 123. 1827.

Stems 25-60 cm. tall, erect, usually simple; leaves 4-10 cm. long, oval to elliptical, deeply cordate-clasping; petals 8 mm. long, yellowish-white; fruit 8-10 cm. long, about 2 mm. thick, ascending, 4-angled, beak about 1.5 mm. long.---Fields, roadsides and waste places. Naturalized from Europe into Canada and northern United States. Our records scattered over Colorado, none as yet from the southwestern quarter of the state, at 4000-7000 feet.

34. Cardamine (Tourn.) L. BITTERCRESS

Perennial or annual herbaceous plants; leaves entire or pinnately compound; flowers in racemes or corymbs; sepals equal at base; petals white or purple, obovate to spatulate; stamens usually 6; fruit elongated and somewhat compressed parallel with the partition, dehiscent, valves nerveless; seeds wingless, in 1 row

1. Leaves all simple -1. C. cordifolia
1. Leaves (at least some) pinnately compound
  2. Petals 5-6 mm. long; beak of fruit about 2.5 mm. long -2. C. vallicola
  2. Petals 2-4 mm. long; beak of fruit if present not over 1 mm. long
    3. Petals 2-3 mm. long; lateral leaflets of cauline leaves linear to elliptic; stems very leafy; plants annual or biennial -3. C. pennsylvanica
    3. Petals 3-4 mm. long; lateral leaflets of cauline leaves oval to rotund; plants perennial -4. C. umbellata

1. Cardamine cordifolia A. Gray, Amer. Acad. Arts & Sci. Mem. Ser. 2. 4:8. 1849.

C. infausta Greene---Perennial plants with short slender rootstocks; stems 30-75 cm. tall, erect, varies from glabrous to densely pilose; leaves 2-5 cm. long, several, petioled, cordate at base, broadly ovate in shape, repand-dentate; petals 7-12 mm. long, white; fruit 2-4 cm. long, about 1.5 mm. wide, erect on ascending pedicels, beak very short.---Moist or wet ground along streams and near springs. Wyoming to Idaho, south to New Mexico and Arizona. Our records widespread over the western two-thirds of Colorado, except in the northwestern part, at 7000-11,500 feet.

2. Cardamine vallicola Greene, Pitt. 3:116. 1896.

Perennial plants with slender rootstocks; stems 20-40 cm. tall, glabrous, rather sparsely leafy especially above; leaves pinnate with 5-9 leaflets, lateral leaflets ovate, entire or toothed, terminal leaflet larger, somewhat lobed, broadly cuneate to cordate at base, 2-4 cm. wide; petals 5-6 mm. long, white; pod 20-35 mm. long, erect on an ascending pedicel the beak over 2 mm. long. This species is closely related to C. breweri Wats. reported from Colorado. Our plants show numerous intergradations and the writer cannot separate the 2 in this area, in fact the 2 are often found on the same plant.---Moist or wet places. Wyoming to Washington, south to Colorado and California. Our records from northcentral Colorado at 5000-9000 feet.

3. Cardamine pennsylvanica Muhl. ex Willd., Spec. Pl. 3:486. 1800.

Annual or sometimes biennial plants; stems 7-80 cm. tall, glabrous or with a few scattered hairs, simple or branched from the base; leaves rather thin, extending to the racemes, with 5-17 linear to elliptic leaflets, these entire or toothed, seldom over 25 mm. long; petals 2-3 mm. long, white; fruit 1.5-3 cm. long, about 1 mm. wide or less, erect or ascending, beak less than 1 mm. long.---Wet or moist places. Newfoundland to British Columbia, south to Florida and Oregon. Our records from northcentral Colorado at 6500-9000 feet.

Rydberg described C. unijuga Rydb. closely related to the above from Montana and Wyoming. Differs in having only 3 leaflets on stem leaves. Some of our plants may be that species.

4. Cardamine umbellata Greene, Pitt. 3:154. 1897.

Perennial plants with rootstocks; stems 20-50 cm. tall, glabrous or nearly so; leaves pinnate with 3-7 rounded, oval, or oblong divisions or terminal one reniform; flowers in a short corymb, pedicels ascending; petals 3-4 mm. long, spatulate; fruit crowded, 2-3 cm. long, about 1 mm. wide, erect, beak minute. This species intergrades somewhat with the preceding in our plants.---Moist or wet places. Alberta to Alaska, south to Colorado and Oregon. Our few (rather doubtful) records from northcentral Colorado at 8500-9500 feet.

35. Barbarea R. Br. WINTERCRESS

Biennial or perennial herbaceous plants; stems erect; leaves lyrate-pinnatifid; flowers in racemes; sepals erect; petals yellow, clawed; style short, stigma more or less 2-lobed; fruit elongate, linear, at least somewhat 4-angled; seeds in 1 row in each cell, wingless.

1. Basal leaves with 2-6 lateral divisions -1. B. orthoceras
1. Basal leaves with 8-20 lateral divisions -2. B. verna

1. **Barbarea orthoceras** Ledeb., Hort. Dorp. 1824.
B. americana Rydb.; Campe stricta of Manual; C. americana (Rydb.) Cockerell---Biennial plants, glabrous throughout; stems 30-50 cm. tall, rather stout, erect; basal leaves oblong or elliptic in shape, lyrately-pinnatifid with a large terminal division and 2-4 smaller lateral ones; stem leaves lyrate-pinnatifid; petals about 2-5 mm. long, pale yellow; pedicels stout; fruit 3.4-5 cm. long and about 1.5 mm. wide, subterete or slightly compressed, obscurely 4-angled, ascending or appressed, even spreading sometimes, beak 0.5-1.5 mm. long.---Waste places, banks of streams or wet places in general. Labrador to Alaska south to New Hampshire, Colorado and California; Europe. Our records scattered in the western half of Colorado, except in the extreme southern part, at 5000-10,000 feet.

2. **Barbarea verna** (Mill.) Asch., Fl. Brandenb. 1:36. 1864.
B. praecox (J. E. Smith) R. Br.---Waste places. Naturalized from Europe into various parts of the United States. Reported from western Colorado but the record is doubtful.

Family 53. Resedaceae MIGNONETTE FAMILY

Annual or perennial herbaceous plants; stems erect or decumbent; leaves alternate or fascicled, lobed or pinnatifid; flowers perfect, unsymmetrical, small, in narrow racemes; calyx 4- to 7-parted; petals 4-7, distinct, toothed or cleft; stamens 8-30, inserted on 1 side of the flower, hence flower unsymmetrical; ovary superior, 3- to 6-carpellate, 1-celled with 3-6 placentae, styles or stigmas 3-6; fruit a 3- to 6-lobed capsule, horned at apex, opening at top before the seeds mature.

1. Reseda L. MIGNONETTE

Characters of the Family

1. **Reseda lutea** L., Sp. Pl. 449. 1753.
Plants pubescent with short stiff hairs or nearly glabrous; stems 25-60 cm. tall, ascending or decumbent; leaves 5-10 cm. long, oblong in outline, deeply lobed or pinnatifid, the segments linear or oblong, margins undulate; racemes at length elongated; pedicels 3-5 mm. long, ascending; petals 3-4 mm. long, greenish-yellow, 5-6, all but the lowest irregularly cleft; capsule 7-10 mm. long and 3-5 mm. wide, oblong, with 3 or 4 short teeth at apex.---Roadsides and hills. Introduced into various parts of the United States. Our records from northcentral Colorado at 5000-5500 feet where it is locally abundant. Also found recently in El Paso County.

Family 54. Crassulaceae ORPINE FAMILY

Plants herbaceous, annual or perennial, mostly fleshy and succulent; leaves opposite or alternate, simple, usually entire, without stipules; flowers usually perfect, regular, usually cymose but sometimes racemose or solitary; calyx usually 4- to 5-parted or lobed; petals separate or united below, equal in number to the calyx lobes, usually persistent; stamens as many as petals or twice as many; ovaries superior, carpels 3-5, distinct or united at base, styles subulate or filiform; fruit made up of 1-celled follicles dehiscent along the ventral side; seeds few to many.

1. Plants annual; leaves opposite; flowers axillary, minute, the petals not over 2 mm. long; stamens 4 -1. Tillaea
1. Plants perennial; leaves alternate; flowers terminal, larger, the petals 3 mm. long or longer; stamens 8-10 -2. Sedum

1. Tillaea L. PIGMY WEED

Small glabrous aquatic plants; leaves opposite; flowers minute, solitary in the axils of the leaves; sepals usually 4; petals usually 4, distinct or united only at base; stamens usually 4; carpels usually 4, distinct, styles short; seeds few to several in each carpel.

1. **Tillaea aquatica** L., Sp. Pl. 128. 1753.
Tillaeastrum aquaticum (L.) Britt.---Stems up to 10 cm. tall, filiform, branching and decumbent at base, often forming mats; leaves 1-5 mm. long, linear or linear-oblong; flowers sessile or short-pedicelled; sepals 0.5-1.5 mm. long; petals about twice as long as the sepals, greenish-white; fruit about as long as the petals.---In water or on mud, possibly on wet rocks. Newfoundland to Washington, south to Maryland, Mexico and California; Eurasia and Africa. Our few records from central and southcentral Colorado at 7500-9500 feet.

2. Sedum L. STONECROP

Plants succulent, mostly glabrous, perennial; leaves alternate, fleshy, terete or flat, often imbricated; flowers usually perfect but sometimes dioecious or polygamous; calyx 4- to 5-parted; petals as many as sepals, distinct or united at base; stamens twice as many as the petals, the filaments adnate to the perianth at base; carpels 4-5, distinct or united at base, each few- to many-seeded. In addition to the following species, a cultivated species of Sedum with trailing prostrate branches has escaped and become a weed in various parts of the state. No flowering specimens are available for specific identification.

1. Flowers rose-purple to whitish; rootstocks thick and fleshy; leaves usually over 16 mm. long; stems often over 20 cm. tall
   2. Flowers in a spikelike raceme with leafy bracts, petals light rose to whitish; leaves oblanceolate to narrower -1. S. rhodanthum
   2. Flowers in a dense cyme, the flowers unilateral on the short branches, petals rose-purple to greenish-purple; leaves obovate to oblong-obovate (rarely to oblanceolate) -2. S. integrifolium
1. Flowers yellow to white; rootstocks slender; leaves not over 16 mm. long; stems rarely over 20 cm. tall
   3. Leaves linear, terete or nearly so; petals yellow; plants common in Colorado -3. S. stenopetalum
   3. Leaves oblanceolate to obovate, flat; petals yellow or white; plants rare in Colorado
      4. Petals white to faintly pinkish; leaves of flowering stems 5 mm. long or more, not densely imbricated -4. S. cockerellii
      4. Petals bright yellow; leaves of flowering stems about 3 mm. long, densely and conspicuously imbricated -5. S. acre

1. Sedum rhodanthum A. Gray, Am. Journ. Sc. ser. 2, 33:405. 1862.
Clementsia rhodantha (A. Gray) Rose---Plants glabrous with thick rootstocks; stems 10-35 cm. tall, several, simple; leaves 15-30 mm. long, alternate, sessile, linear-oblong to oblanceolate, entire or toothed; flowers in dense terminal spikelike racemes with leaflike bracts (could be considered axillary); sepals distinct; petals 7-10 mm. long, light rose, pink or whitish, linear-lanceolate, about twice as long as the sepals; stamens 10, these opposite the petals and adnate to them; carpels 5, erect.---Moist ground of meadows and along lakes and streams. Montana to Utah, south to New Mexico and Arizona. Our records well scattered over the western half of the state at 9000-13,000 feet.

2. Sedum integrifolium (Raf.) A. Nels. in Coulter and Nelson, Man. 233. 1909.
Rhodiola integrifolia Raf.; R. polygama (Rydb.) Britt. & Rose---Thick, red, fleshy, branched rootstocks present; stems 5-30 cm. tall; leaves 1-2.5 cm. long, alternate, sessile, obovate to oblong-obovate occasionally oblanceolate, acute, entire or dentate above the middle, flat and only moderately thick; flowers polygamous or dioecious, in dense terminal cymes; sepals about 2 mm. long, lanceolate, 4-5 in number; petals distinct, about twice as long as sepals, dark rose-purple or rarely greenish-purple, acute or acuminate; stamens longer than petals, the filaments purplish; carpels 3-8 mm. long, beak usually spreading or recurved.---Moist places in the mountains. Alberta to Alaska, south to Colorado and California. Our records scattered over western Colorado, none as yet from the northwestern part, at 9000-13,000 feet.

3. Sedum stenopetalum Pursh, Fl. Am. Sept. 324. 1814.
Plants tufted, with slender rootstocks; stems glabrous, the flowering ones 5-20 cm. tall; leaves 6-16 mm. long, alternate, sessile, crowded but hardly imbricate except on the sterile shoots, linear, terete or nearly so; flowers in a compact forked cyme, each pedicelled and unilateral; sepals about 4 mm. long, lanceolate, sometimes to ovate; petals 6-9 mm. long, yellow, lanceolate, acute or acuminate; stamens 8-10; carpels about 4 mm. long, erect with divergent tips.---Commonest on rocky slopes and ridges. Saskatchewan and Alberta, south to Nebraska, New Mexico and Arizona. Our many records well scattered over the western two-thirds of Colorado at 5000-12,500 feet.

4. Sedum cockerellii Britt., Bull. N. Y. Bot. Gard. 3:41. 1903.
Reported from Colorado, New Mexico and Arizona. The writer has seen no specimens from this state.

5. Sedum acre L., Sp. Pl. 432. 1753.
Stems densely tufted, those of sterile branches prostrate and matted, glabrous, flowering stems 3-8 cm. tall, erect or ascending; leaves alternate, sessile, ovate, very thick, densely imbricated and less than 3 mm. long at least on flowering shoots; flowers in cymes, these usually 2- to 3-forked, the flowers sessile; petals 5-6 mm. long, yellow, linear-lanceolate; carpels spreading.---Sometimes escaping from cultivation. Apparently this does not happen very often in Colorado, our one record from Clear Creek County at 7500 feet.

Family 55. Saxifragaceae SAXIFRAGE FAMILY

Perennial plants, herbaceous or shrubby; leaves simple (may be lobed), alternate, opposite, or basal; flowers usually perfect, regular or nearly so, solitary, racemose or in cymes, usually with a well developed hypanthium and often with a disk; sepals 4-5, distinct or united; petals 4-5, separate, sometimes wanting; stamens variable in number, usually as many as petals, twice as many or numerous; pistil 1 and compound, or the carpels nearly distinct, the ovary inferior to nearly or quite superior; fruit of follicles, a capsule or berry.

This is a family with several diverse elements and has been divided into 3 or 4 families by some authors. However, when this is done at least 1 of the families remains almost as diverse as the original.

1. Plants woody, prostrate or upright shrubs
  2. Leaves alternate; fruit a berry; spines often present   -1. Ribes
  2. Leaves opposite; fruit a capsule; spines never present
    3. Petals 4 (only rarely 5); stamens either 8 or many, never 10
      4. Stamens 8; ovary inferior only at base (less than 1/2)   -2. Fendlera
      4. Stamens 20 or more; ovary at least 2/3 inferior   -3. Philadelphus
    3. Petals 5; stamens 10
      5. Leaves entire, small (not over 1.5 cm. long), sessile or nearly so; flowers small, petals 2.5-4 mm. long   -4. Fendlerella
      5. Leaves toothed, over 1.5 cm. long, definitely petioled; flowers larger, the petals 8-12 mm. long   -5. Jamesia
1. Plants herbaceous, not woody, at least above the base
  6. Stamens 10
    7. Ovary 2-celled with axile placentae; petals entire, not lobed (may be emarginate at apex or slightly angled)   -6. Saxifraga
    7. Ovary 1-celled with 3 parietal placentae; petals 3- to 7-parted   -7. Lithophragma
  6. Stamens 4 or 5 (may be clusters of staminodia between)
    8. Clusters of staminodia alternating with fertile stamens; flowers solitary on the stems; ovary 1-celled with 3 or 4 parietal placentae; leaves entire   -8. Parnassia
    8. No staminodia present; flowers in clusters; ovary either 2-celled or 1-celled with 2 parietal placentae; leaves toothed or lobed
      9. Stems leafy, the flowers clustered in the axils of the upper leaves; stamens and sepals 4; petals lacking   -9. Chrysosplenium
      9. Stems scapose, the leaves nearly or quite basal, flowers terminal; stamens and sepals 5; petals present (may be small or deciduous)
        10. Ovary 2-celled with axile placentae   -10. Sullivantia
        10. Ovary 1-celled with 2 parietal placentae
          11. Inflorescence a raceme; petals pinnatifid to 3-toothed (only rarely entire); stamens alternate or (in our common species) opposite to the petals   -11. Mitella
          11. Inflorescence a panicle (may be dense); petals entire; stamens alternate to the petals   -12. Heuchera

1. Ribes L. GOOSEBERRY; CURRANT

Shrubs with erect, ascending or prostrate branches, often with bristles and spines; leaves alternate, deciduous, palmately veined and palmately lobed, the stipules none or adnate to the petioles; flowers racemose or rarely solitary, on short axillary branches, pedicels subtended by a bract and usually with 2 bractlets at about the middle; calyx tube adnate to the globose ovary and more or less produced above it; calyx lobes 5 or rarely 4; petals 5 rarely 4, usually smaller than calyx lobes, and alternate to them, erect; ovary inferior, 1-celled, with 2 parietal placentae; styles 2, more or less united, stigmas terminal; fruit a berry crowned by withered remains of the flower; seeds several to many.

1. Plants with stiff bristles or spines (often both) on some or all of the twigs
  2. Calyx tube (above ovary) short, not over 2 mm. long, bowl-shaped or saucer-shaped; twigs often with bristles; branches ascending to procumbent
    3. Leaf blades nearly or quite glabrous, often over 3 cm. wide; berry purple-black   -1. R. lacustre
    3. Leaf blades pubescent, not over 3 cm. wide; berry red   -2. R. montigenum
  2. Calyx tube campanulate to cylindrical, over 2 mm. long; twigs rarely with bristles; branches ascending to erect
    4. Style glabrous throughout; calyx tube 4-8 mm. long, long-pubescent   -3. R. leptanthum
    4. Style hairy toward the base; calyx tube 2.5-8 mm. long but if over 4 mm. long then glabrous
      5. Calyx tube campanulate, 2.5-3.5 mm. long, about as long as the calyx lobes; stamens about twice as long as the petals or longer   -4. R. inerme
      5. Calyx tube cylindrical, 5-8 mm. long, about twice as long as the calyx lobes; stamens about equalling the petals   -5. R. setosum
1. Plants lacking stiff bristles or spines
  6. Calyx tube (above ovary) saucer-shaped or bowl-shaped, not over 2 mm. long
    7. Calyx lobes linear to oblong-ovate, definitely longer than wide; calyx tube 1-1.5 mm. long; stems stout, not trailing or prostrate   -6. R. wolfii
    7. Calyx lobes about as wide as long; calyx tube 0.5-1 mm. long; stems weak, prostrate or reclining   -7. R. coloradense
  6. Calyx tube cylindrical, 3 mm. long or longer
    8. Calyx tube 3-4 mm. long; lower leaf surface with sessile orange or yellow dots under a lens   -8. R. americanum
    8. Calyx tube 5-10 mm. long; lower leaf surface glabrous to glandular-pubescent, no sessile orange or yellow dots present
      9. Calyx a bright conspicuous yellow, glabrous; leaves glabrous, vernation convolute; ovary glabrous   -9. R. aureum
      9. Calyx greenish, whitish, or pinkish, hairy; leaves usually hairy, with plicate vernation; ovary glandular-pubescent
        10. Calyx tube (above the ovary) narrow, 3-4 times longer than wide; leaf blades 1-3 cm. wide, sparingly if at all glandular-pubescent; fruit red, fleshy   -10. R. cereum
        10. Calyx tube broad, about twice as long as wide; leaf blades 3-7 cm. wide, glandular-pubescent to glandular-hirsute; fruit black, almost dry   -11. R. viscosissimum

1. Ribes lacustre (Pers.) Poir. in Lam., Encycl. Suppl. 2:856. 1812.

R. parvulum (Gray) Rydb.; Limnobotrya parvula (A. Gray) Rydb.---Shrubs mostly less than 100 cm. tall, with stems ascending or procumbent; branches often hispid-bristly; spines about 2.5 mm. long, slender, weak, usually clustered or rarely absent; leaves 2.5-5 cm. wide, nearly orbicular, 3- to 5-lobed or cleft, the divisions toothed and incised, cordate at base, nearly or quite glabrous; petioles slender, glandular-pubescent; flowers green or purplish; inflorescence loosely 5- to 12-flowered; pedicels slender, about 4-6 mm. long, bracts small; calyx tube bowl-shaped or saucer-shaped, short, about 1 mm. long; calyx lobes about 1.5 mm. long, short and broad, spreading; petals about half as long as calyx lobes; berry about 6-10 mm. in diameter, purple-black, covered with gland-tipped bristles.---In the mountains. Newfoundland to Alaska, south to Pennsylvania, Michigan, Colorado and California. Our records scattered in the western half of Colorado, mostly from the northcentral and southwestern parts, at 8000-10,000 feet.

2. Ribes montigenum McClatchie, Erythea 5:38. 1897.

R. lentum (Jones) Cov. and Rose; Limnobotrya montigena (McClatchie) Rydb.---Low freely branching shrubs 30-60 cm. tall, rarely to 100 cm., with ascending or procumbent branches; stems more or less bristly; nodal spines varying from very short to long; leaves 1.5-3 cm. wide, suborbicular, more or less 5-lobed or 5-cleft, the divisions incised and toothed, usually densely glandular-pubescent, sometimes sparingly so; petioles hairy; flowers greenish-white to rose-purple; raceme 4- to 8-flowered; pedicels short, about 3 mm. long, bracts small; calyx tube short, about 1.5-2 mm. long, bowl-shaped or saucer-shaped; calyx lobes about 2-3 mm. long, rather broad; petals usually about 1/2 as long as the calyx lobes; berry about 6-10 mm. in diameter, red, covered with gland-tipped bristles.---Open slopes and ridges in the mountains. Montana to British Columbia, south to New Mexico and California. Our records scattered in the western half of Colorado at 7500-11,500 feet.

3. Ribes leptanthum A. Gray, Am. Acad. Arts & Sci. Mem. Ser. 2, 4:53. 1849.

Grossularia leptantha (A. Gray) Cov. & Britt.---Erect shrubs 50-200 cm. tall, branches sometimes bristly; nodal spines about 5-15 mm. long, single or triple, rather stout; leaves 5-20 mm. wide, 3- to 5-cleft about 1/2 way in, orbicular or reniform-orbicular in outline, the lobes usually 3-lobed or dentate, glabrous to pubescent, truncate to cordate at base; racemes 1- to 3-flowered, peduncles short; pedicels very short, bracts small; calyx tube 4-8 mm. long, cylindrical, greenish-white, long-pubescent; calyx lobes somewhat shorter to about as long as the tube, greenish-white; petals shorter than the calyx lobes, white to pink; fruit about 6-8 mm. in diameter, blackish, glabrous or rarely glandular, hispid or bristly.---In the mountains, often in dry places. Colorado, Utah, New Mexico and Arizona. Our records scattered in the mountainous western two-thirds of Colorado, except the northwestern corner, at 5500-12,000 feet.

4. Ribes inerme Rydb., N. Y. Bot. Gard. Mem. 1:202. 1900.

Grossularia inermis (Rydb.) Cov. & Britt.; G. purpusi Koehne; R. vallicola Greene; R. saxosum of Manuals at least in part---Erect shrubs about 100 cm. tall; stems smooth or setose, rarely with a few bristles; nodal spines single or triple at each node, up to 1 cm. long, sometimes lacking; leaves 1-6 cm. wide, orbicular to reniform-orbicular in outline, 3- to 5-lobed, the lobes dentate, glabrous or pubescent; petioles about as long as the blades; flowers 1-4, on short peduncles, bracts very small; calyx tube about 2.5-3.5 mm. long, bell-shaped, glabrous or rarely sparsely pilose; calyx lobes about as long as the tube, glabrous or sparsely pilose especially along the margins, calyx greenish or greenish-white; petals pink or white, about 1/3 as long as the calyx lobes; fruit about 8 mm. in diameter, glabrous, wine-colored. The form with downy branches and pubescent petioles, leaf blades, peduncles and bracts has been named var. pubescens Berger but hardly seems worth maintaining in Colorado plants.---Mostly in the mountains. Montana to British Columbia, south to New Mexico and California. Our records scattered in the western three-fourths of Colorado, mostly in the mountains, at 5000-11,000 feet.

5. Ribes setosum Lindl., Trans. Hort. Soc. London 7:243. 1828.

Grossularia setosa (Lindl.) Cov. & Britt.; R. saximontanum E. Nels.---Reported from Saskatchewan south to Wyoming, Idaho and possibly northern Colorado. No specimens from the state were seen by the writer.

6. Ribes wolfii Rothr., Amer. Nat. 8:358. 1874.

Shrubs 100-150 cm. tall or less sometimes, without bristles or spines; stems not trailing, glabrous, young shoots pubescent; leaves 4-9 cm. wide, suborbicular in outline, cordate at base, 3- to 5-lobed, the lobes irregularly toothed, glabrous above, usually somewhat pubescent on the veins and with minute sessile glands; petioles slender, rather long, glandular-pubescent; racemes erect, 4- to 8-flowered, glandular-pubescent on the rachis, pedicel and bracts; pedicels 5-8 mm. long about as long as bracts; flowers greenish-white to reddish; calyx tube very short, about 1-1.5 mm. long, bowl-shaped or saucer-shaped; calyx lobes 3-4 mm. long, linear, oblong or oblong-ovate; petals shorter than the calyx lobes; ovary glandular-pubescent; berry about 8-12 mm. in diameter, black, glandular-bristly.---In the mountains, often in partial shade. Colorado, Utah, New Mexico and Arizona. Our records scattered in the western half of Colorado, except in the northwestern part, at 6500-11,500 feet.

7. Ribes coloradense Coville, Proc. Biol. Soc. Wash. 14:3. 1901.

Stems prostrate or reclining, no bristles or spines present, young shoots puberulent; leaves 4-8 cm. wide, wider than long, thin, cordate-reniform, mostly 5-lobed, the lobes incised or toothed, glabrous or sometimes minutely pubescent below on the veins, and with sessile glands; petioles long, puberulent, rather stout; racemes erect, 6- to 12-flowered; pedicels 4-15 mm. long, longer than minute bracts (lower bracts rarely leaflike); flowers greenish to purplish; calyx tube 0.5-1 mm. long; calyx lobes 3-4 mm. long, and as broad or broader than long, spreading, glandular-hairy; petals shorter than calyx lobes; ovary glandular-pubescent; berry about 10 mm. in diameter, sparingly glandular-hairy, black without a bloom.---Mountains. Colorado, New Mexico and Utah. Our records scattered over the western half of Colorado, except in the northwestern part, at 8000-11,500 feet.

8. **Ribes americanum** Mill., Gard. Dict. ed. 8 no. 4. 1768.
R. floridum L'Her.---Shrubs 100-150 cm. tall, with more or less erect branches these lacking bristles or spines, young shoots somewhat pubescent and with sessile yellow or orange dots; leaves 3-8 cm. wide, chiefly 3- to 5-lobed, the divisions usually acute and serrate-dentate, suborbicular or reniform-orbicular in outline, cordate to broadly cuneate at base, usually glabrous above, more or less pubescent below at least on the veins and with sessile orange or yellow dots; petioles slender, about as long as the blades; racemes drooping, about as long as the leaves, 5- to 10-flowered; bracts linear to linear-lanceolate, usually longer than the pedicels; calyx tube 3-4 mm. long, tubular but enlarged at base, yellowish-green to greenish-white; calyx lobes same color as tube and a little longer; petals about 2/3 as long as the calyx lobes; berry about 6-10 mm. in diameter, black, glabrous.---Woods and thickets. Nova Scotia to Alberta, south to Virginia and New Mexico. Apparently rare, our few records from the vicinity of Denver at 5000-5500 feet.

9. **Ribes aureum** Pursh, Fl. Am. Sept. 164. 1814.
R. longiflorum and R. longifolium of Manuals; Chrysobotrya aurea (Pursh) Rydb.---Erect shrubs 100-300 cm. tall, somewhat taller in shaded areas, spineless, glabrous throughout or leaves and inflorescences pubescent; leaves up to 5 cm. wide, reniform-orbicular to obovate in outline, 3- to 5-lobed, the lobes subobtuse, often somewhat toothed, cordate to cuneate at base; petioles shorter to about as long as the blades; racemes leafy-bracted, 2- to 15-flowered, rarely solitary, on short pedicels; flowers fragrant; calyx tube 5-10 mm. long, cylindrical; calyx lobes 5-8 mm. long, oblong, spreading in anthesis, the whole calyx bright yellow; petals short, about 2 mm. long, erose, yellow or tipped with red; fruit 6-10 mm. wide, globose, glabrous, black or sometimes red to yellowish. This is a variable species in this region.---Plains, foothills and mountains, rather prevalent along roadsides. South Dakota, Saskatchewan and Washington, south to New Mexico and California. Our records well scattered over the entire state at 3500-8000 feet.

10. **Ribes cereum** Dougl., Trans. Hort. Soc. London 7:512. 1830.
R. pumilum Nutt. of Manuals; R. inebrians Lindl.---Shrubs 50-200 cm. tall, much-branched, without spines or bristles, young twigs pubescent; leaves 1-3.5 cm. wide, reniform-orbicular, cordate to truncate at base, 3- to 5-lobed, the lobes rather shallow and rounded, crenate, puberulent and more or less glandular-hairy to almost glabrous; petioles sometimes as long as the blades; racemes few-flowered, pendulous, puberulent and usually glandular; bracts longer than the pedicels, oblong, rhombic or cuneate-obovate, entire, toothed or lobed at apex; calyx tube 5-8 mm. long and 3-4 times longer than wide, cylindrical but dilated at base, usually pink, pubescent; calyx lobes about 2 mm. long, ovate; petals about 1/2 as long as the calyx lobes, whitish; ovary glandular-pubescent; style glabrous or hairy above; berry 6-8 mm. in diameter, red, usually glabrous. R. inebrians is not distinct from the above species in Colorado plants. Some plants of northwestern Colorado have very densely glandular leaves.---Dry slopes, ridges and plains. Montana to British Columbia, south to New Mexico and California. Our records scattered over Colorado at 4000-11,000 feet.

11. **Ribes viscosissimum** Pursh, Fl. Am. Sept. 163. 1814.
Shrubs 80-150 cm. tall, branches without bristles or spines but young twigs puberulent and glandular-pubescent; leaves 3-7 cm. wide, reniform-orbicular in outline, cordate at base, 3- to 5- shallowly lobed, the divisions irregularly crenate-dentate and rounded at apex, glandular-pubescent (almost glandular-hirsute) on both sides; petioles short and stout, dilated at base, glandular-hairy; racemes spreading or ascending, few-flowered; pedicels not over 1 cm. long, about equalling the spatulate or oblanceolate bracts, both glandular hairy; calyx tube 5-7 mm. long and about 2.5-3.5 mm. wide, cylindrical to almost campanulate, greenish to pink; calyx lobes nearly or quite as long as the tube (rarely not much over 1/2 as long), greenish-white to pink; petals half as long as the calyx lobes; ovary glandular-hairy; fruit about 1 cm. in diameter, black, glandular-pubescent, almost dry.---Mountains, often in shade. Montana to British Columbia, south to Colorado and California. Our one record from Routt County at 6700 feet.

2. Fendlera Engelm. & Gray   FENDLERBUSH

Shrubs with intricate branching; leaves opposite, nearly sessile, without stipules, deciduous, entire or nearly so; flowers solitary or 2 or 3 together, perfect, rather large; calyx lobes 4, longer than the 8-ribbed calyx tube; petals 4, white, clawed, blades ovate-deltoid and erose; stamens 8, filaments flattened and 2-forked at apex, the lobes prolonged beyond the anthers; ovary inferior at base, broad, 4-celled, styles 4, stigmas minute, nearly terminal; capsule over 1/2 superior, septicidal; seeds large, few in each cell.

1. **Fendlera rupicola** A. Gray, Pl. Wright. 1:77. 1852.
F. tomentella Thornb.; F. falcata Thornb.---Branching shrubs to 200 cm. tall, twigs and branches reddish to straw-colored, glabrate to pubescent, becoming gray, the bark shredding; leaves 5-40 mm. long, very variable in size, shape and pubescence, linear to oblong, margins often revolute, glabrate to strigose both sides, sometimes whitish below; calyx lobes 3-6 mm. long in flower, elongating finally, longer than the tube; petals 12-20 mm. long; capsule body 9-15 mm. long. A rather variable species especially in relative pubescence of the leaves. Varieties have been separated on this character but Colorado plants exhibit all types of intergradation.---Often on rocky slopes. Colorado, south to Texas and Arizona. Our records from westcentral and southwestern Colorado at 4000-8000 feet.

3. Philadelphus L. MOCKORANGE; SYRINGA

Freely branching shrubs; leaves opposite, entire or toothed; flowers perfect, rather showy, solitary or in few-flowered cymes terminating the leafy branches; calyx 4- or rarely 5-parted, the tube adnate to most of the ovary; petals 4, rarely 5, white or ochroleucous; stamens many (20-60); ovary at least 2/3 inferior, 4-celled, (rarely 3 or 5), styles 3-5, distinct or united; capsule loculicidal, more or less leathery or woody; seeds numerous.

1. Petals 11-17 mm. long; leaves mostly 17-35 mm. long; calyx tube glabrate to sparsely pubescent, the hairs usually thicker on the angles -1A. P. microphyllus microphyllus
1. Petals 9-11 mm. long; leaves mostly 10-16 mm. long; calyx tube glabrate to sparsely pubescent, the hairs not usually confined to the angles -1B. P. microphyllus occidentalis

1. Philadelphus microphyllus Gray, Mem. Am. Acad. II. 4:54. 1849.
Much branched shrubs, 80-200 cm. tall; young branches appressed-pubescent; bark reddish-brown to tan, usually exfoliating the second year; petioles 1-3 mm. long; leaf blades 8-35 mm. long, ovate to lanceolate, appressed-pubescent to nearly glabrate; flowers solitary to in 3's; calyx lobes 3-5 mm. long; calyx tube 2-3 mm. long, glabrate to appressed pubescent, enlarging in fruit; petals 9-17 mm. long, white to cream-colored; styles united or sometimes free near apex. Represented in Colorado by the two following sub-species, which intergrade rather commonly except in northern Colorado.
1A. Philadelphus microphyllus microphyllus
P. microphyllus A. Gray; P. microphyllus ssp. typicus Hitchc.---Leaves mostly 17-35 mm. long; petals 11-17 mm. long; calyx tube glabrate or sparsely pubescent usually on the angles.---Dry hills and cliffs. Colorado, south to Texas and Arizona. Our records mostly from southern Colorado, a few from the westcentral part, at 5500-8000 feet.
1B. Philadelphus microphyllus occidentalis (Nelson) Hitchc., (ssp.) Madrono 7:51. 1943.
P. occidentalis A. Nels.; P. minutus Rydb.; P. nitidus A. Nels.---Leaves mostly 10-16 mm. long; petals 9-11 mm. long; calyx tube sparsely pubescent to glabrate the hairs usually not confined to the angles.---Dry canyons and hillsides. Wyoming, Utah and Colorado. Our records scattered over the western half of Colorado at 5000-8000 feet.

4. Fendlerella Heller

Low shrubs with bark somewhat shreddy; leaves opposite, entire, nearly sessile; flowers perfect, small, in small compound cymes; calyx lobes 5, shorter than the turbinate-campanulate tube; petals 5, white; stamens 10, alternately longer and shorter, filaments dilated below the slender apex; ovary conical, inferior at base (about 1/2), 3-celled, styles 3; capsule narrowed at base, 3-valved, septicidal; seeds 1 in each cell.

1. Fendlerella utahensis (S. Wats.) Heller, Bull. Torrey Club 25:626. 1898.
Diffusely branched shrubs to 100 cm. tall; twigs strigose; leaves numerous, fascicled on older twigs, oblong or oblong-spatulate, strigose, 5-15 mm. long, inconspicuously 3-nerved; cymes several-flowered, terminating leafy branches; calyx lobes 1-1.5 mm. long, the tube as long or longer, strigillose, united to base of ovary; petals 2.5-4 mm. long, oblong or elliptic; capsule about 3.5-4 mm. long.---Canyons and rocky cliffs. Reported for southern Utah and adjacent Arizona and California. Our records from northwestern Colorado at 5000-8000 feet.

5. Jamesia Torr. & Gray CLIFFBUSH; WAXFLOWER

Shrubs with erect or ascending stems and exfoliating bark; leaves opposite, deciduous, toothed; inflorescence cymose; flowers perfect, rather large; calyx 5-lobed, the tube hemispheric or turbinate, united to lower part of ovary; petals 5, white; stamens 10, filaments narrow, the alternate ones shorter; ovary conical, about half inferior, finally 1-celled but partially 3- to 5-celled at first; styles 3-5, distinct, stigmas obtuse; fruit a half-inferior capsule, 3- to 5-valved, with slender beaks; seeds numerous.

1. Jamesia americana T. & G., Fl. N. Amer. 1:593. 1840.
Edwinia americana (T. & G.) Heller---Shrubs about 50-200 cm. tall, young twigs pubescent, branches with reddish bark, this exfoliating; leaves 1.5-6 cm. long, ovate, oval or rarely suborbicular, serrate, green and glabrate above, pale or white-tomentulose below, rather short-petioled; flowers in rather small to moderate terminal cymes, the branches, pedicels and calyx white-canescent; calyx becoming 4-6 mm. long at maturity, lobes about equal to the tube; petals 8-12 mm. long, obovate to oblong-cuneate, somewhat hairy within; body of capsule 4-4.5 mm. long.----Canyons, slopes and cliffs. Wyoming and Utah to New Mexico and Arizona. Our records from northcentral, central and southcentral Colorado at 5500-10,000 feet.

6. Saxifraga L. SAXIFRAGE

Herbaceous perennial plants; stems leafy or scapose; leaves alternate or basal; flowers rather small, perfect, usually regular, solitary, paniculate or in simple or compound cymes; calyx 5-lobed or 5-parted, free or adnate in some degree to the base of the ovary; petals 5 (rarely wanting), with or without claws; stamens 10; ovary 2-celled with axile placentae, the carpels completely united or only at base, varying from superior to partly inferior, styles 2; follicles (or carpels of capsule) beaked, these beaks usually divergent; seeds numerous.

1. Foliage leaves all basal; flowers borne on scapes naked below the inflorescence; petals white; calyx tube short or absent, never as long as the lobes (but read second number 1)
   2. All the flowers represented by clusters of bulblets or with 1 terminal flower only; petals often very irregular -1. S. vreelandii
   2. No bulblets present; petals regular or nearly so
      3. Filaments clavate, wider above; leaves suborbicular to reniform, as wide as long -2. S. arguta
      3. Filaments not clavate, wider below if anywhere; leaves ovate, elliptic or narrow, longer than wide
         4. Cymules wholly or mainly aggregated into heads or a spikelike panicle (at least at first); leaves 2-6 cm. long, margins entire to crenate or dentate; plants common -3. S. rhomboidea
         4. Cymules in moderate to narrow panicles; leaves 6-20 cm. long, remotely glandular-denticulate; plants uncommon in Colorado
            5. Panicle narrow, the peduncles remaining very short; petals clawless -4. S. montanensis
            5. Panicle becoming wide; petals with broad clawlike bases -5. S. oregana
1. Foliage leaves not all basal, stems leafy below the inflorescence (may have only 1 or 2 leaves in S. adscendens and S. caespitosa); petals white, rose-purple or yellow, calyx tube very short to long (in 2 species with 1 to few stem leaves calyx tube is as long or longer than the lobes)
   6. Petals rose-purple, with long claws nearly as long as the blades; leaves round-reniform, doubly crenate but hardly lobed -6. S. jamesii
   6. Petals white to yellow, claws absent or short; leaves various, if round-reniform then definitely lobed
      7. Leaves suborbicular to reniform, little if any longer than wide, 5- to 7-palmately lobed; petals white; calyx tube over 1.5 mm. long
         8. Flowers below the terminal one replaced by clusters of bulblets, the whole spikelike; stems more or less glandular-pubescent -7. S. cernua
         8. No bulblets present, flowers in an open cyme; stems glabrous or nearly so -8. S. debilis
      7. Leaves narrower than suborbicular, definitely longer than wide, entire or 3-toothed or 3-lobed at apex; petals white or yellow; calyx tube long or very short to nonexistent
         9. Plants with long, filiform naked stolons, these leafy and rooting at the tip; petals yellow -9. S. flagellaris
         9. Plants not stoloniferous; petals white to yellow
            10. Calyx tube long, usually becoming as long or longer than the lobes; stem leaves 1 to few, at least some of leaves 3-toothed or 3-lobed at apex; petals white, not over 3.5 mm. long
               11. Leaves entire to merely 3-toothed at apex -10. S. adscendens
               11. Leaves (at least lower) 3-lobed or 3-cleft at apex -11. S. caespitosa
            10. Calyx tube very short or absent; stem leaves several to many, all leaves entire; petals yellow (or white in S. bronchialis), 4 mm. long or more
               12. Petals white, often spotted with orange or purple dots; leaves spine-tipped and spiny-ciliate -12. S. bronchialis
               12. Petals yellow, not dotted; leaves not spine-tipped or spiny-ciliate
                  13. Petals 7-13 mm. long, without claws, elliptic to oblong in shape; leaves crowded but hardly rosulate on the caudices, mostly over 10 mm. long; plants 6-20 cm. tall -13. S. hirculus
                  13. Petals 5.5-7 mm. long, abruptly short-clawed at base, oval to suborbicular in shape; leaves rosulate on the caudices, seldom over 10 mm. long; plants 3-8 cm. tall -14. S. chrysantha

1. Saxifraga vreelandii (Small) Fedde in Just, Bot. Jahesb. 33 pt. II:613. 1906.
   Spatularia vreelandii Small---High mountains. British Columbia to Colorado. No specimens from this state seen by the writer.
2. Saxifraga arguta D. Don, Trans. Linn. Soc. 13:356. 1822.
   Miscanthes arguta (D. Don) Small---Acaulescent plants; leaves with blades 3-9 cm. in diameter, long-petioled, suborbicular to orbicular-reniform, crenate-dentate with gland-tipped teeth, cordate at base, glabrous or nearly so; scapes 20-50 cm. tall, glabrous or nearly so at least below, but commonly glandular-pubescent above and in inflorescence; inflorescence an open panicle or cymules; pedicels and upper branches commonly purplish and glandular-pubescent; sepals 1.5-3 mm. long, oblong to lanceolate, usually purplish-tinged, reflexed; petals 2.5-3.5 mm. long, white with yellowish tinge below, blades suborbicular with narrow claws; stamens clavate and often abortive; ovary almost completely superior; capsule 4-8 mm. long, the carpels usually united for about 1/2 their length.---Moist ground. Montana to Alaska, south to New Mexico and California. Our records scattered over the western half of Colorado, except in the northwestern part, at 9000-11,000 feet.
3. Saxifraga rhomboidea Greene, Pitt. 3:343. 1898.
   Miscanthes rhomboidea (Greene) Small; M. austrina (A. Nels.) Rydb.; S. austrina A. Nels.---Acaulescent plants; leaves 2-6 cm. long (including the rather short petioles which are seldom longer than the blades), in a basal rosette, blades rhombic-ovate to ovate, abruptly or narrowly cuneate, obtuse, sometimes entire but usually crenate or dentate, glabrous except margins more or less ciliate; scapes 10-30 cm. tall, glandular-pubescent (rarely glabrate); inflorescence at first rather capitate, later somewhat interrupted-thrysiform; sepals about 1.5-3.5 mm. long, oval to ovate, erect; petals 2.5-4 mm. long, white, oblong-obovate, often emarginate at apex and clawed at base; stamens not clavate; ovary united to calyx tube at base; capsule 3-4.5 mm. long, the carpels joined about 1/2, often purplish.---Rather moist ground. Montana, south to New Mexico. Our records scattered in the western half of Colorado, except in the northwestern part, at 5500-13,000 feet.
4. Saxifraga montanensis Small, Bull. Torr. Club 23:367. 1896.
   Micranthes brachypus Small; S. subapetala E. Nels.---This plant is closely related to S. rhomboidea Greene and probably is only a form of that species. The cymules are in looser panicles and the leaves are over 6 cm. long.

5. Saxifraga oregana Howell, Erythea 3:34. 1895.

Miscanthes arnoglossa Small---This plant has cymules in moderate to fairly narrow panicles, the clusters on fairly long peduncles 1-5 cm. long; leaves 6-20 cm. long. Intergrades somewhat with S. rhomboidea Greene in Colorado and may not be distinct.

6. Saxifraga jamesii Torr., Ann. Lyc. N. Y. 2:204. 1827.

Telesonix jamesii (Torr.) Raf.; Boykinia jamesii (Torr.) Engler---Plants caespitose with woody caudices and thick scaly rootstocks; stems 10-15 cm. tall, glandular-pubescent; radical leaves 2-3.5 cm. wide, long-petioled, round-reniform, deeply and doubly crenate, cordate to broadly cuneate at base; lower stem leaves similar with petioles shorter, upper subsessile; inflorescence a crowded paniculate cyme; calyx tube campanulate, adnate to the base of the ovary; sepal lobes 4-5 mm. long; petals red-rose to purple, long-clawed, with orbicular beaks, the whole about 7-10 mm. long; carpels united up to and including part of the styles, partly inferior; blades of fruit not divergent.---Among rocks. Limited to Colorado. Our records from central and northcentral Colorado at 9000-12,000 feet.

7. Saxifraga cernua L., Sp. Pl. 403. 1753.

S. simulata Small---No caudex developing; stems 6-15 cm. tall, somewhat gregarious, more or less glandular-pubescent; leaves with blades 8-10 mm. wide, reniform in outline, shallowly or prominentaly 5- to 7-lobed palmately, scattered on the stems, the lower and basal cauline leaves with rather long petioles; flowers represented by axillary clusters of bulblets except the terminal one; calyx tube 1.5-3 mm. long; calyx lobes 2-4 mm. long, oblong-lanceolate to oblong-ovate; petals 6-8 mm. long, white, cuneate or panduriform, clawless; stamens with filaments not clavate; ovary inferior at base; capsule with the carpels united to well above the middle.---Among rocks. Greenland to Alaska, south to New Mexico and Utah; Eurasia. Our records scattered in the high mountains of Colorado at 10,000-14,000 feet.

8. Saxifraga debilis Engelm. Proc. Acad. Phila. 1863:62. 1863.

No caudex developing; stems 4-14 cm. tall, glabrous or nearly so, loosely tufted and weak; leaf blades 6-17 mm. wide, suborbicular or reniform, thin, usually 5-palmately lobed, stem leaves lobed or entire, basal and lower cauline leaves long-petioled; inflorescence of 1-3 flowers; calyx tube 2.5-4 mm. long; calyx lobes 2-3 mm. long, shorter than the tube, ovate or oblong-ovate; petals 3-7 mm. long, white, oblong, somewhat clawed at base; filaments not clavate; ovary partly inferior (calyx tube all adnate); capsule 5-7 mm. long.---Among rocks. Montana, south to Colorado and Utah. Our records scattered in the mountains of Colorado at 9500-14,000 feet.

9. Saxifraga flagellaris Willd. in Sternb., Rev. Saxifr. 25. 1810.

Leptasea flagellaris (Willd.) Small---Stems 6-15 cm. tall, glandular-pubescent at least above; plants bearing long, filiform naked stolons from near base, these leafy and rooting at tip; leaves of caudices crowded-rosulate, blades oblong-spatulate, entire, glandular-ciliate, spine-tipped, stem leaves similar and rather numerous clear to top; inflorescence of 1 to few flowers; sepals 4-5.5 mm. long, ovate to oblong-ovate, glandular-pubescent, obtuse; petals 6-9 mm. long, yellow, obovate, short-clawed; filaments not clavate; ovary somewhat inferior at base; capsule about 5-6 mm. long, carpels united above the bases.---High mountains, often among rocks. Greenland to Alaska, south to New Mexico and Arizona, Eurasia. Our records scattered in the high mountains of Colorado, mostly in the southern part, at 10,000-13,500 feet.

10. Saxifraga adscendens L., Sp. Pl. 405. 1753.

Muscaria adscendens (L.) Small; S. oregonensis (Raf.) A. Nels.---Stems 2.5-8 cm. tall, usually sparingly branched above, glandular-pubescent, from a slender caudex, the plant appearing annual or biennial; leaf blades 5-10 mm. long, oblong-spatulate to obovate, in a basal rosette and a few cauline ones present, hispid-ciliate, some entire but some 3-toothed at apex; inflorescence a compact, leafy-bracted cyme; calyx tube longer than the lobes; calyx lobes 1.5-2 mm. long, oblong-obovate to ovate, somewhat glandular-pubescent; petals 2.5-3.5 mm. long, white, oblong-obovate, 3-nerved; filaments not clavate; ovary about 1/2 inferior; capsules about 3.5-5.5 mm. long, carpels united to above the middle, the calyx tube usually reticulate when pressed mature, due to the seeds within.---Among rocks. Alberta to British Columbia, south to Colorado and Utah; Europe. Our records scattered in the higher mountains of Colorado at 10,000-14,000 feet.

11. Saxifraga caespitosa L., Sp. Pl. 404. 1753.

Muscaria caespitosa (L.) Haw.; M. delicatula Small---Dwarf caespitose plants; stems 3-6 cm. tall, sparingly glandular-pubescent; leaves crowded on the caudices, 5-10 mm. long, 3-lobed at apex the lobes linear, oblong or lanceolate and obtuse, sparingly glandular-ciliate; stem leaves 1 or 2, the upper unlobed and bractlike; stems 1- to 4-flowered; calyx tube 2 mm. long or longer, turbinate at base; lobes of calyx 1.5-3.5 mm. long, oblong to oblong-lanceolate, obtuse or acutish; petals 3 mm. long or more, somewhat longer than the calyx lobes, white, oblong to obovate, clawless; filaments not clavate; ovary adnate at base to calyx tube; capsule 3 mm. long or more, carpels united to above the base.---Mountains. Circumpolar, south to Colorado, Arizona and Oregon. Our records scattered in the higher mountains of Colorado at 11,000-14,000 feet.

12. Saxifraga bronchialis L., Sp. Pl. 400. 1753.

S. austromontana Wieg.; Leptasea austromontana (Wieg.) Small; S. bronchialis ssp. austromontana (Wieg.) Piper---Rather matted plants, caudices rather creeping; stems 6-15 cm. tall, glabrous or finely glandular-pubescent; leaves crowded at the base, 5-16 mm. long, parchmentlike, linear-oblong to narrowly lanceolate, ciliate (almost spiny-ciliate) and spine-tipped, often spreading; leaves of flowering stem smaller, rather remote; inflorescence an open and rather corymbose panicle; sepals 1.5-3 mm. long, ovate to ovate-lanceolate, obtuse, distinct for most of the way, glabrous or glandular-pubescent; petals 4-6 mm. long, white, often spotted with orange or purplish dots, oblong to oblong-lanceolate, clawless; filaments not clavate; ovary only slightly united to calyx at base; capsule 5-9 mm. long, carpels united to above the middle.---Rocky places. Alberta to British Columbia, south to New Mexico, Utah and Washington. Our numerous records well scattered in the western half of Colorado at 7500-13,000 feet.

13. Saxifraga hirculus L., Sp. Pl. 402. 1753.
    Leptasea hirculus (L.) Small---Stems 6-20 cm. tall, glabrous to pubescent; leaves hardly rosulate, crowded on the caudices but also scattered on the flowering stems, blades elliptic-linear to oblong-linear, glabrous, entire, obtuse, cuneate, lower short-petioled; flowers mostly solitary; sepals 3-5 mm. long, oblong to oblong-ovate, ciliate, obtuse, finally reflexed; petals 7-13 mm. long, yellow, elliptic or oblong, without claws; filaments not clavate; ovary nearly or quite superior; capsule 9-16 mm. long, carpels united to above the middle.---Moist places. Circumboreal, south to Colorado. Our records from central and northcentral Colorado at 9500-13,000 feet.
14. Saxifraga chrysantha Gray, Proc. Amer. Acad. 12:83. 1877.
    Leptasea chrysantha (A. Gray) Small---Stems 3-8 cm. tall, glandular-pubescent to glabrate; leaves entire, rosulate on the caudices and a few scattered on the flowering stems, these latter smaller, oblong-ovate to spatulate, obtuse, glabrous or nearly so; flowers solitary to 3 on a stem; sepals 2-3.5 mm. long, ovate or oblong-ovate, obtuse, somewhat glandular-ciliate, soon reflexed; petals 5.5-7 mm. long, golden-yellow, oval to suborbicular, abruptly narrowed to a short claw; stamens not clavate; ovary superior or very nearly so; capsule 3 mm. long or more, the carpels united to above the middle.---Mountains, often in rocky places. Colorado and New Mexico. Our records from the mountains of northcentral, central and southern Colorado at 11,000-14,000 feet.

7. Lithophragma Nutt.   WOODLAND STAR

Perennial herbaceous plants with slender, bulblet-bearing rootstocks and leafy stems, the most of the leaves basal however; leaves petioled with blades suborbicular or reniform and palmately lobed or cleft; petioles with stipulelike dilations at base; flowers in racemes, mostly few-flowered; sepals 5, partly united into a tube which is more or less united to the base of the ovary; petals 5, white or rose, 3- to 7-parted, clawed at base; stamens 10, included; ovary more or less inferior at base, 1-celled, with 2-3 parietal placentae, styles 2-3; capsule 3-valved at apex; seeds many.

1. Leaves of stem with clusters of bulblets in all or some of the axils; flowers often partly or completely replaced by bulblets; upper part of stipules rounded -1. L. bulbifera
1. Bulblets seldom if ever present on the stem or inflorescence; upper part of stipules usually pointed
   2. Calyx tube campanulate, the base somewhat rounded, whole calyx 3-4 mm. long, adnate to the ovary for about 1/5 its length -2. L. tenella
   2. Calyx tube turbinate, cuneate at base, the whole calyx 4-8 mm. long, adnate to the ovary for about 1/2 its length or more -3. L. parviflora

1. Lithophragma bulbifera Rydb., N. Am. Fl. 22(2):86. 1905.
   Tellima bulbifera (Rydb.) A. Nels.---Stems slender, 10-25 cm. tall, glandular-pubescent; leaf blades 1-2 cm. wide, ternately divided to near base, the divisions cleft and toothed, lower leaves long-petioled, the upper ones short-petioled, usually bearing clusters of short-hirsute, dark red bulblets in their axils; stipules with apex rounded; flowers 3-6, sometimes partly or wholly replaced by bulblet clusters, pedicels 3-5 mm. long; calyx 2.5-4 mm. long, the tube united at its base to the base of the ovary, glandular-pubescent; petals 4-8 mm. long, 3-parted but the lower divisions cleft again, sometimes with 2 lobes; ovary slightly inferior.---Dry, sometimes partly shaded places. South Dakota to British Columbia, south to Colorado and California. Our records western and northcentral Colorado at 7000-9500 feet.
2. Lithophragma tenella Nutt. ex T. & G., Fl. N. Amer. 1:584. 1840.
   Tellima tenella (Nutt.) Walp.; L. australis Rydb.---Stems slender, 10-25 cm. tall, more or less glandular-puberulent; leaves ternately divided to near base, the divisions deeply cleft and toothed again, the whole blade 1.5-2 cm. wide, petioles long on basal leaves and shorter upward, none of the leaves bulblet-bearing; stipules pointed at top; flowers 3-12, on pedicels 2-5 mm. long; calyx 3-4 mm. long, the lobes short, the tube adnate to the base of ovary, glandular-puberulent; petals 3-5 mm. long, deeply 3- to 5-cleft (rarely 7) into linear lobes; ovary slightly inferior.---Slopes and meadows. Alberta to California, south to New Mexico. Our records scattered in the western half of Colorado at 7000-9000 feet.
3. Lithophragma parviflora (Hook.) Nutt. ex T. & G., Fl. N. Amer. 1:584. 1840.
   Tellima parviflora Hook.---Stems 10-30 cm. tall, glandular-puberulent and sometimes hirsutulous; leaf blades divided to near or to base into 3-5 divisions, these 1-3 cm. long and once or twice ternately cleft into oblong or linear divisions, lower long-petioled, upper shorter; stipules with apex rounded or pointed; racemes with 3-8 flowers, pedicels 2-5 mm. long; hypanthium turbinate, cuneate at base with shorter sepals about 4-8 mm. long, adnate to the ovary for about 1/2 its length; petals deeply cleft into 3-5 narrowly oblong divisions.---Often in rocky or gravelly places. Alberta to British Columbia, south to Colorado and California. Our one record from Routt County at 6700 feet.

8. Parnassia L.

Glabrous scapose perennial plants with short rootstocks; leaves mostly basal, only 1 bractlike one on the scape, entire; flowers solitary; calyx with a short tube, this free or adnate to the ovary, 5-lobed; petals 5, white or nearly so; stamens 5, alternate with the petals and also with the 5 clusters of gland-tipped staminoidea; ovary superior or slightly inferior, 1-celled, with 3 or 4 parietal placentae, these directly under as many sessile stigmas; fruit a capsule, loculicidal at apex; seeds numerous, winged.

1. Petals fimbriate at base; stem leaf at or above middle; basal leaves reniform-cordate -1. P. fimbriata
1. Petals entire, not at all fimbriate; stem leaf usually below middle; basal leaves oval to ovate, cuneate at base -2. P. parviflora

1. Parnassia fimbriata Konig, Ann. Bot. 1:391. 1804.
P. rivularis Osterh.---Scapes 10-30 cm. tall; scape leaf cordate, more or less clasping, inserted at or above the middle of scape; basal leaves long-petioled, this usually 5-15 cm. long, blades 1-4 cm. wide, reniform to broadly cordate, with about 7 main veins; sepals 3-6 mm. long, lanceolate to oval, acute or obtuse; petals 10-12 mm. long or sometimes less, obovate, clawed and fimbriate-margined below; each staminodium with 5-9 lobes, united below into a fleshy scale.---Moist or wet ground. Alberta to Alaska, south to Colorado and California. Our records scattered in the western half of Colorado, mostly from the northcentral, central and southwestern parts at 8500-12,000 feet.

2. Parnassia parviflora DC., Prodr. 1:320. 1824.
Scapes 10-30 cm. tall; scape leaf ovate, sessile, usually below the middle; basal leaves with petioles 1-5 cm. long, blades 7-20 mm. wide, oval or ovate, cuneate at base; sepals 5-8 mm. long, oblong; petals 6-10 mm. long, elliptic or oval, clawless, entire, not at all fimbriate; staminoidal scales with 5-7 filiform filaments these capitate at apex and resembling small stamens.---Moist or wet ground. Labrador to Alberta, south to Colorado and Utah. Our records scattered in the western half of Colorado, none as yet from the extreme western part, at 6500-10,000 feet.

9. Chrysosplenium L. GOLDEN SAXIFRAGE

Low somewhat succulent glabrous perennial plants with rootstocks; leaves alternate; flowers small, in small clusters in the axils of the upper leaves; calyx lobes 4 or 5, the tube adnate to the lower part of the ovary; petals lacking; stamens 4-8, inserted on a disk; ovary partly inferior, 1-celled, 2-lobed, 2 styles present, placentae parietal; fruit a capsule with numerous seeds.

1. Chrysosplenium tetrandrum Th. Fries, Bot. Notiser 1858:193. 1858.
Perennial plants with slender stoloniferous rootstocks; stems 1.5-8 cm. long, becoming branched above; leaves 4-12 mm. wide, all alternate, lower fairly long-petioled, shorter-petioled upward, blades thick, reniform, crenate with 3-7 broad teeth at apex, almost lobed on radical leaves, upper somewhat larger than lower; sepals usually 4; stamens 4.---Moist or wet places. A circumpolar species extending south to northcentral Colorado, our record from about 11,000 feet.

10. Sullivantia T. & G.

Perennial acaulescent herbaceous plants with small rootstocks; leaves alternate and mostly basal, only 1 or 2 on the lower part of the stems, blades reniform to orbicular, palmately lobed, long-petioled; flowers in panicled cymes; calyx united, sepals 5, erect, shorter than hypanthium; petals 5, persistent, clawed, white or whitish; stamens 5, alternate to petals; ovary fully half-inferior, 2-celled with thickened axile placentae, carpels united to the beaks, 2 styles present; capsule 2-beaked, opening between the beaks; seeds winged or wing-margined.

1. Sullivantia purpusi (Brand) Rosendahl, Minn. Studies in Pl. Sc. No. 6. p. 407. 1929.
S. hapemanii of Manuals in part---Stems 20-30 cm. tall, slender, erect or ascending, sparingly glandular-puberulent especially above; petioles of basal leaves 2-4 times longer than blades, blades 2-3 cm. long and wider, 9- to 12-lobed, the lobes irregularly and sharply dentate; inflorescence glandular, with long and spreading branches, bracteate, the 2 or 3 lower ones 3-lobed; calyx 3-4 mm. long in anthesis, tube glandular below, a little longer than the sepals which are 3-nerved; petals 3-4 mm. long.---Wet rocks. Apparently limited to westcentral and southwestern Colorado. Our few records from about 7000-7500 feet.

11. Mitella L. MITREWORT; MITERWORT

Low slender perennial plants with scaly rootstocks and scapiform flowering shoots; leaves round-cordate, basal; inflorescence racemose; calyx 5-cleft, the tube more or less adnate to the ovary; petals 5, pinnatifid, 3-cleft, 3-toothed or rarely entire; stamens 5, opposite or alternate to the petals, sometimes inserted on a disk that covers the top of the ovary; ovary more or less inferior, styles 2, very short, 1-celled with parietal or almost basal placentae; fruit a capsule; seeds numerous. The species treated below have been withdrawn from the genus Mitella by some authors and further separated into 2 genera.

1. Stamens opposite the petals and alternate to the calyx lobes; petals pinnatifid with more than 3 filiform divisions; pedicels over 1.5 mm. long; calyx lobes greenish -1. M. pentandra
1. Stamens alternate to the petals and opposite to the calyx lobes; petals from entire to cleft into 3 filiform divisions; pedicels less than 1.5 mm. long; calyx lobes whitish
   2. Calyx 3-5 mm. long; petals fully twice as long as the calyx -2. M. stauropetala
   2. Calyx 1.5-3 mm. long; petals only half longer than the calyx -3. M. stenopetala

1. Mitella pentandra Hook., Bot. Mag. 56: pl. 2933. 1829.
Pectianthia pentandra (Hook.) Rydb.---Plants sparingly pubescent with short white hairs or glandular-puberulent; scapes 10-30 cm. tall, slender; leaves 3-6 cm. wide, long-petioled,

blades cordate to round-reniform, coarsely crenate and with 9-11 more or less distinct round shallow lobes; raceme rather lax, flowers often paired; calyx 2.5-3.5 mm. long, the lobes greenish and reflexed; calyx tube longer than the lobes,deep saucer-shaped or shallowly campanulate, completely adnate to the ovary; petals rather irregular, pectinately-pinnatifid with central rachis and filiform divisions; stamens opposite the petals and alternate to the calyx lobes; ovary mostly inferior,covered by a distinct disk.---Moist or wet places, often in shaded places. Alberta to Alaska, south to Colorado and California. Our records scattered in the western half of Colorado at 9000-12,000 feet.

2. Mitella stauropetala Piper, Erythea 7:161. 1899.
Ozomelis stauropetala (Piper) Rydb.---Scapes 30-50 cm. tall, glandular-puberulent throughout or somewhat hirsute-pubescent below; leaves rather long-petioled; leaf blades round-reniform, simply or doubly crenate and more or less shallowly lobed, sparingly hirsute both sides; raceme rather elongated, usually secund, pedicels short, not over 1 mm. long; calyx 3-5 mm. long, the lobes whitish, ascending or somewhat reflexed, the tube about as long as the lobes, campanulate-turbinate, adnate to the lower half of the ovary; petals about twice as long as the sepals, 3-parted for half their length into filiform divisions; stamens alternate to the petals and opposite the sepals; ovary half-inferior, no disk covering it.---Moist or wet places. Montana to Washington, south to Colorado and Oregon. Reported definitely from this state and included on that report.

3. Mitella stenopetala Piper, Erythea 7:161. 1899.
Ozomelis stenopetala (Piper) Rydb.; M. parryi (Piper) A. Nels.---Scapes 15-50 cm. tall, puberulent or glandular-puberulent especially above; leaves rather long-petioled, blades rounded-reniform, crenate and more or less shallowly lobed; racemes secund, flowers very short-pedicelled, these to 1 mm. long; calyx 1.5-3 mm. long, the lobes whitish, tube about as long as the lobes, campanulate, adnate to the lower half of the ovary; petals ternately cleft above the middle into filiform divisions, filiform or rarely entire; stamens alternate to the petals and opposite the sepals; ovary half-inferior, no disk covering it.---Moist places. Wyoming, Utah and Colorado. Our records from the western half of Colorado, mostly in the northcentral and southwestern parts, at 8500-12,000 feet.

12. Heuchera L. ALUMROOT

Perennial herbaceous plants with scaly rootstocks; stems scapose, the leaves mostly basal, long-petioled with palmately veined and orbicular or broadly ovate blades; inflorescence paniculate, diffusely branched to narrow and spikelike; calyx with 5 often unequal lobes, the tube adnate at least 1/2 its length to the ovary base; petals entire, small, often more or less clawed (rarely wanting); stamens 5, opposite the sepals; ovary partly inferior, 1-celled, with 2 parietal placentae, with 2 elongated styles; capsule opening between the 2 more or less divergent beaks; seeds many.

1. Flowers 5-10 mm. long, definitely irregular; calyx tube (on longest side) 4-5 mm. long; scapes and petioles more or less hirsute with hairs about 1 mm. long   -1. H. richardsonii
1. Flowers 2-5 mm. long, nearly regular; calyx tube not over 2 mm. long; scapes if hirsute at all with hairs less than 1 mm. long
    2. Stamens of mature flowers exserted and longer than the petals; inflorescence spikelike or capitate
                                                                                     -2. H. bracteata
    2. Stamens included, definitely shorter than the petals; inflorescence spikelike or capitate only in H. nivalis
        3. Calyx tube campanulate, 1.5-2 mm. long; scapes glandular-puberulent to glabrous but not at all hirsute; styles about 1.5 mm. long, the stigmas about twice as wide as the styles   -3. H. hallii
        3. Calyx tube saucer-shaped, funnel-shaped or shallowly cup-shaped, 0.7-1 mm. long; scapes glandular-puberulent but also more or less short-hirsute, the hairs often to 0.5 mm. long; styles seldom over 0.5 mm. long, the stigmas scarcely wider than the styles
            4. Ovary flat-topped, completely covered to the styles by a disk; inflorescence dense and spikelike, rarely over 6 cm. long; flowers not over 2.5 mm. long; plants seldom over 20 cm. tall, usually found above timberline; seeds verrucose, not spiny   -4. H. nivalis
            4. Ovary conical on top, disk lacking or covering only the margins of the ovary; inflorescence rarely dense and spikelike, seldom less than 6 cm. long; flowers 2-5 mm. long; plants usually over 20 cm. tall, sometimes alpine but ranging down to 5000 feet in Colorado; seeds short-spiny
                                                                                     -5. H. parvifolia

1. Heuchera richardsonii R. Br. in Frankl., Journey 766. 1823.
H. hispida of Manuals---Flowering stems 20-60 cm. tall, more or less hirsute with the hairs about 1 mm. long; basal leaves long-petioled these more or less hirsute with hairs about 1 mm. long, and glandular-puberulent; leaf blades 2-11 cm. wide, round-cordate with about 9 shallow lobes, these toothed, upper surface glabrous or nearly so, lower hirsute with long whitish hairs especially along veins, margins ciliate; panicle rather narrow and few-flowered; flowers 6-10 mm. long, greenish, definitely zygomorphic; calyx tube oblique, 4-5 mm. long on longest side, 1-3 mm. on shortest; sepals erect with a rather wide sinus especially on short side of flower; petals as long or slightly longer than the sepals, spatulate, the blades glandular-denticulate on margins; stamens slightly exserted on most, about equal to sepals on a few; styles little if any exserted; capsules ellipsoidal, the beaks included; seeds black, about 0.75 mm. long, densely echinate.---Woods and slopes. Canada, south to Wisconsin and Colorado. Our few records from central Colorado at 7000-8000 feet.

2. Heuchera bracteata (Torr.) Ser. in DC., Prod. 4:52. 1830.
Flowering stems 10-40 cm. tall, glandular-puberulent or glabrate; leaf petioles long, finely glandular-puberulent, sometimes hirsute but hairs short or even glabrate; leaf blades 2-4 cm.

wide, ovate-reniform, cordate to truncate, shallowly to fairly deeply 5- to 7-lobed, lobes dentate with teeth usually acute, upper surface glandular-puberulent to glabrous, lower glandular-puberulent, sometimes margins ciliate; inflorescence spikelike or sometimes capitate, strongly secund; flowers 2.5-5 mm. long, greenish, nearly regular, glandular-puberulent; calyx tube 0.7-1 mm. long on longer side but attached to ovary for 1-1.5 mm. more; calyx lobes 1-1.5 mm. long; petals oblanceolate or spatulate, equalling or slightly exceeding the sepals; stamens at maturity exserted and longer than petals; styles finally exserted; capsule ovoid; seeds brownish-black, 0.7-0.8 mm. long, echinate.---Mountains and foothills. Wyoming and Colorado. Our records from northcentral, northwestern and probably central Colorado at 5500-9000 feet.

3. Heuchera hallii A. Gray, Proc. Acad. Phila. 1863:62. 1864.

Flowering stems 10-30 cm. tall, glandular-puberulent to glabrate; leaf petioles long, glandular-puberulent to glabrate; leaf blades 1-3 cm. wide, mostly orbicular, usually cordate with a shallow sinus and rather deeply 5- to 7-lobed, lobes coarsely dentate, glabrous or very minutely glandular-puberulent but margins ciliate, often bristly-serrate, the teeth with whitish aristate projections; inflorescence narrowly cylindrical; flowers 4-5 mm. long, greenish-white to pink-yellow, nearly regular, glandular-puberulent to hirtellous; calyx tube campanulate, the free portion about 1.5-2 mm. long; calyx lobes 1.5-2 mm. long with rounded tips; petals oblanceolate, longer than the calyx lobes, sometimes twice as long; stamens included; styles included, beaks of carpels conical, tapering into styles about 1.5 mm. long, stigmas capitate about twice as wide as the styles; capsule ovoid; seeds dark brown with thin sharp spines.---Mountains, often on rocky slopes or cliffs. Limited to Colorado, our records mostly from the central part of the state at 7500-10,000 feet.

4. Heuchera nivalis Rosend., Butters and Lakeld., Minn. Studies Plant Sc. 2:170. 1936.

Flowering stems 10-20 cm. long, rarely to 25, glandular-puberulent and also hirsute; petioles long, somewhat glandular-puberulent; leaf blades 1.5-3 cm. wide, orbicular, shallowly 5- to 7-lobed these crenate-dentate, cordate at base, upper surface glandular-puberulent to glabrous, lower glandular-puberulent, sometimes hirtellous, margins ciliate; inflorescence dense and spikelike, not over 6 cm. long, only the lowest cymules remote if at all; flowers 2-2.5 mm. long, wider than long, greenish to yellowish; calyx tube about 0.7 mm. long, saucer-shaped; calyx lobes about as long as the tube; petals somewhat longer than the sepals, ovate-lanceolate to ovate; stamens included; top of ovary flat, covered clear to the styles by a disk; fruit 3-5 mm. long, ellipsoidal; seeds 0.8-1 mm. long, brownish-black, merely verrucose, not spiny. This plant is related to the next one and probably should be made a variety of it.---High mountains. Apparently limited to Colorado, our records from the northcentral, central and southcentral parts of the state at 9500-12,500 feet.

5. Heuchera parvifolia Nutt. ex T. & G., Fl. N. Amer. 1:581. 1840.

H. utahensis Rydb.; H. flavescens Rydb.---Flowering stems 20-65 cm. tall, glandular-puberulent and also hirsute with white, rather conspicuous hairs, the longest usually to 0.5 mm. long; petioles long, usually glandular-puberulent and often hirsute; leaf blades about 2-6 cm. wide, broadly ovate, orbicular or reniform, cordate, more or less deeply 5- to 9-lobed, the lobes usually crenate-dentate, upper surface glabrous or glandular-puberulent, lower surface usually glandular-puberulent, margins ciliate; inflorescence 6-20 cm. long, varying from rather narrow, especially at first, to loose, usually much interrupted below; flowers 2-5 mm. long (usually 2.5 mm. or more), greenish to yellowish; calyx tube (where unattached) forming a shallow cup or saucer-shape or funnel-shape, about 0.7-1 mm. long, sepals about as long, nearly regular; petals oblanceolate to rhombic-obovate, usually somewhat exceeding the sepals; stamens included; top of ovary conical, rarely covered with a disk on its margins, styles 0.2-0.5 mm. long, stigmas scarcely wider than styles; capsule 3-7 mm. long, ovoid or ellipsoid; seeds 0.6-1 mm. long, brownish-black, with short conical spines. This is a highly polymorphic species but the varieties proposed will not work out for Colorado plants, often 2 or more of these occurring in 1 plant.---Foothills and mountains. Montana and Idaho, south to New Mexico, Arizona and Nevada. Our records scattered in the western two-thirds of Colorado at 5000-12,000 feet, mostly below 10,000 feet.

Family 56. Rosaceae ROSE FAMILY

Herbaceous plants, shrubs or trees; leaves alternate (except in the genus Coleogyne), simple or compound, usually with stipules; flowers regular, commonly perfect, solitary or clustered; sepals partly united, 5 (or 4); petals 5 (or 4) rarely none, distinct; stamens 5 to many, inserted with the petals on the calyx tube or on a disk that lines the calyx tube; ovaries 1 to many, simple and distinct and free from the calyx tube, or carpels united into a 2- to 5-celled ovary which is nearly or completely inferior; fruit various, a follicle, achenes, drupelets (1 to many), a drupe, or a pome. This is an economically important family and includes many of our very important fruit-producing plants such as apple, pear, peach, plum, cherry, blackberry, raspberry, apricot, almond and strawberry. Many of the plants are also cultivated as ornamentals such as the roses. The fruit characteristics in the rose family are very diverse and have resulted in some workers dividing the group into several families based mainly on fruit differences. However, these groups are still united closely by similarities in floral structure. Also at least one of the remaining entities remains very diverse in characters of the fruit. Therefore, the family is considered in its broad aspect here.

1. Ovary inferior, enclosed in and adnate to the calyx tube; fruit a pome; plants woody
  2. Leaves pinnately compound -1. Sorbus
  2. Leaves simple (can be lobed)
    3. Flowers solitary or 2-3 together; styles 2; leaves narrowly oblanceolate, entire or serrulate, sessile or nearly so -2. Peraphyllum

3. Flowers in racemes or corymbs terminating leafy branches; styles 2-5; leaves varying from lanceolate to orbicular, always toothed, short- to long-petioled
    4. Flowers umbellike (central axis very short); petals over 2 cm. long; fruit over 2 cm. wide; thorns never present   -3. Malus
    4. Flowers in racemes or corymbs (central axis definite); petals seldom over 1.5 cm. long; fruit less than 2 cm. wide; thorns present or absent
        5. Branches armed with stout spines; winter buds nearly spherical, rounded at apex; flowers in corymbiform cymes; fruit with as many cells as the styles, the carpels with bony walls at maturity   -4. Crataegus
        5. Branches unarmed; winter buds elongated and pointed; flowers usually racemose; fruit with twice as many cells as styles, carpel walls not bony   -5. Amelanchier
1. Ovary superior (nearly enclosed by the hypanthium disk in Alchemilla); calyx tube either not enclosing the pistils or if so then not adnate to them; fruit not a pome; plants woody or herbaceous
  6. Plants woody, definitely shrubby or treelike (stems sometimes prostrate or matted)
    7. Leaves opposite (sometimes rather crowded), entire; flowers with four sepals and no petals   -6. Coleogyne
    7. Leaves alternate or fascicled, never opposite, either entire, toothed, lobed or compound; flowers various but if apetalous then with 5 sepals present
        8. Petals absent; pistil 1, becoming an achene, included in an elongated calyx tube and with an elongated plumose style; leaves simple   -7. Cercocarpus
        8. Petals present; pistils more than 1 or if 1 then not in an elongated tube and without a plumose style; leaves simple or compound
            9. Carpels enclosed in the calyx tube which becomes fleshy in fruit, and open only at top (but not adnate to bony achenes); leaves pinnately compound and stems usually with prickles; petals large (over 1.5 cm. long) and rose-colored   -8. Rosa
            9. Carpels not enclosed in a fleshy, flask-shaped calyx tube (may be included in a saucer-shaped or campanulate one); leaves simple to compound but if pinnately so then prickles absent; petals small or if large (over 1.5 cm. long) then never rose-colored
                10. Leaves compound with 3 or more definite leaflets; pistils many
                    11. Petals bright yellow; sepals 5 with 5 alternating bractlets present; no bristles or prickles present; fruit of dry achenes   -9. Potentilla
                    11. Petals white, never bright yellow; sepals 5, no bractlets present; bristles or prickles present; fruit fleshy, of druplets   -10. Rubus
                10. Leaves simple (may be deeply lobed); pistils 1 to many
                    12. Leaves entire or toothed, but not lobed
                        13. Pistil 1, becoming a fleshy drupe (plum or cherry)   -11. Prunus
                        13. Pistils 3 to many, becoming dry follicles or achenes
                            14. Low, depressed undershrubs; fruit of numerous achenes or of 3-5 follicles
                                15. Pistils many, becoming achenes with long, plumose styles; leaves crenate toothed; petals 8-15 mm. long   -12. Dryas
                                15. Pistils 3-5, becoming follicles, the styles not elongated and plumose; leaves entire; petals 1.5-2 mm. long   -13. Spiraea
                        14. Upright shrubs; fruit of 5 achenes   -14. Holodiscus
                    12. Leaves palmately or pinnately lobed
                        16. Calyx lobes 5, alternating with as many bractlets (appearing to have 10 sepals); fruit of many achenes with elongated, plumose styles   -15. Fallugia
                        16. Calyx lobes 5, no bractlets present; fruit various (can be of 4-10 achenes with plumose styles)
                            17. Pistils many, becoming druplets, more or less juicy   -10. Rubus
                            17. Pistils 1 to 10, becoming dry achenes or follicles
                                18. Fruit a capsule, the parts folliclelike and dehiscent; leaf blades shallowly lobed with broad lobes, the whole blade over 3 cm. long; usually more or less stellate-hairy, never white tomentose   -16. Physocarpus
                                18. Fruit of 1-10 achenes; leaf blades deeply lobed or cleft, the divisions linear or oblong, the whole blade not over 3 cm. long, not stellate-hairy, white-tomentose below
                                    19. Pistil solitary, style not plumose; plants in the western 2/3 of Colorado   -17. Purshia
                                    19. Pistils 4-10, styles becoming elongated and plumose in fruit; plants limited to southwestern corner of Colorado   -18. Cowania
  6. Plants herbaceous above ground (the perennial rootstock may be woody and persistent at base of plant)
    20. Leaves simple, entire or crenate-toothed
        21. Pistils many, becoming achenes with long, plumose styles; leaves crenate-toothed; petals 8-15 mm. long   -12. Dryas
        21. Pistils 3-5, becoming follicles, the styles not elongating or becoming plumose; leaves entire; petals 1.5-2 mm. long   -13. Spiraea
    20. Leaves compound or deeply divided
        22. Calyx tube with hooked prickles above; pistils 1-2 to a flower   -19. Agrimonia
        22. Calyx tube lacking hooked prickles; pistils 5 to many (except in Ivesia with 1-6)
            23. Petals present; calyx lobes 5 (or appearing 10); stamens 5 to many
                24. Calyx lobes 5, alternating with bractlets (hence appearing to have 10 sepals); pistils 1 to many, becoming dry achenes (may be on a fleshy receptacle)
                    25. Stamens 5; pistils 1-20; petals small and not showy, little if any larger than the sepals

26. Styles terminal or subterminal on the achenes; pistils 1-6; flowers capitate -20. Ivesia
26. Styles basal or subbasal on the achenes; pistils 5-20; flowers not capitate
  27. Leaves palmately compound with 3 leaflets, these oblanceolate or wider, not lobed; plants 5-15 cm. tall; petals yellow -21. Sibbaldia
  27. Leaves 2- to 4-ternately divided into linear lobes; plants 10-30 cm. tall; petals white -22. Chamaerhodos
25. Stamens 10 or more; pistils 10 to very many; petals usually longer than the sepals and often very showy
  28. Basal leaves with only 3 leaflets
    29. Petals white; plants spreading by long aerial stolons; plants acaulescent; receptacle becoming fleshy and juicy (and edible) at maturity -23. Fragaria
    29. Petals light yellow to bright yellow; no aerial stolons present; plants usually more or less caulescent; receptacle not becoming fleshy -9. Potentilla
  28. Basal leaves with more than 3 leaflets
    30. Styles articulated at base and deciduous from the achenes, neither plumose or jointed near the middle -9. Potentilla
    30. Styles not articulated at the base or deciduous but forming a part of the mature fruit; middle of style often with an abrupt bend or joint, style often plumose -24. Geum
24. Calyx lobes 5, no bractlets present; pistils many, becoming fleshy druplets -10. Rubus
23. Petals lacking; calyx lobes 4 (or appearing 8); stamens 4 -25. Alchemilla

1. Sorbus L. MOUNTAIN ASH

Shrubs; leaves alternate, pinnately compound, deciduous; flowers perfect, in compound, terminal corymbose cymes; calyx tubes urn-shaped or turbinate, sepals 5; petals 5, white, short-clawed; stamens many; ovary inferior, adnate to calyx tube, styles 3-5, distinct, cells of ovary as many as styles, each with 2 ovules; fruit a small, red berrylike pome, carpels with cartilaginous or papery walls, each cell with 1-2 seeds.

1. Sorbus scopulina Greene, Pitt. 4:130. 1900.
Shrubs 4 or 5 m. tall; young twigs villous-pubescent; leaflets 11-15, 3-7 cm. long, glabrous, elliptic, lanceolate or oblong-lanceolate, sharply serrate, cuneate or rounded at base, acute or acuminate, paler below; cyme large, 9-15 cm. wide, flat-topped, usually about 40-flowered, branches of cyme hairy; sepals about 1.5 mm. long; petals 4-6 mm. long, oval; fruit globose, 6-10 mm. in diameter, red or orange-red.---Rocky canyons and slopes. Alberta to British Columbia, south to South Dakota, New Mexico and Washington. Our records scattered in the western half of Colorado at 6000-10,000 feet.

2. Peraphyllum Nutt. SQUAW APPLE

Low shrubs with gray bark and rigid branches; leaves simple, sessile or nearly so, deciduous, alternate but fascicled at the ends of the branchlets, with minute, deciduous stipules; flowers solitary or 2 or 3 together, perfect, regular; calyx tube subglobose, adnate to the top of the ovary, sepals 5, persistent; petals 5, obovate, spreading; stamens about 20; ovary inferior, of 2 carpels but 4-celled by a false partition from the back; 2 styles present; fruit a fleshy pome, 4-seeded (1 to each cell), the carpel walls cartilaginous.

1. Peraphyllum ramosissimum Nutt. in T. & G., Fl. N. Amer. 1:474. 1840.
Shrubs to 2 m. tall, intricately branched; leaves 2-4 cm. long, narrowly oblanceolate, cuneate at base, acute at apex, sparingly appressed-silky when young, glabrous in age, entire or serrulate, the teeth glandular and reddish brown, often deciduous at tip, becoming somewhat coriaceous in age; petioles short; sepals triangular, reflexed, pubescent inside and on margins; petals about 7-10 mm. long, pale pink; fruit globose, 10-15 mm. in diameter, glabrous, yellowish to reddish brown, bitter tasting.---Hills and slopes. Colorado to Oregon, south to California. Our records scattered in the western one-third of Colorado, except the extreme northern part, at 5500-8000 feet.

3. Malus Mill. APPLE

Deciduous trees or shrubby somewhat at first; twigs rarely if ever spinescent, young growth woolly-hairy; leaves simple, alternate; flowers white to pink, showy, in umbellike racemes; calyx 5-lobed, tomentose; petals obovate to suborbicular; stamens many; ovary completely inferior, 3- to 5-celled, with 2-5 styles connate at base; fruit a pome usually over 2 cm. wide.

1. Malus pumila Mill., Gard. Dict. ed. 8. 1768.
Malus malus (L.) Britt.; P. malus L. in part---Leaves 2.5-10 cm. long, ovate to elliptic, pubescent both sides, later glabrate above, crenate-serrate; petals 2 cm. long or more, sub-globose, impressed at base and apex.---Woods, thickets and slopes. Introduced from Europe as a fruit tree and occasionally escaping. Our two records from northcentral Colorado apparently are not from cultivated plants.

4. Crataegus L. HAWTHORN; HAW

Shrubs or small trees, usually armed with spines or thorns; leaves deciduous, alternate, petioled, simple, serrate to lobed; inflorescence of cymose corymbs, terminating short leafy branches; calyx tube urn-shaped, campanulate or cup-shaped, adnate to the ovary, calyx lobes 5, reflexed after anthesis; petals 5, white or pink, soon falling; stamens 5-25, on the margin of the calyx tube; ovary inferior, of 1-5 carpels, 1- to 5-celled, with 2 ovules in each cell, only 1 maturing; styles 1-5, distinct; fruit a pome, usually subglobose, usually red, yellow or black, containing 1-5 bony-walled, 1-seeded carpels. Two specific keys, 1 for flowering condition, the other for fruiting condition follow.

I. Crataegus--in flower
  1. Leaves elliptic to lanceolate, about twice as long as wide or more, narrowly cuneate at base; spines usually few, seldom over 3 cm. long
    2. Stamens about 20; leaves crenate-serrate, the teeth small; leaves small, usually not over 4 cm. long; pedicels glabrous -1. C. saligna
    2. Stamens about 10; leaves irregularly and rather sharply serrate, the teeth moderate; leaves usually over 4 cm. long; pedicels glabrous to hairy -2. C. rivularis
  1. Leaves ovate to rhombic-ovate, less than twice as long as wide, usually broadly cuneate to truncate at base; spines usually many, usually over 3 cm. long
    3. Branches of the corymb and pedicels glabrous at flowering time; leaves glabrous beneath; anthers pink, rose or purple when fresh -3. C. erythropoda
    3. Branches of the corymb and pedicels more or less villous or pubescent; leaves commonly pubescent on the veins beneath; anther color usually white, cream or yellow (can be pink, rose or purple, however)
      4. Petioles with dark red raised glands; teeth of leaves conspicuously dark red-tipped; leaves and petioles becoming glabrous or nearly so when mature -4. C. chrysocarpa
      4. Petioles without dark red raised glands; teeth of leaves somewhat glandular at tip but seldom with conspicuous dark red tips; leaves and petioles commonly somewhat pubescent at maturity -5. C. succulenta

II. Crataegus--in fruit
  1. Nutlets not pitted or deeply concave ventrally; fruit never bright red or scarlet when mature
    2. Fruit blue-black when mature, 5-8 mm. in diameter; thorns rather short, seldom over 2.5 cm. long; leaves rhombic-oblanceolate to oblong-lanceolate -1. C. saligna
    2. Fruit golden-yellow to reddish-orange, 8-12 mm. in diameter; thorns rarely less than 3 cm. long; leaves broadly ovate to broadly rhombic -4. C. chrysocarpa
  1. Nutlets pitted or deeply concave on the ventral surface; fruit usually scarlet or bright red but sometimes blackish or orange when mature
    3. Nutlets 5; pedicels of fruit usually glabrous
      4. Leaf blades ovate, less than twice as long as wide, often distinctly lobed, rather broadly cuneate at base; spines usually numerous, often over 3 cm. long; fruit about 7-8 mm. in diameter -3. C. erythropoda
      4. Leaf blades elliptic to lanceolate, about twice as long as wide, scarcely if at all lobed, narrowly cuneate at base; spines usually few and seldom over 3 cm. long; fruit about 10 mm. in diameter -2. C. rivularis
    3. Nutlets 2-4; pedicels of fruit often villous -5. C. succulenta

1. Crataegus saligna Greene, Pitt. 3:99. 1896.
  Small trees about 2-6 m. tall with rather slender, glabrous twigs; spines usually less than 2.5 cm. long, rather blackish; leaves not over 4 cm. long, thick, rather glossy, margins crenate-serrate, the teeth small, or entire near base, not lobed at least on flowering branches, glabrous below, glabrate or appressed hairy above, cuneate at base to a short petiole, acute to obtuse at apex, oblong-lanceolate to rhombic-lanceolate, about 2.5-3 times longer than wide; corymbs 5-flowered or more, the branches and pedicels glabrous; calyx glabrous; stamens about 20, anthers yellow when fresh; styles 5; fruit blue-black, globose, 5-8 mm. in diameter; nutlets 5, not pitted or deeply concave ventrally.---Canyons and banks. Colorado. Our records from the western one-third of the state at 6000-7500 feet.

2. Crataegus rivularis Nutt. in T. & G., Fl. N. Amer. 1:464. 1840.
  C. wheeleri of Rydbs. Flora---Small trees about 3-6 m. tall, with ascending branches; twigs glabrous; thorns mostly not over 3 cm. long, blackish; leaves mostly over 4 cm. long, and about twice as long as wide, elliptic-rhombic to lanceolate, cuneate at base to a moderately long petiole (10-15 mm. here), obtuse to short-acuminate at apex, margins irregularly and rather sharply serrate except near base, usually not lobed at least in flowering branches, glabrous and paler below, scattered and appressed hairy above, at least when young; branches of corymbs, pedicels and calyx glabrous to villous; stamens about 10, anthers whitish, pink, rose or purple when fresh; styles usually 5; fruit crimson to blackish, about 1 cm. in diameter when mature; nutlets 5, pitted or deeply concave ventrally.---Valleys and banks. Wyoming to Idaho, south to New Mexico, Arizona and Nevada. Our records scattered in the western one-third of Colorado at 5500-8500 feet.

3. Crataegus erythropoda Ashe, N.C. Agri. Exp. Stat. 175:113. 1900.
  C. cerronis A. Nels.---Small trees, 2-5 m. tall, widely branched; twigs glabrous; spines about 2-4 cm. long, purple-brown; leaves 3-7 cm. long, less than twice as long as wide, ovate, rhombic-ovate or elliptic-ovate, acute or acuminate, cuneate at base, usually rather broadly so, rather decurrent on the moderate to short petiole (about 10-15 mm. here) irregularly serrate, often incised, glabrous below, glabrous to sparingly appressed hairy above; branches of corymb, pedicels and calyx tube glabrous from the first; sepals usually somewhat hairy at margins or within; stamens about 10 or less, anthers pink, rose or purple when fresh; styles 5; fruit red to dark brown or blackish, about 7-8 mm. in diameter, subglobose; nutlets 5, pitted or deeply concave ventrally.---Banks and valleys. Wyoming to New Mexico and Arizona. Our records scattered in the western half of Colorado at 5500-8000 feet.

4. Crataegus chrysocarpa Ashe, N. C. Agri. Exp. Stat. Bul. 175:110. 1900.

C. doddsii Ramaley---Small trees, 2-7 m. tall; twigs hairy or glabrous especially in age; thorns 3-7 cm. long, numerous; leaves 3-5 cm. long, nearly as wide, orbicular or rounded-obovate, sometimes broadly rhombic, acute at apex, broadly cuneate to almost truncate at base, margins sharply or doubly serrate, the teeth dark brown-tipped, with shallow lobes present, glabrous or nearly so when mature; petioles 1-2 cm. long, finally glabrous or nearly so, narrowly winged from the blade, and with dark red, raised glands, these rather conspicuous; branches of corymbs and pedicels more or less villous or pubescent at flowering time; stamens 5-10, anthers whitish; fruit golden-yellow, orange or reddish, about 8-9 mm. long, usually not quite so wide; nutlets 3-4 (rarely 2), not pitted or deeply concave ventrally.---Canyons, valleys and banks. Manitoba, south to Nebraska and Colorado. Our few records from northcentral Colorado at 5500-6000 feet. However a specimen from the southwestern part of the state may be this species.

5. Crataegus succulenta Schrad. ex Link, Handb. 2:78. 1831.

C. occidentalis Britt.; C. colorado Ashe; C. coloradoides Ramaley; C. coloradensis A. Nels.---Tall shrubs or small trees up to 7 m. tall; twigs of the season glabrous to variously hairy, rather stout; thorns about 3-7 cm. long, slender; leaf blades about 4-7 cm. long and nearly as wide, oval, obovate to nearly orbicular, irregularly sharply serrate and shallowly lobed, teeth and lobes not glandular at tips, acute to rounded at apex, glabrate and rather glossy above, more or less villous-pubescent beneath at least near the veins; petioles 1/5 to 1/2 as long as the blades; stamens 20 or less, ours usually not over 10, anthers white to rose or purple; fruit globose or sub-globose, 7-11 mm. in diameter, bright red to purplish; nutlets 2-4, pitted or deeply concave on the ventral surface. As treated here, a complex of closely related forms, considered to be a part of the northeastern C. succulenta Schrad. The C. coloradensis A. Nels. with glabrous twigs may be a good variety and needs further study, but the other 3 synonyms listed above undoubtedly refer to the same plant.---Hillsides, canyons and slopes. Eastern Canada to Saskatchewan, south to Pennsylvania, Nebraska and Arizona. Our records from northcentral and central Colorado at 5000-7000 feet.

5. Amelanchier Medic. SERVICE-BERRY

Shrubs or very small trees; twigs unarmed; leaves simple, alternate, deciduous, petioled; inflorescence of racemes on short leafy branchlets, flowers perfect, regular, bracteate; calyx tube campanulate, becoming globose, more or less adnate to the ovary; calyx lobes 5; petals 5, white; stamens many, inserted on the rim of the calyx tube; ovary inferior with 2-5 styles and 2-5 cells, each cell with 2 ovules; fruit a pome, 2- to 5-celled but a partition growing in from the back between the paired seeds forming 2 false cells in every true one.

1. Top of ovary glabrous; calyx, pedicels, twigs and leaves glabrous from the first   -1. A. pumila
1. Top of ovary tomentose, this usually persisting; calyx, twigs and leaves pubescent at first (only rarely glabrous)
   2. Leaves permanently puberulent or finely tomentulose at least below (only very rarely glabrous), 5-30 mm. long, becoming subcoriaceous; styles usually 3-4 (sometimes 2 and rarely 5)   -2. A. utahensis
   2. Leaves at maturity glabrous or nearly so, 20-50 mm. long, only moderately firm; styles usually 5 (but often 4 and rarely 3)   -3. A. alnifolia

1. Amelanchier pumila Nutt. ex T. & G., Syn. Mon. 3:145. 1847.

A. polycarpa Greene; A. alnifolia pumila (Nutt.) A. Nels.; A. glabra Greene---Shrubs 1-3 m. tall; twigs and buds glabrous; leaves 1-5 cm. long when mature, glabrous from the start, oval to suborbicular, rather thick, deeper green above, paler below, apex obtuse, base rounded to subcordate, coarsely serrate to middle or somewhat below, lower 1/4 entire-margined; calyx and pedicels glabrous, calyx lobes about 3 mm. long; petals 8-12 mm. long, widest at or above middle; styles 5 (or 4); top of ovary glabrous; fruit dark purple and glaucous, 8-9 mm. in diameter.---Hills, mesas and mountains. Montana to Oregon, south to Colorado and California. Ours well scattered in the western half of Colorado at 5890-9500 feet.

2. Amelanchier utahensis Koehne, Gattung. Pomac. 25. 1890.

A. bakeri Greene; A. crenata Greene; A. elliptica A. Nels.; A. mormonica Schneid.; A. oreophila A. Nels.; A. prunifolia Greene; A. rubescens Greene---Shrubs to 3 or 4 m. tall; twigs and buds usually pubescent; leaves 5-30 mm. long, suborbicular, oval, ovate or obovate, hairy both surfaces even at maturity, the hairs short and curly, rarely glabrous, sub-coriaceous at maturity, apex rounded or acute, base rounded, truncate or sometimes cuneate, margins coarsely serrate-dentate to middle or below, rarely almost entire; pedicels lanate (rarely glabrous) usually even in fruit; calyx tube tomentulose or even sometimes glabrous, its lobes about 3 mm. long usually tomentulose both sides; petals 6-12 mm. long, linear-oblanceolate; styles usually 3-4, sometimes 2 and rather rarely 5; top of ovary more or less tomentulose; fruit 6-10 mm. in diameter, purplish-black.---Rocky slopes, canyons and stream banks. Montana to Oregon, south to New Mexico and California. Our records scattered over the western half of Colorado at 5000-9500 feet.

3. Amelanchier alnifolia Nutt., Journ. Acad. Phila. 7:22. 1834.

Shrubs or small trees 1-4 m. tall; twigs more or less pubescent at first, soon glabrous; buds somewhat villous; leaves 2-5 cm. long, of rather firm texture, oval to suborbicular, tomentulose below when very young, soon glabrous or somewhat pubescent below along veins, mostly obtuse, rounded to subcordate at base, margins coarsely serrate or dentate to the middle or below, rarely entire; pedicels lanate; calyx tube floccose at first, soon glabrous, the lobes about 1.5 to 3 mm. long, more or less pilose within; petals 6-10 mm. long; styles usually 5, rather often 4 and occasionally 3; ovary top tomentose; fruit 10-15 mm. in diameter, purple to black when ripe. This species intergrades somewhat with the preceding in the pubescence and size of the leaves, also somewhat in the number of styles.---Banks, slopes and mountains. Manitoba to Yukon, south to Michigan, Colorado and Idaho. Our records scattered over the western half of Colorado at 5000-10,000 feet.

6. Coleogyne Torr. BLACKBRUSH

Shrubs with diffuse, rigid branches, these often opposite and spinescent at the apex; leaves crowded but opposite, deciduous, coriaceous, entire; flowers solitary, terminal on the young branchlets; calyx tube very short, calyx lobes 4; petals usually none; stamens numerous, inserted near base of an elongated, sheathing tube (disk) that encloses the ovary; pistil 1, superior, 1-ovuled, 1-celled, with a lateral, bent and twisted style; fruit an achene.

1. Coleogyne ramosissima Torr., Pl. Fremont. 8. 1853.
Densely branched shrubs up to 2 m. tall with ashy-gray branches, these opposite and tangled; leaves 5-15 mm. long and about 1-1.5 mm. wide, clavate or linear, strigose with the hairs attached at middle; sepals 5-8 mm. long, yellowish, greenish or purplish, the outer 2 lanceolate, the inner ovate or obovate; tubular sheath about 4-6 mm. long, toothed at apex, glabrous without, densely white-villous within; achene about 3 mm. long, glabrous, with a bent, twisted, exserted style, this very densely villous at base.---Mesas and hillsides, in dry often rocky ground. Colorado to California, south to Arizona. Our records from westcentral Colorado at 4500-5000 feet.

7. Cercocarpus H. B. K. MOUNTAIN MAHOGANY

Shrubs or trees with hard, dark colored wood; leaves alternate and usually fascicled, simple, stipulate; flowers solitary or in small fascicles; calyx tube free from the ovary, salverform, the tube persistent, the limb abruptly widening and campanulate; sepals 5; petals absent; stamens 15 or more, inserted on the limb of the calyx tube in 2 or 3 rows; pistil 1, superior, style terminal, 1-ovuled, ovary inserted at the bottom of the calyx tube, sessile; fruit a coriaceous, terete, villous achene, included in the elongated calyx tube and with an elongated, plumose style.

1. Leaves entire, more or less revolute, thick and coriaceous, linear to oblong-lanceolate; flowers sessile or nearly so
   2. Leaves slightly to moderately revolute, much of the lower surface exposed, oblong-lanceolate; styles on mature fruit 4-7 cm. long; plants often over 2 m. tall   -1. C. ledifolius
   2. Leaves definitely revolute, very little of the lower surface exposed, linear or very narrowly linear-oblong; styles on mature fruit 2-4 cm. long; plants not over 2 m. tall   -2. C. intricatus
1. Leaves toothed at least near apex, not revolute, rather thick but not coriaceous, obovate to oblanceolate; flowers on definite short pedicels
   3. Leaves green above, usually glabrate, obovate   -3. C. montanus
   3. Leaves permanently densely pilose above, hardly green, oblanceolate to obovate   -4. C. argenteus

1. Cercocarpus ledifolius Nutt. ex T. & G., Fl. N. Amer. 1:427. 1840.
Rather large shrubs or small trees to 7 or 8 m. tall with 1 to several trunks, these up to 18 cm. in diameter with furrowed, reddish-brown bark; leaves 1-3 cm. long and 5-10 mm. wide, thick, evergreenlike, resinous and aromatic, oblong-lanceolate, sometimes lanceolate or oblanceolate, margins more or less revolute but most of underside exposed, usually glabrous above in age, green and shining, resinous, whitish-tomentose below; petioles 3-5 mm. long; flowers solitary or in 2's or 3's, sessile; calyx tube 5-7 mm. long, with turbinate-campanulate limb, white-villous to nearly tomentose outside, glabrous within; sepals 1.5-2 mm. long, ovate, spreading, hairy both sides; styles in fruit 4-7 cm. long; achene 8-10 mm. long.---Hills and canyons. Montana to Washington, south to Colorado, Arizona and California. Our records from northwestern and westcentral Colorado at 6500-9000 feet.
2. Cercocarpus intricatus S. Wats., Am. Acad. Arts & Sci. 10:346. 1875.
C. ledifolius var. intricatus (S. Wats.) M. E. Jones---Shrubs 30-150 cm. tall, with intricately branched, more or less spinescent twigs, bark gray; leaves 5-15 mm. long, seldom over 2.5 mm. wide, thick, evergreenlike, resinous and aromatic, linear or very narrowly oblong, green and shining above, glabrate in age, margins strongly revolute, with very little of the under surface exposed, this tomentulose-villous at least when young; petioles 1-2 mm. long; flowers solitary, nearly or quite sessile; calyx tube about 4 mm. long, the limb turbinate-campanulate; styles in fruit 2-4 cm. long; achene about 6-7 mm. long. Apparently intergrades with the preceding in other areas but appears distinct in Colorado.---Hills and canyons. Utah to California, south to Arizona. Our records from extreme western Colorado at 4500-8500 feet.
3. Cercocarpus montanus Raf., Atlant. Journ. 146. 1832.
C. parvifolius Nutt.; C. flabellifolius Rydb.---Shrubs up to 300 cm. tall, bark thin and grayish-brown, twigs more or less villous when young; leaves 2-5 cm. long and 1.5-2.5 cm. wide, oval to broadly obovate, coarsely serrate above the middle, sometimes somewhat below, texture moderately thin, rounded at apex, cuneate at base, upper surface green and more or less pilose especially when young, white or whitish beneath and finely tomentose; petioles 3-6 mm. long; flowers 1-3, on definite pedicels 2-5 mm. long; calyx tube villous to long-sericeous, about 5-10 mm. long, limb campanulate; sepals about 1-1.5 mm. long; style in fruit 6-10 cm. long; achene about 8-10 mm. long.---Hills and slopes often on stony ground. South Dakota and Montana, south to Kansas, New Mexico and Arizona. Our records scattered over Colorado at 4000-8500 feet.
4. Cercocarpus argenteus Rydb., N. Am. Fl. 22:422. 1913.
C. montanus var. argenteus (Rydb.) F. L. Martin ---This rather doubtful species is reported from Colorado, south to Texas and New Mexico. It has not been seen by the writer.

8. Rosa L. ROSE

Shrubs usually with prickly (and often also bristly) stems; leaves alternate, odd-pinnate with adnate stipules; flowers perfect, large and showy, solitary or corymbose; calyx tube globose or urceolate, contracted at the mouth; calyx lobes 5 (rarely 4); petals rose-colored, the same number as the sepals; stamens many, on a silky disk that nearly closes the opening of the calyx tube; pistils several to many, distinct, superior but included in the calyx tube, the styles reaching the mouth or exserted; pistils becoming bony achenes, included. This is a variable genus inclined to hybridize, and two points of view are possible. Rydbergs concept (N. A. Fl. 22:483-532. 1918) recognizes as entities many variations and gives them specific recognition. Erlanson (Bot. Gaz. 96:197-259. 1934) lumps these forms into a relatively few Linnean species, most of which are complexes. This latter viewpoint is followed here as the only logical procedure although further study is certainly necessary. The key will be most usable checking the plants in the field since herbarium specimens rarely are accompanied by the data needed.

1. Infrastipular prickles present (these are 1 or 2 prickles situated on the twig just below the node. They may be the only ones present or in any case are longer or more conspicuous than the scattered prickles)
   2. Flowers often solitary, sometimes 2 or 3, on lateral shoots 3-10 cm. long; petals sometimes over 2.5 cm. long
      3. Stems with bristles clear to the apex, stout or slender, not over 50 cm. tall   -1. R. engelmannii
      3. Stems with bristles only at the very base if present at all, stout, usually over 100 cm. tall
          -2. R. nutkana
   2. Flowers 1-15 (some usually well of 3) on lateral shoots usually over 10 cm. long; petals 1-2.5 cm. long
          -3. R. woodsii
1. Infrastipular prickles absent (stem may have scattered prickles)
   4. Flowers terminal on long, basal shoots as well as on lateral shoots; flowering period long; stems often dying back to ground over winter   -4. R. arkansana
   4. Flowers all from lateral shoots from the old wood; flowering period short; stems not usually dying back over winter
      5. Stems unarmed, without prickles and with bristles only at the base; leaflets 5-11; petals not over 2.5 cm. long   -3. R. woodsii
      5. Stems with bristles or prickles (or both) to apex; leaflets 5-7 (rarely 9); petals often over 2.5 cm. long
         6. Plants 30-200 cm. tall; flowers 1-3, early flowering   -5. R. acicularis
         6. Plants less than 30 cm. tall; flowers usually solitary, flowering late   -4. R. arkansana

1. Rosa engelmannii S. Wats., Garden & Forest 2:376. 1889.
Stems 30-50 cm. tall, low with arching branches, densely bristly and with infrastipular prickles present on some of the twigs; stems usually not dying back to the ground; leaflets 1-3 cm. long, 5-7, oval or elliptic, serrate, somewhat puberulent especially below, often glandular-puberulent; flowers solitary or a few corymbose on short lateral shoots from the old wood; petals about 2.5 cm. long.---Woods and thickets. Montana to Colorado. Our few records scattered in the western half of Colorado, mostly from the central part, at 8000-10,000 feet.

2. Rosa nutkana Presl., Epim. Bot. 203. 1851.
R. aciculata Cockerell; R. macdougalii Holz.; R. melina Greene; R. oreophila Rydb.; R. pandorana Greene; R. underwoodii Rydb.; R. manca Greene; probably R. bakeri Rydb.---Stems usually over 100 cm. tall, much branched, bristles if present at base only, but infrastipular prickles and often other scattered weak prickles present, not dying down to ground over winter; leaflets 1.5-5 cm. long, 5-7, ovate, oval to obovate, serrate, the teeth sometimes irregular or double, glabrous or puberulent below; flowers on lateral shoots 3-10 cm. long from the old wood, sometimes solitary but often 2 or 3 corymbose; petals 1.5-3 cm. long, usually over 2.5 cm. long.---Plains and hills, widely distributed in western Canada and western United States. Our records scattered over Colorado at 3500-10,000 feet.

3. Rosa woodsii Lindl., Ros. Monogr. 21. 1820.
R. fendleri Crepin; R. maximiliana Nees; R. neomexicana Cockerell; R. praetincta Cockerell; R. macounii Greene---Shrubs usually 100-300 cm. tall, sometimes only to 50 cm. tall; stems with infrastipular prickles present or absent, bristles and prickles present or often absent, not dying back to the ground over winter; leaflets 1-3 cm. long, 5-11, oval, elliptic or obovate, serrate, glabrous or somewhat puberulent below; flowers from lateral shoots from the old wood, these shoots usually over 10 cm. long and some usually bearing 4 or more flowers, rarely solitary; petals 1-2.5 cm. long.---Plains, hills and mountains. Widely distributed in western Canada and western United States. Our records widely scattered over Colorado at 3500-9000 feet.

4. Rosa arkansana Porter in Porter & Coult., Syn. Fl. Colo. 38. 1874.
R. pratincola Greene; R. suffulta Greene; R. alcea Greene---Shrubs with stems 10-50 cm. tall, infrastipular prickles absent but stems usually bristly to tip, often dying back to the ground over winter; leaflets 1-6 cm. long, 5-7, elliptic or obovate, coarsely serrate, glabrous to pubescent; flowers sometimes on lateral shoots from the old wood, but in part corymbose on long suckers from the base of the plant; petals 1.5-2.5 cm. long.---Plains and hills. Widely distributed in western Canada and western United States. Our records from the eastern half of Colorado and also the southwestern corner, at 3500-9000 feet.

5. Rosa acicularis Lindl., Ros. Monog. 44. 1820.
R. bourgeauiana Crepin; R. sayi Schwein.---Stems 30-200 cm. tall, infrastipular prickles absent, but covered with bristles or weak prickles, not dying down to the ground over winter; leaflets 1.5-5 cm. long, 5-7 (rarely 9), elliptic or oval, serrate, the teeth sometimes double, glabrous or nearly so above, somewhat pubescent below; flowers on short lateral shoots from the old wood, these shoots less than 7 cm. long, bearing 1-3 flowers; petals 2.5-3 cm. long.---

Plains, valleys and hills. Widely distributed in Canada and northern United States. Our records scattered in Colorado except the northeastern and northwestern corners at 4500-10,000 feet.

9. Potentilla L.

Perennial herbaceous plants (rarely annual or shrubby) with palmately or pinnately compound leaves, these with stipules; flowers solitary to cymose, usually the latter; calyx tube campanulate, saucer- or cup-shaped; calyx lobes 5, with 5 alternating bractlets; petals 5, usually yellow to ochroleucous; stamens 10 or more, inserted on an annular disk near the base of the receptacle; pistils superior, many, rarely less than 11, borne on an elevated receptacle; fruit of achenes with styles terminal or nearly so, lateral or basal, deciduous. This is a difficult genus and is badly in need of a general revision. Many of the segregates probably will not hold. However, some of the entities listed here as synonyms will doubtlessly be found to have at least subspecific rank.

1. Woody shrubs; styles lateral on hairy achenes -1. P. fruticosa
1. Plants herbaceous, woody only at the very base; styles terminal or lateral on glabrous achenes
  2. Styles lateral (sometimes near base of achene)
    3. Flowers solitary from the axils of the leaves crowded at the base of acaulescent stems, or on the long slender stolons; leaflets white-silky or white-tomentose at least below; petals bright yellow
      4. Leaflets green and sparsely silky or glabrate above -2. P. anserina
      4. Leaflets silvery-white above -2A. P. anserina sericea
    3. Flowers in clusters, terminal on the caulescent stems; leaflets glabrous to hairy but never white beneath; petals white to lemon-yellow when fresh
      5. Flowers in open cymes, the branches divergent; petals lemon-yellow when fresh -3. P. glandulosa
      5. Flowers in dense, aggregated cymes, the branches short and appressed or ascending; petals white to cream-colored when fresh
        6. Leaflets 9-13, regularly decreasing in size from apex to base, with rudimentary leaflets often interspersed, lateral leaflets nearly orbicular, rarely over 2 cm. long; petioles of basal leaves seldom over 1/2 as long as the rachis -4. P. fissa
        6. Leaflets 5-9 (very rarely 11), usually rather irregularly decreasing in size from apex to base, the rudimentary leaflets absent or only at base, lateral leaflets ovate, obovate or rhombic (sometimes suborbicular on lower ones), over 2 cm. long; petioles of basal leaves over 1/2 as long as the rachis, often as long
          7. Leaves densely hairy; stem prominently villous, reddish-tinged; cyme essentially capitate; plants usually growing east of the Continental Divide in Colorado -5A. P. arguta arguta
          7. Leaves sparingly hairy; stem only moderately villous or reddish; cyme merely condensed, not really capitate; plants usually growing west of the Continental Divide in Colorado -5B. P. arguta convallaria
  2. Styles terminal or very nearly so
    8. Plants annual or biennial, lacking perennial rootstocks and without rosettes of basal leaves; cyme many-flowered, conspicuously leafy, its leaves hardly bractlike although smaller than other leaves; styles somewhat thickened and glandular at base; petals seldom surpassing the sepals in length
      9. Achenes conspicuously swollen on 1 side near the base; leaves all pinnately 7- to 11-foliate -6. P. paradoxa
      9. Achenes not swollen near base; some or all the leaves 3-foliate
        10. Lower leaves pinnately 5-foliate below, 3-foliate above -7. P. rivalis
        10. All leaves 3-foliate, above and below
          11. Stamens about 15-20; petals nearly as long as the sepals; stem usually hirsute below -8. P. monspeliensis
          11. Stamens about 10; petals definitely shorter than the sepals (often about 1/2 as long); stem not hirsute below
            12. Stems erect or strongly ascending, often glandular-hairy; leaflets obovate to broader, 2-4 cm. long -9. P. biennis
            12. Stems with weak and decumbent branches, not glandular-hairy; leaflets obovate to oblong, 1-1.5 cm. long -10. P. leucocarpa
    8. Plants perennial with stout crowns which are often thick and woody and bear the persistent leaf bases of the previous year or a cluster of basal leaves of the season; cymes usually few-flowered, often with bracts but these not leaflike; styles filiform or glandular-thickened at base; petals usually surpassing the sepals
      13. Basal leaves with 3 leaflets -11. P. nivea
      13. Basal leaves with 5 or more leaflets (only rarely 3 in P. quinquifolia)
        14. Basal leaves palmately compound, the rachis if present very short; leaflets 5-7
          15. Basal leaves with a reduced pair of leaflets remote on the petiole from the others -12. P. subjuga
          15. Remote, reduced leaflets absent
            16. Leaflets of the basal leaves divided 1/2 or more to the midrib into linear, lanceolate or narrowly oblong segments
              17. Stems 5-20 cm. tall; leaflets either densely white-tomentose beneath or glandular-puberulent
                18. Leaflets densely white-tomentose beneath; inflorescence not glandular -13. P. quinquifolia
                18. Leaflets not densely tomentose beneath; inflorescence glandular-puberulent -14. P. subviscosa
              17. Stems 30-50 cm. tall; leaflets strigose to slightly tomentose below, not glandular-puberulent

19. Leaflets glandular, with brownish, strigose hairs — -15. P. brunnescens
19. Leaflets not glandular, no brownish hairs present
 20. Anthers 0.7-1 mm. long; sepals 5-8 mm. long; leaflets with 9-15 divisions, some below the middle — -16. P. pectinisecta
 20. Anthers 0.5-0.6 mm. long; sepals 2-5 mm. long; leaflets with 3-7 divisions, these above the middle — -17. P. diversifolia
16. Leaflets of basal leaves merely toothed or (as sometimes in P. gracilis) if divided 1/2 way to the midrib the segments broad
 21. Leaflets greenish above, white-tomentose below, the contrast conspicuous
  22. Plant 5-20 cm. tall; leaflets 5, small, not over 3 cm. long, usually toothed only above the middle — -18. P. concinna
  22. Plant 20-60 cm. tall; leaflets usually over 5, larger, often over 3 cm. long, toothed below the middle also — -19. P. pulcherrima
 21. Leaflets not conspicuously bicolored, greenish both sides, only slightly if ever tomentose below
  23. Anthers ovate to lanceolate, about 1 mm. long — -20. P. gracilis
  23. Anthers oval or nearly round, about 0.5 mm. long — -17. P. diversifolia
14. Basal leaves pinnately compound on a definite rachis; leaflets often over 7
 24. Basal leaves with 5 digitate leaflets and a reduced remote pair on the petiole — -12. P. subjuga
 24. Basal leaves pinnate, no reduced, remote pair of leaflets present on the petiole
  25. Styles shorter than the mature achene, glandular-thickened at base; leaflets deeply cut — -21. P. pennsylvanica
  25. Styles exceeding the mature achene, filiform, not glandular-thickened at base; leaflets not usually deeply cut (may be toothed)
   26. Basal leaves with leaflets deeply incised to pinnatifid or dissected; stems not over 30 cm. tall
    27. Basal leaves with 5 or rarely 7 leaflets, these tomentose or gray-silky below — -22. P. rubricaulis
    27. Basal leaves with 7-17 leaflets, these glabrous or appressed strigose below — -23. P. plattensis
   26. Basal leaves coarsely to shallowly toothed, not deeply incised to dissected; stems sometimes over 30 cm. tall
    28. Leaflets almost equally white-silky or white-tomentose above and below
     29. Basal leaves with 5-7 leaflets; stems about 10-20 cm. tall — -17. P. diversifolia
     29. Basal leaves with 7-13 leaflets; stems often over 20 cm. tall — -24. P. hippiana
    28. Leaflets glabrous or merely green-strigose at least above
     30. Leaflets white-tomentose below (definitely bicolored above and below) — -19. P. pulcherrima
     30. Leaflets glabrous, silky-villous or slightly tomentose on lower surface when young, not definitely bicolored
      31. Leaflets of basal leaves 5-7, glabrous below except midrib and margins — -25. P. rupincola
      31. Leaflets of basal leaves 9-15, silky-villous beneath
       32. Stems 40-70 cm. tall; leaflets 3-6 cm. long, toothed on sides also; sepals 6-7 mm. long — -26. P. ambigens
       32. Stems 20-30 cm. tall; leaflets 1.5-2 cm. long, toothed only near apex; sepals about 4 mm. long — -27. P. crinita

1. **Potentilla fruticosa** L., Sp. Pl. 495. 1753.
Dasiophora fruticosa (L.) Rydb. also Dasiphora fruticosa---Freely branching shrubs 30-100 cm. tall; branches with brown, shreddy bark; leaves pinnately compound with 3-7, usually 5 leaflets, these 0.5-2 cm. long, oblong to linear-oblong, entire-margined, cuneate and acute, usually whitish beneath, green above, more or less silky and margins usually somewhat revolute; flowers solitary or in loose terminal cymes; calyx tube, pedicels and young twigs silky-villous; calyx tube shorter than lobes which are 4-5 mm. long and ovate; bractlets narrower and longer than the calyx lobes; petals 5-15 mm. long, longer than the sepals, yellow, orbicular; stamens about 20-25; achenes densely hairy, styles near or below the middle.---Valleys and slopes, often in rocky or moist ground. Labrador to Alaska, south to New Jersey, Arizona and California; Eurasia. Our records widely scattered in the western half of Colorado at 7000-11,500 feet.

2. **Potentilla anserina** L., Sp. Pl. 495. 1753.
Argentina anserina (L.) Rydb.---Main stems very short, from a cluster of roots, and producing long runners up to 50 cm. long; basal leaves 8-20 cm. long, pinnate with about 9-31 larger leaflets interspersed with smaller ones, rachis and petioles hairy; larger leaflets 1-4 cm. long, oblong or oblanceolate, usually acute, deeply serrate, green and sparsely silky or glabrate above, white-silky or tomentose below; leaves of runners smaller; flowers solitary from axils of basal leaves or those of stolons, pedicels rather long, about 3-10 cm.; calyx tube shorter than lobes, these about 5 mm. long, ovate, bractlets sometimes toothed or divided, about as long as the sepals; petals 6-10 mm. long, yellow; stamens about 20-25; styles lateral, rather persistent; achenes rather corky and grooved on upper end, glabrous, numerous.---Rather moist open places. Newfoundland to Alaska, south to New York, Arizona and California; Eurasia. Our records in the western half of Colorado, few from the extreme western part, at 4500-9500 feet.

2A. **Potentilla anserina sericea** (L.) Hayne, (var.) Arzneigew 4.31. 1816.
Argentina argentea Rydb.; Potentilla anserina var. concolor Seringe---Differs from the species in having leaves silvery-whitish above as well as below. Possibly merely an ecological form.---Western Canada, south to New Mexico and California. Our few records from northcentral Colorado at 8500 feet.

3. **Potentilla glandulosa** Lindl., Bot. Reg. 19. pl. 1583. 1833.

Drymocallis glabrata Rydb.; D. foliosa Rydb.---Stems 30-50 cm. tall, several from a branching caudex, rather sparingly villous; basal leaves pinnate, slender-petioled, leaflets 7-9, these obovate-rhombic to ovate or even nearly orbicular, entire at base, coarsely and doubly serrate above, glandular-atomiferous and sparingly hairy; stem leaves 3- to 5-foliate, reduced; cyme flat-topped, rather open with divergent branches; calyx tube hairy to nearly glabrous, shorter than the ovate to lanceolate sepals which are 5-7 mm. long, bractlets narrower and shorter than the sepals; petals longer than the sepals, bright yellow; style basal or nearly so; achenes glabrous. This is the plant called ssp. glabrata (Rydb.) Keck.---Hills and mountains. The subspecies from Montana to Washington, south to Wyoming, Utah and Nevada. Our one record from northwestern Colorado at 9000 feet.

4. **Potentilla fissa** Nutt. in T. & G., Fl. N. Amer. 1:446. 1840.

Drymocallis fissa (Nutt.) Rydb.---Stems 20-30 cm. tall, rarely to 40 cm., with rather long stoloniferous branches to the caudex and with shaggy, often brownish hairs; basal leaves pinnate, short-petioled, the petiole rarely over 1/2 the length of the rachis, leaflets 9-13, decreasing regularly in size from apex to base with rudimentary leaflets sometimes interspersed, lateral leaflets rarely over 2 cm. long, leaflets nearly orbicular except terminal sometimes rhombic, incised, serrate, somewhat hairy or subglabrate above; stems leaves similar but reduced; inflorescence in a narrow cyme; calyx tube hairy, shorter than the triangular sepals which are 5-7 mm. long (rarely to 10 mm.), bractlets narrower and shorter than the sepals; style basal or nearly so; achenes glabrous.---Hills and mountains, often on rocky slopes. South Dakota and Wyoming, south to New Mexico. Our records in or near northcentral, central and southcentral Colorado at 5000-10,000 feet.

5. **Potentilla arguta** Pursh, Fl. Amer. Sept. 2:736. 1814.

Stems 30-100 cm. tall, stout and erect, with shaggy, usually brownish hairs, stoloniferous branches none or rarely present; basal leaves rather long-petioled, these rarely less than 1/2 the leaf rachis, often as long, leaflets 7-9, rarely to 11, often irregularly decreasing in size from apex to base of blade, rudimentary leaflets if present only at the very base, ovate, obovate to rhombic, sometimes suborbicular on lower ones, upper leaflets over 2 cm. long, coarsely incised-serrate, densely to sparingly hairy; stems leaves similar but smaller; sepals about 5 mm. long in flower, but enlarging to 15 mm. in fruit, longer than calyx tube, bractlets smaller; petals at least somewhat longer than the sepals, white or whitish when fresh. The two varieties listed below sometimes intergrade in Colorado.

5A. **Potentilla arguta arguta**

P. arguta ssp. typica Keck; Drymocallis arguta (Pursh) Rydb.; D. agrimonioides (Pursh.) Rydb.---Stems anthocyanous and prominently brown-villous; leaves densely hairy; cyme dense, essentially capitate.---Meadows and valleys. New Brunswick to Mackenzie, south to New Jersey and Colorado. Our records from northcentral, central, southcentral and southwestern Colorado at 5500-9000 feet.

5B. **Potentilla arguta convallaria** (Rydb.) Keck, (ssp.) Carneg. Inst. Pub. 520:39. 1940.

Drymocallis convallaria Rydb.; D. corymbosa Rydb.---Stems only slightly anthocyanous or villous; leaves sparingly hairy; cyme merely condensed, not capitate.---Meadows and valleys. Alberta to British Columbia, south to Colorado, Arizona and Oregon. Our records from northcentral and western Colorado at 6500-8500 feet.

6. **Potentilla paradoxa** Nutt. ex T. & G., Fl. N. Amer. 1:437. 1840.

Annual or biennial plants; stems 20-50 cm. tall, spreading, ascending or erect, leafy, sparsely pubescent; leaves pinnately 7- to 11-foliate, leaflets 1-2 cm. long, obovate or oblong, deeply crenate or even incised, sparsely hairy to glabrous; flowers in leafy-bracted cymes; calyx tube shorter than sepals which are 3-5 mm. long, ovate-triangular, the bractlets as long or somewhat longer; petals about as long or somewhat longer than sepals, yellow, obovate-cuneate; stamens 15-20; style terminal, short, thickened-glandular below (roughened from small irregular lumps, these rather shiny-glandular); achenes glabrous, with conspicuous corky swelling on lower side.---Rather low ground. New York to Washington, south to New Mexico. Our few records from northcentral Colorado at 4500-5500 feet.

7. **Potentilla rivalis** Nutt. ex T. & G., Fl. N. Amer. 1:437. 1840.

Annual or biennial plants; stems 15-60 cm. tall, erect, simple or branched below, finely villous-hirsute and more or less glandular; lower leaves pinnately 5-foliate, upper 3-foliate; leaflets obovate with coarse teeth, often incised, somewhat villous to glabrate; inflorescence a leafy cyme, pedicels short and branches ascending; calyx tube shorter than sepals which are 2.5-4 mm. long, ovate and acute, bractlets oblong not quite as long as the sepals; petals definitely shorter than the sepals, yellow; stamens about 10; style terminal or nearly so, short and thickened-glandular below; achenes glabrous, no swelling present.---Valleys. Illinois to Washington, south to Colorado, Arizona and California. Our records in or near northcentral, central and southcentral Colorado at 4500-7500 feet.

8. **Potentilla monspeliensis** L., Sp. Pl. 499. 1753.

Annual or biennial plants; stems rather stout, erect or ascending, often several from the base, hirsute especially near the base; leaves usually 3-foliate, palmate or somewhat pinnate (only rarely 5-foliate), more or less hirsute; leaflets 2.5-5 cm. long, obovate, with broad, deep teeth; cyme leafy-bracted, pedicels short; calyx tube shorter than sepals, sepals about 4-5 mm. long, ovate-triangular, bractlets narrower but about as long; petals nearly as long as the sepals, light yellow, obovate; stamens about 15-20; style nearly terminal, short and thickened-glandular; achenes usually longitudinally rugulose when mature, glabrous. The correct name for this plant may be P. norvegica L.---Waste places and rich ground. Nearly throughout North America; Eurasia. Our records scattered over Colorado, except in the extreme eastern part, at 4500-8500 feet.

9. Potentilla biennis Greene, Fl. Francisc. 65. 1891.

P. lateriflora Rydb.---Annual or biennial plants; stems 20-50 cm. tall, 1 to several, erect or nearly so, pubescent and glandular-hairy, often tinged with dark red; leaves palmately 3-foliate; leaflets 2-4 cm. long, broadly obovate to obovate, coarsely crenate, the teeth usually irregular in size, more or less pubescent; flowers in dense leafy cymes often appearing racemose; calyx and pedicels pubescent and glandular-puberulent; calyx tube shorter than sepals which are about 2.5-3 mm. long, bractlets about as long; petals shorter than the sepals, yellow, obovate-cuneate; stamens about 10; style terminal, short, thickened-glandular below; achenes glabrous, no swelling present.---Waste places. Saskatchewan to British Columbia, south to Colorado, Arizona and Lower California. Our records scattered in the western half of Colorado at 6500-10,000 feet.

10. Potentilla leucocarpa Rydb. in Britt. & Br., Ill. Fl. 2:212, fig. 1924. 1897.

P. millegrana of Manuals---Annual or biennial plants; stems slender, divaricately branching, weak and decumbent, softly pubescent to glabrate; leaves 3-foliate, usually palmately so, more or less pubescent; leaflets about 1-1.5 cm. long, obovate to oblong, coarsely and deeply toothed; cyme spreading and leafy-bracted; calyx tube shorter than sepal lobes which are about 2.5-4 mm. long, bractlets about as long; petals definitely shorter than the sepals, light yellow; stamens about 10; styles short, thickened-glandular, terminal or nearly so; achenes glabrous, no swelling at base.---Valleys and mountain slopes. Illinois to Washington, south to New Mexico and California. Our records scattered in the western two-thirds of Colorado, scarce in the extreme western part, at 4500-9500 feet.

11. Potentilla nivea L., Sp. Pl. 499. 1753.

P. uniflora Ledeb. as to Colo. plants---Perennial plants with short, thick, branched rootstocks covered with dark brown remnants of old leaves; stems 5-20 cm. tall, caespitose, growing in mats, more or less villous or tomentose; leaves mostly basal, 3-foliate, nearly or quite palmately compound; leaflets 1-3 cm. long, oblong, obovate to rhombic-obovate, from coarsely crenate to deeply cut, silky-strigose to glabrate above, densely white-tomentose below; flowers solitary or few in a small cyme; calyx more or less tomentose, the tube shorter than the lobes which are 4-6 mm. long, bractlet shorter and narrower; petals about as long or somewhat longer than the sepals, yellow, obovate; stamens about 20; style terminal, rather short and somewhat glandular below; achenes glabrous.---High mountains. Greenland to Alaska, south to Quebec and Colorado; Eurasia. Our records scattered in the higher mountains of Colorado at 9500-13,000 feet.

12. Potentilla subjuga Rydb., Bull. Torr. Club 23:397. 1896.

Perennial plants from caespitose caudices; stems 10-30 cm. tall, tufted, more or less silky-villous; basal leaves many, palmately 5 (rarely 3)-foliate and an additional smaller pair of leaflets on what appears to be the petiole; leaflets 1-4 cm. long, oblong or oblanceolate to obovate, deeply incised, sparingly silky above, denser below; stem leaves reduced, usually trifoliate; sepals 5-6 mm. long, strigose, the bractlets not as long by about 1/4 to 1/3; petals longer than the sepals, yellow; styles filiform, terminal; achenes glabrous.---High mountains. Apparently limited to Colorado. Our records from the northcentral, central and southwestern parts at 11,500-12,500 feet.

13. Potentilla quinquifolia Rydb., Mem. Bot. Columb. Coll. 2:76. 1898.

P. divisa Rydb.---Perennial plants with short rootstocks; stems 5-20 cm. tall, spreading to ascending, more or less pubescent; basal leaves many, with 5 (rarely 3) digitate leaflets, these 1-3 cm. long, obovate to oblanceolate in outline but deeply cleft 1/2 to midrib or more, into oblong or lanceolate divisions, silky-villous, sometimes greenish above, densely white-tomentose below; flowers few to several; sepals about 4 mm. long, silky-villous, bractlets somewhat shorter; petals about 5-6 mm. long, yellow; stamens about 20, anthers about 0.6-0.8 mm. long; styles long and filiform; achenes glabrous.---Hills, valleys and mountains. Saskatchewan to British Columbia, south to Colorado and Utah. Our records scattered in the Colorado mountains at 6000-13,000 feet.

14. Potentilla subviscosa Greene, Bull. Torr. Club 8:97. 1881.

Perennial plants from short caudices and taproots; stems not over 20 cm. tall, several to many, prostrate or spreading, more or less hirsute-glandular especially near top; leaves mostly basal, the plant subacaulescent; leaves digitately compound, of 5 leaflets (sometimes almost pinnate but rachis remaining very short); leaflets 1-4 cm. long, obovate, cuneate, deeply cleft into oblong divisions, the middle one often 3-parted to near base, puberulent and more or less glandular, rather sparingly hirsute especially on veins and near margins; sepals about 4 mm. long, glandular, bractlets shorter; petals about 5-6 mm. long, yellow; style near apex, rather elongate and filiform; pistils many, but often few maturing, these glabrous and marked with irregular and somewhat branched raised nerves on surface.---Meadows and forests. New Mexico to Arizona, south into Mexico. Our one record from southcentral Colorado at about 8000 feet.

15. Potentilla brunnescens Rydb., Bull. Torr. Club 28:173. 1901.

Perennial plants with short rootstocks; stems 40-50 cm. tall, strict, spreading-villous, tinged with brown or purplish-brown especially above; basal leaves palmately 5- to 7-foliate, the petioles long and villous; leaflets 2.5-7 cm. long, broadly oblanceolate in outline, long-strigose both sides the hairs usually definitely brownish, pectinately divided to near the midrib into linear segments; stem leaves smaller and short-petioled; cyme corymbose; sepals 4-6 mm. long, villous-strigose, the bractlets about 1/2 as long; petals 5-6 mm. long, yellow; styles filiform, inserted near apex of the glabrous achene.---Mountains Wyoming and Colorado. Our few records from northcentral and northwestern Colorado at about 8000-8500 feet.

16. Potentilla pectinisecta Rydb., Bull. Torr. Club 24:7. 1897.

P. bakeri Rydb.---Perennial plants; stems 30-60 cm. long, from heavy, erect rootstocks; leaves mostly basal, with about 7 palmately arranged leaflets, these about 3-5 cm. long, oblanceolate or obovate in outline, each with 9-15 linear or oblong lobes cut in about 2/3 of the way to the midrib, densely silky-hairy on both sides, the hairs appressed; cymes dense in flower but elongating in fruit, not leafy-bracted unless at very base; sepals 5-8 mm. long, silky, bractlets shorter and narrower; petals about 6-8 mm. long; stamens about 20, anthers ovate-cordate, about 0.7-1 mm. long; styles long and filiform; achenes glabrous.---Valleys and slopes. Montana to Oregon, south to Colorado and California. Our records from northcentral, northwestern and westcentral Colorado at 7500-9500 feet.

17. Potentilla diversifolia Lehm., Stirp. Pug. 2:9. 1830.

P. glaucophylla Lehm., P. dissecta var. glaucophylla (Lehm.) S. Wats.; P. intermittens Rydb.---Perennial plants; stems 10-40 cm. tall from woody rootstocks; leaves mostly basal, 5- sometimes to 7-foliate, digitate or sometimes pinnate, the leaflets 1-5 cm. long, oblanceolate, 3- to 7-toothed or incised above the middle, glabrous to silky-strigose especially below but scarcely if at all whitened; cymes few- to several-flowered; calyx sparsely silky, the lobes 2-5 mm. long, bractlets narrower and shorter; petals 5-8 mm. long, yellow; stamens about 20, anthers about 0.5-0.6 mm. long and almost as wide; styles long and filiform; achenes glabrous.---High mountains. Alberta to British Columbia, south to South Dakota, New Mexico and California. Our records from the mountains of Colorado at 7000-12,500 feet.

18. Potentilla concinna Richards. in Frankl., Journ. App. 739. 1823.

P. bicrenata Rydb.---Perennial plants from thick woody rootstocks; stems 5-20 cm. long, more or less tomentose; leaves palmately compound, basal mostly with 5 (rarely 7) leaflets, these 5-30 mm. long, obovate to oblong-cuneate, merely crenate or serrate near apex or sometimes to the middle, greenish and somewhat silky-strigose above, densely white-tomentose below; cymes few- to several-flowered; sepals 3-5 mm. long, silky-villous, bractlets narrower and shorter; petals 4-8 mm. long, yellow; stamens about 20, anthers 0.6-0.8 mm. long; styles long and filiform; achenes glabrous.---Dry hills, valleys and mountain slopes. Saskatchewan and Alberta, south to Colorado and Utah. Our records widely scattered over the western half of Colorado at 6000-12,500 feet.

19. Potentilla pulcherrima Lehm., Nov. Stirp. Pug. 2:10. 1830.

P. filipes Rydb.---Perennial plants; stems 20-60 cm. tall; basal leaves on long petioles; leaflets 5-11, usually 7, digitate or pinnate, 0.5-7 cm. long, obovate to oblanceolate, serrate with teeth cut less than 1/2 the way to the midrib, green and glabrate or sparsely silky-strigose above, conspicuously white-tomentose below, hence bicolored; cyme rather dense in flower, usually many-flowered; sepals 4-8 mm. long, sparsely strigose, bractlets shorter; petals 6-8 mm. long, yellow; stamens about 20, anthers about 0.7-0.8 mm. long, elliptic or ovate; styles filiform, not thickened or glandular at base; achenes glabrous. A rather well marked species in Colorado, distinguished by its bicolored leaves, but dwarf forms are rather hard to tell from P. concinna Rich. The writer cannot distinguish the following named forms: P. propinqua Rydb. and P. viridior Rydb.---Hills, ridges and valleys in the mountains. Manitoba to British Columbia, south to South Dakota, New Mexico, Arizona and Nevada. Our records scattered over the western two-thirds of Colorado at 6000-11,000 feet.

20. Potentilla gracilis Dougl. ex Hook., Bot. Mag. 57: pl. 2984. May 1830.

P. nuttallii Lehm.; P. jucunda A. Nels.---Perennial plants; stems 40-80 cm. tall; leaves digitately 5- to 7-foliate; leaflets about 3-10 cm. long, oblanceolate to obovate-oblanceolate, crenate to serrate or sometimes incised about 1/2 to the midrib, but segments broad, sometimes glabrate but usually more or less silky-strigose both sides, more below, rarely somewhat tomentose but both sides greenish; cymes variable, often many-flowered; calyx strigose-silky; sepals 4-7 mm. long, bractlets shorter; petals 7-10 mm. long, yellow; stamens about 20, anthers about 1 mm. long, ovate to lanceolate; styles long and filiform; achenes glabrous. Our plant has been called ssp. nuttallii (Lehm.) Keck.---Meadows and slopes. The subspecies from Alberta to Alaska, south to South Dakota, Colorado and California. Our records from the western half of Colorado, mostly in the north, at 5000-11,000 feet.

21. Potentilla pennsylvanica L., Mant. 76. 1767.

Perennial plants; stems from about 10 cm. up to 80 cm. tall, upright or decumbent, silky-pilose; leaves pinnately compound but the leaflets sometimes crowded, about 5-9 in number, these 1-5 cm. long, deeply pinnatifid the lobes oblong or linear, silky-strigose and usually gray-green above, densely strigose and more or less tomentose below; cymes 2- to many-flowered; sepals 3-6 mm. long, bractlets narrower and from shorter to about equalling the sepals; petals 3-7 mm. long, yellow; stamens 15-20; styles not much longer than the glabrous achenes, thickened and glandular at base. The following named forms are treated here as part of the species complex: P. platyloba Rydb., P. atrovirens Rydb., P. strigosa Pall., P. bipinnatifida Dougl. and P. pseudosericea Rydb.---Widely distributed in western Canada and the western part of the United States as treated here. Our records from the western two-thirds of Colorado, not so common in the extreme western part, at 4500-9500 feet.

22. Potentilla rubricaulis Lehm., Stirp. Pug. 2:11. 1830.

P. tenerrima Rydb.; P. minutifolia Rydb.; P. saximontana Rydb.; P. rubripes Rydb.---Perennial plants; stems 10-20 cm. tall, rarely more or less, usually caespitose, glabrous to silky-pubescent, often tinged with red; leaves mostly basal, pinnately compound with 5 or rarely 7 leaflets, these 1-5 cm. long, obovate or oblanceolate in outline, pinnately cleft, the divisions over 1/2 way into the midrib, white-tomentose or gray-silky below, glabrate to hirsute or silky-villous above; flowers usually few to several; calyx more or less silky-strigose or villous, the lobes about 4 mm. long, the bractlets shorter and narrower; petals about 6-8 mm. long, yellow; stamens about 20, anthers 0.5-0.8 mm. long; styles long and filiform; achenes glabrous.---As treated here rather widely distributed in western North America. Our records widely scattered in the western half of Colorado at 7000-13,000 feet.

23. Potentilla plattensis Nutt. ex T. & G., Fl. N. Amer. 1:439. 1840.

P. nelsoniana Rydb.; P. pinnatisecta A. Nels. at least in part---Perennial plants; stems 10-20 cm. tall (sometimes to 30 cm.), usually several, erect, ascending or prostrate, glabrate or villous; leaves mostly basal, pinnately compound, of 7-17 leaflets, these obovate-oblong, incised to near the midrib into linear or narrowly oblong lobes, glabrous to appressed strigose; cymes few- to several-flowered; calyx strigose, the sepals 4-5 mm. long, bractlets narrower and definitely shorter; petals about 5-7 mm. long, yellow, an orange spot often present near base; stamens about 20, anthers 0.6-0.8 mm. long; styles long and filiform; achenes glabrous.---Hills and valleys. Saskatchewan to New Mexico and Arizona. Our records in or very near northcentral, central and southcentral Colorado at 5000-10,000 feet.

24. Potentilla hippiana Lehm., Stirp. Pug. 2:7. 1830.

Perennial plants; stems 10-50 cm. long, sometimes even less, more or less silky; leaves pinnately compound, leaflets 7-13, rarely 5, 1-5 cm. long, oblanceolate to cuneate-oblong, coarsely crenate or serrate but never cleft to 1/2 into midrib, white-silky to gray-tomentose about equally on both sides; flowers usually few to several; sepals 4-7 mm. long, bractlets definitely shorter; petals 5-8 mm. long, yellow; stamens about 20, anthers 0.5-0.8 mm. long; styles long and filiform; achenes glabrous. The species as streated here is a complex with several intergrading forms. The writer cannot distinguish the following and the description includes them all: P. effusa Dougl., P. filicaulis (Nutt.) Rydb. and P. coloradensis Rydb.--- As treated here rather widely distributed in the western part of Canada and the United States. Our records scattered in the western two-thirds of Colorado, few in the extreme western part, at 5000-9500 feet.

25. Potentilla rupincola Osterhout, Bull. Torr. Club 26:256. 1899.

Perennial plants from branching caudices; stems 20-30 cm. tall, erect or nearly so, glabrous except for a few stiff hairs; basal leaves rather numerous, pinnately compound with 5-7 leaflets, these 1-6 cm. long, cuneate, sharply serrate, often almost incised except near base, glabrous or nearly so; stem leaves smaller; inflorescence a paniculate or diffuse cyme, more or less tomentose; calyx tube tomentose, sepals about 3-4 mm. long, bractlets about half as long; petals about as long as the sepals, yellow; stamens 18-20; pistils about 6; style filiform, attached near apex; achenes glabrous, but embedded in the wool of receptacle.---Mountains. Apparently limited to Colorado, our few records from the northcentral part at about 7500 feet.

26. Potentilla ambigens Greene, Erythea 1:4. 1903.

Perennial plants with thick, woody taproots; stems 40-70 cm. tall, rather stout, silky-villous; basal leaves up to 20 cm. long, irregularly pinnately compound; leaflets 9-15, 3-6 cm. long, coarsely serrate, often decurrent on the rachis, especially the upper, silky-villous below especially on veins, somewhat tomentulose when young, more glabrous above but the color contrast not striking; stem leaves smaller with fewer leaflets; flowers in a rather narrow cyme; sepals about 6-7 mm. long, lanceolate, strigose, bractlets as long or longer than sepals; petals about 8 mm. long, yellow; style filiform, near apex; achenes glabrous.---Mountains. Wyoming to New Mexico. Our few records from northcentral and southwestern Colorado at 9000-10,000 feet.

27. Potentilla crinita A. Gray, Am. Acad. Arts & Sci. Mem. Ser. 2, 4:41. 1849.

Perennial plants with short-branched caudices; stems 15-30 cm. tall, several, ascending, silky-pilose; basal leaves many, pinnately 11- to 17-foliate; leaflets 1.5-2 cm. long, crowded, oblong or oblong-oblanceolate, toothed near apex only, nearly glabrous above, silky-villous below; stem leaves few and reduced; sepals 4-5 mm. long, strigose, lanceolate or ovate-lanceolate, the bractlets shorter; petals somewhat longer than sepals, yellow; styles filiform, inserted near apex of glabrous achene.---Dry slopes, hills and valleys. Colorado to Utah, south to New Mexico and Arizona. Our few records from southwestern Colorado at about 7000 feet.

10. Rubus L. BLACKBERRY; RASPBERRY

Low shrubs with stems usually arching but sometimes trailing, woody or nearly herbaceous, often with sharp prickles; leaves alternate, simple or compound, with conspicuous stipules; inflorescence various, flowers perfect; calyx tube short, saucer-shaped, without bracts; calyx lobes 5, persistent; petals 5, deciduous, usually white; stamens many; pistils superior, many, on an enlarged receptacle, the style terminal or nearly so; fruit a collection of these pistils, each usually fleshy and ripening into a 1-seeded drupelet, these separate but coherent, and falling separate from or with the receptacle.

1. Stems armed with prickles or stiff bristles
    2. Stems with stiff bristles only; fruit red; inflorescence racemose, on the open order, not closely aggregated at the summit -1. R. strigosus
    2. Stems with prickles (broad at base and very stiff); fruit black or purple; inflorescence corymbose
        3. Mature fruit black (but with a bloom); pedicel prickles not very stout or broad-based, commonly nearly straight; terminal leaflets shortly or abruptly pointed; plants of eastern Colorado -2. R. occidentalis
        3. Mature fruit purple; pedicel prickles very stout and broad-based, usually hooked; terminal leaflets narrowly and very gradually long-pointed; plants, if present at all, in western Colorado -3. R. leucodermis
1. Stems without prickles or stiff bristles
    4. Leaves compound with 3 leaflets; stems weak, herbaceous or essentially so, with tight bark
        5. Petals white, 4-6 mm. long; plants with prostrate, rooting, stolonlike stems -4. R. pubescens
        5. Petals rose-colored, 8-13 mm. long; stems all erect and tufted -5. R. acaulis
    4. Leaves simple (lobed only); stems stout and woody the bark finally peeling or flaking
        6. Flowers in clusters of 2-9; leaves at maturity 6-30 cm. wide -6. R. parviflorus
        6. Flowers solitary; leaves rather small, about 4-6 cm. wide
            7. Peduncles and petioles glandular-hairy; petals 10-12 mm. long; leaves usually not over 3 cm. long; plants of western Colorado -7. R. exrubricundus
            7. Peduncles and petioles not glandular-hairy; petals 15-30 mm. long; leaves over 3 cm. long; plants of eastern Colorado -8. R. deliciosus

1. **Rubus strigosus** Michx., Fl. Bor. Amer. 1:297. 1803.
   R. melanolasius of Rydberg's Manual---Stems woody, erect, often glaucous, armed with stiff bristles, these hardly thick enough to be called true prickles, glabrous or glandular-pubescent when young; leaves 3- to 5-foliate, mostly 3-foliate on the flowering stems; leaflets lanceolate-ovate to narrowly ovate, usually acuminate, usually rounded or cuneate at base, terminal leaflets 3-8 cm. long, prominently doubly-serrate, green and sparingly puberulent or glabrous above, thin gray-pubescent to white-tomentose below; flowers in small terminal and axillary racemes, usually 4-7 together; sepals 7-10 mm. long, usually more or less glandular-pubescent, narrowly ovate, acuminate, soon reflexed; petals 5-6 mm. long, about as long as the erect stamens, white, narrowly oblong; fruit about 10-12 mm. wide, red, hemispheric, druplets falling from receptacle but not cohering together very well.---Mountains, often on open slopes. Newfoundland to Alaska, south to North Carolina, New Mexico and California. Our records scattered in the western half of Colorado at 5500-11,000 feet.

2. **Rubus occidentalis** L., Sp. Pl. 493. 1753.
   Stems woody, armed with stout straight or somewhat recurved prickles, these commonly few, stems often arching and taking root at the tips, glaucous when young; leaves 3-foliate on flowering stems, sometimes 3-foliate or nearly 5-foliate on basal shoots; petioles sparsely prickly; leaflets ovate to broadly ovate, terminal one widest, prominently stalked, laterals very short-petiolulate, leaflets green, glabrous to puberulent above, white-tomentose below especially in shade plants, doubly serrate and often somewhat lobed, terminal leaflets 5-10 cm. long; inflorescence of 3-7 closely clustered flowers in a corymb, this terminal or some lateral near apex, pedicels with prickles or bristles but no glands; sepals 7-8 mm. long, tomentose both sides, ovate-lanceolate, long-acuminate or even attenuate; petals somewhat shorter than sepals, white; fruit to 15 mm. wide, black with a bloom, hemispheric or nearly so.---Borders of woods and fields. Eastern part of Canada and the United States, west to Colorado. Our few records from the northcentral part of the state at 5000-5500 feet.

3. **Rubus leucodermis** Dougl. ex T. & G., Fl. N. Amer. 1:454. 1840.
   Montana to British Columbia, south to New Mexico and California. This species is closely related to the preceding and should be looked for in western Colorado.

4. **Rubus pubescens** Raf., Med. Rep. III. 2:333. 1811.
   R. americanus (Pers.) Britt.---Unarmed perennial plants, suffruticose only at base hence practically herbaceous; sterile stems about 10-100 cm. long, procumbent and acting like stolons, rooting at nodes and ends, glabrous or nearly so, no prickles or bristles; flowering stems short and erect, arising from base; leaves trifoliate (rarely with 5 leaflets); leaflets 3-6 cm. long, ovate, oval to oblong, green and glabrous to sparingly pubescent both sides, acuminate or acute at apex, cuneate often broadly so at base, sharply and doubly serrate, lateral leaflets sessile, terminal short-petiolulate; flowers solitary on the flowering shoots or with 1-3 axillary ones crowded near the ends; sepals 4-5 mm. long, acuminate, lanceolate; petals about 4-6 mm. long, white; fruit 5-12 mm. long, red, globular or conical, druplets adherent to receptacle.---Moist woods and swamps. Canada, south to New Jersey, Iowa, Colorado and Washington. Our few records from northcentral and central Colorado at 7000-7700 feet.

5. **Rubus acaulis** Michx., Fl. Bor. Amer. 1:298. 1803.
   Herbaceous unarmed perennial plants with scaly creeping rootstocks; stems 3-12 cm. tall, perhaps even taller; leaves 3-foliate (may be reduced to 1 leaflet on some of lower); leaflets 2-3 cm. long, oblong to ovate or obovate, commonly cuneate at base, mostly rounded or obtuse at the apex, irregularly crenate-serrate, petiolule short, nearly glabrous to sparingly appressed pubescent; flowers commonly solitary; sepals 7-10 mm. long, narrowly lanceolate, at length reflexed; petals 10-20 mm. long, rose-colored; pistils many; fruit 1 cm. or less wide.---Meadows, woods, and bogs. Newfoundland to Alaska, south to Minnesota and Colorado. Our one record from northcentral Colorado at 9000 feet.

6. **Rubus parviflorus** Nutt., Gen. 1:308. 1818.
   Bossekia parviflora (Nutt.) Greene; Rubacer parviflorum (Nutt.) Rydb.---Stems to 1 or 2 m. tall, woody, erect or upright, bark of older stems more or less flaking or shredding, young shoots glabrous to glandular-hairy, without bristles or prickles; leaves simple, blades usually over 10 cm. long when mature on some of canes, broadly ovate, cordate at base, palmately 3- to 5-lobed, the lobes serrate to dentate, the sinuses wide and rather shallow (mostly about 1/3 of the way in), glabrous or nearly so above, pubescent to nearly glabrous below; inflorescence a cymose cluster of 2-9 flowers; sepals 1-1.5 cm. long, broadly ovate, with abrupt tips finally spreading or reflexed; petals 15-30 mm. long, white; fruit 15-20 mm. wide, red. A rather variable species.---Open woods and open slopes. Ontario to Alaska, south to Michigan, Mexico and California. Our records from the western half of Colorado, few in the extreme western part, at 7000-10,000 feet.

7. **Rubus exrubricundus** L. H. Bailey, Gentes Herbarium 5:917. 1945.
   Upright or ascending shrubs about 1-1.5 m. tall; stems unarmed, the younger pubescent; leaves broadly ovate, simple, more or less 3- to 5-lobed, these shallow and serrate, base of leaf rather cordate, glabrous to sparingly pubescent above, pubescent on the veins at least below, some of the hairs glandular; flowers solitary, terminating a short, leafy branch, the peduncles glandular-hairy; sepals about 10-12 mm. long, acuminate; petals 10-12 mm. long, white, rather showy; fruit small and dry.---Canyons, valleys and cliffs. New Mexico and Arizona. Our one record from westcentral Colorado at about 6500 feet.

8. **Rubus deliciosus** Torr., Ann. Lyc. N. Y. 2:196. 1828.
   Bossekia deliciosa (James) A. Nels.; Oreobatus deliciosus (Torr.) Rydb.---Shrubs to 1.5 m. tall, the branches woody, erect or ascending, older bark flaky, unarmed, young twigs pubescent; leaves 3 to 6 cm. wide, simple, more or less 3- to 5-lobed, these shallow and dentate, cordate at base, orbicular-reniform in outline, soon glabrous above, glabrous to somewhat puberulent below; inflorescence a solitary flower terminal on a short leafy branchlet; sepals 10-15 mm. long, lanceolate-ovate, acuminate or with foliaceous tips, sometimes glandular-pubescent; petals 1.5 to 3 cm. long, white, showy; fruit about 1 cm. broad or more, dark purple, receptacle

dry and druplets with little juice.---Foothills and canyons often in rocky ground. Wyoming and Colorado. Our records from northcentral, central, southcentral and southeastern Colorado at 4500-9000 feet.

11. Prunus L. CHERRY; CHOKECHERRY; PLUM

Trees or shrubs; leaves alternate, simple, deciduous; flowers umbellate, corymbose, racemose or solitary in the leaf axils; calyx tube campanulate or turbinate, free from the ovary, with 5 lobes; petals 5, spreading, inserted on a disk on the edge of the calyx tube; stamens many; pistil 1, superior, ovary 1-celled, style terminal; fruit a drupe. In addition to the species treated in the key, P. angustifolia Marsh might possibly be in southeastern Colorado, and P. persica (L.) Batsch. (the peach) occasionally escapes and persists.

1. Flowers in narrow racemes on leafy branches, many in a cluster — -1. P. virginiana
1. Flowers in umbellate clusters, the clusters sessile or nearly so from lateral buds, few (3-6)
   2. Calyx lobes tomentose inside and on margins; branches usually spiny; young twigs and petioles more or less hairy and leaves often pubescent along midrib even in age; fruit a plum with a ventral groove and a flat stone — -2. P. americana
   2. Calyx lobes glabrous both sides; branches never spiny; young twigs and petioles glabrous, and leaves glabrous at least in age; fruit a cherry without a ventral groove, the stone subglobose
      3. Low and often prostrate shrubs, seldom over 100 cm. tall; leaves glaucescent beneath, entire toward the base; buds often 2 or 3 in the leaf axil (hence often 2 or 3 separate clusters of flowers or leaf shoots together); petals glabrous; fruit 12-18 mm. in diameter — -3. P. besseyi
      3. Shrubs with upright or ascending stems, usually over 100 cm. tall; leaves green both sides, serrated nearly or quite to the base; buds only 1 in a leaf axil; petals hairy outside near base; fruit 6-7 mm. in diameter — -4. P. pensylvanica

1. **Prunus virginiana** L., Sp. Pl. 473. 1753.
P. melanocarpa (A. Nels.) Rydb.---Shrubs or trees up to 10 m. tall; twigs glabrous, reddish-brown, with gray longitudinal lenticels, unarmed; leaves 4-10 cm. long, broadly to narrowly elliptical or obovate, abruptly acute or acuminate at apex, rounded to subcordate at base, margins finely serrate the teeth incurved or appressed at apex, glabrous both sides, paler below; petioles 1-2 cm. long, usually but not always glandular; inflorescence a raceme appearing with the leaves, many-flowered; sepals broad and short, calyx glabrous; petals 3-6 mm. long, white; fruit dark purple or black. Our plant has been called var. melanocarpa (A. Nels.) Sarg.---Hills, valleys and banks. Canada, south to Georgia, New Mexico and California. Our records scattered over Colorado at 4500-9000 feet.

2. **Prunus americana** Marsh., Arb. Amer. p. 111. 1785.
P. ignota A. Nels.---Tall shrubs or small trees to 5 m., frequently in thickets, branches more or less spiny, these spines as sharp ends of short, leafy twigs; young twigs usually pubescent, the hairs often somewhat persisting especially on 1 side; leaves 6-10 cm. long when mature, elliptic, oblong-elliptic, sometimes ovate or obovate, green and glabrous above, paler and glabrous below but very often pubescent along the midribs or the lower margins, sharply and sometimes doubly serrate, cuneate or rounded at base, rather abruptly acuminate; stipules linear or lobed; flowers 18-25 mm. across, 3-4, in sessile umbels from lateral buds, appearing before or with the leaves; calyx tube about 2-3 mm. long, glabrous, the lobes shorter or about as long as tube, glabrous on back, tomentose inside and on margins; petals 6-10 mm. long, white or sometimes pink; fruit 1.5-2.5 mm. long, subglobose to rather elongated, varying from red to reddish-orange.---Slopes and valleys, often along streams. Massachusetts to Manitoba, south to Florida and New Mexico. Our records from the eastern half of Colorado, scarce in the extreme eastern part, at 3500-6000 feet.

3. **Prunus besseyi** Bailey, Bull. Cornell Agri. Exp. Sta. 70:261. 1894.
P. prunella Daniels---Dwarf bushy shrubs to 1.5 m. tall, erect, spreading or prostrate; branches glabrous; leaves 3-6 cm. long at maturity, elliptic-lanceolate, oblong-obovate to elliptic, cuneate at base, abruptly short-acuminate to obtuse at apex, glabrous both sides, green above, glaucescent below, appressed serrate except near base; petioles 4-8 mm. long, sometimes glandular; stipules glandular-serrate to laciniate; flowers in sessile umbellate clusters of 3-4, appearing with or before the leaves; sepals about 2 mm. long, calyx glabrous; petals about 4-7 mm. long, white; fruit 12-18 mm. in diameter, usually black or reddish-black, sometimes red, subglobose.---Sandy hills and plains. South Dakota to Wyoming, south to Kansas and Colorado. Our records mostly from the northeastern quarter of Colorado at 3500-6500 feet.

4. **Prunus pensylvanica** L. f., Suppl. Pl. 252. 1871.
Shrubs or small trees to 5 m. tall; twigs reddish, glabrous, unarmed, buds (hence flowers and leaf clusters and also leaves) with a tendency to cluster 3 or 4 out near end of the longer twigs; leaves 3-8 cm. long or longer, oblong-lanceolate to oval especially at first, margins finely serrate, acute or acuminate at apex, rounded or cuneate at base, glabrous both sides or sometimes sparingly long-hairy below especially along midrib when young; stipules linear, glandular-serrate; flowers 9-12 mm. wide, in clusters of 3-6 from lateral buds, corymbose or umbellate, the inflorescence nearly sessile to peduncled; calyx tube 2-3 mm. long, glabrous, the lobes about 2 mm. long, glabrous on outside near base; fruit 6-7 mm. wide, globose, red. Our plant has been called subspecies corymbulosa (Rydb.) Wight.---Woods, hills and banks. The subspecies from Newfoundland to Hudson Bay, south to Indiana and Colorado. Our records from northcentral, central and southcentral Colorado at 5500-8500 feet.

12. Dryas L. MOUNTAIN AVENS; ALPINE AVENS; DRYAD

Low depressed, matted undershrubs; leaves alternate, simple, petioled; flowers solitary on naked peduncles; calyx tube saucer-shaped, little developed, calyx lobes 8-10, persistent, bractlets not present; petals 8-10; stamens many, inserted on the limb of the calyx tube; ovaries superior, numerous, sessile, styles terminal and persistent; fruit a 1-seeded, indehiscent achene with the style elongated and plumose.

1. Dryas octopetala L., Sp. Pl. 501. 1753.
Branches horizontal; leaves 6-25 mm. long, oblong to elliptic-oblong, obtuse at apex, rounded or subcordate at base, glabrate and dull above, white-tomentose beneath, veins conspicuous, impressed above, margins coarsely crenate and somewhat revolute; scapes 5-20 cm. long, tomentose and more or less black-hairy; calyx tube tomentose and black-hairy; sepals 4-7 mm. long, linear or linear-lanceolate; petals 8-15 mm. long, elliptic or obovate, white; styles in fruit 2-3 cm. long; achenes about 2.5-4 mm. long. Porsild (Canadian Field Nat. 61:175-192. 1947) considered our Rocky Mountain plant to be D. hookeriana Juz., and true D. octopetala L. limited on this continent to Alaska and vicinity.---High mountains. Greenland to Alaska, south to Colorado and Washington. Our records in the higher mountains of Colorado at 11,000-14,000 feet.

13. Spiraea L.

Shrubs or depressed undershrubs with prostrate branches, growing on rocks in mats; leaves alternate, no stipules; flowers variously clustered, in ours in a spikelike raceme; calyx tube hemispheric or campanulate, free from the ovaries, calyx lobes 5; stamens about 20; pistils superior, 3-5, distinct; style filiform, terminal; fruit of follicles, dehiscent along the ventral suture and often splitting part way down the dorsal. Our species has been segregated into another genus Petrophytum on the basis of dehiscence of follicle. However, McVaugh (Flora Nevada 22:12. 1942) points out that species of Spiraea also split dorsally part way.

1. Spiraea caespitosa Nutt. in T. & G., Fl. N. Amer. 1:418. 1840.
Petrophytum caespitosum (Nutt.) Rydb.---Season's shoots very short; leaves 5-12 mm. long, about 2-4 mm. wide, crowded and rather rosulate, coriaceous, entire, spatulate, densely appressed silky-hairy, 1-ribbed; peduncles 3-20 cm. long, silky, with several bractlike reduced leaves; inflorescence short, 1-5 cm. long, dense and spikelike, flowers small and numerous; calyx tube 0.5-1 mm. long, silky hairy, sepals 1-1.5 mm. long, silky; petals 1.5-2 mm. long, oblanceolate, white, pilose inside near base; follicles about 2 mm. long, dehiscent ventrally and usually about 1/2 way from apex on dorsal side.---On rocks. South Dakota, Montana and Idaho, south to New Mexico and California. Our few records from western Colorado at 6500-8000 feet.

14. Holodiscus Maxim. OCEAN SPRAY; MOUNTAIN SPRAY; ROCK SPIREA

Shrubs, much branched, no spines or prickles present; leaves alternate, simple, toothed, deciduous, lacking stipules; inflorescence a terminal panicle; calyx tube saucer-shaped, calyx lobes 5, valvate in bud, erect in fruit; petals 5, white, pinkish or cream-colored; stamens about 20, inserted on a disk lining the calyx tube; pistils 5, superior, distinct, styles terminal; fruit of short-stipitate and villous achenes.

1. Leaf blades toothed only at top, never at or below the middle, obovate to spatulate, widest above the middle; inflorescence narrow and compact   -1. H. microphyllus
1. Leaf blades toothed to middle or below, elliptic to ovate, widest at or below the middle; inflorescence usually open   -2. H. dumosus

1. Holodiscus microphyllus Rydb., Bull. Torr. Club 31:559. 1904.
Sericotheca microphylla Rydb.---Low shrubs, 15-150 cm. tall, spreading, older bark red but becoming gray and exfoliating, young twigs glabrescent to villous-pubescent; leaves 5-20 mm. long, obovate to spatulate, 2- to 3-toothed on each side and only above the middle, cuneate to a very short petiole, glabrescent to villous above, pubescent to villous below; inflorescence 3-10 cm. long, usually narrow and compact; sepals 1-1.5 mm. long; petals 1.5-2 mm. long. Our plant has been called variety typicus Ley.---Mountains. The variety from Wyoming to Idaho, south to Colorado and California. Our few records from the western one-third of Colorado at 8000-9500 feet.
2. Holodiscus dumosus (Nutt.) Heller, Cat. N. Am. Pl. 4. 1898.
Shrubs 50-300 cm. tall, spreading, bark of older twigs dark red, later becoming gray and exfoliating, young twigs villous; leaf blades 1-4.5 cm. long, elliptic or elliptic-ovate, cuneate, with about 3-6 usually simple teeth on each side extending to below the middle, glabrous or nearly so above, from pubescent to villous-tomentose below; petioles rather short, definitely shorter than blades; inflorescence 5-20 cm. long, 3-10 cm. wide, compound especially in lower part; sepals 1.5-2 mm. long, triangular ovate; petals about 2 mm. long, oval. Two varieties are listed by Ley (Bull. Torr. Club 70:275-288. 1943) for Colorado, var. typica Ley and var. australis (Heller) Ley. The author cannot differentiate between the 2 in Colorado plants.---Mountains, mostly on rocky slopes. Wyoming and Utah, south to Texas, Mexico and Arizona. Our records scattered in the western two-thirds of Colorado at 5500-10,000 feet.

15. Fallugia Endl. APACHE PLUME

Low much branched shrubs, the branches whitish to straw-colored, bark finally flaky; leaves alternate but fascicled, simple but pinnately cleft or divided into linear divisions, these revolute on margins; small stipules present; flowers terminal, solitary or few in a corymb; calyx tube hemispheric, calyx lobes 5, imbricate, with 5 alternating bractlets; petals 5, white, orbicular, spreading; stamens numerous on the margin of the calyx tube; pistils superior, many on a conical receptacle, villous, not adnate to the calyx tube, styles terminal, ovules 1; fruit of oblong, coriaceous, villous achenes with the styles becoming elongate, plumose and persistent.

1. Fallugia paradoxa(D. Don.) Endl., Genera Pl. 1246. 1840.
    F. acuminata (Woot.) Rydb.---Shrubs 50-150 cm. tall, young twigs pubescent; leaves 1-2 cm. long, with 3-7 linear lobes, pubescent, villous, or in age glabrate above, rusty-scaly below; calyx tube 3-5 mm. long, silky-villous; sepals 4-8 mm. long, long acuminate-caudate, sometimes this tail somewhat 3-toothed at apex, villous outside, glabrous or nearly so within; petals 11-15 mm. long; styles of achenes 2.5-4 cm. long, often purplish, the achene about 3 mm. long.---Hills, slopes and canyons. Colorado to California, south to Texas and Mexico. Our records from southern Colorado, mostly in the southcentral part at 4500-8500 feet.

16. Physocarpus Maxim. NINEBARK

Shrubs with exfoliating bark; leaves simple, broadly ovate to reniform, palmately lobed, alternate, usually with at least some stellate hairs, and with membraneous deciduous stipules; inflorescence of terminal corymbs, the flowers perfect; calyx tube campanulate or cup-shaped; calyx lobes 5; petals 5, white; stamens 20-40, inserted on a disk at the mouth of the calyx tube; pistils superior, 1-5, more or less united; style terminal, stigma capitate; fruit a capsule, the individual parts folliclelike and dehiscent on both sutures; seeds 2-4.

1. Carpels 3-5, united only at the base, turgid, 7-9 mm. long -1. P. intermedius
1. Carpels 1 or if 2 then united for at least 1/2 their length, turgid or flattened, 3-6 mm. long
    2. Leaf blades not over 15 mm. long, with petioles not over 6 mm. long; carpels 1; filaments alternately long and short; low shrub, not over 1 m. tall, with stellate leaves (densely so below) -2. P. alternans
    2. Leaf blades (at least some) over 15 mm. long, some of petioles over 6 mm. long; carpels normally 2; filaments all alike or nearly so; shrubs often over 1 m. tall, the leaves glabrous or nearly so on both sides
        3. Mature carpels flattened and definitely keeled; styles erect -3. P. malvaceus
        3. Mature carpels turgid; styles divergent -4. P. monogynus

1. Physocarpus intermedius (Rydb.) Schneid., Handb. Laubh. 1:807. 1906.
    P. intermedius (Rydb.) A. Nels.; Opulaster intermedius Rydb.---New York to South Dakota, south to Missouri and Colorado. Closely related to P. monogynus and may not be distinct. Some of our specimens from northcentral and central Colorado may be this species.
2. Physocarpus alternans (M. E. Jones) J. T. Howell, Proc. Calif. Acad. 20:130. 1931.
    Opulaster alternans (M. E. Jones) Heller---Low shrubs up to about 1 m. tall, much branched; branchlets brown and pubescent; leaves about 10 mm. long, sometimes to 15 mm., green and more or less stellate above, pale and densely stellate-tomentose below, with 3 rounded lobes these more or less crenate, usually cordate at base; bracts oblanceolate; corymbs pubescent, about 1- to 6-flowered; calyx stellate, the tube about 2-2.5 mm. long, the sepals about as long as the tube, obtuse; petals about 2.5-3 mm. long, white, glabrous; stamens with alternating longer and shorter filament; carpels 1, 3-5 mm. long, densely stellate.---Rocky slopes and cliffs. Utah and Nevada. Our one record from the northwestern corner of the state at about 8000 feet.
3. Physocarpus malvaceus (Greene) Kuntze, Rev. Gen. 219. 1891.
    Opulaster malvaceus (Greene) Kuntze---Montana to British Columbia, south to Wyoming, Utah and Oregon. The species could well be in northern or western Colorado.
4. Physocarpus monogynus (Torr.) Coult., Contrib. U. S. Nat. Herb. 2:104. 1891.
    Opulaster monogynus (Torr.) Kuntze---Shrubs 50-200 cm. tall; leaf blades more or less deeply 3- to 5-lobed, divisions doubly crenate, glabrous or nearly so on both sides; bracts usually membranous and deciduous, rarely foliaceous and more or less persistent; calyx stellate, the sepals sometimes thinly so on back and the tube sparsely so, the lobes about 3 mm. long; petals 3-5 mm. long; carpels 2 (rarely 1 or 3) adnate to the middle or above, the styles somewhat spreading; fruit about 3-4 mm. long, densely stellate. As treated here this is a complex containing P. ramaleyi A. Nels.; P. pubescens (Rydb.) A. Nels.; P. bracteatus Rydb. and P. glabratus Rydb.---Mountains, often in canyons or rocky slopes. South Dakota to Nevada, south to Texas and Arizona. Our records from the western two-thirds of Colorado, few in the extreme western part, at 5500-10,000 feet.

17. Purshia DC. ANTELOPE-BRUSH; BITTERBRUSH

Intricately branched shrubs with branches spreading or sprawling; leaves alternate but crowded and appearing fascicled, simple but deeply 3-cleft; flowers solitary, terminating short branchlets; calyx tube turbinate or funnelform; calyx lobes 5; petals 5, yellow; stamens about 20-25 in 1 row, inserted on the margin of the calyx tube; pistil superior, solitary (rarely 2), with a short style, sessile at bottom of calyx tube; fruit pubescent, attenuate each end, an achene.

1. **Purshia tridentata** (Pursh) DC., Linn. Soc. London Trans. 12:158. 1817.
   Kunzia tridentata (Pursh) Spreng.---Shrubs to 300 cm. tall, with brown or gray bark; leaves 5-30 mm. long, cuneate-obovate, short-petioled, glabrate or slightly tomentose above, white-tomentose below, margins revolute; calyx tube funnelform, tomentose, often glandular too; calyx lobes about 2-3 mm. long, oblong or elliptic; petals 5-9 mm. long, spatulate-obovate; fruit 8-12 mm. long, exserted.---Hills and slopes. Montana to British Columbia, south to New Mexico and California. Our records scattered in the western half of Colorado at 5000-8000 feet.

18. Cowania D. Don. CLIFFROSE

Shrubs; leaves alternate, simple, coriaceous, usually glandular-dotted and resinous, lobed; flowers rather large, the terminal on short branches; calyx tube hemispheric to turbinate, calyx lobes 5, persistent, no bractlets present; petals 5, spreading, obovate, stamens numerous, on the rim of the calyx tube; pistils 4-10, superior, sessile, not adnate to calyx tube, villous, style terminal; fruit of narrowly-oblong, coriaceous achenes partly included in the calyx tube, the styles persistent, plumose and elongated.

1. **Cowania mexicana** Don., Trans. Linn. Soc. 14:575. 1825.
   C. stansburiana Torr.---Much branched shrubs to 3.5 m. tall, with young twigs reddish-brown or greenish, glandular and also puberulent, older twigs with gray bark, finally shredding somewhat; leaves 6-15 mm. long, obovate in outline, pinnately 3- to 5-lobed, the lobes linear to broader, some usually toothed or again lobed, revolute-margined, green and glandular-punctate above, more or less whitish-tomentose below, crowded on the branches, cuneate at base, petiole lacking (or margined if present), texture coriaceous, resinous; stipules fused with the base of the blade and persistent after its fall; flowers solitary at the end of leafy, short branchlets, pedicels glandular; calyx tube 4-7 mm. long, turbinate to campanulate-turbinate, glandular; calyx lobes 4-6 mm. long, hairy and glandular outside, usually glabrous or nearly so inside; petals 6-10 mm. long, white, cream-colored or yellow; pistils at bottom of calyx tube; styles 4-6 cm. long in fruit; achenes 5-8 mm. long, glabrous at maturity, longitudinally ribbed.---Hills and canyons. Colorado to Nevada, south to Mexico and California. Our records from westcentral and southwestern Colorado at 4500-7500 feet.

19. Agrimonia L. AGRIMONY

Perennial herbaceous plants with rootstocks; leaves alternate, odd-pinnate, with leaflets alternately large and small; stipules conspicuous; flowers regular, perfect, in slender, spikelike racemes; calyx tube hemispheric to obconic, constricted at throat and enclosing the fruit, with hooked prickles above; calyx lobes 5, more or less connivent in fruit; petals 5, clawless, yellow; stamens 5-15; pistils 1 or 2, superior, styles terminal, stigmas 2-lobed; fruit of 1-2 achenes.

1. Bristles all erect or ascending; sepals acute; leaves usually definitely pubescent below   -1. *A. striata*
1. Lower bristles strongly reflexed at maturity; sepals acuminate; leaves sparsely if at all hairy below
                                                                                             -2. *A. gryposepala*

1. **Agrimonia striata** Michx., Fl. Bor. Amer. 1:287. 1803.
   A. brittoniana Bickn.; *A. brittoniana occidentalis* Bickn. in Rydb. Fl. Colo.---Stems 30-100 cm. tall, hirsute and somewhat glandular; larger leaflets 3-10 cm. long, 7-13, more or less hispid or scabrous above, more or less pubescent beneath and glandular dotted, lanceolate to rhombic-obovate, serrate; calyx tube with bristles not reflexed; calyx lobes about 1.5 mm. long; petals about 3 mm. long; calyx tube in fruit about 5 mm. long.---Open or wooded areas. Nova Scotia to British Columbia, south to West Virginia, New Mexico and Arizona. Our scattered records in northcentral, central and southwestern Colorado at 5000-9000 feet.
2. **Agrimonia gryposepala** Wallr., Beitr. Bot. 1:49. 1842.
   This species is widespread in North America and should be looked for in Colorado.

20. Ivesia Torr. & Gray

Perennial herbaceous plants with thick erect rootstocks; leaves mostly basal, pinnately compound, the leaflets divided into narrow lobes; flowers cymose, crowded and capitate; calyx tube campanulate; calyx lobes 5, alternate to 5 bractlets; petals 5, yellow, usually clawed; stamens 5 with filiform filaments; pistils 1-6, superior, styles subterminal; fruit of separate achenes surrounded by bristles from the receptacle. This genus is related to Potentilla and combined with it by a few authors. However Keck (Lloydia 1. 75-142. 1938) argues for the separation of the genus, and our 1 species is very different.

1. **Ivesia gordonii** (Hook.) T. & G. in Newberry, Pacif. Rail. Rep. 6(3):72. 1857.
   Horkelia gordonii Hook.---Stems 10-25 cm. tall, erect or ascending, subscapose, glabrous to glandular-puberulent; leaves 3-15 cm. long, with 10-25 pairs of leaflets, these 2-8 mm. long, divided to base into 2-5 oblong or linear segments, glabrous to pubescent and viscid-glandular; bractlets linear, about half as long as the calyx lobes, these latter 2.5-4 mm. long; petals about as long as the bractlets, sometimes as long as the calyx lobes, yellow (often bleached white).---Mountains, often in rocky ground. Montana to Washington, south to Colorado and California. Our records from the western half of Colorado at 7500-10,500 feet.

21. Sibbaldia L.

Dwarf, tufted, caespitose perennial plants; leaves 3-foliate; flowers cymose, no scape-like peduncle; calyx tube campanulate to saucer-shaped; calyx lobes 5, with 5 alternating bractlets; petals 5, yellow, minute; stamens 5 with filiform filaments; pistils superior, 5-20 on a dry, small receptacle; fruit of achenes with short subbasal, deciduous styles. This genus has been combined with Potentilla by some authors. However, in our limited area it seems distinct enough and is kept separate pending a monographic treatment of the group.

1. Sibbaldia procumbens L., Sp. Pl. 284. 1753.
Densely caespitose plants with short, decumbent, creeping rootstocks; flowering stems 5-15 cm. tall, leafy-bracted near inflorescence, more or less strigose; basal leaves on long, slender petioles about as long as the flowering stems; leaflets 1-2 cm. long, obovate to oblanceolate, cuneate at base, truncate and 3- to 5-lobed at apex, sparsely pubescent both surfaces; flowers rather few, densely cymose; calyx tube shorter than sepals, these 2-3 mm. long, bractlets narrow and shorter; petals shorter than the sepals, oblong; stamens inserted close to base of calyx tube.---High mountains. Circumpolar and alpine in the Northern Hemisphere. Our records scattered in the high mountains of Colorado at 10,000-13,500 feet.

22. Chamaerhodos Bunge

Perennial or biennial herbaceous plants with taproots; leaves 2- to 4-ternately divided; flowers small, in cymes; calyx tube cup-shaped or somewhat campanulate, short, the base with a ciliate disk; sepals 5; no bractlets present; petals 5, somewhat clawed, obovate-cuneate, white; stamens 5, opposite the petals, inserted in the sinuses of the sepals; pistils 5-10 or rarely more, superior, on a dry, villous receptacle, not adnate to calyx tube, styles basal; achenes with deciduous styles.

1. Chamaerhodos nuttallii (T. & G.) Pickering in Rydb., N. Am. Fl. 22:377. 1908.
C. erecta of Manual---Stems 10-30 cm. tall, erect, paniculately branching above, leafy, hirsute-villous and glandular; basal leaves numerous and in a rosette, 2- to 4-ternately divided into linear or oblong divisions; stem leaves rather numerous and similar to the basal ones but smaller and less divided; inflorescence many-flowered, often rather flat-topped with ascending branches; calyx tube 1-1.5 mm. long, hispid, sepals about as long or somewhat longer, hispid with transparent glasslike hairs, lanceolate; petals usually somewhat longer than sepals; fruit about 1.3 mm. long, brownish-black, ovoid-pyriform in shape.---Plains and mountains. Saskatchewan to Alaska, south to South Dakota, Colorado and Utah. Our records in or very near northcentral, central and southcentral Colorado at 6500-9500 feet.

23. Fragaria L. STRAWBERRY

Acaulescent, perennial, stoloniferous, herbaceous plants with short scaly rootstocks; leaves long-petioled, tufted at base, palmately trifoliate, with membranous stipules; flowers in cymes on the scapes; hypanthium almost flat, calyx deeply 5-lobed into 5 sepals and also with 5 alternating bractlets; petals 5, white, broad and rounded; stamens many; receptacle conical or hemispheric, becoming enlarged, red and juicy in fruit with very numerous superior pistils scattered over it; styles short, attached near middle of ovary; fruit of achenes. This genus needs a revision in our area and the nomenclature used here is tentative.

1. Achenes partly buried in deep pits of the receptacle; pubescence of scapes (and commonly of petioles) appressed or ascending; leaves often glaucous beneath, subsessile to short-petiolulate   -1. F. ovalis
1. Achenes in very shallow pits, hence superficial on the receptacle; pubescence of scapes and petioles soon spreading or reflexed; leaflets not glaucous, sessile to subsessile   -2. F. americana

1. Fragaria ovalis (Lehm.) Rydb., Bull. Torr. Club 33:143. 1906.
F. glauca (S. Wats.) Rydb.; F. pauciflora Rydb.; F. prolifica Baker & Rydb.; F. pumila Rydb.---as treated here---Leaflets subsessile to short-petiolulate, often glaucous below; pubescence of scapes and petioles appressed or ascending; achenes partly buried in deep pits of the ripe fruit, hence not superficial.---Open or partly shaded slopes. Widely distributed in the western part of Canada and the United States. Our records scattered over the western two-thirds of Colorado at 5000-11,500 feet.
2. Fragaria americana (Porter) Britton, Bull. Torr. Club 19:222. 1892.
F. bracteata Heller---Leaflets sessile to subsessile (rarely short-petiolulate), seldom really glaucous below; pubescence of scapes (and commonly the leaf petioles) soon spreading or reflexed (at least on some); achenes in very shallow pits on ripe fruit, hence superficial. This plant may not be distinct from the preceding and certainly intergrades with it in this state. The fruit character may work with ripe fresh fruit. The glaucous character is uncertain as is the sessile condition of leaflets. The pubescent characters work fairly well on most, but some intergrades occur--even both types may be found on the same plant.---Open or partly shaded slopes. Widely distributed in North America. Our records scattered over the western two-thirds of Colorado at 4500-9500 feet.

24. Geum L. AVENS

Herbaceous perennial plants; leaves mostly basal, the cauline small, usually lyrate-pinnate; flowers solitary or in a few-flowered cyme or corymbose, rather large; calyx tube campanulate or saucer-shaped; calyx lobes 5, persistent, with 5 alternate bractlets; petals

5; stamens many; pistils superior, many, on a clavate or hemispheric receptacle, style terminal; fruit of many distinct achenes, these topped with the persistent, elongated, often plumose jointed styles or style bases.

1. Stems leafy, the lower ones not much different from the basal ones; leaves with 5-9 large leaflets, terminal one definitely larger; styles in fruit conspicuously geniculate above the middle, the upper section soon deciduous, the lower strongly hooked at apex
    2. Petals flesh-colored to only yellow-tinged; sepals not reflexed; upper part of the styles in fruit 4-5 mm. long -1. G. rivale
    2. Petals bright yellow; sepals very soon reflexed; upper part of styles in fruit 1-2 mm. long
        3. Lower part of styles hispid near base, not glandular-puberulent; upper part of styles hairy, the hairs about 0.5 mm. long or more -2. G. strictum
            4. Leaflets of stem leaves not decurrent on the rachis
            4. Leaflets of stem leaves decurrent on the rachis -2A. G. strictum decurrens
        3. Lower part of styles not hispid near base but glandular-puberulent; upper part of style glabrate or with short hairs less than 0.5 mm. long -3. G. macrophyllum
1. Stems subscapose, all their leaves greatly reduced; basal leaves with 9-33 divisions, the terminal not much larger than the upper lateral ones; styles not conspicuously geniculate, straight or curved near apex, not jointed or slightly so
    5. Petals white, pink or only yellow-tinged; sepals 8-12 mm. long with bractlets usually longer than sepals; calyx tube rounded at base; style elongated and plumose in fruit; plants rather conspicuously villous or hirsute
        6. Leaflets deeply lobed to 1/2 way into the base; bractlets longer than sepals -4. G. ciliatum
        6. Leaflets merely toothed at apex; bractlets slightly shorter to slightly longer than the sepals -4A. G. ciliatum griseum
    5. Petals bright yellow; sepals 3-6 mm. long, with bractlets shorter than sepals; calyx tube cuneate at base; styles not especially elongating in fruit, glabrous; plants glabrate, not conspicuously villous or hirsute -5. G. turbinatum

1. Geum rivale L., Sp. Pl. 501. 1753.
    Stems 25-60 cm. tall, nearly or quite simple, more or less hirsute and glandular-pilose especially above; basal leaves lyrate-pinnate; leaflets obovate or the terminal rounded-reniform, 2-10 cm. long, doubly serrate to deeply incised, uneven in size, some very small leaflets alternating; stem leaves ternate; inflorescence 1- to 4-flowered, nodding in early anthesis, more erect in fruit; sepals 8-12 mm. long, longer than calyx tube, lanceolate to ovate-lanceolate, not reflexed, densely pilose, rather purplish; bractlets less than half as long as sepals, narrowly-linear; petals 6-10 mm. long, flesh-colored or sometimes yellow-tinged, purple-veined, clawed, flabelliform, emarginate; receptacle short-hirsute, more or less stalked in fruit; lower style internode about 7-9 mm. long, hirsute below and more or less glandular-pubescent, upper internode 4-5 mm. long, hirsute, articulated to lower by a curved and geniculate joint and rather early deciduous; achenes about 4 mm. long, hirsute.---Low ground. Newfoundland to British Columbia, south to New Jersey, Missouri and New Mexico; Eurasia. Our records in the western half of Colorado, except in the extreme western part, at 7500-10,000 feet.
2. Geum strictum Ait., Hort. Kew.2:217. 1789.
    Stems 30-80 cm. tall, erect or ascending, hirsute especially below, finely pubescent above and in the inflorescence; basal leaves lyrate-pinnate, rachis more or less hirsute, with 5-9 larger leaflets and alternating smaller ones; leaflets cuneate-obovate to somewhat narrower, the terminal sometimes rounded, variously cleft or divided and double-toothed, more or less pubescent or glabrate; lower stem leaves similar to basal, upper stem leaves 3-foliate, short-petioled; inflorescence few-flowered; calyx tube short, calyx lobes 4-6 mm. long, acuminate, ovate to lanceolate, reflexed; bractlets about a third as long as the sepals, linear; petals 5-8 mm. long, yellow; lower internode of style 4-5 mm. long in fruit, glabrous or hispid at base, not at all glandular-puberulent, upper internode about 1-1.5 mm. long, hirsute with hairs about 0.5 mm. long, strongly geniculate and jointed to lower portion; achenes 3-4 mm. long, appressed-pubescent and hispid toward the apex.---Meadows. Newfoundland to British Columbia, south to Pennsylvania, Missouri, Mexico and California. Our records from the western two-thirds of Colorado at 5000-9500 feet.
2A. Geum strictum decurrens (Rydb.) Kearney & Peebles, (var.) Journ. Wash. Acad. Sci. 20:481. 1939.
    G. decurrens Rydb.---This variety has leaflets of stem leaves decurrent on rachis.---In the southern Rocky Mountain area. Our one record from southwestern Colorado at about 9000 feet.
3. Geum macrophyllum Willd., Enum. Pl. 557. 1809.
    G. oregonense as treated by Rydb.---Stems 30-100 cm. tall, erect, bristly-hairy; basal leaves interruptedly lyrate-pinnate with 5-7 principal leaflets, these oval or obovate with alternating smaller ones, terminal one largest, sometimes round to reniform, leaflets often incised and serrate-dentate; stem leaves 3- to 5-foliate, short-petioled; flowers few to 15, in corymbose cymes; calyx tube short, sepals 3-5 mm. long, reflexed; bractlets not over 1/3 as long as the sepals, linear; petals 4-7 mm. long, yellow, obovate; lower internode of style 4-5 mm. long, glandular-puberulent, upper internode 1-2 mm. long, sparingly hairy at base, these hairs definitely less than 0.5 mm. long, the internode strongly geniculate and jointed to lower. Our plant has been called var. perincisum (Rydb.) Raup.---Meadows. Throughout most of the cooler parts of North America. Our records scattered in the western half of Colorado at 5000-10,000 feet.
4. Geum ciliatum Pursh, Fl. Am. Sept. 352. 1814.
    Sieversia ciliata (Pursh) G. Don.---Rootstocks thick; stems 20-50 cm. tall, villous and more or less glandular-hairy; leaves mostly basal, 10-20 cm. long, interruptedly pinnate,

white-pilose especially when young; leaflets 9-19, rather crowded, largest near apex and gradually smaller, cuneate-obovate in outline, more or less 3- to 5-cleft into linear or cuneate segments, these toothed, the sinuses of the leaflet lobes usually reaching the base; stem leaves usually 1- or 2-pinnatifid; cymes 1- to 5-flowered, flowers often nodding; calyx tube short, pilose and glandular; sepals 8-12 mm. long, pilose and glandular; bractlets usually longer than the sepals, calyx and bractlets usually strongly red-purple tinged; petals 9-15 mm. long, white, pink or yellowish, oval or broadly elliptic; style little if any articulated but elongating in fruit to 2.5-4 cm. long, plumose except right at apex; achenes about 3 mm. long. The correct name of this plant may be G. triflorum Pursh.---Hills and mountains, often on moist slopes. Alberta to British Columbia, south to New Mexico and Washington. Our records scattered in the mountains of Colorado at 7000-11,000 feet.

4A. Geum ciliatum griseum (Greene) Kearney & Peebles, (var.) Journ. Wash. Acad. Sci. 29:481. 1939.

Sieversia grisea (Greene) Rydb.---Leaflets merely toothed at apex, not deeply lobed; bractlets are shorter than to slightly exceeding the sepals.---Mountains. In at least the southern range of the species and to be expected in Colorado.

5. Geum turbinatum Rydb., Bull. Torr. Club 24:91. 1897.

Sieversia turbinata (Rydb.) Greene; Acomastylis turbinata (Rydb.) Greene---Rootstocks thick; stems 5-30 cm. tall, rather slender, glabrate or more or less pubescent; basal leaves 5-15 cm. long, pinnatifid or pinnate, glabrate to silky-strigose, segments 11-33, oblanceolate to obovate, 3- to 5-cleft or toothed; stem leaves 1-3, pinnatifid; calyx tube turbinate, cuneate at base, about 4-5 mm. long; sepals 3-6 mm. long, ovate-triangular, often purplish-tinged; bractlets somewhat shorter than sepals, but over 1/2 as long; petals 6-8 mm. long, bright yellow, broadly obovate or obcordate; styles about 3-5 mm. long, glabrous or nearly so, not greatly elongated or jointed; achenes 2.5-3 mm. long, strigose. The correct name for the above may be G. rossii (R. Br.) Ser.---High mountains. Montana to Nevada, south to New Mexico and Arizona. Our records from the higher mountains of Colorado at 9000-12,500 feet.

25. Alchemilla L.

Perennial herbaceous plants with short rootstocks; stems erect or ascending, silky-canescent; leaves silky-canescent beneath, the basal with blades digitately divided to near base into 5-7 segments, long-petioled, stipules brown and scarious; stem leaves 3-cleft, sessile or short-petioled; stipules connate; flowers in small glomerules; hypanthium campanulate or obconic, with a thick disk almost closing the mouth; sepals 4, subtended by 4 bractlets; petals none; stamens 4, inserted outside the disk between the calyx lobes; pistil usually 1, style basal, stigma capitate; fruit an achene, enclosed or nearly enclosed in the hypanthium.

1. Alchemilla alpina L., Sp. Pl. 123. 1753.
High mountains of Europe; Greenland; Island of Miquelon south of Newfoundland. Reported for Custer County, Colorado, by Fernald (Rhodora 52:47-48. 1950) but the record is somewhat uncertain.

Family 57. Leguminosae PEA FAMILY; LEGUME FAMILY

Herbaceous plants, shrubs or trees; leaves alternate, stipulate, usually compound and leaflets usually entire; flowers mostly perfect, commonly irregular; calyx of more or less united sepals, these 4-5 (or 2-lipped in Lupinus); corolla of 5 petals (rarely 1 or none), commonly papilionaceous (wide upper petal called a banner, 2 lateral petals called wings and 2 lower petals joined along edges to form a keel but claws separate), but sometimes regular; stamens 4 to 10, commonly diadelphous (9 united by their filaments, 1 free or nearly so); ovary 1, superior, 1-celled; fruit a legume (1 row of seeds and dehiscing along 2 sutures), sometimes indehiscent, sometimes 2-celled by intrusion of 1 or both sutures (as in some species of Astragalus).

1. Stamens 3-4; leaves simple without stipules; sepals showy, larger than the petals; fruit globose and spiny -1. Krameria
1. Stamens 5-10; leaves usually compound (rarely unifoliate with distinct stipules); sepals not showy, smaller than the petals (petals sometimes absent); fruit rarely spiny and then not globose
  2. Leaves bipinnate; corolla regular or only slightly irregular; filaments separate or united only at very base
    3. Flowers somewhat irregular, petals 5 mm. long or more; flowers in racemes -2. Hoffmanseggia
    3. Flowers regular, petals not over 3 mm. long; flowers in capitate, spherical heads
      4. Stems and usually the fruits with prickles; petals united to about the middle or above
        5. Fruit 4-angled, not at all jointed; stems prostrate; at least lower leaves with 4-8 pairs of pinnae, the leaflets 10-17 pairs to a pinna -3. Schrankia
        5. Fruit flattened, jointed; stems erect or ascending; leaves with 1-3 pair of pinnae, the leaflets 1-8 pairs to a pinna -4. Mimosa
      4. Stems and fruits without prickles; petals distinct or only slightly united at bases -5. Desmanthus
  2. Leaves once-compound or rarely simple; corolla irregular, usually strongly so (absent in Parryella); filaments separate or (more commonly) variously united
    6. Petals absent; filaments separate; fruit 1-seeded; plants of southwestern Colorado -6. Parryella
    6. Petals (at least 1) present; filaments separate or united; fruit 1- to many-seeded; plants usually generally distributed
      7. Stamens 5, monadelphous; flowers in very dense spikes -7. Petalostemon

7. Stamens 9 or 10, separate or variously united; flowers variously arranged but seldom in very dense spikes
   8. Petals 1 (banner); shrubs with purplish flowers -8. Amorpha
   8. Petals 5; plants herbaceous with variously colored flowers or if shrubs (as in Robinia or Dalea) then with rose, yellow or white flowers
      9. Filaments separate or united only at very base
         10. Corolla not papilionaceous, only slightly irregular; anthers opening by terminal pores -9. Cassia
         10. Corolla papilionaceous, very irregular; anthers not opening by terminal pores
            11. Leaves pinnate with numerous leaflets, stipules small; flowers white to ochroleucous -10. Sophora
            11. Leaves palmate with 3 leaflets, stipules large and foliaceous; flowers yellow -11. Thermopsis
      9. Stamens monadelphous or diadelphous, some or all united well above the base
         12. Fruit covered with hooked prickles, body 12-18 mm. long, not coiled; leaves with more than 3 leaflets -12. Glycyrrhiza
         12. Fruit not covered with hooked prickles, (or if so then smaller and coiled and leaves 3-foliate), body of various lengths; leaves 3- to many-foliate
            13. Anthers strongly differentiated, some very small and versatile, alternating with larger basifixed ones; leaves palmately 4- to 15-foliate -13. Lupinus
            13. Anthers uniform or nearly so (may shed pollen at different times); leaves pinnate or if palmate usually with only 3 leaflets (rarely simple)
               14. Leaflets and usually the calyx glandular-punctate (under strong lens); stamens usually monadelphous
                  15. Leaves palmately compound; flowers white to blue or purple; ovary with 1 or rarely 2 ovules; petal wings and keel free from the staminal tube -14. Psoralea
                  15. Leaves pinnately compound (except in 1 species with yellow corolla); flowers white, yellow or rose; ovules 2-3 (fruit only 1-seeded); petal wings and keel adnate to the staminal tube -15. Dalea
               14. Leaflets and calyx not glandular-punctate; stamens diadelphous (except in Onobrychis)
                  16. Woody shrubs or trees with stipular spines present -16. Robinia
                  16. Herbaceous plants without stipular spines
                     17. Stamens monadelphous; fruit pectinately bristly on margins (and not at all coiled) -17. Onobrychis
                     17. Stamens diadelphous; fruit not bristly on margins (or if so then coiled)
                        18. Leaf rachis produced into a bristle or tendril, this simple or branched
                           19. Styles hairy along the inner side but no apical tuft present; staminal tubes not oblique at summit -18. Lathyrus
                           19. Styles with tuft of hairs at apex; staminal tubes usually oblique at summit -19. Vicia
                        18. Leaf rachis ending in a leaflet (or leaves rarely simple)
                           20. Twining or trailing vines; corolla keel strongly curved or coiled
                              21. Leaves 3-foliate; styles bearded along the inner sides; flowers about 6 mm. long, purple -20. Strophostyles
                              21. Leaves 5- to 7-foliate; styles glabrous; flowers about 12 mm. long, brown-purple -21. Apios
                           20. Stems not twining or trailing (stems may be decumbent but plants not vines); keels not strongly curved or coiled
                              22. Fruit a loment (of 2 or more 1-seeded indehiscent segments constricted between from the edges) -22. Hedysarum
                              22. Fruit not a loment (may sometimes be somewhat constricted from the faces, however)
                                 23. Margins of leaflets toothed, only 3 leaflets present
                                    24. Leaves palmately compound; flowers in heads or very short spikelike racemes, never yellow -23. Trifolium
                                    24. Leaves pinnately compound (terminal leaflet stalked); flowers in oblong to elongated racemes (or if head-like then yellow in color)
                                       25. Fruits strongly curved or coiled; inflorescence capitate or oblong and very dense, not over 5 cm. long (if yellow then capitate, if oblong then purple) -24. Medicago
                                       25. Fruits straight or nearly so; inflorescence a lax raceme over 5 cm. long (flowers white or yellow) -25. Melilotus
                                 23. Margins of leaflets entire, usually over 3 leaflets present (rarely only 1)
                                    26. Filaments all (or 5 alternating ones) broad and flat or dilated above; flowers solitary or 2-3 in subumbellate clusters; corolla yellow to rose -26. Lotus
                                    26. Filaments all filiform or very narrowly-flattened; flowers in capitate or racemose inflorescences, usually over 3; corolla variously colored but rarely yellow
                                       27. Leaves pinnate with more than 3 leaflets or rarely unifoliate

28. Calyx closely subtended by a pair of small caducous bractlets; corolla orange-red when fresh; spreading rootstocks present; usually found only in the San Luis Valley -27. Swainsona
28. Calyx without bractlets (bracts may be present at base of pedicels); corolla various colored but never orange-red when fresh; spreading rootstock rarely present; plants widely distributed
   29. Fruit with ventral or upper sutures partly or completely intruded; corolla keel with a prominent beak at end; plants acaulescent, the scapes erect or ascending -28. Oxytropis
   29. Fruit without intrusion of either suture or both ventral and dorsal intruded, or dorsal (lower) only intruded; corolla keel not beaked at end (or if so the plant caulescent with decumbent stems); stems erect or prostrate -29. Astragalus
27. Leaves palmately 3-foliate
   30. Leaflets densely silky-hairy both sides; flowers 1 to few; plants rarely found above 8000 feet in Colorado -29. Astragalus
   30. Leaflets glabrous to sericeous, never densely silky both sides; flowers usually several to many; plants rarely found below 9500 feet in Colorado, usually alpine -23. Trifolium

1. Krameria Loefl. RATANY

Perennial herbaceous plants or shrubs; stems straggling, grayish-strigose; leaves simple, entire, alternate, no stipules present, grayish-strigose; flowers axillary, or in leafy-bracted racemes, the pedicel with 2 opposite foliaceous bracts; sepals 4-5, large and showy, unequal, colored; petals 4 or 5, smaller than the sepals, the upper 3 long-clawed and united by their claws, the lower 2 reduced to broad fleshy, sessile glandlike appendages; stamens 3-4, more or less united; fruit globose or nearly so, indehiscent, thick-walled, spiny, 1-seeded.

1. Plants herbaceous; stems prostrate or nearly so; peduncles not glandular-hairy; spines of the fruit scabrous but not barbed; plants of southeastern Colorado -1. K. lanceolata
1. Plants shrubby; stems not prostrate; peduncles glandless; spines usually barbed; plants of southwestern Colorado -2. K. parvifolia

1. **Krameria lanceolata** Torr., Ann. Lyc. N. Y. 2:168. 1827.
K. secundiflora of authors not of DC.---Herbaceous plants with perennial woody roots; stems prostrate or somewhat ascending; leaves 8-20 mm. long, linear, sessile; sepals 8-10 mm. long, purplish, glabrous inside, silky-strigose outside; upper petal blades about 1 mm. long, their united claws 3-4 mm. long, claws greenish-yellow, blades purplish, lower 2 petals about 2 mm. long; fruit 6-8 mm. in diameter, body woolly-hairy, with many scabrous spines about 2-4 mm. long.---Sandy plains and hills. Kansas, south to Arkansas, Mexico and Arizona. Our few records from the southeastern corner of Colorado at 4000-4500 feet.

2. **Krameria parvifolia** Benth., Bot. Voy. Sulph. 6. 1844.
K. glandulosa Rose & Painter---Much branched shrubs 30-70 cm. tall with canescent twigs, these rigid and somewhat spinose at ends; leaves 4-12 mm. long, linear, sessile, densely to sparsely strigose; peduncles stipulate-glandular as well as strigose, bracts foliaceous, often glandular; sepals 6-7 mm. long, purplish, glabrous inside, strigose outside; upper petals 4-5 mm. long, united for about 1/2 their length; lower 2 petals about 2-3 mm. long; fruit globular, about 6-7 mm. in diameter, rather densely strigose, the spines with scattered barbs, rarely barbless. Our plant has been called var. glandulosa (Rose & Painter) Macbride.---Dry plains and hills. Texas to California, south into Mexico. Our one record from southwestern Colorado at about 6000-7000 feet.

2. Hoffmanseggia Cav. RUSHPEA

Herbaceous perennial plants; leaves bipinnate with small leaflets, often glandular-punctate; stipules small, readily deciduous; inflorescence of terminal or lateral racemes; calyx tube short; petals 5, yellow, nearly equal, not at all papilionaceous; stamens 10, distinct, anthers longitudinally dehiscent; fruit flat, dehiscent.

1. Leaflets dotted below with conspicuous black glands, the fruit and flowers similarly dotted; fruit lunate, widest above the middle -1. H. jamesii
1. Leaflets, flowers and fruit not conspicuously black-dotted (may be glandular); fruit little if any wider above the middle
   2. Petals with long claws, these glandular; calyx and pedicels with stipitate glands; fruit straight or slightly falcate -2. H. densiflora
   2. Petals with short claws, these not glandular; no stipitate glands present anywhere; fruit strongly falcate -3. H. drepanocarpa

1. **Hoffmanseggia jamesii** T. & G., Fl. N. Amer. 1:393. 1840.
Suffruticose at base with woody rather spindle-shaped roots; stems 15-40 cm. tall, branching at the woody base, appressed-pubescent, more or less glandular; pinnae 5-7; leaflets 8-20, 4-5 mm. long, in pairs, ovate to oblong, conspicuously black glandular-dotted beneath; racemes 5-10 cm. long, 5- to 15-flowered; calyx about 6-8 mm. long, with linear-lanceolate lobes, and very short tube, black-dotted; petals as long or a little longer than the calyx, black-dotted; fruit 2-2.5 cm. long, asymmetrically lunate, widest above middle, black-dotted and with tufts of stellate hairs.---Plains and hills. Kansas and Colorado, south to Texas and Arizona. Our records scattered in the eastern half of Colorado at 3500-5500 feet.

2. **Hoffmanseggia densiflora** Benth. in A. Gray, Pl. Wright. 1:55. 1852.
Stems 10-30 cm. tall, from deep-seated creeping rootstocks, glabrate or puberulent; petioles and rachis of leaves often with stipitate glands; pinnae 5-9; leaflets 3-8 mm. long, 10-22, in pairs, oblong, glabrous or puberulent; inflorescence 5-15 cm. long, pedicels and calyx with

stipitate glands and somewhat tomentose; calyx 6-8 mm. long, tube short; petals somewhat longer than calyx, long-clawed, these bearing stipitate glands; pods 2-4 cm. long, nearly straight to slightly falcate, glabrous or puberulent, more or less stipitate-glandular.---Plains and hills, in this area weedlike. Kansas, south to Texas, Mexico and California. Our few specimens from southeastern Colorado at 4000-4500 feet.

3. Hoffmanseggia drepanocarpa A. Gray, Pl. Wright. 1:58. 1852.
Plants from thick woody taproots, glandless throughout; stems 10-20 cm. tall, finely pubescent; pinnae 7-11; leaflets 8-20, in pairs, oblong, appressed-pubescent; racemes few-flowered; calyx 5-8 mm. long, tube very short, lobes linear, appressed-pubescent; petals slightly longer than calyx, short-clawed, glandless; fruit 3-4 cm. long, 6- to 11-seeded, strongly falcate, curved in a part of a circle, appressed-puberulent.---Plains and hills. Colorado, south to Texas, Mexico and Arizona. Our few records from the southeastern quarter of Colorado at 4000-5500 feet.

3. Schrankia Willd. SENSITIVE BRIER

Perennial herbaceous plants or shrubs; stems arching to procumbent, armed with recurved prickles, angled; leaves bipinnate, usually sensitive; leaflets small, numerous, entire; stipules linear or setaceous; flowers perfect or polygamous, in axillary, peduncled, globose heads; calyx 4- to 5-lobed; corolla regular, of 4-5 partly united petals, white, pink or light red; stamens usually 8-12, distinct or nearly so, anthers all alike; fruit linear, not stipitate, prickly, dehiscent, several-seeded.

1. Leaflets with lower surface bearing prominently raised veins, the lateral veins prominent too
   -1. S. nuttallii
1. Lower surface of leaflets with veins not prominently raised, at least the lateral ones   -2. S. occidentalis

1. Schrankia nuttallii (DC.) Standley, Field Mus. Nat. Hist. Bot. 8:13. 1930.
Leptoglottis nuttallii DC.; Morongia uncinata (Willd.) Britt.---Stems 60-120 cm. long, more or less puberulent; stipules 4-6 mm. long; petioles and rachis prickly; pinnae 4-8 pairs; leaflets 8-15 pairs, 4-8 mm. long, elliptic to oblong, cuspidate at apex, thick and prominently veined beneath; peduncles 3-9 cm. long, prickly; fruit 3-10 cm. long, 4-angled, not flattened, beaked.---Dry plains and hills. Virginia to Colorado, south to Florida and Texas. Our few records from the southeastern corner of the state at 4000-4500 feet.

2. Schrankia occidentalis (Woot. & Standl.) Standl., Field Mus. Nat. Hist. Bot. 8:13. 1930.
Leptoglottis occidentalis (Woot. & Standl.) Britt. & Rose---Stems 40-120 cm. long, puberulent; stipules about 2-3 mm. long; rachis and petioles sparingly prickly; pinnae 4-7 pairs; leaflets 4-6 mm. long, 10-16 pairs, oblong or linear-oblong, main veins rather obscure to fairly prominent but lateral veins obscure; peduncles 3-7 cm. long, unarmed or sparingly prickly; fruit 6-9 cm. long, 2-3 mm. wide, angled but compressed somewhat, beaked.---Dry plains and hills. Texas and New Mexico. Our few records from the southeastern corner of the state at 4000-4500 feet.

4. Mimosa L.

Shrubs with stems prickly; leaves bipinnate, the leaflets entire and often sensitive; flowers perfect or polygamous, small, in capitate heads; calyx small, its teeth short, mostly 4- to 5-toothed; corolla regular, of 4-5 petals united to middle or above; stamens 8-10, distinct, exserted; fruit flattened, segmented (rarely reduced to 1 joint), often with prickles, usually several-seeded.

1. Mimosa borealis A. Gray, Mem. Am. Acad. II. 4:39. 1849.
M. fragrans A. Gray---Shrubs 50-200 cm. tall; stems branching, the branches often horizontal but plants more or less upright; twigs glabrous or puberulent, the prickles mostly solitary and opposite the petiole base; leaves small, slender-petioled, of 1-3 pair of pinnae; leaflets 3-8 pairs, oblong to oval, 2-6 mm. long, obtuse or rounded at ends; peduncles about 1-2 cm. long; fruit 2-5 cm. long, segments 6-8 mm. wide, linear-oblong, glabrous, 1- to 7-seeded and constricted between seeds, unarmed or prickly on edges (often on same plant).---Canyons and rocky hills. Oklahoma, south to Texas and New Mexico. Our records mostly from the southeast corner of the state, one from Lincoln County, at 4000-5000 feet.

5. Desmanthus Willd. BUNDLE FLOWER

Perennial herbaceous or somewhat shrubby plants, no spines or prickles present; leaves bipinnate with numerous, entire leaflets and usually with small stipules; flowers regular, perfect or lower staminate, sessile in peduncled heads or spikes, greenish or whitish; calyx 5-lobed; petals 5, alike, distinct or slightly united at very base; stamens 5 or 10, distinct or nearly so; fruit elongated, straight or curved, several-seeded, dehiscent.

1. Pods oblong, falcate, 1.5-2.5 cm. long, 4-6 mm. wide; stamens 5; stipules over 2 mm. long -1. D. illinoensis
1. Pods linear, straight, 3-7 cm. long, about 3 mm. wide; stamens 10; stipules wanting or 1-2 mm. long
   -2. D. cooleyi

1. Desmanthus illinoensis (Michx.) MacM., Metasperm. Minn. 388. 1892.
Acuan illinoensis (Michx.) Kuntze---Stems erect or ascending, glabrous or nearly so; leaves with 14-30 pairs of pinnae; leaflets 2.5-4 mm. long, many, linear or linear-oblong, glabrous or ciliate; stipules setaceous, to 4 or 8 mm. long; calyx campanulate, about 1 mm. long with very

short lobes; petals about 2 mm. long; stamens 5; fruit in compact heads, each about 1.5-2.5 cm. long and about 4-6 mm. wide, strongly falcate and slightly spirally twisted.---Banks, roadsides and plains. Ohio to South Dakota, south to Florida and New Mexico. Our records from eastern Colorado at 3500-4000 feet.
2. Desmanthus cooleyi (Eaton) Trel., Ark. Geol. Survey Rpt. 1888, 4:178. 1891.
D. jamesii T. & G.---Stems 20-50 cm. long, decumbent or ascending, sometimes nearly upright, angled, glabrous or puberulent; leaves 2-4 cm. long, with 2-5 pairs of pinnae, an orbicular gland between the lower pair; leaflets 2.5-4 mm. long, 8-15 pairs, oblong or linear, oblique at base, the vein excentric, acutish or mucronate, ciliate at least when young; stipules absent or 2 mm. long; peduncles about 2 cm. long or shorter; heads several- to many-flowered; calyx about 1.5 mm. long with short lobes; petals about 3 mm. long, valvate; stamens 10; fruits solitary or few, 3-7 cm. long and about 3 mm. wide, linear, straight or nearly so.---Plains and hills. Nebraska, south to Texas and Arizona. Our few records from the southeastern corner of Colorado at 4000-4500 feet.

6. Parryella T. & G. DUNEBROOM

Perennial low shrubby plants; stems much branched; leaves alternate, pinnately compound with many entire, very narrow stipitate leaflets, glandular-punctate; flowers small, yellowish or yellowish-green in terminal spikelike racemes, these often branched; calyx 5-toothed, the teeth shorter than the tube; corolla absent entirely; stamens 9-10, separate, attached to base of calyx; fruit 1-seeded, small, glandular-dotted.

1. Parryella filifolia T. & G., Proc. Amer. Acad. 7:397. 1868.
Plants low, rushlike; stems about 70-100 cm. tall, glabrous or thinly strigose; leaves 6-12 cm. long or more; leaflets 11-45, 5-15 mm. long, linear-filiform or possibly wider, but not over 2 mm. wide, glabrous or thinly strigose; stipules and stipels minute; racemes 2-13 cm. long; calyx 2.5-4 mm. long, ciliate on lobes, usually glabrous below; fruit 4-6 mm. long, excluding the hairy style.---Dry hills, often on sandy soil. New Mexico and Arizona. Our one record from the southwestern corner of Colorado at about 4500 feet.

7. Petalostemon Michx. PRAIRIE CLOVER

Herbaceous perennials, often deep-rooted; stems leafy; leaves odd pinnate with leaflets entire and more or less glandular-dotted; flowers in terminal dense spikes, bracts present; calyx campanulate the teeth short; corolla indistinctly papilionaceous, the banner broader and long-clawed, inserted at bottom of calyx, the other 4 petals subequal, their short claws attached to the top of the staminal tube; stamens 5, monadelphous; fruit 1-seeded, small, enclosed in the calyx, usually somewhat compressed, indehiscent, membranous.

1. Corolla white or ochroleucous
  2. Calyx tube glabrous or nearly so, (lobes villous-ciliate only); spike 1-7 cm. long, not over 1 cm. thick -1. P. candidus
  2. Entire calyx conspicuously silky-villous; spike 4-15 cm. long, over 1 cm. thick -2. P. compactus
1. Corolla rose to purple
  3. Leaflets 9-17, linear to oblong, 6-12 mm. long; whole plant densely short-villous -3. P. villosus
  3. Leaflets 3-5, linear, 8-20 mm. long; plant usually glabrous to moderately villous
    4. Stems erect or ascending; leaflets definitely glandular-dotted; plants glabrous to villous -4. P. purpureus
    4. Stems (at least outer ones) prostrate-assurgent; leaflets obscurely if at all glandular-dotted; plants villous -4A. P. purpureus pubescens

1. Petalostemon candidus Michx., Fl. Bor. Amer. 2:49. 1803.
P. oligophyllus (Torr.) Rydb.---Plants with woody bases; stems 30-80 cm. tall, several, branched, ascending or decumbent below, glabrous, striate; leaflets 7-9, linear, oblong to oblanceolate or elliptic, glandular-dotted beneath, glabrous; spike oblong, dense, 1-7 cm. long, bracts about as long to exceeding the buds; calyx 3 mm. long, the tube glabrous, 10-ribbed, about 2 mm. long, lobes about 1 mm. long, villous-ciliate; corolla white, blade of banner broader than long, the claw longer, blades of other petals oblong, the claws shorter; fruit about 3 mm. long, glabrous. Our plants seem to be var. oligophyllus (Torr.) Hermann.---Plains and hills. Saskatchewan, south to Mississippi, Texas, Mexico and Arizona. Our records well scattered in the eastern half of Colorado at 3500-7000 feet.
2. Petalostemon compactus (Spreng.) Swezey, Neb. Flow. Pl. 6. 1891.
Plants with woody taproots; stems 40-80 cm. tall, several, glabrous, striated; leaves 2.5-6 cm. long, numerous and rather crowded, petioled; leaflets 5-7, 8-25 mm. long, lanceolate-oblong to oblong or linear-oblong, glabrous, glandular-dotted beneath; spike dense, 4-15 cm. long, cylindrical, on long peduncles (mostly over 8 cm. here); bracts long-attenuate, about as long as the flowers, silky; calyx about 3-4 mm. long, silky-hairy, lobes about as long as the tube; corolla white or ochroleucous, blade of banner broad, 1.5-2 mm. long, its claw longer, other petals about 2-3 mm. long, the claws shorter than the blades; fruit about 3 mm. long, sparingly pubescent.---Sandy hills and plains. Nebraska and Wyoming, south to Colorado. Our records in the eastern half of the state, few in the extreme southern part, at 3500-6000 feet.
3. Petalostemon villosus Nutt., Gen. 2:85. 1818.
Plants with woody taproots or caudices; stems 30-60 cm. tall, decumbent or ascending, several, densely short-villous with white hairs; leaves about 3-5 cm. long, crowded, with 9-17 leaflets, these linear to oblong, 6-12 mm. long, densely whitish-villous; spikes 2.5-8 cm. long, cylindrical, short-peduncled, dense; bracts longer than the calyces, lanceolate, villous; calyx about 2-3 mm. long, 10-ribbed, densely white-villous, lobes shorter than the tube; corolla

light rose to rose-purple, blade of banner about 2 mm. long, the claw a little longer, other petals with blades about 2 mm. long and short-clawed; fruit about 3 mm. long, villous.---Sandy hills and plains. Minnesota to Saskatchewan, south to Missouri, Texas and Colorado. Our records in the eastern half of Colorado, except the southern part, at 3500-6000 feet.

4. Petalostemon purpureum (Vent.) Rydb., Mem. N. Y. Bot. Gard. 1:238. 1900.

P. mollis Rydb.; P. purpureum var. mollis (Rydb.) A. Nels.---Taproots woody; stems several, erect or ascending, branched above or simple, striate, glabrous to villous; leaves petioled, leaflets 3-5, mostly 5, 8-20 mm. long, linear, mostly folded, glabrous to villous, glandular-dotted below; spikes 1-5 cm. long, dense, oblong to cylindrical; bracts abruptly acuminate, villous at base, glabrate near middle and sparingly villous on tip; calyx 2.5-4 mm. long, the lobes shorter than the tube, silky-villous; corolla rose-purple, blade of banner cordate, about 2 mm. long, its claw about 3 mm. long; other petals with oblong blades and shorter claws; fruit about 3 mm. long, pubescent.---Plains and hills. Indiana to Saskatchewan, south to Texas and Arizona. Our records well scattered in the eastern half of Colorado at 3500-7500 feet.

4A. Petalostemon purpureum pubescens (A. Nels.) comb. nov. (var.)

P. pubescens A. Nels.---This plant differs from the species in hairy stems prostrate-assurgent (at least the outer ones of the cluster); leaflets are obscurely if at all glandular and the plant is always definitely villous. Intergrades with the typical form.---Apparently limited to southern Colorado. Our few records from Las Animas County at about 5000-6000 feet.

8. Amorpha L. FALSE INDIGO

Shrubs with gland-dotted foliage; leaves odd-pinnately compound with entire leaflets; inflorescence of spikelike racemes, with narrow deciduous bracts and small flowers; calyx 5-lobed, turbinate, slightly oblique; petals 1 only (banner); stamens 10, united only at the very base; ovary 2-ovuled; fruit small, indehiscent or very tardily dehiscent, oblique, 1- to 2-seeded.

1. Low shrubs rarely over 100 cm. tall; leaves nearly sessile (petiole shorter than width of lowest leaflet); leaflets 8-20 mm. long; calyx lobes about 1/2 to as long as the tube
    2. Glabrous or nearly so from the first; calyx lobes about 1/2 as long as the tube; racemes solitary or a very few together -1. A. nana
    2. Plants densely canescent; calyx lobes about as long as the tubes; racemes usually several to many in a cluster -2. A. canescens
1. Tall shrubs seldom less than 200 cm. tall; leaves petioled (the petiole longer than the width of lowest leaflet); leaflets 20-40 mm. long; calyx lobes less than 1/2 as long as the tube
    3. Racemes usually of several and clustered, 10-50 cm. long; leaflets usually narrowly elliptical -3A. A. fruticosa angustifolia
    3. Racemes usually single, 15-25 cm. long (sometimes with a second shorter one at base); leaflets oblong or elliptic -3B. A. fruticosa occidentalis

1. Amorpha nana Nuttall in Fraser, Cat. 1813.

A. microphylla Pursh---Low erect shrubs 30-90 cm. tall, glabrous or nearly so throughout; leaves 3-10 cm. long, with short petioles, numerous; leaflets 15-31, 8-14 mm. long, oblong, oval or slightly obovate, rounded or abruptly narrowed at bases, rounded or emarginate and mucronate at tips, firm and slightly reticulate, green both sides; inflorescence single or of a few erect densely flowered racemes 3-8 cm. long; calyx lobes about 1/2 as long as tube, lanc-acuminate; standard purple; fruit about 5 mm. long, straight dorsally, with a short, erect or slightly oblique beak, densely punctate-dotted.---Hills and prairies. Saskatchewan, south to Iowa, Kansas and New Mexico. Our record from central Colorado at about 7200 feet.

2. Amorpha canescens Pursh, Fl. Am. Sept. 467. 1814.

Low shrubs 30-100 cm. tall, gray-canescent throughout; leaves nearly sessile, petioles usually shorter than width of the lowest leaflet; leaflets 7-20 mm. long (averaging about 15 mm.), 15-45, crowded, elliptical, oblong or ovate usually rounded at base, mucronate at tip; inflorescence terminal, usually with several to many spikelike racemes 8-25 cm. long; calyx villous, canescent, its lobes about equalling the tube, all about equal; standard about 5 mm. long, blue-purple; fruit villous-canescent.---Plains and hills. Michigan to Saskatchewan, south to Arkansas, Texas and New Mexico. Our records from eastern Colorado at 3500-4500 feet.

3. Amorpha fruticosa L., Sp. Pl. 713. 1753.

The following 2 varieties are in or near Colorado.

3A. Amorpha fruticosa angustifolia Pursh, (var.) Fl. Am. Sept. 466. 1814.

Amorpha angustifolia (Pursh) Boynton; A. fruticosa of Manuals in part---Branched shrubs 2-4 m. high or rarely higher; branches striate, young stems with appressed hairs (sometimes sparse); leaves 7-20 cm. long; petioles 1-2 cm. long; leaflets 2-4 cm. long, 9-25, elliptical or obovate, narrowed or cuneate at base, mucronate, acute or emarginate at apex, minutely puberulent, petiolules 2-3 mm. long; inflorescence usually of several clustered, erect, closely-flowered spikes 10-15 cm. long; calyx sparingly short-pilose, lobes much shorter than tube, upper lobes blunt, lower ones acute; standard 4.5-5 mm. long, purplish; fruit 7-8 mm. long, rather curved, glabrous but with resin dots.---Banks of streams and ditches. Wisconsin to Saskatchewan, south to Texas and Mexico. Our records scattered in the eastern half of Colorado at 3500-6000 feet.

3B. Amorpha fruticosa occidentalis (Abrams) Kearney & Peebles, (var.) Journ. Wash. Acad. Sci. 29:483. 1939.

Reported from Wyoming to California, south to Mexico. The writer has not located any Colorado records.

9. Cassia L. SENNA; PARTRIDGE PEA

Annual or perennial herbaceous plants; stems caulescent; leaves pinnately compound, no terminal leaflet present; flowers perfect, solitary or in axillary clusters of 2-4; calyx 5-lobed, these nearly equal and longer than the tube; corolla not papilionaceous and not strongly irregular; stamens 5-10, distinct, all fertile and alike or some reduced to staminodia, or some enlarged, anthers opening by terminal pores; fruit linear, flat, dehiscent, seeds many.

1. Cassia fasciculata Michx., Fl. Bor. Am. 1:262. 1803.
   C. chamaecrista of Manuals; Chamaecrista fasciculata (Michx.) Greene---Reported for Maine to South Dakota, south to Florida and New Mexico. This plant should be in the eastern half of Colorado but no records have been found.

10. Sophora L.

Perennial herbaceous plants with creeping rootstocks; leaves odd-pinnately compound, with numerous, entire leaflets; stipules small; flowers in terminal racemes; calyx campanulate, 5-lobed; petals papilionaceous; stamens 10, separate to the base or nearly so; pods nearly terete in section, usually moniliform, indehiscent or tardily dehiscent, 1- to several-seeded, stipitate.

1. Sophora sericea Nutt., Gen. Pl. 1:280. 1818.
   Stems 10-30 cm. tall, more or less silky-canescent; leaflets 15-23, elliptic, obovate to oblong, canescent but often glabrate above; pedicels short; calyx tube about 5 mm. long, gibbous at base, lobes short, not over 2 mm. long, triangular; corolla about 14-16 mm. long, white to ochroleucous, keel with 2 subulate beaks; fruit 1-5 cm. long.---Plains, hills and roadsides. South Dakota to Wyoming, south to Texas and Arizona. Our records scattered over Colorado but mostly from the eastern half of the state, at 3500-6000 feet.

11. Thermopsis R. Br. GOLDENPEA

Perennial herbaceous plants with creeping rootstocks; stems erect or ascending; leaves palmately compound with 3 leaflets; stipules large; flowers racemose, large (1-1.5 cm. long) yellow; calyx campanulate, of 4 or 5 sepals; corolla papilionaceous; stamens 10, distinct, anthers about all alike; fruit usually flattened, straight to incurved, sessile (in the calyx) or nearly so, 4-7 cm. long at maturity. The mature or nearly mature fruit is necessary to be sure of the species. Even then some intergradations occur in Colorado plants.

1. Fruit arcuate to curved in a half circle, divaricate to reflexed at maturity; plants rarely over 30 cm. tall; leaflets seldom over 3 cm. long
   -1. T. rhombifolia
1. Fruit straight or slightly curved, erect, ascending or occasionally divaricate; plants over 30 cm. tall; leaflets 3-6 cm. long
   2. Fruit appressed to rachis, straight; stipules linear-lanceolate to lanceolate(but occasionally ovate)
   -2. T. montana
   2. Fruit not appressed but ascending to divaricate; straight to slightly curved; stipules ovate-lanceolate to ovate
   -3. T. divaricarpa

1. Thermopsis rhombifolia Nutt. ex Rich., in Frankl. Narr. First Journ. App. 737. 1823.
   T. arenosa A. Nels.---Stems 10-30 cm. tall, rarely taller, varying from appressed-pubescent to glabrate; stipules broadly ovate to ovate-lanceolate above; leaflets 2-3 or sometimes 4 cm. long, obovate, oblanceolate or elliptic, glabrous to appressed-pubescent especially beneath; fruit divaricate to reflexed, arcuate to curved in a half circle or more, pubescent to glabrous at maturity. The glabrate form (T. arenosa A. Nels.) seems more common in Colorado but all intergradations are found.---Dry plains, hills and slopes often on sandy ground. North Dakota to Alberta, south to Nebraska and Colorado. Our records from the eastern half of Colorado, but a few in the San Luis Valley, at 5000-9000 feet.
2. Thermopsis montana Nutt. ex T. & G., Fl. N. Amer. 1:388. 1840.
   Stems 30-60 cm. tall, glabrate to hairy; stipules linear-lanceolate to lanceolage, sometimes ovate; leaflets 3-6 cm. long, rather narrowly elliptic to oblanceolate, sometimes obovate-oblong, glabrous or nearly so above, more or less appressed to rachis, hairy.---Mountains, often in valleys and meadows. Montana to Oregon, south to Colorado and Nevada. Our records scattered in the western half of Colorado, not many in the northern part, at 6500-10,500 feet.
3. Thermopsis divaricarpa A. Nels., Bot. Gaz. 25:275. 1898.
   T. pinetorum Greene---Stems 30-60 cm. tall, glabrate to pubescent; stipules ovate-lanceolate to ovate; leaflets 3-6 cm. long, obovate-oblong, rhombic-oblanceolate to elliptic, glabrate or sparsely pubescent especially below; fruit straight or slightly curved, ascending to divaricate, not reflexed, pubescent to glabrate. This species intergrades somewhat with T. montana Nutt. in this state.---Valleys and slopes, often in partial shade. Wyoming and Utah, south to New Mexico and Arizona. Our records scattered in the western half of Colorado at 5500-10,500 feet.

12. Glycyrrhiza (Tourn.) L. LICORICE

Perennial, herbaceous plants with thick, sweet roots; stems erect and leafy; leaves odd-pinnately compound, with leaflets entire, numerous and glandular-dotted; flowers perfect, in axillary spikelike racemes; calyx teeth subequal; corolla papilionaceous; stamens diadelphous (9-1) with alternating anthers smaller and larger; fruit indehiscent, sessile, several-seeded, covered with hooked prickles, 1-celled.

1. **Glycyrrhiza lepidota** Pursh, Fl. Amer. Sept. 480. 1814.

Stems 30-100 cm. tall, glabrous or puberulent, minutely glandular-dotted; leaves with 11-19 leaflets 2-5 cm. long, these lanceolate or oblong, acute or mucronate, glabrous to puberulent below; stipules deciduous, lanceolate; calyx about 4-6 mm. long, narrowly campanulate, the teeth longer than the tube; corolla yellowish-white, 10-13 mm. long; fruit 12-18 mm. long, resembling a cocklebur.---Meadows, valleys and roadsides. Ontario to Washington, south to New York, New Mexico and California. Our records scattered over Colorado, few in the extreme eastern part, at 4000-8500 feet.

13. Lupinus L. LUPINE

Annual or perennial herbaceous plants; stems leafy or subscapose; leaves digitately compound with 4-15 entire leaflets; flowers perfect, in terminal racemes or spikes; calyx 2-lipped, the 2 upper and the 3 lower more or less united; corolla papilionaceous, sometimes bicolored, banner broad, commonly with reflexed margins, wings commonly connivent by their edges and enclosing the mostly falcate pointed keel; stamens 10, monadelphous, anthers of 2 kinds, 5 large ones alternating with 5 smaller; fruit at least somewhat flattened, dehiscent; seeds 2 to many. This is one of the most, if not the most confusing genus of plants in Colorado. The following treatment is preliminary and arbitrary and the group needs a complete revision. Many of the entities listed here are highly variable but to give each variant a separate name, as has been done, does not help the matter. The author has selected out forms that have been given the widest sanction of usage and appear to be most distinct. Further study may add new entities or result in a different selection of specific names.

1. Plants annual or possibly biennial; ovules and seeds 2
  2. Racemes dense in flower, usually less than 2 cm. long; pods not noticeably constricted between the seeds; plants mostly west of Continental Divide in Colorado
    3. Plants acaulescent or nearly so, the leafy stem scarcely 1 cm. long, hence the leaves all basal or nearly so; lower lip of calyx more than twice as long as the upper (the latter often obsolete)
        -1. L. brevicaulis
    3. Plants caulescent, the leafy stem 3 cm. long or more, branched; lower and upper lip of calyx subequal
      4. Calyx and pedicels with spreading hairs   -2. L. kingii
      4. Calyx and pedicels with appressed hairs   -2A. L. kingii argillaceus
  2. Racemes in flower rather loose and usually over 3 cm. long; pods somewhat constricted between the seeds; plants mostly east of the Continental Divide in Colorado   -3. L. pusillus
1. Plants perennial; ovules and seeds normally more than 2
  5. Plants low, less than 15 cm. high, acaulescent or nearly so; racemes usually 1-4 cm. long, shorter than the leaves; leaflets 1-2 cm. long   -4. L. caespitosus
  5. Plants over 20 cm. tall (unless depauperate), stems definite, usually long; racemes usually over 4 cm. long and longer than the leaves; leaflets 2-7 cm. long
    6. Hairs of the stems and petioles, at least the longer ones, spreading or ascending
      7. Leaves silky-hairy both sides, leaflets narrowly oblanceolate; flowers mostly ochroleucous (but may be purple)   -5. L. barbiger
      7. Leaves glabrous or glabrate above, leaflets obovate to oblanceolate; flowers blue or purple
        -6. L. ammophilus
    6. Hairs of stems and petioles appressed or subappressed
      8. Calyx at base short-spurred, this spur over 1 mm. long; leaflets silky-hairy above
        -7. L. aduncus
      8. Calyx not spurred but sometimes strongly gibbous; leaflets glabrous to hairy above
        9. Leaflets glabrous or glabrate above
          10. Flowers 12 mm. long or more, conspicuously bicolored, the banner with a conspicuous darker spot   -8. L. plattensis
          10. Flowers less than 12 mm. long, sometimes bicolored but not strikingly so
            11. Flowers small, 5-7 mm. long   -9. L. parviflorus
            11. Flowers 8-12 mm. long   -10. L. argenteus
        9. Leaflets hairy on upper surface
          12. Flowers 7-10 mm. long   -11. L. greenei
          12. Flowers 12 mm. long or longer   -12. L. humicola

1. **Lupinus brevicaulis** Wats. in King, Geol. Expl. 40th Par. 5:53. 1871.

L. scaposus Rydb.---Plants low annuals seldom over 10 cm. tall, acaulescent or nearly so, the leafy stem not over 1 cm. long, densely villous; leaves crowded basal; leaflets 5-9, 8-18 mm. long, spatulate to oblanceolate-obovate, glabrous or nearly so above, sparingly villous below and ciliate; peduncles shorter than the leaves, racemes subcapitate, dense, not over 2 cm. long in flower; calyx villous, upper lip from nearly obsolete to 2 mm. long, lower lip 4-6 mm. long; flowers 6-8 mm. long, blue or purple; fruit 9-11 mm. long, ovate to oblong, villous, seeds 2, pod not constricted between seeds.---Hills and valleys often on sandy soil. Colorado to Oregon, south to New Mexico and California. Our few records from western Colorado at about 5000-6500 feet.

2. **Lupinus kingii** Wats., Proc. Am. Acad. 8:534. 1873.

Annual or possibly biennial plants; stems 5-20 cm. tall, branched near base, erect to widely spreading, silky-villous with spreading often tawny hairs; leaflets 1-3 cm. long, oblanceolate to oblong-lanceolate, silky-villous both sides; raceme dense and subcapitate, seldom over 2 cm. long, peduncles variable but flower cluster shorter to somewhat longer than the leaves; calyx ciliate, the lips subequal, about 4-7 mm. long; corolla about 7-10 mm. long, purplish or blue; fruit ovate or rhombic-ovate, about 1 cm. long, not noticeably constricted between the 2 seeds, villous.---Dry ground. Colorado, Utah, New Mexico, and Arizona. Our records from southern and western Colorado at 5000-8000 feet.

2A. **Lupinus kingii argillaceus** (Woot. & Standl.) Smith, (var.) Bull. Torr. Club 46:325. 1919.

L. argillaceus Woot. & Standl.---The hairs are appressed and inconspicuous, especially on the pedicels and calyx.---Reported for western Colorado.

3. **Lupinus pusillus** Pursh, Fl. Amer. Sept. 2:468. 1814.

L. intermontanus Heller---Annual plants with cotyledons clasping and broad; stems 5-20 cm. tall, branching near the base, the stems decumbent near the base, loosely-villous, hairs rather stiff; leaflets 5-8, 2-3 cm. long, oblong-oblanceolate, glabrous or nearly so above, sparingly long-hairy beneath; racemes 3-5 cm. long, on short peduncles, not exceeding the leaves, pedicels short (about 2 mm.); calyx villous, lower lip about 5-6 mm. long, upper somewhat shorter; corolla 8-12 mm. long, purplish, bluish or almost white; fruit 20-25 mm. long, lanc-oblong, constricted somewhat between the 2 seeds, villous-pubescent.---Plains often in sandy ground. Saskatchewan to Washington, south to Kansas, New Mexico and Arizona. Our records scattered in the eastern half of Colorado at 3500-8000 feet, but also reported from the western part of the state.

4. **Lupinus caespitosus** Nutt. in T. & G., Fl. N. Amer. 1:379. 1840.

L. watsonii Heller---Perennial acaulescent or subacaulescent plants 5-15 cm. tall, silky-hirsute; leaves basal; leaflets 5-7, 1-2 cm. long, oblanceolate, silky-hairy both sides; racemes usually 1-4 cm. long, dense and short, normally much shorter than the leaves; calyx rather unequal, lower lip about 6 mm. long, upper about 3-4 mm. long; corolla 7-8 mm. long, pale blue, lilac or whitish; fruit 8-15 mm. long, hairy, 2- to 4-seeded. Recently made a subspecies of L. lepidus Dougl. by L. Detling.---Dry plains and hills. Montana to Oregon, south to Colorado and Utah. Our records from northcentral, western and central Colorado at 8000-10,000 feet.

5. **Lupinus barbiger** S. Wats., Proc. Amer. Acad. 8:528. 1873.

L. bakeri Greene; L. dichrous Greene---Perennial plants; stems 40-80 cm. tall, with spreading hairs, often short and long hairs intermixed; leaflets 5-7, 3-7 cm. long, narrowly oblanceolate, silky-hairy on both sides; racemes 8-30 cm. long; calyx only slightly to strongly gibbous but not projecting backward much; corolla 10-13 mm. long, ochroleucous or less commonly purplish; fruit 2-3 cm. long, seeds 3-7.---Meadows and slopes. Colorado, Utah and Arizona. Our records mostly from westcentral and southwestern Colorado, one from Douglas County appears to be this species, at 5500-9000 feet.

6. **Lupinus ammophilus** Greene, Pitt. 4:136. 1900.

Perennial plants; stems 20-60 cm. tall, spreading villous or hirsute, the hairs up to 2 mm. long, often tawny; leaflets 7-11, 2-4 cm. long, obovate to oblanceolate, glabrous or glabrate above, rather sparingly spreading villous or hirsute below; racemes 6-20 cm. long; calyx more or less gibbous at base; corolla 9-12 mm. long, blue or purple; fruit 2-4 cm. long, 4- to 6-seeded.---Banks and valleys, especially in sandy soil. Colorado, Utah and New Mexico. Our records from western and northcentral Colorado at 6000-8500 feet.

7. **Lupinus aduncus** Greene, Pitt. 4:132. 1900.

L. argophyllus (A. Gray) Cockll.; L. helleri Greene---Perennial plants; stems 30-100 cm. tall, appressed hairy; leaves many, leaflets 7-9, 2-4 cm. long, linear-oblanceolate to narrowly oblanceolate, appressed silky-hairy both sides; racemes about 5-25 cm. long; calyx spurred or strongly saccate, this extending backward for about 1 mm. or more; corolla about 8-11 mm. long, dark-blue; fruit 2-3 cm. long, 4- to 6-seeded.---Plains, canyons, valleys and banks. Wyoming and Utah, south to New Mexico and Arizona. Our records from western and southcentral Colorado at 6000-10,000 feet.

8. **Lupinus plattensis** S. Wats., Proc. Amer. Acad. 17:369. 1882.

Perennial plants; stems 20-50 cm. tall, appressed-hairy; leaflets 7-9, 2-4 cm. long, narrowly to broadly oblanceolate, glabrous or glabrate above, appressed-hairy below; racemes about 6-15 cm. long; calyx more or less gibbous but this not projecting backward as a spur or sac; corolla 12-14 mm. long, conspicuously bicolored, blue in color but banner with a conspicuous darker spot; fruit 2.5-3.5 cm. long, 3- to 7-seeded.---Plains and hills. Nebraska and Wyoming, south to Kansas and Colorado. Our records from the eastern half of Colorado, mostly along the foothills, at 5500-8000 feet.

9. **Lupinus parviflorus** Nutt. ex H. & A., Bot. Beech. Voy. Suppl. 336. 1841.

L. floribundus Greene---Perennial plants; stems 30-80 cm. tall, appressed-hairy; leaflets 5-11, 3-6 cm. long, linear-oblanceolate to oblanceolate, glabrous or glabrate above, more or less strigose below; racemes 7-30 cm. long; calyx more or less gibbous but not projecting backward as a spur or sac; corolla 5-7 mm. long, light to dark blue or violet, rarely almost whitish; fruit 1.5-2.5 cm. long, 3- to 4-seeded.---Hills, valleys and mountains. South Dakota to Montana, south to New Mexico and Utah. Our records scattered in the western half oc Rolorado at 6500-9500 feet.

10. **Lupinus argenteus** Pursh, Fl. Amer. Sept. 468. 1814.

Perennial plants; stems 30-100 cm. tall, appressed-hairy; leaflets 5-10, 2-7 cm. long, linear-oblanceolate to oblanceolate, strigose below, glabrous to glabrate above; racemes 5-20 cm. long; calyx weakly to strongly gibbous at base on 1 side but not projecting backward as a sac; corolla light to dark blue or purple, 8-12 mm. long; fruit 2-3.5 cm. long, about 3- to 6-seeded. Closely related to the above are several forms that have been given specific names, such as L. alpestris A. Nels.; L. rubricaulis Greene and L. spathulatus Rydb. The writer cannot distinguish these in Colorado and they are arbitrarily included in the above, although extreme forms certainly do differ in general appearance.---Hills, valleys, mountains and plains. North Dakota to Montana, south to Nebraska, New Mexico and Arizona, perhaps more widely distributed. Our records well scattered in the western two-thirds of Colorado at 5000-11,000 feet.

11. **Lupinus greenei** A. Nels., New Man. Bot. Rocky Mtn. 274. 1909.

L. oreophilus Greene---Perennial plants; stems 30-60 cm. tall, appressed-hairy; leaflets 7-10, 2-5 cm. long, linear-oblanceolate to rather narrowly oblanceolate, appressed-hairy on both sides, usually silvery colored; racemes usually rather short, often not over 10 cm. long; calyx more or less gibbous but this not projecting backward; corollas about 7-10 mm. long, blue, purple or violet, varying from light to dark; fruit about 1.5-2.5 cm. long, 3- to 6-seeded.---Plains, hills, mountains and banks. Wyoming to Nevada, south to Colorado and Arizona, perhaps more widely distributed. Our records scattered in the western two-thirds of Colorado, none as yet from the southern part of the state however, at 5000-9000 feet.

12. Lupinus humicola A. Nels., Bull. Torr. Club 25:204. 1898.
Perennial plants; stems 30-60 cm. tall, appressed-hairy; leaflets 7-12, 3-7 cm. long, oblong-oblanceolate to linear-oblanceolate, appressed-hairy both sides; racemes about 8-20 cm. long; calyx more or less gibbous but this not extending backward into a spur or sac; corollas 12 mm. long or longer, blue to purple; fruit about 2-4 cm. long, seeds 3-6.---Hills and mountains. Montana to Colorado. Our few records from northcentral Colorado, with a doubtful one from Elbert County, at 6000-7000 feet.

14. Psoralea L. SCURFPEA

Perennial herbaceous plants with rootstocks or thick tuberlike roots; leaves 3- to 7-foliate, with entire or toothed leaflets, these glandular-punctate; flowers perfect, in peduncled spikes or racemes; calyx campanulate nearly equal or unequal; corolla papilionaceous, the keel incurved; stamens monadelphous or diadelphous, 9 or 10, anthers uniform; ovary sessile or nearly so, 1- to 2-ovuled; fruit ovoid, usually indehiscent, 1-seeded.

1. Flowering stems from thick tuberous, rounded or fusiform taproots; flowers 10 mm. long or more, in short spikelike, subcapitate heads 15 mm. wide or more; leaflets 5-7; fruit regularly or irregularly circumscissle near the middle
   2. Stems 40-60 cm. tall, branched throughout and leafy; plants of eastern Colorado   -1. P. cuspidata
   2. Stems not over 30 cm. tall, simple and rather sparingly branched, most of leaves usually basal; plants of eastern and western Colorado
      3. Stems and petioles villous-hirsute with spreading hairs   -2. P. esculenta
      3. Stems and petioles appressed pubescent or strigose, with few if any spreading hairs
         4. Leaflets obovate, 1-3 cm. long; plants of western Colorado
            5. Flowers 15-20 mm. long; stems above ground short, usually less than 6 cm. long, never over 15 cm. long   -3. P. megalantha
            5. Flowers 10-14 mm. long; stems above ground 15-30 cm. long   -4. P. aromatica
         4. Leaflets narrowly oblanceolate to narrower, 2-5 cm. long; plants of eastern Colorado
                                                                                                 -5. P. hypogaea
1. Flowering stems from creeping rootstocks, not at all tuberous; flowers not over 10 mm. long, variously arranged but if in spikelike heads then these not over 15 mm. wide; leaflets commonly 3 (except in P. digitata); fruit indehiscent
   6. Plants conspicuously silvery-white (except rarely the upper surface of leaflets) from dense silky hairs; leaflets often 5; flowers sessile or nearly so
      7. Leaflets linear or linear-lanceolate, usually 5   -6. P. digitata
      7. Leaflets oblong-oblanceolate to obovate, usually 3   -7. P. argophylla
   6. Plants glabrous or merely strigose on stems and lower surface of leaflets, not silvery-white; leaflets rarely over 3; flowers pedicellate, often strongly so
      8. Fruit tapering gradually to the beak; flowers in rather loose racemes, the pedicels 4-8 mm. long; leaflets always narrowly linear   -8. P. linearifolia
      8. Fruit abruptly beaked; flowers in spikelike or narrow racemes, the pedicels not over 4 mm. long; leaflets narrowly linear to obovate
         9. Fruit subglobose; corolla whitish (tip of the keel may be purplish); calyx lobes not over 1/2 as long as the tube   -9. P. lanceolata
         9. Fruit ovoid; corolla purple or blue; calyx lobes about as long or longer than the tube
            10. Calyx 3-5 mm. long in flower, much inflated in fruit; fruit hairy, the beak about 2 mm. long; leaflets usually 5   -6. P. digitata
            10. Calyx 2-2.5 mm. long, only slightly enlarging in fruit; fruit glabrous (although glandular), with the beak much less than 2 mm. long; leaflets usually 3   -10. P. tenuiflora

1. Psoralea cuspidata Pursh, Fl. Amer. Sept. 741. 1814.
Pediomelum cuspidatum (Pursh) Rydb.---Roots tuberous and farinaceous, elongated, fusiform or ellipsoidal; stems erect or ascending, 10-60 cm. long, branching, stout, sparingly strigose to glabrate below; leaves 5-foliate, not basal; leaflets 2-4 cm. long, elliptic to broadly-oblanceolate or obovate, glandular-punctate, glabrous above, strigose beneath; flowers in dense headlike spikes, bracts 1-1.5 cm. long; calyx about 1 cm. long, strigose and punctate with lobes about 3-6 mm. long, the lowest longest; corolla bluish, 15-20 mm. long; fruit ovoid, strigose, beaked, the body about 6 mm. long.---Hills and plains usually in sandy soil. Our one record from northeastern Colorado at about 3700 feet.

2. Psoralea esculenta Pursh, Fl. Am. Sept. 475. 1814.
Pediomelum esculentum (Pursh) Rydb.---Roots tuberous, farinaceous and deep-seated; stems 10-30 cm. tall, stout, villous, hirsute with long whitish hairs; lower petioles definitely longer than leaflets, these 5, each 2-6 cm. long, oval, oblong or obovate, entire, glabrous above, strigose beneath; flowers in short, dense, peduncled spikes 2-8 cm. long, bracts nearly as long as the flower; calyx tube about 5 mm. long, the lobes longer; corolla about 12-15 mm. long, somewhat bluish-tinged; fruit 5-6 mm. long, oblong or ovoid, long-beaked, body enclosed in the calyx tube.---Plains and hills. Northwestern Territories, south to Louisiana and Texas. Our records from northeastern Colorado at 3500-5000 feet.

3. **Psoralea megalantha** Woot. & Standl., Contrib. U. S. Nat. Herb. 16:140. 1913.

   Pediomelum megalanthum (Woot. & Standl.) Rydb.; Psoralea mephitica of Manuals---Roots tuberous, farinaceous and deep-seated; stems very short above ground, the stems and petioles canescently-strigose or appressed-pubescent (a few hairs may spread), petioles definitely longer than leaflets; leaflets 5-7, 1-3 cm. long, obovate, white-strigose both sides but less so above; peduncles stout, sericeous; flowers in subcapitate spikelike racemes about 2 cm. long, bracts nearly as long as the corolla; calyx tube 5-6 mm. long, the lobes longer, nearly as long as corolla; corolla about 15-20 mm. long, tinged with blue.---Hills and mesas. Colorado and Utah, probably south into New Mexico and Arizona. Our records from western Colorado, especially the westcentral part, at 4500-6000 feet.

4. **Psoralea aromatica** Payson, Bot. Gaz. 60:379. 1915.

   Roots tuberous and deep-seated, branching caudices arising from these; stems 15-30 cm. long, decumbent, sparingly branched, sparsely appressed-hairy; petioles longer than the leaflets; leaflets usually 5, obovate, about 1.5-2 cm. long, sparingly appressed-pubescent; flowers subcapitate in clusters these usually about 1.5-2 cm. long and about as wide; bracts shorter than the calyx; calyx with lower lip nearly as long as the corolla, sparingly pubescent except on the margins of the lobes; corolla about 10-14 mm. long, rose-purple to much lighter.---Dry hills and slopes. Apparently limited to western Colorado, our records from western Montrose County at about 5000-5500 feet.

5. **Psoralea hypogaea** Nutt. in T. & G., Fl. N. Amer. 1:302. 1838.

   Pediomelum hypogaeum (Nutt.) Rydb.---Roots tuberous, farinaceous and deep-seated; stems very short above the ground level, stems and petioles appressed-pubescent with whitish hairs, the petioles usually longer than the leaflets; leaflets 5-7, 2-5 cm. long, linear-lanceolate to linear-oblong, sometimes narrowly oblanceolate, entire, glabrous or nearly so above, strigose beneath; flowers in short, dense, capitate spikes, bracts nearly as long or as long as the corolla; calyx about 9-12 mm. long, the tube about 4 mm. long, lobes as long or longer, the lower one longer and wider than the other 4; corolla about 12 mm. long, tinged with blue; fruit body about 4-5 mm. long, long-beaked this longer than body, appressed-hirsute above at least, body enclosed in calyx tube.---Dry plains and slopes. Nebraska to Colorado, south to Texas and New Mexico. Our records scattered in the eastern half of Colorado at 4500-5500 feet.

6. **Psoralea digitata** Nutt. in T. & G., Fl. N. Amer. 1:300. 1838.

   Psoralidium digitatum (Nutt.) Rydb.---Roots not tuberous, plants from creeping rootstocks; stems 30-60 cm. tall, simple below, branched above, leafy, canescent to appressed-puberulent; leaflets usually 5 at least on lower leaves, 2-3.5 cm. long, linear or oblong-linear, sparingly strigose to glabrate above, densely white-strigose to moderately strigose below, obtuse or mucronate; flowers in lax, interrupted, spicate racemes; calyx about 3-5 mm. long in flower, silky-hairy, lobes as long or somewhat longer than tube, sharp acute to acuminate, enlarging and inflated in fruit; corolla 7-10 mm. long, blue or purple; fruit 6-7 mm. long, ovoid, flattish, tapering to a moderately short beak.---Plains and hills, often on sandy soil. South Dakota to Colorado, south to Texas. Our few records from eastern Colorado at about 3500-7500 feet.

7. **Psoralea argophylla** Pursh, Fl. Am. Sept. 475. 1814.

   Psoralidium argophyllum (Pursh) Rydb.---Creeping thickened rootstocks present; stems erect, widely branched, 25-60 cm. tall, densely silvery-pubescent with appressed hairs; petioles from shorter to longer than leaflets; leaflets 3-5, 5-35 mm. long, obovate, elliptic-lanceolate to oblong-oblanceolate, densely white-silky both sides, however less dense and more green above (rarely somewhat glabrate); flowers in interrupted short spikes, bracts somewhat longer than the corolla; calyx tube about 2 mm. long, upper lobes somewhat longer, lower one much longer than others (6-10 mm.); corolla about 6-7 mm. long, purple; fruit body about 6-7 mm. long, ovate with a straight beak, hairy.---Plains and hills, often on sandy soil. Saskatchewan to Alberta, south to Wisconsin, Missouri and New Mexico. Our records scattered in the eastern half of Colorado, mostly in the extreme eastern part, at 3500-5500 feet.

8. **Psoralea linearifolia** T. & G., Fl. N.Amer. 1:300. 1838.

   Psoralidium linearifolium (T. & G.) Rydb.---Long creeping rootstocks present; stems 40-100 cm. tall, erect, widely branched, sparingly strigose or glabrous; leaves sessile or short-petioled; leaflets usually 3, 2-6 cm. long and 2-3 mm. wide, linear, glandular-punctate, finely rugose, glabrous or nearly so; flowers loosely scattered, the pedicels 4-8 mm. long; calyx about 4 mm. long, the lobes about 1/2 the entire distance or somewhat shorter, somewhat unequal, strigose; corolla 7-8 mm. long, blue; fruit 7-9 mm. long, including a long beak (2-3 mm. long), glandular-punctate but glabrous, ovoid.---Plains and hills. Nebraska to Colorado, south to Texas. Our one record from northeastern Colorado at 3500 feet.

9. **Psoralea lanceolata** Pursh, Fl. Amer. Sept. 475. 1814.

   Psoralidium lanceolatum (Pursh) Rydb.; Psoralea micrantha A.Gray; Psoralidium micranthum (A. Gray) Rydb.---Long-creeping rhizomes present, these in sand becoming 7-10 m. long; stems 15-40 cm. tall, erect or ascending, glabrous or sparingly strigose, glandular-punctate; leaves 3-foliate, petioles as long or shorter than the leaflets; leaflets 1.5-5 cm. long, variable, from narrowly linear to obovate-oblanceolate, acute to mucronate, glabrous or sparingly strigose below especially near veins, conspicuously glandular-punctate; flowers in short dense spikelike racemes 10-25 mm. long; calyx about 2-2.5 mm.long, campanulate, sparingly strigose, glandular, the lobes nearly equal, not over 1/2 as long as the tube; corolla 5-6 mm. long, white or purple-tinged near tip especially on keel; fruit about 4-5 mm. long, globose, abruptly short-beaked, glabrate to densely-villous, densely glandular.---Plains and hills, usually on sandy soil. Saskatchewan to Alberta, south to Missouri, Texas, Arizona and Washington. Our records scattered over Colorado, especially abundant in the eastern half of the state, at 3500-8500 feet.

10. Psoralea tenuiflora Pursh, Fl. Amer. Sept. 475. 1814.
   Psoralidium tenuiflorum (Pursh) Rydb.---Long-creeping rootstocks present; stems erect, 20-60 cm. tall, busy-branching, more or less strigose, glandular-dotted; petioles mostly shorter than leaflets; leaves digitately 3-foliate (sometimes 5 on basal ones); leaflets 1-4 cm. long, strigose below, glabrate above, glandular-dotted both sides, obtuse and mucronate at apex, linear-oblanceolate, oblong-oblanceolate to obovate; flowers in slender racemes 1-4 cm. long, interrupted, the flowers solitary or to 3 at a node, pedicels 2-4 mm. long, the bracts shorter; calyx 2-2.5 mm. long, glandular, lobes about equal to each other and about as long as the tube; corolla about 4-7 mm. long, blue to purple; fruit 5-8 mm. long, ovoid, with a short straight beak, body glabrous but densely glandular dotted. The form with leaflets broadly oblanceolate to obovate has been called var. bigelovii (Rydb.) Macbride---Plains and hills. North Dakota and Montana, south to Kansas and Arizona. Our records mostly from the eastern half of Colorado, a few from the southwestern part of the state, at 4000-7000 feet.

15. Dalea Juss. INDIGO BUSH

Annual or perennial herbaceous plants or shrubs; leaves odd-pinnately compound, or digitately 3-foliate (in 1 species), with small entire glandular-punctate leaflets; stipules minute; flowers perfect, in spikes or spikelike racemes; calyx campanulate, its 5 lobes equal or subequal; petals 5, clawed, standard usually shorter, wings and keel with claws adnate to the tube of the filaments; stamens 9 or 10, monadelphous, anthers alike; fruit short, often included in the calyx, sessile or short-stipitate, indehiscent, 2- to 3-ovuled but usually 1-seeded.

1. Stems and leaves glabrous (only the calyx silky-hairy)
  2. Plants shrubby; inflorescence a 2- to 10-flowered short headlike spike; corolla rose to yellowish
     -1. D. formosa
  2. Plants herbaceous; inflorescence a many-flowered elongated spike, not headlike; corolla white to pink
    3. Spike oblong, densely flowered; leaflets 13-41 to a leaf; plants annual   -2. D. leporina
    3. Spike narrow-elongated, laxly flowered; leaflets 5-9 to a leaf; plants perannial   -3. D. enneandra
1. Stems and at least lower leaflet surface densely silky- or villous-hairy
  4. Calyx lobes ovate or ovate-lanceolate, shorter than the tube; spikes laxly flowered; corolla red to purple (or rarely white) when fresh   -4. D. lanata
  4. Calyx lobes subulate-filiform, longer than the tube; spikes densely flowered; corolla yellow when fresh (often drying to rose or purple)
    5. Leaves palmately compound with 3 leaflets   -5. D. jamesii
    5. Leaves pinnately compound with 5-7 leaflets
      6. Stems 30-80 cm. tall, ascending or erect; upper leaves reduced, the spikes rather long-peduncled (peduncle about 6-10 cm. long), spikes 2-7 cm. long; leaflets averaging about 15 mm. long
         -6. D. aurea
      6. Stems 10-25 cm. long, decumbent to ascending; upper leaves not reduced, the spike sessile or short-peduncled (these about 1-2 cm. long), spikes 1-3 cm. long; leaflets averaging about 10 mm. long
        7. Leaflets silky-sericeous both sides   -7. D. nana
        7. Leaflets green and glabrate above   -7A. D. nana carnescens

1. Dalea formosa Torr., Ann. Lyc. N. Y. 2:177. 1827.
   Parosela formosa (Torr.) Vail---Plants shrubby or suffrutescent; stems 25-60 cm. tall, divaricately branched, glabrous and sparingly glandular; leaves small, rarely over 1 cm. long, pinnately 7- to 11-foliate; leaflets 1.5-2.5 mm. long, oblong, spatulate to cuneate-oblong, rather thick and conduplicate especially when dry, glabrous, glandular-dotted beneath; inflorescence a headlike spike, 2- to 10-flowered, with bracts ovate, glabrous on back and silky on margins, glandular; calyx long-villous, tube 3-4 mm. long, campanulate, ribbed longitudinally and glandular, lobes 5-10 mm. long, setaceous and plumose-appearing; corolla rose-colored or the banner sometimes yellowish; blade of banner 2.5-4 mm. long, ovate, claw about as long; blade of wings 4-5 mm.; blade of keel 5-6 mm. long, claws somewhat shorter and inserted below the middle of the staminal tube; fruit hairy and glandular-dotted above.---Dry ground. Colorado to New Mexico, Mexico and Arizona. Our few records from southeastern and southcentral Colorado at about 4000-5000 feet.
2. Dalea leporina (Ait.) Kearney and Peebles, Journ. Wash. Acad. Sci. 29:483. 1939.
   Dalea alopecuroides Willd.; D. alba Michx.; Parosela alopecuroides (Willd.) Rydb.; P. dalea (L.) Britt. of Manuals---Annual plants; stems 20-60 cm. tall, erect, branched, glabrous; leaves pinnately compound with 15-41 leaflets, these 3-8 mm. long, glabrous, oblanceolate to linear-oblong, obtuse at apex, long-cuneate at base; inflorescence dense, terminal, oblong or cylindrical spike, 2-7 cm. long, bracts ovate or lanceolate, deciduous; calyx lobes long-acuminate, narrow, plumose; corolla 2.5-4 mm. long, white or tinged with rose. Our plant would be var. alba (Michx.) comb. nov.---Plains and hills. The variety from Minnesota to Nebraska, south to Illinois, Texas and Mexico. Our records from northcentral and southcentral Colorado at 5000-8800 feet.
3. Dalea enneandra Nutt., Fras. Cat. 1813.
   Parosela enneandra (Nutt.) Britton; P. lasianthera in Rydb.'s Fl. Colo.---Perennial plants from rather woody taproots; stems 40-120 cm. tall, mostly single, glabrous, branching above and rather flat-topped, ascending or spreading; leaves 1-3 cm. long on shorter petioles, glabrous, pinnately compound of 5-9 leaflets, these linear or narrowly oblong, 5-10 mm. long, conspicuously glandular-dotted below; inflorescence of rather short-peduncled lax spikes, 5-12 cm. long, with bracts broadly ovate and with scarious margins, glabrous but glandular, short-acuminate and folded, enclosing the calyx; calyx tube 3-3.5 mm. long, densely silky, lobes somewhat

longer, filiform, conspicuously plumose; corolla white or pinkish-tinged; blade of banner ovate-cordate, about 3 mm. long, glandular dotted below, claw about as long; blade of wing 3-4 mm. long, short-clawed, inserted near middle of staminal tube; blade of keel about 5 mm. long, short-clawed, inserted above the middle of the staminal tube; fruit villous above. At least some of the plants have a "sour" odor when fresh.---Plains and hills. Iowa to Nebraska, south to Mississippi, Texas and Colorado. Our records from along the eastern border at 3500-4000 feet.

4. Dalea lanata Spreng., Syst. Veg. 3:327. 1826.
Parosela lanata (Spreng.) Britt.---Stems decumbent, 30-60 cm. long, branched at base, densely pubescent or very short-villous; leaflets 7-13, 4-12 mm. long, obovate to narrowly obovate, densely pubescent to very short-villous; spikes slender, rather lax, 2-10 cm. long, many-flowered; bracts long-acuminate; calyx about 3-3.5 mm. long, the lobes lanceolate and somewhat shorter than the tube; corolla about 4-5 mm. long, purple, red or rarely whitish.---Dry ground. Kansas and Colorado, south to Texas and New Mexico. Our one record from the southeastern corner of the state at about 4000 feet.

5. Dalea jamesii (Torr.) T. & G., Fl. N. Amer. 1:308. 1838.
Parosela jamesii (Torr.) Vail; P. porteri A. Nels.---Perennial plants with thick woody branching caudices; stems 5-20 cm. long, numerous, ascending or decumbent, densely silky; leaves digitately 3-foliate, on petioles about as long as the leaflets, these obovate to obovate-oblanceolate, 8-15 mm. long, densely silky-canescent both sides, obscurely if at all glandular; inflorescence of very short-peduncled dense spikes 1.5-6 cm. long and about 15-20 mm. wide, bracts about as long as calyx, ovate to lanceolate, often purple-tinged; calyx 8-10 mm. long, densely silky, the tube about 2.5-3 mm. long, lobes much longer, filiform and setaceous, plumose-appearing; corolla yellow, usually drying rose or purplish; blade of banner about 3 mm. long, broadly ovate, claw about as long; blades of wing 4-4.5 mm. long; blades of keel 5-8 mm. long, shorter-clawed, these inserted somewhat below the middle of the staminal tube; fruit villous.---Dry plains and hills. Kansas to Colorado, south to Mexico. Our records from central and southern Colorado at 4500-9000 feet.

6. Dalea aurea Nutt. in Pursh, Fl. Amer. Sept. 740. 1814.
Parosela aurea (Nutt.) Britt.---Perennial plants, somewhat woody at very base; stems 30-80 cm. tall, several, erect or ascending, silky-canescent, leafy but upper leaves reduced and scattered; leaves 2-5 cm. long, pinnately 5- (rarely 7-) foliate, petioles shorter; leaflets oblong, oblanceolate to obovate, more or less silky-pubescent especially below, minutely glandular-dotted below, 1-2 cm. long; peduncles up to 10 cm. long; flowers in dense spikes 2-7 cm. long and about 1.5 cm. thick; bracts about as long as calyx, silky; calyx 8-10 mm. long, the lobes narrow and somewhat pungent, somewhat longer than the tube and plumose-looking, densely silky; corolla yellow; blade of banner flabelliform, about 3-4 mm. long, contracted to a claw about as long; keel with blades about 5-6 mm. long and short-clawed, attached above middle of long staminal tube; wings shorter than keel, short-clawed; fruit hairy.---Plains and hills. South Dakota to Wyoming, south to Texas, Mexico and Arizona. Our records from the eastern half of Colorado, mostly in the extreme eastern part, at 3500-6000 feet.

7. Dalea nana Torr. in A. Gray, Amer. Acad. Arts & Sci. Mem. ser. 2, 4:31. 1849.
Parosela nana (Torr.) Heller---Perennial plants, herbaceous or somewhat woody at very base; stems 10-25 cm. tall, branching at base, decumbent at base or ascending, silky-canescent; leaves many, 2-3 cm. long, on shorter petioles, with 5 (rarely 7) leaflets, these obovate to oblanceolate, 5-15 mm. long, silky-sericeous both sides, obscurely glandular below; inflorescence an oblong dense spike 1-3 cm. long and 1-1.5 cm. wide, on short peduncles; bracts about as long as the calyx, ovate, short-acuminate, hairy; calyx about 6-8 mm. long, silky-sericeous, the tube about 2-3 mm. long, angled, campanulate, the lobes longer, setaceous and plumose-appearing; corolla yellow, fading rose-colored; blade of banner reniform, 2-2.5 mm. long, claw somewhat longer; blade of wings and keel 2.5-3.5 mm. long, short-clawed, inserted somewhat below the middle of the staminal tube; fruit villous.---Dry plains and hills. Kansas, south to Texas, Mexico and Arizona. Our one record from Crowley County in southeastern Colorado at about 4300 feet.

7A. Dalea nana carnescens (Rydb.) Kearney & Peebles, (var.) Journ. Wash. Acad. Sci. 29:483. 1939.
Parosela carnescens Rydb.; P. rubescens in C. & N. Manual---Leaflets green and glabrate above; stems usually stouter. Intergrades with the species in southern Colorado.

16. Robinia L. LOCUST

Trees or shrubs, usually with spinelike stipules; leaves odd-pinnate, leaflets entire with stipels; flowers large and showy, in rather dense racemes; calyx campanulate, 5-lobed but the upper 2 more united than the rest; corolla papilionaceous; stamens 10, diadelphous (9-1) or the single filament free at base, and more or less united with the staminal tube at middle; fruit narrow, flattened, short-stipitate, 2-valved, several-seeded.

1. Corolla white; leaflets, twigs, and fruit glabrous; plants cultivated or escaped   -1. R. pseudoacacia
1. Corolla rose-pink; leaflets strigose; young twigs puberulent; fruit hairy, usually glandular; plants native (but often cultivated too)   -2. R. neomexicana

1. Robinia pseudoacacia L., Sp. Pl. 722. 1753.
Trees to 10 m. or more; twigs glabrous, with stipular spines usually present; leaflets 9-19, 2-5 cm. long, petiolulate, ovate or oval, glabrous, entire, rounded at base, rounded, emarginate or mucronate at apex; racemes loose and dropping; corolla 14-25 mm. long, white and fragrant; fruit 5-10 cm. long, glabrous, 4- to 7-seeded.---Usually around dwellings or along fence rows as an escape from cultivation. Pennsylvania to Utah, south to Georgia and Oklahoma. (Probably more widely distributed.) Our records from northcentral and western Colorado at 4500-5500 feet but to be expected anywhere in cultivated areas.

2. **Robinea neomexicana** A. Gray, Mem. Amer. Acad. ser. 2, 5:314. 1855.
  R. luxurians (Dieck.) Rydb.; R. subvelutina Rydb.---Trees or large shrubs to 8 m. tall; twigs with the stipular spines straight or somewhat curved, 4-20 mm. long, young twigs puberulent, becoming reddish-brown, finally gray; leaflets 9-19, 2-3 cm. long, elliptic to oblong, sometimes oval, strigose-pubescent, rounded at both ends and mucronate at apex; racemes about 5-10 cm. long, the pedicels, peduncle, and calyx glandular-hispid and puberulent; calyx tube about 6-8 mm. long, lobes shorter; corolla 20-25 mm. long, rose-pink; fruit 6-10 cm. long and about 7-10 mm. wide, 4- to 8-seeded, usually glandular-hispid as well as hirsutulous.---Along streams and valleys. Colorado to Nevada, south to Texas, Mexico and Arizona. Our records from central and southcentral Colorado at 5500-7000 feet.

17. Onobrychis Scop. SAINFOIN

Herbaceous plants or undershrubs; stems erect; leaves pinnate with entire leaflets; stipules rather small and scarious, stipels absent; flowers perfect, in spikes or racemes; calyx tube campanulate, the 5 lobes subequal and 1-sided, subulate; corolla papilionaceous, rose-colored; stamens 10, monadelphous but 1 filament free from the others at the base of the staminal tube only (not at apex), anthers uniform; fruit flat, sessile in calyx, suborbicular, not jointed, indehiscent, pectinately bristly on margins (except down 1 side) and reticulated, 1- to 2-seeded.

1. **Onobrychis viciaefolia** Scop., Fl. Carn. 2:76. 1772.
  O. onobrychis (L.) Rydb.; O. sativa Lam.---Perennial herbaceous plants; stems 30-50 cm. tall, sparsely pubescent to glabrate; leaves about 10 cm. long, the petiole shorter; leaflets linear-oblong to elliptic, 11-23, inconspicuously glandular if at all, glabrous or sparsely hairy on margins and midrib below; calyx tube irregular, about 2-2.5 mm. long on lobed side, lobes on 1 side, about 3-5 mm. long, calyx appressed-pilose; corolla 8-10 mm. long, the banner and keel subequal, the wings much shorter; fruit about 5-6 mm. long, coarsely reticulated and puberulent.---Cultivated and occasionally escaping, at least in the Rocky Mountain area. Our rather few records from northcentral, central, and southcentral Colorado at 5000-8000 feet.

18. Lathyrus L. PEAVINE

Mostly perennial herbaceous plants with erect to trailing or climbing stems; leaves pinnately compound, the rachis ending in a short, straight, bristlelike prolongation or in a prehensile tendril; stipules conspicuous; flowers solitary or in axillary racemes; calyx oblique its teeth about equal; petals papilionaceous, the wings appearing adherent to folds of the keel; stamens diadelphous (9 and 1); style usually curved, flattened, hairy along the inner side for about half its length; fruit usually more or less flattened; seeds usually several.

1. Flowers white (may have darker veins) fading to yellow or tan; calyx not over 9 mm. long; corolla usually not over 20 mm. long
    2. Leaflets 2-4 (rarely 6); corolla not over 14 mm. long; calyx 5-7 mm. long; leaf tendrils never prehensile; leaflets 4-8 cm. long -1. L. arizonicus
    2. Leaflets usually 6 or more; corolla usually over 14 mm. long; calyx 7-9 mm. long; tendrils often prehensile; leaflets 2-5 cm. long -2. L. leucanthus
1. Flowers blue or purple at least in part (may dry or fade to dingy white); calyx often well over 9 mm. long; corolla often over 20 mm. long
    3. Tendrils not prehensile; leaflets linear to linear-oblanceolate, not over 3 cm. long; stems short, rarely over 20 cm. tall -3. L. polymorphus
    3. Tendrils prehensile, at least on some leaves; leaflets linear to as wide as ovate, often over 3 cm. long; stems usually over 20 cm. tall
        4. Keel of corolla definitely shorter (2-4 mm.) than the wings; calyx glabrous or the teeth merely ciliate, the teeth usually longer than the tube -4. L. pauciflorus
        4. Keel and wings of corolla equal or very nearly so; calyx sparsely hairy, the teeth shorter than the tube -5. L. eucosmus

1. **Lathyrus arizonicus** Britt., Trans. N. Y. Acad. Sci. 8:65. 1889.
  Stems 10-40 cm. tall; leaves with 2-6 leaflets, these 4-8 cm. long, linear to elliptic-oblong, glabrous or sparsely hairy; tendrils bristlelike, unbranched, not at all prehensile; corolla 11-14 mm. long, white, ageing to tan or yellow; calyx 5-7 mm. long, the teeth shorter than the tube.---Colorado to Utah, south to New Mexico and Arizona. Our records from southcentral and southwestern Colorado at 6000-9500 feet, with a doubtful record at 11,000 feet.
2. **Lathyrus leucanthus** Rydb., Bull. Torr. Club 28:37. 1901.
  L. laetivirens Greene---Stems 15-50 cm. tall; leaves with 4-10 leaflets, these 2-5 cm. long, linear-elliptic to ovate; tendrils variable, from bristlelike and unbranched to prehensile and branched; corolla 13-22 mm. long, white ageing to tan or yellow; calyx 7-9 mm. long, the teeth shorter than the tube.---Slopes and woods. Wyoming to Utah, south to New Mexico and Arizona. Our records scattered throughout the western two-thirds of Colorado at 6500-11,000 feet.
3. **Lathyrus polymorphus** Nutt., Gen. N. Am. Pl. 2:96. 1818.
  L. ornatus Nutt.; L. incanus (Sm. & Rydb.) Rydb.; L. stipulaceous (Pursh) Butters & St. John; L. decaphyllus Rydb.---Stems 10-20 cm. tall, rarely taller; leaves with 4-12 leaflets, these 2-3 cm. long, linear, linear-elliptic or linear-oblanceolate; tendrils bristlelike, straight or somewhat curved but not prehensile; corolla 20-30 mm. long, blue or purple at least in part; calyx 7-11 mm. long, the teeth shorter than the tube. The typical plants are essentially glabrous and the villous-hirsute ones have been called ssp. incanus (Sm. & Rydb.) Hitchc.---Plains,

hills and thickets. South Dakota to Wyoming, south to Texas and Colorado. Our records scattered in the western half of Colorado at 4000-7500 feet, the hairy subspecies apparently more common.
4. Lathyrus pauciflorus Fernald, Bot. Gaz. 19:335. 1894.
L. utahensis Jones; L. schaffneri Rydb.---Stems 20-80 cm. tall; leaves with 6-13 leaflets, these 3-8 cm. long, linear to ovate; tendrils simple to forked but prehensile; corolla 18-27 mm. long, orchid to purple at least in part; calyx 8-10 mm. long, the teeth usually longer than the tube.---Woods and slopes. Idaho to Washington, south to Colorado, Arizona and California. Our few records from southwestern Colorado at 7000-8500 feet.
5. Lathyrus eucosmus Butters and St. John, Rhodora 19:160. 1917.
L. decaphyllus of Manuals---Stems 15-50 cm. tall; leaves with 6-8 leaflets, these 2.5-6 cm. long, narrowly oblong to oblong-elliptic; tendrils simple or branched but prehensile at least on the upper leaves; corolla 20-30 mm. long, rose to purple at least in part; calyx 10-15 mm. long, the teeth shorter than the tube.---Plains, prairies and hills. Colorado to Utah, south to New Mexico and Arizona; Mexico. Our records scattered over Colorado, mostly in the southern two-thirds of the state at 4500-8500 feet.

19. Vicia L. VETCH

Annual or perennial herbaceous plants; stems weak and climbing or trailing; leaves pinnately compound, tendril-bearing at apex; stipules conspicuous; flowers axillary, solitary or in racemes; calyx with a somewhat oblique tube and unequal teeth; petals papilionaceous, wings adherent to the keel; stamens diadelphous (of 9-1), anthers uniform; style slender with a tuft or ring of hairs at apex; fruit strongly compressed, dehiscent by 2 sutures, few- to several-seeded.

1. Racemes many-flowered, dense
  2. Flowers about 15 mm. long, 5 times as long as wide; limb of banner (flaring part) less than 1/2 as long as the claw; calyx strongly gibbous, the lobes very narrow, about as long as the tube   -1. V. villosa
  2. Flowers about 9-13 mm. long, less than 4 times as long as wide; limb of banner about as long as the claw; calyx not gibbous at base (is oblique) its lobes long-triangular, shorter than the tube   -2. V. cracca
1. Racemes 1- to 10-flowered, lax
  3. Racemes 1- to 2-flowered, the peduncle produced beyond the upper flower   -3. V. producta
  3. Racemes 3- to 10-flowered, the peduncle not produced beyond the flowers   -4. V. americana

1. Vicia villosa Roth, Tent. Fl. Germ. 2:182. 1789.
Annual or biennial plants, villous throughout with spreading, somewhat tangled hairs; leaflets about 8-10, linear to oblong-linear, obtuse, acute or mucronate; racemes rather 1-sided, many-flowered; calyx tube strongly gibbous at base, with subulate lobes 2-4 mm. long, and as long as the tube; petals about 12-15 mm. long, forming a narrow flower at least 5 times as long as broad, standard (banner) with flaring limb less than 1/2 as long as the broad claw; fruit glabrous.---Cultivated and occasionally escaping in the United States especially in the western part. Our records from northcentral and southcentral Colorado at 5000-7500 feet.
2. Vicia cracca L., Sp. Pl. 735. 1753.
Reported for Greenland to Alberta, south to Wisconsin and California. No records as yet from Colorado.
3. Vicia producta Rydb., Bull. Torr. Club 28:500. 1901.
Perennial plants; stems 20-40 cm. long, slender, decumbent or ascending, sparsely hairy and branched, somewhat angled; leaflets 6-10, 5-15 mm. long, oblong to linear, obtuse and mucronate; peduncles 2-4 cm. long, 1- or 2-flowered, produced beyond the upper flower; calyx tube about 2 mm. long, teeth about 1 mm. long; corolla about 7-8 mm. long, ochroleucous but tipped or tinged with purple.---Hills and mountains. Colorado and possibly west and south. Our few records from northern, western and southern Colorado at 5000-8000 feet.
4. Vicia americana Muhl. ex Willd., Sp. Pl. 3:1096. 1803.
V. americana var. oregana (Nutt.) A. Nels.; V. oregana Nutt.; V. trifida Dietr.; V. dissitifolia (Nutt.) Rydb.; V. caespitosa A. Nels.; V. linearis (Nutt.) Greene; V. linearis var. caespitosa A. Nels.; V. pumila Heller; V. sparsifolia Nutt.---Perennial plants; stems from 10-80 cm. long, glabrous or somewhat pubescent, usually trailing or climbing; leaflets 6-14, 10-40 mm. long, varying from linear to oval, glabrous to pubescent; racemes lax, 3- to 10-flowered; flowers 15-20 mm. long, light to dark purple. As treated here the species is a complex with the extremes very unlike. The writer has never been able to separate any of the entities in Colorado since intergradations are so numerous, often 2 represented on the same plant.---Widely distributed over the United States. Our records widely distributed in Colorado at 3500-10,500 feet.

20. Strophostyles Ell. WILDBEAN

Twining or trailing herbaceous annual plants; leaves pinnately 3-foliate, stipules and stipels present; flowers perfect, capitate on long, axillary peduncles; calyx 5-toothed, these subequal or the 2 upper teeth partly united; corolla papilionaceous, keel strongly curved; stamens diadelphous (9-1); fruit linear, flattened, straight.

1. Strophostyles leiosperma (T. & G.) Piper, Contrib. U. S. Nat. Herb. 22:668. 1926.
Strophostyles pauciflora (Benth.) Wats.---Stems slender, retrorsely hirsute-pubescent; leaflets 2-4 cm. long, lanceolate or linear-oblong, entire, obtuse at apex, rounded at base, appressed-pubescent; peduncles long, flowers 2-6, capitate-umbellate; calyx 2.5-4 mm. long, hairy; petals 6-7 mm. long, pale-purplish; fruit 2-4 cm. long and about 4-5 mm. wide, strigose;

seeds 3-4 mm. long, brown-purplish.---Banks and valleys. Minnesota to Colorado, south to Louisiana and New Mexico. Our one record from northeastern Colorado at 3500 feet.

21. Apios Moench. GROUNDNUT; POTATO BEAN

Twining or trailing herbaceous perennial plants with tuberous roots; leaves pinnately compound, stipules small; flowers perfect, in racemes; calyx campanulate, 2-lipped, lowest one longer; corolla papilionaceous, wings adherent to the keel, this twisted and curved; stamens diadelphous (9-1), anthers alike; fruit sessile, narrow, flat, straight or somewhat curved, dehiscent, many-seeded.

1. Apios americana Medic. in Vorles, Churpf. Phys. oekon. Gesellsch. 2:355. 1787.
   A. tuberosa Moench.; A. apios (L.) MacMill.---Stems about 30-200 cm. long, pubescent to glabrous; leaflets 5-9, each ovate to lanceolate, entire, acute or acuminate; racemes axillary on rather short peduncles (these seldom longer than the racemes); calyx 2-4 mm. long; corolla about 8-12 mm. long, brownish-purple; fruit about 5-12 cm. long, linear, straight or slightly curved.---Banks and valleys. New Brunswick to Minnesota, south to Florida, Texas and Colorado. Our few records from northcentral Colorado at 5000-5500 feet.

22. Hedysarum L. SWEET VETCH

Herbaceous perennial plants; leaves odd-pinnate, petioled but with leaflets nearly sessile, stipulate but no stipels present, mostly finely puncticulate above; inflorescence racemose, axillary, peduncled; calyx bracteolate, campanulate, 5-toothed, these nearly equal, pubescent; corolla papilionaceous, the keel longest; stamens diadelphous (9-1); fruit a loment with flat, single-seeded, indehiscent segments deeply indented between above and below, usually stipitate.

1. Segments of fruit definitely winged, 9-14 mm. long, the reticulations polygonal; calyx lobes unequal, not longer than the tube, the upper triangular; veinlets of leaflet conspicuous   -1. H. occidentale
1. Segments of fruit not at all winged, 5-7 mm. long, the reticulations transversely elongated with few cross-veins; calyx lobes about equal, longer than the tube, none of them tringular; only the main vein of leaflet noticeable   -2. H. boreale

1. Hedysarum occidentale Greene, Pitt. 3:19. 1896.
   H. marginatum Greene; H. uintahensis A. Nels.---Stems 30-70 cm. tall, several to many from woody roots, somewhat strigose, striate; leaflets 11-19, 10-25 mm. long, ovate, elliptic to oblong-lanceolate, sparsely pubescent both sides or glabrate above, main vein and veinlets rather conspicuous; inflorescence 6-13 cm. long, flowers pendant; calyx pubescent, the teeth unequal, 1-3.5 mm. long, upper short and triangular, lower subulate but none longer than the tube; corolla 16-20 mm. long, reddish-purple to dark rose, auricles of wings united, linear, about as long as the claw; fruit stipitate, pendant, of 1-4 elliptic segments, these 9-14 mm. long and 7-13 mm. wide, conspicuously wing-margined, usually strigose, reticulations polygonal.---Valleys and banks. Wyoming and Colorado. Our records in the southwestern quarter of Colorado at 7500-11,500 feet.

2. Hedysarum boreale Nutt., Gen. N. Am. Pl. 2:110. 1818.
   H. pabulare A. Nels.; H. carnosulum Greene---Stems 25-60 cm. tall, several to many from woody roots, longitudinally grooved, finely strigose; leaflets 9-15, 1-2 cm. long, linear-oblong to elliptic or obovate, strigose below and sparsely strigose or glabrous above, only main vein noticeable; flowers erect or lower somewhat reflexed; calyx pubescent, teeth nearly equal, 3-5 mm. long, longer than the tube, all subulate; corolla 12-19 mm. long, carmine to rose-purple, auricles of wings not united, much shorter than the claw; fruit pendant to divaricate, usually stipitate, segments 2-5, 5-7 mm. wide and 6-8 mm. long, each orbicular or nearly so, strigose, not wing-margined, reticulations transversely elongated.---Canyons, valleys and banks. Montana, south to Colorado and Utah. Our records scattered over Colorado, but few from the extreme eastern part, at 4000-9500 feet.

23. Trifolium L. CLOVER

Herbaceous plants; stems caulescent or acaulescent; leaves palmately trifoliate or sometimes 3- to 4-foliate; leaflets more or less toothed in most species; flowers perfect, in heads or short spikelike racemes, these frequently subtended by a basal involucre, flowers often reflexed after anthesis; calyx 5-cleft, usually equal or subequal; corolla papilionaceous, often withering-persistent, claws more or less united to staminal tube; stamens diadelphous (9-1), anthers all alike; fruit small, indehiscent or tardily dehiscent shorter than or little longer than the persistent calyx; seeds 1 to few.

1. Flowers yellow;plants annual   -1. T. procumbens
1. Flowers white, ochroleucous, pink, rose or purple; plants rarely annual
   2. Heads of flowers sessile, subtended by 1 or 2 short-petioled leaves   -2. T. pratense
   2. Head of flowers peduncled, subtended if at all by highly modified bracts these not at all leaflike
      3. Head of flowers subtended by a conspicuous involucre of bracts, these often united at least at base and over 3 mm. long
         4. Calyx becoming much inflated and reticulate in fruit, together forming a densely packed globose head 2 or 3 times larger than in flower, each calyx silky-hairy; plants of cultivated fields or adjacent waste places, rare in Colorado   -3. T. fragiferum
         4. Calyx not becoming inflated, the head in fruit little larger than in full flower, each calyx glabrous or strigose; native plants often rather common

5. Leaflets, peduncles, and calyx sparsely to densely pubescent; leaflets linear-lanceolate (to rarely as wide as lanceolate), entire -4. *T. dasyphyllum*
5. Whole plant glabrous; leaflets obovate, oval, oblong (or sometimes as narrow as oblanceolate or lanceolate), margins more or less toothed
   6. Stems acaulescent or nearly so; plants not over 15 cm. tall, growing at higher altitudes (over 9000 feet in Colorado); involucre white-scarious with purple veins, the bracts usually separate nearly to the base; calyx teeth as long or shorter than their tube -5. *T. parryi*
   6. Stems caulescent; plants often over 15 cm. tall, of lower altitudes (not over 8000 feet in Colorado); involucre green, the bracts definitely united; calyx teeth definitely longer than the tube -6. *T. fendleri*
3. Head of flowers without an involucre or with an inconspicuous one, the bracts rarely over 3 mm. long (the flowers of the head may be bracted individually)
  7. Plants caespitose (usually matlike), acaulescent, not over 15 cm. tall (usually much less); high altitude plants (except *T. gymnocarpon* and its var.) usually found over 10,000 feet in Colorado
    8. Heads 1- to 3-flowered; calyx glabrous; plants 2-6 cm. tall with flowers 16-20 mm. long -7. *T. nanum*
    8. Heads 5- to many-flowered; calyx more or less hairy; plants over 6 cm. tall or if less then flowers 6-8 mm. long
     9. Leaflets narrow, linear-lanceolate to lanceolate; involucre bracts present (may be small); calyx lobes about twice as long as the tube -4. *T. dasyphyllum*
     9. Leaves wider, elliptic-oblong to oval or obovate; involucre bracts absent or very minute; calyx lobes as long or somewhat longer than the tube, not twice as long
      10. Leaflets glabrous or glabrate; flowers 13-20 mm. long, soon reflexed; plants of high altitudes (over 11,000 feet in Colorado) -8. *T. brandegei*
      10. Leaflets strigose at least below; flowers 6-13 mm. long, little if at all reflexed; plants of lower altitudes (not over 8,500 feet in Colorado)
       11. Flowers 6-8 mm. long; plants 3-6 cm. tall; leaflets 5-10 mm. long -9. *T. gymnocarpon*
       11. Flowers 9-13 mm. long; plants 6-10 cm. tall; leaflets 10-25 mm. long -9A. *T. gymnocarpon subacaulescens*
  7. Plants not caespitose or matlike (*T. repens* may root at the nodes), caulescent, usually over 15 cm. tall; plants of low to medium altitude (rarely over 10,000 feet in Colorado)
    12. Flowers 4-8 mm. long; plants in cultivation, around fields, roads, and waste places, not native
     13. Stems creeping, often rooting at the nodes; flowers white or merely pink-tinged; calyx 3.5-5 mm. long the teeth with a red or purple spot in the sinuses; leaflets usually with a white spot -10. *T. repens*
     13. Stems erect or ascending, not rooting; flowers pink or rose-tinged (sometimes nearly white); calyx 3-4 mm. long, no red or purple spot in the sinuses of the teeth; leaflets without a white spot -11. *T. hybridum*
    12. Flowers 12-20 mm. long; native plants
     14. Leaflets glabrous; calyx 4-7 mm. long, the teeth about as long as the tube, nearly glabrous or sparingly hairy; banner obtuse or emarginate -12. *T. macilentum*
     14. Leaflets strigose below at least on the veins (often on 1/2 the blade only); calyx 7-10 mm. long, the teeth about twice as long as the tube, villous-plumose; banner acute
      15. Plants from slender, branched roots; flowers white to rose-colored; leaflets mostly over 2 cm. long -13. *T. longipes*
      15. Plants from thick, woody, fusiform roots; flowers rose to salmon-colored; leaflets usually less than 2 cm. long -13A. *T. longipes rushbyi*

1. Trifolium procumbens L., Sp. Pl. 772. 1753.
    Widespread in the United States at least in the eastern and western parts. To be expected in this state.
2. Trifolium pratense L., Sp. Pl. 768. 1753.
    Perennial or biennial plants; stems 10-50 cm. tall, erect or ascending, several from a woody root, more or less pubescent; petioles rather long; leaflets 2-5 cm. long, obovate, oval, ovate or sometimes oblong, entire to finely dentate, sparsely pubescent and often spotted near the middle; head dense, about 2-3 cm. wide, globose or nearly so, sessile or very short-pedunculate, subtended by 1 or 2 leaves, flowers sessile; calyx 5-10 mm. long, tube greenish-white, the teeth narrow and somewhat longer than tube, villous; corolla about 12-20 mm. long, rose to red-purple; fruit 1- to 2-seeded.---In cultivation and escaping especially along roadsides and ditches. Widespread in Canada and the United States. Our records scattered in the western half of Colorado at 4500-9500 feet but to be expected anywhere in the state.
3. Trifolium fragiferum L., Sp. Pl. 772. 1753.
    Perennial plants; stems branching, creeping and rooting at the nodes, glabrous or nearly so; petioles long; leaflets 6-30 mm. long, obovate, finely serrulate to subentire, glabrous or sparingly long-hairy at base; peduncles long, exceeding the leaves; head about 10-14 mm. wide in flower, globose, with an involucre; calyx 3-4 mm. long, silky-hairy, the slender teeth about as long as the tube, becoming much inflated in fruit, increasing the size of head 2 or 3 times, becoming reticulate-veined and often reddish-tinged; corolla about 5-7 mm. long, pink to rose; fruit 1-seeded.---Moist meadows. Introduced sparingly in this country from Europe and occasionally escaping. Our records from northcentral and westcentral Colorado at 4500-5500 feet.
4. Trifolium dasyphyllum T. & G., Fl. N. Amer. 1:315. 1838.
    Caespitose, acaulescent perennial plants, 5-15 cm. tall, from woody branched caudex; leaf petioles rather long; leaflets 1-3 cm. long, linear-lanceolate to sometimes as wide as lanceolate, cuneate, acuminate, sparsely to densely strigose especially below, entire; peduncles fairly long, the heads borne above the leaves, heads globose, 1.5-2.5 cm. wide, with 10-30 flowers, involucre present but variable from long (8 mm.) and subulate, whitish with green nerves, to scarious-truncate and less than 1 mm. long, pedicels short and pubescent (about 1 mm. long), flowers reflexed or not reflexed in age; calyx 6-10 mm. long, strigose, the lobes subulate,

subequal, about twice as long as the tube; corolla 10-15 mm. long, purple to pink, usually bicolored, with wings and keel darker; fruit 3- to 5-seeded. As treated here the species is a complex, including T. lividum Rydb., T. scariosum A. Nels., and T. uintense Rydb. Also T. stenolobum Rydb., T. bracteolatum Rydb., and T. attenuatum Greene are included in the above concept. Further study needs to be made on this group.---High mountains. Throughout the southern Rocky Mountains. Our records scattered in the higher mountains of the state at 9500-13,000 feet.

5. Trifolium parryi Gray, Am. Journ. Sc. ser. 2. 33:409. 1862.

T. montanense Rydb.; T. salictorum Greene---More or less caespitose perennial plants with branching woody caudex; leaves mostly radical; stems 7-15 cm. tall, glabrous; petioles rather long; leaflets 10-25 mm. long, obovate, oval to oblanceolate, serrulate or subentire, glabrous; peduncles stout, lifting the head above the leaves, heads large, about 1.5-3 cm. wide, involucre present, lobed or separated, rather scarious and usually dark blue-veined, over 5 mm. long, flowers 6-20 on short pedicels about 1 mm. long or less, bracted at base; calyx 5-7 mm. long, glabrous, the teeth somewhat shorter than the tube, widely subulate, unequal; corolla 13-20 mm. long, rose-purple or reddish-purple; fruit 3- to 7-seeded.---High mountains. Wyoming and Utah, south to Colorado, but probably more widely distributed. Our records scattered in most of the higher mountains of the state at 9500-12,500 feet.

6. Trifolium fendleri Greene, Pitt. 3:221. 1897.

T. oxydon of Rydb.'s Manuals---Perennial plants from slender roots; stems 8-50 cm. tall, longitudinally striate, glabrous, caulescent; leaves glabrous, bright green, long-petioled; lower leaflets smaller than upper, leaflets about 1 cm. long, upper to about 3 cm. long, obovate, oblong to lanceolate or oblanceolate, margins sharply denticulate to serrulate; heads 1.5-3 cm. wide, globular in shape, long-peduncled, borne above the leaves, with greenish involucres, pedicels slender; calyx 5-8 mm. long, thin, glabrous, the teeth longer than the tube; corolla 8-14 mm. long, white to pink or rose; fruit 2- to 4-seeded.---Meadows and slopes. Colorado, Utah, New Mexico and Arizona. Our records mostly from the southwestern quarter of Colorado, mostly in the central and southcentral parts of the state, at 7000-9000 feet.

Some of the plants from southern Colorado have narrowly oblanceolate leaflets and small reddish-rose or dark rose flowers. These plants have been called T. lacerum Greene which is doubtfully distinct from T. arizonicum Greene. The matter needs further study.

7. Trifolium nanum Torr., Ann. Lyc. N. Y. 1:35. 1824.

Caespitose perennial plants 2-6 cm. tall, erect or spreading from woody branched crowns of roots, acaulescent, glabrous; petioles slender, longer than the leaflets; leaflets 6-15 mm. long, glabrous, narrowly obovate, linear-oblanceolate to oblong, slightly serrate to nearly or quite entire; heads of 2-3 flowers, peduncled, often borne somewhat above the leaves, with 2 or 3 small inconspicuous whitish cupulate involucres, these usually not over 1-1.5 mm. long, flowers ascending on slender pedicels 1-2 mm. long; calyx glabrous, tube 3-4 mm. long, campanulate, teeth about 2 mm. long, lanceolate to deltoid; corolla 16-20 mm. long, reddish-purple or rose-purple, rarely whitish; fruit 5- to 10-seeded.---High mountains. Montana, south to New Mexico and Utah. Our records from the higher mountains of Colorado at 11,500-14,000 feet.

8. Trifolium brandegei Watson, Proc. Am. Acad. 11:130. 1876.

Perennial plants with woody branching roots, 6-15 cm. tall, acaulescent and caespitose, lax and ascending, glabrous or very sparingly long-hairy; petioles long; leaflets 7-30 mm. long, glabrous or glabrate, oval, obovate to elliptic-oblong, entire to sharply denticulate; flowers about 6-15 in a loose head, soon reflexed, on rather stout peduncles which lift the head well above the leaves, involucre absent or nearly so; calyx 6-10 mm. long, glabrous or with scattered long hairs, green or purple, the teeth lanceolate or ovate-lanceolate, taper-pointed at ends, as long or longer than the tube; corolla 13-20 mm. long, purple; fruit 2- to 7-seeded.---High Mountains. Colorado and New Mexico. Our records from southcentral and southwestern Colorado at 11,000-12,500 feet.

9. Trifolium gymnocarpon Nutt. in T. & G., Fl. N. Amer. 1:320. 1838.

Dwarf caespitose perennial plants from a branching woody caudex; leaves mostly radical; stems 3-6 cm. tall, base covered with brown stipules; leaves bluish-green, on fairly long petioles; leaflets 5-10 mm. long, oval or elliptic-oblong, sharply serrate, glabrous above with veins more or less indented, strigose below, sometimes 5 in number; peduncles 1-4 cm. long, strigose, about as long as leaves; head globose or hemispheric, of 5-12 flowers, these little if at all reflexed, involucre very small or absent, pedicels short, about 1 mm. long, strigose; calyx about 4-6 mm. long, strigose, teeth equal or slightly longer than the tube, widely subulate; corolla 6-8 mm. long, ochroleucous to rose-tinted; fruit 1- to 2-seeded, hairy.---Hills and slopes, usually in dry ground. Wyoming south to Colorado. Our records from northcentral and western Colorado at 5500-8500 feet.

9A. Trifolium gymnocarpon subacaulescens (Gray) A. Nels., (var.) in C. & N. Man. 279. 1909.

T. subacaulescens Gray; T. nemorale Greene---This variety is taller than the species, 6-10 cm. tall (instead of 3-6 cm.); leaflets are longer, 10-25 mm. (instead of 5-10 mm.) and flowers are longer, 9-13 mm. long (instead of 6-8 mm.). Martin (Bull. Torr. Club 73:366-369. 1946) united the above with the species proper. However, our Colorado plants seem reasonably distinct.---Colorado, New Mexico and Arizona. With the species in Colorado, our records at 6000-8000 feet.

10. Trifolium repens L., Sp. Pl. 767. 1753.

Perennial plants, glabrous throughout or with scattered hairs; stems branching from base, 5-30 cm. long, caulescent, creeping and often rooting at nodes, rarely erect; petioles long; stipules ovate-lanceolate or oval, membranous; leaflets 1-2 cm. long, obovate, denticulate, serrulate, sometimes subentire, with a white blotch near base; heads 18-25 mm. across, globose, peduncles long, often conspicuously so, involucre absent or nearly so, pedicels 2-4 mm. long; calyx tube about 2 mm. long, whitish, teeth narrow, somewhat shorter than the tube, greenish and with a red or purple spot at the sinus, whole calyx 3.5-5 mm. long; corolla 4-8 mm. long,

reflexed in age, white or merely pink-tinged; fruit 4- to 5-seeded.---Cultivated and escaping. Widely distributed in North America. Our records from the western two-thirds of Colorado at 5000-11,000 feet but to be expected anywhere especially in cultivated areas.
11. Trifolium hybridum L., Sp. Pl. 766. 1753.
Perennial (possibly annual or biennial) plants, glabrous throughout or very nearly so; stems 15-50 cm. tall, erect or ascending, caulescent, branching; petioles long; stipules membranous; leaflets 1.5-3.5 cm. long, obovate, ovate to elliptic, serrulate to sometimes subentire; heads 2-2.5 cm. across, globose, long-peduncled, involucre absent or very small, pedicels about 4-8 mm. long; calyx 3-4 mm. long, the tube whitish, the teeth greenish, narrow, about as long as tube, no red or purple spot at sinus; corolla 6-8 mm. long, reflexed in fruit, rose-tinged, pink to nearly white; fruit 2- to 4-seeded.---Cultivated and escaping. Now widely distributed in this country. Our records widely scattered in the western two-thirds of Colorado at 4500-9500 feet but to be expected anywhere.
12. Trifolium macilentum Greene, Pitt. 3:223. 1897.
Perennial plants; stems 15-40 cm. tall, erect or ascending, glabrous, from woody roots; leaves cauline and radical, lower long-petioled; leaflets large, 2-6 cm. long, thin, oval, ovate to lanceolate, unevenly serrate, glabrous; heads large, about 2-3 cm. wide, globose on rather long often deflexed peduncles, flowers sub-sessile, soon reflexed, involucres absent or nearly so; calyx about 4-7 mm. long, very thin, nearly glabrous or sparingly pubescent, teeth setaceous, more or less flexuous, about as long as the tube; corolla 15-20 mm. long, pink to rose; fruit about 4-seeded. The specimens with hairy calyx may be T. latifolium (Hook.) Greene.---Meadows. Colorado and Utah. Our few records from westcentral and southwestern Colorado at 8000-9500 feet but the species has been reported up to 10,500 feet.
13. Trifolium longipes Nutt. in T. & G., Fl. N. Amer. 1:314. 1838.
T. rydbergii Greene---Perennial plants from slender branched roots; stems 8-40 cm. tall, erect or ascending; leaves radical and cauline; petioles rather long; leaflets mostly over 2 cm. long (2-4.5 cm.), lanceolate, oval to oblong-lanceolate, sharply serrate, glabrous above, sparsely strigose beneath, the hairs often on 1/2 of the blade; heads long-peduncled, exceeding the leaves, peduncles villous above, head globose or turbinate, the flowers sessile, reflexing in age, involucre absent; calyx 7-10 mm. long, teeth setaceous, unequal, about twice as long as the tube, with long hairs; corolla 12-18 mm. long, white, ochroleucous to rose-colored; fruit 3- to 6-seeded.---Meadows and moist valleys. Montana to Washington, south to Colorado and California. Our records from the western half of Colorado at 6000-11,000 feet.
13A. Trifolium longipes rushbyi (Greene) comb. nov., (var.)
T. rushbyi Greene---Differs from species in having rather heavy thick woody fusiform roots; flowers are rose or salmon-colored and leaflets run shorter (usually less than 2 cm. long). The variety intergrades somewhat with the species except in the underground parts.---Colorado to Arizona and probably westward. Our few records from western Colorado at 7000-8000 feet.

24. Medicago L. MEDIC

Annual or perennial herbaceous plants; stems leafy; leaves pinnately 3-foliate, with toothed leaflets; stipules adnate; flowers small, in axillary racemes or heads, perfect; calyx tube campanulate, the lobes nearly equal; corolla yellow to purple, papilionaceous, keel obtuse, shorter than wings or banner; stamens diadelphous (9-1), anthers uniform; fruit indehiscent, 1- to few-seeded, sometimes prickly, strongly curved or spirally coiled.

1. Plants perennial; stems erect or ascending; corolla violet; fruit spirally coiled in 2 or 3 spirals, several-seeded, no hooked prickles or tubercles on margins   -1. M. sativa
1. Plants annual; stems prostrate to somewhat ascending; corolla yellow; fruit coiled in 1 plane and 1-seeded or if in 2-3 spirals and several-seeded then with hooked prickles or tubercles on margins
    2. Fruit coiled in 1 plane, 1-seeded, pubescent, but not prickly, becoming blackish at maturity; racemes ovoid, many   -2. M. lupulina
    2. Fruit coiled in a spiral, several seeded, glabrous but prickly on edges, straw-colored at maturity; racemes subcapitate, few   -3. M. hispida

1. Medicago sativa L., Sp. Pl. 778. 1753.
Perennial plants from elongated taproots; stems 30-100 cm. tall, branched, ascending, sometimes decumbent at base, glabrous to sparsely hairy; leaves petioled; leaflets 10-30 mm. long, oblanceolate to oblong, sometimes to obovate, denticulate toward apex, obtuse or truncate at tip, more or less pubescent; racemes 1-5 cm. long, oblong, dense; corolla 7-10 mm. long, violet or purple; fruit coiled in 2 or 3 spirals, more or less pubescent, somewhat reticulated, several-seeded.---Cultivated and escaping. Well distributed in the United States. Our records from northcentral, central, southcentral and southwestern Colorado at 5000-8000 feet.
2. Medicago lupulina L., Sp. Pl. 779. 1753.
Annual plants or sometimes acting as perennials; stems 30-60 cm. long, branched at base, procumbent, sparingly pubescent; leaves petioled; leaflets 5-15 mm. long, obovate to nearly orbicular, denticulate above, sparingly hairy; flowers in short, headlike or spikelike racemes not over 12 mm. long; corolla about 1.5-2 mm. long, yellow; fruit 1-seeded, reniform, the acuminate tip coiled but only in 1 plane, strongly reticulate, pubescent, becoming blackish at maturity.---Fields and waste places. Now generally distributed over the United States. Our records from northcentral, central, southcentral and southwestern Colorado at 5000-8000 feet but to be expected anywhere in cultivated areas.
3. Medicago hispida Gaertn., Fruct. et Sem. 2:349. 1791.
M. denticulata Willd.---Adventive from Europe and extensively naturalized in the United States. It should be found in Colorado.

25. Melilotus Juss. SWEETCLOVER

Annual or biennial herbaceous plants; stems leafy and branched; leaves pinnately 3-foliate, not glandular-punctate, with dentate leaflets; inflorescence of racemes, these axillary; flowers small; calyx campanulate, 5-lobed, these nearly equal; corolla papilionaceous, wings adnate to the obtuse keel and free from staminal tube; stamens diadelphous (9-1), the anthers uniform; fruit membranous, scarcely if at all longer than the calyx, indehiscent, 1- to 2-seeded, ovoid, reflexed.

1. Corolla white, banner considerably longer than the wings, (over 0.5 mm. longer); fruit not rugose, inconspicuously reticulate -1. Melilotus alba
1. Corolla yellow (may dry cream-colored), banner little if any longer than the wings (rarely over 0.5 mm. longer); fruit rugose -2. Melilotus officinalis

1. **Melilotus alba** Desr. in Lam., Encyl. 4:63. 1797.
   Stems 70-200 cm. tall, erect or ascending, glabrous or finely pubescent above; leaves petioled, rather distant; leaflets 1-2.5 cm. long, variable in shape, obovate, oblanceolate or oblong-ovate, sometimes elliptic, denticulate except near base, rounded or truncate at apex, glabrous or sparingly hairy; racemes numerous, 5-15 cm. long, narrow, flowers with short pedicels; calyx about 1.5-2.5 mm. long, the lobes and tube subequal; corolla white (may dry cream-colored), 4-6 mm. long, the banner over 0.5 mm. longer than the wings; fruit about 3 mm. long, ovoid, glabrous, inconspicuously reticulated, not rugose.---Roadsides and waste places. Widely cultivated and escaping in this country. Our records scattered over the state at 4500-7500 feet.

2. **Melilotus officinalis** (L.) Lam., Fl. Franc. 2:594. 1778.
   Stems 70-200 cm. tall, erect or spreading, glabrous or slightly pubescent; leaves petioled; leaflets 10-20 mm. long, obovate, oblanceolate to oblong-lanceolate, denticulate except at base, glabrous or sparingly hairy, rounded at apex; racemes 5-15 cm. long, numerous, lax and narrow, pedicels short; calyx about 2-2.5 mm. long, the tube and lobes subequal; corolla 4-6 mm. long, yellow (may dry cream-colored), the banner not longer than the wings or seldom over 0.5 mm. longer; fruit about 3-3.5 mm. long, ovoid, rugose.---Waste places and roadsides. Widely cultivated and escaping in this country. Our records scattered over the state at 4000-7500 feet.

26. Lotus L. DEERVETCH; TREFOIL

Annual or perennial herbaceous plants; stems leafy; leaves pinnately compound but sometimes appearing palmate by the shortening of the rachis; leaflets entire, mostly 3; stipules minute and glandlike; inflorescence solitary or in few-flowered umbellike clusters; calyx lobes 5, nearly equal; corolla papilionaceous, yellow to white (fading orange-red or pink), wings and keel not united to staminal tube; stamens diadelphous (9-1), anthers all alike; fruit linear, sessile, several-seeded.

1. Flowers whitish, fading to pink, 4-7 mm. long; plants annual; fruit glabrous -1. L. purshianus
1. Flowers yellow to orange, fading to reddish, 10-15 mm. long; plants perennial from a thick woody root; fruit strigose -2. L. wrightii

1. **Lotus purshianus** (Benth.) Clements & Clements, Rocky Mt. Fl. 183. 1914.
   L. americanus (Nutt.) Bisch.; Acmispon americanus (Nutt.) Rydb.---Reported from Minnesota to British Columbia, south to New Mexico, Arizona and California. No specimens have been seen for Colorado by the writer.

2. **Lotus wrightii** (A. Gray) Greene, Pitt. 2:143. 1890.
   Anisolotus wrightii (A. Gray) Rydb.---Perennial plants from woody roots; stems 20-40 cm. tall, many, erect or ascending, grayish-strigose, branched throughout; leaves apparently sessile or nearly so, with 3-6 leaflets, these crowded on a short rachis and appearing digitate; leaflets 6-20 mm. long, linear, linear-oblong to nearly oblanceolate, strigose; flowers solitary or 2-3 together in the leaf axils, usually subsessile but sometimes on 1- to 3-flowered peduncles, bracts simple; calyx 6-8 mm. long, the subulate teeth somewhat shorter than to as long as the tube; corolla 10-15 mm. long, yellow or orange, turning reddish-orange in age, the keel obtuse or nearly so; fruit 20-25 mm. long and 2-3 mm. wide, linear, not strongly flattened, strigose.---Hills and rocky slopes, sometimes in shaded ground. Colorado, Utah, New Mexico and Arizona. Our records mostly from the southwestern quarter of Colorado, with one from Douglas County, at 5000-8000 feet.

27. Swainsona Salisb. SWAINSON PEA

Herbaceous perennial plants from underground rhizomes; stems erect, strictly branched above; leaves odd-pinnately compound, leaflets entire; stipules small; flowers showy in numerous axillary, pedunculate racemes; calyx with 5 sepals, somewhat 2-lipped with bracts at base of pedicels and 2 bractlets beneath the calyx; corolla papilionaceous; stamens diadelphous (9-1), anthers uniform; fruit stipitate, bladderlike with thin papery walls, globose or oval, 1-celled, the seed-bearing suture somewhat intruded, indehiscent or tardily and irregularly dehiscent; seeds about 1.5-2 mm. long, numerous, brownish-black.

1. **Swainsona salsula** (Pall.) Taubert in Engler and Prantl., Pflanzenfam. III. 3:281. 1894.
   Sphaerophysa salsula (Pall.) DC.---Stems 30-100 cm. tall, closely strigose-puberulent or glabrous, somewhat angular or ridged; leaves about 5-7 cm. long with about 9-21 leaflets, these

elliptic, oblanceolate to obovate-oblong, 5-10 mm. long, glabrous above, strigose-puberulent below; pedicels short, about 2-4 mm. long, bracts small and the 2 bractlets very small (to 1 mm. long); calyx about 3-4 mm. long, campanulate, the teeth about 1 mm. long; corolla 12-15 mm. long, orange-red, the banner and keel yellowish near base, flowers drying purplish; fruit about 1.5-2 cm. long, strigose-pubescent, the stipe 5-6 mm. long, borne horizontally.---Fields and roadsides. Introduced from Asia into Colorado, Utah, and Arizona probably with alfalfa seed. Our records from southcentral Colorado, in the San Luis Valley, at about 7500 feet, but recently recorded from Weld County.

28. Oxytropis DC. LOCO; OXYTROPE; CRAZYWEED

Contributed (except for specific descriptions) by C. L. Porter. Perennial scapose or sometimes caulescent herbaceous plants; leaves alternate, odd-pinnately compound; stipules strongly or weakly adnate to the petioles; hairs simple and basifixed, or pick-shaped (dolabriform) and attached at or below the middle; flowers in racemes or spikes, white, ochroleucous, pink or purplish, papilionaceous, the keel with a prominent ascending beak; stamens diadelphous, the anthers all alike; pods membranous to woody in texture, sometimes inflated, sessile or nearly so, oblong-lanceolate to subspherical in outline, partially or nearly completely 2-celled by the intrusion of the ventral (upper) suture, the dorsal (lower) suture not intruded; seeds reniform, several to many in each fruit.

1. Pods pendulous; stipules only slightly adnate by their bases to the petioles; plants caulescent
　　　-1. O. deflexa
1. Pods erect or spreading, never pendulous; stipules adnate to the petioles; plants scapose or subscapose
    2. Inflorescence not glandular-viscid
        3. Leaves with verticillate leaflets　　　　　　　　　　　　　　　　　　　　　　　　-2. O. splendens
        3. Leaves strictly pinnate, the leaflets paired or scattered but never verticillate
            4. Racemes 1- to 5-flowered, subcapitate in anthesis and not becoming much elongated in fruit; corolla purple or pinkish, never white or yellowish except in rare albino forms
                5. Fruiting calyx inflated and completely enclosing the ripe fruit; plants of mostly middle or lower elevations
　　-3. O. multiceps
                5. Fruiting calyx not inflated nor enclosing the ripe fruit; plants of alpine situations
                    6. Pods papery and inflated, ellipsoid　　　　　　　　　　　　　　　-4. O. podocarpa
                    6. Pods coriaceous at maturity, cylindric or oblong, not inflated　-5. O. parryi
            4. Racemes 6- to many-flowered, usually elongating in fruit; corolla purple or lavender to white or ochroleucous
                7. Corolla purplish, pinkish, or blue, except in rare albino forms or hybrids
                    8. Pods inflated, ovoid, 7-10 mm. wide, with a prominent and abrupt beak
　　-6. O. obnapiformis
                    8. Pods not inflated, oblong or cylindric, 5 mm. wide or less, gradually narrowed into a beak
                        9. Calyx with spreading hairs; inflorescence subcapitate, rarely becoming much elongated in fruit; pubescence of foliage composed entirely of basifixed hairs　-7. O. lagopus
                        9. Calyx with appressed hairs; inflorescence becoming elongated in fruit; pubescence of foliage composed at least in part of 2-armed (dolabriform) hairs　-8. O. lambertii
                7. Corolla white or ochroleucous, the keel often purple-tipped, the whole corolla rarely tinged with purple in hybrid populations
                    10. Flower mostly 12-15 mm. long, the corolla ochroleucous; pods chartaceous; foliage green; plants of middle and higher elevations　　　　　　　　-9. O. campestris glabrata
                    10. Flowers mostly 18-25 mm. long, the corolla white with a purple-tipped keel; pods coriaceous at maturity; foliage gray or silvery; plants of middle and lower elevations
　　-10. O. sericea
    2. Inflorescence glandular-viscid　　　　　　　　　　　　　　　　　　　　　　　　　　-11. O. viscida

1. Oxytropis deflexa (Pall.) DC., Astrag. 96. 1802.
    Astragalus deflexus Pall.; Oxytropis retrorsa Fernald---Plants 10-40 cm. tall, caulescent or subacaulescent; leaflets 25-41, 5-20 mm. long, lanceolate, oblong to ovate, loosely villous to silvery-pubescent; stipules only slightly adnate to the petiole; racemes many-flowered, rather short in flower but elongating in fruit; calyx appressed-villous with some of the hairs usually blackish, tube about 3-4 mm. long, the teeth about 3 mm. long; corolla 6-9 mm. long, sordid white, sometimes purplish near apex; fruit pendulous, oblong, the suture little intruded. The typical form of the species is usually definitely caulescent in habit with loosely villous-pilose pubescence on the foliage. Northern and alpine forms of the species which are subscapose (rarely with more than a single internode) and decidedly silvery-pubescent, with smaller leaves, are var. sericea T. & G., Fl. N. Am. 1:342. 1838.---Circumboreal, extending south to New Mexico. Our records from the western half of Colorado, except the extreme western part, at 6000-10,000 feet.
2. Oxytropis splendens Dougl. in Hook., Fl. Bor. Am. 1:147. 1834.
    Oxytropis splendens var. richardsoni Hook.; Aragallus splendens Greene; A. richardsonii Greene; A. caudatus Greene; A. galioides Greene; Oxytropis richardsoni Wooton & Standl.; Astragalus splendens richardsonii Tidestrom---Plants 10-35 cm. tall, acaulescent or nearly so; leaflets numerous, verticillate, 5-25 mm. long, linear-lanceolate to oblong-elliptic, more or less silky-hairy; stipules adnate to the petiole; racemes many-flowered, becoming somewhat elongated; calyx densely villous, tube about 5 mm. long, teeth 2-3 mm. long; corolla 10-15 mm. long, dark blue to dark purple; fruit erect or spreading, ovoid, suture deeply intruded.---Often in aspen groves at middle elevations. Ranging from Minnesota and North Dakota to Saskatchewan, and southward in the Rocky Mountains to New Mexico. Our records from northcentral, central and southcentral Colorado at 8000-10,000 feet.

3. Oxytropis multiceps Nutt. in T. & G., Fl. N. Amer. 1:341. 1838.
   Oxytropis multiceps var. minor A. Gray; Aragallus multiceps Heller; Oxytropis minor Cockerell---Plants 2-10 cm. tall, acaulescent or nearly so, densely caespitose; leaflets 5-9, 3-10 mm. long, lanceolate to oblong, densely silky with appressed or sub-appressed hairs; stipules adnate to the petioles; racemes 1- to 3-flowered; calyx silky-villous, becoming inflated in fruit, tube 6-10 mm. long, teeth 2-3 mm. long; corolla 17-25 mm. long, purple to rose-purple; fruit erect or spreading, ovoid to ellipsoid, suture partly intruded, wholly included in the inflated calyx.---Dry, gravelly well-drained slopes at middle and lower elevations. Ranging from western Nebraska to Wyoming and Colorado. Our records from northcentral and central Colorado at 5000-9500 feet.
4. Oxytropis podocarpa A. Gray, Proc. Am. Acad. 6:234. 1866.
   Oxytropis arctica inflata Hook.; Oxytropis hallii Bunge; Spiesia inflata Britt.; Aragallus inflatus A. Nels.; A. hallii Rydb.---Plants 2-6 cm. tall, acaulescent or nearly so, caespitose; leaflets 11-25, 3-8 mm. long, linear to oblong, silky hairy; stipules adnate to the petiole; racemes 1- to 3-flowered; calyx villous, tube 5-7 mm. long, teeth 2-3 mm. long; corolla 10-15 mm. long, blue to purple; fruit erect or spreading, ellipsoid or ovoid, becoming large and inflated, suture only slightly intruded.---Arctic and alpine, from Labrador and Baffin Island to Yukon Territory, and southward in the Rocky Mountains to Colorado. Our one record from central Colorado at 12,500 feet.
5. Oxytropis parryi A. Gray, Proc. Am. Acad. 20:4. 1884.
   Aragallus parryi Greene; Astragalus parryanus Tidestrom---Plants 1-10 cm. tall, possibly to 15 cm., acaulescent or nearly so; leaflets 13-21, 3-8 mm. long, oblong-lanceolate, grayish silky-strigose; stipules adnate to the petioles; racemes 1- to 5-flowered, short and not elongating in fruit; calyx cinereous-pubescent, some of the hairs black, tube 3-5 mm. long, teeth 2-3 mm. long; corolla about 15 mm. long, purple; fruit erect or spreading, cylindric or oblong, not inflated but coriaceous, suture well intruded.---Alpine regions. In the Rocky Mountains and westward, Wyoming and Idaho to northern New Mexico and California. Our records from northcentral, central and southcentral Colorado at 9000-11,000 feet.
6. Oxytropis obnapiformis C. L. Porter, Madrono 9:133. 1947.
   Plants 10-30 cm. tall, acaulescent or very nearly so; leaflets 11-25, 5-30 mm. long, oblong-lanceolate, grayish with hairs somewhat appressed; stipules adnate to the petioles; racemes many-flowered, elongating in fruit; calyx strigose to villous, tube 7-8 mm. long, teeth about 2-3 mm. long; corolla 15-20 mm. long, purplish; fruit erect or spreading, ovoid, abruptly beaked, inflated, suture intruded about 1/2 across.---Dry sandy areas at middle elevations. Northcentral Wyoming and northwestern Colorado (Moffat County). Our records at 6000-6500 feet.
7. Oxytropis lagopus Nutt., Journ. Acad. Phila. 7:17. 1834.
   Oxytropis argentata Pursh, not Pers.; Aragallus blankinshipii A. Nels.; Aragallus cellinus sensu Rydb. not A. Nels.---Dry sandy or gravelly hills at middle elevations. Montana and Idaho to southeastern Wyoming near the Colorado line, but not known definitely from Colorado although to be expected in northern Larimer County.
8. Oxytropis lambertii Pursh, Fl. Am. Sept. 740. 1814.
   Aragallus dispar A. Nels.; Aragallus involutus A. Nels.; Aragallus patens Rydb.; Aragallus angustatus Rydb.; Aragallus bigelovii Rydb.; Oxytropis bilocularis A. Nels.---Plants 10-30 cm. tall or sometimes taller, acaulescent or nearly so; leaflets 7-17, 5-40 mm. long, usually linear to oblong but sometimes nearly orbicular, silvery-sericeous to silvery-strigose, at least some of the hairs dolabriform; stipules adnate to the petiole; racemes many-flowered, elongating in fruit; calyx silvery-silky, tube about 5-9 mm. long, teeth about 2-4 mm. long; corolla 12-25 mm. long, purplish-rose to pinkish-blue; fruit erect or spreading, oblong to cylindric, gradually narrowed to a beak, suture intruded about 1/2 across. A variable species which hybridized with O. sericea Nutt. to give robust plants of a general O. sericea aspect but having a lavender tinge in the corolla.---Prairies and mountains, usually in drier situations, at lower to middle elevations. Minnesota to Saskatchewan and southward to Arizona, New Mexico, Texas, and Oklahoma. Our records well scattered over Colorado at 3500-9000 feet.
9. Oxytropis campestris (L.) DC. glabrata Hook., (var.) Fl. Bor. Am. 1:147. 1834.
   O. campestris var. gracilis Barneby; Aragallus gracilis A. Nels.---Plants 20-50 cm. tall, acaulescent or nearly so; leaflets 17-31, 1-2.5 cm. long, oblong-lanceolate, appressed hairy; stipules adnate to the petioles; racemes many-flowered, becoming elongated in fruit; calyx appressed-villous, the hairs often blackish, tube 5-7 mm. long, teeth 3-4 mm. long; corolla about 12-15 mm. long, ochroleucous to yellowish; fruit erect to spreading, oblong-ovate, suture strongly intruded.---Usually in rocky or gravelly slopes at middle and higher elevations. British Columbia and Alberta southward to Oregon and in the Rocky Mountains to Colorado. Our records from northcentral and central Colorado at 9000-10,000 feet.
10. Oxytropis sericea Nutt. in T. & G., Fl. N. Amer. 1:339. 1838.
    O. lamberti ochroleuca A. Nels; Aragallus albiflorus A. Nels. not Oxytropis albiflorus Bunge; Aragallus pinetorum Heller; Aragallus saximontanus A. Nels.; Aragallus majusculus Greene; Aragallus albiflorus var. condensatus A. Nels.; Aragallus pinetorum var. veganus Ckll.---Plants 15-40 cm. tall, acaulescent or nearly so; leaflets 11-21, 15-25 mm. long, oblong to lanceolate or sometimes linear, gray or silvery with appressed hairs; stipules adnate to the petioles; racemes many-flowered, elongating in fruit; calyx appressed silky-hairy with some blackish hairs present, tube 7-9 mm. long, teeth 2-3 mm. long; corolla mostly 18-25 mm. long, white with purple-tipped keel; fruit erect or spreading, oblong, narrowed gradually to a beak, suture intruded about 1/2 across.---Open gravelly or well-drained slopes and hills at lower to middle elevations. Saskatchewan and British Columbia to Wyoming, Colorado, and New Mexico; occasional in Utah. Our records scattered over Colorado at 4000-9000 feet.

11. **Oxytropis viscida** Nutt. ex T. & G., Fl. N. Amer. 1:341. 1838.
O. monticola Gray; O. hudsonicus Fernald; O. gaspensis Fernald & Kelsey; O. ixodes Butters & Abbe; Aragallus hudsonicus Greene; A. viscidulus Rydb.; A. viscidulus depressus Rydb.; A. grayanus Tidestrom---A species readily recognized by the glandular-viscid inflorescence and sometimes the foliage as well; also by its rather blue-purple corolla.---Alpine or subalpine. Hudson Bay to Alaska, south to Colorado and California. Apparently rare in this state, our record from Summit County.

29. Astragalus L. VETCH; MILKVETCH

Contributed (except for specific descriptions) by C. L. Porter. Perennial or sometimes annual, acaulescent or caulescent herbs, in ours, from taproots or rhizomes, the stems sometimes forming a woody and branched caudex at the base, pubescence simple and basifixed or of pick-shaped (malpighiaceous) hairs which are attached at or below the middle; stipules persistent, adnate to the stem, free or connate; leaves alternate, odd-pinnate, trifoliate, or sometimes simple or with a single leaflet, the terminal leaflet sometimes confluent with the rachis or all the leaflets reduced; flowers few to many, white, ochroleucous, yellowish, or purple, in axillary racemes, these sometimes subcapitate or subspicate; calyx with a campanulate to cylindrical tube and 5 subequal lobes or teeth; banner straight to abruptly arched; wings equalling or usually shorter than the banner; keel-petals equalling or usually shorter than the others, rounded on the ends in ours; stamens diadelphous (9 united and 1 free), the anthers all alike; style glabrous and stigma terminal; pods diverse, sessile to long-stipitate, membranaceous to woody or fleshy in texture, sometimes inflated, dehiscent or indehiscent, 1-celled or partially or completely 2-celled by the intrusion of the dorsal suture; seeds reniform, 1 to many in each pod. (Tragacantha Kuntze, Phaca L., and numerous segregates proposed by Rydberg.)

KEY TO THE SUBGENERA AND SOME OF THE SPECIES OF ASTRAGALUS OF COLORADO

1. Leaves all digitately 3-foliate — -1. Orophaca
1. Leaves pinnate, simple, or unifoliolate
   2. Low mat-forming plants with spinulose-tipped leaflets, very small flowers, and 1- to 2-seeded pods — -2. Kentrophyta
   2. Low or tall plants of various aspect, but the leaflets never spinulose-tipped
      3. Tall plants resembling milkweeds, with single, large leathery leaflets and long-stipitate, woody or coriaceous, plum-shaped pods — -31. A. asclepiadoides
      3. Not as above, the leaflets pinnate or else the plants low and pulvinate
         4. Pods 1-celled or partially 2-celled by the intrusion of the dorsal suture, the septum, if any, not united to the ventral suture
            5. Fruit triangular, sagittate, or obcordate in median cross-section, not papery
               6. Legume sessile, strictly 1-celled, the dorsal suture not intruded but sometimes sulcate — -3. Batidophaca
               6. Legume often stipitate, partially 2-celled by a slight or deep intrusion of the dorsal suture — -4. Tium
            5. Fruit rounded or variously flattened but not triangular, sagittate, or obcordate in cross-section unless papery
               7. Pods papery in texture
                  8. Legume bladdery and inflated, the body 8-30 mm. wide, sessile or subsessile -5. Phaca
                  8. Legume papery, sometimes inflated, the body 6 mm. wide or less, sessile or stipitate — -6. Atelophragma
               7. Pods with a leathery or woody texture
                  9. Legume inflated — -28. A. preussii
                  9. Legume not inflated
                     10. Fruit with a low intrusion of the dorsal suture, fleshy when green, becoming woody in age; tall glabrous plants — -7. Jonesiella
                     10. Fruit without any intrusion of the dorsal suture
                        11. Legume with 2 grooves running lengthwise on the upper (ventral) side — -8. Diholcos
                        11. Legume not 2-grooved ventrally
                           12. Pods fleshy when green, becoming woody at maturity; leaflets filiform or very slender and not articulated to the rachis
                              13. Stems erect; pods erect, somewhat laterally flattened, both sutures prominent — -52. A. toanus
                              13. Stems sprawling; pods pendulous, somewhat dorsally flattened, the ventral suture usually not prominent — -43. A. pectinatus
                           12. Pods sometimes fleshy when green but the leaflets then broader, always articulated to the rachis
                              14. Calyx cylindrical; low cinereous or villous plants with flowers mostly 15-30 mm. long — -9. Xylophacos
                              14. Calyx campanulate; low or tall plants with smaller flowers if the plants are low and cinereous
                                 15. Pods thick in texture, sessile, ovoid to ellipsoid in outline somewhat dorsally flattened; corolla white to yellowish — -10. Cnemidophacos
                                 15. Pods thin in texture, or if firm then either linear in outline or laterally flattened; corolla white, ochroleucous, or purplish

```
   16.  Pods flattened laterally, both sutures prominent                                    -11. Homalobus
   16.  Pods terete or dorsally flattened, both sutures not prominent
      17.  Banner of corolla strongly arched; pods nearly terete, sessile or shot-stipitate
         18.  Pods 5-8 mm. long                                                             -12. Microphacos
         18.  Pods 10-20 mm. long                                                           -13. Pisophaca
      17.  Banner of corolla not strongly arched; pods dorsally flattened, long-stipitate   -14. Lonchophaca
4. Pods completely 2-celled, the septum complete and united to the ventral suture
   19.  Fruit fleshy, plumlike, the walls at least 3 mm. thick, indehiscent or very tardily dehiscent, not
        inflated                                                                            -15. Geoprumnon
   19.  Fruit papery or woody, or if fleshy then the walls thinner, mostly dehiscent, sometimes inflated
      20.  Legume inflated
         21.  Pods papery, ovate or lanceolate in outline, with a pronounced beak, glabrous in ours; corolla
              purplish                                                                      -16. Diplocystium
         21.  Pods somewhat thicker-walled, nearly spherical, merely mucronate, not beaked, tomentose;
              flowers white or ochroleucous                                                 -70. A. anisus
      20.  Legume not inflated
         22.  Pods ovoid or oblong in outline, nearly circular in cross-section, leathery or woody, rigid
                                                                                            -17. Eustragalus
         22.  Pods linear, somewhat laterally flattened, thin-textured, scarcely rigid      -18. Hamosa
```

1. OROPHACA

Low, caespitose, and usually pulvinate perennial plants from a branched woody caudex, the older branches covered by stipules and old petioles. Leaves digitately 3-foliate, the leaflets oblong to oblanceolate or cuneate, silvery-canescent. Flowers 6-18 mm. long, on very short axillary racemes or appearing sessile among the stipules, pinkish or purplish when fresh but often drying ochroleucous, the calyx campanulate to cylindric, the banner moderately to strongly arched. Pods subglobose to ovoid, coriaceous, 1-celled, hardly exceeding the calyx. Seeds few.

Key to the Species

```
1. Corolla 15-18 mm. long; flowers subsessile                        -1. A. hyalinus
1. Corolla 6-10 mm. long; flowers usually on short peduncles         -2. A. sericoleucus
```

1. **Astragalus hyalinus** Jones, Proc. Calif. Acad. II. 5:648. 1895.
   Phaca argophylla Nutt. Not Astragalus argophyllus Nutt.; Orophaca argophylla (Nutt.) Rydb.---Stems not over 2 cm. long; leaflets 5-10 mm. long; calyx 7-9 mm. long; corolla 15-18 mm. long; fruit about 7-10 mm. long.---Dry sandy or clayey plains and hills at lower to middle elevations. North Dakota and eastern Wyoming, to western Nebraska and northeastern Colorado. Our one record from Yuma County at 3500 feet.
2. **Astragalus sericoleucus** Gray, Amer. Journ. Sci. II. 33:410. 1862.
   Phaca sericea Nutt. Not Astragalus sericeus DC. nor Lam.; Orophaca sericea (Nutt.) Britt.---Stems 4-20 cm. long but rarely over 10 cm.; leaflets 5-20 mm. long; calyx 3.5 mm. long; corolla 6-10 mm. long; fruit 4-6 mm. long. The very condensed form of the species with nearly sessile leaves and stems reduced to root-crowns in var. aretioides Jones, Contr. West. Bot. 8:13. 1898. Forms with long-petioled leaves and narrow leaflets often 15 mm. long are var. tridactylicus (Gray) Jones, l. c. 10:69. 1902 (Astragalus tridactylicus Gray, Proc. Amer. Acad. 6:527. 1865).---Dry plains and foothills at lower to middle elevations. Western Nebraska and southern Wyoming, to northeastern Colorado. Our records from the northeastern and north-central parts at 3500-7000 feet.

2. KENTROPHYTA

Perennial plants from a strong taproot, the branches sprawling or prostrate and forming loose or dense mats. Leaves pinnate, the leaflets about 5, linear-subulate or lanceolate, stiff, and spinulose-tipped. Flowers small, ochroleucous or purple, usually in 2-flowered racemes among the leaves, the banner arched. Pods ovoid or elliptic, laterally compressed or somewhat rounded, but both sutures usually prominent, 1-celled. Seeds few (1-3).

Key to the Species

```
1. Flowers usually ochroleucous; leaflets mostly 8-12 mm. long; plants of dry areas at lower to middle elevations
                                                                     -3. A. kentrophyta
1. Flowers usually purple; leaflets mostly 3-6 mm. long; plants of moister granitic soils in timbered areas at
   middle and higher elevations                                      -4. A. tegetarius
```

3. **Astragalus kentrophyta** A. Gray, Proc. Acad. Phila. 1863:60. 1863.
   Astragalus montanus Jones not L.; A. kentrophyta var. elatus S. Wats.; A. impensus Wooton & Standl.; A. montanus impensus Jones; Kentrophyta coloradensis Rydb.; K. impensa Rydb.; K. montana Nutt.; K. ungulata Rydb.; K. viridis Nutt.---Stems 10-40 cm. long; leaflets about 8-12 mm. long; calyx about 2.5-4 mm. long; corolla 4-6 mm. long, usually ochroleucous; fruit 4-7 mm. long.---A highly variable species of hot, dry sandy soils at lower elevations. Saskatchewan to southeastern Oregon and southward to New Mexico and Nevada. Our few records scattered in the western two-thirds of Colorado at 4500-9500 feet.

4. **Astragalus tegetarius** S. Wats., Bot. King's Expl. 76. 1871.
   Astragalus tegetarius var. rotundus Jones; A. aculeatus A. Nels.; A. montanus tegetarius Jones; Kentrophyta rotunda Rydb.; Kentrophyta wolfii Rydb.; Kentrophyta minima Rydb.---Stems 1-20 cm. long; leaflets about 4-10 mm. long; calyx about 2-4.5 mm. long; corolla 4-7 mm. long, usually purple; fruit 3-7 mm. long.---Often in open timbered areas on moist granitic soils at middle and higher elevations. Montana and Idaho, south to Colorado and California. Our rather few records from northcentral and central Colorado at 8500-10,000 feet

3. BATIDOPHACA

Low, caespitose, or decumbent-spreading perennial plants with pinnate leaves and rather broad leaflets. Flowers in short spikelike or capitate racemes, white, ochroleucous, yellowish, or reddish-purple, 6-15 mm. long, the calyx campanulate, the banner strongly arched. Pods sessile, 1-celled, membranaceous to leathery, somewhat lunate or obliquely ovate in outline, tapering at both ends, the ventral (upper) suture acute and straight or arched upwards, the dorsal (lower) suture rounded or inflexed at the middle, more strongly arched than the upper, the cross-section at the middle of the pod obcordate or triangular in shape. Seeds several to numerous.

Key to the Species

1. Stipules large and membranaceous, connate around the stem; plants decumbent-spreading   -5. **A. humistratus**
1. Stipules narrow and distinct or nearly so; plants decumbent to ascending or erect
    2. Flowers all in peduncled racemes
        3. Pods villous or hirsute, the hairs mostly not appressed
            4. Leaves and stems pilose; corolla white or ochroleucous, the keel purple-tipped; banner about 15 mm. long; pods 25-30 mm. long   -6. **A. parryi**
            4. Leaves and stems strigose; corolla purplish; banner about 10 mm. long; pods 10-25 mm. long   -7. **A. desperatus**
        3. Pods strigose, the hairs appressed
            5. Stems ascending, the internodes 5 mm. long or less; leaflets narrowly elliptical, mostly acute, never retuse at the apex; pods about 2 cm. long   -8. **A. naturitensis**
            5. Stems decumbent-spreading, mostly 10-20 cm. long, the internodes about 2 cm. long; leaflets oval to obovate, obtuse or retuse at the apex; pods about 7 mm. long   -9. **A. sparsiflorus**
    2. Flowers of two kinds, some well developed and on peduncled racemes, others apparently cleistogamous and sessile or subsessile in the leaf-axils   -10. **A. lotiflorus**

5. **Astragalus humistratus** Gray, Pl. Wright. 2:43. 1853.
   Tium humistratum Rydb.; Batidophaca humistrata Rydb.---Stems decumbent-spreading, about 25-45 cm. long; leaflets 11-19, 7-18 mm. long, oblong to lanceolate; stipules large and membranous, connate around the stem; flowers in dense racemes; calyx 4-7 mm. long; corolla 8-12 mm. long, purple to white; fruit 10-20 mm. long, strigose.---Dry gravelly mesas and hills, often among junipers, at lower to middle elevations. Southern Colorado, southward to Chihuahua and Arizona. Our one record from Archuleta County at 7000 feet.

6. **Astragalus parryi** Gray, Amer. Journ. Sci. II. 33:410. 1862.
   Astragalus succumbens T. & G. not Dougl.; Xylophacos parryi Rydb.; Batidophaca parryi Rydb.---Stems decumbent, 10-30 cm. long; leaflets 19-25, 5-15 mm. long, elliptic to oval; calyx 8-12 mm. long; corolla 12-15 mm. long, ochroleucous or whitish with purple-tipped keel; fruit 10-30 mm. long, villous.---Usually on cut banks and roadsides, middle elevations. Southern Wyoming to central Colorado. Our records from northcentral, central and southcentral parts at 4500-10,000 feet, one from Grand County on the Western Slope.

7. **Astragalus desperatus** Jones, Zoe 2:243. 1891.
   Tium desperatum Rydb.; Batidophaca desperata Rydb.---Stems spreading, 5-20 cm. long; leaflets 9-13, 5-10 mm. long, lanc-elliptic to broader; calyx 4-7 mm. long; corolla 9-15 mm. long, purplish; fruit 10-25 mm. long, villous-hirsute.---Dry mesas along the Colorado River at lower elevations, western Colorado, eastern Utah and northern Arizona. Our records from Mesa County at 4500-6000 feet.

8. **Astragalus naturitensis** Payson, Bot. Gaz. 60:377. 1915.
   Astragalus arietinus var. stipularis Jones---Stems ascending, 2-6 cm. long; leaflets 9-11, 4-7 mm. long, narrowly elliptic to oblong-elliptic; calyx 3.5-8 mm. long; corolla 10-15 mm. long; fruit about 1.5-2 cm. long; strigose. The typical form of the species has a calyx 6-8 mm. long, a bicolored corolla (banner white, apical portion of wings and keel reddish), the banner 14-15 mm. long, and is found in Montrose and Montezuma Counties. The var. deterior Barneby, Leafl. West. Bot. 5:88. 1948, has a calyx 3.5-5 mm. long, a smaller corolla 10-11 mm. long, which is entirely ochroleucous, and is found in the Mesa Verde region. Both forms occur on dry rocky mesas, ledges, and detrital slopes at 5000-7000 feet elevation in southwestern Colorado.

9. **Astragalus sparsiflorus** Gray, Proc. Am. Acad. 6:205. 1864.
   Tium sparsiflorum Rydb.; Batidophaca sparsiflora Rydb.---Stems probably decumbent-spreading, 10-20 cm. long; leaflets 11-21, 4-10 mm. long, oval, obovate to oblanceolate; calyx 2.5-4.5 mm. long; corolla 6-7 mm. long, whitish to ochroleucous, often purple-veined; fruit 7-18 mm. long. A coarse form of the species which is more ascending and with larger pods is var. majusculus Gray, Proc. Am. Acad. 6:206. 1864 (Batidophaca variegata Rydb.).---Dry bushy slopes with mountain mahogany and scrub oak at middle elevations, along the east flanks of the Rocky Mountains of Colorado. Our rather few records from the central part of the state at 5000-8000 feet.

10. Astragalus lotiflorus Hook., Fl. Bor. Am. 1:152. 1831.
    Phaca lotiflora T. & G.; Batidophaca lotiflora Rydb.; Phaca cretacea Buckl.; Astragalus elatiocarpus Sheld.; Astragalus batesii A. Nels.---Stems 2-10 cm. long; leaflets about 5-13, 5-15 mm. long, elliptic to narrowly oblong; earlier flowers on elongated peduncles, calyx 5.5-7 mm. long, corolla 8-9 mm. long, ochroleucous to whitish, the keel often purple-tipped; later flowers sessile or subsessile in the axils of the leaves, apparently cleistogamous with short calyx; fruit 15-25 mm. long, short-villous. Apparently the early flowers may be lacking on young plants.---Dry plains at lower to middle elevations. Minnesota, Saskatchewan and Montana to Texas and New Mexico. Our records scattered over the state but mostly from the eastern half, at 3500-8000 feet.

4. TIUM

Leafy-stemmed perennial plants with pinnate leaves having several to many leaflets. Flowers in axillary racemes, soon reflexed, the calyx campanulate, the corolla ochroleucous or white in ours, the banner not strongly arched, mostly 15-20 mm. long. Pods thin-coriaceous, oblong to linear in shape, mostly 20-30 mm. long, straight or curved, in ours with a stipe which exceeds the calyx tube, the cross-section triangular, sagittate, or obcordate, the dorsal suture sulcate and forming a partial septum within. Seeds several to many.

Key to the Species

1. Foliage conspicuously villous with spreading hairs -11. A. drummondii
1. Foliage strigose with appressed hairs, or glabrous
    2. Pod curved downward, the ventral suture convex externally; leaflets linear, mostly 1-2 mm. wide -12. A. schmollae
    2. Pod straight, or curved upward and the ventral suture then concave externally; leaflets broader
        3. Leaflets mostly 20-30 mm. long, not crowded; pods straight; calyx white-hairy -13. A. racemosus
        3. Leaflets mostly 15 mm. long or less, crowded; pods usually curved upward, the ventral suture concave externally; calyx with some conspicuous black hairs -14. A. scopulorum

11. Astragalus drummondii Dougl. in Hook., Fl. Bor. Am. 1:153. Pl. 57. 1831.
    Tium drummondii Rydb.---Stems erect or somewhat spreading, 25-60 cm. tall or more; leaflets 15-31, 7-25 mm. long, elliptic to narrowly oblong, villous; calyx 7-8 mm. long, black-villous; corolla 16-20 mm. long; fruit 18-40 mm. long, straight or curved.---Dry plains and foothills at lower to middle elevations. Saskatchewan and Alberta, to northern New Mexico and Utah. Our records scattered in the western three-fourths of Colorado, mostly east of the Continental Divide, at 4500-9000 feet.
12. Astragalus schmollae C. L. Porter, Madrono 8:100. Pl. 9, fig. 4-7. 1945.
    Stems erect or nearly so, 45-60 cm. tall; leaflets about 11-13, 10-30 mm. long, linear, strigose; calyx 6-7 mm. long; corolla 12-15 mm. long; fruit 25-40 mm. long, curved downward.---Dry mesas, in pinyon-juniper forest, at 6000-8000 feet. Known only from Mesa Verde in southwestern Colorado.
13. Astragalus racemosus Pursh, Fl. Am. Sept. 740. 1814.
    Astragalus galegoides Nutt.; Astragalus racemosus var. brevisetus Jones; Astragalus racemosus var. longisetus Jones; Tium racemosum Rydb.; Tium platycarpum Rydb. (a form with relative broad pods); Astragalus racemosus var. typicus C. L. Porter---Stems erect or ascending, 25-70 cm. long, possibly longer; leaflets 17-23, 15-30 mm. long, elliptic or linear, oblong, glabrous above and more or less strigose below; calyx 6-9 mm. long, white-hairy; corolla 15-17 mm. long; fruit 20-35 mm. long, straight or nearly so.---Dry plains and foothills at lower elevations. North Dakota and eastern Wyoming, southward to the Texas panhandle and northeastern New Mexico. Our records scattered in the eastern half of Colorado at 3500-6000 feet.
14. Astragalus scopulorum T. C. Porter in Porter & Coulter, Syn. Fl. Colo. 24. 1874.
    Astragalus subcompressus Gray; Astragalus rarus Sheld.; Tium scopulorum Rydb.---Stems ascending, 25-60 cm. tall; leaflets 11-27, 7-15 mm. long, elliptic to narrowly oblong, glabrous above, more or less strigose below; calyx 7-10 mm. long, more or less black-hairy; corolla 15-20 mm. long; fruit 20-25 mm. long, curved upward.---Open slopes, canyons, or meadows at middle elevations. Mountains of Colorado and adjacent New Mexico and Utah. Our records from the southwestern quarter of Colorado (except one from Grand County) at 6500-9000 feet.

5. PHACA

Mostly low (3 dm. or less), leafy-stemmed perennial plants, in ours, with pinnate leaves having several to many leaflets, or sometimes these reduced to linear phyllodes. Flowers in axillary racemes which seldom elongate greatly in fruit, the calyx campanulate, the corolla white, ochroleucous, or only tinged with purple in ours, 7-20 mm. long, the banner strongly arched or nearly straight. Pods papery or bladdery and much inflated, balloon-like, sessile or stipitate, the body ovoid or broadly lanceolate in outline, 1-celled, the dorsal suture not intruded, indehiscent or tardily dehiscent. Seeds several to many.

Key to the Species

1. Pods conspicuously mottled with splashes of red or orange; leaflets usually long-linear or reduced to linear phyllodes, or the leaflets sometimes normal and oblong -15. A. ceramicus

1. Pods seldom conspicuously mottled with red or orange, usually not mottled at all; leaflets normal, oblong to broadly obovate, never narrowly linear or reduced to phyllodes
   2. Stems 5 cm. high or less; leaves (including petioles) 1-2 cm. long, ashy-gray, the leaflets very crowded and somewhat imbricate; plants of the badlands of the White River drainage   -16. A. lutosus
   2. Stems taller; leaves (including petioles) 5-15 cm. long, green, the leaflets not crowded nor imbricate
      3. Mature pods 50-60 mm. long; flowers 18-20 mm. long   -17. A. megacarpus
      3. Mature pods 13-35 mm. long; flowers 7-15 mm. long
         4. Pod with a stipe 4-7 mm. long and exceeding the calyx-tube; flowers 13-15 mm. long
            -18. A. oophorus
         4. Pod sessile or subsessile; flowers 7-8 mm. long
            5. Leaflets oblong to narrowly elliptic, mostly 3 mm. wide or less; pods with a very short beak or merely mucronate   -19. A. cerussatus
            5. Leaflets obovate or broadly elliptic, mostly more than 4 mm. wide; pods tapering into a beak
              -20. A. wetherilli

15. Astragalus ceramicus Sheld., Minn. Bot. Stud. 1:19. 1894.
Phaca picta Gray; Astragalus pictus Gray, not Steud.; Astragalus pictus var. foliosus Gray; Astragalus foliosus Sheld.; Astragalus pictus var. magnus Jones---Stems 10-25 cm. long; leaflets 5-11, 10-30 mm. long, linear, or terminal reduced to a long linear phyllode and laterals often absent; calyx 3-5 mm. long; corolla 7-10 mm. long, whitish, ochroleucous or pinkish, the keel often darker tinged; fruit 15-35 mm. long, ovoid to ellipsoid, glabrous, conspicuously mottled with red or orange; stipe about 3-4 mm. long. The typical form of the species is characterized by having several well developed oblong to linear leaflets, the terminal one similar to the lateral ones, and somewhat smaller pods 10-20 mm. long. The more common form having few or no linear or filiform lateral leaflets, the terminal one reduced to a slender prolongation of the rachis, and pods mostly 15-30 mm. long is var. imperfectus Sheld., Minn. Bot. Stud. 1:19. 1894 (Astragalus longifolius (Pursh) Rydb.).---Dry sandy soils at lower elevations. The species from northern New Mexico and southwestern Colorado through the Navajo Basin to southern Utah and northern Arizona. The variety from western South Dakota and western Nebraska to Idaho and New Mexico. Our records scattered over Colorado at 3500-8000 feet.
16. Astragalus lutosus Jones, Contr. W. Bot. 13:7. 1910.
Phaca lutosa Rydb.---Stems less than 5 cm. tall; leaflets crowded, 19-27, 2-7 mm. long, oval, oblong or obovate; calyx 4-7 mm. long; corolla 8-9 mm. long, white; fruit 20-30 mm. long, elliptic, finely strigose; stipe if present very short.---Dry calcareous shales at lower elevations. Known only from the drainages of the White River, Rio Blanco County, Colorado, and adjacent Utah. Our few records at about 7000 feet.
17. Astragalus megacarpus (Nutt.) Gray, Proc. Am. Acad. 6:215. 1864.
Phaca megacarpa Nutt.; Astragalus megacarpus var. parryi Gray; Astragalus megacarpus var. prodigus Sheld.---This species occurs in semidesert areas across southern Wyoming to Utah north of the Uinta Mountains, and was listed for Gunnison County, Colorado, by M. E. Jones (Rev. Astrag. 120. 1923.), but the meager collection of Jones has smaller flowers and is probably A. oophorus Wats. Astragalus megacarpus is to be expected along the Wyoming-Colorado lines in northwestern Colorado.
18. Astragalus oophorus Wats., Bot. King's Expl. 73. 1871.
Astragalus artipes Gray; Astragalus megacarpus var. caulescens Jones; Astragalus oophorus var. caulescens Jones; Phaca artipes Rydb.; Phaca oophora Rydb.---Stems 10-30 cm. tall, sometimes taller; leaflets 9-21, 7-15 mm. long, oval, obovate to elliptic-oblong; calyx 6-9 mm. long; corolla 13-17 mm. long, cream-colored or ochroleucous, often rose-tinged near end; fruit 20-35 mm. long, ovoid or ellipsoid, glabrous, sometimes mottled with red; stipe longer than the calyx.---Dry semidesert regions at lower elevations. From western Colorado to southern Nevada and northern Arizona. Our records from northwestern and westcentral Colorado at 5000-7000 feet.
19. Astragalus cerussatus Sheld., Minn. Bot. Stud. 1:139. 1894.
Phaca cerussata Rydb.---Stems 10-25 cm. long; leaflets 11-17, 5-12 mm. long, oblong or narrowly elliptic; calyx 2.5-4 mm. long; corolla 5-8 mm. long, whitish or purplish-tinged; fruit 15-20 mm. long, oval with beak short or pod merely cuspidate, pilose; stipe very short if present.---Rocky slopes and hillsides at lower to middle elevations. Central Colorado, southward to northern New Mexico. Our few records from the central and southcentral parts at 5500-8500 feet.
20. Astragalus wetherilli Jones, Zoe. 4:34. 1893.
Phaca wetherilli Rydb.---Stems 10-40 cm. tall; leaflets 9-17, 5-10 mm. long, obovate to broadly elliptic; calyx 4.5-7 mm. long; corolla 8-10 mm. long, whitish to purplish-tinged; fruit 20-22 mm. long, ovoid and tapering to tip, short strigose; stipe shorter than calyx.---Bushy slopes at lower elevations. Known only from westcentral Colorado, our records from 4500-6000 feet.

6. ATELOPHRAGMA

Leafy-stemmed perennial plants from slender rhizomes or woody taproots, with pinnate leaves and several to many leaflets. Flowers in axillary racemes, the calyx campanulate, the corolla purple to white or ochroleucous, 7-12 mm. long, the banner not strongly arched. Pods sessile to long-stipitate, papery, sometimes a little inflated, 6 mm. wide or less (measured from valve to valve), 1-celled or sometimes partially 2-celled by the inflexed or sulcate dorsal suture. Seeds several to many.

Key to the Species

1. Plants from slender creeping rhizomes; stems slender, decumbent; stipules connate
    2. Banner and keel of corolla subequal; stipe of pod equaling the calyx-tube; dorsal suture of pods sulcate
      -21. A. alpinus
    2. Banner of corolla definitely longer than the keel; stipe of pod less than 1 mm. long; dorsal suture of pods not sulcate
        3. Corolla white, only the keel purple-tipped; entire calyx about 4 mm. long; keel about 6 mm. long; herbage green   -22. A. leptaleus
        3. Corolla purplish throughout; entire calyx 6-7 mm. long; keel 8-10 mm. long; herbage gray or silvery
            -23. A. plumbeus
1. Plants from woody taproots; stems mostly stiff and erect; stipules distinct
    4. Corolla ochroleucous, only the keel purple-tipped; stipe of pod 6-8 mm. long, well-exceeding the calyx
        -24. A. aboriginum
    4. Corolla purplish throughout; pods sessile or with a stipe only 3 mm. long and about equaling the calyx
        5. Pods 16-18 mm. long, with a stipe about 3 mm. long   -25. A. occidentalis
        5. Pods 10-12 mm. long, sessile or subsessile
            6. Leaflets acute at each end; fruiting racemes mostly 3-6 cm. long, the fruits crowded
                -26. A. bodini
            6. Leaflets rounded at each end; fruiting racemes mostly 7-15 cm. long, the fruits distant
                -27. A. eucosmus

21. Astragalus alpinus L., Sp. Pl. 760. 1753.
    Phaca astragalina DC.; Colutea astragalina Poir.; Phaca andina Nutt. ex T. & G., as synonym; Astragalus pauciflorus Hook.; Astragalus astragalinus Sheld.; Astragalus giganteus Sheld. in part, not Wats.; Tium alpinum Rydb.; Phaca alpina Piper, not L.; Astragalus andinus Jones---Plants from creeping rhizomes; stems 5-25 cm. long; leaflets 11-23, 4-10 mm. long, oval to elliptic-oblong, pilose both sides or glabrous above; calyx 3-4 mm. long; corolla 7-12 mm. long, keel as long as banner, light to dark purple (wings may be whitish); fruit 10-13 mm. long, pendant; stipe about 2 mm. long.---Meadows and stream banks at middle and higher elevations. Circumboreal, subarctic and alpine in Europe, Asia and North America. Our records scattered in the mountains of Colorado at 7000-11,500 feet.
22. Astragalus leptaleus Gray, Proc. Am. Acad. 6:220. 1864.
    Phaca pauciflora Nutt.?; Astragalus pauciflorus Gray, not Pall. nor Hook.; Phaca leptalea Rydb.---Plants from slender rhizomes; stems 7-20 cm. long; leaflets 15-25, 5-12 mm. long, oblong to linear, sparingly strigose at least below; calyx 4-5 mm. long; corolla 8-10 mm. long, white or the keel purplish this latter about 6 mm. long; fruit 8-10 mm. long, sessile.---Meadows and valleys, at middle elevations. Mountains of Colorado, in southcentral Idaho, and southwestern Montana. Our few records from northcentral and central Colorado at 7500-10,000 feet.
    A related species, keying out about here, is present in southcentral Colorado. It is S. brandegei Porter and differs in having characteristic flagelliform peduncles.
23. Astragalus plumbeus Barneby, Leafl. West. Bot. 5:195. 1949.
    Plants from creeping rhizomes; stems 2-10 cm. long; leaflets 13-21, 4-8 mm. long, oblong to elliptic, gray-green; calyx 5-7 mm. long; corolla 9-12 mm. long, purplish; fruit 7-10 mm. long; stipe short.---Rocky slopes above timberline. Known only from Lake, Pitkin and Summit Counties of central Colorado at 11,000-13,000 feet.
24. Astragalus aboriginum Rich., Bot. App. Frankl. Journ. 746. 1823. (Spelled aboriginorum).
    Astragalus vaginatus Rich.; Astragalus richardsonii Sheld.; Homalobus aboriginum Rydb.; Atelophragma aboriginum Rydb.---Plants from woody taproots; stems 10-30 cm. long; leaflets 7-13, 8-20 mm. long, linear to oblong, villous or strigose at least below; calyx 5-8 mm. long; corolla 8-11 mm. long, ochroleucous with keel purple-tipped; fruit 15-30 mm. long; stipe 6-8 mm. long. The typical form of the species has densely villous-hirsute herbage. Plants with sparse and appressed pubescence, often associated with the typical form, may be referred to var. glabriusculus (Hook.) Rydb., Contr. U. S. Nat. Herb. 3:492. 1896 (Phaca glabriuscula Hook., Fl. Bor. Am. 1:144. 1831).---Rocky or clayey mountain slopes, often in calcareous soils, at middle elevations. Quebec, Manitoba and Yukon Territory, southward in the mountains to Colorado and Nevada. Our few records from northcentral, central and southcentral Colorado (a doubtful one from southwestern) at 7000-10,000 feet.
25. Astragalus occidentalis (Wats.) Jones, Contr. W. Bot. 8:17. 1898.
    Astragalus robbinsii var. occidentalis Walts.; Astragalus macounii Rydb.; Astragalus labradoricus var. occidentalis Jones; Atelophragma macounii Rydb.; Atelophragma occidentale Rydb.---Plants from caespitose rootstocks; stems 20-60 cm. long; leaflets 9-17, 10-30 mm. long, oval to elliptic, glabrous above and sparingly strigose below; calyx 3-4 mm. long; corolla 9-11 mm. long, purple; fruit 16-18 mm. long; stipe about 3 mm. long.---Moist meadows and stream banks at higher elevations. Alberta and British Columbia, south in the mountains to central Colorado and Idaho. Our two records from Gunnison and Summit Counties at 9500-11,500 feet.
26. Astragalus bodini Sheld., Minn. Bot. Stud. 1:122. 1894.
    Astragalus debilis sensu Jones, Rev. Astrag. 88. 1923, in part, not Gray, 1863; Phaca bodini Rydb.---Plants from woody taproots; stems 25-50 cm. long; leaflets 9-25, 6-20 mm. long, elliptic to oblong-elliptic, strigose at least below; calyx 4-5 mm. long; corolla 8-10 mm. long, purple; fruit about 10 mm. long, sessile or very nearly so.---Moist open meadows and stream banks at lower to middle elevations. Central Idaho, southward to western Nebraska, Wyoming, Utah and Colorado. Our rather few records from northcentral and southcentral Colorado at 7500-8500 feet.

27. Astragalus eucosmus Robinson, Rhodora 10:33. 1908.
    Phaca elegans Hook.; Phaca elegans minor Hook.; Phaca parviflora Nutt., not Astragalus parviflorus Nutt.; Astragalus oroboides Gray, not Hornem.; Astragalus oroboides var. americanus Gray; Astragalus elegans Sheld., not Bunge; Astragalus elegans var. curtiflorus Rydb.; Astragalus minor Jones, not Clos.---Plants from woody taproots; stems 25-60 cm. long; leaflets 13-15, 10-20 mm. long, elliptic or oblong, glabrous above, strigose below; calyx 3-4 mm. long; corolla 6-8 mm. long, purple; fruit 9-15 mm. long, sessile or nearly so.---Moist meadows and stream banks, often in partial shade, at lower to middle elevations. Arctic North America from Baffin Island and Alaska southward to northern Maine, northern Ontario, Montana, Wyoming and Colorado. Our records from northcentral and central Colorado at 7500-9000 feet.

7. JONESIELLA

Rather coarse, tufted, leafy-stemmed sparingly pubescent to glabrous perennial plants with pinnate or sometimes large unifoliolate leaves. Flowers in racemes, on long axillary peduncles, or sometimes capitate on short peduncles, the calyx campanulate, the corolla ochroleucous or purple, the banner not strongly arched, 15-20 mm. long. Pods thin-coriaceous to woody at maturity, often fleshy when green, sometimes inflated, subsessile to long-stipitate, the body 15-35 mm. long, in ours elliptical, ovoid, or oblong in outline and rounded in cross-section, the dorsal suture intruded as a thin line or a low, thickened ridge within. Seeds several to many.

Key to the Species

1. Leaflets several to many, thin, 10-30 mm. long; flowers ochroleucous or purple
    2. Pods thin-walled, never fleshy, inflated; flowers purple   -28. A. preussii
    2. Pods thick-walled, fleshy when green, hardly inflated; flowers ochroleucous, the keel often purple-tipped
        3. Body of pod tapering from the middle to the narrow base, thus cuneate at the base   -29. A. recedens
        3. Body of pod rounded at the base, not tapering   -30. A. pattersoni
1. Leaflet single, leathery, those in the upper part of the stems mostly 30-60 mm. long; flowers purple
                                                                                           -31. A. asclepiadoides

28. Astragalus preussii Gray, Proc. Am. Acad. 6:222. 1864.
    Astragalus preussii var. latus Jones; Astragalus preussii var. arctus Sheld.; Phaca preussii Rydb.; Rydbergiella preussii Rydb.; Rydbergiella arcta Rydb.; Jonesiella arcta Rydb.---Stems 15-30 cm. long, possibly longer; leaflets 11-15, 6-15 mm. long; calyx 9-11 mm. long; corolla 17-22 mm. long, purple; fruit 20-30 mm. long, thin-walled and inflated. The Colorado collections seen may be referred to var. eastwoodae (Jones) Jones, which has short-stipitate, oblong pods and narrow, acute leaflets, the stems shorter and more decumbent than in the species.---Dry sandy or clayey plains and hills at lower elevations. Drainages of the Colorado and lower Green Rivers from Colorado to Nevada and in California west of the Colorado River to the Mexican boundary. Our few records from Garfield, Mesa and Montrose Counties at 4500-5500 feet.
29. Astragalus recedens (Greene ex Rydb.) C. L. Porter, Univ. Wyo. Pub. 16(1):19. 1951.
    Jonesiella recedens Greene---Stems 25-50 cm. tall; leaflets 15-23, 10-30 mm. long; calyx 7-9 mm. long; corolla 14-20 mm. long, ochroleucous or keel purple-tipped; fruit 20-30 mm. long, ellipsoid and tapering to base, thick-walled.---Dry plains at lower elevations. Specimens seen from southern Colorado and northern New Mexico. Also reported from Utah by Rydberg. Our one Colorado record from Archuleta County at 6000 feet.
30. Astragalus pattersoni Gray in Brandegee, Bull. U. S. Geol. Surv. Terr. 2:285. 1876.
    Phacopsis pattersoni Rydb.; Rydbergiella pattersoni Fedde & Sydow; Jonesiella pattersoni Rydb.---Stems 30-70 cm. tall; leaflets 13-19, 10-25 mm. long, elliptic to oblong; calyx 7-11 mm. long; corolla 15-20 mm. long, whitish to cream-colored; fruit 20-25 mm. long, oblong-ellipsoid, rounded at base, thick-walled. The species has pods which are typically oblong and about 8 mm. wide. Plants with oval pods which are usually wider near the middle are referred to var. praelongus (Sheld.) Jones, but intergradation between the species and the variety is common.---Dry, usually alkaline and seleniferous shales at middle and lower elevations. Colorado to Texas, westward to southern Nevada and northern Arizona. Our records scattered in the western half of Colorado at 4500-8000 feet.
31. Astragalus asclepiadoides Jones, Zoe 2:238. 1891.
    Jonesiella asclepiadoides Rydb.---Stems 25-50 cm. tall; leaflets reduced to 1, this 20-60 mm. long, suborbicular, cordate; calyx 10-15 mm. long; corolla 17-21 mm. long, purple (often recorded in descriptions as white); fruit 25-40 mm. long, ellipsoid-oblong to lanc-ellipsoid, long-stipitate.---Dry alkaline plains and foothills at lower elevations. Colorado, to eastern Utah. Our records from Garfield, Mesa, Delta and Montrose Counties at 4500-6500 feet.

8. DIHOLCOS (BISULCATI)

Leafy-stemmed perennial plants, often in large clumps. Leaves pinnate with numerous leaflets. Flowers in dense, spikelike racemes, commonly reflexed in age, the corolla purple, ochroleucous, or white. Pods included in the calyx or more commonly well exserted, with a short or long stipe, somewhat coriaceous in texture, 1-celled, with 2 grooves on the ventral (upper) side running lengthwise of the pod, and somewhat dorsally compressed. Seeds few to many.

Key to the Species

1. Calyx campanulate, not inflated, strigose; leaflets ovate to oblong; corolla purple, ochroleucous, or white; pods exserted from the calyx at maturity
   2. Body of pod 11 mm. long or more; corolla usually purple, rarely pale or nearly white, over 10 mm. long
     -32. **A. bisulcatus**
   2. Body of pod 10 mm. long or less; corolla ochroleucous or white, the keel often purple-tipped, less than 10 mm. long
     3. Pods cross-reticulate, blunt at apex -33. **A. haydenianus**
     3. Pods not cross-reticulate, acute at apex -34. **A. diholcos**
1. Calyx inflated-urceolate, shaggy-villous; leaflets linear; corolla white or ochroleucous, the keel purple-tipped; pods included in the calyx at maturity -35. **A. oocalycis**

32. Astragalus bisulcatus (Hook.) Gray, Pacific R. R. Rept. 12:42. 1860.
    Phaca bisulcata Hook.; Diholcos decalvans Rydb.; Diholcos bisulcatus Rydb.---Stems 30-90 cm. tall; leaflets 1-2.5 cm. long, linear-oblong to oval; calyx 4.5-7 mm. long, not inflated, strigose; corolla 12-15 mm. long, usually purple, but sometimes reddish-purple or paler to nearly white; fruit 11-15 mm. long, not cross-reticulate; stipe 3-4 mm. long.---In dry alkaline meadows and hills, on seleniferous soils, at lower and middle elevations. Manitoba, Saskatchewan, and Alberta southward to New Mexico. Our records scattered over Colorado at 3500-8000 feet.

33. Astragalus haydenianus A. Gray, Bull. U. S. Geol. and Geogr. Surv. Terr. 2:235. 1876.
    Astragalus haydenianus var. major Jones; Astragalus haydenianus var. nevadensis Jones; Astragalus grallator S. Wats.; Astragalus demissus Greene, not Boiss. & Heldr.; Astragalus jepsoni Sheld.; Astragalus scobinatulus Sheld.; Homalobus grallator Rydb.; Diholcos haydenianus Rydb.; Astragalus bisulcatus var. haydenianus Jones---Stems 30-60 cm. tall; leaflets 1-3 cm. long, linear-oblong to oblong; calyx 3-5 mm. long, not inflated, strigose; corolla 6-8 mm. long, whitish to cream-colored, keel often purplish at apex; fruit about 7-9 mm. long, cross-reticulate, rather blunt at apex; stipe about 3 mm. long. Apparent hybrids between this and the preceding species have been collected in Grand County.---Rather dry meadows and hills at lower and middle elevations. Southern Wyoming and Utah to New Mexico and Nevada. Our records from the western half of Colorado at 5500-8500 feet.

34. Astragalus diholcos Tidestr., Proc. Biol. Soc. Wash. 50:17-22. 1937.
    Diholcos micranthus Rydb., not Astragalus micranthus Nutt. nor Desv.; Astragalus haydenioides C. L. Porter---Stems 20-40 cm. tall; leaflets 7-20 mm. long, linear-oblong to lanc-oblong; calyx 3-4.5 mm. long, not inflated, strigose; corolla 6-7 mm. long, cream-colored; fruit about 8-11 mm. long, not cross-reticulate, acute at apex; stipe about 3-5 mm. long.---Occasional in dry alkaline areas at middle elevations. Southern Wyoming to northern New Mexico. Our rather few records from the western half of the state at 5000-6500 feet.

35. Astragalus oocalycis Jones, Contr. West. Bot. 8:10. 1898.
    Astragalus urceolatus Greene; Cnemidophacos urceolatus Rydb.; Diholcos oocalycis Rydb.---Stems 30-60 cm. tall; leaflets 2-5 cm. long, linear; calyx 9-11 mm. long, inflated, shaggy-villous; corolla 14-16 mm. long, cream-colored to ochroleucous, the keel purple-tipped; fruit included in the calyx, about 6 mm. long, cross-reticulate; stipe 1-2 mm. long.---Dry alkaline areas at lower and middle elevations. Southern Colorado and northern New Mexico. Our few records from southwestern Colorado, in La Plata and Archuleta Counties, at 6500-7000 feet.

9. XYLOPHACOS (ARGOPHYLLI*)

Mostly rather low, short-caulescent or acaulescent perennial or winter annual plants from a branched caudex or taproot, with pinnate leaves and usually broad leaflets. Flowers sub-capitate or racemose, often on long peduncles, ochroleucous or purple or pinkish in ours, mostly 15-30 mm. long, the calyx-tube cylindric, the banner moderately arched. Pods lance-ovoid to oblong, more or less beaked, often curved upward, 1-celled, sessile, often somewhat dorsally compressed with the dorsal suture variously sulcate but little if any intruded, the texture spongy or fleshy and usually becoming coriaceous or woody at maturity, sometimes densely woolly. Seeds several to many.

Key to the Species

1. Pods shaggy-villous or woolly with long hairs; corolla ochroleucous with purple-tipped keel -36. **A. purshii**
1. Pods glabrous, strigose, or short-pilose but not shaggy-villous; corolla white, ochroleucous, or purplish
   2. Pods spongy in texture -37. **A. chamaeleuce**
   2. Pods coriaceous or woody in texture, sometimes succulent or fleshy when young
      3. Hairs of leaves mostly or entirely pick-shaped, attached at or below the middle (dolabriform)
         4. Pods readily deciduous at maturity, about 30 mm. long, mostly arcuate and dorsally compressed -38. **A. amphioxys**
         4. Pods persistent, 17-25 mm. long, straight or nearly so, subterete or 4-angled, rarely dorsally compressed -39. **A. missouriensis**
      3. Hairs of leaves simple and basifixed (use strong magnification)
         5. Pods 25-45 mm. long, the walls fleshy when young and becoming coriaceous, at least 1 mm. thick; plants of the eastern slope -40. **A. shortianus**
         5. Pods mostly 15-25 mm. long, rarely longer, the walls thinner; plants of the western slope
            6. Pods strigose to villous-hirsute; plants of western Colorado -41. **A. argophyllus**
            6. Pods glabrous; plants of southern Colorado -42. **A. iodopetalus**

*See Barneby, R. C. - A Revision of the Argophylli, Amer. Midl. Nat. 37:421-516. March 1947. Several Colorado collections are cited which are not in the Rocky Mountain Herbarium.

36. Astragalus purshii Dougl. ex Hook., Fl. Bor. Am. 1:152. 1834.
Phaca mollissima Nutt. ex T. & G. Not Astragalus mollissimus Torr.; Phaca purshii (Dougl.) Piper; Xylophacos purshii (Dougl.) Rydb.; Astragalus purshii var. interior Jones; Xylophacos incurvus Rydb. Not Astragalus incurvus Desf.---Stems less than 10 cm. long, the plant subacaulescent; leaflets 7-13, 5-12 mm. long, lanc-elliptic to oval, silky-villous with simple hairs; calyx 8-15 mm. long; corolla 15-20 mm. long, ochroleucous but keel purple-tipped; fruit 15-20 mm. long, ovoid, densely white-villous.---Prairies, valleys, and hillsides, often in sagebrush, at middle and lower elevations. Saskatchewan, Alberta, and British Columbia southward to the Dakotas, northern Colorado, central Nevada, and northeastern California. Our records from northcentral, northwestern and westcentral Colorado (mostly from Moffat County) at 5500-9500 feet.

37. Astragalus chamaeleuce Gray in Ives, Rep. Colo. River, Bot. 10. 1860.
Phaca pygmaea Nutt.; Xylophacos pygmaeus (Nutt.) Rydb.; Astragalus pygmaeus (Nutt.) Jones, Not Astragalus pygmaeus Pall.; Astragalus cicadae Jones; Astragalus chamaeleuce var. laccoliticus Jones---Stems short, the plant subacaulescent; leaflets 5-11, 4-10 mm. long, broadly to narrowly obovate, white-strigose with simple hairs; calyx 6-12 mm. long; corolla 17-20 mm. long, light purple to lavender-pink; fruit 25-40 mm. long, ovoid or oblong-ovoid, walls very thick and spongy, mottled with reddish color, strigose.---Clayey hills and badlands at middle elevations. Basins of the Green and Grand Rivers in southern Wyoming, Utah, and western Colorado. Our records from northwestern and westcentral Colorado at 4500-7000 feet.

38. Astragalus amphioxys Gray, Proc. Am. Acad. 13:366. 1878.
Xylophacos amphioxys (Gray) Rydb.; Astragalus crescenticarpus Sheld.; Astragalus selenaeus Greene; Xylophacos aragalloides Rydb.---Stems 3-6 cm. long; leaflets 9-15, 4-12 mm. long, elliptic to obovate, white-strigose with at least some hairs dolabriform; calyx 10-13 mm. long; corolla 21-28 mm. long, rose-purple or tinged with magenta or metallic green; fruit 25-35 mm. long, oblong-ellipsoid, strigose. Colorado plants of this species are referable to var. vespertinus (Sheld.) Jones (Astragalus vespertinus Sheld.), which has flowers having the banner 22-28 mm. long and more than twice the length of the entire calyx, the typical form of the species having somewhat smaller flowers, the banner 13-23 mm. long and less than twice the length of the entire calyx.---Clayey soil of mesas and valleys in the drainage of the Green, and upper Colorado rivers in western Colorado, southern Utah, and northern Arizona. Our records from western Colorado at 4500-7500 feet.

39. Astragalus missouriensis Nutt., Gen. N. Am. Pl. 2:99. 1818.
Xylophacos missouriensis (Nutt.) Rydb.---Stems 5-12 cm. long; leaflets 9-15, 5-10 mm. long, oval to elliptic, strigose with at least some hairs dolabriform; calyx 8-10 mm. long; corolla 15-20 mm. long, rose-purple; fruit 15-25 mm. long, oblong, strigose. The typical form of the species has nearly straight pods which are persistent on the plant at maturity, thus differing from A. amphioxys which has arcuate pods which are deciduous. Intermediate forms from southern Colorado, where the ranges of these two species meet, having persistent pods which are somewhat arcuate and dorsally compressed, have been designated as A. missouriensis var. amphibolus Barneby.---Plains, foothills, and bluffs, in dry areas at lower to middle elevations. Alberta and Saskatchewan to western Oklahoma, western Texas, Colorado, and New Mexico. Our records scattered over Colorado (none as yet from the northwestern part) at 3500-8000 feet.

40. Astragalus shortianus Nutt. ex T. & G., Fl. N. Amer. 1:331. 1838.
Xylophacos shortianus (Nutt.) Rydb.---Stems usually less than 10 cm. long; leaflets 7-17, 8-20 mm. long, oval, broadly elliptic or obovate, silky-strigose with simple hairs; calyx 11-12 mm. long; corolla 18-24 mm. long, rose-purple; fruit 25-45 mm. long, oblong but curved, strigose.---Plains and foothills, often in granitic soils, at lower to middle elevations. On the eastern flanks of the Rocky Mountains from southern Wyoming to northern New Mexico. Our records from central and northcentral Colorado (mostly from the eastern slope but apparently from Grand and Eagle Counties) at 5000-8000 feet.

41. Astragalus argophyllus Nutt. ex T. & G., Fl. N. Amer. 1:331. 1838.
Xylophacos argophyllus (Nutt.) Rydb.---Stems 6-12 cm. long; leaflets 13-21, 5-15 mm. long, lanceolate to narrowly elliptic, long-strigose with simple hairs; calyx 9-11 mm. long; corolla 20-25 mm. long, pale purple to nearly whitish; fruit 20-25 mm. long, ovoid, strigose to villous-hirsute. Colorado plants of this species are referred to var. pephragmenoides Barneby, which differs from the typical form of the species in being more xerophytic with smaller flowers which are pale purple and with looser pubescence on the herbage and pods.---Dry hillsides and sagebrush mesas at middle elevations. The variety ranging from southern Idaho to western Colorado and on the Kaibab Plateau of Arizona. One one record from northwestern Colorado at 7000 feet.

42. Astragalus iodopetalus Greene ex Barneby, Amer. Midl. Nat. 37:471. 1947.
Xylophacos iodopetalus (Greene) Rydb.---Stems 10-20 cm. tall; leaflets 19-27, 5-15 mm. long, elliptic to obovate, silky-pilose below with simple hairs; calyx 11-13 mm. long; corolla 18-23 mm. long, purple; fruit 20-30 mm. long, glabrous, cross-reticulate.---Mountainsides and canyons, often among scrub-oaks, along the Gunnison and tributaries of the San Juan rivers in southern Colorado and adjacent New Mexico. Our records from Montrose, Gunnison and Archuleta Counties at 5500-7500 feet.

10. CNEMIDOPHACOS

Leafy-stemmed perennial plants, often many-stemmed from the base, with pinnate leaves having several to many leaflets. Flowers in many-flowered, spikelike racemes, mostly erect or ascending, the calyx campanulate, the corolla ochroleucous, yellowish, or white in ours, the banner not strongly arched. Pods coriaceous, sometimes fleshy when green, sessile, ovoid to ellipsoid in outline, rounded or somewhat dorsally flattened in cross-section, erect or ascending, 1-celled. Seeds several to many.

Key to the Species

1. Leaflets filiform, 2-6 cm. long; pods fleshy when green -43. **A. pectinatus**
1. Leaflets linear or narrowly oblong, never filiform, 1-2.5 cm. long; pods coriaceous, never fleshy -44. **A. confertiflorus**

43. Astragalus **pectinatus** (Hook.) Dougl. in Hook., Fl. Bot. Am. 1:142. 1830.
 as synonym
 Phaca pectinata Hook.; Ctenophyllum pectinatum Rydb.; Cnemidophacos pectinatus Rydb.---Stems 25-60 cm. long; leaflets 9-17, 20-60 mm. long, narrowly linear to filiform, sparingly strigose; calyx 7-10 mm. long; corolla 17-20 mm. long; fruit 10-25 mm. long, fleshy when immature, becoming pendulous.---Common on prairies and plains and in foothills at lower to middle elevations. Saskatchewan to Kansas, Colorado, and Utah. Our records scattered in the eastern half of Colorado at 4000-6500 feet.
44. Astragalus **confertiflorus** A. Gray, Proc. Am. Acad. 13:368. 1878.
 Cnemidophacos confertiflorus Rydb.---Stems 10-30 cm. long; leaflets 11-15, 7-25 mm. long, linear to narrowly oblong, silvery-canescent; calyx 6-7 mm. long; corolla 9-13 mm. long; fruit 10-15 mm. long, ascending. The form with yellowish flowers 12-15 mm. long instead of 7-11 mm. long is var. **flaviflorus** (Kuntze) Jones, Rev. Astrag. 241. 1923 (Tragacantha flaviflora Kuntze, and Cnemidophacos flavus Rydb.).---Dry alkaline, seleniferous hills at middle elevations. Utah and Wyoming southward to New Mexico and northern Arizona. Our records from western Colorado at 5000-7000 feet.

11. HOMALOBUS

Perennial plants, sometimes caespitose and pulvinate, the leaves pinnate with few to several leaflets, these sometimes reduced to filiform phyllodes, or the leaves sometimes simple (unifoliolate). Flowers in racemes, white to purple, the banner usually strongly arched. Pods elliptic to linear in outline, sessile or stipitate, mostly flat and laterally compressed with both sutures prominent, 1-celled. Seeds few to many.

1. Plants pulvinate-caespitose; leaves simple or pinnately 3-foliate
   2. Banner of corolla 5-9 mm. long -45. **A. spatulatus**
   2. Banner of corolla 10-15 mm. long
      3. All leaves simple (unifoliolate); pods glabrous, 9-15 mm. long -46. **A. simplicifolius**
      3. Upper leaves 3- to 5-foliolate; pods 15-31 mm. long -47. **A. detritalis**
1. Plants not pulvinate-caespitose; leaves several-foliolate, at least some of them
   4. Corolla 5-10 mm. long, ochroleucous to pale lavender; pods pendulous
      5. Pods 15-20 mm. long, sessile, somewhat rounded in cross-section, the valves convex; inflorescence usually well-exserted beyond the foliage
         6. Plants not rushlike; leaflets not reduced to filiform phyllodes -48. **A. decumbens**
         6. Plants rushlike; leaflets of upper leaves reduced to filiform phyllodes -49. **A. diversifolius**
      5. Pods 8-15 mm. long, with at least a short stipe, flat; inflorescence not well exserted beyond the foliage
         7. Plants, or at least the stipules, blackening in drying; corolla about 5 mm. long; stipe of pod usually exceeding the calyx; pods 8-12 mm. long -50. **A. tenellus**
         7. Plants not blackening in drying; corolla about 8 mm. long; stipe of pod usually included in the calyx; pods 10-15 mm. long -51. **A. wingatanus**
   4. Corolla 15-25 mm. long, ochroleucous or purple; pods erect or pendulous
      8. Pods erect and rounded in cross-section, turgid; corolla about 15 mm. long, ochroleucous -52. **A. toanus**
      8. Pods pendulous and flat; corolla 13-25 mm. long, purple
         9. Pods long-stipitate, the stipe exceeding the calyx; corolla about 13 mm. long -53. **A. coltoni**
         9. Pods sessile; corolla 18-25 mm. long -54. **A. rafaelensis**

45. Astragalus **spatulatus** Sheld., Minn. Bot. Stud. 1:22. 1894.
 Astragalus caespitosus Gray, not Pall.; Homalobus canescens Nutt., not A. canescens DC.; Astragalus simplicifolius caespitosus Jones; Astragalus simplicifolius spatulatus Jones; Astragalus simplex Tidestr.; Homalobus caespitosus Nutt.---Plants acaulescent, pulvinate-caespitose, 2-8 cm. tall; leaves mostly unifoliate, rarely 3- to 5-foliate, blade 10-30 mm. long, linear or linear-oblanceolate, silky-canescent; flowers 3-7; calyx about 4 mm. long; corolla 5-9 mm. long, bluish-purple rarely cream-colored; fruit 6-12 mm. long, short-oblong.---Dry sandy or gravelly plains and hills at lower to middle elevations. Saskatchewan to northern Colorado, Utah, and southeastern Idaho. Our records from northern Colorado, from Weld to Moffat County at 5000-9000 feet.
46. Astragalus **simplicifolius** (Nutt.) Gray, Proc. Am. Acad. 6:231. 1864.
 Phaca simplicifolia Nutt.; Homalobus simplicifolius Rydb.; Astragalus lingulatus Sheld.; Astragalus exilifolius A. Nels.---Plants acaulescent and short-caespitose, 1-4 cm. tall; leaves unifoliate, 5-40 mm. long, linear or narrowly linear-oblanceolate, silvery-canescent; flowers 1-2; calyx 3.5-6 mm. long; corolla 10-15 mm. long, purple or rose, to ochroleucous or nearly white; fruit 6-15 mm. long.---Dry sandy or gravelly hills and bluffs, at middle elevations. Wyoming and northern Colorado, southward to Canon City. Our few records from northwestern Colorado at 6000-9000 feet.
47. Astragalus **detritalis** Jones, Contr. West. Bot. 13:9. 1910.
 Homalobus detritalis Rydb.---Plants acaulescent, pulvinate-caespitose, 2-6 cm. tall; leaves 3- to 5-foliate or lower unifoliate, 10-20 mm. long, usually narrowly oblanceolate but sometimes nearly linear, silky-canescent; flowers 3-6; calyx about 7 mm. long; corolla 10-15 mm.

long, purplish; fruit 15-31 mm. long.---Barren clayey detrital slopes at middle elevations. Infrequent in northeastern Utah and the valley of the White River in Rio Blanco County, Colorado. Our one record at 5400 feet.

48. Astragalus decumbens (Nutt. ex T. & G.) Gray, Proc. Am. Acad. 6:229. 1864.
Homalobus decumbens Nutt. not Astragalus decumbens Boiss., 1867.---Plants with caespitose or branching caudex, not at all pulvinate; stems 10-40 cm. tall, leafy; leaflets 7-25, 4-50 mm. long, linear to oval, glabrate to sparingly strigose below; calyx 3-5 mm. long; corolla 7-12 mm. long, white to cream-colored but often variously tinged with purple especially the apex of the keel; fruit about 15-30 mm. long, linear. Our state has the following 3 varieties:

1. Leaflets mostly narrow, often folded, acute, seldom more than 1 cm. long, usually grayish-strigose; plants low and tufted, mostly 15 cm. high or less
   -var. decumbens
1. Leaflets broad, mostly oblong to elliptic, often obtuse, mostly more than 1 cm. long, green; plants loosely tufted and usually taller
   2. Leaflets crowded, the terminal one not much longer than the others and articulated to the rachis
      -var. oblongifolius
   2. Leaflets distant on the rachis, the terminal one usually longer than the others and not distinctly articulated to the rachis
      -var. decurrens

48A. A. decumbens decumbens (Nutt.) Cronquist, (var.) Leafl. West. Bot. 3:252. 1943.
Homalobus decumbens Nutt.; Astragalus divergens Blank.; Astragalus campestris var. decumbens Jones; Homalobus microcarpus (A. Gray) Rydb.; Homalobus camporum Rydb.---Dry hills and banks at middle elevations. Montana and Idaho to Colorado and Utah. Our records from the western half of Colorado (except southcentral part) at 6000-10,000 feet.

48B. A. decumbens oblongifolius (Rydb.) Conquist, (var.) Leafl. West. Bot. 3:253. 1943.
Homalobus hylophilus Rydb.; Homalobus oblongifolius Rydb.; Astragalus hylophilus A. Nels.; Astragalus hylophilus var. oblongifolius Macbr.; Astragalus campestris var. hylophilus Jones; Astragalus convallarius hylophilus Tidestr.---Common in open timbered areas and in sagebrush hills at middle elevations. Montana and Idaho through South Dakota to Colorado and Utah. Our records from northcentral, northwestern and westcentral Colorado at 6500-9000 feet.

48C. A. decumbens decurrens (Rydb.) Cronquist, (var.) Leafl. West. Bot. 3:253. 1943.
Homalobus decurrens Rydb.; Astragalus rydbergii Macbr.---Usually in timbered areas at middle elevations. Specimens seen only from Colorado. Our records from the western half of the state (with few from the extreme western part) at 7000-10,000 feet.

49. Astragalus diversifolius Gray, Proc. Am. Acad. 6:230. 1864.
Homalobus junceus Nutt., not Astragalus junceus Ledeb.; Astragalus junciformis A. Nels.; Astragalus convallarius Greene; Homalobus salidae Rydb.?; Astragalus campestris var. diversifolius Macbr.---Plants usually with creeping rootstocks, not at all pulvinate, rather rushlike in appearance; stems 15-50 cm. tall, leafy but upper leaves reduced to rachises; leaflets 3-9, 5-20 mm. long, linear-filiform to narrowly oblong, strigose; calyx 4-5 mm. long; corolla 7-10 mm. long, whitish, ochroleucous, pink or rarely rose-purple; fruit 20-50 mm. long, linear.---Dry slopes, often in sagebrush, at middle elevations. Idaho and Montana to Nevada, Colorado, and northeastern Arizona. Our records from the western half of Colorado (except the southcentral part) at 5000-9000 feet.

50. Astragalus tenellus Pursh, Fl. Am. Sept. 2:473. 1814.
Homalobus multiflorus T. & G.; Astragalus nigrescens Gray; Homalobus stipitatus Rydb.; Homalobus clementis Rydb.---Plants with caespitose caudex but not pulvinate, blackening on drying; stem 20-50 cm. tall, leafy; leaflets 7-21, 7-20 mm. long, linear or oblong, glabrous to sparingly strigose; calyx 3-6 mm. long; corolla 6-10 mm. long, purplish to whitish-ochroleucous with keel purple-tipped; fruit 8-10 mm. long, ellipsoid-oblong.---Alkaline banks and meadows, often in somewhat moist situations, or drier hills, mostly at middle elevations. Manitoba and Yukon Territory, southward to Colorado and Nevada. Our records scattered in the western three-fourths of Colorado at 5000-11,000 feet.

51. Astragalus wingatanus Wats., Proc. Am. Acad. 18:192. 1883.
Astragalus wingatensis Jones; Homalobus wingatanus Heller; Astragalus dodgeanus Jones; Astragalus acerbus Sheld.---Plants with caespitose caudex but not pulvinate, not blackening on drying; stems 15-60 cm. tall, leafy; leaflets 5-15, 4-15 mm. long, linear-oblong, oblanceolate or oval, more or less strigose below; calyx 2.5-4 mm. long; corolla 5-8 mm. long, white or tinged with purple; fruit 8-20 mm. long, oblong or elliptic-oblong.---Dry hills and meadows at middle elevations. Colorado and adjacent Utah and New Mexico. Our records from westcentral and southwestern Colorado at 4500-7000 feet.

52. Astragalus toanus Jones, Zoe 3:296. 1893.
Cnemidophacos toanus Rydberg; Astragalus linifolius Osterhout; Ctenophyllum linifolium Osterhout---Plants not pulvinate-caespitose; stems 30-80 cm. long; leaflets 2-5 or wanting especially the terminal one, 10-20 mm. long, linear, glabrate; calyx 6-8 mm. long; corolla 15-18 mm. long, purple but often pale colored especially on drying; fruit 12-17 mm. long, lanc-oblong, ascending or erect at maturity.---Dry adobe hills, at middle elevations. Western Colorado to Utah and Nevada. Our few records from the westcentral part at 4500 feet.

53. Astragalus coltoni Jones, Zoe 2:237. 1891.
Homalobus coltoni Rydb.---Plants with caespitose rootstocks but not at all pulvinate; stems 30-50 cm. long; leaflets 7-15, 5-12 mm. long, linear to linear-oblong, strigose both sides; calyx about 5 mm. long; corolla 13-15 mm. long, rose-purple to dark purple; fruit 20-30 cm. long, oblong, pendulous; stipe about 8-10 mm. long. Colorado specimens may be referred to var. moabensis Jones, Contr. West. Bot. 8:11, 1898, which has pods somewhat shorter than those of the species and with stipes also somewhat shorter.---Dry open hills at middle elevations. Utah, western Colorado, and northwestern New Mexico. Our few records from Mesa and Montrose Counties at 6000-7500 feet.

A plant keying out about here but with whitish to yellowish flowers has recently been collected in southcentral Colorado. It seems to be A. ripleyi Barneby described originally from adjacent New Mexico.

54. **Astragalus rafaelensis** Jones, Rev. Astrag. 146. 1923.
Cnemidophacos rafaelensis Rydberg---Dry clayey or adobe hills at middle elvations. Utah and probably western Colorado. (Known from Dinosaur National Monument Headquarters near Jensen, Utah, close to Moffat County, Colorado.)

12. MICROPHACOS

Slender, leafy-stemmed perennial plants, often sprawling or prostrate with pinnate leaves having several oblong to linear leaflets. Flowers small, in loose, elongate racemes, rose to purple, the corolla strongly arched. Pods coriaceous, sessile, 1-celled, 5-8 mm. long, 2-3.4 mm. wide, cross-reticulate, inversely boat-shaped in cross-section, the ventral (upper) suture prominent. Seeds few.

Key to the Species

1. Leaflets oblong-linear, truncate or retuse at the apex; pods usually about 7 mm. long   -55. **A. gracilis**
1. Leaflets narrowly linear, never retuse at the apex; pods usually about 5 mm. long   -56. **A. parviflorus**

55. **Astragalus gracilis** Nutt., Gen. 2:100. 1818.
Astragalus microlobus A. Gray; Microphacos microlobus Rydb.; Microphacos gracilis Rydb.; Astragalus parviflorus microlobus Jones---Stems 20-50 cm. long; leaflets usually 11 or more, 5-15 mm. long, oblong-linear, truncate or retuse at apex; calyx about 2-3 mm. long; corolla 6-8 mm. long; fruit about 7-9 mm. long.---Plains and low hills at lower to middle elevations. Southeastern Wyoming, adjacent Nebraska, through Colorado and western Oklahoma to northern New Mexico. Our records scattered in the eastern half of Colorado at 3500-9000 feet.

56. **Astragalus parviflorus** (Pursh) Nutt. in A. Gray, Proc. Am. Acad. 6:202. 1864.
as synonym.
Dalea parviflora Pursh; Astragalus gracilis A. Gray, not Nutt.; Phaca parvifolia Nutt.; Microphacos parviflorus Rydb.---Stems 30-60 cm. long; leaflets 5-9, 10-25 mm. long, narrowly linear, not retuse at apex; calyx about 2-2.5 mm. long; corolla 5-6 mm. long; fruit about 5-6 mm. long.---Plains and foothills. Western Dakotas, Kansas, Nebraska, Oklahoma, eastern Colorado and Wyoming. Our few records from extreme eastern Colorado at 3500-4000 feet.

13. PISOPHACA

Slender to fairly robust, leafy-stemmed perennial plants, sometimes sprawling or flexuous, with pinnate leaves and several oblong to elliptic leaflets. Flowers small to medium-sized (5-18 mm. long), in congested or elongate, many-flowered racemes, the calyx tube campanulate, the corolla pale to dark purplish, the banner rather strongly arched. Pods thin-coriaceous to coriaceous, sessile to short-stipitate, 1-celled, 10-20 mm. long, linear, oblong, or lanceolate, rounded or slightly flattened dorsally in cross-section. Seeds several.

Key to the Species

1. Corolla 12-18 mm. long; pods oblong or oblong-ovate
  2. Leaves and pods villous or pilose   -57. **A. puniceus**
  2. Leaves glabrous above, strigose beneath; pods glabrous   -58. **A. hallii**
1. Corolla 5-8 mm. long; pods linear, narrowly oblong, or lanceolate
  3. Pods glabrous, thin textured, 10-15 mm. long   -59. **A. proximus**
  3. Pods strigose, firm textured, 15-20 mm. long   -60. **A. flexuosus**

57. **Astragalus puniceus** Osterhout, Muhlenbergia 1:140. 1906.
Xylophacos puniceus Rydb.; Astragalus gracilentus var. exsertus Jones; Pisophaca punicea Rydb.---Stems decumbent at base, 25-50 cm. long; leaflets 13-25, 5-15 mm. long, oblong to narrowly obovate; calyx 5-7 mm. long; corolla 11-18 mm. long; fruit 16-20 mm. long, oblong and curved, pilose or villous.---Foothills and mesas, at lower to middle elevations. Southeastern Colorado, western Oklahoma, and northeastern New Mexico. Our few records from Las Animas County at 6000 feet.

58. **Astragalus hallii** Gray, Proc. Am. Acad. 6:224. 1864.
Astragalus gracilentus var. hallii Jones; Homalobus hallii Rydb.; Pisophaca hallii Rydb.; A. shearii Rydb.---Stems decumbent at least at base, 30-50 cm. long; leaflets 13-23, 5-20 mm. long, elliptic or oblong; calyx 6-7 mm. long; corolla 12-18 mm. long; fruit 15-25 mm. long, oblong-elliptic to oblanceolate, glabrous.---At middle elevations and sometimes higher, in the mountains of Colorado. Our records from the northcentral, central and southcentral parts at 7000-10,000 feet.

59. **Astragalus proximus** (Rydb.) Wooton & Standley, Contr. U. S. Nat. Herb. 19:366. 1915.
Homalobus proximus Rydb.; Pisophaca proxima Rydb.---Stems erect or nearly so, 20-50 cm. long; leaflets 5-11, 5-20 mm. long, linear or linear-oblong; calyx 2-3 mm. long; corolla 5-6 mm. long; fruit 10-15 mm. long, narrowly oblong, glabrous.---Dry plains and hills at elevations of about 4500-6500 feet. Southwestern Colorado and adjacent New Mexico. Our few records from La Plata and Archuleta Counties at 6000-7000 feet.

60. Astragalus flexuosus Dougl. ex Hook., Fl. Bor. Am. 1:141. 1831.
Phaca flexuosa Hook.; Astragalus fendleri Gray; Astragalus flexuosus var. fendleri Jones; Homalobus fendleri Rydb.; Homalobus flexuosus Rydb.; Pisophaca flexuosa Rydb.; Pisophaca ratonensis Rydb.; Pisophaca elongata Rydb.; Pisophaca sierrae-blancae Rydb.; Astragalus famelicus Sheld.?; Astragalus gracillentus var. fallax Jones?---Stems more or less decumbent at least at base, rarely erect, 20-50 cm. long; leaflets 11-25, 4-18 mm. long, linear to oblong-oblanceolate; calyx 3-4.5 mm. long; corolla 8-10 mm. long, sometimes whitish in color; fruit 15-20 mm. long, linear to narrowly oblong, strigose.---A highly variable species as indicated by the synonymy. The typical form of the species has pods which are linear to narrowly oblong and about 3-4 mm. wide throughout most their length. The var. diehlii (Jones) Barneby differs in having pods which are lanceolate in shape, and 4-5 mm. wide near the base, tapering to a point from near the base. The variety occurs in eastern Utah, western Colorado and northern New Mexico.---Prairies and plains at lower to middle elevations. Manitoba and Alberta southward to Kansas and New Mexico. Our records scattered over the western two-thirds of the state (none from the northwestern part as yet) at 4500-10,000 feet.

14. LONCHOPHACA

Rather tall, rushlike perennial plants from a short woody caudex, the leaves pinnate and the leaflets linear or filiform. Flowers white, ochroleucous, or purplish, 10-20 mm. long, in elongate racemes, the calyx campanulate, the banner not strongly arched. Pods linear, usually long-stipitate, tapering at both ends, 2.5-4 cm. long, thin-coriaceous, 1-celled, without any intrusion of the sutures, dorsally compressed, the cross-section rhombic to elliptic, both sutures rather prominent. Seeds several to many.

Key to the Species

1. Stipe of pod exceeding the calyx tube; flowers white or ochroleucous   -61. A. lonchocarpus
1. Stipe of pod shorter than the calyx tube; flowers ochroleucous or purplish
   2. Corolla ochroleucous, about 20 mm. long   -62. A. osterhoutii
   2. Corolla purplish, about 10-12 mm. long   -63. A. duchesnensis

61. Astragalus lonchocarpus Torr., Pacif. R. R. Rep. 4:80. 1857.
Phaca macrocarpa Gray not DC.; Homalobus macrocarpus Rydb.; Lonchophaca macrocarpa Rydb.; Astragalus macer A. Nels.; Lonchophaca macra Rydb.---Stems 30-80 cm. long; leaflets 3-9, 10-40 mm. long, linear-filiform to linear; calyx 6-10 mm. long; corolla 15-20 mm. long, white to cream-colored; fruit about 30-40 mm. long; stipe longer than the calyx tube.---Dry sandy, clayey, or shaley soil, slopes and meadows, at middle elevations. Colorado and northern New Mexico, westward to Utah and Nevada. Our records from southern and western Colorado at 4000-7500 feet.

A related plant with fruit laterally compressed has recently been collected in Conejos County in southcentral Colorado. It seems to be A. ripleyi Barneby described originally from northern New Mexico.

62. Astragalus osterhoutii Jones, Rev. Astrag. 251. 1923.
Lonchophaca osterhoutii Rydb.---Stems 50-100 cm. long; leaflets 9-13, 15-30 mm. long, linear; calyx 7-8 mm. long; corolla 18-20 mm. long, ochroleucous; fruit about 40 mm. long; stipe shorter than calyx tube.---Shaley slopes at middle elevations. Grand County, Colorado. Our records at about 7000-9500 feet.

63. Astragalus duchesnensis Jones, Contr. West. Bot. 13:9. 1910.
Sandy mesas at middle elevations. Northeastern Utah and northwestern Colorado. (Reported by Jones from Chepeta Wells and White River in Colorado, but no specimens from Colorado have been seen.)

15. GEOPRUMNON

Low, sprawling perennial plants with pinnate leaves and white, ochroleucous, or purplish flowers in dense axillary racemes. Calyx deeply campanulate. Corolla about 15-22 mm. long. Pods ellipsoid or subglobose, plumlike, thick and fleshy, becoming spongy in age, indehiscent or tardily dehiscent, completely 2-celled by the intrusion of the lower (dorsal) suture which is fused with the ventral (upper) suture. Seeds several in each cell.

Key to the Species

1. Pod glabrous, subglobose or broadly ellipsoid
   2. Corolla purple; leaflets oblong to linear   -64. A. crassicarpus
   2. Corolla white or ochroleucous, the keel purple-tipped; leaflets oval or obovate   -65. A. succulentus
1. Pod pubescent, obliquely ovoid and abruptly acuminate   -66. A. plattensis

64. Astragalus crassicarpus Nutt. in Fraser, Cat. 1. 1813.
Astragalus carnosus Pursh; Astragalus caryocarpus Ker; Astragalus succulentus Lindl; Geoprumnon crassicarpum (Nutt.) Rydb.---Stems 10-40 cm. long; leaflets 15-21, 5-18 mm. long, oblong to linear; calyx 7-13 mm. long; corolla 13-20 mm. long, purple; fruit 15-20 mm. long, subglobose, glabrous.---Prairies and plains at lower elevations. Manitoba and Saskatchewan southward to Tennessee, Missouri, New Mexico and Montana. Our records from the eastern half of Colorado at 3500-5500 feet.

65. Astragalus succulentus Rich. in Franklin, Journ. 746. 1823.
Astragalus prunifer Rydb.; Astragalus mexicanus Rydb.; Geoprumnon succulentum (Rich.) Rydb.---Stems 10-40 cm. long; leaflets 13-19, 5-20 mm. long, oval, obovate or elliptic-oblong; calyx 8-13 mm. long; corolla 18-23 mm. long, white to ochroleucous but keel purple-tipped; fruit 15-25 mm. long, subglobose, glabrous.---Prairies and plains, often in dry sandy soil, at lower to middle elevations. Saskatchewan and Alberta southward to Nebraska and Colorado. Our records scattered in the eastern half of Colorado at 3500-7000 feet with a rather doubtful plant from Park County.

66. Astragalus plattensis Nutt. in T. & G., Fl. N. Amer. 1:332. 1838.
Astragalus caryocarpus Torr.; Geoprumnon plattense (Nutt.) Rydb.---Stems 10-30 cm. long; leaflets 15-25, 4-15 mm. long, elliptic, oblong or obovate; calyx 8-10 mm. long; corolla 13-16 mm. long, purplish; fruit 15-20 mm. long, ovoid, pubescent.---Prairies and plains at lower elevations. Minnesota and South Dakota to Alabama and Texas. Our one record from the southeastern corner of the state at 4000 feet.

16. DIPLOCYSTIUM

Low or tall caulescent perennial plants with pinnate leaves and broad leaflets. Flowers on peduncled racemes, the calyx tube campanulate, the corollas purplish in ours, the banner 13-17 mm. long. Pods sessile, papery or parchmentlike in texture, inflated, ovate to lanceolate in ours, 2-celled by the intrusion of the lower (dorsal) suture which meets the upper (ventral) suture. Seeds several.

67. Astragalus lentiginosus Dougl. in Hook., Fl. Bor. Am. 1:151. 1831.
Stems 15-40 cm. long; leaflets 13-21, 8-15 mm. long, obovate, elliptic to oval; calyx 4-9 mm. long; fruit 10-30 mm. long. The following 2 varieties are present in Colorado and are separated as follows:

1. Racemes typically short and dense, 2-6 cm. long in fruit; pods obliquely ovoid, rarely arcuate, only the tip upturned   -67A. A. lentiginosus var. diphysus
1. Racemes typically lax, 2-12 cm. long in fruit; pods lanceolate or narrowly ovoid-acuminate, usually arcuate throughout their length   -67B. A. lentiginosus var. palans

67A. Astragalus lentiginosus diphysus (Gray) Jones, (var.) Proc. Calif. Acad. Sci. II. 5:673. 1895. (for the most part)
Astragalus diphysus Gray; Astragalus macdougali Sheld.; Cystium diphysum Rydb.---Arid plains, dunes, and slopes, at lower elevations. Southwestern Colorado and central New Mexico westward to northern Arizona. Our one record from Montezuma County at 6500 feet.

67B. Astragalus lentiginosus palans (Jones) Jones, (var.) Contr. West. Bot. 8:4. 1898.
Astragalus palans Jones; Astragalus amplexus Payson; Tium palans Rydb.; Tium amplexum Rydb.---Dry slopes along the Colorado River and its tributaries; Colorado to southern Utah and northern Arizona. Our records from western Colorado at 4500-6500 feet.

17. EUASTRAGALUS

Perennial plants, in ours, leafy-stemmed or subscapose, with pinnate leaves having several to many leaflets. Flowers subcapitate or racemose, often on long peduncles, the calyx cylindrical to campanulate. Corolla ochroleucous, yellowish, or purplish, mostly 10-20 mm. long, the banner not strongly arched. Pods sessile or nearly so, coriaceous to woody or rarely chartaceous, mostly ovoid to oblong, rarely sub-spherical in outline, with or without a prominent beak, mostly erect or ascending, completely 2-celled by the intrusion of the dorsal suture to meet the ventral suture. Seeds several to many.

Key to the Species

1. Leaves villous or silky, often densely so; plants subacaulescent, the internodes few and short
    2. Leaves mostly 10-20 cm. long with 10 or more pairs of leaflets; pods longer than broad, prominently beaked; corolla purplish
        3. Pods densely villous or shaggy; leaflets progressively smaller from base to apex of leaf; plants of the western slope   -68. A. bigelovii
        3. Pods glabrous; leaflets not reduced upward; plants of the plains east of the continental divide   -69. A. mollissimus
    2. Leaves mostly 6 cm. long or less with less than 8 pairs of leaflets; pods nearly spherical, merely mucronate at the apex; corolla probably white   -70. A. anisus
1. Leaves strigose or glabrate; plants with well developed leafy stems
    4. Flowers ochroleucous or yellowish
        5. Pods about 10 mm. long and 3-4 mm. in diameter; flowers yellowish, 15-18 mm. long, ascending   -71. A. sulphurescens
        5. Pods 15-20 mm. long and 5-6 mm. in diameter; flowers ochroleucous or pale greenish, 12-15 mm. long, spreading or reflexed   -72. A. canadensis
    4. Flowers purplish except for occasional albino forms
        6. Inflorescence subcapitate; calyx long-pilose, the hairs mostly black; pods 5-6 mm. wide, dorsally flattened, long-pilose; flowers fragrant; stems slender, not stiff   -73. A. agrestis
        6. Inflorescence racemose, mostly 3-5 cm. long; calyx strigose with appressed hairs which are mostly white; pods 3-4 mm. wide, strigose; flowers not fragrant; stems stouter and stiff   -74. A. striatus

68. **Astragalus bigelovii** Gray, Pl. Wright. 2:42. 1853.
   Plants subacaulescent with short stems; leaflets 21-35, 5-10 mm. long, oval, ovate or broadly-elliptic, silky-villous; calyx 11-14 mm. long; corolla about 20-21 mm. long, rose-purple; fruit 11-20 mm. long, oblong and strongly arcuate, villous. Colorado specimens are referable to var. thompsonae (Wats.) Jones, Contr. West. Bot. 8:23. 1898, which differs from the species in having white instead of yellowish or rusty pubescence and smaller leaflets.---Dry and usually loose sandy soil at lower elevations. The variety from western Colorado to Utah and Arizona, the species from western Texas to Arizona and northern Mexico. Our records from extreme western Colorado at 4500-8000 feet.

69. **Astragalus mollissimus** Torr., Ann. Lyc. N. Y. 2:178. 1827.
   Phaca villosa James, not Nutt.; Astragalus simulans Cockerell; not Phaca mollissima Nutt.---Plants subacaulescent with stems less than 10 cm. long; leaflets 21-31, 10-25 mm. long, oval, obovate to broadly elliptic, silky-villous; calyx 10-15 mm. long; corolla 17-21 mm. long, rose-purple; fruit 13-20 mm. long, narrowly oblong, glabrous.---Dry plains and foothills at lower elevations. Southwestern South Dakota, western Nebraska and southeastern Wyoming, southward to northwestern Texas and eastern New Mexico. Our records from the eastern half of Colorado at 3500-5500 feet.

70. **Astragalus anisus** Jones, Zoe 4:34. 1893.
   Plants subacaulescent with decumbent stems 5-10 cm. long; leaflets 9-15, 4-9 mm. long, obovate to oval, silky-canescent; calyx 8-11 mm. long; corolla 15-20 mm. long, apparently whitish; fruit 15-20 mm. long, nearly spherical, strigose.---Dry sagebrush slopes at lower or middle elevations. Known only from two collections: the type from near Pueblo, Pueblo County, Colorado (possibly an error in locality data), and near Gunnison, Gunnison County, Colorado, at 4500-7700 feet.

71. **Astragalus sulphurescens** Rydb., Bull. Torr. Club 28:36. 1901.
   Astragalus crandallii Gandoger---Stems 20-40 cm. long; leaflets 13-19, 12-30 mm. long, oblong to broadly elliptic, glabrous above, sparingly strigose below; calyx 6-10 mm. long; corolla 15-18 mm. long, light yellow; fruit 7-10 mm. long, oblong, strigose the hairs usually black.---Open or wooded slopes at middle elevations. Foothills and mountains of Colorado east of the Continental Divide. Our records from northcentral and central Colorado at 5000-10,000 feet.

72. **Astragalus canadensis** L., Sp. Pl. 757. 1753.
   Phaca canadensis MacMill. as synonym; Astragalus canadensis forma monticola Gand.; Astragalus carolinianus of many American authors, not L.---Stems 30-100 cm. tall; leaflets 15-31, 10-35 mm. long, oval to oblong, glabrous above and strigose below; calyx 6-9 mm. long, possibly longer; corolla 12-15 mm. long, pale greenish to ochroleucous; fruit 10-18 mm. long, oblong, glabrous or glabrate. The typical form of the species consists of plants 4-10 dm. high having pods 15-20 mm. long, glabrous, and with the dorsal suture mostly rounded. Plants with stems 3.5-5 dm. high having pods about 12 mm. long, somewhat strigose and with the dorsal suture sulcate may be referred to var. mortoni (Nutt.) Wats. (Astragalus mortoni Nutt., Phaca mortoni Piper, Astragalus pachystachys Rydb., Astragalus spicatus Nutt., not Pall.).---Meadows and stream banks at lower to middle elevations. Quebec to British Columbia and southward to Virginia, northern Texas, and Utah. The variety from Montana and Washington to northern Colorado, Nevada, and California. Our records in the western half of Colorado (except the northwestern part) at 5000-8000 feet.

73. **Astragalus agrestis** Dougl. ex G. Don, Gen. Hist. 2:258. 1832.
   Astragalus hypoglottis Hook.; Astragalus goniatus Nutt.; Astragalus virgultulus Sheld.; Astragalus hypoglottis var. bracteatus Osterh.; Astragalus agrestis var. bracteatus Jones---Stems 10-30 cm. long; leaflets 11-21, 5-20 mm. long, linear, oblong or elliptic, sparingly pilose at least below; calyx 7-12 mm. long; corolla 12-20 mm. long, purple or rarely whitish; fruit 7-10 mm. long, ovoid, long-pilose. Occasional plants of this species may have pale or white flowers (A. virgultulus Sheld.) but these probably represent mere albino forms which are to be expected in any population having colored flowers.---Meadows, prairies, and mountains, usually in moist, shady situations, at middle elevations. Saskatchewan, Alberta, and Washington southward to the Dakotas and New Mexico. Scattered over Colorado, except in the extreme eastern part, at 4500-10,000 feet.

74. **Astragalus striatus** Nutt. ex T. & G., Fl. N. Amer. 1:330. 1838.
   Astragalus laxmanni Nutt. not Jacq.; Astragalus adsurgens Hook. not Pall.; Astragalus adsurgens robustior Hook.; Astragalus hypoglottis var. robustus Hook.; Astragalus nitidus Dougl. ex Hook. as synonym; Phaca adsurgens Piper; Astragalus adsurgens var. pauperculus Blank.; Astragalus nitidus var. robustior Jones---Stems 10-35 cm. long; leaflets 13-25, 5-20 mm. long, oblong or elliptic, silky-strigose to glabrous, especially above; calyx 6-9 mm. long; corolla 14-18 mm. long, rose-purple or rarely whitish; fruit 7-10 mm. long; oblong, strigose. Occasional white-flowered albino forms (what Jones called A. nitidus var. robustior) occur in normal purple-flowered populations.---Sandy or gravelly soils, open prairies and hills at lower to middle elevations. Manitoba to Washington and southward to Nebraska and New Mexico. Our records mostly from the eastern half of the state at 5000-8000 feet, with one plant from Mineral County at 11,000 feet.

18. HAMOSA

Low green and leafy-stemmed annual or caespitose silvery and subscapose perennial plants, in ours. Leaves pinnate with several leaflets. Flowers in short axillary racemes, the calyx campanulate, the corolla purplish or sometimes white, the banner 5-15 mm. long, not strongly arched. Pods firm, thin-coriaceous, not inflated, sessile, linear to oblong or lanceolate, curved upward from near the base or from the middle, 15-35 mm. long in ours, laterally compressed with the dorsal (lower) suture sulcate and intruded within to meet the ventral (upper) suture, the pods thus completely 2-celled, obcordate or sagittate in cross-section. Seeds several.

Key to the Species

1. Plants perennial and caespitose, subacaulescent; foliage and pods silvery-canescent; flowers 10-15 mm. long; pods about 15 mm. long, 3-4 mm. wide, arcuate near the middle -75. A. calycosus
1. Plants annual, definitely caulescent; foliage and pods glabrate or strigulose but never silvery-canescent; flowers about 5-6 mm. long; pods about 20-35 mm. long, 3 mm. wide, arcuate near the base -76. A. nuttallianus

75. Astragalus calycosus Torr. in S. Wats., Bot. King's Expl. 66. 1871.
Astragalus scaposus Gray; Astragalus candicans Greene, not Pall.; Astragalus calycosus var. scaposus Jones; Astragalus brevicaulis A. Nels.; Hamosa calycosa Rydb.; Hamosa scaposa Rydb.---Perennial subacaulescent plants; leaflets 3-9, 2-10 mm. long, oblong, oblanceolate to obovate, silvery-canescent; calyx 6-7 mm. long; corolla 10-15 mm. long; fruit about 15 mm. long, silvery-canescent.---Dry plains and foothills at middle elevations (about 6000-7000 feet). Idaho to southwestern Wyoming, western Colorado and northwestern New Mexico, westward to Nevada, Arizona and eastern California. Our one record from southwestern Colorado at 6200 feet.

76. Astragalus nuttallianus DC., Prodr. 2:289. 1825.
Astragalus micranthus Nutt., not Desv.; Astragalus nuttallianus var. enneajugus Jones; Astragalus nuttallianus var. quadrilateralis Jones; Hamosa emoryana Rydb. and Astragalus emoryanus Cory; Hamosa imperfecta Rydb. and Astragalus imperfectus Cov.---Annual, caulescent plants; stems 5-25 cm. long; leaflets 7-19, 3-15 mm. long, lanc-oblong to elliptic, strigose or glabrous above; calyx 3-6 mm. long; corolla 5-8 mm. long; fruit 17-35 mm. long, glabrous or sparingly strigose. Plants having strigose pods, instead of glabrous as in the species, have been designated as var. trichocarpus T. & G., Fl. N. Am. 1:334. 1838 (Hamosa austrina Small). An insignificant variety.---Dry plains, hills, and mesas, at lower elevations. Arkansas and Texas to western Colorado, California, and northern Mexico. Our records from Mesa, Montrose, and Montezuma Counties at 4500-5500 feet.

Family 58. Geraniaceae GERANIUM FAMILY

Herbaceous plants; leaves normally opposite but usually crowded at base, the blades palmately or pinnately lobed or divided; stipules present; flowers perfect, regular, cymose or subumbellate; calyx of 5 imbricated sepals; petals separate, 5, deciduous, white to rose-purple; stamens 10, sometimes the 5 alternating ones reduced to staminodia, filaments distinct or more or less united toward the base; ovary superior, of 5 carpels united by thin styles to a central column, from which they break at maturity, each carpel with 2 ovules but only 1-seeded.

1. Leaves pinnately divided; styles in fruit pubescent inside and becoming spirally coiled; stamens with anthers 5 -1. Erodium
1. Leaves palmately lobed or divided; styles glabrous inside, recoiled in fruit but never spirally so; stamens usually 10 -2. Geranium

1. Erodium L'Her. HERONBILL; STORK'S BILL

Annual or sometimes perennial-appearing herbaceous plants, acaulescent at first, later elongating; leaves pinnately divided, often long-petioled; sepals usually awn-tipped; petals of later flowers often reduced in size; stamens 5, alternating with 5 staminodia; style column elongated, the styles pubescent within and when free becoming spirally coiled, carpels narrow, sharp at base.

1. Erodium cicutarium (L.) L'Her. ex Ait., Hort. Kew. 2:414. 1789.
Leaf blades 3-12 cm. long, pinnately divided, with the segments oblong to ovate and pinnatifid or incised; peduncles and pedicels more or less hirsute; sepals 3-7 mm. long, with short tips bearing 1 or 2 bristles; petals somewhat longer than the sepals, pink or rose-colored; style column 2.5-5 cm. long, the carpel body 4-5 mm. long.---Plains, mesas and slopes often in fields and waste places. Nova Scotia to British Columbia, south to Central America; Europe. Our records scattered over Colorado, none as yet from the extreme eastern part, at 4500-7500 feet.

2. Geranium L. CRANESBILL

Annual or perennial herbaceous plants; leaves palmately lobed, cleft or parted; sepals usually awn-tipped; petals imbricate, alternating with 5 glands; stamens 10 (rarely 5); styles glabrous inside, recoiled at maturity but not twisted. This is a difficult genus and this treatment will be better in identifying a living plant. The color of the fresh flowers and the growth habit (caespitose or with a single stem) are important.

1. Plants annual or biennial; petals 2-7 mm. long
    2. Sepals conspicuously awned or subulate-tipped, the tips 0.7-3 mm. long; seeds reticulated
        3. Beak and branches of style column about 2 mm. long; petals pink or whitish; inflorescence compact; fruiting pedicels little if any longer than the calyx -1. G. carolinianum
        3. Beak and branches of style column 4 mm. long or more; petals rose-purple; inflorescence loose; fruiting pedicels much longer than the calyx -1A. G. carolinianum longipes
    2. Sepals awnless or with minute callous tips; seeds smooth or nearly so -2. G. pusillum

1. Plants perennial; petals 10 mm. long or more
    4. Plants more or less caespitose, the several stems together from a branched caudex, the stems erect or often later ascending or decumbent; pedicels with or without glandular hairs
        5. Pedicels with glandular hairs
            6. Petioles of the basal and lower stem leaves glandular-pubescent   -3. G. parryi
            6. Petioles of basal and lower leaves glabrous to retrorsely pubescent but not at all glandular hairy   -4. G. fremontii
        5. Pedicels without glandular hairs
            7. Petals oblanceolate to narrowly obovate, pale pink or lavender; branches of style column 3-4 mm. long; plants in extreme western Colorado (if present)   -5. G. marginale
            7. Petals obovate or obcordate usually broadly so, lavender to deep rose; branches of style column 5-8 mm. long; plants rather widely distributed in Colorado
                8. Petals pilose inside for about 1/2 their length; mature style column about 2.5-3 cm. long   -6. G. caespitosum
                8. Petals pilose inside for about 1/3 their length; mature style column 2-2.5 cm. long   -4A. G. fremontii cowenii
    4. Plants usually single stemmed, this erect from a simple caudex; pedicels always with glandular hairs
        9. Petals pilose on the inner surfaces for about 1/2 their length, white or purple-tinged only; pedicels always in pairs, the glandular hairs purple-tipped   -7. G. richardsonii
        9. Petals pilose inside for about 1/4 their length or less, rose-purple; pedicels sometimes in pairs but often in 3's or 4's, the glandular hairs tawny-tipped   -8. G. nervosum

1. Geranium carolinianum L., Sp. Pl. 682. 1753.
   Annual or biennial plants; stems 17-35 cm. tall, erect or more commonly branched at base, more or less glandular-pubescent; leaf blades 2.5-6 cm. wide, orbicular to reniform in shape, deeply cleft into 5-9 segments, these again toothed or cleft; inflorescence a compact cluster; sepals about 6-8 mm. long, ovate, awn-tipped, this tip 1-2 mm. long; petals about as long as the sepals, pink to whitish; stamens 10-11; style column 12-18 mm. long, the beak and branches together about 2 mm. long; carpel body 3-3.5 mm. long, with erect rather stiff hairs; seeds reticulated.---Meadows and waste places. Canada to Mexico. The species should be in Colorado but no specimens have been seen by the writer.
1A. Geranium carolinianum longipes Wats., (var.) Bot. King's Expl. 50. 1871.
   G. bicknellii Britt.---Inflorescence loose; petals rose-purple; beak and branches of style over 4 mm. long. Fernald (Rhodora 37:295-301. 1935.)maintained G. bicknellii Britt. as a species and made our western plant G. bicknellii var. longipes (Wats.) Fernald.---Our few records in the western half of Colorado, in Routt, Boulder, Jefferson and Archuleta Counties, at 5500-7000 feet.
2. Geranium pusillum Burm. f., Spec. Bot. Geran. 27. 1759.
   Annual or perhaps biennial plants; stems 10-60 cm. long, branching at base, puberulent; leaves reniform to orbicular, divided into 5-7 divisions, these toothed or lobed at apex; sepals 2.5-4 mm. long, awnless; petals about as long or somewhat longer than the sepals, pale-purple or violet; stamens with anthers 5; style column 6-9 mm. long, the beak and branches about 1 mm. long; carpel bodies about 2 mm. long, finely pubescent; seeds smooth.---Fields and waste places. Ontario to British Columbia, south to North Carolina, Nebraska and Utah. Our few records from central and northcentral Colorado at 4500-7000 feet.
3. Geranium parryi (Engelm.) Heller, Cat. N. Am. Pl. ed. 2. 7. 1900.
   G. pattersonii Rydb.---Perennial plants with branched caudices; stems 10-45 cm. tall, tufted, erect, glandular throughout, lower part and petioles of basal leaves glandular-pilose and glandular-villous but with some long non-glandular hairs; petioles long; leaves 2-7 cm. wide, strigose above and below, deeply 3- to 5-parted, the divisions incised and lobed; pedicels usually in 2's, sometimes in 3's, copiously glandular-hairy; sepals 6-10 mm. long, the awn tip 0.5-1.5 mm. long; petals 12-18 mm. long, pale rose to deep rose-purple, obovate, emarginate, pilose inside for about 1/4 their length; mature style column 1.5-3 cm. long, the branches 4-7 mm. long; carpel bodies about 5 mm. long.---Foothills and mountains, often in canyons or rocky ground. Wyoming, Utah, Colorado and Arizona. Our records scattered in the western half of Colorado, except in the northwestern part, at 5000-8000 feet.
4. Geranium fremontii Torr. ex Gray, Pl. Fendl. 26. 1849.
   G. caespitosum in Rydberg's Manual---Perennial plants with branched caudices; stems 20-50 cm. tall, tufted, erect at first, lower part and petioles of basal leaves strigose to glabrate, never glandular-hairy; petioles long; leaves 3-8 cm. wide, 5- to 7-parted, the divisions rather oblong, incised-lobed and toothed, sparsely appressed pubescent above and below; pedicels in 2's, rather densely glandular-hairy; sepals 7-12 mm. long, the awn tip about 1 mm. long; petals 1-1.5 cm. long, pale purple or rarely white, dark veined, obovate, pilose inside for about 1/4 their length; style column about 2.5-3 cm. long when mature, the branches 4-5 mm. long; carpel bodies about 4-5 mm. long.---Plains, foothills and mountains, often in canyons. Wyoming, Colorado, New Mexico and Arizona, possibly further west. Our records scattered in the western two-thirds of Colorado, except the extreme western part, at 5000-9500 feet.
4A. Geranium fremontii cowenii (Rydb.) comb. nov. (var.)
   G. cowenii Rydb.---Perennial plants with branched woody caudices; stems 20-40 cm. tall, tufted, erect, hairy but not glandular-hairy; petioles long; leaves 3-6 cm wide, roundish, strigillose both surfaces, deeply 5-parted, the divisions again deeply 3- to 5-parted; pedicels non-glandular (sometimes slightly so in bud stage); sepals 6-11 mm. long, the awn on apex 1-2 mm. long; petals obcordate to broadly obcordate, lavender to pale rose, pilose inside for about 1/3 their length; mature style column 2-2.5 cm. long, the style branches 5-7 mm. long; carpel bodies about 5 mm. long.---Canyons and slopes. Colorado and possibly Wyoming. Our few records from central, northcentral, and northwestern Colorado at 6000-7000 feet.

5. **Geranium marginale** Rydb. ex Hanks & Small, N. Am. Fl. 25:16. 1907.

A species of Utah that has been reported from Colorado. The writer has seen no specimens from the state and it should be looked for in the western part.

6. **Geranium caespitosum** James apud Gray, Pl. Fendl. 25. 1849.

G. atropurpureum Heller---Perennial plants with caudices woody and usually branched; stems 10-90 cm. tall, tufted, erect or later ascending or procumbent, frequently rooting at nodes, pubescent but not at all glandular-hairy; petioles long; leaf blades 2-4 cm. wide, 5-parted, divisions rhombic and again 3-lobed, appressed-pubescent both sides; peduncles and pedicels pubescent but not at all glandular-hairy; sepals 8-12 mm. long, awn-tipped, this tip 1-2 mm. long; petals 12-18 mm. long, usually deep rose-purple, sometimes paler, pilose inside for about 1/2 their length; mature style column 2-3 cm. long, the branches 5-8 mm. long; carpel bodies 4-5 mm. long.---Slopes and canyons, often in partial shade. Colorado to Utah, south to Texas and Mexico. Our records scattered in the western half of Colorado, except the northwestern part, at 5000-9500 feet.

7. **Geranium richardsonii** Fisch. & Trautv., Ind. Sem. Hort. Petrop. 4:37. 1837.

G. gracilentum Greene---Perennial plants, the caudices usually simple; stems 30-90 cm. tall, solitary or few, erect, glabrous to pubescent, the upper part often glandular; petioles long; leaf blades 3-15 cm. broad, deeply 3- to 5-parted, the rhombic segments incised or lobed, sparsely strigose, at least on veins; peduncles and pedicels glandular-villous with mostly purple-tipped hairs; sepals 6-12 mm. long, awn-tipped, this awn 1.5-2.5 mm. long; petals 10-18 mm. long, white or pink usually with purplish veins, occasionally drying rose-colored, pilose inside for about 1/2 their length; style column when mature 2-2.5 cm. long, glandular, the branches 3-5 mm. long; carpel bodies 2.5-4.5 mm. long---Mountains and foothills, often in partial shade or in rather moist ground. Saskatchewan to British Columbia, south to South Dakota, New Mexico and California. Our records well scattered over the western half of Colorado at 5500-12,000 feet.

8. **Geranium nervosum** Rydb., Bull. Torr. Club 28:34. 1901.

G. strigosum Rydb.---Perennial plants with simple caudices; stems 30-75 cm. tall, solitary, erect, lower stem and petioles of basal leaves with non-glandular hairs; petioles long; leaves 3-10 cm. wide, deeply 5- to 7-parted, the divisions rhombic to ovate, lobed and incised again, strigillose to pubescent both surfaces; pedicels usually paired but frequently in 3's or 4's, densely glandular-pubescent; sepals 7-10 mm. long, the awn tip 1-2 mm. long; petals 1.5-2 cm. long, rose-colored, paler at base with darker veins; mature style column 2.5-3.5 cm. long, the branches 4-5 mm. long; carpel bodies 4-6 mm. long.---Mountains, often in partial shade or moist soil. Alberta to British Columbia, south to South Dakota, Colorado and California. Our records scattered over the western half of Colorado at 6500-9500 feet.

Family 59. Oxalidaceae WOODSORREL FAMILY

Herbaceous plants, annual or perennial, caulescent or acaulescent, often with slender or bulblike rootstocks; leaves alternate or basal, palmately 3-foliate; flowers perfect, regular, in umbellike or dichotomous cymes; sepals 5; petals 5, separate or united only at very base; stamens 10, united at base by their filaments, in 2 series alternating longer or shorter; ovary superior, 5-lobed and 5-celled, styles separate or united; fruit a loculicidal capsule.

1. Oxalis L. WOODSORREL

Characters of the Family

1. Plant acaulescent with a bulblike rootstock; petals violet — -1. O. violacea
1. Plant caulescent, rootstocks if present not at all bulblike; petals yellow
    2. Flowers umbellate or solitary; fruiting pedicels usually horizontal or reflexed (although the capsule is erect); plants rarely with rootstocks or stolons; stipules oblong, rather firm and definite
        — -2. O. stricta
    2. Flowers cymose on well developed plants; fruiting pedicels spreading or ascending; plants producing long slender horizontal rootstocks; stipules nearly obsolete
        3. Upper surface of leaflets glabrous; stems glabrate to pubescent the hairs appressed or ascending
            — -3. O. europaea
        3. Upper surface of leaflets with scattered hairs; stems villous with spreading hairs
            — -3A. O. europaea bushii

1. **Oxalis violacea** L., Sp. Pl. 434. 1753.

Inoxalis violacea (L.) Small---Acaulescent plants from scaly bulbs or bulblike rootstocks, the bulb scales more or less 3-ribbed; leaves 1/2 as long as the scapes or more; leaflets 1-2 cm. wide and not as long as wide, broadly obcordate, glabrous; scapes mostly 10-20 cm. long, glabrous; cymes 3- to 10-flowered; sepals 4-6 mm. long; petals 14-20 mm. long.---Rich ground. Maine to Florida and extending westward into eastern or southern Colorado according to report. Our two records from southcentral and southwestern Colorado at about 7000-9000 feet. Our plants are not typical when compared with those from the eastern United States.

2. **Oxalis stricta** L., Sp. Pl. 435. 1753.

Xanthoxalis stricta (L.) Small---Annual or perannial plants from slender taproots or rarely from elongated rootstocks; stems often branching at base, erect or ascending, occasionally decumbent; stipules oblong, rather firm; leaflets 10-18 mm. wide, obcordate or obovate, glabrous above, glabrous or somewhat strigose below; umbels 2- to 3-flowered rarely 1- or 4-flowered, the pedicels strigose, more or less deflexed in fruit; capsules puberulent with some long viscid hairs intermixed.---Woods, roadsides and cultivated ground. Present throughout the most of North America. Our records scattered in the western two-thirds of Colorado at 4500-8500 feet.

3. **Oxalis europaea** Jord. in Schultz, Arch. Fl. Franc. et Allem. 309. 1854.
Perennial plants with long-creeping, slender rootstocks; stems 5-30 cm. tall, usually erect or nearly so, simple or branched, glabrate or with ascending hairs; leaflets 10-30 mm. broad, obcordate or obovate, glabrous above and glabrate beneath; flowers usually cymose or vigorous plants, the pedicels strigose, ascending or spreading; capsule glabrate or with scattered spreading viscid hairs.---Roadsides, moist places and open woods. Quebec to South Dakota, south to Georgia and Colorado. Our few records from northcentral Colorado at 5000-6000 feet.

3A. **Oxalis europaea bushii** (Small) Widgaud, (var.) Rhodora 27:135. 1935.
Xanthoxalis bushii Small; X. coloradensis Rydb.---Stems villous with spreading hairs; leaflets scattered-hairy above.---Nova Scotia to South Dakota, south to Georgia and Colorado. Our few records from central and northcentral Colorado at 4500-8000 feet.

Family 60. Linaceae FLAX FAMILY

Plants herbaceous, glabrous or puberulent, annual or perennial; leaves alternate, simple, entire, sessile; flowers perfect, regular, in racemes or cymose-paniculate; sepals 5, imbricated, persistent; petals 5, usually blue or yellow, separate, very fugaceous; stamens 5, alternate to the petals, sometimes alternating with 5 rudiments, the filaments united at their bases; ovary superior, of 5 united carpels, styles distinct or united below; capsule 10-celled, each of the 5 carpels with a false septum, each of the 10 cells with 1 seed.

1. Linum L. FLAX

Characters of the Family

This is a puzzling genus and is badly in need of thorough revision. The yellow flowered "species" are based mainly on differences in the length of petals. The petals are so fugaceous that mature ones are seldom preserved on specimens and they undoubtedly shrink in drying. Most of our forms seem to group around the L. rigidum concept and perhaps should be considered as varieties only in this state, if actually different from it.

1. Petals blue (or rarely white); stigmas elongate or at least longer than wide; sepals glandless (may be toothed)
    2. Annual plants, if present outside cultivation then around fields and in waste places; inner sepals ciliate or toothed on margins; stigmas much elongated -1. **L. usitatissimum**
    2. Perennial plants, truly native; inner sepals entire; stigmas only slightly longer than wide -2. **L. lewisii**
1. Petals yellow or orange-yellow; stigmas capitate; inner sepals (at least) with marginal glands
    3. Styles distinct; sepals not exceeding the capsule -3. **L. kingii**
    3. Styles united at least at base; sepals usually longer than capsule
        4. Petals 11-17 mm. long
            5. Plant glabrous or nearly so -4. **L. rigidum**
            5. Plant definitely puberulent on stems and pedicels -5. **L. puberulum**
        4. Petals either longer or shorter than 11-17 mm. long
            6. Petals 17-20 mm. long; sepals 9-15 mm. long -6. **L. berlandieri**
            6. Petals 5-9 mm. long; sepals 6-8 mm. long
                7. Outer sepals with lateral ribs faint near base; branches with angles blunt, glabrous or nearly so -7. **L. australe**
                7. Outer sepals with lateral ribs prominent throughout; branches with angles sharp (almost winglike), nearly glabrous to puberulent -8. **L. compactum**

1. **Linum usitatissimum** L., Sp. Pl. 277. 1753.
Annual plants; stems 20-80 cm. tall, often tufted, simple or branching above, glabrous; leaves 1-3 cm. long, ascending, linear to narrowly linear-lanceolate, 3-nerved, acute or acuminate; sepals about 7-9 mm. long, acuminate, the inner ciliate or toothed, 3-nerved at base; petals 10-15 mm. long, blue; styles distinct or nearly so, the stigmas elongated; capsule 6-8 mm. long, equalling or somewhat longer than the sepals.---Waste places and old fields, sometimes escaped from cultivation. Throughout cultivated North America. Our few records from northcentral Colorado at about 5000 feet but the specimens may be of cultivated plants.

2. **Linum lewisii** Pursh, Fl. Am. Sept. 210. 1814.
L. pratense (Norton) Small---Perennial plants; stems 10-70 cm. tall, often branching at base, glabrous, striate in age at least; leaves about 5-20 mm. long, ascending, crowded, linear, often somewhat involute; sepals 2-5 mm. long, entire, the inner shorter, acute, short-acuminate to mucronate; petals 1-2 cm. long, blue; styles distinct or nearly so, stigmas somewhat longer than wide but not elongated; capsules 5-8 mm. long, about twice as long as the sepals.---Plains and hills, often locally abundant. Saskatchewan to Alaska, south to New Mexico and California. Our records scattered over Colorado, except in the northeastern and eastcentral parts, at 4500-9500 feet.

3. **Linum kingii** Wats. in King, Geol. Expl. 40th Par. 5:49. 1871.
Cathartolinum kingii (S. Wats.) Small---Perennial plants from almost woody bases; stems and branches 5-30 cm. high, somewhat tufted, glabrous and more or less glaucous; leaves about 1-2 cm. long, numerous, thick, linear-oblanceolate to linear; flowers in fastigiate, corymblike cymes; outer sepals about 4 mm. long at maturity, ovate, sparingly glandular-toothed, inner sepals shorter, broader and more abruptly pointed; petals yellow; capsules about 4 mm. long, as long or slightly longer than the sepals; styles distinct.---In the mountains. Wyoming and Utah. Our one record from western Colorado at 8500 feet.

4. Linum rigidum Pursh, Fl. Am. Sept. 210. 1814.
   Cathartolinum rigidum (Pursh) Small---Annual or perhaps perennial plants; stems 10-50 cm. tall, simple or branched, branches angled, glabrous or nearly so; leaves 6-30 mm. long, linear to narrowly linear-lanceolate, readily deciduous after anthesis, at least the upper glandular-serrate; flowers in rather open cymules; outer sepals about 6-8 mm. long, lanceolate, acute to short awn-pointed, glandular-toothed, inner sepals like outer but shorter; petals about twice as long as the sepals, yellow; styles mostly united; capsule about 4-5 mm. long, ovoid, the false septa somewhat thickened at base; styles united at base.---Plains and hills. Manitoba to Alberta, south to Texas and Colorado. Our records scattered in the western half of the state and in southwestern Colorado at 3500-7000 feet.

5. Linum puberulum (Engelm.) Heller, Plant World 1:22. 1897.
   Cathartolinum puberulum (Engelm.) Small---Like the preceding and closely related to it. The plant is densely puberulent and not glabrous.---Dry plains and hills. Colorado, Utah, New Mexico and Arizona. Our records, mostly doubtful, are scattered over Colorado at 4000-8000 feet.

6. Linum berlandieri Hook., Curtis's Bot. Mag. 63 Pl. 3480. 1836.
   Cathartolinum berlandieri (Hook.) Small; Linum arkansanum Osterhout; L. rigidum var. berlandieri (Hook.) T. & G.---Closely related to L. rigidum. Petals are 17-20 mm. long; sepals 9-15 mm. long.---Plains. Wyoming to Texas and Colorado. Our records from eastcentral and southeastern Colorado at 4000-4500 feet.

7. Linum australe Heller, Bull. Torr. Club 25:627. 1898.
   Cathartolinum australe (Heller) Small---Closely related to L. rigidum. The petals are 5-9 mm. long.---Plains. Colorado, south to Mexico. Our records, many doubtful, scattered over Colorado, except in the extreme western part, at 4500-9000 feet.

8. Linum compactum A. Nels., Bull. Torr. Club 31:241. 1904.
   Cathartolinum compactum (A. Nels.) Small---Closely related to L. rigidum and doubtfully distinct. Petals are 5-9 mm. long; branches are often puberulent.---Plains. North Dakota and Montana, south to Kansas and Wyoming. Our few and mostly doubtful records from northeastern Colorado at 3500-4500 feet.

Family 61. Balsaminaceae JEWEL-WEED FAMILY

Herbaceous caulescent plants; nodes of stems more or less swollen, the stem translucent when fresh; leaves simple, alternate, petioled; flowers perfect, irregular, axillary and somewhat clustered; sepals 3, the 2 lateral small, the posterior one petaloid, spurred and saccate; petals 5, or by union of 2 sets only 3; stamens 5, alternate to petals, the anthers more or less connivent around the stigma; ovary superior, of 5 carpels, 5-celled; styles none, stigmas 5-toothed or lobed; fruit a capsule, elastically dehiscent into 5 spirally-coiled valves, expelling the seeds.

1. Impatiens L. JEWELWEED; TOUCH ME NOT

Characters of the Family

1. Impatiens capensis Meerb., Afb. Zedz. Gewass. t. 10. 1775.
   I. biflora Walt.---Stems 60-150 cm. tall, glabrous, often tinged with red, branched; leaves 2-10 cm. long, ovate or oval, green to purplish, lighter below, rather coarsely toothed; inflorescence 2- to 6-flowered, racemose; pedicels with narrow, small bracts above the middle; saccate sepal 15-18 mm. long, about 2/3 as broad, orange to pinkish, usually darker spotted, the spur recurved; petals 6-7 mm. long and wider than long, colored like the saccate sepal; fruit about 2 cm. long.---Moist ground. Newfoundland to Alberta, south to Florida, Alabama and Idaho. Our one record from northcentral Colorado at 5400 feet.

Family 62. Limnanthaceae FALSE MERMAID FAMILY

Annual herbaceous plants of wet places; leaves alternate, pinnately compound; stipules not present; flowers regular, perfect, solitary, axillary, peduncled; hypanthium small and saucer-shaped; sepals 3; petals 3, distinct; stamens 6, distinct, the ones opposite the petals shorter; ovary superior, 2- to 3-celled, deeply 2- to 3-lobed, these nearly distinct but styles united into 1 below; fruit of 2-3 semi-drupaceous nutlets, each 1-seeded.

1. Floerkea Willd. FALSE MERMAID

Characters of the Family

1. Floerkea proserpinacoides Willd., Neue Schr. Ges. Nat. Freunde Berlin 3:449. 1801.
   F. occidentalis Rydb.---Slender glabrous plants; stems 4-30 cm. tall; leaves 1-6 cm. long, leaflets 5 or 3, these 5-25 mm. long, oblong to nearly linear, sometimes 2-cleft; peduncles 1-3 cm. long; sepals 2-3 mm. long or even longer in fruit; petals about 1/2 as long as the sepals, white; fruit 2-4 mm. long, with tubercles.---Moist or wet places. Quebec to Washington, south to Tennessee, Missouri and California. Our few records from western Colorado, (Routt and Gunnison Counties) at 6500-11,500 feet.

Family 63. Zygophyllaceae CALTROP FAMILY

Perennial or annual herbaceous or shrubby plants; leaves opposite, compound, with entire leaflets, stipulate; flowers perfect, regular, axillary; sepals 5 (sometimes 4 or 6), usually distinct; petals 5 (or 4 or 6), distinct; stamens twice the number of petals, in 2 series, distinct; ovary superior, of 2-12 carpels, 2- to 12-celled, styles united; fruit a capsule or divided into few to several smooth or spinescent nutlets (as many or twice as many as carpels).

1. Woody shrubs; fruit woolly-hairy; plants strong-scented -1. Larrea
1. Herbaceous annual or perennials; fruit glabrous to variously hairy but never woolly; plants not strong-scented
  2. Leaflets 2, succulent; plants perennial with creeping roots -2. Zygophyllum
  2. Leaflets 6-10, not succulent; plants annual
    3. Fruits separating into 5 nutlets, these with strong spines, each containing 2 or more seeds in separate compartments -3. Tribulus
    3. Fruits separating into 8-12 nutlets, these not spiny (many have tubercles), each 1-seeded -4. Kallstroemia

1. Larrea Cav. CREOSOTE BUSH

Woody evergreen, strong scented shrubs; branches distichous; leaves with usually 1 pair of leaflets, rarely several pairs present, sessile or nearly so; stipules persistent; peduncles 1-flowered, terminal; sepals 5, deciduous; petals 5, yellow, longer than the sepals; ovary short-stipitate, pilose, 5-celled, styles united but sometimes separable; fruit 5-angled, separating into 5 indehiscent carpels, villous or tomentose.

1. Larrea divaricata Cav., Anal. Hist. Nat. 2:119. 1800.
L. tridentata (DC.) Cov.; Covillea tridentata (DC.) Vail; C. glutinosa (Engelm.) Rydb.---Dry plains and hills. Texas to Utah and California, south to Mexico. This plant has been found rather close to the Colorado line and should be looked for in the southern part of the state.

2. Zygophyllum L. BEAN CAPER

Perennial, somewhat shrubby plants with deep strong creeping roots; stems terete, thick, smooth; leaves with 2 leaflets, these succulent; flowers solitary in the leaf axils, copper-colored, salmon-colored or yellow; calyx 5-parted; petals 5; ovary sessile, 4- to 5-carpelled, 4- to 5-celled, with a disk at base; fruit 4- to 5-valved with 1 seed in each cell.

1. Zygophyllum fabago L., Sp. Pl. 385. 1753.
Glabrous plants; stems branching from the base to form a bushlike plant; leaflets succulent, the larger 1-3.5 cm. long, elliptic-oval, rounded at apex and oblique at base; sepals 5-10 mm. long, nearly distinct, green but with whitish almost scarious margins; petals about 10 mm. long but seem to be about as long as sepals; stamens longer than petals; fruit oblong, immature on our specimens but stated to be 2-5 cm. long when mature.---Waste places and fields. Introduced from Asia into several places in the western part of the United States. Our few records from southcentral Colorado at 7500 feet but also reported from the western part of the state.

3. Tribulus L. BURNUT; PUNCTURE VINE

Herbaceous plants, stems diffuse and prostrate; leaves pinnate, 1 of each pair usually smaller; flowers axillary and peduncled; sepals 5; petals 5, yellow, deciduous; ovary sessile, 5-lobed, 5-celled, surrounded at base by a 10-lobed disk; stigmas 5; fruit spiny, separating at maturity into 5 bony carpels, each again divided into 3-5 1-seeded compartments.

1. Tribulus terrestris L., Sp. Pl. 387. 1753.
Annual plants; stems diffuse and trailing, short-hirsute, somewhat swollen at the nodes; leaves 2-5 cm. long, the leaflets of 5-7 pair, each 3-13 mm. long, oblong or elliptic, oblique at base, acute to obtuse, silky-strigose below and more or less so above especially on the midveins; sepals about 3-4 mm. long, deciduous, strigose; petals about 4 mm. long; fruit to 8-9 mm. long, each carpel with 2 stout spines to 6 mm. long and 2 or more smaller ones, tuberculate, somewhat hispid and often puberulent.---Waste places and fields, often in sandy soil. Naturalized from Europe and now widely distributed into the United States at least in the western part. Our records scattered in the eastern half of Colorado at 4000-6500 feet.

4. Kallstroemia Scop. CALTROP

Annual herbaceous plants; stems diffuse and procumbent; leaves pinnate, 1 of each pair usually smaller; flowers solitary, peduncled, axillary; sepals 5 (or 6), mostly persistent; petals the same number as sepals but longer, caducous, orange or yellow; ovary sessile, 8- to 12-celled; stigma capitate; fruit 8- to 12-angled, roughened or tuberculate but not spiny, separating at maturity into 8-12 bony indehiscent usually 1-seeded nutlets.

1. Beak of fruit 1-2 mm. long at maturity, glabrous; plants glabrous to appressed puberulent, few if any longer hairs present; petals 4-5 mm. long -1. K. brachystylis
1. Beak of fruit 3-4 mm. long at maturity, appressed puberulent at least at base; plants appressed puberulent but also definitely hirsute; petals 5 mm. long or more -2. K. hirsutissima

1. **Kallstroemia brachystylis** Vail, Bull. Torr. Club 24:206. 1897.
New Mexico to California and Mexico. This plant has been reported from southern Colorado but no specimens seen.
2. **Kallstroemia hirsutissima** Vail in Small, Fl. Southeast. U. S. 670. 1903.
Annual plants; stems 15-70 cm. long, much branched and trailing, appressed pubescent and hirsute; leaves 2-3.5 cm. long, short-petioled; leaflets 3-4 pairs, 8-20 mm. long, elliptic to oblong, strigose at least below; sepals 4-6 mm. long, linear-lanceolate, hirsute with spreading hairs; petals 5-7 mm. long sometimes to 12; fruit 6-8 mm. long (counting beak) the body canescent-strigose, the beak 3-4 mm. long, as long or nearly as long as body, conic at base, appressed hairy near base.---Sandy ground. Kansas to Colorado, south to Texas and Mexico. Our records from southeastern and central Colorado at 4000-6500 feet.

Family 64. Rutaceae RUE FAMILY

Small trees, shrubs or nearly herbaceous plants, more or less strong-scented, the herbage glandular-punctate; leaves alternate, simple or compound; flowers perfect or polygamous, regular, usually cymose; sepals 4-5, rarely 6; petals distinct, as many as the sepals; stamens as many or twice as many as the petals, distinct; ovary superior, 2- to 3-celled, of 2-3 united carpels; fruit a samara or leathery capsule.

1. Leaves palmately 3-foliate; fruit an indehiscent samara -1. **Ptelea**
1. Leaves simple; fruit a 2-lobed capsule opening at apex -2. **Thamnosma**

1. Ptelea L. HOPTREE

Spineless shrubs or small trees; leaves palmately 3-foliate; flowers polygamous, in compound cymes, often corymbose; sepals 4 or 5 (rarely 6); petals same number as sepals; stamens same number as petals, abortive in pistillate flowers; ovary 2-celled (rarely 3-celled), inserted on a disk; style short, stigma 2- to 3-lobed; fruit a mostly 2-celled indehiscent samara, the wing surrounding the body.

1. **Ptelea baldwinii** T. & G., Fl. N. Amer. 1:215. 1838.
P. crenulata Greene; P. angustifolia Benth. at least in our Manuals---Shrubs or small trees to 6 m. tall; lateral leaflets 1.5-6 cm. long, sessile or very nearly so, narrowly lanceolate to oval, acute to acuminate, cuneate or obliquely rounded at base; terminal leaflet sessile or short-petiolate, somewhat longer; all leaflets crenulate to entire, punctate and more or less puberulent especially on veins below; branches of inflorescence and pedicels usually puberulent; sepals about 1-3 mm. long; petals about 4-7 mm. long; samara 1-2 cm. long, retuse at both ends, nearly orbicular, the body about 1/3 of the width, the whole surface usually strongly reticulated. Our plant is very doubtfully distinct from the eastern P. trifoliata L.---Woods, slopes and canyons. Florida, Texas, Colorado and California, south into Mexico. Our records scattered in the southern half of Colorado at 5000-9000 feet.

2. Thamnosma Torr. & Frem. DESERT RUE

Shrubs or herbaceous plants woody at the base, strongly-scented; leaves simple, entire, linear to narrower; flowers perfect, in racemes or cymose clusters; sepals 4; petals 4, a cup-like disk present; stamens 8, inserted on the disk; ovary of 2, rarely 3 carpels, 2- or 3-lobed, 2- or 3-celled with 5-6 ovules in each cell, sessile or stipitate; style filiform, stigma capitate; fruit a leathery 2- to 3-celled, 2- to 3-lobed capsule opening at the apex.

1. **Thamnosma texana** (A. Gray) Torr., Bot. Mex. Bound. Surv. 42. 1859.
Reported from Colorado, south to Texas and Mexico. The writer has never seen material from this state.

Family 65. Polygalaceae MILKWORT FAMILY

Herbaceous plants; leaves simple, entire, alternate, opposite or whorled; stipules lacking; inflorescence various; flowers very irregular, perfect; sepals 5, imbricate, distinct, the 2 inner more or less petaloid; petals commonly 3, the lower often beaked and crested (called a keel), all united into a tube which is more or less adnate to the stamens; stamens 6 or 8, filaments united into a tube, anthers opening by apical pores, becoming 1-celled; ovary superior, usually 2-celled; style simple, dilated or lobed above; fruit a 2-celled capsule, laterally compressed; seeds 1 in each cell, usually carunculate at the hilum.

1. Polygala L. MILKWORT

Characters of the Family in Our Area

1. Plants more or less spiny, somewhat shrubby at base; keel with a beak but no crest
    2. Shrubby intricately branched plants usually over 15 cm. tall (to 90 cm.); flowers 4-5 mm. long; leaves rarely over 3 mm. wide -1. **P. acanthoclada**
    2. Plants woody only at very base, not over 15 cm. tall; flowers 7-10 mm. long; leaves 3-6 mm. wide -2. **P. subspinosa**

1. Plants not at all spiny, annual or perennial, hardly woody at base; keel with a fimbriate crest
   3. Plants annual; most of leaves verticillate; flowers about 2 mm. long   -3. <u>P. verticillata</u>
   3. Plants perennial; only the lower leaves verticillate if any; flowers about 3-3.5 mm. long   -4. <u>P. alba</u>

1. Polygala acanthoclada A. Gray, Am. Acad. Arts and Sci. Proc. 11:73. 1876.
   P. acanthocarpa of Manuals---Dry areas. Colorado to California, south to Arizona. The writer has never seen a specimen from this state but the plant has been reported from the southwestern part.
2. Polygala subspinosa S. Wats., Amer. Nat. 7:299. 1873.
   Perennial undershrubs, woody at branching bases; stems many, 5-15 cm. tall, erect or ascending, pale green, puberulent to rarely nearly glabrous, with indurated, spiny-tipped branches; leaves 1-3 cm. long and 3-6 mm. wide, mucronate at the rounded apex, cuneate at base, sparsely puberulent or glabrous; flowers pink-purple and yellow in a few-flowered raceme, bracts small and scarious; smaller sepals 4-6 mm. long **but** wings 7-10 mm. long, no fimbriate crests present; capsule 6-7 mm. long, oval, reticulate, **sparsely** hispidulous on margins.---Hills and slopes. Colorado to Nevada, south to New Mexico **and** Arizona. Our records from western Colorado, mostly in Mesa County, at 4500-5500 feet.
3. Polygala verticillata L., Sp. Pl. 706. 1753.
   Reported from Ontario south to Florida and Texas. This plant has been listed for eastern Colorado but no specimens have been located.
4. Polygala alba Nutt., Gen. Pl. 2:87. 1818.
   Perennial plants; stems 10-35 cm. tall, numerous from the root, erect or ascending, glabrous, angled; leaves 4-30 mm. long, scattered above, usually some whorled at base, linear or lowest oblanceolate or even wider, glabrous, cuspidate; flowers in dense, narrow spikelike racemes 2-8 cm. long, the flowers white with a greenish center, crest often purplish; smaller sepals 1-1.5 mm. long, the wings about 2.8-3.5 mm. long; keel of petals about 3 mm. long, with a crest of 4 lobes on each side; capsule 2.5-3 mm. long.---Dry plains and slopes. North Dakota to Washington, south to Mexico. Our records from eastern Colorado at 3500-5000 feet.

Family 66. Euphorbiaceae SPURGE FAMILY

Herbaceous plants often with milky sap; leaves simple, alternate, opposite or whorled; stipules present or absent; flowers unisexual, monoecious or dioecious; calyx present or absent; a calyxlike involucre sometimes present; corolla of separate petals present or absent; stamens 1 to many; ovary superior, mostly 3-celled (sometimes 1-4), ovules 1-2 to a cell, styles as many as cells, distinct or partly connate, often many-cleft; fruit a capsule, each carpel usually dehiscent by 2 elastic valves but sometimes very tardily if at all dehiscent.

1. Plants canescent with stellate hairs   -1. <u>Croton</u>
1. Plants glabrous or pubescent with simple hairs, or hairs affixed at middle (appearing 2-branched)
   2. Flowers appearing perfect, the cluster consisting of a single pistillate flower and few to many staminate flowers the whole surrounded by a gamophyllous involucre bearing 1-5 often petaloid-appendaged glands on the rim (no sepals present); leaves alternate, opposite or whorled   -2. <u>Euphorbia</u>
   2. Flowers plainly unisexual, the staminate and pistillate clearly separated, no such involucre with <u>glands</u> present (sepals are present); leaves alternate
      3. Petals present (at least on staminate flowers); leaves entire; some or all of hairs malpighiaceous, but never stiff; stamens 8-10, united in a column at base   -3. <u>Ditaxis</u>
      3. Petals absent on all flowers; leaves toothed; hairs if present simple and stiff; stamens 2-5, not united at base
         4. Pistillate flowers and capsules short-pedicelled; capsules and whole plant with bristly hairs; stems rarely over 30 cm. tall   -4. <u>Tragia</u>
         4. Pistillate flowers and capsules sessile; capsules and plants glabrous; stems usually over 30 cm. tall   -5. <u>Stillingia</u>

1. Croton L. CROTON

Herbaceous, heavy-scented stellate-pubescent plants; leaves alternate, entire, or repand; flowers monoecious or dioecious, in racemose axillary or terminal clusters; calyx 4- to 6-lobed (usually 5); petals present or absent, often small and alternating with glands; stamens 5 or more; ovary mostly 3-celled, 1 ovule in each cell; styles 3 or often 1- to 4-cleft; seeds smooth or minutely pitted, carunculate.

1. Croton texensis (Klotzsch) Muell. Arg. in DC., Prodr. 15(2):692. 1866.
   Annual plants; stem 20-60 cm. tall, canescent-stellate, dichotomously branching or spreading; leaves linear, lanceolate or oblong (staminate plants usually with narrower leaves); plants dioecious; petals apparently lacking; staminate flowers in racemes, sepals about 1.5 mm. long, stamens 8-12; pistillate flowers 2-4 together, sometimes solitary, sepals about 2 mm. long; capsule 4-6 mm. long, subglobose, stellate and somewhat warty.---Prairies and plains, often in sandy ground. Illinois to Wyoming, south to Arkansas, Mexico and Arizona. Our records scattered over the eastern half of Colorado at 3500-5500 feet.

2. Euphorbia L. SPURGE

Annual or perennial herbaceous plants; leaves simple, alternate, opposite or whorled; flowers monoecious, borne in structures called cyathia, these simulating a single flower; pistillate flower solitary in center of cyathium, pedicelled, without calyx or corolla, ovary

3-celled, styles 3, usually bifid; staminate flowers with no calyx or corolla, in 5 fascicles, 1 to several in a fascicle, the fascicles opposite the lobes of the involucre, each pedicel of 1 stamen often with a minute scale at base; groups of 1 pistillate and several staminate flowers surrounded by a hypanthium or calyxlike involucre, bearing on its margin 1-5 nectariferous glands alternating with the involucre lobes, petaloid appendages often extending out from beneath these glands; capsule 3-celled, 3-seeded, usually nodding.

1. Glands of the involucre with petaloid appendages growing out below them or if these absent, then the leaves are all strictly opposite and with unequal bases
  2. Leaves all strictly opposite, usually strongly unequal at base; stipules well developed (although small); bracts never petallike
    3. Ovaries and capsule hairy; stems and leaves hairy
      4. Plants perennial; capsule over 2 mm. long   -1. E. lata
      4. Plants annual; capsule less than 2 mm. long
        5. Leaves sharply serrate; seeds pitted or ridged, acute at apex   -2. E. stictospora
        5. Leaves entire; seeds smooth or nearly so, not acute at apex   -3. E. micromera
    3. Ovaries and capsule glabrous; leaves and stems usually glabrous
      6. Plants perennial; some of leaves broadly ovate to orbicular
        7. Stipules of opposite leaves united in the center into 1 scale; capsule 1.7-2.3 mm. long; seeds less than 2 mm. long   -4. E. albomarginata
        7. Stipules of opposite leaves distinct; capsule 2.3-2.5 mm. long; seeds 2 mm. long or longer   -5. E. fendleri
      6. Plants annual; leaves ovate or narrower (except in E. serpens)
        8. Stipules of opposite leaves united in the center into 1 scale; leaves ovate-orbicular to oblong   -6. E. serpens
        8. Stipules of opposite leaves distinct; leaves ovate to narrowly linear
          9. Leaves linear to narrowly linear, equal or very nearly so at base, always entire
            10. Appendages conspicuous, over 4 times wider (appearing longer) than the glands   -7. E. missurica
            10. Appendages small and narrow, rarely extending beyond the glands
              11. Capsule 1.3-1.4 mm. long, sharply angled; staminate flowers 5-10 in an involucre; involucres not over 1 mm. wide   -8. E. revoluta
              11. Capsule 2-2.5 mm. long, bluntly-angled; staminate flowers many in an involucre; involucres 1.5 mm. wide or more   -9. E. parryi
          9. Leaves oblong-linear to ovate, the narrower ones at least always definitely unequal at base (wider ones too except sometimes in E. micromera), entire to serrate
            12. Appendages absent or a minute rudiment present; capsule about 1.3 mm. long; leaves entire, 2-7 mm. long   -3. E. micromera
            12. Appendages present, 1/2 as wide as the glands or wider; capsule 1.4-2.6 mm. long; leaves entire to serrate, some usually over 7 mm. long
              13. Stems pilose; leaves serrate   -10. E. serrula
              13. Stems glabrous; leaves entire to serrate
                14. Seed plump, about 2/3 as wide as long, smooth; leaves always entire   -11. E. geyeri
                14. Seeds narrower, about 1/2 as wide as long, smooth to ridged; leaves often serrate
                  15. Seeds with definite, regular unbranched transverse ridges including the angles; leaves more commonly entire   -12. E. glyptosperma
                  15. Seeds smooth, punctate or irregularly rugose, never with regular transverse ridges and the angles not ridged; leaves almost always serrate   -13. E. serpyllifolia
  2. Leaves alternate or opposite, equal at base; stipules minute and glandlike or none; bracts (in the commonest species) with broad white margins
    16. Leaves linear to oblong, not over 8 mm. wide, short-petioled, all opposite; floral leaves not white-margined   -14. E. hexagona
    16. Leaves ovate to oblong-ovate, well over 10 mm. wide, sessile, the lower usually scattered; floral leaves with conspicuous white margins   -15. E. marginata
1. Glands of the involucre without petaloid appendages; leaves opposite or alternate, essentially equal at bases
  17. Glands deeply cupped; stems never branching into a 3- to several-rayed, symmetrical umbellike inflorescence; leaves mostly opposite throughout
    18. Leaves ovate-lanceolate to broader, coarsely toothed, usually about twice as long as wide   -16. E. dentata
    18. Leaves linear to lanceolate, entire or shallowly toothed, usually 5-10 times longer than wide   -16A. E. dentata cuphosperma
  17. Glands flat or convex, never deeply cupped; inflorescence a symmetrical 3- to several-rayed umbel; stem leaves alternate (opposite only in or just below the inflorescence)
    19. Perennial by vigorous horizontal roots or rootstocks, hence plants growing in patches; seed surface smooth
      20. Leaves linear, not over 3 mm. wide; plants 12-30 cm. tall; an escape from cultivation   -17. E. cyparissias
      20. Leaves linear to oblong, 4-12 mm. wide; plants usually 30-60 cm. tall; present as a serious weed   -18. E. esula

    19. Annual, biennial or perennial by vertical roots, not growing in patches; seed surface smooth, pitted or wrinkled
      21. Leaves serrulate; glands elliptic, entire, not crescent-shaped or horned at the ends; seeds brown or purplish-black, reticulated or wrinkled; involucres usually about 1 mm. long; capsule usually 2.5-3 mm. long, verrucose near ends   -19. E. dictyosperma

21. Leaves entire (may dry finely wrinkled); glands crescent-shaped, often horned at the ends; seeds gray or greenish-gray, smooth or shallowly pitted; involucres about 2 mm. long or more; capsule about 3-4.5 mm. long, not verrucose
    22. Annual plants; glands with slender divergent horns at the ends as long as the body   -20. E. crenulata
    22. Perennial by vertical thickened roots; glands crescent-shaped but horns if present very short
                                                               -21. E. robusta

1. Euphorbia lata Engelm. in Emory, U. S. & Mex. Bound. Surv. 2(1):188. 1859.
    Chamaesyce lata (Engelm.) Small---Perennial plants; stems 10-15 cm. long, ascending or erect, with short somewhat appressed hairs; leaves 5-20 mm. long, opposite, ovate to narrow, often appearing linear and usually more or less falcate, more or less oblique at base, with entire more or less revolute margins, hairs short, somewhat appressed; stipules distinct, narrow and undivided, hairy; involucres 1.7-2 mm. wide; glands about 0.5 mm. long, oblong; appendages absent or very narrow and whitish; capsule 2.5 mm. long, hairy; seeds about 2 mm. long, and about 1/2 as wide, surface smooth but depressed on faces.---Plains. Kansas and Colorado, south to Texas and New Mexico. Our few records from southeastern Colorado at about 4000 feet.
2. Euphorbia stictospora Engelm. in Torr., Bot. U. S. & Mex. Bound. 187. 1859.
    Chamaesyce stictospora (Engelm.) Small---Annual plants; stems 5-25 cm. long, prostrate or ascending, short-villous; leaves 3-10 mm. long, suborbicular to oblong-linear, sharply serrate, base oblique, crispy-hairy both sides or glabrate above; stipules distinct or united, hairy; involucre 0.7-1 mm. wide; glands 0.15-0.3 mm. long, oblong to suborbicular; appendages narrow, white; capsules 1.4-1.9 mm. long, strigose; seeds 1.2-1.4 mm. long, over 1/2 as wide as long, surface with shallow pits or with low subregular transverse ridges.---Dry ground. South Dakota to Wyoming, south to Mexico. Our records scattered over the eastern half of Colorado at 4000-5500 feet.
3. Euphorbia micromera Boiss. ex Engelm., Proc. Amer. Acad. 5:171. 1861.
    Utah and Nevada, south to Texas and Arizona. This plant has been recorded in Utah close to the southwestern part of Colorado and is to be expected in our state.
4. Euphorbia albomarginata T. & G., Rep. Expl. & Surv. Miss. R. to Pac. Ocean 2:174. 1855.
    Chamaesyce albomarginata (T. & G.) Small---Utah to California, south to Oklahoma, Texas and Arizona. Reported in the northeastern corner of Arizona and to be expected anywhere in southern Colorado.
5. Euphorbia fendleri T. & G., Pac. R. R. Rept. 2(4):175. 1855.
    Chamaesyce fendleri (T. & G.) Small; C. greenei in Rydb's. Floras---Perennial plants from taproots, woody in age, glabrous throughout; stems up to 15 cm. long, several to many, erect or decumbent; leaves 3-11 mm. long, opposite, ovate-orbicular to ovate-lanceolate, entire, oblique at base; stipules distinct, mostly entire, narrowly linear; involucres 1.25-1.75 mm. wide; glands about 1 mm. long and not as wide, reddish, concave or convex; appendages about as wide as the glands, white; capsule 2.25-2.5 mm. long; seeds 2-2.5 mm. long, about 1/2 as wide, surface smooth or slightly wrinkled. Our plant has been called var. typica L. C. Wheeler.---Hills and plains. Nebraska to California, south to Oklahoma and Mexico. Our records scattered over the state at 3500-7000 feet.
6. Euphorbia serpens H. B. K., Nov. Gen. et Sp. 2:52(Quarto), 41(Folio) 1817.
    Chamaesyce serpens (H. B. K.) Small---Annual plants, glabrous throughout; stems 5-30 cm. long or even longer, prostrate; leaves 2-7 mm. long, opposite, ovate-orbicular to oblong, entire, base oblique at least on larger leaves; stipules of opposite leaves united into a white membranous scale with more or less lacerate margins; involucres about 1 mm. wide; glands oblong, very narrow; appendages absent or if present little wider than the glands; capsule 1.2 mm. long; seeds about 1 mm. long, 1/2 or more as wide, surface smooth or nearly so.---Prairies and plains. Ontario, south to Tennessee and Colorado; also casual in the eastern states; South America. Our few records from northcentral Colorado at about 5000 feet.
7. Euphorbia missurica Raf., Atlantic Journ. 1:146. 1832.
    E. petaloidea Engelm.; Chamaesyce petaloidea (Engelm.) Small---Annual plants, glabrous throughout; stems 5-40 cm. long, decumbent to erect; leaves 10-30 mm. long, opposite, linear, entire, apex truncate and emarginate or mucronate, base symmetrical or nearly so; stipules distinct or somewhat united, entire to parted; involucre 1.7-1.9 mm. wide; glands 0.4-0.7 mm. long, subcircular to oblong; appendages usually 4 or more times as wide (or long) as the glands, white or pink, conspicuous; capsule 2-2.5 mm. long; seeds 1.5-2 mm. long, plump, 2/3 as wide as long or wider, surface smooth. Our plant would be var. intermedia (Engelm.) Wheeler.---Sandy plains. The variety from Minnesota to Montana, south to Texas and New Mexico. Our records mostly scattered over the eastern half of Colorado (one in Montrose County) at 3500-6000 feet.
8. Euphorbia revoluta Engelm. in Emory, U. S. Mex. Bound. Survey 2(1):186. 1859.
    Chamaesyce revoluta (Engelm.) Small---Annual plants, glabrous throughout; stems 3-20 cm. tall; largest leaves 1-2.6 cm. long, opposite, narrowly linear, margins entire and revolute, equal or nearly so at base; stipules distinct, entire; involucre 0.9-1 mm. wide; glands 0.15-0.3 mm. wide, subcircular; appendages from nearly absent to nearly as wide as the glands; capsule 1.3-1.4 mm. long; seeds 1-1.3 mm. long, over 1/2 as wide as long, surface nearly smooth or with 2 transverse ridges.---Hills and plains. Colorado south to Mexico. Our few records from central to northcentral Colorado at 5000-5500 feet.
9. Euphorbia parryi Engelm., Amer. Nat. 9:350. 1875.
    Chamaesyce parryi (Engelm.) Rydb.; C. flagelliformis (Engelm.) Rydb.---Colorado to Nevada, south to Texas, Mexico and Arizona. Reported for southwestern Colorado but not located by the writer.
10. Euphorbia serrula Engelm. in Emory, U. S. & Mex. Bound. Surv. 2(1):188. 1859.
    Texas to Arizona, south into Mexico. Reported in New Mexico near the Colorado line.

11. Euphorbia geyeri Engelm. in Engelm. & Gray, Bost. Journ. Nat. Hist. 5:260. 1845.
Chamaesyce geyeri (Engelm.) Small---Annual plants, glabrous throughout; stems 6-35 cm. long, mostly prostrate, occasionally ascending or even erect; leaf blades 4-10 mm. long, opposite, oblong to ovate-oblong or elliptic-oblong, margins entire, base oblique; petioles 1-2 mm. long; stipules distinct; involucres 0.8-1.1 mm. wide; glands oval to round; appendages from 1/2 to 2 times longer than the glands, white; capsule about 2 mm. long; seeds 1.3-1.6 mm. long, about 2/3 as wide as long, plump, coat smooth and whitish.---Plains and hills, often on sandy ground. Wisconsin to North Dakota, south to Illinois, Texas and New Mexico. Our one record from Morgan County in northeastern Colorado at 4200 feet.

12. Euphorbia glyptosperma Engelm. in Emory, U. S. & Mex. Bound.Surv. 2(1):187. 1859.
Chamaesyce glyptosperma (Engelm.) Small---Annual plants, glabrous throughout; stems 5-35 cm. long, mostly prostrate, occasionally ascending; leaf blades 3-15 mm. long, opposite, oblong, narrowly oblong to ovate, entire or serrulate especially toward apex, unequal at base; stipules distinct, subulate; involucres 0.6-0.9 mm. wide; glands elliptic or oblong, mostly depressed in middle; appendages 1-1.5 as wide as the glands, white; capsule 1.4-1.7 mm. long; seeds 0.7-0.9 mm. long, about 1/2 as wide, the faces with 4-5 transverse, regular ridges, surface white to tan---Sandy plains. New Brunswick to British Columbia, south to Indiana, Texas and California. Our records scattered over the state, none from the northwestern or southeastern parts as yet, at 4000-8000 feet.

13. Euphorbia serpyllifolia Pers., Syn. Pl. 2:14. 1806.
E. rugulosa (Engelm.) Greene; E. serpyllifolia var. rugulosa Engelm.; Chamaesyce rugulosa (Engelm.) Rydb.; C. serpyllifolia (Pers.) Small; C. albicaulis Rydb.---Annual plants; stems 5-35 cm. long, prostrate or erect, glabrous; leaves 3-14 mm. long, opposite, oblong, ovate, obovate to linear-oblong, glabrous, usually serrulate at least toward the apex, unequal at base; stipules distinct, entire or few-parted; involucres 0.8-1.2 mm. wide, glabrous; glands oblong; appendages white, narrow; capsule 1.5-1.9 mm. long, glabrous; seeds 1-1.4 mm. long and about 1/2 as wide, smooth to somewhat punctate or irregularly rugose, clay-brown to white. Our plant has been called var. genuina Boiss.---Dry plains. The variety from Alberta to British Columbia, south to Michigan, Texas and Mexico. Our records scattered over Colorado, except in the northwestern corner, at 4000-8000 feet.

14. Euphorbia hexagona Nutt.; Spreng., Syst. 3:791. 1826.
Zygophyllidium hexagonum (Nutt.) Small---Annual plants, yellowish-green in color; stems 20-50 cm. tall, slender, erect, usually sparingly hairy, branched with very slender ascending branches; leaves 7-35 mm. long, all opposite, short-petioled, linear, oblong or lanceolate, usually with appressed hairs even at base; stipules absent or very minute; involucres 1.5-3 mm. long, 1 to few in the upper forks, hairy; glands 5, with triangular-ovate, rather conspicuous greenish-yellow to white appendages 2-3 times wider than glands; capsule about 4 mm. wide, 3-lobed, glabrous; seeds ovoid, papillose.---Sandy plains and valleys. Minnesota to Montana, south to Texas and New Mexico. Our records from northcentral and northeastern Colorado at 3500-5500 feet.

15. Euphorbia marginata Pursh, Fl. Am. Sept. 2:607. 1814.
Dichrophyllum marginatum (Pursh) Klotzsch & Garcke; Lepadenia marginata (Pursh) Niewl.---Annual plants; stems 30-70 cm. tall, stout, erect, glabrous or pilose; leaves ovate, obovate or oblong-ovate, sessile, entire, the floral leaves whorled or opposite and with conspicuous white margins; stipules small and deciduous; lower leaves scattered, bases even; inflorescence umbel-like, usually with 3 rays; involucres about 3-4 mm. long, campanulate, usually pubescent, 5-lobed; glands about 1 mm. wide with conspicuous white appendages; capsule 5-6 mm. wide, usually pubescent, with 3 rounded lobes; seeds about 3-4 mm. long, ovoid-globose, ash-colored, tuberculate, often reticulate-tuberculate.---Dry plains and valleys. Montana to Mexico and introduced eastward. Our records mostly from the eastern half of Colorado (one from Montrose County) at 4000-7000 feet.

16. Euphorbia dentata Michx., Fl. Bor. Am. 2:211. 1803.
Poinsettia dentata (Michx.) Small---Annual plants; stems 20-60 cm. tall, erect or ascending, branching, pubescent; leaves 10-60 mm. long, sometimes longer, opposite (only the lower sometimes alternate), ovate-lanceolate to orbicular-oblong, sometimes narrower, coarsely toothed, pubescent, cuneate to a petiole; stipules glandlike; involucres 2.5-3 mm. long, clustered at the ends of the branches, with 1-4 short-stalked glands; no appendages present; capsule 3-5 mm. long, glabrous, smooth; seeds ovoid-globose, inconspicuously angled, irregularly tuberculate, ash-colored to darker.---Dry soil. Pennsylvania to South Dakota, south to Louisiana and Mexico. Our records scattered over the eastern half of Colorado at 3500-5500 feet.

16A. Euphorbia dentata cuphosperma Engelm., (var.) in Torr., Bot. U. S. & Mex Bound. 190. 1859.
E. cuphosperma (Engelm.) Boiss.; Poinsettia cuphosperma (Boiss.) Small---In this variety the leaves are narrower, linear to lanceolate, entire or shallowly toothed; the capsule is glabrous to strigose; the seeds supposed to have 4 sharp angles. Intergrades somewhat with the species in Colorado.---Plains and prairies. South Dakota to Colorado, south to Texas and Arizona. Scattered with the species in Colorado (one record from Montrose County) at 3500-6000 feet.

17. Euphorbia cyparissias L., Sp. Pl. 461. 1753.
Tithymalis cyparissias (L.) Lam.---Plants perennial by horizontal roots; stems 20-30 cm. tall, erect, mostly clustered, glabrous, rather scaly below and leafy above; leaves 12-30 mm. long, scattered, alternate except the whorled ones at the base of the umbels, sessile, linear to nearly filiform, entire; stipules absent; bracts reniform or nearly orbicular; inflorescence of umbels terminating the stems; involucres about 2 mm. long; glands crescent-shaped; no appendages present; capsule about 3 mm. in diameter, glabrous, subglobose; seeds about 2 mm. long, oblong, smooth.---Waste places and around dewellings. Escaping from cultivation in various places in this country. Our few records from northcentral Colorado at about 5000 feet.

18. Euphorbia esula L., Sp. Pl. 461. 1753.

E. virgata Wald. & Kit.---Plants perennial by horizontal roots; stems 30-60 cm. tall, mostly erect, glabrous, scaly below, branched above, main stem and branches topped by a several-rayed umbel; leaves 3-6 cm. long on main stems, smaller on branches, about 4-8 mm. wide, all scattered except the whorled ones at the base of the umbels, linear to oblong or lanceolate-oblong, sessile, entire; stipules absent; bracts broadly ovate or sub-reniform; inflorescence yellow-green; involucres about 2 mm. long, the lobes erect, with about 4 crescent-shaped glands; no appendages present; capsule 4-5 mm. wide, glabrous, somewhat roughened; seeds about 2 mm. long, oblong to ovoid, smooth. There has been a difference of opinion as to the correct name for this plant. Morton (Rhodora 39:45-50. 1937) claimed our western plant should be E. virgata Wald. & Kit. But Croizat (Am. Midl. Nat. 33:231-243. 1945) maintained that our American plant is E. intercedans Podtera.---Waste places, roadsides and fields as a serious weed. At various places in eastern United States, west to Colorado. Our records scattered over the state in cultivated areas at 5000-6500 feet.

19. Euphorbia dictyosperma Fisch. & Meyer, Index Sem. Hort. Petrop. 2:37. 1836.

E. arkansana Engelm. & Gray; Tithymalus arkansanus (Engelm. & Gray) Kl. & Garke; T. arkansanus var. coloradensis (Norton) Rydb.; T. missouriensis (Norton) Small---(This is a variable species, the following description applying only to the plants of Colorado.) Annual plants (possibly sometimes biennial); stems 10-50 cm. tall, erect, branching above, glabrous; leaves 1-3 cm. long, rather few, lower leaves scattered, oblong or oblong-spatulate, serrulate, glabrous, sessile or lowest short-petioled, upper leaves and floral leaves opposite, sessile, rounded-ovate to oblong; stipules absent; inflorescence umbelliform; involucres about 1 mm. long; glands elliptic, not horned; no appendages present; capsule 2.5-3 mm. long, glabrous but verrucose; seeds about 1.5-2 mm. long, ellipsoid or ovoid, brown or purplish-black, finely reticulated or wrinkled. The correct name for this plant may be E. spathulata Lam.---Plains, often in sandy soil. Widely distributed in the United States, especially in western part, and south into Mexico. Our records scattered over Colorado at 4500-6000 feet.

20. Euphorbia crenulata Engelm. in Torr.,Bot. Mex. Bound. 192. 1859.

E. manca A. Nels.; Tithymalus manca (A. Nels.) Heller---Annual, perhaps sometimes biennial plants; stems 10-20 cm. tall, several from the base, erect or decumbent at base, branched above into an umbel; leaves of stem 8-30 mm. long, scattered, obovate to spatulate, cuneate at base, entire or crenulate, obtuse at apex, sessile or short-petioled; floral leaves opposite or ternate, sessile or connate at base, deltoid or broadly ovate; stipules absent; inflorescence once- or twice-trichotomous; involucre about 2 mm. long; glands large, crescent-shaped with slender, diverging horns as long as the body; appendages absent; capsule about 3-4 mm. long, smooth, glabrous; seeds 2-3 mm. long, ash-colored or greenish-gray, with ridges or pits, or nearly or quite smooth.---Valleys and hills. Colorado to Oregon, south to Arizona and California. Our one record from southwestern Colorado at 7000 feet.

21. Euphorbia robusta (Engelm.) Small in Britt.& Brown, Ill. Fl. 2:381. 1897.

E. montana Engelm.; Tithymalus robustus (Engelm.) Small; T. philorus Cockerell---Perennial plants with vertical roots; stems 10-30 cm. tall, clustered, erect to decumbent, usually glabrous, topped by a 3- to 5-rayed umbel, the inflorescence greenish; stem leaves 10-16 mm. long, scattered, ovate or oblong-ovate, thick, entire, obtuse, usually sessile, a whorl of leaves subtending the umbel; bracts opposite, triangular-ovate or subreniform; stipules absent; involucres about 2-2.5 mm. long, with 4 crescent-shaped glands; no appendages present; capsule about 4-4.5 mm. long, glabrous, smooth; seeds about 2-3 mm. long, ovoid, gray, smooth or with shallow pits.---Mountains and hillsides. South Dakota to Montana, south to New Mexico and Arizona. Our records scattered over the state at 4000-9000 feet.

3. Ditaxis Vahl.

Perennial plants, woody in the caudex, with purplish sap, this staining some of the leaves drying, with at least some malpighiaceous hairs present; leaves alternate, simple, entire, strongly veined; flowers monoecious rarely dioecious, borne in bracteate axillary racemes; staminate flowers usually crowded at ends of racemes; sepals 4 or 5; petals 4 or 5, alternate to the calyx lobes or lacking in the pistillate flowers; stamens 8-10, filaments united into a column, and anthers in 2 whorls; ovary 3-celled with a solitary ovule in each cell; styles 3, each bifid; capsule 3-lobed.

1. Flowers in axillary clusters; leaves petioled -1. D. humilis
1. Flowers in elongated axillary racemes; leaves sessile -2. D. mercurialina

1. Ditaxis humilis (Engelm. & Gray) Pax. in Engl. & Prantl., Nat. Pfl. Fam. 3. Abt. 5, 45. 1890.

Stems 10-30 cm. tall, slender, branched, with appressed malpighiacious hairs; leaves 1-3 cm. long, oblong, obovate or oblanceolate, narrowed into definite petioles, margins entire or very obscurely toothed; flowers in axillary clusters; staminate flower with petals somewhat longer than sepals; pistillate flowers with reduced or apparently normal petals; capsule 4-6 mm. broad, appressed-hairy.---Plains. Kansas south to Louisiana, Texas and New Mexico. Our one record from southeastern Colorado at 3500 feet.

2. Ditaxis mercurialina (Nutt.) Coult., Mem. Torr. Club 5:213. 1894.

Stems 10-60 cm. tall, strict, simple or branching at base, with appressed malpighiaceous hairs; leaves 2-5 cm. long, ovate, oblong or oblong-lanceolate, sessile, margins undulate, glabrate to hairy; flowers in terminal or more commonly in axillary racemes; staminate flowers with lanceolate or linear-lanceolate ciliate sepals and spatulate-oblong petals; pistillate flowers with larger calyx, the segments spreading and lanceolate, lacking petals; capsule 6-9 mm. broad, more or less silky-hairy.---Dry ground. Kansas south to Texas and Arizona. Our one record from southeastern Colorado at about 4000 feet.

4. Tragia L. NOSEBURN

Perennial herbaceous plants clothed with stinging hairs; leaves alternate, petiolate, stipulate; flowers monoecious borne in bracteate racemes, these terminal or lateral but not axillary; staminate flowers above, sepals 3-5, no petals, stamens 3-5; pistillate flowers 1 or 2, sepals 3-8 (usually 5-6), ovary usually 3-celled, 1 ovule to each cell, styles 3, more or less united below; capsule 3-lobed, bristly-hairy.

1. Tragia nepetaefolia Cav., Incon. Pl. 6:37. pl. 557., f.l., 1801.
Light green plants, bristly with stinging hairs; stems 10-30 cm. long, slender, usually much branched; leaves 1-4 cm. long, lanceolate to triangular-lanceolate, coarsely and sharply serrate or in some almost laciniate; staminate flowers very small, 2 to many, with 3-4 sepals and 3-5 stamens; pistillate flowers usually 1 or 2 together, with 5 sepals; capsule about 6-8 mm. thick, depressed; seeds globose. Our plant has gone by several names. T. stylaria Muell. is included in the above concept.---Dry ground. Missouri to California, south to Texas and Arizona. Our records in northcentral, central and southern Colorado at 4000-6000 feet.

5. Stillingia L. QUEENS DELIGHT; QUEENS-ROOT

Perennial herbaceous plants; stems leafy; leaves simple, toothed, alternate or somewhat opposite; stipules absent but 2 glands present at base of petioles; plants monoecious; flowers in terminal spikes, the staminate several in the axils of the bractlets and the pistillate solitary in the axils of the lower bractlets; staminate flowers with calyx 2- to 3-lobed, with no petals, and 2-3 exserted stamens; pistillate flowers with calyx 3-lobed, no petals and ovary 3-celled with 3 styles united at base; capsule 3-lobed, 1 seed in each cell.

1. Stillingia sylvatica Garden, ex L. Mant. 19. 1767.
S. salicifolia (Torr.) Raf.---Stems 30-100 cm. tall, glabrous, simple or branched above; leaves 4-8 cm. long, lanceolate or narrowly elliptic, acute, serrulate; capsule 10-15 mm. wide.---Sandy ground. Virginia to Kansas, south to Florida, Texas, and New Mexico. Our one record from southeastern Colorado at about 4000 feet.

Family 67. Callitrichaceae WATER-STARWORT FAMILY

Aquatic caulescent herbaceous plants; leaves simple, opposite, entire, at least the submerged ones narrow, often near end of branches, without stipules; flowers inconspicuous, solitary and axillary, polygamous; calyx and corolla lacking but 2 saccate bracts often present; stamen 1, filament filiform and anther cordate; ovary superior, 4-celled, each cell with 1 ovule, 2 styles present; fruit 4-lobed, often separating at maturity into 4 indehiscent, 1-seeded nutlets.

1. Callitriche L. WATER-STARWORT

Characters of the Family

1. Leaves all linear, 1-veined, submersed; mature carpels separating for more than half their length; fruit 1-2 mm. wide; no bracts present .... -1. C. hermaphroditica
1. Leaves, at least the floating ones, spatulate to obovate, 3-veined; mature carpels not separating; fruit about 1 mm. wide; 2 bracts present below the flowers (may be very small)
   2. Fruit about as wide as high .... -2. C. heterophylla
   2. Fruit about 2/3 as wide as high .... -3. C. palustris

1. Callitriche hermaphroditica L., Cent. I. Pl. No. 90. Feb. 1755.
C. autumnalis L.---Submerged aquatic annual plants; leaves 5-20 mm. long, crowded, linear, 1-nerved; bracts if present very small; fruit 0.8-1.5 mm. in diameter, orbicular or nearly so, little if any winged, mature carpels separating.---In water. New England states to Oregon, south to Louisiana and Colorado. Our few records from northcentral, northwestern and westcentral Colorado at about 6500-9000 feet.
2. Callitriche heterophylla Pursh, Fl. Am. Sept. 3. 1814.
A species closely related to and intergrading with C. palustris L. The fruit is about as wide as long, the lobes not at all winged on the outer margins, although somewhat ridged.---Ponds and slow streams. Widely distributed in the United States especially in the eastern part. Our few records scattered in the western half of Colorado at 5000-10,500 feet.
3. Callitriche palustris L., Sp. Pl. 969. 1753.
Annual plants, aquatic or growing in the mud; submerged leaves 10-15 mm. long, sessile, linear, 1-nerved; floating leaves 5-10 mm. long, rather rosulate, petioled, obovate or spatulate, 3-nerved; fruit 1 mm. wide or less and about 1/3 longer than wide, the lobes more or less winged especially above, the mature carpels not separating, with 2 bracts present.---In shallow water. Widely distributed in the northern hemisphere and throughout most of the United States. Our records scattered, mostly in the northern half of Colorado at 5000-11,500 feet.

Family 68. Anacardiaceae SUMAC FAMILY

Shrubs, small trees or sometimes climbing plants, the sap usually acrid, resinous or milky; leaves alternate, pinnately or palmately 3- or more-foliate; flowers regular, perfect or polygamous, very small, in axillary or terminal panicles; sepals commonly 5 (4-6) with a glandular disk at base; petals distinct, commonly 5 (4-6); stamens as many as and alternate to the petals; ovary superior, 1-celled and 1-ovuled, styles usually 3; fruit dry but drupelike, 1-seeded.

1. <u>Rhus</u> L. SUMAC

Characters of the Family

1. Leaves pinnately compound with 11-21 leaflets; inflorescence terminating leafy twigs -1. <u>R. glabra</u>
1. Leaves simple or with 3 leaflets; inflorescence axillary
  2. Leaves simple -2A. <u>R. trilobata simplicifolia</u>
  2. Leaves 3-foliate
    3. Leaflets obovate, usually less than 3 cm. long, the terminal sessile or on a short stalk much less than 1 cm. long; fruit red or orange-red, hairy -2. <u>R. trilobata</u>
    3. Leaflets ovate to rhombic-ovate, over 3 cm. long, middle one on a stalk usually over 1 cm. long; fruit yellowish-white, usually glabrous -3. <u>R. radicans</u>

1. Rhus glabra L., Sp. Pl. 1:265. 1753.
    <u>R. cismontana</u> Greene; <u>Schmaltzia glabra</u> Small---Large shrub becoming 1-2 m. tall; branches glabrous or sometimes those of inflorescence puberulent; leaflets 11-21, 5-12 cm. long, lanceolate to elliptic-lanceolate, serrate, from deep to light green above, lighter, usually somewhat glaucescent below, accuminate, base somewhat unequal, lateral subsessile, terminal sessile to long-petiolulate; inflorescence a thrysoid panicle about 10-20 cm. long, of many flowers; sepals 1-2 mm. long; petals about 1 mm. long, whitish or greenish-yellow; fruit about 4 mm. long, globular, covered with a dense layer of short red hairs.---Valleys and slopes. New Hampshire to British Columbia, south to Georgia, Mexico and Nevada. Our records scattered in the western two-thirds of Colorado at 5500-7500 feet.
2. Rhus trilobata Nutt. ex T. & G., Fl. N. Amer. 1:219. 1838.
    <u>R. oxycanthoides</u> (Greene) Rydb.; <u>Schmaltzia trilobata</u> (Nutt.) Small---Shrubs 50-200 cm. tall, ill-scented; twigs slender usually puberulent at least when young; leaflets 3, 1-3 cm. long, all nearly or quite sessile, thin to rather thick, green and glabrous above, minutely pubescent below or sometimes glabrate, obovate, the terminal larger; inflorescence dense, compound and spikelike, the flowers appearing before the leaves; sepals about 1-1.5 mm. long; petals 2-3 mm. long, yellowish; fruit about 6-7 mm. long, red or orange-red, with short glandular hairs and some longer simple hairs.---Plains and hills. Alberta to California, south to Iowa and Mexico. Our many records well scattered over Colorado at 3500-9000 feet.
2A. Rhus trilobata simplicifolia (Greene) Barkley, (var.) Ann. Mo. Bot. Gard. 24:410. 1937.
    Like the species but most of the leaves with 1 leaflet, this oval or ovate, nearly or quite as broad as long.---Oklahoma to Utah, south to Mexico. Our few records from western and southwestern Colorado at 4500-5500 feet.
3. Rhus radicans L., Sp. Pl. 266. 1753.
    <u>R. rydbergii</u> Small; R. toxicodendron in part; <u>Toxicodendron rydbergii</u> (Small) Greene---Shrubs, or vinelike plants with aerial rootlets; branches glabrous or puberulent; leaves pinnately 3-foliate, the leaflets 3-20 cm. long, ovate or rhombic-ovate, usually coarsely serrate or dentate, sometimes entire, glabrous above, glabrous to pubescent below, rather thick and bright green; flowers in dense axillary panicles; sepals about 1 mm. long; petals about 3 mm. long, yellowish-white; fruit about 5-6 mm. in diameter, whitish or yellowish-white, usually glabrous. This is Poison Ivy, to which many people are susceptible.---Open woods, hillsides, roadsides, thickets and plains. Nova Scotia to British Columbia, south to Florida and Arizona. Our records in northcentral, central and southern Colorado, but probably more widely distributed, at 4500-8500 feet.

Family 69. Celastraceae STAFFTREE FAMILY;
BURNINGBUSH FAMILY; BITTERSWEET FAMILY

Woody shrubs or vines; leaves alternate or opposite, simple; stipules present or absent; flowers regular, usually perfect, small; calyx deeply parted, of 4 or 5 sepals; petals 4 or 5, distinct; stamens as many as petals or twice as many, inserted on or below the margins of a large disk which lines the calyx base; ovary 1, superior, 2- to 5-celled, style short or wanting, stigma 2- to 5-lobed; fruit a capsule or a follicle, the seeds with arils or caruncles, these minute to very conspicuous.

1. Leaves opposite, serrulate above the middle; sepals and petals 4; carpels 2 -1. <u>Pachystima</u>
1. Leaves alternate, entire; sepals and petals usually 5; carpels usually solitary -2. <u>Forsellesia</u>

1. <u>Pachystima</u> Raf. BOXLEAF; MOUNTAIN LOVER

Low depressed evergreen shrubs; leaves opposite, short-petioled, serrulate or dentate; flowers solitary or clustered in the leaf axils; sepals 4, petals 4; stamens 4; ovary 2-celled, 2 ovules in each cell; capsule 2-valved; seeds with a white aril.

1. Pachystima myrsinites (Pursh) Raf., Sylv. Tellur. 42. 1838.
    Leafy densely branched glabrous shrubs about 20-50 cm. tall; branches 4-ridged; leaves 6-40 mm. long, oval, elliptic to oblong, margins somewhat revolute, appearing thickened, serrate

above the middle; sepals about 0.7 mm. long, greenish; petals about twice as long as the sepals, deep red to brownish-red.---Wooded hills and slopes of mountains. Alberta to British Columbia, south to New Mexico and California. Our records well scattered in the western half of Colorado at 6000-11,000 feet.

2. Forsellesia Greene GREASEBUSH

Small, deciduous intricately branched shrubs; branches greenish, angled and more or less spinescent; leaves alternate, entire; flowers solitary or subsolitary, axillary; sepals 5 (rarely 4-6); petals 5 (rarely 4-6); stamens 5-8; ovary with 1 carpel, 1-celled with 1-2 ovules; fruit a follicle, aril of seed small.

1. Stipules present (but small); leaves pubescent, the petioles not over 1.5 mm. long; sepals 1.5-3 mm. long
    2. Young branches grayish-yellow; petals somewhat constricted below the apex; plants of western Colorado
        -1. F. meionandra
    2. Young branches grayish-green; petals not so constricted; plants present if at all in southeastern Colorado
        -2. F. planitierum
1. Stipules absent; leaves glaucous, the petioles about 2 mm. long; sepals 5-8 mm. long   -3. F. spinescens

1. Forsellesia meionandra (Koehne) Heller, Cat. N. Am. Pl. ed II. 130. 1900.
    Low spinescent, leafy plants not over 60 cm. high; young branches about 1 mm. thick, grayish-yellow, older branches gray with splitting bark; leaves 7-15 mm. long and 3-4 mm. wide, oblanceolate, gray-green, pubescent, acute; petioles 1 mm. long; stipules less than 0.5 mm. long, subulate; pedicels with scarious bracts at base; sepals 5, 3 mm. long and 2 mm. wide; disk more or less cup-shaped; petals 5, 4-6 mm. long and slightly more than 1 mm. wide, oblanceolate, somewhat constricted below apex; stamens 5-7; follicle 4 mm. long and 3 mm. wide.---Hills. Colorado and Utah. Our records from westcentral and southwestern Colorado at 4500-7000 feet with one doubtful record from Clear Creek County.
2. Forsellesia planitierum Ensign, Am. Midl. Nat. 27:509. 1942.
    Western Oklahoma and Texas. The type of this species came from very near the Colorado line (Baca County) and the plant should be sought in that area.
3. Forsellesia spinescens (Gray) Greene, Erythea 1:206. 1893.
    Texas to Arizona. This plant has been reported in Colorado and should be looked for in the southern part.

Family 70. Aceraceae MAPLE FAMILY

Shrubs or trees; leaves opposite, simple or palmately to pinnately compound; no stipules present; flowers regular, perfect, polygamous or dioecious, borne in fascicles or racemes; sepals 4 or 5, sometimes more; petals wanting or same number as sepals and distinct; stamens 4 to 12, borne on a ring-shaped disk or the disk sometimes wanting; ovary superior of 2 carpels, 2-lobed and 2-celled; fruit of 2 united samaras, the wings large and reticulate.

1. Acer L. MAPLE

Characters of the Family

1. Leaves pinnately compound of 3 or more leaflets (the end leaflet on an elongated rachis)
    2. Young twigs greenish, not glaucous but usually short-pubescent; leaflets 3 (rarely 5), glabrous beneath or villous on midrib   -1A. A. negundo interius
    2. Young twigs purplish or violet with a glaucous bloom, usually glabrous; leaflets usually 5-7, usually pubescent beneath   -1B. A. negundo violaceum
1. Leaves simple and lobed to palmately compound with 3 leaflets
    3. Leaves rather thick and leathery, usually pubescent beneath, the lobes with a few large, coarse and rather blunt teeth; petals lacking; inflorescence nearly sessile   -2. grandidentatum
    3. Leaves thin, glabrous, the lobes with numerous acute teeth; petals present; inflorescence long-stalked
        4. Leaf blades lobed but these not extending to the base, hence leaves not at all compound
            -3. A. glabrum
        4. Leaf blades 3-parted or divided nearly or quite to the base, hence some or most of leaves palmately compound
            5. Leaf not over 3 cm. wide; plants of western and northwestern Colorado   -3A. A. glabrum tripartitum
            5. Leaf over 4 cm. wide; plants of southern, central and southwestern Colorado
                -3B. A. glabrum neomexicanum

1. Acer negundo L., Sp. Pl. 1056. 1753.
    Rulac negundo (L.) Hitch.---Trees with trunk usually irregular; leaves pinnately 3- to 5-foliate, the leaflets 5-10 cm. long, ovate to lanceolate, usually becoming glabrous in age, entire to coarsely serrate or dentate, often 3-lobed; flowers dioecious, appearing with the leaves or a little before, staminate fascicled, pistillate in a dropping raceme; calyx 5-lobed; petals absent; stamens 4-6; fruit pendent, each samara 2-5 cm. long, glabrous. Represented in Colorado by the following 2 varieties.
1A. Acer negundo interius (Britt.) Sar., (var.) Bot. Gaz. 67:239. 1919.
    Negundo interius (Britt.) Rydb.---This variety has branches greenish, short-pubescent or rarely glabrous; leaflets 3 rarely 5, glabrous beneath or villous on midrib.---Along streams and canyons. Manitoba to Alberta, south to Missouri and Arizona. Our records scattered over the western two-thirds of Colorado at 4500-7500 feet.

1B. Acer negundo violaceum Jaeg. & Beissn., (var.) Ziergeholze Gart. Park. Ed. 3.6. 1889.

A. nuttallii (Nieuwl.) Lyon; Negundo nuttallii (Nieuwl.) Rydb.---This variety has branches purplish or violet, with a glaucous bloom; leaflets usually 5-7, usually pubescent beneath.---Along streams, valleys and canyons. Massachusetts to Montana, south to Ohio, Kansas and Colorado. Our one record from northwestern Colorado at 6000 feet.

2. Acer grandidentatum Nutt. ex. T. & G., Fl. N. Amer. 1:247. 1838.

Shrubs and trees to 10 m. tall, bark thin and dark brown, separating into plates; older twigs gray, young twigs light brown to reddish; leaves 4-10 cm. wide, simple, 3- to 5-lobed, sinuses rounded and the lobes again shallowly lobed, glabrous above, usually pubescent beneath; flowers polygamous and small; corolla absent; stamens 7-8; fruit with wings 1-2.5 cm. long, glabrous or nearly so.---Valleys, canyons and slopes. Montana to Idaho, south to Texas and Arizona; Mexico. Our one record from southwestern Colorado at 7000 feet.

3. Acer glabrum Torr., Ann. Lyc. N. Y. 2:172. 1828.

Glabrous shrubs 2-6 m. tall, usually rounded in shape; twigs usually gray; leaf blades 2-6 cm. long, 3-7 cm. wide, broadly cordate, 3- to 5-lobed to 3-parted, truncate to subcordate at base, middle lobe largest, all lobes acute with margins doubly and irregularly serrate; flowers monoecious or dioecious; petals shorter than the sepals; samaras each 2-3 cm. long, narrowly to widely divergent. The species intergrades with var. neomexicanum (Greene) Kearney & Peebles in Colorado, sometimes both simple and compound leaves present on one plant.---Mountains, usually on moist slopes or along streams. Wyoming to Idaho, south to Nebraska and Utah. Our records well scattered over the western two-thirds of Colorado at 5000-10,500 feet.

3A. Acer glabrum tripartitum (Nutt.) Pax., (var.) Engler's Botan. Jahrb. 7:218. 1886.

A. tripartitum Nutt.---In this variety the leaf blades are deeply 3-parted, some divided nearly or quite to the base into 3 leaflets, the whole leaf small, rarely over 3 cm. wide. A well marked variety in this state.---Wyoming to California, south to Utah. Our one record from northwestern Colorado at 8000 feet.

3B. Acer glabrum neomexicanum (Greene) Kearney & Peebles, (var.) Journ. Wash. Acad. Sc. 29:486. 1939.

In this variety the leaf blades are deeply 3-parted or divided nearly or quite to the base into 3 leaflets, the whole leaf 4 cm. wide or more. The variety is of doubtful value in Colorado but plants with compound leaves are scattered with the species here. Our records from 7000-9500 feet.

Family 71. Sapindaceae SOAPBERRY FAMILY

Shrubs or trees; leaves alternate, pinnately compound; flowers polygamous, small, regular, in panicles; sepals 4-5, imbricate in 2 rows; petals 4-5, distinct, inserted under a fleshy disk; stamens 8-10, inserted on the disk; ovary superior, 2- to 4-lobed, with same number of cells as lobes, with 1 ovule in each cell, styles 1, stigmas 2- to 4-lobed; fruit berrylike.

1. Sapindus L. SOAPBERRY

Characters of the Family

1. Sapindus drummondii Hook. & Arn., Bot. Beechey Voy. 281. 1840.

Shrubs or small trees to 8 m. tall, bark thin, furrowed; twigs gray in age, yellow-greenish and puberulous when young; leaflets 9-19, 4-8 cm. long, a terminal one present, lanceolate, falcate, oblique at base, glabrous above, more or less pubescent beneath especially on the main veins, margins entire; sepals about 2 mm. long, obtuse; petals about twice as long as the sepals, rhombic-lanceolate, ciliate, often lacerate at apex, whitish to cream-colored; fruit about 1.5 cm. broad, globose, yellow but drying black. This plant has recently been called S. saponaria var. drummondii (Hook. & Arn.) Benson.---Hillsides and valleys. Arkansas to Kansas, south to Louisiana and Arizona. Our records from southeastern Colorado at 4000-4500 feet.

Family 72. Rhamnaceae BUCKTHORN FAMILY

Shrubs or small trees; leaves simple, alternate, opposite or fascicled; flowers small, perfect or polygamous, regular or nearly so; sepals 4 or 5, the tube lined with a disk; petals distinct, 4 or 5; stamens the same number and opposite the petals, both on the disk; ovary free from or coalescent to the fleshy disk, 2- to 3-celled, each cell with 1 ovule; fruit a capsule or a berrylike drupe.

1. Fruit dry, capsulelike; petals 5, long-clawed, 2 mm. long or more; ovary adnate to disk for lower part (partly inferior) -1. Ceanothus
1. Fruit fleshy, drupelike; petals 4, rarely 5, short-clawed, about 1-1.5 mm. long; ovary little if any adnate to disk (essentially superior) -2. Rhamnus

1. Ceanothus L. NEW JERSEY TEA

Shrubs, the branches often spinescent; leaves alternate; flowers small, perfect, in crowded panicles, cymes or umbels; sepals 5, calyx adnate by disk to lower part of ovary; petals 5, hooded and clawed; stamens 5; ovary usually 3-lobed and 3-celled with a 3-cleft style; fruit capsulelike but separating at maturity into 3 nutlets, these finally dehiscent on inner edge, discharging the seeds.

1. Branches spinescent; leaves both entire and pubescent beneath — -1. **C. fendleri**
1. Branches unarmed; leaves toothed or if entire then glabrate below
   2. Leaves entire, seldom over 2 cm. long — -2. **C. martinii**
   2. Leaves toothed, over 2 cm. long
      3. Leaves thick, evergreen, glossy-varnished above and strongly aromatic, broadly elliptical or ovate, some usually over 2.5 cm. wide — -3. **C. velutinus**
      3. Leaves moderate to thin, deciduous, not glossy-varnished above, not strongly aromatic, narrowly elliptical or narrower, rarely over 2 cm. wide — -4. **C. ovatus**

1. Ceanothus fendleri Gray, Pl. Fendl. 29. 1849.
C. subsericeus Rydb.---Low shrubs 30-80 cm. tall; twigs canescent, spinose branchlets present, the plants only rarely nearly unarmed; leaves 10-25 mm. long, seldom over 8 mm. wide, narrowly elliptical or oblong, usually nearly glabrous above, densely gray-tomentulose or silky-canescent below, entire; flowers white, in umbellike clusters arranged in terminal racemes; calyx about 1.5 mm. long; petals 2.5-3 mm. long; fruit about 4-5 mm. wide, subglobose, 3-lobed.---Woods and slopes. Colorado to Utah, south to Texas, Mexico and Arizona; also reported farther north. Our records scattered over the western two-thirds of Colorado, except the northwestern corner, at 5500-9000 feet.

2. Ceanothus martinii M. E. Jones, Contrib. West. Bot. 8:41. 1898.
Utah, Nevada and Arizona. This plant should be sought in western Colorado, and has been reported in Boulder County but the record seems doubtful.

3. Ceanothus velutinus Dougl.; Hook.,Fl. Bor. Am. 1:125. 1830.
Shrubs 1-3 m. tall, usually spreading, unarmed, usually in dense patches, twigs olive-green becoming gray in age; leaf blades 4-8 cm. long, some usually over 25 mm. wide, usually about twice as long as wide or less than twice as long, broadly elliptical or ovate, obtuse or rounded at apex, rounded or subcordate at base, dark green, glabrous and glandular-varnished above, odoriferous, paler and velutinous or canescent below, closely glandular-serrulate; petioles rather short; inflorescence paniculate, dense, the flowers white; sepals about 1.5 mm. long; petals about 2-2.5 mm. long; fruit about 3-6 mm. long, subglobose or triangular, 3-lobed near summit at least, glandular.---Hillsides and mountain slopes. South Dakota to British Columbia, south to Colorado and California. Our records scattered over the western half of Colorado, mostly in the northern part, at 6500-9000 feet.

4. Ceanothus ovatus Desf., Hist. Arbr. & Arbris. 2:381. 1809.
C. pubescens (T. & G.) Rydb.; C. mollissimus Torr.---Shrubs 20-100 cm. tall, unarmed, with twigs puberulent to glabrate; leaf blades 2-6 cm. long and rarely over 2 cm. wide, thin to moderately thick, oblong-elliptic, ovate-lanceolate to elliptic-lanceolate, rounded or acute at base, obtuse or acute at apex, glandular-serrate, glabrous above or pubescent on the veins, usually pubescent beneath; flowers white, in corymbose clusters; calyx about 1-1.5 mm. long; petals about 2-2.5 mm. long; fruit about 4-6 mm. wide, nearly globose, somewhat lobed.---Plains and hills, often on sandy soil. New England states to Manitoba, south to Florida and New Mexico. Our records from northcentral and central Colorado at 5000-7500 feet.

2. Rhamnus L. BUCKTHORN

Shrubs or small trees with or without spines; leaves alternate, opposite or fascicled; flowers perfect or unisexual, axillary, solitary to clustered, small and greenish; sepals 4 or 5, with a disk lining the calyx tube; petals 4 or 5 or absent; stamens 4-5, inserted with the petals on the disk; ovary free or almost free from the calyx and disk, 2- to 4-celled, each cell 1-ovuled; fruit a berrylike drupe with 2-4 nutlets.

1. Winter buds without bud scales; flowers in peduncled umbels, perfect, 5-merous — -1. **R. betulaefolia**
1. Winter buds scaly; flowers solitary or in sessile umbels, unisexual, 4-merous
   2. Cultivated plants, rarely if ever escaping; some of branches usually spiny; leaves about twice as long as wide — -2. **R. cathartica**
   2. Plants native to west and southwestern Colorado; branches never spiny; leaves about 3 times as long as wide — -3. **R. smithii**

1. Rhamnus betulaefolia Greene, Pitt. 3:16. 1896.
Utah and Nevada south to Arizona and New Mexico. This species has been reported close to southwestern Colorado and may be growing in that part of the state.

2. Rhamnus cathartica L., Sp. Pl. 193. 1753.
Shrubs or low trees 1-8 m. tall (occasionally lower when trimmed in cultivation); branches often opposite and ending in sharp thorns, buds with scales; leaves 2-6 cm. long, usually opposite or approximately opposite, oval to ovate or broadly elliptical, irregularly crenate, glabrous or minutely puberulent; flowers 1-4, in the axils of the leaves, unisexual; sepals 4, about 2.5-3 mm. long; petals 4, about 1-1.5 mm. long; fruit about 7-9 mm. wide, globose, glabrous.---Around dwellings and in cultivated areas. Escaping in various parts of Canada and the United States. Our one record from northcentral Colorado at 5100 feet but the plant might have been cultivated.

3. Rhamnus smithii Greene, Pitt. 3:17. 1896.
Unarmed shrubs about 2-3 m. tall; twigs glabrous or pubescent; bud scales present; leaves 2-7 cm. long, alternate or approximately opposite, elliptical to ovate-oblong or oblong-lanceolate, apex acute or obtuse, base rounded or cuneate, serrulate to crenulate, green or yellowish-green but often paler below, glabrous to pubescent both surfaces; flowers unisexual, 1-3 in the axils; sepals 4, about 2 mm. long; petals 4, about 1 mm. long; fruit about 8 mm. long and almost as wide, glabrous, blackish. Our plant has been called ssp. typica Wolf.---Valleys and hillsides. The subspecies from Colorado and New Mexico, probably farther west. Our plants from southwestern and western Colorado at about 7000 feet.

Family 73. Vitaceae GRAPE FAMILY

Woody vines, climbing or trailing, usually with tendrils; stems swollen at the nodes; leaves alternate, petioled, simple or compound with long petioles rather dilated at base; flowers small, greenish, perfect or unisexual, in cymose panicles; calyx entire or 4- to 5-toothed; petals 4 or 5, separate or coherent, valvate, caducous; stamen as many as and opposite to the petals; disk often present above the ovary; ovary superior, of 2 carpels or sometimes 2-6, 2- to 6-celled, 1-2 ovules in each cell, style 1; fruit a berry.

1. Leaves simple; hypogynous disk present around the ovary or its base; fruit juicy -1. Vitis
1. Leaves palmately 5- to 7-foliate; disk not present; fruit with thin pulp -2. Parthenocissus

1. Vitis L. GRAPE

Leaves simple, palmately veined and more or less palmately lobed; flowers dioecious or polygamo-dioecious, rarely perfect; hypogynous disk present; calyx minute; petals more or less coherent at apex until they fall; fruit juicy.

1. Leaves bright green above, glabrous except along veins below, usually longer than wide, young growth glabrous or soon glabrous; plants of eastern Colorado -1. V. vulpina
1. Leaves gray-green above, more or less pubescent or tomentose below at least to midseason, usually wider than long, young growth tomentose or floccose; plants of eastern to southwestern Colorado
   2. Vigorously climbing plants; leaves mostly prominently lobed (cut about 1/3 or more to the midrib); berries 12-15 mm. wide -2. V. doaniana
   2. Plants not vigorously climbing if at all; leaves not lobed or shallowly lobed (not cut in much over 1/4 to midrib); berries 8-12 mm. wide
      3. Leaf surface permanently tomentose or thickly floccose; plants if present in southwestern Colorado -3. V. arizonica
      3. Upper leaf surface glabrate in age; plants of eastern and southeastern Colorado -4. V. longii

1. **Vitis vulpina** L., Sp. Pl. 203. 1753.
   Climbing vines; young growth glabrous or soon becoming so; stems terete with thin twig-node diaphragms about 1-2 mm. wide or even less; leaves 8-18 cm. long, usually not as wide as long, rather thin, not coriaceous, cordate-ovate, margin cut into large and small sharp teeth, often 3-lobed especially on ground shoots, upper surface bright green and glabrous or nearly so, under surface lighter green and usually retaining hairs on veins and in axils of veins; fruit cluster 6-12 cm. long; berry 6-10 mm. thick, black with a heavy bloom; seeds about 6 mm. long.---Banks, roadsides, fields and edges of woods. New Brunswick to Manitoba, south to Tennessee, Texas and Colorado. Our records scattered in the eastern half of Colorado, mostly along the foothills, at 3500-7000 feet.

2. **Vitis doaniana** Munson ex Viala, Une Mission Viticole, 101. 1889.
   Oklahoma to Texas and New Mexico. Recorded as close to southeastern Colorado and may be found there.

3. **Vitis arizonica** Engelm., Am. Nat. 2:321. 1868.
   Texas, New Mexico, Arizona and Utah. This plant should be looked for in southern Colorado.

4. **Vitis longii** Prince, Treat. Vine, 184. 1830.
   Much branched but little climbing plants, the tendrils short or even absent; young growth pubescent, tomentose or floccose, this more or less retained, diaphragms of twig-nodes 1-3 mm. thick; leaves 7-12 cm. long, usually broader than long, thick, broadly cordate, margins indistinctly or shallowly lobed, large and small teeth present, upper side glabrate in age except along veins, under side retaining more or less of pubescence or tomentum, gray-green on both sides; fruit clusters short, 3-6 cm. long; berries about 8-12 mm. thick, black with a heavy bloom; seeds 5-6 mm. long.---Canyons and valleys. Kansas to Colorado, south to Texas and New Mexico. Our few records from extreme eastern Colorado at 3500-4500 feet.

2. Parthenocissus Planch. VIRGINIA CREEPER

Leaves palmately compound with 5-7 leaflets; flowers perfect or polygamo-dioecious; no disk present; sepals 5; petals 5, spreading at anthesis; fruit thin-fleshed.

1. **Parthenocissus vitacea** (Knerr) Hitchc., Sp. Fl. Manhattan 26. 1894.
   Psedera vitacea (Knerr) Greene; P. inserta Kerner---Aerial rootlets lacking; tendrils without sucking disks; leaflets 4-10 cm. long, lanceolate or oval, acuminate, serrate to almost laciniate, with large, lanceolate, often flaring teeth, green both sides; cymes about 5 cm. broad; fruit 5-7 mm. in diameter, blue-black. Our plant may be only a variety of the eastern P. quinquifolia (L.) Planch.---Banks and woods. Ohio to Wyoming, south to New Mexico and Arizona. Our records in the western two-thirds of Colorado, mostly in the northern part, at 4500-7500 feet.

Family 74. Malvaceae MALLOW FAMILY

Herbaceous plants with more or less mucilaginous juice, often pubescent with stellate or forked hairs; leaves alternate, simple, petioled, generally palmately veined or lobed, stipulate; flowers regular, perfect; calyx of 5 more or less united sepals, often subtended by a calyxlike involucre; petals 5, hypogynous, more or less united at base to each other and to the staminal column; stamens many, monadelphous, anthers 1-celled, reniform; ovary superior, of several carpels, each 1-celled, usually united to each other; styles united below, branched above; fruit a capsule, the carpels loculicidal or separating from each other at maturity.

1. Fruit a capsule, the 5 valves not separating from each other or from the axis at maturity (scarcely any central axis present after dehiscence); our species an annual with deeply lobed leaves — -1. Hibiscus
1. Fruit of 5-20 carpels united in a ring around a central axis, usually separating from this at maturity but sometimes connate and remaining attached by a thread; our plants either perennials or if annuals the leaves are not deeply lobed
   2. Style branches longitudinally stigmatic, not at all capitate at ends
      3. Petals crimson when fresh, 2-3 cm. long; bractlets of involucel 8-10 mm. long; calyx about 16-20 mm. long in anthesis — -2. Callirrhoe
      3. Petals white, yellow or rose-purple, less than 2 cm. long; bractlets absent or small; calyx about 4-6 mm. long in anthesis
         4. Stamens in 1 series at the apex of the column; involucel of small narrow bracts; plants annual; leaf blades shallowly if at all lobes — -3. Malva
         4. Stamens more or less united into 2 whorls; involucel absent; plants perennial; leaf blades (at least the upper) deeply parted — -4. Sidalcea
   2. Style branches widening upward to capitate or truncate stigmas
      5. Carpels sharply differentiated into a more or less reticulated, indehiscent basal part and a smooth, dehiscent apical part, the 2 parts separated by a pronounced ventral notch; petals conspicuously salmon-red or brick-red — -5. Sphaeralcea
      5. Carpels completely smooth or completely reticulated; petals white, yellow, pink, rose-purple or red but never brick-red
         6. Ovules and seeds 1 to a carpel; carpels indehiscent or dehiscent only at the apex, more or less reticulated; petals white to ochroleucous when fresh; plants densely whitish-stellate — -6. Sida
         6. Ovules and seeds 2 or more to a carpel; carpels dehiscent to the base or nearly so, not reticulated; petals variously colored, never really whitish; plants sparingly if at all whitish-stellate
            7. Involucel none; column antheriferous only at or near the apex; petals much less than 15 mm. long — -7. Abutilon
            7. Involucel of 3 persistent bractlets; column antheriferous to far below apex; petals large, about 20 mm. long or more — -8. Iliamna

1. Hibiscus L. ROSEMALLOW

Annual herbaceous plants; leaves stipulate, deeply lobed; flowers axillary, peduncled; calyx prominently inflated in fruit; bractlets 3-5; petals showy; staminal column truncate or 5-toothed at summit, anther-bearing below the apex; ovary 5-celled, styles 5, stigmas capitate or peltate; fruit a loculicidal, 5-valved capsule, the valves remaining together after dehiscence; seeds several in each carpel.

1. Hibiscus trionum L., Sp. Pl. 697. 1753.
   Annual plants; stems branching from base, depressed, hirsute to pubescent; leaves ovate or suborbicular in outline, digitately 5- to 7-lobed or parted, the lobes toothed or incised; bractlets 2-4 mm. long, linear; calyx prominently nerved, inflated in fruit, hirsute-pubescent; petals 1-4 cm. long, whitish or pale-yellow with a purple or brownish spot near base (flowering period very short for 1 flower); capsule about 1-1.5 cm. long, globose-ovoid, hairy; seeds roughened.---Fields and waste places. A European species introduced in the United States especially eastward. Our few records widely scattered over Colorado in cultivated areas at 4000-5500 feet.

2. Callirrhoe Nutt. POPPYMALLOW

Perennial herbaceous plants; leaves with lobed or cleft blades; flowers terminal and axillary, showy; involucel present, of 3 bractlets; petals not notched at apex; stamens not in 2 series; style branches filiform, longitudinally stigmatic; carpels 10-20, 1-celled, 1-seeded; beaked.

1. Callirrhoe involucrata (T. & G.) A. Gray, Mem. Am. Acad. (II). 4:16. 1849.
   Roots thick and farinaceous; stems procumbent, more or less hirsute; leaf blades rounded in outline, palmately 5- to 7-divided nearly to the base, the divisions deeply cleft into lanceolate to linear divisions or oblong on lower leaves, more or less pubescent with simple or stellate hairs especially below; bractlets of involucel linear, about 8-10 mm. long, about 1/2 as long as the calyx; petals 2-3 cm. long, crimson but drying or fading to purple, truncate and somewhat erose at apex; carpels more or less rugose, reticulated.---Dry plains. Minnesota to Utah, south to Texas and New Mexico. Our records scattered in the eastern half of Colorado at 3500-6000 feet.

3. Malva L. MALLOW

Annual or perennial herbaceous plants; leaf blades orbicular or reniform, more or less lobed; flowers solitary or in small clusters; involucel present; petals obcordate or emarginate; styles filiform, longitudinally stigmatic; fruit depressed and disklike, carpels numerous, indehiscent, separating when mature, beakless.

1. Corolla not over 6 mm. long, scarcely exceeding the calyx, claw of the petal glabrous; carpels rugose-reticulate -1. **M. parviflora**
1. Corolla 10-15 mm. long, about twice as long as the calyx, claw of petal glabrous or hairy; carpels smooth to rugose-reticulate
    2. Leaves 6-20 cm. wide, strongly crisped and crinkly on the margins; plants upright; carpels rugose-reticulate; plants rare as escapes from cultivation -2. **M. crispa**
    2. Leaves 2-6 cm. wide, little if any crisped on margins; stems procumbent; carpels smooth or faintly reticulate; plants rather common as weeds -3. **M. neglecta**

1. Malva parviflora L., Amoen. Acad. 3:416. 1756.
    Annual plants; stems 20-90 cm. tall, erect or ascending, glabrous or sparingly hairy; leaf blades 2-10 cm. wide, reniform or orbicular, shallowly 5- to 7-lobed, these crenate, glabrous to sparingly hairy; petals about 4-5 mm. long, scarcely if any longer than the calyx, pink to lilac; carpels rugose-reticulated.---Waste places. Introduced from Europe in various parts of the United States. Our record from northcentral Colorado at about 5000 feet.
2. Malva crispa L., Sp. Pl. 970. 1753.
    Annual plants; stems 30-100 cm. tall, erect; leaf blades 8-20 cm. wide, round or reniform in outline, distinctly (but shallowly) lobed, double crenate and crisp on the margins, sparsely strigose both sides; corolla about 1 cm. long and twice as long as the calyx, white to purplish; carpels reticulated.---Waste places and around dwellings, escaping or persisting after cultivation. Our few records from northcentral Colorado at 5000-5500 feet.
3. Malva neglecta Wallr., Syll. Ratisb. 1:140. 1824.
    M. rotundifolia of Manuals-not L.---Annual (or appearing biennial) plants from thickened roots; stems 10-30 cm. long, procumbent and branching at base, more or less pubescent especially when young, often stellate; leaves 2-6 cm. wide, rounded-reniform, with 5-9 shallow crenate lobes or almost lobeless, more or less pubescent both sides; petals about 1 cm. long, twice as long as the calyx, pale blue to whitish; carpels pubescent or glabrate in age, slightly if at all reticulated on the back and sides.---A weed in waste places, gardens and fields. Naturalized from Europe in various parts of the United States. Our records scattered in Colorado, none as yet from the extreme eastern part, at 4500-7000 feet.

4. Sidalcea Gray CHECKERMALLOW; FALSE MALLOW

Perennial herbaceous plants with woody roots; leaves palmately lobed or cleft; inflorescence in terminal, bracted racemes; involucel none; petals rather showy; stamens in 2 more or less distinct series, an outer and inner; style branches filiform, longitudinally stigmatic; carpels 5-9, indehiscent but at maturity separating from the axis, 1-seeded.

1. Petals white to yellowish; stems usually glabrous below; leaves glabrous above, nearly so beneath -1. **S. candida**
1. Petals rose-purple; stems usually pubescent below; leaves pubescent above and below -2. **S. neomexicana**

1. Sidalcea candida Gray, in Mem. Acad. n. ser. 4:20. 1849.
    Stems 40-90 cm. tall, simple, erect, usually glabrous to the inflorescence; leaves 4-15 cm. wide, glabrous above, a few stiff hairs below, basal leaves orbicular, 7-lobed, these coarsely dentate; cauline leaves cut more than 1/2 to base into 1-7 lanceolate-linear mostly entire segments; inflorescence racemose, the rachis and pedicel stellate-pubescent; calyx 4-6 mm. long in anthesis, densely stellate-pubescent; petals 10-15 mm. long, white or yellowish.---Wet meadows or along streams. Wyoming to Nevada, south to New Mexico and Utah. Our records scattered in the western two-thirds of Colorado at 4500-11,000 feet.
2. Sidalcea neomexicana Gray, in Mem. Am. Acad. n. ser. 4:23. 1849.
    Stems 20-90 cm. tall, 1 to several, simple or branched, erect to decumbent at base, hirsute to nearly glabrous; basal leaves 1-6 cm. wide, orbicular, from crenate to shallowly 5- to 9-lobed; cauline leaves 5- to 9-parted into 3- to 5-lobed segments, both leaf types ciliate and pubescent on both surfaces; inflorescence a raceme; calyx about 5 mm. long, more or less hirsute; petals 12-18 mm. long, purple-rose, rarely to white.---Meadows and valleys. Wyoming to Idaho, south to Mexico. Our records scattered in the western three-fourths of Colorado at 5000-9000 feet.

5. Sphaeralcea St. Hil. GLOBE MALLOW

Perennial, stellate-pubescent plants; leaf blades palmately lobed or merely toothed; inflorescence racemose or paniculate; involucel usually present, of 3 subulate deciduous bractlets; corolla large, usually grenadine-red or grenadine-pink; carpels not dehiscent to base, differentiated into a smooth dehiscent apical portion and an indehiscent, reticulated basal portion, often remaining attached to the axis by a stout thread, uniformly stellate-pubescent; ovules and seeds 1-3 in each carpel.

1. Indehiscent reticulate part of carpels forming 2/3 or more of the whole; leaves 3-parted or divided
    2. Stems and leaves silvery-lepidote; upper leaf blades entire, lower 3-divided -1. **S. leptophylla**
    2. Stems and leaves canescent or more coarsely pubescent; all leaves deeply 3-cleft (often appearing 5-cleft)
        3. Primary lateral divisions of the leaf blade 2/3 as long as the middle lobe or longer; dehiscent part of carpel forming less than 1/3 of the whole; plants erect or prostrate, short to tall
            4. Leaf blades dissected but ultimate divisions usually over 4 mm. wide (on widest ones); stems usually decumbent, seldom over 20 cm. tall -2. **S. coccinea**
            4. Leaf blades dissected into narrow divisions the ultimate ones seldom over 3 mm. wide; stems often erect, often over 25 cm. tall -2A. **S. coccinea dissecta**

3. Primary lateral divisions of the leaf blade less than 2/3 as long as the midlobe; dehiscent part of carpel forming about 1/3 of the whole; plants often erect and over 25 cm. tall   -2B. _S. coccinea elata_
1. Indehiscent, reticulate part of carpels smaller, seldom over 1/2 the length, usually less; leaves not divided or if lobed these not extending 1/2 way into the midrib (except in S. digitata - doubtful in Colorado)
   5. Indehiscent part of the carpel usually rugose or muricate dorsally   -3. _S. subhastata_
   5. Indehiscent part of carpel smooth or nearly so dorsally
      6. Leaf blades about as long as wide
         7. Leaf blades cleft more than 1/2 way to midvein   -4. _S. digitata_
         7. Leaf blades without lobes or the lobes not extending more than 1/2 way into the midrib
            8. Pubescence dense, grayish; leaf blades usually thick and finely crenate; carpels broadly ovate, usually acute or cuspidate at apex   -5. _S. parvifolia_
            8. Pubescence usually sparse and plant bright green; leaves usually thin and coarsely toothed; carpels nearly orbicular, obtuse or slightly mucronate at apex   -6. _S. munroana_
      6. Leaf blades definitely longer than wide
         9. Leaf blades narrow, about 4 times longer than wide   -7. _S. angustifolia_
         9. Leaf blades usually more than 1/2 as wide as long
            10. Pubescence yellowish; leaf blades shallowly lobed; column of stamens 6-8 mm. long; hairs of stem very short (not over 0.2 mm. long)
               11. Petioles 1/2 as long as the blade or longer; blades 2/3 as wide as long, usually subcordate or truncate   -8. _S. incana_
               11. Petioles usually less than 1/2 as long as the blades; blades 1/3 to 2/3 as wide as long, cuneate at base   -8A. _S. incana cuneata_
            10. Pubescence grayish; leaf blades often deeply cleft; column of stamens 4-6 mm. long; hairs of stem usually longer (usually 0.6 mm. long or more)   -9. _S. fendleri_

1. Sphaeralcea leptophylla (Gray) Rydb., Bull. Torr. Club 40:59. 1913.
    Malvastrum leptophyllum Gray---Perennial plants with long taproots and woody crowns or with long horizontal roots, silvery-lepidote throughout, the appressed scales with short marginal hairs; stems 30-50 cm. tall, several to many; leaves often fascicled below on stem, upper entire, linear or narrowly oblanceolate, lower 3-divided into linear lobes not over 3 mm. wide; inflorescence racemiform, few-flowered; involucel rather persistent, often to fruiting time; calyx 4.5-7 mm. long, the lobes as long or longer than the tube; petals 9-15 mm. long, grenadine; reticulated indehiscent part of carpel 2/3 to 3/4 of whole, the fruit 3-3.5 mm. high; seeds 1 to a carpel, glabrous or sparsely pubescent.---Dry ground, often on rocky hills or mesas. Colorado to Utah, south to Texas and Mexico. Our records from western Colorado at about 4000-4500 feet.
2. Sphaeralcea coccinea (Pursh) Rydb., Bull. Torr. Club 40:58. 1913.
    Malvastrum coccineum Gray---Perennial plants with woody bases, often in patches; stems 10-25 cm. tall, usually decumbent, usually densely stellate; leaf blades 3-cleft to base, the lateral divisions again deeply parted, the ultimate segments of the widest ones about 4 mm. wide or wider, usually more densely stellate below; involucel usually lacking; calyx at anthesis 5-10 mm. long, the lobes 1-2 times as long as the tube, villous, usually densely so; petals 10-20 mm. long, grenadine; carpels in fruit 3-3.5 mm. high, indehiscent portion about 3/4 of length, very coarsely and prominently reticulate, ovules and seeds 1 to a carpel; seeds glabrous or sparingly pubescent.---Dry plains and hills. Manitoba and Alberta, south to Iowa, Texas and Arizona. Our records scattered over Colorado at 3500-9000 feet.
2A. Sphaeralcea coccinea dissecta (Nutt.) Kearney, (ssp.) Univ. Calif. Pub. in Bot. 19(1):96. 1935.
    S. dissecta (Nutt.) Rydb.; Malvastrum dissectum (Nutt.) A. Nels.---This subspecies differs from the species as follows: plants usually more conspicuously whitish-pubescent; stems often more nearly erect and taller; leaf blades more dissected, the ultimate divisions narrower, not over 3 mm. wide. Intergrades to some degree with the species in Colorado.---Wyoming, south to Texas and Arizona. Our records from the western third of Colorado at 4500-8500 feet.
2B. Sphaeralcea coccinea elata (Baker) Kearney, (ssp.) Univ. Calif. Pub. in Bot. 19(1):97. 1935.
    S. elata (Baker) Rydb.; Malvastrum elatum (Baker) A. Nels.; M. dissectum var. cockerelli A. Nels.---In this variety the leaves are lobed, with the ultimate segments moderately broad, 4 mm. wide or more as in the species; the stem is 25-60 cm. long, more frequently erect or ascending; leaves have the middle lobe elongated somewhat, the lateral lobes only 1/2 to 2/3 as long; inflorescence is more apt to be elongated and not crowded as is so common in the species.---Texas to Utah, south to Arizona. Our few records from southwestern Colorado at 6000-7500 feet.
3. Sphaeralcea subhastata Coult., Contrib. U. S. Nat. Herb. 1:32-33. 1890.
    The ssp. connata Kearney is recorded from New Mexico and Arizona and is to be looked for in southwestern Colorado.
4. Sphaeralcea digitata (Greene) Rydb., Bull. Torr. Club 40:58. 1913.
    New Mexico, Arizona and Utah. To be expected in southwestern Colorado.
5. Sphaeralcea parvifolia A. Nels., Proc. Biol. Soc. Wash. 17:94-95. 1904.
    S. marginata York; S. arizonica Heller---Plants perennial with large roots and woody crowns; stems 20-100 cm. tall, several, erect or ascending, grayish-canescent; leaves 2-5 cm. long (on larger blades), thickish, broadly ovate to rotund or reniform, apex obtuse, lobeless or shallowly 3-lobed, margins toothed, more or less stellate-canescent both sides; involucels narrow, small and not conspicuous; calyx 4-8 mm. long at anthesis, lobes as long or longer than tube; petals 8-15 mm. long, grenadine; carpels 3-5 mm. high, broadly ovate and not as wide as high, indehiscent portion reticulate on face, about 1/4 the length of the carpel, dehiscent portion smooth, acute or short cuspidate. Some of our western specimens tend to intergrade with S. munroana (Dougl.) Spach. in rather sparse pubescence and broad almost obtuse carpels.---

Dry mesas, slopes and forest openings. Colorado to California, south to New Mexico and Arizona. Our records from western Colorado at 4500-6500 feet.
6. Sphaeralcea munroana (Dougl.) Spach. in Gray, Proc. Am. Acad. 22:292. 1887.
Montana to British Columbia, south to Wyoming, Utah and California. This plant should be expected in western Colorado.
7. Sphaeralcea angustifolia (Cav.) Don, Hist. Dichl. Pl. 1:465. 1831.
S. cuspidata of Manuals---Perennial plants with thick woody crowns, rather densely stellate-canescent; stems 30-150 cm. tall, erect or nearly so; leaf blades 5-10 cm. long (about 1/4 to 1/5 as wide as long on most), lanceolate or linear-lanceolate, usually acute or short-acuminate, cuneate and with short basal lobes to subcordate and lobeless in some forms; involucels short, narrow and inconspicuous; calyx at anthesis about 5-9 mm. high, the lobes usually deltoid-ovate and acute; petals about 7-12 mm. long, grenadine or grenadine-pink; carpels 3.5-6.5 mm. long, about 1/2 to 3/5 as wide, the indehiscent portion 1/10 to 2/5 the entire length, reticulate on sides, dehiscent portion smooth, usually mucronate or cuspidate at apex. Our plant is var. cuspidata Gray.---Plains. The variety from Kansas to Colorado, south to Texas, Mexico and California. Our records from the southeastern quarter of Colorado at 3500-7000 feet.
8. Sphaeralcea incana Torr. in Gray, Mem. Am. Acad., n. ser. 4:23. 1849.
Texas to Arizona, south to Mexico. This species could be in southern Colorado.
8A. Sphaeralcea incana cuneata Kearney, (ssp,) Univ. Calif. Pub. Bot. 19:59. 1935.
New Mexico and Arizona. This plant may be found in southwestern Colorado.
9. Sphaeralcea fendleri Gray, Pl. Wright. 1:21-22. 1752.
Perennial plants 60-150 cm. tall, with woody crowns, sparsely to densely stellate-canescent; leaf blades 2-6 cm. long, oblong-ovate to broadly ovate, base usually cuneate, apex acute, obtuse or mucronate, shallowly to deeply 3-lobed, the lateral lobes about 1/4 as long as the middle one, margins of lobes crenate or crenate-dentate or cleft, usually green above, lighter below; bractlets small and narrow; calyx 4-6 mm. high at anthesis, the lobes about as long as the tube; petals 8-13 mm. long, grenadine or grenadine-pink; carpels 4-5 mm. high, about 3/5 as wide, indehiscent portion about 1/5 to 1/3 of the whole carpel, finely and often faintly reticulate on sides; dehiscent portion smooth, usually cuspidate.---Banks and slopes, often in forest openings. Colorado, south to Texas, Mexico and Arizona. Our few records from southwestern Colorado at 6000-7000 feet.

6. Sida L.

Perennial herbaceous plants, pubescent with stellate-canescent or lepidote hairs; flowers axillary, usually solitary, white or ochroleucous often fading to pink or purple; bractlets 1-3 or apparently none; carpels 5 to many, styles filiform with capitate stigmas; carpels in fruit 1-seeded, indehiscent or finally dehiscent only part way from apex.

1. Leaves reniform or orbicular, wider than long, not at all hastate at base, rounded at the apex; involucel of 1-2 subulate bractlets present; carpels indehiscent -1. S. hederacea
1. Leaves lanceolate or oblong-lanceolate, hastate at base, acute to acuminate at apex; involucel absent; carpels dehiscent at apex -2. S. lepidota

1. Sida hederacea (Dougl.) Torr. in Gray, Mem. Am. Acad. n. ser. 4:23. 1849.
Disella hederacea (Dougl.) Greene---Elongated or rather short rootstocks present; stems 10-40 cm. long, decumbent or prostrate, white-stellate; leaves 1-5 cm. wide, broadly deltoid, reniform or suborbicular, dentate, the teeth rather regular in size, rounded at apex, very oblique at base, white-stellate; involucels about 3 mm. long, very narrow; calyx about 6 mm. long; petals about 1 cm. long, whitish to ochroleucous often fading or drying to pink; carpels 6-10, more or less reticulate on the sides.---Low banks and flats. Oklahoma to Washington, south to Texas, Mexico and California. Our one record from southeastern Colorado at 4000 feet.
2. Sida lepidota A. Gray, Pl. Wright. 1:18. 1852.
Disella sagittaefolia (A. Gray) Greene---Caespitose plants; stems 10-40 cm. long, spreading-ascending or decumbent, more or less scurfy-stellate; leaf blades 2-5 cm. long, lanceolate or oblong-lanceolate, hastate at base, sinuate-dentate, more or less scurfy-stellate especially when young; calyx about 5-6 mm. long; petals about 10-12 mm. long, yellowish-white or tinged with rose or purple. Our plant is var. sagittaefolia Gray.---Plains. The variety from Colorado, south to Texas, Mexico and California. Our one record from southeastern Colorado at about 4000 feet.

7. Abutilon Mill. INDIAN MALLOW

Annual or perennial herbaceous plants; leaves entire, toothed or lobed, rounded-cordate; flowers axillary; no involucel present; petals yellow to pink or red; styles filiform, or clavate with terminal capitate stigmas; carpels 4-30, 2- to 9-seeded, tardily or not at all separating from each other, dehiscent dorsally and ventrally, more or less beaked.

1. Plants annual; carpels usually 12-15 with long divergent awns at apex -1. A. theophrasti
1. Plants perennial; carpels 4-6, rounded or cuspidate at apex, no long awns present
    2. Carpels cuspidate or mucronate at apex; stems spreading or ascending; leaves seldom over 3 cm. long -2. A. parvulum
    2. Carpels rounded at apex, not cuspidate or mucronate; stems usually erect; leaves (largest) 4 cm. long or more -3. A. incanum

1. **Abutilon theophrasti** Medik., Malvenfam. 28. 1787.
   A common weed in the eastern part of the United States. It should be expected in this state.
2. **Abutilon parvulum** A. Gray, Pl. Wright, 1:21. 1852.
   Colorado to California, south to Texas and Mexico. Generally reported for Colorado but no specimens from the state have been located by the writer.
3. **Abutilon incanum** (Link) Sweet, Hort. Brit. 53. 1827.
   A. texense Torrey & Gray---Perennial plants with large rootstocks; stems 30-150 cm. tall, erect or nearly so, rather stout, branched, short-stellate; leaves 4-10 cm. long, ovate, cordate at base, abruptly acuminate or acute, dentate or somewhat lobed, rather thick, usually densely tomentulose above and below; flowers solitary in axils or sometimes in leafy panicles; calyx 2-4 mm. long in flower; petals 5-7 mm. long, orange-yellow; carpels muticous or nearly so, stellate-puberulent, of 5-8 carpels.---Dry slopes. Texas to Arizona, south to Mexico. Our one record from southeastern Colorado at about 4500 feet.

8. Iliamna Greene WILD HOLLYHOCK

Perennial plants from woody caudices, sparsely stellate-pubescent; stems tall and leafy; leaves large, rather shallowly 3- to 7-cleft; flowers large, in interrupted spicate or corymbose racemes; involucel of 3 distinct, narrow, persistent bracts; petals large, white, pink or rose-purple; staminal column coarsely stellate-hirsute below; carpels not reticulate, dehiscent to the base on back and about 2/3 down ventrally, attached to the axis by a stout thread, persisting for some time after seed dispersal, with short-stellate hairs and also long hairs on back; ovules and seeds 2-4 (usually 3) in each carpel.

1. **Iliamna rivularis** (Dougl.) Greene, Leafl. Bot. Obs. & Crit. 1:206. 1906.
   Sphaeralcea rivularis (Dougl.) Torr.; Phymosia rivularis (Dougl.) Rydb.---Stems 60-200 cm. tall, erect; leaves 5-15 cm. wide, cordate to reniform in outline, deeply 5- to 7-lobed (resembling a maple leaf), the lobes toothed, deep green above, paler below; pedicels 5-15 mm. long; calyx 5-15 mm. long in anthesis, stellate-puberulent to stellate-hirsute; petals 18-30 mm. long, pinkish-white or rose-purple; carpels 6-15 mm. high, 2- to 3-seeded. Two related species are listed for Colorado, I. crandallii (Rydb.) Wiggans and I. grandiflora (Rydb.) Wiggans. They are separated mainly on relative size of parts but so much intergradation occurs in this respect in Colorado plants that the writer has included them in the above description and tentatively must consider them as synonyms.---Banks, slopes, meadows and along streams. Alberta to British Columbia, south to New Mexico, Arizona and Nevada. Our records scattered in the western third of Colorado at 7500-11,500 feet.

Family 75. Elatinaceae WATERWORT FAMILY

Low herbaceous, semiaquatic annual plants; stems slender, creeping and rooting at the nodes; leaves opposite or verticillate, simple, entire, with stipules; flowers small, axillary, regular, perfect; sepals 2-3, imbricated; petals the same number as sepals, distinct; stamens 2-3; ovary superior, 2- to 5-celled, with 2-5 styles, ovules many; fruit a septicidal capsule with central or basal placenta.

1. Elatine L. WATERWORT

Characters of the Family

1. Leaves emarginate at apex, linear or spatulate, thin; seeds nearly horizontal, attached along an axile placenta -1. E. triandra
1. Leaves rounded, not emarginate at apex, obovate, thick; seeds erect from a basal placenta -2. E. americana

1. **Elatine triandra** Schkur., Bot. Handb. 1:345. 1791.
   Canada and rather irregularly distributed through the United States to Mexico. Reported for Colorado but no specimens from the state have been found by the writer.
2. **Elatine americana** (Pursh) Arn., Edinb. Journ. Sci. 1:430. 1830.
   Stems 1-4 cm. long; leaves 2-6 mm. long, obovate, rounded at apex; sepals and petals usually 2; capsule globose or subglobose, about 1 mm. across; seeds 0.5-1 mm. long, definitely sculptured with longitudinal and cross bars, more or less curved.---In mud and shallow water. Quebec to British Columbia, south to Virginia, Texas, Arizona and California. Our one record from near Denver at about 5300 feet.

Family 76. Tamaricaceae TAMARIX FAMILY

Large shrubs or small trees; branches long and slender, covered when young with small, appressed, imbricated scalelike leaves; flowers in slender terminal spikes or these grouped in paniculate clusters, regular and perfect; sepals 4 or 5; petals distinct, 4 or 5; stamens as many or twice as many as the petals, borne with them on a fleshy disk; ovary superior, 1-celled with basal placenta; styles 3-5; fruit a many-seeded capsule with 3-5 valves; seeds usually with a tuft of hairs at one end.

1. Tamarix L. TAMARIX; SALT CEDAR

Characters of the Family

1. Tamarix gallica L., Sp. Pl. 270. 1753.
Shrubs or low trees usually around 3-8 m. tall; leaves about 1 mm. long, acute; sepals about 0.5 mm. long, deltoid; petals about 1.5 mm. long, white to light rose; fruit about 3-4 mm. long, widest near base and tapering to apex.---Roadsides, waste places and river bottoms. A European species, naturalized at various places in the United States, especially in the southern part. Our records from extreme western and southeastern Colorado at 3500-6000 feet., with some recent collections from the northcentral part.

Family 77. Frankeniaceae FRANKENIA FAMILY

Low shrubs; leaves opposite or whorled, thick, entire with a membrane connecting the bases of the opposite, very short petioles; flowers small, perfect, regular, solitary or cymose-clustered, axillary and terminal, sessile; sepals 4-5, united into a tube, persistent, equal; petals as many as sepals, distinct, clawed; stamens as many as sepals or more (usually 6 in ours); ovary superior, 1-celled with 2-4 parietal placentae; styles 2-4, partly united; fruit a capsule, enclosed in the calyx.

1. Frankenia L.

Characters of the Family

1. Frankenia jamesii Torr. in Gray, Proc. Am. Acad. 8:622. 1873.
Erect shrubs about 20-60 cm. tall, branches rather fascicled, scabro-puberulent; leaves 5-10 mm. long, crowded and fascicled in the axils of the primary leaves, linear, nearly or quite glabrous, with revolute margins; calyx about 5 mm. long with lobes about 1 mm. long, narrow cylindrical; petals white, the blades about 4-6 mm. long, gradually tapering to the long claws, truncate and somewhat erose at apex; stamens usually 6. This plant has the general appearance of a Leptodactylon and often masquerades under that name in herbaria, despite the polypetalous corolla.---Plains and hills. Texas to Colorado, south to New Mexico. Our records from the Arkansas River Valley (Fremont, Pueblo and Otero Counties), and also from southwestern Colorado, at 4000-5500 feet.

Family 78. Hypericaceae ST. JOHNSWORT FAMILY

Herbaceous plants, sometimes woody at the base; leaves opposite, simple, entire, punctate with translucent or dark colored glandular dots, lacking stipules; flowers perfect, regular, cymose; sepals 5 (rarely 4); petals distinct, free, 5 (rarely 4), yellow; stamens few to many, often in 3-5 clusters with the filaments united below; ovary superior, 1-celled, with 3-5 parietal placentae or 3- to 5-celled with axile placentae; fruit a capsule.

1. Hypericum L. ST. JOHNSWORT

Characters of the Family

1. Petals 7-15 mm. long, with glandular-black dots on the margins; plants perennial
   2. Calyx lobes ovate, strongly black-dotted along the margins; stamens not united into groups; leaves oval to elliptic -1. H. formosum
   2. Calyx lobes lanceolate, slightly if at all black-dotted (punctate however); stamens united into 3-5 sets; leaves linear to oblong -2. H. perforatum
1. Petals 2-3.5 mm. long, no black dots present; plants annual -3. H. majus

1. Hypericum formosum H. B. K., Nov. Gen. et Sp. 5:196. 1821.
H. formosum var. scouleri (Hook.) Coult.; H. scouleri Hook.---Perennial herbaceous plants with horizontal rootstocks; stems 20-60 cm. tall, simple at base, sometimes branched above, erect; leaves 1-3 cm. long, oval or elliptic, or sometimes oblong-obovate, obtuse, black-dotted below at least along the margins; flowers in cymes; sepals 2-5 mm. long, ovate, black-dotted at least on margins, obtuse to acuminate; petals 7-14 mm. long, with black dots or glands on margins; stamens separate or nearly so, many (over 20); capsule about 8 mm. long. H. scouleri Hook. cannot be maintained even as a variety in Colorado.---Mountains and hills, sometimes along streams or in meadows. Wyoming to California, south to Arizona. Probably farther north into Canada. Our records scattered over the western half of Colorado at 6000-9000 feet.
2. Hypericum perforatum L., Sp. Pl. 785. 1753.
Perennial plants from somewhat woody bases; stems 30-70 cm. tall, erect, much branched; leaves 1-3 cm. long, linear to oblong, obtuse, sessile, more or less blackish-dotted; cymes terminal, many-flowered, often very large; sepals 3-4 mm. long, lanceolate, acute, glandular-punctate but slightly if at all black-dotted; petals 10-15 mm. long, obovate to oblong-oblanceolate, black-dotted on the margins; stamens many in 3-5 groups; styles 3; capsules about 4-8 mm. long.---Roadsides, fields and waste places. Naturalized from Europe in various parts of North America. Our one record from northcentral Colorado at 5500 feet.
3. Hypericum majus (A. Gray) Britt., Mem. Torr. Club 5:225. 1894.
Annual plants; stems 15-80 cm. tall; erect, simple below and somewhat branched above, 4-lined;

leaves 1.5-5 cm. long, lanceolate to oblong-lanceolate (rarely to ovate-lanceolate), sessile, obtuse, 5- to 7-veined (or sometimes 3 on narrowest ones); cymes several- to many-flowered, the bracts subulate; sepals 3-5 mm. long, lanceolate, not glandular; petals 2-3.5 mm. long, not black-dotted; stamens 5-20; styles 3, distinct; capsule conical-cylindrical.---Moist ground. Maine to British Columbia, south to New Jersey, Colorado and Washington. Our few records from central and northcentral Colorado at 5500-7500 feet.

Family 79. Violaceae VIOLET FAMILY

Herbaceous plants; leaves alternate or basal, simple, stipulate; flowers solitary or somewhat clustered, perfect, irregular; sepals 5, persistent, equal or unequal; petals 5, distinct, unequal, the lower spurred or saccate at the base; stamens 5, anthers connivent or united; ovary superior, 1-celled, with 3 parietal placentae, 1 style and stigma; fruit a 3-valved capsule.

1. Sepals auricled at base; lower petal produced at base into a spur or deep sac; plants caulescent or acaulescent; leaves various but if as narrow as linear-lanceolate then over 3 mm. wide and flowers yellow (others with flowers yellow, white, blue or purple)   -1. Viola
1. Sepals not auricled at base; lower petal merely gibbous or concave at base; plants caulescent; upper leaves linear, rarely over 3 mm. wide and flowers whitish   -2. Hybanthus

1. Viola L. VIOLET

Perennial herbaceous, caulescent or acaulescent plants; leaves from linear-lanceolate to broader; peduncles axillary or from base of plant, 1-flowered; flowers often of 2 kinds, the spring ones showy, the later ones cleistogamous with petals rudimentary or none; showy flowers with sepals auricled at the base; petals 5, the lowest one with a spur or deep sac at base; stamens distinct but more or less coherent, the 2 lower furnished at base with nectar-bearing appendages which project into the spur or sac of the petal; capsule valves bear the seeds along the middle, and after dehiscence they fold lengthwise and eject the seeds with some force.

1. Plants annual; stipules large and leaflike, palmately deeply laciniate or pectinate; petals blue or purple -1. V. rafinesquii
1. Plants perennial; stipules not large and leaflike, entire or spinulose-serrate; petals yellow, white to purple
  2. Petals yellow on the faces (except for darker veins); plants caulescent (sometimes subcaulescent in V. sheltonii)
    3. Leaves palmately divided -2. V. sheltonii
    3. Leaves undivided, entire, toothed or shallowly lobed
      4. Leaf blades round-reniform, usually cordate at base; stems prostrate to ascending -3. V. biflora
      4. Leaf blades linear-lanceolate to ovate but never round-reniform and very rarely even subcordate at base; stems erect to ascending
        5. Dwarf plants seldom over 5 cm. tall above ground level; leaves seldom over 2 cm. long, some of them coarsely toothed or shallowly lobed; capsules puberulent -4. V. purpurea
        5. Plants usually over 5 cm. tall; leaves over 2 cm. long, entire or repand-denticulate; capsules glabrous
          6. Basal leaves 2-5 cm. long, rarely to 7 cm.; plants usually ascending
            7. Leaves linear-lanceolate to oblong-lanceolate, usually acute or subacute, cuneate at base -5. V. nuttallii
            7. Leaves ovate to lanceolate, usually obtuse at apex and some truncate or subcordate at base -5A. V. nuttallii vallicola
          6. Basal leaves 5-9 cm. long; plants usually erect -5B. V. nuttallii linguaefolia
  2. Petals blue, purple or white on the faces (may have darker veins and be somewhat yellowish at very base); plants caulescent or acaulescent
    8. Petals white on the faces (may have purple veins, and be reddish-purple on back)
      9. Plants acaulescent, leaves and flowers all basal
        10. Plants stoloniferous; leaves orbicular-cordate; capsules about 9 mm. long -13A. V. palustris brevipes
        10. Plants not stoloniferous; leaves reniform-cordate; capsules 10-15 mm. long -6. V. renifolia
      9. Plants caulescent
        11. Plants with underground rhizomatous stolons; some leaves wider than long, more pubescent below than above, margins coarsely serrate, ciliate -7. V. rugulosa
        11. Plants without underground stolons or these short; leaves all as long or longer than wide, more pubescent above than below (if pubescent at all), margins finely serrate and glabrous
          12. Stems 15-35 cm. tall, puberulent -8. V. canadensis
          12. Stems 10-15 cm. tall, glabrate -8A. V. canadensis scopulorum
    8. Petals purple or violet on the faces
      13. Plants caulescent (stems above ground may be short)
        14. Plants 5-25 cm. tall, glabrous or puberulent; leaves 10-40 mm. long, some usually subcordate at base; plants mostly found at less than 10,000 feet in Colorado -9. V. adunca
        14. Plants 1-6 cm. long, glabrous; leaves 5-16 mm. long usually truncate or cuneate at base, rarely cordate; plants mostly above 10,000 feet in Colorado -9A. V. adunca bellidifolia
      13. Plants acaulescent, the leaves and flowers all from the base of the plant
        15. Leaves divided or parted
          16. Style club shaped, beakless, the stigma near the center of a cavity; lateral petals beardless -10. V. pedata

16. Style capitate at apex, with conical beak on lower side, the stigma on the end; lateral petals bearded -11. **V. pedatifida**
15. Leaves entire or toothed, not lobed
   17. Upper leaf surface hirtellous; spur over 4 mm. long to nearly as long as the petals; rootstocks very slender (not over 1.5 mm. in diameter), and stolons absent -12. **V. selkirkii**
   17. Upper leaf surface glabrous or nearly so; spur shorter, not over 4 mm. long, much shorter than the petals; rootstocks thick (over 1.5 in diameter) or if thin then stolons developing above ground
      18. Rootstocks slender, almost filiform (not over 1.5 mm. in diameter); stolons developing later in the season from the leaf axils; petals 6-10 mm. long -13. **V. palustris**
      18. Rootstocks thick (over 1.5 mm. in diameter); no stolons present; petals often over 10 mm. long
         19. Spurred petals thickly bearded within; cleistogamous flowers borne on erect peduncles; sepals obtuse; all leaves broadly ovate to reniform, gradually rounding into a short obtuse apex -14. **V. nephrophylla**
         19. Spurred petal glabrous or slightly hairy within; cleistogamous flowers borne on prostrate peduncles; sepals acute to obtuse; upper leaves ovate to deltoid with the sides extending straight or concave out to a definite acute or subacute apex
            20. Spurred petal glabrous, usually rounded at apex -15. **V. papilionacea**
            20. Spurred petal slightly hairy, usually emarginate at apex -16. **V. retusa**

1. Viola rafinesquii Greene, Pitt. 4:9. 1899.
   Annual caulescent plants, 10-25 cm. tall or sometimes shorter, often branching from base; stems subglabrous (pubescent on lower part); lower leaves 3-10 mm. long, suborbicular, upper 10-20 mm. long, oblong-lanceolate to ovate, entire or crenate-dentate, cuneate, margins minutely pubescent; stipules 5-14 mm. long, foliaceous, palmately laciniate or pectinate, the terminal segment elongated; sepals 6 mm. long, lanceolate, acute, ciliate; petals 4-9 mm. long, cream-colored, blue or purple-spotted; spur very short; peduncles 1.5-3 cm. long; capsule 6 mm. long or less, glabrous; seeds light brown. Our plant has been called V. kitaibeliana var. rafenesquii (Greene) Fernald.---Waste ground, fields and slopes. New York to Colorado, south to Georgia and Texas. Our few records from northcentral Colorado at about 5200-5400 feet.

2. Viola sheltonii Torr., Pacif. Rail. Rep. 4:67. 1857.
   Rootstocks ascending, deep-seated; plants 5-15 cm. tall, subacaulescent to caulescent, hirsutulous to nearly glabrous; leaves 1-4 cm. long, palmately 3-divided the divisions again palmately or pedately 3-parted and cleft; petioles 3-10 cm. or more long; stipules lanceolate to ovate, scarious; sepals 4-8 mm. long, linear, lanceolate, acuminate; petals 6-12 mm. long, yellow, strongly veined, tinged with brown below, laterals slightly bearded; spur hardly 1/3 length of petals; peduncles 5-15 cm. long; capsule 6-8 mm. long, globose, brown, glabrous; seeds 2 mm. long, buff-colored.---Dry rocky hillsides. Montana to Washington, south to Colorado and California. Our records from western Colorado at 5300-9000 feet.

3. Viola biflora L., Sp. Pl. 936. 1753.
   Rootstocks ascending; plants 5-20 cm. tall, caulescent; stems prostrate or ascending, slender, generally 2- (sometimes 3-) leaved and 2-flowered; leaves round-reniform, narrowly cordate at base, rounded at apex, wider than long, somewhat hirtellous above and margins crenulate; stipules ovate; sepals 4 mm. long, linear-oblong; petals 6-8 mm. long, yellow streaked with purplish-brown; spur short-conical; peduncles surpassing the leaves; capsule about 6 mm. long; seeds about 1 mm. long.---In wet moss or in alluvial soil along streams. Alaska south to Colorado; Eurasia. Our rather few records from northcentral and central Colorado at 5500-10,000 feet.

4. Viola purpurea Kell., Proc. Cal. Acad. 1:55. 1854.
   V. venosa (S. Wats.) Rydb.; V. atriplicifolia Greene---Rootstocks long, vertical or nearly so; plants caulescent; stems short, several to many, ascending, mostly buried in the ground but arising from 2-6 cm. from the ground level, rarely more, glabrate or puberulent; leaves 1-2 cm. long, rarely to 3, variable, from nearly round to ovate or to ovate-lanceolate, from entire especially on the narrow leaves to coarsely round-toothed or very shallowly lobed, more or less puberulent, the purplish veins conspicuous below; stipules lanceolate, entire; sepals lanceolate; petals 6-10 mm. long, yellow, tinged with purple on back of upper ones and purple-veined on lower petals, usually beardless; spur short; cleistogamous flowers from upper axils; capsule 4-5 mm. long, globose, minutely pubescent, often purplish. Our plants are var. venosa (S. Wats.) Brainerd.---Dry hillsides or openings in forests. The variety from Montana to Washington, south to Colorado, Arizona and California. Our few records from the northwestern corner of Colorado at 7500-8500 feet.

5. Viola nuttallii Pursh, Fl. Am. Sept. 1:174. 1814.
   Rootstocks ascending and rather fleshy; plants 5-25 cm. tall, caulescent, pubescent to nearly glabrous, the stems numerous; leaves 2-7 cm. long, oblong-lanceolate to linear-lanceolate, acute to subacute, some puberulent, entire to repand-denticulate, cuneate at base; stipules narrow, entire; sepals 2.5-3 mm. long, lanceolate to linear; petals 8-12 mm. long, yellow, at times tinged with reddish-purple on outside, slightly bearded; spur short; stigma bearded; capsule 7 mm. long, subglobose, brown, glabrous; seeds 2.5-3 mm. long.---Open plains and foothills, often on sandy, gravelly or rocky ground. Manitoba and Saskatchewan, south to Kansas, Colorado and Arizona. Our records scattered over the state, but mostly along the eastern foothills, at 4000-9000 feet.

5A. Viola nuttallii vallicola (A. Nels.) St. Johns., (var.) Fl. Southeast. Wash. and Adj. Idaho 262. 1937.
   V. vallicola Nels.---Differs from the species in having leaves ovate to lanceolate, obtuse, rounded at base or the basal leaves subcordate. However, both types of leaves are often found on the same plant in Colorado.---Saskatchewan to British Columbia, south to Colorado and Washington. Our records scattered over the state at 3500-9500 feet.

5B. **Viola nuttallii linguaefolia** (Nutt.) Jepson, (var.) Fl. Calif. 2:521. 1936.
V. linguaefolia Nutt.---Rootstocks ascending, slender; plants 10-30 cm. high, caulescent or subcaulescent, erect, hirtellous to nearly glabrous; leaves 5-9 cm. long, narrowly oblong to ovate-lanceolate, mostly obtuse, entire or repand-denticulate, margins ciliate, veins hirtellous; petioles long, twice the length of the blades; stipules lanceolate, acuminate; sepals 5-8 mm. long, linear-lanceolate, acuminate, margins usually ciliate; petals 10-15 mm. long, yellow, veined with purple, laterals somewhat bearded; spur very short; peduncle generally as long as or surpassing the leaves; capsule globose, usually glabrous; seeds 2-3 mm. long. Intergradations with the species are not uncommon in Colorado.---Valleys and slopes, often in partial shade. Montana to Washington, south to Colorado and California. Our records scattered in the western half of Colorado, mostly in the northern part, at 6000-9500 feet.

6. **Viola renifolia** Gray, Proc. Amer. Acad. 8:288. 1870.
Rootstocks ascending, commonly slender but becoming thick in old plants; plants 5-10 cm. tall, acaulescent, pubescent to subglabrous, stolons absent; leaves 3-6 cm. long, rather thin, reniform-cordate, the later sometimes blunt-pointed, margins distinctly crenate-serrate; petioles longer than blades; stipules linear; sepals half as long as petals, lanceolate, pale-margined; petals 6-10 mm. long, white, the lower frequently with purplish veins, obovate-spatulate, beardless or laterals with small tufts of hair; spur short; peduncles from shorter to longer than petioles; capsule 10-15 mm. long, ellipsoidal, often purplish; seeds about 1-5 mm. long. Our plant has been called var. brainerdii (Greene) Fernald.---Moist ground. Newfoundland to British Columbia, south to New York, Colorado and Washington. Our few records from central and northcentral Colorado at 6500-9000 feet.

7. **Viola rugulosa** Greene, Pitt. 5:26. 1902.
V. canadensis var. rydbergii (Greene) House; V. rydbergii Greene---Rootstocks ascending; plants 15-40 cm. tall, caulescent, pubescent to glabrate, stoloniferous, some leaves usually over 5 cm. wide (to 10 cm.), cordate-reniform, abruptly short-pointed, serrate, hirsutulous below; stipules entire, lanceolate, acuminate; sepals 3-7 mm. long, narrowly lanceolate; petals 6-12 mm. long, violet or nearly white, lateral petals bearded; spur short; capsule 6-10 mm. long, ovoid to subglobose; seeds 1.5-2 mm. long, brown.---Moist partly shaded ground. Manitoba to British Columbia, south to Iowa, Colorado and Washington. Our records scattered over the western two-thirds of Colorado at 5000-8500 feet.

8. **Viola canadensis** L., Sp. Pl. 936. 1753.
V. canadensis var. neomexicana (Greene) House---Rootstocks ascending; plants 15-35 cm. tall, caulescent, puberulent; leaves rarely over 5 cm. wide, cordate-ovate, acuminate or acute, serrate, teeth incurved, veins of under surface sometimes pubescent especially on upper leaves; stipules lanceolate, entire; sepals 4-7 mm. long, narrowly lanceolate; petals 6-14 mm. long, violet or nearly white, purple-veined (ours purplish only with age), laterals bearded; spur short; capsule 4-6 mm. long, ovoid or subglobose, mostly glabrous, brown; seeds about 2 mm. long, brown.---Moist, mostly shaded ground. New Brunswick to British Columbia, south to South Carolina, and Arizona. Our records scattered in the western half of Colorado at 5500-11,000 feet.

8A. **Viola canadensis scopulorum** Gray, (var.) Bot. Gaz. 11:291. 1886.
V. scopulorum (Gray) Greene---Stems tufted, about 10-15 cm. tall, glabrate; leaves about 2 cm. wide.---Colorado and possibly New Mexico and Arizona. Our records scattered in the western half of Colorado, except in the extreme northern and northwestern parts, at 5000-11,500 feet.

9. **Viola adunca** Smith, Rees Cycl. 37:pl. 63. 1817.
V. montanensis Rydb.; V. retroscabra Greene---Rootstocks ascending, a long taproot present; plants 4-25 cm. tall, (about 1/3 are of the dwarf, alpine type), caulescent, glabrous to puberulent, stems many; leaves 1-4 cm. long, often as broad, round-ovate to subcordate-ovate, crenulate, cordate to cuneate at base; stipules nearly entire to spinulose-serrate; sepals 5-7 mm. long, linear-lanceolate; petals 7-12 mm. long, violet-purple or paler, the laterals bearded; spur 4-7 mm. long, straight or hooked; capsule 6-7 mm. long, ellipsoid; seeds 1.5-2.0 mm. long. Three subspecies have been listed for Colorado, ssp. typica Baker, ssp. ashtonae Baker, and ssp. radicosa Baker.---Moist shaded hills and valleys. New Brunswick to Alaska, south to Vermont, New Mexico and California. Our records scattered in the western half of Colorado, except the northwestern part, at 5500-12,000 feet.

9A. **Viola adunca bellidifolia** (Greene) comb. nov. (var.)
V. bellidifolia Greene---Rootstocks erect; plants 2-6 cm. tall, with short obscure stems but still caulescent, glabrous throughout; leaves 5-16 mm. long, round-ovate, obtuse, truncate to subcordate to broadly cuneate at base, entire or crenulate; stipules linear with a few aristatelike teeth; sepals 3-5 mm. long, oblong-lanceolate; petals 5-10 mm. long, violet-purple, lateral petals slightly bearded; spur half as long as petals, often curved; peduncles generally surpassing the leaves; capsule 5 mm. long, subglobose; seeds 1.5 mm. long. Considered to be a distinct species by Baker (Madrono 5:223. 1940) but in this state intergrades rather commonly with V. adunca.---Montana to British Columbia, south to Colorado and California. Our records scattered in the higher mountains of the state at 8000-13,500 feet, mostly above 10,000 feet.

10. **Viola pedata** L., Sp. Pl. 933. 1753.
Massachusetts to Iowa, south to Florida and Texas. A specimen from Wolf Creek Pass in Mineral County, Colorado, is this species but the labels may have become mixed.

11. **Viola pedatifida** Don., Gard. Dict. 1:320. 1831.
Rootstocks ascending; plants 7-20 cm. high, not stoloniferous, acaulescent, minutely pubescent or glabrous; leaves 1-7 cm. long, 2- to 3-parted, the divisions ternately cleft into linear, obtuse segments, margins and midrib more or less hirsutulous; stipules ovate-lanceolate; sepals 6-10 mm. long, linear or lanceolate; acute or obtuse; petals 10-20 mm. long, violet, laterals bearded; spur short; peduncles of petaliferous flowers taller than the leaves, of apetalous flowers shorter but erect; capsule 10-15 mm. long, light gray; seeds 2 mm. long, light brown.---Fields, plains and valleys, often in forest openings. Ohio to Saskatchewan, south to Oklahoma and Arizona. Our records scattered in the western half of Colorado, except the extreme western part, at 5000-11,000 feet.

12. **Viola selkirkii** Pursh ex Goldie, in Edinb. Phil. Journ. 6:324. 1822.
Rootstocks slender and creeping; plants 4-10 cm. high, without stolons, acaulescent, nearly glabrous; leaves 1-4 cm. long, thin, broadly cordate-ovate, short-acute or obtuse, crenate, upper surface hirtellous; stipules ovate, often toothed at apex; sepals 4-6 mm. long, lanceolate or ovate-lanceolate, acute to acuminate; petals 5-10 mm. long, pale violet, beardless; spur nearly as long as petals, the end enlarged and round; cleistogamous flowers borne on erect peduncles; capsule 4-8 mm. long, subglobose; seeds 1.5 mm. long, pale buff.---Moist shaded ground. Greenland to British Columbia, south to Pennsylvania, New Mexico and Washington. Our few records from Douglas County in central Colorado at 6000-9000 feet.

13. **Viola palustris** L., Sp. Pl. 934. 1753.
Rootstocks slender, scaly, creeping or horizontal; plants 5-15 cm. tall, stoloniferous, acaulescent, glabrous throughout, scapes usually longer than the leaves; leaves 2-4 cm. long, generally wider than long, thin, ovate, cordate to orbicular or reniform, margins crenulate; stipules lanceolate, sometimes with subulate teeth near apex; sepals 5-6 mm. long, obtuse, ovate to ovate-lanceolate, narrowly white-margined; petals 6-10 mm. long, pale lilac to nearly white with darker veins, the lateral petals bearded; spur short; capsule 4-6 mm. long; seeds 1.5 mm. long, dark brown.---Moist places. Labrador to Alaska, south to New Hampshire, Colorado and California. Our records from central and northcentral Colorado at 9000-11,000 feet, but also reported from the southern part of the state.

13A. **Viola palustris brevipes** M. S. Baker, (ssp.) Madrono 3:235. 1936.
V. blanda of Manuals---Like the species but only 5-8 cm. tall; basal leaves 3.7-5 cm. long; flowers white or sometimes the lower petal with darker veins, lateral petals beardless.---Montana and Idaho, south to Colorado. Our records from northcentral and northwestern Colorado at 6500-9500 feet.

14. **Viola nephrophylla** Greene, Pitt. 3:144. 1896.
V. cognata Greene---Rootstocks generally ascending, moderately thick; plants 5-22 cm. tall, without stolons, acaulescent, nearly or quite glabrous; leaves rather firm, broadly cordate-ovate to reniform, crenate-serrate, obtuse or bluntly short-pointed; stipules ovate-lanceolate; sepals ovate to lanceolate; petals 10-15 mm. long, violet, lighter at base with purple veining, the spurred petal bearded and emarginate; spur rounded, about 1/4 the length of the petal; cleistogamous flowers erect on slender, short peduncles; capsule 8-10 mm. long, green, glabrous; seeds about 2 mm. long. V. cognata Greene may be distinct (Leafl. West. Bot. 5:173-177. 1949).---Moist open woods or low meadows. Newfoundland to British Columbia, south to Connecticut, Wisconsin, New Mexico and California. Our records well scattered over the western two-thirds of Colorado at 4500-10,000 feet.

15. **Viola papilionacea** Pursh, Fl. Am. Sept. 1:173. 1814.
V. pratincola Greene---Rootstocks thick, ascending; plants 7-30 cm. tall, not stoloniferous, acaulescent, usually glabrous; leaves deeply cordate-ovate, tapering to a subacute apex, crenulate; stipules linear-lanceolate; sepals ovate-lanceolate, acute or obtuse; petals 10-15 mm. long, violet, laterals somewhat bearded, the spurred one glabrous, rounded at apex; spur short; peduncles as long or longer than leaves; cleistogamous flowers borne on prostrate peduncles; capsule 6-10 mm. long, ellipsoidal, green or purplish; seeds 1.5-2 mm. long.---Moist fields and slopes, often along streams. Maine to North Dakota, south to Georgia and Colorado. Our few records from northern Colorado at 3500-8000 feet.

16. **Viola retusa** Greene, Pitt. 4:6. 1899.
V. missouriensis Greene---Rootstocks ascending, stout, no stolons present; plants 4-15 cm. tall, acaulescent, glabrous throughout; leaves 2-4 cm. long, in the Spring broadly ovate to cordate-deltoid, acute or obtuse, finely serrate; aestival leaves wider, reniform, often abruptly acuminate, base cordate to truncate-decurrent; stipules ovate; sepals 5-6 mm. long, acute; petals 10-15 mm. long, violet, upper obovate, laterals somewhat bearded, spurred petal slightly hairy, retuse; spur short; peduncles somewhat exceeding leaves; cleistogamous flowers borne on prostrate peduncles; capsule 10-12 mm. long, ellipsoid, green; seeds 2 mm. long, brown. This plant is closely related to the 2 preceding species and may not be distinct.---Borders of streams, often in shade. Illinois to Colorado, south to Missouri and Texas. Our few records from the eastern half of Colorado at 4500-6000 feet.

2. Hybanthus Jacq. GREEN VIOLET

Perennial herbaceous, caulescent plants; leaves opposite, alternate or fascicled; flowers solitary and axillary; sepals not auricled; petals very unequal, the lowest one larger and gibbous at base; lower 2 stamens glandular at base; capsule opening elastically.

1. **Hybanthus verticillatus** (Ortega) Baill., Hist. Pl. 4:345. 1873.
Calceolaria verticillata (Ortega) Kuntze---Stems 10-40 cm. tall, branched, tufted from woody bases, glabrous to slightly pubescent; leaves 1-4 cm. long, alternate, fascicled, or the lower opposite, entire or remotely serrate, upper linear, lower sometimes as wide as oblong; stipules subulate or lacking; flowers nodding; sepals about 2 mm. long; petals, the shorter about 2-3 mm. long, the longer 4-6 mm. long, whitish; capsule 4-5 mm. long, glabrous; seeds large, about 2 mm. long, dark brown to turning black when mature.---Dry plains and hills. Kansas to Colorado, south to Texas and Arizona. Our records from northcentral, central, southcentral and southeastern Colorado at 4000-6000 feet.

Family 80. Loasaceae LOASA FAMILY

Annual or perennial plants, herbaceous or suffrutescent, with stiff hairs, these barbed but not stinging; stems usually becoming whitish and shining with pale exfoliating bark; leaves alternate, simple but often deeply pinnatifid; no stipules present; flowers perfect, regular in terminal inflorescences or solitary in the forks of the branches; calyx tube completely adnate to the ovary, only the 5 lobes free; petals 5-10, distinct or very nearly so, white or cream-colored to yellow; stamens 10-200, inserted with the petals on the rim of the calyx tube, the outer often petaloid; ovary inferior, 1-celled, usually with 3 parietal placentas, the ovules many in 1 or 2 rows on each placenta; fruit a capsule, opening regularly or irregularly at the top; seeds often winged.

1. **Mentzelia L.** STICKLEAF; BLAZING STAR

Characters of the Family

The treatment of the species of this genus in the main follows that of Darlington (Ann. Mo. Bot. Gard. 21:103-226. 1934), this being the latest monograph of the group. She divides our plants into many species but these are often connected by numerous intergrades. As far as this state is concerned the whole genus needs to be realigned and many of the entities reduced to varieties or to synonymy.

1. Petals 2-6 mm. long (rarely to 8 mm.), only 5 present; plants annual; stems narrow, usually not over 4 mm. in diameter; placentae filiform; seeds in 1 row, pendulous, not flattened or winged
   2. Seeds grooved on the angles, minutely muricate, appearing smooth to the naked eye; at least the upper leaves usually ovate or ovate-oblong
      3. Petals 3-4 mm. long
         4. Stems loosely branched; capsules 15-25 mm. long   -1. M. dispersa
         4. Stems densely and compactly branched; capsules 10-13 mm. long   -1A. M. dispersa compacta
      3. Petals 5-6 mm. long   -1B. M. dispersa latifolia
   2. Seeds not grooved on the angles (or rarely on 1 angle), obviously tuberculate; leaves linear-oblanceolate
      5. Petals 2-4 mm. long; calyx lobes less than 3 mm. long   -2. M. albicaulis
      5. Petals 5-8 mm. long; calyx lobes 3-5 mm. long   -2A. M. albicaulis veatchiana
1. Petals 8-80 mm. long, 5-10 present; plants perennial or biennial (perhaps annual only in M. pterosperma); stems stouter, usually well over 4 mm. in diameter; placentae broad; seeds in 1 or 2 rows, usually horizontal (pendulous only in M. oligosperma); seeds usually flattened and winged
   6. Seeds 1-4 to a capsule, pendulous, never winged; placentae rugose but not extending horizontally between the seeds; stamens 20-25; leaves, at least lower ones, petioled   -3. M. oligosperma
   6. Seeds 50-80, horizontal, usually winged; placentae with horizontal lamellae extending out between the seeds; stamens 30 or more; leaves rarely petioled
      7. Petals straw-colored, cream-colored to whitish (sometimes very light yellowish) when fresh
         8. Petals 5-8 cm. long (rarely as small as 4 cm. long); capsules 3-4 cm. long; sepals in anthesis about 20-40 mm. long; flowers closed in daylight; seeds not winged   -4. M. decapetala
         8. Petals 1.6-4 cm. long (rarely to 5 cm.); capsules 1.5-3 cm. long; sepals in anthesis 10-20 mm. long; flowers open in daylight; seeds winged on margins
            9. Petals 20-40 mm. long; sepals sometimes over 12 mm. long
               10. Calyx tube naked at base or subtended by entire bracts; leaves usually less than 7 cm. long; plants usually less than 50 cm. tall   -5. M. nuda
               10. Calyx tube subtended by 2 or more toothed or pinnatifid bracts; leaves often over 7 cm. long; plants often over 50 cm. tall   -5A. M. nuda stricta
            9. Petals 16-20 mm. long; sepals seldom over 12 mm. long   -5B. M. nuda rusbyi
      7. Petals bright yellow to golden-yellow when fresh (greenish-yellow in M. lutea)
         11. Leaves entire or rarely toothed   -6. M. integra
         11. Leaves sinuate-dentate to pinnatifid (at least some)
            12. Flowers greenish-yellow; only 1 collection known - from near Canon City, Colorado   -7. M. lutea
            12. Flowers bright yellow to golden; plants usually widely distributed in Colorado
               13. Upper leaves entire, narrowly linear; seeds margined to narrowly winged; capsules 8-15 mm. long   -8. M. multicaulis
               13. All leaves pinnatifid or lobed, the upper not both narrowly linear and entire; seeds margined to broadly winged; capsules 7-30 mm. long
                  14. Petals 15-20 mm. long; mature capsule usually over 15 mm. long; stems 30 cm. tall or more; leaves mostly over 5 cm. long
                     15. Leaves divided to near the midrib forming linear or linear-oblong divisions   -9. M. laciniata
                     15. Leaves sinuate-dentate to pinnatifid but seldom divided over 1/2 way in to the midrib and then the lobes ovate
                        16. Stems decumbent at base; leaves sinuate-dentate; capsules 25-30 mm. long; petals acute; seeds margined but not winged   -10. M. chrysantha
                        16. Stems erect or nearly so at base; leaves pinnatifid; capsules 15-20 mm. long; petals obtuse or acute; seeds winged   -11. M. multiflora
                  14. Petals 8-15 mm. long; mature carpels usually less than 15 mm. long (except in M. pumila); stems rarely over 30 cm. tall; leaves mostly less than 5 cm. long
                     17. Petals obtuse at apex; plants annual, possibly beiennial; stems 10-20 cm. tall; leaves spatulate or orbicular, sinuate-dentate, rarely entire, not at all pinnatifid   -12. M. pterosperma
                     17. Petals acute at apex; plants perennial or biennial; stems 20-40 cm. tall; leaves oblong-oval to linear, mostly pinnatifid
                        18. Capsules 7-10 mm. long when mature; leaves pinnatifid into narrowly linear segments about 1-1.5 mm. wide   -13. M. humilis

18. Capsules 13-20 mm. long when mature; lobes of leaves over 1.5 mm. wide (except sometimes in M. densa)
  19. Stems simple at base, branched toward the summit; lobes of leaves short, broadly oblong   -14. M. pumila
  19. Stems branched at base; lobes of leaves narrowly linear-lanceolate   -15. M. densa

1. Mentzelia dispersa Wats., Proc. Am. Acad. 11:137. 1876.
    Acrolasia integrifolia (S. Wats.) Rydb.---Annual plants; stems 10-30 cm. tall, slender, erect, glabrous or puberulent above; lower leaves narrowly lanceolate, often entire, middle ovate-lanceolate, sinuate-dentate or rarely pinnatifid, upper leaves ovate, entire, all leaves sessile; flowers axillary but approximate toward ends of branches; calyx tube about 6-8 mm. long, the lobes about 2 mm. long; petals 5, 3-4 mm. long, yellow, obovate or spatulate; filaments filiform; capsule 15-25 mm. long and about 2-2.5 mm. wide, narrowly linear-clavate; placentae filiform, seeds in 1 row on each, pendulous; seeds angled and grooved along the angle, minutely tuberculate, appearing smooth to the naked eye, wingless.---Mostly sandy ground. Montana to Washington, south to Colorado and California. Our records from the northwestern quarter of Colorado at 4500-7500 feet.

1A. Mentzelia dispersa compacta (A. Nels.) Macbride, (var.) Contrib. Gray Herb. n.s. 56:26. 1918.
    M. compacta A. Nels.; Acrolasia compacta (A. Nels.) Rydb.---This variety has low compactly branched (not sparsely branched) stems; capsules are 10-13 mm. long.---Wyoming to Washington, south to Colorado and California. Scattered with the species in Colorado at 4500-5500 feet.

1B. Mentzelia dispersa latifolia (Rydb.) Macbride, (var.) Contrib. Gray Herb. n.s. 56:26. 1918.
    M. latifolia (Rydb.) A. Nels.; Acrolasia latifolia Rydb.---This variety has longer petals than the species, these 5-6 mm. long.---Wyoming to Washington, south to Colorado and California. Our records from northcentral and northwestern Colorado at 6000-8000 feet.

2. Mentzelia albicaulis Dougl. ex Hook., Fl. Bor. Am. 1:222. 1834.
    Annual plants; stems 10-40 cm. tall, slender, white and shining, glabrous or sparsely pubescent; leaves sessile, lower 3-5 cm. long, linear-oblanceolate, dentate or entire, middle pinnatifid with linear lobes (rarely all leaves entire or obscurely toothed); flowers axillary, sessile; calyx tube 10-15 mm. long, the lobes 2-2.5 mm. long; petals 5, 2-4 mm. long, yellow, obovate or spatulate; filaments filiform; capsule 10-15 mm. long and about 2 mm. wide, linear-cylindrical; placentae filiform; seeds in 1 row on each, pendulous, tuberculate-granulate, irregularly angled, the angles not grooved (only rarely on 1), wingless. This complex has been broken down into several entities largely on the basis of size of floral parts and relative lobing of the leaves. Only 1 variety is recognized here and even it intergrades commonly with the species as described above.---Dry slopes and banks often in sandy soil. Wyoming to Washington, south to New Mexico and California. Our records scattered over the western three-fourths of Colorado at 4500-7000 feet.

2A. Mentzelia albicaulis veatchiana (Kell.) Urb. & Gilg., (var.) Nova Acta Acad. C.L.C.G. Nat. Cur. 76:28. 1900.
    Differs from the species as follows: leaves usually all pinnatifid; calyx lobes 3-5 mm. long; petals 5-8 mm. long. Most plants will separate nicely but some intergrades do occur. The above includes Darlington's 2 varieties, gracilis and stenophora.---Montana to British Columbia, south to Arizona and California. Our records scattered in the western two-thirds of Colorado at 4500-6500 feet.

3. Mentzelia oligosperma Nutt. in Sims, Bot. Mag. 42. pl. 1760. 1815.
    Perennial herbaceous plants, rough with many-barbed hairs; stems 20-70 cm. tall, erect, branched, brittle, white or yellowish; leaves 1-6 cm. long, ovate to ovate-lanceolate, apex acute, base cuneate, irregularly toothed or incised, upper sessile, lower usually petioled; flowers in 1- to 3-flowered cymes, terminal; calyx tube 5-6 mm. long, narrowly cylindrical, sessile, the lobes 7-9 mm. long, lanceolate or linear, finally deciduous; petals 5, 8-15 mm. long, yellow or orange; stamens about 20-25; capsule 8-16 mm. long, cylindrical, narrow; placentae broad and irregularly plicated-rugose; seeds usually few, pendulous, not winged, longitudinally undulate-striate.---Slopes and flats, often on rocky, gravelly or sandy ground. Arkansas to Wyoming, south to Texas and New Mexico. Our records scattered in the eastern half of Colorado, except the northeastern part, at 4000-5500 feet.

4. Mentzelia decapetala (Pursh) Urb. & Gilg., in Ber. Deut. Bot. Ges. 10:263. 1892.
    Nuttallia decapetala (Pursh) Greene; Touteria decapetala (Pursh) Rydb.---Perennial or possibly biennial plants from large roots; stems 30-100 cm. tall, stout, erect; leaves 5-15 cm. long, sessile or lower somewhat petioled, lanceolate or oblong-lanceolate, interruptedly sinuate-pinnatifid; flowers solitary or very few together; calyx tube 2-3 cm. long with toothed or laciniate bracts at base and part way up, the calyx lobes 2-4 cm. long; petals 5-8 cm. long rarely as short as 4 cm., whitish to very pale yellow; capsule 3-5 cm. long, oblong; seeds about 3 mm. long, margined but not winged, 50-80, horizontal in 2 rows on broad placentae, these forming horizontal lamellae between the seeds.---Plains, foothills and canyons. Alberta south to Iowa and Mexico. Our records from the eastern half of Colorado, mostly along the eastern foothills, at 3500-7000 feet.

5. Mentzelia nuda (Pursh) T. & G., Fl. N. Amer. 1:535. 1840.
    Nuttallia nuda (Pursh) Greene; Touteria nuda (Nutt.) Eat. & Wr.---Perennial, possibly biennial plants from enlarged roots; stems 30-50 cm. tall, often branched at base; leaves 5-8 cm. long, sometimes longer, lanceolate to linear-lanceolate, or oblanceolate, sinuate-dentate to pinnatifid; calyx tube about 10-18 mm. long, 1 or 2 bractlets usually at base, these entire, calyx lobes about 8-15 mm. long; petals 2-4 cm. long (rarely to 5 cm.), very light yellow to whitish; stamens over 30, outer filaments petaloid; capsule 15-30 mm. long; seeds 3-4 mm. long, broadly winged, 50-80, horizontal in 2 rows, on broad placentae, these forming horizontal lamellae between the seeds.---Plains and hills, often in sandy or gravelly soil. Nebraska and Colorado, south to Texas and New Mexico. Our records in northcentral and southcentral Colorado at 5500-7500 feet.

5A. **Mentzelia nuda stricta** (Osterh.) comb. nov. (var.)
   M. stricta (Osterh.) Stevens; Nuttallia stricta (Osterh.) Greene; Touteria stricta Osterh.---Plants branched above but not from base, often over 50 cm. tall; leaves often over 7 cm. long; calyx tube subtended by 2 or more toothed bracts; petals 20-35 mm. long. Intergrades with the species and with var. rusbyi in Colorado.---Plains and foothills often in sandy ground. South Dakota to Montana, south to Texas and New Mexico. Our records well scattered in the eastern half of Colorado and also in the southwestern part, at 3500-6500 feet.

5B. **Mentzelia nuda rusbyi** (Woot.) comb. nov. (var.)
   M. rusbyi Wooton; Nuttallia rusbyi (Woot.) Rydb.; Touteria rusbyi (Woot.) Rydb.---Differs from the species as follows: plants do not branch at base; calyx subtended by toothed or pinnatifid bracts; sepals 10-12 mm. long; petals 16-20 mm. long. Intergrades with the species and the other variety in this state.---Plains and mountains, often on moist slopes. Wyoming, south to New Mexico and Arizona. Our records scattered in the western two-thirds of Colorado at 5000-8500 feet.

6. **Mentzelia integra** (Jones) Tidestrom, Contrib. U. S. Nat. Herb. 25:362. 1925.
   Nuttallia integra (Jones) Rydb.; Touteria integra (Jones) Rydb.---Reported from Utah, Arizona and New Mexico. The plant should be looked for in southwestern Colorado.

7. **Mentzelia lutea** Greene, Pitt. 3:99. 1896.
   Nuttallia lutea Greene; Touteria lutea (Greene) Rydb.---A species of very doubtful validity, collected only from Fremont County, Colorado. The writer has not seen a specimen of the plant.

8. **Mentzelia multicaulis** (Osterh.) Darlington, Ann. Mo. Bot. Gard. 21:156. 1934.
   M. pumila var. multicaulis (Osterh.) Nelson; Nuttallia multicaulis Osterh.; Touteria multicaulis Osterh.---Perennial plants; stems 20-40 cm. tall, numerous, much branched; leaves 2-5 cm. long, lanceolate, pinnatifid with remote entire divisions; upper leaves linear and entire; flowers at apex of branches, usually many; calyx tube 6-10 mm. long, lobes about 8-10 mm. long; petals 5, 1-2 cm. long, golden-yellow; stamens over 30, outer filaments petaloid; capsule 8-15 mm. long; seeds margined or very narrowly winged, 50-80, horizontal, in 2 rows on broad placentae, these forming horizontal lamellae between the seeds.---Dry open ground. Limited to Colorado, our records from the northwestern quarter of the state, also from the southwestern corner, at 5000-8500 feet. However, some of the records are somewhat doubtful.

9. **Mentzelia laciniata** (Rydb.) Darlington, Ann. Mo. Bot. Gard. 21:173. 1934.
   Nuttallia laciniata (Rydb.) Woot. & Standl.; Touteria laciniata Rydb.---Biennial or perennial plants; stems 30-40 cm. tall, branched above, erect; leaves 5-12 cm. long at least the lower ones, narrowly lanceolate, deeply pinnatifid to near midrib, forming obtuse linear or linear-oblong divisions; flowers terminal, bracted, these commonly laciniate; calyx tube 7-8 mm. long, the lobes 8-10 mm. long; petals 15-20 mm. long, golden-yellow, acute; stamens 30 or more, outer filaments petaloid; capsule 15-20 mm. long and about 5-8 mm. wide; seeds winged, over 50, horizontal on broad placentae, these forming lamellae between the seeds.---Dry slopes. Colorado and New Mexico. Our records scattered in the southwestern quarter of Colorado at 5000-8000 feet.

10. **Mentzelia chrysantha** Engelm. ex Brandegee, Bull. U. S. Geol. Surv. Terr. 2:237. 1876.
    Nuttallia chrysantha (Engelm.) Greene; Touteria chrysantha (Engelm.) Rydb.---Biennial plants; stems 30-60 cm. tall, decumbent toward the base, branched; leaves 2-10 cm. long or lower to 15 cm., ovate-lanceolate, lanceolate, oblanceolate or upper ovate, sinuate-dentate or upper ones entire; calyx tube about 10-15 mm. long, the lobes about 10-12 mm. long; petals 15-20 mm. long, golden-yellow, acute; stamens over 30, the outer petaloid; capsule 25-30 mm. long; seeds narrowly margined but not winged, 50-80, horizontal in 2 rows on broad placentae, these forming horizontal lamellae between the seeds.---Slopes and canyons. Wyoming to Colorado. Our few records from central and northcentral Colorado at 5500-7000 feet.

11. **Mentzelia multiflora** (Nutt,) Gray, Mem. Am. Acad. 4:48. 1849.
    M. speciosa Osterhout.; Nuttallia multiflora (Nutt.) Greene; N. speciosa (Osterh.) Greene; N. sinuata Rydb.; Touteria multiflora (Nutt.) Rydb.; T. speciosa Osterh.; T. sinuata Rydb.---Perennial or possibly biennial plants; stems 30-80 cm. tall, branched above; leaves 5-15 cm. long, narrowly lanceolate, all sinuately-pinnatifid usually not over 1/2 in to the midrib, the lobes ovate; flowers corymbose; calyx tube 12-16 mm. long, narrowed at base, the lobes about 8-12 mm. long; petals 15-20 mm. long, yellow, usually obtuse or cuspidate at apex, less commonly short-acute; stamens 30 or more, the outer filaments petaloid; capsule 15-20 mm. long, rather narrowed and acute at base; seeds with margins winged, over 50, horizontal in 2 rows on broad placentae, these forming horizontal lamellae between the seeds. Closely related to M. pumila (Nutt.) T. & G. and often considered as a variety of that species.---Dry often sandy ground. Wyoming to California, south to Texas and Mexico. Our records scattered over Colorado, except in the extreme eastern part, at 5000-9500 feet.

12. **Mentzelia pterosperma** Eastw., Proc. Cal. Acad. Sci. II. 6:290. 1896.
    Nuttallia pterosperma (Eastw.) Greene---Annual plants with taproots; stems 10-20 cm. tall, divaricately branched from the base; lower leaves 4-5 cm. long, petiolate, spatulate to orbicular, upper leaves about 3 cm. long and 2 cm. wide, sessile, sinuate-dentate; flowers at ends of branches; calyx tube about 6 mm. long, the lobes about 6 mm. long; petals 10-15 mm. long, golden-yellow, obtuse; stamens 30-40, the outer petaloid; capsules 8-15 mm. long and 7-8 mm. wide, campanulate-cylindrical, round at base; seeds 4 mm. wide, winged, 50-80, horizontal, in 2 rows on broad placentae, these forming horizontal lamellae between the seeds.---Dry valleys and mesas, often on sandy ground. Colorado and Utah. Our few records from westcentral Colorado the altitude not known.

13. **Mentzelia humilis** (Gray) Darlington, Ann. Mo. Bot. Gard. 21:155. 1934.
    Nuttallia humilis (Gray) Rydb.---Perennial or possibly biennial plants; stems 20-40 cm. tall, caespitose; leaves 2-4 cm. long and about 1 cm. wide, oblong-lanceolate to oval-oblong, pinnately divided or deeply pinnatifid into linear, obtuse segments about 1-1.5 mm. wide; flowers terminal; calyx tube 3-5 mm. long, somewhat campanulate-cylindrical, the lobes 4-5 mm. long;

petals 8-15 mm. long, yellow, narrow, acute at apex; stamens 50 or more, the outer petaloid; capsules 7-10 mm. long, about 5-6 mm. wide; seeds 50 or more, flat, winged, horizontal in 2 rows on broad placentae, these forming horizontal lamellae between the seeds.---Dry places often on steep slopes or cliffs. Colorado and Utah, south to Texas and New Mexico. Our records from western Colorado and extending east to Eagle and Grand Counties, at 4500-8500 feet.

14. Mentzelia pumila (Nutt.) T. & G., Fl. N. Amer. 1:535. 1840.

Nuttallia pumila (Nutt.) Greene---Biennial plants; stems 20-30 cm. tall, tortuous, low but mostly branched toward apex; leaves obovate-lanceolate to linear-lanceolate, sinuate-dentate to pinnatifid, the lobes short and broadly oblong; flowers terminal, solitary or in clusters up to 3; calyx tube 5-15 mm. long subtended by 1 or 2 linear or setaceous bracts, calyx lobes 7-10 mm. long; petals 9-15 mm. long, yellow, acute; stamens 60 or more, the outer filaments petaloid; capsule 13-20 mm. long; seeds broadly winged, 50 or more, horizontal in 2 rows on broad placentae, these forming horizontal lamellae between the seeds.---Dry plains and slopes. Wyoming and Utah, south to Texas and Arizona. Our records scattered over Colorado at 4000-8000 feet but the ones in the eastern half of the state are somewhat doubtful.

15. Mentzelia densa Greene, Pitt. 3:99. 1896.

M. multiflora var. densa (Greene) A. Nels.; Nuttallia densa Greene; Touteria densa (Greene) Rydb.---Perennial plants; stems 20-30 cm. tall, divaricately branched from the base forming a hemispherical tuft; leaves narrowly linear-lanceolate to oblanceolate, sinuate-pinnatifid into linear-lanceolate lobes about 1.5-2 mm. wide; flowers solitary or up to 3 at ends of branches; calyx tube about 3-5 mm. long the lobes about 4-6 mm. long; petals about 10-15 mm. long, golden-yellow, acute; stamens over 30, the outer filaments petaloid; capsule 13-15 mm. long when mature; seeds over 50 in 2 rows, horizontal on broad placentae, these forming horizontal lamellae between the seeds.---Dry often sandy ground. Apparently limited to Colorado, our one record from the westcentral part at 7000 feet.

Family 81. Cactaceae CACTUS FAMILY

Fleshy-stemmed perennial plants, usually spiny and xerophytic with mucilaginous juice; cushionlike organs (areoles) present from which the spines, branches or flowers arise; stems of 1 or of more than 1 joint, these flattened, globose or cylindrical, often ribbed or tuberculate; leaves wanting or small and early deciduous; flowers perfect, regular and solitary; perianth segments several to many, more or less united, inserted on a hypanthium, outer segments, (calyx) several to many; inner segments (corolla) usually many; stamens many in several series, inserted within the hypanthium tube; ovary inferior, 1-celled, with numerous ovules on several parietal placentae; styles 1 but stigmas several; fruit fleshy or dry, berrylike, and several- to many-seeded. This is an interesting but difficult family, due largely to the scarcity of good herbarium specimens. To properly understand the group a good deal of field study is needed and the plants should be transplanted into gardens or greenhouses. Much of the descriptive material in this treatment is copied. Boissevain and Davidson (Colorado Cacti, 71 pages, Abbey Garden Press 1940) has been very helpful.

1. Areoles bearing minute sharp-pointed barbed bristles (glochids) as well as spines; young stems with a small caducous leaf at the base of each areole; stems composed of joints, these flattened or cylindrical
 -1. Opuntia
1. Areoles not bearing glochids; stem leafless from the first; stems not composed of joints (except as clustered near ground level)
  2. Flowers borne between the tubercles or at the base of grooves on the tubercle at some distance from the spiniferous areoles; tubercles distinct and not in ribs, somewhat grooved above; fruit without spines
 -2. Mamillaria
  2. Flowers borne at the apex of the tubercles contiguous to a spiniferous areole; tubercles usually coalescing to form vertical or spiral ribs but if this is obscure the tubercles not grooved above; fruit with or without spines
    3. Hypanthium and fruit spiny; flowers at the sides of the plant or definitely below the apex; hypanthium forming a distinct tube above the ovary
 -3. Echinocereus
    3. Hypanthium and fruit not spiny; flowers produced at the top of the plant; hypanthium not forming a distinct tube above the ovary
 -4. Echinocactus

1. Opuntia Mill. PRICKLY PEAR

Shrubs or herbaceous perennial plants with jointed stems, the joints flattened or terete, sometimes tuberculate but never ribbed; leaves small, fleshy but scalelike, caducous; areoles furnished with small barbed bristles called glochids as well as longer spines; flowers borne on the same areoles as the spines; tube of hypanthium short, little if any prolonged beyond the ovary, bearing scales resembling the leaves; fruit fleshy and edible to dry, indehiscent, bearing glochids and often spines; seeds compressed or angled, light colored.

1. Joints terete in cross-section
  2. Shrubby plants; spines covered with a delicate sheath; joints with prominent laterally compressed tubercles 2-3 cm. long
    3. Plants 100 cm. tall or more; flowers purple to rose-pink; plants mostly east of the Continental Divide in Colorado
 -1. O. arborescens
    3. Plants seldom over 30 cm. (rarely to 50 cm.) tall; flowers yellow; plants west of the Divide in Colorado
 -2. O. davisii
  2. Plants not shrubby; spines without a sheath; joints without tubercles or these not as conspicuous as above

4. Joints brittle, the upper easily detached, 1.5-4 cm. long; areoles less than 1 cm. apart; flowers about 5 cm. long -3. O. fragilis
4. Joints not brittle (at least on older plants), 4-10 cm. long; areoles 1-1.5 cm. apart; flowers about 6 cm. long -4. O. rutila
1. Joints flattened in section (at least some of them)
   5. Fruit juicy and edible
      6. Joints 12-20 cm. long; areoles 2.5-4 cm. apart -5. O. phaeacantha
      6. Joints 8-12 cm. long; areoles 1.5-2.5 cm. apart -6. O. humifusa
   5. Fruit dry at maturity
      7. At least some of the joints rounded or turgid and not definitely flattened
         8. Joints brittle, the upper easily detached, 1.5-4 cm. long; areoles less than 1 cm. apart; flowers about 5 cm. long -3. O. fragilis
         8. Joints not brittle (at least on older plants) 4-10 cm. long; areoles 1-1.5 cm. apart; flowers about 6 cm. long -4. O. rutila
      7. All joints definitely flattened (except very young seedlings)
         9. Joints about 5 cm. long (occasionally to 7 cm.), spines only on the upper half -7. O. schweriniana
         9. Joints 7.5 cm. long or more, spines usually on lower half too (except sometimes in O. rhodantha)
           10. Areoles over 1 cm. apart
              11. Spines 6-12 at an areole, 4-12 cm. long, pointing in all directions; joints armed over entire surface -8. O. hystricina
              11. Spines 1-4 at an areole, 2-6 cm. long, mostly pointing outward; joints often spiny only on upper half -9. O. rhodantha
           10. Areoles less than 1 cm. apart
              12. Some of the spines (on lower part of older joints especially) elongated to 12 cm. long and bristlelike -10. O. trichophora
              12. None of spines bristlelike, all stiff and needlelike -11. O. polyacantha

1. Opuntia arborescens Engelm. in Wislizenus, Mem. Tour. North Mex. 90. 1848.
   O. imbricata (Haw.) DC.---Shrubby, 1-2 m. tall, rarely taller, trunk 2.5-6 cm. in diameter, with a woody skeleton; joints about 5-15 cm. long and 2-2.5 cm. in diameter, verticillate, cylindrical, narrowed toward base, strongly tubercled, these laterally compressed and 20-30 mm. long, with woolly areoles at upper end; spines 6-20, 1-3 cm. long, white to brown, covered with grey-white papery sheaths; flowers 4-6 cm. long, purple to rose-pink, often green outside; fruit about 2.5-3.5 cm. long, yellow, dry, tuberculate with very deciduous spines; seeds about 3-4 mm. long, yellowish-grey.---Dry plains and hills. Colorado to Texas and Mexico. Our records from the southern half of Colorado, mostly east of the Continental Divide, at 4000-7500 feet.
2. Opuntia davisii Engelm. & Bigel., Proc. Am. Acad. 3:305. 1856.
   Low spreading shrubs 10-30 cm. tall (rarely to 50 cm.), the branches sometimes procumbent; joints cylindric, 10-15 cm. long and about 1-2 cm. in diameter, slender at base, covered with laterally compressed tubercles about 2-3 cm. long, woolly areoles at upper end, with usually 1 longer (1-2.5 cm.) deflexed spine and several shorter spines and bristles, the spines brownish but covered with straw-colored, papery sheaths; flowers 6 cm. long, yellow; fruit about 2.5 cm. long, yellow, hardened, tubercled and with somewhat deciduous spines and bristles; seeds about 3-4 mm. long, yellowish-grey. The correct name for our plant may be O. whipplei Engelm. & Bigel.---Dry ground. Texas to Colorado and perhaps west to California. Our records from southwestern Colorado at 6000-7000 feet.
3. Opuntia fragilis (Nutt.) Haw., Suppl. Pl. Succ. 82. 1819.
   Stems low and spreading or subdecumbent; joints 1.5-4 cm. long, often nearly globular to ovoid but sometimes flattened, the terminal ones easily breaking off (often carried away on the pelts of animals); areoles close, less than 1 cm. apart, filled with white wool; spines 3-7, 1-3 cm. long; flowers about 5 cm. long, pale yellow with orange center or sometimes pale pinkish; fruit dry, tubercled and more or less spiny; seeds about 5-7 mm. long, yellowish, irregular.---Dry plains and hills. Wisconsin to British Columbia, south to Texas and Arizona. Our records scattered over Colorado at 4500-7500 feet but most of the eastern ones rather doubtful.
4. Opuntia rutila Nutt. in T. & G., Fl. N. Amer. 1:555. 1840.
   Stems ascending or diffuse-spreading; joints 4-10 cm. long, flat and ovate or obovate but some of them cylindrical; distance between areoles 1-1.5 cm., the areoles white-woolly; spines 1-6, 1-3 cm. long, all point upward or outward; flowers about 6 cm. long, pink to yellowish; fruit 2.5-3.5 cm. long, dry, with bristles and occasionally spiny; seeds about 5-6 mm. long, yellowish, irregular. Difficult to distinguish from the preceding and perhaps not distinct from it.---Dry plains and hills. Wyoming to Nevada, south to Colorado and Utah according to report. Our records from western Colorado at 5500-6500 feet.
5. Opuntia phaeacantha Engelm. in Gray, Mem. Amer. Acad. Ser. 2. 4:52. 1849.
   O. camanchica Engelm. & Bigel.---Stems spreading or prostrate; joints 12-20 cm. long and narrower than long, obovate, flat; areoles 2.5-4 cm. apart, often only the upper ones spine-bearing; spines 1-4, 3-4 cm. long (occasionally to 6 cm.), brown or variegated; flowers 5-7 cm. long, yellow, orange or pink; fruit 3-6 cm. long, clavate or ovoid, purplish-red, juicy and edible; seeds 4 mm. long, wingless.---Dry plains and hills. Colorado to Utah, south to Texas and Mexico. Our records scattered in the southern half of Colorado at 4000-7500 feet.
6. Opuntia humifusa Raf., Med. Bot. 2:247. 1830.
   O. rafinesquei Engelm.; O. mesacantha and O. tortispina of our Manuals---Plants prostrate or ascending; joints 8-12 cm. long, flattened, orbicular to obovate; areoles 1.5-2.5 cm. apart; spines 1-3 (sometimes to 5), 1-3 cm. long (sometimes to 6 cm.), often only on upper part of joint, white to brown; flowers 6-8 cm. long, yellow-orange or pink; fruit 4-5 cm. long, juicy and edible, ovoid to clavate; seeds 5 mm. long, irregular, yellowish.---Dry often sandy ground. Wisconsin to South Dakota, south to Mississippi and New Mexico. Our records from the eastern half of Colorado at 4000-5500 feet.

7. Opuntia schweriniana K. Schuman, Monatsschr. Kakteenk. 9:148. 1899.
This is a plant closely related to O. polyacantha. It has smaller joints, about 5 cm. long, and the spines are scantier and shorter, only the upper areoles having 1 or 2 stouter erect ones; flowers yellow to pinkish; fruit dry and spiny. Doubtfully distinct in this area, possibly a good variety.---Dry hills. Apparently limited to Colorado, our one doubtful record from the southern part of the state.

8. Opuntia hystricina Engelm. & Bigel. in Engelm., Proc. Amer. Acad. 3:299. 1856.
This is a plant closely related to O. polyacantha but the areoles are over 1 cm. apart, 1-2 central spines are present, these 4-12 cm. long; petals are yellow or pink; fruit dry and spiny. Probably not distinct as a species.---Dry places. Colorado, New Mexico and Arizona. Our one somewhat doubtful record from southwestern Colorado.

9. Opuntia rhodantha Schumann, Gesamtb. Kakteen. 735. 1898.
O. xanthostemma Sch.---This plant is related to O. polyacantha. The joints are 5-12 cm. long, but the areoles are over 10 mm. apart (usually 15-20 mm.); the spines are 2-6 cm. long; flowers are yellowish, salmon colored or rose-purple; fruit is dry and spiny. Doubtfully distinct as a species.---Dry ground. Colorado and Utah. Our records from the extreme western part of Colorado at 4500-7500 feet.

10. Opuntia trichophora (Engelm.) Britt. & Rose, Smiths. Misc. Coll. 50. 535. 1908.
This plant is closely related to O. polyacantha and was originally described as a variety of that species. The areoles are less than 10 mm. apart like that species but the spines are often elongated to 12 cm. and bristlelike or hairlike especially on lower areoles of lower joints; flowers yellow or pink; fruit about 2 cm. long, dry and spiny. Doubtfully distinct as a species.---Dry ground. Oklahoma to Colorado, south to Texas and New Mexico. Our few records from southern Colorado.

11. Opuntia polyacantha Haw., Suppl. Pl. Succ. 82. 1819.
Low spreading plants usually forming clumps, joints 7.5-10 cm. long, flat, obovate or orbicular; areoles small, closely set, less than 10 mm. apart, with glochids and 3-9 spines, those on sides appressed, often 1 or 2 elongated and erect ones to 3 cm. long, white, brown or variegated, no definite central spines present; flowers 4-7 cm. long, pale yellow or tinged with orange or pink; fruit about 2 cm. long, dry, globular to oblong, spiny; seeds 5-6 mm. long, irregular, yellowish or whitish. This species apparently forms the center of an intergrading group of dry-fruited Opuntias in this area.---Dry plains and hills. North Dakota to Alberta, south to Texas, Arizona and Washington. Our records scattered over Colorado at 4500-6000 feet.

2. Mamillaria Haw. PINCUSHION CACTUS; BALL CACTUS

Small or low plants with solitary to numerous, mostly globose or cylindrical 1-jointed stems; tubercles distinct not coalescing in ribs but spirally arranged, the tubercles, except for very earliest ones, mostly grooved on the upper sides and the areoles on the ends; spines straight or hooked, no glochids present; flowers from apex or from sides, from between the tubercles or on the tubercles at the base of the grooves; flowers funnelform to somewhat campanulate, diurnal, hypanthium naked or with a few scales; fruit fleshy, without spines.

1. Flowers yellowish-green; mature fruit scarlet; central spines none or sometimes 1 -1. M. missouriensis
1. Flowers pink-purple; mature fruit green; central spines usually over 1 -2. M. vivipara

1. Mamillaria missouriensis Sweet, Hort. Brit. 171. 1827.
M. similis Engelm.; Coryphantha missouriensis (Sweet) Britt. & Rose; C. similis (Engelm.) Britt. & Rose; Cactus similis (Engelm.) Britt. & Rose; C. missouriensis (Sweet) Kuntze; Neobesseya missouriensis (Sweet) Britt. & Rose---Plants solitary or in small clusters, 2.5-8 cm. in diameter, globose, tubercles prominent, over 5 mm. long; radial spines 10-20, gray with brownish tips, pubescent; central spines none or sometimes 1; flowers yellow, outer perianth segments ciliate; fruit globose ovoid, fleshy, scarlet, appearing in the spring of the following year; seeds black.---Dry plains and hills. North Dakota to Montana, south to Texas and Colorado. Reported as occurring throughout Colorado but our few records from the western and southern parts at 4500-7500 feet.

2. Mamillaria vivipara (Nutt.) Haw., Syn. Pl. Succ. Suppl. 72. 1819.
Coryphantha radiosa (Engelm.) Rydb.; C. vivipara (Nutt.) Britt. & Rose; Cactus viviparus Nutt.; C. radiosus (Engelm.) Coult.---Plants solitary or in clusters, 3-8 cm. high, somewhat cylindric to globose, with prominent tubercles these around 4-5 mm. long or longer; areoles large, radial spines about 16-18, white to brownish; central spines usually 3-6, sometimes solitary and stouter, brown or blackish; flowers pink to salmon-pink or pale purple the outer perianth segments ciliate; fruit green at maturity or sometimes somewhat brownish-red, juicy, globular to ellipsoid, smooth or with several scales scattered over the surface; seeds light brown.---Dry plains and hills. Manitoba to Alberta, south to Texas and New Mexico. Our records from southern, central and northcentral Colorado at 4500-8500 feet.

3. Echinocereus Engelm. HEDGEHOG CACTUS; STRAWBERRY CACTUS

Plants globose to cylindrical, single or caespitose; stems 1-jointed, erect or strongly ascending, tubercled, these joined into more or less vertical ribs, the spine-bearing areoles on these ribs; spines all straight, no glochids present; flowers borne just above the lateral spine-bearing areoles, hence lateral on the stem; flowers funnelform or subcampanulate, variously colored, hypanthium prolonged above the ovary, spiny; fruit spiny with large and easily detachable spine clusters, thin skinned, often edible; seeds black, tuberculate.

1. Flowers yellowish-green, 2-3 cm. long; ribs of stem usually 13 or 14 the stem seldom over 7 cm. high
   -1. E. viridiflorus
1. Flowers scarlet or purple, 5-9 cm. long; ribs of stem seldom over 12 or 13, the stem often over 7 cm. high
   2. Flowers scarlet or orange-red; fruit red when ripe
      3. Spines angled (usually definitely so); ribs of stem 5-8 (usually 7)   -2. E. triglochidiatus
      3. Spines terete, not angled; ribs 8-11 (usually 9)   -3. E. coccineus
   2. Flowers rose-purple to light purple; fruit red or green
      4. Radial spines 5-6 mm. long; central spines wanting or if 1 or 2 present, then resembling the radial
         ones   -4. E. reichenbachii
      4. Radial spines about 1-2 cm. long; central spine present, 2.5-4 cm. long and curved upward, hence not
         like the radial ones   -5. E. fendleri

1. Echinocereus viridiflorus Engelm. in Wislizenus, Tour. North. Mex. 91. 1848.
   Plants single or in small clusters, 2.5-7.5 cm. tall, subglobose to cylindrical, ribs often spiralled, 13 or 14; radial spines about 16; central spines absent or 2 or 3, all spines short, white to dark brown; flowers 2-3 cm. long, borne laterally or near base of plant, yellowish-green (outer perianth segments with reddish-brown midrib); fruit about 1 cm. long, greenish, bearing woolly areoles with spines; seeds about 1 mm. long.---Dry plains and hills. South Dakota to Wyoming, south to Kansas and New Mexico. Our records mostly from the eastern half of Colorado (one from Saguache County) at 4000-8000 feet.
2. Echinocereus triglochidiatus Engelm. in Wislizenus, Mem. Tour. North. Mex. 93. 1848.
   E. gonacanthus (Engelm. & Bigel.) Lem.; E. paucispinus (Engelm.) Rumpl.---Stems caespitose, with few to many simple cylindrical or ovoid sections, these 10-15 cm. high and 5-6 cm. wide, about 5-8 ribbed (usually 7); lateral spines 3-8, to 3 cm. long, gray to black, stout, usually angled; central spines absent or 1 present; flowers 5-8 cm. long, scarlet, spines on ovary and tube few, white-felted; fruit about 3 cm. long, spiny at first, bright red, juicy and edible; seeds about 1 mm. long. Closely related to the next and doubtfully specifically distinct from it.---Dry often rocky or sandy hills. Colorado to Texas and New Mexico. Our records from central, westcentral and southern Colorado at 5000-8000 feet.
3. Echinocereus coccineus Engelm. in Wislizenus, Mem. Tour. North. Mex. 94. 1848.
   E. roemeri and E. aggregatus of Manuals---Plants usually caespitose, often forming large mounds, joints 5-10 cm. high and about 3-5 cm. wide, simple, cylindrical; ribs 8-11, usually about 9; radial spines 7-12, 1-2 cm. long usually whitish; central spines 1-3, longer and stouter than the radial ones, usually yellowish or gray-white, terete; flowers 5-7.5 cm. long, crimson or orange-red; ovary and tube with short bristly spines; fruit about 2 cm. long, red, juicy, edible with rather deciduous spines; seeds about 1 mm. long. A form from western Colorado with spines absent or rudimentary has been called var. inermis (Schumann) Boiss. and David.---Dry plains and hills. Colorado, Utah, New Mexico and Arizona. Our records from southern and western Colorado at 4000-8000 feet.
4. Echinocereus reichenbachii (Terscheck) Haage jr., Index Kew.2:813. 1893.
   E. caespitosus Engelm.---Stems 2.5-15 cm. tall, about 3-8 cm. in diameter, single or somewhat clustered, simple or rarely branched, ovoid or cylindrical; ribs 12-13; radial spines about 12-18, about 5-6 mm. long; central spines wanting or 1 or 2, like the radial ones, white to brown, not interlocking; flowers about 6-8 cm. long, deep rose-purple or light purple; fruit about 1 cm. long, ovoid, green, woolly and spiny; seeds about 1-1.5 mm. long.---Dry ground. Kansas to Colorado, south to Texas. Definitely reported from southeastern Colorado and included on the strength of that report.
5. Echinocereus fendleri (Engelm.) Rumpler in Forster, Handb. Cact. ed. 2. 801. 1886.
   Plants about 10 cm. long and 5-6 cm. wide, in small clusters or sometimes solitary, ovoid or somewhat cylindrical; ribs 9-12; radial spines 5-10, about 1-2 cm. long; central spines 1, 2.5-4 cm. long, curved upward, dark colored; flowers 6-9 cm. long, purple, borne on upper part of plant, the tube spiny; fruit about 2.5-3 cm. long, ovoid or globose, red, edible, spiny; seeds 1-1.5 mm. long.---Dry plains and hills. Utah, New Mexico and Arizona. Definitely reported for southwestern Colorado but no specimens seen.

4. Echinocactus Link & Otto   HEDGEHOG CACTUS; BARREL CACTUS

Plants small to rather large, not jointed, caespitose or simple (usually the latter); tubercles confluent in more or less vertical or spiral continuous ribs; spines straight, curved or hooked, no glochids present; flowers vertical or nearly so, arising from special areoles just above an immature spine-producing areole at the stem apex; flowers funnelform to campanulate, the hypanthium not forming a distinct tube above the ovary, with the scales naked or with a tuft of short wool in their axils; fruit usually with scales but spineless, dry or fleshy, more or less dehiscent. The above has been divided into several genera, with some justification when considering a limited area like Colorado.

1. Ovary (outside) and fruit scaly, these (when lifted up) showing a cluster of woolly hairs in their axils;
   flowers purple to lavender in color, 3.5-5 cm. long; plants often cylindrical   -1. E. whipplei
1. Ovary and fruit naked or with a few scales these naked in their axils; flowers cream-colored, yellowish to
   pink, seldom over 3.5 cm. long; plants usually globose
   2. Plants strongly tubercled, usually over 6 cm. tall; central spines about 8; flowers usually pink, not over
      2.5 cm. long, plants more widely distributed in Colorado   -2. E. simpsonii
   2. Plants only moderately tubercled, seldom much over 6 cm. tall; central spines none or sometimes 1; flowers
      yellow to greenish-white, usually over 3 cm. long; plants apparently limited to Montezuma County
                                                                                             -3. E. mesa-verdae

1. Echinocactus whipplei Engelm. & Bigel. in Engelm., Proc. Am. Acad. Arts and Sci. 3:271.1856.
E. glaucus K. Sch.; E. subglaucus Rydb.; Sclerocactus whipplei (Engelm. & Bigel.) Britt. & Rose; S. franklinii Evans---Stems 10-25 cm. long and 7-15 cm. wide, usually solitary or sometimes in small clusters, obovoid, globose or cylindrical; ribs usually 13-15, generally spiralled, more or less tubercled; lateral spines 7-15, somewhat flattened, straight or recurved; central spines 1-4, up to 3-5 cm. long, at least the lower one hooked, spines whitish to black; flowers 3.5-5 cm. long and nearly as wide, from near center of plant on the ribs, purple or lavender; fruit 1-1.5 cm. long, oblong, dry, red, few-scaled, the hairs shorter than scales and hidden by them.---Dry often sandy ground. Colorado, Utah and Arizona. Our records from westcentral and southwestern Colorado at 4500-5500 feet.
2. Echinocactus simpsonii Engelm., Trans. St. Louis Acad. 2:197. 1863.
E. simpsonii var. minor Engelm.; Pediocactus simpsonii (Engelm.) Britt. & Rose---Plants 4-15 cm. in diameter, depressed, globular to turbinate usually single, strongly tubercled but these borne in spiralled obscure ribs; young areoles very woolly, or in age nearly naked; radial spines 15-25, whitish; central spines about 5-8, 0.4-3 cm. long, whitish to brown; flowers small, about 3 cm. across, massed in center and surrounded by brown or whitish wool, funnelform in the tube, rather campanulate above, usually pink, verging to yellowish or whitish; stamens and stigmas yellow or yellowish-green; fruit few-scaled at apex, these naked in axils, dry, smooth, splitting on 1 side; seeds black, tuberculate, keeled on back. This plant resembles our species of Mamillaria.---Dry plains in the foothills, mesas and mountains. Montana to Washington, south to New Mexico and Utah. Our records scattered over the western half of Colorado at 6000-10,000 feet.
3. Echinocactus mesae-verdae (Boiss. & David.) L. Benson, Leafl. West. Bot. 6:163. 1951.
Coloradoa mesae-verdae Boiss. & David.---Plants about 6 cm. high, solitary or caespitose, globose or depressed, with 13-17 ribs of confluent tubercles; areoles densely woolly when young; radial spines 8-10, about 1 cm. long, whitish with brownish base, spreading; central spine usually wanting; flowers about 3.5 cm. long, yellow to greenish-white, the tube naked or with a few scales these not hairy in their axils; fruit naked, no scales or hairs, irregularly dehiscent, perianth persistent at top; seeds large (4 mm. long), black.---Dry slopes. Apparently limited to southwestern Colorado.

Family 82. Cistaceae ROCKROSE FAMILY

Shrubs or herbaceous plants, slightly woody at the base; leaves simple, alternate, entire; flowers solitary or several racemose, regular, perfect, often of 2 kinds, some with large fugaceous petals and many stamens, others with petals small or lacking, with 3-10 stamens and cleistogamous; sepals 3-5; petals distinct, 5 (3 in the genus Lechia) or wanting; ovary superior, 1-celled (partly 3-celled in Lechia) with 3 parietal placentae; styles 1, stigma capitate or 3-lobed; fruit a capsule; seeds few.

Another genus Lechia L. could be in eastern Colorado. L. intermedia Leggett is recorded for northwestern Nebraska. The 2 genera can be separated as follows:

1. Petals 5 and fugaceous or wanting, yellow   Helianthemum
1. Petals 3, persistent, marcescent, not yellow   Lechia

1. Helianthemum Pers. FROSTWEED

Characters of the Family

1. Helianthemum bicknellii Fernald, Rhodora 21:36. 1919.
H. majus and Crocanthemum majus of Manuals; C. bicknellii (Fern.) Britt.---Stems 20-60 cm. tall, erect, slightly woody at base, simple at first, later with short ascending branches, canescent-stellate, especially at first; leaves 1.5-3.5 cm. long, oblong-lanceolate or oblanceolate, acute or obtuse, stellate both sides but sparser and green above, dense and whitish below, short-petioled; petaliferous flowers several in terminal cymes; sepals 5, canescent, 3 inner ones broad, the 2 outer ones narrow and bractlike; petals 7-9 mm. long, yellow, oval; apetalous flowers appearing later, clustered in the axils of the leaves, nearly sessile.---Dry ground. Nova Scotia to South Dakota, south to South Carolina, Texas and Colorado. Our few records from central Colorado at 7000-7500 feet.

Family 83. Elaeagnaceae OLEASTER FAMILY

Shrubs or trees, scurfy-pubescent with small stellate scales; leaves alternate or opposite, simple, entire; flowers axillary, solitary or in clusters, regular, perfect, polygamous or dioecious; hypanthium of pistillate flowers tubular or urn-shaped, appearing adnate to and enclosing the ovary, the upper part 4-lobed, these deciduous; petals lacking; stamens 4 or 8; disk present, annular or lobed; ovary appearing inferior, 1-celled, 1-ovuled, 1 style; fruit drupelike, dry or fleshy.

1. Leaves alternate; stamens 4; flowers perfect or polygamous; fruit dry and mealy, covered with scurfy scales
                                                                                          -1. Elaeagnus
1. Leaves opposite; stamens normally 8; flowers dioecious; fruit succulent, juicy, not covered with scales
                                                                                          -2. Shepherdia

1. Elaeagnus L. OLEASTER; SILVERBERRY

Shrubs or trees; leaves alternate; flowers perfect or polygamous; stamens 4; fruit scaly, dry and mealy.

1. Young branches uniformly silvery-scurfy; leaves oblong-lanceolate to narrower, not often over 12 mm. wide; plants usually near habitations or in settled areas -1. E. angustifolia
1. Young branches brown-scurfy (at least somewhat); leaves elliptic-ovate to oblong-ovate, often over 12 mm. wide; plants truly native -2. E. commutata

1. **Elaeagnus angustifolia** L., Sp. Pl. 121. 1753.
Shrubs or often trees to 7 m. tall, especially in cultivation; branches often thorny, silvery-scurfy when young; leaves 3-8 cm. long, oblong-lanceolate to linear-lanceolate, silvery-scurfy beneath, green or somewhat scurfy above; flowers 1-3; fruit about 1 cm. long, ellipsoid, silvery-scurfy.---Cultivated and sometimes escaping, mostly in low ground along streams or valleys. Our records scattered in the agricultural areas of Colorado, probably many of them from cultivated plants, at 4500-6500 feet.

2. **Elaeagnus commutata** Bernh., Allg. Thuer. Gartenz. 2:137. 1843.
E. argentea Pursh---Shrubs or small trees 2-5 m. high; twigs brown-scurfy or at least somewhat brownish; leaf blades 2-10 cm. long and usually over 12 mm. wide, elliptic-ovate to oblong-ovate.---Banks and hillsides. Quebec to Yukon, south to Minnesota, Wyoming and Utah. Recorded from southern Wyoming and to be expected in northern Colorado. Our one surprising record from the southern part of the state but the plant may have been cultivated or the labels mixed.

2. Shepherdia Nutt. BUFFALOBERRY

Shrubs; leaves opposite; flowers dioecious; stamens normally 8; fruit succulent, not scaly.

1. Leaves green above, brown-scurfy (at least somewhat) below; low, thornless shrubs, seldom over 3 m. tall -1. S. canadensis
1. Leaves silvery-scurfy both sides; tall usually thorny shrubs or small trees, often over 3 m. tall -2. S. argentea

1. **Shepherdia canadensis** (L.) Nutt., Gen. Pl. 2:241. 1818.
Lepargyraea canadensis (L) Greene---Shrubs 1-3 m. tall, branches thornless, brown-scurfy when young; leaves ovate or oval, dark green and nearly or quite glabrous above, mixed silver- and brown-scurfy beneath; flowers in clusters at the nodes; fruit 4-6 mm. long, oval, red or yellowish.---Moist usually shaded slopes. Newfoundland to Alaska, south to New York, New Mexico and Oregon. Our records scattered in the western half of Colorado at 7500-11,500 feet.

2. **Shepherdia argentea** (Pursh) Nutt., Gen. Pl. 2:241. 1818.
Lepargyraea argentea (Nutt.) Greene---Shrubs or small trees 2-7 m. tall; branches usually thorny, silvery-scurfy when young; leaves 2-5 cm. long, oblong, obtuse, cuneate at base, silvery-scurfy both sides; flowers fascicled at the nodes; fruit about 4-6 mm. long, ovoid-round to ellipsoid, scarlet to golden, sour but edible.---Banks and valleys along streams or low meadows. Manitoba to Alberta, south to Kansas, New Mexico and California. Our records scattered in the western two-thirds of Colorado at 4500-7500 feet.

Family 84. Lythraceae LOOSESTRIFE FAMILY

Herbaceous plants; leaves opposite or alternate, simple, entire; flowers perfect, regular, solitary, in axillary clusters or cymes; calyx with 4 or 5 lobes often with as many accessory teeth, the tube free from the ovary but enclosing it closely; petals distinct, 4 or 5, borne on the throat of the calyx; stamens 4-12, on the calyx; ovary superior, 2- to 6-celled or soon becoming 1-celled, styles 1, stigma capitate; fruit a 1- to 4-celled capsule, sometimes bursting irregularly; seeds many.

1. Plants annual; capsule bursting irregularly; calyx tube campanulate; petals less than 2 mm. long -1. Ammannia
1. Plants perennial; capsule 2-valved; calyx tube cylindrical; petals over 2 mm. long -2. Lythrum

1. Ammannia L.

Annual glabrous plants; stems 4-angled; leaves opposite, sessile; flowers 1-several in the axils; sepals 4, with alternating smaller teeth, the tube campanulate and 4-angled; petals 4, readily deciduous; stamens 4-8; ovary 2- to 4-celled; capsule bursting irregularly.

1. **Ammannia coccinea** Rottb., Pl. Hort. Havn. Descr. 7. 1773.
Stems 10-50 cm. tall, erect, branching below; leaves 3-7 cm. long, linear or linear-lanceolate, acute, cordate-auriculate at base; calyx 3-4 mm. long, the lobes short; petals about 1-1.5 mm. long, purple, obovate.---Wet ground or shallow water. New Jersey to Washington, south to Florida and California; South America. Our one record from near Denver, Colorado, at about 5200 feet.

2. Lythrum L. LOOSESTRIFE

Perennial herbaceous plants; stems angled; leaves alternate or opposite below; flowers solitary in the upper axils; calyx lobes 4-6, with alternating teeth, the tube cylindrical and grooved; petals 4-6; stamens 8-12; ovary 2-celled; fruit 2-valved.

1. Lythrum alatum Pursh, Fl. Am. Sept. 334. 1814.
Stems 30-100 cm. tall, usually much branched, glabrous; leaves 2-5 cm. long, sessile, lanceolate to oblong, acute or obtuse at apex; calyx about 4-6 mm. long, the lobes short; petals about 4-6 mm. long, purple, obovate to oblanceolate, ascending; flowers dimorphic with stamens of 2 lengths.---Low moist open ground. Ontario to British Columbia, south to Kentucky, Texas and Colorado. Our records from northcentral and northeastern Colorado at 3500-6000 feet.

Family 85. Onagraceae EVENING PRIMROSE FAMILY

Herbaceous or rarely woody plants, caulescent or acaulescent; leaves alternate, opposite or basal; stipules none; flowers perfect, mostly regular, in axillary or terminal racemes; hypanthium adnate to the ovary and usually prolonged beyond it; sepals 4 or 2; petals distinct, 4 or 2, inserted at the summit of the hypanthium; stamens as many or twice as many as the petals and borne with them; ovary inferior, usually 4-celled; style 1, stigma capitate, 4-lobed or discoid; fruit usually a capsule but sometimes nutlike and indehiscent.

1. Sepals and petals 2; stamens 2; fruit bearing hooked hairs -1. Circaea
1. Sepals and petals 4; stamens usually 8; fruit not bearing hooked hairs
  2. Seeds with a tuft of hair at 1 end
    3. Hypanthium tube 2-3 cm. long, inflated just above the ovary, then long-tubular with a row of 8 scales inside about the middle; flowers scarlet -2. Zauschneria
    3. Hypanthium tube less than 1 cm. long or absent entirely, not inflated just above the ovary, no scales present; flowers white, pink or rose-purple -3. Epilobium
  2. Seeds without a tuft of hair
    4. Fruit indehiscent, nutlike, 1 or sometimes 1- to 4-seeded, not over 1 cm. long; petals somewhat unequal
      5. Hypanthium tubes filiform, 8-10 mm. long; filaments with no basal scales present; ovary 1-celled -4. Stenosiphon
      5. Hypanthium tubes cylindrical or obconical, never filiform, usually not over 8 mm. long; filaments often with a basal scale present below each; ovary (not fruit!) 4-celled -5. Gaura
    4. Fruit dehiscent or if woody and tardily deciduous then not so small; seeds usually many; petals usually about equal
      6. Ovary and fruit 2-celled; hypanthium tube not at all prolonged beyond the ovary; flowers minute, the petals not over 1.5 mm. long -6. Gayophytum
      6. Ovary and fruit 4-celled; hypanthium tube prolonged beyond the ovary (may be as short as 1 mm. or as long as 15 cm.); flowers large or small but petals over 1.5 mm. long -7. Oenothera

1. Circaea L. ENCHANTER'S NIGHTSHADE

Low slender perennial herbaceous plants with rootstocks and small tubers; leaves opposite, simple, unlobed, cauline, thin, petioled, the petiole about as long as the blade; flowers small, in paniculately arranged racemes; hypanthium only slightly prolonged beyond the ovary, deciduous, with a ringlike disk within; sepals 2, reflexed; petals 2, whitish, notched at apex; stamens 2, alternate to the petals; ovary 1- to 2-celled, each cell with 1 ovule; fruit nutlike, indehiscent, obovoid, 1- to 2-seeded, bearing hooked hairs.

1. Leaves cordate at base, sharply and coarsely serrate; plants not over 20 cm. tall; racemes with minute bracts, at least when young -1. C. alpina
1. Leaves rounded or truncate at base, undulate-denticulate; plants usually over 20 cm. tall; racemes bractless -2. C. pacifica

1. Circaea alpina L., Sp. Pl. 9. 1753.
Stems 5-20 cm. tall, simple or sometimes branched above, glabrous or pubescent above, the hairs often glandular; leaves 2-5 cm. long, ovate, mostly cordate at base, sharply but coarsely dentate; pedicels subtended, at least when young, with minute bracts; fruit about 2 mm. long.---Usually in shade. Labrador to Alaska, south to Georgia, Colorado and Washington; Eurasia. Our records from northcentral and central Colorado at 6000-8000 feet.
2. Circaea pacifica Aschers. & Magnus, Bot. Zeit. 29:392. 1871.
Reported for Montana to British Columbia, south to Colorado and California. A specimen from Boulder County may be this species.

2. Zauschneria Presl. HUMMINGBIRD-TRUMPET; FIRECHALICE

Perennial herbaceous plants; leaves opposite (except those of inflorescence), more or less fascicled, sessile or nearly so, toothed; inflorescence spicate, the flowers scarlet, large and showy; hypanthium globose-inflated above the ovary, then narrowed into a long tube bearing within a transverse row of 8 appendages; sepals 4; petals 4, notched; stamens 8, the alternate ones shorter, anthers versatile; ovary 4-celled, stigmas 4-lobed or capitate; capsule many-seeded; seeds narrowed at base, comose at apex.

1. **Zauschneria garrettii** A. Nels., Proc. Biol. Soc. Wash. 20:36. 1907.
Reported for Wyoming, Idaho and Utah. The species should be looked for in northern or northwestern Colorado.

3. Epilobium L. WILLOWWEED; WILLOW HERB

Mostly herbaceous plants, annual or perennial; leaves opposite or alternate, sessile or short-petioled, entire or toothed; hypanthium either not prolonged beyond the ovary or more commonly prolonged as a short tube; sepals 4; petals 4, usually notched at apex, purplish, pink or white, sometimes yellow; stamens 8, the alternating ones shorter; stigmas oblong or 4-lobed; fruit an alongated, cylindrical, fusiform or clavate, 4-celled, 4-sided, loculicidal capsule; seeds with a tuft of hair at upper end. This is a puzzling genus in Colorado and needs an up-to-date monograph. This treatment follows that of Munz in several state floras. The presence or absence of turions is usually impossible to tell with most herbarium material.

1. Hypanthium not prolonged beyond the ovary; flowers large, the petals 1-2 cm. long, spreading; stigmas 4-lobed
    2. Styles pilose at base, longer than the stamens; leaves 5-15 cm. long, membranous, lateral veins definite, confluent in marginal loops; racemes many-flowered, elongate, not leafy   -1. E. angustifolium
    2. Styles glabrous; shorter than the stamens; leaves 2-6 cm. long, thick and fleshy, not veiny; racemes few-flowered, short and leafy   -2. E. latifolium
1. Hypanthium prolonged beyond the ovary (may be short); flowers smaller, the petals not over 1 cm. long, ascending; stigmas oblong, not lobed
    3. Annual plants; stems with exfoliating epidermis; plants of dry situations
        4. Stems 30-90 cm. tall, glabrous (except in upper part); leaves mostly alternate with fascicles in axils; hypanthium over 1 mm. long   -3. E. paniculatum
        4. Stems 5-30 cm. tall, puberulent throughout; leaves mostly opposite, without fascicles; hypanthium scarcely 1 mm. long   -4. E. minutum
    3. Perennial plants; stems not exfoliating; plants usually of moist places
        5. Leaves linear or linear-oblong, margins entire or nearly so, more or less revolute, sessile or nearly so
            6. Leaves glabrous, 3-7 mm. wide; stems slender   -5. E. palustre
            6. Leaves definitely pubescent, at least above, 1-3.5 mm. wide; stems stout   -6. E. leptophyllum
        5. Leaves linear-lanceolate to ovate or oval, margins not revolute, entire, serrulate, sessile or petioled
            7. Rootstocks bearing turions (globose or ovoid winter buds with fleshy overlapping scales), the scales sometimes rather loose (complete material needed here,in case of doubt try both numbers); seeds usually more or less papillose under strong magnification
                8. Petals 5-10 mm. long; leaves 5-12 cm. long; stems 30-90 cm. tall   -7. E. glandulosum
                8. Petals 2-5 mm. long; leaves 1-5 cm. long; stems 10-40 cm. tall (rarely to 60 cm.)
                    9. Stems glabrous below, pubescent or glandular above but not with decurrent, longitudinal lines of hair from the leaf bases   -8. E. brevistylum
                    9. Stems with decurrent, longitudinal lines of hair from the leaf bases
                        10. Leaves lanceolate-linear, often decurrent at base, not crowded, the margins more commonly irregularly dentate   -9. E. halleanum
                        10. Leaves, at least the basal,ovate with rounded bases, often crowded, the margins more commonly entire   -10. E. saximontanum
            7. Rootstocks without turions although sometimes with rather fleshy rosettes present; seeds smooth or papillose
                11. Stems 30-100 cm. tall, freely branched above; leaves sessile or short-petioled; capsule 4-6 cm. long; coma of seeds white
                    12. Leaves ovate to elliptical-lanceolate; petals about 4 mm. long   -11. E. adenocaulon
                    12. Leaves narrowly lanceolate; petals 5-6 mm. long   -11A. E. adenocaulon occidentale
                11. Stems 5-30 cm. tall, simple or nearly so; leaves definitely petioled (not long); capsules 2-5 cm. long; coma of seeds white to dingy
                    13. Stems densely caespitose, sigmoidly bent, 5-15 cm. tall; leaves 1-2 cm. long; capsules 2-4 cm. long; petals purple or rose
                        14. Capsules linear, slender, not over 1 mm. thick and 2-4 cm. long; seeds smooth, about 1 mm. long; leaves oblong-ovate to lanceolate   -12. E. alpinum
                        14. Capsules subclavate, 1.5-2 mm. thick and 2-2.5 cm. long; seeds papillose (under magnification), 1.5-2 mm. long; leaves broadly ovate   -13. E. clavatum
                    13. Stems not densely caespitose, erect, not sigmoidly curved, 10-30 cm. tall; leaves 1.5-5 cm. long; capsules 3-5 cm. long; petals white or white with pink tips (in all but E. hornemanii)
                        15. Petals rose or violet, 5-8 mm. long; seeds papillose (under strong magnification)   -14. E. hornemanni
                        15. Petals white or with pink tips only, about 3 mm. long; seeds smooth or papillose
                            16. Leaves obtuse, mostly opposite in pairs; fruiting pedicels erect; seeds smooth   -15. E. lactiflorum
                            16. Leaves acute, mostly alternate; fruiting pedicels arched or ascending; seeds papillose   -16. E. ciliatum

1. **Epilobium angustifolium** L., Sp. Pl. 347. 1753.
Chamaenerion angustifolium (L.) Scop.; C. spicatum (Lam.) S. F. Gray---Perennial herbaceous plants; stems 50-250 cm. tall, erect, mostly unbranched, glabrate below, puberulent above; leaves 5-15 cm. long, alternate, lanceolate to linear-lanceolate, sessile or nearly so, nearly entire, lighter below, veins rather prominent, the lateral looped before reaching margins; flowers in long terminal racemes, very showy; hypanthium not prolonged above the ovary; sepals 8-12 mm. long; petals 8-18 mm. long, lilac-purple to rose (rarely to white); stigma with 4 linear or narrowly oblong lobes, the style longer than the stamens, pilose at base; capsule 5-8 cm. long;

seeds 1-1.4 mm. long, coma long and dingy. Our Colorado plant has been called var. platyphyllum (Daniels) Fernald.---Edges of woods, often in burnt over ground, usually in rather moist soil. Quebec to Alaska, south to New Mexico and California. Our records scattered in the western half of Colorado at 6500-11,000 feet.

2. Epilobium latifolium L., Sp. Pl. 347. 1753.

Chamaenerion latifolium (L.) Sweet---Perennial plants with caespitose rootstocks; stems several, 10-60 cm. tall, depressed or ascending, glabrous below, puberulent above; leaves 2-6 cm. long, lanceolate to elliptic-ovate, subopposite, thick, glaucous, subsessile, veins inconspicuous, entire or minutely denticulate; flowers in terminal racemes, showy; hypanthium not prolonged; sepals 8-15 mm. long; petals 15-25 mm. long, rose-colored, purple or white, darker veined; stigma irregularly 4-lobed, the lobes oblong, styles shorter than the stamens, glabrous; capsule 5-8 cm. long; seeds not seen.---Moist ground. Greenland to Alaska, south to South Dakota, Colorado and California; Eurasia. Our records in the western half of Colorado except the northwestern part, at 7500-12,500 feet.

3. Epilobium paniculatum Nutt. ex T. & G.,, Fl. N. Amer. 1:490. 1840.

E. adenocladon (Haussk.) Rydb.; E. laevicaule Rydb.---Annual plants; stems 30-100 cm. tall, erect with shreddy epidermis below, simple at base and paniculately branched above, glabrous throughout or slightly puberulent above; leaves 2-5 cm. long, linear-lanceolate to linear, remotely denticulate, usually alternate with fascicles of leaves in the axils; hypanthium 2-3 mm. long; sepals 2-3 mm. long; petals 5-6 mm. long, rose or lilac; fruiting pedicels 3-15 mm. long; capsule about 2-3 cm. long, linear to linear-clavate; seeds large, about 2 mm. long, with tawny or dingy coma, surface minutely papillate.---Rather dry often sandy plains and hills. South Dakota to British Columbia, south to New Mexico and California. Our records scattered in the western half of Colorado at 5500-9000 feet.

4. Epilobium minutum Lindl. ex Hook., Fl. Bor. Am. 1:207. 1834.

Reported for Montana to British Columbia, south to Arizona and California. This plant may be in western Colorado.

5. Epilobium palustre L., Sp. Pl. 348. 1753.

E. wyomingense A. Nels.---Perennial plants, the rootstocks bearing turions; stems 20-40 cm. tall, slender, erect, simple or branched, glabrous below, minutely puberulent above; leaves 3-5 cm. long or more, 3-7 mm. wide, in pairs, thin, linear-oblong, acute, cuneate, glabrous; petals 2-4 mm. long, white; capsule 4-7 cm. long, linear, more or less puberulent; seeds numerous, smooth, coma white. Our plant is var. grammadophyllum Haussk.---Moist or wet ground. The variety from Newfoundland to Alaska, south to Minnesota and Colorado. Our few records from northwestern and northcentral Colorado (the latter ones somewhat doubtful) at 5500-8500 feet.

6. Epilobium leptophyllum Raf., Precis des Decouv. 41. 1814.

E. lineare of Manuals---Perennial plants with small inconspicuous turions; stems 35-100 cm. tall, rather stout, more or less branched especially above, with short incurved hairs; leaves 1.5-4.5 cm. long, linear, margins entire or nearly so, somewhat revolute, pubescent with incurved hairs at least on veins below, sessile or nearly so; hypanthium about 1 mm. long; sepals 2.5-3 mm. long; petals 3-5 mm. long, pink or white; fruiting pedicels 10-25 mm. long; capsule 3-5 cm. long, linear, pubescent with incurved hairs; seeds 1.4-1.6 mm. long, surface smooth to minutely papillose, coma dingy.---Moist or wet ground. Quebec to Alberta, south to West Virginia and Colorado. Our records from northern Colorado, mostly northcentral, at about 5000 feet.

7. Epilobium glandulosum Lehm., Pugillus 2:4. 1830.

Reported from Canada, south to Colorado and California. No specimens from the state have been seen by the writer.

8. Epilobium brevistylum Barbey in Brewer & Wats., Bot. Calif. 1:220. 1876.

Perennial plants with compact turions; stems 20-60 cm. tall, erect, glabrous below, crisp-pubescent to somewhat glandular above, simple or subsimple; leaves 2-4 cm. long sometimes to 6 cm., mostly opposite, ovate to elliptic-lanceolate, sometimes to linear-lanceolate, sessile, denticulate; hypanthium about 2 mm. long; sepals 2-3 mm. long; petals 3-5 mm. long, rose-purple or paler; fruiting pedicels 5-15 mm. long; capsule 4-6 cm. long, linear, sparingly pubescent; seeds about 1.2-1.5 mm. long, papillose, coma dingy.---Moist or wet ground. Montana to Washington, south to Colorado and California. Our records thinly scattered in the western half of Colorado at 7500-9000 feet.

9. Epilobium halleanum Haussk., Mon. Epil. 261. 1884.

E. drummondii Haussk.---Perennial plants with turions; stems 10-40 cm. tall, erect, simple or nearly so, subglabrous below, with longitudinal lines of hairs from leaf bases; leaves 1.5-4 cm. long, linear-lanceolate to lanceolate, mostly opposite, entire to serrulate, erect, sessile or nearly so, often clasping, acute; hypanthium 1-1.5 mm. long; sepals 2-3 mm. long; petals 2-4 mm. long, white to purple; fruiting pedicels 3-5 mm. long; capsule 2-5 cm. long; seeds 1.2-1.4 mm. long, papillose, coma dingy. Some of the plants of northwestern Colorado we have been calling the above may be E. leptocarpum var. macounii Trel.---Moist or wet places. Montana to British Columbia, south to Colorado and California. Our records thinly scattered in the western half of Colorado at 7500-10,500 feet.

10. Epilobium saximontanum Haussk., Bot. Zeitschr. 29:119. 1879.

E. ovatifolium Rydb.---Perennial plants with turions; stems 10-40 cm. tall, single or clustered, simple or branched, with decurrent longitudinal lines of hairs from the leaf bases; leaves 2-4 cm. long, opposite or upper alternate, lower ones ovate to ovate-lanceolate, upper ones ovate to elliptical, acute, sessile or short-petioled, serrulate to entire; hypanthium 1-1.5 mm. long; sepals 2 mm. long; petals 3-5 mm. long, white, pink or purple; fruiting pedicels 5-10 mm. long; capsule 4-6 cm. long, more or less pubescent; seeds about 1-1.3 mm. long, papillose, coma white to dingy.---Moist or wet places. Alberta, south to Colorado and Arizona. Our records thinly scattered over the western half of Colorado at 9000-11,000 feet.

11. Epilobium adenocaulon Hausskn., Oesterr. Bot. Zeitschr. 29:119. 1879.
Perennial plants without turions but with rosettes; stems 30-100 cm. tall, glabrous below, glandular-pubescent above, simple or sparingly branched below, usually freely branching above; leaves 3-6 cm. long, ovate to elliptic-lanceolate, obtuse to acute, rounded into very short petioles or even sessile at base, nearly opposite, glabrous or nearly so; hypanthium about 1-1.5 mm. long; sepals 2-3 mm. long; petals about 4 mm. long, white to pale rose-red; fruiting pedicels about 7-10 mm. long; capsule 4-6 cm. long, slender, glabrate in age; seeds about 1 mm. long, surface papillose, coma white.---Moist or wet ground. New Brunswick to British Columbia, south to West Virginia and California. Our records scattered rather irregularly in the western two-thirds of Colorado at 5000-10,000 feet.

11A. Epilobium adenocaulon occidentale Trel., (var.) Ann. Rept. Mo. Bot. Gard. 9:95. 1891.
E. occidentale (Trel.) Rydb.---Differs from the species in having narrower leaves, these narrowly lanceolate; petals longer, these 5-6 mm. long.---British Columbia, south to Utah and California. Our one record from northcentral Colorado at about 5000 feet.

12. Epilobium alpinum L., Sp. Pl. 348. 1753.
E. anagallidifolium Lam.---Perennial stoloniferous plants lacking turions; stems 5-15 cm. tall, caespitose, slender, sigmoidally bent and nodding at the apex, glabrous or pubescent in longitudinal lines; leaves 1-2 cm. long, mostly opposite, oblong-ovate to lanceolate, obtuse, entire or nearly so, on short petioles; hypanthium 1-1.5 mm. long; sepals 2-2.5 mm. long; petals 4-5 mm. long, lilac, rose or purple; fruiting pedicels 4-15 mm. long; capsule 2-4 cm. long, linear, sparingly hairy; seeds 1 mm. long, surface smooth, coma dingy.---Wet ground or rocky slopes. Labrador to Alaska, south to Colorado and California; Eurasia. Our records scattered in the higher mountains of Colorado at 8500-12,500 feet.

13. Epilobium clavatum Trel., Ann. Rep. Mo. Bot. Gard. 2:111, pl. 48. 1891.
Perennial plants without turions; stems 5-15 cm. tall, caespitose, sigmoid, more or less crisp-pubescent the hairs in more or less definite longitudinal lines above; leaves 1-2 cm. long, mostly opposite, ovate, entire to serrulate with short winged petioles; hypanthium about 1.5 cm. long; sepals 2.5-4 mm. long; petals 3.5-6 mm. long, rose or purple; fruiting pedicels 8-20 mm. long; capsules 2-2.8 cm. long and about 1.5-2 mm. wide, subclavate; seeds about 1.5-2 mm. long, minutely papillose, coma dingy to whitish.---Mountains. Montana to British Columbia, south to Colorado and Oregon. Our record from northwestern Colorado at 10,800 feet.

14. Epilobium hornemanni Reichenb., Ic. Crit. 2:73. 1824.
Perennial plants without turions; stems 10-30 cm. tall, erect, slender, simple, glabrous except for crisp-pubescence in decurrent longitudinal lines from the leaf bases, somewhat glandular above; leaves 1.5-4 cm. long, mostly opposite, ovate to elliptic-ovate, mostly obtuse, short-petioled, margins subentire to remotely denticulate; hypanthium 1-1.5 mm. long; sepals 2.5-4 mm. long; petals 5-8 mm. long, rose or violet; capsule 4-5 cm. long, linear, glabrate; seeds about 1 mm. long, smooth or papillose, coma dingy.---Wet places. Greenland to Alaska, south to New Hampshire, Colorado and California; Eurasia. Our few records scattered in the higher mountains of Colorado at 9500-12,000 feet.

15. Epilobium lactiflorum Hausskn., Oesterr. Bot. Zeitschr. 29:89. 1879.
Perennial plants without turions; stems 10-30 cm. tall, not densely caespitose or sigmoidally curved, glabrous or nearly so; leaves 2-4 cm. long, mostly opposite in pairs, oblong-ovate to elliptic, obtuse, short-petioled; pedicels erect in fruit; hypanthium short; petals about 3 mm. long, white or pink-tipped; capsules 3-5 cm. long; seeds about 1 mm. long, smooth.---Moist ground. New Hampshire to Alaska, south to Colorado and California; Eurasia. Our few records from northcentral Colorado at 9500-10,500 feet.

16. Epilobium ciliatum Raf., Med. Repos. N. Y. hex. II. v. 361. 1808.
E. americanum Hausskn.; E. perplexans Trel.---Newfoundland to Washington, south to New York, Colorado and California. No specimens from the state have been seen by the writer.

4. Stenosiphon Spach

Perennial herbaceous caulescent plants; stems erect; leaves alternate and sessile, narrow; flowers in narrow terminal spikes; hypanthium much prolonged beyond the ovary into a filiform tube; sepals 4; petals 4, white, clawed, rather unequal; stamens 8, declined, no scales at base of filaments; ovary 1-celled, stigma 4-lobed with a cuplike border below; fruit indehiscent, 8-ribbed, 1-seeded.

1. Stenosiphon linifolium (Nutt.) Britt., Mem. Torr. Club 5:236. 1894.
Nebraska to Colorado, south to Arkansas, Texas and Mexico. This plant may well be in eastern Colorado.

5. Gaura L. BUTTERFLY WEED

Annual, biennial or perennial, caulescent, herbaceous plants; leaves alternate, lobeless, narrow; flowers white or pink, more or less irregular, in spikes, spicate or subcapitate racemes; hypanthium obconical or cylindrical; sepals 4, deciduous; petals 4, clawed; stamens usually 8, often with a scale at the base of each filament; ovary 4-celled, 1 ovule in each cell, styles filiform, declined, stigmas 4-lobed and provided at base with a cuplike border, or discoid; fruit nutlike, nearly or quite indehiscent, 1- to 4-seeded.

1. Anthers oval, 0.5-1 mm. long; sepals 1.5-3 mm. long; petals 1.5-2 mm. long
   2. Ovary and fruit glabrous -1. G. parviflora
   2. Ovary and fruit pubescent -1A. G. parviflora lachnocarpa
1. Anthers linear, 2-5 mm. long; sepals 5-11 mm. long; petals 3-10 mm. long
   3. Fruit with a slender stipe (the stipe not over 1 mm. in diameter) 3-6 mm; hypanthium about 2 mm. long -2. G. villosa

3. Fruit without a stipe or with a short very stout one over 1 mm. wide; hypanthium 5 mm. long or more
   4. Stems erect, 40-70 cm. tall; stem leaves 5-10 cm. long; hypanthium 7-12 mm. long, sepals 9-11 mm. long; petals 8-10 mm. long
      5. Stems villous and short-pubescent; inflorescence with at least some glandular hairs; plants mostly of southern Colorado   -3. G. neomexicana
      5. Stems short-pubescent only; inflorescence with all hairs non-glandular; plants mostly of northern Colorado   -3A. G. neomexicana coloradensis
   4. Stems ascending, rarely erect, seldom over 30 cm. tall; stem leaves 1-3.5 cm. long; hypanthium 5-8 mm. long; sepals 5-8 mm. long; petals 3-6 mm. long
      6. Stems and leaves more or less hairy
         7. Leaves (at least lower ones) lanceolate and some sinuate-dentate   -4. G. coccinea
         7. Leaves all linear and entire   -4A. G. coccinea parvifolia
      6. Stems and leaves glabrous   -4B. G. coccinea glabra

1. **Gaura parviflora** Dougl. ex Hooker, Fl. Bor. Am. 1:208. 1834.
Plants probably biennial or winter-annual; stems 20-200 cm. tall, erect, simple below, commonly widely branched above, soft-villous and also with short, commonly glandular hairs; leaves 3-10 cm. long, soft-pubescent, the cauline almost or quite sessile, lanceolate to ovate-lanceolate, acute or acuminate, entire or remotely sinuate-denticulate; flowers in terminal, long, slender spikes with caducous bracts 2-5 mm. long; hypanthium 1.5-3 mm. long, puberulent; sepals 1.5-3 mm. long, mostly glabrous; petals 1.5-2 mm. long, mostly pink or rose; anthers about 0.5-1 mm. long, reddish, oval; stigmas with 4 short lobes little exserted from the basal cup; fruit 6-10 mm. long, about 2 mm. thick, sessile, somewhat fusiform, 4-angled above, narrowed but not into a stipe, glabrous. This plant has been called var. typica Munz, and its form with the hypanthium glabrous forma glabra Munz.---Waste places, roadsides and plains. The variety from South Dakota to Washington, south to Missouri and Mexico. Our records from the eastern half and the southwestern part of Colorado at 3500-7000 feet.

1A. **Gaura parviflora lachnocarpa** Weatherby, (var.) Rhodora 27:14. 1925.
Differs from the species in having the capsule and ovary with divergent hairs (not glabrous).---Missouri to Utah, south into Mexico; South America. Our one record from Elbert County, Colorado, at about 6500 feet.

2. **Gaura villosa** Torr., Ann. Lyc. N. Y. 2:200. 1828.
Perennial, usually almost suffrutescent plants; stems 30-100 cm. tall, several from the base, ascending-erect, branching, canescent to villous or both; leaves 3-7 cm. long, canescent-villous, lower ones spatulate to lanceolate or ovate, mostly sinuate-serrate, subsessile or short-petioled, upper ones narrow and smaller; inflorescence a slender, long spike with small caducous bracts; hypanthium about 2 mm. long, funnelform, cinereous-strigulose; sepals 6-10 mm. long; petals about 5-8 mm. long, white turning reddish; anthers 2-3 mm. long, linear; stigma lobes short; capsule 7-9 mm. long, cinereous-strigulose, oblong, shraply angled, narrowed both ways with a very slender stipelike base 3-6 mm. long. This plant has been called var. typica Munz.---Plains and hills. The variety from Kansas, south to Texas and New Mexico. Our few records from southeastern Colorado at 4500-5500 feet.

3. **Gaura neomexicana** Wooton, Bull. Torr. Club 25:307. 1898.
Biennial or perennial herbaceous plants; stems 40-70 cm. tall, 1 to several from woody roots, erect, simple or branched above, villous as well as short-pubescent; leaves of stem 5-10 cm. long, mostly lanceolate, strigose, subentire to sinuate-denticulate, narrowed to a short petiole; inflorescence spicate with some of hairs glandular, bracts lanceolate, short, caducous; hypanthium 7-12 mm. long, pubescent; sepals 9-11 mm. long; petals 8-10 mm. long, pink turning rose; anthers 3-4 mm. long, linear; stigma lobes rather long, surpassing the basal cup; fruit 7-10 mm. long, fusiform, obtusely 4-angled, not stiped. The above has been called var. typica Munz and in central Colorado it intergrades commonly with the other variety.---Hills and valleys. Colorado and New Mexico. Our few records from Douglas and Archuleta Counties at 6000-7000 feet.

3A. **Gaura neomexicana coloradensis** (Rydb.) Munz, (var.) Bull. Torr. Club 65:114. 1938.
G. coloradensis Rydb.---The stems are short-pubescent only; inflorescence has no glandular hairs.---Apparently limited to northcentral Colorado, our records at 5000-6500 feet.

4. **Gaura coccinea** Nutt. ex Pursh, Fl. Am. Sept. 733. 1814.
G. marginata Lehm.---Perennial herbaceous plants; stems 10-30 cm. tall, several, simple or branched, ascending, canescent-strigose; leaves 1-3.5 cm. long, oblong to lanceolate (or linear on upper), repand-dentate to entire especially on upper, more or less canescent; spikes not peduncled, floral bracts persistent, linear or lanceolate; hypanthium 5-8 mm. long, narrow; sepals 5-8 mm. long, reflexed in anthesis; petals 3-6 mm. long, whitish, pink or red, the latter especially in age; anthers 2-5 mm. long, linear; stigma lobes short, suborbicular; fruit 5-10 mm. long, body 4-angled and tapering to tip, constricted to a gradually narrowing stout terete base. This plant has been called var. typica Munz. A large number of intergradations occur in this area between this and the other 2 varieties.---Plains and hills. Manitoba to Alberta, south to Texas and California. Our records from the eastern half and southwestern quarter of Colorado at 3500-8500 feet.

4A. **Gaura coccinea parvifolia** (Torr.) T. & G., (var.) Fl. N. Amer. 1:518. 1840.
G. parvifolia Torr.---Stems and leaves more or less pubescent, but all leaves entire, linear and seldom over 15 mm. long.---Kansas, Colorado and New Mexico. Distributed with the typical form in this state, at 4000-7000 feet.

4B. **Gaura coccinea glabra** (Lehm.) T. & G., (var.) Fl. N. Amer. 1:518. 1840.
G. glabra Lehm.---Leaves mostly sinuate-dentate as in the typical form of the species but the stems and leaves are glabrous.---Distribution generally and in Colorado about as in the typical form. Our records at 4500-7500 feet.

6. Gayophytum A. Juss. BABY'S BREATH; GROUNDSMOKE

Annual caulescent plants; stems slender; leaves alternate or lower opposite, entire, narrow, subsessile or short-petioled; flowers small, in the upper axils; hypanthium not produced above the ovary; sepals 4, often reflexed, deciduous; petals 4, white to rose-colored; stamens 8 but those opposite the petals much reduced and usually sterile; stigma capitate; fruit a 2-celled, 4-valved, linear or clavate capsule; seeds many, in 1 row in each cell, not comose.

1. Plants branched mostly at the very base, not so much above; capsule not strongly torulose if at all, erect; pedicels lacking or usually not over 2 mm. long
   2. Seeds glabrous -1. G. racemosum
   2. Seeds appressed-canescent -2. G. helleri
1. Plants freely branching mostly above the base; capsule torulose, erect or deflexed; pedicels definite, usually 2 mm. long or more
   3. Petals about 0.5 mm. long (less than 1 mm.); pedicels 4-7 mm. long, longer than the capsule which is 2-5 mm. long; plants glabrous -3. G. ramosissimum
   3. Petals about 1-1.5 mm. long; pedicels usually less than 4 mm. long, shorter than the capsule which is 5-12 mm. long; plants usually somewhat pubescent above
      4. Seeds glabrous -4. G. nuttallii
      4. Seeds strigose-canescent -5. G. lasiospermum

1. **Gayophytum racemosum** T. & G., Fl. N. Amer. 1:514. 1840.
   Stems 10-30 cm. tall, usually branched mostly at base, not so much above, strigulose or nearly glabrous; leaves 1-3 cm. long, linear to linear-oblanceolate, not much reduced above; pedicels from almost lacking to 2 mm. long; sepals less than 1 mm. long; petals about 1 mm. long, reddish in age; capsule 6-14 mm. long, erect, not strongly torulose if at all; seeds glabrous. Our plant has been called var. typicum Munz.---Often sandy ground. The variety from Montana to Washington, south to Colorado, Arizona and California. Our records from the western half of Colorado, none as yet from the northwestern part, at 5000-9500 feet.
2. **Gayophytum helleri** Rydb., Bull. Torr. Club 40:65. 1913.
   Plants 10-20 cm. tall, subsimple or more commonly branching at the very base; leaves 1-3 cm. long, linear to linear-oblanceolate; pedicels lacking or not over 3 mm. long; sepals about 1 mm. long; petals about 1-1.5 mm. long, white or turning red; capsule 6-10 mm. long, subterete, little if any torulose, erect; seeds appressed-canescent. Our plant has been called var. glabrum Munz.---Hills and mesas. The variety from Idaho to Washington, south to Colorado and California. Our one record from northwestern Colorado at 8000 feet.
3. **Gayophytum ramosissimum** T. & G., Fl. N. Amer. 1:513. 1840.
   Stems 15-50 cm. tall, diffusely branched mostly above the base, glabrous or rarely slightly strigose in the inflorescence; leaves 1-3 cm. long, rather reduced upward, the upper bractlike, linear or lanceolate-linear; pedicels 4-7 mm. long; sepals about 0.5 mm. long; petals 0.5 mm. long (sometimes to 1 mm. in intergrading forms); capsule 2-5 mm. long, shorter than the pedicels, torulose, mostly spreading-deflexed; seeds glabrous.---Hills and valleys often on sandy slopes. Montana to Washington, south to Colorado, Arizona and California. Our records scattered in the western half of Colorado at 6500-7500 feet.
4. **Gayophytum nuttallii** T. & G., Fl. N. Amer. 1:514. 1840.
   G. intermedium Rydb.---Stems about 20-50 cm. tall, diffusely branched but above the base, usually appressed-pubescent in the upper part; leaves 1-3 cm. long, lanceolate-linear or linear, upper bractlike; pedicels 1-3 mm. or sometimes to 5 mm. long; sepals 1-1.5 mm. long; petals 1-1.5 mm. long, reddish in age; capsule 5-12 mm. long, usually exceeding the pedicels, erect or deflexed, torulose when mature; seeds glabrous.---Hills and mountains often on dry sandy slopes. South Dakota to Washington, south to New Mexico and California; South America. Our records scattered in the western half of Colorado at 5000-9500 feet.
5. **Gayophytum lasiospermum** Greene, Pitt. 2:164. 1891.
   Stems 15-50 cm. tall, diffusely branched throughout, more or less puberulent above; leaves 1-3 cm. long, linear to lanceolate-linear, upper ones bractlike; pedicels 2.5-4 mm. long, possibly longer; sepals about 1 mm. long; petals about 1-1.5 mm. long, white but turning to rose; capsule 5-9 mm. long, torulose, spreading to deflexed; seeds strigose-canescent. Our plant has been called var. typicum Munz.---Sandy ground. Montana to Washington, south to Wyoming and California. Our record from the northwest corner of the state at 8000 feet.

7. Oenothera L. EVENING PRIMROSE; SUNDROPS

Annual, biennial or perennial, caulescent or acaulescent herbaceous plants; leaves basal or alternate; flowers yellow to white (often aging orange, rose, red or purple); hypanthium prolonged beyond the ovary, deciduous; sepals 4, reflexed in anthesis; petals 4; stamens 8, equal, or the alternating ones opposite the petals shorter, anthers mostly versatile; stigmas variable, from 4 linear lobes to capitate or discoid; capsule membranous to woody, straight to curved, 4-celled, 4-valved, sometimes tardily dehiscent; seeds many, not comose. This genus seems to consist of several rather defined groups that have been called genera. The author follows Munz in keeping these groups united.

1. Stigmas with 4 linear lobes; flowers mostly opening in the evening
   2. Capsule winged at least on the upper part, ovoid and rather short for its thickness; seeds corky or wing-angled at apex; petals yellow when young; plants acaulescent or nearly so
      3. Seeds in 1 row in each cell, with corky tubercles; hypanthium 5-15 cm. long; petals 3-5 cm. long (subgenus Megapterium) -1. O. brachycarpa

3. Seeds in 2 rows in each cell, not corky; hypanthium 4-8 cm. long; petals 1-3 cm. long (subgenus Lavauxia) -2. O. flava
2. Capsule terete, round-angled or sharp-angled but never winged, usually elongated; seeds not corky or wing-angled at apex (may be sharp-angled); petals white or yellow when young; plants acaulescent or caulescent
    4. Capsule sharply angled, not over 1 cm. long; seeds in several rows in each cell, angled; flowers white when young (subgenus Gauropsis) -3. O. canescens
    4. Capsule terete or round-angled, over 1 cm. long; seeds in 1 or 2 rows in each cell; flowers white or yellow when young
        5. Flowers yellow when young; seeds sharply angled, in 2 rows in each cell; plants caulescent (subgenus Onagra)
            6. Hypanthium tube 8-15 cm. long; sepals and petals 4-6 cm. long -4. O. jamesii
            6. Hypanthium tube 3-4.5 cm. long; sepals and petals less than 4 cm. long
                7. Petals and sepals 25-40 mm. long -5. O. hookeri
                7. Petals and sepals not over 20 mm. long -6. O. strigosa
        5. Flowers white when young (yellow in 1 annual, acaulescent species doubtfully in Colorado); seeds not sharply angled, in 1 or 2 rows; plants acaulescent or caulescent
            8. Seeds with a deep furrow along the raphe; capsule ovoid or ovoid-lanceolate; plants usually acaulescent or nearly so (subgenus Pachylophus)
                9. Petals yellow when fresh; plants annual -7. O. primiveris
                9. Petals white when young; plants perennial
                    10. Capsule sessile
                        11. Capsule distinctly tubercled
                          12. Plants acaulescent
                            13. Herbage glabrous or nearly so -8A. O. caespitosa caespitosa
                            13. Herbage pubescent -8B. O. caespitosa jonesii
                          12. Plants caulescent -8C. O. caespitosa eximia
                      11. Capsule not tubercled -8D. O. caespitosa montana
                  10. Capsule distinctly pedicelled -8E. O. caespitosa marginata
        8. Seeds lacking such a furrow; capsule oblong, cylindrical or subcylindrical; plants caulescent
            14. Capsule membranous, somewhat enlarged above the base; seeds in 2 rows in each cell, with shallow pits in regular rows on the surface (subgenus Raimannia)
                15. Plants perennial with creeping rootstocks; hypanthium with a conspicuous tuft of hairs in the throat; petals 7-15 mm. long; capsule 8-20 mm. long -9. O. coronopifolia
                15. Plants annual or winter annual; hypanthium not long-hairy in the throat; petals 15-40 mm. long; capsule 20-40 mm. long -10. O. albicaulis
            14. Capsule woody, somewhat narrowed toward the apex; seeds in 1 row in each cell, not pitted (subgenus Anagra)
                16. Buds, sepals and often the upper part of the plants with rather conspicuous long, white, pilose hairs; annual or perennial plants with taproots
                    17. Whole upper part of plants conspicuously pilose; leaves sinuate-dentate, 1.5-3 cm. long -11. O. engelmanni
                    17. Long-pilose hairs limited to the flower, especially the sepals; leaves sinuately-pinnatifid at least above, 3-6 cm. long -12. O. trichocalyx
                16. No long pilose hairs present, the buds and sepals glabrous, strigose or pubescent (rarely with a few long appressed hairs); plants perennial with long-creeping rootstocks (except in number 12)
                    18. Plants glabrous or nearly so; mostly west of the Continental Divide in Colorado
                        19. Leaves pinnatifid; buds and sepals usually with at least a few long-pilose hairs; plants with taproots (glabrate forms of) -12. O. trichocalyx
                        19. Leaves subentire or merely sinuate-dentate; buds and sepals glabrous or nearly so, no long-pilose hairs present; plants with creeping rootstocks -13. O. pallida
                    18. Plants glandular-pubescent or strigose at least above; mostly east of the Continental Divide in Colorado
                        20. Hypanthium tube, sepals and often the upper part of the plant glandular-pubescent, usually strigose only on the lower leaf surface -14. O. nuttallii
                        20. Hypanthium tube, sepals and plants in general never glandular-pubescent, usually definitely strigose -15. O. latifolia
1. Stigmas capitate, discoid or slightly 4-lobed or 4-toothed; flowers mostly opening in the daytime
    21. Ovary with a long, narrow upper sterile part grading into the basal, enlarged fertile part, the upper part simulating a hypanthium; plants perennial and acaulescent (subgenus Taraxia) -16. O. heteranthera
    21. Ovary without such an elongated sterile part; plants annual or perennial, caulescent (or if somewhat acaulescent then annual)
        22. Hypanthium tube 25-50 mm. long; stamens all nearly equal in length (subgenus Salpingia)
            23. Stems and calyx tube villous-hirsute; stems usually over 20 cm. long; leaves usually over 15 mm. long on some plants -17. O. greggii
            23. Stems and calyx tube not villous-hirsute; stems not over 20 cm. long; leaves not over 15 mm. long
                24. Hypanthium tube and sepals strigose-canescent, not glandular; plants mostly in eastern Colorado -18. O. lavandulaefolia
                24. Hypanthium tube and sepals glandular-pubescent, not strigose-canescent; plants mostly in western Colorado -18A. O. lavandulaefolia glandulosa
        22. Hypanthium tube 1-15 mm. long; stamens of 2 lengths
            25. Stigmas discoid, somewhat shallowly 4-lobed; plants perennial the base often woody; hypanthium tube 5-15 mm. long (subgenus Calylophis) -19. O. serrulata

25. Stigmas capitate, often spherical; plants annual or biennial, never woody at base; hypanthium tube less than 5 mm. long
    26. Capsule distinctly pedicelled, cylindrical to clavate; leaves mostly near base of plants; petals yellow when young (subgenus Chylismia)
        27. Capsule linear, not tapering much to base, 1-1.5 mm. wide; stems and leaves often villous; styles not longer than the petals -20. O. multijuga
        27. Capsule clavate, tapering to base, 2-2.5 mm. wide; stems and leaves never villous; styles often longer than the petals -21. O. scapoidea
    26. Capsule sessile or very nearly so, cylindrical; leaves scattered; petals yellow to white (subgenus Sphaerostigma)
        28. Flowers yellow when young, borne in the axils of foliage leaves
            29. Plants with several, naked, fine stems bearing a leafy inflorescence at apex; capsule 5-8 mm. long, nearly straight; petals about 1.5-2.5 mm. long -22. O. andina
            29. Plants with stems leafy from base; capsule 15 mm. long or more, usually strongly curved; petals 2.5-3.5 mm. long -23. O. contorta
        28. Flowers white when young, in terminal spikes
            30. Capsule cylindrical, not tapering from base to apex, 25-50 mm. long; hypanthium tube 2.5-3 mm. long; petals and sepals 2.5-3 mm. long -24. O. chamaenerioides
            30. Capsule somewhat enlarged near base and tapering to apex, 18-25 mm. long; hypanthium tube about 2 mm. long; petals and sepals about 2 mm. long -25. O. minor

1. Oenothera brachycarpa A. Gray, Pl. Wright. 1:70. 1852.
Lavauxia brachycarpa (Gray) Britt.---Acaulescent or subacaulescent perennial plants with heavy, thick caudex, closely cinereous throughout; leaves 4-15 cm. long and 1-3 cm. wide, clustered, thick, lanceolate to ovate-lanceolate to oblong-lanceolate, entire to sinuate-pinnatifid, petioled; hypanthium 5-15 cm. long; sepals 3-5 cm. long, the tips 1-4 mm. long; petals 3-5 cm. long, yellow when young but drying an orange-red; stigma of 4 linear lobes; capsules 2.5-3 cm. long and 6-8 mm. thick, ovoid or cylindrical-ovoid, winged from near summit almost to base; seeds in 1 row in each cell, dark brown, more or less corky-tubercled. Our plant has been called var. wrightii (A. Gray) Leveille.---Plains, hills, often on clay banks. The variety from Kansas and Idaho, south to Texas and Mexico. Our records from northwestern, northcentral and central Colorado at 5000-8500 feet.

2. Oenothera flava (A. Nels.) Garrett, Sp. Fl. of Wasatch Region, 4th ed. 1927.
Lavauxia flava A. Nels.---Acaulescent or subacaulescent perennial plants with deep taproots, subglabrous to pubescent throughout; leaves 3-20 cm. long and 1-2 cm. wide, crowded, linear to oblanceolate, deeply and irregularly runcinate-pinnatifid, petioled; hypanthium about 4-8 cm. long; sepals 1-2.5 cm. long, the tips 3-6 mm. long; petals 1-3 cm. long, yellow but in age reddish; stigmas of 4 linear lobes; capsules 1-3.5 cm. long, ovoid, 4-winged, especially at apex; seeds in 2 rows in each cell, numerous, winged near summit, granular but not tuberculate.---Hills, valleys and mountains. Canada, south to North Dakota, Mexico and California. Our records scattered in the western two-thirds of Colorado at 5000-10,000 feet.

3. Oenothera canescens Torr. & Frem., Rep., 315. 1845.
Gaurella canescens (Torr. & Frem.) A. Nels.; G. guttata (Geyer) Small---Bushy, perennial, herbaceous plants diffusely branching at base; stems 10-20 cm. tall, decumbent or ascending, appressed-canescent; leaves 5-15 mm. long, lanceolate or linear-lanceolate, appressed-canescent, almost sessile, repand-denticulate to almost entire, often fascicled in the axils; flowers axillary on the upper part of the stem; hypanthium 5-10 mm. long; sepals 8-12 mm. long, coherent; petals 8-15 mm. long, white to pink with red spots or stripes; stamens subequal; stigma with 4 linear lobes; capsule 6-8 mm. long, ovoid-pyramidal, sharply 4-angled, somewhat beaked; seeds in several rows in each cell.---Plains and hills, often in dried up ponds. Nebraska to Wyoming, south to Texas and New Mexico. Our records scattered in the eastern half of Colorado at 4000-5500 feet.

4. Oenothera jamesii T. & G., Fl. N. Amer. 1:493. 1840.
Onagra jamesii (T. & G.) Small---Oklahoma, south to Texas and Mexico. Reported for Colorado and may be in the eastern part of the state.
    A similar plant, O. longissima Rydb., has recently been collected in southwestern Colorado. It will key out to the above but the leaves are narrower, not over 2 cm. wide.

5. Oenothera hookeri T. & G., Fl. N. Amer. 1:493. 1840.
Onagra hookeri (T. & G.) Small; Oenothera hirsutissima (A. Gray) Rydb.---Biennial or possibly short-lived perennial plants with taproots; stems 30-150 cm. tall, erect, reddish, more or less canescent and hirsute, simple or branched; leaves 5-15 cm. long, lanceolate to oblanceolate or the lower spatulate, sinuate-dentate, petioled, more or less hirsute and canescent; inflorescence rather elongated with leafy bracts; hypanthium about 2.5-4.5 cm. long, reddish; sepals about 2.5-4 cm. long, the tips free in the bud; petals 2.5-4 cm. long, yellow but turning reddish-rose in age; stamens about equal in length; stigmas of 4 linear lobes; capsules 2-5 cm. long, obtusely angled, tapering upward; seeds reddish-brown, in 2 rows in each cell. Munz (El Aliso 2: 1-47, 1949) separated this species into several subspecies. The following 3 are recorded from Colorado, ssp. hirsutissima (Gray) Munz, ssp. angustifolia (Gates) Munz and ssp. hewettii Cockll.---Valleys, roadsides and slopes. Colorado to Utah, south to Mexico; probably farther north. Our records from the southern half of Colorado, few in the extreme eastern part, at 4000-10,000 feet.

6. Oenothera strigosa (Rydb.) Mack. & Bush, Fl. Jackson Co. Mo. 139. 1902.
O. biennis var. hirsutissima A. Gray; Onagra strigosa Rydb.---Biennial plants more or less grayish-strigose throughout; stems 30-100 cm. tall, erect, usually simple, more or less hirsute and also strigose, often reddish-tinged; lower leaves 3-10 cm. long, spatulate to oblanceolate, obtuse; cauline leaves lanceolate, acute, repand-denticulate, shorter petioled; inflorescence long, leafy-bracted; hypanthium 3-4 cm. long; sepals 10-15 mm. long; petals 12-20 mm. long, yellow often fading to whitish or pink in age; stamens subequal; stigma with 4 linear lobes;

capsule 25-35 mm. long, subcylindrical and tapering upward, obtusely angled; seeds in 2 rows in each cell, reddish-brown. A form in southern Colorado has slightly smaller flowers, and the hairs although similar are sparser. It would fit O. procera Woot. & Standl., a species of doubtful validity.---Plains, hills and valleys, often in sandy soil. Minnesota to Washington, south to Kansas and Oregon. Our records scattered over the state, with few in the extreme eastern part, at 3500-8500 feet.

7. Oenothera primiveris A. Gray, Pl. Wright. 2:58. 1853.
Utah to Nevada, south to Texas, Mexico and California. Not reported directly from this state but to be expected.

8. Oenothera caespitosa ex Fraser, Cat. n. 53. 1813; Sims. Bot. Mag. 39:pl.1593. 1813.
Pachylophus caespitosus (Nutt.) Raim.---Caespitose perennial plants from thick taproots, usually acaulescent (not in 1 variety); leaves 3-12 cm. long and 1-2 cm. wide, subentire to pinnatifid, glabrous, ciliate or pubescent, petioled; hypanthium 5-8 cm. long or more; sepals 2.5-3.5 cm. long; petals 2.5-4 cm. long, whitish but soon aging pink or rose; stigma of 4 linear lobes; capsule 1-3 cm. long, ovoid or ovoid-lanceolate, beaked, sessile or short-pedicelled, often tubercled on the angles; seeds in 2 rows in each cell, dark brown, minutely roughened. The varieties found in Colorado intergrade freely. However, typical plants of each are quite distinct from the others.

8A. Oenothera caespitosa caespitosa
O. caespitosa var. typica Munz---From Canada south to South Dakota, Wyoming and Oregon. May be in northern Colorado.

8B. Oenothera caespitosa jonesii Munz, (var.) Am. Journ. Bot. 18:731. 1931.
Eastern Utah. To be looked for in western Colorado.

8C. Oenothera caespitosa eximia (A. Gray) Munz, (var.) Am. Journ.Bot. 18:731. 1931.
Pachylophus eximius A. Gray; P. exiguus (Rydb. Fl. Colo.-misprint)---Plants distinctly caulescent with a stem 10-30 cm. long; leaves pilose especially on margins and veins, subentire to sinuate-dentate; capsule 2.5-4 cm. long when mature, sessile, definitely tubercled on the angles.---Colorado and New Mexico. Our records from central, southcentral and southeastern Colorado at 4000-7500 feet.

8D. Oenothera caespitosa montana (Nutt.) Durand, (var.) Bot. Basin of Great Salt Lake of Ut. 164. 1859.
Pachylophus montanus (Nutt.) A. Nels.; P. macroglottis Rydb.; P. hirsutus Rydb. and P. caulescens Rydb. in part---Plants acaulescent; leaves canescent-pubescent on the margins and often on the veins or on lower surface, otherwise glabrous; capsule about 2 cm. long, sessile, with sinuate cross-ridges or cross-veins on the angles but no tubercles.---Montana to Oregon, south to New Mexico and Arizona. Our records from the western two-thirds of Colorado at 5000-10,000 feet.

8E. Oenothera caespitosa marginata (Nutt.) Munz, (var.) Am. Journ. Bot. 18:733. 1931.
Pachylophus marginatus (Nutt.) Rydb.; P. hirsutus Rydb. in part---Plants usually caulescent; leaves usually sinuate-pinnatifid, villous throughout but especially on the leaf margins and veins; capsule 3-4 cm. long, linear-cylindrical, the ridges with low tubercles, definitely pedicelled.---Colorado, Idaho and Washington, south to New Mexico and California. Our records scattered in the western half of Colorado, mostly from the extreme western part of the state, at 4500-8500 feet.

9. Oenothera coronopifolia T. & G., Fl. N. Amer. 1:495. 1840.
Anogra coronopifolia (T. & G.) Britt.---Perennial caulescent plants with slender rootstocks; stems 5-25 cm. tall, 1 to few in a cluster, simple or branched, erect or nearly so, finely strigose, usually canescent and with scattered stiff hairs; lower leaves about 1.5-2.5 cm. long, oblong-lanceolate to oblanceolate in outline, obtuse, coarsely sinuate-serrate, early deciduous; cauline leaves oblong-lanceolate in outline, deeply and regularly sinuate-pinnatifid or pectinate, the lobes narrow; flowers in upper axils; hypanthium 1.5-3 cm. long, hairy and with long conspicuous hairs in throat; sepals 8-15 mm. long, tips slightly free in the nodding bud; petals 7-15 mm. long, white soon aging pink or rose; stamens subequal; stigma with 4 linear lobes; capsule 8-20 mm. long, oblong-fusiform, somewhat 4-sided; seeds in 2 rows in each cell.---Plains and valleys, often along roads. South Dakota to Utah, south to Kansas and Arizona. Our records scattered over Colorado, none as yet from the southeastern corner, at 3500-8500 feet.

10. Oenothera albicaulis Pursh, Fl. Sept. Am. 2:733. 1816.
Anogra perplexa Rydb.; A. albicaulis (Pursh) Britt.---Annual or winter-annual caulescent plants; stems up to 45 cm. tall, variable in habit, simple and erect or with additional decumbent or ascending branches from base, strigulose and more or less cinerous, often with scattered long hairs especially above; basal rosette leaves 1-5 cm. long, spatulate to oblanceolate to obovate, obtuse, often entire but sometimes toothed below; cauline leaves lanceolate to linear-oblanceolate in outline, usually pectinate-pinnatifid into narrow lobes; flowers solitary in the axils; hypanthium 1.5-3.5 cm. long, hairy but not long-pilose at throat; sepals 14-25 mm. long, without free tips in the nodding bud; petals 1.5-4 cm. long, about as wide, white or pink in age; stamens subequal; stigma with 4 linear lobes; capsule 2-4 cm. long, cylindrical, sessile, ribbed; seeds in 2 rows in each cell.---Plains and hills, often on sandy soil. South Dakota to Montana, south to Texas, Mexico and Arizona. Our records scattered in Colorado at 4000-8000 feet.

11. Oenothera engelmanni (Small) Munz, Am. Journ. Bot. 18:316. 1931.
Anogra engelmanni (Small) Woot. & Standl.---Annual or possibly perennial, caulescent plants; stems 20-60 cm. tall, erect, upper portion conspicuously pilose with long spreading hairs; leaves 2-6 cm. long, sessile, lanceolate to oblong-lanceolate, coarsely sinuate-dentate, canescent with longer hairs on veins and margins; flowers in upper leaf axils; hypanthium 1.5-3 cm. long, pilose and strigillose, no long hairs in throat; sepals 10-15 mm. long, pilose and strigillose; petals 1-1.5 cm. long, white or soon rose; stamens subequal; stigma with 4

linear lobes; capsule 2.5-4.5 cm. long, standing at about right angles to stem, somewhat 4-angled, sessile, cylindrical but tapering somewhat from base; seeds brown, in 1 row in each cell.---Valleys and slopes. Texas and New Mexico. Our one record from the southeastern corner of Colorado at about 4200 feet.

12. Oenothera trichocalyx Nutt. ex T. & G., Fl. N. Amer. 1:494. 1840.

Anagra trichocalyx (Nutt.) Small; A. rhizomata A. Nels.; A. violacea A. Nels.---Biennial or perennial caulescent plants with taproots; stems usually 15-40 cm. tall, usually branching at base and often above, finely strigose throughout to glabrate especially below, with scattered long hairs above; cauline leaves 3-6 cm. long, usually lanceolate or oblong, evenly and sinuately pinnatifid into lanceolate lobes, sessile; basal leaves petioled and often subentire; flowers axillary; hypanthium 2-3 cm. long, more or less pilose and strigose, not long-pilose in throat; sepals 10-15 mm. long, both strigose and pilose-hirsute especially in bud, tips hardly free in the nodding bud; petals 10-19 mm. long, white becoming reddish in age; stamens subequal; stigma with 4 linear lobes; capsule 1.3-3 cm. long, cylindrical, slightly narrowed to apex, straight or contorted; seeds brown, in 1 row in each cell. Occasionally glabrous or nearly glabrous forms of the above occur in Colorado.---Dry often sandy plains and hills. Wyoming, Colorado and Utah. Our records from western, central and southcentral Colorado at 4500-9000 feet.

13. Oenothera pallida Lindl., Bot. Reg. 14:1142. 1828.

Perennial caulescent plants with creeping rootstocks; stems 20-50 cm. tall, erect, branching, glabrous or nearly so; leaves 2-6 cm. long, lanceolate to lanceolate-linear, subentire to sinuate-dentate, sessile or short-petioled, usually glabrous; flowers in upper axils; hypanthium 2-3.5 cm. long, usually glabrous, not long-hairy at throat; sepals 12-18 mm. long, glabrous or nearly so; petals 1.2-3 cm. long, white soon reddish; stamens subequal; stigmas with 4 linear lobes; capsules subcylindric, often somewhat curved, tapering gradually to apex; seeds brown or blackish, in 1 row in each cell. Our plants have been called var. typica Munz.---Valleys and mesas. The variety from Idaho to Washington, south to New Mexico and Arizona. Our few records from western Colorado at 6500-7500 feet.

14. Oenothera nuttallii Sweet, Hort. Brit. ed 2, 199. 1830.

Anagra nuttallii (Sweet) A. Nels.---Perennial caulescent plants with underground rootstocks; stems 40-100 cm. tall, erect, often branched above, glabrous with white exfoliating epidermis; leaves 2-7 cm. long, pale green, oblong-linear to lanceolate, usually entire, sometimes remotely denticulate, glabrous above, strigose beneath, sessile or petioled; hypanthium 2-4 cm. long, glandular-pubescent outside, not long-hairy in throat; sepals 2-3 cm. long, glandular-pubescent, short tips free in the nodding buds; petals 1.5-2.5 cm. long, whitish but soon pink or rose; stamens subequal; stigma with 4 linear lobes; capsule 2-3 cm. long, erect, tapering toward apex; seeds dark, in 1 row in each cell.---Plains, valleys and slopes, often in sandy soil. Saskatchewan to British Columbia, south to Minnesota and Colorado. Our records from northcentral, northeastern, central and eastcentral Colorado at 4000-8500 feet.

15. Oenothera latifolia (Rydb.) Munz, Am. Journ. Bot. 18:371. 1931.

Anagra latifolia Rydb.; A. cinerea Rydb.---Perennial caulescent plants with underground creeping rootstocks; stems 10-60 cm. tall, 1 to several, branching from the base or above, strigose-canescent; leaves 1-4 cm. long, ovate, oblong to lanceolate, denticulate or shallowly sinuate-dentate, sessile or petioled, strigose-canescent both sides; hypanthium 15-30 mm. long, strigose-canescent, not long-hairy in throat; sepals 12-25 mm. long, tips free in the nodding bud; petals 15-25 mm. long, white but aging reddish; stamens subequal; stigma with 4 linear lobes; capsule 1.5-6 cm. long, cylindrical, divaricate or spreading; seeds brown, 1 row in each cell. Plants occur on the eastern plains with a tendency toward a glabrous condition and thus intergrade with O. pallida Lindl.---Plains, hills and valleys. South Dakota to Utah, south to Oklahoma and New Mexico. Our records scattered over Colorado, none from the extreme western part, at 3500-10,000 feet.

16. Oenothera heterantha Nutt., Journ. Phila. Acad. 7:22. 1834.

Taraxia subacaulis of Manuals---Acaulescent perennial plants, glabrous or nearly so except on leaf margins; leaf blades 3-15 cm. long, lanceolate to oblanceolate or wider, entire to sinuate; petioles long; hypanthium proper only 1-2 mm. long but the sterile part of the ovary 3-10 cm. long and simulating a hypanthium; sepals 5-8 mm. long; petals 7-10 mm. long, yellow when fresh, often appears whitish when dried; stamens alternating longer and shorter; stigma somewhat discoid; capsule 12-15 mm. long, oblong-ovoid, somewhat 4-angled; seeds straw-colored. ---Plains and valleys. Wyoming to Washington, south to Colorado and California. Our few records from northwestern Colorado at about 6500 feet.

17. Oenothera greggii A. Gray, Pl. Fendl. 46. 1848.

Suffrutescent or suffruticose plants, at least woody at base; stems 10-30 cm. tall, not tufted, glandular-puberulent and villous-hirsute; basal leaves oblanceolate to obovate, short-petioled; cauline leaves 10-25 mm. long, narrowly ovate to lanceolate, margins entire to remotely denticulate, commonly wavy, more or less glandular-puberulent and hirsute; flowers 1 to few, single from upper leaf axils; hypanthium 25-40 mm. long, widening upward, glandular-puberulent and villous-hirsute; sepals 6-18 mm. long, greenish-yellow, sometimes spotted or tinged with dark purple-red; petals 12-30 mm. long, yellow when fresh, fading to orange; stamens subequal; stigma disklike, rather squarish; capsule 10-20 mm. long, nearly cylindrical; seeds 1-2 mm. long, angled. Our plant has been called var. lampasana (Bockl.) Munz.---Valleys and hills. The variety from Nebraska, south to Texas and Arizona. Our few records from the southeastern corner of Colorado at about 4500 feet.

18. Oenothera lavandulaefolia T. & G., Fl. N.Amer. 1:501. 1840.

Galpinsia lavandulaefolia (T. & G.) Small---Perennial plants from woody caudex, suffrutescent at base; stems 5-20 cm. tall, caespitose, grayish-strigose; leaves 5-15 mm. long, linear to narrowly oblanceolate, sessile or nearly so, grayish-strigose, crowded; flowers single in upper axils; calyx tube 2.5-5 cm. long, strigose-canescent; sepals 8-12 mm. long, strigose-canescent; petals 13-22 mm. long, yellow, reddish in age; stamens subequal; stigma discoid; capsule 8-20 mm. long, cylindrical; seeds brownish. The above plant has been called var. typica Munz.---Plains and hills. The variety from Nebraska to Wyoming, south to Texas, New Mexico and Utah. Our records from the eastern half of Colorado and also in the westcentral and southwestern parts, at 3500-6000 feet.

18A. Oenothera lavandulaefolia glandulosa Munz, (var.) Am. Journ. Bot. 16:705. 1929.

Differs from the typical form of the species by having hypanthium and calyx lobes glandular-pubescent (not strigose-canescent).---Colorado to Nevada, south to Texas and Arizona. Our records from western and southeastern Colorado at 5000-7500 feet.

19. Oenothera serrulata Nutt., Gen. Pl. 1:246. 1818.

Meriolix serrulata and M. melanoglottis of Manuals---Perennial plants, from herbaceous to suffrutescent; stems 10-50 cm. tall, caespitose and branched to single-stemmed and simple, canescent especially above, erect to nearly decumbent; leaves 1-6 cm. long, linear, lanceolate or oblanceolate, entire to sharply dentate, sessile or short-petioled; flowers in axils of upper leaves; hypanthium 5-15 mm. long, funnelform; sepals 3-10 mm. long; petals 3-13 mm. long, yellow; stamens alternating longer and shorter; stigma disklike, somewhat 4-lobed; capsule 1-2.5 cm. long, cylindrical, sessile; seeds dark. Our plants have been called var. typica Munz.---Dry plains and hills. The variety from Manitoba to Alberta, south to Texas and Arizona. Our records from the eastern half of Colorado at 3500-7500 feet.

20. Oenothera multijuga S. Wats., Am. Nat. 7:300. 1873.

This species resembles the next, differing in the characters given in the key. The var. orientalis Munz has some of the leaves part way up the stem, these often with few if any lateral pinnules; petals 1.5-2 mm. long; capsules 12-20 mm. long. The variety is reported from eastern Utah, northeastern Arizona and westcentral to southwestern Colorado, our few records from about 5500 feet.

21. Oenothera scapoidea Nutt. ex T. & G., Fl. N. Amer. 1:506. 1838-1840.

Chylismia scapoidea (Nutt.) Small---Annual plants; stems 10-45 cm. tall, simple or branching at base, erect or spreading, glabrous to pubescent or glandular-pubescent; leaves 1-4 cm. long, mostly near base of plant, simple or with a few segments separated and along the petiole, ovate to oblong-ovate to oblanceolate, obtuse, entire to irregularly dentate; flowers in racemes, with spreading pedicels; hypanthium 1.5-3 mm. long; sepals about 2-6 mm. long; petals 2-4 mm. long, yellowish, frequently with reddish dots at base; stamens alternating longer and shorter; stigma capitate; capsule 10-25 mm. long, more or less clavate and narrowing toward base; seeds brownish. Several varieties of this species have been reported from Colorado but the writer has never been able to separate them satisfactorily, except perhaps var. eastwoodae Munz.---Plains and hills, often in draws. Wyoming to Oregon, south to Colorado and Utah. Our records from western Colorado at 4500-7000 feet.

22. Oenothera andina Nutt. ex T. & G., Fl. N. Amer. 1:512. 1840.

Sphaerostigma andinum (Nutt.) Walp.---Canada, south to Colorado and California. No records from the state seen by the writer.

23. Oenothera contorta Dougl. ex Hook., Fl. Bor. Am. 1:214. 1834.

Sphaerostigma flexuosum (A. Nels.) Rydb.; S. contortum var. flexuosum A. Nels.---Annual plants; stems 5-10 cm. tall, slender, leafy, glabrate to pubescent, often reddish; leaves 5-25 mm. long, linear to linear-oblanceolate; flowers few to many in the axils of leafy bracts; hypanthium 1-2 mm. long; sepals 1.5-3 mm. long; petals 2.5-3.5 mm. long, yellow or reddish in age; stamens alternately longer and shorter; stigmas capitate; capsules 15-40 mm. long, linear, often torulose or curved; seeds brownish.---Rather dry ground. Wyoming to Washington, south to Texas and California. Our one record from the northwestern corner of Colorado at 5000 feet.

24. Oenothera chamaenerioides Gray, Pl. Wright. 2:58. 1853.

Utah and Nevada, south to Texas and California. May be in southern or western Colorado.

25. Oenothera minor (A. Nels.) Munz, Bot. Gaz. 85:238. 1928.

Sphaerostigma tortum (Lev.) A. Nels.; S. tortum var. eastwoodae A. Nels.; S. minutiflora (S. Wats.) Rydb.---Wyoming to Idaho, south to Utah and Nevada. Should be looked for in northern and northwestern Colorado.

Family 86. Haloragidaceae WATERMILFOIL FAMILY

Perennial aquatic or marsh-dwelling plants; stems wholly or partly submerged; leaves commonly in whorls, simple to finely dissected; flowers regular, minute, perfect or unisexual, sessile in the axils of the leaves or in terminal spikes; sepals 2-4 or lacking; petals distinct, 2-4, or lacking; stamens 1-8; ovary inferior, 1- to 4-celled, with 1-4 styles; fruit indehiscent, drupelike or nutletlike (Haloragaceae).

1. Submerged leaves finely pinnatifid; stems slender, not erect; flowers usually imperfect with 4-8 stamens (on staminate flower) -1. Myriophyllum
1. All leaves simple; stems stout and erect; flowers perfect, with 1 stamen -2. Hippuris

1. Myriophyllum L. WATERMILFOIL; PARROTFEATHER

Stems slender, usually floating; leaves alternate or whorled, the submerged ones (and often all) finely pectinately and pinnately dissected into capillary divisions; flowers monoecious or polygamous, in axillary or in interrupted spikes; upper flowers usually staminate with short hypanthium and 2-4 sepals; petals 2-4 when present, and 4-8 stamens; lower flowers pistillate with 4 minute sepals, 4 small petals or none; ovary 2- to 4-celled, stigmas 2-4, plumose; fruit bony, splitting into 2-4 1-seeded nutlets.

1. Floral bracts pinnatifid, mostly much longer than the flowers -1. M. verticillatum
1. Floral bracts serrate or entire, mostly shorter than the flowers -2. M. exalbescens

1. Myriophyllum verticillatum L., Sp. Pl. 992. 1753.
Stems up to 200 cm. long, simple or sparingly branched, not drying whitish; leaves about 1-4 cm. long with 9-13 pairs of divisions, verticillate in 4's or sometimes in 3's or 5's; flowers whorled in an interrupted spike 4-12 cm. long, bracts mostly much longer than the flowers, pinnatifid; fruit 2-3 mm. long, rounded on back.---Still water and slow streams. Newfoundland to British Columbia, south to Delaware, Nebraska and Utah. Our one record from Gunnison County, the altitude uncertain but probably around 9000-10,000 feet.
2. Myriophyllum exalbescens Fernald, Rhodora 21:120. 1919.
M. spicatum of our Manual---Stems up to 100 cm. long, simple or branching, drying whitish; leaves 1-3 cm. long with 7-11 pairs of divisions, verticillate, commonly in 4's sometimes 3's or 5's; flowers whorled in an interrupted terminal spike 2-10 cm. long, bracts barely as long as the fruit, serrate to entire; fruit 2-3 mm. long, rounded on back.---Still water and slow streams. Greenland to Washington, south to Florida, Arizona and California. Our records scattered over the western two-thirds of Colorado at 5000-8500 feet.

2. Hippuris L. MARESTAIL

Stems simple, usually partly submerged, erect from creeping rootstocks; leaves narrow, entire, simple, in whorls of 6 or more; flowers axillary and perfect; calyx lobes minute or lacking; petals wanting; stamens 1; ovary 1-celled with 1 ovule, the style filiform and lying in a groove in the anther; fruit 1-seeded, indehiscent, drupelike.

1. Hippuris vulgaris L., Sp. Pl. 4. 1753.
Stems 20-60 cm. tall, glabrous; leaves 1-3 cm. long, linear, sessile; stamen with a short thick filament and large anther; fruit about 2 mm. long.---Swamps and ponds. Greenland to Alaska, south to New York, New Mexico and California; South America; Eurasia. Our records scattered over the western two-thirds of Colorado at 5500-10,000 feet.

Family 87. Araliaceae GINSENG FAMILY

Perennial herbaceous plants; leaves alternate but basal, compound and long-petioled; flowers regular, perfect or polygamous, in umbels; sepals 5, often obsolete; petals 5, distinct; stamens 5, alternate to the petals; ovary inferior, cells and styles 4-6, usually 5, ovules solitary in each cell; fruit a berry or drupe.

1. Aralia L. WILD SARSAPARILLA; WILD GINSENG

Characters of the Family

1. Plants scapose; scapes with 2-7 umbels -1. A. nudicaulis
1. Plants leafy-stemmed; scapes with many umbels -2. A. racemosa

1. Aralia nudicaulis L., Sp. Pl. 274. 1753.
Acaulescent herbaceous plants with rootstocks often long and branched; leaves glabrous or nearly so, with petioles up to 40 cm. long, with 3 digitate primary divisions, each one pinnately 3- to 5-foliate; leaflets 5-13 cm. long, ovate, oval or elliptic, acuminate at apex, unequal at base, margins finely serrate; scapes usually shorter than the petioles, with 2-6 umbellate stalks at summit, each with a 20- to 50-flowered umbel at apex; calyx about 2 mm. long; petals 1.5-3 mm. long, whitish or greenish; fruit 3-6 mm. in diameter, purplish-black at maturity.---Shaded ground. Newfoundland to British Columbia, south to Georgia, Colorado and Washington. Our records from northcentral and central Colorado at 6000-8000 feet.
2. Aralia racemosa L., Sp. Pl. 273. 1753.
Eastern Canada, south to Georgia and Arizona. This plant should be looked for in eastern and southern Colorado.

Family 88. Umbelliferae CARROT FAMILY; PARSNIP FAMILY

Annual or perennial herbaceous plants; stems commonly hollow; leaves alternate or basal, mostly compound with sheathing petioles; flowers small, perfect or polygamous, regular, in simple or compound umbels (rarely almost capitate), the rays often with an involucre and the umbels of flowers often with an involucel; sepals small or obsolete; petals 5, distinct, the tips usually inflexed; stamens 5, alternate to the petals and inserted on an epigynous disk; ovary inferior, 2-celled, with 1 ovule in each cell, styles 2 and often swollen at base to form a "stylopodium"; fruit of 2 parts called "mericarps", united by their faces, "commissures", each mericarp with 5 ribs, 1 on the back, 2 on extreme sides and the other 2 between, oil tubes usually present in the walls in the interval between the ribs and on the commissural side; mericarps splitting open at maturity, each 1-seeded, usually suspended from the summit of a slender prolongation of the axis, the "carpophore". This is a difficult family largely because reasonably mature fruit is almost always necessary. At the same time certain floral characteristics such as color, and size of calyx teeth are usually desirable. Flowers and well-developed fruit may be on the same plant or on adjoining plants.

1. Fruit linear-oblong, linear-fusiform or linear-clavate, at least 4 times as long as broad, the apex (in all but 1 species) sharp-pointed, the base with retrorse bristles (in all but 1 species) -1. Osmorhiza
1. Fruit narrowly oblong to orbicular, less than 4 times as long as wide, apex hardly sharp pointed, glabrous, pubescent or with hooked bristles
   2. Ovary and fruit with hooked or barbed bristles
      3. Fruit bristly on the ribs, naked between the rows; flowers white; leaves pinnately decompound, the ultimate divisions linear or narrowly lanceolate -2. Daucus
      3. Fruit bristly all over, these not in rows; flowers greenish-yellow to greenish-white; leaves palmately compound, the leaflets obovate, ovate or oval -3. Sanicula
   2. Ovary and fruit glabrous or pubescent, no hooked or barbed bristles present
      4. Fruits (and usually the ovary) round in cross-section or flattened laterally, the ribs not winged (except in Ligusticum, Aletes, Oreoxis and possibly Sium)
         5. Low acaulescent or short-caulescent plants, subscapose or with naked, unbranched peduncles from a cluster of basal leaves
            6. Stylopodium present, conic to low-conic
               7. Flowers white, on pedicels 3-8 mm. long; involucels inconspicuous or wanting; ribs of fruit narrowly winged -4. Ligusticum
               7. Flowers greenish-yellow, on pedicels 1-2 mm. long; involucels conspicuous; ribs of fruit filiform, not at all winged -5. Podistera
            6. Stylopodium lacking entirely at maturity
               8. Plants from globose or ovoid tubers; mericarps with a corky projection extending down the middle of the commissure; calyx teeth obsolete -6. Orogenia
               8. Plants from slender or thick taproots; mericarps lacking any such projection on the commissure; calyx teeth conspicuous
                  9. Carpophores absent; ribs conspicuously corky-winged and (in cross-section) each wing thick, usually ovate or obovate in shape and much longer (really wider) than the width of the mericarp wall -7. Oreoxis
                  9. Carpophores present; ribs inconspicuously to narrowly corky-winged but the ribs (in cross-section) never longer than the width of the mericarp wall
                     10. Oil tubes typically 3 in the intervals; plants hirtellous-scabrous below the umbels and in the rays; fruit granular-scabrous -8. Musineon
                     10. Oil tubes solitary in the intervals; plants glabrous or puberulent at base of umbels, not scabrous; fruit smooth -9. Aletes
         5. Tall caulescent plants with several to many stem leaves
            11. Plants subcaespitose, not over 50 cm. tall; fruit glandular-roughened; stylopodium lacking on mature fruit -10. Harbouria
            11. Plants not caespitose, often over 50 cm. tall; fruit smooth (ribs crenate in Conium); stylopodium present (on all but Zizia)
               12. Stylopodium lacking from mature fruit; flowers with both bright yellow petals and prominent calyx teeth; central flower of each umbellet sessile or nearly so -11. Zizia
               12. Stylopodium present, conic or depressed-conic; flowers not both yellow and with prominent calyx teeth (yellow in Foeniculum with calyx teeth obsolete); central flower of each umbellet not sessile
                  13. Plants stoloniferous; ribs of fruit obscure in the thick corky pericarp; leaves heteromorphic, the upper leaves more deeply cut than the lower; involucre bractlets conspicuous, narrow but foliaceous; plants growing in water or very wet places -12. Berula
                  13. Plants not stoloniferous; ribs filiform, winged or corky but pericarp not thick or corky; leaves more or less alike on the plant; involucel bractlets not conspicuous and foliaceous (except in Sium); plants not in water or wet places (except Cicuta and Sium)
                     14. Fruit with low but prominent corky ribs; plants of water or very wet places
                        15. Involucres of conspicuous, subfoliaceous bracts; involucel of numerous usually foliaceous bractlets; ribs of fruit subequal in cross-section -13. Sium
                        15. Involucres wanting or of a few inconspicuous bracts; involucels of usually few inconspicuous bractlets; ribs of fruit unequal in cross-section, the lateral ones definitely thicker than the dorsal -14. Cicuta
                     14. Fruit with filiform ribs or narrowly winged ribs (in Ligusticum), never with wide corky ribs; plants rarely growing in water or wet places
                        16. Ribs of fruit narrowly winged -4. Ligusticum
                        16. Ribs filiform, never winged
                           17. Plants glaucous, with a strong anise odor; flowers yellow; involucel and involucre wanting -15. Foeniculum
                           17. Plants rarely glaucous, lacking an anise odor; flowers white to pink; involucre and involucel usually present
                              18. Stems spotted; ribs of fruit undulate and crenate; plants over 1 m. tall in favorable habitats; oil tubes very small and numerous -16. Conium
                              18. Stems not spotted; ribs of fruit not undulate or crenate; plants rarely over 1 m. tall; oil tubes 1 in the intervals
                                 19. Calyx teeth conspicuous; leaves pinnate or occasionally bipinnate; fruit 2-3 mm. long; involucel bractlets linear, 1-4 mm. long; native plants -17. Perideridia
                                 19. Calyx teeth obsolete; leaves pinnately dissected; fruit 3-4 mm. long; bractlets setaceous, 5-10 mm. long or lacking; plants escaped from cultivation -18. Carum

4. Fruits flattened dorsally, some (at least the lateral) or all the ribs winged
  20. Stylopodium present
    21. Plants annual, with a strong anise odor; involucel usually wanting; leaves pinnately dissected into filiform divisions not over 0.5 mm. wide   -19. Anethum
    21. Plants perennial or biennial, no strong anise odor present; involucel present or absent; leaves compound into lanceolate or ovate leaflets over 0.5 mm. wide
      22. Flowers yellow; plants of waste places and roadsides; fruit 5-8 mm. long, the dorsal ribs inconspicuous   -20. Pastinaca
      22. Flowers white, pink or purple; plants seldom in waste places; fruit various but if 5-8 mm. long, then the dorsal ribs prominent
        23. Outer petals of the umbel radiant (enlarged) and often 2-cleft; mature fruit 8-12 mm. long; stout, coarse plants, typically over 80 cm. tall, the leaves ternate with 3 leaflets   -21. Heracleum
        23. Outer petals like the others, never 2-cleft; mature fruit 3-8 mm. long; plants various, leaves only rarely with 3 leaflets and then plants not over 80 cm. tall
          24. Dorsal ribs of fruit filiform and inconspicuous; calyx teeth conspicuous; involucel lacking; leaves once-pinnate   -22. Oxypolis
          24. Dorsal ribs of fruit (in ours) conspicuous and more or less winged; calyx teeth minute or obsolete; involucel usually present (absent in Angelica pinnata); leaves usually more decompound than once-pinnate
            25. Leaflets incised or parted into narrow segments, hence not very distinct; fruit 2-3 mm. wide; stylopodium conic   -23. Conioselinum
            25. Leaflets serrate, dentate or sparingly lobed, always very distinct and set apart; fruit 3-6 mm. wide; stylopodium low-conic   -24. Angelica
  20. Stylopodium absent entirely
    26. Dorsal ribs prominent, at least 1 of them winged
      27. Calyx teeth prominent, linear-lanceolate, acuminate   -25. Pteryxia
      27. Calyx teeth absent or if present then ovate or deltoid, little if any acuminate
        28. Peduncles conspicuously hirtellous-pubescent at the base of the umbel; plants subacaulescent to caulescent   -26. Pseudocymopterus
        28. Peduncles glabrous or merely scabrous below the umbels; plants subacaulescent to acaulescent   -27. Cymopterus
    26. Dorsal ribs filiform or obsolete, not at all winged
      29. Fruit with a corky projection extending down the middle of the commissure on each mericarp; plants 5-15 cm. tall, from ovoid or globose tubers   -6. Orogenia
      29. Fruit without such a corky projection on the commissure; plants often over 15 cm. tall, usually from taproots
        30. Peduncles conspicuously hirtellous-pubescent at the base of the umbels   -26. Pseudocymopterus
        30. Peduncles glabrous or pubescent throughout
          31. Calyx teeth prominent, linear-lanceolate and acuminate; plants acaulescent; flowers always yellow   -25. Pteryxia
          31. Calyx teeth obsolete or minute; plants acaulescent or caulescent; flowers yellow, white or purple   -28. Lomatium

1. Osmorhiza Raf.   SWEET CICELY; SWEETROOT

Caulescent, pubescent to somewhat hispid perennial plants; stems branching from thick fascicled roots; leaves ternate or ternate-pinnate; inflorescence of loose compound umbels; involucre wanting or of 1 or more foliaceous bracts; involucel lacking or of several narrow, foliaceous, reflexed bractlets; pedicels spreading or divaricate; stylopodium conical; fruit linear-oblong to linear-fusiform or clavate, obtuse to tapering at apex, somewhat laterally flattened, the mericarps nearly round in section, bristly-hispid or glabrous, ribs filiform; oil tubes obscure or lacking; seeds subterete in section, the face concave or sulcate; carpophore 2-cleft above. The long tapering antrorsely hispid base allows the fruit to penetrate clothing, and the hair of animals and cling there. This assists in seed distribution.

1. Fruit glabrous, no hispid hairs, the base obtuse, not long-caudate; involucels usually lacking   -1. O. occidentalis
1. Fruit with appressed, hispid hairs near the long-tapering base; involucels present or absent
  2. Involucels present and definite; styles 2-3 mm. long; fruit 18-22 mm. long   -2. O. longistylis
  2. Involucels absent or very small and caducous; styles not over 0.5 mm. long; fruit 10-20 mm. long
    3. Fruit clavate, obtuse or abruptly short-pointed at apex, 10-15 mm. long; rays and pedicels often widely divergent to even reflexed   -3. O. obtusa
    3. Fruit linear-oblong, tapering to a definite beak, 12-20 mm. long; rays and pedicels spreading-ascending   -4. O. chilensis

1. **Osmorhiza occidentalis** (Nutt.) Torr., Bot. Mex. Bound. Surv. 71. 1859.
Washingtonia occidentalis (Nutt.) Coult. & Rose; Glycosma occidentalis Nutt.---Stems 30-100 cm. tall, usually hairy at the nodes at least; leaflets 2-10 cm. long, oblong-lanceolate to ovate, serrate and usually incised and lobed, glabrous or somewhat pilose; involucre usually wanting; involucel usually wanting; rays 5-12, ascending or somewhat spreading; pedicels 3-8 mm. long, spreading or ascending; flowers yellowish; styles about 1 mm. long; fruit 12-20 mm. long, linear-fusiform, obtuse at base, glabrous.---Hillsides and valleys. Alberta to British Columbia, south to Colorado and California. Our records from the western one-third of Colorado at 6000-10,000 feet.

2. **Osmorhiza longistylis** (Torr.) DC., Prodr. 4:232. 1830.
   Washingtonia longistylis (Torr.) Britt.---Stems 30-100 cm. tall, glabrate; leaflets 3-10 cm. long, ovate or ovate-lanceolate, coarsely serrate, incised or lobed, more or less short-pilose especially on veins; involucre of 1 or more foliaceous bracts; involucel of several bractlets like the bracts, reflexed; rays 3-6, spreading-ascending; pedicels 5-8 mm. long, spreading-ascending; flowers white; styles 2-3 mm. long; fruit 18-22 mm. long, oblong, tapering to acute apex, long-tapering to caudate base, sparingly hispid.---Woods often in damp ground. Quebec to Alberta, south to Georgia and Colorado. Our rather few records from northcentral, central and southcentral Colorado at 5000-6500 feet.
3. **Osmorhiza obtusa** (Coult. & Rose) Fernald, Rhodora 4:154. 1902.
   Washingtonia obtusa Coult. & Rose---Stems 15-70 cm. tall, glabrate to hispidulous especially on younger portions; leaflets 1.5-5 cm. long, broadly lanceolate to ovate, coarsely serrate, incised or lobed, sparingly hispid especially on veins; involucre wanting or of 1 foliaceous bract; involucel wanting; rays 2-5, widely divergent to reflexed; pedicels 10-30 mm. long, 2-5, widely divergent; flowers greenish-white; styles minute; fruit 10-15 mm. long, clavate, obtuse or abruptly short-acute at apex, tapering to a caudate base, hispid below.---Woods. Labrador to British Columbia, south to Colorado and Arizona. Our records scattered over the western two-thirds of Colorado at 6000-11,500 feet.
4. **Osmorhiza chilensis** Hook. & Arn., Bot. Beechey Voyage 26. 1830.
   O. nuda Torr.; O. divaricata Nutt.; Washingtonia divaricata Britt.; O. intermedia (Rydb.) Blankenship---Stems 30-100 cm. tall, more or less hispid especially on younger portions; leaflets 2-6 cm. long, ovate-lanceolate to orbicular, coarsely serrate, incised or lobed, somewhat hispidulous especially on veins; involucre usually wanting; involucel wanting; rays 3-8, spreading-ascending; pedicels 5-30 mm. long, spreading-ascending; flowers greenish-white; styles 0.2-0.5 mm. long; fruit 12-20 mm. long, linear-oblong, tapering or somewhat abruptly ending in a short beak, long-caudate at base, densely hispid below.---Woods. Newfoundland to British Columbia, south to Colorado, Arizona and California; South America. Our records scattered in the western half of Colorado at 6000-9000 feet.

2. Daucus L. CARROT

Annual or biennial caulescent plants with taproots; leaves pinnately decompound with small and narrow ultimate divisions; inflorescence of compound umbels; involucre of conspicuous, foliaceous, pinnately divided or pinnate bracts (rarely entire); involucels of linear and entire or rarely pinnate bractlets; flowers white or the central purplish; petals obcordate; stylopodium conical; fruit dorsally flattened, primary ribs filiform, bristly, the secondary ribs winged, the wings divided into a row of barbed or hooked bristles; oil tubes solitary under the secondary ribs, 2 on the commissure; face of seed plane to shallowly concave, flattened dorsally.

1. Bracts of involucre pinnately divided into short linear or lanceolate divisions; rays of umbel usually 4-40 mm. long; fruit broadest below the middle; central flowers whitish like the rest   -1. D. pusillus
1. Bracts of involucre pinnately divided into elongate filiform divisions; rays of umbel 30-75 mm. long; fruit broadest right at middle; central flowers of each umbellet usually pink to purple   -2. D. carota

1. **Daucus pusillus** Michx., Fl. Bor. Amer. 1:164. 1803.
   South Carolina to British Columbia, south to Florida and California. This plant should be in Colorado but no record is known to the writer.
2. **Daucus carota** L., Sp. Pl. 242. 1753.
   Plants biennial; stems 15-80 cm. tall, solitary, glabrous to retrorsely hispid; leaves oblong in outline, ultimate divisions linear or lanceolate, entire or few-cleft, glabrous or hispid on veins or margins, divisions of cauline leaves often more elongate; involucres 3-30 mm. long, pinnately divided or rarely entire, usually reflexed; involucel entire or rarely pinnate, the bractlets equalling or exceeding the flowers; flowers whitish, the central one of each umbellet usually pink or purple; fruit 3-4 mm. long, ovoid.---Waste places; Eurasia and introduced widely throughout the world. Our few records from central and southcentral Colorado at about 6000 feet.

3. Sanicula L. SNAKE ROOT

Erect biennial or perennial caulescent plants from woody roots; leaves palmately 3- to 5-parted or appearing to be 7-parted; inflorescence of compound umbels, the umbellets capitate with pedicels lacking or very short, the flowers perfect or staminate; involucre foliaceous; involucel similar to involucre but smaller; stylopodium lacking; fruit subglobose to ovoid, sessile or short-stiped, somewhat flattened laterally, covered with hooked bristles, ribs lacking; oil tubes large, solitary in the intervals, 2 on the commissural side.

1. Fruit sessile, not at all stipitate; styles moderate to long, in fruit exceeding the bristles; staminate flowers in the same umbellet as the perfect flowers   -1. S. marilandica
1. Fruit short-stipitate; styles very short, in fruit shorter than the bristles; staminate flowers often on separate umbellets   -2. S. canadensis

1. **Sanicula marilandica** L., Sp. Pl. 235. 1753.
   Stems 30-100 cm. tall, glabrous, usually solitary; leaves palmately parted the divisions oval or obovate, toothed or incised, the lateral ones bifid; petioles of basal leaves long; pedicels (of fertile flowers) lacking; flowers greenish-white; styles long, exceeding the bristles of the fruit, usually recurved; fruits 4-6 mm. long, ovoid, sessile.---Woods and thickets. Newfoundland

to British Columbia, south to Florida and New Mexico. Our records from northcentral and central Colorado at 5000-7500 feet.
2. Sanicula canadensis L., Sp. Pl. 235. 1753.
   Vermont to South Dakota, south to Florida and Texas. Should be looked for in eastern Colorado.

4. Ligusticum L. LOVAGE

Scapose to caulescent, glabrous to puberulent perennial plants from fibrous root crowns, these from taproots; stems stout or slender, simple or branching; leaves thin to subcoriaceous, ternate or ternate-pinnately compound, the leaflets pinnately dissected; inflorescence of loose, compound umbels; involucre wanting or sometimes of 1 deciduous bract; involucel wanting or of several narrow bractlets; rays spreading-ascending; pedicels spreading-ascending; flowers white; calyx teeth minute to evident; stylopodium low-conic, styles short; fruit oblong, slightly flattened laterally or nearly round in cross-section, glabrous, ribs narrowly winged; oil tubes small, 3-6 in the intervals, 6-10 on the commissure; seeds dorsally flattened, channeled under the intervals, the face concave to plane; carpophores 2-cleft to the base.

1. Leaflets ovate or oblong, usually over 3 mm. wide; plants stout, stems usually 4 mm. in diameter or more, never subscapose
   2. Leaflets ovate, 2.5-5 cm. long   -1. L.porteri
   2. Leaflets oblong, 5-7 mm. long   -1A. L.porteri brevilobum
1. Leaflets linear, 1-3 mm. wide; plants slender, stems usually less than 4 mm. in diameter, often subscapose
   3. Plants caulescent; pedicels 8-12 mm. long; fruit 5-7 mm. long   -2. L.filicinum
   3. Plants subcaulescent (only 1 cauline leaf); pedicels 3-8 mm. long; fruit 3-5 mm. long
                                                                                                         -2A. L. filicinum tenuifolium

1. Ligusticum porteri Coult. & Rose, Rev. N. Am. Umbell. 86. 1888.
    L. affine A. Nels.; L. simulans C. & R.---Stems 50-100 cm. tall, stout, caulescent, freely branched, glabrous or puberulent in inflorescence; leaves 15-30 cm. long, ovate in outline, 1- to 3-ternately pinnate; leaflets 2.5-5 cm. long, ovate, incised; rays 11-24; pedicels 5-12 mm. long; fruit 5-8 mm. long.---Valleys and slopes, often in woods. Wyoming to Nevada, south to Mexico. Our records well scattered in the western half of Colorado at 7500-11,500 feet.
1A. Ligusticum porteri brevilobum (Rydb.) Math. & Const., (var.) Bull. Torr. Club 68:123. 1941.
    L. brevilobum Rydb.---Limited to central and southeastern Utah and to be looked for in adjacent Colorado.
2. Ligusticum filicinum S. Wats., Proc. Am. Acad. 11:140. 1876.
    Montana and Idaho, south to Wyoming and Utah. The species proper may be in northern Colorado.
2A. Ligusticum filicinum tenuifolium (S. Wats.) Math. & Const., (var.) Bull. Torr. Club 68:123. 1941.
    L. tenuifolium S. Wats.---Subscapose plants; stems 20-60 cm. tall, very slender, glabrous or scaberulous in the inflorescence; leaves oblong to oval in outline, ternate, then pinnately decompound; leaflets pinnately divided; cauline leaf usually reduced and solitary, sessile; rays 5-15; pedicels 3-8 mm. long; fruit 3-5 mm. long.---Woods. Montana to Oregon, south to Colorado and Utah. Our rather few records from the northwestern quarter of Colorado at 8500-10,000 feet.

5. Podistera S. Wats.

Low caespitose, caulescent, perennial, glabrous plants, from long thickened roots; leaves pinnate, the leaflets deeply 2- to 3-lobed, the lobes cleft into linear segments; inflorescence of subcompact, compound umbels exceeding the leaves; involucre wanting; involucel of several toothed bractlets longer than the flowers and fruit; rays few, short, stout, spreading; pedicels very short; flowers greenish-yellow; calyx teeth ovate, conspicuous; stylopodium conic; fruit oval, slightly flattened laterally, glabrous, ribs filiform; oil tubes 2-3 in the intervals and about 4 on the commissure; seed face slightly concave; carpophore 2-cleft to base.

1. Podistera eastwoodae (Coult. & Rose) Math. & Const.,Bull. Torr. Club 69:247. 1942.
    Ligusticum eastwoodae Rose; Ligusticella eastwoodae Coult. & Rose---Plants 7-30 cm. tall; leaves oblong in general outline; leaflets 10-15 mm. long, ovate in outline; involucel bractlets 4-6 mm. long, oval or obovate, sometimes entire but some toothed at apex, almost lobed; rays 5-8; pedicels 1-2 mm. long; fruit 3-4 mm. long.---Mountains. Colorado, Utah and New Mexico. Our records from the southeastern quarter of Colorado at 10,000-11,500 feet.

6. Orogenia S. Wats. INDIAN POTATO

Low, glabrous acaulescent or very short-caulescent, perennial plants from globose or ovoid tubers; leaves once- or twice-ternate or rarely simple, the leaflets narrow, elongate, mostly entire; inflorescence a loose, compound umbel; involucre none; involucel of a few minute, narrow bractlets or none; rays few, unequal, spreading; pedicels short or obsolete; flowers white; calyx teeth obsolete; stylopodium lacking, styles short and spreading; fruit oblong to oval, nearly round in section, glabrous, dorsal ribs filiform and rather prominent, a corky rib running down middle of commissure; oil tubes several in intervals (but reported as having 3 under each rib) and several on the commissure; carpophore none.

1. Orogenia linearifolia S. Wats., Bot. King's Expl. 120. 1871.
    Plants 5-15 cm. tall from a tuber about 5-12 mm. in diameter; leaflets 1.5-7 cm. long, linear or lanceolate; pedicels about 1 mm. long; fruit 3-4 mm. long.---Mountain slopes and valleys.

Montana to Washington, south to Colorado and Utah. Our rather few records from the western third of Colorado at 6500-9000 feet.

7. Oreoxis Raf.

Low, caespitose, acaulescent, perennial plants from slender, elongated roots; leaves pinnate or bipinnate with the ultimate divisions linear; inflorescence of subcompact, compound umbels, exceeding the leaves; involucre usually wanting; involucel usually dimidiate, entire or toothed, usually exceeding the flowers; rays few; pedicels short and spreading or obsolete; flowers whitish or yellowish; calyx teeth prominent; stylopodium lacking, style slender; fruit oblong to ovoid-oblong, slightly flattened laterally, the ribs corky-winged, glabrous or slightly pubescent; oil tubes small, 1 to several in the intervals, 2 to several on the commissure, often an accessory one in each wing; seed face plane or slightly concave, carpophore lacking.

1. Bractlets obovate, toothed at apex, usually purplish -1. O. bakeri
1. Bractlets linear, entire, greenish
   2. Plants puberulent; each wing of fruit obovate or ovate in section; plants generally distributed -2. O. alpina
   2. Plants glabrous or puberulent only in inflorescence or near it on the peduncles; each wing of fruit broadly linear in section; plants apparently limited to the Pikes Peak region of central Colorado -3. O. humilis

1. Oreoxis bakeri Coult. & Rose, Contr. U. S. Nat. Herb. 7:144. 1900.
Cymopterus bakeri (Coult. & Rose) Jones---Plants 1-12 cm. tall, base of umbels and rays slightly puberulent; leaves 5-50 mm. long, narrowly oblong in outline, usually twice-pinnate, ultimate divisions more or less distinct; involucel dimidiate, the bractlets 3-5 mm. long, obovate, usually 3-toothed at apex; rays 3-5 mm. long, 3-8; flowers yellow or whitish; fruit 2-4 mm. long, ovoid-oblong, usually purplish, wings mostly linear-oblong and rounded to subacute at apex in cross-section; oil tubes 3-4 in the intervals.---High mountains. Colorado, Utah and New Mexico. Our few records from southwestern and southcentral Colorado at about 12,000-12,500 feet.
2. Oreoxis alpina (A. Gray) Coult. & Rose, Contr. U. S. Nat. Herb. 7:144. 1900.
Cymopterus alpinus A. Gray---Plants 2-10 cm. tall, more or less puberulent especially on peduncles; leaves oblong in outline, 1- to 2-pinnate, the first division divided into 1-7 linear divisions, these 2-30 mm. long and 1-2 mm. broad; involucre absent or sometimes of a small, linear bract; bractlets of involucel 2-5 mm. long, linear; rays 2-7 mm. long, 3-6; flowers yellow; fruit 3-6 mm. long, pubescent when young, with broad wings each obovate or ovate in section; oil tubes usually 1 in intervals. Some of our plants intergrade with O. humilis Raf.---High mountains. Wyoming to Utah, south to New Mexico and Arizona. Our records from the western half of Colorado at 9500-13,500 feet.
3. Oreoxis humilis Raf. Bull. Bot. Seringe 1:217. 1830.
Cymopterus humilis (Raf.) Tide. & Kitt.---Plants glabrous or sparsely puberulent, the hairs limited to the inflorescence and the peduncle; wings of fruit broadly linear in section. Doubtfully distinct from O. alpina (A. Gray) C. & R., as represented in this area.---Apparently limited to central Colorado. Our few records from 12,000-13,000 feet.

8. Musineon Raf.

Low, erect, short-caulescent or acaulescent perennial plants from thick taproots; leaves once-pinnate or once-ternate, the divisions linear or oblong, distinct or confluent; inflorescence of loose or subcompact compound umbels; involucre lacking or sometimes rather conspicuous; involucel of a few linear distinct or nearly distinct bractlets; rays few, spreading; pedicels short, spreading; flowers yellow or white; calyx teeth ovate and conspicuous; stylopodium lacking, the styles slender; fruit ovoid to narrowly oblong, laterally flattened and somewhat constricted at the commissure, glandular-scabrous, ribs acute and prominent; oil tubes usually 3 in the intervals and 2-6 on the commissure, sometimes 1 in each rib; seed subround in section with plane or concave face; carpophore entire, bifid at apex or deeply 2-cleft.

1. Plants acaulescent; ultimate divisions of the leaves narrowly linear -1. M. tenuifolium
1. Plants caulescent; ultimate divisions oblong (and lobed)
   2. Stems glabrous -2. M. divaricatum
   2. Stems scabrous -2A. M. divaricatum hookeri

1. Musineon tenuifolium Nutt.; T. & G., Fl. N. Amer. 1:642. 1840.
Daucophyllum tenuifolium (Nutt.) Rydb.; D. linearis Rydb.---Plants 6-30 cm. tall, acaulescent, erect, subcaespitose; leaves 1.5-10 cm. long, oblong or narrowly-oblong in outline, 1- to 3-pinnate, the ultimate divisions distinct, linear, not over 1 mm. wide; peduncles exceeding the leaves, more or less hirtellous at base of umbel; involucre lacking or of 1-2 inconspicuous or conspicuous bracts; involucel 1-3 mm. long, rather inconspicuous, subdimidiate; rays 8-30; pedicels 1-2 mm. long; fruit 2-5 mm. long and 1-2 mm. wide, ovoid to narrowly oblong, more or less granular; oil tubes mostly 3 in the intervals, 2-4 on the commissure. In characters of fruit and bractlets our plants seem to intergrade with M. lineare (Rydb.) Math. a species of Utah.---Hills and plains. South Dakota and Wyoming, south to Nebraska and Colorado. Our few records from northcentral and northeastern Colorado at 5000-6000 feet.

2. **Musineon divaricatum** (Pursh) Nutt.; T. & G., Fl. N. Amer. 1:642 in part. 1840.
   M. pedunculatum A. Nels.---Plants caulescent, spreading or erect, stems glabrous or somewhat scabrous, usually dichotomously branching; leaves ovate-oblong in outline, 1- to 2-pinnate or ternate-pinnate, the leaflets 5-15 mm. long, and 3-10 mm. wide, oblong, pinnately lobed; bractlets 2-4 mm. long, more or less distinct, shorter than the yellow flowers; rays 10-20; pedicels 1-3 mm. long; fruit 3-6 mm. long, ovoid to oblong, constricted at apex, glabrous or minutely scaberulous; oil tubes 3-4 in the intervals, 4-6 on the commissure.---Dry hills and plains. South Dakota and Montana, south to Nebraska and Colorado. Our records from the northeastern quarter of Colorado at 4500-6000 feet.
2A. **Musineon divaricatum hookeri** T. & G., (var.) Fl. N. Amer. 1:642. 1840.
   M. trachyspermum Nutt.; M. angustifolium Nutt.; M. hookeri (T. & G.) Nutt.---Like the species but the stems scaberulous at the base of the umbel and often throughout; leaves narrower in outline; fruit scaberulous to densely scabrous; oil tubes rarely solitary in the intervals. Intergrades very commonly with the species and hardly worth maintaining as a variety in this state.---Saskatchewan, south to North Dakota, Colorado and Nevada. Our records from northcentral and northeastern Colorado (also reported from the central part of the state) at 5000-5500 feet.

9. Aletes Coult. & Rose

Low, caespitose, acaulescent, perennial plants from slender roots; leaves pinnate to bipinnate, the leaflets linear to ovate, entire to toothed or pinnately lobed; inflorescence a loose, compound umbel, sometimes on a peduncle shorter than the leaves; involucre usually lacking; involucel of narrow bractlets; rays spreading or reflexed; pedicels short and spreading; flowers yellow; calyx teeth deltoid-ovate, conspicuous; stylopodium lacking, the styles slender and spreading; fruit oblong, slightly flattened laterally or subterete (the carpels round or somewhat dorsally flattened), ribs obscure or somewhat corky-winged; oil tubes solitary in the intervals, 2 on the commissure, sometimes a small one at the apex of each rib; carpophore 2-cleft to the base, sometimes deciduous.

1. Peduncles mostly shorter than the leaves; ribs of fruit obscure; plants known only from extreme northern Colorado -1. A. humilis
1. Peduncles longer than the leaves; ribs of fruit conspicuous, somewhat corky-winged; plants of central and southern Colorado
   2. Rays 4-8; bractlets distinct or nearly so -2. A. macdougali
   2. Rays 8-15; bractlets connate at base -3. A. acaulis

1. **Aletes humilis** Coult. & Rose, Contr. U. S. Nat. Herb. 7:107. 1900.
   Plants 2-10 cm. tall, glabrous or slightly puberulent; leaves 1.5-4 cm. long, oblong in outline, 1- to 2-pinnate, the leaflets linear to ovate-oblong, acute and more or less confluent; peduncles short, usually shorter than the leaves; bractlets 2-4 mm. long, linear, acute, distinct; rays 4-6, subequal; pedicels about 2 mm. long; fruit 3-4 mm. long and about 2 mm. wide, ovoid-oblong, ribs inconspicuous; oil tubes 1 in the intervals.---Mountains. Apparently limited to northcentral Colorado, our records from 7500 feet.
2. **Aletes macdougali** Coult. & Rose, Contr. U. S. Nat. Herb. 7:107. 1900.
   Oreoxis macdougali (Coult. & Rose) Rydb.---Plants 6-25 cm. tall, glabrous; leaves 2-8 cm. long, oblong in outline, pinnate or sometimes twice-pinnate, leaflets linear-oblong and entire or broader and incised; peduncles longer than the leaves; involucel of several, linear, distinct bractlets, these 2-3 mm. long; rays 4-8; pedicels about 1 mm. long; fruit 3-8 mm. long, oblong, ribs corky-winged, broadly linear in section; oil tubes 1 in the intervals.---Canyons and mesas. Colorado, Utah and Arizona. Our few records from the southwestern corner of Colorado at about 6500-7000 feet.
3. **Aletes acaulis** (Torr.) Coult. & Rose, Rev. N. Am. Umbell. 126. 1888.
   A. obovata Rydb.---Plants 5-35 cm. tall, mostly puberulent; leaves 2-10 cm. long, oblong in outline, 1- to 2-pinnate, the leaflets 4-15 mm. long, lanceolate to orbicular, pinnately lobed and spinulose-dentate, sometimes puberulent below; inflorescence longer than the leaves; bractlets of involucel 2-3 mm. long, lanceolate or linear, connate at base; rays 8-15, subequal, spreading or reflexed; pedicels short; fruit 4-7 mm. long, oblong, with corky obtuse wings, an accessory oil tube in the apex of each rib.---Hills and mountains. Colorado to Texas and Mexico. Our few records from northcentral and central Colorado at 5500-7000 feet.

10. Harbouria Coult. & Rose

Plants subcaespitose, caulescent, perennial from long taproots; stems glabrous or nearly so, erect, slender and branching; leaves mostly basal, pinnately decompound, the ultimate divisions linear and distinct; inflorescence of loose to subcompact compound umbels, the terminal usually in pairs; involucre wanting or of 1-2 inconspicuous bracts; involucel of several linear bractlets no longer than the pedicels; flowers yellow; calyx teeth evident but small; rays spreading, subequal; pedicels spreading; stylopodium lacking, styles short; fruit ovoid, laterally flattened, constricted at the commissure, granular-roughened, ribs prominent, obtuse, corky, subequal; oil tubes usually solitary in the intervals, 1-3 on the commissure; seeds rounded in section but plain on the face; carpophores entire.

1. **Harbouria trachypleura** (A. Gray) Coult. & Rose, Rev. N. Am. Umbell. 125. 1888.
   Plants 8-50 cm. tall, hirtellous just below umbel and often in umbel, otherwise glabrous, with long brown fibers from apex of root; leaves ovate-oblong in outline, ultimate divisions 2-30 mm. long; rays 8-30; pedicels 1-4 mm. long; fruit 3-6 mm. long, often only 1 mericarp developing

and this curved or bowed.---Hills and mountains. Wyoming and Utah to Mexico. Our records from northcentral, central and southcentral Colorado at 5500-10,500 feet.

11. Zizia Koch

Caulescent, glabrous or nearly glabrous perennial plants from fascicles of fleshy roots; stems erect, simple or branching; leaves simple (on basal ones) or ternately compound with leaflets toothed or lobed, rather broad; inflorescence of loose compound umbels, terminal or lateral; involucre wanting; involucel of a few narrow bractlets; rays rather few, spreading or spreading-ascending; pedicels short and stout, spreading, the central flower of each umbellet sessile or nearly so; flowers yellow; caylx teeth prominent; stylopodium none; fruit oblong or oval, flattened laterally, glabrous, the ribs all filiform; oil tubes solitary in the intervals, 2 on the commissure; seed face plane to somewhat concave, rather sulcate, under the oil tubes; carpophore 2-cleft about 1/2 way down.

1. Basal leaves simple, oval or broadly ovate, cordate at base -1. Z. aptera
1. Basal leaves compound, the leaflets lanceolate to ovate, rarely cordate at base -2. Z. aurea

1. Zizia aptera (A. Gray) Fernald, Rhodora 41:441. 1939.
Z. cordata (Walt.) Koch of Manuals---Plants 30-60 cm. tall; basal leaves 4-8 cm. long, simple, cordate, oval or broadly ovate, crenate-dentate; cauline leaves ternately divided, the divisions oval to lanceolate, coarsely serrate and often lobed; fruit 2-4 mm. long and about 2 mm. broad.---Meadows and open woods, usually in moist ground. Canada, south to Georgia, Utah and Oregon. Our records from northcentral, central and southcentral Colorado at about 8500-9000 feet.
2. Zizia aurea (L.) Koch, Nova Acta Acad. Leop.-Carol. 12:129. 1825.
New Brunswick to Saskatchewan, south to Florida and Texas. This plant should be looked for in eastern Colorado.

12. Berula Hoffm. WATER PARSNIP

Slender, erect, stoloniferous, caulescent, branching, glabrous, aquatic, perennial plants, from fascicled-fibrous roots; leaves pinnate or decompound, the leaflets subentire to toothed or lobed, heteromorphic the upper leaves more deeply cut than the lower, the submerged leaves often decompound; inflorescence of loose compound terminal and axillary umbels exceeding the leaves; involucre of conspicuous narrow bracts; involucel of rather conspicuous narrow bractlets; rays subequal, 6-15, erect-ascending or spreading; pedicels spreading; flowers white; calyx teeth minute; stylopodium conical; fruit oval to suborbicular, glabrous, flattened laterally, ribs obscure in the thick corky pericarp; oil tubes numerous and closely surrounding the seed cavity; seed round in cross-section, the face plane; carpophore 2-cleft to base, the halves adnate to the mericarp.

1. Berula erecta (Huds.) Coville, Contr. U. S. Nat. Herb. 4:115. 1893.
Stems 20-80 cm. tall; leaves 10-30 cm. long, narrowly oblong in general outline, leaflets variable in size on upper and lower leaves; pedicels 2-5 mm. long; fruit 1.5-2 mm. long.---Moist or wet ground. Ontario to British Columbia, south to Florida, Mexico and California; Europe. Our records scattered in Colorado, mostly in the northern half of the state at 3500-9000 feet.

13. Sium L. WATER PARSNIP

Caulescent, glabrous, perennial plants, aquatic or growing in swamps, with fascicles of fusiform roots; stems stout, erect or branching; leaves pinnately compound or rarely simple, the leaflets narrow, serrate or incised; inflorescence of loose compound umbels, these terminal or lateral; involucre of conspicuous, linear, entire or incised, unequal bracts; involucel of conspicuous, entire bractlets; rays rather few, subequal, spreading-ascending, pedicels spreading; flowers white; calyx teeth minute or obsolete; stylopodium depressed or rarely conic, styles short and reflexed; fruit oval or orbicular, slightly flattened laterally and somewhat constricted on the commissure, glabrous, the ribs prominent and corky, subequal; oil tubes 1-3 in the intervals, 2-6 on the commissure; seed round in section and flattened on the face; carpophore 2-cleft to base, the halves adnate to the mericarp.

1. Sium suave Walt., Fl. Car. 115. 1788.
S. cicutaefolium of Manuals---Stems 50-100 cm. tall; leaves oblong or ovate in outline, leaflets 1-4 cm. long, lanceolate or linear; involucre of 6-10 bracts, these 3-15 mm. long, reflexed; involucel of 4-8, linear-lanceolate bractlets, these 1-3 mm. long; rays 10-20; pedicels 3-5 mm. long; fruit 2-3 mm. long. This plant resembles Cicuta in younger stages. The veins of the leaflets run out toward the apex of the teeth, however, not toward the notches between the teeth---Moist or wet ground. Newfoundland to British Columbia, south to Virginia and California. Our records scattered over the western half of Colorado at 6500-8000 feet.

14. Cicuta L. WATER HEMLOCK

Stout, erect, caulescent, glabrous, perennial plants from tuberous bases, these bearing fibrous or elongated-tuberous roots; stems branching; leaves once-pinnate to thrice-pinnate; inflorescence of loose compound umbels; involucre none or of a few narrow bracts; involucel usually present, of several narrow bractlets; rays numerous, spreading-ascending; pedicels spreading; flowers white or greenish; calyx teeth evident; stylopodium depressed or low-conical; fruit oval or orbicular, flattened laterally, constricted at the commissure, glabrous, ribs rather prominent, obtuse, corky, the lateral ribs of adjacent carpels separated, hence all ribs about equal around the fruit; oil tubes rather small, solitary in the intervals and 2 on the commissure; seeds round in cross-section, somewhat flattened on face; carpophore 2-cleft to base, deciduous.

1. Cicuta douglasii (DC.) Coult. & Rose, Contr. U.S. Nat. Herb. 7:95. 1900.
C. occidentalis Greene; C. vagans Greene; C. cinicola A. Nels.---Roots or base of stem developing cross-partitions in older plants; stems 60-120 cm. tall; leaflets 3-10 cm. long, linear-lanceolate to ovate-lanceolate, toothed to incised; pedicels 3-8 mm. long; fruit 2-4 mm. long. This is a poisonous plant that resembles species of Sium and Angelica. In Cicuta the secondary veins of the leaflets end in the notches between the teeth, in the others they tend to go out to the tip of the teeth.---Moist or wet ground. Alberta to Alaska, south to New Mexico, Mexico and California. Our records scattered over Colorado at 3500-8000 feet.

15. Foeniculum Adans. FENNEL

Caulescent, glabrous and glaucous, perennial or biennial, strongly odoriferous plants from taproots; stems slender, erect, branching; leaves pinnately decompound, the ultimate divisions filiform; inflorescence of loose compound umbels, these terminal and axillary; involucre wanting; involucel wanting; rays numerous and ascending; pedicels spreading; flowers yellow; calyx teeth obsolete; stylopodium conic, style very short and recurved; fruit oblong, slightly flattened laterally, glabrous, ribs rather prominent; oil tubes solitary in the intervals, 2 on the commissure; seed flattened dorsally in section, the face plane or slightly concave; carpophore 2-cleft to base.

1. Foeniculum vulgare Mill., Gard. Dict. ed. 8. Foeniculum No. 1. 1768.
Adventive throughout the United States and to be expected here.

16. Conium L. POISON HEMLOCK

Erect caulescent glabrous biennial plants from stout taproots; stems spotted, branching; leaves pinnately decompound, the ultimate divisions pinnately incised; inflorescence composed of loose, terminal and axillary compound umbels; involucre of numerous, lanceolate, rather inconspicuous bracts; involucel bractlets resembling the bracts, shorter than the pedicels; rays spreading-ascending; pedicels spreading to spreading-ascending; flowers white; calyx teeth obsolete; stylopodium depressed-conical, the styles reflexed; fruit broadly ovoid, flattened laterally, glabrous, the ribs prominent, obtuse, crenulate to undulate; oil tubes very small and numerous; seed faces deeply to narrowly sulcate; carpophore entire.

1. Conium maculatum L., Sp. Pl. 243. 1753.
Plants 50-300 cm. tall; leaves 15-30 cm. long on longest ones, broadly ovate in general outline, petioles dilated; rays 15-25 mm. long; involucre bracts ovate to ovate-lanceolate, acuminate; bractlets similar to bracts, with midribs; pedicels 4-6 mm. long; fruit 2-2.5 mm. long, the ribs prominent in dry fruit.---Waste places, valleys and low ground in general. Widely distributed in Eurasia, Africa and North America. Our records scattered in the western half of Colorado at 5000-9000 feet.

17. Perideridia Reichenb. YAMPA

Caulescent, glabrous, perennial plants from thickened, fascicled roots, these tuberlike; stems erect and branching; leaves thin, pinnate or occasionally bipinnate, the ultimate divisions linear to linear-lanceolate; inflorescence of loose compound umbels, these terminal and lateral, longer than the leaves; involucre of several setaceous bracts or wanting; involucel of 1 to several short, narrow bractlets; rays spreading-ascending; pedicels spreading; flowers white or pink; calyx teeth conspicuous; stylopodium low-conic, styles short and recurved; fruit orbicular to suborbicular, laterally flattened, glabrous, with ribs filiform; oil tubes solitary in the intervals, 2 on the commissure; seeds round in section; carpophore 2-cleft to base.

1. Perideridia gairdneri (H. & A.) Mathias, Brittonia 2:244. 1936.
Carum gairdneri (H. & A.) A. Gray; Atenia gairdneri H. & A.; C. garrettii A. Nels.; A. garrettii (A. Nels.) Rydb.---Stems 30-100 cm. tall from solitary or fascicles of fusiform tubers; leaves oblong to ovate in outline, the ultimate divisions 2-15 cm. long, entire or sometimes toothed or lobed; bractlets of involucel 1-4 mm. long, scarious or green; rays 8-20; pedicels 3-7 mm. long; fruit 2-3 mm. long.---Valleys and meadows. Alberta to British Columbia, south to New Mexico and California. Our few records from the northwestern quarter of Colorado at 6500-8000 feet.

18. Carum L. CARAWAY

Caulescent, glabrous, biennial or perennial plants from taproots; stems slender and branching; leaves pinnately dissected, the ultimate divisions narrow to filiform, entire or toothed, cauline leaves with dilated petioles; inflorescence of loose compound umbels, terminal and lateral; involucre lacking or of a few filiform bracts; involucel lacking or of a few filiform bractlets; rays few, spreading-ascending, unequal; flowers white or pinkish;

calyx teeth obsolete; stylopodium low-conic, styles short and spreading; fruit oblong to oblong-oval, laterally flattened, glabrous, ribs filiform but prominent; oil tubes solitary in the intervals and 2 on the commissure; seeds rounded in cross-section, the faces plane; carpophore 2-cleft to the base.

1. Carum carvi L., Sp. Pl. 263. 1753.
Stems 30-100 cm. tall; leaves oblong or oval in outline, the ultimate divisions filiform-linear on upper leaves, wider on other leaves; rays 7-14; pedicels 3-13 mm. long; fruit 3-4 mm. long.---Waste places as an escape from cultivation. Variously distributed in Canada and the United States. Our records from the western half of Colorado, except in the extreme western part of the state, at 5000-9500 feet.

19. Anethum L. DILL

Plants glabrous or glaucous, caulescent and annual, strongly odoriferous, from subfusiform roots; stems slender, erect, branching; leaves pinnately decompound, with the ultimate divisions filiform; inflorescence of loose compound umbels, these terminal and lateral; involucre usually wanting; involucel usually wanting; rays numerous, spreading-ascending; pedicels spreading; flowers yellow; calyx teeth obsolete; stylopodium conic, styles short and reflexed; fruit ovoid, flattened dorsally, glabrous, ribs narrowly winged, the laterals much broader; oil tubes solitary in the intervals, 2-4 on the commissure; seeds dorsally flattened in section, the face plane or slightly concave; carpophore 2-cleft to base.

1. Anethum graveolens L., Sp. Pl. 263. 1753.
Commonly cultivated in this country and often persisting or escaping. We have no record of this occurring in Colorado as yet.

20. Pastinaca L. PARSNIP

Tall, stout, caulescent, biennial or perennial plants from fleshy taproots; leaves large, petiolate, pinnately compound with serrate to lobed leaflets; inflorescence of loose compound flat-topped umbels; involucre usually wanting; involucel usually wanting; rays and pedicels spreading-ascending; flowers yellow; calyx teeth minute or obsolete; stylopodium depressed-conic; fruit oval or ovate, strongly flattened dorsally, glabrous, dorsal ribs inconspicuous, lateral thin-winged; oil tubes large, solitary in the intervals and 2-4 on the commissure, extending the full length of the mericarps; carpophore 2-cleft to the base.

1. Pastinaca sativa L., Sp. Pl. 262. 1753.
Stems 30-100 cm. tall; leaves large, leaflets 5-10 cm. long, oblong to ovate; rays 2-10 cm. long, 15-25, unequal; pedicels 5-10 mm. long; fruit 5-8 mm. long.---Roadsides and waste places. Europe, but now widely distributed in North America. Our records scattered over the western two-thirds of Colorado at 5000-8000 feet.

21. Heracleum L. COW PARSNIP

Tall, stout, caulescent, biennial or perennial plants from taproots or fascicled roots; leaves petioled, large, orbicular in outline, ternately compound with 3 leaflets, the upper cauline leaves with conspicuous dilated sheaths at bases of petioles; inflorescence of loose, compound, terminal and lateral flat-topped umbels; flowers white, the outer ones radiant with enlarged and often 2-cleft petals; involucre wanting or of 5-10 deciduous bracts; involucel of numerous narrow, entire bractlets; calyx teeth minute or obsolete; stylopodium more or less conical, the styles short; fruit orbicular to elliptic, usually pubescent, strongly flattened dorsally, dorsal ribs inconspicuous, the lateral broadly winged; oil tubes large, visible as brownish bands on the surface extending only part way from apex toward base of the mericarps, solitary in the intervals, 2-4 on the commissure; seeds dorsally flattened; carpophore 2-cleft to the base.

1. Heracleum lanatum Michx., Fl. Bor. Am. 1:166. 1803.
Stems 80-200 cm. tall, woolly-villous especially above; leaflets 10-40 cm. long, ovate to orbicular, cordate at base, coarsely serrate and variously lobed; rays of umbel rather unequal in length, 15-20; fruit 8-12 mm. long, somewhat pubescent.---Moist, often partly shaded ground. Newfoundland to Alaska, south to Georgia, Arizona and California; Asia. Our records scattered in the western two-thirds of Colorado at 5000-10,500 feet.

22. Oxypolis Raf. COWBANE

Glabrous, caulescent, perennial plants from fleshy-fascicled tuberlike roots; stems erect, slender; leaves thin, simply pinnate, the leaflets 5-13; inflorescence of loose compound umbels, these terminal and axillary; involucre wanting; involucel wanting; rays strictly ascending; pedicels ascending or spreading; flowers white to purple; calyx teeth conspicuous; stylopodium conic, styles slender and spreading; fruit oblong to oval, strongly flattened dorsally, glabrous, dorsal ribs filiform, not at all winged, lateral ribs broadly winged and nerved dorsally, giving the carpels the appearance of having 5 dorsal ribs instead of 3; oil tubes solitary in the intervals, 2-6 on the commissure; seeds flattened dorsally, the face plane; carpophores 2-cleft to the base.

1. **Oxypolis fendleri** (A. Gray) A. Heller, Bull. Torr. Club 24:478. 1897.
Stems 50-100 cm. tall; leaves about 8-15 cm. long, oblong in general outline, leaflets mostly 7-9, 3-5 cm. long, ovate to narrowly lanceolate, toothed or rarely incised; rays 5-14; pedicels 3-10 mm. long, ascending; fruit 3-5 mm. long.---Moist ground, often on stream banks. Wyoming, south to New Mexico and Utah. Our records from the western half of Colorado at 8000-11,500 feet.

23. Conioselinum Hoffm. HEMLOCK-PARSLEY

Plants caulescent, perennial, glabrous except in inflorescence, from taproot or a cluster of fleshy roots; stems rather slender, simple or sparingly branched; leaves thin, ternate-pinnately decompound, the leaflets pinnately incised; inflorescence of loose compound umbels, these terminal and lateral; involucre of 1 to several very narrow bracts or wanting; involucel of several narrow bractlets often connate at base; rays rather numerous, spreading-ascending; pedicels spreading-ascending; flowers white; calyx teeth obsolete; stylopodium conic, styles short; fruit oval, flattened dorsally, glabrous, dorsal ribs low and corky, the lateral ones more broadly thin-winged; oil tubes small, 1-2 in the intervals, 2-4 on the commissure; seeds dorsally flattened, the face plane or slightly concave; carpophore 2-cleft to the base or nearly so.

1. **Conioselinum scopulorum** (A. Gray) Coult. & Rose, Contr. U. S. Nat. Herb. 7:151. 1900.
C. coloradense Osterh.---Stems 30-90 cm. tall; leaves about 10-20 cm. long, lanceolate or ovate in outline, 1- to 2-pinnate or 1- to 2-ternate-pinnate; leaflets 20-65 mm. long, ovate; bractlets 2-8 mm. long, a little shorter than the flowers; rays 10-20; pedicels 4-12 mm. long; fruit 4-6 mm. long, only slightly compressed dorsally.---Mountains, often on rocky, shaded ground. Wyoming, south to New Mexico and Arizona. Our records scattered in the western half of Colorado at 7000-10,000 feet.

24. Angelica L.

Stout to rather slender, caulescent, perennial plants from taproots; leaves ternate-pinnately or pinnately compound, with rather broad toothed or lobed leaflets; petioles of cauline leaves often inflated at base; inflorescence of loose compound umbels, these terminal or lateral; involucre none or of 1 or more conspicuous bracts; involucel none or of narrow bractlets; rays many or few, equal or unequal, frequently webbed at base (margins dilated and united), spreading-ascending; pedicels slender, spreading; flowers white, pink or purple; calyx teeth minute or obsolete; stylopodium low-conical; fruit oblong-oval to orbicular, pubescent or glabrous, strongly flattened dorsally, dorsal ribs inconspicuous to narrowly winged or corky-winged, lateral ribs thin-winged or corky-winged; oil tubes numerous and adhering to the seed which is free in the pericarp, or few and remaining in the pericarp which adheres to the seed; seeds flattened dorsally, the commissural face plane or concave; carpophore 2-cleft to the base.

1. Stems over 100 cm. tall; fruit 7-8 mm. long at maturity, oil tubes numerous, adhering to the seed which is loose in the pericarp at maturity; whole umbels globular in shape; ovaries glabrous -1. **A. ampla**
1. Stems less than 100 cm. tall; fruit 4-7 mm. long, oil tubes solitary (rarely in pairs) in the intervals and remaining in the pericarp wall which is adherent to the seeds; whole umbels rather flat-topped; ovaries glabrous, scabrous or hispidulous
    2. Ovaries and immature fruit glabrous; bractlets of involucels linear-lanceolate to lanceolate, usually over 1 mm. wide; flowers usually purplish-brown -2. **A. grayi**
    2. Ovaries and mature fruit scabrous to hispidulous; bractlets of involucels lacking, filiform or narrowly linear, not over 1 mm. wide; flowers white to pink
        3. Involucel none; leaves pinnate or very incompletely bipinnate; stems slender (seldom over 7 mm. thick at base) -3. **A. pinnata**
        3. Involucel present of narrow bractlets; leaves ternate-pinnate, only rarely once-pinnate; stems stout (usually over 7 mm. thick at base) -4. **A. roseana**

1. **Angelica ampla** A. Nels., Bull. Torr. Bot. Club 25:375. 1898.
Stems usually over 1.5 m. tall, stout (well over 1 cm. in diameter), hollow, purplish, glabrous or sometimes sparingly puberulent near or in the inflorescence; leaves ternate, then twice-pinnate, deltoid in general outline; leaflets 3-20 cm. long, ovate or ovate-lanceolate, sessile, crowded, serrate or incised; petioles of cauline leaves conspicuously dilated; involucre wanting; involucel of a few short, filiform bractlets; rays 30-45, spreading and subequal; pedicels 5-12 mm. long, spreading; both umbel and umbellets globular; flowers white; fruit 7-8 mm. long, oblong-oval, glabrous, the ribs very narrowly winged; oil tubes small, continuous about the seed and adherent to it; seed loose in the pericarp, the face concave.---Moist or wet ground, often in partial shade. Wyoming and Colorado. Our records scattered over the western half of Colorado, except in the extreme south, at 7500-9500 feet.

2. **Angelica grayi** Coult. & Rose, Contrib. U. S. Nat. Herb. 7:154. 1900.
Leaves and inflorescence more or less scabrous; stems rather low, 20-60 cm. tall, stout (mostly over 1 cm. thick at base); leaves pinnate to bipinnate or ternate-pinnate, the middle division larger; leaflets 1-5 cm. long, ovate to lanceolate, sessile or nearly so, serrate to sometimes lobed; cauline leaves with conspicuously dilated sheaths; involucre wanting or of foliaceous bracts; involucel of bractlets 5-18 mm. long, usually over 1 mm. wide, linear-lanceolate to lanceolate; rays many, spreading-ascending, the whole umbel rather flat on top; pedicels 2-6 mm. long, spreading-ascending; flowers purplish-brown (perhaps rarely whitish); fruit 4-5 mm. long, glabrous even when young, oval, dorsal ribs narrowly winged, the laterals

broader winged; oil tubes solitary in the intervals, 2-4 on the commissure; seed face plane or nearly so, the seed remaining united to the pericarp.---Higher mountains. Wyoming and Colorado. Our records from the western half of Colorado, few from the extreme western part of the state, at 8000-13,000 feet.

3. **Angelica pinnata** S. Wats., Bot. King's Expl. 126. 1871.

A. leporina S. Wats.---Plants glabrous or the leaves and inflorescences sometimes scabrous; stems 25-90 cm. tall, rather slender (about 5-7 mm. thick near base); leaves once-pinnate or incompletely bipinnate; leaflets 3-9 cm. long, lanceolate to ovate-lanceolate, sessile or nearly so, serrate to rarely entire, some of lower often lobed or nearly divided; cauline leaves with rather conspicuously dilated petioles; involucre wanting or of sheathlike bracts; involucel wanting; rays 6-25, ascending or spreading-ascending, unequal, the whole umbel somewhat flattened above; pedicels 3-8 mm. long, spreading and ascending; flowers white or pinkish; fruit 3-6 mm. long, nearly orbicular, glabrate but hispidulous when young, dorsal ribs narrowly winged, lateral ribs broader winged, these nearly as wide as the body; oil tubes solitary in the intervals, sometimes in pairs, 2-4 on the commissure; seed face slightly concave, the seed remaining attached to the pericarp.---Moist or wet ground. Wyoming to Utah, south to New Mexico. Our records from the western half of Colorado at 5000-9500 feet.

4. **Angelica roseana** Henderson, Contrib. U. S. Nat. Herb. 5:201. 1899.

Plants 30-85 cm. tall, stems stout, foliage and inflorescence scaberulous or rarely glabrate; leaves 5-30 cm. long, deltoid in outline, 1- to 3-ternate or ternately-pinnately divided, the leaflets ovate to lanceolate, spinulose-dentate; peduncles stout; involucre wanting or occasionally of a few bracts; involucel of a few linear or filiform bractlets 1.5-16 mm. long; rays 15-35, unequal; pedicels 3-11 mm. long, spreading-ascending; flowers white or pink; fruit 4-7 mm. long and 3-4 mm. wide, oblong, scaberulous, dorsal ribs narrowly winged, lateral broader winged; oil tubes 1 in intervals.---Mountains, often on talus. Montana to Idaho, south to Colorado and Utah. Our one record from central Colorado at about 9000 feet.

25. Pteryxia Nutt.

Caespitose acaulescent or subacaulescent, essentially glabrous, perennial plants from long slender taproots; stems little if any branched, low but erect; leaves bipinnate to ternate-pinnately decompound, the ultimate divisions narrow; inflorescence of loose to subcompact compound umbels, these terminal and sometimes lateral, equalling or exceeding the leaves; involucre lacking or rarely of 1 or 2 linear bracts; involucel of narrow bractlets, equalling or exceeding the flowers; rays spreading or ascending; pedicels spreading, sometimes almost obsolete; flowers yellow; calyx teeth prominent and often unequal; stylopodium lacking; fruit ovoid to oblong, flattened dorsally, glabrous at least in age, lateral ribs thin-winged, dorsal ribs similar but often shorter or nearly obsolete; oil tubes small, 1 to several in the intervals, several on the commissure; seeds dorsally flattened in section, the faces plane or concave; carpophore 2-cleft to the base.

1. Leaves narrowly oblong, bipinnate, not over 3.5 cm. wide
   2. Rays conspicuously unequal, up to 6 cm. long; dorsal wings of mericarps reduced or obsolete
      -1. **P. anisata**
   2. Rays nearly equal, up to 3 cm. long; dorsal wings well developed, some as wide as the laterals
      -2. **P. hendersoni**
1. Leaves broadly ovate to ovate-oblong in general outline, pinnately or ternate-pinnately decompound, over 3.5 cm. wide
      -3. **P. terebinthina**

1. **Pteryxia anisata** (A. Gray) Math. & Const., Bull. Torr. Club 69:248. 1942.

Pseudocymopterus anisatus (Gray) C. & R.; Pseudocymopterus aletifolius Rydb.; Pseudopteryxia anisata (Gray) Rydb.; Pseudopteryxia aletifolia Rydb.---Plants 10-35 cm. tall, acaulescent; leaves narrowly oblong in outline, bipinnate, the ultimate divisions 1-6 mm. long, rigid, acute, appearing as narrow lobes of a subcuneate leaflet; bractlets 3-15 mm. long, conspicuous, entire, longer than the flowers; rays 6-9; pedicels 1-10 mm. long; fruit 4-6 mm. long, lateral wings narrower than the body, the dorsal much narrower than the laterals or obsolete.---Hills and mountains, usually on dry or rocky soil. Apparently limited to Colorado. Our records thinly scattered in the western half of the state at 6500-10,000 feet.

2. **Pteryxia hendersoni** (Coult. & Rose) Math. & Const., Bull. Torr. Club 69:248. 1942.

Pseudopteryxia hendersonii (C. & R.) Rydb.; Pseudocymopterus hendersoni C. & R.; Pseudocymopterus anisatus of Mans. at least in part---Plants 5-40 cm. tall, acaulescent; leaves not over 2.5 cm. wide, narrowly oblong in outline, bipinnate, the ultimate divisions 1-15 mm. long, linear, acute; bractlets 2-12 mm. long, dimidiate, linear-lanceolate, entire or bifid; rays 3-30 mm. long, 4-8, slightly unequal; pedicels 1-3 mm. long; fruit 4-7 mm. long, ovoid-oblong, wings shorter than the body, dorsal equal to or shorter than the lateral.---Mountains. Montana to Idaho, south to New Mexico and Nevada. Our few records from the western half of Colorado at 6500-10,000 feet.

3. **Pteryxia terebinthina** (Hook.) Coult. & Rose, Contrib. U. S. Nat. Herb. 7:171. 1900.

P. calcarea (M. E. Jones) C. & R.---Plants acaulescent or shortly caulescent; stems 10-60 cm. tall; leaves gray-green, ovate-oblong to broadly ovate in general outline, pinnately or ternate-pinnately decompound, the ultimate divisions broadly linear; bractlets of involucel 2-6 mm. long, usually entire; rays short, 7-24, unequal; pedicels 1-8 mm. long; fruit 5-8 mm. long, dorsal wings about as broad as the laterals. Our plant is var. calcarea (M. E. Jones) Mathias.---Hills and valleys, often on rocky ground. The variety from Montana and Idaho, south to Colorado and Nevada. Our few records from northcentral and northwestern Colorado at 5500-10,000 feet.

26. Pseudocymopterus Coult. & Rose

Plants subacaulescent to caulescent, perennial from long, slender taproots; stems slender, erect; leaves thin, 1- to 3-pinnate, leaflets or ultimate divisions filiform to lanceolate; inflorescence of loose, compound umbels, these terminal and sometimes lateral, exceeding the leaves, conspicuously hirtellous-pubescent at base of umbel; involucre usually lacking; involucel of ovate to filiform bractlets connate at base, longer or shorter than the flowers; rays spreading-ascending or spreading; pedicels spreading; flowers yellow to purple; calyx teeth evident, ovate; stylopodium lacking; fruit ovoid or ovoid-oblong, dorsally flattened, glabrous, lateral ribs broadly winged, the dorsal ribs similar to much reduced; oil tubes 1 to several in the intervals, several on the commissure; seeds dorsally flattened in section, the faces plane; carpophores 2-cleft to the base.

1. Pseudocymopterus montanus (A. Gray) Coult. & Rose, Rev. N. Am. Umbell. 74. 1888.
   P. sylvaticus A. Nels.; P. tenuifolius (Gray) Rydb.; P. multifidus Rydb.; P. tidestromii C. & R.; P. versicolor Rydb.; P. purpureus (C. & R.) Rydb.---Stems 20-80 cm. tall, variable; leaves ovate-oblong to broadly ovate in outline; petioles often with scarious or purplish margins; involucel bractlets 2-8 mm. long; rays 5-25; pedicels 1-5 mm. long; fruit 3-7 mm. long.---Mountains. Wyoming to Utah, south to Texas, Mexico and Arizona. Our records scattered in the western two-thirds of Colorado at 7000-12,000 feet.

27. Cymopterus Raf.

Acaulescent or subcaulescent, glabrous or pubescent perennial plants from elongated thickened taproots; subterranean stems often elongated into pseudoscapes bearing at the summit the leaf clusters and peduncles; leaves thin or often somewhat fleshy, variable, from pinnate or bipinnate to pinnately or ternate-pinnately decompound, ultimate divisions entire, toothed or lobed, usually narrow; inflorescence of rather loose to definitely compact compound umbels, these terminal; involucre lacking or of scarious or foliaceous bracts; involucel of conspicuous bractlets or these small and linear, rarely obscure; rays few, spreading; pedicels often very short; flowers white, yellow or purple; calyx teeth small or obsolete; stylopodium lacking; fruit ovoid to oblong, dorsally flattened, glabrous to pubescent, the lateral and usually at least 1 of the dorsal ribs broadly winged; oil tubes 1 or more in the intervals, 2 or more on the commissure; seeds somewhat flattened in section, the face concave to sulcate; carpophore 2-cleft to the base.

1. Wings of fruit narrowed at base (in cross-section), thicker and corky-spongy near middle or near end; leaves not fleshy and not glaucescent; pedicels less than 1 mm. long or lacking entirely; bractlets of involucels conspicuously foliaceous
   2. Peduncles shorter than or at least not exceeding the leaves; bracts usually lacking; flowers usually white; central umbellet pedicellate; rays not over 10 mm. long -1. C. acaulis
   2. Peduncles longer than the leaves; bracts present; flowers usually yellow; central umbellet sterile and sessile; rays often over 10 mm. long
      3. Leaflets about as wide as long; pseudoscapes never developing -2. C. newberryi
      3. Leaflets definitely longer than wide; pseudoscapes 4-20 cm. high -3. C. fendleri
1. Wings of fruit not narrowed at base (in cross-section), never corky-thickened out from base (either wider at base or this about the same width as rest of wing); leaves somewhat fleshy, pallid or glaucescent; pedicels 2 mm. long or more (in all but C. multinervatus doubtful in Colorado); bractlets small to foliaceous
   4. Peduncles when mature shorter than or not exceeding the leaves; wings of fruit (best seen in cross-section) conspicuously enlarged at base; plants of the eastern plains to the foothills -4. C. montanus
   4. Peduncles exceeding the leaves at maturity; wings of fruit little if any larger at base; plants of western Colorado (only C. bulbosus found east of the Continental Divide and then only rarely)
      5. Bractlets of involucels conspicuously many-nerved, often purplish; pedicels obsolete or less than 1 mm. long -5. C. multinervatus
      5. Bractlets few nerved if at all, never purplish; pedicels 2-12 mm. long
         6. Involucre present, conspicuous; involucel bractlets conspicuous, whitish-scarious; seed face (in cross-section) plane or slightly concave
            7. Rays not over 15 mm. long; fruit broadly ovoid, the wings 2-3 times as wide as the body -6. C. purpurascens
            7. Rays often over 15 mm. long; fruit ovoid-oblong to oblong, the wings seldom as wide as the body -7. C. bulbosus
         6. Involucre bracts lacking; involucel of several linear, rather inconspicuous bractlets; seed face (in cross-section) concave to very deeply concave
            8. Plants developing a very conspicuous pseudoscape, this usually fleshy and 5-16 cm. long
               9. Flowers purple; pedicels 2-3 mm. long; fruit 2-3 mm. wide, the lateral wings as long or shorter than the body -8. C. planosus
               9. Flowers yellow (only rarely purplish); pedicels 3-8 mm. long; fruit 3-6 mm. wide, the lateral wings broader than the body -9. C. longipes
            8. Plants with an inconspicuous pseudoscape, this not fleshy and not over 2 cm. long -10. C. purpureus

1. Cymopterus acaulis (Pursh) Raf., Herb. Raf. 40. 1833.
   C. parryi (C. & R.) M. E. Jones; C. lucidus Osterh.; C. leibergii C. & R.---Pseudoscape definite but not very elongated (up to 7 cm. long); plants 3-30 cm. tall; leaves 1-9 cm. long, ovate to oblong-obovate in outline, not fleshy or especially glaucescent; leaflets entire to pinnately lobed; peduncles usually shorter than or equalling the leaves; involucre wanting or rarely very minute; involucel usually of linear, thin, sometimes scarious-margined bractlets;

fertile rays 3-5; pedicels about 1 mm. long or less; flowers white; fruit 3-8 mm. long, ovoid or ovoid-oblong, wings narrowed at base, enlarging outwards, narrow than to about equalling the body; seed face nearly plane to slightly concave.---Dry open ground. Minnesota, Saskatchewan and Oregon, south to Colorado. Our records scattered over Colorado at 4000-7500 feet.
2. Cymopterus newberryi (S. Wats.) M. E. Jones, Zoe 4:47. 1893.
Utah and Arizona. Should be looked for in western and southwestern Colorado.
3. Cymopterus fendleri A. Gray, Mem. Am. Acad. II. 4:56. 1849.
Pseudoscape present but short; plants 4-30 cm. tall; leaves 1-7 cm. long, ovate to oblong-obovate in outline; leaflets longer than broad, entire to pinnately lobed; peduncles usually at least as long as the leaves, often longer; involucre a low sheath or sometimes 1-3 narrow bracts; involucel from subscarious to foliaceous, of linear to ovate-oblong, often 3-toothed bractlets; fertile rays usually 3-5, central umbellet sessile and sterile; pedicels obsolete; flowers yellow; fruit 5-13 mm. long, ovoid or ovoid-oblong, wings narrowed at base and broadened above, about 1/2 as wide as the body; seed face more or less concave.---Dry usually gravelly soil. Utah south to New Mexico, Mexico and Arizona. Our few records from westcentral and southwestern Colorado at 4500-6500 feet.
4. Cymopterus montanus (Nutt.) T. & G., Fl. N. Amer. 1:624. 1840.
Phellopterus montanus Nutt. ; P. macrocarpus Osterh.---Pseudoscape present; plants 5-30 cm. tall; leaves 1.5-8 cm. long, ovate-oblong in general outline, somewhat fleshy and pallid; leaflets entire to pinnately lobed, these lobes usually obtuse and mucronate; peduncles shorter than or merely equalling the leaves; involucre either wanting, a sheath, or of conspicuous bracts; involucel of conspicuous, ovate-oblong bractlets, these white with a brown-green central nerve; fertile rays 3-6; pedicels 2-5 mm. long; flowers white or purple; fruit 5-12 mm. long, ovoid to ovoid-oblong, wings conspicuously enlarged at base, narrowed to apex, about twice as broad as the body; seed face concave.---Dry hills, valleys and plains. South Dakota, south to Texas and New Mexico. Our records scattered in the eastern half of Colorado, one from Saguache County, at 4000-7500 feet.
5. Cymopterus multinervatus (Coult. & Rose) Tide., Proc. Biol. Soc. Wash. 48:41. 1935.
Phellopterus multinervatus C. & R.---Utah and Nevada, south to Texas, Mexico and California. This plant should be expected in the southern part of Colorado.
6. Cymopterus purpurascens (A. Gray) M. E. Jones, Zoe 4:277. 1893.
Phellopterus purpurascens (A. Gray) Coult. & Rose---Acaulescent or subcaulescent plants with pseudoscape, 3-15 cm. high, from long slender taproots, glabrous; leaves 1.2-5 cm. long, ovate-oblong, bipinnate, pinnate or occasionally ternate-pinnate, pallid and somewhat fleshy, ultimate divisions rounded and confluent; peduncles equalling or exceeding the leaves; involucre of conspicuous white bracts, usually connate below; involucel of conspicuous bractlets similar to the bracts, white, with 1-5 conspicuous nerves, equalling or exceeding the flowers; rays 4-15 mm. long, 4-10; pedicels 5-8 mm. long; flowers purplish; fruit 8-18 mm. long, broadly ovoid, wings thin, 2 or 3 times as broad as body, narrow at base; oil tubes 3 or 4 in the intervals.---Dry ground. Idaho and Nevada, south to Arizona and California. Our one record from the southwestern corner of Colorado at about 6000 feet.
7. Cymopterus bulbosus A. Nels., Bull. Torr. Club 26:241. 1899.
Phellopterus bulbosus (A. Nels.) C. & R.; P. camporum Rydb.; P. purpurascens var. eastwoodea (Jones) C. & R.; P. utahensis (M. E. Jones) Woot. & Standl.---Pseudoscape present and definite; plants 5-30 cm. tall; leaves ovate-oblong in outline, somewhat fleshy, pallid; leaflets entire to pinnately lobed; peduncles usually longer than leaves; involucre more or less conspicuous, from a mere sheath to white 1- to 3-nerved bracts; involucel similar to involucre; fertile rays 3-8; pedicels 3-12 mm. long; flowers purplish; fruit 7-17 mm. long, ovoid-oblong to oblong, wings narrow, only slightly enlarging to base, narrower to broader than body; seed face almost plane to very shallowly concave.---Dry hills, mesas and plains. Wyoming, south to Texas and New Mexico. Our records mostly from the western one-third of the state, but also in central and southcentral Colorado, at 4500-8000 feet.
8. Cymopterus planosus (Osterh.) Math., Brittonia 2:245. 1936.
Aulospermum planosum Osterh.---Pseudoscape conspicuous; plants 10-30 cm. tall; leaves 1-6 cm. long, usually oblong in general outline, somewhat fleshy, glaucescent; leaflets entire to pinnately lobed; peduncles longer than leaves; involucre wanting; involucel of several linear bractlets; fertile rays 3-4; pedicels 2-3 mm. long; flowers purple; fruit 5-7 mm. long, not over 3 mm. wide, oblong, wings remaining narrow to base, as wide or narrower than the body; seed face deeply concave.---Dry ground. Limited to Colorado, our records from the northwestern quarter of the state at 6500-8500 feet.
9. Cymopterus longipes S. Wats., Bot. King's Expl. 124. 1871.
Aulospermum longipes (S. Wats.) C. & R.; A. angustum Osterh.; Cogswellia lapidosa (M. E. Jones) Rydb.---Pseudoscape very definite; plants 3-30 cm. tall; leaves 1.5-8 cm. long, ovate-oblong in outline, somewhat fleshy and glaucescent; leaflets pinnately lobed; peduncles equalling or exceeding the leaves; involucre wanting; involucel of several linear bractlets; fertile rays 3-8; pedicels 3-8 mm. long; flowers usually yellow; fruit 5-9 mm. long, oblong to ovoid-oblong, wings remaining rather narrow to base, lateral broader than body, dorsal more or less obsolete; seed face very deeply concave.---Dry hills and valleys. Wyoming and Idaho, south to Colorado and Utah. Our few records from northwestern Colorado at 6500-9500 feet.
10. Cymopterus purpureus S. Wats., Am. Nat. 7:300. 1873.
Aulospermum purpureum (Wats.) C. & R.; Coriophyllus purpureus (Wats.) Rydb.; C. betheli (Osterh.) Rydb.---Pseudoscapes inconspicuous; plants 10-25 cm. tall; leaves 2-13 cm. long, ovate to broadly ovate-oblong in outline, somewhat fleshy and glaucescent; leaflets somewhat remote, toothed or lobed; peduncles taller than the leaves; involucre lacking; involucel dimidiate, of linear bractlets; fertile rays 5-12; pedicels 4-11 mm. long; flowers purple or yellow (rarely whitish); fruit 6-12 mm. long, oblong or ovoid-oblong, wings narrow at base, somewhat broader than body; seed face concave.---Dry hills and plains. Colorado and Utah, south to New Mexico and Arizona. Our records from western Colorado at 4500-7000 feet.

28. Lomatium Raf. BISCUITROOT

Acaulescent to caulescent perennial plants, glabrous to pubescent, from fleshy tubers, fleshy taproots or slender roots; stems tall or low, simple or branching; leaves mostly basal, variously compound or decompound; inflorescence of loose compound umbels these usually solitary and terminal, equalling or exceeding the leaves; involucre none or inconspicuous; involucel of filiform to obovate bractlets, rarely lacking; rays few to many, spreading to strict, those in center of umbel often shorter with sterile umbellets; pedicels slender or stout, long or short, central flower often sterile and sessile; flowers yellow, cream-colored, sometimes purple or white; calyx teeth obsolete or small; stylopodium lacking; fruit linear to orbicular, dorsally flattened, glabrous to pubescent, lateral ribs with thin or corky wings, the dorsal ribs filiform or obsolete; oil tubes 1 or more in the intervals, small or large, 2 or none on the commissure; seeds flattened dorsally in section, the face plane or slightly concave; carpophores 2-cleft to the base.

1. Fruit narrowly oblong, usually at least 3 times longer than wide (never less than 2 times)
  2. Plants strictly acaulescent; leaves once- or twice-pinnate; plants glabrous; oil tubes 3-5 in the intervals -1. **L. nuttallii**
  2. Plants short-caulescent; leaves once- or twice-ternate, each division then 2- to 4-pinnate (hence leaf very decompound); plants glabrous, scabrous or hairy; oil tubes usually less than 3 in the intervals
    3. Plants glabrous or scabrous; ovary and young fruit glabrous; plant from subglobose or somewhat elongated tubers -2. **L. leptocarpum**
    3. Plants villous to tomentose; ovary and young fruit sparingly villous; plants from slender or swollen taproots -3. **L. macrocarpum**
1. Fruit oblong to orbicular, rarely over twice as long as wide (never as much as 3 times)
  4. Leaves twice-ternate, the ultimate divisions over 3 cm. long; fruit often as broad as long -4. **L. simplex**
  4. Leaves pinnate, twice-pinnate or more decompound, the ultimate divisions in any case not over 2 cm. long; fruit definitely longer than wide
    5. Leaves once- or twice-pinnate (lower leaflets may sometimes be lobed), oblong, rarely over 7 cm. long
      6. Leaves scabrous; pedicels 6-17 mm. long; fruit 8-10 mm. long when mature, the wings about as wide as the body -5. **L. eastwoodae**
      6. Leaves glabrous; pedicels 1-2 mm. long; fruit 5-8 mm. long, the wings less than 1/2 as wide as the body -6. **L. concinnum**
    5. Leaves not once- or twice-pinnate, either 3- to 4-pinnate or once- or twice-ternate with the divisions again 2- to 4-pinnate (hence very decompound), often over 7 cm. long
      7. Ovary and young fruit pubescent -7. **L. foeniculaceum**
      7. Ovary and young fruit glabrous
        8. Bractlets of involucel obovate to linear-lanceolate; flowers white; plants of eastern Colorado, west to the eastern foothills -8. **L. orientale**
        8. Bractlets linear, filiform or subulate; flowers purple or yellow; plants of western Colorado, west of eastern foothills
          9. Low plants 12-20 cm. tall; leaves not over 6 cm. long; fruit 5-8 mm. long -9. **L. juniperinum**
          9. Taller plants usually 20 cm. tall or more; leaves usually well over 10 cm. long; fruit 6-16 mm. long
            10. Wings of fruit corky-thickened; leaves more or less puberulent, over 15 cm. long; plants 30-100 cm. tall -10. **L. dissectum**
            10. Wings of fruit thin and membranous; leaves glabrous, rarely over 15 cm. long; plants lower, not over 40 cm. tall -11. **L. grayi**

1. **Lomatium nuttallii** (A. Gray) Macbr., Contr. Gray Herb. 56: 35. 1918.
Cynomarathrum nuttallii (A. Gray) C. & R.---Wyoming to Nevada, south to Nebraska and New Mexico. Should be in Colorado but no specimens from the state seen by the writer.
2. **Lomatium leptocarpum** (T. & G.) Coult. & Rose, Contr. U. S. Nat. Herb. 7:213. 1900.
Cogswellia leptocarpa (Nutt.) M. E. Jones; C. ambigua (Nutt.) M. E. Jones; C. bicolor (S. Wats.) M. E. Jones---Plants 15-50 cm. tall, short-caulescent from tuberous roots, glabrous or scaberulous; leaves 9-14 cm. long, broadly obovate in outline, once- or twice-ternate, then again 2- to 4-pinnate, the ultimate divisions filiform or linear, 5-40 mm. long; involucel of several linear acute bractlets; rays 4-15, rather erect and strict, unequal; pedicels 2-7 mm. long; flowers yellow; fruit 10-15 mm. long and 2-5 mm. wide, glabrous from first, narrowly oblong, wings thin, less than 1/2 the width of the body; oil tubes 1 in the intervals, 2-4 on the commissure.---Hills and plains. Colorado to Idaho, south to Arizona and California. Our records from northwestern Colorado at 7500-10,000 feet.
3. **Lomatium macrocarpum** (H. & A.) Coult. & Rose, Contr. U. S. Nat. Herb. 7:217. 1900.
Cogswellia macrocarpa in Manuals---Plants 10-50 cm. tall, short-caulescent, densely tomentose, villous or glabrate, from slender or swollen taproots; leaves 3-12 cm. long, oblong to obovate in general outline, ternate, then again 2- to 3-pinnate, the ultimate divisions 1-7 mm. long, oblong to linear; involucel of dimidiate, linear-lanceolate bractlets, becoming reflexed; rays 5-25, spreading; pedicels 1-14 mm. long, spreading; flowers white, yellow or purple; fruit 9-20 mm. long and 2-8 mm. wide, narrowly oblong, when young sparingly villous, wings narrower than the body; oil tubes usually 1 in the intervals, 2-6 on the commissure.---Hills and plains. North Dakota, Manitoba and British Columbia, south to Nevada and California. Our rather few records from the northwestern quarter of Colorado (one with a doubtful label from La Plata County) at 6500-8500 feet.
4. **Lomatium simplex** (Nutt.) Macbr., Contr. Gray Herb. 56:34. 1918.
L. platycarpum (Torr.) C. & R.; Cogswellia platycarpa (Torr.) M. E. Jones; C. simplex of Rydb's. Manual---Plants 15-50 cm. tall, caulescent or acaulescent, densely puberulent, clustered from slender taproots; leaves 11-20 mm. long, obovate in general outline, biternate

or middle division pinnate, ultimate divisions 3 cm. or more long, linear, usually glabrous above; involucel of linear or filiform bractlets; rays 8-17, spreading or ascending; pedicels 1-9 mm. long; flowers yellow; fruit 7-14 mm. long and 7-10 mm. wide, broadly oblong to suborbicular, glabrous, wings as broad or even broader than the body; oil tubes solitary in the intervals, 2 on the commissure.---Hills and mountains. Montana to Washington, south to Colorado, Utah and Oregon. Our records from the northcentral and western part of Colorado at 5500-8500 feet.

5. **Lomatium eastwoodae** (Coult. & Rose) Macbr., Contr. Gray Herb. 56:35. 1918.
Cynomarathrum eastwoodae C. & R.---Plants 10-15 cm. tall from subwoody caudex, acaulescent; leaves 3-7 cm. long, scabrous, narrowly oblong in outline, 1- to 2-pinnate; leaflets 2-4 mm. long, remote, oblong-lanceolate; involucel of a few linear, entire bractlets; rays 4-6, ascending, unequal; pedicels 6-17 mm. long; flowers probably yellow; fruit 8-10 mm. long and about 6 mm. wide, oblong, glabrous, the wings about as wide as the body; oil tubes 3 or 4 in the intervals, 6-8 on the commissure.---Hills and plains. Apparently limited to westcentral Colorado, our one record at 4500 feet.

6. **Lomatium concinnum** (Osterhout) Mathias, Ann. Mo. Bot. Gard. 25:276. 1937.
Cogswellia concinna Osterh.---Plants 12-25 cm. tall, short-acaulescent, purplish at least below, from long slender taproots, glabrous; leaves 2-7 cm. long, oblong, bipinnate, the lower leaflets often lobed, ultimate divisions usually rounded at apex; peduncles exceeding the leaves; involucel of conspicuous, dimidiate, ovate-lanceolate, foliaceous bractlets at least as long as the flowers; rays 8-13, spreading, subequal; pedicels 1-2 mm. long; flowers yellow; fruit 5-8 mm. long and 4-5 mm. broad, glabrous, ovate, the wings less than 1/2 the width of the body; oil tubes 4 or 5 in the intervals. A related species L. latilobum (Rydb.) Math. has been reported from western Colorado by Barneby (Leaft. West. Bot. 5:64. 1947). It differs in having the leaves usually once-pinnate instead of twice-pinnate.---Hills and plains. Limited to westcentral and southwestern Colorado. Our few records at 5500-7000 feet.

7. **Lomatium foeniculaceum** (Nutt.) Coult. & Rose, Contr. U. S. Nat. Herb. 7:222. 1900.
Cogswellia foeniculacea (Nutt.) C. & R.; C. villosa (Raf.) Schultes---Plants 10-50 cm. tall, acaulescent, villous to glabrate, from thickened taproots; leaves 5-13 cm. long, ovate to oblong in outline, 3- to 4-pinnate, the ultimate divisions 2-4 mm. long, linear, crowded; involucel of dimidiate, lanceolate, entire or lobed bractlets, connate below and scarious-margined; rays 8-24, spreading or erect, subequal; pedicels 2-13 mm. long; flowers yellow; fruit 7-10 mm. long and 4-6 mm. wide, ovate-oblong, pubescent, the wings narrower than the body; oil tubes 3 in the intervals, 4 on the commissure.---Dry plains and hills. Manitoba, south to Missouri, Oklahoma and Wyoming. Our record from southeastern Colorado at 5500 feet but probably of wider distribution.

8. **Lomatium orientale** Coult. & Rose, Contr. U. S. Nat. Herb. 7:220. 1900.
Cogswellia orientale (C. & R.) M. E. Jones---Plants 10-40 cm. tall, short-caulescent, soft-pubescent or puberulent, caulescent, from long slender taproots; leaves 4-11 cm. long, ovate to oblong in outline, tripinnate, the ultimate divisions 1-12 mm. long, linear; involucel of obovate or linear-lanceolate bractlets, these distinct and scarious-margined; rays 6-20, subequal; pedicels 3-9 mm. long; flowers white; fruit 3-10 mm. long and 3-7 mm. wide, ovate-oblong, glabrous from first, wings rather thin, narrower than the body; oil tubes 1-4 in the intervals, 2-8 on the commissure.---Dry plains and hills. Minnesota to Montana, south to Missouri and Colorado. Our records from the northeastern quarter of Colorado at 5000-9000 feet.

9. **Lomatium juniperinum** (M. E. Jones) Coult. & Rose, Contr. U. S. Nat. Herb. 7:235. 1900.
Cogswellia juniperina M. E. Jones---Wyoming and Idaho, south to Utah. This species may be in northwestern Colorado.

10. **Lomatium dissectum** (Nutt.) Math. and Const., Bull. Torr. Club 69:246. 1942.
Leptotaenia multifida Nutt.; L. eatoni C. & R.---Plants 30-90 cm. tall or more, caulescent or rarely acaulescent, from stout roots; leaves 15-35 cm. long, puberulent or rarely glabrate, deltoid-orbicular, ternate, then 2- to 4-pinnate, the ultimate divisions 2-20 mm. long, linear-oblong; involucre wanting or of a few setaceous or even foliacious bracts; involucel of several linear, entire bractlets; rays many, spreading, subequal; pedicels 4-20 mm. long; flowers purple or yellow; fruit 12-16 mm. long and 6-10 mm. wide, glabrous from the first, oblong-oval, wings very thick and corky, much narrower than the body; oil tubes obscure. Our plants are var. multifidum (Nutt.) Math. & Const.---Hills and mountains. The variety from Alberta to British Columbia, south to Colorado, Arizona and California. Our records from the western half of Colorado at 6000-8500 feet.

11. **Lomatium grayi** Coult. & Rose, Contr. U. S. Nat. Herb. 7:229. 1900.
Cogswellia grayi C. & R.; Leptotaenia filicina M. E. Jones---Plants 20-60 cm. tall, glabrous throughout or scaberulous sometimes, acaulescent or short-caulescent, from a thickened taproot; leaves 10-20 cm. long, broadly obovate in outline, 1- to 2-ternate or quinate, then again 2- to 3-pinnate, ultimate divisions 1-7 mm. long, linear to filiform; involucel of filiform, entire or occasionally toothed bractlets; rays 7-22, spreading; pedicels 6-22 mm. long; flowers yellow; fruit 7-16 mm. long and 5-8 mm. wide, glabrous from the first, ovate-oblong to oblong, wings narrower than or equalling the body; oil tubes solitary in the intervals or rarely 2 or 3, 2-4 or rarely 6 on the commissure.---Dry plains and hills. Wyoming to Washington, south to Colorado and Nevada. Our records from the western one-third of Colorado at 6000-9500 feet.

Family 89. **Cornaceae** DOGWOOD FAMILY

Shrubs or herbaceous perennial plants; leaves opposite or whorled, simple, entire, without stipules; inflorescence capitate or cymose but often flat-topped, sometimes with a petaloid involucre present; flowers perfect, regular, rather small; sepals 4-5, very small or almost absent; petals 4-5, distinct; stamens the same number as the petals and alternate to them; ovary inferior, usually 2-celled, usually with 1 ovule per cell, styles 1, stigmas minute or capitate; fruit a 1- or 2-seeded drupe.

1. Cornus L. DOGWOOD

Characters of the Family

1. Plants herbaceous, not over 25 cm. tall; leaves 4-6 in a single whorl near apex of plant; inflorescence capitate, subtended by 4 petaloid bracts; fruit red -1. C. canadensis
1. Woody shrubs over 25 cm. tall; leaves many and all opposite; inflorescence a flat-topped cyme, never capitate, no petaloid bracts present; fruit white -2. C. stolonifera

1. Cornus canadensis L., Sp. Pl. 118. 1753.
Herbaceous plants from woody rootstocks; stems 5-20 cm. tall; leaves 4-6, apparently whorled in 1 series near apex but usually a pair of smaller or scalelike ones below, leaf blades commonly 3-6 cm. long, ovate or obovate, acute to short-acuminate, cuneate at base, sparsely strigose above, glabrous and lighter in color below; flowers many in a capitate cyme subtended by 4 large white or yellowish bracts, these about 10-15 mm. long and nearly as wide; sepals about 0.4 mm. long; petals 1.5 mm. long, yellowish to purple; fruit about 8 mm. in diameter, globose, red.---In woods. Greenland to Alaska, south to Colorado and California. Our records from northcentral and central Colorado at 7500-11,000 feet.

2. Cornus stolonifera Michx., Fl. Bor. Am. 1:92. 1803.
C. instoloneus A. Nels.; Svida instolonea A. Nels.; S. interior Rydb.; S. stolonifera var. riparia Rydb.---Woody shrubs sometimes to 4 m. tall, the branches often procumbent and rooting at apex; young twigs strigillose to more or less tomentose; leaves variable in length, commonly 5-9 cm. long, lanceolate, elliptic to ovate, acute to acuminate, cuneate, glabrous or nearly so above, strigillose below with spreading hairs along the veins or sometimes hairs spreading on whole lower surface; flowers many in a rather flat-topped cyme; sepals about 0.5 mm. long; petals 2-3 mm. long, white; fruit 7-9 mm. in diameter, whitish. This plant has been called a subspecies of C. sericea L.---Hills, slopes and banks. Newfoundland to Alaska, south to Mexico. Our records well scattered in the western two-thirds of Colorado at 4500-10,000 feet.

Family 90. Ericaceae HEATH FAMILY

Woody shrubs or perennial herbs (none of ours treelike), with or without green chlorophyll; leaves simple, alternate, (may appear whorled or opposite sometimes, or actually be opposite), entire or toothed, evergreen to deciduous or much reduced; flowers usually perfect, regular, symmetrical or nearly so; calyx with 5 or sometimes 4 distinct or united sepals; corolla with 5 or sometimes 4 petals, united to nearly or quite separate; stamens as many as and alternate to petals or twice as many, free from the corolla or nearly so, anthers usually opening by terminal or terminal-appearing pores or chinks, frequently bearing 2 awnlike appendages; ovary superior or inferior, 3- to 10-celled, placentae usually axile, style 1, stigma 1, entire or merely lobed; fruit a capsule or berrylike; seeds small.

1. Plants lacking chlorophyll; leaves reduced and scalelike, not evergreen
    2. Petals distinct; anthers awnless -1. Monotropa
    2. Petals united; anthers with deflexed awns -2. Pterospora
1. Plants with chlorophyll; leaves not reduced to scales, evergreen (in all but Vaccinium and Rhododendron)
    3. Petals distinct or very nearly so; plants herbaceous or somewhat woody at the very base only
        4. Stems 1-flowered -3. Moneses
        4. Stems more than 1-flowered
            5. Stems leafy (though short); flowers corymbose; filaments dilated near base; styles very short or lacking -4. Chimaphila
            5. Stems leafy at base only; flowers in elongated racemes; filaments not especially dilated at base; styles (in most species) over 3 mm. long -5. Pyrola
    3. Petals very definitely united; shrubs (may be undershrubs)
        6. Ovary inferior; fruit berrylike; leaves deciduous -6. Vaccinium
        6. Ovary superior; fruit berrylike or a capsule; leaves evergreen (in all but Rhododendron)
            7. Corolla rotate with 10 pouches enclosing the anthers in the bud; fruit a capsule -7. Kalmia
            7. Corolla urceolate to open campanulate, no pouches present; fruit a capsule or berrylike
                8. Leaves small, not over 1 cm. long and 2 mm. wide; flowers in terminal crowded racemes appearing umbellate; fruit a capsule -8. Phyllodoce
                8. Leaves longer and wider; flowers axillary or in terminal racemes or panicles, not umbellate; fruit a capsule or berrylike
                    9. Leaves deciduous; fruit a dry capsule (not bright red); corolla 10-15 mm. long -9. Rhododendron
                    9. Leaves evergreen; fruit berrylike (usually bright red or scarlet); corolla not over 8 mm. long
                        10. Sepals definitely united at base, the tube becoming fleshy in fruit and enclosing the capsule; anthers not awned; flowers solitary in the axils of the leaves -10. Gaultheria
                        10. Sepals nearly or quite distinct, not becoming fleshy (the ovary becoming a berry); anthers with a recurved dorsal awn; flowers terminal in short racemes or panicles -11. Arctostaphylos

1. Monotropa L. INDIANPIPE; PINESAP

Plant lacking chlorophyll, white to reddish when fresh, often turning blackish in drying, arising from a balllike cluster of matted roots; stems fleshy with reduced scalelike leaves; flowers perfect, solitary or in a raceme; sepals 2-5, separate and bractlike; petals 3-6, erect, scalelike, often saccate at base, distinct; stamens 6-12, twice as many as the petals, anthers more or less reniform, 2-valved; ovary superior, 4- to 5-celled, style erect, stigma rather disklike, a disk present at base of ovary, this 8- to 12-toothed; fruit a capsule; seeds numerous.

1. Flowers 1 to a stem, 15-20 mm. long -1. M. uniflora
1. Flowers several to many to a stem, not over 14 mm. long -2. M. hypopitys

1. Monotropa uniflora L., Sp. Pl. 387. 1753.
Deep woods. Newfoundland to British Columbia and south to Florida and California. No record as yet from Colorado but to be expected.
2. Monotropa hypopitys L., Sp. Pl. 387. 1753.
Hypopitys latisquama Rydb.; H. multiflora of C. & N. Manual---Plants pink or reddish, sometimes yellowish; stems 10-30 cm. tall, usually more or less pubescent, moderately stout; scales (leaves) 1-1.5 cm. long on upper, lower smaller, rather thick, usually broadly ovate; flowers in a raceme, 5-merous in upper ones, often 3- or 4-merous in lower; sepals about 8-11 mm. long, ciliate; petals 10-12 mm. long, cuneate or obovate, pubescent, saccate at base; capsule 5.5-7 mm. long, globose to ovoid-globose, usually 4- or 5-celled; style about 4-5 mm. long. Kearney and Peebles recognize a variety in our western plant which is larger and more inclined to be red or pink- M. hypopitys var. latisquama (Rydb.) Kearney and Peebles, Journ. Wash. Acad. Sc. 29:487. 1939.---Woods. Montana to British Columbia south to Mexico and California. Apparently uncommon in Colorado, collected in Douglas and Mineral Counties at about 7000-8000 feet.

2. Pterospora Nutt. PINEDROPS

Roots densely matted, forming round mats often 5 cm. in diameter, plants without chlorophyll, purplish-brown or chestnut-colored; stems stout; leaves reduced or scalelike; flowers perfect, many, pendulous in elongated racemes; calyx of 5 slightly united sepals; corolla urceolate, globose, the petals united with short spreading lobes; stamens 10, included, anthers longitudinally dehiscent, bearing deflexed awns; ovary superior, depressed-globose, 5-celled; style stout; capsule 5-lobed; seeds numerous, broadly winged at apex.

1. Pterospora andromedea Nutt., Gen. Pl. 1:269. 1818.
Stems 20-80 cm. tall, simple, stout, glandular-hairy; leaves 1-3.5 cm. long, thick, lanceolate to linear-lanceolate, crowded near base of stem; racemes erect, with bracts narrow, often longer than the pedicels; calyx 3-6 mm. long; corolla 6-8 mm. long, white; capsules 8-12 mm. wide, spheroidal.---In rich woods. Quebec to British Columbia south to Pennsylvania, Mexico and California. Scattered over the western two-thirds of Colorado at 7000-9500 feet. but no records as yet from the northwest part.

3. Moneses Salisb. WOODNYMPH

Glabrous perennial herbaceous plants with slender rootstocks; stems short, the plants scapose, scapes 1-flowered, the flower perfect, nodding; sepals 5 (rarely 4); petals 5 (rarely 4) distinct, widely spreading; stamens 10 (rarely 8), anthers opening by pores at the 2-horned base (appearing as apex); disk not present; ovary superior, style straight, stigma peltate with 5 narrow lobes, broader than the style, ovary 4- to 5-celled; loculicidal capsule 5-angled, depressed- globose, the valves not cobwebby on margins; seeds many.

1. Moneses uniflora (L.) A. Gray, Man. 273. 1848.
Stems 1-3 cm. long above ground, the scape 5-12 cm. tall; leaves about 1-4, orbicular or oval, rounded at apex, cuneate at base to a petiole 5-30 mm. long, crenate; sepals about 3 mm. long, ovate, obtuse, ciliolate on margins; petals about 1 cm. long, white, ovate; capsule 7-8 mm. wide. ---In deep woods. Greenland to Alaska, south to Pennsylvania, Arizona and Oregon; Eurasia. Scattered over the mountainous area of Colorado at 8500-11,500 feet.

4. Chimaphila Pursh PIPSISSEWA

Herbaceous or suffruticose perennials with creeping rootstocks; stems leafy but short; leaves evergreen, coriaceous, subverticillate; flowers perfect, in corymbs or racemes; sepals 5; petals 5, distinct; stamens 10, filaments expanded near base, anthers opening by pores at basal (appearing apical) 2-horned end; ovary superior, 5-celled, 5-lobed, style very short, stigma peltate, flat; capsule loculicidally dehiscent from top, the valves not cobwebby on margins; seeds many.

1. Chimaphila umbellata (L.) Bart., Veg. Mat. Med. U. S. 1:17. 1817.
C. occidentalis Rydb.---Plants 10-30 cm. tall; leaves 3-9 cm. long, broadly oblanceolate to nearly obovate with margins serrate except toward cuneate base, apex acute; sepals 5-6 mm. long, broadly ovate to deltoid; petals about 5-6 mm. long, oval, concave and ciliolate; capsule 6-7 mm. in diameter, subglobose. Our plant is var. occidentalis (Rydb.) Blake. Blake (Rhodora 19:237-244, 1917) makes var. acuta (Rydb.) Blake from N. Mexico and Arizona and this could be in

Colorado. However, his key differences especially as to leaves and inflorescences will not work for us as some of our northern Colorado plants are intergrades or key to var. acuta.---Dry woods. The var. occidentalis from Montana to British Columbia south to Colorado and California. Scattered over the mountainous part of Colorado at 8000-11,500 feet.

5. Pyrola L. PYROLA

Perennial herbaceous plants with slender rootstocks; leaves basal, evergreen, plant more or less scapose; flowers perfect, in racemes, several, nodding; calyx 5-parted; petals 5, distinct or very slightly united; stamens 10, anthers opening by pores at the somewhat 2-horned base (appearing as apex); disk usually not present; ovary superior, 5-celled, stigma 5-lobed or 5-rayed; fruit a 5-lobed, loculicidal capsule, the valves with cobwebby margins when opening; seeds numerous.

1. Style straight, narrow right out to the peltate stigma which is much wider; stamens not declined but the filaments straight or nearly so
   2. Style short, 1-2 mm. long; raceme not 1-sided; hypogynous disk not present   -1. P. minor
   2. Style 4-5 mm. long; racemes 1-sided, hypogynous disk present at base of ovary   -2. P. secunda
1. Style deflexed at base then curved upward, gradually thickened to a truncate collar, the stigma smaller than this; stamens curved upwards
   3. Petals pink to rose-purple; sepals often definitely longer than wide
      4. Leaves suborbicular to round-reniform, usually cordate or subcordate at base   -3. P. asarifolia
      4. Leaves elliptic, obovate to suborbicular, the base rounded to broadly cuneate
                                                               -3A. P. asarifolia purpurea
   3. Petals white or greenish-white; sepals about as wide as long
      5. Leaves mottled with white above along the veins, usually acute at apex   -4. P. picta
      5. Leaves not mottled above, usually rounded or obtuse at apex
         6. Leaves 1-3 cm. long, usually shorter than their petioles   -5. P. chlorantha
         6. Leaves 3-6 cm. long, longer than their petioles   -6. P. elliptica

1. Pyrola minor L., Sp. Pl. 396. 1753.
Erxlebenia minor (L.) Rydb. ---Stems above ground very short, the scape 10-20 cm. tall; leaves 1-4 cm. long on petioles about 1-2 cm. long, from very near base of plant, rather thin, oval or suborbicular, crenulate rounded or somewhat acute; racemes short, not especially 1-sided; sepals about 1.5 mm. long, broadly triangular; petals about 3-4 mm. long, white or rose-colored, orbicular, no tuberclelike nectaries present; stamens straight; style about 1-2 mm. long, erect, straight; stigma peltate much broader than the styles, no hypogynous disk present at base of ovary; capsule about 5-6 mm. wide, depressed-globose.---In woods. Labrador to Alaska south to New England, Colorado and California; Eurasia. Scattered in the mountainous part of Colorado at 8000-12,000 feet.

2. Pyrola secunda L., Sp. Pl. 396. 1753.
Ramischia secunda (L.) Garcke.---Stems often slightly woody above ground, leaf-bearing portion 1-10 cm. high, scapes 10-20 cm. tall; leaves 1-5 cm. long, on petioles 1-2 cm. long, few, rather thin, ovate to oval, acute or sometimes rounded at apex, finely serrulate to crenulate; racemes 1-sided; sepals about 1-1.5 mm. long, oval or elliptic; petals 4-5 mm. long, each with 2 tubercles inside, oblong or elliptic, greenish-white; stamens straight; style about 4-5 mm. long, erect, stigma peltate, much broader than the style; ovary with a 10-lobed hypogynous disk present; capsule about 4 mm. wide, depressed-globose.---Woods. Labrador to Alaska, south to New Jersey, New Mexico and California; Eurasia. Scattered over the mountainous part of Colorado at 5500-10,000 feet.

3. Pyrola asarifolia Michx., Fl. Bor. Am. 1:251. 1803.
Stems above ground very short, the scapes 10-25 cm. tall, sometimes taller; leaves 2-5 cm. long, suborbicular to round-reniform, usually cordate or subcordate at base, thick, dark green and shiny above, usually brownish beneath, rather obscurely crenulate, on petioles 2-7 cm. long; racemes not especially 1-sided; sepals about 1/4 to 1/3 the length of the petals, ovate to ovate-lanceolate, often definitely longer than wide; petals about 5 mm. long, pink to rose-purple; stamens curved upward; style about 7-9 mm. long, deflexed at base then curved upward, thickened gradually upward to a truncate collar, stigma narrower than this collar; fruit 7-8 mm. wide, depressed-globose.---Wet woods and swamps. Nova Scotia to Yukon, south to Massachusetts and New Mexico. Scattered in the mountainous part of Colorado at 8000-11,000 feet.

3A. Pyrola asarifolia purpurea (Bunge) Fernald, (var.) Rhodora 51:103. 1949.
P. asarifolia var. incarnata (DC.) Fernald; P. uliginosa Torr.---Differs from species in shape of leaves, these elliptic to obovate or suborbicular with base broadly cuneate or rounded. Intergrades with species, in several cases both types of leaves being on same plant! But the extremes are different.---Scattered over the western half of the state at 6500-12,000 feet.

4. Pyrola picta Smith in Rees., Cycl. 29:No. 8. 1814.
Stems (leaf-bearing portion) 1-5 cm. long, the scapes 10-20 cm. tall; leaf-blades 1.5-6 cm. long, ovate or rounded-ovate, denticulate or entire, upper surface green but white-mottled along the veins, lower pale or tinged with red, on petioles 1-4 cm. long; racemes not especially 1-sided, 1-3 scales present below; sepals about 1.5 mm. wide and long, ovate-deltoid, acute; petals about 4-6 mm. long, greenish-white or brownish inside; stamens curved upward; style little if any longer than the petals, deflexed at base, then curved upwards, gradually thickened to a truncate collar, the stigma smaller than this.---Woods. Montana to British Columbia, south to New Mexico and California. Apparently rare in Colorado, our one specimen from Ouray County at 7800 feet.

5. Pyrola chlorantha Swartz, Svenska Vetensk. Akad. Handl. 31:190. 1810.
Stems above ground very short, the scape 10-20 cm. tall; leaves 1-3 cm. long on petioles usually longer than the blade (often longer and shorter on same plant), near ground, usually

suborbicular but sometimes orbicular, reniform to elliptic or oblong-ovate (often on same plant), rounded to broadly cuneate at base, obscurely crenulate; racemes short, not especially 1-sided; sepals triangular, about 1.5 mm. long and about as wide; petals 5-7 mm. long, greenish-white; stamens curved upwards; style deflexed at base then curved upwards, equal to or somewhat longer than the petals, gradually thickened upward to the collar, stigma narrower than this collar; fruit about 6 mm. wide depressed-globose. The correct name for this plant may be P. virens Schweigg.---In wet woods. Labrador to British Columbia, south to Maryland, Arizona and California; Europe. Scattered in the mountainous part of Colorado at 7500-12,000 feet.
6. Pyrola elliptica Nutt., Gen. Pl. 1:273. 1818.
 Closely related to P. chlorantha and intergrading with it. Doubtfully distinct. Supposed to have leaves 3-6 cm. long, blades shorter than petioles. A specimen from southwestern Colorado may be this species.

6. Vaccinium L. BLUEBERRY; WHORTLEBERRY

Low shrubs; leaves alternate, deciduous; flowers perfect, solitary or few together, rather small; sepals 5 or 4, very small, persistent; gamopetalous corolla more or less urceolate, 5- or 4-lobed or toothed, white to rose; stamens twice as many as the lobes of the corolla, anthers included, prolonged upward into tubes, opening by terminal pores or chinks, often awned on the back; ovary inferior, 4- to 5-celled or 8- to 10-celled by false partitions, ovules several to many in each cavity, ovary crowned at top by an epigynous disk, style straight, stigma small; fruit berrylike, sweet and edible, many-seeded.

1. Branches terete (many wrinkled irregularly when dry); fruit blue with a bloom   -1. V. caespitosum
1. Branches strongly angled longitudinally; fruit red, black or blue-black
  2. Fruit red; grooves of branches usually glabrous; leaves usually less than 12 mm. long   -2. V. scoparium
  2. Fruit black or blue-black; grooves of branches usually puberulent; leaves usually over 12 mm. long
                                                                                        -3. V. myrtillus

1. Vaccinium caespitosum Michx., Fl. Bor. Am. 1:234. 1803.
 Plants 5-30 cm. tall, branchlets glabrous or obscurely puberulent, terete; leaves 12-25 mm. long, obovate to oblong-oblanceolate, glabrous, serrulate, obtuse or acute, cuneate and nearly sessile; flowers mostly solitary in axils of leaves, pedicels shorter than the corolla; calyx lobes nearly obsolete; corolla 3-5 mm. long, pink or white; fruit 6-8 mm. wide, blue with a bloom.---Labrador to Alaska, south to Maine and Colorado. In the mountains of Colorado especially in the north, at 8500-12,000 feet.
2. Vaccinium scoparium Leiberg, Mazama 1:196. 1897.
 V. erythrococcum Rydb.---Plants 10-20, or rarely to 30 cm. tall; branches strongly angled, the grooves commonly glabrous; leaves 1-1.5 cm. long, glabrous, ovate, oval or lanc-ovate, acute or obtuse at apex, usually broadly cuneate at base and nearly sessile, margins serrulate; flowers usually solitary and axillary; calyx lobes nearly obsolete; corolla about 3-4 mm. long; fruit about 5 mm. wide, red.---Alberta to British Columbia, south to California and Colorado. Our records from the northwestern, northcentral and central part of the state at 8500-11,500 feet.
3. Vaccinium myrtillus L., Sp. Pl. 349. 1753.
 V. oreophilum Rydb.---Plants 10-30 cm. tall; branchlets strongly angled, the grooves usually puberulent; leaf blades 1-2.5 cm. long, ovate, oval or oblong-elliptic, glabrous, serrulate, acute at apex, broadly cuneate at base, nearly sessile; flowers mostly solitary in the leaf axils; calyx lobes nearly obsolete; corolla about 4 mm. long, pink or white; fruit 5-8 mm. wide, black or blue-black.---Alberta to British Columbia south to New Mexico and Arizona. Scattered over the mountainous part of Colorado at 8000-12,000 feet.

7. Kalmia L. LAUREL; KALMIA

Low diffusely branching shrubs; leaves evergreen, opposite (in ours), coriaceous entire; flowers perfect, in terminal corymbs or umbels; calyx deeply 5-parted, persistent, leathery; petals united, corolla rotate, 5-lobed with 10 pouches below the lobes and their sinuses; stamens 10, anthers awnless, opening by terminal pores, at first enclosed in the corolla pouches; ovary superior, 5-celled, style slender; fruit a globose septicidal capsule, 5-valved. (The expansion of the opening corolla curves the filaments outward and backward under a tension. Irritation as by an insect causes these elastic filaments to release the anthers with considerable force.)

1. Kalmia polifolia Wang., Beob. Ges. Nat. Freunde Berlin 2:130. 1788.
 K. microphylla (Hook.) A. Heller---Shrubs to 20 cm. high; leaves 1-1.5 cm. rarely to 2.5 cm. long, oval to elliptic-oblong, entire and more or less revolute on margins and sometimes appearing narrower, acute or obtuse at apex, cuneate at base, nearly or quite sessile, dark green above, glaucous-white beneath; sepals about 2.5-3 mm. long, ovate, concave; corolla 10-15 mm. across, about 8-10 mm. long, lilac-purple to light rose. Our plant is var. microphylla (Hook.) Rehd.---Cold bogs. The variety from Alberta to Alaska south to Colorado and California. Our few records from northcentral Colorado at 9000-11,500 feet.

8. Phyllodoce Salisbury MOUNTAIN HEATH

Evergreen low branching shrubs; stems with short glandular bristles; leaves crowded, alternate, thick, 2-grooved below and appearing revolute, linear, almost sessile; flowers perfect, racemose-crowded and appearing in umbels, rose-colored (in ours), pedicels glandular each with 2 bracts at base; sepals normally 5, persistent; corolla 5-lobed, campanulate to urn-shaped; stamens included, 10 (rarely less), anthers dehiscent by oblique terminal chinks, awnless; ovary superior, 5-celled, style filiform, stigma minutely capitate or obscurely 5-lobed; capsule globose, septicidally dehiscent from the summit, with many seeds.

1. Phyllodoce empetriformis (Smith) D. Don, Edinb. New Phil. Journ. 17:160. 1834.
Swamps and moist places. Alberta to Alaska, south to California and Colorado. Recorded in this state from Clear Creek County but no specimen could be located. It would be in subalpine or alpine areas if present.

9. Rhododendron L. RHODODENDRON; AZALEA

Shrubs with erect branches; leaves alternate, petioled, entire, evergreen or deciduous (in ours); flowers perfect, in terminal umbellike racemes or (in ours) solitary or in pairs from lateral buds; calyx 5-lobed, foliaceous and persistent; corolla open-campanulate or rotate, 5-lobed, these slightly unequal but corolla not 2-lipped; stamens 10 (in ours), anthers opening by terminal pores; ovary superior, 5-celled (in ours) style slender with capitate or peltate stigma this 5-lobed; fruit a septicidal capsule.

1. Rhododendron albiflorum Hook., Fl. Bor. Am. 2:43. 1834.
Azaleastrum albiflorum (Hook.) Rydb.; A. warrenii A. Nels.---A shrub described as not over 30 cm. tall or with us reaching 2 m., bark of branches gray, twigs brown; leaves 1-2 cm. long, elliptic, ovate, oval or obovate, more or less hirsute and glandular-ciliate; flowers solitary or several crowded together; sepals narrowly elliptic to narrowly obovate or oblanceolate, covered with glandular hairs; corolla 10-15 mm. long, white or pink-tipped, with orbicular lobes.---Woods. Alberta to British Columbia, south to Oregon and Colorado. Has been collected several times in western Jackson and eastern Routt County at 9000-11,000 feet.

10. Gaultheria L. CREEPING WINTERGREEN

Depressed procumbent undershrubs (ours); leaves alternate, evergreen; flowers perfect, axillary and solitary; calyx 5-cleft or 5-parted, persistent; corolla campanulate, 5-toothed or lobed; stamens 10, included, filaments dilated below, anthers opening by terminal pores, sometimes 2-awned; ovary superior, depressed, 5-lobed and 5-celled, seated on a thin undulate-toothed disk; calyx becoming fleshy and at length enclosing the capsule forming a globose, berrylike fruit.

1. Gaultheria humifusa (Graham) Rydb., Mem. N. Y. Bot. Gard. 1:300. 1900.
Forming loose mats, the procumbent stems 5-20 cm. long, glabrous or somewhat pilose or puberulent; leaves 1-2 cm. long, orbicular or oval, nearly entire to crenulate or serrulate, sometimes setose-serrulate especially toward the apex, obtuse or somewhat acute at apex, rounded at base to a short petiole; pedicels about 2-3 mm. long with 3-5 small but thick bractlets; calyx glabrous, about as long as the corolla; corolla about 3-5 mm. long, white; anthers obscurely 4-pointed at apex, not awned; fruit 5-6 mm. wide, scarlet.---Woods. Alberta to British Columbia, south to California and Colorado. In this state rather uncommon in the central and northcentral part at 10,000-11,500 feet.

11. Arctostaphylos Adans. BEARBERRY; KINNIKINICK; MANZANITA

Shrubs, the branches prostrate to erect, often with thin exfoliating polished looking bark; leaves evergreen, coriaceous; flowers perfect, in terminal racemes or panicles, these often short; sepals 5, nearly or quite distinct, persistent; corolla urceolate with 5 short recurved lobes; stamens 10 included, filaments dilated and usually hairy at base, anthers opening by terminal pores with 2 recurved dorsal awns; hypogynous disk present at base of ovary; ovary superior, 4- to 10-celled, a single ovule in each cell, style slender; fruit berrylike with 4-10 seedlike nutlets, irregularly separating or united into a single stone.

1. Stems erect or only somewhat spreading; plants 80 cm. tall or more; most of leaves widest below the middle; fruit not bright red; western Colorado only -1. A. patula
1. Stems decumbent or spreading; plants not over 60 cm. tall; most of leaves widest above the middle; fruit red; western Colorado and over the state
   2. Stems with bark soon rough and shreddy (before stems become 3 mm. in diameter); leaves seldom over 2 cm. long; branches strongly rooting, rarely over 20 cm. off the ground; all over mountainous Colorado -2. A. uva-ursi
   2. Stems with bark remaining smooth (until over 3 mm. in diameter); leaves usually 2 cm. long; branches rooting but usually not conspicuously so, the plant 30-60 cm. tall from ground, limited to western Colorado -3. A. nevadensis coloradensis

1. Arctostaphylos patula Greene, Pitt. 2:171. 1891.
A. platyphylla (A. Gray) Rydb.; A. pinetorum Rollins---Erect shrubs 1-2 m. tall, forming dense mats; stems erect or somewhat spreading, intricately and widely branched, with dark brownish-red bark, smooth (at least on all but real old and large ones), branchlets glandular-puberulent; leaves 2-4 cm. long, ovate to nearly orbicular, obtuse or acutish at apex, rounded to truncate or even subcordate at base, thick, margins entire, glabrous or glandular, petioles glandular-puberulent; inflorescence paniculate, bracts and rachises somewhat glandular-puberulent; bracts about 1-3 mm. long, subulate to lanceolate, acute; pedicels glabrous 2-6 mm. long; sepals 1.5-2 mm. long, orbicular to broadly ovate, rounded at apex; corolla 5-8 mm. long, white to rose; ovary glabrous; fruit about 8-10 mm. wide, depressed-globose, glabrous, creamy-white

to yellowish-brown, nutlets variously separable. Dr. W. A. Weber has found what appears to be hybrids between this and A. nevadensis var. coloradensis on the Uncompahgre Plateau.---Open ground and forest openings. Colorado to Oregon, south to California. In western Colorado at 7000-9000 feet.

2. Arctostaphylos uva-ursi (L.) Spreng., Syst. Veg. 2:287. 1825.

Main stems prostrate, these rooting, forming mats, bark exfoliating before reaching 3 mm. diameter, the erect branches up to 15 cm. long; branchlets with tomentulum or viscid-villous hairs; leaves 1-2.5 cm. long, oval to obovate or oblanceolate or oblong-spatulate, rounded to subacute at apex, cuneate at base, margins entire and often somewhat revolute, glabrous and shining; flowers in short few-flowered racemes, rachis often pubescent; bracts small; pedicels 3-4 mm. long, glabrous; sepals about 1 mm. long, broadly ovate, acute; corolla 4-5 mm. long, white to pink; ovary glabrous; fruit 6-10 mm. wide, bright red, globose, nutlets separable. Fernald and Macbride (Rhodora 16:21-213. 1914) recognized 3 plants in North America.

1. Branchlets minutely tomentulose, commonly somewhat viscid, soon losing all the hairs
-A. uva-ursi (L.) Spreng.
2. Branchlets with dense canescent tomentum, this persistent at least for several years, not viscid
-A. uva-ursi var. coactilis Fern. & Macbr.
3. Branchlets viscid-villous commonly intermixed with stipitate black glands
-A. uva-ursi var. adanotricha Fern. & Macbr.

They cited the first two from Colorado. But Adams (Elisha Mitchell Sci. Soc. Journ. 56:16. 1940) cited the last one from Colorado.---Woods. Greenland to Alaska, south to Virginia, New Mexico and California. Rather widely distributed in the mountainous part of the state at 6000-10,000 feet.

3. Arctostaphylos nevadensis coloradensis (Rollins) comb. nov. (var.)

A. coloradensis Rollins---Shrubs 30-60 cm. high forming dense beds; stems spreading or decumbent, often rooting, intricately branching, bark red, smooth, finally exfoliating, branchlets glandular-puberulent; leaves 2-3 cm. long, obovate to nearly oblanceolate, cuneate at base, rounded or somewhat cuspidate at apex, glabrous or somewhat glandular; petioles 4-7 mm. long, glandular-puberulent; inflorescence congested-paniculate; rachis glandular-puberulent; bracts about 2-3 mm., lanceolate the lower somewhat foliacious; sepals about 1-1.5 mm. long, orbicular; corolla 5-7 mm. long, pink to light rose; fruit about 7-8 mm. wide, globose, bright red; nutlets more or less coalescent. This plant is closely related to the western A. nevadensis A. Gray and further study may indicate that the 2 are the same entity.---Hillsides and forest openings. Known only from western Colorado at 7500-8500 feet but a collection from Routt County appears to be this species.

Family 91. Plumbaginaceae PLUMBAGO FAMILY

Perennial, caespitose, acaulescent or subacaulescent plants; leaves rosulate but alternate, narrow, sessile; scape with 1 head, this spherical to hemispheric, subtended by bracts, the outer reflexed and united into a sheath below the head; head of much condensed cymules and with perfect flowers; calyx of 5 partly united sepals; petals 5, regular, united; stamens 5, opposite the petals and adnate to their bases; ovary superior or nearly so with 5 styles slightly united at base, ovary 1-celled, 1-ovuled; fruit a 1-seeded utricle.

1. Armeria Willd. THRIFT

Characters of the Family

1. Armeria labradorica Wallr., Beitr. Heft 2:185. 1844.

A. vulgaris Willd. and Statice armeria L. in part---Plants caespitose; leaves 3-10 cm. long and to 1.8 mm. wide, similar, narrowly linear, acute; scape 3-20 cm. tall, glabrous to pubescent, sheath shorter than the head; head 1-2 cm. wide; petals pink, purple or whitish.---On mountains and sea coasts. Hudson Bay, Labrador, Newfoundland and Quebec. Collected on Hoosier Pass between Park and Summit Counties at 12,000-13,000 feet, representing an amazing range extension.

Family 92. Primulaceae PRIMROSE FAMILY

Plants herbaceous, annual or perennial, caulescent or acaulescent; leaves simple, entire or toothed, alternate, opposite or whorled; inflorescences various; flowers perfect, regular; sepals commonly 5, more or less united; petals usually 5, gamopetalous but sometimes cleft to base (wanting entirely in Glaux); stamens as many as petals and opposite them, inserted on the corolla tube (except in Glaux); ovary superior or partly inferior (in Samolus), 1-celled with a free central placenta, style and stigma 1; fruit a capsule.

1. Ovary partly inferior, its base definitely adnate to the calyx tube; plant caulescent -1. Samolus
1. Ovary superior, not at all adnate to the calyx tube; plant caulescent or acaulescent
  2. Corolla lacking, the calyx lobes petaloid -2. Glaux
  2. Corolla present
    3. Plants acaulescent or nearly so, the leaves radical or clustered near base of plant; scapes terminated by an umbellike inflorescence (rarely reduced to 1 flower)
      4. Corolla lobes reflexed; stamens exserted, connivent in a cone and somewhat monadelphous -3. Dodecatheon

4. Corolla lobes erect or merely spreading, stamens included, distinct
   5. Flowers red, lilac or purple (rarely white), showy with corollas over 8 mm. long, their tubes exceeding or equalling the calyx   -4. Primula
   5. Flowers white or pinkish, not showy, the corollas less than 8 mm. long, their tubes shorter than the calyx   -5. Androsace
3. Plants caulescent, with leafy stems; flowers axillary, solitary or in racemes
   6. Plants perennial with rootstocks; capsules longitudinally dehiscent; flowers yellow, in racemes or if solitary their corollas over 6 mm. long   -6. Lysimachia
   6. Plants annual; capsules circumscissle; flowers white, pink or scarlet, solitary with corollas less than 6 mm. long
      7. Flowers on peduncles 1 cm. long or more; leaves opposite or whorled; corolla normally scarlet, longer than the calyx   -7. Anagallis
      7. Flowers sessile or nearly so; leaves alternate (may be opposite below); corolla white to pink, shorter than the calyx   -8. Centunculus

1. Samolus L. WATERPIMPERNEL; BROOKWEED

Glabrous perennial caulescent, herbaceous plants; leaves alternate, entire; flowers in terminal racemes or panicles; calyx campanulate, 5-cleft; corolla nearly campanulate, 5-lobed or 5-parted; stamens 5, on the tube of the corolla, filaments short, staminodia inserted in sinuses of the corolla well above the fertile stamens; ovary partly inferior; capsule glabose or ovoid, 5-valved from the summit; seeds many.

1. Samolus floribundus H. B. K., Nov. Gen. 2:224. 1817.
   In wet places. Newfoundland to British Columbia, south to Florida, Texas and California; reported from Mexico and South America. We have no specimens from Colorado as yet but the plant should be here.

2. Glaux L. SEA MILKWORT

Somewhat succulent caulescent herbaceous perennials; leaves opposite, entire, sessile; flowers small, axillary, sessile or nearly so; calyx 5-lobed, campanulate, petaloid; corolla lacking; stamens 5, inserted at the base of the calyx and alternate to its lobes, anthers cordate; ovary superior; capsule 5-valved at the top, globose-ovoid and beaked, few-seeded.

1. Glaux maritima L., Sp. Pl. 207. 1753.
   Plant glabrous and rather pale green or glaucouslike; stems erect or ascending, 5-30 cm. tall from slender rootstocks; leaves 4-12 mm. long, oval to linear-oblong; calyx 3-4 mm. long, white or pinkish with lobes oval and rounded at apex.---In saline or subsaline moist soil. Newfoundland to Alaska, south to New Jersey, Nebraska and California. Our records sparsely scattered over the western three-fourths of the state, mostly in the north, at 4800-8500 feet.

3. Dodecatheon L. SHOOTING STAR

Scapose perennial herbaceous plants with short rootstocks; leaves in a basal rosette; flowers umbellate or solitary, showy, terminating the scapes; calyx 5-parted (sometimes 4-parted), reflexed in the flower; corolla 5-parted (sometimes 4-parted) the tube short, throat dilated and the narrow divisions reflexed; stamens usually 5, filaments short and flat, united at least in younger flowers, inserted on the throat of the corolla; ovary superior with filiform, exerted style; capsules many-seeded.

1. Anthers over 3 times longer than the filament tube (anthers usually 6-7 mm. long, tube about 1.5 mm.); leaves usually over 6 cm. long, the margins often sinuate   -1. D. radicatum
1. Anthers only 3 times longer than the filament tube or much less (anthers usually about 5-5.5 mm. long, tube usually about 2 mm. long); leaves seldom over 6 cm. long, margins usually entire   -2. D. pulchellum

1. Dodecatheon radicatum Greene, Erythea 3:37. 1895.
   D. multiflorum Rydb.; D. philoscia A. Nels.; D. sinuatum Rydb.---Glabrous plants; scapes 15-40 cm. tall, 1- to 20-flowered; leaves 4-20 cm. long, usually over 6 cm., from narrowly oblanceolate to elliptic, entire or sinuate on margins; calyx 4-10 mm. long; corolla about 12-20 mm. long, purple-rose, a dark wavy line in the throat, yellow in the tube; filament tube 1-2 mm. long, usually around 1.5 mm.; anthers 4-8 mm. long, usually around 6-7 mm. and over 3 times longer than the filament tube in any case; capsules opening by short teeth at apex. Intergrades some with D. pulchellum, but relative length of anther and filament tube combined with leaf size will separate almost all specimens.---Wet meadows and shaded slopes. South Dakota and Wyoming south to New Mexico and Arizona. Scattered over the mountainous part of Colorado at 5500-11,000 feet, apparently not common in the west.

2. Dodecatheon pulchellum (Raf.) Merr., Journ. Arn. Arb. 29:212. 1948.
   D. pauciflorum (Durand) Greene---Scapes 10-40 cm. tall; leaves 3-10 cm. long but commonly not over 6 cm., glabrous, oblanceolate or linear-oblanceolate, entire; flowers 1-10; calyx about 5-6 mm. long; corolla about 12-18 mm. long, purple-rose, a dark wavy line in the throat, yellow in the tube; filament tube 1.5-2 mm. long, usually about 2 mm.; anthers 4.5-6 mm. long, not over 3 times as long as the filaments in any case and usually less; capsules opening at apex by the short teeth.---Moist or wet meadows and shaded banks. Saskatchewan to British Columbia, south to Colorado, Arizona and California. Scattered over the western two-thirds of this state at 6000-10,500 feet.

4. Primula L. PRIMROSE; COWSLIP

Perennial scapose herbaceous plants; leaves all in a basal rosette; flowers solitary or umbellate (in ours); calyx tubular, funnelform or campanulate usually angled, 5-lobed; corolla funnelform or salverform, the tube as long or longer than the calyx, showy; stamens 5, distinct, on the tube or throat of the corolla; ovary superior, style filiform, stigma capitate; fruit a 1-celled capsule, 5-valved at the apex; seeds many.

1. Lower surface of leaves whitish-farinose; involucre bracts gibbous at base; corolla lobes each again deeply cleft, the sinuses extending about 1/4 the length of the lobe -1. P. incana
1. No part of plant farinose; involucre not gibbous at base; lobves of corolla merely notched at apex
  2. Plant over 10 cm. tall; inflorescence 3- or more-flowered, the flowers usually over 18 mm. long -2. P. parryi
  2. Plants shorter; flowers usually solitary or sometimes 2 on a scape, flowers smaller -3. P. angustifolia

1. Primula incana Jones, Proc. Cal. Acad. Sc. Ser. 2, 5:706. 1895.
P. americana Rydb.---Perennial, scapose plants; leaves 1.5-8 cm. long, 0.5-2 cm. broad, shallowly denticulate to sub-entire, obtuse or nearly so, farinose below; scape 10-45 cm. tall, strict; involucre bracts 6-10 mm. long, linear to linear-oblong, flat, gibbous at base; calyx 6-8 mm. long, farinose to some degree, nearly equalling corolla tube, lobes oblong, obtuse or somewhat acute; corolla tube about 7 mm. and limb about 3 mm. long, lilac, lobes obcordate, the lobe about 1/4 way in limb length; capsule only slightly exceeding the calyx.---Swamps and wet meadows. Saskatchewan to Alberta south to Colorado and Utah. Scattered in the central mountains, mostly in central Colorado, at 7000-10,000 feet.
2. Primula parryi Gray, Am. Journ. Sc. 34:257. 1862.
Perennial, scapose plants; leaves 6-30 cm. long, spatulate-oblong to narrowly oblanceolate, obtuse or acute, rather fleshy, often somewhat puberulent, not farinose; scape 8-40 cm. tall, erect; involucre bracts 3-12 mm. long, lanceolate; calyx 7-15 mm. long, its lobes 5-8 mm. long, lanceolate-acuminate; corolla 12-20 mm. long, deep red (or purple in drying), the tube about as long or a little longer than the limb, the limb 1.5-3 cm. broad, the lobes notched; capsule 7-11 mm. long.---Wet or moist ground often along streams. Montana to Idaho, south to Colorado and Arizona. Scattered at high elevations over Colorado at 9500-13,000 feet.
3. Primula angustifolia Torr., Ann. Lyc. N. Y. 1:34 pl. 3, fig. 3. 1824.
Plants 0.5-7 cm. tall; leaves 1.5-5 cm. long, 2-7 mm. wide, lanceolate-spatulate to linear-lanceolate, obtuse, entire, not farinose (but slightly short glandular-hairy on some), rather thick; involucre bracts 1-7 mm. long, lanceolate, scape 1- or sometimes 2-flowered; calyx 5-8 mm. long, lobes 2-3 mm. long, acute or acuminate; corolla tube 5-8 mm. long, limb 5-6 mm. long and 1-2 cm. broad, purple or occasionally white; capsule ovoid.---Alpine areas of the southern central Rocky Mountains. Scattered in the higher mountains of Colorado at 8000-14,000 feet.

5. Androsace L. ROCKJASMINE

Small annual, biennial or perennial plants with basal leaves, these petioled or very nearly sessile; inflorescence umbellate, open or contracted; calyx 5-lobed to 5-parted, persistent; corolla 5-lobed, salverform or funnelform, often shorter than the calyx; stamens 5, included, filaments short; ovary superior, style short, stigma capitate; capsule 5-valved, few- to many-seeded; seeds minutely pitted.

1. Caespitose perennials; umbels subcapitate, the pedicels not over 5 mm. long; corolla showy, the lobes 2-4 mm. long, longer than the sepals; leaves dimorphic, the inner ones broader; capsule few-seeded -1. A. carinata
1. Annuals or occasionally short lived perennials; umbels never capitate, the pedicels over 5 mm. long; corolla not showy, the lobes seldom over 2 mm. long, shorter than or slightly longer than the sepals (except in fruit); leaves approximately alike; capsule 5- to many-seeded
  2. Sepals broadly triangular, each 3-nerved, flat; calyx tube not 5-angled; capsule much exceeding the calyx at maturity -2. A. filiformis
  2. Sepals narrowly triangular to subulate, not 3-nerved, carinate or inrolled from sides; calyx tube definitely 5-angled; capsule included in the calyx
    3. Involucre bracts ovate to lanceolate-ovate
      4. Scapes 1-25; umbel 2- to 10-flowered -3. A. occidentalis
      4. Scapes mostly solitary; umbel 1- to 4-flowered -3A. A. occidentalis simplex
    3. Involucre bracts narrowly lanceolate to subulate
      5. Plants mostly with 1 well-developed scape, lateral scapes small or absent; pedicels with glandular hairs -4A. A. septentrionalis glandulosa
      5. Plants usually with several to many well-developed scapes; pedicels and scapes glabrous or hairy, not glandular
        6. Calyx lobes 1/4 to 1/3 the length of the tube; scapes 7-25 cm. tall; below 10,000 feet in Colorado and glabrous except for minute puberulence on base of scapes and on leaves -4B. A. septentrionalis subulifera
        6. Calyx lobes 1/3 to 1/2 the length of the tube; scapes 1-10 cm. long; often above 10,000 feet in Colorado, if below then plant definitely puberulent
          7. Plant glabrous except minute puberulence at base of scape and on leaves; plants not over 5 cm. tall, above 10,000 feet in Colorado -4C. A. septentrionalis subumbellata
          7. Plants more or less puberulent all over, often on the sepals too; plants 2-10 cm. tall, above and below 10,000 feet in Colorado -4D. A. septentrionalis puberulenta

1. **Androsace carinata** Torr., Ann. Lyc. Nat. Hist. N. Y. 1:30, 31. 1824.

Drosace carinata (Torr.) A. Nels.---Caespitose perennial plants, 1-5 cm. tall from slender taproots; scapes more or less arachnoid-pilose, solitary from close rosulate leaf clusters, leaves dimorphic, outer linear-lanceolate to lanceolate, the inner lanceolate to oblanceolate, glabrous to arachnoid-pilose or silky but ciliate; involucre bracts lanceolate to linear-lanceolate; umbel subcapitate, 2- to 5-flowered, the pedicels 2-4 mm. long; calyx broadly campanulate, lobes ovate or lanceolate, shorter than the tube; corollas showy, the lobes 2-4 mm. long, white to cream-colored with an orange, yellow or pinkish "eye"; capsules few-seeded.---High mountains. Colorado, Utah and probably northward. Our plants in southcentral, central and northcentral Colorado at 10,500-13,000 feet.

2. **Androsace filiformis** Retz., Obs. Bot. fasc. 2:10-11. 1781.

A. capillaris Greene---Annual or short lived perennial plants, glabrous or sparingly glandular-hairy; scapes 5-13 cm. tall, several to many, the lateral ones often arching or flexuous; leaves 1-3 cm. long, ovate or ovate-lanceolate abruptly narrowed to a definite petiole, denticulate; bracts minute, subulate; pedicels 1-5 cm. long, subequal, filiform, usually many; calyx tubes hemispheric, the lobes 3-nerved, foliaceous, broadly triangular, becoming papery in age; corolla white, minute at first, finally exceeding the calyx; capsule much exceeding the calyx.--- Wet margins of lakes and streams. Montana to Washington, south to Colorado and Utah; Asia. Our records from northcentral Colorado at 6500-9500 feet.

3. **Androsace occidentalis** Pursh, Fl. Am. Sept. 1:137. 1814.

Annual plants; scapes 1-10, 3-10 cm. tall, lateral ones often arching or flexuous; leaves ovate-lanceolate to linear, entire or sparingly toothed, only slightly petiolate, margins and upper surface with short stiff hairs; bracts lanceolate-ovate to ovate, acute; pedicels 1- to 10-flowered, unequal but over 5 mm. long; calyx tube whitish contrasting with green lobes, 5-angled, subcampanulate to obpyramidal, lobes rounded, not flat, 1/3 to 1/2 the length of the tube, calyx more or less puberulent; corolla shorter than calyx, white, minute; capsule included in calyx tube.---Rather dry ground. Mississippi River valley and in Rocky Mountains from Montana and British Columbia south to New Mexico and Arizona. Our Colorado records from northcentral and western parts at 4500-6000 feet.

3A. **Androsace occidentalis simplex** (Rydb.) Robbins, (forma) Am. Midl. Nat. 32:153. 1944.

A. simplex Rydb.---A form with solitary scapes, and 1- to 4-flowered umbels. Not too well marked in our plants.---Our records from northern Colorado at 5000-6000 feet.

4. **Androsace septentrionalis** L., Sp. Pl. 142. 1753.

Annual or short-lived perennial plants with slightly thickened root crown; scapes 2-30 cm. tall, 1 to many, glabrous, puberulent or glandular-hairy; leaves linear to linear-oblanceolate, entire to denticulate, somewhat rounded below but not abruptly narrowed to a definite petiole, glabrous to puberulent; bracts subulate to narrowly-lanceolate; umbel compact to diffuse, pedicels usually over 2 cm. long; calyx campanulate, the tube 5-ridged, whitish-green, the lobes ridged or rounded, usually greenish, subulate to narrowly lanceolate, sometimes foliose in fruit, 1/4 to 1/2 as long as the tube, calyx often puberulent; corolla shorter or slightly exceeding the calyx, white to pinkish; fruit globose, included in the calyx tube. The species proper is a northern form not found in Colorado. The varieties intergrade rather commonly, especially var. subulifera and var. puberulenta.

4A. **Androsace septentrionalis glandulosa** St. John, (var.) Victoria Mem. Mus. Memoir 126:47. 1922.

A. glandulosa Woot. & Standl.---Scapes 1 or a few short lateral ones present; pedicels and scapes glandular-hairy.---Mountains. Colorado, New Mexico and Arizona. Central and southcentral Colorado at 7000-10,000 feet.

4B. **Androsace septentrionalis subulifera** Gray, (var.) Syn. Fl. N. Amer. 2:60. 1878.

A. diffusa Small; A. pinetorum Greene---Several to many scapes present, glabrous or nearly so; calyx lobes 1/4 to 1/3 the length of the tube.---In wooded areas. Wyoming to British Columbia, south to New Mexico and Arizona. Widely scattered and rather abundant over the western two-thirds of Colorado at 5000-9500 feet.

4C. **Androsace septentionalis subumbellata** A. Nels.; (var.) Bull. Wyo. Exp. Sta. 28:149. 1896.

A. subumbellata (A. Nels.) Small---Scapes several to many, glabrous or nearly so, seldom over 5 cm. tall; calyx lobes 1/3 to 1/2 the length of the tube.---Mountains. Alberta to British Columbia, south to Colorado and California. Scattered in the mountainous part of Colorado at 8000-12,000 feet, more common at the higher elevations.

4D. **Androsace septentrionalis puberulenta** (Rydb.) Knuth., (var.) Engl., Pflanzenreich. 4:216. 1905.

A. puberulenta Rydb.---Several to many scapes present; plants 2-10 cm. tall, more or less puberulent; caylx lobes 1/3 to 1/2 the length of the tube.---Hills and mountains. Alberta to British Columbia, south to New Mexico and Arizona. Scattered and rather common in the mountainous two-thirds of Colorado at 7000-12,500 feet.

6. Lysimachia L. LOOSESTRIFE

Perennial herbaceous caulescent plants (ours with rootstocks); leaves opposite or appearing whorled, entire; flowers solitary or in terminal or axillary racemes; sepals 5-7, nearly distinct; corolla of 5-7 petals (rarely more or less), deeply parted, rotate to funnelform; stamens 5-7, often appearing somewhat united below, sometimes alternating with staminodia; ovary superior; capsules few- to many-seeded. Includes the genera Steironema and Naumburgia following Allen (Rhodora 22:193-194. 1920).

1. Flowers solitary; corolla lobes obovate; leaves on conspicuously ciliate petioles -1. **L. ciliata**
1. Flowers in racemes; corolla lobes linear; leaves sessile -2. **L. thyrsiflora**

1. Lysimachia ciliata L., Sp. Pl. 147. 1753.
Steironema ciliatum (L.) Raf.---Stems glabrous, 30-100 cm. tall; leaves 5-12 cm. long, ovate, lanceolate or oblong-ovate, acute to acuminate at apex, rounded truncate or subcordate at base, glabrous except the ciliate margin; petioles about 1-2 cm. long, conspicuously ciliate; flowers solitary, axillary, on slender nodding peduncles; sepals about 5-6 mm. long, linear-lanceolate, acuminate; corolla 7-12 mm. long, yellow, deeply 5-parted, the lobes obovate, erose-denticulate; stamens 5, with narrowly triangular, long-acuminate staminodia between bases of filaments; capsules 10- to 20-seeded.---Moist ground. Canada south to Georgia, New Mexico and Washington. Widely distributed over the state at 3500-8000 feet.
2. Lysimachia thyrsiflora L., Sp. Pl. 147. 1753.
Naumburgia thyrsiflora (L.) Duby.---Stems 25-60 cm. tall, simple, erect, glabrous or slightly hairy; leaves 5-15 cm. long, glabrous, sessile, lanceolate or linear-lanceolate, acute at apex, cuneate at base, lighter beneath, smaller below and finally reduced to ovate scales; flowers in axillary crowded racemes 1-2.5 cm. long on long slender peduncles; sepals about 2-2.5 mm. long, linear-lanceolate, purple-spotted (in ours); corolla 3-4 mm. long, deeply 5- to 7-parted (rarely more or less) into linear, yellow, purple-spotted divisions; stamens 5-7, rarely if ever with teeth or staminodia between the filament bases; capsule few-seeded.---In water or moist ground. Nova Scotia to Alaska, south to Pennsylvania, Colorado and California; Eurasia. Our few specimens from northcentral Colorado at 5000-7500 feet.

7. Anagallis L. POORMANS WEATHERGRASS; PIMPERNEL

Annual or rarely perennial, caulescent, herbaceous plants; stems spreading or decumbent; leaves mostly opposite or whorled, entire; flowers axillary, peduncled; calyx deeply 5-cleft into spreading persistent lobes; corolla rotate, deeply 5-parted or lobed, the lobes rotate or spreading; stamens 5, filaments sometimes hairy, adnate to base of corolla; ovary superior; capsule circumscissle, subglobose, with many seeds.

1. Anagallis arvensis L., Sp. Pl. 148. 1753.
Waste places. Newfoundland to Alaska, south to Florida, Texas and California. We have no record for Colorado but the plant should be expected.

8. Centunculus L. CHAFFWEED; FALSE PIMPERNEL

Small, annual, caulescent, herbaceous plants; leaves alternate (or lower opposite) entire; flowers minute, solitary, axillary, sessile or nearly so; calyx 4- (or 5-) parted, the lobes narrow; corolla 4- (or 5-) cleft, the tube subglobose, the lobes spreading; stamens 4 or 5, inserted in the throat of the corolla, anthers cordate; ovary superior; capsule subglobose, circumscissle; seeds many.

1. Centunculus minimus L., Sp. Pl. 116. 1753.
Moist soil. Minnesota to British Columbia, south to Florida, Texas and Arizona; South America and Europe. No records from Colorado but it should be expected.

Family 93. Oleaceae OLIVE FAMILY

Undershrubs, shrubs or trees, woody, at least at base; leaves opposite (at least below) deciduous, simple or pinnately compound, without stipules; flowers small, regular, perfect or unisexual; calyx lobes 4 to 15; corolla absent or if present gamopetalous, rotate and 5-lobed (in ours), or more rarely of 1 or 2 separate petals; stamens 2-4, adnate to base of corolla when this is present; ovary superior, 2-celled, style 1, stigma capitate or 2-lobed; fruit a samara, capsule or drupe.

1. Corolla present, conspicuous (over 7 mm. long); plants woody only at base, not over 35 cm. tall; fruit a capsule
　　-1. Menodora
1. Corolla absent or reduced to 1 or 2 minute petals; plants definite shrubs ordinarily over 100 cm. tall; fruit a drupe or samara
　　2. Fruit a samara; leaves variable but of the ovate type, some usually well over 2 cm. wide; twigs large, young ones usually over 2 mm. in diameter　　-2. Fraxinus
　　2. Fruit a drupe; leaves variable but of the oblanceolate type, rarely any over 2 cm. wide; twigs moderate, the young ones rarely as much as 2 mm. in diameter　　-3. Forestiera

1. Menodora H. & B. MENODORA

Shrubs or undershrubs, woody only at very base; leaves opposite or alternate especially above, simple, entire, sessile; flowers in subcorymbose cymes (in ours), perfect; calyx deeply cleft with 7-12 lobes (in ours); corolla 5-lobed (rarely 6-lobed), subrotate (in ours) and bright yellow; stamens 2; ovules 4 (rarely 2) in each cell; capsule deeply 2-parted, circumscissle.

1. Menodora scabra (Engelm.) A. Gray, Am. Journ. Sc. ser. 2, 14:44. 1852.
Undershrubs, woody at very base, nearly herbaceous; stems 7-35 cm. tall, not spinescent (resembling species of flax), sparsely to moderately scabro-puberulent; leaves 1-3 cm. long, opposite, or alternate especially above, linear to lanceolate or lower broader, scabro-puberulent especially on margins and midrib; calyx 4-10 mm. long, lobes narrow often unequal; corolla 7-12 mm. long, lobes longer than the tube; capsule 5-7 mm. long, 7-12 mm. wide.---Dry ground. Texas

to southern Utah and Arizona. Our few records from Fremont and Pueblo Counties at 5300-5700 feet.

2. Fraxinus L. ASH

Trees or shrubs; leaves deciduous, simple or rarely 3-foliate (in ours); flowers polygamous (in ours) in panicles; calyx minute, 4-lobed or 4-toothed; corolla none (in ours); stamens 1-3 (usually 2); fruit a 1-seeded samara. Besides our native species any of the common ash trees with pinnately compound leaves may be planted in Colorado. (F. americanus L; F. pennsylvanica Marsh.; F. pennsylvanica var. lanceolata Sarg.) These may persist for years in abandoned areas or possibly escape at times.

1. Fraxinus anomala Torr. ex Wats. in King, Geol. Expl. 40th Par. 5:283. 1871.
Shrubs or small trees about 2-8 m. tall, twigs 4-angled; leaves 2-5 cm. long, simple (or sometimes 3-foliate), glabrate, ovate to rhombic-elliptic, cuneate to subcordate at base, rounded to acute or acuminate at apex, entire or somewhat crenate; samaras 12-25 mm. long and about 8-10 mm. wide, oblong, cuneate at base, emarginate to acute at apex, winged all around.---Canyons and slopes. Colorado to California south to New Mexico and Arizona; perhaps Texas. Westcentral and southwestern Colorado at 4500-6000 feet.

3. Forestiera Poir. FORESTIERA; ADELIA

Shrubs; leaves opposite, simple; flowers inconspicuous, dioecious or polygamo-dioecious, in small fascicles or panicles, appearing before the leaves; calyx minute or none; corolla wanting or rarely of 1 or 2 small petals; stamens 2-4; fruit a 1-seeded drupe.

1. Forestiera neomexicana A. Gray, Am. Acad. Arts & Sci. Proc. 12:63. 1876.
Adelia neomexicana (A. Gray) Kuntze---Glabrous upright shrubs 1-3.5 m. tall (or perhaps taller according to report) branches somewhat spinescent; leaves 1-4 cm. long, spatulate-oblong, to obovate and oblanceolate or rhombic-elliptic, cuneate at base, entire or serrulate-crenate, obtuse or obtusely-acuminate, short-petioled, often fascicled; drupe 6-8 mm. long, black or blue-black.---Dry valleys. Colorado to California, south to western Texas and Arizona. Westcentral and southwestern Colorado at 4500-7000 feet.

Family 94. Menyanthaceae BUCKBEAN FAMILY

Perennial herbaceous plants with creeping rootstocks; leaves alternate but basal, long-petioled and 3-foliate (in ours); flowers perfect, regular, racemose or paniculate; calyx deeply 5-parted; corolla gamopetalous short-funnelform, 5-lobed or 5-cleft; stamens 5, inserted on the corolla tube and alternate to its lobes, anthers sagittate; ovary superior, 1-celled with 2 parietal placentae, 1 style, stigma 2-lobed; fruit capsulelike but indehiscent or finally rupturing irregularly, seeds few. This family is often united with Gentianaceae but is kept separate here following Lindsey (Am. Journ. Bot. 25:480-485. 1938.).

1. Menyanthes L. BUCKBEAN; BOGBEAN

Characters of the Family

1. Menyanthes trifoliata L., Sp. Pl. 145. 1753.
Leaflets 5-10 cm. long, oval to elliptic, thick, glabrous, entire, each cuneate to a sessile base; petioles 5-25 cm. long, sheathing at base; calyx very short; corolla about 10-15 mm. long, white or tinged with rose or purple, bearded within.---In water or very moist ground. Newfoundland to Alaska, south to Pennsylvania, Colorado and California. Apparently rare in Colorado our 4 records from La Plata, Garfield, Boulder and Larimer Counties at 8500-9200 feet.

Family 95. Gentianaceae GENTIAN FAMILY

Mostly glabrous, herbaceous plants with colorless, bitter juice; leaves normally opposite or verticillate, cauline sessile, often somewhat connate, basal sometimes narrowed to a petiolelike base, simple, entire, lacking stipules; flowers perfect, regular (at least the corolla) solitary or in cymose clusters; calyx of 2 to 5 (sometimes more) more or less united sepals; corolla of 4 to 5 (rarely more) more or less united petals, the lobes commonly convolute in the bud; stamens the same number and alternate to the corolla lobes, the filaments partly adnate to the tube; ovary superior, 1-celled with 2 parietal placentae (these often broad, making ovules appear to be in a solid ring); style 1 or none, stigma usually 2-lobed; fruit a 2-valved capsule; seeds usually many.

1. Stigmas as 2 decurrent bands on the sutures of the ovary; styles none; corolla rotate   -1. Pleurogyne
1. Stigmas on end of style or sessile at apex of ovary, never in decurrent bands down the sides; corolla rotate, salverform, cylindrical, funnelform or campanulate
   2. Corolla lobes with 1 or 2 conspicuous fringed glands and pits toward the base; corolla rotate, the lobes over twice as long as the short tube   -2. Swertia
   2. Corolla lobes without conspicuous glands or pits at base (lobes may be fringed at margin or at base in Gentiana); corolla salverform, campanulate, funnelform or cylindrical, lobes variable, seldom over twice as long as the tube

3. Corolla salverform, pink, white or yellowish; anthers twisting spirally after shedding pollen   -3. <u>Centaurium</u>
3. Corolla campanulate, funnelform, or cylindrical; anthers not especially twisting
    4. Styles very short or none; lobes seldom over 1/2 the length of the corolla (never 3/4 and often much less than 1/2)   -4. <u>Gentiana</u>
    4. Styles definite (nearly as long as the ovary); lobes about 3/4 the length of the corolla   -5. <u>Eustoma</u>

1. Pleurogyne Eschol. FELWORT

Slender annual glabrous plants; leaves opposite; flowers solitary or in narrow racemes or panicles; calyx deeply 4- to 5-parted (nearly divided) the lobes narrow, often unequal; corolla rotate, deeply 4- to 5-parted, lobes convolute, acute, with a pair of scalelike appendages at base; stamens on the short corolla tube, hence appearing basal; anthers appearing versatile but sagittate at 1 end and filament attached between the lobes; style none, stigmas very peculiar, being decurrent lines on the sutures of the ovary; capsules narrowly oblong or lanceolate, not stipitate; seeds many.

1. Pleurogyne rotata (L.) Griseb., Gent. 309. 1839.
   <u>P. fontana</u> A. Nels.---Stems 10-40 cm. tall, slender, mostly simple; leaves 2-3 cm. long, linear or basal oblong-oblanceolate; sepals 7-12 mm. long, linear; corolla lobes 7-12 mm. long, whitish or tinged with blue, elliptic-oblong to ovate-lanceolate, acute, about as long or somewhat shorter than the sepals.---Mountain bogs. Greenland to Alaska and south to Colorado. Our records from northcentral, central and southcentral Colorado at 8500-10,500 feet.

2. Swertia L. SWERTIA

Perennial (often short-lived) caulescent plants; stems rather tall and erect; leaves opposite or whorled (occasionally alternate); flowers in panicles (cymose); calyx 4- or 5-parted or divided; corolla rotate, 4- or 5-lobed or divided, each lobe bearing 1 or 2 conspicuous fringed glands; stamens inserted on the base of the corolla; style from very short or none to definite, stigma 2-lobed; capsule coriaceous, more or less flattened, commonly surrounded by the marcescent corolla; seeds rather large, marginal or winged.

1. Corolla normally blue or purple (may be streaked or spotted lighter or darker), the lobes normally 5; style none; glands 2 to a petal; stem leaves sometimes alternate   -1. <u>S. perennis</u>
1. Corolla greenish-white or yellowish-green (may be purple-dotted), the lobes normally 4; style present; glands 1 or 2 to a petal; stem leaves opposite or whorled
    2. Glands 2 to a petal; anthers 3-4 mm. long; plant typically over 60 cm. tall; leaves not white-margined   -2. <u>S. radiata</u>
    2. Glands 1 to a petal; anthers 1-1.5 mm. long; plant below 60 cm. tall; leaves definitely white-margined
       3. Leaves whorled, 3 or more at a node; fovea (glands) lobed at apex; crown of a few cilia; pedicels in ours usually over 2 cm. long   -3. <u>S. albomarginata</u>
       3. Leaves opposite; fovea (glands) lobed at base; crown of a definite scale; pedicels less than 2 cm. long   -4. <u>S. coloradensis</u>

1. Swertia perennis L., Sp. Pl. 1:226. 1753.
   <u>S. scopulina</u> Greene; <u>S. congesta</u> A. Nels.; <u>S. palustris</u> A. Nels.---Stems single from a short rootstock, 10-30 cm. tall, glabrous; leaves glabrous, mostly basal, obovate, oblanceolate, lanceolate, or elliptic, basal petioled, stem leaves sessile, opposite or alternate; inflorescence thrysoid, the flowers in the axils of the upper reduced leaves or bracts, on pedicels of varying lengths; calyx 5-8 mm. long, deeply 5-parted (rarely 4 or 6); corolla solid blue, or streaked or spotted darker or lighter, 5-parted (rarely 4 or 6) the lobes 8-12 mm. long, longer than the calyx, 2 glands at the base of each lobe, these fringed, less than 1 mm. long; anthers 1-2 mm. long.---Moist ground. Canada, extending southward to Colorado; Europe and Asia. Scattered in the mountains (no records from westcentral or northwestern part) at 9000-12,500 feet.

2. Swertia radiata (Kellogg) Kuntze, Rev. Gen. Pl. 2:430. 1891.
   <u>Frasera speciosa</u> (Griseb.); <u>F. speciosa</u> var. <u>scabra</u> M. E. Jones; <u>F. scabra</u> (Jones) Rydb.; <u>F. stenosepala</u> Rydb.; <u>F. speciosa</u> Dougl.; <u>F. angustifolia</u> Rydb.; <u>Tesseranthium scabrum</u> (Jones) Rydb.; <u>T. stenosepalum</u> (Rydb.) Rydb.; <u>T. speciosum</u> (Dougl.) Rydb.; <u>T. angustifolium</u> (Rydb.) Rydb.---Plants 30-150 cm. tall; stems stout, erect, single from a stout taproot, glabrous to puberulent; radical and lower cauline leaves 10-30 cm. long, the basal longer, obovate to narrowly oblong, puberulent and cauline leaves whorled in 3's to 7's, lanceolate to linear, sessile; flowers on an elongated many-flowered thryse, in the axils of leaflike bracts; calyx deeply parted, the lobes 12-30 mm. long, linear; corolla deeply parted, the lobes 15-25 mm. long, often shorter than the calyx, greenish-white, frequently dotted with purple, 2 fringed glands on each lobe, these 3-7 mm. long; anthers 3-4 mm. long. This plant varies in size of parts. However, the tall form (about 1 m. tall) is the characteristic and conspicuous plant in Colorado.---In mountains. South Dakota to Washington, south to New Mexico and California; Mexico. Scattered in the mountainous two-thirds of Colorado at 7000-10,000 feet.

Some of our plants have glabrous leaves (but puberulent bracts). They have been called S. radiata var. macrophylla (Greene) St. John, but since most specimens have no real leaves present the variety has little value in checking herbarium material.

3. Swertia albomarginata (Wats.) Kuntze, Rev. Gen. Pl. 2:431. 1891.
   <u>Frasera albomarginata</u> Wats.; <u>Leucocraspedum albomarginatum</u> (Wats.) Rydb.---Plants 20-60 cm. tall, glabrous; stems single or few from a taproot; leaves 4-10 cm. long, the basal longer, whorled in 3's or 4's, linear to oblanceolate, inclined to be conduplicate, and appearing undulate on some, light green with white margins; panicle broad with lower branches elongate and thus corymbose, pedicels usually over 2 cm. long, bracts opposite and subulate; calyx deeply

4-parted, the lobes 3-5 mm. long, narrow; corolla 4-parted, greenish-white or greenish-yellow, the lobes 7-10 mm. long, cuspidate-acuminate, gland single, elongated, fringed, emarginate at apex; anthers 1-1.5 mm. long.---Dry ground. Colorado to California, south to Arizona. Our few records from extreme southwestern Colorado at 6500-7500 feet.

4. Swertia coloradensis Rogers, Madrono 10:108. 1949.

Stems 10-20 cm. long, several to many, caespitose, ascending or spreading from a larger taproot, puberulent; leaves 5-20 cm. long, opposite, thick, minutely puberulent to glabrate, basal leaves longer and tapering to petioledlike bases, white-margined, linear-oblanceolate to linear-lanceolate; inflorescence corymbose; calyx about 7-12 mm. long, deeply 4-parted, lobes linear-lanceolate; corollas cream-colored, not spotted, its lobes 8-10 mm. long, often convolute near the apex, fovea lobed at base, crown of a definite scale; anthers 1-2 mm. long.---Dry rocky slopes and knolls. Known to the author only from southeastern Colorado at 4000 feet, but the cited specimens at 5000-5500 feet.

3. Centaurium Hill. CENTAURY

Plants annual or biennial; stems branched, leafy; leaves opposite, sessile; flowers in terminal cymes; calyx 4- or 5-parted, its lobes narrow and keeled; corolla 4- or 5-parted or lobed, salverform or somewhat funnelform, pink, rose or yellowish; stamens inserted on the throat of the corolla, anthers commonly exserted, twisting spirally after shedding their pollen; style filiform, deciduous, stigma 2-lobed.

1. Corolla lobes 7-8 mm. long, at least 3/4 as long as the tube　　　-1. C. calycosum
1. Corolla lobes about 3-5 mm. long, about 1/2 as long as the tube　　　-2. C. exaltatum

1. Centaurium calycosum (Buckl.) Fernald, Rhodora 10:54. 1908.

C. arizonicum (A. Gray) Heller; Erythraea arizonica (Gray) Rydb.---Stems angled, 20-40 cm. tall, simple or somewhat racemosely branched; leaves about 2-5 cm. long, narrowly oblong, lanceolate or linear, the basal sometimes spatulate; pedicels longer than the calyx; calyx lobes about 7-8 mm. long; corolla tube about 10-15 mm. long, lobes about 8-10 mm. long, 4 or 5, narrowly-oblong (on ours). Our plants may be distinct enough to be called Centaurium calycosum var. arizonicum (Gray) Tidestrom.---River valleys. Texas to Nevada south to northern Mexico. Our few records from westcentral Colorado at 5000-6000 feet.

2. Centaurium exaltatum (Griseb.) W. F. Wight, Contrib. U. S. Nat. Herb. 11:449. 1906.

Nebraska to Washington, south to Utah and California. This species should be in Colorado but no specimens could be found by the writer.

4. Gentiana L. GENTIAN

Annual or perennial plants; stems mostly erect, simple or sparingly branched; leaves opposite and sessile; flowers terminal or axillary, solitary or in cymose clusters, these often forming narrow leafy panicles; calyx 4- or 5-cleft to nearly divided; corolla cylindrical, campanulate, funnelform or salverform, with 4 or 5 lobes, lobes shorter than the rest of corolla, often with teeth or plaited folds in the sinuses; stamens inserted on the corolla tube at the middle or below; style short or none, stigmas 2; capsule sessile or stipitate; seeds many.

1. Corolla plicate and folded in the sinuses, these more or less extending upward into teeth or lobes between the true corolla lobes; plants perennial or biennial (sometimes annual in G. prostrata and G. fremontii)
    2. Flowers solitary and terminal, not over 18 mm. long; plants rarely over 10 cm. tall
        3. Capsules at maturity long-stipitate and trumpet-shaped, long-exserted from the corolla tube (stipe to 2 cm. long); leaves and calyx lobes with conspicuous white margins　　　-1. G. fremontii
        3. Capsules at maturity short-stipitate, not exserted; leaves and calyx lobes slightly if at all white-margined　　　-2. G. prostrata
    2. Flowers usually more than 1, over 20 mm. long; plants rarely less than 10 cm. tall
        4. Corolla nearly closed at apex, the lobes nearly obsolete, apex of plaits incurved　　　-3. G. andrewsii
        4. Corolla open at apex, lobes present, spreading or ascending
            5. Corolla yellowish-white, spotted or streaked with purple, over 3 cm. long; seeds lamellose-rugose　　　-4. G. romanzovii
            5. Corolla blue to purple, less than 3 cm. long (except in G. parryi); seeds smooth or nearly so
                6. Flowers 1-6 to a stem, in terminal or subterminal clusters; corollas campanulate to funnelform, 3-4 cm. long; floral bracts ovate to ovate-lanceolate, often somewhat scarious　　　-5. G. parryi
                6. Flowers several to many, some usually axillary and well below the apex of the stem; corolla funnelform, 2-3 cm. long; floral bracts lanceolate or narrower, not scarious
                    7. Calyx very irregular, cleft on 1 or both sides, resembling a spathe, lobeless or with subulate teeth much less than 1/3 the length of the tubes　　　-6. G. forwoodii
                    7. Calyx nearly regular, not spathaceous, with well developed lobes at least 1/3 as long as the tubes
                        8. Bracts lanceolate to linear, seldom as long as the flower; upper leaves ovate to lanceolate; corolla usually over 2.5 cm. long　　　-7. G. affinis
                        8. Bracts linear, usually longer than the flowers; upper leaves linear or linear-lanceolate; corolla usually 2-2.5 cm. long　　　-8. G. bigelovii
1. Corollas not plicate or folded in the sinuses, no extra lobes extending up between the true corolla lobes; plants annual (all but G. barbellata)
    9. Corolla lobes usually 4, with erose or fimbriate margins; corollas usually over 2 cm. long, not fringed in the throat at the base of lobes; flowers solitary or terminating long peduncles of branches (these usually naked for at least 2 cm.)

10. Plants perennial; flowers closely invested by bractlike pair of upper leaves; stems not over 15 cm. tall
-9. G. barbellata
10. Plants annual; flowers on naked peduncles (naked for at least 2 cm. below the flowers) hence not invested by bractlike leaves; stems usually over 15 cm. tall
-10. G. thermalis
9. Corolla lobes usually 5, margins entire or nearly so; corollas not over 2 cm. long, fringed more or less in the throat at base of lobes; flowers several to many, clustered on rather short pedicels (except in G. tenella)
11. Flowers solitary on a naked peduncle 2-10 cm. long; plants not over 10 cm. tall -11. G. tenella
11. Flowers several to many in short-peduncled clusters; plants usually over 10 cm. tall
12. Calyx lobes conspicuously unequal, the outer 2 large and foliaceous usually enclosing the smaller inner ones (outer at least 1/3 longer and twice as wide as inner) -12. G. heterosepala
12. Calyx lobes more or less unequal but not conspicuously unlike as above
13. Flowers numerous and crowded, short-pedicelled or on ascending peduncles forming a narrow dense spikelike inflorescence; corolla usually white to greenish-yellow, rarely blue; middle leaves usually longer than their internodes -13. G. strictiflora
13. Flowers not numerous or especially crowded, distinctly pedicelled, the inflorescence not dense and spikelike; corolla blue or bluish, rarely nearly white; middle leaves usually shorter than their internodes -14. G. plebeia

1. Gentiana fremontii Torr. in Frem., Rept. Exped. Rocky Mt. 94. 1845.
Chondrophylla fremontii (Torr.) A. Nels.---Annual or biennial plants; stems 3-10 cm. tall, erect, simple or several from base; leaves about 4-6 mm. long, round-ovate on basal to oblong or lanceolate on cauline, pale green, conspicuously white-margined; flowers solitary and terminal; calyx 4- or 5-lobed, the tube about 5-7 mm. long, lobes about 2 mm. long, scarious-margined these extending down into tube from sinus; corolla 8-15 mm. long, the lobes about 1/5 as long as the tube, salverform with broad emarginate plaits in the sinuses, whitish to greenish-purple; capsule long-stipitate (to 2 cm. here), trumpet-shaped when dehiscent, much exserted from corolla when mature.---Banks and slopes. Alberta, south to New Mexico and California. Our rather few records from northcentral, central and southcentral Colorado at 7500-10,000 feet.
2. Gentiana prostrata Haenke, Jacq. Coll. Bot. 2:66. 1788.
Chondrophylla americana (Engelm.) A. Nels.---Annual or biennial plants; stems 1-10 cm. long, erect, simple or branching at base and the branches often decumbent; leaves 3-5 mm. long, ovate or obovate on basal to ovate or oblong on cauline, slightly if at all white-margined; flowers solitary and terminal; calyx 6-10 mm. long, 4- or 5-lobed, these about 1.5-2 mm. long, slightly if at all scarious-margined; corolla 10-18 mm. long, rather salverform, blue, the lobes about 1/4 as long as the tube with broad plaits in the sinuses; capsules short-stipitate, not exserted from the corolla. Our plants are var. americana Engelm.---Higher altitudes in the mountains. The variety from Alberta to Alaska south to Colorado and California. Our few records scattered in the Colorado mountains at 9500-12,000 feet.
3. Gentiana andrewsii Griseb. in Hook., Fl. Bor. Am. 2:55. 1834.
Dasystephana andrewsii (Griseb.) Small---Perennial plants; stems 20-60 cm. tall, glabrous, erect; leaves 5-15 cm. long, ovate to lanceolate, glabrous or rough-margined; flowers in a terminal cluster and in the axils of upper leaves; calyx tube 8-12 mm. long, lobes about 5-10 mm. long, ovate or lanceolate, obtuse, commonly spreading; corolla 2-4 cm. long, oblong to cylindrical-clavate, nearly closed at throat, blue or perhaps sometimes white, the lobes nearly obsolete, apex of plaits somewhat incurved; capsule short-stipitate; seeds winged.---Wet or moist places. In Canada extending south to Georgia, Nebraska and Colorado. Our records from northcentral Colorado at 5000-5500 feet.
4. Gentiana romanzovii Ledeb., Nouv. Mem. Soc. Nat. Mosc. 1:215. 1829.
Dasystephana romanzovii (Ledeb.) Rydb.---Perennial plants with short rootstocks; stems 5-20 cm. long often tufted; basal leaves 3-10 cm. long, several to many, linear-oblanceolate to oblanceolate, thickish, cauline leaves linear to narrowly-oblong, somewhat sheathing-connate at the sessile base; flowers 1- to 3-clustered, subtended by leaflike bracts; calyx 15-25 mm. long, the narrow rather unequal lobes about 1/2 the entire length or somewhat less; corolla about 35-45 mm. long, yellowish to yellowish-white, spotted with purple and streaked with blue, funnelform, the lobes triangular, 4-5 mm. long, plaited in the sinuses; capsule stipitate; seeds lamellose.---High altitudes or arctic. Montana to Alaska, south to New Mexico and Utah. Scattered over the mountainous parts of Colorado at 10,500-13,000 feet.
5. Gentiana parryi Engelm., Trans. St. Louis Acad. Sci. 2:218. 1863.
Dasystephana parryi (Engelm.) Rydb.; G. bracteosa Greene; G. parryi var. bracteosa (Greene) A. Nels.---Perennial plants; stems 10-40 cm. tall, often numerous from a thick root; leaves somewhat glaucous, varying from ovate to oblong-lanceolate or even linear-lanceolate; flowers 1-6 in a terminal or subterminal cluster, subtended by 2 or 3 leaflike bracts which are ovate to ovate-lanceolate and somewhat scarious, often concealing the calyx; calyx 14-24 mm. long, the lobes usually about 1/2 as long as the tube, narrow lanceolate or linear; corolla 3-4 cm. long, blue to purple, banded with green outside, campanulate to funnelform, the lobes about 5 mm. long, with plaits in the sinuses forming lobes about 2/3 as long as the corolla lobes; capsule stipitate; seeds smooth.---High altitudes in the mountains. Wyoming and Utah south to New Mexico and Arizona. Rather common in the mountainous part of Colorado at 9000-12,000 feet.
6. Gentiana forwoodii A. Gray, Proc. Am. Acad. 19:86. 1883.
Dasystephana forwoodii (A. Gray) Rydb.---Perennial plants; stems 10-45 cm. tall, usually several; leaves 1.5-3 cm. long, rather numerous, ovate-lanceolate to oblong-lanceolate; flowers short-pedicelled in the upper axils forming a spikelike inflorescence; calyx 4-8 mm. long, irregular and more or less cleft and spathelike, the lobes lacking or of subulate short teeth much less than 1/3 as long as the tube; corolla 2-3 cm. long, blue-purple, funnelform, with lobes about 5-6 mm. long, plaited in the sinuses, the apical appendages nearly as long as the

corolla lobes; capsule stipitate; seeds smooth.---Hills and mountains. Alberta to Colorado and Arizona. Our few records from northcentral and central Colorado at 6000-8500 feet.
7. Gentiana affinis Griseb. in Hook., Fl. Bor. Amer. 2:56. 1838.
Dasystephana affinis (Griseb.) Rydb.---Perennial plants; stems 10-30 cm. tall, several; leaves rather numerous, ovate, oblong or lanceolate; flowers usually several to many, and some usually axillary well below the apex of the stem, the leaflike bracts lanceolate or linear, seldom as long as the flowers; calyx about 8-10 mm. long the lobes linear to oblong-lanceolate, unequal but often about as long as the tube; corolla 2.5-3 cm. long, funnelform, blue or purple, the lobes about 4 mm. long, ovate and acute, plaited in the sinuses, the appendages at the apex about 2/3 as long as the corolla lobes; capsule stipitate, seeds smooth.---Mountains and hills often in low meadows. Saskatchewan to British Columbia, south to Colorado, Arizona and California. Scattered over the mountainous part of Colorado at 7000-9000 feet but reported up to 12,000 feet.
8. Gentiana bigelovii A. Gray, Proc. Am. Acad. 19:87. 1883.
Dasystephana bigelovii (A. Gray) Rydb.---This plant is much like G.affinis and intergrades with it commonly. The bracts are linear, usually longer than the flowers, and the upper leaves are linear or lanceolate-linear, corolla is shorter (20-25 mm. long).---Mountains. Colorado, New Mexico and Arizona. Scattered in the northcentral, central and southcentral parts of Colorado at 7000-8500 feet.
9. Gentiana barbellatus Engelm., Trans. Acad. St. Louis 2:216. 1863.
Anthopogon barbellatas (Engelm.) Rydb.---Perennial glabrous plants from slender but fleshy roots; stems 5-15 cm. tall; basal leaves 3-5 cm. long, oblanceolate to spatulate, rather thick; stem leaves 2-6 pairs, oblong to linear, sessile; flowers often solitary but sometimes several, a pair of leaflike bracts close to the flower (usually closer than 1 cm., certainly not over 2 cm. away); calyx 10-20 mm. long, lobes rather unequal but usually about as long as the tube, green to purplish; corolla 2.5-5 cm. long, showy, blue or blue-purple, rather funnelform, lobes variable in length, usually about 3/4 to as long as the tube, oblong, erose-dentate at very apex, fimbriate on the sides nearly to sinuses, sinuses not plaited; filaments more or less bearded below; capsule stipitate.---Higher mountains. Wyoming to Utah, south to New Mexico and Arizona. In the higher mountains of Colorado at 10,000-11,500 feet.
10. Gentiana thermalis Kuntze, Rev. Gen. Pl. 2:427. 1891.
Anthopogon thermalis (Kuntze) Rydb.; Gentiana elegans A. Nels.; G. elegans var. unicaulis A. Nels.---Annual glabrous plants; stems usually 20-30 cm. tall but often larger or smaller (we have one 2 cm. tall), simple or branched; stems and branches terminated by a single flower, peduncles naked for at least 2 cm., mostly more, no leaflike bracts near the flower; basal leaves 1-3 cm. long, spatulate to obovate or oblanceolate, obtuse, long-cuneate at base, cauline leaves oblong to linear, sessile; calyx usually 1.5-5 cm. long, angled, green or spotted or blotched with purple, the long-acuminate rather unequal lobes about 1/2 to 1/3 the entire length; corolla ordinarily about 2-5 cm. long, very showy, bluish to purple, rather funnelform, lobes about 1/2 to 1/3 the entire length, lobes obovate to obovate-oblong, dentate at apex, fimbriate on upper part of sides, not plaited in sinuses; filaments not hairy; capsule stipitate.---Moist meadows. Mackenzie to Idaho, south to New Mexico and Arizona. Rather common in the mountainous part of Colorado at 7500-12,500 feet.
11. Gentiana tenella Rottb., Act. Hafn. 10:436. 1776.
G. monantha A. Nels.; Amarella monantha (A. Nels.) Rydb.---Annual plants, not over 10 cm. tall, glabrous; stems simple or few-branched from base, true stems very short, only about 1-3 cm. long with few pairs of leaves; leaves about 5-10 mm. long, oblong to spatulate; peduncle long and naked and longer than the true stem, 1-flowered; sepals about 1/2 to 3/4 as long as the corolla, 4 or 5, oblong-lanceolate, equal to unequal; corolla 10-12 mm. long, blue or bluish, not plaited in the throat but with a fimbrate crown, the 4 or 5 lobes shorter than the tube.---Mountains. Arctic regions south to Colorado, Arizona and California. Our few records scattered in the mountainous parts of Colorado at 8500-11,500 feet.
12. Gentiana heterosepala Engelm., Trans. Acad. St. Louis 2:215. 1863.
Amarella heterosepala (Engelm.) Greene---Annual plants; stems 10-30 cm. tall, erect, mostly simple; basal leaves 1-3 cm. long, oblanceolate to spatulate, stem leaves to 4 cm. long, middle ones usually not as long as their internodes, lanceolate; flowers 1-3 in the axils of the upper 4-6 leaves, on slender pedicels, also some terminal; outer 2 calyx lobes about 8-15 mm. long, large and foliaceous (at least 1/3 longer and twice as wide as the inner) usually enfolding and enclosing the linear-lanceolate inner ones; corolla about 10-15 mm. long, blue or purplish, funnelform, lobes acute, about 1/4 the total length, not plaited in the sinuses but corolla with a fringed crown in throat at base of lobes; capsule slightly if at all stipitate; seeds smooth.---Mountains. Colorado to Utah, south to New Mexico and Arizona. Our few records mostly from southwestern Colorado at 7500-10,000 feet.
13. Gentiana strictiflora (Rydb.) A. Nels., Bot. Gaz. 34:26. 1902.
Amarella strictiflora (Rydb.) Greene---Annual plants; stems 10-40 cm. tall, strict, simple or branched; leaves numerous, basal about 2-3 cm. long, spatulate or oblanceolate, stem leaves lanceolate, middle ones usually longer than their internodes; flowers axillary and terminal, numerous and crowded, short-pedicelled or on ascending peduncles, the whole appearing distinctly spikelike; calyx 5-12 mm. long, the lobes somewhat unequal, much longer than the tube which is about 1-2 mm. long; corolla 8-15 mm. long, white to greenish-yellow sometimes with blue lobes, occasionally all blue, funnelform to cylindric, lobes about 1/4 the length or less, not plaited in the sinuses but with at least a few setae on a crown in the throat at the base of the lobes (rarely no setae); capsule little if any stipitate; seeds smooth.---Moist meadows and open woods. Saskatchewan to Alaska, south to New Mexico, Arizona and California. Scattered in the mountains, our records from northcentral and central Colorado at 7000-10,500 feet.

14. Gentiana plebeia Cham., Linnaea 1:181. 1826.

Amarella plebeia (Cham.) Greene---Annual plants; stems 5-40 cm. tall, simple with ascending branches, the middle internodes usually much longer than the leaves; basal leaves 8-30 mm. long, spatulate or oblanceolate-obovate, cauline leaves often longer than the basal, lanceolate to oblong-lanceolate; flowers 1-4 in the upper axils, on slender pedicels, more numerous and crowded at summit; calyx variable, in our plants 5-10 mm. long, the lobes somewhat unequal, about 2/3 to 3/4 as long as the whole; corolla 10-18 mm. long, funnelform, blue or bluish, rarely paler, lobes acute, about 1/4 as long as the tube, not plaited in sinuses but corolla with a fringed crown in throat at the base of lobes; capsule slightly if at all stipitate; seeds smooth. The author cannot separate the above from G. scopulorum Greene which is supposed to have longer calyx lobes. All intergradations occur. The reduced alpine form with few flowers has been called G. plebeia var. holmii Wettst.---Mountains and meadows. Saskatchewan to Alaska, south to New Mexico and perhaps California. Scattered through the mountainous part of Colorado at 7000-11,000 feet.

5. Eustoma Salisb. PRAIRIE GENTIAN

Annual or perennial plants; stems erect or nearly so, leafy; leaves opposite, sessile or clasping; flowers in cymose, few-flowered panicles; calyx 5-parted into narrow keeled lobes (rarely 4 or 6); corolla campanulate, 5-parted, (sometimes 4 or 6), the lobes well over twice as long as the tube, erose-dentate, convolute in the bud, deep purple or rarely white; stamens adnate to the corolla tube, anthers sagittate.

1. Eustoma russelianum (Hook.) Griseb. in DC., Prodr. 9:51. 1845.

E. andrewsii A. Nels.---Annual or perennial plants with rosettes and vertical taproots; stems 20-60 cm. tall, more or less glaucous; leaves 2-4 cm. long, elliptic-oblong to lanceolate, mostly 3-nerved; calyx 14-20 mm. long, the tube about 1/5 the total length; corolla 3-4 cm. long, the tube about 1/4 the length, lobes elliptic-obovate; anthers basifixed, sagittate; style stout, about as long or somewhat shorter than the ovary, stigma lobes about 2-3 mm. wide, oval or round.---Meadows. Nebraska to Colorado, south to Texas and New Mexico. Scattered in the eastern part of Colorado west to the foothills at 3500-5500 feet.

Family 96. Apocynaceae DOGBANE FAMILY

Perennial caulescent plants, herbaceous or slightly woody at base, with milky acrid juice; leaves simple, entire, opposite or alternate, without stipules; flowers perfect, regular, mostly in corymbose cymes; calyx deeply 5-parted; corolla of 5 partly united petals, these petals convolute in the bud; stamens 5, attached separately to the corolla tube alternate to its lobes; ovaries superior, of 2 separate carpels, each 1-celled with 1 parietal placenta, the 2 carpels joined above by their styles and with 1 common stigma; fruit of 2 (or only 1 developing) follicles; seeds often comose.

1. Leaves alternate; seeds naked; anthers not connivent around the stigma -1. Amsonia
1. Leaves opposite; seeds comose; anthers connivent around the stigma -2. Apocynum

1. Amsonia Walt. AMSONIA

Leaves alternate; corolla salverform, the tube somewhat enlarging upward and longer than its lobes, lead-purple, blue or whitish; anthers included, not connivent, unappendaged; seeds naked, many.

1. Amsonia jonesii Woodson, Ann. Mo. Bot. Gard. 15:414. 1928.

A. latifolia M. E. Jones; apparently A. texana of Manuals---Glabrous plants from thickened often woody roots; stems 20-50 cm. tall; leaf blades 3-6 cm. long, thick, ovate to oblong-ovate to nearly lanceolate, acute or obtuse, glaucous, petioles about 4-5 mm. long; inflorescence dense, many-flowered, pedicels 3-5 mm. long; calyx lobes about 1.5-2 mm. long; corolla glabrous externally or nearly so, the tube 6-8 mm. long, lobes 3-6 mm. long; follicles 5-10 cm. long, glabrous, not constricted, terete.---Rocky areas. Colorado, Utah and Arizona. Our few records from westcentral and southwestern Colorado at about 5000-6000 feet.

2. Apocynum L. DOGBANE; INDIAN HEMP

Plants spreading by rhizomelike horizontal roots; leaves opposite; flowers in terminal or lateral cymes; corollas campanulate to cylindrical, bearing 5 distinct appendages within, these adnate to the tube and opposite the lobes; stamens converging about the pistils, the anthers connivent to the stigma; ovary surrounded at base by 5 distinct nectaries; fruit usually of two terete follicles; seeds numerous, comose. A very variable genus especially in leaf shape. The relative length of the calyx and corolla as used in the key works for most specimens but some seem to intergrade even there.

1. Corolla 2-3 times as long as the calyx, usually over 4 mm. long, the lobes spreading or recurved, commonly with pinkish veins; leaves wide spreading or drooping
   2. Corolla about 3 times as long as the calyx, usually over 5 mm. long, lobes often recurved; leaves drooping
      3. Leaves more or less pilose-tomentose beneath -1. A. androsaemifolium
      3. Leaves glabrous beneath -1A. A. androsaemifolium glabrum

    2. Corolla about 2 to 2½ times as long as the calyx, not over 5 mm. long, lobes merely spreading; leaves usually spreading, not drooping
        4. Lower leaf surface somewhat hairy -2A. **A. medium lividum**
        4. Lower leaf surface glabrous -2B. **A. medium floribundum**
1. Corolla less than 2 times as long as the calyx, often only slightly longer (about 2 times longer in A. suksdorfii), the lobes erect or only slightly spreading, not at all pinkish-veined; leaves ascending or only slightly spreading
    5. Calyx lobes 1-1.5 mm. long, about 1/2 as long as the corolla -3. **A. suksdorfii**
    5. Calyx lobes 1.5-3 mm. long, (usually over 2 mm.) more than 1/2 as long as the corolla
      6. Leaves of main stem (not lower though) petioled, cuneate to rounded at base; bracts of the inflorescence usually inconspicuous and scarious
        7. Entire plant tomentulose -4A. **A. cannabinum pubescens**
        7. Entire plant glabrous -4B. **A. cannabinum glaberrimum**
      6. Leaves of main stem nearly or quite sessile, cordate to subrounded at base; bracts of inflorescence often sub-foliaceous
        8. Corolla about as wide as long -5. **A. sibericum**
        8. Corolla definitely longer than wide -5A. **A. sibericum salignum**

1. Apocynum androsaemifolium L., Sp. Pl. ed. 2., 311. 1762.
    Stems 20-50 cm. tall, erect or ascending, glabrous or nearly so, rather dichotomously branched alternately; leaves 2-10 cm. long, rather shortly petiolate, drooping, ovate to oblong-lanceolate, essentially glabrous above, more or less pilose-tomentulose beneath; cymes few- to several-flowered; calyx 2-4 mm. long; corolla 4-12 mm. long, usually about 3 times as long as the calyx, white to pinkish-veined, campanulate with the lobes spreading or reflexed; follicles 6-15 cm. long, pendulous.---Woods and fields. Newfoundland to British Columbia, south to Georgia and Arizona. Probably scattered over the state but our records from the western three-fourths at 5000-10,000 feet.

1A. Apocynum androsaemifolium glabrum Macoun, (var.) Cat. Can. Pl. 2:317. 1884.
    A. ambigens Greene; A. scopulorum Greene---Foliage wholly glabrous.---Intergrades some with species and intermixed with it in Colorado.

2. Apocynum medium Greene, Pitt. 3:229. 1897.
    Stems 20-60 cm. tall, erect or ascending, glabrous to puberulent, branched; leaves 5-10 cm. long, petiolate to subsessile, usually spreading, ovate to oblong-lanceolate, cuneate to subcordate at base, glabrous or nearly so above, glabrous to somewhat hairy below; calyx lobes 1.5-3 mm. long; corolla 3-5 mm. long, campanulate, cylindrical to urceolate, lobes spreading, white with pinkish veins; follicles 7-15 cm. long, pendulous.
    A northern species represented in Colorado by 2 varieties.

2A. Apocynum medium lividum (Greene) Woodson, (var.) Ann. Mo. Bot. Gard. 17:115. 1930.
    A. lividum Greene---Lower leaf surfaces somewhat hairy.---Scattered over Colorado mostly in the western part at 5000-8000 feet.

2B. Apocynum medium floribundum (Greene) Woodson, (var.) Ann. Mo. Bot. Gard. 17:112. 1930.
    A. cannabinum var. lividum of C. & N. Manual---Plants wholly glabrous.---Probably scattered over Colorado but our few specimens from northcentral, southwestern parts, and eastern parts at 3400-7500 feet.

3. Apocynum suksdorfii Greene, Pitt. 5:65. 1902.
    Stems 40-70 cm. tall, erect or somewhat ascending, glabrous, branching; leaves 4 to 8 cm. long, petiolate or lower subsessile, ascending or only slightly spreading, oblong-ovate, glabrous; calyx lobes 1-1.5 mm. long, glabrous; corolla 2-4 mm. long, cylindrical, white, about twice as long as the calyx, the lobes erect or only slightly spreading; follicles 9-15 cm. long, pendulous.---Colorado to Washington, south to Arizona and California. Our few records widely scattered over the state at 3700-5000 feet.

4. Apocynum cannabinum L., Sp. Pl. 213. 1753.
    Stems 30-90 cm. tall, erect or ascending, glabrous, branched; leaves 2-10 cm. long, petiolate or the lower subsessile, cuneate to rounded at the base, ascending or only slightly spreading, ovate to lanceolate, glabrous above, more or less pubescent or tomentulose below; calyx lobes 2-3 mm. long; corolla 2-5 mm. long, white, cylindrical to urceolate, scarcely longer than the calyx to about twice as long, lobes erect or only slightly spreading; follicles 12-20 cm. long, pendulous. Represented in Colorado by 2 varieties.

4A. Apocynum cannabinum pubescens (R. Br.) A. DC. in DC., (var.) Prodr. 8:440. 1844.
    Entire plant tomentulose. Generally distributed in the United States. Surely present in Colorado but no specimens encountered by the author.

4B. Apocynum cannabinum glaberrimum A. DC. in DC., (var.) Prodr. 8:439. 1844.
    Entire plant glabrous.---Probably scattered throughout the state but our records from the western two-thirds at 4700-7500 feet.

5. Apocynum sibericum Jacq., Hort. Vindob. 3:37. 1770.
    A. hypericifolium Ait.---Stems 20-80 cm. tall, erect or ascending, glabrous, branching; leaves 1.5-14 cm. long, sessile or subsessile especially on the main stem (those on branches often short-petiolate), ascending or slightly spreading, oblong or oblong-lanceolate to oval, rounded to cordate at base; calyx lobes 1.5-3 mm. long; corolla 2-5 mm. long, white, cylindrical to urceolate, scarcely longer to nearly twice as long as the calyx, the lobes erect or slightly spreading; follicles 4-10 cm. long, pendulous. This northern and eastern species may be present in Colorado.

5A. Apocynum sibiricum salignum (Greene) Fernald, (var.) Rhodora 37:328. 1935.
    Corolla definitely longer than wide (instead of about as wide as long as in species).---Manitoba to British Columbia, south to Texas and California. Scattered over all of Colorado at lower altitudes of 3500-7500 feet.

Family 97. Asclepiadaceae MILKWEED FAMILY

Perennial herbaceous plants with milky sap; leaves simple, usually entire, opposite, whorled or irregularly approximate; flowers perfect, regular, of very specialized structure, usually in umbels; calyx 5-parted or 5-divided; corolla of united petals, 5-lobed, these often reflexed; stamens 5, inserted on the base of the corolla and closely connivent about the pistils into a tube which is blended above with the style column, a "crown" of separate or united, often hoodlike appendages usually present between the corolla and the stamens, anthers tipped with a scarious membrane which is inflexed over the style, with longitudinal slits between the anthers, pollen in pear-shaped masses, suspended in pairs from the summit, 1 from each anther; pistils 2, with distinct superior ovaries enclosed in the staminal ring, styles distinct below but united by the common stigmatic disk, ovaries each 1-celled with 1 parietal placenta; fruit of follicles, usually just 1 developing in each flower; seeds silky-comose.

1. Stems twining; leaves sagittate or hastate at base — -1. Sarcostemma
1. Stems not twining; leaves various but never sagittate or hastate at base — -2. Asclepias

1. Sarcostemma R. Br.

Stems twining; leaves opposite; flowers in lateral umbels; calyx 5-parted or cleft; corolla campanulate-rotate, deeply 5-lobed the lobes more or less twisted; crown double, the outer about as long as the anther column.

1. Sarcostemma crispum Benth., Pl. Hartw. 291. 1841.
   Funastrum crispum (Benth.) Schlechter; S. lobata Waterfall---Stems sometimes to 100 cm. long or more, twining, more or less appressed-puberulent; leaves 3-10 cm. long, narrowly to broadly lanceolate, sagittate to hastate at base, petioled, more or less puberulent on surfaces and often crisped on margins; peduncles shorter than the leaves, 5- to 10-flowered, on pedicels nearly as long as the peduncles; calyx lobes 4-5 mm. long; corolla lobes 6-9 mm. long, greenish with brown tinge, ovate; outer crown white.---Canyons and rocky slopes. Texas to Arizona and Mexico. Our few specimens from southern Colorado at 4000-5500 feet.

2. Asclepias L. MILKWEED

Stems commonly erect, never twining; flowers in umbelliform cymes, terminal or lateral; calyx and corolla divisions mostly reflexed in anthesis; filament column bearing a circle of 5 hoods, these often with an incurved hornlike crest within.

1. Corolla lobes erect or merely spreading, not reflexed, over 7 mm. long; leaves linear to lanceolate
   2. Inflorescences on very short peduncles (immediately subtended by the leaves); hoods rose-colored; leaves rounded or broadly cuneate at base; mostly in eastern Colorado — -1. A. capricornu
   2. Inflorescences definitely peduncled; hoods dark brown-purple; leaves cuneate at base; mostly in western Colorado — -1A. A. capricornu occidentalis
1. Corolla lobes conspicuously reflexed in the mature flower (not in bud), less than or more than 7 mm. long; leaves various, from linear to orbicular
   3. Corolla lobes orange to red in color; stems hirsute to rough-pubescent — -2. A. tuberosa
   3. Corolla lobes whitish, greenish, yellowish, rose or purplish, never orange or red; stems glabrous, puberulent, short-pubescent or tomentose
      4. Hoods not crested or horned within
         5. Mass of anthers and stigma nearly globose, about as wide as long; hoods truncate-emarginate or toothed at apex; leaves narrowly linear to linear-filiform
            6. Hoods truncate-emarginate at apex, hence appearing to have 2 teeth, no smaller tooth between; lower umbels on peduncles usually 1 cm. long or longer; leaves not over 3 mm. wide, usually alternate; fruit on a deflexed pedicel — -3. A. engelmanniana
            6. Hoods with 2 broad teeth at apex and a shorter hornlike tooth in their sinus; umbels sessile or on short peduncles rarely over 1 cm. long; leaves usually over 3 mm. wide, usually opposite; fruit on an ascending pedicel — -4. A. stenophylla
         5. Mass of anthers and stigma definitely longer than wide, hoods rounded at apex, not toothed or truncate-emarginate; leaves usually lanceolate or broader — -5. A. viridiflora
      4. Hoods crested or horned within
         7. Flowers large, the corolla lobes 7-13 mm. long; leaves ovate-lanceolate to orbicular, not over twice as long as wide (except sometimes in A. speciosa)
            8. Leaves oblong to lanceolate-ovate, over 1½ times longer than broad, usually tapering from middle to an obtuse or acute apex; fruit white-tomentose, often with soft-spinose processes; corolla lobes pink to greenish-purple — -6. A. speciosa
            8. Leaves ovate-orbicular, broadly obovate, oblong-oval to orbicular, not over 1½ times longer than wide, not tapering to apex; fruit glabrous to puberulent, never soft-spiny; corolla lobes greenish-white to greenish-yellow
               9. Corolla lobes 10-14 mm. long; plants small, seldom over 20 cm. tall; follicles 3-5 cm. long; plants of extreme western Colorado — -7. A. cryptoceras
               9. Corolla lobes 7-10 mm. long, plants over 20 cm. tall; follicles 5-10 cm. long; plants mostly of the eastern half of Colorado
                  10. Stems and leaves canescent-tomentose, only in age glabrate; petioles usually over 8 mm. long on some leaves — -8. A. arenaria
                  10. Stems and leaves puberulent to glabrous; petioles rarely over 8 mm. long — -9. A. latifolia

7. Flowers small to medium, the corolla lobes 3-6 mm. long; leaves ovate-lanceolate or oblong-lanceolate to linear-filiform, over twice as long as wide (only very rarely less than twice as long as wide in A. hallii)
   11. Leaves narrowly-linear to linear-filiform, less than 4 mm. wide (seldom over 3 mm.), opposite, whorled or scattered
      12. Leaves opposite; stems busy-branched; follicles erect from a reflexed pedicel   -10. A. macrotis
      12. Leaves scattered-alternate to whorled; stems not bushy-branched (may branch at very base into several stems); follicles erect on erect or nearly erect pedicels
         13. Leaves linear-filiform, seldom over 1 mm. wide (never over 2 mm.), irregularly alternate or obscurely whorled; stems tufted, seldom over 20 cm. tall; peduncles rather short, often not as long as the pedicels   -11. A. pumila
         13. Leaves narrowly-linear, some over 1 mm. wide (usually some over 2 mm.), whorled; stems 1 to several but hardly tufted, rarely less than 20 cm. tall; peduncles longer than the pedicels   -12. A. subverticillata
   11. Leaves linear-lanceolate to ovate-lanceolate, some over 4 mm. wide
      14. Dwarf plants not over 30 cm. tall; leaves not over 15 mm. wide
         15. Leaf blades less than 5 cm. long, ciliate with curved hairs, these hairs also on midrib; umbels sessile or nearly so, all crowded near end of stem   -13. A. uncialis
         15. Leaf blade over 5 cm. long (on longest), hairs if present not confined to margins and midrib, hence blade not ciliate; umbel short-peduncled, some remote from end of stem   -14. A. brachystephana
      14. Taller plants normally over 30 cm. tall; some of leaves usually over 15 mm. wide
         16. Hoods 4.5-6 mm. long, surpassing the horn within and much longer than the anthers; column below anthers short, less than 1 mm. long   -15. A. hallii
         16. Hoods 2-3 mm. long, surpassed by the horn within, and in ours rarely longer than the anthers; column below anthers about 1 mm. long   -16. A. incarnata

1. **Asclepias capricornu** Woodson, Ann. Mo. Bot. Gard. 32:370. 1945.
Stems 20-60 cm. tall, usually several from a woody rootstock, decumbent or ascending, rough-puberulent; leaves 4-15 cm. long, alternate or 2 or 3 approximate, lanceolate to linear, rounded or nearly so at base; umbels solitary or corymbose, terminal on short peduncles; calyx lobes about 3-4 mm. long; corolla lobes about 7-9 mm. long, greenish to cream-colored, ovate or oval, acute, spreading or erect; hoods light rose, shorter than the corolla lobes, bearing a crest within but not a horn; follicles 5-10 cm. long, more or less minutely spiny, erect on deflexed pedicels. This is the typical form and has been called ssp. capricornu Woodson.---Dry ground. Kansas to Nevada, south to Texas and Arizona. Our records from the southern half of Colorado at 4000-7500 feet.

1A. **Asclepias capricornu occidentalis** Woodson, (ssp.) Ann. Mo. Bot. Gard. 32:371. 1945.
Like the typical form but leaves supposed to be cuneate at base; inflorescences definitely peduncled; hoods dark purple-brown; follicles smooth. Intergrades in leaf characters with the typical form of the species in Colorado.---This is the western form of the species. Colorado to California south to Texas. Our records from southern and western Colorado at 4500-8000 feet.

2. **Asclepias tuberosa** L., Sp. Pl. 217. 1753.
Milky sap scanty; stems 30-80 cm. tall, hirsute to rough-pubescent, simple or branched above; leaves numerous, scattered and appearing alternate, sessile or subsessile, broadest below middle, linear-lanceolate to oblong-lanceolate, rarely wider, acute, broadly cuneate to truncate or cordate at base, margin often more or less revolute; umbels several and mostly terminal; sepals about 1-2 mm. long, reflexed in flower; corolla lobes 5-8 mm. long, greenish-orange to reddish-orange; linear-oblong, obtuse, reflexed in flower; hoods about 4.5-6 mm. long, with a horn somewhat shorter, bright orange or yellow; follicles 7-10 cm. long, finely pubescent, nearly erect on decurved pedicels. Our plants have been called ssp. interior Woodson.---Meadows and canyons. The subspecies from Ohio to Utah, south to lower Mississippi Valley and Arizona. Rather uncommon in Colorado, our records from the southern half of the state from 5500-7000 feet.

3. **Asclepias engelmanniana** Woodson, Ann. Mo. Bot. Gard. 28:207. 1941.
Acerates auriculata Engelm.----Stems 20-80 cm. tall, rather stout, usually simple, glabrous and somewhat glaucous, leafy; leaves 5-15 cm. long, alternate and scattered and sometimes appearing whorled, narrowly linear or almost filiform-linear, margins scabrous and often revolute; umbels many-flowered, 1 to many, axillary, the peduncles rather short, about 5-20 mm. long; calyx lobes about 2-3 mm. long, reflexed in flower; corollas greenish-yellow or tinged with dull purple, the lobes 4-5 mm. long, reflexed in flower, oblong; hoods about 4 mm. long, yellowish with usually purplish keel, truncate and emarginate at apex, margins involute and auriculate at base, no horns present; follicles fusiform, about 6-10 cm. long, erect on deflexed pedicels.---Plains and prairies, often on rocky ground. Nebraska to Colorado, south to Texas and Arizona. Scattered over Colorado (no records yet from northwestern part) at 3500-7500 feet.

4. **Asclepias stenophylla** Gray, Proc. Am. Acad. 12:72. 1876.
Aceratas angustifolia (Nutt.) Dec.---Stems 25-60 cm. tall, usually several from a thick root, simple, puberulent or glabrate; leaves about 5-13 cm. long, opposite or somewhat alternate, narrowly linear, glabrous, margins revolute and somewhat scabrous as is the midrib; umbels several, axillary, short-peduncled or subsessile; calyx lobes 2-3 mm. long, reflexed in flower; corolla yellowish-green, the lobes about 5 mm. long, oblong, reflexed in flower; hoods about 3 mm. long but hardly equalling the anthers, whitish, erect, 2-toothed at truncate apex, another tooth in the sinus between, inner margins with small, erose, truncate, scalelike processes, no horn present; follicles slender, about 7 cm. long or longer, erect on an ascending pedicel.---Dry plains and prairies. Nebraska to Colorado south to Texas. Our records thinly scattered over the eastern half of the state at 3500-6000 feet.

5. **Asclepias viridiflora** Raf., Med. Repos. N. Y. 5:360. 1808.
Acerates viridiflora (Raf.) Eaton---Stems 15-60 cm. tall, decumbent or ascending, tomentose to glabrate, simple; leaves about 3-8 cm. long, sometimes longer, alternate or opposite,

short-petioled, oval, ovate, oblong, lanceolate, oblong-lanceolate to narrowly linear, obtuse, cuspidate to acute at apex, rarely retuse, rather thick, margins often undulate; umbels many-flowered, 1 to several, axillary, sessile or nearly so; calyx lobes about 2-3 mm. long, reflexed in flower; corolla greenish-white, its lobes 4-6 mm. long, oblong, reflexed in flower; hoods about 3-4 mm. long, yellowish-green to purple-tinged, narrowly oblong to oblong-lanceolate, roundish-entire at apex and somewhat auriculate at base, no horn present; follicles 5-10 cm. long, erect on deflexed pedicels, puberulent. Vail (Bull. Torr. Cl. 25:33-35. 1898.) recognized var. ivesii Britt. with leaves lanceolate and var. linearis A. Gray with leaves elongated-linear. Our plants show all kinds of intergradations even on one specimen.---Dry plains, often in rocky ground. Massachusetts to Saskatchewan south to Florida and Mexico. Scattered over the eastern half of the state from 3500-6000 feet. A collection from Park County at a higher altitude may be this species.

A related plant with the upper leaves forming a sort of involucre around the terminal umbel has been recently collected in southeastern Colorado and identified as A. involucrata Engelm.

6. Asclepias speciosa Torr., Ann. Lyc. N. Y. 2:218. 1828.

Stems 40-150 cm. tall, simple, canescent-tomentose to glabrate; leaves 8-20 cm. long, opposite, shortly-petiolate, oblong to ovate-lanceolate, canescent-tomentose to glabrate, thick; umbels solitary at upper nodes and terminal, peduncles rather stout, densely tomentose; calyx lobes about 4-5 mm. long, reflexed in flower; corolla lobes about 9-13 mm. long, pink to greenish-purple or very light rose, spreading-reflexed; hoods about 1 cm. long or more, longer than the corolla lobes, pink or whitish, spreading, sharply tapering at apex, a horn present; follicles 7-12 cm. long, white-tomentose, from smooth to spiny, erect on deflexed pedicels.---Plains and valleys. Saskatchewan to British Columbia, south to Kansas and Arizona. Widely scattered over Colorado at 3500-7500 feet.

7. Asclepias cryptoceras S. Wats. in King, Geol. Expl. 40th. Par. 5:283. 1871.

Stems 10-25 cm. tall, glabrous, decumbent, several from a woody root; leaves 3-7 cm. long, 2-4 pairs, opposite, short-petioled or subsessile, ovate-orbicular, mostly subcordate at base, mucronate, glabrous or ciliate, more or less glaucous; umbels 2-3 axillary and sessile, or 1 and terminal; calyx lobes about 6-8 mm. long, reflexed in flower; corollas greenish-yellow, lobes about 10-14 mm. long, reflexed in flower, ovate, acute; hoods about 6-7 mm. long with an included horn, pink or flesh-colored, saccate-ovate; follicles ovate or ovate-oblong, 3-5 cm. long, nearly glabrous and reflexed on the pedicels.---Plains, hills and slopes. Wyoming to Oregon, south to Arizona and California. Comes into western Colorado, our records at 4500-6500 feet.

8. Asclepias arenaria Torr., Bot. Mex. Bound. Surv. 162. 1859.

Stems 20-60 cm. tall, ascending or decumbent, tomentose or in age glabrate, rather stout; leaves 5-10 cm. long, opposite, rather short-petioled, the blades obovate or oblong-oval or the lower often ovate, truncate to retuse and cuspidate at apex with base rounded, truncate or subcordate, margins somewhat undulate, tomentose; umbels dense, many-flowered, all lateral, sessile or short-peduncled; calyx lobes about 4-5 mm. long, reflexed in flower; corolla greenish-white, lobes 7-10 mm. long, oval-oblong, reflexed in flower; hoods about 4-5 mm. long, truncate at apex, somewhat lobed at inner margins, horn horizontally exserted from a broad base which is included; follicles 5-10 cm. long, erect from decurved pedicels, puberulent.---River plains and hills often in sandy soil. Nebraska to Colorado, south to Texas and Mexico. Our records from the eastern plain area of Colorado at 3500-6000 feet.

9. Asclepias latifolia (Torr.) Raf., Atl. Journ. 146. 1832-1833.

Stems 30-80 cm. tall, simple, erect or ascending, stout, minutely puberulent to glabrous; leaves 7-15 cm. long, opposite, sessile or short-petioled, orbicular or rounded-oval, cordate or subcordate at base, mucronate and often emarginate at apex, conspicuously nerved, thick; umbels 2-4, many-flowered, in upper leaf axils or rarely terminal, peduncles short and stout; calyx lobes 2-4 mm. long, reflexed in flower; corolla greenish to greenish-cream, lobes 7-10 mm. long, reflexed in flower, oblong-ovate; hoods about 3-4 mm. long, yellowish-cream to whitish, truncate at apex, the horn projecting from a crest, this crest flattened and broad; follicles ovoid, smooth, 5-7 cm. long, erect on deflexed pedicels.---Plains. Nebraska to Utah, south to Texas and Arizona. Should be expected anywhere in the state but our rather few records from eastern Colorado at 3500-4500 feet.

10. Asclepias macrotis Torr., Bot. U. S. & Mex. Bound. 164, pl. 45. 1859.

Stems 15-20 cm. tall, much-branched from a woody root, more or less puberulent; leaves 2-8 cm. long, opposite, narrowly linear with revolute margins and appearing almost filiform, glabrate; umbels terminal and lateral, 3- to 5-flowered, short-peduncled or sessile, pedicels short; corolla purplish or greenish, the lobes about 4 mm. long, reflexed in flower, ovate; hoods ovate at base, contracted above to an incurved subulate apex, horn broad, short and blunt, the apex barely exserted; follicles ovate-lanceolate to lanceolate, about 2.5-5 cm. long, erect from reflexed short pedicels, glabrous.---Hills and slopes, often in the rocky ground. Texas to New Mexico and south to Mexico. Our few records from southeastern Colorado at 4500-5000 feet.

11. Asclepias pumila (A. Gray) Vail in Britt. & Brown, Ill. Fl. 3:12. 1898.

Stems 5-25 cm. tall, tufted from a woody root, glabrous or nearly so; leaves 2-5 cm. long, very numerous, crowded irregularly or obscurely whorled, linear-filiform, margins revolute, glabrous or somewhat scabro-puberulent; umbels 2- to several-flowered, from upper axils and terminal, rather short-peduncled and somewhat corymbose; calyx lobes about 2 mm. long, reflexed in flower; corolla greenish-white, the lobes 3-5 mm. long, oblong, reflexed in flower; hoods about 1.5 mm. long, whitish, erect, each bearing a horn within; follicles narrowly fusiform, 3-8 cm. long, finely puberulent, erect on erect or nearly erect pedicels.---Dry plains and hills. South Dakota to Montana, south to Arkansas and New Mexico. Our records from the eastern half of Colorado at 3500-5500 feet.

12. **Asclepias subverticillata** (Gray) Vail, Bull. Torr. Club 25:178. 1898.
    A. galioides of Manuals---Stems 15-100 cm. tall, erect or nearly so from an extensive root system, usually with dwarf microphyllous axillary branches, glabrous or minutely pubescent; leaves 2-13 cm. long, whorled on the main stem, shortly-petiolate, linear, glabrous or nearly so; inflorescences usually solitary at upper nodes; calyx lobes about 1-1.5 mm. long, reflexed in flower; corolla lobes 3-5 mm. long, white to cream-colored or sometimes greenish-purple, reflexed in flower; hoods about 1.5-2 mm. long, nearly erect, more or less toothed at base, bearing a horn within; follicles narrowly fusiform, 5-10 cm. long, smooth or minutely puberulent, erect on erect pedicels.---Dry plains and pastures. Kansas to Idaho, south to Texas and Mexico. Scattered over Colorado but mostly in the southeastern and western parts, at 3400-7000 feet.
    A related species A. verticillata L., of the eastern part of the U. S. has been reported from eastern Colorado. Woodson (Ann. Mo. Bot. Gard. 31:365. 1944) said it differs from the above by lacking axillary shoots. The writer has not seen this plant from Colorado.

13. **Asclepias uncialis** Greene, Bot. Gaz. 5:64. 1880.
    Stems 2.5-10 cm. tall, usually several, puberulent; leaves 3-5 cm. long, opposite or approximate, lanceolate to narrowly linear-lanceolate or rarely ovate-lanceolate, acute or acuminate, cuneate to a short but definite petiole, white-ciliate with incurved hairs; umbels 3- to 4-flowered, sessile, crowded near ends of stems; calyx lobes 1.5-2 mm. long, reflexed in flower; corolla dull greenish-purple to dull rose, lobes 3-4 mm. long, ovate, reflexed in flower; hoods about as wide as long, about 1.5 mm. long, whitish, with triangular wings about as long as the main body, the horn represented by a broadly ovate plate somewhat shorter than the body; follicles on deflexed pedicels.---Dry plains and hills. Wyoming to New Mexico and Arizona. Our few records from the eastern half of Colorado at 4000-5300 feet.

14. **Asclepias brachystephana** Engelm. in Torr., Bot. Mex. Bound. Surv. 163. 1859.
    Kansas to Wyoming, south to Mexico. This species should be in Colorado but the writer could not locate a specimen from this state.

15. **Asclepias hallii** Gray, Proc. Am. Acad. 12:69. 1876.
    Stems 30-40 cm. tall, rather stout, puberulent to glabrate; leaves 5-12 cm. long, opposite or at least approximate in pairs, ovate-lanceolate, oblong-lanceolate to lanceolate, rather acute at apex, rounded to cuneate at base, puberulent to glabrate; umbels few to several, terminal and in upper leaf axils, on long peduncles which often make a corymbose cluster; calyx lobes about 2-3 mm. long; corolla greenish-white, tinged with red or purple, the lobes 5-6 mm. long, oblong, reflexed in mature flower; hoods 4.5-6 mm. long, oblong, hastate-gibbous at base, surpassing the sickle-shaped incurved horn which widens below apex; follicles lanceolate, about 8 cm. long, tomentulose, in age glabrate.---Slopes and hills, often in gravelly ground. Apparently limited to Colorado, our records from the southern mountainous half of the state at 7500-9200 feet.

16. **Asclepias incarnata** L., Sp. Pl. 215. 1753.
    Stems 60-200 cm. tall, simple or branched above, leafy, glabrous or somewhat pubescent; leaves 7-18 cm. long, opposite or at least approximate in pairs, oblong-lanceolate to linear-lanceolate, cuneate to subcordate at base; umbels usually several, corymbose at summit of stem or on branches; calyx lobes about 1.5-2 mm. long, reflexed in flower; corolla pink to deep rose-purple rarely white, the lobes 4-6 mm. long, oblong, reflexed in flower; hoods about 2-3 mm. long with a longer incurved horn; follicles 5-10 cm. long, erect on erect pedicels, glabrous or minutely puberulent.---Moist or wet ground. New Brunswick to Manitoba south to Florida and New Mexico. Our records from the northeastern quarter of Colorado at 3400-5400 feet, but the plant should be expected anywhere in the eastern part of the state.

Family 98. Convolvulaceae MORNING-GLORY FAMILY

Plants herbaceous, mostly with twining or trailing stems (in Cuscuta without chlorophyll); leaves alternate, simple but often deeply lobed (in Cuscuta reduced to minute scales); flowers perfect, regular, usually showy, solitary or in cymes, the pedicels articulated; sepals mostly 5, distinct or nearly so, imbricated; corolla sympetalous, 5-plaited or 5-lobed, twisted in the bud; stamens 5, borne on the corolla and included, alternate to its lobes; ovary superior, 2- to 3- (rarely 1-) celled usually with 2 ovules in each cell, styles 1 or 2; fruit a capsule, dehiscent or indehiscent, 1- to 4-seeded.

1. Plants lacking green chlorophyll, parasites on the stems of various host plants, no true roots present when mature; leaves reduced to scales　-1. **Cuscuta**
1. Plants with chlorophyll, not parasites, true roots present; leaves well developed
    2. Styles 2, separate or nearly so to the ovary, each 2-cleft; corolla small, not over 1 cm. long; stems erect or ascending　-2. **Evolvulus**
    2. Styles 1 or if 2-cleft then only at the apex and the divisions entire; corolla larger over 1 cm. long; stems usually climbing or prostrate
        3. Stigmas 1, globose or nearly so, entire or lobed; corolla white, pink, rose, red, blue or purple　-3. **Ipomoea**
        3. Style 2-cleft at apex, the 2 stigmas ovate to narrowly linear, corollas white to pink　-4. **Convolvulus**

1. Cuscuta L. DODDER

Plants parasitic, lacking chlorophyll, rootless; stems yellowish, filiform, twining; leaves reduced to scales; flowers small, sessile or nearly so, in few- to many-flowered cymose clusters, mostly 5-merous; appendages commonly present at base of corolla opposite the stamens, these scalelike and more or less toothed or fimbriate; ovary 2-celled, with 2 styles, the stigmas usually capitate; capsule bursting irregularly or circumscissle near base. Soon attaching themselves by haustoria to various plant hosts, often to cultivated crops like clover and alfalfa. Most species grow readily on various host species. The seed is often found mixed with that of cultivated crops, such as clover or alfalfa. For that reason other species may be found in Colorado that are not treated here. For example, C. epithymum Murr., C. applanata Engelm., C. potosina var. globifera Schaffner, C. plattensis Nels. and C. salina Engelm. have been found in adjacent states.

1. Stigmas linear-elongated; styles equal; capsule circumscissle in a definite line -1. C. approximata
1. Stigmas capitate or peltate, not elongated; styles equal or unequal; capsule not opening by circumscission (except in C. umbellata)
   2. Sepals nearly or quite distinct, closely invested with bracts -2. C. cuspidata
   2. Sepals definitely united, not immediately invested with bracts
      3. Corolla appendages lacking entirely -3. C. occidentalis
      3. Corolla appendages present opposite the stamens, more or less fringed
         4. Most of flowers with 3 or 4 sepals, petals and stamens; corolla lobes erect or slightly spreading
            5. Corollas membranous, when withered, remaining at top of capsule, the lobes obtuse, not inflexed at tip; sepals obtuse, about 1/2 as long as the corolla tubes -4. C. cephalanthi
            5. Corollas thickened-fleshy, when withered not remaining at top of capsule, the lobes acute with inflexed tips; sepals acute, only slightly if at all shorter than the corolla tubes -5. C. coryli
         4. Most flowers with 5 sepals, petals and stamens; corolla lobes erect, spreading or reflexed
            6. Capsules opening by circumscission near base; calyx lobes acute or acuminate -6. C. umbellata
            6. Capsules splitting irregularly or coming away entirely from the receptacle (no regular line of cleavage); calyx lobes obtuse
               7. Corolla fleshy-papillate, its lobes commonly erect with inflexed tips; flowers 3-5 mm. long -7. C. indecora
               7. Corolla not fleshy-papillate, its lobes not both erect and inflexed at tips (may be either one); flowers not over 3 mm. long (except 2-4 in C. gronovii)
                  8. Corolla lobes ovate, no longer than wide, obtuse; capsules ovoid, longer than wide
                     9. Scales (corolla appendages) truncated or bifid at apex, mostly shorter than the corolla tube; styles not over 1/4 as long as the capsule; seeds about 2-3 mm. long -8. C. megalocarpa
                     9. Scales usually not truncated or bifid at apex, usually nearly as long or as long as the corolla tube; styles usually about 1/3 or more the length of the capsule; seeds about 1.5 mm. long -9. C. gronovii
                  8. Corolla lobes triangular or lanceolate, longer than wide, acute; capsules globose or depressed-globose, no longer than wide
                     10. Flowers about 1.5 mm. long (or in fruit to 2 mm.); calyx lobes widely overlapping in the sinuses protruding to form angles -10. C. pentagona
                     10. Flowers 2-3 mm. long; calyx lobes only slightly overlapping but not protruding to form angles -11. C. campestris

1. Cuscuta approximata Babington, Ann. & Mag. Nat. Hist. 13:253. 1844.
    C. gracilis Rydb.; C. anthemi Nels.---Flowers 3-4 mm. long, sessile in few- to several-flowered glomerules; calyx lobes triangular-ovate, more or less pointed, overlapping in the sinuses, somewhat fleshy, about as long as the corolla tube; corolla campanulate, soon globose, lobes spreading, triangular-ovate, obtusish, about equalling the tube to shorter; scales not as long as the corolla tube, shallow-fringed above; styles separate, stigmas nearly as long as the styles, narrowly linear, often reddish; capsule wider than high, depressed-globose, circumscissle in a definite line near base. Our plants are var. urceolata (Kuntze) Yuncker.---On legumes and other plants. The variety from Eurasia and Africa; established in various parts of the western United States. Our few records in northcentral and northwestern Colorado, on Medicago and Argemone, at 5000-6500 feet.

2. Cuscuta cuspidata Engelm., Bost. Journ. Nat. Hist. 5:224. 1847.
    Flowers about 4 mm. long, pedicelled or subsessile; 1 or more ovate to orbicular bracts closely subtending the calyx; calyx lobes distinct or very slightly united, sepals about 1/2 as long as the corolla tube, orbicular to ovate, obtuse or acute; corolla funnelform or subcylindrical, the lobes shorter than the corolla tube and narrowly ovate, acute or obtuse, usually spreading; scales fringed, shorter than the corolla tube; styles longer than the ovary, separate, stigmas capitate; capsule globose, carrying the withered corolla at the apex; seeds about 1.4 mm. long.---On species of Compositae and Leguminosae. Indiana to North Dakota and Utah, south to Louisiana, Texas and Colorado. Our few records scattered in the eastern one-third of Colorado at 4000-5500 feet.

3. Cuscuta occidentalis Millspaugh in Mill. & Nutt., Field Mus. Nat. Hist. Bot. 5:204. 1923.
    C. californica var. breviflora Engelm.---Flowers about 3 mm. long, glabrous or often glandular, mostly sessile in small compact clusters; calyx about 1-1.5 mm. long, shorter than the corolla tube, the lobes ovate-lanceolate, acuminate, rather fleshy; corolla globose-campanulate, the lobes about as long or somewhat shorter than the tube, lanceolate, acuminate, spreading, giving a starlike appearance from above; scales absent; styles separate, somewhat longer than the globose ovary, stigmas capitate; capsule globose; seeds about 1.5 mm. long.---On various hosts. Colorado to the Pacific States. Our one record from westcentral Colorado, on a species of Compositae, at 5500 feet.

4. Cuscuta cephalanthi Engelm., Am. Journ. Sc. & Arts, 43:336. 1842.
    On various plants such as species of Compositae and Labiatae. Maine to Oregon, south to Virginia and Texas and westward. This plant has been reported for contiguous states and should be looked for in Colorado

5. Cuscuta coryli Engelm., Am. Journ. Sci. & Arts 43:337. 1842.
   On various woody and herbaceous plants. Rhode Island to Montana, south to Virginia, Texas and Arizona. Not reported for this state but the plant should be expected here since Colorado is within its recorded range.
6. Cuscuta umbellata H. B. K., Nov. Gen. et Sp. Pl. 3:121. 1818.
   Flowers 2-3 mm. long, smooth, rarely puberulent, on pedicels shorter or longer than the flowers, in dense compound cymes; calyx lobes about as long as the corolla tube, triangular-ovate, acute to acuminate at apex, definitely joined at base; corolla broadly campanulate, lobes about as long as the tube, reflexed, lanceolate, acute to acuminate; scales moderately fringed, about as long as the corolla tube; styles longer than the ovary, stigmas capitate; capsule globose, circumscissle finally by a definite ring at base; seeds about 1 mm. long.---On a large number of herbaceous plants. Texas to Colorado and Arizona; Mexico and South America. Our one record from southern Colorado, on genus Salsola, at 5000 feet.
7. Cuscuta indecora Choisy, Soc. Phys. Nat. Hist. Geneve Mem. 9:278. 1841.
   Flowers 3-5 mm. long, fleshy, papillate, on pedicels longer than the flowers, in cymose clusters; calyx lobes shorter than or equalling the corolla lobes, triangular-ovate, acute to obtuse; corolla broadly campanulate, lobes shorter than the tube, triangular, acute with inflexed tips; scales about as long as the corolla tube, deeply fringed; styles about as long as the ovary, separate, stigmas capitate; capsule globose; seeds about 1.7 mm. long. Our plants are var. neuropetala (Engelm.) Hitchc.---On a large number of herbaceous hosts especially on the families Compositae and Leguminosae. The variety from Illinois to California, south to Florida and Mexico; West Indies; South America. Our records from the eastern half of Colorado, mostly on Compositae and Leguminosae, at 4500-6000 feet.
8. Cuscuta megalocarpa Rydb., Bull. Torr. Club 28:501. 1901.
   C. curta (Engelm.) Rydb.---Stems coarse; flowers 2-3 mm. long on short pedicels mostly shorter than the flowers, in cymose panicles, these becoming somewhat globular in fruit; calyx lobes about 1/2 as long as the corolla tube, ovate, obtuse, definitely joined at base; corolla campanulate, the lobes triangular usually becoming spreading or reflexed in fruit; scales more or less fringed or divided; styles short, about 1/4 the length of the globose-conical ovary, separate, stigmas capitate; capsule 3-6 mm. long and somewhat longer than wide, globose-conical; seeds 2-2.8 mm. long.---On various plants, these mostly herbaceous. Minnesota to Wyoming, south to New Mexico. Our few records from central and southcentral Colorado at 7000-7500 feet.
9. Cuscuta gronovii Willd. in R. & S., Syst., 6:205. 1820.
   C. gronovii var. vulgivaga Engelm.---On a large number of woody and herbaceous plants. Nova Scotia to Manitoba, south to Florida, Texas and Arizona. This plant has been reported from Colorado and should be present but the writer has seen no specimens from the state.
   A specimen from the southwestern corner of the state has been provisionally identified as the related species C. denticulata Engelm. It has a mostly 1-seeded capsule and the styles are always much shorter than the ovary.
10. Cuscuta pentagona Engelm., Am. Journ. Sci. & Arts, 43:340. 1842.
    On many herbaceous plants. Massachusetts to California, south to Florida. This plant was reported from Colorado but the writer has seen no specimens from the state.
11. Cuscuta campestris Yuncker, Mem. Torr. Bot. Club 18:138. 1932.
    C. pentagona var. calycina Engelm.; C. arvense of Manuals---Flowers about 2-3 mm. long, on pedicels mostly shorter than the flowers, in compact, globular clusters; calyx lobes about as long as the corolla tube, oval or short-orbicular, mostly overlapping at sinuses but not protruding as angles; corolla lobes about as long as the corolla tube, triangular, acute, tips often inflexed, lobes usually spreading; scales at least as long as the corolla tube, fringed; styles 2, stigmas capitate; capsule usually depressed-globose; seeds about 1.5 mm. long.---On various herbaceous plants, often on alfalfa and clover. Virginia to Saskatchewan, south to California and New Mexico; South America. Our records scattered over Colorado at 3500-6000 feet.

2. Evolvulus L.

Perennial herbaceous plants, silky-pubescent or silky-hirsute; stems not climbing, branching; leaves alternate, entire, narrow; flowers solitary and axillary, not very showy; sepals nearly equal; corolla funnelform-campanulate, purple, blue or light rose, 5-angled; ovary 2-celled, styles 2, distinct, each 2-cleft, stigmas long and filiform; capsule 2- to 4-valved, 1- to 2-seeded.

1. Evolvulus nuttallianus R. & S., Syst. Veg. 6:198. 1820.
   E. pilosus Nutt.; E. argenteus Pursh---Stems 10-25 cm. long, several from a somewhat woody base, erect or ascending, branched below; leaves 8-20 mm. long, subsessile, linear, oblong or oblanceolate; pedicels shorter than sepals, reflexed in fruit, bracted; sepals 4-5 mm. long, long-acuminate, linear-lanceolate; corolla about 8-12 mm. wide, purple-rose to rose, rotate or broadly funnelform.---Sandy plains and hills. North Dakota to Montana, south to Tennessee, New Mexico and Arizona. Our records scattered over Colorado, except in the northwestern corner, at 3500-5000 feet.

3. Ipomoea L. MORNING GLORY

Herbaceous annuals or perennials; stems prostrate or climbing; leaves entire to lobed; flowers solitary or in few-flowered clusters; outer sepals commonly larger than the inner ones; corolla mostly funnelform, the limb entire or very shallowly lobed; style 1, unbranched, stigma 1, capitate or nearly so, entire or lobed.

1. Perennial plants; stems not at all twining; leaves linear-lanceolate or oblong-lanceolate, not cordate at base, never lobed
   2. Corolla white or cream-colored; stems prostrate (except near base)   -1. I. longifolia
   2. Corolla purple, rose or pink-purple; stems erect to ascending, not prostrate   -2. I. leptophylla
1. Annual plants; stems twining; leaves ovate-lanceolate to broadly ovate, cordate at base or lobed
   3. Leaves entire; sepals acute; corolla 4-6 cm. long   -3. I. purpurea
   3. Leaves 3-lobed; sepals caudate-acuminate at apex; corolla 2.5-4 cm. long   -4. I. hederacea

1. Ipomoea longifolia Benth., Pl. Hartw. 16. 1839.
   Reported from Oklahoma, south to Arizona and Mexico. The plant should be looked for in eastern and southern Colorado.
2. Ipomoea leptophylla Torr. in Frem., Rep. 94. 1845.
   Perennial plants with deep-seated enormous roots often weighing 25 pounds, glabrous throughout; stems 30-120 cm. long, several, erect to ascending, rarely ascending-prostrate; leaves 5-15 cm. long, linear to linear-lanceolate, entire, acute, cuneate; petioles rather short; peduncles short, 1- to 4-flowered; sepals 6-10 mm. long, ovate to broadly ovate, unequal; corolla 4.5-8 cm. long, pink-purple to rose, funnelform; capsule 2-celled, about 16-25 mm. long.---Dry often sandy plains and banks. South Dakota to Montana, south to Texas and New Mexico. Our records from the eastern half of Colorado at 3500-5500 feet.
3. Ipomoea purpurea (L.) Lam., Tabl. Encyl. 1:466. 1797.
   Cultivated ground and waste places, often as an escape from cultivations. Nova Scotia to Colorado, south to Florida and Texas. This plant has been reported from Colorado but the writer has never located a specimen from the state.
4. Ipomoea hederacea (L.) Jacq., Collect. 1:124. 1786.
   Fields and thickets in dry, often sandy soil. Virginia to Kansas, south to Florida and tropical America. This plant should be looked for in eastern and southern Colorado.

4. Convolvulus L. BINDWEED; MORNING GLORY

   Plants perennial, trailing or twining herbaceous vines; leaves usually more or less lobed or auriculate at base; flowers axillary, solitary or clustered, rather showy; sepals nearly equal or the outer larger, often with a pair of bracts at base; corolla broadly funnelform, the limb entire to 5-lobed; styles 1, filiform, stigmas 2, ovate-oblong to linear; ovary with 4 ovules; capsule globose or subglobose, 4 seeds in the 2 cells (or by abortion in 1 cell), 2- to 4-valved.

1. Calyx subtended by a pair of ovate bracts longer than the sepals (10-20 mm. long); corollas over 3 cm. long   -1. C. sepium
1. Bracts well away from the calyx, linear or narrower, shorter than the sepals (not over 5 mm. long); corollas less than 3 cm. long
   2. Basal lobes of leaves (on some at least) with linear or lanceolate teeth or lobes; plants canescent-sericeous or cinereous; corolla lobes definite and pointed; plants apparently limited to southern Colorado, hardly weeds   -2. C. incanus
   2. Basal lobes of leaves entire or sparingly dentate; plants glabrous or pubescent; corolla lobes hardly if at all noticeable, not pointed; plants generally distributed over Colorado, very serious weeds   -3. C. arvensis

1. Convolvulus sepium L., Sp. Pl. 153. 1753.
   C. americanus (Sims) Greene; C. interior House and C. repens L. appear not to be specifically distinct---Plants twining or climbing, sometimes prostrate, glabrous or more usually sparsely hairy but occasionally densely pubescent; stems up to 300 cm. long; leaves 3-10 cm. long, petioled, hastate with the basal lobe angled, or sagittate with basal lobe rounded or slightly pointed; peduncles from shorter to longer than the leaves, 1-flowered, with a pair of ovate or cordate-ovate bracts about 1-3 cm. long at the summit near the flower; sepals smaller than the bracts; corollas about 4-6 cm. long, white to pink; stigmas oblong or nearly oval. Tryon (Rhodora 41:415-423. 1939.) listed var. americanus Sims from Colorado.---Fields, roadsides and waste places. Introduced weed in most of North America. Our records scattered over the state, especially in cultivated areas, at 3500-8500 feet.
2. Convolvulus incanus Vahl., Symb. 3:23. 1794.
   Stems mostly procumbent or scrambling over bushes, cinereous or canescent-sericeous; leaves about 2-5 cm. long, variable, from linear-sagittate to oblong-ovate and hastate, basal lobes of some toothed or cleft and middle division more or less toothed, canescent-sericeous; peduncles 1- to 3-flowered, with a pair of small bracts (these up to 4 mm. long and linear-filiform) well away from the flower; corollas 1-2 cm. long, white to tinged with rose especially in the middle, the lobes definite and pointed; stigmas linear to filiform.---Dry hills and plains. Nebraska and Colorado, south to Texas and Arizona. Our records from the southeastern quarter of Colorado at 3500-6000 feet.
   A wide-leafed plant has been called C. hermannioides A. Gray and has been reported for southeastern Colorado. The writer cannot separate it from the above.
3. Convolvulus arvensis L., Sp. Pl. 153. 1753.
   C. ambigens House---Stems prostrate from horizontal roots, often in dense colonies, glabrous to densely-pubescent; leaves 2-5 cm. long, glabrous to short-pubescent, very variable in shape (ovate to linear) but usually oblong-ovate with base more or less sagittate to hastate (often merely truncate); peduncles 1- to 2-flowered, with 2 bracts, these linear and about 2-4 mm. long, well away from the flower; corollas 1.5-2.5 cm. long, white to pink, usually with longitudinal darker bands outside, the 5 lobes barely if at all distinguishable; stigmas linear to filiform. This is probably our worst weed. Brown (Ia. State Coll. Journ. Sc. 20:269-276.

1946.) grew from same lot of seeds plants with leaves ovate-oblong and truncate to linear and hastate, with all intergradations between. He concluded no varieties can be maintained.---Waste places and fields. Now naturalized in most of North America. Our records scattered over the entire state, especially in cultivated areas, at 4000-8000 feet.

Family 99. Polemoniaceae PHLOX FAMILY

Annual or perennial herbaceous plants or somewhat woody at base; leaves opposite or alternate, simple and entire to compound; flowers perfect, regular, solitary or in various types of inflorescences; calyx persistent, of 5 (or 4) partly united sepals; corolla of 5 (or 4) united petals, from rotate, campanulate or funnelform to salverform, its lobes convolute in the bud; stamens 5 (rarely 4), equal or unequal, filaments attached to the corolla tube alternate to the petals; ovary superior, 3-celled, styles 1 but stigmas 3- or rarely 2-lobed; fruit a loculicidal capsule, few- to many-seeded; seeds often emitting spiral threads from their mucilaginous coats when moistened.
    The writer wishes to acknowledge special help given on this family by Dr. Edgar Wherry who identified many of the specimens and provided preliminary lists and keys. The writer is responsible for any errors.

1. Calyx tube remaining unbroken to maturity of capsule; leaves all alternate, opposite only near base of stem, or lacking entirely
  2. Leaves lacking, represented by persistent cotyledons and bracts around the head of flowers -1. Gymnosteris
  2. Ordinary leaves present
    3. Leaves simple and entire; sepal-junction membrane gibbous   -2. Collomia
    3. Leaves compound or pinnatifid (at least most of them); sepal-junction membrane absent or if present not gibbous
      4. Annual plants not over 10 cm. tall; leaves pinnatifid into setaceous or spinulose divisions; corolla small, less than 8 mm. long   -3. Navarretia
      4. Perennial plants normally over 10 cm. tall; leaves pinnately compound or if pinnatifid the divisions not setaceous or spinulose; corolla larger, over 8 mm. long   -4. Polemonium
1. Calyx tube splitting along the sepal-junction membrane at maturity of capsule; leaves all opposite although bracts may be alternate (except in Gilia)
  5. Leaves simple and entire, opposite (bracts may be alternate)
    6. Perennial plants; corolla tube cylindrical to base, over 6 mm. long; calyx scarcely accrescent, the sepals uniform; seeds unchanged when moistened   -5. Phlox
    6. Annual plants; corolla tube flaring at base, less than 6 mm. long; calyx somewhat accrescent, the sepals somewhat unequal; seeds becoming sticky when moistened   -6. Microsteris
  5. Leaves more or less lobed or parted, opposite or alternate (sometimes entire in Gilia but alternate)
    7. Leaves pinnately lobed or parted (rarely entire), alternate (or lower only opposite); plants rarely woody at base   -7. Gilia
    7. Leaves palmately lobed or parted; opposite or upper appearing somewhat alternate; plants woody at base (unless annual)
      8. Plants annual   -8. Linanthus
      8. Plants perennial, woody at base or in caudex
        9. Junction membrane of calyx broad; woody tissue well developed (plants definitely woody above base); leaves often somewhat alternate above with acerose divisions; corollas usually over 14 mm. long   -9. Leptodactylon
        9. Junction membrane of calyx narrow or lacking; plants woody only in crown; leaves all strictly opposite, the divisions subacerose; corollas not over 14 mm. long   -10. Linanthastrum

1. Gymnosteris Greene

Small annual plants with simple stems; proper leaves none, the foliage represented by persistent connate-perfoliate cotyledons near base of stem and the bracts; flowers few, in terminal heads, the 4 or 5 bracts scarious at base, partly united as an involucre; calyx membranous, not folded in sinuses, not breaking by the expanding fruit, its lobes more or less unequal; corollas salverform or somewhat funnelform; anthers borne in the corolla throat; seeds mucilaginous but not developing spiral threads when wet, membranous-margined or narrowly winged.

1. Gymnosteris parvula (Rydb.) Heller, Muhl. 1:3. 1904.
    G. nudicaulis of Manuals---Plants 2-3 cm. tall, rarely taller, glabrous; bracts 5-15 mm. long, ovate-lanceolate; calyx 3-4 mm. long with ovate, acute teeth, these herbaceous, rest of calyx scarious; corolla white or pinkish, yellowish or bronze, its tube about 4-6 mm. long, expanding at base in age to enclose the developing fruit, lobes about 1.5-3 mm. long; capsules dehiscent.---Plains and mesas, often in sandy soil. Wyoming to Oregon, south to Colorado and California. Our records in northcentral and northwestern Colorado at 7000-8500 feet.

2. Collomia Nutt.

Annual caulescent plants; leaves sessile, alternate, or opposite below, entire; flowers in subcapitate terminal clusters with foliaceous bracts; calyx obpyramidal or cup-shaped, its sepals equal or nearly so, with outwardly projecting folds (gibbous) in their sinuses especially in age, the tube not ruptured by the developing capsule; corolla funnelform to salverform; stamens unequally inserted on the corolla tube; seeds mucilaginous and developing spiral threads when wet.

1. Corolla white to salmon-pink or apricot-colored, usually over 20 mm. long, the limb 1 cm. wide or more; calyx lobes ovate-lanceolate, obtuse -1. **C. grandiflora**
1. Corolla whitish to rose or lilac-purple, not over 20 mm. long, the limb seldom over 6 mm. wide; calyx lobes subulate or linear-lanceolate, acute -2. **C. linearis**

1. **Collomia grandiflora** Dougl. ex Lindl., Bot. Reg. 14, pl. 1174. 1828.

Stems 15-70 cm. tall, erect or branched above, canescent-puberulent especially above; leaves 2.5-8 cm. long, linear to linear-lanceolate; bracts ovate, somewhat viscid; flowers in few- to many-flowered clusters on the end of the stem or its branches; calyx more or less chartaceous especially in fruit, the lobes herbaceous, ovate-lanceolate, obtuse; corolla about 2-3 cm. long, white when fresh to salmon- or apricot-colored especially in age, usually at least 3 times as long as the calyx.---Gravelly or rocky ground. Montana to British Columbia, south to Arizona and California. Our few records from the southwestern corner of Colorado at 7000-8000 feet.

2. **Collomia linearis** Nutt., Gen. Pl. 1:126. 1818.

Stems 5-30 cm. tall, usually simple but sometimes branched, erect, puberulent or glandular above; leaf blades 1.5-6 cm. long, linear to linear-lanceolate; bracts ovate or lanceolate-ovate, often paler at base or tinged with red; flowers in few- to many-flowered clusters terminal on the stem or branches; calyx rather chartaceous in fruit, the lobes subulate or narrowly lanceolate, acute; corolla 6-15 mm. long, rose, lilac-purple to nearly white, usually about twice as long as the calyx.---Usually open ground, often on sandy soil. Quebec to British Columbia, south to Minnesota, Colorado, Arizona and California. Our records well scattered over the western two-thirds of Colorado at 5000-10,000 feet.

3. Navarretia R. & P.

Small annual plants; leaves alternate, setaceously or spinulosely pinnatifid or the lower subentire; flowers in crowded, bracted terminal clusters; calyx tube not accrescent, not broken by the expanding fruit, scarious but not folded or gibbous in the sinuses, the lobes spiny, mostly with 4 or 5 unequal lobes; corolla white to yellow, salverform or funnelform, lobes 4 or 5; stamens 4 or 5, inserted at the throat of the corolla; capsules regular, dehiscent by valves or splitting irregularly; seeds separate or agglutinated in masses, usually mucilaginous and emitting spiral threads when wetted.

1. Corolla white, usually with 4 lobes; styles 2-cleft; herbage not glandular; capsules splitting irregularly -1. **N. minima**
1. Corolla yellow, usually with 5 lobes; styles 3-cleft; herbage glandular-puberulent; capsules regularly dehiscent -2. **N. breweri**

1. **Navarretia minima** Nutt., Journ. Acad. Nat. Sci. Phila. II,1:160. 1848.

Gilia minima (Nutt.) A. Gray---Stems 1-6 cm. high, simple or branched near base, glabrous or minutely puberulent; leaves 7-25 mm. long, glabrous or nearly so, the pinnate divisions linear-subulate and acerose; bracts similar to leaves but smaller; calyx teeth usually as long or longer than the corolla, 4, subulate-acerose, entire or spiny-toothed; corolla about 3-7 mm. long, white, lobes 4; stamens 4; styles 2-cleft or rarely entire; capsule with hyaline walls, splitting irregularly; seeds, 1-8 agglutinated together.---Dry often sandy ground. Nebraska to Washington, south to Arizona and California. Our records in northcentral and northwestern Colorado at 6500-8500 feet.

2. **Navarretia breweri** (Gray) Greene, Pitt. 1:137. 1887.

Gilia breweri Gray---Stems 1-10 cm. tall, simple or branched below, glandular-puberulent; leaves 4-12 mm. long, the pinnate divisions linear-subulate, acerose, glandular-puberulent; bracts similar to leaves but smaller; calyx teeth usually as long or longer than the corolla, subulate-acerose; corolla about 6-8 mm. long, yellow, lobes 5; stamens 5; styles 3-cleft; capsules with rather firm walls, splitting regularly; seeds not agglutinated.---Open slopes and flats, often in sandy ground. Wyoming to Oregon, south to Arizona and California. Our records from western Colorado at 6500-8500 feet.

4. Polemonium L. JACOB'S LADDER; SKUNK LEAF

More or less caulescent, perennial herbaceous plants; leaves alternate, pinnately compound or pinnately parted, leaflets sessile; flowers solitary to variously clustered; calyx campanulate, 5-cleft, not scarious below the sinuses; corolla funnelform, rotate to nearly cylindrical; filaments equally inserted, often hairy at base; ovules 3-several in each cell; fruit 3-valved; seeds often angled or narrowly winged.

1. Leaflets (on all or at least the basal leaves) appearing verticillate on the leaf rachis, the leaflets terete in general makeup, more than 2-ranked; corollas over 2 cm. long, the lobes about 1/3 as long as the tube -1. **P. viscosum**
1. Leaflets 2-ranked on all the leaves, flat; corollas often less than 2 cm. long, the lobes exceeding or subequal to the tube
    2. Stems erect, each single from the end of a rootstock, simple or branched above, usually over 30 cm. tall; leaves predominantly cauline
        3. Upper leaflets not confluent; inflorescence a narrow thyrse; upper cauline leaves reduced and sparse -2. **P. occidentale**
        3. Upper leaflets confluent; inflorescence a corymbose cyme; cauline leaves not especially reduced above -3. **P. foliosissimum**
    2. Stems spreading, the lower part of stem or caudex divergent-branching, stems less than 30 cm. tall; leaves predominantly basal

4. Corolla over 15 mm. long; calyx over 8 mm. long -1. P. viscosum
4. Corolla not over 12 mm. long; calyx not over 7 mm. long -4. P. delicatum

1. Polemonium viscosum Nutt., Journ. Acad. Nat. Sc. Phila. II, 1:154. 1848.
   P. confertum A. Gray; P. grayanum Rydb.; P. brandegei (Gray) Greene; P. speciosum Rydb.; P. mellitum (Gray) A. Nels.---Plants spreading from more or less divergently branched caudices; stems 5-30 cm. tall, rarely taller, glandular-puberulent to viscid-pilose, sparingly leafy; leaflets 1-10 mm. long, usually verticillate and rather terete, linear to oval, 30-60 present; flowers in subcapitate to elongated racemes; calyx 8-16 mm. long, the segments 1/3 as long as the tube, densely glandular-puberulent; corolla 17-40 mm. long, the lobes definitely shorter than the tube, blue, yellow or white. Several subspecies have been proposed (Univ. of Calif. Pub. Bot. 23:262-267. 1950).---In the mountains. Alberta and British Columbia, south to New Mexico, Arizona and Oregon. Our records well scattered over the mountainous part of Colorado at 6000-13,500 feet.
2. Polemonium occidentale Greene, Pitt. 2:75. 1890.
   Stems 30-90 cm. tall, strict, simple, glandular-puberulent above, glabrate to sparsely pubescent below; cauline leaves reduced upwards in size and number, leaflets 7-40 mm. long, 15-27 in number 2-ranked, lanceolate to ovate-oblong, acute or acuminate; inflorescence a narrow thryse; calyx 6-8 mm. long, segments longer than the tube, glandular-puberulent; corolla 10-15 mm. long, the lobes about twice as long as the tube, blue or violet; stamens shorter than the corolla. Our plant has been called ssp. typicum Wherry and recently was made a subspecies of P. caeruleum L. (Univ. Calif. Pub. Bot. 23:225. 1950).---Valleys and open woods, usually in moist places. Saskatchewan to the Yukon, south to Colorado, Utah and California. Our records from the western half of Colorado at 7500-10,000 feet.
3. Polemonium foliosissimum Gray, Syn. Fl. N. Amer. 2(1):151. 1886.
   P. archibaldae; P. molle Greene; P. robustum Rydb.; P. grande Greene---Stems 30-90 cm. tall, erect, simple or terminally branched, glabrous to villous, often glandular above; leaves little reduced upwards; leaflets about 12-25 mm. long, 2-ranked, commonly confluent near tip, lanceolate to oblong-lanceolate; inflorescence a corymbose cyme; calyx 6-10 mm. long, segments as long as or longer than the tube, often glandular; corolla 8-20 mm. long, the lobes about twice as long as the tube, purplish to white; stamens included. The various subspecies and forms described intergrade in Colorado.---Mountains. Wyoming and Idaho, south to New Mexico and Nevada. Our records scattered in the western half of Colorado at 5000-10,500 feet.
4. Polemonium delicatum Rydb., Bull. Torr. Club 28:29. 1901.
   P. scopulinum Greene---Plants rather caespitose with numerous basal leaves and few cauline ones; stems 5-20 cm. tall, slender, spreading from a caudex or a more or less divergently branched and slender rootstock; leaflets 5-20 mm. long, 11-23 in number, 2-ranked, ovate-lanceolate to elliptical; calyx 3-6 mm. long, segments longer than the tube, glandular; corolla 6-12 mm. long, lobes usually longer than the tube, blue or violet; stamens somewhat included. The various subspecies described intergrade in this area.---Mountains. Colorado, Utah, New Mexico, and Arizona. Our records well scattered over the western half of Colorado at 8500-14,000 feet.

5. Phlox L. PHLOX

Plants perennial, herbaceous or somewhat woody; leaves opposite, entire; inflorescence cymose, corymbiform or paniculate, sometimes reduced to a solitary flower; sepals 5, uniform, the lobes about equal, calyx little if any accrescent, splitting at sepal junction from pressure of developing fruit; corolla usually showy, typically salverform, the tube cylindrical, lobes 5; stamens 5, included, unequally inserted on corolla tube; seeds unchanged when moistened.

1. Shoots tending to be elongated and little branched, usually over 10 cm. long; inflorescence compound, several to many-flowered; habit erect to somewhat caespitose
   2. Corolla tubes 20-25 mm. long; plants of southwestern and southern Colorado -1. P. stansburyi
   2. Corolla tubes 10-19 mm. long; plants not limited to southwestern or southern Colorado
      3. Junction membrane of calyx gibbous-carinate; glandular hairs sometimes present; plants widely distributed in mountainous 2/3 of Colorado -2. P. longifolia
      3. Junction membrane of calyx flat or nearly so; glandular hairs never present; plants of southwestern, western or northwestern Colorado
         4. Corolla tubes 14-17 mm. long, the lobes 7-10 mm. long; calyx 11-15 mm. long, plants 12-20 cm. tall; present in southwestern Colorado -3. P. caryophylla
         4. Corolla tubes about 12 mm. long, the lobes about 4 mm. long; calyx 7-11 mm. long; plants usually over 20 cm. tall; present if at all in northwestern Colorado -4. P. grahami
1. Shoots tending to be short and branched, in most species usually less than 10 cm. long; inflorescence of 1-5 flowers; habit caespitose, often pulvinate
   5. Largest leaves over 2 mm. wide
      6. Longest leaves averaging 30-32 mm. long; flowers on pedicels, the longer ones 10-30 mm. long -5. P. patula
      6. Longest leaves averaging 20 mm. long or less; flowers sessile or on pedicels 1-12 mm. long
         7. Nodes few, well spaced; leaves not especially cartilaginous or white and thickened on margins -6. P. variabilis
         7. Nodes several to many, very short; leaves thickened and white-cartilaginous on the margins -7. P. alyssifolia
   5. Largest leaves 0.5-2 mm. wide
      8. Leaves coarsely ciliate at base with stiff hairs, commonly glandular-puberulent; plants of mountains -8. P. caespitosa

8. Leaves finely ciliate, if at all, near base with flaccid or crinkly weak hairs, seldom glandular (and then only in a plains species); plants of mountains and plains
   9. Junction membrane of calyx decidedly carinate; leaf texture firm, the tips more or less acerose
          -9. P. austromontana
   9. Junction membrane of calyx flat (or slightly carinate in P. diffusa with leaf texture not firm); leaf texture usually flaccid to moderately firm
      10. Longest leaves averaging 10-30 cm. long and 1-2 mm. wide (some over 15 mm. long and 1.5 mm. wide); calyx 7-11 mm. long; corolla tube 12-15 mm. long
         11. Inflorescence 1-flowered; habit compact-caespitose; leaves narrowly oblong to linear; plants of the mountains and high foothills
               -10. P. multiflora
         11. Inflorescence 1- to 5-flowered (usually over 2); habit open-caespitose; leaves subulate to linear; plants of the plains
               -11. P. andicola
      10. Longest leaves averaging 2-12 mm. long and 0.5-1.5 mm. wide (none over 15 mm. wide); calyx 3-8 mm. long; corolla tube 6-13 mm. long
         12. Stems and leaves copiously arachnoid-hairy; leaves elliptic to lanceolate, hardly subulate, 3-5 mm. long, closely imbricate
               -12. P. bryoides
         12. Stems and leaves glabrate to sparsely pilose or pubescent, not copiously arachnoid; leaves subulate or nearly so, 3-12 mm. long, crowded or somewhat remote
            13. Junction membrane of calyx more or less carinate; decumbent stems well developed; nodes more or less spaced
               -13. P. diffusa
            13. Junction membrane of calyx flat, not at all carinate; decumbent stems little developed; nodes crowded
               -14. P. hoodii

1. **Phlox stansburyi** (Torrey) Heller, Bull. Torr. Club 24:478. 1897.
   Habit erect and openly caespitose; stems sparingly branched, shoots usually over 10 cm. long (10-20 cm.), more or less hairy and often glandular; leaves 2-3 cm. long and about 2-3 mm. wide, linear or linear-lanceolate, rather thick in texture; flowers several to many; calyx about 10 mm. long, more or less carinate in junction-membrane below the sinuses in the tube, lobes about as long as the tube; corolla pink or rose, tube 20-25 mm. long, lobes 7-8 mm. long. Our plant has been called ssp. eustansburyi Brand.---Dry areas. The subspecies from Colorado to California, south to New Mexico and Arizona. Our one record from southwestern Colorado at 6500 feet.

2. **Phlox longifolia** Nutt., Acad. Nat. Sc. Phila. Journ. 7:41. 1834.
   P. puberula (E. Nels.) A. Nels.; P. cernua E. Nels.---Base more or less woody, often prostrate with ascending or erect branches; stems 10-30 cm. tall, sometimes rather openly caespitose, erect, from glabrous to pubescent or glandular-pubescent, the shoots simple or sparingly branched, the nodes remote; leaves about 2-8 cm. long, narrowly linear; inflorescence several- to many-flowered, the flowers pedicellate; calyx 7-11 mm. long, glandular, pubescent or glabrous, the lobes somewhat shorter than the tube, junction-membrane between sepals broad and gibbous-carinate; corolla white to rose, its tube 12-19 mm. long, its lobes 6-10 mm. long; seeds 2.5-4.5 mm. long with large embryos. Wherry has named 5 subspecies that are present in Colorado. These are ssp. calva Wherry; ssp. cortezana (A. Nels.) Wherry; ssp. humilis (Dougl.) Wherry; ssp. longipes (Jones) Wherry; and ssp. typica Wherry. The subspecies intergrade very much but the extremes do look different.---Plains and hills. Wyoming to British Columbia, south to New Mexico and California. Our records well scattered over the western two-thirds of Colorado at 4500-8500 feet.

3. **Phlox caryophylla** Wherry, Not. Nat. No. 146:4. 1944.
   Plants about 12-20 cm. high, woody at base, open-tufted, pilose-pubescent with more or less remote nodes; leaves 3-5 cm. long and 2-3 mm. wide, linear to oblong, short-acuminate; inflorescence of 3-12 flowers on fairly long pedicels up to 2.5 cm. long; calyx 11-15 mm. long, broadly scarious-membranous below the sinuses, these flat or somewhat wrinkled but not carinate, lobes slightly shorter than the tube; corolla bright purple or pink at least in the lobes, the tube 14-17 mm. long, the lobes 7-10 mm. long.---Sparsely wooded slopes. Colorado and New Mexico. Our few records from southwestern Colorado at 6500-7000 feet.

4. **Phlox grahami** Wherry, Britt. 5:63. 1943.
   Described from the northeastern corner of Utah and to be looked for in adjacent Colorado.

5. **Phlox patula** A. Nels., Univ. Wyo. Pub. in Bot. 1:48. 1924.
   Low caespitose plants somewhat woody at base, these branches spreading assurgent or prostrate; season's stems suberect, 6-15 cm. tall, nodes few, glabrous or pubescent on peduncles; leaves 3-6 pair present, longest averaging 30 mm. long, larger ones 2-3 mm. wide, (internodes 1 cm. or more long), glabrous, linear; flowers 1 to several, single on branches and on definite peduncles 10-30 mm. long; calyx about 10-14 mm. long, glabrous or nearly so, membranous between the prominent costa, this not carinate but flat or essentially so, calyx lobes as long or longer than the tube; corolla white to pink, the tube 12-15 mm. long, a little longer than the calyx, the lobes 8-10 mm. long, oval-suborbicular; seeds 2-3 mm. long, the embryos small. Our material intergrades with P. multiflora A. Nels.---Foothills. Colorado. Our records from central and northcentral Colorado at 5000-7000 feet.

6. **Phlox variabilis** Brand in Eng., Pflanzenreich, Polemon. 87. 1907.
   Phlox kelseyi ssp. variabilis (Brand) Wherry---Habit dwarf and caespitose; stems 2-10 cm. tall, nodes crowded and stems concealed; longest leaves 10-15 mm. long and 2 mm. wide or more, oblong or linear-oblong, ciliate and often glandular, margins whitish but not very thickened; flowers sessile or nearly so, solitary or very few; calyx 6-8 mm. long, lobes rather thick, membrane between not carinate, usually glandular; corolla white to blue, tube 8-12 mm. long (in ours, but 8-18 mm. recorded), lobes 4-8 mm. long; seeds 2-3 mm. long, embryos small.---High mountains. Colorado and Utah. Our few records from northcentral and southwestern Colorado, the altitude uncertain but probably above 10,000 feet.

7. **Phlox alyssifolia** Greene, Pitt. 3:27. 1896.
Saskatchewan, south to South Dakota and Utah. The ssp. vera Wherry is to be expected in northern Colorado.
8. **Phlox caespitosa** Nutt., Acad. Nat. Sci. Phila. 7:41. 1834.
P. condensata (Gray) E. Nels.---Low caespitose, often pulvinate plants, about 3-10 cm. tall, the season's stems often branched, erect or spreading; leaves 5-14 mm. long and 1-2 mm. wide, much imbricated, the internodes short, rigid, linear or oblong-linear, apiculate, usually flat, more or less thickened on the margins and coarsely-ciliate at least near base, surface glabrous to (more commonly) short glandular-puberulent; flowers sessile or nearly so, solitary on end of branches; calyx about 6-8 mm. long, hispid-ciliate and usually glandular-puberulent, the lobes about as long as the tube, narrowly lanceolate and pungent, tube scarious below sinuses, these not carinate, costa medium to strong; corolla white or light blue, the tube about 10 mm. long, definitely longer than the calyx, the lobes about 4-7 mm. long; seeds 2-3 mm. long with small embryos. Our plants would be in 3 subspecies, ssp. condensata (Gray) Wherry; ssp. eucaespitosa Brand and ssp. pulvinata Wherry.---Mountains. Montana to Oregon, south to New Mexico and California. Our records scattered in the higher mountains of Colorado at 10,500-13,000 feet.
9. **Phlox austromontana** Coville, Contrib. U. S. Nat. Herb. 4:151. 1893.
Habit dwarf-tufted and caespitose; stems 5-10 cm. long, canescent-pubescent, nodes crowded; leaves about 1-1.5 cm. long and 0.5-1.5 mm. wide, acerose, rather firm, more or less hairy but not glandular; flowers 1 to few, subsessile; calyx 6-9 mm. long, the teeth shorter than the tube, the junction-membrane decidedly carinate; corollas white to purple, the tubes 8-14 mm. long, lobes about 5-7 mm. Our plant has been called ssp. vera Wherry.---Dry areas in the mountains. The subspecies from Wyoming to Oregon, south to New Mexico and California. Our few records from western Colorado at about 6500-7000 feet.
10. **Phlox multiflora** A. Nels., Bull. Torr. Club 25:278. 1898.
P. depressa (E. Nels.) Rydb.---Compact caespitose plants much branched at base; season's stems 2-10 cm. high, nearly erect, glabrous to more or less pubescent, nodes more or less crowded; leaves 1-3 cm. long and 1-2 mm. wide, linear to narrowly-oblong, glabrous or nearly so but finely ciliate near base, apiculate; flowers solitary on the branches, on pedicels 2-25 mm. long; calyx 8-11 mm. long, glabrous to pubescent, lobes subulate, about as long as the tube, scarious membrane below sinuses broad, flat and not carinate, costa conspicuous; corolla white, pink or light blue, the tube 12-14 mm. long, exceeding the calyx, the lobes 5-9 mm. long; seeds 2-3 mm. long the embryos small. Represented in Colorado by 3 subspecies, ssp. costata (Rydb.) Wherry, ssp. depressa (E. Nels.) Wherry and ssp. typica Wherry.---Plains and mountains. Montana to Idaho, south to Colorado and Utah. Our records scattered in the western two-thirds of Colorado at 6000-11,500 feet.
11. **Phlox andicola** (Britt.) E. Nels., Rev. W. N. A. Phlox, Wyo. Agr. Col. Ann. Rept. p. 11. 1899.
P. planitiarum A. Nels.---Plants more or less caespitose in colonies, spreading by rhizomes; shoots 5-10 cm. tall, branching about ground level, with whitish bark, with nodes not especially crowded, more or less arachnoid at nodes; largest leaves 1-3 cm. long and 1-2 mm. wide, linear to subulate, pungent, sparsely pubescent below more strongly so above, sometimes glandular-puberulent, finely ciliate near base with rather crinkly hairs; inflorescence 1- to 5-flowered, often over 1; calyx 7-11 mm. long, costa prominent, scarious membrane below sinuses broad and flat, not carinate, lobes about as long as the tube, subulate and pungent; corolla white, the tube 12-15 mm. long, definitely longer than the calyx, the lobes 5-8 mm., sometimes to 10 mm. long; seeds 2-3 mm. long with small embryos. The 2 intergrading subspecies, ssp. typica Wherry and ssp. planitiarum (A. Nels.) Wherry are present in Colorado.---Open plains. South Dakota to Montana, south to Nebraska and Colorado. Our records from the southeastern quarter of Colorado at 3500-5200 feet.
12. **Phlox bryoides** Nutt., Pl. Gamb. 153. 1848.
Densely caespitose and pulvinate plants, forming cushionlike mats less than 10 cm. tall (ours about 4-6 cm.); leaves 2-5 mm. long, elliptic, narrowly oblong to lanceolate, densely imbricated in 4 ranks, concave with inflexed margins, mucronate but hardly acerose, beset with arachnoid hairs; flowers solitary and sessile at ends of branches; calyx 3-6 mm. long, lanate, the teeth somewhat subulate about as long as or shorter than the tube; corollas white, tube about 6-10 mm. long, longer than the calyx, lobes about 3-5 mm. long; seeds 2-3 mm. long, with small embryos.---Dry hills, ridges and slopes often on rock outcrops. Montana, south to Nebraska and Utah. Our records scattered over the state, mostly in the northern part, at 5000-8000 feet.
13. **Phlox diffusa** Benth., Pl. Hartw. 325. 1849.
Stems with branches often decumbent, less than 10 cm. tall, glabrous to sparsely hairy; leaves 6-12 mm. long and 0.5-1.5 mm. wide, linear to linear-subulate; flowers few; calyx 5-9 mm. long, junction-membrane more or less carinate; corolla white, tube 9-13 mm. long, lobes about 5-7 mm. long.---Hills and mountains. British Columbia to Oregon, south to Wyoming. One specimen from western Colorado seems to be this species.
14. **Phlox hoodii** Rich. in Frankl., Narr. J. Pol. Sea, App. 733. 1823.
P. canescens T. & G.; P. glabrata (E. Nels.) Brand---Caespitose and tufted, the decumbent stems poorly developed; stems short, 3-10 cm. tall, sparsely pubescent or glabrate to arachnoid-canescent, with crowded nodes; larger leaves 3-12 mm. long and usually 0.5-1 mm. wide, gray-green, canescent to arachnoid-canescent rarely glabrate, imbricated, subulate; flowers solitary and sessile at ends of branches; calyx 5-8 mm. long, membrane below sinuses rather wide and nearly or quite flat, not carinate, the costa conspicuous, calyx teeth about as long as the tube, resembling the leaves; corolla white, the tube 6-12 mm. long, equalling or somewhat exceeding the calyx, the lobes 4-5 mm. long; seeds 2-3 mm. long, the embryos small. Represented in Colorado by 4 subspecies, ssp. canescens (T. & G.) Wherry; ssp. genuina Wherry; ssp. glabrata

(E. Nels.) Wherry and ssp. muscoides (Nutt.) Wherry.---Dry hills and plains. Saskatchewan and Idaho, south to New Mexico and California. Our records scattered over Colorado, except in the extreme eastern part, at 5000-9000 feet.

6. Microsteris Greene

Small, branched annual plants; leaves opposite (bracts may be alternate); flowers minute, axillary; calyx tubular, 5-cleft, more or less accrescent, splitting along line below sinuses from pressure of developing fruit, lobes more or less unequal; corolla tube flaring out at base, glabrous inside or nearly so; seeds viscid when wet.

1. Microsteris humilis (Dougl.) Greene, Pitt. 3:301. 1898.
    M. micrantha (Kellogg) Greene---Stems usually branched, 3-15 cm. tall, pubescent and usually glandular; leaf blades 6-40 mm. long, linear to lanceolate, or lower oblanceolate; calyx glandular, from shorter than to somewhat longer than the corolla; corolla 4-7 mm. long, white but limb usually purplish-tinged or rose; stamens included in the corolla throat. Closely related to and perhaps not distinct from M. gracilis (Dougl.) Greene.---Hills and valleys. Montana to British Columbia, south to Colorado and California. Our records scattered over the western three-fourths of Colorado at 4500-9000 feet.

7. Gilia R. & P.

Annual or biennial, less often perennial herbaceous plants; leaves all alternate or lower only opposite, usually more or less pinnately lobed or dissected but sometimes entire; flowers variously arranged; calyx teeth equal or subequal, the tube somewhat campanulate, scarious-membranous below each sinus and splitting finally by the pressure of the developing fruit; corolla funnelform, salverform, rotate or somewhat campanulate; stamens usually equally inserted; seeds usually many, sometimes mucilaginous when wet. As treated here the genus is rather limited. Many workers prefer to include Linanthastrum, Leptodactylon, Navarretia, and often Linanthus and Microsteris as sections under the genus Gilia. The writer follows Wherry in the treatment of these difficult groups and keeps them as separate genera.

1. Corollas rotate, the lobes longer than the tube; plants definitely woody at base -1. Gilia acerosa
1. Corollas funnelform to salverform, the lobes shorter than the tube; plants not definitely woody at base (except in G. congesta and G. roseata)
    2. Flowers congested in capitate or spicate inflorescences, the pedicels short or wanting; corolla salverform
        3. Plants perennial or possibly biennial
            4. Corolla 9-14 mm. long; styles about 1/2 as long as the corolla tube; inflorescence capitate or spicate; anthers elliptic and flattened, often longer than the filaments
                5. Plants shrubby, with many stems, from a definitely woody branching base; corolla 12-14 mm. long; inflorescence capitate; plants limited to the western borders of Colorado
                    -2. G. roseata
                5. Plants not shrubby, stems 1 or few from a taproot, not woody at base; corolla 9-12 mm. long; inflorescence capitate or spicate; plants not limited to extreme western Colorado (no collection as yet from there)
                    6. Inflorescence spikelike; calyx short glandular-hairy; plants not over 10,000 feet in Colorado -3. G. spicata
                    6. Inflorescence globose and capitate; calyx hairs long-silky; plants mostly above 11,000 feet in Colorado -3A. G. spicata capitata
            4. Corolla 6-8 mm. long; styles as long or longer than the corolla tube; inflorescence capitate; anthers rounded, not flattened, definitely shorter than the filaments
                7. Leaves mostly pinnately divided or parted -4. G. congesta
                7. Leaves entire or only a very few pinnatifid -4A. G. congesta burleyana
        3. Plants annual
            8. Leaf blades all entire -5. G. gunnisoni
            8. Leaf blades, at least the basal ones, pinnatifid or deeply toothed
                9. Anthers included in the corolla tube; corolla 3-5 mm. long; styles glabrous throughout -6. G. polycladon
                9. Anthers exserted; corolla 5-9 mm. long; styles hairy at least near base -7. G. pumila
    2. Flowers in open panicles or rather narrow thrysiform panicles but not spicate or capitate, at least some on definite pedicels; corolla salverform to funnelform
        10. Corolla 15-50 mm. long
            11. Inflorescence an open panicle inclined to be flat-topped; plants annual or possibly biennial (may be short-lived perennial in G. montezumae)
                12. Corolla tube 15-25 mm. long, the lobes 3-5 mm. long
                    13. Biennial or short-lived perennial plants; corolla purplish; plants apparently limited to southwestern Colorado, west of the Continental Divide -8. G. montezumae
                    13. Annual (or possibly biennial) plants; corolla white to blue-tinged; plants east of the Continental Divide -9. G. laxiflora
                12. Corolla tube 28-50 mm. long, the lobes 8-10 mm. long -10. G. longiflora
            11. Inflorescence a narrow thrysiform panicle not flat-topped; plants definitely biennial (intergrades rather common from here to second number 10)
                14. Calyx tube 3/4 to 4/5 the length of the cylindrical calyx
                    15. Corolla white to pink, little expanding upward (when pressed little if any over 2.5 mm. wide near apex) -11A. G. candida vera
                    15. Corolla red to deep pink, expanding upward (usually over 2.5 mm. wide at apex) -11B. G. candida collina

14. Calyx tube united less than 3/5 the length of the subcampanulate calyx
   16. Corolla tube slender with slight expansion upward (when pressed little if any over 2 mm. wide at apex), white, yellow or pink
      17. Corolla 20-30 mm. long; calyx tube about 1/3 to 1/2 the length of the calyx   -12. G. attenuata
      17. Corolla 35-60 mm. long; calyx tube about 1/2 to 3/5 the length of the calyx   -13. G. tenuituba
   16. Corolla tube with marked expansion upward (over 3 mm. wide near apex when pressed), red in color (at least when fresh)
      18. Calyx tube less than 1/2 the total calyx length which is 6-9 mm.
         19. Stamens included; corollas 20-30 mm. long   -14A. G. aggregata euaggregata
         19. Stamens exserted; corollas 30-45 mm. long   -14B. G. aggregata formosissima
      18. Calyx tube 1/2 to 3/5 the total length of the calyx which is 4-7 mm.
         20. Stamens included; corolla 20-25 mm. long; plants 20-30 cm. tall   -15. G. arizonica
         20. Stamens more or less exserted; corolla 25-40 mm. long; plants 30-100 cm. tall   -16. G. texana
10. Corolla 4-15 mm. long
   21. Plants perennial or definitely biennial; stems usually over 20 cm. tall; anthers often strongly exserted (exserted over 1 mm.)
      22. Corollas 10-15 mm. long; stamens only slightly exserted   -17. G. haydeni
      22. Corollas 6-10 mm. long; stamens conspicuously exserted
         23. Lobes of leaves about 3-10 mm. long; corolla tube about twice as long as the calyx; calyx 2-3 mm. long; plants widely distributed in Colorado   -18. G. calcarea
         23. Some of leaf lobes over 10 mm. long (often to 20 mm.); corolla tube shorter than or only slightly longer than the calyx; calyx 4-5 mm. long; plants apparently limited to Archuleta County in Colorado   -19. G. polyantha
   21. Plants annual; stems seldom over 20 cm. tall; anthers only slightly if at all exserted (exserted less than 1 mm.)
      24. Corollas 7-10 mm. long   -20. G. sinuata
      24. Corollas 4-6 mm. long   -21. G. leptomeria

1. Gilia acerosa (Gray) Britt., Man. Fl. N. U. S., 761. 1901.
   Giliastrum acerosum (Gray) Rydb.; Gilia rigidula var. acerosa Gray---Suffruticose perennial plants, definitely woody at base; stems 8-25 cm. tall, branching from base, glandular-puberulent; leaves pinnatifid with 3-5 linear, acerose divisions about 4-8 mm. long, glandular; flowers open-cymose, rather flat-topped; calyx about 5-7 mm. long, glandular, scarious below sinuses, teeth lanceolate, somewhat shorter than the tube; corolla about 1 cm. long and about as wide (when open), nearly or quite rotate, blue or purple, lobes broadly ovate, longer than the tube; anthers bright yellow, exserted.---Open plains and hills. Kansas and Colorado, south to Texas, Mexico and Arizona. Our records from southeastern Colorado at about 4000 feet.

2. Gilia roseata Rydb., Bull. Torr. Club 31:633. 1904.
   Perennial plants from a definitely woody base, this branching and shrubby; stems 10-25 cm. tall, more or less villous or crisp-hairy, usually more or less tinged with purple; leaves pinnately parted into linear lobes, these about 1-2 cm. long, pungently pointed; flowers in capitate clusters, these corymbosely arranged, the bracts shorter than the flowers; calyx 5-7 mm. long, more or less rose-tinged, crisp-hairy, the tube broadly scarious below the sinuses, the lobes much shorter than the tube, pungently-pointed, this point over 0.5 mm. long; corolla 12-14 mm. long, salverform, whitish, the tube longer than the calyx, the lobes 4-5 mm. long, about 1/2 as long as the tube; anthers elliptical, longer than or nearly as long as the filaments; styles included, about 1/2 as long as the corolla tube.---Canyons and hills, usually in dry ground. Colorado and Utah. Our few records from westcentral and northwestern Colorado at 6000-6500 feet.

3. Gilia spicata Nutt., Journ. Acad. Nat. Sci. Phila. 2 ser. 1:156. 1848.
   G. spicata var. deserta A. Nels.---Plants biennial or perennial from a taproot; stems 10-35 cm. long, usually 1, simple or branched at the very base, arachnoid-villous; leaves usually over 3 cm. long (basal ones), entire and linear or irregularly pinnately parted into linear lobes, these short-mucronate; flowers in capitate clusters these crowded into a spikelike thyrsus, the bracts longer than flowers; calyx about 5 mm. long, tube broadly scarious below the sinuses and short glandular-hairy, the lobes not as long as the tube, short spiny-mucronate, this tip not over 0.5 mm. long; corolla 9-12 mm. long, whitish to cream-colored when fresh often turning brownish on pressing, salverform, the tube longer than the calyx and much longer than its lobes; anthers elliptic, longer than the filaments; style included, about 1/2 as long as the corolla tube.---Dry hills and plains. South Dakota and Wyoming, south to Kansas, New Mexico and Utah. Our records scattered in the eastern half of Colorado, mostly along the eastern foothills, at 5000-7500 feet.

3A. Gilia spicata capitata A. Gray, (var.) Proc. Am. Acad. 8:274. 1870.
   G. globularis Brand; G. cephaloidea of Mans. as to Colo. plants---Differs from the species in having a globose, capitate inflorescence and with calyx hairs long and silky. The flowers seem to be a little darker brown on pressed material but are probably cream or whitish when fresh. Perhaps better regarded as a separate species especially since altitudinal ranges do not even approach each other. However, some lower altitude plants do approach the variety.---High mountains. Colorado. Our few records from central Colorado at 12,000-14,000 feet.

4. Gilia congesta Hook., Fl. Bor. Amer. 2:75. 1838.
   G. iberidifolia Benth.; G. congesta iberidifolia (Benth.) Brand---Perennial from a somewhat woody branching caudex; stems 10-30 cm. tall, erect or nearly so, simple to near apex at least, arachnoid-villous with jointed hairs; leaves mostly pinnately divided or parted into 3-5 linear divisions 5-30 mm. long, these mucronate-spiny at apex; flowers in 1 or several corymbose-capitate cymes with bracts shorter than the flowers; calyx about 3-4 mm. long, very broadly scarious below the sinuses, the lobes somewhat shorter than the tube, spiny-mucronate; corolla about 6-8 mm. long, white, salverform, the tube a little longer than the calyx, lobes

about 1/3 as long as the tube; anthers rounded on definitely longer filaments; styles as long or longer than the corolla tube.---Dry hills and plains. South Dakota to Oregon, south to Nebraska, Utah and California. Our records from northern and western Colorado at 4000-8500 feet, also from Park County at 9700 feet.

4A. Gilia congesta burleyana (A. Nels.) Const. & Roll., (var.) Am. Journ. Bot. 23:440. 1936.
G. spergulifolia Rydb. in part; G. burleyana A. Nels.; G. frutescens Rydb.---Differs from the species in having all or nearly all the leaves entire but some intergradations occur in Colorado.---Montana to Colorado and Utah. Our records from southwestern Colorado at 5000-6000 feet.

5. Gilia gunnisoni T. & G., U. S. Rept. Expl. Miss. Pacific 2(2):128. 1855.
Annual plants; stems 10-25 cm. tall, divaricately branched, sparingly hirsutulous-villous and glandular-puberulent; leaves 1-4 cm. long, linear-filiform, entire, sparingly hirsutulous-villous and glandular-puberulent; flowers clustered in small terminal, leafy-bracted heads; calyx 3-4 mm. long, more or less hairy, the acerose-tipped lobes shorter than the tube; corolla white, tube 2-4 mm. long, lobes about 2 mm. long; anthers exserted; styles sparsely hairy near base. This species intergrades with G. pumila Nutt. to some degree and may be only a variety of it.---Dry often sandy ground. Colorado and Utah, south to New Mexico and Arizona. Our few rather doubtful specimens are from southern Colorado.

6. Gilia polycladon Torr., U. S. & Mex. Bound. Bot. 146. 1859.
Annual plants; stems 5-15 cm. tall, few to several from a taproot, spreading or erect, sparingly leafy, puberulent or sparsely pubescent, sometimes slightly glandular; leaves pinnatifid with 3-7 teeth or short oblong lobes these seldom over 5 mm. long; flowers cymose-glomerate at the ends of the branches, the clusters leafy-bracteate; calyx about 3-4 mm. long, scarious below the sinuses, the lobes subulate, about as long as the calyx tube; corollas salverform, white, the tube 3-5 mm. long, little longer than the calyx, the lobes about 1-1.5 mm. long; anthers included; styles glabrous throughout.---Dry areas. Colorado and Utah, south to Texas and California. Our records from westcentral Colorado at 4500-5500 feet.

7. Gilia pumila Nutt., Acad. Nat. Sc. Phila. Journ., ser. 2. 1:156. 1848.
Annual plants; stems 6-20 cm. long, erect but branched, crisp-hairy, almost woolly especially above, sparingly leafy; leaves 2- to 4-parted into divergent linear cuspidate lobes these usually at least 5 mm. long, or sometimes some leaves entire; flowers cymose-glomerate at the ends of the branches, the clusters leafy-bracteate; calyx about 4-5 mm. long, the lobes subulate, about as long or shorter than the tube, scarious below sinuses; corolla white to light rose or light blue, salverform, the tube about 7-9 mm. long, about twice as long as the calyx, corolla lobes about 1.5-2 mm. long; anthers rather definitely exserted, more or less declined; styles sparsely hairy at least near base.---Dry ground. Wyoming to Utah, south to Texas and Arizona. Our records mostly from western and southern Colorado at 4000-6500 feet.

8. Gilia montezumae Tide. & Dayton, Bull. Torr. Club 55:73. 1928.
Short-lived perennial or biennial plants; stems about 30-55 cm. tall, stout, solitary, soon branching, sparsely crisp-pubescent; basal leaves 5-8 cm. long, pinnatifid, lobes lanceolate, upper leaves mostly reduced; inflorescence many-flowered, elongated but loose and open; calyx 5-7 mm. long, glandular, teeth usually shorter than the tube; corolla 15-25 mm. long, purplish, tubular-funnelform; stamens not exserted.---Apparently limited to the type area in southwestern Colorado, at about 9000-9500 feet.

9. Gilia laxiflora (Coult.) Osterh., Bull. Torr. Club 24:51. 1897.
Annual (appearing so on most) or possibly biennial plants; stems 10-40 cm. tall, erect, branching, glabrous or glandular-puberulent above; leaves all pinnatifid with filiform or narrowly linear divisions or upper entire to few-lobed; inflorescence open and inclined to be flat-topped, the flowers on slender pedicels often over 1 cm. long; calyx about 5-6 mm. long, scarious below sinuses, the teeth about 1/3 the total length, subulate and slender-spinulose at apex; corolla white or tinged with blue, salverform, the tube 15-25 mm. long, the lobes 3-5 mm. long; anthers included.---Plains, hills and mesas. Colorado, south to Texas and New Mexico. Our records mostly in the eastern half of Colorado (one from Chaffee County) at 4000-7000 feet.

10. Gilia longiflora (Torr.) G. Don., Hist. Dichl. Pl. 4:245. 1837.
Annual or possibly biennial plants; stems glabrous or glandular-puberulent above, erect, usually 1 from a taproot, branching above; leaves pinnatifid with rather few filiform or narrowly-linear divisions, or upper ones sometimes entire; flowers open-paniculate, usually somewhat flat-topped, the flowers on slender pedicels often over 1 cm. long; calyx 5-7 mm. long, scarious below the sinuses, the teeth shorter than the tube, subulate and spinulose at apex; corolla white to pink or rose-lilac in age, salverform, the tube 28-50 mm. long, the lobes about 8-10 mm. long; anthers included.---Open, often sandy soil. Nebraska to Utah, south to Texas and Arizona. Our records scattered in the eastern half of Colorado, also from southcentral and the southwestern parts, at 3500-8000 feet.

11. Gilia candida Rydb., Bull. Torr. Club 28:29. 1901.
Biennial plants; stems 25-60 cm. tall, usually 1 from a taproot, glandular; leaves pinnately divided into linear, acutish and more or less spinulose-tipped segments; flowers many in a narrow thyrsus; calyx usually over 7 mm. long, cylindrical, united for 3/4 to 4/5 its length, the lobes short and rounded to an abrupt subulate-aristate tip; corolla 30-45 mm. long, white, pink or more rarely red, usually narrow.---Open ground. Colorado and New Mexico. Our records (including both subspecies) mostly from northcentral, central and southcentral Colorado at 5000-9500 feet.

11A. Gilia candida vera Wherry, (ssp.) Bull. Torr. Club 73:197. 1946.
G. scariosa Rydb.---Corolla white or pale pink, the tube expanding upward very little. Fairly distinct in this state.

11B. Gilia candida collina (Greene) Wherry, (ssp.) Bull. Torr. Club 73:197. 1946.
Callisteris collina Greene---Corolla deep pink to red, the tube definitely expanding upward. Intergrades in Colorado with G. texana (Greene) Woot. & Standl.
12. Gilia attenuata (Gray) A. Nels., Bull. Torr. Club 25:278. 1898.
G. aggregata var. attenuata Gray---Biennial plants; stems 30-70 cm. tall, usually 1 from a taproot, somewhat glandular-pubescent; leaves pinnately divided into linear, mucronate-spinulose divisions; flowers many in a narrow thrysus; calyx 5.5-8 mm. long, subcampanulate, united 1/3 to 1/2 its length, the lobes rather long-awned; corolla 20-30 mm. long, funnel-salverform, white to pink or somewhat yellowish, the tube hardly expanded upward (little if any over 2 mm. wide just at apex of tube); filaments included or almost so. This species intergrades somewhat in this state with G. texana (Greene) Woot. & Standl. and G. candida Rydb.---Dry ground. Colorado to Washington. Our records from northern and western Colorado at 6000-9000 feet.
13. Gilia tenuituba Rydb., Bull. Torr. Club 40:472. 1913.
Callisteris violacea Greene---Biennial plants; stems 25-60 cm. tall, from a taproot, more or less glandular-puberulent and hirsutulous-villous; leaves pinnately divided into linear cuspidate divisions, these crisp-hairy and more or less glandular-puberulent; flowers many in a narrow thrysus; calyx 5-6 mm. long, subcampanulate, glandular, tube about 1/2 to 3/5 the length of the whole calyx, lobes awn-tipped; corolla 35-60 mm. long, flesh-colored or whitish, the tubes cylindrical only slightly expanded upward (not over 3 mm. wide near apex when pressed); stamens included.---Dry ground. Colorado to California. Our few records from north-central Colorado at 8000-8500 feet.
14. Gilia aggregata (Pursh) Spreng., Syst. Veg. 1:626. 1825.
Biennial plants; stems 15-60 cm. tall, erect, usually only 1 from a taproot, more or less hairy and glandular; leaves pinnately divided into linear mucronate divisions; flowers many in a narrow thrysus; calyx 6-9 mm. long, subcampanulate, the tube less than 1/2 the length, the lobes subulate from a more or less broad base, awn-tipped; corolla 20-45 mm. long, deep red (except in mutants) more or less mottled with yellow, funnel-salverform, the tube rather stout and expanding upward (when pressed over 3 mm. wide near the top); stamens included or exserted. Represented in Colorado by the following 2 subspecies, both of which intergrade to some degree with G. texana (Greene) Woot. & Standl.---Hills and plains. Montana to British Columbia, south to New Mexico and California. Our records (including both subspecies) from the western half of Colorado at 4500-9000 feet.
14A. Gilia aggregata euaggregata Brand (ssp.) in Englers, Pflanzenreich IV. 115. 1907.
G. pulchella Dougl.---Stamens included; corollas 20-30 mm. long. The 2 subspecies do not intergrade very much with each other in this state.
14B. Gilia aggregata formosissima (Greene) Wherry, (ssp.) Bull. Torr. Club 73:198. 1946.
Stamens exserted; corolla 30-45 mm. long.
15. Gilia arizonica (Greene) Rydb., Bull. Torr. Club 40:472. 1913.
Utah to Nevada, south to Arizona. This plant may be in southwestern Colorado.
16. Gilia texana (Greene) Woot. & Standl., Contrib. U. S. Nat. Herb. 16:161. 1913.
Biennial plants; stems 30-100 cm. tall usually over 40 cm., single to few from a taproot, strict or ascending, more or less sparsely pubescent; leaves pinnately divided into narrowly linear, mucronate divisions; flowers many in a narrow thrysus; calyx 4-7 mm. long, subcampanulate, the tube 1/2 to 3/5 the total length (rarely a little more), the lobes deltoid to oblong but more or less abruptly contracted to a subulate tip; corolla 25-40 mm. long, red, funnel-salverform, the tube rather stout, expanding upward (when pressed 3 mm. wide or more near apex); stamens more or less exserted.---Dry ground. Colorado to California, south to Mexico. Our records scattered over the western two-thirds of Colorado, except the northern part, at 6000-11,000 feet.
17. Gilia haydeni A. Gray, Proc. Am. Acad. 11:85. 1876.
G. crandallii Rydb.; G. subnuda haydeni (Gray) Brand---Biennial plants (some look annual or almost perennial); stems 20-50 cm. tall, erect or ascending, glabrous to pubescent and glandular especially above; basal leaves oblanceolate, toothed or shallowly pinnately lobed, cauline leaves toothed or entire, linear; inflorescence an open panicle, somewhat flat-topped; calyx 3-5 mm. long, scarious below the sinuses, the lobes subulate, shorter than the tube, usually glandular-puberulent; corolla 10-15 mm. long, rose-colored, the tube several times longer than the calyx, whole corolla funnelform to somewhat salver-form; anthers slightly exserted. G. subnuda Torr. and G. pentstemoides M. E. Jones are related entities probably not different from the above and the specific name may have to be changed.---Plains and mesas. Colorado, Utah and New Mexico. Our records from westcentral and southwestern Colorado at 4500-7500 feet.
18. Gilia calcarea M. E. Jones, Contrib. West. Bot. 8:36. 1898.
G. pinnatifida Nutt.---Biennial or perennial plants; stems 10-50 cm. tall, usually single from a taproot, simple, or branching above the base, erect or ascending, glandular-puberulent and often viscid; leaves pinnatifid into linear or narrowly oblong segments about 3-8 mm. long, glandular; inflorescence open-paniculate, often compound from branches bearing flowers; calyx 2-3 mm. long, glandular, united for 2/3 or more of its length, scarious below the sinuses; corolla 6-9 mm. long, whitish to blue or purplish, salverform, the tube about twice as long as the calyx, tube often yellowish; stamens conspicuously exserted. This species as treated here is variable in general appearance.---Open often sandy ground. Nebraska and Wyoming, south to New Mexico. Our records scattered over Colorado, few in the extreme eastern and western parts, at 4500-11,000 feet.
19. Gilia polyantha Rydb., Bull. Torr. Club 31:634. 1904.
Perennial (possibly biennial) plants from a taproot; stems 30-60 cm. tall, only 1, unbranched at very base, branching upward, leafy, glandular-puberulent; leaves 3-4 cm. long, pinnatifid, the lobes over 10 mm. long (on same at least), narrowly linear (the lobes no wider

than leaf rachises), spinulose-mucronate; flowers many in a rather narrow thrysus, often several of these on branches; calyx 4-5 mm. long, lobes about 1.5 mm. long, broadly triangular and abruptly cuspidate-pungent; corolla about 10 mm. long, white, sometimes purple-dotted, the tube 4.5-6.5 mm. long, from shorter than to slightly longer than the calyx, somewhat trumpet-shaped to nearly salverform; stamens noticeably exserted.---Dry ground. Apparently known only from the type locality in Archuleta County, Colorado, at about 7000 feet.

20. Gilia sinuata Dougl. ex Benth., in DC. Prodr. 9:313. 1845.

G. inconspicua (Smith) Dougl.---Annual plants; stems 10-30 cm. tall, glandular-puberulent and often more or less woolly when young, often branching near base; leaves mostly near base, pinnatifid or upper leaves may be entire; flowers loosely panicled or somewhat crowded; calyx about 3-4 mm. long, scarious below sinuses, the subulate teeth shorter than the tube; corolla 7-10 mm. long, whitish to purplish or rose-violet, narrowly funnelform to somewhat salverform, the tube definitely exceeding the calyx; anthers exserted but not strongly so (not over 1 mm.). ---Dry hills and mesas. Wyoming to Washington, south to Texas and California. Our records scattered over the western two-thirds of Colorado at 4500-8000 feet.

21. Gilia leptomeria A. Gray, Am. Acad. Arts & Sci. Proc. 8:278. 1870.

Annual plants; stems 6-20 cm. tall, erect or ascending, few to many from base and branching, more or less glandular-puberulent; leaves mostly radical, these oblong-linear to oblanceolate, pinnately lobed or toothed, cauline leaves may be entire; flowers corymbose-paniculate; calyx about 2-4 mm. long, scarious below the sinuses, the teeth shorter than the tube; corolla 4-6 mm. long, white to rose, narrowly funnelform to somewhat salverform, the tube longer than the calyx; anthers exserted but not strongly so (not over 1 mm.). Related forms have been called G. sedifolia Brand; G. minutiflora Benth.; G. subacaulis Rydb., and G. tweedyi Rydb., and perhaps some may be distinct on further study of this group.---Dry soil. Wyoming to Washington, south to New Mexico and California. Our records well scattered over the western two-thirds of Colorado at 4500-7500 feet.

8. Linanthus Benth.

Erect low annual plants; leaves opposite, palmately divided into 3-5 filiform or narrowly linear divisions; flowers axillary or terminal, usually solitary; calyx tube scarious below the sinuses, rupturing by the pressure of the developing fruit, the lobes acerose; corolla with a funnelform throat and short tube; stamens equally inserted and included in the corolla throat; capsule 3-celled; seeds 2-5 in each cell, not emitting spiral threads when wetted.

1. Linanthus harknessii (Curran) Greene, Pitt. 2:225. 1892.

Stems 5-20 cm. tall, usually simple below, glabrous or pubescent; leaf segments 4-20 mm. long, glabrous or pubescent (ours somewhat strigose especially on margins); flowers solitary on filiform pedicels; calyx 2.5-3.5 mm. long, glabrous to strigillose; corolla 1-1½ times as long as the calyx, white, a hairy ring present inside; stamens somewhat unequal in length; seeds producing mucilage when wetted. Our plants have been called var. septentrionalis (Mason) Jepson & Bailey and may be worthy of specific rank.---Dry often sandy ground. The variety from Montana to Oregon, south to Colorado and California. Our records from the northwestern quarter of Colorado at 6500-11,000 feet.

9. Leptodactylon Hook. and Arn.

Much branched undershrubs, definitely woody at base; leaves palmately parted into acerose divisions, opposite at least at very base of plants, opposite or appearing alternate above, with smaller leaves fascicled in the axils of the primary ones; flowers showy, solitary and sessile or few in a cluster, white to yellowish (may be purplish in throat in 1 species); calyx with rather short teeth, broadly scarious-membranous below the sinuses, the tube rupturing from the developing fruit; corolla salverform with funnelform throat, the tube longer than the calyx; stamens short, inserted in or below the throat, anthers included; seeds not developing mucilage or spiral threads when wet.

1. Plant depressed-pulvinate, not over 7 cm. tall; sepals, petals and stamens 4   -1. L. caespitosum
1. Plants not pulvinate, over 7 cm. tall; sepals, petals and stamens 5
    2. Leaves all strictly opposite, the divisions often over 8 mm. long; plants woody only at base; corollas 18-25 mm. long, the lobes 9-12 mm. long
       -2. L. watsonii
    2. Some of upper leaves subopposite or alternate, seldom over 8 mm. long; plants woody well above the base; corollas 15-20 mm. long, the lobes 6-8 mm. long
       -3. L. pungens

1. Leptodactylon caespitosum Nutt., Journ. Acad. Phila. II. 1:157. 1848.

Gilia caespitosa of Manuals---Densely pulvinate-caespitose plants, the much-branched woody base hardly emerging from the ground; stems not over 7 cm. tall; leaves crowded, alternate, subopposite or opposite below, the divisions about 5-7 mm. long, acerose-subulate; calyx 6-8 mm. long, teeth shorter than the tube; corolla about 12-16 mm. long, white to yellowish, lobes 4, about 4 mm. long.---Dry hills and plains. Nebraska to Nevada. Our few records from northern Colorado at about 5000 feet.

2. Leptodactylon watsonii (A. Gray) Rydb., Colo. Ag. Exp. St. Bul. 100:279. 1906.

Gilia watsonii Gray---Suffruticose plants, woody at branched caudex; stems 10-20 cm. tall, glandular-puberulent or finally smooth; leaves all opposite with 3-7 narrowly linear-acerose divisions 7-15 mm. long, these more or less glandular-puberulent; calyx about 6-8 mm. long, tubular, the lobes less than 1/2 as long as the tube, this membranous below sinuses and glandular-ciliate at apex; corolla 18-25 mm. long, whitish to cream-colored but purple in the throat, with lobes about 9-12 mm. long.---Rocky cliffs and hills. Colorado and Utah. Our few records from western Colorado at about 6000-6500 feet.

3. **Leptodactylon pungens** (Torr.) Rydb., Fl. Colo. 279. 1906.
   L. brevifolium Rydb.; *L. pungens* (Torr.) Nutt.; *Gilia pungens* (Torr.) Benth.---Suffruticose plants but the stems rather woody well above the base; stems several to many, branching, densely leafy, glabrate to viscid-puberulent; leaves alternate or subopposite above, lower sometimes opposite, with 3-7 acerose-linear subulate divisions 3-9 mm. long, glabrate to glandular-puberulent; calyx about 7-15 mm. long, tubular, glabrate to glandular-puberulent, the teeth acerose, much shorter than the tube, this membranous below sinuses and ciliate at apex; corolla 15-20 mm. long, white or cream-colored, sometimes purplish in the throat, lobes about 6-8 mm. long. Wherry (Am. Midl. Nat. 34:381-387. 1945) lists 2 subspecies in or near Colorado. These intergrade to some degree.

1. Principal leaves 5-7 mm. long - ssp. brevifolium (Rydb.) Wherry
2. Principal leaves 7.5-15 mm. long - ssp. eupungens (Brand) Wherry

    Plains, valleys and hills, often on sandy ground. Montana to Washington, south to New Mexico and California. Our records well scattered over the western three-fourths of Colorado at 4500-9000 feet. The short-leaved plants are in the extreme western part of the state.

10. Linanthastrum Ewan

    Perennial herbaceous plants with tufted stems from a woody root-crown; leaves opposite or nearly so (appearing whorled), the blades palmately 3-7 parted into narrow divisions; flowers in upper axils, scarcely exserted from a tuft of leaves, in capitate clusters or loose cymes; calyx campanulate, its lobes somewhat shorter than the tube, not or very narrowly scarious-membranous below the sinuses, the tube splitting from pressure of the developing fruit; corolla white, rose or blue or sometimes cream-colored, funnelform to rather salverform, the tube not surpassing the calyx; stamens inserted just below the throat; seeds 2-4 in each locule.

1. Leaves 7- to 3-lobed, rather soft; flowers numerous, on pedicels rarely over 2 mm. long
       -1. **Linanthastrum nuttallii**
1. Leaves 3- to 1-lobed, rather firm; flowers sparse, on pedicels rarely less than 5 mm. long
       -2. **Linanthastrum floribundum**

1. **Linanthastrum nuttallii** (Gray) Ewan, Journ. Wash. Acad. Sc. 32:139. 1942.
   Siphonella nuttallii (A. Gray) Hel.; *Gilia nuttallii* A. Gray; *Leptodactylon nuttallii* (A. Gray) Rydb.---Stems 12-25 cm. tall, several, simple or branched from a woody branching crown, pubescent; leaves 7- to 3-parted, the segments 1-1.5 cm. long, narrowly linear, glabrous or glabrate, sparingly ciliate; flowers numerous, pedicels to 2 mm. long; calyx 7-10 mm. long, very narrowly scarious below the sinuses; corolla 7-14 mm. long, the tube commonly yellowish or cream-colored, the lobes whitish, glabrous within, pubescent without; capsule about 5 mm. long.---Rocky outcrops, borders of meadows and hillsides. Wyoming to Washington, south to New Mexico and California. Our records scattered in the higher mountains of Colorado at 6500-11,500 feet. The report from southeastern Colorado appears to be the result of a mididentification.

2. **Linanthastrum floribundum** (Gray) Wherry, Am. Midl. Nat. 34:386. 1945.
   Siphonella floribunda (Gray) Jepson---Wherry maintained this as a species, differing from the preceding in having leaves 3- to 1-parted, rather firm; flowers sparse, on pedicels 5-20 mm. long.---Valleys and slopes. Our records from the mountains of central and southern Colorado at about 8000-10,000 feet.

Family 100. Hydrophyllaceae WATERLEAF FAMILY

    Plants herbaceous, annual or perennial; leaves alternate, rarely opposite or basal, simple or pinnate, without stipules; flowers perfect, regular or nearly so, 5-merous, mostly in cymes, these often elongated, 1-sided and racemelike but flowers sometimes solitary and axillary; sepals more or less united, often appendaged in each sinus; corolla gamopetalous, mostly campanulate or funnelform, often appendaged inside; stamens 5, inserted on the corolla near its base, alternate to the petals; pistil of 2 carpels, ovary superior, mostly 1-celled, with 2 parietal placentae, sometimes 2-celled, styles more or less united; fruit a capsule; seeds often pitted or reticulated.

1. Leaves entire, less than 5 mm. wide; plants annual; flowers solitary and axillary or in 1- to 3-flowered terminal cymes; styles distinct to base or nearly so   -1. **Nama**
1. Leaves mostly sinuate to pinnate (if entire over 5 mm. wide); plants annual to perennial (few if any annuals have entire leaves); flowers various, solitary to many (if leaves entire then flowers in scorpioid spikelike racemes); styles usually united to above the middle
   2. Flowers solitary on axillary or terminal peduncles; plants annual; corollas not longer than the sepals
      3. Sepals with reflexed auricles in the sinuses, these auricles about 1.5 mm. long   -2. **Nemophila**
      3. Sepals without reflexed auricles in the sinuses (rarely a minute sepaloid appendage present)
             -3. **Ellisia**
   2. Flowers in several to many-flowered terminal or axillary inflorescences; plants annual to perennial; corollas usually longer than the sepals
      4. Corolla lobes convolute in the bud; ovary 1-celled with placentae broad and fleshy in age, expanding, membranous and saclike; flowers in capitate clusters, not especially scorpioid   -4. **Hydrophyllum**

4. Corolla lobes imbricate in the bud; ovary 1-celled or 2-celled by the meeting of narrow placentae; flowers rarely in capitate clusters, usually in scorpioid, spikelike racemes   -5. Phacelia

1. Nama L.

Herbaceous annual plants; leaves alternate and entire; flowers borne singly in the axils or in lateral or terminal cymes; calyx divided nearly to base; corolla tubular to narrowly campanulate, pubescent outside; filaments glabrous, stamens included; styles 2, united less than 1/2 their length; ovary 1-celled but often appearing 2-celled by the growth of the placentae, pubescent; capsules usually loculicidal; seeds many.

1. Corolla 8-15 mm. long; seeds not pitted (may be minutely reticulated)   -1. N. hispidum
1. Corolla about 5 mm. long (never over 7 mm.); seeds with large pits   -2. N. dichotomum

1. **Nama hispidum** Gray, Proc. Am. Acad. 5:339. 1861.
Oklahoma and Colorado, south to Texas and Mexico. Reported for Colorado but no specimens seen by the writer. Should be looked for in the southern part of the state.
2. **Nama dichotomum** (Ruiz & Pavon) Choisy, Mem. Soc. Phys. Geneve 6:113. 1833.
N. angustifolium (A. Gray) A. Nels.; Marilaunidium angustifolium (A. Gray) Kuntze---Colorado south to New Mexico and Arizona, and Mexico; South America. Reported for southern Colorado but no specimens seen by the writer.

2. Nemophila Nutt. BABY-BLUE EYES

Annual herbaceous plants; stems somewhat succulent and brittle, diffuse; leaves alternate, variously lobed or pinnately divided; flowers solitary on axillary or terminal peduncles; calyx divided about to base, auricled in the sinuses; corollas white to blue-purple, campanulate, scales 2 to a filament; styles 1, bifid at apex, ovary 1-celled; seeds usually few.

1. **Nemophila breviflora** Gray, Proc. Amer. Acad. 10:315. 1875.
Stems 5-20 cm. long, weak, sharply-angled, with minute reflexed prickles, otherwise glabrous; leaves 5-20 cm. long, the lower ovate-deltoid, pinnately divided into 3-6 oblong-lanceolate rather remote divisions; flowers on rather short peduncles; calyx lobes about 3 mm. long, reflexed auricles about 1.5 mm. long; corolla shorter than the calyx; capsule 3-5 mm. long; seeds usually only 1, 2-4 mm. in diameter.---Rich soil of canyons and hillsides, often in shade. Montana to British Columbia, south to Colorado and California. Our records from western and northern Colorado at 6500-9000 feet.

3. Ellisia L.

Annual plants; lower leaves opposite, the others alternate, pinnately divided; flowers solitary in the upper leaf axils; calyx cleft nearly to base, not really appendaged in the sinuses (may be sepaloid teeth present on some specimens), enlarging in fruit; corolla campanulate, white to bluish, from shorter than to as long as the calyx, the lobes shorter than the tube, with scalelike, minute appendages within; stamens included; ovary 1-celled, ovules on fleshy placentae attached at top and bottom of cell, styles cleft at apex less than 1/2 the length; seeds usually 4, regularly reticulated.

1. **Ellisia nyctelea** L., Sp. Pl. ed. 2, 1662. 1763.
Macrocalyx nyctelea (L.) Kuntze---Stems 10-40 cm. tall or often ever shorter, simple or diffusely branched, angled, retrorsely hirsute; leaves 2-8 cm. long, divisions 7-13, oblong, acute or obtuse, more or less hispidulous; calyx lobes 3-5 mm. long, lanceolate or ovate-lanceolate; stamens included; capsule hispid; seeds 2-3 mm. in diameter, globose, dark brown. Constance (Rhodora 42:38. 1940) said the western plants with depressed and more pubescent habit and shorter styles intergrade too much to justify Ellisia nyctelia var. coloradensis Brand.---Valleys, shady places and rich soil. Atlantic States west to Saskatchewan and New Mexico. Our records well scattered over Colorado at 4000-9000 feet.

4. Hydrophyllum L. WATER LEAF

Plants perennial, ours with short horizontal rootstocks; leaves alternate, mostly basal, long-petioled, pinnate or pinnatifid; flowers in more or less scorpioid cymes, these often congested; calyx deeply parted, the sinuses not appendaged; corolla tubular-campanulate, the tube short, lobes convolute in bud, each with a linear appendage within opposite the lobes; stamens exserted, the filaments hairy at middle or base; ovary 1-celled with fleshy placentae attached at bottom and top, ovules 4, styles 2-cleft near apex; capsule 2-valved; seeds 1-4.

1. Flowers in close capitate clusters, peduncles shorter than the petioles of the subtending leaves; leaf blades usually shorter than their petioles; leaflets entire or toothed or lobed only at the apex; anthers short-oblong, not over 1 mm. long   -1. H. capitatum
1. Flowers in rather lax capitate clusters, the peduncles longer than the petioles of the subtending leaves; leaf blades longer than their petioles; leaflets incised or toothed usually to near base; anthers linear-oblong, over 1 mm. long   -2. H. fendleri

1. **Hydrophyllum capitatum** Dougl. ex Benth., Trans. Linn. Soc. 17:273. 1836.
Rhizomes very short, bearing a fascicle of fleshy fingerlike roots; plants 10-45 cm. tall but stems short, spreading-hirsute; leaves 5-12 cm. long on longer petioles, ovate to oval in

outline, blades pinnately parted or divided with 5-7 obovate to oblong or lanceolate divisions, these sometimes cleft again, lobes and divisions obtuse, acute or mucronate; cymes 1 to several, globose, on peduncles 1-5 cm. long, definitely shorter than the leaves, usually recurved in age; calyx lobes 3-4 mm. long, ciliate and appressed hirsute; corolla 5-9 mm. long, the lobes about 1/2 the entire length, purplish-blue to white; anthers short-oblong, 0.7-1 mm. long.---Hillsides, often among bushes. Alberta to British Columbia, south to Colorado and Oregon. Our records in western Colorado, mostly in the westcentral and northwestern parts, at 6000-9500 feet.

2. **Hydrophyllum fendleri** (Gray) Heller, Pl. World 1:23. 1897.

Rhizomes rather short, bearing fleshy-fibrous roots; plants 20-90 cm. tall, the stems retrorsely hispid or hirsute; leaf blades 6-30 cm. long, oblong or oval in outline, pinnatifid with 9-13 main divisions, these ovate to lanceolate, acuminate, coarsely serrate or incised; cymes lax in fruit, more or less capitate, peduncles 2-13 cm. long, often as long or longer than the subtending leaves; calyx lobes 4-6 mm. long, ciliate and strigose, often hispid; corolla 6-10 mm. long, white to violet, lobes about 1/2 length; anthers 1-2 mm. long, linear-oblong.---Along streams and valleys, often among bushes. Wyoming to Washington, south to New Mexico and Utah. Our records scattered over the western two-thirds of Colorado, except the northwestern corner, at 5500-10,500 feet.

5. Phacelia Juss. SCORPION WEED

Plants annual or perennial, mostly hirsute or hispid; leaves alternate (or lower sometimes opposite), simple and entire to pinnatifid or pinnately compound; flowers often showy, mostly in 1-sided false racemes (really cymes); calyx 5-lobed or 5-parted, without appendages in the sinuses; corolla 5-lobed, funnelform, tubular, campanulate or nearly rotate, usually with vertical folds or plaits in the tube; ovary 1-celled or nearly 2-celled by union of placentae, styles 2-cleft or 2-divided; fruit 2-valved; seeds reticulated or roughened.

1. Stamens included (shorter than the corolla); corolla more or less tubular; plants annual
  2. Seeds transversely corrugated on the ventral side; corollas 2.5-4.5 mm. long
    3. Flowers white, deciduous; leaves often pinnately divided to the midrib   -1. P. ivesiana
    3. Flowers light yellow (sometimes purplish-tinted too), marcescent; leaves unlobed to serrate-lobed   -2. P. submutica
  2. Seeds not corrugated but minutely pitted; corollas 4.5-8 mm. long
    4. Lower leaves pinnately divided, upper pinnatifid to incised; corollas about 5 mm. long; seeds about 4 mm. long or more   -3. P. denticulata
    4. All leaves undulate or entire; corollas 6-8 mm. long; seeds 1-1.5 mm. long   -4. P. demissa
1. Stamens exserted; corolla more or less campanulate (except in P. integrifolia); plants annual, biennial or perannial
  5. Stems hispid or hirsute, not glandular; leaves simple and entire or with 1-5 entire-margined lobes or leaflets at base, the veins conspicuous; filaments more or less hairy   -5. P. heterophylla
  5. Stems often glandular, not hispid or hirsute or if so then leaves toothed or lobed with more than 5 divisions (these main divisions seldom entire), veins usually not especially conspicuous; filaments glabrous
    6. Plants perennial; inflorescence of spikelike racemes, these not scorpioid, crowded into a narrow spikelike panicle; plants not glandular-hairy
      7. Leaf divisions linear or narrowly oblong, the ultimate ones averaging 2-4 mm. wide, surface often sericeous especially when young   -6. P. sericea
      7. Leaf divisions oblong, the ultimate divisions averaging over 4 mm. wide, surface not sericeous   -7. P. idahoensis
    6. Plants annual (possibly biennial in a few); inflorescence of scorpioid parts, rarely in a spikelike panicle (except in P. glandulosa); plants usually more or less glandular-hairy
      8. Seeds transversely corrugated on the ventral side; leaves sinuate, crenate or pinnately lobed
        9. Seeds 3.5-4.5 mm. long; corolla 7-12 mm. long   -8. P. corrugata
        9. Seeds not over 2.5 mm. long; corolla about 6 mm. long   -9. P. intermedia
      8. Seeds without transverse corrugations on the ventral side; leaves usually pinnately divided or bipinnate (except in P. integrifolia)
        10. Corolla tubular; leaves usually deeply crenate, sometimes pinnately cleft   -10. P. integrifolia
        10. Corolla campanulate; leaves pinnately divided, pinnately cleft or bipinnate
          11. Corolla lobes toothed or fimbriate, white   -11. P. neomexicana
          11. Corolla lobes entire (or if toothed not white), usually blue-purple or violet
            12. Leaflets (or pinnae) lobed or pinnatifid (on most); stamens 2-3 mm. longer than the corolla; plants apparently limited to Jackson and Grand Counties   -12. P. formosula
            12. All leaflets entire or merely toothed; stamens 4-7 mm. longer than the corolla; plants of western and southern Colorado, so far not in Jackson County
              13. Seeds 3.5-4 mm. long, stems glabrous to glandular-puberulent   -13. P. splendens
              13. Seeds 2.5-3.5 mm. long; stems rather densely villous or hispidulous and glandular   -14. P. glandulosa

1. **Phacelia ivesiana** Torr. in Ives, Colo. River Explor. Exped. Bot. 21. 1860.

P. campestris A. Nels.---Annual plants from taproots; stems about 5-15 cm. tall, erect or widely spreading, more or less hirsutulous and also glandular-puberulent; leaves 1-3 cm. long, oblong to oblanceolate, petioled, from deeply toothed to pinnately divided, hirsutulous and usually glandular; flowers in racemes, not especially congested, lower pedicels often reflexed or recurved in fruit; calyx segments narrowly oblong to oblong-lanceolate, about as long as the corolla, united below; corolla 2.5-4 mm. long, tubular-funnelform, limb white, tube whitish to yellow; stamens not exserted, filaments glabrous; styles not exserted; seeds 1-1.5 mm. long, transversely corrugated on the ventral side. Represented in Colorado by var. typica Howell.---Sandy soil of plains, mountain slopes and hills, often among bushes. The variety from Wyoming to Nevada, south to New Mexico and California. Our few records from extreme western Colorado at 4500-7000 feet.

2. Phacelia submutica J. T. Howell, Proc. Calif. Acad. Sci. 4th ser. 25:370. 1944.

Annual plants; stems 2-8 cm. long, often branched at base, more or less pilose-hirsute; leaves 5-15 mm. long, scattered, elliptic-oblong to obovate, obtuse and cuneate, more or less hirsutulous, entire to serrate-lobed; racemes with flowers rather crowded; calyx lobes 3.5-5 mm. long, enlarging in fruit, hirsute and glandular; corolla 3.5-4.5 mm. long, light yellow, sometimes tinged with purple, tubular; stamens included, filaments glabrous; seeds 1.5-2 mm. long, subovate to ovate-oblong with 6-12 transverse corrugations.---Dry ground. Apparently known only from the type locality in Mesa County, Colorado, at about 4700 feet.

3. Phacelia denticulata Osterhout, Torreya 16:70. 1916.

Annual plants, viscid and glandular, slightly hispid; stems 10-40 cm. tall, erect, usually simple; leaves 3-7 cm. long, oblong-lanceolate in outline, lower at least, pinnately divided, or upper pinnately cleft, pinnae irregularly pinnately cleft; racemes to 4-5 cm. long in fruit, scorpioid, 1 to several; sepals about 1/2 as long as the corolla; corolla 4-5 mm. long, white to pale blue or violet, the lobes with irregular toothed or fimbriate margins; stamens barely included in corolla, filaments glabrous; styles included; capsules 5 mm. long; seeds 4-5 mm. long, flattened-elliptical, with a prominent ridge somewhat excavated between, minutely pitted. ---Slopes and valleys. Wyoming and Colorado. Our records in northcentral, central and southcentral Colorado east of the Continental Divide at 5000-8500 feet.

4. Phacelia demissa Gray, Proc. Amer. Acad. 10:326. 1875.

P. knightii A. Nels.---Annual plants; stems 3-20 cm. tall, branched from base and above, usually dichotomously, glandular-pubescent; leaves 10-25 mm. long, ovate to orbicular-ovate, rounded at apex, broadly cuneate to nearly truncate at base, entire to repand-dentate; flowers in short racemes, terminal and some appearing axillary; calyx about 1/2 to 2/3 as long as corolla; corolla 4-7 mm. long, the limb lavender-violet to purple, the tube somewhat yellowish; stamens included, with few hairs on filaments; styles included; seeds pitted, about 1-1.5 mm. long. Our plants would be var. typica Howell.---Barren arid places. The variety from Wyoming, south through Utah to Arizona and probably New Mexico. Our record from Montezuma County at about 5500 feet.

5. Phacelia heterophylla Pursh, Fl. Am. Sept. 1:140. 1814.

P. biennis A. Nels.---Plants perennial or possibly biennial; stems 10-80 cm. tall, stout, single to several, more or less hispid or hirsute and tomentulose, beneath this coating of hairs sometimes also strigose and canescent; leaves 1.5-9 cm. long, scattered but somewhat crowded basal, varying from entire to lobed or divided at base, these divisions entire, whole blade varying elliptic to elliptic-lanceolate (or even orbicular to linear-lanceolate) with more or less hirsute-hispid hairs, these usually appressed to somewhat sericeous and beneath this either puberulent or canescent-tomentose, leaves very variable; inflorescence virgate or spreading, the segments scorpioid; calyx from about as long as corolla to 3/4 its length, hirsute-hispid and tomentulose; corolla 4-8 mm. long, white, pink, rose or light purple; stamens exserted, about twice as long as the corolla, filaments villous. Our plant has been called var. typica Dundas. The above description includes that of P. leucophylla Torr.; P. nervosa Rydb.; P. alpina Rydb., which the writer cannot separate in this state.---Hills, valleys and mountains. The variety from Montana to Washington, south to New Mexico and California. Our records well scattered over the western two-thirds of Colorado at 4500-11,000 feet.

6. Phacelia sericea (Graham) A. Gray, Proc. Amer. Acad. 10:323. 1875.

Perennial plants; stems 10-30 cm. tall, rather stout, more or less caespitose from a branching caudex, strigose-canescent to sericeous, sometimes sparingly so; leaves 3-10 cm. long, tending to crowd down near base of plant, pinnately parted with many linear or narrowly oblong divisions, these often toothed, incised or pinnatifid, the ultimate divisions averaging 2-4 mm. wide, more or less sericeous especially when young; inflorescence of short spikelike racemes crowded into a narrow rather spikelike panicle, not clearly scorpioid; calyx about 5-6 mm. long, the lobes linear, much longer than the tube; corolla 5-7 mm. long, violet-blue, rarely white, campanulate; stamens long-exserted, about twice as long as the corolla, filaments glabrous. Intergrades in some degree with P. idahoensis Henderson.---Mountains, often on banks and slopes. Alberta to British Columbia, south to Colorado, Arizona and California. Our records scattered in the mountains of the western half of the state at 8000-13,000 feet.

7. Phacelia idahoensis Henderson, Bull. Torr. Club 22:48. 1895.

P. ciliosa Rydb.---Perennial plants; stems 20-50 cm. tall, weakly caespitose if at all, more or less strigose or villous-pubescent, hardly sericeous; leaves 5-15 cm. long, scattered, somewhat crowded near base, pinnately cleft to parted into oblong divisions, these entire or cleft, but divisions averaging over 4 mm. wide, more or less strigose-pubesent, not sericeous; inflorescence of short spikelike racemes, hardly scorpioid, crowded into rather dense spikelike panicles, these often interrupted; calyx about 4-5 mm. long, the lobes linear, much longer than the tube; corolla 5-6 mm. long, violet, blue or purple, campanulate; stamens exserted, in ours twice as long as the corolla or more, filaments glabrous. The correct name for this plant may be P. sericea var. biennis (Nels.) Brand.---Mountains. Alberta to British Columbia, south to Colorado and Nevada. Our records scattered in the western half of Colorado at 6500-12,000 feet.

8. Phacelia corrugata A. Nels., Bot. Gaz. 34:26. 1902.

Annual or possibly biennial plants; stems 15-50 cm. tall, simple or branched from base, moderately pubescent and somewhat glandular; leaves 1.5-10 cm. long, ovate to narrowly oblong, from sinuate to pinnately lobed; inflorescence of scorpioid racemes to 15 cm. long in fruit; calyx 1/2 or less the length of the corolla, hispid; corolla 7-12 mm. long, deep blue or violet, campanulate, the lobes entire or nearly so, rather wide; stamens exserted 3.5 mm. or more, the filaments glabrous; seeds 3.5-4.5 mm. long, finely pitted, excavated on either side of salient ridge, ventrally corrugated on ridge and margins.---Dry ground. Colorado, Utah and Arizona. Our records from western Colorado at 4500-6000 feet.

9. Phacelia intermedia Wooton, Bull. Torr. Club 25:457. 1898.
Texas to Arizona and Utah. This plant should be looked for in southern or southwestern Colorado.

10. Phacelia integrifolia Torr., Ann. Lyc. N. Y. 2:222, t. 3, 1828.
The var. typica Voss is reported from Kansas to Utah, south to Texas and Mexico, but none are cited for Colorado. The writer has seen no specimens from this state although the plant should be here.

11. Phacelia neomexicana Thurber ex Torr., in Bot. Mex. Bound. Surv. 143. 1859.
P. alba Rydb.---Annual plants, hispid to villous and glandular especially above; stems 15-70 cm. tall, erect or ascending, simple or branching throughout; leaves 2-4.5 cm. long, ovate, irregularly bipinnate; inflorescence rather dense and branching, often corymbiform, the racemes scorpioid, about 2 cm. long in flower to 9 cm. in fruit; calyx about 2/3 as long as the corolla; corolla 4-5 mm. long, white with dentate or fimbriate lobes; stamens exserted for about 3 mm., filaments glabrous; styles exserted; seeds 2.5-3 mm. long, finely pitted, excavated on each side of the ventral ridge. Represented in Colorado by var. alba (Rydb.) Brand, but a few specimens appear rather purple-flowered and suggest the presence of the species proper.---Plains, hills and valleys. The variety from Wyoming to Mexico. Our records from northcentral, central, southcentral and southwestern Colorado at 5000-9000 feet.

12. Phacelia formosula Osterhout, Bull. Torr. Club 46:54. 1919.
Annual plants; stems 15-25 cm. tall, erect, branching from base or simple, somewhat hispid or hispidulous and usually glandular; leaves 5-7 cm. long, narrowly ovate or elliptic, pinnately divided, the pinna more or less pinnatifid; inflorescence of rather close panicles; calyx 1/2 as long as the corolla; corollas about 6 mm. long, blue or violet, campanulate, the lobes entire; stamens exserted for 2-3 mm., filaments glabrous; seeds 3-4 mm. long, pitted, excavated ventrally on each side of the ridge.---Apparently limited to northcentral Colorado at 8500-9500 feet. The type is from Jackson County but a specimen from Grand County appears to be this species.

13. Phacelia splendens Eastwood, Zoe 4:9. 1893.
P. glandulosa ssp. splendens (Eastw.) Brand---Annual plants; stems 10-30 cm. tall, simple or branching from base, glabrous, or glandular-puberulent; leaves 1.5-6 cm. long, ovate, pinnately divided or parted; inflorescence of scorpioid racemes; calyx about 1/2 as long as the corolla; corolla 7-9 mm. long, campanulate, blue, lobes entire or nearly so; stamens exserted for 7-8 mm., filaments glabrous; seeds about 3.5-4.5 mm. long, pitted, excavated on each side of the ridge.---Dry ground. Colorado and Utah. Our records from westcentral and southwestern Colorado at 4500-6000 feet.

14. Phacelia glandulosa Nutt., Journ. Acad. Phila. n. ser. 1:160. 1848.
P. bakeri (Brand) Macbride---Annual or biennial plants; stems 10-30 cm. tall, simple, branching from base, viscid-glandular, often canescent-pilose or densely hispidulous; leaves 1.5-12 cm. long, lanceolate to narrowly elliptical or oblong, once-pinnately divided; inflorescence usually crowded-spicate, the racemes not loosening much in age; calyx about 1/2 as long as the corolla; corolla 5-7 mm. long, purple, blue or violet, campanulate, lobes entire to somewhat toothed; stamens exserted for 4-7 mm., filaments glabrous; seeds 2.5-3.25 mm. long, pitted, sometimes deeply excavated on each side of the ridge. The suggested differences between the above and P. bakeri will not work out for our plants.---Hills, valleys and mountains. Montana to Idaho, south to Colorado and Utah. Our records well scattered over the western half of Colorado at 6500-12,000 feet.

Family 101. Boraginaceae BORAGE FAMILY

Annual or perennial, herbaceous to shrubby plants, usually bristly-hairy; leaves simple, usually alternate, rarely opposite or whorled; flowers perfect, regular (or irregular in Lycopsis and Echium), solitary or cymose, the cymes glomerate-racemose or spicate, frequently scorpioid, the bracts to 1 side or opposite the flowers; calyx usually deeply 5-lobed, more or less irregular; corolla gamopetalous, 5-lobed with folds or saccate-intruded appendages common in the throat; stamens 5, borne on the corolla tube alternate to its lobes; ovary superior, of 2 carpels, usually with 4 ovules, entire or lobed, commonly breaking up into 4 one-seeded nutlets (appearing to consist of 4, 1-ovuled carpels), style entire or 2-cleft, at apex of ovary or down between the lobes; fruit commonly of 4 nutlets these often on an upward prolongation of the receptacle (gynobase). Mature fruit is desirable and usually a necessity in checking plants in this family.

1. Styles arising from the apex of the 2- to 4-lobed ovary; stigma annulate, usually surpassed by a 2-lobed appendage   -1. Heliotropium
1. Styles arising from between the lobes and attached to the receptacle (gynobase); stigmas various but never as above
  2. Nutlets with uncinate, glochidiate or barbed prickles on back, margins or at apex
    3. Nutlets with dorsal surface rather uniformly covered with prickles, no definite margins present
                                                                                             -2. Cynoglossum
    3. Nutlets with a definite margin, the prickles confined to this (back may be muricate or tuberculate but no prickles present)
      4. Nutlets stellately spreading, attached at the apical end   -3. Pectocarya
      4. Nutlets erect, incurved or weakly divergent, attached at or below the middle

    5. Plants annual; pedicels erect or nearly so; styles surpassing the nutlets; subulate gynobase about as long as the nutlets -4. <u>Lappula</u>
    5. Plants perennial or biennial; pedicels reflexed in fruit; styles usually shorter than the nutlets; pyramidal gynobase about half as long as the nutlets -5. <u>Hackelia</u>
2. Nutlets without hooked or barbed prickles
  6. Corolla irregular with an oblique limb (not strongly so), the lobes somewhat unequal
    7. Stamens equal, included, filaments very short -6. <u>Lycopsis</u>
    7. Stamens unequal, the longer exserted on long filaments -7. <u>Echium</u>
  6. Corolla regular or very nearly so
    8. Calyx in fruit much enlarged, becoming conspicuously veiny, folded and flattened; stems procumbent, angled with stiff retrorse bristles on the angles -8. <u>Asperugo</u>
    8. Calyx in fruit little if any enlarging, not becoming veiny, folded and flattened; stems various but not as above
      9. Nutlet attachment surrounded by a swollen ring, leaving a distinct pit on the gynobase; plants of fields and waste places -9. <u>Anchusa</u>
      9. Nutlet attachment flat or somewhat concave, not surrounded by a ring and not leaving a distinct pit on the gynobase; plants of various habitats (some in waste grounds and fields)
        10. Corolla normally blue (aberrant white-flowered plants occasionally are found), or reddish in the bud stage
          11. Nutlets with an oblique dorsal face encircled by an upturned flange or rim, this often irregularly toothed; depressed-pulvinate plants seldom over 7 cm. tall, of alpine areas in Colorado -10. <u>Eritrichium</u>
          11. Dorsal face of nutlets (if present) not encircled by an upturned flange or rim; plants not depressed-pulvinate, usually over 7 cm. tall, most species growing below alpine areas in Colorado
            12. Corolla lobes convolute in the bud; nutlets basally attached to a flat gynobase; corolla salverform -11. <u>Myosotis</u>
            12. Corolla lobes imbricate in the bud; nutlets obliquely attached to a convex gynobase; corolla with a tube and usually campanulate throat, not salverform -12. <u>Mertensia</u>
        10. Corolla white, greenish-white, yellow or orange
          13. Nutlets attached by their very base to a flat gynobase; stigmas usually bifid or geminate
            14. Corolla lobes acute or acuminate, erect; styles long-exserted, protruding as the buds open; anthers sagittate; flowers greenish-white or yellowish-green -13. <u>Onosmodium</u>
            14. Corolla lobes rounded or obtuse, spreading or ascending; styles included or short-exserted when flower is fully opened; flowers usually yellow to orange -14. <u>Lithospermum</u>
          13. Nutlets attached somewhat above the base to a broadly to narrowly pyramidal gynobase; stigmas solitary and simple
            15. Nutlets attached along an open or nearly closed ventral groove or slit, this sometimes opening in triangular fashion at base; lower leaves usually alternate; annual, biennial or perennial plants -15. <u>Cryptantha</u>
            15. Nutlets lacking a ventral groove, this replaced by a keel; lower leaves alternate or opposite; annual plants
              16. Cotyledons 2-lobed; corolla orange or yellow, the tube definitely longer than the calyx -16. <u>Amsinckia</u>
              16. Cotyledons not lobed; corolla white, the tube short and rarely longer than the calyx -17. <u>Plagiobothrys</u>

1. <u>Heliotropium</u> L. HELIOTROPE

Annual or perennial herbaceous plants; leaves alternate, entire; flowers regular, blue to more commonly white, in scorpioid racemes or axillary; calyx of 5 slightly united sepals; corolla salverform to funnelform, 5-lobed, spreading in the flower, plaited in the bud; styles 1, arising from apex of ovary, stigmas annular or conic, ovary 2- (or appearing) 4-celled, 2- to 4-grooved or very slightly 2- to 4-lobed; fruit separating into 2 or 4 nutlets, the styles falling with the nutlets.

1. Plant strigose on stems, leaves, calyx and corolla tube; flowers (at least some) axillary (or with leaflike bracts); corollas 10-15 mm. long; fruit separating into 2 nutlets -1. <u>H. convolvulaceum</u>
1. Plant glabrous everywhere; flowers in terminal bractless racemes; corollas 6-8 mm. long; fruit separating into 4 nutlets -2. <u>H. spathulatum</u>

1. Heliotropium convolvulaceum (Nutt.) A. Gray, Mem. Am. Acad. 6:403. 1857.
    <u>Euploca convolvulacea</u> Nutt.---Annual plants; stems 10-40 cm. tall, low, freely branched, strigose; leaves 1-4 cm. long, ovate to lanceolate, strigose-canescent; flowers lateral and terminal, but mostly solitary and opposite to leaflike bracts; calyx lobes about 3 mm. long, linear-subulate to lanceolate; corolla 10-15 mm. long, white, salverform to funnelform, the tube strigose; fruit separating into 2 nutlets, these about 3-3.5 mm. long, pubescent.---Sandy plains and hills. Nebraska to Utah, south to Texas and Mexico. Our records in the eastern half of Colorado, except the extreme southern part, at 3500-5500 feet.

2. Heliotropium spathulatum Rydb., Bull. Torr. Club 30:262. 1903.
    <u>H. curassavicum</u> var. <u>obovatum</u> DC.---Perennial or possibly annual plants; stems diffuse, rather fleshy, glabrous, branching; leaves rather fleshy and succulent, oblanceolate-spatulate or linear-oblong, glabrous; flowers in unilateral, scorpioid racemes, usually 2

racemes present from a common branch; calyx lobes about 2 mm. long, lanceolate to ovate-lanceolate; corolla about 6-8 mm. long, white or tinged with blue, salverform; fruit separating into 4 nutlets each about 2.5-3 mm. long.---Valleys often in moist saline soil. Montana to Washington, south to Colorado and Nevada. Our records thinly scattered over the eastern two-thirds of Colorado at 3500-8000 feet.

2. Cynoglossum L. HOUNDSTONGUE

Annual or biennial, caulescent, rather tall herbaceous plants; leaves alternate and entire; flowers regular in scorpioid racemes, these panicled, several to many, terminal; calyx 5-parted, enlarged and the lobes spreading or reflexed in fruit; corolla dull red to reddish-purple or rarely white, salverform or funnelform, the tube short and throat closed by 5 scales; stamens included; ovary deeply 4-lobed; fruit separating into 4 equally divergent nutlets, dorsally covered with short barbed prickles, attached at upper end to a rather low conical gynobase, splitting away at maturity but hanging attached to the style.

1. Cynoglossum officinalis L., Sp. Pl. 134. 1753.
Stems 40-80 cm. tall, erect, stout, softly pilose and somewhat canescent; basal leaves 10-30 cm. long, petioled, oblong or oblong-lanceolate, pilose, upper leaves lanceolate, sessile; calyx lobes about 3 mm. long, ovate-lanceolate to ovate; corolla about 7-10 mm. wide and about 6-9 mm. long; fruit pyramidal, the nutlets about 7 mm. long, flattened on upper outer face.---Fields, roadsides and waste places. Native to Europe, now widely introduced in the United States west to Montana, Utah and Arizona. Our records thinly scattered in cultivated areas of Colorado at 5000-9000 feet.

3. Pectocarya DC. COMBSEED

Small branching annual plants; leaves alternate or subopposite, narrow, entire; flowers regular and minute, scattered, solitary in the axils; calyx deeply 5-parted, spreading in fruit; corolla white or whitish, salverform to tubular with crests nearly closing the throat; anthers included; ovary 4-lobed, stigma 1, capitate; nutlets attached to a depressed receptacle by their apical ends, the attachment without an annular ring surrounding the scar, widely divergent, margin undulate, laciniate, with uncinate prickles on this margin at least near apex; fruiting pedicels coarse and stiff, nodding or reflexed.

1. Pectocarya penicillata (H. & A.) A. DC., Prodr. 10:120. 1846.
A northern and western species recorded from southern Wyoming. Should be looked for in northern or northwestern Colorado.

4. Lappula Moench STICKSEED; BEGGARS TICKS

Rather small annual plants; leaves alternate, entire; flowers regular, small, in racemes, these abundantly bracteate, the bracts although smaller resemble the stem leaves; calyx of 5 narrow lobes, slightly if at all accrescent; corolla blue or white, salverform or funnelform; ovary 4-lobed; fruiting pedicels erect or nearly so; fruit separating into 4 nutlets, the dorsal surface of each with an upturned rim bearing glochidiate prickles, attached to a gynobase taller than wide and about as long as the nutlets, attached all along the ventral keel, not leaving a pit on the gynobase; style usually longer than the nutlets.

1. Nutlets with marginal prickles in at least 2 rows
  2. Attached fruit about 3 mm. high; nutlets with dorsal granulations uniform; introduced plants
                                                                                                                      -1. L. echinata
  2. Attached fruit about 4-5 mm. high; nutlets with dorsal granulations high toward middle, lower toward the sides; native plants
    3. Plants bushy-branched from base, green, pubescence short, sparse and appressed; fruit (attached) about 5 mm. high   -2. L. cenchrusoides
    3. Plant strict and usually branching above, canescent, the hairs rather long and loose; attached fruit about 4 mm. tall   -3. L. fremontii
1. Nutlets with marginal prickles definitely in 1 row
  4. Margin of 2 or more of the nutlets obese, completely inflated, bearing a row of very short, terete bristles seated upon its usually rounded edge, (bristles not appearing by the confluence of their bases to form part of the margin)
    5. Only 3 of the nutlets obese as described in first number 4 (the other more like second number 4)
                                                                                -4. L. texana
    5. All the nutlets alike, obese (and like first number 4)   -4A. L. texana coronata
  4. Margins of the nutlets consisting of a row of distinct appendages or the margin more or less cup-shaped through the obvious basal union of the appendages, basal portion of appendages sometimes inflated but this always appearing as lobes to the margin (hence never with margins obese and inflated with terete prickles seated on it)
    6. Appendages of nutlet margin distinct, not at all united at base; nutlets alike   -5. L. redowskii
    6. Appendages of at least 3 nutlet margins with bases more or less confluent and dilated; nutlets often unlike (1 being like first number 6)   -5A. L. redowskii desertorum

1. Lappula echinata Gilib., Fl. Lith. 1:25. 1781.
L. lappula (L.) Karst.---Branching mostly above the base; stems 25-60 cm. tall, hispid to appressed pubescent; racemes leafy-bracted; corollas about 3 mm. long, blue, little longer than the calyx lobes; fruit about 3-3.5 mm. long, nutlets alike, margins armed with 2 rows of distinct

prickles.---A weed in fields and waste ground. Introduced from Europe and widely distributed in Canada and United States. Our few records scattered in the northwestern quarter of Colorado at 6500-7500 feet.
2. Lappula cenchrusoides A. Nels., Bull. Torr. Club 26:243. 1899.
    A species apparently known only from southeastern Wyoming. To be expected in adjacent Colorado.
3. Lappula fremontii (Torr.) Greene, Pitt. 4:96. 1899.
    L. erecta A. Nels.---Saskatchewan, south to Wyoming and Utah. To be looked for in northern Colorado.
4. Lappula texana (Scheele) Britt., Mem. Torr. Bot. Club 5:273. 1894.
    L. cupulata (A. Gray) Rydb.; L. heterosperma Greene---Stems 10-50 cm. long, branched in most of ours from very base, hirsute to appressed pilose; racemes leafy-bracted; corolla 2.5-3 mm. long, little longer than calyx lobes, light blue; fruit about 3-4 mm. long, 3 of nutlets with margins obese and swollen, bearing a row of terete bristles about 0.5 mm. long, the swollen margins obviously not formed from the swollen bases of these bristles, fourth nutlet with 1 row of longer bristles not seated on an inflated margin.---River valleys and waste places. South Dakota to Idaho, south to Texas, Arizona and Mexico. Our records scattered over Colorado, except in the northeastern part, at 4000-8000 feet.
4A. Lappula texana coronata (Greene) Nelson & Macbride, (var.) Bot. Gaz. 61:41. 1916.
    L. heterosperma var. homosperma A. Nels.---Differs from species in having all nutlets alike, with short bristles seated on an obese, inflated margin.---Alberta to Idaho, south to Colorado and Arizona. Our records northcentral, central and southwestern Colorado at 4000-8000 feet.
5. Lappula redowskii (Hornem.) Greene, Pitt. 2:182. 1891.
    L. occidentalis (S. Wats.) Greene; L. montana Greene; L. calycosa Rydb.---Plants 10-60 cm. tall, strict and erect, branching only above or branching at base with ascending branches, hirsute to strigose; racemes leafy-bracted; corolla about 2.5-3 mm. long, little longer than the calyx lobes, whitish to light blue; fruit about 3 mm. long, nutlets alike, the margins little if any inflated, bearing 1 row of bristles, these about 1 mm. long, not dilated or only moderately so at base and nearly or quite distinct.---Fields, waste places, plains and hills, often in dry or sandy ground. North Dakota to Washington, south to Texas and California; Eurasia; South America. Our records scattered over Colorado at 4000-9000 feet.
5A. Lappula redowskii desertorum (Greene) Johnston, (var.) Contrib. Arn. Arb. 3:93. 1932.
    L. collina Greene; L. foliosa A. Nels.; L. cucullata A. Nels.; L. leucotricha Rydb.---Differs from species in nutlets often unlike (1 being like species) the others with bases more or less dilated and confluent but these always appearing as lobes of the margins (margin itself not inflated). This variety intergrades somewhat with the species in Colorado.---Montana to Washington, south to New Mexico and California; South America. Our records from western Colorado (a doubtful one in the central part of the state) at 4500-8000 feet.

5. Hackelia Opiz

Coarse biennial or perennial, caulescent, herbaceous plants; stems rather strict; leaves spatulate, oblanceolate to linear-lanceolate; flowers regular, in terminal scorpioid racemes, bractless or with minute bracts present, not leafy-bracted except at very base; calyx with 5 narrow lobes, slightly if at all accrescent; corolla blue or white, salverform or funnelform, tube short, throat closed by scales; pedicels of fruit recurved; ovary 4-lobed; fruit separating into 4 nutlets, these attached to the pyramidal gynobase which is wider than high and about 1/2 as long as the nutlets, the scar not leaving a pit on the gynobase, nutlets attached by a deltoid or ovate scar, not along the ventral keel, dorsal surface of nutlet with an upturned rim bearing glochidiate prickles, the surface smooth to prickly; style usually shorter than the nutlets.

1. Stems hispid, the hairs with pustulose bases; stems with leaves crowded toward base, few-leaved above; plants apparently limited to southwestern corner of Colorado   -1. H. gracilenta
1. Stems hirsute, pilose, strigose or pubescent, the hairs without pustulate bases; stems leafy throughout; plants not limited to southwestern Colorado
    2. Corolla minute, 1.5-2 mm. wide; nutlets usually prickly on back (as well as margins)   -2. H. deflexa
    2. Corolla larger; nutlets usually without prickles on the back
        3. Corolla 2-4 mm. wide; nutlets with marginal prickles often united at base sometimes for 1/2 their length; fruit 3-5 mm. wide   -3. H. leptophylla
        3. Corolla 4-12 mm. wide; nutlets with prickles of margin not at all united at base; fruit 5-6 mm. wide
                                                                                            -4. H. floribunda

1. Hackelia gracilenta (Eastw.) Johnston, Contrib. Gray Herb. 68:46. 1923.
    Lappula gracilenta Eastw.---Stems 25-60 cm. tall, hispid the hairs from pustulate bases; leaves 2-8 cm. long, tending to crowd toward the base of the plant, these basal leaves spatulate or oblanceolate, stem leaves lanceolate; corolla 6-8 mm. wide, blue to whitish-blue; fruit 4-5 mm. wide, the prickles on margins slightly united at base.---Canyons and slopes. Apparently limited to southwestern Colorado, our records from 7000-8000 feet.
2. Hackelia deflexa (Wahl.) Opiz in Bercht., Fl. Bohm. 2(2):147. 1839.
    Lappula americana (Gray) Rydb.---Manitoba and British Columbia, south to Wyoming and Idaho. This plant may be in northern Colorado.
3. Hackelia leptophylla (Rydb.) Johnston, Contrib. Gray Herb. 68:46. 1923.
    Lappula leptophylla Rydb.; L. scaberrima Piper; L. angustata Rydb.; L. besseyi Rydb.---Stems 50-100 cm. tall, hirsute; leaves strigose-hirsute to hispidulous-pubescent; corolla 2-4 mm. wide, blue; fruit 3-5 mm. wide, prickles of margin often partly united (often 1/2) with

surfaces of nutlets not at all prickly.---Hills, valleys and canyons. South Dakota and Montana, south to Nebraska and Colorado. Our records scattered in the western half of Colorado at 6000-9500 feet.

4. Hackelia floribunda (Lehm.) Johnston, Contrib. Gray Herb. 68:46. 1923.
Lappula floribunda (Lehm.) Greene; L. subdecumbens (Parry) A. Nels.---Stems 30-150 cm. tall, strigose, pilose to softly hirsute; leaves 5-15 cm. long, narrowly oblanceolate, strigose, pilose to softly hirsute; corolla 4-12 mm. wide, blue to whitish; fruit 5-6 mm. wide, prickles on margin only, not at all united at base.---Dry hillsides and valleys, often among bushes. Manitoba to British Columbia, south to New Mexico, Arizona and California. Our records scattered over the western half of Colorado at 6500-10,000 feet.

6. Lycopsis L. BUGLOSS

Annual caulescent, hispid, herbaceous plants; leaves alternate and entire; flowers slightly irregular, blue or bluish, in dense spikelike, leafy-bracted, scorpioid racemes; calyx 5-parted; corolla salverform, the tube distinctly bent near middle, the lobes subirregular and oblique, throat closed by hairy scales; stamens included, with very short filaments; ovary 4-lobed, these separating in fruit, stigmas 1; nutlets wrinkled, erect, attached by the base to a flat gynobase, the attachment surrounded by a tumid ring leaving a pit upon the gynobase.

1. Lycopsis arvensis L., Sp. Pl. 139. 1753.
A weed native to Europe and widely introduced into North America. Reported for Colorado but no specimen seen.

7. Echium L.

Biennial or possibly perennial, hispid, herbaceous plants; leaves alternate, entire; flowers blue to violet-purple, in leafy-bracted scorpioid, spikelike racemes; calyx 5-parted; corolla tubular-funnelform, irregular, unequally 5-lobed, the throat not appendaged; stamens unequal at least the longer ones exserted on long filaments; ovary 4-lobed, these separating in fruit, style 2-cleft at apex; nutlets erect, rugose, attached by their bases to a flat gynobase, the scar flat or somewhat concave, not leaving a pit.

1. Echium vulgare L., Sp. Pl. 139. 1753.
Stems 20-70 cm. tall, erect, simple or at length branched; leaves 2-15 cm. long, linear or oblong; calyx 5-6 mm. long in anthesis; corolla 15-20 mm. long, pale blue or purplish.---Fields, roadsides and waste places. Native to Europe and introduced in the eastern part of the United States, west to South Dakota and New Mexico. Our one record from southeastern Colorado at 4200 feet.

8. Asperugo L. CATCHWEED

Annual rough-hispid plants; stems procumbent, with stiff bristly hairs; leaves alternate or upper sometimes opposite, entire; flowers regular, 1-3 together on short, recurved pedicels in the upper leaf axils; calyx campanulate, unequally 5-cleft, much enlarged and reticulate-veiny in fruit, lobes incised-dentate, the teeth often appearing as extra lobes in the sinuses; corolla tubular-campanulate, 5-lobed; stamens included; ovary 4-lobed, these separating in fruit; nutlets grandular-tuberculate, keeled laterally, attached above the middle to an elongate-conic receptacle, the scar not leaving a pit.

1. Asperugo procumbens L., Sp. Pl. 138. 1753.
Stems 20-60 cm. long, slender, retrorsely short-hispid; leaves 1-4 cm. long, oblong, elliptic, lanceolate or spatulate; fruiting calyx 8-12 mm. wide; corolla small, about 2-3 mm. long, blue, purple or perhaps purplish-red.---Waste places. Introduced from Europe in various parts of Canada and the United States, more abundant eastward. Our records from northern and western Colorado at 5000-7500 feet.

9. Anchusa L. BUGLOSS; ALKANET

Perennial herbaceous caulescent plants with hispid or villous stems and leaves; flowers regular, bracteate, in scorpioid racemose cymes; calyx 5-cleft; corolla blue or purple, funnelform or salverform, lobes spreading, the throat closed by appendages; stamens included; ovary 4-lobed, these separating in fruit into rugose or granulate nutlets, inserted by their bases on a flat gynobase, the attachment surrounded by an annular rim leaving a pit upon gynobase.

1. Anchusa officinalis L., Sp. Pl. 133. 1753.
Perennial plants with taproot; stems 30-80 cm. tall; basal leaves 8-20 cm. long, oblanceolate, stem leaves lanceolate; calyx 5-7 mm. long, the lobes lanceolate or narrowly triangular, about as long as the tube; corolla about 1 cm. long, dark blue; nutlets about 2-3 mm. long.---Roadsides and waste places. Native to Eurasia and introduced into eastern United States as far west as Utah. Our records from central and northcentral Colorado at 4500-6500 feet.

10. Eritrichium Schrad. ALPINE FORGET-ME-NOT

Depressed-pulvinate perennial plants; leaves small, entire, densely clustered on short branches; flowers regular, in few-flowered racemes; calyx 5-parted, the lobes erect or ascending in fruit; corolla blue, rarely white, funnelform to nearly rotate with tube short, crested in the throat; stamens included; ovary 4-lobed, stigmas 1; fruit separating into 4 nutlets, these smooth, attached by their middle to a stout gynobase, but not leaving a pit on it, gynobase about 1/2 as long as the nutlets, nutlets with an obliquely truncated apex, this portion surrounded by an entire or toothed margin.

1. Oblique truncated face of nutlets with margins entire or obscurely toothed -1. E. elongatum
1. Oblique truncated face of nutlets with margins definitely toothed -1A. E. elongatum argenteum

1. Eritrichium elongatum (Rydb.) Wight, Bull. Torr. Club 29:408. 1902.
Plants villous often silvery looking, forming a tuft about 2-4 cm. tall (not counting flowering branches); leaves 5-10 mm. long, narrowly ovate to oblong or oblanceolate; flower cluster compact when sessile among leaves or sometimes racemelike when borne on a leafy flowering branch up to 7 cm. long; calyx lobes about 2 mm. long, linear; corolla variable in size, from 1-7 mm. across (usually about 4-5 mm.), bright blue (rarely white); nutlets with an entire margin to the truncated oblique portion, rarely with a few obscure teeth.---Alpine. Montana to Oregon, south to New Mexico and Utah. Our records scattered in the high mountains of Colorado at 10,500-13,000 feet.
1A. Eritrichium elongatum argenteum (Wight) Johnston, (var.) Contrib. Gray Herb. 70:53. 1924.
E. argenteum Wight---Differs from species in having the oblique truncated face of the nutlets with margins toothed. Some of our plants appear to have obscure teeth and tend to intergrade with the species.---High mountains. Colorado, Utah and Wyoming. Our records from central and northcentral Colorado at 11,500-13,000 feet.

11. Myosotis L. FORGET-ME-NOT

Annual or perennial plants, caulescent but leaves tending to crowd down to base; leaves alternate, entire, narrow; flowers regular, in slender bractless, unilateral racemes; calyx 5-cleft, lobes erect or spreading, some hairs on the basal part uncinate or gland-tipped; corolla blue to white, salverform, tube short, throat crested; stamens included; ovary 4-lobed, stigma solitary and simple; fruit separating into 4 smooth nutlets, these basally attached to the flat gynobase, the scar of attachment flat, not pitlike.

1. Densely tufted perennial plants; corolla bright blue, conspicuous, 5-8 mm. wide -1. M. alpestris
1. Annual or biennial plants not densely tufted; corolla white to very light blue, inconspicuous, 1-2 mm. wide
-2. M. verna

1. Myosotis alpestris Schmidt., Fl. Boehm. 3:26. 1794.
Perennial plants; stems 10-30 cm. tall, densely tufted, hirsute, nearly erect, with some loose papery sheaths usually present at base; basal leaves linear-oblanceolate to spatulate, hirsute; stem leaves 2-7 cm. long, oblong-linear to linear-lanceolate; calyx lobes about 2 mm. long, erect in fruit; corolla 5-8 mm. wide, bright blue.---Moist places in the mountains. Alberta and Alaska, south to Colorado and Oregon; Europe. Our records from northwestern Colorado at 6000-11,500 feet.
2. Myosotis verna Nutt., Gen. 2:Add. 1818.
M. virginica of Manuals--- Maine to Ontario, south to Florida and Texas, reappearing in British Columbia, Idaho and Oregon. The plant may be in Colorado.

12. Mertensia Roth. BLUEBELLS

Perennial caulescent herbaceous plants; leaves entire, alternate; flowers regular, in scorpioid cymes or paniculate; calyx 5-parted, or deeply cleft; corolla tubular-funnelform or somewhat trumpet-shaped, usually campanulate above the tube, usually blue in age and reddish in the bud; stamens included or somewhat exserted; ovary 4-lobed, stigmas entire or very nearly so; fruit separating into 4 (or sometimes fewer) nutlets, these attached about 1/4 to 1/2 way up to a convex gynobase, not leaving a pit on it, the nutlets rugose. This is a difficult genus in our area. M. lanceolata, M. viridis, M. fusiformis and M. bakeri with their varieties show numerous intergradations. The writer is following Williams (Am. Mo. Bot. Gard. 24:17-159. 1937), as the latest broad treatment with the realization that the result is often unsatisfactory in Colorado.

1. Cauline leaves with distinct lateral veins, large, to 15 cm. long (usually over 6 cm.); plants ordinarily over 40 cm. tall, mostly in moist or shaded habitats
    2. Leaves strigillose on upper surface -1. M. franciscana
    2. Leaves glabrous or somewhat papillose above, not at all hairy
        3. Calyx lobes 3-4 mm. long, triangular-acute; corolla tubes 4-6 mm. long, the limb somewhat longer
-2. M. arizonica
        3. Calyx lobes 1.5-3 mm. long, obtuse at apex; corolla tubes 6-8 mm. long, the limb about the same length
-3. M. ciliata
1. Cauline leaves without distinct lateral veins, the middle cauline rarely over 6 cm. long; plants usually less than 40 cm. tall, often in rather dry open habitats (lateral veins sometimes present in M.oblongifolia but plant not over 30 cm. tall and of open dry habitats)
    4. Filaments attached in the corolla tube, the anthers not projecting (or only very slightly) into the throat; anthers usually longer and wider than the filaments; styles shorter than or as long as the calyx
        5. Leaves glabrous (rarely faintly strigillose near base below); plants apparently limited to northcentral and northeastern parts of Colorado close to the Wyoming border -4. M. humilis

5. Leaves strigillose above, glabrous below; plants not limited to northcentral and northeastern Colorado
  6. Backs of calyx lobes and stems glabrous; plants of subalpine or alpine zones (above 10,000 feet) in Colorado; middle cauline leaves usually acute -5. M. alpina
  6. Backs of calyx lobes and stems pubescent; plants usually below 10,000 feet in Colorado; middle cauline leaves often rounded at apex -6. M. brevistyla
4. Filaments attached near the throat of the corolla and the anthers projecting into the throat; anthers often shorter and narrower than the filaments; styles longer than the calyx
  7. Calyx lobes not over 1/2 the length of the whole calyx; plants rarely above 9000 feet in Colorado
    8. Leaves pubescent on both surfaces -11A. M. lanceolata pubens
    8. Leaves glabrous, or pubescent only above
      9. Leaves glabrous both sides; plants of northcentral Colorado -11B. M. lanceolata brachyloba
      9. Leaves pubescent above; plants of southern and northwestern Colorado -11C. M. lanceolata fendleri
  7. Calyx lobes 2/3 or more the length of the whole calyx; plants sometimes above 9000 feet in Colorado
    10. Leaves pubescent on both surfaces
      11. Corolla tube shorter than or as long as the limb (expanded portion); cauline leaves often over 4 cm. long; plants not over 9000 feet in Colorado -11D. M. lanceolata secundorum
      11. Corolla tube longer than the limb; cauline leaves seldom over 4 cm. long; plants usually over 9000 feet in Colorado
        12. Leaves more or less unilateral, rarely over 3 cm. long -8A. M. viridis cana
        12. Leaves not unilateral, some over 3 cm. long
          13. Calyx usually pubescent on back; pubescence of upper leaf surface appressed; plants usually drying sordid brown and corollas drying purple -7. M. bakeri
          13. Calyx usually glabrous on back; pubescence of upper leaf surface spreading; plants usually drying green and corollas drying blue -7A. M. bakeri osterhoutii
    10. Leaves pubescent only above or glabrous both surfaces
      14. Filaments shorter than their anthers; calyx divided quite to the base; styles usually not reaching the anthers; anthers straight; plants often above 10,000 feet in Colorado
        15. Leaves glabrous both surfaces -8B. M. viridis dilatata
        15. Leaves strigose above
          16. Plants usually erect; leaves not unilateral, usually quite glaucous
            17. Pubescence of pedicels appressed -8. M. viridis
            17. Pubescence of pedicels spreading -8C. M. viridis cynoglossoides
          16. Plants usually ascending; leaves usually unilateral, not at all glaucous -8D. M. viridis parvifolia
      14. Filaments longer than the anthers; calyx not divided quite to base; styles usually reaching or surpassing the anthers; anthers usually curved; plants seldom above 10,000 feet in Colorado
        18. Limb of the corolla definitely shorter than the tube; plants limited to northwestern Colorado -9. M. oblongifolia
        18. Limb of corolla longer than or subequal to tube; plants not limited to northwestern Colorado
          19. Root fusiform; calyx pubescent on the back; leaves usually strigillose above, the hairs directed toward the nearest margin of the leaf; plants of southwestern Colorado -10. M. fusiformis
          19. Root not fusiform; calyx glabrous on the back; leaves glabrous above or if strigose the hairs directed toward the apex; plants not in southwestern Colorado -11. M. lanceolata

1. **Mertensia franciscana** Heller, Bull. Torr. Club 26:549. 1899.
   M. alba Rydb.; M. pratensis Heller---Stems 10-100 cm. tall; basal leaves to 10 cm. long, elliptical, base rounded or subcordate, petiole longer than the blade; cauline narrowly ovate to almost lanceolate, the lowest petiolate, strigillose above, nearly glabrous below; branches of cyme elongating; calyx about 3 mm. long, divided almost to base, the lobes narrow, acute and ciliate; corolla limb 4-9 mm. (often about 6 mm.) long, subequal to tube.---Wet meadows and along streams. Colorado to Nevada, south to New Mexico and Arizona. Our records from southcentral, southwestern and westcentral Colorado at 6000-13,000 feet.
2. **Mertensia arizonica** Greene, Pitt. 3:197. 1897.
   The var. grahami Williams is apparently limited to western Colorado but no specimens were seen by the writer.
3. **Mertensia ciliata** (James) G. Don., Gen. Hist. 4:372. 1838.
   M. picta Rydb.; M. pallida Rydb.; M. ciliata var. longipedunculata A. Nels.; A. ciliata var. punctata (Greene) A. Nels.; A. ciliata var. polyphylla (Greene) A. Nels.; M. punctata Greene; M. polyphylla Greene---Stems to 100 cm. tall, erect or ascending, usually in clumps or many stems from 1 rootstock; basal leaves to 12 cm. long, 3-6 cm. wide, oblong to ovate or almost lanceolate, ciliate, petioles often as long or longer than the blades; cauline leaves variable, lanceolate to ovate, acute (or sometimes obtuse), attenuate to subcordate at base, ciliate, often papillate above, often glaucous, thin; inflorescence from axils of leaves, peduncles becoming elongated, pedicels short, glabrous or papillate; calyx lobes 1.5-3 mm. long, glabrous on back, ciliate on margins, rather obtuse, divided nearly to base; corolla tube 6-8 mm. long, limb about the same; style about length of the corolla.---Moist or wet places. Montana to Oregon, south to New Mexico. Our records well scattered in the western half of Colorado at 5500-13,000 feet.
4. **Mertensia humilis** Rydb., Bull. Torr. Club 36:681. 1909.
   Stems 4-20 cm. tall, ascending or erect; basal leaves ovate to oblong-lanceolate, petioled; cauline leaves oblong-lanceolate, sessile or nearly so; all leaves glabrous or nearly so, pustulose, faintly strigillose near base below, rather thick; inflorescence congested; calyx 2.5-5 mm. long, glabrous on back, lobes ciliate; corolla tube 2-6 mm. long, often exceeding the calyx,

limb 3-6 mm. long; anthers longer than the filaments, included in tube; style as long as calyx.---High plains and hills, often on stony slopes. Wyoming and Colorado. Our records from northcentral Colorado at about 8000-8500 feet.
5. Mertensia alpina (Torr.) G. Don, Gen. Hist. 4:372. 1838.
M. tweedyi Rydb.; M. obtusiloba Rydb.; M. brevistyla obtusiloba (Rydb.) A. Nels.---Stems 10-20 cm. tall, glabrous; basal leaves oblong-lanceolate to linear, petiole winged and sometimes as long as the blade; cauline leaves lanceolate to oblanceolate, often narrowly so, strigillose above, glabrous below; all leaves without definite lateral veins; pedicels rarely over 5 mm. long, strigillose in some of ours; calyx divided almost to base, ciliate; corolla tube 3-4 mm. long, limb 3-4 mm. long; anthers inserted in tube and not projecting beyond it, fornices prominent; style short, about as long as the calyx.---Alpine. Montana to Idaho, south to New Mexico. Our records in central and northcentral Colorado at 11,000-14,000 feet.
6. Mertensia brevistyla S. Wats., U. S. Geol. Expl. 40th Par. 5:239. 1871.
Stems 10-40 cm. tall, erect or ascending, more or less pubescent, from a fusiform root; leaves 2-6 cm. long, broadly lanceolate, oblong to obovate-oblong, densely strigillose above, glabrous below, obtuse or acute; inflorescence congested at first, later panicled, pedicels strigillose; calyx 2-5 mm. long, divided almost to base, strigillose on face as well as margins; corolla tube 2-4 mm. long, limb 4-6 mm. long; anthers longer than the filaments, included in the tube; styles shorter than calyx lobes.---Mountains. Wyoming and Idaho, south to Colorado and Utah. Our records from the western half of Colorado, except in the extreme southern part, at 6500-10,500 feet.
7. Mertensia bakeri Greene, Pitt. 4:90. 1899.
M. bakeri var. lateriflora (Greene) A. Nels.; M. nivalis (S. Wats.) Rydb.; M. myosotifolia Heller; M. lateriflora Greene---Stems to 25 cm. tall, erect or ascending, often several from 1 rootstock, pubescent; basal leaves petioled; cauline to 4 cm. long, linear-lanceolate to oblong-lanceolate, canescent on both sides, sessile; calyx to 4 mm. long, divided almost to base, pubescent on back and edges (may be sparse on back); corolla tube up to 8 mm. long, usually longer than calyx, limb up to 7 mm. long; style as long or longer than the anthers.---Mountains. Colorado, Utah and New Mexico. Our records scattered in the higher mountains of Colorado at 8000-13,000 feet.
7A. Mertensia bakeri osterhoutii Williams, (var.) Ann. Mo. Bot. Gard. 24:120. 1937.
Like species but pubescence spreading on leaves; calyx lobes glabrous or sparingly pubescent on back; corolla dries a clearer blue (not so purple), and leaves a brighter green.---Limited to Colorado. Our records from the central and northcentral parts at 7500-11,000 feet.
8. Mertensia viridis A. Nels., Bull. Torr. Club 26:244. 1899.
M. papillosa lineariloba (Rydb.) A. Nels.; M. ovata Rydb.; M. lineariloba Rydb.; M. parryi Rydb.; M. perplexa Rydb.---Stems 5-35 cm. tall; basal leaves with winged petioles, lanceolate to oblanceolate, sometimes wider, strigillose above, glabrous below; cauline leaves to 5 cm. long, sessile, oblong-lanceolate to ovate, strigillose above, glabrous below; inflorescence rather crowded; calyx 2-6 mm. long, divided almost to base, lobes ciliate, but glabrous on back; corolla tube 3-9 mm. long, limb 4-9 mm. long, about same length as tube, sometimes shorter or longer; filaments longer or shorter than anthers.---In the mountains, mostly alpine. Montana, south to Colorado and Utah. Our records scattered in the higher mountains of Colorado at 6500-14,000 feet, mostly above 10,000 feet.
8A. Mertensia viridis cana (Rydb.) Williams, (var.) Ann. Mo. Bot. Gard. 24:115. 1937.
M. canescens Rydb.; M. cana Rydb.---Like the species but leaves inclined to be unilateral, linear to narrowly ovate, more or less densely canescent on both sides; calyx lobes usually glabrous on back but sometimes slightly pubescent.---Colorado and Utah. Our records from central, westcentral and northcentral Colorado at 9000-12,000 feet.
8B. Mertensia viridis dilatata (A. Nels.) Williams, (var.) Ann. Mo. Bot. Gard. 24:113. 1937.
Differs from species in having the leaves glabrous on both surfaces.---Wyoming, Colorado and Utah. Our few records from northcentral Colorado at about 9000-10,000 feet.
8C. Mertensia viridis cynoglossoides (Greene) Macbride, (var.) Contrib. Gray Herb. 48:13. 1916.
M. cynoglossoides Greene; M. muriculata Greene---Differs from species in having the hairs of the pedicels spreading.---Apparently limited to Colorado. Our records from Boulder, Lake, and Gunnison Counties at 7000-10,000 feet.
8D. Mertensia viridis parvifolia Williams, (var.) Ann. Mo. Bot. Gard. 24:115. 1937.
Differs from species in having stems ascending, not erect; leaves are usually unilateral on the stems, not at all glaucous.---Limited to Colorado. Our records from the northcentral part, with a rather doubtful one from westcentral, at 7500-12,500 feet.
9. Mertensia oblongifolia (Nutt.) G. Don, Gen. Hist. 4:372. 1838.
M. nevadensis A. Nels.; M. coronata A. Nels.; M. intermedia Rydb.; M. foliosa A. Nels.; M. tubiflora Rydb.; M. nutans Howell; M. praecox Smiley---Stems 10-30 cm. tall; from an elongated rootstock, erect or ascending, glabrous or nearly so; cauline leaf blades 2-10 cm. long, lanceolate-oblong to ovate, glabrous but usually pustulate above, 1-nerved or lateral veins developing especially on larger basal leaves; pedicels pustulose, rarely strigose; calyx 3-7 mm. long, lobes much longer than the tube, linear to linear-oblong, acute or obtuse, glabrous dorsally but ciliate; corolla tube variable, 5-10 mm. long, slightly longer than the limb; anthers shorter than the filaments which are attached about at mouth of tube. Our plant is var. nevadensis (A. Nels.) Williams.---Mesas, hills and mountains. The variety from Montana to Washington, south to Wyoming, Utah and California. Our one record from northwestern Colorado at about 7500-8000 feet.
10. Mertensia fusiformis Greene, Pitt. 4:89. 1899.
M. papillosa var. fusiformis (Greene) A. Nels.---Stems apparently erect, glabrous, few from fusiform rootstocks; basal leaves 6-12 cm. long, elliptical to oblong-ovate, strigose above, hairs pointing toward margins, glabrous below, on definite petioles; cauline leaves oblong-lanceolate to wider, nearly sessile at least above, obtuse, pubescence that of basal leaves;

inflorescence congested; calyx 3-6 mm. long, lobes ciliate and usually pubescent on back; corolla tube 4-7 mm. long, limb 5-7 mm.; filaments longer or shorter than anthers.---Hillsides and mountains. Colorado and Utah. Our records from the western third of Colorado at 6500-11,000 feet.

11. **Mertensia lanceolata** (Pursh) A. DC., Prodr. 10:88. 1846.
M. linearis Gr.; M. papillosa Greene---Stems 10-45 cm. tall, usually several to a rootstock; basal leaves petioled, ovate-lanceolate, glabrous below and strigillose to pustulate above; cauline leaves usually lanceolate to oblong-lanceolate; pedicels glabrous or strigillose; inflorescence finally panicled but congested at first; calyx 2-9 mm. long, ciliate, glabrous on back, a definite tube present; corolla tube 3-6.5 mm. long, subequal to or shorter than the limb; anthers shorter and narrower than filaments.---Dry plains and slopes. Saskatchewan, south to New Mexico. Our records scattered over Colorado, few in the extreme eastern and western parts, at 5000-10,000 feet.

11A. **Mertensia lanceolata pubens** (Macbr.) Williams, (var.) Ann. Mo. Bot. Gard. 24:98. 1937.
Like the species but leaves pubescent on both sides; calyx divided less than 1/2 or about 1/2 way to base.---Colorado and New Mexico. Our records from northcentral, central and southcentral Colorado at 5000-9500 feet.

11B. **Mertensia lanceolata brachyloba** (Greene) A. Nels., (var.) Coult. & Nelson's Man. Ry. Mt. Bot. 422. 1909.
Like the species but stems and leaves entirely glabrous and somewhat pustulate; calyx campanulate, the lobes shorter than the tube.---Apparently limited to northcentral Colorado, our records from 5500-7500 feet.

11C. **Mertensia lanceolata fendleri** Gray, (var.) Proc. Am. Acad. 10:53. 1875.
Leaves rather larger than species, glabrous on lower surface, upper short-strigillose; calyx divided about to middle, strigillose on back and edges.---Colorado and New Mexico. Our records from southcentral, central and northwestern Colorado at 6000-9000 feet.

11D. **Mertensia lanceolata secundorum** Ckll., (var.) Torreya 18:180. 1918.
M. media Osterh.---Leaves linear to lanceolate, strigose above, strigose to densely hispid below; calyx glabrous to strigose (usually ciliate and glabrous on back), lobed over 1/2 to base.---Wyoming and Colorado. Our records from northcentral, central and southcentral Colorado at 5500-9500 feet.

13. Onosmodium Michx.   FALSE GROMWELL; MARBLESEED

Perennial herbaceous plants; stems stout, leafy, hairy; leaves alternate, entire, strongly nerved; flowers regular, in terminal leafy-bracted scorpioid racemes or spikes; calyx with 5 narrow lobes; corolla greenish-white to yellowish, tubular, with 5 erect lobes, throat not appendaged; stamens included, anthers sagittate; ovary 4-lobed, style exserted; fruit separating into 4 nutlets (sometimes not all 4 maturing) these white, shiny, smooth or rarely sparingly pitted, attached by their bases to a flat gynobase, the scar small and flat.

1. **Onosmodium molle** Michx., Fl. Am. Bor. 1:133. 1803.
O. occidentale Mack.---Stems 30-80 cm. tall, stout, branched above, with rough somewhat spreading hirsute-hispid hairs; leaves about 4-6 cm. long, ovate-lanceolate to oblong-elliptic, coarsely hirsute-strigose, 5- to 9-ribbed; calyx 7-12 mm. long; corolla 12-20 mm. long; nutlets 3.5-4.5 mm. long. Our plant is var. occidentalis (Mack.) Johnston.---Plains and slopes. The var. from Minnesota to Saskatchewan, south to Illinois, New Mexico and Utah. Our records in the eastern half of Colorado at 4500-7000 feet.

14. Lithospermum L.   GROMWELL; PUCCOON

Plants annual or perennial, caulescent and herbaceous; leaves alternate, entire, narrow, stiff-hairy; flowers regular, white, orange or yellow, in bracted spikes or racemes (early flowers conspicuous but most of the seed may be produced by later inconspicuous cleistogamous flowers); calyx persistent, 5-lobed; corolla mostly salverform or funnelform, throat either appendaged or pubescent, lobes rounded, often toothed; stamens included; ovary 4-lobed; stigmas capitate or 2-lobed; nutlets whitish, shiny and smooth (brown and wrinkled in L. arvense), attached by their bases to a flat gynobase, not leaving a pit on the gynobase.

1. Plants annual or biennial; flowers white; nutlets gray or brown, roughened   -1. **L. arvense**
1. Plants perennial; flowers greenish to yellow; nutlets white, smooth
  2. Corolla less than 10 mm. long, the tube seldom much longer than the calyx, green or pale-yellow in color; nutlets 4-6 mm. long   -2. **L. ruderale**
  2. Corolla 10 mm. long or more, the tube definitely exceeding the calyx, yellow (L. incisum has small cleistogamous flowers later in the season)
    3. Flowers not heterostyled; stamens all borne near summit of corolla tube; corolla usually over 20 mm. long its lobes toothed or fimbriate; later flowers cleistogamous and much smaller   -3. **L. incisum**
    3. Flowers strongly heterostyled, styles of 2 lengths; stamens borne about at middle or near top of corolla tube; corolla not over 20 mm. long, its lobes entire or nearly so; smaller cleistogamous flowers absent
      4. Corolla orange-yellow; calyx 6-13 mm. long; nutlets 3-4 mm. long; plants of northeastern Colorado   -4. **L. caroliniense**
      4. Corolla yellow; calyx 4-6 mm. long; nutlets 2-3 mm. long; plants not limited to northeastern Colorado
        5. Floral leaves greatly reduced, scarcely if at all exceeding the calyx lobes and simulating them; corolla throat with slightly swollen but not invaginate appendages (on fresh flower); pollen grains broadest at middle; plants not limited to eastern Colorado   -5. **L. multiflorum**

5. Floral leaves not greatly reduced, much surpassing the calyx lobes and not at all simulating them (hence inflorescence leafy); corolla throat with invaginate appendages; pollen grains constricted at middle or broadest at one end; plants (if present at all) of eastern border of Colorado   -6. **L. canescens**

1. Lithospermum arvense L., Sp. Pl. 132. 1753.
    Plants annual or biennial; stems 20-70 cm. tall, simple or branched at base, strigose; leaves 1-8 cm. long, linear, linear-lanceolate or narrowly oblanceolate, canescent-strigose; flowers in a leafy panicle; calyx about 5 mm. long; corolla white or whitish, tubular-funnelform, tube about 5 mm. long, lobes about 1-1.5 mm. long, entire; nutlets 2-3 mm. long, gray to brownish, rugose; cleistogamous flowers lacking.---Waste places and fields. Introduced from Europe into various parts of North America. Our record from northcentral Colorado at 5800 feet.
2. Lithospermum rudarale Dougl. in Lehm., Pug. 2:28. 1830.
    L. pilosum Nutt.; L. torreyi Nutt.; L. lanceolatum Rydb.---Perennial plants; stems 20-60 cm. tall, several, ascending from a woody root, canescent, pilose or strigose; leaves 3-8 cm. long, linear or linear-lanceolate, strigose, crowded out toward and almost concealing the flowers; flowers in a very leafy panicle, along short, rather spreading branches especially when in fruit; calyx about 5 mm. long; corolla greenish or pale yellow, campanulate-funnelform, variable in size but the tube about 5-8 mm. long, often no longer than the calyx, lobes short (about 1.5 mm.) entire or nearly so; nutlets about 4-6 mm. long, white and shiny; cleistogamous flowers lacking. Flowers monomorphic.---Plains, hills and canyons. Alberta to British Columbia, south to Colorado and California. Our records from the western half of Colorado, except in the south-central part, at 6000-9500 feet.
3. Lithospermum incisum Lehm., Asperif. 2:303. 1818.
    L. angustifolium Michx.; L. mandanense Spreng.; L. linearifolium Goldie; L. breviflorum Engelm. & Gray; L. asperum A. Nels.; L. oblongum Greene; L. albicans Greene---Perennial plants from a thick woody root; stems 10-50 cm. tall, usually several, erect or ascending, strigose to somewhat hirsute; leaves 10-50 mm. long, linear to linear-oblong, strigose; inflorescence of terminal leafy racemes; calyx 6-10 mm. long; corolla 10-30 mm. long, yellow, the tube seldom over 2.5 mm. wide when pressed, salverform, limb 9-18 mm. wide, lobes fimbriate to toothed; nutlets 3-4 mm. long, white and shining; cleistogamous flowers present, in fruit usually with recurved pedicels. Flowers monomorphic.---Dry plains and slopes. Ontario to British Columbia, south to Illinois, Texas, Mexico and Arizona. Our records well scattered over Colorado at 3500-8500 feet.
4. Lithospermum caroliniense (Walt.) MacMill., Metasp. Minn. Valley 438. 1892.
    L. hirtum (Muhl.) Lehm.; L. gmelini (Michx.) Hitchc.; L. croceum Fernald ---Perennial plants from a woody thick root; stems 25-60 cm. tall, usually clustered, stout, often branched above, hispid-strigose; leaves 2.5-6 cm. long, lanceolate to oblong, strigose-hispid, little reduced above, the floral leaves definitely longer than the sepals and not at all simulating them; flowers in dense leafy racemes often on corymbose branches; calyx 7-13 mm. long; corolla 12-20 mm. long, yellow to orange-yellow, funnelform with dilated throat, the lobes short and entire or nearly so; nutlets 3-4 mm. long, smooth and white; cleistogamous flowers absent. Flowers heterostyled and dimorphic.---Plains, slopes, woods and thickets, often in sandy soil. Ontario to Montana, south to Florida and Texas; Mexico. Our few records from northeastern Colorado at 3500-4500 feet.
5. Lithospermum multiflorum Torr. ex A. Gray, Proc. Am. Acad. 10:51. 1874.
    Perennial plants from a thick woody root containing a purplish dye (this often staining the mounting paper); stems 30-60 cm. tall, more or less tufted, often virgately branched above, strigose-hispid; leaves 2-6 cm. long, linear or linear-lanceolate, appressed-strigose above, hirsute beneath, becoming smaller and bractlike near flowers, scarcely if at all longer than the calyx lobes and simulating them; flowers racemose, short-pedicelled, often on several ascending corymblike branches; calyx lobes about 4-6 mm. long; corolla yellow, tubular-funnelform, the tube about 8-13 mm. long, the lobes short (about 2 mm.), rounded, not fimbriate; nutlets about 3 mm. long, white and shining; cleistogamous flowers absent. Flowers heterostyled and dimorphic. ---Hills, canyons and mountain slopes. Wyoming to Mexico. Our records scattered in the western half of Colorado, except the northwestern part, at 5500-11,000 feet.
6. Lithospermum canescens (Michx.) Lehm., Asperif, 2:305. 1818.
    Ontario to Saskatchewan, south to Alabama and Kansas. Reported for eastern Colorado but no specimens seen by the writer.

15. Cryptantha Lehm.

Annual, biennial or perennial herbaceous plants, usually bristly-hairy (mostly painful to handle with bare hands); leaves entire, linear to spatulate, alternate or opposite below; flowers regular, in scorpioid cymes, these in a continuous or glomerate cluster, fasciculately or cymosely disposed; calyx lobed to middle or base; corolla from minute to evident or almost showy, white to yellow, salverform to funnelform, the tube equalling or exceeding the calyx, the throat with crests; stamens included, filaments short or absent; ovary 4-lobed; fruit separating into 4 nutlets, these sometimes heteromorphic, attached laterally to a subulate or narrowly pyramidal gynobase through an elongated medial ventral groove, not leaving a pit on the gynobase, more or less erect or vertical, unmargined or with a more or less developed marginal wing, surface smooth or variously roughened. A difficult genus and some of the species listed here probably should be combined especially in the section Oreocarya, the biennial and perennial forms. The mature nutlets are almost always necessary in identifying the species.

1. Annual plants with slender stems; flowers white, inconspicuous (rarely over 2 mm. wide)
    2. Calyx circumscissile at maturity, basal part persisting, upper half falling away   -1. **C. circumscissa**
    2. Calyx not circumscissile

3. Margins of nutlets winged -2. C. pterocarya
3. Margins of nutlets rounded or merely angled, not at all winged
    4. Nutlets of 2 kinds, 1 differing definitely in size, color, surface markings or shape from the others, some of nutlets tuberculate (except in C. pattersoni)
        5. Three similar nutlets smooth and shiny or very slightly tuberculate indeed -3. C. pattersoni
        5. Three nutlets definitely tuberculate
            6. Spikes with foliaceous bracts subtending some of the flowers along the spike -4. C. minima
            6. Spikes naked or with a few bracts near base
                7. Odd nutlet spinular-tuberculate (tubercules ending in short spines - use strong lens); midrib of calyx lobes conspicuously thickened; plants apparently limited to westcentral Colorado -5. C. crassisepala
                7. Odd nutlet granulate, the tubercules rounded at ends; midrib of calyx lobes only moderately thickened; plants not limited to westcentral Colorado -6. C. kelseyana
    4. Nutlets all alike or essentially so (sometimes 1 or more fails to develop), nutlets all smooth (except C. ambigua and C. recurvata)
        8. Nutlets grandular and tuberculate at least toward the apex
            9. Nutlets and calyx bent and recurved; nutlet 1 -7. C. recurvata
            9. Nutlets and calyx straight, not recurved; nutlets 4 -8. C. ambigua
        8. Nutlets smooth and more or less shiny
            10. Ventral groove or nutlet eccentric, not in middle -9. C. affinis
            10. Ventral groove of nutlet central (unless nutlet was immature and crushed in pressing)
                11. Nutlets broadly ovate -10. C. torreyana
                11. Nutlets oblong-ovate to lanceolate
                    12. Styles about 2/3 to 3/4 as long as the nutlets, only 1 or rarely 2 or 3 nutlets developing -11. C. gracilis
                    12. Styles as long or very nearly as long as the nutlets, usually 4 nutlets developing
                        13. Margins of nutlets acute-angled at least above -12. C. watsoni
                        13. Margins of nutlets rounded or obtuse -13. C. fendleri
1. Biennial or perennial plants with rather coarse stems; flowers often very showy, white to yellow, the limb (when fresh) 5-12 mm. wide
  14. Tube of corolla 6-14 mm. long, usually distinctly longer than the calyx lobes in anthesis; plants of western Colorado
      15. Nutlets smooth; corolla yellow -14. C. flava
      15. Nutlets papillose to rugose; corolla white to light yellow (crests in throat may be yellow and corolla may dry yellowish)
          16. Nutlets densely and uniformly muriculate, not at all rugose -15. C. fulvocanescens
          16. Nutlets more or less rugose as well as tuberculate on dorsal surface
              17. Corolla tube 12-14 mm. long; lower leaf surface with hairs pustulate at base -16. C. longiflora
              17. Corolla tube 6-12 mm. long; hairs on lower leaf surface not pustulate (may be on upper)
                  18. Corolla tube 6-7 mm. long; scars of nutlets surrounded by an elevated margin and closed -17. C. bakeri
                  18. Corolla tube 7-12 mm. long; scars of nutlets with elevated margins but open
                      19. Margins of nutlets not in contact; plants 5-12 cm. tall; leaves 1.5-3 cm. long; corolla tubes 10-12 mm. long -18. C. paradoxa
                      19. Margins of nutlets in contact; plants 10-30 cm. tall; leaves 3-8 cm. long; corolla tubes 7-10 mm. long -19. C. flavoculata
  14. Corolla tube not over 5 mm. long, shorter than or merely as long as the calyx lobes; plants of eastern and western Colorado
      20. Dorsal surface of nutlets smooth, their margins widely separated
          21. Stems and upper leaf surfaces definitely and densely hairy -20. C. jamesii
          21. Stems and upper leaf surfaces glabrous or very nearly so -20A. C. jamesii pustulosa
      20. Dorsal surface of nutlets roughened-rugose, tuberculate or muricate, their margins in contact or only slightly separated
          22. Inner surface of nutlets smooth or nearly so, not definitely roughened (dorsal surface roughened)
              23. Inflorescence spicate with conspicuous bracts greatly exceeding the flowers; stems usually solitary and unbranched -21. C. virgata
              23. Inflorescence open or narrow but hardly spicate, no elongated floral bracts present that greatly exceed the flowers; stems usually more than 1, or more branched
                  24. Inflorescence very broad and rounded in outline; nutlets not wing-margined; plants found east of the Continental Divide in Colorado -22. C. thrysiflora
                  24. Inflorescence rather narrow and crowded; nutlets narrowly wing-margined; plants found west of the Continental Divide in Colorado -23. C. stricta
          22. Inner surface of nutlets rugose, tuberculate or muricate, in any case definitely roughened
              25. Plants of the eastern plains or mountains east of the Continental Divide in Colorado
                  26. Dorsal surface of nutlets uniformly muricate (small projections rather pointed at apex), not at all rugose or tuberculate; limb of corolla 5-6 mm. wide; stems 8-15 cm. tall, caespitose; leaves silky-strigose, the hairs uniform -24. C. cana
                  26. Dorsal surface of nutlets rugose and tuberculate (may be muricate also); limb of corolla about 7-10 mm. wide; stems 15-40 cm. tall, not caespitose; leaves never uniformly silky-strigose
                      27. Inflorescence very broad and rounded in outline -22. C. thrysiflora
                      27. Inflorescence rather narrow and somewhat glomerate, not broad and rounded in outline -25. C. bradburiana
              25. Plants growing west of the Continental Divide in Colorado

28. Nutlets uniformly muricate dorsally, not at all tuberculate or rugose -26. C. nana
28. Nutlets with dorsal surface more or less rugose and tuberculate (may be muricate also)

    (Note: the following 6 species intergrade)

    29. Corolla tube 5 mm. long or more (usually about 6-7 mm.); groove of nutlets closed, with conspicuous elevated tuberculate margins -17. C. bakeri
    29. Corolla tubes less than 5 mm. long; groove of nutlet various, open or closed, but if closed the margins not elevated
        30. Upper surface of leaves uniformly appressed strigose, without pustules (may be pustulate below)
            31. Ventral groove of nutlet open near base, an elevated tuberculate margin present; plants densely caespitose -27. C. osterhoutii
            31. Ventral groove of nutlet closed or nearly so, no elevated margin present; plants less evidently if at all caespitose -28. C. sericea
        30. Upper surface of leaves with 2 distinct kinds of hairs, some of these pustulate at base
            32. Stems 5-10 cm. tall, densely caespitose; sepals in fruit about 6-7 mm. long; styles scarcely if at all exceeding the nutlets (not over 0.5 mm. longer) -29. C. caespitosa
            32. Stems 10-50 cm. tall, not usually densely caespitose if at all; sepals in fruit over 7 mm. long; styles 1-2 mm. longer than the nutlets
                33. Sepals in fruit 2-3 mm. longer than the nutlets, the latter about 5 mm. long; tube of corolla about 4 mm. long -30. C. elata
                33. Sepals in fruit about 5 mm. longer than the nutlets, these about 2.5-3 mm. long; corolla tube about 2.5-3 mm. long -31. C. aperta

1. Cryptantha circumscissa (H. & A.) Johnston, Contrib. Gray Herb. 68:55. 1923.
Greeneocharis circumscissa (H. & A.) Rydb.; Piptocalyx circumscissus Torr.---Annual plants; stems 2-10 cm. tall, rather simple and erect or bushy-branched, strigose or hirsute; leaves 3-15 mm. long, oblanceolate, strigose or hirsute, obscurely pustulate; flowers axillary or in obscure leafy racemes; fruiting calyx 2.5-4 mm. long, circumscissile at maturity the upper part falling away, basal cupulate part persisting, lobes linear-lanceolate, more or less hirsute; corolla about 1-2 mm. wide, whitish; nutlets 4, 1.2-1.7 mm. long, homomorphous or nearly so, smooth or obscurely muricate, ovate to oblong-lanceolate; style about as long as the nutlets. Our plant has been called var. genuina Johnston.---Dry plains. The variety from Wyoming to British Columbia, south to Colorado and California. Our few records in northcentral and northwestern Colorado at 6500-8500 feet.

2. Cryptantha pterocarya (Torr.) Greene, Pitt. 1:120. 1887.
Annual plants; stems 10-40 cm. tall, with erect or ascending branches, finely strigose or short-hirsute; leaves 1-2.5 cm. long, linear to lanceolate, strigose or hirsute, more or less pustulate; spikes usually clustered, naked or inconspicuously bracted below, pedicels short; fruiting calyx 2.5-3 mm. long, lobes ovate or lanceolate, appressed-hirsute or hispid; corolla about 1 mm. wide or less, whitish, inconspicuous; nutlets 4, body muricate on back, homomorphous and all winged, or axial one wingless and the other 3 with broad entire or lobed wings; style longer than the body proper of the nutlet, usually surpassed by the wings. In the var. cycloptera (Greene) Macbr. the nutlets are all winged.---Dry ground. Colorado to Nevada, south to Texas and California. Our records from western Colorado at 4500-8000 feet.

3. Cryptantha pattersoni (Gray) Greene, Pitt. 1:120. 1887.
Plants annual; stems 8-15 cm. tall, loosely branched, hirsute and usually more or less strigose; leaves 1-3 cm. long, oblanceolate or even spatulate on lower, hirsute and more or less pustulate; spikes solitary or few together, naked; fruiting calyx 4-5 mm. long, lobes slightly asymmetrical, linear-lanceolate, hirsute; corolla 1-1.5 mm. broad, white, inconspicuous; nutlets usually 4, heteromorphous but not conspicuously so, odd one slightly larger, about 1.9 mm. long, ovate, smooth or sparsely and inconspicuously tuberculate, more persistent and spreading than the others, other nutlets about 1.6 mm. long, smooth or nearly so, groove medial; style about as long or slightly shorter than the 3 similar nutlets.---Dry plains and hills. Wyoming and Colorado. Our records from the western half of Colorado at 4500-8000 feet.

4. Cryptantha minima Rydb., Bull. Torr. Club 28:31. 1901.
Plants annual; stems 10-20 cm. tall, erect or widely spreading, usually numerous and branching, finely strigose and coarsely hirsute; leaves 1-3 cm. long, oblanceolate, hirsute or hispid, little reduced above and continuing into the inflorescence as foliaceous bracts; calyx in fruit 5-7 mm. long, asymmetrical, the linear-lanceolate lobes with midribs becoming strongly thickened; corolla 1-1.5 mm. wide, white, inconspicuous; nutlets 4, dimorphic, odd nutlet 2-3 mm. long, persistent, finely papillate-granulate, ovate, other nutlets 1.2-1.5 mm. long, ovate, strongly tuberculate; style equalling or surpassing the 3 nutlets, shorter than the odd one.---Plains, valleys and hills. Saskatchewan, south to Texas and Colorado. Our records mostly from the eastern half of Colorado (a few from the extreme westcentral part), at 3500-7000 feet.

5. Cryptantha crassisepala (T. & G.) Greene, Pitt. 1:112. 1887.
Plants annual; stems 5-15 cm. tall, erect or widely spreading, branched and loosely ascending, hirsute; leaves 1-3 cm. long, oblanceolate, hirsute; spikes naked or few-bracted below; fruiting calyx 6-7 mm. long, somewhat unsymmetrical, the lobes linear-lanceolate, midrib strongly thickened and indurated; corolla 1-1.5 mm. wide, white or whitish, inconspicuous; nutlets usually 4, dimorphic, odd one 2-2.5 mm. long, persistent, finely granulate, these granules spinulose-muricate (use a strong lens), other nutlets 1.2-1.5 mm. long, oblong-ovate, coarsely tuberculate; style as long or somewhat longer than the 3 nutlets, shorter than the odd one. This species intergrades with C. minima Rydb. and C. kelseyana Greene in Colorado.---Dry plains and hills, often in waste ground. Colorado and Utah, south to Texas and Mexico. Our records from western Colorado at 4500-5500 feet.

6. **Cryptantha kelseyana** Greene, Pitt. 2:232. 1892.
   Plants annual; stems 5-25 cm. tall, 1 to several, spreading or ascending, hirsute and more or less hispid-strigose; leaves 1-3 cm. long, linear-oblanceolate, hirsute; spikes naked or with a few bracts near base; fruiting calyx 4-6 mm. long, somewhat asymmetrical, lobes linear and midrib thickened; corolla 1-2 mm. wide, white or whitish, inconspicuous; nutlets 4, dimorphic, odd nutlet 2-2.6 mm. long, lanceolate-ovate, smoothish or minutely grandular, other 3 nutlets 1.8-2.3 mm. long, oblong-ovate to narrowly ovate, coarsely tuberculate; style about as long as the 3 nutlets, shorter than the odd one. The above intergrades with C. minima Rydb. and C. crassisepala (T. & G.) Greene in Colorado.---Dry often sandy plains and hills. Saskatchewan, south to Colorado and Utah. Our records scattered over Colorado, mostly in the northern half of the state, at 4500-8500 feet.
7. **Cryptantha recurvata** Coville, Contrib. U. S. Nat. Herb. 4:165. 1893.
   Annual plants; stems 5-30 cm. tall, branched, usually strigose; leaves 8-20 mm. long, few, oblanceolate to linear-oblanceolate, appressed hispid and minutely pustulate; spikes naked; fruiting calyx about 3-4 mm. long, bent and recurved, very asymmetrical, sessile, lobes linear, hirsute; corolla about 2 mm. long, whitish, inconspicuous; only 1 nutlet maturing, this curved, granulate-muricate, groove medial; style shorter than nutlet.---Dry ground. Utah to Oregon, south to Arizona and California. Our few records from westcentral Colorado at 4500-5000 feet.
8. **Cryptantha ambigua** (Gray) Greene, Pitt. 1:113. 1887.
   C. multicaulis A. Nels.---Plants annual; stems 6-25 cm. tall, usually loosely branched from base, hirsute and more or less short-strigose; leaves 1-3 cm. long, linear to narrowly lanceolate, usually hispid-hirsute, the hairs usually appressed; spikes naked or bracted in lower flowers, usually short; fruiting calyx 4-7 mm. long, the lobes linear or lanceolate-linear, midrib only slightly thickened; corolla 1-2 mm. wide, white, inconspicuous; nutlets 4, 1.6-2 mm. long, homomorphous, broadly ovate, grandular and tuberculate or sometimes tending to be smooth toward base, margins rounded, groove medial; styles as long or nearly as long as the nutlets.---Dry, often sandy ground. Montana to Washington, south to Colorado and California. Our few specimens from central, northcentral and northwestern Colorado at 7500-9000 feet.
9. **Cryptantha affinis** (Gray) Greene, Pitt. 1:119. 1887.
   C. confusa Rydb.---Annual plants; stems 10-30 cm. tall, usually sparsely branched, hispid or short-hirsute; leaves 1-4 cm. long, narrowly to broadly lanceolate, rather few, short-hirsute and usually somewhat pustulate; spikes solitary or clustered, commonly leafy-bracted at base; fruiting calyx 2.5-4 mm. long, lobes lanceolate, hirsute to appressed-hispid; corolla 1-2 mm. broad, whitish, inconspicuous; nutlets 4, 1.8-2.5 mm. long, homomorphous, smooth or nearly so, obliquely compressed and the groove evidently eccentric; style shorter than or rarely equal to nutlets.---Dry ground. Montana to Washington, south to Wyoming and California. Our one specimen from northcentral Colorado at 8500 feet.
10. **Cryptantha torreyana** (Gray) Greene, Pitt. 1:118. 1887.
    C. calycosa (Torr.) Rydb.; C. flexuosa A. Nels.---British Columbia and Alaska, south to Wyoming and California. This plant has been reported for Colorado and is to be looked for in the northwestern part.
11. **Cryptantha gracilis** Osterh., Bull. Torr. Club 30:236. 1903.
    Annual plants; stems 10-20 cm. tall, branched, the branches erect or ascending, short-hispid or hirsute; leaves 1-3 cm. long, rather few, linear or narrowly oblanceolate, short-hispid with ascending hairs, usually pustulose; spikes solitary or clustered, naked; fruiting calyx 2-3 mm. long, lobes lanceolate, tawny hispid-villous; corolla not over 1 mm. wide, whitish, inconspicuous; nutlets 1-3, 1.5-2 mm. long, homomorphous or nearly so, lanceolate, smooth; style about 2/3 to 3/4 as long as the nutlets.---Dry ground. Colorado to Idaho, south to Arizona and California. Our records from western Colorado at 4500-6000 feet.
12. **Cryptantha watsoni** (Gray) Greene, Pitt. 1:120. 1887.
    Small annual plants; stems 6-30 cm. tall, solitary, sparsely to loosely branched, short-hispid; leaves 1-4 cm. long, linear to oblanceolate, hispid; spikes sometimes leafy-bracted below; fruiting calyx 2-4 mm. long, the lobes lanceolate, scarcely thickened at midrib; corolla about 1 mm. broad, white or whitish; nutlets 4, 1.5-2 mm. long, homomorphous or nearly so, lanceolate, smooth and usually shiny, the margins sharp-angled at least above, the groove medial; style nearly as long or as long as the nutlets.---Hills and mountain slopes. Montana to Washington, south to Colorado and Nevada. Our few records from central and northcentral Colorado (west of the Continental Divide) at 6500-7500 feet.
13. **Cryptantha fendleri** (Gray) Greene, Pitt. 1:120. 1887.
    C. ramulosissima A. Nels.---Plants annual; stems 10-50 cm. tall, usually simple below and branching above but with a central axis, more or less hispid; leaves 2-5 cm. long, narrowly oblanceolate, appressed hispid; spikes sparsely if at all bracteate; fruiting calyx 4-7 mm. long, slightly asymmetrical, the lobes linear or lanceolate-linear, with midrib thickened; corolla about 1 mm. wide, white, inconspicuous; nutlets usually 4, 1.5-2 mm. long, homomorphous, smooth and somewhat shiny, lanceolate, acuminate, angles rounded and obtuse, groove medial; style as long or slightly longer than the nutlets.---Plains and valleys, often in sandy soil. Alberta and Saskatchewan, south to Nebraska, New Mexico, Arizona and Nevada. Our records scattered over the western two-thirds of Colorado at 5000-9000 feet.
14. **Cryptantha flava** (A. Nels.) Payson, Ann. Mo. Bot. Gard. 14:259. 1927.
    Oreocarya flava A. Nels.---Perennial plants from a woody root; stems 15-30 cm. tall, usually few to many, whitish-strigose below, yellowish-hirsute above; leaves linear to linear-oblanceolate, appressed-strigose below, rather uniformly strigose above, more or less pustulate; inflorescence narrow and rather crowded; sepals in fruit about 6 mm. longer than the nutlets; corolla definitely yellow, the tube 9-11 mm. long, about twice as long as the calyx, limb 7-9 mm. across; usually less than 4 nutlets maturing, 3-4 mm. long, homomorphous, margins in contact, smooth dorsally and ventrally, groove nearly closed, margin not elevated.---Hills and

plains, often in sandy soil. Wyoming and Utah, south to New Mexico and Arizona. Our records from western Colorado at 4500-7500 feet.

15. Cryptantha fulvocanescens (Gray) Payson, Ann. Mo. Bot. Gard. 14:319. 1927.

Oreocarya fulvocanescens (Gray) Greene; O. nitida Greene---Perennial and more or less caespitose plants; stems 8-20 cm. tall, few to many, appressed-hirsute below, strigose and hirsute above; leaves 2-5 cm. long, tending to cluster near base of plant, oblanceolate to linear-oblanceolate, uniformly silky-strigose, more or less pustulate below; inflorescence rather narrow, bracts rather small; sepals in fruit 10-12 mm. long, much longer than the nutlets; corolla white, the tube 9-11 mm. long, exceeding the sepals by 2-4 mm., limb 6-9 mm. wide; usually 1-2 nutlets maturing, these about 4 mm. long, margins in contact, dorsal and ventral surfaces densely and uniformly muricate, not at all rugose and tuberculate, groove short, nearly closed, margins not elevated.---Dry, often rocky ground. Colorado and Utah, south to New Mexico and Arizona. Our records from western Colorado at 4500-6500 feet.

16. Cryptantha longiflora (A. Nels.) Payson, Ann. Mo. Bot. Gard. 14:326. 1927.

Oreocarya longiflora A. Nels.---Perennial or possibly biennial plants; stems 10-30 cm. tall, several, unbranched, hispid and puberulent beneath; leaves 2-7 cm. long, mostly basal, spatulate to narrowly obovate, strigose and more or less hispid, the hairs pustulate on both sides; inflorescence rather broad and loose, occupying most of the length of the stem; sepals 10-12 mm. long in fruit, greatly exceeding the nutlets; corolla white, dimorphic, the tube 12-14 mm. long, limb about 7-10 mm. wide; fruit usually with less than 4 nutlets maturing, these with margins in contact or nearly so, all surfaces more or less rugose and definitely tuberculate, groove straight, closed or narrow, margin slightly elevated.---Plains and mesas. Colorado and Utah. Our records from western Colorado, mostly in the westcentral part, at 4500-8000 feet.

17. Cryptantha bakeri (Greene) Payson, Ann. Mo. Bot. Gard. 14:331. 1927.

Oreocarya bakeri Greene; O. eulophus Rydb.---Perennial plants; stems 12-30 cm. tall, 1 to several, hirsute and hispid above; leaves 4-6 cm. long, somewhat clustered at base of plant, oblanceolate to spatulate, strigose above, strigose below with pustulate hairs; inflorescence more or less glomerate in a narrow thryse, occupying over 1/2 the stem length, foliar bracts few; sepals in fruit about 8-10 mm. long, about twice as long as the nutlets; corolla white, the tube 6-7 mm. long, usually a little longer than the sepals, limb 7-8 mm. wide; 2-4 nutlets maturing, these with margins not in close contact, dorsal surfaces rugose and tuberculate, ventral mainly tuberculate, groove closed with elevated tuberculate margins.---Plains and hills. Colorado and Utah. Our records scattered in the western third of Colorado at 6000-8000 feet.

18. Cryptantha paradoxa (A. Nels.) Payson, Ann. Mo. Bot. Gard. 14:330. 1927.

Oreocarya paradoxa A. Nels.---Densely caespitose perennial plants with much branched caudex; stems 5-12 cm. tall, slender, more or less strigose and hirsute above; leaves tending to crowd at base of stem on end of caudex branches, spatulate or oblanceolate, strigose and pustulate on back; inflorescence not congested, foliar bracts present but rather inconspicuous; calyx 4-5 mm. long in anthesis, enlarging in fruit, lobes linear-lanceolate, hirsute or weakly hispid; corolla white, tube 10-12 mm. long; usually 4 nutlets maturing, 2-3 mm. long, lanceolate, densely tuberculate and rugose; style definitely exceeding the nutlets.---Dry hills. Colorado and Utah. Our records from western Colorado at about 5000-5500 feet.

19. Cryptantha flavoculata (A. Nels.) Payson, Ann. Mo. Bot. Gard. 14:334. 1927.

Oreocarya flavoculata A. Nels. O. cristata Eastw.---Perennial plants, more or less caespitose; stems 10-30 cm. tall, few to many, strigose and hispid; leaves 2-6 cm. long, linear-lanceolate to spatulate, strigose and weakly hirsute, pustulate below at least, often silky looking above; inflorescence rather narrow, continuous or interrupted, bracts not conspicuous; sepals in fruit 7-10 mm. long, about twice as long as the nutlets; corolla white or pale yellow in the tube, tube 7-10 mm. long, 1.5-3 mm. longer than the calyx, limb 7-12 mm. wide; fruit with 4 nutlets commonly maturing, these 2.5-3.5 mm. long, margins mostly in contact or nearly so, dorsal surface more or less rugose, tuberculate and muricate, ventral surface tuberculate, groove open with elevated margins. This species intergrades with C. paradoxa (A. Nels.) Payson in Colorado.---Dry ground. Wyoming to Nevada, south to Colorado, Utah and California. Our records from western Colorado, as far east as Eagle County, at 5000-6500 feet.

20. Cryptantha jamesii (Torr.) Payson, Ann. Mo. Bot. Gard. 14:242. 1927.

Oreocarya multicaulis (Torr.) Greene; O. cinerea Greene; O. suffruticosa (Torr.) Greene---Perennial plants; stems 10-30 cm. tall, branching from the base and usually upward too, hirsute and more or less cinereous; leaves 3-10 cm. long, linear-oblanceolate or linear, appressed hairy to cinereous, rarely hirsute, more or less pustulate; inflorescence of usually open rather elongated cymes; sepals in fruit 5-6.5 mm. long, lanceolate, exceeding the nutlets by 3-4 mm.; corolla white, the tube 2.5-3 mm. long, no longer than the calyx, limb 6-8 mm. wide; fruit with 1-4 nutlets maturing, these 2-2.5 mm. long, margins well separated, ventral and dorsal surfaces smooth or slightly rugose near margins only on back, scar closed, no elevated margin. The following varieties have been reported from Colorado, var. cinerea (Greene) Payson, var. multicaulis (Torr.) Payson and var. typica Payson. Although extreme specimens do appear to be different, these varieties are connected by too many intergrades to be maintained in Colorado plants.---Dry ground. South Dakota to Nevada, south to Texas and California. Our records mostly scattered in the eastern two-thirds of Colorado at 3500-8000 feet.

20A. Cryptantha jamesii var. pustulosa (Rydb.) comb. nov.

Cryptantha pustulosa (Rydb.) Payson; Oreocarya pustulosa Rydb.---Perennial plants; stems 25-50 cm. high, branching at base, glabrous or very nearly so; leaves 3-10 cm. long, linear-oblanceoate, rather numerous, glabrous and without pustules above, pustulate and sometimes short-hairy below; cymes rather elongated and lax in age, bracts inconspicuous; sepals about 4 mm. long, lanceolate, strigose and with appressed short bristles; corolla white, tube 2.5-3 mm. long, hence shorter than calyx, limb 5-6 mm. broad; 1-4 nutlets maturing, these 2.5-3 mm.

long, margins acute and separated, all surfaces smooth and glossy.---Canyons and slopes. Colorado and Utah. Our few records from southcentral Colorado at about 8000 feet.
21. Cryptantha virgata (Porter) Payson, Ann. Mo. Bot. Gard. 14:270. 1927.
Oreocarya virgata (Porter) Greene; O. spicata Rydb.---Biennial plants from a taproot; stems 20-60 cm. tall, usually solitary, stout, unbranched, hirsute and also hispid with long sharp bristles; leaves 3-10 cm. long, narrowly oblanceolate, hirsute and hispid, pustulate hairs on both surfaces; inflorescence long, leafy and spicate, the leaflike bracts much longer than the cymules; sepals in fruit about 5-6 mm. longer than the nutlets; corolla white, the tube about 3-4 mm. long, the limb about 7-10 mm. wide; all 4 nutlets commonly maturing, these 2.5-3.5 mm. long, homomorphous, ovate, dorsally more or less rugose and tuberculate, ventrally smooth or very nearly so, margins in contact, ventral groove nearly or quite closed.---Dry hills, and plains and mountains. Wyoming and Colorado. Our records mostly from central and northcentral Colorado at 5000-9000 feet.

A species keying down to the above, but perennial and caespitose and with nutlets less than 2.5 mm. long with an open scar, has been recently described from Saguache County. It is C. weberi Johnston.

22. Cryptantha thyrsiflora (Greene) Payson, Ann. Mo. Bot. Gard. 14:283. 1927.
Oreocarya thrysiflora Greene---Perennial or sometimes biennial plants; stems 20-40 cm. tall, 1 to several, stout, setose-hispid; leaves 4-10 cm. long, oblanceolate, coarsely strigose-hispid, pustulate especially below; inflorescence broad and round-topped, bracts rather large but not conspicuous in the wide inflorescence; sepals in fruit about 6-8 mm. long, much longer than the nutlets; corolla white, tube 3-4 mm. long, limb about 7-8 mm. wide; often less than 4 nutlets maturing, these about 3-3.5 mm. long, homomorphous, margins in contact, rugose and smooth to tuberculate ventrally, groove narrow, margin not elevated.---Hills, valleys and plains. Nebraska and Wyoming, south to New Mexico. Our records well scattered over the eastern two-thirds of Colorado, east of the Continental Divide, at 4500-9500 feet.

23. Cryptantha stricta (Osterh.) Payson, Ann. Mo. Bot. Gard. 14:264. 1927.
Oreocarya stricta Osterh.---Plants perennial; stems 10-30 cm. tall, solitary or a few from a taproot, erect, hispid and strigose; leaves 1.5-5 cm. long, tending to cluster near the base, oblanceolate, hirsute and strigose, the hairs pustulate; inflorescence rather narrow and crowded, bracts small; sepals 6-8 mm. long in fruit, exceeding nutlets by 4-5 mm.; corolla white or yellowish when dry, tube 3-4 mm. long, as long or shorter than the sepals, limb 6-10 mm. broad; nutlets narrowly winged, with margins in contact, dorsal surface rugose and often tuberculate, ventral surface smooth or nearly so, groove nearly closed, no elevated margins present.---Dry ground. Colorado. Our records from the northwestern part at 6500-9000 feet.

24. Cryptantha cana (A. Nels.) Payson, Ann. Mo. Bot. Gard. 14:316. 1927.
Oreocarya cana A. Nels.---Caespitose perennial plants; stems 8-15 cm. tall, silvery-strigose; leaves 1.5-6 cm. long, narrowly oblanceolate, tending to cluster on the caudex, whitish from uniform looking silky-strigose hairs; inflorescence narrow, short and dense, usually restricted to upper 1/2 of stem; sepals in fruit 5-6 mm. long, about twice as long as the nutlets; corolla white, tube about 3 mm. long, about as long as the sepals, limb 5-6 mm. broad; only 1 nutlet usually maturing, this 2-3.5 mm. long, margins in contact (if more than 1 develops) dorsal and ventral surfaces densely muricate, the dorsal more strongly so, not at all tuberculate or rugose, groove narrowly triangular, open at base, no elevated margins.---Dry hills and plains. Nebraska, Wyoming and Colorado. Our records from the northwestern fourth of Colorado at 5000-5500 feet.

25. Cryptantha bradburiana Payson, Ann. Mo. Bot. Gard. 14:307. 1927.
Oreocarya affinis Greene; O. glomerata (Pursh) Greene; O. perennis in part---Biennial plants, possibly short-lived perennials, from a taproot; stems 15-35 cm. tall, simple or branched from base, coarsely hispid; leaves 2-5 cm. long, in rosettes the first year, radical spatulate or oblanceolate to nearly ovate, cauline narrower and longer, setose and subtomentose, pustulate; inflorescence rather narrow and somewhat glomerate, leafy-bracted below; calyx hirsute and setose, the lobes about 4 mm. long and lengthening in fruit, linear-lanceolate, acute; corolla white, tube shorter than or just equalling the sepals, limb about 1 cm. wide; nutlets usually 4, margins acute and in contact, dorsally rugose-tuberculate, ventrally more or less tuberculate or rugose.---Dry hills and plains. North Dakota and Alberta, south to Nebraska and Colorado. Our few records from northeastern Colorado at about 5000 feet.

26. Cryptantha nana (Eastw.) Payson, Ann. Mo. Bot. Gard. 14:312. 1927.
Oreocarya nana Eastw.---Perennial more or less caespitose plants; stems 5-15 cm. tall, hispid to hirsute; leaves 1-4 cm. long, spatulate to oblanceolate, coarsely strigose and hirsute; inflorescence narrow; sepals in fruit 7-10 mm. long, about twice as long as the nutlets; corolla white, the tube about 3 mm. long, no longer than the calyx, limb 6-10 mm. wide; 1-4 nutlets maturing, margins in contact, the dorsal and usually the ventral surface densely and uniformly muricate, not at all rugose or tuberculate, groove somewhat open especially at base. Our plant has been called var. typica Payson.---Dry hills and mesas. The variety from Colorado and Utah. Our records from western Colorado, mostly the westcentral part, at 4500-7000 feet.

27. Cryptantha osterhoutii (Payson) Payson, Ann. Mo. Bot. Gard. 14:329. 1927.
Oreocarya osterhoutii Payson---Densely caespitose perennial plants with branched caudex; stems 2-7 cm. tall, slender, with appressed white hairs at base, strigose and hirsute above; leaves 1-1.5 cm. long, spatulate to oblanceolate, strigose and appressed hirsute-pustulate below, strigose above; inflorescence reduced but rather open, foliar bracts inconspicuous; calyx 2-4 mm. long in anthesis, lobes linear-lanceolate, strigose-hirsute; corolla white, tube about 3 mm. long, about as long as the sepals; nutlets usually only 1-2 maturing, about 3 mm. long, lanceolate to ovoid, dorsal tuberculate and somewhat rugose, ventral surface tuberculate, scar open; styles slightly longer than nutlets.---Dry ground. Colorado and Utah. Our few records from westcentral Colorado at 4500-5000 feet.

28. **Cryptantha sericea** (Gray) Payson, Ann. Mo. Bot. Gard. 14:286. 1927.
Oreocarya sericea (A. Gray) Greene; O. argentea Rydb.---Perennial plants; stems 15-45 cm. tall, 1 to many, hirsute to hispid, the hairs above often tawny; leaves 2.5-10 cm. long, often clustered near base of plant, these spatulate to oblanceolate, strigose and appressed-hirsute below, pustulate, uniformly silky-strigose above, not pustulate; inflorescence rather narrow at first, broadening later, foliar bracts conspicuous at least at first; sepals 7-9 mm. long in fruit, about twice as long as the nutlets; corolla white, tube about 4 mm. long, about equalling the sepals, limb 5-8 mm. wide; 4 nutlets commonly maturing, these about 3.5-5 mm. long, margins in contact, more or less rugose and densely tuberculate on dorsal and ventral surface, scars closed but slightly open at base, no elevated margins. Our plants have been called var. typica Payson.---Dry hills and plains. Wyoming, Colorado and Utah. Our records scattered in the northwest quarter of the state at 4500-7500 feet.

29. **Cryptantha caespitosa** (A. Nels.) Payson, Ann. Mo. Bot. Gard. 14:281. 1927.
Oreocarya caespitosa A. Nels.---Known only from southern Wyoming and to be looked for in northern Colorado.

30. **Cryptantha elata** (Eastw.) Payson, Ann. Mo. Bot. Gard. 14:285. 1927.
Oreocarya elata Eastw.---Perennial plants, probably short-lived; stems 30-50 cm. tall, rather stout, 1 to several, strigose and hirsute; radical leaves 1-3 cm. long, spatulate, cauline leaves 2-4 cm. long, linear-oblanceolate, all leaves densely pustulate and appressed-hirsute as well as strigose above, strigose and pustulate below; inflorescence long, upper cymes tending to elongate and inflorescence becoming wider at apex, bracts inconspicuous; calyx about 4 mm. long, elongating in fruit, lobes in fruit exceeding the nutlets by 2-3 mm., lanceolate, hirsute-hispid; corolla white, tube about 4 mm. long, limb rather broad; all 4 nutlets commonly maturing, these about 4-5 mm. long, ovate-lanceolate, densely tuberculate and more or less rugose dorsally and ventrally, scar straight, closed or nearly so, no elevated margins present; style exceeding the nutlets.---Dry ground. Known only from the type locality near Grand Junction, Colorado, at about 4500 feet.

31. **Cryptantha aperta** (Eastw.) Payson, Ann. Mo. Bot. Gard. 14:295. 1927.
Oreocarya aperta Eastw.---Perennial plants with woody root; stems 10-20 cm. tall, several; leaves tending to cluster at base of plant, sparsely strigose and pustulate-setose both sides, radical leaves about 3 cm. long, spatulate to oblanceolate, cauline leaves oblanceolate; inflorescence branched at base with many spreading or aggregated spikes; sepals in fruit 8-10 mm. long; corolla white, the tube 2.5-3 mm. long, limb about 6 mm. broad; 4 nutlets commonly maturing, 2.5-3 mm. long, margins in contact, acute, lanceolate, dorsal surface tuberculate and more or less rugose and muriculate, ventral surface irregularly roughened, scar closed, no elevated margin present.---Dry ground. Apparently limited to the type locality near Grand Junction, Colorado, at about 4500 feet.

16. Amsinckia Lehm. FIDDLENECK

Annual pungent-bristly herbaceous plants; stems erect or with spreading branches, leafy; leaves alternate, entire or nearly so; flowers regular, yellowish to orange, in naked or sparsely bracted scorpioid spikes; calyx 5-lobed, persistent; corolla salverform, with tube longer than the calyx, limb narrow, no appendage in the throat; ovary 4-lobed, separating into 4 nutlets, heterostyled, stigmas simple; nutlets on a pyramidal gynobase, attached by a caruncular scar borne upon or at basal end of a ventral keel, surface tuberculate and often rugose.

1. **Amsinckia rugosa** Rydb., Fl. Rocky Mts. 729, 1066. 1917.
Stems about 20-50 cm. tall, erect or branches slightly decumbent, hispid and more or less strigose; leaves narrowly lanceolate or oblong-lanceolate, hispid, the hairs sometimes appressed; racemes not leafy-bracted; sepals about 8-9 mm. long in fruit, linear; corolla 5-7 mm. long; nutlets about 3 mm. long, rugose with tesellate ridges interspersed with very small tubercles.---Dry ground. Idaho and Washington, south to Utah and Nevada. One specimen from northwestern Colorado at 8000 feet seems to be this species.

Another species with longer corolla (7-13 mm.) and with nutlets rugose but not tesellate has been found in Clear Creek County, probably as a casual introduction. It seems to be the far western A. intermedia Fisch. and Meyer or A. douglasiana A. DC.

17. Plagiobothrys Fisch. & Mey. POPCORN FLOWER

Annual, usually appressed-hairy (at least in part) herbaceous plants; at least the lower leaves opposite, upper may be alternate; flowers regular, in scorpioid, short-pedicelled racemes, some leafy-bracted; calyx 5-parted, both tube and lobes persistent; corolla white, rather salverform with short tube; stamens included; ovary 4-lobed; fruit separating into 4 nutlets, these rugose, erect or incurved, attached near base or middle to a depressed gynobase much shorter than nutlet, not leaving a pit on the gynobase, nutlets with a well developed ventral keel.

1. Nutlets with minute hooked or forked hairs on the back (use good lens); calyx lobes with midrib enlarged and more or less indurated in fruit, usually unequal; calyx base narrowly conical   -1. P. nelsonii
1. Nutlets without hooked or forked hairs (may be slightly tuberculate); calyx lobes with midrib scarcely indurate or thickened in fruit, usually equal; calyx base broadly conical
    2. Calyx lobes 2-3 mm. long in fruit, twice as long as the nutlets; plants often glabrous -2. P. orthocarpus
    2. Calyx lobes 1-2 mm. long in fruit, 1-1.5 times as long as nutlets; plants hairy   -3. P. scopulorum

1. Plagiobothrys nelsonii (Greene) Johnston, Contrib. Gray Herb. 68:77. 1923.
   Allocarya nelsonii Greene---Plants profusely branching from the base; stems 5-20 cm. tall, diffuse, hispid-hirsute with appressed hairs; leaves 8-30 mm. long, linear to linear-oblanceolate, strigose; raceme moderately dense, bracts from oblong and not more than twice as long as the flower to somewhat leaflike and linear-elongated; calyx about 1.5 mm. long in flower but the lobes enlarged to 2 mm. or more in fruit, the midrib enlarging and becoming somewhat indurated, lobes usually becoming more or less unequal; corolla about 1.5 mm. long, inconspicuous; nutlets about 1.5 mm. long, with minute hooked or forked hairs on the back.---River banks and plains, often in sandy soil. Saskatchewan to Oregon, south to Wyoming and Nevada. Our few records from northcentral and northwestern Colorado at 5000-7000 feet.
2. Plagiobothrys orthocarpus (Greene) Johnston, Contrib. Gray Herb. 68:78. 1923.
   Allocarya orthocarpa Greene---Washington, south to Colorado and Nevada. Reported for northcentral Colorado but no specimens seen by the writer.
3. Plagiobothrys scopulorum (Greene) Johnston, Contrib. Gray Herb. 68:79. 1923.
   Allocarya scopulorum Greene---Stems 5-20 cm. tall, more or less branching from base, ascending or spreading, with stiff appressed hairs; leaves 1-5 cm. long, linear, strigose; racemes rather lax, the bracts when present resembling the leaves; calyx about 1-1.5 mm. long in flower, the lobes about 1-2 mm. long in fruit, 1-1.5 times as long as the nutlets; corolla about 1.5 mm. long, inconspicuous; nutlets about 1.5-1.8 mm. long, no hooked or forked hairs present.---Often on moist or sandy soil. Saskatchewan to Washington, south to Nebraska and Nevada. Our records scattered in the western two-thirds of Colorado at 4500-8500 feet.

Family 102. Verbenaceae VERVAIN FAMILY

Annual or perennial herbaceous plants; leaves opposite or whorled, simple, entire to dissected, lacking stipules; flowers perfect, mostly small, more or less irregular, in spikes or heads; calyx of 4 or 5 more or less united sepals, usually bracted; corolla usually at least slightly 2-lipped, of 4 or 5 partly united petals; stamens 4, didynamous, attached to the corolla and alternate to its lobes (when lobes are 4); ovary superior, 2- or appearing 4-celled, ovules 1 in each cavity, ovary not deeply lobed externally, styles 1, terminal, stigmas 1- or 2-lobed; fruit of 2-4 nutlets, each 1-seeded, separating at maturity.

1. Calyx and corolla 5-lobed; nutlets 4; spikes terminal on stems or branches   -1. Verbena
1. Calyx 2- to 4-lobed, corolla 4-lobed; nutlets 2; spikes lateral   -2. Phyla

1. Verbena L. VERVAIN

Annual or perennial plants; leaves entire, toothed or dissected; flowers in terminal spikes or short headlike clusters; calyx tube usually tubular, teeth somewhat unequal; corolla salverform, the tube often curved, more or less bilabiate, 5-lobed; ovary appearing 4-celled and 4-ovuled, stigma 2-lobed, only 1 lobe stigmatic; fruit separating into 4 nutlets. The 6 species listed here with showy large flowers in short and broad spikes apparently intergrade among themselves, especially V. ciliata Benth. and V. wrightii A. Gray. However, most of our material keys down fairly well in Perry's treatment (Ann. Mo. Bot. Gard. 20:239-362. 1933).

1. Flowers not especially showy, corolla tube not over 6 mm. long; calyx 2.5-5 mm. long; inflorescence of elongated narrow spikes (not so very long in V. bracteata at first); leaves usually not lobed (except in V. bracteata); sterile style lobe adjacent to stigmatic surface but usually not protruding beyond it (Section Verbenacea).
   2. Plants branching at base, the branches decumbent or ascending; bracts conspicuous, at least twice as long as the calyx; leaves 3-parted or pinnatifid, not over 4 cm. long   -1. V. bracteata
   2. Plants rarely branching at base, stems erect or nearly so; bracts inconspicuous, less than 1-1/2 times longer than the calyx; leaves serrate-dentate to incised-serrate (except V. hastata), sometimes hastate lobed at base but then 5-15 cm. long
      3. Leaf blades thin, not reticulated, often hastately lobed at base; fruiting spikes narrow, less than 7 mm. wide, subtended at base only by inconspicuous bracts, spikes usually many   -2. V. hastata
      3. Leaf blades thick and rugose-reticulated, not hastate at base; fruiting spikes thick, over 7 mm. wide, subtended at or near base by leaflike bracts, spikes 1 to several
         4. Bracts 1-2 mm. longer than their calyx; leaves oblong-elliptical to ovate-lanceolate, short-petioled and usually acute; plants mostly of southcentral Colorado   -3. V. macdougalii
         4. Bracts about as long as the calyx; leaves ovate, oval or suborbicular, sessile or nearly so, acute to obtuse; plants mostly of eastern and northcentral Colorado   -4. V. stricta
1. Flowers showy, corolla tube 8 mm. long or more; calyx 7-13 mm. long; inflorescence of short, broad spikes (elongating some in fruit); leaves commonly lobed to bipinnatifid (except in V. gooddingii); sterile style lobe adjacent to stigmatic surface and protruding well beyond it (Section Glandularia)
   5. Corolla about twice as long as the calyx, limb 11-15 mm. wide   -5. V. canadensis
   5. Corolla tube as long or 1-1/2 times as long as the calyx, limb 6-10 mm. wide
      6. Leaves coarsely dentate or less commonly very shallowly lobed or incised   -6. V. gooddingii
      6. Leaves 3-cleft, incised-pinnatifid or bipinnatifid, usually the latter
         7. Floral bracts as long or longer than the calyx; calyx not glandular; corolla limb about 8-10 mm. wide (when fresh)   -7. V. bipinnatifida
         7. Floral bracts shorter than the calyx; calyx usually somewhat glandular; corolla limb about 6-8 mm. wide
            8. Calyx teeth 2-3 mm. long   -8. V. ambrosifolia
            8. Calyx teeth short, less than 2 mm. long
               9. Plants with prostrate-compact habit; margins of leaf divisions strongly revolute; calyx glandular but hardly viscid   -9. V. ciliata

9. Plants with stems ascending to erect; margins of leaf divisions not strongly revolute; calyx somewhat viscid
-10. **V. wrightii**

1. Verbena bracteata Lag. & Rodr., An. Cienc. Nat. 4:260. 1801.
   V. bracteosa Michx.---Stems 10-50 cm. long, usually several from a common base, diffusely branched, these branches decumbent or ascending and more or less hirsute; leaves 1-4 cm. long, cuneate-spatulate to cuneate-obovate in outline, pinnately parted to lobed, usually of 3 divisions, the middle largest and incisely toothed or cleft, more or less hirsute on both upper and lower surfaces; spikes sessile, conspicuously bracted, these much longer than the calyx (ours 2 or more times as long); calyx 3-4 mm. long; corolla light blue to purple, the tube slightly longer than the calyx, corolla limb 2.5-3 mm. broad; nutlets about 2 mm. long, reticulate above, striate below.---Roadsides, fields and waste places. Widespread in North America. Our records well scattered over Colorado at 3500-7500 feet.
2. Verbena hastata L., Sp. Pl. 20. 1753.
   Stems 30-100 cm. tall, erect or nearly so, branched above only, rough-pubescent; leaves 5-15 cm. long, rather thin, not at all rugose, lanceolate to ovate-lanceolate, definitely petioled, coarsely or almost incised-serrate often hastately 3-lobed at base, acuminate at apex, rough-pubescent above and below; spikes straight, usually numerous in an upright panicle, not leafy right at base, each spike commonly not over 7 mm. wide, bracts commonly shorter than the calyx; calyx 2.5-3 mm. long; corolla blue to purple, the tube somewhat longer than the calyx, limb 3-4.5 mm. wide; nutlets about 2 mm. long, nearly smooth or faintly striate.---Valleys, thickets, plains and pastures. Widespread in Canada and the United States. Our records mostly from the northeastern quarter of Colorado at 3500-5000 feet.
3. Verbena macdougalii Heller, Bull. Torr. Club 26:588. 1899.
   Stems 30-80 cm. tall, erect, usually simple at base, commonly branching above, pilose-hirsute to pubescent; leaves oblong-elliptical to ovate-lanceolate, short-petiolate, usually acute, coarsely and irregularly serrate-dentate, rugose above, rather thick, pilose-pubescent to hirtellous; spikes 1 to several, short-peduncled with leaves at or near base, each spike usually over 7 mm. thick, bracts 1-2 mm. longer than the calyx; calyx 4-5 mm. long; corolla blue to purple, the tube scarcely longer than the calyx, limb about 6 mm. wide when fresh; nutlets 2.5 mm. long, reticulate above, striate below.---Valleys and plains. Wyoming south to New Mexico and Arizona. Our records from central and southern (mostly southcentral) Colorado at 6500-8500 feet.
4. Verbena stricta Vent., Hort. Cels. 53, pl. 53. 1800.
   Stems 30-100 cm. tall, erect or nearly so, simple or branched above, densely pilose to hirsute; leaves 5-10 cm. long, ovate to suborbicular or oval, sessile or nearly so, sharply serrate, mostly doubly so, sometimes nearly incised-serrate, acute or obtuse, hirsute-villous, thickish-rugose above; spikes solitary or several, short-pedunculate with leaves close to base, each spike usually over 7 mm. thick, bracts about as long as the calyx; calyx 4-5 mm. long; corolla tube slightly longer than the calyx, limb 8-9 mm. wide when fresh; nutlets about 2.5 mm. long, reticulate above, striated below.---Dry valleys, hills and plains. Common in the central part of the United States, extending eastward to Pennsylvania and west to the Rocky Mountains. Our records scattered in the eastern half of Colorado, except in the extreme southern part, at 3500-5500 feet.
5. Verbena canadensis (L.) Britt., Mem. Torr. Club 5:276. 1894.
   Stems 15-40 cm. tall, decumbent or ascending, more or less branched, glabrate to hirsute; leaves ovate to elongate-ovate, from incised to incised-pinnatifid or 3-cleft, glabrate to appressed hirsute; spikes short and broad in anthesis, bracts shorter than the calyx; calyx becoming 10-13 mm. long, glandular; corolla reddish-purple or lilac, the tube about twice as long as the calyx, the limb about 11-15 mm. wide when fresh; nutlets about 3 mm. long, more or less reticulate.---Hills, plains and valleys. North Carolina to Colorado, south to Florida and Texas. One specimen from southwestern Colorado at 6500 feet may be this species.
6. Verbena gooddingii Briq., Ann. Conserv. et Jard. Bot. Geneve 10:103. 1907.
   Utah to California, south to Mexico. This plant has been reported for Colorado but no specimens located by the writer.
7. Verbena bipinnatifida Nutt., Journ. Acad. Nat. Sci. Phila. 2:123. 1821.
   Stems 10-40 cm. tall, more or less diffusely branching from base, loosely ascending, hispid-hirsute; leaves bipinnately parted to triparted with divisions more or less bipinnatifid, margins of segments sometimes revolute; spikes short and broad in anthesis, the bracts longer than the calyx; calyx 8.5-10 mm. long in fruit, not glandular; corolla rose to purple, the tube about 1½ times as long as the calyx, limb 8-10 mm. wide; nutlets about 3 mm. long, more or less reticulated.---Plains, and slopes. South Dakota, south to Alabama and Arizona. Our records from central and southeastern Colorado at about 4000-6000 feet.
8. Verbena ambrosifolia Rydb. in Small, Fl. Southeast. U. S. ed 1:1011. 1903.
   Stems 15-40 cm. tall, loosely decumbent-ascending, somewhat hirsute, more or less diffusely branched from base; leaves bipinnatifid with ultimate segments lanceolate, appressed-hirsute; spikes broad and short (in flower), bracts shorter than the calyx; calyx 8-9 mm. long in fruit, glandular, the teeth 2-3 mm. long; corolla rose to purple, the tube 1-1/3 - 1-1/2 times as long as the calyx, the limb 6-8 mm. wide; nutlets 2.5-3 mm. long, more or less reticulated.---Plains and slopes. Oklahoma, south to Texas, Arizona and Mexico. Our records from northcentral, central, southcentral and southeastern Colorado at 4000-6000 feet.
9. Verbena ciliata Benth., Pl. Hartw. 21. 1839.
   Stems about 5-25 cm. tall, several from the base, prostrate to decumbent, branched, hirsute; leaves subbipinnate or trifid with divisions more or less deeply incised, ultimate divisions linear-oblong, often revolute, hirsute-strigose; spikes broad and short in anthesis, bracts shorter than the calyx; calyx 7-8 mm. long, more or less glandular, the calyx teeth less than 2 mm. long; corolla rose to purple, the tube 1-1/3 to 1-1/2 times as long as the calyx, the

limb about 6-8 mm. broad; nutlets 2.5-3 mm. long, more or less reticulated.---Dry plains and slopes. Texas to Arizona and south into Mexico. Our records from central and southern Colorado at 4000-7000 feet.

10. Verbena wrightii A. Gray, Syn. Fl. N. Amer. 2(1):337. 1878.

Stems 20-60 cm. tall, usually several from base, decumbent-ascending to erect, branched, more or less hispid-hirsute; leaves bipinnatifid or trifid with divisions more or less incised, hirtellous to hirsute; spikes broad and short in anthesis, bracts shorter than the calyx; calyx 7-9 mm. long in fruit, glandular and more or less viscid, teeth less than 2 mm. long; corolla rose to purple, the tube 1-1/3 to 1-1/2 times as long as the calyx, the limb 6-8 mm. wide; nutlets 2.5-3 mm. long, more or less reticulated.---Plains and hills. Colorado, south to Texas and Arizona. Our records from southwestern Colorado at about 6500-7000 feet.

2. Phyla Lour. FOGFRUIT

Perennial herbaceous plants, glabrous or with forked hairs; stems creeping or procumbent, often rooting at the nodes; leaves opposite, rather narrowly cuneate at base, toothed; flowers bracted in short dense headlike spikes, these on rather long axillary peduncles; calyx flattened, 2- to 4-toothed and 2-lipped; corolla 4-lobed, more or less 2-lipped and irregular, the tube cylindrical; ovary 2-celled, stigma oblique or recurved; fruit separating into 2 nutlets.

1. Leaves thick and rigid, oblanceolate-cuneate, widest above the middle, only the midrib noticeable, with 1 to 4 pair of teeth these tending to crowd out near apex, more or less canescent-strigose   -1. P. cuneifolia
1. Leaves thin, not rigid, rhombic-lanceolate to ovate-oblong, the widest part at or below the middle, pinnately veined, the side veins noticeable, more than 4 pair of teeth present, these extending to the middle or below, leaf green   -2. P. lanceolata

1. Phyla cuneifolia (Torr.) Greene, Pitt. 4:47. 1899.
   Lippia cuneifolia (Torr.) Steud.---Stems 20-100 cm. long, puberulent almost canescent; leaves 1-3 cm. long, thick and rigid, oblanceolate, long-cuneate to base, only the midrib conspicuous and noticeable, with 1-4 pair of teeth above the middle and near apex, more or less canescent-strigose; peduncles from shorter than the leaves to 5 cm. long; heads globose in flower, elongating in fruit; corollas about 4 mm. long, white to pink or purplish-rose.---Dry plains. Nebraska to Wyoming, south to Texas and Arizona. Our records scattered over the eastern half of Colorado at 4000-6000 feet.

2. Phyla lanceolata (Michx.) Greene, Pitt. 4:47. 1899.
   Lippia lanceolata Michx.---New Jersey to Minnesota, south to Florida, Texas, Arizona and southern California. This plant was reported for central Colorado but no specimens were located by the writer.

Family 103. Labiatae MINT FAMILY

Often aromatic, usually herbaceous, annual or perennial plants, but sometimes shrubs or undershrubs; stems usually square and 4-angled; leaves simple, usually toothed or lobed, opposite; flowers perfect, usually definitely irregular, variously disposed; calyx of 5 united sepals, commonly more or less 2-lipped, the upper 3 teeth more or less joined, the lower more free or sometimes the 5 teeth about equal; corolla gamopetalous, usually evidently 2-lipped, the upper 2 petals almost joined to form an erect lip, the lower 3 usually spreading, or sometimes the 5 lobes of the corolla nearly alike; stamens 4, usually in 2 unequal pairs, or 2, often lying under the upper lip; ovary superior, more or less 4-lobed, 4-celled, each cell with 1 ovule, styles 1, usually bifid at apex, arising from between the lobes, or at apex of the ovary when the lobes are more united; fruit of 4 little seedlike nutlets each formed from a lobe of the ovary, each 1-seeded. When studying the staminal arrangement of this family the procedure is to split the corolla down the middle of the lower lip and then open it out. When 2 pairs of stamens are present the upper pair will thus be the inner pair since the upper lip lies in the middle of the spread corolla. The 2 outer stamens will constitute the lower pair and are commonly longer or shorter than the other 2.

1. Stamens, with anthers, 2 (small rudimentary staminodes sometimes also present)
   2. Each stamen with the connective between the 2 anther sacs elongated (usually longer than true filament) only 1 branch usually bearing an anther sac; flowers in terminal spikelike racemes   -1. Salvia
   2. Stamens not as above, anther sacs not widely separated by a connective; flowers in clusters, these usually axillary, never in spikelike racemes
      3. Corolla regular or nearly so   -2. Lycopus
      3. Corolla distinctly 2-lipped, definitely irregular
         4. Low woody shrubs clothed with feltlike tomentum   -3. Poliomintha
         4. Herbaceous annual or perennial plants, not woody, glabrous to hairy but not as above
            5. Flowers in dense subglobose clusters, these terminal and solitary or several and axillary, forming interrupted spikes; calyx teeth about equal   -4. Monarda
            5. Flowers seldom over 6 together; in axillary clusters but these not dense nor subglobose, usually many clusters present; calyx more or less 2-lipped   -5. Hedeoma
1. Stamens, with anthers, 4
   6. Calyx teeth 10, hooked at apex; stems densely white-woolly   -6. Marrubium
   6. Calyx teeth 5 or less, never hooked at apex; stems usually not white-woolly
      7. Calyx 2-lipped, the lips entire, upper bearing an erect crest well below the apex   -7. Scutellaria
      7. Calyx not 2-lipped or if so then the lips lobed or toothed, no erect crest present

8. Upper lip of corolla with a deep slit between its 2 petals, this running down to the top of the calyx tube or below (hence corolla appearing to have 1 large lower lip only, this 5-lobed); style not basal   -8. Teucrium
8. Upper lip of corolla entire or merely 2-lobed or corolla nearly regular (but no deep slit on top of corolla); styles basal
   9. Corolla regular or nearly so, obscurely 2-lipped
      10. Anther sacs parallel; flowers either in terminal spikes or in dense axillary clusters   -9. Mentha
      10. Anther sacs divergent; flowers in terminal globose clusters   -10. Monardella
   9. Corolla very definitely 2-lipped (the upper lip commonly erect and entire or notched, the lower spreading and 3-lobed)
     11. Bracts spinose-toothed (like a holly leaf); upper calyx tooth ovate, twice as broad as the others   -11. Moldavica
     11. Bracts, if present, not spinose-toothed (do not mistake leaves for bracts); calyx regular, slightly irregular or if 2-lipped not as above
       12. Anther sacs parallel or nearly so
         13. Leaves 3- to 5-palmately cleft   -12. Leonurus
         13. Leaves entire to toothed
           14. Leaves entire; flowers in dense, short, terminal glomerules; upper lip of corolla flat, not concave or galeate   -13. Pycnanthemum
           14. Leaves toothed; flowers in dense glomerules, some of them axillary, or in spikelike racemes; upper lip of corolla concave or galeate
              15. Leaves ovate to ovate-deltoid, definitely petioled; flowers many, densely crowded in a continuous or interrupted spike; upper (inner) pair of stamens longer than the lower   -14. Agastache
              15. Leaves lanceolate to oblong-lanceolate, sessile; flowers in a rather loose, spikelike raceme, but not very numerous; upper (inner) pair of stamens shorter than the lower   -15. Dracocephalum
       12. Anther sacs widely divergent or placed end to end
         16. Flowers, at least some, in clusters in the axils of the upper leaves, sometimes in only 1 terminal, leaf-subtended cluster
           17. Flowers subtended by short setaceous, conspicuously hirsute-ciliate bracts (in addition to subtending leaves); stems erect and plants perennial with rootstocks   -16. Clinopodium
           17. Flowers without such bracts; stems decumbent (or if erect then plants annual)
              18. Stems erect; leaves ovate, rounded to cuneate at base, never cordate   -17. Galeopsis
              18. Stems creeping and decumbent; leaves orbicular or very broadly ovate, at least some very definitely cordate at base
                 19. Upper (inner) pair of stamens longer than the lower; plants perennial; calyx teeth about 1/3 as long as their tube   -18. Glecoma
                 19. Upper (inner) pair of stamens shorter than the lower; plants annual or biennial; calyx teeth about 1/2 as long as the tube   -19. Lamium
         16. Flowers in dense or somewhat interrupted spikes, these leafy if at all, only at base
           20. Calyx definitely 2-lipped; flowers in dense uninterrupted spikes, bracts ovate to reniform, not at all leaflike   -20. Prunella
           20. Calyx teeth equal or nearly so, not at all 2-lipped; flowers in rather interrupted spikes, the bracts when present rather leaflike (except in size)
              21. Upper (inner) pair of stamens longer than the lower; plant strongly aromatic (catnip-odor); stems branched; leaves ovate, definitely petioled   -21. Nepeta
              21. Upper (inner) pair of stamens shorter than the lower; plants little if any aromatic; stems usually simple; leaves oblong to ovate-oblong, sessile or nearly so   -22. Stachys

1. Salvia SAGE

    Annual or perennial herbaceous plants; stems leafy; leaves entire to toothed; flowers in terminal bracteate spikelike racemes; calyx 2-lipped, the upper lip entire or 3-toothed, lower with 2 lobes; corolla tubular, strongly 2-lipped, upper lip concave, entire or 2-notched, lower lip longer than upper, 3-toothed and spreading; stamens 2, connective between the anther sac well developed and usually longer than the true filament, bearing 1 sac terminal on 1 branch, the other branch deflexed and destitute of a sac; nutlets smooth.

1. Plants annual; stems not over 30 cm. tall; corolla tube included in the calyx; plants common   -1. S. reflexa
1. Plants perennial; stems 30-100 cm. tall; corolla tube exserted; plants rare in Colorado
   2. Corollas 15-30 mm. long; upper lip of calyx entire; leaves linear to linear-lanceolate, all short-petioled   -2. S. azurea
   2. Corollas 10-14 mm. long; upper lip of calyx 3-toothed; leaves lanceolate to oblong or ovate, lower definitely petioled, upper sessile   -3. S. sylvestris

1. Salvia reflexa Hornem., Enum. Pl. Hort. Hafn. 1:34. 1807.
    S. lanceolata of Manuals---Plants annual; stems 10-30 cm. tall, puberulent to nearly glabrous, branched from base; leaves 2-6 cm. long, oblong-lanceolate to oblong, sometimes as narrow as linear, entire to undulate or remotely serrate; calyx 4-8 mm. long, enlarging some in fruit; corolla tube included in the calyx, the whole corolla 8-12 mm. long, light purple or whitish tinged with blue.---Prairies, plains, fields and roadsides. Wisconsin to Wyoming, south to Illinois, Texas, Arizona and Mexico. Our records scattered over Colorado, except the northwestern part, at 3500-8000 feet.

2. **Salvia azurea** Michx. ex Lam., Journ. Hist. Nat. 1:409. 1792.
S. pitcheri Torr.---Perennial plants; stems 40-100 cm. tall or more, stout, strigose; leaves 3-12 cm. long, linear or linear-lanceolate, short-petioled, toothed or entire, strigillose; calyx 6-8 mm. long, canescent; corolla 15-30 mm. long, blue or whitish, upper lip bearded on back. Our plant is ssp. pitcheri (Torr.) Epling or possibly var. grandiflora Benth.---Dry plains and hills. The subspecies from Missouri to Nebraska, south to Arkansas and Texas. Our few records from southeastern Colorado at 4000-4500 feet.

3. **Salvia sylvestris** L., Sp. Pl. 24. 1753.
Perennial plants; stems 30-90 cm. tall, rather stout, puberulent to densely short-villous, branched above; lower leaves 5-10 cm. long, oblong, petioled; upper leaves short-petioled or sessile, subcordate at base, ovate to lanceolate, densely puberulent, crenate; flowers appearing racemose, crowded, with reduced ovate to ovate-lanceolate often colored bracts; calyx about 6 mm. long, tinged with purple, puberulent, upper lip 3-toothed, lower 2-cleft; corolla 10-14 mm. long, violet-blue, the tube exserted from the calyx.---Roadsides, fields and waste ground. Adventive from Eurasia and appearing as a weed in various parts of the United States. Rare in Colorado.

2. Lycopus L. BUGLEWEED; WATER HOREHOUND

Plants not odoriferous, perennial and herbaceous; stems leafy, erect; leaves sessile or nearly so, serrate to pinnatifid; flowers white to lavender, small, in dense sessile axillary clusters; calyx campanulate, regular or nearly so with 4 or 5 teeth; corolla nearly regular, with 4 lobes, or 1 of the lobes emarginate; stamens 2, the other 2 if present very rudimentary; anther sacs 2, parallel; nutlets 3-angled.

1. Calyx lobes obtuse to barely acute; mature nutlets as long or longer than the calyx -1. L. uniflorus
1. Calyx lobes acuminate to awn-tipped; mature nutlets about 2/3 as long as the calyx
  2. Leaves pinnatifid (at least the lower); calyx lobes awn-tipped; ridges of nutlets entire -2. L. americanus
  2. Leaves serrate or the lower rarely incised; calyx lobes merely acuminate; ridges of nutlets rugose-tuberculate -3. L. lucidus

1. **Lycopus uniflorus** Michx., Fl. Bor. Am. 1:14. 1803.
Stems 10-50 cm. tall, slender, puberulent, from stolons that are often tuber-bearing; leaves 3-8 cm. long, lanceolate to ovate-lanceolate, pale beneath, serrate to somewhat incised; calyx about 2 mm. long, the lobes subobtuse and broadly triangular; corolla longer than the calyx, funnelform, whitish; nutlets as long as or exceeding the calyx.---Moist soil. Newfoundland to Alaska, south to Virginia and Oregon. Our one record from central Colorado at 6000 feet.

2. **Lycopus americanus** Muhl.; Bart., Fl. Phila. Prodr. 15. 1815.
L. sinuatus Ell.---Stems 20-70 cm. tall, glabrous or puberulent; leaves 3-10 cm. long, lanceolate to ovate in outline, pinnatifid or incised at least the lower; calyx 2-3 mm. long, the teeth rather rigid, about as long as the tube, tipped with a short awn; corolla barely exceeding the calyx, white or lavender; ridges of nutlets entire, mature nutlets about 2/3 as long as the calyx.---Moist or wet ground. Newfoundland to British Columbia, south to Florida and California. Our records from the western two-thirds of Colorado, except the northwestern part, at 4500-7500 feet.

3. **Lycopus lucidus** Turcz. ex Benth. in DC., Prodr. 12:178. 1848.
L. asper Greene; probably L. velutinus Rydb.---Stems 20-80 cm. tall, glabrate or pubescent; leaves 3-10 cm. long, oblong-lanceolate to narrowly lanceolate, sharply serrate; calyx 2.5-4 mm. long the teeth triangular-subulate to ovate-lanceolate, acuminate or long-acute at apex, teeth at least as long as the tube; corolla as long as or to 1 mm. longer than the calyx, whitish; nutlets about 2/3 as long as the calyx, ridges rugose-tuberculate.---Moist or wet ground. Minnesota to British Columbia, south to Kansas, Arizona and California; Asia. Our records mostly from northcentral and central Colorado at 4500-7000 feet.

3. Poliomintha A. Gray ROSEMARY MINT

Shrubs, clothed with minute, feltlike tomentum; stems leafy; leaves entire, linear or linear-oblong, thickish; flowers pale blue, rose or purple, in small axillary cymules; calyx cylindrical, 15-veined, pilose on the tube, teeth subequal and more or less connivent; corolla with hairy rings in the throat, 2-lipped, the upper erect, emarginate, lower lip spreading, 3-cleft, the middle lobe broader and emarginate; stamens 2, attached above the middle of the corolla tube and ascending under the upper lip; nutlets smooth, oblong.

1. **Poliomintha incana** (Torr.) A. Gray, Proc. Am. Acad. 8:296. 1870.
Sandy ground. Texas to Utah and Arizona. This plant may be present in western or southwestern Colorado.

4. Monarda L. HORSEMINT; BEEBALM

Perennial or annual herbaceous plants; stems leafy; leaves toothed to entire, petioled; flowers in dense glomerules which may be terminal and solitary, or several in an interrupted spike, subtended by an involucre of bracts; calyx tubular, 13- to 15-veined, the teeth about equal, mostly pubescent in the throat; corolla strongly bilabiate, upper lip erect or sickle-shaped, entire or notched, lower lip 3-lobed, spreading; stamens with anthers 2, rudimentary ones present or absent, anthers narrow, 2-celled, sacs divergent; nutlets smooth.

1. Heads solitary; upper lip of corolla usually erect and straight, the stamens exserted beyond it
  2. Petioles more than 1 cm. long; stems usually branched in upper parts   -1. M. fistulosa
  2. Petioles not more than 8 mm. long; stems usually simple   -1A. M. fistulosa menthaefolia
1. Heads 2 or more, forming an interrupted spike; upper lip of corolla falcate, the stamens usually not exserted
  3. Calyx teeth deltoid, about as broad as long, more or less acuminate but not at all aristate
                                                         -2. M. punctata
  3. Calyx teeth slender, aristate   -3. M. pectinata

1. Monarda fistulosa L., Sp. Pl. 1:22. 1753.
    M. mollis L.---Prairies, hills and plains. Eastern United States, extending west to Nebraska and Texas. This species has been reported close to the eastern borders of Colorado and some of our collections are intergradations between the species and its variety.
1A. Monarda fistulosa menthaefolia (Graham) Fernald, (var.) Rhodora 46:494. 1944.
    M. menthaefolia Graham; M. ramaleyi A. Nels.; M. comata Rydb.; M. stricta Wooton---Perennial plants; stems usually 30-75 cm. tall, occasionally taller, usually unbranched, usually pubescent above and glabrous below; leaves ovate-lanceolate, serrate, upper surface hairy or glabrous, lower pubescent, nearly tomentose or glabrous, often definitely lighter; petioles 2-5 mm. long but rarely to 12 mm.; glomerules 1.5-2.5 cm. wide; outer bracts foliar, often pinkish; calyx 7-10 mm. long rarely longer, the teeth 1 mm. long, acuminate, orifice hirsute; corolla 26-33 mm. long, rarely to 38 mm., lavender or rose-purple, usually twice the length of calyx, expanding upward; stamens exserted beyond upper lip.---Valleys, hills and plains. Saskatchewan to Alberta, south to Texas and Arizona. Our records well scattered over the western two-thirds of Colorado at 5000-9000 feet.
2. Monarda punctata L., Sp. Pl. 22. 1753.
    The subspecies occidentalis Epling has been reported in Kansas very close to the Colorado border.
3. Monarda pectinata Nutt., Journ. Acad. Phil. ser. 2. 1:82. 1847.
    M. nuttallii A. Nels.---Annual plants; stems commonly 15-40 cm. tall, branched from base, with retrorse hairs; leaves, the largest 20-50 mm. long, oblong-lanceolate to oblong, both surfaces glabrous or sparingly puberulent, margins remotely serrate or subentire; petioles 2-25 mm. long; glomerules 15-25 mm. wide, exclusive of corolla; outer bracts foliaceous, inner oblong-elliptical, acuminate to bristle-tipped, midnerve conspicuous, the 1 or sometimes 2 pairs of lateral veins less so, margins with stiff bristles, outer surface glabrous or nearly so, inner glabrous; calyx tube 6-8 mm. long, orifice hirsute, teeth 2-6 mm. long, slender, not rigid, hirsute; corolla light rose, pink to whitish, its tube 8-14 mm. long including throat which is 3-5 mm. long, lips about equal, usually shorter than the tube, lower lip often darker spotted.---Plains and hills, often on sandy ground. Nebraska to Utah, south to Texas, Arizona and Mexico. Our records scattered over Colorado, except in the northwestern part, at 3500-8500 feet.

        5. Hedeoma Pers.   MOCK-PENNYROYAL; FALSE-PENNYROYAL

    Annual or perennial herbaceous plants, usually aromatic; stems leafy; leaves linear to oval, entire, sessile or nearly so; flowers very short-pedicelled, in axillary remote cymules, sometimes present even to base of plant; calyx teeth nearly as long or as long as the tube, dissimilar and more or less 2-lipped, the upper 3 more united and shorter than the 2 lower ones, all teeth hispid-ciliate and calyx hispid in throat, the tube more or less gibbous near base; corolla bluish to blue-purple, 2-lipped, the tube tubular in shape, upper lip 2-lobed or entire and erect, lower 3-lobed and more or less spreading; stamens 2, but 2 very rudimentary ones sometimes also present, stamens exceeding the corolla tube, anther cells divergent; ovary deeply 4-parted with basal style; nutlets smooth, oblong.

1. Plants annual; calyx tube 2-3 mm. long, teeth nearly equal in length, upper 3 recurving somewhat after flowering and leaving the throat open
                                                                                   -1. H. hispida
1. Plants perennial, the caudices often woody; calyx tube 5-7 mm. long, lower teeth longer than upper (often twice as long), all teeth converging after flowering and closing the throat   -2. H. drummondii

1. Hedeoma hispida Pursh, Fl. Am. Sept. 414. 1814.
    Annual plants; stems 10-30 cm. tall, simple or branching chiefly at base, with long deflexed hairs at least above; leaves spreading, linear to linear-elliptical, hispid-ciliate, otherwise glabrous to hispidulous; cymules 1- to 6 or more-flowered, clusters numerous and rather crowded, borne well along the entire length of the stem; calyx tube 2-3 mm. long, definitely gibbous near base, the pouch forming 2/3 to 3/4 the length of the tube, teeth definitely 2-lipped, the upper 3 teeth somewhat recurved, all teeth nearly equal in length, after flowering leaving the throat open; corolla about 6 mm. long, bluish-purple, the tube narrowly funnelform.---Plains and valleys, often on sandy soil. Ontario to Alberta, south to Connecticut, Louisiana and Colorado. Our records from the northeastern quarter of Colorado at 3500-5500 feet.
2. Hedeoma drummondii Benth., Lab. Gen. et Sp. 368. 1836.
    H. sancta Small; H. ovata Nels.; H. camporum Rydb.; prob. H. nana in Rydb. Fl. Colo.---Perennial plants; stems 8-25 cm. tall, arising from usually woody caudices, rather widely branched, more or less pubescent above with recurved hairs; leaves 1-2 cm. long, linear, elliptic-oblong to oval or ovate; cymules 1- to 6-flowered, mostly above the middle of the plant, sometimes to very base; calyx tubes 5-7 mm. long, more or less gibbous near base, teeth not strongly bilabiate (sinuses more or less equally deeply cut) but lower 2 teeth definitely longer than upper 3, often to twice as long, all teeth converging and closing the orifice after flowering; corolla about 6-12 mm. long, rose to purple.---Hills and plains, usually in dry ground. North Dakota to Montana, south to Texas, Arizona and Mexico. Our records scattered over Colorado at 3500-7500 feet.

6. Marrubium L. HOARHOUND

Perennial herbaceous plants; stems leafy, white-woolly; leaves broad, rugose, white-woolly especially below; flowers whitish, crowded in dense, subglobose clusters in the axils of the upper leaves, often forming interrupted spikes; calyx tubular with 10 spreading, hooked, teeth; corolla small, the tube little longer than the calyx, 2-lipped, the upper lip erect, notched, lower lip spreading and 3-lobed, the middle lobe larger; stamens 4, paired, anthers with 2 divergent sacs; nutlets more or less roughened.

1. Marrubium vulgare L., Sp. Pl. 583. 1753.
    Stems 20-100 cm. tall, erect, rather stout; leaf blades suborbicular to ovate or broadly oval; calyx 4-7 mm. long; corolla 5-10 mm. long.---Waste places, roadsides and fields. Naturalized from Europe and widely distributed in the United States. Our records scattered over Colorado at 5000-7500 feet.

7. Scutellaria L. SKULLCAP

Perennial plants with rhizomes or stolons present; leaves entire or toothed; flowers blue, in terminal and axillary racemes or solitary and axillary; calyx rather pouch-shaped or campanulate, with 2 entire equal and rounded lips, the upper one bearing a conspicuous crest; corolla tubular, 2-lipped, the lateral lobes more or less joined with the upper lip to form a galea, this including the style and stamens; stamens 4 in pairs, anthers 2-celled in upper pair, 1-celled in lower; nutlets tuberculate or papillose.

1. Flowers in racemes; rhizomes slender; petioles of middle leaves -1. S. lateriflora
1. Flowers solitary in the axils; rhizomes stout; middle leaves sessile or with shorter petioles
    2. Corolla tube and galea over 22 mm. long; middle leaves sessile, entire, narrowing to base
        -2. S. brittonii
    2. Corolla tube and galea shorter; middle leaves short-petioled, crenate-serrate, truncate to subcordate at base -3. S. galericulata

1. Scutellaria lateriflora L., Sp. Pl. 598. 1753.
    Perennial plants spreading by slender rhizomes; stems 20-80 cm. tall, channeled, branching above, glabrate or usually appressed hairy on upper part, sometimes glandular; leaf blades 3-7 cm. long, thin, deltoid-ovate, acute, rounded-truncate at base, crenate-serrate, upper surface glabrous, lower paler and often appressed hairy on veins; petioles 5-25 mm. long; flowers in racemes usually 3-8 cm. long, with bracts present; calyx 1.5-2.5 mm. long; corolla blue, short-pubescent, galea and tube 5-7 mm. long.---Along streams and swamps. Newfoundland to British Columbia, south through United States except the extreme southern part. Apparently uncommon in Colorado our one record from the northeastern part at 3500 feet.
2. Scutellaria brittonii Porter, Bull. Torr. Bot.Club 21:177. 1894.
    S. virgulata A. Nels.; S. brittonii var. virgulata (A. Nels.) Rydb.---Perennial plants spreading by thickened usually yellowish and knotty rhizomes; stems about 15-30 cm. tall, usually branched at base, glabrous or with retrorsely-appressed to upcurved hairs, sometimes with glandular hairs; lower leaves 8-15 mm. long, usually oval, short-petioled (3-5 mm.); upper leaves 17-35 mm. long, sessile, narrowly oblong to ovate-elliptical, entire, glabrous or pubescent and glandular; flowers axillary, pedicels 3-4 mm. long; calyx 4.5-6.5 mm. long to longer at maturity; corolla deep violet-blue, tube and galea 23-32 mm. long, tube curved.--- Hills and valleys. Wyoming to New Mexico. Our records from in or near northcentral, central and southcentral Colorado at 5000-10,000 feet.
3. Scutellaria galericulata L., Sp. Pl. 599. 1753.
    Perennial marsh plants spreading by slender stolons; stems 30-60 cm. tall, simple or branched, usually pubescent at least on angles with short curly hairs, sometimes glandular-hairy; petioles 1-3 mm. long or sometimes to 4 mm.; leaves 3-6 cm. long, oblong-ovate, truncate to subcordate at base, acute, crenate-serrate, upper surface glabrous or nearly so, lower paler puberulent rarely nearly glabrous; flowers solitary in the axils of the upper leaves on short pedicels 2-2.5 mm. long; corolla blue, tube and galea 14-21 mm. long.---Moist or wet ground along ditches, streams and lakes. Newfoundland to Alaska, south to New England, Wisconsin, Arizona and California; Eurasia. Our records scattered in Colorado, few from the extreme western or eastern parts, at 5000-8000 feet.

8. Teucrium L. GERMANDER

Perennial herbaceous plants; stems leafy; leaves serrate to pinnately parted; flowers white, purple, pale blue or lilac, solitary in axils of leaves or in bracteate terminal crowded spikes; calyx campanulate with 10 veins and 5 equal or unequal lobes; corolla tube short, upper lip short, deeply parted to top of calyx tube or below, lower lip conspicuous and spreading, middle lobe larger, lateral lobes small (coralla appearing to have 1 lower lip with 5 lobes on it in our species); stamens 4, paired, exserted from the deep cleft at top of corolla; nutlets glabrous, usually wrinkled, attached laterally with the style not basal.

1. Leaves serrate but not at all parted; calyx teeth shorter than the tube; corolla usually light rose to purplish; stems usually over 25 cm. tall -1. T. canadense
1. Leaves pinnately parted; calyx teeth about twice as long as their tube; corolla usually white or whitish; stems rarely over 25 cm. tall -2. T. laciniatum

1. Teucrium canadense L., Sp. Pl. 564. 1753.
T. occidentale Gray---Plants from creeping rootstocks; stems 30-60 cm. tall, erect, branching mostly in the inflorescence if at all, pubescent to villous-hirsute; leaves 4-9 cm. long, ovate, oblong to oval, sometimes lanceolate, serrate, villous-pubescent at least below, some of hairs glandular; calyx 5-7 mm. long, villous with some of hairs glandular, teeth unequal, shorter than the tube; corolla 7-15 mm. long, light rose to purplish, sometimes even cream-colored. Our plants seem to be var. occidentalis (Gray) McClintock & Epling.---Usually moist ground of swamps, thickets or stream edges. The species widely distributed in North America. Our records scattered in Colorado, except in the northwestern part, at 3500-7500 feet.

2. Teucrium laciniatum Torr., Ann. Lyc. N. Y. 2:231. 1828.
Melosmon laciniatum (Torr.) Small---Plants from caespitose caudices and rather woody roots; stems 7-20 cm. tall, branching from base, usually unbranched above, glabrous or sparsely hairy; leaves pinnately parted nearly to the midrib into usually entire linear lobes, these sometimes lobed again, whole leaf 1.5-5 cm. long, glabrous or nearly so; calyx 8-13 mm. long in flower, the teeth equal or nearly so, about twice as long as the tube, glabrous or nearly so; corolla 12-20 mm. long, white or possibly pale blue or lilac.---Plains, hills and valleys. Oklahoma to Colorado, south to Texas and New Mexico. Our records from southcentral and southeastern Colorado at 4000-6500 feet.

9. Mentha L. MINT

Perennial aromatic herbaceous plants with creeping rhizomes; leaves serrate, short-petioled; flowers white to pink, small, in dense clusters in the axils of the upper leaves or in narrow dense terminal spikes; calyx campanulate, with 5 equal lobes shorter than the tube; corolla campanulate to funnelform, obscurely 2-lipped, upper lip deeply to shallowly notched, lower lip more deeply 3-lobed, but corolla nearly regular; stamens 4, about equal or upper (inner) pair shorter, anthers with 2 parallel sacs; nutlets smooth.

1. Flowers in terminal spikes -1. M. spicata
1. Flowers in axillary clusters -2. M. arvensis

1. Mentha spicata L., Sp. Pl. 576. 1753.
Occasionally cultivated and perhaps escaping in moist or shady places. Two similar species may be found in this state, cultivated or escaped. The three can be separated as follows:

1. Spikes less than 1 cm. thick; leaves sessile or nearly so M. spicata L.
1. Spikes more than 1 cm. thick; leaves petioled
   2. Leaves lanceolate, narrowed at base M. piperita L.
   2. Leaves ovate, rounded or truncate at base M. citrata Ehrh.

2. Mentha arvensis L., Sp. Pl. 577. 1753.
M. penardi (Briq.) Rydb.; M. glabrior (Hook.) Rydb.; M. borealis in Rydb. Fl. Colo.; M. canadensis L.---Stems 10-40 cm. tall, glabrous to pubescent; leaves 1-5 cm. long, oblong-ovate to oblong-lanceolate but sometimes broader or narrower, rounded to cuneate at base, crenate-serrate to sharply serrate; flowers in dense axillary clusters; calyx about 2-3 mm. long; corolla about 3-6 mm. long, pink to light rose, rarely white. The writer cannot distinguish the separated varieties in Colorado plants (Rhodora 46:331-335. 1944).---Moist or wet ground. A circumpolar species extending south in North America to Pennsylvania, New Mexico and California. Our records scattered over Colorado, except the extreme eastern part, at 4500-9500 feet.

10. Monardella Benth.

Perennial herbaceous plants; stems leafy; leaves entire; flowers commonly rose-purple, borne in terminal globose bracteate glomerules, the bracts leaflike but usually colored; calyx tubular, narrow, about 13-nerved, 5-toothed, these narrow, erect and subequal; corolla tube usually exserted from calyx, subbilabiate, the 2 lips nearly equal, upper lip 2-lobed, the lower 3-lobed; stamens 4, somewhat exserted, the upper (inner) pair shorter or equal to the lower, anthers 2-celled, the sacs divergent; nutlets smooth.

1. Monardella odoratissima Benth., Lab. Gen. & Sp. 332. 1834.
Monardella parvifolia Greene; Madronella parvifolia (Greene) Rydb.; M. sessilifolia Rydb.; M. oblongifolia Rydb.---Plants from woody often contorted and decumbent stems with splitting bark, branches about 20-30 cm. tall, erect or ascending, usually unbranched, pubescent often canescent; leaves 1-3 cm. long, usually lanceolate, sometimes to ovate or oblong, glabrous to pubescent, usually entire, sessile or short petioled; glomerules 1-3 cm. wide, bracts ovate to round, variously shaded with purple, ciliate; calyx 6-10 mm. long; corolla about 8-20 mm. long; nutlets about 2 mm. long.---Canyons, valleys and slopes. Montana to Washington, south to New Mexico and California. Our records from westcentral and southwestern Colorado at 7000-10,000 feet.

11. Moldavica (Tourn.) Adans. DRAGONHEAD

Annual, biennial or possibly perennial herbaceous plants; stems erect, leafy; leaves petioled, sharply serrate or lower incised; flowers blue to rose-pink, crowded in dense terminal (sometimes axillary too) clusters with conspicuous spinulose-pectinate bracts; calyx campanulate-tubular, 5-toothed, the upper tooth enlarged; corolla tube enlarging above, 2-lipped, the upper lip spreading and 3-lobed, the middle lobe larger and sometimes notched at apex; stamens 4, paired the upper (or inner) pair longer, anthers with 2 divergent sacs; nutlets smooth.

1. Moldavica parviflora (Nutt.) Britt. in Britt. & Br., Ill. Fl. ed. 2.3:115. 1913.
Dracocephalum parviflorum Nutt.---Stems 20-60 cm. tall, more or less pubescent; leaves 1-6 cm. long, lanceolate to oblong, sometimes ovate especially the lower, coarsely serrate, the upper with spinulose teeth, more or less puberulent at least below; bracts ovate or oblong, pectinate with awn-pointed teeth; calyx 9-15 mm. long; corolla bluish to light rose, scarcely longer than the calyx.---Valleys, hills and mountains. New York to Alaska, south to New Mexico and Arizona. Our records well scattered over the western two-thirds of Colorado at 4500-10,000 feet.

12. Leonurus L. MOTHERWORT

Biennial or perennial herbaceous plants; stems erect, leafy; leaves long-petioled, the blades 3-cleft or parted with the segments sometimes incised; flowers pink, purple, red or white, in dense axillary clusters near apex of plant; calyx tubular-campanulate, 5-veined, the 5 teeth nearly equal, triangular-aristate; corolla strongly 2-lipped, the upper lip erect and slightly concave, entire, lower lip spreading, 3-lobed, the middle lobe larger, truncate or emarginate; stamens 4, paired, the upper (inner) shorter, ascending under the upper lip, anthers with 2 mostly parallel sacs; nutlets smooth, truncate at apex.

1. Leonurus cardiaca L., Sp. Pl. 584. 1753.
Perennial plants; stems 30-100 cm. tall, strigose-pubescent; leaves palmately 3- to 5-cleft, the lobes entire or toothed, lower leaves broad ovate, upper often narrower; calyx about 6-8 mm. long, the teeth spreading or reflexed; corolla 6-10 mm. long, pale purple, rose to whitish, densely hairy without, especially on back of upper lip.---Waste places. Now widespread throughout most of temperate North America. Our rather few records from northcentral and south-central Colorado at 4500-9000 feet.

13. Pycnanthemum Michx. MOUNTAIN MINT

Perennial herbaceous plants; stems leafy; leaves entire or toothed; flowers in dense terminal clusters, subtended by a pair of leaflike bracts; calyx cylindrical, with 5 lobes; corolla tube enlarging somewhat upward and longer than the calyx, 2-lipped, the upper erect and notched, the lower spreading and 3-lobed; stamens 4, paired, the lower pair somewhat longer, all exserted, anthers 2-celled, the sacs parallel; nutlets smooth or tipped with short hairs.

1. Pycnanthemum virginianum (L.) Durand & Jackson ex Robins.& Fern., Man. 707. 1908.
Maine to North Dakota, south to Tennessee and Texas. This species has been reported in eastern Colorado but is probably not here.

14. Agastache Clayton GIANT-HYSSOP

Perennial herbaceous plants; stems leafy and tall, arising from creeping rootstalks; leaves ovate or deltoid-ovate, crenate-serrate, petioled; flowers in dense sessile clusters which are disposed in spikes or panicles; calyx tubular or campanulate, 5-toothed, the teeth about equal or upper somewhat longer, often whitish or colored other than green; corolla purplish, rose or whitish, 2-lipped, upper lip erect, concave, 2-lobed, the lower somewhat spreading and 3-cleft; stamens 4, anther sacs parallel or nearly so; nutlets smooth or hairy at apex.

1. Calyx teeth 1-2 mm. long, blue or violet-tinged; corollas blue, their tubes about 6.5-7.5 mm. long; lower leaf surface whitish from minute, appressed, feltlike hairs -1. A. foeniculum
1. Calyx teeth 2-5 mm. long, white to rose; corollas white to rose or violet, the tubes 8-13 mm. long; lower leaf surface glabrous to pubescent (may be paler green but not whitish from feltlike hairs)
   2. Upper lip of corolla prominently thrust forward, 2-2.5 mm. long; stamens all parallel, both pairs thrust similarly out of the corolla tube -2. A. pallidiflora
      3. Calyx green with whitish teeth; corolla white or pallid -2. A. pallidiflora
      3. Calyx rose or green with rose-colored teeth; corolla rose -2A. A. pallidiflora neomexicana
   2. Upper lip of corolla not prominently thrust forward, 1-1.5 mm. long; lower pair of stamens ascending under the upper lip of corolla, upper pair thrust down and crossing the lower -3. A. urticifolia

1. Agastache foeniculum Kuntze, Rev. Gen. 511. 1891.
A. anethiodora (Nutt.) Britt.---Stems 50-100 cm. tall, rarely taller, little branched, glabrous but pubescent in the inflorescence; leaf blades about 5-8 cm. long, ovate or deltoid-ovate, green and glabrous above, lower surface densely covered with minute appressed feltlike hairs hence more or less whitened and glaucous appearing; verticels of flowers crowded or somewhat separate especially the lower, the spike 4-8 cm. long and 1.5-2 cm. wide; calyx tube 5-7 mm. long, teeth 1-2 mm. long, deltoid to narrowly deltoid, violet or blue-tinged, often the upper part of tube colored; corolla blue, the tube 6.5-7.5 mm. long; lower stamens ascending under the lip of the corolla, the upper stamens thrust down and crossing the lower.---Plains and slopes. Ontario to Alberta, south to Wisconsin and Colorado. Our records from northcentral and central Colorado at 7000-8000 feet.

2. **Agastache pallidiflora** (Heller) Rydb., Bull. Torr. Club 33:150. 1906.
Stems 40-60 cm. tall or sometimes to 100 cm.; leaf blades about 2.5-4 cm. long or sometimes longer, nearly glabrous to pubescent or hirtellous but always green both sides though somewhat paler below; verticels of flowers sessile, usually crowded or lowest remote, in a cylindrical spike about 4-8 cm. long and about 20 mm. thick in fruit; calyx green with whitish deltoid to subulate teeth 2-4 mm. long, the tube about 5-8 mm. long; corolla white to pallid, the tube 9-13 mm. long, the upper lip 2-2.5 mm. long and rather prominently thrust forward; stamens parallel, both pairs thrust similarly out of the tube. This plant has been called ssp.typica Lint & Epling.---Valleys and slopes. Colorado, south to New Mexico and Arizona. Our records from southcentral and southwestern Colorado at 7500-9000 feet.

2A. **Agastache pallidiflora neomexicana** (Briq.) Lint & Epling, (ssp.) Am. Midl. Nat. 33:220. 1945.
Differs from the typical form of the species in having calyx rose or green with rose teeth; corolla rose-colored.---With the species. Our few records from southwestern Colorado at 9000-9500 feet.

3. **Agastache urticifolia** Kuntze, Rev. Gen. 511. 1891.
Stems 60-200 cm. tall, often over 1 m., glabrous or nearly so; leaves 3.5-8 cm. long, mostly ovate to deltoid-ovate, usually glabrous above, lower glabrous to pubescent, both sides green but lower usually paler; verticels of flowers 4-15 cm. long on fairly robust plants, mostly 20-30 mm. wide, compact or somewhat remote especially below; calyx about 4-7 mm. long, green or rose-colored at least in teeth and near apex of tube, tube about 4-7 mm. long, teeth 2.5-5 mm. long, deltoid-lanceolate; corolla white to rose or violet, the tube about 8-13 mm. long, the upper lip about 1-1.5 mm. long, not prominently thrust forward; lower stamens ascending under the upper lip of the corolla, the upper stamens thrust down and crossing the lower.---Valleys and slopes. British Columbia, south to Colorado and California. Our records from northcentral and western Colorado at 6000-10,000 feet.

15. Dracocephalum L. FALSE DRAGONHEAD

Perennial herbaceous plants; stems erect, leafy; leaves sessile, rather narrow, toothed; flowers purple to rose-pink, in terminal showy spikelike racemes; calyx campanulate-tubular, more or less inflated, the 5 teeth about equal, short; corolla with the tube dilated upward, 2-lipped, the upper erect and slightly concave, entire, lower lip spreading and 3-lobed the middle lobe emarginate; stamens 4, paired, the upper (inner) pair shorter, ascending under the upper lip, anthers with 2 parallel sacs; nutlets smooth.

1. **Dracocephalum nuttallii** Britt. in Britt. & Brown, Ill. Fl. Ed. 2.3:117. 1913.
Physostegia parviflora Nutt.---Saskatchewan to British Columbia, south to South Dakota and Oregon. This plant has been reported from Colorado, possibly on the basis of a cultivated or escaped specimen.

16. Clinopodium L. WILD-BASIL

Perennial herbaceous plants; stems leafy, erect, hirsute; leaves oval to narrowly ovate, subentire; flowers white to purple but usually rose-colored, in dense cymules, these terminal and hemispheric or forming an interrupted spike, the lower cluster axillary; calyx tubular, 2-lipped, upper lip with 3 shorter teeth; corolla 2-lipped, the upper somewhat erect, somewhat concave, entire or emarginate, the lower lip spreading and 3-lobed; stamens 4, paired, the upper probably shorter, anthers with 2 divergent sacs; nutlets smooth, ovate.

1. **Clinopodium vulgare** L., Sp. Pl. 587. 1753.
Rootstocks present; stems 10-50 cm. tall, simple or nearly so, more or less hairy; leaf blades 1-4 cm. long; flowers in dense axillary and terminal clusters, sometimes the clusters reduced to a terminal one but subtended by leafy bracts; calyx 8-9 mm. long, villous-hirsute, lobes subulate-aristate; corolla 8-12 mm. long, white or purple.---Waste places. Nova Scotia to Manitoba, south to North Carolina and Arizona; Eurasia. Our one record from northwestern Colorado at 6500 feet.

17. Galeopsis L. HEMPNETTLE

Annual herbaceous plants; stems erect, branched; leaves opposite, petioled; flowers in 1 to several clusters in the upper leaf axils or in 1 terminal cluster subtended by leaves, no bracts present; calyx campanulate or tubular-campanulate, 5-ribbed, with 5 equal or nearly equal spinulose teeth; corolla 2-lipped, upper lip erect and entire, concave, lower 3-cleft, and spreading; stamens 4, anthers 2-celled, these spreading or appearing so, upper (inner) pair of stamens shorter than the lower; ovary deeply 4-parted; nutlets smooth, ovoid.

1. **Galeopsis tetrahit** L., Sp. Pl. 579. 1753.
Stems 30-90 cm. tall, rather coarse, swollen below the nodes, with long, jointed, glass-like hairs; leaves 3-15 cm. long, cuneate or rounded at base, coarsely dentate or serrate; calyx about 7-9 mm. long, the teeth about as long as the tube; corolla 10-20 mm. long, pink or pale purple, variegated with white.---Waste places. Newfoundland to Alaska, south to various parts of the United States; Eurasia. Our few records from Eagle and Pitkin Counties at 8500-9000 feet.

18. Glecoma L. GROUND IVY; GILL-OVER-THE-GROUND

Low creeping perennial herbaceous plants; stems leafy; leaves long-petioled, orbicular or reniform, crenate; flowers rather large, blue or purple, in small axillary clusters; calyx tubular, somewhat unequally 5-toothed; corolla tube exserted, enlarged above, 2-lipped, the upper lip erect, 2-lobed or emarginate; stamens 4, in pairs, the upper (inner) pair longer, anther sacs divergent; nutlets smooth.

1. Glecoma hederacea L., Sp. Pl. 578. 1753.
Stems 20-50 cm. long, pubescent; leaves 1-4 cm. broad, green both sides; calyx 5-7 mm. long, oblique; corolla 12-20 mm. long.---Waste places, lawns and gardens. Newfoundland to Washington, south to Georgia and California; Eurasia. Our few records from northcentral and central Colorado at 5000-6000 feet.

19. Lamium L. DEAD NETTLE

Annual or biennial herbaceous plants; stems leafy, branching; leaves petioled, toothed, orbicular to ovate; flowers purplish-red, in remote axillary and terminal clusters, the subtending leaves sessile, clasping or petioled; calyx campanulate, about 5-nerved with 5 equal or somewhat unequal teeth about as long as the tube; corolla with the tube longer than the calyx, strongly 2-lipped, upper lip erect, concave and usually entire, more or less hairy, lower lip spreading, 3-lobed, the middle lobe enlarged and notched, the lateral lobes small; stamens 4, the upper (inner) pair shorter, anthers with 2 divergent sacs, ascending under the upper lip; nutlets smooth or tuberculate.

1. Upper leaves sessile or clasping -1. L. amplexicaule
1. Upper leaves petioled -2. L. purpureum

1. Lamium amplexicaule L., Sp. Pl. 579. 1753.
Stems 10-50 cm. tall, slender, glabrous, usually decumbent; lower leaf blades orbicular, ovate or cordate, crenate, long-petioled; upper leaves similar but sessile and clasping; corolla 10-16 mm. long, purplish or red.---Waste places and cultivated ground. Introduced from Europe and found in most parts of temperate North America. Our one record from northcentral Colorado at about 5500 feet.
2. Lamium purpureum L., Sp. Pl. 579. 1753.
This eastern species has been reported for northcentral Colorado (Rhodora 28:112. 1926.).

20. Prunella L. SELFHEAL; HEALALL

Perennial herbaceous plants; stems leafy; leaves petioled, subentire; flowers violet or purple, in dense terminal bracteate spikes the bracts sheathing and foliose; calyx 2-lipped, the upper truncate bearing 3 cusps, lower 2-lobed, closing at maturity; corolla 2-lipped, the upper lip arched and somewhat galeate, the lower lip 3-lobed, the middle lobe broader; stamens 4, lying under the upper lip, in pairs, the upper (inner) pair shorter, lower and longer pair 2-toothed or 2-pronged at apex, 1 tooth bearing the 2-celled anther, the other sterile, anthers divergent; nutlets smooth, ovate.

1. Prunella vulgaris L., Sp. Pl. 600. 1753.
Stems 5-30 cm. tall, procumbent, ascending or erect, glabrous or pubescent; leaves 2-7 cm. long, ovate, oblong or lanceolate, cuneate at base; spikes 1.5-5 cm. long, sometimes longer in fruit, bracts reniform to ovate, cuspidate and ciliate; calyx 5-10 mm. long, green or purplish-tinged on teeth; corolla 8-15 mm. long.---Slopes, roadsides, fields and in shade, usually in rather moist ground. Widespread in the cooler parts of North America; Eurasia. Our records widely scattered in the western half of Colorado at 5000-10,000 feet.

21. Nepeta L. CATNIP

Perennial herbaceous very aromatic plants; stems leafy and branched; leaves broad, truncate or subcordate at base, rather coarsely toothed; flowers white to purplish-spotted, in dense cymes these disposed in an interrupted terminal spike; calyx tubular or campanulate, the tube somewhat constricted above, the 5 teeth deltoid-subulate, nearly equal in length but lower 3 more joined at base; corolla tube longer than calyx, 2-lipped, the upper lip erect, notched, lower lip spreading and 3-lobed, the middle lobe larger; stamens 4, paired, the upper (inner) pair longer, exserted, anthers with 2 divergent sacs; nutlets smooth.

1. Nepeta cataria L., Sp. Pl. 570. 1753.
Stems 50-100 cm. tall, erect, canescent-tomentose; leaves 2-8 cm. long, ovate, coarsely crenate, green above, white-tomentose below; calyx 4-7 mm. long; corolla 7-12 mm. long, white or purple-spotted.---Waste places, roadsides and around dwellings. Native to Europe but now widely distributed in North America. Our records scattered over the state at 3500-6500 feet.

22. Stachys L. BETONY; WOUNDWORT; HEDGENETTLE

Perennial herbaceous plants, with rootstocks; stems erect, leafy; leaves oblong, sessile or nearly so, toothed; flowers in clusters usually in 3's in the axils of leaflike bracts and disposed in interrupted spikes; calyx turbinate or campanulate, enlarging somewhat at maturity, the 5 teeth nearly equal and more or less spinulose at apex; corolla white, pallid or purple-rose, 2-lipped, upper lip erect, entire or notched, lower lip spreading, 3- or rarely 2-lobed; stamens 4, the upper (inner) pair shorter, attached near middle of corolla tube, anthers 2-celled the sacs divergent and all fertile, filaments naked, all ascending under the upper lip; ovary deeply 4-parted with basal style.

1. **Stachys palustris** L., Sp. Pl. 580. 1753.
   S. scopulorum Greene; S. teucriformis Rydb.; S. teucrifolia Rydb.---Stems 15-80 cm. tall, simple or sparingly branched, with spreading hirsute or villous hairs especially above; leaves about 4-8 cm. long, villous-hirsute to villous with spreading hairs; calyx 5-9 mm. long; corolla about 10-15 mm. long, the tube about as long as the calyx or slightly longer, mottled with darker spots. Our plant has been called ssp. pilosa (Nutt.) Epling.---Moist open soil. The species widespread in North America. Our records scattered in the western two-thirds of Colorado at 4500-9500 feet.

Family 104. Solanaceae POTATO FAMILY

Mostly herbaceous plants but some woody; leaves alternate or rarely opposite, mostly simple; flowers perfect, usually regular and usually cymose; calyx of 5 (rarely 4 or 6) more or less united sepals; corolla gamopetalous, 5-lobed, these lobes often obscure, rotate, campanulate, funnelform or salverform; stamens as many as the corolla lobes and alternate with them, inserted on the corolla tube; ovary superior, mostly 2-celled and many-ovuled, these on central placentae, style terminal, stigma entire or slightly 2-lobed; fruit a berry or a capsule.

1. Fruit a dehiscent capsule; corollas funnelform to nearly salverform, usually over 2 cm. long
   2. Corollas 5 cm. long or longer, all solitary in the axils or forks of the branching stems; capsules normally spiny   -1. Datura
   2. Corollas not over 4 cm. long, at least some in terminal racemes or panicles; capsules not spiny
      3. Capsules circumscissile near apex, completely included in the calyx; corolla slightly irregular   -2. Hyoscyamus
      3. Capsules opening by longitudinal valves, not completely included in calyx; corolla strictly regular   -3. Nicotiana
1. Fruit a more or less fleshy berry, not dehiscent; corollas rotate to rotate-campanulate or open-campanulate, never funnelform, usually less than 2 cm. long
   4. Plants woody shrubs (upright or scrambling), more or less spiny; leaves entire   -4. Lycium
   4. Plants herbaceous, spiny or not spiny (if woody at base as in Solanum dulcamara the plant spineless and the leaves hastate or 3-lobed); leaves various
      5. Calyx inflated and concealing the fruit or enlarging and enclosing the fruit except at very top; plants never spiny; anthers longitudinally dehiscent throughout
         6. Calyx closely fitted to the fruit, thin and obscurely veiny, lobes not closing at apex, hence the top of the fruit exposed   -5. Chamaesaracha
         6. Calyx bladdery-inflated and conspicuously veiny in fruit, lobes closing or connivent over the top of the berry
            7. Sepals of calyx nearly distinct, auriculate at base; ovary 3- to 5-celled; corolla blue to almost white and plant annual   -6. Nicandra
            7. Sepals united to near apex, not auriculate; ovary mostly 2-celled; corolla usually yellowish or greenish-yellow (if purple or violet the plant perennial)   -7. Physalis
      5. Calyx not enlarging or inflated and not at all enclosing the fruit (except in 2 species with sharp prickles); anthers opening by terminal pores or slits, not dehiscent throughout   -8. Solanum

1. Datura L. THORNAPPLE; JIMSON WEED

Coarse weedlike, herbaceous, ill-scented plants; stems stout, mostly erect, branched; leaves large, petioled, repand to pinnately lobed; flowers solitary in the axils or forks of the branching stem, very large and showy, white to lavender or violet; calyx cylindric or prismatic-funnelform, 5-lobed; corolla funnelform with a spreading 5-lobed plaited border; stamens 5, included, filaments adnate to near middle of tube; ovary 2-celled or sometimes falsely 4-celled, style slender, stigmas 2-lobed; fruit a large, globose or ovoid capsule, normally spiny, 4-valved or irregularly dehiscent; seeds flat.

1. Corollas 15-20 cm. long; calyx 7-12 cm. long; leaves and stems canescent-puberulent; capsules nodding, globose, breaking irregularly   -1. D. meteloides
1. Corollas 5-11 cm. long; calyx 3-6 cm. long; leaves and stems glabrous or sparsely puberulent; capsules erect, ovoid, regularly dehiscent into 4 valves   -2. D. stramonium

1. **Datura meteloides** Dunal. in DC., Prodr. 13(1):544. 1852.
   Annual or perennial plants; stems 30-100 cm. tall, erect, pruinose looking with fine gray pubescence; leaves about 10-20 cm. long on largest, ovate, repand-dentate to nearly entire, often uneven at base, canescent-puberulent; calyx 7-12 cm. long, lobes lanceolate, about 20 mm. long; corolla 15-20 cm. long, whitish to violet; fruit nodding, irregularly breaking open, subglobose.---Valleys, plains and hills. Colorado, south to Texas, Mexico and California; South America. Our certain records from southwestern Colorado at 6000 feet.

2. **Datura stramonium** L., Sp. Pl. 179. 1753.
   D. tatula L.---Annual plants; stems 30-100 cm. tall, glabrous or very nearly so; leaves 10-20 cm. long, ovate or oblong, repand to coarsely sinuately toothed, nearly lobed, glabrous or sparingly puberulent; calyx 3-6 cm. long, lobes about 4-7 mm. long, triangular-lanceolate; corolla 5-11 cm. long, white to violet; fruit erect, regularly dehiscent into 4 valves, ovoid.---Waste places and cultivated ground, often around barnyards. Widespread throughout North and South America, Eurasia and Africa. Our records from northcentral Colorado at 5000-6000 feet.

2. Hyoscyamus L. HENBANE

Annual or biennial herbaceous plants; stems leafy; leaves alternate, lobed or pinnatifid; flowers solitary in the upper axils and in terminal racemes, these more or less 1-sided; calyx urn-shaped to campanulate, 5-cleft, enlarging and becoming more or less reticulate in fruit, enclosing the capsule; corolla greenish-yellow to whitish with dark rose or purple veins, funnelform, with slightly oblique 5-lobed limb; stamens exserted, anthers opening longitudinally; ovary 2-celled, stigma capitate; fruit a capsule, circumscissile above the middle, included in calyx.

1. Hyoscyamus niger L., Sp. Pl. 179. 1753.
Stems about 30-100 cm. tall, stout, viscid short-villous; leaves 6-20 cm. long, oblong, ovate to lanceolate in outline, irregularly lobed, cleft or pinnatifid, sessile or upper clasping, viscid and short-villous; calyx 2-2.5 cm. long in fruit; corollas about 1.5-2 cm. long; capsule about 10-14 mm. long.---Waste places around dwellings and along roadsides. Nova Scotia to Montana, south to New York and Colorado; naturalized from Europe and probably more widespread in North America than as above. Our records from southcentral and southwestern Colorado at 7500-9000 feet, but recently collected in Grand County.

3. Nicotiana L. TOBACCO

Annual or perennial herbaceous plants; stems erect, leafy, viscid-puberulent; leaves sessile or petioled, entire or repand; flowers in terminal racemes or panicles; calyx campanulate or tubular-campanulate, 5-toothed or 5-lobed; corolla tubular-funnelform to nearly salverform, the 5 lobes spreading; stamens included, anthers opening longitudinally; ovary 2-celled, stigma capitate; fruit a capsule, dehiscent by 2-4 valves near apex; seeds many and small.

1. Leaves petioled, not cordate or auriculate-clasping at base; corolla externally glabrous or very sparsely hairy -1. **N. attenuata**
1. Leaves cordate and sessile or auriculate-clasping at base; corolla copiously pubescent externally -2. **N. trigonophylla**

1. Nicotiana attenuata Torr. ex S. Wats., Bot. Kings Exped. 276. 1871.
Annual plants; stems 30-60 cm. tall, often branched; leaves 3-10 cm. long, ovate to lanceolate or the upper linear; flowers in paniculate racemes; calyx about 6-9 mm. long, the lobes triangular-lanceolate; corolla 2-4 cm. long, white to greenish-white; fruit about 1 cm. long, glabrous or very sparsely pubescent externally.---Dry often sandy ground. Montana to British Columbia, south to Texas, Mexico and California. Our records from the western half of Colorado, none as yet from the southcentral part, at 5000-8000 feet.
2. Nicotiana trigonophylla Dunal in DC., Prodr. 13(1):562. 1852.
Texas to California, south into Mexico. May be found in southern Colorado.

4. Lycium L. MATRIMONY VINE; WOLFBERRY; DESERT-THORN

Shrubs or woody vinelike, usually spiny plants; leaves short-petioled, alternate, commonly fascicled, entire; flowers regular, solitary or few in the leaf axils; calyx campanulate or tubular, usually 5-toothed or cleft, often enlarging somewhat in fruit; corolla rotate-campanulate to tubular-funnelform or somewhat salverform, greenish to purplish, with 5 or rarely 4 lobes; stamens 4 or 5, adnate to corolla tube; ovary 2-celled, stigma entire or shortly 2-lobed; fruit a fleshy but often rather dry berry.

1. Corolla purple, lavender, lilac or light rose, the tube 3-7 mm. long; berry orange-red or salmon-red, not glaucescent; stems sparingly branched, these spreading, arching or recumbent -1. **L. halimifolium**
1. Corolla greenish or tinged with purple, the tube about 15-20 mm. long; berry red to reddish-blue, glaucescent; more or less upright shrubs with intricate branching -2. **L. pallidum**

1. Lycium halimifolium Mill., Gard. Dict. ed. 8. 1768.
L. vulgare (Ait.) Dunal---Sparingly branched, spreading, recumbent or climbing shrubs, 1-6 m. tall, usually spiny; stems glabrous; leaves 2-6 cm. long, light green, oblong, lanceolate or elliptic, glabrous; calyx about 4 mm. long, the lobes about 1/2 as long as the tube but often split deeper and irregular; corolla rotate-campanulate or short-funnelform, purple to lavender, lilac or light rose, the tube 3-7 mm. long, the lobes somewhat shorter than the tube; stamens about as long as the corolla lobes, exserted; fruit about 1 cm. long, ovoid, fleshy, salmon-red.---Thickets, waste places, often around dwellings or along roadsides and ditches. Throughout most of the United States and Mexico. Our records from northcentral, westcentral and central Colorado at 4500-8000 feet.
2. Lycium pallidum Miers, Ann. & Mag. Nat. Hist. ser. 2, 14:131. 1854.
Intricately branched shrub, 1-2 m. tall, upright but with spreading branches, spiny; stems glabrous or sparingly pubescent; leaves 1-4 cm. long, glaucous, green, glabrous, oblong-spatulate, oblanceolate or elliptic, rarely broader; calyx 5-8 mm. long, the lobes as long or somewhat longer than the tube; corolla 15-20 mm. long, narrowly funnelform, greenish or tinged with purple; stamens usually slightly exserted; fruit about 1 cm. long, red to reddish-blue due to glaucescence, ovoid.---Dry hills and plains. Colorado to Utah, south to Texas, Mexico and Arizona. Our records from southern Colorado at 5000-7000 feet.

5. Chamaesaracha A. Gray

Perennial, low, herbaceous plants; stems leafy, decumbent or prostrate, branched; leaves alternate, entire, repand or pinnatifid, sessile to petioled, the petioles margined; flowers solitary on slender peduncles or sometimes 2-3 at a node, the peduncles recurved in fruit; calyx campanulate, 5-lobed, not becoming bladderlike but closely investing the fruit; corolla rotate, yellow, greenish-white or purplish, 5-angled; filaments adnate to base of corolla, anthers opening longitudinally; fruit a berry; seeds flattened and more or less wrinkled.

1. Leaves, stems and calyx (at least when young) with short stellate hairs, not viscid, longer hairs when present usually branched; leaves mostly sessile or subsessile, oblong-lanceolate to linear   -1. C. coronopus
1. Leaves, stems, and calyx densely viscid-puberulent most of these hairs simple, longer hairs when present simple or very rarely branched; leaves mostly petioled, oblanceolate to ovate (some usually wider than oblong-lanceolate)   -2. C. conioides

1. Chamaesaracha coronopus (Dunal) A. Gray, Bot. Calif. 1:540. 1876.
Stems about 10-20 cm. long, scurfy especially when young, with short, flat, stellate hairs but not at all viscid, also long flat, segmented and usually branching hairs often present; leaves mostly sessile or subsessile, oblong-lanceolate to linear, with hairs resembling those of stem, rather sparse on older leaves especially; calyx 3-5 mm. long, with short stellate hairs and a few long flat segmented usually branched hairs often present; corolla about 8-13 mm. wide, greenish-white or tinged with purple.---Dry hills, valleys and plains. Kansas to Utah, south to Mexico. Our records from the southern half of Colorado at 4000-6000 feet.

2. Chamaesaracha conioides (Moric.) Britt., Mem. Torr. Bot. Club 5:287. 1895.
Stems about 10-15 cm. long, viscid-puberulent and also more or less villous with long, flat, segmented but rarely branched white hairs; leaves mostly distinctly petioled, oblong-lanceolate, oblanceolate, obovate-spatulate to ovate, with hairs similar to those of the stems; calyx 3-5 mm. long, hairy like stem but the long flat segmented hairs usually more abundant; corolla about 8-13 mm. wide, pale yellow or purplish.---Dry plains and hills. Kansas to Colorado, south to Mexico. Our records from southeastern Colorado at 3500-4500 feet.

6. Nicandra Adans.   APPLE-OF-PERU

Annual herbaceous plants; stems leafy, glabrous; leaves alternate, petioled, sinuate-dentate to lobed; flowers showy, blue to almost white, solitary in the axils of the leaves; calyx 5-angled, deeply 5-parted, becoming much enlarged and strongly reticulated in fruit, the segments cordate or sagittate at base; corolla broadly campanulate, obscurely 5-lobed; stamens included, inserted on base of corolla tube, anthers opening longitudinally; ovary 3- to 5-celled, style slender, stigmas 3- to 5-lobed; fruit a berry, this nearly dry and surrounded by the calyx; seeds many.

1. Nicandra physalodes (L.) Pers., Synops. Plant. part 1:219. 1805.
Stems 20-100 cm. tall, angled; leaves 6-20 cm. long, ovate or oblong to ovate-lanceolate; calyx about 1 cm. long in flower; corolla 1.5-4 cm. long; calyx in fruit 1.5 cm. long or more.---Waste ground and fields, often as an escape from gardens. Nova Scotia, Ontario and Idaho, south to Florida and Colorado; South America. Probably more widely distributed in North America. Our records from eastern and central Colorado at 4000-6000 feet.

7. Physalis L.   GROUND CHERRY

Annual or perennial herbaceous plants; stems leafy; leaves alternate, entire, sinuately toothed or pinnatifid; flowers rather large, yellow, whitish or purple, solitary from the axils of leaves, or in 2-6's; calyx campanulate, 5-toothed or 5-lobed, becoming greatly inflated and bladderlike in fruit, the surface veiny, papery and 5-angled; corolla rotate, open-campanulate or rotate-campanulate, obscurely 5-lobed, often darker-spotted in center; stamens inserted near base of corolla, anthers opening by longitudinal slits; fruit a berry; seeds few to many. A puzzling genus because of the intergradations between the species listed here.

1. Corolla violet or purple, rotate   -1. P. lobata
1. Corolla yellow or yellowish, usually brownish in center, open-campanulate
    2. Stems and leaves cinereous with some minute forked or stellate hairs on leaves and stems (use a good lens)   -2. P. fendleri
    2. Stems and leaves with hairs all simple (or if branched then with long hairs confined to lower surface of leaf), or glabrous entirely
        3. Lower leaf surface with long branching hairs   -3. P. pumila
        3. Lower leaf surface glabrous or with various types of unbranched hairs
            4. Leaves and stems glabrous or with a few appressed hairs, never viscid
                5. Leaves lanceolate, oblanceolate to linear, about 4-5 times longer than wide   -4. P. longifolia
                5. Leaves ovate, ovate-lanceolate to ovate-oblong, usually not over 3 times longer than wide   -5. P. subglabrata
            4. Leaves and stems conspicuously pubescent, many of the hairs spreading, sometimes viscid
                6. Hairs rather sparse, not viscid, rather long, flat and segmented, especially on the stem, no other type of hairs present; leaves lanceolate, ovate-lanceolate, oblanceolate to spatulate   -6. P. lanceolata
                6. Hairs dense, more or less viscid, long flat hairs may be present, but other types of hairs also present and more common; leaves ovate to broader

7. Plants annual; corollas 3-10 mm. wide; leaves usually definitely oblique at base
   8. Leaves definitely cordate at base, sinuately toothed to base (at least on some), the teeth large
                                                                                               -7. *P. pruinosa*
   8. Leaves truncate to slightly cordate at base, sinuately-crenate, the teeth small   -8. *P. pubescens*
7. Plants perennial from creeping rootstocks; corolla 12-20 mm. in diameter; leaves usually nearly equal at base
   9. Leaf blades commonly 5 cm. long or more, pubescence of both short glandular hairs and long, flat, segmented hairs   -9. *P. heterophylla*
   9. Leaf blades commonly less than 5 cm. long, glandular-hairy with few if any long, flat, segmented hairs   -10. *P. hederaefolia*

1. **Physalis lobata** Torr., Ann. Lyc. N. Y. 2:226. 1828.
Quincula lobata (Torr.) Raf.---Perennial plants often in patches; stems spreading or prostrate, finally diffusely branched, younger parts more or less scurfy-granuliferous; leaves rather fleshy, variable, oblanceolate, spatulate to oblong, repand to sinuately-pinnatifid; flowers commonly in pairs; corolla 15-25 mm. wide, rotate, dark violet or purple, with a central-rayed white-woolly star; calyx about 1.5-2 cm. long at maturity, 5-angled in fruit, sunken at base, ovoid.---Plains and hills, often in waste ground as a weed. Kansas to Colorado, south to Texas, Mexico and Arizona. Our records well scattered over the eastern half of Colorado at 3500-6000 feet.
2. **Physalis fendleri** A. Gray, Proc. Am. Acad. 10:66. 1874.
Perennial plants from deep fleshy rootstocks; stems about 15-60 cm. tall, variable, sometimes low and branched, sometimes elongated and branching above, finely puberulent the hairs short but some stellate, more or less cinereous; leaves variable, from cordate or deltoid to ovate-lanceolate, repand to sinuate-toothed, short-puberulent and some of the hairs stellate; flowers usually solitary; corolla about 1-2 cm. wide, dull yellow with a brownish center; calyx in fruit ovoid, about 2-3 cm. long.---Dry plains and hills, often in rocky ground. Colorado to Utah, south to New Mexico, Mexico and Arizona. Our records from the southern half of Colorado at 4500-7500 feet.
3. **Physalis pumila** Nutt., Trans. Am. Phil. Soc., ser. 2. 5:193. 1837.
Illinois to Colorado, south to Arkansas and Texas. Reported from eastern Colorado but not located by the writer.
4. **Physalis longifolia** Nutt., Trans. Am. Phil. Soc. ser. 2. 5:193. 1837.
Perennial plants from rather thick rootstocks; stems 20-80 cm. tall, rather stout, branched above, glabrous or with a very few flat long hairs especially above; leaf blades about 4-5 times longer than wide, lanceolate, oblanceolate to linear, entire to repand; flowers usually solitary; corolla 1-2 cm. broad, yellow with a brown center; fruiting calyx ovoid, about 3-5 cm. long at maturity.---Valleys and plains. Iowa to Montana, south to Arkansas, Mexico and Arizona. Our records from the eastern half of Colorado, also in the western and southwestern part, at 3500-6500 feet.
5. **Physalis subglabrata** Mack. & Bush, Trans. St. Louis Acad. Sci. 12:86. 1902.
Perennial plants; stems 25-80 cm. tall, glabrous or very sparingly hairy above with long, flat, segmented hairs; leaves ovate, ovate-lanceolate to ovate-oblong, glabrous or with a very few hairs similar to those on the stem, entire or undulate; flowers usually solitary; corolla yellowish with a brownish center, 1-2 cm. wide.---Valleys, plains and hills, often in waste ground or fields. Vermont to Washington, south to Florida and Texas. Our few records from northern, southwestern and western Colorado at 5000-7000 feet.
6. **Physalis lanceolata** Michx., Fl. Bor. Amer. 1:149. 1803.
P. polyphylla Greene---Perennial plants, rootstocks often slender and creeping, often growing in patches; stems 10-50 cm. tall, erect or spreading, rather sparingly hairy with rather long, flat, segmented hairs, no others present, not at all viscid; leaves oblanceolate, spatulate, lanceolate to oblong-ovate, with hairs similar to those of stem especially on lower surface at margins and along the veins; flowers usually solitary, dull yellow or greenish-yellow with brownish center, about 1-2 cm. wide; fruit about 2-4 cm. long, variable in size.---Plains and hills. Illinois to South Dakota, south to Arkansas and Arizona. Our records from the eastern half of Colorado, also in the southwestern part, at 3500-7000 feet.
7. **Physalis pruinosa** L., Sp. Pl. 184. 1753.
Massachusetts to Kansas, south to Florida and Missouri; probably adventive westward. This plant has been reported for eastern Colorado.
8. **Physalis pubescens** L., Sp. Pl. 183. 1753.
P. neomexicana Rydb. probably---Annual plants; stems 25-60 cm. tall, rather stout, finely pubescent and slightly if at all viscid, erect; leaf blades broadly ovate to orbicular, sinuate-crenate with small teeth, usually unevenly truncate to slightly cordate at base, finely pubescent; flowers usually solitary; corolla usually about 6-10 mm. across, yellowish with a dark center; fruiting calyx about 2-3 cm. long.---Valleys and plains. Pennsylvania to Colorado, south to Florida and Central America. Our one record from northcentral Colorado at 5100 feet.
9. **Physalis heterophylla** Nees, Linnaea 6:463. 1831.
Perennial plants from creeping rootstocks; stems 25-80 cm. tall, erect or decumbent, viscid with short glandular hairs and also more or less villous with long flat jointed hairs; leaves large, usually over 5 cm. long, thick, broadly ovate and cordate at base to sometimes oblong, with hairs similar to stem, subentire to sinuately toothed; flowers usually solitary; corolla yellow with darker center, about 1.5-2 cm. wide; calyx in fruit variable, 2.5-4 cm. long.---Plains and hills, often on sandy or disturbed ground. New Brunswick to Saskatchewan, south to Florida and Arizona. Our records from the northeastern quarter of Colorado at 5000-7000 feet.
10. **Physalis hederaefolia** A. Gray, Proc. Am. Acad. 10:65. 1874.
P. comata Rydb.; *P. rotundata* Rydb.---Perennial plants with rootstocks; stems erect or spreading, often diffusely branched, viscid-puberulent to short-pubescent and often a very few long flat segmented hairs present above; leaves rarely as long as 5 cm., reniform, cordate, orbicular or ovate, sometimes rhombic-ovate, nearly entire to repand or sinuately toothed, with

hairs similar to those of stem; corolla about 1-2 cm. wide, usually solitary, yellowish or greenish-yellow with a darker center; calyx in fruit 2-4 cm. long.---Plains and hills, often on dry or rocky ground. Colorado to Utah, south to Texas, Mexico and California. Our records from the eastern half of Colorado at 4000-5000 feet.

8. Solanum L. NIGHTSHADE

Annual or perennial herbaceous plants; stems leafy, sometimes twining; leaves alternate, petioled, entire to bipinnatifid or pinnately compound; flowers in cymes, racemes, umbels, panicles or solitary; calyx campanulate to rotate with 5 lobes; corolla rotate, 5-angled or 5-lobed, plicate; stamens with short filaments and large anthers which are somewhat coherent in most species and form a cone-shaped ring around the style, the anthers opening by apical pores or short slits; ovary mostly 2-celled, stigma small; fruit a many-seeded berry with the calyx persistent or enclosing it.

1. Calyx large and investing the fruit; stamens unequal, 1 much enlarged and darker; annual plants with sharp prickles
   2. Leaves glandular-puberulent, sparsely stellate; corolla violet   -1. *S. heterodoxum*
   2. Leaves not at all glandular, copiously stellate; corolla yellow   -2. *S. rostratum*
1. Calyx not especially enlarging, not investing the fruit except at very base; stamens all alike; plants without prickles or if with them then perennials with creeping rootstocks
   3. Plants with stellate hairs; stems and leaves usually with prickles; perennial plants with creeping rootstocks
      4. Leaves greenish, not densely scurfy-white, margins sinuate-toothed or lobed   -3. *S. carolinense*
      4. Leaves densely scurfy-white, margins undulate to sinuate-toothed   -4. *S. elaeagnifolium*
   3. Plants glabrous to hairy but not stellate; annual or perennial plants lacking long creeping rootstocks
      5. Plants climbing; perennial plants with the stems somewhat woody below; leaves hastate or 2-lobed near base   -5. *S. dulcamara*
      5. Plants not climbing; annual or herbaceous perennial plants; leaves entire to lobed but not hastate or 2-lobed near base
         6. Plants perennial with globose tubers; leaves pinnately compound; corollas 12-18 mm. wide   -6. *S. jamesii*
         6. Plants annual, no tubers; leaves entire to pinnatifid; corollas 6-12 mm. wide
            7. Leaf blades (at least some) deeply pinnatifid with acute, triangular segments; berry green at maturity   -7. *S. triflorum*
            7. Leaf blades entire to sinuate-dentate or sinuately-lobed, never deeply pinnatifid; berry black or yellow
               8. Stems and leaves glabrate, puberulent or strigose; berry black at maturity   -8. *S. nigrum*
               8. Stems and leaves viscid-villous; berry yellow at maturity   -9. *S. sarachoides*

1. Solanum heterodoxum Dunal, Hist. Solan. 235. 1813.
Annual plants, glandular-pubescent and a few stellate hairs usually present on the leaves, copiously armed with yellow prickles especially on the petioles, stems, and calyces; stems branched, 15-80 cm. tall; leaves 2-12 cm. long, irregularly pinnatifid or bipinnatifid, segments ovate to obovate and obtuse; inflorescence cymose; calyx enlarging and closely investing the fruit, armed with long, straight, straw-colored, very sharp spines; corolla 1.5-4 cm. broad, 5-cleft and slightly irregular, violet; stamens unequal, the lowest anther spreading, larger and violet-colored, other 4 stamens similar, smaller and yellow; fruit appearing spiny from investing spiny calyx.---Dry ground. Colorado, south to Texas, Mexico and Arizona. Our few records from the eastern half of Colorado at 5000-6500 feet.

2. Solanum rostratum Dunal, Solan. Syn. 234. 1813.
Androcera rostrata (Dunal) Rydb.---Annual plants, the stems and leaves not glandular but copiously stellate-pubescent, often also puberulent, armed with long straight prickles especially on the petioles, stems, and calyces; stems 20-70 cm. tall, widely branching; leaf blades 4-12 cm. long, once- or twice-pinnatifid, the segments broad and obtuse; inflorescence cymose; calyx enlarging and closely investing the fruit, armed with long, straight very sharp straw-colored spines; corolla 18-28 mm. wide, slightly irregular, yellow; stamens unequal, the lower one spreading, larger and darker, the other 4 equal and yellow; fruit appearing spiny from the investing calyx.---Valleys and plains, often in waste places, roadsides and fields. North Dakota to Wyoming, south to Mexico; introduced eastward. Our records mostly from the eastern half of Colorado (but a few from western) at 3500-6000 feet.

3. Solanum carolinense L., Sp. Pl. 184. 1753.
Native in the southern states, now scattered as a weed over temperate North America. Not located as yet in Colorado.

4. Solanum elaeagnifolium Cav., Icon. Pl. 3:22. 1794.
Perennial plants from creeping deep rootstocks, stems, leaves, and calyx with prickles usually present but in any case whitish with a dense scurflike stellate effect; stems 30-100 cm. tall, branched; leaves 4-10 cm. long, oblong to lanceolate or linear, entire to coarsely sinuate-dentate; flowers cymose; calyx not enlarging and investing the fruit at maturity but persistent at its base; corollas 20-30 mm. wide, purple, violet or whitish, 5-lobed; anthers all alike; berry 8-15 mm. wide, globose, yellow to black.---Dry ground. Kansas and Colorado, south to tropical America. Our records from eastern Colorado, mostly along the mountains, at 5000-6000 feet.

5. Solanum dulcamara L., Sp. Pl. 185. 1753.
Native to Europe and now distributed in North America, mostly in the eastern part. It should be expected in Colorado.

6. **Solanum jamesii** Torr., Ann. Lyc. N. Y. 2:227. 1828.
Perennial plants with rootstocks bearing nearly globose tubers; stems 10-30 cm. tall, glabrous to sparingly pilose, erect or spreading; leaves pinnately compound, the 5-9 leaflets lanceolate to ovate-oblong, subentire, glabrate or with a few scattered hairs; inflorescence cymose, few- to several-flowered; calyx not investing the fruit although somewhat enlarging, not at all spiny; corolla 12-18 mm. across, normally white; anthers all alike.---Mountains. Colorado to Utah, south to Texas and Arizona. Our records from southcentral and southwestern Colorado at 6000-7500 feet. A record from northern Colorado probably is from a cultivated plant.
   A related tuber-bearing species has apparently been collected once in "S. Colorado". It is S. fendleri A. Gray with corolla lobes broadly ovate and not over 1/2 the total length of the corolla (instead of lanceolate or ovate-lanceolate and at least 1/2 as long as the corolla).
7. **Solanum triflorum** Nutt., Gen. Pl. 1:128. 1818.
Annual plants; stems 10-90 cm. long, branching at base, spreading, glabrous or appressed-hairy; leaves deeply pinnatifid to pinnately lobed, the segments triangular and acute, the sinuses wide and rounded, rather sparsely appressed-hirsute; inflorescence of 1- to 3-flowered cymes; calyx not enlarging and not investing the fruit but persisting at its base, not at all spiny; corolla 8-12 mm. wide, white; anthers all alike; berry 9-15 mm. wide, green at maturity, globose.---Waste ground and fields. Ontario to British Columbia, south to Kansas, Arizona and California. Our records scattered over all of Colorado at 3500-9000 feet.
8. **Solanum nigrum** L., Sp. Pl. 186. 1753.
   S. interius Rydb.---Annual plants; stems 10-80 cm. tall, branched, glabrous to sparsely pubescent or strigose; leaves 2-8 cm. long, ovate to oblong-ovate or lanceolate, entire to irregularly dentate or lobed; inflorescence an umbelliform cyme; calyx not at all spiny, not enlarging and not investing the fruit but persistent at its base; corolla 6-10 mm. wide, white; anthers all alike; berry 5-8 mm. wide, black at maturity. Stebbins and Paddock (Madrono 10:70-81. 1949) separated a closely related species, S. americanum Mill. from the above and some or all of our plants may be that entity.---Waste ground and fields. Apparently introduced from Europe and now widespread in the United States. Our records from the eastern half of Colorado at 3500-5500 feet.
9. **Solanum sarachoides** Sendt. ex Mart., Fl. Bras. 10:18. 1846.
   S. villosum Mill.; S. nigrum var. villosum Mill. of Manuals for Colorado Plants.---Stems and leaves viscid-villous; berries yellow at maturity; pedicels expanding in fruit near its junction with the calyx; calyx enlarging in fruit.---Waste places and fields. Native to South America but now widely distributed in North America. Our few records from northcentral and central Colorado at 5000-8500 feet.

Family 105. Scrophulariaceae FIGWORT FAMILY

Annual or perennial mostly herbaceous plants; stems mostly round; leaves commonly opposite but sometimes alternate or whorled, simple, entire to pinnately parted, without stipules; flowers perfect, more or less irregular; calyx of 5 or 4 more or less united sepals; corolla gamopetalous, usually 2-lipped, sometimes almost regular (or sometimes absent in Besseya); stamens inserted on the corolla tube and more or less alternate to its lobes, commonly 4, in unequal pairs, a staminode often present, or sometimes only 2 stamens present or (in Verbascum) all 5 stamens present; ovary superior, 2-celled (rarely 1-celled near apex), styles 1, stigmas entire or 2-lobed; fruit a 2-valved capsule, usually many-seeded, with central placentae.

1. Anther-bearing stamens 5; corolla nearly regular   -1. Verbascum
1. Anther-bearing stamens 2 or 4 (rarely 5 in a few species of Penstemon with irregular corolla); corolla usually irregular
   2. Anther-bearing stamens normally 2
      3. Cells of the anthers separated by a broad connective wider than the sacs; corolla nearly regular
                                                                                             -2. Gratiola
      3. Anther cells contiguous at apex, sometimes confluent; corolla irregular (in all but Veronica)
         4. Two sterile stamens present, each 2-lobed at the apex; sepals 5   -3. Lindernia
         4. Sterile stamens absent or if present, not 2-lobed at the apex; sepals usually 4
            5. Stems leafy with opposite leaves at least below the inflorescence; corolla nearly regular;
               stamens not conspicuously exserted at anthesis, if at all   -4. Veronica
            5. Stems scapelike, the cauline leaves bractlike and alternate; corolla absent or when present
               definitely irregular; stamens conspicuously exserted at anthesis   -5. Besseya
   2. Anther-bearing stamens 4
      6. Corolla with narrow spur at base   -6. Linaria
      6. Corolla without a spur
         7. Stamens 5, of these 4 anther-bearing and a fifth sterile and lacking an anther, sometimes much
            reduced in size; leaves opposite or whorled
            8. Annual plants; corolla not over 6 mm. long, with middle lobe of lower lip deeply concave and
               enclosing the stamens; sterile stamen reduced, minute and glandlike   -7. Collinsia
            8. Perennial (or possibly biennial) plants; corolla over 6 mm. long, the middle lobe of lower lip
               not concave nor enclosing the stamens; sterile stamen scalelike or elongate
               9. Sterile stamen short and scalelike, nearly as wide as long; corolla 5-10 mm. long, greenish
                  to greenish-brown, not at all showy   -8. Scrophularia
               9. Sterile stamen slender and elongated, little if any shorter than the anther-bearing ones;
                  corolla usually over 10 mm. long, white to variously colored but not greenish to greenish-brown, almost always showy
                  10. Calyx deeply 5-parted; lower corolla lip without a palate; seeds without a cellular-
                      reticulate outer coat   -9. Penstemon

10. Calyx obscurely and shallowly 5-lobed; lower corolla lip with a bearded palate near base; seeds with a loose cellular-reticulated coat -10. Chionophila
7. Stamens 4, all anther-bearing, no reduced or rudimentary ones present; leaves opposite, basal or alternate
11. Anther sacs separate and dissimilar, outer 1 versatile, inner pendulous by its apex and mostly smaller (sometimes rudimentary); leaves alternate, often more or less dissected or lobed
12. Upper corolla lip very much longer than the lower; plants perennial (in all but 1 species); calyx tubular, cleft above and below, the lateral lobes usually toothed or cleft; flowers or bracts often red -11. Castilleja
12. Upper corolla lip little if any longer than the lower; plants annual; calyx with 4 equal or subequal lobes, or cleft to base on sides or appearing as 1 sepal; flowers and bracts purplish to yellowish to white, never red
13. Calyx gamosepalous with 4 equal or nearly equal lobes; leaves entire or 3-cleft -12. Orthocarpus
13. Calyx cleft to base on sides with 2 sepals, or only upper sepal present and not gamosepalous; leaves 3- to 5-parted -13. Cordylanthus
11. Anther sacs either similar in shape and attachment, or confluent, never as above; leaves usually opposite or basal and usually entire or toothed (may be alternate or dissected in Pedicularis)
14. Upper lip of corolla conspicuously arched upward in the middle (galea), the corolla conspicuously bilabiate; leaves toothed or variously lobed
15. Plants annual; leaves opposite and sharply dentate or serrate; calyx 4-toothed, becoming bladder-like and veiny, completely enclosing the fruit but not filled by it; fruit symmetrical, both cells dehiscing equally -14. Rhinanthus
15. Plants perennial; leaves usually alternate, either crenate or variously lobed and dissected; calyx cleft on lower and sometimes on upper side becoming distended but not bladderlike and completely enclosing the fruit; fruit asymmetrical, curved, opening chiefly or wholly on 1 side -15. Pedicularis
14. Upper lip of corolla not conspicuously arched, the corolla nearly regular to weakly bilabiate (except in some species of Mimulus); leaves entire or undulate (except in some species of Mimulus), (Note: Mimulus never has an arched upper corolla lip)
16. Anther cells wholly confluent, appearing as a single cell; capsule 1-celled above; corolla minute, 2-2.5 mm. long; plants subscapose -16. Limosella
16. Anther cells clearly 2 (may be somewhat confluent at apex only); capsule 2-celled throughout; corolla 5 mm. long or more; plants usually caulescent
17. Calyx parted nearly or quite to base; corolla white or yellowish, 5-7 mm. long -17. Bacopa
17. Calyx with sepals united over half their length into a tubular or campanulate tube, merely 5-toothed or lobed; corolla various but rarely less than 7 mm. long and then bright yellow
18. Corolla with lower lip external and overlapping in the bud, corolla rose-purple, not at all strongly 2-lipped; anther cells parallel; calyx not prismatic or strongly 5-ribbed; leaves linear and entire -18. Gerardia
18. Corolla with upper lip external and overlapping in the bud, the corolla usually yellow (may be red or blue but seldom rose-purple), often definitely 2-lipped; anther cells divergent; calyx prismatic with 5 longitudinal ribs or plaits; leaves various but seldom both linear and entire -19. Mimulus

1. Verbascum L. MULLEIN

Biennial or possibly perennial herbaceous plants; stems tall and strict, glandular-hairy (at least above) or with branched woolly hairs, leafy; leaves alternate, stem leaves sessile or decurrent, entire or toothed; flowers in spikes or spikelike racemes, yellow or white; calyx 5-parted; corolla rotate, 5-lobed, only slightly irregular, the upper lobe exterior in the bud; stamens 5, all fertile, exserted, some or all the filaments pilose; capsules globular to ovoid-oblong, septicidal, usually 2-cleft at apex; seeds longitudinally rugose, not winged.

1. Leaves green and glabrous; hairs, if present anywhere, simple with some glandular; inflorescence a lax interrupted raceme with pedicels definite; all filaments with violet hairs -1. V. blattaria
1. Leaves yellowish-green or whitish; stems and leaves woolly with branching non-glandular hairs; inflorescence of crowded flowers, spikelike; the 3 upper filaments yellow-hairy, 2 lower glabrous -2. V. thapsus

1. Verbascum blattaria L., Sp. Pl. 178. 1753.
Stems 40-120 cm. tall, slender (usually not over 1 cm. thick), glabrous below and glandular above with simple hairs; leaves of stem 2-12 cm. long, green, glabrous, dentate, incised or lobed, oblong to ovate, sessile or clasping; first year leaves in a rosette; inflorescence a lax, interrupted raceme up to 50 cm. long, pedicels becoming well over 1 cm. long; calyx lobes 5-6 mm. long, lanceolate, shorter than mature fruit; corolla 2.5-3 cm. wide or more, yellow to whitish; all filaments with knobbed purple hairs; capsules 7-8 mm. long; seeds 0.8-0.9 mm. long, dark gray.---Fields, roadsides, valleys and waste places. Naturalized from Europe and widespread in North America. Our few records from northcentral and eastern Colorado at 4000-5500 feet.
2. Verbascum thapsus L., Sp. Pl. 177. 1753.
Stems 30-200 cm. tall, very stout, usually over 1 cm. thick, densely woolly with branched nonglandular hairs, simple or with some erect branches; leaves of stem about 10-40 cm. long, elliptic to oblanceolate, decurrent, margins dentate or crenate, acutish, yellowish-green from woolly branching hairs; leaves of the first year in a close rosette, often very large; flowers in a dense spikelike inflorescence terminating the stems but sometimes with short branches present, often up to 30 cm. or more long; calyx lobes about 6-10 mm. long, lanceolate to ovate, about as long as the fruit; corolla about 1.5-2.5 cm. wide, yellow; 3 upper stamens hairy, 2 lower ones glabrous; capsule 6-10 mm. long, stellate; seeds 0.4-0.5 mm. long, cylindrical, truncate at both ends, brownish-gray. A related species, V. phlomoides L., with leaves not

decurrent, has been reported in Colorado and is to be expected in cultivated areas of the state.---Fields, roadsides, valleys and waste places. Naturalized from Europe and common in temperate North America. Our records scattered in Colorado at 4500-9000 feet.

2. Gratiola L. HEDGE HYSSOP

Annual herbaceous plants; leaves opposite, sessile, entire, to denticulate; flowers solitary on axillary peduncles; calyx deeply 5-parted, the divisions narrow, with a pair of bractlets closely subtending it and similar to its lobes; corolla tubular-funnelform, tube greenish-yellow, the lobes white, nearly regular and only indistinctly 2-lipped; fertile stamens 2, included, with anther sacs separated by a broad connective larger than the sacs; 2 sterile rudimentary stamens usually present; capsule 4-valved, both loculicidal and septicidal; seeds many, striate and reticulated.

1. Gratiola neglecta Torr., Cat. Pl. N. Y. 89. 1819.
    G. virginiana of Manuals for our plants at least---Plants glabrate or more or less glandular-puberulent; stems 7-20 cm. tall, simple or branched; leaves 1-5 cm. long, linear to oblong; some pedicels becoming over 10 mm. long, slender; sepals about 5 mm. long; corolla 8-12 mm. long on earlier flowers, with clavate hairs on upper side of throat; capsules ovoid, seeds about 0.5 mm. long.---Moist or wet ground. Quebec to British Columbia, south to Georgia, Texas and California. Our records scattered in the western two-thirds of Colorado at 4500-8000 feet.

3. Lindernia Allioni FALSE PIMPERNEL

Annual or biennial, caulescent herbaceous plants; leaves opposite, glabrous, entire or nearly so, sessile; flowers slender-peduncled, solitary in the axils, not bracteate near calyx; sepals 5, deeply 5-parted to distinct; corolla white to purplish, 2-lipped, upper lip shorter, erect, 2-cleft, lower lip spreading and 3-lobed; fertile stamens 2, anther sacs divergent; sterile stamens 2, each 2-lobed at apex; capsule septicidal; seeds wrinkled, not winged.

1. Lindernia anagallidea (Michx.) Pennell, Am. Acad. Nat. Sci. Phila. Mon. 1, p. 152. 1935.
    Ilysanthes inaequalis (Walt.) Pennell---Plants 5-20 cm. tall; stems slender; leaves 5-15 mm. long, obovate, oblong or ovate, rounded-clasping at base; corolla 4-8 mm. long.---Moist or wet ground as edges of streams or ponds. New Hampshire to Washington, south to Florida, Texas and California. Our few records from northcentral, southcentral and central Colorado at about 5000-7500 feet.

4. Veronica L. SPEEDWELL

Annual or perennial herbaceous caulescent terrestrial or aquatic plants; leaves opposite, or alternate above, sessile or short-petioled, entire or toothed; flowers small, solitary to racemose or spicate, terminal or axillary; calyx of 4 slightly united sepals (rarely 5); corolla only slightly irregular, rotate, 4-lobed; stamens 2, anther sacs confluent at summit; style slender, stigma capitate; capsule flattened, usually notched or 2-lobed at apex, loculicidal; seeds few to many.

1. Racemes all axillary, main stem never terminating in an inflorescence; leaves opposite throughout; plants aquatic or semiaquatic
    2. Leaves all short-petioled   -1. V. americana
    2. Leaves, at least the upper ones, sessile and clasping the stem
        3. Capsules much wider than long, strongly 2-lobed; sepals shorter than the capsules, equal; leaves linear to lanceolate; pedicels filiform, reflexing in fruit   -2. V. scutellata
        3. Capsules only moderately if at all wider than long, never strongly 2-lobed (may be notched at apex); sepals nearly if not as long as the capsules, slightly unequal; leaves lanceolate to oblong-ovate; pedicels rather stout, ascending-spreading
            4. Sepals acute to acuminate; capsules not evidently if at all notched; leaves oblong-ovate, mostly widest above the middle   -3. V. anagallis-aquatica
            4. Sepals obtuse to acutish; capsules evidently notched; leaves lanceolate, mostly widest below the middle   -4. V. salina
1. Racemes terminating the stem, or flowers solitary and axillary, never in axillary racemes; leaves opposite or alternate above; plants of wet places but seldom truly aquatic
    5. Perennial plants; only the upper leaf axils flower-bearing (the subtending leaves bractlike and reduced), the inflorescence appearing to form a definite raceme
        6. Capsules wider than long, notched at apex for about 1/4 the length; lower leaves petioled   -5. V. serpyllifolia
        6. Capsules longer than wide, merely emarginate at apex; lower leaves sessile   -6. V. wormskjoldii
    5. Annual plants; most leaf axils flower-bearing, no reduced bracts, inflorescence appearing to be of axillary flowers only
        7. Pedicels definitely longer than the sepals, often longer than the subtending leaves; corolla over 4 mm. wide   -7. V. persica
        7. Pedicels shorter than the sepals, much shorter than the leaves; corolla 2-3 mm. wide
            8. Leaves ovate or oval, petioled (or only the upper nearly sessile); hairs obscurely if at all glandular   -8. V. arvensis
            8. Leaves oblanceolate, spatulate or oblong, sessile (or very lowermost sometimes petioled); hairs usually conspicuously gland-tipped   -9. V. peregrina

1. Veronica americana Schwein. ex Benth. in DC., Prodr. 10:468. 1846.
V. americana crassula Rydb.---Perennial plants; stems 10-60 cm. long, erect or decumbent, branched or unbranched, glabrous; leaves 2-8 cm. long, all opposite, the blades lanceolate to ovate, widest nearest base, mostly acute or acutish near apex, glabrous; flowers in axillary racemes, these about 10- to 25-flowered, on pedicels 5-13 mm. long; sepals 2-3 mm. long, acute; corolla 4-5 mm. across, blue or nearly white; capsule about as wide as long, notched at apex.--- In water or very wet ground, often in springs or along streams. Widely distributed in North America. Our records widely scattered in the western two-thirds of Colorado at 5000-10,000 feet.
2. Veronica scutellata L., Sp. Pl. 12. 1753.
Perennial plants; stems 10-50 cm. long, ascending or decumbent, glabrous or sparingly hairy; leaves 1-7 cm. long, all opposite, the blades linear to lanceolate, entire or remotely denticulate, all sessile and acute, glabrous or nearly so; flowers in axillary racemes, these 5- to 20-flowered, the rachis zigzag and the pedicels spreading and several times longer than the fruit; sepals 1-2 mm. long, acute; corolla 4-8 mm. wide, blue to purple, possibly sometimes whitish; fruit 2-3 mm. long, wider than long and strongly 2-lobed, on reflexed filiform pedicels. ---Moist or very wet places. Newfoundland to Yukon, south to Virginia and California. Our few records from northcentral and northwestern Colorado at about 9500 feet.
3. Veronica anagallis-aquatica L., Sp. Pl. 12. 1753.
V. anagallis in Rydbs. Fl. Colo.---Perennial plants; stems 20-90 cm. long, glabrous or nearly so, branched, mostly decumbent; leaves 2-10 cm. long, mostly widest above the base, all opposite, oblong-ovate to orbicular, mostly serrate, usually glabrous; racemes about 30- to 60-flowered, all axillary; pedicels 5-8 mm. long, rather stout, usually ascending; sepals about 4-5 mm. long, mostly longer than the capsule, acute to acuminate; corolla about 4-5 mm. across, blue to blue-lilac; capsule 3-4 mm. long and about as wide, not notched or scarcely so at apex. Our plants tend to intergrade with V. salina Schur.---Moist or very wet places as in slow streams. Widely distributed in North America, South America and Eurasia. Our records scattered over the southern half of Colorado at 4000-5500 feet.
4. Veronica salina Schur., Enum. Pl. Transsil. 492. 1866.
V. catenata Pennell; V. connata Pennell, prob. not Raf. 1830; V. connata glaberrima Pennell---Perennial plants; stems 5-30 cm. long, glabrous; leaves 2-6 cm. long, all opposite, widest usually below the base, lanceolate, crenate to nearly entire; racemes about 15- to 30-flowered, all axillary; pedicels 3-6 mm. long, usually spreading, rather stout; sepals 2.5-3.5 mm. long, about as long or somewhat shorter than the capsule, obtuse to acutish; corolla about 3-5 mm. across, pale blue or white; capsule about 3 mm. long and about 3.5 mm. wide, decidedly emarginate at apex. Many of our plants are intergrades with V. anagallis-aquatica L.---Moist or wet places or in water. Widely distributed in the northern hemisphere. Our records scattered over Colorado at 3500-7500 feet.
5. Veronica serpyllifolia L., Sp. Pl. 12. 1753.
Perennial plants from rootstocks; stems creeping at base, 5-30 cm. tall, ascending above, glabrous to pubescent, the hairs mostly spreading; leaves 5-15 mm. long, opposite below, shading up into alternate bracts, lower ones petioled, ovate-oblong or oval, entire or crenulate; flowers in what appears to be a terminal raceme, from reduced leaves or bracts, lower leaves not bearing flowers; pedicels equal to or somewhat longer than the sepals; sepals about 2-5 mm. long; corolla about 3 mm. long and about 3-4 mm. wide, blue, rarely white; capsule 3-5 mm. long, wider than long, notched at apex for about 1/4 the length; seeds flat, numerous, smooth. Our plant has been called var. humifusa (Dickson) Vahl.---Meadows and open woodlands. Newfoundland to Alaska, south to South Carolina and Central America. Our records scattered in the western half of Colorado at 5500-11,500 feet.
6. Veronica wormskjoldii Roem. & Schult., Syst. Veg. 1:101. 1817.
Probably V. alpina in Manuals as to our plants---Perennial plants with rootstocks; stems 10-40 cm. tall, strict, glandular-villous, pubescent or glabrate especially below; leaves 1-3 cm. long, opposite at least to the inflorescence, sessile, ovate to oblong, entire to crenulate; flowers crowding out in what appears to be a terminal raceme with bracts subtending them, these different from the ordinary leaves in size; pedicels mostly 2-5 mm. long, little if any longer than the sepals; sepals 3-6 mm. long, glandular-villous; corolla about 4-6 mm. long and 4-6 mm. broad, dark blue, glabrous within; capsule about 4-6 mm. high, glandular-hairy, elliptic-obovate, longer than wide, merely emarginate at apex; seeds numerous, flattened. Our plants are closely related to V. alpina L., and may be the same.---Moist or wet meadows and draws. Greenland to Alaska, south to New Hampshire, New Mexico and California. Our records from the western half of Colorado at 8500-12,500 feet.
7. Veronica persica Poiret, Encyc. Meth. Bot. 8:542. 1808.
V. buxbaumii Tenore in our Manuals---Annual plants; stems diffusely branching, spreading or ascending, finely pubescent; leaves 8-15 mm. long, opposite especially below or alternate above, short-petioled, blades ovate, oval or suborbicular, coarsely serrate to nearly incised; flowers axillary from ordinary leaves, no reduced bracts above; pedicels longer than the sepals often longer than the subtending leaves; sepals 2-5 mm. long, narrowly ovate, enlarging in fruit; corolla 6-9 mm. wide the petals longer than the sepals, blue; capsule about 6-8 mm. wide and shorter than wide, the notch at apex wide, the 2 halves spreading, often acutish at apex, style about 1.5-2 mm. long; seeds over 1.3 mm. long, few, roughened, convex-arched.--- Fields, lawns, roadsides and waste places often in partial shade. Newfoundland to Alaska, south to Florida, Texas and Colorado; Europe. Our few records from northcentral and northwestern Colorado at 5000-8000 feet.
8. Veronica arvensis L., Sp. Pl. 13. 1753.
A weedy plant naturalized from Europe and now scattered in North America. This plant should be expected in Colorado, in fact it has been reported from this state.

9. Veronica peregrina L., Sp. Pl. 14. 1753.

V. xalapensis H.B.K.---Annual plants, glabrous or more or less glandular-pubescent throughout; stems 8-35 cm. tall, erect or ascending, often branched; leaves spatulate, oblanceolate to oblong, rather thick, lower leaves opposite, sometimes petioled, upper leaves alternate and strictly sessile, all entire to denticulate; flowers solitary and axillary, most leaf axils bearing flowers, hence no reduced bracts above; pedicels shorter than the sepals; sepals about 2-4 mm. long, lanceolate; corolla 2-3 mm. wide, whitish to bluish; capsule about 2-4 mm. long, strongly flattened, nearly orbicular, notched at apex, with minute style, glabrous to glandular hairy; seeds less than 1 mm. long, many, flat, smooth. All our plants seem to belong to var. xalapensis (H.B.K.) Pennell.---Banks, fields and gardens, usually in moist ground. Widespread in North and South America. Our records scattered in the western two-thirds of Colorado at 4500-8500 feet.

5. Besseya Rydb. KITTEN-TAILS

Perennial subscapose herbaceous plants; leaves alternate and mostly basal; stem leaves sessile, bractlike and smaller, basal blades cordate-ovate to oblong, toothed, short-petioled; flowers in terminal spikes or spikelike racemes, conspicuously bracted; sepals usually 4 and almost distinct but sometimes 1-3 and variously united; corolla wanting or when present irregular and 2-lipped, violet-purple, yellow or white, upper lip entire and concave, lower lip shorter and more or less 3-lobed; stamens 2, exserted, anther sacs parallel or nearly so, adnate to corolla or when latter is lacking inserted on a hypogynous disk; seeds several, flat.

1. Corolla lacking; filaments conspicuously colored; plants limited to extreme northern Colorado   -1. B. cinerea
1. Corolla present; filaments not very conspicuously colored; plants not limited to northern Colorado (if there at all)
   2. Corolla violet-purple, the upper lip with middle lobe about 1/2 as long as the whole lip; plants 5-15 cm. tall with 4-6 bractlike leaves below the inflorescence; alpine plants   -2. B. alpina
   2. Corolla white to yellow (or slightly purplish-tinged) the upper lip with middle lobe less than half as long as the lip; plants over 15 cm. tall, with over 6 bractlike leaves; plants seldom alpine
      3. Corolla yellow, lower lip with lobes not over 1/5 the length of the lip; lateral sepals united about 1/2 their length or more; plants of southwestern Colorado, west of the Continental Divide   -3. B. ritteriana
      3. Corolla white to pinkish or purplish-tinged, the lower lip with lobes over 1/2 the length of the lip; lateral sepals united less than 1/3 their length; plants east of the Continental Divide in Colorado   -4. B. plantaginea

1. Besseya cinerea (Raf.) Pennell, Proc. Acad. Nat. Sci. Phila. 85:104. 1933.

B. gymnocarpa (A. Nels.) Rydb.; B. wyomingensis (A. Nels.) Rydb.; Synthyris gymnocarpa (A. Nels.) Heller; S. wyomingensis (A. Nels.) Heller; S. wyomingensis gymnocarpa (A. Nels.) A. Nels.---Plants soft-pubescent or puberulent to glabrate; leaves 2-5 cm. long, basal oblong, elliptic to ovate, truncate to cuneate at base, crenate-serrate; scapes 6-30 cm. tall, several bractlike leaves present below the inflorescence; calyx 2-lobed, united anteriorly and placed on anterior side of capsule; corolla wanting; filaments conspicuously colored.---Hills and mountains, often on rocky slopes. South Dakota to Alberta, south to Colorado and Utah. Our records from northcentral Colorado (a doubtful one from Park County) at 5000-7500 feet.

2. Besseya alpina (Gray) Rydb., Bull. Torr. Club 30:279. 1903.

Synthyris alpina Gray---Plants more or less woolly at first, becoming glabrous; basal leaf blades 2-5 cm. long, cordate-ovate to elliptic, crenate-serrate; scapes 5-15 cm. long or rarely to 20 cm. with 4-6 bractlike leaves below the inflorescence; sepals lanceolate, white-villous especially near the edge; corolla about 6-8 mm. long, the middle lobe of lower lip about half as long as the lip, violet-purple; filaments not especially conspicuously colored.---High mountains. Wyoming and Utah to New Mexico. Our records scattered in the higher mountains of Colorado at 11,500-14,000 feet.

3. Besseya ritteriana (Eastw.) Rydb., Bull. Torr. Club 30:280. 1903.

B. reflexa (Eastw.) Rydb.; Synthyris reflexa Eastw.; S. ritteriana obtusa A. Nels.; S. flavescens A. Nels.---Plants more or less pubescent to glabrate; basal leaves 5-12 cm. long, elliptic to oblong, cuneate, crenate; scapes 20-30 cm. tall, occasionally shorter, with several to many bractlike leaves below the inflorescence; calyx with lateral lobes less than 1/2 the length of the calyx; lower lip of corolla with lobes only about 1/5 as long as the lip, lemon-yellow; filaments not especially conspicuously colored.---Mountains, often on moist banks or meadows. Limited to westcentral and southwestern Colorado, our records from 7000-11,500 feet.

4. Besseya plantaginea (James) Rydb., Bull. Torr. Club 30:280. 1903.

Synthyris plantaginea (James) Benth.---Plants more or less tomentose especially at first; basal leaves 5-15 cm. long, ovate to ovate-oblong, sometimes lanc-oblong, broadly to narrowly cuneate at base, crenate; scapes usually 20-40 cm. tall, but sometimes shorter, with several to many bractlike leaves below the inflorescence; sepals with lateral lobes united at base for less than 1/3 their length; corolla about 5-8 mm. long, the lower lip with lobes over 1/3 the length of the lip, white to purplish-tinged; filaments not especially conspicuously colored.---Hills and mountains, often on moist wooded slopes. Wyoming to New Mexico. Our records from northcentral, central and southcentral Colorado, east of the Continental Divide, at 6000-10,000 feet.

6. Linaria (Bauhin) Miller   TOADFLAX

Annual or perennial herbaceous plants; stems erect or nearly so, leafy, simple or sparingly branched; leaves entire, sessile, alternate or opposite, occasionally whorled; flowers rather showy in terminal racemes, these often paniculate; calyx of 5 partly united sepals; corolla strongly bilabiate, yellow or violet-purple, tube with a spur at base and throat with a prominent palate, upper lip 2-lobed, lower 3-lobed; stamens 4, in pairs, included; capsule opening near summit by pores or chinks; seeds numerous.

1. Flowers yellow (or partly orange), usually over 15 mm. long (excluding the spur); plants perennial
   2. Stem leaves cordate-clasping, upper ovate to ovate-lanceolate, some over 1 cm. wide   -1. L. dalmatica
   2. Stem leaves not at all clasping, linear or linear-lanceolate, not over 5 mm. wide   -2. L. vulgaris
1. Flowers whitish to violet, less than 15 mm. long; plants annual or biennial
   3. Seeds tuberculate; corolla over 10 mm. long, excluding the spur which is 5-9 mm. long   -3. L. texana
   3. Seeds smooth or very nearly so; corolla less than 10 mm. long, excluding the spur which is 2-6 mm. long
                                                                                                                                      -4. L. canadensis

1. **Linaria dalmatica** (L.) Mill., Gard. Dict. ed. 8. no. 13. 1768.

Perennial plants, apparently sometimes developing horizontal rootstocks; stems 80-120 cm. tall, robust, glabrous and more or less glaucous; leaves about 3-8 cm. long, ovate, ovate-lanceolate or even lanceolate below, sessile and cordate-clasping at base, alternate but crowded and sometimes appearing opposite; flowers yellow but often purplish-red at apex in the bud, in racemes; sepals 6-8 mm. long; corolla about 17-35 mm. long, exclusive of the spur which is about 13-20 mm. long; seeds reported to be wingless. Our plant is the small-flowered one that has been called var. macedonica Fanzl.---Roadsides especially near dwellings but spreading to valleys and sagebrush flats. Apparently escaping from cultivation in various parts of Colorado, our records from 5000-6500 feet.

2. **Linaria vulgaris** Mill., Gard. Dict. ed. 8, no. 1. 1768.

L. vulgaris Hill; L. linaria (L.) Karst.---Perennial plants with rootstocks, forming dense patches; stems 20-80 cm. tall, glabrous or puberulent above; leaves 2-7 cm. long, seldom over 3 mm. wide, linear or rarely linear-lanceolate, mostly alternate; flowers yellow with an orange palate, in rather dense racemes, these often paniculate-clustered; sepals about 3 mm. long; corolla about 15-20 mm. long or more, exclusive of spur which is about 10-20 mm. long; seeds rugose, winged.---Pastures, fields, roadsides and often near dwellings. Widely distributed in North America. Our records from the western half of Colorado at 6000-8500 feet.

3. **Linaria texana** Scheele, Linnaea 21:761. 1848.

L. canadensis of our Manuals---Plants annual or biennial; stems erect but prostrate shoots often present, glabrous; leaves of stems 1-3 cm. long, linear, scattered, those on shoots shorter and wider and usually opposite or whorled; flowers light blue to violet-purple (perhaps rarely white), in slender racemes; sepals 2-3 mm. long; corolla 10-13 mm. long, exclusive of spur which is 5-9 mm. long; seeds densely tuberculate, wingless.---Hills and rocky slopes and ravines. Missouri, South Dakota and Washington, south to Virginia and Mexico. Our records from northcentral and central Colorado at 5000-7500 feet.

4. **Linaria canadensis** (L.) DuMont de Cours, Bot. Cult. 2:96. 1802.

This plant of the eastern and also the western United States has not been recorded from Colorado but has been collected in western Kansas and western Nebraska.

7. Collinsia Nutt.   BLUE-EYED MARY

Annual caulescent herbaceous plants; leaves opposite or whorled; flowers solitary or whorled; calyx of 5 united, subequal sepals; corolla 2-lipped, white to pale blue; stamens 4 in 2 pairs, anther cells confluent at apex; staminodium glandlike, near base of corolla.

1. **Collinsia parviflora** Dougl. in Lindl., Bot. Reg. 13: pl. 1082. 1827.

C. tenella (Pursh) Piper---Stems 5-40 cm. tall, ascending or erect, glabrate to puberulent; leaves 0.7-5 cm. long, entire to serrulate, glabrate-puberulent to glandular, often purplish below, oblong to lanc-linear but lower ovate to oblong-ovate, petioled; flowers 2-5 in a whorl above, solitary below; pedicels 3-15 mm. long, puberulent to glandular; calyx 3-7 mm. long, somewhat membranous below, puberulent to glabrate, from 1/2 to the full length of corolla; corolla 4-6 mm. long, dorsal region gibbous, crested in throat; filaments rather stout, glabrous; stigma 2-lobed; capsule 3-4 mm. long, slightly exceeding the calyx lobes; seeds 1-2 mm. long, normally 4, reddish-brown.---Shaded hillsides and valleys. Ontario to British Columbia, south to Michigan, Colorado, Arizona and California. Our records from the western two-thirds of Colorado at 5000-9500 feet.

8. Scrophularia L.   FIGWORT

Caulescent, coarse perennial herbaceous plants; stems erect or nearly so, 4-angled; leaves opposite, petioled, toothed to incised, ovate; flowers in terminal, paniculate cymes; calyx with 5 rather short lobes; corolla 2-lipped, greenish to greenish-brown, the 2 upper lobes nearly erect, 2 lateral ones ascending and lower spreading or reflexed; stamens 5, only 4 bearing anthers, the fifth reduced to a scale on the roof of the corolla tube; stigma capitate or truncate; capsule septicidal; seeds rugose, many.

1. **Scrophularia lanceolata** Pursh, Fl. Amer. Sept. 419. 1814.

S. occidentalis (Rydb.) Bickn.; S. serrata Rydb.---Stems 50-200 cm. tall, more or less puberulent or glabrate below and glandular in the inflorescence; leaves 5-15 cm. long, cuneate to truncate, often with clusters of smaller leaves in the axils; inflorescence narrow; corolla 6-10 mm. long; sterile filament greenish-yellow, flabellate; capsule ovoid-conical.---Hills, valleys and plains. New England to Washington, south to Virginia, New Mexico and California. Our records well scattered over the western two-thirds of Colorado at 4500-9500 feet.

9. Penstemon Mitchell BEARDTONGUE

(Treatment contributed by C. Wm. T. Penland)

    Herbaceous or rarely suffrutescent perennial plants, usually erect and tufted, but occasionally low and creeping; leaves opposite, entire or toothed, the upper ones sessile and often clasping; calyx equally 5-parted; flowers usually showy, in open or contracted panicles; corolla tubular, usually ventricose, bilabiate, the upper lip 2-lobed, the lower lip 3-lobed; fertile stamens 4, in 2 pairs, their filaments arched; anthers 2-celled, the dehiscence longitudinal but sometimes only partial; fifth stamen represented by a conspicuous, sterile filament (the staminode) attached to the upper side of the corolla, and usually widened and bearded at the apex; style elongate, slender, with a capitate stigma; fruit a 2-celled, septicidal capsule with numerous angular seeds.

    The generic name, formerly more often spelled Pentstemon, alludes to the anomaly of the fifth stamen. In several of the species, the staminode has been induced to become fertile, that is, to produce anthers and pollen grains, by the early removal of the fertile stamens. The common name "Beardtongue" is applicable to all our species, except for a half-dozen or so which have glabrous staminodes.

    Included in the genus are a number of our most spectacular wild-flowers. The various species occupy different habitats all the way from the plains upward into alpine regions. P. unilateralis, P. strictus, P. virens and P. teucrioides have all been observed at times to form extensive, more or less solid sheets of color on disturbed soils of abandoned fields and elsewhere. The interesting P. ambiguus may be similarly abundant in its native habitat on the sandy plains. As a whole, the genus has supplied a few cultivated varieties, and seems worthy of further experimentation in this direction.*

1. Flowers bright red, usually scarlet
  2. Calyces and pedicels definitely glandular-pubescent; anther sacs dehiscent across their joined apices and distally for less than 1/2 their length   -1. P. bridgesii
  2. Calyces and pedicels glabrous or pubescent, but at most very obscurely glandular; anther sacs dehiscent from their free ends half-way or more toward their juncture
    3. Corolla strongly bilabiate, lower lip reflexed, upper projecting or spreading
      4. Anther sacs glabrous or very minutely pubescent
        5. Lower lip of corolla bearded inside with yellow hairs   -2. P. barbatus
        5. Lower lip of corolla glabrous inside or with very few white hairs   -2A. P. barbatus torreyi
      4. Anther sacs with few to many long hairs   -2B. P. barbatus trichander
    3. Corolla obscurely bilabiate, the lips about equally erect or spreading
      6. Stems glabrous; leaves glabrous or nearly so   -3. P. eatoni
      6. Stems and leaves puberulent   -3A. P. eatoni undosus
1. Flower white, pink, blue, lavender, or purple**
  7. Anther sacs pubescent along their sides (away from the line of dehiscence), often sparsely so
    8. Pubescence of anthers sparsely long-villous to lanate, the hairs usually much longer than the length of the anther sacs, and sometimes nearly hiding the anthers; anther sacs opening throughout their length or essentially so
      9. Corolla pale blue, the tube nearly as long as the throat; at least the lower peduncles elongate and somewhat divergent; lower stem leaves finely puberulent and somewhat glaucous; basal leaf blades commonly oval or oblong-spatulate   -4. P. comarrhenus
      9. Corolla deep blue (bluish-lilac in subspecies 5B), the tube much shorter than the throat; lower peduncles short and more or less appressed; the lower stem leaves essentially green and glabrous (except in subspecies 5A); basal leaf blades commonly linear-lanceolate to spatulate
        10. Sepals 3-5 mm. long, obtuse or acute (sometimes acuminate and then to 6 mm. long), with narrow scarious margins
          11. Basal leaves and stems glabrous, usually not glaucous; leaves mostly lanceolate, occasionally linear-lanceolate   -5. P. strictus
          11. Basal leaves and stems (at least near base) puberulent, often somewhat glaucous; leaves linear or linear-lanceolate   -5A. P. strictus angustus
        10. Sepals 6-9 mm. long, abruptly acuminate-tipped, with conspicuous scarious margins   -5B. P. strictus strictiformis
    8. Pubescence of anthers of short, straight, or sometimes flexuous hairs shorter than the length of the anther sacs, often sparse; anthers opening throughout or partially

---

*Acknowledgment is made for extended use of the several publications of Dr. F. W. Pennell, and Dr. D. D. Keck on this genus. This applies especially to out-of-state distributions, synonymy, and citations.

**Several of the blue-flowered species have repeatedly been observed to produce pink and albino individuals.

12. Stem and leaves densely short-pubescent, usually cinereous but often greenish; corolla limb not strongly bilabiate, the tube not or only slightly ventricose -6. P. fremontii
12. Stems and leaves usually glabrous and green (pubescent in P. alpinus forma riparius); corolla usually strongly bilabiate and ventricose
   13. Corolla 22-38 mm. long, with few to many long hairs within or (infrequently) glabrous
      14. Sepals 2-4 mm. long, the apex broadly rounded, with relatively short, acute tip or none -7. P. glaber
      14. Sepals 4-7 mm. long, usually ovate, and with an acuminate tip as long as or exceeding the body
         15. Staminode shallowly or scarcely notched at apex, usually bearded near apex and a short distance proximad with yellow hairs; corolla 24-32 mm. long; stem glabrous (except in f. riparius) -8. P. alpinus
         15. Staminode deeply notched at tip, glabrous, or with very few hairs at apex; corolla 30-38 mm. long; stem puberulent at least below -8A. P. alpinus brandegei
   13. Corolla 15-21 (-24 in P. saxosorum ?) mm. long, glabrous within throat
      16. Calyx strongly glandular-puberulent; corolla sparsely but definitely glandular externally -9. P. mensarum
      16. Calyx and corolla glabrous externally, or very obscurely glandular-puberulent
         17. Anther sacs opening throughout, the hairs sparse and much shorter than width of sacs; upper bracts, pedicels, and calyces sparsely glandular; inflorescence commonly 1/3 or less the height of plant -10. P. saxosorum
         17. Anther sacs opening partially; hairs frequently numerous and equalling or exceeding width of sacs; upper bracts, pedicels, and calyces essentially glabrous; inflorescence commonly more than 1/3 height of plant -11. P. cyanocaulis
7. Anther sacs glabrous along their sides (or only microscopically puberulent with high magnification)
   18. Leaves, at least some of them, more or less regularly toothed, though sometimes merely denticulate (variable species may also be found under next 18 in key)
      19. Corolla 37-50 mm. long -12. P. cobaea
      19. Corolla under 35 mm. long
         20. Corolla white (occasionally pale lavender), not long-bearded but densely glandular-puberulent within -13. P. albidus
         20. Corolla not white (lavender, lilac, purplish, deep blue, or very dark purple), lond-bearded within but not, or only sparsely, glandular-puberulent
            21. Leaves glabrous or essentially so (in P. jamesii sometimes puberulent)
               22. Corolla under 17 mm. long, blue to violet-blue -14. P. virens
               22. Corolla 17 mm. or over, other-colored
                  23. Calyx 4-6 mm. long -15. P. gracilis
                  23. Calyx 7-15 mm. long
                     24. Corolla with two longitudinal grooves on under side of tube*; plant of forested regions of state -16. P. whippleanus
                     24. Corolla not 2-grooved on lower side, round in cross-section; plant of dry plains only in southeastern section of state -17. P. jamesii
            21. Leaves densely puberulent and usually canescently villous as well -18. P. eriantherus
   18. Leaves all entire, or only remotely undulate-toothed
      25. Leaves all narrowly linear or narrowly oblanceolate, not over 5 mm. wide, and mostly under 20 mm. long (but up to 40 mm. long in 22B)
         26. Staminode glabrous, much shorter than the corolla tube -19. P. ambiguus
         26. Staminode bearded, as long as or longer than the corolla tube
            27. Stems glabrous, or only minutely puberulent; corolla glabrous outside, white, greenish-white, or purple
               28. Corolla purple -20. P. laricifolius
               28. Corolla white, greenish-white, or lightly pink-tinged outside -20A. P. laricifolius exilifolius
            27. Stems markedly pubescent; corolla glandular-pubescent outside, blue, violet-blue, or purplish
               29. Corolla rounded below (not 2-ridged within on lower side), the throat abruptly and strongly enlarged above tube proper
                  30. Staminode densely long-bearded most of its length -21. P. linarioides
                  30. Staminode sparsely short-bearded, hairs mostly in an apical tuft -21A. P. linarioides coloradoensis
               29. Corolla flattened beneath and 2-ridged within on lower side, the throat not abruptly and strongly enlarged above tube proper
                  31. Calyx lobes usually with prominent scarious margins; leaves glabrous at least apically
                     32. Leaves narrowly oblanceolate to elliptic, obovate, or spatulate
                        33. Stems ascending-erect; leaves gradually tapering to base -22. P. crandallii
                     33. Stems decumbent; leaves mostly abruptly petiolate -22A. P. crandallii procumbens
                  32. Leaves linear -22B. P. crandallii glabrescens
                  31. Calyx lobes usually without prominent scarious margins; leaves from cinereous-pubescent to greenish puberulent, or (some specimens of P. caespitosus) glabrate

---

*With dried specimens, this important character is best determined by boiling the flower and then floating it out in a vessel of water.

34. Leaves essentially linear, usually not over 1.5 mm. wide -23. P. teucrioides
34. Leaves narrowly oblanceolate or spatulate, more than 1.5 mm. wide
    35. Calyx-tube 2-3 mm. long; leaves pubescent throughout, the pubescence coarse and somewhat spreading -24. P. retrorsus
    35. Calyx-tube barely 1 mm. long; leaves pubescent throughout, or blades essentially glabrous, the pubescence fine and apparently appressed -25. P. caespitosus
25. Leaves various, but some at least other than linear, and regularly over 5 mm. wide and 20 mm. long
  36. Stems prostrate, or decumbent-ascending; matlike plants of alpine habitats -26. P. harbourii
  36. Stems erect
    37. Stamens well-exserted from corolla -27. P. cyathophorus
    37. Stamens included by the corolla, or at most barely visible in side view
      38. Plants pubescent in some degree, either throughout or at least in the inflorescence (stems, bracts, pedicels)
        39. Corolla glabrous on outside; rest of inflorescence glabrous or pubescent
          40. Calyx 2.5-3.5 mm. long; flowers somewhat congested but individual pedicels evident; basal rosette leaves absent -28. P. watsoni
          40. Calyx 3-6 mm. long; flowers in dense fascicles, with individual pedicels obscured; basal rosette leaves present
            41. Corolla 6-10 mm. long; flowers usually declined; scarious margins of sepals not conspicuous -29. P. procerus
            41. Corolla 10-16 mm. long; flowers usually horizontal; scarious margins of sepals usually conspicuous (except in 30A)
              42. Sepals relatively wide, with abrupt, slender acumination, the scarious margins broad and strongly erose -30. P. rydbergii
              42. Sepals relatively narrow, acuminate-attenuate, the scarious margins narrow and slightly or scarcely erose -30A. P. rydbergii aggregatus
        39. Corolla distinctly glandular-pubescent outside; rest of inflorescence also glandular-pubescent
          43. Corolla white (infrequently pale lavender), with violet guide lines; plant of eastern plains -13. P. albidus
          43. Corolla not white, if whitish, then plants not of eastern plains
            44. Leaves of lower stem wholly glabrous, bright or dark green
              45. Corolla 10-17 mm. long, blue to violet-blue; calyx less than 5 mm. long -14. P. virens
              45. Corolla 17-28 mm. long, color variable but not clear blue (dull white to very dark purple); calyx 7-15 mm. long -16. P. whippleanus
            44. Leaves of lower stem either pubescent throughout or at least on veins and margins near base, usually dull or grayish green
              46. Corolla somewhat flattened beneath, with 2 strongly marked longitudinal grooves on under side and 2 corresponding ridges within
                47. Basal rosette leaves absent; plants only near northern border of state -31. P. radicosus
                47. Basal rosette usually well-developed; plants only in southcentral part of state -32. P. oliganthus
              46. Corolla tube more or less rounded, without marked grooves and ridges
                48. Staminode very prominently exserted, bearded with hairs reaching 5 mm. or longer; corolla mostly 24-36 mm. long; stems and leaves puberulent and also villous-canescent -18. P. eriantherus
                48. Staminode included, or if exserted, the hairs not over 3 mm. long; corolla 9-24 mm. long; stems and leaves puberulent in some degree, but not additionally villous or canescent
                  49. Corolla 8-14 mm. long
                    50. Calyx 3-5 mm. long; leaves lanceolate, to oblanceolate-spatulate; plants only of northwestern part of state -33. P. humilis
                    50. Calyx over 5 mm. long; leaves linear to linear-lanceolate; plants of southwestern part of state -34. P. parviflorus
                  49. Corolla 14-24 mm. long
                    51. Anther sacs not dehiscing throughout their entire length; basal leaves linear to linear-lanceolate; plants of southeastern plains -35. P. auriberbis
                    51. Anther sacs dehiscing throughout; basal leaves narrowly to broadly oblanceolate or spatulate; plants of southwestern part of state
                      52. Anther sacs widely divaricate but not explanate; staminode mostly included
                          53. Basal leaves broadly oblanceolate to spatulate -36. P. moffatii
                          53. Basal leaves narrowly oblanceolate -36A. P. moffatii paysonii
                      52. Anther sacs widely divaricate and explanate; staminode usually exserted -17A. P. jamesii ophianthus
      38. Plants glabrous throughout (but some specimens of P. angustifolius scabrid-pubescent)
        54. Staminode not strongly widened toward apex, glabrous or bearded; leaves dark green (at least not strongly glaucous) and thin
          55. Corolla 6-16 mm. long
            56. Corolla 11-16 mm. long; calyx 2.0-3.5 mm. long -28. P. watsoni

56. Corolla 6-10 mm. long; calyx 3-6 mm. long   -29. P. procerus
55. Corolla 17-30 mm. long
   57. Plants 10-25 cm. tall; of alpine habitats, 10,000 feet or above; corolla immediately enlarged above calyx, with scarcely any tube proper   -37. P. hallii
   57. Plants 20-120 cm. tall; of lower elevations, 6,000-10,000 feet; corolla tube gradually enlarged above calyx
      58. Staminode glabrous or nearly so; calyx 3-5 mm. long; anther sacs about 2 mm. long, with broad line of contact; plants very common along foothills and mountains of eastern slope
                                                                                             -38. P. unilateralis
      58. Staminode bearded with short, flat hairs distally; calyx 5-7 mm. long; anther sacs about 3 mm. long, with short line of contact; plants known only from type collection in Jackson County
                                                                                            -8B. P. alpinus magnus
54. Staminode strongly widened toward apex, mostly with prominent dense bearding as well; leaves firm-thickened or fleshy-thickened, glaucous
   59. Corolla 35 mm. or more long; plants of southeastern part of state   -39. P. grandiflorus
   59. Corolla under 35 mm. long; plants widely distributed in Colorado
      60. Basal leaves broadly oblanceolate to obovate-spatulate or orbicular, usually broader than those of the upper stem; plants of western slope
         61. Inflorescence more or less secund, peduncles elongate, flowers not congested in the fascicles
                                                                                                  -40. P. lentus
         61. Inflorescence not secund, peduncles short, flowers in congested fascicles -41. P. osterhoutii
      60. Basal leaves linear to narrowly- or somewhat broadly-oblanceolate, usually narrower than those of the upper stem; plants mainly of eastern slope, but 42 only in Moffat County, and 43A rarely represented in Montezuma County
         62. Sepals lanceolate to lanc-acuminate; inflorescence congested, commonly not secund; corolla glabrous within or occasionally sparsely bearded, prevailingly cerulean or deeper blue in color, but may be pinkish or lavender
            63. Corolla 10-14 (-15) mm. long, lobes of upper lip larger than those of the lower or subequal; plants only in extreme northwestern part of state   -42. P. arenicola
            63. Corolla 15-20 mm. long, lobes of upper lip markedly smaller than those of the lower; plants of the eastern slope or in extreme southwestern part of state
               64. Bracts narrowly ovate-lanceolate, mostly tapering gradually to the tip
                                                                              -43. P. angustifolius
               64. Bracts broadly ovate, mostly tapering abruptly to an acute tip or a caudate acumination
                                                                            -43A. P. angustifolius caudatus
         62. Sepals ovate to ovate-lanceolate, often with an acute or short-acuminate tip; corolla bearded within, sometimes sparsely; inflorescence rather open and usually strongly secund; corolla pinkish-purple but often purplish-blue   -44. P. secundiflorus

1. Penstemon bridgesii A. Gray, Proc. Am. Acad. Arts andSci. 7:379. 1868.
P. rostriflorus Kellogg; P. bridgesii rostriflorus Schelle---Stems 29-60 cm. tall, few to several, often branched above, from a somewhat woody base, mostly glabrous below, glandular-pubescent above; leaves entire, glabrous, the basal linear-oblanceolate or narrowly elliptic, 4-8 mm. wide, 2-6 cm. long, the upper similar but with shorter petiolelike bases, and often longer; inflorescence rather few-flowered and open; calyx glandular-pubescent, 3-5 mm. long, the lobes ovate to ovate-lanceolate, with acute or short acuminate tip, and without a scarious margin; corolla scarlet-red, 22-30 mm. long, glandular-pubescent outside and somewhat so within, the tube widened but not ventricose, strongly bilabiate, the upper lip erect-projecting, the lower lip reflexed; anthers essentially glabrous, the line of dehiscence extending across the juncture of the sacs and for less than halfway along their sides, distal portion of sacs remaining closed; staminode glabrous, slender, slightly enlarged distally.---Southwestern Colorado, southwestward to Lower California. In Colorado, only in Montezuma and Dolores Counties, on rocky or sandy areas, commonly with scrub oak, pinyon, and junipers; 6000-8000 feet.

2. Penstemon barbatus (Cav.) Roth., Cat. Bot. 3:49. 1806.
Chelone barbata Cav.---No Colorado collections of the species are known. Description of the following subspecies applies here, except that the species has the corolla bearded within the throat with woolly, yellow hairs, instead of being glabrous. The range of the species is from Mexico, northward as far as southern Utah.

2A. Penstemon barbatus torreyi (Benth.) Keck, (ssp.) Journ. Wash. Acad. Sci. 29:491. 1939.
P. torreyi Benth.; P. barbatus var. torreyi (Benth.) A. Gray---Stems 40-130 cm. tall, solitary or few, slender, glabrous or slightly puberulent below; leaves entire, the basal puberulent to glabrate, oblong-oblanceolate or spatulate, 1-4 cm. wide, 4-10 cm. long, often reddish especially beneath, the cauline linear to lanceolate, glabrous, 3-7 mm. wide, 4-12 cm. long; inflorescence elongate, loosely flowered; calyx glabrous to obscurely glandular, 3-6 mm. long, the lobes ovate with acute or short-acuminate tip and narrow, scarious margins; corolla scarlet, glabrous throughout but sometimes with few white hairs within at base of lower lip, 26-36 mm. long, the tube enlarged upward but scarcely ventricose, the limb strongly bilabiate, lobes of upper lip little separated, erect-projecting and inclosing stamens, the lower lip mostly reflexed; anthers glabrous, the sacs opening partially, widely joined at their bases; staminode glabrous, slender, only slightly enlarged distally.---Southwestern and southcentral Colorado to northern New Mexico and Arizona. In Colorado, from El Paso County west to Garfield County and southward, on open, rocky or gravelly soils of brushy or wooded slopes of the lower mountains on both sides of the Continental Divide; 6000-9600 feet.

2B. Penstemon barbatus trichander (A. Gray) Keck, (ssp.) Journ. Wash. Acad. Sci. 29:491. 1939.
P. barbatus var. trichander A. Gray; P. trichander (A. Gray) Rydb.---Differs from the two preceding in that the anthers are sparsely to moderately lanate-villous. Southwestern Colorado and adjoining Utah, Arizona, and New Mexico. On dry, sandy or gravelly soils of hillsides and mesas; 5500-9000 feet.

3. **Penstemon eatoni** A. Gray, Proc. Amer. Acad. Arts & Sci. 8:395. 1872.

Stems 30-60 cm. tall, solitary or few, simple, erect, essentially glabrous; leaves mostly entire, essentially glabrous, the basal oblanceolate, spatulate or oblong-elliptic, 1-4 cm. wide, 4-10 cm. long, the cauline sessile, ovate to cordate-clasping, 1.5-4 cm. wide, 3-8 cm. long; inflorescence narrow, the flowers somewhat crowded in fascicles; calyx sparsely puberulent or glabrous, 3-6 mm. long, the lobes ovate-lanceolate, acute to short acuminate, with narrow scarious margin; corolla scarlet-red, 23-29 mm. long, glabrous throughout, tubular, the throat a slight and gradual enlargement from the tube proper, obscurely bilabiate, the lobes nearly equal and equally erect to slightly spreading; anthers barely included by corolla, minutely puberulent to glabrate, the sacs opening from their free ends only part way toward their juncture; staminode included, glabrous to minutely puberulent at apex, slender, very slightly widened upward.---Southwestern Colorado to central Arizona and California. In Colorado, apparently rare, only in Montezuma County, in open areas or among junipers on rocky slopes; 5500-7000 (?) feet.

3A. **Penstemon eatoni undosus** (M. E. Jones) Keck, (ssp.) Journ. Wash. Acad. Sci. 29:491. 1939.

P. eatoni var. undosus M. E. Jones; P. coccinatus Rydb.---Differs from the species in having stems and leaves finely puberulent instead of glabrous. Southwestern Colorado to northern Arizona and southern Utah. This is the common representative of the species in Colorado, from Montrose to Montezuma Counties, on rocky, dry hills and slopes of the juniper-pinyon association; 5500-7000 feet.

4. **Penstemon comarrhenus** A. Gray, Proc. Am. Acad. Arts & Sci. 12:81. 1876.

Stems 40-100 cm. tall, few to several, puberulent to glabrate; leaves entire, somewhat glaucous, the basal oblong-oblanceolate, obtuse, tapering to a more or less winged petiole, usually several in a rosette, puberulent, 1-2.5 cm. wide, 5-12 cm. long, the cauline linear-oblanceolate to lanc-linear, attenuate, sessile but narrowed some above base, 3-6 mm. wide, 3-10 cm. long; inflorescence much elongated, typically loosely flowered at full anthesis, somewhat secund, ultimate bracts minute; calyx glabrous or obscurely puberulent, 4-5 mm. long, the lobes ovate, obtuse to abruptly acute-tipped, with prominent erose-scarious margins; corolla glabrous throughout, usually pale blue, frequently purplish-blue, 30-35 (-38) mm. long, the tube narrow, 1/3 or more of the length of the corolla and much paler, definitely ventricose above middle on lower side, ampliate also above, definitely bilabiate, the upper lip projecting-erect, the lower spreading-reflexed; anthers villous-woolly, the sacs opening almost throughout; staminode glabrous or with a few hairs near apex, widened upward.---Southwestern Colorado to Utah and northeastern Arizona. In Colorado rather infrequent, from Garfield, south to Montezuma and Archuleta Counties, on mesas and gravelly or sandy sagebrush slopes; 5200-9000 feet.

5. **Penstemon strictus** Benth. in DC., Prodr. 10:324. 1846.

Stems 20-70 (-100) cm. tall, solitary or many, mostly strict or rather stout, glabrous or obscurely puberulent below, occasionally somewhat glaucous; leaves entire, glabrous, the basal narrowly oblanceolate to more broadly so, petioled, mostly acute, 0.5-1.5 (-2) cm. wide, 3-15 cm. long, the cauline sessile, oblanceolate to lanceolate, ovate-lanceolate, or lanc-linear, attenuate, clasping, usually under 1.5 cm. wide and under 10 cm. long; inflorescence elongate, slender, mostly secund, leafy-bracted only below, branches and flowers ascending-erect; calyx glabrous, 3-5 (-6) mm. long, the lobes ovate or oblong-ovate, rounded or with short, acute tip, scarious-margined; corolla deep blue, tube and lower throat often much lighter, sometimes violet, 22-30 mm. long, glabrous throughout, tube often elongate but not as long as the ventricose-rounded throat, definitely bilabiate, upper lip arched-spreading, the lower spreading-reflexed; anthers somewhat lanate to rather sparsely bearded with long flexuous hairs longer than the width of the sac, the sacs opening nearly throughout but not to the line of juncture; staminode widened upward and slightly notched at apex, glabrous to more commonly, sparsely long-bearded. Noteworthy are some specimens from the southern part of the state which have longer sepals than those farther north (reaching 6 mm., whereas in the northern part of the state the sepals rarely surpass 4 mm.). In these forms the sepals are lanc-acuminate, but the body is narrower than in subspecies 5B; otherwise they agree with the species as described. Further, one collection (Mineral County, 10-11,000 feet) is only 10 cm. tall and with but 3-5 flowers per stem, simulating P. hallii.---Southern Wyoming to northern New Mexico, northeastern Arizona, and Utah. In Colorado, very common across the whole western one-half of the state; apparently very rare on the eastern slope except in the northcentral and southcentral counties. On dry, chiefly gravelly, rocky, or sandy-loam soils of mesas, bluffs, hillsides, and mountain slopes with sagebrush, other scrub or timber; 6000-11,000 feet.

5A. **Penstemon strictus angustus** Pennell, (ssp.) Contr. U. S. Nat. Herb. 20:356. 1920.

Differs from the species in having basal leaves and at least the lower part of the stem puberulent, and more commonly glaucous, while the corolla may reach 32 mm. in length. The narrow leaves are often matched by specimens of the species, and the latter may also show a small amount of pubescence on the lower stem, leaf petioles and margins. The often longer corolla tube and pubescent basal parts suggest a close relation to P. comarrhenus as well.---Colorado specimens recognized here range from Mesa south to Montezuma, and east to Archuleta Counties. In rocky soil with sagebrush, scrub oaks, and pines; 6400-9000 feet. Also reported from adjacent New Mexico, Arizona and Utah.

5B. **Penstemon strictus strictiformis** (Rydb.) Keck, (ssp.) Journ. Wash. Acad. Sci. 29:490. 1939.

P. strictiformis Rydb.---Differs from the species in having longer (6-10 mm.), broader sepals. The somewhat acuminate tip is shorter than the broadly ovate, or ovate-lanceolate body (about 4 mm. wide, instead of the usual 2-3 mm. for the species). A sepal length of 7-10 mm. is usually stated, but few of ours measure as much as 9 mm. The subspecies is also believed to be marked by fuller, paler corollas (in those seen fresh, bluish-lilac to pale violet); by staminodes which are always moderately or sparsely bearded with long hairs; and by more glaucous herbage and sepals than in the species.---As circumscribed above, the subspecies in Colorado is probably limited to adobe, shaly, or partly sandy soils, chiefly of the juniper-pinyon association, in Montezuma, La Plata, Archuleta, and Mineral (?) Counties; 6500-8000 feet.

6. Penstemon fremontii T. & G. in A. Gray, Proc. Amer. Acad. Arts & Sci. 6:60. 1862.
   P. glaber var. fremontii M. E. Jones---Stems 10-45 cm. tall, 1 to several, erect, rather stoutish, mostly densely puberulent throughout, sometimes glabrate in upper inflorescence; leaves entire, mostly densely cinereous-puberulent, but often sparsely so and green, the lower spatulate to oblong-oblanceolate, 1-2.5 cm. wide, 3-10 cm. long, with mostly winged petioles, the cauline broadly linear to ovate-lanceolate, and partly clasping, 1-2 cm. wide, 3-5 cm. long; inflorescence cylindrical, mostly congested, leafy-bracted below; calyx minutely puberulent to essentially glabrous, 4-5 mm. long, the lobes ovate to ovate-lanceolate, acute to short-acuminate, margin scarious and somewhat erose; corolla dark bluish-purple to violet-purple, 17-22 mm. long, glabrous throughout or sparsely and minutely puberulent outside, the throat nearly tubular to somewhat ventricose-ampliate, the limb not markedly bilabiate, the lobes subequal; anthers pubescent with rather sparse, slender, stiffish hairs shorter than the width of the sac, the sacs opening throughout but scarcely to the line of juncture; staminode included, widened upward, sparsely bearded most of its length.---Northwestern Colorado and adjacent Wyoming and Utah. In Colorado, in Routt, Moffat, Rio Blanco, and Garfield Counties, in clay, or sandy-clay loams of banks, knolls, or draws; 5500-6500 feet.
7. Penstemon glaber Pursh, Fl. Amer. Sept. 738. 1814.
   P. gordoni Hook.; P. erianthus Nutt.---As recognized, this species has apparently not been collected in Colorado. The essential difference between it and the following lies in the calyx as defined in the key. The range of P. glaber is from North Dakota to western Nebraska and eastern Wyoming.
8. Penstemon alpinus Torr., Ann. Lyc. N. Y. 1:35. 1824.
   Chelone alpina Spreng.; P. glaber var. alpinus A. Gray; P. riparius A. Nels.---Stems 10-80 cm. tall, few to many, stout, mostly assurgent, glabrous to puberulent; leaves entire, glabrous to puberulent, often rather glaucous and thickish, exceedingly variable in size through the altitudinal range, the basal mostly from poorly-defined petioles, oblanceolate, obtuse to acute, up to 10 cm. long, the cauline mostly much broader, lanceolate to cordate-clasping, acute to acuminate, 5-13 cm. long; inflorescence congested, secund, leafy-bracted at base; calyx glabrous, 5-10 mm. long, the lobes ovate or ovate-lanceolate, with prominent erose, scarious margin and mostly abrupt-acuminate tip equalling or exceeding the body; corolla deep blue or bluish-purple (sometimes pale), mostly 22-32 mm. long (-38 mm. in subspecies 8A), glabrous outside, usually sparsely white-bearded within, ventricose below, the lips, especially the lower, sharply divergent or deflexed; anthers with sparse, short hairs, sacs opening nearly throughout but not across their juncture; staminode included, widened toward the apex which is more or less notched, bearded near apex with sparse or moderately dense hairs, or glabrous.---Southeastern Wyoming, south to northern New Mexico. In Colorado, quite common, primarily in gravelly soils of disturbed areas, along the eastern slope from the base of the mountains to 11,000 feet. Puberulent to pubescent specimens are referred to forma riparius (A. Nels.) Pennell.
8A. Penstemon alpinus brandegei (Porter) n. comb. (ssp.)
   P. brandegei var. cyananthus Porter; P. brandegei Porter ex Rydb.---This entity is doubtfully maintained on the basis of the key characters given. In addition, the stems are always robust, 30-70 cm. tall; the stem leaves are broadly cordate-clasping, up to 3 cm. wide and often over 10 cm. long. There appears to be much intergradation with the species. As recognized, the subspecies ranges from southcentral Colorado to northern New Mexico in the same habitats as the species; 5500-7500 feet.
8B. Penstemon alpinus magnus (Pennell) n. comb. (ssp.)
   P. magnus Pennell---Stems 30-50 cm. tall, stoutish, assurgent, glabrous; leaves entire, glabrous or with very sparse pubescence, glaucous (?), rounded, the lower oblanceolate, blade and petiole scarcely distinguishable, about 1.5 cm. wide, 6-12 cm. long, the upper lanceolate with broadly attached base, acute or acuminate; inflorescence somewhat crowded, secund, leafy-bracted below; calyx glabrous, 5-7 mm. long, the lobes ovate to oblong-ovate, with scarious, erose margins and barely acute or short-acuminate tip; corolla blue (?), 25-30 mm. long, glabrous outside, the throat ventricose-widened from slender throat; anthers essentially glabrous, the sacs not opening throughout; staminode included, widened distally and there bearded with short, flat hairs.---Collected but once, in "low open grounds," near Teller, Jackson County at 8000 feet. Only additional collections from the type locality will establish certain status of this entity, but it seems from the evident close relationship to P. alpinus to warrant no more than subspecific rank. As described above, the chief differences from P. alpinus are the glabrous anthers and the longer body of the calyx lobes with correspondingly shorter acuminations. In addition, the anther sacs appear to be about 3 mm. long as contrasted with 2-2.5 mm. for the species.
9. Penstemon mensarum Pennell, Contr. U. S. Nat. Herb. 20:380. 1920.
   Stems 40-100 cm. tall, 1 to few, erect, glabrous up to the glandular inflorescence; leaves entire, glabrous, the basal with long elliptic, to oblong-or lanc-elliptic blades, narrowed to nearly discrete petioles, mostly acute, 1-1.5 cm. wide, 6-20 cm. long, the cauline sessile, oblanceolate or oblong-lanceolate to ovate-lanceolate, acute or attenuate, cordate-clasping above; inflorescence elongate, slender, the rather short branches appressed; calyx glandular-pubescent, 3-4 (-5) mm. long, the lobes ovate-lanceolate or oblong-ovate, acute, with no or very narrow, scarious margin; corolla dark blue or blue tinged with purple, 14-18 (-20) mm. long, glandular-pubescent externally, glabrous within, the throat rounded and moderately ventricose below, not abruptly inflated, narrowed some toward orifice, the lips divergent-spreading, the lower longer; anthers sparsely short-hairy, the hairs straight, not as long as the width of sac, the sacs opening partially (nearly to line of juncture); staminode included, gradually widened upward, somewhat notched at apex, bearded most of its length, rather densely so at apex.---Apparently only in Mesa and Delta (?) Counties in Colorado, among oaks or aspens; 7200-9500 feet.

10. **Penstemon saxosorum** Pennell, Contr. U. S. Nat. Herb. 20:349. 1920.
Stems 10-60 (-80) cm. tall, usually several, erect, glabrous, or minutely glandular in the inflorescence; leaves entire, glabrous, the basal from linear-oblanceolate to broader, 0.4-1.5 cm. wide, 3-14 cm. long, the cauline narrowly to broadly oblanceolate to lanceolate, the upper ovate-lanceolate and somewhat broadly clasping; inflorescence elongate, narrow and congested, not leafy-bracted or only below, pedicels and upper bracts sparsely glandular; calyx usually very obscurely glandular, 5-7 (-10) mm. long, the lobes ovate, with long acuminate tips and prominent, somewhat erose, scarious margin; corolla deep blue to purplish-blue, 18-24 mm. long (some, "smaller than typical" down to 16 mm. long), very obscurely glandular near base of tube, glabrous within, the throat rounded and inflated below, but not abruptly, and scarcely (or not ?) narrowed to the orifice, definitely bilabiate, the upper lip projecting-erect, the lower widely spreading; anthers bearded sparsely with hairs shorter than the width of the sac, the sacs opening almost throughout but not across the juncture; staminode widening upward, the apex retuse, bearded near the apex with numerous hairs, and then sparsely hairy for about 1/2 the length.---Northern Colorado and some adjacent counties of southern Wyoming. In Colorado, from Larimer, west to Moffatt Counties, of both continental slopes, on gravelly or rocky hillsides in sagebrush or open timber; 8200-9500 feet. The representative collections of this species are somewhat heterogeneous, the delimitation from P. subglaber, P. mensarum (larger flowered specimens), and P. cyanocaulis, difficult. It is probable that wider collections may result in a quite different interpretation than that given here.

11. **Penstemon cyanocaulis** Payson, Bot. Gaz. 60:380. 1915.
Stems 15-60 cm. tall, few to many, erect or somewhat assurgent, glabrous throughout or obscurely puberulent; leaves entire, or somewhat crenate-undulate, glabrous or slightly puberulent, the basal spatulate to oblanceolate, mostly obtuse, 1-2.5 cm. wide, 3-9 cm. long, the cauline sessile, oblong-oblanceolate to lanceolate, with rounded or acute tip, mostly longer than the internodes, 1-1.5 cm. wide, 3-7 cm. long; inflorescence elongate, 1/3 to more than 1/2 the height of the stem, leafy-bracted only below; calyx essentially glabrous, 4-6 mm. long, the lobes ovate or ovate-lanceolate, with acute or short-acuminate tip and scarious margin; corolla deep blue to blue-violet, 15-21 mm. long, glabrous or obscurely glandular outside, glabrous within, definitely and somewhat ventricosely inflated, but not constricted toward orifice, the limb not markedly bilabiate, somewhat oblique, lobes sub-equal, both lips spreading-divergent, the upper more so; anthers sparsely or moderately densely hairy with mostly straight, slender hairs, shorter than the length of the sacs, the sacs opening partially; staminode widened upward, bearded chiefly at the apex with sparse to rather dense hairs.---Westcentral Colorado to adjacent Utah. In Colorado, Mesa and Montrose Counties, on dry, rocky hills and in gulches of open or brushy country; 5300-9000 feet.

12. **Penstemon cobaea** Nutt., Trans. Amer. Phil. Soc. N.S.5:182. 1837.
P. hansonii A. Nels.; P. cobaea ssp. typicus Pennell---Stems 30-60 cm. tall, few to several, stout, puberulent below, glandular-pubescent above; leaves toothed (at least the upper), the basal petioled, oblanceolate, mostly glabrous, 6-12 cm. long, the cauline sessile, oblong- to ovate-lanceolate or cordate-clasping, usually pubescent, 1.5-4 cm. wide, 5-10 cm. long; inflorescence rather short, with few to many flowers, open; calyx glandular-pubescent, 10-14 mm. long, the lobes ovate or oblong-lanceolate, acute, without an appreciable scarious margin; corolla purplish to white, 40-50 mm. long, glandular-pubescent outside, glandular but not villous within, the tube abruptly and widely inflated, the limb scarcely bilabiate, its lobes rounded and spreading; anthers glabrous, opening throughout and explanate; staminode exserted, sparsely bearded near apex.---Nebraska to Oklahoma and Texas. In Colorado, one collection known, and this from Baca County along a highway.

13. **Penstemon albidus** Nutt., Gen. Pl. N. Am. 2:53. 1818.
P. teretiflorus Nutt.; Chelone albida Spreng.; P. viscidulus Nees---Stems 15-40 cm. tall, solitary to several, erect, puberulent below, glandular-pubescent above; leaves entire or more commonly some at least serrate-denticulate, usually puberulent, the basal oblanceolate with poorly defined petioles, rounded or acute, 0.4-2 cm. wide, 3-9 cm. long, the cauline sessile, lanceolate to ovate-lanceolate, mostly acute, 0.4-1.5 cm. wide, 3-7 cm. long; inflorescence elongate, of few to several, mostly distinct, rather compact or loose fascicles; calyx glandular-pubescent, 5-8 mm. long, the lobes lanceolate; corolla white with violet guide-lines, infrequently pale lilac throughout, 15-20 mm. long, glandular-pubescent outside and inside, tube and throat slender, the latter gradually widening, the lips divergent-spreading; anthers glabrous, sacs explanate; staminode included, somewhat widened toward apex, moderately bearded for about 1/2 its length.---Alberta to Manitoba, and southward to Oklahoma, Texas and New Mexico. In Colorado, very common on hills and sandy plains from the base of the mountains to the eastern border; 3500-7000 feet.

14. **Penstemon virens** Pennell ap. Rydb., Fl. Rocky Mts. 773, 1066. 1917.
Stems 10-35 (-40) cm. tall, few to many, slender, erect, forming matlike clumps, pubescent in lines below, glandular-pubescent above; leaves entire or sometimes denticulate, mostly bright and often shiny green, glabrous, basal rosette leaves mostly oblanceolate to elliptic, acute, petioled, 0.5-1.5 cm. wide, 3-10 cm. long, the cauline lanceolate to ovate-lanceolate, somewhat acuminate, the upper often subcordate and clasping, 0.5-1 cm. wide, 2-4 cm. long; inflorescence rather narrow, leafy-bracted only at base, the few to several fascicles well separated but each somewhat densely flowered; calyx glandular-pubescent, 2.5-4.5 mm. long, the lobes ovate to lanc-ovate, acute or acuminate, narrowly scarious-margined; corolla light to dark blue, often with a mixture of violet, 10-15 (-18 rarely) mm. long, glandular-pubescent outside, lightly bearded within, the throat widened laterally, only slightly ampliate, not strongly 2-ridged within on lower side, definitely bilabiate, the upper lip erect-recurved, the lower spreading and exceeding the upper; anthers glabrous, opening throughout, not explanate; staminode slender, not widening upward, bearded densely only near apex, a few sparse hairs proximad of the apex.---Colorado and adjacent southcentral Wyoming. In Colorado, quite common across

the mountainous central part along the eastern slope (apparently only in Grand County of the western slope). On gravelly and rocky, wooded or brushy slopes of lower and middle mountains; 5300-10,000 feet.

15. Penstemon gracilis Nutt., Gen. N. Amer. Pl. 2:52. 1818.

Chelone gracilis Spreng.; P. glaucus R. Grah.; P. digitalis var. glaucus Trautv., and P. digitalis var. gracilis Trautv.; P. pubescens var. gracilis A. Gray---Stems 25-55 cm. tall, solitary or few, glabrate to puberulent below, glandular-pubescent above; leaves more or less regularly and finely toothed or nearly entire, glabrous or nearly so, the basal oval to oblanceolate, obtuse, 0.5-2 cm. wide, 3-8 cm. long, the cauline linear-lanceolate, attenuate and somewhat clasping, 0.5-1 cm. wide, 3-6 cm. long; inflorescence narrow, rather open to congested, the lower branches elongated and appressed; calyx glandular-pubescent, 4-6 mm. long, the lobes ovate to ovate-lanceolate, acuminate, with narrow, scarious margin; corolla pale lavender or pale violet, 15-23 mm. long, glanular-pubescent outside, bearded within, the tube rather slender, but more or less ampliate, 2-ridged within on lower side, the limb definitely bilabiate, the lower lip reflexed and projecting beyond the upper; anthers glabrous, opening throughout and across the juncture, not explanate; staminode barely exserted, widened upward, densely bearded for most its length.---From Alberta south to New Mexico, and eastward to Ontario and Wisconsin. In Colorado, rather infrequent, in meadowlands, and with scrub oaks and pines at the base of the mountains on the eastern slope; 5300-7500 feet.

16. Penstemon whippleanus A. Gray, Proc. Amer. Acad. Arts & Sci. 6:73. 1862.

P. glaucus var. stenosepalus A. Gray; P. arizonicus Heller; P. stenosepalus Howell; P. puberulus Woot. & Standl.; P. metcalfei Woot. & Standl.; P. pallescens Osterh.---Stems 10-50 (-70) cm. tall, usually many in clumps from woody root-crown, slender to stout, glabrous below or sometimes minutely puberulent, glandular-pubescent above; leaves entire, or the basal and sometimes the upper denticulate, dark green, glabrous, the basal oval, ovate, elliptic, or broadly oblong-ovate, acute, mostly tapering abruptly into well-defined petioles 1 to several times length of blade, 1-3 cm. wide, 1-8 (-15) cm. long, the cauline mostly oblong to lanc-acuminate, with cordate-clasping base, frequently caudate-curved, 1-3 cm. wide, 3-8 cm. long; inflorescence somewhat capitate with few flowers (alpine forms), or, more commonly, with 2-5 densely-flowered fascicles, the lowermost well separated; calyx glandular-pubescent, 7-15 mm. long, the lobes lanceolate, long-attenuate, margins not scarious; corolla prevailingly light wine-purple to almost black-purple in color, but other shades common, including a dull white, or this with admixture of yellow, green, brown, or dull dark blue, 17-28 mm. long, glandular-pubescent outside, bearded rather sparsely within on lower lip, or sometimes glabrous, the tube rather abruptly inflated to throat, but then not much further enlarged, strongly bilabiate, the upper lip somewhat arched-spreading, the lower projecting-spreading much beyond the upper; anthers glabrous, the sacs opening throughout, explanate; staminode well exserted, widening upward slightly, bearded with a tuft of hairs near the apex, or occasionally glabrous. ---Southeastern Idaho, western and southern Wyoming, through Utah and Colorado, to Arizona and New Mexico. In Colorado, quite common throughout the mountains from timberline down to ponderosa pine. In loamy, gravelly soils of meadows, valleys, and wooded slopes; 8000-12,500 feet.

17. Penstemon jamesii Benth., in DC. Prod. 10:325. 1846.

P. similis A. Nels.; P. jamesii ssp. typicus Keck---Stems 10-45 cm. tall, few, erect, puberulent below (ours), glandular-pubescent above; leaves entire to regularly serrate, glabrous or puberulent, those below mostly narrowly oblanceolate, 0.5-1 cm. wide, 2-8 cm. long, those above linear-lanceolate to lanceolate, 0.5-1.5 cm. wide, 2-10 cm. long; inflorescence elongate, many-flowered, somewhat crowded, secund; calyx glandular-pubescent, 8-12 mm. long, the lobes lanceolate or ovate-lanceolate, attenuate, scarious margined only at base; corolla "orchid" or "blue-lavender" (not seen fresh), 25-32 mm. long, glandular-pubescent outside, prominently long white-bearded within, the throat abruptly inflated-ventricose, limb strongly bilabiate, the upper lip projecting-erect, the lower spreading-reflexed; anthers glabrous, dehiscent throughout, explanate; staminode only slightly widened upward, prominently tufted-bearded apically with tortuous, spreading hairs, and near the middle with a dense group of straight, parallel, backwardly projecting hairs.---Southeastern Colorado to eastern New Mexico and southwestern Texas. In Colorado, apparently only in Las Animas County, "In field. Frequent here." About 5700 feet.

17A. Penstemon jamesii ophianthus (Pennell) Keck, (ssp.) Bull. Torr. Club 65:240. 1938.

P. ophianthus Pennell; P. pilosigulatus A. Nels.; P. jamesii ssp. breviculus Keck---The subspecies may be distinguished from the species by the corolla length, which is 14-22 mm. in the former, 25-32 mm. in the latter. In addition, the Colorado specimens have entire, or only remotely repand leaves. The subspecies ranges from western Colorado into adjacent New Mexico, Arizona and Utah. In Colorado, from San Miguel to Montezuma Counties, on dry, rocky slopes of the juniper-pinyon belt; 5300-6700 feet. A number of sheets show corollas running from 14-16 mm. in length, which was one of the basic characters for separating ssp. breviculus Keck. Since recent collections show intermediate length in the corolla, the subspecies is not maintained here.

18. Penstemon eriantherus Pursh, Fl. Amer. Sept. 2:737. 1814.

P. cristatus Nutt.; Chelone eriantherea Steud.; Chelone cristata Spreng.; P. saliens Rydb.; P. eriantherus ssp. saliens (Rydb.) Pennell---Stems 10-40 cm. tall, solitary to few, erect or assurgent-erect, puberulent and viscid-villous throughout; leaves entire to remotely or regularly and saliently serrate, densely puberulent to canescent-villous, the basal spatulate to oblanceolate, 1-2 cm. wide, 2-8 cm. long, the cauline sessile, lanceolate to oblong and somewhat clasping, 0.5-2.5 cm. wide, 3-5 cm. long; inflorescence dense and leafy-bracted; calyx markedly viscid-pubescent, 9-11 mm. long, the lobes lanc-acuminate, without a scarious margin; corolla purplish, 20-35 mm. long, glandular-pubescent outside, scarcely so within, long-bearded within on lower lip, the tube abruptly inflated to a wide throat, limb strongly bilabiate, the upper lip arched-projecting, the lower spreading or reflexed; anthers glabrous, the sacs dehiscent throughout, widely contiguous, explanate; staminode slightly widened toward apex, conspicuously exserted and bearded with long yellow hairs.---North Dakota to Nebraska and northern Colorado, thence northwestward to western Washington, British Columbia and Alberta. In Colorado, only in northern Larimer and Weld Counties, on plains and rocky hills; 5000-7500 feet.

19. Penstemon ambiguus Torr., Ann. Lyc. N. Y. 2:228. 1828.
Leiostemon purpureus Raf.; P. ambiguus var. foliosus Benth.; Leiostemon ambiguus Greene---Stems 20-50 cm. tall, diffusely and slenderly branched from a woody base, glabrous or scabrid-puberulent; leaves filiform, about 1 mm. wide, or the lower linear and up to 3 mm. wide, mostly 1.5-3 cm. long, glabrous to scabrid-puberulent; inflorescence loose but whole plant very floriferous; calyx glabrous, 3-4 mm. long, the lobes ovate to ovate-lanceolate with acute or acuminate tip and scarious margin; corolla white, with a striking sheen on the inside (upper) surface of the lobes, pink to dull rose on tube and reverse of lobes, color imprint of contiguous overlapping lobes often visible on back of lobes when fresh, 15-25 mm. long, the tube slender, decurved and scarcely widened into a throat, pubescent within, the limb scarcely bilabiate, but Phlox-like, oblique on tube, facing upward; anthers glabrous, explanate, on filaments shorter than the corolla tube; staminode included, filiform, glabrous.---Colorado and Kansas southward to Mexico, thence northward to Utah and Nevada. Mostly infrequent, except locally, on sand hills and other sandy areas of the eastern one-third of Colorado; 3500-6000 feet.

20. Penstemon laricifolius Hook. et Arn., Bot. Beech. Suppl. 376. 1840.
P. exilifolius var. desertus A. Nels.; P. laricifolius ssp. typicus Keck---Typical P. laricifolius is apparently found only in Wyoming, but possibly also in Idaho. Its corolla is longer than in the following subspecies, and is purple in color.

20A. Penstemon laricifolius exilifolius (A. Nels.) Keck, (ssp.) Bull. Torr. Club 64:381. 1937.
P. exilifolius A. Nels.; P. laricifolius var. exilifolius (A. Nels.) Payson---Stems 10-20 cm. tall, many, erect from branching, woody, root-crown, forming cushions, minutely puberulent to glabrous; leaves filiform, involute, essentially glabrous, the basal in dense tufts, 15-25 mm. long, the cauline similar in shape and size; inflorescence narrow, few- to many-flowered, scarcely dense; calyx glabrous, 4-6 mm. long, the lobes ovate-lanceolate, attenuate, prominently scarious-margined at base; corolla white, greenish-white, or sometimes light rose outside, 11-15 mm. long, glabrous outside, bearded within on lower lip, evidently bilabiate, the throat rounded below (not 2-ridged within), somewhat abruptly inflated, lips mostly spreading; anthers glabrous, the sacs opening throughout and through the line of juncture, not explanate; staminode included or barely exserted, widened upward, bearded strongly 1/3 to 1/2 its length.---Albany County in southeastern Wyoming to Larimer County in northcentral Colorado. Dry, rocky plains and hills; 7000-9000 feet.

21. Penstemon linarioides A. Gray, in Torr., Bot. Mex. Bound. 112. 1859.
P. linarioides ssp. typicus Keck---Only the following one of several subspecies is to be found within the state. The species occurs in New Mexico and Arizona. It is mostly taller than the subspecies below and has the staminode more densely bearded with longer hairs over most of its length.

21A. Penstemon linarioides coloradoensis (A. Nels.) Keck, (ssp.) Bull. Torr. Club 64:375. 1937.
P. coloradoensis A. Nels.---Stems 15-25 (-36) cm. tall, many, erect, simple or branched from base, from an often stout and elongate, woody caudex, puberulent throughout; leaves linear or linear-lanceolate, often crowded toward base, this appearance often enhanced by additional sterile shoots, puberulent throughout, cinereous to grayish-glabrate, mucronate, rather uniform in size from top to base of stem, the upper somewhat reduced, mostly 1-2.5 mm. wide, 1-2 (-3) cm. long; inflorescence narrow, elongate, secund, scarcely crowded, often apparently naked due to elongation of upper internodes; calyx glandular-puberulent, 5-7 mm. long, the lobes ovate-lanceolate, with acute or acuminate tips, scarious-margined at base; corolla reddish-lavender to bluish-purple, 14-17 mm. long, glandular-pubescent externally, lightly bearded within, the throat abruptly and strongly inflated from tube, campanulate (not folded or ridged below), the lips rather equally spreading; anthers glabrous, opening throughout and through the juncture, not explanate; staminode included, barely widened distally, bearded with a rather dense tuft of hairs near apex, very sparsely bearded proximally.---Southwestern Colorado to northwestern New Mexico. In Colorado, from San Miguel to Montezuma and Archuleta Counties, on dry sandy, or rocky slopes in sagebrush, chaparral, or woodland; 6200-8500 feet.

In addition to the above subspecies, Penstemon linarioides A.Gray may be represented in Colorado by Penstemon linarioides var. viridis Keck, Bull. Torr. Club 64:375. 1937. The variety differs from the subspecies in having essentially glabrous leaves, and in having elsewhere fine, erect, retrorsely-spreading hairs, instead of flattened, scalelike, closely appressed hairs. Found in Utah and Arizona, and doubtfully recognized from Archuleta County in Colorado; one specimen from an open, dry flat; 6600 feet.

22. Penstemon crandallii A. Nels., Bull. Torr. Club 26:354. 1899.
P. crandallii ssp. typicus Keck---Stems 8-15 cm. tall, many, ascending to erect, from slender or very coarse, woody, spreading rootstocks, puberulent; leaves linear-oblanceolate to spatulate, mucronate, mostly glabrate, at least toward the apex, sometimes more or less puberulent throughout, basal and cauline leaves scarcely distinguished, frequently crowded, 2-5 mm. wide, 10-30 mm. long; inflorescence glandular-puberulent, mostly leafy, flowers few or somewhat crowded, secund; calyx puberulent, 4-7 mm. long, the lobes ovate-lanceolate, long-attenuate, mostly with a prominent scarious margin on the body; corolla light to dark bluish-purple, 15-22 (-24) mm. long, glandular-pubescent outside, bearded within the throat, nearly tubular to ampliate above, 2-ridged within on the lower side, definitely bilabiate, the lower lip projecting-spreading beyond the erect upper lip; anthers glabrous, the sacs opening throughout and across the line of juncture, not explanate; staminode slender, scarcely enlarged upward, densely bearded its full length.---Central Colorado, southwestward into Utah. In Colorado, not uncommon on clay-loam soils of open, brushy hillsides and mountain slopes, from Teller, Gilpin, and Eagle Counties southwestward; 5900-10,000 feet.

22A. Penstemon crandallii procumbens (Greene) Keck, (ssp.) Bull. Torr. Club 64:369. 1937.

P. procumbens Greene; P. xylus A. Nels.---Differs from the species by characters given in the key. It is believed that the leaves mark a real separation, but that the decumbent stems are probably characteristic only of plants in the open and at higher altitudes. Only flowerless plants have been seen fresh. Apparently occurring only in Gunnison County in Colorado; 7300-9500 feet.

22B. Penstemon crandallii glabrescens (Pennell) Keck, (ssp.) Bull. Torr. Club 64:369. 1937.

P. glabrescens Pennell---Differs from the species in having the leaves narrowly linear instead of linear-oblanceolate, and the leaves are often more densely crowded. As recognized, the subspecies may also grow taller (up to 30 cm.); the leaves may be longer (up to 35 mm.); and the rootstock may be much larger (up to 1.5 cm. in diameter). Southwestern Colorado and adjacent New Mexico. On dry hills, and open areas among brush and trees; 5500-10,000 feet.

23. Penstemon teucrioides Greene, Pl. Baker. 3:23. 1901.

Stems 4-12 cm. tall, numerous, from creeping woody rootstocks, slender, ascending or erect, puberulent; leaves crowded, linear or very narrowly oblanceolate, involute, cinereous-puberulent, sharply mucronate, mostly about 1 mm. wide and 8-15 mm. long; inflorescence narrow, few to many-flowered, rather leafy; calyx viscid-puberulent, 3-5 mm. long, the lobes linear-lanceolate, with no, or scarcely any, scarious margin; corolla bluish or bluish-purple, 14-19 mm. long, glandular-pubescent outside, bearded within, the tube very slender, slightly enlarged upward, 2-ridged within on the lower side, strongly bilabiate, the lower lip spreading or reflexed, in the latter case surpassed by the scarcely spreading upper lip; anthers glabrous, dehiscent throughout, widely opened, not explanate; staminode slender throughout, bearded most of its length.---In Colorado only, in Park, Chaffee, Gunnison, Saguache, and Hinsdale Counties. On dry, clay, rocky or sandy loam soils of sagebrush slopes; 7200-10,000 (? 11,000) feet.

24. Penstemon retrorsus Payson ex Pennell, Contr. U. S. Nat. Herb. 20:373. 1920.

Stems 10-20 cm. tall, mostly erect, from woody rootstock, cinereous throughout; leaves cinereous with dense, retrorse pubescence, the basal crowded, oblanceolate or spatulate, 2-3 mm. wide, 8-15 mm. long, the upper seemingly reduced abruptly and bractlike in some specimens, in others much larger than the basal, oblanceolate, and up to 4 mm. wide and to 2.5-3.5 cm. long; inflorescence narrow, somewhat crowded or rather loose, not densely leafy, rather distinct from the more densely leafy lower portion of stem; calyx densely cinereous-pubescent, 3-5 mm. long, the tube nearly as long or longer than the lanc-linear lobes, which have no, or very inconspicuous, scarious margins; corolla bluish-purple (? not seen fresh), 15-20 mm. long, glandular-pubescent outside, moderately bearded within, nearly tubular, the throat flattened and 2-ridged within on the lower side, the 2 lips subequally projecting, or the lower somewhat longer and more spreading; anthers glabrous, opening throughout; staminode slender, bearded most of its length.---Apparently only in Montrose County, Colorado; adobe hills at 5700-6000 feet.

25. Penstemon caespitosus Nutt. ex A. Gray, Proc. Am. Acad. Arts & Sci. 6:66. 1862.

P. caespitosus ssp. typicus Keck---Stems not over 5 cm. tall, numerous, prostrate-assurgent, herbaceous, from creeping rootstocks, usually puberulent, forming mats up to 4 feet in diameter; leaves mostly linear-oblanceolate, but from almost linear to spatulate, 2-5 mm. broad, 7-20 mm. long, usually cinereous-puberulent throughout, but in some areas green and glabrate; inflorescence few-flowered, these with short branches, the bracts scarcely distinguished from the leaves; calyx puberulent, 5-7 mm. long, the lobes attenuate and with no or very narrow, scarious margins at base only; corolla light blue or purplish, 12-18 mm. long, more or less pubescent outside, sparsely villous within throat, the tube most noticeably inflated above, on the lower side somewhat flattened and 2-ridged within, strongly bilabiate, the upper lip erect and exceeded by the projecting-spreading lower lip; anthers glabrous, dehiscent throughout, essentially explanate; staminode slender, strongly bearded virtually its entire length.---Western Wyoming, northwestern Colorado, and northeastern Utah. In Colorado, very common in clay and rocky, clay-loam soils of sagebrush country; 6000-8800 feet.

Recently a species tending to key out to the above has been found in the northwestern corner of the state at 6500 feet. It is Penstemon acaulis Williams and differs from the other narrow-leaved species by having stems none or essentially so, all the leaves being in compact rosettes at the surface of the soil.

26. Penstemon harbourii A. Gray, Proc. Amer. Acad. Arts & Sci. 6:71. 1862.

P. bakeri Greene---Stems 5-15 cm. long, slender, assurgent to weakly erect, from creeping rootstocks, puberulent below, viscid-villous above; leaves entire, rather dullish green, retrorsely puberulent but not densely so, and often glabrate on the blades, no basal rosette, those of the lower stem spatulate to oblanceolate, somewhat petiolate, about 0.5 cm. wide, 1-2 cm. long, those of the upper stem ovate to lanc-ovate, gradually narrowed to an apparent petiole, or sessile to subsessile with a broad base, acute or obtuse; inflorescence rather short (2-4 cm. long) and leafy, mostly few-flowered and somewhat crowded and secund; calyx canescently viscid-pubescent, 6-9 mm. long, the lobes lanc-linear and without a scarious margin; corolla lilac-purple, 15-20 mm. long, glandular-pubescent externally, densely bearded within on lower lip, definitely bilabiate, the throat slender, 2-ridged within on lower side, the lips subequally projecting-spreading; anthers glabrous, opening throughout and across the juncture of the sacs, explanate; staminode widened distally and densely bearded nearly its entire length.---Only in Colorado, on scree and gravel slopes of high alpine summits across the state, from the Medicine Bow range in the north, south and southwestward to the La Plata and San Juan Mountains; 10,500-13,500 feet.

27. Penstemon cyathophorus Rydb., Bull. Torr. Club 31:643. 1905.

Stems 20-60 cm. tall, solitary or few, erect, stout, glabrous and glaucous; leaves entire, fleshy, glabrous and glaucous, the basal broadly oblanceolate or spatulate, obtuse or acute, with scarcely a proper petiole, 1-2 cm. wide, 2-6 cm. long, the cauline sessile, ovate or orbicular, with broadly attached bases, blades of opposite pairs meeting and overlapping around stem, mostly acute, gradually diminishing into the bracts; inflorescence cylindrical, crowded, the conspicuous bracts shorter than the flowers, as broad as or broader than long, with acute tip and more or less erose, scarious margins; calyx glabrous, 4-7 mm. long, the lobes lanceolate or ovate-lanceolate, attenuate, with conspicuous erose-scarious margins; corolla pale violet-blue, 9-15 mm. long, glabrous throughout, the tube funnelform, scarcely ventricose, the limb hardly bilabiate, the lobes subequal, lips slightly spreading; anthers glabrous, the sacs parallel, opening throughout, the filament attached opposite middle of length of sacs; staminode somewhat exserted, widened upward, densely bearded near apex.---Grand and Jackson Counties in northern Colorado and Carbon and Sweetwater Counties of adjacent southern Wyoming. Rocky, clay loam soils of sagebrush hills and flats; 7000-8500 feet.

28. Penstemon watsoni A. Gray, Syn. Fl. 2(1):267. 1878.
P. fremontii var. parryi A. Gray ap. Wats.; P. phlogifolius Greene; P. watsoni ssp. typicus Keck---Stems 30-70 cm. tall, usually many, forming clumps, erect from woody root-crown, puberulent to glabrate; leaves entire, green, mostly glabrous, basal rosette not developed, the lower cauline oblanceolate to lanceolate, short-petioled, acute, 0.5-1 cm. wide, upper cauline lanceolate to broadly ovate-lanceolate, cordate-clasping, acuminate, 1-3 cm. wide, 2-6 cm. long; inflorescence elongate, of few to many, loose to rather densely flowered fascicles, the lower on elongate but erect branches, lowermost fascicles in leaf axils, upper bracts strongly reduced; calyx glabrous, 2-3.5 mm. long, the lobes broadly triangular above the middle, rounded to base below, the tip acute or short acuminate, margins scarious and somewhat erose; corolla light blue, deep blue, or blue-purple with reddish throat, 11-16 mm. long, glabrous except for light bearding within on lower lip, the throat gradually dilated from tube, not strongly 2-ridged within, moderately bilabiate, lobes subequal, the upper lip arched-erect, the lower spreading-reflexed; anthers glabrous, the sacs opening throughout but not explanate; staminode not or just exserted, scarcely widened upward, bearded with long yellow hairs 1/3 to 1/2 its length, densely so near tip.---Southwestern Wyoming and northwestern Colorado to Idaho, Nevada, and northwestern Arizona. In Colorado, from Grand and Eagle Counties to the western border of the state from Moffat to Montrose Counties. On dry, gravelly or loamy slopes, with grass, sagebrush or other scrub; 7000-8500 feet.

29. Penstemon procerus Dougl. ex R. Grah., Edinb. N. Phil. Journ. 7:348. 1829.
P. micranthus Nutt.; Lepteiris parviflora Raf.; P. confertus var. violaceous Trautv.; P. confertus caeruleo-purpureus A. Gray; P. confertus var. procerus Cov.; P. procerus var. micranthus Jones; P. procerus ssp. pulvereus Pennell; P. procerus ssp. typicus Keck---Stems 10-30 (-45) cm. tall, few to many, erect or upcurved at base, minutely puberulent throughout or glabrate; leaves entire, mostly glabrous, but often puberulent below especially on petioles and midribs, basal rosette usually not well developed, the lower leaves narrowly oblanceolate or elliptic, petioled, 3-10 mm. wide, 2-6 cm. long, the upper oblanceolate, oblong-lanceolate, to narrowly lanceolate, clasping, 5-10 mm. wide, 2-5 cm. long; inflorescence capitate to elongate-cylindric, 1 to several fascicles of closely packed flowers, only lowermost fascicles leafy-bracted; calyx glabrous or minutely puberulent, 3-6 mm. long, the lobes oblong-ovate or obovate, long-attenuate, with erosulate to nearly entire scarious margins; corolla deep blue-purple, more or less declined, 6-10 mm. long, glabrous outside, rather densely bearded within, tube and throat narrow and scarcely separable, limb obscurely bilabiate, the lips spreading; anthers glabrous, the sacs opening throughout, essentially explanate; staminode included, slightly enlarged apically, with a tuft of yellow hairs near tip.---Alaska, southward through eastern Washington and Montana, to southcentral Colorado. In Colorado, rather common in the central mountainous portion as far south as Saguache County. Loamy soils of high meadows, or bordering timbered or brushy slopes; 8300-12,000 feet.

30. Penstemon rydbergii A. Nels., Bull. Torr. Club 25:281. 1898.
P. erosus Rydb.; P. lacerellus Greene; P. latiusculus Greene; P. rydbergii ssp. typicus Keck---Stems 15-60 cm. tall, few to many, slender to stoutish, erect or upcurved at base, glabrate to puberulent below, always pubescent in some degree above (either generally, or only in lines running downward from intervals between leaf-pairs), rather canescently-hirsute than finely puberulent; leaves entire, glabrous or minutely puberulent (mostly on petioles, midribs and margins of lower), the basal elliptic to oblanceolate, narrowed to mostly indistinct petioles, obtuse or acute, 0.5-1.5 cm. wide, 3-10 cm. long, the cauline oblanceolate to oblong-lanceolate, or ovate-lanceolate, some clasping, 0.5-2.5 cm. wide, 3-12 cm. long; inflorescence elongate, slender, of 2 to several dense fascicles, usually well separated at least below, only lowermost fascicle conspicuously leafy-bracted, bracts within fascicles with broad fimbriate-scarious margins; calyx glabrous, or occasionally pubescent (?), 4-7 mm. long, the lobes ovate-lanceolate, or roughly obovate-spatulate, with very broad erose-lacerate margins, and an abrupt acumination (appendage) shorter than the body; corolla dark violet-purple, 10-15 mm. long, mostly horizontal or ascending, glabrous outside, bearded within, the tube gradually widening upward, 2-ridged within on the lower side, limb not markedly bilabiate, the lobes divergent-spreading; anthers glabrous, dehiscent throughout and essentially explanate; staminode included, slightly widened distally, densely golden-bearded near the apex.---Southcentral Wyoming to New Mexico, Utah, and Arizona. In Colorado, across the mountainous western half of the state, in loamy soils of high meadows, aspen thickets, and grassy slopes; 7600-11,000 feet.

30A. Penstemon rydbergii aggregatus (Pennell) Keck, (ssp.) Am. Midl. Nat. 33:158. 1945.
P. aggregatus Pennell---The subspecies differs from the species in having the calyx lobes lanceolate or linear-lanceolate, with moderately or scarcely erose-scarious margins, and a relatively longer acumination, which seems more of a continuation of the body; the corolla appears to run lighter in color in the subspecies, and is somewhat longer (our specimens are 13-16 mm. long, but are stated to run 15-18 mm. long elsewhere). In addition, our specimens, as recognized, almost all have a pubescent calyx, which is very rarely true for the species. Our specimens are found only in the northcentral counties of the state at elevations and in habitats similar to the species. A recognition of the last three entities is not always satisfactory; particularly, specimens from Middle Park seem to show intergradations, for which hybridization may be accountable.

31. Penstemon radicosus A. Nels., Bull. Torr. Club 25:280. 1898.

P. lineolatus Greene---Stems 15-40 cm. tall, many, slender, erect, puberulent throughout, also somewhat glandular above, forming clumps from a woody root-crown; leaves entire, somewhat gray-green, puberulent throughout or glabrate on upper surface, no basal rosette, stem leaves linear-lanceolate to elliptic, or oblanceolate below, mostly acute, the lower with poorly defined petiole, the upper sessile, not often clasping, most of them 3-8 mm. wide, 2-4 cm. long; inflorescence rather short and compact, but scarcely dense; calyx glandular-pubescent, 4.5-6.5 mm. long, the lobes lanceolate or ovate-lanceolate, attenuate, or acuminate tipped, with scarious and somewhat erose margins; corolla pale blue, 16-21 mm. long, glandular-pubescent outside, bearded within below, whole corolla rather slender, the throat gradually enlarged, 2-ridged within on lower side, definitely bilabiate, the upper lip arched-reflexed, the lower spreading and projecting beyond the upper; anthers glabrous, opening throughout and through the juncture, not explanate; staminode included, slightly widened upward, bearded about half its length, densely so near apex.---Northcentral Colorado (North Park) through western Wyoming to Nevada, Idaho and Montana. The few Colorado collections known are from rocky, sagebrush areas of Jackson County; 8000-9000 feet.

32. Penstemon oliganthus Woot. & Standl., Contr. U. S. Nat. Herb. 16:172. 1913.

P. griffinii A. Nels.---Stems 20-50 cm. tall, rather few, slender, erect, puberulent at least below, glandular-pubescent in the inflorescence; leaves entire, puberulent or glabrous, basal rosette leaves well developed, these mostly oblanceolate to elliptic, rounded to acute, 5-10 mm. wide, 2-6 cm. long, the cauline linear-oblanceolate to lanc-linear, much reduced above; inflorescence elongate, narrow, few flowered, open, branches elongate but erect; calyx sparsely glandular-pubescent, 4-6 mm. long, the lobes ovate- or oblong-lanceolate, acuminate, with prominent erose, scarious margins; corolla pale blue or purplish, much lighter beneath and on tube, 15-25 mm. long, pubescent outside, densely golden-bearded within, tube and throat rather narrow, sometimes the latter abruptly inflated, throat 2-ridged within on lower side, definitely bilabiate, the lower lip projecting, exceeding the upper spreading or reflexed lip; anthers glabrous, opening throughout, not explanate; staminode not exserted, scarcely widening upward, densely bearded most its length.---Central Colorado to New Mexico and eastern Arizona. In Colorado, rather infrequent, in dry meadows or on grassy, rocky, wooded or brushy slopes, from Park County southwestward to Mineral and Conejos Counties; 7400-10,500 feet.

33. Penstemon humilis Nutt. ex A. Gray, Proc. Amer. Acad. Arts & Sci. 6:69. 1862.

P. collinus A. Nels.; P. puberulus Jones; P. brevis A. Nels.; P. humilis ssp. typicus Keck---Stems 10-30 cm. tall, numerous, slender, erect, in dense clumps usually, puberulent throughout, also glandular above; leaves entire (all our specimens), margins subrevolute, grayish or light green, puberulent throughout, the basal lanceolate to oblanceolate-spatulate, acute, petiole often definite, 0.5-1.5 cm. wide, 2-6 cm. long, the cauline oblanceolate, lanc-oblong, or lanc-linear, 0.5-0.7 cm. wide, 1.5-2.5 cm. long; inflorescence rather slender, of few mostly separated fascicles, scarcely densely flowered; calyx glandular-pubescent, 3-5 mm. long, the lobes ovate to ovate-lanceolate, acute or short acuminate, scarcely scarious-margined; corolla deep sky-blue to violet-blue, 9-14 mm. long, glandular-pubescent externally, sparingly bearded within on lower lip, the throat flattened beneath, scarcely 2-ridged within on lower side, definitely bilabiate, the lower lip projecting a little beyond the upper; anthers glabrous, opening throughout, explanate; staminode mostly included, not widened distally, densely bearded only at apex, then sparsely about 1/2 its length.---Western Wyoming to northwestern Colorado and westward to Idaho and California. In Colorado, only in Rio Blanco, Moffat and Routt (?) Counties, on gravelly or sandy sagebrush hills and slopes; 6000-9000 feet.

34. Penstemon parviflorus Pennell, Contr. U. S. Nat. Herb. 20:340. 1920.

Stems 15-20 cm. tall, cinereous-puberulent, above also glandular; leaves entire, those of the lower stem finely canescent, those of the upper stem glandular-pubescent, linear to linear-lanceolate, 3-5 mm. wide, 3-5 cm. long; inflorescence narrow, of several loose fascicles; calyx glandular-pubescent, the lobes 8 mm. long, linear-lanceolate; corolla purplish blue (?), 12 mm. long, glandular-puberulent externally, pubescent within at base of lower lip, the throat inflated and rounded, definitely bilabiate, the upper lip arched, the lower lip spreading; anthers minutely puberulent, the sacs opening almost throughout, not explanate; staminode included or barely exserted, widened distally, densely bearded nearly its entire length.---Known only from the type collection from Montezuma County. Apparently very infrequent as indicated by its absence from the many collections made near the type locality in recent years. Its nearest relative is stated to be P. auriberbis Pennell, of the eastern slope. (Plant not seen; the description is adapted from those of Dr. Pennell and Dr. Keck.)

35. Penstemon auriberbis Pennell, Contr. U. S. Nat. Herb. 20:339. 1920.

Stems 10-35 cm. tall, 1 to several, simple, erect or assurgent, puberulent below, glandular-pubescent above; leaves entire, or the upper not infrequently denticulate, mostly puberulent, sometimes glabrate, the basal linear to linear-lanceolate, often very numerous, mostly 2-5 mm. wide, 3-6 cm. long, the cauline mostly broader, sessile, with somewhat clasping base, long-attenuate, 3-7 mm. wide, 4-8 cm. long; inflorescence rather narrow, of few flowers and somewhat secund, or denser with separated fascicles; calyx glandular-pubescent, 6-9 mm. long, the lobes lanceolate-attenuate with narrow scarious margins; corolla lavender to purplish-blue, 16-24 mm. long, glandular-pubescent outside, long-bearded within the lower lip, the tube moderately to markedly inflated upward, the lips widely spreading; anthers glabrous, opening almost throughout (not quite to the line of juncture); staminode exserted, widened distally, densely bearded for most its length.---Eastcentral Colorado to northern New Mexico. Common in Colorado on dry, sandy, sandy-loam, or "hard" soils, from the Platte-Arkansas divide southward at the base of the mountains and on the plains to the southern border; 4500-7500 feet.

36. Penstemon moffatii Eastw., Zoe 4:9. 1893.

P. moffatii ssp. typicus Keck---Stems 10-30 cm. tall, solitary or few, erect or ascending, puberulent throughout, glandular above; leaves entire or somewhat undulate, mostly puberulent throughout, the main basal spatulate or broadly oblanceolate, the broader portion oval, ovate, or narrower, abruptly narrowed to a poorly defined petiole, the larger basal 4-8 cm. long, 1-2.5 cm. wide, the cauline in taller specimens cordate-clasping above, oblongish or oblanceolate but widened again at base and clasping, up to 1.5 cm. wide and 4.5 cm. long (in smaller specimens all dimensions reduced); inflorescence elongated, half or more height of stem, in larger specimens stoutish, branches elongate, leafy-bracted below; calyx glandular-pubescent, 5-8 mm. long, lobes oblong- or linear-lanceolate, mostly attenuate, acute and hardly scarious-margined; corolla blue to purple, 15-20 mm. long, moderately glandular-pubescent outside, sparsely bearded within, tube and throat slender, the widening moderate, not strongly bilabiate, lips subequally erect-spreading, lobes subequal; anthers glabrous (except for microscopic puberulence), the sacs opening throughout, not explanate; staminode included or barely exserted, widened upward somewhat, sparsely bearded for about 1/2 its length.---Westcentral Colorado and adjacent Utah. In Colorado, Garfield and Mesa Counties, on clay or sandy-clay soils of dry, rocky slopes and valleys, with scrub or juniper; 4300-5100 feet.

36A. Penstemon moffatii paysonii (Pennell) Keck, (ssp.) Bull. Torr. Club 65:243. 1938.
P. paysonii Pennell---Differs from the species by the characters given in the key. Due to the apparent intergradation between the two, it seems possible that additional collections may still further reduce this category. As recognized, the subspecies occurs only in Colorado, in Garfield, Mesa, and Montrose Counties. On dry, rocky, or sandy clay slopes and knolls, with brush or juniper; 4500-5800 feet.

37. Penstemon hallii A. Gray, Proc. Amer. Acad. Arts & Sci. 6:70. 1862.
Stems 10-20 cm. tall, usually several to many, erect, glabrous below, sparsely glandular-pubescent above; leaves entire, glabrous, all of them oblanceolate, often narrowly so, the upper occasionally linear-lanceolate, obtuse or acute, 3-8 mm. wide, 2-5 cm. long; inflorescence mostly rather capitate and few-flowered, but often elongate, slender and many-flowered, secund; calyx glandular-puberulent, 5-9 mm. long, the lobes broadly ovate with wide, scarious, erose margins, the tips rounded abruptly acute or acuminate; corolla bluish- to violet-purple, 18-24 mm. long, minutely glandular to glabrous outside, glabrous within, the throat abruptly inflated from almost no tube proper, moderately ventricose below, the lips almost equal and equally projecting, the lower often reflexed; anthers apparently glabrous, actually microscopically puberulent, with a few more evident hairs at juncture of sacs near summit of filament, the sacs opening throughout and across the line of juncture, not explanate; staminode gradually widened upward, often exserted, short-bearded throughout its length, densely so near apex.---In rocky, or gravelly soils, near and above timberline in the Colorado mountains from Gray's and Pike's Peaks, southwestward to the San Juan and La Plata Mountains; 10,000-13,000 feet.

38. Penstemon unilateralis Rydb., Bull. Torr. Bot. Club 33:150. 1906.
P. secundiflorus A. Gray, not Benth.---Stems 20-60 (-100) cm. tall, solitary or many, strict, or somewhat upcurved from base, slender to quite stout, glabrous throughout; leaves entire, glabrous, green to somewhat glaucous, the basal narrowly to more widely oblanceolate, petioles mostly ill-defined, rounded to acute, 0.3-1 cm. wide, 2-10 cm. long, the cauline often larger, mostly lanceolate but frequently linear-lanceolate to somewhat ovate-lanceolate and clasping, attenuate, acute, 0.4-1.8 cm. wide, 3-10 cm. long; inflorescence slender, elongate, secund, upper part often curved-nodding in bud, fascicles from few to very densely flowered, branches and flowers ascending-erect; calyx glabrous, 3-5 cm. long, the lobes ovate or oblong-ovate, with mostly rounded (to acute) tip, and narrow, scarious margin, which is subentire to erosulate; corolla predominantly blue or bluish, but often suffused with violet and streaked with violet guide lines, nearly pure pink or lavender-pink forms not uncommon, 17-26 mm. long (late flowers still shorter), glabrous throughout or bearded within with sparse, long hairs, the throat abruptly inflated, ventricose-rounded, strongly bilabiate, the upper lip arched-erect, the lower divergent-reflexed; anthers glabrous, the sacs opening throughout and across the juncture, not explanate; staminode widened distally, usually glabrous, occasionally with sparse, short hairs near apex.---Southeastern Wyoming to northern New Mexico. In Colorado, very common along the whole eastern slope. In parks, or on slopes of lower mountains and foothills, on gravelly or sandy loam soils, particularly of disturbed areas; 6000-10,000 feet.

39. Penstemon grandiflorus Nutt., Fraser's Catal. 2. 1813.
P. bradburii Pursh---Stems 60-120 cm. tall, stout, glabrous and somewhat glaucous, leaves entire, glabrous and glaucous, the lower obovate or broadly oblanceolate, narrowed into broad petioles, obtuse or acute, 1-4.5 cm. wide, 3-15 cm. long, the cauline broadly ovate, oval, or orbicular, cordate-clasping, 2-4 cm. wide, 2-4 cm. long; inflorescence narrow, leafy-bracted below, branches shorter than the bracts; calyx glabrous, 8-11 mm. long, the lobes ovate-lanceolate, with narrow, entire, scarious margins; corolla lavender-blue to sky-blue, 40-50 mm. long, glabrous throughout, the throat much inflated, rather abruptly so above, the lips somewhat spreading; anthers glabrous, the sacs opening throughout and across the juncture, not explanate; staminode much widened distally, densely short-bearded near apex.---Texas to Wyoming, North Dakota and Wisconsin. In Colorado, apparently very rare, in Baca County and "eastern Colorado", probably on sandy plains.

40. Penstemon lentus Pennell, Contr. U. S. Nat. Herb. 20:359. 1920.
Stems 20-30 cm. tall, few to several, erect, from branching, somewhat woody caudex, glabrous throughout; leaves entire, glabrous and glaucous, thickened-fleshy, the basal broadly oval, or ovate to oblanceolate-spatulate, rather abruptly narrowed to a poorly defined petiole, 1-3.5 cm. wide, 3-9 cm. long, obtuse or acute, the cauline sessile, ovate to ovate-lanceolate, clasping, mostly acute and mucronate, about 2 cm. wide, 2-4 cm. long; inflorescence elongate, 1/3 or more of stem length at full anthesis, the lowermost branches equalling or longer than the bracts, but appressed, secund, the fascicles not appearing distinct nor crowded; calyx glabrous, 5-7 (-9) mm. long, the lobes broadly ovate to ovate-lanceolate, with acute or acuminate tip and conspicuous scarious margins which may extend nearly to the tip; corolla dark blue, light purple to bluish-purple, 18-24 mm. long, glabrous except for sparse beard within on lower side, tube and throat about equal in length, the throat definitely enlarged but rather gradually so, rounded below, the lips spreading, the lower much larger; anthers glabrous, the sacs opening throughout

and through the juncture, not explanate; staminode not or only slightly exserted, much widened toward apex, densely to sparsely bearded along 2 marginal lines near the apex, then sparsely hairy proximally for nearly half its length.---Southwestern Colorado to southeastern Utah and northeastern Arizona. In Colorado, from Montrose to Montezuma Counties, in dry, clayey, or adobe soils of hills and mesas; 5800-7800 feet.

41. Penstemon osterhoutii Pennell, Contr. U. S. Nat. Herb. 20:358. 1920.

Stems 30-70 (-80) cm. tall, solitary to few, erect, stout, glabrous and glaucous; leaves entire, fleshy-thickened, glabrous and glaucous, the basal with ovate, or oblong-ovate to oblanceolate blades, narrowed to poorly defined petioles, rounded, acute or even emarginate, mucronate, 1-4 cm. wide, 3-12 cm. long, the cauline oval, oblong-ovate, or ovate-lanceolate, cordate-clasping, rounded or acute, mucronate, 1.5-2.5 cm. wide, 2-5 cm. long; inflorescence elongate, 1/3 to 1/2 height of stem, the broad bracts conspicuous well above the base, the lower bracts exceeding the peduncles, the few to several fascicles densely flowered usually and well separated; calyx glabrous, 5-7 mm. long, the lobes ovate to ovate-lanceolate, attenuate or acuminate, with conspicuous, scarious margins almost to tip; corolla light purple to violet-blue, 14-20 mm. long (-24 spread out), glabrous except for sparse beard within in lower lip, the throat from nearly tubular to definitely, but not abruptly enlarged from the subequal tube, the lips spreading, lobes rounded, the lower somewhat to much larger than the upper; anthers glabrous, the sacs opening throughout and across the juncture, but not explanate; staminode barely exserted, cuneately widened at the apex, densely short-bearded at apex and along 2 marginal lines for 1/3 to 1/2 its length, the hairs much shorter than the width of the staminode.---Northwestern Colorado and adjacent Utah. In Colorado, from Eagle to Moffat and Garfield Counties, in gulches, canyons, and on sagebrush slopes; 5500-7200 feet. This species is closely related to Penstemon pachyphyllus A. Gray, which has a range farther west, from Nevada to northern Arizona. The staminode of the latter is linear-widened (about 1 mm.), and densely bearded with hairs much longer than the width of the staminode. Specimens of P. osterhoutii through most of its range have the staminodes 2-3 mm. wide. A more complete series of specimens might relegate P. osterhoutii to subspecific rank.

42. Penstemon arenicola A. Nels., Bull. Torr. Club 25:280. 1898.

Stems 10-30 cm. tall, 1 to many, simple, assurgent-erect, glabrous throughout; leaves entire, glabrous, the basal oblanceolate, with mostly poorly-defined petioles, rounded or acute, mucronate, 2-7 cm. long, 0.4-1.5 cm. wide, the cauline sessile, oblanceolate, oblong-lanceolate, or somewhat ovate-lanceolate, mostly 3-7 cm. long, 0.5-1.5 cm. wide; inflorescence narrow, cylindrical, usually 1/2 or more height of plant, with closely-spaced and densely-flowered fascicles, the bracts oblong- to ovate-lanceolate, prominently whitish-margined; calyx glabrous, 4-6 mm. long, the lobes lanceolate or lanc-acuminate, narrowly scarious-margined to the tip; corolla deep blue or blue-purple, tube and throat scarcely differentiated (base of tube nearly of same diameter as throat), limb not markedly bilabiate, the lobes subequal, often the 2 erect-spreading, upper lobes larger than the 3 spreading lower ones; anthers glabrous, the sacs opening throughout and across the juncture, not explanate; staminode included, dilated-widened upward (as large as in P. angustifolius which has a larger corolla), densely golden-bearded (sometimes purple) near apex and more sparsely so in 2 lateral rows proximally for about 1/2 its length.---Southwestern Wyoming and adjacent Utah and Colorado. In Colorado, only in northwestern Moffat County, in sandy areas; probably 6000-7500 feet.

43. Penstemon angustifolius Nutt. ex Pursh, Fl. Amer. Sept. 738. 1814.

P. caeruleus Nutt.; Chelone angustifolia Steud.; Chelone coerulea Spreng.---Stems 10-25 cm. tall, solitary or few, simple, erect, mostly glabrous and glaucous, sometimes scabrid-pubescent; leaves entire, mostly glabrous but sometimes scabrid-puberulent, glaucous, the basal ones mostly linear-lanceolate, linear-oblanceolate, or broader, with poorly defined petioles, mostly 0.5-1 cm. wide, 3-8 cm. long, middle and upper stem leaves sessile, linear-lanceolate to broadly ovate-lanceolate and clasping, always some of them broader than the basal ones and attenuate, up to 2 cm. wide, 4-10 cm. long; inflorescence mostly dense, cylindrical, often interrupted, the bracts lanceolate to narrowly ovate-lanceolate, and mostly gradually tapering from the base; calyx glabrous or sometimes puberulent, 4-8 mm. long, the lobes lanceolate to lanc-acuminate, with narrow scarious margin at base; corolla pale to deep sky-blue but occasionally in bud, with age, and at other times with pinkish or purplish cast, often with several shades in 1 inflorescence, 14-20 mm. long, glabrous throughout, or occasionally with very few hairs within on lower lip, the tube gradually widening upward, not markedly ventricose, the lobes widely spreading; anthers glabrous, the sacs opening throughout, divaricate but not explanate; staminode strongly widened upward, bearded 1/2 or more of its length, rather densely so near the apex.---North Dakota and eastern Montana, southward to northern New Mexico. In Colorado, from Sedgwick to Larimer Counties and then southward at the base of the mountains to northern New Mexico; apparently very infrequent in the southernmost counties, where its place is largely taken by the subspecies caudatus. Mostly of sandy areas; 3500-7500 feet.

43A. Penstemon angustifolius caudatus (Heller) Keck, (ssp.) Journ. Wash. Acad. Sci. 29:490. 1939.

P. caudatus Heller; P. angustifolius var. caudatus (Heller) Rydb.; P. secundiflorus var. caudatus (Heller) A. Nels.; P. angustifolius ssp. venosus Keck---Differs from the foregoing by characters given in the key. The name is accepted here only because typical P. angustifolius appears not to occur in the southern range of the subspecies, while the latter does not occur in the northern part of the range of the species. The broader, and often caudate, bracts upon which P. caudatus was based, show such complete intergradation with the narrower ones of the species, that the assignment of many specimens in the overlapping ranges is wholly arbitrary. In the subspecies, the plants are generally more robust, being usually 20-40 cm. tall, and having upper stem leaves very broadly ovate-lanceolate. In addition, the color of the corolla in the subspecies is often lighter than in the species, running into various shades of pinkish-blue, magenta, or lilac, and suggesting a delicate balance of the hydrogen ion concentration in the cell sap for expression of the blue or pink anthocyan pigments. Among the specimens here recognized are a few which show prominently veined bracts and often leaves as well. Their sporadic appearance seems to indicate nothing above the rank of form.---Eastern Kansas to eastern Colorado and southward to northern New Mexico and western Oklahoma. In Colorado, from Weld County eastward and southward, on the sand hills and other sandy areas of the plains; 3000-7500 feet. One specimen also from Montezuma County of the western slope.

44. Penstemon secundiflorus Benth. in DC., Prodr. 10:325. 1846.
P. secundiflorus ssp. lavendulus Pennell; P. versicolor Pennell---Stems 10-40 cm. tall, 1 to many, glabrous and usually glaucous, erect or upcurved at base; leaves entire, firm-thickened to sub-fleshy, mucronate, glabrous and glaucous to dark green, the basal typically rather narrowly oblanceolate, 4-10 mm. wide, 2-7 cm. long, those of shoots up to 2.5 cm. wide and 4-10 cm. long, the cauline lanceolate to ovate-lanceolate and cordate-clasping, mostly gradually diminishing from the middle stem, acute to short-acuminate, mucronate, upper pairs commonly shorter than the internodes, 0.5-2.5 cm. wide, 2-5 (-7) cm. long; inflorescence elongate, rather slender, secund, fascicles loose to moderately dense, even the lower bracts rarely as wide as, and never wider than the upper stem leaves; calyx glabrous, 4-6 mm. long, the lobes ovate to ovate-lanceolate, acute to short-acuminate, prominently scarious-margined (often pink), frequently ribbed; corolla commonly light to pinkish-purple but often darker purple, 15-24 mm. long, though late flowers often as short as 14 mm., glabrous except for sparse bearding within on lower side, throat from relatively narrow (only slightly larger than tube) to rather widely expanded, scarcely ventricose, the lips widely flaring-spreading; anthers glabrous, the sacs opening throughout and across the juncture, not explanate; staminode included or barely exserted, much widened toward apex, densely bearded with golden hairs about 1/2 its length.---Southern Wyoming through central Colorado to central New Mexico. In Colorado, a common species across the state along the eastern base of the mountains as well as in North, Middle, and South Parks. On gravelly or rocky slopes, mesas, and hills of grassland, brushy or wooded areas; 5000-9500 feet.

45. Penstemon buckleyi Pennell, Proc. Acad. Nat. Sci. Phila. 73:486. 1921.
This species is reported to occur in the southeastern part of the state (Pennell, Acad. Nat. Sci. Phila. Monogr. 1:267, 1935). It is distinguished from numbers 43 and 43A above as follows: "Corolla pale lavender, its lobes 3-4 mm. long; sterile filament slightly bearded distally with yellowish hairs; seeds 3-4 mm. long; bracts ovate, in fruit conspicuously reticulate-ridged." None of our specimens examined appear to check with the description in all points. It is believed that only field studies will serve adequately to distinguish between this and the two proposed subspecies of P. angustifolius. The record is made here for that purpose.

10. Chionophila Benth. SNOWLOVER

Perennial caulescent herbaceous plants; leaves opposite or somewhat alternate on stem, but mostly in a basal tuft, entire; flowers in spikelike racemes; calyx funnelform, obscurely 5-lobed; corolla 2-lipped, the upper erect and retuse, lower 3-lobed with a bearded base forming a palate, the whole corolla somewhat horizontally flattened especially near apex (when fresh); anther bearing stamens 4, the anther sacs divaricate and confluent, sterile filament short and glabrous; seeds with a loose cellular-reticulated coat.

1. Chionophila jamesii Benth. in DC., Prodr. 10:331. 1846.
Stems 5-10 cm. tall, rarely to 15 cm., glabrous or minutely pubescent; basal leaves thick, spatulate or oblanceolate, stem leaves linear; inflorescence dense, mostly secund, bracteate; calyx 7-8 mm. long; corolla about 10-12 mm. long, greenish-white to cream-colored.---High mountains, mostly on gravelly moist slopes. Wyoming and Colorado. Our records scattered in the higher mountains of Colorado at 11,000-13,500 feet.

11. Castilleja Mutis PAINTED CUP; PAINT BRUSH

Annual or perennial plants, herbaceous or somewhat woody at base only, often partial root-parasites; leaves alternate, sessile, entire or pinnatifid; flowers very irregular, in terminal bracted spikes, the bracts usually petaloid; calyx tubular, laterally flattened, 4-lobed, more deeply cleft above and below than on the sides; corolla long and narrow, strongly 2-lipped, the upper (galea) elongated, entire, laterally compressed and often keeled, lower lip very short, 3-toothed; stamens 4, in pairs, ascending under and enclosed by the galea, anther cells unequal, the outer attached by its middle, the inner pendulous and attached at its apex; stigmas entire or 2-lobed; capsule loculicidal; seeds many, reticulated.

1. Roots annual; stems usually short and erect -1. C. minor
1. Roots perennial, more or less woody; stems usually slender and more or less decumbent at base
  2. Lower lip of corolla prominent, 1/2 to 2/3 as long as the upper (galea); galea short, 1/2 as long as the corolla tube or less; bracts green or yellow, rarely red
    3. Corolla about 2 cm. long, the lip more or less saccate; plants of high mountains above 8000 feet -2. C. puberula
    3. Corolla 4-5 cm. long, lip petaloid and spreading, not saccate; plants of plains and foothills below 8000 feet -3. C. sessiliflora
  2. Lower lip of corolla relatively small, usually less than 1/3 the length of the galea, never over 1/2; galea over 1/2 as long as the corolla tube, at times even exceeding it; bracts various, often reddish
    4. Calyx cut much deeper below than above; bracts divided into linear lobes, often not as conspicuous as the calyx
      5. Corolla 1.5-2.5 cm. long; bracts yellow or rarely yellowish-red tipped; calyx yellowish -4. C. flava

5. Corolla 3-5 cm. long; bracts red or scarlet; calyx red or scarlet   -5. C. linariaefolia
 4. Calyx equally or subequally cut above and below; bracts divided to entire, broad and conspicuous
   6. Plants with stems and leaves more or less densely tomentose; seed coats dark and often pubescent
     7. Bracts crimson, entire or cleft near apex; corollas over 2.5 cm. long; inflorescence short and broad;
        plants widely distributed in Colorado   -6. C. integra
     7. Bracts dull yellowish, deeply lobed; corollas less than 2.5 cm. long; inflorescence narrow; plants
        limited to southcentral Colorado   -7. C. lineata
   6. Plants glabrous to variously hairy but not at all tomentose; seed coats light colored, never pubescent
     8. Alpine plants growing above 11,000 feet in Colorado; stems not over 20 cm. tall
       9. Bracts entire or short-lobed at apex, these lobes not linear, greenish-yellow or somewhat streaked
          with red; plants generally distributed in Colorado   -8. C. occidentalis
       9. Bracts cleft into linear lobes, crimson, rose or lilac-purple; plants reported only from south-
          western corner of Colorado   -9. C. haydeni
     8. Plants not alpine, rarely above 11,000 feet in Colorado; stems over 20 cm. tall on normal plants
        (a low-altitude form of C. chromosa may be only 10 cm. tall)
       10. Leaves (at least upper ones) deeply cleft into linear, divaricate lobes; stems and leaves usually
           with both short and long hairs   -10. C. chromosa
       10. Leaves entire or upper occasionally shallowly lobed near apex; usually only 1 kind of hairs
           present (if at all)
         11. Inflorescence yellow, bracts, calyx and corolla never tinged with red -11. C. septentrionalis
         11. Inflorescence more or less red or rose, either the bracts, calyx or corolla at least reddish-
             tinged (occasionally a plant with yellowish inflorescence is found among red ones)
           12. Bracts entire or very shallowly lobed, often rose-colored   -12. C. rhexifolia
           12. At least some of the bracts rather deeply cleft, never really rose-colored
                                                                                -13. C. miniata

1. Castilleja minor Gray, Bot. Calif. 1:573. 1876.
   C. exilis A. Nels.; C. stricta Rydb.---Annual plants; stems 30-80 cm. tall, strict, slender to stout, usually unbranched, more or less glandular-pubescent to hirsute; leaves 4-8 cm. long, 3-nerved, entire, linear to lanceolate, glandular-pubescent; inflorescence becoming much elongated; bracts similar to leaves, only the uppermost scarlet at apex; calyx about 2 cm. long, deeper cleft above than below, the 2 lateral lobes toothed at apex; corolla little if any longer than the calyx, greenish-yellow or sometimes pinkish on margins; galea about 1/2 the length of the corolla tube; lower lip 3-cleft with obtuse lobes short (about 1-2 mm. long).---Usually in moist or wet soil. Montana to Washington, south to New Mexico, Mexico and Arizona. Our few records from westcentral Colorado at about 5500 feet.

2. Castilleja puberula Rydb., Bull. Torr. Club 31:644. 1904.
   C. flavoviridis L. Kelso; C. brachyantha var. subinflata E. H. Kelso; C. brachyantha of authors---Perennial plants; stems about 8-15 cm. tall, several, erect or ascending from a woody caudex, finely puberulent to short-tomentose in the inflorescence; leaves 2-3 cm. long, linear and entire.or upper with linear lateral lobes, finely puberulent; inflorescence rather short; bracts 3-lobed, shorter than the flowers, tinged with green, yellow or red, tomentose; calyx 1-1.5 cm. long, cleft about half way above and below or deeper below, lateral lobes cleft at summit, usually greenish or yellowish; corolla about 2 cm. long, green to yellowish; galea about 6-8 mm. long, about 1/2 as long as the corolla tube; lower lip about 3-5 mm. long, 3-lobed, about 1/2 to 2/3 the length of the galea.---High mountains. Limited to northcentral and central Colorado. Our records from 10,000-13,000 feet.

3. Castilleja sessiliflora Pursh, Fl. Am. Sept. 738. 1814.
   C. sessiliflora bethelii Cockerell; C. grandiflora Sprengel---Perennial plants; stems 10-40 cm. tall, several to many from a woody caudex, ascending, finely pubescent to somewhat tomentose; lower leaves 3-5 cm. long, linear, entire, puberulent, to pubescent, upper leaves usually cleft into linear lobes; inflorescence at first short, then elongating; bracts similar to upper leaves, green; calyx cleft about 1/2 below, about the same or somewhat deeper above, the lateral lobes then cleft about 1/2 their length, greenish or somewhat yellowish at apex; corolla 4-5 cm. long, much exceeding the bracts and calyx, greenish-yellow often pinkish or purplish tinged near apex especially; galea short, about 1/5 the length of the corolla; lower lip with 3 prominent spreading lobes, about 2/3 the length of the galea.---Hills, plains and valleys, often on gravelly or sandy soil. Illinois to Saskatchewan, south to Texas and Arizona. Our records from the eastern half of Colorado at 3500-7500 feet.

4. Castilleja flava Watson, Bot. King's Exped. 230. 1871.
   C. brachyantha Rydb.; C. breviflora Gray 1862 not Berth. 1846. C. curticalix Nelson & Macbride---Perennial plants; stems 20-50 cm. tall, several to many from a woody caudex, erect or ascending, simple or branched above, pubescent or sometimes glabrate; leaves 2-6 cm. long, lower linear and entire, upper usually cleft into 3-5 linear lobes, short-pubescent; inflorescence elongated, pubescent; bracts 3- to 5-cleft, the lobes of lower usually longer than the flowers, yellowish-tipped or rarely yellowish- red-tipped; calyx about 15 mm. long, more or less hairy, cleft about 1/2 way above, less deeply so above, lateral segments short-cleft; corolla 1.5-2.5 cm. long, yellow; galea about 7-10 mm. long, shorter than the tube; lower lip short, rarely over 1/4 the length of the galea, 3-cleft.---Dry plains, hills and mountains. Wyoming to Idaho, south to Colorado and Nevada. Our records from northcentral and northwestern Colorado at 7000-9000 feet.

5. Castilleja linariaefolia Benth. in DC., Prodr. 10:532. 1846.
   Perennial plants; stems 30-80 cm. tall, erect from a woody caudex, glabrous to crisp-pubescent especially below; leaves 4-10 cm. long, narrowly to broadly linear, glabrous to sparingly puberulent, entire or deeply divided above into linear lobes; bracts red or scarlet, divided into linear lobes, usually not as conspicuous as the calyx; calyx scarlet or red, finely pubescent to tomentose, deeper cleft below, usually about 1/2 as deep above, lobes

laterally bifid; corolla 3-5 cm. long, greenish-yellow; galea about as long or slightly shorter than the corolla tube; lower lip much reduced, less than 1/5 the length of the galea.---Hills, mountains, often on rocky slopes. Wyoming to Oregon, south to New Mexico and California. Our records well scattered in the western half of Colorado at 5000-9500 feet.

6. Castilleja integra Gray in Torr., Bot. Mex. Bound. 119. 1859.

C. gloriosa Britt.; C. tomentosa Gray---Perennial plants; stems 10-40 cm. tall, several from a woody root, tomentose throughout; leaves 4-6 cm. long, entire or upper sometimes cleft, narrowly lanceolate to linear, canescent-pubescent; inflorescence short; bracts conspicuous, broader than leaves, entire or cleft near apex, crimson, usually tomentose at base; calyx 2.5-3.5 cm. long, rather inflated in fruit, crimson near apex, deeply cleft, usually deeper above, lateral lobes divided about 1/2 their length; corolla about 3-4 cm. on most but some shorter and longer, longer than calyx, greenish or margins crimson; galea about 10-15 mm. long, shorter than corolla tube, often about 1/2 its length; lower lip short, about 1/5 as long as galea, short-lobed.---Dry hills, mesas and plains often on gravelly or rocky soil. Colorado, south to Texas, Mexico and Arizona. Our records scattered in Colorado, few from the extreme eastern and western parts, at 4500-9500 feet.

7. Castilleja lineata Greene, Pitt. 4:151. 1900.

C. scabrida Eastw.---Perennial plants; stems 10-40 cm. tall, several, erect or ascending from a woody caudex, grayish-tomentose; leaves 2-4 cm. long, linear, usually with 1 or 2 pairs of lateral lobes, tomentose; inflorescence narrow and elongating in age; bracts broader than the leaves, 3- or more cleft, middle lobe broader, dull yellowish in color; calyx about 18-20 mm. long, yellowish or greenish, subequally cleft to middle above and below and deeply cleft laterally; corolla about 20 mm. long, greenish; galea about 4-7 mm. long, shorter than corolla tube; lower lip short, not much over 1 mm. long.---Mountains and hills. Colorado and New Mexico. Our records from southcentral and southwestern Colorado at 7000-10,000 feet.

8. Castilleja occidentalis Torrey, Ann. Lyc. N. Y. 2:230. 1828.

Perennial plants; stems 5-20 cm. tall, several from a woody caudex, erect or ascending, glabrous below and more or less pilose above especially in the inflorescence; leaves 1.5-4 cm. long, lanceolate, or linear-lanceolate, glabrous or puberulent, 3-nerved, entire or upper sometimes with a pair of lateral lobes; inflorescence short and dense in flower, more or less villous; bracts 1.5-2 cm. long, obovate and entire or often with a pair of lateral lobes near apex, greenish-yellow or somewhat streaked with red; calyx about 1.5-2 cm. long, subequally cleft above and below, the lobes short-cleft; corolla usually exceeding the calyx, greenish; galea shorter than the corolla tube; lower lip about 1.5-3 mm. long.---High mountains. Colorado, Utah and New Mexico. Our records from the higher mountains of the state at 10,500-13,000 feet.

9. Castilleja haydeni (Gray) Cockerell, Bull. Torr. Club 17:37. 1890.

Perennial plants; stems 10-15 cm. tall, clustered on a woody caudex, purplish, finely pubescent or glabrate below; leaves 2-8 cm. long, linear and entire or upper with 1 or 2 pair of spreading lobes, glabrate to finely puberulent; inflorescence rather short at first; bracts cleft, often ciliate and villous on vein, crimson, rose-red or lilac-purple; calyx about 2 cm. long, about equally cleft above and below, lobes divided halfway down; corolla 2-2.5 cm. long, equalling or exceeding the calyx; galea about 7-8 mm. long, shorter than the tube; lower lip about 2-3 mm. long.----A little known species from high mountains. Colorado and New Mexico. Our one record from southwestern Colorado at 11,000 feet.

10. Castilleja chromosa A. Nels., Bull. Torr. Club 26:245. 1899.

C. collina A. Nels.; C. dubia A. Nels.; C. buffumii A. Nels.---Perennial plants; stems 10-40 cm. tall, simple or branched, several, decumbent at base to erect, 2 kinds of hairs present, 1 fine and the other long and pilose; lower leaves 3-7 cm. long, simple or somewhat cleft into narrow lobes the lateral somewhat divaricate, the middle sometimes cleft again, upper leaves cleft; inflorescence rather short in flower; bracts usually tipped with red, sometimes with yellow, lobed like upper leaves; calyx subequally cleft above and below or slightly deeper cut above, less deeply cut laterally, reddish or yellowish; corolla about 2 cm. long, greenish-yellow; galea about 1 cm. long, about as long as the tube; lower lip short, about 2-3 mm. long.---Dry plains, mesas and hills. Wyoming to Oregon, south to New Mexico and California. Our records from the western half of Colorado at 5000-9000 feet.

11. Castilleja septentrionalis Lindley, Bot. Reg. pl. 925. 1825.

C. sulphurea Rydb.; C. luteovirens Rydb.; C. brunnescens Rydb.---Perennial plants; stems 30-50 cm. tall, usually glabrous below, more or less pilose and somewhat viscid in the inflorescence, erect or somewhat decumbent at base, more or less clustered; leaves 3-5 cm. long, lanceolate to linear-lanceolate, 3-nerved or upper broader and 5-nerved, glabrous or nearly so, entire or rarely the upper 3-lobed at apex; inflorescence short or rarely elongating, yellow; bracts broad, usually about as long as the flowers or longer, entire or commonly with short lateral lobes, puberulent to villous, yellow; calyx about 1.5-2 cm. long, subequally cut above and below, the lateral lobes 2-cleft; corolla 2-2.5 cm. long; galea about 7-10 mm. long, shorter than the tube; lower lip about 2-4 mm. long, usually about 1/3 the length of the galea.---Mountains. Eastern Canada to Alberta, south to New England, New Mexico and Utah. Our records scattered in the western two-thirds of Colorado at 6500-12,000 feet.

12. Castilleja rhexifolia Rydb., Mem. N. Y. Bot. Gard. 1:356. 1900.

C. lauta A. Nels.; probably C. obtusiloba Rydb.; C. humilis Rydb.---Perennial plants; stems 20-60 cm. tall, erect or nearly so, glabrous or more or less puberulent; leaves 3-6 cm. long, linear-lanceolate to ovate, 3-nerved, entire to shallowly 3-lobed above, usually sparingly pubescent; inflorescence short and dense, often crisp-pubescent; bracts thin and large, entire or shallowly lobed, crimson, scarlet, rose or occasionally yellowish; calyx about 2 cm. long, about equally cut above and below, lateral lobes shallowly cut; corolla 2-3 cm. long, yellow or tinged with red; galea about 7-10 mm. long, shorter than the tube; lower lip about 2-4 mm. long. Closely related to the preceding species if really distinct.---Open woods and mountain slopes, often in moist soil. Montana to Oregon, south to Colorado and Utah. Our records from the western half of Colorado at 7500-13,000 feet.

13. Castilleja miniata Douglas in Hook., Fl. Bor. Am. 2:106. 1838.
C. confusa Greene; C. crista-galli Rydb.; C. lancifolia Rydb.; C. trinervis Rydb.---Perennial plants; stems 30-60 cm. tall, erect or decumbent at base, glabrous to pubescent; leaves 3-5 cm. long, linear-lanceolate to ovate, entire or rarely cleft above, 3-nerved, glabrous to pubescent; inflorescence short in flower, elongating in fruit; bracts red, rarely yellow, mostly lobed or toothed; calyx about 1.5-2 cm. long, subequally cut above and below, lateral lobes less deeply cut; corolla about 2-3 cm. long, often tinged with red; galea and corolla tube usually about equal, tube sometimes longer, hence galea about 10-15 mm. long; lower lip 2-4 mm. long. This species is very close to C. rhexifolia Rydb. and the writer cannot separate it from C. linearis Rydb. in Colorado.---Mountains. Montana to Alaska, south to New Mexico and California. Our records from the western two-thirds of Colorado at 7500-12,000 feet.

12. Orthocarpus Nutt. OWLCLOVER

Annual caulescent herbaceous plants; leaves alternate, sessile or nearly so, entire or 3-cleft; flowers yellow, white or purple, in terminal leafy-bracted spikes; calyx tubular-campanulate, with 4 equal or subequal lobes; corolla bilabiate, tube slender, upper lip (galea) erect or slightly arched, lower lip about as long as the upper, entire to more or less 3-saccate at apex; stamens 4, didynamous, ascending under the upper lip, 2 anther sacs dissimilar, outer one attached by its middle, inner pendulous from its apex; styles slender, stigma entire; capsule oblong or narrowly elliptic, loculicidal; seeds many, reticulate, ridged or alveolar.

1. Corolla yellow, 10-15 mm. long, tip of galea obtuse not inflexed, lower lip not much wider than the galea; plants general over Colorado -1. O. luteus
1. Corolla purple or white, 13-20 mm. long (usually 15 mm. or more), tip of galea with a mucronate, inflexed apex, lower lip much wider than the galea; plants limited to southwestern Colorado -2. O. purpureo-albus

1. Orthocarpus luteus Nutt., Gen. Pl. 2:57. 1818.
Stems 10-30 cm. tall, erect, usually simple or branching near the top, more or less pubescent; leaves 1-4 cm. long, linear to linear-lanceolate, usually entire or rarely 3-cleft; inflorescence many-flowered, from 1 to 10 cm. or more long; bracts 10-15 mm. long, 3- to 5-cleft, rather leaflike; calyx 5-8 mm. long, the lobes 1-2 mm. long; corolla 10-15 mm. long, golden-yellow, apex of galea obtuse, not inflexed, lower lip about as long and not much wider than the galea; seeds 1-1.25 mm. long.---Plains, hills and mountains. Manitoba to British Columbia, south to New Mexico and California. Our records well scattered in the western two-thirds of Colorado at 5500-10,000 feet.

2. Orthocarpus purpureo-albus A. Gray in King, Geol. Expl. 40th Par. 5:458. 1871.
Stems 10-40 cm. tall, erect, simple or branching above, glandular-pubescent, purplish; leaves 1.5-3.5 cm. long, linear-lanceolate to almost filiform, entire or apparently more commonly 3-cleft into filiform lobes; inflorescence few- to many-flowered, 2-15 cm. long or more; bracts 10-20 mm. long, 3-lobed; calyx 6-10 mm. long, lobes acute to acuminate; corolla about 13-20 mm. long, white or purplish, apex of galea mucronate and inflexed, lower lip much wider and usually slightly shorter than the galea; seeds 2-2.5 mm. long.---Mesas, slopes and valleys. Colorado, New Mexico and Arizona. Our records in or very near southwestern Colorado at 6500-8000 feet.

13. Cordylanthus Nutt. BIRDBEAK; CLUBFLOWER

Annual caulescent plants; stems branching; leaves alternate, 3- to 5- (or more) parted into narrowly linear to filiform divisions; flowers solitary or in headlike spikes at ends of branches, with foliose bracts; calyx spathelike, cleft to base on sides and appearing as 1-2 leaflike sepals; corolla tubular, bilabiate with nearly equal lips, upper lip narrow and entire, lower entire to 3-toothed; stamens 4, in pairs, anther cells unequal, 1 pendulous from 1 end, the other inserted about at its middle, enclosed in the galea; ovary with style hooked near apex, stigma entire; fruit flattened, loculicidal; seeds few.

1. Calyx with only 1 part, not subtended by small bractlets (has larger bracts); stems glandular-puberulent, sometimes villous, leaves 3-cleft into narrowly linear divisions -1. C. kingii
1. Calyx with 2 parts, subtended by 2-4 small bractlets (as well as larger bracts); stems glabrous to puberulent, not glandular or villous; leaves 3- or more-parted into filiform divisions
    2. Corolla over 2 cm. long; stems usually glabrous to sparsely puberulent, usually over 30 cm. tall -2. C. wrightii
    2. Corolla less than 2 cm. long; stems usually cinereous-puberulent, seldom over 30 cm. tall -3. C. ramosus

1. Cordylanthus kingii Wats., Bot. Kings Exp. 233. 1871.
Adenostegia kingii (Wats.) Greene---Stems 10-30 cm. tall, diffusely branched, glandular-puberulent or glandular-villous; leaves 3-cleft into narrowly linear divisions; bracts leaflike, no bractlets present; flowers glomerate and sessile; calyx about 12-15 mm. long, monophyllous; corolla 15-20 mm. long, purplish.---Hills and ridges. Colorado to Nevada. Our few records from southwestern Colorado at about 6500-7000 feet.

2. Cordylanthus wrightii A. Gray in Torr., Bot. Mex. Bound. 120. 1859.
Adenostegia wrightii (Gray) Greene---Stems 30-50 cm. tall, paniculately branched, glabrous or inconspicuously puberulent, rarely definitely puberulent; leaves 3- to 5-parted into filiform divisions; flowers in headlike spikes or rarely solitary; bracts leaflike; calyx about 2-3 cm.

long, of 2 parts; corolla 20-30 mm. long, yellow to purple.---Dry hills and valleys. Colorado and Utah, south to Texas, Mexico and Arizona. Our records from westcentral and southwestern Colorado at 5000-8000 feet.

3. Cordylanthus ramosus Nutt.; Benth. in DC., Prodr. 10:597. 1846.
Adenostegia ramosa (Nutt.) Greene---Stems 15-30 cm. tall, much branched, cinereous-puberulent; leaves dissected into filiform divisions; flowers rarely solitary but mostly in headlike spikes; bracts leaflike; calyx about 15-25 mm. long, of 2 parts; corolla 15-20 mm. long, yellow. Intergrades somewhat with the preceding in length of flower.---Dry plains and hills. Montana to Oregon, south to Colorado and California. Our records from the western one-third of Colorado at 5500-8000 feet.

14. Rhinanthus L. RATTLEWEED; YELLOWRATTLE

Annual, caulescent, herbaceous plants; stems erect, 4-angled; leaves opposite, sessile, rather thick and rigid, scabrous, linear to lanceolate, sharply serrate or dentate; flowers in a rather dense, leafy-bracted spikelike raceme; calyx 4-toothed, becoming bladderlike and veiny in fruit but compressed; corolla yellow at least the tube, bilabiate, the upper lip arched and minutely 2-toothed below apex, lower lip shorter with 3 spreading lobes; stamens 4, didynamous, ascending under the upper lip, anthers pilose, anther sacs equal, parallel; capsule orbicular-compressed, loculicidal; seeds several, winged.

1. Rhinanthus rigidus Chab., Herb. Boissier Bul. 7:516. 1899.
Stems 20-70 cm. tall, often pilose in longitudinal lines, simple or branching; leaves linear to lanceolate; corolla 7-9 mm. long. A related species R. kyrollae Chab. has been reported from this area but may not be distinct.---Meadows and valleys. Alberta to Alaska, south to Colorado, Arizona and Washington. Our records from the western half of Colorado, mostly in the central and southern parts, at 7000-9000 feet.

15. Pedicularis L. LOUSEWORT; WOODBETONY

Perennial herbaceous plants, apparently partial root-parasites but with chlorophyll present, caulescent or subacaulescent; stems branched if at all only at base; leaves alternate, opposite or basal, toothed to bipinnatifid; flowers in terminal bracted spikes or spikelike racemes; calyx 2- to 5-lobed, cleft on lower side and sometimes on upper; corolla strongly bilabiate, the lower lip 3-lobed, these mostly spreading or reflexed, upper lip compressed on the sides, arched and often beaked at the apex (galea); stamens 4, didynamous, ascending under the galea, anther sacs parallel; capsules compressed, oblique or curved, loculicidal; seeds many, pitted, striate or ribbed.

1. Galea prolonged into a slender recurved beak (curved outward and upward) this beak over 8 mm. long (galea and beak resembling the head and trunk of an elephant) -1. P. groenlandica
1. Galea beakless or with a shorter and straight or incurved beak not over 5 mm. long
   2. Leaves merely toothed or pinnately lobed, the sinuses not extending over 2/3 of the way to the midrib
      3. Galea produced into an incurved beak 3-5 mm. long; corolla white -2. P. racemosa
      3. Galea beakless (many have small teeth near apex); corolla rarely white
         4. Leaves pinnately lobed, the sinuses extending about 1/2 to 2/3 of the way to the midrib
            5. Plants subacaulescent; anthers aristate-acuminate at base, the awns projecting like teeth from the hood of the galea; corolla over 25 mm. long -3. P. centranthera
            5. Plants caulescent; anthers not aristate at base; corolla not over 25 mm. long -4. P. canadensis
         4. Leaves doubly crenate toothed, not at all lobed -5. P. crenulata
   2. Leaves pinnately deeply parted or divided to or nearly to the midrib (in the latter case the midrib narrowly winged) but the sinuses always much over 2/3 of the distance into the midrib
      6. Corolla large, 25-30 mm. long, the galea beakless but with 2 lateral teeth just below apex (flowers sordid yellow or rarely streaked with red) -6. P. grayi
      6. Corolla smaller, less than 25 mm. long, whitish, yellow, rose or purple, galea beaked or beakless and lacking teeth (except P. scopulorum with corolla rose or purple)
         7. Corolla rose or purple, galea beakless but with 2 broadly triangular teeth near apex -7. P. scopulorum
         7. Corolla whitish to yellowish, galea either beaked, or beakless and lacking teeth
            8. Galea tapering into a short beak 1-2 mm. long; leaves narrow, the primary divisions seldom over 1 cm. long (the leaf rarely over 2 cm. wide) -8. P. parryi
            8. Galea beakless, and truncate at apex; leaves broader, primary divisions usually over 2 cm. long (leaf over 4 cm. wide) -9. P. paysoniana

1. Pedicularis groenlandica Retz., Fl. Scand. Prodr. ed. 2, 145. 1795.
Elephantella groenlandica (Retz.) Rydb.---Stems 15-60 cm. tall, glabrous or nearly so; leaves basal and cauline, lanceolate in outline, pinnately parted nearly to midrib but this narrowly winged, segments crenulate or incised, often doubly so; inflorescence becoming 10-30 cm. long, spicate, dense; bracts narrow, lower more or less leaflike, upper small; calyx 5-7 mm. long, 5-toothed, oblique; corolla about 1 cm. long (not counting beak) reddish-purple or dark rose, galea rather short and abruptly curved downward but extending into a long conspicuous upturned beak 8-15 mm. long (whole galea resembling the head and trunk of an elephant).---Swamps and wet meadows. Greenland to Alaska, south to New Mexico and California. Our records from the western half of Colorado at 8000-13,000 feet.

2. Pedicularis racemosa Dougl. ex Hook., Fl. Bor. Amer. 2:108. 1838.
Stems 20-50 cm. tall, usually several in a clump, glabrous or nearly so; leaves 3-8 cm. long, mostly cauline, lanceolate or linear, not divided but merely doubly crenate; flowers rather few in a rather short loose spikelike raceme or lower in upper leaf axils; bracts leaflike and conspicuous below, shorter above; calyx appearing 2-lobed, 4-6 mm. long; corolla 12-20 mm. long, white, galea about 5-7 mm. long but curved downward and produced into slender uncurved beak 3-5 mm. long, lower lip of corolla long and wide (compared to other species).---Mountains, often on shaded slopes. Alberta to British Columbia, south to New Mexico and California. Our records from the western half of Colorado at 9000-13,000 feet.

3. Pedicularis centranthera A. Gray in Torr., U. S. & Mex. Bound. Bot. 120. 1859.
Stems about 10-15 cm. tall, glabrous; leaves 4-15 cm. long, mostly crowded near base, parted about 2/3 of the way to the midrib or less, divisions oval or ovate-oblong, doubly toothed, these callous tipped; flowers few; bracts linear or nearly so; calyx about 1.5-2 cm. long, lobes linear or linear-lanceolate; corolla 3-3.5 cm. long, purple or yellowish, galea slightly curved with short lobes.---Mountains and valleys. Colorado, Utah, New Mexico and Arizona. Our records from the western part of Colorado at 6000-7500 feet.

4. Pedicularis canadensis L., Mant. 86. 1767.
Stems 10-45 cm. long, usually more than 1, pubescent or glabrate below; leaves 4-12 cm. long, cauline and basal, pinnately lobed not over 2/3 of the way to the midrib, the lobes obtuse and incised or dentate; inflorescence spicate, short in flower, elongating in fruit to 10 cm. long or more; upper bracts usually small and inconspicuous, the lower bracts usually foliaceous; calyx about 3-5 mm. long, cleft below, the lobes minute; corolla 18-25 mm. long; yellowish to reddish, perhaps sometimes whitish, galea about 7-9 mm. long, arched and incurved, not beaked but with 2 slender teeth below the apex.---Woods and meadows, often on sandy or loam soil. Quebec to Manitoba, south to Florida, Texas and Mexico. Our records from central and southcentral Colorado at 7000-10,000 feet.

5. Pedicularis crenulata Benth. in DC., Prodr. 10:568. 1846.
Stems 10-30 cm. long, usually more than 1, more or less villous or pubescent, erect from a usually decumbent base; leaves 3-7 cm. long, basal and cauline, linear to oblong-linear, doubly crenate; inflorescence 2-10 cm. long, spicate, dense; bracts leaflike; calyx 8-12 mm. long, cleft below; corolla about 2-2.5 cm. long, rose, purplish or possibly whitish, galea about 10 mm. long, curved near apex, not beaked but with 2 small teeth near end.---Moist meadows. Wyoming to Nevada, south to Colorado. Our records from the western half of Colorado at 7000-9500 feet.

6. Pedicularis grayi A. Nels., Biol. Soc. Wash. Proc. 17:100. 1904.
P. procera Gray---Stems 40-100 cm. tall, glabrous or pubescent above and in inflorescence; leaves 20-60 cm. long, basal and cauline, glabrous or somewhat pubescent at first, pinnately divided or parted to or nearly to midrib, segments pinnatifid into serrate or incised lobes; inflorescence about 15-40 cm. long, spicate, many-flowered; bracts linear from an ovate-lanceolate base, the lower ones pinnatifid and often longer than the flowers; calyx 10-15 mm. long, 5-lobed, these lanc-linear; corolla about 25-30 mm. long, sordid yellow or sometimes streaked with red, galea 9-15 mm. long, curving downward and cucullate, not at all beaked but with 2 lateral teeth just below apex. The correct name for our plant may be P. procera A. Gray.---Mountain slopes and valleys, usually in shade. Wyoming and Utah, south to New Mexico and Arizona. Our records scattered in the western half of Colorado at 7000-10,500 feet.

7. Pedicularis scopulorum A. Gray, Syn. Fl. 2:308. 1878.
Stems 10-20 cm. tall, glabrous except in the villous or lanate rachis of spike; leaves 3-8 cm. long, basal and cauline, pinnately parted not quite to midrib (hence rachis winged) the divisions lanceolate-toothed to incised; inflorescence spicate, short in flower, becoming 3-5 cm. long; bracts rather leaflike below but not conspicuous, shorter above; calyx about 8-10 mm. long, arachnoid-villous, the lobes triangular-subulate; corolla 15-20 mm. long, rose to purple, galea about 10 mm. long, curved near end, with 2 broadly triangular teeth near apex, not beaked.---High mountains. Colorado and possibly north to Montana. Our records from northcentral, westcentral, central and southcentral Colorado at 10,000-14,000 feet.

8. Pedicularis parryi A. Gray, Am. Journ. Sc. ser. 2, 34:250. 1862.
Stems 10-40 cm. tall, glabrous; leaves 5-12 cm. long, basal and cauline, pinnately parted to near the midrib this narrowly winged, divisions toothed or incised; inflorescence 3-20 cm. long, spicate but flowers short-pedicellate; bracts narrow, usually with narrow lateral divisions, lower often longer than flowers, upper shorter; calyx about 7-12 mm. long, becoming somewhat inflated, with longitudinal darker lines; corolla 12-20 mm. long or sometimes longer, whitish to yellow (especially when pressed), galea about 6-8 mm. long, curved downward near apex, with a short straight beak 1-2 mm. long this merging rather gradually into the galea.---Mountains. Montana, south to New Mexico and Arizona. Our records scattered in the higher mountains of the state at 9000-12,000 feet.

9. Pedicularis paysoniana Pennell, Bull. Torr. Club 61:447. 1934.
P. bracteosa for Colorado plants---Stems 30-90 cm. tall, glabrous below the inflorescence; leaves 5-15 cm. long, mostly cauline, basal long-petioled, cauline shorter-petioled, bipinnately or pinnately divided, the first division extending to the midrib, the primary segments irregularly toothed or pinnatifid; inflorescence 6-30 cm. long, spicate, densely flowered; bracts lanceolate, slightly caudate, 1-2 cm. long; calyx 8-10 mm. long, lobes lanceolate to linear, villous and rarely glandular, the upper sepal shorter than others; corolla about 2 cm. long, yellow or yellowish, galea 8-10 mm. long, erect, curved near apex, this truncate and without teeth, not beaked.---Moist mountains and valleys, often in shade. Montana south to Colorado and Utah. Our records from the higher mountains of Colorado at 9000-12,000 feet.

16. Limosella L. MUDWORT

Small, subscapose, somewhat succulent, glabrous or nearly glabrous annual plants or possibly perennial by stolons; leaves rosulate at base of plant, narrow and entire; flowers small, solitary on long slender peduncles; calyx campanulate, 5-lobed; corolla white to purplish, nearly regular, rotate-campanulate, 5-cleft; stamens 4, anther sacs confluent; styles short, stigmas capitate; capsule 2-celled at base, 1-celled above; seeds many.

1. Limosella aquatica L., Sp. Pl. 631. 1753.
    L. tenuifolia in Manuals for western plants---Leaf blades 5-30 mm. long, elliptic or oval, sometimes oblanceolate or sometimes even bladeless; pedicels 5-30 mm. long; calyx campanulate, with triangular teeth; corolla 2-2.5 mm. long.---Muddy flats and shallow water. Widely distributed in North America and Eurasia. Our records scattered, mostly in central and north-central Colorado, at 4500-9500 feet.

17. Bacopa Aubl. WATERHYSSOP

Perennial succulent herbaceous plants; stems creeping or floating, rooting at the nodes; leaves opposite, entire or toothed, sessile, with broad blades; flowers white or yellowish, solitary, peduncled and axillary; calyx 5-parted, upper lobe largest; corolla 5-lobed, the lobes spreading, only slightly irregular; stamens 4, somewhat didynamous, included, the anther sacs equal; capsule shorter than the calyx.

1. Bacopa rotundifolia (Michx.) Wettst. in Engler & Prantl., Naturl. Pflanzenfam. IV:76. 1891.
    Monniera rotundifolia Michx.; Macuillamia rotundifolia (Michx.) Raf.; Hydranthelium rotundifolium (Michx.) Pennell---Stems to 40 cm. long, branched or simple, more or less pilose; leaves 1-3 cm. long, entire or slightly undulate, obovate or orbicular, palmately veined; calyx about 4-5 mm. long, with outer sepals nearly round; corolla 5-7 mm. long, white or yellowish.---Mud or shallow water, Indiana to Montana, south to Tennessee, Texas and Colorado. Our few records from near Denver, Colorado, at about 5300 feet but the plant has also been reported from the southwestern part of the state.

18. Gerardia L.

Annual caulescent herbaceous plants; stems erect, slender; leaves entire, mainly opposite, linear or lanc-linear; flowers solitary in the axils of the leaves or in racemes; calyx tube campanulate, the limb 5-lobed; corolla rose-purple, the tube funnelform or campanulate, 5-lobed, somewhat irregular and 2-lipped but not at all strongly so; stamens 4, in 2 pairs, anther cells equal, filaments more or less hairy; styles elongated, stigmas fused as a line on each side of the style apex; capsule loculicidal; seeds reticulate.

1. Capsules globose, 3-7 mm. long; pedicels 7-20 mm. long in flower; calyx tube 2.5-4 mm. long; corolla 10-15 mm. long -1. G. tenuifolia
1. Capsules oblong to ovoid-oblong, 7-11 mm. long; pedicels 4-13 mm. long in flower; calyx tube 4-6 mm. long; corolla 18-25 mm. long -2. G. aspera

1. Gerardia tenuifolia Vahl., Symb. Bot. 3:7. 1794.
    G. besseyana Britt.; Agalinis besseyana Britt.---Plants tending to blacken on drying; stems 10-60 cm. tall, nearly glabrous to somewhat scabrous, branches few and ascending; leaves about 2-5 cm. long with axillary fascicles usually developed; racemes elongated, 6- to 20-flowered, pedicels 7-20 mm. long in flower; calyx tube 2.5-4 mm. long, broadly campanulate, the lobes about 1-2 mm. long; corolla 10-15 mm. long; filaments glabrous to somewhat lanate; anthers sparingly hairy; capsules 3-7 mm. long, globose.---Dry loam or sandy soil, often in partial shade. Maine to North Dakota, south to Florida, Texas and Colorado. Our records from the northeastern quarter of Colorado at 3500-5500 feet.
2. Gerardia aspera Dougl., Benth. in DC., Prodr. 10:517. 1846.
    Agalinus aspera (Dougl.) Britt.---Plains and hills. Minnesota to North Dakota, south to Illinois and Oklahoma. Reported from western Nebraska and Kansas and may be expected in eastern Colorado.

19. Mimulus L. MONKEYFLOWER

Annual or perennial herbaceous plants; stems leafy or scapose; leaves opposite or basal, sessile or petioled, entire to dentate; flowers solitary and axillary or in terminal leafy racemes, usually showy; calyx tubular or campanulate, 5-angled and usually 5-toothed or 5-lobed, these more or less unequal and shorter than the tube; corolla usually yellow (but also red or blue), bilabiate to nearly regular, 5-lobed with a cylindrical or funnelform tube this with a pair of longitudinal ridges on the lower side of the throat; stamens 4, in pairs, the anther sacs divergent; style slender, stigma more or less 2-lobed; capsule loculicidal; seeds many.

1. Corolla 3.3-5.5 cm. long, crimson, red, pink, rose or lilac; anthers bearded
    2. Stems erect; corolla pink or rose -1. M. lewisii
    2. Stems prostrate; corolla red or crimson -2. M. eastwoodiae
1. Corolla usually not over 3.3 cm. long (if over then blue or yellow with glabrous anthers), variously colored; anthers glabrous or bearded

3. Pedicels 2-3 mm. long, less than half as long as the calyx; upper part of style glandular-puberulent; corolla usually reddish-purple or purple and 1.5-2 cm. long; anthers bearded -3. **M. nanus**
3. Pedicels 4 mm. long or more, as long or longer than the calyx; upper part of style glabrous to puberulent, never glandular; corolla various but never both reddish-purple and 1.2-2 cm. long; anthers bearded or glabrous
    4. Corolla blue or purple, 2.5-3.5 cm. long, the throat nearly closed by a prominent palate -4. **M. ringens**
    4. Corolla usually yellow, reddish, pink or rose, of various lengths (but if even approaching blue or purple then not over 15 mm. long), throat open (except in 2 species with yellow corollas)
        5. Calyx teeth decidedly unequal, the upper longer than the others (often 2 or 3 times); corolla yellow (may be red-spotted); anthers glabrous
            6. Throat of corolla open, the whole corolla 7-18 mm. long; some of calyx teeth often obscure or obsolete
                7. Fruiting pedicels 10-20 mm. long, less than twice as long as the subtending bracts; calyx 5-10 mm. long; corolla 8-12 mm. long -5A. **M. glabratus fremontii**
                7. Fruiting pedicels 25-50 mm. long, more than twice as long as the subtending bracts; calyx 10-15 mm. long; corolla 12-18 mm. long -5B. **M. glabratus utahensis**
            6. Throat of corolla partly or nearly closed by a prominent palet, the whole corolla usually over 20 mm. long; calyx teeth usually all 5 definite -6. **M. guttatus**
        5. Calyx teeth equal or nearly so; corolla yellow, red, pink, rose or lilac; anthers glabrous or hairy
            8. Mature calyx strongly inflated; corolla yellow and about 7-14 mm. long; leaves with definite petioles these often longer than the blades -7. **M. floribundus**
            8. Mature calyx little or not at all inflated (may be distended somewhat by the developing fruit inside); corolla various but seldom both yellow and 7-14 mm. long; leaves sessile or with short petioles, much shorter than the leaves
                9. Perennial plants; anthers bearded; corolla over 1.5 cm. long -8. **M. moschatus**
                9. Annual plants; anthers glabrous; corolla 6-10 mm. long
                    10. Corolla yellow, 5-6 mm. long; calyx teeth glabrous; stigma lobes unequal -9. **M. suksdorfii**
                    10. Corolla red or rose-colored, rarely yellow, 6-10 mm. long; calyx teeth usually ciliate; stigma lobes equal -10. **M. rubellus**

1. **Mimulus lewisii** Pursh, Fl. Am. Sept. 2:427. 1814.
Perennial plants from a rootstock; stems 30-70 cm. tall, mostly simple, more or less viscid-pubescent; leaves 2-7 cm. long, rather thin, oblong-lanceolate to oblong-ovate, irregularly dentate to nearly entire, 3- to 5-nerved, sessile; flowers few to several, pedicels usually longer than the subtending leaves; calyx 15-25 mm. long, teeth nearly equal; corolla 3-5.5 cm. long, rose-red to pink often blotched or lined with darker red; anthers more or less bearded. ---Along streams and lakes. Minnesota to British Columbia, south to Colorado and California. Our few records from northcentral and northwestern Colorado at 7000-11,000 feet.
2. **Mimulus eastwoodiae** Rydb., Bull. Torr. Club 40:483. 1913.
Utah, Nevada and Arizona. Has been reported in southeastern Utah and northeastern Arizona and may be in adjacent Colorado.
3. **Mimulus nanus** Hook.& Arn., Bot. Beechey's Voy. 378. 1840.
Eunanus tolmiei Benth.---Montana to Oregon, south to Colorado and California. The writer has never been able to locate a specimen from this state.
4. **Mimulus ringens** L., Sp. Pl. 634. 1753.
Perennial plants with rootstocks; stems 30-100 cm. tall, glabrous, erect or nearly so, 4-angled or 4-winged; leaves 2.5-10 cm. long, glabrous, oblong to oblong-lanceolate, serrate, tapering to the sessile base; pedicels about 2-3.5 cm. long, stout; calyx 1.2-1.7 cm. long, teeth equal, slender and subulate, tube strongly angled; corolla 2.5-3.5 cm. long, blue or purple, lobes unequal, throat nearly closed by a palate; anthers glabrous; styles glabrous.---Moist or wet ground. Canada, south to Florida and Texas. Our record from near Denver, Colorado, at about 5300 feet.
5. **Mimulus glabratus** H.B.K., Nov. Gen. et Sp. 2:370. 1818.
Perennial plants; stems 10-40 cm. long, creeping and freely rooting at nodes, glabrous or nearly so; leaves broadly ovate to oval or orbicular, glabrous or nearly so, irregularly dentate or sometimes entire, lower petioled, upper sessile; flowers axillary on upper part of stem, pedicels longer than the calyx; calyx 5-15 mm. long, teeth broad, obtuse and lateral ones sometimes obsolete, upper teeth larger (about 2-3 times) than others; corolla 7-18 mm. long, open in throat, yellow; anthers glabrous; styles glabrous. Represented in this state by two varieties.
5A. **Mimulus glabratus fremontii** (Benth.) Grant, (var.) Ann. Mo. Bot. Gard. 11:190. 1924.
M. geyeri Torr.---Pedicels in fruit 10-20 mm. long, less than twice as long as the subtending bracts; calyx 5-10 mm. long; corolla 7-12 mm. long.---Mud or in shallow water. Manitoba, south to Mexico. Our records scattered in the eastern half of Colorado at 3500-7000 feet.
5B. **Mimulus glabratus utahensis** Pennell, (var.) Acad. Nat. Sc. Phila. Mon. 1, p. 123. 1935.
Fruiting pedicels 25-50 mm. long, more than twice as long as the subtending bracts; calyx 10-15 mm. long; corolla 12-18 mm. long.---Mud and shallow water. Colorado to Nevada. Our few records from western Colorado at 5500-6000 feet.
6. **Mimulus guttatus** DC., Cat. Pl. Hort. Monsp. 127. 1813.
M. langsdorfii Donn; M. hallii Greene; M. puberulus Greene---Annual or perennial plants; stems 5-55 cm. tall, usually more or less erect but sometimes trailing, little branched, glabrous to pubescent especially above; leaves 8-10 cm. long, variable in shape, broadly ovate to oblong-lanceolate or spatulate, coarsely and irregularly dentate, glabrous to pubescent, lower petioled, upper sessile; flower racemose to solitary, pedicels twice the length of the calyx; calyx 8-17 mm. long, inflated in fruit, teeth short-triangular, the upper tooth definitely longer (about twice); corolla 2-4 cm. long rarely less, yellow, usually spotted with red, a

palate nearly closing the throat; anthers glabrous; styles glabrous or puberulent. The above description includes M. tilingi Regel and M. corallinus Greene, which the writer cannot separate from M. guttatus in this state.---Wet ground or in shallow water. Montana to Alaska, south to Mexico and California. Our records scattered in the western half of Colorado at 5500-12,000 feet.

7. Mimulus floribundus Dougl. in Lindl., Bot. Reg. 13: pl. 1125. 1828.

M. membranaceus A. Nels.---Annual plants; stems 3-30 cm. tall, viscid-villous to nearly glabrous, more or less branched, erect to diffusely spreading and reclining; leaves 4-40 mm. long, rather thin, ovate to ovate-lanceolate, dentate, on petioles longer or somewhat shorter than the blades, viscid-villous to nearly glabrous; flowers on pedicels longer than the calyx; calyx 4-7 mm. long in flower, longer and inflated in fruit, teeth short-triangular, equal, tube strongly angled; corolla 7-14 mm. long, yellow, dotted or streaked with red in the throat, open in the throat; anthers glabrous; styles glabrous.---Moist or wet ground. Wyoming to British Columbia, south to Mexico and California. Our records in the western half of Colorado at 5000-8500 feet.

8. Mimulus moschatus Dougl. in Lindl., Bot. Reg. pl. 1118. 1828.

Perennial plants, more or less viscid-pubescent; stems 4-30 cm. long, creeping or decumbent and rooting at lower nodes; leaves 1-4 cm. long, ovate, entire or sparingly denticulate, on petioles definitely shorter than the blades; flowers few, pedicels longer than the calyx; calyx 8-10 mm. long, teeth lanceolate, subequal or nearly equal; corolla 14-20 mm. long, yellow, usually red-striped in throat; anthers bearded; styles glabrous.---Moist or wet ground. Montana to British Columbia, south to Colorado and California. Our few records from northcentral and northwestern Colorado at 6500-8500 feet.

9. Mimulus suksdorfii A. Gray. Syn. Fl. N. Amer. ed. 2. 2(1):450. 1886.

Annual plants; stems 1.5-7 cm. tall, sparsely viscid-puberulent and more or less tinged with red; leaves 3-12 mm. long, oblong, oblanceolate or linear, entire or irregularly toothed, sessile or lower petioled, more or less tinged with red especially below; flowers 1 to many on pedicels 5-7 mm. long and mostly longer than the calyx; calyx 5-6 mm. long, 5-angled, with 5 equal or subequal teeth, these broad; corolla 5-6 mm. long, yellow; anthers glabrous; styles glabrous.---Moist or wet ground. British Columbia, south to Colorado, Arizona and California. Our few records from northcentral Colorado at 7500-8000 feet.

10. Mimulus rubellus A. Gray in Torr., Bot. Mex. Bound. 116. 1859.

M. gratioloides Rydb.---Annual plants; stems 3-15 cm. tall, simple or branched from base, glabrous to glandular-puberulent; leaves oblong, lanceolate to linear, obtuse, sessile or nearly so, margin entire to irregularly toothed; flowers scattered, pedicels 1-2 cm. long; calyx 4-9 mm. long, teeth short, less than 1 mm. long, equal, usually ciliate; corolla 6-10 mm. long, yellow, rose-red or sometimes white, throat funnelform; anthers glabrous.---Mountains. Wyoming to Washington, south to New Mexico and California. Our records from the western two-thirds of Colorado at 5500-9000 feet.

Family 106. Lentibulariaceae BLADDERWORT FAMILY

Small aquatic or semiaquatic herbaceous plants; leaves capillary-divided, usually bearing small inflated bladders (that float the plant and trap small water organisms); flowers perfect, irregular, solitary or racemose, on slender scapes, the pedicels with 2 bractlets; calyx 2-lipped, the lips entire; corolla gamopetalous, yellow, 2-lipped, the upper lip usually erect and entire, the lower lip 3-lobed, a projecting palet often more or less closing the throat, spurred at the base; stamens 2, anthers 1-celled, filaments adnate to base of corolla tube; ovary superior, 1-celled with a free central placenta, style 1, short and thick, stigma 2-cleft; fruit a capsule, 2-valved or irregularly dehiscent; seeds numerous.

1. Utricularia L. BLADDERWORT

Characters of the family

1. Corolla about 12 mm. broad, the spur hornlike, curved and conspicuous; leaves usually pinnately divided, the ultimate divisions usually over 3 mm. long on some -1. U. vulgaris
1. Corolla about 6 mm. broad, the spur reduced to a protuberance, almost lacking; leaves dichotomously divided, the ultimate segments usually less than 3 mm. long -2. U. minor

1. Utricularia vulgaris L., Sp. Pl. 18. 1753.
Stems 30-100 cm. long, branched, submerged or free-floating, leafy; leaves 2- to 3- times pinnately divided, bearing numerous bladders these about 2-5 mm. long; scapes 7-30 cm. long; racemes 2- to 12-flowered; corolla 10-15 mm. long, spur hornlike, shorter than lower lip but conspicuous and somewhat curved.---Shallow water of ponds and streams. Throughout most of North America. Our records scattered over the state, few in the eastern half, at 3500-12,000 feet.

2. Utricularia minor L., Sp. Pl. 18. 1753.
Stems rather short, usually less than 30 cm. long; leaves 2- to 4- times dichotomously forked, the ultimate divisions flat and less than 3 mm. long, bladders about 1-2 mm. long; scapes 3-15 cm. tall, 2- to 9-flowered; corollas 5-8 mm. long, spur very short and saccate, almost wanting.---Shallow water. Greenland to Alaska, south to New Jersey and California. Our few records from northcentral and westcentral Colorado at 5500-9000 feet.

Family 107. Orobanchaceae BROOMRAPE FAMILY

Herbaceous perennial plants, lacking green chlorophyll, parasitic on roots of other plants; stems fleshy; leaves alternate, reduced to simple scales; flowers perfect or rarely dioecious, irregular, in loosely fastigiate to spicate inflorescences; calyx of 4-5 more or less united and equal or nearly equal sepals; corolla gamopetalous, somewhat bilabiate and definitely oblique, 5-lobed, persistent and withering; stamens 4, didynamous, mostly included, filaments adnate to the corolla tube; ovary 1-celled, with 4 parietal placentae, styles 1, stigma peltate; fruit a 2-valved capsule; seeds many.

1. Orobanche L.

Characters of the Family

1. Flowers with bracts close to the calyx, many, in dense spicate inflorescences or these sometimes branched below; pedicels none or shorter than the flowers; calyx lobes more or less unequal
    2. Anthers more or less woolly-hairy; corolla 20-35 mm. long -1. O. multiflora
    2. Anthers glabrous or slightly pubescent; corolla 15-20 mm. long
        3. Corolla lobes rounded, not pointed; anthers glabrous; plants of western Colorado -1A. O. multiflora arenosa
        3. Corolla lobes acute; anthers often sparingly pubescent; plants found east of Continental Divide in Colorado -2. O. ludoviciana
1. Flowers not subtended by bracts, few, in loose inflorescences; pedicels usually much longer than the flowers; calyx lobes equal or very nearly so
    4. Plants with 1-3 pedicels and flowers, these pedicels over twice as long as the true stems; calyx lobes definitely longer than the tube -3. O. uniflora
    4. Plants with several to many pedicels and flowers, pedicels usually somewhat shorter than the true stems; calyx lobes as long as or shorter than the tube
        5. Corolla purplish to pink; plants purplish -4. O. fasciculata
        5. Corolla yellow; plants yellowish -4A. O. fasciculata lutea

1. Orobanche multiflora Nutt., Journ. Acad. Nat. Sci. Phila. II. 1:179. 1848.
    Myzorrhiza multiflora (Nutt.) Rydb.---Stems 5-30 cm. tall, gray viscid-pubescent especially above; flowers nearly sessile or lower short-pedicelled, many, in dense spicate or thyrsoid inflorescences; calyx 10-17 mm. long, lobes somewhat unlike, longer than the tube; corolla 20-35 mm. long, dark purple-rose especially inside, the lobes rounded or slightly mucronate at apex; anthers woolly-hairy at least in part. This plant would be var. typica Munz.---Plains and woods. The variety from Colorado, south to Texas and Mexico. Our records scattered in the eastern half of Colorado at 4000-6000 feet.
1A. Orobanche multiflora arenosa (Suksdf.) Munz, (var.) Bull. Torr. Club 57:623. 1930.
    Anthers glabrous and corollas shorter than in the typical form of the species.---Washington, south to Colorado and California. Our records from western Colorado, (Montrose County) at about 5500 feet.
2. Orobanche ludoviciana Nutt., Gen. N. Am. 2:58. 1818.
    Myzorrhiza ludoviciana (Nutt.) Rydb.---Stems 5-20 cm. tall, usually branched, solitary or clustered, viscid-pubescent; flowers many in spicate inflorescences but lower sometimes on pedicels up to 1 cm. long; calyx 8-12 mm. long, lobes somewhat unequal and longer than the tube, lanc-linear; corolla 15-20 mm. long, usually purplish at least on upper lip, rarely yellowish, the lobes acute; anthers glabrous or slightly pubescent especially after anthesis. Our plant has been called var. genuina G. Beck, but intergrades are rather common to var. cooperi (A. Gray) G. Beck.---Plains and slopes, often on sandy ground. The typical variety from Montana to British Columbia, south to Texas and California. Our records from the eastern half of Colorado at 4500-8000 feet.
3. Orobanche uniflora L., Sp. Pl. 633. 1753.
    Thalesia sedi (Suksdorf) Rydb.---Stems short and nearly subterranean, bearing 1-3 scapes 2-7 cm. long, at least 2 or 3 times as long as the stem, glandular, pubescent; scapelike pedicels each 1-flowered, no bracts borne near calyx; calyx 6-12 mm. long, campanulate, 5-lobed, these almost all alike, the lobes longer than the tube; corolla 15-22 mm. long, usually straw-colored, yellow or tinged with lavender; anthers glabrous. Our plants seem to be var. sedi (Suksdorf) Achey.---Plains and hills. The variety from Montana to British Columbia, south to Colorado and California. Our few records from central and northcentral Colorado at 5000-6000 feet.
4. Orobanche fasciculata Nutt., Gen. N. Am. 2:59. 1818.
    Thalesia fasciculata (Nutt.) Britt.---Stems 3-15 cm. long, mostly underground, glandular-pubescent; scapelike pedicels 3-15 cm. long, usually somewhat shorter than stem, occasionally somewhat longer, glandular-pubescent, 1-flowered, no bracts near calyx; calyx 6-10 mm. long, campanulate, the lobes alike or nearly so, as long or shorter than the tube; corolla 15-30 mm. long, the lobes usually rounded, purplish to pink; anthers glabrous or pubescent. This is the plant called var. typica Achey.---Plains, hills and slopes. The variety from Michigan to British Columbia, south to Illinois, Texas and California. Our records scattered over the state at 3500-10,000 feet.
4A. Orobanche fasciculata lutea (Parry) Achey, (var.) Bull. Torr. Club 60:449. 1933.
    Thalesia lutea (Parry) Rydb.---Plants and corolla yellow, the latter 3-8 mm. wide at the throat, the lobes usually acute.---About the range of the typical form. Reported for northcentral, central and southcentral Colorado.

Family 108. **Martyniaceae** UNICORNPLANT FAMILY

Annual or perennial coarse herbaceous, viscid-pubescent plants; stems branching; leaves simple, petioled, opposite, or alternate on upper part of stem, the blades large and cordate; flowers few in terminal racemes, large and showy, irregular, perfect; calyx inflated somewhat, unequally 5-cleft with 1-2 bractlets at base; corolla gamopetalous, more or less 2-lipped, funnelform-campanulate, inflated, 5-lobed, these spreading; stamens 4, didynamous, anthers gland-tipped; ovary superior, 1-celled with 2 parietal placentae these expanded in center, style slender, stigmas 2; fruit a capsule, somewhat fleshy ending in a long incurved, hooked beak; seeds numerous, tuberculate.

1. **Martynia** L. UNICORNPLANT

Characters of the Family

1. Martynia louisianica Mill., Gard. Dict. no. 3. 1768.
M. louisiana Mill. of Manuals; Proboscoidea louisiana (Mill.) Woot. & Standl.---Annual plants; stems simple and short or with spreading branches 20-100 cm. long; leaves 5-30 cm. wide, suborbicular, entire or sinuate; calyx up to 2 cm. long; corolla up to 5 cm. long and nearly as wide, cream-colored or yellowish, spotted with purple, or sometimes clear reddish-violet; rudimentary 5th stamen often present; fruit body up to 10 cm. long, the curved horns longer than the body, hairy.---Waste places. Delaware to Colorado, south to Virginia and Mexico; also in California. Our records scattered in the eastern half of Colorado at 3500-4500 feet.

Family 109. **Rubiaceae** MADDER FAMILY

Annual or perennial, usually herbaceous plants; stems leafy, mostly 4-angled; leaves simple, entire, opposite or appearing verticillate from large leaflike stipules; flowers regular, mostly perfect (in a few species dioecious), mostly cymose; calyx with the tube completely adnate to the ovary, the limb obsolete or toothed; corolla gamopetalous and 3- to 5-lobed, rotate to funnelform; stamens as many as the corolla lobes and alternate to them, inserted on the tube; ovary inferior, 2-celled, ovules 1 in each cell, style 1 or 2; fruit dry, 2-lobed and separating into 2 indehiscent carpels, these often beset with bristles.

1. Corolla rotate; leaves appearing whorled (4 or more at a node) from the presence of leaflike stipules (2 leaves on some nodes of G. bifolium) -1. Galium
1. Corolla funnelform-salverform; leaves opposite only 2 at a node, or whorled
  2. Leaves opposite, 2 at a node -2. Kelloggia
  2. Leaves whorled, usually 8 at a node -3. Asperula

1. **Galium** L. BEDSTRAW

Annual or perennial plants, herbaceous or somewhat woody at base; stems 4-angled, often winged, usually weak and reclining on other plants; leaves opposite but appearing whorled because of the leaflike stipules; flowers mostly white, perfect or unisexual, in axillary or terminal cymules or clusters or sometimes solitary and axillary; calyx limb obsolete; corolla rotate, usually 4-lobed; styles 2, short, stigma capitate; fruit with long hairs, these often uncinate, or glabrous.

1. Plants dioecious; fruit with straight hairs; stems somewhat woody above the base; plants perennial -1. G. coloradoense
1. Flowers perfect; fruit usually either glabrous or with curved or uncinate hairs; stems not at all woody; plants annual or perennial
  2. Plants annual; fruit with uncinate hairs (if leaves 5-8 in a whorl then stem retrorsely hispid)
    3. Leaves 6-8 to a whorl; flowers in 1- to 10-flowered cymes; stems retrorsely hispid
      4. Fruit 3-5 mm. wide at maturity; leaves commonly over 2 cm. long -2. G. aparine
      4. Fruit 2-3 mm. wide at maturity; leaves usually less than 2 cm. long -2A. G. aparine echinospermum
    3. Leaves 2-4 to a whorl; flowers solitary; stems glabrous or hispid but not retrorsely so
      5. Flowers sessile or very nearly so; leaves ovate or oblong; stems usually hispidulous -3. G. proliferum
      5. Flowers pedicellate; leaves linear or linear-oblong; stems glabrous -4. G. bifolium
  2. Plants perennial; fruit glabrous to hairy, the hairs not usually uncinate (except in G. triflorum with stems smooth and leaves 5-6 in a whorl)
    6. Leaves 3-nerved; stems erect; flowers in dense many-flowered panicles; fruit with short or long hairs, not uncinate -5. G. boreale
    6. Leaves 1-nerved; stems usually weak, reclining or prostrate; flowers in few- to several-flowered inflorescences; fruit glabrous or with uncinate hairs
      7. Fruit uncinate-hairy; leaves appearing in 6's or sometimes 5's at a node -6. G. triflorum
      7. Fruit glabrous; leaves appearing in 4's at a node
        8. Stems usually less than 12 cm. long, low and prostrate, forming mats, smooth or nearly so on angles; leaves glabrous or very nearly so; pedicels rather stout (by comparison) -7. G. brandegei
        8. Stems usually over 12 cm. long, loosely spreading or even erect, not forming mats, usually retrorsely scabrous or hispid on angles; leaves hispidulous or scabrous especially on margins and veins; pedicels capillary, very slender -8. G. trifidum

1. **Galium coloradoense** W. F. Wright, Zoe 5:54. 1900.

Perennial plants; stems 20-35 cm. tall, somewhat woody at base, glabrous, erect or at least branches erect; leaves 1-3 cm. long, linear, 1-ribbed, rather thick, 4 in a whorl, acute-cuspidate; flowers dioecious, in 1- to 15-flowered cymes, the staminate flowers more numerous; corolla white to yellowish; fruit with long straight hairs.---Dry plains and hills. Colorado, Utah and Arizona. Our records from the western fourth of Colorado at 4500-8000 feet.

2. **Galium aparine** L., Sp. Pl. 108. 1753.

Annual plants; stems 20-100 cm. long, weak, prostrate-scrambling and spreading, sometimes short and erect, angles retrorsely hispid; leaves 2-7 cm. long, 6-8 in a whorl, oblanceolate to linear-oblong, mucronate-acute, margins and middle veins hispid, 1-nerved; flowers perfect in 1- to 3-flowered cymes, in upper leaf axils; corolla white to greenish-white; fruit 3-5 mm. wide, with short uncinate bristles.---Plains and valleys often along streams. Widely distributed in North America; naturalized from Europe. Our records scattered in the western half of Colorado at 5000-9500 feet.

2A. **Galium aparine echinospermum** (Wallr.) Farwell, (var.) Rep. Mich. Acad. Sci. 19:261. 1917.

G. aparine var. vaillantii (DC.) Koch; G. vaillantii DC.---Leaves are shorter (5-20 mm. long) in this variety and fruit is 2-3 mm. wide at maturity (not 3-5 mm.). Intergrades with the species in Colorado although the extremes are certainly different.---About the same distribution as the species, both generally and in Colorado. Our records at 5000-6000 feet.

3. **Galium proliferum** A. Gray, Pl. Wright. 2:67. 1853.

Utah, south to Texas, Mexico and Arizona. May well be in southern or western Colorado.

4. **Galium bifolium** Wats., Bot. King's Exped. 134. 1871.

Annual plants; stems 5-15 cm. tall, erect or ascending, glabrous; leaves 6-15 mm. long, appearing 2-4 at a node, thin, acutish, 1-nerved, when 4 present the alternate pair usually smaller; flowers perfect, solitary but on definite axillary peduncles; corolla white; fruit with uncinate hairs.---Meadows and moist ground in general. Montana to British Columbia, south to Colorado and California. Our records from northcentral and northwestern Colorado at 6500-8500 feet.

5. **Galium boreale** L., Sp. Pl. 108. 1753.

Perennial plants; stems 20-70 cm. long, erect or nearly so, glabrous or nearly so; leaves 2-5 cm. long, in 4's at a node, linear to broadly lanceolate, obtuse or acutish, 3-nerved, margins sometimes ciliate; flowers perfect, in dense terminal, many-flowered panicles; corolla white or ochroleucous; fruit with short appressed incurved hairs to villous-hirsute with straight spreading hairs, possibly glabrous at times according to report but none seen in this area.---Rocky slopes and valleys along streams. Quebec to Alaska, south to Pennsylvania, Texas and California; Eurasia. Our records widely scattered over the western two-thirds of Colorado at 5000-10,000 feet.

6. **Galium triflorum** Michx., Fl. Bor. Am. 1:80. 1803.

G. flaviflorum Heller---Perennial plants from slender creeping rootstocks; stems 20-80 cm. long, rather slender, reclining, glabrous or very slightly scabrous; leaves 2-7 cm. long, appearing 5-6 at a node, oblanceolate to lanceolate-ovate, cuspidate, 1-nerved, somewhat scabrous on midrib and margins; peduncles usually 2- to 3-flowered, the flowers perfect; corolla whitish to greenish or purplish; fruit with soft uncinate hairs.---Mostly in damp, shaded ground. Newfoundland to Alaska, south to Florida and California. Our records from the western half of Colorado at 6500-9000 feet.

7. **Galium brandegei** A. Gray, Am. Acad. Arts & Sci., Proc. 12:58. 1877.

Perennial plants; stems 5-12 cm. long, low, weak, forming dense very leafy mats, glabrous to minutely hispidulous; leaves 3-12 mm. long, commonly 4 at a node, obovate to oblong-lanceolate, 1-nerved, this faint, usually glabrous, obtuse or somewhat mucronate at apex; flowers perfect, usually solitary on short, rather thickened pedicels; corolla white; fruit small and glabrous.---Usually moist ground. Wyoming, south to New Mexico and California. Our records from the western half of Colorado, none from the southern part as yet, at 6500-9500 feet.

8. **Galium trifidum** L., Sp. Pl. 105. 1753.

Perennial plants; stems 10-30 cm. long, slender and weak, ascending or prostrate, slightly scabrous or hispidulous on the angles; leaves 4-16 mm. long, commonly 4 at a node, oblong to linear-spatulate, 1-nerved, thin, obtuse or acutish, midrib and margins usually scabrous; flowers perfect, in slender axillary 1- to 3-flowered scabrous peduncles; flowers white or whitish; fruit glabrous. Our plants intergrade very commonly with G. subbiflorum (Wieg.) Rydb. which is supposed to have stem-angles and pedicels smooth or nearly so. The writer sees no value in maintaining the latter even as a variety in this state.---Moist ground. Labrador to Alaska, south to New York, Nebraska, Arizona and Colorado. Our records from the western half of Colorado, mostly from the northern part, at 6500-9500 feet.

2. Kelloggia Torr.

Plants perennial and herbaceous from slender rootstocks; leaves opposite with small stipules, sessile, narrowly lanceolate to lanceolate; flowers perfect, small, white or pinkish, in open cymose panicles; calyx teeth minute; corolla funnelform-salverform, usually 4 lobes present; stamens exserted; style filiform, stigmas 2, clavate; fruit with uncinate bristles.

1. **Kelloggia galioides** Torr. in Wilkes, U. S. Expl. Exped. 17:332. 1874.

Wyoming to Washington, south to Arizona and California. Not reported directly from Colorado but close to the western part.

3. Asperula L. WOODRUFF

Perennial herbaceous plants; stems 4-angled; leaves whorled usually in 8's; flowers in paniculate cymes, terminal or from upper axils; calyx limb obsolete; corollas funnelform, 4-lobed; styles 2-cleft; fruit hispid.

1. Asperula odorata L., Sp. Pl. 103. 1753.
   Stems erect, glabrous; leaves 2-4 cm. long, oblong-oblanceolate, acute; flowers white or pink, corollas 3-6 mm. long.---Waste places usually as an escape. Native to Europe and in various parts of this country especially in the eastern states. Our one record from near Denver, Colorado, at about 5300 feet.

Family 110. **Plantaginaceae** PLANTAIN FAMILY

Annual or perennial acaulescent herbaceous plants; leaves alternate but all basal, simple, blades often broad and prominently ribbed; flowers small, regular, usually perfect but sometimes unisexual, in terminal long-peduncled bracted spikes; calyx of 4 persistent sepals, all alike or 2 longer, often scarious-margined; corolla gamopetalous, campanulate or tubular, with 4 erect or spreading lobes, scarious at least on margins; stamens 2 or 4, distinct, attached to the tube of corolla, anthers versatile, opening lengthwise; ovary superior, 1- to 4-celled, with 1 style and 1 filiform stigma; fruit a circumscissile capsule (pyxis).

1. Plantago L. PLANTAIN

Characters of the Family

1. Leaves linear to filiform, rarely over 1 cm. wide; plants annual; scapes or leaves more or less silky-pubescent or somewhat cinereous-pubescent
   2. Stamens 2; corolla lobes (in some flowers) erect and closing over the capsule; flowers more or less dioecious or polygamous; leaves linear-filiform, about 1 mm. wide   -1. P. elongata
   2. Stamens 4; corolla lobes permanently spreading or reflexed; flowers all perfect; leaves linear, but well over 1 mm. wide
      3. Bracts, at least the lower ones, 2 or more times as long as the calyx
         4. Bracts all drying green or gray; plants light green   -2. P. spinulosa
         4. Bracts all drying dark brown or black; plants dark green   -3. P. aristata
      3. Bracts shorter than or not over twice as long as the calyx   -4. P. purshii
1. Leaves lanceolate to broader, rarely less than 1 cm. wide; plants perennial (or occasionally biennial); scapes and leaves glabrous to hairy but not silky-pubescent or cinereous-pubescent
   5. Leaf blades broadly ovate, abruptly contracted into the petiole; seeds 6-20 to a capsule (over 2 to a cell); spikes usually over 6 cm. long
      6. Capsule circumscissile at the middle, ovoid in shape   -5. P. major
      6. Capsules circumscissile much below the middle, elongate-ovoid in shape   -6. P. asiatica
   5. Leaf blades lanceolate, oblanceolate to elliptic, gradually tapering into the petiole; seeds 2-4 to a capsule (not over 2 to a cell); spikes of various lengths but often less than 6 cm. long
      7. Leaves narrowly oblong-lanceolate to narrowly lanceolate; capsule circumscissile at middle; seeds 2, concave on the faces; spikes short, oblong or ovoid and commonly less than 3 cm. long at first (may elongate in fruit)   -7. P. lanceolata
      7. Leaves lanceolate to elliptic; capsule circumscissile near base or well below the middle; seeds 2-4, slightly if at all concave on faces; spikes long or short, commonly over 3 cm. long and cylindrical when in flower
         8. Crown of plants conspicuously woolly-hairy at base of petioles; spikes usually over 5 cm. long; leaves thick   -8. P. eriopoda
         8. Crown of plants not woolly or only slightly so; spikes usually less than 5 cm. long; leaves moderate to thin   -9. P. tweedyi

1. Plantago elongata Pursh, Fl. Amer. Sept. 729. 1814.
   P. myosuroides Rydb.---Annual plants; leaves more or less cinereous-puberulent, linear-filiform, 3-8 cm. long and about 1 mm. wide; scapes 3-10 cm. high, longer than the leaves; spikes 1-8 cm. long, loosely flowered; bracts about as long as the calyx, triangular-ovate; flowers subdioecious or ploygamous; sepals about 1.5-2 mm. long; corolla lobes about 0.5 mm. long, on some flowers becoming erect and closing over the capsule; stamens 2; capsule ovoid to oblong-ovoid, circumscissile just below the middle; seeds about 4, nearly flat on both sides.---Usually moist ground. Montana to Alberta, south to Oklahoma, Utah and Oregon. Our records thinly scattered over the western two-thirds of Colorado at 5000-7000 feet.
2. Plantago spinulosa Decne.ex DC., Prodr. 13:713. 1852.
   Annual plants, light green in color; leaves 2.5-15 cm. long and 1-4 mm. wide, linear, glabrate to villous, entire; scapes 5-10 cm. tall; spikes about 3-14 cm. long, short to long-cylindrical, sometimes interrupted; bracts 6-15 mm. long, aristate, over twice as long as the calyx; flowers perfect; sepals 2-3 mm. long; corolla lobes about 1-1.5 mm. long, spreading and remaining so; stamens 4; capsule oblong, circumscissile at the middle; seeds 2, usually somewhat concave on face.---Plains and hills. Saskatchewan and Alberta, south to Texas and New Mexico. Our records from the eastern half of Colorado, except the extreme southern part, at 4000-6000 feet.
3. Plantago aristata Michx., Fl. Bor. Am. 1:95. 1803.
   New York to South Dakota, south to Florida and Texas. To be expected in eastern Colorado.
4. Plantago purshii Roem. & Schult., Syst. Veg. 3:120. 1818.
   P. xerodea Morris---Annual plants, woolly or silky throughout; leaves 3-10 cm. long and 1-4 mm. wide, linear, entire; scapes 4-30 cm. long, erect or ascending; spikes 1-2 cm. long but variable in length, dense and cylindrical; bracts variable in length but rarely over twice as long as the calyx, linear-subulate; flowers perfect; sepals about 2-3 mm. long; corolla lobes 1-2 mm. long, spreading and remaining so; stamens 4; capsule oblong, circumscissile about

the middle; seeds 2, usually concave on face.---Plains and slopes, often on sandy soil. Saskatchewan to British Columbia, south to Texas and California. Our records scattered over the state at 3500-7000 feet.

5. Plantago major L., Sp. Pl. 112. 1753.

Perennial plants, glabrous or somewhat pubescent; leaves 5-35 cm. long, thick, ovate to oval, abruptly contracted at base into the petiole, 5- to 7-ribbed, entire to somewhat toothed; scapes 8-60 cm. long; spikes variable but usually 5-30 cm. long, rather dense; bracts about as long as calyx or somewhat longer at base of spike; flowers perfect; sepals about 2-3 mm. long; corolla lobes about 1 mm. long, spreading and remaining so; stamens 4; capsule ovoid, circumscissile about at the middle; seeds 5-18, not concave on the face.---Waste places, common in lawns. Widely distributed in North America; naturalized from Europe. Widely scattered in Colorado, our records at 3500-9000 feet.

6. Plantago asiatica L., Sp. Pl. 113. 1753.

P. nitrophila A. Nels.---Closely related to P. major and probably only a variety of it. Differs in having the leaves usually thinner; capsules elongate-ovoid, circumscissile much below the middle.---Waste places. Widely distributed especially in the western half of North America. Our few records from northcentral and southcentral Colorado at 5000-7500 feet.

7. Plantago lanceolata L., Sp. Pl. 113. 1753.

Biennial or perennial plants; leaves 4-30 cm. long and mostly over 1 cm. wide, entire or denticulate, more or less pubescent and woolly-hairy at base, 3- to 7-ribbed; scapes 10-60 cm. long; spikes 1-3 cm. long, dense at first and short-ovoid or ovoid-oblong, somewhat elongated in fruit; bracts broad, about as long as the calyx; flowers usually perfect; sepals 2-3 mm. long; corolla lobes about 1.5 mm. long, spreading and remaining so; stamens 4; capsule oblong, circumscissile about at the middle; seeds 2, deeply concave on the sides.---Lawns, fields and waste places. Widely distributed in North America; naturalized from Europe. Our records scattered over the state at 4500-7000 feet.

8. Plantago eriopoda Torr., Ann. Lyc. N. Y. 2:237. 1828.

Perennial plants, the rootstock rather long and conspicuously reddish-woolly among the leaf bases, other parts glabrate to pubescent; leaves 6-20 cm. long, oblanceolate, oblong-lanceolate to narrowly ovate, 3- to 9-ribbed, entire or repand-dentate, gradually cuneate to petiole; scapes 10-40 cm. tall; spikes 4-12 cm. long, usually over 5 cm., dense above; bracts about as long as the calyx; flowers perfect; sepals about 2.5-3 mm. long; corolla lobes 1-2 mm. long, spreading and remaining so; stamens 4; capsule ovoid-oblong, circumscissile below the middle; seeds 2-4, not concave on face.---Plains, hills and mountains. Nova Scotia to Alberta, south to Minnesota, New Mexico and California. Our records scattered over Colorado, few in extreme eastern or western parts, at 3500-10,000 feet.

9. Plantago tweedyi Gray, Syn. Fl. N. Amer. 2(1):390. 1886.

Perennial plants, glabrous to more or less pubescent especially on the scape but not woolly-hairy at base of petioles; leaves 3-7 cm. long, rather thin, lanceolate-spatulate to lanceolate, or even narrowly ovate, tapering gradually to the petiole, 3- to 5-nerved, entire or nearly so; scapes 10-20 cm. high; spikes 2-5 cm. long, rarely over 5 cm. even in fruit, rather dense; bracts about as long or somewhat shorter than the calyx; flowers perfect; sepals about 1.5-2.5 mm. long; corolla lobes about 1 mm. long, spreading and remaining so; stamens 4; capsule ovoid, circumscissile near the base; seeds 2-4, not concave on the face.---Slopes and mountains. Montana, south to New Mexico and Utah. Our records from northcentral and southcentral Colorado at 8000-11,000 feet.

Family 111. Caprifoliaceae HONEYSUCKLE FAMILY

Shrubs, trees or vines, if appearing herbaceous then at least woody near base; leaves simple or pinnately compound, opposite; flowers perfect, regular or more or less irregular, mostly cymose; calyx 3- to 5-lobed or toothed; corolla of united petals, funnelform, tubular to rotate, sometimes gibbous at base, 4- to 5-lobed, sometimes 2-lipped; stamens 5, inserted on the tube of the corolla and alternate to its lobes (4 in Linnaea); ovary inferior, 1- to 6-celled, stigmas capitate or 2- to 5-lobed or cleft; fruit berrylike, capsulelike or drupelike.

1. Leaves pinnately compound -1. Sambucus
1. Leaves simple (may be lobed)
   2. Styles 3-lobed or 3-cleft at apex; corolla rotate to very short campanulate; flowers in compound more or less flat-topped cymes -2. Viburnum
   2. Styles not branched, stigmas capitate; corolla funnelform, salverform, tubular or campanulate; flowers in pairs or in irregular close clusters
      3. Low evergreen plants, only slightly woody at base, with prostrate, creeping stems; flowers in pairs on an elongated, slender terminal peduncle; stamens 4; fruit dry -3. Linnaea
      3. Deciduous, definitely woody, upright or spreading plants; flowers in clusters or if in pairs then axillary; stamens usually 5; fruit fleshy
         4. Corolla more or less irregular, gibbous on 1 side at base; ovary 2- to 3-celled (the partitions sometimes incomplete); berry never white (ours red to black), often containing more than 2 seeds; flowers borne in close pairs (rarely 3) on axillary peduncles; leaves usually over 3 cm. long -4. Lonicera
         4. Corolla regular, the tube not gibbous or only slightly so at base; ovary 4-celled; berry white (in all but S. orbiculatus doubtful in Colorado), 1- to 2-seeded; flowers in clusters or solitary, not always 2 on a common peduncle; leaves in most species less than 3 cm. long -5. Symphoricarpos

1. Sambucus L. ELDER

Shrubs or small trees; twigs with large pith; leaves odd-pinnate; leaflets lanceolate to ovate and serrate; flowers in broad compound cymes; calyx lobes minute, generally 5; corolla white to yellowish, rotate or saucer-shaped, regular, 5-lobed; stamens 5, inserted at the base of the corolla; styles short, 3- to 5-cleft; ovary 3- to 5-celled, 1 ovule in each cell; fruit a berrylike drupe, containing 3-5 seedlike nutlets.

1. Inflorescence flat-topped or umbrella-shaped with elongated compound rays, the axis seldom extending beyond the lowest branches, the whole over 8 cm. wide; fruit purple-black or blue; pith of older twigs white
   2. Fruit dark blue with a glaucous bloom, usually over 5 mm. wide; plants of western Colorado
                                              -1. S. coerulea
   2. Fruit purplish-black, not glaucous, 4-5 mm. wide; plants of eastern Colorado   -2. S. canadensis
1. Inflorescence short-pyramidal, the axis extending beyond the lowest branches, not over 7 cm. wide; fruit black, yellowish, red or orange; pith of older twigs brownish
   3. Fruit black; cymes 5-7 cm. wide or more   -3. S. melanocarpa
   3. Fruit red, yellowish or orange; cymes 3-5 cm. wide   -4. S. pubens

1. **Sambucus coerulea** Raf., Alsogr. Amer. 48. 1838.
S. glauca Nutt. of Manuals---Large shrubs or small trees 2-6 m. tall; stems glabrous, pith white in color; leaves light green, glabrous, with 5-9 leaflets these 4-8 cm. long, rarely longer, oblong or oblong-lanceolate (ours sometimes elliptic-oblong) to lanceolate, acuminate (rather shortly so on some), serrate, usually oblique at base; inflorescence 8-15 cm. wide rarely wider, flat-topped or umbrella-shaped the axis not extending above the lowest branches, flowers white; berry 5-6 mm. in diameter, dark blue and glaucous, thus decidedly blue when fresh. Probably S. neomexicana Woot. is a synonym of the above.---Creeks, valleys and bases of cliffs. Alberta to British Columbia, south to Arizona and California. Our few records from western Colorado at 5500-8000 feet.

2. **Sambucus canadensis** L., Sp. Pl. 269. 1753.
Shrubs 1-3 m. tall, glabrous or nearly so throughout; older stems with whitish pith; leaflets 5-11, usually 7, ovate, oval or elliptic, acuminate or acute, sharply serrate; inflorescence 10-20 cm. wide, flat-topped or umbrella-shaped, axis not extending above the lower branches; flowers white; berry 4-5 mm. in diameter, not glaucous, deep purple or black.---Roadsides and valleys. Nova Scotia to Manitoba, south to Florida and Texas. Our records in northcentral and central Colorado, probably as escapes from cultivation, at about 5000-5500 feet.

3. **Sambucus melanocarpa** A. Gray, Proc. Am. Acad. 19:76. 1883.
Shrubs up to 3 m. tall, rarely taller; young branches usually pubescent, pith white or light brown at first, darker on older twigs; leaflets 5-7, these 5-15 cm. long, scurfy-puberulent or sparsely villous below, rarely glabrate, oval, ovate or oblong-lanceolate, serrate, acuminate; cyme 5-7 cm. wide or more, short-pyramidal, ovoid or hemispherical but central axis extending above lowest branches; flowers white; fruit black at maturity, about 6 mm. in diameter.---Along streams and on mountain slopes. Alberta to British Columbia, south to New Mexico and California. Our records in the western half of Colorado at 7500-10,000 feet.

4. **Sambucus pubens** Michx., Fl. Bor. Am. 1:181. 1803.
S. racemosa of Authors for American plants; S. microbotrys Rydb.---Shrubs 60-400 cm. tall; young branches usually glabrous, sometimes more or less pubescent, pith of older twigs brown or reddish-brown; leaves commonly with 5-7 leaflets, these 5-12 cm. long, glabrous or more or less pubescent especially below, ovate-lanceolate to oval, sharply serrate, acuminate, more or less oblique at base; cyme 4-5 cm. wide and usually slightly longer, short-pyramidal, the central axis extending beyond the lowest branches; flowers white to yellowish; fruit about 5 mm. in diameter, orange-red to red, sometimes yellowish.---Along streams and on moist slopes. Newfoundland to British Columbia, south to Pennsylvania, New Mexico and California. Our records scattered in the western half of Colorado at 8000-12,000 feet.

2. Viburnum L.

Upright shrubs or small trees; leaves simple, lobed or toothed; flowers in compound flat-topped cymes; calyx 5-toothed; corolla rotate to short-campanulate, regular, 5-lobed; stamens 5; styles short, 3-lobed or 3-cleft; ovary 1- to 3-celled, 1 ovule to a cell; fruit a 1-seeded drupe.

1. Some of leaves 3-lobed at apex; cyme small, not over 3 cm. across, on definite peduncles 1-2 cm. long; fruit red   -1. V. pauciflorum
1. None of leaves 3-lobed; cyme rather large, over 4 cm. across, sessile; fruit blue-black   -2. V. lentago

1. **Viburnum pauciflorum** Pylaie; T. & G., Fl. N. Amer. 2:17. 1841.
Shrubs 1-2 m. tall; twigs glabrous, gray to brown, buds with 2 connate outer scales; leaves variable, 4-10 cm. long, from elliptic and pinnately veined with margins coarsely to serrate-dentate to suborbicular or broadly obovate and with 3 lobes at apex and 3 palmate veins, at least some of the latter leaves present, glabrous above, more or less pubescent on margins and veins beneath; cyme small, 1.5-3 cm. across, rather few-flowered, these all perfect and small, peduncles 1-2 cm. long; fruit 8-10 mm. in diameter, round to ovoid, red, stone flat. The correct name for our plant may be V. edule (Michx.) Raf.---Woods. Newfoundland to Alaska, south to New Hampshire, Minnesota, Colorado and Oregon: Asia. Our few records from northcentral and central Colorado at 7000-9000 feet.

2. **Viburnum lentago** L., Sp. Pl. 268. 1753.

Shrubs or small trees, up to 10 m. tall; twigs glabrous or somewhat pubescent; buds with 1 pair scales visible; leaves 4-10 cm. long, ovate to elliptic, sharply serrate, pinnately veined, glabrous or slightly pubescent beneath; cyme 5-12 cm. across, sessile, flowers many, all perfect; fruit 10-15 mm. long, ellipsoid or oval, blue-black but with a whitish bloom.---Along streams and in rich woods. Hudson Bay to Manitoba, south to Georgia and Colorado. Our one record, perhaps from a cultivated plant or an escape from cultivation, from northcentral Colorado at about 5500 feet.

3. Linnaea L. TWINFLOWER

Plants somewhat woody at base; stems creeping, slender, forming loose mats; leaves simple, evergreen and thickish, obovate to nearly orbicular, crenulate; flowers nodding, borne in pairs at the apex of elongated terminal peduncles; calyx teeth 5, subulate-lanceolate, deciduous from the mature fruit; corolla nearly regular, broadly funnelform, 5-lobed, white to pink; stamens 4, unequal in length; ovary 3-celled, 1 cell with a perfect ovule, other cells with ovules abortive; style slender, stigma capitate; fruit 1-seeded, coriaceous dry and indehiscent.

1. **Linnaea borealis** L., Sp. Pl. 631. 1753.

L. americana Forbes---Stems often to 100 cm. long, younger twigs pubescent; leaves 8-20 mm. long, usually crenate above the middle, somewhat pubescent, short-petioled; peduncles 3-12 cm. long, slender, glandular, with 2 linear-subulate bracts at summit, the pedicels 4-20 mm. long, divergent, glandular, 2-bracteate near apex and with 2 ovate scales near ovary; calyx segments 1.5-5 mm. long; corolla 8-15 mm. long; fruit ovoid. Our plant has been called var. americana (Forbes) Rehder.---Moist woods. The species from Greenland to Alaska, south to New Jersey, New Mexico and California; Eurasia. Our records from the western half of Colorado at 8500-11,000 feet.

4. Lonicera L. HONEYSUCKLE

Erect shrubs; leaves simple, entire, deciduous; flowers in pairs rarely in 3's, on axillary peduncles, often very showy; calyx teeth 5, very short; corolla tubular or funnelform, from nearly regular to strongly 2-lipped, 5-lobed; stamens 5; ovary 2- to 3-celled but the partitions between the cells sometimes incomplete, ovules several in each cell; style slender-elongated, stigma capitate and usually exserted; fruit berrylike, usually few-seeded.

1. Bracts and bractlets minute or wanting (seldom over 2 mm. long); fruit orange, red or yellow; corolla usually over 15 mm. long; leaves obtuse at apex -1. **L. utahensis**
1. Bracts and bractlets large and foliaceous (over 4 mm. long); fruit black; corolla less than 15 mm. long; leaves acute or acuminate at apex -2. **L. involucrata**

1. **Lonicera utahensis** S. Wats. in King, Geol. Expl. 40th Par. 5:133. 1871.

British Columbia, south to New Mexico and California. Not reported from Colorado but should be expected in the western part of the state.

2. **Lonicera involucrata** (Richards.) Banks in Spreng. Syst. Veg. 1:759. 1825.

Distegia involucrata (Richards.) Cockerell---Upright shrubs 50-300 cm. tall, glabrate to pubescent; leaves 5-15 cm. long, obovate, oval, ovate to ovate-oblong, acute or acuminate, rounded to broadly cuneate at base, on rather short petioles; flowers in pairs on a more or less elongated axillary peduncle, bracts and bractlets foliaceous, enlarging and becoming reddish, closely investing the fruit; corolla 12-15 mm. long, narrowly funnelform, slightly gibbous at base, yellow, glandular-hairy; berries black, the 2 distinct, about 8 mm. in diameter or larger.---Wooded banks and slopes. Quebec to Alaska, south to Michigan, Mexico and California. Our records scattered in the western half of Colorado at 7000-11,500 feet.

5. Symphoricarpos Duhamel. SNOWBERRY; CORALBERRY

Shrubs, the older bark commonly exfoliating and shredding; leaves simple, deciduous, entire to sinuate or even lobed, short-petioled; flowers rather small, pink or white, in axillary or terminal clusters, or solitary in the upper axils; calyx teeth 4 or 5, short; corolla regular or nearly so, campanulate, funnelform or salverform, 4- to 5-lobed; stamens 4 or 5; ovary 4-celled, 2 containing abortive ovules; style 1, stigma capitate; fruit a 2-seeded berrylike drupe.

1. Corolla shortly campanulate, often less than 7 mm. long, the lobes nearly as long to longer than the tube, villous in the throat, more or less ventricose on 1 side of tube
    2. Fruit red; styles about 2 mm. long and pilose; corolla 3-4 mm. long; veins of leaves conspicuously impressed above -1. **S. orbiculatus**
    2. Fruit white; styles glabrous or if pilose then 4-8 mm. long; corolla 4-9 mm. long; veins not especially impressed on upper surface of leaves
        3. Styles 4-8 mm. long, sometimes pilose, definitely exserted and longer than the corolla lobes on most flowers; anthers usually longer than the corolla lobes; leaves over 3 cm. long, often toothed or lobed -2. **S. occidentalis**
        3. Styles about 2-3 mm. long, always glabrous and shorter than the corolla tube; anthers shorter than the corolla lobes; leaves less than 3 cm. long, entire or somewhat sinuate
            4. Young twigs glabrous; leaves usually glabrous beneath; fruits usually 10-15 mm. in diameter at maturity, in terminal and axillary clusters; plants over 1 m. tall -3. **S. rivularis**

4. Young twigs shortly crisp-pubescent; leaves usually densely pubescent below; fruits 6-10 mm. in diameter, solitary or in pairs in the upper leaf axils; plants seldom over 1 m. tall -4. **S. albus**
1. Corolla elongate-campanulate, tubular-funnelform or salverform, 7 mm. long or more (occasionally 6 mm. in S. **vaccinioides**), the lobes 1/3 or less than 1/3 as long as the tube, usually not villous right in the throat (often pilose within below the throat of the tube), tube not at all ventricose
   5. Corolla salverform, 11-13 mm. long, only 1 basal nectary within; styles 5-7 mm. long, usually pilose above; anthers sessile; leaves lanceolate or oblanceolate, usually less than 5 mm. wide (mostly about 3 times longer than wide) -5. **S. longiflorus**
   5. Corolla tubular-funnelform to campanulate, often less than 11 mm. long, 5 basal nectaries present inside; styles 3-4.5 mm. long, glabrous; anthers on filaments as long or nearly as long; leaves ovate-elliptic to wider, over 5 mm. wide (rarely over 2 times longer than wide)
      6. Young twigs glabrous
         7. Corolla elongate-campanulate (when pressed 2.5 mm. wide or more at middle mark), 7-9 mm. long -6. **S. tetonensis**
         7. Corolla tubular-funnelform (when pressed about 2 mm. wide at middle mark), 11-13 mm. long -7. **S. oreophilus**
      6. Young twigs hairy
         8. Corollas elongate-campanulate (the tube usually 3 mm. wide or more at middle mark when pressed), 6-8 mm. long -8. **S. vaccinioides**
         8. Corollas tubular-campanulate (the tube when pressed usually less than 3 mm. wide at middle mark), over 8 mm. long
            9. Young twigs pubescent with short, straight, spreading hairs; anthers reaching only to base of corolla lobes; leaves oval to round -9. **S. rotundifolius**
            9. Young twigs tomentulose-puberulent with short, curved hairs; anthers reaching to middle or above on corolla lobes; leaves oval to narrower
               10. Erect shrubs; leaves scarcely paler beneath, veins on upper surface prominent; petioles 2-4 mm. long; nutlets 5-7 mm. long; plants not limited to southern Colorado -10. **S. utahensis**
               10. Trailing shrubs; leaves paler beneath, veins on upper surface obscure; petioles 1-2 mm. long; nutlets 4-5 mm. long; plants limited to southern Colorado -11. **S. palmeri**

1. Symphoricarpos orbiculatus Gray, Journ. Linn. Soc. Bot. 14:12. 1873.
   New York to South Dakota, south to Florida and Mexico. Reported definitely for Colorado and may be present in the eastern part of the state.
2. Symphoricarpos occidentalis Hook., Fl. Bor. Am. 1:285. 1833.
   Erect shrubs 50-150 cm. tall, often forming dense colonies; twigs puberulent to rarely glabrous; leaves 3-10 cm. long, oval, entire or undulate-crenate to lobed, obtuse, thickish, green above and glabrous to short-pilose especially along veins, lower surface paler and usually thinly pubescent; petioles 4-10 mm. long; flowers sessile in short dense axillary and terminal many-flowered spicate clusters; bracts and bractlets oval, ciliate; calyx teeth 0.7-0.8 mm. long, ciliate; corolla 6-9 mm. long, campanulate, pink to light rose, the lobes about as long or somewhat longer than the tube, the tube densely villous inside; anthers exserted and usually longer than the corolla lobes; style 4-8 mm. long or more, definitely exserted and longer than the corolla lobes on most flowers, pilose to glabrous; fruit 6-8 mm. long, nearly globose, pale greenish-white.---Hillsides, valleys and banks, often along streams. Michigan to British Columbia, south to Illinois and New Mexico. Our records scattered over Colorado, except the southwestern part, at 3500-8500 feet.
3. Symphoricarpos rivularis Suksdorf, Werdenda 1:41. 1927.
   S. racemosus var. laevicarpus Fernald; S. albus of Manuals---Erect branching shrub about 1-3 m. tall, sometimes shorter; young twigs glabrous or very nearly so; leaves 2-3 cm. long, oval, acute or obtusish, entire or sometimes sinuate, upper surface green and glabrous, more or less paler below and glabrous or sparsely hairy; petioles 2-4 mm. long; flowers often numerous in short-peduncled terminal or sometimes axillary racemes; bracts and bractlets glabrous; calyx teeth about 0.5 mm. long, glabrous; corolla 4-7 mm. long, shortly campanulate, definitely ventricose on lower side, rose-pink to whitish, lobes about as long as the tube, whitish-villous inside; anthers shorter than the corolla lobes; styles about 2 mm. long, glabrous; fruit up to 10-15 mm. long, subglobose to ellipsoid, white, often borne in glomerules.---Hillsides, valleys and river banks. Montana to Alaska, south to Colorado and California. Our records from the northwestern quarter of Colorado at about 6000-7000 feet.
4. Symphoricarpos albus (L.) Blake, Rhodora 16:118. 1914.
   S. racemosus var. pauciflorus Robbins; S. pauciflorus (Robbins) Britt.---Erect shrubs 20-80 cm. tall; young twigs crisp-pubescent with short curved hairs; leaves 1-3 cm. long, oval, ovate to nearly orbicular, acute or obtuse, entire or somewhat sinuate, bright green above and usually glabrate, paler or glaucous below, short-pilose at least on veins and margins; petioles 2-3 mm. long; flowers short-pedicelled, 1-5 in the upper leaf axils; bracts lanceolate, bractlets deltoid; calyx lobes about 0.5 mm. long, glabrous; corolla 5-6 mm. long, short campanulate, pink, the lobes somewhat shorter than the tube, this somewhat ventricose at base and densely villous in the throat; anthers shorter than the corolla lobes; style 2-3 mm. long but not exceeding the corolla tube, glabrous; fruit 6-10 mm. long, depressed-globose, white.---Banks, cliffs and hillsides. Quebec to Alberta, south to Virginia and Colorado. Our records from northcentral and central Colorado at 5500-7500 feet.
5. Symphoricarpos longiflorus Gray, Journ. Linn. Soc. Bot. 14:12. 1873.
   Rather low spreading shrubs 50-100 cm. tall; young twigs glabrous or sparsely short-pilose; leaves 6-15 mm. long and seldom over 5 mm. wide, lanceolate to oblanceolate, sometimes oval, entire, obtuse to acute, glabrous to short-pilose, pale green or glaucous; petioles 1-3 mm. long; flowers solitary or in pairs in the upper axils or in terminal few-flowered racemes; calyx lobes 0.7-1 mm. long, glabrous or somewhat pubescent; corolla 11-13 mm. long, salverform,

symmetrical, pink to rose, tube narrow, lobes 1/3 to 1/5 as long as the tube, glabrous within, with only 1 basal nectary; anthers sessile, about 1/4 as long as the corolla lobes; style 5-7 mm. long, usually pilose above; fruit 8-10 mm. long, ellipsoid, white.---Slopes and valleys. Oregon, south to Texas and California. Our few records from extreme western Colorado at about 5000 feet.

6. Symphoricarpos tetonensis A. Nels., Bull. Torr. Club 31:246. 1904.

Upright much branched shrub 1-1.5 m. tall; young twigs glabrous and more or less glaucous; leaves 1-3 cm. long, oval, entire or occasionally with a few irregular teeth, acute, glabrous, paler beneath; petiole 2-4 mm. long; flowers mostly in pairs in upper axils, short-pedicelled; bracts small and glabrous; calyx lobes about 1 mm. long, glabrous; corolla 7-9 mm. long, cylindrical-campanulate, ochroleucous especially in age, to pink or whitish, lobes about 1/3 as long as the tube, this glabrous inside or slightly hairy at base; anthers about 1/2 as long as the corolla lobes; styles 4 mm. long, glabrous; fruit 8-10 mm. long, ellipsoid, white.--- Plains and hills. Montana to Idaho, south to Colorado and Nevada. Our records scattered in the western half of Colorado at 6000-9000 feet.

7. Symphoricarpos oreophilus Gray, Journ. Linn. Soc. Bot. 14:12. 1873.

Erect shrubs 1-1.5 m. tall, much branched; young twigs glabrous; leaves 1-3 cm. long, oval, entire or serrate-dentate, usually acutish, usually glabrous, somewhat paler beneath; petioles about 2 mm. long; flowers mostly in axillary pairs or in terminal few-flowered spikelike inflorescences; bracts 1 mm. long, oval; calyx lobes 0.5-1 mm. long; glabrous; corolla 11-13 mm. long, tubular-funnelform, symmetrical, rose, lobes about 1/4 to 1/3 as long as the tube, this glabrous or sparsely pilose within; anthers about 2/3 as long as the corolla lobes; style about 3 mm. long, glabrous; fruit 8-10 mm. long, ovoid or ellipsoid, white.--- Hillsides and valleys, often on river banks. Colorado and Nevada, south to Texas and Mexico. Our records scattered over the western two-thirds of Colorado at 5500-10,000 feet.

8. Symphoricarpos vaccinioides Rydb., Mem. N. Y. Bot. Gard. 1:371. 1900.

S. rotundifolius var. vaccinioides (Rydb.) A. Nels.---Shrubs to 1.5 m. tall, much branched; twigs when young puberulent or densely grayish-pubescent with curved hairs; leaves 1-2.5 cm. long, oval to oval-ovate or ovate-elliptic, entire or slightly dentate, acute or acutish, puberulent, green above and more or less paler beneath; petioles 2-4 mm. long; flowers solitary or in pairs in upper axils, on short pedicels; bracts lanceolate, puberulent; calyx lobes about 1 mm. long, puberulent; corolla 6-8 mm. long, cylindrical-campanulate, symmetrical, pink or light rose, lobes about 1/3 the length of the tube, this pilose within at base; anthers about 2/3 as long as the corolla lobes; style about 4 mm. long, glabrous; fruit about 6-10 mm. long, ellipsoid, white.---Hillsides and valleys. Montana to British Columbia, south to Colorado and California. Our records from the western half of Colorado at 6000-10,500 feet.

9. Symphoricarpos rotundifolius Gray, Plant. Wright. 2:66. 1853.

Erect but straggling shrubs not over 1 m. tall; young twigs densely pubescent with straight hairs; leaves 1-3 cm. long, suborbicular to broadly oval or ovate, obtuse to roundish at apex, rarely acutish, usually entire, usually puberulous above and short-pilose below, somewhat lighter below; petioles 1-3 mm. long; flowers almost sessile in the axils; bracts shorter than the ovary; calyx lobes about 0.5-1 mm. long, more or less ciliate; corolla 9-10 mm. long, tubular-funnelform, symmetrical, light pink to rose, the lobes about 1/4 to 1/3 as long as the tube, this pilose inside at lower part; anthers little if any longer than the corolla tube; style 3-4 mm. long, glabrous; fruit about 10 mm. long, ovoid or ellipsoid, white.---Slopes and valleys. Colorado, New Mexico and Arizona. Our records from western and southcentral Colorado at about 4000-7000 feet.

10. Symphoricarpos utahensis Rydb., Bull. Torr. Club 26:544. 1899.

S. oreophilus var. utahensis (Rydb.) A. Nels.---Shrubs 1-2 m. tall; young twigs crisp-puberulent; leaves 1.5-4 cm. long, oval or ovate, obtuse or acutish, usually entire, puberulent above and below, only slightly paler below; petioles 2-4 mm. long; flowers in small clusters or short spikes from the axils of the upper leaves; corolla 9-12 mm. long, tubular-funnelform, symmetrical, pink or light rose, lobes about 3 mm. long, pubescent inside; anthers about 2/3 as long as the corolla lobes; style about 3 mm. long, glabrous; fruit about 8-10 mm. long, ellipsoid, white.---Mountainsides, valleys and canyons. Wyoming to Utah, south to Colorado and Arizona. Our records from westcentral and southwestern Colorado (but reported from the central part also) at 8000-10,000 feet.

11. Symphoricarpos palmeri G. N. Jones, Journ. Arnold Arb. 21:243. 1940.

Shrubs 1-3 m. tall, the branches more or less trailing; young twigs tomentose-puberulent to somewhat pubescent-villous, the hairs usually curved; leaves 1-3 cm. long, oval or ovate, acute to acutish, margins often somewhat sinuate, upper surface dark green and short-pilose, lower short-pilose, paler with rather prominent veins; petioles 1-3 mm. long; flowers solitary or in pairs, axillary, short-pedicelled; bracts lanceolate, bractlets oval; calyx teeth about 0.3-0.5 mm. long, glabrous or ciliate; corolla 9-12 mm. long, tubular-funnelform, symmetrical, pink to dull light rose, lobes about 1/4 to 1/3 as long as the tube, this pilose below within; anthers about as long as the middle of the corolla lobes or shorter, but longer than the tube; style 2-4 mm. long, glabrous; fruit 6-8 mm. long, ellipsoid, glabrous. Intergrades with S. rotundifolius in southwestern Colorado.---Banks, slopes and valleys. Colorado, south to Texas and Arizona. Our few records from southcentral and southwestern Colorado at 5500-7500 feet.

Family 112. Adoxaceae MOSCHATEL FAMILY

Herbaceous glabrous plants with scaly or tuberiferous rootstocks; leaves basal and also 1 pair opposite on the stem, ternately compound; flowers small and inconspicuous, perfect, regular, in terminal, capitate clusters; calyx tube adnate to ovary, its limb 2- to 3-lobed;

corolla gamopetalous, rotate, 4- to 6-lobed; stamens twice as many as corolla lobes, borne in pairs on the corolla tube, alternate to its lobes, filaments short, anthers 1-celled; ovary inferior, 3- to 5-celled, 1 ovule to each cell, styles deeply 3- to 5-parted; fruit a greenish dry drupe with 3-5 nutlets.

1. Adoxa L. MUSKROOT; MOSCHATEL

Characters of the Family

1. Adoxa moschatellina L., Sp. Pl. 367. 1753.
Stems 5-15 cm. tall, erect, simple, slender; basal leaves 1-4, long-petioled, segments broadly ovate to orbicular, thin, glabrous, 3-cleft or parted, lobes obtuse or mucronate, the pair of stem leaves short-petioled; flowers 3-8, in a head 6-8 mm. in diameter; corolla about 2-3 mm. long, greenish, on terminal flower 4- to 5-lobed, on other flowers 5- to 6-lobed.--- Woods, especially on rocky slopes. Arctic America, south to Wisconsin, Iowa and Colorado. Our records from the mountains of the state, none as yet from the northwestern part, at 8000-13,000 feet.

Family 113. Cucurbitaceae GOURD FAMILY

Annual or perennial herbaceous vines; stems leafy, trailing or climbing by means of tendrils; leaves alternate, petioled, broad, simple and palmately veined and often palmately lobed or palmately compound; flowers monoecious, axillary, solitary to variously clustered, regular; calyx of 4-6 (usually 5) more or less united sepals; corolla of 4-6 (usually 5) petals, these more or less united; stamens commonly 3, the anthers more or less united or coherent, 1 anther 1-celled, the other 2 anthers 2-celled, or only 1 anther present; ovary inferior, 1- to 3-celled, style with stigma variously lobed; fruit a pepo, either spiny or with a tough rind; seeds flattened.

1. Flowers all solitary; corolla campanulate, large (over 5 cm. long), yellow or orange-yellow; plants perennial from thick roots; stems usually prostrate; leaves usually entire or merely angled; fruit not at all spiny
-1. Cucurbita
1. Flowers (staminate ones) racemose or paniculate; corolla rotate, white to yellowish-green, (not over 1 cm. long); plants annual; stems usually climbing; leaves definitely palmately lobed or compound; fruit spiny
2. Leaves palmately compound (or some deeply parted); anthers completely fused and appearing as a horizontal ring, opening all around
-2. Cyclanthera
2. Leaves palmately lobed, the lobes not over 1/2 to the base; anthers more or less united but evidently more than 1, opening longitudinally
-3. Echinocystis

1. Cucurbita L. GOURD

Coarse rough perennial plants forming large colonies in age from a large thickened root; stems mostly prostrate on the ground; leaves thick, large, triangular-ovate, entire to denticulate or occasionally somewhat lobed; flowers solitary and axillary; calyx campanulate, 5-lobed; corollas campanulate, 5-lobed, yellow, large (over 5 cm. long); stamens 3, the anthers connivent into an often twisted body; styles 1, stigmas 3-5, each 2-lobed or 2-branched; pepo with a thick rind, 3- to 5-celled or appearing 1-celled at maturity; seeds large, many.

1. Cucurbita foetidissima H. B. K., Nov. Gen. et Sp. 2:123. 1817.
Pepo foetidissimus (H.B.K.) Britt.---Plants commonly but not always ill-smelling; root fusiform, perpendicular, often to 25 cm. thick; patches becoming 10 m. across or even larger; leaves truncate or cordate at base, the larger 10-30 cm. long, scabro-pubescent and grayer beneath than above; petioles stout; corolla 6-15 cm. long, rough-pubescent; fruit 5-10 cm. across, globose or ovoid-globose, striped or mottled dark green and light green.---Dry plains and hills, common along railroad tracks in this area. Missouri to California, south to Texas and Mexico. Our records scattered over the eastern half of Colorado, more abundant in the southern part, at 3500-6000 feet.

2. Cyclanthera Scrad.

Annual glabrous climbing herbaceous plants; stems slender; leaves thin, palmately compound or some deeply palmately parted; flowers small, greenish or white, monoecious; staminate flowers in small racemes or panicles, pistillate solitary, in the same axil as the staminate; calyx cup-shaped, 5-toothed; corolla rotate, 5-parted; anthers appearing as 1, this annular; ovary beaked, 1- to 3-celled, styles short, stigmas large; fruit ovoid, somewhat asymmetric, acuminate, awned with long, smooth spines, bursting irregularly; seeds few.

1. Cyclanthera dissecta (T. & G.) Arn., London Journ. Bot. 3:280. 1841.
Stems grooved, branching, often 1-2 m. long; leaves palmately 3- to 7-parted to (more commonly) 3- to 7-foliate, the middle leaflet especially stalked, leaflets 2-7 cm. long, more or less scabrous both sides, toothed to incised; corolla about 3 mm. long; fruit about 2-3 cm. long.--- Woods and thickets. Kansas, south to Louisiana, Arizona and Mexico. Our few records from eastern Colorado at 4000 feet.

3. Echinocystis Torr. & Gray MOCK CUCUMBER

Annual climbing herbaceous plants; leaves palmately lobed, thin; staminate flowers numerous, racemose or paniculate; pistillate flowers solitary or few, from the same axils as the staminate flowers; calyx of 5-6 sepals; corolla deeply 5- to 6-parted, rotate; stamens 2-3, the anthers more or less united; ovary 2- to 3-celled, style short, stigmas lobed or hemispheric; fruit 1- to 2-celled, densely soft-spiny becoming papery, spongey and fibrous within, irregularly dehiscent at the apex; seeds 1-4 in each cavity.

1. Echinocystis lobata (Michx.) T. & G., Fl. N. Amer. 1:542. 1840.
   Micrampelis lobata (Michx.) Greene---Stems glabrous or nearly so, climbing often to a height of 4-8 m., angular and grooved; leaf blades about 3-8 cm. long, 3- to 7-lobed, these triangular, scabrous on both sides; staminate flowers numerous the lobes 4-7 mm. long, light yellowish-green; fruit 3-5 cm. long, ovoid or globose-ovoid.---Bushes, banks and waste places in general. New Brunswick to Manitoba, south to Pennsylvania, Texas and Arizona. Our records from the eastern half of Colorado at 3500-7500 feet.

Family 114. Campanulaceae BELLFLOWER FAMILY

Annual or perennial herbaceous plants, often with milky or acrid juice; leaves alternate, simple, without stipules; flowers perfect, regular or irregular, solitary or in inflorescences; sepals mostly 5 (3-5), partially united; corolla gamopetalous, irregular or regular, 5-lobed; stamens 5, alternate to the corolla lobes (at least in regular flowers) and inserted with the corolla, often united; ovary at least partly inferior, of 2-5 carpels, 2- to 5-celled, style 1, stigma 1, stigma capitate to 2- to 5-lobed; fruit a capsule; seeds minute, numerous. This family is often divided and the Lobeliaceae segregated on the basis of the irregular corolla and united anthers.

1. Corolla regular; anthers and filaments distinct; capsule opening by lateral pores formed by the rolling up of small lids (except in Heterocodon)
   2. Flowers sessile, dimorphic, the lower ones cleistogamous; plants annual
      3. Flowers in spikes; capsule opening by pores at apex or near the middle; corolla usually over 6 mm. long -1. Triodanis
      3. Flowers solitary; capsule opening irregularly; corolla not over 6 mm. long -2. Heterocodon
   2. Flowers on definite peduncles or pedicels, not in spikes, all alike; plants perennial -3. Campanula
1. Corolla irregular; anthers and filaments united; capsule opening from apex by valves or irregularly
   4. Corolla tube slit down on upper side nearly to base; perennial or biennial plants; flowers (in our 2 species) 20 mm. long or more -4. Lobelia
   4. Corolla tube not slit down to near base; plants annual; flowers about 8-12 mm. long -5. Porterella

1. Triodanis Raf. VENUS'S LOOKING-GLASS

Annual plants; stems angled, these continuous from the decurrent leaf bases, leafy; at least the upper leaves sessile or clasping; inflorescence spicate, flowers at lower part regularly cleistogamous with undeveloped corollas; ordinary flowers with 5 sepals; corolla usually lavender-blue, regular, rotate to funnelform, deeply lobed; filaments and anthers distinct, former ciliate at base; ovary inferior; capsule opening by pores at apex or near middle, 1- or 3-celled.

1. Flower bracts lanceolate to linear, usually 6-8 times as long as wide; capsules of cleistogamous flowers with more or less spreading tips; middle and upper leaves elliptic to oblanceolate or linear -1. T. leptocarpa
1. Flower bracts ovate to reniform, not over 1.5 times as long as wide, often wider than long; capsules of cleistogamous flowers with appressed tips; middle and upper leaves elliptic to ovate or ovate-cordate
   2. Middle leaves cordate-clasping; bracts usually strongly cordate-clasping; pores of capsule oval to rounded, 0.5-1.5 mm. wide; seeds smooth or evenly muriculate -2. T. perfoliata
   2. Middle leaves not clasping; bracts sessile and only somewhat clasping; pores of capsule linear, 0.2-0.4 mm. wide; seeds low-tuberculate in longitudinal lines -3. T. holzingeri

1. Triodanis leptocarpa (Nutt.) Nieuwl., Am. Midl. Nat. 3:192. 1914.
   Specularia leptocarpa (Nutt.) A. Gray---Stems erect, 10-35 cm. tall, retrorsely hispidulous on the angles or nearly glabrous, simple or branched from the base; leaves lanceolate or linear-lanceolate, elliptic or linear, sessile but not clasping, entire or sparingly denticulate; floral bracts about 6-8 times as long as wide, lanceolate to linear; calyx lobes about 8-10 mm. long, subulate, spreading; corolla 6-8 mm. long, rotate or broadly funnelform.---Draws, rocky slopes and banks. Minnesota to Montana, south to Iowa, Colorado and Texas. Our one record from northcentral Colorado at about 5500 feet.
2. Triodanis perfoliata (L.) Nieuwl., Am. Midl. Nat. 3:192. 1914.
   Specularia perfoliata (L.) A. DC.---Stems 20-50 cm. tall or more, usually erect, pilose or hispid-scabrous at least at base; middle leaves sessile, ovate-cordate and clasping, crenate, lower from broadly elliptic or obovate, glabrous above, ciliate on margins, more or less hairy beneath; bracts ovate to reniform, sessile and usually cordate-clasping, entire to crenate-serrate, wider than long, as large as leaves; calyx lobes of ordinary flowers 4-8 mm. long; corollas of ordinary flowers 8-12 mm. long, bluish-lavender; cleistogamous capsules 4-6 mm. long, of ordinary flowers about 10 mm. long, pores 0.8-1.5 mm. wide, broadly elliptic to round, the lid with narrow central rib and broad scarious margins; seeds smooth or muriculate.---Hillsides and dry woods, sometimes in disturbed ground. Ontario to British Columbia, south to Florida,

Mexico and California. Our records in or very near northcentral and central Colorado at 5000-8500 feet.

3. **Triodanis holzingeri** McVaugh, Wrightia 1:45. 1945.

Stems erect, 25-60 cm. tall, hispid or hirsute at base, upper part hispid to scabrous; leaves and bracts glabrous above, scabrous below chiefly on veins, or glabrous, middle leaves about 16-24 mm. long and nearly as wide, ovate to elliptic, sessile but not really clasping at base, lower leaves oblanceolate to obovate; bracts broad, ovate, sessile and slightly clasping, both leaves and bracts crenate; calyx lobes of ordinary flowers 4-6 mm. long; corollas of ordinary flowers 5-9 mm. long, blue to purple; capsules about 6-8 mm. long, pores 0.2-0.4 mm. wide, linear, the central rib with very narrow scarious margins; seeds minutely tuberculate in longitudinal lines.----Prairies, plains, valleys and canyons. Missouri to Wyoming, south to Tennessee, Texas and Arizona. Our one record from southeastern Colorado at about 4000 feet.

2. Heterocodon Nutt.

Annual plants; stems slender, delicate, diffusely branched and spreading; leaves alternate, broadly oval to orbicular, sessile and clasping-cordate at base, coarsely dentate and hispid-ciliate; flowers of 2 kinds, the earlier cleistogamous with rudimentary corollas, solitary; later flower-calyx 5-lobed, these rather foliaceous; corollas regular, short-campanulate, 5-lobed, blue; capsules 3-celled, 3-angled, bursting irregularly; seeds numerous.

1. **Heterocodon rariflorum** Nutt., Trans. Am. Phil. Soc. N. S. 8:255. 1843.

Stems about 5-30 cm. high; leaves about 3-10 mm. long and wide; cleistogamous flowers with calyx lobes 2-2.5 mm. long; later flowers with calyx lobes somewhat longer; corolla 3-6 mm. long; seeds narrowly ellipsoid.---Shaded, grassy often gravelly ground. Idaho to British Columbia, south to California. Our one record (the label not very complete) from northwestern Colorado at about 6500 feet.

3. Campanula L. BELLFLOWER; HAREBELL

Perennial herbaceous plants with rootstocks; leaves sometimes dimorphic, entire to toothed; flowers usually blue or violet and showy, solitary or in cymose racemes or panicles; calyx of 5 narrow sepals; corolla regular, campanulate to funnelform, 5-lobed; stamens distinct; ovary 3- to 5-celled, stigmas 3-5 (cells and stigmas usually 3); capsule inferior or partly so, opening by small lateral pores.

1. Corolla white to pale blue, 5-8 mm. long; stems retrorsely scabrous on the angles (like Galium); sepals 1.5-2 mm. long; inflorescence paniculate, the flowers on long filiform pedicels -1. C. aparinoides
1. Corolla normally blue or purple, 8-23 mm. long; stems not retrorsely scabrous on angles (may be hispidulous near base); sepals 3-18 mm. long; flowers solitary or if racemose or paniculate the flowers not on long filiform pedicels
    2. Anthers 1.5-2.5 mm. long; flowers 1, erect; capsule erect, opening by pores near its summit; calyx lobes 3-4 mm. long; corolla 7-10 mm. long -2. C. uniflora
    2. Anthers 4-6.5 mm. long; flowers 1 or more, erect or nodding; capsule erect or nodding, opening by pores near base of summit; calyx lobes 5-18 mm. long; corolla 10-23 mm. long
        3. Bases and petioles of rosette leaves and lower cauline leaves ciliate, the hairs 0.3-0.7 mm. long, plants glabrous otherwise; flowers usually 1 only, erect; capsule erect, opening by pores near apex; sepals 7-18 mm. long, often callus-toothed -3. C. parryi
        3. Plants glabrous or hispidulous, leaf bases and petioles glabrous or at most hispidulous on margins, the hairs not over 0.2 mm. long; flowers usually more than 1 (except at high altitudes and in unfavorable habitats), erect or nodding; capsule nodding, opening by pores near the base; sepals 5-7 mm. long, entire -4. C. rotundifolia

1. **Campanula aparinoides** Pursh, Fl. Am. Sept. 159. 1814.

New Brunswick to Saskatchewan, south to Georgia and Colorado. Reported from near Denver but no specimens seen by the writer.

2. **Campanula uniflora** L., Sp. Pl. 163. 1753.

Stems up to 10 cm. tall, simple, 1-flowered, glabrous or nearly so; lower leaves 2-3 cm. long, spatulate, upper linear or linear-oblong, entire or sparingly dentate; flower erect; calyx lobes about 3-4 mm. long, erect; corolla 7-10 mm. long, narrowly campanulate, blue; anthers 1.5-2.5 mm. long; capsule erect, opening by pores near the summit.---High mountains. Greenland to Alaska, south to Colorado; Eurasia. Our records from northcentral and central Colorado at 10,000-13,000 feet.

3. **Campanula parryi** A. Gray, Syn. Fl. N. Amer. ed. 2, 2(1):395. 1886.

Stems 10-30 cm. tall, glabrous except sometimes near base; lower leaves 2-6 cm. long, spatulate to oblanceolate, entire or denticulate, coarsely ciliate with white hairs 0.3-0.7 mm. long near base and on the petiole; upper leaves linear or lanc-linear; flowers erect and ordinarily 1 (or rarely with 1-4 additional flowers on subordinate lateral branches); sepals 7-18 mm. long, rather subulate, erect; corolla 10-23 mm. long, broadly funnelform, blue to purple; anthers 4-6.5 mm. long; capsule erect, opening by valves or pores near the summit.---Mountains. Wyoming to Utah, south to New Mexico and Arizona. Our records scattered in the mountains of Colorado at 7000-12,000 feet.

4. **Campanula rotundifolia** L., Sp. Pl. 163. 1753.

C. petiolata A. DC.---Stems 10-40 cm. tall, erect or diffuse, glabrous to hispidulous especially below; basal leaves petioled, ovate-cordate to ovate to elliptic, often early deciduous, lower stem leaves linear-lanceolate to linear, radical and lower leaves often hispidulous-ciliate near bases and on petioles, the hairs short, rarely over 0.2 mm. long; upper leaves linear to nearly filiform; flowers sometimes solitary especially at high altitudes but usually several and sometimes many, erect or some usually more or less nodding; sepals subulate, 5-7 mm. long; corolla 12-20 mm. long, campanulate, blue; anthers 4-6.5 mm. long; capsule nodding, opening by pores at base.---Hills, valleys and mountains. Boreal North America, south to Mexico and California; Eurasia. Our records from the western two-thirds of Colorado at 5000-13,000 feet.

4. Lobelia L. LOBELIA

Stems more or less erect; leaves variously toothed to subentire; inflorescence racemose to paniculate; calyx of 5 sepals; corolla irregular, with a dorsal fissure extending from the apex nearly to the base, 5-lobed, more or less 2-lipped; anthers united into a tube, the lower 2 smaller and hairy-tufted, filaments flat, united above; ovary inferior or partly inferior, 2-celled, stigmas 2-lobed with a ring of hairs below the apex; capsule loculicidally 2-valved at apex; seeds many; (corolla, stamens, and styles withering-persistent on the fruit).

1. Corolla red or crimson; filament tube over 15 mm. long -1. L. cardinalis
1. Corolla blue to nearly white; filament tube not over 15 mm. long
  2. Corolla 20-26 mm. long; filament tube 12-15 mm. long; auricles of calyx conspicuous, about 2-5 mm. long -2. L. siphilitica
  2. Corolla not over 16 mm. long; filament tube not over 5 mm. long; auricles absent or if present inconspicuous
    3. Pedicels with 2 conspicuous bractlets near middle -3. L. kalmii
    3. Pedicels with 2 inconspicuous bractlets near base -4. L. spicata

1. Lobelia cardinalis L., Sp. Pl. 930. 1753.
L. splendens Willd.---Plants perennial by offsets; stems 30-100 cm. tall, erect, unbranched, glabrous or sparsely pubescent; leaves 6-15 cm. long, linear, linear-lanceolate to ovate, usually toothed but occasionally entire, glabrous or sparsely hairy; inflorescence a spikelike raceme, few- to many-flowered, bracts rather leaflike, a pair of bractlets present near flower; calyx lobes 7-16 mm. long, no auricles in sinuses or rarely minute ones present; corolla 20-40 mm. long, deep crimson or red; capsule about 6-8 mm. long, about 1/2 inferior; seeds about 0.8-1.2 mm. long. Our plant has been called ssp. graminea (Lam.) McVaugh.---Moist places. The species widely distributed in United States, Mexico and Central America. Our records from the eastern third of Colorado, mostly from the southeastern part at 3500-5000 feet.
2. Lobelia siphilitica L., Sp. Pl. 931. 1753.
Perennial plants by offsets, rootstock short; stems 20-100 cm. tall, erect, unbranched, glabrous or nearly so, angled from the decurrent leaf bases; leaves of stem 1.5-4 cm. long, rather few, obovate, oblong, ovate or ovate-lanceolate, subentire to more or less serrate, merging gradually into the bracts, glabrous or nearly so; inflorescence a rather spikelike raceme, few- to rarely many-flowered, bracts and bractlets present; calyx lobes often 8-11 mm. long, foliaceous, with foliaceous auricles 2.5 mm. long in the sinuses; corolla 20-26 mm. long, bright blue to whitish near base and striped white inside; capsule about 7-10 mm. long, 1/2 to 2/3 inferior; seeds about 0.8-1 mm. long. Our plant has been called var. ludoviciana A. DC.---Moist or wet ground. The variety from Wisconsin to Manitoba, south to Texas and Central America. Our records from the eastern half of Colorado, none as yet from the extreme southern part, at 3500-6000 feet.
3. Lobelia kalmii L., Sp. Pl. 930. 1753.
Newfoundland to Hudson Bay, south to Pennsylvania, Minnesota and Montana. Reported from northcentral Colorado but the record is questionable.
4. Lobelia spicata Lam., Encyc. 3:587. 1791.
L. hirtella (A. Gray) Greene---Minnesota to Alberta, south to Indiana and Nebraska. Reported from western Nebraska very near the northeastern corner of our state.

5. Porterella Torr.

Annual plants of wet soil, often partially submerged, glabrous; leaves linear-subulate to lanceolate, entire or nearly so; hypanthium obconic to obturbinate in flower, cylindric or turbinate in fruit; sepals 5, narrow, cut nearly to the hypanthium; corolla usually blue with white or yellow eye, strongly irregular, but not split on upper side; filaments and anthers connate, 2 anthers shorter than others; ovary inferior or nearly so, 2-celled; capsule 2-celled and 2-valved, opening near apex.

1. Porterella carnosula (H. & A.) Torr., Rep. U. S. Geol. Surv. Terr. 5:488. 1872.
P. eximia A. Nels.; Laurentia eximia A. Nels.---Wyoming to Oregon, south to Utah and Nevada. This plant may be in western or northern Colorado.

Family 115. Valerianaceae VALERIAN FAMILY

Annual or perennial herbaceous plants; leaves opposite, entire to pinnately divided, lacking stipules; flowers perfect or unisexual, somewhat irregular, in cymes; calyx teeth obsolete or of 3-15 usually pappuslike structures, the tube adnate to the ovary; corollas gamopetalous, tubular, funnelform or salverform, white, yellowish-white, greenish-white to tinged with rose, blue or purple, usually with 5 lobes; stamens commonly 3, borne on the corolla tube; ovary inferior, 1- to 3-celled (but if more than 1-celled, then 2 cells not bearing ovules), styles 1, entire or 2- to 3-lobed at apex; fruit 1-seeded; indehiscent, achenelike.

1. *Valeriana* L. VALERIAN

Characters of the Family

The calyx lobes are persistent, inrolled in the flower but elongating and conspicuous in fruit.

1. Plants from conical taproots; leaves thick with veins nearly parallel to the midrib; corolla rotate, not over 3 mm. long .... -1. **V. edulis**
1. Plants from elongated, usually horizontal rootstocks; leaves thin with spreading lateral veins; corolla various, usually not rotate, usually over 3 mm. long
   2. Corolla funnelform, over 4 mm. long .... -2. **V. capitata**
   2. Corolla rotate to short-campanulate, not over 3.5 mm. long .... -3. **V. occidentalis**

1. Valeriana edulis Nutt., T. & G., Fl. N. Amer. 2:48. 1841.
   V. trachycarpa Rydb.; V. furfurescens A. Nels.; V. ceratophylla (Hook.) Piper---Plants with large vertical taproots, these stout and erect, often branching below; stems 10-60 cm. high, erect, glabrous or nearly so; leaves thick, glabrous to ciliate, with several conspicuous veins nearly parallel with the midrib; basal leaves oblanceolate or spatulate, long-petioled, entire or pinnately parted with few divisions; cauline leaves 1-3 pairs, sessile or nearly so, usually pinnately parted into 3-7 divisions, these usually linear or lanceolate; panicle elongated or open especially in age; corolla 1.5-3 mm. long, rotate, yellowish to whitish; fruit 2.5-5 mm. long, glabrous or pubescent. Our plants have been called ssp. edulis.---Hillsides and meadows, often in moist ground. British Columbia, south to Mexico. Our records scattered over the western half of Colorado at 7000-12,000 feet.
2. Valeriana capitata Pallas ex Link, Jahrb. d. Gewachsk. 1:66. 1820.
   V. pubicarpa Rydb.; V. puberulenta Rydb.; V. acutiloba Rydb.---Plants with elongated, usually horizontal rootstocks and rather slender roots; stems 15-45 cm. tall, erect, glabrous or more or less glandular-pubescent especially above and at nodes; leaves moderate to thin, with rather conspicuous lateral veins, these not parallel to midnerve, glabrous to ciliate; basal leaves mostly spatulate, oblanceolate or obovate, usually entire, long-petioled; cauline leaves sessile or nearly so, 2-4 pairs present, sometimes entire but some pinnately 2- to 4-lobed into entire divisions; panicles short and dense, at least at first; corolla 4-8 mm. long, funnelform, white to pink; fruit 2.5-5 mm. long, glabrous to pubescent. Our plant has been called V. capitata ssp. acutiloba (Rydb.) F. G. Meyer.---Hillsides and mountains, often in moist ground. The subspecies from Wyoming, south to New Mexico and Arizona. Our records in the western two-thirds of Colorado at 7500-13,000 feet.
   The plants with ovate leaves and longer corollas from southcentral Colorado have been considered to be a different species, V. arizonica Gray (V. ovata Rydb.).
3. Valeriana occidentalis Heller, Bull. Torr. Club 25:269. 1898.
   V. micrantha E. Nels.---Plants with elongated, usually horizontal rootstocks and rather slender fibrous roots; stems 30-80 cm. tall, erect, glabrous or sparingly puberulent but more or less hairy at nodes; leaves moderate to thin, with rather conspicuous lateral veins; basal leaves petioled, elliptic to lanceolate or spatulate, usually entire; cauline leaves sessile or on short petioles, ovate to lanceolate, usually 3- to 9-pinnately lobed, the divisions lanceolate to elliptic, usually entire; panicles rather compact when young but becoming looser in age; corolla 3-3.5 mm. long, or even shorter, rotate to campanulate, whitish; fruit 4-5 mm. long, glabrous to pubescent.---Meadows and hillsides, usually in moist ground. Montana to Oregon, south to Colorado and Arizona. Our records from the western half of Colorado, except the southcentral part of the state, at 7500-10,000 feet.

Family 116. *Dipsacaceae* TEASEL FAMILY

Herbaceous plants; stems stout, prickly, leafy; leaves opposite, simple, prickly on midrib below; flowers perfect, irregular, borne in dense involucrate heads, each flower with a bractlet (chaff), bracts and bractlets narrow and rather rigid; calyx cup-shaped, 4-toothed; corolla 4-lobed, irregular; stamens 4, distinct; ovary inferior, 1-celled, style elongate, stigma entire; fruit an achene.

1. *Dipsacus* L. TEASEL

Characters of the Family

1. Dipsacus sylvestris Huds., Fl. Angl. 49. 1762.
   Stems 80-200 cm. tall, with prickles on the angles; leaves 20-60 cm. long, sessile or upper perfoliate, lanceolate, oblong to oblanceolate, crenate-serrate or upper entire; heads 5-8 cm. long, ovoid or becoming somewhat cylindrical; involucre bracts linear, prickly on margin, curving upward, as long or longer than the head; bractlets of the head with a stiff straight awn; corolla about 8-12 mm. long, white to lilac.---Roadsides and waste places, often along ditches and edges of forests. Maine to Utah, south to Virginia and Colorado; naturalized from Europe. Our few records from central Colorado at 6000-8000 feet.

Family 117. Compositae THISTLE FAMILY; COMPOSITE FAMILY

Plants herbaceous or shrubby; leaves opposite or alternate, rarely whorled, entire to variously dissected but never truly compound; flowers borne in a head (rarely this 1-flowered), on a common receptacle, this surrounded by an involucre (the bracts sometimes called phyllaries); flowers perfect, pistillate, functionally staminate (an abortive pistil may be present) or neutral, often with bracts (chaffy scales) interspersed among them; calyx limb (pappus) obsolete, of scales, bristles, awns, etc., crowning the top of the ovary; corolla gamopetalous, either regular, tubular and 5-toothed (rarely 2- to 4-toothed) or bilabiate or ligulate (flattened and strap-shaped very soon above the base and usually 2- to 5-toothed at apex); regular flowers called disk-flowers, the ligulate flowers called ray-flowers; stamens usually 5, inserted on the corolla, united by their anthers or sometimes by their filaments; ovary inferior, 1-celled, 1-ovuled, styles usually 2-branched, these stigmatic inside; fruit an achene, often bearing the pappus, this often aiding in distribution of the fruit. This is a very large family and considered difficult. Two sets of keys are given here. One consists of a key to the 11 tribes represented in this state, followed by keys to the genera under each tribe. The other is an artificial key to the genera direct, not considering the tribes at all.

I. Key to the tribes of Compositae
   (followed by a key to the genera of each tribe)

1. Corollas not all ligulate, some or all of them tubular; milky juice usually absent; leaves alternate, basal or opposite
   2. Anther sacs not caudate or long-tailed at base (usually obtuse or acute or sometimes somewhat sagittate); ligulate marginal flowers often present
      3. Style branches half-round in cross-section, with long-acuminate tips and with long hairs on the outside from apex to below the cleft, stigmatic surface extending from base to tip of branches; no ligules present; flowers never yellow -Tribe 1. Vernonieae
      3. Style branches various but not as above; ligules present or absent; flowers often yellow
         4. Stigmatic lines not extending to the tips of the style branches, these clavate and very short-hairy; ligules absent; flowers never yellow -Tribe 2. Eupatorieae
         4. Stigmatic lines extending to the tips of the branches, these usually acute or acuminate (if blunt then long-hairy); ligules usually present; flowers (at least disk-flowers) usually yellow
            5. Anthers connate; heads rarely unisexual; pappus usually present; ligules usually present
               6. Style branches (of perfect flowers) with long hairs on upper outside surface, mainly on a flattened lanceolate or triangular appendage, branches glabrous inside, never with a ring of long hairs; leaves alternate or basal; receptacle never chaffy; pappus usually of capillary hairs -Tribe 3. Astereae
               6. Style branches (of perfect flowers) with a tuft of long hairs at apex or toward the apex, if appendaged then with long hairs both inside and outside, or as a ring of hairs below the appendage; leaves alternate, basal or opposite; receptacle often chaffy; pappus various (may be of capillary hairs)
                  7. Pappus not of capillary bristles (except in a few genera with either the receptacle chaffy or with oil or resinous dots present); receptacle chaffy or naked; oil glands present or absent; involucre bracts usually in 2 or more rows, often very imbricated
                     8. Involucre bracts scarious on the margins; pappus wanting or a mere crown; leaves alternate, aromatic -Tribe 4. Anthemideae
                     8. Involucre bracts not scarious; pappus usually present; leaves alternate or opposite, seldom very aromatic
                        9. Receptacle not chaffy (naked, fimbriate or hairy) -Tribe 5. Helenieae
                        9. Receptacle chaffy -Tribe 6. Heliantheae
                  7. Pappus of capillary bristles; receptacle never chaffy; oil glands not present; involucre bracts in 1 row of equal length -Tribe 7. Senecioneae
            5. Anthers nearly or quite distinct; heads often unisexual; pappus none or vestigial; ligules absent -Tribe 8. Ambrosineae
   2. Anther sacs caudate at base; ligules never present
      10. Anthers not appendaged at apex; pistillate corollas filiform, not deeply cleft; receptacle not bristly; heads dioecious or marginal flowers pistillate and fertile -Tribe 9. Inuleae
      10. Anthers with elongated appendages at apex; corollas not filiform but deeply cleft; receptacle usually bristly; flowers usually either all perfect and fertile or the marginal neutral -Tribe 10. Cynareae
1. Corollas all ligulate; plants usually with milky juice; leaves alternate or basal -Tribe 11. Cichorieae

Tribe 1. Vernonieae

One genus   -1. Vernonia

Tribe 2. Eupatorieae

1. Leaves opposite or whorled
   2. Leaves whorled, 3 or more at a node, 5-20 cm. long   -2. Eupatorium
   2. Leaves opposite, 2 at a node, rarely over 5 cm. long
      3. Achenes 5-angled or 5-ribbed; involucre bracts subequal or in 2 series   -2. Eupatorium
      3. Achenes 10-ribbed or 10-angled; involucre bracts imbricated in several series of different lengths
                                                                                           -3. Brickellia
1. Leaves alternate (may be crowded)
   4. Flowers rose-purple; heads in narrow spikes or spikelike racemes, rarely spikelike panicles; involucre bracts not striated, often partly colored
                                                                                           -4. Liatris

4. Flowers white, cream-colored, ochroleucous or sometimes purplish-tinged; heads in open branching cymes or corymbs, never spikelike; involucre bracts striated, rather chartaceous and whitish, never colored (other than green or white)
    5. Pappus plumose, the side-cilia about 0.4-0.5 mm. long, obviously plumose even without a lens; anthers tending to separate   -5. Kuhnia
    5. Pappus merely barbellate or (in B. brachyphylla) subplumose with side-cilia only about 0.2-0.3 mm. long, not appearing very plumose to the naked eye; anthers united   -3. Brickellia

Tribe 3. Astereae

1. No ligules of any kind present on the heads
  2. Plants dioecious; woody at least at base   -6. Baccharis
  2. Plants not dioecious, usually herbaceous, rarely woody (except in Chrysothamnus)
    3. Pappus of 2-8 slender but almost paleaceous, stiff caducous awns   -9. Grindelia
    3. Pappus (at least inner if double) of more than 8 capillary bristles, usually not caducous
      4. Plants annual
        5. Lower leaves toothed or lobed; heads small the involucres about 3 mm. high   -14. Conyza
        5. All leaves entire; involucres 5-8 mm. high   -16. Aster
      4. Plants perennial, usually woody at least at base
        6. Pappus double, outer of short narrow squamellae   -15. Erigeron
        6. Pappus single (bristles may be unequal in length but not in 2 sets)
          7. Leaves 1- to 3-ternately lobed   -15. Erigeron
          7. Leaves entire or toothed, not at all ternately toothed
            8. Heads usually narrowly turbinate to narrowly cylindrical; involucre bracts in more or less clearly defined vertical rows; plants woody shrubs (if undershrubs the bracts clearly vertical)   -7. Chrysothamnus
            8. Heads campanulate or hemispheric; involucre bracts not in vertical rows; herbaceous or woody only at base   -12. Haplopappus
1. Ligules present (may be minute to conspicuous)
  9. Ligules yellow or distinctly yellowish when fresh (may fade to whitish on old or poorly pressed specimens)
    10. Pappus distinctly double, inner of long capillary bristles, outer of short scales (sometimes bristle pointed)   -8. Chrysopsis
    10. Pappus single, of bristles, awns or scales but all alike (these may be of different lengths but never in 2 distinct sets as above)
      11. Pappus of 2-8 slender caducous awns   -9. Grindelia
      11. Pappus of numerous capillary bristles or narrow scales, these usually persistent
        12. Pappus of oblong to narrow scales   -10. Gutierrezia
        12. Pappus of capillary bristles
          13. Heads usually small (involucres seldom over 6.5 mm. high) and numerous, often in a panicle; plant always herbaceous; involucre bracts rarely herbaceous at apex   -11. Solidago
          13. Heads usually large (involucres seldom less than 6.5 mm. high) and usually few, variously disposed but if small and panicled then the plant woody at base; plants herbaceous to woody; involucre bracts often distinctly herbaceous at apex   -12. Haplopappus
  9. Ligules white, pink, blue, rose, violet or purple when fresh, never yellow
    14. Pappus (at least of disk flowers) of several to many rigid bristles; achenes pubescent with hairs 2-forked or glochidiate at apex (use strong lens)   -13. Townsendia
    14. Pappus (at least of disk flowers) of many long capillary bristles, at least in part; achenes glabrous or pubescent with simple hairs
      15. Ligules very inconspicuous, rays shorter than their tubes and scarcely if at all exceeding their pappus; central perfect flowers few; plants annual   -14. Conyza
      15. Ligules usually conspicuous, when short then longer than their pappus and tubes; central perfect flowers usually several to many; plants annual, biennial or perennial
        16. Involucre bracts relatively long and narrow, herbaceous to non-herbaceous but loss of green color either uniform or more pronounced toward the tips, bracts usually subequal but sometimes definitely imbricated but then scarcely if at all herbaceous; pappus often double; style appendages short, commonly obtuse to merely acutish at apex but sometimes sharply acute; achenes mostly 2-nerved; ligules often narrow and very numerous   -15. Erigeron
        16. Involucre bracts relatively broad, at least with usually expanded tips, herbaceous to non-herbaceous but loss of green color more pronounced near the base, bracts commonly regularly imbricated, sometimes subequal but then usually herbaceous nearly throughout; pappus not double; style appendages relatively long, usually acute to acuminate; achenes mostly several-nerved; ligules commonly broad and relatively few   -16. Aster

Tribe 4. Anthemideae

1. Definite rays present (may be small)
  2. Receptacle with chaff (at least among the upper or middle flowers); ligules normally white
    3. Heads in close terminal corymbs; achenes flattened; ligules not over 6 mm. long, nearly as wide, only 4-5 normally present to a head   -17. Achillea
    3. Heads solitary at the ends of branches, never closely corymbose; achenes not compressed; ligules 6-15 mm. long, longer than wide, over 5 present
      4. Receptacle conical, bearing stiff narrow chaff near apex; rays 6-10 mm. long; involucres 4-6 mm. high   -18. Anthemis
      4. Receptacle convex, bearing blunt, membranous scalelike chaff throughout; rays 10-15 mm. long; involucres 8-10 mm. high   -30. Leucampyx
  2. Receptacle naked; ligules white to yellow

5. Heads in definite corymbs; ligules reduced, not over 5 mm. long -19. Tanacetum
5. Heads solitary at the ends of the stems or rather long branches, not at all corymbose; ligules 7 mm. long or more
   6. Leaves merely toothed or coarsely incised; ligules light yellow or if white then 12 mm. long or more -20. Chrysanthemum
   6. Leaves bi- or tri-pinnatifid into linear filiform divisions; rays white, 7-10 mm. long -21. Matricaria
1. Definite rays absent (pistillate marginal flowers usually present but corollas reduced to a tube, this sometimes oblique at apex but not at all ligulate)
   7. Heads solitary on the stem or on rather long branches, or sometimes corymbose or capitate; anthers with rounded tips; receptacle convex to conical; plants herbaceous or woody only at the base
      8. Plants annual; heads solitary on rather long branches; leaves green and glabrous -21. Matricaria
      8. Plants perennial, usually with a somewhat woody base; heads corymbose or capitate, very rarely solitary; leaves mostly canescent, in any case hairy
         9. Involucre bracts in 2-3 series; plants sometimes aromatic but never with a minty odor; heads either in capitate clusters or if corymbose then the plant has bipinnatifid leaves or leaves lobed at apex only -19. Tanacetum
         9. Involucre bracts in 4-5 series; plants with minty odor when fresh; heads corymbose and leaves undivided or pinnately lobed near the base only -20. Chrysanthemum
   7. Heads spicate, racemose or (more commonly) panicled; anthers with pointed tips; receptacle flat to convex; plants herbaceous or (in some very common species) shrubby -22. Artemisia

Tribe 5. Helenieae

1. Involucre bracts bearing conspicuous translucent resinous, glandular dots, these often yellow or orange; plants annual
   2. Leaves linear, entire, opposite; pappus of 4-5 scales, rarely a few awns in addition -23. Pectis
   2. Leaves pinnately or bipinnately dissected or divided, opposite or alternate; pappus of 8-15 scales, each soon divided into 5-10 bristles -24. Dyssodia
1. Involucre bracts without such dots; plants annual or perennial
   3. Ray flowers none or ligules minute, not over 1 mm. long (marginal tubular flowers may be enlarged but not ligulate in Chaenactis)
      4. Corollas of disk flowers white, flesh-colored or rose
         5. Leaves entire or nearly so, never lobed or dissected
            6. Leaves all basal, oblong, ovate or suborbicular; pappus scales spatulate, not acuminate at apex -25. Chamaechaenactis
            6. Leaves cauline, linear or lanceolate; pappus scales ovate or lanceolate, acuminate-attenuate at apex -26. Palafoxia
         5. Leaves, at least in part, 1-3 times pinnately divided or dissected
            7. Involucre bracts thin, scarious with whitish or yellowish margins and tips; pappus scales about 10-20 -27. Hymenopappus
            7. Involucre bracts herbaceous, lacking scarious-colored margins and tips; pappus scales about 4-10 -28. Chaenactis
      4. Corollas of disk flowers from light to dark yellow
         8. Leaves entire, serrulate or merely sinuate, not at all deeply lobed or divided, opposite
            9. Leaves sessile, slightly connate at base, linear to lanceolate; rays present but minute; pappus none -32. Flaveria
            9. Leaves petioled, not at all connate, hastate to ovate; rays entirely lacking; pappus present as a crown of lacerate-ciliate scales (or awns in addition) -29. Pericome
         8. Leaves deeply lobed or divided, at least in part, usually alternate (or lower opposite)
            10. Plants more or less floccose-tomentose; pappus of about 10-20 scales; plants perennial or (in 1 species) biennial -27. Hymenopappus
            10. Plants not tomentose; pappus of 8 scales; plants annual -36. Bahia
   3. Ray flowers present, large or small but over 1 mm. long
      11. Rays white; receptacle with chaffy scales between the disk flowers -30. Leucampyx
      11. Rays yellow, orange, rose or reddish-purple; receptacle naked (or with subulate or setiform chaff in Gaillardia and sometimes a few scales between rays and disk flowers in Helenium)
         12. Rays rose or reddish-purple, no yellow present; plants annual
            13. Receptacle with subulate or setiform chaff among the disk flowers, convex; leaves usually toothed or lobed, rarely entire; plants not glandular -31. Gaillardia
            13. Receptacle naked, flat; leaves entire; plants glandular above -26. Palafoxia
         12. Rays yellow or orange, at least in part; plants annual or perennial
            14. Leaves opposite and serrulate, slightly connate at base; rays not over 2 mm. long; pappus none -32. Flaveria
            14. Leaves alternate or if opposite then deeply lobed or divided, never connate at base; rays usually over 2 mm. long; pappus present (in all but Bahia dissecta)
               15. Rays persistent on the achenes and becoming papery; plants woolly or tomentose
                   16. Rays 3-6; achenes linear, striated but slightly angled, glabrous or nearly so; leaves spatulate, lanceolate or oblanceolate -33. Psilostrophe
                   16. Rays 7 or more (usually over 10); achenes obpyramidal, 5-angled, densely hirsute; leaves or leaf divisions usually linear-oblanceolate to linear -35. Hymenoxys
               15. Rays not persistent on the achenes or becoming papery; plants glabrous to hairy but not woolly or tomentose
                 17. Receptacle with subulate or setiform chaff between the disk flowers; disk flowers with lobes reddish-purple and bearing moniliform hairs -31. Gaillardia
                 17. Receptacle naked, or rarely with a few scales between the disk flowers and the ray flowers; disk flower lobes not reddish-purple nor with moniliform hairs

18. Involucre bracts spreading or reflexed; leaves entire or toothed and some cauline; rays golden-yellow or orange-yellow -34. <u>Helenium</u>
18. Involucre bracts erect; leaves various, often lobed, if entire then commonly all basal; rays yellow
  19. Leaves alternate or crowded at base
    20. Achenes 5-angled, only 2 or 3 times as long as wide, densely hirsute -35. <u>Hymenoxys</u>
    20. Achenes 4-angled, 4 or more times longer than wide, glabrate or sparingly hairy -36. <u>Bahia</u>
  19. Leaves opposite and mostly cauline -36. <u>Bahia</u>

Tribe 6. <u>Heliantheae</u>

1. Rays absent entirely
  2. Receptacles high-conical, becoming 3-5 cm. long -48. <u>Rudbeckia</u>
  2. Receptacles flat, convex or hemispheric, never even approaching 3 cm. long
    3. Involucre bracts imbricated but all essentially alike except for length, not in 2 sets; heads solitary on a scape (this may be short)
      4. Central flowers maturing achenes; plants 15-25 cm. tall; involucres 12-15 mm. high -37. <u>Enceliopsis</u>
      4. Central flowers perfect but sterile; plants only 2-5 cm. tall; involucres about 5 mm. high -38. <u>Parthenium</u>
    3. Involucre bracts in 2 distinct sets, the outer herbaceous, the inner colored or more or less scarious and differing in shape; heads usually several to many on leafy branches
      5. Leaves alternate, ovate, not lobed; pappus lacking -39. <u>Parthenice</u>
      5. Leaves opposite (at least below), either lobed or divided or if not lobed then much narrower than ovate; pappus of 2-4 retrorsely hispid or barbed awns
        6. Inner involucre bracts united to the middle or above, forming a cup -53. <u>Thelesperma</u>
        6. Inner bracts not united or if so only at very base -54. <u>Bidens</u>
1. Rays present (may be small and inconspicuous)
  7. Involucre bracts (all or inner set only) with margins enclosing or enfolding the ray achenes
    8. Involucre bracts in a single series, all enclosing ray achenes; leaves alternate; rays yellowish, inconspicuous (about 1 mm. long) -40. <u>Madia</u>
    8. Involucre bracts in 2 distinct sets, only the inner enclosing the ray achenes; leaves opposite; rays white to pinkish, conspicuous (6 mm. long or more) -41. <u>Melampodium</u>
  7. Involucre bracts not enclosing or enfolding the ray achenes
    9. Rays white, whitish, pink, purple or rose, never yellow, golden or orange
      10. Leaves alternate; plants perennial; rays rose, purple or brownish-purple
        11. Leaves entire -42. <u>Echinacea</u>
        11. Leaves pinnately parted or divided -49. <u>Ratibida</u>
      10. Leaves opposite; plants annual or rarely appearing somewhat perennial; rays white, whitish or pink
        12. Leaves dissected; involucre bracts in 2 rather distinct sets, these unlike in color or texture; pappus of retrorsely hispid or barbed awns
          13. Achenes conspicuously beaked; rays usually pink -43. <u>Cosmos</u>
          13. Achenes not at all beaked; rays whitish to yellowish-white, not pink -54. <u>Bidens</u>
        12. Leaves merely toothed, not at all lobed; involucre bracts not in 2 distinct sets unlike in color or texture (may be in 2 series, the outer shorter); pappus lacking or of teeth or scales, never of retrorsely hispid or barbed awns
          14. Pappus lacking or rarely of minute teeth; leaves lanceolate and sessile (except sometimes the lower ones) -44. <u>Eclipta</u>
          14. Pappus (at least of disk flowers) of definite scales; leaves ovate, all petioled -45. <u>Galinsoga</u>
    9. Rays yellow, golden or orange (may be purple or red at base or in fine longitudinal lines but essentially yellow)
      15. Leaves all alternate or crowded-basal, none opposite
        16. Disk flowers perfect but sterile, the ovaries abortive and style undivided at apex; leaves pinnatifid; ray flowers fertile
          17. Involucre bracts more or less imbricated but otherwise essentially alike; plants often over 100 cm. tall -56. <u>Silphium</u>
          17. Involucre bracts in 2 sets, the outer different in shape or texture; plants rarely over 70 cm. tall
            18. Plants hirsute, the leaves not canescent below; pappus of ray flowers a crown of scales; inner bracts conspicuous and firm -46. <u>Engelmannia</u>
            18. Plants puberulent, the leaves whitish-canescent below; pappus of ray flowers absent; inner bracts foliaceous and thin, margins somewhat scarious -47. <u>Berlandiera</u>
        16. Disk flowers perfect and fertile, maturing their achenes and with a style divided at apex in anthesis; leaves seldom pinnatifid; ray flowers fertile or sterile
          19. Receptacles conical or columnar, usually conspicuously so; leaves (in most species) lobed or dissected
            20. Achenes 4-angled, not definitely flattened; leaves either entire or if parted or divided then on a plant 100 cm. tall or more -48. <u>Rudbeckia</u>
            20. Achenes flattened; leaves parted or divided but the plants not over 80 cm. tall -49. <u>Ratibida</u>
          19. Receptacles flat or merely convex; leaves (in most species) not lobed or dissected
            21. Achenes flat, the edges thin and margined or winged (at least on the disk flowers); pappus of disk flowers of long slender awns, at least in part
              22. Ray flowers maturing achenes; achenes of disk flowers with corky wings; no intermediate scales present between the pappus awns -50. <u>Verbesina</u>

                22. Ray flowers sterile; none of achenes really winged; small scales present between the pappus awns
                                                                                                        -59. Helianthella
            21. Achenes slightly if at all flattened, not margined or winged (at least on the disk flowers); pappus of
                disk flowers none, of scales or awnlike scales, rarely of slender awns
                23. Ray flowers maturing achenes; plants with thick balsamiferous taproots; leaves always alternate or
                    basal; pappus, if present, persistent
                    24. Pappus none; stems scapose, cauline leaves when present much reduced; basal leaves cordate-
                        ovate and entire or pinnately divided   -51. Balsamorhiza
                    24. Pappus a crown of scales or awnlike scales; stems more or less leafy, not scapose; leaves not
                        lobed, ovate-lanceolate or narrower   -52. Wyethia
                23. Ray flowers sterile; taproots if present not thick or balsamiferous; leaves often opposite at
                    least in part; pappus if present, deciduous
                    25. Pappus none   -60. Viguiera
                    25. Pappus of scales or awnlike scales (readily deciduous but conspicuous)   -61. Helianthus
    15. Leaves opposite, at least in part
        26. Involucre bracts in 2 distinct sets, the outer narrow and herbaceous, the inner broader, membranous or
            scarious-margined and differing in color from the outer
            27. Inner involucre bracts connate to the middle or above   -53. Thelesperma
            27. Inner bracts distinct or united only at base
                28. Pappus of 2-4 retrorsely barbed awns   -54. Bidens
                28. Pappus none or of 2 very short teeth not at all barbed   -55. Coreopsis
        26. Involucres not distinctly double (bracts may be in 2 rows but these essentially alike except sometimes in
            length)
            29. Disk flowers perfect but sterile, ovaries abortive and styles remaining undivided   -56. Silphium
            29. Disk flowers maturing achenes, styles divided in anthesis
                30. Rays persistent on the achenes and becoming papery; ray flowers maturing achenes and achenes
                    of disk flowers not winged
                    31. Leaves entire; disk flower achenes compressed; rays 10-16 mm. long; plants not over 25
                        cm. tall   -57. Zinnia
                    31. Leaves (at least lower) dentate; achenes 3- to 4-angled; rays 15-30 mm. long; plants
                        usually over 50 cm. tall   -58. Heliopsis
                30. Rays deciduous; ray flowers sterile except in Verbesina with disk achenes corky-winged
                    32. Ray flowers maturing achenes; disk achenes compressed and corky-winged; pappus of disk
                        flowers of 2 long slender awns, no intermediate scales present   -50. Verbesina
                    32. Ray flowers sterile; disk achenes, if compressed at all never corky-winged; pappus of
                        disk flowers none, of scales or of scales and awns together
                        33. Achenes flattened with margined edges; pappus of disk flowers of long slender awns
                            (at least in part), the scales if present not deciduous   -59. Helianthella
                        33. Achenes slightly if at all flattened; pappus of disk flowers none or of scales or
                            awnlike scales, these very deciduous
                            34. Pappus none   -60. Viguiera
                            34. Pappus of scales or awnlike scales (very deciduous)   -61. Helianthus

                                        Tribe 7. Senecioneae

1. Leaves alternate or crowded at base of plant, none really opposite
    2. Plants dioecious or nearly so; leaves all large (10 cm. long or more) and basal, appearing after the
       flowering stems; flowers whitish   -62. Petasites
    2. Plants not dioecious; leaves almost always partly cauline, when the plant is scapose none of the leaves
       even near 10 cm. long; flowers yellow or orange-red
        3. Woody shrubs; involucre bracts 4-6, equal in length but strongly overlapping from the sides; rays
           absent   -63. Tetradymia
        3. Herbaceous plants woody if at all only at the base; involucre bracts over 6, not or only slightly
           overlapping from sides; rays present in most but not all species   -64. Senecio
1. Leaves opposite (some may be basal)
    4. Involucre bracts 8 or more, little if any overlapping from the sides; plants herbaceous   -65. Arnica
    4. Involucre bracts 4-5, overlapping laterally; plants woody at the base   -66. Haploesthes

                                        Tribe 8. Ambrosineae

1. Heads all unisexual; pistillate heads with 1-4 flowers enclosed in a nutlike or burlike involucre usually
   with spines or sometimes with tubercles
    2. Involucre bracts of staminate heads distinct, receptacles elongated; fruit a bur, usually over 1 cm.
       long, covered by uncinate bristles   -67. Xanthium
    2. Involucre bracts of staminate heads more or less united, receptacles flat or merely convex; fruit rarely
       over 1 cm. long, if even approaching that length then spines not hooked at apex
        3. Fruiting involucres with tubercles or short spines in a single series, only 1 flower present
                                                                                                        -68. Ambrosia
        3. Fruiting involucres with several to many spines in 2 or more series, 1-4 flowers present
                                                                                                        -69. Franseria
1. Heads usually with both pistillate and staminate flowers (sometimes all staminate but never all pistillate);
   pistillate flowers few and marginal, not in burlike involucres
    4. Achenes densely long-villous; leaves or their lobes linear-filiform, not over 1 mm. wide, some of leaves
       deeply pinnatifid; plants semi-woody   -70. Oxytenia
    4. Achenes not long-villous; none of leaves linear-filiform and never deeply pinnatifid into linear-filiform
       lobes; plants herbaceous
        5. Achenes with pectinate or toothed winged margins; pistillate flowers subtended by large scarious chaff
           simulating inner involucre bracts   -71. Dicoria

5. Achenes without pectinate or toothed wings; large scarious chaff present or absent -72. Iva

Tribe 9. Inuleae

1. Pappus none; receptacles chaffy -73. Evax
1. Pappus of capillary bristles; receptacles naked
   2. Heads unisexual, the plants dioecious (some heads of pistillate flowers, others of staminate flowers with abortive ovaries and undivided styles - hence functionally staminate)
      3. Pappus bristles of pistillate flowers united at base and falling together; pappus bristles of staminate flowers usually clavate at the tip; plants usually less than 30 cm. tall; basal leaves commonly in a rosette, the stem leaves reduced and different in shape, leaves usually tomentose both sides -74. Antennaria
      3. Pappus bristles of pistillate flowers separate at base and falling separately; pappus bristles of staminate flowers not clavate at tip; plants usually over 30 cm. tall; leaves about alike, no basal rosette, leaves usually green and glabrate above -75. Anaphalis
   2. Heads with marginal flowers pistillate, the central flowers perfect and fertile
      4. Heads subdioecious, the pistillate heads usually with a few perfect flowers in the center; other heads of perfect flowers but in any case all the perfect flowers are functionally staminate with abortive ovaries and undivided styles; leaves usually green and glabrate above and not glandular; plants perennial with creeping rootstocks -75. Anaphalis
      4. Heads all alike, of marginal pistillate flowers and central perfect ones, these maturing their achenes and with divided styles at anthesis; leaves either about equally tomentose above and below or if green above then glandular; plants annual or biennial (possibly perennial) with taproots -76. Gnaphalium

Tribe 10. Cynareae

1. Leaves spinescent, usually with spinescent teeth or lobes, rarely entire with spinulose tips
   2. Pappus of plumose bristles; receptacles densely bristly -77. Cirsium
   2. Pappus various but if of bristles these barbellate but never plumose; receptacles bristly or not bristly
      3. Heads proper 1-flowered, but many heads crowded into a globose common head; pappus cup-shaped with toothed edges or somewhat scalelike -78. Echinops
      3. Heads many-flowered, these not strictly globose; pappus of bristles (except in Carthamus)
         4. Pappus of scales or flattened bristles; flowers orange or orange-yellow -79. Carthamus
         4. Pappus of bristles; flowers purple or purplish
            5. Receptacles densely bristly; leaves glabrate or very sparsely tomentose -80. Carduus
            5. Receptacles honeycombed but scarcely bristly; leaves often white-tomentose -81. Onopordon
1. Leaves not spinescent, entire, toothed or pinnately lobed
   6. Involucre bracts with uncinate spines; lower leaves very large, usually cordate at base -82. Arctium
   6. Involucre bracts spineless or spines not hooked at apex; lower leaves not especially large, not cordate at base
      7. Achenes obliquely attached by 1 side; weedy plants of fields and waste ground at lower altitudes; stems usually over 30 cm. tall; pappus various but never of plumose bristles -83. Centaurea
      7. Achenes attached by their base; plants never weedy, alpine in Colorado; stems not over 20 cm. tall; pappus of plumose bristles -84. Saussurea

Tribe 11. Cichorieae

1. Pappus of plumose bristles (at least in part)
   2. Achenes definitely beaked at apex; involucres 2 cm. long or longer
      3. Receptacles chaffy with thin and narrow scales; involucre bracts imbricated (outer short); leaves pinnatifid to dentate -85. Hypochoeris
      3. Receptacle naked; involucre bracts in 1 series of equal length; leaves entire -86. Tragopogon
   2. Achenes not at all beaked at apex; involucres not over 2 cm. long
      4. Rays rose- to flesh-colored (at least when fresh); plants with branching, rushlike stems and leaves reduced above -87. Stephanomeria
      4. Rays yellow; plants with stems not especially branching, not rushlike -88. Ptilocalais
1. Pappus never of plumose bristles, none, of scales or simple bristles (bristles may be scabrous or barbellate but never plumose)
   5. Pappus none -89. Lapsana
   5. Pappus of scales or bristles present (may be both)
      6. Pappus double, the outer a row of scales, the inner of capillary bristles -90. Krigia
      6. Pappus of 1 type, either of scales or bristles
         7. Pappus a short-toothed crown or of short blunt scales; flowers normally blue or purple -91. Cichorium
         7. Pappus of long capillary bristles or elongated bristlelike scales; flowers seldom blue or purple
            8. Achenes more or less flattened; stems leafy; heads in paniculate or umbellate inflorescences
               9. Involucres cylindric or ovoid-cylindric; achenes beaked (weakly so in L. spicata); flowers yellow or blue -92. Lactuca
               9. Involucres broadly campanulate to hemispheric; achenes not at all beaked; flowers yellow -93. Sonchus
            8. Achenes not flattened; stems leafy or scapose; heads solitary to many, variously clustered
               10. Rays yellow or orange-red when fresh
                  11. Heads only 1 to a stem or scape
                     12. Stems more or less leafy
                        13. Pappus white or whitish; involucre bracts somewhat thickened at base or on the midrib -94. Crepis

533

                13. Pappus some shade of brown; involucre bracts not thickened   -95. Hieracium
            12. Stems scapose, the leaves all basal or radical
                14. Pappus bristles broader and flattened near base, really very narrow bristlelike scales (use
                    strong lens)   -96. Nothocalais
                14. Pappus bristles not at all flattened
                    15. Achenes beaked at apex
                        16. Achenes 10-ribbed or nerved, not spinulose-muricate; involucre bracts usually
                            imbricate in several graduated series   -97. Agoseris
                        16. Achenes 4- to 5-ribbed (may be smaller nerves too), spinulose-muricate especially
                            near apex; principal bracts of the involucre in a single series, the outer ones
                            much shorter   -98. Taraxicum
                    15. Achenes not beaked at apex   -94. Crepis
        11. Heads 2 or more to a main stem
            17. Pappus bristles readily deciduous, more or less united below and falling together, only a few of
                the stout outer ones may be persistent   -99. Malacothrix
            17. Pappus bristles persistent or tardily deciduous and then falling separately (achene beak, when
                present, may break and pappus then falls together)
                18. Pappus white or whitish; involucre bracts somewhat thickened at base or on the midrib
                    -94. Crepis
                18. Pappus tan to brown; involucre bracts not thickened   -95. Hieracium
    10. Rays white, purple, rose-colored, or pink
        19. Achenes beaked at apex; heads solitary on long leafless scapes   -97. Agoseris
        19. Achenes not beaked at apex (may narrow somewhat upward); heads usually 2 to many, rarely on leafless
            scapes
            20. Flowers white to ochroleucous   -95. Hieracium
            20. Flowers pink, rose or purple
                21. Heads in spikelike panicles; leaves lanceolate or oblanceolate to broader and over 5 cm. long
                    -100. Prenanthes
                21. Heads terminating the branches or racemose, never in spikelike panicles; leaves either linear
                    or linear-lanceolate or if broader then not over 4 cm. long   -101. Lygodesmia

II. Artificial key to the genera of Compositae

1. Corollas all ligulate; plants usually with milky juice; leaves alternate or basal
    2. Pappus of plumose bristles, at least in part
        3. Achenes definitely beaked at apex; involucres 2 cm. long or more
            4. Receptacles chaffy with thin, narrow scales; involucre bracts imbricated (outer short); leaves
                pinnatifid to dentate   -85. Hypochoeris
            4. Receptacles naked; involucre bracts in 1 series of equal length; leaves entire   -86. Tragopogon
        3. Achenes not at all beaked at apex; involucres not over 2 cm. long
            5. Rays rose- to flesh-colored (at least when fresh); plants with branching, rushlike stems and
                leaves reduced above   -87. Stephanomeria
            5. Rays yellow; stems not especially branching, not rushlike, leaves not reduced above
                -88. Ptilocalais
    2. Pappus none, of scales or of simple bristles (bristles may be scabrous or barbellate but never plumose)
        6. Pappus none   -89. Lapsana
        6. Pappus of scales or bristles present (may be both)
            7. Pappus double, the outer a row of scales, the inner of capillary bristles   -90. Krigia
            7. Pappus of 1 type, either of scales or bristles
                8. Pappus a short-toothed crown or of short blunt scales; flowers normally blue or purple
                    -91. Cichorium
                8. Pappus of long capillary bristles or of elongated bristlelike scales; flowers not commonly blue
                    or purple
                    9. Achenes more or less flattened; stems leafy; heads in paniculate or umbellate inflorescences
                        10. Involucres cylindric or ovoid-cylindric; achenes beaked (weakly so in L. spicata);
                            flowers yellow or blue   -92. Lactuca
                        10. Involucres broadly campanulate to hemispheric; achenes not at all beaked; flowers
                            yellow   -93. Sonchus
                    9. Achenes not flattened; stems leafy or scapose; heads solitary to many
                        11. Rays yellow or orange-red when fresh
                            12. Heads 1 to a stem or scape
                                13. Stems more or less leafy
                                    14. Pappus white or whitish; involucre bracts somewhat thickened at base or
                                        on midrib   -94. Crepis
                                    14. Pappus some shade of brown; involucre bracts not thickened
                                        -95. Hieracium
                                13. Stems scapose, the leaves all basal or radical
                                    15. Pappus bristles broadened and flattened near base, really very narrow
                                        bristlelike scales (use strong lens)   -96. Nothocalais
                                    15. Pappus bristles not at all flattened
                                        16. Achenes beaked at apex
                                            17. Achenes 10-ribbed or 10-nerved, not spinulose-muricate;
                                                involucre bracts usually imbricated in several graduated
                                                series   -97. Agoseris
                                            17. Achenes 4- to 5-ribbed (may be smaller nerves in addition),
                                                spinulose-muricate especially near apex; principal bracts of
                                                the involucre in a single series, the outer much shorter
                                                -98. Taraxicum

                16. Achenes not beaked at apex   -94. Crepis
        12. Heads 2 or more to a stem
                18. Pappus bristles readily deciduous, more or less united below and falling together, only a few
                    of the stout outer ones may be persistent   -99. Malacothrix
                18. Pappus bristles persistent or tardily deciduous and then falling separately (achene beak when
                    present may break and pappus then falls with it)
                        19. Pappus white or whitish; involucre bracts somewhat thickened at base or on the midrib
                                                                                                  -94. Crepis
                        19. Pappus tan to brown; involucre bracts not thickened   -95. Hieracium
    11. Rays white, pink, rose or purple
        20. Achenes beaked at apex; heads solitary on long, leafless scapes   -97. Agoseris
        20. Achenes not beaked at apex (may narrow upwards somewhat); heads usually 2 to many, rarely on leafless
            scapes
            21. Flowers white to ochroleucous   -95. Hieracium
            21. Flowers pink, rose or purple
                22. Heads in spikelike panicles; leaves lanceolate or oblanceolate to broader and then over
                    5 cm. long   -100. Prenanthes
                22. Heads terminating the branches or racemose, never in spikelike panicles; leaves either
                    linear or linear-lanceolate or if broader then not over 4 cm. long   -101. Lygodesmia
1. Corollas not all ligulate, some or all of them tubular; milky juice usually absent; leaves opposite, alternate
   or basal
    23. Corollas all tubular, no rays present (or the rays vestigial and minute-sometimes the marginal disk
        flowers are enlarged and simulate rays but are never ligulate)
        24. Heads unisexual, the pistillate heads with 1-4 flowers enclosed in a nutlike or burlike involucre
            usually with spines or sometimes tubercles present
            25. Involucre bracts of staminate heads distinct, receptacle elongated; fruit usually over 1 cm.
                long, covered with bristles hooked at apex   -67. Xanthium
            25. Involucre bracts of staminate heads more or less united, receptacle flat or merely convex;
                fruit rarely over 1 cm. long, if even approaching that length the spines not hooked at apex
                26. Fruiting involucres with tubercles or short spines in a single series, only 1 flower
                    present   -68. Ambrosia
                26. Fruiting involucres with several to many spines in 2 or more series, 1-4 flowers present
                                                                                                  -69. Franseria
        24. Heads bisexual (with perfect or staminate and pistillate flowers together) or if unisexual the
            pistillate flowers not in modified and specialized heads
            27. Stamens not united by their anthers; flowers always unisexual, the pistillate corollas none or
                much reduced
                28. Achenes densely long-villous; leaves or their lobes linear-filiform, not over 1 mm. wide,
                    some of leaves deeply pinnatifid; plants semi-woody   -70. Oxytenia
                28. Achenes not long-villous; none of leaves linear-filiform and never deeply pinnatifid into
                    linear-filiform lobes (entire, toothed or incised); plants herbaceous
                    29. Achenes with pectinate or toothed winged margins; pistillate flowers subtended by
                        large scarious chaff simulating inner involucre bracts   -71. Dicoria
                    29. Achenes without pectinate or toothed wings; large scarious chaff subtending the
                        pistillate flowers present or absent   -72. Iva
            27. Stamens with anthers connate (or anthers rarely distinct in a few bisexual-flowered species),
                some of flowers usually bisexual
                30. Involucre bracts with conspicuous, translucent, usually orange or yellow dots; pappus of
                    8-15 paleae, each soon dissected into 5-10 bristles   -24. Dyssodia
                30. Involucre bracts without such dots; pappus various but rarely as above
                    31. Pappus (at least in part) of capillary bristles, these smooth, scabrous, barbellate
                        or plumose
                        32. Leaves opposite or whorled, some or all on the stem
                            33. Corollas yellow or pale yellow; involucre bracts in 1 series, or 2 series
                                but all equal in length
                                34. Pappus bristles united at base to scales, hence in groups; leaves
                                    ovate or triangular, cordate, truncate or hastate at base
                                                                                                  -29. Pericome
                                34. Pappus bristles all separate to base, hence not united in groups;
                                    leaves lanceolate or narrower, cuneate at base   -65. Arnica
                            33. Corollas white, ochroleucous, flesh-colored, blue or purple, not at all
                                yellow; involucre bracts in 2 to several series, outer usually shorter
                                35. Leaves whorled, 3 or more at a node, 5-20 cm. long   -2. Eupatorium
                                35. Leaves opposite, 2 at a node, rarely over 5 cm. long
                                    36. Achenes 5-angled or 5-ribbed; involucre bracts subequal or in 2
                                        series   -2. Eupatorium
                                    36. Achenes 10-ribbed or 10-angled; involucre bracts imbricated in
                                        several series of different lengths   -3. Brickellia
                        32. Leaves alternate (at least on lower part) or basal (and actually alternate)
                            37. Heads proper 1-flowered, these heads crowded in a globose involucrate head;
                                leaves spinulose   -78. Echinops
                            37. Heads more than 1-flowered, not crowded into a globose structure; leaves
                                spinulose or not spinulose
                                38. Receptacle with dense bristles or narrow chaffy scales between the
                                    disk flowers; leaves often very spinescent (like a thistle)
                                    39. Leaves spinescent, usually with spinescent teeth or lobes, rarely
                                        entire with spinulose apex

535

40. Pappus bristles plumose; receptacles densely bristly -77. Cirsium
40. Pappus various but if of bristles then not plumose; receptacle with bristles or scales
   41. Pappus of narrow scales; flowers orange or orange-yellow -79. Carthamus
   41. Pappus of bristles; flowers purplish -80. Carduus
39. Leaves not spinescent (entire, denticulate or pinnately lobed)
   42. Involucre bracts with hooked spines; lower leaves very large (often over 15 cm. wide), usually cordate at base -82. Arctium
   42. Involucre bracts without spines or spines not hooked at apex; lower leaves not especially large, not cordate at base
      43. Achenes obliquely attached by 1 side; weedy plants of fields and disturbed areas at lower altitudes; stems usually over 30 cm. tall; pappus various, never of plumose bristles -83. Centaurea
      43. Achenes attached at very base; plants never weedy, alpine; stems not over 20 cm. tall; pappus in part of plumose bristles -84. Saussurea
38. Receptacle naked or at most short-hairy, never with dense bristles or scales; leaves various but seldom spinescent
   44. Heads unisexual, plants dioecious (staminate flowers may have styles but ovary is abortive and does not develop)
      45. Involucre bracts in 1 series; basal leaves very large, 10-30 cm. long -62. Petasites
      45. Involucre bracts imbricated in several series; basal leaves usually much smaller, if present
         46. Anthers not caudate at base; plants shrubs or woody at base; leaves usually toothed or lobed; involucre bracts not strongly scarious-margined; plants not tomentose -6. Baccharis
         46. Anthers caudate at base; plants herbaceous; leaves entire; involucre bracts strongly scarious at least on margins; plants more or less tomentose
            47. Pappus bristles of pistillate flowers united at base and falling together; pappus bristles of staminate flowers usually clavate at the apex; plants usually less than 30 cm. tall, basal leaves commonly in a basal rosette, the stem leaves reduced and different in shape; leaves usually tomentose both sides -74. Antennaria
            47. Pappus of pistillate flowers separate to base and falling separately; pappus bristles of staminate flower not clavate at apex; plants usually over 30 cm. tall, the leaves about all alike, no basal rosette; leaves usually green and glabrate above -75. Anaphalis
   44. Heads with the central flowers at least perfect
      48. Involucre bracts scarious or hyaline at least at the broad margins
         49. Heads subdioecious, pistillate heads usually with a few perfect flowers in center, other heads of perfect flowers but in any case all perfect flowers functionally staminate with abortive ovaries and undivided styles; leaves usually green and glabrate above and not glandular; plants perennial with creeping rootstocks -75. Anaphalis
         49. Heads all alike, of marginal pistillate flowers and central perfect ones, these with divided styles and maturing their achenes; leaves either about equally tomentose above and below or if green above then glandular; plants annual or biennial (possibly sometimes perennial) with a taproot -76. Gnaphalium
      48. Involucre bracts slightly if at all scarious
         50. Involucre bracts in a single series (edges may overlap somewhat), a few very short ones may be at base
            51. Plant scapose, normal leaves radical, 10-30 cm. long, only much reduced leaves on the stem; flowers whitish, subdioecious -62. Petasites
            51. Plants not scapose some of leaves on the stem, radical leaves if present usually much smaller; flowers, at least central ones, yellow or orange
               52. Woody shrubs; involucre bracts 4-6 to a head -63. Tetradymia
               52. Herbaceous plants; involucre bracts over 6 to a head
                  53. Heads with the outer flowers pistillate and the inner perfect; plants annual -14. Conyza
                  53. Heads with all the flowers perfect; plants perennial (in all but Senecio vulgaris rare in Colorado)
                     54. Style branches with a tuft of long hairs near the truncate apex; involucre bracts in 1 series only (a few short ones may be present) -64. Senecio
                     54. Style branches with a flattened appendage with hairs aggregated on the upper part but not as a tuft at the apex which is not truncate; involucre bracts on close examination of 2 or more series of equal length -15. Erigeron
         50. Involucre bracts of 2 or more series, these often of different lengths
            55. Disk flowers yellow, orange or brownish
               56. Pappus of 2-8 slender but almost paleaceous, stiff, caducous awns -9. Grindelia
               56. Pappus of more than 8 capillary bristles, usually not caducous
                  57. Plants annual
                     58. Lower leaves toothed or lobed; heads small, the involucres about 3 mm. high -14. Conyza
                     58. All the leaves entire; involucres 5-8 mm. high -16. Aster
                  57. Plants perennial, often woody at least at base
                     59. Pappus double, outer of narrow squamellae -15. Erigeron
                     59. Pappus single (bristles may be unequal in length)
                        60. Leaves 1- to 3-ternately lobed -15. Erigeron
                        60. Leaves entire or toothed, not at all ternately lobed
                           61. Heads usually narrowly turbinate to narrowly cylindrical; involucre bracts in more or less clearly defined vertical rows; plants woody shrubs (if undershrubs the bracts clearly vertical) -7. Chrysothamnus
                           61. Heads campanulate to hemispheric; involucre bracts not in vertical rows; plants herbaceous or woody only at base -12. Haplopappus

55. Disk flowers white, ochroleucous, rose-purple to purple
    62. Involucre bracts tipped with long spines; anther sacs caudate at base   -81. Onopordon
    62. Involucre bracts not tipped with long spines; anther sacs not caudate at base
        63. Pappus double, inner of capillary bristles, outer of short bristles or squamellae
                                                                                                                                                                       -1. Vernonia
        63. Pappus single, bristles may be uneven but not in 2 distinct sets
            64. Flowers rose-purple; heads in narrow spikes or in spikelike racemes, rarely spikelike panicles; involucre bracts not striated, often partly colored   -4. Liatris
            64. Flowers white, cream-colored, ochroleucous or sometimes purplish-tinged, usually in open branching cymes or corymbs, never in spikes or spikelike inflorescences; involucre bracts striated, rather chartaceous and whitish but never petaloid
                65. Pappus plumose, with side-cilia about 0.4-0.5 mm. long, the bristles obviously plumose to the naked eye; anthers tending to separate   -5. Kuhnia
                65. Pappus bristles barbellate (or in a rare species) sub-plumose with side-cilia only about 0.2-0.3 mm. long, hence not appearing very plumose to the naked eye; anthers united   -3. Brickellia
31. Pappus none or if present not as capillary bristles
    66. Receptacles with bristles or chaffy scales among the flowers (scales may be limited to a row between the central and marginal flowers)
        67. Receptacles densely bristly   -83. Centaurea
        67. Receptacles with chaffy scales
            68. Leaves definitely spiny on margins; plants cultivated, possibly escaping   -79. Carthamus
            68. Leaves not spiny; plants not cultivated
                69. Involucre bracts in 2 distinct sets, the outer herbaceous, the inner colored and more or less scarious, differing in shape and texture; leaves often opposite
                    70. Leaves alternate, ovate and toothed; pappus none   -39. Parthenice
                    70. Leaves opposite (at least below), either lobed or divided or if not lobed then much narrower than ovate; pappus of 2-4 retrorsely hispid or barbed awns
                        71. Inner involucre bracts united to the middle or above, forming a cup   -53. Thelesperma
                        71. Inner involucre bracts not united or only at the very base   -54. Bidens
                69. Involucre bracts not in 2 unlike sets, either in 1 row or if imbricated the bracts all essentially alike in texture and shape (outer may be shorter); leaves alternate or crowded-basal
                    72. Involucre bracts in 1 row, their margins incurved from the sides, each bract enclosing a marginal flower; short yellow rays present (about 1 mm. long)   -40. Madia
                    72. Involucre bracts in 1 or more series but none enclosing the flowers; no rays of any kind present
                        73. Low woolly annual plants not over 15 cm. tall; heads more or less glomerate, leafy-bracted   -73. Evax
                        73. Taller perennial plants usually over 15 cm. tall (except in Parthenium), glabrous to hispid-canescent, never woolly; heads solitary on long peduncles or scapes
                            74. Receptacles high-conical, becoming 3-5 cm. long; stems not scapose, some of leaves cauline; some of leaves lobed   -48. Rudbeckia
                          74. Receptacles convex, never even approaching 3 cm. long; stems scapose or subscapose; leaves not lobed
                              75. Central flowers maturing achenes; plants 15-25 cm. tall; involucres 12-15 mm. high; plants of western Colorado   -37. Enceliopsis
                              75. Central flowers perfect but sterile; plants 2-5 cm. tall; involucres about 5 mm. high; plants of eastern Colorado   -38. Parthenium
    66. Receptacles naked or very short-hairy (sometimes pitted, the pits with raised margins)
        76. Pappus none
            77. Leaves opposite, some cauline, somewhat connate at base   -32. Flaveria
            77. Leaves alternate or basal (and really alternate)
                78. Heads solitary on the stems or rather long branches, or sometimes corymbose or capitate; anthers with rounded tips; receptacles convex to conic; plants herbaceous or woody only at the base
                    79. Plants annual; heads solitary on rather long branches; leaves green and glabrous   -21. Matricaria
                    79. Plants perennial, usually with a somewhat woody base; heads corymbose or capitate, rarely solitary; leaves mostly silvery-canescent, sometimes glabrate
                        80. Involucre bracts in 2-3 series; plants sometimes aromatic but never with a mintlike odor; heads either capitate (rarely solitary) or if corymbose then the leaves bipinnatifid or lobed at the apex only   -19. Tanacetum
                        80. Involucre bracts in 4-5 series; plants with strong mintlike odor when fresh; heads corymbose and leaves not lobed or somewhat once-pinnate at the base only   -20. Chrysanthemum
                78. Heads spicate, racemose or (more commonly) panicled; anthers with pointed tips; receptacles flat to convex; plants herbaceous or (as in some very common species) shrubby   -22. Artemisia
    76. Pappus of some sort present
        81. Heads strictly dioecious, heads on some plants entirely pistillate, on others with perfect but sterile and functionally staminate flowers; plants usually shrubby; pappus of long flattened bristles   -6. Baccharis
        81. Heads not dioecious, all or at least central flowers perfect and fertile; plants not shrubby; pappus otherwise
            82. Leaves opposite, some cauline

83. Corollas pink or rose; leaves entire, linear or lanceolate -26. Palafoxia
83. Corollas pale yellow or yellow; leaves divided or if entire then ovate or deltoid
  84. Leaves divided, some usually alternate -36. Bahia
  84. Leaves entire, serrulate or merely sinuate, not at all divided and strictly opposite
    -29. Pericome
82. Leaves alternate or basal (and essentially alternate)
  85. Pappus of 2-8 caducous awns; plants often strongly glutinous -9. Grindelia
  85. Pappus of a crown or scales (these may be awn-pointed); plants seldom glutinous
    86. Pappus a crown with margins entire or of short squamellae united into a crown; involucre bracts with definite scarious margins
      87. Plants annual; heads solitary on the rather long branches; all the flowers of the head perfect; leaves green and glabrous -21. Matricaria
      87. Plants perennial; heads corymbose or capitate, rarely solitary; some of marginal flowers pistillate only, the others perfect; leaves mostly silvery-canescent, sometimes glabrate
        -19. Tanacetum
    86. Pappus of paleae not united into a crown; involucre bracts often without scarious margins
      88. Corollas white, flesh-colored or rose
        89. Leaves entire or nearly so, never lobed or dissected
          90. Leaves all basal, oblong, ovate or suborbicular; pappus scales spatulate, not acuminate at apex -25. Chamaechaenactis
          90. Leaves mostly cauline, linear or lanceolate; pappus scales ovate or lanceolate, acuminate-attenuate at apex -26. Palafoxia
        89. Leaves, at least in part, 1-3 times pinnately divided or dissected
          91. Involucre bracts thin, scarious and whitish or yellowish at margins and tips; pappus scales about 10-20 -27. Hymenopappus
          91. Involucre bracts herbaceous, not scarious or colored at margins and tips; pappus scales about 4-10 -28. Chaenactis
      88. Corollas light yellow to yellow
        92. Plants more or less floccose-tomentose; pappus of about 10-20 scales; plants perennial or in 1 species biennial -27. Hymenopappus
        92. Plants not tomentose; pappus of 8 scales; plants annual -36. Bahia
23. Ligulate marginal flowers present (rays may be short)
  93. Pappus of capillary bristles, at least in part
    94. Rays yellow, golden or orange
      95. Receptacles long-bristly; involucre bracts ending in spines -83. Centaurea
      95. Receptacles naked or short-hairy, sometimes pitted with raised margins but never with long bristles; involucre bracts never ending in spines
        96. Involucre bracts with conspicuous translucent resin dots; pappus of 8-15 squamellae, each dissected in bristles above but united and entire at base -24. Dyssodia
        96. No such dots present; pappus otherwise
          97. Leaves opposite, at least below, some cauline
            98. Involucre bracts 4 or 5, overlapping from the sides; plants woody at base
              -66. Haploesthes
            98. Involucre bracts 8 or more, slightly if at all overlapping from sides; plants always herbaceous -65. Arnica
          97. Leaves alternate or basal (and still essentially alternate)
            99. Pappus double, outer of short scales (sometimes bristle-pointed), inner of capillary bristles -8. Chrysopsis
            99. Pappus of bristles or awns but all alike (may be of different lengths but not in 2 sets)
              100. Pappus of 2-8 caducous awns -9. Grindelia
              100. Pappus of many capillary bristles
                101. Proper involucre bracts in 1 series, all equal (sometimes with very short outer bractlets at base); style tips truncate -64. Senecio
                101. Involucre bracts in more than 1 series, usually more or less unequal and graduated; style tips not truncate
                  102. Heads usually small (involucres seldom over 6.5 mm. high) and numerous, often in a panicle; plants always herbaceous; involucre bracts rarely herbaceous at apex -11. Solidago
                  102. Heads usually large (involucres seldom less than 6.5 mm. high) and usually few or solitary but if small and panicled then the plant woody at base; plants herbaceous or woody; involucre bracts often distinctly herbaceous at apex
                    -12. Haplopappus
    94. Rays white, pink, blue, rose, violet or purple when fresh, never yellow
      103. Receptacles densely bristly; outer flowers enlarged and falsely radiate -83. Centaurea
      103. Receptacles naked, no long bristles; true rays present (sometimes small)
        104. Scapose plants, basal leaves ovate to deltoid, tomentose beneath, 10-30 cm. long; heads sub-dioecious; involucre bracts equal in 1 series -62. Petasites
        104. Plants not scapose, or if so then the basal leaves otherwise; heads not at all dioecious; involucre bracts rarely in 1 series only
          105. Pappus (at least of disk flowers) of several to many rigid bristles; achenes pubescent with hairs 2-forked or glochidiate at apex (use strong lens)
            -13. Townsendia
          105. Pappus (at least of disk flowers) of many long capillary bristles, at least in part; achenes glabrous or pubescent with simple hairs

106. Rays very inconspicuous, rays shorter than their tube and scarcely if at all exceeding their pappus; central perfect flowers few; plants annual -14. Conyza
106. Rays usually conspicuous, when short then longer than their tube and pappus; central perfect flowers usually several to many; plants annual, biennial or perennial
    107. Involucre bracts relatively long and narrow, herbaceous to non-herbaceous but loss of green color either uniform or more pronounced toward the tips, bracts usually subequal but sometimes definitely imbricated but then scarcely if at all herbaceous; pappus often double; style appendages short, commonly obtuse to merely acutish at apex but sometimes sharply acute; achenes mostly 2-nerved; rays often narrow and very numerous -15. Erigeron
    107. Involucre bracts relatively broad, at least the tips usually expanded, herbaceous to non-herbaceous but loss of green color more pronounced near the base, bracts commonly regularly imbricated, sometimes subequal but then usually herbaceous nearly throughout; pappus never double; style appendages relatively long, usually acute to acuminate; achenes mostly several-nerved; rays commonly broad and relatively few -16. Aster
93. Pappus none or when present not of capillary bristles
  108. Receptacles with long bristles or chaffy scales among the disk flowers or as a row of scales between the outer ray flowers and the central disk flowers
    109. Receptacle with bristles or subulate or setiform fimbrillae, no chaffy scales present
      110. Rays not conspicuous, really not present, the marginal disk flowers enlarged to simulate a ray; central flowers without moniliform hairs; involucres ovoid or globose the bracts not spreading (except sometimes at apex) -83. Centaurea
      110. Rays conspicuous, commonly over 8 mm. long; central flowers with moniliform hairs on the corolla lobes; involucres rotate to saucer-shaped, bracts strongly reflexed in age -31. Gaillardia
    109. Receptacle with chaffy scales
      111. Involucre bracts (all or inner set) with margins from the sides enclosing or enfolding the ray flowers and achenes
        112. Involucre bracts in a single series, all enclosing ray achenes; leaves alternate; rays yellowish, inconspicuous (about 1 mm. long) -40. Madia
        112. Involucre bracts in 2 distinct sets, only the inner enclosing the ray achenes; leaves opposite; rays white to pink, conspicuous (6 mm. long or more) -41. Melampodium
      111. None of involucre bracts enclosing or enfolding the ray flowers or ray achenes
        113. Rays white, whitish, pink, purple or rose, never yellow, golden or orange
          114. Leaves opposite, some cauline
            115. Leaves dissected; involucre bracts in 2 rather distinct sets, these unlike in color and texture; pappus of retrorsely hispid or retrorsely barbed awns
              116. Achenes conspicuously beaked; rays usually pink -43. Cosmos
              116. Achenes not at all beaked; rays whitish to greenish-white, not pink -54. Bidens
            115. Leaves merely toothed, not at all lobed or dissected; involucre bracts not in 2 distinct sets (may be in 2 series but not unlike in color or texture); pappus lacking or of teeth or scales, never of retrorsely hispid or barbed awns
              117. Pappus lacking or rarely of minute teeth; leaves lanceolate and sessile (except sometimes the lower ones) -44. Eclipta
              117. Pappus (at least of disk flowers) of definite scales; leaves ovate, all petioled -45. Galinsoga
          114. Leaves alternate or basal (and essentially alternate)
            118. Rays white (rarely rose), not over 15 mm. long; involucre bracts scarious or with definite scarious margins; leaves dissected, divided or parted
              119. Heads in close terminal corymbs; achenes flattened; rays not over 6 mm. long, nearly as wide, only 4 or 5 present on a head -17. Achillea
              119. Heads solitary at the ends of the branches, never definitely and closely corymbose; achenes not compressed; rays 6-15 mm. long, longer than wide, over 5 normally present
                120. Receptacle conic, bearing stiff narrow palea near apex; rays 6-10 mm. long; involucres 4-6 mm. high -18. Anthemis
                120. Receptacle convex, bearing blunt membranous scalelike palea throughout; rays 10-15 mm. long; involucres 8-10 mm. high -30. Leucampyx
            118. Rays purple or rose, 2 cm. long or more; involucre bracts not at all scarious; leaves entire -42. Echinacea
        113. Rays yellow, golden or orange at least in part
          121. Chaff of receptacle, when present, reduced to a few scales between the ray and disk flowers -34. Helenium
          121. Chaff of receptacle of many scales, usually among the disk flowers, less commonly as a definite series between the ray and disk flowers
            122. Leaves opposite at least in part, at least some cauline
              123. Involucres distinctly double, outer bracts narrow and herbaceous, inner broader, membranous or scarious-margined, distinctly differing in shape, texture and color from the outer
                124. Inner involucre bracts connate to the middle or above -53. Thelesperma
                124. Inner involucre bracts distinct or united only at the base
                  125. Pappus of 2-4 retrorsely barbed awns -54. Bidens

125. Pappus none or of 2 very short teeth, not at all barbed   -55. Coreopsis
123. Involucres not distinctly double (bracts may be in 2 series but these essentially alike except sometimes in length)
    126. Disk flowers perfect but sterile, ovaries remaining abortive and styles undivided
        -56. Silphium
    126. Disk flowers perfect and maturing achenes, the styles dividing in anthesis
        127. Rays persisting on the achenes and becoming papery; ray flowers maturing achenes and disk achenes not winged
            128. Leaves entire; achenes of disk flowers compressed; rays 10-16 mm. long; plants not over 25 cm. high   -57. Zinnia
            128. Leaves (at least lower) dentate; achenes 3- to 4-angled; rays 15-30 mm. long; plants usually over 50 cm. tall   -58. Heliopsis
        127. Rays deciduous; ray flowers not maturing achenes (except in Verbesina with disk achenes corky-winged)
            129. Ray flowers maturing achenes; disk achenes compressed and corky-winged; pappus of disk flowers of 2 long slender awns, no intermediate scales present
                -50. Verbesina
            129. Ray flowers sterile; disk achenes if compressed at all, never corky-winged; pappus of disk flowers lacking, of scales or of awns and scales together
                130. Achenes flattened with margined edges; pappus of disk flowers of long slender awns (in part), scales if present not deciduous
                    -59. Helianthella
                130. Achenes slightly if at all flattened; pappus of disk flowers none, of scales (or awnlike scales) these very deciduous
                    131. Pappus none   -60. Viguiera
                    131. Pappus of scales or awnlike scales (very deciduous and can be missed)   -61. Helianthus
122. Leaves alternate or basal (and essentially alternate), none cauline and opposite
    132. Disk flowers perfect but sterile, ovaries abortive and styles undivided at apex; leaves pinnatifid; ray flowers maturing achenes
        133. Involucre bracts more or less imbricated but all essentially alike; plants often over 100 cm. tall   -56. Silphium
        133. Involucre bracts in 2 distinct sets, these differing in shape or texture; plants not over 70 cm. tall
            134. Plants hirsute, the leaves not at all canescent below; pappus of ray flowers a crown of scales; inner involucre bracts coriaceous and firm   -46. Engelmannia
            134. Plants puberulent, the leaves whitish-canescent below; pappus of ray flowers absent; inner bracts of involucres foliaceous and thin, the margins somewhat scarious
                -47. Berlandiera
    132. Disk flowers perfect and fertile, maturing their achenes and with styles divided at apex at anthesis; leaves various but not usually pinnatifid; ray flowers sterile to fertile
        135. Receptacle conical or columnar, usually conspicuously so; leaves (in most species) lobed or dissected
            136. Achenes 4-angled, not definitely flattened; leaves either entire, or if parted or divided then on a plant 100 cm. high or more   -48. Rudbeckia
            136. Achenes flattened; leaves parted or divided and the plants not over 80 cm. high
                -49. Ratibida
        135. Receptacle flat or merely convex; leaves (in most species) not lobed or dissected
            137. Achenes flat, the edges thin and margined or winged (at least on disk flowers); pappus of disk flowers of long slender awns (at least in part)
                138. Ray flowers maturing achenes; achenes of disk flowers with corky wings; no intermediate scales present between the pappus awns   -50. Verbesina
                138. Ray flowers sterile; none of achenes corky-winged; small scales present between the pappus awns   -59. Helianthella
            137. Achenes slightly if at all flattened, not margined or winged (at least on the disk flowers); pappus of disk flowers none, of scales or of awnlike scales, rarely of slender awns
                139. Ray flowers maturing achenes; plants with thick balsamiferous taproots; leaves always alternate or basal; pappus if present persistent
                    140. Pappus none; stems scapose, cauline leaves when present much reduced; basal leaves cordate-ovate and entire or pinnately divided
                        -51. Balsamorhiza
                    140. Pappus a crown of scales or of awnlike scales; stems more or less leafy, not scapose; leaves not lobed but ovate-lanceolate or narrower
                        -52. Wyethia
                139. Ray flowers sterile; taproot if present not thick and balsamiferous, leaves often opposite in part; pappus if present very deciduous
                    141. Pappus none   -60. Viguiera
                    141. Pappus of scales or awnlike scales (very deciduous and may be missed)
                        -61. Helianthus
108. Receptacles naked or very short-hairy (may be pitted with edges of pits raised) never any bristles or scales present among the flowers
    142. Involucre bracts with conspicuous, large, translucent resinous dots, these usually yellow or orange
        143. Leaves pinnately divided; involucre bracts often partly united; pappus of 8-15 squamellae each soon dissected into 5-10 bristles   -24. Dyssodia
        143. Leaves simple; involucre bracts not at all united; pappus of scales each bearing only 1 awn at apex or none at all

144. Involucre bracts in 1 series; leaves opposite; foliage rather lemon-scented when fresh -23. Pectis
144. Involucre bracts imbricated in 2 or more series; leaves alternate; foliage with odor but not lemon-scented -10. Gutierrezia
142. Involucre bracts without such dots
    145. Rays yellow, golden or orange
        146. Leaves opposite, at least some cauline
            147. Leaves simple, somewhat connate at base; rays 1-2 mm. long, usually only 1 to a head -32. Flaveria
            147. Leaves 3-5 parted, not at all connate at base; rays 2-4 mm. long, over 1 to a head -36. Bahia
        146. Leaves alternate or basal (and then essentially alternate)
            148. Pappus of 2-8 rigid, caducous awns, not at all scalelike -9. Grindelia
            148. Pappus none, of a short crown, or of paleae (these may be awned from apex)
                149. Pappus wanting or of short squamellae united into a crown; leaves variously divided or parted
                    150. Rays essentially lacking, some of marginal flowers with oblique raylike limbs much less than 6 mm. long; plants perennial -19. Tanacetum
                    150. Rays present, definite, 7-12 mm. long; plants annual
                        151. Involucre bracts herbaceous, not scarious on margins; plants glandular above -36. Bahia
                        151. Involucre bracts with broad scarious margins; plants completely glabrous -20. Chrysanthemum
                149. Pappus present as separate scales; leaves not lobed (except in some species of Hymenoxys)
                    152. Leaves lobed or divided -35. Hymenoxys
                    152. Leaves entire or toothed
                      153. Rays persistent on the achenes and becoming papery; plants woolly or tomentose
                        154. Rays 3-6; achenes linear, striated but slightly angled, glabrous or nearly so; leaves spatulate, lanceolate or oblanceolate -33. Psilostrophe
                        154. Rays 7 or more (usually over 10); achenes obpyramidal, 5-angled, densely hirsute; leaves usually linear-oblanceolate to linear -35. Hymenoxys
                    153. Rays not persistent on the achenes nor becoming papery; plants glabrous to hairy but never woolly or tomentose
                      155. Involucre bracts distinct and definitely graduated in several series, scarious-margined; ray flowers not over 8 to a head -10. Gutierrezia
                      155. Involucres either in 2 series with the outer united at base, or all bracts distinct and nearly equal in length, in any case not graduated in several series, usually not scarious-margined; ray flowers often over 8 to a head
                        156. Receptacle with subulate or setiform chaff between the disk flowers; disk flowers with corolla lobes reddish-purple and bearing moniliform hairs -31. Gaillardia
                        156. Receptacle naked; disk flower corolla lobes not reddish-purple nor with moniliform hairs
                            157. Involucre bracts spreading or reflexed; leaves entire or toothed and some cauline; rays golden-yellow or orange-yellow -34. Helenium
                            157. Involucre bracts erect; leaves usually entire and basal; rays yellow
                                158. Achenes 5-angled, only 2 or 3 times as long as wide, densely hirsute; plants generally distributed in Colorado -35. Hymenoxys
                                158. Achenes 4-angled, 4 or more times as long as wide, glabrate or sparingly hairy; plants of western or southwestern Colorado -36. Bahia
    145. Rays white, flesh-colored, rose or reddish-purple
        159. Pappus none or of a crown on the achene, the margin usually entire; rays white
            160. Leaves bi- to tri-pinnatifid into linear-filiform lobes; rays 7-10 mm. long; pappus a crown -21. Matricaria
            160. Leaves merely toothed or coarsely incised; rays usually over 12 mm. long; pappus absent -20. Chrysanthemum
        159. Pappus present, of distinct and rather conspicuous scales; rays white, flesh-colored, rose or reddish-purple
            161. Receptacle with subulate or setiform chaff among the disk flowers, convex -31. Gaillardia
            161. Receptacle naked, flat
                162. Leaves lobed; heads falsely radiate, the outer disk flowers enlarged and simulating rays -28. Chaenactis
                162. Leaves entire, not at all lobed; true rays present
                    163. Involucre bracts subequal; plants glandular and viscid; pappus of 6-10 lanceolate or ovate-lanceolate paleae; plants annual -26. Palafoxia
                    163. Involucre bracts imbricated, outer shorter; plants not glandular; pappus of many rigid bristles (at least of disk flowers); plants usually distinctly perennial -13. Townsendia

1. <u>Vernonia</u> Schreb. IRONWEED

Perennial herbaceous plants; stems leafy; leaves alternate, simple, entire or toothed; heads corymbose-paniculate; involucres campanulate to cylindrical, bracts imbricated in several series, purplish to greenish; receptacle flat, naked; ray flowers none; disk flowers purple or rose, rarely whitish; anthers coherent, sagittate at base; achenes 8- to 10-ribbed, truncate at apex; pappus double, inner of long capillary bristles and the outer of short bristles or small paleae, purple to tawny.

1. Vernonia marginata (Torr.) Raf., Atl. Journ. 146. 1832.
<u>V. jamesii</u> T. & G.---Stems 40-80 cm. tall, glabrous to puberulent; leaves linear or lanc-linear, entire or minutely serrulate, 1-nerved with very indistinct lateral veins, punctate especially below; inflorescence flat-topped; involucres about 1 cm. high, bracts ovate or lanc-ovate, short-acuminate; achenes glabrous.---Plains and meadows. Kansas and Colorado, south to Texas and New Mexico. Our one record from the southeastern corner of Colorado at about 4000 feet.

Recently a related species, Vernonia fasciculata Michx. has been collected in the northeastern corner of the state. It differs in having linear-lanceolate to lanceolate, sharply serrate-dentate leaves.

2. <u>Eupatorium</u> L. THOROUGHWORT

Perennial plants, herbaceous or woody only at the base; stems usually erect, leafy; leaves opposite or verticillate, simple, often punctate; inflorescence usually cymose-paniculate; involucres oblong-cylindrical, campanulate or hemispheric, bracts imbricated in 2 to several series; receptacle flat to conical, naked; ray flowers none; disk flowers white, blue or purple; achenes 5-angled or 5-ribbed, truncate at the apex; pappus a single row of long capillary scabrous bristles.

1. Leaves verticillate, 3 or more at a node, 5-20 cm. long, cuneate at base; plants limited to east of the Continental Divide in Colorado ....-1. <u>E. maculatum</u>
1. Leaves opposite, 2 at a node, 2-5 cm. long, usually truncate to cordate at base; plants not limited to east of the Continental Divide ....-2. <u>E. herbaceum</u>

1. Eupatorium maculatum L., Amoen. Acad. 4:288. 1755.
<u>E. bruneri</u> A. Gray---Herbaceous plants; stems 50-150 cm. tall, more or less purple-spotted, puberulent or scabrous at least above, sometimes nearly canescent; leaves 5-20 cm. long, most commonly in whorls of 4's or 5's, sometimes 3's or 6's, ovate to lanceolate, tapering to apex and to base, short-petioled or nearly sessile, sharply serrate to crenate-serrate, glabrous or scabrous above, glabrate to canescent below; inflorescence or its component parts flat-topped; involucres about 7-9 mm. high, broadly oblong, bracts in 4 or 5 series the outer shorter, obtuse, striated, pink; flowers 9-15 to a head, usually deep purple, pink or rose; achenes 3-4.5 mm. long, more or less glandular-dotted.---Moist soil in meadows or along streams and ponds. Newfoundland to British Columbia, south to Pennsylvania and New Mexico. Our records along the eastern foothills in northcentral and central Colorado at 5000-7500 feet.

2. Eupatorium herbaceum (A. Gray) Greene, Pitt. 4:279. 1901.
<u>E. arizonicum</u> (Gray) Greene---Plants herbaceous, woody only in the caudex; stems 30-80 cm. tall, rather yellowish-green, puberulent; leaves opposite, about 2-5 cm. long, ovate to triangular-ovate, truncate or subcordate at base, acute or acuminate, rather obscurely toothed with small blunt teeth, minutely scabrous; petioles 4-12 mm. long; inflorescence corymbiform-paniculate; involucres 3-4 mm. long, campanulate, minutely puberulent, bracts in about 2 rows, subequal in length or sometimes a few shorter ones present, lanceolate, long-acute, more or less striated; flowers about 10-20 to a head, white or faintly purplish-tinged; achenes 1.5-2.5 mm. long, more or less hairy.---Canyons and rocky slopes. Colorado and Utah, south to New Mexico and Arizona. Our records from southwestern Colorado, but one from the central part of the state, at 5500-8000 feet.

A related species, E. texense (T. & G.) Rydb., with involucre bracts in 2 definitely unequal series, is reported from Colorado. The writer cannot distinguish it from the above in our material.

3. <u>Brickellia</u> Ell. BRICKELLBUSH

Perennial herbaceous or shrubby plants; stems leafy; leaves opposite or alternate, simple; heads in panicles, racemes, cymes, corymbs or rarely solitary; involucres campanulate to cylindrical, bracts imbricated in several series, striated, more or less chartaceous; receptacles flat or shallowly convex, naked; ray flowers none; disk flowers white, ochroleucous or flesh-colored; anthers united; achenes 10-ribbed; pappus of many long capillary bristles, these scabrous, barbellate or subplumose with side-cilia 0.2-0.3 mm. long.

1. Leaves, at least the cauline, on definite petioles 4-70 mm. long, blades usually with many teeth, usually deltoid-ovate, often with cordate or truncate base and over 12 mm. long
    2. Outer involucre bracts with loose herbaceous caudate-attenuate or subulate tips; flowers about 20-40 to a head; achenes about 5 mm. long ....-1. <u>B. grandiflora</u>
    2. Outer involucre bracts acute to rounded, never caudate-attenuate at tips; flowers about 10-20 to a head; achenes about 3 mm. long ....-2. <u>B. californica</u>
1. Leaves sessile or on short petioles not over 3 mm. long, usually entire or few-toothed, either linear to narrowly ovate and of various lengths, or if ovate with truncate or subcordate base then not over 12 mm. long

3. Pappus bristles subplumose (side-cilia about 0.2-0.3 mm. long); involucre bracts abruptly acuminate to a short stiff terete point; achenes rather densely strigose (obscuring the surface)   -3. B. brachyphylla
3. Pappus bristles minutely barbed, never even approaching plumose; involucre bracts not abruptly acuminate to a point; achenes sparingly scabro-strigose (surface not concealed)
    4. Involucres 10-14 mm. high, the bracts with tips appressed or nearly so; flowers about 40-50 to a head; leaves linear to elliptic-lanceolate; achenes about 5-6 mm. long   -4. B. oblongifolia
    4. Involucres 7-9 mm. high, the outer bracts with spreading tips; flowers 10-12 to a head; leaves ovate to oblong-ovate; achenes about 3-3.5 mm. long   -5. B. scabra

1. Brickellia grandiflora (Hook.) Nutt., Trans. Amer. Phil. Soc. Ser. 2, 7:287. 1841.
    B. grandiflora var. minor Gray; B. grandiflora var. peteolaris Gray; B. ambigens (Greene) A. Nels.; Coleosanthus grandiflorus (Hook.) Kuntze; C. umbellatus Greene---Plants herbaceous or somewhat woody at very base; stems 30-100 cm. tall, paniculately branched above, puberulent to glabrate; leaves 2-10 cm. long, alternate, deltoid-ovate or upper deltoid-lanceolate, truncate, subcordate or sometimes broadly cuneate, rather long-acute, coarsely dentate, rather thin, somewhat hairy to almost glabrous; petioles 10-70 mm. long; heads paniculate-cymose, in rather congested clusters,, usually nodding; involucres about 8-10 mm. long, inner bracts obtuse or acute, outer broader and at least some with loose caudate-attenuate or subulate, herbaceous tips, puberulent or ciliate; flowers about 20-40 to a head, greenish-cream to cream-colored; achenes about 5 mm. long, the angles scabro-strigose; pappus bristles minutely barbed, whitish. ---Canyons and rocky slopes. Montana to Washington, south to Arkansas, New Mexico and California. Our records scattered in the western two-thirds of Colorado at 5000-10,000 feet.
2. Brickellia californica (T. & G.) A. Gray, Mem. Amer. Acad. Ser. 2, 4:64. 1849.
    B. tenera A. Gray; B. wrightii A. Gray; Coleosanthus californicus (T. & G.) Kuntze; C. tenera (A. Gray) Kuntze; C. albicaulis Rydb.---Plants shrubby at least at base (often much branched from a short woody base); stems 25-100 cm. tall, puberulent to hirsute-scabrous; leaves 1.2-5 cm. long, opposite or alternate, deltoid-ovate, often cordate or truncate at base, rounded or short-acute at apex but variable, more or less crenate-dentate, usually rather thick, scabro-puberulent; petioles 4-20 mm. long; heads glomerate-paniculate, terminal on the branches or axillary; involucres 6-9 mm. long, bracts subacute to rounded and appressed at tips; flowers 10-20 to a head, usually greenish-cream; achenes about 3 mm. long, the angles scabrous; pappus bristles merely scabrous, whitish to brownish.---Canyons and rocky slopes. Colorado to California, south to Texas and Mexico. Our records scattered in the western half of Colorado, few in the north, at 5500-7000 feet.
3. Brickellia brachyphylla A. Gray, Pl. Wright. 1:84. 1852.
    Coleosanthus brachyphyllus (Gray) Kuntze---Stems about 25-40 cm. tall, from a somewhat woody caudex, simple or branched above, puberulent; leaves 1-4 cm. long, opposite or alternate, lanceolate or narrowly ovate, acute, entire or slightly serrate, more or less puberulent and glandular-dotted; petioles absent or not over 3 mm. long; heads in elongated racemes or panicles; involucres about 6-8 mm. long, bracts narrowly oblong, abruptly acuminate or almost aristate into a short bristle, the tips appressed; flowers about 8-10 to a head, apparently cream-colored; achenes 3-4 mm. long, rather densely strigose (compared to our other species); pappus subplumose (side-cilia about 0.2-0.3 mm. long), white to tan.---Hills and canyons. Colorado, south to Texas and Arizona. Our few records from southern Colorado at 5500-7500 feet.
4. Brickellia oblongifolia Nutt., Trans. Am. Phil. Soc. Ser. 2, 7:288. 1841.
    B. linifolia Eaton; Coleosanthus linifolius (Eaton) Kuntze---Stems 20-50 cm. tall, herbaceous or woody at base, caespitose from a woody caudex, often branching above, short-pubescent or glandular; leaves 1.2-4 cm. long, alternate, linear to elliptic-lanceolate or narrowly oblong, usually obstuse and entire, more or less puberulent; heads 1 to few on the slender branches; involucres 10-14 mm. high, bracts linear to oblong-lanceolate, tips acute and appressed; flowers 40-50 to a head, usually greenish-cream; achenes 5-6 mm. long, scabro-striate along the angles; pappus minutely barbed, whitish. Our plant has been called var. linifolia (D.C. Eat.) Robinson.---Dry hills and valleys. Colorado to Nevada, south to New Mexico and California. Our records from westcentral and southwestern Colorado at 4500-9000 feet.
5. Brickellia scabra (A. Gray) A. Nels. ex Robinson, Gray Herb. Mem. 1:43. 1917.
    Shrubs or subshrubs woody at the base; stems 40-80 cm. tall, with slender ascending branches, glandular-puberulent or scabro-puberulent; leaves 8-12 mm. long, reduced in size on the flowering branches, alternate, ovate to oblong-ovate, broadly cuneate to subcordate, subentire to rather sparingly dentate, more or less scabrous; petioles lacking or up to 3 mm. long; heads 1-2 on the branches; involucres about 7-9 mm. long, outer bracts with green and spreading tips; flowers about 10-12 to a head, ochroleucous to purple-tinged; achenes about 3-3.5 mm. long, the angles scabrous; pappus minutely barbed, sordid-white to whitish.---Dry ground. Wyoming to Nevada, south to New Mexico and Arizona. Our records scattered in the western one-fourth of Colorado at 4500-7500 feet.

4. Liatris Schreb. BLAZING STAR; GAYFEATHER

    Perennial herbaceous plants from ovoid to globular corms or sometimes from elongated rootstocks; stems simple or slightly branched; leaves alternate or mostly basal, simple, narrow, entire, more or less punctate; heads in spikes or narrow racemes; involucres oblong, ovoid or hemispheric, bracts imbricated in several series, not striated, margins narrowly to broadly petaloid and ciliate or erose; ray flowers none; disk flowers rose-purple, rarely whitish; anthers united; receptacles flat or slightly concave, naked; achenes cylindrical but tapering to the base, 10-ribbed, more or less hairy; pappus of bristles in 1-2 series, from plumose to barbellate.

1. Pappus bristles plumose, the side-cilia about 0.4-0.6 mm. long, distinctly visible to the naked eye; leaves 1.5-7 mm. wide
   2. Heads 4- to 8-flowered, longer than wide, sessile or nearly so; bracts appressed except at very tip -1. **L. punctata**
   2. Heads 20- to 30-flowered, usually nearly as broad as long, on short but distinct peduncles; bracts usually squarrose or spreading -2. **L. squarrosa**
1. Pappus bristles barbellate, the side cilia about 0.2 mm. long or less, hardly visible without a lens; leaves (at least basal) 7-20 mm. wide
   3. Heads 9- to 15-flowered; involucres about 8-10 mm. high, oblong to campanulate; inflorescence of many heads these sessile or very nearly so in a long spike -3. **L. lancifolia**
   3. Heads 40- to 70-flowered; involucres 12-20 mm. high, broadly campanulate to hemispheric; inflorescence of few to several heads in a rather short raceme, the peduncles about 1-2 cm. long -4. **L. ligulistylis**

1. **Liatris punctata** Hook., Fl. Bor. Am. 1:306. 1834.
    Lacinaria punctata (Hook.) Kuntze---Stems 15-80 cm. tall, from an elongated mostly branched rootstock, glabrous or nearly so; leaves 8-15 cm. long and 1.5-6 mm. wide, numerous and rather evenly distributed, grading into the bracts above, linear, rather rigid, punctate, glabrous but often coarsely ciliate and thickened on the margins; heads sessile or very nearly so, in spikes, rather crowded; involucres about 1-1.6 cm. high, oblong or cylindrical, bracts thick, appressed except for very apex, outer acuminate or cuspidate, inner acute or acuminate, often ciliate; flowers 4-8 to a head, purple or rose, rarely white; achenes 6-7 mm. long, hairy; pappus plumose the side-cilia about 0.4-0.6 mm. long. Two varieties have been recognized from this area, var. nebraskana Gaiser and var. typica Gaiser. These show so many intergradations in Colorado plants that the writer cannot separate them.---Dry plains and hills. Iowa to Saskatchewan, south to Texas and New Mexico. Our records well scattered over the eastern half of Colorado at 3500-8000 feet. There is one report from Gunnison County perhaps the result of an error.

2. **Liatris squarrosa** (L.) Michx., Fl. Bor. Am. 2:92. 1803.
    L. glabrata Rydb.; Lacinaria squarrosa of Manuals for our plants---Stems 15-60 cm. tall, from rounded corms up to 4 cm. in diameter, glabrous; leaves 10-20 cm. long, linear, rigid, evenly distributed, the upper grading into the leaflike bracts, punctate, glabrous; involucres 14-18 mm. long, cylindrical, bracts foliaceous, linear to narrowly ovate, acuminate or cuspidate, recurved at anthesis, glabrous; flowers 20-30 to a head, rose-purple; achenes 4-6 mm. long, appressed hairy; pappus plumose with side-cilia about 0.4-0.6 mm. long. Our plant has been called var. glabrata (Rydb.) Gaiser.---Sandy plains and hills. The variety from South Dakota to Colorado, south to Missouri and Texas. Our records from eastern Colorado, possibly all from Yuma County, at about 3500 feet.

3. **Liatris lancifolia** (Greene) Kittell in Tide. & Kitt., Fl. Ariz. & N. Mex. 370. 1941.
    Lacinaria lancifolia Greene---Stems 50-100 cm. tall from a globose corm, glabrous; leaves 10-30 cm. long and about 7-9 mm. wide, numerous, punctate, lower leaves linear to linear-oblanceolate, upper reduced and grading into the bracts above; heads many, sessile or very nearly so, in a dense spike; involucres about 8-10 mm. high, oblong to campanulate, bracts erect, ovate, oval or oblong, obtuse or inner acutish, glabrous or at least the outer ciliate, punctate; flowers 9-15 to a head, purple or light rose; achenes about 3-4 mm. long, hairy; pappus barbellate the side-cilia about 0.2 mm. long or less, whitish but often rose-tinged at apex.---Valleys and river plains. South Dakota to Colorado, south to New Mexico. Our few records from northeastern Colorado at 3500-4000 feet.

4. **Liatris ligulistylis** (A. Nels.) K. Sch.; Just, Bot. Jahresb. 29(1):569. 1903.
    Lacinaria ligulistylis A. Nels.---Stems 10-60 cm. tall, from shallow rounded corms, usually glabrate below and whitish pubescent above; basal leaves 8-15 cm. long and about 1-2 cm. wide, lanceolate-oblong, usually with a margined petiole, upper leaves linear and becoming reduced and bractlike, all leaves glabrous to pubescent but always ciliate; heads few to several, in a rather short raceme with stout peduncles 1-3 cm. long, the terminal head often enlarged; involucres 12-20 mm. high, broadly campanulate to hemispheric, bracts erect, orbicular to oblong or spatulate, the outer wider, erose and rounded at apex, glabrous; flowers 40-70 to a head, purple to rose; achenes 5-6 mm. long, hairy; pappus barbellate, the side-cilia about 0.2 mm. long or less, becoming gray-purple.---Hills and valleys. Wisconsin to Manitoba, south to New Mexico. Our records scattered in the western two-thirds of Colorado, not common in the extreme western part, at 6000-7500 feet.

5. Kuhnia L. FALSE BONESET

    Perennial herbaceous plants, usually with a vertical taproot; stem leafy; leaves alternate (rather unevenly so at least above), simple, usually more or less resinous-dotted; inflorescence of solitary to corymbose-paniculate heads; involucres turbinate-campanulate to subcylindric, bracts imbricated in few to several series, striated and chartaceous; receptacle flat, naked; ray flowers none; disk flowers creamy-white to purplish; anthers cohering at least at base; achenes columnar, 10- to 20-ribbed; pappus of 10-20 equal plumose bristles, the side-cilia about 0.4-0.5 mm. long.

1. Leaves linear, rarely lanceolate, often less than 5 mm. wide, entire or with a pair of basal teeth; heads solitary or in loose clusters of 2-5, on peduncles 5-30 mm. long; plants mostly west of the Continental Divide in Colorado -1. **Kuhnia chlorolepis**
1. Leaves linear-lanceolate to oblong, usually over 5 mm. wide on the largest, mostly toothed or laciniate especially on the lower leaves; heads mostly in dense clusters of 3-8 on peduncles 3-15 mm. long; plants mostly east of the Continental Divide in Colorado -2. **Kuhnia eupatorioides**

1. **Kuhnia chorolepis** Woot. & Standl., Contrib. U. S. Nat. Herb. 16:177. 1913.

Stems 30-70 cm. tall, more or less branched, minutely pubescent; leaves often less than 5 mm. wide, linear to narrowly oblong or lanceolate, entire or rarely with a basal pair of teeth, sessile or subsessile, scabrous above, punctate and usually hairy at least on the veins below; inflorescence loose and open, heads solitary on the branches or in loose clusters, the peduncles 5-30 mm. long; involucres 8-12 mm. high, bracts nearly glabrous to pubescent; achenes about 5 mm. long, usually 10-ribbed. This species is closely related to K. rosmarinifolia Vent. and may be a variety of it. In eastern Colorado our 2 species intergrade commonly.---Dry hills and valleys. Colorado, south to Texas and Mexico. Our records scattered over the western half of the state, scarce in the northern part, at 5000-8500 feet.

2. **Kuhnia eupatorioides** L., Sp. Pl., ed. 2, 2:1662. 1763.

K. hitchcockii A. Nels.; K. gooddingi A. Nels.; K. reticulata A. Nels.; K. glutinosa of Manuals---Stems 30-80 cm. tall, usually few to several from 1 root, finely pubescent to puberulent; leaves over 5 mm. wide (on larger), linear-lanceolate to oblong-lanceolate, mostly dentate to laciniate especially below, firm, rather prominently veined, scabrous above and more or less pubescent below at least on the veins, subsessile or on short petioles; inflorescence from a compact corymb at end of stem to many heads in a bushy branched effect, mostly in compact clusters of 3-8 on peduncles 3-15 mm. long; involucres 8-14 mm. high, bracts more or less hirsutulous and resin-dotted; achenes about 5 mm. long, usually 10-ribbed. Our plant has been called var. corymbosa T. & G.---Dry prairies and hills. The variety from Ohio to Montana, south to Alabama and New Mexico. Our records scattered over the eastern half of Colorado at 3500-6500 feet.

6. Baccharis L. GROUNDSEL TREE

Shrubs or perennial plants woody only at the base; stems leafy; leaves alternate, simple, entire or toothed; heads usually numerous, panicled or corymbose, functionally dioecious; involucres campanulate, the bracts imbricated in several series; receptacles flat, naked but commonly pitted; rays none; disk flowers yellowish to whitish, corollas of staminate flowers tubular, 5-cleft, a style present but achenes not developing, corollas of pistillate flowers reduced to a slender tube; anthers united; achenes small, 5- to 10-ribbed; pappus of pistillate flowers of copious capillary bristles, of staminate flowers of shorter, scabrous, often twisted bristles.

1. Plants only 30-60 cm. tall, woody only at the base; upper leaves linear-subulate, entire, 1-nerved
 -1. **B. wrightii**
1. Plants 100 cm. tall or more, woody throughout (except B. glutinosa); leaves linear-oblong or wider, usually toothed, more or less 3-nerved at base
    2. Plants woody only at base; achenes 5-nerved; pappus rather rigid, scant, not especially elongating in fruit, not exceeding the stigmas -2. **B. glutinosa**
    2. Plants woody throughout; achenes 10-nerved; pappus copious, soft, elongating in fruit, much longer than the stigmas
        3. Involucre bracts all acute; involucres 5-10 mm. wide -3. **B. salicina**
        3. Involucre bracts (at least outer) obtuse; involucres 3-5 mm. wide -4. **B. emoryi**

1. **Baccharis wrightii** A. Gray, Pl. Wright. 1:101. 1852.

Plants 30-60 cm. tall, herbaceous or woody at the base only; stems diffusely branched, angled, glabrous; leaves 7-30 mm. long, linear or the upper linear-subulate, entire, 1-nerved; heads solitary and terminal on the slender branches; involucres 6-10 mm. high, bracts lanceolate, acute or acuminate, green on the back with margins scabrous; achenes 8- to 10-ribbed or perhaps only 5-ribbed at times, scabrous; pappus bristles copious, in more than 1 series, soft, elongating in fruit and longer than the stigmas.---Dry often saline soil. Kansas to Colorado, south to Texas, Mexico and Arizona. Our records from southeastern Colorado at 4500-5500 feet.

2. **Baccharis glutinosa** Pers., Syn. Pl. 2:425. 1807.

Colorado, south to Texas, Mexico and California; South America. Reported for southern Colorado but no specimens from the state seen.

3. **Baccharis salicina** T. & G., Fl. N. Amer. 2:258. 1842.

Shrubs 1-4 m. tall; branches striate-angled, green and glabrous, often with resinous exudations; leaves 2-5 cm. long, oblong or oblong-lanceolate, sometimes linear-lanceolate, sparingly toothed to sometimes entire, somewhat 3-nerved at base; heads in paniculate clusters; involucres 4-8 mm. high, the pistillate longer, bracts ovate to lanceolate, mostly acute; achenes 10-nerved, glabrous; pistillate pappus soft and copious, elongating in fruit and much surpassing the stigmas.---Often in saline ground. Kansas to Colorado, south to Texas and New Mexico. Our records from the southern half of Colorado at 3500-5500 feet.

4. **Baccharis emoryi** A. Gray in Torr., Bot. U. S. & Mex. Bound. 83. 1859.

May not be distinct from the preceding. It has smaller heads, 3-5 mm. wide and the outer bracts inclined to be obtuse.---Utah to California, south to Texas and Arizona. A few of our southern Colorado plants seem to be this species.

7. Chrysothamnus Nutt. RABBIT BRUSH

Shrubs or sometimes woody only at base; stems leafy; leaves alternate, linear to filiform, entire, commonly resinous and aromatic, sessile or appearing so; heads small to medium, variously disposed, usually in panicles or cymes; involucres cylindrical, bracts imbricated in various lengths and in more or less distinct vertical ranks, mostly with a central rib, coriaceous or herbaceous, sometimes herbaceous only at the tip; receptacles small, naked; rays none; flowers 4-7 to a head, all perfect and alike; achenes slender, terete or slightly angled, glabrous to pubescent; pappus of numerous slender capillary bristles, usually white. This is a variable and rather difficult genus and many species have been proposed. The monograph of Hall & Clements (Phy. Meth. in Tax. Carnegie Inst. Wash. Pub. 326. p. 157-324. 1923) has proven reasonably satisfactory for our Colorado plants and is followed closely in this treatment.

1. Twigs covered with feltlike or pannose tomentum (of long weak hairs more or less infiltrated with resin - may be mistaken for the bark itself; scrape with a knife edge)
   2. Heads in leafy terminal racemes (these may be somewhat branching and subpaniculate); outer involucre bracts commonly prolonged into slender herbaceous tips or appendages; involucres 10-15 mm. high and achenes villous
      3. Flowers 10-20 to a head; leaves 1.5-3 mm. wide, 3-nerved (only 1 conspicuous) -1A. C. parryi parryi
      3. Flowers 5-7 to a head; leaves about 1 mm. wide, 1-nerved
         4. Leaves gray-tomentose, uppermost usually elongated and overtopping the inflorescence; flowers pale yellow -1B. C. parryi howardi
         4. Leaves green, not at all tomentose, upper usually not projecting beyond the inflorescence; flowers yellow -1C. C. parryi attenuatus
   2. Heads cymose, often corymbose, at the ends of branches; outer involucre bracts obtuse to acute and lacking herbaceous or elongated tips; involucres 6-10 mm. high or if to 11.5 mm. then the achenes glabrous
      5. Achenes densely pubescent; involucres 6-10 mm. high
         6. Involucres puberulent or tomentose, at least the outer bracts; plants often less than 60 cm. tall
            7. Rather low shrubs 20-60 cm. tall; corollas 6.5-8 mm. long (when fresh or boiled) -2A. C. nauseosus nauseosus
            7. Taller shrubs 60-200 cm. tall; corollas 8-11 mm. long -2B. C. nauseosus speciosus
         6. Involucre bracts all glabrous; plants usually over 60 cm. tall
            8. Leaves over 1 mm. wide, some usually with 3-5 nerves -2C. C. nauseosus graveolens
            8. Leaves 1 mm. wide or less, 1-nerved
               9. Corolla lobes (on fresh or boiled flowers) 0.5-1 mm. long; inflorescence typically rounded or pyramidal -2D. C. nauseosus pinifolius
               9. Corolla lobes 1-2 mm. long; inflorescence typically cylindrical to pyramidal -2E. C. nauseosus consimilis
      5. Achenes glabrous; involucres 10-11.5 mm. high -2F. C. nauseosus bigelovi
1. Twigs glabrous to puberulent, not as described above
   10. Involucres 9-13 mm. high, bracts strongly keeled, arranged in very distinct vertical rows; achenes glabrous or very minutely pubescent but not 10-striated (may be 4- to 8-angled)
      11. Leaves linear, sledom over 2 mm. wide; corollas (fresh or boiled) 10-14 mm. long -3. C. pulchellus
      11. Leaves oblanceolate to spatulate, often over 2 mm. wide; corollas 7-9 mm. long -4. C. depressus
   10. Involucres 5-8 mm. high, the bracts only moderately keeled, the vertical rows not sharply defined; achenes hairy or if glabrous then 10-striated
      12. Achenes glabrous or very nearly so, longitudinally 10-striated, about 5 mm. long -5. C. vaseyi
      12. Achenes usually densely hairy, not at all striated (sometimes sparsely hairy but never striated), about 3-4 mm. long
         13. Involucre bracts, at least some, attenuated or abruptly narrowed to a subulate tip; corollas (fresh or boiled) 4-4.5 mm. long; leaves not over 1.2 mm. wide, never more than 1-nerved
            14. Leaves 1-1.2 mm. wide, mostly 2.5-3.5 cm. long -6A. C. greenei greenei
            14. Leaves less than 1 mm. wide and usually less than 2 cm. long -6B. C. greenei filifolius
         13. Involucre bracts obtuse or acute; corollas 4.5-7 mm. long; leaves often over 1.2 mm. wide and often over 1-nerved
            15. Leaves glabrous above and below; shrubs often over 50 cm. tall
               16. Leaves 2-8 mm. wide; shrubs 50-240 cm. tall
                  17. Involucre bracts thin, without a distinct sub-apical spot; plants not of alkaline soil -7A. C. viscidiflorus viscidiflorus
                  17. Involucre bracts, at least the outer, thick with a conspicuous greenish or brownish spot near apex; plants of alkaline soils -7B. C. viscidiflorus linifolius
               16. Leaves not over 2 mm. wide; shrubs 10-50 cm. tall
                  18. Leaves linear, 1-2 mm. wide -7C. C. viscidiflorus pumilus
                  18. Leaves linear-filiform, 1 mm. wide or less -7D. C. viscidiflorus stenophyllus
            15. Leaves (at least the upper) definitely puberulent; shrubs 10-50 cm. tall or only rarely taller
               19. Involucre bracts with a thickened green spot at apex -7E. C. viscidiflorus elegans
               19. Involucre bracts without a thickened green apex
                  20. Leaves 2.5-6 mm. wide, 3- to 5-nerved -7F. C. viscidiflorus lanceolatus
                  20. Leaves not over 1 mm. wide, 1-nerved -7G. C. viscidiflorus puberulus

1. Chrysothamnus parryi (A. Gray) Greene, Erythea 3:113. 1895.
Shrubs commonly 20-60 cm. tall; twigs covered with a white or green pannose tomentum; leaves 2-8 cm. long and about 1-3 mm. wide, narrowly to broadly linear, 1- to 3-nerved, glabrous to tomentulose; heads in terminal leaf racemes, these sometimes somewhat branching; involucres 9-13 mm. high, bracts loosely pubescent at least on the margins; flowers 5-20 to a head, 8-11 mm. long with lobes 0.5-2.5 mm. long; achenes 5-6 mm. long, 4-angled, densely appressed-villous. Represented in Colorado by 3 subspecies.
1A. Chrysothamnus parryi parryi
C. wyomingensis A. Nels.; C. parryi ssp. typicus H. & C.---Leaves 3-8 cm. long and 1.5-3 mm. wide, broadly linear, 3-nerved but only 1 prominent, glabrous or nearly so; involucre bracts in obscure vertical rows, not strongly keeled; flowers 10-20 to a head.---Dry hills, valleys and plains. Wyoming to Nevada, south to Colorado and Utah. Our records well scattered in the western half of Colorado at 6000-11,000 feet.

1B. Chrysothamnus parryi howardi (Parry) H. & C., (ssp.) Phylog. Meth. Tax. 201. 1923.
    C. howardii (Parry) Greene---Leaves 2-4 cm. long and about 1 mm. wide, narrowly linear, 1-nerved, gray-tomentose, the uppermost usually overtopping the raceme; involucre bracts in rather well defined vertical ranks, keeled; flowers 5-7 to a head, pale yellow.---Dry hills and mesas. Wyoming to Utah, south to Nebraska and Colorado. Our records from northcentral Colorado at about 8500 feet, but also reported from the central and southcentral parts.
1C. Chrysothamnus parryi attenuatus (Jones) H. & C., (ssp.) Phylog. Meth. Tax. 201. 1923.
    C. newberryi Rydb.---Leaves 2-4 cm. long and about 1 mm. wide, narrowly linear, 1-nerved, not tomentulose, upper ones not projecting beyond the raceme; involucre bracts keeled; flowers 5-7 to a head, yellow.---Hills and valleys. Idaho, south to New Mexico and Arizona. Our few records from northcentral and central Colorado at 5500-9500 feet but also reported from the southwestern part.
2. Chrysothamnus nauseosus (Pallas) Britt. in Britt. & Brown, Ill. Flora 3:326. 1898.
    Shrubs usually 20-200 cm. tall; twigs covered with a closely packed gray-green or white feltlike tomentome (appears like twig surface until scraped with a sharp blade); leaves filiform to broadly linear, not twisted, 1- to 3-nerved, from glabrous to tomentose; heads in terminal rounded or round-topped cymes, these often compound; involucre bracts usually 20-25, in vertical ranks; flowers usually 5 to a head; corollas 6.5-11 mm. long, the lobes 0.5-2 mm. long; achenes 5-5.5 mm. long, 5-angled, glabrous or more commonly densely strigose or appressed villous. This is a very variable species in such characters as height of plant and shape of leaf, the extremes widely different. The 6 subspecies listed here for Colorado do intergrade to some degree.
2A. Chrysothamnus nauseosus nauseosus
    C. nauseosus ssp. typicus H. & C.; C. nauseosus of Manuals; C. collinus Greene; C. frigidus Greene; C. frigidus var. concolor A. Nels.; C. pallidus A. Nels.---Shrubs commonly 20-60 cm. tall; leaves 2-5 cm. long and 0.5-1.5 mm. wide, narrowly linear; involucres 6.5-8 mm. long, bracts scarcely keeled, outer ones at least tomentulose; corollas 6.5-8 mm. long, the lobes 1.2-2 mm. long; achenes densely hairy.---Hills and plains. Saskatchewan to British Columbia, south to Colorado and Utah. Our records scattered over Colorado at 5000-9000 feet.
2B. Chrysothamnus nauseosus speciosus (Nutt.) H. & C., (ssp.) Phylog. Meth. Tax. 211. 1923.
    C. speciosus Nutt.; C. pulcherrimus A. Nels.; possibly C. formosus Greene---Shrubs commonly 60-200 cm. tall; leaves 2-6 cm. long and about 1 mm. wide (sometimes to 3 mm.), 1-nerved; involucres 8-10 mm. long, bracts strongly keeled in obvious vertical rows, at least the shorter outer ones tomentulose on back; corollas 8-11 mm. long, the lobes 0.8-2 mm. long; achenes densely hairy.---Valleys and hills. Montana to Washington, south to Colorado and California. Our records scattered in the western half of Colorado at 6000-8000 feet.
2C. Chrysothamnus nauseosus graveolens (Nutt.) H. & C., (ssp.) Phylog. Meth. Tax. 214. 1923.
    C. graveolens (Nutt.) Greene; C. virens Greene---Shrubs commonly 60-150 cm. tall; leaves 4-6 mm. long and 1-2 mm. wide, some usually 3- to 5-nerved; involucres 6-8 mm. long, bracts keeled, in rather well defined vertical rows, glabrous at least on back; corollas 7-9 mm. long, the lobes 0.5-1.5 mm. long; achenes densely hairy.---Dry plains and hills. North Dakota to Idaho, south to New Mexico and Arizona. Our records well scattered over the western two-thirds of Colorado at 4500-10,000 feet.
2D. Chrysothamnus nauseosus pinifolius (Greene) H. & C., (ssp.) Phylog. Meth. Tax. 215. 1923.
    C. pinifolius Greene; C. confinus Greene; C. patens Rydb.---Shrubs commonly 60-150 cm. tall; leaves 3.5-6 cm. long and 1 mm. wide or less, 1-nerved; involucres 6-9 mm. long, bracts keeled, in vertical rows, glabrous; corollas 6.5-9 mm. long, the lobes 0.5-1 mm. long; achenes densely hairy.---Dry hills and plains, often in alkaline soil. Colorado and New Mexico. Our records scattered in the western half of Colorado at 5500-9000 feet.
2E. Chrysothamnus nauseosus consimilis (Greene) H. & C., (ssp.) Phylog. Meth. Tax. 215. 1923.
    C. consimilis Greene; C. oreophilis A. Nels.---Wyoming to Oregon, south to New Mexico and California. This plant should be here but no specimens have been seen by the writer.
2F. Chrysothamnus nauseosus bigelovi (A. Gray) H. & C., (ssp.) Phylog. Meth. Tax. 217. 1923.
    C. bigelovii (Gray) Greene---Shrubs commonly 30-100 cm. tall; leaves 1-2.5 cm. long and 0.5-1 mm. wide, 1-nerved; involucres 10-11.5 mm. high, bracts tomentose or sometimes ciliate; corollas 9-10 mm. long, the lobes 0.8-1.5 mm. long; achenes glabrous.---Dry plains and hills. Colorado, south to Texas and Arizona. Our records from central and southcentral Colorado at about 8000 feet.
3. Chrysothamnus pulchellus (A. Gray) Greene, Erythea 3:107. 1895.
    Shrubs 30-50 cm. tall, possibly taller; twigs greenish to whitish, glabrous; leaves 1-4 cm. long and not over 2.5 mm. wide, linear or linear-oblong, 1-nerved, glabrous but minutely ciliolate; heads several to numerous, in cymes; involucres 10-13 mm. long, bracts rather definitely keeled on back, in vertical ranks, greenish-spotted near apex; flowers about 5 to a head; corollas about 10-14 mm. long, the lobes 1.5-2 mm. long; achenes about 6-7 mm. long, 4-angled, usually glabrous. Our plants have the leaves ciliolate and apparently would be ssp. baileyi (Woot. & Standl.) H. & C.---Dry hills. The species from Kansas to Utah, south to Texas and New Mexico. Our one record from southeastern Colorado at about 4000 feet.
4. Chrysothamnus depressus Nutt., Journ. Phila. Acad. II. 1:171. 1847.
    Shrubs or undershrubs commonly 10-30 cm. tall; twigs densely cinereous or scabrous; leaves 8-20 mm. long and 1-4 mm. wide, narrowly oblanceolate to spatulate, 1-nerved, finely puberulent; heads in small terminal cymes; involucres 9-12 mm. long, bracts keeled or boat-shaped, arranged in 5 sharply defined vertical rows, attenuated to a short awn or mucro, minutely rough-puberulent; flowers about 5 to a head; corollas 7-9 mm. long, the lobes 1-2.3 mm. long; achenes 5-5.5 mm. long, 4- to 8-angled or ribbed, glabrous or obscurely pubescent toward the apex.---Plains, hills and mountains. Colorado to Nevada, south to New Mexico and California. Our records from the western one-fourth of Colorado at 6500-8000 feet.

5. **Chrysothamnus vaseyi** (A. Gray) Greene, Erythea 3:96. 1895.
Shrubs commonly 10-30 cm. tall; twigs pale green or whitish, glabrous; leaves 1-2.5 cm. long and 1-2.5 mm. wide, linear or very narrowly oblanceolate, 1-nerved, flat to twisted, glabrous; heads in small compact cymes; involucres 5.5-7 mm. high, bracts obscurely keeled, in poorly defined vertical rows, glabrous or ciliate, outer often thicker and greenish-spotted near apex; flowers 5-7 to a head; corollas 5.5-6.5 mm. long, the lobes 1.5-2 mm. long; achenes about 5 mm. long, 10-nerved, glabrous to minutely granular, glabrous or nearly so.---Dry plains, valleys and hills. Wyoming and Utah, south to New Mexico. Our records from the southern half of Colorado at 6000-8500 feet.

6. **Chrysothamnus greenei** (A. Gray) Greene, Erythea 3:94. 1895.
Shrubs commonly 10-35 cm. tall; twigs green at first but soon white and shining, glabrous; leaves 1-3.5 cm. long and 1 mm. wide or less, linear to linear-filiform, 1-nerved, glabrous or scabro-ciliate; heads in terminal loose to compact cymes; involucres 5-7 mm. high, bracts in poorly defined rows, abruptly narrowed to a subulate tip or outer ones gradually attenuated, glabrous; flowers about 5 to a head; corollas 4-4.5 mm. long, the lobes 0.8-1.3 mm. long; achenes about 3 mm. long, densely appressed villous. Two subspecies are present in Colorado.

6A. **Chrysothamnus greenei greenei**
C. greenei spp. typicus H. & C.; C. pumilus var. acuminatus A. Nels.---Shrubs commonly 10-20 cm. tall; leaves usually 2-3.5 cm. long and about 1 mm. wide, narrowly linear; heads in loose cymes, mostly on distinct peduncles.---Dry hills and plains. Colorado to Nevada, south to New Mexico. Our records in the southwestern quarter of Colorado at 5000-8500 feet.

6B. **Chrysothamnus greenei filifolius** (Rydb.) H. & C., (ssp.) Phylog. Meth. Tax. 191. 1923.
C. filifolius Rydb.---Shrubs commonly 20-35 cm. tall; leaves usually 1-2 cm. long and less than 1 mm. wide, linear-filiform; heads in compact cymes, sessile or subsessile.---Dry hills and plains. Colorado to Nevada, south to New Mexico and Arizona. Our records from the western half of Colorado, none as yet from the extreme western part, at 7000-8500 feet.

7. **Chrysothamnus viscidiflorus** (Hook.) Nutt., Trans. Am. Phil. Soc. II. 7:324. 1840.
Shrubs 10-240 cm. tall; twigs glabrous or puberulent; leaves 1-6 cm. long and 1-10 mm. wide, narrowly linear to oblong-lanceolate, plane or twisted, 1- to 5-nerved, glabrous or puberulent; heads in a terminal rounded or flat-topped cyme; involucres 5-8 mm. high, bracts about 15, in poorly defined vertical rows, sometimes with a greenish or brownish thickened spot near apex, glabrous to puberulent; flowers about 5 to a head; corollas 4.5-7 mm. long, the lobes 1-2 mm. long; achenes 3-4 mm. long, 5-angled, densely to sparsely villous to sericeous. Our 7 subspecies are reasonably distinct.

7A. **Chrysothamnus viscidiflorus viscidiflorus**
C. viscidiflorus ssp. typicus H. & C.; C. stenolepis Rydb.; C. glaucus A. Nels.; C. serrulatus (Torr.) Rydb.---Shrubs 50-120 cm. tall; leaves 2-5 cm. long and 2-5 mm. wide, broadly linear to narrowly lanceolate, 1- to 3-nerved, plane or twisted, glabrous on faces; involucres 5-7 mm. high, bracts not keeled, without an apical spot.---Dry plains and hills. Montana to Washington, south to Colorado, Arizona and California. Our records from the northwestern quarter of Colorado at about 6000-7000 feet.

7B. **Chrysothamnus viscidiflorus linifolius** (Greene) H. & C., (ssp.) Phylog. Meth. Tax. 184. 1923.
C. linifolius Greene---Shrubs commonly 80-240 cm. tall; leaves 2-5 cm. long and 4-8 mm. wide, lanceolate to oblong-lanceolate, mostly 3-nerved, not twisted, glabrous on the faces; involucres 5-6 mm. high, bracts not keeled, at least the outer with a subapical green or brown spot.---Dry or alkaline lowlands. Wyoming and Utah, south to New Mexico. Our records from central and western Colorado at 5000-9000 feet.

7C. **Chrysothamnus viscidiflorus pumilus** (Nutt.) H. & C., (ssp.) Phylog. Meth. Tax. 182. 1923.
C. pumilus Nutt.---Shrubs commonly 10-50 cm. tall; leaves 2-4 cm. long and 1-2 mm. wide, linear, plane or twisted, 1- to 3-nerved, glabrous on faces; involucres 5-6 mm. high, bracts not keeled, no apical spot present.---Dry plains and hills. Montana to Washington, south to Colorado and California. Our records scattered in the western half of Colorado, mostly in the central and northwestern parts, at 5500-9500 feet.

7D. **Chrysothamnus viscidiflorus stenophyllus** (A. Gray) H. & C., (ssp.) Phylog. Meth. Tax. 183. 1923.
C. stenophyllus (A. Gray) Greene---Shrubs commonly 10-30 cm. tall; leaves 1-3 cm. long, 1 mm. wide or less, linear or linear-filiform, often twisted, 1-nerved, glabrous on faces; involucres 4-6 mm. high, bracts not keeled, without subapical spot.---Dry ridges and slopes. Montana, south to New Mexico and California. Our one record from northwestern Colorado at 6500 feet.

7E. **Chrysothamnus viscidiflorus elegans** (Greene) H. & C., (ssp.) Phylog. Meth. Tax. 183. 1923.
C. elegans Greene---Shrubs commonly 10-40 cm. long; leaves 1.5-3 cm. long and 1-2 mm. wide, linear, usually definitely twisted, mostly 3-nerved, puberulent at least on upper; involucres about 5 mm. high, keeled, with a thickened green or brown subapical spot present at least on some.---Dry plains and valleys. Wyoming to Nevada, south to New Mexico and Arizona. Our records scattered in the northwestern quarter of Colorado at 6000-9500 feet.

7F. **Chrysothamnus viscidiflorus lanceolatus** (Nutt.) H. & C., (ssp.) Phylog. Meth. Tax. 181. 1923.
C. lanceolatus Nutt.---Shrubs 20-50 cm. tall, rarely taller; leaves 1.5-4 cm. long and 2.5-6 mm. wide, broadly linear to linear-lanceolate, usually not at all twisted, 3- to 5-nerved, at least upper ones rough-pubescent; involucres 5-6.5 mm. high, bracts scarcely keeled, with subapical spots.---Dry plains and hills. Montana to Washington, south to Colorado and Nevada. Our records from the northern and western parts of Colorado at 7000-10,500 feet.

7G. **Chrysothamnus viscidiflorus puberulus** (D.C.Eaton) H. & C., (ssp.) Phylog. Meth. Tax. 182. 1923.
C. puberulus (D.C.Eat.) Greene---Shrubs commonly 20-50 cm. tall; leaves 1.5-4 cm. long

and 1 mm. wide or less, often twisted or rolled, 1-nerved, densely pubescent; involucres about 6 mm. high, bracts not keeled, no subapical spot present.---Dry plains and hills. Montana to British Columbia, south to Colorado and California. Reported for central and western Colorado.

8. Chrysopsis Nutt. GOLDEN ASTER

Plants perennial and herbaceous; stems leafy, branching at least at base; leaves alternate, entire; heads medium-sized, rather showy, usually few and cymose; involucres campanulate to hemispheric, bracts imbricated in several series; receptacles usually flat, naked and pitted; ray flowers golden-yellow, pistillate, usually many; disk flowers perfect and fertile; achenes compressed; pappus double, the outer series of short scales, these sometimes bristle-pointed, inner series of much longer capillary bristles. This is a very difficult genus and the treatment here is provisional waiting a revision of the genus. The 5 species listed are rather polymorphic and the extremes are certainly different looking, but connected by numerous intergradations. See Cronquist (Bull. Torr. Club 74:149. 1947).

1. Involucres pubescent but the glands obscure or none
   2. Plants canescent; leaves usually not petioled except sometimes at base of plant   -1. C. foliosa
   2. Plants grayish-green, hardly canescent; leaves distinctly petioled at least at middle and lower part of stem   -2. C. villosa
1. Involucres sparingly to definitely glandular
   3. Heads leafy-bracted; leaves sessile, 2-4 cm. long   -3. C. fulcrata
   3. Heads not leafy-bracted; leaves petioled, seldom over 2.5 cm. long
      4. Stems and involucres sparsely glandular   -4. C. hispida
      4. Stems and involucres densely glandular   -5. C. viscida

1. Chrysopsis foliosa Nutt., Trans. Amer. Phil. Soc. n. ser. 7:316. 1840.
C. mollis Nutt.; C. hirsutissima Greene; C. foliosa amplifolia (Rydb.) A. Nels.; C. foliosa imbricata A. Nels.---Stems 20-60 cm. tall, spreading-ascending, canescent-strigose or canescent-hirsute; leaves 2-7 cm. long, elliptic, obovate or oblanceolate, whitish-canescent with strigose-silky hairs both sides, sessile or lower ones more or less petioled; heads 1 to few, at the ends of the stems or branches, often leafy-bracted; involucres 6-9 mm. high, strigose-hirsute but without glands or these very obscure.---Dry ground. Minnesota to Washington, south to New Mexico and Arizona. Our records scattered in the western half of the state at 4500-7000 feet.
2. Chrysopsis villosa (Pursh) Nutt. ex DC., Prodr. 5:327. 1836.
C. bakeri Greene; C. arida A. Nels.---Stems 10-50 cm. tall, spreading or ascending; leaves 2-5 cm. long, varying from linear to obovate, strigose but not canescent, middle and lower petioled; heads solitary to several near ends of branches, leafy-bracted, nearly sessile to long-peduncled; involucres 7-10 mm. high, strigose but slightly if at all glandular.---Dry ground. Minnesota to Saskatchewan, south to Texas and Arizona. Our records scattered over Colorado at 4000-10,000 feet.
3. Chrysopsis fulcrata Greene, Bull. Torr. Club 25:119. 1898.
C. resinolens A. Nels.; C. resinolens obtusata A. Nels.---Stems 20-60 cm. tall, spreading or ascending, more or less hirsute and glandular; leaves 2-4 cm. long, oblong, oblong-oblanceolate to elliptic, usually green and not canescent, hirsute or strigose, often ciliate, sessile; heads 1 to several at the ends of branches and stems definitely leafy-bracted; involucres 8-10 mm. high, appressed-hirsute and definitely glandular.---Dry ground. Montana, south to Texas and Arizona. Our few records from the northern half of Colorado at 5500-10,000 feet, possibly as high as 12,000 feet.
4. Chrysopsis hispida (Hook.) DC., Prodr. 7:279. 1838.
Stems 20-30 cm. tall, ascending or spreading, hirsute and more or less glandular; leaves rarely over 2 cm. long, oblanceolate, oblong to obovate, moderately to densely strigose-hirsute, often nearly canescent, narrowed to petiolelike bases; heads 1-several, near ends of branches, not especially leafy-bracted at base; involucres about 7-8 mm. high, hirsute and also moderately glandular.---Dry ground. Saskatchewan to British Columbia, south to Arizona and California. Our rather few records widely scattered over Colorado at 4000-10,000 feet.
5. Chrysopsis viscida (A. Gray) Greene, Erythea 2:105. 1894.
Closely related to the preceding and may not be distinct. The involucres are densely glandular and very viscid.---Dry ground. Colorado to Texas and Arizona. Our few records from the eastern half of Colorado at 4000-7500 feet.

9. Grindelia Willd. GUMWEED; RESINWEED

Biennial or perennial herbaceous plants; stems leafy; leaves alternate, simple, entire to pinnatifid, more or less resinous-dotted; heads medium in size, solitary or often corymbose; involucres hemispheric to campanulate, the bracts imbricated in several series, the tips often subulate and spreading or recurved, mostly gummy-resinous; receptacles flat or slightly convex, naked but more or less pitted and the edges of the pits often developing projections; ray flowers present or absent, yellow, pistillate and fertile; disk flowers yellow, perfect; anthers united; achenes rather short and thick, sometimes compressed; pappus of 2-8 slender caducous stiff and almost paleaceous awns.

1. Heads discoid, no rays present
   2. Perennials plants; involucre bracts with tips much thickened and coriaceous
      3. Stems 50-150 cm. tall; leaves conspicuously resinous-punctate; stigmas linear-lanceolate; plants of western Colorado   -1. G. fastigiata

3. Stems 25-60 cm. tall; leaves less conspicuously resinous-punctate; stigmas oblong-lanceolate to oblong; plants of the eastern half of Colorado
    4. Main middle and upper cauline leaves ovate to broadly oblong, obtuse or nearly so, 1½ - 3 times longer than wide   -2. G. inornata
    4. Main middle and upper cauline leaves oblong to oblong-oblanceolate, acute or acutish, over 3 times longer than wide   -2A. G. inornata angusta
2. Annual or biennial plants; involucre bracts only slightly thickened and coriaceous
    5. Upper and middle cauline leaves oblong to oblanceolate, 5-10 times longer than wide, often entire; mature achenes deeply ribbed; pappus awns serrulate; plants mostly found in southwestern Colorado   -3. G. aphanactis
    5. Upper and middle cauline leaves oval to broadly oblong, about 1½ - 3 times longer than wide, rarely entire; mature achenes smooth or slightly striate; pappus awns smooth or remotely serrulate; plants mostly found in southeastern Colorado   -6A. G. squarrosa nuda
1. Heads radiate
    6. Involucre bracts (at least middle and upper) with appressed or nearly erect tips, these not revolute or coriaceous
        7. Leaves all or mostly laciniate-dentate or pinnatifid   -4. G. laciniata
        7. Leaves entire to serrate or dentate near apex   -5. G. arizonica
    6. Involucre bracts with upper part strongly spreading to recurved, subulate
        8. Middle and upper leaves crenate-serrate, the teeth obtuse at apex
            9. Middle and upper leaves ovate-oblong to broadly oblong, about 2-4 times longer than broad   -6. G. squarrosa
            9. Middle and upper leaves oblong, oblanceolate or linear-oblong, mostly 5-8 times longer than broad   -6B. G. squarrosa serrulata
        8. Leaves entire, incised, or if toothed with teeth acute or acuminate at apex
            10. Involucre bracts slightly to only moderately resinous; plants of southwestern Colorado
                11. Leaves firmly membranaceous; involucre bracts with tips rather abruptly and closely recurved; leaves entire to dentate, not pinnatifid   -7. G. decumbens
                11. Leaves only moderately membranaceous; involucre bracts loosely recurved; leaves dentate but often more or less pinnatifid   -7A. G. decumbens subincisa
            10. Involucre bracts conspicuously resinous; plants of eastern and northcentral Colorado, so far not found in the southwestern part of the state
                12. Pappus awns 4-8, closely serrulate; plants of northcentral Colorado
                    13. Stems usually several from the base, slender; main cauline leaves 1.5-6 cm. long, upper cauline leaves narrowed to base   -8. G. subalpina
                    13. Stems usually 1 from the base, stout; main cauline leaves 6-15 cm. long, upper clasping or nearly so   -8A. G. subalpina erecta
                12. Pappus awns 2-4, rather remotely serrulate; plants not limited to northcentral Colorado
                    14. Middle and cauline leaves about 6-10 times longer than broad, upper usually at least 5 times longer than broad; plants usually of the northeastern quarter of Colorado, sometimes farther south, however
                        15. Leaves closely and evenly crenate-serrulate to closely serrulate, radical and basal rarely incised-serrate or pinnatifid   -6B. G. squarrosa serrulata
                        15. Leaves entire to remotely serrulate, lower leaves often incised or pinnatifid   -9. G. perennis
                    14. Middle and cauline leaves mostly 2½ - 5 times as long as wide, upper at most 4 times longer than broad; plants of the southeastern quarter of Colorado (not as yet north of Douglas County)
                      16. Leaves subcoriaceous and thickened, entire to remotely dentate or denticulate with short broad teeth   -10. G. revoluta
                    16. Leaves merely firmly membranaceous, sharply toothed, the teeth sometimes acuminate-setulose   -11. G. acutifolia

1. Grindelia fastigiata Greene, Pitt. 3:102. 1896.
Perennial plants; stems 30-150 cm. tall, several, stramineous, glabrous or nearly so; leaves 2-7 cm. long, oblanceolate to oblong-lanceolate, serrate or the upper entire, definitely punctate; heads 7-12 mm. high, usually longer than wide; involucre bracts in several series, thickened and coriaceous, the outer and middle squarrose, more or less resinous; rays absent; achenes 3.5-5 mm. long, brown; pappus awns 2-3, 3-5.5 mm. long, more or less serrulate.---Dry hills and valleys. Colorado and Utah. Our records from western Colorado at 4500-7000 feet.
2. Grindelia inornata Greene, Pitt. 3:102. 1896.
Plants perennial; stems 25-60 cm. tall, several, subcorymbosely or paniculately branched above, stramineous, glabrous; leaves 2-7 cm. long, ovate to broadly oblong, obtuse or nearly so, dentate with teeth subspinulose or sharply acute at tips, glabrous; heads 8-12 mm. high, campanulate-cylindrical, mostly broader than long; involucre bracts to 10 mm. long, in 5-6 series, upper part subterete, thickened and strongly reflexed, glabrous and more or less resinous; ray flowers none; achenes about 3-5 mm. long, stramineous to pale brown, smooth or slightly striated; pappus awns 2-3, 3-4.5 mm. long, serrulate to setulose.---Dry hills and valleys. Limited to central Colorado, our records at 5000-6000 feet.
2A. Grindelia inornata angusta Steyermark, (var.) Ann. Mo. Bot. Gard. 21:496. 1934.
Leaves oblong to oblong-oblanceolate, over 3 times longer than wide, acute or acutish.---Apparently limited to the central area of Colorado, our records at 5000-6500 feet.
3. Grindelia aphanactis Rydb., Bull. Torr. Club 31:647. 1904.
G. fastigiata of Manuals in part---Biennial plants; stems 25-40 cm. tall, 1 to several, branching at least in the inflorescence; leaves oblong to oblanceolate, 5-10 times as long as wide, slightly to moderately resinous-punctate, entire to crenate-serrate or denticulate or the basal leaves often pinnatifid; heads 7-20 mm. high and 10-28 mm. wide; involucre bracts in 5-6

series, upper part of bracts loosely reflexed and with subulate tips but these only slightly thickened and coriaceous, moderately to slightly resinous; ray flowers none; achenes 2.3-3 mm. long, mostly brown. deeply furrowed; pappus awns 2-3, 3-5.5 mm. long, somewhat serrulate.--- Moist or dry meadows, fields and roadsides. Colorado and Utah, south to Texas and Arizona. Our records from northcentral, central and southern Colorado at 5000-8000 feet.

4. Grindelia laciniata Rydb., Fl. Rocky Mts. 848 & 1066. 1917.

Reported for southeastern Utah and adjacent Arizona, close to the Colorado line, and it should be looked for in the southwestern corner.

5. Grindelia arizonica Gray, Proc. Am. Acad. 17:208. 1882.

G. decumbens of Manuals in part---Plants apparently perennial; stems 15-45 cm. tall, several, corymbosely branched above; leaves 3-7 cm. long, oblanceolate mostly 5-8 times longer than wide, entire to serrate or denticulate on apical half, membranaceous, scarcely resinous-punctate; heads 8-10 mm. high and about as wide, campanulate to hemispheric; involucres moderately to slightly resinous, the bracts in 4-5 series, the tips acute, erect and appressed; achenes 3-3.5 mm. long; pappus awns 2-3, not at all serrulate. Our plant has been called var. stenophylla Steyermark.---Dry fields, hills and plains along rivers. The variety from Colorado and New Mexico. Our records from southwestern and southcentral Colorado at 6500-9500 feet.

6. Grindelia squarrosa (Pursh) Dunal, Mem. Mus. Par. 5:50. 1819.

Plants biennial or possibly perennial; stems 25-100 cm. tall, 1 to several, corymbosely branched above; leaves 3-7 cm. long, ovate-oblong to broadly oblong, about 2-4 times longer than wide, closely and evenly crenate-serrate with obtuse teeth, thick and subcoriaceous, abundantly resinous-punctate; heads 1.2-2 cm. wide, wider than high; involucres 7-9 mm. high, conspicuously resinous, the bracts in 5-6 series, tips subcoriaceous and terete, descending or reflexed; rays 8-10 mm. long, lemon-yellow to bright yellow; achenes 2.3-3 mm. long, striate about the angles; pappus awns 2-3 rarely more, subentire to rather remotely serrulate.---Prairies, plains, roadsides and fields. Minnesota to Wyoming, south to Iowa, Texas and Arizona; introduced east and west. Our records scattered in Colorado, except in the extreme southern part, at 4500-5000 feet.

6A. Grindelia squarrosa nuda (Wood) Gray, (var.) Syn. Fl. N. Amer. 1(2):118. 1884.

G. nuda Wood---Leaves oval to broadly oblong, 1½ - 3 times longer than wide; crenate to crenate-serrate with obtuse teeth; rays none.---Plains, hills, roadsides and fields. Kansas to Colorado, south to Texas and New Mexico; introduced eastward. Our records from the southeastern quarter of Colorado at about 4500 feet.

6B. Grindelia squarrosa serrulata (Rydb.) Steyermark, (var.) Ann. Mo. Bot. Gard. 21:227. 1934.

G. serrulata Rydb.---Stems 15-60 cm. tall; leaves 3-6.5 cm. long, linear-oblong to oblanceolate or oblong, about 5-8 times as long as wide, closely crenate-serrate with obtuse teeth or serrulate with teeth rather sharp; rays commonly 10-14 mm. long.---Dry open places. Wyoming to Utah, south to New Mexico and Arizona; introduced eastward and westward. Our records scattered over the state, none as yet from the southeastern part, at 3500-8500 feet.

7. Grindelia decumbens Greene, Pitt. 3:102. 1896.

Plants perennial; stems 15-45 cm. tall, several to many, slender, corymbosely branched above; leaves 2-8 cm. long, broadly to narrowly oblong or oblong-oblanceolate, 3-7 times longer than wide, entire to dentate, firmly membranous; heads 8-11 mm. high, usually somewhat wider than high; involucres lightly to moderately resinous, bracts in about 5 series, tips subterete and slender, upper part abruptly reflexed; rays 7-10 mm. long, yellow; achenes 3-3.5 mm. long; pappus awns 2-4, usually only moderately serrulate.---Dry hills and plains. Apparently limited to Colorado, our records from the southwestern and westcentral parts at 6500-8000 feet.

7A. Grindelia decumbens subincisa (Greene) Steyermark, (var.) Ann. Mo. Bot. Gard. 21:503. 1934.

G. subincisa Greene---Leaves 2-6 cm. long, about 6-15 times longer than wide, entire, denticulate to pinnatifid; rather thin; involucre bracts reflexed but not abruptly so.---Hills and valleys. Colorado and New Mexico. Our records from southwestern Colorado at 6500-7000 feet.

8. Grindelia subalpina Greene, Pitt. 3:297. 1898.

Plants biennial or perennial; stems 15-60 cm. tall, usually several from base, corymbosely branched above; leaves 1.5-6 cm. long, oblanceolate-oblong to lanceolate-oblong, about 3-7 times longer than wide, submembranous, serrate to dentate or the basal incised-serrate, the upper sometimes entire and mostly narrowed to a sessile base; heads 8-11 mm. high and wider than high; involucres moderately to slightly resinous, the bracts in 5-6 series, tips subulate and upper part reflexed; rays about 10-14 mm. long, yellow; achenes 2.5-3.5 mm. long, smooth or slightly striated on the angles; pappus awns 4-8, closely serrulate-setose.---Dry open slopes and valleys. Wyoming and Colorado. Our few records from northcentral and central Colorado at 5500-9000 feet.

8A. Grindelia subalpina erecta (A. Nels.) Steyermark, (var.) Ann. Mo. Bot. Gard. 21:501. 1934.

G. erecta A. Nels.; G. texana of Manuals---Stems 30-65 cm. tall, usually 1; main cauline leaves 6-15 cm. long, upper shorter and subclasping.---Open plains and hills. Wyoming and Colorado. Our rather few records from northcentral and central Colorado at 6000-9500 feet.

9. Grindelia perennis A. Nels., Bull. Torr. Club 26:355. 1899.

Plants biennial; stems 10-50 cm. tall, 1 to several, corymbosely branched; leaves 3-6 cm. long, narrowly oblong to oblanceolate, mostly 5-8 times longer than wide, entire to remotely serrate the lower often pinnatifid, rather thick, resinous-punctate; heads 7-20 mm. wide; involucre bracts in 4-5 series, the upper part revolute and subulate, reflexed; rays 7-15 mm. long, bright orange to golden-yellow; achenes 2.5-3.3 mm. long, lightly striated, brown to straw-colored; pappus awns 2-4, sometimes more, only remotely serrulate.---Fields and prairies, especially along rivers or dried up ponds. Saskatchewan to British Columbia, south to Minnesota, South Dakota, Colorado and California; introduced in Central America. Our records scattered over the western half of Colorado at 5000-8500 feet.

10. Grindelia revoluta Steyermark, Ann. Mo. Bot. Gard. 21:496. 1934.
Plants annual or biennial; stems 20-80 cm. tall, branched, stramineous, glabrous; leaves 3-7.5 cm. long and 6-25 mm. wide, oblong or oblong-oblanceolate, 2½ - 5 times longer than wide, entire or rather remotely serrate with rather sharp teeth, subcoriaceous; heads 2-3.5 cm. wide; involucres 4-10 mm. high, campanulate-hemispheric or depressed hemispheric, very resinous, the bracts with subulate reflexed tips; rays 4-11 mm. long, yellow; achenes 3.5-4 mm. long, oblong; pappus awns 2-4, 3.5-5 mm. long, setulous-serrulate.---Dry open ground. Apparently limited to central and southcentral Colorado, our records from about 6000-6500 feet.
11. Grindelia acutifolia Steyermark, Ann. Mo. Bot. Gard. 21:498. 1934.
Plants apparently perennial; stems several, branching above, glabrous; leaves 2.5-8 cm. long, oblong, lanceolate, oblanceolate or ovate-lanceolate, about 4-5 times longer than wide on cauline ones, sharply serrate or dentate, mostly glabrous, upper sessile and somewhat clasping; heads 1.2-1.5 cm. high and slightly wider; involucre bracts in 5-6 series, moderately resinous, upper part subulate and definitely recurved; rays about 9-11 mm. long; achenes 4-5 mm. long, oblong; pappus awns 2-3, remotely serrate.---Dry open slopes and plains. Colorado and New Mexico. Our few records from southcentral Colorado (western Las Animas County) at 6000 feet.

10. Gutierrezia Lag. SNAKEWEED; BROWNWEED

Herbaceous perennial or suffrutescent, more or less glutinous plants; stems leafy; leaves alternate, narrow and entire; heads rather small and usually many, crowded in terminal corymbs; involucres cylindrical to campanulate, bracts imbricated, chartaceous, scarious-margined with small green tips; receptacles flat, naked commonly pitted; ray flowers few, pistillate and fertile, yellow; disk flowers usually perfect and fertile, yellow, the corollas 5-lobed; anthers united; achenes small, oblong to obovoid; pappus of several to many oblong to narrower scales, present at least on the disk flowers.

1. Resin dots of leaves small and not bordered by hyaline margins
   2. Heads cylindrical, about 1 mm. thick; rays 1 (rarely 2); disk flowers 1 or 2   -1. G. lucida
   2. Heads turbinate, usually over 1.5 mm. thick; rays 3-8; disk flowers 1-8
      3. Involucres usually over 2 mm. thick; rays 3-8; disk flowers 3-8   -2. G. sarothrae
      3. Involucres 1-1.5 mm. thick; rays 4-5; disk flowers 1-3   -2A. G. sarothrae microcephala
1. Leaves with large conspicuous resin dots, each bordered by a hyaline margin   -3. G. lepidota

1. Gutierrezia lucida Greene, Fl. Francisc. 361. 1897.
G. glomerella Greene---Plants 20-60 cm. tall, woody at least at the base; stems branched, glabrous to pubescent; leaves narrowly linear; heads numerous, crowded near the end of the branches; involucres about 3 mm. long and 1 mm. wide, cylindrical; ray flowers usually 1 to a head; disk flowers 1-2.---Dry open plains and hills. Colorado to Nevada, south to Texas, Mexico and California. Our records from western Colorado at 4500-7000 feet.
2. Gutierrezia sarothrae (Pursh) Britt. & Rusby, N.Y. Acad. Sci. Trans. 7:10. 1887.
G. juncea Greene; G. linearis Rydb.; G. euthamiae T. & G.; G. longifolia Greene;
G. diversifolia Greene; G. divaricata (Nutt.) T. & G.; G. scoparia Rydb.; G. fasciculata Greene---Plants 10-70 cm. tall, bushy, shrubby or woody only at the base, glabrous to scabro-puberulent; stems branching; leaves linear to linear-filiform, glabrous to puberulent; heads many, usually in clusters near the ends of the branches; involucres 3-6 mm. long, usually 2 mm. wide or more, turbinate; rays 3-8 to a head; disk flowers 3-8. This is a variable species but the various forms do not segregate out in any logical fashion in this area.---Open dry plains and hills. Saskatchewan, south to Kansas, Mexico and California. Our records widely scattered over Colorado, (none as yet from the northeastern corner) at 4000-10,000 feet.
2A. Gutierrezia sarothrae microcephala (DC.) Benson, (var.) Am. Journ. Bot. 30:631. 1943.
Involucres 1-1.5 mm. thick; rays 4-5 to a head; disk flowers 1-3. Hardly distinct from the species.---Dry plains and hills. Idaho, south to Texas, Mexico and Arizona. Our few records from southern Colorado at 6000-8000 feet.
3. Gutierrezia lepidota Greene, Pitt. 4:57. 1899.
A little known species related to G. sarothrae and probably not distinct. The leaves have large resin dots bordered by hyaline margins.---Known only from westcentral Colorado.

11. Solidago L. GOLDENROD

Perennial, herbaceous, caulescent or subacaulescent plants; leaves alternate, entire or merely toothed; heads small, usually in racemiform, cymose or corymbose clusters, often secund on the branches of the inflorescence; involucres campanulate, oblong or cylindrical, usually narrow, the bracts imbricated in several series, usually thin and dry, sometimes with herbaceous tips; receptacles flat or somewhat convex, small, alveolate to fimbriolate but no chaff present; ray flowers pistillate, small, yellow; disk flowers mostly perfect, yellow; achenes terete or angular, usually 5- to 10-nerved; pappus of numerous, scabrous, white or tawny capillary bristles. This is a difficult genus badly in need of a general revision. The nomenclature has become involved and the established names are retained in most cases.

1. Heads in corymbs or flat-topped cymes; plants with at least one of the following 4 characters- (1). Involucre bracts striated (2). Leaves more or less punctate (3). Stems caespitose from a branched caudex (4). Ray flowers more numerous than the disk flowers
   2. Involucre bracts longitudinally striate; leaves and stems hairy; leaves over 5 mm. wide, with 1 longitudinal nerve   -1. S. rigida

2. Involucre bracts not striate; stems and leaves glabrous (or leaves scabrous on margins); leaves less than 5 mm. wide, with 3 longitudinal nerves (2 laterals often faint)
   3. Stems not over 20 cm. tall, caespitose from a branched caudex; rays 1-3   -2. S. petradoria
   3. Stems over 20 cm. tall, not caespitose from a branched caudex; rays 6-20
      4. Stems 30-50 cm. tall; plants of eastern Colorado; outer involucre bracts broader than the inner which are usually obtuse   -3. S. graminifolia
      4. Stems 50-100 cm. tall; plants mostly of western Colorado; outer involucre bracts about the same width as the inner which are acute   -4. S. occidentalis
1. Heads racemose or very numerous and panicled, the inflorescence usually of racemose branches; plants rarely if ever with any of the 4 characters listed above
   5. Leaves 1-nerved; heads not at all secund and stems usually glabrous at least below
      6. Involucre bracts linear-lanceolate or lanceolate, acute or acuminate; leaves villous-ciliate at least near their bases   -5. S. ciliosa
      6. Involucre bracts oblong or linear-oblong, obtuse; leaves not ciliate
         7. Leaves all entire; plants usually over 40 cm. tall; achenes glabrous or nearly so   -6. S. pallida
         7. Leaves (at least basal) toothed; plants 10-40 cm. tall; achenes strigose-hirsute
            8. Plants 10-15 cm. tall, usually growing above 9000 feet in Colorado; involucres 4-5 (rarely to 5.5) mm. high   -7. S. decumbens
            8. Plants 20-40 cm. tall, usually below 9000 feet in Colorado; involucres 5-8 mm. high   -7A. S. decumbens oreophila
   5. Leaves triple-nerved (1 pair of lateral veins decidedly more prominent than the others); heads usually secund on the branches of the inflorescence (when this is obscure the stem is definitely pubescent); stems glabrous to hairy
      9. Involucres 2-3 mm. high (rarely to 3.5 mm.)
         10. Leaves glabrous or pubescent only on the veins; stems glabrous or sparingly pubescent above   -8. S. canadensis
         10. Leaves densely canescent-pubescent at least below; stems densely puberulent   -8A. S. canadensis gilvocanescens
      9. Involucres 3-6.5 mm. high
         11. Stems glabrous or very sparingly pubescent in or near the inflorescence
            12. Plants 40-200 cm. tall; leaves lanceolate, usually serrate, abundant on the stem; achenes always pubescent   -9. S. gigantea
            12. Plants 20-40 cm. tall; leaves oblanceolate to spatulate, entire or sometimes serrate, stem leaves rather few; achenes glabrous to pubescent   -10. S. missouriensis
         11. Stems definitely pubescent at least above and usually throughout
            13. Involucre bracts ovate or oval, obtuse or acute; heads usually not definitely secund on the branches of the inflorescence
               14. Involucre bracts acute; cauline leaves usually oval, ovate or occasionally obovate   -11. S. mollis
               14. Involucre bracts rounded at apex; cauline leaves usually oblanceolate, sometimes obovate   -12. S. nana
            13. Involucre bracts narrower, lanceolate to linear-subulate, acute to attenuate at apex, rarely obtuse; heads usually definitely secund
               15. Leaves definitely pubescent at least below; stems densely hairy throughout
                  16. Stems 30-60 cm. tall; leaves rather few, dimorphous to some degree on complete specimens (basal longer than middle and upper and petioled); bracts acute to acuminate   -13. S. sparsiflora
                  16. Stems 60-100 cm. tall; leaves numerous and nearly uniform (upper leaves like lower only smaller); bracts acute to obtuse   -14. S. altissima
               15. Leaves pubescent below only on the midrib and main veins, occasionally very sparsely hairy all over; stems glabrous at base
                  17. Leaves subentire or sparingly serrate above the middle   -15. S. lepida
                  17. Leaves coarsely and sharply serrate   -15A. S. lepida fallax

1. Solidago rigida L., Sp. Pl. 880. 1753.
   S. rigida humilis Porter; Oligoneuron canescens Rydb.---Stems 30-100 cm. tall, simple at least below, rather stout, densely pubescent; leaves oblong to oval, crenate to entire, thick and rigid, canescent both sides, 1-nerved but the lateral veins rather prominent, basal long-petioled, upper sessile; heads in rather dense corymblike cymes; involucres 5-7 mm. high, campanulate, the bracts oblong and obtuse, longitudinally striated, outer pubescent; rays about 6-10; achenes several-nerved, glabrous or somewhat hairy above.---Prairies, hills and valleys. Ontario to British Columbia, south to Georgia, Texas and New Mexico. Our records mostly from the northeastern quarter of Colorado, but also in Archuleta County, at 3500-7500 feet.

2. Solidago petradoria Blake in Tide., Contrib. U. S. Nat. Herb. 25:540. 1925.
   Petradoria pumila (Nutt.) Greene---Stems 10-20 cm. tall, tufted and caespitose from a branching caudex, glabrous; leaves linear-oblanceolate, firm and rigid, 3-nerved, punctate and more or less resinous, glabrous, reduced in size upwards; heads in a flat-topped cyme; involucres about 5-6 mm. high, subcylindric, the bracts in more or less distinct vertical rows, somewhat keeled, spotted greenish or brownish near tips; rays 1-3, short; disk flowers few; achenes 5- to 10-striated, glabrous.---Dry ridges and rocky knolls. Wyoming to Oregon, south to Texas and California. Our records from the western one-fourth of Colorado at 4500-8000 feet.

3. Solidago graminifolia (L.) Salisb., Prodr. 199. 1796.
   Euthamia camporum Greene---Stems 30-50 cm. tall, leafy, striated, glabrous or nearly so; leaves linear, entire, 3-ribbed, but lateral ones often faint, more or less resinous-punctate, glabrous; heads in small rounded cymose clusters at the ends of the branches, aggregated into a more or less corymbose inflorescence; involucres 3-5 mm. long, oblong or campanulate, the

outer bracts acute or obtuse and wider than the inner, these latter often obtuse; rays small and inconspicuous but more numerous than the disk flowers; achenes hairy. Our plants are var. camporum (Greene) Fernald.---Low plains and meadows. The variety from Manitoba to Alberta, south to Kansas and Colorado. Our records from northeastern, central and southcentral Colorado at 3500-7500 feet.

4. Solidago occidentalis (Nutt.) T. & G., Fl. N. Amer. 2:226. 1842.

Euthamia occidentalis Nutt.---Stems 50-100 cm. tall, leafy, striate, glabrous; leaves linear or lanc-linear, entire, 3-ribbed, more or less resinous-punctate, glabrous; heads in small rounded cymose clusters at ends of branchlets, aggregated in a more or less corymbose inflorescence; involucres about 3-5 mm. high, oblong or campanulate, outer bracts acute or obtuse, about as wide as the acute inner ones; rays small and inconspicuous but more numerous than the disk flowers; achenes sparingly to densely hairy.---Usually moist ground. Alberta to British Columbia, south to New Mexico and California. Our records mostly from the western half of Colorado, but a few from northeastern, at 3500-7500 feet.

5. Solidago ciliosa Greene, Pitt. 3:22. 1896.

S. scopulorum (A. Gray) A. Nels.; S. rubra Rydb.; S. laevicaulis Rydb.; S. corymbosa of Manuals at least in part---Stems 10-70 cm. tall, more or less decumbent at base, often reddish-tinged, glabrous or loosely pubescent above, rarely throughout; basal leaves spatulate-obovate to oblanceolate, crenate-serrate at least near apex, 1-nerved, glabrous to somewhat hairy and villous-ciliate at least near base, petioled; stem leaves lanceolate or linear-lanceolate, 1-nerved, upper sessile; heads racemose in a narrow thyrsus but not at all secund; involucres 4-6 mm. long, campanulate, the bracts lanceolate or linear-lanceolate, acute or acuminate, usually ciliolate; rays usually fewer than the disk flowers; achenes more or less strigose. Our plants show some intergradations with S. decumbens Greene.---Hills and mountains. Alberta to British Columbia, south to Colorado, Arizona and California. Our records scattered in the western half of Colorado, scarce in the extreme southern part, at 6500-13,000 feet.

6. Solidago pallida (Porter) Rydb., Bull. Torr. Club 33:153. 1906.

S. speciosa var. pallida Porter---Stems 30-80 cm. tall, glabrous; leaves obovate, ovate to lanceolate, entire, usually paler below, 1-nerved, upper ones narrower and reduced; heads in a narrow panicle, not at all secund; involucres 5-8 mm. high, bracts oblong to linear-oblong, obtuse; rays usually fewer than the disk flowers; achenes glabrous or nearly so.---Hills and plains. Colorado, possibly also east and north. Our records from central and northcentral Colorado at 5500-6000 feet.

7. Solidago decumbens Greene, Pitt. 3:161. 1897.

Stems 10-15 cm. long, decumbent at base, often dark reddish, glabrous or sparsely puberulent above; basal leaves spatulate-obovate to oblanceolate, obtuse or rounded at apex, more or less distinctly serrate near tip, 1-nerved, petioled; upper leaves similar but smaller, glabrous or scabrous on margins, rarely pubescent all over; heads racemose in a narrow, often interrupted thyrsus, not at all secund; involucres 4-5 mm. long (rarely to 5.5) campanulate, bracts oblong to linear-oblong, obtuse; rays usually fewer than the disk flowers; achenes strigose-hirsute. Closely related to S. glutinosa Nutt.---Higher mountains. British Columbia, south to Colorado, Arizona and Oregon. Our records scattered in the higher mountains of Colorado at 9000-12,000 feet.

7A. Solidago decumbens oreophila (Rydb.) Fernald, (var.) Rhodora 38:202. 1936.

S. oreophila Rydb.---Stems 20-40 cm. tall; involucres 5-8 mm. long. Doubtfully distinct from the species, probably being merely a lower altitude form.---Mountains. British Columbia, south to Colorado. Our records scattered in the mountains of Colorado at 7000-9000 feet or sometimes at higher elevations.

8. Solidago canadensis L., Sp. Pl. 878. 1753.

Stems 40-120 cm. tall, more or less pubescent at least above; leaves narrowly lanceolate, acuminate, cuneate, serrate or upper entire, 3-nerved, glabrous or pubescent on veins beneath, sessile; heads in a panicle, often flat-topped, secund on the branches; involucres 2-3 mm. long, rarely to 3.5, the bracts linear or lanc-linear, acute; rays usually fewer than the disk flowers; achenes sparingly hairy.---Slopes and plains, sometimes among bushes. Newfoundland to British Columbia, south to Florida, Texas and Arizona. Our records scattered over the state at 4500-8500 feet.

8A. Solidago canadensis gilvocanescens Rydb., (var.) Contrib. U. S. Nat. Herb. 3:162. 1895.

S. gilvocanescens (Rydb.) Smyth---Stems densely pubescent; leaves densely canescent-pubescent at least below.---Plains and hills, often on sandy ground. Minnesota to Montana, south to New Mexico. Our few records from northern Colorado at 4000-6500 feet.

9. Solidago gigantea Ait., Hort. Kew. ed. 1. 3:211. 1789.

S. serotina Ait.; S. pitcheri Nutt.---Stems 40-200 cm. tall, erect, glabrous or glaucous right up to the inflorescence; leaves lanceolate, acuminate, cuneate, 3-nerved, glabrous or nearly so but often ciliate on the margins, sessile or very short-petioled; heads in a broad pyramidal panicle, secund, lower branches often elongated and recurved; involucres 3.5-5 mm. high, bracts linear or linear-lanceolate, obtuse or acute; rays usually fewer than the disk flowers; achenes pubescent. Our plant would be var. leiophylla Fernald.---Meadows, valleys and plains. The variety from Newfoundland to British Columbia, south to Georgia, Texas, Utah and Oregon. Out records from the eastern half of Colorado and also in the northern part, at 3500-7000 feet.

10. Solidago missouriensis Nutt., Journ. Phila. Acad. Nat. Sci. 7:32. 1834.

S. glaberrima Martens; S. concinna A. Nels.; S. viscidula Rydb.---Stems 20-40 cm. tall, sometimes even taller, erect; basal leaves oblanceolate to spatulate, petioled; cauline leaves lanceolate to linear, acute, 3-nerved, scabro-ciliate but surfaces glabrous or nearly so; heads usually secund, in a broad round or flat-topped panicle; involucres 3-5 mm. high, bracts oblong to linear-oblong, obtuse or acute; rays usually fewer than the disk flowers; achenes hairy to nearly glabrous.---Mountains, plains and hills. Michigan to British Columbia, south to Tennessee and Arizona. Our records scattered over the state at 4000-10,000 feet.

11. Solidago mollis Bartl., Ind. Sem. Hort. Goett. 5. 1836.
Stems 15-40 cm. tall, strict, leafy, canescent throughout; leaves oval, oblong or sometimes oblanceolate, entire to dentate, 3-nerved, rigid, lower petioled, upper sessile; heads scarcely secund, in a thrysoid panicle; involucres 3-6.5 mm. long, bracts ovate or oblong, usually at least some acute, sometimes obtuse; ray flowers usually fewer than disk flowers; achenes sparingly hirsutulous.---Dry hills and plains. Manitoba and Saskatchewan, south to Texas and New Mexico. Our records scattered in the eastern half of Colorado, mostly along the eastern foothills, at 4500-7000 feet.

12. Solidago nana Nutt., Trans. Amer. Phila. Soc. ser. 2. 7:327. 1840.
S. pulcherrima A. Nels.; S. radulina Rydb.---Stems 10-80 cm. tall, canescent-pubescent; leaves oblanceolate to obovate, entire to sparingly crenulate, usually rounded at apex, 3-nerved but the laterals often obscure, canescent; heads secund or not secund, in a panicle; involucres 4-6 mm. long, the bracts oblong or oblong-ovate, obtuse; rays usually less than the disk flowers; achenes hairy.---Dry plains and mountains. Alberta, south to Nebraska, Arizona and Nevada. Our records scattered in the western half of Colorado at 5000-8000 feet.

13. Solidago sparsiflora A. Gray. Proc. Amer. Acad. 12:58. 1877.
S. trinervata Greene---Plants variable in general appearance; stems 30-60 cm. tall, cinereous-pubescent; leaves oblanceolate to lanc-linear, entire or sparingly serrate, 3-nerved, puberulent and scabro-ciliate; heads usually secund on the branches of a panicle; involucres 3-5 mm. high, bracts lanceolate to linear, acute or acuminate; rays usually fewer than disk flowers; achenes hispidulous.---Plains and hills. South Dakota to Wyoming, south to Texas and Arizona. Our records scattered over Colorado, except in the extreme eastern part, at 5000-9000 feet.

14. Solidago altissima L., Sp. Pl. 878. 1753.
S. polyphylla Rydb.; S. canadensis scabriuscula Porter---Stems 70-100 cm. tall, from short-pubescent to hirsute throughout; basal leaves lanceolate to elliptic, serrate; cauline leaves lanceolate to elliptic, sharply serrate above the middle, acute, cuneate, thick, 3-nerved, scabrous above and pubescent below; heads secund on recurved branches of a narrow or wide panicle; involucres 3-5 mm. high, the bracts linear, acute or less commonly obtuse; rays usually fewer than the disk flowers; achenes pubescent-hirsute. Our Colorado plants have been called var. procera (Ait.) Fernald.---Often along streams. Newfoundland to Alberta, south to Florida and New Mexico. Our records scattered in Colorado at 3500-9000 feet.

15. Solidago lepida DC., Prodr. 5:339. 1836.
S. elongata Nutt.; S. serra Rydb.---Stems 30-80 cm. tall, erect, glabrous or minutely puberulent near the panicle; leaves lanceolate, acute to acuminate, cuneate, subentire or somewhat serrate above the middle, 3-nerved, more or less pubescent to glabrate; heads usually secund on the branches of a narrow panicle; involucres 3-5 mm. high, the bracts linear or linear-subulate, acute but attenuate at apex; rays small, usually fewer than the disk flowers; achenes pubescent. This plant would be var. elongata (Nutt.) Fernald.---Dry hills and valleys. The variety from Quebec to Alberta, south to New York and Colorado, possibly westward. Our records over Colorado, none as yet from the extreme eastern part, at 7000-9500 feet.

15A. Solidago lepida fallax Fernald, (var.) Rhodora 17:9. 1915.
Leaves coarsely and sharply serrate.---Newfoundland to British Columbia, south to Maine, Michigan, Utah and Washington. Our records scattered in Colorado, except the extreme eastern part, at 5000-9500 feet.

12. Haplopappus Cassini GOLDENWEED

Plants annual or perennial, herbaceous or shrubby, usually resinous or glandular; leaves alternate, entire to variously lobed; heads small to large, variously disposed, sometimes solitary; involucre bracts usually gradually imbricated but not in vertical rows; receptacle alveolate but not truly chaffy; rays usually present, yellow; disk flowers perfect and usually fertile, several to many; achenes cylindrical to turbinate; pappus copious of numerous graduated scabrous capillary bristles. Also spelled Aplopappus.

1. Heads discoid, no ray flowers present
    2. Leaves with spinulose, bristle-tipped teeth, not lobed at base   -1. H. nuttallii
    2. Leaves with margins entire, sometimes lobed at base but lobes entire
        3. Involucre bracts about 18 mm. long, conspicuously acuminate to cuspidate; leaves usually oblong-lanceolate, over 6 mm. wide; heads usually 1 to a stem   -14A. H. fremontii monocephalus
        3. Involucre bracts 4-13 mm. long, obtuse to acute or sometimes acuminate; leaves narrowly linear to oblong but usually not over 6 mm. wide; heads usually more than 1 to a stem
            4. Young stems tomentose; involucres 10-13 mm. long, the bracts only slightly imbricated if at all; leaves usually over 3 mm. wide   -2. H. macronema
            4. Young stems glabrous or hairy, never tomentose; involucres 4-10 mm. long, the bracts imbricated definitely with 3 or more lengths; leaves often less than 3 mm. wide
                5. Involucres 8-10 mm. long, the bracts all abruptly acuminate to cuspidate; achenes glabrous; corolla tube gradually widening upward into the throat   -3. H. engelmannii
                5. Involucres 4-8 mm. long, the outer bracts obtuse, inner obtuse or acute; achenes silky-hairy; corolla tube narrow upwards then abruptly expanding into the throat
                    6. Involucres 4-5 mm. high and 2-3 mm. wide; primary leaves usually with smaller fascicled ones in their axils   -4. H. pluriflorus
                    6. Involucres 6-8 mm. high and 4-5 mm. wide; primary leaves lacking smaller fascicled ones in their axils   -5. H. drummondii
1. Heads with ray flowers present (these may be rather short)
    7. Leaves definitely toothed or lobed; pappus brown to reddish, rarely white
        8. Annual or possibly biennial plants

9. Involucres 12-18 mm. long; leaves oblong to oval-ovate, usually over 15 mm. wide; achenes glabrous
-6. H. ciliatus
9. Involucres not over 10 mm. long; leaves narrowly oblong to narrower, not over 10 mm. wide; achenes villous to canescent
10. Plants 30-100 cm. tall; principal leaves 3-8 mm. wide; ligules 9-14 mm. long when fresh; involucre bracts usually somewhat squarrose -7. H. phyllocephalus
10. Plants 5-20 (rarely to 30) mm. tall; leaves 1-3 mm. wide; ligules 7-8 mm. long when fresh; involucre bracts not at all squarrose -8. H. gracilis
8. Perennial plants from a definite caudex or taproot
11. Heads large, the involucres 10-15 mm. high; achenes over 5 mm. long when mature; leaves usually entire
12. Involucre bracts green and herbaceous (except for margins) to very base, abruptly acute; ray flowers often over 40 to a head -9. H. clementis
12. Involucre bracts pale and cartilaginous at base, usually acute; ray flowers less than 40 to a head -10. H. integrifolius
11. Heads smaller, the involucres 5-8 mm. high; achenes 2-4 mm. long when mature; leaves usually toothed or lobed
13. Heads broad, the involucres 13-18 mm. wide (when pressed flat); longest leaves 5-10 cm. long, never lobed
14. Involucre bracts all equal or nearly equal in length; heads 1 to several in a raceme, not corymbose; achenes 2-3 mm. long -11. H. uniflorus
14. Involucre bracts definitely imbricated in 3-4 lengths; heads 1 to many, often in corymbs; achenes 3-4 mm. long -12. H. lanceolatus
13. Heads narrow, the involucres 8-12 mm. wide; leaves 1.5-6 cm. long, often lobed
15. Leaves densely scabrous to tomentose
16. Peduncles short, leafy up to or nearly to the heads
17. Leaves more or less tomentose at least above -13A. H. spinulosus spinulosus
17. Leaves not at all tomentose -13B. H. spinulosus australis
16. Peduncles obvious, naked or with much reduced leaves -13C. H. spinulosus scabrellus
15. Leaves glabrous or nearly so -13D. H. spinulosus glaberrimus
7. Leaves entire, undulate or very obscurely toothed; pappus variously colored but sometimes white
18. Involucres 11-22 mm. high; pappus light brown, of comparatively few rigid bristles
19. Heads subtended by numerous scarcely reduced upper leaves; involucre bracts abruptly acuminate to cuspidate -14. H. fremontii
19. Heads naked at base or on a few-bracted peduncle, the upper leaves much reduced; involucre bracts acute or obtuse, sometimes with a small mucro
20. Involucre bracts obtuse (or with a small mucro); involucres 15-18 mm. high; achenes glabrous
-15. H. croceus
20. Involucre bracts acute; involucres 11-15 mm. high; ahcenes glabrous to villous
21. Involucre bracts green and herbaceous (except for edges) to very base, abruptly acute; ray flowers often over 40 to a head -9. H. clementis
21. Involucre bracts pale and cartilaginous at base, gradually acute; ray flowers less than 40 to a head -10. H. integrifolius
18. Involucres 6-11 mm. high; pappus white to sordid (only rarely light brown), bristles soft and usually comparatively copious
22. Leaves dotted with conspicuous resin pits, blades 1-2.5 mm. wide; definite woody shrubs usually over 40 cm. tall -16. H. linearifolius
22. Leaves not conspicuously resin dotted or pitted, the blades usually over 2.5 mm. wide; plants herbaceous or low subshrubs woody at the base (if woody not over 20 cm. tall)
23. Plants over 15 cm. tall; heads usually more than 1 to a stem (not in H. uniflorus); plants never woody at base; leaves usually over 5 mm. wide; style appendages slender and acute or acuminate
24. Plants spreading by rootstocks; stems rather equably leafy from near base to apex; achenes usually glabrous; involucres 10-11 mm. high; pappus white -17. H. parryi
24. Plants with taproots (caudex may branch some) but no spreading rootstocks present; stems with leaves conspicuously reduced upwards in size and number; achenes sericeous; involucres 6-10 mm. high; pappus sordid
25. Bracts of the involucre all equal or subequal in length; heads usually 1, rarely 2-3 in a raceme but not corymbose; achenes 2-3 mm. long -11. H. uniflorus
25. Bracts of the involucre imbricated in 3 or 4 lengths; heads 1 to many, often in corymbs; achenes 3-4 mm. long -12. H. lanceolatus
23. Plants not over 15 cm. tall; heads solitary on the stems; plants often woody at base; leaves often less than 5 mm. wide; style appendates thick and obtuse
26. Heads on nearly naked peduncles, the leaves conspicuously reduced in size near apex of stems; subshrubs woody near base; rays 6-15 to a head
27. Involucre bracts ovate to ovate-lanceolate, 7-9 mm. long, at least some of them narrowed to an acute or acuminate apex, not regularly imbricated; plan+s rarely over 10 cm. tall; leaves 3 cm. long or less -18. H. acaulis
27. Involucre bracts broadly oblong to oblong-oval, 10-11 mm. long, rounded-obtuse at apex, imbricated in 3-4 lengths; plants often over 10 cm. tall; leaves 3-8 cm. long
-19. H. armerioides
26. Heads on leafy peduncles, some of leaves clustered close to the head, the leaves slightly if at all reduced upward on the stems; plants not really woody near base; rays 15-30 to a head
28. Involucre bracts lanceolate to linear-lanceolate, tapering to an acute apex; stems and leaves more or less glandular-hairy; leaves 5-12 mm. wide; plants often spreading by rootstocks -20. H. lyallii
28. Outer involucre bracts broadly oblong, rounded-obtuse or cuspidate at apex; stems and leaves never glandular; leaves 1.5-5 mm. wide; plants never widely spreading by rootstocks -21. H. pygmaeus

1. **Haplopappus nuttallii** T. & G., Fl. N. Amer. 2:242. 1842.
    Sideranthus grindelioides of Manuals---Herbaceous perennial plants or subshrubs with woody caudex, often caespitose; stems 10-30 cm. tall, nearly glabrous to tomentose; leaves 2-3.5 cm. long and 4-10 mm. wide, oblong to oblong-spatulate, dentate or serrate with spinulose teeth, thick, strigose to glabrate, sessile; heads 1 to several in a cyme; involucres 7-10 mm. high and 8-10 mm. wide, campanulate, bracts rather few, imbricated in about 3 series, acute, mostly green; rays none; corolla of disk flowers gradually enlarged above; achenes 2.5-3 mm. long, sericeous-canescent; pappus reddish-brown to tan.---Dry plains, mesas and hills. Saskatchewan and Alberta, south to Nebraska, New Mexico, Arizona and Nevada. Our records scattered in the western half of Colorado at 4500-7500 feet.

2. **Haplopappus macronema** A. Gray, Proc. Am. Acad. 6:542. 1866.
    Macronema discoideum Nutt.; M. obtusum Rydb.---Subshrubs 15-40 cm. tall, with numerous short branches from near base, forming a rounded bush; twigs leafy, white-tomentose when young, this often sparse or wanting near heads; leaves 1-3 cm. long and 3-6 mm. wide, oblong or oblanceolate, entire or somewhat undulate, glandular-scabrous; heads solitary to several; involucres 9-13 mm. high and 8-12 mm. wide, campanulate, bracts not imbricated and all about the same length, acute to somewhat obtuse, herbaceous to chartaceous, glandular-scabrous where exposed; ray flowers none; disk flowers 10-25, corollas tubular-funnelform, slender, villous; pappus dull white to sordid straw-colored. Our plant has been called ssp. typicus Hall.---Mountains. The subspecies from Idaho, south to Colorado and California. Our records from the higher mountains of Colorado, mostly in the central part, at 9000-12,000 feet.

3. **Haplopappus engelmannii** (Gray) Hall, Gen. Haplopappus 93. 1928.
    Oonopsis engelmannii (A. Gray) Greene---Plants herbaceous, from a suffrutescent base, a branching caudex or a rootstock; stems 10-30 cm. tall, bark pale or brown below, glabrous; leaves 3-7 cm. long and 1-3 mm. wide, narrowly linear, entire often rather involute, rigidly erect, glabrous or scabro-ciliate; heads few to many, cymose-glomerate; involucres 8-10 mm. high and about 6-8 mm. wide, turbinate-campanulate, bracts imbricated in 3 or more lengths, oblong or outer lanceolate, all abruptly cuspidate, mostly chartaceous but green below the tip, glabrous to short-ciliate; ray flowers none; disk flowers 15-20 to a head, corolla tubes widening gradually upwards to throat; achenes glabrous, nearly prismatic; pappus scanty, rigid and brown.---Dry plains. Colorado and Kansas. Our records scattered in the eastern half of Colorado, except in the extreme northern part, at 4000-6000 feet.

4. **Haplopappus pluriflorus** (A. Gray) Hall, Gen. Haplopappus 237. 1928.
    Isocoma pluriflora (T. & G.) Greene in Manuals; I. wrightii (A. Gray) Rydb.---Colorado, south to Texas, Mexico and Arizona. Reported for southern Colorado but no specimens located.

5. **Haplopappus drummondii** (T. & G.) Blake, Contrib. U. S. Nat. Herb. 23:1491. 1926.
    Colorado, south to Texas and Arizona. Reported for southwestern Colorado but no specimens seen.

6. **Haplopappus ciliatus** (Nutt.) DC., Prodr. 5:346. 1836.
    Prionopsis ciliata Nutt.---Plants annual or possibly biennial; stems 50-100 cm. tall, leafy sparingly branched above, glabrous; leaves 3-8 cm. long and 15-40 mm. wide, oval, ovate or oblong, spinulose-dentate, glabrous, sessile; heads few, in open cymes or rarely solitary; involucres 12-18 mm. high and wider than high, broadly hemispheric, bracts well imbricated, the outer more or less spreading or squarrose, linear-lanceolate, acuminate, firm and rather chartaceous but exposed parts green; ray flowers many, rays 12-20 mm. long; disk flowers many; achenes 2-4 mm. long, glabrous; pappus not copious, bristles rather rigid, reddish-brown in age at least, tardily deciduous.---Hills and valleys. Missouri to Colorado, south to Texas and New Mexico. Our one record from the southeastern corner of Colorado at about 4000 feet.

7. **Haplopappus phyllocephalus** DC., Prodr. 5:347. 1836.
    Sideranthus annuus Rydb.---Annual or possibly biennial plants with a straight taproot; stems 25-100 cm. tall, leafy, glandular-scabrous to villous or tomentose; leaves 3-5 cm. long and 3-13 mm. wide, oblong to spatulate-oblong or oblong-lanceolate, toothed or lobed these bristle-tipped, glandular-scabrous to villous; heads 2 to many, cymose or paniculate; involucres 6-8 mm. high and 10-15 mm. wide, the bracts imbricated in 2-4 lengths, lanceolate, attenuate to an awn, squarrose from middle, scarious on margins, herbaceous in middle; rays 25-40, 9-14 mm. long; achenes 2 mm. long or less, turbinate, densely villous-pubescent; pappus pale brown to reddish-brown, rather copious. Our plant has been called ssp. annuus (Rydb.) Hall.---Plains and hills, often on sandy ground. The subspecies from Kansas and Colorado, south to Texas. Our records from the northeastern quarter of Colorado at 3500-5500 feet.

8. **Haplopappus gracilis** (Nutt.) A. Gray, Mem. Amer. Acad. II. 4:76. 1849.
    Sideranthus gracilis of Manuals---Plants annual from a taproot; stems 5-30 cm. tall, branching from base and usually throughout, strigose, leafy; leaves 1-3 cm. long and 1-3 mm. wide, somewhat unequal in size upwards, linear or oblanceolate, dentate to pinnately divided the tooth or lobe ending in a bristle, strigose; heads solitary or terminating short branches; involucres 6-7 mm. high and 8-10 mm. wide, bracts imbricated definitely in 4-6 lengths, linear-lanceolate, acuminate to an awn, middle and tip green, edges whitish-scarious; rays 18-28, 7-8 mm. long; achenes about 2.5 mm. long, narrowly turbinate, canescent; pappus reddish-brown to whitish.---Dry hills and valleys. Colorado, south to Texas, Mexico and California. Our records from southwestern and westcentral Colorado at 5000-10,000 feet.

9. **Haplopappus clementis** (Rydb.) Blake, Contrib. U. S. Herb. 25:543. 1925.
    Pyrrocoma clementis Rydb.---Perennial herbaceous plants from a small taproot; stems 10-40 cm. long, ascending or decumbent, somewhat villous especially above; leaves 5-15 cm. long and 8-12 mm. wide, reduced in size upwards, oblanceolate to oblong-lanceolate, entire to sharp-toothed, midnerve conspicuous, glabrous or sparsely puberulent; heads solitary or few; involucres about 11-12 mm. high and 20-25 mm. wide, broadly hemispheric, the bracts in several series but not definitely imbricated, oblong to narrowly oblanceolate, abruptly acute, herbaceous to base except for hyaline margins; rays 40-60, 10-14 mm. long; achenes about 6 mm. long,

4-angled, nearly prismatic, more or less villous; pappus not copious, rather stiff, light brown.---Mountains. Wyoming, Colorado and Utah. Our records from central and western Colorado at 9500-12,500 feet.

10. Haplopappus integrifolius A. Gray, Syn. Fl. 1:128. 1884.

Pyrrocoma integrifolia of Manuals---Plants perennial and herbaceous with taproot and compact caudex; stems 20-60 cm. tall, glabrous or nearly so; leaves 7-14 cm. long and 10-30 mm. wide, rather reduced in size above, oblong to oblanceolate-oblong, entire, undulate or upper with a few sharp teeth, midnerve conspicuous, glabrous to scabrous, petioled; heads usually several in an open raceme, sometimes solitary; involucres 12-15 mm. high and 15-20 mm. wide, broadly hemispheric, bracts in several series but not imbricated, the outer usually about as long as the inner, gradually acute, pale and cartilaginous at base and edges, otherwise greenish; rays 20-40, 12-18 mm. long; achenes 6-7 mm. long, 4-angled and nearly prismatic, glabrous to pubescent; pappus of coarse brown bristles. Our plant has been called ssp. typicus Hall.---Meadows. The subspecies from Montana and Idaho, south to Wyoming and Utah. A few plants of northern, southern and western Colorado appear to be this species but show intergradation with H. croceus A. Gray.

11. Haplopappus uniflorus (Hook.) T. & G., Fl. N. Amer. 2:241. 1842.

Pyrrocoma uniflora (Hook.) Greene; P. inuloides (Nutt.) Greene---Plants perennial from a stout and deep taproot and with a simple or sometimes branching caudex, this with a coarse fibrous coat but not woody; stems 10-40 cm. tall, more or less reddish-tinged, often with a loose tomentum; basal leaves usually over 5 mm. wide, narrowly to broadly lanceolate to oblanceolate, entire to dentate or even laciniate, glabrous, sparsely villous-tomentose to slightly ciliate; cauline leaves much reduced in size and distribution, sessile; heads solitary or rarely 3 together; involucres 6-8 mm. high and 13-18 mm. wide, hemispheric, glabrous to tomentose, bracts subequal or sometimes the outer shorter, appressed; rays 25-45, 7-9 mm. long; achenes 2-3 mm. long, subcylindrical, sericeous; pappus sordid-white, bristles numerous and soft. Our plants have been called ssp. typicus Hall.---The subspecies from Saskatchewan and Idaho, south to Colorado and California. Our records scattered in the western one-third of Colorado at 8000-11,000 feet.

12. Haplopappus lanceolatus (Hook.) T. & G., Fl. N. Amer. 2:241. 1842.

Pyrrocoma lanceolata (Hook.) Greene; P. vaseyi (Parry) Rydb.; P. lagopus Rydb.---Perennial herbaceous plants from a deep taproot or short-branched caudex; stems 20-50 cm. tall, ascending and upwardly curving, not as leafy above as below, glabrous to slightly tomentose; leaves 5-15 cm. long and 5-35 mm. wide, narrowly to broadly lanceolate, entire to sharply dentate, glabrous to tomentulose, petioled to nearly sessile; heads 1 to many, corymbose, racemose or paniculate; involucres 7-10 mm. long and 12-18 mm. wide (pressed), hemispheric, the bracts imbricated in 3-4 series, lanceolate to lanceolate-oblong, acute to acuminate, green at apex and pale coriaceous at base; rays 15-45, 5-10 mm. long; achenes 3-4 mm. long, subcylindrical, densely sericeous; pappus sordid, rather soft.---Meadows, plains, hills and mountains. Saskatchewan and Alberta, south to Nebraska and Nevada. Our records from northcentral, central and southcentral Colorado at 7500-8500 feet.

13. Haplopappus spinulosus (Pursh) DC., Prodr. 5:347. 1836.

Perennial herbaceous plants from a woody caudex, the branches sometimes also woody at base; stems 20-60 cm. tall, erect or ascending, glabrous to tomentose, usually glandular; leaves 1.5-6 cm. long and 2-10 mm. wide, oblong-spatulate to linear-spatulate, dentate with teeth bristle-tipped or once- to twice-pinnately parted with the lobes bristle-tipped, glabrous to tomentose; heads solitary on leafy or naked branches; involucres 5-8 mm. long and 8-12 mm. wide, hemispheric, the bracts many and imbricated in 4-6 series, linear, acute or bristle-tipped, usually spreading or reflexed in age, chartaceous but with green centers; rays 15-50, 8.5-10 mm. long; achenes 2-2.5 mm. long, narrowly turbinate, appressed pubescent; pappus brown, rather scanty. Represented here by four subspecies.

13A. Haplopappus spinulosus spinulosus

H. spinulosus ssp. typicus Hall; Sideranthus spinulosus (Pursh) Sweet.---Upper leaves, at least, tomentose, leaves about equable in size and distribution.---Prairies and plains. Minnesota to Alberta, south to Texas, New Mexico and Mexico. Our records mostly scattered in the eastern half of Colorado, one from Archuleta County, at 3500-6500 feet.

13B. Haplopappus spinulosus australis (Greene) Hall, (ssp.) Gen. Haplopappus 77. 1928.

Sideranthus australis (Greene) Rydb.; S. puberulus Rydb.---Stems leafy upwards; leaves puberulent but not at all tomentose.---Dry hills and valleys. Colorado, south to Texas, Mexico and Arizona. Our records scattered in the southern half of Colorado at 4000-7500 feet.

13C. Haplopappus spinulosus scabrellus (Greene) Hall, (ssp.) Gen. Haplopappus 74. 1928.

Eriocarpum scabrellum Greene---Leaves hairy; peduncles naked or with much reduced leaves. ---Plains, hills and mesas. Colorado, south to Texas, Mexico and Arizona. Our records from western Colorado at 4500-5000 feet.

13D. Haplopappus spinulosus glaberrimus (Rydb.) Hall, (ssp.) Gen. Haplopappus 77. 1928.

Sideranthus glaberrimus Rydb.; S. spinulosus glaberrimus(Rydb.) A. Nels.---Leaves glabrous or nearly so. Dry hills and plains. Iowa to South Dakota, south to Texas and New Mexico. Our one record from central Colorado at 7500 feet.

14. Haplopappus fremontii A. Gray, Proc. Acad. Phil. 1863:65. 1864.

Oonopsis foliosa (A. Gray) Greene---Plants perennial, the stems many and herbaceous from a woody caudex or crown; stems 15-30 cm. tall, glabrous or slightly tomentulose above; leaves 5-10 cm. long and 5-15 mm. wide, linear-lanceolate to oblong-lanceolate, entire, glabrous; heads several and cymose or solitary; involucres 15-22 mm. long and 20-30 mm. wide, hemispheric, bracts imbricated but nearly equal in length, broadly lanceolate, conspicuously acuminate to cuspidate, the outer greenish; rays 15-25, 8-15 mm. long; achenes 5-7 mm. long, narrowly turbinate, glabrous or with a few short hairs near apex; pappus scanty, rigid, brown. This has been

called ssp. typicus Hall.---Plains and hills. The subspecies from Kansas and Colorado. Our records scattered in the southeastern quarter of Colorado at 4000-6000 feet.
14A. Haplopappus fremontii monocephalus (A. Nels.) Hall, (ssp.) Gen. Haplopappus 87. 1928.
Oonopsis monocephala A. Nels.---Heads solitary; involucres about 18 mm. high; ray flowers none.---Plains and hills. Colorado. Our few records from Las Animas County at 5500-6000 feet.
15. Haplopappus croceus A. Gray, Proc. Acad. Phila. 1863:65. 1864.
Pyrrocoma crocea (Gray) Greene---Plants perennial and herbaceous, from a deep thick taproot; stems 25-60 cm. tall, several to many, glabrous to tomentose especially above; lower leaves 10-25 cm. long and 2-4 cm. wide, reduced in size upwards, lanceolate, oblanceolate to elliptic, entire or undulate, glabrous or puberulent, petioled; heads usually solitary; involucres 15-18 mm. high and 20-30 mm. wide, broadly hemispheric, bracts imbricated or of nearly equal length, broadly oblong to ovate, broadly rounded at apex or sometimes mucronate, herbaceous except for the pale cartilaginous base; rays 25-70, 13-30 mm. long; achenes about 6 mm. long, 4-angled, glabrous; pappus rather scanty, rigid, light brown to tan. Our plant has been called ssp. typicus Hall.---Mountains. Wyoming and Utah, south to New Mexico and Arizona. Our records scattered in the western half of Colorado at 6000-11,000 feet.
16. Haplopappus linearifolius DC., Prodr. 5:347. 1836.
Stenotopsis interior (Cov.) Rydb.---Colorado to California. The ssp. interior (Coville) Hall has been reported for western Colorado.
17. Haplopappus parryi A. Gray, Am. Journ. Sci. II. 33:239. 1862.
Oreochrysum parryi (Gray) Rydb.---Herbaceous perennial plants spreading by slender horizontal rootstocks; stems 15-50 cm. tall, rarely even less at high altitudes, finely pubescent at least above; leaves 3-20 cm. long and 8-25 mm. wide, oblanceolate to oblanceolate-obovate or upper lanceolate, entire, glabrous or puberulent; heads 2 to many, open-cymose to glomerate; involucres 10-11 mm. long and 8-10 mm. wide, campanulate to hemispheric, bracts little imbricated, oblong and obtuse to lanceolate and acute, outer mostly herbaceous, others green near apex, pale chartaceous otherwise; rays 12-20, 5-8 mm. long; achenes 3.5-4.5 mm. long, more or less fusiform, 4-angled, glabrous or sometimes with scattered hairs; pappus copious, soft and white.---Mountains. Wyoming, south to New Mexico and Arizona. Our records scattered in the mountains of Colorado at 8000-11,500 feet.
18. Haplopappus acaulis (Nutt.) Gray, Proc. Am. Acad. 7:353. 1868.
Stenotus acaulis Nutt.---Perennial subshrubs from a branching woody caudex, forming mats; stems 4-12 cm. tall, subscapose with the leaves definitely reduced above, somewhat scabrous; leaves 1-3 cm. long and 2-7 mm. wide, narrowly to broadly oblanceolate, the veins prominent, glabrous to scabrous; heads solitary on the scapelike peduncle; involucres 7-9 mm. long and 8-12 mm. wide, hemispheric to campanulate, bracts in 3 series but not strongly imbricated, ovate to ovate-lanceolate, at least some acute or acuminate, pale chartaceous or green in the center; rays 6-15, 7-12 mm. long; achenes about 4 mm. long, more or less compressed, usually silky-villous but occasionally glabrous; pappus soft, white to pale brown. Our plant has been called ssp. typicus Hall.---Dry hills and mountains, often on rocky ridges. The subspecies from Montana and Idaho, south to Colorado and California. Our records from northwestern and northcentral Colorado at 5500-9000 feet.
19. Haplopappus armerioides (Nutt.) Gray, Syn. Fl. N. Amer. 1:132. 1884.
Stenotus armerioides Nutt.---Subshrubs; stems 5-15 cm. tall, rarely to 20 cm., caespitose on the branches of a stout woody caudex, forming mats, usually glabrous, leaves reduced above; leaves 3-8 cm. long and 3-12 mm. wide, linear-oblanceolate, entire, veins prominent, chiefly basal, glabrous or margins scabrous, often resinous; heads usually solitary on subscapose peduncles; involucres 10-11 mm. long and 10-12 mm. wide, broadly campanulate, bracts imbricated in 3-4 series, broadly oval to broadly oblong, rounded-obtuse or mucronate, pale and coriaceous but outer green near apex; rays 8-15, 10-12 mm. long; achenes about 4-5 mm. long, compressed, densely silky-villous; pappus rather copious, soft and white.---Dry hills and slopes, often on rocky ridges. Montana, south to Nebraska, New Mexico and Arizona. Our records mostly from the western one-third of Colorado, also in Weld County, at 5000-8000 feet.
20. Haplopappus lyallii A. Gray, Proc. Acad. Phila. 1863:64. 1864.
Tonestus lyallii (A. Gray) A. Nels.---Plants herbaceous, with several erect stems from a branched caudex or sometimes with rootstocks; stems 3-15 cm. tall, leafy, puberulent the hairs mostly gland-tipped; leaves 2-7 cm. long and 5-12 mm. wide, spatulate-oblong, usually entire, 1-nerved, erect, glandular-puberulent both sides; heads solitary, mostly naked at very base; involucres 9-10 mm. high and 8-12 mm. wide, broadly campanulate, bracts nearly equal, lanceolate to linear-lanceolate, tapering to an acute tip, herbaceous to chartaceous, glandular-puberulent; rays 20-25, 7-8 mm. long; disk flowers 80-100; achenes 3-4 mm. long, sparsely puberulent; pappus whitish and soft.---High mountains. Alberta and British Columbia, south to Colorado and Oregon. Our records from northern Colorado at 11,000-11,500 feet.
21. Haplopappus pygmaeus (T. & G.) Gray, Am. Journ. Sc. II. 33:239. 1862.
Tonestus pygmaeus (T. & G.) A. Nels.---Perennial herbaceous plants; stems 2-6 cm. tall, several to many from a taproot and a close underground crown, forming a cushion, crisp-puberulent to tomentose, nearly equably leafy above and below; leaves 2-5 cm. long and 1.5-5 mm. wide, linear-oblanceolate to oblong-oblanceolate, entire, pubescent at least on the margins; heads solitary, leafy-bracted; involucres 9-10 mm. long and 8-14 mm. wide, campanulate to hemispheric, bracts about equal in length, outer broadly oblong, obtuse or broadly cuspidate, herbaceous, the inner narrower and chartaceous; rays 15-30, 6-8 mm. long; achenes 3-4 mm. long, subcylindrical but compressed, sparsely villous; pappus rather scanty, soft and whitish.---Higher mountains. Wyoming to New Mexico. Our records scattered in the higher mountains of Colorado at 11,000-12,500 feet.

13. Townsendia Hook.

Herbaceous perennial, rarely annual or biennial, caulescent or acaulescent plants; leaves alternate, mostly narrow and broadest above the middle, entire, often crowded, sessile or petioled; heads medium to large, sessile to peduncled; involucres of many imbricated appressed bracts, these mostly lanceolate with scarious margins; receptacles flat, naked; ray flowers numerous in 1 series, pistillate and fertile, the rays white, rose or violet; disk flowers perfect, yellow; achenes of disk flowers compressed, with 2-forked or glochidiate short hairs; pappus of disk flowers a single series of several to many long rather rigid scabrous bristles, pappus of ray flowers similar or of paleae, often reduced.

1. Involucre bracts with rigid-acuminate points
  2. Pappus of disk flowers coroniform-united but with 2 stout awns (rarely 3 or 4 shorter awns present)
          -1. T. eximia
  2. Pappus of disk flowers of many stiff rather long bristles   -2. T. grandiflora
1. Involucre bracts obtuse or merely acute, never rigid-acuminate
  3. Plants strictly acaulescent, the heads sessile among the crowded basal leaves
    4. Leaves glabrous or soon glabrate; involucre bracts thick but not scarious on margins, often obtuse at apex; pappus of ray flowers much reduced; plants above 11,000 feet in Colorado   -3. T. rothrockii
    4. Leaves more or less permanently hairy; involucre bracts more or less scarious-margined, acute; pappus of ray flowers usually not much reduced from that of disk flowers; plants below 10,000 feet in Colorado
      5. Leaves with hirsute-appressed hairs; involucres usually less than 1 cm. high, margins of bracts rather narrowly scarious; plants apparently limited to Grand County, Colorado   -4. T. leptotes
      5. Leaves with fine appressed hairs; involucres 1-2 cm. long, margins of bracts definitely scarious; plants not limited to Grand County in Colorado   -5. T. sericea
  3. Plants caulescent or subacaulescent at least at maturity, the stems or leafy peduncles short but definite
    6. Pappus of ray flowers similar to that of the disk flowers and as long or nearly so (well over 1/2); leaves narrowly linear with hirsute-appressed hairs; plants apparently limited to Grand County, Colorado   -4. T. leptotes
    6. Pappus of ray flowers usually of united scales or short bristlelike scales from very short to about 1/2 as long as those of the disk flowers; leaves linear to spatulate, glabrous to cinereous-strigillose; plants not limited to Grand County, Colorado
      7. Leaves glabrous or very sparingly hairy; rays blue, light purple or lilac, 8-12 mm. long, plants apparently limited to southwestern Colorado   -6. T. glabella
      7. Leaves definitely cinereous-strigillose; rays white, rose or sometimes lilac, 6-8 mm. long; plants present but not limited to southwestern Colorado   -7. T. incana

1. Townsendia eximia A. Gray, Mem. Amer. Acad. n.s.4:70. 1849.
   T. vreelandii Rydb.---Plants perennial or possibly biennial, caulescent; stems 8-30 cm., usually 2 or more from a caudex, simple or sparingly branched, ascending; leaves 2-6 cm. long, oblanceolate to spatulate or upper lanceolate, numerous, nearly glabrate but often appressed hairy on veins and margins; involucres about 12-16 mm. high, the bracts in 3-4 series, ovate-lanceolate and rigidly acuminate, margins scarious and lacerate-ciliate; rays about 10-15 mm. long, many, blue or purple; achenes with glochidiate hairs; ray pappus of coroniform-united short paleae; disk flowers with pappus like ray flowers but in addition with 2 stout awns, sometimes with several shorter and slender ones but never many.---Mountainsides and valleys. Colorado and New Mexico. Our records from southcentral Colorado at 8500-10,000 feet.

2. Townsendia grandiflora Nutt., Trans. Am. Phil. Soc. n.s.7:306. 1841.
   Plants caulescent; stems 2-20 cm. tall (usually over 5 cm.), branching from base; leaves 2-5 cm. long, linear-oblanceolate, canescent to nearly glabrous; involucres about 10-15 mm. long, the bracts in about 3 series, ovate-lanceolate, rigidly acuminate, margins scarious and lacerate-ciliate; rays 10-20 mm. long, rose-purple or pinkish; achenes with glochidiate sparse hairs; ray pappus reduced to a crown of paleae, disk pappus of many stiff bristles longer than the achenes.---Plains and hills. South Dakota to Wyoming, south to Nebraska and New Mexico. Our records from the eastern half of Colorado but mostly along the eastern foothills, at 5000-7500 feet.

3. Townsendia rothrockii A. Gray in Wheeler Rept. 6:148. 1878.
   Plants strictly acaulescent; leaves about 1.5-2 cm. long, spatulate to broadly oblanceolate, rosulate, glabrous or nearly so; heads sessile; involucres about 7-9 mm. high, bracts in 3-4 series, oblong or ovate, thick-margined, obtuse or acutish, ciliate, purplish; achenes with rather long glochidiate hairs; ray pappus up to 1 mm. long, disk pappus about 4-5 mm. long.---High mountains. Limited to Colorado. Our records from Park and Gunnison Counties at about 12,000-12,500 feet.

4. Townsendia leptotes (Gray) Osterh., Muhl. 4:69. 1908.
   Perennial plants with branching caudex, acaulescent or with short leafy stems up to 3 cm. long; leaves 2-4 cm. long, surpassing the heads, narrowly linear, numerous, with appressed hirsute or strigose hairs; heads usually sessile; involucres about 8-10 mm. long, bracts in 3-4 series, broadly linear, margins narrowly scarious; rays several to many, lavender to blue-purple; achenes with sparse glochidiate hairs; ray pappus like disk pappus, this of many bristles.---Hills and mountains. Apparently limited to Grand County, Colorado. Our records from 7000-7500 feet.

5. Townsendia sericea Hook., Fl. Bor. Am. 2:16. 1834.
   T. exscapa (Rich.) Porter; T. wilcoxiana Wood; T. intermedia Rydb.---Depressed acaulescent perennial plants from a woody and usually more or less branched caudex; leaves 1-5 cm. long, linear-oblanceolate to linear, flat to thickened and nearly terete from the involute margins, crowded, sericeous-strigose to sparsely appressed pubescent; involucres 1-2 cm. high, bracts

in 4-6 series, linear-lanceolate, obtuse to long-acute, margins scarious and lacerate-ciliate; rays 6-12 mm. long, white to purple-tinged; achenes with glochidiate hairs; ray pappus from reduced short paleae to like disk pappus, disk pappus of many bristles. The plants with leaves sparsely pubescent have been called T. exscapa (Rich.) Porter but too many intergradations occur in this area to allow for such a separation.---Plains and hills. Saskatchewan and Alberta, south to Texas and Arizona. Our plants well scattered over Colorado at 4000-10,000 feet, often coming in blossom in January or February.

6. Townsendia glabella A. Gray, Proc. Am. Acad. 16:86. 1880.
   Caespitose perennial, subacaulescent plants; stems 2-6 cm. long, from a thick, woody often branched caudex; leaves 2-4 cm. long, linear-oblanceolate to spatulate, rosulate on the crown or crowded on the short branches, rather thick, soon becoming glabrous; heads solitary on naked peduncles 1-5 cm. long; involucres 7-10 mm. long, bracts in 2-3 series, oblong, acute, margins narrowly scarious and lacerate-ciliate; rays 8-12 mm. long, blue, light purple to lilac; achenes sparsely glochidiate-hairy; ray pappus irregular but short, to about 1/3 as long as the disk pappus, this of many long bristles.---Hills and mountains. Limited to southwestern Colorado. Our records from 6500-8500 feet.

7. Townsendia incana Nutt., Trans. Am. Phil. Soc. n.s.7:305. 1841.
   T. arizonica A. Gray; T. strigosa Nutt.; T. diversa Osterh.---Plants perennial or sometimes appearing to be winter annual or biennial, caulescent to very short-caulescent; stems 3-20 cm. long, caespitose with several from the caudex, cinereous-strigillose; leaves 1-4 cm. long, linear to linear-oblanceolate, tending to crowd out near the heads, cinereous-strigillose; heads 1-3; involucres 7-10 mm. high, bracts in 2-4 series, oblong-lanceolate to broadly lanceolate, obtuse to acute, margins scarious and erose-ciliate; rays about 6-8 mm. long, white to lilac; achenes with glochidiate hairs; ray pappus of stiff bristles or bristlelike paleae, varying from very short to about 1/2 as long as the disk pappus, this of many stiff bristles. The length of the flowering stems and the relative length of the ray and disk pappus varies so much in our plants that the writer cannot segregate them on that basis with any reasonable satisfaction.---Hills and mountains, often on gravelly ground. Wyoming and Utah, south to New Mexico and Arizona. Our records mostly in the western one-third of Colorado, but a few in the central and southcentral parts, at 4500-8000 feet.

14. Conyza L.

Annual herbaceous plants; stems leafy; leaves alternate, rather narrow; heads numerous, paniculate; involucres campanulate or subcylindrical; involucre bracts narrow, in 1-3 series; pistillate marginal flowers numerous with filiform corollas, the ligules absent or when present inconspicuous and shorter than the tube, scarcely if at all exceeding the pappus in length, whitish; central perfect flowers few; achenes flattened; pappus of 1 series of capillary bristles.

1. Involucres, leaves and stems glandular or viscid; leaves often toothed or lobed
   2. Ligules absent entirely; stems not arachnoid; achenes 0.5-0.8 mm. long -1. C. coulteri
   2. Ligules short but present on marginal flowers; stems arachnoid-villous; achenes 1-1.4 mm. long -2. C. schiedeana
1. Plants without glandular hairs; leaves, at least the upper, usually entire -3. C. canadensis

1. Conyza coulteri A. Gray, Proc. Amer. Acad. 7:355. 1868.
   Eschenbachia coulteri (A. Gray) Rydb.---Stems 30-60 cm. tall, erect, hirsute-villous and viscid; lower leaves 2-5 cm. long, spatulate, to oblanceolate, dentate or laciniate, hirsute and glandular-puberulent; upper leaves oblong or linear-lanceolate; heads numerous in an elongated panicle; involucre about 3-4 mm. high, bracts subequal; pistillate flowers without ligules; disk flowers 5-7; achenes 0.5-0.8 mm. long.---Along valleys and streams. Colorado, south to Texas, Mexico and California. Our one record from central Colorado at about 6000 feet.

2. Conyza schiedeana (Less.) Cronquist, Bull. Torr. Club 70:632. 1943.
   Erigeron schiedeanus Less.---Stems 20-60 cm. tall, erect, arachnoid-pilose and glandular; leaves 2-5 cm. long, lanceolate to oblong-linear, sparingly dentate or lower laciniate, definitely glandular and more or less pilose; heads many in a narrow raceme or panicle; involucres 4-5 mm. long, glandular and more or less villous, bracts subequal to somewhat imbricated; pistillate flowers all with short ligules less than 2 mm. long and shorter than the tube, slightly if at all longer than the pappus; disk flowers 6-10; achenes 1-1.4 mm. long.---Hills and valleys. Colorado, south to New Mexico, Mexico and Arizona. Our few records from central and southern Colorado at 7500-9500 feet.

3. Conyza canadensis (L.) Cronquist, Bull. Torr. Club 70:632. 1943.
   Leptilon canadense (L.) Britt.; Erigeron canadensis L.---Stems 20-120 cm. tall, erect, glabrous to hirsute; lower leaves 2-10 cm. long, spatulate or oblanceolate, entire or incised, usually hirsute or at least ciliate, petioled; upper leaves linear and usually entire, sessile; heads numerous in an elongated panicle or few in depauperate plants; involucres 2-4 mm. high, glabrous or sparingly hirsute, bracts in 2-3 series; ligules numerous but shorter than the pappus, white.---Dry hills, plains, waste places and roadsides. Naturalized from Europe and widely distributed in the Western Hemisphere. Our records scattered over Colorado at 4000-7500 feet.

15. Erigeron L.   FLEABANE; WILD-DAISY

Herbaceous plants, usually perennial but sometimes annual or biennial; stems leafy or sometimes scapelike; leaves alternate or basal, entire or pinnatifid; heads usually solitary or few, commonly borne on naked or subnaked peduncles; involucres hemispheric or shallower, the bracts narrow, varying from herbaceous and subequal to scarcely herbaceous and evidently imbricated, the loss of green either uniform or more prominent toward the apex; receptacles flat or nearly so, naked; pistillate marginal flowers more or less numerous, bearing evident and often very narrow ligules these as long as or longer than the pappus or ligules rarely absent, rays white to pink, blue or purple; disk flowers perfect, usually numerous, yellow; achenes mostly 2-nerved; pappus of few to many capillary and usually brittle bristles, commonly with a few outer shorter bristles or squamellae.

(Note: The 3 species of Conyza are included here)

1. Leaves (at least some) pinnately or palmately lobed, clustered at base of plant; stems scapose or subscapose
  2. Leaves ternately once- or twice-lobed, not pinnatifid
    3. Caudex stout, simple or somewhat branched, the branches short and stout; leaves 1- to 3-ternate
      4. Leaves mostly 2- to 3-ternate -1A. E. compositus glabratus
      4. Leaves mostly once-ternate -1B. E. compositus discoideus
    3. Caudex diffuse, branched in several to many slender rhizomelike branches; leaves once-ternate
      -2. E. vagus
  2. Leaves pinnatifid
    5. Pistillate flowers present, with well developed ligules; plants not limited to western Colorado
      -3. E. pinnatisectus
    5. Pistillate flowers absent, hence no ligules present; plants present, if at all, in western Colorado only -4. E. mancus
1. Leaves mostly entire, sometimes toothed, rarely lobed and then stem leafy; stems scapose to leafy
  6. Ligules absent or short, not over 4 mm. long at maturity
    7. Pappus of disk flowers evidently double, of long inner capillary bristles, the outer of shorter but conspicuous narrow squamellae, these usually setose; involucres 2-6 mm. high
      8. Perennial plants with short branching caudex; stems 2-30 cm. tall -5. E. aphanactis
      8. Annual or biennial plants, no definite caudex present, at least no branching caudex; plants often over 30 cm. tall
        9. Long pappus-bristles lacking entirely from the outer pistillate (usually rayed) flowers (present on disk flowers)
          10. Leaves many; plants mostly 60-150 cm. high; hairs of stem long and spreading
            -6. E. annuus
          10. Leaves sparse; plants mostly 30-70 cm. tall; hairs of stem usually short and appressed
            -7. E. strigosus
        9. Long pappus bristles present on both central and marginal flowers
          11. Earliest heads on leafy peduncles; plants without long stoloniferous branches
            -8A. E. divergens divergens
          11. Earliest heads on long naked peduncles; plants later producing long leafy stolons or stoloniform branches -8B. E. divergens cinereus
    7. Pappus all of long bristles or sometimes with a few inconspicuous shorter setae but never evidently double; involucres 2-10 mm. high
      12. Ligules either entirely absent on marginal, pistillate flowers or very short, not over 2 mm. long, little if any exceeding the pappus; central perfect flowers few to several; involucres 2-5 mm. high
        13. Stems, leaves and involucres glandular or viscid; leaves often toothed or lobed
          14. Ligules absent entirely from the pistillate, marginal flowers; stems not at all arachnoid; achenes 0.5-0.8 mm. long -Conyza coulteri
          14. Ligules (short) present on marginal flowers; stems arachnoid-pilose; achenes 1-1.4 mm. long -Conyza schiedeana
        13. No glandular hairs present, leaves (at least upper) usually entire -Conyza canadensis
      12. Ligules present on at least the outer pistillate flowers, 2-4.5 mm. long, usually longer than the pappus; central perfect flowers usually many; involucres 4-10 mm. high
        15. Tubular, filiform, essentially rayless pistillate flowers present between the perfect central flowers and the outer rayed flowers; inner involucre bracts usually long-attenuate at apex, almost caudate, often glandular; inflorescence corymbiform (or solitary); cauline leaves narrowly lanceolate to broader, rarely linear
          16. Plants usually 30 cm. tall or taller, with several to many heads
            -9A. E. acris asteroides
          16. Plants less than 30 cm. tall, with few to solitary heads -9B. E. acris debilis
        15. Rayless pistillate flowers absent; inner involucre bracts merely sharply acute to acuminate, never glandular; inflorescence racemiform (or solitary); cauline leaves usually linear
          -10. E. lonchophyllus
  6. Ligules present and at least reasonably conspicuous, over 4 mm. long at maturity (avoid immature heads or heads with ligules shattered in age)
    17. Involucre bracts definitely imbricated in several rows of different lengths and achenes 4-14 nerved (rarely 3) (Note: Read second No. 17); plants perennial with branching caudex; heads usually solitary
      18. Achenes glabrous, 8- to 14-nerved; old leaf bases persisting and becoming conspicuously fibrous -11. E. canus
      18. Achenes more or less hairy, 3- to 5-nerved; old leaf bases, if persistent, not becoming strongly fibrous
        19. Involucres villous-hirsute with crinkled, spreading hairs, longest bracts about 6-9 mm. long -12. E. pulcherrimus

19. Involucres, more or less strigose with appressed hairs, longest bracts about 3-6 mm. long
   20. Ligules about 9-18 mm. long; leaves usually longer than the internodes of the stem, often spreading
          -13A. E. utahensis tetrapleuris
   20. Ligules about 4-8 mm. long; leaves usually shorter than the internodes and erect
          -13B. E. utahensis sparsifolius
17. Involucre bracts usually equal or subequal and achenes 2-nerved (involucre bracts may be imbricated on some and achenes over 2-nerved on others but not both together); plants annual, biennial or perennial with caudex; heads solitary to many
   21. Long pappus bristles lacking entirely on the outer pistillate (usually ligulate) flowers; rays often inconspicuous
      22. Leaves many; plants mostly 60-150 cm. high; hairs of stem long and spreading
          -6. E. annuus
      22. Leaves sparse; plants mostly 30-70 cm. tall; hairs of stem usually short and appressed
          -7. E. strigosus
   21. Long pappus bristles present on all the flowers of the head; rays usually conspicuous
      23. Annual, biennial or short-lived perennial plants, without rootstocks or well developed woody or branching caudices (Note: in case of doubt try both 23's)
         24. Pappus double, inner of longer capillary bristles, outer of short setae or setose squamellae; plants often producing long leafy stolons or stoloniferous branches later in the season
            25. Stems with hairs appressed; long sparsely leafy stolons always developing later in the season
              -14. Erigeron flagellaris
            25. Stems with spreading hairs; long leafy stolons sometimes developing, sometimes none at all
               26. Earliest heads on leafy peduncles; plants without long stoloniferous branches
                 -8A. E. divergens divergens
               26. Earliest heads on long naked peduncles; plants later developing long leafy stolons or stoloniferous branches
                 -8B. E. divergens cinereus
         24. Pappus single, of long capillary bristles essentially equal in length; no long stolons or stoloniferous branches produced
            27. Stems with short incurved hairs, usually branching throughout; plants annual or occasionally biennial; leaves linear or oblanceolate; rays about 30-70 to a head
               28. Lower part of stems slender, mostly 1-2 mm. thick; plants usually intricately branched
                 -15A. E. bellidiastrum bellidiastrum
               28. Lower part of stems mostly 2.5-5 mm. thick; plants usually only moderately branched
                 -15B. E. bellidiastrum robustus
            27. Stems with long spreading hairs, rarely nearly glabrous, not branching at base; plants biennial to short-lived perennial; leaves broadly oblanceolate or wider; rays very numerous, mostly over 150 to a head
               -16. E. philadelphicus
      23. True perennial plants, usually with rootstocks or well developed caudices, these often woody or branching
         29. Involucres woolly-villous with multicellular hairs
            30. Stems 20-60 cm. tall, with leaves little if any reduced above, cauline ones ovate to lancovate, over 1 cm. wide; heads often over 1
               -17. E. elatior
            30. Stems 2-20 cm. tall, cauline leaves reduced in size and number, narrower, less than 1 cm. wide; heads always solitary
               31. Hairs of involucre with conspicuous black or dark purple crosswalls; pappus unequal but not definitely double
                 -18. E. melanocephalus
               31. Hairs of involucres with clear crosswalls or lowermost crosswalls sometimes bright reddish-purple; pappus double, inner of long bristles, outer of short setose squamellae
                 -19. E. simplex
         29. Involucres not woolly-villous, glabrous, glandular or variously hairy (if hairs multicellular then not woolly)
            32. Cauline leaves little reduced in size and number upwards, ample, usually lanceolate or broader; stems erect, often over 35 cm. tall
               33. Hairs of involucre with black crosswalls near their base; pappus not double or very inconspicuously so
                 -20. E. coulteri
               33. Hairs of involucres without black crosswalls; pappus double (outer row shorter and of setae or setose-squamellae) except in E. peregrinus
                 34. Achenes 4- to 7-nerved (usually 5); pappus usually not double; ligules 2-4 mm. wide
                   -21. E. peregrinus
                 34. Achenes 2-nerved (rarely 2-4); pappus double; ligules not over 2 mm. wide (rarely to 2.5 mm. in E. caespitosus)
                   35. Leaf surfaces glabrous (margins may be ciliate); stems glabrous below, glabrous, glandular or sparsely hairy above
                     36. Cauline leaves not ciliate on margins, little if any longer than the internodes
                       -22. E. superbus
                     36. Cauline leaves ciliate, usually longer than the internodes
                       37. Uppermost leaves lanceolate; involucre bracts commonly with a few hairs
                         -23A. E. speciosus speciosus
                       37. Uppermost leaves ovate; involucre bracts without hairs
                         -23B. E. speciosus macranthus
                   35. Leaf surfaces hairy (at least on the main veins); stems hairy throughout
                     38. Stems curved or decumbent at base, 5-30 cm. tall; ligules 30-100, 1-2.5 mm. wide; involucres 4-7 mm. high
                       -24. E. caespitosus
                     38. Stems erect, usually over 30 cm. tall; ligules 100-150, about 1 mm. wide; involucres 6-9 mm. high
                       -25. E. subtrinervis
            32. Cauline leaves much reduced in size or number upwards, mostly linear, lanceolate or oblanceolate; stems often spreading or decumbent at base, mostly less than 35 cm. tall

39. Achenes 4- to 7-nerved; pappus not definitely double; ligules often over 2.5 mm. wide
   40. Stems and leaves glabrous to sparsely villous; basal leaves broadly oblanceolate to spatulate, the blade over 5 mm. wide, definite and abruptly tapering to the petiole; rays usually rose-purple or darker   -21. E. peregrinus
   40. Stems and leaves gray-strigose; basal leaves linear or narrowly oblanceolate, not over 5 mm. wide, the blade if distinguishable tapering very gradually to the petiole; rays usually pink or white   -12. E. pulcherrimus
39. Achenes 2- or rarely 3-nerved; pappus usually definitely double (in all but E. brandegei and E. nematophyllus); ligules rarely over 2.5 mm. wide
   41. Hairs of stem appressed or ascending, or rarely wanting (in a few species hairs spreading just under the heads)
      42. Involucres finely glandular, no nonglandular hairs present; leaves usually glabrous, usually broadly oblanceolate or broader   -26. E. leiomeris
      42. Involucres with sparse to dense nonglandular hairs, glandular hairs often present also; leaves usually more or less hairy, linear to oblanceolate
         43. Involucres strigose to hirsute-strigose, the hairs appressed
            44. Ligules 15-55 to a head, 1.3-2.3 mm. wide, often white; stems 4-15 cm. tall; pappus single (shorter setae may intermingle with longer)   -27. E. nematophyllus
            44. Ligules 125-175, about 1 mm. wide, rarely white; stems usually over 15 cm. tall; pappus double, outer layer short and setose   -28A. E. glabellus glabellus
         43. Involucres hirsute, villous to hispid, the hairs spreading
            45. Caudex with several to many slender rhizomatous branches, without a well defined taproot or long central axis; base of stems conspicuously purple   -29. E. ursinus
            45. Caudex simple or with short, stout branches, a taproot present (in all but E. glabellus); base of stems not conspicuously purple (except in E. eatoni)
               46. Plants without an evident taproot but with numerous fibrous roots; ligules 125-175 to a head, about 1 mm. wide   -28A. E. glabellus glabellus
               46. Plants with an evident taproot and no fibrous roots; ligules 20-100, often over 1 mm. wide
                  47. Petioles or margins of at least the basal leaves with some coarse spreading hairs, unlike the other leaf hairs; leaves not 3-nerved
                     48. Stems and usually the leaves glandular; ligules rarely white   -30. E. vetensis
                     48. Stems and leaves not glandular; ligules usually white   -31. E. engelmanni
                  47. Petioles of the basal leaves without such coarse spreading hairs; basal leaves often 3-nerved
                     49. Leaves 3-nerved, at least basal ones; stems usually conspicuously reddish-purple at base; outer pappus bristles setose, fine and inconspicuous   -32. E. eatoni
                     49. Leaves not veined; stems not conspicuously reddish-purple at base; outer pappus of conspicuous thick setae or of squamellae   -33. E. ochroleucus
   41. Hairs of stem widely spreading, sometimes rather scanty
      50. Pappus an irregular crown, terminating in bristles   -34. E. brandegei
      50. Pappus of distinct bristles, not joined at base
         51. Stems and usually the leaves glandular, and basal leaves linear-oblanceolate; heads solitary   -30. E. vetensis
         51. Stems and leaves either not glandular, or if so then leaves oblanceolate or broader; heads solitary to many
            52. Plants with well developed caudices but no taproots (roots fibrous); ligules often over 100 to a head, about 1 mm. wide
               53. Involucre bracts glandular; stems usually glandular, curved at base
                  54. Involucres more or less hirsute as well as glandular; uppermost leaves more or less long-hairy, commonly not glandular   -35A. E. formosissimus formosissimus
                  54. Involucres nearly or completely without nonglandular hairs; uppermost leaves glandular, sometimes also sparsely hairy   -35B. E. formosissimus viscidus
               53. Involucre bracts not glandular; stems not glandular, usually erect at base   -28B. E. glabellus pubescens
            52. Plants with a taproot as well as caudex (no fibrous root system present); ligules not over 100 to a head, usually over 1 mm. wide
               55. Leaves appressed hairy, basal linear to linear-oblanceolate, not over 5 mm. wide   -33. E. ochroleucus
               55. Leaf hairs (at least some) spreading, basal leaves oblanceolate to spatulate, over 5 mm. wide
                  56. Involucres canescent with short white hairs, the bracts slightly to strongly imbricated; stems not glandular   -24. E. caespitosus
                  56. Involucres hirsute, with long hairs, the bracts equal or nearly so; stems sometimes glandular
                     57. Outer pappus of inconspicuous setae; ligules typically white   -36A. E. pumilus pumilus
                     57. Outer pappus of evident squamellae; ligules typically blue, very rarely white   -36B. E. pumilus concinnoides

1. Erigeron compositus Pursh, Fl. Am. Sept. 2:535. 1814.
   Perennial plants, the caudex simple or with stout and short branches; stems 5-15 cm. tall, scapiform; leaves crowded at base, ternately 1- to 3-lobed or dissected into linear or linear-oblong segments, more or less glandular and with at least a few hispid-hirsute spreading hairs; cauline leaves few, reduced, usually entire; heads solitary; involucres 5-10 mm. high, more or

less glandular and hirsute, bracts subequal; pistillate flowers about 20-60, rays up to 12 mm. long and 2 mm. wide but usually smaller, white, pink or rose-blue, often much reduced or absent; achenes 2-nerved, hairy; pappus single, of 12-20 coarse bristles. The following 2 varieties intergrade very commonly in Colorado.
1A. Erigeron compositus glabratus Macoun, (var.) Cat. Can. Pl. 2:231. 1884.
   E. multifidus Rydb.---Leaves 2- to 3-ternate.---Mountains, often in rocky ground. Greenland to Alaska, south to South Dakota, Colorado, Arizona and California. Our records scattered over the western half of the state at 6000-11,500 feet.
1B. Erigeron compositus discoideus A. Gray, (var.) Am. Journ. Sci. II. 33:237. 1862.
   E. trifidus Hook.; E. trifidus var. discoideus A. Nels.---Leaves mostly 1-ternate.---Mountains, often in rocky ground. Range of the preceding. Our records from central and northcentral Colorado at 6500-9000 feet.
2. Erigeron vagus Payson, Univ. Wyo. Publ. Bot. 1:179. 1926.
   Perennial plants with diffuse caudex of several to many slender rhizomelike branches; stems 1-6 cm. tall, scapiform; leaves 1-ternate, the segments short and rounded, crowded at base, spreading hirsute and usually more or less glandular; heads solitary; involucres 5-7 mm. high, glandular and more or less villous or hirsute, bracts subequal; rays 25-35, 4-7 mm. long and about 1-2 mm. wide, white or pink; achenes 2-nerved, slightly hairy; pappus single, of about 20 bristles.---Rocky slopes in the mountains. Colorado to Oregon, south to Utah and California. Our few records from westcentral, southwestern and southcentral Colorado at 8000-12,500 feet.
3. Erigeron pinnatisectus (A. Gray) A. Nels., Bull. Torr. Club 26:246. 1899.
   Perennial plants with stout, simple or more commonly branched caudex; stems 4-12 cm. tall, rather scapiform, somewhat glandular and with sparse spreading hairs; leaves clustered at base, pinnatifid into linear segments, petioles ciliate-bristly, otherwise glabrous; cauline leaves few, reduced, usually linear and entire; heads solitary; involucres 5-8 mm. high, glandular and usually more or less hirsute, bracts subequal, usually purplish at apex and margins; rays 40-70, 7-12 mm. long and 1-2 mm. wide, blue or purple; achenes 2-nerved, hairy; pappus single, of 25-30 usually unequal bristles.---High mountains. Wyoming to New Mexico. Our records scattered in the higher mountains of Colorado at 9000-13,000 feet.
4. Erigeron mancus Rydb., Fl. Rocky Mts. 902. 1917.
   Apparently limited to high altitudes in the La Sal mountains of eastern Utah. It may be in Colorado.
5. Erigeron aphanactis (A. Gray) Greene, Fl. Fran. 389. 1897.
   Perennial plants with short-branching caudex and taproot; stems 2-20 cm. tall, hirsute; basal leaves numerous, linear-oblanceolate to spatulate, petioled; cauline leaves few and reduced, all leaves hirsute, the petiole margins often with stiffer longer hairs; heads 1 to several; involucres 4-6 mm. long, hirsute and often glandular, bracts subequal to slightly imbricated; pistillate flowers without ligules or with very short ones; achenes 2-nerved, sparsely hairy; pappus double, inner of 7-20 long bristles, outer of narrow squamellae.---Dry valleys and hills. Colorado to Oregon, south to New Mexico, Arizona and California. Our few records from westcentral, southwestern and southcentral Colorado at 4500-8000 feet.
6. Erigeron annuus (L.) Pers., Syn. Pl. 2:431. 1807.
   A weedy species of Canada and northern United States. It should be found in Colorado.
7. Erigeron strigosus Muhl. ex Willd., Sp. Pl. 3:1956. 1803.
   E. ramosus (Walt.) B.S.P.---Plants annual or possibly biennial; stems 30-70 cm. tall, finely strigose to hirsute, sometimes glabrous; basal leaves mostly oblanceolate to elliptic, entire or toothed; cauline leaves rather few and reduced above, linear to oblanceolate, entire to toothed; heads several to many in an open rather flat-topped inflorescence; involucres 2-5 mm. high, finely glandular and more or less hairy, bracts subequal to somewhat imbricated; rays about 50-100 or sometimes lacking, up to 6 mm. long and 1 mm. wide, white or sometimes pink or blue; achenes 2-nerved, hairy; pappus double, inner of 10-15 fragile bristles, but lacking on the pistillate marginal flowers, outer of setose squamellae.---Waste ground and fields. Canada and United States; introduced into Europe. Our records from northcentral and northwestern Colorado at 5000-8500 feet.
8. Erigeron divergens T. & G., Fl. N. Amer. 2:175. 1841.
   Plants usually biennial but sometimes annuals or short-lived perennials with taproot and only rarely caudex; stems 10-70 cm. tall, branching, with short spreading hairs; leaves with spreading or appressed hairs; basal leaves oblanceolate to spatulate, entire or coarsely toothed, petioled; cauline leaves linear to oblanceolate, numerous or reduced on the peduncles; heads 1 to several; involucres 3-6 mm. long, glandular and more or less hirsute, bracts subequal or somewhat imbricated; rays about 75-150, 5-10 mm. long and 0.5-1.2 mm. wide, but occasionally much reduced, rose-purple, rose to white; achenes 2 or rarely obscurely 4-nerved, sparsely hairy; pappus double, inner of 5-12 long fragile bristles, outer of short usually conspicuous setose squamellae. Our 2 varieties are reasonably distinct.
8A. Erigeron divergens divergens
   E. divergens T. & G.; E. wootoni Rydb.; E. divergens var. typicus Cronq.---Earliest heads on leafy peduncles; plants not producing long stoloniform branches.---Dry often sandy ground. South Dakota to British Columbia, south to Texas, Mexico and California. Our records scattered over Colorado, except the extreme eastern part, at 4500-8000 feet.
8B. Erigeron divergens cinereus A. Gray, (var.) Pl. Wright. 1:91. 1852.
   E. cinereus A. Gray; E. nudiflorus Buckl.; E. commixtus Greene; E. colo-mexicanus A. Nels.; E. divergens nudiflorus (Buckl.) A. Nels.---Earliest heads on long naked or nearly naked peduncles; plants producing long leafy stolons later in the season.---Dry often sandy soil. Kansas to Nevada, south to Texas and Arizona. Our records scattered over Colorado, mostly in the southern part, at 4000-8000 feet.

9. Erigeron acris L., Sp. Pl. 863. 1753.

Plants biennial or perennial with short, simple or slightly branched caudex; stems 5-80 cm. tall, glabrous to more or less hirsute, often glandular; leaves glabrous to more or less hirsute and glandular; basal leaves oblanceolate to spatulate, entire to remotely serrulate; cauline leaves well developed, narrowly lanceolate to ovate; heads solitary to many; involucres about 4-10 mm. high, glandular or hirsute, bracts subequal to somewhat imbricated, inner long-attenuated, almost caudate; pistillate flowers numerous, of 2 kinds, the outer with short rays about 2.5-4.5 mm. long and 0.2-0.5 mm. wide, whitish or pinkish, the inner flowers lacking rays; disk flowers normal; achenes 2-nerved, sparsely hairy; pappus of 25-35 bristles, sometimes with a few shorter inconspicuous ones. Two varieties are present in Colorado.

9A. Erigeron acris asteroides (Andrz.) DC., (var.) Prodr. 5:290. 1836.

E. asteroides Andrz.; E. droebachensis O. Muell.; E. yellowstonensis A. Nels.; E. lapiluteus A. Nels.---Plants robust, the stems 30-80 cm. tall; heads several to many.---Woods and mountains. Labrador to Alaska, south to Michigan, Colorado and California. Our records from northcentral, central and western Colorado at 7500-10,000 feet.

9B. Erigeron acris debilis A. Gray, (var.) Syn. Fl. N. Amer. 1:220. 1884.

E. jucundus Greene---Stems not over 30 cm. tall; heads solitary to few.---Woods and mountains. Colorado to California. Our records from the western half of Colorado at 7500-12,000 feet.

10. Erigeron lonchophyllus Hook., Fl. Bor. Am. 2:18. 1834.

E. minor (Hook.) Rydb.---Biennial or short-lived perennial plants with little if any caudex; stems 5-50 cm. tall, erect, more or less hirsute; leaves more or less hirsute to glabrate, often ciliate; basal leaves oblanceolate to spatulate, entire; cauline leaves usually linear; little reduced upwards; heads solitary or racemose; involucres 4-9 mm. long, more or less hirsute but not glandular, bracts usually somewhat imbricated, inner acute or acuminate and often purplish near apex; rays many, about 2-3 mm. long and less than 0.5 mm. wide; achenes 2-nerved, sparsely hairy; pappus of 20-30 long bristles, sometimes with a few short inconspicuous outer setae.---Moist ground. Quebec to Alaska, south to North Dakota, Utah and California. Our records scattered over the western half of Colorado at 6000-10,000 feet.

11. Erigeron canus A. Gray, Mem. Am. Acad. 4:67. 1849.

Wyomingia cana (A. Gray) A. Nels.---Perennial plants with stout taproot and a usually branched caudex crowded with old leaf bases, these becoming fibrous; stems 5-30 cm. tall, usually erect or nearly so, grayish-strigose; leaves grayish-strigose, basal tufted, oblanceolate to linear-oblanceolate, cauline reduced above, mostly linear; heads usually solitary; involucres 5-7 mm. high, canescent-hirtellous with short spreading hairs, bracts strongly imbricated; rays 25-40, 6-12 mm. long and 0.8-1.4 mm. wide, blue or white; achenes 8- to 14-nerved, glabrous; pappus double, inner of about 20-25 long bristles, outer of short setae.---Dry plains and hills. South Dakota to Wyoming, south to Nebraska, New Mexico and Arizona. Our records mostly from the eastern half of Colorado, also in the southwestern part, at 5000-9000 feet.

12. Erigeron pulcherrimus Heller, Bull. Torr. Club 25:200. 1898.

E. wyomingia Rydb.; Wyomingia pulcherrima (Heller) A. Nels.; W. cinerea A. Nels.---Perennial plants with taproot and much branched caudex; stems 5-35 cm. high, gray-strigose; basal leaves 3-7 cm. long and usually over 1.5 mm. wide on some, linear to narrowly linear-oblanceolate, tufted, rather strigose; cauline leaves reduced in size upwards; heads solitary; involucres about 6-9 mm. high, more or less villous with multicellular hairs, bracts imbricated in several rows, the outer definitely shorter; rays about 25-60, 8-15 mm. long and 2-4 mm. wide, pink, white to bluish; achenes usually 4-nerved, hairy; pappus of 30-50 sordid bristles, sometimes clearly double. Our plants have been called var. wyomingia (Rydb.) Cronq.---Dry plains and hills. The variety from Wyoming and Utah, south to New Mexico. Our plants from westcentral Colorado at 4500-6000 feet.

13. Erigeron utahensis A. Gray, Proc. Am. Acad. 16:89. 1880.

Perennial plants with stout woody taproot and branching caudex; stems 10-50 cm. tall, erect, somewhat gray-green in color, more or less strigose especially at base; leaves more or less strigose, usually gray-green, basal and lower cauline linear-oblanceolate to linear, cauline reduced above, erect to spreading; heads solitary to several; involucres 3-6 mm. high, more or less strigose and often finely glandular, bracts strongly imbricated; rays about 10-40, 4-18 mm. long and about 1-2.7 mm. wide, pink, blue or white; achenes 4- or rarely 6-nerved, hairy; pappus double, inner of long bristles, outer of rather conspicuous short setae. Two varieties are here.

13A. Erigeron utahensis tetrapleuris (A. Gray) Cronquist, (var.) Brittonia 6:272. 1947.

E. tetrapleuris (A. Gray) Heller---Leaves often spreading and usually longer than the internodes; rays mostly 8-18 mm. long.---Dry ground. Colorado, Utah and Arizona. Our records from extreme western Colorado at 4500-6500 feet.

13B. Erigeron utahensis sparsifolius (Eastw.) Cronquist, (var.) Brittonia 6:273. 1947.

E. sparsifolius Eastw.---Leaves usually erect, shorter than the internodes; rays usually about 4-8 mm. long.---Dry ground. Colorado, Utah and Arizona. Our records from westcentral and southwestern Colorado at 5000-7000 feet.

14. Erigeron flagellaris A. Gray, Mem. Am. Acad. II. 4:68. 1849.

Plants usually biennial but possibly short-lived perennials, caudex not present; flowering stems 5-30 cm. tall, erect, sparsely leafy or subnaked, some of the main stems or branches of the flowering stems become long, trailing and sparsely leafy, often rooting at the tips or bearing a head, strigose or strigose-hirsute with hairs antrorsely appressed; basal leaves oblanceolate, entire, petioled; cauline leaves linear to linear-oblanceolate, strigose; heads solitary; involucres 3.5-5 mm. high, glandular and more or less hirsute, the hairs usually appressed, bracts equal or subequal; rays 50-100, 5-10 mm. long and about 1 mm. wide, white or sometimes blue or pink; achenes 2-nerved, sparsely hairy to glabrous; pappus double, inner of about 10-15 bristles, outer of rather inconspicuous short setae. Our plant has been called var. typica Cronq.---Open banks and slopes. The variety from South Dakota to British Columbia, south to Texas and Arizona. Our records scattered in Colorado, except in the northwestern and eastern parts, at 5000-9500 feet.

15. Erigeron bellidiastrum Nutt., Trans. Am. Phil. Soc. II. 7:307. 1840.

Plants annual or perhaps occasionally biennial, no caudex; stems 10-50 cm. tall, moderately to intricately branched, hirsute with incurved hairs that are broadest near their base; leaves numerous, linear to oblanceolate, entire or rarely toothed or even lobed, lower sometimes petioled, with hairs similar to those of stem; heads usually many, at ends of the branchlets; involucres 3-5 mm. long, hirsute with incurved hairs and more or less glandular, bracts thick, subequal or imbricated; rays 30-70, 4-6 mm. long and about 1 mm. wide, white to pink; achenes 2-nerved, more or less hairy; pappus single, of about 15 fragile bristles. Two varieties are in Colorado.

15A. Erigeron bellidiastrum bellidiastrum

E. bellidiastrum var. typicus Cronquist---Plants slender with stems mostly 1-2 mm. thick near base, usually branching throughout.---Mostly sandy ground. South Dakota to Nevada, south to Oklahoma, New Mexico and Utah. Our records scattered over Colorado, mostly in the eastern half, at 3500-8000 feet.

15B. Erigeron bellidiastrum robustus Cronquist, (var.) Brittonia 6:256. 1947.

Plants robust, the stems mostly 2.5-5 mm. thick near base, only moderately branching.---Mostly in sandy soil. Nebraska to Colorado, south to Texas. Our few records from southeastern Colorado at about 4000 feet.

16. Erigeron philadelphicus L., Sp. Pl. 863. 1753.

Biennial or short-lived perennial plants from a simple or nearly simple caudex; stems 20-70 cm. tall, glabrous or with long spreading hairs; basal leaves narrowly to broadly oblanceolate rarely wider, crenate to lobed, short-petioled; upper leaves crenate, dentate or entire, sessile and clasping at base; heads 1 to many; involucres 4-6 mm. long, bracts subequal, nearly glabrous to villous along the midrib; rays mostly over 125, 5-10 mm. long and 0.2-0.6 mm. wide, white to rose-purple; achenes 2-nerved, sparsely hairy; pappus single.---Woods and fields. Throughout Canada and the United States. Our records from southern Colorado at about 6500 feet.

17. Erigeron elatior (A. Gray) Greene, Pitt. 3:163. 1897.

Plants perennial from a stout short caudex; stems 20-60 cm. tall, amply and equally leafy, more or less spreading-hirsute, also glandular above; leaves entire, villous-hirsute, upper ones occasionally glandular; basal leaves oblanceolate to elliptic, petioled; cauline leaves ovate to lanc-ovate, upper clasping and relatively wider than the others; heads solitary or occasionally 2 or 3; involucres 7-12 mm. high, woolly-villous with long hairs having reddish-purple crosswalls, bracts about equal; rays about 75-150, 10-20 mm. long and 0.8-1.6 mm. wide, pink or rose-purple; achenes 2-nerved, hairy; pappus double, inner of about 15-20 long bristles, outer of conspicuous setose squamellae.---Along streams and in moist places in the mountains. Colorado and Utah. Our records scattered in the western half of Colorado at 9500-12,500 feet.

18. Erigeron melanocephalus A. Nels., Bull. Torr. Club 26:246. 1899.

Perennial plants with somewhat branching caudex; stems 5-15 cm. tall, villous with hairs having dark crosswalls; leaves largely basal and reduced in size and numbers upwards, glabrous or somewhat hirsute especially on margins; basal leaves oblanceolate to spatulate, petioled; cauline leaves linear or narrowly linear-oblanceolate, few; heads solitary; involucres 5-9 mm. high, densely woolly-villous, the hairs with conspicuous black or dark purple crosswalls, bracts equal or nearly so; rays about 50-70, 7-11 mm. long and about 1.4-2 mm. wide, white or pink; achenes 2-nerved, slightly hairy; pappus of 20-25 unequal bristles but not in 2 definite sets. Closely related to number 19.---Higher mountains. Wyoming to Utah, south to New Mexico. Our records scattered in the higher mountains of Colorado at 11,500-12,500 feet.

19. Erigeron simplex Greene, Fl. Fran. 387. 1897.

E. leucotrichus Rydb.---Perennial plants with simple or slightly branching caudex; stems 4-20 cm. tall, 1 to few, more or less villous; leaves glabrous to somewhat hirsute especially on margins, mostly crowded basal; basal leaves oblanceolate to spatulate, petioled; cauline leaves few and much reduced in size especially in width; heads solitary; involucres 5-8 mm. high, rather woolly-villous the hairs having clear crosswalls or occasionally basal walls are reddish-purple, bracts about equal; rays about 50-125, 7-11 mm. long and 1.2-2.5 mm. wide, blue or pink, rarely white; achenes 2-nerved, sparsely hairy; pappus double, inner of 10-15 long bristles, outer of rather conspicuous short setose squamellae.---Higher mountains. Montana to Oregon, south to New Mexico and Arizona. Our records scattered in the higher mountains at 9500-13,500 feet.

20. Erigeron coulteri Porter in Porter & Coulter, Fl. Colo. 61. 1874.

Perennial plants with slender rhizomes or short-branching caudex; stems 10-60 cm. tall, rather equally leafy to top, more or less spreading hirsute; leaves sparsely to densely hirsute; lower leaves broadly oblanceolate to elliptic, petioled; upper leaves broadly lanceolate to oblong, entire to dentate, usually clasping; heads 1 or occasionally 2 or 3; involucres 7-10 mm. high, more or less villous-hirsute, the hairs spreading and with crosswalls, these black near base of hair, bracts equal; rays 50-100, 9-25 mm. long and 1.2-1.7 mm. wide, white; achenes 2-nerved, hairy at least on nerves and near apex; pappus of about 20-25 bristles, rarely with a few short inconspicuous outer setae.---Mostly in meadows or on streambanks. Wyoming to Washington, south to New Mexico and California. Our records scattered in the western half of Colorado, except the northwestern part, at 9000-12,000 feet.

21. Erigeron peregrinus (Pursh) Greene, Pitt. 3:166. 1897.

E. callianthemus Greene; E. salsuginosus of Manuals---Plants perennial from a rhizome or short caudex; stems 10-70 cm. tall, glabrous to more or less villous especially just below the heads, leafy in large forms, subscapose in smaller; basal leaves oblanceolate to spatulate, petioled; cauline leaves linear to ovate, reduced above, glabrous to ciliate or more or less villous, often subclasping; heads usually solitary; involucres 7-11 mm. high, bracts equal or slightly imbricated, attenuated at apex, glandular; rays 30-80, 8-25 mm. long and 2-4 mm. wide, rose-purple to dark purple, rarely lighter; achenes 4- to 7-nerved (usually 5), sparsely hairy; pappus single, of about 20-30 long bristles, or sometimes with a few outer setae present. Our plants are ssp. callianthemus (Greene) Cronq.---Moist meadows or along streams. The species from Canada and Alaska, south to New Mexico and California. Our records scattered in the higher mountains of Colorado at 9500-12,500 feet.

22. Erigeron superbus Greene ex Rydb., Colo. Ag. Exp. Sta. Bull. 100: 351, 364. 1906.

Perennial plants from a simple or slightly branched caudex, often with rhizomes; stems 15-60 cm. tall, glabrous below, glandular in the inflorescence; leaves not much reduced in size above but very few and mostly shorter than the internodes, more or less 3-nerved, glabrous or upper glandular, lower sometimes very slightly ciliate; basal leaves oblanceolate to oval, petioled; upper leaves lanceolate, ovate or oblong; heads 1-6; involucres 6-9 mm. high, glandular or sometimes with a few long hairs, bracts about equal; rays mostly 40-80, 12-20 mm. long and about 1-2 mm. wide, blue or rose, rarely white; achenes 2-nerved, hairy; pappus double, inner of 20-25 long bristles, outer of short, sometimes scanty setae.---Mountains. Wyoming and Utah, south to Texas and Arizona. Our records scattered in the higher mountains of Colorado at 7000-10,000 feet.

23. Erigeron speciosus (Lindl.) DC., Prodr. 5:284. 1836.

Perennial plants with more or less branching caudex; stems 15-80 cm. tall, glabrous or somewhat hairy above; leaves commonly 3-nerved, entire, ciliate, the cilia sometimes few and only on a part of the margins, glabrous on faces or rarely with a few hairs on main veins; basal leaves oblanceolate to broadly spatulate, petioled; cauline leaves narrowly lanceolate, oblong to broadly ovate, the upper not markedly reduced; heads 1-10; involucres 6-9 mm. high, glandular or with a few long nonglandular hairs, bracts equal or nearly so; rays about 75-150, 9-18 mm. long and about 1 mm. wide, blue or very rarely lighter; achenes 2- to 4-nerved, hairy; pappus double, inner of about 20-30 long bristles, outer of short setae. Two varieties are in Colorado.

23A. Erigeron speciosus speciosus

E. salicinus Rydb.; E. speciosus var. typicus Cronquist---Uppermost leaves mostly lanceolate; involucre bracts usually with at least a few long hairs along with short glandular ones.---Mountains. Alberta to British Columbia, south to South Dakota, New Mexico and Mexico. Our records scattered in the western half of Colorado at 6500-10,000 feet.

23B. Erigeron speciosus macranthus (Nutt.) Cronquist, (var.) Bull. Torr. Club 70:269. 1943.

E. macranthus Nutt.; E. vreelandii Rydb.---Upper leaves usually narrowly to broadly ovate; involucre bracts without long nonglandular hairs.---About the same range as the preceding variety, both generally and in Colorado. Our records from 5500-9500 feet.

24. Erigeron caespitosus Nutt., Trans. Am. Phil. Soc. II. 7:307. 1840.

E. subcanescens Rydb.---Perennial plants with taproot and stout usually branched caudex; stems 10-25 cm. tall, several, curved at base, hirtellous or canescent with short spreading hairs; leaves short-hirsute, basal oblanceolate, narrowly linear-oblanceolate to spatulate, entire, 3-nerved; cauline leaves more or less reduced upward in size and distribution; heads 1 to few; involucres 4-7 mm. high, more or less canescent and also glandular, bracts slightly to strongly imbricated; rays about 30-100, 4-15 mm. long and about 1-2.5 mm. wide, white to light rose; achenes 2-nerved, hairy; pappus double, inner of 15-25 long bristles, outer of short rather evident setose squamellae.---Dry open hills and plains often in rocky ground. Saskatchewan to Alaska, south to Nebraska, New Mexico and Arizona. Our records from north-central and northwestern Colorado at 6000-10,000 feet.

25. Erigeron subtrinervis Rydb., Mem. Torr. Club 5:238. 1894.

E. incanescens Rydb.---Perennial plants with more or less branching caudex; stems 15-90 cm. tall, leafy, with spreading hairs; leaves more or less 3-veined, entire, ciliate and hairy at least on the veins; lower leaves oblanceolate, petioled; cauline leaves narrowly lanceolate to ovate, not markedly reduced above; heads 1 to many; involucres 6-9 mm. high, glandular and more or less hirsute, bracts equal or nearly so; rays about 100-150, about 7-18 mm. long and 1 mm. wide, blue or rose-purple; achenes 2- to 4-nerved, hairy; pappus double, inner of 20-30 bristles, outer short and setose. Our plant has been called ssp. typicus Cronq.---Mountains. The subspecies from South Dakota to Wyoming, south to Nebraska and Utah. Our plants from the western half of Colorado, but scarce in the extreme western part, at 6000-10,500 feet.

Since this treatment was prepared a related species has been collected in northwestern Colorado. It has been identified as E. uintahensis Cronq. and differs in having short glandular hairs on the stems and leaves.

26. Erigeron leiomeris A. Gray, Syn. Fl. N. Amer. 1:211. 1884.

Perennial plants with taproot or central axis and branching caudex; stems 5-15 cm. tall, usually sparsely strigose, rarely glabrous; leaves glabrous or nearly so at maturity; basal leaves spatulate to narrowly oblanceolate, clustered, petioled; cauline leaves definitely reduced in size and distribution; heads solitary; involucres 4-6 mm. high, finely glandular, bracts slightly to moderately imbricated; rays about 15-60, 5-11 mm. long and about 1.5-2.5 mm. wide, deep blue or sometimes white; achenes 2-nerved, hairy; pappus double, inner of about 15-25 long bristles, outer of short inconspicuous setae.---Higher mountains often in rocky ground. Wyoming to Idaho, south to New Mexico and Nevada. Our few records from the western one-third of Colorado at about 10,000 feet.

27. Erigeron nematophyllus Rydb., Bull. Torr. Club 32:124. 1905.

Perennial plants with taproot and branched caudex, this chaffy from old leaf bases; stems 4-15 cm. tall, moderately strigose to nearly glabrous; leaves sparsely strigose to nearly glabrous; basal leaves tufted, linear to narrowly linear-oblanceolate, often ciliate near base; cauline leaves few and reduced; heads solitary; involucres 4-6.5 mm. high, moderately strigose to hirsute-strigose the hairs all appressed, bracts more or less imbricated; rays about 15-50, 3-8 mm. long and 1-2.3 mm. wide, white or pinkish; achenes 2-nerved, hairy; pappus single, of 15-25 long bristles but sometimes with a few intermingling short setae.---Open places in the hills and plains. Wyoming and Colorado. Our few records from northcentral and northwestern Colorado at 6500-8000 feet.

28. Erigeron glabellus Nutt., Gen. Pl. 2:147. 1818.

Biennial or perennial plants with simple or slightly branching caudex and a fibrous root system, no taproot present; stems 10-70 cm. tall, erect or curved at base, with appressed or spreading hairs; leaves more or less hirsute or strigose; basal leaves oblanceolate, entire or irregularly toothed, petioled; cauline leaves lanceolate to linear, reduced in size upwards; heads 1-10; involucres 5-9 mm. high, hirsute or strigose but not glandular, bracts subequal or slightly imbricated; rays about 100-175, 7-15 mm. long and about 1 mm. wide, blue or pink, occasionally white; achenes 2-nerved, hairy; pappus double, inner of long bristles, outer of short sometimes scanty setae. Two subspecies are in Colorado.

28A. Erigeron glabellus glabellus

E. asper Nutt.; E. earlei Rydb.; E. glabellus ssp. typicus Cronquist---Hairs of stems appressed or closely ascending; crosswalls of hairs usually not conspicuous if present.---Open ground, often in meadows. Manitoba and Saskatchewan, south to South Dakota, Colorado and Utah. Our records from the western half of Colorado, except the extreme western part, at 5000-8000 feet.

28B. Erigeron glabellus pubescens (Hook.) Cronquist, (ssp.) Bull. Torr. Club 70:273. 1943.

E. consobrinus Greene---Hairs of stems spreading; crosswalls of hairs often very conspicuous.---Open ground. Wisconsin to Alaska, south to Colorado. Our records well scattered over the western half of Colorado, except in the extreme western part, at 5500-10,000 feet.

29. Erigeron ursinus D.C. Eat., Bot. Kings Exp. 148. 1871.

Perennial plants, the caudex with several to many slender almost rhizomatous branches these bearing fibrous roots, no taproot or central axis; stems 5-25 cm. tall, usually curved and strongly reddish-purple at base, strigose or appressed hirsute; leaves glabrous but ciliate, occasionally sparsely strigose; basal leaves oblanceolate, petioled; cauline leaves linear or lanceolate, reduced upwards in size and distribution; heads solitary; involucres 5-7 mm. high, glandular and more or less hirsute or villous, bracts nearly equal; rays 30-100, about 6-15 mm. long and 1-2 mm. wide, blue to pink-purple; achenes 2-nerved, hairy; pappus double, inner of about 10-20 long bristles, outer of short and evident setae or setose squamellae.---Mountain meadows. Montana to Idaho, south to Colorado, Arizona and Nevada. Our records from northcentral, northwestern and central Colorado at 8500-13,000 feet.

30. Erigeron vetensis Rydb., Bull. Torr. Club 32:126. 1905.

E. glandulosus Porter---Perennial plants with stout branching caudex and taproot; stems 5-25 cm. tall, more or less glandular and also sparingly hirsute with spreading hairs; leaves more or less glandular, sometimes sparsely hirsute, some or all the margins ciliate with long, stiff spreading hairs especially near base of blade; basal leaves oblanceolate, often narrowly so; cauline leaves reduced upwards in size and distribution; heads solitary; involucres 4-8 mm. high, glandular and usually sparsely hirsute, bracts subequal or slightly imbricated; rays about 30-90, 5-15 mm. long and 1-1.7 mm. wide, blue, sometimes pink or pink-purple, rarely white; achenes 2-nerved, hairy; pappus double, inner of 18-25 long bristles, outer of short often rather obscure setae.---High mountains often in wooded places. Wyoming, south to New Mexico. Our records mostly in northcentral, central and southern Colorado at 5500-9000 feet.

31. Erigeron engelmanni A. Nels., Bull. Torr. Club 26:247. 1899.

Perennial plants with taproot and branching caudex; stems 5-25 cm. tall, spreading or erect, often decumbent at base, hirsute with appressed or ascending hairs; basal leaves linear-oblanceolate, often narrowly so, appressed hairy, the margins especially near base of blade and on the petiole with long stiff spreading hairs unlike the other hairs of leaf; cauline leaves reduced in size and number; heads 1 to few; involucres 4-8 mm. high, more or less hirsute or hirsute-hispid, usually more or less glandular also, bracts subequal; rays about 30-100, 4-14 mm. long and about 0.7-2.2 mm. wide, usually white, occasionally pink or blue; achenes 2-nerved, hairy; pappus double, inner of about 12-20 long bristles, outer of short setae or narrow squamellae. Our plant has been called ssp. typicus Cronq.---Dry often sandy or rocky ground. Wyoming to Idaho, south to Colorado and Utah. Our records well scattered over Colorado, except in the southeastern part, at 3500-9000 feet.

32. Erigeron eatoni A. Gray, Proc. Am. Acad. 16:91. 1880.

E. microlonchus Greene---Perennial plants with taproot and crown or somewhat branching caudex; stems 5-25 cm. tall, decumbent and usually reddish at base, sparsely to moderately strigose or strigose-hirsute with appressed or ascending hairs; leaves with hairs like the stems; basal leaves tufted and conspicuous, linear to oblanceolate, entire, usually 3-nerved; cauline leaves reduced upwards in size and distribution; heads usually 1, sometimes up to 4; involucres 5-7 mm. high, glandular and more or less hirsute, bracts more or less imbricated; rays about 20-50, about 4-10 mm. long and 1-3 mm. wide, white or occasionally blue or purplish-rose; achenes 2- or sometimes 3-nerved, sparsely hairy; pappus of about 15-25 long bristles with a few very inconspicuous short outer setae. Our plant has been called ssp. typicus Cronq.---Open ground. The subspecies from Wyoming and Idaho, south to Colorado and Arizona. Our records scattered in the western half of Colorado, mostly from the western one-fourth of the state, at 6000-9000 feet.

33. Erigeron ochroleucus Nutt., Trans. Am. Phil. Soc. II. 7:311. 1840.

Saskatchewan and Alberta, south to Nebraska and Wyoming. To be looked for in northern, especially northeastern Colorado.

34. Erigeron brandegei A. Gray, Syn. Fl. 1:210. 1884.

Apparently a rare species, the type from southwestern Colorado, possibly a form of E. pumilus Nutt.

35. Erigeron formosissimus Greene, Bull. Torr. Club 25:121. 1898.

Perennial plants with simple or somewhat branching caudex, root system fibrous, not a taproot; stems 10-40 cm. tall, usually curved at base, more or less glandular at least above and often with long spreading nonglandular hairs, glabrous to hirsute below; lower leaves oblanceolate, spatulate or oval, usually glabrate, rather abruptly narrowed to the petioles; cauline leaves lanceolate to oblong or linear, reduced above, upper with glandular or nonglandular hairs or both; heads 1-6; involucres 5-8 mm. high, glandular or viscid, often hirsute also, bracts about equal; rays 75-150, 7-15 mm. long and about 1 mm. wide, blue rarely pink or white; achenes 2-nerved, hairy; pappus double, inner of about 15-25 long bristles, outer of short often scanty setae. Two varieties are here.

35A. Erigeron formosissimus formosissimus

E. formosissmus var. typicus Cronquist---Upper leaves more or less long-hairy, commonly not glandular; involucres more or less hirsute as well as glandular.---Open ground, often in meadows. Alberta, south to New Mexico and Arizona. Our records mostly in the northwestern quarter of Colorado at 5000-10,500 feet.

35B. Erigeron formosissimus viscidus (Rydb.) Cronquist, (var.) Bull. Torr. Club 70:272. 1943.

E. viscidus Rydb.; E. eximius Greene; E. smithii Rydb.; E. rubricundus Greene---Upper leaves glandular, sometimes with nonglandular hairs also; involucres without nonglandular hairs or these very sparse.---South Dakota and Wyoming, south to Mexico. Our records scattered in the western half of Colorado, except the northwestern part, at 5000-11,000 feet.

36. Erigeron pumilus Nutt., Gen. Am. 2:147. 1818.

Perennial plants with taproot and caudex, this short- or long-branched; stems 5-30 cm. tall, hirsute with spreading hairs, often glandular also; leaves entire, more or less hirsute, the hairs sometimes slightly appressed; basal leaves oblanceolate to narrowly linear-oblanceolate; cauline leaves more or less reduced upwards in size and distribution; heads 1 to several; involucres 4-7 mm. high, more or less hirsute or somewhat glandular, bracts subequal; rays about 50-100, 5-15 mm. long and mostly 0.6-1.5 mm. wide, white, rose or purple; achenes 2-nerved, more or less hairy; pappus double, inner of 7-27 long bristles, outer of rather inconspicuous setae or of evident squamellae. Two subspecies are in Colorado.

36A. Erigeron pumilus pumilus

E. pumilus ssp. typicus Cronquist---Rays usually white; outer pappus of inconspicuous setae.---Dry plains, often among sagebrush. Saskatchewan south to Kansas, Colorado and Arizona. Our records mostly scattered in the eastern half of Colorado at 3500-8000 feet, all east of the Continental Divide.

36B. Erigeron pumilus concinnoides Cronquist, (ssp.) Brittonia 6:181. 1947.

E. concinnus T. & G.; E. condensatus (D.C.Eat.) Greene---Rays often rose to violet, but frequently white in our area; outer pappus of conspicuous squamellae.---Dry plains and hills. Wyoming and Idaho, south to New Mexico and California. Our records scattered in the western one-fourth of Colorado at 4500-7500 feet.

16. Aster L. ASTER

Mostly perennial herbaceous plants but sometimes annual, biennial or shrubby; leaves alternate, entire, toothed or pinnatifid, sometimes bipinnatifid; heads medium to large, solitary to variously disposed, usually corymbose or paniculate; involucres usually hemispheric, campanulate or turbinate, bracts usually graduated in several series, the tips usually herbaceous or foliaceous; receptacles flat or convex, naked or alveolate and fimbriate but no chaffy scales present; rays white, pink, violet or purple, rarely wanting; achenes hairy or glabrous; pappus of equal or subequal capillary bristles. This genus is badly in need of a revision and many of the proposed segregates must be reduced to synonymy. The treatment here, especially in the sections Euaster and Machaeranthera is arbitrary.

1. Leaves, at least the lower, once- or twice-pinnatifid
   2. Green tips of involucre bracts short, not spreading; involucres not over 6 mm. long; rays not over 7 mm. long -1. A. parvulus
   2. Green tips long (as long as whitish part), spreading or squarrose; involucres 8-12 mm. long; rays over 7 mm. long -2. A. tanacetifolius
1. Leaves entire or merely toothed
   3. Rays reduced to a short tube or so small that they slightly if at all exceed the pappus; plants annual; leaves entire
      4. Involucre bracts linear, widest at or below the middle and tapering to an acute apex -3. A. brachyactis
      4. Involucre bracts oblanceolate, oblong-oblanceolate or spatulate-oblong, widest above the middle especially on outer ones, obtuse, abruptly acute or cuspidate at apex -4. A. frondosus
   3. Rays conspicuous, longer than the pappus, usually at least as long as the width of the disk; plants annual, biennial or perennial; leaves entire to toothed
      5. Plants annual or biennial (rarely appearing to be short-lived perennials with taproot); involucre bracts in several series, the tips reflexed (or at least usually spreading); leaves toothed, the teeth tipped with spinulose bristles (rarely entire but with a spinulose bristle at tip)
         6. Leaves densely cinereous-puberulent; stem leaves linear, usually entire
            7. Involucre bracts canescent, obscurely if at all glandular -5. A. canescens
            7. Involucre bracts not canescent but densely glandular especially near the tips -6. A. leucanthemifolius
         6. Leaves glabrous or variously hairy but never cinereous-puberulent; stem leaves usually wider than linear, usually toothed
            8. Stems strigose-puberulent but obscurely if at all glandular below the inflorescence -7. A. rubrotinctus
            8. Stems definitely glandular-villous or glandular-hispidulous
               9. Dwarf plants rarely over 20 cm. tall; plants usually above 10,000 feet in Colorado -8. A. pattersoni

9. Taller plants 20-100 cm. tall; plants usually below 10,000 feet in Colorado　　　-9. A. bigelovii
5. Plants perennial with rootstock or branching caudex; involucre bracts seldom with reflexed or spreading tips (but see choice No. 29); leaves usually entire and apex not spinulose (except in a few species)
　10. Plants with woody branched caudex, or lower part of stem definitely woody when caudex is not definite; plants either small with stems less than 15 cm. tall and leaves not over 1 cm. long, or else larger and with bracts and leaves tipped with callous points or spines (or spinulose-toothed)
　　　11. Heads small, the involucre about 7 mm. high; plants less than 15 cm. tall; leaves entire, not over 1 cm. long, not callous- or spinulose-tipped　　　-10. Aster arenosus
　　　11. Heads larger, the involucres 8 mm. high or more; plants usually over 15 cm. tall; leaves usually over 2 cm. long, entire and callous- or spinulose-tipped or with spinulose teeth
　　　　　12. Achenes densely silky-hairy; involucre bracts all narrow, the outer not passing into the leaves; heads distinctly pedicelled; leaves spinulose-toothed to entire
　　　　　　　13. Leaves spinulose-toothed; stems seldom over 10 cm. tall　　　-11. A. coloradoensis
　　　　　　　13. Leaves entire (spinulose at apex); stems usually over 10 cm. tall
　　　　　　　　　14. Involucres 8-12 mm. high; disk 1.2-2 cm. wide　　　-12. A. xylorrhiza
　　　　　　　　　14. Involucres 13-16 mm. high; disk 2-3 cm. wide　　　-13. A. venustus
　　　　　12. Achenes glabrous; outer involucre bracts broad and foliaceous, passing into the leaves; heads subsessile; leaves spinulose and toothed　　　-14. A. horridus
　10. Plants without woody branching caudex, stems not woody below (rarely caudex appears branching but in such a case the following applies); plants either over 15 cm. tall or leaves over 1 cm. long (or both), leaves not callus or spinulose-tipped or with spinulose teeth (ending in a callus bristle in A. ericoides, A. commutatus and A. falcatus)
　　　15. Involucre bracts dry and chartaceous, with thin or scarious tips (at least the inner ones), strongly keeled on back
　　　　　16. All the involucre bracts acute; stems tall, often over 50 cm. tall; leaves green above
　　　　　　　　　　　　　　　-15. A. engelmanni
　　　　　16. All or at least the outer bracts obtuse at apex; stems 30-50 cm. tall; leaves glaucous or whitish-green above
　　　　　　　17. Inner bracts acute (outer only obtuse); pedicels usually glabrous　　　-16. A. glaucodes
　　　　　　　17. Inner as well as outer bracts obtuse; pedicels often pubescent
　　　　　　　　　　　　　　　-16A. A. glaucodes formosus
　　　15. Involucre bracts with distinct, rather thickened herbaceous tips or green throughout (often foliaceous), usually not keeled on back
　　　　　18. Involucres and usually the peduncles glandular
　　　　　　　19. Stems glabrous; leaves linear to linear-oblanceolate, seldom over 5 mm. wide, over 3 cm. long　　　-17. A. pauciflorus
　　　　　　　19. Stems more or less hairy; leaves various but if narrower than lanceolate or oblanceolate then usually not over 3 cm. long
　　　　　　　　　20. Leaves 3 cm. long or longer, stem ones clasping at base; involucres 8-11 mm. high
　　　　　　　　　　　21. Stems 40-250 cm. tall; lower leaves not at all petioled; rays very numerous, often over 30 present on each head　　　-18. A. novae-angliae
　　　　　　　　　　　21. Stems 20-40 cm. tall; lower leaves tapering to a petiole; rays about 15-25 to a head　　　-19. A. integrifolius
　　　　　　　　　20. Leaves 2-3 cm. long (rarely longer), stem ones not clasping at base; involucres 6-8 mm. high
　　　　　　　　　　　22. Involucre bracts linear or linear-lanceolate, acuminate or attenuate at apex, in 3 rows but all nearly equal in length　　　-20. A. campestris
　　　　　　　　　　　22. Involucre bracts oblanceolate, obtuse to acute at apex, in 3-4 series, the outer definitely shortest
　　　　　　　　　　　　　23. Leaves scabrous-hirsutulous, mostly spreading or reflexed -21. A. kumleini
　　　　　　　　　　　　　23. Leaves glabrous except for bristly-ciliate margins, ascending
　　　　　　　　　　　　　　　　　　　　　-22. A. fendleri
　　　　　18. Involucres and peduncles glabrous to hairy but not at all glandular
　　　　　　　24. Outer bracts (at least some) foliaceous, equalling or surpassing the inner in length
　　　　　　　　　25. Inflorescence a long, narrow, leafy panicle with erect or stiffly ascending branches and numerous heads　　　-23. A. oregonus
　　　　　　　　　25. Inflorescence few-headed or if heads many then inflorescence open and cymose-paniculate with reduced leaves
　　　　　　　　　　　26. Middle stem leaves 1 cm. wide or more, mostly less than 7 times as long as broad
　　　　　　　　　　　　　27. Involucre bracts linear, acute to acuminate, outer foliaceous ones linear or lanceolate and very acute　　　-24A. A. foliaceus frondeus
　　　　　　　　　　　　　27. Involucre bracts oblong to ovate, obtuse to acutish, outer foliaceous ones broadly lanceolate to ovate, apex rounded, obtuse or acutish
　　　　　　　　　　　　　　　　　　　-24B. A. foliaceus canbyi
　　　　　　　　　　　26. Middle stem leaves mostly less than 1 cm. wide and more than 7 times as long as broad
　　　　　　　　　　　　　28. Stems decumbent, 10-20 cm. high; plants above 10,500 feet in Colorado
　　　　　　　　　　　　　　　　　　　-24C. A. foliaceus apricus
　　　　　　　　　　　　　28. Stems erect, 20-30 cm. tall; plants below 10,500 feet in Colorado
　　　　　　　　　　　　　　　　　　　-25. A. occidentalis
　　　　　　　24. Outer bracts neither truly foliaceous nor equalling or surpassing the inner ones
　　　　　　　　　29. Involucre bracts tipped with a callus point, squarrose; rays white or somewhat pinkish
　　　　　　　　　　　30. Stems excurrent with ascending recurved branches; heads numerous, more or less secund; rootstocks caespitose
　　　　　　　　　　　　　31. Involucre bracts narrowly oblanceolate, the inner very acute
　　　　　　　　　　　　　　　　　　　-26. A. ericoides

31. Involucre bracts broadly oblanceolate to spatulate, obtuse   -27A. *A. commutatus polycephalus*
30. Stems branched or sometimes excurrent but then branches not recurved; heads solitary on branches or few, not secund; rootstocks creeping
    32. Involucre bracts definitely imbricated in 3-5 distinct series; stems hispid or hirsute
        -27B. *A. commutatus crassulus*
    32. Involucre bracts nearly equal in length or the outer over half the height of the involucre; stems strigose or strigillose
        33. Involucre bracts slightly imbricated, thick, very squarrose, definitely strigose; leaves densely strigose   -27. *A. commutatus*
        33. Involucre bracts all equal, not at all imbricated, thin, only slightly squarrose, glabrate; leaves sparingly strigose or glabrate   -28. *A. falcatus*
29. Involucre bracts acute to obtuse, not callus-pointed, slightly if at all squarrose; rays white, pink, rose, purple or violet
    34. Dwarf plants less than 20 cm. tall; leaves not over 5 cm. long; involucre bracts definitely hairy on back; rays violet   -29. *A. alpinus*
    34. Plants usually over 20 cm. tall; leaves over 5 cm. long; involucre bracts glabrous on back or sparingly hairy; rays white, rose, violet or purple
        35. Pappus bristles definitely enlarged at apex (use good lens); bracts of involucre more or less keeled on back; rays white to ochroleucous   -30. *A. ptarmicoides*
        35. Pappus bristles not enlarged at apex; bracts of involucre not at all keeled; rays various, from white to purple
            36. Outer involucre bracts oblanceolate, spatulate or obovate-oblong, rounded and obtuse at apex (may be cuspidate); stems uniformly pubescent at least above; rays rose to purple
                -31. *A. adscendens*
            36. Involucre bracts linear, lanc-linear to oblong, acute or acutish (sometimes oblanceolate but then stem pubescent in lines); stems glabrous to pubescent; rays white to purple
                37. Plant glabrous throughout (except sometimes leaf edges and involucre bracts ciliate)
                    38. Involucres 4-5 mm. high; rays white; bracts green nearly throughout
                        -32. *A. porteri*
                    38. Involucres about 8-9 mm. high; rays blue; bracts whitish below with a rhombic green spot near apex
                      39. Leaves abruptly reduced in size on the branches   -33. *A. laevis*
                      39. Leaves of branches only gradually reduced from those of stems
                          -33A. *A. laevis geyeri*
                37. Plants hairy on the stem, at least above and in the inflorescence
                    40. Heads solitary or few in a flat-topped inflorescence, rarely many and corymbose
                      41. Stems pubescent above in definite longitudinal lines; leaves narrowly linear; rays usually white   -34. *A. junciformis*
                      41. Stems uniformly pubescent above (may be in lines below); leaves linear to oblong; rays blue, violet or rose-purple   -25. *A. occidentalis*
                    40. Heads numerous in panicles, not especially flat-topped
                      42. Veinlets of leaves conspicuous forming little netlets (areolae) about as long as wide   -35. *A. praealtus*
                      42. Veinlets of leaves inconspicuous or if forming definite areolae these definitely longer than wide
                          43. Outer bracts of involucre usually not definitely wider than the inner, the green tips not enlarged, less than 1/5 as wide as the length of the bract; rays mostly blue   -36A. *A. hesperius hesperius*
                          43. Outer bracts usually broader than the inner, with enlarged green tips, these at least 1/5 as wide as the length of the bract; rays mostly white or pink   -36B. *A. hesperius laetivirens*

1. Aster parvulus Blake in Tide., Contrib. U. S. Herb. 25:563. 1925.
Machaeranthera parviflora A. Gray---Plants annual; stems 15-50 cm. tall, glabrous or short and sparsely glandular; leaves once-pinnatifid with linear segments, sparingly if at all glandular; heads 1 to few on the branches; involucres 4-6 mm. high, hemispheric, bracts lanceolate, acute, the green tips shorter than the white base and not or scarcely spreading; rays 5-6 mm. long, purplish or possibly whitish; achenes terete, short-hairy; pappus bristles about equal.--- Plains, hills and mesas. Utah, south to New Mexico and Arizona. Our few records from western and southern Colorado at 5500-7500 feet.

2. Aster tanacetifolius H.B.K., Nov. Gen. et Sp. 4:95. 1820.
Machaeranthera tanacetifolia (H.B.K.) Nees.; M. coronopifolia (Nutt.) A. Nels.---Plants annual; stems 10-50 cm. tall, often branched from base, glandular-puberulent and more or less villous; leaves 1-4 cm. long, once- or twice-pinnatifid, the segments linear or oblong-oblanceolate, ending in callus bristles; heads few, solitary on branches but collectively corymbose or paniculate; involucres 8-12 mm. high, hemispheric, bracts imbricated in several series, narrowly linear, chartaceous-whitish at base, the green tips linear-subulate, usually spreading, as long or longer than the whitish base, ending in a bristle; rays usually over 1 cm. long, blue-purple to purple; achenes appressed hairy; pappus of numerous rather firm bristles. ---Plains and hills, often on sandy ground. South Dakota to Alberta, south to Texas, Mexico and Arizona. Our records scattered in Colorado at 4000-9500 feet.

3. Aster brachyactis Blake in Tide., Contrib. U. S. Nat. Herb. 25:564. 1925.
A. angustus T. & G.; Brachyactis angustus (T. & G.) Britt.---Plants annual; stems 20-60 cm. tall, branching, leafy, glabrous or sparingly puberulent above; leaves 3-10 cm. long, linear or narrowly linear, entire, acute, more or less ciliate; heads numerous, paniculate; involucres 5-8 mm. high, campanulate, bracts in 2-3 series but almost equal in length, linear or linear-oblong, widest at or below the middle, tapering to an acute and somewhat callus tip, with a

midrib, glabrous or slightly ciliate; rays lacking or possibly sometimes present but shorter than the pappus; achenes appressed pubescent; pappus copious, white.---Wet, often saline ground. Wisconsin to Alberta, south to Missouri and Utah. Our records scattered in the western two-thirds of Colorado, except in the extreme southern part, at 5000-8500 feet.

4. Aster frondosus (Nutt.) T. & G., Fl. N. Amer. 2:165. 1841.
Brachyactis frondosa (Nutt.) A. Gray---Plants annual; stems 20-50 cm. tall, leafy, glabrous or nearly so; leaves 3-10 cm. long, linear or lanc-linear, entire, acute or obtuse, glabrous but ciliate; heads numerous, paniculate on the branches; involucres 5-8 mm. long, campanulate, bracts in 2-3 series, almost equal in length, oblanceolate to oblong-oblanceolate, widest above the middle and apex obtuse, abruptly acute to cuspidate, with a distinct midrib, glabrous or slightly ciliate; rays lacking or very small and shorter than the pappus; achenes appressed pubescent; pappus copious, white.---Moist, often saline soil. Wyoming to Oregon, south to New Mexico and California. Our few records from the northern half of Colorado at 4500-5500 feet.

5. Aster canescens Pursh, Fl. Amer. Sept. 547. 1814.
Machaeranthera canescens (Pursh) A. Gray---Plants biennial from a stout taproot, possibly annual or short-lived perennial; stems 10-40 cm. tall, simple or branched, glabrate or more or less cinereous-canescent; heads rather small, usually numerous and corymbose; involucres 6-9 mm. high, hemispheric, bracts imbricated in several rows, lower part whitish-chartaceous, upper part green and attenuated to a salient point, ascending or somewhat spreading to reflexed, canescent but not glandular; rays about 7-10 mm. long, deep blue to purple; achenes appressed-hairy, often very sparsely so; pappus bristles numerous and rather firm.---Open, often sandy ground. Saskatchewan to British Columbia, south to New Mexico and California. Our records scattered in the western two-thirds of Colorado at 5000-8500 feet.

6. Aster leucanthemifolius Greene, Erythea 3:119. 1895.
Machaeranthera pulverulenta (Nutt.) Greene---Plants biennial with stout taproot, possibly annual or short-lived perennial; stems 20-50 cm. tall, simple or branching throughout, glabrate or more or less canescent; leaves 2-7 cm. long, linear or linear-lanceolate, entire or sometimes sparingly toothed, cinereous-canescent; heads rather small, usually numerous and more or less corymbose; involucres about 6-8 mm. high, hemispheric, bracts imbricated in several series, basal part whitish-chartaceous, upper part green and attenuated at apex more or less spreading or reflexed, densely glandular especially near tip; rays 7-10 mm. long, purple; achenes sparingly appressed hairy; pappus bristles numerous, rather firm.---Plains, hills and mountains. Montana to Idaho, south to Colorado and Nevada. Our records scattered in the northern half of Colorado at 5000-8000 feet.

7. Aster rubrotinctus Blake in Tide., Contrib. U. S. Nat. Herb. 25:564. 1925.
Machaeranthera rubricaulis Rydb.---Plants biennial with taproot, or possibly annual or short-lived perennial; stems 15-60 cm. tall, simple or branched, often tinged with red, glabrous to strigose or puberulent but very sparingly if at all glandular; leaves 3-8 cm. long, oblanceolate or spatulate to linear-oblanceolate, usually more or less saliently dentate but sometimes linear and entire, puberulent to glabrous but never cinereous; heads in corymbs or panicles; involucres about 7-12 mm. high, hemispheric, bracts imbricated in several series, narrow, lower part whitish-chartaceous, upper green and narrowing at apex, usually more or less reflexed; rays about 7-12 mm. long, blue to violet or purple; achenes usually appressed hairy, mostly sparingly so; pappus bristles numerous and rather firm. The following related species, Machaeranthera glabella Greene, M. ramosa A. Nels., M. selbyi Rydb. and M. spectabilis Greene are supposed to differ in having the green tips of the involucre bracts much shorter than the whitish basal part. The writer cannot make any such distinction in our plants.---Mesas, hills and plains. Colorado, Utah, New Mexico and Arizona. Our records well scattered over the western two-thirds of Colorado at 5000-11,000 feet.

8. Aster pattersoni A. Gray, Proc. Amer. Acad. 8:272. 1878.
Machaeranthera pattersoni (A. Gray) Greene---Plants biennial with taproot, possibly short-lived perennials; stems 10-20 cm. tall, 1 to several, glandular and usually villous; leaves 4-6 cm. long, spatulate to oblanceolate, usually saliently toothed at least near apex, glabrate but hispid-ciliate near base, not cinereous; basal leaves numerous and petioled; heads solitary or few; involucres 8-11 mm. high, hemispheric, bracts imbricated in several series, base whitish-chartaceous, upper part green, attenuated at apex and reflexed; rays 8-15 mm. long, purple or violet; achenes sparingly appressed hairy; pappus bristles numerous and rather firm.---Mountains. Apparently limited to Colorado, our few records probably from the central or northcentral part at 10,000-14,000 feet but data on label inadequate.

9. Aster bigelovii A. Gray, U. S. Rept. Expl. Miss. Pacif. 4:97. 1857.
Machaeranthera bigelovii (A. Gray) Greene; M. aspera Greene---Plants annual or biennial, with taproot; stems 20-100 cm. tall, usually branching, definitely glandular-puberulent to glandular-hispid, at least above and usually throughout; leaves about 5-10 cm. long, oblong to lanceolate or oblanceolate, rarely linear above, usually saliently dentate often runcinate-dentate, scabrous, glandular-pruinose to glabrous, not at all cinereous; heads usually large and showy, corymbose or paniculate; involucres about 10-15 mm. high, rarely shorter, hemispheric, bracts imbricated in several series, narrow, lower part whitish-chartaceous, the upper part green, narrowing at apex, usually reflexed or spreading; rays 8-15 mm. long, purple or violet; achenes glabrous to sparingly hairy; pappus bristles many, rather firm. The following related species, Machaeranthera fremontii Rydb. and M. cichoriacea Greene are supposed to differ in having the green tips of the involucre bracts much shorter than the basal portion. The writer cannot make any such distinction in our plants.---Plains, hills and mountains. Colorado, New Mexico and Arizona. Our records well scattered over the western half of Colorado at 6500-10,000 feet.

10. Aster arenosus Blake, Journ. Wash. Acad. Sci. 30:471. 1940.
A. ericaefolius Roth.; Leucelene arenosa Heller; L. serotina (Greene) Rydb.; L. hirtella (A. Gray) Rydb.; L. alsinoides Greene; L. ericoides (Torr.) Greene---Plants perennial with

woody branching caudex and horizontal or oblique creeping rootstocks; stems 6-12 cm. tall, numerous in tufts, sparsely to densely strigose and often more or less glandular; leaves rarely over 1 cm. long, linear to linear-oblanceolate or lower spatulate, entire, hispid-ciliate, reduced in size and distribution near heads; heads solitary on slender branches; involucres 5-7 mm. high, turbinate or campanulate, bracts in about 3-7 series, imbricated, lanceolate or oblong-lanceolate, with midrib, green with margins scarious, more or less appressed puberulent and ciliate especially near apex; rays short, white; disk flowers yellow when fresh, losing this color when dried and reported in the manuals as white; achenes appressed hairy at least near apex; pappus copious, white. The preceding applies to the normal spring form. Later in the season the secondary branches may elongate and flower, the whole plant having a different aspect.---Dry plains and hills often in sandy or rocky ground. Wyoming to California, south to Texas and Mexico. Our records well scattered over Colorado at 4000-8000 feet.

11. Aster coloradoensis A. Gray, Proc. Amer. Acad. 11:76. 1876.

Xylorrhiza coloradensis (A. Gray) Rydb.; X. brandegei Rydb.---Perennial plants with thick woody taproot and short but definite branched caudex; stems 4-10 cm. tall, canescent; leaves 1-4 cm. long spatulate-oblanceolate to narrowly oblong, coarsely dentate the teeth spinulose, densely canescent; heads large, solitary on short peduncles; involucres 7-9 mm. high, hemispheric, bracts imbricated in 2-3 series, acuminate to a salient point, more or less keeled especially at base; rays 8-10 mm. long, violet-purple to rose-violet; achenes short, densely hairy; pappus bristles rather stiff, tan or fulvous.---Mountains. Apparently limited to Colorado, our records from the central, westcentral and southwestern parts at 9000-11,000 feet.

12. Aster xylorrhiza T. & G., Fl. N. Amer. 2:158. 1841.

Xylorrhiza glabriuscula Nutt.; X. villosa Nutt.---Perennial plants with thick woody taproot and woody branching caudex; stems 10-20 cm. tall, clustered, more or less short-villous; leaves 2-5 cm. long and 2-6 mm. wide, linear-oblanceolate to linear, entire, tipped with a callus point, more or less villous; involucres 8-12 mm. high, hemispheric, bracts in 2-3 series but nearly equal in length, spinulose-acuminate, short-villous on back and margins; rays 8-12 mm. long, white, pinkish or possibly rose; achenes appressed silky-hairy; pappus bristles rather stiff, tan or fulvous. Some of our plants have the outer bracts shorter than the inner, although similar in shape, and resemble A. parryi Gray, a species usually not credited to Colorado.---Rocky or clayey open ground. Wyoming to Nevada, south to Colorado. Our records from westcentral and northwestern Colorado at 5000-6500 feet.

13. Aster venustus M. E. Jones, Zoe 2:247. 1891.

Xylorrhiza venusta (M. E. Jones) Heller---Perennial plants with thick woody taproot and branching caudex; stems 20-50 cm. tall, clustered, more or less villous; leaves 3-6 cm. long and usually about 5-10 mm. wide, reduced above, spatulate or oblanceolate, cuspidate to a callus point, entire, villous; heads solitary on long almost naked peduncles; involucres 13-16 mm. high, hemispheric, bracts nearly equal in length but in 2-3 series, spinulose-acuminate, short-villous on back and margins; rays 15-25 mm. long, white to light rose or light purple; achenes appressed silky; pappus bristles rather stiff, tan to fulvous.---Dry plains and hills. Colorado and Utah. Our records from western Colorado at 4500-7000 feet.

14. Aster horridus (Woot. & Standl.) Blake, Journ. Wash. Acad. Sc. 27:379. 1937.

Herrickia horrida Woot. & Standl.---New Mexico and possibly southern Colorado.

15. Aster engelmanni (D.C. Eaton) A. Gray, Syn. Fl. N. Amer. 1:199. 1884.

Eucephalus engelmanni (D.C. Eat.) Greene---Perennial plants with rootstocks; stems 50-150 cm. tall, leafy, glabrous or sparsely puberulent; leaves 5-10 cm. long, ovate-oblong to broadly lanceolate, entire or toothed, rather thin and reticulate-veiny, green above and lighter below; heads few to many, corymbose; involucres 8-12 mm. high, campanulate or hemispheric, bracts imbricated in several series, acute or acuminate, keeled on back, whitish-chartaceous at base, thin and scarious at margins and apex, often purplish tinged or some of outer green near apex or sometimes throughout, glabrous to pubescent, more or less ciliate; rays conspicuous, usually about 1-2 cm. long, white or possibly violet; achenes hairy or glabrate; pappus copious, white or whitish.---Mountains. Alberta to British Columbia, south to Colorado and Nevada. Our records from western and northern Colorado at 8000-10,500 feet.

16. Aster glaucodes Blake, Proc. Biol. Soc. Wash. 35:174. 1922.

A. glaucus T. & G.; Eucephalus glaucus Nutt.---Perennial plants with rootstocks; stems 30-50 cm. tall, branched, glabrous, pale green or glaucous; leaves 3-7 cm. long, lanceolate, oblong or sometimes linear, rather firm, reticulate-veiny, glabrous but minutely ciliate, pale green or glaucous; heads few to many, corymbose at the ends of branches or sometimes paniculate; involucres 5-7 mm. high, campanulate or hemispheric, bracts imbricated in several series, keeled, chartaceous-whitish near base, thin at margins and apex, outer obtuse, inner acute, often purplish-tinged at apex, glabrous or minutely ciliate; rays about 7-12 mm. long, white to violet; achenes hairy or glabrate; pappus copious, white or whitish.---Mountains. Wyoming to Utah, south to Colorado and Arizona. Our records well scattered in the western half of Colorado at 6500-9500 feet.

16A. Aster glaucodes formosus (Greene) Kittell, (var.) Fl. Ariz. and N. Mex. 404. 1941.

A. glaucous formosus A. Nels.; Eucephalus formosus Greene---Pedicels often hairy; involucre bracts all obtuse.---Reported from southern Colorado.

17. Aster pauciflorus Nutt., Gen. Pl. 2:154. 1818.

Perennial plants; stems 15-50 cm. tall, in clumps, branching above, glabrous below and glandular only near the heads; leaves 3-10 cm. long, linear to linear-oblanceolate or basal as wide as oblanceolate and petioled, entire, glabrous and rather fleshy; heads solitary on the branches, these rather corymbose; involucres 5-8 mm. high, hemispheric, bracts in 2 or 3 series, linear-lanceolate, acute, glandular; rays about 15-25, about 4-6 mm. long, blue to whitish; achenes glabrous or sparsely hairy on the angles; pappus bristles many and equal, white-tawny.---Usually in saline soil. Saskatchewan, south to Texas, Mexico and Arizona. Our few records from the western half of Colorado at 6000-8000 feet.

18. **Aster novae-angliae** L., Sp. Pl. 875. 1753.
    Perennial plants with rootstocks; stems 40-150 cm. tall, rather stout and strict, very leafy, branched above, hirsute and glandular; leaves 3-12 cm. long, lanceolate or oblong, entire, firm, all sessile and half clasping by an auriculate or broadly cordate base; inflorescence leafy and corymbose; involucres 8-11 mm. high, hemispheric or campanulate, bracts more or less imbricated, whitish near base with attenuated usually squarrose tip, glandular; rays many, about 9-12 mm. long, rose to violet-purple; achenes appressed silky-hairy; pappus bristles numerous, tan to fulvous.---Low ground. Quebec to Alberta, south to South Carolina and New Mexico. Our few records from northcentral and central Colorado at 5000-6000 feet.
19. **Aster integrifolius** Nutt., Trans. Amer. Phil. Soc. n.s.7:291. 1840.
    Reported from Montana to Washington, south to Colorado and California. The writer has not seen material from this state.
20. **Aster campestris** Nutt., Trans. Am. Phil. Soc. 7:293. 1841.
    Perennial plants with decumbent or creeping rootstocks; stems 20-40 cm. tall, simple or branched above, often brownish or purplish, glabrate to glandular especially above; leaves 2-3 cm. long, oblong or linear-oblong, entire, somewhat 3-nerved, glandular-puberulent to glabrate, sessile but not clasping the stem; heads solitary or racemose on the branches; involucres 6-8 mm. high, hemispheric, bracts in 3 rows but nearly equal in length, linear or linear-lanceolate, acuminate or attenuate at tip, glandular-puberulent; rays about 6-10 mm. long, dark blue, purplish or light violet; achenes appressed hairy; pappus bristles many, sordid.---Plains, valleys and hills. Alberta to British Columbia, south to Colorado. Our few records from central and northcentral Colorado at 8500-9500 feet.
21. **Aster kumleini** Fries in Rydb., Fl. Colo. 354. 1906.
    A. oblongifolius var. _rigidulus_ A. Gray---Perennial plants; stems 20-50 cm. tall, branching above, often straw-colored, glandular-scabrous or glandular-puberulent; leaves 2-3 cm. long, oblong to oblong-lanceolate, mostly spreading or reflexed, scabro-pubescent, all sessile; heads 1 to several on the branches; involucres 6-7 mm. high, hemispheric, bracts oblanceolate, obtuse or acute in 3-4 series, the outer definitely shorter and more or less squarrose, glandular-puberulent; rays 6-8 mm. long, bluish-violet; achenes more or less hairy; pappus tawny or whitish-tawny.---Dry plains and hills. Wisconsin to Colorado, south to Illinois and Texas. Our few records from southcentral Colorado at about 7000 feet.
22. **Aster fendleri** A. Gray, Mem. Amer. Acad. n.s. 4:66. 1849.
    Perennial plants with thick rather woody rootstocks; stems 10-30 cm. tall, rather rigid, glandular and sparsely hispidulous; leaves 1.5-3 cm. long, rarely longer, linear, entire, firm, glabrous on both sides but hispid-ciliate; heads usually few and racemose; involucres 5-8 mm. high, campanulate or turbinate, bracts imbricated in several series, oblanceolate, obtuse especially the outer to acute, whitish at base the green apex glandular; rays 6-10 mm. long, violet; achenes appressed hairy; pappus bristles numerous, tan to whitish.---Plains and hills, often in sandy soil. Kansas to Colorado, south to Oklahoma and New Mexico. Our one record from northcentral Colorado at 5200 feet.
23. **Aster oregonus** (Nutt.) T. & G., Fl. N. Amer. 2:163. 1841.
    A. fulcratus Greene; _A. eatonii_ (A. Gray); probably _A. lonchophyllus_ Greene---Perennial plants with rootstocks; stems 30-100 cm. tall, usually reddish-tinged, pubescent at least above; leaves 5-15 cm. long and 4-10 mm. wide, usually over 7 times longer than wide, usually entire, glabrous or scabrous on margins, middle leaves linear to lanceolate; heads many in a rather long narrow leafy panicle with erect or stiffly ascending branches; involucres 5-9 mm. high, hemispheric to campanulate, bracts loose, the outer about as long as the inner, at least some foliaceous and completely green, obtuse to acute; rays 5-12 mm. long, white to rose; achenes more or less pubescent; pappus bristles many, white or whitish-tawny.---Moist ground. Alberta to British Columbia, south to New Mexico and California. Our few records from northcentral Colorado at 5000-8500 feet, but more widely distributed in the state according to report.
24. **Aster foliaceus** Lindl. in DC., Prodr. 5:228. 1836.
    Perennial plants with rootstocks; stems 5-100 cm. tall, usually several together, commonly reddish, nearly glabrous to pubescent the hairs often in lines; lower leaves obovate to oblanceolate, usually petioled; stem leaves about 5-12 cm. long, lanceolate to oblong-ovate, mostly entire, glabrous to pubescent, upper leaves often grading into the involucre bracts; heads 1 to few, medium to large; involucres 7-12 mm. high, about as long as the disk, hemispheric, bracts loose or appressed, outer foliaceous and completely green, usually about as long as the inner; rays about 1-1.5 cm. long, rose to purple; achenes usually pubescent; pappus bristles many, white or tawny. The 3 varieties intergrade somewhat in this state.
24A. **Aster foliaceus frondeus** A. Gray, (var.) Syn. Fl. N. Amer. 1:193. 1884.
    A. ciliomarginatus Rydb.; _A. frondeus_ (A. Gray) Greene---Stems 30-100 cm. tall; leaves commonly over 10 cm. long and well over 1 cm. wide, usually less than 7 times longer than wide, rather thin, glabrous or the petioles ciliate; outer involucre bracts linear or lanceolate and very acute, inner linear.---Moist ground, often along streams. Wyoming to Washington, south to New Mexico and California. Our records scattered in the western half of Colorado at 6000-9500 feet.
24B. **Aster foliaceus canbyi** A. Gray, (var.) Syn. Fl. N. Amer. 1:193. 1884.
    A. canbyi Vasey; _A. burkei_ (A. Gray) Howell; _A. tweedyi_ Rydb.; _A. phyllodes_ Rydb.---Stems 25-100 cm. tall, glabrous to pubescent especially above; leaves often over 10 cm. long and well over 1 cm. wide, usually less than 7 times as long as wide, broadly oblanceolate to elliptic, rather thick; outer involucre bracts widest above the middle, oblong-obovate, obtuse to abruptly acute or cuspidate, inner oblong to ovate or acute.---Moist woods and meadows. Wyoming to Washington, south to New Mexico and California. Our records from northcentral, central and western Colorado at 5500-9500 feet.

24C. **Aster foliaceus apricus** A. Gray, (var.) Syn. Fl. N. Amer. 1:193. 1884.

A. apricus (A. Gray) Rydb.---Stems 5-20 cm. tall, decumbent or ascending, 1 to several, pubescent at least above; middle leaves about 3-8 cm. long and less than 1 cm. wide, usually over 7 times longer than wide, oblong or oblanceolate, glabrous or ciliate; outer involucre bracts broadly linear, obtuse or acute, inner linear, acute or acuminate, often purplish at tip and margins.---High mountains. Wyoming to British Columbia, south to Colorado and California. Our records from northcentral, westcentral and central Colorado at 11,000-12,000 feet.

25. **Aster occidentalis** (Nutt.) T. & G., Fl. N. Amer. 2:164. 1841.

A. fremontii (T. & G.) A. Gray; probably A. corymbiformis Rydb.---Perennial plants from rather slender rootstocks; stems 15-50 cm. tall, usually 2 or 3 together, usually reddish or brownish, not especially leafy above, pubescent at least above, the hairs uniform around the stem; leaves 4-10 cm. long, linear, lanceolate or oblong or lower oblanceolate, usually entire, usually glabrous; heads 1 to few in a nearly naked cyme or cymose panicle; involucres 6-9 mm. high, hemispheric, bracts little imbricated, the outer about as long as the inner and often wholly green, linear to oblong, acute or occasionally obtuse, appressed to loose, usually glabrous on back and often ciliate; rays 6-15 mm. long, blue to violet or sometimes rose-purple; achenes hairy; pappus bristles many, white to tawny. Our plant has been called var. typicus Cronq.---Meadows, banks and hillsides. Alberta and British Columbia, south to Colorado and California. Our records scattered in the mountains of Colorado at 8500-9500 feet.

26. **Aster ericoides** L., Sp. Pl. 2:875. 1753.

A. hebecladus DC.; A. multiflorus Ait.; A. stricticaulis (T. & G.) Rydb.---Perennial plants with short rootstocks; stems about 20-60 cm. tall, simple below, branched above with ascending-recurved branches, scabrous to rough pubescent; leaves 1-5 cm. long, linear or linear-oblong, obtuse to acute with a callus bristle at apex, entire, rigid, hirsute-strigose to glabrate; heads numerous and secund on the recurved branches, nearly sessile; involucres 3-5 mm. high, turbinate to hemispheric, bracts imbricated in about 3 series, ciliate and more or less strigose, outer bracts narrowly oblanceolate, obtuse and cuspidate with a sharp callus point, inner acute; rays about 3-4 mm. long, white or rarely pinkish; achenes puberulent; pappus bristles many, tawny.---Plains, valleys and meadows. Maine to Montana, south to Georgia and Mexico. Our records from northcentral, central and southcentral Colorado at 5000-7500 feet.

27. **Aster commutatus** (T. & G.) A. Gray, Syn. Fl. N. Amer. 1:185. 1884.

Perennial plants with rather long rootstocks; stems about 20-60 cm. tall, branching throughout or at least above the middle, the branches spreading and not especially recurved, strigose; leaves 1-4 cm. long, linear, entire, apex with a callus point, densely strigose; heads 1 to few on the branches, not especially secund; involucres 5-8 mm. high, turbinate to hemispheric, bracts only slightly imbricated, thick, outer ones oblanceolate or spatulate, rounded and cuspidate at tip with a callus point, strigose, inner often acute; rays 4-5 mm. long, white; achenes appressed pubescent usually rather sparsely so; pappus bristles numerous, tawny. This species has been united with A. falcatus Lindl. (Bull. Torr. Club 74:144. 1947) but in Colorado the two seem reasonably distinct and are kept separate here.---Plains, hills and banks. Minnesota to British Columbia, south to Kansas, New Mexico and Arizona. Our few records from northcentral and central Colorado at 5000-7500 feet.

27A. **Aster commutatus polycephalus** (Rydb.) Blake, (var.) Contrib. U. S. Nat. Herb. 25:560. 1925.

A. polycephalus Rydb.---Stems few branched below, the branches often recurved; heads often secund on the branches; involucre bracts broadly oblanceolate to spatulate.---Alberta, south to Texas and Arizona. Our few records from northcentral Colorado at 6000-7000 feet.

27B. **Aster commutatus crassulus** (Rydb.) Blake, (var.) Contrib. U. S. Nat. Herb. 25:560. 1925.

A. crassulus Rydb.; A. exiguus (Fernald) Rydb.---Stems hispid or hirsute; involucre bracts imbricated rather definitely in 3-5 series.---Saskatchewan, south to South Dakota, Colorado and California; perhaps farther east. Our records scattered in the western half of Colorado at 5000-7000 feet.

28. **Aster falcatus** Lindl. in DC., Prodr. 5:241. 1806.

A. cordineri A. Nels.---Perennial plants from rather long rootstocks; stems about 20-40 cm. tall, branching from the base or above, the branches not especially recurved, glabrate or sparingly strigose; leaves 2-5 cm. long, linear, entire, ending in a callus point, crowded, glabrate to sparingly strigose; heads 1 to few on the branches, not secund; involucres 4-6 mm. high, hemispheric to turbinate, bracts oblanceolate, acute to a callus point but tips little if any reflexed, rather thin, glabrate to sparingly strigose; rays 3-5 mm. long, white to pinkish; achenes appressed hairy; pappus bristles many, whitish or tawny.---Plains, hills and valleys. Arctic America south to Colorado. Our records from northcentral, central and southcentral Colorado at 5000-7500 feet.

29. **Aster alpinus** L., Sp. Pl. 872. 1753.

A. culminis A. Nels.---Canada south to Colorado; Eurasia. No good specimens seen from the state.

30. **Aster ptarmicoides** (Nees.) T. & G., Fl. N. Amer. 2:160. 1841.

Unamia alba (Nutt.) Rydb.---Perennial plants with short rootstocks; stems 20-50 cm. tall, often tufted, simple to the inflorescence, glabrous, scabrous or hispidulous; leaves 5-15 cm. long, linear or linear-oblanceolate, 3- to 5-ribbed, firm in texture, rough margined or ciliate, sessile or the lowest petioled; heads few to several, rarely many, corymbose at the end of the stem; involucres 4-6 mm. high, campanulate to hemispheric, bracts oblong-lanceolate to oblong-linear, imbricated in 2-4 series, keeled on back, at least the outer green throughout, glabrous or nearly so; rays about 5-7 mm. long, white to ochroleucous; achenes glabrous; pappus bristles rather rigid and dilated at apex, white.---Banks and bluffs, often on rocky ground. New England to Saskatchewan, south to Colorado. Our records from northcentral and central Colorado at 6000-7500 feet.

31. **Aster adscendens** Lindl. in DC., Prodr. 5:231. 1836.
A. nuttallii T. & G.; A. violaceus Greene; A. nelsonii Greene; A. underwoodii Rydb.; A. griseus Greene; A. subgriseus Rydb.; A. griseolus Rydb.; probably A. vallicola Greene and A. armeriaefolius Greene---Perennial plants with short to long rootstocks; stems 10-80 cm. tall, sparsely to densely pubescent, if densely so the hairs usually uniform and appressed, or rarely glabrous; cauline leaves 2-12 cm. long, linear to lanceolate, entire, glabrate to densely pubescent; lower leaves often elliptic to oblanceolate, petioled; heads few to many in racemes, panicles or cymes, occasionally solitary; involucres 5-8 mm. high, hemispheric, bracts definitely imbricated in 3-4 series, the outer spatulate-oblanceolate or obovate-oblong, narrowed to the whitish base, apex rounded and obtuse or cuspidate, greenish, inner bracts narrower, acute or acuminate, glabrous or ciliate, rarely sparsely pubescent on back, rays 6-10 mm. long, blue, violet or pink, rarely white; achenes more or less hairy; pappus bristles many, tawny to white. This species is exceedingly diverse in general aspect but each possible form seems to be connected by numerous intergrades. Recently this species has been made a subspecies under A. chilensis Nees. (Am. Midl. Nat. 29:429-468. 1943).---- Plains, marshes, banks, hills and woodlands. Saskatchewan to Washington, south to New Mexico and California. Our records scattered in the western half of Colorado at 5000-9500 feet.

32. **Aster porteri** A. Gray, Proc. Amer. Acad. 16:89. 1881.
Perennial plants with rootstocks; stems 20-40 cm. tall, simple or branched above, glabrous; leaves 5-10 cm. long, the middle ones usually not over 5 mm. wide, linear or lower linear-oblanceolate, glabrous or sometimes hirsute-ciliate; heads many, in corymbs or panicles; involucres 4-5 mm. high, hemispheric, bracts only slightly imbricated, rigid, linear-subulate, with green nearly erect acute tips, green midrib and whitish-chartaceous base; rays 4-8 mm. long, white; achenes minutely pubescent; pappus bristles many, whitish-tawny.---Mountains. Colorado and New Mexico. Our records from northcentral and central Colorado at 5500-9500 feet.

33. **Aster laevis** L., Sp. Pl. 876. 1753.
Perennial plants with rootstocks; stems 30-120 cm. tall, rather stout, glabrous and more or less glaucous; leaves about 5-20 cm. long, the leaves on the branches conspicuously reduced in size, ovate to oblong or lanceolate, usually somewhat serrate, glabrous or somewhat ciliate, lower short-petioled, upper sessile with auriculate or partly clasping base; heads usually many, in panicles; involucres 8-9 mm. high, hemispheric to campanulate, bracts imbricated in several series, acute, whitish-chartaceous at base, tips green and rhombic, appressed, glabrous or ciliate; rays about 8-10 mm. long, blue or violet; achenes glabrous or nearly so; pappus bristles many, tawny.---Dry or rather moist ground. New England to Saskatchewan, south to Louisiana and New Mexico. Our records from northcentral, central and southcentral Colorado at 5000-7000 feet.

33A. **Aster laevis geyeri** A. Gray, (var.) Syn. Fl. N. Amer. 1:183. 1884.
A. geyeri (Gray) Howell; A. subsalignus Rydb.---Leaves only gradually reduced from the stems to the branches.---South Dakota to Alberta, south to Colorado. Our records in northcentral, central and southern Colorado at 5000-9500 feet.

34. **Aster junciformis** Rydb., Bull. Torr. Club 37:142. 1910.
Probably A. junceus of Manuals---Perennial plants with slender rootstocks; stems 15-50 cm. tall, simple below, hairy in longitudinal lines or glabrous below; leaves 4-8 cm. long, narrowly linear, entire, glabrous or scabro-ciliate, sessile; heads 1 to few, corymbose; involucres 5.5-8 mm. long, bracts oblong to linear and acute or outer oblanceolate and obtuse, glabrous on back, ciliate; rays 6-8 mm. long, white or rarely light rose or lilac; achenes nearly glabrous to sparsely hairy; pappus bristles many, sordid-white.---Meadows and swamps. Ontario to British Columbia, south to New York, Wisconsin, Colorado and Washington. Our few records from northcentral Colorado at 7500-8500 feet but also reported from the central and southcentral parts of the state.

35. **Aster praealtus** Poir., Encyc. Suppl. 1:493. 1810.
A. salicifolius and A. paniculatus of Manuals---Perennial plants with rootstocks; stems 40-100 cm. tall, densely paniculately branched above, glabrous below, pubescent in longitudinal lines above; leaves about 6-13 cm. long, lanceolate or elliptic-lanceolate, entire, thick and firm, veinlets conspicuously reticulated the nets about as wide as long, glossy, glabrous or scabro-puberulent; heads forming a broad panicle; involucres 5-7 mm. high, hemispheric or campanulate, bracts in 3-4 series, oblong-lanceolate to linear-lanceolate, acute, firm, green near apex, pale at margins and base; rays 6-15 mm. long, bluish-purple; achenes more or less hairy; pappus bristles many, tawny or whitish.---Often in low ground. Michigan to British Columbia, south to Kentucky, Arkansas and Arizona. Our one record from northcentral Colorado at 6000 feet.

36. **Aster hesperius** A. Gray, Syn. Fl. N. Amer. 1(2):192. 1884.
Perennial plants with rootstocks; stems 30-100 cm. tall or more, usually branching above, commonly glabrous below; leaves 5-15 cm. long, linear to broadly lanceolate, entire or sometimes toothed, sometimes prominently veined but netlets irregular and longer than broad, glabrous or scabro-hirsute above; heads usually many in a long panicle; involucres 5-8 mm. high, hemispheric to campanulate, bracts imbricated in several series, linear-lanceolate, acute, outer sometimes wholly green, otherwise green at apex and whitish below; rays 6-14 mm. long, white, pink or blue; achenes pubescent; pappus bristles white or tawny. Two varieties are here.

36A. **Aster hesperius hesperius**
A. fluviatilis Osterh.; A. coerulescens of Mans.---Outer bracts of involucres about as wide as the inner, the green tips not especially enlarged; rays mostly blue.---Valleys, plains and hills, often along ditches or streams. Wisconsin to Alberta, south to Texas, Arizona and California. Our records scattered over Colorado at 3500-9000 feet.

36B. **Aster hesperius laetivirens** (Greene) Cronq., (var.) Leafl. West. Bot. 6:44. 1950.

A. laetivirens Greene; *A. osterhoutii* Rydb.---Outer involucre bracts usually wider than the inner, with wide green tips; rays mostly white or pink.---Along streams and ditches, in valleys and mountains. North Dakota to Saskatchewan, south to New Mexico and Utah. Our records from northern Colorado at 5000-9500 feet but this variety has been reported as well distributed over the western half of the state.

17. Achillea L. YARROW; MILFOIL

Perennial herbaceous plants with slender rootstocks; stems erect, leafy, usually villous or pilose; leaves alternate, finely dissected into numerous short linear to oblong divisions, these seldom over 1 mm. wide; heads small, numerous, in dense corymbose terminal panicles; involucres campanulate to hemispheric, bracts imbricated in 3-4 series the outer much shorter, appressed, thin dry and scarious; receptacles nearly flat to conical, with membranous chaff; ray flowers pistillate and fertile, rays few, small, about as wide as long, normally white; disk flowers about 15-70, perfect and fertile, whitish; anthers obtuse at base, not caudate; achenes oblong to obovate, compressed; pappus none.---Some authors have considered all our plants to be forms of *A. millefolium* L. On the basis of cytological and experimental evidence other workers maintain that we have a distinct native species, *A. lanulosa* Nutt. On the basis of ordinary morphological differences used to separate taxonomic entities the writer cannot distinguish between the two plants in Colorado. Our native plant apparently sometimes invades cultivated ground and waste areas. The key presented here is only tentative.

1. Plants of plains to subalpine situations; stems over 25 cm. tall; margins of involucre bracts not at all blackish (may be brownish)
    2. Ultimate leaf segments ovate to lanceolate; rays about 2-3 mm. long; introduced plants usually around fields, roads and dwellings; leaves usually glabrate to sparsely villous   -1. **A. millefolium**
    2. Ultimate leaf segments linear; rays about 3-6 mm. long; plants native; leaves usually densely long-villous   -2A. **A. lanulosa lanulosa**
1. Plants of subalpine and alpine situations; stems rarely over 25 cm. tall; margins of involucre bracts often blackish   -2B. **A. lanulosa alpicola**

1. **Achillea millefolium** L., Sp. Pl. 899. 1753.

A weedy plant naturalized from Europe. Widely distributed in the eastern part of the United States and reported for this state. A few specimens from northcentral Colorado may be this species.

2. **Achillea lanulosa** Nutt., Journ. Acad. Nat. Sci. Phila. 7:36. 1834.

Stems 10-60 cm. tall, usually copiously villous with long hairs; leaves 3-10 cm. long, linear to narrowly oblong in outline, finely bipinnatifid, ultimate segments linear or linear-lanceolate and acicular, almost spinulose, more or less long-villous; involucre bracts with greenish keel and usually brownish or blackish margins; rays 2-6 mm. long, usually white but rarely rose to red-purple. The two subspecies intergrade somewhat in this area.

2A. **Achillea lanulosa lanulosa**

A. lanulosa ssp. *typica* Keck---Stems over 25 cm. tall; involucre bracts with margins tan to brownish and not especially conspicuous.---Plains and mountains. Alberta, south to New Mexico and California. Our records widely scattered in the western two-thirds of Colorado at 5000-10,000 feet, rarely to 10,500 feet.

2B. **Achillea lanulosa alpicola** (Rydb.) Keck, (ssp.) Carnegie Inst. Wash. Pub. No. 520:300.1940.

A. alpicola Rydb.; *A. subalpina* Greene ---Stems seldom over 25 cm. tall; involucre bracts with margins blackish to blackish-brown, conspicuous.---High mountains. Western part of North America. Our records scattered in the higher mountains of Colorado at 11,000-12,000 feet.

18. Anthemis L. CAMOMILE; DOG FENNEL

Annual or possibly sometimes biennial herbaceous plants with rank or fetid odor; stems branching, leafy; leaves alternate, bi- or tripinnatifid into linear or filiform cuspidate ultimate segments; heads solitary at the ends of the branches; involucres saucer-shaped or hemispheric, the bracts in 2 to several series, outer more or less shorter, appressed, scarious-margined; receptacles conic, bearing narrow chaff near the apex; ray flowers about 10-15, pistillate and fertile or neutral, rays white; disk flowers perfect and fertile, yellow; anthers united, not caudate at base; achenes subcylindric, 10-ribbed and often glandular roughened; pappus none.

1. Rays neutral, lacking styles; stems glabrous or somewhat pubescent above only   -1. **A. cotula**
1. Rays pistillate and fertile, with styles present; stems more or less villous-pubescent throughout   -2. **A. arvensis**

1. **Anthemis cotula** L., Sp. Pl. 894. 1753.

Maruta cotula (L.) DC.---Stems 20-60 cm. tall, glabrous or more or less pubescent above; leaves about 3-5 cm. long, glabrous to more or less hairy; involucres 4-6 mm. high, bracts oblong to oblong-lanceolate, midrib darker and margins broadly scarious, more or less tomentose; rays about 6-10 mm. long, at length reflexed, ray flowers neutral; chaff stiff and awllike.---Fields and waste ground. Naturalized from Europe and now spread throughout most of Canada and United States. Our few records scattered in Colorado in cultivated areas at 5000-9500 feet.

2. **Anthemis arvensis** L., 894. 1753.
Stems 20-40 cm. tall, usually much branched, villous-hairy; leaves about 3-5 cm. long, sparsely villous to nearly glabrous; involucres about 4-6 mm. high, bracts oblong to oblong-lanceolate, midrib darker and margins scarious, more or less hairy; rays about 6-12 mm. long, spreading or somewhat reflexed, ray flowers pistillate and fertile; chaff lanceolate.---Fields and waste places. Naturalized from Europe into Canada and United States. Apparently uncommon in the Rocky Mountain area. Our two records from cultivated areas in the eastern half of Colorado at 4000-5000 feet.

19. Tanacetum L. TANSY

Herbaceous perennial or suffruticose plants, often aromatic; leaves alternate, often crowded at base of plant, simple and toothed or lobed at apex, or pinnately to palmately divided or 2- to 3-pinnatifid, usually silvery-canescent or punctate; heads corymbose, in small cymes, in subcapitate clusters or rarely solitary; involucres hemispheric to broadly campanulate, bracts in 2-3 series, not very unequal in length, scarious-margined; receptacles convex to conic, naked; marginal flowers pistillate and fertile, rays lacking or very short oblique yellow ones present; disk flowers many, perfect and fertile; anthers united, not caudate at base, obtuse or mucronate at apex; achenes subcylindric, 3- to 10-ribbed or angled; pappus wanting or of short paleae united into a crown.

1. Short rays present; heads many; plants 50-100 cm. tall -1. **T. vulgare**
1. Rays wanting; heads few; plants 10-20 cm. tall -2. **T. nuttallii**

1. **Tanacetum vulgare** L., Sp. Pl. 844. 1753.
Stems 50-100 cm. tall, glabrous or nearly so; leaves 1- to 3-pinnatifid, the divisions serrate, glabrous or nearly so; flowers many in a corymbose panicle; involucres 4-6 mm. high; rays very short, yellow; achenes about 1 mm. long, 3- to 5-angled.---Roadsides and waste areas, probably escaped from cultivation. Rather generally distributed in the United States but apparently uncommon in the Rocky Mountain area. Our few records from northcentral Colorado at about 5000 feet.
2. **Tanacetum nuttallii** T. & G., Fl. N. Amer. 2:415. 1843.
Plants loosely caespitose with woody caudex; stems 10-20 cm. tall, decumbent at base, silvery-canescent to short-sericeous; leaves tending to crowd toward base but also scattered on stems, densely silvery-white, broadly cuneate, obtusely 3- to 5-lobed at broad apex, those of flowering stems oblong or linear and entire especially above; heads usually 3-5 in a corymbose cyme but clustered at apex of stems, at least some of heads peduncled; involucres 3-5 mm. high, bracts yellowish-green or yellowish-brown; rays none; achenes about 2 mm. long, 10-ribbed.---Dry hills and mesas. Montana and Idaho, south to Wyoming and Utah. Our few records from northwestern Colorado at 6500-7000 feet.

20. Chrysanthemum L. OXEYE DAISY

Annual or perennial herbaceous plants; stems leafy; leaves alternate, toothed, coarsely incised or pinnatifid; heads solitary and terminal on stems or long branches, or corymbose; involucres rather saucer-shaped, bracts imbricated in several series, appressed, definitely scarious-margined; receptacles flat or convex, naked; ray flowers pistillate and fertile, rays conspicuous and white or pale yellow or lacking entirely; disk flowers many, yellow; anthers united, not caudate at base; achenes, at least of disk flowers 5- to 10-ribbed or angled; pappus none or rarely of a short crown.

1. Rays none or less than 8 mm. long; heads numerous and corymbose -1. **C. balsamita**
1. Rays present, over 8 mm. long; heads solitary at ends of long naked peduncles or branches
    2. Leaves toothed or incised-pinnatifid, basal petioled and cauline sessile or nearly so; involucre bracts with narrow scarious margins, the scarious apex not over 1.5 mm. wide; rays white, normally over 12 mm. long -2. **C. leucanthemum**
    2. Leaves 2- to 3-pinnatifid, all essentially alike; involucre bracts with broader scarious margins, the scarious apex about 2-3 mm. wide; rays pale yellow, 8-12 mm. long -3. **C. coronarium**

1. **Chrysanthemum balsamita** L., Sp. Pl. ed. 2, 1252. 1763.
Balsamita major Desf.---Perennial aromatic and mint scented plants with horizontal rootstocks; stems 50-175 cm. tall, branched, puberulent or canescent; leaves obovate, oblanceolate or oblong, crenate-serrate or crenate-dentate, obtuse, punctate, basal large and petioled; stem leaves 2-5 cm. long, often with a pair of lateral lobes near base, sessile; heads numerous in terminal corymbs, on the stems and branches; involucres about 6-16 mm. wide, bracts narrow, obtuse, pubescent; rays none or rarely present and short and white; pappus none or a short crown.---Sparingly escaping from gardens. Native to Eurasia and escaping in the eastern part of the United States. Our few records from northern and southcentral Colorado at 6500-8000 feet.
2. **Chrysanthemum leucanthemum** L., Sp. Pl. 888. 1753.
Leucanthemum leucanthemum (L.) Rydb.---Perennial plants; stems 25-80 cm. tall, glabrous or nearly so; basal leaves obovate to spatulate, coarsely and irregularly toothed or incised; stem leaves oblanceolate to linear, toothed to sub-pinnatifid, base lacerate, sessile at least above; heads on long naked peduncles; involucres 6-10 mm. high, broad and flattish, bracts lanceolate or oblong-lanceolate, very light green with darker green midrib, rather narrowly scarious on margins, this becoming wider near apex and with a brown line just below the apex;

rays normally 12-15 mm. long, white; achene ribs whitish, blackish between. Our plant is the var. pinnatifidum Lec. & Lam.---Meadows, roadsides and fields. Native of Europe and widespread in the United States but rather uncommon westward. Our records scattered over Colorado at 5000-8000 feet.

3. Chrysanthemum coronarium L., Sp. Pl. 890. 1753.
Annual plants; stems 25-100 cm. tall, glabrous; leaves about similar, broadly obovate, 2- to 3-pinnatifid, the ultimate segments narrowly oblong to linear, callus-tipped, glabrous; heads on moderately long naked peduncles; involucres 8-11 mm. long, broad and flattish, bracts ovate to oblong, with darker midrib and with broad scarious margins without brownish line inside this band; rays 8-12 mm. long, pale yellow.---Introduced from Europe and occasionally escaping in the United States. Our one record from central Colorado at 5500 feet.

21. Matricaria L. FALSE CAMOMILE; MAYWEED

Annual glabrous plants; stems branching, leafy; leaves alternate, bipinnatifid or tripinnatifid into linear-filiform segments; heads solitary at the ends of the branches; involucres saucer-shaped, bracts in 2-3 series but subequal in length, obtuse, with broad to moderate scarious margins; receptacles convex to conic, naked; ray flowers pistillate and fertile, rays white or lacking; disk flowers perfect and fertile, numerous; anthers united, obtuse at apex, not caudate at base; achenes unsymmetrical, with 3-5 ribs on inner side and smooth or rugose on outer; pappus a crown with entire margins or minute to obsolete.

1. Rays present and conspicuous; receptacles convex and rounded at apex -1. M. inodora
1. Rays lacking; receptacles conic and rather pointed at apex -2. M. matricarioides

1. Matricaria inodora L., Fl. Suec. ed. 2. 297. 1755.
Chamomilla inodora (L.) Gillib.---Stems 20-50 cm. tall; leaves 2- to 3-pinnatifid into linear-filiform divisions; involucres about 4 mm. high, bracts with green midnerve and moderate to narrow brownish scarious margins; receptacles convex; rays 15-25, 7-10 mm. long, spreading, white; disk flowers 5-lobed; achenes with 3 ribs on inner face, rugose on outer; pappus crown well developed.---Fields and waste places. Introduced from Europe and escaping in various parts of North America. Our one record from southcentral Colorado at about 7500-8000 feet.

2. Matricaria matricarioides (Less.) Porter, Mem. Torr. Bot. Club 5:341. 1894.
Chamomilla suaveolens (Pursh) Rydb.---Stems 10-40 cm. tall; leaves 2- to 3-pinnatifid with short linear-filiform divisions; involucres about 3 mm. high, bracts green-nerved with moderately broad scarious margins; receptacles conic and often becoming pointed at apex; rays absent; disk flowers 4-lobed; achenes with 4 ribs on inner face, smooth on outer; pappus minute or obsolete. At least some of our plants have a rather pleasant pineapple odor when fresh.--- Roadsides and waste places, often on moist or sandy ground. Montana to Alaska, south to Arizona, and California; naturalized eastward and in Eurasia. Our records from northcentral, northwestern and central Colorado at 7500-9500 feet.

22. Artemisia L. SAGEBRUSH; WORMWOOD

Annual or perennial herbaceous or shrubby plants, usually aromatic; leaves alternate, from entire to variously lobed or dissected; heads small, commonly panicled, sometimes in racemes or spikelike clusters; involucres ovoid to campanulate or hemispheric, bracts imbricated in 2-4 series, dry in texture, at least the inner scarious or with scarious margins; receptacles flat to convex, naked or hairy; marginal flowers pistillate and fertile or wanting, no definite rays ever present; disk flowers perfect, fertile or sterile; anthers united, not caudate at base but with lanceolate or subulate tips; achenes rather short, usually glabrous; pappus none.

1. Plants shrubs or undershrubs, sometimes low but always definitely woody at least at base; leaves usually silvery-canescent both sides
  2. Leaves (except sometimes the upper) 2-3 times pinnately parted or dissected; receptacles often long-hairy
    3. Receptacles with long woolly hairs between the flowers; leaves silky-canescent or silvery-canescent on upper as well as lower surface
      4. Ultimate leaf segments narrowly linear; plants 10-50 cm. tall; native plants -1. A. frigida
      4. Ultimate leaf segments lanceolate, oblanceolate or narrowly oblong; plants usually over 50 cm. tall; introduced plants -2. A. absinthium
    3. Receptacles not hairy; leaves green (glabrous to sparsely puberulent) above -3. A. abrotanum
  2. Leaves entire, palmately toothed or palmately divided from apex; receptacles glabrous
    5. Branches very spiny; corollas copiously hairy; achenes villous -4. A. spinescens
    5. Branches not spiny; corollas not copiously hairy; achenes glabrous
      6. Leaves linear to linear-oblanceolate and entire (rarely with a few teeth or lobes), over 1 mm. wide -5. A. cana
      6. Principal leaves toothed to divided, of various shapes, or if entire then linear-filiform and seldom over 0.5 mm. wide
        7. Leaves 3-parted or sometimes entire but the leaves or their divisions linear-filiform and rarely over 0.5 mm. wide -6. A. filifolia
        7. Leaves toothed or divided, but leaves or their divisions linear to broader, usually over 0.5 mm. wide
          8. Principal leaves cleft or divided into linear or narrowly spatulate divisions over 5 times longer than wide
            9. Inflorescence a raceme; a few of the marginal flowers lacking stamens, shorter than the others; plants 10-15 cm. tall -7. A. pedatifida

9. Inflorescence a panicle; all the flowers of the head perfect and alike; plants 20-60 cm. tall
 -8. A. tripartita
8. Principal leaves 3- to 5-toothed at apex or cleft into relatively shorter and broader divisions less than 5 times longer than wide
 10. Heads very numerous in rather open panicles; shrubs 50-400 cm. tall (even taller); leaves merely toothed at apex   -9A. A. tridentata tridentata
 10. Heads fewer in spikelike racemes (rather open panicles in A. bigelovii but plant rarely over 35 cm. tall); undershrubs usually under 40 cm. tall; leaves toothed to divided
  11. Involucres 5-6 mm. high; flowers usually 8 or more to a head   -9B. A. tridentata rothrockii
  11. Involucres rarely over 4 mm. high; flowers 3-9 to a head
   12. Involucres glabrous or glabrescent (except sometimes for outer short bracts), yellowish or yellowish-brown   -9C. A. tridentata nova
   12. Involucres tomentose to canescent, not yellowish or yellowish-brown
    13. Involucres 3-4 mm. high and about 3 mm. wide; flowers 5-9 to a head, all corollas alike with stamens present   -9D. A. tridentata arbuscula
    13. Involucres 2-3 mm. high and 2-2.5 mm. wide; flowers 1-5 to a head, 1-2 of the outer with much shorter corollas and usually lacking stamens -10. A. bigelovii
1. Plants annual or perennial and herbaceous, stems often thickened at base but never decidedly woody; leaves usually not silvery-canescent both sides
 14. Receptacles densely hairy between the flowers
  15. Leaf divisions oblong, lanceolate or oblanceolate, ultimate segments 1.5-4 mm. wide; plants 40-100 cm. tall, mostly in waste ground as an escape   -2. A. absinthium
  15. Leaf divisions linear or linear-filiform, ultimate segments about 1 mm. wide; plants seldom over 40 cm. tall, native
   16. Heads many, usually in panicles; plants woody and rather matlike at base; stems very leafy
                  -1. A. frigida
   16. Heads 1-20, in racemes; plants not at all woody or matlike at base; stems sparsely leafy
    17. Heads 5-20; perfect flowers 15-30 to a head; basal leaves mostly 2-pinnatifid
                  -11. A. scopulorum
    17. Heads 1-5; perfect flowers 30-100 to a head; basal leaves 1-pinnately divided or cleft
                  -12. A. pattersoni
 14. Receptacles not evidently hairy between the flowers
  18. At least some leaves 2- or 3-pinnatifid
   19. Plants annual or biennial with taproot
    20. Inflorescence spikelike (although really composed of many short spikes); disk flowers maturing achenes   -13. A. biennis
    20. Inflorescence a narrow panicle; disk flowers sterile   -14. A. caudata
   19. Plants perennial with branching caudex or rootstocks
    21. Leaves silky-pubescent, silky-canescent, pilose, villous or glabrous, never woolly or tomentose
     22. Achenes developing only on marginal pistillate flowers (ovaries of central flowers much smaller than outer from the first); styles of central flowers rarely 2-branched at apex; central flowers 10-30; involucres 2-4 mm. long
      23. Involucres 2-3 mm. high and about as wide; heads in elongated panicles
                  -15. A. pacifica
      23. Involucres 3-4 mm. high and 4-5 mm. broad; heads often in spikelike panicles
       24. Involucre bracts glabrous or nearly so   -16. A. borealis
       24. Involucre bracts densely hairy   -17. A. spithamaea
     22. Achenes maturing on all the flowers of the head; styles all 2-branched at apex (branches may lie together in young flowers); central flowers 30-75; involucres 3-5.2 mm. high
      25. Involucres about 3 mm. high and 4-5 mm. broad; corollas of central flowers about 2 mm. long or less
       26. Leaves glabrous or sparingly pilose below, not especially bicolored
                  -18. A. parryi
       26. Leaves definitely whitish and silky-strigose below, green above and definitely bicolored   -19. A. franserioides
      25. Involucres about 4-5.2 mm. high and 5-7 mm. broad; corollas of central flowers 2.5-3.5 mm. long   -20. A. norvegica
    21. Leaves definitely tomentose or woolly, at least below
     27. At least some of leaf segments obtuse at apex; flowers 50 or more to a head; involucres 2.5-3 mm. high and about 5 mm. broad   -19. A. franserioides
     27. Leaf segments acute; flowers 20-50 to a head; involucres 3-4 mm. high and about as wide
      28. Involucres 3.5-4 mm. high; leaves usually bright green above, primary divisions again divided (hence truly bipinnately divided) -21. A. michauxiana
      28. Involucres 3-3.5 mm. high; leaves usually more or less tomentose above, primary lobes entire, toothed or merely lobed   -24C. A. ludoviciana incompta
  18. Leaves entire, toothed to pinnately divided but the segments merely toothed or lobed, hence leaf never really bipinnatifid or bipinnately divided
   29. Leaves glabrous, silky-canescent or silky-villous, not at all woolly or tomentose; achenes developing only on marginal flowers, abortive on central flowers
    30. Leaves mostly entire or the lower 1- to 3-cleft, often over 2 mm. wide
     31. Heads 3-4 mm. wide; leaves 2-10 mm. wide, glabrous -22. A. dracunculus dracunculus
     31. Heads 2-3 mm. wide; leaves 1-3 mm. wide, glabrous or hairy
                  -22B. A. dracunculus glauca
    30. Leaves pinnatifid, the ultimate segments often less than 2 mm. wide

   32. Involucre bracts glabrous or nearly so   -16. A. borealis
   32. Involucre bracts densely hairy   -17. A. spithamaea
 29. Leaves densely tomentose or woolly at least below; all flowers of the head developing achenes
   33. Principal leaves entire or merely lobed (the sinuses seldom extending more than 1/2 into midrib); involucres 3-5 mm. high (Note: The leaf is rarely divided in A. longifolia but the involucres are 4-5 mm. high)
     34. Involucres 4-5 mm. long; leaves linear or linear-lanceolate, usually entire   -23. A. longifolia
     34. Involucres 3-4 mm. long; leaves usually lanceolate or broader (sometimes linear in A. ludoviciana typica), the principal leaves usually toothed or lobed
       35. Leaves 3-10 cm. long, mostly lanceolate, oblanceolate or linear
                                        -24A. A. ludoviciana ludoviciana
       35. Leaves 1-2 cm. long, mostly elliptic to obovate   -24B. A. ludoviciana albula
   33. Principal leaves parted or divided; involucres 2.5-3.5 mm. high
     36. Leaves 1-3 cm. long, divisions linear-filiform, 0.5-1 mm. wide (rarely to 1.5 mm.); leaves commonly fascicled   -25. A. carruthii
     36. Leaves usually over 3 cm. long, the divisions linear to lanceolate, 2-4 mm. wide; leaves usually not fascicled
       37. Panicles very narrow, spikelike or racemelike, 1-2 cm. broad, not very leafy; some of primary leaf divisions usually toothed or lobed; flowers 21-40 to a head
                                          -24C. A. ludoviciana incompta
       37. Panicles broader than racemelike, usually about 3-10 cm. wide, leafy; primary leaf divisions entire; flowers 11-30 to a head   -24D. A. ludoviciana mexicana

1. **Artemisia frigida** Willd., Sp. Pl. 3:1838. 1804.
Perennial herbaceous plants or decidedly woody at base and an undershrub; stems 10-40 cm. tall, branching from base, young twigs finely canescent; basal leaves crowded, twice-pinnatifid into linear or linear-filiform divisions 5-15 mm. long, silvery-canescent but often turning brown in age; upper leaves less dissected; heads many, in a narrow leafy panicle, sometimes in a reduced raceme; involucres 2-3.5 mm. high and 4-6 mm. wide, villous, canescent or tomentose; marginal flowers about 10-17, lacking anthers, fertile, corollas definitely shorter; disk flowers 20-50, perfect and fertile; receptacles densely villous; achenes glabrous.---Dry plains, hills and mountains. Canada and Alaska, south to Texas and Arizona; Siberia. Our records scattered over Colorado mostly in the western half, at 4500-10,000 feet.

2. **Artemisia absinthium** L., Sp. Pl. 848. 1753.
Perennial herbaceous plants but often woody at base; stems 40-100 cm. tall, simple below, glabrate to finely canescent; basal leaves 3-7 cm. long, long-petioled, 2- to 3-pinnately divided into oblong, oblanceolate or lanceolate divisions, silky-canescent sometimes less so above; upper leaves with fewer lobes; heads nodding, many, in a large panicle; involucres 2-3.5 mm. high and 3-5 mm. wide, bracts canescent; marginal flowers 9-20, lacking stamens but fertile, with shorter corollas; disk flowers 20-50, perfect and fertile; receptacles long-hairy; achenes glabrous.---Waste places and around dwellings. Native in Europe and northern and eastern United States and introduced at various other places. Our records from northcentral and central Colorado at 5000-6000 feet.

3. **Artemisia abrotanum** L., Sp. Pl. 845. 1753.
Shrubs, woody at the base only in our area, pleasantly scented; stems 50-200 cm. tall, much branched, glabrous or somewhat canescent-strigillose on young twigs; leaves 2-6 cm. long, numerous, 2-3 times pinnately dissected into linear-filiform segments, green and glabrous or sparsely puberulent above, lightly tomentose beneath; upper leaves less dissected; heads many in a large elongated panicle, the branches very leafy below; involucres 2.5-3 mm. high, outer bracts canescent, inner sparsely if at all hairy; marginal flowers 5-15, lacking stamens but fertile, with shorter corollas; disk flowers 10-20, perfect and fertile; receptacles glabrous; achenes glabrous.---Escaping from cultivation. Native to Eurasia, now present in various parts of North America. Our one record from northcentral Colorado at about 5000 feet.

4. **Artemisia spinescens** D.C. Eaton in King's Expl. 180. 1841.
Picrothamnus desertorum Nutt.---Rounded, spiny shrubs 5-50 cm. tall, definitely woody at least at base; stems crowded, white-tomentose or short-villous; principal leaves 5-20 mm. long including petioles, flabellate in outline, palmately 3- to 5-divided, the divisions again cleft into linear-spatulate lobes, villous; upper leaves less dissected; heads nodding, solitary to several in short racemes, in axils or at the ends of short branches; involucres about 2-3.5 mm. high and 3-4 mm. wide, bracts densely villous; marginal flowers 2-6, without stamens but fertile, with shorter corollas; disk flowers 5-13, perfect but sterile, corollas long-hairy; receptacles glabrous; achenes densely villous.---Dry plains and slopes. Montana to Oregon, south to New Mexico and California. Our records from western Colorado at 4500-8000 feet.

5. **Artemisia cana** Pursh, Fl. Am. Sept. 521. 1814.
A. cana viscidula Osterh.; A. viscidula (Osterh.) Rydb.---Shrubs 30-200 cm. tall; stems branching throughout, twigs with gray or yellowish-green tomentum or canescent; principal leaves 1-4 cm. long, linear or lanc-linear, entire or occasionally with a few irregular teeth or lobes, silvery-canescent; upper leaves little reduced; heads many in a narrow leafy panicle, mostly in small glomerules; involucres 4-5 mm. high and 3-4 mm. wide, sometimes smaller, bracts canescent or tomentose; flowers 5-15, all perfect and fertile; receptacles glabrous; achenes granduliferous.---Plains, hills and valleys. Saskatchewan to British Columbia, south to Nebraska, New Mexico and California. Our records scattered over the western half of Colorado at 5000-10,000 feet.

6. **Artemisia filifolia** Torr., Ann. Lyc. N. Y. 2:211. 1828.
Shrubs or sometimes undershrubs woody at least near base, 30-150 cm. tall; stems freely branching, twigs canescent or tomentulose; leaves 3-8 cm. long, less than 0.5 mm. wide, filiform and entire or more commonly ternately divided into filiform divisions, often fascicled,

canescent or minutely tomentulose; heads numerous in a narrow panicle; involucres 1-2 mm. high and about as wide, bracts canescent or tomentulose; marginal flowers 2-3, lacking stamens but fertile; disk flowers 1-6, perfect but sterile; receptacles glabrous; achenes glabrous.---Dry plains. Nebraska and Wyoming to Nevada, south to Texas and Mexico. Our records well scattered in the eastern half of Colorado at 3500-5500 feet.

7. Artemisia pedatifida Nutt., Trans. Phil. Soc. II. 7:399. 1841.

Subshrubs definitely woody below; stems 5-15 cm. tall, erect, finely canescent; basal leaves 1-2 cm. long, 1- to 2-ternately divided into linear or narrowly spatulate rather short divisions with fine dense almost tomentose hairs; upper leaves smaller and rather few; heads few in a spikelike or racemelike inflorescence; involucres 3-4 mm. high and 3-4 mm. wide, bracts tomentose; marginal flowers about 4-7, pistillate and fertile, with definitely shorter corollas; disk flowers 5-9, perfect but sterile; receptacles glabrous; achenes glabrous.---Dry mesas, ridges and plains. Montana, Wyoming and Idaho. Our one record from northwestern Colorado at 5500 feet.

8. Artemisia tripartita Rydb., Mem. N. Y. Bot. Gard. 1:432. 1900.

A. trifida Nutt.; A. tridentata ssp. trifida (Nutt.) H. & C.---Shrubs or undershrubs 20-50 cm. tall, sometimes to 60 cm.; branches canescent; leaves 1.5-4 cm. long, cuneate to flabelliform in outline, deeply cleft into linear or linear-oblanceolate divisions about 1 cm. long, the divisions sometimes again lobed, silvery-canescent; upper leaves often entire; heads paniculate; involucres about 3-4 mm. high and 2-3 mm. wide, bracts canescent; flowers 5-8 to a head, all perfect and fertile; receptacles glabrous; achenes granduliferous. This plant is clearly related to the A. tridentata complex but seems distinct enough in our area to be treated as a separate species.---Dry plains and mesas. Montana to British Columbia, south to Colorado and California. Our few records from northern Colorado at 8000-9000 feet.

9. Artemisia tridentata Nutt., Trans. Am. Phil. Soc. ser. 2.7:398. 1841.

Shrubs or undershrubs, 10-400 cm. tall, much branched, these more or less silvery-canescent; leaves 5-35 mm. long, narrowly to broadly cuneate, 3- to 5-lobed or cleft at apex, but the sinuses seldom extending half way into the base, the divisions sometimes again toothed or lobed somewhat, silvery-canescent; upper leaves sometimes entire; heads few to many, in spikelike, racemose or paniculate inflorescences; involucres about 3-6 mm. high and 2-5 mm. wide, bracts glabrous, canescent or tomentose; flowers 2-15, all perfect and fertile; receptacles glabrous; achenes granuliferous. This is a complex and the various entities have been given specific rank. However, some intergradation exists in this area and it has proven more satisfactory to follow Hall & Clements treatment. (Phylog. Meth. Tax. 31-156. 1923).

9A. Artemisia tridentata tridentata

A. tridentata ssp. typica H. & C.---Shrubs 40-400 cm. tall or even taller; leaves 1-3 cm. long, 3-toothed at apex (rarely 5); heads many in a definite, rather broad panicle; involucres about 4 mm. long and 2.5-3 mm. wide, bracts tomentose or canescent, at least the outer; flowers 4-8.---Dry plains and hills. Montana to British Columbia, south to New Mexico, Mexico and California. Our records well scattered in the western half of Colorado at 4500-8500 feet.

9B. Artemisia tridentata rothrockii (A. Gray) H. & C., (ssp.) Phylog. Meth. Tax. 138. 1923.

A. rothrockii A. Gray; A. spiciformis Osterh.---Shrubs or undershrubs 10-70 cm. tall; leaves 1-3.5 cm. long, 3-toothed or 3-cleft at apex, the teeth or lobes sometimes again toothed; heads rather few in spikelike or very narrow inflorescences; involucres 5-6 mm. high and 3-5 mm. broad, bracts canescent to glabrate, sometimes greenish-yellow; flowers 8-15.---Dry plains and mountains. Wyoming to Washington, south to Colorado and California. Our records from northcentral, central and western Colorado at 8500-9000 feet.

9C. Artemisia tridentata nova (A. Nels.) H. & C., (ssp.) Phylog. Meth. Tax. 137. 1923.

A. nova A. Nels.---Undershrubs normally 10-30 cm. tall; leaves 0.5-1.5 cm. long, 3-toothed at apex; heads rather few in a narrow, racemiform panicle; involucres about 3-4 mm. long and about 2 mm. wide, bracts glabrous or short outer ones sometimes canescent, greenish-yellow to yellowish-brown; flowers 2-6.---Dry plains, mesas and hills. Montana to California, south to New Mexico and Arizona. Our rather few records from western Colorado at 7000-8000 feet.

9D. Artemisia tridentata arbuscula (Nutt.) H. & C., (ssp.) Phylog. Meth. Tax. 138. 1923.

A. arbuscula Nutt.---Low, almost undershrubs, 10-30 cm. tall; leaves 5-15 mm. long, 3- to 5-lobed or cleft at apex, the segments sometimes again lobed; heads rather few in a spikelike inflorescence; involucres 3-4 mm. high and about 3 mm. wide, bracts canescent to short-tomentose; flowers 5-9.---Dry plains, mesas and hills. Wyoming to Washington, south to Colorado and California. Our few records from northwestern Colorado at 7000-8000 feet.

10. Artemisia bigelovii A. Gray, Pacif. R. R. Rept. 4:110. 1857.

Subshrubs definitely woody at base, 20-30 cm. tall; stems erect, twigs silvery-canescent; leaves about 1-1.5 cm. long, mostly 3-toothed at apex but many entire especially upper, linear-cuneate, silvery-canescent; heads rather numerous in a narrow sometimes spikelike panicle; involucres about 2-3 mm. long and about 1.5-2.5 mm. wide, bracts short-tomentose or canescent; marginal flowers 1-2, without stamens but fertile, with corollas definitely shorter; disk flowers 1-3, perfect and fertile; receptacles glabrous; achenes glabrous.---Rocky banks, hills and canyons. Colorado and Utah, south to Texas and Arizona. Our few records from central and southcentral Colorado at 4500-5000 feet.

11. Artemisia scopulorum A. Gray, Proc. Acad. Phila. 1863:66. 1863.

Perennial herbaceous plants with caespitose or slender-branching caudex; stems 10-25 cm. tall, several to many, simple, glabrate, canescent or lightly tomentose; basal leaves 2-7 cm. long, mostly 2-parted or divided into linear or oblanceolate-oblong segments about 1 mm. wide, crowded, petioled, silky-canescent; upper leaves reduced and less parted; heads 5-12 or rarely more, spicate or racemose, often crowded at apex; involucres about 4 mm. high and 4-7 mm. wide, bracts with black or dark brown scarious margins, copiously villous; marginal flowers 6-13, lacking stamens but fertile; disk flowers 15-30, perfect and fertile, corollas about 2-2.5 mm. long; receptacles hairy; achenes glabrous.---High mountains. Montana, south to New Mexico and Utah. Our records scattered in the higher mountains at 11,000-13,000 feet.

12. **Artemisia pattersoni** A. Gray, Syn. Fl. N. Amer. 1:453. 1886.

Perennial herbaceous plants with long or short rootstocks; stems 8-20 cm. tall, erect and unbranched, tomentulose to glabrate; basal leaves 2-4 cm. long (counting petiole) spatulate or oblong-spatulate in outline, once-pinnately parted into narrow divisions 1-2 mm. wide, these sometimes crowded and tufted near apex of leaf, silky-canescent; upper leaves reduced sometimes entire; heads 1-5 in racemes, horizontal or nodding; involucres about 5 mm. high and 5-8 mm. wide, bracts ovate to lanceolate, villous; marginal flowers 7-27, lacking stamens but fertile, the corollas definitely shorter; disk flowers 32-100, perfect and fertile; receptacles copiously villous; achenes glabrous.---High mountains. Colorado and New Mexico. Our records from northcentral and central Colorado at 11,000-13,000 feet.

13. **Artemisia biennis** Willd., Phytogr. 11. 1794.

Annual or biennial plants with taproot; stems 30-100 cm. tall, simple to the inflorescence, glabrous, often red-tinged; basal leaves twice-pinnately parted into lanceolate, sharply toothed or cleft segments, crowded, glabrous; upper leaves mostly once-pinnate; heads many in dense spikes these crowded into a spikelike or rarely open panicle, leafy throughout; involucres 2-3 mm. high and 2-3.5 mm. wide, glabrous; marginal flowers 6-25, lacking stamens but fertile; disk flowers 15-40, perfect and fertile, corollas about 1 mm. long; receptacles glabrous; achenes glabrous.---Moist places. Widespread in Canada and the United States as a weed, native in the western part. Our records mostly in northcentral, central and southcentral Colorado at 5000-7500 feet.

14. **Artemisia caudata** Michx., Fl. Bor. Am. 2:129. 1803.

A. forwoodii Wats.; A. campestris ssp. caudata (Michx.) H. & C.---Herbaceous biennial plants, the taproot small to rather large but never perennial; stems 30-60 cm. tall, single or occasionally 2 or 3, somewhat hairy when young, leafy especially below and more or less red-tinged; basal leaves 3-8 cm. long, twice-pinnately divided into narrowly linear segments about 1 mm. wide or sometimes thrice-pinnate, canescent or silky-hairy rarely glabrate, petioled; upper leaves reduced; heads many in an elongated panicle with many ascending branches; involucres 2-3 mm. high and 2-3 mm. wide, bracts glabrous or sparingly short-hairy; marginal flowers 5-20, lacking stamens but fertile; disk flowers 10-25, perfect but not maturing achenes, corollas 1.8-2.5 mm. long; receptacles glabrous; achenes glabrous.---Plains and hills. New Brunswick to Saskatchewan, south to Florida and New Mexico. Our few records from northcentral and central Colorado at 5000-7500 feet.

15. **Artemisia pacifica** Nutt., Trans. Am. Phil. Soc. ser. 2.7:401. 1841.

A. camporum Rydb.; A. campestris ssp. pacifica (Nutt.) H. & C.; probably A. scouleriana (Besser) Rydb.---Perennial herbaceous plants; stems 30-60 cm. tall, several to many, erect from a spreading base, glabrous to slightly hairy, often reddish-tinged; basal leaves 4-7 cm. long, twice-pinnately divided into linear segments about 1 mm. wide or these sometimes narrowly oblong and over 1 mm. wide, more or less silky-canescent, petioled; upper leaves sessile and less dissected; heads many in a narrow leafy panicle with numerous ascending branches; involucres 2-3 mm. high and about 2-3 mm. wide, glabrous or sparingly short-hairy; marginal flowers 5-20, lacking stamens but fertile; disk flowers 10-15, perfect but ovaries abortive, corollas 1.8-2.5 mm. long; receptacles glabrous; achenes glabrous.---Plains and hills. South Dakota to British Columbia, south to Nebraska, New Mexico and Arizona. Our records scattered in the northern three-fourths of Colorado, mostly in the central area, at 4500-9000 feet.

16. **Artemisia borealis** Pallas, Reise 3:755. 1776.

A. campestris ssp. borealis (Pallas) H. & C.; A. maccallae Rydb.---Greenland to Yukon, south to Vermont and Washington. Reported from Colorado but doubtfully present.

17. **Artemisia spithamaea** Pursh, Fl. Am. Sept. 522. 1814.

A. campestris ssp. spithamaea (Pursh) H. & C.---Perennial herbaceous plants; stems 10-40 cm. tall, several, crowded on the crown, silky-hairy; principal leaves once or twice ternately or pinnately divided into short linear or linear-oblanceolate segments, silky-hairy; upper leaves less divided or entire; heads usually several to many in a spikelike or narrow panicle, erect or nodding, sessile or subsessile; involucres about 4 mm. high and 4-5 mm. wide, densely villous; marginal flowers 5-20, lacking stamens but fertile; disk flowers 15-30, perfect but sterile, corollas about 3 mm. long; receptacles glabrous; achenes glabrous.---Mountains. Greenland to Alaska, south to Quebec, Colorado and Oregon. Our few records scattered in the higher mountains of Colorado at 11,500-12,000 feet.

18. **Artemisia parryi** A. Gray, Proc. Am. Acad. 7:361. 1868.

A. saxicola var. parryi A. Nels.---Perennial herbaceous plants with rootstocks; stems 10-40 cm. tall, simple below, glabrous; basal and lower leaves 4-8 cm. long counting petioles, twice-pinnately divided into narrowly linear segments, crowded, glabrous or nearly so; upper leaves smaller and less dissected; heads mostly nodding, many, in an elongated raceme or panicle; involucres about 3 mm. long and 4-5 mm. wide, bracts elliptic, obtuse, glabrous or sparsely villous; marginal flowers about 8, lacking stamens but fertile, corollas smaller; disk flowers 20-50, perfect and fertile, corollas about 2 mm. long; receptacles glabrous; achenes glabrous.---High mountains. Colorado and Utah. Our few records from southern Colorado at 9000-10,000 feet.

19. **Artemisia franserioides** Greene, Bull. Torr. Club 10:42. 1883.

Perennial herbaceous plants with rootstocks; stems 30-100 cm. tall, often reddish-tinged, tomentose at first to glabrate in age; leaves 4-7 cm. long, pinnately divided into 5-9 elliptic or lanceolate divisions, these again cleft or divided into entire or toothed, usually obtuse segments, glabrous or puberulent above, finely tomentose or short silky-villous beneath; upper leaves sometimes less dissected; heads secund and nodding on branches, in a narrow panicle 1-4 cm. wide; involucres 2.5-3 mm. high and about 4-6 mm. wide, bracts tomentose to nearly glabrate; marginal flowers 6-13, lacking stamens but fertile; disk flowers 45-90, perfect and fertile, corollas 1.5-2 mm. long; receptacles glabrous; achenes glabrous.---Mountains. Colorado, south to Mexico. Our records from central and southern Colorado at 6000-10,000 feet.

20. Artemisia norvegica Fries, Novit. Fl. Suec. ed. 1:56. 1817.
   A. saxicola Rydb.---Perennial herbaceous plants with caespitose caudex or short rootstocks; stems 10-15 cm. tall, clustered, sometimes ascending at base, often reddish-tinged, loosely villous or rarely glabrate; lower leaves 2-12 cm. long, twice-pinnately divided into linear or lanceolate, acute divisions, these rarely divided again, varying from copiously villous to glabrous, petioled; upper leaves smaller and less dissected; heads usually not very numerous, in a narrow to rather open panicle; involucres 4-5.2 mm. high and 4-10 mm. wide, bracts with blackish or dark brown margins, usually villous; marginal flowers 6-20, lacking stamens but fertile; disk flowers 30-75, perfect and fertile, corollas 2.5-3.5 mm. long; receptacles glabrous; achenes glabrous and granuliferous. Our plants are ssp. saxitilis (Besser) H. & C.---High mountains. The subspecies from Alberta to Alaska, south to Colorado and California; Siberia. Our records from northcentral and central Colorado at 11,000-13,000 feet.
21. Artemisia michauxiana Bess. in Hook., Fl. Bor. Am. 1:324. 1833.
   A discolor Dougl.; A. vulgaris ssp. discolor (Dougl.) H. & C.---Alberta to British Columbia, south to Wyoming and Washington. This plant has been reported for northern Colorado.
22. Artemisia dracunculus L., Sp. Pl. 849. 1753.
   Perennial herbaceous plants with rather thick nearly woody rootstock; stems 30-70 cm. tall, commonly reddish-tinged, glabrate to hairy; leaves 2-8 cm. long and 1-10 mm. wide, linear, lanceolate or oblong, entire or sometimes 1- to 3-cleft, glabrous to silky-canescent, sessile; upper leaves only slightly reduced; heads nodding, numerous in a leafy-bracted panicle, the branches ascending at least near base; involucres 2-3 mm. high and 2-4 mm. wide, glabrous or sparingly hairy; marginal flowers 6-30, lacking stamens but fertile; disk flowers 10-30, perfect but sterile, corollas 1.5-2 mm. long; receptacles glabrous; achenes glabrous. Intermediate forms between the 2 subspecies are present.
22A. Artemisia dracunculus dracunculus
   A. aromatica A. Nels.; A. dracunculus ssp. typica H. & C.---Leaves 2-10 mm. wide, glabrous; involucres 3-4 mm. wide.---Mountains. Our few records from southwestern and westcentral Colorado at about 9500-10,000 feet, but reported also from Summit County.
22B. Artemisia dracunculus glauca (Pallas) H. & C., (ssp.) Phylog. Meth. Tax. 116. 1923.
   A. glauca Pallas; A. dracunculoides Pursh---Leaves 1-3 mm. wide, glabrous to hairy; involucres 2-2.5 mm. wide.---Plains and mountains. Wisconsin to British Columbia, south to Texas and Mexico; Siberia. Our records scattered over Colorado, none as yet from the extreme eastern part, at 5000-9000 feet.
23. Artemisia longifolia Nutt., Gen. 2:142. 1818.
   A. natronensis A. Nels.; A. vulgaris ssp. longifolia (Nutt.) H. & C.---Saskatchewan to Alberta, south to Nebraska and Oregon. This species has been listed for Colorado.
24. Artemisia ludoviciana Nutt., Gen. 2:143. 1818.
   Perennial herbaceous plants; stems 25-100 cm. tall, tomentose to nearly glabrous; leaves very variable, from narrow and entire to broad and divided, sometimes even bipinnatifid, tomentose below, glabrate to tomentose above; heads in a spikelike to open panicle; involucres 2.5-4 mm. high and 2-4 mm. wide, glabrate to tomentose; marginal flowers 5-12, lacking stamens but fertile; disk flowers 6-30, perfect and fertile, corollas 1.5-3 mm. long; receptacles glabrous; achenes glabrous. The four subspecies intergrade somewhat in this state.
24A. Artemisia ludoviciana ludoviciana
   A. ludoviciana ssp. typica Keck; A. gnaphalodes Nutt.; A. purshiana Bess.; A. rhizomata A. Nels.; A. diversifolia Rydb.; A. silvicola Osterh.; A. brittonii Rydb.; A. pudica Rydb.; A. pabularis (A. Nels.) Rydb.; A. gnaphalodes var. diversifolia A. Nels.; A. mexicana silvicola (Osterh.) A. Nels.; A. cuneata Rydb.; A. argophylla Rydb.; A. vulgaris ssp. ludoviciana (Nutt.) H. & C.; A. vulgaris ssp. gnaphalodes (Nutt.) H. & C. ---Stems 30-100 cm. tall, tomentose to nearly glabrous; leaves 3-10 cm. long, linear to lanceolate, oblanceolate or elliptic, sometimes cuneate, entire, few-toothed or lobed especially near apex, the lobes entire, densely tomentose both sides or loosely floccose to green and glabrate above; heads in an elongated, usually compact panicle; involucres 3-4 mm. high and 2-3 mm. wide.---Plains and mountains. Ontario to Alberta, south to Arkansas, New Mexico and California; introduced in eastern United States. Our records well scattered over the state at 3500-8500 feet.
24B. Artemisia ludoviciana albula (Woot.) Keck, (ssp.) Proc. Calif. Acad. Sc. 25:446. 1946.
   A. albula Woot.---Stems 30-100 cm. tall, usually white-tomentose; leaves mostly 1-2 cm. long, mostly obovate to elliptic, toothed or lobed, margins often narrowly revolute, white-tomentose both sides or merely tomentulose above and bicolored; heads racemose on the branches of an open panicle; involucres about 3 mm. high and 2 mm. wide.---Plains and hills. Colorado, south to Texas, Mexico and California. Our records scattered in the eastern half of Colorado at 3500-7500 feet but also reported from Delta County.
24C. Artemisia ludoviciana incompta (Nutt.) Keck, (ssp.) Carn. Inst. Wash. Pub. 520:327. 1940.
   A. incompta Nutt.; A. discolor incompta (Nutt.) Gray; A. vulgaris ssp. flodmanii (Rydb.) H. & C.---Stems 25-50 cm. tall, more or less floccose or tomentose; leaves 2-8 cm. long, green and glabrate above, white-tomentose beneath or often tomentose throughout; lower leaves parted or divided into linear or lanceolate segments these often again toothed or lobed; upper leaves less dissected or even entire; heads in a narrow spikelike or racemelike panicle; involucres 3-3.5 mm. high and 2.5-4 mm. wide.---Hills and mountains. Montana to Idaho, south to Colorado and California. Our few records scattered in the western half of Colorado at 7000-10,000 feet.
24D. Artemisia ludoviciana mexicana (Willd.) Keck, (ssp.) Proc. Calif. Acad. Sci. 25:452. 1946.
   A. mexicana Willd.; A. underwoodii Rydb.; A. vulgaris ssp. mexicana (Willd.) H. & C.---Stems 30-80 cm. tall, floccose to tomentose; leaves about 3-10 cm. long, pinnately deeply lobed or parted into spreading entire segments, or sometimes ternately parted, tomentose below, glabrate to floccose above; upper leaves entire and often narrowly revolute; heads in a rather

dense leafy panicle; involucres 2.5-3.5 mm. high and 2-3 mm. wide.---Hills and mountains. Missouri to Colorado, south to Mexico. Our records from the western half of Colorado at 6000-7000 feet.

25. Artemisia carruthii Wood ex Carruth., Trans. Kan. Acad. Sci. 5:51. 1877.

A. wrightii A. Gray; A. coloradensis Osterh.; A. wrightii coloradensis (Osterh.) A. Nels.; A. bakeri Greene; A. mexicana bakeri (Greene) A. Nels.; A. kansana Britt.; A. vulgaris ssp. wrightii (Gray) H. & C.---Perennial herbaceous plants; stems 20-60 cm. tall, very leafy, more or less floccose or tomentose; leaves 1-3 cm. long, fascicled leaves often present, pinnately divided into linear-filiform segments about 0.5-1 mm. wide, margins revolute, these segments sometimes again toothed, tomentose below, green and glabrate to tomentose above; heads in a narrow or spikelike panicle; involucres 3-3.5 mm. high and 1.5-3 mm. wide, tomentulose; marginal flowers 6-14, lacking stamens but fertile; disk flowers 6-20, perfect and fertile, corollas 1.5-3 mm. long; receptacles glabrous; achenes glabrous.---Plains and mountains. Kansas to Utah, south to Texas, Mexico and Arizona. Our records scattered over Colorado, mostly in the southern half of the state, at 5500-9500 feet.

23. Pectis L. FETID-MARIGOLD

Low branching glabrous usually scented annual plants; stems slender, leafy; leaves opposite, linear, entire, dotted with pellucid glands, usually ciliate near base with a few stiff bristles; heads several to many, cymose; involucres cylindrical or turbinate; bracts in 1 series, round-carinate on back, gibbous at base, glandular-dotted above; receptacles small, naked; ray flowers pistillate and fertile, rays few, small but definite, yellow; disk flowers perfect and fertile, rather few, yellow; anthers united, not caudate at base; achenes slender, terete or somewhat angled; pappus of 4-5 scales, rarely a few awns besides.

1. Pectis angustifolia Torr., Ann. Lyc. N. Y. 2:214. 1828.

Diffuse, dichotomously branching, lemon-scented plants; stems about 5-20 cm. tall; leaves narrowly linear and rather fleshy; involucres about 4-5 mm. high, bracts 8-10, each with a large apical gland; rays about 3-4 mm. long.---Dry, often sandy ground. Nebraska to Colorado, south to Texas and Mexico. Our records scattered in the eastern half of Colorado at 3500-10,000 feet.

24. Dyssodia Cav. DOGWEED

Annual, mostly strong-scented plants; stems leafy, branched; leaves opposite or alternate, pinnately or bipinnately dissected or divided into linear or filiform segments, marked with conspicuous translucent glands; heads solitary or somewhat paniculate; involucres turbinate-hemispheric to broadly campanulate, bracts in about 1-2 series, about equal in length, more or less united, with glandular spots, outer shorter bracts usually present; receptacles naked or puberulent; ray flowers pistillate and fertile, rays inconspicuous, 1-6 mm. long, yellow; anthers united, not caudate at base; achenes angled or somewhat compressed; pappus of 8-15 paleae, these each erose at apex or dissected into 5-12 bristles.

1. Pappus scales dissected into 5-12 bristles, the whole about 3 mm. long; rays inconspicuous, about 1-1.5 mm. long; involucres 7-10 mm. high; leaf divisions spinulose-tipped   -1. D. papposa
1. Pappus scales truncate and erose at apex, no bristles present, the scale about 0.3-0.4 mm. long; rays rather conspicuous, about 3-6 mm. long; involucres 4-6 mm. high; leaf divisions not at all spinulose-tipped
                                                                                                 -2. D. aurea

1. Dyssodia papposa (Vent.) Hitchc., Trans. Acad. Sc. St. Louis 5:503. 1891.

Boebera papposa (Vent.) Rydb.---Stems 8-30 cm. tall; leaves mostly opposite, pinnately- to bipinnately-divided into narrowly linear segments, these spinulose-tipped; involucres 7-10 mm. long, outer bracts linear, green, inner oblong, scarious at apex and brownish-green or purplish-tinged, united at base only; rays few and inconspicuous, about 1 mm. long; achenes somewhat compressed; pappus of 8-15 short paleae but these each dissected into 5-12 capillary bristles, the whole about 3 mm. long.---Roadsides, fields and waste ground. Illinois to Montana, south to Louisiana and Arizona. Our records mostly from the eastern half of Colorado in cultivated areas at 3500-7500 feet.

2. Dyssodia aurea (A. Gray) A. Nels., Coult. & Nels. Man. 563. 1909.

Thymophylla aurea (A. Gray) Greene; Lowellia aurea A. Gray---Stems 5-15 cm. tall; leaves alternate or lower opposite, pinnately parted into linear-filiform segments these not spinulose-tipped; involucres 4-6 mm. high, bracts oblanceolate, green below, brown-scarious above, seemingly united to above the middle; rays 3-6 mm. long; achenes angled; pappus of oblong paleae about 0.3-0.4 mm. long, truncate and erose at apex.---Dry ground. Kansas and Colorado, south to Texas and Mexico. Our records mostly from the southeastern quarter of Colorado at 4000-6000 feet.

25. Chamaechaenactis Rydb.

Perennial caespitose herbaceous plants from branching caudex; leaves all basal, simple, oblong, ovate or suborbicular, entire or nearly so, coriaceous, strigose below, sparingly hirsute and punctate above; heads solitary; involucres turbinate, bracts about 12, in 2 rather distinct series, inner longer with scarious margins and reddish tips; receptacles naked; rays none; disk flowers perfect and fertile, flesh-colored; anthers united, not caudate at base; achenes clavate or obpyramidal, 4-angled, densely villous; pappus of about 8 hyaline scales, these spatulate with thick midribs.

1. **Chamaechaenactis scaposa** (Eastw.) Rydb., Bull. Torr. Club 33:156. 1906.
   <u>Actinella carnosa</u> A. Nels.---Scapes 3-8 cm. tall; leaves 8-12 mm. long; involucres about 13-16 mm. long, bracts obtuse, densely white-hairy.---Dry plains and mesas. Wyoming and Utah, south to Colorado and Arizona. Our few records from westcentral and southwestern Colorado at 4500-5500 feet.

26. Palafoxia Lag.

Annual glandular plants; stems leafy, usually solitary from a taproot; leaves alternate or lower opposite, narrow and entire, petioled; heads corymbose or paniculate; involucres turbinate to campanulate, bracts subequal, in 1-3 series, herbaceous but more or less colored at least at apex; receptacles small and flat, naked; ray flowers pistillate and fertile, rays broad, 3-cleft, rose-colored, or absent entirely; anthers united, not caudate at base; achenes obpyramidal, 4-angled; pappus of disk flowers of 6-10 lanceolate or ovate paleae, these usually over 4 mm. long, acuminate-attenuate at apex.

1. Rays present   -1. <u>P. sphacelata</u>
1. Rays absent   -2. <u>P. macrolepis</u>

1. **Palafoxia sphacelata** (Nutt.) Cory, Rhodora 48:86. 1946.
   <u>Othake sphacelata</u> (Nutt.) Rydb.; <u>Polypteris hookeriana</u> of Manuals for Colorado plants---Stems about 15-60 cm. tall, more or less strigose-hirsutulous especially above; leaves 3-8 cm. long, linear-lanceolate to lanceolate or sometimes linear; involucres 8-11 mm. high; rays 1-2 cm. long, deeply 3-cleft.---Sandy ground. Kansas and Colorado, south to Texas and Mexico. Our records scattered in the eastern half of Colorado at 3500-6000 feet.
2. **Palafoxia macrolepis** (Rydb.) Cory, Rhodora 48:86. 1946.
   <u>Othake macrolepis</u> Rydb.; O. texanum var. macrolepis (Rydb.) Ammerman---Stems 15-40 cm. tall, strigose-puberulent especially above; leaves 3-5 cm. long, linear to lanceolate; involucres 8-10 mm. high; rays absent. Possibly a discoid form of the preceding.---Plains. Wyoming and Colorado. Our few records from southeastern Colorado at 4000-5500 feet.

27. Hymenopappus L'Her.

Perennial or biennial herbaceous more or less floccose-tomentose plants; leaves alternate, often crowded near base, mostly once- or twice-pinnatifid; heads solitary or in corymbose cymes; involucres hemispheric to campanulate, bracts 6-12, in 1-2 subequal series, appressed and more or less petaloid-scarious on margins; receptacles small, naked; ray flowers none; disk flowers perfect and fertile, usually many, cream-colored to yellow; anthers united, not caudate at base; achenes obpyramidal, 4- to 5-angled, hairy; pappus of 10-20 minute hyaline scales sometimes obscured by the long apical achene hairs. This genus is badly in need of a revision and the following treatment is tentative, with numerous intergradations present between the species.

1. Plants biennial with leafy stems; plants limited to east of the Continental Divide in Colorado
    2. Involucre bracts conspicuously colored (usually yellow) at apex; corolla lobes about equalling the throat in length   -1. <u>H. corymbosus</u>
    2. Involucre bracts inconspicuously colored; corolla lobes shorter than the throat   -2. <u>H. tenuifolius</u>
1. Plants perennial with branching caudex, leaves mostly basal or cauline; plants generally distributed in Colorado
    3. Pappus about 1 mm. long or more, not hidden by the achene hairs; corolla throat often 3-4 times as long as the lobes (sometimes only twice as long); stems scapose, the cauline leaves much reduced in size and number
        4. Corolla throat about twice as long as the lobes
            5. Stems sparingly grayish-tomentose, glabrate in age; achenes loosely villous   -3. <u>H. cinereus</u>
            5. Stems permanently white-tomentose; achenes silky-hairy   -4. <u>H. arenosus</u>
        4. Corolla throat about 3-4 times longer than the lobes   -5. <u>H. lugens</u>
    3. Pappus minute and hidden by the hairs of the achene, possibly lacking entirely sometimes; corolla throat about twice as long as the lobes; stems often leafy with the cauline leaves not much reduced
        6. Ultimate segments of leaves 1-3 mm. long, crowded, permanently canescent to white-tomentose; stems 10-20 cm. long; plants apparently limited to northwestern Colorado   -6. <u>H. luteus</u>
        6. Ultimate leaf segments 3-20 mm. long, remote, glabrate; stems 25 cm. tall or more; plants widely scattered in Colorado   -7. <u>H. filifolius</u>

1. **Hymenopappus corymbosus** T. & G., Fl. N. Amer. 2:372. 1842.
   Reported from Kansas, Oklahoma, Texas and New Mexico. This plant is to be expected in southeastern Colorado and a few specimens almost check out here.
2. **Hymenopappus tenuifolius** Pursh, Fl. Am. Sept. 2:742. 1814.
   Biennial plants, caudex usually simple; stems 30-80 cm. tall, leafy, usually single, loosely floccose; leaves tending to crowd somewhat near base of plant, twice-pinnatifid into filiform segments, loosely floccose; heads several to many; involucres 7-8 mm. high, bracts with yellowish to whitish tips and margins but the yellow not very conspicuous; corolla throat 1-2 times longer than the lobes, cream-colored to light yellow; achenes densely pubescent; pappus about 1 mm. long or more, not hidden by the achene hairs.---Dry plains and hills. North Dakota to Wyoming, south to Texas and New Mexico. Our records well scattered in the eastern half of Colorado at 3500-7000 feet.

3. Hymenopappus cinereus Rydb., Bull. Torr. Club 27:634. 1900.

H. polycephalus Osterh.; H. filifolius var. cinereus (Rydb.) I. M. Johnston---Perennial plants with branching caudex; stems 15-30 cm. tall, mostly subscapose, sparingly grayish-floccose or glabrate; leaves mostly basal, bipinnatifid into linear segments, usually sparingly floccose; heads few; involucres 6-7 mm. high, bracts more or less yellowish at apex and margins; corolla throat about twice as long as the lobes; achenes loosely villous; pappus about 1 mm. long, not hidden by the achene hairs.---Dry hills and plains. Alberta, south to North Dakota, New Mexico and Utah. Our records scattered over the western two-thirds of Colorado at 4500-7000 feet.

4. Hymenopappus arenosus Heller, Bull. Torr. Club 25:200. 1898.

Perennial plants with branching caudex; stems 20-30 cm. tall, densely white-floccose to tomentose; leaves mostly basal, 2 or 3 times pinnatifid into narrowly linear segments; stem leaves reduced; heads few; involucres 5-7 mm. long, bracts with whitish or yellowish margins; corolla throat about twice as long as the lobes; achenes long-silky; pappus about 1 mm. long, not hidden by the achene hairs. Closely related to numbers 3 and 5, probably not specifically distinct from them.---Dry hills, valleys and plains. Colorado to New Mexico and Arizona. Our records from the western two-thirds of Colorado, mostly in the southern and none in the north-central parts, at 4500-9000 feet.

5. Hymenopappus lugens Greene, Pitt. 4:43. 1899.

H. macroglottis Rydb.; H. parvulus Greene; H. scaposus Rydb.---Perennial plants with more or less branching caudex; stems 10-50 cm. tall, sparingly to moderately floccose; leaves mostly basal, twice-pinnatifid into linear segments, sparingly to definitely tomentose; stem leaves reduced; heads 1 to few; involucres 6-9 mm. high, bracts with scarious and somewhat colored margins; corolla throat about 3-4 times as long as the lobes; achenes silky-hirsute; pappus about 1 mm. long or more, not hidden by the achene hairs.---Dry hills, ridges and plains. Colorado to Nevada, south to New Mexico and California. Our records mostly in the western half of Colorado at 5000-8000 feet.

6. Hymenopappus luteus Nutt., Trans. Am. Phil. Soc. II. 7:374. 1841.

Low subscapose perennial plants with branching caudex; stems 10-20 cm. tall, woolly at base; leaves mostly basal, twice-pinnatifid into short oblong or linear segments 1-3 mm. long, permanently tomentose or canescent; stem leaves few and rather reduced in size; heads few; involucres 5-6 mm. high, bracts with pale yellow tips, tomentose; corolla throat about twice as long as the lobes or less; achenes villous-hirsute; pappus minute and hidden by the hairs of the achene. Doubtfully distinct from the next.---Dry hills, valleys and plateaus. Reported from southwestern Wyoming. Our records from the northwestern corner of Colorado at about 6500 feet.

7. Hymenopappus filifolius Hook., Fl. Bor. Amer. 1:317. 1834.

Perennial plants from branching caudex in older plants at least; stems 25-50 cm. tall, floccose; leaves tending to crowd near base but cauline not much reduced, twice-pinnatifid into filiform segments 3-20 mm. long, sparingly floccose; heads rather few to many; involucres 5-7 mm. high, bracts with yellowish tips; corolla throat usually about twice as long as the lobes, at least longer; achenes silky-villous; pappus usually less than 1 mm. long, in any case almost or completely hidden by the long achene hairs.---Dry plains, hills and mesas. Saskatchewan, south to Kansas, New Mexico, Nevada and Washington. Our records scattered over Colorado, few in the extreme southern part, at 3500-8000 feet.

28. Chaenactis DC. FALSE YARROW

Annual, biennial or perennial plants, woody if at all only at the very base; stems leafy; leaves alternate, more or less 1- to 3-pinnate; heads solitary to corymbose; involucres campanulate to hemispheric, bracts in about 2 series but subequal, herbaceous; receptacles flat, alveolate but naked; marginal flowers often enlarged but not ligulate, flowers perfect and fertile, whitish or flesh-colored; achenes terete but somewhat compressed, tapering to base; pappus of hyaline linear-lanceolate to obovate scales, usually 4-10 present.

1. Pappus of 4 scales; plants annual; heads 8-10 mm. high -1. C. stevioides
1. Pappus of 8-10 scales (5 may be shorter); plants perennial or biennial (possibly annual in C. douglasii); heads 10-18 mm. high
    2. Perennial plants with branching caudex or with rootstocks; older plants often with more than 1 leaf crown from base
        3. Stems erect and corymbose, extending well beyond the basal leaf cluster, not subscapose -2. C. humilis
        3. Stems prostrate or caespitose, terminating in the leaf cluster, subscapose
            4. Leaves glabrate at least in age; involucres and peduncles densely hispid or hirsute -3. C. alpina
            4. Leaves lanate; involucres and peduncles lanate -3A. C. alpina leucopsis
    2. Biennial or short-lived perennial plants (possibly annual) with no well developed caudex or rootstocks present; single leafy crown present
        5. Plants sparingly floccose, in age glabrate, the leaves usually hispid or villous; stems stout, usually over 20 cm. tall -4. C. douglasii
        5. Plants copiously lanate, pilose or floccose; stems slender, rarely over 20 cm. tall
            6. Leaves ovate or elliptic in outline; plants green or red-tinged -4A. C. douglasii achilleaefolia
            6. Leaves linear-lanceolate; plants white or gray -5. C. angustifolia

1. Chaenactis stevioides Hook. and Arn., Bot. Beech. 353. 1840.

Annual plants; stems 10-35 cm. tall, few from the base; leaves 2-5 cm. long, pinnate or occasionally bipinnate into linear obtuse segments about 2-10 mm. long; peduncles short, glandular; heads 8-10 mm. long; involucres 5-7 mm. high, bracts linear, acute, glandular; corollas 4-5 mm. long, white; achenes 4-5 mm. long, gray-strigose; pappus of 4 scales.---Dry places. Idaho to California, south to New Mexico and Arizona. Our records from western Colorado at 4500-6500 feet.

2. **Chaenactis humilis** Rydb., N. Am. Fl. 34:72. 1914.
Wyoming, Idaho, Colorado and Utah. Reported from western Colorado.
3. **Chaenactis alpina** (A. Gray) M. E. Jones, Proc. Acad. Sc. Calif. II. 5:699. 1895.
C. pedicularia Greene---Perennial plants with thick roots and short rootstocks, each terminated by a caespitose leaf cluster; stems subscapose, 5-20 cm. tall; leaves 2-6 cm. long, elliptic to narrower in outline, pinnate, the segments obscurely divided, lanate at least when young, glabrate in age; peduncles 1-5 cm. long, short-tomentose; heads 15-17 mm. long; involucres 10-14 mm. high, bracts oblong to linear, obtuse to acute, canescent; corollas 5-6 mm. long, whitish or flesh-colored, glandular-puberulent to puberulent; achenes 7-8 mm. long, hirsute to strigose; pappus scales 10, oblanceolate, about 2.5-3 mm. long, 5 somewhat shorter.---High mountains. Montana to Oregon, south to Colorado and Idaho. Our records from the higher mountains of Colorado, except the southcentral part, at 8000-13,000 feet.
3A. **Chaenactis alpina leucopsis** (Greene) Stockwell, (var.) Contrib. Dudley Herb. 3:114. 1940.
Reported for southwestern Colorado but not located in this study.
4. **Chaenactis douglasii** (Hook.) H. & A., Bot. Beech. Suppl. 354. 1840.
Biennial or possibly annual plants; stems 20-35 cm. tall, erect; leaves 2-10 cm. long, deltoid or ovate in outline, 2- to 3-pinnate, sparingly to copiously pilose to sparingly tomentose; basal rosette of leaves present at least in withered condition; heads corymbose, 14-18 mm. long, on glandular peduncles 2-10 cm. long; involucres 12-16 mm. high, bracts oblong to linear, acute, glandular; corollas about 6 mm. long, whitish to light rose, glandular-puberulent below; achenes about 6 mm. long, somewhat 4-sided, sparingly glandular and strigose; pappus 3-4.5 mm. long, usually of 10 scales, 5 shorter, linear to narrowly oblanceolate.---Plains and hills, often on sandy or rocky ground. Montana to British Columbia, south to Colorado and Nevada. Our many records scattered in the western half of Colorado at 5500-10,500 feet.
4A. **Chaenactis douglasii achilleaefolia** (H. & A.) A. Nels., (var.) Coulter & Nelson Man. 557. 1909.
Stems less than 20 cm. tall, base often rather woody; leaf segments crowded, whitish-tomentose or lanate.---With the species generally and in Colorado, our records from 6000-10,000 feet.
5. **Chaenactis angustifolia** Greene, Leafl. Bot. Obs. and Crit. 2:223. 1912.
Plants perennial; stems 5-25 cm. tall, erect, corymbose and usually single, canescent; leaves 2-6 cm. long, linear-lanceolate in outline, 1- to 2-pinnate, permanently canescent; heads 2-6, 14-16 mm. long, on short peduncles; involucres about 10 mm. high, bracts linear to lanceolate, acute; corollas 6-7 mm. long, whitish to flesh-colored; achenes 6-7 mm. long, stout-clavate, strigose, marginal usually corky; pappus of 5 obtuse scales about 6-7 mm. long and 5 short acute scales.---Dry rocky or sandy slopes. Montana to Idaho, south to Colorado and Utah. Our one record from central Colorado at about 6500 feet, but the plant should be looked for in the northern and western parts.

29. Pericome A. Gray

Perennial herbaceous plants; stems tall, leafy and branching; leaves opposite, blades hastate to triangular, acuminate to caudate-acuminate at apex, often cordate at base, 3- to 5-nerved, petioled; heads in rather large cymes; involucres turbinate-campanulate, bracts in 1 row, narrow, with hyaline margins and thickened midribs, more or less united; receptacles conical, naked; ray flowers none; disk flowers perfect and fertile, pale yellow; anthers united, subsagittate at base; achenes linear-oblong, compressed, villous-ciliate on the callus margins; pappus a crown of lacerate-ciliate paleae sometimes with 1 or 2 longer awns in addition.

1. Stems not glandular; leaves 5-12 cm. long, hastate to truncate at base, long caudate-acuminate at apex
   -1. P. caudata
1. Stems glandular; leaves 2-4 cm. long, truncate or cordate, seldom really hastate at base, not long caudate-acuminate at apex
   -1A. P. caudata glandulosa

1. **Pericome caudata** A. Gray, Pl. Wright. 2:82. 1853.
Stems 80-150 cm. tall, minutely puberulent; leaves 5-12 cm. long, hastate or deltoid, entire or sinuate, truncate or hastate at base, caudate-acuminate at apex, minutely puberulent and punctate; petioles 1-3 cm. long; involucres 5-6 mm. high; crown of pappus about 1 mm. long, awns when present about 4 mm. long.---Canyons and rocky hills. Colorado to California, south to Mexico. Our records from central and southcentral Colorado at 7000-9000 feet.
1A. **Pericome caudata glandulosa** (Goodm.) comb. nov. (var.)
P. glandulosa Goodm.---Stems glandular and puberulent; leaves 2-4 cm. long, broadly ovate or cordate, rarely deltoid, acuminate, glandular both sides; petioles about 1 cm. long. Intergrades commonly with the species in this state.---Hills. Oklahoma and Colorado. Our few records from the southeastern corner of the state at 4000-4400 feet.

30. Leucampyx A. Gray WILD COSMOS

Perennial more or less floccose herbaceous plants; stems sparingly leafy; leaves alternate, mostly basal, 2- to 3-pinnately parted or divided into linear segments; heads loosely corymbose; involucres hemispheric, bracts in 2-3 series but subequal, broad with wide scarious tips and margins; receptacles convex, with broad membranous chaffy paleae half enclosing the achenes; ray flowers pistillate and fertile, rays conspicuous, 3-toothed at apex, white or cream-colored; disk flowers perfect and fertile; anthers united, not sagittate at base; ray achenes cuneate, 3-angled, disk achenes 4- to 5-angled, all achenes more or less incurved, striate, muriculate; pappus a minute crown or none.

1. Leucampyx newberryi A. Gray in Porter & Coult., Syn. Fl. Colo. 77. 1874.
Stems 20-60 cm. tall; leaves 10-15 cm. long; involucres 8-10 mm. high, bracts obovate and obtuse; rays 10-15 mm. long; achenes about 4-5 mm. long, blackish.---Canyons and slopes, often in rocky ground. Colorado and New Mexico. Our records in central, southcentral and southwestern Colorado at 6500-9000 feet.

31. Gaillardia Foug. BLANKETFLOWER

Annual or perennial herbaceous plants; stems leafy at base or throughout; leaves alternate or crowded basal, entire, toothed or pinnatifid; heads peduncled, solitary on stem or branches; involucres rotate to saucer-shaped, bracts in 2-3 series, ovate or lanceolate, strongly reflexed in fruit; receptacles convex, alveolate, bearing subulate or setiform fimbrillae; ray flowers usually neutral, rarely pistillate, rays usually present and conspicuous, yellow to reddish-purple; disk flowers perfect and fertile, lobes bearing moniliform hairs, yellow to purple; anthers united, auricled at base but not caudate; achenes obpyramidal, at least partly hairy; pappus of lanceolate paleae, these awn-pointed or with an awn twice as long as the body.

1. At least some of the leaves pinnately parted; corolla lobes of disk flowers broadly triangular and short; pappus awns about 1/2 as long as their basal scale; plants perennial with involucres 8-10 mm. high
-1. G. pinnatifida
1. Leaves entire, toothed or lobed, very rarely pinnately parted; corolla lobes of disk flowers triangular-acuminate; pappus awns as long or longer than their basal scale; plants either annual or if perennial then the involucres 10-20 mm. high
    2. Plants annual; involucres 8-10 mm. high; disk 12-16 mm. wide; pappus awns about as long as the basal scales; rays usually reddish, yellow only at apex -2. G. pulchella
    2. Plants perennial, possibly biennial; involucres 10-20 mm. high; disk 15-30 mm. wide; pappus awns about twice as long as their scales; rays usually yellow, reddish only near base -3. G. aristata

1. Gaillardia pinnatifida Torr., Ann. Lyc. N. Y. 2:214. 1828.
Plants of various aspects, perennial but young plants appearing biennial, usually with a caudex; stems 15-60 cm. tall, more or less appressed hairy; leaves mostly on lower part of stem, 3-8 cm. long, oblanceolate in outline, at least some pinnatifid-parted, more or less hairy with moniliform hairs; disk about 15-20 mm. wide; involucres 8-10 mm. high, bracts lanceolate and acuminate; rays about 10-15 mm. long, yellow but often reddish at base, usually with reddish-purple veins; disk flowers reddish-purple at least on the broadly triangular lobes; achenes about 3 mm. long, densely silky-hairy; pappus about 3-6 mm. long, the awned apex about 1/2 as long as the body.---Dry plains and hills. Colorado and Utah, south to Texas and Mexico. Our records scattered in the southern half of Colorado at 4000-6500 feet.
2. Gaillardia pulchella Foug., in Mem. Acad. Sci. Paris 1786:5. 1788.
Annual plants; stems 20-50 cm. tall, branched, pubescent or appressed pubescent with moniliform hairs, leafy; leaves 2-8 cm. long, oblong or oblanceolate, entire to coarsely toothed or lobed, resinous-dotted, sparingly hairy and more or less ciliate with moniliform hairs; disks 12-16 mm. wide; involucres about 8-10 mm. high, bracts lanceolate and acute; rays about 8-20 mm. long, usually reddish with yellow apex, sometimes all reddish or all yellowish; disk flowers purple at least on the triangular-acuminate lobes; achenes 2-2.5 mm. long, white-hirsute at least below; pappus about 3-6 mm. long, the apical awn about as long as the basal scale.---Plains and hills. Virginia, Missouri and Colorado, south to Florida and Arizona. Our rather few records from the southeastern quarter of Colorado at 3500-6000 feet.
3. Gaillardia aristata Pursh, Fl. Am. Sept. 2:573. 1814.
Perennial plants but often appearing almost annual when young; stems 25-60 cm. tall, hirsute or villous with jointed hairs, leafy; leaves 5-20 cm. long, oblanceolate to linear-lanceolate, entire to coarsely toothed or pinnately lobed, occasionally pinnately-parted, more or less hirsute-pubescent with jointed hairs; disks 15-30 mm. wide; involucres 1-2 cm. high, bracts lanceolate and long-acuminate; rays about 1-2.5 cm. long, yellow or reddish-purple at base; disk flowers reddish-purple at least on the triangular-acuminate lobes; achenes about 4 mm. long, with long silky brown or white hairs, at least near base; pappus 3-7 mm. long, the apical awn about twice as long as the basal scale.---Hills and plains. Saskatchewan to Alberta, south to Colorado, Utah and Oregon. Our records mostly from the western half of Colorado, none in the westcentral and southwestern parts, at 5000-9000 feet.

32. Flaveria Juss.

Annual plants; stems leafy; leaves opposite, narrow, sessile or slightly connate; heads in close subsessile clusters at the ends of the stems or branches; involucres cylindric, bracts about 2-5, equal or nearly so but sometimes with 1 or 2 smaller outer ones; receptacles small, naked; ray flowers pistillate and fertile, usually 1 to a head, rays yellowish and very inconspicuous; disk flowers usually 3-5, perfect and fertile, yellow; anthers united, not caudate at base; achenes linear-oblong, 8- to 10-ribbed, glabrous; pappus none.

1. Flaveria campestris J. R. Johnston, Proc. Am. Acad. 39:287. 1903.
F. angustifolia of Manuals---Stems 20-60 cm. tall, glabrous or villous at the nodes; leaves 2.5-6.5 cm. long, linear to lanceolate, serrulate or the upper nearly entire, more or less distinctly 3-ribbed, glabrous; clusters of heads leafy; involucres about 5-6 mm. high, longer inner bracts mostly 3; rays about 1-2 mm. long.---Rather low, mostly alkaline soil. Missouri to Colorado, south to Texas and New Mexico. Our records from southeastern and westcentral Colorado at 4000-4500 feet but also reported from the southcentral part.

33. Psilostrophe DC. PAPERFLOWER

Perennial or rarely biennial herbaceous or sometimes suffrutescent, more or less woolly or tomentose plants; leaves alternate, entire or lobed; heads corymbose; involucres cylindrical to campanulate, bracts connivent in a single series, sometimes a few small scarious ones within; receptacles small, naked; ray flowers pistillate and fertile, rays 3-lobed, yellow, becoming papery in age and persistent on the achenes; disk flowers perfect and fertile, yellow; anthers united, not caudate at base; achenes linear, striate; pappus of about 4-6 scales.

1. Pappus scales obtuse, ovate to broadly oblong, about 1/3 as long as the disk corollas; rays over 8 mm. long
-1. P. bakeri
1. Pappus scales acute, lanceolate to subulate, about 1/2 as long as the disk corollas; rays less than 8 mm. long
-2. P. tagetina

1. Psilostrophe bakeri Greene, Pl. Baker. 3:29. 1901.
Plants with taproot and woody caudex; stems 10-20 cm. tall, branching, long-villous or loosely floccose; lower leaves 5-10 cm. long, spatulate or oblanceolate, entire or rarely more or less lobed, densely villous to floccose; upper leaves oblong to linear-oblanceolate; involucres 7-9 mm. high; rays about 4-6, 8-15 mm. long; pappus scales usually about 1/3 as long as the disk corollas, sometimes nearly half as long, ovate to broadly oblong, rounded or obtuse, denticulate.---Dry plains and hills. Colorado and Utah. Our records from western Colorado at 4500-6000 feet.

2. Psilostrophe tagetina (Nutt.) Greene, Pitt. 2:176. 1891.
Dry plains and hills. Texas to Arizona, south to Mexico and possibly north to southern Colorado.

34. Helenium L. SNEEZEWEED

Perennial herbaceous plants; stems leafy; leaves alternate, entire or toothed, sometimes decurrent; heads corymbose; involucres nearly or quite rotate, bracts in 2 or 3 equal or subequal series, linear, subulate to lanc-subulate, spreading or becoming reflexed; receptacles flat, convex or conic, naked or sometimes with a few chaffy scales between the ray and disk flowers; ray flowers pistillate and fertile, rays conspicuous, yellow, golden-yellow or orange; disk flowers perfect and fertile, numerous, usually yellow, anthers united, not caudate at base; achenes turbinate or obpyramidal, 4- to 10-ribbed; pappus of 5-8 thin and acuminate, sometimes even awn-tipped paleae.

1. Stem leaves sessile or short-petioled but not at all decurrent; rays 15-30 mm. long; pappus 2-3 mm. long
-1. H. hoopesii
1. Stem leaves decurrent forming wings on the stem; rays 6-10 mm. long; pappus about 1-1.5 mm. long
-2. H. autumnale

1. Helenium hoopesii A. Gray, Proc. Acad. Nat. Sci. Phila. 1863:65. 1863.
Dugaldia hoopesii (A. Gray) Rydb.---Stems 30-100 cm. tall, more or less puberulent, pubescent or tomentose when young, soon glabrate; basal leaves 10-30 cm. long, oblanceolate to oblong-oblanceolate, entire, wing-petioled; upper leaves lanceolate to lanc-ovate, sessile but not decurrent; involucres 7-10 mm. high; rays 15-30 mm. long, golden-yellow to orange; pappus about 2-3 mm. long.---Valleys, hills and mountains. Wyoming to Oregon, south to New Mexico and California. Our records well scattered over the western half of Colorado at 6000-11,500 feet.

2. Helenium autumnale L., Sp. Pl. 886. 1753.
H. montanum Nutt.---Stems 30-100 cm. tall, more or less puberulent; leaves 5-10 cm. long, lanceolate, denticulate to subentire, decurrent; involucres about 7-9 mm. high; rays 6-10 mm. long, yellow; pappus about 1-1.5 mm. long.---Meadows. Quebec to British Columbia, south to Florida, Texas and Arizona. Our records from the northeastern quarter of Colorado at 3500-8500 feet.

35. Hymenoxys Cass. ACTINEA

(Contributed by K. F. Parker)

Caespitose perennials or rarely annuals, from a strong taproot; caudex simple or multicipital, herbaceous or woody; plants caulescent or scapose; stems simple or greatly branched, ribbed, especially beneath the heads; leaves alternate, simple, forming basal rosettes, or both basal and sparingly to densely cauline, entire, short-lobed or pinnatifid into 2-30 narrowly linear lobes, usually glandular-punctate with small, impressed or sunken glands; heads radiate, rarely discoid, small to large, disks (excluding rays) 3-35 mm. broad, solitary on the stem or on the branches, often corymbose; involucres campanulate, hemispherical; involucral bracts in 2 definite or 3 indefinite series, outer distinct or united 1/4 to 1/2 their length, inner similar to outer or dissimilar and rigid, subequal or longer, always distinct to base; receptacles conical or convex, naked; ray flowers in 1 series, pistillate, fertile, corollas bright yellow, but whitish and sharply reflexed in age, showy; disk flowers many, perfect, fertile; anthers united, not caudate at base; achenes 1.5-5 mm. long, turbinate or obpyramidal, 5-angled, densely hirsute to silky-villous; pappus of 5-8 paleaceous scales, these ovate to narrowly lanceolate, obtuse, acute, acuminate, or midrib prolonged into a short awn, whitish.

1. Involucral bracts similar, in 2 to 3 indefinite series, subequal, none united, distinct to the base; heads solitary; plants caulescent or scapose
  2. Plants strictly scapose; scapes always simple, leafless; leaves all basal
    3. Leaves all entire, not lobed
      4. New leaves crowded on the short, stout caudex, not imbricated along the base of the stem; pappus scales 2.5-4 mm. long   -1. H. acaulis
      4. New leaves densely imbricated on the slender elongated branches of the caudex and base of the stem; pappus scales mostly 1-2.5 mm. long   -2. H. scaposa
    3. Leaves entire or with 2-5 short lobes   -2A. H. scaposa scaposa
  2. Plants caulescent; stems simple or often branched, 1 to many leaves; leaves both basal and 1 to many cauline
    5. Leaves all entire, not lobed nor pinnatifid; heads 8-20 mm. broad   -1D. H. acaulis ivesiana
    5. Leaves (at least some) pinnatifid into 2 to many linear segments; heads 1.5-3.5 mm. broad
      6. Stems simple; some basal leaves always entire, some divided into 2-5 linear segments   -3. H. brandegei
      6. Stems usually branching; basal leaves divided into 3 to many linear segments   -4. H. grandiflora
1. Involucral bracts dissimilar, rigid, in 2 definite series, outer thickened at base, united $\frac{1}{4}$ to $\frac{1}{2}$ their length, inner coriaceous, longer than outer at maturity, distinct to base; heads few to many, corymbose; caulescent plants with very leafy stems
  7. Plants perennial from a woody, simple or multicipital caudex; basal leaves imbricated into several to many tufts; plants branching mostly in the inflorescence
    8. Leaves 0.5-2.5 mm. broad; basal leaf axils usually densely woolly   -5. H. richardsonii
    8. Leaves 2-11 mm. broad; basal leaf axils not woolly   -6. H. helenioides
  7. Plants annual; basal leaves forming a large, single, quickly-withering rosette; plants greatly branching throughout   -7. H. odorata

1. **Hymenoxys acaulis** (Pursh) Parker, Madrono 10:159. 1950.
Gaillardia acaulis Pursh; Actinella acaulis (Pursh) Nutt.; Actinea acaulis (Pursh) Spreng.; Actinella depressa Torr. & Gray var. pygmaea A. Gray; Tetraneuris simplex A. Nels.; Tetraneuris incana A. Nels.; Tetraneuris eradiata A. Nels.; Tetraneuris septentrionalis Rydb.; Actinea osterhoutii A. Nels.---Caespitose perennials from a short thick multicipital caudex; acaulescent with unbranched scapes (or few-branched in var. ivesiana), plants dwarfs or the scapes 2-50 cm. high, densely villous to glabrate; leaves 3-6 cm. long, 2-8 (or 10) mm. broad, all basal, or sometimes a single reduced leaf on scape (more in var. invesiana), densely to sparingly villous to glabrous, densely glandular punctate to epunctate, linear-oblanceolate to nearly linear, entire; heads solitary, 7-20 mm. broad, radiate except in 1 form; involucral bracts distinct, in 2-3 indefinite subequal series, 4-8 mm. high; ray flowers 5-19, corollas 6-15 mm. long; achenes 2.5-4.5 mm. long; pappus scales 5-7 (rarely 9), 2.5-4.5 mm. long, lanceolate or ovate-lanceolate, obtuse, acute or short-awned. This complex species is a tetraploid (2n=60) which accounts for its extreme variability. It is segregated into varieties on the basis of the type of leaf pubescence (not quantity), the presence or absence of punctate glands, the position of the leaves, and whether the plant is caulescent or acaulescent.---Dry, often rocky slopes, plains, hills and mountains. Widely distributed from south central Canada southward to California and eastward to Texas; flowering from April to September, mostly in June; at 3500-12,000 feet elevation.

Key to the Colorado Varieties of HYMENOXYS ACAULIS

1. Leaves densely or sparingly silky-sericeous, often silvery; rays usually present, sometimes lacking   -1A. H. acaulis acaulis
1. Leaves densely or sparingly lanate villous to glabrate; rays always present
  2. Plants acaulescent with unbranched scapes; leaves all basal
    3. Leaves all densely and conspicuously glandular punctate   -1B. H. acaulis arizonica
    3. Leaves not all glandular punctate, both punctate and epunctate or entirely epunctate   -1C. H. acaulis caespitosa
  2. Plants caulescent with simple or few branched stems; leaves both basal and cauline   -1D. H. acaulis ivesiana

1A. **H. acaulis acaulis**
Only part of the complex with silky, appressed leaf pubescence; all others have villous leaf pubescence. Widely distributed on the plains from southcentral Canada to Texas. Our records scattered over Colorado at 3500-9600 feet, particularly common in the eastern and entire central portion. Rayless colonies are common in southern Pueblo County.

1B. **H. acaulis** (Pursh) Parker **arizonica** (Greene) Parker, (var.) Madrono 10:159. 1950.
Tetraneuris arizonica Greene; Actinea acaulis (Pursh) Spreng. var. lanata Macbr. forma arizonica (Greene) Macbr.; Actinea arizonica (Greene) A. Nels.; Actinea acaulis (Pursh) Spreng. subsp. arizonica (Greene) Blake---Leaves always covered on both surfaces with small punctate glands. Common in the mountains and plains from southern Idaho to northern Arizona, westward to southern California. Our records scattered over Colorado, at 5000-11,000 feet, mostly in the western and central part of the state.

1C. **H. acaulis** (Pursh) Parker **caespitosa** (A. Nels.) Parker, (var.) Madrono 10:159. 1950.
Actinea integrifolia Torr.; Actinella lanata Nutt.; Tetraneuris brevifolia Greene; Tetraneuris acaulis (Pursh) Greene var. caespitosa A. Nels.; Actinella epunctata A. Nels.; Tetraneuris crandallii Rydb.; Tetraneuris lanigera Daniels; Actinea acaulis (Pursh) Spreng. var. lanata Macbr.; Actinea acaulis (Pursh) Spreng. var. lanigera Blake---An extremely variable entity. Altitude seems to have little effect on these variations. Leaves from collections from the Grand Mesa at 10,500 feet are glabrate while those from Monarch Pass are densely

woolly. Some leaves, at least the lowest and shortest, are always glandless in this variety, while all are conspicuously punctate in var. arizonica. From sagebrush hillsides to above timberline in southern Wyoming, Colorado, Utah, and northern New Mexico. Our records scattered over Colorado, except the eastern part, at 5400-14,000 feet.

1D. H. acaulis (Pursh) Parker invesiana (Greene) Parker, (var.) Madrono 10:159. 1950.
Tetraneuris ivesiana Greene; Tetraneuris mancosensis A. Nels.; Tetraneuris intermedia Greene; Actinea leptoclada (A. Gray) Kuntze var. ivesiana (Greene) Macbr.; Tetraneuris pilosa Greene---It is impractical to maintain this variety as a separate species. Typically it is distinct from var. arizonica since it bears cauline leaves and sometimes the stems are branched. It is not unusual, however, for an unbranched stem to bear only 1 reduced leaf or often none and thus the plant is indistinguishable from var. arizonica. Frequently too, var. arizonica may bear a small leaf on the scape. Pinyon-juniper-yellow pine associations of western Colorado, southeastern Utah, northern New Mexico, and northcentral Arizona. Our records from westcentral Colorado and southwestern Colorado, at 5000-9500 feet.

2. Hymenoxys scaposa (DC.) Parker, Madrono 10:159. 1950.
Cephalophora scaposa DC.; Actinella scaposa Nutt.; Actinea scaposa Kuntze; Tetraneuris angustata Greene; Tetraneuris angustifolia Greene---Scapose perennials, caudex usually very strongly developed, woody, with many elongated branches covered with imbricated bases of old and new leaves; scapes 10-36 cm. high, densely to sparingly strigose; leaves all densely crowded at the base, imbricated 1-5 cm. along stem base, 2-11 cm. long, 1-14 (mostly 1-2) mm. broad, villous to glabrate, linear to linear-oblanceolate, entire (rarely with 1-5 short lobes), glandular-punctate; heads solitary, 9-19 mm. broad; involucral bracts in 2-3 indefinite series, 4-6 mm. high, distinct, subequal, villous; ray flowers 12-25, corollas 7-20 mm. long, yellow-orange, turning whitish, veins purplish brown beneath; achenes 2-3.5 mm. long; pappus scales 5-7, 1-2.5 mm. long, oval to lanceolate, obtuse, acute, or midrib prolonged into a short awn. Species represented by 2 varieties.

2A. H. scaposa scaposa
Plants with poorly developed caudex, never bushy nor woody; leaves narrowly linear or broader, 2-14 mm. broad, entire or with 1-5 short lobes. Common in Mexico, Texas, New Mexico to Kansas, rare in Colorado. Known only from one collection in central Colorado (Chaffee County) at 8000 feet.

2B. H. scaposa (DC.) Parker linearis (Nutt.) Parker, (var.) Madrono 10:159. 1950.
Actinella scaposa (DC.) Nutt. var. linearis Nutt.; Tetraneuris fastigiata Greene; Tetraneuris stenophylla Rydb.; Actinea scaposa (DC.) Kuntze var. linearis (Nutt.) Robins.---Plants with a many branched, woody caudex, bushy; leaves all narrowly linear,1-2 (rarely 3) mm. broad, entire, never lobed, imbricated in distinct fascicled tufts around the base of each stem. Common on sandy plains from Mexico to Kansas. Scattered over eastern Colorado, at 3500-5000 feet.

3. Hymenoxys brandegei (Porter) Parker, Madrono 10:159. 1950.
Actinella grandiflora Torr. & Gray var. glabrata Porter; Actinella brandegei Porter ex A. Gray; Actinea brandegei (Porter) Kuntze; Rydbergia brandegei (Porter) Rydb.---Low perennials with a fleshy taproot and a stout branching caudex; stems unbranched, 6-20 cm. high, woolly above, glabrate below; leaves 4-11 cm. long, mostly basal, few cauline, some entire, and some with 2-3 (rarely 5) linear lobes on the upper part of the leaf, glabrate or sparingly villous; heads solitary, 1.5-2 cm. broad; peduncles 0.5-5 cm. long; involucral bracts in 2-3 indefinite series, distinct to base, cinereous, all but the tips often hidden by dense wool, outer bracts 7-13 mm. high, inner shorter; ray flowers 13-19, corollas 11-22 mm. long, with about 8-11 indistinct veins; achenes 2-3 mm. long; pappus scales 5-6, 3-5 mm. long, lanc-attenuate, often lead-colored.---Uncommon; above timberline on the alpine mountain peaks of southern Colorado, northern New Mexico, and Arizona where it replaces H. grandiflora. Our records from southcentral Colorado, at 11,000-13,000 feet.

4. Hymenoxys grandiflora (T. & G.) Parker, Madrono 10:159. 1950.
Actinella grandiflora T. & G.; Actinea grandiflora (T. & G.) Kuntze; Rydbergia grandiflora (T. & G.) Greene---Stout, low-growing perennials with a fleshy taproot and usually a simple caudex; stems stout, 3-30 cm. high, 1 to several, simple or usually branched below, densely floccose-woolly or villous; leaves basal and cauline, 2-12 cm. long, 2-3 ternately or quinately, somewhat palmately divided into 2-30 linear segments, floccose-lanate to glabrate; heads solitary, largest in the genus, 1.5-3.5 cm. broad; peduncles 0.5-4 cm. high; involucral bracts numerous, in 2-3 indefinite series, distinct to the base, 10-16 cm. long, subequal or outer longer, densely woolly; ray flowers 19-54, corollas 26-34 mm. long, 8-12 veins; achenes 3-5 mm. long; pappus scales 5-8, 3.5-7 (rarely 8) mm. long, whitish, attenuate.---Common on high peaks and alpine meadows from southwestern Montana and eastcentral Idaho south to Colorado and Utah. Throughout Colorado, except in extreme southcentral part, at 9000-14,000 feet.

5. Hymenoxys richardsonii (Hook.) Ckll., Bull. Torrey Club 31:468. 1904.
Picradenia richardsonii Hook.; Actinella richardsonii (Hook.) Nutt.; Actinea richardsonii (Hook.) Kuntze; Hymenopappus ligulaeaflorus A. Nels.; Picradenia ligulaeflora (A. Nels.) A. Nels.; Hymenoxys macrantha (Ckll.) Rydb.; Hymenoxys macounii (Ckll.) Rydb.---Bushy, caespitose perennials; caudex woody, many branched, covered by dead leaf bases usually with densely matted, woolly axils; stems numerous, 6-45 cm. high, sparingly to greatly branched above, scattered pubescence, granuliferous; leaves forming basal tufts around each stem, also densely cauline, 2-15 cm. long, entire, or mostly divided into 3-7 linear segments, fleshy, pubescent to glabrate; heads 1 to many, in flat-topped corymbs, 4-19 mm. broad; peduncles 1-6 cm. long; involucral bracts in 2 definite, dissimilar series, outer 6-16, 5-8 mm. high, united ¼ to ½ their length, lanceolate, acute to acuminate, carinate to strongly keeled, inner distinct to base, as long or longer than the outer, cuneate-obovate, cuspidate, coriaceous, rounded, upper margins erose, minutely floccose; ray flowers 9-14, corollas 8-23 mm. long; achenes 2.5-4 mm. long; pappus scales 5-6 (rarely 7), 1.5-4.5 mm. long, acuminate or short-awned.

---Common on dry rocky or clay soil, of plains and mountain slopes. Widely distributed from southwestern Canada to Texas at 1500-12,000 feet.

5A. H. richardsonii richardsonii

Short plants, 6-13 (rarely 19) cm. high; heads few, 1-5 to each stem, rarely more, 7-19 mm. broad; outer involucral bracts 8-16 (at least some heads with 10), 5-8 mm. high; ray flowers 9-14; pappus scales 2.5-4.5 (mostly 3-3.5) mm. long; flowering from May to late July, uncommon later.---Common on plains and mountains of southern Alberta and Saskatchewan southward to Montana, Wyoming, northeastern Utah, and Colorado; occasionally in Idaho, North and South Dakota and northern New Mexico. Found in western and northcentral Colorado at 4500-10,000 feet.

5B. H. richardsonii (Hook.) Ckll. floribunda (A. Gray) Parker, (var.) Madrono 10:159. 1950.

Actinella richardsonii (Hook.) Ckll. var. floribunda A. Gray; Picradenia floribunda (A. Gray) Greene; Picradenia earlei Ckll.; Picradenia intermedia A. Heller; Hymenoxys metcalfei Ckll.; Hymenoxys olivacea Ckll.---Taller (13-45 cm. high), and bushier than var. richardsonii with the caudex branches 5-10 cm. above ground in old plants; heads more numerous, 5-90 to each stem, smaller, 4-10 mm. broad; outer involucral bracts fewer, 6-9, usually shorter, 3-6 mm. long, typically carinate with a keeled ridge on the back; pappus scales mostly shorter, 1.5-3 mm. long; flowering later, usually from July to September.---The northern limit of this variety is Colorado and Utah, thence southward to northcentral Arizona, New Mexico, and occasional in southwestern Texas. More common in Colorado than var. richardsonii. Occurs on mountain slopes, mostly in the southwestern and southcentral part of the state at 6500-12,000 feet.

6. Hymenoxys helenioides (Rydb.) Ckll., Bull. Torrey Club 31:481. 1904.

Picradenia helenioides Rydb.; Dugaldia helenioides (Rydb.) A. Nels.---Tall perennials with a simple or multicipital caudex; stems stout, 15-50 cm. high, scurfy, branched above, finely pubescent to glabrate; leaves flat, 2-19 cm. long, 2-11 mm. broad, linear-oblanceolate, entire or with 2-5 segments, finely glandular-punctate, puberulent; heads few or several, 12-18 mm. broad, corymbose; peduncles 1-6 cm. long; involucral bracts in 2 definite unlike series, 6-8 mm. high, outer 9-17, carinate in age, united 1/4-1/3 of their length, acuminate, inner 1-2 mm. shorter when young but equal or slightly longer at maturity, oblanceolate with a long mucronate tip; ray corollas 10-20 mm. long; achenes 3 mm. long; pappus scales 5-7, 2.5-3.3 mm. long, ovate-lanceolate, acuminate. This species, with the broad leaf segments, apparently intergrades with H. richardsonii.---Along creeks or mountain ridges. Rare, known only from 3 locations; southwestern and southcentral Colorado and in central Utah on Castle Ridge in Carbon County. Found in Colorado in Costilla County and Hinsdale County, at 7200-9000 feet.

7. Hymenoxys odorata DC., Prodr. 5:661. 1836.

Actinella odorata (DC.) A. Gray; Actinea odorata (DC.) Kuntze; Hymenoxys chrysanthemoides DC. var. excurrens Ckll.; Hymenoxys davidsonii (Greene) Ckll.; Hymenoxys multiflora (Buckl.) Rydb.; Hymenoxys cockerellii Woot. & Standl.; Actinella multiflora (Buckl.) Clements & Clements---Hardy, bushy annuals; stems 8-50 cm. high, greatly branched throughout, pubescent or puberulent, glandular-granuliferous; basal leaves forming a withering rosette, and cauline densely covering the stem, 2-10 cm. long, pinnatifid into 3-13 narrowly linear divisions, punctate, more or less pubescent; heads many, corymbose, small, 3-12 mm. broad; peduncles 3-14 cm. long; involucral bracts in 2 definite, dissimilar series, 3-5 mm. high, outer 8-13, united 1/4 to 1/3 of their length, thickened at the base, lanceolate, acute, punctate, inner distinctly longer than outer, pressed inward in fruit; ray flowers 8-13, corollas 5-11 mm. long; achenes 1.5-2 mm. long; pappus scales 5, rarely 6, 1.5-2.3 mm. long.---Common on disturbed soil and overgrazed ranges from southeastern Colorado and western Kansas, southward to Mexico, westward to southeastern California. From eastcentral to southeastern Colorado, at 4000-5500 feet.

36. Bahia Lag.

Annual or perennial herbaceous plants; leaves alternate or opposite, entire to variously divided or dissected; heads corymbose; involucres campanulate, hemispheric, turbinate or obconic, bracts in 2 or 3 series, herbaceous to scarious or colored in part; receptacles mostly flat, naked or alveolate; ray flowers pistillate and fertile, rays from inconspicuous to definite, or in 1 species lacking, yellow; disk flowers perfect and fertile, yellow; anthers united, not caudate at base; achenes narrow, 4-angled; pappus of several paleae, these sometimes with the thickened midrib excurrent as an awn, or in 1 species pappus none.

1. Leaves entire, 1 cm. broad or more; involucres 7-10 mm. high; rays 7-10 mm. long
    2. Leaves chiefly basal, stems scapiform; leaf blades mostly oval to lanceolate   -1. B. nudicaulis
    2. Stems leafy throughout, not scapiform; leaf blades mostly oblong   -2. B. oblongifolia
1. Leaves variously parted or divided (upper leaves sometimes entire but narrow); involucres 5-7 mm. high; rays various but often less than 7 mm. long (except in B. dissecta)
    3. Pappus none; plants normally over 25 cm. tall; rays 7-9 mm. long   -3. B. dissecta
    3. Pappus present; plants rarely over 25 cm. tall, usually much less; rays absent or up to 4 mm. long
        4. Rays absent; plants annual   -4. B. neomexicana
        4. Rays present, 2-4 mm. long; plants perennial more or less woody at base
            5. Achenes hirsutulous especially toward the base; pappus scales lanceolate, acute or acuminate, the nerve reaching the apex and usually excurrent as a short awn   -5. B. woodhousei
            5. Achenes sessile-glandular only; pappus scales obovate rarely ovate, blunt at apex, the nerve disappearing before reaching the tip   -6. B. oppositifolia

1. Bahia nudicaulis A. Gray, Proc. Amer. Acad. 19:27. 1883.

Platyschkuria integrifolia (A. Gray) Rydb.---Perennial plants with woody caudex; stems scapiform, 10-40 cm. tall, cinereous-puberulent and more or less glandular above; leaves alternate, basal 2-7 cm. long, ovate, oval-ovate to lanceolate, entire, 3-nerved, petioled, cinereous-puberulent; stem leaves much reduced; heads 1 to few; involucres 7-10 mm. high, bracts in about 2 series nearly equal in length, lanceolate or oblanceolate, sometimes oblong, acute to obtuse, herbaceous and very narrowly scarious-margined if at all, glandular, outer at least 3-nerved; rays 8-20, 7-10 mm. long; achenes oblong-pyramidal, angled, rather sparingly hairy; pappus of 8-14 scales, these about 2 mm. long, lanceolate, midrib extending to apex or somewhat excurrent.---Dry plains and slopes. Wyoming, Colorado and Utah. Our records from westcentral and southwestern Colorado at 4500-7000 feet.

2. Bahia oblongifolia A. Gray, Proc. Amer. Acad. 19:27. 1883.
Platyschkuria oblongifolia (Gray) Rydb.---Colorado, Utah, New Mexico and Arizona. Reported for southwestern Colorado.

3. Bahia dissecta (A. Gray) Britt., Trans. N. Y. Acad. Sci. 8:68. 1889.
Amauriopsis dissecta (A. Gray) Rydb.---Annual plants with taproot; stems 25-100 cm. tall, puberulent or glabrate below, definitely glandular in the inflorescence; leaves alternate, 2-3 times ternately divided into oblong or linear segments, more or less puberulent; involucres 5-7 mm. high, glandular-pubescent, bracts in 2-3 series, oblong-oblanceolate to oblong-obovate, abruptly acuminate, herbaceous, outer at least 3-nerved; rays 12-20, 7-9 mm. long, cuneate; achenes narrowly obpyramidal, angled, usually glandular; pappus none.---Along streams and valleys. Wyoming to Mexico. Our records scattered in the western two-thirds of Colorado at 6500-9500 feet.

4. Bahia neomexicana A. Gray, Proc. Am. Acad. 19:27. 1883.
Cephalobembix neomexicana (A. Gray) Rydb.; Achyropappus neomexicana A. Gray.---Annual plants; stems 8-20 cm. tall, slender, branched from near base, glabrate, hirsutulous or strigose, more or less glandular above; leaves opposite or the upper alternate, pinnately 3- to 7-divided into linear-filiform segments, sparingly hirsutulous and impressed punctate; involucres 5-7 mm. high, bracts obovate to oblanceolate, herbaceous but more or less scarious on apex and margins, this usually purplish-tinged, hirsutulous to puberulent; rays none; achenes narrowly obpyramidal, 4-angled, sparsely hairy; pappus scales about 8, about 1-1.5 mm. long, obovate, rounded at apex.---Open often sandy soil. Colorado to Mexico. Our few records from central and southcentral Colorado at 7500-8000 feet.

5. Bahia woodhousei A. Gray, Proc. Amer. Acad. 19:28. 1883.
Picradeniopsis woodhousei (Gray) Rydb.---Perennial plants more or less woody at base; stems 8-20 cm. tall, branching below, appressed hairy; leaves mostly opposite, 3-parted into linear segments, strigose-puberulent, impressed punctate; involucres 5-6 mm. high, bracts oblong, obtuse, herbaceous, strigose-puberulent, outer 3-nerved; rays few, about 2-4 mm. long, oblong, sometimes ochroleucous; achenes narrowly obpyramidal, angled, hirsutulous; pappus scales narrowly lanceolate, acute or acuminate, the stout nerve reaching the apex and usually excurrent as an awn.---Plains and hills. Colorado, south to Texas and Arizona. Our few records from eastcentral Colorado at about 5000 feet.

6. Bahia oppositifolia (Nutt.) DC., Prodr. 5:656. 1836.
Picradeniopsis oppositifolia (Nutt.) Rydb.---Perennial plants more or less woody at base; stems 8-25 cm. tall, much branched below, canescent-puberulent; leaves nearly all opposite, 3- to 5-parted into linear or narrowly oblong segments or upper entire, canescent-strigose and impressed punctate; involucres 5-7 mm. high, canescent, bracts in about 2 series, outer more or less keeled and herbaceous, inner somewhat scarious-margined; rays few, 2-4 mm. long, oval; achenes narrowly obpyramidal, angled, sessile-glandular but not hairy; pappus scales obovate sometimes ovate, the nerve disappearing before reaching the blunt apex.---Dry plains and hills. North Dakota to Montana, south to Texas and Arizona. Our records scattered over the eastern half of Colorado at 4000-7500 feet.

37. Enceliopsis (A. Gray) A. Nels.

Perennial scapose or subscapose plants from stout tuber-form roots and short woody caudex; leaves basal, in tufts on the short caudex branches; heads solitary, nodding in fruit; involucres hemispheric or flattened-hemispheric, bracts imbricated in 2-3 series the outer shorter; receptacles somewhat convex, chaffy; rays none; disk flowers perfect and fertile, yellow; anthers united, sagittate at base; achenes flattened but not winged, villous; pappus of 2 short subulate teeth bordered between by membranous confluent paleae, all very short and essentially lacking.

1. Enceliopsis nutans (Eastw.) A. Nels., Bot. Gaz. 47:433. 1909.
Leaves 2-5 cm. long, oval, obtuse, 3- to 5-ribbed, hispid, canescent to strigose, on margined petioles; scapes 1-5, 15-25 cm. tall, hirsute the hairs more or less reflexed; heads 2-4 cm. wide; involucres 12-15 mm. high, bracts lanceolate, hirsute; achenes about 1 cm. long, obovate.---Dry hills and slopes. Colorado and Utah. Our records from westcentral Colorado at 4500-5500 feet.

Another species E. nudicaulis (A. Gray) A. Nels., with stems and leaves white-tomentose and rays present has been collected in northeastern Utah within a mile of Colorado.

38. Parthenium L. FEVERFEW

Acaulescent caespitose herbaceous perennial plants; leaves all basal and crowded on the branches of the caudex, spatulate to linear, entire, silvery-canescent; heads solitary, sessile, subsessile or short-peduncled among the leaves; involucres hemispheric, bracts rather few in about 2-3 series, broad; receptacles convex, chaffy; rays none, a few marginal flowers pistillate and fertile with corollas truncate at apex, or with rays 1-2 mm. long; disk flowers perfect but sterile; anthers united, entire at base; achenes flattened and somewhat winged above; pappus of 2 rather thick teeth extending upward from the winged achene edges.

1. Parthenium alpinum (Nutt.) T. & G., Fl. N. Amer. 2:285. 1840.
Plants 2-5 cm. tall; leaves 1-3 cm. long, long-villous in the axils; chaff of receptacle pubescent at apex; involucres about 5 mm. long; rays none; achenes about 4 mm. long, appressed puberulent; pappus scales as long or shorter than the corollas. Our Colorado plant is P. alpinum var. tetraneuris (Barneby) Rollins (P. tetraneuris Barneby).---Among rocks and on barren flats or ridges. Wyoming and Colorado. Our one record from central Colorado (Fremont County) at 5500 feet.

Recently another species, P. ligulatum (Jones) Barneby, has been found in Rio Blanco County, Colorado. It has rays 1-2 mm. long and sessile heads.

39. Parthenice A. Gray

Annual herbaceous plants; stems leafy, paniculately branching; leaves alternate, ovate, toothed, long-petioled; heads numerous, in panicles; involucres hemispheric, bracts in 2 distinct sets, the outer 5 somewhat herbaceous, inner 6-8 somewhat larger, orbicular, more or less scarious, subtending the marginal flowers; receptacle convex, chaffy, the paleae linear-oblong or spatulate, some or all subtending the outer flowers of the head; rays none but marginal flowers 6-8, pistillate and fertile with oblique corolla tubes; disk flowers many, perfect but sterile, greenish-white; anthers united, entire at base; achenes flattened but not winged, incurved and apiculate at apex, glabrous, falling with the paleae; pappus wanting.

1. Parthenice mollis A. Gray, Pl. Wright. 2:85. 1853.
Reported for Arizona, New Mexico and southern Colorado but no specimens from this state have been located.

40. Madia Molina TARWEED

Annual glandular-viscid, heavy-scented plants; stems leafy; leaves alternate, entire, linear; heads many, glomerate near ends of branches; involucres campanulate to ovoid, angled by the narrow backs of the bracts, bracts in a single series, the margins inflexed from sides and enclosing the ray achenes; receptacles flat or convex, chaffy, the paleae as a single series enclosing the disk flowers as a kind of inner involucre; ray flowers about 2-5, pistillate and fertile, rays cuneate, 3-lobed, inconspicuous, yellowish; disk flowers about 8-12, perfect and fertile, yellowish; anthers united, not sagittate at base; achenes angled or flattened on the ray flowers; pappus none.

1. Madia glomerata Hook., Fl. Bor. Amer. 2:24. 1834.
Stems 20-60 cm. tall, simple or branched above, hirsute and glandular especially above; leaves about 3-6 cm. long, narrowly linear; involucres about 6-8 mm. long and 3-5 mm. wide, glandular; rays about 1 mm. long; ray achenes more or less curved, 1-nerved on each face.--- Open ground, often along roadsides and in partial shade. Saskatchewan to Washington, south to Colorado and California. Our records scattered in the western half of Colorado at 5500-9000 feet.

41. Melampodium L. BLACKFOOT

Perennial herbaceous plants with taproots and caudex; stems leafy; leaves opposite, entire to sinuately lobed; heads terminal on the stem and branches; involucres campanulate, bracts in 2 sets, the outer 4 or 5 herbaceous, broad, flat and partially united, the inner hooded, each embracing a ray achene and deciduous with it; receptacles convex or conical, chaffy; ray flowers pistillate and fertile, rays white to somewhat pink, spreading, conspicuous; disk flowers perfect but sterile, with undivided style; anthers united, entire at base; achenes broadening upward and more or less incurved; pappus wanting.

1. Melampodium cinereum DC., Prodr. 5:518. 1836.
M. leucanthum T. & G. as to Colorado plants---Stems 10-30 cm. tall, more or less woody at base and from a rather woody caudex, canescent or cinereous-strigose; leaves 3-5 cm. long, linear or the lower oblanceolate, entire or sometimes sinuate-lobed, cinereous-strigose; involucres about 5 mm. long, cinereous, outer bracts ovate and appressed; rays 6-13 mm. long and about 5 mm. wide, white above, more or less purple-veined below. The form with long rays with veins prominent below has been called M. leucanthum but all intergradations occur in Colorado.---Dry plains and hills. Kansas to Colorado, south to Texas and Arizona. Our records from the southeastern quarter of Colorado at 4000-7000 feet.

42. Echinacea Moench PURPLE-HEADED CONEFLOWER

Herbaceous perennial plants from thick and vertical or horizontal rootstocks; stems leafy; leaves alternate, rather narrow, entire; heads solitary on long peduncles; involucres depressed hemispheric, bracts in 2-3 series, narrow, herbaceous or reflexed; receptacles conic or hemispheric at first, elongating in fruit, chaffy, the paleae longer than the disk flowers and spinulose at apex; ray neutral or with rudimentary styles and ovaries, several, rays showy, purple or rose, spreading or reflexed; disk flowers perfect and fertile; anthers united, not sagittate at base; achenes 4-angled; pappus a toothed crown or essentially lacking.

1. Echinacea angustifolia DC., Prodr. 5:554. 1836.
Brauneria angustifolia (DC.) Heller---Stems 25-60 cm. tall, erect, hirsute at least below; basal leaves petioled, cauline short-petioled or sessile above, oblong-lanceolate to nearly

linear, rough and tuberculate-hirsute; heads 1.5-3 cm. long; involucres 6-11 mm. high, bracts lanceolate; rays 2-3.5 cm. long, spreading; achenes 5 mm. long.---Plains and hills. Minnesota to Wyoming, south to Texas. Our rather few records from eastern Colorado at 3500-4500 feet.

43. Cosmos Cav.

Plants annual; stems leafy, slender; leaves opposite, dissected into narrow lobes; heads terminal on the stems or branches; involucres hemispheric, bracts in 2 distinct and unlike series, thin, outer narrower and herbaceous, inner with definite scarious margins; receptacles flat or nearly so, chaffy; ray flowers neutral, rays few, rose, pink or whitish; disk flowers perfect and fertile, yellow; anthers united, not caudate at base; achenes fusiform, conspicuously beaked; pappus of 2-4 retrorsely hispid awns.

1. Rays less than 12 mm. long -1. C. parviflorus
1. Rays 15-30 mm. long -2. C. bipinnatus

1. Cosmos parviflorus (Jacq.) H. B. K., Nov. Gen. et Sp. 4:241. 1820.
Stems 30-80 cm. tall, glabrous; leaves bi- or tri-pinnatifid into filiform divisions; involucres 6-7 mm. long; rays 7-12 mm. long; achenes 7-10 mm. long including the beak which is about 1/3 to 1/2 as long as the body; pappus awns about 2 mm. long.---Plains and hills. Colorado, south to Texas, Mexico and Arizona. Our record from southcentral Colorado at about 6800 feet. Recently collected in the southwestern part of the state.
2. Cosmos bipinnatus Cav., Icon. et Descr. 1:10. 1791.
Differs from the preceding in having rays much longer, 1.5-3 cm. long.---Common in cultivation and occasionally escaping. Our specimens from northcentral and central Colorado at about 5500 feet.

44. Eclipta L.

Annual herbaceous plants; stems leafy; leaves opposite, simple, narrow; heads peduncled in the upper axils; involucres hemispheric to broadly campanulate, bracts imbricated in about 2 series, equal or outer longer; receptacles flat or convex, chaffy with bristlelike paleae; ray flowers pistillate and fertile, numerous, rays short, white or whitish; disk flowers perfect and mostly fertile, the corollas usually 4-toothed; achenes short, thick, 3- or 4-angled or compressed, truncate at apex; pappus none or possibly a few very short teeth present.

1. Eclipta alba (L.) Hassk., Pl. Jav. Rar. 528. 1848.
Widely distributed in the warmer regions of the world and extending north in this continent to Massachusetts, Nebraska, Arizona and California. It may be in this state.

45. Galinsoga R. & P. QUICKWEED

Annual plants; stems leafy, branching; leaves opposite, ovate or ovate-lanceolate, serrate, petioled; heads small, peduncled, terminal or axillary; involucres hemispheric or broadly campanulate, bracts in 2 series, nearly equal or outer shorter, ovate and obtuse; receptacles conic, chaffy; ray flowers pistillate and fertile, rays small and white; disk flowers perfect and fertile, yellow; anthers united, minutely sagittate at base; achenes angled or flattish; pappus of awn-pointed scales, sometimes very narrow or absent from the ray flowers. The 2 species are very similar and intergrade somewhat in our material.

1. Pappus of ray flowers wanting or of a few short bristles; disk flower pappus fimbriate but not aristate at apex; ray achenes glabrous or finely pilose on 1 side; nodes and peduncles finely appressed pilose -1. G. parviflora
1. Pappus of ray flowers of linear scales; disk flower pappus usually aristate at apex; ray achenes densely hispid on inner faces; nodes and peduncles with dense, coarse spreading hairs -2. G. ciliata

1. Galinsoga parviflora Cav., Icon. 3:41. 1794.
Stems 10-80 cm. long, more or less appressed hairy; leaves coarsely and bluntly serrate, 3-nerved; involucres about 3-4 mm. high, bracts usually glabrous; rays about 1-1.5 mm. long; achenes of ray flowers glabrous or sparsely pilose on inner face; ray pappus absent or reduced to short bristles, disk pappus linear-lanceolate, margins fimbriate but not aristate at apex.---In yards and waste places, usually in rather damp shaded ground. Native to South America and Central America but widely introduced in the United States. Our records from northcentral and central Colorado at 5000-6000 feet.
2. Galinsoga ciliata (Raf.) Blake, Rhodora 24:35. 1922.
G. aristulata Bickn.; G. parviflora hispida DC.---Similar in general appearance to the preceding. Hairs on nodes and peduncles coarse and spreading; ray achenes densely hispid on inner face; ray pappus of linear scales about as long as the corolla tube, disk pappus usually aristate at apex, margins fimbriate.---Yards and waste places, usually in shade. Distribution of the preceding both generally and in Colorado.

46. Engelmannia T. & G. ENGELMANN DAISY

Perennial, herbaceous, hirsute or hispid plants; stems leafy; leaves alternate, pinnatifid; heads corymbose; involucres hemispheric or campanulate, bracts in 2 or 3 sets, the outer linear, middle suborbicular with linear tips, inner oval or obovate; receptacles flat, chaffy; ray flowers pistillate and fertile, rays 8-10, conspicuous, golden-yellow; disk flowers perfect but sterile, with undivided styles; anthers united, minutely toothed at base; achenes flat but not winged, obovate, 1-nerved on each side; ray pappus an irregularly toothed or lobed crown, disk pappus much reduced.

1. Engelmannia pinnatifida T. & G., Fl. N. Amer. 2:283. 1841.
 Stems 25-70 cm. tall, usually branched above; leaves 5-15 cm. long, upper sessile, lower short-petioled; involucres 8-13 mm. high; rays about 7-10 mm. long.---Plains and hills. Kansas to Colorado, south to Louisiana and Arizona. Our records from the southeastern part of Colorado at 3500-4500 feet.

47. Berlandiera DC.

Perennial finely canescent-tomentulose plants; stems somewhat leafy; leaves alternate but rather basal, lyrate-pinnatifid at least in part; heads solitary and long-peduncled on the stems and branches; involucres broad and depressed-hemispheric, bracts imbricated in about 3 series, outer short and broad, inner obovate and reticulated when mature; receptacles flat or nearly so, chaffy; ray flowers 5-12, pistillate and fertile, yellowish; disk flowers perfect but sterile, with undivided styles; anthers united, entire or minutely toothed at base; ray achenes flattened but not winged, adnate at base to inner involucre bracts and chaff of the inner flowers, the whole falling together; pappus none.

1. Berlandiera lyrata Benth., Pl. Hartw. 17. 1839.
 Stems 10-40 cm. tall, canescent-puberulent; leaves 5-15 cm. long, segments mostly crenate, terminal one largest, canescent-tomentulose especially below, upper often green and glabrate; involucres 12-15 mm. high; rays 1-1.5 cm. long, yellow above, brownish-red beneath at least in longitudinal lines; achenes obovate.---Dry plains, hills and mesas. Arkansas to Kansas, south to Texas, Mexico and Arizona. Our records from the southeastern corner of Colorado at 4000-4500 feet.

48. Rudbeckia L. CONEFLOWER

Herbaceous perennial plants or often appearing to be annual; stems more or less leafy; leaves alternate, entire, toothed, lobed or pinnatifid; heads often showy, terminal on the stems or branches, long-peduncled; involucres hemispheric, bracts in about 2 or more series, herbaceous to foliose, loose and spreading; receptacles convex, conic or cylindrical in fruit, chaffy; ray flowers neutral, rays absent or present and conspicuous, yellow, or partly or wholly brownish-red; disk flowers perfect and fertile; anthers united, not caudate at base; achenes 4-angled; pappus a low border, a toothed crown or lacking.

1. Leaves entire; stems 30-70 cm. tall; rays present -1. R. hirta
1. Leaves (at least lower) lobed or divided; stems usually over 100 cm. tall; rays present or absent
 2. Rays present; disk ovoid in fruit, less than 2.5 cm. long, dull yellowish at least when young
 -2. R. laciniata
 2. Rays lacking; disk becoming cylindrical in fruit, 3-5 cm. long, purplish-black from the first
 -3. R. montana

1. Rudbeckia hirta L., Sp. Pl. 907. 1753.
 R. flava Moore---Plants perennial or sometimes appearing annual; stems 30-70 cm. tall, erect, simple or sparingly branched, hirsute and often purple-dotted; lower leaves 5-10 cm. long, oblong-lanceolate or oblanceolate, entire, thick; upper leaves becoming linear and subsessile; involucres 12-18 mm. high, bracts oblong-linear, spreading or reflexed, hirsute; rays 1.5-3 cm. long, yellow rarely darker at base; disk globose-ovoid, blackish or brownish; no pappus present. Out plant has recently been considered to be R. serotina Nutt. (Rhodora 50:172-176. 1948).---Plains and hills. Quebec to British Columbia, south to Florida, Texas and Colorado. Our records from in or rather near northcentral, central and southcentral Colorado at 5000-9000 feet. The plant should also be present in the eastern part of the state.
2. Rudbeckia laciniata L., Sp. Pl. 906. 1753.
 R. ampla A. Nels.---Perennial plants; stems 100-200 cm. tall, branched above, glabrous; basal leaves large, long-petioled, pinnately divided into 3-7 ovate or lanceolate segments, these cleft and coarsely serrate, glabrate above, glabrous or somewhat hairy above; upper leaves 3-parted or entire; involucres about 1-2 cm. high, bracts oblong to ovate-oblong, often very unequal and foliose, becoming reflexed, glabrous to sparsely hairy; rays 3-5 cm. long, conspicuous, yellow; disk dull yellowish, becoming ovoid in fruit; pappus a very short crown or border.---Moist ground, usually along streams. Maine to Saskatchewan, south to Florida and Arizona. Our records well scattered in the western two-thirds of Colorado at 5000-9000 feet.
3. Rudbeckia montana A. Gray, Proc. Amer. Acad. 17:217. 1882.
 Perennial plants; stems about 100-200 cm. tall, glabrous and often glaucous; lower leaves large, long-petioled, pinnately parted or divided into 3-9 segments, these oblong-lanceolate or rhombic-ovate in outline but usually deeply cleft, glabrous or nearly so; upper leaves ovate to lanceolate, lobed to entire; involucres about 15-40 mm. high, bracts linear or linear-lanceolate, glabrous or sparingly short-hairy; rays none; disk purplish-black, cylindrical and at length 3-5 cm. long; pappus an irregularly margined almost toothed crown.---Mountains. Apparently limited to Colorado, our records from the western third, mostly from the westcentral part, at 6500-9500 feet.

49. Ratibida Raf. PRAIRIE CONEFLOWER

Perennial herbaceous plants from taproots; stems leafy; leaves alternate, pinnately parted or divided; heads terminal and long-peduncled; involucre bracts in 2 series, inner not over 1/2 as long as the outer, herbaceous; receptacles globose to columnar, the chaff with apex incurved and hairy at least at first; ray flowers neutral, rays conspicuous, yellow or brownish at base, rarely wholly brown-purple; disk flowers perfect and fertile, yellow or yellowish-gray; anthers united, sagittate at base; achenes rather stout and broad, flattened; pappus of 1-2 short and rather awnlike teeth, sometimes of several very deciduous scales, sometimes practically absent.

1. Receptacle globular or oblong-ellipsoid, 6-13 mm. high; rays 3-8 mm. long; peduncles about 1-5 cm. long
-1. R. tagetes
1. Receptacle soon cylindric or columnar, 10-40 mm. high; rays usually 8-30 mm. long; peduncles 6-25 cm. long
-2. R. columnifera

1. Ratibida tagetes (James) Barnh., Bull. Torr. Club 24:410. 1897.
Lepachys tagetes (James) A. Gray---Stems 15-40 cm. tall, erect but branched, strigose-hirsute, sometimes more or less glandular; basal leaves from entire to pinnately or bipinnately cleft; middle leaves pinnately or bipinnately parted or divided into linear or linear-lanceolate segments, more or less strigose-hirsute both sides; peduncles 1-5 cm. long or rarely longer; heads 6-13 mm. long and about 8-10 mm. wide; involucre bracts about 3-5 mm. long, reflexed; rays 5-7, 4-6 mm. long, rarely to 8 mm., reflexed, yellowish to bluish-purple; achenes 2-2.5 mm. long, ovate-oblong, winged; pappus of numerous lacerate setae on a crown or essentially lacking.---Plains and hills. Kansas and Colorado, south to Texas and Arizona. Our records in southeastern Colorado west to Archuleta County and north to Denver at 3500-7500 feet.

2. Ratibida columnifera (Nutt.) Woot. & Standl., Contrib. U. S. Nat. Herb. 19:706. 1915.
R. columnaris (Sims.) D. Don; Lepachys columnaris (Sims.) T. & G.---Stems 25-80 cm. tall, 1 or more from a root, more or less branched above, hirsute-strigose and more or less glandular; leaves pinnately divided or deeply parted into 5-13 linear to oblong divisions, these sometimes lobed again; peduncles 6-25 cm. long; heads 10-40 mm. long and about 7-10 mm. wide; outer involucre bracts 4-12 mm. long, linear; rays 8-30 mm. long, yellow to variously bluish-purple or rose-purple; achenes 2 mm. long, winged; pappus of 2 toothlike projections at the edge of a crown.---Plains and hills. Minnesota to British Columbia, south to Tennessee, Colorado and Arizona. Our records scattered in the eastern half of Colorado, also in the southwestern part of the state, at 3500-7000 feet.

50. Verbesina L. CROWNBEARD

Annual plants from a taproot; stems leafy; leaves alternate or sometimes opposite, simple, coarsely toothed to incised; heads peduncled on stems and branches; involucres hemispheric, bracts in several series but nearly equal in length, narrow, herbaceous or foliose; receptacles convex, chaffy; ray flowers pistillate and fertile, rays conspicuous, yellow or orange; disk flowers perfect and fertile, numerous; anthers united, not caudate at base; achenes of disk flowers flattened and corky-winged, ray achenes often wingless; pappus of ray flowers apparently none, disk pappus of 2 slender long awns.

1. Verbesina encelioides (Cav.) Benth. and Hook. ex Gray, Bot. Calif. 1:350. 1876.
Ximenesia encelioides Cav.---Stems 25-60 cm. tall, branching, cinereous or canescent-tomentulose; leaves ovate to rhombic-lanceolate, variously serrate or laciniate-dentate especially near base, with dilated auricles sometimes present at base of petiole, minutely strigose but greenish above, more or less whitish-strigose below; involucres about 7-10 mm. high; rays 12-20 mm. long, golden-yellow to orange-yellow. Our plants lack a definite auriclelike appendage at the petiole base and have been called var. exauriculata Robins. & Greenm. However, some leaves have an indefinite auricle and intergrade to the species.---Plains and valleys, often along ditches and roads. Kansas to Montana, south to Texas, Mexico and California. Our records scattered over Colorado at 4000-9000 feet.

51. Balsamorhiza Hook. BALSAM-ROOT

(Contributed by W. A. Weber)

Perennial herbaceous plants from stout taproots; stems with basal leaves, scapose, the flowering stems with a few greatly reduced leaves; leaves entire and cordate or variously pinnately dissected; heads usually solitary; involucre bracts in 2-4 series, subequal; receptacles broadly convex, chaffy; ray flowers pistillate and fertile, rays yellow; disk flowers perfect and fertile, yellow; ray achenes quadrangular, disk achenes trigonal; pappus none.

1. Leaves cordate-ovate, entire, velvety white-pubescent or tomentose   -1. B. sagittata
1. Leaves pinnately dissected, green, hispidulous and glandular   -2. B. hispidula

1. Balsamorhiza sagittata (Pursh) Nutt., Trans. Am. Phil. Soc. II. 7:350. 1841.
Basal leaves 20-40 cm. long and 5-15 cm. wide, cordate-ovate, entire, acute, silvery-tomentulose to velutinous; heads solitary, sometimes with a few reduced heads below the terminal one, 6-11 cm. wide including rays; involucre bracts ovate-lanceolate to lanceolate, acuminate to

attenuated, densely white-tomentose; rays 2-4 cm. long, yellow; mature achenes 7-8 mm. long, glabrous.---Hillsides, plateaus and parks. South Dakota to British Columbia, south to Colorado and California. Our records scattered in the western half of Colorado, except the southcentral part, at 6000-9000 feet.

2. Balsamorhiza hispidula W. M. Sharp, Ann. Mo. Bot. Gard. 22:137. 1935.

Basal leaves 10-40 cm. long, pinnately divided, green on both surfaces, hispidulous and resinous; heads 4.5-6 cm. wide including the rays, solitary; involucre bracts linear-lanceolate, acute to acuminate; rays 2.5-3 cm. long, yellow; achenes 6-7 mm. long, glabrous.---High desert grasslands and sagebrush plateaus. Wyoming to Idaho, south to Colorado and Nevada. Our few records from the northwestern corner of the state at 6500-8000 feet.

52. Wyethia Nutt. MULES-EARS

(Contributed by W. A. Weber)

Perennial herbaceous plants with stout taproots; stems leafy, with or without enlarged basal leaves; leaves alternate, lanceolate or ovate-lanceolate, entire or denticulate; heads large, solitary or few; involucre bracts in 2-4 series, the outer often foliaceous and longer than the disk; receptacles broadly convex, chaffy; ray flowers pistillate and fertile, rays yellow; disk flowers perfect and fertile, yellow; achenes 6-15 mm. long, ray achenes trigonal, disk quadrangular in cross-section; pappus a crown of unequal, laciniate, persistent scales, often prolonged into awns on the angles of the achenes.

1. Leaves linear to linear-lanceolate; basal leaves absent, reduced or similar in size to the stem leaves
   2. Involucre bracts moderately imbricated, more or less erect, the outer series pubescent mostly on the margins, the inner series more uniformly clothed with finer, more appressed hairs; plants of northwestern Colorado   -1. W. scabra
   2. Involucre bracts very closely imbricated, the outer conspicuously recurved-spreading, all clothed with a very dense coat of fine, appressed hairs; plants of southwestern Colorado   -1A. W. scabra canescens
1. Leaves lanceolate to ovate-lanceolate; basal leaves present and usually much larger than the cauline
   3. Plants entirely glabrous, as if resinous-varnished; upper stem leaves clasping   -2. W. amplexicaulis
   3. Plants densely hirsute to glabrate, never completely glabrous; upper stem leaves petioled   -3. W. arizonica

1. Wyethia scabra Hook., Lond. Journ. Bot. 6:245. 1847.

Stems 15-40 cm. tall, very pale and white, very hispid or scabrous; enlarged basal leaves absent, leaves 3-15 cm. long and 3-17 mm. wide, linear to linear-lanceolate, veins pale, the lateral ones confluent toward the leaf margin giving a 3-nerved appearance, sessile, hispid or scabrous; heads solitary; outer involucre bracts attenuate from rather broad bases.---Dry plains and hills. Wyoming to Utah. Our few records from the northwestern corner of Colorado at 5000 feet.

1A. Wyethia scabra canescens W. A. Weber, (var.) Amer. Midl. Nat. 35:425. 1946.

Involucre bracts very closely imbricated, the outer conspicuously recurved-spreading, all clothed with a very dense coat of fine appressed hairs, apex bearing enlarged hairs.---Colorado, Utah, south to New Mexico and Arizona. Our one record from the southwestern corner of Colorado at about 6000 feet.

2. Wyethia amplexicaulis (Nutt.) Nutt., Trans. Amer. Phil. Soc. II. 7:352. 1840.

Plants glabrous; leaves deep green, leathery and glossy, basal very large, 20-40 cm. long or longer and 5-15 cm. wide, oblong-lanceolate, entire or denticulate, acute or acuminate, pinnately veined, narrowed gradually to a short petiole; stem leaves smaller, sessile and usually clasping the stem; heads large, several or occasionally solitary, the terminal one largest; outer involucre bracts subequal, foliaceous, ovate-lanceolate, acute to acuminate, somewhat dilated above. Hybrid swarms are to be expected where this species and the next one overlap in distribution. Intermediate specimens suggesting hybridization have been found in Routt County, Colorado.---Moist draws, open grasslands and woods. Montana to Washington, south to Colorado and Nevada. Our records from the northwestern quarter of Colorado at 8000-11,000 feet.

3. Wyethia arizonica A. Gray, Proc. Am. Acad. 8:655. 1873.

Similar to the preceding but plants densely hirsute-pubescent, becoming scabrous or sometimes glabrate; leaves are smaller and narrower, with slender petioles, only the uppermost sessile or somewhat clasping.---Dry hills and mountain slopes. Colorado to Utah, south to New Mexico and Arizona. Our records scattered in the western third of Colorado, mostly from the southern half of that area, at 6000-9000 feet.

53. Thelesperma Less. GREENTHREAD

Annual, biennial or perennial, glabrous, herbaceous plants; leaves opposite, usually finely pinnately parted or divided into linear or narrower divisions, rarely entire; heads on long peduncles; involucres hemispheric to campanulate, bracts in 2 very unlike series, outer narrow herbaceous and often spreading, inner broad, scarious-margined, connate to about the middle or higher; receptacles flat, with white-scarious, broad, 2-nerved chaff; ray flowers none or present and showy, neutral, rays yellow; disk flowers perfect and fertile; anthers united, not caudate at base; achenes oblong to linear, slightly if at all compressed; pappus of 2 retrorsely hispid awns or lacking entirely.

1. Rays normally absent or very short and inconspicuous; pappus of 2 awns 1.5-2.5 mm. long   -1. T. megapotamicum
1. Rays normally present and conspicuous (at least on some of the heads); pappus of 2-5 short awns less than 1 mm. long or lacking entirely

2. Disk flowers with ovate lobes much shorter than the expanded throat; leaves mostly basal, the plants sub-
      scapose; leaf divisions often over 2 mm. wide   -2. **T. subnudum**
   2. Disk flowers with lanceolate or narrower lobes longer than the expanded throat; leaves scattered on the
      stems; leaf divisions rarely over 2 mm. wide   -3. **T. trifidum**

1. **Thelesperma megapotamicum** (Spreng.) Kuntze, Rev. Gen. Pl. 3(2):182. 1898.
      T. gracile (Torr.) A. Gray---Perennial plants from a deep root; stems 30-80 cm. tall, branched and leafy; leaves 1- to 2-divided into narrowly linear rather rigid segments or upper leaves entire; outer involucre bracts 4-6, oblong or ovate, obtuse, about 1/4 as long as the inner; inner bracts about 8-12 mm. high, connate to above the middle, the lobes narrowly scarious-margined; rays lacking; disk flowers yellowish or possibly brownish, the corolla throat abruptly expanded and shorter than the lobes; outer achenes more or less papillose on back; pappus of 2 retrorsely hispid awns about 1.6-2.2 mm. long, longer than the width of the achene. ---Plains and hills. Nebraska, Wyoming and Utah, south to Texas, Mexico and Arizona; South America. Our records scattered in the eastern half of Colorado at 3500-6500 feet.
2. **Thelesperma subnudum** A. Gray, Proc. Amer. Acad. 10:72. 1874.
      Perennial plants from an often rather woody root; stems 10-30 cm. tall, subscapose; leaves clustered at base of plant, pinnately or bipinnately divided into linear or broadly linear segments; heads about 1-1.5 cm. wide (exclusive of rays); outer involucre bracts oblong to oblong-lanceolate, narrowly scarious-margined, about 1/4 to 1/2 as long as the inner; inner bracts about 7-10 mm. high, united less than 1/2 their length, lobes rather conspicuously scarious-margined; rays normally present on some of the heads of the plant, conspicuous, yellow; disk flowers yellow or possibly brown, with ovate corolla lobes much shorter than the expanded throat; achenes short-hairy at apex; pappus a 4- to 5-toothed crown or lacking entirely.---Dry plains and hills. Colorado to Utah, south to New Mexico and Arizona. Our records from southern and westcentral Colorado at 4500-8000 feet.
3. **Thelesperma trifidum** (Poir.) Britt., Trans. N. Y. Acad. Sci. 9:182. 1890.
      T. tenue Rydb.---Annual, biennial or perennial plants; stems 30-80 cm. tall, leafy; leaves bipinnately divided into linear-filiform segments no broader than the rachis, the leaf rather flaccid; outer involucre bracts linear or linear-subulate, usually 1/2 or more as long as the inner; inner bracts about 8-12 mm. high, connate up to the middle or below, lobes scarious-margined; rays normally present, conspicuous, yellowish; disk flowers yellow or possibly brown, the corolla lobes lanceolate or narrower, longer than the expanded throat; outer achenes more or less papillose; pappus of 2 short, rather triangular awns less than 1 mm. long. Two related species, T. ambiguum A. Gray and T. intermedium Rydb., cannot be distinguished from the above in this area by the writer.---Dry plains, valleys and hills. Nebraska to Colorado, south to Texas. Our records scattered over the eastern half of Colorado at 3500-8000 feet.

54. **Bidens** L. BEGGARTICKS; SPANISH NEEDLES

Annual, or rarely appearing somewhat perennial, herbaceous plants; leaves at least the lower opposite, entire to dissected; heads solitary or variously paniculate; involucres usually campanulate or subhemispheric, bracts distinct or slightly united at base, of 2 definite sets, outer herbaceous, inner membranous or scarious-margined; receptacles flat, chaffy; ray flowers neutral, rays yellow or whitish, sometimes lacking; disk flowers perfect and fertile; anthers united, not caudate at base; achenes linear, linear-fusiform or cuneate, compressed and 4-angled, rarely nearly terete; pappus of 2-4 retrorsely barbed awns, or rarely wanting.

1. Leaves serrulate to coarsely serrate, sometimes incised especially near base of blade, never divided
   2. Achenes with 3 awns; some of leaves short but definitely wing-petioled; ray flowers never present
                                                                                                     -1. **B. comosa**
   2. Achenes with 2 or 4 awns; leaves sessile or connate at base, rarely subsessile; ray flowers often present
      3. Chaff of receptacle yellow at apex (but with red stripes often present); rays lacking or when present
         not over 1.5 cm. long; heads often nodding; achenes often with paler margins   -2. **B. cernua**
      3. Chaff of receptacle reddish at tip; rays present and 1.5-3 cm. long; heads rarely nodding; achenes
         without paler margins   -3. **B. laevis**
1. Leaves once to twice pinnately dissected
   4. Rays conspicuous, 10-25 mm. long, yellow   -4. **B. polylepis**
   4. Rays lacking or inconspicuous, not over 7 mm. long, yellow to white
      5. Leaves with ultimate segments narrowly-linear to filiform, not over 2.5 mm. wide
         6. Heads small and narrow, to 5 mm. high and 3 mm. wide; outer involucre bracts 3-5 mm. long; disk
            flowers mostly 8-13 to a head; ultimate leaf divisions filiform or linear-filiform, seldom over
            1 mm. wide   -5. **B. heterosperma**
         6. Heads larger, 6-10 mm. high and 4-10 mm. wide; outer involucre bracts 5-7 mm. long; mostly over 13
            disk flowers present; ultimate leaf divisions narrowly linear, often over 1 mm. wide
                                                                                                -6. **B. tenuisecta**
      5. Leaves with ultimate segments lanceolate, oblong or deltoid-lanceolate, over 2.5 mm. wide
         7. Achenes all narrowly cuneate to obovate-cuneate, 2-awned; leaves pinnately divided, the first
            divisions serrate, not lobed; outer involucre bracts over 1 cm. long, usually much longer than inner
            8. Outer involucre bracts 10-16, usually 1-2 cm. long; inner involucre bracts about 7-9 mm. long,
               about equalling the disk flowers   -7. **B. vulgata**
            8. Outer involucre bracts 5-8, often over 2 cm. long; inner involucre bracts about 5-7 mm. long,
               usually shorter than the disk flowers   -8. **B. frondosa**
         7. Achenes linear, 2- to 4-awned; leaves with first main divisions again divided, parted or lobed;
            outer involucre bracts 3-8 mm. long, shorter or only slightly longer than the inner

9. Achenes strongly dimorphic, inner long and linear, tapering to the apex, outer linear-cuneate, much shorter and broader, and truncate at apex; outer involucre bracts 5-8 mm. long; 2-3 awns usually present on achenes
-9. **B. bigelovii**
9. Achenes essentially alike in shape, linear and tapering to apex (outer may be shorter than inner, however); 3-4 awns usually present on achenes
-10. **B. bipinnata**

1. Bidens comosa (Gray) Wieg., Bull. Torr. Club 24:436. 1897.
Stems 30-80 cm. tall, glabrous, branching; leaves 4-10 cm. long, elliptic-lanceolate, at least lower short wing-petioled, serrate and often incised-lobed at very base of blade; head about 1-2 cm. high exclusive of outer bracts; outer involucre bracts very conspicuous, 2-5 cm. long, 2-4 times longer than the disk, foliose, linear to lanceolate, often toothed; inner bracts short, lanceolate, membranous; ray flowers none; achenes 5-10 mm. long, cuneate, olive-brown to dark purplish; pappus awns usually 3, about 4-6 mm. long, retrorsely barbed.---Wet or damp ground. Quebec to North Dakota, south to North Carolina, New Mexico and Utah. Our records in the eastern half of Colorado at 3500-5500 feet.
2. Bidens cernua L., Sp. Pl. 832. 1753.
Stems 20-70 cm. tall, erect and sparingly if at all branched at base, glabrous or sparingly hairy; leaves linear to lanceolate or sometimes broadly oblanceolate, serrate rarely somewhat incised, sessile or connate at base; heads about 6-15 mm. high, often nodding; outer involucre bracts foliaceous, usually much surpassing the head, linear-lanceolate and unequal; inner bracts ovate-lanceolate to obovate-oblanceolate, membranous, yellow but usually red-streaked; chaff yellowish at apex and throughout but usually with longitudinal red lines; rays absent or when present rarely over 1.5 cm. long, yellow; achenes 5-7 mm. long, cuneate, 4-angled, apex and margins usually paler often yellowish; pappus awns 2 or 4, about 2-3 mm. long, retrorsely barbed.---Moist or wet ground. New Brunswick to British Columbia, south to North Carolina, New Mexico and California; Eurasia. Our records in or near northcentral, central and southcentral Colorado at 5000-7500 feet.
3. Bidens laevis B. S. P., Prelim. Cat. N. Y. 29. 1888.
New Hampshire to Florida, west to California; South America. Should be looked for in southern Colorado.
4. Bidens polylepis Blake, Proc. Biol. Soc. Wash. 35:78. 1922.
Illinois to Colorado, south to Iowa and Texas. Reported from Colorado but no specimens seen by the writer.
5. Bidens heterosperma A. Gray, Pl. Wright. 2:90. 1853.
Colorado, south to New Mexico, Mexico and Arizona. Listed for southern Colorado but no records seen by the writer.
6. Bidens tenuisecta A. Gray, Mem. Amer. Acad. Ser. 2, 4:86. 1849.
Stems 20-80 cm. tall, more or less branched especially above, glabrous or sparsely hairy; leaves at least basal petioled, usually 2- to 3-pinnately dissected into narrowly linear ultimate segments 0.5-2.5 mm. wide; heads about 6-10 mm. high and 4-10 mm. wide; outer involucre bracts 6-12, about 5-7 mm. long, linear; inner bracts subequal; rays inconspicuous, 4-6 mm. long, yellow, or lacking; achenes linear but tapering to apex, inner 8-15 mm. long, outer shorter and broader, different; pappus of 2 or rarely 3 retrorsely barbed awns 1.5-3 mm. long, at least on inner achenes.---Moist ground. Idaho, south to Texas and Mexico and Arizona. Our records from the western half of Colorado, but none as yet from the northwestern part of the state, at 5000-8000 feet.
7. Bidens vulgata Greene, Pitt. 4:72. 1899.
Stems 30-150 cm. tall, branching, glabrous or nearly so; leaves petioled, mostly 3- to 5-pinnately parted the segments lanceolate, acuminate and serrate, mostly glabrous; heads about 1.2-1.8 cm. high and 1.5-2.8 cm. wide; outer involucre bracts 10-16, about 1-2 cm. high, rather foliaceous; inner bracts 7-9 mm. long, about equalling the disk flowers; rays inconspicuous, less than 3.5 mm. long, yellow; achenes 6-12 mm. long, obovate-cuneate to oblong-cuneate; pappus awns 2, barbed, about 3-4 mm. long.---Moist or wet ground. Nova Scotia to Alberta, south to North Carolina and California; Europe. Our records from northcentral and southcentral Colorado at 5000-7500 feet.
8. Bidens frondosa L., Sp. Pl. 832. 1753.
Stems 30-100 cm. tall, usually branching, glabrous or sparingly hairy; leaves petioled, pinnately divided or compound into 3-5 divisions or leaflets, these lanceolate, serrate, acuminate and more or less petiolulate; heads about 8-12 mm. high and about as wide; outer involucre bracts 5-8, often over 2 cm. long, conspicuous and foliaceous, linear or linear-spatulate; inner bracts about 5-7 mm. long, usually slightly shorter than the disk flowers; rays none or minute, not over 3.5 mm. long, yellow; achenes 6-10 mm. long, narrowly cuneate, truncate at apex, all achenes essentially alike; pappus of 2 retrorsely barbed awns 3-4.5 mm. long.---Moist or wet places. Newfoundland to Saskatchewan and Washington, south to West Virginia, Texas and California; Europe. Our records mostly in the northeastern quarter of Colorado at 5000-6000 feet.
9. Bidens bigelovii A. Gray in Torr., Bot. Mex. Bound. 91. 1859.
Stems 40-80 cm. tall, usually branched, almost or quite glabrous; leaves petioled, 2- or 3-pinnate, the ultimate segments oblong to cuneate; heads about 6-9 mm. high and about as wide; outer involucre bracts 6-9, 5-7 mm. high, linear; inner bracts lanceolate and often shorter than outer; rays absent or small, not over 5-7 mm. long, whitish; achenes strongly dimorphic, inner 8-12 mm. long, narrowly linear, tapering to apex, outer linear-cuneate and truncate at apex, much shorter and broader (ours papillose and lighter in color); pappus awns 2-3, retrorsely barbed.---Along streams in moist soil. Colorado, south to Texas, Mexico and Arizona. Our records from central and southeastern Colorado at 4000-7500 feet.
10. Bidens bipinnata L., Sp. Pl. 832. 1753.
Stems 30-100 cm. tall, erect, usually branching, glabrous or nearly so; leaves petioled, usually 2- to 3-pinnate, the ultimate segments deltoid-lanceolate, oblong or cuneate; heads

about 5-7 mm. high and 4-6 mm. wide; outer involucre bracts 7-10, 3-5 mm. long, linear; inner bracts only slightly longer than outer; rays absent or small, not over 3 mm. long, yellowish-white; achenes 7-18 mm. long, the outer often smaller but all essentially similar in shape, linear tapering to the apex; pappus awns 3-4, retrorsely barbed.---Damp soil and waste land. Rhode Island to California, south to Florida and Mexico; South America; Australia; Eurasia. Our few somewhat doubtful records from central Colorado at 7000-8000 feet.

55. Coreopsis L. TICKSEED

Annual or perennial herbaceous plants; leaves mostly opposite, simple and entire to pinnately divided; heads solitary on pedunclelike branches; involucres campanulate or hemispheric, bracts in 2 very distinct series but all more or less united at base; outer bracts shorter to nearly as long as the inner, herbaceous and narrow; inner bracts broad, orange or brown, scarious at least on margins; receptacles flat or slightly convex, chaffy; ray flowers neutral, rays conspicuous, yellow or with a purple-brown base; disk flowers yellow or reddish-brown; anthers united, not caudate at base; achenes compressed-winged to wingless; pappus none, a mere border or of 2 very short teeth.

1. Leaves 1- to 2-pinnately divided; achenes wingless -1. C. tinctoria
1. Leaves entire or rarely with 1-2 small lobes but never divided; achenes winged -2. C. lanceolata

1. Coreopsis tinctoria Nutt., Journ. Acad. Phila. 2:114. 1821.
Annual plants; stems 40-80 cm. tall, erect, branched, glabrous; leaves subsessile or very short-petioled, about 5-10 cm. long, 1- or 2-pinnately divided into linear to narrowly linear, mostly entire segments; heads numerous, subcorymbose; involucres glabrous, outer bracts about 8, 1-3 mm. long, much shorter than inner, linear-oblong; inner bracts 5-7 mm. long, ovate to oblong-ovate, brownish-red; rays 6-15 mm. long, yellow but brownish-red at base; achenes about 2-4 mm. long, wingless; pappus of 2 minute awns or none.---In moist ground escaping from cultivation. Minnesota to British Columbia, south to Louisiana and California; Asia. Our few records from northeastern and eastcentral Colorado at 4000-5000 feet.

2. Coreopsis lanceolata L., Sp. Pl. 908. 1753.
Plants perennial; stems 20-50 cm. tall, branching, glabrous; leaves 5-15 cm. long, narrowly oblong to linear-lanceolate or oblanceolate, entire or rarely with 1-2 small basal lobes; heads few; outer involucre bracts from somewhat shorter to as long as the inner, lanceolate or oblong-ovate, usually narrower than inner; inner bracts about 7-12 mm. long; rays about 1-3 cm. long, yellow; achenes about 2-3 mm. long, winged; pappus of 2 very short teeth.---Widely cultivated and occasionally escaping. Michigan to Colorado, south to Florida and New Mexico. Our few records from northcentral and central Colorado at about 5000-5500 feet.

56. Silphium L. ROSIN-WEED; COMPASS PLANT

Perennial, caulescent, herbaceous plants, often with resinous sap; leaves opposite or alternate, entire, toothed or lobed; involucres hemispheric to campanulate, bracts broad and somewhat imbricated; receptacles flat or nearly so, chaffy; ray flowers pistillate and fertile, rays yellow; disk flowers perfect but sterile, yellow; anthers united, entire or somewhat toothed at base but not sagittate; achenes flat and more or less 2-winged; pappus none or the achene wings extending up into 2 pappuslike triangular teeth.

1. Leaves alternate, deeply lobed or parted -1. S. laciniatum
1. Leaves opposite, entire to dentate -2. S. integrifolium

1. Silphium laciniatum L., Sp. Pl. 919. 1753.
Stems 1-3.5 m. tall, coarsely hispid; leaves mostly basal and turned vertically mostly lying north and south, 10-40 cm. long, pinnately lobed or parted; cauline leaves clearly alternate; involucres 2-2.5 cm. high; rays 3-5 cm. long.---Prairies. Michigan to North Dakota, south to Alabama and Texas. Our two records from central Colorado at 6000-7500 feet, perhaps as casual introductions.

2. Silphium integrifolium Michx., Fl. Bor. Am. 2:146. 1803.
Stems 70-150 cm. tall, rough-pubescent; leaves opposite, about 7-12 cm. long, ovate to lanceolate, entire to dentate, scabrous, mostly sessile; involucres about 2 cm. high; rays 2-3 cm. long.---Prairies. Michigan to Colorado, south to Louisiana and Texas. Our few records from northeastern Colorado at 3500 feet.

57. Zinnia L.

Perennial plants with woody root and caudex; leaves opposite, linear, entire; heads showy, terminating the branches; involucres campanulate to nearly cylindrical, bracts imbricated in 2-3 series, obtuse, dry, firm and appressed; receptacles conic to nearly cylindrical, chaffy; ray flowers pistillate and fertile, rays yellow or orange, sometimes turning whitish in age, persistent on the achenes and becoming papery; disk flowers perfect and fertile, brick-red; ray achenes more or less compressed but 3-angled, disk achenes strongly compressed; pappus of a few awns or teeth, or lacking.

1. Zinnia grandiflora Nutt., Trans. Amer. Phil. Soc. n. ser. 7:348. 1840.
Crassina grandiflora (Nutt.) Kuntze---Stems 10-20 cm. tall, branching, puberulent; leaves 1-3 cm. long, more or less 3-nerved, impressed punctate, scabro-hispidulous, somewhat connate at

base; involucre 6-9 mm. high; ray flowers 4-5, at maturity about 10-16 mm. long. A plant with short ligules has been reported for southern Colorado and has been called Z. anomala A. Gray. Specimens so named appear to be a form of the above.---Plains and hills. Kansas to Nevada, south to Texas, Mexico and Arizona. Our records from the southeastern quarter of Colorado at 4000-6000 feet.

58. Heliopsis Pers. OXEYE

Perennial herbaceous plants; stems leafy; leaves opposite, ovate to ovate-lanceolate, toothed, petioled; heads long-peduncled on end of stem and branches; involucres hemispheric or broadly campanulate, bracts in 2-3 series but nearly equal in length, outer herbaceous; receptacles convex or conic, chaffy; ray flowers pistillate and fertile, with conspicuous yellow rays persistent on the achenes; disk flowers perfect and fertile; anthers united, entire or nearly so at base; achenes 3- to 4-angled, truncate at apex; pappus a short laciniate crown or 1-4 teeth, sometimes apparently lacking.

1. Heliopsis scabra Dunal, Mem. Mus. Paris 5:56. 1819.
    Stems 50-100 cm. tall, more or less hispidulous-scabrous; leaves 5-10 cm. long, often cordate at base, strongly 3-veined, coarsely dentate or upper entire, scabro-hispidulous; involucres 8-12 mm. high, short-hairy; rays 15-30 mm. long, oblong.---Dry ground. New York to British Columbia, south to Illinois, Arkansas and New Mexico. Our records from the southeastern quarter of Colorado at 5500-7000 feet.

59. Helianthella T. & G.

(Contributed by W. A. Weber)

Coarse perennial herbaceous plants with stout taproot and woody caudex; stems several, leafy, usually hirsute and scabrous in age, enlarged basal leaves present or absent; leaves, at least the lower opposite, linear, lanceolate, elliptic, oblong or ovate-lanceolate, usually 3-nerved, entire; heads medium to large, terminal on the stem or branches, usually solitary and long-peduncled; involucres hemispheric to flat, bracts more or less imbricated and often foliaceous; receptacles flat or convex, chaffy; ray flowers neutral, rays yellow; disk flowers perfect and fertile, yellow or purple; achenes cuneate-obovate or obcordate, strongly laterally compressed, thin-edged or wing-margined at least when young; pappus a pair of persistent awns or chaffy teeth with a crown of intermediate chaffy scales usually present also.

1. Involucre bracts ovate, strongly graduated; heads numerous, the rays very short; disk flowers purple or brown; plants of southwestern Colorado   -1. H. microcephala
1. Involucre bracts lanceolate, subequal; heads solitary or few, the rays ample; disk flowers yellow; plants not limited to southwestern Colorado
    2. Involucre bracts uniformly pubescent, not distinctly ciliate; leaves acute or obtuse, tapering to short petioles; enlarged basal leaves absent; heads erect; plants apparently limited to northwestern Colorado
                                                                                                     -2. H. uniflora
    2. Involucre bracts conspicuously ciliate, not or little pubescent in the face; leaves acute or acuminate, tapering to long petioles; enlarged basal leaves present; heads nodding; plants not limited to northwestern Colorado
        3. Plants tall and stout; heads 4-5 cm. broad, excluding the rays; leaves leathery, up to 50 cm. long, attenuate at both ends   -3. H. quinquenervis
        3. Plants low and slender; heads 1.5-2 cm. broad excluding the rays; leaves not leathery, less than 10 cm. long, acute   -4. H. parryi

1. Helianthella microcephala (A. Gray) A. Gray, Proc. Amer. Acad. 19:10. 1883.
    H. scabra Payson; Encelia microcephala Gray---Stems 20-60 cm. tall, leafy, appressed hispidulous; basal leaves numerous, linear-spatulate to very broadly spatulate, tapering to long petioles; cauline leaves reduced, narrow, almost sessile; heads few to many, paniculate but inflorescence tending to be flat-topped, heads 1 cm. broad and high; involucres hemispheric, bracts strongly graduated in about 3 rows, broadly oblanceolate, obtuse; chaff-paleae rigid but weak, obtuse; rays 8-10, 10-13 mm. long; disk corollas 6 mm. long, lobes purple or brown; achenes 7-8 mm. long, at maturity 3-4 times as long as the awns, not strongly compressed, faces with dense, long, appressed, white hairs; pappus of awns and well developed intermediate squamellae.---Deserts and dry areas. Colorado, Utah and Arizona. Our records from westcentral and southwestern Colorado at 5500-6500 feet.
2. Helianthella uniflora (Nutt.) T. & G., Fl. N. Amer. 2:334. 1842.
    H. multicaulis D. C. Eaton---Stems 40-120 cm. tall, somewhat hirsute or merely puberulent, becoming scabrous; enlarged basal leaves absent; cauline leaves 12-25 cm. long, lanceolate to elliptic, rarely ovate, in 3-6 opposite pairs, obtuse to acute, 3-nerved from below the middle, short-petioled or sessile; heads erect, usually solitary; involucres 12-18 mm. high and 1.5-2 cm. broad, bracts subequal, lanceolate-attenuate, the outermost frequently elongate and foliaceous, appressed, uniformly cinereous-pubescent drying green; chaff-paleae rigid; rays light yellow, ample; disk flowers yellow; achenes 6-7 mm. long, narrowly obovate, ciliate above, appressed pubescent on lateral faces; pappus awns 2, 3.5-4 mm. long.---High plateaus and valleys. Montana to Washington, south to Colorado and Arizona. Our records from northwestern Colorado at 6500-8000 feet.
3. Helianthella quinquenervis (Hooker) A. Gray, Proc. Amer. Acad. 19:10. 1883.
    Stems 50-150 cm. tall, stout, sparingly hirsute or glabrate; enlarged basal leaves present, up to 50 cm. long; cauline leaves leathery, about 4 pairs present, ovate-lanceolate or

elliptic-lanceolate, acuminate, tapering to a prominent petiole, 2 pairs of lateral veins usually prominent; heads solitary or a few reduced lateral ones present, long-peduncled, nodding; involucres 2 cm. high and 4-5 cm. wide, bracts broadly ovate-lanceolate, acute or acuminate, prominently ciliate, blackening on drying; chaff-paleae soft and scarious; rays 2.5-3 cm. long, pale yellow; disk flowers yellow; achenes 8-10 mm. long, ciliate-margined, appressed pubescent on the faces, blackish-brown; pappus awns 2, slender, half the length of the achene.---Mountains. Montana, south to Mexico. Our records scattered in the western half of Colorado at 7500-11,000 feet.

4. Helianthella parryi A. Gray, Proc. Acad. Sci. Phila. 1863:65. 1863.

Stems 20-50 cm. tall, appressed hirsutulous; basal leaves numerous, spatulate, acute, tapering to long petioles; cauline leaves 1-2 pairs, greatly reduced, 3-nerved; heads solitary or few, 1 cm. high and 1.5-2 cm. broad; involucre bracts subequal in 3 rows, lanceolate, tapering from base to apex, acuminate, ciliate; chaff-paleae thin and weak; rays 8-10, 25-30 mm. long and 5-7 mm. wide, pale yellow; disk flowers 5 mm. long, yellow; achenes 7-8 mm. long, margins and faces appressed pilose, black and shining; pappus awns stout, much shorter than the achenes, squamellae present.---Mountains. Colorado, New Mexico and Arizona. Our records from central, southcentral and southwestern Colorado at 7000-12,000 feet.

60. Viguiera H. B. K. GOLDENEYE

Perennial herbaceous plants; stems leafy; leaves opposite or alternate, narrow, entire or obscurely toothed; heads corymbose or paniculate; involucres hemispheric or campanulate, bracts in 2 or 3 series, narrow and herbaceous, outer somewhat shorter; receptacles more or less conic, chaffy; ray flowers neutral, rays yellow and rather showy; disk flowers perfect and fertile, yellow or brownish; achenes 4-angled or somewhat flattened; pappus none.

1. Viguiera multiflora (Nutt.) Blake, Contrib. Gray Herb. 54:108. 1918.

Gymnolomia multiflora (Nutt.) B. & H.---Stems 25-100 cm. tall, branched, finely pubescent to subglabrous; leaves opposite below and alternate above, lanceolate to linear-lanceolate or sometimes to lanc-ovate, acuminate to obtuse at apex, entire to serrate, cuneate to a short petiole, strigillose both sides; heads in irregular loose panicles; involucres 6-10 mm. high, bracts linear-lanceolate, strigose; rays 7-17 mm. long, yellow and conspicuous; chaff of receptacle hairy above; achenes 2.5-3 mm. long, glabrous, mottled in age.---Plains, hills, roadsides and streambanks. Montana to Nevada, south to New Mexico and California. Our records scattered in the western half of Colorado at 4500-11,000 feet.

61. Helianthus L. SUNFLOWER

Annual or perennial herbaceous plants; stems leafy; leaves opposite or alternate, usually entire or merely toothed, rarely laciniate; heads large, peduncled, usually solitary or few-corymbose; involucres hemispheric or depressed-hemispheric, bracts imbricated in several series, herbaceous; receptacles flat or convex, rarely conic, chaffy; ray flowers pistillate and sterile, rays conspicuous, yellow; disk flowers perfect and fertile, yellow, brown or purplish-brown; anthers united, not caudate at base; achenes more or less 4-angled and somewhat compressed but not at all winged; pappus of 2 readily deciduous scales or awns, sometimes a few small intermediate ones present.

1. Leaf margins conspicuously bristly-ciliate on the margins; rays inconspicuous, not over 1 cm. long; plants limited to southcentral Colorado   -1. H. ciliaris
1. Leaf margins not especially bristly ciliate; rays conspicuous, normally over 1 cm. long; plants widely distributed in Colorado
   2. Lobes of disk flowers red or purple; annual or perennial plants
      3. Cauline leaves linear or linear-lanceolate, not over 1 cm. wide, more than 5 times longer than wide; plants perennial with leaves mostly alternate   -2. H. salicifolius
      3. Cauline leaves lanceolate to ovate, usually over 1 cm. wide and less than 5 times longer than wide; annuals with leaves mostly alternate or perennial with leaves mostly opposite
         4. Involucre bracts glabrous, at least outer, on back (short-ciliate on margins), mostly shorter than the disk; perennial plants with leaves mostly opposite   -3. H. rigidus
         4. Involucre bracts pubescent on back (ciliate or not ciliate), mostly equalling or longer than the disk; annual plants with leaves mostly alternate
            5. Leaves lanceolate to ovate, all cuneate at base; involucre bracts usually not ciliate; disk seldom over 3 cm. wide; chaff toward center of disk conspicuously white-bearded at apex (best seen in a young head)   -4. H. petiolaris
            5. Leaves ovate, lower at least broadly so and subcordate at base; involucre bracts ciliate; disk often over 3 cm. wide; chaff not conspicuously white-bearded (may be short-hairy)-5. H. annuus
   2. Lobes of disk flowers yellow (may have a few reddish longitudinal lines); perennial plants
      6. Midrib of lower leaf surface with stiff widely spreading or even retrorse hairs   -6. H. pumilus
      6. Midrib of lower leaf surface with hairs antrorsely appressed or strongly ascending
         7. Leaves ovate or ovate-oblong; rootstocks often bearing tubers   -7. H. tuberosus
         7. Leaves lanceolate or narrowly elliptic; rootstocks not bearing true tubers
            8. Leaves, at least some, conduplicate and pressing folded; involucre bracts conspicuously canescent on back; stems hispidulous-scabrous especially above   -8. H. maximiliani
            8. Leaves rarely if ever folded and conduplicate; involucre bracts glabrous to sparingly setose or appressed-pubescent; stems glabrous
               9. Leaves deeply and coarsely serrate (teeth often over 2 mm. long on upper side), blades often over 15 cm. long and 4 cm. wide; involucre bracts usually definitely longer than the disk; leaves short-pubescent or canescent (not rough) below, slightly if at all scabrous above   -9. H. grosse-serratus

9. Leaves entire to serrulate or serrate with teeth seldom over 2 mm. deep, blades seldom over 15 cm. long or 4 cm. wide; involucre bracts only slightly longer than the disk; leaves scabro-hispidulous below, scabrous above -10. H. nuttallii

1. Helianthus ciliaris DC., Prodr. 5:587. 1836.
Perennial plants with spreading rootstocks; stems about 40-60 cm. tall, glabrous to sometimes hispid; leaves mostly opposite, 2-6 cm. long and 5-20 mm. wide, lanceolate to rarely linear, entire to laciniate, glaucous and bluish-green, mostly glabrous on surfaces; heads solitary on stems and branches, disk about 1-1.5 cm. wide; chaff more or less hispid, entire or 3-toothed at apex; involucre bracts shorter than the disk, ovate, usually obtuse, erect, appressed, glabrous or subglabrous on surface but usually ciliate; rays inconspicuous, rarely over 1 cm. long, yellow with reddish longitudinal lines; corolla lobes of disk flowers brown or reddish; achenes glabrous; pappus of 2 broadly ovate-acuminate short scales, no intermediate paleae present.---Dry ground, often acting as a weed in cultivated areas. Texas to Arizona, south to Mexico. Our record from the San Luis Valley in southcentral Colorado at 7500 feet.
2. Helianthus salicifolius A. Dietr., Allg. Gartenz. 2:337. 1834.
Nebraska and Colorado, south to Texas. This plant should be in eastern Colorado according to report.
3. Helianthus rigidus (Cass.) Desf., Cat. Hort. Par. ed. 3, 184. 1829.
H. subrhomboideus Rydb.; H. scaberrimus Ell.---Perennial plants with stout, often branched rootstocks; stems 30-80 cm. tall, more or less hirsute-scabrous; leaves mostly opposite, 5-10 cm. long and over 1 cm. wide, variable in shape, from ovate to linear-lanceolate, acute, entire to sometimes serrate, cuneate to a short or long petiole, thick and firm, scabrohirsute both sides; heads solitary at ends of stem and branches, disk 1-2 cm. wide; involucre bracts shorter than the disk, in 3-4 series, ovate or ovate-lanceolate, obtuse or broadly acute, closely appressed, glabrous on back at least on outer ones and rather conspicuously short-ciliate; rays 1.5-2.5 cm. long, yellow or streaked with darker lines; corolla lobes of disk flowers brown-purple or reddish; achenes more or less hairy at apex; pappus of disk flowers of 2 broadly lanceolate, awnlike scales, usually with accessory paleae at base or between.---Plains, prairies and hills. Western Canada, south to New Mexico and introduced eastward. Our records from northcentral, central, southcentral and southwestern Colorado, at 6000-8000 feet.
4. Helianthus petiolaris Nutt., Journ. Acad. Nat. Sci. Phila. 2:115. 1821.
Annual plants; stems about 20-100 cm. tall, erect, hirsute-strigose or scabro-hispid; leaves alternate, 3-15 cm. long and 1-5 cm. wide, narrowly lanceolate to deltoid-ovate, mostly acute, entire or sparingly dentate, mostly cuneate at base to an often long petiole, hispidulous-scabrous; heads solitary at ends of stems and branches, disk about 1-3 cm. wide; involucre bracts about as long or a little longer than the disk, broadly lanceolate, acute or abruptly acuminate, appressed hispidulous, sometimes ciliate but seldom conspicuously so; chaff 3-toothed and hispid near apex, especially in center of young heads; rays about 1.5-2 cm. long, bright yellow; disk corolla lobes red-purple or brownish-red; achenes usually villous or strigose; pappus of 2 broadly lanceolate awnlike scales about 1/2 as long as the corollas.---Plains, hills, roadsides and waste places. Saskatchewan to British Columbia, south to Missouri, Texas and California; introduced eastward. Our records scattered over Colorado at 3500-8000 feet.
5. Helianthus annuus L., Sp. Pl. 904. 1753.
H. aridus Rydb.; H. lenticularis Dougl.---Annual plants; stems 30-200 cm. tall, very hispid and rough; leaves normally alternate, 4-20 cm. long and about 3-15 cm. wide, ovate, rarely ovate-lanceolate, obtuse or acute, dentate or entire, often cordate especially the lower, long-petioled; heads terminating the stem and branches, disk about 3-4 cm. wide but sometimes smaller; involucre bracts about as long as the disk, broadly ovate to lanceolate, acuminate, often attenuate, usually hirsute or ciliate at least near base; chaff 3-toothed and more or less pubescent at apex; rays 15-40 mm. long, yellow; disk corolla lobes reddish-purple or brownish-red; achenes glabrous to finely strigose; pappus of 2 broadly lanceolate awnlike scales. A variable plant that has been broken down into many ill-defined varieties.---Plains, hills, waste places and fields. Native or introduced in the most of Canada and the United States. Our records scattered over Colorado at 4000-8500 feet.
6. Helianthus pumilus Nutt., Trans. Am. Phil. Soc. n. ser. 7:366. 1841.
H. excubitor E. E. Wats.---Perennial plants, roots woody but no rootstocks; stems 30-80 cm. tall, mostly tufted, scabro-hispid; leaves opposite, about 3-10 cm. long and 1.5-2 cm. wide, ovate or sometimes lanceolate, entire or nearly so, thick, strigose-hispid both sides, midrib with stiff-spreading or retrorsely leaning hairs; heads 1-3 on short peduncles terminating the stems, disk 1-1.8 cm. wide; involucre bracts a little shorter than the disk, ovate-lanceolate, acute, erect, usually hirsute; chaff lanceolate, entire and hirsute at apex; rays about 1-2.5 cm. long, yellow; disk flowers yellow or sometimes with a few reddish longitudinal lines; achenes glabrous; pappus of 2 broadly lanceolate awnlike scales, with a few lacerate paleae at base.---Plains and hills. Wyoming and Colorado, possibly eastward. Our records from northcentral, central and southcentral Colorado at 5000-8000 feet.
7. Helianthus tuberosus L., Sp. Pl. 905. 1753.
Perennial plants with thickened rootstocks these bearing tubers; stems 100-250 cm. tall, hirsute or pubescent; lower leaves opposite, 5-25 cm. long, ovate or broadly ovate-lanceolate, serrate, acuminate, base cuneate to subcordate to a short petiole, 3-ribbed, scabrous above, finely pubescent beneath; upper leaves usually alternate; heads several to many, 1-1.5 cm. high; involucres hemispheric, bracts usually longer than the disk, linear-lanceolate, loose or reflexed, more or less appressed pubescent; chaff pubescent, mostly 3-toothed at apex; rays 1.5-3 cm. long, yellow; disk flowers yellow; achenes pubescent.---Moist or alluvial ground.

Nova Scotia to Northwestern Territories, south to Georgia and Nebraska, probably introduced anywhere in North America. Our few records from the eastern half of Colorado at 4000-5000 feet.
8. Helianthus maximiliani Schrad., Ind. Sem. Hort. Gotting. 1835.
H. subtuberosus Bourg.---Perennial plants with very short rootstocks; stems 40-250 cm. tall, stout, solitary to tufted, rough hispidulous-scabrous especially above; leaves usually alternate, 5-15 cm. long and about 1-2 cm. wide, linear-elliptic to lanceolate, acute, strongly conduplicate and some remaining folded after drying, entire or serrate, firm, cuneate to a short winged petiole, scabro-setose and light green both sides; heads usually racemose, disk 2-3 cm. wide; involucre bracts longer than the disk, linear-lanceolate, attenuate at apex, conspicuously canescent or densely appressed pubescent; chaff linear, entire and pubescent at apex; rays 15-40 mm. long, yellow; disk flowers yellow; achenes glabrous; pappus of 2 broadly lanceolate, awnlike scales, no accessory paleae on disk flower.---Prairies, valleys and banks. Manitoba to British Columbia, south to Missouri and Texas. Our few records scattered in Colorado but mostly in the northeastern quarter of the state, at 3500-7000 feet.
9. Helianthus grosse-serratus Martens, Sel. Sem. Hort. Lov. 1840.
Maine to Wyoming, south to Virginia and New Mexico. Not found in this state by the writer.
10. Helianthus nuttallii T. & G., Fl. N. Amer. 2:324. 1842.
H. fascicularis Greene---Perennial plants with slender rootstocks and thickened fleshy roots; stems 30-100 cm. tall, glabrous or slightly scabrous, sometimes with a very few hispid hairs above; lower leaves opposite, upper alternate, occasionally all opposite, 5-15 cm. long and about 1.5-3 cm. wide or sometimes wider, lanceolate, acute or acuminate, entire to denticulate, cuneate to a short winged petiole, scabro-hispidulous both sides especially below; heads solitary or in cymes or panicles, disk 1.5-3 cm. wide; involucre bracts usually longer than the disk, linear-lanceolate, loose, glabrate to only moderately appressed pubescent on surfaces but more or less ciliate; chaff linear, entire or nearly so and pubescent at apex; rays about 2-3 cm. long, yellow; disk flowers yellow; achenes glabrous; pappus of 2 narrowly lanceolate awnlike scales, rarely intermediate paleae present.---Plains, hills and valleys. Saskatchewan to Alberta, south to New Mexico and Arizona. Our records scattered in the western half of Colorado at 4500-8000 feet.

62. Petasites Gaertn. BUTTER BUR; SWEET COLTSFOOT

Herbaceous perennial plants with thick creeping rootstocks; stems scapiform; leaves basal, large, long-petioled, white-tomentose beneath; scape with scalelike leaves; heads racemose or corymbose, subdioecious; involucres campanulate, the bracts in 1 series; receptacle flat or nearly so, naked; corolla whitish, the pistillate heads with disk flowers and ray flowers, substerile heads discoid and bisexual but sterile, or with a few pistillate ray flowers near margins; anthers connate, not caudate at base; achenes narrow, 5- to 10-ribbed; pappus of soft, white, scabrous to barbellate bristles.

1. Petasites sagittata (Pursh) A. Gray in Brew. & Wats., Bot. Calif. 1:407. 1880.
Scapes 20-30 cm. tall, more or less tomentose, very bracted, these 4-7 cm. long; basal leaves 10-30 cm. long, ovate to deltoid, cordate to truncate at base and more or less sagittate or hastate, repand-dentate, green and glabrate above; heads rather corymbose; involucres 6-9 mm. high; rays 3-7 mm. long, smallest on substerile heads.---Wet or moist ground. Labrador to Alaska, south to Minnesota, and Colorado. Our records from the western half of Colorado, except the extreme western part, at 7500-9000 feet.

63. Tetradymia DC. HORSEBRUSH

Rather low shrubs with stiffly much branched stems, usually canescent-tomentose or tomentose throughout at least when young; leaves alternate, narrow, entire, solitary or fascicled, or the primary leaves modified into spines; heads solitary from the upper axils or clustered at the tips of the branches; involucres cylindrical to oblong, bracts 4-6, all equal in length but overlapping from sides, often enlarged and thickened at base; receptacles small, flat, naked; no rays present; disk flowers yellow with elongated tubes; anthers connate, not caudate at base; achenes obovoid to terete, 5-nerved; pappus of copious, whitish, minutely scabrous bristles.

1. Primary leaves converted into spines
   2. Flowers 5-9 to a head; branches densely white-tomentose even in age; involucre bracts 5-6; heads solitary in the upper axils, not at all corymbose -1. T. spinosa
   2. Flowers 4 to a head; branches glabrate in age; involucre bracts usually 4 (sometimes 5); heads clustered-corymbose at ends of branches -2. T. nuttallii
1. Primary leaves foliate and not at all spinulose -3. T. canescens

1. Tetradymia spinosa Hook. & Arn., Bot. Beechey Voy. 360. 1840.
Shrub 50-120 cm. tall; stems white-tomentose, beset with spines 5-15 mm. long; primary leaves transformed into these often recurved spines, secondary leaves fascicled in their axils, linear-clavate, fleshy, glabrous or soon glabrate; heads on stout peduncles, solitary in the upper axils; involucres about 7-10 mm. high, bracts 5-6; flowers 5-9 to a head; achenes with soft hairs, woolly at base, these hairs about as long as the pappus.---Dry hills, valleys and plains. Montana to Oregon, south to Colorado and California. Our records from western Colorado at 4500-7000 feet.
2. Tetradymia nuttallii T. & G., Fl. N. Amer. 2:447. 1843.
Shrub 30-100 cm. tall; stems woolly-canescent when young, glabrate in age, densely spiny; primary leaves mostly converted into these spines, 10-18 mm. long; secondary leaves densely

fascicled in their axils, thickish, oblanceolate to linear-oblanceolate, loosely tomentose with deciduous hairs; heads corymbose-clustered near the ends of the branches; involucres 4-8 mm. long, with 4 or sometimes 5 bracts; flowers 4 to a head; achenes densely white-tomentose to white-villous.---Dry barren hills and plains. Wyoming to Utah. Our record from northwestern Colorado at about 6500 feet.

3. Tetradymia canescens DC., Prodr. 6:440. 1837.

T. linearis Rydb.; T. inermis Nutt.---Shrub 20-100 cm. tall; stems canescent-tomentose, not at all spinose; primary leaves 6-30 mm. long, oblanceolate to linear, often with fascicles of leaves in the axils, gray-canescent to tomentulose; heads clustered near tips of branches; involucres about 6-15 mm. high, bracts 4; flowers 4 to a head; achenes silky-hairy. The form with shorter broader leaves has been called T. inermis Nutt. but intergradations in this area are as common as the extremes.---Dry hills, ridges and plains. Montana to British Columbia, south to New Mexico and California. Our records well scattered in the western half of Colorado at 4500-9000 feet.

64. Senecio L. GROUNDSEL

Annual or perennial plants, herbaceous or sometimes woody at base; stems leafy but the leaves often much reduced above, rarely the stems scapose; leaves alternate, entire, toothed or variously pinnatifid; heads solitary, corymbose or paniculate; involucres cylindrical to campanulate, bracts many, the principal ones in 1 series but often a few short ones at base; receptacles flat or slightly concave, naked but often pitted; ray flowers usually present, pistillate and fertile, rays yellow, rarely orange-red; disk flowers perfect and fertile, usually yellow; anthers connate, not sagittate at base; achenes terete or slightly flattened on marginal, 5- to 10-ribbed; pappus of many capillary, smooth to rough bristles. A large and puzzling genus.

Since this treatment was prepared a discoid plant with low stems and tomentose, entire leaves has been collected in Saguache County. It has been identified as S. hallii Britt.

1. Heads nodding especially in bud
   2. Ray flowers present
      3. Leaves (at least some) pinnatifid; plants floccose-tomentose at least on lower leaf surface
         -1. S. taraxacoides
      3. Leaves entire to sinuate-dentate, not at all pinnatifid; plants glabrous or slightly floccose when young
         4. Plants over 20 cm. tall
            -2. S. amplectens
         4. Plants not over 20 cm. tall
            5. Leaves sharply dentate, the teeth about 1 mm. deep or more, blade obovate to lanceolate, often over 4 cm. long, usually tapering gradually to the petiole, seldom purple below; heads about 15-18 mm. high
               -2A. S. amplectens holmii
            5. Leaves entire or denticulate, the teeth seldom 1 mm. deep, blade obovate to rotund, usually truncate to broadly cuneate at base, seldom over 4 cm. long, usually purple below; heads about 18-20 mm. high
               -3. S. soldanella
   2. Ray flowers lacking
      6. Heads 8-12 mm. high and about 8 mm. wide, narrowly campanulate   -4. S. pudicus
      6. Heads 12-20 mm. high and about 15-20 mm. wide, broadly campanulate
         7. Auricles of upper leaves large and usually toothed; midveins of leaves not long-villous
            -5. S. bigelovii
         7. Auricles of upper leaves small and entire margined; midveins of leaves long-villous
            -5A. S. bigelovii hallii
1. Heads erect or nearly so
   8. Ray flowers absent
      9. Plants annual; leaves (including basal) sinuate-pinnatifid; involucre bracts black-tipped; achenes hairy on the angles
         -6. S. vulgaris
      9. Plants biennial or perennial; leaves entire to incised, usually not pinnatifid (at least on basal); involucre bracts not black-tipped; achenes glabrous or very nearly so
         10. Plants scapose or nearly so; leaves entire or slightly dentate at apex; heads 10-12 mm. high
            -38A. S. werneriaefolius incertus
         10. Plants with evident cauline leaves (although reduced upwards), these more or less pinnatifid or sharply dentate; heads 5-10 mm. high
            11. Stems and leaves densely white-tomentose; all leaves linear or oblong-linear, all pinnatifid
               -36A. S. fendleri lanatus
            11. Stems and leaves glabrous or glabrate; not all leaves (if any) pinnatifid, at least the basal oblanceolate to wider
               12. Heads many; basal leaves obovate to oblanceolate, usually widest above the middle, mostly over 6 cm. long; stems coarse (about 3 mm. or more in diameter near base); none of cauline leaves pinnatifid
                  -7. S. rapifolius
               12. Heads 2 to several; basal leaves reniform, oval, ovate or ovate-oblong, widest at or below the middle, not over 6 cm. long; stems more slender; cauline leaves usually pinnatifid
                  13. Plants not over 15 cm. tall; heads about 6 mm. high, 2-3 to a stem
                     -8. S. fedifolius
                  13. Plants over 15 cm. tall; heads 7-10 mm. high, often over 3 to a stem
                     -9. S. debilis
   8. Ray flowers present
      14. Heads 15-25 mm. high (usually 18-25 mm.), usually solitary; plants not over 15 cm. tall
         -3. S. soldanella

14. Heads not over 15 mm. high (if rarely to 16 mm. the heads not solitary), 1 to many; plants often over 15 cm. tall
   15. Stems about equally leafy above and below (the upper leaves only slightly reduced in size and distribution)
      16. Leaves simple and linear-filiform, or pinnately divided into linear or linear-filiform segments
         17. Stems permanently tomentose at least above; heads broadly campanulate, mostly over 8 mm. wide
            -10. S. longilobus
         17. Stems glabrous or glabrate, never tomentose in age; heads subcylindric or narrowly campanulate, not over 8 mm. wide
            18. Leaves all simple or rarely with a pair of filiform lobes   -11. S. spartioides
            18. Leaves (at least lower) pinnately divided into narrow divisions   -12. S. multicapitatus
      16. Leaves simple, linear-lanceolate or broader, or if lobed the divisions wider than linear-filiform
         19. Leaves pinnatifid
            20. Heads 7-10 mm. high; involucre bracts 5-7 mm. long, usually conspicuously black-tipped
               -13. S. ambrosioides
            20. Heads 10-12 mm. high; involucre bracts 7-10 mm. long, not conspicuously black-tipped
               -14. S. eremophilus
         19. Leaves toothed or entire, rarely dentate-lobed near base but never pinnatifid
            21. Leaves obovate, spatulate or oval, usually broadly obtuse or rounded at apex, not over 6 cm. long   -15. S. carthamoides
            21. Leaves linear-lanceolate, lanceolate or elongated-triangular (rarely obovate) and always tapering to an acute or acuminate apex, the longest over 6 cm. long
              22. Leaf blades (at least some) broadly triangular-lanceolate, truncate, cordate or sometimes very broadly cuneate at base, the blade usually over 3 cm. wide on widest   -16. S. triangularis
              22. Leaf blades usually linear-lanceolate, never triangular-lanceolate, gradually cuneate at base, the blade usually less than 3 cm. wide
                 23. Stems 50-150 cm. tall; heads usually many; leaves sharply serrate; involucres cylindric, 5-8 mm. wide   -17. S. serra
                 23. Stems 20-50 cm. tall; heads usually few to several; leaves usually denticulate or dentate; involucres broadly campanulate, some usually over 8 mm. wide   -18. S. crassulus
   15. Stems few-leaved or the upper leaves definitely reduced in size or distribution
      24. Rays orange-red   -19. S. crocatus
      24. Rays light yellow to bright yellow, no trace of orange-red (at least when fresh)
         25. Plants glabrous (sometimes somewhat tomentose when young but soon glabrate) not at all tomentose when mature except at base of petioles in the leaf axils or in the inflorescence; hairs often long-jointed especially if persisting
            26. Plants dwarf alpines not over 10 cm. tall; heads solitary; leaves reniform or subreniform   -20. S. porteri
            26. Plants over 10 cm. tall, usually over 15 cm.; heads rarely less than 2; leaves various but never reniform
              27. Basal leaves more or less pinnatifid; cauline leaves also pinnatifid
                 28. Achenes densely hirtellous; involucre bracts usually about 13   -21. S. multilobatus
                 28. Achenes glabrous or but slightly hirtellous; involucre bracts usually about 21 to a head   -22. S. uintahensis
              27. Basal leaves entire or merely toothed; cauline leaves not pinnatifid in most species
                 29. Tall bog plants, usually over 70 cm. tall; heads not over 10 mm. high   -23. S. hydrophilus
                 29. Plants of various characters but never with all those mentioned above (mostly less than 70 cm. tall, if over then not bog plants and heads usually over 10 mm. high)
                    30. Rootstocks very short and erect (usually wider than high) with numerous fleshy fibrous roots; hairs when present usually long-jointed; all leaves entire or denticulate, the longest over 8 cm. long
                       31. Leaves entire or sinuate, upper widest at base, tapering to apex; stems hollow, usually flattening on drying; involucre bracts not especially thickened on back or base; terminal head of the inflorescence often sessile or shorter-peduncled than the others; heads usually over 5   -24. S. integerrimus
                       31. Leaves denticulate, upper widest a little above very base; stems solid, not especially flattened; involucre bracts definitely thickened on back and at base; terminal head on a well developed peduncle; heads rarely over 5   -18. S. crassulus
                    30. Rootstocks well developed, horizontal or ascending, woody, not usually densely covered with fleshy roots; hairs when present not long-jointed; leaves various but rarely with the combination of characters listed above
                       32. Cauline leaves (at least some) lyrate to pinnatifid; leaves thin in texture (except sometimes in S. mutabilis) not succulent when fresh
                          33. Achenes hirtellous to hispidulous; leaves rather thick and firm in texture   -25. S. mutabilis
                          33. Achenes glabrous (rarely very sparsely hirtellous on angles); leaves thin
                             34. Basal leaves entire or very obscurely crenate   -26. S. dimorphophyllus
                             34. Basal leaves definitely toothed
                                 35. Basal leaves oblanceolate, cuneate, never cordate at base   -27. S. flavulus

35. Basal leaves ovate-rotund to oblong-ovate, at least some cordate or subcordate at base
-28. S. pseudaureus
32. Cauline leaves entire to toothed, never really lyrate or pinnatifid; leaves thick or firm in texture, more or less succulent when fresh
36. Achenes definitely hirtellous or hirsutulous; leaves often oblanceolate   -25. S. mutabilis
36. Achenes glabrous or very rarely sparsely hirtellous on angles; leaves oblong-lanceolate to wider
37. Lower stem leaves spatulate with broad winged petioles, upper leaves sessile with clasping much enlarged bases   -26. S. dimorphophyllus
37. Stem leaves neither with broad winged petioles or with clasping much enlarged bases
38. Stems stout from a rootstock which is not caespitose; leaves usually sharply dentate with subcartilaginous teeth, sometimes entire   -29. S. wootonii
38. Stems slender from a usually caespitose rootstock; leaves entire or toothed but never sharply dentate with subcartilaginous teeth
39. Lower leaves mostly obovate   -30. S. cymbalarioides
39. Lower leaves mostly oblong-oblanceolate   -31. S. acutidens
25. Plants tomentose (at least when young) sometimes becoming somewhat glabrate in age; none of hairs long-jointed
40. Involucre bracts about 8 to a head; plants 30-80 cm. tall; achenes glabrous   -32. S. atratus
40. Involucre bracts about 13-21 to a head; plants of various heights but often below 30 cm.; achenes glabrous to hairy
41. All or some of the leaves lobed or pinnatifid
42. Achenes hirtellous or hispidulous, at least on the angles
43. Heads 10-12 mm. high; involucre bracts about 21 to a head   -33. S. neomexicanus
43. Heads 8-10 mm. high; involucre bracts 13-21 to a head
44. Stems and leaves glabrous or loosely floccose-tomentose at first, soon glabrate especially on upper leaf surface; heads few to several   -25. S. mutabilis
44. Stems and leaves permanently floccose-tomentose, this rather dense; heads often numerous   -34. S. plattensis
42. Achenes glabrous
45. Plants permanently white-tomentose, only rarely glabrate on upper leaf surface; heads large, 10-12 mm. high, few to several   -35. S. canus
45. Plants glabrous or tomentose at first, soon glabrate; heads 8-10 mm. high, several to many
46. Involucre bracts about 21 to a head; plants limited to northwestern and westcentral Colorado   -22. S. uintahensis
46. Involucre bracts about 13 to a head; plants not limited to northcentral or westcentral Colorado   -36. S. fendleri
41. None of the leaves lobed, all entire or merely toothed
47. Stems scapose or subscapose, cauline leaves either none or bractlike and very few; plants usually less than 20 cm. tall and involucre bracts about 21 to a head
48. Achenes hirtellous; plants white-tomentose and not becoming glabrate; leaves always entire; plants known only from Archuleta County at about 7000 feet   -37. S. molinarius
48. Achenes glabrous; plants becoming more or less glabrate; leaves entire to toothed; plants more widely distributed, often above 8000 feet in Colorado
49. Leaf blades linear to elliptic-oblanceolate, definitely longer than wide
-38. S. werneriaefolius
49. Leaf blades rotund-obovate to broadly spatulate, nearly as broad as long
-39. S. saxosus
47. Stems not at all scapose (cauline leaves reduced upward, however); plants over 20 cm. tall or if less then involucre bracts about 13 to a head
50. Achenes hirtellous or hispidulous; plants rarely less than 20 cm. tall; involucre bracts usually about 21 to a head
51. Tall plants usually over 55 cm. tall; leaves often over 12 cm. long, seldom glabrate above   -40. S. sphaerocephalus
51. Plants not over 50 cm. tall; leaves less than 12 cm. long, usually becoming glabrate above
52. Heads 10 mm. wide or more, 10-12 mm. high; bracts about 21 to a head
-33. S. neomexicana
52. Heads less than 10 mm. wide and usually less than 10 mm. high; bracts 13-21 to a head   -25. S. mutabilis
50. Achenes glabrous; plants often less than 20 cm. tall; involucre bracts about 13 to a head (except in S. canus)
53. Involucre bracts about 21 to a head; petioles of basal leaves shorter than the blades
-35. S. canus
53. Involucre bracts about 13 to a head; petioles of basal leaves often longer than the blades
54. Heads 10-12 mm. high; involucre bracts 7-10 mm. long   -41. S. harbourii
54. Heads 7-10 mm. high; involucre bracts 4-7 mm. long
55. Leaves becoming more or less glabrate in age, usually more or less lobed, at least definitely toothed; plants commonly over 20 cm. tall; basal elongated offsets often present   -36. S. fendleri
55. Leaves persistently tomentose, usually entire; plants seldom over 20 cm. tall; elongated basal offsets never present   -42. S. purshianus

1. Senecio taraxacoides (Gray) Greene, Pitt. 4:119. 1900.
Perennial plants; stems 5-9 cm. tall, rarely to 12 cm., leafy only at base, more or less floccose-tomentose; leaves 3-5 cm. long, oblanceolate to spatulate, at least some more or less pinnatifid often deeply so, the lobes sharp, floccose-tomentose at least below; heads solitary, nodding or at least horizontal on most, 12-15 mm. wide and about as high; involucres glabrous

or sparingly arachnoid, bracts linear-lanceolate, acuminate, a few present about 1/2 as long as the others; rays 8-14 mm. long, yellow.---Alpine peaks. Colorado and New Mexico. Our records from northcentral, central and southcentral Colorado at 11,000-12,500 feet.

2. Senecio amplectens A. Gray, Am. Journ. Sci. II. 33:240. 1862.

Perennial plants; stems 20-60 cm. tall, leafy, glabrous to slightly floccose when young; leaves 7-15 cm. long, oblanceolate to elliptic, denticulate to sharply dentate or upper leaves nearly entire, sessile, glabrous or slightly floccose; heads 1 to few, about 15-20 mm. wide, nodding; involucres broadly campanulate, bracts linear, acuminate or acute, often dark purplish, glabrous or nearly so; rays about 1-2 cm. long, yellow. S. pagosanus Heller is very close if not synonymous with the above.---Moist mountains. Wyoming to Nevada, south to Colorado. Our records scattered in the higher mountains of Colorado at 9500-13,000 feet.

2A. Senecio amplectens holmii (Greene) comb. nov., (var.)

S. holmii Greene---Plants less than 20 cm. tall. Perhaps only a dwarf form.---With the species in Colorado, our records at 10,500-12,500 feet.

3. Senecio soldanella Gray, Proc. Acad. Nat. Sci. Phil. 15:67. 1863.

Perennial plants with stout rootstocks bearing numerous fleshy roots; stems 8-15 cm. tall, more or less purplish, glabrous; leaves few, basal and usually 1-2 cauline, subrotund to oblong-obovate, subcordate, truncate or broadly cuneate to a long petiole, thick, glabrous but somewhat hairy near apex, more or less purplish-red; heads 15-25 mm. high, solitary, erect or slightly nodding; involucres broadly campanulate, glabrous, principal bracts with a few short ones at base; rays 6-18, 6-10 mm. long, yellow.---High mountains, often on talus slopes. Apparently limited to Colorado, our records scattered in the higher mountains at 9500-13,000 feet.

4. Senecio pudicus Greene, Pitt. 4:118. 1900.

S. cernuus A. Gray---Perennial plants; stems 30-80 cm. tall, leafy, often branched above, glabrous; basal leaves about 5-10 cm. long, oblong to narrowly lanceolate, entire to denticulate, rarely coarsely dentate, glabrous, petioled; upper leaves linear, sessile or nearly so; heads 8-12 mm. high and about 7-8 mm. wide, several to many, paniculate, nodding; involucres campanulate, rather narrowly so, bracts linear, acute; rays absent.---Meadows and valleys in the mountains. Colorado, Utah and New Mexico. Our records in the western half of Colorado, except the extreme northern part at 7000-12,000 feet.

5. Senecio bigelovii A. Gray, U. S. Rpt. Expl. Miss. Pacif. 4:111. 1857.

S. choranthus Greene; S. contristatus Greene---Perennial plants; stems 30-100 cm. tall, leafy; basal and lower stem leaves 10-20 cm. long, oblanceolate to lanc-oblong, dentate or denticulate, tapering to a winged petiole; upper stem leaves sessile and more or less clasping, the basal auricle enlarged, from glabrate to woolly-pilose especially along the midrib; heads 12-20 mm. long and about 15-20 mm. wide, 1 to several, racemose or corymbose, nodding; involucre bracts oblong to linear, acute, thin, glabrous; rays none.---Mountain valleys and slopes. Colorado, New Mexico and Arizona. Our records scattered over the western half of Colorado at 7500-11,500 feet.

5A. Senecio bigelovii hallii Gray, (var.) Proc. Phila. Acad. 1863:67. 1864.

S. accedens Greene---Differs from the species in having small, entire auricles on upper leaves; midveins long-villous. Intergrades very commonly with the species in Colorado.---With the species both generally and in this state. Our records at 8000-11,000 feet.

6. Senecio vulgaris L., Sp. Pl. 2:867. 1753.

Introduced from Europe and now widely distributed in Canada and United States. Reported from Clear Creek County, Colorado.

7. Senecio rapifolius Nutt., Trans. Am. Phil. Soc. 7:409. 1841.

Perennial plants with rather well developed rootstocks; stems 20-50 cm. tall, stout, rather leafy, glabrous and often glaucous; basal leaves petioled, 6-10 cm. long, obovate to broadly oblanceolate, sharply and deeply dentate, almost incised-dentate, glabrous, often glaucous; stem leaves oval, oblong or lanceolate, dentate, upper at least sessile, clasping and more or less auricled at base; heads 6-8 mm. high, not nodding, many, panicled and usually corymbiform; involucres campanulate, bracts narrowly oblong, acute, glabrous; rays none.---Valleys and slopes in the mountains. South Dakota to Idaho, south to Colorado. Our few records from northcentral Colorado at about 8500 feet.

8. Senecio fedifolius Rydb., Bull. Torr. Club 27:183. 1900.

S. discoideus of Manuals in part---Reported from central Colorado and apparently limited to that area. No specimens seen by the writer.

9. Senecio debilis Nutt., Trans. Am. Phil. Soc. 7:408. 1841.

S. nephrophyllus Rydb.; S. discoideus of Manuals in part---Perennial plants with rootstocks, glabrous throughout or slightly tomentose in the leaf axils and sometimes on midrib of upper leaf surface; stems 20-50 cm. tall, erect; leaves rather thickish, reduced upward, basal 1-6 cm. long, on petioles often longer than the blades, subreniform to ovate-oblong, entire to unequally dentate; cauline leaves sessile or petioled below, sublyrate to pinnately divided; heads 7-10 mm. high, not nodding, few to several, in corymbose cymes; involucres about 6-7 mm. high; rays absent; achenes glabrous.---Moist or wet ground. Montana to Washington, south to Colorado and Idaho. Our records from northcentral and central Colorado at 8000-9000 feet.

10. Senecio longilobus Benth., Pl. Hartw. 18. 1839.

S. filifolius Nutt.; probably S. ridellii T. & G.---Perennial plants; stems 30-100 cm. tall, suffruticose at base, several together often forming a clump, equably leafy to apex, more or less tomentose when young; leaves pinnately divided into narrowly linear, entire segments or upper entire and linear-filiform; heads 12-15 mm. high and usually over 8 mm. wide, not nodding, several to many, corymbose; involucres broadly campanulate, floccose-tomentose, at least near base; rays 10-14 mm. long.---Valleys and plains. Colorado to Utah, south to Texas and Mexico. Our records from the southern half of Colorado at 4500-7000 feet.

11. Senecio spartioides T. & G., Fl. N. Amer. 2:438. 1843.

Perennial plants; stems 20-60 cm. tall, suffruticose at base, often several from the same root, equably leafy to apex, glabrous or sparsely puberulent; leaves narrowly linear and entire or less commonly with 1 or 2 pair of short lobes at base, rather fleshy, glabrous or sparsely puberulent; heads 8-12 mm. long and less than 8 mm. wide, not nodding, many, in corymbose cymes; involucres cylindrical or narrowly campanulate, principal bracts with shorter ones at base; rays 8-10 mm. long, rather few, yellow.---Valleys and plains. Nebraska to Wyoming, south to Texas and Arizona. Our records scattered in Colorado, except in the northeastern and northwestern parts, at 5000-9000 feet.

12. Senecio multicapitatus Greenm. ex Rydb., Bull. Torr. Club 33:160. 1906.

Perennial plants with thick woody roots; stems 40-100 cm. tall, branched and broomlike, equably leafy to apex, glabrous or nearly so; leaves irregularly pinnately divided into linear-filiform segments or upper entire, glabrous or nearly so; heads 7-9 mm. high and about 3-4 mm. wide, not nodding, numerous in corymbose clusters; involucres subcylindrical or very narrowly campanulate, glabrous, bracts 6-8 mm. long, with shorter ones at base; rays 7-8 mm. long, yellow.---Valleys and plains. Colorado, Utah, New Mexico and Arizona. Our records from the western half of Colorado, mostly from the southwestern quarter, at 4500-8000 feet.

13. Senecio ambrosioides Rydb., Bull. Torr. Club 37:467. 1910.

S. eremophilus and S. macdougallii of Manuals in part---Perennial plants; stems 30-60 cm. tall, 1 to several from a rather woody base, equably leafy to apex; leaves 3-10 cm. long, lanceolate to ovate-lanceolate in outline but more or less laciniately pinnatifid into coarsely and unequally dentate segments, glabrous or nearly so, mostly short-petioled; heads 7-10 mm. high, not nodding, usually numerous, in corymbiform cymes; involucres 5-7 mm. high, subcampanulate, bracts commonly black-tipped, glabrous or nearly so; rays 4-6 mm. long, yellow. This may be merely a small-flowered form of the northern S. eremophilus Rich. and intergrades are present especially in northern Colorado.---Valleys and mountains. Wyoming to Idaho, south to New Mexico and Arizona. Our records in the western half of the state at 5500-11,500 feet.

14. S. eremophilus Richards., App. Frankl. 1st. Journ. 31. 1823.

Reported from northwestern Canada, south to Colorado and Utah. Listed from central Colorado but not located by the writer.

15. S. carthamoides Greene, Pitt. 4:122. 1900.

S. blitoides Greene; S. invenustus Greene---Perennial plants; stems 10-50 cm. tall, tufted, spreading or decumbent at base, equably leafy or becoming leafless below, glabrous or sparingly puberulent; leaves 2-6 cm. long, obovate, spatulate or oval, mostly broadly obtuse or rounded at apex, coarsely and sharply often doubly dentate, often dentate-lobed especially near base, somewhat fleshy, glabrous or sparingly puberulent, sessile and more or less clasping at base; heads 10-15 mm. high, not nodding, usually few to several in a leafy corymb; involucres broadly campanulate, glabrous or sparingly puberulent, bracts definitely thickened at base and on back; rays 6-11 mm. long, yellow to orange.---High mountains, often in rocky or gravelly ground. Colorado and New Mexico. Scattered in the higher mountains of Colorado at 9500-13,000 feet.

16. Senecio triangularis Hook., Fl. Bor. Am. 1:332. 1834.

Perennial or possibly biennial plants; stems 50-150 cm. tall, usually several in a clump, rather stout, equably leafy to apex, glabrous or sparingly floccose near nodes; leaves 5-20 cm. long, broadly triangular-lanceolate to triangular-ovate, acute at apex, cordate, truncate or sometimes very broadly cuneate at base, more or less dentate, thin, glabrous or nearly so; heads about 10-12 mm. high, not nodding, usually many in an open leafy corymbose cyme; involucres campanulate, bracts thin, glabrous on back but hairy at apex; rays about 10-12 mm. long, yellow.---Moist or wet ground. Saskatchewan to Alaska, south to New Mexico and California. Our records from the western half of Colorado at 9000-11,000 feet.

17. Senecio serra Hook., Fl. Bor. Amer. 1:333. 1834.

S. admirabilis Greene; S. serra var. admirabilis (Greene) A. Nels.---Perennial plants with rootstocks; stems 60-120 cm. tall, strict, branching above near inflorescence, about equably leafy to apex, glabrous or sparingly hairy; leaves 5-15 cm. long, lanceolate or linear, long-acute or acuminate at apex, narrowly cuneate at base, sharply serrate, sessile or lower short-petioled; heads 11-14 mm. long and about 5-8 mm. wide, not nodding, usually many, corymbose-paniculate; involucres cylindric, bracts definitely thickened on back and near base, glabrous or glabrate; rays 6-10 mm. long, light yellow.---Meadows and damp ground, often along streams in the mountains. Wyoming to Washington, south to Colorado and California. Our records from the western half of Colorado, except the extreme southern part, at 8500-10,500 feet.

18. Senecio crassulus A. Gray, Proc. Amer. Acad. 19:54. 1883.

S. semiamplexicaulis Rydb.; S. lepathifolius Greene---Perennial plants; stems 20-50 cm. tall, rarely less, rather equably leafy to apex in most plants; leaves 8-15 cm. long on longest, oblong-lanceolate to linear-oblanceolate, sometimes obovate, denticulate, glabrous or nearly so, lower cuneate to a winged petiole, upper sessile and partly clasping; heads 10-15 mm. high, not nodding, few to several, corymbose; involucres broadly campanulate, bracts fleshy thickened on back and near base, blackish-hairy near apex; rays about 7-10 mm. long, yellow.---Mountains, often in meadows. South Dakota to Idaho, south to New Mexico and Utah. Our records scattered in the higher mountains of Colorado at 8000-13,000 feet

19. Senecio crocatus Rydb., Bull. Torr. Club 24:299. 1897.

S. longipetiolatus Rydb.; S. tracyi Rydb.; S. pyrrhochrous Greene---Perennial plants with rather stout rootstocks; stems 20-70 cm. tall, glabrous or nearly so, or somewhat tomentulose in the axils of the bracts; lower leaves 2-8 cm. long, oblong to oblong-ovate, subcordate to cuneate at base, obtuse or nearly mucronate at apex, entire to somewhat crenate-dentate, glabrous or nearly so; cauline leaves reduced above, short-petioled or sessile and clasping, lyrate-pinnatifid to lobed; heads 7-10 mm. high, not nodding, few to many in a corymbose cyme; involucres about 6-8 mm. high, campanulate to hemispheric, glabrous, bracts more or less tinged with purple; rays about 6-8 mm. long, orange-red or saffron, varying somewhat to yellowish. The form with basal leaves less than 2 cm. long has been called var. wolfii Greenm.---Mountain valleys and meadows. Colorado, Utah and Wyoming. Our records scattered in the western half of Colorado, but few from the extreme western part, at 8000-13,000 feet.

20. **Senecio porteri** Greene, Pitt. 3:186. 1897.
Perennial, glabrous, more or less purplish-tinged plants from slender rootstocks; stems 4-10 cm. tall, ascending, scapiform; leaves basal or nearly so, blades 8-15 mm. long on longer petioles, mostly reniform but sometimes broadly ovate, crenate to shallowly lobed or incised; heads solitary, 10-14 mm. high; involucres campanulate, 9-12 mm. long, bracts about 13, linear-lanceolate, acute; rays 8-10, about 8 mm. long, yellow; achenes glabrous.---Higher mountains. Colorado and Oregon. Our one record from northern Gunnison County at 12,000 feet.

21. **Senecio multilobatus** T. & G. ex Gray, Mem. Am. Acad. 4:109. 1849.
Perennial plants from definite ascending or horizontal rootstocks; stems usually 15-30 cm. tall, erect, 1 to several, leafy but leaves reduced above, glabrous or tomentose in the leaf axils; basal leaves 2-10 cm. long and 5-20 mm. wide, petioled, oblanceolate to spatulate-oblong in outline, pinnatifid to lyrate, rarely merely dentate, glabrate; stem leaves pinnatifid, sessile above; heads 8-10 mm. high, not nodding, few to several in a corymbose cyme; involucres campanulate, sparingly calyculate, glabrous, bracts usually 13, 5-8 mm. long; rays 5-8, about 5-7 mm. long, yellow; achenes hirtellous.---Dry plains and slopes. Wyoming to Nevada, south to New Mexico and Arizona. Our records from the western one-third of the state at 4500-8500 feet.

22. **Senecio uintahensis** (A. Nels.) Greenm., Monogr. Senecio I. Teil 24. 1901.
Perennial plants from definite ascending or horizontal rootstocks; stems 15-40 cm. tall, 1 to several, leafy but leaves reduced above, glabrous or slightly tomentose; lower leaves 2-10 cm. long and 5-25 mm. wide, obovate to oblong-oblanceolate in outline, lyrate to pinnatifid, glabrous to slightly floccose-tomentose; upper stem leaves pinnatifid, sessile; heads 8-10 mm. high, not nodding, usually many in a corymbose cyme; involucres campanulate, sparingly calyculate, glabrous, bracts usually about 21, 5-6 mm. long; rays about 8, about 5-7 mm. long, yellow or somewhat orange; achenes glabrous.---Hills, plateaus, valleys and mountain slopes. Wyoming to Oregon, south to Arizona and California. Our records from the western one-fourth of Colorado at 4500-7000 feet.

23. **Senecio hydrophilus** Nutt., Trans. Amer. Phil. Soc. n. s. 7:411. 1841.
Perennial or possibly biennial plants with very short vertical rootstocks crowded with fleshy roots; stems 70-100 cm. tall, strict, hollow, leafy but leaves reduced upwards, glabrous to glaucous; basal leaves 10-30 cm. long, fleshy-coriaceous, oblanceolate, tapering to a long winged petiole, entire or slightly denticulate; upper leaves oblong-lanceolate to linear, subsessile to sessile and clasping; heads about 8-10 mm. high, not nodding, numerous in a branching often large cyme; involucres campanulate, glabrous, bracts 8-12, 5-7 mm. long, more or less black-tipped; rays few, small, yellow; achenes glabrous.---In wet or moist ground, often in partial shade. South Dakota to British Columbia, south to Colorado and Utah. Our records from northcentral Colorado at 7000-8500 feet.

24. **Senecio integerrimus** Nutt., Gen. Pl. 2:165. 1818.
S. dispar A. Nels.; S. perplexus A. Nels.; S. flintii Rydb.; S. columbianus Greene; S. perplexus var. dispar A. Nels.---Perennial or biennial plants with very short thick vertical rootstocks on which are crowded fleshy roots; stems 30-70 cm. tall, sometimes less, hollow and flattened in pressing, glabrous to loosely woolly-pubescent with jointed hairs, these usually partly or completely lost in age, leafy but leaves definitely reduced above; basal leaves 6-10 cm. long, rather thick, oblanceolate to oblong-lanceolate, entire or denticulate, glabrous or somewhat woolly; stem leaves lanceolate to linear-lanceolate, usually entire, often auricled and widest at very base and tapering upwards; heads 10-14 mm. high, not nodding, 6 to many in a corymbose cyme, the center head usually on a shorter thicker peduncle; involucres campanulate, bracts about 7-8 mm. long, usually glabrous but hairy near apex which is usually black; rays about 5-15 mm. long, yellow; achenes glabrous. A rather variable species especially in general appearance and some of the synonyms listed above may prove to have varietal rank. A doubtfully distinct species, S. hookeri T. & G., may be present in Colorado.---Meadows, valleys and moist slopes. Minnesota to British Columbia, south to Nebraska and California. Our records scattered in the western two-thirds of Colorado at 5000-10,000 feet, occasionally even higher.

25. **Senecio mutabilis** Greene, Pitt. 4:113. 1900.
S. cognatus Greene---Perennial plants from definite ascending or horizontal rootstocks; stems 10-50 cm. tall, 1 to several, loosely floccose-tomentulose to nearly glabrate, leafy but leaves reduced upwards; basal leaves 1.5-6 cm. long, obovate to oblanceolate, subentire to dentate, usually rounded at apex, base tapering to a rather long petiole; stem leaves oblanceolate to narrowly lanceolate, entire to pinnatifid, sessile to petioled, all leaves usually more or less tomentose at first, more or less glabrate at maturity; heads 8-10 mm. high, not nodding, few to several in a corymbose cyme; involucres campanulate, sparingly calyculate, glabrous to somewhat tomentose, bracts 13-21, 5-7 mm. long; rays 8-12, about 5-8 mm. long, yellow; achenes hirtellous or hispidulous. A very variable species, as treated here including S. tridenticulatus Rydb., S. oblanceolatus Rydb., S. condensatus Rydb., and S. densus Greene as more glabrate forms.---Plains, hills, valleys and mountains. Manitoba, south to Texas and Arizona. Our records well scattered over Colorado at 4000-9000 feet.

26. **Senecio dimorphophyllus** Greene, Pitt. 4:109. 1900.
S. heterodoxus Greene; S. crocatus of Manuals in part---Perennial plants with definite ascending or horizontal rootstocks; stems 10-30 cm. tall, 1 to several, glabrous or nearly so, leafy but leaves reduced upwards; basal leaves 1-4 cm. long, ovate, obovate to oblong-lanceolate, entire or obscurely crenate, obtuse; stem leaves lanceolate to ovate, some at least more or less pinnatifid or sinuate-toothed, rarely merely crenate, often dilated and clasping at base, mostly sessile, all leaves glabrous or nearly so; heads 8-10 mm. high, not nodding, few

to several in a corymbose cyme; involucres campanulate, calyculate, bracts about 20-21, about 6-7 mm. long, glabrous or slightly tomentulose, often reddish-tipped; rays 10-12, about 6-8 mm. long, yellow; achenes glabrous.---Mountains. Wyoming and Colorado. Our records from the higher mountains of Colorado at 10,000-13,000 feet.

A closely related species, S. oodes Rydb., was described from Colorado material and may be distinct.

27. Senecio flavulus Greene, Pitt. 4:108. 1900.

S. flavovirens Rydb.; S. rydbergii A. Nels.---Perennial plants with definite, ascending or horizontal rootstocks; stems 20-50 cm. tall, erect, glabrous to tomentose near base of petioles, leafy but leaves reduced above; lower leaves 1-6 cm. long, usually oblanceolate, crenate to coarsely and unequally dentate, obtuse or rounded at apex, petioled; stem leaves more or less lyrate-pinnatifid, short-petioled to sessile; heads 7-9 mm. high, not nodding, few in a corymbose cyme; involucres campanulate, calyculate, bracts 13-31, 5-7 mm. long, glabrous except at apex; rays 10-12, about 6-7 mm. long, yellow; achenes glabrous or slightly hirtellous on the angles.---Plains, valleys and hills. British Columbia, south to Colorado and Idaho. Our records thinly scattered over the state, possibly none in the eastern one-third, at 4000-8000 feet.

28. Senecio pseudaureus Rydb., Bull. Torr. Club 24:298. 1897.

Perennial plants with definite ascending or horizontal rootstocks; stems 30-70 cm. tall, erect, glabrous or tomentose in axils of leaves or in inflorescence; leafy but leaves reduced upwards; basal leaves 1-10 cm. long, ovate-rotund to oblong-ovate, crenate to doubly serrate, usually rounded at apex, some more or less cordate at base, long-petioled; stem leaves more or less lyrate-pinnatifid, short-petioled or sessile and clasping, all leaves glabrous or glabrate; heads 8-10 mm. high, not nodding, few to many in a corymbose cyme; involucres campanulate, calyculate, bracts about 21, rarely as few as 13, 6-8 mm. long, glabrous except at apex; rays 10-13, about 6-9 mm. long, yellow; achenes glabrous.---Meadows and valleys. Saskatchewan to British Columbia, south to New Mexico and California. Our records scattered over the western half of Colorado at 5500-9500 feet.

29. Senecio wootonii Greene, Bull. Torr. Club 25:122. 1898.

S. anacletus Greene---Perennial plants from ascending, stout, definite simple or branched rootstocks; stems 20-50 cm. tall, erect, 1 to several, glabrous and usually rather glaucous, cauline leaves few; lower leaves 5-15 cm. long, obovate to oblong-oblanceolate, entire to sinuate-dentate with subcartilaginous teeth, long wing-petioled; stem leaves very few, oblanceolate, sessile and entire above, all leaves rather thick and glabrous; heads 10-12 mm. high, not nodding, several to many in a corymbose cyme; involucres campanulate, calyculate, bracts about 13, 7-9 mm. long, glabrous except at tips; rays 8-10, about 7-8 mm. long, yellow; achenes glabrous.---Mountains, usually in meadows or moist ground. Colorado to Mexico. Our records scattered in the western half of Colorado, except the northwestern part, at 8000-11,000 feet.

30. Senecio cymbalarioides Nutt., Trans. Am. Phil. Soc. n. s. 7:412. 1841.

S. subcuneatus Rydb.---Perennial plants from definite ascending or horizontal rootstocks; stems 10-40 cm. tall, 1 to several, glabrous or floccose-tomentose in axils of leaves, leafy but leaves reduced upwards; basal leaves 1-6 cm. long, mostly broadly ovate to obovate, entire or dentate toward apex, thick and firm, petioled; cauline leaves deeply serrate to entire, petioled or sessile, all leaves glabrous; heads 7-10 mm. high, not nodding, few to several in a corymbose cyme; involucres campanulate, sparingly calyculate, bracts 13-21, 5-8 mm. long, glabrous or rarely tomentose at base; rays 8-12, about 4-8 mm. long, yellow; achenes glabrous.---Meadows and mountain valleys. Alberta to British Columbia, south to New Mexico and Nevada. Our records scattered in the higher mountains of Colorado at 7000-12,500 feet, mostly above 10,000 feet.

31. Senecio acutidens Rydb., Bull. Torr. Club 27:180. 1900.

Perennial plants from definite ascending or horizontal rootstocks; stems 15-25 cm. tall, more or less tufted, glabrous or glabrate except tomentum persisting sometimes in leaf axils, leafy but leaves reduced upwards; basal leaves 3-6 cm. long, mostly oblong-lanceolate, dentate toward apex, thick, petioled; stem leaves oblanceolate to narrower, entire to dentate, all leaves glabrous and somewhat glaucous; heads 8-10 mm. high, not nodding, few to several in a corymbose cyme; involucres campanulate, sparingly calyculate, bracts about 14-21, 5-8 mm. long; rays 8-10, about 5-6 mm. long; achenes glabrous. Probably merely a narrow-leafed form of the preceding.---Meadows and mountain valleys. Wyoming to New Mexico. Our few records from the western half of Colorado at 7500-10,000 feet.

32. Senecio atratus Greene, Pitt. 3:105. 1896.

S. milleflorus Greene; S. atratus var. milleflorus (Greene) Greenm.---Perennial plants with rather stout, often long, ascending rootstocks; stems 20-80 cm. tall, erect or nearly so, white floccose-tomentose or tardily glabrate, leafy but leaves reduced upwards; basal leaves 10-20 cm. long, oblanceolate to oblong-obovate, subentire to dentate, floccose-tomentose both sides, on definite petioles; upper stem leaves lanceolate, sessile; heads 10-12 mm. high, not nodding, many, in a corymbose or round-topped cyme; involucres narrowly campanulate to cylindrical, calyculate, bracts commonly 8, 6-8 mm. long, more or less conspicuously black-tipped, somewhat tomentose at base; rays 3-5, about 5-8 mm. long, yellow; achenes glabrous.---Mountain valleys and slopes. Colorado to Utah, south to New Mexico. Our records scattered in the higher mountains of Colorado at 8000-12,000 feet.

33. Senecio neomexicanus A. Gray, Proc. Am. Acad. 19:55. 1883.

Perennial plants from definite ascending or upright rootstocks; stems 15-50 cm. tall, more or less tomentose at least at first, often rather glabrate in age, leafy but leaves reduced above; basal leaves 2-8 cm. long, oblanceolate to obovate, subentire to lyrate, thick, white-tomentose at first, later more or less glabrate; stem leaves irregularly dentate to pinnatifid, sessile above; heads 10-12 mm. high and about as wide, not nodding several to many in a corymbose cyme; involucres campanulate, sparingly calyculate, tomentose to nearly glabrous, bracts about 21, 6-8 mm. long; rays 10-13, about 8 mm. long, yellow; achenes hirtellous. This species is closely related to S. mutabilis Greene and the records listed below may actually be that plant.---Valleys, plateaus and slopes. New Mexico and Arizona. Our records from westcentral and southwestern Colorado at 7000-9500 feet.

34. Senecio plattensis Nutt., Trans. Am. Phil. Soc. n.s. 7:413. 1841.
S. balsamitae Torr.---Perennial plants from definite, ascending rootstocks; stems 10-50 cm. tall, erect, 1 to several, more or less persistently white-tomentose, tardily if at all glabrate, leafy but leaves reduced above; leaves rather variable, lower 1-8 cm. long, ovate, oblong to lanceolate, crenate, serrate-dentate or sublyrate to pinnatifid at base of plant, rounded at apex, cuneate to subcordate at base, petioled; upper stem leaves linear-lanceolate, toothed to pinnatifid, sessile; all leaves tomentose at first, more or less glabrate in age; heads 8-10 mm. high, not nodding, usually few to several in a corymbose cyme; involucres campanulate, calyculate, glabrous or somewhat tomentose; rays 10-12, 6-9 mm. long, yellow or orange-yellow; achenes hispidulous along the angles.---Plains and hills. Ontario to Saskatchewan, south to Louisiana and Texas. Our records from the eastern half of Colorado at 3500-7000 feet.

35. Senecio canus Hook., Fl. Bor. Am. 1:333. 1834.
Perennial plants from definite ascending or horizontal rootstocks; stems 12-50 cm. tall, usually more than 1, white-tomentose and tardily if at all glabrate, leafy but leaves reduced above; basal leaves 1-8 cm. long, obovate to oblanceolate, entire to dentate, thick, obtuse, rather short-petioled; stem leaves lanceolate to linear, entire, dentate or pinnatifid, upper sessile and more or less clasping, all leaves white-tomentose or occasionally somewhat glabrate above; heads 10-12 mm. high, not nodding, few to several in a corymbose cyme; involucres campanulate, sparingly calyculate, glabrous to tomentose, bracts about 21, about 6-8 mm. long; rays 8-12, about 7-8 mm. long, yellow; achenes glabrous.---Dry hills and mountainsides. Manitoba to British Columbia, south to Nebraska and Colorado. Our records from the northern part of the state at 5000-7500 feet.

36. Senecio fendleri A. Gray, Mem. Am. Acad. n. s. 4:108. 1849.
S. nelsonii Rydb.; S. salicinus Rydb.; S. rosulatus Rydb.; S. canovirens Rydb.--- Perennial plants with definite ascending or horizontal rootstocks, frequently bearing short or elongated offsets; stems 10-60 cm. tall, 1 to several, floccose-tomentose at first, more or less glabrate in age, leafy but leaves somewhat reduced upwards; leaves variable, basal and lower petioled, 4-10 cm. long, linear-oblanceolate to rarely oblong, entire, sinuately toothed or pinnatifid, sometimes crenate or plaited; upper leaves sessile, linear to lanceolate, entire to pinnatifid, all leaves tomentose at first but more or less glabrate especially above; heads 7-10 mm. long, not nodding, several to many in a corymbose cyme; involucres campanulate, sparingly calyculate, tomentose to glabrous, bracts about 13, 4-6 mm. long; rays 7-12, 5-7 mm. long; achenes glabrous.---Hills and mountains. Wyoming to New Mexico. Our records scattered in Colorado, few from the extreme eastern and western parts, at 5000-11,000 feet.

36A. Senecio fendleri lanatus Osterh., (var.) Bull. Torr. Club 31:358. 1904.
S. lanatifolius Osterh.---Stems 10-20 cm. tall, about equally leafy to apex; leaves linear or linear-oblong, all uniformly pinnatifid; rays apparently lacking entirely. This plant is different enough in this state to warrant specific status.---Plains and hills. Nebraska to Colorado. All our records from Eagle County, Colorado, at 7000 feet.

37. Senecio molinarius Greenm., Ann. Mo. Gard. 5:65. 1918.
A little known plant resembling number 38. Plant usually more permanently white-tomentose; leaves always entire; achenes hirtellous.---Reported from Archuleta County, Colorado, but a specimen from Eagle County seems to be the same. Both records from about 7000 feet.

38. Senecio werneriaefolius A. Gray, Proc. Am. Acad. 19:54. 1883.
S. perennans A. Nels.---Perennial plants from definite ascending rootstocks and branching caudex; stems 5-20 cm. tall, 1 to several, floccose-tomentose and tardily glabrate, subscapose; leaves chiefly basal, 1-5 cm. long, linear-oblong to oblong-oblanceolate, entire or toothed at apex, slightly revolute, thick, petioled, tomentose or glabrate in age; stem leaves linear and subulate, bractlike; heads 9-12 mm. high, not nodding, 1 to several in a corymbose cyme; involucres campanulate, sparingly calyculate, tomentose at least at base, bracts usually about 21, 6-8 mm. long; rays 5-12, about 5-7 mm. long, yellow; achenes glabrous.---Foothills, valleys, canyons and mountains. South Dakota to Wyoming, south to Arizona and Utah. Our records scattered in the western half of Colorado, but none as yet from the southwestern corner at 6500-10,000 feet.

38A. Senecio werneriaefolius incertus Greenm., (var.) Ann. Mo. Bot. Gard. 5:62. 1918.
No rays present.---Reported from Clear Creek County, Colorado.

39. Senecio saxosus Klatt., Anal. Naturhist. Hofmus. Wien 9:366. 1894.
S. petrocallis Greene; S. pentodontus Greene; S. turbinatus Rydb.---Perennial plants from definite ascending rootstocks which form a branching caudex above; stems 5-15 cm. tall, solitary to many, white-tomentulose at first, later more or less glabrate, subscapose; leaves chiefly basal, 1-4 cm. long, rotund-obovate and nearly as wide as long, entire or crenate-dentate toward apex, petioled, floccose when young, more or less glabrate in age; stem leaves reduced and bractlike above; heads 10-12 mm. high, not nodding, 1 to several in a corymbose cyme; involucres campanulate, sparingly calyculate, usually slightly floccose at base, bracts about 21, 7-10 mm. long; rays 10-12, 6-8 mm. long, yellow; achenes glabrous.---High mountains. Montana to Idaho, south to Colorado. Our records from most of the higher mountains of the state at 9500-13,000 feet, mostly above 11,500 feet.

40. Senecio sphaerocephalus Greene, Pitt. 3:106. 1896.
S. altus Rydb.---Montana to Idaho, south to Colorado and Nevada. Reported from Clear Creek County, Colorado, but not located by the writer.

41. Senecio harbourii Rydb., Bull. Torr. Club 33:158. 1906.
Caespitose perennial plants, flocculose-tomentose throughout but sometimes sparsely so; stems 10-20 cm. tall, simple or branched, usually rather leafy and these not so very reduced upwards in some plants; leaves about 3-10 cm. long, oblong-oblanceolate to narrowly oblanceolate, entire or crenate-undulate; lower leaves on petioles longer than the blade, upper sessile and often auriculate at base; heads 10-12 mm. high, not nodding, few in a subcorymbose cyme; involucres campanulate, bracts usually about 13, 7-10 mm. long, outer shorter ones few, more or less tomentose; rays about 8, 10-12 mm. long; achenes glabrous.---High mountains. Apparently limited to Colorado. Our records from the northcentral, westcentral and central parts at 10,000-12,000 feet.

42. Senecio purshianus Nutt., Trans. Am. Phil. Soc. n. s. 7:412. 1841.
S. canus purshianus (Nutt.) A. Nels.---Perennial plants from definite, often branched rootstock; stems 10-20 cm. tall, caespitose, white-tomentose, rarely glabrate, leafy, but leaves more or less reduced upwards; basal leaves 2-5 cm. long, narrowly oblanceolate, sometimes oblong-obovate, obtuse, entire to somewhat dentate, petioled, tomentose or sometimes somewhat glabrate; stem leaves lanceolate to linear, entire or dentate, sessile and often clasping on upper ones; heads 7-10 mm. high, not nodding, few to several in a cyme; involucres campanulate, sparingly calyculate, tomentulose to glabrous, bracts about 13, 5-7 mm. long; rays about 8, about 6-8 mm. long, yellow; achenes glabrous.---Hills and mountains. South Dakota to Montana, south to Nebraska and Colorado. Our records from the northcentral and central parts of the state at 10,000-13,000 feet.

65. Arnica L.

Herbaceous perennial plants; stems simple or branched from base; leaves mostly opposite, entire or toothed; heads solitary to many, in cymose clusters; involucres hemispheric to turbinate, bracts equal in length, in 1-2 series; receptacles naked; ray flowers pistillate and fertile, yellow, or absent in 1 species; disk flowers perfect and fertile, yellow; anthers united, not sagittate at base; achenes cylindric to fusiform, 5- to 10-nerved, glabrous to hairy; pappus of white to tawny, barbellate to subplumose, capillary bristles.

1. Rays lacking; heads nodding in the bud -1. A. parryi
1. Rays present; heads erect or nearly so in the bud
   2. Stems with 5-10 pairs of cauline leaves; pappus cream-colored, stramineous or tawny, barbellate (side bristles 0.1-0.2 mm. long)
      3. Involucre bracts obtuse or merely acutish, pilose within at apex
         4. Rootstocks conspicuously elongated (to 30 cm. long), not rooting the first year; heads 5-15 -2. A. chamissonis
         4. Rootstocks short and conspicuously rooting the first year; heads 1-3 -3. A. fulgens
      3. Involucre bracts definitely acute, not pilose within at apex -4. A. longifolia
   2. Stems with 2-4 pairs of cauline leaves; pappus various (often white or subplumose)
      5. Involucre bracts obtuse to merely acutish, pilose or pubescent within at apex; heads 1-3; pappus barbellate
         6. Lower cauline leaves sessile or subsessile; rays 4-6, with apical teeth less than 0.5 mm. long; involucres 9-11 mm. high; heads campanulate-turbinate; pappus white; plants 10-25 cm. tall -5. A. rydbergii
         6. Lower cauline leaves definitely petioled; rays 12-20, with apical teeth 1 mm. long or more; involucres 11-15 mm. high; heads broadly hemispheric; pappus cream-colored to tawny; plants 20-60 cm. tall -3. A. fulgens
      5. Involucre bracts definitely acute to acuminate, not especially pilose or pubescent within; heads often over 3; pappus barbellate, subplumose (side bristles 0.2-0.35 mm. long), or plumose (side bristles 0.35-0.6 mm. long)
         7. Pappus subplumose to plumose, dirty white to tawny
            8. Pappus dirty white, subplumose; heads 7-17; rays 8-10, 10-15 mm. long -6. A. paniculata
            8. Pappus tawny, plumose; heads 1-3; rays 14-18, 18-24 mm. long -7. A. mollis
         7. Pappus barbellate, white to occasionally cream-white, rarely stramineous
            9. Achenes uniformly hirsute or sometimes also uniformly short-stipitate glandular; peduncles just below the head usually densely white-pilose; cauline leaves usually cordate or truncate; involucre bracts 14-18 mm. long -8. A. cordifolia
            9. Achenes glabrate to sparingly short-hirsute to glandular above, glabrate below; peduncles just below the heads rather sparingly puberulent to glandular-hairy, not densely white-pilose; cauline leaves seldom cordate at base; involucre bracts 7-15 mm. long
               10. Involucre bracts 7-9 mm. long and 1-1.5 mm. wide, narrowly lanceolate; heads 5-11; only 2 pair of cauline leaves; achenes 5.5-6.5 mm. long -9. A. gracilis
               10. Involucre bracts 10-15 mm. long and 1-3 mm. wide, lanceolate; heads 1-5; cauline leaves usually 3 or more pairs; achenes 6-8 mm. long -10. A. latifolia

1. Arnica parryi A. Gray, Am. Nat. 8:213. 1874.
Rootstocks slender; stems 30-50 cm. tall, pilose to lanate-villous or puberulent above, the hairs sometimes sparse; cauline leaves 2-3 pairs; lower leaves lanceolate, entire to denticulate, cuneate to a short petiole, more or less villous and glandular above; heads 3-9, turbinate-campanulate, peduncles pilose and glandular-hairy just below; involucre bracts 12-20, 8-15 mm. long, acute, puberulent and glandular especially toward the apex; rays wanting; achenes 4.5-5.5 mm. long, glabrous, hirsute only above or uniformly hirsute, also often glandular; pappus barbellate to subplumose, stramineous to tawny. Our plants have been called ssp. genuina Maguire.---Parks and open woods. Alberta to British Columbia, south to Colorado and California. Our records scattered in the western half of Colorado at 9000-11,500 feet.

2. **Arnica chamissonis** Lessing, Linnaea 6:238. 1831.

A. foliosa Nutt.; *A. macilenta* Greene; *A. celsa* A. Nels.; *A. rhizomata* A. Nels.; *A. tomentulosa* Rydb.; *A. stricta* A. Nels.---Rootstocks long (to 30 cm.) not rooting the first year; stems 25-90 cm. tall, variously hairy and glandular-hairy above; cauline leaves 5-10 pairs, lanceolate, oblanceolate to lanc-oblong, lower leaves petioled, upper sessile, entire to denticulate, variously hairy; heads 5-15, hemispheric-campanulate, moderately to densely villous just below on peduncles; involucre bracts 12-18, 7-10 mm. long, obtuse or somewhat acutish, pilose at apex, more or less glandular outside; rays 12-18, 12-18 mm. long; achenes 4-5.5 mm. long, short-hirsute to glandular to glabrate; pappus stramineous and barbellate or rarely tawny and subplumose. Our plant has been called ssp. foliosa (Nutt.) Macguire.---Meadows and moist places. The subspecies from Manitoba to Northwest Territories, south to New Mexico and California. Our records from the western half of Colorado at 6000-10,000 feet.

3. **Arnica fulgens** Pursh, Fl. Am. Sept. 527. 1814.

A. pedunculata Rydb.; *A. monocephala* Rydb.---Rootstocks short, thick and profusely rooting; stems 20-60 cm. tall, more or less puberulent and glandular-hairy; cauline leaves 4-6 pairs, broadly to narrowly oblanceolate, upper sessile or nearly so, the lower petioled, entire or rarely denticulate, more or less puberulent and glandular to pilose; heads 1-3, broadly hemispheric, peduncles pilose and glandular just below; involucre bracts 14-20, 11-15 mm. long, obtuse or merely acutish, pilose near apex within, villous-puberulent at least near apex within, villous-puberulent at least near base; rays 12-20, 15-25 mm. long; achenes 4.5-5.5 mm. long, hirsute; pappus barbellate, cream-colored to tawny.---Hills and meadows. Saskatchewan to British Columbia, south to Colorado and Nevada. Our records from northcentral and central Colorado at 5500-11,000 feet.

4. **Arnica longifolia** D. C. Eat. in S. Wats., Bot. King's Expl. 5:186. 1871.

Rootstocks usually short; stems 30-60 cm. long, subglabrate below and sparsely glandular to puberulous above; cauline leaves 5-7 pairs, upper lanceolate to lanc-elliptic, connate-sheathing, entire or denticulate, glandular and puberulent, leaves on branches often petioled; heads 3-20, turbinate-campanulate; involucre bracts 11-15, 7-10 mm. long, narrowly to broadly lanceolate, acute to acuminate, pubescent at least below; achenes 4.5-5.5 mm. long, glandular to subglabrous; pappus barbellate, stramineous. Our plant has been called ssp. genuina Macguire. ---Mountains, often in meadows and wet places. The subspecies from Montana and Idaho, south to Colorado and Nevada. Our few records scattered in the western half of Colorado at 8000-10,000 feet.

5. **Arnica rydbergii** Greene, Pitt. 4:36. 1899.

A. tenuis Rydb.---Rootstocks moderately long; stems 10-25 cm. tall, more or less glandular-puberulent and pilose; cauline leaves 3-4 pairs, upper sessile, lanceolate to ovate, lower sessile or subsessile, oblanceolate, obovate to spatulate, entire or toothed, glabrate to glandular-hairy or pilose; heads 1-3, campanulate-turbinate, more or less pilose and glandular hairy on peduncles just below; involucre bracts 10-14, 9-11 mm. long, obtuse or merely acutish, hairy inside near tip, glabrate to thinly subpilose; rays 7-10, 4-16 mm. long, the apical teeth not over 0.5 mm. long; achenes 5-6 mm. long, short-villous; pappus barbellate, white.---Slopes, ridges and meadows in the high mountains. Alberta to British Columbia, south to Colorado and Oregon. Our records from westcentral and in or very near northcentral and central Colorado at 10,000-12,000 feet.

6. **Arnica paniculata** A. Nels., Man. Bot. Rocky Mts. 572. 1909.

Montana, Wyoming and Utah. Reported for southern Wyoming and may be expected in adjacent Colorado.

7. **Arnica mollis** Hook., Fl. Bor. Am. 1:331. 1834.

A. subplumosa Greene; *A. silvatica* Greene; *A. coloradensis* Rydb.; *A. subplumosa silvatica* (Greene) A. Nels.; *A. ovata* Greene---Rootstocks short and branched, conspicuously rooting; stems 20-50 cm. tall, puberulent to pilose and often glandular, sometimes sparsely - hairy; cauline leaves 3 pairs, all sessile or lower petioled, ovate, obovate, elliptic, lanceolate or oblanceolate, subentire to denticulate or serrate, glandular-hairy to pilose or sometimes glabrous; heads 1-3, hemispheric-campanulate, sparsely to moderately pilose on peduncle just below; involucre bracts 18-22, 12-14 mm. long, acute and not especially pilose inside, pilose below and glandular-hairy; rays 14-18, 18-24 mm. long; achenes 6-7 mm. long, more or less hirsute; pappus plumose, tawny.---In the mountains, usually in moist ground. Alberta to British Columbia, south to Colorado and California. Our records from the higher mountains of Colorado at 9500-13,000 feet.

8. **Arnica cordifolia** Hook., Fl. Bor. Am. 1:331. 1834.

A. pumila Rydb.---Rootstocks slender and extensive; stems 20-45 cm. tall, usually simple, glandular-puberulent to loosely pilose; cauline leaves 2-3 pairs, ovate to lanceolate, upper sessile, lower petioled and often orbicular and cordate, subcordate to truncate, entire to dentate, puberulent to somewhat pilose, often glandular; heads 1-3, broadly turbinate to campanulate, densely pilose on peduncle below head or rarely only glandular-hairy; involucre bracts 10-14, 14-18 mm. long, acute to acuminate, pilose at base, puberulent above; rays 9-13, 20-28 mm. long; achenes 6.5-8 mm. long, uniformly hirsute and sometimes also uniformly glandular-hairy; pappus barbellate, white. Our plant has been called ssp. genuina Macguire.---Mountains, often in wooded areas. The subspecies from Alberta to Alaska, south to New Mexico and California. Our records scattered in the western half of Colorado at 6000-10,500 feet.

9. **Arnica gracilis** Rydb., Bull. Torr. Club 24:297. 1897.

Alberta to British Columbia, south to Wyoming, Utah and Oregon. May be in northern or northwestern Colorado.

10. **Arnica latifolia** Bong., Mem. Acad. St. Petersb. VI. 2:147. 1832.

A. ventorum Greene; *A. platyphylla* A. Nels.---Rootstocks short to moderately long; stems 8-60 cm. tall, glabrate, puberulent to short-pilose and often glandular-hairy; cauline leaves 3-4 pairs or rarely 2 pairs, ovate to lanceolate, sometimes cordate at base, crenate-serrulate to dentate, upper sessile, lower sessile or petioled, glabrate, puberulent to pilose and glandular-hairy; heads 1-5, narrowly turbinate to broadly campanulate-hemispheric, usually puberulent to pilose and glandular on peduncles just below; involucre bracts 8-16, 10-15 mm. long, acute to acuminate, puberulent to pilose and glandular or sometimes nearly glabrous; achenes 6-8 mm. long, glabrous to pubescent and often glandular-hairy; pappus barbellate, white.---Mountain meadows and woods. Alaska, south to Colorado and California. Our records from the western half of Colorado, scarce in the extreme southern part, at 6000-12,000 feet.

66. Haploesthes A. Gray

Herbaceous perennial plants or sometimes undershrubs; stems leafy; leaves opposite, linear-filiform and somewhat fleshy; heads cymose-paniculate; involucres short-campanulate, bracts 4-5, equal in length but broad and overlapping; receptacles flat, naked; ray flowers few, pistillate and fertile, rays yellow; disk flowers perfect and fertile; anthers united, not sagittate at base; achenes linear, terete, longitudinally ribbed, glabrous; pappus a single series of rather rigid, scabrous, white bristles.

1. Haploesthes greggii A. Gray, Mem. Amer. Acad. n. s. 4:109. 1849.
   Kansas to Colorado, south to Texas and Mexico. Reported from southeastern Colorado but not seen by the writer.

67. Xanthium L. COCKLEBUR

Weedy annual, herbaceous plants; stems simple or branched; leaves alternate, slender-petioled, blades toothed or lobed; heads unisexual, the plant monoecious. Staminate heads uppermost, many-flowered; involucre bracts separate in 1-3 series; corollas all discoid; anthers distinct, rudimentary style and ovary present; receptacles cylindrical. Pistillate heads burlike, bracts all fused, forming a body usually 2-beaked at apex and bearing stiff, hooked prickles, body 2-celled, mostly 2-flowered; corollas lacking; achenes linear to ovate, more or less compressed, beaked; pappus none. This is a difficult genus with numerous intergradations. The treatment follows Millspaugh and Sherff (N. A. Flora 33(1):37-44. 1922) although the last three species form a complex and perhaps should be united under one name.

1. Leaves lanceolate or lanc-ovate, long-cuneate at base, canescent below; axils of leaves bearing conspicuous 3-forked spines; fruit about 1 cm. long, the prickles doubly-curved (swan's neck) -1. X. spinosum
1. Leaves deltoid to broadly ovate, usually cordate at base, green and scabrous below; leaf axils without spines; fruit over 1 cm. long; prickles straight or once-curved
   2. Prickles densely hispid on lower half; fruit pubescent -2. X. italicum
   2. Prickles glabrous to merely pubescent or glandular-pubescent near base; fruit glabrous or subglabrous
      3. Fruit ovoid or fusiform, rarely over 1.5 cm. long; prickles 3-5 mm. long, not glandular-pubescent at base -3. X. chinense
      3. Fruit cylindric to oblong, often over 1.5 cm. long; prickles 3-7 mm. long, glandular-pubescent toward base -4. X. pennsylvanicum

1. Xanthium spinosum L., Sp. Pl. 987. 1753.
   Stems 25-100 cm. tall, erect to ascending; leaves lanceolate to ovate-lanceolate, acute or acuminate, cuneate, entire or lobed, densely white-strigose or canescent beneath, bearing conspicuous, yellowish, 3-forked spines to 25 mm. long in their axils; fruit about 1 cm. long, oblong-cylindrical, pubescent.---Waste places and fields. A weed throughout most of the warmer and temperate parts of the world. Our records from southeastern Colorado at 3500-5500 feet.
2. Xanthium italicum Moretti, Giorn. Fis. II. 5:326. 1822.
   X. commune Britt.---Stems about 20-80 cm. tall, finally branched, scabrous and purple-dotted; petioles about as long as the blades, these widely ovate, often cordate especially on lower, more or less sinuately lobed and dentate, scabrous both sides; fruit 1-2 cm. long, cylindrical, glandular-pubescent, prickles 3-7 mm. long, hispid from base to middle.---Waste places and fields. Quebec to Saskatchewan, south to Virginia, Mexico and California. Our records scattered over Colorado at 3500-7000 feet.
3. Xanthium chinense Mill., Gard. Dict. ed. 8, No. 4. 1768.
   Ontario to California, south to Florida and Mexico. This plant, if distinct from the next, should be in Colorado.
4. Xanthium pennsylvanicum Wallr., Beitr. Bot. 1:236. 1842.
   Stems 20-90 cm. tall, scabrous and purple-dotted; petioles about as long as the blades, these deltoid-ovate to cordate, dentate or sinuately lobed, scabrous both sides; fruit about 1-2 cm. long, glabrous or subglabrous, prickles 3-7 mm. long, glandular-pubescent and sometimes very sparsely hispid near base.---Waste places and fields. Massachusetts to Washington, south to Florida and California. Our records from western Colorado at 5000-7000 feet but to be expected anywhere in the state.

68. Ambrosia L. RAGWEED

Annual or perennial herbaceous plants; stems leafy; leaves opposite or alternate, mostly lobed or dissected; heads monoecious or rarely dioecious. Staminate heads in terminal spikes or racemes; involucres saucer-shaped or hemispheric, of 5-12 partly united bracts; receptacles flat or somewhat convex, naked or with filiform paleae; corollas discoid, funnelform; anthers not united. Pistillate heads below the staminate, 1-flowered, enclosed by the nutlike involucre, this obovoid and with a single series of tubercles or short spines near apex; corollas wanting; achenes ovoid or obovoid; pappus lacking.

1. Leaves unlobed or palmately 3- to 5-lobed; body of fruit about 5 mm. long -1. A. trifida
1. Leaves pinnatifid to bipinnatifid; body of fruit about 3 mm. long
   2. Annual plants; leaves rather thin, the lower usually bipinnatifid; beak of fruit about 1 mm. long
-2. A. elatior
   2. Perennial plants with rootstocks; leaves rather thick, mostly once-pinnatifid; beak of fruit about 0.5 mm. long -3. A. coronopifolia

1. **Ambrosia trifida** L., Sp. Pl. 987. 1753.
Annual plants; stems usually over 100 cm. tall, scabrous to nearly glabrous; leaves opposite, ovate in outline, mostly 3- to 5-lobed palmately or cleft, rarely all entire or toothed and narrower. Staminate heads numerous, involucres 3 mm. wide; saucer-shaped with 6-8 lobes. Pistillate heads in small clusters at the base of the staminate racemes, subtended by 3-cleft bracts; body of fruit about 5 mm. long, 4-5 ridged, with short conical spines, these sometimes reduced to rudiments. The form with leaves not at all lobed has been called var. integrifolia (Muhl.) T. & G.---Waste places and fields, especially in moist ground. Quebec to British Columbia, south to Florida and New Mexico. Our records scattered in the eastern half of Colorado at 4500-6000 feet.

2. **Ambrosia elatior** L., Sp. Pl. 987. 1753.
A. artemisifolia L.; A. media Rydb.---Annual plants; stems 30-100 cm. tall, strigose, hispid or scabrous; leaves opposite above, lower more commonly alternate, 1- to 2-pinnatifid, the latter especially on lower, scabrous, hispidulous or glabrate above, more or less strigose below, rather thin. Staminate heads numerous; involucres about 3 mm. wide. Pistillate heads in small clusters in the upper leaf axils; body of fruit about 3 mm. long, beak about 1 mm. long, subulate, spines 3-7, rather short, acute. Jones (Am. Midl. Nat. 17:673-700. 1936) gave reasons for rejecting A. artemisifolia L. as a name for our plant.---Waste places and fields. Now widespread throughout Canada and the United States. Our records from the eastern half of Colorado at 4000-6000 feet.

3. **Ambrosia coronopifolia** T. & G., Fl. N. Amer. 2:291. 1842.
A. psilostachya of Manuals for our plant---Perennial plants with horizontal, spreading rootstocks; stems 30-80 cm. tall, canescent-strigose; leaves opposite above, alternate below, pinnatifid, the divisions entire, toothed or again lobed but not usually strictly bipinnatifid, gray-strigose both sides, rather thick. Staminate heads numerous; involucres 3-3.5 mm. wide. Pistillate heads 1-3 in the axils of the upper leaves; body of fruit 3 mm. long, conical beak scarcely over 0.5 mm. long, spines none or 1-4, blunt at apex.---Plains and hills. Michigan to Saskatchewan, south to Illinois, Mexico and Washington. Our records from the eastern half of Colorado at 4000-6500 feet, also from the westcentral part.

69. Franseria Cav. BURSAGE

Annual herbaceous plants, or perennial with creeping horizontal rootstocks; stems leafy; leaves mostly alternate, toothed or pinnatifid; heads monoecious. Staminate heads in terminal spikes or racemes at the ends of branches; involucres with 5-12 more or less united bracts; receptacles flat or somewhat convex, with filiform to spatulate chaff; corollas all discoid, tube short; anthers distinct; pistil rudimentary. Pistillate heads nutlike, solitary or in clusters usually below the staminate, erect; involucres globose or ovoid, closed, beaked, enclosing 1-4 flowers; corollas none; fruit burlike with several to many spines in several series.

1. Fruit 2-4 mm. long, the spines hooked, not over 0.8 mm. long
   2. Leaves tomentulose as well as strigose below; corollas glabrous, not glandular; plants seldom over 30 cm. tall, the heads racemose; leaves seldom over 4 cm. long -1. F. linearis
   2. Leaves not tomentulose below; corollas glandular-puberulent or glandular-dotted; plants often over 30 cm. tall, the heads paniculate; leaves over 4 cm. long -2. F. confertiflora
1. Fruit 4-10 mm. long, the spines usually not hooked, over 0.8 mm. long
   3. Plants annual; leaves green or only slightly paler below, strigose both surfaces; spines of fruit often over 3 mm. long; beak of fruit only 1 -3. F. acanthicarpa
   3. Plants perennial with creeping rootstocks; leaves white-tomentose below; spines of fruit rarely over 3 mm. long; beaks of fruit 2
      4. Chaff of staminate heads equalling or exceeding the corollas, 3- to 5-nerved; leaves short-sericeous and more or less whitish above -4. F. tomentosa
      4. Chaff of staminate heads shorter than the corollas, 1-nerved; leaves strigose to glabrous and greenish above -5. F. discolor

1. **Franseria linearis** Rydb., N. Am. Fl. 33(1):27. 1922.
Gaertneria linearis Rydb.---Perennial plants, often rather woody at base; stems about 20-30 cm. tall, more or less hirsute; leaves 3-4 cm. long, bipinnatifid to tripinnatifid with linear segments, strigose both sides but also whitish-tomentose beneath; heads racemose. Staminate heads with narrowly spatulate paleae; corollas glabrous. Fruit 2-4 mm. long, strigose, 1 beak present, spines 8-15, nearly terete, hooked at apex, less than 1 mm. long.---Dry plains and hills. Apparently limited to central and eastcentral Colorado. Our records at about 6500 feet.

2. **Franseria confertiflora** (DC.) Rydb., N. Am. Fl. 33:28. 1922.
F. tenuifolia Harv. & Gray; Gaertneria tenuifolia (Gray) Kuntze---Perennial plants; stems 30-60 cm. tall, hirsute with appressed or spreading hairs; leaves 5-15 cm. long, bipinnatifid to tripinnatifid into oval to linear segments, coarsely strigose both sides; heads more or less paniculate. Staminate heads with narrowly spatulate paleae; corollas glandular-puberulent or glandular-dotted. Fruit 2-3 mm. long, glandular-puberulent and often hirsute as well, rather

reticulated, beak single, spines about 6-18, less than 1 mm. long, more or less broadened below and hooked at apex.---Plains, valleys and hills. Oklahoma to California, south to Texas and Mexico. Our records from southeastern and southcentral Colorado at 4000-6000 feet.
3. Franseria acanthicarpa (Hook.) Coville, Contr. U. S. Nat. Herb. 4:129. 1893.
Gaertneria acanthicarpa (Hook.) Britt.---Annual plants; stems 10-60 cm. tall, diffusely branched, strigose as well as hirsute; leaves bipinnatifid almost to the midrib, strigose and greenish both sides, slightly if at all paler below. Pistillate heads 1-flowered; fruit 6-10 mm. long, glabrate to sparingly hirsute, reticulate-ridged, beak 1, spines 8-20, about 3-4 mm. long at maturity, flattened, rarely hooked at least on all of the spines.---Plains and valleys, common on sandy ground. Minnesota to Alberta, south to Texas and California. Our records scattered over Colorado at 3500-8000 feet.
4. Franseria tomentosa A. Gray, Mem. Am. Acad. 4:80. 1849.
F. grayi A. Nels.; Gaertneria tomentosa (A. Gray) Kuntze---Perennial plants with creeping rootstocks; stems 25-60 cm. tall, villous-tomentose at least when young, glabrous and chestnut-brown in age; leaves about 4-10 cm. long, simple especially on upper, to pinnatifid, the divisions often again pinnatifid, silky and more or less whitish above, white-tomentose or densely white strigose-villous beneath; inflorescence rather paniculate. Staminate heads with linear-lanceolate, 3- to 5-nerved paleae longer than the flowers. Fruit 6-8 mm. long, glandular-puberulent, 2 beaks present, spines 12-15, about 2-3 mm. long, subulate-conic, not hooked or sometimes somewhat hooked at apex.---Plains and valleys. South Dakota and Wyoming, south to Nebraska and Arizona. Our records from eastern Colorado at 3500-4500 feet.
5. Franseria discolor Nutt., Trans. Am. Phil. Soc. Ser. 2. 7:345. 1840.
F. tomentosa (Nutt.) A. Nels.; Gaertneria tomentosa (Nutt.) Heller---Perennial plants with creeping rootstocks; stems 10-40 cm. tall, sparingly pubescent; leaves irregularly bipinnatifid, strigose or glabrate above, white-tomentose beneath. Staminate heads with spatulate, 1-nerved chaff shorter than the corollas. Fruit 4-6 mm. long, strigose, beaks 2, spines 8-12, about 1-1.5 mm. long, conical, little if any hooked at least on all.---Dry plains, hills, waste ground and fields. Nebraska and Wyoming, south to New Mexico and Arizona. Our records from northcentral, central and northwestern Colorado at 5000-8000 feet but to be expected anywhere in the state in cultivated areas.

70. Oxytenia Nutt. COPPERWEED

Perennial plants woody at base; stems slender and leafy; leaves alternate, pinnately parted into 3-5 long filiform divisions or upper leaves entire; heads numerous, in dense panicles, each with 10-20 staminate flowers and about 5 marginal pistillate ones; involucres of 5 distinct, ovate, acuminate bracts; receptacles convex, chaffy at least on outer staminate flowers. Staminate flowers with white discoid corollas; anthers distinct; pistils rudimentary. Pistillate flowers without corollas; achenes obovate, long-villous; pappus none or a minute scale.

1. Oxytenia acerosa Nutt., Journ. Acad. Phila. Ser. 2. 1:172. 1848.
Stems 100-200 cm. tall, sometimes leafless and rushlike, grayish-strigose especially above and more or less canescent; leaf segments about 1 mm. wide; involucres about 4-5 mm. wide, bracts canescent; achenes about 2 mm. long.---Dry plains and valleys. Colorado to Nevada, south to New Mexico and California. Our records from western Colorado at 4500-6500 feet.

71. Dicoria T. & G.

Annual plants; stems leafy, branched; leaves opposite or alternate, toothed or entire, petioled; heads small and numerous, in leafy panicles, each with 6-12 staminate flowers and 1-2 pistillate flowers, or head all staminate; involucres of 5 bracts but the thin, subscarious, narrow paleae subtending the staminate flowers larger than the bracts. Staminate flowers with discoid corollas; anthers distinct; rudimentary pistil present. Pistillate flowers 1-2; corollas none; achenes flattened, the margins pectinate or toothed; pappus none or rudimentary.

1. Margins of achenes irregularly toothed; pistillate flowers usually 2   -1. D. paniculata
1. Margins of achenes pectinately winged; pistillate flowers usually 1   -2. D. brandegei

1. Dicoria paniculata Eastw., Proc. Calif. Acad. Ser. 2. 6:298. 1896.
Colorado, Utah and New Mexico. Reported from southwestern Colorado but the record actually seems to be from southeastern Utah.
2. Dicoria brandegei A. Gray, Proc. Am. Acad. 11:76. 1876.
Colorado, Utah, New Mexico and Arizona. Reported from southwestern Colorado but not located by the writer.

72. Iva L. POVERTYWEED

Annual or perennial herbaceous plants; stems leafy; leaves opposite at least below; heads solitary and axillary, in bracteate spikes, or paniculate and not leafy-bracted, staminate and pistillate flowers in the same head; involucres hemispheric or cup-shaped, of 3-6 distinct or partly united bracts; receptacles chaffy. Staminate flowers with discoid corollas; anthers distinct. Pistillate flowers 1-8, marginal; corollas wanting or short-tubular; achenes usually obovoid, somewhat compressed; pappus none.

1. Plants annual; petioles over 5 cm. long (on longest), leaf blades toothed or incised, over 6 cm. long; heads in paniculate clusters -1. I. xanthifolia
1. Plants perennial with creeping rootstocks; leaves sessile, entire, rarely over 3 cm. long; heads solitary and axillary -2. I. axillaris

1. Iva xanthifolia Nutt., Gen. Pl. 2:185. 1818.
Cyclachaena xanthifolia (Nutt.) Fresen.---Annual plants; stems 70-180 cm. tall, stout, pubescent to glabrous; leaves mainly opposite, petioles 5-15 cm. long, blades 6-15 cm. long, ovate to ovate-lanceolate, broadly cuneate or cordate at base, coarsely serrate or incised, scabrous above, more or less strigose and often whitish below; heads in axillary and terminal panicles but each sessile, no leafy bracts among the heads; involucres 4-5 mm. wide, bracts 5, distinct. Staminate flowers about 8-20. Pistillate flowers about 5; corollas none; achenes about 3 mm. long.---Waste places and fields, usually in moist ground. Wisconsin to Alberta, south to New Mexico and Arizona; introduced eastward. Our records scattered over Colorado at 4500-7000 feet.

2. Iva axillaris Pursh, Fl. Am. Sept. 743. 1814.
Perennial plants with creeping, horizontal rootstocks; stems 10-50 cm. tall, ascending, simple or branched, strigose, hirsute or glabrate; lower leaves opposite, inflorescence leaves alternate, sessile, about 1-3 cm. long, oblong to obovate, entire, pubescent or glabrate; heads solitary in the axils of floral leaves but the whole appearing racemose; involucres about 5 mm. wide, of 4-5 bracts united at base. Staminate flowers about 12-20. Pistillate flowers about 5-8; corollas short-tubular; achenes about 3 mm. long.---Meadows, plains, roadsides, waste ground and fields. Saskatchewan to British Columbia, south to New Mexico and California. Our records well scattered in Colorado at 3500-7500 feet.

73. Evax Gaertn.

Low, woolly, annual plants; leaves alternate and entire; heads globose, more or less glomerate and leafy-bracted; involucre bracts few, more or less scarious; receptacles convex or conical, chaffy; rays none; outer flowers pistillate and fertile, central flowers few, perfect but usually sterile with undivided style; anthers united, caudate or sagittate at base; achenes obcompressed, glabrous; pappus none.

1. Evax prolifera Nutt. in DC., Prodr. 5:459. 1836.
Filago prolifera (Nutt.) Britt.; Diaperia prolifera Nutt.---Stems about 5-15 cm. tall, simple or branching from the base, also sometimes from the glomerules of heads, loosely tomentose; leaves up to 1 cm. long, spatulate, tomentose; heads in a sessile, leafy-bracted cluster from which later may develop 1-4 branches bearing similar clusters of heads, sometimes even again branching; involucres oblong or cylindrical; chaff of pistillate flowers mostly glabrous and scarious, of central flowers herbaceous and woolly at apex.---Dry plains and hills. South Dakota, south to Texas and Colorado. Our records from the eastern half of Colorado at 4000-7000 feet.

74. Antennaria Gaertn. PUSSYTOES

Low tomentose dioecious perennial plants usually with prostrate or nearly erect stolons; leaves alternate but usually in a basal rosette, small and entire; heads usually in capitate or corymbose clusters; involucres oblong to campanulate, bracts imbricated in several series, scarious the tips often white or colored; receptacles convex to nearly flat, naked. Staminate heads with perfect-appearing tubular flowers these with rudimentary ovary and undivided style; anthers united, caudate at base; pappus of bristles, clavate and flattened in apex (except in A. dimorpha). Pistillate corollas filiform, not at all raylike; style 2 cleft; achenes small, usually terete and glabrous; pappus of copious capillary bristles united at base and deciduous together.

1. Plants low, rarely over 5 cm. tall; heads usually solitary, sessile, subsessile among the leaves or on short stems rarely over 3 cm. long
    2. Leaves 5-10 mm. long, closely silky-tomentose; heads sessile or on stems 1 cm. long or less; plants mostly of southern Colorado -1. A. rosulata
    2. Leaves usually over 10 mm. long, tomentose; heads often on stems over 1 cm. long; plants usually generally distributed in Colorado
        3. Plants surculose-proliferous with spreading leafy stolons; pappus bristles of staminate flowers clavate-thickened upwards, inner involucre bracts of pistillate heads oblong, obtuse or acute; heads usually over 1 -2. A. parvifolia
        3. Plants not surculose-proliferous, lacking stolons; pappus bristles of staminate flowers not at all clavate; inner involucre bracts of pistillate heads linear-lanceolate, acuminate-attenuate at apex; heads solitary -3. A. dimorpha
1. Plants over 5 cm. tall; heads few to many on stems rising well above the basal leaves (stems usually 10 cm. high or more)
    4. Leaves (at least on flowering stems) soon glabrous and green above (the dense tomentum on lower surface may form a marginal line as seen from above)
        5. Involucre bracts whitish near apex -4. A. neglecta
        5. Involucre bracts greenish-brown or brown near apex -5. A. umbrinella
    4. Leaves permanently gray-, yellow-, or more commonly white-tomentose on both sides
        6. Largest basal leaves over 2.5 cm. long; plants lacking horizontal stolons (in 3 out of 5 species)
            7. Involucres 4-5 mm. high; leaves about 3-6 cm. long; stems seldom over 25 cm. tall

8. Plants surculose-proliferous with spreading leafy stolons; leaves narrowly oblanceolate to linear, rarely oblanceolate -6. **A. corymbosa**
8. Plants not surculose-proliferous, stolons if present erect or nearly so; leaves oblanceolate to spatulate -7. **A. oblanceolata**
7. Involucres 6 mm. high or more; leaves usually over 6 cm. long; stems often over 25 cm. tall
9. Plants surculose-proliferous with spreading leafy stolons; leaves obovate-cuneate, about 3-5 cm. long -8. **A. obovata**
9. Plants not surculose-proliferous, stolons if present erect or nearly so; leaves oblanceolate or narrower, 8-15 cm. long
10. Involucre bracts of pistillate heads 6-8 mm. high, in 3-4 series but subequal in length; involucre bracts of staminate heads with white tips longer than the body -9. **A. anaphaloides**
10. Involucre bracts of pistillate heads 8-10 mm. high, in 6-7 unequal series; involucre bracts of staminate heads with brownish tips shorter than the body -10. **A. pulcherrima**
6. Basal leaves about 5-20 mm. long; plants with spreading leafy stolons (in all species)
11. Pistillate heads 8-10 mm. high; involucres 6-9 mm. high; stems seldom over 15 cm. tall -2. **A. parvifolia**
11. Pistillate heads 5-8 mm. high; involucres 5-6 mm. high (rarely longer in **A. microphylla**); stems often over 15 cm. tall
12. Involucre bracts (at least inner) rose-colored, or apex rose-colored or pink
13. Leaves cuneate-spatulate, without distinction of blade and petiole; outer involucre bracts of pistillate heads brown, inner pink or rose -11. **A. concinna**
13. Leaves oblanceolate, sometimes spatulate or even obovate but petioles distinct; all bracts of pistillate heads rose or bright pink at apex -12. **A. rosea**
12. Involucre bracts white to brown at apex, not at all rose or bright pink
14. Involucre bracts green or light brown at base, definitely white above; leaves usually not over 1 cm. long -13. **A. microphylla**
14. Involucre bracts blackish-green or brownish-green (sometimes yellowish) at base, upper part green or brown not or only slightly whitish; leaves often over 1 cm. long
15. Involucre bracts blackish or blackish-green at base, mostly throughout; stems 2-7 cm. tall -14. **A. media**
15. Involucre bracts dark brown to yellowish at base, usually green or brown at apex; stems usually over 7 cm. tall
16. Leaves obtuse to mucronate, broadly spatulate to spatulate-obovate; involucres about 4-5 mm. high -15. **A. reflexa**
16. Leaves acute to abruptly acute, spatulate to narrower; involucres about 5-6 mm. high -5. **A. umbrinella**

1. Antennaria rosulata Rydb., Bull. Torr. Club 24:300. 1897.
A. sierrae-blancae Rydb.---Plants densely matted, acaulescent or nearly so, stems hardly rising above the rosette of leaves, surculose-proliferous but stolons short; stems absent or to 1 cm. long; leaves 5-10 mm. long, oblong-spatulate to spatulate-obovate, obtuse or acutish, closely silvery-tomentose, almost sericeous-strigose or rarely almost glabrous above; heads solitary or 2-3 together; involucres 5-7 mm. long, bracts acute to rounded at apex, outer not scarious or whitish, inner broadly white at apex; pappus bristles of staminate flowers moderately clavate at apex.---Mountains. Colorado, Utah, New Mexico and Arizona. Our records from the southwestern quarter of Colorado (a somewhat doubtful record from Boulder County) at 7000-9000 feet.

2. Antennaria parvifolia Nutt., Trans. Am. Phil. Soc. n. s. 7:406. 1841.
A. aprica Greene---Plants low, usually matted and caespitose; stems seldom over 15 cm. tall, sometimes not over 2 cm.; basal leaves 1-2 cm. long, obovate to cuneate-oblanceolate, acute or obtuse, permanently gray- or white-tomentose both sides; heads 3-8, rarely solitary, glomerate and sessile or subsessile; involucres 6-9 mm. high, pistillate head longest, bracts obtuse or acute, usually brown-spotted at base and with white or pinkish tips; pappus bristles of staminate heads clavate but only moderately so.---Plains, hills and mountains. Manitoba to British Columbia, south to New Mexico and Arizona. Our records from the western three-fourths of Colorado at 5000-11,500 feet.

3. Antennaria dimorpha (Nutt.) T. & G., Fl. N. Amer. 2:431. 1843.
Plants depressed-caespitose in matted tufts with thickish rootstocks covered with long roots; flowering stems 1-3 cm. tall or seemingly lacking sometimes; leaves 1-2 cm. long or even slightly shorter, oblanceolate to spatulate, obtuse to acutish, tapering to a petiolelike base, rather loosely gray- or white-tomentose on both sides; heads solitary; involucres 6-10 mm. high; bracts of pistillate heads longer, linear-lanceolate, attenuate-acuminate at apex, brown or purplish at base or near middle, rather whitish on margins at apex; bracts of staminate head wider, obtuse or broadly acute at apex, brown at base, lighter brown to brownish-white near apex; pappus bristles of staminate heads not at all clavate.---Dry plains and slopes. Montana to British Columbia, south to Nebraska, Utah and California. Our records mostly from western Colorado, one, however, from Grand County, at 6500-8000 feet.

4. Antennaria neglecta Greene, Pitt. 3:173. 1897.
A. marginata Greene---Plants surculose-proliferous; stems 10-30 cm. tall on pistillate plant, about 6-12 cm. tall on staminate; leaves 1-6 cm. long, obovate-cuneate or oblanceolate, not petioled, white-tomentose below, glabrate above; heads about 3-7, crowded at first; involucres of pistillate heads 7-9 mm. high, bracts brownish at base, white near apex, linear or lanceolate, at least some acute; staminate involucres 5-7 mm. high, bracts oblong and obtuse; pappus bristles of staminate heads clavate. The small-leaved **A. marginata** Greene may be distinct.---Fields, pastures and open woods. New Brunswick to Utah, south to Virginia and Arizona. Our records from northcentral, southcentral and southwestern Colorado at 5500-10,000 feet.

5. Antennaria umbrinella Rydb., Bull. Torr. Club 24:302. 1897.

Plants surculose-proliferous; stems 8-15 cm. tall; basal leaves 1-2 cm. long, spatulate to narrowly oblanceolate, acute, abruptly acute or mucronate, densely white-tomentose both sides; stem leaves linear to linear-oblong; heads about 3-8, conglomerate; involucres about 5-6 mm. high, bracts dark brown to yellowish-brown at base, usually brown at apex or inner sometimes whitish on pistillate, obtuse or sometimes acutish; staminate pappus bristles clavate at apex. A dwarf form of high altitudes, with leaves more or less glabrate above has been called A. alpina (L) Gaertn. and may be distinct.---Mountains. Montana to British Columbia, south to Colorado and Arizona. Our records from the higher mountains of Colorado at 9000-11,500 feet.

6. Antennaria corymbosa E. Nels., Bot. Gaz. 27:212. 1899.

A. nardina Greene---Plants surculose-proliferous, stolons slender and flexible; flowering stems 15-30 cm. tall, slender; leaves about 3-3.5 cm. long, thin, 1-nerved or indistinctly 3-nerved, sparsely and appressed tomentose both sides; basal leaves narrowly oblanceolate, acute to mucronate, tapering to the petiole, stem leaves linear; heads corymbose, lowest peduncles longest; involucres 4-5 mm. high, bracts greenish- or brown-spotted near base, white at apex; pistillate bracts elliptic and obtuse or inner narrower and acute, staminate broader and obtuse or truncate; pappus bristles of staminate flowers clavate at apex.---Mountains. Montana to Oregon, south to Colorado and Nevada. Our records scattered in the mountains of Colorado at 8000-10,500 feet.

7. Antennaria oblanceolata Rydb., Mem. N. Y. Bot. Gard. 1:409. 1900.

Plants with branching caudex, not surculose-proliferous, the stolonlike structures, when present, erect or ascending; flowering stems 20-40 cm. tall; leaves 3-6 cm. long, basal oblanceolate to spatulate, rather 3-ribbed, mucronate or rounded at apex, densely white-tomentose; upper leaves linear; heads few in a small corymb; involucres 4-5 mm. high; inner bracts brownish at base with white or straw-colored tips, outer usually entirely brown, acute to obtuse; pappus bristles of staminate heads clavate at apex. This species may not be distinct from A. luzuloides T. & G. with narrower basal leaves.---Mountains. Montana to British Columbia, south to Colorado and California. Our few records from northcentral and northwestern Colorado from 8000-9000 feet.

8. Antennaria obovata E. Nels., Bot. Gaz. 27:213. 1899.

Plants surculose-proliferous, stolons leafy near ends; flowering stems 20-30 cm. tall; basal leaves 3-5 cm. long, obovate, cuneate to a petiole, 3-ribbed, rounded or mucronate at apex, firm in texture, permanently tomentose both sides; stem leaves smaller, oblong-linear to linear; heads about 3-7 in corymbs; involucres 7-8 mm. high, pistillate bracts in several series, outer short and usually obtuse, inner narrow and acute or acuminate, mostly with a brownish spot near middle, brownish at apex; pappus bristles of staminate heads not seen but apparently clavate.---Plains, hills and lower mountains. Manitoba to Alberta, south to South Dakota and Colorado. Our records from northcentral and central Colorado at 5500-8500 feet.

9. Antennaria anaphaloides Rydb., Mem. N. Y. Bot. Gard. 1:409. 1900.

Plants not surculose-proliferous, stolonlike branches, if present, erect or nearly so; stems about 30-40 cm. tall but shorter in poor ground or high altitudes; basal leaves 10-15 cm. long, oblanceolate to linear-oblanceolate, acute, more or less distinctly 3-nerved, loosely tomentose both sides; heads several to many in a corymb; involucres 6-8 mm. high, bracts brown at base with broad white tips much longer than the basal brown part, obtuse or acute in pistillate, rounded or truncate in staminate; pappus bristles of staminate flowers clavate at apex. May not be distinct from the next species.---Hills and mountains. Montana to British Columbia, south to Colorado and Nevada. Our records scattered in the mountains of Colorado at 5500-11,500 feet.

10. Antennaria pulcherrima (Hook.) Greene, Pitt. 3:176. 1898.

Reported for Saskatchewan to Yukon, south to Colorado and Washington. Not seen by the writer from this state.

11. Antennaria concinna E. Nels., Proc. U. S. Nat. Mus. 23:705. 1901.

This plant may be merely a form of the next species. Leaves lack any distinction between petiole and blade; outer involucre bracts tend to be brown instead of rose.---Hills and mountains. Wyoming to Washington, south to New Mexico and Utah. Our records from the western half of Colorado at 7500-11,000 feet.

12. Antennaria rosea (D. C. Eat.) Greene, Pitt. 3:281. 1898.

A. imbricata E. Nels.---Plants surculose-proliferous but sterile branches erect or ascending; stems 10-40 cm. tall; basal leaves 1.5-2 cm. long, narrowly oblanceolate to broadly spatulate but petioles distinct from blade, acute to obtuse, closely tomentose both sides; stem leaves oblong, oblanceolate or linear; heads few to many in close or rather open structures; involucres 5-6 mm. high, sometimes to 7 mm.; pistillate bracts elliptic to linear-oblong, obtuse or acutish, rose or pink at least near spex, staminate bracts more or less pink-tinged or rose; pappus bristles of staminate heads apparently clavate. The rose color of the involucre bracts is an arbitrary character and this species may not be distinct from the preceding and the next one.---Meadows, hills and mountain slopes. Montana to Alaska, south to Colorado and California. Our records from the western half of Colorado at 5500-11,000 feet.

13. Antennaria microphylla Rydb., Bull. Torr. Club 24:303. 1897.

A. arida E. Nels.; A. bracteosa Rydb.; A. parvifolia bracteosa (Rydb.) A. Nels.; A. viscidula A. Nels.; A. arida viscidula A. Nels.---Plants surculose-proliferous; stems 10-40 cm. tall; basal leaves 5-12 mm. long, rarely to 20 mm., spatulate, often broadly so, finely appressed tomentose on both sides; stem leaves oblong, oblanceolate or linear; heads about 3-10, corymbose; involucres 5-8 mm. high; pistillate bracts oblong or lanceolate, obtuse or acute, brown or brownish-green at base, white at apex, staminate bracts broader and rounded at apex; pappus bristles of staminate flowers clavate. Probably not distinct from the preceding. The form with low stems and viscid involucres has been called A. sedoides Greene.---Plains, hills and mountains. Saskatchewan to Alaska, south to New Mexico and Nevada. Our records well scattered in the western half of Colorado at 5000-11,000 feet.

14. **Antennaria media** Greene, Pitt. 3:286. 1898.

Surculose-proliferous plants; stolons slender and rather short; flowering stems 2-7 cm. tall; basal leaves 1-2 cm. long, spatulate-oblanceolate, abruptly acute or acutish, white- or gray-tomentose both sides; stem leaves linear above; heads 3-7, glomerate on short peduncles at ends of stems; involucres about 4 mm. high, bracts dark brownish or brownish-black throughout at least on pistillate heads, oblong, obtuse or inner acute; pappus bristles of staminate flowers clavate. This may be a form of A. umbrinella Rydb. connected through A. alpina (L.) Gaertn.---High mountains. Alberta to Alaska, south to Colorado and California. Our records scattered in the higher mountains of Colorado, mostly from the northern part, at 9500-11,500 feet.

15. **Antennaria reflexa** E. Nels., Bot. Gaz. 27:208. 1899.

A. flavescens Rydb.---Plants surculose-proliferous with leafy stolons; stems 5-15 cm. tall; basal leaves 8-15 mm. long, broadly spatulate to spatulate-obovate, obtuse to somewhat mucronate, densely whitish-tomentose on both sides; stems leaves oblong or linear; heads 4-8, sessile or nearly so, in dense terminal, subcapitate clusters; involucres 4-5 mm. long, bracts brown to brownish-yellow at base, upper part brownish-white or greenish-white, not pure white, truncate, obtuse to acuminate at tip; staminate pappus bristles clavate.---Hills and parks. Montana to Colorado. Our few records from northcentral Colorado at 8500-9500 feet.

75. Anaphalis DC.   PEARL-EVERLASTING

Perennial, more or less tomentose, herbaceous plants; stems erect, leafy from rootstocks; leaves alternate, entire, narrow, sessile; heads corymbose, terminal on stem and upper branches; involucres campanulate to hemispheric, bracts imbricated in several series, outer shorter, pearly-white and scarious, spreading when dry; receptacles mostly convex, naked; plants polygamo-dioecious, the pistillate heads usually with a few perfect flowers in the center. Pistillate flowers discoid, fertile. Perfect flowers sterile with undivided styles; corollas discoid; anthers united, caudate at base. Achenes short, glabrous; pappus of capillary bristles, falling separately, little if any thickened at apex.

1. **Anaphalis margaritacea** (L.) Benth. & Hook., Gen. Pl. 2:303. 1873.

A. subalpina (Gray) Rydb.; A. margaritacea var. occidentalis Greene---Stems 25-60 cm. tall; leaves about 3-10 cm. long, linear to linear-lanceolate or narrowly lanceolate, 1- to 3-nerved, white-tomentose below, usually less so and floccose, or green and glabrous above; heads in close or open corymbose clusters; involucres 6-7 mm. high. The writer cannot distinguish varieties in Colorado plants.---Mountain slopes, often in open woods. South Dakota to British Columbia, south to Colorado and Arizona. Our records scattered in the mountains of the state at 7000-11,000 feet.

76. Gnaphalium L.   CUDWEED; EVERLASTING

Annual, biennial or perennial herbaceous plants with taproots, more or less woolly throughout; stems leafy, erect or diffuse; leaves alternate, narrow and entire; heads glomerate or paniculate; involucres campanulate or hemispheric, bracts imbricated in several series, scarious; receptacles convex to conic, naked and usually pitted; heads all alike; marginal flowers pistillate and fertile, rayless; central flowers fewer, perfect and fertile, discoid; achenes terete or slightly flattened; pappus of capillary bristles, these falling separately.

1. Heads not leafy-bracted; involucres 4-6 mm. high; plants usually over 20 cm. tall
  2. Leaves green and glandular above, tomentose below, strongly decurrent; involucre bracts all acute
                    -1. G. macounii
  2. Leaves gray-tomentose both sides, only moderately if at all decurrent; involucre bracts obtuse or sometimes abruptly pointed
    3. Leaves not at all or only very slightly decurrent; involucre bracts white, obtuse or abruptly pointed and apiculate   -2. G. wrightii
    3. Leaves moderately decurrent or at least with adnate auricles; involucre bracts straw-colored or even yellowish, obtuse at apex   -3. G. chilense
1. Heads in leafy-bracted clusters; involucres 2-4 mm. high; plants rarely to 20 cm. tall
  4. Plants loosely floccose-tomentose; leaves oblong, spatulate or oblanceolate   -4. G. palustre
  4. Plants appressed tomentose (at least the leaves); leaves narrowly oblanceolate to linear
    5. Leaves narrowly oblanceolate, the largest over 3 mm. wide   -5. G. uliginosum
    5. Leaves all linear or narrowly linear, none over 3 mm. wide   -6. G. exilifolium

1. **Gnaphalium macounii** Greene, Ottawa Nat. 15:278. 1902.

G. decurrens Ives---Plants probably always biennial with taproot; stems 40-80 cm. tall, corymbosely branched above, stout; leaves about 5-10 cm. long, lanceolate or linear especially above, obviously decurrent, whitish-tomentose below, green and glandular-pubescent above; heads numerous in close glomerules at the end of the branches, usually bracted at base but these not leaflike; involucres 5-6 mm. high, campanulate-subglobose, bracts acute, straw-colored or somewhat whitish at base, sometimes pale brown at apex.---Open slopes and valleys. Nova Scotia to British Columbia, south to West Virginia, Texas and California. Our records mostly from northcentral, central and southcentral Colorado at 6000-8500 feet.

2. **Gnaphalium wrightii** A. Gray, Proc. Am. Acad. 17:214. 1882.

Biennial or perennial plants with taproots; stems 30-50 cm. tall, upright, much branched, white-tomentose; leaves 2-4 cm. long, oblanceolate or upper linear, sessile but not decurrent or clasping, gray-tomentose both sides; heads many in an open panicle, often somewhat glomerate at ends of branches; involucres 4-6 mm. high, campanulate, bracts pearly-white, obtuse or more commonly acute to abruptly pointed.---Dry ground. Colorado, south to Texas, Mexico and California. Our records from westcentral and northcentral Colorado at 5000-10,000 feet but the plant should be in the central and southern parts also.

3. Gnaphalium chilense Spreng., Syst. Veg. 3:480. 1826.

G. sulphurescens Rydb.---Annual or biennial plants; stems 20-70 cm. tall, usually simple but sometimes branching above, woolly-tomentose or somewhat glabrate at base; leaves about 2-3 cm. long, linear or lanceolate, lower spatulate, mostly erect, stem leaves definitely decurrent, or with broad adnate auricles and appearing so; heads sessile in small glomerules terminating the stem and branches, mostly bracted at base but these not leaflike; involucres 4-5 mm. long, campanulate-subglobose, bracts obtuse, straw-colored or distinctly yellowish, glossy.---Hills and valleys, often in moist ground. Montana to Washington, south to Texas and California. Our records from the western half of Colorado at 5000-9000 feet.

4. Gnaphalium palustre Nutt., Trans. Am. Phil. Soc. Ser. 2. 7:403. 1841.

Annual plants; stems 5-20 cm. tall, becoming diffusely branched from the base or sometimes erect and unbranched, loosely floccose-woolly; leaves 1-2.5 cm. long and about 3-8 mm. wide, oblong, oblanceolate, obovate or spatulate, loosely floccose both sides; heads in small glomerules terminating the stems and branches, subtended by leaflike bracts; involucres 3-4 mm. long, campanulate, bracts linear or oblong, brown or whitish at apex, more or less woolly at base.---Usually moist or wet ground. Alberta to Alaska, south to New Mexico and California. Our records from the western half of Colorado at 5000-8000 feet.

5. Gnaphalium uliginosum L., Sp. Pl. 856. 1753.

Closely related to the G. exilifolium complex. The leaves are narrowly oblanceolate and over 3 mm. wide on largest.---Moist or wet ground. Newfoundland to Oregon, south to Colorado and Utah; Europe. Our few records from northcentral, westcentral and southcentral Colorado at 4500-9000 feet.

6. Gnaphalium exilifolium A. Nels., Bull. Torr. Club 29:406. 1902.

G. grayi Nels. & Macbr.; G. strictum A. Gray---Annual plants; stems 8-25 cm. tall, varying from erect and simple to branching from the base with the branches erect-ascending to diffusely spreading, appressed-tomentose; leaves 1-4 cm. long, seldom over 3 mm. wide and usually less, linear or occasionally linear-oblanceolate, appressed tomentose; heads in glomerules, these in crowded cymes or often spicate, subtended by leaflike bracts; involucres about 3 mm. long, bracts acutish to obtuse, brown below with whitish tips. A variable species but the proposed segregates are connected in this area by many intermediate forms. All our plants may actually be a part of G. uliginosum L.---Moist or wet places. Wyoming to New Mexico and Arizona. Our records scattered in the western two-thirds of Colorado, except the extreme western part, at 4500-10,000 feet.

77. Cirsium Hill THISTLE

Biennial or perennial herbaceous plants with spine-tipped or spine-toothed leaves and involucres; leaves alternate, usually toothed or lobed, rarely entire, often decurrent; heads medium-sized to large, flowers usually pink, reddish or white; involucres broad, bracts many, imbricated in several series, at least some usually spine-tipped; receptacles flat, bristly; rays lacking; disk flowers perfect or sometimes the plant dioecious, corollas tubular and deeply 5-cleft; anthers united, sagittate at base; achenes obovate to oblong, somewhat flattened; pappus of numerous plumose bristles in a single series, united at very base and falling together. This is a variable and difficult genus with numerous intergradations in this state. Many of these intergrades have been called hybrids, often arbitrarily so. Good specimens are relatively rare in herbaria. The genus is badly in need of a revision in North America. The following names have been applied to plants reported from Colorado and may turn out to be synonyms or even valid species upon further study:

Genus Cirsium - C. acuatum Rydb.; C. araneans Rydb.; C. engelmanii Rydb.; C. erosum K. Schumann; C. floccosum Rydb.; C. foliosum DC.; C. griseum K. Schumann; C. hookerianum Nutt.; C. laterifolium Rydb.; C. modestum Rydb.; C. oblanceolatum K. Schumann; C. pulcherrimum K. Schumann; C. spathulifolium Rydb.; C. vernale Cockl.

Genus Carduus - C. araneosus Osterh.; C. canadensis Osterh.; C. crassus Osterh.; C. filipendulus Rydb.; C. osterhoutii Rydb.; C. spathulatus Osterh.

1. At least some of the involucre bracts with conspicuous dilated-fringed tips
   2. Involucre bracts densely arachnoid-pubescent; leaves glabrate below; flowers greenish-yellow
      -1. C. parryi
   2. Involucre bracts not arachnoid-pubescent or only slightly so on the margins; leaves more or less tomentose below; flowers white, red, rose or purple
      3. Corollas white to ochroleucous; involucre bracts without a glutinous dorsal ridge; at least some of leaves sinuately-parted to divided   -2. C. centaureae
      3. Corollas red, purple or rose; involucre bracts with a glutinous dorsal ridge, this sometimes rather obscure; leaves toothed to very shallowly lobed   -3. C. perplexans
1. None of bracts with dilated-fringed tips, usually ending in a spine (if tip dilated then merely erose but not fringed)
   4. Involucre bracts densely arachnoid-pubescent
      5. At least some of the involucre bracts pectinate-ciliate with lateral spines; flowers greenish-yellow; leaves arachnoid-pilose below, not tomentose   -4. C. pallidum
      5. None of bracts pectinate-ciliate; flowers white, rose, purple to red, rarely ochroleucous; leaves usually tomentose below
         6. Involucre bracts squarrose, the outer reflexed   -5. C. neomexicanum
         6. Involucre bracts not squarrose or reflexed, only the tips sometimes spreading

7. Leaves conspicuously decurrent, forming wings on the stem; heads usually over 4 cm. high; plants usually in waste places and fields as an introduced weed -6. C. lanceolatum
7. Leaves little if any decurrent; heads less than 4 cm. high; native plants
    8. Anthers glabrous; plants of montane to alpine areas, generally distributed in Colorado -7. C. scopulorum
    8. Anthers hairy; plants of alpine areas, known only from the southwestern corner of Colorado -8. C. hesperium
4. Involucre bracts not densely arachnoid-pubescent, or only slightly so on the margins
  9. Plants acaulescent -12A. C. drummondii acaulescens
  9. Plants caulescent
    10. Lower leaf surface glabrous to somewhat arachnoid-pubescent, not tomentose
      11. Heads 2-2.5 cm. high; involucre bracts with a glutinous dorsal ridge; flowers apparently dioecious; plants perennial by horizontal underground rootstocks, hence in patches -9. C. arvense
      11. Heads 3-4.5 cm. high; involucre bracts lacking a glutinous dorsal ridge; flowers perfect; plants not commonly with horizontal rootstocks, seldom in large patches
        12. Outer involucre bracts spinulose-ciliate -10. C. eatoni
        12. Outer bracts not spinulose-ciliate
          13. Inner involucre bracts with elongated, attenuate, plane, usually bright red or purple tips -11. C. bipinnatum
          13. Inner bracts usually with more or less dilated and twisted, often erose tips -12. C. drummondii
    10. Lower leaf surface definitely tomentose
      14. Involucre bracts with a glutinous dorsal ridge (this sometimes not very conspicuous)
        15. Plants dioecious; spinulose tips of involucre bracts not over 2 mm. long; heads seldom over 2 cm. high -9A. C. arvense vestitum
        15. Plants with perfect flowers; spinulose tips of involucre bracts often 3 mm. long or at least on the outer; heads over 2 cm. high
          16. Heads less than 4 cm. high
            17. Heads 3-4 cm. high, solitary or few -13. C. undulatum
            17. Heads 2-3 cm. high, several to many
              18. Middle or outer involucre bracts with spines over 5 mm. long, medial glutinous ridge usually definite -14. C. canescens
              18. Spines of bracts less than 5 mm. long, glutinous ridge usually rather obscure -15. C. spathulifolium
          16. Heads over 4 cm. high
            19. Corollas yellow, ochroleucous or white; involucre spines 3-5 mm. long -16. C. plattense
            19. Corollas rose to purple; involucre spines 3-10 mm. long
              20. Involucre spines 6-10 mm. long, stout -17. C. ochrocentrum
              20. Involucre spines 3-4 mm. long, slender -18. C. megacephalum
      14. Involucre bracts lacking a glutinous dorsal ridge
        21. Involucre bracts squarrose, the lower reflexed with apical spines over 4 mm. long -5. C. neomexicanum
        21. Involucre bracts not squarrose or reflexed, only the tips sometimes spreading, apical spines rarely over 4 mm. long
          22. Inner involucre bracts more or less dilated, erose and twisted at apex -12. C. drummondii
          22. Inner bracts with elongated, plane tips -19. C. pulchellum

1. Cirsium parryi (A. Gray) Petrak, Bot. Tidesskr. 31:68. 1911.
Carduus parryi (Gray) Greene---Perennial or possibly biennial plants; stems 30-100 cm. tall, more or less arachnoid-pubescent; leaves lanceolate to oblong-lanceolate, sinuate-dentate to sinuate-lobed, the spines 2-15 mm. long, glabrate above and below; heads solitary or 2-4 clustered at end of stem and branches, occasionally in the upper axils, 20-30 mm. high and about as wide; involucre bracts rather densely arachnoid-hairy, no glutinous dorsal ridge present; at least some of outer bracts spinulose-ciliate, with a terminal spine 1-5 mm. long, inner with dilated-fringed tips, at least on some; corollas light greenish-yellow.---Mountains. Colorado, Utah, New Mexico and Arizona. Our records from the mountains of Colorado, mostly in the southern part, at 8000-11,000 feet.

2. Cirsium centaureae (Rydb.) K. Schum., Just. Bot. Jahresb. 29(1):566. 1903.
C. americanum (Gray) Daniels; Carduus americanum (Gray) Greene; Carduus centaureae Rydb.---Perennial or possibly biennial plants; stems 60-100 cm. tall, sparingly arachnoid-pilose to glabrate; leaves oblong-lanceolate to lanceolate, sinuately-pinnatifid to merely toothed on a few, the spines 1-4 mm. long, glabrate and light green above, more or less tomentose below; heads solitary or 2-3 in subsessile clusters at end of stem and branches, about 20-30 mm. high and about 15-25 mm. wide; involucre bracts glabrate or sparingly arachnoid-pubescent on margins, without a dorsal glutinous ridge; outer bracts abruptly acuminate, the spines 3-6 mm. long with lacerate fringes near apex, inner gradually acuminate; corollas ochroleucous to white.---Mountains. Wyoming, Colorado and Utah. Our records from western and northern Colorado at 6000-9500 feet.

3. Cirsium perplexans Rydb., Bull. Torr. Club 32:132. 1905.
Carduus americanus perplexans (Rydb.) A. Nels.---Biennial or perennial plants; stems 20-60 cm. tall, often purplish, sparingly tomentose; leaves oblanceolate to lanceolate, dentate or very sparingly lobed, spines short, glabrate above, thinly tomentose below, upper ones clasping at base; heads solitary on stem or branches, about 2.5-3.5 cm. high and about as wide; involucre bracts glabrate or nearly so, with glutinous dorsal ridge present but this sometimes

obscure; outer bracts with a short spine, middle and inner with conspicuous erose-dilated tips; corollas rose or red-purple.---Mountains and plains.   Apparently limited to Colorado, our records from the westcentral part of the state at 4500-7000 feet.

4. Cirsium pallidum Woot. & Standl., Contrib. U. S. Nat. Herb. 19:751. 1915.

Perennial or possibly biennial plants; stems 80-200 cm. tall, arachnoid-pilose; leaves lanceolate, spinulose-dentate with spines 1-6 mm. long, glabrate above and more or less arachnoid-pilose below; heads 3-6, congested at end of stem or on short axillary branches, 25-30 mm. high and about 25 mm. wide; involucre bracts densely arachnoid-pubescent, no glutinous, dorsal ridge present; outer bracts gradually attenuated to a terminal spine 5-7 mm. long and some at least pectinate with lateral spines, inner bracts mostly spine-tipped; corollas greenish-yellow.---Mountains. Colorado and New Mexico. Our one record from southcentral Colorado at 11,000 feet.

5. Cirsium neomexicanum A. Gray, Pl. Wright. 2:101. 1853.

Carduus neomexicanus (Gray) Greene---Biennial plants; stems 30-120 cm. tall, arachnoid-tomentose; leaves lanceolate to oblong, sometimes linear, sinuately-lobed or pinnatifid with spines 2-8 mm. long, upper surface tomentose sometimes loosely so and lower surface tomentose; heads solitary to few, at end of stem and branches, 35-60 mm. high and about 30-50 mm. wide; involucre bracts from sparingly to densely arachnoid-pubescent, no dorsal, glutinous ridge present; outer bracts elongated-attenuate and squarrose or the bract itself reflexed, the spine 4-8 mm. long, inner bracts long-attenuate; corollas white to pink.---Plains and hills. Colorado to Nevada, south to New Mexico and California. Our records from westcentral and southwestern Colorado at 4500-6500 feet.

6. Cirsium lanceolatum (L.) Hill, Herb. Brit. 1:80. 1769.

Carduus lanceolatus L.---Biennial plants; stems 60-150 cm. tall, more or less villous or tomentose when young; leaves lanceolate, deeply to moderately pinnatifid with lanceolate or triangular-lanceolate lobes tipped with stout yellowish spines 3-4 mm. long, upper surface strigose-pubescent to hirsute and lower more or less gray-tomentose, leaf strongly decurrent; heads scattered and mostly solitary at ends of branches, 3-5 cm. high and nearly as wide; involucre bracts arachnoid, all attenuated into spines 1-4 mm. long, outer spines longer and inclined to spread, midnerves definite but not glutinous; corollas rose-purple. The correct name for the above may be C. vulgare (Savi) Tenore.---Waste places, pastures, roadsides and fields. Introduced from Europe and widespread in North America. Our few records from northcentral, central and southwestern Colorado at 5000-7500 feet, but to be expected anywhere in cultivated areas.

7. Cirsium scopulorum (Greene) Cockll. in Daniels, Univ. Mo. Stud. Sci. 22:253. 1911.

Carduus scopulorum Greene---Perennial plants; stems 20-60 cm. tall, sparingly arachnoid-pilose; leaves lanceolate to oblong-lanceolate, sinuately-lobed or pinnatifid with spines 2-4 mm. long, upper surface glabrate and green, lower more or less tomentose; heads aggregated, usually in a nodding or subsessile cluster, 30-40 mm. high and about 25-35 mm. wide; involucre bracts densely arachnoid-hairy, a dorsal glutinous ridge present or absent; outer bracts gradually attenuate to a spine 3-5 mm. long, inner bracts acuminate; corollas ochroleucous to white, perhaps sometimes purplish. Our plants may be C. hookerianum Nutt.---Mountains. Montana to Colorado and Utah. Our records from the higher mountains in or near northcentral, westcentral, central and southcentral Colorado at 9000-13,000 feet.

8. Cirsium hesperium (Eastw.) Rydb., Fl. Rocky Mts. 1007. 1917.

Carduus hesperium (Eastw.) Heller; Carduus hookerianus hesperius (Eastw.) A. Nels.---A little known species apparently limited to southwestern Colorado. The writer has not seen it.

9. Cirsium arvense (L.) Scop., Fl. Carn. ed. 2. 2:126. 1772.

Carduus arvensis (L.) Rob.---Perennial plants with horizontal, creeping rootstocks and forming patches; stems 30-100 cm. tall, branching above, glabrous or nearly so; leaves lanceolate or oblong-lanceolate, deeply pinnatifid or rarely entire, with spines 1-6 mm. long, glabrous or sparsely pubescent above and below; heads many, corymbose, apparently dioecious the staminate heads more globose in shape, 15-25 mm. high and less than 25 mm. wide; involucre bracts glabrous or nearly so; outer bracts tipped with a short spine rarely over 1 mm. long, with a glutinous dorsal ridge, inner bracts spineless and elongated at apex; corollas purple to white. The form with leaf margins entire has been called var. integrifolium Wm. & Grab.---Fields, pastures, roadsides and waste places. Introduced from Europe and now widespread in North America. Our records scattered over Colorado, mostly from cultivated areas, at 5000-9500 feet.

9A. Cirsium arvense vestitum Wm. & Grab., (var.) Fl. Sil. Vratislaviae 3:1829.

Lower leaf surface white-tomentose. With the species in this area.

10. Cirsium eatoni (A. Gray) Robinson, Rhodora 13:240. 1911.

Perennial plants; stems 20-60 cm. tall, glabrous or sparsely arachnoid; leaves lanceolate to linear-lanceolate, sinuately-pinnatifid with spines 2-10 mm. long, glabrous or nearly so above and glabrate or somewhat arachnoid-hairy below; heads 3-6, aggregated in sessile or subsessile clusters, 25-35 mm. high and 20-30 mm. wide; involucre bracts glabrate or very sparingly arachnoid-pubescent near margins, no glutinous dorsal ridge present; outer bracts nearly as long as inner, gradually acuminate to a spine 5-10 mm. long, spinulose-ciliate on margins, inner bracts gradually acuminate; corollas rose-purple.---Mountains. Idaho, south to Colorado and Nevada. Our few records scattered in western and central Colorado at 8000-11,400 feet.

11. Cirsium bipinnatum (Eastw.) Rydb., Fl. Rocky Mt. 1010, 1068. 1917.

C. pulchellum ssp. bipinnatum (Eastw.) Petrak; Carduus bipinnatus (Eastw.) Heller---Apparently perennial plants; stems 50-150 cm. tall, glabrous or slightly arachnoid; leaves pinnatifid, the divisions often again pinnate, spinulose, glabrous or slightly floccose below when young; heads scattered, 3-4.5 cm. high and 2-3 cm. wide; involucre bracts with short prickles, glabrous or sparingly hairy but arachnoid-ciliate, no glutinous dorsal ridge present;

inner bracts with elongated, plane, often red or purplish tips; corollas purplish.---Mountains, canyons, parks and valleys. Colorado, Utah, New Mexico and Arizona. Our rather few records scattered in the western half of Colorado at 4500-7500 feet.

12. Cirsium drummondii T. & G., Fl. N. Amer. 2:459. 1843.

C. oreophilum Rydb.; C. coloradense (Rydb.) Cockll.; Carduus drummondii (T. & G.) Cov.---Perennial plants; stems short-caulescent to taller, 10-40 cm. tall or even more, more or less arachnoid-pubescent; leaves oblong to linear-lanceolate, sinuate-pinnatifid, the spines about 1-10 mm. long, glabrate to arachnoid-pubescent above, tomentose to arachnoid or rarely glabrate below; heads 1-8, congested at end of stem or in upper axils, 3-4.5 cm. high and 2-3.5 cm. wide; involucre bracts glabrous but sparingly arachnoid on margins, no glutinous dorsal ridge present; outer bracts gradually attenuate to a spine 2-5 mm. long, rarely less, inner bracts with more or less dilated, twisted, erose and acuminate apex; corollas rose-purple to white.---Mountain slopes and valleys. Saskatchewan to British Columbia, south to Arizona and California. Our records from the western half of Colorado except the extreme southern part, at 7500-9500 feet.

12A. Cirsium drummondii acaulescens (Gray) Macbr., (var.) Contr. Gray Herb. n. s. 53:22. 1918.

C. acaulescens (Gray) K. Schum.; Carduus acaulescens (Gray) Rydb.; Carduus drummondii acaulescens (Gray) Coville---Like the species but strictly acaulescent.---Colorado to Nevada, south to New Mexico. Our records scattered in the western half of Colorado at 7500-10,500 feet.

13. Cirsium undulatum (Nutt.) Spreng., Syst. Veg. 3:374. 1826.

C. tracyi Rydb.; Carduus undulatus Nutt.; Carduus tracyi Rydb.---Plants biennial; stems 30-80 cm. tall, more or less arachnoid-tomentose; leaves ovate-oblong to lanceolate, sinuately-pinnatifid or rarely nearly entire, spines 2-6 mm. long, upper surface more or less floccose-tomentose but loosely so and usually greenish, lower surface tomentose; heads solitary or 2-3 at end of stems and branches, few to the entire stem, 3-4 cm. high and 2-3.5 cm. wide; involucre bracts glabrate or sparingly arachnoid-hairy on margins, a glutinous dorsal ridge present; outer bracts attenuated to a spine 3-5 mm. long or rarely shorter, inner bracts somewhat dilated and usually twisted and erose at apex; corollas rose-purple or less commonly whitish.---Plains and hills. Michigan to British Columbia, south to Texas and Arizona. Our records scattered in the western two-thirds of Colorado at 5000-9500 feet.

14. Cirsium canescens Nutt., Trans. Amer. Phil. Soc. n. s. 7:420. 1841.

C. flodmani (Rydb.) Arthur; Carduus flodmani Rydb.; Carduus undulatus canescens (Nutt.) Porter---Perennial plants with creeping rootstocks apparently sometimes present; stems 20-100 cm. tall, more or less arachnoid-tomentose; leaves oblong to lanceolate, sinuate-pinnatifid to almost entire, spines 1-10 mm. long, sparingly arachnoid-tomentose above, tomentose below; heads 1-2 at ends of stem and branches, usually several to many to a stem, 2-3 cm. long or sometimes longer, about 2-3 cm. wide; involucre bracts glabrate or sparingly arachnoid on margins, with a glutinous dorsal ridge; outer bracts gradually attenuate to a spine 3-10 mm. long, middle bracts with spines usually over 5 mm. long, inner attenuate and only slightly dilated at apex if at all; corollas rose or rose-purple.---Plains, hills and mountain valleys. Saskatchewan, south to Iowa, New Mexico and Utah. Our records from the western half of Colorado at 6500-9000 feet.

15. Cirsium spathulifolium Rydb., Fl. Rocky Mts. 1011. 1917.

Carduus spathulatus Osterh.---Perennial plants, perhaps with horizontal rootstocks on occasion; stems 60-80 cm. tall, rather slender to moderately stout, sparsely tomentose to nearly glabrous; leaves oblong, pinnatifid, the lobes narrowly oblong to ovate with spines 1-5 mm. long, or lower leaves merely toothed, glabrate above, densely white-tomentose below; upper leaves sessile and clasping, lower leaves on winged petioles; heads several on leafy branches, usually about 2-3 cm. high but sometimes longer or shorter; involucre bracts glabrate or somewhat floccose especially on the edges, with an inconspicuous dorsal glutinous ridge, this often brownish-spotted; outer bracts with nearly erect spinulose tips 1-5 mm. long, inner bracts acuminate or long-attenuate but hardly spinulose; corollas purple to white. Closely related and probably not distinct from Carduus griseus Rydb. and Carduus modestus Osterh.---Mountain parks and slopes. Apparently limited to Colorado, our records from the northwestern quarter of the state at 7500-8500 feet.

16. Cirsium plattense (Rydb.) Fernald, Rhodora 13:240. 1911.

Carduus plattensis Rydb.---Biennial plants; stems 30-90 cm. tall, tomentose; leaves lanceolate to lanceolate-oblong, sinuate-dentate to sinuate-pinnatifid with spines 1-8 mm. long, upper surface more or less arachnoid-tomentose, lower tomentose; heads solitary at ends of stem and branches, few, 4-5 cm. long; involucre bracts glabrate or sparingly arachnoid on margins, a glutinous dorsal ridge present; outer bracts abruptly attenuate to a spine 3-5 mm. long, inner bracts usually definitely dilated into a twisted and erose apex; corollas white to ochroleucous.---Plains, hills and valleys. Nebraska and Colorado. Our records from the northeastern quarter of Colorado at 3500-7500 feet.

17. Cirsium ochrocentrum A. Gray, Mem. Amer. Acad. n. s. 4:110. 1849.

Carduus ochrocentrus (Gray) Greene---Biennial plants; stems 30-100 cm. tall, rarely taller, tomentose; leaves lanceolate to oblong-lanceolate, spinulose-dentate to sinuate-pinnatifid, the spines 4-12 mm. long, upper surface arachnoid-tomentose this usually loose in age, lower surface tomentose; heads few, solitary on the stem and branches, 4-6 cm. high and about as wide; involucre bracts glabrate or sparingly arachnoid on margins, with a glutinous dorsal ridge present; outer bracts gradually attenuate to a spine 6-10 mm. long, inner bracts more or less dilated to a twisted and erose apex; corollas purplish to rose.---Plains, hills and valleys. Nebraska to Colorado, south to Texas and Arizona. Our records from the eastern half of Colorado, also in the southwestern corner, at 3500-7500 feet.

18. **Cirsium megacephalum** (Gray) Cockll. in Daniels, Univ. Mo. Stud. Sci. 2(2):254. 1911.
Carduus megacephalus (Gray) A. Nels.---Biennial plants; stems 30-100 cm. tall, rarely taller, arachnoid-tomentose; leaves oblong-lanceolate to lanceolate, dentate to sinuately-pinnatifid, the spines 2-7 mm. long, upper surface tomentose, in age usually loosely so and greenish, lower surface tomentose; heads solitary or 2-3 on short peduncles, 4-7 cm. high and about 3.5-4.5 cm. wide; involucre bracts glabrate or sparingly arachnoid on margins, a dorsal glutinous ridge present; outer bracts gradually acuminate to a spine 3-4 mm. long, rarely to 5 mm., inner bracts more or less dilated to a twisted and erose apex; corollas rose or purple. This and the two preceding species are closely related and may not be distinct.---Plains, hills and mountains. North Dakota to Idaho, south to Texas and Arizona. Our records scattered over the state at 3500-6000 feet.

19. **Cirsium pulchellum** (Greene) Woot. & Standl., Contrib. U. S. Nat. Herb. 19:752. 1915.
Biennial or perennial plants; stems 30-100 cm. tall, sparingly arachnoid especially when young; leaves oblong to lanceolate, sinuately-pinnatifid with spines 1-5 mm. long, upper surface glabrescent to arachnoid, lower tomentose; heads solitary at end of stem and branches, 3-4 cm. high and about 1.2-2.5 cm. wide; involucre bracts glabrate or sparingly arachnoid on margins, no glutinous dorsal ridge present; outer bracts attenuate to a spine 2-5 mm. long, inner with tips elongated but plane and not erose; corollas purplish, pink-purplish to white. ---Mountains and valleys. Colorado, Utah, New Mexico and Arizona. Our records from southwestern Colorado at 6000-7500 feet.

78. Echinops L. GLOBETHISTLE

Perennial coarse, thistlelike, herbaceous plants; stems leafy; leaves alternate, pinnatifid or rarely dentate, lobes or teeth spinulose; involucres 1-flowered, cylindric, outer bracts bristlelike, about 1/2 as long as the inner fimbriate ones; many of these 1-flowered heads crowded into a globose cluster, this with a small common involucre of reflexed bracts, the whole cluster simulating a single head; corollas discoid, white or blue, with cylindrical tube; anthers caudate or sagittate at base; achenes hairy; pappus cup-shaped with toothed edges or somewhat scalelike.

1. **Echinops sphaerocephalus** L., Sp. Pl. 814. 1753.
Stems 100-300 cm. tall, striated, glandular and with villous purplish hairs; leaves 15-40 cm. long, oblanceolate to oblong, upper surface more or less glandular and villous, lower tomentose; cluster of heads 2-5 cm. in diameter. Our plant may be E. exaltus Schrad.---Cultivated and sometimes escaping. Native to Europe. Our one record from central Colorado at about 6000 feet.

79. Carthamus L. SAFFLOWER

Annual herbaceous plants; stems leafy; leaves alternate or somewhat opposite above, spiny-toothed; heads solitary on corymbose branches; involucres campanulate or ovoid, bracts in several series, the outer leaflike and spinulose, the inner narrower, drier and lighter in color, spine-tipped; receptacles chaffy; flowers all discoid, perfect or sometimes the marginal ones pistillate, corollas deeply 5-cleft, orange to orange-yellow; achenes glabrous; pappus of several series of narrow scales.

1. **Carthamus tinctorius** L., Sp. Pl. 830. 1753.
Stems 30-100 cm. tall, glabrous; leaves ovate, clasping, glabrous; heads about 3-4 cm. high, outer leaflike bracts about as long as the head; pappus 3-5 mm. long.---Now being cultivated in Colorado and very likely to escape.

80. Carduus L. BRISTLETHISTLE

Stout annual or biennial herbaceous plants; stems leafy; leaves alternate, pinnately lobed or deeply dentate, the lobes or teeth spinose, strongly decurrent on the stems; heads solitary on the stem and short branches; involucres hemispheric, bracts rather foliaceous, imbricated in several series, lanceolate or linear-lanceolate with strong midrib this excurrent as a spine; receptacles bristly; flowers perfect, all discoid, perfect and fertile, corollas tubular, deeply 5-cleft, purple; anthers sagittate at base; pappus of many scabrous bristles.

1. **Carduus nutans** L., Sp. Pl. 821. 1753.
A weedy plant native to Europe and rather widespread in the eastern part of the United States. It has been reported in northeastern Arizona and may well be in Colorado.

81. Onopordon L. COTTONTHISTLE

Coarse biennial herbaceous plants; stems stout, leafy, usually much branched, usually white-tomentose; leaves alternate, lobed and spinescent, decurrent; heads solitary on the stem and branches; involucres globose, bracts imbricated in many series, tipped with long spines; flowers discoid, pale purple; receptacles flat, honeycombed but not bristly; achenes oblong to obovate, 4-angled, truncate at apex; pappus of many capillary bristles, these barbellate but never truly plumose.

1. **Onopordon acanthium** L., Sp. Pl. 827. 1753.
Stems 90-200 cm. tall; leaves to 30 cm. long, oblong, white-tomentose; heads 2.5-4 cm. high, floccose; pappus brownish.---Waste places, fields and roadsides. Native to Eurasia and

escaping from cultivation at various parts of this continent. Our one record from northcentral Colorado at 7000 feet.

Another species, O. tauricum Willd. has been collected several times in Pueblo County where it seems to be maintaining itself and spreading. The leaves are glandular-puberulent but not at all tomentose.

82. Arctium L. BURDOCK

Large, coarse, mostly biennial, herbaceous plants; stems usually much branched; leaves alternate, broad, petioled; heads rather large, racemose, corymbose or paniculate at the ends of the branches; involucres subglobose to globose, bracts imbricated in many series, tipped with hooked bristles; corollas all discoid, rose or rose-purple rarely whitish; receptacles flat, densely bristly; achenes oblong, more or less compressed and 3-angled, truncate at apex; pappus of numerous short rigid scalelike bristles.

1. Arctium minus (Hill) Bernh., Syst. Verz. Erfurt. 154. 1800.
Stems 50-200 cm. tall, stout, much branched, puberulent; leaves large, oblong-ovate, cordate at least on lower ones, repand, gray-tomentulose below; heads in a leafy panicle, often with racemose branches; involucres 15-25 mm. wide, bracts shorter than the corollas, glabrous or somewhat arachnoid. A related species A. lappa L. with heads 3-4.5 cm. wide and bracts equalling or exceeding the corollas is to be expected in this area.---Waste places usually around dwellings. A weed naturalized from Europe and widely distributed especially in the eastern part of North America. Our records from northcentral and central Colorado at 4500-7000 feet.

83. Centaurea L. STAR-THISTLE; KNAPWEED

Annual or perennial herbaceous plants; stems leafy; leaves alternate, entire to pinnately lobed; heads 1 to several at end of branches; involucres ovoid or globose, bracts imbricated, the tips scarious and entire to pectinately-fringed or spinulose; receptacles flat, densely bristly; corollas discoid, all alike and fertile, or marginal ones neutral and corollas enlarged thus appearing falsely radiate; achenes compressed and obtusely 4-angled, obliquely attached and crowned by a disk or an elevated margin; pappus of several series of bristles, of scales or wanting entirely. In addition to the two species given below, C. cyanus L., the Bachelor's Button cultivated in gardens may escape in Colorado.

1. Involucre bracts tipped with yellow spines 1-2 cm. long; corollas yellow; leaves decurrent
-1. C. solstitialis
1. Involucre bracts entire or denticulate, not at all spinose; corollas purplish; leaves not decurrent
-2. C. picris

1. Centaurea solstitialis L., Sp. Pl. 917. 1753.
Annual plants; stems 30-60 cm. tall, canescent-tomentulose; leaves linear, decurrent, lower pinnatifid, upper entire, canescent-tomentulose; heads 1 to few at the ends of branches; involucres about 1-2 cm. high and wide, bracts tipped with a stiff, yellow, spreading spine 1-2 cm. long, this with 1-3 pair of smaller spines at base; corollas yellow.---Waste places and fields. Native to Europe and occasionally found in much of the United States. Our record from northcentral Colorado at 5000 feet.
2. Centaurea picris Pall., "Tabl. Taur. 58," Willd., Sp. Pl. 3:2302. 1804.
Perennial plants with horizontal creeping rootstocks; stems 30-50 cm. tall, striate, finely puberulent to tomentulose; leaves linear to linear-lanceolate, lower pinnately lobed, middle often dentate, upper entire; heads 1 to few at the ends of the branches; involucres about 1 cm. high and wide, bracts thin and whitish, entire or merely denticulate; outer bracts obtuse, glabrous or nearly so, inner bracts pointed and pilose; corollas rose-purple. Our plant may be C. repens L. A related species, C. americana Nutt., with bracts pectinately-fringed and heads 1-2 cm. wide may be expected in southern Colorado.---Waste places, pastures, fields and roadsides. Native to Eurasia. Michigan to Washington, south to Missouri and California. Our records scattered over the state at 4500-7500 feet.

84. Saussurea DC.

Perennial herbaceous plants; stems leafy; leaves alternate, entire or sinuate-dentate; heads racemose-crowded; involucres turbinate to campanulate, bracts imbricated in several series; receptacles flat, chaffy with long narrow paleae; rays absent; flowers perfect and fertile, corollas with slender tube, inflated throat and 5-toothed margin, purple or violet; anthers united, caudate at base and with setiform, hairy appendages at apex; pappus double, outer of short rigid bristles, inner of stout plumose bristles united at base.

1. Saussurea alpina DC., Ann. Mus. Paris 16:198. 1810.
S. densa (Hook.) Rydb.---Plants loosely arachnoid to glabrate; stems 8-20 cm. tall; leaves 5-10 cm. long, broadly to narrowly lanceolate, upper sessile, lower petioled; involucres 8-10 mm. high, bracts lanceolate to ovate, acute to acuminate, lighter in color near base; outer pappus bristles about 1-2 mm. long, inner plumose bristles about 5-8 times as long as the outer. Our plants may be S. monticola Richards.---High mountains. Canadian Rockies to the Arctic Coast. Our few records from central Colorado, from the area of Hoosier Pass, at 12,500-14,000 feet.

85. Hypochoeris L. CATSEAR

Perennial, scapose, herbaceous plants; leaves mostly in a basal rosette; heads long-peduncled on corymbose branches; involucres campanulate or oblong-cylindrical, the bracts imbricated rather gradually, the outer short; receptacles nearly flat, with thin and narrow chaff; corollas all ligulate, yellow; achenes linear to fusiform, 10-ribbed, tapering at apex into a slender beak longer than the body; pappus of plumose bristles or some of outer ones simple and shorter.

1. Hypochoeris radicata L., Sp. Pl. 811. 1753.
Plants 30-80 cm. tall, scapes simple or branched, bracted; leaves 5-10 cm. long, lyrate-pinnatifid to deeply dentate, hirsute; involucres 15-25 mm. high, glabrous or sparsely hairy.---Waste places, fields and roadsides. Adventive from Eurasia in various parts of this continent. Our one record from northcentral Colorado at 5500 feet.

86. Tragopogon L. GOATSBEARD; SALSIFY

Biennial or perennial herbaceous plants from fleshy taproots; leaves alternate and clasping at base, elongated, linear, rather grasslike, entire, strongly nerved; heads large, solitary on fistulose peduncles; involucres cylindric or narrowly campanulate, with 8-13 bracts in a single series united at base; receptacles flat or nearly so, naked; flowers all ligulate, yellow to rose-purple; achenes linear-fusiform, muricate, longitudinally ribbed, long-beaked (rarely outer flowers with beakless achenes); pappus of plumose bristles connate at base. Intermediate forms are not uncommon in Colorado

1. Rays purple or rose; involucre bracts longer than the rays   -1. T. porrifolius
1. Rays yellow; bracts longer or shorter than the rays
  2. Involucre bracts 8-9, equalling or shorter than the chrome-yellow rays   -2. T. pratensis
  2. Involucre bracts 10-13, definitely longer than the lemon-yellow rays   -3. T. dubius

1. Tragopogon porrifolius L., Sp. Pl. 789. 1753.
Plants commonly 50-100 cm. tall; leaves straight; peduncles strongly fistulose below the heads; involucre bracts 8-10, commonly 3-5 cm. long, longer than the rays; rays purple or rose.---Waste places, fields and roadsides, often near dwellings as an escape from cultivation. Our records well scattered over Colorado at 4000-7500 feet.
2. Tragopogon pratensis L., Sp. Pl. 789. 1753.
Plants commonly 30-80 cm. tall; leaves often twisted at ends; peduncles thickened but hardly strongly fistulose below the heads; involucre bracts 8-9, commonly 2-3 cm. long, equalling or shorter than the rays; rays chrome-yellow (dandelion-yellow).---Fields and waste places. Adventive from Europe and now widely distributed in Canada and the United States. Our records from over Colorado, except the extreme eastern part, at 5000-8000 feet.
3. Tragopogon dubius Scop., Fl. Carn. ed. 2. 2:95. 1772.
Plants 30-90 cm. tall; leaves not especially twisted at ends, more crowded than in preceding species; peduncles more or less enlarged and fistulose below the heads; involucre bracts 10-13, commonly 3-6 cm. long, longer than the rays; rays lemon-yellow. Our plant has been called ssp. major (Jacquin) Vollman.---Waste places and fields. Adventive from Europe. Colorado to Idaho, south to New Mexico and Arizona; also in various states east of our area. Our records scattered in the western two-thirds of Colorado at 4500-10,500 feet.

87. Stephanomeria Nutt. WIRELETTUCE

Annual or perennial herbaceous plants, often rushlike in aspect; stems branching, often rigid; leaves alternate, linear to oblong, entire to pinnatifid, upper reduced and mostly bractlike; heads usually panicled; involucres cylindric, oblong or somewhat campanulate, the inner bracts few to several, equal in a single series, but outer ones shorter, or more rarely regularly imbricated; flowers 3-20, all ligulate, rose or flesh-colored; receptacles flat, naked; achenes columnar, 5-angled or ribbed, truncate at apex, no beak present; pappus of 12-20 bristles these often paleaceous toward base, plumose at least above, sometimes connate into groups.

1. Plants annual
  2. Pappus plumose to base or very near base but never paleaceous; heads subsessile on spiciform branches   -1. S. virgata
  2. Pappus plumose only to about 1.5-2 mm. from base, often paleaceous below; heads paniculate on fairly long peduncles   -2. S. exigua
1. Plants perennial
  3. Pappus brownish-tinged, bristles plumose to about 1 mm. from base, then scabrous   -3. S. pauciflora
  3. Pappus bright white, plumose nearly or quite to the very base   -4. S. tenuifolia

1. Stephanomeria virgata Benth., Bot. Voy. Sulph. 32. 1844.
Ptiloria virgata (Benth.) Greene---Reported from Wyoming to Nevada, south to New Mexico and California. Probably present in western Colorado but not found by the writer.
2. Stephanomeria exigua Nutt., Trans. Amer. Phil. Soc. ser. 2.7:428. 1841.
Ptiloria exigua (Nutt.) Greene---Annual plants; stems 15-50 cm. tall, diffusely branched with ascending or spreading branches, glabrous or glandular-puberulent above; lower leaves narrowly oblong and remotely lobed, upper reduced; heads scattered and paniculate on the ends of the branches; involucres 6-9 mm. long, principal bracts about 5; flowers 5-8; achenes

5-angled, rugose-tuberculate; pappus whitish to brown-tinged, plumose to about 1.5-2 mm. from the base, often paleaceous-dilated below.---Dry hills, plains and valleys. Wyoming, south to New Mexico and California. Our rather few records from extreme western Colorado at 4500-7000 feet.

3. Stephanomeria pauciflora (Torr.) A. Nels., Man. Rocky Mt. 588. 1909.
Ptiloria pauciflora (Torr.) Raf.---Perennial plants; stems 20-50 cm. tall, herbaceous from a woody root, with stiff, ascending branches; leaves linear or narrowly lanceolate, more or less runcinate-pinnatifid, light green in color; heads solitary or few on the branches; involucres 8-10 mm. high, glabrous; achenes striate and often more or less rugose; pappus brownish-tinged, plumose to about 1 mm. from base, then merely scabrous, not at all paleaceous below.---Dry plains, hills and valleys. Nebraska to California, south to Texas and Arizona. Our records scattered in Colorado, except in the northeastern and northwestern parts, at 4500-8000 feet.

4. Stephanomeria tenuifolia (Torr.) Hall, Calif. Univ. Pub. Bot. 3:256. 1907.
Ptiloria tenuifolia (Torr.) Raf.; P. ramosa Rydb.---Perennial plants; stems 10-50 cm. tall, herbaceous, much branched with ascending or spreading branches, glabrous or puberulent below; leaves narrowly linear to oblanceolate, entire to runcinate-pinnatifid, pale green; heads usually solitary and terminal on the branches; involucres 7-10 mm. high, glabrous; achenes striate and often more or less rugose; pappus bright white, plumose to the base and not paleaceous below.---Dry plains, hills and valleys. Montana to Washington, south to Arizona and California. Our records well scattered over Colorado at 3500-8000 feet.

88. Ptilocalais Greene

Perennial glabrous herbaceous plants from fusiform roots; leaves mostly from near base, entire to pinnately lobed; heads solitary at the ends of long branches or peduncles; involucres campanulate to cylindric, of 8-12 linear bracts, these in 2-3 series, outer ones short; receptacles flat, with narrow scarious chaff; flowers all ligulate, yellow; achenes columnar-linear, 8- to 10-ribbed, truncate and beakless at apex; pappus of 15-20 thin oblong to linear-lanceolate, white to sordid tan scales, each terminating in a soft plumose bristle.

1. Ptilocalais nutans (Geyer) Greene, Bull. Calif. Acad. II. 2:54. 1886.
P. tenuifolia Osterh.---Plants 10-50 cm. tall; leaves entire or with short to narrow and long lobes; involucres 10-20 mm. high, bracts acuminate especially the longer; flowers 8-20. May belong to the genus Microseris.---Hillsides, mountains and valleys. Montana to British Columbia, south to Colorado and California. Our records from the western half of Colorado, except the extreme southern part, at 7000-10,000 feet.

89. Lapsana L. NIPPLEWORT

Annual herbaceous plants; leaves alternate, dentate to pinnatifid; heads panicled on slender peduncles; involucres cylindrical, principal bracts in 1 series, nearly equal, a few shorter ones on the outside; receptacles flat and naked; flowers all ligulate, yellow; achenes obovate-oblong, 20- to 30-nerved, narrowed below and at least somewhat curved; pappus none.

1. Lapsana communis L., Sp. Pl. 811. 1753.
Stems 25-80 cm. tall, paniculately branched, more or less hispid-pubescent at least below; lower leaves ovate, repand-dentate and often with 2-6 lobes on the petiole, pubescent to glabrate; upper leaves narrow, mostly entire; involucres 4-6 mm. high, about 8 long bracts present.---Waste places and roadsides. Native to Europe and introduced in the United States, particularly in the eastern part. Our few records from near Denver, Colorado, at 5000-5500 feet.

90. Krigia Schreb. DWARF DANDELION

Slender perennial herbaceous plants; leaves chiefly basal, oblanceolate to spatulate, entire or sinuate-dentate; stem leaves small, alternate, sessile and somewhat clasping at base; heads 1 to few; involucres campanulate with 9-18 thin bracts about equal in length and in 1-2 series; flowers all ligulate, yellow to orange; receptacles flat, naked; achenes oblong to turbinate, truncate and beakless at apex, many-ribbed; pappus double, of 10-15 small oblong scales and several to many inner simple capillary, much longer bristles.

1. Krigia biflora (Walt.) Blake, Rhodora 17:135. 1915.
K. virginica (L.) A. Nels.; Adopogon virginicum (L.) Kuntze---Stems 25-50 cm. tall; basal leaves wing-petioled; cauline leaves about 1-3; involucre 8-10 mm. high, glabrous.---Plains and hills, often in moist partly shaded ground. New York to Minnesota, south to Georgia, Texas and Arizona. Our few records from the southern half of Colorado at 7000-7500 feet.

91. Cichorium L. CHICORY

Perennial herbaceous plants from a deep-seated taproot; stems branching; leaves alternate but mostly basal, those of the stem and branches reduced and bractlike; basal leaves pinnatifid, cauline usually toothed and clasping; heads sessile along the branches or on the tips of thick branchlets; involucres cylindrical, bracts in 2 series, the shorter outer ones somewhat spreading, about 5 in number, inner 8-10 bracts partly enfolding the outer achenes at base; receptacles flat, naked or somewhat fimbrillate; flowers all ligulate, blue or rarely pink to white; achenes obovoid, somewhat ribbed, truncate and beakless at apex; pappus a short-toothed crown or of short blunt scales.

1. **Cichorium intybus** L., Sp. Pl. 813. 1753.
Stems 30-90 cm. tall, more or less hispid; basal leaves 7-16 cm. long, runcinate-pinnatifid, spreading; involucres about 6-10 mm. high, bracts more or less glandular-ciliate at least the outer.---Roadsides, fields and waste places. A common weed in Canada and the United States. Native to Europe. Our rather few records widely scattered in Colorado at 4000-7000 feet.

92. Lactuca L. WILD LETTUCE

Annual or perennial caulescent herbaceous plants with milky juice; leaves alternate, variable, from linear and entire to pinnatifid; heads paniculate; involucres narrowly cylindrical or conical in fruit, bracts imbricated in 3 or more series; receptacles flat or nearly so, naked; flowers several to many, all ligulate, yellow or blue; achenes strongly flattened, abruptly to gradually beaked (beak very short in L. spicata), this beak expanded at apex to a disk; pappus of numerous capillary bristles which fall separately. Mature achenes are almost always necessary in identifying the species.

1. Pappus brown; achenes beakless or very nearly so   -1. L. spicata
1. Pappus white; achenes with long to short beaks
  2. Plants perennial with spreading rootstocks (usually in dense patches); achenes lanceolate or oblong-lanceolate, the beak less than 1/2 as long as the body; flowers blue or violet   -2. L. pulchella
  2. Plants annual or biennial, rarely perennial but never spreading by rootstocks and growing in dense patches; achenes oval or oval-oblong, beak 1/2 as long as the body or longer; flowers yellow but often turning blue-purple in age
    3. Leaf blades spinulose-toothed and often spinulose on midrib below
      4. Heads 12- to 20-flowered; achenes 1- to 3-nerved on each face   -3. L. ludoviciana
      4. Heads 6- to 12-flowered; achenes 5 or more nerved on each face
        5. Leaves pinnatifid   -4. L. scariola
        5. Leaves not at all lobed   -4A. L. scariola integrata
    3. Leaf blades not spinulose on margins or midrib below
      6. Achene body about 5-6 mm. long, the beak about 1/2 as long; leaves mostly narrowly linear and entire, the lower often broader and pinnatifid   -5. L. graminifolia
      6. Achene body about 3-4 mm. long, the beak about as long; leaves spatulate to obovate, the lower pinnatifid (upper may not be lobed)
        7. Involucres 8-12 mm. long, bracts not regularly imbricated (outer short, then a break to longer inner ones on most heads)   -6. L. canadensis
        7. Involucres 12-20 mm. long, bracts rather regularly imbricated   -3. L. ludoviciana

1. **Lactuca spicata** (Lam.) Hitchc., Trans. Acad. St. Louis 5:506. 1891.
Annual or biennial plants; stems 30-200 cm. tall, glabrous; leaves oblong to obovate, pinnatifid to lobed the teeth and lobes not spinulose; involucres about 10-12 mm. high, bracts imbricated; flowers 6-15, blue, yellowish or white; achenes beakless, about 4 mm. long, about 5-nerved on each face, not rugose or setose; pappus light brown or tawny.---Often moist ground as along ditches. Newfoundland to Manitoba, south to North Carolina, Colorado and California. Our few records from northcentral and southwestern Colorado at 5500-7000 feet.

2. **Lactuca pulchella** (Pursh) DC., Prodr. 7:134. 1838.
L. tatarica ssp. pulchella (Pursh) Stebbens---Perennial plants with horizontal spreading rootstocks; stems 30-100 cm. tall, glabrous and often somewhat glaucous; leaves linear-lanceolate, lanceolate to oblong, entire, dentate or even some often pinnatifid but not at all spinulose, glabrous or glaucous; involucres 14-20 mm. high, bracts gradually imbricated; flowers blue or blue-purple; achenes blackish, about 4 mm. long, lanceolate or oblong-lanceolate, several-ribbed on each face, tapering to a short beak less than 1/2 as long as the body; pappus white. May not be distinct from the Eurasian species, L. tatarica (L.) C. A. Mey.---Meadows, plains, roadsides especially in moist ground. Saskatchewan to British Columbia, south to Missouri, New Mexico and California. Our records scattered over Colorado, none as yet from the southeastern corner, at 3500-8000 feet.

3. **Lactuca ludoviciana** (Nutt.) DC., Prodr. 7:141. 1838.
Biennial plants; stems 50-150 cm. tall, glabrous; leaves obovate, oblong to oblong-lanceolate, dentate to pinnatifid, the teeth or lobes more or less spinulose, glabrous; involucres 15-20 mm. high, sometimes as short as 12 mm., bracts regularly imbricated; flowers 12-20, yellow but turning purple-blue in age; achenes dark brown or black, body oval or oval-oblong, 1- to 3-nerved on each face, transversely rugose, beak nearly as long as the body; pappus white.---Banks often in moist ground. Minnesota to Montana, south to Missouri, Texas and Arizona. Our records from northcentral and northeastern Colorado at 3500-8000 feet.

4. **Lactuca scariola** L., Sp. Pl. ed. 2. 1119. 1763.
Annual or possibly biennial plants; stems 30-100 cm. tall, glabrous or bristly hispid below; leaves oblong or oblong-lanceolate, pinnatifid with lobes spinulose-toothed and usually spinulose on the midrib below; involucres about 10 mm. high, bracts regularly imbricated; flowers 6-8, yellow or turning blue-purple in age; achenes gray or brown, about 3 mm. long, oval or oval-oblong, tapering to a beak longer than the body, body setose near apex and rugose only on angles below or not at all, 5-nerved on each face; pappus white. This species is sometimes listed as L. serriola L.---Waste places, fields and roadsides. Native to Europe and now widespread in southern Canada and the United States. Our rather few records scattered, mostly in northcentral Colorado, at 4500-6000 feet.

4A. **Lactuca scariola integrata** Gren. & Godr., (var.) Fl. France 2:320. 1850.
L. integrata (Gren. & Godr.) A. Nels.---Leaves unlobed.---Our few records from northcentral Colorado at 4500-6000 feet.

5. **Lactuca graminifolia** Michx., Fl. Bor. Amer. 2:85. 1803.
South Carolina to Colorado, south to Florida and Arizona. Reported for southern Colorado but not seen by the writer.
6. **Lactuca canadensis** L., Sp. Pl. 796. 1753.
Biennial or annual plants; stems 100-300 cm. tall, glabrous and often glaucescent; leaves spatulate to oblong, sinuately-pinnatifid but not spinulose-toothed, midrib below only rarely with a few sparse bristles; involucres 8-12 mm. long, bracts not regularly imbricated, outer short and inner long; flowers 12-20, yellow; achenes blackish, body 3-4 mm. long, broadly oval, 1- to 3-nerved and transversely rugose on faces, contracted to a beak nearly as long or as long; pappus white.---Waste ground, often in moist places. Nova Scotia to Alberta, south to Florida and New Mexico. Our rather few records from northcentral, central and western Colorado at 5000-6500 feet.

93. Sonchus L. SOWTHISTLE

Annual or perennial, mostly glaucous plants; stems leafy; leaves alternate, subentire to pinnatifid, mostly clasping the stem; heads cymose or umbellate; involucres ovoid or campanulate, bracts regularly imbricated, becoming corky-thickened at base; receptacles flat or nearly so, naked; flowers all ligulate, yellow; achenes strongly flattened, ribbed on each face, beakless; pappus of soft white numerous simple capillary bristles these falling separately.

1. Perennial plants with spreading rootstocks; involucres 15-25 mm. high, often glandular setose
   2. Involucres more or less glandular-setose   -1. S. arvensis
   2. Involucres glabrous   -1A. S. arvensis glabrescens
1. Annual plants; involucres 9-12 mm. high, glabrous
   3. Achenes longitudinally ribbed and also transversely rugose; auricles of leaves acute   -2. S. oleraceus
   3. Achenes longitudinally ribbed but not at all transversely rugose; auricles of leaves rounded
      -3. S. asper

1. **Sonchus arvensis** L., Sp. Pl. 793. 1753.
Perennial plants with horizontal, spreading rootstocks; stems 50-100 cm. tall, less leafy above; lower leaves pinnatifid, the lobes spinulose-dentate, upper pinnatifid or unlobed, glabrous; heads in corymbose panicles or cymes; involucres 15-25 mm. high, more or less glandular-setose; achenes about 2.5-3 mm. long, only moderately compressed, with longitudinal ribs, which are rugose.---Fields, roadsides and waste places. Introduced weed from Europe. Newfoundland to British Columbia, south to New Jersey and Utah. Our records thinly scattered over Colorado in cultivated areas at 5000-6500 feet.
1A. **Sonchus arvensis glabrescens** Guenth.; Grab. & Wimm., (var.) Enum. Stirp. Phan. Siles. pt. 2. 2:220. 1829.
Involucres without any glandular-setose hairs. Intergrades commonly with the species.--- Our one record from western Colorado at 5000 feet.
2. **Sonchus oleraceus** L., Sp. Pl. 794. 1753.
Annual plants; stems 50-150 cm. tall or more; lower leaves pinnatifid with spinulose-dentate lobes, upper leaves usually pinnatifid with acute auricles; heads in corymbose cymes; involucres 10-12 mm. high, glabrous; achenes about 2-3 mm. long, definitely compressed, longitudinally ribbed and also transversely rugose.---Fields, roadsides and waste places. Native to Europe and now widespread in most parts of North America. Our records from northcentral, central, southcentral and southwestern Colorado at 4500-6000 feet.
3. **Sonchus asper** (L.) Hill, Herbarium Brit. 1:47. 1769.
Annual plants; stems 30-150 cm. tall; leaves unlobed to pinnatifid especially on lower, spinulose-dentate, clasping by rounded auricles; heads in corymbose cymes; involucres about 9-12 mm. high, glabrous; achenes about 2-3 mm. long, definitely compressed, longitudinally ribbed but not transversely rugose.---Fields, roadsides and waste places. Native to Europe and now widespread in North America. Our records scattered over Colorado, none as yet from the extreme eastern part, at 5000-7500 feet.

94. Crepis L. HAWKSBEARD

Annual, biennial or perennial herbaceous plants, when perennial the caudex woody and leafy at crown; leaves mostly radical but some usually cauline and alternate, entire, toothed or pinnatifid; heads 1 to many, on branched or scapiform stems; involucres cylindrical to campanulate or turbinate-campanulate, calyculate; outer bracts shorter, few to many, inner bracts in 1-2 series; receptacles flat or nearly so, naked or short-ciliate but never chaffy; flowers all ligulate, yellow; anthers united; achenes terete or nearly so, 10- to 20-ribbed, beakless or sometimes attenuate into a short beaklike structure; pappus of capillary, smooth or barbellate (never plumose), white or yellowish-white bristles usually longer than the involucres.

1. Dwarf plants, 2-7 cm. tall, of alpine areas; plants glabrous or nearly so   -1. C. nana
1. Plants over 8 cm. tall, rarely below 15 cm., seldom as high as alpine; plants rarely completely glabrous
   2. Stems and leaves glabrous or glaucous (or sometimes hispidulous but never tomentose); cauline leaves usually 1 and reduced or none; involucres turbinate-campanulate; plants of moist situations
      3. Involucres more or less glandular-pubescent, at least toward the base
         4. Basal leaves narrowly obovate to oblanceolate, seldom over 3.5 cm. wide and about 4-8 times as long as wide   -2A. C. runcinata runcinata

  4. Basal leaves obovate, usually over 3.5 cm. wide and about 2-4 times as long as wide
                             -2B. C. runcinata hispidulosa
 3. Involucres not at all glandular   -2C. C. runcinata glauca
2. Stems and leaves more or less tomentose (often hispidulous as well); 1-3 cauline leaves usually well developed; involucres narrowly to broadly cylindric-campanulate; plants in dry situations
 5. Involucre bracts rather densely beset with blackish, straight, glandless setae and basal part of stem with stiff, long yellowish setae   -3. C. modocensis
 5. Involucre bracts either without setae or with glandular setae, or if a few glandless setae are present then the basal part of the stem is without setae
  6. Inner involucre bracts glabrous, 5-7 present (8 in subspecies pleuriflora with glabrous bracts)
   7. Involucres with 5-7 inner bracts; flowers 5-10   -4A. C. acuminata acuminata
   7. Involucres with 8 inner bracts; flowers 9-12   -4B. C. acuminata pleuriflora
  6. Inner bracts more or less tomentose (may be setose or glandular as well), 5-13 present
   8. Involucres thick-cylindrical, 5-10 mm. wide at the middle just before maturity, more or less glandular-pubescent or glandular-setose at maturity; plants seldom over 35 cm. tall
    9. Involucre bracts not setose; largest heads with 10-13 inner bracts and 18-30 flowered
                      -5A. C. occidentalis occidentalis
    9. Involucre bracts bearing glandular setae; largest heads with 8 inner bracts, 12-14 flowered
                      -5B. C. occidentalis costata
   8. Involucres narrow-cylindric, 3-5 mm. wide at middle just before maturity, no glands present (setae if present never glandular); plants often over 35 cm. tall
    10. Achenes dark green to light green; lobes of leaves lanceolate to linear, usually entire; plants of northcentral Colorado   -6. C. atribarba
    10. Achenes yellowish, buff or brown; lobes of leaves lanceolate, usually toothed; plants of westcentral and southwestern Colorado   -7. C. intermedia

1. Crepis nana Richards, Bot. App. of Franklin 1st Journ. ed. 1. 746. 1823.
 Youngia nana (Rich.) Rydb.---Perennial plants with taproots or creeping rootstocks, acaulescent or subacaulescent, 2-7 cm. tall; leaves chiefly basal, about as long as the scape, obovate to spatulate, sometimes narrower, entire, repand-dentate or lyrate, glabrous or nearly so; heads several and congested; involucres 10-13 mm. long, cylindrical, bracts ciliate near apex and spongey-thickened near base; outer bracts 5-8, unequal, inner bracts about 10; achenes 4-6 mm. long, golden-brown. Our plant has been called ssp. typica Babcock.---Arctic and alpine. Asia; Arctic North America, south to Colorado and California. Our few records from central and westcentral Colorado at 12,000-13,000 feet.
2. Crepis runcinata T. & G., Fl. N. Amer. 2:487. 1843.
 Perennial plants; stems usually about 20-50 cm. tall, 1-3, glabrous to glaucous sometimes hispidulous especially above; leaves obovate to linear, pinnatifid, toothed or entire, glabrous to glaucous, sometimes hispidulous; heads 1-30 in corymbose-cymes; involucres 8-21 mm. high, turbinate-campanulate, slightly tomentulose to glandular-hispid or glandular-pubescent; outer bracts 5-12, unequal, inner bracts 10-16; flowers 20-50; achenes 3.5-7.5 mm. long, 10- to 13-ribbed, light to very dark brown. Three subspecies are present in Colorado.
2A. Crepis runcinata runcinata
 C. runcinata ssp. typica Babc. & Stebbins; C. denticulata Rydb.; C. glaucella Rydb.; C. alpicola (Rydb.) A. Nels.; C. perplexans Rydb.; C. tomentulosa Rydb.---Basal leaves narrowly obovate to lanceolate, about 5-35 mm. wide and about 5-8 times as long; involucres definitely glandular-pubescent.---Moist meadows and valleys. Minnesota to Manitoba, south to New Mexico and Utah. Our records scattered in Colorado, few from the extreme eastern or western parts, at 4500-10,000 feet.
2B. Crepis runcinata hispidulosa (Howell) Babc. & Stebbens, (ssp.) Carnegie Inst. Wash. Publ. 504:96. 1938.
 Stems glabrous to glandular-hispid; leaves obovate, mostly 3-8 cm. wide and 2-4 times as long; involucres glandular.---Moist banks and meadows. Montana to Washington, south to Colorado and California. Our records from northcentral and northwestern Colorado at about 7000-8500 feet.
2C. Crepis runcinata glauca (Nutt.) Babc. & Stebbens, (ssp.) Carnegie Inst. Wash. Publ. No. 504:98. 1938.
 C. glauca (Nutt.) T. & G.---Leaves vary from narrowly obovate to oblanceolate; involucres glabrous to somewhat tomentose, not at all glandular-hairy.---Moist plains and valleys. Saskatchewan to Idaho, south to New Mexico and Arizona. Our records thinly scattered over the western half of Colorado at 4500-8000 feet.
3. Crepis modocensis Greene, Erythea 3:48. 1895.
 C. scopulorum Coville---Perennial plants; stems 10-45 cm. tall, glabrate to tomentose, with scattered stiff yellowish setae at least below; leaves deeply pinnatifid the segments sometimes again pinnatifid, glabrate to tomentose and usually setose along the midrib and petiole; heads 1-10; involucres 11-16 mm. high and 5-10 mm. wide just before maturity, cylindrical-campanulate, usually tomentose and setose with blackish, straight or nearly straight hairs; outer bracts 8-18, longest about 1/3 to 1/2 as long as the inner ones, inner bracts 8-18, becoming somewhat carinate in fruit; flowers 10-60; achenes 7-12 mm. long, greenish-black to deep reddish-brown.---Rather dry places in the mountains. Montana to British Columbia, south to Colorado and California. Our records scattered in western and northern Colorado at 5000-9000 feet.
4. Crepis acuminata Nutt., Trans. Am. Phil. Soc. n. s. 7:437. 1841.
 Perennial plants; stems 20-65 cm. tall, 1-3, from taproots, tomentose at least near base; leaves pinnately lobed into rather narrow segments, these entire or dentate, densely to rather sparsely tomentose; heads 15-100 in compound corymbose cymes; involucres 9-15 mm. high and 2.5-4 mm. wide just before maturity; outer bracts 5-8, glabrous or rarely tomentulose; flowers 5-12; achenes 5.5-9 mm. long, about 12-ribbed, pale yellow to brownish. Two subspecies are in Colorado.

4A. Crepis acuminata acuminata
   C. acuminata ssp. typica Babc. & Stebbens; C. angustata Rydb. in part---Involucre with 5-7 inner bracts; flowers 5-10.---Dry hills and mountains. Montana to Washington, south to Colorado and Arizona. Our records from the western half of Colorado, except the southcentral part, at 6500-8500 feet.
4B. Crepis acuminata pleuriflora Babc. & Stebbens, (ssp.) Carnegie Inst. Wash. Publ. No. 504:178. 1938.
   Involucres with 8 inner bracts; flowers 9-12.---Colorado, Utah and New Mexico. With the typical form in Colorado but less abundant. The entity seems of doubtful value in our plants.
5. Crepis occidentalis Nutt., Journ. Acad. Phila. 7:29. 1834.
   Perennial plants; stems 8-40 cm. tall, more or less gray-tomentose usually glandular-hairy above; leaves pinnately lobed to parted, segments lanceolate to linear, toothed, more or less gray-tomentose and usually the upper leaves glandular-hairy; heads 1-20 in corymbose cymes; involucres 11-19 mm. high and 5-10 mm. wide just before maturity, cylindrical-campanulate; outer bracts 6-8, longest about 1/3 to 1/2 as long as the inner, inner bracts 8-13, gray-tomentose, slightly to strongly glandular-pubescent or glandular-setose, the latter often blackish; flowers 12-30; achenes 6-10 mm. long, light to dark brown. Two subspecies are in Colorado.
5A. Crepis occidentalis occidentalis
   C. occidentalis ssp. typica Babc. & Stebbens---Upper cauline leaves more or less glandular but not setose; largest heads with 10-13 inner bracts; involucres slightly to strongly glandular-hairy but not setose; flowers 18-30.---Dry plains, hills and mountains. Wyoming to Oregon, south to New Mexico and California. Our records from the western half of Colorado at 4500-8500 feet.
5B. Crepis occidentalis costata (Gray) Babc. & Stebbens, (ssp.) Carnegie Inst. Wash. Publ. No. 504:124. 1938.
   Upper cauline leaves usually bearing glandular setae; largest heads with 8 inner bracts; involucres with conspicuous glandular setae; flowers 12-14.---Dry plains and mountains. Saskatchewan to British Columbia, south to Colorado and California. Our few records scattered over the western half of Colorado at 5500-8500 feet.
6. Crepis atribarba Heller, Bull. Torr. Club 26:314. 1899.
   C. gracilis (Eaton) Rydb.; C. exilis Osterh.; C. angustata Rydb. in part---Perennial plants; stems 15-70 cm. tall, glabrous to tomentose; leaves pinnately lobed or parted into lanceolate to linear often falcate entire to toothed segments, glabrate to tomentose; heads 3-30 in corymbose cymes; involucres 8-14 mm. high and 3-5 mm. wide just before maturity, cylindrical-campanulate, canescent-tomentose or rarely glabrous, sometimes with black glandless setae; outer bracts 5-10, about 1/4 to 1/3 as long as inner, inner bracts 5-13; flowers 6-35; achenes 3-10 mm. long, more or less attenuate at apex, ribs prominent, dark or light green or rarely brown.---Hills and mountains in rather dry ground. Montana to British Columbia, south to Colorado and Nevada. Our records from northcentral and northwestern Colorado at 5000-8000 feet.
7. Crepis intermedia A. Gray, Syn. Fl. 1(2):432. 1884.
   Perennial plants; stems 30-70 cm. tall, densely to sparsely canescent-tomentose; leaves pinnatifid with lanceolate, entire or dentate segments, densely to sparsely canescent-tomentose; heads 10-60 in corymbose-cymes; involucres 10-16 mm. high and about 4-5 mm. wide just before maturity, cylindrical-campanulate, canescent-tomentose this sometimes sparse, occasionally with a few black or green setae near apex; outer bracts 6-8, longest 1/5-1/3 as long as the inner, inner bracts 7-8; flowers 7-12; achenes 5.5-9 mm. long, attenuate to the apex, buff or brown. This species is very difficult to distinguish from the preceding in this state.---Hills and mountains, usually in dry ground. Alberta to Washington, south to New Mexico and California. Our few records from the western half of Colorado at 6000-10,000 feet.

95. Hieracium L. HAWKWEED

Perennial caulescent to acaulescent herbaceous plants; leaves basal or alternate, entire to dentate; heads solitary or in panicles or corymbs; involucres cylindrical to campanulate, of 1-2 series of longer bracts, with shorter ones at base, or sometimes all imbricated; receptacles flat, naked or short-fimbriate; flowers all ligulate, yellow or occasionally white or orange; achenes fusiform or columnar, not flattened, 10- to 15-ribbed, not at all beaked; pappus of simple, capillary, brownish to sordid-white bristles.

1. Achenes tapering upward; involucres 12-14 mm. high   -1. H. fendleri
1. Achenes cylindrical, little if any tapering; involucres 7-10 mm. high
   2. Rays white to ochroleucous; involucres glabrous to sparsely hairy; basal leaves long-hirsute at least on the veins below; stems usually over 30 cm. tall; leaves usually over 8 cm. long   -2. H. albiflorum
   2. Rays yellow; involucres usually densely black-hirsute; basal leaves glabrous or nearly so; stems less than 30 cm. tall; leaves rarely over 7 cm. long   -3. H. gracile

1. Hieracium fendleri Schultz. Bip., Bonplandia 9:173. 1861.
   Heteropleura fendleri (Schultz. Bip.) Rydb.---Plants subscapose, 20-30 cm. tall, setose-hirsute; basal leaves spatulate or obovate, setose-hirsute; stem leaves few and reduced upwards; heads rather few, racemose-paniculate; involucres 12-14 mm. high, puberulent, glabrate or with a few long hairs; rays light yellow; achenes tapering upwards; pappus sordid-white to light tan.---Hills and mountains often in woods. South Dakota, south to New Mexico, Mexico and Arizona. Our records from northcentral, central, southcentral and southwestern Colorado at 5500-10,000 feet.
2. Hieracium albiflorum Hook., Fl. Bor. Amer. 1:298. 1834.
   Stems 30-70 cm. tall, leafy below and nearly naked above, hirsute below and glabrate above; basal leaves about 8-15 cm. long, petioled, blades oblanceolate to oblong, more or less

long-hirsute with tawny hairs especially on the midrib below; stem leaves sessile, lanceolate, less hairy; heads usually numerous in corymbose panicles; involucres 8-10 mm. high, glabrous or sparsely hairy; rays white to ochroleucous; achenes cylindrical, little if any tapering upward; pappus sordid to very light brown.---Hills and mountains, often in woods. Saskatchewan, south to Colorado and California. Our records from northcentral, northwestern, central and westcentral Colorado at 7000-10,500 feet.

3. Hieracium gracile Hook., Fl. Bor. Amer. 1:298. 1834.

Stems 10-30 cm. tall, usually scapose, puberulent to glabrate; leaves about 2-7 cm. long, obovate to oblong-spatulate, glabrous or nearly so; heads usually few and racemose; involucres 7-10 mm. high, usually densely black-hirsute at least near base; rays yellow; achenes cylindric, little if any tapering upward; pappus sordid or tawny.---Hills and mountains. Alberta to Alaska, south to New Mexico and California. Our records from the western half of Colorado in the higher mountains at 7500-12,000 feet.

96. Nothocalais Greene

Perennial herbaceous plants; leaves tufted-basal, narrow, entire or wavy-margined these margins white-tomentose; heads solitary on scapose peduncles; involucres oblong-campanulate, bracts in 2-3 series, nearly equal in length, all appressed; receptacles flat, naked; flowers all ligulate, yellow; achenes cylindric-fusiform, not flattened, 10-ribbed, somewhat narrowed above but beakless; pappus of soft, white, very narrow unequal scales (appearing as bristles to the naked eye).

1. Nothocalais cuspidata (Pursh) Greene, Bull. Calif. Acad. II. 2:55. 1886.

Leaves linear, long-acuminate, thick; scapes 5-35 cm. tall, rather stout, more or less tomentose at least above; involucres 18-25 mm. high, usually glabrous. This species has been placed in Agoseris and Microseris (Rhodora 50:33-35. 1948).---Plains and hills, often on rocky ground. Wisconsin to South Dakota, south to Missouri and Colorado. Our records from the eastern half of Colorado, mostly along the foothills, at 4000-6500 feet.

97. Agoseris Raf. MOUNTAIN-DANDELION

Perennial, mostly scapose, herbaceous plants; leaves usually basal-tufted and sessile, entire to pinnatifid; heads solitary on long scapes; involucres campanulate to oblong-cylindrical, bracts usually imbricated in several series, the outer shorter and broader; receptacles flat, naked; flowers all ligulate, yellow, orange or purple; achenes subfusiform to oblong, not flattened, 10-ribbed, narrowing above to a beak; pappus of numerous, white, simple, capillary bristles. The species are polymorphic and have been multiplied out of all proportion. The genus needs a thorough revision and the following treatment is very arbitrary. Fairly mature achenes and information as to the color of the fresh flowers are necessary.

1. Beak of achene short and stout, nerved throughout, much shorter than the body (scarcely over 1/2 as long)
   2. Leaves definitely pinnatifid   -1A. A. glauca laciniata
   2. Leaves entire or subentire
      3. Leaves linear to linear-lanceolate, seldom over 1 cm. wide   -1B. A. glauca parviflora
      3. Leaves lanceolate or broader, usually about 2 cm. wide   -1C. A. glauca dasycephala
1. Beak of achene slender, not nerved at least above, only slightly shorter to much longer than the body
   4. Involucres of 2 distinct sets of bracts, the inner long, the outer short; beak 2 to 3 times as long as the body of the achene; plants apparently limited to northcentral Colorado   -2. A. rostrata
   4. Involucres rather regularly imbricated; beak never 2 times longer than the body of the achene; plants usually more widely distributed in Colorado
      5. Ligules light yellow when fresh, often drying pinkish or purplish
         6. Beak longer than the body of the achene; plants often over 40 cm. tall   -3. A. elata
         6. Beak never exceeding in length the body of the achene; plants rarely over 40 cm. tall   -4. A. arizonica
      5. Ligules deep orange or brownish-red when fresh, drying to purple   -5. A. aurantiaca

1. Agoseris glauca (Pursh) D. Dietr., Syn. Pl. 4:1332. 1847.

Troximon glaucum Pursh---Plants variable, scapes 10-50 cm. tall, glabrous to woolly especially below; leaves linear, lanceolate, oblanceolate, to oblong, entire, dentate or pinnatifid, pubescent to glabrous; involucres variable in size, 10-25 mm. high, bracts usually lanceolate to linear-lanceolate, acute to acuminate; rays yellow when fresh; achene beak short and stout, much shorter than the body, nerved throughout. This is a polymorphic species as treated here. The following names have been used for various segregates in this region - A. taraxacifolia (Nutt.) Dietr., A. pubescens Rydb.; A. aspera Rydb.; A. villosa Rydb.; A. altissima Rydb.; A. maculata Rydb.; A. turbinata Rydb.; A. attenuata Rydb.; A. pumila (Nutt.) Rydb.; A. scorzoneraefolia (Schrad.) Greene; A. agrestis Osterh.; A. roseata Rydb.; A. rosea (Nutt.) Greene; A. laciniata (Nutt.) Greene, A. leontodon Rydb. Further study may prove that some of these may have specific or subspecific value. The 3 varieties recognized here are usually fairly distinct in Colorado.---Manitoba to British Columbia, south to New Mexico and Arizona. Our records scattered over the western half of Colorado at 7000-10,000 feet.

1A. Agoseris glauca laciniata (D. C. Eaton) Smiley, Univ. Calif. Pub. Bot. 9:404. 1921.

Leaves pinnatifid. The plant varies from rather tall and robust to shorter. The heads vary from small to large.---Our records scattered in the western half of Colorado at 4500-9500 feet.

1B. **Agoseris glauca parviflora** (Nutt.) Rydb., (var.) Contrib. U. S. Nat. Herb. 3:511. 1896.
A. parviflora (Nutt.) Greene; Troximon parviflorum Nutt.---Leaves entire or subentire, linear to linear-lanceolate, seldom over 1 cm. wide. The plant is usually small and the involucres are usually not over 15 mm. long.---Our records scattered in the western half of Colorado at 6000-10,000 feet.

1C. **Agoseris glauca dasycephala** (T. & G.) Jepson, (var.) Man. Fl. Pl. Calif. 1005. 1925.
Leaves lanceolate to broader, usually 2 cm. wide or more, entire to subentire. Plant is usually robust with more or less hairy involucres.---Our records mostly from the northwestern quarter of Colorado at 7500-12,000 feet.

2. **Agoseris rostrata** Rydb., Bull. Torr. Club 32:137. 1905.
Scapes 20-60 cm. tall, more or less villous especially just under the head; leaves narrowly linear-lanceolate to almost linear, usually more or less pinnatifid with linear segments but sometimes entire, glabrous and somewhat glaucous; involucres 25-30 mm. high, bracts in 2 distinct sets, the outer oblong or ovate, obtuse or acute, about 1/2 to 2/3 as long as the inner, inner bracts linear or linear-lanceolate; rays orange or purple; achene beak about 2-3 times as long as the body, not nerved throughout.---Mountains of Colorado. Our few records from northcentral Colorado at 6500-9500 feet.

3. **Agoseris elata** (Nutt.) Greene, Pitt. 2:177. 1891.
Troximon elatum (Nutt.) A. Nels.---Scapes usually robust, 20-50 cm. tall, sometimes shorter or longer, usually villous above; leaves lanceolate to oblanceolate, entire or pinnatifid; involucres 20-25 mm. high, bracts lanceolate to linear-lanceolate, acute, more or less hairy; rays yellow at first, drying to pinkish or purplish; achene beak longer than the body, not nerved throughout.---Hills and mountains. Montana to British Columbia, south to Colorado and California. Our records from the western half of Colorado at 6000-9000 feet.

4. **Agoseris arizonica** Greene, Pitt. 2:176. 1891.
A. leptocarpa Osterh.; Troximon arizonicum Greene---Scapes 10-40 cm. tall, usually less than 20 cm., glabrous or somewhat villous above; leaves linear to linear-oblanceolate, entire, toothed or pinnatifid, glabrate to hairy; involucres 15-25 mm. high, bracts rather regularly imbricated in about 3 series, lanceolate to linear-lanceolate, midribs often purplish, usually glabrous; rays yellow when fresh, drying pinkish or purple; achene beak shorter to as long as the body, not nerved throughout. Related to the preceding species. A. montana Osterh., A. frondifera Osterh. and A. arachnoidea Rydb. are also related to the above and may not be distinct.---Hills, mountains and valleys. Wyoming, south to New Mexico and Arizona. Our records scattered over the western half of Colorado at 5500-8000 feet or higher.

5. **Agoseris aurantiaca** (Hook.) Greene, Pitt. 2:177. 1891.
A. purpurea (Gray) Greene; Troximon purpureum (Gray) A. Nels.; A. gracilens (A. Gray) Kuntze ---Scapes 10-60 cm. tall, glabrous to sparingly lanate; leaves linear, linear-oblong to oblanceolate, entire, dentate to pinnatifid; involucres about 15-20 mm. high, bracts rather regularly imbricated, linear-lanceolate, oblong or lanc-ovate, the outer acuminate to obtuse at apex; rays deep orange or brownish-red when fresh, drying to a purple; achene beak shorter to longer than the body, nerved throughout. A. nana Rydb., A. carnea Rydb., and A. graminifolia Greene are doubtfully distinct. Our plants have been called var. purpurea (Gray) Cronq.---Parks, meadows and banks. Alberta to British Columbia, south to New Mexico and Arizona. Our records from the western half of Colorado at 5500-12,000 feet.

98. Taraxacum Hall DANDELION

Perennial, acaulescent plants with fleshy taproots and milky juice; leaves in a rosette, pinnatifid to toothed, rarely entire; heads solitary on a hollow scape, this elongating in fruit; involucres oblong or campanulate, bracts in 2 series, outer much shorter and often recurved, inner in 1 row, nearly equal; receptacles flat to low-convex, naked; flowers all ligulate, yellow, rays 5, toothed at truncate apex; anthers united; achenes more or less fusiform, not flattened, 4- to 5-ribbed, 5- to 10-nerved, muricate at least above, prolonged into a slender beak; pappus of numerous, unequal, persistent simple, capillary bristles. The plants are partly or completely parthenogenetic which has resulted in the preservation of a multitude of similar but distinguishable forms. More species may be present in Colorado than given here. Mature fruit is necessary to identify these plants.

1. Mature achenes blackish (sometimes reddish near apex only); outer involucre bracts erect or appressed
  -1. **T. lyratum**
1. Mature achenes reddish, brownish or greenish; outer bracts from appressed to reflexed
  2. Achenes reddish or reddish-brown
    3. Outer involucre bracts spreading or recurved, inner bracts with corniculate appendages; leaves thin and definitely pinnatifid   -2. **T. erythrospermum**
    3. Outer involucre bracts appressed, inner not or only slightly corniculate appendaged; leaves thickish, shallowly pinnatifid to entire   -3. **T. eriophorum**
  2. Achenes straw-colored, olive-green or brownish
    4. Outer involucre bracts appressed, inner corniculate appendaged   -4. **T. ceratophorum**
    4. Outer involucre bracts widely spreading to recurved, inner bracts not corniculate appendaged
  -5. **T. officinale**

1. **Taraxacum lyratum** (Led.) DC., Prodr. 7:148. 1838.
T. scopulorum (Gray) Rydb.; Leontodon scopulorum (Gray) Rydb.---Plants 2-15 cm. tall, seldom over 12 cm.; leaves lanceolate to spatulate, entire, dentate to pinnatifid; scapes glabrous to somewhat villous especially when young; heads 6-12 mm. high; outer involucre bracts erect or appressed, inner without corniculate appendages; achenes to 5 mm. long, murications rather short and blunt, blackish to sometimes reddish-tinged toward apex.---High mountains. Greenland

and Canada, south to Colorado and Arizona; Asia. Our few records scattered in the western half of Colorado at 9500-14,000 feet.

2. **Taraxacum erythrospermum** Andrz. ex Besser, Enum. Pl. 75. 1822.

Plants 5-30 cm. tall; leaves usually dissected deeply, nearly to the midrib, or sometimes toothed and entire, glabrous or somewhat pilose; heads 1-2 cm. high; outer involucre bracts widely spreading to recurved, inner bracts corniculate appendaged on all or most; achenes usually muricate only above the middle, reddish or reddish-brown. Our plant has been considered to be the same as the old world species T. laevigatum (Willd.) DC.---Waste places, roadsides and lawns. Nova Scotia to British Columbia, south to West Virginia and New Mexico; Eurasia and Africa. Our records from the westcentral and eastern half of Colorado at 5000-9000 feet but probably widely distributed.

3. **Taraxacum eriophorum** Rydb., Fl. Montana Mem. N. Y. Bot. Gard. 1:454. 1900.

T. angustifolium Greene; T. ammophilum A. Nels.; Leontodon angustifolium (Greene) Rydb.; L. ammophilum (A. Nels.) Rydb.---Plants 3-10 cm. tall, sometimes up to 30 cm.; leaves lanceolate to oblanceolate, mostly shallowly pinnate, toothed or even entire, rather thickish, glabrous to sparsely pilose; heads 1.5-2.5 cm. high; outer involucre bracts appressed or somewhat spreading, inner bracts either not corniculate appendaged or rarely dilated corniculate at apex; achenes 4-5 mm. long, murications rather spinelike, reddish or reddish-brown.---Hills, valleys and mountain meadows. Alberta to Alaska, south to Colorado. Our few records scattered in the higher mountains of Colorado at 7500-13,000 feet.

4. **Taraxacum ceratophorum** (Led.) DC., Prodr. 7:146. 1838.

T. dumetorum Greene; T. leiospermum Rydb.; T. montanum Nutt.; Leontodon dumetorum (Greene) Rydb.; L. leiospermum Rydb.; L. monticola Rydb.---Plants about 7-25 cm. tall; leaves lanceolate to oblanceolate, sinuate-dentate to pinnatifid, rarely entire to more dissected, glabrous to sparsely pilose; scapes when young more or less lanate-pilose; heads 1.5-2.5 mm. high; outer involucre bracts appressed or somewhat spreading, inner bracts, at least some, corniculate appendaged; achenes 4-5 mm. long, muriculations closely crowded and extending almost to the base, straw-colored, greenish-brown to gray-brown.---Valleys and meadows in the mountains. Labrador to Alaska, south to Massachusetts, New Mexico and California; Eurasia. Our records scattered in the higher mountains of the state at 7500-10,000 feet.

5. **Taraxacum officinale** Wiggars, Prin. Pl. Holst. 56. 1780.

T. vulgare (Lam.) Schrank.; T. taraxacum (L.) Karst.; T. mexicanum DC.; Leontodon taraxacum L.; L. mexicanum (DC.) Rydb.---Plants 5-30 cm. tall; leaves oblanceolate, usually deeply pinnatifid, sparsely pilose to glabrous; heads 1.5-5 cm. high; outer involucre bracts recurved-spreading to reflexed, inner bracts only slightly if at all corniculate appendaged; achenes usually muriculate only above the middle, pale gray to olive-brown.---Waste places, lawns, around dwellings, meadows and slopes. Widespread throughout most of the world. Our records scattered over Colorado at 4500-9500 feet., apparently up to 12,000 feet on occasion.

99. Malacothrix DC. DESERT DANDELION

Annual herbaceous plants; leaves alternate and basal, mostly pinnatifid; heads several to many, sometimes solitary; involucres campanulate to turbinate, principal bracts in 1-2 series, with several shorter outer ones; receptacles flat, naked or bristly; flowers all ligulate, yellow; achenes columnar, not flattened, 5- to 10-ribbed, truncate and beakless at apex; pappus of soft, scabrous or minutely barbellate bristles which fall together, no stiff outer persistent bristles present or if so only 1 or 2.

1. **Malacothrix sonchoides** (Nutt.) T. & G., Fl. N. Amer. 2:486. 1843.

M. runcinata A. Nels.---Plants glabrous or nearly so; stems 5-25 cm. tall, branching at base, the side branches widely spreading; leaves oblong or upper linear, dentate to shallowly or deeply pinnatifid; involucres 8-12 mm. high, bracts in about 3 series, outer broader, inner linear-lanceolate; achenes about 2.5-3 mm. long, with 5 stronger ribs. The form with 1-2 outer persistent scales or bristles has been called var. torreyi Gray.---Dry hills, valleys and plains. Nebraska to Idaho, south to New Mexico and California. Our records from western Colorado at 4500-5500 feet.

100. Prenanthus L. RATTLESNAKE ROOT

Perennial, caulescent, herbaceous plants; leaves alternate, entire to dentate, lower petioled, upper sessile and partly clasping; heads in spikelike panicles; involucres narrow, cylindric to campanulate, principal bracts in 1-2 series and nearly equal, with a few shorter outer ones; receptacles flat, naked; flowers all ligulate, rose or purplish; achenes oblong or tapering to base, not compressed, mostly ribbed or nerved, apex not at all beaked; pappus of simple, copious, rather rigid capillary pale to brown bristles.

1. **Prenanthus racemosa** Michx., Fl. Bor. Am. 2:83. 1803.

Nabalus racemosus (Michx.) DC.---Stems 30-100 cm. tall, glabrous and more or less glaucous to the inflorescence; leaves glabrous, lower oval to oblanceolate, petioles long and winged; upper leaves lanceolate to ovate-lanceolate; involucres 10-14 mm. high, bracts linear, loosely hirsute; flowers 8-16, rarely fewer; pappus exceeding the involucres in fruit.---Moist, usually open ground. New Brunswick to Alberta, south to New Jersey, Missouri and Colorado. Our records from northcentral, central, southcentral and southwestern Colorado at 5500-9500 feet.

101. Lygodesmia D. Don. SKELETON PLANT

Annual or perennial herbaceous plants, nearly or quite glabrous; stems usually tough, rigid and rushlike, branching; leaves alternate, entire, toothed or more or less runcinate, usually the upper reduced or often bractlike; heads small to large, terminal on the branches or racemose; involucres cylindric to oblong, of a few equal inner bracts and 1 to several outer shorter ones; flowers 3-12, all ligulate, pink to rose; receptacles flat, naked; achenes linear, subcylindric to fusiform, not flattened, few-ribbed, beakless; pappus of numerous, stiff or soft, simple white or sordid-white bristles.

1. Involucres 4-5 mm. long; plants annual; pappus bright white   -1. L. exigua
1. Involucres 10 mm. long or more; plants annual or perennial; pappus usually sordid white or possibly light brown
  2. Plants annual; involucres 10-16 mm. high   -2. L. rostrata
  2. Plants perennial; involucres 10-25 mm. high
    3. Involucres 18-25 mm. high; leaves 5-10 cm. long; achenes 10 mm. long or more   -3. L. grandiflora
    3. Involucres 10-14 mm. high; leaves rarely over 5 cm. long; achenes 5-7 mm. long   -4. L. juncea

1. Lygodesmia exigua A. Gray, Proc. Amer. Acad. 9:217. 1874.
Prenanthella exigua (Gray) Rydb.---Annual plants; stems 6-20 cm. tall, much branched; lower leaves 1-3 cm. long, obovate to oblanceolate, upper leaves reduced and bractlike; heads solitary at the ends of the branches, 3- to 6-flowered; involucres 4-5 mm. high, about 4-5 inner linear bracts present; ligules pink to rose; achenes 3-4 mm. long; pappus white.---Colorado south to Texas and California. Our one record from the very southwestern corner of Colorado at 5500 feet.
2. Lygodesmia rostrata A. Gray, Proc. Am. Acad. 9:217. 1874.
Annual plants; stems 20-80 cm. tall, erect with ascending branches; leaves 5-20 cm. long, narrowly linear and entire, upper small and subulate; heads solitary at ends of long or short branches; involucres 10-16 mm. high, about 7-9 inner bracts present; flowers 7-9, pink to rose; achenes about 8-10 mm. long; pappus whitish.---Plains and hills. Saskatchewan, south to Nebraska and Colorado. Our records from northcentral and northeastern Colorado at 3500-5500 feet.
3. Lygodesmia grandiflora (Nutt.) T. & G., Fl. N. Amer. 2:485. 1843.
Perennial plants with deep-seated rootstocks; stems 10-40 cm. tall, leafy at least on large plants, 1 to few together; leaves 5-10 cm. long, linear, entire, firm, ascending; heads solitary at the end of the stem or branches; involucres 18-25 mm. high, bracts linear; flowers 6-10, rose or pink, large and showy in anthesis; achenes about 10 mm. long or longer; pappus sordid.---Hills, valleys and plains, usually in dry often gravelly soil. Wyoming to Idaho, south to New Mexico and Arizona. Our records from western Colorado at 4500-8500 feet.
4. Lygodesmia juncea (Pursh) D. Don., Edinb. Phil. Journ. 6:311. 1829.
Perennial plants with deep-seated rootstocks; stems 10-40 cm. tall, much branched, stiff, often swollen by globose galls; lower leaves seldom over 5 cm. long, linear to lanceolate, entire, rigid, upper leaves smaller and scalelike or subulate; heads solitary at ends of branches; involucres 10-14 mm. high, about 5, linear, inner bracts present; flowers mostly 5, rose or pink; achenes 5-7 mm. long; pappus sordid.---Plains, hills and valleys, usually in dry ground. Minnesota to Alberta, south to Missouri and New Mexico. Our records scattered over Colorado, except the northwestern part, at 4000-7500 feet.

NEW COMBINATIONS PUBLISHED IN THIS WORK

Kobresia bellardi (All.) Degland var. macrocarpa (Clokey) H. D. Harrington---based on Kobresia macrocarpa Clokey, N. Am. Fl. 18:5. 1931. (See page 118.)

Luzula subcapitata (Rydb.) H. D. Harrington---based on Juncoides subcapitatum Rydb., Bull. Torr. Club 31:401. 1904. (See page 151.)

Arenaria fendleri A. Gray var. eastwoodiae (Rydb.) H. D. Harrington---based on Arenaria eastwoodiae Rydb., Bull. Torr. Club 31:406. 1904. (See page 235.)

Aconitum columbianum Nutt. var. bakeri (Greene) H. D. Harrington---based on Aconitum bakeri Greene, Pl. Baker. 3:5. 1901. (See page 239.)

Sisymbrium elegans (Jones) Payson var. juniperorum (Payson) H. D. Harrington---based on Sisymbrium juniperorum Payson, Univ. Wyo. Pub. Sci. 1:12. 1922. (See page 280.)

Petalostemon purpureum (Vent.) Rydb. var. pubescens (A. Nels.) H. D. Harrington---based on Petalostemon pubescens A. Nels., Bot. Gaz. 31:395. 1901. (See page 319.)

Dalea leporina (Ait.) Kearney & Peebles var. alba (Michx.) H. D. Harrington---based on Dalea alba Michx. in Roem., Cat. Hort. Turic. 1802. (See page 325.)

Trifolium longipes Nutt. var. rusbyi (Greene) H. D. Harrington---based on Trifolium rusbyi Greene, Pitt. 1:5. 1887. (See page 332.)

Geranium fremontii Torr. var. cowenii (Rydb.) H. D. Harrington---based on Geranium cowenii Rydb., Fl. Colo. 218. 1906. (See page 352.)

Viola adunca Smith var. bellidifolia (Greene) H. D. Harrington---based on Viola bellidifolia Greene, Pitt. 4:292. 1901. (See page 377.)

Mentzelia nuda (Pursh) T. & G. var. stricta (Osterh.) H. D. Harrington---based on Hesperaster strictus Osterh., Bull. Torr. Club 29:174. 1902. (See page 381.)

Mentzelia nuda (Pursh) T. & G. var. rusbyi (Woot.) H. D. Harrington---based on Mentzelia rusbyi Woot., Bull. Torr. Club 25:261. 1898. (See page 381.)

Arctostaphylos nevadensis A. Gray var. coloradensis (Rollins) H. D. Harrington---based on Arctostaphylos coloradensis Rollins, Rhodora 39:463. 1937. (See page 420.)

Cryptantha jamesii (Torr.) Payson var. pustulosa (Rydb.) H. D. Harrington---based on Oreocarya pustulosa Rydb., Bull. Torr. Club 40:480. 1913. (See page 466.)

Penstemon alpinus Torr. ssp. brandegei (Porter) C. Wm. T. Penland---based on Penstemon brandegei Porter ex Rydb., Mem. N. Y. Bot. Gard. 1:343. 1900. (See page 496.)

Penstemon alpinus Torr. ssp. magnus (Pennell) C. Wm. T. Penland---based on Penstemon magnus Pennell, Contrib. U. S. Nat. Herb. 20:346. 1920. (See page 496.)

Pericome caudata A. Gray var. glandulosa (Goodm.) H. D. Harrington---based on Pericome glandulosa Goodm., Rhodora 39:209. 1937. (See page 589.)

Senecio amplectens A. Gray var. holmii (Greene) H. D. Harrington---based on Senecio holmii Greene, Pitt. 4:120. 1900. (See page 611.)

SUMMARY

	Families	Genera	Species	Varieties	Subspecies
Pteridophyta	8	22	59	3	0
Spermatophyta					
Gymnospermae	2	6	17	3	0
Angiospermae					
Monocotyledoneae	17	138	587	38	0
Dicotyledoneae					
Apetalae and Polypetalae	62	283	1162	124	28
Sympetalae	28	244	969	95	60
Total	117	693	2794	263	88

The preceding summary contains only those species, varieties, and subspecies actually appearing in the diagnostic keys. However, of the 2,794 species treated about 242 appear without a specific description since specimens collected in Colorado were not seen.

GLOSSARY

A. A prefix meaning "without," as in apetalous.
Abortive. Imperfectly developed; rudimentary.
Acaulescent. Stemless or apparently so, or the stem subterranean; leaves basal.
Accrescent. Enlarging after flowering, usually the sepals.
Acerose. Needle-shaped, as the leaves of spruce.
Achene. A small, dry, 1-celled, 1-seeded indehiscent fruit, the seed attached to the pericarp at one place.
Acorn. The 1-celled, 1-seeded fruit of oaks; consists of a cuplike base and the nut.
Acrid. Sharp, irritating or biting to the taste.
Acuminate. Tapering to the apex, the sides more or less pinched in before reaching the tip. Compare acute.
Acute. Tapering to the apex with the sides straight or nearly so; usually less tapering than acuminate.
Adnate. The union of unlike parts, as an inferior ovary to the calyx tube. Compare connate.
Adventitious. Developing in an unusual or irregular position.
Aerial. In the air, as roots borne above the ground or water.
Aggregate. Crowded into a dense cluster but not united.
Alpine. The area above timberline.
Alternate. Borne singly and not opposite--in leaves one at a node.
Alveolar. Honeycombed.
Alveolate. Honeycombed. Same as alveolar.
Ament. A spike or spikelike, usually pendulous inflorescence of unisexual flowers. Same as catkin.
Amplicate. Enlarged.
Androgynous. Staminate and pistillate flowers in the same inflorescence, the staminate above. Used chiefly in the genus Carex.
Annual. Completing the life cycle in one growing season. Living only one season as an annual stem.
Annular. In the form of a ring.
Annulate. In the form of a ring.
Anterior. On the front side away from the axis.
Anther. The pollen-bearing part of the stamen.
Antheridium (pl. antheridia). The male organ of reproduction in ferns, corresponding to the anther in seed plants.
Anthesis. Period when the flower is open.
Anthocyanous. With anthocyanin pigments in the cells, these reddish or bluish.
Antrorse. Directed forward or upward as of hairs.
Apetalous. Lacking petals.
Aphyllopodic. Lower leaves bladeless or nearly so; used in sedges.
Apiculate. Ending in an abrupt slender tip which is not stiff.
Appressed. Lying flat or close against something. Often used for hairs.
Approximate. Close together but not united.
Aquatic. Living in water.
Arachnoid. Beset with cobwebby or entangled hairs.
Archegonium (pl. archegonia). The female reproductive organ in ferns, corresponding to the pistil in seed plants.
Arcuate. Arching or moderately curved like a bow.
Areola (pl. areolae). A small space marked out upon or beneath the surface; often used in leaves for the area between small veins. Also spelled Areole.
Areole. See areola.
Aril. An appendage growing at or about the hilum of a seed. Compare caruncle.
Aristate. With an awn or stiff bristle, usually at the apex.
Aristulate. Minutely aristate.
Articulating. With a joint or node separating at maturity by a clean-cut scar.

Ascending. Growing obliquely upward, often curving upward usually at about 40°-60°.
Assurgent. Ascending.
Attenuate. Gradually narrowing to a tip or base, this usually narrow and slender.
Auricle. An ear-shaped lobe or appendage.
Auriculate. With auricles.
Awl-shaped. Tapering gradually upward from a broader base to a sharp point, narrowly triangular; usually small structures.
Awn. A slender bristlelike organ usually at the apex of a structure.
Axile. In the axil, the angle between an organ and its axis.
Axillary. In or related to the axis.

Balsamiferous. Sticky and odoriferous like balsam.
Banner. The upper, usually larger petal in a papilionaceous or "sweetpea type" flower.
Barbed. With rigid short reflexed processes like the barb of a fishhook.
Barbellate. Finely barbed usually down the sides of the structure as well as at the apex.
Basifixed. Attached by the base. Compare versatile.
Beak. A hard or firm point or projection.
Bearded. Furnished with long or stiff hairs.
Berry. A fleshy pulpy fruit with immersed seeds. Rather loosely used.
Biangulate. Two-angled.
Bicolored. Of two rather contrasting colors.
Biconvex. Convex on both sides like a lens.
Bidentate. With two teeth.
Bidentulate. With two small teeth. Compare bidentate.
Biennial. Living for two years.
Bifid. Two-cleft or 2-lobed, usually at the apex.
Bifurcate. Divided into two forks or branches.
Bilabiate. Two-lipped.
Bimultifid. Twice cleft into many lobes or segments.
Bipinnate. Doubly or twice-pinnate, the primary divisions once-again pinnate.
Bipinnatifid. Twice or doubly pinnatifid, the primary divisions once-again pinnatifid.
Biseriate. Occupying two rows.
Bisexual. Having both stamens and pistils, usually used for a flower.
Biturbinate. Rather top-shaped but the widest part not directly at one end.
Bladder. An inflated, thin-walled structure.
Blade. The expanded usually flat portion of a leaf. Compare sheath and petiole. Also used for the expanded portion of a petal. Compare claw.
Bloom. A whitish powdery, glaucous usually waxy covering of a surface. Also used in reference to a flower.
Bract. A more or less modified leaf situated near a flower or inflorescence.
Bracteate. Having bracts.
Bracteolate. With bractlets.
Bracteole. Same as bractlet.
Bractlet. A secondary bract as one on the pedicel of a flower usually smaller than the bracts.
Bristle. A stiff hairlike structure on the order of a pig bristle.
Bud. The rudimentary state of a stem or branch. Also used for an unexpanded flower.
Bulb. A subterranean leaf-bud with fleshy scales like an onion.
Bulblet. A small bulb especially one borne above the ground, as in onion-sets.
Bur. A seed or fruit bearing spines or prickles, these usually hooked or bearded.

Caducous. Falling off unusually early as compared with similar structures in general.

Caespitose. Growing in tufts. Also written cespitose.
Callous. Having a hard texture, often swollen.
Callus. A hard protuberance or callosity. In grasses the indurated downward extension of the lemma, morphologically a part of the rachilla.
Calyculate. Having bracts around the calyx or involucre, these usually smaller.
Calyx. The outer series of the perianth, used especially when it differs in size, shape or color from the inner (or petals).
Calyx tube. That part of the calyx where the sepals are united.
Campanulate. Bell-shaped, rather cup-shaped with a flaring rim.
Canaliculate. Longitudinally channeled or grooved.
Canescent. With gray or white short hairs, short-hoary. Often loosely used to mean any gray or white surface.
Capillary. Very slender and hairlike.
Capitate. In a globular or head-shaped cluster.
Capitellate. Headlike; a diminutive of capitate.
Capsule. A dry dehiscent fruit made up of more than one carpel.
Carinate. Keeled with a longitudinal ridge.
Carpel. A simple pistil formed from one sporophyll, or that part of a compound pistil formed from one sporophyll.
Carpophore. The slender prolongation of the floral axis which in Umbelliferae supports the pendulous carpels.
Cartilaginous. Firm and tough but elastic like cartilage.
Caruncle. An escrescence or appendage at or about the hilum of a seed. Usually fleshy and less tendril-like than an aril.
Caryopsis. A dry, 1-seeded, indehiscent fruit in which the seed is grown fast to the pericarp at all points.
Castaneous. Of a chestnut or dark brown color.
Catkin. An ament.
Caudate. With a slender taillike appendage.
Caudex. (pl. caudices). The persistent, often woody base of an otherwise annual herbaceous stem.
Caulescent. Having a manifest leafy stem above ground. Compare with acaulescent.
Cauline. Of or pertaining to the stem.
Cell. A microscopic structural unit of a plant. When used in connection with a pistil then the same as locule.
Cellular. Made up of small pits or compartments.
Centimeter (abbreviation cm.). Ten millimeters or about 2.54 of an inch.
Chaff. A thin dry scale or bract. One of the bracts between the individual flowers in the head of the Compositae.
Chaffy. Possessing or resembling chaff.
Chartaceous. Having the texture of stiff writing paper or parchment.
Chlorophyll. The green pigment in plants associated with photosynthesis.
Ciliate. Beset with a marginal fringe of hairs.
Ciliolate. Ciliate but the hairs minute.
Circinate. Coiled from the tip downward, resembling the upper end of a violin.
Circumscissle. Dehiscing in a transverse circular line, the top separating like the lid of a pill box.
Clavate. Club-shaped and widest nearer the apex.
Claw. The narrowed base or stalk to some petals.
Cleft. Cut in about 1/2 way to the midvein or base, especially when the sinus is sharp.
Cleistogamous. Fertilized in the bud, the flower never opening.
Coalescent. Union of parts of the same kind.
Cochleate. Coiled or shaped like a snail shell.
Coerulean. Blue or bluish.
Coetaneous. Appearing together or at the same time.

Collar. The area on the outside of a grass leaf at the junction of the blade and the sheath.
Collateral. Situated at the side of something.
Column. A group of united filaments as in Malvaceae. Also the coalesced style and filaments in the Orchidaceae.
Coma. A tuft of hairs especially at the tips of seeds.
Commissure. The surface by which one carpel joins another in Umbelliferae.
Comose. Furnished with a tuft of hairs or coma.
Complete. A flower with sepals, petals, stamens and pistils present.
Compound leaf. A leaf completely separated into two or more leaflets.
Compound ovary. An ovary with two or more carpels.
Compressed-keeled. Flattened laterally, the fold constituting a ridge or keel.
Conduplicate. Folded lengthwise down the middle.
Cone. The dry multiple fruit of pines, spruces, etc., consisting of overlapping scales. Same as strobilus. Also used as a shape "cone-shaped."
Conical. Cone-shaped, attached at the broad end.
Confluent. Running together; blending in one.
Connate. The union of like structures.
Connective. That portion of a stamen that connects the two halves of an anther.
Connivent. Converging; in close contact but not actually united by tissue.
Continuous. Said of a rachis or axis that does not break up at joints at maturity. Compare articulate.
Contorted. Twisted or bent, or twisted on itself.
Contracted. Said of an inflorescence that is narrow and dense with short or appressed branches.
Convex. Rounded on the surface as a lens.
Convolute. Rolled up longitudinally; technically one edge inside the other but loosely used especially in grasses.
Coralloid. Resembling the growth and form of coral.
Cordate. Of a conventional heart-shape, the point apical. Compare obcordate.
Coriaceous. Texture of leather.
Corm. A thickened, vertical solid underground stem. Compare bulb.
Corniculate. Furnished with a small horn or horns.
Corolla. The inner series of the floral envelope; collective name for petals.
Corrugated. Wrinkled or in folds.
Corymb. A flat-topped or convex open inflorescence; technically a contracted raceme.
Corymbiform. Shaped like a corymb.
Corymbose. Borne in corymbs or corymblike.
Cotyledon. The embryo leaf in a seed, often functioning as the first leaf of a seedling.
Crenate. Toothed with teeth rounded at apex.
Crenulate. Crenate with small teeth.
Crest. An elevated ridge or projection on the surface.
Crisp. Margins wavy up and down.
Crispate. Having a crisped appearance.
Crown. An inner appendage to a petal or throat of a corolla. Also used for the persistent base of a tufted perennial plant, especially a grass.
Crustose. Of a hard and brittle texture.
Cucullate. Hooded or hood-shaped; like a cowl.
Culm. The specialized stem of grasses, sedges and rushes.
Cuneate. Wedge-shaped; rather narrowly triangular, the acute angle downward.
Cuspidate. Tipped with an abrupt, short, sharp, firm point. Compare mucronate.
Cyme. A flower cluster, often convex or flat-topped, in which the central or terminal flower blooms earliest.
Cymose. Bearing cymes or cymelike.
Cymule. A small cyme or portion of one.

Deciduous. Falling away, not persistent or evergreen.
Decompound. More than once-compound, the primary divisions again completely separated.
Decumbent. Reclining on the ground but with the end ascending; used for stems.
Decurrent. Extending downward from the point of insertion; said of a leaf decurrent on the stem.
Deflexed. Bent or turned abruptly downward or backward.
Dehiscent. Opening by definite pores or slits to discharge the contents.
Deltoid. Shaped like the Greek letter △, attached at the center of one side.
Dense. Said of inflorescences where the flowers are crowded.
Dentate. Toothed with the teeth directed outward. Sometimes loosely used for any large teeth.
Denticulate. Dentate with small teeth.
Depauperate. Reduced or undeveloped; said of plants dwarfed and stunted.
Depressed. More or less flattened from above.
Diadelphous. Stamens in two often unequal sets.
Dichotomous. Two-forked, the branches equal or nearly so.
Didymous. Twinlike; in equal pairs.
Didynamous. Stamens in two pairs of unequal length.
Diffuse. Loosely or widely spreading.
Digitate. Compound with the parts radiating out from a common point like the fingers on a hand. Same as palmate.
Dimorphic. In two forms.
Dimorphous. With two forms.
Dioecious. Flowers unisexual, the staminate and pistillate borne on separate plants.
Disarticulating. The parts separating at maturity. Compare articulating.
Disc. Same as disk.
Discoid. Resembling a disk. A discoid head in Compositae is one where no ray flowers are present.
Disk. An enlargement or prolongation of the receptacle of a flower around the base of a pistil. In Compositae the central part of the head bearing regular tubular flowers.
Disk flowers. The regular tubular flowers on the heads of Compositae. Compare ray-flowers or ligules.
Dissected. Cut or divided into numerous and usually narrow segments.
Distal. The end opposite the point of attachment.
Distichous. In two vertical ranks, usually conspicuously so.
Distinct. Separate, not at all united to each other. Compare connate.
Diurnal. Occurring in the daytime.
Divaricate. Widely spreading or diverging.
Divided. Deeply lobed, the sinuses extending to the base of the leaf or to the midrib; nearly compound.
Dolabriform. Pick-shaped; said of hairs apparently attached at their middle.
Dorsal. Pertaining to the back or outer surface of an organ.
Drupe. A fleshy indehiscent, 1-seeded fruit, the inner layer of the pericarp stony.
Druplet. A diminutive drupe, as the small parts of a raspberry fruit.

E. A prefix meaning "lacking" or "without."
Ebracteate. Lacking bracts.
Eccentric. Not situated at the central axis; off-center.
Echinate. Provided with prickles.
Ellipsoid. A solid body, elliptic in outline.
Elliptic. Shaped like an ellipse; widest in center and the 2 ends equal. Loosely used.
Elliptical. Same as elliptic.
Emarginate. With a shallow notch at the apex.
Embryo. The rudimentary plant within a seed.
Endemic. Confined to a limited geographical area.
Ensiform. Shaped like a sword, as the leaf of Iris.

Entire. Margins without teeth or lobes.
Ephemeral. Lasting for one day or less.
Epidermis. The outer layer of cells.
Epigynous. Growing on the summit of the ovary or appearing to do so.
Equitant. Leaves that are conduplicate and in two ranks; also 2-ranked leaves, flattened with edges toward and away from the axis.
Erose. Margin irregular as if gnawed.
Erosulate. More or less erose.
Evergreen. Bearing green leaves throughout the year. Compare deciduous.
Ex. A prefix meaning "lacking" or "without."
Excurrent. Running out or beyond, as a nerve of a leaf projecting out beyond the margin.
Excurved. Curving outward or away from the axis.
Exfoliating. Peeling off in thin layers.
Explanate. Spread out flat.
Exserted. Projecting beyond a surrounding organ, as a stamen exserted from a corolla. Compare included.

Falcate. Scythe- or scimitar-shaped, curved sidewise and flat, tapering upwards; asymmetric.
Fan-shaped. Shaped like an opened folding fan; triangular with the upper side convex.
Farinaceous. Starchy; mealy.
Farinose. Covering with a mealy, usually whitish substance.
Fascicled. Borne in close bundles or clusters.
Fastigiate. Erect or near together with a broomlike effect.
Fernlike. Used of a leaf dissected or divided into narrow segments, like many ferns.
Fertile. Capable of producing fruit and seeds; a fertile flower may be pistillate or perfect.
Fibrillose. With fine fibers. Sometimes written fibrillate.
Fibrous. Composed of, or resembling fibers.
Filament. Any threadlike body; used especially for that part of the stamen that supports the anther.
Filamentose. Composed of threads. Also written filamentous.
Filiferous. Producing or bearing threadlike growths.
Filiform. Threadlike; long, slender and terete.
Fimbriate. Margins with a fringe of hairs, these longer and coarser than ciliate.
Fimbrilla (pl. fimbrillae). A single unit of a marginal fringe.
Fistulose. Hollow and cylindrical like onion leaves, often rather enlarged.
Flabellate. Same as fan-shaped.
Flabelliform. Same as fan-shaped.
Flaccid. Lax and weak; without rigidity.
Flange. A projecting edge or rim; edge flaring and conspicuous.
Flexuose. Bent alternately in opposite directions, usually not strongly so.
Flexuous. Same as flexuose.
Floccose. Clothed with loose tufts of woollike hair, this not uniform over the entire surface.
Floret. A small flower, especially one in a dense cluster. Also a special term for a grass flower with its lemma and palea included.
Floriferous. Bearing flowers.
Fluted. With grooves or furrows.
Foliaceous. Leaflike especially in color.
-foliate. Having leaves.
-foliolate. Having leaflets.
Follicle. A dry fruit with one carpel and splitting down one side only.
Fornix (pl. fornices). Small arching crests in the throat of a corolla.
Fovea (pl. foveae). Small depressions or pits.
Free. Not adnate; unlike parts not connected.
Free central placenta. The ovary is 1-celled and the ovules are borne on a central stalk not connected at the top.

Frond. Leaf of a fern. In Lemnaceae the expanded thalluslike stem which functions as a leaf.
Fruit. The ripened ovary and any other structures that enclose it at maturity.
Fruticose. Shrublike; at least somewhat woody. Also written frutescent.
Fugaceous. Falling or fading very early. About the same as caducous.
Fulvous. Dull yellow; yellow tinged with brown or gray.
Furfuraceous. Resembling flakes or grains of bran; scurfy.
Fuscous. Grayish-brown or dusky-brown.
Fusiform. Spindle-shaped, broadest at the middle and tapering both ways.

Galea. A hooded, or helmet-shaped part of the perianth, usually the upper lip of an irregular corolla.
Galeate. Shaped like a galea or helmet.
Gametophyte. The sexual stage in plants which bear sperm and eggs. Used here for the prothallus of ferns.
Gamopetalous. Petals more or less united. Same as sympetalous.
Gamophyllous. The leaves or leaflike organs more or less united one to another.
Gamosepalous. The sepals more or less united.
Geminate. Equal, in pairs like twins.
Gemma (pl. gemmae). A bud or budlike body by which some plants propagate themselves.
Geniculate. Bent abruptly like a knee or stovepipe bend.
Gibbous. Enlarged, humped or swollen on one side.
Glabrate. Becoming glabrous in age.
Glabrescent. About the same as glabrate.
Glabrous. No hairs present at all; also used for smooth.
Gland. A secreting surface or structure, or an appendage having the general appearance of such an organ.
Glandular. Bearing glands. A glandular hair has an enlargement like a hatpin at the apex.
Glaucescent. Tending to be glaucous; somewhat glaucous.
Glaucous. Covered with a whitish or bluish waxy covering, this should rub off readily but the term is sometimes loosely used for any whitish surface.
Glochid. A barbed hair or bristle; usually used for the minute bristles in Opuntia.
Glochidiate. Barbed at the tip.
Glomerate. Crowded, congested or compactly clustered.
Glomerule. A dense crowded cluster, usually of flowers.
Glume. A chafflike bract; used particularly for the two lower empty bracts of a grass spikelet.
Glutinous. Covered with a sticky gluelike or gummy exudation.
Grain. A swollen, seedlike structure as on the fruit of some species of Rumex; also used as a synonym for caryopsis.
Granulate. Composed of, or appearing to be covered by small granules. Same as granulose.
Granule. A minute rounded object.
Granuliferous. Composed of, or covered with very minute granules.
Grasslike. Resembling grasses; usually used for sedges and rushes.
Grenadin. A conspicuous orange- or brick-red color, most characteristic of petals of Sphaeralcea.
Gynaecandrous. With staminate and pistillate flowers on the same spike, the pistillate above; used in the genus Carex.
Gynobase. An enlargement or prolongation of the receptacle, bearing the ovary.

Halbert-shaped. Same as hastate.
Hastate. Arrow-head shaped but with the basal lobes pointing outward instead of backward. Compare sagittate.

Haustoria. Rootlike sucking attachments of parasitic plants like Cuscuta.
Head. A dense cluster of sessile or nearly sessile flowers or fruits on a very short axis; used especially for the involucrate inflorescence in Compositae.
Herb. A plant with no persistent woody stem above ground. Also plants used in seasoning or in medicine.
Herbaceous. Having the characteristic of a herb; also leaflike in color or texture.
Heteromorphous. Of more than one kind or form.
Heterostyled. With more than one length of style.
Hilum. The scar or point of attachment of a seed.
Hirsute. With moderately coarse and stiff hairs.
Hirsutulose. Same as hirsutulous.
Hirsutulous. Somewhat hirsute.
Hirtellous. Minutely hirsute.
Hispid. With stiff and rigid bristles or bristlelike hairs, these usually stiff enough to penetrate the skin.
Hispidulous. Minutely hispid.
Homomorphous. Of only one form or kind.
Hooked. Abruptly curved at tip.
Horn. A stiff tapering appendage somewhat like the horn of a cow.
Hyaline. Thin, dry and transparent or translucent.
Hygroscopic. Altering form or position due to changes in moisture content.
Hypanthium. An enlargement or elongation of the floral axis below the calyx, commonly partly or completely enclosing the pistils; when this occurs the ovary is here considered to be inferior.
Hypogynous. Situated on the receptacle below the ovary; a flower having the petals and stamens so situated.

Imbricate. Partly overlapping like shingles on a roof, either vertically or laterally.
Immersed. Growing submerged in water.
Imperfect flowers. Lacking either stamens or pistils. Compare perfect, unisexual and bisexual.
Incised. Cut sharply and usually irregularly with sharp sinuses, deeper than teeth but seldom as deep as 1/2 way in to the base or midrib.
Included. Not at all protruding from the surrounding organ. Compare exserted.
Incurved. Curved toward the axis or attachment.
Indehiscent. Remaining persistently closed; not opening by definite lines or pores.
Indurated. Hardened and stiffened.
Indusium (pl. indusia). The thin scalelike outgrowth of the fern leaf forming a covering for the young sorus. Sometimes the inrolled margin functions as an indusium.
Inferior ovary. One that is adnate to the hypanthium or calyx tube, appearing to be sunken in the stem, the flower parts appearing to come off from above the ovary; used here in the broad sense.
Inflated. Bladderlike; enlarged with thin walls.
Inflexed. Turned abruptly or bent inwards; incurved.
Infrastipular. Situated below the stipules; used in Rosa when a pair of prickles below the node is enlarged or conspicuous because of the absence or scarcity of other prickles.
Innovation. A basal offshoot from the main stem, shorter and less modified than a rhizome or stolon; in grasses an incomplete young shoot.
Internode. The portion of a stem or other structure between two nodes.
Involucel. A secondary involucre as in Umbelliferae.
Involucrate. With an involucre.
Involucre. A whorl of distinct or united leaves or bracts subtending a flower or an inflorescence.
Involute. Both edges inrolled toward the midnerve on the upper surface; loosely used in grasses for any leaf rolled on the upper surface.

Irregular flower. With inequality in the size, form or union of its similar parts; not radially symmetrical.

Keel. A dorsal projecting usually central rib, like the keel of a boat; also the name for the two anterior united petals of a papilionaceous "sweetpea" flower.

Labiate. Lipped. Belonging to the Labiatae or mint family.
Lacerate. Irregularly cut or cleft as if torn.
Laciniate. Narrowly incised or slashed; margins cut into narrow and usually pointed lobes.
Lacuna (pl. lacunae). An air space in the midst of tissue.
Lamella (pl. lamellae). A thin flat plate or laterally flattened ridge.
Lanate. With long tangled woolly hairs.
Lanceolate. Lance-shaped; several times longer than wide, broadest toward the base and tapering to apex.
Lateral. Borne on the sides of a structure or object.
Lax. Loose; often used for a soft open inflorescence or for soft drooping stems or foliage.
Leaflet. One of the divisions of a compound leaf.
Legume. The characteristic fruit of the Leguminosae family; usually a dehiscent fruit formed from one carpel with two lines of dehiscence. Also used for any plant with this type of fruit.
Lemma. The lower of the two bracts enclosing a grass flower, above the glumes; formerly called flowering glume.
Lenticel. A group of loose corky cells formed beneath the epidermis of woody plants, rupturing the epidermis and admitting gases to the inner tissues.
Lenticular. Lens-shaped; biconvex in shape.
Lepidote. Covered with small scurfy scales like the leaf of Elaeagnus.
Ligulate. Furnished with a ligule; also used for a strap-shape like a ligule.
Ligule. The flattened, usually strap-shaped corolla in the ray flowers of Compositae. Also a hairlike or membranous projection up from the inside of a grass sheath at its junction with the blade.
Limb. The expanded portion of a gamopetalous corolla above the throat.
Linear. Narrow and flat with parallel sides like a grass leaf blade.
Lip. Either the upper or lower division of a bilabiate or 2-lipped corolla. Also the upper (but by twisting of the pedicel appearing to be the lower) petal in Orchidaceae.
Lobe. Any segment of an organ especially if rounded.
Lobed. Bearing lobes; loosely used, but technically cut in not over half way to the base or midvein and the sinuses and apex of segments rounded.
Lobulate. With small lobes.
Locule. The cell or compartment of an ovary or anther.
Loculicidal. A dehiscent fruit splitting down the center of a compartment or locule.
Loment. A legume fruit conspicuously constricted between the seeds.
Lunate. Crescent-shaped like the crescent moon.
Lyrate. Pinnatifid with the terminal segment large and rounded and the lower lobes small.

Macrospore. Same as megaspore.
Malphigiaceous. Straight hairs seemingly attached by their middle, pick-shaped. Same as dolabriform.
Many. Eleven or more. Same as numerous.
Marcescent. Withering but still persistent.
Mealy. A surface covered with minute particles, these usually rounded.
Medial. Refers to the middle of a structure.
Megaspore. The larger of the two kinds of spores, used particularly in the Pteridophytes.

Membranaceous. Same as membranous.
Membranous. Thin, more or less translucent and pliable; loosely used in grasses for any thin structure.
Mericarp. A portion of a fruit that splits away as a seemingly separate unit, especially used for the two carpels in Umbelliferae.
Meter (abbreviation M. or m.). Unit of measurement consisting of 100 centimeters; almost 40 inches.
Micron. A microscopic unit of measurement, 1/1000 of a millimeter.
Microspore. The smaller kind of spore when two types are present; used especially in the Pteridophytes.
Millimeter (appreviation mm.). A small unit of measurement, 1/10 of a centimeter or about 1/25 of an inch.
Moniliform. Cylindrical with rounded contractions at regular intervals, resembling a string of beads.
Monadelphous. Stamens united by their filaments into one set.
Monoecious. Flowers unisexual but the staminate and pistillate ones borne on the same plant.
Monophyllous. Used for leaves in plants where related species have compound leaves, but the leaflets here reduced to 1.
Mosslike. With low thin stems and small thin leaves like a moss plant.
Mucilaginous. Slimy or mucilagelike.
Mucro. A short, small, abrupt toothlike tip; loosely used but not very sharp at extreme apex. Compare cuspidate.
Mucronate. Tipped with a mucro.
Mucronulate. Minutely mucronate, the mucro very small.
Multicellular. Consisting of many cells or small compartments.
Multicipital. With many heads, referring to the crown of a single root or to several caudices.
Multifid. Cleft into many lobes or segments, these usually narrow.
Muricate. Roughened with short hard points.
Muriculate. Very finely muricate.
Muticous. Blunt and without a point.

Naked. Lacking some structure, as an appendage or hairs which might ordinarily be expected to be present.
Nectary. A gland or tissue for secreting nectar; often located in highly specialized structures which may themselves be called nectaries.
Needlelike. Long, slender, rather rigid and more or less sharp at apex like a needle. Usually round or square in cross-section but sometimes flattened.
Nerve. A simple or unbranched vein or slender rib.
Netted. Same as reticulated.
Nerviform. On the order of a nerve.
Neuter. Without functional stamens or pistils. Same as neutral.
Neutral. See neuter.
Node. The place on a stem where leaves or branches normally originate; the place on an axis that bears other structures; any swollen or knoblike structure.
Nodose. Knobby or knotty.
Nodulose. Provided with minute knobs.
Numerous. Eleven or more. Same as many.
Nut. A 1-seeded, indehiscent fruit with a hard wall.
Nutlet. A small nut or nutlike fruit; used especially for the separating lobes of the mature ovary in Boraginaceae, Labiatae and Verbenaceae.

Ob. A prefix signifying inversion.
Obcompressed. Flattened the opposite to the usual way.
Obcordate. Inverted heart-shaped, attached at the point.
Obconical. Inversely cone-shaped, attached at the pointed end.

Oblanceolate. Inversely lanceolate, attached at the tapered end.
Oblique. Sides unequal, especially the base of a leaf.
Oblong. Two to four times longer than wide and the sides parallel or nearly so.
Obovate. Inversely ovate, attached at the narrow end.
Obovoid. A 3-dimensional figure of obovate outline.
Obpyramidal. Inverted pyramidal, attached at the pointed end.
Obsolete. Rudimentary or not at all evident; particularly applied to organs usually present.
Obtuse. Blunt or rounded at the apex.
Ochroleucous. Yellowish-white or cream-colored.
Ocrea (pl. ocreae). A tubular stipule or pair of sheathing confluent elongated stipules. Characteristic in the family Polygonaceae.
Oil tube. Small longitudinal ducts in the walls of the fruit of Umbelliferae, presumably containing volatile oils.
Opposite. Leaves two at a node and situated across the stem from each other.
Orbicular. A 2-dimensional figure circular in outline. Compare spherical.
Oval. Loosely used for broadly elliptical, the width over 1/2 the length; some authors have used it as the same as ovate.
Ovary. That part of the pistil that contains the ovules.
Ovate. Egg-shaped in outline, attached at the wide end.
Ovoid. A 3-dimensional figure, ovate in outline.
Ovule. The structure that develops into the seed.

Palate. A rounded projection on the lower lip of a bilabiate corolla, closing the throat.
Palea (pl. paleae or paleas). A chaffy scale or bract; the inner of the two bracts enclosing the grass flower. Compare lemma.
Paleaceous. Chaffy, thin, small and often translucent.
Palmate. The lobes or divisions attached or running down toward one place at the base. Compare pinnate.
Pandurate. Fiddle-shaped. Same as panduriform.
Panduriform. Same as pandurate.
Panicle. A compound inflorescence with the younger flowers at the apex or center; a compound raceme or corymb.
Paniculate. Borne in a panicle; resembling a panicle.
Pannose. With the texture or appearance of felt or closely woven woolen cloth.
Papery. Thin and usually whitish like paper. Compare chartaceous which is usually thick-papery.
Papilionaceous. Like the "sweetpea" type flower of Leguminosae with standard (banner) wings and keel.
Papilla (pl. papillae). A minute nipple-shaped projection.
Papillose. Bearing papillae.
Pappus. The modified calyx limb in Compositae, forming a crown of various character at the summit of the achene.
Parasite. An organism growing upon and obtaining nourishment from another; usually lacking chlorophyll in plants.
Parasitic. Like a parasite.
Parallel veined. A leaf with the veins running parallel to each other, usually all about the same size (except sometimes the midrib) and the connections between obscure. Characteristic of the leaf of Monocotyledoneae.
Parietal. Borne on or pertaining to the wall or inner surface of an ovary or fruit.
Parted. Lobed or cut in over half-way and usually very near to the base or midrib. The sinuses and segments may be sharp or rounded.
Pectinate. Pinnatifid with the segments narrow and arranged like the teeth of a comb; comblike.

Pedicel. The stalk to a single flower of an inflorescence; also used as a stalk to a grass spikelet.
Pedicelled. With a pedicel. Same as pedicillate.
Pedicillate. Borne on a pedicel.
Peduncle. The stalk to a solitary flower or to an inflorescence.
Pedunculate. Borne upon a peduncle.
Peltate. Shield-shaped, attached to the center or near the center, at least in-a-ways from the margin, on the order of an umbrella.
Pendulous. More or less hanging or declined.
Pentagonal. Five-angled.
Pepo. The fleshy indehiscent fruit characteristic of the Cucurbitaceae. It differs from a berry chiefly by having a hard, more or less thickened rind.
Perennial. A plant lasting for three or more years; a stem not dying back over winter.
Perfect. A flower with both functional stamens and pistils.
Perfoliate. A leaf with the stem apparently passing through it or opposite leaves joined around the stem at their bases.
Perianth. The floral envelope consisting of calyx and corolla however incomplete or modified. Used particularly when the calyx and corolla cannot be readily distinguished.
Pericarp. The ripened wall of the matured ovary, sometimes used for the outer wall of the mature fruit regardless of its origin.
Perigynium. The bract in the pistillate flower of sedges that completely surrounds the pistil; it is often inflated and usually almost or completely joined at the edges.
Perigynous. Situated around but not attached to the ovary or its base directly; a flower with stamens and pistils on the calyx tube and the ovary superior.
Persistent. Remaining attached after like parts ordinarily fall off.
Petal. One of the individual parts of the corolla.
Petaloid. Resembling a petal in some way, usually colored other than green.
Petiolate. With a petiole.
Petiole. The stalk to a leaf blade.
Petiolule. The stalk to a leaflet in a compound leaf.
Phyllary. A special name sometimes used for an involucral bract on the head of Compositae.
Phyllopodic. Lower leaves with well developed blades. Used in sedges.
Pilose. With long, soft straight hairs.
Pinna. (pl. pinnae). One of the first or primary divisions of a pinnately compound or decompound leaf; used especially in ferns.
Pinnate. Compound leaf with the leaflets on two opposite sides of an elongated axis.
Pinnatifid. Pinnately lobed, cleft or parted, usually 1/2 way in to the midrib or more.
Pinnule. The pinnate segment of a pinna in a bipinnate leaf. In a tripinnate leaf the pinnules are again pinnately divided.
Pistil. The seed-producing organ, consisting usually of ovary, style, and stigma.
Pistillate. Provided with pistils, used when stamens are lacking.
Pith. The spongy center of a stem, surrounding or joining to the inner part of the vascular bundles.
Pitted. Marked with small depressions or pits.
Placenta (pl. placentae). Any part of the interior of an ovary that bears ovules.
Plane. With flat surface.
Plano-convex. An object usually a fruit or seed, flat on one side and convex on the other.
Plicate. Folded in plaits, usually lengthwise on the order of a folding fan.
Plumose. Hairs with side hairs along the main axis like the plume of a feather.

Plumule. The stem- and leaf-producing structure of an embryo in the seed.
Pod. Any dry dehiscent fruit, often used as a synonym for legume.
Pollen. The male spores in an anther.
Pollinium (pl. pollinia). A mass of waxy or coherent pollen grains as in Asclepias or Orchidaceae.
Polygamo-dioecious. Polygamous but chiefly dioecious; having bisexual flowers and unisexual flowers on separate individuals.
Polygamo-monoecious. Polygamous but chiefly monoecious; having bisexual flowers and unisexual flowers on the same individual.
Polygamous. Having bisexual flowers and unisexual flowers on the same or different individuals.
Polymorphous. With several forms; variable as to habit.
Polypetalous. With separate petals.
Pome. A fleshy indehiscent fruit with an inferior ovary and more than 1 locule; of the apple type.
Posterior. On the side next to or close to the axis. Compare anterior.
Precocious. Appearing or developing very early; used in Salix where the aments develop before the leaves.
Prickle. A small, usually slender outgrowth of the young bark, coming off with it. Compare spine and thorn.
Prismatic. Shape of a prism, in general cylindric with flat sides.
Procumbent. Lying or trailing on the ground, usually not rooting at the nodes. See prostrate.
Proliferous. Producing bulbs or plantlets from leaves or other offshoots.
Prostrate. Lying flat on the ground, if a stem then may or may not root at nodes.
Prothallus (pl. prothallia). A usually flat thallus-like growth resulting from the germination of a spore upon which are produced sexual organs or new plants.
Promixad. Toward the point of attachment.
Proximal. The end of an organ by which it is attached.
Pruinose. With a waxy powdery usually whitish covering, this usually rubbing off readily; glaucous to a conspicuous degree.
Pseudo-. A prefix meaning false.
Puberulent. With very short hairs; minutely pubescent.
Pubescent. Loosely used for covered with hairs; technically with short soft hairs.
Pulvinate. Cushioned or shaped like a close thick mat or cushion.
Punctate. Dotted with depressions, or with translucent internal glands or colored dots.
Puncticulate. Minutely punctate.
Pungent. Tipped with a sharp rigid point.
Pustulose. Beset with pimplelike or bristlelike elevated areas. Same as pustulate.
Pyramidal. Shape of the conventional pyramid, attached at the broad base.
Pyxis. A capsule with circumscissle dehiscence, the top coming off as a lid.

Quinate. With five nearly similar structures (as leaflets) from a common point.

Raceme. An inflorescence with pedicelled flowers borne along a more or less elongated axis with the younger flowers nearest the apex.
Racemiform. In the form of a raceme.
Racemose. Racemelike or bearing racemes.
Rachilla. A small rachis; applied particularly to the axis of a grass spikelet, and to the secondary axis in sedges.
Rachis. The central elongated axis to an inflorescence or a compound leaf.
Radiate. Spreading from or arranged around a common center. In Compositae meaning with ray-flowers.
Radical. Belonging to the root, or apparently arising from or very near the root. The leaves of dandelion are called radical. Compare rosette.
Radicle. The short structure of an embryo in a seed giving rise to the primary root.
Ray. The branch of an umbel or a similar inflorescence. The ligulate or strap-shaped flower in the Compositae, used especially for marginal flowers different from the central regular ones.
Receptacle. The more or less expanded portion of the flower stalk that bears the organs of a flower, or the collected flowers of a head as in Compositae.
Reclinate. Turned or bent abruptly downward.
Reclining. Lying upon something.
Recumbent. Leaning or reposing upon the ground.
Recurved. Curved outward, downward or backward.
Reflexed. Abruptly bent or turned downward or backward.
Regular. A flower with all the members of each set alike in form, size and color; radially symmetrical.
Reniform. Kidney-shaped, usually attached at the center of the incurved side.
Repand. With a wavy surface or margin, not as deep as sinuate. Same as undulate.
Replum. The septum of certain dry dehiscent fruits, persisting after the valves have fallen away; used in the Cruciferae.
Reticulate. In the form of a network; leaf veins in a network.
Retrorse. Directed backward or downward.
Retuse. A rounded apex with a shallow notch.
Revolute. Rolled backward from each margin upon the lower side. Opposite of involute.
Rhizomatous. Having the characters of a rhizome. Sometimes written rhizomatose.
Rhizome. Any prostrate more or less elongated stem growing partly or completely beneath the surface of the ground; usually rooting at the nodes and becoming upcurved at apex. See rootstock.
Rhombic. Outline of an equilateral oblique-angled figure; 4-sided like a diamond shape.
Rhomboid. A solid figure rhombic in outline.
Rib. A primary or prominent vein, usually of a leaf.
Root. The descending axis of the plant without nodes and internodes, absorbing moisture from the ground.
Rootlet. A small root; often used for the aerial supporting roots put out by some vines.
Rootstock. A rootlike stem or branch under or sometimes on the ground. Like rhizome but loosely used by some to include any elongated underground structure that spreads the plant.
Rosette. A dense basal cluster of leaves arranged in circular fashion like the leaves of the common dandelion.
Rostrate. Having a beak.
Rotate. A wheel-shaped corolla with short tube and wide horizontally flaring limb.
Rudiment. An imperfectly developed, usually minute organ.
Rufous. Reddish-brown.
Rugose. With wrinkled or creased surface.
Rugulose. Minutely rugose.
Runcinate. Sharply incised or pinnatifid with the segments directed backward.
Runner. A very slender or filiform stolon; sometimes limited to those that root only at the tip.
Rushlike. Grasslike in general appearance, the flowers usually not colored or conspicuous.

Saccate. Sac-shaped or pouch-shaped.
Sagittate. Shaped like an arrow-head with the basal lobes directed backward. Compare hastate.
Salverform. A corolla with a long slender tube, abruptly flaring into a circular limb.
Samara. A dry indehiscent winged fruit.
Scaberulent. Slightly scabrous.

Scaberulous. Slightly scabrous.
Scabrous. Rough or harsh to the touch, usually from very short stiff hairs or short sharp projections.
Scale. Any thin scarious body resembling the scale of a fish or reptile; often used for such structures present on the basal or underground portion of the plant. Compare bract.
Scape. A naked flowering stem rising from the ground without proper leaves.
Scapose. Bearing a scape or resembling one.
Scarious. Thin, dry, membranous and more or less translucent, not green.
Scorpioid. Coiled at the apex like the tail of a scorpion, used especially for inflorescences.
Scurfy. Covered with small scalelike or branlike particles
Secund. Borne or directed to one side of the axis.
Seed. The matured ovule, consisting of embryo and its coats, with a supply of food.
Sepal. One of the parts of the outer whorl of the floral envelope or calyx, usually green in color.
Sepaloid. Of the color or texture of a sepal, or resembling one in some way.
Septate. Divided by one or more partitions.
Septicidal. A capsule splitting down the septa and not through the locule. Compare loculicidal.
Septum. (pl. septa.) Any kind of a partition.
Sericeous. Covered with long, straight, soft, appressed hairs giving a silky texture.
Serrate. With sharp teeth directed forward.
Serrulate. Serrate with small teeth.
Sessile. Without a stalk of any kind.
Seta. (pl. setae). A bristlelike hair.
Setaceous. Bristlelike.
Setose. Beset with bristles.
Setiform. Like or on the order of a bristle.
Sheath. A tubular envelope, usually used for that part of the leaf of a sedge or grass that envelops the stem.
Shrub. A woody perennial plant smaller than a tree and usually with several basal stems. Compare tree.
Sigmoid. Doubly curved like the letter S.
Silky. Of silklike appearance caused by long, straight, soft, appressed hairs. See sericeous.
Simple. Of only one piece, not completely divided into separate parts. Compare compound.
Sinuate. Strongly wavy-margined, deeper than undulate or repand.
Sinus. The depression or recess between two adjoining lobes.
Sordid. Dirty in tint.
Sorus. (pl. sori). A cluster of sporangia on a fern frond.
Spadix. A spike with a thick and fleshy axis, usually densely flowered with imperfect flowers.
Spathe. A large bract sheathing or enclosing an inflorescence.
Spatulate. Broad and rounded at apex and tapering at base, like a druggist's spatula; flattened spoon-shaped.
Spermatozoid. A motile, ciliated male reproductive cell.
Spherical. A 3-dimensional solid round in outline, like the earth.
Spicate. Arranged in or resembling a spike.
Spike. An inflorescence with the flowers sessile on a more or less elongated axis, with the younger flowers at the apex.
Spikelet. A small or secondary spike, used particularly in grasses and sedges. In grasses it is a unit subtended by two glumes (these rarely absent) and containing one or more florets.
Spikelike. Resembling a spike, used where the flowers are on a short stalk or on very short panicle branches.

Spine. A sharp-pointed rigid deep-seated outgrowth from the stem, not pulling off with the bark. Compare prickle. Sometimes differentiated from thorn by absence of vascular tissue.
Spinescent. Bearing a spine or ending in a spinelike sharp point.
Spinulose. Minutely spiny; beset with small spines.
Sporangium. (pl. sporangia). The spore-bearing case in Pteridophytes.
Spore. The small reproductive body in Pteridophytes.
Sporocarp. The fruit-cases of certain Pteridophytes containing sporangia or spores.
Sporophyll. A leaf bearing spores, often highly modified.
Sporophyte. The spore-bearing asexual generation. Used especially in Pteridophytes for the conspicuous plant body. Compare gametophyte.
Sprawling. Lying on or leaning upon or over another object.
Spreading. Diverging nearly at right angles; nearly prostrate.
Spur. A hollow, saclike or tubular extension of a floral organ, usually nectariferous.
Squamella. (pl. squamellae). A small chaffy bract or scalelike appendage.
Squarrose. Having the parts or processes (usually the tips) spreading or recurved.
Stamen. One of the pollen-bearing organs of a flower. Made up of filament and anther.
Staminal tube. The united part of the filaments when this occurs.
Staminate. Bearing stamens only.
Staminodium. (pl. staminodia). A sterile stamen or any structure lacking an anther but corresponding to a stamen. Also written staminode.
Standard. Same as banner.
Stellate. Starlike or star-shaped with slender segments or hairs radiating out from a common center.
Sterile. Infertile and unproductive, as a flower without a pistil, a stamen without an anther or a leafy shoot without flowers.
Stigma. That part of the pistil that receives the pollen, usually at or near the apex of the pistil and mostly hairy, papillose or sticky.
Stigmatic. Belonging to or having the characteristics of a stigma.
Stipe. The stalklike support of a pistil (above the other flower parts). Also the name for the petiole of a fern frond.
Stipel. An appendage like a stipule but subtending the leaflet.
Stipitate. Provided with a stipe or with a slender stalklike base.
Stipulate. Provided with stipules.
Stipule. An appendage at the base of the petiole or leaf at each side of its insertion; often more or less united.
Stolon. A trailing shoot above ground, rooting at the nodes. Compare runner.
Stoloniferous. Bearing stolons.
Stoloniform. On the general order of a stolon.
Stomate. (pl. stomata). A small opening on the surface of a leaf through which gaseous exchange takes place. Sometimes written stoma.
Stramineous. Straw-colored.
Striate. Marked with fine longitudinal lines, grooves, furrows or streaks.
Strict. Very straight and upright.
Strigillose. Like strigose but hairs very short.
Strigose. With appressed, stiff, rather short hairs.
Strobilus. (pl. strobili). An inflorescence characterized by imbricated bracts or scales as a pine cone.
Strophiole. An appendage at the hilum of some seeds.
Style. The usually stalklike part of a pistil connecting the ovary and stigma.
Stylopodium. A disklike expansion of the base of the style as in Umbelliferae.

Sub-. A prefix meaning "almost" or "below."
Subtending. Situated closely beneath something often enclosing or embracing it.
Subulate. Awl-shaped; narrowly triangular and tapering to a sharp point.
Succulent. Fleshy and full of juice.
Suffruticose. Low-shrubby; applied to perennials the lower part of the stems woody but the upper part herbaceous.
Sulcate. Grooved or furrowed, especially if the grooves are deep and longitudinal.
Sulcus. (pl. sulci). A furrow or groove.
Superior ovary. An ovary with the perianth inserted below it.
Surculose-proliferous. Producing runners or offsets from the base or from rootstocks.
Suture. A junction or seam of union; a line of dehiscence.
Sympetalous. Petals more or less united. Same as gamopetalous.

Taproot. The primary root continuing the axis of the plant downward.
Tawny. Dull yellowish with a tinge of brown.
Tendril. A slender cauline or foliar outgrowth, commonly coiling at apex and serving as an organ of support.
Terete. Circular in cross-section and more or less elongated. Like cylindrical but may be slightly tapering.
Ternate. Arranged in three's.
Terrestrial. A plant growing in the air with its basal parts in soil. Compare aquatic with parts immersed in water.
Tesellate. Checkered.
Thalloid. Resembling or on the order of a thallus.
Thallus. A vegetative often flattened body not differentiated into stems and leaves.
Thorn. A stiff, hard, sharp-pointed emergence more deeply seated than a prickle. By some differentiated from a spine in having vascular tissue.
Throat. The orifice of a gamopetalous corolla or calyx, at or just below the junction of the tube with the limb.
Thyrse. A contracted, cylindrical or ovoid-pyramidal, usually densely flowered panicle, like a cluster of grapes.
Thyrsoid. Resembling a thyrse.
Tomentose. With a dense woollike covering of matted, intertangled hairs of medium length. Compare lanate and canescent.
Tomentulose. Sparingly or minutely tomentose.
Tomentum. The covering of closely interwoven and tangled hairs in a tomentose surface.
Tortuous. Twisted or bent.
Torulose. Minutely torose as in a small pod constricted between the seeds.
Trailing. Prostrate but not rooting.
Translucent. Transmitting rays of light without being transparent.
Tree. A perennial woody plant of considerable stature at maturity and with one or a few main trunks. Rather loosely used but a fairly well understood concept.
Trichome. A hairlike outgrowth of the epidermis.
Trichotamous. Three-forked with equal or nearly equal branches.
Trifoliate. A compound leaf with three leaflets.
Trigonal. Three-angled.
Trigonous. Three-angled. Same as trigonal.
Tripinnate. Pinnately compound three times; the pinnules again pinnate.
Tripinnatifid. Pinnatifid three times; pinnatifid with the primary segments pinnatifid and these secondary segments again pinnatifid.
Triquetrous. With three salient angles, the sides concave or channeled.

Triternate. Three times ternate; ternate with the three main divisions once and once-again ternate.
Truncate. Squared at the tip or base as if cut off.
Tube. Any hollow cylindrical structure, especially the tubular basal part of a gamopetalous corolla.
Tuber. A thickened, short subterranean stem having numerous buds called eyes; like a potato.
Tubercle. A small rounded structure, often pimplelike.
Tuberculate. Bearing small processes or tubercles.
Tufted. Having a cluster of hairs or other slender outgrowths; stems in a very close cluster.
Tumid. Swollen.
Tunicated. Having concentric coats as an onion bulb.
Turbinate. Top-shaped; inversely conical. About the same as obconical.
Turgid. Swollen or tightly drawn, said of a thin covering expanded by internal pressure.
Turion. A scaly often succulent shoot produced from a bud on an underground rootstock.
Twining. Ascending by coiling around a support.

Umbel. A convex or flat-topped inflorescence the flowers all arising from one point, the younger in the center.
Umbellate. In or like an umbel.
Umbellet. A small or secondary umbel in a compound umbel.
Umbonate. Bearing a stout projection in the center; bossed.
Uncinate. Hooked, near the apex or in the form of a hook.
Undershrub. A small low shrub or a perennial plant woody only at the base.
Undulate. The margin gently wavy. Same as repand.
Unifoliate. A theoretically compound leaf with all but one leaflet suppressed; a simple-appearing leaf in a group with compound leaves.
Unilateral. Arranged on one side.
Uniseriate. Arranged in one row or series.
Unisexual. With either stamens or pistils, not both. Compare bisexual and perfect.
Urceolate. Hollow and cylindrical or ovoid but contracted at or near the mouth like an urn.
Urn-shaped. Same as urceolate.
Utricle. A small thin-walled 1-seeded fruit; any bladderlike body.

Valvate. Opening by valves or provided with valves; also for parts meeting together edge to edge without overlapping.
Valve. One of the parts or segments into which a dehiscent fruit splits.
Vascular bundle. An elongated group of cells specialized for conduction and often support. In a leaf the veins.
Vein. Threads of vascular tissue in a leaf or other organ especially those which branch. Compare nerve.
Velum. The fold of the innerside of the leaf base of Isoetes functioning as an indusium.
Ventral. Belonging to the inner or axis side of an organ; the upper surface of a leaf.
Ventricose. Inflated or swollen unequally as on one side.
Vernation. The particular arrangement of a leaf or its parts in the bud.
Verrucose. Covered with wartlike elevations.
Versatile. An anther attached at or near its middle and turning freely on its support. Compare basifixed.
Verticillate. With three or more leaves or other structures arranged in a circle about a stem or other common axis. Same as whorled.
Villous. With long, soft, somewhat wavy hairs. Compare pilose.
Vine. A plant climbing or scrambling on some support, the stem not standing upright of itself.
Viscid. Glutinous, sticky or gummy to the touch.

Weed. A troublesome or aggressive plant that intrudes where not wanted, especially a plant that vigorously colonizes disturbed areas. To the rangeman a weed is a herbaceous nongrasslike plant on the range.

Whorled. With three or more leaves or other structures arranged in a circle around a stem or some common axis. Same as verticillate.

Wing. Any membranous or thin expansion bordering or surrounding an organ; one of the lateral petals in a papilionaceous corolla.

Winged. Provided with wings.

Winter annual. A plant where the seed germinates in the fall, the seedling surviving the winter and completing its growth in the spring of the next season.

Xerophytic. Adapted to dry or arid habitats.

INDEX

Abies, 27, 184
Abronia, 219
Abutilon, 372
Acer, 365
Aceraceae, 365
Acerates, 433
Achillea, 578
Achyropappus, 595
Acnida, 216
Acomastylis, 314
Aconitum, 239
Acorus, 144
Acrolasia, 380
Actaea, 244
Actinea, 592, 593, 594
Actinea, 591
Actinella, 587, 592, 593, 594
Acuan, 317
Adderstongue Family, 12
Adelia, 425
Adelia, 425
Adenostegia, 508, 509
Adiantum, 18
Adopogon, 632
Adoxa, 523
Adoxaceae, 522
Aegilops, 69
Agalinis, 511
Agastache, 477
Agoseris, 637
Agrimonia, 311
Agrimony, 311
Agropyron, 65
Agrostemma, 226
Agrostideae, 41, 79
Agrostis, 81
Aizoaceae, 222
Alchemilla, 314
Alder, 179
Aletes, 405
Alisma, 37
Alismaceae, 36
Alkali-grass, 50
Alkanet, 457
Allionella, 221
Allionia, 220, 221, 222
Allium, 155
Allocarya, 469
Alnus, 179
Alopecurus, 83
Alpine Avens, 309
Alpine Forget-me-not, 457
Alsine, 231, 232
Alsinopsis, 234, 235
Alumroot, 293
Alyssum, 263
Alyssum, 263
Apinus, 27
Amaranthaceae, 216
Amaranth Family, 216
Amaranthus, 216
Amaranthus, 216
Amarella, 429
Amaryllidaceae, 159
Amaryllis Family, 159
Amauriopsis, 595
Ambrosia, 618
Ambrosineae, 532
Amelanchier, 298
American Mistletoe, 185
Ammannia, 387
Amorpha, 319
Amphilophus, 114
Amsinckia, 468
Amsonia, 430

Amsonia, 430
Anacardiaceae, 363
Anacharis, 37
Anagallis, 424
Anaphalis, 624
Anchusa, 457
Androcera, 484
Andropogon, 114
Andropogoneae, 43, 113
Androsace, 422
Androstephium, 154
Anemone, 241
Anemone, 241
Anemopsis, 181
Anethum, 408
Angelica, 409
Angiospermae, 29
Anisolotus, 333
Anogra, 396, 397
Anteclea, 157
Antelope-brush, 310
Antennaria, 621
Anthemideae, 529
Anthemis, 578
Antheropogon, 102
Anthopogon, 429
Anthoxanthum, 104
Apache Plume, 310
Apios, 329
Aplopappus, 555
Apocynaceae, 430
Apocynum, 430
Apple, 296
Apple-of-peru, 482
Aquilegia, 239
Arabidopsis, 277
Arabis, 270
Araceae, 144
Aragallus, 334, 335, 336
Aralia, 399
Araliaceae, 399
Arceuthobium, 184
Arctium, 630
Arctostaphylos, 419
Arenaria, 233
Argemone, 252
Argentina, 302
Argophylli, 343
Aristida, 96
Armeria, 420
Arnica, 616
Arrhenatherum, 77
Arrowgrass, 35
Arrowgrass Family, 35
Arrowhead, 36
Artemisia, 580
Arum Family, 144
Asclepiadaceae, 432
Asclepias, 432
Ash, 425
Asparagus, 156
Asparagus, 156
Aspen, 165
Asperugo, 457
Asperula, 516
Asplenium, 15, 17
Aster, 570
Aster, 570
Astereae, 529
Astragalus, 336, 334, 335
Atelophragma, 340
Atenia, 407
Athyrium, 16
Atragene, 243
Atriplex, 204

653

Aulospermum, 412
Avena, 76, 77
Aveneae, 40, 73
Avens, 312
Azalea, 419
Azaleastrum, 419
Azolla, 22
Azolla, 22
Baby-blue Eyes, 450
Baby's Breath, 393
Baccharis, 545
Bacopa, 511
Bahia, 594
Ball Cactus, 384
Balsaminaceae, 355
Balsamita, 579
Balsamorhiza, 599
Balsam-root, 599
Baneberry, 244
Barbarea, 283
Barberry, 251
Barberry Family, 251
Barley, 72
Barrel Cactus, 385
Bassia, 210
Bastard Toadflax, 185
Batidophaca, 338, 339
Batrachium, 247
Bean Caper, 356
Bearberry, 419
Beardtongue, 491
Beargrass, 153
Beckmannia, 100
Bedstraw, 515
Beebalm, 473
Beech Family, 180
Bee-plant, 255
Beggars Ticks, 455
Beggarticks, 601
Bellflower, 525
Bellflower Family, 524
Bentgrass, 81
Berberidaceae, 251
Berberis, 251
Berlandiera, 598
Berula, 406
Besseya, 489
Betony, 479
Betula, 179
Bidens, 601
Bilderdykia, 196, 197
Bindweed, 438
Birch, 179
Birch Family, 179
Birdbeak, 508
Bistorta, 198, 199
Bisulcati, 342
Bitterbrush, 310
Bittercress, 282
Bitterroot, 224
Bittersweet Family, 364
Blackberry, 306
Blackbrush, 299
Blackfoot, 596
Bladderfern, 18
Bladder Pod, 269
Bladderwort, 513
Bladderwort Family, 513
Blanketflower, 590
Blazing Star, 379, 543
Blepharoneuron, 91
Blitum, 212
Blowout Grass, 61
Bluebells, 458
Blueberry, 418
Blue-eyed Grass, 160
Blue-eyed Mary, 490

Blue Flag, 159
Bluegrass, 52
Bluestem, 114
Boebera, 586
Bogbean, 425
Bog Orchid, 162
Borage Family, 453
Boraginaceae, 453
Bossekia, 307
Botrychium, 12
Bouteloua, 101
Boxleaf, 364
Boykinia, 290
Brachyactis, 572, 573
Brachypodium, 47
Bracken, 19
Brassica, 281
Brauneria, 596
Braya, 256
Brickellbush, 542
Brickellia, 542
Brigham Tea, 29
Bristle Grass, 111
Bristlethistle, 629
Brodiaea, 154
Bromegrass, 44
Bromus, 44
Brookgrass, 61
Brookweed, 421
Broomrape Family, 514
Brownweed, 552
Buchloe, 103
Buckbean, 425
Buckbean Family, 425
Buckthorn, 367
Buckthorn Family, 366
Buckwheat, 200
Buckwheat Family, 185
Buffaloberry, 387
Buffalograss, 103
Bugleweed, 473
Bugloss, 457
Bugseed, 215
Bulbilis, 103
Bulrush, 140
Bundle Flower, 317
Burdock, 630
Burningbush Family, 364
Burnut, 356
Burreed, 30
Burreed Family, 30
Burrograss, 64
Bursa, 262
Bursage, 619
Butter Bur, 607
Buttercup, 245
Butterfly Weed, 391
Cactaceae, 382
Cactus, 384
Cactus Family, 382
Calamagrostis, 79
Calamovilfa, 80
Calceolaria, 378
California Poppy, 253
Callirhoe, 369
Callisteris, 447
Callitrichaceae, 363
Callitriche, 363
Calochortus, 153
Caltha, 241
Caltrop Family, 356
Calypso, 162
Calypso, 162
Camelina, 267
Camomile, 578
Campanula, 525
Campanulaceae, 524

Campion, 226, 228
Canarygrass, 104
Cannabis, 182
Capnoides, 253
Capparidaceae, 254
Capper Family, 254
Caprifoliaceae, 518
Capsella, 262
Caraway, 407
Cardamine, 282
Cardaria, 259
Carduus, 629, 625, 626, 627, 628
Carex, 118
Carpet-weed, 222
Carpet-weed Family, 222
Carrion Flower, 154
Carrot, 402
Carrot Family, 399
Carthamus, 629
Carthiera, 278
Carum, 407
Caryophyllaceae, 225
Caryopitys, 27
Cassia, 320
Castilleja, 505
Catabrosa, 61
Catchfly, 226
Catchweed, 457
Cathartolinum, 354, 355
Catnip, 479
Catsear, 631
Cat-tail, 30
Cat-tail Family, 29
Caulanthus, 278
Ceanothus, 366
Cedar, 26
Celastraceae, 364
Celtis, 182
Cenchrus, 113
Centaurea, 630
Centaurium, 427
Centaury, 427
Centunculus, 424
Cephalobembix, 595
Cephalophora, 593
Cerastium, 230
Ceratophyllaceae, 236
Ceratophyllum, 236
Cercocarpus, 299
Chaenactis, 588
Chaetochloa, 112
Chaffweed, 424
Chamaechaenactis, 586
Chamaecrista, 320
Chamaenerion, 389, 390
Chamaerhodos, 312
Chamaesaracha, 482
Chamaesyce, 360, 361
Chamomilla, 580
Checkermallow, 370
Cheilanthes, 19
Cheirinia, 276, 277
Chelone, 496, 497, 498, 504
Chenopodiaceae, 203
Chenopodium, 211
Cherry, 308
Chickweed, 231
Chicory, 632
Chimaphila, 416
Chionophila, 505
Chlorideae, 42, 99
Chloris, 101
Chloris, 101
Chokecherry, 308
Chondrophylla, 428
Chorispora, 280
Chrysanthemum, 579

Chrysopsis, 549
Chrysosplenium, 292
Chrysothamnus, 545
Chylismia, 398
Cichorieae, 533
Cichorium, 632
Cicuta, 406
Cinna, 83
Circaea, 388
Cirsium, 625
Cistaceae, 386
Cladothrix, 218
Clammy Weed, 255
Claytonia, 225
Clematis, 242
Clementsia, 284
Cleome, 255
Cleomella, 254
Cliffbrake, 20
Cliffbush, 288
Cliffrose, 311
Clinopodium, 478
Cloakfern, 19
Clover, 329
Clubflower, 508
Clubmoss, 24
Clubmoss Family, 24
Cnemidophacos, 344, 343, 345, 346, 347
Cocklebur, 618
Cockspur, 110
Coeloglossum, 163
Cogswellia, 412, 413, 414
Coleogyne, 299
Coleosanthus, 543
Collinsia, 490
Collomia, 439
Coloradoa, 386
Columbine, 239
Colutea, 341
Comandra, 185
Combseed, 455
Commelina, 145
Commelinaceae, 145
Compass Plant, 603
Compositae, 528
Composite Family, 528
Coneflower, 598
Conioselinum, 409
Conium, 407
Conringia, 282
Convolvulaceae, 435
Convolvulus, 438
Conyza, 561
Copperweed, 620
Coralberry, 520
Corallorhiza, 161
Coral Root, 161
Cordgrass, 100
Cordylanthus, 508
Coreopsis, 603
Coriophyllus, 412
Corispermum, 215
Cornaceae, 414
Corncockle, 226
Cornus, 415
Corydalis, 253
Corydalis, 253
Corylaceae, 179
Corylus, 179
Coryphantha, 384
Cosmos, 597
Cottongrass, 139
Cottonsedge, 139
Cottontop, 106
Cottonwood, 165
Covillea, 356
Cowania, 311

Cowbane, 408
Cowlily, 235
Cow Parsnip, 408
Cowslip, 422
Crabgrass, 106
Cranesbill, 351
Crassina, 603
Crassulaceae, 283
Crataegus, 296
Crazyweed, 334
Creeping Wintergreen, 419
Creosote Bush, 356
Crepis, 634
Cress, 267
Cristatella, 254
Crocanthemum, 386
Croton, 358
Croton, 358
Crowfoot, 245
Crowfoot Family, 236
Crownbeard, 599
Cruciferae, 256
Crunocallis, 225
Cryptantha, 462
Cryptogramma, 15
Ctenophyllum, 345, 346
Cucurbita, 523
Cucurbitaceae, 523
Cudweed, 624
Cupgrass, 106
Currant, 285
Cuscuta, 435
Cutgrass, 105
Cyclachaena, 621
Cyclanthera, 523
Cycloloma, 208
Cymopterus, 411, 404
Cynareae, 533
Cynodon, 99
Cynoglossum, 455
Cynomarathrum, 413, 414
Cynosurus, 62
Cyperaceae, 116
Cyperus, 116
Cypripedium, 162
Cyrtorhyncha, 248
Cystium, 349
Cystopteris, 18
Cytherea, 162
Dactylis, 61
Dalea, 325, 347
Dandelion, 638
Danthonia, 78
Dasiophora, 302
Dasiphora, 302
Dasystephana, 428, 429
Datura, 480
Daucophyllum, 404
Daucus, 402
Dayflower, 145
Dead Nettle, 479
Death Camas, 157
Deervetch, 333
Delphinium, 237
Deschampsia, 75
Descurainia, 274
Desert Dandelion, 639
Desert Plume, 257
Desert-thorn, 481
Desmanthus, 317
Diaperia, 621
Dichrophyllum, 361
Dicoria, 620
Dicotyledoneae, 165
Digitaria, 106
Diholcos, 342, 343
Dill, 408

Diplocystium, 349
Dipsaceae, 527
Dipsacus, 527
Disella, 372
Disporum, 159
Distegia, 520
Distichlis, 61
Ditaxis, 362
Dithyrea, 258
Dock, 200
Dodder, 435
Dodecatheon, 421
Dogbane, 430
Dogbane Family, 430
Dog Fennel, 578
Dogtail, 62
Dogtoothgrass, 99
Dogtooth Violet, 157
Dogweed, 586
Dogwood, 415
Dogwood Family, 414
Dondia, 210
Double Bladder Pod, 258
Douglas Fir, 28
Draba, 263
Dracocephalum, 478, 477
Dragonhead, 476
Dropseed, 89
Drosace, 423
Dryad, 309
Dryas, 309
Drymocallis, 303
Dryopteris, 17
Ducksmeat, 144
Duckweed, 144
Duckweed Family, 144
Dugaldia, 591, 594
Dwarf Dandelion, 632
Dwarf Mistletoe, 184
Dyssodia, 586
Eatonia, 74
Echinacea, 596
Echinocactus, 385
Echinocereus, 384
Echinochloa, 110
Echinocystis, 524
Echinops, 629
Echinopsilon, 210
Echium, 457
Eclipta, 597
Edwinia, 288
Eel-grass, 38
Elaeagnaceae, 386
Elaeagnus, 387
Elatinaceae, 373
Elatine, 373
Elder, 519
Eleocharis, 142
Elephantella, 509
Eleusine, 99
Ellisia, 450
Elm Family, 182
Elodea, 37
Elymus, 69, 66
Elyna, 118
Encelia, 604
Enceliopsis, 595
Enchanter's Nightshade, 388
Enclosed-seeded Plants, 29
Endolepis, 207
Engelmann Daisy, 597
Engelmannia, 597
Enomegra, 253
Ephedra, 29
Epilobium, 389
Epipactis, 164
Equisetaceae, 22

Equisetum, 22
Eragrostis, 58
Erianthus, 113
Ericaceae, 415
Erigeron, 561
Eriocarpum, 558
Eriochloa, 106
Eriocoma, 91
Eriogonum, 185
Eriophorum, 139
Eritrichium, 457
Erocallis, 224
Erodium, 351
Eruca, 281
Erxebenia, 417
Erysimum, 276, 280
Erythraea, 427
Erythronium, 157
Eschenbachia, 561
Eschscholzia, 253
Euastragalus, 349
Eucephalus, 574
Euklisia, 278
Eunanus, 512
Eupatorieae, 528
Eupatorium, 542
Euphorbia, 358
Euphorbiaceae, 358
Euploca, 454
Eurotia, 208
Eustoma, 430
Euthamia, 553, 554
Evax, 621
Evening Primrose, 393
Evening Primrose Family, 388
Everlasting, 624
Evolvulus, 437
Fagaceae, 180
Fagopyrum, 200
Fairybells, 159
Fallugia, 310
False Boneset, 544
Falsebrome, 47
False Buffalograss, 103
False Bugbane, 243
False Camomile, 580
False Dragonhead, 478
False Flax, 267
False Gromwell, 461
False Hellebore, 156
False Indigo, 319
False Mallow, 370
False Melic, 63
False Mermaid, 355
False Mermaid Family, 355
False-penny royal, 474
False Pimpernel, 424, 487
False Solomon's Seal, 158
False Yarrow, 588
Fame-flower, 224
Fawnlily, 157
Felwort, 426
Fendlera, 287
Fendlerbush, 287
Fendlerella, 288
Fennel, 407
Fern Family, 14
Ferns, 12
Fescue, 48
Festuca, 48, 50
Festuceae, 39, 44
Fetid-marygold, 586
Feverfew, 595
Fiddleneck, 468
Figwort, 490
Figwort Family, 485
Filago, 621

Filbert, 179
Filix, 18
Fimbristylis, 143
Fingergrass, 101, 106
Fir, 27
Firechalice, 388
Fishweed, 31
Flatsedge, 116
Flaveria, 590
Flax, 354
Flax Family, 354
Fleabane, 561
Floerkea, 355
Flowering Plants, 25
Foeniculum, 407
Fogfruit, 471
Forestiera, 425
Forestiera, 425
Forget-me-not, 458
Forsellesia, 365
Four-O'clock, 220
Four-O'clock Family, 218
Foxtail, 83
Fragaria, 312
Frankenia, 374
Frankeniaceae, 374
Frankenia Family, 374
Franseria, 619
Frasera, 426
Fraxinus, 425
Fritillaria, 158
Fritillary, 158
Froelichia, 218
Frogs-bit Family, 37
Frostweed, 386
Fumariaceae, 253
Fumitory Family, 253
Funastrum, 432
Funnellily, 154
Gaertneria, 619, 620
Gaillardia, 590, 592
Galeopsis, 478
Galinsoga, 597
Galium, 515
Galpinsia, 397
Garden Rocket, 281
Gaultheria, 419
Gaura, 391
Gaurella, 395
Gayfeather, 543
Gayophytum, 393
Gentian, 427
Gentiana, 427
Gentianaceae, 425
Gentian Family, 425
Geoprumnon, 348, 349
Geraniaceae, 351
Geranium, 351
Geranium Family, 351
Gerardia, 511
Germander, 475
Geum, 312
Giant-hyssop, 477
Gilia, 444, 440, 448, 449
Giliastrum, 445
Gill-over-the-ground, 478
Ginseng Family, 399
Glasswort, 204
Glaux, 421
Glecoma, 478
Globeflower, 245
Globe Mallow, 370
Globethistle, 629
Glyceria, 50
Glycyrrhiza, 320
Gnaphalium, 624
Gnetaceae, 29

Goatgrass, 69
Goatsbeard, 631
Golden Aster, 549
Goldeneye, 605
Goldenpea, 320
Goldenrod, 552
Goldenweed, 555
Goldstargrass, 159
Goodyera, 164
Gooseberry, 285
Goosefoot, 211
Goosefoot Family, 203
Goosegrass, 99
Gourd, 523
Gourd Family, 523
Grama, 101
Gramineae, 38
Grape, 368
Grape Family, 368
Grapefern, 12
Graphephorum, 75
Grass Family, 38
Gratiola, 487
Grayia, 208
Greasebush, 365
Greasewood, 204
Greenbrier, 154
Greeneocharis, 464
Greenthread, 600
Green Violet, 378
Grindelia, 549
Gromwell, 461
Grossularia, 286
Ground Cherry, 482
Ground Ivy, 478
Groundnut, 329
Ground Pine, 24
Groundsel, 608
Groundsel Tree, 545
Groundsmoke, 393
Gumweed, 549
Gutierrezia, 552
Gymnolomia, 605
Gymnospermae, 25
Gymnosteris, 439
Habenaria, 162
Hackberry, 182
Hackelia, 456
Hairgrass, 75
Halerpestes, 247
Halimolobos, 277
Halogeton, 209
Halogeton, 209
Haloragaceae, 398
Haloragidaceae, 398
Hamosa, 350, 351
Haploesthes, 618
Haplopappus, 555
Harbouria, 405
Hardgrass, 52
Harebell, 525
Hare's Ear Mustard, 282
Haw, 296
Hawksbeard, 634
Hawkweed, 636
Hawthorn, 296
Hazel-nut, 179
Healall, 479
Heath Family, 415
Hedeoma, 474
Hedgehog Cactus, 384, 385
Hedge Hyssop, 487
Hedgenettle, 479
Hedysarum, 329
Helenieae, 530
Helenium, 591
Heliantheae, 531

Helianthella, 604
Helianthemum, 386
Helianthus, 605
Helictotrichon, 76
Heliopsis, 604
Heliotrope, 454
Heliotropium, 454
Helleborine, 164
Hemicarpha, 140
Hemlock-parsley, 409
Hemp, 182
Hempnettle, 478
Henbane, 481
Heracleum, 408
Heronbill, 351
Herrickia, 574
Hesperidanthus, 278
Hesperis, 256, 277
Hesperochloa, 50
Heteranthera, 146
Heterocodon, 525
Heterothrix, 277
Heuchera, 293
Hibiscus, 369
Hieracium, 636
Hierochloe, 103
Hilaria, 98
Hilaria, 98
Hippuris, 399
Hoarhound, 475
Hoffmanseggia, 316
Holcus, 77, 115
Hollyfern, 17
Holodiscus, 309
Homalobus, 345, 341, 343, 346, 347, 348
Homolocenchrus, 105
Honeysuckle, 520
Honeysuckle Family, 518
Hop, 182
Hop-sage, 208
Hoptree, 357
Hordeae, 40, 65
Hordeum, 72
Horkelia, 311
Horned Pondweed, 35
Hornwort, 236
Hornwort Family, 236
Horsebrush, 607
Horsemint, 473
Horsetail, 22
Horsetail Family, 22
Houndstongue, 455
Hummingbird-trumpet, 388
Humulus, 182
Hutchinsia, 262
Hybanthus, 378
Hydranthelium, 511
Hydrocharitaceae, 37
Hydrophyllaceae, 448
Hydrophyllum, 450
Hymenopappus, 587, 593
Hymenoxys, 591
Hyoscyamus, 481
Hypericaceae, 374
Hypericum, 374
Hypochoeris, 631
Hypopitys, 416
Hypoxis, 159
Icegrass, 83
Iliamna, 373
Ilysanthes, 487
Impatiens, 355
Indian Corn, 116
Indiangrass, 115
Indian Hemp, 430
Indian Lettuce, 225
Indian Mallow, 372

Indianpipe, 416
Indian Potato, 403
Indigo Bush, 325
Inoxalis, 353
Inuleae, 533
Ipomoea, 437
Iridaceae, 159
Iris, 159
Iris, 159
Iris Family, 159
Ironweed, 542
Isocoma, 557
Isoetaceae, 23
Isoetes, 23
Iva, 620
Ivesia, 311
Jacob's Ladder, 440
Jamesia, 288
Jewelweed, 355
Jewel-weed Family, 355
Jimson Weed, 480
Johnsongrass, 115
Jointfir, 29
Jointfir Family, 29
Jonesiella, 342
Juglandaceae, 179
Juglans, 179
Juncaceae, 146
Juncaginaceae, 35
Juncoides, 150, 151
Juncus, 146
Junegrass, 73
Juniper, 26
Juniperus, 26, 184, 185
Kallstroemia, 356
Kalmia, 418
Kalmia, 418
Kelloggia, 516
Kentrophyta, 337, 338
Kinnikinick, 419
Kitten-tails, 489
Knapweed, 630
Knotweed, 195
Kobresia, 118
Kochia, 209
Koeleria, 73
Koniga, 262
Krameria, 316
Krigia, 632
Kuhnia, 544
Kunzia, 311
Labiatae, 471
Lacinaria, 544
Lactuca, 633
Ladies Tresses, 164
Lady Fern, 18
Lady's Slipper, 162
Lamium, 479
Lappula, 455, 456, 457
Lapsana, 632
Larkspur, 237
Larrea, 356
Lathyrus, 327
Laurel, 418
Laurentia, 526
Lavauxia, 395
Lechia, 386
Leersia, 105
Legume Family, 314
Leguminosae, 314
Leiostemon, 499
Lemna, 144
Lemnaceae, 144
Lentibulariaceae, 513
Leontodon, 638, 639
Leonurus, 477
Lepachys, 599

Lepargyraea, 387
Lepidium, 260
Leptasea, 290, 291
Lepteiris, 501
Leptilon, 561
Leptochloa, 99
Leptodactylon, 448
Leptoglottis, 317
Leptotaenia, 414
Lesquerella, 269
Leucampyx, 589
Leucanthemum, 579
Leucelene, 573
Leucocraspedum, 426
Leucocrinum, 154
Lewisia, 224
Liatris, 543
Licorice, 320
Ligusticella, 403
Ligusticum, 403
Liliaceae, 151
Lilium, 157
Lily, 157
Lily Family, 151
Limnanthaceae, 355
Limnia, 225
Limnobotrya, 285
Limnorchis, 163
Limosella, 511
Linaceae, 354
Linanthastrum, 449
Linanthus, 448
Linaria, 490
Lindernia, 487
Linnaea, 520
Lipfern, 19
Lippia, 471
Listera, 163
Listera, 163
Lithophragma, 291
Lithospermum, 461
Lizardtail Family, 181
Lloydia, 158
Loasaceae, 378
Loasa Family, 378
Lobelia, 526
Loco, 334
Locust, 326
Lolium, 73
Lomatium, 413
Lonchophaca, 348
Lonicera, 520
Loosestrife, 388, 423
Loosestrife Family, 387
Lophocarpus, 36
Loranthaceae, 183
Lotus, 333
Lousewort, 509
Lovage, 403
Lovegrass, 58
Lowellia, 586
Lupine, 321
Lupinus, 321
Luzula, 150
Lychnis, 228
Lycium, 481
Lycopodiaceae, 24
Lycopodium, 24
Lycopsis, 457
Lycopus, 473
Lycurus, 84
Lygodesmia, 640
Lysiella, 162
Lysimachia, 423
Lythraceae, 387
Lythrum, 388
Machaeranthera, 570, 572, 573

Macrocalyx, 450
Macronema, 557
Macuillamia, 511
Madder Family, 515
Madia, 596
Madronella, 476
Maidenhair, 18
Maize, 116
Malacothrix, 639
Malaxis, 163
Malcolmia, 277
Mallow, 369
Mallow Family, 368
Malus, 296
Malva, 369
Malvaceae, 368
Malvastrum, 371
Mamillaria, 384
Manna Grass, 50
Manzanita, 419
Maple, 365
Maple Family, 365
Marbleseed, 461
Marestail, 399
Marijuana, 182
Mariposa Lily, 153
Marrubium, 475
Marsh Marigold, 241
Marsilea, 21
Marsileaceae, 21
Martynia, 515
Martyniaceae, 515
Maruta, 578
Matricaria, 580
Matrimony Vine, 481
Mayweed, 580
Meadow Rue, 243
Medic, 332
Medicago, 332
Melampodium, 596
Melandrium, 228
Melic, 62
Melica, 62
Melic Grass, 62
Melilotus, 333
Melosmon, 476
Menodora, 424
Menodora, 424
Mentha, 476
Mentzelia, 379
Menyanthaceae, 425
Menyanthes, 425
Meriolix, 398
Mertensia, 458
Micrampelis, 524
Microphacos, 347
Microseris, 637
Microsteris, 444
Mignonette, 283
Mignonette Family, 283
Milfoil, 578
Milkvetch, 336
Milkweed, 432
Milkweed Family, 432
Milkwort, 357
Milkwort Family, 357
Millet, 111
Mimosa, 317
Mimulus, 511
Mint, 476
Mint Family, 471
Mirabilis, 220
Miscanthes, 289, 290
Miscanthus, 113
Mistletoe Family, 183
Mitella, 292
Miterwort, 292

Mitrewort, 292
Mock Cucumber, 524
Mockorange, 288
Mock-pennyroyal, 474
Moehringia, 234
Moldavica, 476
Mollugo, 222
Monarda, 473
Monardella, 476
Moneses, 416
Monkeyflower, 511
Monkshood, 239
Monniera, 511
Monocotyledoneae, 29
Monolepis, 215
Monotropa, 416
Montia, 225
Moonwort, 12
Moraceae, 182
Morning Glory, 437, 438
Morning-glory Family, 435
Moschatel, 523
Moschatel Family, 522
Mosquito Fern, 22
Motherwort, 477
Mountain Ash, 296
Mountain Avens, 309
Mountain-dandelion, 637
Mountain Heath, 418
Mountainlily, 154
Mountain Lover, 364
Mountain Mahogany, 299
Mountain Mint, 477
Mountain Sorrel, 195
Mountain Spray, 309
Mouse-ear Chickweed, 230
Mousetail, 239
Mudplantain, 146
Mudwort, 511
Muhlenbergia, 85
Muhly, 85
Mulberry Family, 182
Mules-ears, 600
Mullein, 486
Munroa, 103
Muscaria, 290
Musineon, 404
Muskroot, 523
Mustard, 281
Mustard Family, 256
Myosotis, 458
Myosurus, 239
Myriophyllum, 398
Myzorrhiza, 514
Nabalus, 639
Naiad, 35
Naias, 35
Nail Wort, 228
Najadaceae, 31
Najas, 35
Najas Family, 31
Naked-seeded Plants, 25
Nama, 450
Naumbergia, 424
Navarretia, 440
Needlegrass, 93
Negundo, 365
Nemexia, 154
Nemophila, 450
Neobesseya, 384
Nepeta, 479
Nettle, 183
Nettle Family, 183
New Jersey Tea, 366
Nicandra, 482
Nicotiana, 481
Ninebark, 310

Nipplewort, 632
Nolina, 153
Norta, 280
Noseburn, 363
Nothocalais, 637
Notholaena, 19
Nuphar, 235
Nutallia, 380, 381, 382
Nyctaginaceae, 218
Nymphaea, 235
Nymphaeaceae, 235
Oak, 180
Oakfern, 21
Oat, 76
Oatgrass, 77, 78
Ocean Spray, 309
Ochrocodon, 158
Odostemon, 252
Oenothera, 393
Oleaceae, 424
Oleaster, 387
Oleaster Family, 386
Oligoneuron, 553
Onagra, 395
Onagraceae, 388
Onion, 155
Oniongrass, 62
Onobrychis, 327
Onoclea, 15
Onopordon, 629
Onosmodium, 461
Oonopsis, 557, 558, 559
Ophiglossaceae, 12
Ophrys, 164
Opulaster, 310
Opuntia, 382
Orache, 204
Orchard Grass, 61
Orchidaceae, 160
Orchid Family, 160
Oreobatus, 307
Oreobroma, 224
Oreocarya, 465, 466, 467, 468
Oreochrysum, 559
Oreoxis, 404, 405
Orobanchaceae, 514
Orobanche, 514
Orogenia, 403
Orophaca, 337
Orpine Family, 283
Orthocarpus, 508
Oryzeae, 43, 105
Oryzopsis, 91
Osmorhiza, 401
Othake, 587
Owlclover, 508
Oxalidaceae, 353
Oxalis, 353
Oxeye, 604
Oxeye Daisy, 579
Oxypolis, 408
Oxyria, 195
Oxytenia, 620
Oxytrope, 334
Oxytropis, 334
Ozomelis, 293
Pachylophus, 396
Pachystima, 364
Paeonia, 245
Paint Brush, 505
Painted Cup, 505
Palafoxia, 587
Paniceae, 43, 106
Panicularia, 51
Panicum, 108
Panicum, 108
Papavera, 253

Papaveraceae, 252
Paperflower, 591
Parietaria, 183
Parnassia, 291
Paronychia, 228
Parosela, 325, 326
Parrotfeather, 398
Parryella, 318
Parsnip, 408
Parsnip Family, 399
Parthenice, 596
Parthenium, 595
Parthenocissus, 368
Partridge Pea, 320
Paspalum, 107
Paspalum, 107
Pasque Flower, 241
Pastinaca, 408
Pea Family, 314
Pearl-everlasting, 624
Pearlwort, 232
Peavine, 327
Pectianthia, 292
Pectis, 586
Pectocarya, 455
Pedicularis, 509
Pediocactus, 386
Pediomelum, 323, 324
Pellaea, 20
Pellitory, 183
Pennellia, 277
Pennisetum, 112
Pennisetum, 112
Penny Cress, 262
Penstemon, 491
Peony, 245
Pepo, 523
Peppergrass, 260
Pepperweed, 260
Pepperwort, 21
Pepperwort Family, 21
Peramium, 164
Peraphyllum, 296
Pericome, 589
Perideridia, 407
Peritoma, 255
Persicaria, 199, 200
Petalostemon, 318
Petasites, 607
Petradoria, 553
Petrophytum, 309
Phaca, 339, 337, 340, 341, 342, 343, 344, 345, 347, 348, 350
Phacelia, 451
Phacopsis, 342
Phalarideae, 43, 103
Phalaris, 104
Phegopteris, 21
Phellopterus, 412
Philadelphus, 288
Philotria, 37
Phippsia, 83
Phleum, 84
Phlox, 441
Phlox, 441
Phlox Family, 439
Phoradendron, 185
Phragmites, 62
Phyla, 471
Phyllodoce, 418
Physalis, 482
Physaria, 258
Physocarpus, 310
Physostegia, 478
Picea, 28, 184
Pickerel-weed Family, 146
Picradenia, 593, 594

Picradeniopsis, 595
Pigmy Weed, 283
Pigweed, 211, 216
Pimpernel, 424
Pinaceae, 25
Pincushion Cactus, 384
Pine, 26
Pinedrops, 416
Pine Family, 25
Pinesap, 416
Pink Family, 225
Pinus, 26, 184
Piperia, 163
Pipsissewa, 416
Piptochaetium, 92
Pisophaca, 347, 348
Plagiobothrys, 468
Plantaginaceae, 517
Plantago, 517
Plantain, 517
Plantain Family, 517
Platyschkuria, 594, 595
Pleuraphis, 98
Pleurogyne, 426
Pleurophragma, 279
Plum, 308
Plumbaginaceae, 420
Plumbago Family, 420
Poa, 52
Podgrass, 35
Podistera, 403
Poinsettia, 361
Poison Hemlock, 407
Poison Suckleya, 208
Polanisia, 255
Polemoniaceae, 439
Polemonium, 440
Poliomintha, 473
Polygala, 357
Polygalaceae, 357
Polygonaceae, 185
Polygonatum, 154
Polygonum, 195
Polypodiaceae, 14
Polypodium, 15
Polypody, 15
Polypogon, 84
Polypogon, 84
Polypteris, 587
Polystichum, 17
Pond Lily, 235
Pondweed, 31
Pond-weed Family, 31
Pontederiaceae, 146
Poolmat, 35
Poormans Weatherglass, 424
Popcorn Flower, 468
Poplar, 165
Poppy, 253
Poppy Family, 252
Poppymallow, 369
Populus, 165
Porterella, 526
Portulaca, 223
Portulacaceae, 223
Potamogeton, 31
Potato Bean, 329
Potato Family, 480
Potentilla, 301
Poverty Weed, 215
Povertyweed, 620
Prairie Clover, 318
Prairie Coneflower, 599
Prairie Gentian, 430
Prenanthella, 640
Prenanthus, 639
Prickly Pear, 382

Prickly Poppy, 252
Primrose, 422
Primrose Family, 420
Primula, 422
Primulaceae, 420
Prince's Plume, 257
Prionopsis, 557
Proboscoidea, 515
Prunella, 479
Prunus, 308
Psedera, 368
Pseudocymopterus, 411, 410
Pseudopteryxia, 410
Pseudotsuga, 28, 184
Psilostrophe, 591
Psoralea, 323
Psoralidium, 324, 325
Ptelea, 357
Pteridium, 19
Pteridophyta, 12
Pteris, 19
Pterospora, 416
Pteryxia, 410
Ptilocalais, 632
Ptiloria, 631, 632
Puccinellia, 50
Puccoon, 461
Pulsatilla, 241
Puncture Vine, 356
Purple-headed Coneflower, 596
Purshia, 310
Purslane, 223
Purslane Family, 223
Pussytoes, 621
Pycnanthemum, 477
Pyrola, 417
Pyrola, 417
Pyrrocoma, 557, 558, 559
Quamoclidion, 221
Queens Delight, 363
Queens-root, 363
Quercus, 180
Quickweed, 597
Quillwort, 23
Quillwort Family, 23
Quincula, 483
Rabbit Brush, 545
Radicula, 268
Radish, 280
Ragweed, 618
Ramischia, 417
Ranunculaceae, 236
Ranunculus, 245
Raphanus, 280
Raspberry, 306
Ratany, 316
Ratibida, 599
Rattlesnake Plantain, 164
Rattlesnake Root, 639
Rattleweed, 509
Razoumofskya, 184
Redfieldia, 61
Reed, 62
Reedgrass, 79
Reseda, 283
Resedaceae, 283
Resinweed, 549
Rhamnaceae, 366
Rhamnus, 367
Rheum, 200
Rhinanthus, 509
Rhodiola, 284
Rhododendron, 419
Rhododendron, 419
Rhubarb, 200
Rhus, 364
Ribes, 285

Ricegrass, 91
Ringwing, 208
Robinia, 326
Rockbrake, 15
Rockcress, 270
Rocket Salad, 281
Rockjasmine, 422
Rockrose Family, 386
Rock Spirea, 309
Rorippa, 267
Rosa, 300
Rosaceae, 294
Rose, 300
Rose Family, 294
Rosemallow, 369
Rosemary Mint, 473
Rosin-weed, 603
Rubacer, 307
Rubiaceae, 515
Rubus, 306
Rudbeckia, 598
Rue Family, 357
Rulac, 365
Rumex, 200
Ruppia, 35
Rush, 146
Rush Family, 146
Rushpea, 316
Russian Thistle, 209
Rutaceae, 357
Rydbergia, 593
Rydbergiella, 342
Ryegrass, 73
Sabina, 26
Safflower, 629
Sage, 472
Sagebrush, 580
Sagina, 232
Sagittaria, 36
Sainfoin, 327
Salicaceae, 165
Salicornia, 204
Salix, 167
Salsify, 631
Salsola, 209
Saltbrush, 204
Salt Cedar, 374
Saltgrass, 61
Salvia, 472
Salvinia Family, 22
Salviniaceae, 22
Sambucus, 519
Samolus, 421
Samphire, 204
Sandalwood Family, 185
Sandbur, 113
Sandgrass, 64
Sandlily, 154
Sandpuffs, 219
Sandreed, 80
Sandspurry, 229
Sandverbena, 219
Sandwort, 233
Sanicula, 402
Santalaceae, 185
Sapindaceae, 366
Sapindus, 366
Saponaria, 226
Sarcobatus, 204
Sarcostemma, 432
Saururaceae, 181
Saussurea, 630
Savastana, 103
Saxifraga, 288
Saxifragaceae, 284
Saxifrage, 288
Saxifrage Family, 284

Schedonnardus, 100
Schizachne, 63
Schizachyrium, 114
Schmaltzia, 364
Schoenocrambe, 280
Schrankia, 317
Scirpus, 140, 142, 143
Sclerocactus, 386
Sclerochloa, 52
Scleropogon, 64
Scorpion Weed, 451
Scouring Rushes, 22
Scrophularia, 490
Scrophulariaceae, 485
Scurfpea, 323
Scutellaria, 475
Seablite, 210
Sea Milkwort, 421
Sea Purslane, 223
Sedge, 118
Sedge Family, 116
Sedum, 284
Seed Plants, 25
Seepweed, 210
Sego Lily, 153
Selaginella, 24
Selaginella, 24
Selaginellaceae, 24
Selaginella Family, 24
Selfheal, 479
Senecio, 608
Senecioneae, 532
Senna, 320
Sensitive Brier, 317
Sensitive Fern, 15
Serapias, 164
Sericotheca, 309
Service-berry, 298
Sesuvium, 223
Setaria, 111
Shepherdia, 387
Shepherds Purse, 262
Shooting Star, 421
Sibbaldia, 312
Sida, 372
Sidalcea, 370
Sideranthus, 557, 558
Sieversia, 313, 314
Silene, 226
Silphium, 603
Silverberry, 387
Silvergrass, 113
Sinapis, 281
Siphonella, 449
Sisymbrium, 279, 278, 268
Sisyrinchium, 160
Sitanion, 72
Sium, 406
Skeleton Plant, 640
Skullcap, 475
Skunk Leaf, 440
Sloughgrass, 100
Small Mistletoe, 184
Smartweed, 195
Smelowskia, 274
Smilacina, 158
Smilax, 154
Snakecotton, 218
Snake Root, 402
Snakeweed, 552
Sneezeweed, 591
Snowberry, 520
Snowlover, 505
Soapberry, 366
Soapberry Family, 366
Soapweed, 152
Soapwort, 226

Solanaceae, 480
Solanum, 484
Solidago, 552
Solomon's Plume, 158
Soloman's Seal, 154
Sonchus, 634
Sophia, 275
Sophora, 320
Sorbus, 296
Sorghastrum, 115
Sorghum, 115
Sorghum, 115
Sowthistle, 634
Spanish Bayonet, 152
Spanish Needles, 601
Sparganiaceae, 30
Sparganium, 30
Spartina, 100
Spatularia, 289
Spectacle Pod, 258
Specularia, 524
Speedwell, 287
Spergula, 229
Spergularia, 229
Spermatophyta, 25
Sphaeralcea, 370, 373
Sphaerophysa, 333
Sphaerostigma, 398
Sphenopholis, 74
Spiderflower, 255
Spiderwort, 145
Spiderwort Family, 145
Spiesia, 335
Spike Fescue, 50
Spike-rush, 142
Spikesedge, 142
Spiraea, 309
Spiranthes, 164
Spirodela, 144
Spleenwort, 15
Sporobolus, 89, 86
Sprangletop, 99
Spring Beauty, 225
Spruce, 28
Spurge, 358
Spurge Family, 358
Spurry, 229
Squaw Apple, 296
Squirrel Tail, 72
Stachys, 479
Stafftree Family, 364
Stanleya, 257
Stanleyella, 277
Stargrass, 159
Starlily, 154
Star-thistle, 630
Starwort, 231
Statice, 420
Steironema, 424
Stellaria, 231
Stenophragma, 277
Stenosiphon, 391
Stenotopsis, 559
Stenotus, 559
Stephanomeria, 631
Stickleaf, 379
Stickseed, 455
Stillingia, 363
Stipa, 93
St. Johnswort, 374
St. Johnswort Family, 374
Stonecrop, 284
Stork's Bill, 351
Strawberry, 312
Strawberry Cactus, 384
Streptanthella, 278
Streptanthus, 278, 280

Streptopus, 158
Strophostyles, 328
Suaeda, 210
Suckleya, 208
Sullivantia, 292
Sumac, 364
Sumac Family, 363
Summer-cypress, 209
Sundrops, 393
Sunflower, 605
Svida, 415
Swainsona, 333
Swainson Pea, 333
Swamp Potato, 35
Sweet Alyssum, 262
Sweet Cicely, 401
Sweet Clover, 333
Sweet Coltsfoot, 607
Sweetflag, 144
Sweetgrass, 103
Sweetroot, 401
Sweet Vetch, 329
Swertia, 426
Swertia, 426
Symphoricarpos, 520
Syntherisma, 106
Synthris, 489
Syringa, 288
Talinum, 224
Tamaricaceae, 373
Tamarix, 374
Tamarix, 374
Tamarix Family, 373
Tanacetum, 579
Tansy, 579
Tansy Mustard, 274
Taraxia, 397
Taraxicum, 638
Tarweed, 596
Teasel, 527
Teasel Family, 527
Telesonix, 290
Tellima, 291
Tesseranthium, 426
Tetradymia, 607
Tetraneuris, 592, 593
Teucrium, 475
Thalesia, 514
Thalictrum, 243
Thamnosma, 357
Thelesperma, 600
Thelypodiopsis, 280
Thelypodium, 279, 278, 277
Thermopsis, 320
Thistle, 625
Thistle Family, 528
Thlaspi, 262
Thornapple, 480
Thoroughwort, 542
Three Awn, 96
Thrift, 420
Thymophylla, 586
Tickseed, 215, 603
Tidestromia, 218
Tillaea, 283
Tillaeastrum, 283
Timothy, 84
Tissa, 229, 230
Tithymalis, 361, 362
Tium, 339, 338, 341, 349
Toadflax, 490
Tobacco, 481
Tonestus, 559
Torresia, 103
Touch Me Not, 355
Touteria, 380, 381, 382
Townsendia, 560

Toxicodendron, 364
Toxicoscordium, 157
Tradescantia, 145
Tragia, 363
Tragopogon, 631
Trautvetteria, 243
Trefoil, 333
Tribulus, 356
Trichachne, 106
Tridens, 63
Trifolium, 329
Triglochin, 35
Trillium, 154
Triodanis, 524
Triodia, 63, 64
Triplasis, 64
Tripsaceae, 44, 116
Tripterocalyx, 219
Trisetum, 74
Trisetum, 74
Triticum, 69
Trollius, 245
Troximon, 637, 638
Tumblegrass, 100
Turritis, 271
Twayblade, 163
Twinflower, 520
Twinpod, 258
Twisted-stalk, 158
Twistflower, 278
Typha, 30
Typhaceae, 29
Ulmaceae, 182
Ulmus, 182
Umbelliferae, 399
Umbrella Plant, 185
Umbrella-wort, 220
Unamia, 576
Unicornplant, 515
Unicornplant Family, 515
Urtica, 183
Urticaceae, 183
Utricularia, 513
Vaccaria, 226
Vaccinium, 418
Vagnera, 159
Valerian, 527
Valeriana, 527
Valerianaceae, 526
Valerian Family, 526
Vallisneria, 38
Valota, 106
Velvetgrass, 77
Venus's Looking-glass, 524
Veratrum, 156
Verbascum, 486
Verbena, 469
Verbenaceae, 469
Verbesina, 599
Vernal Grass, 104
Vernonia, 542
Vernonieae, 528
Veronica, 487
Vervain, 469
Vervain Family, 469
Vetch, 328, 336
Viburnum, 519
Vicia, 328
Viguiera, 605
Viola, 375
Violaceae, 375
Violet, 375
Violet Family, 375
Viorna, 242, 243
Virginia Creeper, 368
Virgin's Bower, 242
Vitaceae, 368

Vitis, 368
Vulpia, 49
Wakerobbin, 154
Walbergella, 228
Wallflower, 276
Walnut, 179
Walnut Family, 179
Washingtonia, 401, 402
Water Hemlock, 406
Waterhemp, 216
Water Horehound, 473
Waterhyssop, 511
Water Leaf, 450
Waterleaf Family, 449
Water Lily Family, 235
Watermilfoil, 398
Watermilfoil Family, 398
Water Parsnip, 406
Waterpimpernel, 421
Waterplantain, 37
Waterplantain Family, 36
Water-starwort, 363
Water-starwort Family, 363
Waterweed, 37
Waterwort, 373
Waterwort Family, 373
Waxflower, 288
Wedeliella, 220
Wedgegrass, 74
Wedgescale, 74
Wheat, 69
Wheat Cockle, 226
Wheatgrass, 65
White Sage, 208
Whitetop, 259
Whiteweed, 259
Whitlow-grass, 263
Whitlow-wort, 228
Whortleberry, 418
Widgeonweed, 35
Wild-basil, 478
Wildbean, 328
Wild Cabbage, 278
Wild Candy Tuft, 262
Wild Celery, 38
Wild Cosmos, 589
Wild-daisy, 561
Wild Ginseng, 399
Wild Hollyhock, 373
Wild Lettuce, 633
Wildrice, 105
Wildrye, 69
Wild Sarsaparilla, 399
Willow, 167
Willow Family, 165
Willow Herb, 389
Willowweed, 389
Windflower, 241
Windmill Grass, 101
Winged Pigweed, 208
Wintercress, 283
Winterfat, 208
Wirelettuce, 631
Witchgrass, 108
Wolfberry, 481
Wolftail, 84
Woodbetony, 509
Woodfern, 17
Woodnymph, 416
Woodreed, 83
Woodruff, 516
Woodrush, 150
Woodsia, 17
Woodsia, 17
Woodsorrel, 353
Woodsorrel Family, 353
Wormwood, 580

Woundwort, 479
Wyethia, 600
Wyomingia, 566
Xanthium, 618
Xanthoxalis, 353, 354
Ximenesia, 599
Xylophacos, 343, 338, 344, 347
Xylorrhiza, 574
Yampa, 407
Yarrow, 578
Yellowrattle, 509
Yerba Mansa, 181
Youngia, 635
Yucca, 152
Yucca, 152
Zannichellia, 35
Zauschneria, 388
Zea, 116
Zinnia, 603
Zizania, 105
Zizanieae, 43, 105
Zizia, 406
Zosterella, 146
Zoysieae, 42, 98
Zygadenus, 157
Zygophyllaceae, 356
Zygophyllidium, 361
Zygophyllum, 356